D1802697

SOLAR WIND TEN

Previous Proceedings in the Series of Conferences on Solar Wind

Year	Title/Held in	Editors	Publisher	Publication Information
1998	Solar Wind Nine/ Nantucket, Massachusetts	S. Habbal, R. Esser, J.V. Hollweg, and P.A. Isenberg	AIP Conference Proceedings Volume 471	ISBN 1-56396-865-7
1995	Solar Wind Eight/ Dana Point, California	D. Winterhalter, J.T. Gosling, S.R. Habbal, W.S. Kurth, and M. Neugebauer	AIP Conference Proceedings Volume 382	ISBN 1-56396-551-8 [out of print]
1991	Solar Wind Seven/ Goslar, Germany	E. Marsch and R. Schwenn	Pergamon Press	ISBN 0-08-042049-4
1987	Solar Wind Six/ Estes Park, Colorado	V.J. Pizzo, T. Holzer, and D.G. Sime	National Center for Atmospheric Research (NCAR)	NCAR/TN-306+Proc
1982	Solar Wind Five/ Woodstock, Vermont	M. Neugebauer	NASA	NASA CP-2280
1978	Solar Wind Four Burghausen, Germany	H. Rosenbauer	Max-Planck Institut für Aeronomie	MPAE-W-100-81-31
1974	Solar Wind Three/ Asilomar, Pacific Grove, California	C.T. Russell	Inst. of Geophysics and Planetary Physics, University of California, Los Angeles	
1971	Solar Wind/ Asilomar, Pacific Grove, California	C.P. Sonett, P.J. Coleman, Jr., J.M. Wilcox	NASA	NASA SP-308
1964	The Solar Wind/ Pasadena, California	R.J. Mackin, Jr. and M. Neugebauer	Jet Propulsion Laboratory, California Institute of Technology, Pasadena, California	JPL Technical Report No. 32-630

Related Titles from AIP Conference Proceedings

649 Dusty Plasmas in the New Millennium: Third International Conference on the Physics of Dusty Plasmas
Edited by R. Bharuthram, M. A. Hellberg, P. K. Shukla, and F. Verheest, December 2002, 0-7354-0106-3

608 Space Technology and Applications International Forum - STAIF 2002: Conference on Thermophysics in Microgravity; Conference on Innovative Transportation Systems for Exploration of the Solar System and Beyond; 19th Symposium on Space Nuclear Power and Propulsion; Conference on Commercial/Civil Next Generation Space Transportation
Edited by Mohamed S. El-Genk, February 2002, 0-7354-0052-0;
CD-ROM sold separately: 0-7354-0053-9

598 Solar and Galactic Composition: A Joint SOHO/ACE Workshop
Edited by Robert F. Wimmer-Schweingruber, December 2001, CD-ROM included, 0-7354-0042-3

537 Waves in Dusty, Solar, and Space Plasmas
Edited by F. Verheest, M. Goossens, M. A. Hellberg, and R. Bharuthram, October 2000, 1-56396-962-9

528 Acceleration and Transport of Energetic Particles Observed in the Heliosphere: ACE 2000 Symposium
Edited by Richard A. Mewaldt, J. R. Jokipii, Martin A. Lee, Eberhard Möbius, and Thomas H. Zurbuchen, July 2000, 1-56396-951-3

SOLAR WIND TEN

Proceedings of the Tenth
International Solar Wind Conference

Pisa, Italy *17-21 June 2002*

EDITORS

Marco Velli
Università di Firenze
Firenze, Italy

Roberto Bruno
IFSI CNR, Roma, Italy

Francesco Malara
Università della Calabria
Arcavacata di Rende, Italy

SPONSORING ORGANIZATIONS
Università di Firenze
Università di Pisa
Università della Calabria
IFSI - Istituto di Fisica dello Spazio Interplanetario, CNR
ASI - Agenzia Spaziale Italiana
ESA - European Space Agency
INAF - Istituto Nazionale di Astrofisica

Melville, New York, 2003
AIP CONFERENCE PROCEEDINGS ■ VOLUME 679

Editors:

Marco Velli
Dipartimento di Astronomia e
Scienza dello Spazio
Università di Firenze
Largo E Fermi 2
50125 Firenze
ITALY

E-mail: velli@arcetri.astro.it

Roberto Bruno
Istituto di Fisica dello Spazio Interplanetario
Via Fosso del Cavaliere 100
00133 Roma
ITALY

E-mail: bruno@ifsi.rm.cnr.it

Francesco Malara
Dipartimento di Fisica
Università della Calabria
Ponte B. Bucci, WBO 31C
87036 Arcavacata di Rende
ITALY

E-mail: malara@fis.unical.it

Cover: The eclipse photo is used courtesy of the High Altitude Observatory/National Center for Atmospheric Research, Boulder, Colorado.

L.C. Catalog Card No. 2003109646
ISBN 0-7354-0148-9 Set
ISSN 0094-243X

Printed in the United States of America

CONTENTS

KEYNOTE LECTURES

SESSION I
STRUCTURE OF THE CORONA AND HELIOSPHERE IN 3D: SOLAR MINIMUM AND MAXIMUM

SESSION II
CORONAL HEATING AND THE ACCELERATION OF SOLAR AND SOLAR-TYPE WINDS

SESSION III
WAVES, TURBULENCE, AND KINETIC PHYSICS

SESSION IV
INTERNAL STATE, COMPOSITION, HIGH ENERGY PARTICLES

SESSION V
DYNAMIC ACTIVITY OF THE CORONA AND THE HELIOSPHERE

SESSION VI
NEW MISSIONS, OPPORTUNITIES, AND TECHNIQUES FOR HELIOSPHERIC PHYSICS

APPENDIX ON CD-ROM ONLY

Preface

The Tenth International Solar Wind Conference was held in Italy from June 17 to 21, 2002 in the city of Pisa, hosted by the Universities of Pisa, Firenze and Cosenza and the Institute for Physics of Interplanetary Space of the Italian National Research Council. This was the first time the conference was held in Italy, and with 240 participants it had the largest attendance of any conference since the series began thirty-eight years ago. Scientists from more than twenty countries came together, during what perhaps fittingly turned out to be the hottest week of the summer, to discuss the most recent findings on the origin and acceleration of the solar wind, and its dynamical interactions throughout the heliosphere and the interstellar medium at its boundaries. The meeting took place during the June festival associated with Pisa's patron saint, S. Ranieri, beginning on the night of the Luminara, when the buildings on the Arno riverfront are lit up in a brilliant display of thousands of candles. A taxi overload left some participants to walk all the way from Pisa airport and train station to their hotels in the city; they discovered, fortunately, that Pisa is a safe place even in the middle of the night. The Palazzo dei Congressi auditorium provided a comfortable venue for the presentations, while the gallery space allowed all posters to be displayed throughout the meeting.

It was thought fit to begin this conference with a series of keynote lectures, setting the stage for the more in-depth discussions on recent results in solar wind and heliospheric research. We were fortunate to have with us Giancarlo Noci, who provided a historical background to our understanding of the solar corona; Marcia Neugebauer, initiator of the solar wind meeting series, who lectured on the evolution of our understanding of the solar wind provided by in-situ measurements; Joe Hollweg, who gave an up-to-date review of the problem of solar wind acceleration taking into account the exciting new experimental results of the last five years; and Professor V. Baranov, who pioneered work on the global structure of the heliosphere and the boundary region separating the sun from the interstellar medium.

One of the main scientific goals of the conference was to bring together the unprecedented in situ observations of the solar wind in three dimensions from the Ulysses spacecraft, now extending over the full solar cycle; data ranging from the inner heliosphere to the proximity of the heliopause provided by Helios, ACE, and Voyager, the comprehensive remote sensing observations of the outer solar atmosphere from the SOHO, Yohkoh, and Trace spacecraft, and theoretical and computational models to better understand the origins and dynamics of the solar wind throughout the solar activity cycle. The meeting was divided into six sessions: the first five following clearly defined yet overlapping science objectives, namely, the structure of the solar corona and heliosphere in 3 dimensions and its variation during the solar cycle, the heating of the corona and acceleration of the solar wind, the dynamics of turbulence from macroscopic fluid scales down to the kinetics of dissipation, the internal state of the wind, and dynamic activity of the corona. The organizing committee also agreed to hold a sixth session specifically devoted to perspectives on the future of heliospheric science. Each session was organized by a group of conveners whose task was to select invited and contributed talks from the abstracts submitted in the individual sessions. Specific poster discussion sessions were included in the program, though scientific discussions at the boards could be seen throughout the meeting, sustained in no small part by the excellent coffee, cappuccino, and refreshments provided in the mornings as well as the wine and parmesan aperitifs available in the evening.

The contributions in this volume follow the chronological session order during the conference. All papers were subject to peer review, organized by the editors with the help of conveners and participants, and our thanks go to all those who have assisted in maintaining the high standard of the Solar Wind conference series.

The organizing committee, chaired by Marco Velli, included Giorgio Einaudi and Francesco Pegoraro from the University of Pisa; Roberto Bruno and Angela Rossetti from the Institute for Physics of Interplanetary Space, Consiglio Nazionale delle Ricerche; Francesco Malara from the University of Cosenza; and Luca Del Zanna, Simone Landi, and Claudio Chiuderi from the University of Firenze. We would like to express our gratitude to the conveners in making the meeting a scientific success: J. Gosling, M. Neugebauer, and G. Noci (session I); S. R. Habbal, J. Hollweg, and F. Malara, (session II); R. Bruno, M. L. Goldstein, and A. Mangeney (session III); J. Mazur, E.C. Roelof, and R. Von Steiger (session IV); S. Antiochos, P. Cargill, and P. Riley (session V); and C. Chiuderi, P.C. Liewer, and R. Schwenn (session VI). We would also like to thank session chairs Claudio Chiuderi, Giancarlo Noci, Ed Stone, Marcia Neugebauer, Jack Gosling, Joe Hollweg, Shadia Habbal, Gary Zank, Mel Goldstein, Rudi Von Steiger, Joe Mazur, Steven Spangler, Thomas Zurbuchen, Paulett Liewer, and Rainer Schwenn for keeping the intense meeting schedule on time without sacrificing the lively debates arising from questions and answers after each talk.

Group photo taken on the lawn in front of the Certosa di Pisa (Calci, Pi), on June 20. In the background, view of the Monti Pisani.

The conference was sponsored by the Universities of Firenze, Pisa, and Cosenza; the Institute for the Physics of Interplanetary Space of the Consiglio Nazionale delle Ricerche; the Consiglio Nazionale delle Ricerche, the Istituto Nazionale di Astrofisica; the Italian Space Agency; the European Space Agency; and the City of Pisa. Thanks to this support we were able to provide full or partial assistance to 25 scientists who would have otherwise been unable to attend.

Special thanks also go to Angela Rossetti and Barbara Moscatelli (IFSI-CNR) for overall organization and assistance at the secreterial desk; to Simone Landi, who helped organize the conference web page, abstract book and program; to Chiara Giannotti, who distributed the microphone during questions; to Franco Rappazzo, Luca Del Zanna, Luana Stanganini, Paola Spadolini, and Nadia Ugolini; to Antonella Lorini and Francesca from the Consorzio Turistico Pisa è, and to Silvia Tomasi, whose artistic talents contributed the poster and conference logo. The University of Pisa graciously allowed the conference dinner to be held in the scenic Certosa di Pisa. De Carlo Catering's cuisine helped make it an unforgettable evening, of which the group photo on the adjacent page is a small reminder.

It is customary for the solar wind conferences to end by selecting the organizers of the subsequent meeting. At the conference, both Thomas Zurbuchen from the University of Michigan and Ruth Esser of the University of Tromso offered to host Solar Wind Eleven. The participants were evenly split in their preference, and both volunteering groups were quite happy to have the other host the conference. The choice was finally left to chance, heads or tails, and as a result Thomas Zurbuchen and the University of Michigan will host the next solar wind conference.

As this preface is completed, clouds of war have gathered over the international community, causing divisions between friendly nations which would have been hard to predict only a few months ago. Though research in space was born in military competition in the middle of the last century, interplanetary space exploration via satellite experiments grew to foster a spirit of collaboration and friendship among nations. It is our hope that progress in our understanding of the heliosphere continues in the same vein and that as scientists we continue to build bridges and bring understanding between societies.

The Solar Wind Ten editors,
Marco Velli, Roberto Bruno and Francesco Malara

Keynote Lectures

Observations of the Solar Corona and Solar Wind Sources

Giancarlo Noci

Dipartimento di Astronomia e Scienza dello Spazio - Università di Firenze, Largo E. Fermi 2 - 50125 Firenze, Italy

INTRODUCTION

This talk presents a historical survey of the remote sensing observations of the solar corona, and in particular of those more relevant for our understanding of the solar wind. I have chosen a schematic presentation to emphasize the steps forward which have characterized the progress of our knowledge. The matter is divided in sections, according to the tecniques emploied for the observations, and follows a chronologique sequence, although some overlap in time, between different sections, has been unaivodable. I am convinced that this organization of the matter is the one which best illustrates the development of our knowledge of the solar corona.

ECLIPSE AND CORONAGRAPHIC OBSERVATIONS FROM THE GROUND

- Although the solar corona, before the modern times, could be observed only during a total eclipse of the sun (extremely rare event, for a given location), Plutarch (A. D. 46 - 120) [1] refers to the existence of a weakly luminous halo around the rim of the moon when the sun is completely occulted.
- Kepler describes the sun in the 1567 eclipse as being surrounded by a luminous sphere, and makes reference to Plutarch [2].
- The beginning of scientific observations could be put at 1842. At the eclipse of that year the first objective was to establish whether the chromosphere, prominences and corona were solar or produced in Earth atmosphere.
- 1851 - First photographic image of totally eclipsed Sun, by Berkowsky (quoted by de Vaucouleurs [3]). Photography more and more used in subsequent eclipses.
- 1860 - The solar nature of prominences was established in the 1860 eclipse, by comparing photographs obtained at two sites 250 miles apart by De la Rue and Secchi [4].
- 1868 - First valuable spectroscopic observation in a total solar eclipse: emission lines observed by several scientists in the prominence spectra, which showed the gaseous nature of these phenomena.
- 1869 - An emission line from the corona (green line) first observed during the 1869 eclipse by Harkness and Young [5, 6]. Attributed initially to iron, later to new element 'coronium'. Also continuous emission observed by Harkness.
- 1871 - Fraunhofer absorption lines in the coronal continuum observed by Janssen [7] at the 1871 eclipse. From the stability of the coronal image during the eclipse, Janssen deduces the solar nature of the corona; he then concludes that the coronal radiation has two components: the emission lines, produced by a gaseous atmosphere, and reflected photospheric light (which explained also earlier polariscopic observations), the reflection being caused probably by solid particles present in the solar corona.
  Also, at the 1871 eclipse, Lockyer and Respighi established that the green line emission extended to very large heights [8], which posed a dynamical problem, since the coronal temperature was thought to be much lower than discovered later.
- 1878 - The comparison of the images of the 1878 and 1871 eclipses confirmed the idea, put forward after the 1975 eclipse, that the corona changed with the sunspot cycle.
  From the images of this eclipse it was also noted the resemblance of the coronal streamers with magnetic lines of force.
- 1930 - Invention of coronagraph by Lyot. Beginning of observation of the solar corona outside total eclipses.
- 1942 - Identification of 'coronium' lines by Edlèn [9].

CP679, *Solar Wind Ten: Proceedings of the Tenth International Solar Wind Conference*, edited by M. Velli, R. Bruno, and F. Malara

High temperature of the corona. Its value difficult to determine because cross-sections for ionization calculations too uncetrain. It became clear the existence of a contribution to the visible continuum from free electrons scattering photospheric radiation. L, K and F corona. Electron density from intensity of K corona.

- $\sim$ 1950 - Once the mass of the ions producing the coronal lines was known, it became possible to transfer the line widths (observed with coronagraphs) into kinetic temperatures. The result was a confirmation of the high coronal temperature. The determination of its value gave $1.7 - 2.1 \times 10^6$ K (Dollfus, quoted by van de Hulst [10]).
- Discovery of large temperature gradient in the cromosphere-corona transition region (pointed out by Giovannelli, on the basis of eclipse observations, in 1949 [11]; Giovannelli also pointed out the importance of energy transport by conduction. Very steep temperature gradient in the conduction dominated transition region model of Woolley and Allen (1950) [12]).
- Ground eclipse observations important also in recent years because of low level of stray light and high space resolution achievable. These observations have shown that the solar corona is a very structured (inhomogeneous) atmosphere.

RADIO OBSERVATIONS

The detection of radio emission from the sun permitted a big step forward in our knowledge of the corona because new phenomena could be observed (particularly transient phenomena) and the coverage was continuous.

- 1942 - First detections of solar radioemission: Hey at meter wavelengths [13] and Southworth at centimeter wavelengths [14].
- It was soon clear that solar radioemission consisted of various components, one of which was in agreement with the thermal emission expected from the coronal plasma. Hence it was possible to determine electron temperature and density, but difficulties arised for space resolution owing to diffraction and to curved path of radiation, particularly for extended corona. (High space resolution is achievable at present with large interferometers, but the observation of extended features would require a long time).
- Large non-thermal transient component. Identification of various kinds of transients and determination of associated coronal phenomena (e. g. path of ejections and shocks). Particularly by means of the spectrum analyzer realized by Wild and McCready [15]). Determination of the magnetic field, from polarization, for some transients.
  The solar corona as a very active environment.
- An interesting outcome of the radio observations is a coronal temperature lower than in previous determinations, in the quiet sun. Being the radioemission due to free-free transitions, the temperature obtained with the radio observations is an electron temperature. Hence: T_e (radio)$< T_{ion}$ (visible, line widths).

UV AND X-RAY OBSERVATIONS ON THE DISK - ROCKETS AND OSO SATELLITES

The solar corona can be observed, in the ultraviolet, also on the disk, because the photosphere, being much cooler, does not emit in this spectral region. The observations in this spectral regions are very important because they concern the proper emission of the coronal plasma. Furthermore their interpretation is simplified by the fact that the corona is transparent at these wavelengths.

- Data from rockets began to be obtained in the early sixties [e. g. 16, 17].
- Determination of density-temperature structure (emission measure $\int N_e^2 dh$ as a function of T) [18, 19] and abundances [19]. Here the temperature is obtained by ionization balance calculations.
- Confirmation of large temperature gradient in the chromosphere-corona transition region [20].
- 1960 - Confirmation of temperature problem (Seaton, quoted by Lust et al. [21]): T_e (UV, from ioniz. balance) $< T_{ion}$ (visible, line widths). (T_e (radio) $\simeq T_e$ (UV, from ioniz. balance).) Note that the temperature deduced from the ionization state is an electron temperature, because the ionization equilibrium in corona arises from a balance between electron collision ionization and recombinations.
- Discovery of the importance of dielectronic recombination [22]: T_e (UV, from ioniz. balance)$\simeq T_{ion}$ (visible, line widths).
  However T_e (radio) sensibly lower. Still unresolved discrepancy. Effect of inhomogeneity?
- Discovery of Coronal Holes from rocket data [23], OSO data [24], ground based K-coronameter data [25].
- Coronal holes identified as sources of fast solar wind from rocket X-ray data and *in situ* velocity data [26], and from OSO UV data and geomagnetic data [27, 28].

FOLLOWING YEARS: OBSERVATIONS OF THE EXTENDED CORONA IN THE VISIBLE AND OF THE CORONAL BASE IN THE UV AND X-RAYS, FROM SATELLITES

- The OSO white light coronagraph - Study of coronal variations on time scales of hours or more. Discovery of Coronal Mass Ejections (CME) [29].
- Skylab (1973 - 1974) - Skylab yielded a large amount of data on the solar corona, having on board instruments operating in the UV and X-ray domains and a coronagraph operating in the visible. The main results of the 'Skylab era' can be summarized as:
  Strong confirmation of the inhomogeneity of the solar corona. Deeper understanding of loops and coronal holes. Coronal holes mainly polar, but importance for geomagnetic effects of large elongated holes extending from polar area across equator, mainly in declining phase of activity cycle. Rigid rotation of coronal holes.
  Confirmation of fast solar wind from coronal holes and better understanding of heliospheric structure.
  A review of the Skylab results on coronal holes and fast solar wind can be found in Zirker [30].
- Other solar satellites - Helios, Solwind, Solar Maximum Mission, Yhokoh, TRACE.
  Improved data. For what concerns the extended corona, particularly important the data on CME's from SMM.

OCCULTATION OF RADIO SOURCES, INTERPLANETARY SCINTILLATION, RADAR

Interplanetary scintillation tecniques are very important for their capability of giving information on the solar wind speed outside the ecliptic plane (they were the only source of this information before Ulysses), and close to the sun.

- Fifties - Observations, in the radio domain, of Crab Nebula occultations by the solar corona, for the determination of the coronal electron density [31, 32].
- Sixties - Beginning of using IPS techniques for interplanetary electron density or/and solar wind speed. Evidence of larger velocity for the solar wind from the polar regions than in the ecliptic plane. [33].
- Late sixties and later - IPS determinations of solar wind speed close to the sun [34], even at $r < 2r_{\odot}$ [35].
- Radar measurement of outflow speed at $r < 2r_{\odot}$ [36].

The interplanetary scintillation measurements showed that the solar wind attained high speeds close to the sun.

UV OBSERVATIONS OF THE EXTENDED CORONA

The extended (i. e. above $\sim 1.2r_{\odot}$) corona, previously observed in the visible and radio domains, began to be observed in the UV in 1970. Since the radiation from the coronal base is much brighter ($\sim 10^5$) than that from the extended corona, there is an observational problem also in this spectral domain, and, in fact, as in the visible, the observations were first made during a total eclipse of the sun, and then, outside eclipses, by means of coronagraphs.

The 1970 eclipse observation from a rocket

- Slitless spectrograph flown by Astrophysics Research Unit (Culham), Harvard College Observatory (Cambridge), Imperial College (London), York University (Toronto) [37].
- The observations showed that the UV coronal spectrum was dominated by the Ly-α emission. It was shown that this emission was due to scattering, by residual neutral hydrogen, of the disk Ly-α [38].

Coronagraphic techniques in the UV

- The analysis of the 1970 eclipse data suggested the possibility of determining the solar wind velocity in the extended corona from Doppler dimming (Noci, quoted by Kohl and Withbroe [39]), and gave the start to the realization of a coronagraph operating in the UV at the Center for Astrophysics.
- The Ultraviolet Coronagraph Spectrometer (UVCS) [40] was flown first on rockets (first flight in 1979), then on Spartan. It measured Ly$-\alpha$ intensities and profiles in the extended corona.

SOHO

Five instruments observing the solar corona: EIT, CDS, SUMER, LASCO, UVCS. Their main results, for what concerns the extended corona and the solar wind, are the following:

- Determination of electron temperature in streamers [41] and in coronal holes [42]. The latter in contrast with electron temperature deduced from ionization state of the solar wind. Hence (see also the results shown above) the problem of the electron temperature in the solar corona appears to be a rather complex one.
- Determination of outflow speed above streamer cusp (slow solar wind) [43]. Confirmation of minimum speed on streamer stalk [44].
- Measurement of unexpectedly large kinetic temperatures ($> 10^8$ K) for heavy ions in coronal holes [45]. This is not a discrepancy with the coronagraphic determinations mentioned above (Dollfus), because the latter referred undoubtedly to streamers, much brighter then coronal holes.
- Discovery of asymmetry in the velocity distribution of heavy ions in coronal holes (bimaxwellian, much wider in the direction perpendicular to the radial direction than in the one parallel to it) [46].
- Determination of outflow speed in coronal holes (fast solar wind), for protons and OVI ions; larger for the latter [46].
- Confirmation of previous measurements from rockets [47] of dominant outflow at the base of coronal holes [48].
- Data on the structure of the current sheet above streamers: single or multiple? [49, 50].

ACKNOWLEDGMENTS

The author wishes to acknowledge, for what concerns section 2, the importance of the book of Meadows [51]; some information was found also in the article of Goldberg [52] and in the book of Berry [53]. Thanks are due to Prof. D. Coppini for her help with Plutarch.

REFERENCES

1. Plutarch, *De facie quae in orbe lunae apparet*, 932 B.
2. Kepler, J., *Epitome Astronomiae Copernicanae*, 1618.
3. de Vaucouleurs, G., *Astronomical photography: from the daguerreotype to the electron camera*, Faber and Faber, London, 1961, plate II.
4. de la Rue, W., *Proceedings of the Royal Society*, **13**, 442 (1864).
5. Young, C. A., *American Journal of Sciences and Arts (2nd series)*, **48**, 370 (1869).
6. Harkness, W., *Astronomical and Meteorological Observations made at the United States Naval Observatory during the Year 1867 - Appendix II*, Government Printing Office, Washington, 1870, p. 60.
7. Janssen, J., *Annuaire du Bureau des Longitudes*, **x**, 623 (1879).
8. Respighi, L., *Nature*, **5**, 217 (1872).
9. Edlèn, B., *Zeiscrift fur Astrophysik*, **22**, 30 (1942).
10. van de Hulst, H. C., , in *The Sun*, edited by G. P. Kuiper, University of Chicago Press, Chicago, 1953, p. 273.
11. Giovannelli, R. G., *Monthly Notices Royal Astronomical Society*, **109**, 372 (1949).
12. v. d. R. Woolley, R., and Allen, C. W., *Monthly Notices of the Royal Astronomical Society*, **110**, 358 (1950).
13. Hey, J. S., *Nature*, **157**, 47 (1946).
14. Southworth, G. C., *Journal of the Franklin Institute*, **239**, 285 (1945).
15. Wild, J. P., and McCready, L. L., *Australian Journal of Scientific Research A*, **3**, 387 (1950).
16. Detwiler, C. R., Garrett, D. L., Purcell, J. D., and Tousey, R., *Annales de Géophysique*, **17**, 9 (1961).
17. Hinteregger, H. E., *Journal of Geophysical Research*, **66**, 2367 (1961).
18. Ivanov-Kolodnyi, G. S., and Nikol'skii, G. M., *Soviet Astronomy-AJ*, **5**, 31 (1961).
19. Pottasch, S. R., *Astrophysical Journal*, **137**, 945 (1963).
20. Pottasch, S. R., *Space Science Rewiew*, **3**, 816 (1964).
21. Lust, R., Meyer, F., Trefftz, E., and Biermann, L., , in *The Solar Corona*, edited by J. W. Evans, IAU Symposium n. 16, 1962, p. 21.
22. Burgess, A., *Astrophysical Journal*, **139**, 776 (1964).
23. Burton, W. M., , in *Structure and Development of Solar Active Regions*, edited by K. O. Kiepenheuer, IAU Symposium n. 35, 1968, p. 395.
24. Withbroe, G. L., Dupree, A. K., Goldberg, L., et al., *Solar Physics*, **21**, 272 (1971).
25. Altschuler, M. D., and Perry, R. M., *Solar Physics*, **23**, 410 (1972).
26. Krieger, A. S., Timothy, A. F., and Roelof, E. C., *Solar Physics*, **29**, 505 (1973).
27. Bell, B., and Noci, G., *Bulletin of the American Astronomical Society*, **5**, 269 (1973).
28. Neupert, W. M., and Pizzo, V., *Journal of Geophysical Research*, **79**, 3701 (1974).
29. Tousey, R., *Space Research*, **XIII**, 713 (1973).
30. Zirker, J. B., *Coronal Holes and High Speed Wind Streams*, Colorado Associated University Press, Boulder, 1977.
31. Hewish, A., *Proceedings of the Royal Society A*, **228**, 238 (1955).
32. Vitkevitch, V. V., *Doklady Academii Nauk S. S. S. R.*, **101**, 429 (1955).
33. Dennison, P. A., and Hewish, A., *Nature*, **213**, 343 (1967).
34. Ekers, R. D., and Little, L. T., *Astronomy and Astrophysics*, **10**, 310 (1971).
35. Woo, R., *Astrophysical Journal*, **223**, 704 (1978).
36. James, J. C., , in *Radar Astronomy*, edited by J. V. Evans and T. Hagfors, Mc Graw-Hill, New York, 1968, p. 323.
37. Speer, R. J., Garton, W. R. S., Goldberg, L., et al., *Nature*, **226**, 249 (1970).
38. Gabriel, A. H., Garton, W. R. S., Goldberg, L., et al., *Astrophysical Journal*, **169**, 595 (1971).
39. Kohl, J. L., and Withbroe, G. L., *Astrophysical Journal*, **256**, 263 (1982).
40. Kohl, J. L., Reeves, E. M., and Kirkham, B., , in *New*

Instrumentation for Space Astronomy, edited by G. Vaiana and K. A. van der Hucht, Pergamon, Oxford, 1978, p. 91.

41. Raymond, J. C., Kohl, J. L., Noci, G., et al., *Solar Physics*, **175**, 645 (1997).
42. David, C., Gabriel, A. H., Bely-Dubau, F., et al., *Astronomy and Astrophysics*, **336**, L90 (1998).
43. Sheeley, N. R., Jr., Wang, Y.-M., Hawley, S. H., et al., *Astrophysical Journal*, **484**, 472 (1997).
44. Habbal, S. R., Woo, R., Fineschi, S., et al., *Astrophysical Journal*, **489**, L103 (1997).
45. Kohl, J. L., Noci, G., Antonucci, E., et al., *Solar Physics*, **175**, 613 (1997).
46. Kohl, J. L., Noci, G., Antonucci, E., et al., *Astrophysical Journal*, **501**, L127 (1998).
47. Cushman, G. W., and Rense, W. A., *Astrophysical Journal*, **207**, L61 (1976).
48. Hassler, D. M., Dammasch, I. E., Lemaire, P., et al., *Science*, **283**, 810 (1999).
49. Wang, Y.-M., Sheeley, N. R., Jr., Howard, R. A., et al., *Astrophysical Journal*, **485**, 875 (1997).
50. Noci, G., Kohl, J. L., Antonucci, E., et al., , in *The Corona and Solar Wind near Minimum Activity*, edited by O. Kjeldseth-Moe, ESA SP-404, 1997, p. 75.
51. Meadows, A. J., *Early Solar Physics*, Pergamon, Oxford, 1970.
52. Goldberg, L., , in *The Sun*, edited by G. P. Kuiper, University of Chicago Press, Chicago, 1953, p. 1.
53. Berry, A., *Compendio di Storia dell'Astronomia - Italian translation of D. Gambioli*, Società Editrice Dante Alighieri, Roma, 1907.

In-situ Measurements of the Solar Wind

Marcia Neugebauer

Lunar and Planetary Laboratory, University of Arizona, Tucson, AZ 85721-0092, USA

Abstract. In-situ measurements of the solar wind began with simple ion traps on Soviet spacecraft in 1959. It wasn't, however, until 1962 that the major properties of the solar wind were determined by Mariner 2. Improvements in instrumentation since then include the use of particle detectors, extension of the observations of the solar wind velocity vector to 2 or 3 dimensions, better time and energy resolution, greater variety of measurement locations, multipoint measurements, electron measurements, and the use of mass spectrometers and sample returns. This review summarizes what was learned about the solar wind from each of these improvements in instrumentation.

INTRODUCTION

In-situ measurements of the solar wind were preceded by both remote sensing and theoretical predictions. An excellent review of the status of our knowledge of and guesses about the interplanetary medium at the beginning of the space age was given by Parker [1] at Solar Wind Nine. Attempts to use in-situ measurements to determine whether the interplanetary medium could be described by Parker's supersonic solar wind [2] or by Chamberlain's subsonic solar breeze [3] occupied space physicists from 1959 to 1962. This paper reviews those first attempts and traces the principal avenues of advances in in-situ measurements from 1962 to the present. The final section is a discussion of the strengths and weaknesses of the three methods of studying the solar wind -- in situ measurements, remote sensing, and theory/modeling.

EARLY ATTEMPTS TO MEASURE THE SOLAR WIND

Table 1 summarizes the early attempts to directly measure and characterize what is now known as the solar wind.

Not surprisingly, the first tries were made by the Soviet Union. The Soviets launched ion traps on four interplanetary missions. These were Faraday cups with an inner grid at a voltage of ~200 V to prevent the escape of photoelectrons plus an outer grid at a positive potential to define the minimum energy/charge of the ions that could enter the cup. The total current measured by the Faraday cup was the sum of the ion flux with energy/charge above the voltage on the positive grid, electrons with energies above 200 V which could escape from the cup, and photoelectrons emitted from the negative grid. There were no publishable results from Lunik 1. Lunik 2 was the most successful of the four missions, measuring an intermittent flux of $\sim 2\times 10^{8}$ $cm^{-2}s^{-1}$ with energy/charge > 15 V [4]. The direction of this flux was not determined. The Soviet results were consistent with, but certainly not proof of Parker s solar wind theory. The positive-grid voltages were higher (up to 25 and 50 V) on Lunik 3 and the Venus probe, but only very intermittent fluxes were detected.

TABLE 1. Early attempts to measure the solar wind

Mission	Year	Flux	*V*	Direction	Persistence
Lunik 1	1959				
Lunik 2	1959	√			
Lunik 3	1959	√			
Venus pb	1961	√			
Expl. 10	1961	√	√	√	
Expl. 12	1961				
Ranger 1	1961				
Ranger 2	1961				
Mariner 1	1962				
Mariner 2	1962	√	√	√	√
Expl. 14	1962				

CP679, *Solar Wind Ten: Proceedings of the Tenth International Solar Wind Conference*, edited by M. Velli, R. Bruno, and F. Malara

The first American instrument was a Faraday cup modified by MIT to have a square-wave potential on the positive grid; with such an arrangement the ac current to the cup was a measure of the ion flux, whereas, to first order, photoelectrons provided only a dc current. This instrument was flown on Explorer 10 which, in retrospect, traveled down the flanks of the magnetosheath passing in and out of the magnetosphere where no ion flux was measured. When an ion flux was detected, it came from somewhere within a 60° field of view that included the solar direction. Explorer 10 determined the proton speed, density, and temperature and established that the flow was supersonic [5]. Thus Parker's prediction of a supersonic solar wind was confirmed, but the persistence of the wind was still in doubt.

The next six efforts were made by American groups flying curved-plate analyzers in which the ions were deflected perpendicular to their direction of motion and those ions that exited the tunnel were recorded by an electrometer. The instrument on Explorer 12 had neither the sensitivity nor the correct look direction to detect the solar wind. Rocket failures led to Rangers 1 and 2 never getting above the ionosphere. Mariner 1 was destroyed by Range Safety when it headed for the North Atlantic shipping lanes. The instrument on Explorer 14 was blinded by solar UV whenever it looked within 3° of the Sun so that very little solar-wind data were obtained.

After a hair-raising set of failures, malfunctions, and recoveries [6], Mariner 2 finally made it out into interplanetary space where it measured the solar wind nearly continuously over 3 months as the spacecraft traveled from Earth to Venus and beyond [7]. The Mariner 2 measurements showed that the flow was incident from within 10° of the Sun and it blew continuously. Figure 1 is an example of one of the better spectra obtained by the curved-plate analyzer on Mariner 2. It had measurably large currents in five voltage (energy/charge) channels that defined two spectral peaks — the first due to protons and the second due to alpha particles. Sometimes only two or three channels had above-background currents.

Despite the crudeness of the measurements, Mariner 2 learned a lot about the solar wind. It discovered that the solar wind was organized into a series of approximately weeklong high-speed streams which recurred at the solar rotation rate. The leading edges of the streams were steeper than the trailing edges, the temperature was correlated with the speed, and the density was highest on the leading edges where the fast wind overtook the slower wind in its path. The average density varied with the inverse square of the distance from the Sun. A correlation was found between the speed and the Kp index of geomagnetic activity [8]. Mariner 2 also discovered that the alpha-particle abundance was highly variable. Furthermore, the two-peak spectra were inconsistent with the protons and alphas having the same temperature, but could be fit by assuming that the alphas were four times hotter than the protons. Mariner 2 also carried a magnetometer whose data validated Parker's prediction of a spiral configuration of the interplanetary magnetic field [9]. Mariner 2 data were also used to demonstrate the existence of collisionless MHD shocks [10].

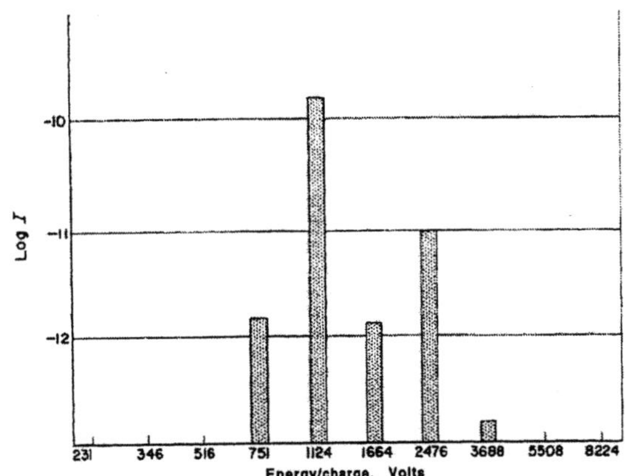

FIGURE 1. One of the better energy/charge spectra obtained by the Mariner 2 curved plate analyzer. The current *I* is given in amperes.

EVOLUTION OF IN-SITU SOLAR WIND MEASUREMENTS

In the four decades since Mariner 2 firmly established the existence of and determined a few of the properties of the solar wind there have been enormous improvements in the instrumentation for in-situ measurements. This section presents a synopsis of some of those improvements and the resulting increase in our knowledge of solar wind physics. The discussion is limited to the measurements of the low-energy charged particles that make up the bulk of the solar wind plasma. Important advances in the capabilities to measure waves and energetic particles are not covered.

One early advance was the use of **particle counters** rather than electrometers as sensors in curved-plate analyzers. A succession of different devices was used to turn the impact of a single ion or electron into a cascade of electrons that could be detected as a pulse. Although ac electrometers are still used with the Faraday cup instruments (e.g., on Voyager and Wind), the detection of individual particles has allowed increased dynamic ranges in ever-smaller instruments.

The extension of the observations into **two and three dimensions** (two angles plus the magnitude of the velocity vector) led to a wealth of discoveries about the solar wind. Systematic angular deflections of the solar wind velocity vector were found, as expected, at the leading edges of high-speed streams where fast plasma pushes aside the ambient slower plasma in its path [11]. The correlation of each component of the vector proton velocity with the corresponding component of the magnetic field led to the discovery of a strong antisolar flux of Alfvén waves, which was especially strong in the high-speed wind [12]. It was quickly discovered that the solar wind was not isotropic, but had different temperatures parallel and perpendicular to the interplanetary magnetic field [13]. Nine proton distribution functions observed by Helios 2 are shown in Figure 2, taken from Marsch et al. [14]. In this figure, speed increases from left to right and the radial distance from the Sun decreases from top to bottom. Dashed lines indicate the projection of the interplanetary magnetic field. This figure illustrates the increase of the temperature, of the anisotropy ($T_{perpendicular}/T_{parallel}$) of the core of the distributions, and of a high-energy tail or a secondary peak with increasing speed and decreasing solar distance. On the basis of early measurements of the anisotropy, Bame et al. [15] interpreted the high perpendicular temperature in the fast solar wind as evidence for interplanetary heating. Later, Tu [16] explained the radial increase of the first adiabatic invariant (proportional to $T_{perpendicular}/B$) measured by Helios 2 on the basis of the turbulent cascade of the energy in the Alfvén waves to higher frequencies.

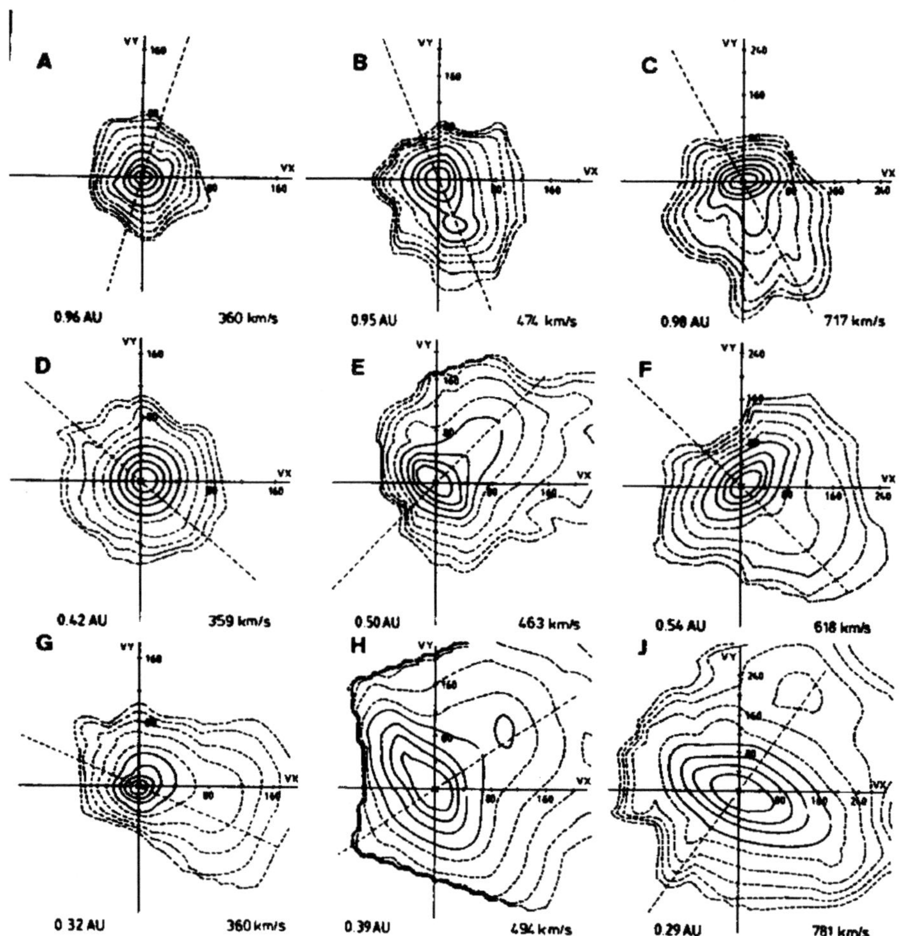

FIGURE 2. Some contours of proton velocity distributions observed by Helios 2. Contour levels correspond to 0.8, 0.6, 0.4, 0.2, 0.1, 0.03, 0.01, 0.003, and 0.001 of the peak phase space density. The dashed line in each frame indicates the projected direction of the interplanetary magnetic field. From [14].

As spacecraft communications capabilities increased, it became possible to improve the **time resolution** of observations of changes in the solar wind. Although some spacecraft have obtained plasma spectra at a cadence as fast as 3 seconds, even the fastest plasma measurements are at least an order of magnitude slower than the measurements of the magnetic field. For example, only under highly unusual circumstances has it been possible to observe the particle population at the frequencies required to study convected structures as small as the ion gyroradius.

The decades since 1962 have also provided opportunities for measuring the solar wind at a variety of **measurement locations**. The early eccentric Earth orbiters with apogees in the solar wind for part of each year have been followed by spacecraft (ISEE 3, ACE, SOHO, and Genesis) that remain continuously in the solar wind by virtue of their halo orbits about the Lagrangian libration point L1. With the Helios missions it was possible to study the solar wind as close as 0.3 AU to the Sun. The outer heliosphere has been sampled by solar-wind instrumentation on a series of planetary missions, most notably Pioneers 10 and 11 and Voyagers 1 and 2. Most recently, Ulysses has obtained the first data from the polar regions of the heliosphere. The community has long advocated and is still working towards extending these observations by going even closer to the Sun than Helios (to 4 solar radii with a solar probe) and much farther from the Sun with an interstellar probe well instrumented to determine the properties of the interstellar medium at a distance of ~200 AU.

Some use has been made of **multipoint** measurements, whereby solar wind structures such as shocks or discontinuities can be studied with an array of spacecraft. In this way it should be possible to distinguish between propagating and convected features. NASA s planned Stereo mission is another step forward in this regard, but arrays of at least four spacecraft would be even better.

The study of **electrons** has also been crucial to our understanding many fascinating aspects of the solar wind. Among the earliest discoveries was that protons and electrons do not have the same temperature; the electron temperature is relatively insensitive to solar wind speed, while the proton temperature is greater/less than the electron temperature in the fast/slow wind [17]. These results established a clear need for multi-fluid models of the solar wind. Although electrons do carry a heat flux away from the Sun, the flux is lower than that originally envisioned [17]. Further investigation revealed several populations of electrons, in addition to the ever-present photoelectrons and secondary electrons emitted from the sunward side of the spacecraft. There is a thermal core population and a hotter halo population which carries most of the heat flux [18]. Especially in the high-speed solar wind, there is a sharp tail, or strahl, of energetic electrons which is strongly focussed along the magnetic field [19]; the strahl electrons are thought to be a collisionless population which has been focussed by the conservation of the magnetic moment. Sometimes there appears to be sort of a double strahl, with electrons streaming in both directions along the field. This occasional counterstreaming of suprathermal electrons has become a popular tool for the identification of transient solar wind from coronal mass ejections [20], but the counterstreaming can also be caused by magnetic connection to shocks or other field-strength increases farther out in the solar system [21].

An improvement in **energy resolution** accompanied the other advances. Contrast, for example, the energy spectrum in Figure 3 with the 5-channel spectrum in Figure 1. At the time the data in Figure 3 were taken the solar wind was sufficiently cold that individual peaks of highly charged heavy ions could be discerned above the instrumental background. Such measurements provided the first estimates of the plasma's ionization state, which is related to the temperature of the solar corona.

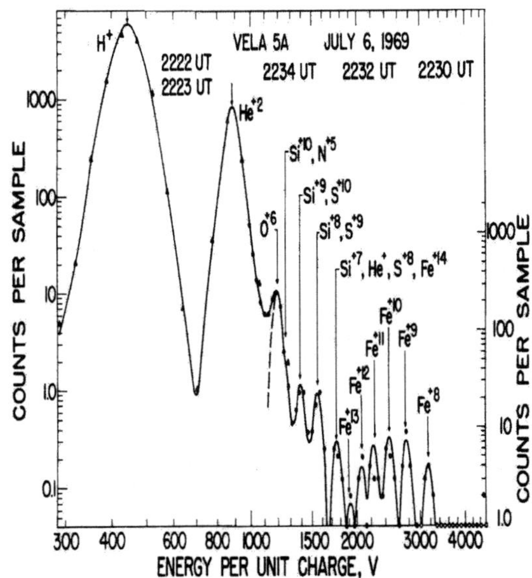

FIGURE 3. An energy/charge spectrum obtained in the slow solar wind by the Vela 5A spacecraft. Arrows indicate the expected positions of some iron and other ions. From [31].

The better energy resolution also led to the discoveries that the alpha particles flowed away from the Sun faster than the protons and that both the proton and alpha particles are often double peaked, as shown in Figure 2.

A problem with ion spectra such as that shown in Figure 3 is that the solar wind is often too hot for the heavy ion peaks to be resolved; at times they can be completely swamped by the alpha particles. Thus the motivation for ion **mass spectrometers**. The first step in that direction was the use of a Wien (velocity) filter following a curved-plate analyzer (e.g., [22]). Later improvements, exemplified by the SWICS instruments on Ulysses and other spacecraft [23], included sequential electrostatic analysis (yielding $mv^2/2q$), time of flight analysis (v), and measurement of total energy ($mv^2/2$) of an ion in a solid state detector. The requirement for triple coincidences (between the start and stop pulses in the time-of-flight section and the total-energy pulse) resulted in a very low background. Figure 4 shows plots of the velocity distribution functions, normalized by the solar wind speed, for three ion species. Besides the solar wind peak at a relative speed of 1, there is a steep drop in interstellar pickup ions at a relative speed of 2, and spectral tails extending to even higher energies; the spectra for the different ion species could not have been separated without a mass spectrometer. This type of instrument has also been used to study the kinetics of some of the heavy ion species, comparing their speeds, temperatures, and anisotropies to those of protons and alphas [24].

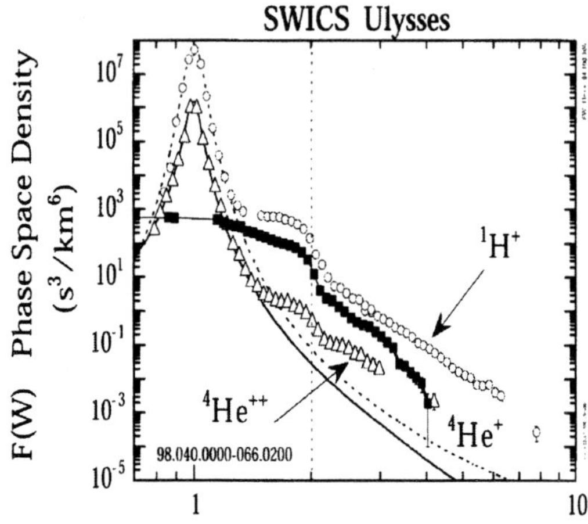

FIGURE 4. Phase space density versus speed relative to the solar wind speed for several species observed by the SWICS instrument on Ulysses. From [26], ©1999, with kind permission of Kluwer Academic Publishers.

Newer spectrometers with improved mass resolution operate on the principle that time of flight in a harmonic oscillator potential (potential varying with the square of the distance) depends only on the mass/charge of an ion and is independent of its velocity [25].

For the purposes of planetary science, even greater precision in the elemental and isotopic abundances of the Sun is required. For example, planetary scientists want to know the solar abundance of the ^{17}O isotope, which has yet to be measured by spacecraft mass spectrometers. Such high precision can only be obtained by **sample return** missions. The first solar wind sample return missions were carried out as part of the Apollo program. Samples collected on the lunar surface returned to Earth by the Apollo astronauts led to determination of the isotopic abundances of helium, neon and argon [27]. The Genesis mission [28], launched in August, 2001, is collecting solar wind samples to be returned to Earth in 2004. Because the Genesis collection time is about 100 times longer than the Apollo collection times and the sample-collecting materials are much purer and more varied, it is expected that Genesis will be able to determine the abundances of most of the elements in the periodic table. The systematics of ion fractionation between the Sun and the solar wind as determined by spacecraft mass spectrometers must be used in interpreting the data from the Genesis samples in terms of solar-system history.

DISCUSSION

In-situ measurements of the solar wind have clearly led to some wondrous discoveries and insights that are totally beyond the realm of what s feasible with remote sensing. By its very nature, remote sensing is limited to the study of features visible in some useful wavelength with an appropriate optical depth; the observations are integrated along the line of sight and usually have limited angular or spatial resolution. Theory and modeling of the solar wind is necessarily oversimplified.

In-situ observations, however, also have their drawbacks or shortcomings. They are limited to the study of time profiles of the properties of the convected plasma at one, or at most a few points. As with remote sensing, there are often problems with the absolute calibration of the instruments (see, for example, Russell et al. [29]). Estimates of the uncertainty in measured plasma density are often ~30%. Determinations of the directions of the ion beams are often off by one or a few degrees. The greatest problem is perhaps with the determination of the plasma temperature. Temperature is usually defined by the second moment of the plasma distribution, which means that the part of the distribution farthest from the peak and closest to the noise level is weighted the most heavily. The quoted value of temperature is highly dependent on how far into the noise, or into the high-energy tail, the summation is taken. The proton temperatures obtained by Genesis were systematically more than 20% greater than the proton temperatures obtained by the nearby ACE spacecraft. The difference is that the autonomous determination by Genesis summed counts over the entire field of view, whereas the ACE calculations were limited to the vicinity of the spectral peak. Accurate calculation of plasma temperature may also involve complex separation or deconvolution of the plasma thermal motions from instrumental smear or other contributions (see, for example, the appendix of Neugebauer et al. [30]). Whereas the data distributed by the US National Space Science Data Center are extremely valuable for a wide range of research projects, their absolute values must be viewed with some skepticism.

I think what is important to keep in mind at the start of this conference is how valuable the interplay can be between in-situ measurements, remote sensing, and theory. One example of the benefits of combining the three methods of studying the solar wind is the mapping of solar wind measured in-situ to its solar sources. Such mapping typically uses theoretical models (either MHD or potential) of the coronal magnetic field which uses remotely sensed magnetograph data as an inner boundary condition. While we will continue to see great advances in each area, the real payoff probably comes from the synergy between the three scientific approaches.

REFERENCES

1. Parker, E.N., "Space physics before the space age," in *Solar Wind Nine, AIP Conference Proceedings 471*, Edited by S.R. Habbal, *et al.*, Amer. Inst. Physics, Woodbury, NY, 1999, p. 3.
2. Parker, E.N., *Astrophys. J.* **128**, 664 (1958).
3. Chamberlain, J.W., *Astrophys. J.* **131**, 47 (1960).
4. Gringauz, K.I., *et al., Soviet Phys. : Doklady* **5**, 361 (1960).
5. Bonetti, A., *et al., J. Geophys. Res.* **68**, 4017 (1963).
6. Wheelock, H.J., Editor. *Mariner: Mission to Venus*, McGraw-Hill, publisher, New York, 1963.
7. Neugebauer, M. and Snyder C.W.,, *J. Geophys. Res* **71**, 4469 (1966).
8. Snyder, C.W., Neugebauer, M., and Rao, U.R., *J. Geophys. Res.* **68**, 6361 (1963).
9. Smith, E.J., "Interplanetary magnetic fields," in *Space Physics*, Edited by D.P. LeGalley and A. Rosen, J. Wiley & Sons, New York, 1964, pp.. 350-396..
10. Sonett, C.P., *et al., Phys. Rev. Lett.* **13**, 153-156 (1964).
11. Lyon, E., *et al., Space Research* **8**, 99 (1968).
12. Belcher, J.W. and Davis, L. Jr., *J. Geophys. Res.* **76**, 3534-3563 (1971).
13. Hundhausen, A.J., Bame, S.J., and Ness, N.F., *J. Geophys. Res.* **72**, 5265 (1967).
14. Marsch, E., *et al., J. Geophys. Res.* **87**, 52 (1982).
15. Bame, S.J., *et al., Geophys. Res. Lett.* **2**, 373 (1975).
16. Tu, C.-Y., *J. Geophys. Res.* **93**, 7-20 (1988).
17. Montgomery, M.D., "Average thermal characteristics of solar wind electrons, NASA SP-308," in *Solar Wind*, Edited by C.P. Sonett, P.J. Coleman, Jr., and J.M. Wilcox, National Aeronautics and Space Administration, Washington, DC, 1972, p. 208.
18. Feldman, W.C., *et al., J. Geophys. Res.* **80**, 4181 (1975).
19. Rosenbauer, H., *et al., J. Geophys.* **42**, 561 (1977).
20. Gosling, J.T., *et al., J. Geophys. Res.* **92**, 8519 (1987).
21. Gosling, J.T., Skoug, R.M., and Feldman W. C.,, *Geophys. Res. Lett.* **28**, 4155 (2001).
22. Coplan, M.A., *et al., IEEE Trans. Geosci. Electron.* **GE-16**, 185 (1978).
23. Gloeckler, G.L., *et al., Astron. Astrophys. Suppl.* **92**, 267 (1992).
24. von Steiger, R., *et al., Space Sci. Rev.* **72**, 71 (1995).
25. Hovestadt, D. *et al., Solar Phys.* **162**, 441 (1995).
26. Gloeckler, G., *Space Sci. Rev.* **89**, 91 (1999).
27. Geiss, J., *et al., J. Geophys. Res.* **75**, 5972 (1970).
28. Burnett, D.S., *et al., Space Sci. Rev.* **in press**, (2002).
29. Russell, C.T. and Petrinec, S.M., *Geophys. Res. Lett.* **19**, 961-963 (1992).
30. Neugebauer, M., *et al., J. Geophys. Res.* **106**, 5635 (2001).
31. Bame, S.J., "Spacecraft observations of the solar wind composition," in *Solar Wind*, Edited by C.P.Sonett, P.J. Coleman Jr., and J.M. Wilcox, NASA, Washington, DC, 1972, p. 535.

Origin of the Fast Solar Wind: From an Electron - Driven Wind to Cyclotron Resonances

Joseph V. Hollweg

Space Science Center, University of New Hampshire, Durham, NH 03824 USA

Abstract. Even before the discovery of the fast solar wind in the mid - 1970s, it was known that even the average solar wind could not be well explained by models in which electron heat conduction was the energy source and the electron pressure gradient was the principal accelerating force. The outward - propagating Alfvén waves discovered around 1970 were thought for a while to provide the sought - after additional energy and momentum, but their wave pressure ultimately failed to explain the rapid acceleration of the fast wind close to the Sun in coronal holes. By the late 1970s, various in situ data were suggesting that protons and heavy ions were being heated and accelerated by the ion - cyclotron resonance far from the Sun. This notion was soon applied to the acceleration region in coronal holes close to the Sun. The models which resulted suggested that the fast wind could be driven mainly by the proton pressure gradient (which is mainly the mirror force if the anisotropy is large), and that the high temperatures and flow speeds of heavy ions could originate within a few solar radii of the coronal base; these models also emphasized the importance of treating the extended coronal heating and solar wind acceleration on an equal footing. By the mid 1990s, SOHO, especially the UVCS (Ultraviolet Coronagraph Spectrometer), provided remarkable data which have given great impetus to studies of the ion cyclotron resonance as the principal mechanism for heating the plasma in coronal holes, and ultimately driving the fast wind. We will discuss the basic ideas behind current research, emphasizing the particle kinetics. We will discuss remaining problems such as the source of the ion - cyclotron resonant waves (direct launching, turbulence, microinstabilities), problems concerning OVI and MgX, the roles of inward - propagating waves and instabilities, the importance of oblique propagation, and the electron heating. Some alternatives, such as shock heating and turbulence - driven magnetic reconnection, will also be reviewed.

ELECTRON - DRIVEN WINDS

Parker's [1] original model of the solar wind was motivated by the observation that electrons conduct heat so effectively that a static corona would have a pressure far from the Sun greatly in excess of the interstellar pressure. Such a corona could not be contained, and thus could not be static. In effect, these early solar wind models were driven mainly by the electron pressure gradient, which was maintained at a high level by the slow decline of the electron temperature, which was itself a consequence of the efficient electron heat conduction out of the hot corona at the base. However, in his 1965 review [2], Parker recognized that such models were not fully successful in accounting for the observed properties of the wind. He stated "The model for the hypothetical conduction corona leads to a temperature falling too rapidly with radial distance from the Sun, indicating that the actual solar corona is probably actively heated for some considerable distance by the dissipation of waves." It now appears he may well have been right. Not only is the electron temperature in coronal holes low, $< 10^6$ K [3], but available evidence points to the action of waves in coronal holes, and in the distant solar wind as well.

Parker's point was driven home in 1966 and 1968 by Hartle and Sturrock [4, 5], who produced the first two - fluid models of the wind, with separate energy equations for the electrons and protons. As in Parker's models, all coronal heating was assumed to take place in a thin layer near the base. This produced a high base temperature, $> 10^6$K, which served as an inner boundary condition. Because protons are poor conductors of heat, their temperature dropped very rapidly to only a few x 10^3 K at 1 AU, in contrast to the observed proton temperatures, T_p, of about 2 x 10^5K in high - speed streams (see Feldman's 1976 article [6] describing the properties of the high - speed wind). The flow speeds predicted by Hartle and Sturrock were much too slow: some 250 km s^{-1} compared to observed values of 700 - 800 km s^{-1} in the fast wind. The authors concluded "Departures [are] attributed to heating by a flux of non - thermal energy".

WAVE - DRIVEN WINDS

In 1969 and 1971, Belcher and co - workers [7, 8]

CP679, *Solar Wind Ten: Proceedings of the Tenth International Solar Wind Conference*,
edited by M. Velli, R. Bruno, and F. Malara

provided strikingly clear evidence that Alfvén waves are ubiquitous in the solar wind, especially in the high - speed streams (see also [9, 10] for earlier indications). The waves were found to propagate predominantly outward from the Sun, suggesting a solar source. Were these the waves mentioned by Parker? Were they Hartle and Sturrock's "flux of non - thermal energy"?

In 1971 Alazraki and Couturier [11] and Belcher [12] realized that the Alfvén waves could contribute to the acceleration of the wind via the wave pressure gradient (also called the ponderomotive force), $-\nabla<\delta \mathbf{B}^2>/8\pi$ (in cgs units which we use throughout), where $\delta \mathbf{B}$ is the magnetic fluctuation, and the angle brackets denote a time (or ensemble) average. In 1973 Hollweg [13] showed how the waves could be incorporated into the energy balance, allowing for wave damping, and he produced some detailed two - fluid models which looked promising when compared with the observed solar wind near 1 AU. Such "wave - driven" solar wind models were soon explored further by many others (see [14] for a review).

These models were incomplete in one major respect: They had to guess how the Alfvén waves might be damped, since they do not readily undergo Landau or transit - time damping [15, 16]. At first it was simply postulated that the waves damp by saturating at some maximum level of $<\delta \mathbf{B}^2>/B_o^2$ ($\mathbf{B}_o$ is the ambient magnetic field); much later Roberts [17] provided evidence that this does not occur. Wave damping via a turbulent cascade was proposed in the early 1980s [18, 19], and that idea has been predominant ever since; more on this later.

Wave - driven models were generally very successful in reproducing observed properties of the wind far from the Sun. But by the early 1990's it was realized that they had difficulties close to the Sun, where observations were indicating that the wind accelerates faster than the wave - driven models could explain. Radio observations (e.g. [20]) suggested rapid acceleration, but they might have been measuring propagating disturbances rather than the solar wind flow (that is now the prevailing view). At about the same time, improved observations of densities in polar coronal holes became available (e.g. [21]). With knowledge of the mass flux (obtained from Ulysses), the coronal velocities could be deduced. They were indeed found to be very fast close to the Sun, implying a sonic critical point inside $3r_s$ (r_s is the solar radius). Wave - driven models could not explain such rapid acceleration, because the ponderomotive force only became significant at greater distances, beyond $3r_s$ [14].

The heavy ions, especially He^{++}, presented further difficulties. They tend to flow faster than the protons by about the Alfvén speed, v_A, and they tend to be roughly mass - proportionally hotter than the protons. The most likely explanation seemed to be heating and acceleration by the cyclotron resonance; see [22 - 27] for some early discussions.

There were already available other hints that cyclotron resonances were at work. Here we refer to the excellent review by Marsch [28]. His Figure 8.31 shows that the proton magnetic moment increases with heliocentric distance r between 0.3 and 1 AU, implying heating perpendicular to $\mathbf{B}_o$; figure 8.25 shows the same for He^{++}. Moreover, Marsch's Figure 8.1 shows proton distribution functions whose "cores" are anisotropic in the sense $T_\perp > T_\parallel$ (T is temperature, and the subscripts indicate directions relative to $\mathbf{B}_o$). Perpendicular heating is again implied, with the cyclotron resonance being the most apparent candidate.

Three - fluid (electrons, protons, He^{++}) models which incorporated the cyclotron resonance succeeded qualitatively but failed in detail [25, 29]. An essential difficulty was the fact that once the ions flow significantly faster than the protons, they tend to drop out of resonance with no further acceleration and heating. In retrospect, these models may also have failed because the cyclotron resonances were assumed to act only far from the Sun.

PROTON - DRIVEN WINDS

By the mid - 80's, some of these ideas were applied to the acceleration region of the solar wind by Hollweg [30] and Hollweg and Johnson [31]. The basic idea was that the Sun launches a flux of low - frequency Alfvén waves which damp via a turbulent cascade to high frequencies where the energy is absorbed by the protons (He^{++} was omitted). The volumetric heating Q was essentially Kolmogorov: $Q = \rho<\delta \mathbf{V}^2>/L_c$, where ρ is density, $\delta \mathbf{V}$ is the wave velocity fluctuation, and L_c is the correlation length transverse to $\mathbf{B}_o$. The wave propagation was handled in the WKB limit. This means that there was an internal inconsistency, since the nonlinear terms cancel for WKB Alfvén waves, and no turbulence can occur (for correct treatments see Dmitruk and Matthaeus, this volume, and [32, 33]. Nonetheless, the model could give the required rapid acceleration close to the Sun. A noteworthy feature of the model was that the protons were substantially hotter than the electrons, with $T_p \approx 4 \times 10^6$K at (3 - 4)$r_s$. It was mainly the proton pressure gradient which led to the rapid acceleration; we therefore call this model "proton - driven". The then available data argued against such high proton temperatures and the model was rejected.

Isenberg [34] extended the model to include He^{++}. Quasilinear theory was used to apportion the cascaded energy between the protons and He^{++}, but it was necessary to assume the form of the power spectrum in the resonant range. The model produced many features of the acceleration region which we now know to be at least qualitatively in accord with the UVCS/SOHO observations of coronal holes (e.g. [35, 36]): protons hotter than electrons, He^{++} (and other ions) more than mass - proportionally hotter than the protons, rapid acceleration driven mainly by the proton pressure gradient, He^{++} already flowing faster than the protons close to the Sun. Again, this model was rejected because of the high T_p, but it anticipated the UVCS/SOHO discoveries of hot and fast coronal protons and ions.

UVCS/SOHO also indicated that 0^{+5} is heated primarily perpendicular to $\mathbf{B}_0$ [37], again implicating cyclotron heating. Moreover, the UVCS/SOHO results [35, 36] show that the ion heating extends far into the corona, even beyond the sonic critical point, in strong contrast to the early models in which all the heating was in a thin shell at the base: coronal heating and solar wind acceleration must be treated together.

Before closing this section, we note that solar wind models with hot protons and ions were investigated years ago by Ryan and Axford [38]. Esser, Habbal, and co - workers [39 - 41] also explored the consequences of hot coronal protons, and showed explicitly that high - speed streams could be produced by the coronal proton pressure gradient, with no additional momentum input from the wave pressure gradient. All of these papers used ad hoc heating functions, however, rather than attempting to model the effects of a specific physical heating (and acceleration) mechanism.

THE CYCLOTRON RESONANCE

Consider a wave with angular frequency ω and wavenumber $k_{||}$ propagating parallel to $\mathbf{B}_0$. A particle with gyrofrequency Ω will see a constant electric field when

$$\omega - k_{||} v_{||} = \pm \Omega; \qquad (1)$$

$v_{||}$ is the particle's velocity along $\mathbf{B}_0$, the (+) refers to waves which are circularly polarized in the sense of the particle's gyration, while the (–) refers to the opposite sense of polarization. Space limits us to consideration only of ion - cyclotron waves, i.e. the (+).

When (1) is satisfied, the particle gains or loses energy secularly, depending on the phase relative to the electric field. In a random field, phase will be random, and the particle will random walk in velocity space. Random walks imply diffusion so a collection of particles will diffuse in velocity space.

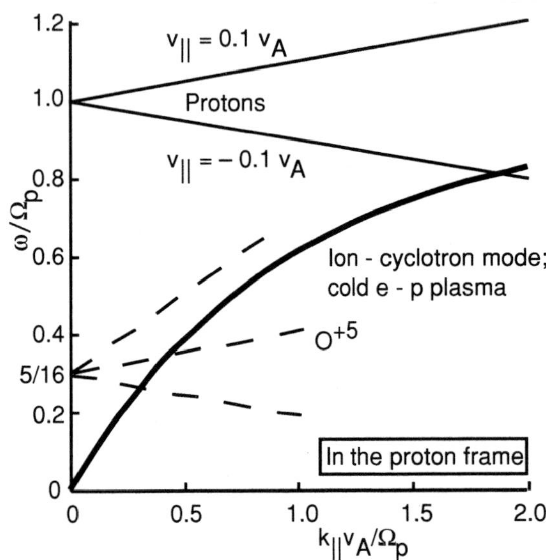

FIGURE 1. The heavy curve is the dispersion relation. The thin lines represent the resonance condition (1). The solid thin lines show that only backward moving protons can resonate. The dashed lines show that both forward and backward moving oxygen can resonate, but that no resonance is possible if the oxygen moves too fast.

The heavy curve in Figure 1 is the dispersion relation for parallel - propagating ion - cyclotron waves in a cold electron - proton plasma, as viewed in the bulk proton frame. The thin solid lines are equation (1) for the protons, for two values of $v_{||}$. The intersection of (1) with the dispersion relation gives ω and $k_{||}$ of the resonant wave. If $v_{||} > 0$, there is no intersection; a proton moving with the wave cannot be in resonance. For $v_{||} < 0$ there can be resonance. Roughly speaking, only half of the proton distribution can resonate with ion - cyclotron waves. If the waves are propagating away from the Sun, only sunward moving protons (as seen in the bulk proton frame) can resonate.

The situation is more favorable for heavy ions, in several respects. The dashed lines show equation (1) for 0^{+5} ($\Omega = 5\Omega_p/16$). The lower two dashed lines show that both forward and backward moving 0^{+5} can be in resonance. However, the upper dashed line shows that 0^{+5} can drop out of resonance if it moves too fast ($v_{||} \gtrsim 0.3\ v_A$). Note too that heavy ions can resonate with waves having lower $k_{||}$ than found for the protons. This too gives ions an advantage, since we expect more wave power at lower $k_{||}$, and because lower - $k_{||}$ waves have higher phase speeds. (He^{++} is abundant enough to

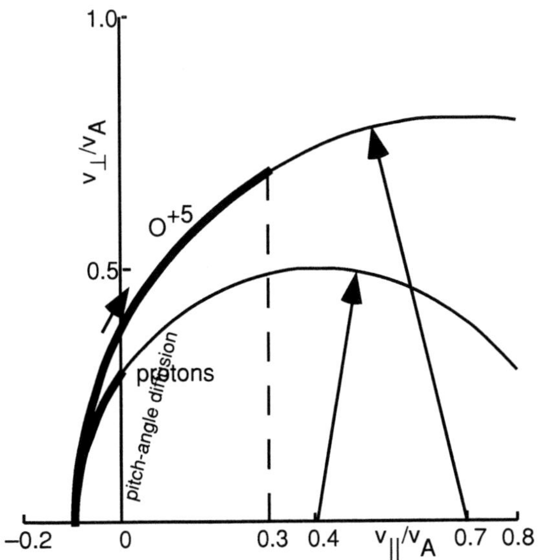

FIGURE 2. Particle motion in $v_{||}$-$v_\perp$ space due to resonance.

affect the dispersion relation significantly, but there is insufficient room for a discussion. See recent reviews [42, 43] as well as [44 - 47]).

Figure 2 shows the ion advantage in a different way. If there is a single wave (or if we ignore dispersion), a particle will conserve energy in the wave frame. In that frame it will move on a circle such that $v_\perp^2 + v_{||}^2 =$ constant. The figure shows two such circles. The O^{+5} circle has a greater radius than the proton circle, qualitatively reflecting the fact that O^{+5} resonates with faster - moving waves than the protons. The dark arcs show the portions of the circles within which the particles can be in resonance. If we imagine a group of particles starting with small values of $v_\perp$, they will diffuse upwards, tending to uniformly fill the dark arcs. This will give perpendicular heating. Note that the ions can attain larger $v_\perp$ than the protons, implying more than mass proportional perpendicular temperatures, in qualitative agreement with the UVCS/SOHO results for O^{+5}. Note too that as the particles diffuse upwards, they also move to the right, increasing $v_{||}$. Resonant heating implies resonant acceleration; simple models [42, 48] suggest that the resonant acceleration cannot be overlooked, even though the magnetic mirror acceleration is usually substantially larger. Finally, note that the mirror acceleration is $v_\perp^2$ dlogB_0/dr; if the ions attain larger values of $v_\perp$, they will experience a greater acceleration from mirroring. (See Isenberg, this volume, for further variations of Figure 2.)

We point out two difficulties with this picture. The first concerns O^{+5}, which is observed to have $T_\perp$ increasing with r out to $3.5r_s$ (the outer limit of the data in [35], while the protons have $T_\perp$ nearly constant. This O^{+5} behavior is difficult to understand. Once oxygen's $v_\perp$ exceeds that of the protons, it experiences a greater mirror acceleration and a greater adiabatic cooling. Moreover, as the oxygen accelerates, it resonates with higher $k_{||}$ waves, which have less power and slower phase speeds. The oxygen can even drop out of resonance when its speed becomes fast enough (Figure 1). These factors work against the fairly steady increase of the oxygen temperature with r. In fact, models such as in [42, 48] behave in the way described above: the temperature at first increases rapidly, mirroring then gives rapid acceleration, and the temperature then drops as adiabatic cooling overwhelms resonant heating.

Isenberg ([49] and this volume) suggests a possible solution involving both outward and sunward - propagating waves. By redrawing Figure 2 with circles corresponding to both propagation directions, it can readily be seen that a proton can resonate with either propagation direction, but not both simultaneously. An oxygen ion can simultaneously resonate with both types of waves. Thus oxygen can undergo second - order Fermi acceleration, while protons cannot. This may be the extra ingredient needed to explain the O^{+5} data.

Mg^{+9} also presents a problem. Even though O^{+5} and Mg^{+9} have very nearly the same charge - to - mass ratios, q/m, the former is definitely more than mass - proportionally hotter than the protons, while the latter is only mass - proportionally hotter [50]. Why species with nearly the same q/m behave so differently is a challenge to the cyclotron heating scenario. Cranmer [43] notes that the second most abundant coronal ion (after He^{++}) is O^{+6}, which has the same q/m, and thus the same Ω, as Mg^{+9}. He suggests that O^{+6} depletes wave power in the vicinity of the O^{+6} and Mg^{+9} resonance, leaving little power for Mg^{+9}, without affecting the power available to O^{+5}. This idea needs detailed evaluation. We note, however, that it will fail if O^{+6} and Mg^{+9} have different $v_{||}$'s; see equation (1).

WHENCE THE CYCLOTRON WAVES?

Although many workers agree that ion - cyclotron waves can explain a variety of observations, there is substantial disagreement on where these high - ω, high - $k_{||}$ waves come from. There are two principal theories.

The models in references [19, 30, 31, 34, 42, 48, 51 - 54], among others, assume that the Sun launches low frequency waves, and that the high - frequency cyclotron waves are produced by a turbulent cascade. This is motivated by in situ observations (e.g. [55]) showing most power at low frequencies, with power law power spectra at higher frequencies, suggestive of turbulent cascades. Radio science data (e.g. [56]) show that density fluctuations in the corona, as close as $r \approx 5r_s$, have similar power spectra.

One question that arises in this context is whether there is sufficient low - frequency power to drive the high - speed wind. Hollweg et al. [57] studied Faraday rotation fluctuations on the signal from Helios. These were caused by coronal magnetic fluctuations, which were associated with Alfvén waves. They concluded that there was enough power to drive the wind. Andreev et al. [58] obtained similar results. However, other observers disagree. For example, Mancuso and Spangler [59] look at Faraday rotation fluctuations on natural radio sources. They conclude that the power in the corona is insufficient to drive the wind. The differences seem to arise because the observed fluctuations depend on the sizes of the scattering elements, and how many there are along the line of sight. This requires knowledge of the correlation scale, which has to be guessed; different authors have made different guesses. It is also fair to say that no studies have properly accounted for dissipation, in extrapolating from the observation region back to the Sun.

Turbulence faces another problem. MHD turbulence tends to produce high $k_\perp$, not high $k_{||}$ (see papers by Dmitruk and Matthaeus, and by Milano et al., this volume). The cascaded energy then dissipates via reconnection, not via cyclotron resonance. If this is what is happening, we have to start from scratch and ask whether reconnection can explain the ion heating seen by UVCS/SOHO. But before we change the paradigm, be aware that in situ data do show wave power at high $k_{||}$, even though there is a preference for high $k_\perp$ (e.g. [60]).

In this context we inform the reader of a neglected paper by Ulrich [61]. He found evidence for upward - propagating Alfvén waves in the chromosphere, with periods in a broad range around five minutes. The upward Poynting flux was about 3 x 10^7 erg cm^{-2} s^{-1}. This is far in excess of the energy requirements of the fast wind (few times 10^5 erg cm^{-2} s^{-1} at the coronal base), but it is comparable to the energy requirements of active regions, which is where the observations were made. Perhaps the fast wind is supplied by five minute waves, but that would still require a turbulent cascade.

An entirely different approach postulates that small reconnection events on the Sun launch the required energy in the kilohertz range [62 - 65]. The waves would be below the cyclotron resonance where they are launched, but would be resonant at greater heights where Ω is less. This is often called the "sweeping mechanism", since the resonant frequency sweeps from high values to lower values as the waves get further from the Sun.

Tu and Marsch [64, 65] presented some detailed calculations based on this idea. They postulated a wave power spectrum which was adjusted to deposit the required energies at various heliocentric distances. They assumed parallel - propagating waves, which is of course not realistic in a structured solar atmosphere. Hollweg [66] asked what would be different if that assumption were relaxed. He noted that obliquely propagating waves would be compressive, and used the model's magnetic power spectrum to estimate the spectrum of density fluctuations in the corona. The estimated spectrum was found to be much larger than density spectra inferred from radio scattering (IPS). This probably means that magnetic spectra such as those used in the sweeping models are not there. However, the oblique waves damp, via Landau and transit - time damping on the electrons, and this might explain why the large density fluctuations are not observed. This needs to be worked out in detail, but our guess is that if the Sun launches high - frequency waves propagating at many angles to $\mathbf{B}_o$, and if the oblique waves damp, then too much electron heating would result, and the electrons would carry more heat flux than observed.

Are there alternatives? Recently there has been a flurry of activity looking at whether instabilities can drive the high - frequency waves. These instabilities are generally highly oblique to $\mathbf{B}_o$, some may be electrostatic, and some may propagate sunward. The driving mechanisms looked at include cross - field currents [67, 68], large but intermittent electron heat flux ([69], and Markovskii and Hollweg, this volume), and proton beams launched by reconnection events [70].

This review would not be complete without mentioning a totally different view. Lee and Wu [71] showed how obliquely - propagating fast shocks can preferentially heat heavy ions such as 0^{+5}. The trouble here is that the shocks need to be strong, with velocity jumps greater than about 0.3 v_A. These velocity jumps should show up as line broadening in the UVCS/SOHO data. In coronal holes v_A can be very large, and the shock - related line broadening could be comparable to or even larger than the observed broadening. Much more detailed modeling needs to be done to assess whether this mechanism can explain the line widths observed by UVCS/SOHO.

CONCLUSIONS

In the years since the discovery of the solar wind, we have progressed from an electron - driven wind, through a wave - driven wind, to a proton - driven wind which is accelerated mainly by the proton pressure gradient (which is equivalent to the mirror force if $T_{p\perp} >> T_{p\|}$). We now know that the coronal heating is not confined to a thin layer at the coronal base; the heating extends throughout the acceleration region, implying that coronal heating and wind acceleration need to be treated on an equal footing. And we now realize that, in coronal holes at least, the heating works mainly on the protons and ions. That the heating is mainly transverse to $\mathbf{B}_0$ has been verified for one ion, 0^{+5}; it seems likely that this will turn out to be a general result. The heating is not Ohmic dissipation, in contrast to what many workers believe to be the case elsewhere in the solar atmosphere. Will it turn out that other regions are heated in the same way as coronal holes?

Even if one accepts the cyclotron heating paradigm, there is still much to be done. What is the source of the waves: turbulent cascade, direct launching, or local instabilities? How important is oblique propagation, which has received very little attention [72]? What is the mix of outward and sunward - propagating waves? Are sunward - propagating waves produced mainly by instabilities, or by reflections? How important is second - order Fermi acceleration? Perhaps the most important area for future development, and probably the most difficult, is full kinetic solutions. This aspect is still in its infancy ([49, 73 - 76] and Isenberg, this volume). Perhaps one of the most difficult features to incorporate will be the self - consistent evolutions of the particle distribution functions and the waves, including instabilities; intuition based on bi - Maxwellians is misleading (Isenberg, this volume).

We again warn the reader that some models based on MHD turbulent cascades ultimately dissipate their energy via reconnection. If true, this would necessitate a complete rethinking of the coronal heating and solar wind acceleration problem.

Finally, we have to say something about the electrons. Their pressure gradient may contribute less to accelerating the wind than that of the protons, but it is not negligible either. They determine the charge states which emerge out of the corona, they are responsible for the electromagnetic radiation emerging from the corona and transition region, and their downward heat conduction has much to do with how the corona joins with the lower layers. For these reasons it is obviously important to get a grasp on the electron energy equation, which still proves elusive: we still do not have a good prescription for the electron heat conduction. The electron energy equation is important for another reason as well. If we knew the level of electron heating we would have one further constraint on the physical processes going on in the corona; recall our discussion of the sweeping models of Tu and Marsch. At present the electrons can not be used in this fashion because their energy equation is not understood.

ACKNOWLEDGEMENTS

The author acknowledges valuable conversations with P. A. Isenberg, M. A. Lee, S. A. Markovskii, and S. R. Cranmer. This work was supported by the NASA Sun - Earth Connections Theory Program under grants NAG5 - 8228 and NAG5 - 11797 to the University of New Hampshire.

REFERENCES

1. Parker, E. N., *Astrophys. J.* **128**, 664 (1958).
2. Parker, E. N., *Space Science Rev.* **4**, 666 (1965).
3. Wilhelm, K., Marsch, E., Dwivedi, B. N. et al., *Astrophys. J.* **500**, 1023 (1998).
4. Sturrock, P. A., and Hartle, R. E., *Phys. Rev. Lett.* **16**, 628 (1966).
5. Hartle, R. E., and Sturrock, P. A., *Astrophys. J.* **151**, 1155 (1968).
6. Feldman, W. C., Asbridge, J. R., Bame, S. J. et al., *J. Geophys. Res.* **81**, 5054 (1976).
7. Belcher, J. W., Davis, L., Jr., and Smith, E. J., *J. Geophys. Res.* **74**, 2302 (1969).
8. Belcher, J. W., and Davis, L., Jr., *J. Geophys. Res.* **76**, 3534 (1971).
9. Coleman, P. J., Jr., *Phys. Rev. Lett.* **17**, 207 (1966).
10. Unti, T. W. J., and Neugebauer, M., *Phys. Fluids* **11**, 563 (1968).
11. Alazraki, G., and Couturier, P., *Astron. Astrophys.* **13**, 380 (1971).
12. Belcher, J. W., *Astrophys. J.* **168**, 509 (1971).
13. Hollweg, J. V., *Astrophys. J.* **181**, 547 (1973).
14. Hollweg, J. V., *Rev. Geophys.* **16**, 689 (1978).
15. Barnes, A., *Astrophys. J.* **154**, 751 (1968).
16. Barnes, A., *Astrophys. J.* **157**, 479 (1969).
17. Roberts, D. A., *J. Geophys. Res.* **94**, 6899 (1989).
18. Tu, C.-Y., Pu, Z.-Y., and Wei, F.-S., *J. Geophys. Res.* **89**, 9695 (1984).
19. Tu, C.-Y., *Solar Phys.* **109**, 149 (1987).
20. Grall, R. R., Coles, W. A., Klinglesmith, M. T. et al., *Nature* **379**, 429 (1995).
21. Guhathakurta, M., and Fisher, R., *Astrophys. J.* **499**, L215 (1998).
22. Abraham-Shrauner, B., and Feldman, W. C., *J. Geophys. Res.* **82**, 618 (1977).
23. Dusenbery, P. B., and Hollweg, J. V., *J. Geophys. Res.* **86**, 153 (1981).
24. Hollweg, J. V., and Turner, J. M., *J. Geophys. Res.* **83**, 97 (1978).
25. Isenberg, P. A., and Hollweg, J. V., *J. Geophys. Res.* **88**, 3923 (1983).
26. Marsch, E., Goertz, C. K., and Richter, K., *J. Geophys. Res.* **87**, 5030 (1982).

27. McKenzie, J. F., and Marsch, E., *Astrophys. Space Sci.* **81**, 295 (1982).
28. Marsch, E., "Kinetic physics of the solar wind plasma", in *Physics of the Inner Heliosphere, 2, Particles, Waves and Turbulence,* edited by R. Schwenn, and E. Marsch, Springer-Verlag, Berlin, 1991, p. 45.
29. Isenberg, P. A., *J. Geophys. Res.* **89**, 6613 (1984).
30. Hollweg, J. V., *J. Geophys. Res.* **91**, 4111 (1986).
31. Hollweg, J. V., and Johnson, W., *J. Geophys. Res.* **93**, 9547 (1988).
32. Dmitruk, P., Milano, L. J., and Matthaeus, W. H., *Astrophys. J.* **548**, 482 (2001).
33. Dmitruk, P., Matthaeus, W. H., Milano, L. J. et al., *Astrophys. J.*, in press (2002).
34. Isenberg, P. A., *J. Geophys. Res.* **95**, 6437 (1990).
35. Kohl, J. L., and al., e., *Astrophys. J.* **501**, L127 (1998).
36. Cranmer, S. R., and al., e., *Astrophys. J.* **511**, 481 (1999).
37. Antonucci, E., Dodero, M. A., and Giordano, S., *Solar Phys.* **197**, 115 (2000).
38. Ryan, J. M., and Axford, W. I., *J. Geophys.* **41**, 221 (1975).
39. Esser, R., and Habbal, S. R., *Geophys. Res. Lett.* **22**, 2661 (1995).
40. Esser, R., and Habbal, S. R., "Modeling high flow speeds in the inner corona", in *Solar Wind Eight,* edited by D. Winterhalter, J. Gosling, S. R. Habbal, W. S. Kurth, and M. Neugebauer, AIP, New York, 1996, p. 133.
41. Esser, R., Habbal, H. R., Coles, W. A. et al., *J. Geophys. Res.* **102**, 7063 (1997).
42. Hollweg, J. V., and Isenberg, P. A., *J. Geophys. Res.*, in press (2002).
43. Cranmer, S. R., *Space Sci. Rev.*, in press (2002).
44. Gomberoff, L., and Elgueta, R., *J. Geophys. Res.* **96**, 9801 (1991).
45. Gomberoff, L., Gratton, F. T., and Gnavi, G., "Acceleration and heating of heavy ions in high speed solar wind streams", in *Solar Wind Eight,* edited by D. Winterhalter, J. T. Gosling, S. R. Habbal, W. S. Kurth, and M. Neugebauer, Amer. Inst. Physics, New York, 1996, p. 319.
46. Gomberoff, L., Gratton, F. T., and Gnavi, G., *J. Geophys. Res.* **101**, 15661 (1996).
47. Gomberoff, L., and Astudillo, H., "Ion heating and acceleration in the solar wind", in *Solar Wind Nine,* edited by S. R. Habbal, R. Esser, J. V. Hollweg, and P. A. Isenberg, Amer. Inst. Phys., New York, 1999, p. 461.
48. Hollweg, J. V., *J. Geophys. Res.* **105**, 15699 (2000).
49. Isenberg, P. A., *Space Sci. Rev.* **95**, 119 (2001).
50. Esser, R., Fineschi, S., Dobrzycka, D. et al., *Astrophys. J.* **510**, L63 (1999).
51. Tu, C.-Y., *J. Geophys. Res.* **93**, 7 (1988).
52. Li, X., Habbal, S. R., Hollweg, J. V. et al., *J. Geophys. Res.* **104**, 2521 (1999).
53. Hu, Y. Q., Habbal, S. R., and Li, X., *J. Geophys. Res.* **104**, 24819 (1999).
54. Hu, Y. Q., Esser, R., and Habbal, S. R., *J. Geophys. Res.* **105**, 5093 (2000).
55. Marsch, E., and Tu, C.-Y., *J. Geophys. Res.* **95**, 8211 (1990).
56. Coles, W. A., and Harmon, J. K., *Astrophys. J.* **337**, 1023 (1989).
57. Hollweg, J. V., Bird, M., Volland, H. et al., *J. Geophys. Res.* **87**, 1 (1982).
58. Andreev, V. E., Efimov, A. I., Samoznaev, L. N. et al., *Solar Phys.* **176**, 387 (1997).
59. Mancuso, S., and Spangler, S. R., *Astrophys. J.* **525**, 195 (1999).
60. Leamon, R. J., Smith, C. W., Ness, N. F. et al., *J. Geophys. Res.* **103**, 4775 (1998).
61. Ulrich, R. K., *Astrophys. J.* **465**, 436 (1996).
62. Axford, W. I., and McKenzie, J. F., "The acceleration of the solar wind", in *Solar Wind Eight,* edited by D. Winterhalter, J. T. Gosling, S. R. Habbal, W. S. Kurth, and M. Neugebauer, Amer. Inst. Phys., New York, 1996, p. 72.
63. McKenzie, J. F., Banaszkiewicz, M., and Axford, W. I., *Astron. Astrophys.* **303**, 45 (1995).
64. Tu, C.-Y., and Marsch, E., *Solar Phys.* **171**, 363 (1997).
65. Marsch, E., and Tu, C.-Y., *Astron. Astrophys.* **319**, L17 (1997).
66. Hollweg, J. V., *J. Geophys. Res.* **105**, 7573 (2000).
67. Markovskii, S. A., *Astrophys. J.* **557**, 337 (2001).
68. Markovskii, S. A., and Hollweg, J. V., *J. Geophys. Res.*, in press (2002).
69. Markovskii, S. A., and Hollweg, J. V., *Geophys. Res. Lett.*, in press (2002).
70. Voitenko, Y., and Goossens, M., *Solar Phys.* **206**, 285 (2002).
71. Lee, L. C., and Wu, B. H., *Astrophys. J.* **535**, 1014 (2000).
72. Hollweg, J. V., and Markovskii, S. A., *J. Geophys. Res.*, in press (2002).
73. Tam, S. W. Y., and Chang, T., *Geophys. Res. Lett.* **26**, 3189 (1999).
74. Tam, S. W. Y., and Chang, T., *Geophys. Res. Lett.* **28**, 1351 (2001).
75. Isenberg, P. A., Lee, M. A., and Hollweg, J. V., *J. Geophys. Res.* **106**, 5649 (2001).
76. Isenberg, P. A., *J. Geophys. Res.* **106**, 29249 (2001).

Gas Dynamic Structure of the Heliosphere: Theoretical Predictions and Experimental Data

Baranov V. B.

Institute for Problems in Mechanics, Russian Academy of Sciences, Moscow, 119526, Prosp. Vernadskogo, 101-1

Abstract. Gas dynamic structure of the heliosphere is determined by the solar wind interaction with the supersonic flow of the partially ionized local interstellar medium (LISM). Historical review of this problem investigation is given here. General picture of the considered flow, a mathematical formulation of the gas-kinetic model and its basic results are presented. It is shown that for forecasting of future experimental results it is necessary to construct the models with physically and mathematically correct theoretical basis. Examples of experimental data which can be explained on the basis of theoretical predictions are given. Results of the Ohm's law analysis for partially ionized hydrogen gas show that a conception of the magnetic field freezing in plasma for the problem of the solar wind interaction with the magnetized LISM flow can be not correct if to take into account the processes of the resonance charge exchange.

INTRODUCTION

Constructing a quantitative theoretical model for the prediction and explanation of experimental data is an important goal in various branches of scientific knowledge. However, such a model is useful if it has a reliable, physically correct theoretical basis (see, for example, [2]); otherwise, an interpretation of the experiments could be wrong. This paper is devoted to a critical survey of the present-day situation in theoretical models of the solar wind interaction with the local interstellar medium (LISM) as well as their role in interpreting a number of observed physical phenomena. A historical review of constructed models in the problem considered is given in Section 2. In Section 3 a qualitative picture of the solar wind interaction with the supersonic LISM flow is considered. A mathematical formulation of the gas-kinetic model by Baranov and Malama [4] is presented in Section 4 and its numerical results are given in Section 5. A number of this model predictions (for example, "hydrogen wall") is analyzed here. Analysis of magnetohydrodynamic (MHD) models is given in Section 6 with point of view of the last results obtained recently in [7].

CRITICAL HISTORICAL REVIEW

Parker [36] was the first who predicted theoretically the solar wind on the basis of the spherically symmetric and one-fluid hydrodynamic equations. In particular, under not very restrictive assumptions concerning a dependence of the temperature on the heliocentric distance r he obtained integral curves for the radial velocity V painted at Figure 1. Here $R_{\odot}$, a and r_c are solar radius, sound velocity and a point, where $V = a$, respectively. Curve 1 determines the supersonic solar wind. Later this phenomenon was confirmed experimentally by means of spacecraft [17, 35]. However, with theoretical viewpoint the solution for the solar wind should be consistent with the solution for the interstellar medium. Such a self-consistent solution for the interaction of the stellar wind with the interstellar gas was constructed by Parker [37] at assumption that an interstellar gas flow and a stellar wind downwind of the termination shock are incompressible and potential fluid.

The axisymmetric hydrodynamic model of the solar wind interaction with the supersonic flow of the LISM (two-shock model) was first suggested by Baranov et al. in [8] in the thin layer approximation where the thickness of the region between the bow and terminal shocks (heliosheath) is assumed to be small as compared with the distance of this region to the Sun in the upwind direction (Figure 2). To use hydrodynamic equations these authors assumed that the LISM is a fully ionized hydrogen gas. Supersonic character of the LISM flow relative to the Sun was experimentally confirmed by Bertaux and Blamont [12] and Thomas and Krassa [42] on the basis of measurements of the scattered solar radiation in Lyman-α on the board of OGO-5 spacecraft, i.e. a motion of the LISM hydrogen atoms with the supersonic bulk velocity ($V_{LISM} \approx 20 km/s$) at the LISM temperature $T_{LISM} \approx 10^4 K$ was discovered by these authors. Therefore, the assumption by Baranov et al [8] that the LISM gas is the fully ionized hydrogen was not correct for the

CP679, *Solar Wind Ten: Proceedings of the Tenth International Solar Wind Conference*, edited by M. Velli, R. Bruno, and F. Malara

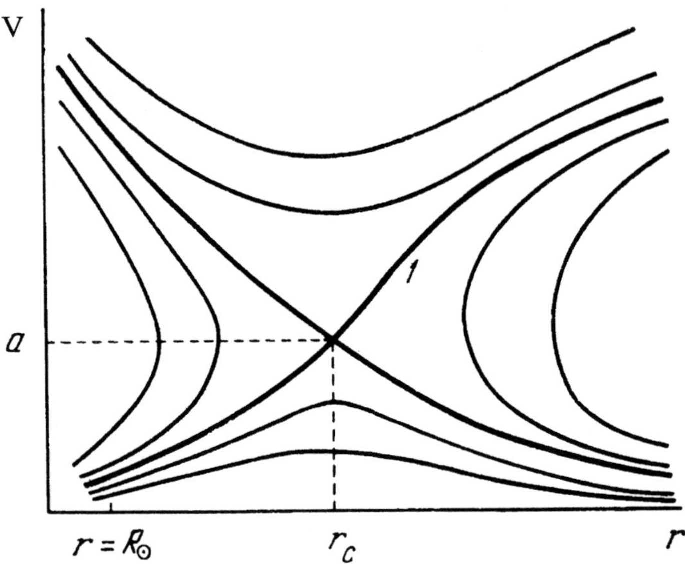

FIGURE 1. Integral curves of spherically simmetric flow from source. Curve 1 is the solar wind.

problem considered although their two-shock model was used for other branches of astrophysics (see, for example, [15]).

The model by Baranov et al. [8] was reanimated for the problem of the solar wind interaction with the LISM by Wallis [43] who assumed, that the LISM is a partially ionized gas, and showed that plasma component (electrons and protons) and H-atoms can influence each other by resonance charge exchange effects. This mutual influence has two aspects. First, the plasma heliosheath becomes a kind of a "filter" for LISM hydrogen atoms, i.e. their parameters are changed after penetrating the solar system. Second, the resonance charge exchange processes can affect plasma parameters, change the heliosheath structure, its size, and its distance from the Sun. These effects were not taken into account till 1985 at the interpretation of measurements of the solar Lyman α scattered radiation (see, for example, [13])

At present there is no doubt that the LISM is partially ionized gas where the main component is partially ionized hydrogen. A revolution in this region was made due to ground observations by Lallement and Bertin [28] and an interpretation of observations by GHRS instrument on Hubble Space Telescope (HST) [29]. They showed that the solar system is embedded in the local interstellar cloud (LIC) moving relative to the Sun with the velocity V_{LIC}=25.7 $\pm$ 0.5km/s and temperature T_{LIC}=7000 K. This is a supersonic flow with the Mach number $M_{LIC}=V_{LIC}/a_{LIC} \sim 2$, where $a_{LIC} = \sqrt{\gamma R T_{LIC}}$,$\gamma$ and R are the specific heat ratio and gas constant, respectively. These parameters of LIC were recently confirmed by the Neutral Gas Experiment [44] on board of the Ulysses spacecraft. Their results were obtained on the basis of helium atom measurements. Helium atoms do not interact with heliosheath due to their small charge exchange cross section with protons.

In 1981 Baranov et al. [9] constructed an simplified axisymmetric model of the solar wind interaction with the supersonic LISM flow self-consistently taking into account mutual influence of plasma component and the LISM H-atoms due to processes of the resonance charge exchange. To estimate effects of H-atoms on the plasma structure of the heliosphere they used the hydrodynamic equations for the plasma component with "source terms" [20], the simplified continuity equation for H-atom only with disappearing H-atoms due to resonance charge exchange (secondary H-atoms did not take into account) and a crude assumption that temperature T_H and bulk velocity V_H of H-atoms are constant. The last assumption was made due to mean free path of H-atoms is comparable with characteristic size of the problem (for example, with the size of the heliosphere) and hydrodynamic momentum and energy equations are not correct in this case. We would like to note here that this simplified model for H-atoms was used in [30] for calculation of 3D MHD problem. However, this model gives rise to good results for plasma component in the vicinity of the upwind direction, but it is absolutely unfit for the tail region and for the interpretation of scattered solar Lyman α experiments.

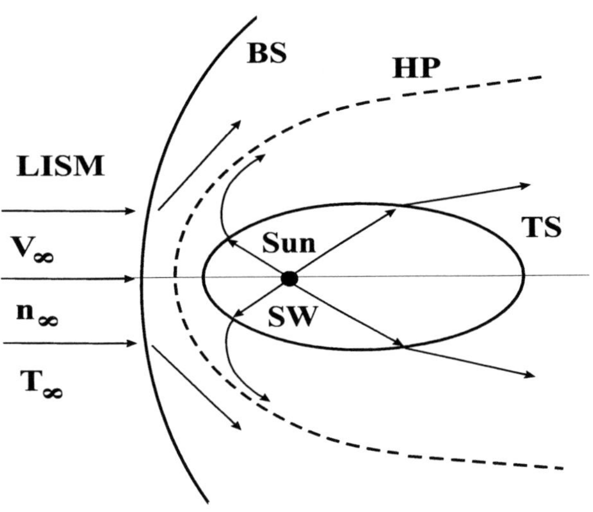

FIGURE 2. Qualitative picture of the first model of the solar wind interaction with supersonic flow of the LISM. Thin layer approximation

QUALITATIVE PICTURE OF THE FLOW CONSIDERED

General picture of the flow caused by the interaction of the solar wind with the supersonic partially ionized hydrogen LISM in Figure 3 is shown. Here the heliopause (HP) is tangential (or contact) discontinuity separating the LISM plasma component and the solar wind. The bow shock (BS) is formed in the supersonic LISM plasma component due to its deceleration at approach to an "obstacle" (here to the HP). The termination shock (TS) is also formed due to deceleration of the solar wind. The heliosheath is the region between the BS and the TS. This region is separated on the outer heliosheath (between BS and HP) and the inner heliosheath (between TS and HP). Of course, geometrical pattern with strong discontinuities (shocks and tangential discontinuity), presented in Figure 3, can be formed only in a hydrodynamic approximation. It can be justified for plasma component rather than for H-atoms, because $Kn = l/L$˜1 for the H-atoms in the problem of the solar wind interaction with the LISM, where Kn, l and L are Knudsen number, mean free path of H-atoms for resonance charge exchange processes and characteristic size of the problem (for example, the size of the heliopause), respectively. In this case a description of H-atom motion in the framework of a hydrodynamic approximation is not correct.

Hydrogen atoms from the LISM penetrating through the four regions presented in Figure 3 are subjected to charge exchange with protons of the plasma component. Since the protons in regions 1 - 4 have different parameters (for example, the temperature and the bulk velocity), the four populations of H-atoms are distinguished by different parameters due to their different places of birth. Population 1 is born in region 1 by the charge exchange of hydrogen atoms from the LISM with supersonic solar wind protons. This population has a large bulk velocity of antisolar direction compared with the thermal velocity. Population 2 is formed as a result of charge exchanges between the hydrogen atoms from the LISM and the solar wind protons thermalized downstream of the TS (in the inner heliosheath). Population 2 of H-atoms has relatively small local bulk velocity and large thermal velocity. Population 3 consists of secondary interstellar H-atoms which are born in region 3 (outer heliosheath) by the charge exchange of interstellar protons thermalized downstream of the BS with the hydrogen atoms from the LISM. Baranov et al. [10] have shown that the number density of this population has no monotonic distribution with heliocentric distance. The maximum of this distribution (the "hydrogen wall") is localized in the vicinity of the HP. Population 3 can also be formed in the disturbed part of the region 4. This disturbed part is formed upwind of the BS due to charge exchange of Populations 1 and 2 of H-atoms and the LISM protons [18]. Population 4 represents primary H-atoms that move from the

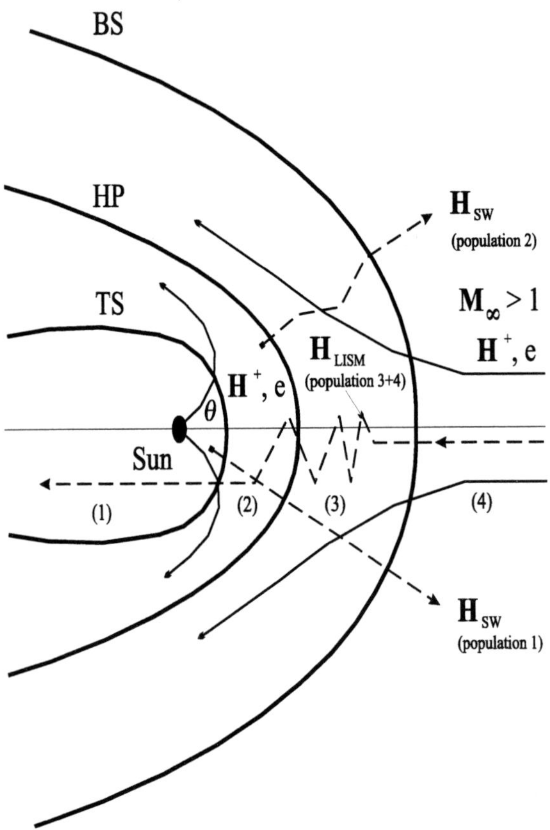

FIGURE 3. Real qualitative physical picture of the flow considered.

undisturbed LISM into the solar system without undergoing charge exchanges. Populations 3 and 4 are measured in the solar scattered Lyman α experiments.

We considered in this Section the qualitative picture of the flow considered. However, it is necessary to construct a mathematical model to obtain quantitative results. As we seen from Introduction, such a model must have a reliable, physically correct theoretical basis to interpret experimental data correctly. Next Section will be devoted to the model by Baranov and Malama (1993) [4], where the plasma component is described by hydrodynamic equations with "source terms" and H-atom component is described kinetically.

MATHEMATICAL FORMULATION OF THE PROBLEM

In this Section we will consider an axisymmetric model of the solar wind interaction with the supersonic flow of the partially ionized LISM suggested and calculated numerically in [4]. They assumed that the plasma component is described by hydrodynamic equations which must take into account the momentum and energy changes due to processes of resonance charge exchange, i.e. these equations must have "source" terms taking into account this effect of H-atoms on the plasma component. For a stationary problem the equations of mass, momentum and energy conservation for ideal gas have the following form (one-fluid approximation for plasma component)

$$\begin{gathered}\nabla \cdot \rho \vec{V} = 0, \\ (\vec{V} \cdot \nabla \vec{V}) + (1/\rho)\nabla p = \vec{F}_1[f_H(\vec{r}, \vec{w}_H), \rho, \vec{V}, p], \\ \nabla \cdot [\rho \vec{V}(\varepsilon + p/\rho + V^2/2)] = F_2[f_H(\vec{r}, \vec{w}_H), \rho, \vec{V}, p], \\ p = (\gamma - 1)\rho\varepsilon, \end{gathered} \tag{1}$$

where p , ρ , $\vec{V}$ and ε are pressure, mass density, bulk velocity and internal energy, respectively; γ is the ratio of specific heats; $\vec{F}_1$ and F_2 are the functionals, describing the change of momentum and energy of the plasma component due to collisions between H atoms and protons which characterize the resonance charge exchange ("source" terms); and $f_H(\vec{r}, \vec{w}_H)$ is the H atom distribution function depending on position-vector $\vec{r}$ and the individual velocity $\vec{w}_H$ of H-atoms. The distribution function must satisfy the Boltzmann equation

$$\vec{w}_H \cdot \frac{\partial f_H(\vec{r},\vec{w})}{\partial \vec{r}} + [(\vec{F}_r + \vec{F}_g)/m_H] \cdot \frac{\partial f_H(\vec{r},\vec{w}_H)}{\partial \vec{w}_H} = \quad (2)$$
$$f_p(\vec{r},\vec{w}_H) \int | \vec{w}'_H - \vec{w}_H | \sigma f_H(\vec{r},\vec{w}'_H) d\vec{w}'_H -$$
$$f_H(\vec{r},\vec{w}_H) \int | \vec{w}_H - \vec{w}_p | \sigma f_p(\vec{r},\vec{w}_p) d\vec{w}_p$$

In equation (2) f_p is the local Maxwellian distribution function of protons with gasdynamic values $\rho(\vec{r})$, $\vec{V}(\vec{r})$ and $T(\vec{r})$ which satisfy to hydrodynamic equations (1), $\vec{w}_p$ is the individual velocity of a proton, $\vec{F}_r$ and $\vec{F}_g$ are the force of the solar radiation pressure and the solar gravitational force, respectively, and σ is the cross section of the resonance charge exchange. If the distribution function f_H is known the "source" terms $\vec{F}_1$ and F_2 in equations (1) can be calculated exactly. We have in this case

$$\vec{F}_1 = \frac{1}{n_p} \int d\vec{w}_H \int d\vec{w}_p \sigma | \vec{w}_H - \vec{w}_p |$$
$$(\vec{w}_H - \vec{w}_p) f_H(\vec{r},\vec{w}_H) f_p(\vec{r},\vec{w}_p),$$
$$F_2 = m_H \int d\vec{w}_H \int d\vec{w}_p \sigma | \vec{w}_H - \vec{w}_p |$$
$$(w_H^2/2 - w_p^2/2) f_H(\vec{r},\vec{w}_H) f_p(\vec{r},\vec{w}_p)$$
$$n_H = \int d\vec{w}_H f_H(\vec{r},\vec{w}_H), n_p = \int d\vec{w}_p f_p(\vec{r},\vec{w}_p) \quad (3)$$

However, Baranov and Malama in [4] used Monte Carlo method to solve numerically the Boltzmann equation (2). The trajectories of H atoms were calculated by the complicated Monte Carlo scheme with "splitting" of trajectories [32] in the field of the plasma gasdynamic parameters. Such an approach allows to calculate the "source" terms in the equations (1) in the framework of a kinetic description of H-atoms if to use formulae (3) (multiple charge exchange are also taken into account by this Monte Carlo method). The solar wind was assumed to be spherically symmetric. That is why the problem was considered as axisymmetric.

To solve the system of equations (1) - (3) the following physical boundary conditions for plasma component were used: the Rankine-Hugoniot relations on the shock waves BS and TS (see Figure 3); the equality of pressures and vanishing normal component of the plasma bulk velocity on the heliopause HP (tangential or contact discontinuity); the velocities, proton (electron) number densities and Mach numbers were given in the undisturbed LISM (index "∞") and at the Earth orbit (index "E").

To calculate H-atom trajectories and "source" terms the distribution function f_H was assumed to be Maxwellian in the undisturbed LISM (at infinity) with the temperature T_∞ , number density $n_{H\infty}$, and velocity V_∞. In so doing, the motion of H atoms is also determined by the solar gravitation force, the force of solar radiation pressure and resonance charge exchange. Any later [6] effects of the photoionization of H atoms near the Sun and H atom ionization due to electron impact in the inner heliosheath were also taken into account. Nonreflecting conditions on the right boundary of the computation region were used. A global iterative method for solution of the problem, suggested in [10] and completed in [4] consists of several steps. First, the trajectories of H atoms were calculated by Monte Carlo method in the field of plasma parameters obtained without "source" terms for fully ionized hydrogen (see, for example, the zero iteration made in [10]). Than the momentum and energy "sources" $\vec{F}_1$ and F_2 in (1) were calculated in this step of iteration using equations (3). In the first iteration, the hydrodynamic equations (1) with these "sources" were solved using the gasdynamic boundary conditions formulated above. Then, the new distribution of plasma parameters was used for the next Monte Carlo iteration for H atoms. The gasdynamic problem was solved again with the new "source" terms of this iteration (the second iteration) and so on. This process of iterations was continuated until the results of two subsequent iterations practically coincide. To solve the gasdynamic part of the problem numerically Baranov and Malama (1993) have used the discontinuity-fitting "second order" technique, which is based on the scheme of Godunov's method [16].

BASIC RESULTS OF GAS-KINETIC MODEL

Formulated in Section 4 model is numerically solved at different parameters given in the undisturbed LISM and at the Earth orbit. Below we will present the basic results separately for plasma component and for H - atoms of several populations for the following values of the specific parameters

$$n_{pE} = 7cm^{-3}, V_E = 450km/s, M_E = 10,$$
$$V_\infty = 25km/s, M_\infty = 2,$$
$$n_{p\infty} = 0.07cm^{-3}, n_{H\infty} = 0.14cm^{-3}, \quad (4)$$
$$\mu = F_r/F_g = 0.75$$

We will also compare the results of our theoretical predictions with the experimental data obtained with spacecraft. In so doing, direct and indirect (remote) methods can be used to detect physical phenomena in space. For example, at present the parameters of inner and outer heliosheaths (regions 2 and 3, respectively, in Figure 3) can be detected only by indirect methods although in future they can be detected by direct methods after crossing Voyager spacecraft the TS or after realization in USA of the Interstellar Probe project. However, direct measurements of the solar wind parameters and the LISM helium atom parameters in region 1 by Voyager and Ulysses spacecraft are now continuing.

Basic results for plasma component

One of the main results of the axisymmetric model, described in Section 4, is the geometrical pattern of the heliosheath. Figure 4 shows the BS, TS, and HP shapes and their heliocentric distances in the xOz plane at the boundary magnitudes of parameters (4). Here the Oz axis coincides with axis of symmetry and is antiparallel to the vector of the LISM's velocity V_∞ (the Sun is in the center of the coordinate system). The Ox axis is normal to the Oz axis.

We see from Figure 4 that heliocentric distances to the TS, HP and BS in the upwind direction are equal about 100 A.U., 180A.U. and 550A.U., respectively. The distance to the TS in the downwind direction is more and equal about 160A.U., i.e. the termination shock are not a sphere. For comparison, Figure 4 demonstrates the geometrical pattern at $n_{H\infty} = 0$. We see that effect of the resonance charge exchange gives rise, first, to decreasing the heliocentric distances of the BS, HP and TS, second, to a smooth shape of the TS (without triple point A and Mach disc MD) and, third, to a subsonic flow in the inner heliosheath.

The following effects in the distant supersonic solar wind due to processes of resonance charge exchange were predicted in [4]: (i) the decrease of the solar wind velocity; (ii) the deviation of the proton number density from the law $1/r^2$; and (iii) the increase of the plasma temperature with the distance from the Sun. However, the last effect is theoretically too large due to using the one-fluid model, where temperatures of electrons and protons are equal. It is interesting to note here that there are direct experimental detections of such effects obtained by Voyager 1/2 observations (see [14]). For example, the deceleration of the solar wind velocity as 30 km/s (about 8% of the average velocity 450 km/s) was estimated from these experiments. It is coinciding with the predictions by Baranov and Malama [4] in the upwind direction. At present indirect observations confirmed theoretical predictions that the outer heliosheath (the region 3) is formed in the problem considered. This conclusion follows, for example, from a detection of the "filter" effect, mentioned above, for the LISM H atoms and its absence for LISM He atoms (see, for example, [44, 27]) due to small cross section of charge exchange between He-atoms and protons.

Basic results for hydrogen atoms

Parameters of four populations of H atoms mentioned in Section 3 were separately calculated on the basis of the model by Baranov and Malama [4]. A basic result is connected with Population 3, i.e. with the secondary LISM H - atoms. The number density of this Population has non-monotonic character with maximum in the vicinity of the HP [10, 4]. This effect was named as "hydrogen wall". The hydrogen wall was experimentally discovered by Linsky and Wood [31] on the basis of interpretation of the Lyman α absorption profile obtained by the GHRS instrument on the HST spacecraft (in details, see, for example, [3]). Basic scheme of such measurements in Figure 5 is painted. Linsky and Wood [31] could only explain the absorption spectrum for the αCen line of sight (52^0 from upwind direction) by including absorption by the heliospheric hydrogen wall (Figure 5). It should be noted here that the largest changes in the hydrogen wall properties (H-atom heating, bulk velocity deceleration and maximum of the H-atom number density) from their ambient values occur in the upwind direction. That is why the star $36OphA$, which is only 12^0 from the upwind direction was observed in [45] with STIS instrument on HST. Their results show that the properties of the heliospheric absorption are consistent with previous measurements of this absorption for the αCen line of sight.

Thus the experimental discovery of the heliospheric hydrogen wall predicted theoretically in [10] is now confirmed by the absorption due to secondary neutral hydrogen atoms formed in the vicinity of the HP. Detailed analysis of the Lyman - alpha absorption profile for different lines of sight shows that hydrogen walls are presented around other stars with sufficiently strong winds [46].

The Lyman-α absorption profile toward Sirius (139^0 from upwind direction) obtained by GHRS instrument on the HST (see Figure 5) was analysed by Izmodenov et al. (1999). They showed that the observed properties of the absorption cannot be explained without taking into account energetic H- atoms (Population 2) in the inner heliosheath, because the hydrogen wall is not easily observable almost in the downwind direction toward Sirius. Detailed study of Lyman α absorption toward six nearby stars was done in [25]

We also believe that the inner heliosheath image in the energetic H-atom (Population 2) fluxes with 1A.U. planned in USA (see, for example [19]) is a very promising method for detecting the inner heliosheath, because such measurements could be carried out regularly. Theoretical calculations of these fluxes and their anisotropy (from upwind to downwind directions) for different assumed ionization fractions in the LISM were obtained in [23].

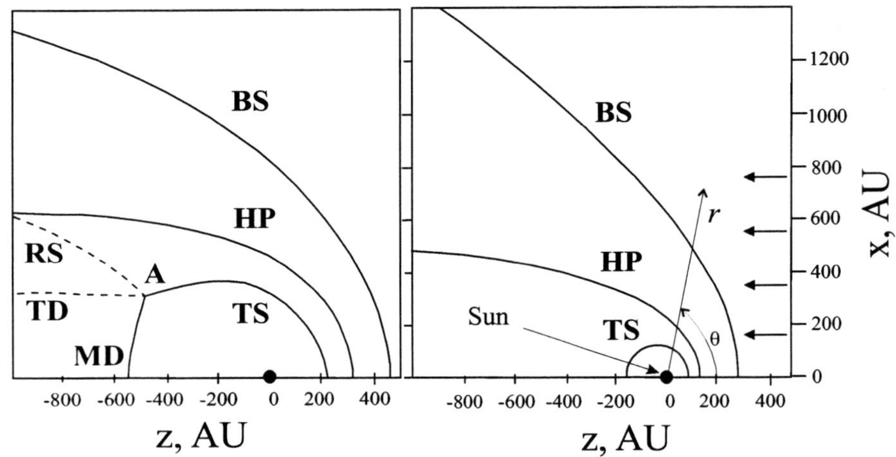

FIGURE 4. Results of calculations. Left picture is results for $n_H = 0$. A is triple point. Right picture is results for $n_H \neq 0$.

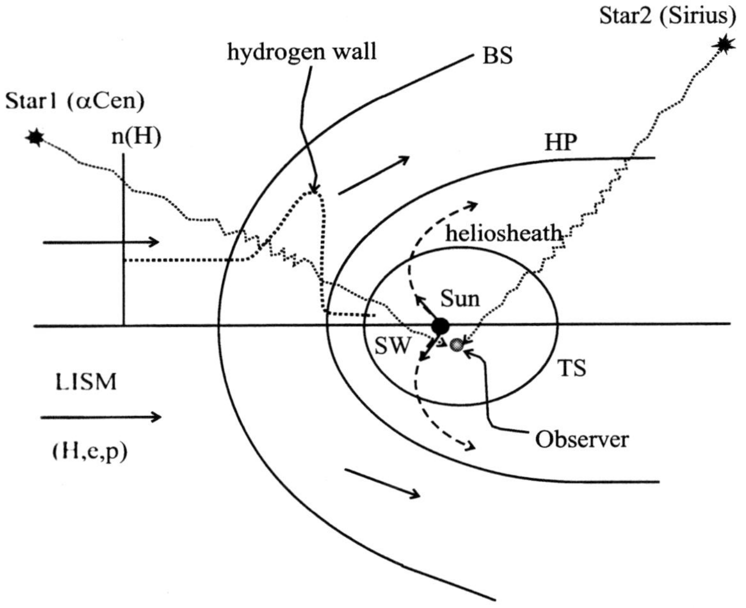

FIGURE 5. Schematic picture of Lyman-alpha absorption experiment (HST).

It is necessary to note here, that the distribution function of H-atoms which is Maxwellian in the LISM becomes quickly non - Maxwellian at crossing interface region between the solar wind and the LISM. This result was first obtained in [11] on the basis of calculations of H-atom moments and was confirmed in [21, 24] on the basis of calculations of H-atom distribution function. Therefore, the hydrodynamic approximation for H-atoms, used by several authors, are not correct in the problem considered. It is naturally because Knudsen number $Kn = l/L \sim 1$ for H-atoms as metioned in Section 3.

ON THE PROBLEM OF THE INTERSTELLAR MAGNETIC FIELD FREEZING FOR THE PARTIALLY IONIZED LISM.

Many authors have been studed the problem of the solar wind interaction with the magnetized LISM without taking into account H-atoms [5, 33, 38, 39, 40, 41] and with taking into account resonance charge exchange processes [30, 1]. The equation for a vector of magnetic field induction $\vec{B}$ in a form

$$\frac{\partial \vec{B}}{\partial t} = \nabla \times (\vec{V} \times \vec{B}) \tag{5}$$

was used in these papers. This equation determines MHD conception of the magnetic field freezing in plasma. However, as we have seen above, the LISM hydrogen atoms and processes of resonance charge exchange connected with these atoms play a determining role in the problem of the solar wind interaction with the LISM. That is why the first group of papers, where were considered MHD problems without taking into account H-atoms, have only a theoretical interest for MHD rather than for problem considered. Recently, the conception of the magnetic field freezing in partially ionized interstellar gas was considered in [7]. Analyzing the Ohm law for partially ionized hydrogen gas and taking into account processes of the resonance charge exchange these authors showed that the last processes give rise to the equation of the magnetic field induction in a form

$$\frac{\partial \vec{B}}{\partial t} = \nabla \times (\vec{V} \times \vec{B}) + \{\nabla \times [\frac{(1-\alpha)^2 c^2}{\sigma B^2}] \tag{6}$$
$$(\frac{\alpha}{1+\alpha} \nabla p \times \vec{B} - \frac{1}{4\pi} \vec{B} \times (\nabla \times \vec{B}) \times \vec{B})\}$$

where α is a degree of ionization ($\alpha = 1$ is for the fully ionized hydrogen). The equation (6) coincides with the equation (5) only at $\alpha = 1$. Estimations in [7] show that the last term on the right side of the equation (6) can be comparable or more with the first term on the right side of this equation, i.e. the results of second group of papers could be reconsidered although the results by Baranov and Fahr [7] are obtained in hydrodynamic approximation. An investigation of this problem must be continuated.

CONCLUSION

In conclusion we would like to discuss effect of other physical phenomena on the axisymmetric model formulated in Section 4.

1. Effect of Galactic Cosmic Rays (GCR) is practically negligible as compared with the effect of resonance charge exchange [34].

2. At present our group has the preliminary results of anomalous cosmic rays (ACR) effect. Effect of ACR gives rise to a change of the TS structure and, therefore, to some changes of distribution of plasma and neutral component parameters in the inner heliosheath.

3. Results obtained on the Ulysses spacecraft showed that the solar wind is not spherically symmetric. Therefore, it is necessary to construct 3D self-consistent model of the solar wind interaction with the LISM. At present the problem is in progress in our group.

4. Our group constructed gas-kinetic time dependent model of the solar wind interaction with the LISM taking into account effects of the solar cycle [47, 26].

5. It is necessary to reconstruct models of the solar wind interaction with magnetized partially ionized LISM with kinetic point of view taking into account processes of resonance charge exchange [7].

Acknowledgments The author is grateful to organizers of the Solar Wind 10 meeting in Pisa for travel support. This work was also sponsored in part by RFBR-projects 01-01-00759 and 01-02-17551, by INTAS grant 01-270 "Physics of the heliosheath plasma flow and structure of the termination shock" (PI - R. Kallenbach). A part of this work was done in the frame of ISSI team "Physics of the heliotail". The author thank International Space Science Institute (ISSI) in Bern for excellent organization of the first team-meeting and for partial support.

REFERENCES

1. Alexashov D.B., Baranov V.B., Barsky E.V. and Myasnikov A.V., Astron. Letters, 2000, v. 26, No 11, pp. 743 - 749.
2. Baranov V.B., Astrophys. Space Sci., 2000, v. 274, pp. 3 - 16.
3. Baranov V.B., Planetary Space Sci., 2002, v. 50, pp. 535 - 539.

4. Baranov V.B. and Malama Yu.G., J.Geophys. Res., 1993, v. 98, No A9, pp. 15,157 - 15,163.
5. Baranov V.B. and Zaitsev N.A., Astron. Astrophys., 1995, v. 304, pp.631 - 637.
6. Baranov V.B. Malama Yu. G., Space Sci. Rev., 1996, v. 78, pp. 305 - 316.
7. Baranov V.B. and Fahr H.-J., J. Geophys. Res., 2002 (accepted for publication).
8. Baranov V.B., Krasnobaev K.V. and Kulikovskii A.G., Dokl. Akad. Nauk SSSR, 1970, v. 194, pp. 41 - 44.
9. Baranov V.B., Ermakov M.K and Lebedev M.G., Sov. Astron. Lett., 1981, v. 7, pp. 206 - 210.
10. Baranov V.B., Lebedev M.G. and Malama Yu.G., Astrophys. J., 1991, v. 375, pp. 347 - 351.
11. Baranov V.B., Izmodenov V.V. and Malama Yu. G., J. Geophys. Res., 1998, v. 103, No A5, pp. 9575 - 9585.
12. Bertaux J.-L. and Blamont J., Astron. Astrophys., 1972, v. 11, pp. 200 - 217.
13. Bertaux J.-L., Lallement R., Kurt V.G. and Mironova E., Astron. Astrophys., 1985, v. 150, pp. 1 - 20.
14. Burlaga L.F., Ness N.F., Belcher J.W., Lazarus A.J. and Richardson J.D., Astrophys. Space Sci, 1995, v.78, pp. 33 - 42.
15. Dyson J., Astrophys. Space Sci., 1975, v. 35, pp. 299 - 312.
16. Godunov S.C., Zabrodin A.V., Ivanov M.Ya., Kraiko A.N. and Prokopov G.P., 1976, Nauka, Moskva (in Russian).
17. Gringauz K., Bezrukich V., Ozerov V. and Ribchinskii R., Dokl. Akad. Nauk SSSR, 1960, v.131, p. 1301.
18. Gruntman M.A., Sov. Astron. Letters, 1981, v. 8, pp. 24 - 26.
19. Gruntman M.A., Proceedings of COSPAR Colloq. "The Outer Heliosphere: The Next Frontiers", Potsdam, Germany, 24 - 28 July, 2000, Eds. K.Scherer et al., p.295.
20. Holzer T., J. Geophys. Res., 1972, v. 77, p. 5407.
21. Izmodenov V.V., Space Sci. Rev., 2001, v. 97., pp. 385 - 388.
22. Izmodenov V.V., Lallement R., and Malama Yu. G., Astron. Astrophys., 1999, v. 342, pp. L13 - L16.
23. Izmodenov V.V., Gruntman M.A., Baranov V.B. and Fahr H.-J., Space Sci. Rev., 2001, v. 97, pp. 413 - 416.
24. Izmodenov, V.V., M. Gruntman, Y.G. Malama, Interstellar Hydrogen Atom Distribution Function in the Outer Heliosphere, J. Geophys. Res. Vol. 106 , No. A6 , p. 10,681-10,690 (2000JA000273).
25. Izmodenov V.V., Wood B., Lallement, R., J. Geophys. Res., 2002, 107(10), doi: 10.1029/2002JA009394.
26. Izmodenov V.V., Malama, Yu.G., in Proceedings of COSPAR 2002, Eds. G.Genta, I.Cairns, K.Scherer, in press, 2003.
27. Lallement R., Space Sci. Rev., 1996, v. 78, pp. 361 - 374.
28. Lallement R. and Bertin P., Astron. Astrophys., 1992, v. 266, pp. 479 - 485.
29. Lallement R., Bertin P., Ferlet R., Vidal-Madjar A. and Bertaux J.-L., Astron. Astrophys., 1994, v. 286, pp. 898 - 908.
30. Linde T., Gombosi T., Roe P., Powell K. and DeZeeuw D., J. Geophys. Res., 1998, v. 109, pp. 1889 - 1904.
31. Linsky J.L. and Wood B.E., Astrophys. J., 1996, v.463, p. 254.
32. Malama Yu.G., Astrophys. Space Sci., 1991, v. 176, pp. 21 - 46.
33. Myasnikov A.V., Preprint IPM RAS, 1997, No 585.
34. Myasnikov A.V., Alexashov D.B., Izmodenov V.V. and Chalov S.V., J. Geophys. Res., 2000, v. 105, No A3, pp. 5167 - 5177.
35. Neugebauer M. and Snyder C., Science, 1962, v. 138, No 3545, pp. 1095 - 1097.
36. Parker E.N., Astrophys. J., 1958, v. 128, pp. 664 - 676.
37. Parker E.N., Astrophys. J., 1961, v.134, pp. 20 - 27.
38. Pogorelov N.V. and Semenov A.Y., Astron. Astrophys., 1997, v.321, pp. 330 - 337.
39. Pogorelov N.V. and Matsuda T, J. Geophys. Res., 1998, v.103, pp. 237 - 245.
40. Pogorelov N.V. and Matsuda T., Astron. Astrophys., 2000, v. 354, pp. 697 - 702.
41. Ratkiewicz R., Barnes A., Molvik G.A., Spreiter J.R., Stahara S.S., Vinokur M. and Venkateswaran S., Astron. Astrophys., 1998, v.335, p.363.
42. Thomas G. and Krassa R., Astron. Astrophys., 1971, v. 11, pp. 218 - 233.
43. Wallis M., Nature, 1975, v. 254, No 5497, pp. 202 - 203.
44. Witte M., Rosenbauer H., Banaszkiewicz M. and Fahr H.-J., Adv. Space Res., 1993, v. 13(6), pp. 121 -130.
45. Wood B.E., Linsky J.L. and Zank G.P., Astrophys J., 2000, v. 537, pp. 304 - 311.
46. Wood B.E., Linsky J.L., Muller H.-R., and Zank G.P., Astrophys. J., 2001, v. 547, pp. L49 - L52.
47. Zaitsev, N.A., Izmodenov V. V., in Proceedings of COSPAR Colloquium on The Outer Heliosphere: The Next Frontiers, Potsdam, Germany, 24 - 28 July 2000; Eds. K. Scherer, H. Fichtner, H.-J. Fahr, E. Marsch, 2001, pp. 65-68.

Session I

Structure of the Corona and Heliosphere in 3D: Solar Minimum and Maximum

The Three-Dimensional Structure of the Solar Wind Over the Solar Cycle

David J. McComas

Southwest Research Institute
San Antonio, Texas 78228, USA

Abstract. Throughout declining and minimum phases of the solar cycle the solar wind displays a simple global structure with fast, tenuous flows emanating from large polar coronal holes filling the interplanetary medium at mid and high latitudes, and slower, denser, and much more variable flows at low latitudes. Approaching solar maximum a complicated mixture of flows from streamers, small coronal holes, and coronal mass ejections extends to higher and higher heliolatitudes, ultimately covering even the poles as the polar coronal holes shrink, fragment, and disappear. The most recent observations have continued through solar cycle 23 maximum with the formation of a mid-sized, circumpolar coronal hole in the northern hemisphere. By April of 2002, Ulysses had again moved back down to ~45° N, however, this hole has not yet grown to nearly the size of those observed in the previous orbit nor pushed fast solar wind down to mid-latitudes. Rather, a complex mixture of solar wind flows is observed below ~70° N in these recent observations. This interval also provided a unique geometry where Ulysses skimmed along, nearly parallel to the boundary of the polar coronal hole over several solar rotations. These times contain substantial intermediate speed solar wind, supporting the previous findings of McComas et al. [2002a] of thin boundary layers (CHBLs) flanking coronal holes.

INTRODUCTION

Over the past decade, the joint NASA/ESA Ulysses mission has made mankind's first two pioneering orbits through the high latitude heliosphere and unlocked many of the secrets of the three-dimensional (3-D) structure of the solar wind over the solar cycle. This paper briefly summarizes a number of these discoveries and goes on to show the most recent solar wind observations from Ulysses. The majority of the observations described here were made with the Solar Wind Observations Over the Poles of the Sun (SWOOPS) instrument [Bame et al., 1992]. The next three sections describe: 1) the 3-D solar wind around solar minimum; this structure is particularly simple and well ordered by heliolatitude; 2) the much more chaotic 3-D solar wind structure observed during the approach to and through solar maximum; and 3) the most recent, post-maximum observations taken by Ulysses, respectively.

THE 3-D SOLAR WIND SPANNING SOLAR MINIMUM

Throughout Ulysses' first complete polar orbit the solar wind displayed a remarkably simple three-dimensional structure, with persistently fast, tenuous and uniform solar wind at high heliolatitudes and slower, more variable, and highly structured wind at low latitudes [e.g., McComas et al., 1998]. Figure 1 (adapted from McComas et al. [2002b]) summarizes Ulysses' two polar orbits in the context of the solar cycle. The bottom panel displays the monthly and smoothed sunspot numbers. The top left panel shows Ulysses' first polar orbit, beginning in February 1992, when the spacecraft swung past Jupiter, extending through December 1997. This interval covered the declining phase of solar cycle 22, through solar minimum, and into the very early rise of cycle 23. This panel also indicates the times of fast solar wind (>700 km s^{-1}, shaded), which was consistently observed at high heliolatitudes.

As Ulysses made its first southward sojourn in 1992-1993, it encountered a strong, recurrent co-rotating interaction region (CIR). With each rotation of the Sun, Ulysses alternately encountered both the fast solar wind from the large southern polar coronal hole and the slower and more variable low-latitude solar wind [Bame et al., 1993]. CIRs form because of the rotation of the Sun, which serves to radially align fast wind with the slower solar wind ahead of it. Such interactions are particularly significant at mid-latitudes when there are large, stable polar coronal holes and a significant tilt or warp in the heliospheric current sheet.

CP679, *Solar Wind Ten: Proceedings of the Tenth International Solar Wind Conference*, edited by M. Velli, R. Bruno, and F. Malara

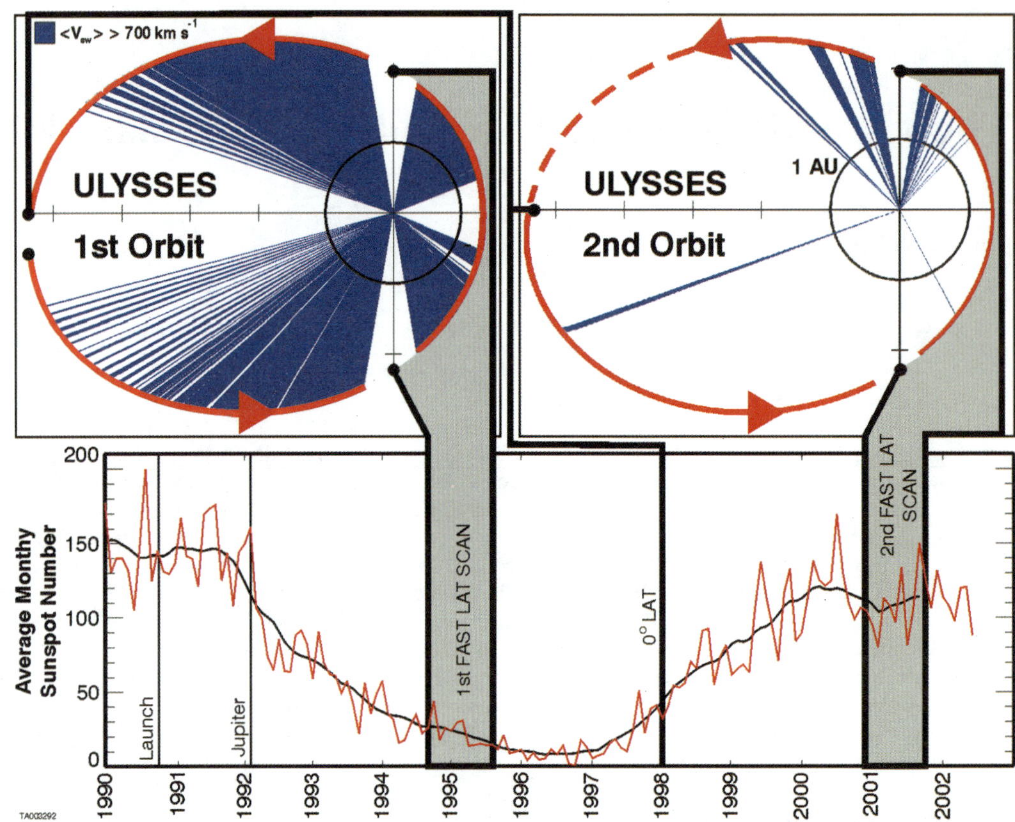

FIGURE 1. Ulysses' first (left) and second (right) polar orbits mapped to the sunspot number (bottom) over the past solar cycle, adapted and extended from McComas et al. [2002b]. Fast solar wind (daily average speed >700 km s^{-1}) is indicated by shading. The highest latitude regions are left blank as Ulysses' orbit did not sample above |80.2°| heliolatitude.

Consistent flow deflections across the recurrent CIR indicated a characteristic north-south tilt, which forms as the three-dimensional structure of a CIR develops and propagates outward [Gosling et al., 1993]. Subsequently, the opposite tilt was observed in the northern hemisphere as Ulysses returned to low latitudes, further confirming this simple physical model [Gosling et al., 1997]. An additional ramification of the consistent tilts of CIRs is that in both hemispheres the forward waves, which often steepen into shocks within a few AU of the Sun, propagate equatorward, while the reverse waves (shocks) propagate poleward. This characteristic asymmetry between forward and reverse shocks was also observed throughout Ulysses' first orbit with reverse shocks being seen up to much higher heliolatitudes than were forward shocks [e.g., Gosling, 1996, and references therein].

Another important aspect of the mid-latitude CIRs observed in Ulysses first orbit was characteristic ion composition variations. In particular, the fast wind from the large polar coronal holes exhibited low ion freezing-in temperatures and a diminished difference between the low-FIP and high-FIP abundances [Geiss et al. 1995]. These results showed that solar wind from this polar coronal hole was fundamentally distinct from the slower, low latitude solar wind, both 1) in the corona, where the electron density rapidly drops and sets the charge state distributions of the solar wind heavy ions and 2) wherever the elemental composition is determined (probably down in the chromosphere).

The solar wind plasma observations covering Ulysses' first complete polar orbit were quantitatively examined by McComas et al. [2000a]. These authors provided statistics on over 20 solar wind parameters in the coronal hole flows and investigated their characteristic variations with radial distance, heliolatitude, and over the solar cycle. They found that the solar wind from the large polar coronal holes was extremely uniform. Once the radial variations were determined and accounted for, only a very weak heliolatitude dependence remained in most of the parameters. One example of this residual latitude variation is the slight increase in solar wind speed with heliolatitude. From 36°-80° the speed is very well characterized by a linearly increasing speed, which is 760 km s^{-1} at 60° and has a slope of +1 km s^{-1} deg^{-1}. If the characteristic speed is an inverse function of a magnetic expansion factor [Wang and Sheeley, 1992], then these observations indicate that throughout the main portion of polar coronal holes, the expansion factor is only slightly increasing with decreasing latitude.

Comparison of the fast solar wind from southern and northern hemispheres showed much smaller differences across the fast latitude scan as compared to the more distant portions of Ulysses' orbit. Since the observations were taken over only ~10 months out of the nearly 6 year orbit, smaller differences from this scan indicate that the observed north-south asymmetries were more likely due to temporal variations over the solar cycle instead of a quasi-permanent hemispheric asymmetry. Thus, observations of ~40% lower mass flux in the northern than the southern hemisphere, indicates that there is much more energy going into the polar solar wind during the declining phase of the solar cycle than near minimum. Further, the average values of the solar wind speed are extremely similar between the hemispheres. This combination suggests that the additional energy needed to accelerate the higher mass flux must have been deposited below critical point down in the lower corona [Holzer and Leer, 1981].

A new class of forward-reverse shock pairs was also identified in the high latitude solar wind on Ulysses. These shock pairs were shown to be driven by "over expansion" of some high latitude coronal mass ejections, which travel out with a speed comparable to the surrounding high speed solar wind [Gosling et al., 1994; 1998]. Over expansion most likely occurs because these CMEs start out with an excess of initial pressure, causing them to expand as they propagate outward. This expansion forms a compressive wave (which eventually steepens into shock) that encircles the CME and is observed as a forward shock on the upstream side and a reverse shock on the downstream side as it passes over the spacecraft. Finally, a rarefaction forms in the center of the CME as the compressions propagate outward, effectively leaving a region depleted in plasma.

The left panel in Figure 2 (from McComas et al., 2002b) summarizes the near minimum configuration of solar wind speed (polar plot), overlaid on a set of coronagraph images, which show the characteristically simple coronal configuration typical of solar minimum. The Ulysses data were taken over its first fast latitude scan, which occurred just before solar minimum (September 1994 – July 1995). Comparison of this figure with a similar plot covering the entire first orbit [McComas et al., 1998] shows the same macroscale structure persisted over much of the declining phase and the near minimum portions of past solar cycle. Thus, a simple picture of the declining phase and near-minimum solar wind structure emerges: fast, tenuous flows emanate from large polar coronal holes to fill the interplanetary medium at mid and high latitudes, while slower, denser, and much more variable flows are present at low latitudes.

ULYSSES FAST-LATITUDE SCANS

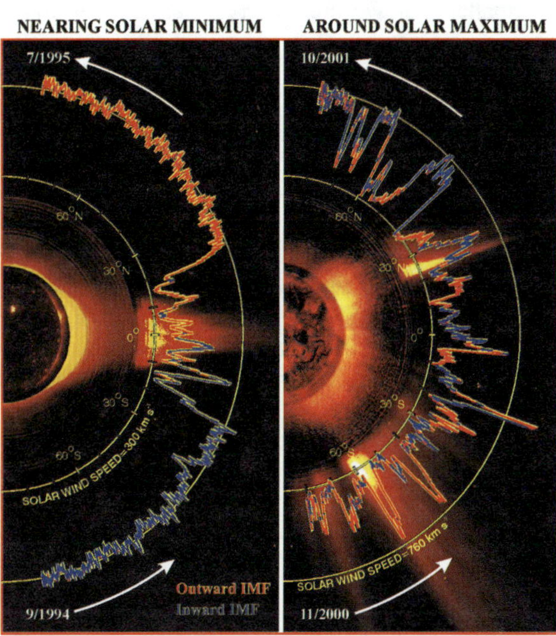

FIGURE 2. Comparison of solar wind speed profiles observed over Ulysses' two fast latitude scans [taken from McComas et al., 2002b]. The IMF polarity is color coded for inward (blue) and outward (red) fields. Each plot has been overlaid on a composite of three concentric solar images from SOHO-EIT (195 A Fe XII, inner), Mauna Loa (middle), and SOHO-LASCO/C2 (outer) for context. The near minimum and near maximum images are from 17 August 1996 and 18 June 2001, respectively.

THE 3-D SOLAR WIND LEADING UP TO SOLAR MAXIMUM

In contrast to the relatively simple conditions described above, the right panel in Figure 2 shows that the three dimensional solar wind is much more complex around solar maximum. At these times Ulysses observed an irregularly structured mixture of mostly slow and intermediate-speed flows. Coronal mass ejections and fast CME-associated transient shocks were much more common during the second orbit and occurred at high as well as low heliolatitudes, although their frequency of occurrence still dropped off with increasing latitude [McComas et al., 2000b]. The *in situ* observations of more CMEs around maximum is consistent with coronagraph observations of about ten times higher rates of CMEs around solar maximum as compared to solar minimum [Webb and Howard, 1994].

Throughout the southern half of Ulysses' second polar orbit (top right panel of Figure 1), there were no flows from large, stable coronal holes. However, short intervals of relatively faster wind from a variety of smaller holes were observed at high southern latitudes using a combination of SWOOPS and Ulysses SWICS data [McComas et al., 2002a]. Flows from these small coronal holes were identifiable from their low coronal freezing-in temperatures and relative depletions in low-FIP ions. While these composition signatures indicate that coronal holes produce a unique type of solar wind, variations in the acceleration process produce a wide range of solar wind speeds from these holes. Further, these authors showed that high-speed wind (>700 km s^{-1}) can be produced in small as well as large holes, although the very highest speed non-transient winds do come from the centers of the largest holes.

Figure 3, relabeled slightly from McComas et al. [2002a], shows the composite variation between fast and slow winds at the edges of the nine small coronal holes examined (inset). These authors found that the rarefaction regions map back to a coronal hole boundary layer (CHBL). In the CHBL oxygen and carbon freezing-in temperatures drop in concert with increasing solar wind acceleration (and ultimately solar wind speed in interplanetary space). In contrast, the Fe/O ratio (low/high FIP) tends to stay depressed throughout the bulk of the CHBL indicating that this boundary layer originates from within the edges of the coronal holes back at the Sun (as indicated schematically in Figure 3).

HIGH LATITUDE SOLAR WIND SHORTLY AFTER MAXIMUM

Just after the first peak in sunspot number, Ulysses rose above ~60° N heliolatitude in its second polar orbit (Figure 1). Extended and solar rotation ordered intervals of high speed wind were again observed, arising from a circumpolar coronal hole that was reforming around the Sun's northern pole [McComas et al., 2002b]. Thus, it appeared that the complex heliospheric structure around solar maximum, was short lived, and that the simple bimodal structure observed throughout Ulysses first orbit represented the vast majority of the solar cycle [McComas et al., 2002b].

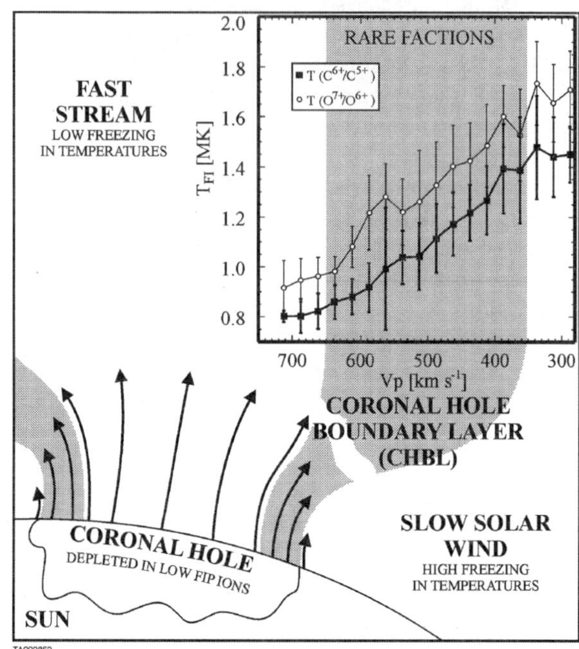

FIGURE 3. Schematic diagram of a coronal hole, from McComas et al. [2002a]. The inset shows composite coronal freezing-in temperatures binned as a function of solar wind speed at the edge of fast streams. These regions map back to a coronal hole boundary layer (CHBL), with steep gradients in the solar wind acceleration (speed) and freezing-in temperatures.

Subsequent SWOOPS observations, however, indicate a somewhat more complex picture. Figure 4 shows the solar wind speed since October 2001, when Ulysses crossed its highest northern heliolatitude, as a function of latitude and Carrington longitude (using constant velocity mapping back to the Sun). The gray bar indicates a solar wind speed ranging from 300-760 km s^{-1}, the typical fast solar wind observed over Ulysses first orbit. The height of the black bar indicates the observed solar wind speed.

Clearly the fast flow from this polar coronal hole has not yet expanded back down to intermediate latitudes. Instead, fast flows are only observed above ~70° at most longitudes, and down to ~55° for a small range of longitudes (centered at ~50° Carrington longitude); He 10830 Å maps of the northern coronal hole back at the Sun indicate that the coronal hole boundary has a small equatorward extension around this longitude at the times that Ulysses observed these fast winds. In addition, while the mid-sized northern polar coronal hole remained relatively stable in size, no similar polar hole had even begun to form in the south.

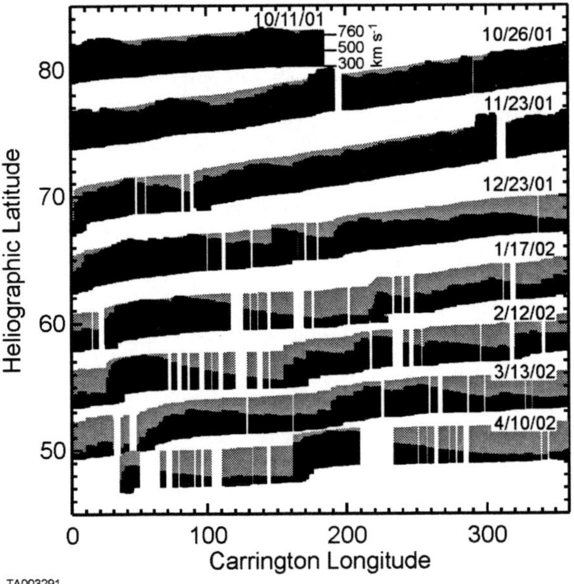

FIGURE 4. Solar wind speed from October 2001 through April 2002, mapped with a constant velocity back to Sun. The speed is indicated by the height of the black band. The grey band indicates speeds from 300-760 km s^{-1}. While Ulysses observed fast wind above ~70°, below that the spacecraft skimmed along the boundary of the coronal hole flow, observing both fast and slow winds.

Detailed comparison of the 10830 maps with Ulysses observations for this interval indicates that Ulysses was skimming the boundary of the polar coronal hole flows for several solar rotations as it dropped back down in heliolatitude. Figure 4 shows that for much of this interval, Ulysses consistently observed intermediate speed solar wind in the 500-700 km s^{-1} range. At these times, Ulysses also observed other unusual plasma properties, such as higher proton temperatures, which is more typical of the fast solar wind. In addition, freezing-in temperatures almost always vary in lock step with the solar wind speed [G. Gloeckler, private communication], suggesting that this interval will display the same characteristic speed and freezing-in temperature variation demonstrated for small coronal holes by McComas et al. [2002a]. These results indicate that as the Ulysses spacecraft skimmed along the edge of the recent mid-sized polar coronal hole, it was sampling a coronal hole boundary layer plasma, similar to the CHBL observations previously documented by McComas et al. [2002a].

CONCLUSIONS

This brief paper has reviewed a number of the notable findings from the Ulysses mission about the three-dimensional structure of the solar wind over the solar cycle. Around solar minimum this structure is simple and bimodal with fast, tenuous solar wind emanating from large polar coronal holes, filling the high latitude regions, and slower, denser, and much more variable wind at low heliolatitudes. This configuration, along with the intrinsic tilt of the solar dipole with respect to the Sun's rotation axis, leads to large recurrent CIRs, which have a characteristic equatorward tilt in both hemispheres. In turn, this geometry preferentially drives the CIR-associated reverse shocks to high heliolatitudes. Finally while there is a relative paucity of CMEs away from solar maximum, those that are observed at high heliolatitudes display a unique configuration, consistent with the over-expansion of structures that are initially overpressured back at the Sun.

In contrast, the 3-D solar wind approaching and through solar maximum is remarkably more complex, with a time-varying structure including numerous CMEs, small coronal holes, and slow to intermediate solar wind observed at all heliolatitudes. The most recent, post-maximum Ulysses observations show fast wind from a single mid-sized polar coronal hole in the northern hemisphere, producing intermediate speed flows all along its edge, and a complex mixture of flows persisting at mid-latitudes.

Small coronal holes near solar maximum generate a variety of solar wind speeds, including flows nearly as fast as those coming from the huge, near minimum polar coronal holes. Both these smaller coronal holes and the mid-sized polar hole observed recently as Ulysses dropped back down from its highest northern heliolatitudes seem to be bounded by coronal hole boundary layers. These observations suggest that CHBLs are ubiquitous structures in the solar wind, bounding flows from all coronal holes. Thus, the global solar wind is not simply comprised of fast and slow flows. Rather, a continuous heating and acceleration process exists, at least within the relatively narrow CHBLs.

Finally, all of the observations described in this study have been from a single 11-year interval (spanning the end of cycle 22 and start of 23). The real 3-D structure of the solar wind surely evolves over the full 22 year cycle and varies from one of these cycles to the next. Thus, continuing observations of the high-latitude solar wind are absolutely essential to understanding the full picture of its three-dimensional structure.

ACKNOWLEDGEMENTS

I gratefully acknowledge valuable discussions with and input from Heather Elliott, George Gloeckler, Jack Harvey, and Nathan Schwadron.

REFERENCES

Bame, S.J., D.J. McComas, B.L. Barraclough, J.L. Phillips, K.J. Sofaly, J.C. Chavez, B.E. Goldstein, and R.K. Sakurai, The Ulysses Solar Wind Plasma Experiment, *Astron. Astrophys. Suppl. Ser. 92*, 237-265, 1992.

Bame, S.J., B.E. Goldstein, J.T. Gosling, J.W. Harvey, D.J. McComas, M. Neugebauer, and J.L Phillips, Ulysses observations of a recurrent high speed solar wind stream and the heliomagnetic streamer belt, *Geophys. Res. Lett., 20*, 2323-2326, 1993.

Geiss, J., et al., The southern high-speed stream: Results from the SWICS instrument on Ulysses, *Science, 268*, 1033, 1995.

Gosling, J.T., Corotating and transient solar wind flows in three dimensions, *Annu. Rev. Astron. Astrophys., 34*, 35, 1996..

Gosling, J.T., S.J. Bame, D.J. McComas, J.L. Phillips, V.J. Pizzo, B.E. Goldstein, and M. Neugebauer, Latitude variation of solar wind corotating stream interaction regions: Ulysses, *Geophys. Res. Lett., 20*, 2789-2792, 1993.

Gosling, J.T., S.J. Bame, D.J. McComas, J.L. Phillips, E.E. Scime, V.J. Pizzo, B.E. Goldstein, and A. Balough, A forward-reverse shock pair in the solar wind driven by over-expansion of a coronal mass ejection: Ulysses observations, *Geophys. Res. Lett., 21*, 237-240, 1994.

Gosling, J.T., S.J. Bame, W.C. Feldman, D.J. McComas, P. Riley, B.E. Goldstein, and M. Neugebauer, The northern edge of the band of solar wind variability: Ulysses at ~4.5 AU, *Geophys. Res. Lett., 24*, 309-312, 1997.

Gosling, J.T., P. Riley, D.J. McComas, and V.J. Pizzo, Over-expanding coronal mass ejections at high heliographic latitudes: Observations and simulations, *J. Geophys. Res., 103*, 1941-1954, 1998.

Holzer, T.E. and E. Leer, Theory of mass and energy flow in the solar wind, *Solar Wind 4*, H. Rosenbauer, ed., 28-41, 1981.

McComas, D.J., A. Balogh, S.J. Bame, B.L. Barraclough, W.C. Feldman, R. Forsyth, H.O. Funsten, B.E. Goldstein, J.T. Gosling, M. Neugebauer, P. Riley, and R. Skoug, Ulysses' return to the slow solar wind, *Geophys. Res. Lett., 25*, 1-4, 1998.

McComas, D.J., B.L. Barraclough, H.O. Funsten, J.T. Gosling, E. Santiago-Muñoz, R.M. Skoug, B.E. Goldstein, M. Neugebauer, P. Riley, and A. Balogh, Solar wind observations over Ulysses' first full polar orbit, *J. Geophys. Res., 105*, 10419-10433, 2000a.

McComas, D.J., J.T. Gosling, and R.M. Skoug, Ulysses observations of the irregularly structured mid-latitude solar wind during the approach to solar maximum, *Geophys. Res. Lett., 27*, 2437-2440, 2000b.

McComas, D.J., H.A. Elliott, and R. von Steiger, Solar wind from high latitude coronal holes at solar maximum, *Geophys. Res. Lett., 29(9)*, 10.1029/2001GL013940, 2002a.

McComas, D.J., H.A. Elliott, J. T. Gosling, D.B. Reisenfeld, R. M. Skoug, B.E. Goldstein, M. Neugebauer, and A. Balogh Ulysses' second fast-latitude scan: Complexity near solar maximum and the reformation of polar coronal holes, *Geophys. Res. Lett., 29(9)*, 10.1029/2001GL014164, 2002b.

Wang, Y.-M. and N.R. Sheeley, The relationship between solar wind speed and the areal expansion factor, Solar wind 7, ed., E. Marsch and R. Schwenn, 125-128, 1992.

Webb, D.F. and R.A. Howard, The solar cycle variation of coronal mass ejections and the solar wind mass flux, *J. Geophys. Res., 99*, 4201, 1994.

Voyager 1 Studies of the HMF to 81 AU During the Ascending Phase of Solar Cycle 23

L. F. Burlaga[a], N. F. Ness[b], Y.-M. Wang and N. R. Sheeley, Jr.[c]

[a]*Laboratory for Extraterrestrial Physics, NASA-Goddard Space Flight Center, Greenbelt, MD 20771, USA*
[b]*Bartol Research Institute, University of Delaware, Newark, DE 19716, USA*
[c]*E. O. Hulburt Center for Space Research, Naval Research Laboratory, Washington, DC 20375, USA*

Abstract. The paper analyzes the magnetic field strength B and polarity observed in the distant heliosphere from 1996 to early 2001 and will be discussed in relation to the variation of B from 1978 through 1996. The observations extend the results of Burlaga et al. [1]. The polarity of the heliospheric magnetic field (HMF) from 1997 to early 2001 is studied in relation to the extrapolated position of the heliospheric current sheet (HCS). These observations of polarity extend the earlier results of Burlaga et al. [2] and Burlaga and Ness [3].

The V1 observations of the heliospheric magnetic field strength B agree with Parker's model of the global heliospheric magnetic field from 1 to 81 AU and from 1978 to 2001, when one considers the solar cycle variations in the source magnetic field strength and the latitude/time variation in the solar wind speed. Parker's model, without adjustable parameters, describe the general tendency for B to decrease with increasing distance R from the Sun, and the solar cycle time variations causing the three broad increases of B around 1980, 1990, and 2000, and the minima of B in 1987 and 1997. The variation of magnetic polarity observed by V1 and V2 was caused by the increasing latitudinal width of the sector zone with increasing solar activity, which in turn was related to the increasing maximum latitudinal extent and the decreasing minimum latitudinal extent of the footpoints of the HCS.

INTRODUCTION

The radial variation of the magnetic field strength to 66 AU was determined by Burlaga et al. [1], using Voyager and Pioneer 11 data, and were found to agree with Parker's model [4, 5] when considering the latitudinal and temporal variations of the source magnetic field strength and the solar wind speed at 1 AU.

The heliospheric magnetic field has regions of positive polarity (field vectors pointing away from the Sun) and negative polarity (field vectors pointing toward the Sun) [6]. During much of the solar cycle, the polarity of the magnetic field is predominantly positive (negative) above the heliospheric current sheet (HCS) and negative (positive) below the HCS. Schulz [7] attributed a 2-sector pattern to a tilt of the nearly planar HCS near the Sun relative to the solar equatorial plane, and a 4-sector pattern to warps in the HCS when it lies close to the equatorial plane.

The HCS tends to extend from the solar corona to the distant heliosphere. Its internal structure can be complex and its shape is severely distorted with increasing distance from the Sun [8]. Nevertheless, the latitudinal extent of the HCS generally changes little (O(10°)) with increasing distance from the Sun out to 40 AU [9]. The extent of the HCS defines the latitudinal boundaries of a zone in which both positive and negative polarities of the magnetic field are observed. This zone can be identified even at large distances, and it can be related to solar magnetic fields. Burlaga and Ness [9] call this region the "sector zone."

During 1989, V1 and V2 observed both positive and negative polarities; both spacecraft were in the sector zone, because the HCS extended to moderate latitudes approaching solar maximum [10, 11]. Ulysses observed a single magnetic polarity at northern latitudes above the HCS during 1996 and opposite polarities when it was within the heliolatitude range covered by the excursion of the neutral line from 1997 through 2000 [12].

CP679, *Solar Wind Ten: Proceedings of the Tenth International Solar Wind Conference*, edited by M. Velli, R. Bruno, and F. Malara

OBSERVATIONS OF B

This section presents new observations of the yearly averages of the Voyager 1 data for the magnetic field strength, $B_{V1}(t)$, from 1997 to 2001.34. In addition, we compare the observations $B_{V1}(t)$ with Parker's model. At 80 AU, the distance of V1 in 2001, the radial and azimuthal components of the heliospheric magnetic field are expected to be ≈ 0.0007 nT and ≈ 0.05 nT, respectively. The observations of $B_{V1}(t)$ are shown as dots in Figure 1. As expected, B decreases with increasing distance from the Sun (which corresponds to increasing time as V1 moves away from the Sun). There are also three relative maxima in B near 1980, 1990, and 2000 and two relative minima in B during 1987 and 1997. The relative maxima and minima in $B_{V1}(t)$ in Figure 1 are related to solar cycle variations in the source field strength (measured by the magnetic field at 1 AU, $B_1(t)$) and the solar wind speed at V1, $V(t, \theta(t))$. Figure 2 shows $B_1(t)$ and $V(t, \theta(t))$ from 1978 through 2000.34 in the bottom and top panels, respectively. The variations of the sunspot number during the same interval are shown in the middle panel. The magnetic field strength at 1 AU, $B_1(t)$, was obtained from the NSSDC Omni data set and some recent data from ACE. The speed $V(t, \theta(t))$ on V1 was not measured directly beyond 10 AU, so we determined it using the Wang and Sheeley [13] method.

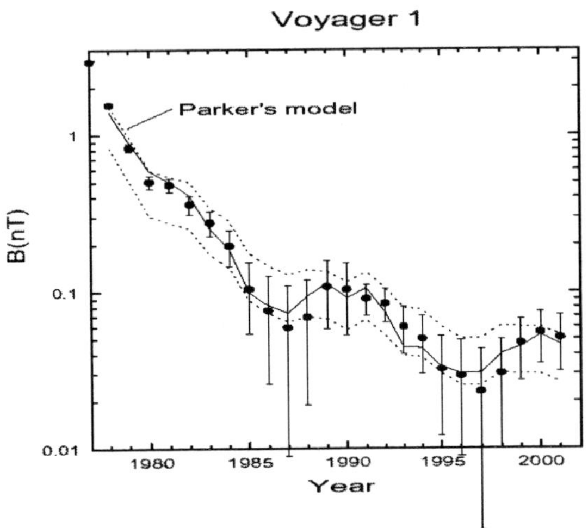

Figure 1. The dots show the yearly averages of the magnetic field strength B measured by Voyager 1 from 1978 to 2001. The solid curve is Parker's model, computed as described in the text. The three local maxima and two local minima in B are associated with solar cycle variations in the source field magnetic field strength and with both solar cycle and latitude variations in the solar wind speed. The top (bottom) dashed curve is the prediction of Parker's model for a solar wind speed of 400 km/s (800 km/s). Standard deviations of the observed averages are indicated also.

A solar cycle variation of $B_1(t)$ is evident in Figure 2; $B_1(t)$, had a minimum strength equal to ≈5 nT and a maximum strength of ≈9 nT. The three intervals with relatively strong fields observed by V1 were associated with relatively strong magnetic fields at 1 AU shortly after the sunspot maxima in ≈ 1980, 1990, and 2000. The intervals with relatively weak fields observed by V1 were associated with relatively weak magnetic fields at 1 AU in 1987 and 1997 when the solar activity was low.

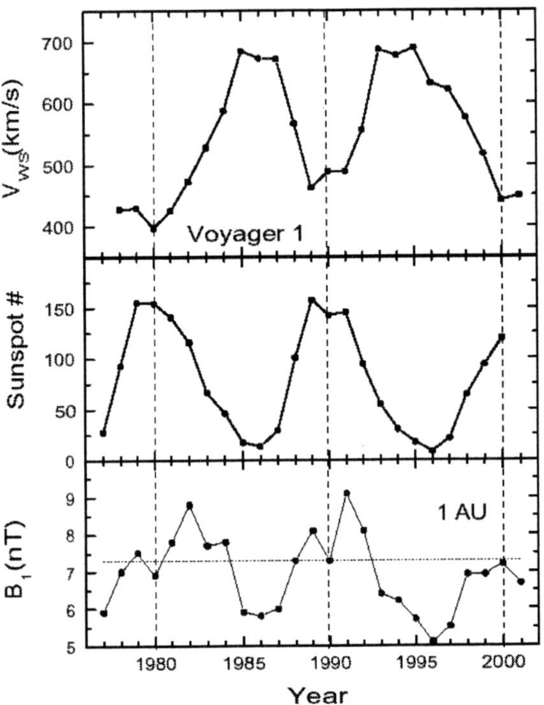

Figure 2. Solar cycle variation of the solar wind speed computed at V1 (top panel) and the magnetic field strength at 1 AU (bottom panel). The sunspot number is shown in the middle panel for reference.

A solar cycle variation of $V(t, \theta(t))$ is evident in Figure 2, but it is complicated by a latitudinal variation that is also present. At 1 AU there are relative minima in V near the solar maxima in 1980, 1990 and 2000, and there are maxima in V near the solar minima in 1986 and 1996. The travel time of the solar wind from 1 AU to 81 AU is approximately 300 days, so that the effects of solar minimum and solar maximum are observed significantly later in the distant heliosphere than at 1 AU.

COMPARISON OF THE OBSERVATIONS OF B WITH PARKER'S MODEL

We now compare the V1 observations of the 1-year averages of the magnetic field strength B with the predictions of Parker's model. Using the observations $B_1(t)$ and $V_1(t)$ at 1 AU and the speed profile $V(t,\theta(t))$ discussed above, the magnetic field strength at V1 that is predicted from Parker's model was calculated. The result is shown by the solid curve in Figure 1. The effects of the speed on the theoretical magnetic field strength are shown by the dotted curves in Figure 1, which show Parker's model for the cases $V(t, \theta(t))$ = 400 km/s (top curve) and 800 km/s (bottom curve). Parker's model, without adjustable parameters, reproduces the basic features of the observed magnetic field strength profile, including: 1) the general tendency to decrease with increasing distance R from the Sun; 2) the three broad maxima around 1980, 1990, and 2000; and 3) the two minima in 1987 and 1997.

The values of B at V1 in ≈1987 and 1997 are lower than the model of Parker predicts for the values of $V(t, \theta(t))$ that we used. It is possible that the departures of the observations from Parker's model in 1987 and 1997 in Figure 1 are associated with the presence of a vortex street at those times.

VARIATIONS OF THE POLARITY OF THE HELIOSPHERIC MAGNETIC FIELD AND THE EXTENT OF THE SECTOR ZONE

We now consider the variation of the magnetic field polarity in the distant heliosphere during the ascending phase of solar cycle 23.

The percentage of positive magnetic polarity measured at V1 and V2 from 1997 (near solar minimum in the distant heliosphere) through 2001.5 (near solar maximum in the distant heliosphere) is shown as a function of time in the top panels of Figure 3, respectively. The uncertainty in these measurements is of the order of 5% - 10%. In the northern hemisphere, V1 observed a decrease in the percentage of positive polarities from ≈100% during 1997 to ≈50%. In the southern hemisphere, V2 observed an increase in the percentage of positive polarities from ≈ 0% during 1997 to ≈ 50% during the same period.

The percentage of positive polarities observed by V1 (V2) is related to the maximum (minimum) latitudinal extent of the HCS, shown in the bottom left (right) panels of Figure 3. When V1 observed ≈100% positive magnetic polarity in the northern hemisphere in 1997, the maximum northern extent of the HCS was very low, ≈10° N. In that year, V2 observed essentially no positive magnetic polarity in the southern hemisphere, and the minimum southern extent of the HCS was ≈5° S. The percentage of positive polarity measured by V1 decreased from 1997 to 2000 as the maximum latitude of the HCS moved northward above the latitude of V1. The percentage of positive polarity measured by V2 from 1997 to 2000 increased (the percentage of negative polarity decreased) as the minimum latitudinal extent of the HCS moved southward below the latitude of V2.

The latitudinal extent of the HCS was propagated radially from the source surface to the distant heliosphere in constructing Figure 3 [11].

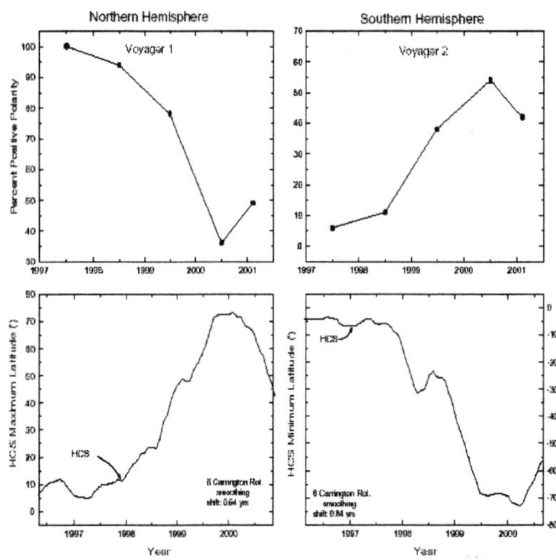

Figure 3. Top left. The percentage of positive polarities observed by Voyager 1 in the years 1997 to 2001. Top right. The percentage of positive polarities observed by Voyager 2 in the years 1997 to 2001. Bottom left. The maximum latitudinal extent of the HCS as a function of time computed from solar observations. Bottom right. The minimum latitudinal extent of the HCS as a function of time computed from solar observations.

We conclude that the variation of magnetic polarity observed by V1 and V2 from 1997 to 2001 was caused by the increasing width of the sector zone with increasing solar activity, which was related to the changing northern and southern latitudinal extent of the footpoints of the HCS.

ACKNOWLEDGMENTS

It was largely through discussions with Dr. M. Acuna and the outstanding efforts of T. McClanahan and S. Kramer that we were able to utilize the measurements of very weak magnetic fields discussed in this paper. Valuable assistance was also provided by Diana Taggart. N.F. Ness acknowledges partial support by Jet Propulsion Laboratory contract 959167. We thank T. Hoeksema (WSO/Stanford) for helpful comments on the manuscript and for providing his results on the positions of the heliospheric current sheet, which are an essential part of this paper, on the World Wide Web. N.R. Sheeley, Jr. and Y.-M. Wang received financial support from NASA and the NRL/ONR 6.1 Accelerated Research Initiative, Solar Magnetism and the Earth's Environment.

REFERENCES

1. Burlaga, L.F., Ness, N.F., Wang, Y.-M., and Sheeley, N.R., *J. Geophys. Res.* **103 (A10)**, 23,727-23,732, (1998).
2. Burlaga, L.F., Ness, N.F., and McDonald, F.B., *J. Geophys. Res.* **100 (A8)**, 14,763-14,771, (1995).
3. Burlaga, L.F., and Ness, N.F., *J. Geophys. Res.* **101 (A6)**, 13,473-13,481, (1996).
4. Parker, E. N., *Astrophys. J.* **128**, 664, (1958).
5. Parker, E.N., *Interplanetary Dynamical Processes*, Publisher, New York, 1963, p. 138.
6. Ness, N.F., and Burlaga, L.F., *J. Geophys. Res.* **106 (A8)**, 15,803-15,817, (2001).
7. Schulz, M., *Astrophys. Space Sci.* **34**, 371, (1973).
8. Behannon, K.W., Burlaga, L.F., Hoeksema, J.T., and Klein, L.W., *J. Geophys. Res.* **94 (A2)**, 1245-1260, (1989).
9. Burlaga, L. F., and Ness, N.F., *J. Geophys. Res,* **98, (A10)** 17,451-17,460, (1993).
10. Hoeksema, J.T., *Adv. Space Res.* **11**, 15, (1991).
11. Hoeksema, J.T., *Space Sci. Rev.* **72,** 137, (1995).
12. Balogh, A., and Smith, E.J., *Space. Sci. Rev.* **97**, 147-160, (2001).
13. Wang, Y.-M., and Sheeley, N.R., Jr., *Astrophys. J.* **355,** 726-732, (1990).

Note: The full discussion of these results is given in: Heliospheric Magnetic Field Strength and Polarity from 1 to 81 AU During the Ascending Phase of Solar Cycle 23, L.F. Burlaga, N.F. Ness, Y.-M. Wang and N.R. Sheeley, Jr., *J. Geophys. Res.*, Submitted March, 2002.

LASCO Observations Of The K-Corona From Solar Minimum To Solar Maximum And Beyond

Michael D. Andrews

Computational Physics Inc., NRL Code 7660, Washington DC 20375 e-mail andrews@louis14.nrl.navy.mil

Russell A. Howard

E.O. Hulburt Center for Space Research, Naval Research Laboratory, Washington DC 20375

Abstract. The LASCO C2 and C3 coronagraphs on SOHO have been recording a regular series of images of the corona since May 1996. This sequence of data covers the period of solar minimum, the increase to solar maximum, and the beginning of the decline toward the next solar minimum. The images have been analyzed to determine the brightness of the K-corona (solar photons Thomson scattered from free electrons). The total brightness of the K-corona is approximately constant from May 1996 through May 1997. The brightness is then seen to increase steadily until early in the year 2000. The structure of the K-corona changes dramatically with solar cycle. The shape as seen in C2 becomes almost circular at solar maximum while the C3 images continue to show equatorial streamers. The magnitude of the solar cycle variation decreases as the height increases. We present data animations (movies) to show the large-scale structure. We have inverted 28-day averages of the white light images to determine radial profiles of electron density. We present these electron profiles, show how they vary as a function of both latitude and time, and compare our observed profiles with other models and observations.

INTRODUCTION

The LASCO coronagraphs [1] have been recording a regular cadence of standard images since May 1996. While there have been a number of short interruptions, the only significant break in this sequence was in 1998 at the time of the SOHO mission interruption.. This image sequence has been processed to generate average images of the K-corona. These images are averages over a period that is usually two weeks.

In the following sections we first briefly discuss the data processing method used to separate the K-corona from the other contributions to the LASCO Images. The time variation of the observed coronal structures is presented. The observed coronal brightness has been used to derive electron density (N_e) as a function of solar height and latitude. The time variation of N_e is presented and these results compared with previously published values.

The time variation of the coronal structures is shown in data animations (movies in mpeg format) that are included on the CD-ROM published with this volume. The CD-ROM also includes 25 figures in gif format that are intended to expand the material in this paper

DATA PROCESSING

An individual LASCO image, I, can be approximated by a sum as follows:

$$I = F + K + S + N. \qquad (1)$$

where F and K are the signals from the F- and K-corona, S is due to stray light and other instrumental effects, and N is noise. For the purpose of this study, stars, planets, and cosmic-rays are considered noise.

CP679, *Solar Wind Ten: Proceedings of the Tenth International Solar Wind Conference*, edited by M. Velli, R. Bruno, and F. Malara

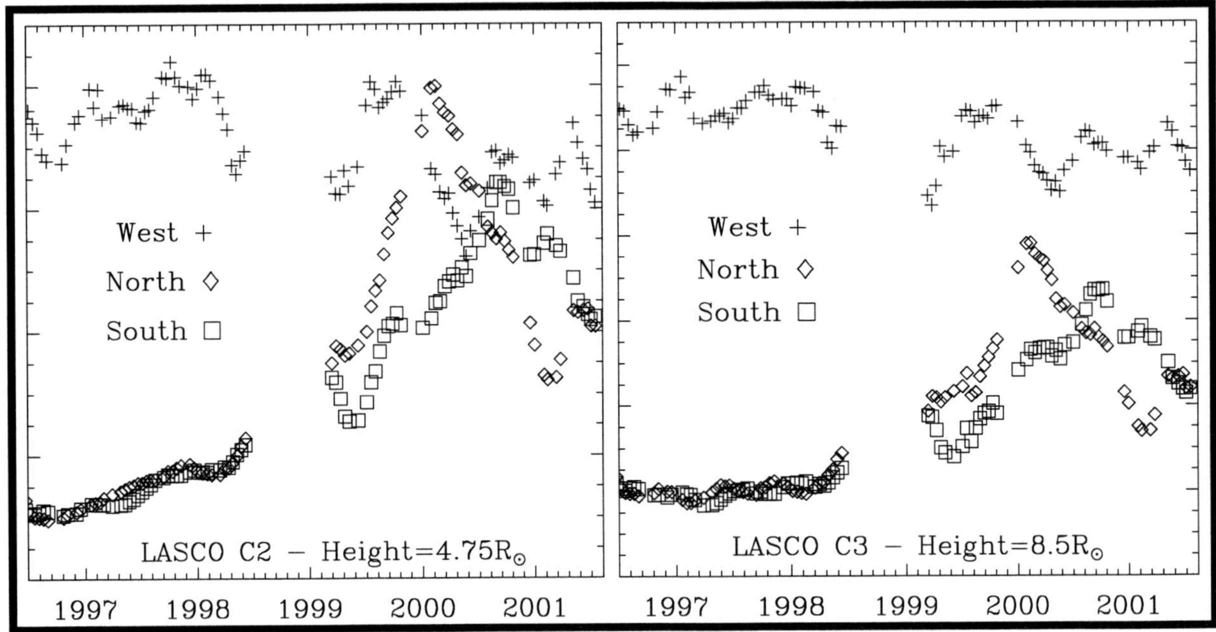

All of the images taken in the "standard" optical configuration are assembled for a period that is typically two weeks. Two images are calculated from these data. A minimum image, M, is formed by collecting the smallest observed value at each pixel over the entire set of images and forming an image from these pixel values.

The images are filtered to remove star and cosmic-rays. An average image, A, is calculated by taking the arithmatic mean using only one image per hour. With the assumption that the F-corona and S are constant over two weeks, these two images can be expressed as

$$M = F + S + k. \tag{2}$$

$$A = F + \langle K \rangle + S + \langle N \rangle, \tag{3}$$

where k is the smallest observed value of the K-corona and <K> and <N> are the average value of the K-corona and noise. The noise level is usually small with two important exceptions. Several planets are so bright that the filtering algorithm does not remove the signal. The result is wide streaks in the average image. The second exception occurs during times of large, energetic particle events. These events can generate a background level in the images that cannot be removed and increases the level of the average images. These particle events also cause a raised level in the minimum images.

The difference between the minimum and average images is then <K> - k. The minimum image is further processed to determine k. This is done by resampling the image to form a rectangulararray in which the X-coordinate is angle and the Y-coordinate is height. A "high-pass" filter is applied to isolate the structures that are narrow in angle. The residual K-corona is seen only where there are persistent streamers. These structures are effectively isolated since they extend over the entire height range in the high-pass filtered image. This estimate of k is added to the difference to form the final K-corona image.

Images from December 1996 are included on the CD-ROM to illustrate this process. The files min96.gif and dif96.gif, included on the CD-ROM, show the minimum image and the raw difference image. The final K-corona image is shown in the file k96.gif. The difference image shows dark lanes in the center of the streamer structures. This is corrected in the K-corona image. The editing process does produce an artifact at low heights in C3 along the southern edge of the east-limb streamer. This is due to the post in the southeast quadrant. (Additional examples of 4-week averaged K-corona images are shown in the files label mmm##.gif on the CD-ROM.)

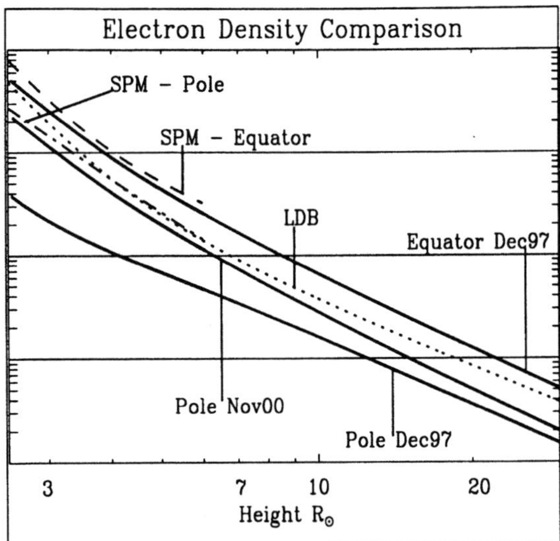

FIGURE 2. Comparison of calculated electron densities. The heavy lines show the electron density profiles reported in this paper. A single profile is shown for the equator along with polar profiles near solar min and max. Also shown are the electron density functions of [3], [4].

TIME VARIATION OF THE K-CORONA

The variation of the K-corona is illustrated in Figure 1 and the movies k2.mpg and k3.mpg on the CD-ROM that accompanies this volume. Figure 1 slows the average coronal brightness as observed by C2 (C3) for 30° wedges over the height range of 4.5-5.0 (7.0-10.0) $R_\odot$. There are three wedges shown centered on 0°, 180°, and 270° solar latitude, e.g., north, south, and west. The plotted values have been smoothed over seven two-week intervals

The equatorial K-corona is approximately 10% brighter during 1996-1998 than it is at later times. The magnitude of this change in brightness decreases with increasing height.

The variation above the poles is much more interesting. The observed brightness at low heights begins a slow, steady increase in early 1997. This continues until the SOHO mission interruption in June 1998. The brightness increase is not observed by C3 until the beginning of 1998. Throughout this period, the coronal brightness in the north and south are approximately equal.

There is no high-quality data until February 1999. The northern corona brightens very rapidly until early in 2000. At that time, the northern corona is brighter than is observed in the equatorial region. The brightening in the south is more gradual with the peak values observed nine months later than in the north. The C3 observations show the same time dependence. However, the magnitude of the brightness increase is smaller. At 4.75 $R_\odot$, the northern corona brightness increases by a factor of eight. The brightening in the south is a factor of six. At 8.5 $R_\odot$, the corresponding factors are about 4 and 3.5.

The shape of the corona as observed by C2 changes dramatically with solar cycle. This time variation is clearly seen in the mpeg movies on the CD-ROM. Near minimum, the structure is clearly dominated by equatorial streamers. The shape becomes almost circular in 2000. While the polar regions do show significant brightening, the C3 images are dominated by equatorial streamers at al times. The equatorial K-corona does not show significant solar cycle variation at heights above ~ 20 $R_\odot$.

ELECTRON DENSITY

We have used the observed brightness profiles to calculate electron density, N_e, profiles. Adjacent two - week K-corona images were combined to generate a four-week average for C2 and C3. The C2 and C3 averages were combined to generate vectors of brightness versus height at a constant polar angle. Each of these vectors is inverted to determine a profile of electron density as a function of distance from the sun [2].

The inversions have been done for seven periods: December 1996 and 1997, March 1998, July 1999, February 2000, November 2000 and 2001. A function of the form

$$N_e = c_1 r^{-2.2} + c_2 r^{-3} + c_3 r^{-4} + c_4 r^{-5} + c_5 r^{-6} \qquad (4)$$

was used for all of the inversions. The density at large heights depends primarily on the leading exponent. The value of –2.2 was selected to duplicate the 1AU value of 7/cm^3 adopted in [3]. The seven sets of data that were processed yielded values of 6-9/cm^3 in the ecliptic at 1AU and values of 0.1-0.5/cm^3 above the pole at a distance of 1AU.

DISCUSSION

Two files are included on the CD-ROM for each of the time periods for which the inversion has been done. The files labeled mmmyy_23.gif show the C2 and C3 K-corona images. The files labeled mmmyy_b_n.gif show the profiles of brightness and the N_e profiles derived by inverting the brightness data.

The profiles labeled east and west are slices through the equatorial region. These profiles show only small variation. Above about 20 $R_{\odot}$, the variation is negligible as can be seen by comparing the data from December 1996, solar minimum conditions, with February 2000 when the polar K-corona was the brightest.

The polar profiles are much more variable. The brightness and density at low heights vary by a factor of almost 10. The magnitude of the variation decrease with increasing height but does remain significant and is estimated to be a factor of perhaps four at a few AU.

Figure 2 presents a comparison of the electron densities calculated using this method and the profiles previous published by [3], [4]. The thick lines are the results of this study. A single equatorial profile is plotted as typical of all of the equatorial profiles. The eqatorial profile is NOT a function of solar activity at least at significant height in the corona. The polar profiles of December 1996 and November 2000 are plotted as the solar minimum and maximum profiles, respectively.

The N_e profile of [3] was determined by radio tracking of coronal disturbances under solar minimum conditions. This profile is significantly lower than our equatorial density. Since the density decreases rapidly with increase solar latitude at solar minimum, this difference could be explained if the events considered in [3] were at an angle from the equator. The electron density of [3] is very similar to these results at an angle of 15° from the equator.

The two profiles in [4] derived from SMM observations from May 1973 to February 1974 in the declining phase of solar cycle 20 are limit to low hsieghts. The equatorial and polar profiles of [4] are very close to these results for the equator and the maximum activity polar profile.

CONCLUSION

We have presented a new method of generating average images of the K-corona. The variation of coronal brightness with solar cycle can be easily seen in the mpeg files included on the CD-ROM that accompanies this volume. The variation of the K-corona with solar cycle is shown to be much larger over the pole than near the equator. The magnitude of the variation decreases with increase solar altitude.

The K-corona images have been combined to calculate monthly averaged electron density profiles as function of distance and solar latitude. These profiles have been shown to be in reasonable agreement with previously published results [3], [4].

ACKNOWLEDGMENTS

The LASCO instruments are on the SOHO space-craft. SOHO is a mission on international cooperation between NASA and ESA. The LASCO project at NRL is supported by NASA under contract S-13631-Y.

REFERENCES

1. Brueckner, G.E., Howard, R.A., Koomen, M.J. et al. *Solar Phys.*, **162**, 357-402, (1995).

2. Hayes, A.P., Vourlidas, A., and Howard, R.A. *Astrophys. J.*, **548**, 1081-1086 (2001).

3. LeBlanc, Y., Dulk, G.A., and Bougeret, J.-L. *Solar Phys.*, **183**, 165-180, (1998)..

4. Saito, K., Poland, A.J., and Munro, R.H. *Solar Phys.*, **55**, 121-134, (1977).

Location of the Termination Shock at Solar Maximum

E. C. Stone and A. C. Cummings

California Institute of Technology, Pasadena, CA 91125 USA

Abstract. During the recent solar maximum, Voyager 1 was beyond 80 AU. Extrapolation of the small gradients of anomalous cosmic rays at solar minimum and the larger gradients at solar maximum indicate that the solar wind termination shock is at $\lesssim$ 92 AU at the beginning of 2002.

INTRODUCTION

Various estimates of the location of the solar wind termination shock suggest a distance in the range of 90±10 AU (see summary in [1] and [2]). During the recent solar maximum in 2001, Voyager 1 and 2 (V1 and V2) were at distances of >80 AU and >63 AU, respectively. During this time the radial mean free path for 1.5 GV anomalous cosmic ray (ACR) O was a factor of ten smaller than during the most recent solar minimum period [3]. The resulting differences in the radial gradients of ACRs between periods of minimum and maximum solar modulation can be used to estimate the distance to the termination shock where ACRs are accelerated.

EXTRAPOLATION OF LARGE SCALE, LONG TERM GRADIENTS

Since they were launched twenty-five years ago, the two Voyager spacecraft have been observing changes in the modulation of ACRs, as did Pioneer 10 and 11 for much of this period. As shown in Figure 1, the Voyager observations now span three periods of maximum solar modulation. With Pioneer and Voyager, it is possible to determine both the latitudinal and radial gradients. As shown in Figure 1, the radial gradient at solar minimum is much smaller than during periods of solar maximum, although at solar maximum in 1990-91 individual gradient measurements are affected by propagating transients. The small radial gradient beyond ~10 AU at solar minimum for the positive solar magnetic polarity (A>0) is also reflected in the nearly identical flux observed by Pioneer 10 in 1978 and 1996 when it was at ~15 and ~64 AU [4, 5].

To illustrate the large scale variation more directly, the ACR O flux is shown in Figure 2 as a function of the heliocentric distance at the time of the observation. Although the observed fluxes are affected by latitudinal as well as radial gradients, latitudinal gradients are small at solar maximum when the solar magnetic field is reversing polarity. The intensity minima marking periods of maximum modulation at the three spacecraft show a strong radial dependence corresponding to a large scale gradient of ~7.5%/AU from ~10 to 80 AU.

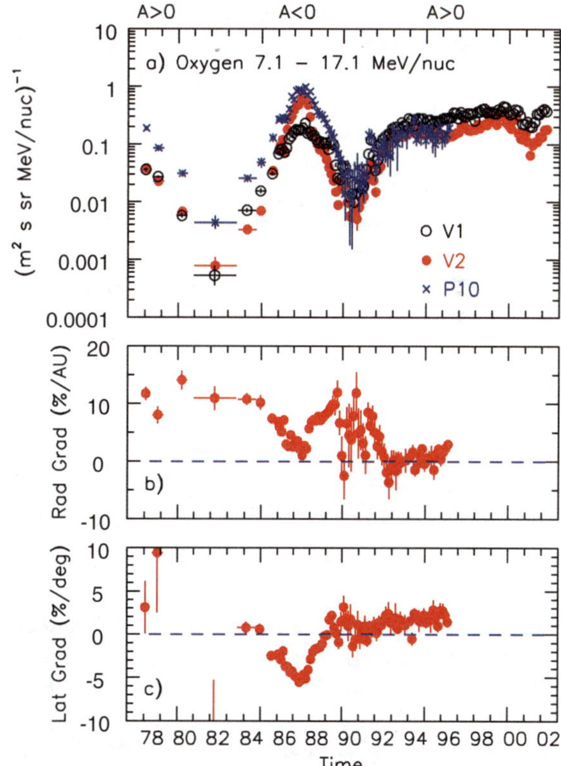

FIGURE 1. a) Intensity of ACR O with 7.1-17.1 MeV/nuc versus time at three spacecraft. b) Radial gradient of ACR O versus time. c) Latitudinal gradient of ACR O versus time.

CP679, *Solar Wind Ten: Proceedings of the Tenth International Solar Wind Conference*, edited by M. Velli, R. Bruno, and F. Malara

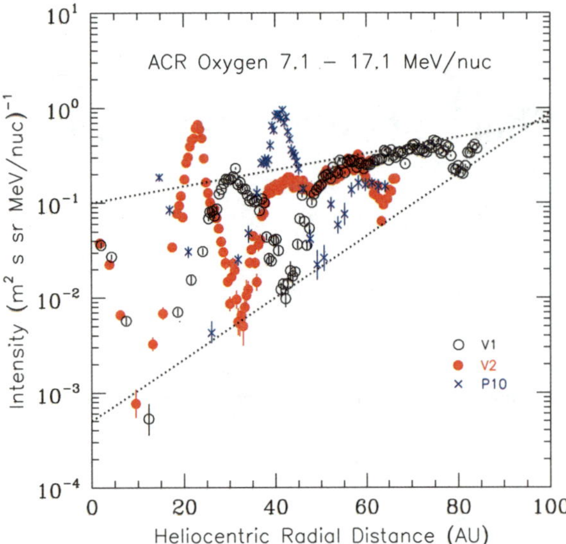

FIGURE 2. ACR O intensities shown in Figure 1 plotted versus radial distance of the observation. The two dotted lines show the intensity variation for gradients of 2%/AU (solar minimum) and 7.5%/AU (solar maximum).

During solar minimum periods, latitudinal as well as radial gradients contribute to the differences in intensities at the three spacecraft. However, the latitude of V1 changed by only 1° while moving from 50 to 80 AU, so the long term intensity change was dominated by the radial gradient of ~2%/AU shown in Figure 2.

The converging intensity gradients in Figure 2 are an indication of the distance of the shock from V1 during periods of solar maximum when the heliosphere is smallest [see, e.g., 6]. Extrapolation of the gradients in Figure 2 indicates a shock location at ~95 AU at solar maximum, assuming that the shock location and source intensity of ACR O at the latitude of V1 did not change between 1999 and 2001.

EXTRAPOLATION OF INSTANTANEOUS, LOCALIZED GRADIENTS

The extrapolation in Figure 2 is based on determinations of the large scale gradients measured over long periods of time and assumes there is no evolution of the radial profile or changes in the shock location. To allow for such changes, radial gradients for specific epochs and radial intervals can be estimated using simultaneous V1 and V2 observations. However, gradient measurements for individual epochs around solar maximum can be significantly affected by outward propagating disturbances that affect V1 and V2 at different times.

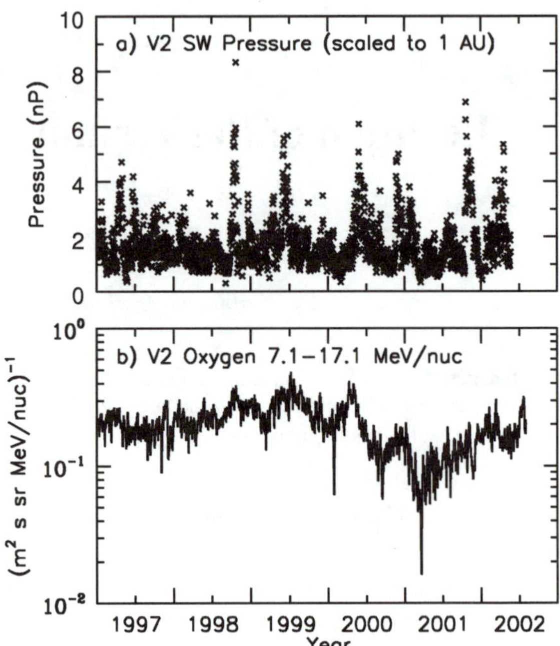

FIGURE 3. a) V2 solar wind pressure scaled by r^2 versus time, where r is the radial distance of the observation in AU. b) Five-day moving averages of the intensity of oxygen with 7.1-17.1 MeV/nuc observed at V2 versus time.

Figure 3 illustrates some of the transient effects at V2. The increases in solar wind dynamic pressure indicate the arrival of merged interaction regions (MIRs). These are barriers to cosmic rays that can also act as a snowplow producing an enhanced intensity [see, e.g., 7]. This effect is especially apparent with the arrival of the large global merged interaction region (GMIR) in 2000.4, but is also seen for some of the other MIRs. To be useful for extrapolating outward, gradients must be measured during periods without such large transient increases. In addition, GMIRs serve as barriers, producing a transient decrease in ACR intensity while propagating between the observation point and the termination shock [8].

As shown in Figure 4, the projected arrival time at 95 AU of the GMIR observed at V2 in 2000.4 was 2000.8. The intensity remained depressed until a subsequent GMIR passed 95 AU about six months later. Subsequent MIRs appear to have had little effect on the high energy ACR O intensity, suggesting that the gradients observed in the most recent five 52-day periods in Figure 4 can be used to extrapolate from V1 out to the shock, as shown in Figure 5. Note that the gradients in Figure 5 are of the form $j \propto r^{\alpha}$, as would result from a diffusion mean free path proportional to r as indicated by Voyager data [3].

Estimates of the radial gradient for three years of solar minimum are also indicated in Figure 5. Yearly averages

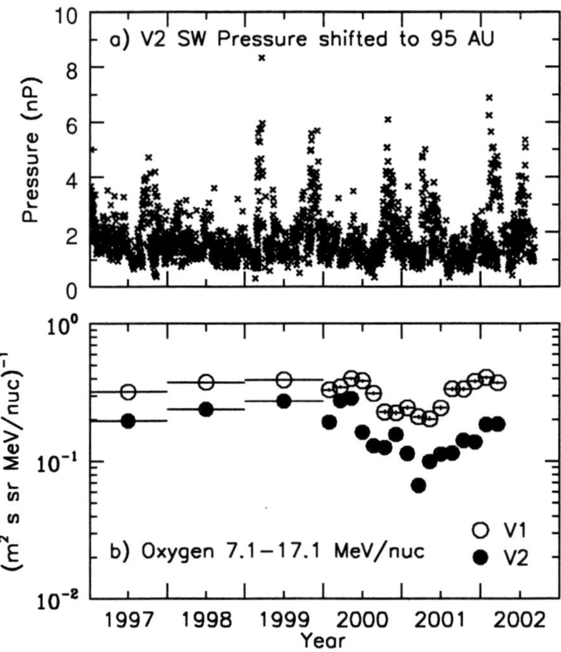

FIGURE 4. a) Same as Figure 3a except the observations have been shifted to 95 AU assuming the solar wind speed measured at V2. b) Intensity of ACR O at V1 and V2 versus time.

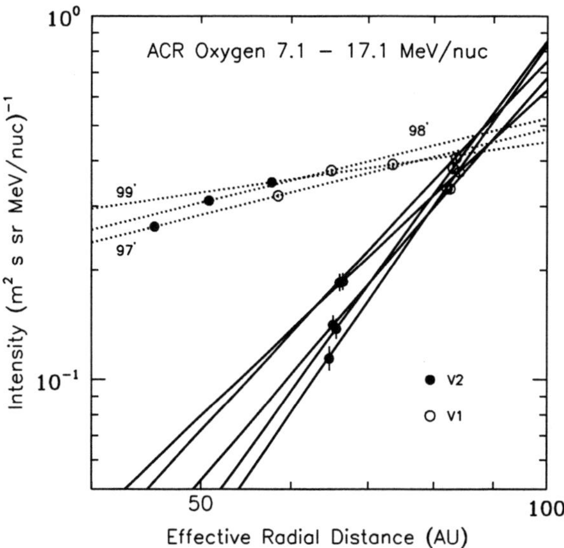

FIGURE 5. Same as Figure 4b, except plotted versus effective radial distance of the observations. The effective radial distances differ from actual radial distances only for 1997, 1998, and 1999 and the procedure is described in the text. The lines connect V1 and V2 data measured at the same time.

have been used in order to reduce the effects of transients on the derived gradients. During solar minimum, the gradient between V1 and V2 is also affected by a latitudinal gradient. To estimate the radial gradient, the V2 intensity has been adjusted to the latitude of V1 by 1.9%/degree, the average latitudinal gradient observed between mid-1994 and mid-1996 when Pioneer 10 was operational (see Figure 1).

The intensities in 1997, 1998, and 1999 are also affected by changes in the location of the termination shock. MHD calculations (Whang and Burlaga, personal communication; see also [6]) indicate that relative to the beginning of 2002, the average location of the shock in 1997, 1998, and 1999 was at increased distances of $\Delta r_s \approx 14.5, 8.2$, and 1.1 AU, respectively. The increased distance results in a corresponding decrease in flux according to $j(r) = j_s(r/r_s^{'})^{\alpha}$, where j_s is the intensity at a shock located at $r_s^{'} = r_s + \Delta r_s$ and r_s is the (unknown) shock location in 2002. From this relationship, it follows that the flux $j(r)$ for a shock at $r_s^{'}$ is the same as the flux $j(r^{'})$ for a shock at r_s and an effective radial location $r^{'}$ if $r^{'} = kr$, where the scaling factor is $k = (r_s/r_s^{'})$. The scaling factor k is relatively insensitive to the exact values of r_s. For $90 \leq r_s \leq 100$ AU, k = 0.87, 0.92, and 0.99 to within 0.01 for 1997, 1998, and 1999, respectively. These factors have been applied to the radial positions of the solar minimum intensities in Figure 5.

As in Figure 2, the gradients in Figure 5 exhibit a strong difference between solar minimum and solar maximum. The intersections of the extrapolated intensities provide estimates of the location (r_s) of the shock in 2002, assuming that the intensity at the shock was the same for all of the epochs. The systematic differences among the five radial profiles for solar maximum profiles give some indication of the variability from epoch to epoch due to systematic uncertainties from residual transient effects. The intersections correspond to a shock location of $\lesssim 92$ AU in early 2002.

CONCLUSION

It is possible that before Voyager 1 reaches 95 AU in 2005, it will encounter the shock. If the shock is beyond that distance, however, Voyager 1 may not reach it before the onset of increased solar wind pressure pushes the shock outward ahead of the spacecraft for several more years.

ACKNOWLEDGMENTS

We thank W. R. Webber for providing Pioneer 10 data. The V2 solar wind data was obtained from J. Richardson and the MIT Space Plasma Group at their website `http://web.mit.edu/space/www/voyager.html`

REFERENCES

1. Stone, E. C., *Science*, **293**, 55–56 (2001).
2. Stone, E. C., and Cummings, A. C., "Estimate of the location of the solar wind termination shock," in *Proc. 27th Internat. Cosmic Ray Conf.*, Hamburg, 2001, vol. 10, pp. 4263–4266.
3. Cummings, A. C., and Stone, E. C., "Inferring energetic particle mean free paths from observations of anomalous cosmic rays in the outer heliosphere at solar maximum," in *Proc. 27th Internat. Cosmic Ray Conf.*, Hamburg, 2001, vol. 10, pp. 4243–4246.
4. Cummings, A. C., and Stone, E. C., *Adv. Space Res.*, **23**, 509–520 (1999).
5. McDonald, F. B., Cummings, A. C., Lal, N., McGuire, R. E., and Stone, E. C., "Cosmic rays in the heliosphere over the solar minimum of cycle 22," in *Proc. 27th Internat. Cosmic Ray Conf.*, Hamburg, 2001, vol. 9, pp. 3830–3833.
6. Whang, Y. C., and Burlaga, L. F., *Geophys. Res. Lett.*, **27**, 1607–1610 (2000).
7. Burlaga, L. F., McDonald, F. B., and Ness, N. F., "Intense Magnetic Fields Observed by Voyager 2 during 1998," in *Proc. 26th Internat. Cosmic Ray Conf.*, Salt Lake City, 1999, vol. 7, pp. 107–110.
8. le Roux, J. A., and Fichtner, H., *J. Geophys. Res.*, **104**, 4708–4730 (1999).

Active-Region Sources of Solar Wind near Solar Maximum

Paulett C. Liewer*, Marcia Neugebauer*# and Thomas Zurbuchen†

*Jet Propulsion Laboratory, California Institute of Technology, Pasadena, CA 91109
*# Lunar and Planetary Laboratory, University of Arizona, Tucson, AZ 85721
†University of Michigan, Space Physics Research Laboratory Ann Arbor, MI 48105

Abstract. Previous studies of the source regions of solar wind sampled by ACE and Ulysses showed that some solar wind originates from open flux areas in active regions. These sources were labeled *active region sources* when there was no corresponding coronal hole in the He 10830 Å synoptic maps. Here, we present results on an investigation of the magnetic topology of these active region sources and a search for corresponding features in EUV and soft X-ray images. In most, but not all, cases, a dark hole or lane is seen in the EUV and SXT image as for familiar coronal hole sources. However, in one case, the soft-X ray images and the magnetic model showed a coronal structure quite different from typical coronal hole structure. Using ACE data, we also find that the solar wind from these active region sources generally has a higher Oxygen charge state than wind from the Helium-10830Å coronal hole sources, indicating a hotter source region, consistent with the active region source interpretation.

INTRODUCTION

The sources of solar wind at solar minimum are understood to some extent: fast wind (V >700 km/s) comes from the large polar coronal holes and slow wind (V <600 km/s) comes from a broad region surrounding the heliospheric current sheet and also from equatorial coronal holes (Neugebauer et. al., 1998).

Recently, Neugebauer et. al. (2002, hereafter Paper I) reported results of a study of the sources and properties of solar wind near solar maximum. The photospheric sources were determined using a magnetic mapping procedure. We found that much wind sampled by ACE and Ulysses near solar maximum originated from open field regions in or adjacent to active regions. When the calculated source was a region of high field strength and there was no corresponding coronal hole in He 10830 Å synoptic map, these sources were labeled *active region sources* to distinguish them from the traditional "coronal hole" sources that show a coronal hole in NSO He 10830 Å synoptic maps. The source region temperature of these active region sources, as determined from the solar wind Oxygen charge state data, was found to be higher than for coronal hole source regions, consistent with an active region source identification. No other property investigated, including velocity, showed a statistically significant difference between coronal hole and active region sources.

While others have concluded that some open flux and solar wind originates in active regions (Levine, 1977; Kojima et. al, 1999), Paper I was the first to use *in situ* data to show that wind from open flux in active regions has distinctive properties (e. g., more variability, higher freezing-in temperature).

Here we report a more detailed study of several of the active regions sources that were part of the statistical study in Paper I. Specifically, we ask if these active regions sources are evident as dark features in EUV and soft X-ray images (see Levine, 1977). We study the topology of the corona above these active region sources using both the images and potential magnetic field models. We also compare the Oxygen charge state of wind from active region and coronal hole sources and find distinct signatures. Using these signatures, we find that we are sometimes able to determine the type of source regions in cases where the magnetic mapping alone was ambiguous

It is known that He 10830Å is an imperfect diagnostic for determining the footpoints of coronal open flux. Because of this, some would argue that open flux sources of solar wind in active regions should also be called coronal holes. However, because the wind from these active-region sources shows different properties than wind from traditional "coronal holes," we use the term *active region source* for these open flux sources in active regions with no corresponding hole in He 10830 Å synoptic maps.

CP679, *Solar Wind Ten: Proceedings of the Tenth International Solar Wind Conference*,
edited by M. Velli, R. Bruno, and F. Malara

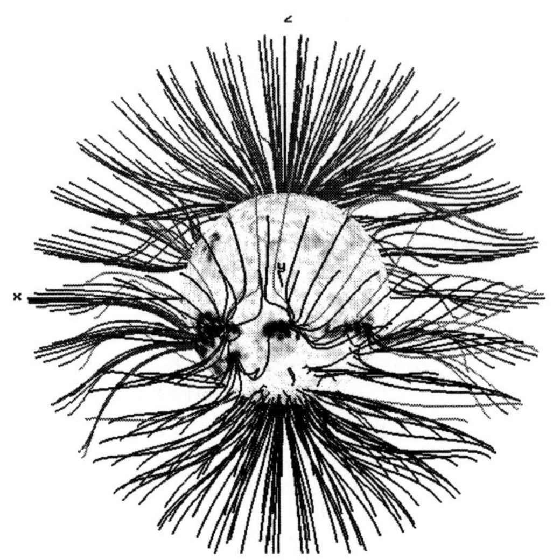

FIGURE 1. CR 1934 magnetogram and magnetic model showing the mapping of open flux from the source surface ($R=2.5R_S$) to the photosphere. Red (blue) is inward (outward) polarity; green lines lie near the current sheet. April 7, 1998.

DETERMINING SOURCE REGIONS

The photospheric sources of solar wind sampled by ACE is determined using a two-step mapping process as in Neugebauer et al. (1998). First, the solar wind speed measured at ACE is used to ballistically map the solar wind back to a source surface, R= 2.5 R_s (solar radii), assuming the wind travels radially at the measured velocity from the source surface to ACE.

In the second step, a potential field source surface model extrapolated from an NSO Kitt Peak synoptic magnetogram is used to trace the solar wind from the source surface to the photosphere, assuming the wind follows the magnetic field lines. Further details can be found in Paper I. We check each mapping against three criteria: (1) Does the potential field model predict open flux where the NSO He 10830Å synoptic maps show coronal holes? (2) Does the polarity of the magnetic field at the spacecraft agree with the polarity predicted by the mapping? (3) Does the solar wind data show evidence of a source region boundary at the locations predicted by the model? Paper I analyzed the mappings for four Carrington rotations where these criteria were generally well satisfied. Here, we further investigate the active region sources from these rotations. While the potential field models show the open flux of the He 10830 Å synoptic coronal holes, the models often show additional small regions of open flux in active regions with no corresponding He 10830 Å coronal hole in the synoptic maps.

Mapping Results for CR1934

Figure 1 shows open flux field lines for the potential field source surface model computed for Carrington rotation 1934 (corresponding to days 77-104 of 1998 at the Sun). The field lines were generated by tracing field lines from the source surface to the photosphere; the lines are uniformly spaced in area at the source surface. Note that open flux comes from both the polar coronal holes and the active region belt.

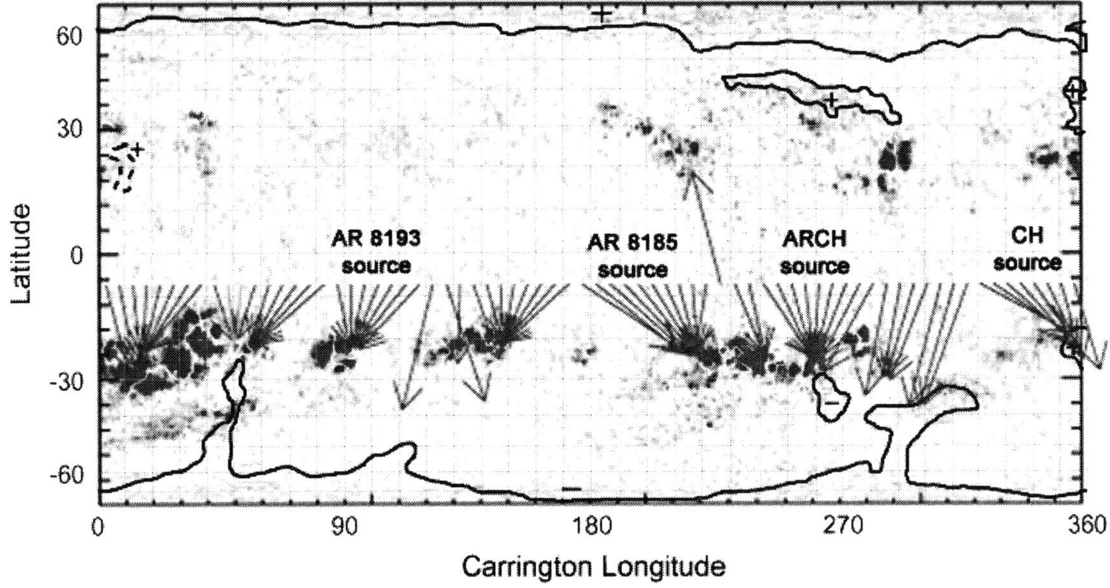

FIGURE 2. Carrington rotation 1934 magnetogram, synoptic coronal holes and arrows mapping ACE solar wind from the source surface (arrow tails) to the photosphere (heads). Several different sources and streams are evident. The two active-region sources (AR), one coronal hole source (CH), and one indeterminate (ARCH) source are discussed in the text.

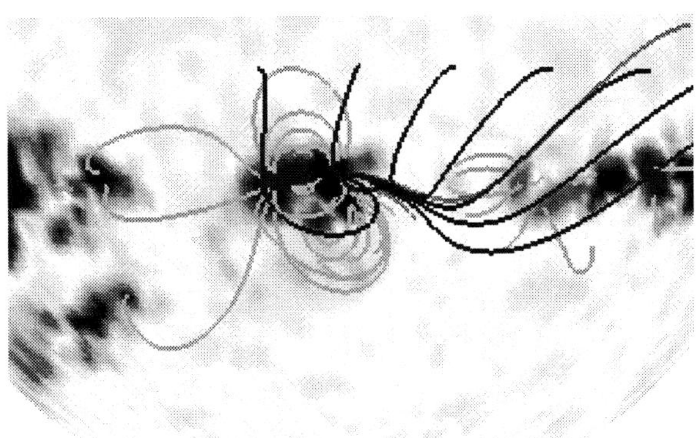

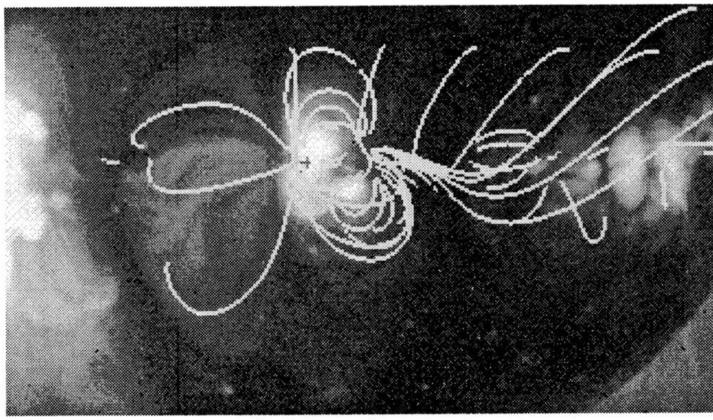

FIGURE 3. Left: Magnetic model of active region AR 8193. The dark lines are the open field lines of ACE solar wind that mapped to AR 8193 (see Fig. 2). The open flux comes from one outer edge of the bipolar AR 8193. The light lines are closed field lines. View as in Fig. 1: April 7, 1998. RIGHT: Yohkoh SXT image with the model magnetic field lines of the left panel superimposed. The open flux is at the edge of the dark lane seen by SXT (April 7, 1998 12:51 UT).

In Figure 2, the arrows showing the potential field mapping of the ACE solar wind from the source surface ($R=2.5R_S$) to the photosphere overlie the synoptic magnetogram. The arrow tail is the solar wind field line location at the source surface (as determined from the ballistic mapping) and the head is at the photospheric source (field line footpoint). Also shown are the coronal holes from the He 10830 Å synoptic maps. The field polarity measured at ACE agrees well with that predicted by the model. Note that most of the sampled wind comes from the active region belt even though the polar coronal holes are still present. Several different sources and streams are evident from the bunching of the arrow heads.

In this rotation, we identify two active regions sources (see Fig. 2). The stream with source surface location between 90°-110° longitude is on open field lines that trace to active region AR 8193. The wind with source surface longitude 180°-220° is on open field lines that map to active region AR 8185. Neither source region has a synoptic 10830 Å coronal hole nearby. On the other hand, the wind from 250°-280° longitude maps to a strong field region with an adjacent synoptic 10830 Å coronal hole. Because the mapping is not accurate enough to discriminate between open flux from the active region or from the adjacent coronal hole, we label this an ARCH source to indicate the uncertainty. Wind from 320°-360° is on open flux from the synoptic 10830 Å coronal hole at ~360° and this is labeled a CH source region. Below, we look at topology of the two AR source and solar wind from all four sources.

Topology of the Active Region Sources

The left panel of Figure 3 shows the magnetic topology of AR 8193 as computed from the potential field model; the view corresponds to the Earth's view of the Sun on April 7, 1998. The dark lines are the open field lines of ACE solar wind that mapped to AR 8193 (see Fig. 2). The light lines are closed field lines. Note that the open flux comes from an outer edge of the bipolar active region, e.g. the edge away from the opposite polarity region of the same active region. The right panel of Figure 3 shows these same field lines superimposed on a Yohkoh SXT image from April 7, 1998. Note that the open flux lines from the active region are at the edge of a dark lane in the SXT image. Dark lanes in Skylab images were also associated with open flux by Levine (1977). Here, the dark lane is much larger than the photospheric region with open flux. For this AR source, there was no coronal hole in either the daily or the synoptic 10830Å maps.

The magnetic topology of the other active region source for this rotation, AR 8185, was similar in that the open flux came from the outer edge of the active region. In this case, a transient coronal hole appears in daily 10830 Å maps for March 24-27. The transient coronal hole is also evident as a dark hole in both the SOHO EIT and Yohkoh SXT images for March 28, 1998 when AR 8185 was near central meridian

We have also looked in detail at two other AR sources from other rotations analyzed in Paper I. AR 8681 in CR 1953 also shows open flux coming from an outer edge of the active region. A very small transient coronal hole is seen in the daily 10830Å maps. A region much larger than this transient coronal hole is seen as a dark lane in the EIT and SXT images for August 29, 1999 when the active region was near the central meridian. For the above three active region sources, the open flux comes from an outer edge of the active region and a dark hole or lane in the SXT and EIT images can be associated with the open flux.

In contrast, the active region in CR 1953 has a different topology (not shown). Here, two neighboring bipolar strong field regions form a dual dipole; the open flux comes from the interior of one polarity region where both the magnetic model and the Yohkoh SXT image show a magnetic separatrix: To one side of the open flux, the loops connect eastward (to the opposite polarity region of the same bipolar regions), and on the other side of the open flux, the loops connect to the west (to the opposite polarity region of the neighboring bipolar region). The separatrix can be well seen in the SXT images of December 18, 1999 as a somewhat dimmer "part" in the loop structure, but no dark feature was evident. Thus, not surprisingly, there appear to be a variety of magnetic topologies that allow open flux from active regions.

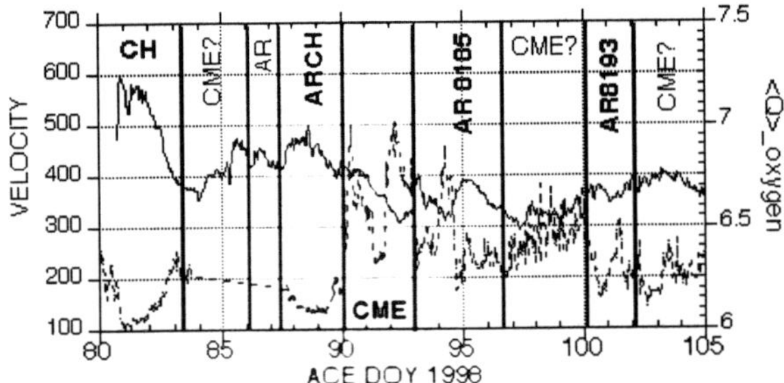

FIGURE 4. ACE solar wind proton velocity (solid) and average Oxygen charge state <Q> (dashed). The vertical lines separate wind from different sources, determined from the mapping (Figure 2). Carrington longitude goes from right to left: Day 80 maps to ~360° and Day 105 maps to 55°.

Solar Wind Charge State Data

Figure 4 shows ACE solar wind data [proton velocity (solid line) and average Oxygen charge state <Q>(dashed line)] for ACE days 80-105 of 1998, covering the four source regions labeled in Figure 2 (as well as the intervening streams). Day 80 maps to ~360° Carrington longitude and Day 105 maps to 55°. These data are typical of the four rotations studied. Note the low Oxygen charge state in the wind from the coronal hole (ACE days 80-83); here, the minimum <Q> is close to 6, similar to wind from polar coronal holes at solar minimum. The speed of the wind from this coronal hole, however, is somewhat low, V<600 km/sec. Note that <Q> is significantly higher and more variable in the wind from the two active regions, AR 8185 and AR 8193. The wind from these active region sources also appears to be composed of multiple substreams as noted in Paper I. Both characteristics are consistent with our active region source identification. The wind from active regions may have been heated on closed field lines which opened by reconnection, allowing the plasma to escape. Such an interpretation is consistent with that of Zurbuchen et al. (2000) who concluded that slow wind with higher charge states emerged from closed coronal magnetic structures.

Note in Figure 4 that the value and behavior of <Q> in the ARCH wind is clearly similar to that of coronal hole wind. Using this information, we can now assume that this wind is from the small coronal hole at 260-270° longitude and not the nearby active region. Thus the charge state data allows us to remove some ambiguities in the mapping procedure.

CONCLUSIONS

We have studied active region sources of solar wind. Our study of the topology shows that in most cases, a dark hole or lane is seen in the EUV and SXT image at the location of the active region source, as for familiar coronal hole sources. In one case, when the magnetic model showed the open flux was associated with a separatrix within a strong field region, the soft-X ray images did not show a dark lane or hole. Here, the separatrix was evident in the X-ray loop structure. Using ACE data, we find that the solar wind from these active region sources generally has a higher Oxygen charge state than the coronal hole wind, indicating a hotter source region, consistent with the active region source interpretation.

ACKNOWLEDGMENTS

We thank Z. Mikic, SAIC, for the potential magnetic field and field line tracing programs used in this work. Much of this work is the result of research performed at the Jet Propulsion Laboratory of the California Institute of Technology under a contract with NASA. The University of Michigan work was also supported by NASA.

REFERENCES

1. Neugebauer, M., et al., Spatial structure of the solar wind and comparison *J. Geophys. Res., 103*, 14587 (1998).
2. Neugebauer, M., Liewer, P. C., Smith, E. J., Skoug, R. M., and Zurbuchen, T. H., Sources of the Solar Wind at Solar Activity Maximum, *J. Geophys. Res.* (2002).
3. Levine, R. H., Large scale solar magnetic fields and coronal holes, in *Coronal Holes and High Speed Wind Streams*, ed. J. B. Zirker, pp. 103-143, Co. Assoc. Univ. Press, Boulder (1977).
4. Kojima, M., K. Fujiki, T. Ohmi, M. Tokamaru, A. Yokobe, and K. Hakamada, Low-speed solar wind from the vicinity of solar active regions, *J. Geophys. Res., 104*, 16993 (1999).
5. Zurbuchen, T. H., S. Hefti, L. A. Fisk, G. Gloeckler, and N. A. Schwadron, Magnetic structure of the slow solar wind: Constraints from composition data, *J. Geophys. Res., 105*, 18327 (2000).

Ubiquitous Open Magnetic Field Lines in the Inner Corona

Richard Woo[1] and Shadia Rifai Habbal[2,3]

[1]Jet Propulsion Laboratory, California Institute of Technology, Pasadena, California 91109
[2]University of Wales, Department of Physics, Aberystwyth, Ceredigion, SY23 3BZ, UK
[3]Harvard-Smithsonian Center for Astrophysics, Cambridge, MA 02138

Abstract. The notion that density structure reflects magnetic field lines makes it possible to deduce information on coronal magnetic fields from density measurements. The purpose of this paper is to summarize the observational evidence for ubiquitous open magnetic field lines in the inner corona from density measurements. Based on both global and filamentary structures, these density measurements explain the unexpected predominance of the radial component of coronal magnetic field discovered in polarimetric observations over three decades ago.

INTRODUCTION

It has become increasingly clear that the Sun's magnetic field is the main source of structure and variability in the solar atmosphere, yet the magnetic field measurements we have are essentially only those of the photosphere. Our knowledge and understanding of the coronal magnetic field has instead come from two indirect sources. Assuming that density reflects magnetic field, we have gleaned the Sun's magnetism from density structure observed in white-light images. Polar plumes, the small-scale, faint, thin, hair-like structures resembling the lines of force of a bar magnet, suggest that polar coronal holes are the source of open field lines (1). The abundance of photospheric magnetograms made on a routine basis has led solar astronomers to extrapolate photospheric measurements into the solar corona using the so-called source surface magnetic field models. Magnetic field modeling represents the other source of information on coronal magnetic fields. Results obtained from these models reinforce the impressions from the density structure of white-light pictures by showing that open magnetic field lines emanate from polar coronal holes, while closed fields are associated with coronal streamers (2). These models also represent the most widely used tool for tracing the distant solar wind observed by spacecraft back to its source at the Sun (3).

For nearly five decades, radio occultation measurements based on a wide variety of radio propagation and scattering phenomena have yielded a wealth of information on coronal density. The extensive but disparate results have only recently been combined to form a unified picture of coronal structure (4). For a long time, progress with these measurements was stymied by the paradigm that the observed density variations were caused solely by turbulence convected along with the solar wind; fine-scale structures which were aligned with the magnetic field and rotated with the Sun were not considered. In the meantime, improvements in spatial resolution and the processing of coronal images revealed increasingly smaller structures that filled the solar corona. An important advance in coronal density studies was the relating of structure observed by the two major remote sensing tools, radio occultation and white-light measurements (5). Synergistic comparisons of the results from these two observing techniques improved our understanding of the distribution of coronal density structure, and by implication, coronal magnetic fields.

Knowing where open magnetic field lines prevail in the inner corona is fundamental not only for understanding the Sun's magnetism, but also for determining how the distant solar wind directly probed by interplanetary spacecraft connects to the Sun. The purpose of this paper is to summarize the evidence from density observations that ubiquitous open field lines permeate the entire inner corona rather than being restricted to coronal holes.

DENSITY IMPRINT OF THE SUN

When the variation of density in the outer corona observed by ranging and white-light measurements was compared with the density variation in the inner corona closest to the Sun (which we refer to as the density 'imprint' of the Sun), a completely unexpected result was found. Hidden in the tenuous dark regions of the outer corona beyond the closed-field regions of the inner corona, and away from the tapered bright coronal streamers, was the density imprint of the Sun (4). How could such a major connection have escaped detection for so long?

CP679, *Solar Wind Ten: Proceedings of the Tenth International Solar Wind Conference*,
edited by M. Velli, R. Bruno, and F. Malara

There are two reasons. First, measurements of the outer corona may not have been sensitive enough to detect the tenuous coronal hole regions. Second, not evident in the white-light pictures because of the steep decline in density with radial distance, but brought to light in the quantitative profiles, is the fact that density changes by one to two orders of magnitude from streamer to coronal hole in the outer corona, but only by a factor of 2–3 in the inner corona (4). These differences reflect the fact that density falls off more slowly in closed-field streamers than in open-field coronal holes. The bright streamers and their conspicuous high-contrast boundaries overwhelm the tenuous and relatively low-contrast (factor of 2–3) imprint of the Sun, even when the measurements are sensitive enough. Consequently, the streamer boundaries in the outer corona are mistaken for the polar coronal hole boundaries, and for over two decades, white-light images have misled us into believing that polar coronal holes diverge and expand superradially into interplanetary space (6).

Instrumental sensitivity is not an issue with direct measurements of the polar solar wind made by Ulysses. Conditioned by three decades of *in situ* measurements confined to near the ecliptic plane where streamers reside, we have been accustomed to seeing large variations in the solar wind plasma parameters. At high latitude, Ulysses found a wind exhibiting relatively small daily variations in all of its solar wind properties (7, 8). Since polar coronal holes were thought to be structureless (9), it was natural to conclude that the fast wind probed by Ulysses must have come from the diverging polar coronal holes. Comparisons with simultaneous white-light measurements of the inner corona have shown that the factor of 2 daily density variations of the fast wind observed by Ulysses beyond 2 AU represented the variations of the imprint of the Sun (10, 11).

The imprint of the Sun found in the outer corona and in the distant solar wind probed by Ulysses could only have been transported there by ubiquitous and approximately radial open field lines emanating from both the quiet Sun and coronal hole. Is there more direct observational evidence for these open fields? The answer is yes, in observations of small-scale density structures.

SMALL-SCALE STRUCTURES

To anyone who has observed a total solar eclipse in person, the corona looks richer than that captured by a photograph. Three decades ago, Serge Koutchmy started processing eclipse pictures by enhancing their density gradients. The result was images that not only showed the boundaries of the large-scale streamers, but many of the small-scale filamentary and raylike structures seen by eye (Figure 1). Why do unprocessed and processed pictures give such different impressions?

The answer lies in the fact that small-scale (filamentary or striated) structures are difficult to see in unprocessed pictures, not because they are faint, but because their density levels are not significantly different from the average level, i.e., $\Delta n/n$ is small. These low-contrast features are revealed in the processed images because of their steep density gradients. Such structures are also more

FIGURE 1. Combined eclipse and SOHO LASCO C2 white-light images of the August 11, 1999 solar eclipse (courtesy of Serge Koutchmy). The upper image is unprocessed, while the lower image has been processed to enhance density gradients. The Yohkoh image of the solar disk is superimposed on the processed image to show that distinct open field lines are seen to be emanating from active regions on the northeast and northwest limbs of the Sun.

readily observed by eclipse watchers because the human eye is adept at distinguishing low-contrast features.

The small-scale structures revealed in white-light pictures represent those individual structures that have the largest density gradients. The corona is actually filled with many more structures that are not seen in the images, either because they are smaller than the spatial resolution or because their density gradients are weak. Highly sensitive and highly sampled Doppler measurements detect all of the structures (5). They are consistent with a continuum of filamentary structures whose scale sizes are described by an inverse power-law spectrum with the smallest filamentary structures being about 1 km at the Sun, more than two orders of magnitude smaller than those observed in white-light images (12). It is these unseen filamentary structures in white-light images that collectively carry the imprint of the Sun into interplanetary space. Do we have any evidence from these ubiquitous but hidden (from white-light images) filamentary structures themselves that they, like the imprint they transport, extend radially from the Sun? The answer is yes, if we look at their density gradients.

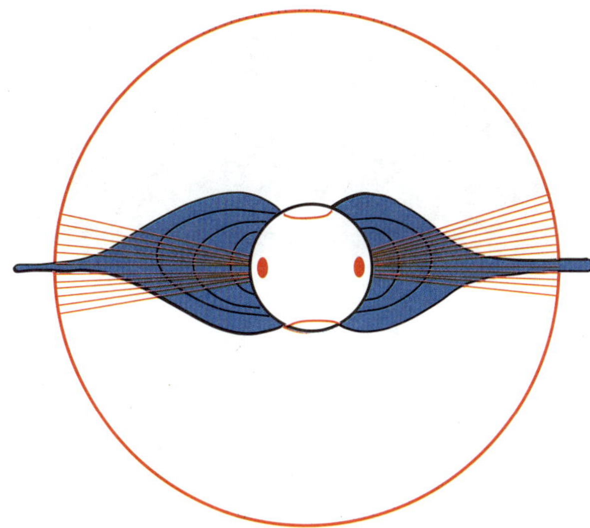

FIGURE 2. Cartoon illustrating how the density gradients that characterize the ubiquitous filamentary structures are weakest in coronal holes (blue lines), higher in the quiet Sun (black lines), and highest in the active regions (red lines).

DENSITY GRADIENTS

Doppler measurements show that density gradients across the fine-scale filamentary structures are lowest in the radial extension of coronal holes, slightly higher in that of the quiet Sun, and highest over active regions of the Sun. Solar eclipse measurements reveal that the density gradients of the filamentary structures are also highest in the brightest regions of the inner corona. Density gradients, therefore, characterize the small-scale filamentary structures by their source region at the Sun, and demonstrate that the structures extend roughly radially outwards from the Sun, carrying with them the imprint of the Sun (Figure 2) (13).

Some of the individual open structures that emanate from the active regions are evident in images processed to enhance density gradients (Figures 1), and are a surprise since active regions are often thought of as exclusively closed field regions. The open field lines can be expected to carry some imprint of active regions, and hence their underlying sunspots, into the solar wind.

It is not possible to image the distant solar wind, but shown in Figure 3 is a synoptic map of solar wind velocity constructed based on Ulysses measurements over a period of 5 solar rotations (3). Solar wind flow would be expected to be slowest over the active regions where the closed fields are strongest. The interaction between fast and slow wind changes the distribution of velocity as the solar wind flows away from the Sun, but the islands of slowest velocity observed by Ulysses are unmistakably the imprint of the active regions in the distant solar wind.

CORONAL MAGNETIC FIELD

Although there are no measurements of magnetic field strength, magnetic field direction was inferred from polarization measurements of the corona over thirty years ago (14). Such measurements showed a coronal magnetic field that was unexpectedly predominantly radial. Unexplained by magnetic field models that extrapolated photospheric fields into the corona, these results were largely forgotten until now (15). It is clear that the radial extension of the density imprint of the Sun, the ubiquitous filamentary structures that are radially oriented, and the predominance of radial magnetic fields, are all manifestations of the same phenomenon, reinforcing the notion that coronal density reflects coronal fields.

Because density associated with closed magnetic field structures is significantly higher than that of open structures, the subjective impression of white-light images is to over-emphasize streamers and their evolution at the expense of the radial extension of open structures. Closed and open structures are treated more equally by polarization measurements, as well as by white-light images processed to enhance density gradients. These latter images portray the truer picture of magnetic field topology: although streamers are mixed regions of open and closed fields, open fields far outnumber the closed fields.

Streamers are shaped by the non-radial component of coronal magnetic field that appears to be associated with the strong and solar cycle dependent photospheric field. On the other hand, the radial field originating from all over the Sun seems to be the coronal counterpart of the

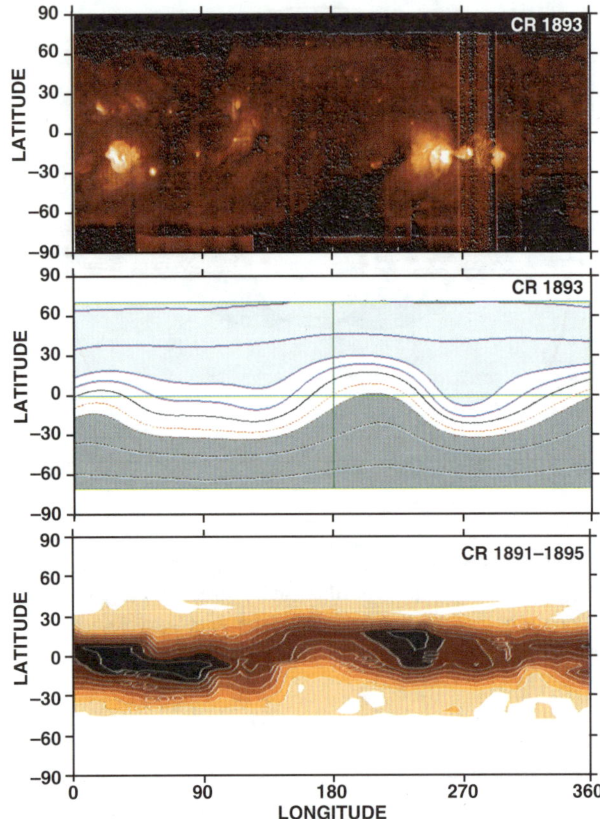

FIGURE 3. Latitude-longitude plots of: Yohkoh soft X-ray measurements for Carrington Rotation CR 1893 (upper panel), source surface magnetic field strength contours produced by the Wilcox Solar Observatory for CR 1893 (middle panel), and solar wind speed contours reproduced from (3) and constructed from solar wind measurements by Ulysses and Wind during CR 1891-1895 (lower panel). The influence of the heliospheric current sheet and the imprint of the active regions are both evident in the flow speed distribution of the distant solar wind.

weak photospheric field that depends weakly on the solar cycle (16). The strong photospheric field is thought to be produced by the classical dynamo, while the weak field by the fast dynamo (17). Ironically, streamers taper to the heliospheric current sheet which occupies only a small volume of interplanetary space. Evidence for both the non-radial streamer field and the radially evolved imprint of the Sun can be found in the distant solar wind engulfing the heliospheric current sheet (Figure 3).

CONCLUSIONS

The new results on coronal magnetic fields are a consequence of unifying extensive density observations made over three decades by white-light, radio occultation and *in situ* measurements. The density measurements cover extensive latitudes and longitudes, span time periods as long as a year, and include global as well as the smallest filamentary structures in the corona. Magnetic field models have shown that magnetic fields are open in coronal holes and predominantly closed everywhere else, a result that has not been validated by coronal magnetic field observations. With the growing evidence from density measurements that ubiquitous open field lines permeate the solar corona, there is a need to take a closer look at the assumptions of these models to better understand why they seem to be missing the radial component of the coronal magnetic field.

ACKNOWLEDGMENTS

This paper describes research carried out at the Jet Propulsion Laboratory, California Institute of Technology, under a contract with the National Aeronautics and Space Administration.

REFERENCES

1. DeForest, C.E., et al., *Solar Phys.* **175**, 393–410(1997).
2. Linker, J.A., et al., *J. Geophys. Res.* **104**, 9809–9830 (1998).
3. Neugebauer, M., et al., *J. Geophys. Res.* **103**, 14587–14599 (1998).
4. Woo, R., and Habbal, S.R., in *Solar Wind Nine*, eds. S.R. Habbal, R. Esser, J.V. Hollweg, and P.A. Isenberg, AIP Conference Proceedings 471, Woodbury, New York, 1999, pp. 71–76.
5. Woo, R., in *Solar Wind Eight*, eds. D. Winterhalter, J.T. Gosling, S.R. Habbal, W.S. Kurth, and M. Neugebauer, AIP Conference Proceedings 382, Woodbury, New York, 1996, pp. 38–43.
6. Munro, R., and Jackson, B., *Astrophys. J.* **213**, 874–886 (1977).
7. McComas, D., et al., *Astrophys. J.* **489**, L103–106 (1997).
8. Balogh, A., Marsden, R.G., and Smith, E.J., *The Heliosphere Near Solar Minimum*, Springer Praxis, London, 2001.
9. Guhathakurta, M., et al., *Astrophys. J.* **458**, 817–831 (1996).
10. Habbal, S.R., and Woo, R., *Astrophys. J.* **549**, L253–L256 (2001).
11. Woo, R., et al., *Astrophys. J.* **538**, L171–L174 (2000).
12. Woo, R., *Nature* **379**, 321–322 (1996).
13. Woo, R., et al., *in preparation* (2002).
14. Eddy, J., et al., *Solar Phys.* **30**, 351–369 (1973).
15. Habbal, S.R., et al., *Astrophys. J.* **558**, 852–858 (2001).
16. Harvey, K.L, in *IAU Colloq. 143, The Sun as a Variable Star: Stellar and Solar Irradiance Variations,* eds. J. Pap, C. Frölich, H. Hudson, and S. Solanki, Cambridge University Press, Cambridge, 1994, 217.
17. Cattaneo, F., and Hughes, D.W., *Astron. Geophys.*, **42**, 3.18–3.22 (2001).

Large-scale structure of the polar solar wind at solar maximum: ULYSSES/URAP observations

Karine Issautier*, Michel Moncuquet* and Sang Hoang*

*Observatoire de Paris, LESIA, CNRS UMR 8109, 92195 Meudon, France

Abstract. We outline the recent in situ radio observations obtained by Ulysses during its second fast latitude scan near the 2001 solar maximum. From $\sim 72^\circ$ N, Ulysses was embedded in a continuous fast wind associated with a northern polar coronal hole. During that period, we enlight the variation of the electron density and thermal temperature with the heliocentric distance and latitude, obtained with the quasi-thermal noise spectroscopy method. Since the scaled mass flux is observed roughly constant, we derive the profile of the electron thermal temperature, assuming a simple power-law model. Indeed, we find that the electron density varies as $R^{-2.0\pm0.1}$ while the thermal temperature varies as $R^{-0.7\pm0.1}$, in good agreement with the results obtained in polar coronal holes in 1995 near solar minimum.

INTRODUCTION

Ulysses is the unique out-of-ecliptic mission that gives us the opportunity since its launch in 1990 to continuously study the 3-D structure of the heliosphere during a complete solar cycle and over a wide range of latitude. During its first pole-to-pole latitude scan in 1994-1995, Ulysses confirmed the rather simple structure of the corona near the solar activity minimum, and showed that two types of solar wind dominate the heliosphere during that period. Indeed, a steady-state fast solar wind is continuously observed at high latitudes as coming from large polar coronal holes, whereas within $\sim \pm 20^\circ$ a more complicated mixture of winds, corresponding to slow, intermediate speed flows, interaction regions and transient events predominates in the ecliptic plane. At high latitude poleward 40°S, the inverse-square-drop-off of the density suggests that the large-scale latitudinal or temporal variations of the solar wind were very small. Thus the observed electron thermal temperature profile of $R^{-0.64}$ is considered to be a genuine radial profile [1].

Ulysses explores its second polar passage during the rising solar cycle to solar maximum. The state of the corona appears to be dramatically different and much more complex than at minimum, as one can see from the SOHO/LASCO coronograph images. In addition, the pole-to-pole radio spectrogram acquired from November 2000 to mid-October 2001 [3] exhibits large-scale variations, and numerous solar radio emissions at high frequency are observed in consistency with the increasing solar activity. During all its journey from pole-to-pole, Ulysses encounters different regimes of wind, slow and intermediate wind from streamers, in addition to sporadic fast wind flows from small coronal holes. The unique long interval of fast solar wind is observed at high northern latitudes above $\sim 72^\circ$N nearly at the end of the polar pass.

SOLAR WIND ELECTRON PROPERTIES

In this paper, we focus on the large-scale electron properties of the steady-state fast solar wind associated with this northern polar coronal hole observed by Ulysses for a few weeks near the 2001 solar maximum. The data were obtained from 4 September 2001 (when the radial distance R was 1.76 AU and the heliolatitude was 72°) to 14 November 2001 (when R was 2.25 AU and the heliolatitude was 76°), beyond the north pole (for which the latitude 80.2° was reached on October 13rd 2001), using the quasi-thermal noise (QTN) method from the Unified Radio and Plasma wave (URAP) experiment [11] as explained below. Indeed, the restricted period mentioned above is based on solar wind speed observations [4] from which we found the largest roughly constant period of high-speed wind, around 750 km/s ±50 km/s.

Thermal Noise Spectroscopy Analysis

The so-called "quasi-thermal noise" is due to the random motion of the ambient particles which excite plasma waves near the plasma frequency f_p. The voltage induced on the wire electric antenna is measured by a sensitive radio receiver. The theoretical interpretation of the QTN

CP679, *Solar Wind Ten: Proceedings of the Tenth International Solar Wind Conference*,
edited by M. Velli, R. Bruno, and F. Malara

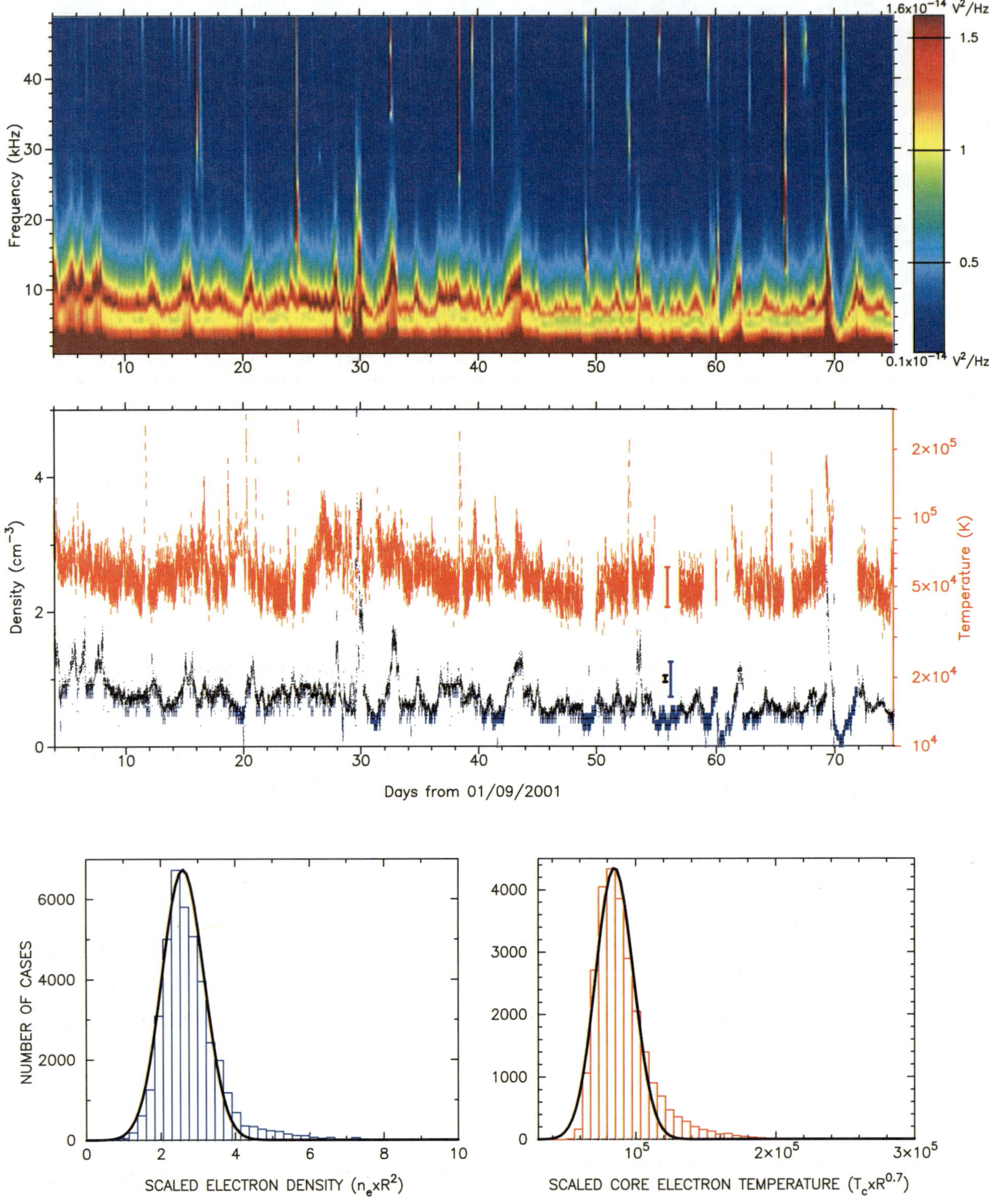

FIGURE 1. (Top) Radio spectrogram from URAP experiment obtained in the northern polar coronal hole near solar maximum. (Middle) From the thermal noise analysis, we deduce the electron density ($\sim 37,000$ black dots, given by the L-M fit and ~ 3000 blue dashes, deduced from the plasma frequency cut-off) and thermal temperature ($\sim 27,000$ red dashes, given by the L-M fit) versus time, together with their corresponding typical uncertainties (near day # 56). (Bottom) Histograms of the electron density and core temperature, normalized to 1 AU using radial variations obtained in the present study. The distributions roughly reveal a single wind population, with a tail of outliers which may correspond to the slow, hot and dense wind coming from the edge of the coronal hole.

spectrum, assuming a sum of maxwellian velocity distribution functions for the core and halo electron populations, yields in situ plasma parameters [5]. In particular, it gives an accurate determination of the electron density (within a few percent) since it is based on the plasma frequency location, therefore nearly independent of gain calibration and photoelectron perturbations, contrary to particle analysers. The overall shape and level of the spectrum also give the thermal temperature T_c, suprathermal parameters and solar wind speed [2]. In practice, we compute a model of quasi-thermal noise spectral, depending on all above-mentioned parameters (but mainly on f_p and T_c), and we fit the model to each measured spectrum by using the Levenberg-Marquardt (L-M) least-squares method [8, p678], which also provides the standard errors on the fitted parameters. This original and simple technique is a powerfull tool to obtain an accurate in situ solar wind plasma diagnostics and can be used to cross-check other plasma sensors.

In the following section, we select the core temperature for which the fitting of the voltage power spectrum to the observations is the best, corresponding indeed to the sigma of the fit better than 2.5%, as we did for the data set obtained near solar minimum. Moreover, to get the more accurate temperature measurements, it is important to take this constraint on the fit into account during solar maximum, since many solar radio emissions often pollute the high-frequency part of the spectrum, where the core temperature is partially sensitive.

Electron Density and Core Temperature

Figure 1 shows the radio spectrogram, displayed as frequency versus time, with intensity coded by a color scale, obtained from 4 september 2001 to 14 November 2001, when Ulysses explored the polar coronal hole of the Sun, from 72°N to 76°N beyond the north pole. One can see the plasma line, revealed by the intense noise in red, which fluctuates between 6 and 14 kHz. Using the thermal noise technique on 128-second radio spectra on Ulysses, we obtain the scatter plot of the corresponding electron density (black and blue), deduced from the plasma frequency ($n_e \propto f_p^2$), and the core electron temperature (red), shown in the middle panel. The red bold vertical bar is the averaged standard error ($\sim$ 20%) on T_c provided by the L-M fit. For very low densities, it is sometimes impossible to accuratly fit to the observed power spectrum: we get indeed too few measurements below f_p, since that frequency is too close to the lowest frequency of the URAP low-frequency receiver (1.25 kHz). In this case, we may use instead the empiric cut-off of f_p and so deduce only the plasma density, shown as blue vertical dashes in the middle panel. The blue bold vertical bar is the uncertainty on such estimated density, and corresponds to twice the difference between two equally spaced frequency channels (0.75 kHz), whereas the black bold vertical bar is the typical uncertainty ($\sim$ 5%) on the density as obtained from the L-M fit.

Statistical features of this polar wind can be enlighted in the bottom panel of Figure 1 from the histograms of the electron density and temperature normalized to 1 AU using radial variations obtained in the present study in the steady-state solar wind. The mean of the electron distribution is around 2.8 whereas the fitted Gaussian distribution to the data is centered at 2.6 (solid line), which is exactly the same value as obtained at solar minimum. On another hand, the histogram of temperature peaks at 8.5×10^4 K , computed from the fit of a Gaussian distribution to the data, and its mean value is around 9.3×10^4 K. Both distributions roughly reveal a single type of flow. However, the departure from a Gaussian distribution of the two histograms, which both present a "tail", shows that the fast wind is somehow "polluted" by other kind of winds, probably coming from the edge of the coronal hole. Even if the solar wind speed reaches the maximum of 800 km/s for this period, the fast solar wind appears to be less fast and much variable than during the solar minimum [4], as also observed in magnetic field data [10].

Radial Profiles

To analyse the basic trend of the electron density variation with the heliocentric distance, we fit a power law to the data set of about 40,000 data points obtained for that period. We get the radial profile for the density, represented in logarithmic scales in Figure 2, processing a "robust" linear regression by minimizing absolute deviation [8, p694]. The reason we cannot perform a least-squares linear regression, as we did in fast high-latitude wind during solar minimum[1], lies in the presence of numerous "outliers", clearly visible in these data. These outlying points, which form the tail of normalized histograms (bottom of Figure 1) are real (not artefact) but, as explained before, out from the fast stationary solar wind we want to focus on. In this case, the classical least-squares method must be disqualified because it is too much sensitive to these outliers, and a so-called robust method has been used instead; however it provides larger errors on the fitted line than least-squares technique used in [1]. The lower solid line on Figure 2 thus shows the robust linear regression result on density data; the power law index, deduced from the slope of the straight line, is -2.0 ± 0.1. The density extrapolated to 1 AU, given by the intercept, is found to be 2.75 cm^{-3}, which is in very good agreement with observations of the fast solar wind from the southern coronal hole around solar minimum.

This suggests that the density profile is independent of latitude or time variations, and corresponds to a spherical expansion, since the corresponding bulk velocity is roughly constant up to 76°N beyond the passage of the north pole. Moreover, the scaled mass flux is also observed [3] to be constant.

From the above finding, we consider this density data set as a rather good sample of stationary high-speed wind in spherical expansion and thus we may describe the thermal temperature with a power law depending on the radial distance only. In the same way as for the density, we find that the electron thermal temperature profile in such a flow varies as $R^{-0.7\pm0.1}$ between 1.76 and 2.25 AU, which is midway between adiabatic and isothermal behavior. The result of the fit is the upper solid line shown in Figure 2. The index of the power law is compatible with that found near solar minimum in 1995 in the southern polar coronal hole ($T_c \propto R^{-0.64}$), whereas the intercept of the temperature at 1 AU is 9×10^4 K, larger to the last solar minimum value of 7.5×10^4 K. The value of the power index is also in good agreement with the results of *Scime et al.* [1994], where the radial profile was deduced from 1 to 5 AU at solar maximum in the ecliptic plane.

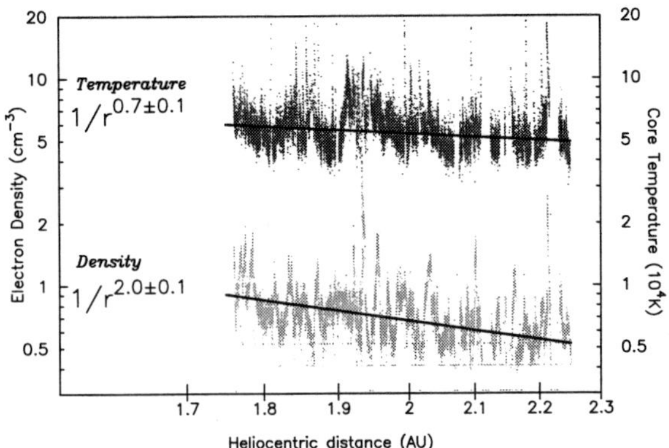

FIGURE 2. For the northern polar coronal hole near solar maximum, radial variations of the electron density (light grey dots) and core temperature (dark grey dots) are obtained by a robust linear fit. The solid black lines are the best fit power laws to the data with the index values shown.

SUMMARY & FINAL REMARKS

The second polar pass of Ulysses at solar maximum also provides an opportunity to measure the radial gradients of solar wind parameters in the fast wind, poleward of 72° N. From plasma thermal noise spectroscopy, we find a core electron temperature profile varying roughly between isothermal and adiabatic up to 2.25 AU. Likewise, the corresponding high-speed wind observed near solar maximum in a polar coronal hole has rather the same features as the single and well-defined flow measured poleward of 40°, near solar minimum during the first Ulysses fast latitude scan. However, slight discrepancies are probably due to the fact that Ulysses was not completely immersed in the polar coronal hole, because of its limited size, and might have measured the edge of the hole, which gives an average of density and temperature (see histograms) 5% and 20% larger respectively than at solar minimum.

It is important to note that either fluid or kinetic models are tempting to try to explain the electron temperature profile [7]. Thus, to correlate observations to any theory, one should accuratly measure the total electron temperature. An improved plasma diagnostics using the thermal noise spectroscopy with kappa velocity distribution function for the electrons is in progress, and would give in situ this fundamental parameter, which could then be directly compared with models and constraint them.

ACKNOWLEDGMENTS

The Ulysses URAP investigation is a collaboration of NASA/GSFC, Observatoire de Paris-Meudon, University of Minnesota, and CETP, Velizy, France. The French contribution is supported by CNES and CNRS.

REFERENCES

1. Issautier K. et al., *J. Geophys. Res.* **103**, 1969-1979 (1998a)
2. Issautier K., Meyer-Vernet N., Moncuquet M., Hoang S., McComas D.J., *J. Geophys. Res.*, **104**, 6691 (1998b)
3. Issautier K., Moncuquet M., Hoang S., *Geophys. Res. Lett.*, submitted (2003)
4. McComas et al., *Geophys. Res. Lett.* **29(9)**, doi 10.1029/2001GL014164 (2002)
5. Meyer-Vernet N. and Perche C., *J. Geophys. Res.* **94**, 2405-2415 (1989)
6. Meyer-Vernet N. and Issautier K., *J. Geophys. Res.* **103**, 29,705-29,717 (1998)
7. Meyer-Vernet N. et al., *Solar Wind 10*, AIP Proceedings (2002)
8. Press, W.H., et al., *Numerical Recipes*, 2nd edition, Cambridge Univ. Press, New-York (1992)
9. Scime et al., *J. Geophys. Res.* **99**, 23,401-23,410 (1994)
10. Smith et al., *Solar Wind 10*, AIP Proceedings (2002)
11. Stone R.G. et al., *Astron. Astrophys. Suppl.* **92**, 291 (1992)

HST Lyman-alpha absorption spectra toward nearby stars as a remote diagnostic of the heliosheath plasma properties

Vlad Izmodenov*, Brian Wood† and Rosine Lallement**

*Lomonosov Moscow State University, Department of Aeromechanics, Faculty of mechanics and mathematics, Vorob'evy gory, Glavnoe Zdanie MGU, 119899, Moscow, Russia
†JILA, University of Colorado, and NIST, Boulder, USA
**Service d'Aeronomie, CNRS, France

Abstract. A self-consistent model of the solar wind interaction with the two-component (plasma and H atoms) interstellar medium is employed to calculate heliospheric Lyman-alpha absorption spectra. It is shown that the heliospheric absorption consists of absorption in the hydrogen wall region and in the heliosheath. Both absorption components are computed in different directions. We perform a parametric study by varying the proton and H atom number densities of the interstellar gas. The theoretical absorption spectra are compared with Hubble Space Telescope observations toward six nearby stars. Results of the comparison and possible constraints on the structure of the heliospheric interface are presented and discussed.

INTRODUCTION

The heliospheric interface is formed by the interaction of the solar wind with the charged component of the interstellar medium (see Figure 1). Interstellar hydrogen atoms are disturbed in the interface by charge exchange with protons. In the heliospheric interface, atoms newly created by charge exchange have the properties of local protons. Since the plasma properties are different in the four regions of the heliospheric interface shown in Figure 1, the H atoms in the heliosphere can be separated into four populations, each having significantly different properties. For example, population 3 consists of the atoms created by charge exchange with relatively hot protons in the region of disturbed interstellar plasma around the heliopause (region 3 in Figure 1).

The Ly-α transition of atomic H is the strongest absorption line in stellar spectra. Thus, the heated and decelerated atomic hydrogen within the heliosphere produces a substantial amount of Ly-α absorption. The absorption was first detected by Linsky and Wood in 1996 [1] in Ly-α absorption spectra of the very nearby star α Cen taken by the Goddard High Resolution Spectrograph (GHRS) instrument on board the Hubble Space Telescope (HST). Since that time, it has been realized that the absorption can serve as a remote diagnostic of the heliospheric interface, and for stars in general their "astrospheric" interfaces.

In this study, we use the Baranov-Malama model [2, 3] of the heliospheric interface to model the heliospheric absorption toward six nearby stars. An advantage of our

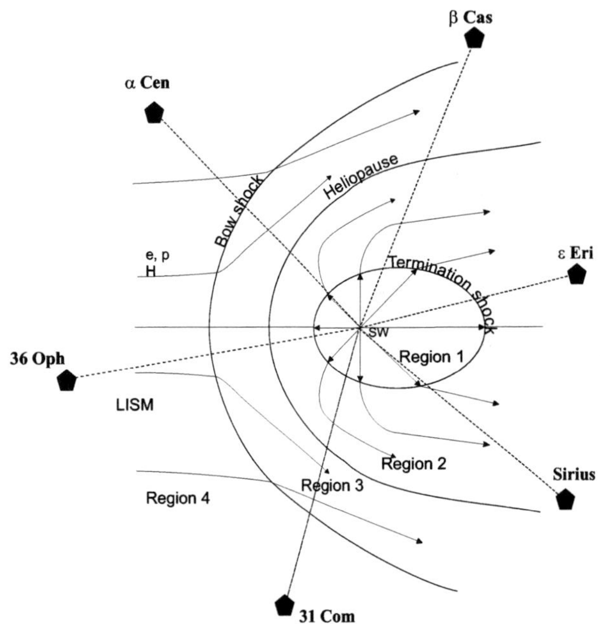

FIGURE 1. The heliospheric interface is the region of the solar wind interaction with LIC. The heliopause is a contact discontinuity, which separates the plasma wind from interstellar plasmas. The termination shock decelerates the supersonic solar wind. The bow shock may also exist in the interstellar medium. The heliospheric interface can be divided into four regions with significantly different plasma properties: 1) supersonic solar wind; 2) subsonic solar wind in the region between the heliopause and termination shock (i.e., the heliosheath) ; 3) disturbed interstellar plasma region (or "pile-up" region) around the heliopause; 4) undisturbed interstellar medium.

CP679, *Solar Wind Ten: Proceedings of the Tenth International Solar Wind Conference*, edited by M. Velli, R. Bruno, and F. Malara

model is the possibility of separation of heliospheric H atoms into four populations, as discussed above. This model advantage allows us to consider separately two types of heliospheric absorption, hydrogen wall absorption (or absorption by population 3) and heliosheath absorption (or absorption by population 4).

We computed the global heliospheric interface structure and distribution of interstellar hydrogen in the interface using the Baranov-Malama model. Our calculations assume the following values for the fully ionized solar wind at 1 AU: $n_{p,E} = 6.5$ cm^{-3}, $V_{p,E} = 450$ km s^{-1}, and $M_E = 10$, where $n_{p,E}$, $V_{p,E}$ and M_E are the proton number density, solar wind velocity, and Mach number, respectively. For the inflowing partially ionized interstellar gas, we assume a velocity of 25.6 km s^{-1} and a temperature of 7000 K. These values are consistent with in situ observations of interstellar helium. Two other input parameters, interstellar proton and H atom number densities, are varied. The values assumed for these parameters for the various models are listed in Table 1. We vary $n_{p,LIC}$ in the range $0.1 - 0.2$ cm^{-3}, while $n_{H,LIC}$ is varied in the range $0.05 - 0.2$ cm^{-3}.

The Ly-α lines of these six stars observed by HST are shown in Figure 2 (left column). All of the spectra except that of 36 Oph A were taken using the Goddard High Resolution Spectrograph (GHRS) instrument. The 36 Oph A data were obtained by the Space Telescope Imaging Spectrograph (STIS), which replaced GHRS in 1997. The spectra are plotted on a heliocentric velocity scale. All show broad, saturated H I absorption near line center and narrower deuterium (D I) absorption about 80 km s^{-1} blueward of the H I absorption. Figure 2 also shows the assumed intrinsic stellar Ly α lines and the best estimates for interstellar absorption (e.g., [4] and references therein). Excess absorption on the red side of the Ly α absorption, when present, is interpreted as heliospheric absorption, while excess absorption on the blue side, when present, is interpreted as astrospheric absorption [*Wood et al.*, 2000]. In this paper we focus on the red side of the Ly α absorption. The left side of the Ly α absorption spectra was discussed, for example, in [5, 6]. The stellar profiles and interstellar absorption estimates are based on previously published work, which is summarized in [4].

COMPARISON OF THEORY AND OBSERVATIONS

The middle and right columns of Figure 2 show the Ly α lines zoomed in on the red side of the HI absorption line, since that is where most of the heliospheric absorption is expected. The Ly α profile after interstellar absorption is shown in the plots as solid lines. The middle column also shows Ly α profiles after interstellar and hydrogen wall absorption, while the right column shows Ly α profiles after interstellar, hydrogen wall, and heliosheath absorption. The hydrogen wall and heliosheath components are shown for all six models discussed above (see Table 1).

TABLE 1. Sets of model parameters

Model	$n_{H,LIC}$	$n_{p,LIC}$	Notation in figure 2
1	0.10	0.10	solid
2	0.15	0.05	dotted
3	0.15	0.10	dashed
4	0.20	0.05	dot-dash
5	0.20	0.10	dot-dot-dot-dash
6	0.20	0.20	long dash

The absorption spectra toward upwind directions (36 Oph and α Cen) are fitted rather well by hydrogen wall absorption only. In comparing the theoretical absorption with data, the most important place for the model to agree with the data is at the base of the absorption, where it is very difficult (if not impossible) to correct any discrepancies by altering the assumed stellar Ly α profile [5]. Toward 36 Oph, model 2 produces slightly more absorption and model 6 produces slightly less absorption at its base compared to the data. For other models the differences between model predictions and HST data at the base can be reduced by small corrections to the assumed stellar profile and interstellar absorption. Toward α Cen, all models show slightly more absorption at the base than the data, except for model 6. It is interesting to note that model 6 fits the α Cen data remarkably well, while it does not fit as well the Ly α spectrum of 36 Oph. Note that these discrepancies with the data just discussed are small enough that they could in principle be corrected by reasonably small changes to the assumed stellar Ly α profile or interstellar absorption, although the better fits in Figure 2 are still to be preferred.

The heliosheath absorption, which is added to the interstellar and hydrogen wall absorption in the right column of Figure 2, does not change the fits much for the upwind lines of sight. Model 2 predicts too much absorption toward both 36 Oph and α Cen and produces the greatest difference with the data at the base of the absorption where it matters most. Consideration of the heliosheath absorption significantly increases the total heliospheric absorption predicted by model 4. The models with the worst agreement with the upwind directions (model 2 and model 4) correspond to the extreme case of small interstellar proton number density, $n_{p,LIC}$.

The observed absorption in the crosswind lines of sight to 31 Com and β Cas shows no need for any heliospheric absorption, and our models do not generally predict significant absorption in the crosswind directions that can be clearly distinguished from interstellar absorp-

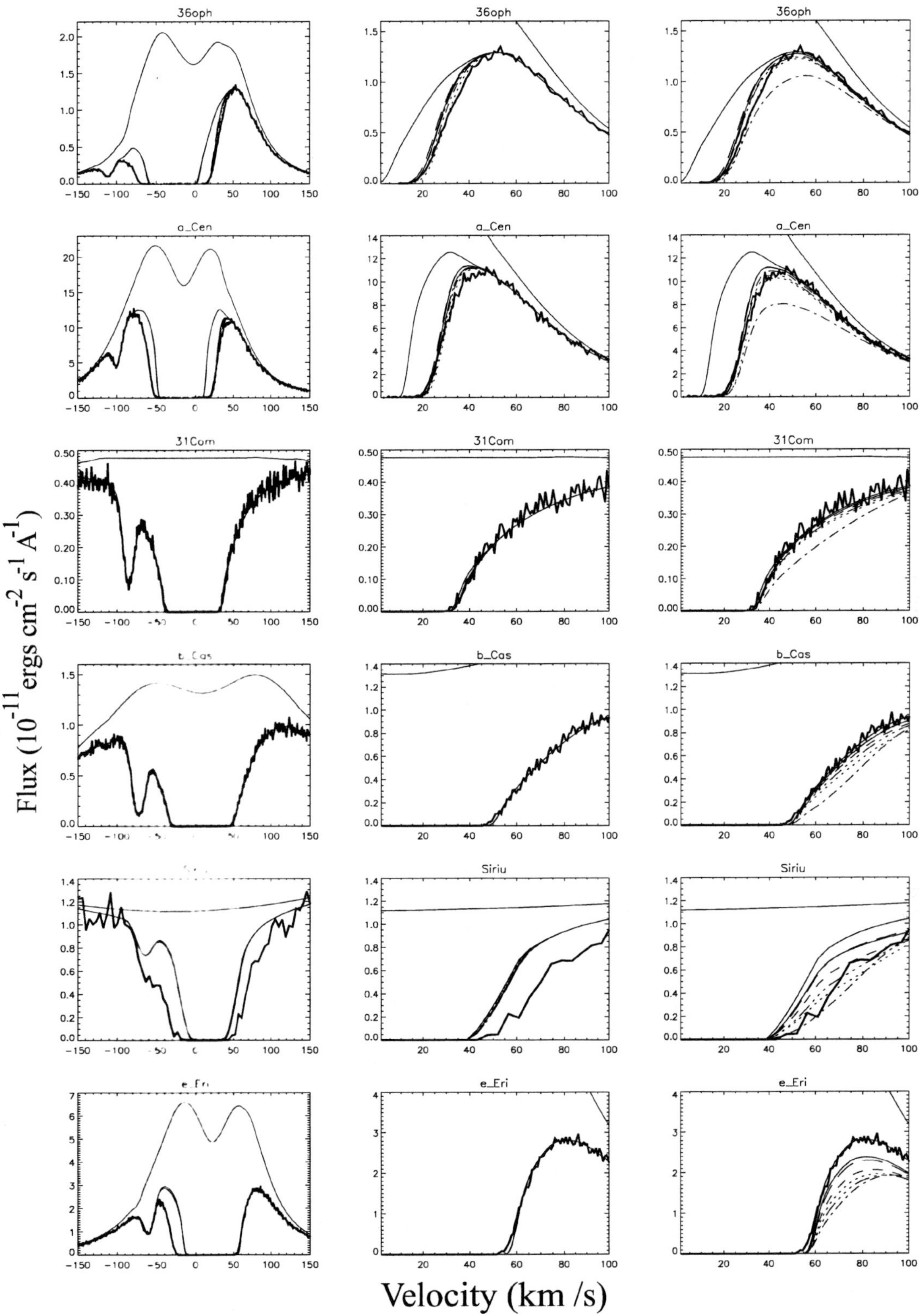

FIGURE 2. Left column: HST Ly α spectra of six stars: 36 Oph, α Cen, 31 Com, β Cas, Sirius, ε Eri, respectively from top to bottom. Each plot shows the observed profile (thick solid line), assumed stellar line profile and interstellar absorption (thin solid lines). Middle column: Reproduction of left column, zoomed in on the red side of Ly-α absorption line. In addition, simulated Ly-α profiles after interstellar and hydrogen wall absorption are shown for models 1-6. Solid lines, when it is different from assumed stellar profile and interstellar absorption, correspond to model 1; dotted lines show results of model 2; dashes correspond to model 3; dot-dash curves correspond to model 4, dot-dot-dot-dash lines correspond to model 5, long dashes correspond to model 6. Right column: same as the middle column, but heliosheath absorption is added to the simulated Ly-α profiles.

tion. Significant heliosheath absorption in model 4 makes the model the worst fit to the data in the crosswind directions. The other models fit the data acceptably well.

Results of the comparison for downwind lines of sight are more puzzling. It is clearly seen that the missing absorption toward Sirius can be easily explained by absorption in the heliosheath. Moreover, the heliosheath absorption predicted by different models is noticeably different. Model 4, which has the worst fit to the data for all other lines of sight, fits the observed Sirius spectrum better than other models.

In contrast to Sirius, for ε Eri the interstellar absorption fits the observed spectrum very well and there is no need for heliospheric absorption. There is no hydrogen wall absorption in this direction (middle column of Figure 2), but all of our models predict too much heliosheath absorption. A similar problem was also found in [5] when they compared their models with the data. The discrepancy may be even more dramatic than our models suggest, because our computational grid extends only 700 AU in the downwind direction, which is not far enough to account for all the absorption. A larger grid would presumably make the ε Eri discrepancies worse for all models, and also change which model fits best for Sirius.

SUMMARY

We have compared H I Ly-α absorption profiles toward six nearby stars observed by HST with theoretical profiles computed using Baranov-Malama model of the heliospheric interface with six different sets of model parameters. Our results are summarized as follows:

1. The absorption produced by the hydrogen wall does not depend significantly on local interstellar H atom and proton number densities for upwind and crosswind directions. In downwind directions the hydrogen wall absorption is sensitive to interstellar densities, but this absorption component is most easily detected in upwind directions. In crosswind and downwind directions the hydrogen wall absorption is hidden in the saturated interstellar absorption and cannot be observed.

2. The heliosheath absorption varies significantly with interstellar proton and H atom number densities. For all models, the heliosheath absorption is more pronounced in crosswind and downwind directions. The heliosheath absorption is redshifted in crosswind directions compared with the interstellar and hydrogen wall absorption components.

3. Comparison of computations and data shows that all available absorption spectra, except that of Sirius, can be explained by taking into account the hydrogen wall absorption only. Considering heliosheath absorption, we find that all models have a tendency to overpredict heliosheath absorption in downwind directions. Toward upwind and crosswind stars the small differences between model predictions and the data can be corrected by small alterations of the assumed stellar Ly-α profile. However, the downwind ε Eri line of sight is a problem, as the models predict too much heliosheath absorption in that direction, and for many, if not most, of the models the discrepancy with the data is too great to resolve by reasonable alterations of the stellar profile.

4. It is puzzling that model 4 provides the best fit to the absorption profile toward Sirius but the worst fit to the other lines of sight. This may be due to our models underestimating heliosheath absorption in downwind directions due to limited grid size, or perhaps the detected excess absorption toward Sirius is not really heliospheric in origin, as suggested by other authors [7]. It is also possible that the difficulties the models have with the downwind lines of sight towards Sirius and ε Eri might be resolved by modifications to the models, perhaps by taking into account the multi-component nature of the heliosheath plasma flow.

ACKNOWLEDGMENTS

This work was supported in part by INTAS Awards 2001-0270, RFBR grants 01-02-17551, 02-02-06011, 01-01-00759, CRDF Award RP1-2248, and International Space Science Institute in Bern.

REFERENCES

1. Linsky, J. L., Wood, B. E., Astrophys. J., 463, 254 (1996).
2. Baranov V.B., Malama Yu.G., J. Geophys. Res., 98, 15,157 (1993).
3. Izmodenov, V., Gruntman, M., Malama, Yu. G., J. Geophys. Res., 106, 10681-10690 (2001).
4. Izmodenov, V., Wood, B., Lallement, R., J. Geophys. Res., submitted (2002).
5. Wood, B. E., Müller H. R., G. P. Zank, Astrophys. J, 537, 304-311 (2000).
6. Izmodenov, V. V., Lallement, R., Malama, Yu. G., Astron. Astrophys. 342, L13-L16 (1999).
7. Hebrard, G., Mallouris, C., Ferlet, R., Koester, D., Lemoine, M., Vidal-Madjar, A., York, D., Astron. Astrophys., 350, 643 (1999)

Open Magnetic Flux: Variation with Latitude and Solar Cycle

Edward J. Smith[1] and Andre Balogh[2]

[1]Jet Propulsion Laboratory, California Institute of Technology, Pasadena, CA 91109, USA

[2]The Blackett Laboratory, Imperial College of Science, Technology and Medicine, London, UK.

Abstract. Recent Ulysses observations of the heliospheric magnetic field (HMF) reveal that the sun's open magnetic flux in the solar wind, as measured by r^2 Br, is independent of heliographic latitude at solar maximum and at solar minimum. It follows that $4\pi\, r^2$ Br at any latitude provides an accurate estimate of the total open flux from the sun. An additional Ulysses result is that long term averages of r^2 Br are very nearly the same at the recent solar maximum as at the preceding minimum. The model of the HMF developed by L. Fisk and his colleagues, which includes several features absent from the well- known Parker model, leads to the prediction that the open flux is relatively constant or perhaps invariant. Motivated by these considerations, Br has been analyzed over the longer interval of four sunspot cycles (20-23) using the archived OMNI field measurements obtained in the ecliptic at 1 AU. Averages of Br, after separation into inward and outward sectors as with Ulysses data, agree within a few percent with previous results based on |Br|, the modulus of the radial component. Averages of Br increase by a factor of about 2 from solar minimum to maximum in cycles 21 and 22 but change only slightly in cycles 20 and 23. The variations in open flux are better correlated with total unsigned flux in the photosphere than with sunspot number, both reaching peak values after sunspot maximum. The open flux decreases at solar minima and secondary decreases occur when the polar cap fields vanish. It appears that the near equality of the open flux at minimum and maximum observed by Ulysses was, in fact, caused by a time variation with the measurements near maximum being obtained while a secondary decrease was occurring. Thus, the open flux is generally variable and only constant during shorter intervals when the solar field is stable.

INTRODUCTION

Ulysses measurements at both solar minimum and maximum show that r^2 Br, i.e., the radial component of the heliospheric magnetic field multiplied by the square of the radial distance at which the observations were made, r, is independent of heliographic latitude (Smith and Balogh, 1995; Smith et al., 2001). This independence (Figure 1) is explained by excess magnetic pressure in the solar wind source regions causing a spreading out of the open magnetic fields (and non-radial solar wind flow) until a pressure balance is achieved (Smith and Balogh, 1995; Suess and Smith, 1996). Values of r^2 Br multiplied by 2π are a measure of the signed open flux in the solar wind, the usual units, nT $(AU)^2$, are equivalent to webers (maxwells) and the total unsigned open flux from the sun is simply $4\pi r^2\, |Br| = F(|Br|)$.

If the latitude invariance is not restricted to the present solar cycle but is a general result, the nearly continuous measurements of Br made in the ecliptic at 1 AU by various spacecraft determines the open flux over previous sunspot cycles. Several independent estimates of open flux have been compared with F(|Br|). Lockwood and Stamper (1999) used a correlation between Br and the aa index of geomagnetic activity to determine F(aa) and compared it with F(|Br|) during solar cycles 20-22. Wang *et al.* (2000) estimated open flux from a potential field source surface model, F(PFSS) between 1971 and 1999. Both approaches revealed a close correspondence between the three estimates of open flux over the last several solar cycles and are consistent with the latitude invariance in Br being a general condition.

CP679, *Solar Wind Ten: Proceedings of the Tenth International Solar Wind Conference*,
edited by M. Velli, R. Bruno, and F. Malara

The availability of the aa index between 1868 and the present enables an extrapolation of the open flux backward in time throughout the previous century. The open flux estimates enabled studies of their correlation with other solar parameters such as total radiance (Lockwood and Stamper (1999) and other variables such as cosmic rays (Lockwood, 2001). Wang *et al.* (2000) found that the open flux in the photosphere evolved systematically from high to low latitudes during the solar cycle.

The average value of Br at Ulysses during solar maximum extrapolated to 1 AU is ≅3 nT approximately the same as at minimum. This result is surprising considering the gross changes in the solar magnetic field between the two phases. However, a model of the HMF developed recently by Fisk and Schwadron (2001) that includes magnetic reconnection at the solar surface implies that the amount of open flux on the sun tends to be conserved. The argument is made that open flux at the surface reconnects predominantly with closed flux causing a migration or diffusion of field lines across the surface without changing the net open flux. The availability of Br measurements in the ecliptic over the past 4 sunspot cycles allows this invariance to be tested more thoroughly.

The previous studies used the modulus of the radial component, |Br|, in order to avoid the compensating effect of the signs in opposing magnetic sectors. However, use of the modulus is equivalent to rectifying the radial component without regard to the sector structure. Rectification is a non-linear process that can add the power in the ever-present HMF fluctuations to the mean and affect the value obtained for the open flux. On the other hand, the Ulysses results have been based on Br in the two sectors separately, i.e., Br(+) and Br(-).

In the following, we compare the signed Br at 1 AU in the ecliptic (OMNI data) with |Br| and use it to reexamine its variation in this and previous sunspot cycles. Details of the variation in Br, the sunspot number and total magnetic flux on the sun are examined to try to understand their relationship and to ascertain whether or not a time variation may have contributed to the apparent invariance of the Ulysses results.

ANALYSIS

The independence of r^2 Br on heliolatitude is demonstrated in Figure 1, solar rotation averages of Ulysses measurements plotted against time with latitude shown along the upper scales. The intervals

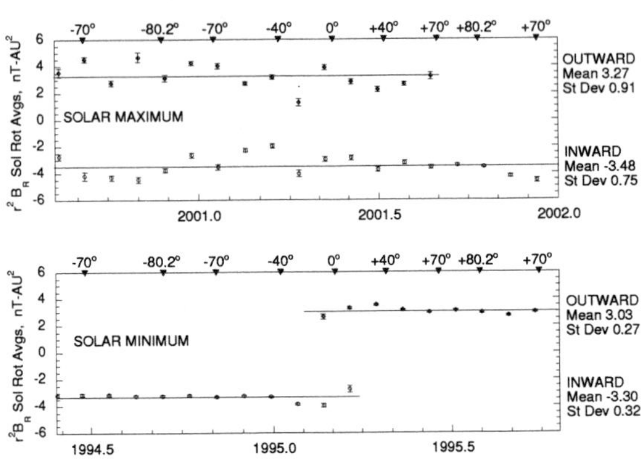

FIGURE 1. Ulysses observations of open flux at sunspot minimum and maximum. Averages and errors of r^2 Br over successive solar rotations are shown as well as overall values for inward and outward sectors and for sunspot minimum and maximum. The two intervals cover the fast latitude scans (or perihelion passages or pole-to-pole transits).

covered are the two fast latitude (or perihelion) scans in 1994-1995 (sunspot minimum, lower panel) and 2000-2002 (maximum, upper panel). Both Br(+) and Br(-) are included with only single polarities present above +/- 20° during *minimum* and a single negative polarity above 70° at *maximum* (indicative of the polar cap polarity reversal that occurs near solar maximum). The averages and standard deviations are more variable at *maximum* because the variable low speed wind extends to essentially all latitudes and the number of coronal mass ejections increases. Nevertheless, no significant latitude variation is evident at *maximum* or at *minimum.* Although the means are slightly higher in the upper panel, the standard deviations imply that all four are consistent with same value for r^2 Br and the open flux in both phases of the sunspot cycle.

Before investigating the behavior of the in-ecliptic Br during past cycles, we address the issue of how mean values are affected by using the modulus rather than separating the measurements into the two sectors before averaging (symbolized by brackets, < >). The result of this comparison, Figure 2, is a plot of <|Br|> vs. the combined average, < Br(+,-)> = <Br(+)> – <Br(-)>. The least squares fit indicates that |Br| is about 2.5% larger on average than Br(+,-) as expected. This difference is probably insignificant in previous or future studies provided that longer term averages (such as shown here) are based on averages over shorter intervals (hourly averages here). The power in the field fluctuations increases with increasing period (decreasing frequency) and the differences due rectification will grow larger if the base data set consists of longer period averages. The succeeding

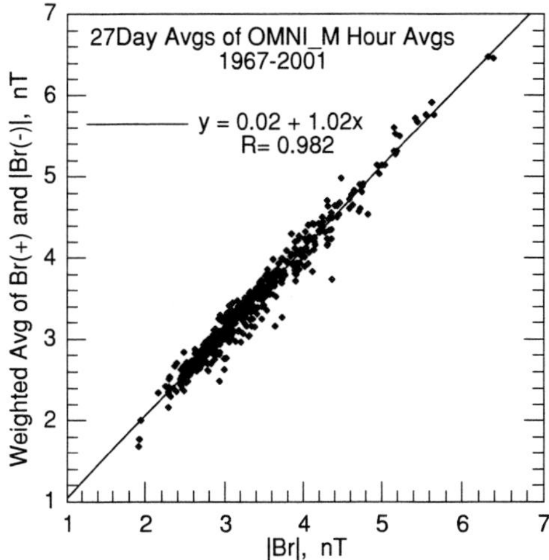

FIGURE 2. The modulus of Br compared with the weighted average of Br(+) and Br(-). Parameters derived from a least squares fit are shown.

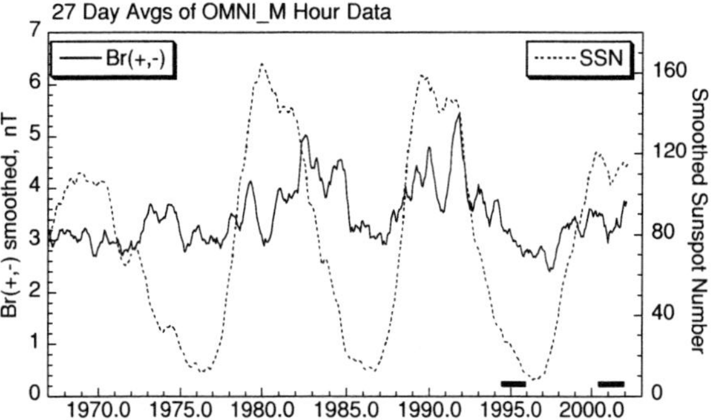

FIGURE 3. Smoothed averages of Br(+,-) and sunspot number during cycles 20-23. The solar rotation averages of Br(+,-) are smoothed further by 7 point running averages. Sunspot numbers are dotted. Horizontal bars indicate when the Ulysses measurements were made.

analyses are based on <Br(+,-)> but are generally consistent with the previous publications.

In-ecliptic values of <Br(+,-)> obtained from the OMNI data set are plotted in Figure 3 along with the sunspot number, R, over the four cycles, 20-23. The 27 day means are filtered by applying 7 point smoothing to emphasize only the longer period variations that may correlate with R. In cycles 21, 22, <Br(+,-)> varies systematically by about a factor of two (2.8 - 5.4 nT). The variation in Br is correlated with R but is more complex. Minima occur near sunspot minimum but the maxima occur later in Br than in R. Furthermore, Br has secondary minima near sunspot maximum suspiciously near the times at which the polar cap magnetic fields vanish and then reappear with the reversed polarity. The increases in Br following sunspot maximum also coincide with the times at which the sun's axial dipole reaches maximum values. The open flux is more closely related to the solar dipole and its changes rather than to the sunspot number.

Cycles 20 and 23 exhibit only modest variations in Br along with a lesser variability in R. Cycle 20 shows a small maximum prior to the minimum in R consistent with cycles 21,22. A slight decrease in 1970 may be correlated with the disappearance of the polar cap fields but is of questionable significance. Of greater relevance to the present study are the recent variations in Br, R and r^2 Br. The intervals during which the Ulysses measurements were made are indicated by the horizontal solid lines. The variation in Br in the ecliptic in cycle 23 is modest (2.6 -3.5 nT) and exhibits another of the decreases, this time in 2001, accompanying the polar cap reversals. Thus, both the limited increase in Br during the present cycle and the timing of the Ulysses pole-to-pole scan evidently kept r^2 Br nearly the same at maximum as at minimum.

The solar magnetic flux is compared more directly with Br in Figure 4. The total unsigned flux, from which the open flux is derived, is shown between 1975 and the present with sunspot maxima and minima indicated (courtesy of K. Harvey). The salient features are the large variation over the sunspot cycle by a factor of about 5 and the appearance of maxima in total flux following the sunspot maxima by about 2 years. The leveling-off of the flux near sunspot maximum is suggestive of the decrease in the polar cap fields at that time. The smoothed means of <Br(+,–)> are shown and the values are converted to open flux on the right hand scale. Solid horizontal lines again show the two Ulysses intervals. The correlation between Br and total flux is better than with R and provides supporting evidence that significant changes in the axial field occurred during the Ulysses measurements making them more nearly equal at *minimum* and *maximum.*

DISCUSSION

The Ulysses observations of nearly constant open magnetic flux at solar minimum and maximum during the recent cycle were influenced by the low values of the sunspot numbers and the related total unsigned flux on the sun. Time variations on the sun associated with the disappearance and reversal of the axial dipole and polar cap fields coincided with the Ulysses measurements and also contributed to low values of the Ulysses observations during maximum. However, the open magnetic flux was shown by Ulysses to be

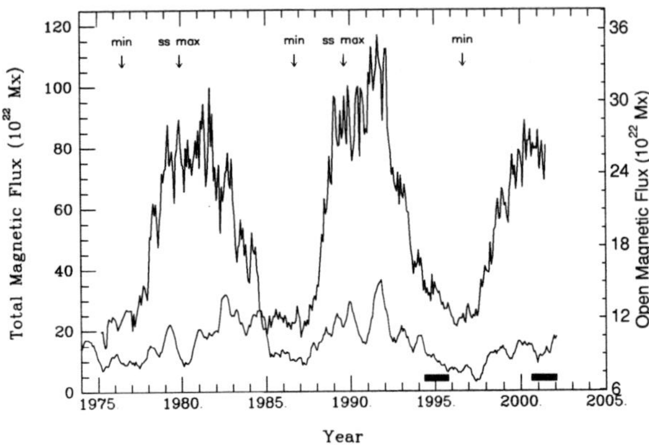

FIGURE 4. Open magnetic flux compared with total flux in the photospheric. The values of <Br(+,-)> are converted to flux units on the right scale. Horizontal bars indicate when the Ulysses observations in Figure 1 were made.

independent of solar latitude during both maximum and minimum. In-ecliptic observations of Br could then be used to extend the investigation of the constancy of the open flux to the previous three cycles. Significant variations in Br and, presumably, the open flux occurred during cycles 21 and 22 but were again at a low level in cycle 20. It seems evident that the open flux is not typically conserved over the sunspot cycle contrary to models which envision a simple rotation of the heliospheric current sheet during the solar cycle with the flux above and below the current sheet being conserved. It may be that when solar conditions are stable the open flux remains unchanged for shorter time intervals.

The interesting correlations between open flux, sunspot number and total photospheric flux are consistent with current understanding of solar cycle variations in the sun's magnetic field represented, for example, by the Babcock- Leighton phenomenological model (Foukal, 1990). As solar activity increases, the sun's field becomes increasingly structured as the strong bipolar fields emerge in association with sunspots. The trailing regions, which have the opposite polarity to the polar cap magnetic fields, give rise to open "unipolar magnetic regions" (and coronal holes) that are stretched as they drift poleward under the simultaneous influence of differential rotation. When these regions arrive at the polar caps, they erode the polar field as a result of magnetic reconnection eventually causing it to disappear. Flux of the opposite sign then builds up to produce the strong dipole field seen after solar maximum.

Since the solar fields, unipolar regions and open flux are ultimately derived from sunspots, a correlation of the open flux with total flux and of both with sunspot number is not surprising. Furthermore, the reconnection of open fields on the sun, indicated by the dip in open flux when the unipolar regions are reversing the polar caps, violates the conditions assumed in the Fisk and Schwadron model where the oppositely-directed fields are alternately open and closed not both open field regions.

ACKNOWLEDGMENTS

Joyce Wolf provided invaluable assistance in the analysis, preparation of the figures and formatting of the text. The research performed at the Jet Propulsion Laboratory was carried out under a contract between the California Institute of Technology and the Aeronautics and Space Administration. Ulysses research at Imperial College is supported by the UK Particle Physics and Astronomy Research Council.

REFERENCES

1. Fisk, L. A. and N. A. Schwadron, *Space Sci. Rev.*, **97**, 21 (2001).
2. Foukal, P. V., *Solar Astrophysics*, John Wiley and Sons, New York, 387 (1990).
3. Lockwood, M. and R. Stamper, *Geophys. Res. Lett.*, **26**, 2461 (1999).
4. Lockwood, M., *J. Geophys. Res.*, **106**, 16,021 (2001).
5. Smith, E. J. and A. Balogh, *Geophys. Res. Lett.*, **22**, 3317 (1995).
6. Smith, E. J., A. Balogh, R. J. Forsyth and D. J. McComas, *Geophys. Res. Lett.*, **28**, 4159 (2001).
7. Suess, S. T. and E. J. Smith, *Geophys. Res. Lett.*, **23**, 3267 (1996).
8. Stamper, R., M. Lockwood and M. N. Wild, *J. Geophys. Res.*, **104**, 28, 325 (1999).
9. Wang, Y.-M and N.R.Sheeley, *Astrophys. J.*, 447, L143 (1995).
10. Wang, Y.- M., J. Lean and N. R. Sheeley, *Geophys. Res. Lett.*, 27, 505 (2000).

The Solar Wind in the Outer Heliosphere at Solar Maximum

John D. Richardson and Chi Wang

Center for Space Research, Massachusetts Institute of Technology, Cambridge, MA 02139

Abstract. This paper reviews solar wind observations in the outer heliosphere, concentrating on the recent data near solar maximum. The speed and temperature tend to be lower at solar maximum, due to the lack of coronal holes. The near-absence of a latitudinal speed gradient at solar maximum allows us to measure the speed decrease of the solar wind and find a value for the H density in the local interstellar medium (LISM) at the termination shock of 0.09 cm^{-3}. The temperature profile is well-matched by a model using pickup ion heating and a speed dependence of the temperature. The density profile at solar maximum is dominated by MIRs; we show one case where converging CME ejecta form a MIR.

INTRODUCTION

In June 2002 Voyager 2 was 68 AU from the Sun and at 24° S heliolatitude. The Voyager 2 trajectory is shown in Figure 1, which illustrates the radial distance of Voyager 2 and its heliolatitude. The state of the heliosphere changes dramatically with the solar cycle. At solar minimum the slow (400 km/s) solar wind is confined to a thin strip of half-width 10-20° near the heliographic equator, but at solar maximum the slow wind fills almost the entire heliosphere. In this paper we describe plasma conditions in the outer heliosphere in the context of their solar cycle variation, use the unique conditions at solar maximum to determine the interstellar neutral density from the slowdown of the solar wind, model the temperature profile using a combination of pickup ions and a dependence of temperature on speed, and show how the merged interaction regions (MIRs) that are prevalent at solar maximum can be formed by transient solar events.

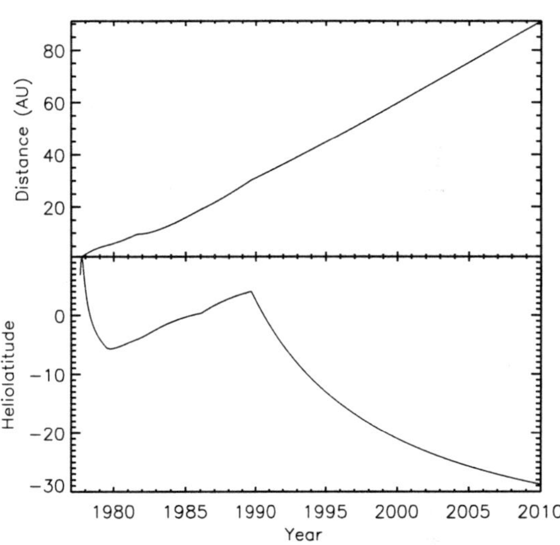

FIGURE 1. The Voyager 2 trajectory showing distance from the SUN in AU (top) and heliolatitude (bottom)

VOYAGER 2 DATA OVERVIEW

Figure 2 shows the solar wind speed, normalized proton density (NR^2), proton temperature, and sunspot number. The hashed regions show the three solar maximums observed by Voyager 2 to date. The speed during solar maximum tends to be lower than at other times, due to the presence of only low-speed solar wind. The temperature is also lower at solar maximum, probably due to the well-known correlation of speed and temperature in the solar wind. The density structure at solar maximum is increasingly dominated by MIRs as Voyager 2 moves farther from the Sun. The solar wind dynamic pressure at 1 AU (IMP and Wind data) and at Voyager 2 are shown in Figure 3. The dynamic pressure of the solar wind is what balances the pressure of the LISM and thus determines the locations of the heliopause and termination shock. In the past two solar cycles, the pressure has been at a minimum near solar maximum, then risen sharply near the end of the solar maximum period. This factor of two increase in pressure drives out the termination shock; as of mid-2002 the increase in pressure has not been observed at 1

CP679, *Solar Wind Ten: Proceedings of the Tenth International Solar Wind Conference*, edited by M. Velli, R. Bruno, and F. Malara

AU, after which it will take roughly a year to reach the termination shock.

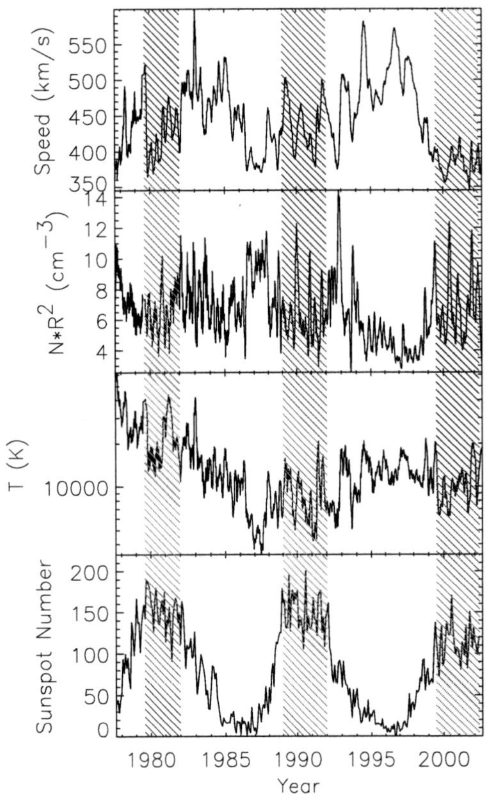

FIGURE 2. The solar wind speed, density normalized to 1 AU, temperature and sunspot number. The shaded regions show solar maxima

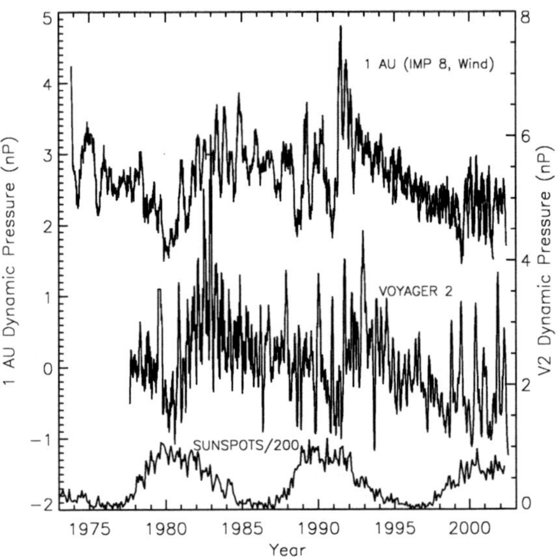

FIGURE 3. The solar wind dynamic pressure observed at 1 AU by IMP 8 and WIND and from 1-65 AU by Voyager 2. The bottom trace is the sunspot number

SOLAR WIND SLOWDOWN

The solar wind interacts with the neutral material from the LISM as it moves outward. These neutrals, mostly H, are ionized and then accelerated to the solar wind speed, with the energy for this acceleration coming from a slowdown of the solar wind. This slowdown depends of the density of the LISM H at the termination shock. The problem with measuring this slowdown is that the solar wind speed varies with time, latitude, and longitude. Generally one needs spacecraft at the same latitudes in the inner and outer heliosphere to determine the speed decrease, since latitudinal speed gradients are large and systematic. At solar maximum, however, the latitudinal speed gradient is small (1). Thus the solar wind speed decrease can be determined using any two spacecraft at different radial distances.

Figure 4 shows 50-day running averages of the solar wind speed and normalized density from 1999 to 2002.2. The bottom panel shows the heliolatitude of Earth, Ulysses, and Voyager 2 during this time period. Earth (and IMP 8) are near the helioequator, Ulysses moves from 20 to 45° S heliolatitude, and Voyager 2 is always at intermediate latitudes. The speeds observed by IMP 8 and Ulysses track each other very well and have nearly the same magnitude. The Voyager 2 speeds, however, are significantly below those observed by the other spacecraft, clear evidence of the slowing of the solar wind from the incorporation of the pickup ions. The density profiles at IMP 8 and Ulysses are also very similar, more evidence that the same solar wind is present at low and medium latitudes. The Voyager 2 density profile is dominated by a large interaction region. To determine the magnitude of the solar wind speed decrease and the LISM H density at the termination shock, a 1-D MHD model was used to propagate the solar wind from IMP 8 to Ulysses and then from Ulysses to Voyager 2 (2). Table 1 shows the average observed solar wind speeds and densities at Earth, Ulysses, and Voyager 2. The density of the termination shock was adjusted to find the value which best fit the data, and that value was 0.09 cm^{-3}.

The model values based on propagating IMP 8 data out to Ulysses and Voyager 2 are also shown. The model and observed speed and density

Table 1: Solar Wind Parameters

	ACE	Uly (5 AU)		V2 (60 AU)	
	data	model	data	model	data
V (km/s)	441	432	428	379	378
N (cm $^{-3}$)	7.0	7.3	7.2	6.1	6.6

are almost identical at Ulysses, giving credence to our assumption that the solar wind is independent of latitude. The model speed matches that observed at Voyager 2, and the densities are also close. The solar wind speed decrease is 53-62 km/s, the largest yet observed. The LISM value of 0.09 cm^{-3} is consistent with other derivations of this number (3).

TEMPERATURE PROFILE

The temperature of the thermal solar wind protons measured by Voyager 2 is shown in figure 5, where we have plotted 50-day running averages. The temperature decreases out to about 30 AU, then increases to 50 AU, then starts to decrease again from 50-65 AU. The decrease in temperature is much less than an adiabatic model would predict (dashed line). Smith et al. (4) describe a model which incorporates stream interactions and pickup ion heating which gives a reasonable fit to the overall shape of the data The large temperature structures with scales of a few AU are not reproduced by the model. The speed and temperature of the solar wind are correlated, and

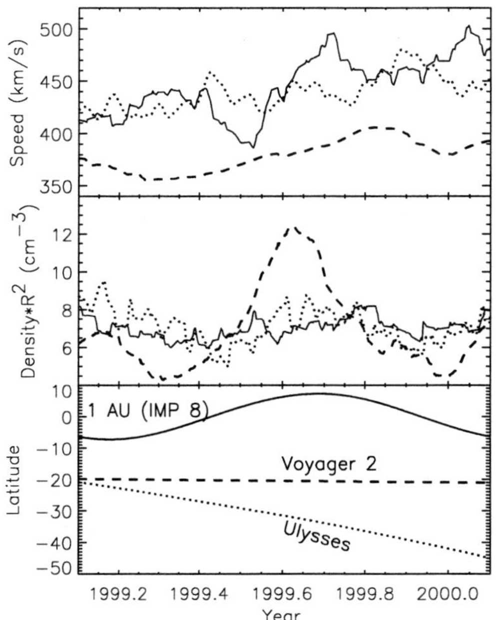

FIGURE 4. The solar wind speed, normalized density observed at IMP 8, Ulysses, and Voyager 2 and the heliolatitudes of these spacecraft

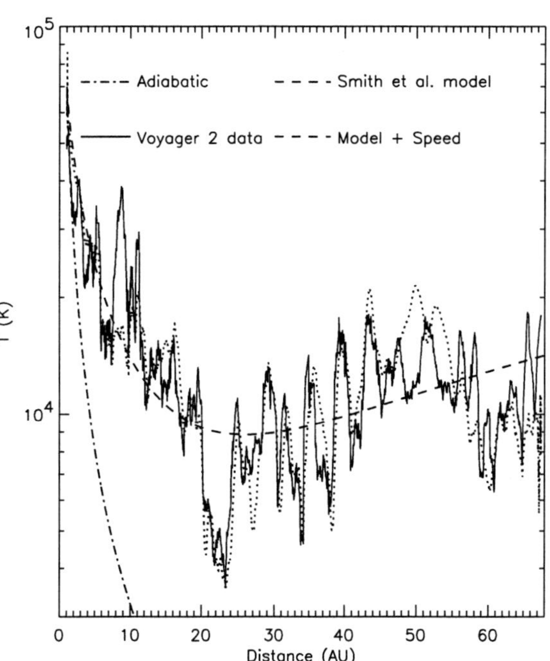

FIGURE 5. The solar wind proton temperature observed by Voyager 2 compared with several model predictions

Richardson (5) investigated whether a combination of the Smith et al. model and a temperature dependence on speed could match the observations. The upper dotted profile shows the result of this work. Much of the finer structure is now matched by the model curve; the correlation coefficient between the model prediction and data is 0.86. In this case the model curve is produced by adding (V - 440 km/s)*80 to the Smith et al. curve, where V is a 100-day running average of the speed. Although this model generally works very well it does not explain all the features of the temperature profile, such as the large rise near 10 AU and the period near 50 AU which corresponds to the 1996 solar minimum.

DEVELOPMENT OF A MIR

MIRs dominate the density structure near solar maximum in the outer heliosphere. These regions of enhanced magnetic field and often plasma density restrict the inward diffusion of energetic particles, resulting in step decreases in cosmic ray intensities. Transient events at the Sun which pile up the plasma and field ahead of them have been suggested as the drivers of these MIRs. In general it is difficult to test this hypothesis since signatures of transient events such as CMEs dissipate by the time they reach the outer heliosphere. However, large alpha particle ratios were used to track one CME from Ulysses to Voyager 2 (6). We found another case where a CME could be tracked from Earth to Ulysses to Voyager 2; these two cases happened to bracket a MIR observed at Voyager 2 (7).

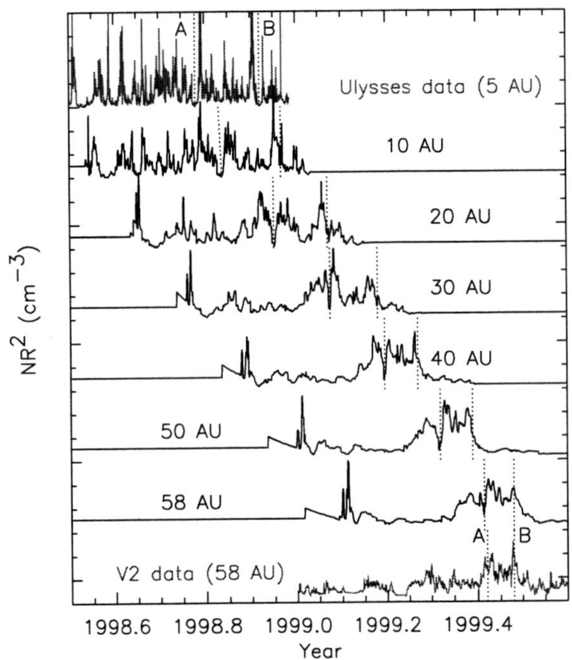

FIGURE 6. From top to bottom, Ulysses density data (which was used as input to the 1-D MHD model), model density profiles at 10, 20, 30, 40, 50 AU and 58 AU, and the Voyager 2 data at 58 AU. The model profiles are time-shifted to roughly align the density decreases associated with the CME ejecta. Tick marks are every 10 cm^{-3}. The vertical dotted lines show the enhanced alpha abundance regions at Ulysses and Voyager 2.

Figure 6 shows the evolution of the solar wind from Ulysses at 5 AU to Voyager 2 at 58 AU. A 1-D MHD model was used to propagate the Ulysses data outward. The locations of the CME ejecta are shown by the two vertical dashed lines. The CMEs were initially about 60 days apart. As they propagate outward they converge, piling up plasma between them, and are about 30 days apart at Voyager 2. The bottom trace shows the measured Voyager 2 density; it is very similar to the model prediction in the trace above. The density in the MIR is about a factor of 2 above the surrounding solar wind density, consistent with the factor of 2 decrease in the spacing of the CME ejecta.

CONCLUSION

The Voyager mission continues to explore the heliosphere at a pace which allows study of solar cycle changes. The combination of Voyager, Ulysses, and Earth-orbiting spacecraft allows us to look differentiate between solar cycle and spatial changes, giving information on the global solar wind outflow and in this case allowing the best determination yet of the effect of pickup ions on the solar wind, in terms of heating and slowing the flow. As we enter the declining phase of the solar cycle we expect to see how the stream structure observed last solar cycle at 35 AU has evolved, providing new science for SW11.

ACKNOWLEDGMENTS

This work was supported under NASA grant NAG5-11623 and NASA contract 959203 from the Jet Propulsion Laboratory to the Massachusetts Institute of Technology.

REFERENCES

1. McComas, D. J., J. T. Gosling, R. M. Skoug, Ulysses observations of the irregularly structured mid-latitude solar wind during the approach to solar maximum, *Geophys. Res. Lett.* **27**, 2437-2440 (2000).
2. Wang, C., J. D. Richardson, Determination of the solar wind slowdown near solar maximum, *J. Geophys. Res.*, **106**, in press (2002).
3. Gloeckler, G., G. A. Fisk, and J. Geiss, Anomalously small magnetic field in the local interstellar cloud, *Nature*, **386**, 374 (1997).
4. Smith, C. W., W. H. Matthaeus, G. P. Zank, N. F. Ness, S. Oughton, J. D. Richardson. Heating of the low-latitude solar wind by dissipation of turbulent magnetic fluctuations, *J. Geophys. Res.*, **106**, 8253-8272 (2001).
5. Richardson, J. D., The speed-temperature relation from 1-65 AU, in preparation for submission to GRL (2002).
6. Paularena, K. I., C. Wang, R. von Steiger, and B. Heber, An ICME observed by Voyager 2 at 58 AU and by Ulysses at 5 AU, *Geophys. Res. Lett.* **28**, 2755-2758 (2001).
7. Richardson, J. D., K. I. Paularena, C. Wang, and L. F. Burlaga. The life of a CME and the development of a MIR: From the Sun to 58 AU, *J. Geophys. Res.*, **106**, 10.1029/2001JA000175 (2002).

Time-dependent tomography of hemispheric features using interplanetary scintillation (IPS) remote-sensing observations

B.V. Jackson, P.P. Hick and A. Buffington

Center for Astrophysics and Space Sciences, University of California at San Diego, LaJolla, CA, U.S.A.

M. Kojima, M. Tokumaru, K. Fujiki, T. Ohmi and M. Yamashita

Solar-Terrestrial Environment Laboratory, Nagoya University, Japan

Abstract. We have developed a Computer Assisted Tomography (CAT) program that modifies a time-dependent three-dimensional kinematic heliospheric model to fit interplanetary scintillation (IPS) observations. The tomography program iteratively changes this global model to least-squares fit IPS data. The short time intervals of the kinematic modeling (~1 day) force the heliospheric reconstructions to depend on outward solar wind motion to give perspective views of each point in space accessible to the observations, allowing reconstruction of interplanetary Coronal Mass Ejections (CMEs) as well as corotating structures. We show these models as velocity or density Carrington maps and remote views. We have studied several events, including the July 14, 2000 Bastille-day halo CME. We check our results by comparison with additional remote-sensing observations, and observations from near-Earth spacecraft.

INTRODUCTION

In solar physics, there have been numerous attempts to reconstruct coronal structure and the heliosphere in three dimensions. These techniques have been developed for coronal mass ejections (CMEs) to understand better the physical principles of their initiation. Using slightly differing techniques others (1, 2, 3) have analyzed views from the Earth using Thomson-scattering data to obtain three-dimensional results.

Since the 1960's interplanetary scintillation (IPS) measurements have been used to probe solar wind features with ground-based meter-wavelength radio observations (4, 5). Observations from the UCSD (6) and Nagoya (7) multi-site scintillation array systems have been used to determine velocities in the interplanetary medium since the early 1970's. The IPS intensity scintillation observations, that arise from small-scale (~200 km) density variations, highlight heliospheric disturbances of larger scale that vary from one day to the next and are often associated with geomagnetic storms on Earth (8).

We have developed a Computer Assisted Tomography (CAT) program that modifies a time-dependent three-dimensional kinematic heliospheric model to fit IPS observations. The tomography program iteratively changes this global model to least-squares fit IPS data. Three-dimensional results for IPS data covering a wide range of elongations have been obtained using a heliospheric model that incorporates both outward solar wind flow and solar rotation (9, 10, 11, 12, 13). Here scintillation strength is caused by small-scale density variations that is in turn scaled to bulk density and solar distance by means of a power law. These previous IPS tomographic programs all assumed that the kinematic heliospheric model remains unchanged over the duration of the observations. Thus the observed heliospheric structures do not change other than by outward radial expansion within this time period.

The new tomographic modeling technique described here relaxes the assumption that heliospheric structure remains constant over time. In this newest extension a global kinematic model is formed at regular time intervals, and the iterative process provides the three-dimensional heliospheric parameters that fit observed data. The next section describes the tomographic program that has been developed. The third section compares and calibrates the kinematic models based on the IPS data to Earth-based *in situ* measurements. The fourth section displays and discusses the kinematic model values as a remote observer would view them. We conclude in the last section.

TOMOGRAPHIC ANALYSIS

The IPS technique relies on several assumptions to relate changes in scintillation level and velocity integrated along each line of sight to local changes in the scintillation level and velocity. In weak scattering (assumed here exclusively) the Born approximation holds, and the diffraction pattern is a sum of contributions from each thin scattering layer perpendicular to the line of sight (14). The analysis

CP679, *Solar Wind Ten: Proceedings of the Tenth International Solar Wind Conference,*
edited by M. Velli, R. Bruno, and F. Malara

proceeds much as in (11), but using evenly placed time steps in the analysis.

Radio source scintillation-level observations have been obtained from several tens of sources measured each day by the STELab Kiso radio telescope from 1997 to the present. This analysis used data from a relatively short time interval during July 2000. The value of the disturbance factor g is defined as

$$g = m/\langle m \rangle, \quad (1)$$

where m is the fractional scintillation level ΔI/I, the ratio of source intensity variation to intensity and $\langle m \rangle$ is the mean level of ΔI/I for the source at that elongation. Scintillation level measurements from the STELab radio facility analyses are available at a given sky location as an intensity variation of the source signal strength. For each source, data are automatically edited to remove any obvious interference discerned in the daily observations. To yield g-levels in real time, the white noise P_{WN} is subtracted from the scintillation signal spectrum $P(f)$, and then system gain corrections are determined by automatically calibrating with the white noise level at the high frequency end of the power spectrum. To obtain m, the white noise is subtracted from the scintillation signal,

$$m = \int_{f_1}^{f_2} (P(f) - P_{WN})/P_{WN}\, df \cdot \quad (2)$$

At UCSD, g-values for a source are determined in real time from m by a least square fit to the axially symmetric solar wind model. We assume that it is sufficient to fit 8 daily measurements in order to obtain a value of $\langle m \rangle$ for a given source. With the STELab 327 MHz analyses weak scattering results are usually obtained from sources outward from 11.5° elongation. However, since ample data are available, our following analyses use a 17.5° limit to be certain to be in the weak scattering regime.

The scintillation level weighting factor along the line of sight $W_C(z)$ can be approximated in weak scattering as in (11) at the 327 MHz frequency of the STELab IPS observations. The scintillation level m is related to the small-scale density variations along the line of sight by

$$m^2 = \int dz\, \Delta N_e(z)^2\, W_C(z). \quad (3)$$

Here, $\Delta N_e(z)$ are the small-scale density variation values at distance z along the line of sight. The density values along the line of sight are not *a priori* known, but we assume that the small-scale variations scale with a power law of heliospheric density,

$$\Delta N_e = A_C R^{PWR} N_e^{PWN}, \quad (4)$$

where A_C is a proportionality constant, PWR is a power of the radial falloff (13) and PWN is the power of the density. In the present analysis, the program fits the value of A_C and the values of PWR and PWN to best fit the data over the interval chosen. For the time period presented here, $A_C = 1$, and the two powers PWR and PWN are –3.5 and 0.7, respectively, to best fit *in situ* density over a ten-day time interval centered on the time the Bastille-day CME reaches Earth.

Generally, valid IPS velocity data are available from the same radio sources as observed in scintillation level each day. IPS velocities are based on observations from up to four scintillation arrays operated STELab, Japan. To use these data our tomography program assumes that the line of sight IPS velocity follows a similar line of sight weighting relationship to that of the intensity scintillation. We approximate the velocity observed at Earth as in (11) [and see (12), for a more complete formulation and validity tests].

The UCSD tomography program (11) applies corrections to a kinematic model, modifying the model until there is a least squares best fit match with the observations. Density (rather than the small-scale density variation) is used and propagated outward in the UCSD kinematic model. The density and velocity are projected outward from a reference surface (source surface) below the lowest lines of sight. Consistent approximately with *in situ* spacecraft observations, the solar wind motion is assumed to be radial outward from this surface. Thus, for example, when faster solar wind catches up with slower wind, the resultant solar wind speed is continued after merging by assuming both mass and mass flux are conserved within the latitudinal band resolved by the model. At the reference surface the velocity structure of the model is smoothed using a Gaussian filter weighted according to the angular distance of the adjacent resolution elements on this surface. Since the resolution of rectangular Carrington coordinate maps increase in longitude with increasing latitude, this filter is used to even the spatial resolution over the whole map.

In the kinematic model described here, the heliosphere can change over time intervals as short as one day. This assumption essentially limits the tomographic reconstruction to rely on outward solar wind flow to form the perspective views. For each observed line of sight at a given time, the position along this line in the model is calculated. The model g-levels along each line of sight defined by the densities are summed using the weighting mentioned in Eq. 3. These model values are then compared with the observed g-levels, and this comparison is used to change the model. For one solar rotation typically 500 to 1000 lines of sight can be used to determine model density from the scintillation-level measurements and velocity. This implies 20 to 40 crossed line of sight components contribute input to latitude and longitude positions each day subject to the Gaussian spatial filter described earlier, and a similar Gaussian filter that combines data from one day to the next. This implies a possibility of determining the density and velocity for 20 to 40 latitude and longitude locations each day. In practice, lines of sight often extend over several consecutive time steps. The amount and quality of the available observations and the heliographic coordinate resolution and temporal data cadence dictate this resolution even more strongly.

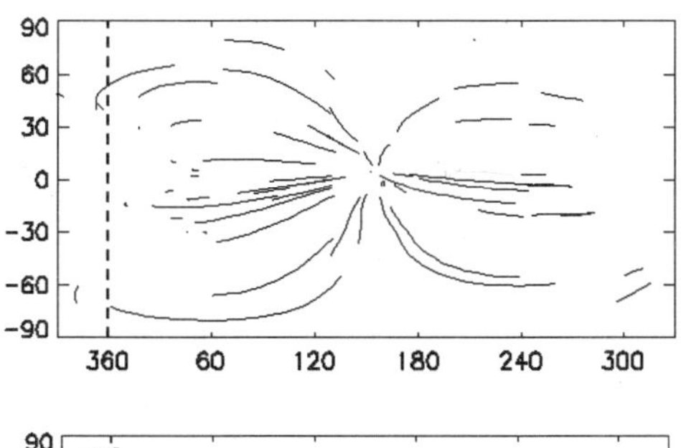

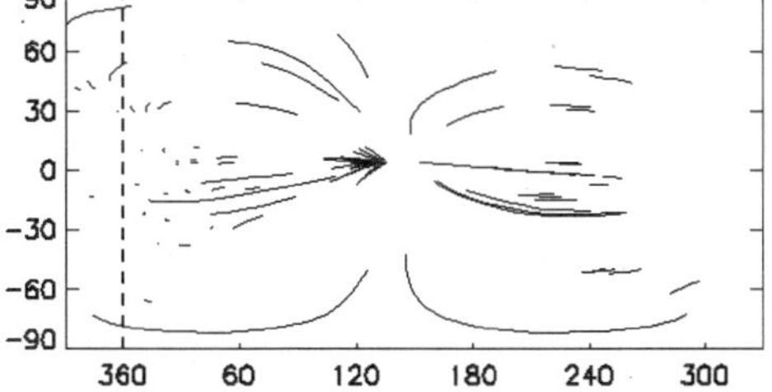

FIGURE 1. Consecutive-day (July 13 and July 14, 2000 latitude and longitude line of sight projections onto the source surface. Lines of sight extend outward from Earth for 2 AU beginning near the projected sub-Earth point at the center of the map. Some lines of sight complete their projection on adjacent days. Perspective views are realized from the different weights on the source surface maps at each latitude and longitude point.

For the UCSD time dependent tomographic program using STELab data, 20° by 20° heliographic latitude and longitude resolution is used and a one day cadence. The regions near the Earth are those most frequently crossed by different lines of sight while those far from it, over the solar poles and especially to the south, are not. This is shown in Fig. 1 for two consecutive days during the Bastille-Day event.

For several different perspective lines of sight to produce changes in the modeled values, we require more than one line of sight crossing on the source surface be present within a 20° by 20° heliographic interval for changes to be made at that position. If the model cannot be updated at some location, these coordinate positions are left blank in the final result. The reference surface maps are smoothed at each iteration using a Gaussian spatial filter that incorporates equal solar surface areas and a Gaussian temporal filter. These spatial and temporal filters can be varied to ensure convergence. Filter changes by large percentages have a significant effect on the result. Filter parameters were set to a 1/e width of 13.5° and 0.85 days, for the 20° by 20° and 1-day model digitization, respectively during the July, 2000 interval shown here. The tomography program iterates to a solution, generally converging to an unchanging model within a few iterations. Convergence is monitored using techniques as described in (11).

IN-SITU COMPARISON

Tomographic model densities and velocities are available in three dimensions and can be extrapolated to any heliocentric distance, for example to 1 A.U. Here they compare directly to the measured results from *e.g.* the Advanced Composition Explorer (ACE) spacecraft near Earth. We smooth the ACE data into 18-hour averages, consistent with the approximate spatial resolution present from the longitudinal and temporal binning of the tomography data. The densities mapped to 1 AU are shown as a time series for rotation 1965 in Fig. 2. The correlation for rotation 1965 in model to ACE *in situ* values is 0.6 and 0.9 respectively for velocity and density over the 10-day period centered on the Earth arrival time of the July 14 CME.

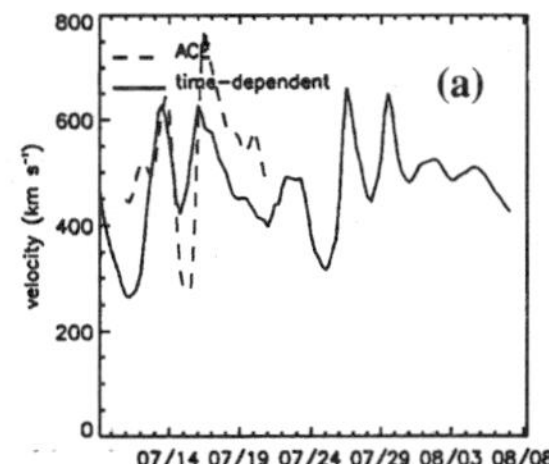

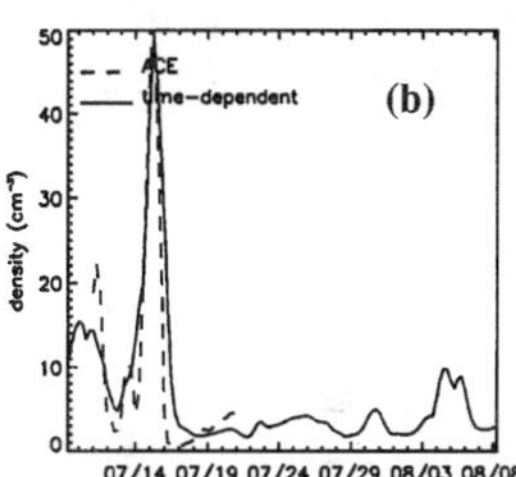

FIGURE 2. Rotation 1965. **a)** 10-day velocity time series from the three dimensional time-dependent model projected to 1 AU compared to the velocity time series from the ACE spacecraft (dashed line). **b)** Model and ACE density correlation.

DISCUSSION

Since few other *in situ* observations exist with which to compare these results, the only guarantee in the current analysis is that the three-dimensional model constructed remotely by the IPS analysis over a large portion of the heliosphere agrees with *in situ* data near Earth. However, we can also view the model's shape for these events and see if they match remotely sensed data from the LASCO coronagraphs.

Fig. 3a shows a LASCO C2 coronagraph image of the July 11 halo CME compared with two views of the density modeled as the CME is about to reach 1 AU. The reconstruction shows that this CME moves mostly to the east and north of the Earth as also indicated in the coronagraph image. Similarly, Fig. 3b shows the Bastille-day CME compared with two views of the reconstructed density as the CME is about to hit Earth. Given the expanse of heliosphere that the CMEs have traversed to reach 1 AU, the comparisons with LASCO near-Sun observations are excellent. The results of the present 3-dimensional reconstruction are in good agreement for the Bastille-day CME with an alternate reconstruction analysis by (15).

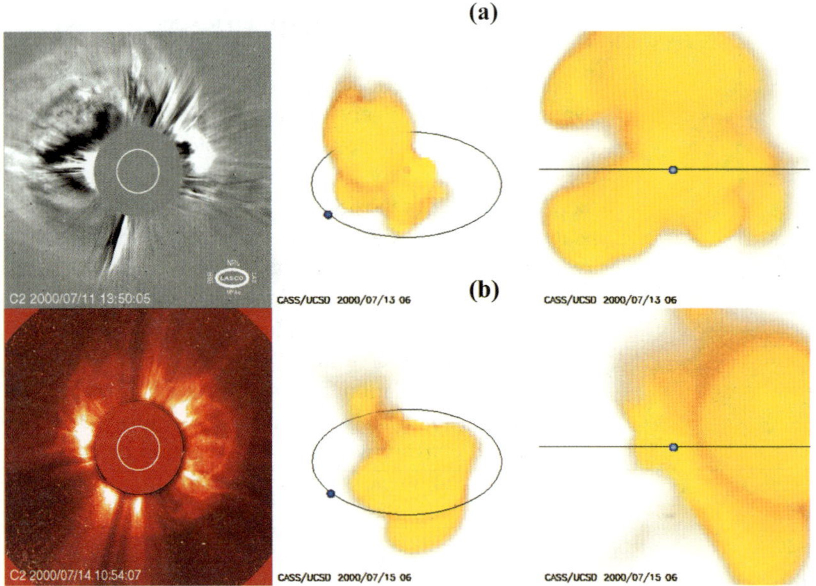

FIGURE 3. LASCO C2 images and two views of the reconstruction of the halo CMEs in July, 2000 with N_e >30e$^-$ cm^{-3} normalized to 1 AU shown. Views (left to right) are 3° across from 1 AU; 55° across from 3 AU, 30° above the ecliptic plane 45° west of the Sun-Earth line; and 100° across at 1.1 AU on the Sun-Earth line. **a)** July 11, 2000 CME in LASCO reconstructed July 13 at 6 UT. **b)** July 14 CME in LASCO reconstructed July 15 at 6 UT.

CONCLUSION

In comparison with *in situ* data at Earth, the tomographic analysis gives superior results to previous corotating analyses (11, 12). This is true even though the spatial resolution of the present model is dramatically decreased from the corotating model to insure convergence. We reconstruct as complete as possible a global three-dimensional model to obtain a good fit to observations at Earth, even though these global models amount to only a few tens of data points per day. In real-time analysis, data drop-outs and noise make the task of forecasting CME arrival using this technique with the present STELab arrays even more problematic.

We expect that only when new and bigger IPS systems are available will the technique provide a more refined tomographic analysis to accurately forecast CME arrival to within a few hours. Other large array systems at different Earth longitudes will also be helpful. The Solar Mass Ejection Imager (SMEI) will allow even more complete sky coverage in density when data from it becomes available, but the SMEI analyses alone cannot as completely determine the velocities required to complete a global solar wind model.

The kinematic model currently fit by the tomography can be improved significantly by using a technique where the boundary conditions (source surface) for a 3D-MHD model are adjusted to give a best fit to the three-dimensional tomographic analysis. One attempt is shown for corotating tomography in (16).

ACKNOWLEDGEMENTS

The work of B.V. Jackson, P.P Hick and A. Buffington was supported at the UCSD by AFOSR grant AF49620-01-1-0054, NSF grant ATM 98-199947 and NASA grant NAG5-8504.

REFERENCES

1) Munro, R.H., Topical Conference on Solar and Interplanetary Physics, Tucson, Arizona, January 12-15, 10 (1977).
2) Crifo, F., J.P. Picat and M. Cailloux, *Solar Phys.*, **83**, 143 (1983).
3) MacQueen, R.M., *Solar Phys.*, **145**, 169 (1993).
4) Hewish, A., P.F. Scott and D. Wills, *Nature*, **203**, 1214 (1964).
5) Houminer, Z., *Nature Phys. Sci.*, **231**, 165 (1971).
6) Coles, W. A. and J.J. Kaufman, *Radio Science*, **13**, 591 (1978).
7) Kojima, M. and T. Kakinuma, *J. Geophys. Res.*, **92**, 7269 (1987).
8) Gapper, G.R., A. Hewish, A. Purvis and P.J. Duffet-Smith, *Nature*, **296**, 633 (1982).
9) Jackson, B.V., P.L. Hick, M. Kojima and A. Yokobe, *Adv. Space Res.*, **20**, (1), 23 (1997).
10) Kojima, M., K. Asai, P.L. Hick, B.V. Jackson, M. Tokumaru, H. Watanabe and A. Yokobe, in: Robotic Exploration close to the Sun: Scientific Basis, edited by S.R. Habbal, *AIP Conference Proceedings* **385**, 97 (1997).
11) Jackson, B.V., P.L. Hick, M. Kojima and A. Yokobe, *J. Geophys. Res.*, **103**, 12,049 (1998).
12) Kojima, M., M. Tokumaru, H. Watanabe, A. Yokobe, K. Asai, B.V. Jackson and P.L. Hick, *J. Geophys. Res.*, **103**, 1981 (1998).
13) Asai, K., M. Kojima, M. Tokumaru, A. Yokobe, B.V. Jackson, P.L. Hick and P.K. Manoharan, *J. Geophys Res.*, **103**, 1991 (1998).
14) Tatarski, V.I., *Wave propagation in a turbulent medium*, McGraw-Hill, New York (1961).
15) Tokumaru, M., M. Kojima, K. Fujiki and M. Yamashita, (this conference), June 17-21 (2002).
16) Hayashi, K., K. Fujiki,, M. Kojima and M. Tokumaru, (this conference), June 17-21 (2002).

Understanding the Solar Sources of In Situ Observations

Pete Riley, Zoran Mikic, and Jon Linker

Science Applications International Corporation, San Diego, California.

Thomas H. Zurbuchen

University of Michigan, Ann Arbor, Michigan.

Abstract. The solar wind can, to a good approximation be described as a two-component flow with fast, tenuous, quiescent flow emanating from coronal holes, and slow, dense and variable flow associated with the boundary between open and closed magnetic fields. In spite of its simplicity, this picture naturally produces a range of complex heliospheric phenomena, including the presence, location, and orientation of corotating interaction regions and their associated shocks. In this study, we apply a two-step mapping technique, incorporating a magnetohydrodynamic model of the solar corona, to bring in situ observations from Ulysses, WIND, and ACE back to the solar surface in an effort to determine some intrinsic properties of the quasi-steady solar wind. In particular, we find that a "layer" of ~35,000 km exists between the Coronal Hole Boundary (CHB) and the fast solar wind, where the wind is slow and variable. We also derive a velocity gradient within large polar coronal holes (that were present during Ulysses' rapid latitude scan) as a function of distance from the CHB. We find that v = 713 km/s + 3.2 d, where d is the angular distance from the CHB boundary in degrees.

INTRODUCTION

Early in situ observations in the vicinity of Earth established that the solar wind consisted primarily of two components: slow (~300 km/s), dense, and cool wind; and fast (~700 km/s), tenuous, and hot wind (1). The relatively quiescent properties of the high-speed wind observed by Ulysses (2,3,4) suggest that this is the basic equilibrium state of the solar wind. The origin of the slow solar wind and its variability, on the other hand, remain poorly understood.

More recently, the idea that there might exist more than two basic types of solar wind has been raised within a number of contexts (e.g., 5, 6). While there may be merit to such views, in this study we limit ourselves to the bimodal viewpoint for simplicity. To support this, in Figure 1 we show Ulysses plasma and composition measurements from launch through the present. Density (logarithmic) runs along the x-axis, speed runs along the y-axis, and temperature (logarithmic) runs along the z-axis.

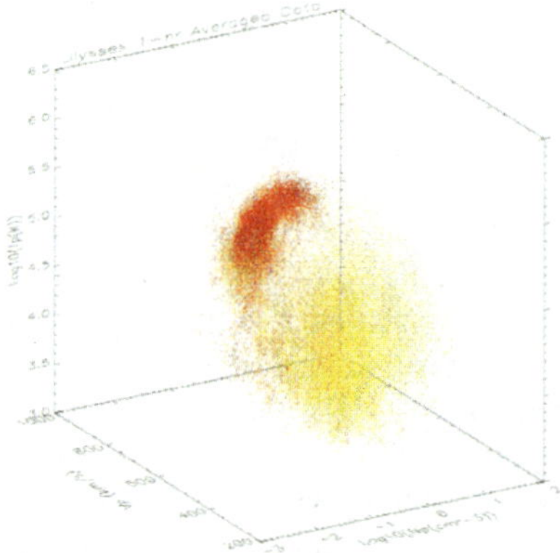

FIGURE 1. Ulysses 1-hr averaged measurements of density (x), speed (y), and proton temperature (z), color-coded according to the ratio O7/O6 (red indicating the lowest values and yellow indicating the highest values).

CP679, *Solar Wind Ten: Proceedings of the Tenth International Solar Wind Conference*, edited by M. Velli, R. Bruno, and F. Malara

The points have been color-coded according to the ratio of O7 to O6 (7). While there is room in such a plot to hide additional intrinsic components, clearly the dominate flow can be described as bimodal.

In this report, we map in situ data back to the Sun to uncover some of the intrinsic properties and gradients associated with the solar wind. We begin with a brief summary of the mapping procedure, then summarize our main results, and conclude with a discussion of their implications.

MAPPING PROCEDURE

A number of techniques have been developed to connect in situ observations with various solar features. Typically this involves some kind of mapping procedure to extrapolate the solar wind observations from the point of measurement back to the Sun. These range from the simplest "constant speed," or "ballistic" mapping (e.g., 8) where it assumed that the plasma maintains constant radial velocity from the Sun to the point of observation) to more sophisticated global, time-dependent MHD simulations that can trace field lines all the way from the solar surface to the point of observation (9). We have investigated the errors associated with several of these approaches and find that a combination of the ballistic mapping, together with a realistic MHD model for the coronal magnetic field typically provides the best compromise (10). The ballistic mapping is used to bring the data from the point of observation (1-5 AU) to $30R_s$ where the dominant variation is with respect to longitude. Between 1 and $30R_s$, on the other hand, latitudinal variations are most significant and the MHD model is used to trace along magnetic field lines back to the solar surface.

In figure 2 we use a global MHD model to illustrate the errors introduced by the ballistic mapping procedure. The two curves compare the MHD solution at $30R_s$ with the MHD solution at 1 AU, which has been ballistically mapped back to $30R_s$. Differences of 5 -10 are apparent, and represent the limit in longitudinal accuracy for the mapping process. While one might suppose that 3-D MHD solutions driven by solar observations would provide a more accurate way to map the in situ data, relatively small errors of ~50 km/s in the solution at 1 AU lead to longitudinal offsets of ~8 . Although the heliospheric MHD models have been very successful as a tool to investigate the large-scale structure of the solar wind (e.g., 9,11) improved specification of the boundary conditions will be necessary before they can be used as a reliable mapping tool.

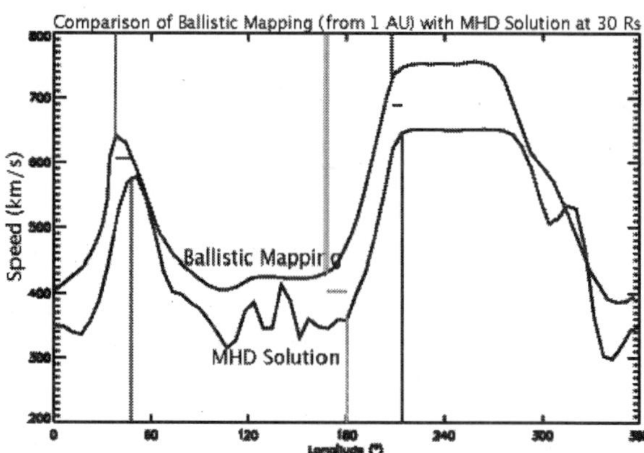

FIGURE 2. Comparison of the ballistic mapping with the MHD solution at $30R_S$. Vertical lines and short horizontal bars indicate differences in longitude for 3 distinct points in the stream profile.

RESULTS

We illustrate the mapping technique in Figure 3, which combines data and modeling results. The closed regions represent coronal holes, as computed by the coronal MHD solution for this time period (Carrington Rotation (CR) 1953 – beginning August 18th, 1999). These are often similar, but not identical to the coronal holes computed from Kitt Peak He I 1083 nm observations (typically being a superset) and are color-coded according to the underlying (highly smoothed) observed photospheric fields. Superimposed onto this are the trajectories of ACE and Ulysses. At the time ACE was located at 1 AU and 6.6 degrees above the heliographic equator. Ulysses was located at 4.66 AU and ~32.2 *S*. The starting point of the trajectories (already mapped to $30R_S$) are straight-line traces moving from right to left (with increasing time) and all points map back to origins within coronal holes (by definition). These mappings have been color-coded according to the measured polarity of the interplanetary magnetic field. Given the approximations in the mapping procedure and the relatively high level of solar activity during this time period, the strong correlation between the mapped in situ polarities and the polarity of the coronal holes is quite remarkable, providing strong support that the mapping procedure has been successful.

Once the plasma data have been mapped back to the solar surface, we can investigate its relationship with various solar features. For this study, we focus on the speed of the solar wind and its distance to the computed Coronal Hole Boundary (CHB). In Figure 4 we have calculated the minimum distance to the CHB for each plasma measurement. These observations were made at the minimum of the solar cycle, when large, long-lived polar coronal holes were present (CR 1913, beginning August 22nd, 1996). During this time Ulysses was completing its first sampling of the northern polar coronal hole and was returning toward the ecliptic.

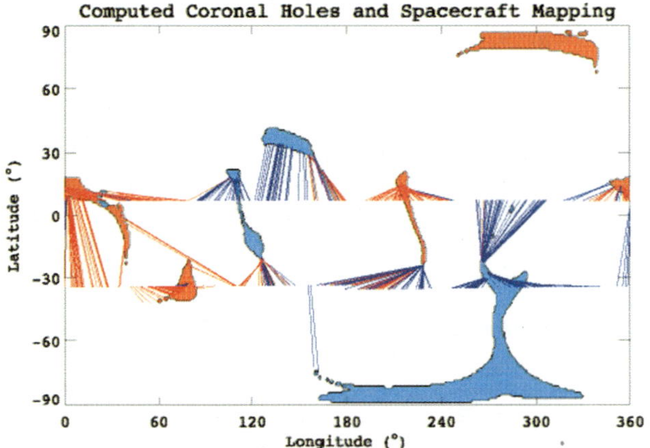

FIGURE 3. ACE (closest to equator) and Ulysses trajectories mapped back to the Sun for CR 1953. Coronal holes are colored according to the observed photospheric field. Trajectories are color-coded according to the measured in situ polarity of the IMF (red/blue corresponds to outward/inward polarity).

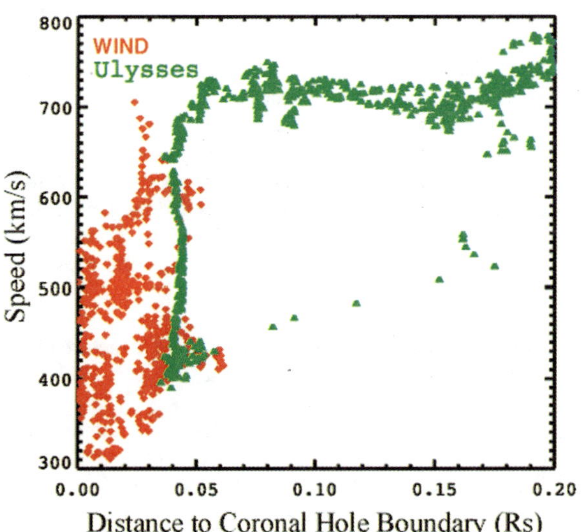

FIGURE 4. Mapped speed as a function of distance to the CHB for CR 1913. Ulysses measurements are shown in green and WIND measurements are shown in red.

We infer a "boundary layer" of ~$0.05R_S$, corresponding to ~35,000 km over which the solar wind is slow and variable, and beyond which the wind is fast and relatively steady. This scale size matches well with the average size of a supergranule (32,000 km). While this profile tends to be common during solar minimum conditions, near solar maximum it is rarely, if ever present. As an example we return to CR 1953 and in Figure 5 we again display speed as a function of the distance to the CHB.

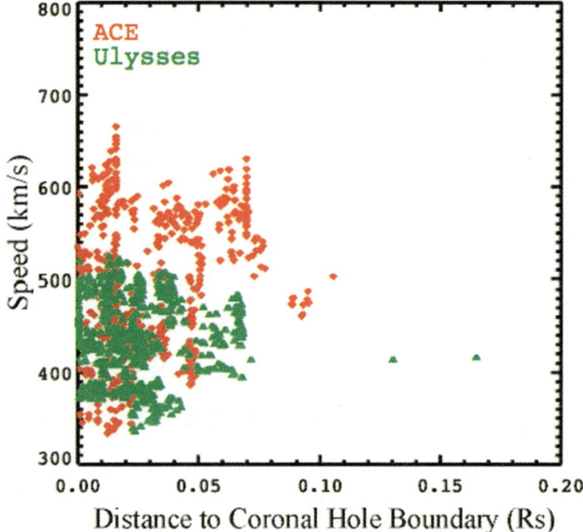

FIGURE 5. Mapped speed as a function of distance to the CHB for CR 1953. Ulysses measurements are shown in green and ACE measurements are shown in red.

Although there is a tendency for the wind speed to increase with distance away from the CHB, there is no obvious sign of a "layer" of slow wind, at least as far as 0.07 R_S. However, as can be seen from Figure 3 during this time period, there were no extensive coronal holes that were intercepted by the trajectory of either spacecraft.

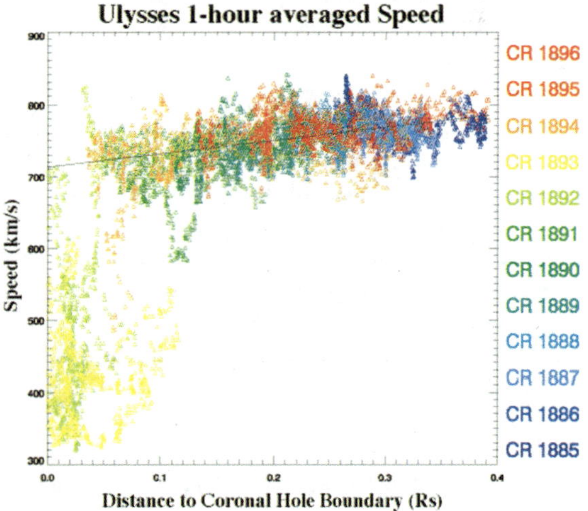

FIGURE 6. Solar wind speed versus distance to the coronal hole boundary for 12 Carrington rotations occurring during Ulysses rapid latitude scan. The solid line represents a least-squares fit to speeds above 600 km/s.

To investigate possible gradients deeper within coronal holes, we mapped data from 12 solar rotations during Ulysses' so-called rapid latitude scan (which occurred during the declining phase of the solar cycle) and computed the distance to the CHB. These results are shown in Figure 6. The distance to the CHB can be interpreted either in units of R_S or Sun-centered radians ($1R_S$=1 radian). Thus the x-axis covers a range of 23 , indicating that Ulysses became deeply immersed within the coronal holes. Again, note that most of the slow wind tends to fall within ~$0.05R_S$ of the CHB. Beyond this, the wind tends to be fast and relatively steady. A least squares fit to data above 600 km/s emphasizes the relatively constant positive gradient in speed with respect to distance to the CHB.

DISCUSSION

The least squares fit to the data in Figure 6 gives the following relationship between distance to the CHB (d) and speed:

$$v = 713 kms^{-1} + 3.2d(\ ) \qquad (1)$$

Or, recasting it in units of solar radii (or equivalently radians):

$$v = 713 kms^{-1} + 183.7d(R_s) \qquad (2)$$

These results may be contrasted with previous work (4) which found a mean high-speed flow of 703 km/s and a latitudinal gradient of 0.950 km/degree. Our results suggest a gradient that is ~3.4 times larger. We suggest that equation (1) is closer to the true speed gradient, at least within large polar coronal holes and that the lower gradient in the previous study (4) reflects the approximate correlation of distance into a coronal hole with increasing heliographic latitude. Since even large polar CHBs can vary significantly in latitude, as a function of longitude, however, the intrinsic gradient is masked, and the apparent disagreement is thus a "dilution" effect. By the same token, the CHBs computed in the MHD models, are only approximately correct and equations (1) and (2) should be viewed as lower limits to the true gradient.

ACKNOWLEDGMENTS

The authors gratefully acknowledge the support of the national Aeronautics and Space Administration (SEC-GI, SEC-TP, SR&T, and LWS programs). We also thank the National Science Foundation at the San Diego Supercomputer Center for providing computational support.

REFERENCES

1. Neugebauer, M., and C. W. Snyder, Mariner 2 observations if the solar wind. I. Average properties, *J. Geophys. Res. 71*, 4469, 1966.

2. Riley, P., S. J. Bame, B. L. Barraclough, W. C. Feldman, J. T. Gosling, G. W. Hoogeveen, D. J. McComas, J. l. Phillips, B. E. Goldstein, and M. Neugebauer, Ulysses solar wind plasma observations at high latitudes, *Adv. Space Res., 20,* 15, 1997.

3. McComas, D. J., S. J. Bame, B. L. Barraclough, W. C. Feldman, H. O. Funsten, J. T. Gosling, P. Riley, R. Skoug, A. Balogh, R. J. Forsyth, B. E. Goldstein, and M. Neugebauer, Ulysses rerun to the slow solar wind, *Geophys, Res. Lett., 25,* 1, 1998.

4. McComas D. J., B. L. Barraclough, H. O. Funsten, J. T. Gosling, E. Santiago Munoz, R. M. Skoug, B. E. Goldstein, M. Neugebauer, P. Riley, A. Balogh, Solar wind observations over Ulysses' first full polar orbit, *J. Geophys. Res., 105,* 10419, 2000.

5. Neugebauer, M., P. C. Liewer, E. J. Smith, R .M. Skoug, T. H. Zurbuchen, Sources of the Solar Wind at Solar Activity Maximum, submitted to *J. Geophys. Res.*, 2002.

6. McComas D. J., H. A. Elliott, and R. von Steiger, Solar wind from high-latitude coronal holes at solar maximum, *Geophys. Res., Lett.*, in press, 2002.

7. von Steiger R., J. Geiss, and G. Gloeckler, Composition of the Solar Wind, in Cosmic Winds and the Heliosphere edited by J. R. Jokipii, C. P. Sonett and M. S. Giampapa, Tucson, Arizona, The University of Arizona Press, pp. 581-616, 1997.

8. Nolte J. T. and E. C. Roelof, Large-Scale Structure of the Interplanetary Medium, *Solar Physics 33,* 483, 1973.

9. Riley P., J. A. Linker, and Z. Mikic, An empirically-driven global MHD model of the solar corona and inner heliosphere *J. Geophys. Res., 106,* 15889, 2001.

10. Neugebauer M., R. J. Forsyth, A. B. Galvin, K. L. Harvey, J. T. Hoeksema, A J. Lazarus, R. P. Lepping, J. A. Linker, Z. Mikic, J. T. Steinberg, R. von Steiger, Y.-M. Wang, and R. F. Wimmer-Schweingruber, Spatial structure of the solar wind and comparisons with solar data and models, *J. Geophys. Res., 103,* 14587, 1998.

11. Riley, P., J. A. Linker, and Z. Mikic, Modeling the heliospheric current sheet: Solar-cycle variations, *J. Geophys. Res., 107,* 10.1029/2001JA000299, 2002.

Space Weather at 75 AU

R. A. Mewaldt

California Institute of Technology, Pasadena, CA 91125, USA

Abstract. Recent outer-heliosphere observations are reviewed from a space weather point of view by comparing the nature of solar wind, solar particle, and cosmic ray variations at the Voyagers and 1 AU. While the Sun still controls the interplanetary medium at 75 AU, the nearby boundaries of the heliosphere exert a strong influence on the environment.

INTRODUCTION

The agents of space weather at 1 AU, including transients such as coronal mass ejections (CMEs) and shocks, variations in the interplanetary magnetic field, solar and interplanetary particle events, and solar UV and x-rays, all make their connection to 1 AU across a distance of "only" 150 million km, allowing them to act on time scales ranging from 8 minutes to a few days. As a result, it is usually possible to make a direct connection between events observed on the Sun (be it a CME, flare, or passage of an active region or coronal hole) and subsequent effects on the near-Earth environment. In the outer solar system, say beyond 30 AU, the time scales are slowed by a factor of >30, and the intensities of solar wind and solar particles are diminished by a factor of >1000. As we will see, this change in distance scale has a rather dramatic effect on the manner by which the Sun exerts its control on the interplanetary environment, and on the nature and magnitude of variations in that environment.

In order to encourage an interest in space weather in the outer heliosphere, the reader is invited to imagine that he/she resides on a Kuiper Belt Object (KBO) orbiting at ~75 AU. Imagine that this society launched two Voyagers in the 1950's as part of their space program, using a gravity assist at Jupiter in 1979 to reach Saturn, Neptune, and Uranus, and to carry on towards the boundaries of the heliosphere. This KBO society also recently positioned two spacecraft (SOHO and ACE) at L1 in order to monitor the Sun and understand how it affects the outer heliosphere.

This paper will survey outer-heliosphere observations from a space-weather point of view. It will compare the nature of solar-wind transients at 1 AU and beyond 50 AU, follow solar energetic particles on their six-month journey from the Sun to 70 AU, look at the nature of cosmic-ray intensity variations in the inner and outer heliosphere, assess the nature of life near the termination shock in a high-radiation zone, and compare particle intensities near and far from the Sun.

SOLAR WIND TRANSIENTS

The five-year period from 1998 through 2002, including the maximum of solar cycle 23, has provided the best opportunity yet to observe and understand the effects of solar variations on the near-Earth environment because of the array of new spacecraft and instruments now in operation. Similarly, Voyager now has a first opportunity to explore the effects of solar maximum beyond 60 AU from the Sun. During 1998 thru 2001 there were ~4000 CMEs observed by SOHO [1], about 5% of which could be identified as interplanetary CMEs (ICMEs) observable at Earth [2]. In addition, ACE observed ~200 interplanetary shocks at L1 during this period [3].

Daily average solar-wind parameters at 1 AU during this time period (top three panels of Figure 1) reveal considerable high-frequency time structure, much of it associated with CMEs. Contrast this with Voyager-2 observations during the same period (Figure 2), where the solar wind speed rarely varies by >10% from one day to the next, where the hundreds of individual ICMEs observed at Earth no longer stand out, and where the dominant solar wind structures are merged interaction regions (MIRs) that occur only infrequently but last for weeks at a time (see, e.g., [4]).

CP679, *Solar Wind Ten: Proceedings of the Tenth International Solar Wind Conference*, edited by M. Velli, R. Bruno, and F. Malara

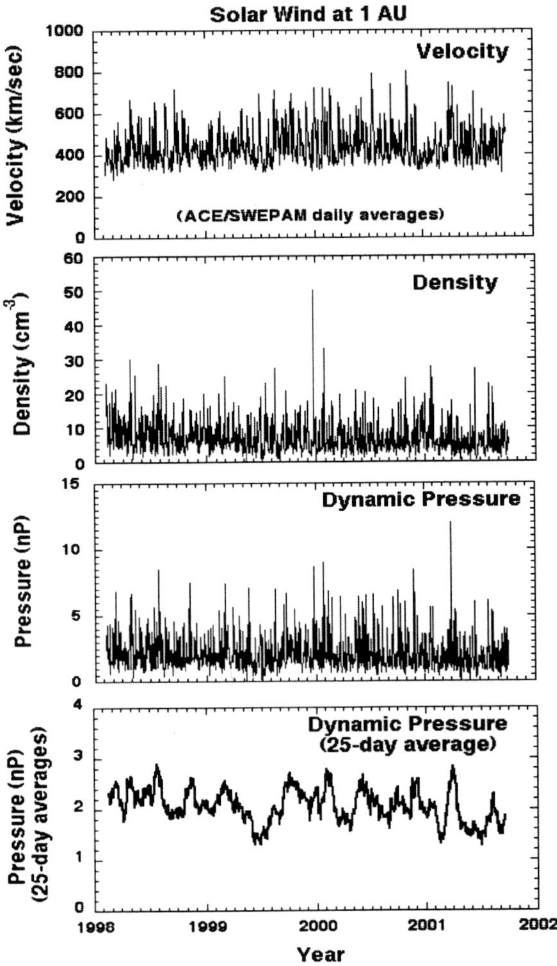

FIGURE 1. Daily average (top 3 panels) and 25-day average solar wind data at 1 AU from ACE/SWEPAM.

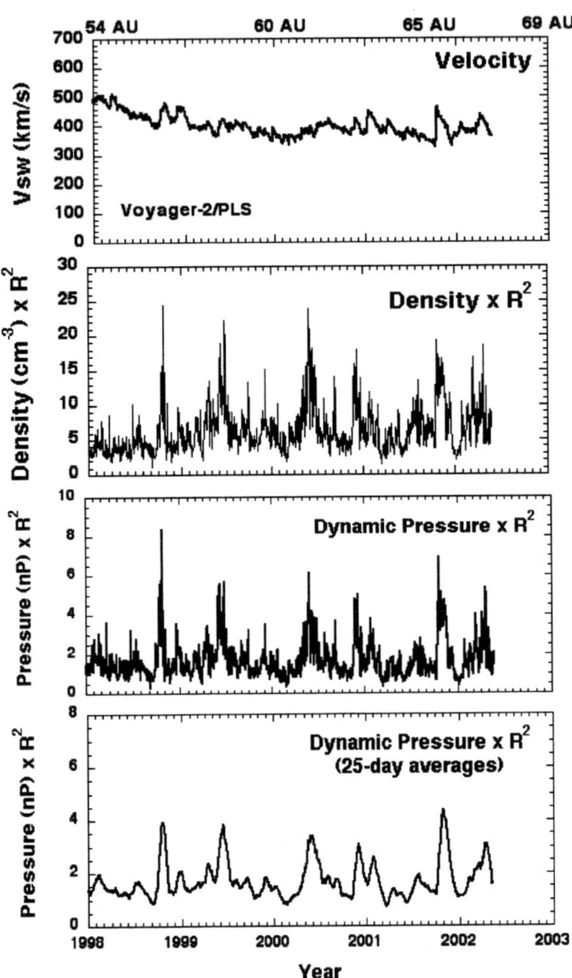

FIGURE 2. Daily average and 25-day average solar-wind data from the PLS instrument on Voyager-2.

Merged interaction regions (MIRs) result from the entrainment of slower wind by faster, upstream wind (e.g., [5]). During solar maximum the highest velocity solar-wind structures are associated with ICMEs, and the fastest and largest ICMEs eventually become the core of MIRs in the outer heliosphere. In contrast to the many ICMEs and shocks observed at 1 AU, only ~6 MIRs and ~18 shocks were observed by Voyager 2 from 1998 through mid-2002 (J. Richardson, private communication; see Figure 2). From January 1998 to June 2002 Voyager-1 moved from ~69 to ~85 AU at an average heliographic latitude of +33.6°, while Voyager-2 moved from ~54 to ~62 AU at an average latitude of -21°.

From daily-average data at 1 AU (top three panels of Figure 1) it is difficult to recognize the structure that will emerge at Voyager-2. However, with 25-day running averages (bottom of Figure 1) it is evident that 1-AU solar wind does include time structure similar to that observed in the outer heliosphere.

Bastille-Day Event in the Outer Heliosphere

The solar fireworks of July 14, 2000 (Bastille Day event) led to the largest space weather events at Earth of this solar maximum. News of the Bastille Day event did not reach Voyager-2 until 6 months later, when an interplanetary shock was observed [6], followed by a Forbush decrease in the cosmic-ray intensity and an increase in the low energy particle intensities that reached a maximum about a month later [7, 8]. Although the arrival of the Bastille Day shock is more difficult to identify at Voyager-1 because the plasma instrument is no longer working, there is a Forbush decrease in the cosmic ray intensity starting in early February, 2001 that is comparable to that at Voyager-2, followed by an increase in the ~1 MeV particle intensity peaking at ~5/1/01.

SOLAR ENERGETIC PARTICLES

The Great Race

In the Bastille-Day event the intensity of ~1 MeV protons at Voyager-2 peaks well after the arrival of the shock. This is common in the outer heliosphere [9], while in the inner heliosphere low-energy shock-accelerated particles typically peak at the shock. However, there are also instances in the outer heliosphere where most of the particles arrive before the shock, as in 1998 [7]. Simulations by Rice et al. [9] show that particles can arrive late if the shock has weakened and can no longer inject and accelerate particles by the time it reaches Voyager-2.

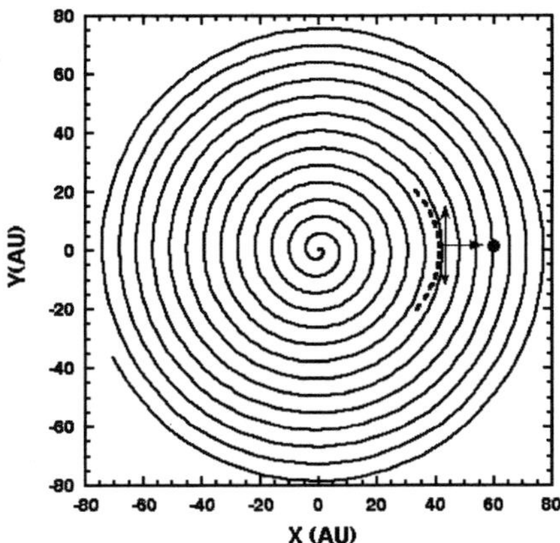

FIGURE 3. The nominal IMF is shown out to 80 AU, including a shock assumed incapable of injecting particles beyond 40 AU and a spacecraft at 60 AU.

The illustration in Figure 3 assumes that the shock stops accelerating particles at 40 AU. Particles observed at 60 AU will have been transported from inside 40 AU to 60 AU by a combination of diffusion and convection. Indeed, particles inevitably move by both processes, but it is not obvious which approach is best to win the race from the last point of acceleration to the spacecraft. In the inner heliosphere 1-MeV particles generally diffuse in radius considerably faster than they are convected. In the outer heliosphere diffusion is much more difficult for low-energy particles because the magnetic field lines are so much longer (the length of the Parker spiral from 0 to R AU is ~$R^2/2$ for a nominal solar wind speed of ~400 km/sec). Particles that move primarily by convection arrive late because the shock speed exceeds the convective speed. Particles that move primarily by diffusion may arrive either early or late, depending on whether they can advance faster by moving along the Parker spiral faster than by convection [7]. It is also possible that particles can advance more quickly by diffusing across field lines (much like Rosie Ruiz "won" the 1980 Boston marathon by running the first few km, leaving the course and taking a subway across town, and then rejoining the race just before the finish). Decker et al. [10] have performed simulations of particle events such as this and compared them with the observed time intensity profiles and anisotropies.

Solar Particle Intensities in the Outer Heliosphere

Because of the difficulty of tracing individual SEP events from 1 AU into the outer heliosphere it is instructive to take a more global view. Figure 4 shows 25-day running averages of the ~2 MeV particle intensities from ACE and the two Voyagers. Note that at the Voyagers there are only a few, very broad intensity maxima. While these are difficult to match with 1-AU events using daily-average data, there is more hope with 25-day averages. The brackets labeled 1 to 4 in Figure 4 represent an attempt to relate particles that originated in the same series of events at the Sun, taking into account propagation delays. The first of these includes the events of April/March 1998, the second is centered on the Bastille Day event, the third includes the events of Nov. 2000, and the fourth the events in April 2001. A fifth period visible in late 2001 at 1 AU is just starting to appear at Voyager-2.

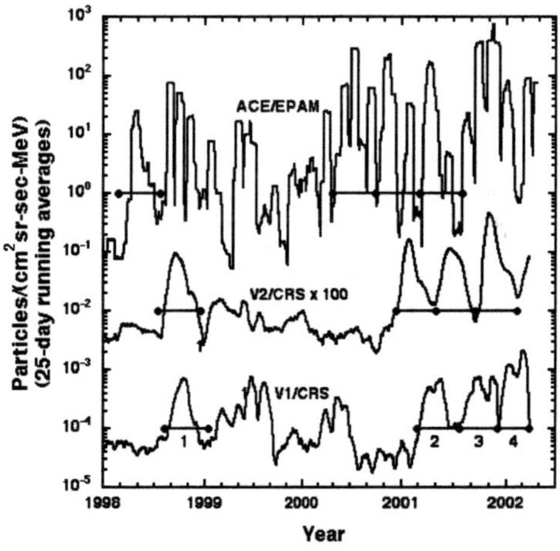

FIGURE 4. Intensity of SEP protons at 1-AU (1.2–5 MeV) and the two Voyagers (1.8-3.3 MeV). Intervals during which the intensities were compared are indicated (see Figure 5).

Integrated particle fluences from four selected periods (see Figure 4) are plotted versus radial distance in Figure 5. In all cases particle intensities in the outer heliosphere are smaller than would be expected from simple scaling by r^{-2}; with suggested slopes ranging from $r^{-2.4}$ to $r^{-2.8}$. Intensities at the two Voyagers appear to be consistent with the same radial dependence in spite of the >50° latitude separation. Decker et al. [13] obtained similar results for protons accelerated in co-rotating interaction regions (CIRs).

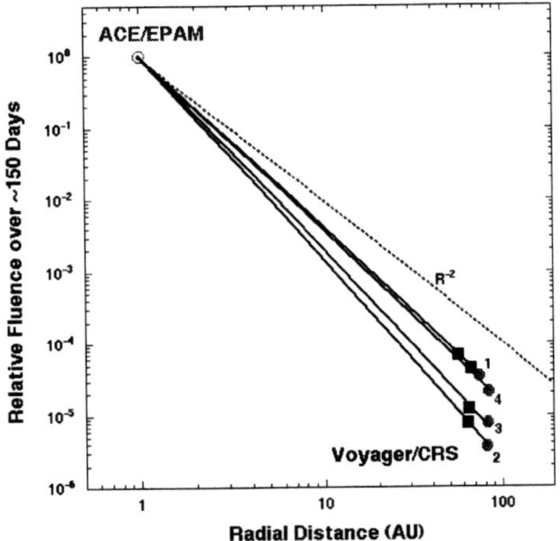

FIGURE 5. Measured fluence of ~2 MeV protons at ACE, Voyager-1, and Voyager-2 for the periods in Figure 4.

There are several factors that affect how SEP intensities vary with distance from the Sun. The most important of these is probably adiabatic energy loss, coupled with a typically falling SEP energy spectrum ($dJ/dE \sim E^b$ with $b \approx -4$ to -2). Solar particle events generally originate at low heliographic latitudes (<30°). If particles escape to higher latitudes by cross-field diffusion the low-latitude intensities will decrease faster than r^{-2}.

Other factors work in the opposite direction. Although the Voyagers are connected to the near-Sun region at all heliographic longitudes over a 25-day period, and thus integrate over all SEP events that occur during these 150-day periods, observations at L1 are well connected to only ~120° of solar longitude at a time – particles accelerated on the back side of the Sun (viewed from Earth) tend to escape into the outer heliosphere before they can be observed at L1. An r^{-2} radial profile might be expected if the vast majority of the particles originate inside 1 AU, as appears to generally be the case at higher energies (>10 MeV/nucleon). Continued acceleration beyond 1 AU will tend to flatten the radial dependence. It is clear that the interpretation of SEP intensities in the outer heliosphere is a very interesting problem that could benefit from theoretical modeling and simulations taking into account all relevant processes.

150-day Periodicities in the Outer Heliosphere

The Voyager data in Figure 4 appear to have a periodicity of ~150 days during 2000-2002. In addition, the same time structure is seen (shifted in time) in ~150-day averages of the frequency of interplanetary shocks observed by ACE (not shown here for lack of space). It is well known that a variety of solar phenomena exhibit ~150 day periodicities, including x-ray flares, SEP events, and the solar wind (see. e.g., [12]). Other periodicities have also been reported (e.g., [13, 14]). Hill et al. [15] reported periodicities of ~150 days in the outer-heliosphere ACR intensity during 1998-1999. Evidence that MIRs act as barriers to produce snow-plow-like increases in the intensity of ACRs in the outer heliosphere [7, 16] provides a mechanism for coupling the time structure of MIRs (Figure 2) to that of ACRs. It is beyond the scope of this paper to investigate whether the interplanetary periodicities apparent in 2000-2002 are statistically significant, as in Hill et al. [15]. However, it appears that the interplanetary medium introduces a low-pass filter into the very chaotic, high-frequency activity that characterizes solar maximum near the Sun. As a result, the outer heliosphere may be a preferred location to investigate this mysterious rhythm of solar activity.

High-Energy SEPs in the Outer Heliosphere.

Although it is desirable to extend the analysis in Figure 7 to higher energies, this proves to be difficult, as illustrated by the time history of 7.5 - 57 MeV protons observed by Voyager-1 (Figure 6). Particle intensities in the early 1980's were dominated by the SEP events of solar cycle 21. However, by the next solar maximum, withVoyager-1 at ~40 AU, only a few SEP events are evident, including the well-known March 1991 event. From 1995 to 2000 it very difficult to observe SEP events at Voyager-1 in this energy range because of a steady "background" that emerges in 1992 and remains through 2000. This background, due to ACR hydrogen, increases in intensity as the Voyagers approach the termination shock and is also subject to diminishing solar modulation effects. Anomalous H is also evident during the 1987 solar minimum at much lower intensity (Fig. 6). Stone [17] reviews estimates of the termination shock location based on approaches that include extrapolating the ACR gradients [18].

Measured and calculated solar-minimum spectra for cosmic-ray protons are shown in Figure 7, where ACR H is the dominant proton component from ~5 to ~140 MeV at Voyager-1. The theoretical curves are from a model by Cummings, Stone and Steenberg (here-in-after CS&S) that was fit to ACR species at Voyager-1&2 [19]. Also shown is the deduced ACR H spectrum at the termination shock for a relatively weak shock (compression ratio = 2.4).

It is reasonable to ask whether ACR intensities at the termination shock will constitute a radiation hazard for the Voyagers. The proton spectrum that CS&S

deduce for the termination shock (Fig. 7) is very similar in shape and intensity to that observed near Earth during the peak of the Bastille Day event [20] – the largest SEP event observed at Earth since 1989 - if the Bastille-Day spectrum (averaged over 8-hours of maximum intensity) is divided by 10^4. In contrast, the Voyagers may experience maximum particle intensities near the termination shock for a few years, leading to a total fluence somewhat less than that of the Bastille Day event at 1 AU. So, while the termination-shock radiation level (if experienced at Earth) would exceed NOAA SEP alert levels for several years (in the case of a weak shock), it will probably not present a hazard to the Voyagers. For a strong shock the levels will be somewhat lower [19].

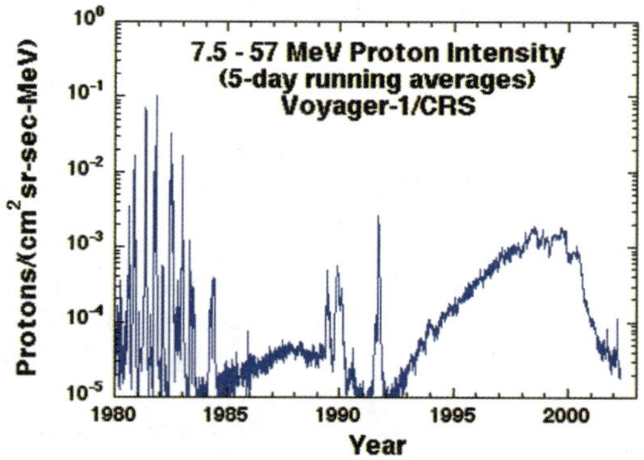

FIGURE 6. Intensity of 7.5-57 MeV protons at Voyager-1. Broad peaks in 1987 and 1998-99 are due to ACR hydrogen. Individual SEP events are difficult to observe after 1991.

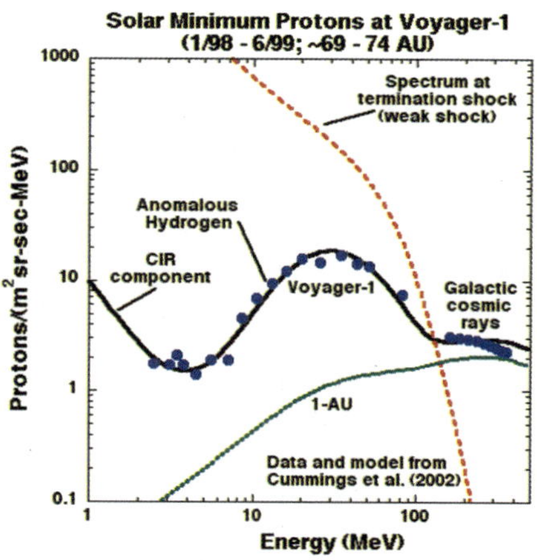

FIGURE 7. Solar-minimum proton spectrum at Voyager-1 compared to model and termination shock spectra [19].

Comparative Particle Intensities

A comparison of long-term average spectra for oxygen nuclei at 1 and 75 AU is shown in Figure 8. Most of the 1-AU spectra were measured by ACE from 10/97 to 6/00 [21]. Inner-source and interstellar pickup ions are also shown [22]. The solar wind and SEP spectra were scaled to 75 AU using simple $1/r^2$ scaling, although results in Figure 5 suggest a steeper radial gradient for energetic particles. Note that all components of solar origin, including inner-source pickup ions, have greatly decreased in importance at 75 AU, while those of interstellar origin (ACRs, GCRs, and interstellar pickup ions) have all increased in importance.

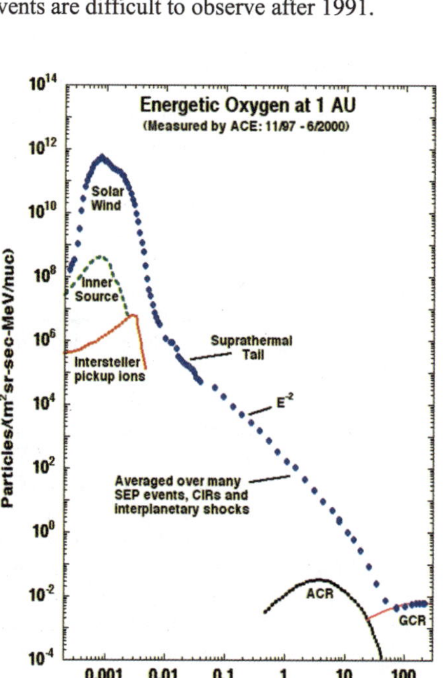

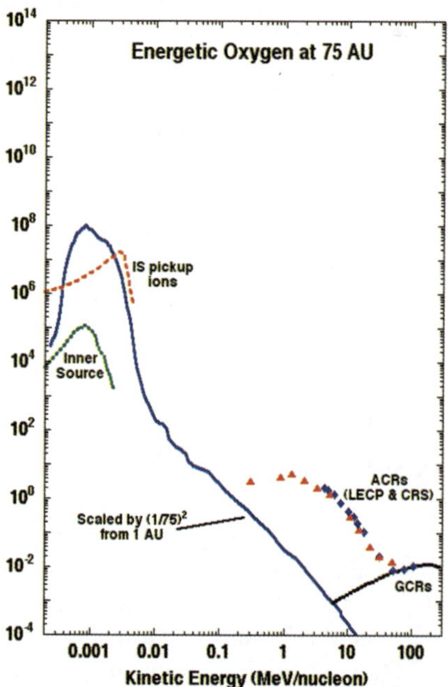

FIGURE 8. Long-term average particle intensities measured at 1-AU (left panel, [21] and scaled to 75 AU (right panel). Pickup ion spectra are from Gloeckler [22]. The ACR and GCR measurements at 75 AU are from [19]. Note that above ~75 keV/nuc interstellar components dominate at 75 AU.

SUMMARY

The examples shown here illustrate that while the Sun is still in control at 75 AU, it cannot exert this control on a day-to-day basis. Solar variations take six months or more to reach 75 AU, and in doing so most of the short-term variations become entrained in broad, long-lived structures. At this distance SEP intensities are attenuated by a factor of $>10^4$ and anomalous and galactic cosmic rays are the dominant high-energy particle radiation. While the Sun still modulates cosmic rays on time scales that range from weeks to the 22-year solar cycle, the range of these variations is much smaller than at 1 AU, and it is evident that the nearby boundaries of the heliosphere exert a powerful influence on the environment.

Postscript: We now return briefly to the Kuiper Belt Society, who, having analyzed the data from the Voyagers, ACE, and SOHO, was debating the future direction of their space program. There were those that favored a sample return mission to Pluto, since it was hypothesized to have been the cradle of life in the solar system as it was carried into solar system on interstellar grains. However, in this planetary program lost out to a two-part space weather program. It was clear to KBO space physicists that the most important space weather threat at 75 AU was neither solar wind transients nor solar particles - rather it was the much higher radiation environment of the termination shock and ISM beyond, and the long-term stability of the shield provided by the heliosphere. Worried that if the "day the solar wind almost disappeared" ever lasted for a year or more, the heliosphere would suddenly shrink, placing their KBO in the ISM, they advocated sending a probe close to the Sun to study how the solar wind is accelerated. In addition, they realized that the heliosphere would also be compressed in size if it encountered a moderate-density interstellar cloud [23], and therefore advocated sending a probe upstream of the heliopause to explore the environment at the boundaries of the heliosphere and beyond, and to monitor approaching density enhancements in the local ISM. Of course, they called this two-part space-weather program "Living with the Stars".

ACKNOWLEDGMENTS

I am grateful to the PLS, LECP, and CRS teams for the use of unpublished Voyager data and to the SWEPAM, MAG, and EPAM teams and ACE Science Center for unpublished ACE data. I also thank A. C. Cummings, R. B. Decker, and J. D. Richardson, E. C. Roelof and E. C. Stone for helpful discussions. This work was supported by NASA under NAS5-6912.

REFERENCES

1 LASCO CME catalog, http://cdaw.gsfc.nasa.gov/
2. Cane. H. V., personal communication, 2002.
3. Personal communication from the MAG and SWEPAM teams on ACE.
4. Richardson, J. D., in The Outer Heliosphere: The Next Frontiers, K. Scherer et al., eds, COSPAR Colloquia Series #11, Pergamon, Amsterdam, 301-310 (2001).
5. Burlaga, L. F., Interplanetary Magnetohydrodynamics, Oxford University Press, New York (1995).
6. Burlaga, L. F., Ness, N. F., Richardson, J. D., and Lepping, R. P., Solar Physics, **204**, 399-411 (2002).
7. Decker, R. B., E. C. Roelof, and S. M. Krimigis, in *Acceleration and Transport of Energetic Particles Observed in the Heliosphere*, edited by R. A. Mewaldt et al., AIP Conf. Proc. 528, New York, 161-164 (2000).
8. McDonald. F. B. et al., Proc. of ICRC 2001, **9**, 3637-3640 (2001).
9. Rice, W. K. M., Zank, G. P., Richardson, J. D., and Decker, R. B., Geophys. Res. Lett. **27**, 509-512 (2000).
10. Decker, R. B., Roelof, E. C., and Krimigis, S. M., "Solar Energetic Particle Propagation from 1 to 72 AU during 1998", in preparation (2002).
11. Decker, R. B., Paranicas, C., Krimigis, S. M., Paularena, K. I., and Richardson, J. D., in The Outer Heliosphere, The Next Frontiers, K. Scherer et al., eds, COSPAR Colloquia Series #11, Pergamon, Amsterdam, 321-324 (2001).
12. Cane, H. V., Richardson, I. G., and von Rosenvinge, T. T., Geophys. Res. Lett. **25**, 4437-4440 (1998).
13. Dalla, S., and Balogh A., Geophys. Res. Lett. **27** (2), 153-156 (2000).
14. Richardson, J. D., Paulerena, K. I., Belcher, J. W., and Lazarus, A. J., Geophys. Res. Lett. **21**, 1559-1560 (1994).
15. Hill, M. E., Hamilton, D. C., and Krimigis, S. M., J. Geophys. Res., **106**, No. A5, 8315-8322, (2001).
16. Stone, E. C., and Cummings, A. C., this conference (2002).
17. Stone, E. C., Science, **293**, 55-56 (2001).
18. Stone, E. C., and Cummings, A. C., Proc. 26th Internat. Cosmic Ray Conf. **7**, 500-503 (1999).
19. Cummings, A. C., Stone, E. C., and Steenberg, C, "Composition of Anomalous Cosmic Rays and Outer Heliospheric Ions", to be published in Astrophys. J. (2002).
20. Tylka, A. J., Cohen, C. M. S., Dietrich, W. F., Maclennan, C. G., McGuire, R. E., Ng, C. K., and Reames, D. V., Astrophys. J. Lett. **558**, L59-L63 (2001).
21. Mewaldt, R. A., et al., in Solar and Galactic Composition, edited by R. F. Wimmer-Schweingruber, AIP Conf. Proc. 598, New York, pp. 165-170 (2002).
22. Gloeckler, G., Fisk, L. A., Zurbuchen, T. H., and Schwadron, N. A., in Acceleration and Transport of Energetic Particles Observed in the Heliosphere, edited by R. A. Mewaldt et al., AIP Conf. Proc. 528, New York, AIP, pp. 221-228 (2000).
23. Zank, G. P., and Frisch, P. C., Astrophys. J. **518**, (2) 965-973 (1999).

Modeling heavy ions and atoms throughout the heliosphere

Hans-R. Müller*† and Gary P. Zank†

*Bartol Research Institute, University of Delaware, Newark, DE 19716, USA
†Institute for Geophysics and Planetary Physics, University of California, Riverside, CA 92521, USA

Abstract. Most investigations addressing the global structure of the heliosphere, including explicitly the interaction between the solar wind and the partially ionized local interstellar medium (LISM), have focused on hydrogen since it is the most abundant particle species. We use kinetic models that include heavy elements such as He, C, N, O, and others, to study the heliospheric distribution of neutrals and the singly charged ions of these species, besides H. Our model describes the evolution of interstellar heavy neutral atom distributions throughout the heliosphere, and we include the interaction of heavy particles with neutral hydrogen and protons through charge exchange (i.e., the creation of pickup ions), while the heavy particles are subject to photoionization and gravity. We use improved, recently published charge exchange cross-sections as well as recently identified LISM boundary conditions. A realistic description of the basic heavy element distribution and filtration at heliospheric boundaries will provide an important theoretical basis for interpreting observations of pickup ions made by Ulysses and ACE.

INTRODUCTION

The interaction of the ionized solar wind with the partially ionized local interstellar medium (LISM) defines the heliosphere with its characteristic boundaries, the termination shock and the heliopause. The dominant particle species are neutral hydrogen (H) and its atomic constituents, protons and electrons. Consequently, numerical models to study the global heliosphere focus on the self-consistent modeling of the interaction of solar wind hydrogen plasma with the LISM hydrogen wind which consists of both neutral H and hydrogen plasma. The consideration of neutral H in the heliosphere is an important ingredient of heliospheric physics since the coupling of neutral H to the plasma via charge exchange affects the plasma distribution profoundly, which in turn feeds back to the neutral H distribution.

The solar wind and the LISM contain elements heavier than hydrogen as well. The direct solar wind includes heavy ions in various high charge states. In addition, there are sources for neutral heavy atoms in the inner heliosphere, such as the "inner source" (likely released from dust) and planetary atmospheres. However, the LISM provides a much higher density of neutral atoms. These interstellar atoms can penetrate through the heliospheric boundaries into the inner heliosphere, where they, together with the corresponding pickup ions, serve as messengers from the LISM which is currently out of reach for *in-situ* measurements.

In order to gain an understanding of the LISM through the measurement of neutral heavy atoms and their singly charged heavy ions, it is necessary to know what changes the interstellar neutral atom population undergoes on its path through the heliosphere. The dominant interactions are the charge exchange of a neutral atom with the plasma proton background, $A + p \rightarrow A^+ + H$, creating a singly charged heavy pickup ion, and the reverse process, $A^+ + H \rightarrow A + p$ which creates a neutral heavy atom. These two processes let the hydrogen background influence the heavy interstellar elements in the heliosphere. In addition, heavy atoms will experience photoionization close to the Sun, as well as gravity.

MODEL

In this study, we first model a complete hydrogen heliosphere, with an assumed steady solar wind with density $n_p = 5\text{cm}^{-3}$, velocity $v = 400\text{km s}^{-1}$, and temperature $T = 10^5\text{K}$ at 1 AU. The interstellar hydrogen velocity is set to $v = 26\text{km s}^{-1}$. The density and temperature are less well known, and we use two sets of values, namely, a low-density, high-ionization fraction background model 1 with $n_H = 0.14$, $n_p = 0.1\text{cm}^{-3}$ and $T = 8000\text{K}$, and the alternative background model 2 with $n_H = 0.216$, $n_p = 0.047\text{cm}^{-3}$ and $T = 7000\text{K}$. The hydrogen distributions are modeled with a multifluid code [1], and representative plots for plasma temperature and neutral H number density are given in Figures 1 and 2. Both cases have a pronounced hydrogen wall between heliopause and bow shock, with neutral densities reaching peaks of 2.3 times above their respective interstellar neutral

CP679, *Solar Wind Ten: Proceedings of the Tenth International Solar Wind Conference*, edited by M. Velli, R. Bruno, and F. Malara

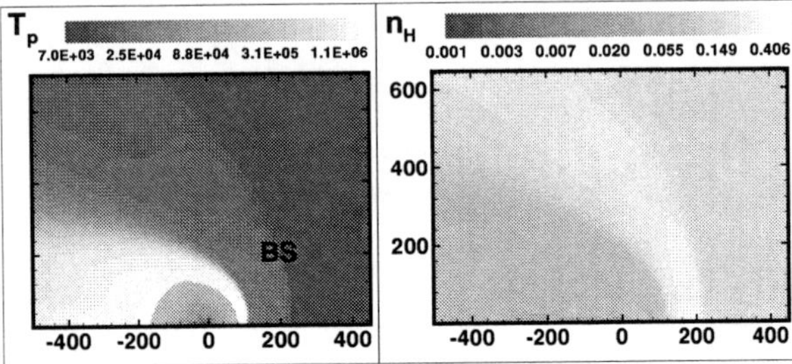

FIGURE 1. Hydrogen background 1, shown through plasma temperature (left, in K) and neutral density (right, in cm^{-3}).

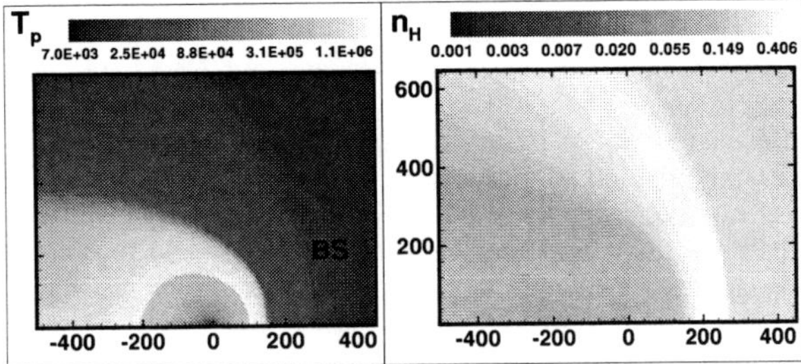

FIGURE 2. Same as Fig. 1, for hydrogen background 2.

densities. While the locations of termination shock and heliopause shift in response to the different interstellar ionization fraction, the filtration ratios at the termination shock (the amount of neutral H reaching the TS, compared to the interstellar density of neutral H) are similar (0.40 and 0.46, respectively). In absolute terms, of course, there is much more neutral H present in the background model 2 throughout the heliosphere, and correspondingly more pickup protons are produced.

With the hydrogen as a background, we model the flow of interstellar neutral heavy atoms and singly charged heavy ions into and around the heliosphere. We use a kinetic direct Boltzmann solver (modified from [2, 3, 4]) that tracks the two species through trajectories of macroparticles while exchanging charge with the hydrogen background, and while being affected by photoionization and gravity.

The focus of this study are the elements N and O, as well as the noble gases He, Ne, and Ar. Pickup processes with these elements in the heliosphere are relevant for processes leading to anomalous cosmic rays. All these elements are present in the ISM both in neutral and ionized form. We additionally study carbon which is almost completely ionized in the LISM. For the quantitative boundary values, we use the number densities given in Table 1. They are derived from model 17 of Slavin & Frisch [5] who model the nearby ISM taking into account measured column densities along different sightlines and models for the radiation environment of the Local Interstellar

TABLE 1. Interstellar model boundary parameters.

	neutral density	singly ionized	doubly ionized
	cm^{-3}	cm^{-3}	cm^{-3}
H*	0.140	0.100	
H†	0.216	0.047	
H	0.230	0.110	
He	0.017	0.017	3.5×10^{-4}
C	7.7×10^{-8}	1.5×10^{-4}	5.4×10^{-6}
N	1.7×10^{-5}	1.3×10^{-5}	8.8×10^{-9}
O	1.6×10^{-4}	6.6×10^{-5}	2.3×10^{-8}
Ne	4.8×10^{-6}	2.8×10^{-5}	1.2×10^{-5}
Ar	1.8×10^{-7}	5.0×10^{-7}	3.3×10^{-7}

* Background model 1
† Background model 2

Cloud. As can be seen from Table 1, among the six elements mentioned here only Ne and Ar have a significant density of charge state larger than 1.

The Slavin & Frisch [5] calculations leading to model 17 also give an estimate for the interstellar H densities, which we did not use here for our background model. Their neutral hydrogen value is similar to background 2, and the proton density is similar to the one of background 1. Therefore, while the LISM hydrogen boundary values assumed in the present study are not self-consistent with the rest of the elemental densities, the two H background models used here straddle the Slavin & Frisch H values.

RESULTS

Figures 3–5 show typical heavy atom (left) and ion (right) distributions resulting from the two-dimensional kinetic models. Direct interstellar ions are excluded from entering the heliosphere. The interior heliosphere is left with a very low density of ions created from neutrals via charge exchange (pickup ions). The heliopause, created by hydrogen, bounds this depleted cavity. The stagnating flow of H plasma upwind of the heliopause translates into a deceleration and density enhancement of the heavy ions as well [6].

The charge exchange cross section of (He, H) and (O, H) are very different, both in the way they depend on the particle energy, and in magnitude, with the helium exchange cross section lower than that of oxygen. This fact finds its most prominent expression in the appearance of a neutral oxygen wall [7] upwind of the heliopause (Figure 4), whereas neutral helium traverses the heliospheric boundaries practically unimpeded, with only a focusing downwind of the Sun due to gravity that is not balanced

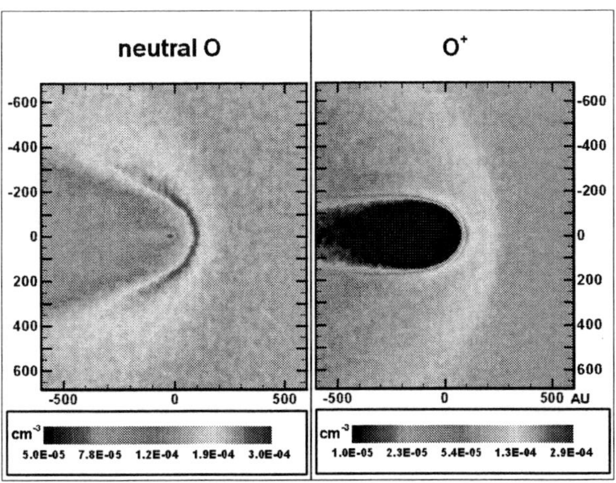

FIGURE 3. Neutral and singly ionized oxygen (model against H background 1), shown through neutral density (left) and ion density (right, in cm^{-3}).

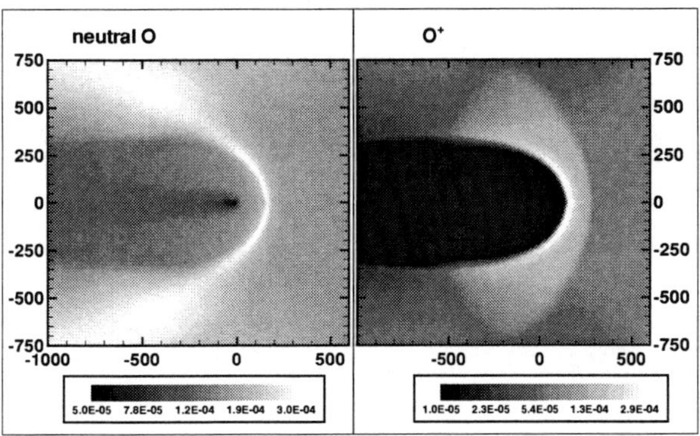

FIGURE 4. Same as Fig. 3, for hydrogen background 2.

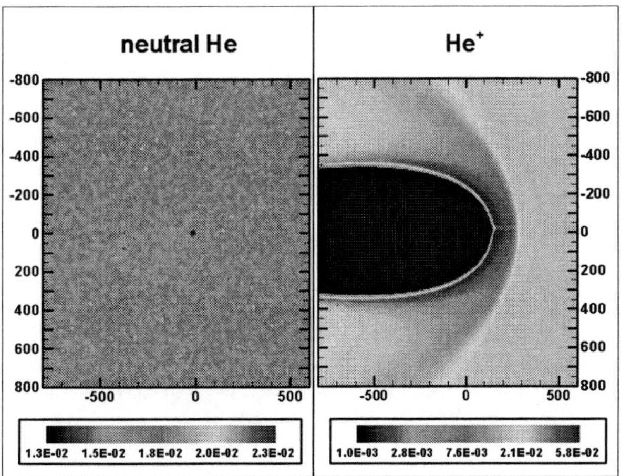

FIGURE 5. Neutral and singly ionized helium (model against H background 2), shown through neutral density (left) and ion density (right, in cm^{-3}).

TABLE 2. Model results for H backgrounds 1 and 2.

	filtration upwind TS		**filtration downwind TS**		**amplification in wall**	
# bg	1	2	1	2	1	2
H	0.40	0.46	0.20	0.22	2.26	2.32
He	0.99	0.98	0.87	0.94	1.10	1.01
C	0.95	0.96	0.48	0.52	1.03	1.00
N	0.96	1.07	0.58	0.58	1.32	1.38
O	1.17	1.08	0.54	0.64	1.75	1.96
Ne	0.87	0.95	0.60	0.58	1.02	1.00
Ar	0.90	0.95	0.51	0.46	0.99	0.99

by radiation pressure (Figure 5). A similar focusing occurs also for oxygen. The oxygen wall feeds back to the oxygen plasma so that the ion enhancement upwind of the heliopause is even more pronounced than for helium.

When comparing the oxygen 2D density maps for two distinct H background models in Figures 3 and 4, the size differences are obvious and relate to the H background (Figures 1 and 2). In addition, because of the larger H densities in background model 2, the corresponding neutral oxygen wall has a larger maximal value. Figure 6 shows the number density profiles along the LISM flow symmetry axis (with Sun at 0) for all 6 modeled element species using H background 2. The bold lines represent the neutral number density, while the thin dashed lines are pickup and interstellar heavy ion densities. The locations of termination shock TS, heliopause HP, and bow shock BS of the heliosphere are marked in the oxygen panel. While on the order of 10^7 particles in the simulations produce reasonable statistics in most parts of the profiles, the scarcity of pickup ions results in substantial statistical noise for their part in the ion density curves.[1]

As is evident from Figure 6, only oxygen, and to a lesser extent nitrogen due to its charge exchange cross section dependence on energy similar to that of oxygen, has a wall-like neutral density enhancement upwind of the heliopause. The neutral helium density is basically constant on the chosen scales. Ne, Ar, as well as C do not show an enhancement in the form of a wall, but exhibit depletion close to the Sun on the upwind side, and refilling of the ionization cavity due to focusing on the downwind side. For the singly ionized heavy interstellar particles, there is a density enhancement upwind of the heliopause, as already discussed above.

Table 2 summarizes some results obtained from the two sets of heavy elements models. The filtration ratios for the upwind termination shock location are important for all calculations of pickup ion densities in the inner

[1] The simulations with H background 1 have been done with fewer particles, resulting in a higher level of graininess in Figure 3.

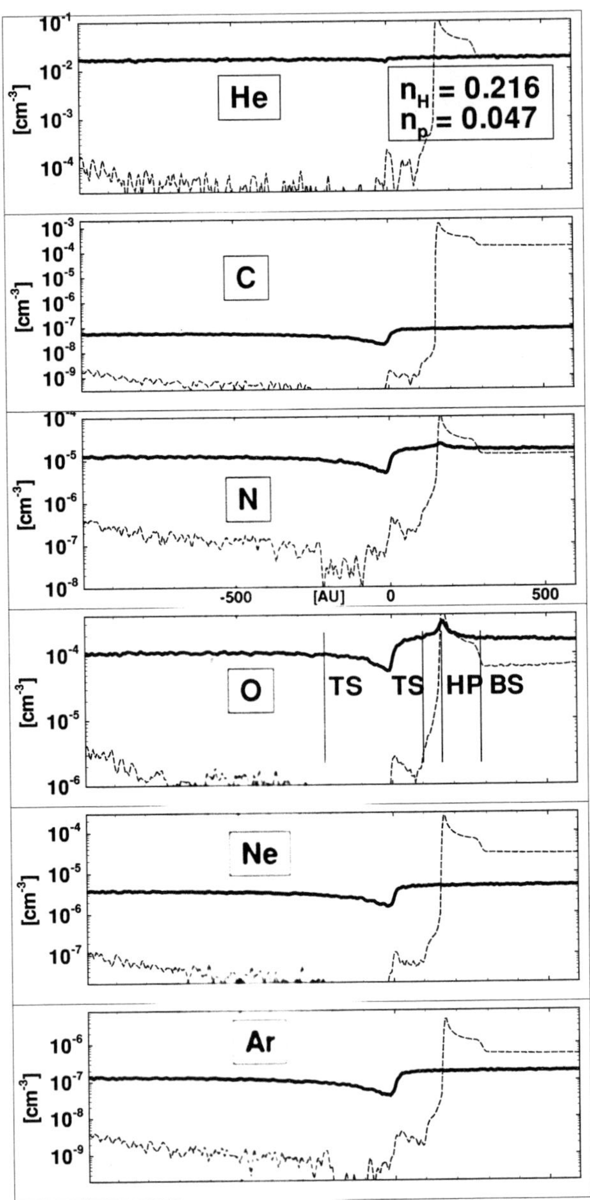

FIGURE 6. Number densities of heavy neutrals (thick solid lines) and heavy ions (thin dashed lines) along the stagnation axis, for H background 2.

heliosphere, and as input for ACR acceleration processes at the TS. The filtration ratios for the downwind TS are given as an illustration of the depletion of neutral atoms while they traverse the region of the supersonic solar wind. Lastly, the amplification in the wall is an expression of the peak density of neutral heavy atoms just upstream of the heliopause.

The results given in Table 2 support the discussions above. He, C, and N do not experience appreciable filtration, and Ne and Ar only a modest amount. The models predict even a small enhancement for neutral oxygen. However, this result might not hold up when additional ionization processes, such as electron impact ionization, are considered in future models. The downwind T-S values are about the same across the elements, with the exception of He which owes its better survival to the lower photoionization rate assumed here. The neutral atom walls upwind of the heliopause behave as discussed above, with the exception of a slight, unexpected overdensity of helium for background model 1.

Table 2 also allows for an estimate of how sensitive such heavy element calculations are to the assumed hydrogen background. Comparing adjacent columns of Table 2 to each other, there is an overall agreement, indicating somewhat low sensitivity to the H background. As expected due to their larger charge exchange cross sections, nitrogen, and even more so oxygen, do react to changes in the H background, with upwind filtration and amplification values varying here by $\sim 10\%$. In general, the filtration ratio seems to depend on the H background on that level, with the notable exception of helium.

In summary, the results emphasize the importance of the charge exchange cross sections to the filtration and distribution of interstellar neutral heavy atoms in the heliosphere. Elements weakly coupled to H experience only modest losses when traversing the heliosphere, and filtration is not sensitive to the details of the H background. Elements coupled more strongly to H, such as O and N, are sensitive to the background H distribution, and react more strongly in the form of neutral walls and changed filtration ratios.

ACKNOWLEDGMENTS

The authors gratefully acknowledge support by NASA contract NAG5-11621 and NSF contract ATM-0296114 awarded to the University of California at Riverside.

REFERENCES

1. Zank, G. P., Pauls, H. L., Williams, L. L., and Hall, D. T., *J. Geophys. Res.*, **101**, 21639 (1996).
2. Zank, G. P., Müller, H. R., and Lipatov, A. S., *AIP Conference Proceedings*, **505**, 811–814 (1999).
3. Müller, H. R., Zank, G. P., and Lipatov, A. S., *J. Geophys. Res.*, **105**, 27419–27438 (2000).
4. Lipatov, A. S., and Zank, G. P., *Solar Syst. Res.*, **34**, 169–172 (2000).
5. Slavin, J. D., and Frisch, P. C., *Astrophys. J.*, **565**, 364–379 (2002).
6. Fahr, H. J., Osterbart, R., and Rucinski, D., *Astron. Astrophys.*, **294**, 587–600 (1995).
7. Izmodenov, V. V., Malama, Y. G., and Lallement, R., *Astron. Astrophys.*, **317**, 193 (1997).

Heliospheric Plasma and Current Sheet Structure

N. U. Crooker

Center for Space Physics, Boston University, Boston, MA 02215, USA

Abstract. Recent analyses using suprathermal electrons as sensors of true magnetic polarity indicate that the steady-state concepts of the heliospheric current sheet and its encasing plasma sheet often break down at solar wind time scales of less than a day owing to transient and turbulent effects. These create magnetic configurations with localized current and plasma sheets and sometimes cause the current to split off from the surface separating fields of true opposite polarity.

INTRODUCTION

The heliospheric community generally agrees on the following definitions for the topics of this paper. The heliospheric current sheet (HCS) is a surface that separates magnetic field lines of opposite polarity. It forms as the solar wind draws the Sun's dipolar field lines into space and constitutes the heliomagnetic equatorial plane. The heliospheric plasma sheet (HPS) is a high-density layer that ensheathes the HCS.

While these definitions certainly hold at the global scale, they describe steady state conditions and often do not apply at the mesoscale on downward, where transient outflows and turbulence create complications [1]. Mesoscale here means of the scale of interplanetary coronal mass ejections (ICMEs), which cover solar wind time scales of about 1 day. Moreover, regarding the heliospheric plasma sheet, there is as yet no consensus on its diagnostic parameters and scale size. This paper reviews the problems of understanding the mesoscale and smaller-scale structure of the HCS and HPS and reports on progress resulting from use of suprathermal electrons as a tool for sensing true magnetic polarity.

SUPRATHERMAL ELECTRONS AS SENSORS OF TRUE POLARITY

Because suprathermal ($E > 80$ eV) electrons continually carry heat flux away from the Sun, their direction relative to the magnetic field gives the true polarity of any given field line as it leaves the Sun, even if the field line locally turns back on itself in the heliosphere [2, 3]. Electrons stream parallel to the field on lines with away polarity and antiparallel to the field on lines with toward polarity.

The usefulness of electrons as polarity sensors was first realized by Kahler and Lin [4, 5] using higher-energy ($E > 2$ keV) electrons. They found the following HCS dichotomy: Sometimes in situ field reversals occur without true polarity reversals, and sometimes true polarity reversals occur without in situ field reversals. While the former are easily understood in terms of fields locally turned back on themselves, creating localized current sheets, the meaning of the latter has remained elusive. This paper reports on initial progress in understanding both aspects of the dichotomy in terms of fields turned back on themselves in mesoscale transient outflows. The reported insights derive from an attempt to identify the HPS at true polarity reversals, as described in the next section.

HELIOSPHERIC PLASMA SHEET

Several parameters with variations over a range of scale sizes have been offered as diagnostic signatures of the heliospheric plasma sheet. These are reviewed below, after which follows a discussion of the variability of the smallest-scale signatures and a summary on HPS usage.

Borrini et al. [6] and Gosling et al. [7] first identified the heliospheric plasma sheet, though not by that name, using superposed epoch analysis. They interpreted the region of high density surrounding the HCS as the heliospheric extension of the coronal streamer belt. The sharp density (n) peak derived in the analysis coincided with a broad, days-long proton temperature (T) depression. Together these parameters yield low entropy, since entropy is proportional to $T/n^{\gamma-1}$, where γ is a constant (the ratio of specific heats). Thus low entropy can be used as a diagnostic parameter.

Burlaga et al. [8] pointed out that the increase in entropy at stream interfaces at the leading edge of high-speed streams is followed by a corresponding decrease on the trailing edges and, citing a number of forerunner

CP679, *Solar Wind Ten: Proceedings of the Tenth International Solar Wind Conference*,
edited by M. Velli, R. Bruno, and F. Malara

studies, called the resulting low-entropy region between streams the "heliospheric plasma sheet." Burton et al. [9] found the same entropy pattern and showed that it matched the composition pattern that distinguishes what was originally slow wind from what was originally fast wind [10]. They thus confirmed that although entropy is not a streamline constant, as it would be if the solar wind expanded adiabatically, its time variations still serve as markers of plasma origins. In this context, if low entropy is treated as the diagnostic HPS parameter, the HPS is a days-long structure and becomes synonymous with what was originally slow wind.

Sometimes the scale size of a low-entropy HPS is shorter, on the order of several hours to a day. This was particularly true for the solar maximum period analyzed by Neugebauer et al. [11]. In addition, they found low-entropy HPSs that lacked HCSs. These occurred where flows from different solar sources converged, as do flows from the two polar coronal holes at the HCS, but in these cases the sources had the same magnetic polarity. From this point of view the HPS is a source boundary rather than a sheath for a current sheet.

Based upon high-density signatures surrounding the HCS in the inner heliosphere and their similarity to the structure of coronal streamers obtained from radio occultation measurements, Bavassano et al. [12] described the HPS as an hours-long feature immersed in a days-long halo. This view incorporates the large-scale slow wind structure discussed above but focuses on the shorter scale for the HPS itself.

Winterhalter et al. [13] drew the attention of the community to the smallest-scale HPS in a paper entitled "The Heliospheric Plasma Sheet." They searched for HCS cases with the sharpest magnetic field reversals and found them embedded in minutes-long spikes of β, their diagnostic parameter, where β is the ratio of gas pressure to magnetic pressure. Follow-on studies with suprathermal electron data, however, have shown that high-β HPSs can have transient characteristics [14] and can occur at localized current sheets as well as at the HCS [2, 15]. In this context the general term "plasma sheets" seems more appropriate than "HPSs," especially since high-β plasma sheets seem to be more a property of current sheets in general than a property specific to the HCS.

High-β Plasma Sheet Variability

Figure 1 illustrates the variability encountered when searching for the HPS as defined by high beta. Three true polarity reversals identified in suprathermal electron data are marked with dashed vertical lines. They align with reversals in the longitude angle ϕ_B of the magnetic field in the top panel, from the toward sector to the away sector (hatched area) and back again, as expected at the HCS. The first of the three reversals coincides with a sharp spike in β in the bottom panel, an excellent example of the kind of HPS identified by Winterhalter et al. [13]. What complicates the pattern is the lack of comparable spikes at the remaining reversals and the occurrence of comparable β spikes elsewhere. The second reversal coincides with a broader, less intense peak in β and the third with essentially no β signature. Some of the β spikes in the first half of the plot align with ϕ_B reversals that lack electron signatures, indicating crossings of localized current sheets, as mentioned at the end of the previous section.

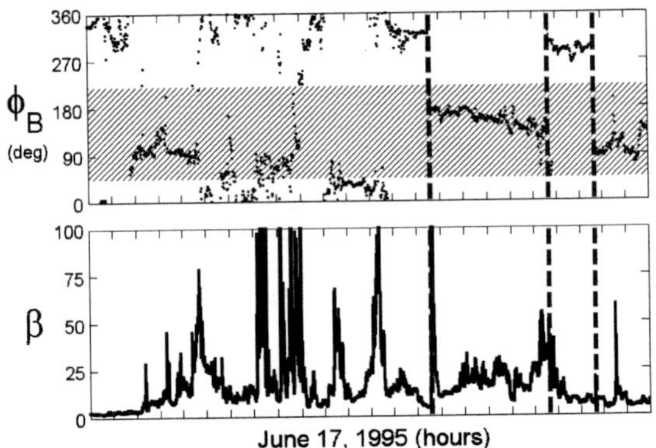

FIGURE 1. Time variations of magnetic longitude ϕ_B and β, the ratio of total gas (electron plus ion) pressure to magnetic pressure, calculated with data from the MFI and SWE instruments on the Wind spacecraft, courtesy of R. P. Lepping and K. Ogilvie. The dashed lines mark true polarity reversals apparent in electron data obtained from the 3DP instrument, courtesy of R. Lin.

Some survey studies have begun to quantify the variability in high-β plasma sheets apparent in Figure 1. This section reports on surveys that test current sheets for plasma sheet occurrence, true polarity reversals for plasma sheet occurrence, and plasma sheets for current sheet and polarity reversal occurrence.

A. D. Coleman [15], under the guidance of A. J. Lazarus, searched for current sheets in magnetic field data across which the field reversed polarity, independent of whether the current sheets were the HCS, with a true polarity reversal, or just localized current sheets. He found that 74% were encased in some type of high-β plasma sheet. Coupled with the remarkable finding of Szabo et al. [16], discussed in the next section, that most field reversals do not coincide with true polarity reversals, one can conclude that high-β plasma sheets are common properties of current sheets, most of which, however, are not the HCS.

Another survey, currently underway, is assessing the probability of encountering a high-β plasma sheet at the HCS in 35 successive crossings of the global sector boundary late in the declining phase of the solar cycle, when the sector structure was well-ordered. Preliminary results indicate that even under these optimal conditions the steady-state concept of a high-β-plasma-sheet-encased HCS applies, at most, about half the time.

A complementary survey covering a similar time period is assessing the probability that a pronounced high-β structure (hourly average of proton $\beta > 10$) is an HPS. Preliminary results indicate that most contain field reversals without true polarity reversals, supporting the idea that high-β plasma sheets are more closely associated with localized current sheets than with the HCS.

High-β plasma sheets fall into the larger category of pressure balance structures, which cover a wide range of scale sizes [17]. In the range of minutes up to a few hours, a preliminary log-log histogram of the durations of events with proton $\beta > 1$ shows a straight-line fall-off, consistent with turbulent cascades from convected plasma structures of solar origin [18]. Thus the high degree of variability in high-β plasma sheets is consistent with a strong presence of transient and turbulent structures.

What is the HPS?

The heliospheric community has yet to settle on what is meant by "heliospheric plasma sheet." As reviewed above, the term has been applied to structures with solar wind time scales ranging from minutes to days and with diagnostic parameters ranging from high density to low entropy to high β. Its meaning is commonly understood only in the most general sense as a sheath for the HCS, and at short time scales it often is not even that, either because it is missing or because the field reversal embedded in it is often not a true polarity reversal. Moreover, as a source boundary, it need not even contain a current sheet. Logically, since "HPS" is a steady-state concept, the term should apply to something that is always there, as proposed for the global-scale low-entropy regions between high-speed streams. But usage, not logic, prevails, and the term "slow wind" has already gained wide usage for that plasma regime. Current HPS usage seems to favor the small-scale high-β structures, but this may change in view of the increasing evidence of their variability and association with localized current sheets. In the long run, owing to these ambiguities, the term "HPS" may disappear from our vocabulary.

HELIOSPHERIC CURRENT SHEET

Recent progress in understanding the structure of the heliospheric current sheet has been made primarily through the use of suprathermal electrons as polarity sensors. For example, Kahler et al. [3] showed that intrasector field reversals, those occurring in what appears to be the middle of a sector, are not isolated islands of opposite polarity. Half result from fields turned back on themselves, and the remaining half are associated with interplanetary coronal mass ejections (ICMEs). Further, Kahler et al. [19] found that fields turned back on themselves constitute about 7% of data collected in the ecliptic plane at 1 AU and that the orientation of these fields borders on the ortho-Parker-spiral direction.

Perhaps the most surprising result to come from studies using suprathermal electrons is the finding of Szabo et al. [16], mentioned in the previous section, that not only some but most field reversals lack true polarity reversals, reflecting the first aspect of the HCS dichotomy discussed in the second section. This finding strongly impacts conclusions drawn from earlier studies of the fine structure of the HCS, which treated every field reversal as an HCS crossing [e.g., 20, 21]. One might now conclude that the HCS is much simpler than previously thought. This conclusion would be premature, however, because it fails to take into account the second aspect of the HCS dichotomy, that not all true polarity reversals constitute HCS crossings. Sometimes the boundary between fields of true opposite polarity separates from the HCS, leaving patches of it current-free. As discussed below, what turns out to be simple is not the HCS but the polarity reversal boundary.

Mismatched Field and Polarity Reversals

In the search for high-β HPSs at successive sector boundary crossings reported in the previous section, one reason so few were found is that some cases were dismissed for lack of field reversals at the true polarity reversals, reflecting the second aspect of the HCS dichotomy. Figure 2 gives two examples.

As in Figure 1, the dashed vertical lines in Figure 2 mark the true polarity reversals identified in the electron data. In contrast to Figure 1, however, they are not aligned with clear field reversals. In the top panel, there is no change at all in ϕ_B at the polarity reversal. The field stays in the away sector while the electron data give an incontrovertible signature of toward polarity. After about 8 hours, ϕ_B dips into the toward sector twice but remains primarily in the away sector until a total of 21 hours after the true polarity reversal. At that point ϕ_B clearly returns to the toward sector and stays there. Since the electron data continue to indicate toward polarity after that point, the two data sets are back in agreement after a 21-hour mismatch.

The second panel of Figure 2 shows a similar pattern, this time with a 15-hour mismatch. In this case the field actually reverses at the true polarity reversal but then quickly returns to the away sector, although

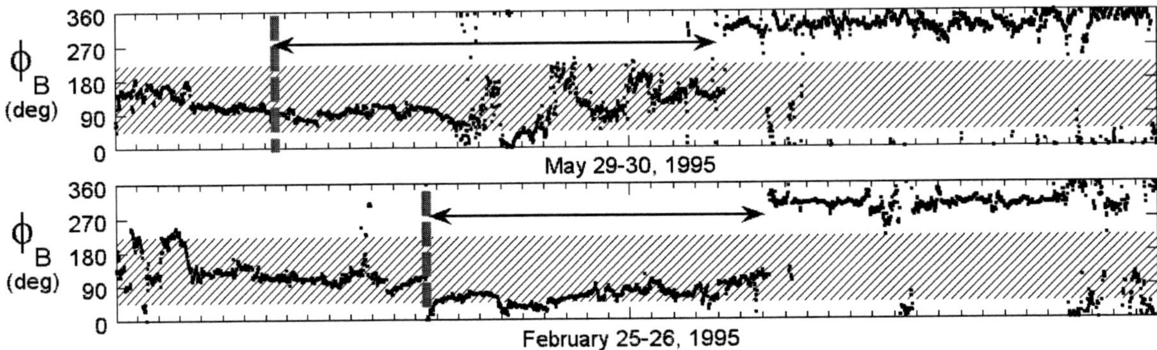

FIGURE 2. Same as top panel of Figure 1. Ticks mark hours, and double-headed arrows mark intervals between true polarity reversals and in situ field reversals.

just barely. It hovers for hours around the ortho-Parker-spiral direction, as discussed above in general for fields that turn back on themselves [19], before moving back to the Parker spiral direction and then clearly reversing 15 hours after the true polarity reversal. Strictly speaking this case does not have a polarity reversal without a field reversal; but, in the larger context, the briefness of the return to the toward sector, the clear field reversal much later, and the similarity not only to the case in the top panel but to several other cases during the first half of 1995 argue for its inclusion as a case of mismatched field and polarity reversals. A complete analysis of these events will be published elsewhere.

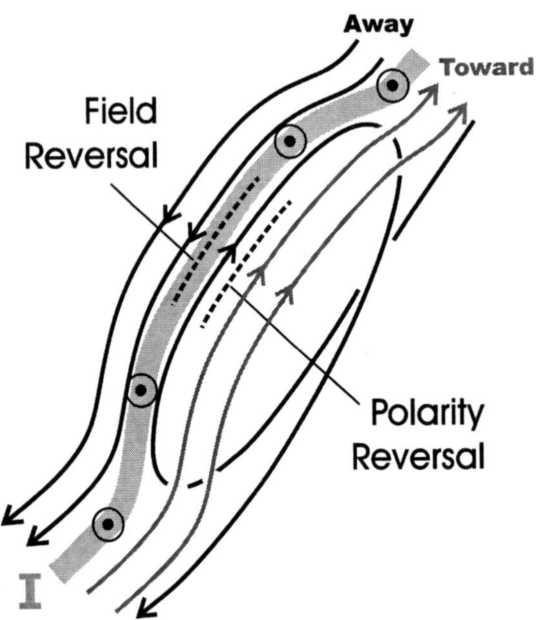

FIGURE 3. Magnetic configuration in which the reversal between field lines with toward (black) and away (gray) polarity is displaced from the local magnetic field reversal owing to a field line with away polarity that coils back on itself. The current I (gray band) in the HCS consequently separates from the polarity reversal boundary and passes around the coil.

What kind of structure can account for mismatched field and true polarity reversals? It must have fields turned back themselves and thus be transient. Figure 3 shows an example. A field line with away polarity coils back on itself so that the top portion of the coil locally points toward the Sun. It is thus parallel to the adjacent field lines with toward polarity so that no current **I** is required in that section of the polarity reversal boundary. On the other side of the coiled field line lie uncoiled field lines with the same away polarity but opposite direction to the top of the coil. Hence **I** must flow there to accommodate the local field reversal. The net effect is that the HCS is locally displaced from the polarity reversal boundary.

Figure 3 thus demonstrates that not all field reversals without true polarity reversals are necessarily signatures of localized current sheets. Those that pair with polarity reversals that lack field reversals, a pairing that is required for current continuity, must be treated as the HCS. These must then be subtracted from the larger pool of field reversals without true polarity reversals before one can address the question of the degree to which the HCS is simpler than previously thought.

For clarity Figure 3 depicts all fields aligned with the Parker spiral, more like the fields across the true polarity reversal in the top panel of Figure 2 than in the bottom panel. The hovering about the ortho-Parker-spiral direction in the bottom panel, characteristic of fields turned back on themselves, might be a signature of interchange reconnection in the Fisk model of footpoint circulation [22, N. Schwadron, private communication], a possibility that will be considered in a future publication.

The configuration in Figure 3 is only one of several that can explain mismatched field and polarity reversals. For example, the coiled field line could be extended into a series of singly coiled field lines or into a flux rope, in which case the signature would result from a skimming trajectory. A trajectory through the heart of a flux rope inclined to the ecliptic plane and attached to the Sun at only one end will also give a signature of mismatched reversals [23]. All the field lines in a flux rope have a single polarity so that any true polarity change will be at its boundary. The field reversal, on the other hand, takes place gradually as a smooth rotation across the rope. In the case of a flux rope, the current of the HCS is distributed throughout the structure. Finally, when considered in three dimensions, mismatched reversals can result from interchange reconnection in the legs of ICMEs [24].

Heliospheric Polarity Reversal Sheet

While at the global scale the HCS can be considered as a boundary between field lines of opposite polarity, at the mesoscale this is not always true. Sometimes the current separates from that boundary, and sometimes the current extends well beyond it. The current flows where it must in order to satisfy the given magnetic field configuration, which can be complicated in transient structures. Discussing the noncoincidence of the HCS and the boundary across which the polarity of the magnetic field reverses is facilitated with the introduction of a new term, the "heliospheric polarity reversal sheet" or HPRS. Unlike the HCS, the HPRS is a true surface or sheet. It separates field lines which have either one polarity or the other, depending upon their direction as they leave the Sun. It is the sector boundary in the truest sense of that term, but unfortunately "sector boundary" is often used in a more general way. The familiar global view of the HCS as a twirling ballerina skirt [25] applies to the HPRS on all scales.

ACKNOWLEDGMENTS

Preliminary results have been reported from ongoing research projects with collaborators S. W. Kahler, D. E. Larsen, and L. Kepko and students C. Huang and S. Lamassa. This research was supported by the National Science Foundation under grant ATM-011970 and by the National Aeronautics and Space Administration under grant NAG5-10856.

REFERENCES

1. Crooker, N. U., G. L. Siscoe, S. Shodhan, D. F. Webb, J. T. Gosling, and E. J. Smith, *J. Geophys. Res.*, **98**, 9371-9381 (1993).
2. Crooker, N. U., M. E. Burton, G. L. Siscoe, S. W. Kahler, J. T. Gosling, and E. J. Smith, *J. Geophys. Res.*, **101**, 24,331-24,341 (1996).
3. Kahler, S. W., N. U. Crooker, and J. T. Gosling, *J. Geophys. Res.*, **101**, 24,373-24,382 (1996).
4. Kahler, S., and R. P. Lin, *Geophys. Res. Lett.*, **21**, 1575-1578 (1994).
5. Kahler, S., and R. P. Lin, *Sol. Phys.*, **161**, 183-195 (1995).
6. Borrini, G., J. T. Gosling, S. J. Bame, W. C. Feldman, and J. M. Wilcox, *J. Geophys. Res.*, **86**, 4565-4573 (1981).
7. Gosling, J. T., G. Borrini, J. R. Asbridge, S. J. Bame, W. C. Feldman, and R. T. Hansen, *J. Geophys. Res.*, **86**, 5438-5448 (1981).
8. Burlaga, L. F., W. H. Mish, and Y. C. Whang, *J. Geophys. Res.*, **95**, 4247-4255 (1990).
9. Burton, M. E., M. Neugebauer, N. U. Crooker, R. von Steiger, and E. J. Smith, *J. Geophys. Res.*, **104**, 9925-9932 (1999).
10. Geiss, J., G. Gloeckler, and R. von Steiger, *Space Sci. Rev.*, **72**, 49-60 (1995).
11. Neugebauer, M., P. C. Liewer, E. J. Smith, R. M. Skoug, and T. H. Zurbuchen, *J. Geophys. Res.*,in press (2002).
12. Bavassano, B., R. Woo, and R. Bruno, *Geophys. Res. Lett.*, **24**, 1655-1658 (1997).
13. Winterhalter, D., E. J. Smith, M. E. Burton, N. Murphy, and D. J. McComas, *J. Geophys. Res.*, **99**, 6667-6680 (1994).
14. Crooker, N. U., M. E. Burton, E. J. Smith, J. L. Phillips, and A. Balogh, *J. Geophys. Res.*, **101**, 2467-2474 (1996).
15. Coleman, A. D., Senior Thesis, Dept. Physics, MIT, Cambridge, MA (1998).
16. Szabo, A., D. E. Larson, and R. P. Lepping, The heliospheric current sheet on small scale, in *Solar Wind Nine*, edited by S. Habbal et al. , New York, Amer. Inst. Phys., 1999, pp. 589-592.
17. Burlaga, L. F., J. D. Scudder, L. W. Klein, and P. A. Isenberg, *J. Geophys. Res.*, **95**, 2229-2239 (1990).
18. Roberts, D. A., *J. Geophys. Res.*, **95**, 1087-1090 (1990).
19. Kahler, S. W., N. U. Crooker, and J. T. Gosling, *J. Geophys. Res.*, **103**, 20,603-20,612 (1998).
20. Villante, U., R. Bruno, F. Mariani, L. F. Burlaga, and N. F. Ness, *J. Geophys. Res.*, **84**, 6641-6652 (1979).
21. Behannon, K. W., F. M. Neubauer, and H. Barnstorf, *J. Geophys. Res.*, **86**, 3273-3287 (1981).
22. Fisk, L. A., *J. Geophys. Res.*, **101**, 15,547-15,553 (1996).
23. Crooker, N. U., A. H. McAllister, R. J. Fitzenreiter, J. A. Linker, D. E. Larson, R. P. Lepping, A. Szabo, J. T. Steinberg, A. J. Lazarus, Z. Mikic, and R. P. Lin, *J. Geophys. Res.*, **103**, 26,859-26,868 (1998).
24. Crooker, N. U., J. T. Gosling, and S. W. Kahler, *J. Geophys. Res.*, ***107***(2), 10.1029/2001JA000236 (2002).
25. Smith, E. J., B. T. Tsurutani, and R. L. Rosenberg, *J. Geophys. Res.*, **83**, 717-724 (1978).

A systematic 22-year pattern in solar wind

K. Mursula*, T. Hiltula* and B. Zieger†

*University of Oulu, Department of Physical Sciences, P.O.Box 3000, FIN-90014, University of Oulu, Finland
†Geodetic and Geophysical Research Institute, Sopron, Hungary

Abstract.
It has been shown recently [1] that the average solar wind speed at 1 AU is faster (slower) in Spring than in Fall around positive (negative, respectively) helicity minima. This implies a related 22-year variation in the solar hemisphere from which a faster solar wind stream is received at 1 AU during each season. We have earlier studied [2] the effective latitudinal gradients of the solar wind speed in the two magnetic hemispheres around the heliographic equator by comparing observations at the Earth's orbit in Spring and Fall. We found that there is a large effective gradient of about 10 km/s/deg in the southern magnetic hemisphere around each solar minimum. However, no statistically significant effective gradient was found in the northern magnetic hemisphere. Here we discuss the related properties of the solar wind proton temperature and show that the temperature and speed behave very similarly in the two hemispheres. In particular, there is a large effective gradient of about 2700-5400 K/deg in the southern magnetic hemisphere while no significant gradient exists in the northern hemisphere. This supports the earlier result [2] that the streamer belt is systematically displaced toward the northern magnetic hemisphere. The displacement of the streamer belt implies a new, persistent north-south asymmetry related to the solar magnetic cycle which needs to be explained by realistic solar dynamo models.

INTRODUCTION

Solar wind (SW) can be roughly divided into two main components: the slow, cool and dense solar wind and the fast, hot and sparse solar wind. The former is related to closed solar magnetic field lines and the heliospheric current sheet, while the latter originates from the open magnetic field lines of solar coronal holes [3,4]. Polar coronal holes start growing after solar maxima and attain their largest extension towards the heliographic equator during the declining phase of the solar cycle [5,6]. Simultaneously, large latitudinal gradients in SW speed arise close to the equator between the fast, hot SW of coronal holes and the slow, cool SW of the streamer belt. Then, if the heliospheric current sheet (HCS) is sufficiently tilted, the high SW streams from coronal holes can be observed even at the Earth's orbit. Fig. 1 shows the annual averages of SW (proton) temperature and speed for 1965-2000 calculated from the OMNI data base, demonstrating the well known fact that SW speed and temperature at 1 AU are maximized during the late declining phase of the solar cycle [7,8,9].

Due to the tilt of the solar rotation axis with respect to the ecliptic, the Earth achieves its highest northern (southern) heliographic latitudes in September (March, respectively), enhancing the fraction of fast, hot SW at these times. Thus, two semiannual maxima are expected in SW speed and temperature. However, it was found earlier [1] that the two semiannual maxima in SW speed are systematically unequal, leading to a dominant annual (rather than semiannual) variation around solar minima. Moreover, it was found [1] that the phase of the annual variation changes from one solar minimum to another so that the annual SW speed maximum is observed in Spring, i.e., at the highest southern heliographic latitudes, during solar minima with a positive magnetic polarity (between cycles 20-21 and 22-23) and in Fall dur-

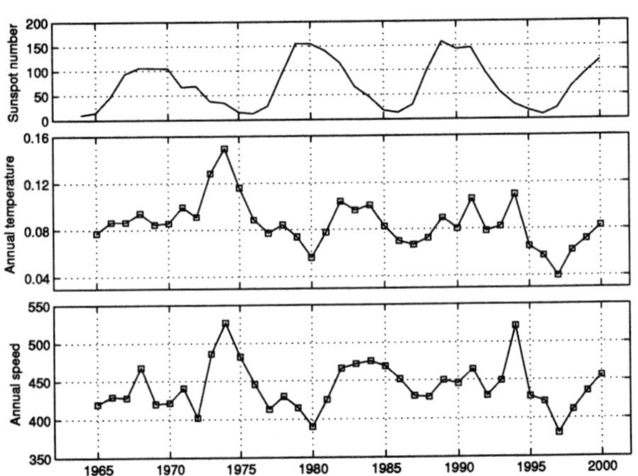

FIGURE 1. Top: annually averaged sunspot numbers; middle: annually averaged SW temperatures (in million K) for 1965-2000; bottom: annually averaged SW speeds (in km/s) for 1965-2000.

CP679, *Solar Wind Ten: Proceedings of the Tenth International Solar Wind Conference*, edited by M. Velli, R. Bruno, and F. Malara

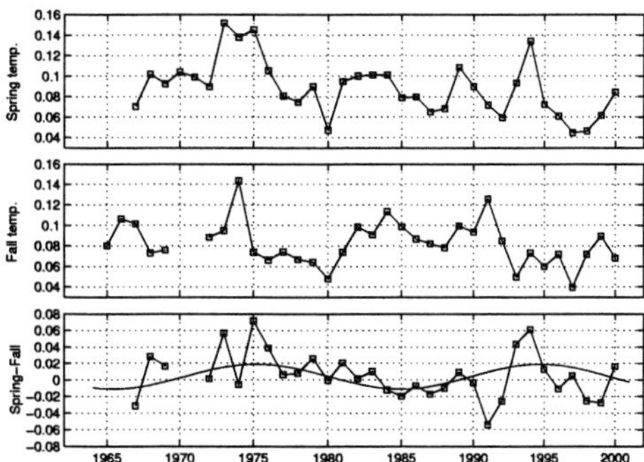

FIGURE 2. SW temperature (in million K); top: in Spring; middle: in Fall; bottom: Spring-Fall difference. The curve is the best fitting sinusoid.

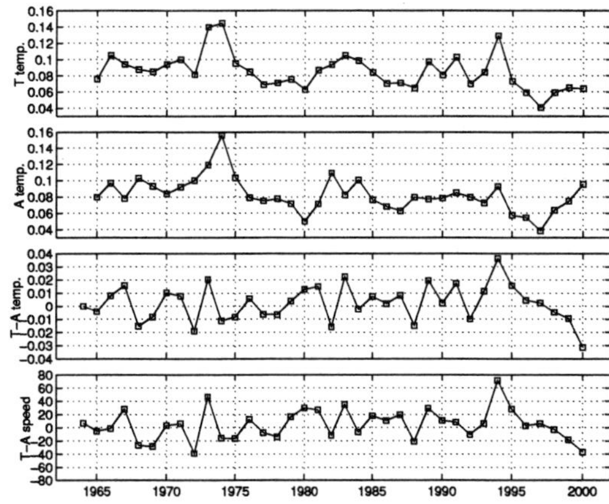

FIGURE 3. Annual SW temperatures (in million K); top: toward sector; second: away sector; third: toward-away difference. Bottom: toward-away difference for annual SW speed.

ing negative minima (between cycles 19-20 and 21-22). We have studied if the same is true for SW temperature. Fig. 2 depicts the SW temperature for Spring and Fall separately, as well as their difference. (Here we use 3-month averages of February to April for Spring and August to October for Fall). It is seen (Fig. 2, bottom panel) that Spring clearly dominates during positive polarity times and Fall weakly during negative polarity times. The resulting effective 22-year variation is depicted by the best fitting sinusoid with a period of about 19.6 years and amplitude of about 15000 K. There is an overall offset of about 4000 K in the Spring-Fall difference.

SW IN TWO MAGNETIC HEMISPHERES

During a positive polarity minimum there is a dominance of the away (A) sector of the interplanetary magnetic field (IMF) in Fall while the toward (T) sector dominates in Spring [10]. The situation is reversed one solar cycle later during a negative polarity minimum. Accordingly, the polarity dependent Spring-Fall difference in SW properties depicted in Fig. 2 could, in principle, result from an intrinsic difference between the northern and southern coronal holes so that the southern magnetic hemisphere (T sector) would always produce a faster and hotter solar wind than the northern magnetic hemisphere (A sector).

However, we have shown [2] that there is no systematic difference in SW speed between the two magnetic hemispheres. The same is found here to be true for the SW temperature, as demonstrated in Fig. 3. Here we have divided the SW data into two groups according to the direction of the simultaneously measured IMF. (We use the plane division of IMF so that, e.g., the T sector corresponds to Bx > By). Fig. 3 depicts the annual averages of SW temperature in the two magnetic hemispheres, as well as their difference. The T-A difference is also given for SW speed for comparison. The annual SW temperatures from the two magnetic hemispheres follow each other as well as the overall annual average (see Fig. 1) quite reliably. Despite some evidence for an overall slightly faster and hotter solar wind from the T sector, the difference between T and A sector is neither statistically significant over the whole interval nor follows any specific pattern with respect to the solar magnetic cycle. Therefore the Spring-Fall difference and its 22-year cycle depicted in Fig. 2 is not due to differences between the two magnetic hemispheres. On behalf of SW speed, this conclusion is in a good agreement with Ulysses results [11].

SPRING-FALL DIFFERENCE IN TWO IMF SECTORS

As another alternative explanation for the annual variation it was suggested in [1] that SW speed at the Earth's orbit is north-south asymmetric across the heliographic equator during times when fast SW streams exist close to the ecliptic. This was verified for SW speed in [2] by studying SW speed in Spring and Fall in the two IMF sectors separately. Fig. 4 presents the SW temperature in the T sector in Spring and Fall, as well as the Spring-Fall difference. Note first that, according to the expected latitudinal variation in the late declining phase, the highest SW temperatures in Spring and Fall are larger than the corresponding annual averages depicted in Fig. 1. This is particularly true for Spring in mid-1970's and mid-1990's. The Spring-Fall difference for SW temperature in the T sector depicts a clear variation related to the

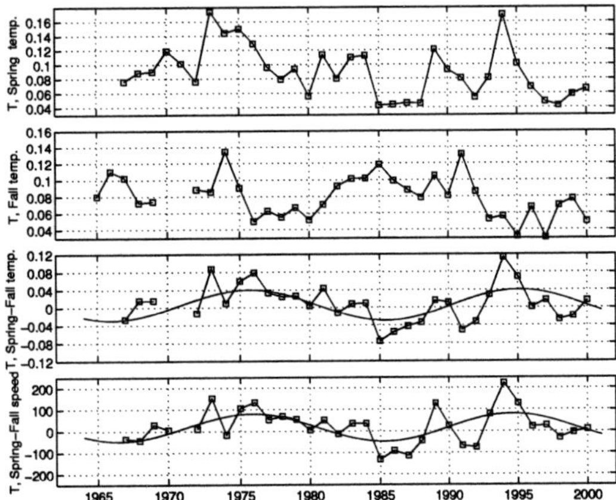

FIGURE 4. SW temperatures (in million K) in the toward sector; top: Spring; second: Fall; third: Spring-Fall difference. Bottom: Spring-Fall difference for SW speed in the toward sector. Curves give the best fitting sinusoids.

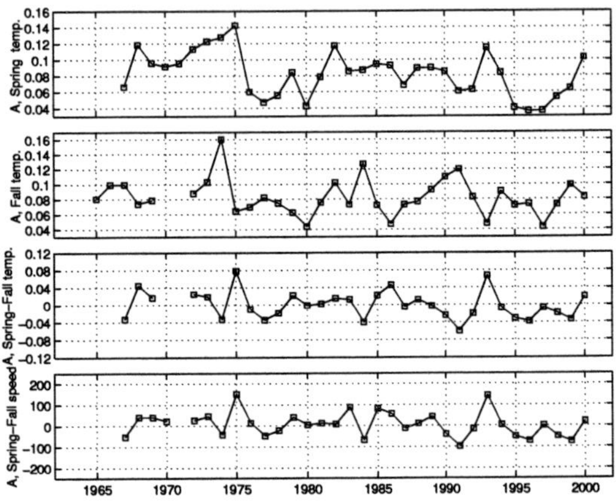

FIGURE 5. SW temperatures (in million K) in the away sector; top: Spring; second: Fall; third: Spring-Fall difference. Bottom: Spring-Fall difference for SW speed in the away sector.

magnetic cycle. The best fitting sinusoid has a period of about 19.4 years, close to the average length of the modern magnetic cycles of about 20 years. The amplitude of the best fitting sinusoid is about 35000 K but the largest values are much larger, more than 100000 K. Note that the period and phase of the best fitting sinusoid for SW temperature are very similar to those of the SW speed depicted in Fig. 4 (bottom panel) [2]. Note also that the relative amplitude of the 22-year variation is clearly larger, about 35 % (roughly 35000K/100000K) for SW temperature than for SW speed, about 15 % (roughly 60/400 in km/s).

Fig. 5 depicts the SW temperature in Spring and Fall for the A sector. As observed earlier for SW speed [2], the magnetic cycle in the Spring-Fall difference in the A sector is far less evident than for the T sector. The phase of the best fitting sinusoid (not shown) is, as expected, roughly opposite to the phase of the corresponding sinusoid in the T sector but the amplitude is not statistically significant. Note that despite the small amplitudes, the Spring-Fall differences for SW temperature and speed are greatly similar.

DISCUSSION

During positive polarity minima, the T (A) sector is the favoured IMF sector at high southern (northern) heliographic latitudes, i.e., in Spring (Fall). Accordingly, in Spring the solar wind of the T sector comes preferably from the relatively higher southern heliomagnetic latitudes. On the other hand, in Fall the T sector is disfavoured and the corresponding SW comes preferably from the low heliomagnetic latitudes. Therefore, as noted in [2], taking the Spring-Fall difference in the T sector gives, in addition to the SW variation in the southern magnetic hemisphere across the heliographic equator, a rough estimate for the effective SW gradient with heliomagnetic latitude in the southern magnetic hemisphere during positive polarity times (and the gradient with negative sign during negative polarity times). Similarly, the Spring-Fall difference in the A sector also gives, in addition to the heliographic variation in the northern magnetic hemisphere, the heliomagnetic gradient with negative sign (gradient with positive sign) in that hemisphere during positive (negative, resp.) polarity times.

Accordingly, the strong quasi-22-year cycle in the Spring-Fall difference in the T sector (see Fig. 4) proves that there is a large effective latitudinal (heliographic and heliomagnetic) gradient in SW temperature in the southern magnetic hemisphere around the ecliptic. Using the amplitude of the sinusoid (about 35000 K) and the Spring-Fall latitude difference of the Earth's orbit (now effectively about 13 deg) we get a rough estimate for the average effective gradient of about 2700 K/deg in the southern magnetic hemisphere. During some solar minima the gradient can even be twice as large. Note also that the latitudinal gradient and its evolution over the solar magnetic cycle is very similar for the SW temperature and speed (see the two lowest panels of Fig. 4).

On the other hand, the Spring-Fall difference of SW temperature in the A sector (see Fig. 5) does not depict a significant quasi-22-year cycle. This means that, contrary to expectations based on a symmetric streamer belt, the effective heliomagnetic (and heliographic) latitudinal gradient is insignificantly small in the northern magnetic hemisphere. Accordingly, the latitudinal gradients around the ecliptic are systematically different in the two magnetic hemispheres.

As noted in [2], a symmetric case where the minimum speed locus of the streamer belt coincides with HCS and the heliographic equator, and where the latitudinal gradients in the two hemispheres are the same is excluded. The present results demonstrate that the same is true for the solar wind temperature. It was shown recently that HCS was displaced southward by about 10 degrees in 1994-95 [12,13]. Supposing that the SW distribution would strictly follow the HCS and was similarly displaced southward would, however, lead to small (large) SW gradients in the T (A, resp.) sector, contrary to observations at this time (see Figs. 4 and 5) and to the general pattern during positive polarity times. Accordingly, the HCS and the streamer belt were separated at least at this time. On the other hand, if the streamer belt is displaced toward the northern magnetic hemisphere by a few (e.g., 4-6) degrees, we would find a significant effective gradient in the southern magnetic hemisphere and a small or negligible effective gradient in the northern magnetic hemisphere. This is in agreement with the present observations. The different effective latitudinal gradients in the SW parameters across the heliographic equator in the northern and southern magnetic hemispheres imply a new, persistent north-south asymmetry which is related to the solar magnetic cycle.

CONCLUSIONS

We have shown here that the effective latitudinal gradients of SW temperature across the heliographic equator are different in the northern and southern magnetic hemisphere, as earlier observed for SW speed [2]. As expected, there is a large, statistically significant effective gradient of about 2700-5400 K/deg in the southern magnetic hemisphere around each solar minimum. However, no systematic and statistically significant effective gradient is found in the northern magnetic hemisphere. This difference in the effective gradients implies a persistent displacement of the minimum SW temperature locus of the streamer belt toward the northern magnetic hemisphere in the same way as for the minimum SW speed locus. At least during one minimum in 1994-95, the streamer belt and the heliospheric current sheet have been oppositely displaced.

We have also demonstrated that the average SW temperatures from the two magnetic hemispheres are roughly equal, as they are for the SW speeds at the Earth's orbit [2] and at Ulysses [11]. We suggest that the polar coronal holes extend closer toward the heliographic equator in the southern magnetic hemisphere, shifting the streamer belt toward the northern magnetic hemisphere and leading to larger effective latitudinal gradients in the southern magnetic hemisphere at 1 AU. The displacement of the streamer belt and the different effective SW gradients in the northern and southern magnetic hemispheres imply a persistent north-south asymmetry in the Sun which lasts at least for the whole 40-year interval of directly measured solar wind. The solar wind is probably the first solar parameter which depicts a long-term north-south asymmetry clearly related to the magnetic cycle. This asymmetry also suggests for a persistent north-south asymmetry in large-scale solar magnetic fields which needs to be explained by realistic solar dynamo models.

ACKNOWLEDGMENTS

Financial support by the Academy of Finland is greatfully acknowledged. We are also grateful to NSSDC for OMNI data.

REFERENCES

1. Zieger, B., and K. Mursula, *Geophys. Res. Lett., 25*, **841** (1998).
2. Mursula, K., T. Hiltula, and B. Zieger, *Geophys. Res. Lett.*, 2002, in print.
3. Krieger, A. S., A. F. Timothy and E. C. Roelof, *Solar Phys., 29*, **505** (1973).
4. Sheeley, N. R., Jr., J. W. Harvey, and W. C. Feldman, *Solar Phys.*, **49**, 271 (1976).
5. Newkirk, G., Jr., and L. A. Fisk, *J. Geophys. Res.*, **90**, 3391 (1985).
6. Rickett, B. J., and W. A. Coles, *J. Geophys. Res.*, **96**, 1717 (1991).
7. Gosling, J. T., and S. J. Bame, *J. Geophys. Res.*, **77**, 12 (1972).
8. Bame, S. J., J. R. Asbridge, W. C. Feldman, and J. T. Gosling, *Astrophys. J.*, **207**, 977 (1976).
9. Mursula, K., and B. Zieger, *J. Geophys. Res.*, **101**, 27077 (1996).
10. Rosenberg, R. L., and P. J. Coleman, *J. Geophys. Res.*, **74**, 5611 (1969).
11. McComas, D. J., et al., *J. Geophys. Res.*, **105**, 10419 (2000).
12. Simpson, J. A., M. Zhang, and S. Bame, *Astroph. J.*, **465**, L69 (1996).
13. Smith, E. J., J. R. Jokipii, J. Kota, R. P. Lepping, and A. Szabo, *Astroph. J.*, **533**, 1084 (2000).

Solar Wind Characteristics from Soho-Sun-Ulysses Quadrature Observations

G. Poletto* and S. T. Suess†

*Osservatorio Astrofisico di Arcetri, Largo E. Fermi 5, 50125 Firenze, Italy
†NASA Marshall Space Flight Center, NSSTC, Huntsville, AL 35812

Abstract. Coronal and solar wind observations of the same plasma, first observed remotely in the corona and later, *in situ*, provide the best way to determine the evolution of plasma as it is being accelerated from the corona out to interplanetary distances. We have used this technique to derive solar wind characteristics from the analysis of data acquired by SOHO and Ulysses when the SOHO-Sun-Ulysses included angle is 90 degrees: that is, when SOHO, the Sun and Ulysses are in *quadrature*. We summarize here the results obtained from the study of the December 1998 quadrature, when we focussed on the behavior of slow wind from low-latitude regions, and anticipate some results from the June 2000 quadrature, which focussed on establishing a relationship between coronal and wind abundances of different elements and whose analysis is in progress. We conclude by illustrating briefly the objectives of future quadrature studies.

INTRODUCTION

In the pre-SOHO era, the only possibility to observe the same plasma first remotely in the corona and, later, *in situ*, occurred when the Helios 1 and 2 probes were moving away from the Earth, and conveniently located with respect to the P78 coronagraphs [1]. Today, when SOHO, the Sun and Ulysses are in *quadrature*, that is, when the SOHO-Sun-Ulysses included angle is 90 degrees, we have the opportunity of again making observations of the same plasma parcel at different positions as it moves from the corona to the interplanetary medium.

We have made several SOHO-Sun-Ulysses quadrature campaigns and plan to make more in future. These occur twice a year, with the latitude of the radial to Ulysses, at the time of the quadrature, depending on the position of Ulysses along its orbit. Hence, the position of Ulysses dictates the scenario of the campaign, and, in part, its objective: slow/fast wind analyses depending on whether Ulysses is at low/high latitudes. Making observations over several years, we also have a means of studying the variation, with solar cycle, of the characteristics of the slow (high) speed streams. Figure 1 shows when past, and planned, quadrature campaigns occur and the quadrant and latitude of Ulysses, as a function of the phase of the solar cycle.

A quadrature campaign relies heavily on SOHO LASCO and UVCS coronal observations. LASCO provides the overall coronal context above 2 solar radii and the UVCS spectrograph acquires data over a limited range of altitudes, usually between $\approx$ 1.5 and 4.5 solar radii. The UVCS slit is set normal to the solar radius, with the radial to Ulysses crossing through its center. The grating positions, as well as the spatial and spectral resolution, are chosen to fit the objective of the campaign. Coronal parameters are derived from data of these two experiments and *in situ* parameters are provided by Ulysses SWOOPS and SWICS experiments. Data from other sources, SUMER and CDS on SOHO or Sacramento Peak National Observatory FeXIV maps and Wilcox Solar Observatory magnetic field maps, or the Ulysses magnetometer, have been occasionally used to complement the UVCS/LASCO and SWOOPS/SWICS data sets.

At the time of this writing, we have led 8 quadrature campaigns, making observations at latitudes ranging from 10 to $\approx$ 80 degrees, sampling both low and high latitude coronal plasma. In the next section we illustrate the results we obtained from observations made in 1998, at a latitude of 17 degrees, as an example of the opportunities offered by quadrature data. After this, we anticipate a few results from the June 2000 campaign which focussed on elemental abundances in the corona and solar wind, at a time when the radial to Ulysses was crossing through an evolving streamer complex. A brief description of anticipated results from upcoming campaigns concludes the paper.

CP679, *Solar Wind Ten: Proceedings of the Tenth International Solar Wind Conference*, edited by M. Velli, R. Bruno, and F. Malara

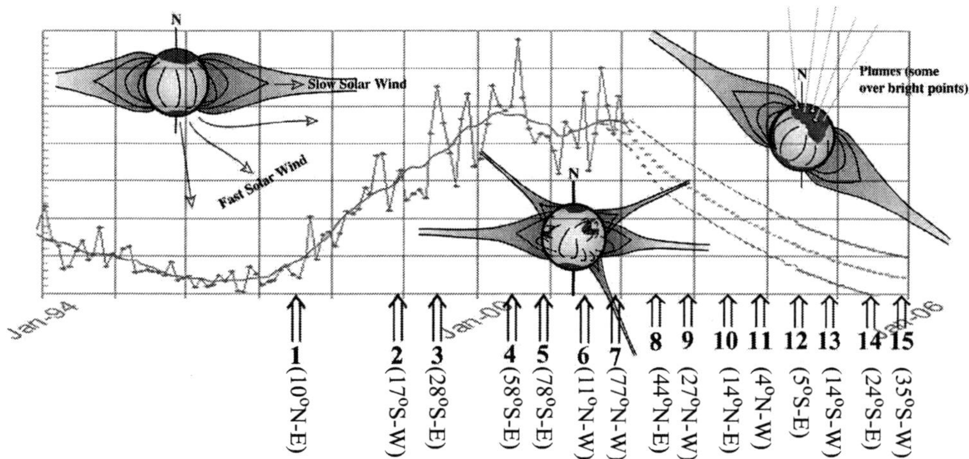

FIGURE 1. Cartoon showing past and future quadratures campaigns (arrows), as a function of the phase of the solar cycle.

SLOW WIND FROM THE FALL 1998 QUADRATURE CAMPAIGN

The 1998 quadrature occurred when the radial to Ulysses, at a southern latitude of 17.4 degrees, crossed through a small coronal hole, before being immersed in a streamer complex for the rest of the campaign. At Ulysses, the speed of the solar wind was low, as expected from low-latitude streams, but connecting the *in situ* to the coronal plasma allowed us to separate the coronal hole wind from the wind emerging from the streamer complex. We were able to derive the plasma (and heavy ion) outflow speed, at 3.5 and 4.5 $R_{\odot}$, from spectra taken by UVCS for both the coronal hole and streamer sources.

We used Doppler dimming to compute the values of the outflow speed that allowed us to reproduce both the observed ratio of the intensities of the O VI 1032 and 1037 Å doublet lines (which can be considered a crude proxy for the heavy ion speed: [2]) and the individual O VI line intensities, as well as the intensity of the H Lyman-α line. We refer the reader to [3] for a detailed description of the method. Here it suffices to note that the line intensities depend on the electron density and temperature, the proton (or ion) outflow speed, the parallel and perpendicular temperature of the ion originating the line, the heavy ion abundance and the intensity of the exciting chromospheric radiation. Electron densities were derived from pB LASCO measurements, perpendicular temperatures were derived from line widths, electron temperatures were been taken from literature, and disk intensities during the quadrature were either available or extrapolated from measurements made at other times.

We want to point out, however, that the parallel temperature cannot be inferred from observations and that the value of the abundance of heavy elements is usually assumed *a priori*. Moreover, the interplay between the element abundance and the poorly-known electron temperature (higher abundances leading to higher line intensities, higher electron temperatures leading to lower line intensities) introduces an ambiguity that usually cannot be resolved. The role of Ulysses data has been crucial in constraining the values of these parameters by providing values of mass flux and Oxygen abundances in the solar wind: that is, by providing us with more conditions that need to be satisfied.

The outflow speeds we derived are shown in Figure 2. The proton and heavy ion speeds in plasma emerging from low latitude regions is lower than the polar plasma speed. Moreover, independent of the region where it originates, low-latitude plasma is accelerated at greater heights and through a more extended region than fast wind from large coronal holes. For the small, low-latitude hole we analyze, the ratio between polar and low latitude wind speed at 1 AU is about 1.5. The same ratio at coronal levels is on the order of 2.5, implying that equatorial wind accelerates throughout a more extended altitude range than polar wind.

The fall 1998 quadrature also gave some indication on another issue, still not well understood. It has long been known that the slow wind is highly variable, but the reason for this is not yet clear. When UVCS was traversing the small coronal hole, the plasma speed turned out to be slightly higher than the plasma speed measured when the wind was emerging from the streamer complex. This is illustrated in Figure 2, which shows that a 20% variability in the speed of slow wind can be ascribed to the presence of open field regions interspersed amidst streamers. These could be responsible for the fastest wind.

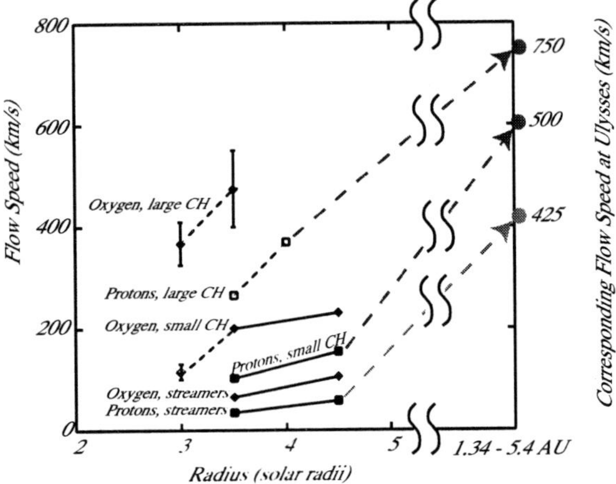

FIGURE 2. Outflow speed at 3.5 and 4.5 $R_{\odot}$ for: full symbols, squares -proton flow speed for the low latitude streamer and coronal hole observed during the 1998 fall quadrature; full symbols, diamonds -as above, for O VI ions; open symbols, squares, polar proton flow speed from [4] minimum corona; open symbols, diamonds, as above for polar holes and from [5] for low-latitude holes. The range of values given by the authors -which correspond to different choices of the parallel temperature and other parameters- are shown in the figure. These would be constrained if the Oxygen mass flux were known. The flow speed at Ulysses during the quadrature, and a typical high flow speed, are also given.

QUADRATURE OBSERVATIONS AND ELEMENTAL ABUNDANCES

Elemental abundances play a key role in studies aimed at relating coronal and solar wind plasmas. Similar coronal and wind abundances provide a clue in the search for the site where solar wind originates. This has been used by [6], for instance, to suggest that slow wind originates from streamer boundaries. Quadrature observations, where the source of the solar wind may be easily identified, offer an unique opportunity for checking on the coronal vs. solar wind elemental abundances. As a further benefit the distribution of element abundance vs. altitude in the corona, or the temporal fluctuations of element abundances at a given location in the corona may be tested with quadrature data.

A quadrature campaign, using SUMER, CDS and UVCS, was held in June 2000, with the purpose of analyzing elemental abundances. At that time, the radial to Ulysses crossed through an evolving streamer complex. The sub-Ulysses point skimmed along the neutral line fon the Wilcox Solar Observatory source surface map for several days. This provides only a crude indication, however, because of the rapid changes in the coronal configuration over the campaign. At Ulysses, the solar wind speed was typically around 350 km/s. Ulysses was at about 3.25 AU, so it took the plasma about 17-18 days to reach the spacecraft. Hence, data taken on 20 to 6 of July are representative of the coronal plasma at the time of our observations.

UVCS took spectra that include lines from many ions, including O VI, Si XII, Fe X, XII and XIII, Ar XII, plus H Lyman$-\alpha$, Lyman$-\beta$ and Lyman$-\gamma$ lines, at altitudes ranging from 1.6 to 2.2 $R_{\odot}$. Absolute Oxygen abundances can be derived once the collisional and radiative components of the H and of the O VI 1032 and 1037 Å lines are identified, under the assumption that plasma is in ionization equilibrium ([6]). In a streamer complex, at the altitudes we consider, this hypothesis is tenable. The Fe absolute abundance can be calculated from the predicted vs. observed ratio of the iron lines to the collisional components of H lines. This way one gets an estimate of the Fe/O ratio in the corona, that may be compared with *in situ* values of the same quantity.

It is well known that fast and slow wind plasma have different FIP effect. Being Fe and O, respectively, low and high FIP elements, a measure of the Fe/O ratio is a proxy for the plasma FIP effect and may offer a further clue in the quest for the slow wind origin. Preliminary estimates of Fe/O along the radial to Ulysses, at 1.6 $R_{\odot}$, for three days of the campaign, lead to coronal values that are within 10% of those measured by Ulysses. The same ratio, when measured by the ACE spacecraft shows a high variability that was not seen during the quadrature either in SWICS or UVCS data. This, however, may be ascribed to different factors, including the long integration time for SWICS, to derive good values of the ratio, and both the long integration time and the line-of-sight effect for UVCS. However, the agreement between the coronal and the *in situ* abundance ratios seems to imply that whatever mechanism is responsible for FIP effects is ineffective beyond $\approx$ 1.5 solar radii and that the precise connection between coronal and interplanetary FIP measurements can be made.

PLANS AND ANTICIPATED RESULTS FROM FUTURE CAMPAIGNS

As to what we can expect from future studies, we refer to Figure 1. The first 8 quadratures campaigns are completed. The next one, in fall 2002, with Ulysses at low latitude ($\approx$ 26 degrees) will focus on the study of variations in the physical parameters of the corona and solar wind, as a consequence of flares. The low latitude of the spacecraft, and the phase of the activity cycle, not so distant from maximum, leave hope for a flare occurrence during the two week campaign interval. SOHO-Ulysses data will then be complemented by HESSI data.

As to future studies, we list a couple of issues. We haven't analyzed, yet, CME data. Some CMEs have al-

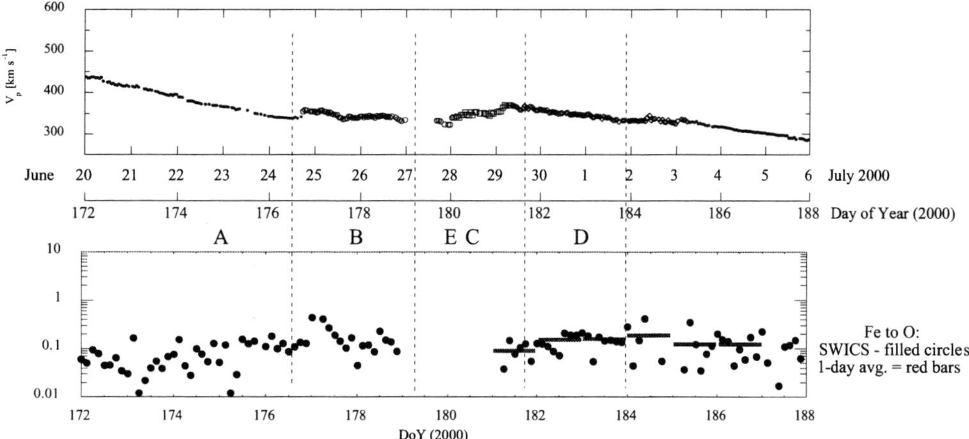

FIGURE 3. Top: wind speed for the June 20 to July 6, 2000, time interval, from the Ulysses SWOOPS experiment. Bottom: Fe/O ratio, over the same time span, from Ulysses SWICS experiment. The bars, which represent 1-day averages over SWICS data points, are drawn for the days when the coronal ratio has been evaluated from UVCS data.

ready been observed in past campaigns (e.g. campaigns 4, 5, 7 and 8) and we may expect more in the next quadratures. We expect to be able to correlate abundance variation in the CMEs with abundances measured *in situ* and to analyze the temperature changes and the ensuing variations in the ion populations at those times. Another topic is the boundary crossing between streamers and coronal holes. Ulysses analyses have shown that composition changes abruptly across the boundary (see, e.g. [7]): but not much has been done on this subject at coronal levels. Whether we will really have the opportunity to measure this depends entirely on the unpredictable solar morphology at the time and position of future quadratures.

In this brief report we meant to give an instance of the kind of research activity that quadrature configurations have allowed us to pursue and plan. For the December 1998 campaign this meant the results that smaller coronal holes have delayed solar wind acceleration compared to large polar coronal holes and that a precise connection can be made between the FIP effect in the solar wind and at 1.6 solar radii. We like to end the report by thanking the Ulysses and SOHO community for the support given throughout the campaigns we lead: without their collaboration we would have been unable to work on the exciting data we had the privilege to collect.

ACKNOWLEDGMENTS

The work of G. P. has been partially supported by ASI (Italian Space Agency) and MURST (Italian Ministry for the University and the Scientific and Technological Research). The work of S.T. Suess has been supported by the Ulysses/SWOOPS experiment team. SOHO and Ulysses are missions of international cooperation between ESA and NASA.

REFERENCES

1. Sheeley-Jr., N. R., Howard, R. A., Koomen, J. J., and *others*, *Journal of Geophysical Research*, **90**, 163–175 (1985).
2. Noci, G., Kohl, J. L., and Withbroe, G. L., *Astrophysical Journal*, **315**, 706 (1987).
3. Poletto, G., Suess, S. T., and *others*, *Journal of Geophysical Research* (2002).
4. Cranmer, S. R., Kohl, J. L., Noci, G., and *others*, *Astrophysical Journal*, **511**, 481 (1999).
5. Miralles, M. P., Cranmer, S. R., Panasyuk, A. V., Romoli, M., and Kohl, J. L., *Astrophysical Journal*, **549**, L257 (2001).
6. Raymond, J., Kohl, J. L., and *others*, *Solar Physics*, **349**, 956–960 (1997).
7. Wimmer-Schweingruber, R. F., Steiger, R. V., and Paerli, R., *Journal of Geophysical Research*, **102**, 407–417 (1997).

Velocity Dispersion Of Energetic Particles Observed By SOHO/CELIAS/STOF

M. Hilchenbach, H. Sierks

Max-Planck-Institut für Aeronomie, 37191 Katlenburg-Lindau, Germany

B. Klecker

Max-Planck-Institut für extraterrestrische Physik, 85740 Garching, Germany

K. Bamert

Physikalisches Institut der Universität Bern, 3012 Bern, Switzerland

R. Kallenbach

International Space Science Institute, 3012 Bern, Switzerland

Abstract. Since the launch of the Solar and Heliospheric Observatory (SOHO) on Dec. 2, 1995, the Suprathermal Time-Of-Flight (STOF) energetic particle sensor of the Charge, Element and Isotope Analysis System (CELIAS) has observed gradual and impulsive solar particle events from solar activity cycle minimum to maximum. The instrument CELIAS/ STOF has the capability of detecting energetic ions and determine the mass, energy and charge of each particle. We report on the measurements of the detected energetic particle events from 1996 to 2001, i.e. the velocity dispersion of energetic He ions in the energy/charge range between 30 and 800 keV/nuc. For energetic particle events we analyse the velocity dispersion of the suprathermal He and O ions (78 and 43 events, resp.) and estimate the onset time of the solar event and set limits on the propagation length along the connecting magnetic field line.

INTRODUCTION

The gradual and impulsive solar energetic particle events can be studied presently by a fleet of spacecrafts carrying in-situ instrumentation, i.e. ULYSSES, WIND, SOHO or ACE. The instruments cover the energy regime from the solar wind, the suprathermal, energetic and high-energy ions. The origin of impulsive solar energetic particle events is attributed to the acceleration in solar flares and they have characteristic abundance enhancements such as high $^3He/^4He$ or Fe/O ratios of about 1 /1-4/. The acceleration time is much shorter than the timescales for the propagation to the spacecraft, i.e. to 1 AU. The dispersion of the velocity distribution of the energetic ions is then dominated by propagation effects in the interplanetary medium /5-7/. The large gradual solar energetic particle events had also been attributed to originate from solar flares. However, in the recent years it became more and more obvious from observations that the large gradual events are accelerated in shock waves driven out of the corona by coronal mass ejections (CMEs) and not in solar flares /8-10/. Recently different classes of solar proton

CP679, *Solar Wind Ten: Proceedings of the Tenth International Solar Wind Conference*,
edited by M. Velli, R. Bruno, and F. Malara

events were reported, i.e. the electrons and protons were found to differ in onset times or path length /11-13/. In the following, we will report on the observations by CELIAS/HSTOF onboard SOHO, in the time interval from 1996 to 2001. For selected events, we attempt to examine the connection of the particle event to solar X-ray events and set limits on the propagation length of the ions along the magnetic field lines.

INSTRUMENTAL

The CELIAS instrument package is onboard of SOHO, a 3 axis stabilized satellite, launched on Dec 2, 1995 and orbiting the sun in a Halo orbit near the Lagrange Point L1 at about 0.99 AU. HSTOF is a section of the CELIAS/STOF instrument. HSTOF has its boresight set at 37° west of the Sun-SOHO line and a field of view (FOV) 2° in and 17° off the ecliptic plane. HSTOF identifies the mass and energy of each incident energetic particle by its speed measured by a time-of-flight (TOF) unit and the residual energy deposited in a pixilated solid-state detector. In front of the TOF unit is a flat-field electrostatic energy/charge (E/Q) filter, that cuts off ions of E/Q < 80 keV/e. The geometrical factor of HSTOF is 0.22 cm^2 sr. A detailed description of the instrument can be found in /14/.

OBSERVATIONS

In Fig.1 the method and the geometry of the energetic particle velocity dispersion is illustrated. Velocity dispersion results because faster ions travelling along the connecting interplanetary field line are detected earlier than slow ions, assuming all are accelerated in the same solar event. The velocity of the ions is proportional to the root of energy/ nucleon. The time t required for ions with the velocity v to propagate along a magnetic field line of length L is given by $t = L / (v \cdot \mu)$, where μ is the cosine of the particle's pitch angle with respect to the magnetic field line /15/. The time t is plainly the difference between the time of arrival at the detector and the time of the solar event onset

An example of the observed velocity dispersion of energetic He ions is shown in Fig. 2. The broad line indicates the events selected to calculate the onset time of the solar event. Plotting the foremost arrival or detection time of the energetic ions versus the reciprocal velocity allows the calculation of the time of the solar event onset (intercept) and L/μ (slope) from a linear fit. L/μ is the maximum travel distance or length of the magnetic field line as μ is always < 1.

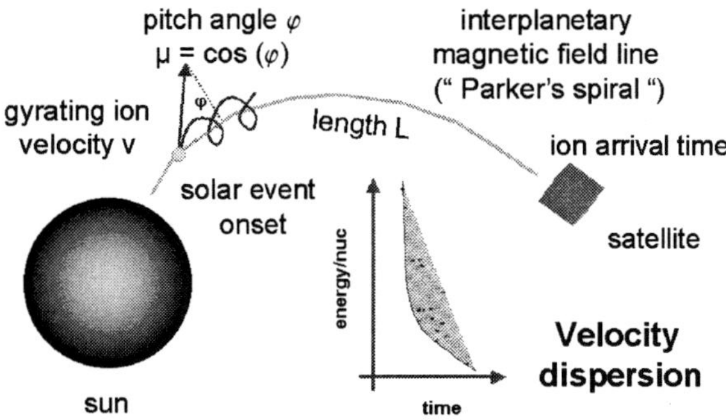

FIGURE 1. Velocity dispersion of solar energetic ions gyrating around and travelling along the interplanetary magnetic field lines.

The minimum value of L/μ is the radial distance between the solar event site and the detector, i.e. 1 AU. With the linear fit

$$\text{arrival time} = \text{solar event onset} + (L/\mu) * v^{-1} \quad (1)$$

the onset time of the preceding solar event (June 4, 1999, 5 UT) and the ion path along the interplanetary magnetic field L (L < 2.5 AU or L/μ = 2.5 AU) is calculated. Solar X-ray data from GEOS Satellite Data /16/ are shown for the respective time period, the arrow indicates the calculated event onset time from the ion velocity distribution.

In Fig. 3 the velocity dispersion of a solar energetic particle event on May 1, 2000 is plotted. The maximum path length is L = 1.24 AU or about the "canonical" value expected from the interplanetary magnetic field lines in the ecliptic plane along "Parker's spiral". While the energetic He ions show clear velocity dispersion, there are lower energetic ions detected during the same observation time interval. The CELIAS/MTOF solar wind monitor did observe a high speed solar wind, but did not identify a shock in the interplanetary medium /17/. The source of these ions is not uniquely identified and a plausible explanation is an energetic particle event behind the limb of the solar disk, i.e. X-rays of this event are not detected by the GEOS Satellites. Superimposed energetic particle events could not always be identified as clearly as on May 1, 2000. This introduces some systematic error into the statistical analysis of the velocity distribution of energetic ions of solar atmospheric origin, i.e. if the source region of the energetic particles is on the far side of the sun and not observable by remote or in-situ instruments.

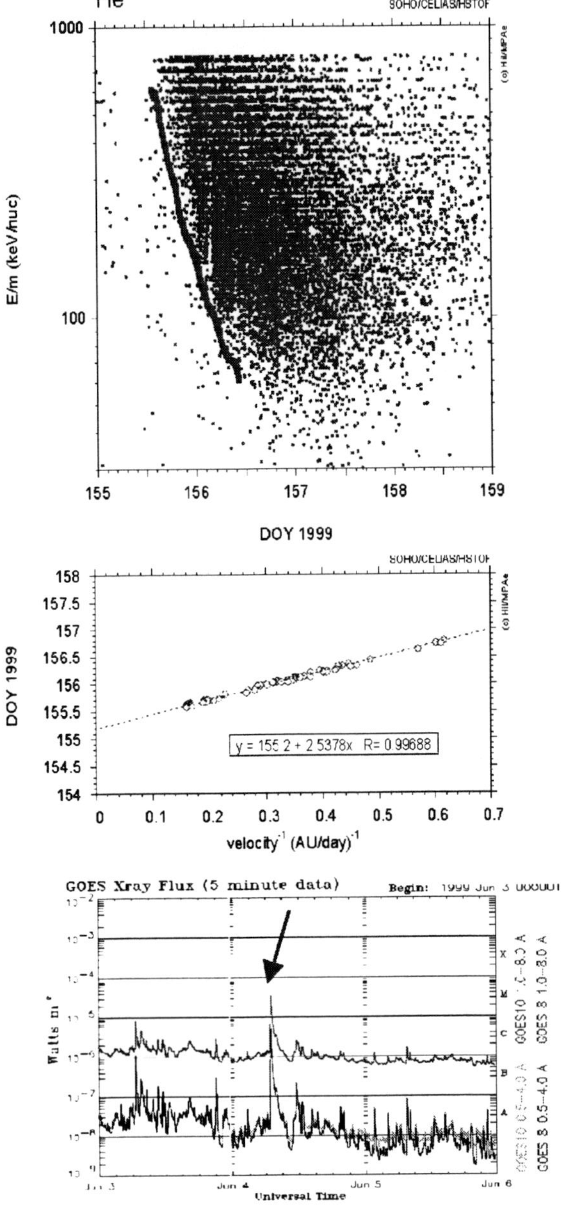

FIGURE 2. Velocity dispersion of the suprathermal He ions originating from solar energetic particle events on June 4, 1999. The interplanetary field line connection length between the acceleration region and the detector is < 2.5 AU, i.e. about twice the "canonical" length expected along "Parker's spiral".

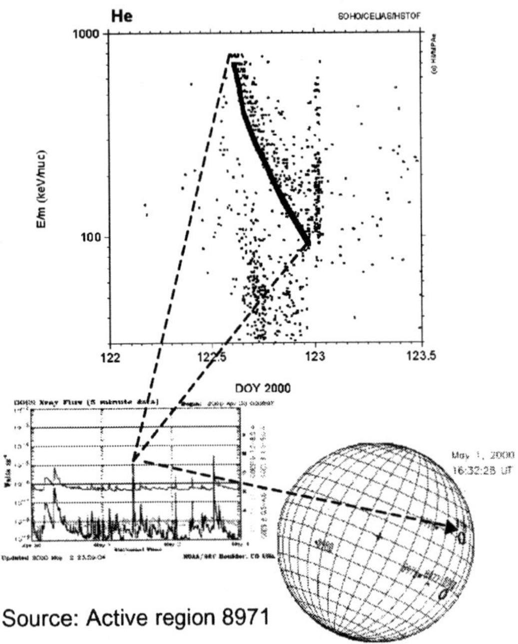

FIGURE 3. Multiple energetic particle events superimpose within the same detection interval (May 1, 2000). While the high energy He ions show a clear velocity dispersion (with L/μ about 1.24 AU) and are most likely originating from the solar active region 8971, the source of the low energy He ions is unidentified, likely their source is located beyond the limb of the solar disk.

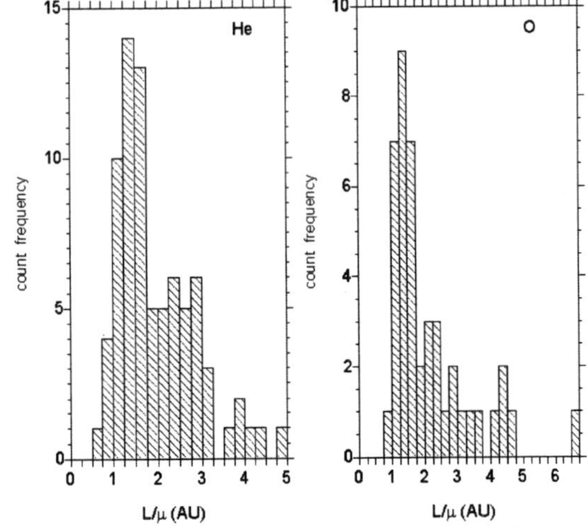

FIGURE 4. Count frequency distribution of L/μ or the maximum length of path length L for 78 He and 43 O events analysed between 1996 and 2001. More than 90% of the suprathermal particle events are registered during the active cycle of the solar activity cycle. Both distributions are peaked at about 1.25 AU, the length expected along the interplanetary magnetic field spiral.

In Fig. 4 the count frequency histograms show the result of the analysis of the maximum path length (L/μ) for 78 He and 43 O energetic solar particle events between 1996 and 2001. Both distributions are clearly peaked at the "canonical" value of 1.25 AU (binning about 0.25 AU) as derived for the ideal interplanetary magnetic field in the ecliptic plane or Parker's spiral. For a sample of 13 out of 78 energetic He ion events the source region was identified from the NOAA data set /16/. The observed events showed no north / south preference and the active solar source regions are distributed between solar latitude - 30° and + 30°. The observed events are, as expected, preferentially observed if the active source region is in the western hemisphere of the solar disk, only a minor fraction might be connected to the eastern solar hemisphere (< 15%).

SUMMARY

The velocity dispersion of energetic ions was used to determine the event onset time and the maximum path length of the magnetic field lines (L/μ). No statistical significant difference between the velocity dispersion of energetic He or O ions was observed and the L/μ most probably value is 1.25 AU, i.e. the suprathermal ions move along Parker's spiral.

ACKNOWLEDGEMENTS

The authors would like to thank all individuals who contributed to the success of the CELIAS / STOF sensor on SOHO.

REFERENCES

1. Mason, G. M., Reames, D. V., von Rosenvinge, T. T., Klecker, B., & Hovestadt, D. , The heavy-ion compositional signature in He- 3-rich solar particle events, ApJ, 303, 849, 1986.

2. Reames, D. V., Richardson, I. G., & Wenzel, K. -. Energy spectra of ions from impulsive solar flares, ApJ, 387, 715 ,1992.

3. Reames, D. V., Barbier, L. M., von Rosenvinge, T. T., Mason, G. M., Mazur, J. E., & Dwyer, J. R. Energy Spectra of Ions Accelerated in Impulsive and Gradual Solar Events, ApJ, 483, 515, 1997.

4. Mason, G. M., Dwyer, J. R., & Mazur, J. E. New Properties of He-3-rich Solar Flares Deduced from Low-Energy Particle Spectra, ApJ, 545, L157, 2000.

5. Mason, G. M., Ng, C. K., Klecker, B., & Green, G. Impulsive acceleration and scatter-free transport of about 1 MeV per nucleon ions in (He-3)-rich solar particle events, ApJ, 339, 529, 1989.

6. Mazur, J. E., Mason, G. M., Dwyer, J. R., Giacalone, J., Jokipii, J. R., & Stone, E. C. Interplanetary Magnetic Field Line Mixing Deduced from Impulsive Solar Flare Particles, ApJ, 532, L79, 2000.

7. Giacalone, J., Jokipii, J. R., & Mazur, J. E. Small-scale Gradients and Large-scale Diffusion of Charged Particles in the Heliospheric Magnetic Field, ApJ, 532, L75 , 2000.

8. Lee, M. A. & Ryan, J. M. Time-dependent coronal shock acceleration of energetic solar flare particles, ApJ, 303, 829, 1986.

9. Gosling, J. T., The solar flare myth, JGR, 98, 18949, 1993.

10. Reames, D. V., Kahler, S. W., & Ng, C. K. Spatial and Temporal Invariance in the Spectra of Energetic Particles in Gradual Solar Events, ApJ, 491, 414, 1997.

11. Krucker, S., D.E. Larson & R.P. Lin, On the origin of impulsive electron events observed at 1 AU, ApJ, 519, p.864-875, 1999

12. Krucker, S. & R.P. Lin, Two classes of solar proton events derived from onset time analysis, ApJ, 542, L61-L64, 2000

13. Hilchenbach M., H. Sierks, B. Klecker, K. Bamert, and R. Kallenbach, Solar energetic particle events observed by SOHO/CELIAS/STOF, in: Proceedings of the 27th International Cosmic Ray Conference, Hamburg 2001, vol. 10, p. 3144, 2001.

14. Hovestadt, D., Hilchenbach, M., Bürgi,A., Klecker, B., Grünwaldt, H. et al., CELIAS-charge, element and isotope analysis system for SOHO, Solar Physics, 162, 441, 1995.

15. Reames, D. V., Ng, C. K., & Tylka, A. J. Initial Time Dependence of Abundances in Solar Energetic Particle Events, ApJ, 531, L83, 2000.

16. NOAA, GOES Satellite Data, http://sec.noaa.gov/ Data/goes.html and http://www.solar.ifa.hawaii.edu/ ARMaps/archive.html

17. Ipavich, Real time solar wind monitor, http://umtof.umd.edu/pm

Solar Wind Sources and Flow Structure Over the 1995-2000 Period

N.A.Lotova, V.N.Obridko

IZMIRAN, Troitsk, Moscow Region, 142190 Russia
tel.: (7-095)3340902, (7-095)3340282 / fax: (7-095)3340124
e-mail: nlotova@izmiran.troitsk.ru / solter@izmiran.troitsk.ru

K.V.Vladimirsky

Lebedev Physical Institute, Moscow, 117924 Russia
tel.: (7-095)9382251 / fax: (7-095)1326274
e-mail: vlad@sci.lebedev.ru

Abstract. Evolution of the large-scale stream structure of the solar wind flow is studied in the main acceleration zone at 10 to 40 solar radii from the Sun. Three independent sets of the experimental data were used: observations of the radio wave scattering using the large radio telescopes of the Lebedev Physical Institute, white solar corona images obtained with the SOHO spacecraft, and solar magnetic field strength computed from J.Wilcox Solar Observatory data. The positions of the transonic region of the solar wind flow derived from the radio astronomical observations data were used as a parameter reflecting the intensity of the solar wind acceleration process. Correlation studies of these data with the magnetic field strength in the solar corona permit us to reveal several different types of the solar wind streams. The 1995--2000 data show important changes in the solar corona magnetic fields and corresponding changes of the solar wind flow.

INTRODUCTION

Evolution of the solar wind sources and flow stream structure is studied on the basis of regular radio scattering observations with compact natural sources [Lotova et al., 1985; 1995; 1998; 2000a,b; 2002a,b]. Large radio telescopes of the Lebedev Physical Institute, Pushchino, were used (DCR-1000 at 103 MHz and RT-22 at 22.2 GHz). Apparent movement of the sources permits us to obtain in lasting about a month series of daily observations a radial dependence of the scattering. Characteristic, repeating shape of this dependence reveals an area of increased scattering, which is identified as a transition, transonic region [Lotova et al., 1985]. Internal boundary of this area R_{in} is used as a natural characteristic of the solar wind acceleration process intensity: an increased level of acceleration corresponds to a closer to the Sun location of the transonic region.

Transonic area studies are very important because it is an erea of basic solar wind acceleration [Bird and Edenhofer, 1990]. An insight into the mechanism of the solar wind streams formation was obtained from correlation plots of R_{in} values combined with the magnetic field intensity $|B_R|$ at the source area of the solar wind, at the start of the acceleration process (Figure 1a-d). As can be seen from Fig. 1, correlation relations R_{in}, $|B_R|$ fall into several branches on each diagram. Different types of the solar wind streams correspond to each of these branches, differing in the source conditions and in the acceleration progress. The nature of these differences was revealed by the more detailed study of both magnetic field and flow structures, by calculations giving the magnetic field topology, and by direct flow observations with the SOHO coronographs. Investigation of the stream structure of the flow seems to be one of the most important problems in studies of the real mechanisms of the supersonic solar wind formation. It is enough to remind that the existence of an analogous structure of the flow in nonconducting media is totally impossible;

CP679, *Solar Wind Ten: Proceedings of the Tenth International Solar Wind Conference*,
edited by M. Velli, R. Bruno, and F. Malara

interaction of a submerged jet with the surrounding liquid is characterized there by a conic structure of the flow with an opening angle of 25° [Landau and Lifshitz, Hydrodynamics]. In near-solar plasmas, streams of magnetized plasmas with an opening angle of about 5° or less can be observed. The cause for such a violent difference are the well-known effects of frozen-in fields, but the details of these processes are very complicated and not yet studied up to now.

SOURCE AREA DATA

Solar magnetic fields were calculated in a narrow area R ~ (1-2.5)R_s disregarding possible variations introduced by the medium. Parameters of the solar wind plasma in this area, density of free electrons ≤10^8 cm^{-3} [Guhathakurta and Sitter, 1999], support this approach. Real hard point is incompleteness of the initial data. J.Wilcox Solar Observatory results for R = R_s, solar surface, were used. These Zeeman measurements give a modulus, not a vector of the field, some interpolation methods were used to overcome this trouble and deficient density of the measuring data [Hoeksema et al., 1982,1983; Obridko and Shelting, 1992, 1999; Lotova et al., 2002]. The results of calculations were presented in two forms. The $|B_R|$ used in Fig. 1 is the radial component of the field vector at the specified points of the sphere R = 2.5 R_s; angular coordinates determined so that R_{in}, $|B_R|$ pairs correspond to the same streamline of the flow. The second form is topology of the magnetic field in the neighbourhood of the same points, open type with field lines going into the space, or closed with loop-shape field lines. The difference between these two types is very important one concerning the frozen-in field phenomena.

Solar magnetic fields are doubtless decisive factor in formation of the solar wind streams. Nevertheless, optical observations of the white solar corona structure, INTERNET data obtained with LASCO C3 and LASCO C2 coronographs of the SOHO spacecraft, were of invaluable help in understanding the nature of some types of the solar wind streams. In comparisons of the single-momented optical observations with the R_{in} data obtained in long series of the radio astronomical observations the temporal and space coordinates were matched with the moments of the R_{in} determinations, which ensures that both radio and optical data belong to the same stream of the solar wind.

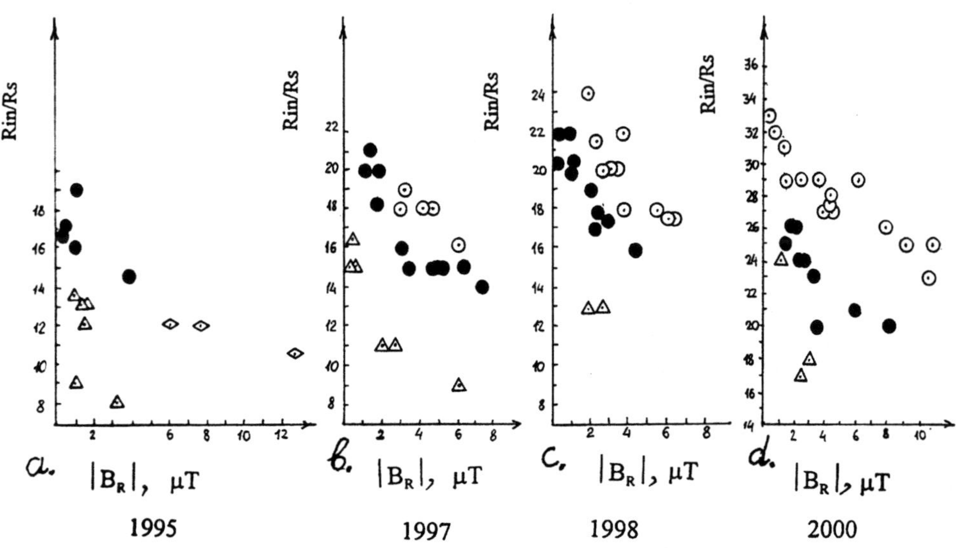

Fig 1 (a-d). Correlation diagrams: position of the inner boundary of the transonic region R_{in} vs. magnetic field $|B_R|$

BASIC RESULTS

The use of three independent diagnostic methods permitted us to isolate four characteristic types of the solar wind streams differing in the source conditions and in the progress of acceleration, two high-speed and two low-speed ones. Repeatedly presented in observations are high-speed streams designated in Fig.

1 with triangles. They are characterized by weak or moderate field strength and open field structure in the source area. In the white corona structure this group of streams originates in local coronal holes or in side lobes of the streamers.

Quite different type of high-speed streams was observed in [Lotova et al., 200b] (diamonds in Fig. 1). They start from polar coronal holes, from strong open magnetic field and are only observable at the minima of solar activity, when polar coronal holes reach medium solar latitudes.

The filled circles in Fig. 1 designate low-speed streams. This group corresponds to the closed loop-like structure of the magnetic field lines and to the main body of the streamers.

The slowest type of the streams, which corresponds to the farthest from the Sun position of the transonic area, is designated in Fig. 1 by open circles. For this type of the streams, a mixed type of the field-line structure is characteristic, in which the field lines going into the space alternate with the loop-like lines. In the white corona structure, wide areas of intense, structureless luminosity are observed.

On the whole, the correlation plots in Fig. 1 characterize evolution of the process of the solar wind stream formation during the first half of the solar activity cycle no. 23. The completeness of the results is to a great extent due to availability of the SOHO data for these years. The year-to-year differences in the flow structure are well pronounced. In the year of solar activity minimum, 1995, the high-speed streams prevail due to the open configuration of the magnetic field. In 2000, at the maximum of solar activity, on the contrary, low-speed streams dominate. The dominating role of the magnetic field structure is clearly pronounced. One can see that the field topology, rather than strength, is here the crucial factor.

ACKNOWLEDGEMENTS

The authors are extremely grateful to the J.Wilcox Observatory staff for INTERNET solar magnetic field data and to the SOHO staff for the INTERNET white corona data.

REFERENCES

Bird M.K. and Edenhofer P. (1990): in Physics of the Inner Heliosphere, R.Schwenn, E.Marsch (Eds.), Springer, Vol. 1, p. 80, 1990.

Guhathakurta M. and Sitter E. (1999): in Proc. 9th Int. Solar Wind Conf., Sh. R. Habbal et al., (Eds.), New York, p. 79, 1999.

Hoeksema J.T. et al. (1982): J. Geophys. Res.,Vol. 87, p. 1033, 1982.

Hoeksema J.T. et al. (1983): J. Geophys. Res., Vol. 88, p. 9910, 1983.

Lotova N.A., Blums D.F., and Vladimirskii K.V. (1985): Astron. Astrophys.. vol. 150, p. 266, 1985.

Lotova N.A. et al. (1995): Astron. Zh.., Vol. 72, p. 757, 1995 (in Russian).

Lotova N.A., Obridko V.N., and Vladimirskii K.V. (1998): Astron. Zh., Vol. 75, p. 626, 1998 (in Russian).

Lotova N.A., Vladimirskii K.V., and Obridko V.N. (2000a): Phys. Chem Earth ©, Vol. 25, p. 121, 2000a.

Lotova N.A., Obridko V.N., and Vladimirskii K.V. (2000b): Astron. Astrophys., Vol. 357, p. 1051, 2000b.

Lotova N.A., Obridko V.N., and Vladimirskii K.V. (2002a): Astron Zh. (in Russian) (in press).

Lotova N.A., Obridko V.N., Bird M.K., and Janardhan P. (2002b): Solar Phys. (in press).

Obridko V.N. and Shelting B.D. (1992): Solar Phys., vol. 137, p. 167, 1992.

Obridko V.N. and Shelting B.D. (1999): Solar Phys., vol. 184, p. 187, 1999.

Development of Multidimensional MHD Model for the Solar Corona and Solar Wind

E. C. Sittler Jr.[1], L. Ofman[1,2], S. Gibson[3], M. Guhathakurta[4], J. Davila[1], R. Skoug[5], A. Fludra[6] and T. Holzer[3]

1. *NASA/Goddard Space Flight Center, Greenbelt, MD*
2. *The Catholic University of America, Washington, D. C.*
3. *NCAR/High Altitude Observatory, Boulder, CO*
4. *NASA Headquarters, Washington, D. C.*
5. *Los Alamos National Laboratory, NM*
6. *Rutherford Appleton Laboratory, UK*

Abstract. We are developing a time stationary self-consistent 2D MHD model of the solar corona and solar wind that explicitly solves the energy equation, using a semi-empirical 2D MHD model of the corona to provide an empirically determined effective heat flux q_{eff} (i.e., the term effective means the possible presence of wave contributions). But, as our preliminary results indicate, in order to achieve high speed winds over the poles we also need to include the empirically determined effective pressure P_{eff} as a constraint in the momentum equation, which means that momentum addition by waves above 2 R_S are required to produce high speed winds. At present our calculations do not include the P_{eff} constraint. The estimates of P_{eff} and q_{eff} come from the semi-empirical 2D MHD model of the solar corona by Sittler and Guhathakurta (1999a,2002) which is based on Mk-III, Skylab and Ulysses observations. For future model development we plan to use SOHO LASCO, CDS, EIT, UVCS and Ulysses data as constraints for our model calculations. The model by Sittler and Guhathakurta (1999a, 2002) is not a self-consistent calculation. The calculations presented here is the first attempt at providing a self-consistent calculation based on empirical constraints.

INTRODUCTION

Modeling of the solar wind in coronal streamers was first attempted by Pneuman & Kopp (1971). They used an isothermal, steady state solution with an iterative technique to calculate the solar wind and the magnetic field. The iterative technique was expanded to model the steady state thermally conductive solar wind flow (e.g., Cuperman, Ofman & Dryer 1990; Stewart & Bravo, 1997). Time dependent 2D MHD models of the solar wind in a streamer have been developed by Steinolfson, Suess & Wu (1982), Wang et al. (1993), Mikic and Linker (1994), Linker and Mikic (1995), Suess et al. (1996), Wang et al. (1998). The first three-fluid 2D model was developed by Ofman (2000). Most models assume polytropic energy equations, or an ad-hoc heating function. Chen & Hu (2001) included Alfvén waves as the driving force of the wind. The stability of a 3-streamer model was investigated analytically by Wiegelmann, Schindler & Neukirch (2000). Sittler and Guhathakurta (1999b) developed a semi-empirical steady state model of the solar wind in a multipole 3-streamer structure, with the model constrained by Skylab observations. However, the semi-empirical model is not a self-consistent calculation. Here, we use the time-dependent 2D MHD equations and a multipolar initial magnetic field to obtain the self-consistent solar wind solution in a three-streamer structure. We investigate polytropic, ad hoc heating term with heat conduction and empirical heating solutions, and discuss the implications of the various approximations.

2D MHD EQUATIONS AND MODEL

The normalized resistive MHD equations are

CP679, *Solar Wind Ten: Proceedings of the Tenth International Solar Wind Conference*, edited by M. Velli, R. Bruno, and F. Malara

$$\frac{\partial \rho}{\partial t}+\nabla\cdot\left(\rho\vec{V}\right)=0, \quad (1)$$

$$\rho\left[\frac{\partial \vec{V}}{\partial t}+\left(\vec{V}\cdot\nabla\right)\vec{V}\right]=-\nabla p-\frac{GM_S\rho}{r^2} +\frac{1}{c}\vec{J}\times\vec{B}+\vec{F}_v \quad (2)$$

$$\frac{\partial \vec{B}}{\partial t}=-c\nabla\times\vec{E}, \quad (3)$$

$$\vec{E}=-\frac{1}{c}\vec{V}\times\vec{B}+\eta\vec{J}, \quad (4)$$

$$\nabla\times\vec{B}=\frac{4\pi}{c}\vec{J}, \quad (5)$$

$$\frac{\partial T}{\partial t}=-(\gamma-1)T\,\nabla\cdot\vec{V}-\vec{V}\cdot\nabla T+ (\gamma-1)(H_c/\rho+H_i), \quad (6)$$

where the heat conduction by Spitzer along magnetic field lines is

$$H_c=\nabla\left(\xi T^{5/2}\frac{\nabla T\cdot\vec{B}}{B^2}\right)\cdot\vec{B}, \quad (7)$$

Here we note that Spitzer's heat conduction does not apply over the poles beyond 2 R_S. We have used uniform initial temperature. The density and the velocity were initialized by the isothermal Parker's (1963) solar wind. The boundary conditions were open at 5 R_S, and extrapolated variables with fixed $\vec{B}=\vec{B}_0$ at r = 1R_S. The resolution was 320 X 300, and $\eta=10^{-4}$. We have investigated three cases for the energy equation: (A) Polytropic model with γ=1.05, no heat conduction. (B) Heat conduction with ad hoc heat input required for the solar wind given by

$$H_i=H_0(r-1)e^{-r/\lambda}, \quad (8)$$

where H_0 = 2.0 and λ = 0.7 R_S. (C) Effective heat flux q_{eff} obtained from the semi-empirical model of Sittler and Guhathakurta (1999, 2002) on open field lines with the corresponding heat input

$$H_{i,eff}=-\vec{B}\cdot\nabla\frac{q_{eff}}{B}, \quad (9)$$

for which we note that the heat conduction is not explicitly included in case (C). Sittler and Guhathakurta (1999a) have developed a magnetic field model given by Equation (19) in that paper which uses a multipole expansion for which η_M = 0.282, η_D = 0.109, and η_Q = 0.108 are ratios of the monopole, dipole and quadropole terms relative to the octupole term, respectively. Using the radial component of the magnetic field at 1 AU they estimate the parameter B_0 = 12 G which is the strength of the octupole term. This magnetic field model was used in Equation (9) to calculate q_{eff} and $H_{i,eff}$ for their initial estimate for case (C). To eliminate the analytical current sheet that reaches r = 1R_S, and the disconnected field lines near the equator we set η_M=0 in the magnetic field used to initialize the self-consistent MHD model. Below we present the numerical results for cases (A)-(C).

Results

For case (A) we used the following initial parameters: γ = 1.05, B_0 = 12 G, T_0 = 1.6 MK, $n_0 = 10^8$ cm^{-3}, with the resulting normalization V_A = 2617 km/s, τ_A = 267.4 s . For case (B) we set γ = 5/3, B_0 = 12 G, T_0 = 1.4 MK, n_0 = 5x10^8 cm^{-3}, with the resulting normalization V_A = 1170.6 km/s, τ_A = 598.0 . The heat conduction coefficient was $\xi = 1.1^{-3}$. Here we will not show the solutions for case (A) since they are similar to that for case (B), and rather will only show the solutions for case (B), while noting differences with case (A).

In Figure 1a we show a color display of the plasma temperature as a function of radial distance (i.e., 1 $\leq$ r $\leq$ 5) and co-latitude from pole to pole for case (B). Color is used to indicate intensity. A similar format is used in Figure 1b where we show j^2/ρ. Both these figures show the presence of three helmet streamers and current sheets where in the closed field line regions the temperature can exceed 2 MK. When compared to case (A) the tops of the helmet streamers are similar and the current intensity within the current sheets is greater for case (B) when compared to case (A). These self-consistent solutions support the predictions by Sittler and Guhathakurta (1999b) that there should be three current sheets because the octupole term dominates near the Sun. Also, the wind speed at 5 R_S is 300 km/s for case (B), while for case (A) it is 250 km/s. In the model solutions by Sittler and Guhathakurta (1999a, 2002) the flow speed exceeds 400 km/s over the poles at 5 R_S. It is clear that these solutions cannot produce high speed winds over the poles as observed. Finally, in Figures 2a and 2b we show the field line topology and solar wind velocity vectors, respectively. It shows the flows are radially outward above 2-3 R_S for all latitudes. Except for the magnetic field these self-consistent solutions do not provide realistic solutions for the density, flow velocity and plasma temperature.

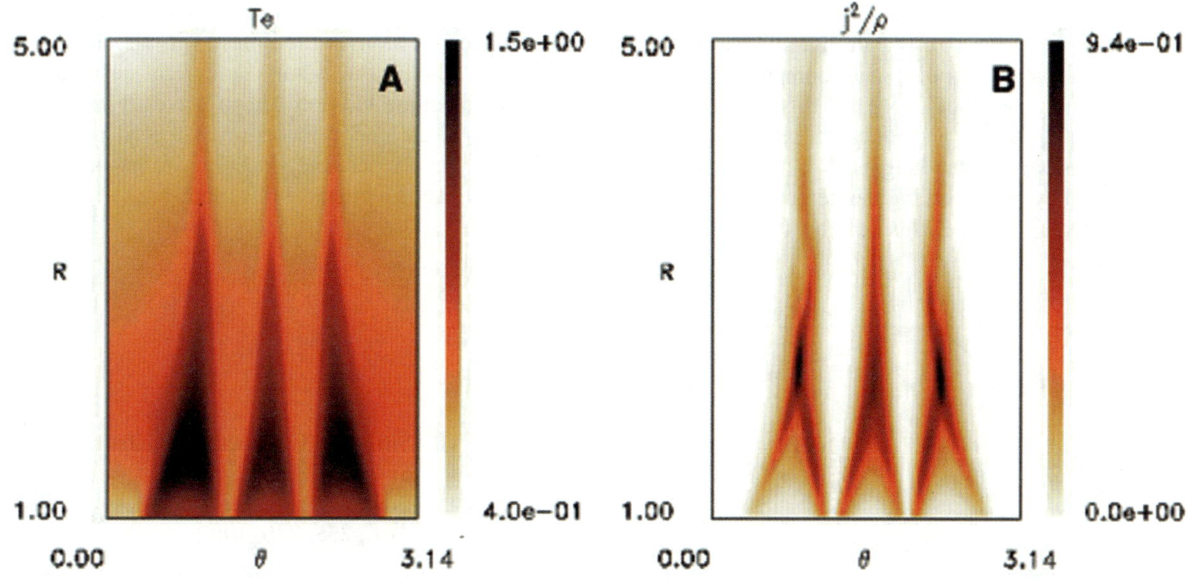

Figure 1. Color intensity maps of plasma temperature in panel A and current density squared divided by the mass density ρ in panel B. The abscissa is the co-latitude in radians and ordinate in radial distance in solar radii. The temperature is in units $1.4x10^6$ °K, while j^2/ρ is in units $1.7x10^{17}$ $statamp^2/(gm\text{-}cm)$.

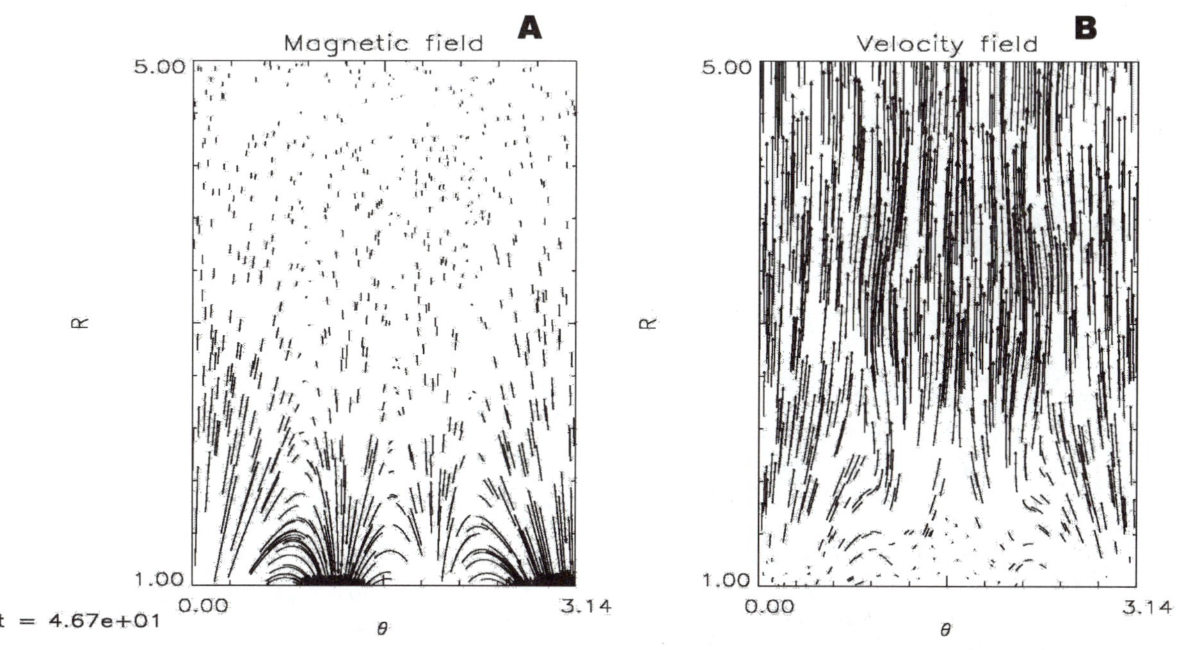

Figure 2. In panel A we show the magnetic field vectors as a function of co-latitude θ in radians and radial distance in solar radii. In panel B we show the flow velocity vectors using the same format as in panel A.

For case (C) we inserted the q_{eff} from the model by Sittler and Guhathakurta (1999a, 2002) into Eq. (9) (i.e., semi-empirical heating term) and dropped the heat conduction term in the energy equation. For case (C) we set $\gamma = 5/3$, $B_0 = 12$ G, $T_0 = 1.4$ MK, $n_0 = 5x10^8$ cm^{-3}, with the resulting normalization $V_A = 1170.6$

km/s, τ_A=598.0. The normalization for q_{eff} given in erg/cm^2/s was 1.13x10^7. In Figure 3 we show the flow speed over the poles and compare it to a Parker solution. Inside of 2 R_S the flow speed exceeds the Parker solution and then drops below the Parker solution at greater heights. We believe this result is caused by not including momentum addition terms due to waves. We feel this will be corrected once P_{eff} is

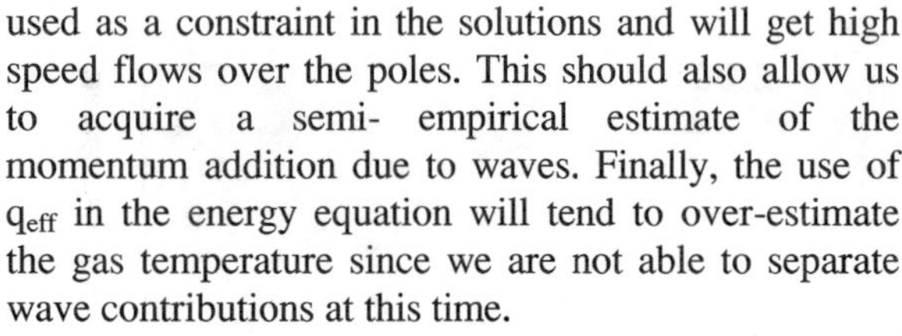

used as a constraint in the solutions and will get high speed flows over the poles. This should also allow us to acquire a semi- empirical estimate of the momentum addition due to waves. Finally, the use of q_{eff} in the energy equation will tend to over-estimate the gas temperature since we are not able to separate wave contributions at this time.

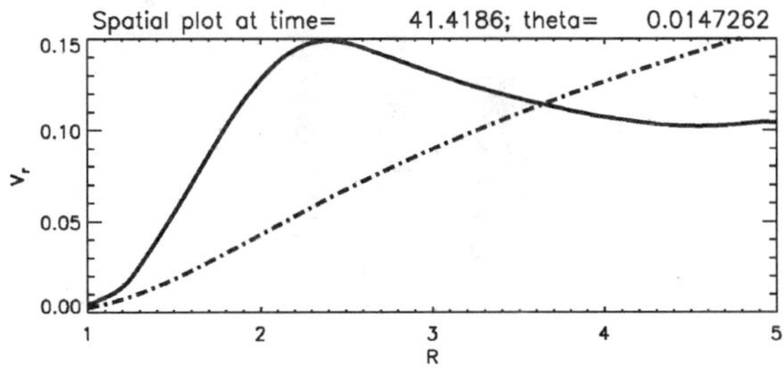

Figure 3. The flow speed over the poles for case (C), solid line, in units of V_A=1171 km/s. For reference we show Parker solution as dash-dot line.

ACKNOWLEDGMENTS

This work was done by L. Ofman as a Research Professor at the Catholic University of America, under the NASA SOHO Guest Investigator program.

REFERENCES

1. Chen, Y. and Y. Q. Hu, *Sol. Phys.*, **199**, 371, 2001.
2. Cuperman, S.,L. Ofman and M. Dryer, *ApJ.*, **350**, 846, 1990.
3. Linker, J. A. and Z. Mikic, *ApJ.*, **438**, L45, 1995.
4. Mikic Z. and J. A.Linker, *ApJ.*, **430**, 898, 1994.
5. Ofman L., *Geophys. Res. Lett.*, **27**, 2885, 2000.
6. Parker, E. N., *Interplanetary Dynamical Processes*, New York, Interscience, 1963.
7. Pneuman, G.W. & Kopp, R.A., *Sol. Phys.*, **18**, 258, 1971.
8. Sittler, E. C. Jr. and M. Guhathakurta, *ApJ.*, **523**, 812, 1999a.
9. Sittler, E. C. Jr. and M. Guhathakurta, *Solar Wind 9*, 401, 1999b.
10. Sittler, E. C. Jr. and M. Guhathakurta, *ApJ.*, **564**, 1062, 2002.
11. Steinolfson, R. S., S. T. Suess and S. T. Wu, *ApJ.*, **255**, 730, 1982.

NOZOMI observation of transient, non-spiral magnetic field in interplanetary space associated with limb CMEs

T. Nakagawa*, A. Matsuoka† and NOZOMI/MGF team

*Tohoku Institute of Technology, Taihaku-ku, Sendai, 982-8577, Japan
†Institute of Space and Astronautical Science, Yoshinodai, Sagamihara, 229-8510, Japan

Abstract. The magnetic fields of interplanetary objects that were ejected as coronal mass ejections (CMEs) from the limb of the Sun were observed by the NOZOMI spacecraft at 1.38 AU above the east limb of the Sun. The solar wind magnetic field whose launch time coincided with ejection of the limb CME was different from that estimated from ACE observations near the Earth, suggesting that NOZOMI encountered transient magnetic structures, or that the heliospheric magnetic field re-structured after ejection of the CMEs. The corresponding interplanetary magnetic field showed non-spiral magnetic field, enhancements of magnetic field, or magnetic discontinuities, but no common structure was yet found.

INTRODUCTION

Acceleration of the solar wind is the most important problem in the study of the solar wind. Evolution of the speed of the solar wind as a function of the distance from the Sun remains uncertain. We do not have enough information on the velocity, density, and temperature of solar wind streams at various distances from the Sun observed simultaneously.

A comparison of the solar wind speed in interplanetary space with its initial speed at the departure from the Sun would give basic information to the solar wind profile. To do this, we have to find some signatures to distinguish solar wind streams. Although there is no established method to mark a stream, we can start with coronal mass ejections (CMEs) and some unusual, perhaps non-spiral, interplanetary magnetic field.

The initial speed of a CME is measured most accurately on the limb of the Sun as it is launched into interplanetary space in a direction perpendicular to the line of sight of the imagers in the vicinity of the Earth. In order to detect interplanetary counterparts of the limb ejections, a spacecraft must be located in planetary orbit above the limb of the Sun.

Schwenn [1] used plasma measurements from Helios 1 and 2 together with white-light-coronagraphs. They showed that flare-related ejecta with shocks and disturbed plasma had no post acceleration in interplanetary space, while slowly rising transients had post acceleration beyond several solar radii. Sheeley *et al.* [2] compared interplanetary shocks detected by Helios 1 with CMEs. Burlaga *et al.* [3] studied an association of a magnetic cloud observed by Helios 1 with a low-

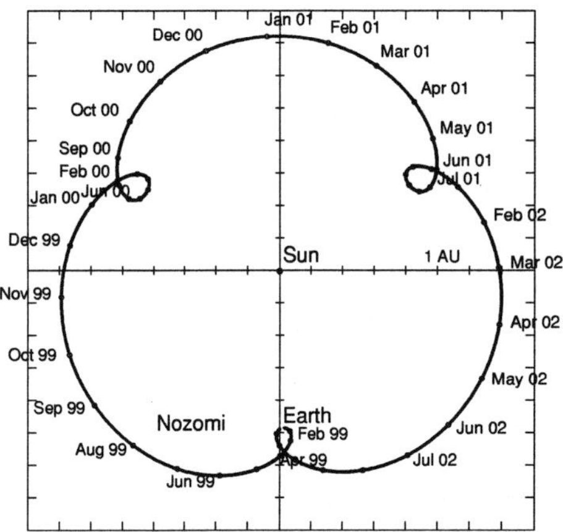

FIGURE 1. Orbit of Nozomi in ecliptic plane with Sun-Earth line fixed.

latitude CME. Richardson *et al.* [4] used ICE energetic ion data combined with IMP 8 magnetic field to compare with SMM coronagraphs. Lindsay *et al.*[5] investigated plasma and magnetic field data from Pioneer Venus Orbiter and Helios 1. Gopalswamy *et al.* [6] [7] presented a model to predict CME arrival time at 1AU, but they had to struggle against effect of projection of CMEs onto the line of sight of the imager.

The Japanese Mars explorer Nozomi has brought us another opportunity to detect interplanetary counterparts of limb CMEs. Figure 1 shows the separation of Nozomi from the Sun-Earth line. Nozomi went as far as $83^o - 97^o$ east from the Sun-Earth line at a radial distance of 1.38 AU from the Sun during the period from November 1, 1999, to November 30, 1999. The magne-

CP679, *Solar Wind Ten: Proceedings of the Tenth International Solar Wind Conference*, edited by M. Velli, R. Bruno, and F. Malara

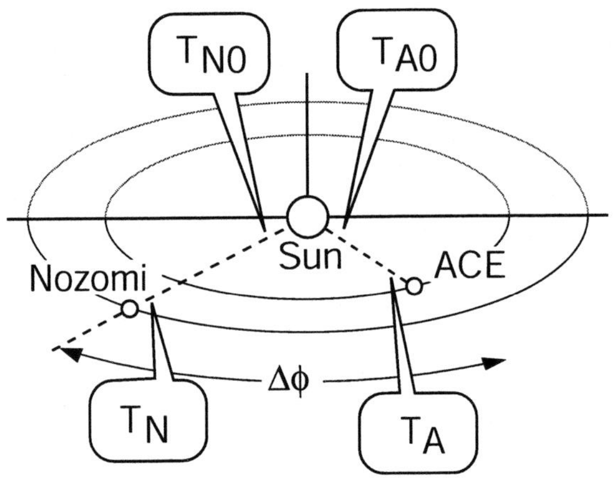

FIGURE 2. The time T_N of solar wind observation by NOZOMI, related with the time T_{N0} of launch toward NOZOMI, the time T_{A0} of launch of the plasma toward ACE, and the time T_A of solar wind observation by ACE.

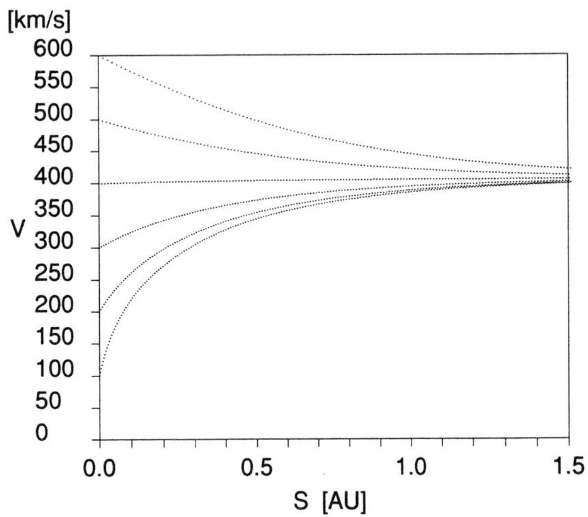

FIGURE 3. Model calculation of solar wind velocity V as a function of distance S from the Sun.

tometer [8] [9] was operating during the period. Imaging spacecraft such as Yohkoh and SOHO were also operating simultaneously with Nozomi. We can compare the interplanetary magnetic field observed by Nozomi with limb CMEs in the SOHO/LASCO CME Catalog (http://cdaw.gsfc.nasa.gov/CME_list/) presented by Gopalswamy *et al.* [10].

ESTIMATION OF LAUNCH TIME AND ARRIVAL TIME

In order to examine the magnetic field of the solar wind whose launch time coincided with a time of ejection of a limb CME, we need to relate the time of launch with the time of arrival at NOZOMI. The relationship was estimated in 2 ways: i) estimation of launch time from arrival time, and ii) estimation of arrival time from launch time.

Launch time of the solar wind

Figure 2 illustrates the method of estimation of the time T_{N0} of launch of the solar wind encountered by NOZOMI at T_N. As NOZOMI had only limited opportunities to measure solar wind speed, the launch time was estimated with an aid of ACE/SWEPAM observation.

The solar wind observed by ACE at time T_A must have left the Sun at around $T_{A0} = T_A - \frac{R_A}{V_A}$, where R_A is the distance between ACE and the Sun, and V_A is the average of the wind speed measured at ACE. Here we use 1-hour averages of the solar wind speed. The launch time T_{N0} of a wind stream toward NOZOMI from the same source is calculated as $T_{N0} = T_{A0} - \frac{\Delta\phi}{\omega_{SUN}}$ where ω_{SUN} and $\Delta\phi$ are the angular velocity of the spin of the Sun, and the separation between the two spacecraft in heliospheric latitude, respectively. The solar wind launched toward Nozomi at T_{N0} was to be observed at $T_N = T_{N0} + \frac{R_N}{V_N}$, where R_N is the heliospheric distance of NOZOMI. Assuming that the solar wind speed from the source remained the same, $V_N \sim V_A$, we can relate T_N and T_{N0} as $T_N = T_{N0} + \frac{R_N}{V_A}$.

Arrival time of CMEs

The arrival time of limb CMEs at NOZOMI at 1.38 AU from the Sun was estimated in a separate way. Gopalswamy *et al.* [6] [7] showed statistically that there is a linear relationship between initial speed V_0 of ejection and its initial acceleration a_0 as

$$a_0 = c_0 - c_1 V_0, \tag{1}$$

where $c_0 = 2.193[\text{m/s}^2]$ and $c_1 = 0.54 \times 10^{-5}[/\text{s}]$ for limb CMEs [7]. This relationship suggests some dragging force acting on CMEs, and it is likely that the dragging force is acting beyond a significant distance from the Sun. If we extend the relationship as

$$\frac{dV(t)}{dt} = c_0 - c_1 V(t), \tag{2}$$

we can solve the differential equation to get the velocity as a function of time t

$$V(t) = w + (V_0 - w)\exp(-c_1 t) \tag{3}$$

and the distance

$$S(t) = wt + \frac{1}{c_1}(V_0 - w)(1 - \exp(-c_1 t)) + 1R_{\text{SUN}}, \tag{4}$$

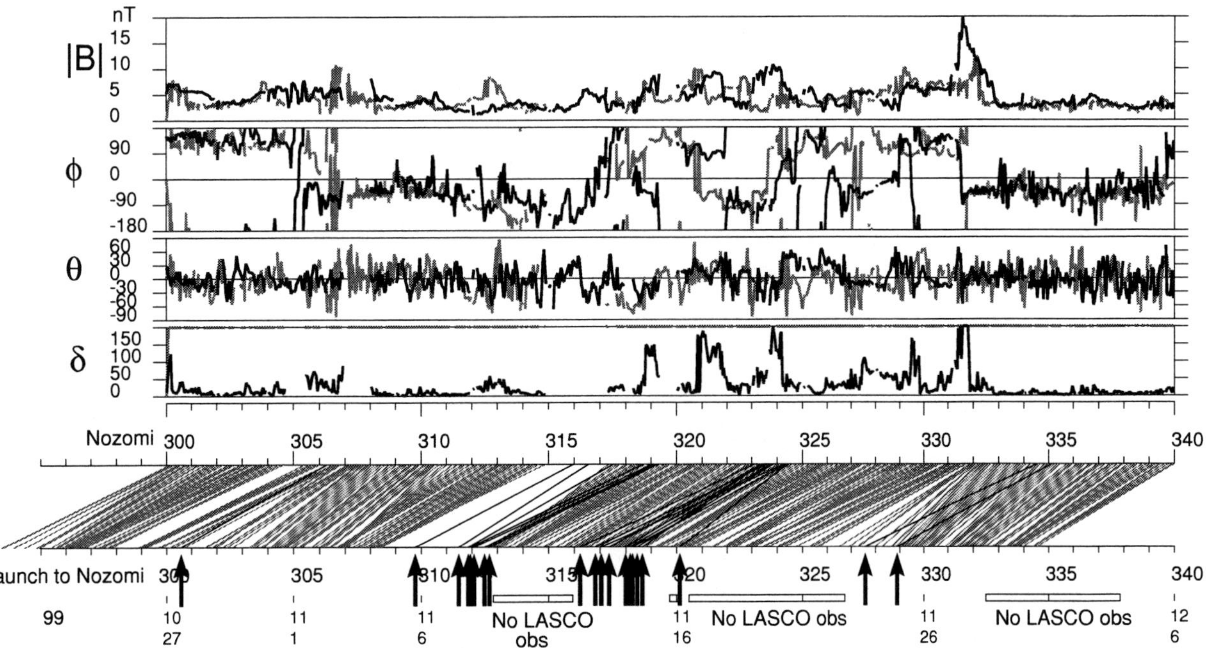

FIGURE 4. Hourly averaged magnetic fields obtained by NOZOMI (dark) and ACE (light), covering the period from November 1, 1999 to November 30, 1999. ACE data are shifted to the dates on which the solar wind of the same source was observed by NOZOMI. From top to bottom, the magnitude $|\mathbf{B}|$ of the magnetic field, azimuthal angle ϕ and inclination θ of the field vector from the ecliptic plane, and deviation $\delta \equiv |\mathbf{B}_{\text{NOZOMI}} - \mathbf{B}_{\text{ACE}}|^2$ of the NOZOMI magnetic field vector $\mathbf{B}_{\text{NOZOMI}}$ from the ACE observation $\mathbf{B}_{\text{ACE}}$. The bottom diagram shows the time T_N of NOZOMI observation related with the estimated launch time T_{N0}. Light lines indicate estimation based on ACE observation, while dark lines indicate results from model calculation. Arrows indicate time of CMEs that appeared on east limb at low latitude and looked likely to be ejected toward NOZOMI. The limb CMEs are adapted from the SOHO/LASCO CME catalog (http://cdaw.gsfc.nasa.gov/CME_list/).

where $w = \frac{c_0}{c_1}$ is the speed to which $V(t)$ converges. Figure 3 illustrates the evolution of solar wind velocity $V(t)$ as a function of the distance $S(t)$ from the Sun. The speeds of CMEs converge to a value $w = 406[\text{km/s}]$ as they propagate in interplanetary space after their launch at various initial speeds. Substituting the initial speed V_0 of each CME into equation (4) we can calculate the time of arrival and the speed at $S = 1.38$ AU.

INTERPLANETARY MAGNETIC FIELD EJECTED AS CMES

Figure 4 shows hourly averaged magnetic fields obtained by NOZOMI (dark) and ACE (light), covering the period from November 1, 1999 to November 30, 1999. ACE data are shifted to the time T_N and corrected for the heliocentric distance. The bottom diagram shows the relationship between the time T_{N0} of launch of solar wind or CMEs and the time T_N of encounter by NOZOMI estimated from ACE observation (light lines) and model calculation (thick lines). Arrows in the bottom diagram indicate CMEs that would hit NOZOMI.

The magnetic field observed by NOZOMI agreed with the result from ACE shifted to NOZOMI position until November 10, 1999, indicating that the heliospheric magnetic field structure was stable and the solar wind speed was unchanged. A disagreement started on November 13, 1999 at NOZOMI position, and lasted for 2 weeks. The period of disagreement between NOZOMI and ACE magnetic field corresponds to the successive release of CMEs on the east limb which started on November 7, 1999.

The limb CMEs launched on November 7 - 8, 1999, were related to the directional disagreement between NOZOMI and ACE weak magnetic fields on November 13 - 14 according to the assumption of unchanged solar wind speed (light lines), while they were related to the disagreement of intense fields observed on November 15, 1999 (at NOZOMI position) according to equation (4) (thick lines). The propagation speed of CMEs estimated from equation (4) was slower than the solar wind speed estimated from ACE observation. It seems likely that the magnetic field on the solar surface changed its structure after the release of the CMEs and the wind speed also varied. On November 13, 1999, there were 2-hour periods of non-spiral field at 0:40-2:40. On November

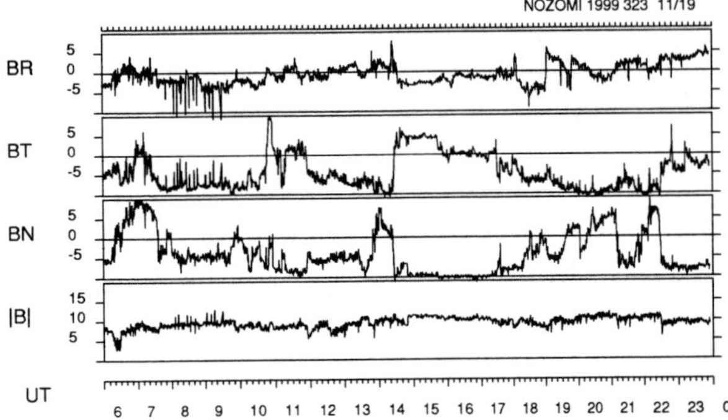

FIGURE 5. Interplanetary magnetic field observed by NOZOMI on November 19, 1999 in RTN coordinates. Non-spiral field and number of tangential discontinuities are recognized.

15, 1999, an intense, non-spiral magnetic field structure bounded by discontinuities was observed by NOZOMI. At present, it is difficult to say which non-spiral field corresponds to the limb CMEs.

The CMEs which left the Sun during the period from November 12 - 16, 1999, were related to the NOZOMI observation on November 19, 1999, according to equation (4) (thick lines), while some of the arrival times were estimated to be 2 days earlier on the assumption of the unchanged solar wind speed (light lines). On both days, i.e., November 17 and November 19, 1999, NOZOMI observed enhancements of magnitude $|\mathbf{B}|$ of the magnetic field and deviation δ from ACE magnetic field. Figure 5 shows the interplanetary magnetic field observed by NOZOMI on November 19, 1999, in RTN coordinate system. The magnetic field showed non-spiral directions, with number of discontinuities. Many of them had characteristic of tangential discontinuities.

The two CMEs observed on November 23 and November 24, 1999, coincided with the sector boundary observed by NOZOMI on November 27, 1999, according to the assumption of unchanged wind speed (light lines). The arrival dates calculated from equation (4) was 3 days behind, where no special signature was found but average spiral field was observed. Around this sector boundary crossing, slower wind was followed by faster wind, and ejecta from wide range were compressed to be observed in a short time in interplanetary space. There is another possibility that these CMEs were just missed by the point observation by NOZOMI.

SUMMARY

The period of disagreement between NOZOMI and ACE magnetic field observations, which indicates loss of stability of the heliospheric structure of magnetic field and solar wind speed, coincided with successive release of CMEs on the east limb of the Sun. The interplanetary magnetic field whose launch time coincide with limb CMEs often showed non-spiral magnetic field, enhanced magnetic field, or discontinuities, but no common structure is yet found.

ACKNOWLEDGMENTS

The authors are indebted to Charles W. Smith for ACE/MAG data, David J. McComas for ACE/SWEPAM data, and N.Gopalswamy, S.Yashiro and G.Michalek for SOHO/LASCO CME Catalog. The CME catalog is generated and maintained by the Center for Solar Physics and Space Weather, The Catholic University of America in cooperation with the Naval Research Laboratory and NASA. SOHO is a project of international cooperation between ESA and NASA.

REFERENCES

1. Shwenn, R., Direct correlations between coronal transients and interplanetary disturbances, *Space Science Rev.*, **34**, 85-99 (1983).
2. Sheeley, N. R. Jr., Howard, R. A., Koomen, M. J., Michels, D. J., Shwenn, R., Muȟlhaüser, K. H., and Rosenbauer, H., Coronal mass ejections and interplanetary shocks, *J. Geophys. Res.*, **90**, 163-175 (1985).
3. Burlaga, L. F., Klein, L., Sheeley, N. R. Jr., Michels, D. J., Howard, R. A., Koomen, M. J. Shwenn, R., and Rosenbauer, H., A magnetic cloud and a coronal mass ejection, *Geophys. Res. Lett.*, **9**, 1317-1320 (1982).
4. Richardson, I. G., Farrugia, C. J., Winterhalter, D., Solar activity and coronal mass ejections on the western hemisphere of the Sun in mid-August 1989: Association with interplanetary observations at the ICE and IMP8 spacecraft, *J. Geophys. Res.*, **99**, 2513-2529 (1994).
5. Lindsay, G. M., Luhmann, J. G., Russell, C. T., and Gosling, J. T., Relationship between coronal mass ejection speeds from coronagraph images and interplanetary characteristics of associated interplanetary coronal mass ejections, *J. Geophys. Res.*, **104**, 12515 (1999).
6. Gopalswamy, N., Lara, A., Lepping, R., Kaiser, M. L., Berdichevsky, D., and Cyr, C. St., Interplanetary acceleration of coronal mass ejections, *Geophys. Res. Lett.*, **27** , 145 (2000).
7. Gopalswamy, N., Lara, A., Yashiro, S., Kaiser, M. L., and Howard, R. A., Predicting the 1-AU Arrival times of coronal mass ejections, *J. Geophys. Res.*, **106** , 29207 (2001).
8. Yamamoto, T., and Matsuoka, A., Planet-B magnetic fields investigation, *Earth, Planets. Space*, **50**, 189 (1998).
9. Nakagawa, T., Matsuoka, A., and NOZOMI/MGF team, NOZOMI observation of the interplanetary magnetic field in 1998, *Adv. Space Res.*, **29**-3, 427 (2002).
10. Gopalswamy, N., Yashiro, S., Kaiser, M. L., Howard, R. A., and Bougeret, J.-L., Characteristics of coronal mass ejections associated with long wavelength type II radio bursts, *J. Geophys. Res.*, **106**, 29219 (2001).

Radial Variation of Magnetic Flux Ropes: Case Studies with ACE and NEAR

C. T. Russell[1], T. Mulligan[1] and B. J. Anderson[2]

[1]*IGPP and ESS, University of California, Los Angeles, CA 90095, USA*
[2]*Applied Physics Laboratory, John Hopkins University, Laurel, MD 20723, USA*

Abstract. Multiple spacecraft observations are used to examine the spatial variation of magnetic flux ropes in the interplanetary medium near 1 AU. When the observing spacecraft are close and radially aligned they observe nearly the same structure. When they are radially aligned and far apart they reveal the expansion of the ropes with heliocentric distance. When the spacecraft are azimuthally separated and the separation vector is along the rope axis, this axis is shown to be bent. The observed radial expansion is consistent with earlier Helios results.

INTRODUCTION

While our studies with a non-force-free model of magnetic ropes clearly shows that the ropes are often nearly force-free, our data also indicates that these magnetic ropes expand as they move outward from the Sun. This expansion can be inferred from the "temporal" asymmetry in the magnetic field profile. Often the ropes exhibit a much faster rise in field strength as they cross the spacecraft than the later fall in magnitude. Since our model [1] and those of other groups when applied to a single spacecraft's data assumes cylindrical symmetry, we resort in our model inversion to use an "expansion" parameter to account for this asymmetry. An alternate approach of normalizing the components of the field by the field strength is less desirable because the rise and fall in field strength signals the pressure gradient in the field that balances the curvature force of the rope. Thus such normalization prevents the true force balance in the rope to be discerned.

Even with a single spacecraft one can confirm that flux ropes expand with radial distance using a statistical approach if that spacecraft is in an elliptical heliocentric orbit and covers a range of radial distances from the Sun. Wind, ACE and Pioneer Venus could not provide such data. However, Helios 1 and 2 could do so and were used to show that flux rope diameters scaled as $R^{0.78 \pm 0.10}$ where R is the distance from the Sun. In our studies we have been able to obtain examples of two spacecraft measurements of the same flux rope and herein we exploit those data to verify this statistical relationship in individual cases.

We use the cylindrically symmetric form of our model [1] in which the axial field falls off from the center as a negative exponential of the distance from the center of the rope to a power. The power and the scale length of the fall off are both parameters of the fit. The axial field is proportional to one minus a negative exponential raised to a power. Again the power and the scale length of the exponential fall off are independently fitted in the inversion. Other parameters include the impact parameter of the rope crossing, the direction of the rope axis and an expansion factor.

In this study we attempt to capture a rope at two locations R_1 and R_2 as sketched in Figure 1 where the rope has radii, r_1, and r_2. We believe that these flux ropes are connected back to the sun as sketched in Figure 2. This connectivity places constraints on how the tube expands as discussed later.

VERIFYING THE PROCEDURE

On January 7, 1998 NEAR was heading toward a gravity assist with the Earth when an ICME crossed it and the ACE spacecraft when they were separated by 0.04 AU radially and 0.02 AU azimuthally. The magnetic field data and the best independent fits to the data and the currents are shown in Figure 3. The directions of the axes of the rope obtained from the two fits differ by only 6° in cone angle (62° vs 56°) and 5° in clock angle (15° vs 10°). Both fits returned a flux content of 10 TWb and a left-handed rope. The impact parameters are 0.15% for Wind and 0% for NEAR. Both ropes are force-free except near the outer edge of the rope. While the peak parallel currents are similar the rope at Wind has a much

CP679, *Solar Wind Ten: Proceedings of the Tenth International Solar Wind Conference*, edited by M. Velli, R. Bruno, and F. Malara

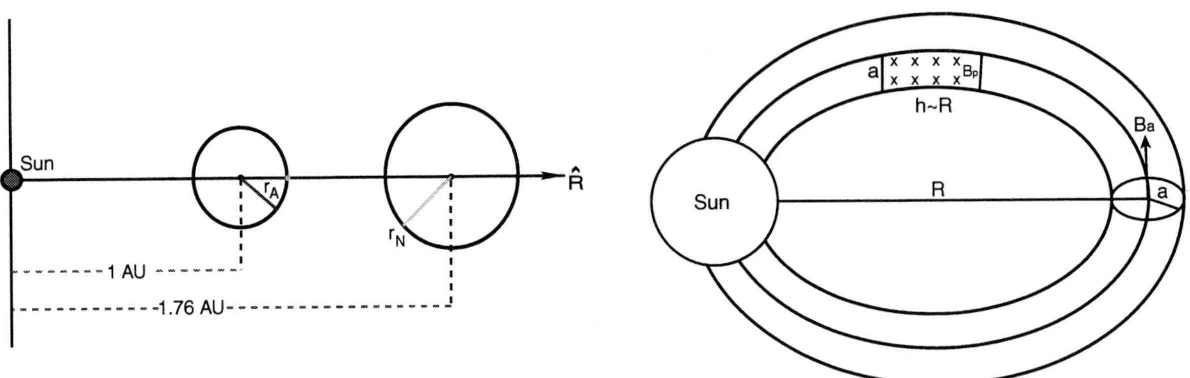

FIGURE. 1. Snapshot of rope cross sections at two radial distances from Sun.

FIGURE 2. Schematic illustration of connection of rope to the Sun shown in the plane containing the axial field.

TABLE 1. Characteristic and Integrated Parameters of Flux Rope Model Fits for the July 15-17, 2000 Magnetic Cloud

	Independent Fits		Simultaneous Fits		Free Expansion Fits	
Property	ACE	NEAR	ACE	NEAR	ACE	NEAR
Handedness	L	L	L	L	L	L
Clock Angle (θ)	53°	48°	50°	50°	50°	50°
Cone Angle (Φ)	92°	76°	83°	83°	83°	83°
Radius, AU	0.24	0.43	0.25	0.43	0.25	0.43
Flux, 10^{12} Wb	95	137	108	130	118	158
Total Current, 10^6 A	72	97	66	79	97	85
Twist (α*), μA/Wb	0.76	0.72	0.60	0.60	0.72	0.63
N_T	0.28	0.47	0.22	0.41	0.28	0.41
Impact Parameter, AU	0.10	0.16	0.11	0.17	0.10	0.16
Expansion Factor γ	0.66	0.35	0.50	0.50	0.73	0.73
B_a Expansion law, R^{-x}	2	2	2	2	1.4	1.4
B_p Expansion law, R^{-x}	2	2	2	2	1.2	1.2
B_a Spread	0.71	0.87	0.71	0.71	0.81	0.81
B_p Spread	0.75	0.06	0.53	0.53	0.53	0.53
B_a Exponential Power	2.4	2.7	2.5	2.5	2.5	2.5
B_p Exponential Power	1.7	2.8	2.1	2.1	2.0	2.0

lower parallel current near the center. If the current normalized by the field strength (which is fairly constant here) is constant, we refer to this as being in the Taylor state. Thus the rope is closer to the Taylor state at NEAR than at Wind.

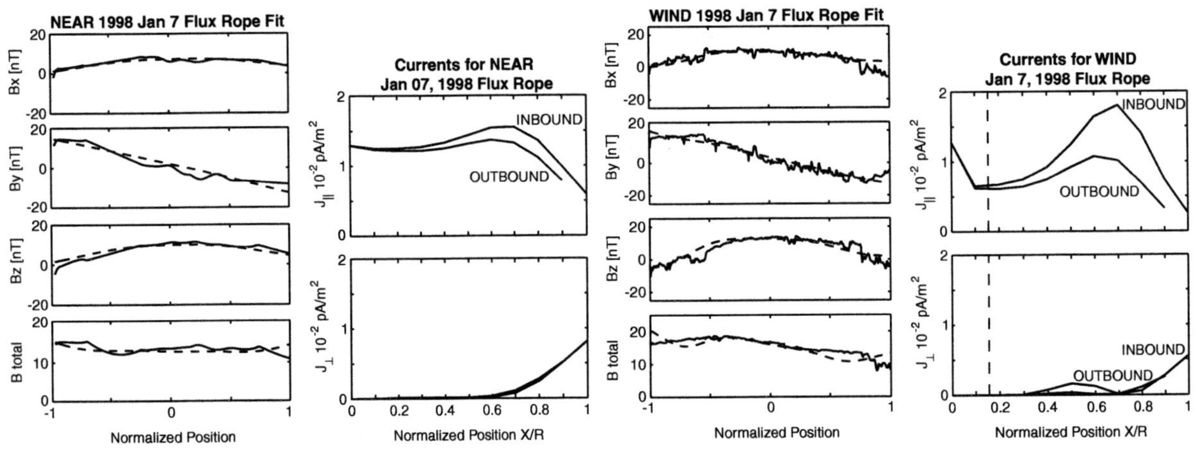

FIGURE 3. Field and currents seen by Wind and NEAR on January 7, 1998. Dashed line gives best fit model from which currents are derived.

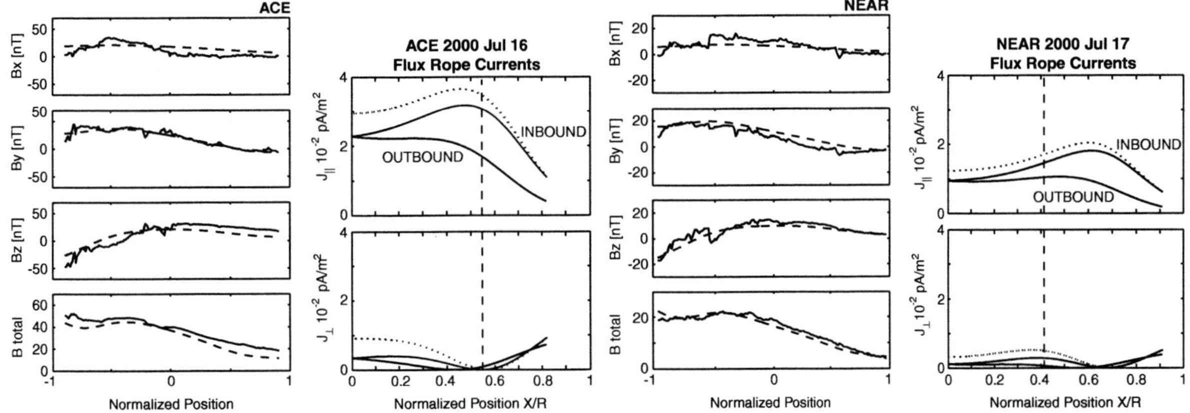

FIGURE 4. Fields and currents seen by ACE and NEAR for Bastille day rope. Dashed line in best fit to field from which currents derived. Dotted line gives current with rope expansion removed.

THE BASTILLE DAY EVENT

In mid July 2000 when NEAR was aligned almost radially with the Earth, a very strong ICME termed the Bastille day event crossed ACE and later NEAR. The magnetic field data and derived currents are shown at ACE and NEAR in Figure 4, using a fit in which the same rope was allowed to expand arbitrarily as it moved from ACE to NEAR. The dotted lines show a snapshot of the rope at a fixed time. The radius has changed from 0.25 AU to 0.43 AU over 1.76 AU. Thus the radius varies as $R^{0.96}$, slightly higher than the Helios number $R^{0.78 \pm 0.10}$ [2,3]. Table 1 gives the parameters of the fitting procedure first by treating each fit as totally independent, secondly by adopting an expected fall off and fitting simultaneously, and lastly by fitting with an arbitrary fall off. All three approaches give similar results.

THE AUGUST 13, 2000 EVENT

On August 13, 2000 NEAR was 15° east of the Earth-Sun line and at 1.72 AU from the Sun when an ICME crossed ACE and then NEAR. Figure 5 shows the observed and fitted magnetic field profiles for this event and the currents. The rope appears to be nearly force-free in the central regions but decidedly non-force-free at the edges. Here we show the "free expansion" fit which allows the magnetic field and scale of the rope to have an arbitrary fall off. The size of the rope is found to vary as $R^{0.84}$ a value quite consistent with Helios [2]. Table 2 gives the parameters for this event as in Table 1 for the Bastille day event.

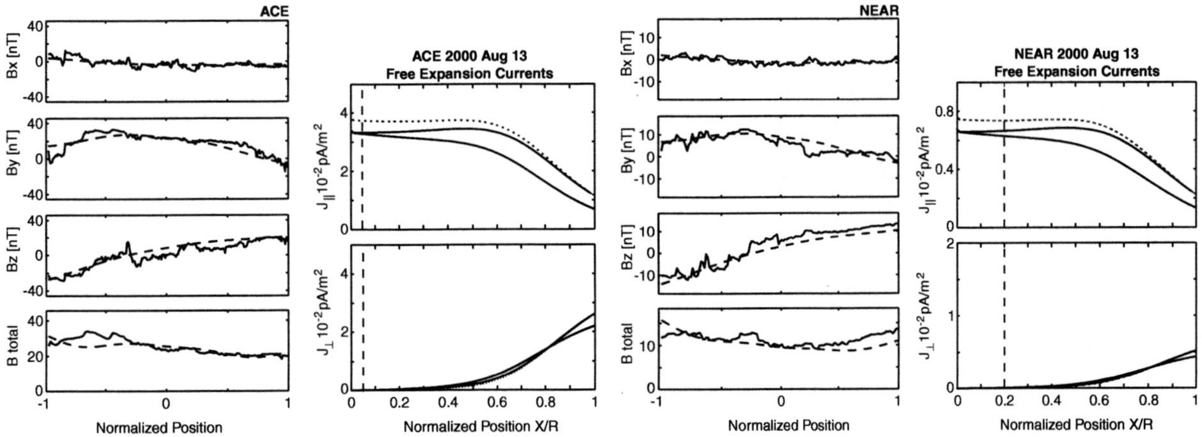

FIGURE 5. Fields and currents seen by ACE and NEAR for August 13, 2000 event. Comments of Figure 4 caption apply.

TABLE 2. Characteristic and Integrated Parameters of the Flux Rope Model Fits for the August 13-15, 2000 Magnetic Cloud

	Independent Fits		Simultaneous Fits		Free Expansion Fits	
Property	ACE	NEAR	ACE	NEAR	ACE	NEAR
Handedness	L	L	L	L	L	L
Clock Angle (θ)	83°	62°	78°	68°	78°	68°
Cone Angle (Φ)	102°	97°	105°	98°	105°	98°
Radius, AU	0.17	0.29	0.17	0.29	0.17	0.29
Flux, 10^{12} Wb	38	41	37	36	41	40
Total Current, x 10^6 A	141	69	47	47	79	79
Twist (α*), μA/Wb	3.71	1.70	1.29	1.32	1.92	1.95
N_T	1.38	1.10	0.47	0.85	0.72	1.26
Impact Parameter, AU	0.01	0	0	0.01	0.01	0.06
Expansion Factor γ	0.34	0	0.17	0.17	0.23	0.23
B_a Expansion law, R^{-x}	2	2	2	2	1.8	1.8
B_p Expansion law, R^{-x}	2	2	2	2	1.3	1.3
B_a Spread	0.81	0.87	0.87	0.87	0.83	0.83
B_p Spread	3.2	0.77	0.79	0.79	0.89	0.89
B_a Exponential Power	9.6	2.2	3.2	3.2	3.0	3.0
B_p Exponential Power	1.7	0.89	1.1	1.1	1.2	1.2

SUMMARY

The similarity of the two independent inversions of the January 7 event as observed by Wind and NEAR suggest that our independent inversions are accurate or at least consistent. In Figures 6 and 7 we show the geometry that we obtain from these independent cylindrically symmetric fits for the July and August 2000 events. The July event clearly shows the expansion of the rope with radial distance. For the August event we have propagated the ACE solution with its velocity and expansion parameters to NEAR's heliocentric location to illustrate that the bend in the rope axis appears to be significantly less than that predicted by our dipolar extrapolation of the axial magnetic field.

Figure 2 can be used to estimate the expected radial expansion of a cylindrically symmetric force-free flux rope. First in a cylindrically symmetric force-free rope we expect the outward magnetic pressure of the rope that we approximate by the axial field squared to be balanced by the cross terms in the magnetic stress which we approximate by the product of the axial and poloidal fields. Then we note that the conservation of magnetic flux tells us that the axial field varies as the radius of the rope squared and the poloidal field varies as the product of the radius of the rope and the distance to the Sun. This latter relationship allows us to relate the radius of the rope to the heliocentric distance and show that the expected variation of the rope size with radial distance is roughly proportional to R. This is in fact close to the number we see.

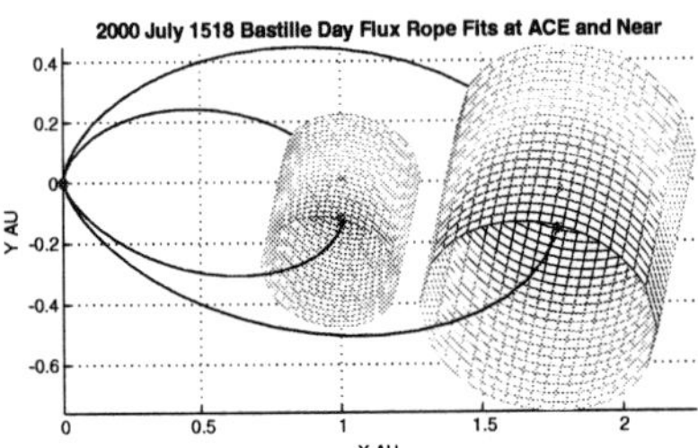

FIGURE 6. Cylindrical fit to Bastille day rope.

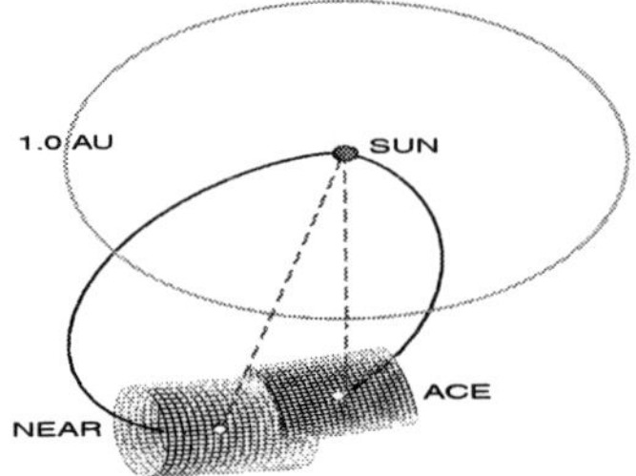

FIGURE 7. Cylindrical fit to August 13, 2000 event.

ACKNOWLEDGMENTS

This work was originally supported by the NEAR project by the National Aeronautics and Space Administration under a grant administered by the Johns Hopkins University, JHU #782797. Presently this work is being supported by NAG5-11583 and NAG5-10834.

REFERENCES

1. Mulligan, T. and Russell, C. T., *J. Geophys. Res.*, **106**, 10,581-10,596 (2001).
2. Bothmer, V. and Schwenn R., *Space Sci. Rev.*, **70**, 215-220, (1994).
3. Bothmer, V. and Schwenn, R., *Annales Geophys.*, **16**, 1-24, (1998).

The Limitation of Bessel Functions for ICME Modeling

C. T. Russell and T. Mulligan

IGPP and ESS, University of California, Los Angeles, CA 90095, USA

Abstract. Most inversions of the structure of magnetic ropes in ICMEs have assumed that the rope can be approximated as a force-free structure in the Taylor state in which the current is not only parallel to the magnetic field but everywhere has the same constant of proportionality to the field strength. The solution of this problem is a magnetic field that is describable by Bessel functions: J_o for the axial component and J_1 for the poloidal component. The Taylor state approximation has a maximum twist that is exceeded by about half the observed flux ropes. Moreover, many flux ropes are not force-free. The vast majority of non-force-free ropes have an excess of pressure pushing outward. Thus these ropes are either expanding or are balanced by non-magnetic forces. Thus while the Bessel function approach may be useful for determining the orientation of rope axes, its limited ability to correctly measure twist and its inability to assess any magnetic force imbalance mitigate against its usefulness in studies of ICME genesis and evolution.

INTRODUCTION

Most work on inverting the structure of ICMEs and especially on their most simple form, the magnetic cloud, has been performed using a Bessel function [1,2]. The justification for this is that in a Tokamak the plasma often relaxes to a state in which the parallel current per unit magnetic flux is constant across the diameter of the machine. This so-called Taylor state [3] is thought to occur due to reconnection and turbulence in the plasma. While often true in a fusion plasma, this need not be true in a space plasma. In this paper we first review the physics of force-free magnetic flux ropes and show the solution for the Taylor state. We then show the greater flexibility afforded by a non-force-free and non-Taylor-state solution. We apply this solution to flux ropes encountered by Pioneer Venus at 0.72 AU and evaluate the applicability of the Taylor state and force-free assumptions.

PHYSICS OF FORCE-FREE FLUX ROPE

In a force-free flux rope there is no net magnetic stress

$$\mathbf{J} \times \mathbf{B} = 0 \qquad (1)$$

We can rewrite (1) as the sum of a magnetic pressure gradient force and the divergence of the magnetic stress tensor.

$$-\nabla\left(\frac{B^2}{2\mu_o}\right) + \frac{1}{\mu_o}\nabla \bullet (\mathbf{B}\,\mathbf{B}) = 0 \qquad (2)$$

Thus in a rope the outward force of the magnetic pressure is balanced by the inward force of the twisted magnetic field.

A force-free rope has no current perpendicular to **B**. Thus we may write

$$\nabla \times B = \mu_o J = \alpha B \qquad (3)$$

There is no constraint on the functional form of α. It may vary with distance from the center of the rope. If it is constant, then we take the curl of (3) to obtain

$$\nabla(\nabla \bullet B) - \nabla^2 B = \alpha(\nabla \times B) \qquad (4)$$
$$\nabla^2 B = -\alpha(\nabla \times B)$$

The solutions of this second order equation are the zeroth and first-order Bessel functions [4]

$$B_a = B_o\, J_o\, (\alpha r) \qquad (5)$$
$$B_p = B_o\, J_1\, (\alpha r)$$
$$B_r = 0$$

If α is constant, the rope is said to be in the Taylor state

In the Taylor state $\frac{J}{B}\left(=\frac{\alpha}{\mu_o}\right)$ is constant.

CP679, *Solar Wind Ten: Proceedings of the Tenth International Solar Wind Conference*, edited by M. Velli, R. Bruno, and F. Malara

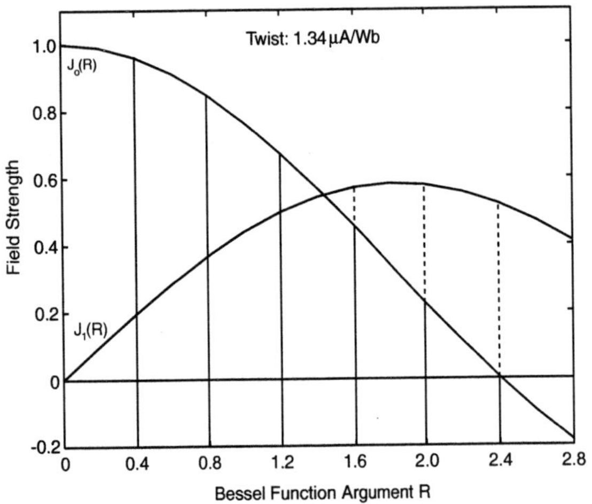

FIGURE 1. Magnetic field of a Taylor state rope.

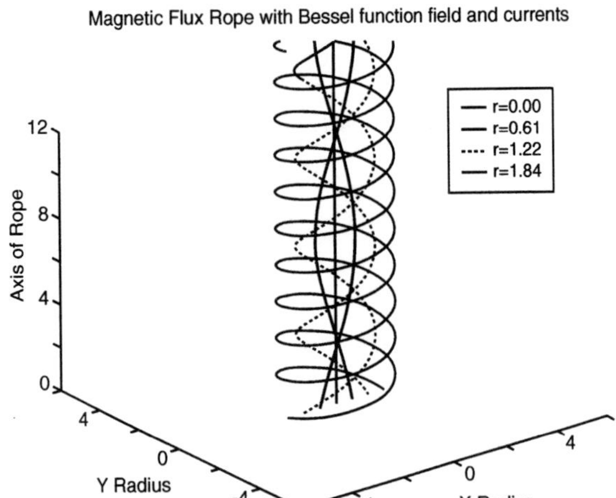

FIGURE 2. A Taylor state flux rope.

TAYLOR-STATE FORCE-FREE ROPE: THE BESSEL FUNCTION SOLUTION

Figure 1 shows the J_0 and J_1 Bessel functions that represent the variations of the magnetic field with R in a Taylor-state, force-free magnetic flux rope. For a typical flux rope seen in the vicinity of the Earth or Venus, the α-parameter would be about 1.34 μA/Wb. The field strength would be about 20 nT in the center of the rope and the diameter of the rope would be about 0.2 AU. Since there is only one Bessel function solution, once the central field strength and the radius are determined, there is very little flexibility in fitting the field variation. If the field has very little twist, one can assume that the Bessel function solution has a limited range, e.g. 0 to 0.4 in Figure 1. If it is very twisted around the axis, then the fit can be extended out to 2.4. However, the functional variation with rope radius is not variable. Another way to visualize this is given in Figure 2 where field lines in a Taylor-state, Bessel function flux rope are drawn. The solutions available to the modeler consist only of choosing the twist of the outer most field line. Once chosen there is a fixed path from the central (straight) field line to the most twisted one.

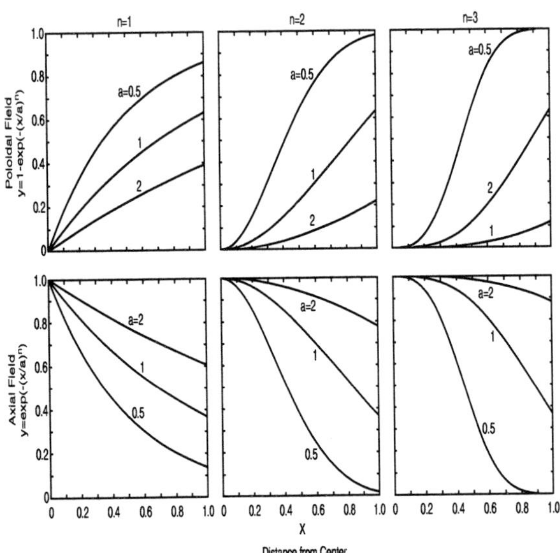

FIGURE 3. Poloidal and axial fields in a non-force-free solution for varying exponents and scales.

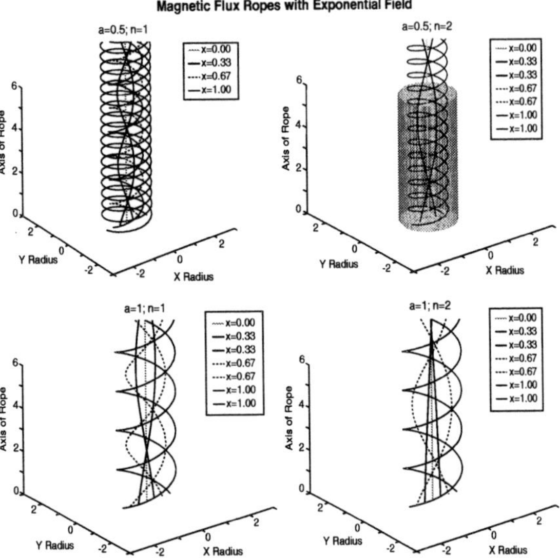

FIGURE 4. Different flux ropes with varying n and a.

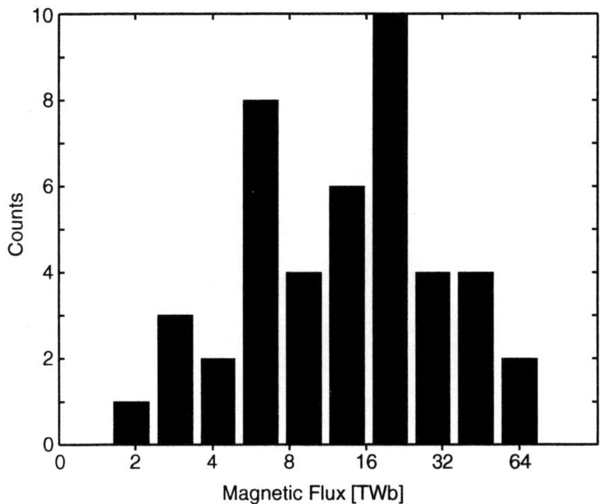

FIGURE 5. Occurrence rate of different flux content in PVO ropes.

NON-FORCE FREE CYLINDRICALLY SYMMETRIC MODEL

We express the poloidal and axial fields in terms of components

$$B_a = B_1 \left[\exp\{-(r/\sigma_a l)^m\}\right]$$
$$B_p = B_o \left[1-\exp\{-(r/\sigma_p l)^n\}\right]$$

that increase and decrease from the center of the rope respectively as exponentials raised to powers n, and m, each with their own independent field strength, exponential scale length and exponent [5]. Here l is the radius of the outside edge of the rope. The two parameters of the Bessel function fit have been replaced by six in the exponential fit.

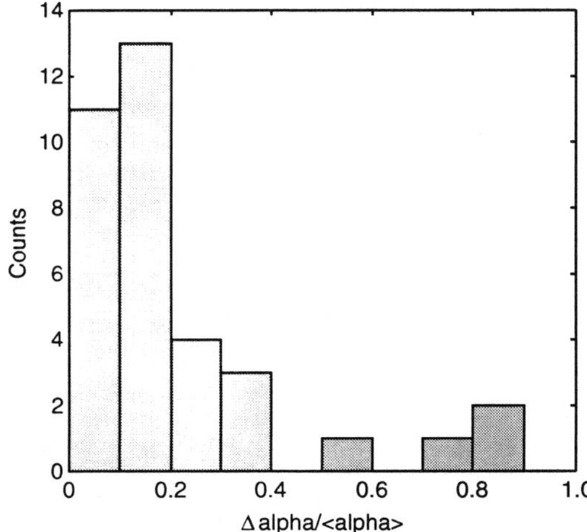

FIGURE 7. Distribution of ropes with varying fractional alpha change.

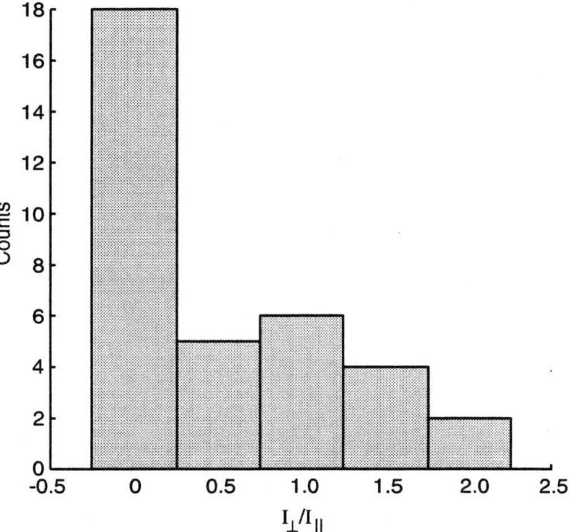

FIGURE 6. Occurrence rate of ropes with varying perpendicular to parallel current strengths.

In practice we have used an "expansion" factor δ to account for asymmetry in the time profile, most probably due to the expansion of the rope as it moves across the observer.

Both models also solve for the orientation of the rope, clock and cone angles, and the impact parameter, the distance of the satellite from the central axis of the rope at closest approach.

Figure 3 shows the magnetic field profiles for the poloidal and axial components of the field for different exponents, n, and different scale lengths of the exponentials, a. In this model the poloidal and axial fields are fit independently and can have different scale lengths and different exponents. This flexibility allows the rope to be non-force-free and if force-free to not be in the Taylor state. Examples of ropes that can be made are shown in Figure 4.

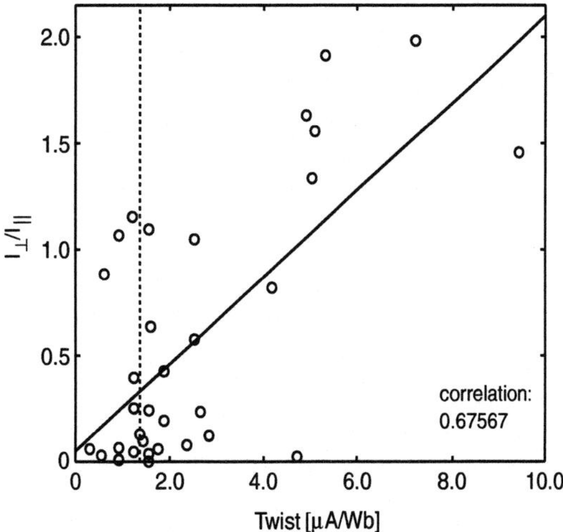

FIGURE 8. Correlation of non-force-free character with the magnitude of alpha.

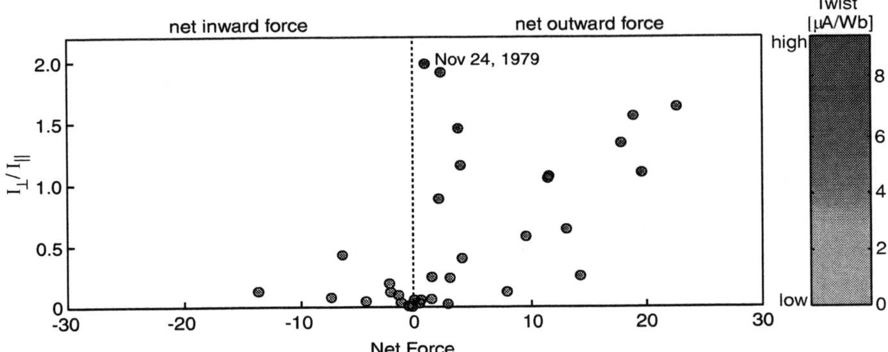

FIGURE 9. Ratio of perpendicular to parallel current strength versus net force in arbitrary units.

APPLICATION TO PIONEER VENUS DATA

Pioneer Venus spent an entire solar cycle in orbit about Venus at 0.7 AU from the Sun. Its 24-hour orbit took it well into the solar wind much of the time making it very well suited for ICME studies. We have applied the non-force-free model to the ICMEs identified during that time. Figure 5 shows the inferred flux content of these ropes making allowance for the (separately) inferred average oval cross-section of these ropes. The median flux content is about 10^{13} Wb. Figure 6 shows the number of ropes with varying ratios of integrated perpendicular to parallel current where the absolute value of the current has been used in the ratio. Thus these ropes could have zero net force but have non-force-free regions within them. About half of the ropes have significant non-force-free regions. Alpha gives the parallel current per unit magnetic flux and is constant in the Taylor state. Most of the ropes have at most small deviations from the Taylor state. This is illustrated in Figure 7 that shows the distribution of the peak to minimum variation in alpha normalized by the average alpha. Examination of the ropes with the largest variations in alpha indicate that some of these may be due to poor fits to the rope structure. If we plot the quantification of the non-force-free nature of the ropes $I_\perp/I_{||}$ versus alpha or twist we obtain the values shown in Figure 8. Clearly those ropes with the strongest parallel currents (per unit field strength) statistically are the ropes with the greatest non-force-free character. The dashed line shows the alpha value for a typical flux rope inferred from the Bessel function fit. Since the Bessel function has no $I_\perp$, only the values near the base of this line could be accurately portrayed with this model.

In Figure 9 we study the net force in the flux ropes. A minority of ropes have net inward force. These ropes all have a nearly force-free character. The majority of ropes that have a net force, have an outward force and that net outward force is greater, the larger is the non-force-free fraction of the current. The alpha parameter is color coded. Those ropes with high alpha are generally also not force-free but there seems to be no correlation of net force and alpha.

SUMMARY

The Bessel function model of a cylindrically symmetric flux rope is adequate only for force-free ropes in the Taylor state. Magnetic clouds seen by Pioneer Venus in the 1980's were often not force-free and not in the Taylor state. When ropes had an imbalance of force it was generally outward. Ropes are in general not cylindrically symmetric. We need multiple observations of flux ropes at separations of 0.1 to 0.4 AU to better understand their true geometry.

ACKNOWLEDGMENTS

This work was supported by the National Aeronautics and Space Administration under research grants NAG5-10834 and NAG5-11583.

REFERENCES

1. Burlaga, L. F., *J. Geophys. Res.*, **93**, 7217-7231 (1988).
2. Lepping, R. P., et al., *J. Geophys. Res.*, **95**, 11,957-11,965 (1990).
3. Taylor, J. B., *Rev. Mod. Phys.*, **58**, 741 (1986).
4. Lundquist, S., *Ark., Fys.* **2**, 361 (1950).
5. Mulligan, T., and Russell, C. T., *J. Geophys. Res.*, **106**, 10,581-10,596 (2001).

A revised theory of the charge-exchange coupling between plasma-gas counterflows in the heliosheath

H.J. Fahr

Institute for Astrophysics and Space Research, University of Bonn, Auf dem Huegel 71, D-53121 Bonn, Germany

Abstract. Various hydrodynamic models meanwhile were presented in the literature giving views of the interaction of the heliospheric plasma bubble with the counterflowing partially ionized interstellar medium in gradually increasing degrees of consistency. In these models the solar and interstellar hydrodynamic flows of neutral atoms and protons are coupled by mass-, momentum-, and energy- exchange terms due to charge exchange collisions between ionized and neutral particle species. In a simplified case by which we describe the main physics of the penetration of an H-atom flow through the well known plasma wall ahead of the heliopause, we show that the exchange terms used in the up-to-now hydrodynamic treatments unavoidably lead to an O-type critical point at the sonic point of the H-atom flow. At this point no continuation of the integration of the hydrodynamic set of differential equations is possible. As we show the way out of this problem is given by a more accurate formulation of the momentum exchange term for quasi- and sub-sonic H-atom flows. With a momentum exchange term derived from basic kinetic Boltzmann principles we instead arrive at a characteristic equation with an X-type critical point allowing for a contiunuous solution from supersonic to subsonic flow conditions. Under these new auspices the already often treated problem of the penetration of interstellar H-atoms into the inner heliosphere has to be revised since under these newly derived, more effective plasma - gas friction forces substantially different results are to be expected in this context.

CHARGE-EXCHANGE COUPLING OF DIFFERENTIAL PLASMA - GAS MOTIONS

In problems like the mutual interaction of plasma and H-atom gas flows in the heliosheath due to the large local Knudsen numbers Kn , in principle a kinetic treatment of charge-exchange induced coupling processes is required. In such kinetic approaches the distribution function $f_H(\vec{r},\vec{v})$ of the H-atom gas for the stationary case is described by a Boltzmann-Vlasov integro-differential equation (see e.g. Ripken and Fahr, 1983, Osterbart and Fahr, 1992, Baranov and Malama, 1993, Pauls and Zank, 1996, Fahr, 1996), McNutt et al. (1998,1999), Bzowski et al. (1997, 2000) given by:

$$v\frac{df_H(\vec{r},\vec{v})}{ds} = \tag{1}$$

$$= f_p(\vec{r},\vec{v})\int^3 f_H(\vec{r},\vec{v}')v_{rel}(\vec{v},\vec{v}')\sigma(v_{rel})d^3v'$$

$$- f_H(\vec{r},\vec{v})\int^3 f_p(\vec{r},\vec{v}')v_{rel}(\vec{v},\vec{v}')\sigma(v_{rel})d^3v'$$

where $\vec{r}$ and $\vec{v}$ are the relevant phase-space variables, ds is the increment of the line element on the associated dynamical particle trajectory, v_{rel} denotes the relative velocity between collision partners of velocities $\vec{v}$ and $\vec{v}'$, and $\sigma(v_{rel})$ is the velocity-dependent charge exchange cross-section.

Changing over from Equ.(1) to a set of hydrodynamic moment equations by introducing the moments $\Phi_0 = m_p$; $\Phi_1 = m_p\vec{v}$; $\Phi_2 = \frac{1}{2}m_p v^2$, then first leads to the following equation for the average mass exchange: $\langle\Phi_0\rangle = Q_0^+ - Q_0^- = 0$, i.e. as evident no net mass gain will occur under pure charge exchange reactions. To evaluate the exchange terms for the higher moments i = 1, 2 some knowledge of the distribution functions for protons, f_p, and H-atoms, f_H, is needed. Writing these distribution functions as functions of the lowest hydrodynamic moments themselves using shifted Maxwellians with isotropic temperatures T_p and T_H (see e.g. Holzer, 1972, Ripken and Fahr, 1983, Fahr and Ripken, 1984, Isenberg, 1986, Pauls and Zank, 1996, Fahr, 1996, Lee, 1997) then permits to present the above expressions in the forms:

$$\langle\Phi_1\rangle = \sigma_{rel}\langle v_{rel}\rangle m_p n_p n_H(\vec{V}_H - \vec{V}_p) \tag{2}$$

and:

$$\langle\Phi_2\rangle = \sigma_{rel}\langle v_{rel}\rangle m_p n_p n_H \cdot \left[\frac{1}{\gamma-1}\left(\frac{P_p}{\rho_p} - \frac{P_H}{\rho_H}\right) - \frac{1}{2}(\vec{V}_H - \vec{V}_p)^2\right], \tag{3}$$

where n_H, V_H, P_H and n_p, V_p, P_p are density, bulk velocity, and pressure of the H-atoms and of the protons, respectively, $\sigma_{rel} = \sigma(\langle v_{rel}\rangle)$ is the actual charge exchange

CP679, *Solar Wind Ten: Proceedings of the Tenth International Solar Wind Conference*, edited by M. Velli, R. Bruno, and F. Malara

cross section, and $\langle v_{rel} \rangle$ is the double-Maxwellian average of the relative velocity between protons and H-atoms as given e.g. by Holzer (1972):

$$\langle v_{rel} \rangle = \sqrt{\frac{128}{9\pi}\left(\frac{P_p}{\rho_p} + \frac{P_H}{\rho_H}\right) + (\overrightarrow{V_H} - \overrightarrow{V_p})^2}. \quad (4)$$

THE HYDRODYNAMIC EQUATIONS

The problem manifest ahead of the heliopause resembles the passage of an H-atom gas through a fixed quasistatic plasma structure downstream of an outer interstellar shock (see Baranov and Malama, 1993, Zank, 1999, Fahr et al., 2000). In a one-dimensional first order approach along the stagnation line (z-axis!) this plasma, due to its low sonic Mach number, can be approximated as incompressible and nearly stagnating. For the H-atom flow one then obtains the following set of equations:

$$\frac{d}{dz}(\rho_H V_H) = 0, \quad (5)$$

$$\rho_H V_H \frac{d}{dz} V_H = -\frac{d}{dz} P_H - \sigma_{rel} V_{rel} n_p \rho_H V_H, \quad (6)$$

$$\frac{d}{dz}\left[V_H\left(\frac{\rho_H V_H^2}{2} + \frac{\gamma P_H}{\gamma - 1}\right)\right] = \quad (7)$$

$$\sigma_{rel} V_{rel} n_p \rho_H \cdot \left[\frac{1}{\gamma - 1}\left(\frac{P_p}{\rho_p} - \frac{P_H}{\rho_H}\right) - \frac{V_H^2}{2}\right],$$

where γ is the polytropic index both of the protons and the H-atoms. $\sigma_{rel} = \sigma(V_{rel})$ is the relevant charge exchange cross section for protons and H-atoms interacting with an average relative velocity V_{rel} equal to:

$$V_{rel} = \sqrt{\frac{128}{9\pi}\left(\frac{P_p}{\rho_p} + \frac{P_H}{\rho_H}\right) + V_H^2}. \quad (8)$$

With the mass flow constant :

$$C_0 = \rho_0 V_{H0} = \rho_H V_H, \quad (9)$$

and the use of the normalized coordinate ζ defined by $z = \zeta D$ and of the auxiliary quantity $\Lambda = D/\lambda = D\sigma_{rel} n_p$ ($\zeta = 0$ and $\zeta = 1$ mark the two borders of the plasma wall) one then obtains the following characteristic equation:

$$\frac{d}{d\zeta} V_H = \frac{V_{rel}\Lambda\left(\Delta_p P_p - P_H + \frac{1}{2} C_0 V_H (\gamma + 1)\right)}{\gamma P_H - C_o V_H} \quad (10)$$

and the differential equation for the pressure:

$$\frac{d}{d\zeta} P_H = -V_{rel}\Lambda C_0 - C_0 \frac{d}{d\zeta} V_H. \quad (11)$$

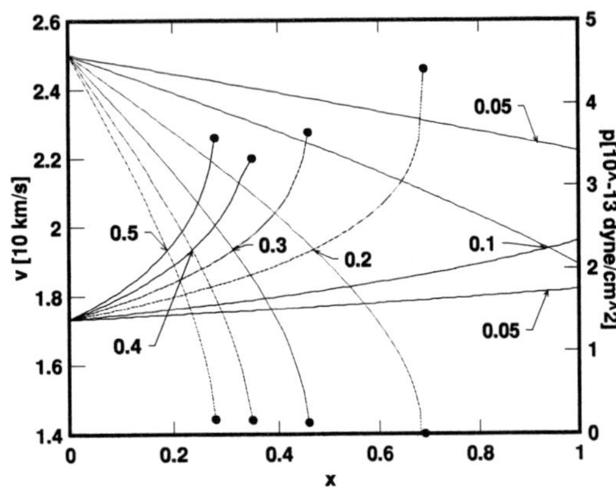

FIGURE 1. *Shown as a function of the plasma wall coordinate $x = z/D$, D being the linear extent of the plasma wall, is the H-atom bulk velocity V_H in units of [km/s] (left ordinate) and the H-atom pressure P_H in units of $[10^{-13} dyne/cm^2]$ (right ordinate) calculated for various values of the parameter $\Lambda = D\sigma_{rel} n_p$. The H-atom flow in all cases enters the plasma wall at $x = 0$ with a velocity of $V_H(0) = 25$ km/s. Curves reaching the singular point x_c where $\gamma P_{H,c} = C_0 V_{H,c}$ stop at this point.*

Starting the integration at $\zeta = 0$ with supersonic H-atom flow velocities, i.e.with $V_{H0}^2 \geq \gamma P_{H0}/\rho_{H0}$, one at first, i.e. for small values of ζ, obtains reasonable results (see Figure 1) .

Arriving, however, at some critical point $\zeta = \zeta_c \geq 0$ where locally $\gamma P_{Hc} = C_0 V_{Hc}$ is valid the integration of the above system of equations cannot be continued anymore. The singularity at $\zeta = \zeta_c$ hereby cannot be avoided by taking care that at this point both the numerator and the denominator of Equ.(10) vanish, since putting $\gamma P_{Hc} = C_0 V_{Hc}$ and introducing this into the numerator, the following requirement had to be fulfilled at ζ_c:

$$V_{Hc} = \sqrt{\gamma \frac{P_p}{\rho_p}\left(\frac{2}{2 - \gamma(\gamma + 1)}\right)} \quad (12)$$

As evident this requirement, however, cannot be fulfilled by real values of V_{Hc}!

THE CHARGE-EXCHANGE INDUCED PLASMA-GAS FRICTION

As explicitly derived in Fahr (2002) the charge-exchange induced momentum loss rate is calculated with the expression:

$$\langle Q_1^-(\overrightarrow{v_H})\rangle =$$

$$\frac{-2}{\sqrt[2]{\pi}} n_p n_H m_p \left(\frac{2KT_p}{m_p}\right) \cdot \int_o^\infty \int_0^\pi x \cos\vartheta \, \sigma(v_{rel}) \cdot \quad (13)$$

$$\sqrt{x_H^2 + x^2 - 2x_H x \cos\vartheta} \exp(-x^2) x^2 dx \sin\vartheta d\vartheta,$$

On the other hand the charge-exchange induced momentum gain rate has to be evaluated from the following expression:

$$\langle Q_1^+(\vec{v_H})\rangle = \frac{2}{\sqrt{\pi}} n_p n_H m_p \sqrt{\frac{2KT_p}{m_p}} \vec{v_H} \int_0^\pi \int_0^\infty \sigma(v_{rel}) \cdot \tag{14}$$

$$\sqrt{x_H^2 + x^2 - 2x_H x \cos\vartheta} \exp(-x^2) x^2 dx \sin\vartheta_H d\vartheta_H$$

The net momentum exchange rate resulting from the sum of the two above expressions evaluates to (see Fahr, 2002):

$$\langle Q_1^+(\vec{x}_H)\rangle_H + \langle Q_1^-(\vec{x}_H)\rangle_H = \Pi[C(M_H) - \tag{15}$$

$$\sqrt{\pi}(-9g_1 M_H - 2g_2 M_H + 5\alpha g_1 M_H + 2\alpha g_1 M_H^3)],$$

where the function $C(M_H)$ is defined by:

$$C(M_H) = \frac{\pi}{8M_H^2} \int_0^\infty \Sigma(x) \exp(-(x^2 + M_H^2)) \cdot \tag{16}$$

$$[2xM_H \cosh(2xM_H) - \sinh(2xM_H)]\, dx.$$

Here the following notations have been used:

$\alpha = T_H/T_p$; $M_H^2 = \frac{\rho_H V_H}{\gamma P_H}$; $g_1 = \frac{1}{15}(1 + \frac{B}{\sqrt{\sigma_{rel}}})$; $g_2 = g_1 - \sqrt{\pi}\frac{2B}{\sqrt{\sigma_{rel}}}\frac{1}{X_{rel}}$

where:

$X_{rel} = \sqrt{\frac{64}{9\pi}(1+\alpha) + M_p^2}$

and the charge exchange cross section as taken from Maher and Tinsley (1977) is given by:

$\sigma_{rel} = [A + B\log(X_{rel})]^2$

Now it can be shown that expression (15) valid for moderate and small Mach numbers M_H further evaluates to:

$$Q_{1,sub} = \frac{7\Pi}{3} g_1 M_H - \tag{17}$$

$$\Pi\sqrt{\pi}(-9g_1 M_H - 2g_2 M_H + 5\alpha g_1 M_H + 2\alpha g_1 M_H^3).$$

whereas, compared to the above expression, in the up to now literature always the following expression was used:

$$Q_{1,super} = -\Pi M_H \sqrt{\frac{4\pi}{9}(1+\alpha) + \alpha M_H^2} \tag{18}$$

Figures 2 and 3 show the two different representations, $Q_{1,sub}$ and $Q_{1,super}$, respectively, of the momentum transfer rate between H-atoms and protons as function of the Mach number M_H, and they clearly demonstrate that at

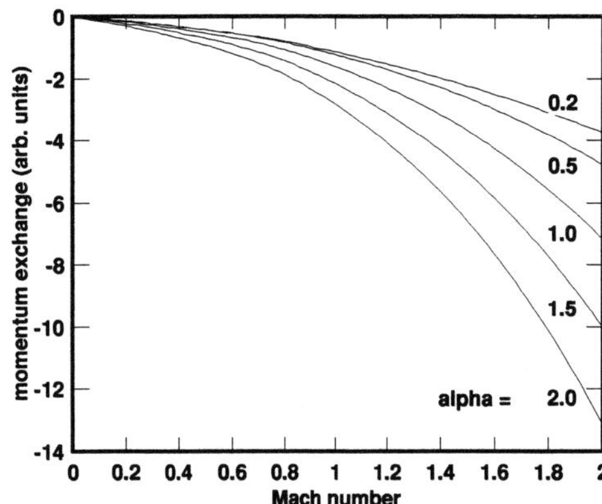

FIGURE 2. *Shown as function of the H-atom flow Mach number M_H is the net momentum exchange rate $Q_{1,sub}$ for various values of the temperature ratio $\alpha = T_H/T_p$.*

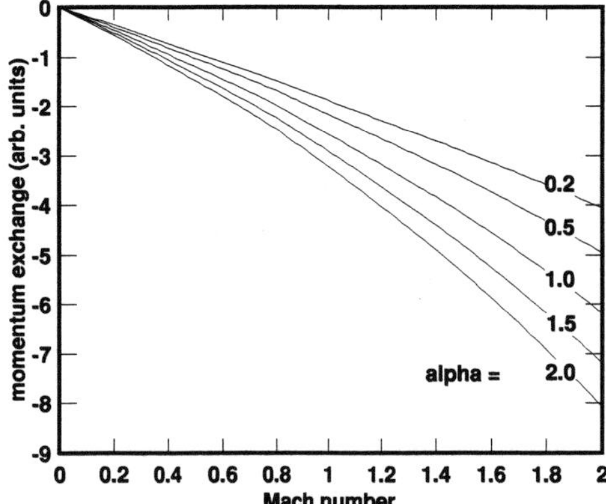

FIGURE 3. *Shown as function of the H-atom flow Mach number M_H is the net momentum exchange rate $Q_{1,super}$ for various values of the temperature ratio $\alpha = T_H/T_p$.*

equal Mach numbers M_H the revised rates $Q_{1,sub}$ always are larger than those $Q_{1,super}$ taken in conventional theories. This evidently means that at moderate and small Mach numbers the plasma-gas friction force was substantially underestimated.

THE CHARACTERISTIC EQUATION

With the expression given by (17) and with $\gamma 1 = \gamma - 1$ we now obtain instead of Equ.(10) the characteristic equation in the following form:

$$\frac{dV_H}{dz} = \frac{\sigma V_{rel} n_p \left(\Delta_p P_p - P_H - \frac{1}{2} C_0 V_H \gamma 1\right) + \gamma V_H Q_{1,sub}}{\gamma P_H - C_0 V_H} \tag{19}$$

The critical point condition for a vanishing numerator of Equ.(19) now writes in the following form:

$$1+\frac{1}{2}\gamma(\gamma-1) \geq \tag{20}$$

$$\frac{\gamma^2 \frac{2}{\pi}[\frac{7}{3}g_1 - \sqrt{\pi}(-9g_1 - 2g_2 + 5\alpha g_1 + \alpha g_1 \frac{\gamma^2}{2})]}{\sqrt{\frac{64}{9\pi}(1+\alpha) + M_p^2}}$$

As an example, adopting numerical values of the parameters after Fahr (2000), for $T_P = 20000$ K, $\alpha = 0.4$ and $V_H = 26$ km s^{-1}, $\gamma = 5/3$ we have $V_{rel} = 41.5$ km s^{-1}, $\sigma_{rel} = 6.4 \cdot 10^{-15}$ cm^{-2}, $M_P = 1.43$ and, when putting in values for g_1 and g_2 and the cross-sectional constants A and B given for the charge exchange reaction between H-atoms and protons by Maher and Tinsley (1977) one obtains the request that:

$$\frac{14}{9} = 1.56 \geq 1.39$$

clearly meaning that this requirement can be fulfilled and a continuous integration of the set of differential equations now is possible.

CONCLUSIONS

As evident from works by Baranov and Malama (1993) Baranov et al. (1998), Fahr (2000) or Fahr et al. (2000) the Mach numbers M_H of the relative flows between the LISM proton plasma and the LISM H-atom fluid in the heliosheath region in all cases are smaller than 2. Consequently and strictly speaking, in this region the specific momentum exchange term in the form of Equ.(17) instead of Equ. (18) had to be used. Since the effective momentum exchange rate described by this expression at identical thermodynamical conditions is greater than that described by the conventionally used term given by Equ.(18) one can presume that the adaption of the LISM H-atom flow to the nearly stagnating LISM proton plasma in this interface region ahead of the heliopause is operating faster, i.e. the LISM H-atoms will be much more effectively decelerated down to the strongly reduced LISM proton flow velocity.

The use of an inadequate momentum exchange term may thus lead to the erroneous claim for much higher LISM proton or H-atom densities to reach the same amount of reduced LISM H-atom bulk flow velocity. Hence more reliable quantitative interpretations of the H-atom flow through the heliosphere in terms of needed LISM parameters (see papers by Scherer and Fahr, 1996, Scherer et al., 1997, 1999, Izmodenov et al., 1997, Izmodenov, 2000) should be obtained with the application of the new term given in Equ. (17). LISM parameters claimed on the basis of these earlier interpretations may thus need substantial revisions.

REFERENCES

. Baranov, V.B., and Y.G.Malama, J.Geophys.Res.,157, 1993
. Baranov, V.B., and Y.G.Malama, Space Sci.Rev., 78, 305, 1996
. Baranov, V.B., Izmodenov, V.,and Y.G.Malama, J.Geophys.Res.,103, 9575,1998
. Bzowski, M., Fahr, H.J., Rucinski, D., and Scherer, H., Astron.Astrophys., 326, 396, 1997
. Bzowski, M., Fahr, H.J., and Rucinski, D., Astrophys.J., 544, 496, 2000
. Fahr, H.J., Solar Physics, 30, 193, 1973
. Fahr, H.J., Space Sci.Rev., 78, 199, 1996
. Fahr, H.J., Astrophys.Space Sci., 274, 35, 2000
. Fahr, H.J., Astrophys.Space Sci., 2002, in press
. Fahr, H.J., and Ripken, H.W., Astron.Astrophys., 139, 551, 1984
. Fahr, H.J., and Rucinski, D., Astron.Astrophys., 350, 1071, 1999
. Fahr, H.J., Kausch, T., and Scherer, H., Astron.Astrophys., 357, 268, 2000
. Holzer, T.E., J.Geophys. Res., 77, 5407, 1972
. Holzer, T.E., and Leer, E.G., Astrophys.Space Sci., 24, 335, 1973
. Izmodenov, V.V., Lallement, R., and Malama, Y.G., Astron.Astrophys., 317, 1 93, 1997
. Izmodenov, V.V., Astrophys.Space Sci., 274, 55, 2000
. Isenberg, P.A., J.Geophys.Res., 91, 9965, 1986
. Khabibrakhmanov, I.K., Summers, D., Zank, G.P., and Pauls, H.L., Astrophys. J., 469, 921, 1996
. Lee, M.A., in "Stellar Winds", ed.by R.Jokipii et al., pp.857-886, 1997
. Maher, L., and Tinsley, B., J.Geophys.Res., 82, 689, 1977
. McNutt, R.L., Lyon, J., and Goodrich, C.C., J.Geophys.Res., 103, 1905, 1998
. McNutt, R.L., Lyon, J. Goodrich, C.C., and Wildberg, M., Solar Wind Nine, CP471, 823, 1999
. Osterbart, R., and Fahr, H.J., Astron.Astrophys., 264, 260, 1992
. Pauls, H.L., and Zank, G.P., J.Geophys.Res., 101, 17081, 1996
. Pauls, H.L., Zank, G.P., and Williams, L.L., J.Geophys.Res., 100, 21595, 19 95
. Ripken, H.W., and Fahr, H.J., Astron.Astrophys., 122, 181, 1983
. Scherer, H., and Fahr, H.J., Astron.Astrophys., 309, 957, 1996
. Scherer, H., Fahr, H.J., and Clarke, J.T., Astron.Astrophys., 325, 745, 199 7
. Williams, L.L., Zank, G.P., and Matthaeus, W.H., J.Geophys.Res., 100, 17059 , 1995
. Whang, Y.C., Astrophys.J., 468, 947, 1996
. Whang, Y.C., J.Geophys.Res., 103, 17419, 1998
. Zank, G.P., Space Science Rev., 89, 413, 1999

Three-Dimensional MHD Simulation of Solar Wind Structure

D. A. Roberts*, M. L. Goldstein* and A. Deane†

*Code 692, NASA Goddard Space Flight Center, Greenbelt, MD 20771, USA
†Institute of Physical Science and Technology, University of Maryland, College Park, MD, USA

Abstract. We examine the 3-D structure of field lines in the expanding solar wind using an MHD code that allows the imposition of streams, a current sheet, flux tubes, microstreams, waves, and quasi-2-D turbulence. We find the natural development of closed field lines and lines with more complex geometry in the current sheet region. We show that microstreams cause an increased divergence of the field lines. The various initially strange looking heliospheric field lines we find have observational support in early work by K. Schatten and others, who found very similar structures using simple projection methods, as well in more recent work examining the nature of current sheet crossings. Our simulations exhibit most of the solar wind field configurations suggested by Schatten in his Solar Wind Two review.

1. INTRODUCTION

The solar wind magnetic field, as observed at one or a few points with interplanetary spacecraft, exhibits nearly impenetrable complexity, especially near the interplanetary current sheet. Even a large collection of spacecraft would not fully reveal the field configuration. Simulations provide an alternate avenue to discerning the three-dimensional structure, and comparisons with spacecraft measurements provide tests of whether or not we have a good representation. This paper presents a brief overview of recent results of our MHD simulations of the solar wind, emphasizing the global aspects that result from a variety of more-or-less realistic time-dependent boundary conditions. Of course, MHD simulations at attainable resolutions have inherent limitations. We only simulate large- to moderate scales (hours to days in the spacecraft frame), and the lack of kinetic effects precludes a proper treatment of, e.g., reconnection. Nonetheless, we believe that such features as global field lines and larger-scale fluctuation amplitudes are well represented here.

2. METHOD

Our MHD simulations use a spherical 3-D mesh and Flux-Corrected-Transport to deal with sharp gradients. The equations are given in *Goldstein, et al.* [this volume]; see also [1]. The code runs on parallel machines using MPI.

Typical simulation grid resolutions are 70^3 to 150^3 and none of the qualitative results here changed as the resolution increased. The inflow boundary is supersonic and superAlfvénic, and is where we impose waves; quasi-2-D fluctuations; stream shear layers; microstreams; a tilted, rotating current sheet; and pressure balanced structures. The outer radial boundary has outflow conditions, and the sides of the box have either outflow (direct extrapolation or continuation of the linear gradient) or periodic conditions, the latter typically being more stable when possible.

3. OBSERVATIONS

K. Schatten produced 2-D pictures of fields by assuming a stationary wind and using kinematic projections [2]. The resulting two-dimensional vector fields revealed kinks, loops, and reversals that suggested the relevance of a variety of possible models in different situations. Such structures have become more prevalent in the literature associated with complex heliospheric current sheet crossings and various solar ejecta (see [3] and references therein). Many studies have shown that heliospheric current sheet (HCS) crossings involve a rotation with a very small normal component to the magnetic field (see, e.g., [4]) Here we demonstrate that the typical current sheet field rotation and some of the complexity can result from very simple boundary conditions involving a tilted, rotating current sheet in an initially uniform flow. We also show that a "field line random walk" [5] is easily driven by microstreams as also discussed by *Suess et al.* [6].

The first figure shows the typical interplanetary field configuration near a current sheet, namely, the field does

CP679, *Solar Wind Ten: Proceedings of the Tenth International Solar Wind Conference*,
edited by M. Velli, R. Bruno, and F. Malara

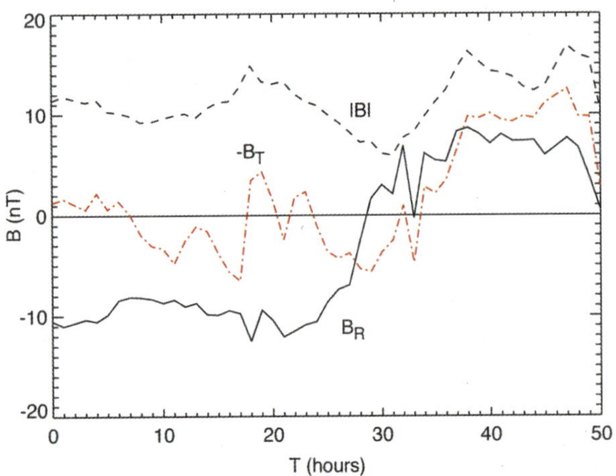

FIGURE 1. Helios 2 hourly-averaged observations of a sector crossing showing the common feature of the displacement in time of the zeros of the radial and tangential magnetic fields. This phenomenon is common in the Helios data, and it is consistent with the results reported by Smith [4].

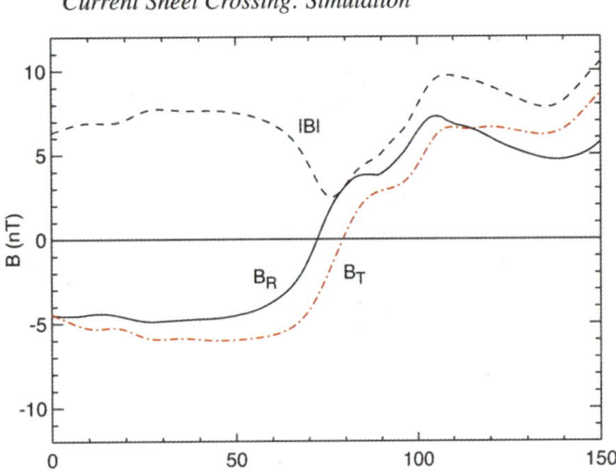

FIGURE 2. Time series from a point ("spacecraft") in our MHD simulation of a tilted, rotating current sheet. Note the similarity to Figure 1. The component normal the plane of the vector rotation is very small, again consistent with observations (e.g., [4]).

not simply change sign, but rather undergoes a rotation. A signature of this is the displacement in time of the zeros of the radial and tangential components, consistent with the typical case found by *Smith* [4].

4. SIMULATIONS

The second figure shows time series at a point for a typical current sheet crossing in a simulation of a rotating current sheet with no waves or shear imposed. Note the characteristic displacement of the zeros of the magnetic field. The normal component of the field (not shown) is very small, consisent with the near zero values often observed [4].

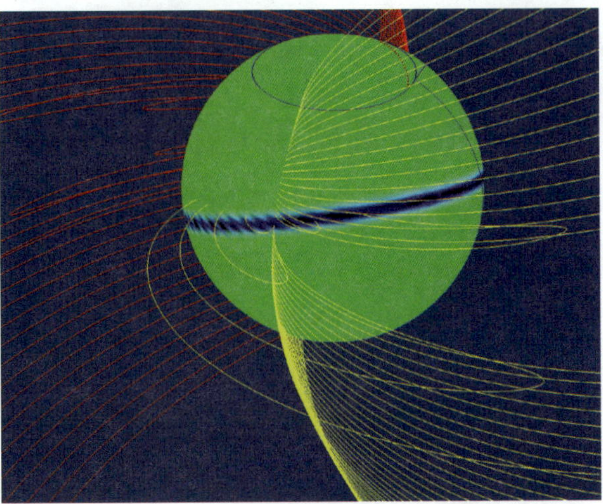

FIGURE 3. The three-dimensional configuration corresponding to Figure 1 from 0.5 to 2.5 AU with the inflow field magnitude with a central null shown as color contours, and the field shown by yellow and red lines. The natural way to have a transition with a rotation is to form loops that are closely confined to the current sheet region.

Figure 3 illustrates the 3-D configuration associated with Figure 2. We see the formation of loops when a current sheet field is imposed at the boundary. This is not due to reconnection inside the box, but rather to the inconsistency of the imposed ideal field reversal conditions with real MHD flows. In particular, the self-consistent fields cannot maintain the initial magnetic-kinetic pressure balance. leading to flows transverse to the current sheet. In this situation, the natural configuration seems to be one with a small normal component and loops that come from the decoupled phasing of the zeros of the radial and tangential magnetic fields as a function of radial distance from the Sun. We have considered many input boundary conditions, such as either imposing or extrapolating the radial and/or normal field from the box, all with the same result. Note that there are magnetic stresses in the resulting configuration such that the field tension tends to oppose the flow, but the stresses are small and simulations show the field is frozen in at least to tens of AU (as far as we have gone).

When waves or other perturbations are imposed, the current sheet fields can become more complex. Spiral fields that initially look like they violate div(**B**) = 0 can result from a variety of causes. In Figure 4 we show a case involving a slow time dependence in the boundary rotation rate (vaguely mimicking solar irregularities). Here the phasing between the radial and tangential mag-

FIGURE 4. The loops of the previous figure can metamorphose into more complex patterns when other features are added to the boundary conditions. Here, a slow linear increase in the effective solar rotation rate leads to spiral patters; these are also seen with other types of boundary variability such as the imposition of quasi-two dimensional fluctuations or waves. The images show normalized (by a function of r) magnetic field magnitude surfaces.

netic fields changes with time. This produces a near-null point where there is a small field component perpendicular to the plane embedded in a spiral. We also show an analytically generated spiral with div(**B**) = 0 that demonstrates that the fields can spiral as long as they vary appropriately in three dimensions. While our boundary conditions are somewhat artificial, we believe that real solar boundary conditions are very likely to generate complex fields of which those shown here are a simple subset. Moreover, since the field at the center of the spiral is a near-null, these field configurations may be one possible configuration associated with interplanetary "magnetic holes" (see [7] and references therein), although the scale of the structures here is large due to the slow changes that cause them.

A more complex case is illustrated in Figure 6 where waves and shear interact with current sheet loop fields to produce configurations that have many kinks and loops. These configurations may contribute to the complexity seen in various observations (e.g., [3] and references therein).

Field line wandering due to microstreams is illustrated in Fig. 7. The initial waves were nearly planar, but transverse wave vectors and perturbed fields are generated by small-scale sheer from corotating wind speed irregularities with amplitudes consistent with the variability in typical fast wind. Further studies will demonstrate the relative importance of this and other suggested origins for the field line meandering such as the motion of the field footpoints on the Sun. This origin of meandering was discussed previously [6], but we believe this is the first simulation of this effect. Note that the scale on which the random walk occurs is determined by the transverse scale of the pattern on the surface, and that field lines originating in relatively uniform, small region of the flow (not shown) tend to stay together.

5. SUMMARY

We have shown, through MHD simulation, that the interplanetary field can become quite complex despite simple boundary conditions. The rotation of the magnetic field across sector boundaries is very likely due to the formation of loop fields near the Sun. Many complex patterns of the magnetic field can occur, especially near the current sheet; this was expected, but now we can simulate fairly realistic cases in 3-D. Field lines can "random walk" due to irregularities in the flow. This may be more important than solar surface field wanderings, but this must be checked. Finally, spiral fields are possible configurations in MHD flows, and may be associated with magnetic holes.

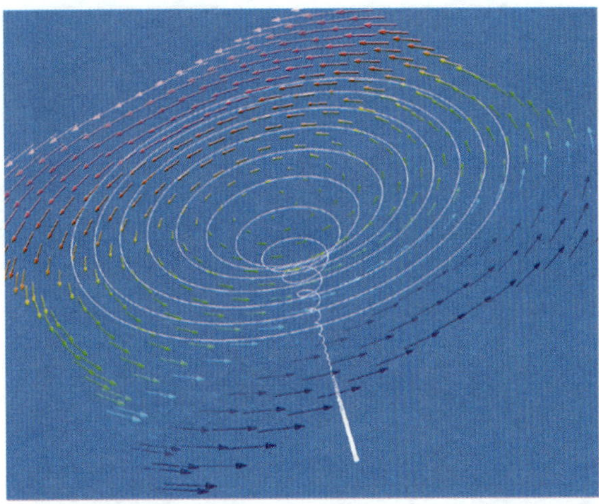

FIGURE 5. The nearly planar spiral pattern of the fields in the previous figure is consistent with a divergenceless field because of variations in the third dimension. Above we show field lines for an analytical solution of $\nabla \cdot (\mathbf{B}) = \mathbf{0}$. The "tail" of the spiral is flattened to be near the sector boundary in the previous simulation.

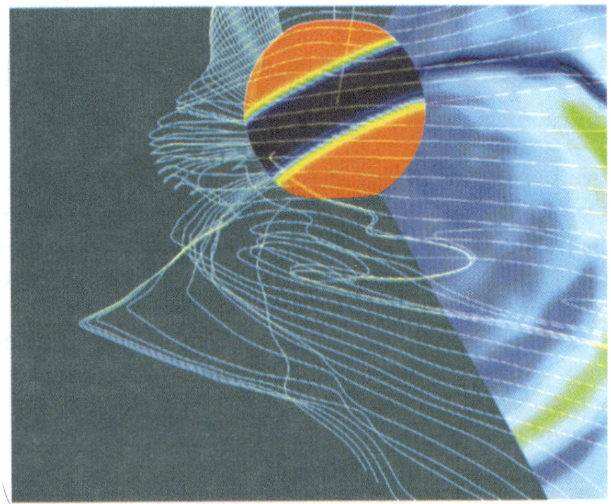

FIGURE 6. Waves, streams (slow surrounded by fast, as shown in the inflow contours), and a current sheet can lead to highly complex loops that do not completely return to the Sun. Lines outside the current sheet region simply wave gently in response to the imposed plane waves (nearly uniform in angular variables) because of the large transverse correlation length (see [8]).

ACKNOWLEDGMENTS

We would like to acknowledge the support of James Fischer, Manager for High Performance Computing at the Goddard Space Flight Center, for helping to arrange use of computing resources at the NASA Advanced Supercomputing Division.

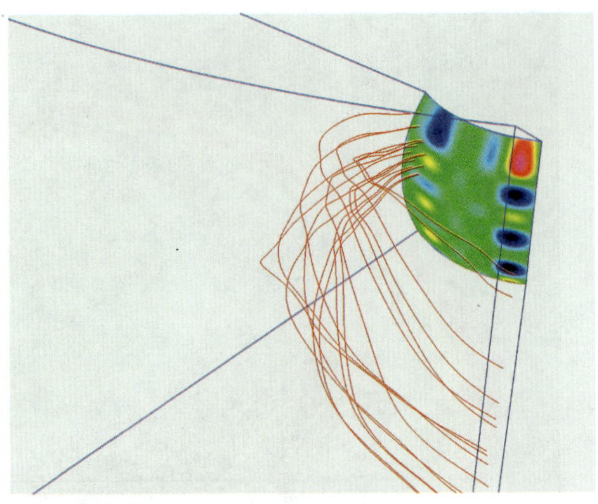

FIGURE 7. A stationary, corotating, radial flow pattern on the inflow boundary of the simulation leads to the wandering ("random walk") of (red) field lines shown above. The flow pattern is shown by the color contours on the input surface, with slow flow surrounded by fast flow.

REFERENCES

1. Goldstein, M. L., Roberts, D. A., Deane, A. E., Ghosh, S., and Wong, H. K., *J. Geophys. Res.*, **104**, 14,437–14,452 (1999).
2. Shatten, K. H., "Large-Scale Properties of the Interplanetary Magnetic Field," in *Solar Wind*, edited by C. P. Sonnett, P. J. Coleman, and J. M. Wilcox, Scientific and Technical Information Office, NASA, Washington, DC, 1972, vol. SP-308, pp. 65–92.
3. Crooker, N. U., Burton, M. E., Siscoe, G. L., Kahler, S. W., Gosling, J. T., and Smith, E. J., *J. Geophys. Res.*, **101**, 24,331–24,341 (1996).
4. Smith, E. J., *J. Geophys. Res.*, **106**, 15,819–15,831 (2001).
5. Jokipii, J. R., and Parker, E. N., *Phys. Rev. Lett.*, **21**, 44 (1969).
6. Suess, S. T., McComas, D. J., Bame, S. J., and Goldstein, B. E., *J. Geophys. Res.*, **100**, 12,261–12,273 (1995).
7. Zurbuchen, T. H., Hefti, S., Fisk, L. A., Gloeckler, G., Schwadron, N. A., Smith, C. W., Ness, N. F., Skoug, R. M., McComas, D. J., and Burlaga, L. F., *J. Geophys. Res.*, **106**, 16,001–16,010 (2001).
8. Matthaeus, W. H., Gray, P. C., D. H. Pontius, J., and Bieber, J. W., *Phys. Rev. Lett.*, **75**, 2136 (1995).

Evidences for Low-speed Streams from Small Coronal Hole

Tomoaki Ohmi*, Masayoshi Kojima*, Keiji Hayashi*, Atushi Yokobe*, Munetoshi Tokumaru*, Ken'ichi Fujiki* and Kazuyuki Hakamada†

*Solar-Terrestrial Environment Laboratory, Nagoya University, Toyokawa 442-8507, Japan
†Department of Natural Science and Mathematics, Chubu University, Kasugai 487-8501, Japan

Abstract. Using the WIND spacecraft data, we have studied properties of the locally bunched low-speed stream which was found in association with active regions by tomographic analysis of interplanetary scintillation observations. The source region of this low-speed stream was inferred to be a small coronal hole at vicinity of active regions by tracing potential magnetic field lines. The following WIND spacecraft observations support this inference of coronal hole origin. (1) Observed magnetic fields have properties of coronal hole origin: IMF polarity is the same as that of the coronal hole, and a magnetic neutral sheet was not observed in the stream. (2) Variations of velocity and density in the stream are as steady and uniform as those in typical high-speed wind. In addition, we have found that the relative He abundance N_α/N_p in this low-speed stream has 0.032, which is more than two times higher than that in low-speed wind in the heliospheric plasma sheet (0.013) and very near to that of high-speed wind from the large coronal hole (0.040). However, proton mass flux density and freeze-in temperature from the ratio of O^{7+}/O^{6+} are about 1.5 times higher than those in the coronal hole high-speed streams. These results imply that the low-speed steam is originated from a small coronal hole with high mass flux density and is strongly heated in the lower corona.

INTRODUCTION

It has been reported that a small coronal hole is a source of a low-speed stream [1, 2, 3, 4]. An inverse correlation between the solar wind speed and the expansion factor of a magnetic flux tube indicates that the lower speed wind flows out from a more largely diverging region [5, 6, 7]. According to the potential-field model, these large diverging regions are at the boundaries of the large coronal holes and in the small coronal holes [8, 9]. In the solar activity mimimum phase, comparison study between the spacecraft measurements and several coronal magnetic field models shows that the low-speed streams originate not only from coronal hole boundaries but also from small coronal holes at low latitudes [10].

Low-speed regions (≤ 350 km s^{-1}) are often observed bunched into a compact area associated with active regions [11]. In recent years, the tomographic analysis for the interplanetary scintillation (IPS) observations has been developed to obtain the solar wind velocity accurately with high spatial resolution [12, 13, 14]. By using this analysis, the origin of low-speed streams observed in association with active regions at solar activity minimum phase was investigated, and it was found that these low-speed streams originated from open field regions located at vicinity of one polarity side of active regions [15].

Our focus in this study is to confirm whether the origin of the the locally bunched low-speed stream found by [15] is a small coronal hole or not. This can be verified by *in situ* measurements when a spacecraft traverses these streams: If the compact low-speed streams are originated from a small coronal hole, a magnetic polarity change will not be observed when the spacecraft traverses this low-speed region. In addition, the measured flow parameters within these streams will be steadier and more uniform than those within other low-speed streams. In order to study in this manner, plasma and magnetic field data measured by the WIND spacecraft are investigated. In this paper, we report these results bliefly, and details of this study and discussions will be submitted to the *Journal of Geophysical Research* as a full paper.

LOW-SPEED WIND FROM SMALL OPEN FIELD REGION

Figure 1 shows the synoptic maps for the solar wind velocity and magnetic structure at the Sun for Carrington rotation number (CR) 1896 (May 16 to June 12 in 1995). The solar wind velocity distribution at 2.5 solar radii (R_S) (Figure 1a) is obtained from IPS tomographic analysis. The distribution of solar wind velocity is constructed on a projection surface with assumptions of radial and constant velocity. For comparing the coronal structures, the projection surface is set at 2.5 R_S. Figure 1b and 1c show the synoptic maps of photospheric magnetic fields ob-

CP679, *Solar Wind Ten: Proceedings of the Tenth International Solar Wind Conference*, edited by M. Velli, R. Bruno, and F. Malara

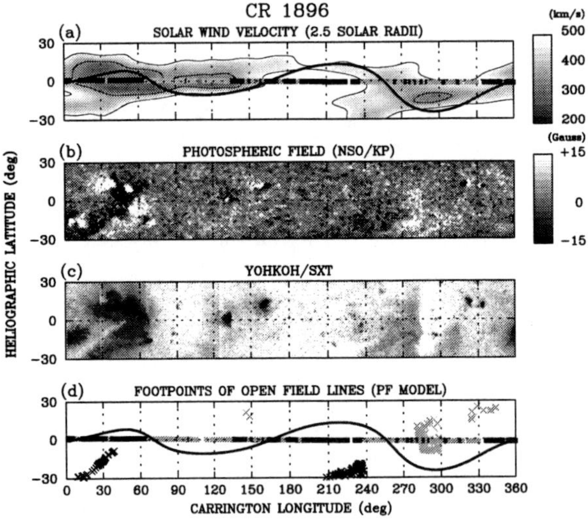

FIGURE 1. Synoptic maps for (a) solar wind velocity projected at 2.5 R_S, (b) photospheric magnetic field, (c) *Yohkoh* soft X-ray, and (d) open field footpoints at ths Sun for Carrington rotation number (CR) 1896 (May 16 to June 12 in 1995). The contour lines in the velocity map are for 500 and 350 km s^{-1}. The heavy solid line shown in (a) and (d) represents a magnetic neutral line derived from the potential-field model. The black and gray sections of the WIND trajectory are negative and positive magnetic polarities measured with the WIND, respectively.

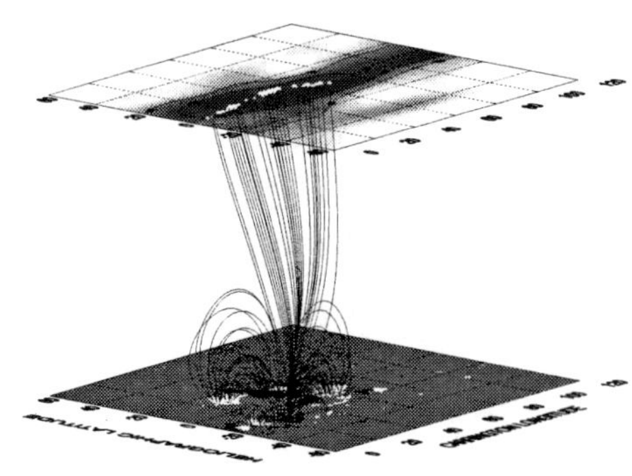

FIGURE 2. Magnetic potential-field lines from the photosphere to the source surface at 2.5 R_S for CR 1896. To avoid plotting too many magnetic field lines, the plotted open field lines are restricted to those which originate from the lower latitude boundaries of the open field regions. Closed loops in the corona are shown for photospheric magnetic field lines stronger than 15 G.

served at the National Solar Observatory at Kitt Peak (NSO/Kitt Peak) and soft X-ray images from the *Yohkoh* SXT observations. Figure 1d shows the magnetic field regions on the photosphere from which magnetic fields are open to interplanetary space. Open field regions and the magnetic neutral lines (MNL) are estimated with a potential-field analysis developed by [6] from the synoptic data of photospheric magnetic field observed at the NSO/Kitt Peak. Black and gray points represent the open field regions with negative and positive polarities, respectively. To verify this estimations, the polarities of interplanetary magnetic field (IMF) measured with the WIND were compared. A measured magnetic polarities were mapped back to the surface of 2.5 R_S by using a constant speed method with measured solar wind speed.

At longitudes of 10°–60° in Figure 1, there are (a) low-speed regions ($\leq$ 350 km s^{-1}), (b) strong complex magnetic structures, and (c) narrow low-intensity soft X-ray region. This narrow low-intensty structure agree with the computed open field regions. Figure 1d shows good agreement between the observed and calculated magnetic polarities, and magnetic polarity change was not observed in this regions.

In order to investigate the magnetic field structure in the regions where the low-speed winds were observed, we calculated the potential-field lines in the corona (Figure 2), and found that the open field lines fanned out largely from the narrow open field regions into the interplanetary space where the low speed streams are observed.

PROPERTIES OF SOLAR WIND FROM SMALL CORONAL HOLE

By tracing the potential magnetic field lines, the source region of the compact low-speed stream is inferred to be a small open field region. (The term of "compact" means "bunched into a compact area" hereafter.) In general the coronal hole and solar wind from there have following typical properties: Coronal hole is the regions of low temperature, low density and unipolar open magnetic field [16]. Variations of velocity and density in the streams from the coronal hole are steady and uniform [17]. The helium to proton density ratio N_α/N_P in the coronal hole stream is relatively higher than that in the heliospheric plasma sheet with high-density low-speed streamer [17, 18]. To verify weather the compact low-speed streams are originated from the small coronal hole, we have investigated the plasma data measured with the WIND spacecraft.

Figure 3 shows the IPS velocity map and the plasma parameters from the WIND observations. To compare with the spacecraft measurements, the projection surface of IPS tomography is set at 1 AU. The compact low-speed region at 2.5 R_S which is located around the 45° longitude in CR 1896 (Figure 1) is moved around the 300°–330° longitude of CR 1897 (June 12 - July 10, 1995) at 1 AU. The IPS velocities extracted along the

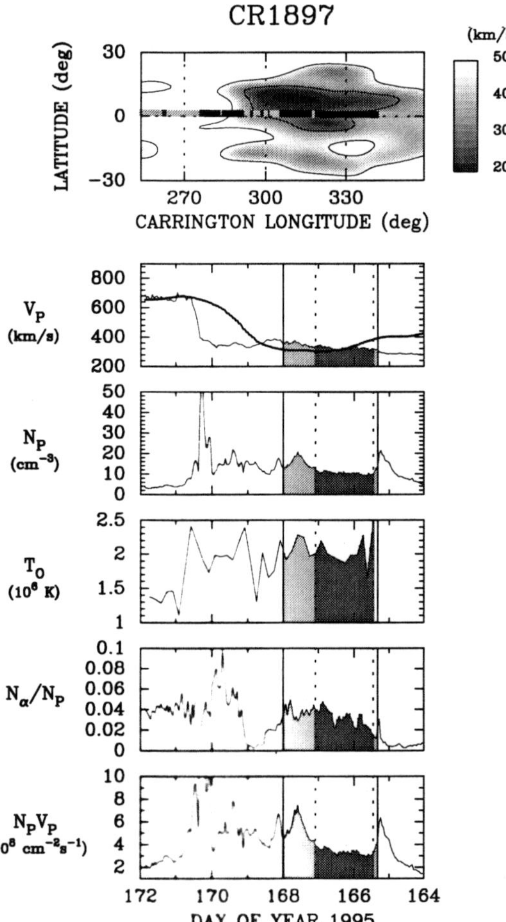

FIGURE 3. Comparison the IPS velocity with the solar wind parameters measured by the WIND spacecraft for CR 1897 (June 12 - July 10, 1995). IPS velocity map is made at 1 AU. The IPS velocities extracted along the WIND trajectory are shown as a heavy solid line in the velocity plot of WIND data.

WIND trajectory are shown as a heavy solid line in the velocity plot of WIND data.

In the gray shaded period between straight solid lines (7 UT on DOY 165 – 1 UT on DOY 168) in the WIND data plot, the compact low-speed streams are observed with the unipolar magnetic field. The velocities of compact low-speed streams obtained from the IPS observations are in good agreement with measured bulk velocities by the WIND spacecraft. It is very interesting of this compact low-speed streams that the density variation is steadier and more uniform than that of other low-speed streams around the heliospheric plasma sheet (HPS). In the the dark gray shaded period, the steady and uniform streams were observed from the near center of the small equatorial coronal hole (seCH). Another interesting is that the He abundance of N_α/N_P was as high as 0.032. This is slightly lower than that in the large equatorial

TABLE 1. Averaged properties of solar wind for CR 1897. Here the velocities are in km s^{-1}, the densities in cm^{-3}, the temperatures in 10^6 K, and the fluxes in 10^8 cm^{2}s^{-1}.

	Slow (HPS)	Slow (seCH)	Fast (leCH)
V_p	343±22	323±9	665±13
N_p	11.8±3.9	10.2±0.7	3.8±0.6
T_O	1.92±0.30	1.99±0.19	1.38±0.07
N_α/N_P	0.013±0.013	0.031±0.008	0.040±0.004
N_PV_P	4.41±1.62	3.30±0.24	2.49±0.34

coronal hole (leCH) streams (N_α/N_P=0.040) but higher than that in the HPS (N_α/N_P=0.013). The freeze-in temperature from the charge states of O^{7+}/O^{6+} is about 2.0 MK, and the proton mass flux density is 1.5 times as large as that in the leCH streams.

SUMMARY AND CONCLUSIONS

In order to study whether the compact low-speed streams ($\leq$ 350 km s^{-1}) were originated from small equatorial coronal hole, the solar wind plasma and magnetic field data obtained from the WIND measurements were investigated. As a result, we have found the evidences that they have properties of coronal hole origin: The IMF polarity is the same as that of the coronal hole, and the variations of velocity and density in the stream are as steady and uniform as those in typical coronal hole streams. The averaged properties of solar wind found in this study are summarized in Tabale 1. In this table, the "Slow" wind is defined as the streams with velocities less equal 400 km s^{-1}, while the velocities of "Fast" wind is greater equal 600 km s^{-1}, and the data at the compressive interacting regions are excluded.

The helium abundance in the seCH streams was steady and lower than that in the leCH streams. The similar results to ours have been obtained by Neugebauer [19] that the averaged value of N_α/N_P correlates with the average speed of the flow. The temperature in the seCH estimated from the charge states of O^{7+}/O^{6+} was about 2.0 MK, which is higher than that in the typical coronal holes (see also [20]), and the proton flux density was higher than that in the fast solar wind. It is well known that heat addition in the subsonic region in the lower corona increases the mass flux and works against the solar wind acceleration [21, 9]. Our results therefore suggest that the solar wind originated from the small coronal hole is strongly heated in the lower corona and emanates the low-speed streams.

ACKNOWLEDGMENTS

We would like to thank the National Solar Observatory at Kitt Peak for use of their synoptic magnetic field data. The NSO/Kitt Peak data used here are produced cooperatively by NSF/NOAO, NASA/GSFC, and NOAA/SEL. We would like to thank the use of WIND plasma and magnetic field data. Hourly averages of the WIND data were obtained from the World Wide Web through NDADS (NSSDC (National Space Science Data Center) Data Archives and Distribution System) and WIND-SWE Data Page at MIT. We would like to thank *Yohkoh* team for the soft X-ray solar images. We wish to acknowledge engineering support of our IPS observations from Y. Ishida, K. Maruyama and N. Yoshimi. This work was partially supported by the Scientific Research Fund of the Japan Society for the Promotion of Sience (grant 12440130).

REFERENCES

1. Nolte, J. T., Krieger, A. S., Timothy, A. F., Gold, R. E., Roelof, E. C., Vaiana, G., Lazarus, A. J., Sullivan, J. D., and McIntosh, P. S., *Solar Phys.*, **46**, 303 (1976).
2. Neugebauer, M., *Space Sci. Rev.*, **70**, 319 (1994).
3. Ohmi, T., Kojima, M., Yokobe, A., Tokumaru, M., Fujiki, K., and Hakamada, K., *J. Geophys. Res.*, **106**, 24,923 (2001).
4. Arge, C. N., Harvey, K. L., Hudson, H. S., and Kahler, S. W., this volume (2002).
5. Wang, Y.-M., and Sheeley Jr., N. R., *Astrophys. J.*, **355**, 726 (1990).
6. Hakamada, K., and Kojima, M., *Solar Phys.*, **187**, 115 (1999).
7. Hakamada, K., Kojima, M., Tokumaru, M., Ohmi, T., Yokobe, A., and Fujiki, K., *Solar Phys.*, **207**, 173 (2002).
8. Levine, R. H., Altschuler, M. D., Harvey, J. W., and Jackson, B. V., *Astrophys. J.*, **215**, 636 (1977).
9. Wang, Y.-M., *Astrophys. J.*, **437**, L67 (1994).
10. Neugebauer, M., et al., *J. Geophys. Res.*, **103**, 14,587 (1998).
11. Watanabe, H., Kojima, M., Kozuka, Y., and Yamauchi, Y., in *Solar Wind Eight*, edited by D. Winterhalter et al., AIP Conference Proceedings 382, New York, 1995, p.117.
12. Asai, K., Kojima, M., Manoharan, P. K., Jackson, B. V., Hick, P. L., Tokumaru, M. and Yokobe, A., *J. Geophys. Res.*, **103**, 1991 (1998).
13. Jackson, B. V., Hick, P. L., Kojima, M., and Yokobe, A., *J. Geophys. Res.*, **103**, 12,049 (1998).
14. Kojima, M., Tokumaru, M., Watanabe, H., Yokobe, A., Asai, K., Jackson, B. V., and Hick, P. L., *J. Geophys. Res.*, **103**, 1981 (1998).
15. Kojima, M., Fujiki, K., Ohmi, T., Tokumaru, M., Yokobe, A., and Hakamada, K., *J. Geophys. Res.*, **104**, 16,993 (1999).
16. Zirker, J. B., ed., *Coronal Holes and High Speed Solar Wind Streams*, Colorado Associated University Press, Boulder, 1977.
17. Bame, S. J., Asbridge, J. R., Feldman, W. C., and Gosling, J. T., *J. Geophys. Res.*, **82**, 1487 (1977).
18. Gosling, J. T., Borrini, G., Asbridge, J. R., Bame, S. J., Feldman, W. C., and Hansen, R. T., *J. Geophys. Res.*, **86**, 5438 (1981).
19. Neugebauer, M., in Solar Wind Seven, edited by E. Marsh and R. Schwenn, Pergamon Press, Oxford, 1992, p.69.
20. Geiss, J., Gloeckler, G., and von Steiger, R., *Space Sci. Rev.*, **72**, 49 (1995).
21. Leer, E., and Holzer, T. E., *J. Geophys. Res.*, **85**, 4681 (1980).

Solar Cycle Dependence of High-Latitude Solar Wind

K. Fujiki*, M. Kojima*, M. Tokumaru*, T. Ohmi*, A. Yokobe* and K. Hayashi*

**Solar-Terrestrial Environment Laboratory, Nagoya University, 3-13 Honohara, Toyokawa, Aichi Japan*

Abstract.
How has the high-latitude solar wind velocity changed over the solar activity cycle? We analyzed interplanetary scintillation data during the years 1985-2001 (excluding the few years around solar maximum) and obtained the following results: (1) the solar wind in the polar region did not change its speed even during the phases of rising and declining solar activity, (2) the N-S asymmetry of the high-latitude solar wind speed is a stable structure from 1987 to 1998, (3) the latitudinal velocity gradient at high latitude becomes steeper with increasing solar activity.

INTRODUCTION

It is well known that the velocity structure of the solar wind changes drastically over the course of the solar cycle. In solar minimum, the solar wind is generally characterized by two velocity components (bimodal), fast and slow solar wind (Woch et al., 1997). Ulysses found in its first rapid latitude scan that the high-latitude solar wind had a speed in range of 700-800 km/s and that there was a small but noticeable gradual increase of the solar wind toward higher latitudes. In the latitude scan the solar wind velocity at the northern high latitude was faster than that at the southern high latitude (Goldstein et.al.,1996). As the solar activity increases the slow solar wind region extends to higher latitudes. Then fast solar wind region is greatly reduced in solar maximum (McComas et al., 2000).

Ground-based interplanetary scintillation (IPS) observations provide valuable estimates of the three-dimensional velocity structure of the inner heliosphere on a continuous basis. We use the interplanetary scintillation observations of natural radio sources obtained with the Solar-Terrestrial Environment Laboratory system at Nagoya University in Japan. In this work, we investigate the variation of the solar wind structure through solar cycle using IPS data.

IPS OBSERVATION

The IPS observations at a frequency of 327 MHz were obtained using four remote stations. Observations are obtained 8 hours a day from each station except during winter (mid-December to March) when heavy snow lies in the antenna reflector. Velocity maps are obtained in each Carrington rotation. The IPS observation is always affected by line-of- sight bias. To reduce this effect, we apply the computer-assisted tomography (CAT) technique (Kojima et al., 1998).

Each V-map is obtained using IPS data for three solar rotations to improve image quality. We average over the longitudinal structure in this study because of our focus on the latitudinal variations.

SOLAR WIND STRUCTURE THROUGH THE SOLAR CYCLE

Figure 1 is stack map in years from 1985 to 2001. It is clearly seen that solar wind structure is bimodal except for a few years around solar maximum. In solar maximum, the equatorial slow wind region extends to higher latitudes and the area of the fast solar wind is greatly reduced.

Velocity at the pole

Figure 2 shows the variation of velocity of the fast solar wind around the north pole. We calculated a mean velocity for the latitude range of 80°-90° in the V-map derived using the CAT analysis. The CAT procedure sometimes underestimates the velocity around the poles because the data coverage (number of line-of-sight) around the pole is insufficient. To remove this effect, we corrected the mean velocity around the pole by model calculation. The mean velocity is 789±68 km/s which agrees well with the velocity estimated by extrapolating the Ulysses observations to the polar region. Almost all data points scatter in the one σ (68 km/s) belt and there is no systematic velocity change with change in the area of

CP679, *Solar Wind Ten: Proceedings of the Tenth International Solar Wind Conference*,
edited by M. Velli, R. Bruno, and F. Malara

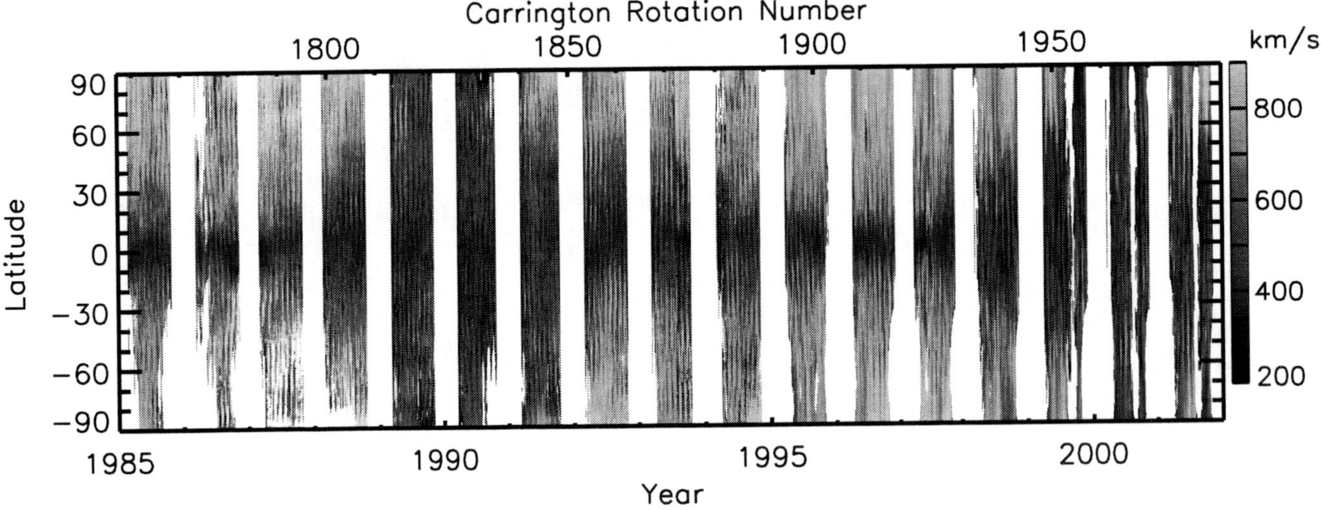

FIGURE 1. Stack map from 1985 to 2001. White regions are data gaps.

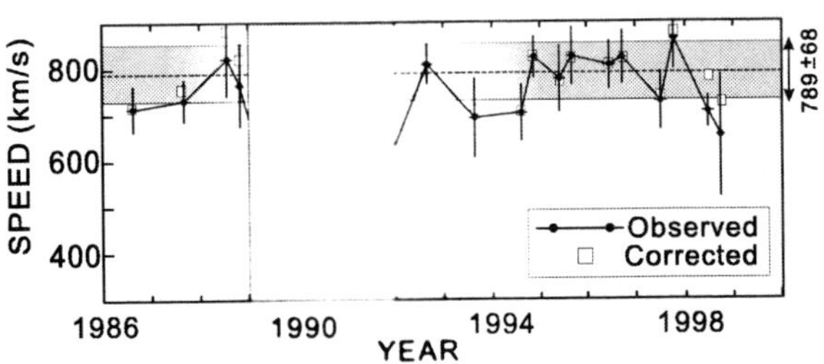

FIGURE 2. Solar wind speed at the northern high latitudes. The latitude range of N80°-N90° in each V-map are averaged. Shaded area is the solar maximum when the fast solar wind disappeared around the poles.

high-speed flow.

N-S asymmetry

We compared the solar wind velocities between the northern and southern high latitudes (Fig. 3). For this comparison we used mean velocity over the latitudes 70°-80° where the data coverage is better than the latitude range of 80°-90°. Then the mean velocity is averaged over each year. As a result, we found that the solar wind velocity at northern high latitudes is usually higher than that at southern high latitudes. Ulysses detected hemispherical differences of 13 km/s at a latitude of 8 0° (Goldstein et al., 1996). Ulysses sampled a velocity along its trajectory in latitude and longitude, while our analyses were made by averaging velocities over all longitudes and in the latitude range 70°-80°.

Figure 3 and Figure 1 (bin of 1991) also show that the recovery of the fast wind around the north pole precedes that around the south pole by several months after the

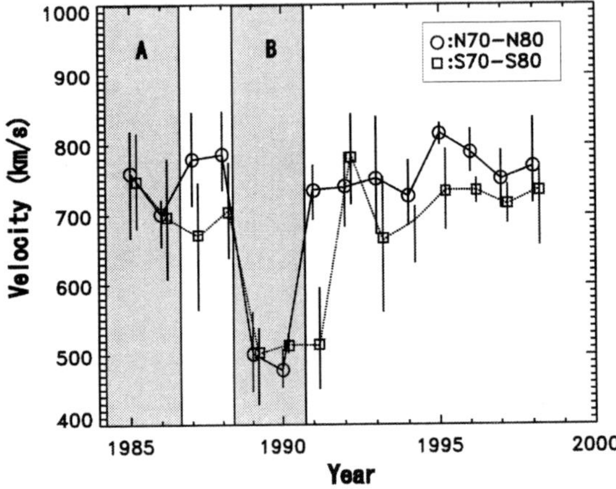

FIGURE 3. N-S asymmetry of the solar wind. The latitude range of 70°-80° in each V-map are averaged. Circle and square show north and south hemisphere, respectively.Shaded area A is a period when the data coverage of high latitudes is insufficient. Shaded area B is the solar maximum.

22nd solar maximum. A similar trend is observed in the 23rd solar maximum (Fujiki et al.in this book).

Velocity gradient in the fast wind region

We analyzed the latitudinal velocity gradient of the fast solar wind (Fig. 4). At high latitudes, there are several fine structures in the velocity map derived from the CAT analysis, such as a lower speed island and abnormally high-speed regions. In order to remove the lower speed island from the data set, we first checked the velocity distribution at each latitude along a meridian, and then search the mode of the velocity of high-speed wind. As a result, the velocity gradient is low at the solar mini-

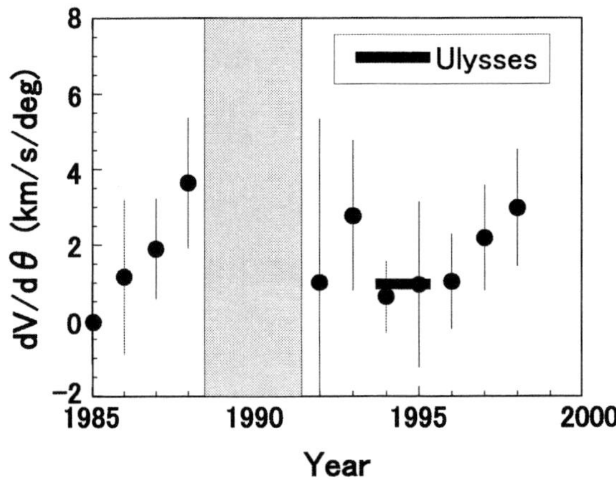

FIGURE 4. Velocity gradient in the high speed region. Thick line shows Ulysses result measured in the first fast latitude scan.

mum and increases with solar activity.

SUMMARY

We analyzed the variation of the high-latitude solar wind structure through the solar cycle (1985-2001). The solar wind structure derived from IPS observations, when augmented with the successful CAT technique, agrees well with Ulysses measurements.

Analysis method used in this study is not applicable during solar maximum because the solar wind structure becomes quite complex in this period. First results of a study relating to solar maximum are reported in Fujiki et al. (2002).

First results of a study relating to solar maximum are reported in Fujiki et al.(2002).

ACKNOWLEDGMENTS

We would like to thank to the Japan Science Society which supported financially for this presentation in Pisa, Italy.

REFERENCES

1. Goldstein, B. E., M. Neugebauer, J. L. Phillips, S. Bame, J. T. Gosling, D. McComas, Y.-M. Wang, N. R. Sheeley, and S. T. Suess, Ulysses plasma parameter: latitudinal, radial, and temporal variations, *Astron. Astrophys.*, **316**, 296-303, 1996
2. Fujiki, K., M. Kojima, M. Tokumaru, A. Yokobe, T. Ohmi, K. Hayashi, D. J. McComas, and H. A. Elliott, Solar wind velocity structure around the solar maximum *Proc. of SW10*(this book)
3. Kojima, M., M. Tokumaru, H. Watanabe, A. Yokobe, K. Asai, B. V. Jackson, and P. L. Hick, Heliospheric tomography using interplanetary scintillation observations, 2, Latitude and heliospheric distance dependence of solar wind structure at 0.1-1 AU, *J. Geophys Res.*, **103**, 1981-1989, 1998
4. McComas D. J., B. L. Barraclough, H. O. Funsen, J. T. Gosling, E. Santiago-Muñoz, R. M. Skoug, B. E. Goldstein, M. Neugebauer, P. Reley, and A. Balogh, Solar wind observation over Ulusses' first full poler orbit, *J. Grophys. Res.*, **105**, 10419-10422, 2002.
5. Woch, J., W. I. Axford, U. Mall, B. Wilken, S. Livi, J. Geiss, G. Gloecker, and R. J. Forsyth, SWICS/Ulysses observation: The three-dimensional structure of the heliosphere in the declining/minimum phase of the solar cycle, *Geophys. Res. Lett.*, **24**, 2885-2888, 1997

Basic Results of MHD Tomography Analysis Method for IPS Observation

Keiji Hayashi*, Ken'ichi Fujiki*, Masayoshi Kojima* and Munetoshi Tokumaru*

*Solar-Terrestrial Environment Laboratory, Nagoya University, Toyokawa,442-8507,Japan

Abstract. We present the tomography analysis method using MHD simulation code to derive three-dimensional solar wind structures from IPS observations. By incorporating MHD simulation, this new tomography method, hereafter called MHD tomography, can treat the nonlinear MHD process in solar wind and will improve the result of the reconstruction of global solar wind structure from IPS measurement data. The practical calculation is done as iteration procedure. At the first step of iteration, MHD simulation is carried out with given provisional boundary conditions, and the numerical three-dimensional solar wind is made. At the next step, the IPS observations are simulated in this three-dimensional solar wind. The discrepancy between the actual IPS measurement and the numerically reproduced ones are traced back along streamlines onto the inflow boundary surface. Then, the velocity distribution on the boundary surface is modified so that the differences may be reduced. The modified boundary distribution is used in the next iteration, and the iteration is continued until the three-dimensional solar wind structure matching the LOS-integrated IPS observational data is obtained. The basic result of this analysis is shown, and the possible applications of this MHD-based and observation-based solar wind are demonstrated.

INTRODUCTION

The observation of the interplanetary scintillation (IPS) is, at present, a unique ground-based observation technique for the solar wind. Solar-Terrestrial Environment Laboratory (STEL) at Nagoya University, Japan has the multi-site observational facilities for IPS that can observe the velocity and density at $r = 0.2 \sim 1$ AU and makes more than 400 observations during one month. The solar wind quantities derived from the IPS observation is, however, line-of-sight (LOS) integrated properties and do not directly represent the quantities at the particular positions.

Jackson et al. [1] and Kojima et al. [2] developed the inversion algorithm called the computer assisted tomography (CAT) and succeeded in reconstructing the spatial distribution of the solar wind from the LOS-integrated IPS observational data. However, the nonlinear process of the hydrodynamics (HD) nor magnetohydrodynamics (MHD) had not been treated exactly in the CAT analysis.

Recently, we have developed the MHD tomography analysis method, in which the MHD simulation code is embedded [3] (hereafter Paper I). The MHD code in this analysis method enables the CAT analysis to treat the MHD process of the solar wind and reconstruct the MHD-based three-dimensional structure of the solar wind from the IPS observational velocity data. At the same time, the CAT analysis allows the MHD code to deal with the situation detected by IPS observation. We anticipate that this mutual relation between the observation analysis and the numerical simulation will enhance the study of the solar wind. In this paper, the basic result of MHD tomography is briefly shown, and the simulation of disturbance propagation and the extrapolation of the solar wind at $r > 1$ AU are demonstrated as the examples of the applications of the MHD-based and IPS-observation-based solar wind.

METHOD

Figure 1 is the flowchart of MHD tomography analysis to find the solution of the three-dimensional solar wind structure best matching IPS observational velocity data, with the thick arrows showing the recursion part. The algorithm is based on the analysis by Kojima et al. [2], but the additional procedures are introduced in order to carry out the MHD simulation. The details are described in Paper I.

In this presented analysis, the inner boundary of the MHD simulation is the 50 solar radii sphere. This choice is determined from the two factors. The one is the domain of sensitivity of IPS observation ($0.2AU < r < 1AU$ for STELab IPS observation at 327 Mhz), and the other is the requirement from the MHD simulation method used in this analysis that the solar wind must be super-Alfvénic on the inner boundary and in the domain of computation. We address here that we can choose the inner boundary more close to the Sun, for example, 10 or 20 R_s sphere,

CP679, *Solar Wind Ten: Proceedings of the Tenth International Solar Wind Conference*,
edited by M. Velli, R. Bruno, and F. Malara

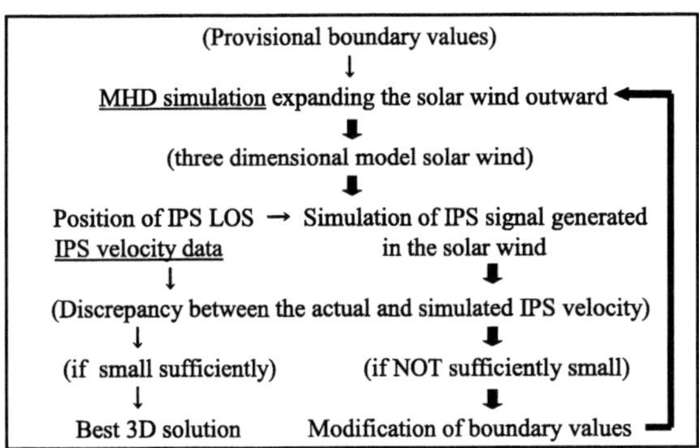

FIGURE 1. Flowchart of MHD tomography analysis.

in case that solar wind $r < 50R_s$ are to be reconstructed.

The MHD code embedded in this analysis method derives three-dimensional global solar wind variables at $r > 50R_s$ using the distribution of solar wind variables at $50R_s$ as the initial values. Since the boundary velocity at $50R_s$ is modified by referring the observational IPS velocity data, the IPS data contribute as the constraints of the conditions in the MHD simulation. To begin the MHD simulation, the other variables, density, temperature and magnetic field at $50R_s$ must be prepared. We derived the empirical functions $n(V)$ and $T(V)$ from the Helios data to determine the density and temperature on the inner boundary at $50R_s$. We used the observational photospheric magnetic field data and calculated the magnetic field at $50R_s$ by means of the potential field model. Therefore, all variables at $50R_s$ are determined from the observational data.

Writing the time-dependent MHD equations in the matrix-vector form,

$$\frac{\partial \vec{W}}{\partial t} = -\mathbf{M}_r \frac{\partial \vec{W}}{\partial r} - \mathbf{M}_\theta \frac{1}{r}\frac{\partial \vec{W}}{\partial \theta} - \mathbf{M}_\phi \frac{1}{r\sin\theta}\frac{\partial \vec{W}}{\partial \phi} + \vec{S}, \quad (1)$$

the equations the MHD code are to solve can be written as

$$\frac{\partial \vec{W}}{\partial r} = \mathbf{M}_r^{-1}\left(-\mathbf{M}_\theta \frac{1}{r}\frac{\partial \vec{W}}{\partial \theta} - \mathbf{M}_\phi \frac{1}{r\sin\theta}\frac{\partial \vec{W}}{\partial \phi} + \vec{S}\right) \quad (2)$$

because the solar wind to be simulated ($> 50R_s$) is always super-Alfvenic and the solar wind quantities can be calculated using the solar wind parameters at the upwind. We used the two-step centered-differencing method to increment the radius and derive the solar wind variables at $r > 50R_s$. The practical calculation is done using conservative variables.

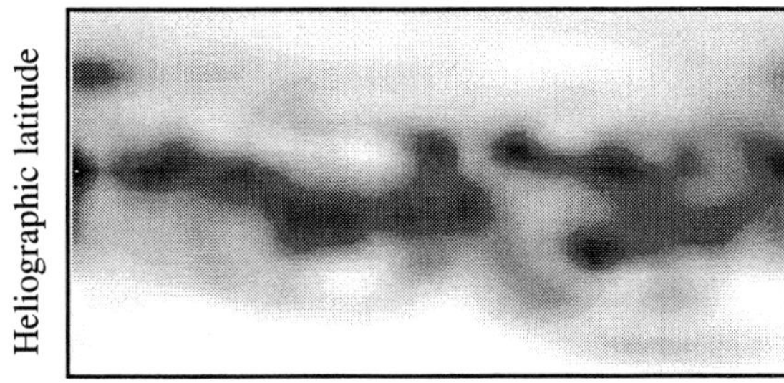

FIGURE 2. The boundary velocity map of the quiet solar wind reconstructed by the MHD tomography analysis. The darker gray represents the slower wind, and the brighter does the faster. The range of derived velocity is approximately from 200 to 800 km/s.

BASIC RESULT AND APPLICATIONS

The gray scale of Figure 2 shows the solar wind velocity at $50R_s$ derived by the MHD tomography analysis for the period of Carrington rotation number 1939, as an example. That is, the MHD-based three-dimensional solar wind with this boundary distribution is best matching the LOS integrated IPS observation data made during this period. The other variables, density, temperature and magnetic field are also calculated together with plasma flow speed from 50 R_s to 1 AU for the CAT analysis. As shown later, the MHD code can expand the steady solar wind from 1 AU to several AUs. In Paper I, we showed good positive correlations of density, temperature, flow speed and magnetic field between the derived three-dimensional solar wind and the direct measurement by spacecrafts, Ulysses, IMP8, ACE, WIND and ISEE3.

Simulation of interplanetary disturbance propagation for space weather prediction

The MHD simulation of interplanetary disturbance propagation is one of the most suitable applications where the IPS-MHD tomography analysis shows its ability. Because the time-evolution of the interplanetary disturbance depends on not only the properties of the trigger event like flare and coronal mass ejection (CME) but also the structure of the background quiet solar wind, to know the structure of the quiet solar wind in advance will be greatly helpful for the space weather prediction.

The time-dependent MHD simulation of the disturbance propagation can be initiated by giving the mimic CME-associated material injection at the inner numerical boundary surface into the quiet solar wind and by solving the response of the system. In our case, the IPS-based

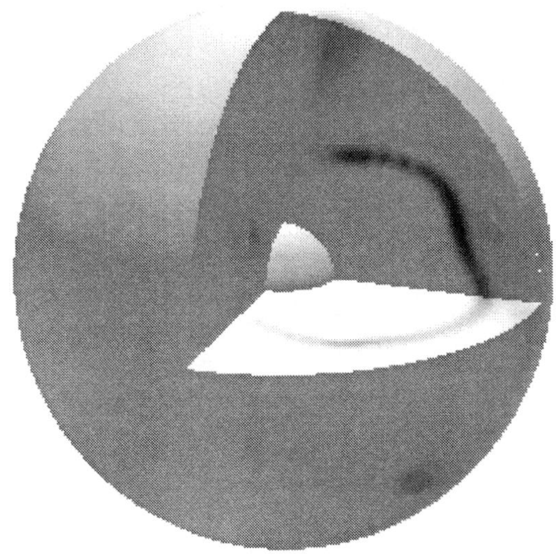

FIGURE 3. Cross sections view of the density enhancements propagating in the interplanetary space. The numerical perturbation was given on the inner boundary sphere at $50R_S$.

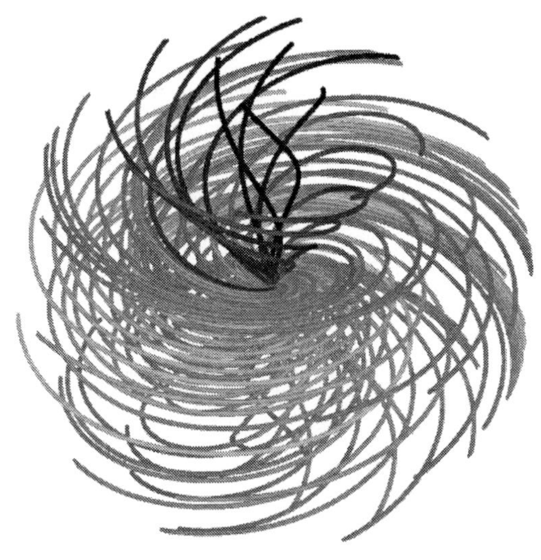

FIGURE 4. Line of interplanetary magnetic field or streamlines viewed in the rotating frame at $50R_s < r < 6$ AU. The thickness of lines represents the local flow speed.

and MHD-based steady solar wind can be used. Figure 3 shows the cross-section view of the solar wind plasma density enhancement of the interplanetary disturbance simulated in this way. The shape of disturbance shock front depends on the ambient background solar wind, which is now obtained from the IPS observational data through the tomography analysis. In the demonstrated case, the artificial mass injection was symmetry, and the bent shape of the shock, in other word, the dependence of shock upon directions were obtained.

Extrapolation of solar wind properties at $r > 1$ AU

The MHD simulation code embedded in the MHD tomography analysis can expand the solar wind outward using the results of our MHD tomography analysis for $r < 1$ AU and make extrapolations of the solar wind quantities at $r > 1$ AU. This extrapolation will be greatly helpful for the study of the distant solar wind from the Sun, for example, the interactions with planets and interstellar medium. Because the solar wind evolution along distance depends both its longitudinal and latitudinal variations, our method that can reconstruct the global solar wind structure is a unique method to make observation-based extrapolation of the global solar wind at the regions distant from the Sun.

Figure 4 shows the streamlines within 6 AUs where the streamlines are defined in the frame rotating with the Sun. The grayscale represents the flow speed, and the streamlines drawn start with the uniform interval in longitude at $r = 50R_s$. It is observed that the longitudinal intervals between streamlines vary with distances. This variation of the streamline intervals represents the nonlinear evolution of the solar wind with distances.

The solar wind properties can be extracted from this three-dimensional numerical solar wind. For the extrapolation at several AUs, considering the travel time of the solar wind from 1 AU, the numerical quiet solar wind derived from the IPS data made during the previous Carrington rotation is used.

Figure 5 demonstrates the comparison of the analyzed data with near-Earth data (at 1 AU) and Ulysses data (at about 4.5 AU) during period of CR 1848 as an example. The positive correlations at these two different heliocentric distances show that both reconstruction of the solar wind at $r < 1AU$ and extrapolation for $r > 1AU$ work with reliable accuracy.

SUMMARY

In this paper, we briefly described the algorithm of the MHD tomography and demonstrated the velocity map of the solar wind reconstructed with the MHD tomography from the IPS observational data. Then, as an example of the application of the analysis result, the time-dependent MHD simulation of the disturbance propagation that may be applicable to the space weather prediction study is shown. Additionally, the extended global solar wind at

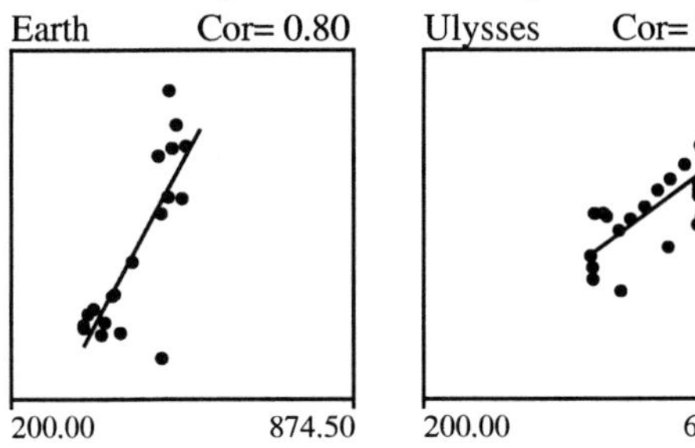

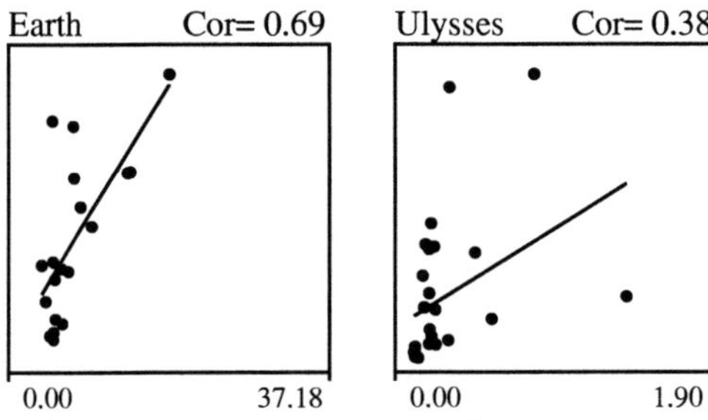

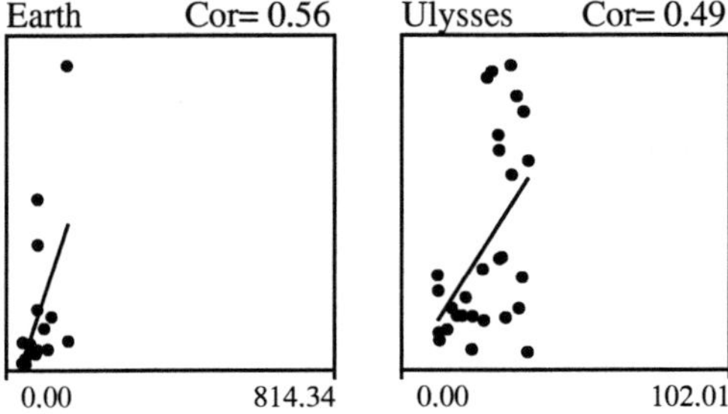

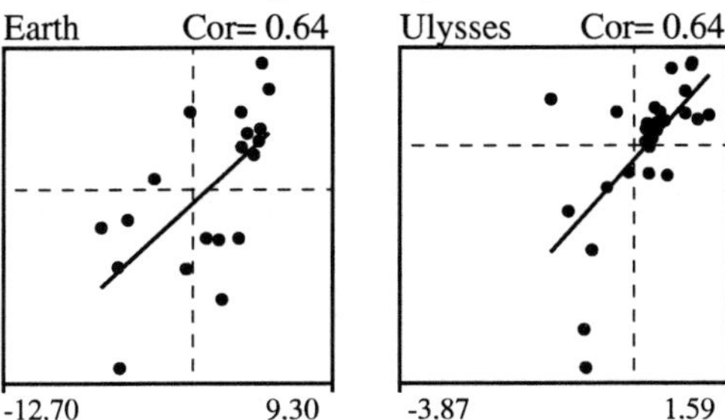

FIGURE 5. Correlation maps of analyzed (ordinate) and measured (abscissa) solar wind variables at Earth (1 AU ; left) and Ulysses spacecraft (4.5 AU ; right) during CR 1848. From the top, daily averaged values of the radial component of plasma velocity, plasma number density, plasma temperature and azimuthal component of azimuthal magnetic field are plotted, together with the square-fitting lines. Correlation coefficients are written on the top of each box. The range of values along x- and y-axis are identical and denoted at the bottom.

$r > 1$ AU are shown, together with the comparisons with Ulysses and near-Earth data.

The presented MHD tomography assumes that the solar wind is steady and super-Alfvénic in the domain of analysis. However, transient events such as interplanetary disturbances exist in reality, and even the global structure varies in time. In addition, to combine the trans-Alfvénic regions ($1R_s < r < 10R_s$) with the super-Alfvénic regions may be useful for the study of the Sun-Earth connection of space weather and the solar wind heating/acceleration mechanism. To handle these, we are studying "time-dependent" MHD tomography.

ACKNOWLEDGMENTS

This work is supported by the IPS project of the Solar-Terrestrial Environment Laboratory, Nagoya University.

REFERENCES

1. Jackson,V.B., Hick,P.L., Kojima,M. and Yokobe,A., *J. Geophys. Res.***103,** 12049–12067, 1998.
2. Kojima,M., Tokumaru,M., Watanabe,H. and Yokobe,A., *J. Geophys. Res.***103,** 1981–1989, 1998.
3. Hayashi,K, Kojima,M., Tokumaru,M, and Fujiki,K., *submitted J. Geophys. Res.*, 2002.

Using Galactic Cosmic Ray Observations for Determination of the Heliosphere Structure During Different Solar Cycles

Lev I. Dorman

Israel Cosmic Ray Center and Emilio Segre' Observatory, affiliated to Tel Aviv University, Technion and Israel Space Agency, Israel; IZMIRAN, Russian Academy of Science, Troitsk, Russia;

Abstract. Our studies of the neutron and muon components data from many observatories with different cut-off rigidities during solar cycles 19-22 have made it possible to investigate the hysteresis character of the relationships between the variations in solar activity and in galactic cosmic ray intensity. We use here a special model described the connection between solar activity and CR convection-diffusion global modulation with taking into account time-lag of plasma processes in the Heliosphere relative to the active processes on the Sun. We supposed different dimension of the modulation region and for each dimension was determined the correlation coefficient between variations of expected and observed CR intensities. We found that the maximum of correlation coefficient occurred for even cycles for about two-three times in the shorter time than for odd cycles. We came to conclusion that this difference is caused by CR drift effects: during even cycle drift effect from minimum to maximum of SA produced the small increasing of CR global modulation additional to the caused by convection-diffusion mechanism, and after maximum of SA - about the same decreasing of CR modulation. This gives sufficient decreasing of observed time-lag between CR and SA in even solar cycles. For odd solar cycles we have inverse situation: drift effect from minimum to maximum of SA produced the additional decreasing of CR global modulation caused mainly by convection-diffusion mechanism, and after maximum of SA - increasing of CR modulation. This gives sufficient increasing of observed time-lag between CR and SA in odd solar cycles. By comparison of expected results with observed for particles of different energy we determine the relative role of convection-diffusion and drift mechanisms in formation of CR global modulation in the Heliosphere.

COSMIC RAY LONG-TERM MODULATION, TIME-LAG AND CONVECTION-DIFFUSION MECHANISM

Short Historical Information on the Problem of CR Long-Term Modulation

The investigation of the hysteresis phenomenon in the connection between long-term variations in cosmic ray (CR) intensity observed at the Earth and solar activity (SA), started about 40 years ago [1-6]. In the middle of sixties many scientists came to conclusion that the: dimension of the modulation region (or Heliosphere, as it is now known) is about 5 AU, and not more than 10-15 AU [7-11]. The radius r_0 of the CR modulation region was found to be very small both by analysis of the intensity of coronal green line in some helio-latitude regions ($r_0 \approx 5$ AU), or by investigation the CR modulation as caused by sudden jumps in solar activity ($r_0 \approx 10\text{-}15$ AU). In [12-15] the hysteresis phenomenon was investigated on the basis of neutron monitor (NM) data for about one solar cycle in the frame of convection-diffusion model of CR global modulation in the Heliosphere taking into account the time lag of processes in the interplanetary space relative to processes on the Sun. It was shown that the dimension of the modulation region should be much bigger, about 100 AU. These

CP679, *Solar Wind Ten: Proceedings of the Tenth International Solar Wind Conference*, edited by M. Velli, R. Bruno, and F. Malara

investigations were continued on the basis of CR and SA monthly average data for about four solar cycles in [16, 17]. Though many authors have worked on this problem, the time lag of processes in the interplanetary space relative to processes on the Sun has not been taken into account (see review in [18]). The method, described below, considers that CR intensity observed on the Earth at moment *t* is caused by solar processes occurring for many months before *t*. In a recent paper [19] CR and SA data for solar cycles 19-22 was considered again taking into account drift effects according to [20]. It was shown that including drift effects (depending on the sign of solar polar magnetic field and determined by difference of total CR modulation at A>0 and A<0, and with amplitude proportional to the value of tilt angle between interplanetary neutral current sheet and equatorial plane) is very important: it became possible to explain the great difference in time-lags between CR and SA in hysteresis phenomena for even and odd solar cycles.

Hysteresis Phenomenon and Model of CR Global Modulation in the Frame of Convection-Diffusion Mechanism

It was shown in [12] that the time of propagation through the Heliosphere of particles with rigidity bigger than 10 GV (to whom NM are sensitive) is not longer one month. This time is at least about one order of magnitude smaller than the observed time-lag in the hysteresis phenomenon. It means that the hysteresis phenomenon on the basis of NM data can be considered as quasi-stationary problem with parameters of CR propagation changing in time. In this case according to [20, 21]

$$n(R, r_{obs}, t)/n_o(R) \approx \exp\left(-a\int_{r_{obs}}^{r_o} \frac{u(r,t)dr}{D_r(R,r,t)}\right), \quad (1)$$

where $n(R, r_{obs}, t)$ is the differential rigidity CR density, $n_o(R)$ is the differential rigidity density spectrum in the local interstellar medium out of the Heliosphere, $a \approx 1.5$, $u(r,t)$ is the effective solar wind velocity (taking into account also shock waves and high speed solar wind streams), and $D_r(R,r,t)$ is the radial diffusion coefficient in dependence of the distance r from the Sun of particles with rigidity R at the time t. According to [13, 14] the connection between $D_r(R,r,t)$ and solar activity can be described by the relation

$$D_r(R,r,t) \propto r^{\beta}(W(t - r/u))^{-\alpha}, \quad (2)$$

where $W(t - r/u)$ is the sunspot number in the time $t - r/u$. By the comparison with observation data it was determined in [13, 14] that parameter $0 \le \beta \le 1$ and $\alpha \approx 1/3$ in the period of high solar activity $(W(t) \approx W_{max})$ and $\alpha \approx 1$ near solar minimum $(W(t) << W_{max})$. Here we suppose, in accordance with [15], that

$$\alpha(t) = 1/3 + (2/3)(1 - W(t)/W_{max}), \quad (3)$$

where W_{max} is the sunspot number in the maximum of solar activity cycle.

According to (1) the value of the natural logarithm of observed CR intensity global modulation at the Earth's orbit, taking into account (2) and. (3), will be

$$\ln\left(n(R,r_E,t)_{obs}\right) = A(R,X_o,\beta,t_1,t_2) - B(R,X_o,\beta,t_1,t_2) \times F\left(t,X_o,\beta,W(t-X)\Big|_{X_E}^{X_o}\right), \quad (4)$$

where

$$F\left(t,X_o,\beta,W(t-X)\Big|_{X_E}^{X_o}\right) = \int_{X_E}^{X_o} \left(\frac{W(t-X)}{W_{max}}\right)^{\frac{1}{3}+\frac{2}{3}(1-W(t-X)/W_{max})} X^{-\beta} dX \quad (5)$$

$X = r/u$, $X_E = 1AU/u$, $X_o = r_o/u$ (X_E and X_o are in units of average month). Let us note that regression coefficient $A(R,X_o,\beta,t_1,t_2)$ determined the CR intensity out of the Heliosphere, and $B(R,X_o,\beta,t_1,t_2)$ characterized the effective diffusion coefficient in the interplanetary space; these coefficients can be determined by correlation between observed values $\ln(n(R,r_E,t)_{obs})$ and the values of F, calculated according to (5) for different values of X_o and β. In [16] three values of $\beta = 0; 0.5; 1$ have been considered; it was shown that $\beta = 1$ strongly contradicts CR and SA observation data, and that $\beta = 0$ is the most reliable value. Therefore, we will consider here only this value.

EVEN-ODD CYCLE EFFECT AND ROLE OF DRIFTS

Cosmic Ray Time-Lags in Odd and Even Solar Cycles

To determine $X_{o\max}$, corresponding to the maximum value of the correlation coefficient for (4), we compare 11 months moving averages of the Climax NM (H = 3400 m, cut-off rigidity $R_c = 2.99\,GV$) for solar cycles 19-22 and onset of cycle 23 with expectation according to (4) and. (5). For each time-lag, $X_o = r_o/u$ =1, 2, 3, ... 60 av. months, we determined the correlation between observed and expected CR intensities. The Climax NM data correspond to an effective rigidity of primary CR of about 10-15 GV. For higher energy particles (about 30-40 GV) we used Huancayo ($R_c = 12.92\,GV, H = 3400m$) and Haleakala ($R_c = 12.91\,GV$, $H = 3030m$) NM data from January 1953 to August 2000. A big difference in $X_{o\max}$ for odd and even solar cycles was found.

How Drift Effects Influenced on the Time-Lag in Odd and Even Cycles?

We assume that observed long-term CR modulation is caused by two processes: the convection-diffusion mechanism (e.g. [21-23]) independent of the sign of the solar magnetic field, and the drift mechanism (e.g. [20, 24-25]) what gave opposite effects with changing sign of solar magnetic field. For the convection-diffusion mechanism we use the model described in detail in [19], shortly given above by (1)-(5). We considered three Approaches of drift effects: First, we assume a constant value of drift modulation between two reversals of solar magnetic field with negative sign at A>0 and positive sign at A<0, and in the short period of reversal we suppose linear transition through 0 from one polarity cycle to other (in this case for convection-diffusion part of CR modulation we obtained sufficient differences in CR maximums near SA minimums in contradiction with observations); Second, we correct the first by reducing the value of drift modulation near SA minimums; Third, we assume that the drift effect is proportional to the value of the tilt-angle T with negative sign at A>0 and positive sign at A<0, and in the period of reversal we again suppose linear transition through 0 from one polarity cycle to other (see Figures 1-4 in [20]; we assume that average of curves for A>0 and A<0 in these figures characterized convection-diffusion modulation, and difference of these curves – double drift modulation). Data on tilt-angles for solar cycles 19 and 20 are not available. We used relation between sunspot numbers W and T to made homogeneous analysis of the period 1953-2000. Based on data for 18 years (May 1976- September 1993), we found that there are very good relation between T and W; for 11 months smoothed data T = 0.349W + 13.5° with correlation coefficient 0.955. We used 11 months smoothed data of W (shown in Figure 1) and determined the amplitude A_{dr} of drift effects as drift modulation at W11M = 75 (average value of W11M for 1953-1999). The reversal periods were: August 1949 ± 9 months, December 1958 ± 12 months, December 1969 ± 8 months, March 1981 ± 5 months, and June 1991 ± 7 months. We determined correlation coefficients between the expected integrals F for different values of X_o = 1, 2, 3, ... 60 av months with the observed LN(CL11M) and LN(HU/HAL11M), as well as with corrected for the drift effects according to the 1-st, 2-nd and 3-rd Approaches with A_{dr} from 0.15% up to 4%. An example for correction of observed CR intensity on the drift effects (to obtain only convection-diffusion modulation) is shown for period January 1953-November 2000 in Figure 1.

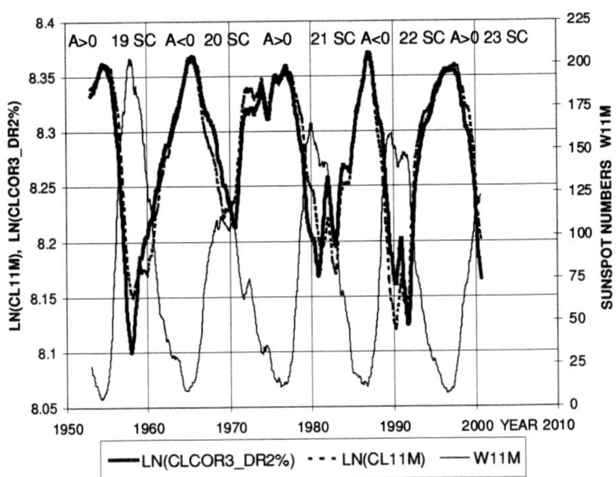

FIGURE 1. An example of CR data correction on drift effects in 1953-2000 (19-22 cycles and onset of 23 cycle): LN(CL11M) – observed natural logarithm of Climax NM counting rate smoothed for 11 months, LN(CLCOR3_DR2%) – corrected on assumed drift effect according to the 3-rd Approach with A_{dr} =2% at W11M=75. Interval between two horizontal lines corresponds 5% of CR intensity variation.

Estimation of Role of Drift Effects in Long-Term Modulation and Dimension of CR Modulation Region (Heliosphere)

In Fig. 2 the dependences of $X_{o\max}$ on A_{dr} are shown for Climax NM.

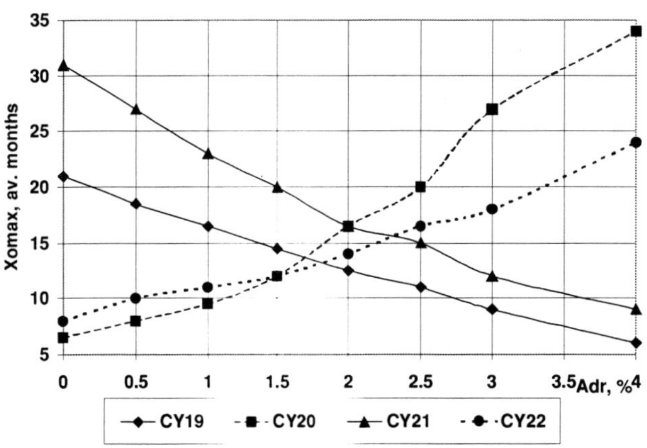

FIGURE 2. Dependences $X_{o\max}(A_{dr})$ for Climax NM.

From Figure 2 it can be seen that the region of crossings of $X_{o\max}(A_{dr})$ for odd and even cycles is: $13 \le X_{o\max} \le 16.5$, $1.7\% \le A_{dr} \le 2.3\%$. For Huancayo/Haleakala NM this region is: $13 \le X_{o\max} \le 18$, $0.23\% \le A_{dr} \le 0.43\%$, Thus we came to conclusion that the amplitude of the drift effect is about 2.0% for Climax NM and about 0.33% for Huancayo/Haleakala NM. We came also to conclusion that for primary CR with rigidity 10-15 GV a relative contribution of drift effects is about 20-25%. For CR with rigidity 35-40 GV a relative role of drift effects is about 2-3 times smaller. For $X_{o\max}$ we obtained for both 10-15 and 35-40 GV about 15 av. months, what corresponds $r_o \approx 100$ AU (with an average solar wind speed 400 km/s).

ACKNOWLEDGEMENTS

This research is partly supported by INTAS grant 0810.

REFERENCES

1. Dorman, L. I., *Cosmic Ray Variations*, Moscow, Gostekhteorizdat, 1957.

2. Forbush, S. E., *J. Geophys. Res.*, **63**, 651 (1958).

3. Neher, H. V. and Anderson, H. R, *J. Geophys. Res.*, **67**, 1309-1315 (1962).

4. Simpson, J. A., *Pontificiae Academiae Scientiarum Scripta Varia* (Vatican), **25**, 323-352 (1963).

5. Dorman, L. I., *Progress in Physics of Cosmic Ray and Elementary Particles*, **7**, 1-320 (1963).

6. Dorman, L. I., *Cosmic Ray Variations and Space Research*, Moscow, Nauka, 1963.

7. Quenby, J. J., *Proc. 9th ICRC*. **1**, 3-13 (1965).

8. Charakhchyan, A. N., and Charakhchyan, T. N., *Canad. Journ. of Phys.*, **46**, No. 10, part 4, 879-882 (1968).

9. Charakhchyan, A. N., and Charakhchyan, T. N., *Proc. 12th ICRC*, **5**, 1984-1991 (1971).

10. Stozhkov, Yu. I. and Charakhchyan, T. N., *Geomagnetism and Aeronomy*, **9**, 803-808 (1969).

11. Pathak, P. N. and Sarabhai V., *Planet.. Space Sci.*, **18**, 81-94 (1970).

12. Dorman, I. V. and Dorman, L. I., *Cosmic Rays*, **7**, 5-17 (1965).

13. Dorman, I. V. and Dorman, L. I., *J. Geophys. Res.*, **72**, 1513-1520 (1967).

14. Dorman, I. V. and Dorman, L. I., *J. Atmosph. and Terr. Phys.*, **29**, 429-449 (1967).

15. Dorman, L. I., *Variations of Galactic Cosmic Rays*, Moscow, Moscow State University Press, 1975.

16. Dorman, L. I. et al, *Proc. 25 th ICRC*, **2**, 69-72 (1997).

17. Dorman, L. I. et al., *Proc 26th ICRC*, **7**, 194-197 (1999).

18. Belov, A., *Space Sci. Reviews*, **93**, 79-105 (2000).

19. Dorman, L. I., *Adv. Space Res.*, **27**, 601-606 (2001).

20. Burger, R. A., and Potgieter, M. S., *Proc. 26th ICRC*, **7**, 13-16 (1999).

21. Dorman, L. I., *Proc. 6th ICRC*, **4**, 328-334 (1959).

22. Parker, E. N., *Interplanetary Dynamical Processes* . New York, Interscience. Publ., 1963.

23. Dorman, L. I., *Proc. 9-th ICRC*, **1**, 292-295 (1965).

24. Jokipii, J.R., and Davila J.M., *Astrophys. J.*, **248**, Part 1, 1156-1161 (1981).

25. Ferreira, S. E. et al., *Proc. 26th ICRC*, **7**, 77-80 (1999).

On the Deceleration of CMEs in the Corona and Interplanetary Medium deduced from Radio and White-Light Observations

M. J. Reiner*, M. L. Kaiser† and J.-L. Bougeret**

*Catholic University and NASA/Goddard Space Flight Center, Greenbelt, MD 20771, USA
†NASA/Goddard Space Flight Center, Greenbelt, MD 20771, USA
**Observatoire de Paris, Meudon, France

Abstract. It is generally acknowledged that CMEs, especially the faster ones, must deceleration somewhere between the high corona and interplanetary medium. However, the detailed characteristics and spatial region over which this deceleration occurs is still largely unknown. Simultaneous radio and white-light observations can provide information on the speed profiles of the CME or CME/shock in the high corona and/or interplanetary medium. These observations are consistent with constant deceleration for most CMEs. From the theoretical or modelling point of view, it is usually assumed that the deceleration of CMEs is due to a drag term, which is usually taken to be proportional to the relative speed of the CME compared to that of the solar wind or to this relative speed squared. We point out that such a form for the drag force on the CME inevitably leads to speed profiles that are inconsistent with both the radio and white-light observations.

INTRODUCTION

It is well known that CME speeds measured in the coronagraph are generally significantly higher than the speeds implied either by their transit times to spacecraft far from the sun or by the in-situ measured shock speeds [19, 18, 5]. However, the spatial regime over which the deceleration occurs and the detailed characterization of the corresponding speed profiles are not well established, due in part to the fact that the coronagraph measurements extend only out to $\sim 30R_{\odot}$.

Many CMEs, generally the faster ones, can drive shocks that, in turn, can generate low-frequency radio emissions [3, 1]. Since the density in the corona and interplanetary medium falls off with increasing heliocentric distance, these type II radio emissions will decrease in frequency as the CME/shock moves farther from the sun. Thus these frequency-drifting radio emissions can provide an alternative method of measuring CME dynamics. Such radio measurements have the advantage that they measure the radial speed or 'true" speed of the shock. Therefore, they can be particularly useful for deducing the dynamics for halo, Earth-directed CMEs, where projection effects are important [7]. Furthermore, the low-frequency radio observations can extend the distance range to well beyond $\sim 30R_{\odot}$ [8, 2, 10].

In those cases where there is significant deceleration implied by the time-sequence of white-light coronagraph images, it is generally found that, at least for most CMEs above $\sim 2R_{\odot}$, the fits to the height-time data are generally consistent with constant deceleration [6]. Modellers, on the other hand, argue that once the CME is released from the sun it is subject to a drag force as it propagates through the interplanetary plasma. The analytical form of this drag force is unknown. However, modellers generally assume that, in analogy with objects falling through the atmosphere, it is proportional to the density and the relative speed squared. In this paper we demonstrate that such a form for the drag force on the CME is not consistent with the observed radio and white-light data.

RADIO AND WHITE-LIGHT OBSERVATIONS OF A CME/SHOCK

In order to describe and understand the dynamics of CMEs, it is necessary to make time sequenced, remote observations of signatures of the CME as it propagates through the solar corona and/or interplanetary medium. One way of doing this is to analyze the time sequence of images made by white-light coronagraphs. An alternative way of measuring the dynamics of CMEs is to use the observed frequency drift of the radio emissions generated by the shock driven ahead of the CME [12, 17]. Both these methods are complementary. The white-light observations are more direct but have the difficulty that they are only plane-of-sky observations (which require

CP679, *Solar Wind Ten: Proceedings of the Tenth International Solar Wind Conference*, edited by M. Velli, R. Bruno, and F. Malara

unknown projection effect corrections) and that they only extend out to $30R_{\odot}$. The radio observations, on the other hand, measure the radial motion and can extend all the way to 1 AU and beyond, but are less direct in that the observations are made in the frequency domain and therefore require a density model in order to convert to the spatial regime. The limitations of each of these methods suggest that the best approach therefore is to use the combined radio and white-light data and to determine the CME dynamics by requiring consistency between the simultaneous radio and white-light observations [16].

White-light observation for the January 14, 2002 CME

On January 14, 2002 the SOHO/LASCO coronagraph observed a white-light CME propagating outward from SW solar limb. It first appeared in LASCO C2 at 05:40 UT at a height of $2.7R_{\odot}$; it was visible in C3 at least to 09:00 UT, when it was at $26R_{\odot}$. We estimated the leading edge height of the CME in each LASCO image and display the corresponding height-time data by the dots in Figure 1a. The CME initially accelerated until about 06:15 UT. However, after 06:15 UT the height-time data can be fit either with a straight line (corresponding to a constant speed of 1463 km/s) or with a modest deceleration (see below). The solar origin of this CME is not known (there were no SOHO/EIT images), however the SOHO/MDI images suggest a possible active site near the western limb. Hence this height-time data may be close to the true height-time relationship for this CME.

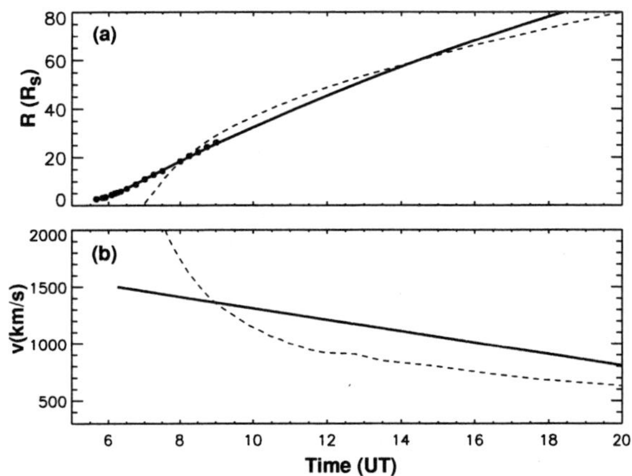

FIGURE 1. (a) Height-time plot for the CME observed on January 14, 2002. (b) Corresponding velocity-time profile (see text)

Radio observation for the January 14, 2002 CME

The frequency-drifting type II emissions generated by this January 14 CME are shown in dynamic spectrum in Figure 2. A GOES M4.4 LDE X-ray, with peak intensity at 06:27 UT, was associated with this radio event. As is usually the case, this CME associated flare was accompanied by a very intense and complex type III-like radio burst [11, 14], which is the dominant (overexposed) feature shown in Figure 2. The weaker, much more slowly drifting, diffuse radio emissions in Figure 2, observed from $\sim 07:30$ to $\sim 10:00$ UT, are the type II radio emissions that were generated as the CME/shock propagated through the high corona and interplanetary medium. To more readily discern the relationship between these frequency-drifting radio emissions and the CME dynamics, we have plotted the dynamic spectrum as inverse frequency versus time [9, 10]. The reason for doing this is that in the interplanetary medium the plasma density falls off as $1/R^2$, where R is the heliocentric distance, so that the observed radio frequency must scale as 1/R. Thus plotting the radio dynamic spectrum as 1/f versus time is essentially equivalent to plotting it as R versus time. The dashed black straight lines in Figure 2 then correspond to the CME/shock moving through the interplanetary medium at a constant speed. We get two straight lines, one with half the slope of the other, corresponding to radio emissions at either the fundamental or harmonic of the plasma frequency (as labelled). We assumed that the diffuse type II emissions were at the harmonic. The solid black curves in Figure 2, which provide a somewhat better fit to the observed frequency drift of the type II emissions, corresponds to the shock moving with a constant deceleration of 14 m s^{-2} (see below). The height-time and velocity-time relation corresponding to this fit is shown by the solid curves in Figure 1. These curves are clearly consistent with the observed white-light LASCO data for this near-limb CME.

Speed profile for the January 14, 2002 CME

Possible speed profiles for the CME/shock can be determined from the measured parameters of the shock when it was observed in-situ. We then try to select the "true" speed profile by requiring that the corresponding height-time profile simultanouesly fit both the white-light CME height-time data and the frequency-drifting radio data [15].

The CME-driven shock for this event was observed at Wind at 05:25 UT on January 17, 2002, corresponding to a transit time to Earth of 71 hours, and therefore a transit speed of 587 km s^{-1}. From the values of the plasma

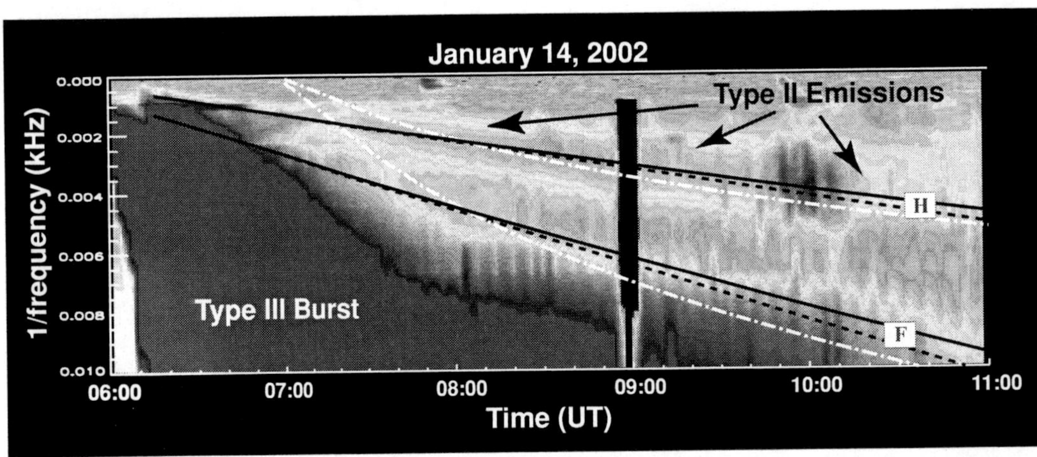

FIGURE 2. Dynamic spectrum of the Wind/WAVES radio emissions observed from 06:00 to 11:00 UT on January 14, 2002 in the frequency band from 100 kHz to 14 MHz. This is a plot of the intensity of the radio emissions as a function of inverse frequency (vertical axis) and time (horizontal axis).

parameters on either side of the shock, assuming a gas hydrodynamic shock, we estimated a shock speed of 403 km s^{-1}. The fact that this shock speed is significantly lower than the transit speed suggests that the CME/shock must have decelerated between the sun and Earth. Some of this discrepancy could also be due to the fact that at Earth we observed the flank of this western-directed shock. However, we ignore this additional complication here; it is not critical to our argument.

Consistent with the low-frequency type II radio data [12], we assume that the CME/shock accelerated at a constant rate, a, until time, t_a, i.e., $v = v_o + at$, after which it moved at a constant speed, v_{1AU}, to 1 AU. The determination of the acceleration profile, i.e., a and t_a, then depends on the initial speed v_o, the measured shock speed v_{1AU} at 1 AU, and the transit time to 1 AU. Since in the present case the transit time to 1 AU and the shock speed at 1 AU are known, for any given initial CME speed, v_o, we get a unique solution for the CME speed profile from the sun to 1 AU. Specifically, for the January 14, 2002 event, if we assume an initial CME speed of 1500 km s^{-1} at 06:15 UT, when the CME/shock was at $\sim 5.3 R_\odot$, then the known transit time (71 hours) and the speed of the CME shock at 1 AU (403 km s^{-1}) requires an acceleration of a = -14 m s^{-2} and a deceleration time of $t_a = 22$ hours. The corresponding speed profile, as a function of time (to $\sim$100 $R_\odot$), is shown by the solid curve in Figure 1b and, as a function of heliocentric distance (to 1 AU), by the solid curve in Figure 3a. The height-time relationship calculated from this speed profile is shown by the solid curve in Figure 1a and the acceleration relationship, in Figure 3b. The height-time profile is clearly consistent with the LASCO height-time data measured for this CME.

Now, in order to compare this CME speed profile with the frequency-drifting type II emissions in Figure 2, we next need to convert the corresponding height-time relationship in Figure 1a to a frequency-time relationship. To do this we assumed a $1/R^2$ falloff of the plasma density in the interplanetary medium. Then, we adjusted the value of the density at 1 AU in order that this corresponding frequency versus time curve provide a "best fit" to the type II frequency drift over the entire frequency range of the type II emissions shown in Figure 2. This procedure gave the solid black curves in Figure 2. This fit, which corresponds to the best overall simultaneous fit to the radio and white-light data, corresponded to a density, normalized at 1 AU, of 15 cm^{-3}.

This procedure can be repeated for other assumed initial speeds for the CME, in each case solving for the deceleration and deceleration time, given the contraints imposed by the known transit time and the shock speed at 1 AU. In each case, we determined the "best fit" to the frequency drifting type II data by adjusting the density at 1 AU. Each case will yield a somewhat different curve on the $1/f$ versus time dynamic spectrum. For example, an initial speed of 1800 km/s, would require an acceleration of a $= -23$ m s^{-2} and a deceleration time, t_a = 6.25 hours, with density at 1 AU of 24 cm^{-3}. However, this solution (not shown) does not provide as good a simultaneous fit to the white-light and radio data.

CME DECELERATION VIA A DRAG FORCE

In the above example we have shown that the speed profile of the CME or CME/shock deduced from the frequency drift of the type II radio emissions and the CME height-time measurements is consistent with constant deceleration for a finite time period. On the other hand, modellers typically assume that the deceleration of CMEs in the solar corona and interplanetary medium is caused by a drag force [4]. This drag force, F_D, is commonly assumed to be proportional to v^2, i.e., $F_D \propto \rho A v^2$, where ρ is the plasma density of the medium (corona or interplanetary medium), A is the effective area, and v is the speed relative to the medium. If this drag force is the primary cause of CME deceleration, then it is clear that this will lead to a CME deceleration that is not constant in time. To illustrate this we suppose that, after the initial ejection of the CME, this drag force and gravity (which gives only a small contribution) are the only forces acting on the CME. Then we have,

$$m\frac{d^2R}{dt^2} = G\frac{M_\odot m}{R^2} - \frac{C_D\rho_o A_o}{2}(v - v_{sw})^2, \quad (1)$$

where C_D is the drag coefficient, m is the CME mass, v_{sw} is the solar wind speed, G is the gravitational constant and $M_\odot$ is the solar mass. In writing Eq. (1) we assumed that $\rho = \rho_o/R^2$ and that $A = A_o R^2$. Eq. (1) was then solved for the speed profile. We did this by starting with the known shock speed at 1 AU and then integrating Eq. (1) inward towards the sun. The values of the various parameters for this CME are unknown, so we simply started with typical values, which were then adjusted until the solution approached $1R_\odot$ after 71 hours. As typical values we have taken $\rho_o = 50 \times 10^{-20}$ kg m^{-3}, $C_D = 0.5$, $m = 1.0 \times 10^{12}\ kg$, $A = 0.6 \times 10^{21}\ m^2$, $v_{sw} = 3.5 \times 10^5\ m\ s^{-1}$, $v = 4.0 \times 10^5\ m\ s^{-1}$,

The solution to Eq. (1) is shown by the dashed curves in Figures 1 and 3 and by the dot-dashed white curves in Figure 2. In comparing this solution with the velocity profile determined from the fits to the white-light and radio data, we see that this solution is fundamentally different. There is no way that we can adjust the parameters so that the solution would provide a reasonable fit either to the white-light height-time data or to the frequency drift of the type II radio emissions. These results then suggest that the commonly assumed analytical expression for the drag force is inconsistent with the CME dynamics implied by both the observed radio and white-light observations, i.e. it will lead to a CME velocity profile that is inconsistent with the data. Either this is not the correct form for the actual drag force acting on the CME or the drag force is not what causes the CME to decelerate. It is also clear that using a drag term proportional to the velocity will also yield a nonconstant deceleration, which is inconsistent with observations.

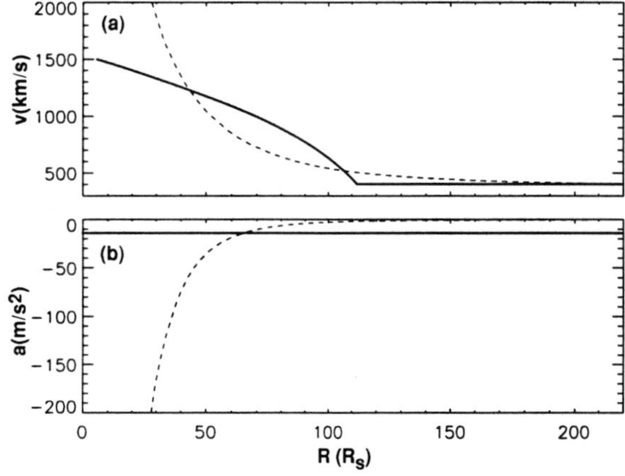

FIGURE 3. (a) Velocity-distance relationship for the January 14, 2002 CME. (b) Acceleration-distance relationship.

CONCLUSION

In this paper, we have demonstrated that quite reasonable fits to the dynamics of CMEs, implied by simultaneous white-light and radio data, are obtained by assuming constant deceleration in the high corona and interplanetary medium. We have further demonstrated that the commonly used analytical form for the drag force can not account for the observed CME deceleration as implied by the white-light and radio data. Therefore, in the spirit of Galileo Galilei, it is proposed here that a number of CMEs be dropped from La Torre di Pisa and that their drag forces be directly measured.

REFERENCES

1. Bale, S. D., et al., *Geophys. Res. Lett.* **26**, 1573 (1999).
2. Cane, H. V., et al., *Sol. Phys.* **78**, 187 (1982).
3. Cane, H. V., Sheeley Jr., N. R., Howard, R. A., *J. Geophys. Res.* **92**, 9869 (1987).
4. Chen, J. and Garren, D. A., *Geophys. Res. Letts.* **21**, 2319, (1993).
5. Gopalswamy, N., et al., *Geophys. Res. Lett.* **27**, 145 (2000).
6. Gopalswamy, N., et al., *J. Geophys. Res.* **106**, 29219 (2001).
7. Howard, R. A., et al., *Astrophys. J.* **263**, L101 (1982).
8. Malitson, H. H., et al., *Astrophys. Lett.* **14**, 111 (1973).
9. Reiner, M. J., et al., *Proc. 31st ESLAB Symp.: ESA SP-415* 183 (1997).
10. Reiner, M. J., et al., *J. Geophys. Res.* **103**, 29664 (1998).
11. Reiner, M. J. and Kaiser, M. L., *Geophys. Res. Letts.* **26**, 397, (1999).
12. Reiner, M. J., et al., *Proc. 9th Solar Wind Conference*, 653 (1999).
13. Reiner, M. J., et al., *Astrophys. J.* **529**, L53 (2000).
14. Reiner, M. J., et al., *Astrophys. J.* **530**, 1049 (2000).
15. Reiner, M. J., et al., *Solar Phys.* **204**, 123 (2001).
16. Reiner, M. J., et al., *Astrophys. J.*, submitted (2002).
17. Reiner, M. J., et al., in preparation (2002).
18. St. Cyr, O. C., et al., *J. Geophys. Res.* **105**, 18169 (2000).
19. Schwenn, R. *Space Sci. Rev.* **44**, 139 (1986).

Ulysses Observations of Changes in the Solar Polar Regions Around Solar Maximum

Geraint H. Jones and André Balogh

Space and Atmospheric Physics Group, The Blackett Laboratory, Imperial College, London SW7 2BW, United Kingdom.

Abstract. During 2000-2001, the Ulysses spacecraft executed its second rapid pass from very high southern to very high northern heliolatitudes. Unlike the previous fast latitude scan, the recent one occurred near the maximum in solar activity. The solar magnetic field reversal, as evidenced by heliospheric magnetic field data, occurred between Ulysses's highest southern latitudes and highest northern latitudes. We examine the magnetic field structure of the solar polar regions by analyzing data returned by the magnetometer aboard Ulysses before and after the maximum southern and northern heliolatitude excursions. An attempt is made to infer the location of the heliospheric current sheet at high latitudes from these observations. The reversal of the magnetic field at the solar wind source surface is inferred to have occurred between November 2000 and August 2001, with the most likely time of reversal being early within that period.

INTRODUCTION

Since early 1992, the ESA/NASA Ulysses spacecraft has been in a 6.2-year near-polar orbit which takes it to maximum heliolatitudes of ± 80.2° (figure 1). The eccentricity of Ulysses's orbit leads to a very rapid scan around perihelion from high southern to high northern latitudes in only 10 months. Its first fast latitude scan occurred in 1994/1995, while approaching a minimum in the solar activity cycle. During 2000/2001, Ulysses repeated the scan, during a period coincident with solar maximum. The latter period has provided the first direct measurements of solar polar magnetic fields around the time of magnetic field polarity reversal. It is hoped that the data presented here will aid in the understanding of the polarity reversal process, and may help in the refinement of potential field models that depend on lower-latitude photospheric field measurements alone.

DATA ANALYSIS

The polarity of the heliographic magnetic field (HMF) at Ulysses was determined using standard techniques. With the use of SWOOPS plasma data [1], the instantaneous Parker spiral direction [2] was calculated from:

$$\phi = \arctan\frac{(R_h - R_s)\Omega\sin\theta}{v_{sw}} \qquad (1)$$

where θ=colatitude, Ω=solar rotation rate, R_h=heliocentric distance, R_s=radius of solar wind source surface (taken to be 2.5 R_{sun}), and v_{sw} the solar wind velocity measured at Ulysses. The Carrington sidereal solar rotation period of 25.38 days was employed. Hourly averaged magnetic field data from Ulysses's magnetometer [3] were compared to the Parker spiral winding angle, and the polarity of the HMF determined from the angle between the actual azimuth angle and the Parker spiral angle. The source location of the solar wind was then estimated using standard ballistic mapping techniques. Heliolatitude was taken to be constant between the solar source surface and the spacecraft. The polarity of the solar wind source surface could then be estimated, either as being positive, i.e. outwards (here shown as white), or negative, i.e. inwards (grey). The timing of the departure of the plasma from the Sun was used to assign a Carrington rotation to the data values most relevant for later comparison with Earth-based observations.

CP679, *Solar Wind Ten: Proceedings of the Tenth International Solar Wind Conference*, edited by M. Velli, R. Bruno, and F. Malara

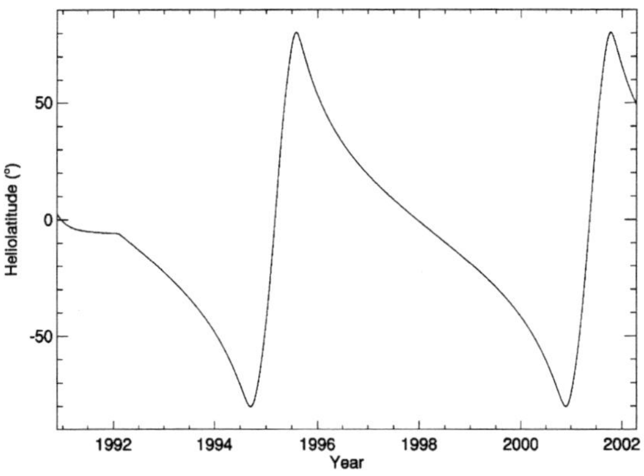

MAGNETIC POLARITY DISTRIBUTION OVERVIEW

FIGURE 1. The heliolatitude of Ulysses from launch to April 2002.

The spatial distribution of source surface magnetic polarities estimated from Ulysses data is given in figures 2 and 3. Each of the eight views in the two figures shows the inferred HMF polarity mapped back to the solar wind source surface, looking down on the rotational poles. Northern hemisphere views are shown in the upper panels; southern hemisphere views in the lower panels. Each phase of the mission from equator to pole is shown, beginning with the first excursion to high southern latitudes after the Jovian encounter in early 1992. The dichotomy in magnetic polarity around solar minimum is clear, as shown by the almost uniform polarities in the two hemispheres to late 1997.

RESULTS

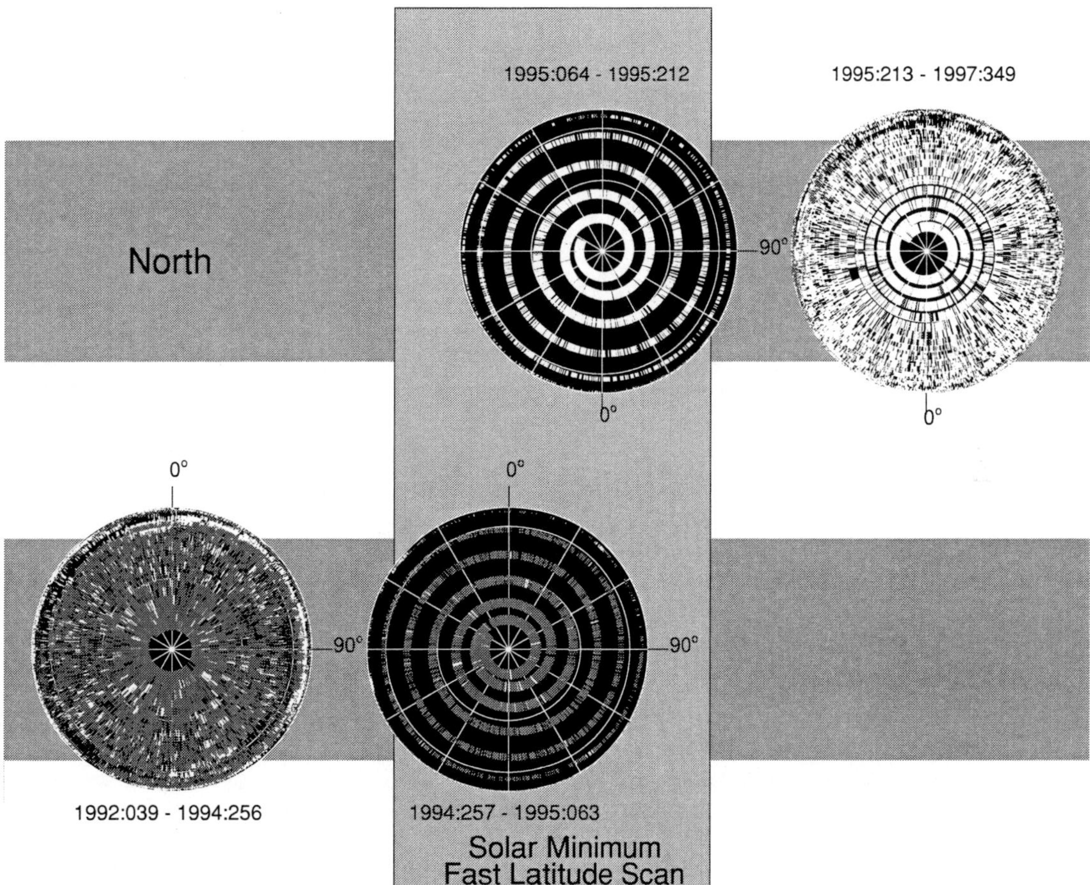

FIGURE 2. Polarity of the HMF at Ulysses, mapped back to the solar wind source surface. See text for description.

The short-lived high-latitude field reversals are not believed to be associated with the heliospheric current sheet, but rather with large-scale Alfvénic fluctuations in the magnetic field [4]. As expected, as solar maximum approached, this clear delineation of magnetic polarity by hemisphere was replaced by a far more complex mixture of the two polarities. As Ulysses ascended to high southern latitudes during 1998-2000, the "old" northern polarity (white) continued to be detected - this would become the "new" southern polarity.

The Southern Polar Region near Solar Maximum

Almost a full solar rotation was detected at maximum southern latitude during which the polarity remained that of the "old" southern polarity (grey) [5]. This suggests that Ulysses's increasing latitude kept pace with the increase in inclination of the heliospheric current sheet (HCS) during the ascent to high southern latitudes, and then overtook the sheet near 80° south.

If the HCS remained a single, coherent structure without significant warps, this single-polarity solar rotation constrains the minimum time of polarity reversal at the solar wind source surface to the period at highest southern latitudes (~2000:332). However, although it is almost certain that the HCS inclination was near its maximum around then, only a small warp in the sheet would have been required to produce the observations described whilst most of the sheet had actually tipped past the south pole.

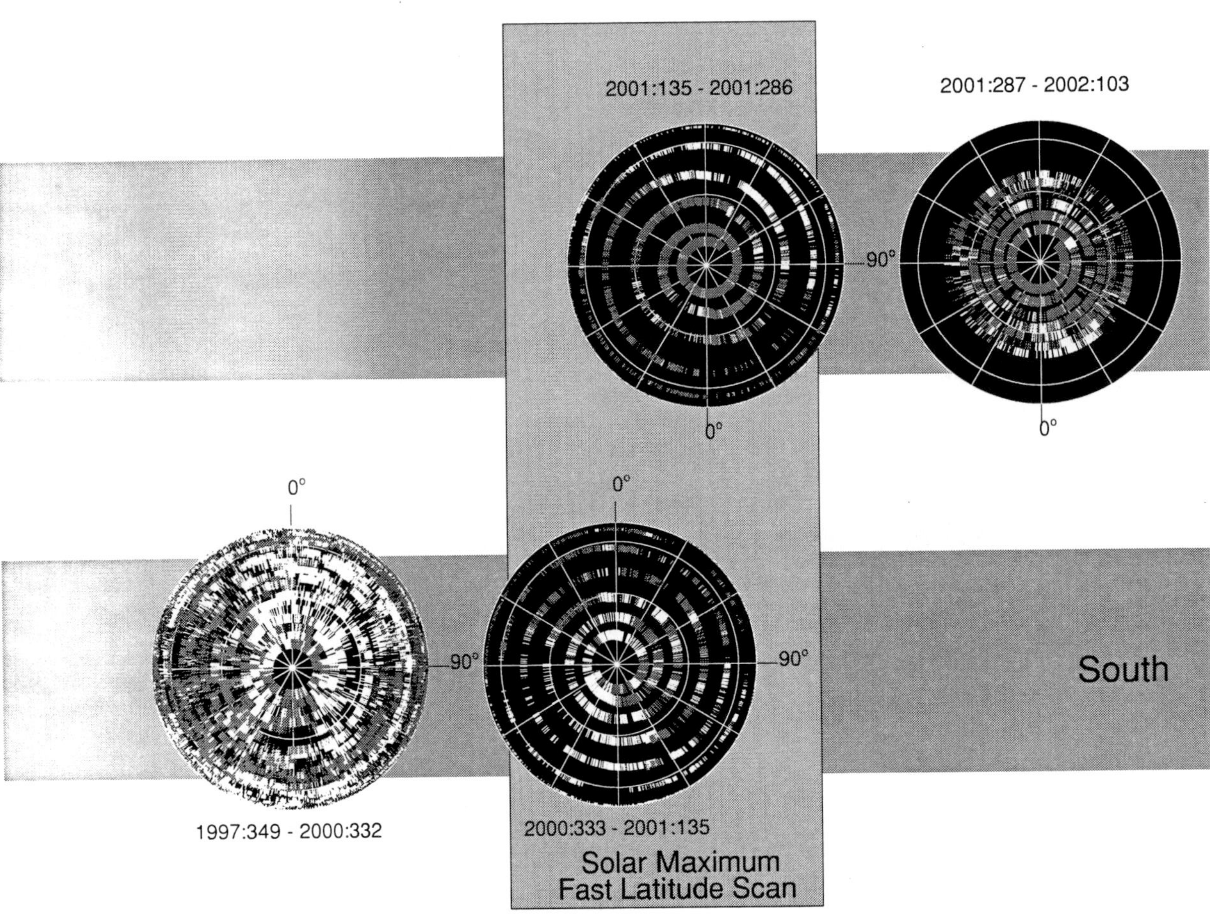

FIGURE 3. Polarity of the HMF at Ulysses, mapped back to the solar wind source surface. See text for description.

The Northern Polar Region near Solar Maximum

Northwards of heliolatitude ~67° north, from ~2001:236 onwards, there was little sign of the "old" northern polarity (white). On the return to lower latitudes, the sector structure may have been detected around 2002:018, at a heliolatitude of ~62° north. However, this possible sector was not clear, and it was two full solar rotations later when the first unambiguous sector structure was detected, at a heliolatitude of ~50° north. As the new northern polarity was clearly dominant at this time, this constrains the time of polarity reversal at the source surface to earlier than 2001:236. Quite striking is the shift in Carrington longitude covered most recently by the "new" southern polarity in the northern hemisphere. This is very different to the longitude range within which it was last seen during the ascent to high northern latitudes.

SUMMARY

The data presented here represent the first direct measurements of the solar polar magnetic fields at solar maximum. As expected, the polarity distribution was much more complex than that near solar minimum. There were few signs of non-dipolar field components during the solar maximum fast latitude scan. The polarity distribution detected by Ulysses indicates that the magnetic reversal in the solar wind source surface occurred between November 2000 and August 2001, and is likely to have been early within that time range. Refinements to the analysis techniques employed here continue to be made. Possible inadequacies in the use of ballistic mapping techniques are to be considered. The synthesis of the Ulysses data and in-ecliptic in-situ measurements from near-Earth and interplanetary spacecraft, and comet ion tail observations (e.g. [6]), will help construct a fuller picture of the inner heliosphere's structure around solar maximum.

ACKNOWLEDGMENTS

Ulysses research at Imperial College is supported by the UK Particle Physics and Astronomy Research Council. We are grateful to D. J. McComas for the use of Ulysses SWOOPS instrument data.

REFERENCES

1. Bame, S. J., et al., *Astron. Astrophys. Suppl, Ser.* **92**, 237-265 (1992)

2. Parker, E. N., *Interplanetary Dynamical Processes*, New York:Wiley-Interscience, (1963)

3. Balogh, A., et al., *Astron. Astrophys. Suppl. Ser.* **92**, 221-236 (1992)

4. Balogh, A., et al., *Geophys. Res. Lett.* **26**, 631-634 (1999)

5. Smith, E. J., Balogh, A., Forsyth, R. J., and McComas, D. J., *Geophys. Res. Lett.* **28**, 22, 4159-4162 (2001)

6. Brandt, J. C., Jones, G. H., McComas, D. J., Snow, M., Trujillo, T., and Burghard, A., Observations of cometary plasma tail and the heliosphere near solar maximum, *Proceedings of The Asteroids, Comets, Meteors 2002 Conference*, ESA SP-500, Noordwijk, Netherlands: ESA Publications Division, (in press)

The Helios Faraday Rotation Data Archive

M.K. Bird, H. Volland*, G.S. Levy, C.T. Stelzried, B.L. Seidel† and A.I. Efimov, V.E. Andreev, L.N. Samoznaev**

*Radioastronomisches Institut, Universität Bonn, 53121 Bonn, Germany
†Jet Propulsion Laboratory, California Institute of Technology, Pasadena, CA 91109, USA
**Inst. for Radio Engineering & Electronics, Russian Academy of Science, Moscow, 101999, Russia

Abstract. The Helios Faraday Rotation (FR) Experiment, a passive radio science investigation requiring no on-board hardware other than the existing spacecraft radio subsystem, was designed to study the dynamic and quiescent structure of the magnetic fields and electron density in the solar corona. Measurements of coronal Faraday rotation were derived from the linearly polarized S-band downlink carrier signal, which probed otherwise inaccessible regions of the corona in the radial range from 2 to 15 solar radii during the regularly recurring solar conjunctions. More than 1250 hours of Helios FR data were recorded over the duration of the Helios 1 (1974-84) and Helios 2 (1976-80) missions. The time scales of FR variations provide information on various physical phenomena: (a) slowly-varying rise and fall associated with the changing ray path offset, combined with the rotation of the quasi-static corona; (b) ubiquitous random oscillations with higher fluctuation amplitude at smaller solar offset distances, probably caused by coronal Alfvén waves; (c) occasional nearly discontinous jumps in the polarization angle, most likely caused by transient events such as coronal mass ejections (CMEs). The Helios FR data, aspects of which have been reported in more than forty publications to date, have now been systematically collected in a data archive for public dissemination. A brief review of the main results of the Helios FR Experiment are presented, together with some suggestions for possible use of the archive for continued solar wind research.

INTRODUCTION

The primary goal of the Helios Faraday Rotation experiment was to investigate the magnetic field $\vec{B}$ in the solar corona (Volland et al., 1977, 1984; Levy et al., 1980; Bird and Edenhofer, 1990). The passive radio science experiment required no on-board hardware other than the existing spacecraft radio communications subsystem. Helios polarization angle (FR) data, recorded using the linearly polarized S-band downlink carrier signal during times of superior conjunction, have been made available in an online archive. Selected results gleaned from the unique Helios FR data set with relevance for the coronal magnetic field and MHD-wave environment, as well as some possible future uses of the archive, are presented.

EXPERIMENT DESCRIPTION

A significant component of linear polarization is necessary for recording FR measurements with spacecraft radio signals. Only Pioneers 6 and 9 had been used for this purpose prior to Helios (Levy et al., 1969; Stelzried et al., 1970). In the more recent era, the Galileo spacecraft was optimally equipped with dual-frequency linear polarization (Howard et al., 1990), but prospects for conducting FR experiments were dashed prior to solar conjunction when the high gain antenna could not be deployed. The Helios FR data volume thus remains the largest collection of radio sounding data sensitive to the coronal magnetic field in the region of the solar wind transition to supersonic velocities.

Linear polarized electromagnetic waves propagating through a magnetized plasma undergo 'Faraday rotation' of their plane of polarization, which can be used to deduce the magnetic structure of the propagation medium. For the case of the cylindrical despun Helios antenna, the electric vector of the *transmitted* linearly polarized carrier signal was oriented perpendicular to the ecliptic. Measurements of the coronal contribution to signal Faraday rotation (Ω) were obtained by tracking the *received* signal polarization angle q, referenced to the local horizon, and transforming this value back to the ecliptic frame from the known parallactic angle p, as shown schematically in Fig. 1. Coronal FR is related to the longitudinal component of the magnetic field along the ray path $B_s = \vec{B} \cdot \hat{s}$ and electron density N_e by

$$\Omega = \frac{K}{f^2} \int_H^{\oplus} N_e(s)\, B_s(s)\, ds \quad \text{radians} \tag{1}$$

where $K = 2.36 \times 10^4$ in both MKS and cgs units, f is the radio frequency (2.296 GHz), and the integral in Eq. (1)

CP679, *Solar Wind Ten: Proceedings of the Tenth International Solar Wind Conference*, edited by M. Velli, R. Bruno, and F. Malara

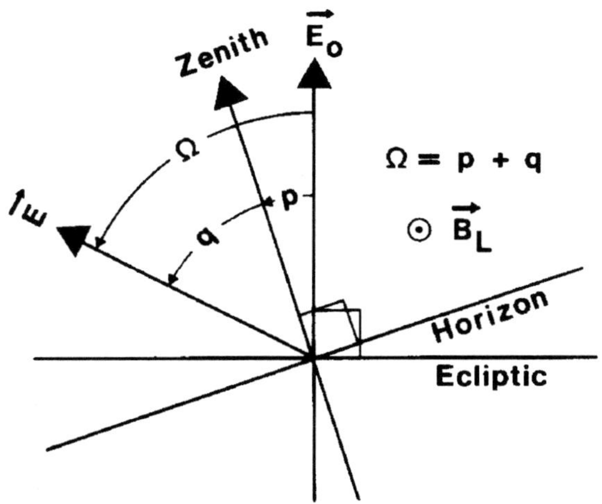

FIGURE 1. *Polarization angle reference in local and ecliptic systems, as viewed from an observer looking toward Helios.*

is taken along the ray path from Helios (H) to the ground station on Earth (⊕), assumed to be along a straight line. The quantity Ω can be plus or minus depending on the integrated product of electron density and component of magnetic field along the ray path (quasi-longitudinal approximation). The direction of rotation is called positive (counterclockwise) when the magnetic field points toward the observer, as sketched in Fig. 1, and negative (clockwise) when directed away from the observer.

The short-period, zero-inclination, heliocentric orbits of the two identical Helios spacecraft were ideally designed for conducting radio science investigations of the solar corona during the recurring solar occultations. Helios 1, with an original design lifetime of 18 months, was launched 10 Dec 1974 and underwent a total of 10 solar conjunctions before its demise in 1985. The official end of mission (EOM) was declared as 15 Mar 1986.

Fig. 2 shows an ecliptic plane view of the Helios 1 orbit in a Sun/Earth fixed system during the first year after launch (left panel) and for all orbits through the end of September 1981 (right panel). The shaded wedge-shaped area denotes the region of space where interplanetary spacecraft attain a solar elongation within 3°, roughly the region where coronal Faraday rotation of an S-band radio signal becomes measureable. Dots are placed at 10 day intervals along the trajectory. Although most solar occultations proceeded diametrically along a solar radial from West to East limb, an occultation near spacecraft aphelion, such as the very first solar conjunction shown in Fig. 2 (left panel), could extend over several weeks.

The counterpart of Fig. 1 for Helios 2 is very similar, except that the orbital perihelion was lowered to 0.29 AU from the 0.31 AU of Helios 1. The associated change in the orbital period caused the clockwise orbital precession rate to decrease from ~14° per year (Fig. 1, right) to ~6° per year for Helios 2. From its launch on 15 Jan 1976, Helios 2 was tracked regularly until it ceased operating on 3 Mar 1980 (formal EOM: 8 Jan 1981).

AREAS OF INVESTIGATION

The time scales of FR variations, observed during virtually all occultations, provide information on various physical phenomena in the solar corona:

Slowly-varying background. The slow rise and fall in FR is associated with the changing ray path offset, combined with the rotation of the quasi-static corona. Fig. 3 shows such slow FR variations observed at solar offsets $R \simeq 3\ R_\odot$ near and during the overlap interval of two tracking stations (Volland et al., 1977). Pätzold et al. (1987) utilized the absolute FR measurements from 1975-76 for a determination of the mean coronal magnetic field $B(R)$ during solar minimum. A more general study using all FR data to study $B(R)$ at other solar cycle phases would be a valuable extension of this work.

Random oscillations. This ubiquitous phenomenon, with higher FR fluctuation amplitude at smaller solar offset distances, is probably caused by coronal Alfvén waves (Hollweg et al., 1982; Efimov et al., 1993, 2000; Andreev et al., 1996, 1997a,b). Fig. 4 shows simultaneous recordings at two tracking stations during an interval of quite obvious wavelike FR fluctuations. A preliminary study determined that ca. 30% of the velocities determined from two-station cross correlations were directed *radially inward toward the Sun* (Bird et al., 1992). A more exhaustive study using the entire FR data set should be undertaken to confirm these results.

FR discontinuities. These occasional abrupt jumps in the polarization angle are assumed to be caused by transient events such as coronal mass ejections (CMEs). A search for white-light CMEs during the 1979 Helios solar conjunctions yielded 5 distinct events which produced obvious abrupt FB signatures associated with their passage through the signal ray path (Bird et al., 1985). These 'FR transients' were first seen at solar maximum (Levy et al., 1969), even before their optical counterparts were widely publicized from the Skylab observations. Not restricted to epochs of solar activity maximum, an example from 1976 presented by Bird et al. (1977) is shown in Fig. 5. Some recent work has associated the FR transients with interplanetary sector boundaries (Woo, 1997), rather than CMEs (Pätzold and Bird, 1998). The Helios FR archive offers by far the largest data source for pursuit of this controversy.

In summary, the Helios Faraday rotation experiment provided valuable clues about the quiet and disturbed magnetic structure of the solar corona in the critical

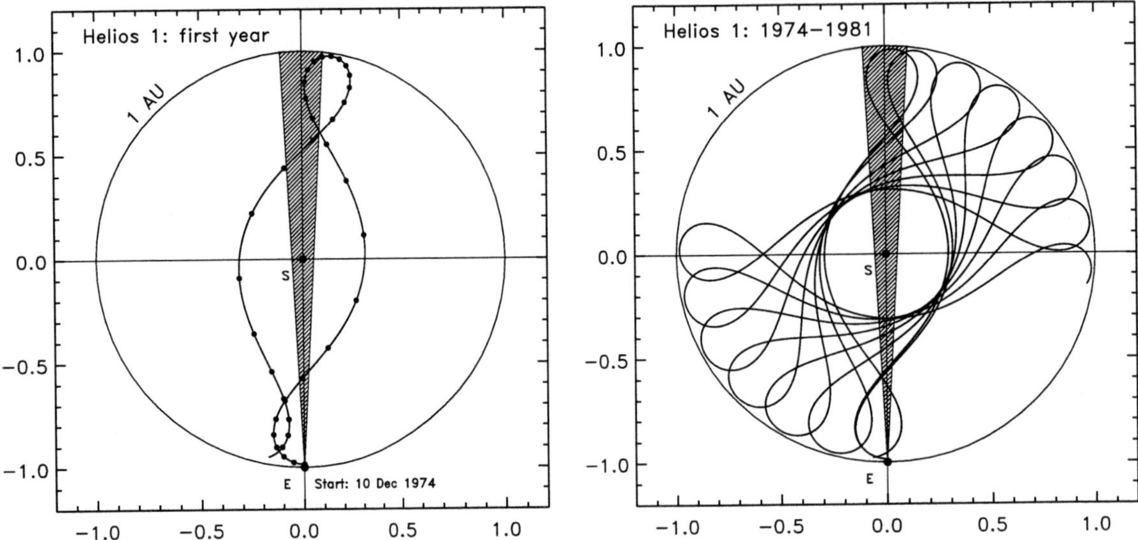

FIGURE 2. *Helios 1 orbit in a Sun/Earth fixed system during the first year (left) and through the end of 1981 (right).*

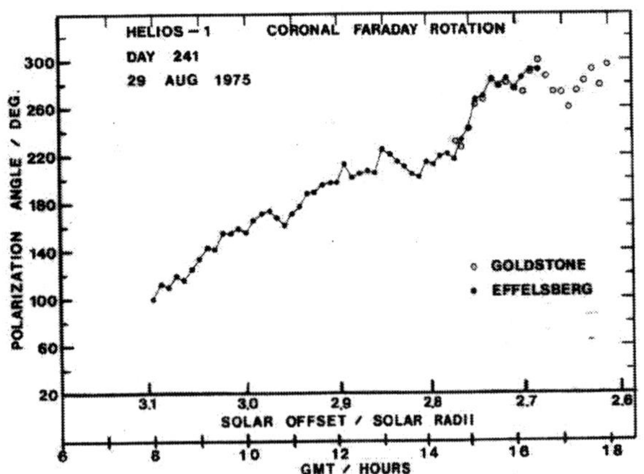

FIGURE 3. *Overlapping FR measurements at Effelsberg and Goldstone during the second solar conjunction of Helios 1 (from Volland et al., 1977).*

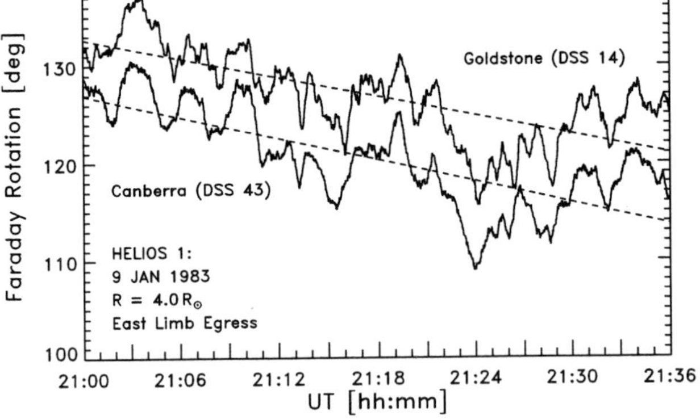

FIGURE 4. *Simultaneous two station Faraday rotation measurements. The constant offset between the values of FR at DSS 14 and 43 is due to different local ionospheric contributions.*

radial range $2\ R_{\odot} < R < 15\ R_{\odot}$. The unexpectedly long lifetime of the Helios probes enabled studies to extend from deepest solar activity minimum into the following maximum and beyond.

FR DATA ARCHIVE STRUCTURE

The FR data consist of ground-received, time-tagged measurements of the locally referenced polarization angle using automatic tracking polarimeters. This angle is transformed to the ecliptic plane (see Fig. 1) to determine the coronal Faraday rotation given by Eq. (1) as a function of (R,Θ,Φ), the heliocentric spherical coordinates of the ray path's point of solar closest approach (the "solar offset"). Significant coronal FR at levels easily distinguishable from the diurnal variation of the Earth's ionosphere is typically obtained for $R < 15R_{\odot}$. It is important to continually monitor the FR during an occultation (see Fig. 3), avoiding gaps between ground station transitions. This is because any isolated measurement of polarization angle is known only to modulo 180°. It follows that the actual value of coronal FR can become ambiguous by $\pm n \times 180°$ if the tracking is interrupted too long. This condition was actually realized on a number of occasions during the many solar conjunctions (e.g., Fig. 5).

Helios FR data exist for the years 1975-1984 during most of the periods when the ray path offset was below $15\ R_{\odot}$. Only the large Goldstone station (DSS 14) of the

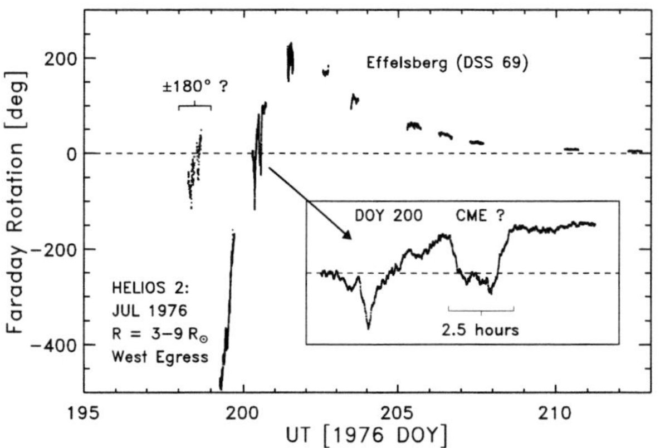

FIGURE 5. *FR measurements during the 1976 solar conjunction of Helios 2 (occultation egress, west solar limb). The inset shows the data of DOY 200 at an enlarged time scale featuring a Faraday rotation transient event (possible CME).*

NASA Deep Space Network (DSN) and the 100 m Effelsberg Radio Telescope were used during the first Helios mission years. The experiment could be conducted on a global basis starting in 1977 when the DSN added automatic tracking polarimeters at the 64-m stations in Canberra (DSS 43) and Madrid (DSS 63). These polarimeters were removed from the DSN ground stations in 1984, effectively ending the Helios FR experiment a few months before Helios 1 operations were terminated.

Data from the Helios FR archive, comprised of 227 tracking passes with 1277 hours of FR data as well as short statistical summaries of each tracking pass, are now available for downloading at the following URL:

http://www.astro.uni-bonn.de/~mbird/helios_html/

ACKNOWLEDGMENTS

The Helios Faraday Rotation Experiment could not have been performed without valuable contributions from the Helios Project Team, the German Space Operations Center (GSOC, Oberpfaffenhofen), the Max-Planck-Institute für Radioastronomie (MPIfR, Bonn), the NASA Deep Space Network (DSN), and the Radio Science Support Team at Jet Propulsion Laboratory (JPL, Pasadena). We wish to extend special thanks to Dr. H. Porsche for his continued supportive efforts and to Dr. M. Pätzold for many helpful suggestions. The work in Bonn was supported by the Deutsche Forschungsgemeinschaft (DFG) and the Deutsches Zentrum für Luft- und Raumfahrt (DLR), known during the Helios days as the Deutsche Forschungs- und Versuchsanstalt für Luft- und Raumfahrt (DFVLR).

REFERENCES

1. Andreev, V.E., A.I. Efimov, L.N. Samoznaev, et al., 1996, Faraday rotation fluctuation spectra observed during solar occultation of the Helios spacecraft, in *Solar Wind Eight* [AIP Conf. Proc. 382], D. Winterhalter et al. (eds.), AIP Press, Woodbury, NY/USA, 34-37.
2. Andreev, V.E., M.K. Bird, A.I. Efimov, et al., 1997a, Intensity of coronal Alfvén waves from polarization radio sounding, *Astron. Lett.* **23**, 194-199.
3. Andreev, V.E., A.I. Efimov, L.N. Samoznaev, et al., 1997b, Characteristics of coronal Alfvén waves deduced from Helios Faraday rotation data, *Solar Phys.* **176**, 387-402.
4. Bird, M.K., P. Edenhofer, Remote sensing of the solar corona, 1990, in *Physics of the Inner Heliosphere*, R. Schwenn, E. Marsch (eds.), Springer-Verlag, Berlin, 13-97.
5. Bird, M.K., H. Volland, R.A. Howard, et al., 1985, White-light and radio sounding observations of coronal transients, *Solar Phys.* **98**, 341-368.
6. Bird, M.K., H. Volland, A.I. Efimov, et al., 1992, Coronal Alfvén waves detected by radio sounding during the Helios solar occultations, in *Solar Wind Seven*, E. Marsch, R. Schwenn (eds.), Pergamon Press, Oxford, 147-150.
7. Bird, M.K., H. Volland, G.S. Levy, et al., 1977, Faraday rotation transients observed during Helios solar occultation, in *Study of Travelling Interplanetary Phenomena/1977*, [AFGL-TR-77-0309], M.A. Shea et al. (eds.), 63-75.
8. Efimov, A.I., I.V. Chashei, V.I. Shishov, et al., 1993, Faraday rotation fluctuations during radio transillumination of circumsolar plasma, *Astron. Lett.* **19**, 57-61.
9. Efimov, A.I., L.N. Samoznaev, V.E. Andreev, et al., 2000, Quasi-harmonic Faraday-rotation fluctuations of radio waves in the outer solar corona, *Astron. Lett.* **26**, 544-552.
10. Hollweg, J.V., M.K. Bird, H. Volland, et al., 1982, Possible evidence for coronal Alfvén waves, *J. Geophys. Res.* **87**, 1-8.
11. Howard, H.T., V.R. Eshleman, D.P. Hinson, et al., 1992, Galileo radio science investigations, *Space Sci. Rev.* **60**, 565-590.
12. Levy, G.S., T. Sato, B.L. Seidel, et al., 1969, Pioneer 6: measurement of transient Faraday rotation phenomena observed during solar occultation, *Science* **166**, 596-598.
13. Levy, G.S., H. Volland, M.K. Bird, et al., 1980, Faraday Rotation Experiment, in *The HELIOS Solar Probes: Science Summaries*, [NASA-TM 82005], J.H. Trainor (ed.), 85-94.
14. Pätzold, M., M.K. Bird, 1998, The Pioneer 6 Faraday rotation transients - On the interpretation of coronal Faraday rotation data, *Geophys. Res. Lett.* **25**, 2105-2108.
15. Pätzold, M., M.K. Bird, H. Volland, et al., 1987, The mean coronal magnetic field determined from HELIOS Faraday rotation measurements, *Solar Phys.* **109**, 91-105.
16. C.T. Stelzried, Levy, G.S., T. Sato, et al., 1970, K.H. Schatten, J.M. Wilcox, 1970, The quasi-stationary coronal magnetic field and electron density as determined from a Faraday rotation experiment, *Solar Phys.* **14**, 440-456.
17. Volland, H., M.K. Bird, G.S. Levy, et al., 1977, Helios-1 Faraday rotation experiment: Results and interpretations of the solar occultations in 1975, *J. Geophys.* **42**, 659-672.
18. Volland, H., G.S. Levy, M.K. Bird, et al., 1984, Das Faraday-Rotations-Experiment, in *10 Jahre HELIOS*, H. Porsche (ed.), DLR, Oberpfaffenhofen, 118-121.
19. Woo, R., 1997, Evidence for the reversal of magnetic polarity in coronal streamers, *Geophys. Res. Lett.* **24**, 97-100.

Changing of the Modulation Region Structure With Particle Rigidities According to Data of Hysteresis Phenomenon (Cycle 22)

Lev I. Dorman

Israel Cosmic Ray Center and Emilio Segre' Observatory, affiliated to Tel Aviv University, Technion and Israel Space Agency, Israel; IZMIRAN, Russian Academy of Science, Troitsk, Russia;

Abstract. Our studies of neutron monitor observatory data with different cut-off rigidities during solar cycle 22 have made it possible to investigate the hysteresis properties of the relationship between the variations in solar activity and galactic cosmic ray intensity. Hysteresis arises due to the delay of interplanetary processes (responsible for cosmic ray modulation) with respect to the initiating solar processes, which correspond to some effective solar wind and shock wave propagation velocity. It allows determination of the effective dimension of the modulation region as a function of the effective energy of galactic cosmic rays. We extend previous investigations made in the framework of the convection-diffusion model by taking into account drifts that change sign in periods of solar magnetic field reversal. From comparisons with experimental data on long-term cosmic ray variation in cycle 22, we determine the role of convection-diffusion and drifts in global modulation, and the effective dimension of the modulation region in dependence of particle rigidities.

INTRODUCTION

A short historical introduction to the research of the lag between long-term variations of cosmic ray (CR) intensity observed at Earth and solar activity (SA) is given in [1]. Analysis made in [1] leads to the conclusion that observed long-term CR modulation is caused by two processes: a convection-diffusion mechanism that does not depend on the sign of the solar magnetic field (SMF), and a drift mechanism (e.g. [2-7]) which gives opposite effects depending on the sign of the SMF. In [1] the relative role of convection-diffusion and drifts in the long-term CR modulation on the basis of a comparison of observations in odd and even cycles of SA was considered: it was shown that the time–lag $X_{o\max}$ between CR and SA in the odd cycles 19, 21 decreases with increasing of the amplitude of the drift effect A_{dr}, but in the even cycles 20, 22, $X_{o\max}$ increases with increasing A_{dr}. To determine $X_{o\max}$ and A_{dr} separately, in [1] was assumed that for a first approximation $X_{o\max}$ and A_{dr} are about the same in odd and even solar cycles. In the present paper we try to solve the problem of determining A_{dr} and $X_{o\max}$ only on the basis of data in solar cycle 22. We will therefore correct the observed cosmic ray long-term variation in cycle 22 for drift effects with different values of the amplitude A_{dr}; for each A_{dr} we determine the correlation coefficient $R(X_o, A_{dr})$ of corrected cosmic ray long-term variation according to a convection-diffusion model for different values of the time-lag X_o (from 0 to 60 av. months with monthly steps). Then we determine the value of $X_{o\max}(A_{dr})$ when $R(X_o, A_{dr})$ reaches the maximum value $R_{\max}(X_{o\max}, A_{dr})$. For each A_{dr} we will determine $R_{\max}$ and $X_{o\max}$. It is natural to assume that the most reliable value of A_{dr} will correspond to the biggest $R_{\max}(X_{o\max}, A_{dr})$ value, i.e. when the correction for drift effects is the best (in the frame of the model used for drift effects for long-term CR variations). This way will also determine the

CP679, *Solar Wind Ten: Proceedings of the Tenth International Solar Wind Conference*, edited by M. Velli, R. Bruno, and F. Malara

most reliable value for $X_{o\max}$ characterizing the dimension of the cosmic ray modulation region in the Heliosphere.

COSMIC RAY LONG-TERM VARIATION CAUSED BY CONVECTION-DIFFUSION

Because the basic convection-diffusion quasi-stationary model of CR-SA hysteresis phenomenon was described in details in [1], we will give only the final equations used in this paper. According to this model the expected contribution to cosmic ray long-term variations caused by the convection-diffusion mechanism at the Earth's orbit is:

$$\ln(n(R,r_E,t))_{obs} = A(X_o,\beta,t_1,t_2) - B(X_o,\beta,t_1,t_2) \times F\left(t,X_o,\beta,W(t-X)\right)_{X_E}^{X_o}, \quad (1)$$

where

$$F\left(t,X_o,\beta,W(t-X)\right)_{X_E}^{X_o} = \int_{X_E}^{X_o} \left(\frac{W(t-X)}{W_{\max}}\right)^{\frac{1}{3}+\frac{2}{3}(1-W(t-X)/W_{\max})} X^{-\beta} dX, \quad (2)$$

with $X = r/u$, $X_E = 1\,\mathrm{AU}/u$, $X_o = r_o/u$ (X_E and X_o are in units of average months = (365.25/12) days = 2.628×10^6 s). As it was mentioned in [1], the regression coefficient $A(R,X_o,\beta,t_1,t_2)$ determined the CR intensity out of the Heliosphere, and $B(R,X_o,\beta,t_1,t_2)$ characterized the effective diffusion coefficient in the interplanetary space; these coefficients can be determined from correlation of observed values $\ln(n(R,r_E,t))_{\rm obs}$ with values of $F\left(t,X_o,W(t-X)\right)_{X_E}^{X_o}$ calculated according to (2) for different values of X_o in the time interval $t_1 \le t \le t_2$. Here r is the distance from the Sun, r_o is the assumed radius of modulation region, $r_E = 1AU$, W is the monthly sunspot number, and $W_{\max}$ is the sunspot number at the maximum of SA.

COSMIC RAY LONG-TERM VARIATION CAUSED BY DRIFTS

According to the main idea of the drift mechanism (see [2-7]), we assume that the drifts are proportional to the value of tilt angle T and changed sign during periods of the SMF polarity reversal. We used data on tilt-angles for the period May 1976-September 1993. On the basis of these data we determined the correlation between T and W for 11 month-smoothed data as

$$T = 0.349W + 13.52° \quad (3)$$

with correlation coefficient 0.955±0.013. Important for the Cycle 22 reversal periods were: March 1981±5 months, and June 1991±7 months. As example, the drift effect according to this model for the period January 1985-December 1996 is shown in Figure 1 for the 11-month-smoothed data of W according to (3) and A_{dr} =1% at W=75.

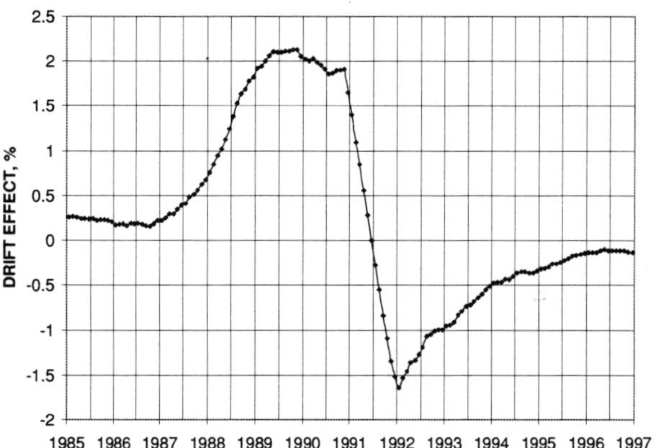

FIGURE 1. An example of assumed drift modulation in Cycle 22 for A_{dr} =1% at W=75.

NEUTRON MONITOR DATA

Results for Climax NM

According to the procedure described above we correct the 11-month-smoothed data on the drift effect for different values of A_{dr} from 0% (no drift effect) up to 4% at W=75. The dependence of the correlation coefficient on the value of the expected time-lags is shown in Figure 2. For each value of A_{dr} in Figure 2, $X_{o\max}(A_{dr})$ can be easy determined when the correlation coefficient reaches a maximum value $R_{\max}$.

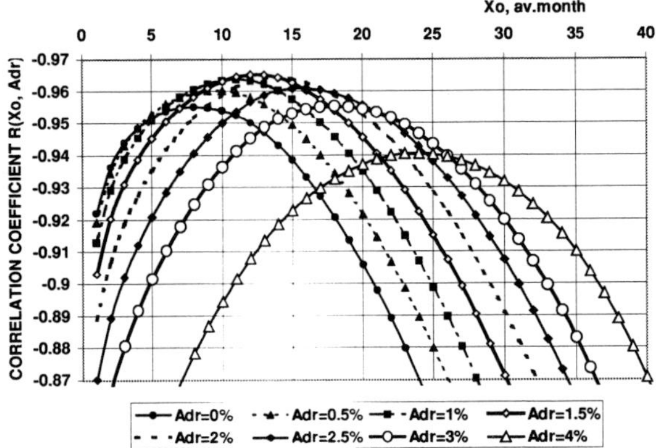

FIGURE 2. Correlation coefficient $R(X_o, A_{dr})$ according to 11-month-smoothed data of Climax NM (N39,W106; height 3400 m, 2.99 GV) in Cycle 22 for different A_{dr} from 0% up to 4% at W=75.

The functions $R_{\max}(A_{dr})$ and $X_{o\max}(A_{dr})$ are shown in Figure 3.

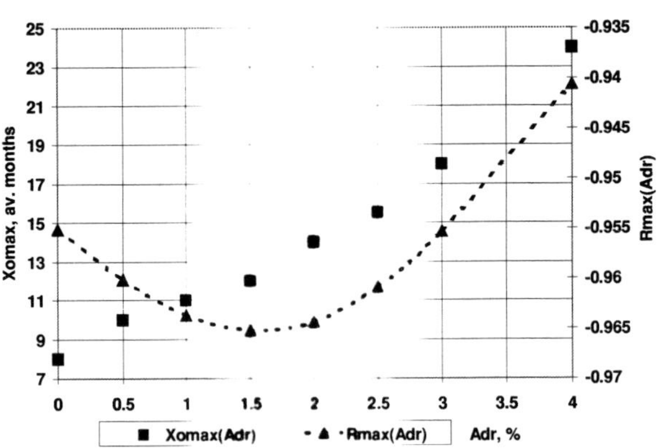

FIGURE 3. Functions $R_{\max}(A_{dr})$ and $X_{o\max}(A_{dr})$ for Climax NM data in Cycle 22.

The function $R_{\max}(A_{dr})$ can be approximated with correlation coefficient 0.9985±0.0007 by parabola

$$R_{\max}(A_{dr}) = aA_{dr}^2 + bA_{dr} + c \ , \qquad (4)$$

where a = 0.004065±0.000079, b = -0.01253±0.00024, and c = -0.9551±0.0185. From (4) we can determine $A_{dr\max}$ when $R_{\max}$ reaches the biggest value:

$$A_{dr\max} = -b/2a = 1.54 \pm 0.04\% \ . \qquad (5)$$

With this information, we can now correct the Climax NM data of Cycle 22 for drifts, with the most reliable amplitude $A_{dr\max}$ according to (5) and the function $R(X_o, A_{dr\max})$ is shown in Figure 4.

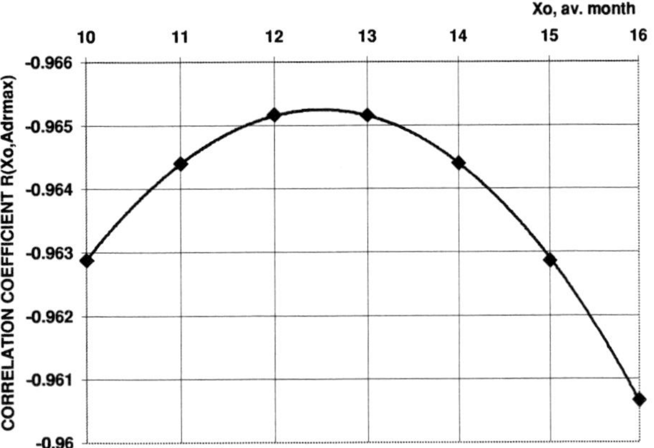

FIGURE 4. The function $R(X_o, A_{dr\max})$ for Climax NM data in Cycle 22

From Figure 4 can be seen that the function $R(X_o, A_{dr\max})$ can be approximated with a correlation coefficient 0.99994±0.00003 by a parabola:

$$R(X_o, A_{dr\max}) = dX_o^2 + eX_o + f \ , \qquad (6)$$

where d=0.000377±0.000002, e=-0.00942±0.00004, and f = -0.906±0.004. By (6) we can determine the most reliable value of $X_{o\max}$ corresponding to $A_{dr\max}$:

$$X_{o\max} = -e/2d = 12.5 \pm 0.1 \text{ av. month.} \qquad (7)$$

At obtained values of $A_{dr\max}$ and $X_{o\max}$ the connection between expected and observed CR intensity is characterized by correlation coefficient $R_{\max}(X_{o\max}, A_{dr\max}) = 0.9652$ (see Figure 4).

Results for Kiel NM Data

The function $R_{\max}(A_{dr})$ for Kiel NM (sea level; 2.32 GV) data can be approximated with a correlation coefficient 0.9992±0.0004 by (4) with regression coefficients a=0.0095±0.0001, b=-0.0250±0.0004, c=-0.960±0.014, what gives, according to (5), $A_{dr\max}$ = 1.32±0.04%. Next, we determine $R(X_o, A_{dr\max})$ that can be approximated with a correlation coefficient 0.99988±0.00006 by (6) with a regression coefficients d=0.000466±0.000003, e=-0.01191±0.00007, f=-0.897±0.005, that gives, according to (7), $X_{o\max}$ = 13.4±0.2 av. months. The obtained values for $A_{dr\max}$

and for $X_{o\max}$ are about the same as for the Climax NM. In this case the correlation between the predicted and observed CR intensity is characterized by a coefficient of $R_{\max}(X_{o\max}, A_{dr\max}) = 0.977$.

Results for Tyan-Shan NM Data

The Tyan-Shan NM (43N, 77E, near Alma-Ata; 3.34 km, 6.72 GV) is sensitive to more energetic particles than the Climax NM and the Kiel NM. For the Alma-Ata NM the function $R_{\max}(A_{dr})$ was approximated with correlation coefficient of 0.9996±0.0002 by (4), with regression coefficients a=0.0149±0.0015, b=-0.019±0.002, c=-0.957±0.009, that gives $A_{dr\max}$ = 0.634±0.012%. Next, we determined $R(X_o, A_{dr\max})$ that can be approximated with a correlation coefficient of 0.9997 by (6) with a regression coefficients d=0.000388, e=-0.00845±0.00005, f=-0.917, that gives, according to (7), $X_{o\max}$ = 10.9±0.2 av. months. In this case the correlation between the predicted and observed CR intensity is characterized by a coefficient of $R_{\max}(X_{o\max}, A_{dr\max}) = 0.963$.

Results for Huancayo/Haleakala NM Data

The Huancayo NM (12S, 75W; 3.4 km, 12.92 GV)/ Haleakala NM (20N, 156W; 3.03 km, 12.91 GV) is sensitive to primary CR particles of 35–40 GV which is about 2-3 times larger than for the Climax and Kiel NM. For Huancayo/ Haleakala NM the function $R_{\max}(A_{dr})$ was approximated with a correlation coefficient of 0.9998 by (4) with regression coefficients a=0.0621, b=-0.0165, c =-0.978, which gives $A_{dr\max}$ =0.133±0.002%. Next, we determined $R(X_o, A_{dr\max})$ that can be approximated with a correlation coefficient 0.99998±0.00001 by (6) with regression coefficients d=0.000406, e=-0.00842, f=-0.935±0.002, that gives $X_{o\max}$ =10.38±0.05 av. months according to (7). In this case the correlation between the predicted and observed CR intensity is characterized by $R_{\max}(X_{o\max}, A_{dr\max}) = 0.979$.

DISCUSSION AND CONCLUSIONS

The taking into account drift effects (see Figure 1) gives an important possibility, using data only for solar cycle 22, to determine the most reliable amplitude $A_{dr\max}$ (at W=75) and the time-lag $X_{o\max}$ (the effective time of the solar wind moving with frozen magnetic fields from the Sun to the boundary of the modulation region on the distance $r_o \approx uX_{o\max}$). We found that with an increasing effective CR primary particle rigidity from 10–15 GV (Climax NM and Kiel NM) up to 35–40 GV (Huancayo/Haleakala NM) are decreased both the amplitude of drift effect $A_{dr\max}$ (from about 1.5% to about 0.15%) and time-lag $X_{o\max}$ (from about 13 av. months to about 10 av. months). It means that in Cycle 22, for the total long term modulation of CR with rigidity 10-15 GV, the relative role of the drift mechanism was ≈1/4 and the convection-diffusion mechanism about 3/4; for rigidity 35-40 GV these values were ≈ 1/10 for the drift mechanism, and about 9/10 for the convection-diffusion mechanism. If we assume that the average velocity of the solar wind in the modulation region was about the same as the observed average velocity near the Earth's orbit in 1965-1990: u=4.41x10^7=7.73 AU/av.month, the predicted dimension of modulation region in Cycle 22 will be 100 AU for CR with rigidity of 10-15 GV and about 80 AU for CR with rigidity of 35-40 GV. It means that at distances more than 80 AU the magnetic field in solar wind is too weak to influence intensity of 35-40 GV particles.

ACKNOWLEDGEMENTS

This research was partly supported by Israel Cosmic Ray Center and Emilio Segre' Observatory (Tel Aviv University), and by INTAS grant 0810.

REFERENCES

1. Dorman, L. I., "Using galactic cosmic ray observations for determination of the Heliosphere structure during different solar cycles" *This issue*, Paper SI-36.
2. Jokipii, J.R., and Davila J.M., *Ap.. J.*, **248**, Part 1, 1156-1161 (1981).
3. Jokipii, J. R., and Thomas, B., *Ap.. J.*, **243**, 1115-1122, 1981.
4. Lee, M. A., and Fisk, L. A. , *Ap. J.*, **248**, 836-844, 1981.
5. Kota, J., and Jokipii, J. R., *Proc. 26th ICRC*, **7**, 9-12, 1999.
6. Burger, R. A., and Potgieter, M. S., *Proc. 26th ICRC*, **7**, 13-16 (1999).
7. Ferreira, S. E .S., Potgieter, M. S., and Burger, R. A., *Proc. 26th ICRC*, **7**, 77-80 (1999).

A Solar Wind Source Tracking Concept for Inner Heliosphere Constellations of Spacecraft

J.G. Luhmann[1], Yan Li[1], C.N. Arge[2], Todd Hoeksema[3] and Xuepu Zhao[3]

(1) Space Sciences Laboratory, University of California Berkeley, (2) CIRES, University of Colorado and NOAA-SEC, (3) Stanford University

Abstract. During the next decade, a number of spacecraft carrying in-situ particles and fields instruments, including the twin STEREO spacecraft, ACE, WIND, and possibly Triana, will be monitoring the solar wind in the inner heliosphere. At the same time, several suitably instrumented planetary missions, including Nozomi, Mars Express, and Messenger will be in either their cruise or orbital phases which expose them at times to interplanetary conditions and/or regions affected by the solar wind interaction. In addition to the mutual support role for the individual missions that can be gained from this coincidence, this set provides an opportunity for evaluating the challenges and tools for a future targeted heliospheric constellation mission. In the past few years the capability of estimating the solar sources of the local solar wind has improved, in part due to the ability to monitor the full-disk magnetic field of the Sun on an almost continuous basis. We illustrate a concept for a model and web-based display that routinely updates the estimated sources of the solar wind arriving at inner heliospheric spacecraft.

DISPLAY TOOLS FOR EVALUATING SOLAR WIND SOURCES

Several web-based space weather tools for corona and solar wind monitoring have been developed over the last few years that take advantage of regular full-disk magnetograph observations and the Potential Field Source Surface Model (e.g. Altschuler et al., 1969; Wang and Sheeley, 1992). One of these, at SSL-UCB, keeps track of the coronal hole and coronal magnetic field configuration, including changes in the open and closed regions between full-disk observations (currently available at the URL http://sprg.ssl.berkeley.edu/mf_evol). The other is part of a Rapid Prototyping Center activity at NOAA-SEC where it is used to make predictions of solar wind speeds and magnetic field polarities at the Earth (URL http://solar.sec.noaa.gov/ws/index.html).

These applications are possible because of the typically daily construction of "updated" photospheric magnetic field synoptic maps that occurs at Wilcox Solar Observatory, and at the NOAA-SEC for other magnetic observatories including Mt. Wilson, Kitt Peak, and the GONG network. The SOHO-MDI full-disk data are also used for this purpose, and moreover provide the possibility of updates at 96 minute cadence - or better during limited campaign periods. Figures 1a-b show some of the standard displays from the aforementioned websites. The programs to generate these run automatically on standard platforms. Looking ahead to STEREO and the Heliospheric Sentinels, this capability can be exploited toward identification of likely sites of solar wind sources for each spacecraft location.

While the Potential Field Source Surface Model has well-understood limitations, it provides an excellent approximation to the observed coronal holes (e.g. Levine, 1982), and projected interplanetary magnetic field polarities {3}. Such mappings of the inferred coronal hole flows to the source surface, and then into the heliosphere, were first demonstrated in the era of SkyLab {7,4}. In these mappings, the coronal potential field model open field lines are regarded as solar wind streamlines in the corona. Kinematic extrapolations from the source surface, typically assumed at ~2.5 solar radii, provide the connection to heliospheric sites where in-situ plasma and field measurements are made by spacecraft.

CP679, *Solar Wind Ten: Proceedings of the Tenth International Solar Wind Conference*,
edited by M. Velli, R. Bruno, and F. Malara

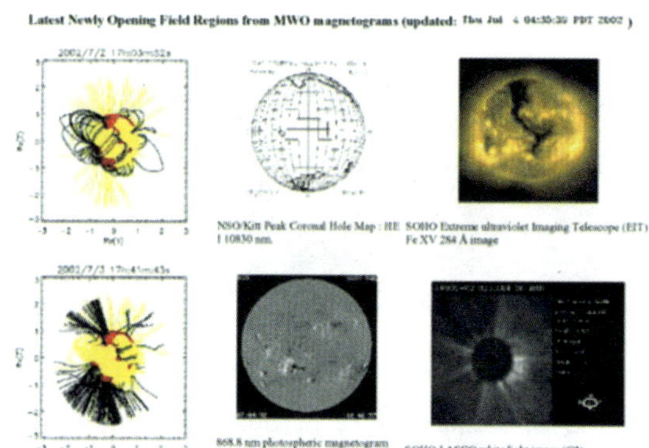

FIGURE 1A. Sample page from the website http://sprg.ssl.berkeley.edu/mf_evol that tracks changes in the coronal magnetic fields based on magnetograph observations and the potential field source surface model.

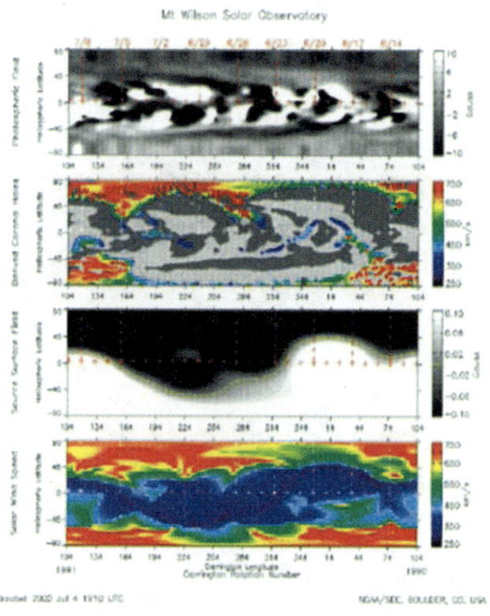

FIGURE 1B. Sample page from the website http://solar.sec.noaa.gov/ws/index.html that predicts solar wind speeds and interplanetary field polarities based on magnetograph observations and the potential field source surface model.

Arge and Pizzo {2} and Fry et al. {8} recently applied the Wang and Sheeley {11} method of estimating the solar wind velocity from the divergence of coronal open flux tubes, and then used a modified kinematic mapping corrected for the effects of stream interactions to obtain quantitative predictions of solar wind speed and interplanetary field polarity. Another application to Ulysses data analysis by Neugebauer et al. {6} compared the results of similar mappings with those from a global MHD model. They found that for the moderate solar activity interval studied, the inferred sources of the measured solar wind obtained with both models were very similar.

Although the MHD models are based on more physically rigorous treatments (e.g. Usmanov, 2000), the good agreement on the solar wind source locations obtained with the much less computationally demanding potential field source surface model approach establishes its credibility as a first-order source mapping tool. Moreover, the latter is much more amenable to the higher spatial resolution coronal field modeling required during active periods of the solar cycle.

The displays shown in Figures 2 through 7 illustrate the types of visualizations possible with only modest extensions of the framework developed for the above-mentioned websites. The advantage of these displays is that they are all 3-dimensional. They can be shown as rotating Sun movies, to give the viewer a sense of the longitudinal variation of the solar wind sources at a particular time, or with new applications like Java 3D, can be manipulated to any orientation by the website viewer using their mouse.

The general philosophy represented by these displays is the use of available synoptic data from magnetographs and other full-disk imagers, together with the Potential Field Source Surface model, to infer the solar connections of interplanetary features. Setting up the mechanisms now, using solar observations of the type we expect to have available to us in 2006, will prepare us to exploit the observations from STEREO and its partners in the heliosphere, including Messenger at Mercury, Nozomi at Mars, and especially the near-Earth spacecraft ACE and WIND.

Description of Figures

The figures in this brief report show various combinations of 3D information from observations and the potential field source surface model of the coronal hole sources of solar wind. The example of Carrington Rotation 1932 (22 Jan-17 Feb, 1998) is used for illustration. Mt. Wilson Observatory synoptic maps provide both the photospheric magnetic fields shown on the globe of the photosphere, as well as the basis for the coronal field model. They were chosen because of their particularly well-characterized polar field corrections (e.g. see Arge and Pizzo, 2000). The other synoptic maps shown are from the EIT experiment on SOHO. These are available through the NRL LASCO EIT website. The 195 Angstrom maps are used because they often exhibit dark areas similar to the coronal hole footprints obtained with the potential field source surface model.

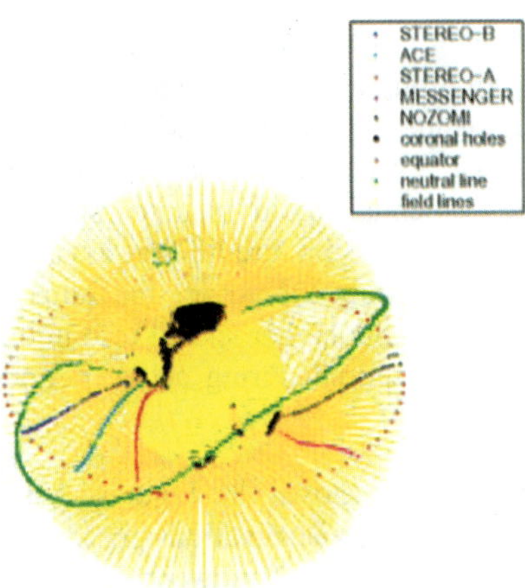

FIGURE 2. Illustration of the solar wind source tracking tool concept described in this report. Hypothetical spacecraft locations are mapped back to the source surface, and then to their coronal hole footpoints along interplanetary and coronal magnetic field lines. The coronal hole pattern on the inner boundary of the potential field source surface model is included, together with some other open coronal field lines for context. Only the coronal part of the mapping is shown here.

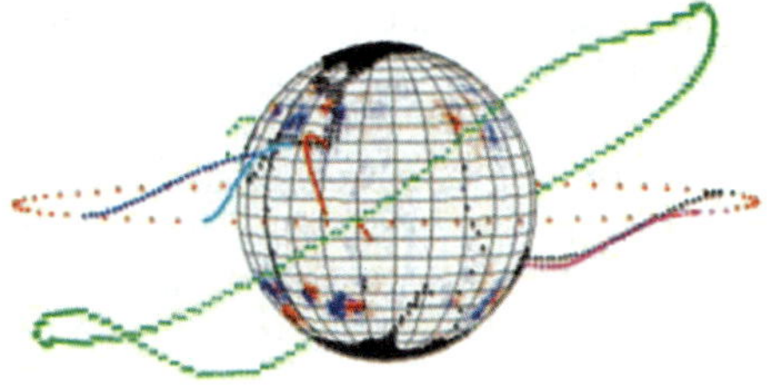

FIGURE 3. Like Figure 2, but including a synoptic map of the photospheric magnetic field from the Mt. Wilson Observatory website at UCLA for context, and eliminating all but the mapped coronal field lines that connect the spacecraft to the Sun. The relationship of the connection points to active regions is the aim of this display.

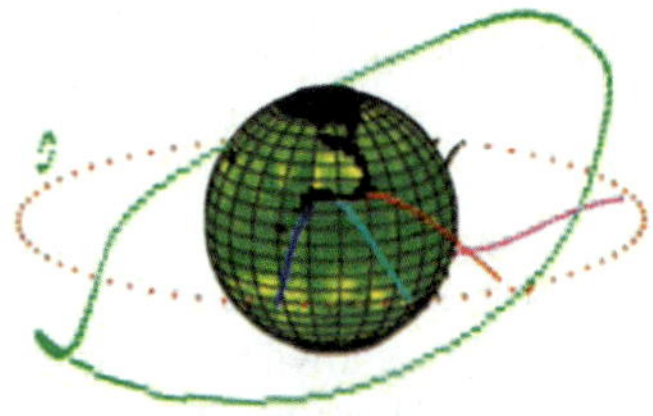

FIGURE 4. Like Figure 3, but including a synoptic map of 195 Angstrom EUV flux from the SOHO EIT website at NRL for context. The EIT maps show the coronal holes as dark features, for comparison with the calculated coronal holes.

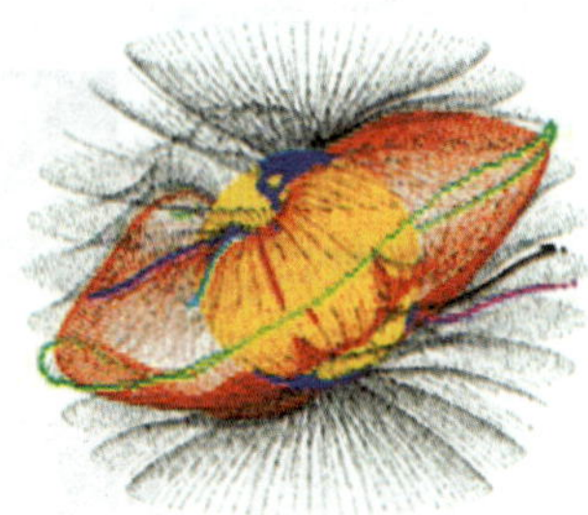

FIGURE 5. Like Figure 2, but including the last closed field lines of the coronal helmet streamer belt for context. This display emphasizes the location of the mapped points relative to the closed/open field line boundaries.

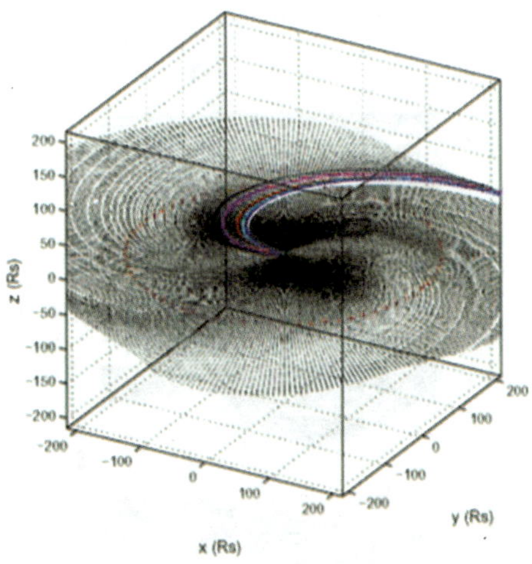

FIGURE 6. Illustration of a subsequent mapping of magnetic field lines from the coronal model source surface out to 1 AU using a modified kinematic approach, More sophisticated techniques, especially MHD models, can be substituted here for more physical realism.

The fine dotted lines are coronal magnetic field lines from the spherical source surface (Rss=2.5 solar radii) potential field model. The green dotted line represents the trace of the heliospheric neutral sheet on the source surface, while red-dotted circles indicate either the source surface equator or Earth's orbit at 1 AU. The remaining colored dotted lines are those that map to near-term inner heliospheric spacecraft at hypothetical locations in their orbits. These mappings can be used to infer the arrival of wind from a particular feature such as a polar coronal hole extension, and the context of the spacecraft with respect to the stream structure and magnetic sector boundaries. Our extrapolation from the spacecraft

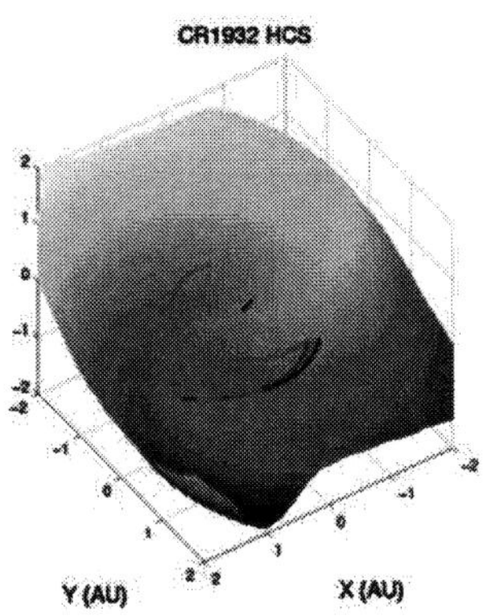

FIGURE 7. Heliospheric current sheet inferred from the potential field source surface coronal model and modified kinematic interplanetary mapping. The 1 AU circle and spiral line segments indicate the Earth's orbit and mapped interplanetary field lines intersecting the 3 dimensional opaque structure used to render the current sheet.

location back to the source surface uses a modified kinematic approach with ad-hoc stream interactions similar to those applied by Fry et al.{8} and Arge and Pizzo {2}. In this case they are introduced into a radial solar wind flow with an initially heliomagnetic latitude-dependent speed. The results from more rigorous approaches can of course be substituted in these visualizations. The solar wind extrapolations also allow 3D rendering of the prevailing heliospheric current sheet configuration with respect to the mapped field lines, akin to those of Riley et al. {9, also this volume}.

Altogether, this source tracking toolkit can be regularly updated in a space-weather forecasting fashion or used for retrospective analyses. With multipoint (e.g. SOHO and STEREO) mission images on the horizon, the ability to illustrate and examine the inferred 3D solar wind structure will be a key part of tying the in-situ measurements into our "picture" of the inner heliosphere.

ACKNOWLEDGMENTS

The visualizations described here could not have been done without open access to solar data provided by Mt. Wilson Observatory (Roger Ulrich and coworkers at UCLA), and the SOHO EIT investigators (Nathan Rich at NRL, especially). This work was supported in part by the Space Weather Program at NSF as part of award 164766 to Boston University, and by the Solar MURI project at UC Berkeley, sponsored by the Air Force Office of Scientific Research.

REFERENCES

1. Altschuler, M.D. and G. Newkirk, Jr., Magnetic fields and the structure of the solar corona, *Solar Physics*, **9**, 131, (1969).

2. Arge, C.N. and V.J. Pizzo, Improvement in the prediction of solar wind conditions using near-real time solar magnetic field updates, *J. Geophys. Res.* **105**, 10465, (2000).

3. Hoeksema, J.T., J.M. Wilcox and P.H. Scherrer, Structure of the heliospheric current sheet: 1978-1982, *J. Geophys. Res.* , **88**, 9910, (1983).

4. Levine, R.H., The relation of open magnetic structures to solar wind flow, *J. Geophys. Res.*, **83**, 4193, (1978).

5. Levine, R.H., Open magnetic fields and the solar cycle, I. Photospheric sources of open magnetic flux, *Solar Phys.* **79**, 203, (1982).

6. Neugebauer, M., et al., Spatial structure of the solar wind and comparisons with solar data and models, *J. Geophys. Res.*, **103**, 14,587, (1998).

7. Nolte, J.T., et al., Coronal hole sources of solar wind, *Solar Phys.*, **46**, 303, (1976).

8. Fry, C.D.; W. Sun, C.S. Deehr, M. Dryer, Z. Smith, S.I. Akasofu, M. Tokumaru, M. Kojima, Improvements to the HAF solar wind model for space weather predictions, *J. Geophys Res.*, **106**, 20,985, (2001).

9. Riley, P., J.A. Linker and Z. Mikic, An empirically driven global MHD model of the solar corona and inner heliosphere, *J. Geophys. Res.*, **106**, 15,889, (2001).

10. Usmanov, A.V., A global MHD solar wind model with WKB Alfven waves: Comparison with Ulysses data, *J. Geophys. Res.*, **105**, 12,675, (2000).

11. Wang, Y-M. and N.R. Sheeley, Jr., Solar wind speed and coronal flux tube expansion, *Ap. J.*, **355**, 726, (1990).

12. Wang, Y-M. and N.R. Sheeley, Jr., On potential field models of the solar corona, *Ap.J.*, **392**, 310, (1992).

Parameterizing the Wind 3DP Heat Flux Electron Data

S.W. Kahler*, N.U. Crooker† and D.E. Larson**

*Space Vehicles Directorate, Air Force Research Laboratory
†Center for Space Physics, Boston University
**Space Sciences Laboratory, University of California

Abstract. Solar wind heat flux (HF) electrons are valuable as tracers of the interplanetary magnetic field (IMF) topology, distinguishing positive from negative solar polarities and indicating the presence of magnetically closed CMEs when the flows are counterstreaming. All past applications of heat fluxes to determine field topologies have been based on visual inspection of color spectrograms of electron pitch angle distributions (PADs). However, HF PADs can take a range of shapes and amplitudes, which challenges the visual analysis. We now take a quantitative approach to HF analysis by parameterizing the HF PADs of the UC Berkeley 3DP data with a Fourier harmonic analysis. We have calculated the harmonic cosine coefficients A_0 through A_4 for a five-year period of the Wind 3DP data set with a 10-min time resolution. With these data we intend to derive quantitative criteria for unidirectional and bi-directional flows and other possible diagnostics of interplanetary field dynamics or configurations. Some initial considerations and results of the 3DP parameterization are presented.

1. INTRODUCTION

Solar-wind heat flux electrons have provided a powerful tool for determining the solar magnetic polarities of the interplanetary magnetic field (IMF) [1, 2]. Those electrons, with energies $E \geq 80$ eV, stream antisunward parallel to positive polarity field lines and antiparallel to negative polarity field lines, independently of the local field directions. By determining whether the pitch angle distribution (PAD) is concentrated at 0° or 180°, i.e., parallel or antiparallel to the field, respectively, we can establish the solar polarities of the IMF. It is possible to detect fields locally turned back to the Sun or, with bidirectional electron (BDE) flows, the closed fields of ICMEs (Figure 1).

The first plots of HF flow directions were based on a solar-pointing coordinate system and required a visual comparison of the observed IMF direction with the HF flow direction to estimate the PAD (e.g., [3, 4]). HF plots from current spacecraft are done in a coordinate system fixed on the IMF direction, which immediately yields the PAD. While simplifying the analysis, it does not avoid the considerable variation among PADs (e.g., [5, 6]) which sometimes leaves the net HF flow direction or the presence of bidirectionality [3] in doubt. In addition, the PADs are usually plotted with a color table which may or may not be normalized for the total number of electrons observed in the PADs. The recent reports of HF depletions at 90° PA [6, 7] suggest another possible HF diagnostic of IMF structure, but the identification and magnitude of these features is compromised when we are limited to a subjective analysis based on the color-coded PAD contour plots.

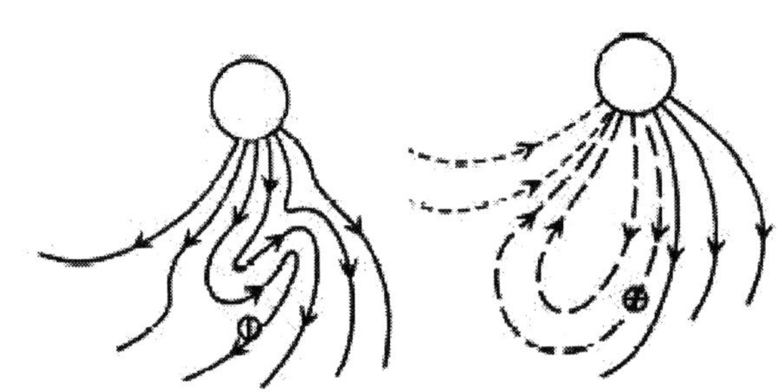

FIGURE 1. Schematic illustrations of open fields of the IMF locally turned back to the Sun (left) and a closed field topology of an ICME (right). The polarity of the turned back fields can be inferred from the primary direction of the HF PAD, and the closed fields are indicated by bidirectional HF PADs.

2. ANALYSIS

To provide quantitative diagnostics for the different types of PADs, we are beginning to analyze the 3DP [8] HF electron PAD data in terms of Fourier cosine harmonics using the following least-squares fit to the

CP679, *Solar Wind Ten: Proceedings of the Tenth International Solar Wind Conference*, edited by M. Velli, R. Bruno, and F. Malara

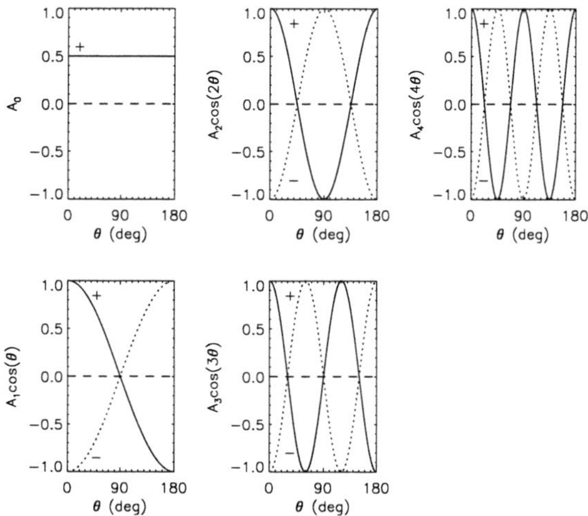

FIGURE 2. Graphical presentations of the terms of the fits to equation (1). The A_2 and A_4 terms are important for bidirectionality and depletions, respectively. Dotted lines show profiles of negative values.

13-point HF PAD intensities

$$\log I(\theta) = A_0 + A_1\cos(\theta) + A_2\cos(2\theta) + A_3\cos(3\theta) + A_4\cos(4\theta)$$

where θ is the PA, which ranges from 0° to 180°. Because of the large dynamic range of the HF intensities, we do a logarithmic rather than a linear fit to the intensity I. Plots of the five harmonic terms as functions of θ are shown in Figure 2. Note that a bidirectional flow [3] should have a significant positive A_2 term and a depletion [6, 7] a significant negative A_4 term. We have now calculated the five harmonic components for all the WIND 3DP HF data from 1995 through 1999 in 10-min intervals for electron energies of 125, 250, and 500 eV.

To examine the signatures of the various A_n we selected 3 days with very different characteristic PADs (Figure 3) : 1 Jan 1995 has a poorly defined and relatively flat PAD; 3 Aug 1996 has a well defined 0° PAD and 25 Jun 1998 has a BDE PAD, with peaks at 0° and 180°, most of the day.

2.1. A_2 and the BDE Diagnostic

A_2 is the key parameter for detecting BDE flows. Figure 4 shows the plot of log I(θ) for various ratios of A_2/A_1 when A_2 and $A_1 > 0$. We see that bidirectionality is not apparent until $A_2/A_1 \sim 1$. Selecting an arbitrary value of A_2/A_1 to define periods of bidirectionality will be a critical choice for our anticipated 3DP survey. We note that Richardson and Reames [9] chose $A_2/A_1 > 0.8$ for their survey of bidirectional energetic ($\sim$ 1 MeV) ion

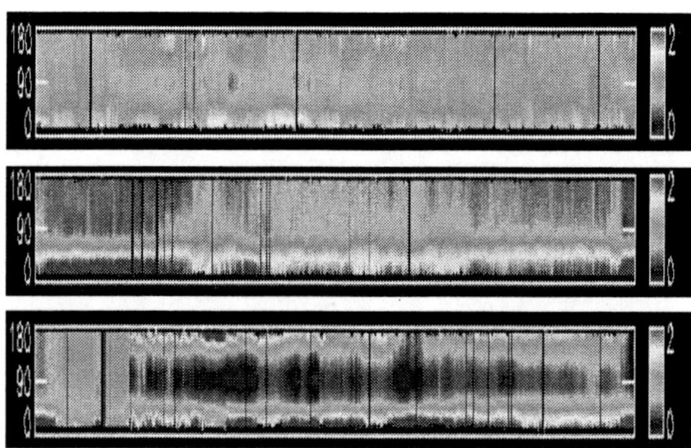

FIGURE 3. Three days with very different 3DP 260 eV electron PADs: top, flat PADs of 1 Jan 1995; middle, directed PADs of 3 Aug 1996; bottom, BDE PADs of 25 Jun 1998. The PA scales extend from 0° on the bottom to 180° at the top.

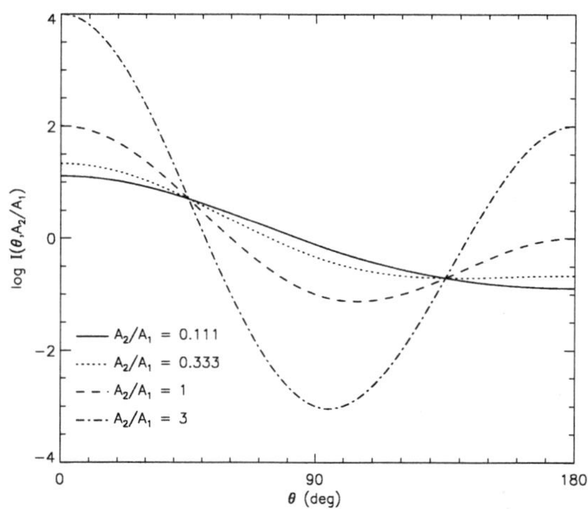

FIGURE 4. Plot of log I(θ) for four different ratios of $|A_2|/|A_1|$ with other A_n set to 0.

flows.

Our goal here is to get a general impression of the parametric variations among the three days with the very different PADs as determined by eye. We plotted log($|A_2|$) against log($|A_1|$) for the 10-min intervals of each of the three selected days and at each of the three electron energies used in the harmonic analysis. Figure 5 shows the results. The diagonal lines of the plots enable us to compare the magnitudes of A_2 with A_1. We see first that the plots are very similar for each of the three selected electron energies, suggesting that the particular energy range chosen as the standard for analysis is not critical. We also find that A_2, nearly always positive throughout each period, exceeds 0.2 for most of the BDE

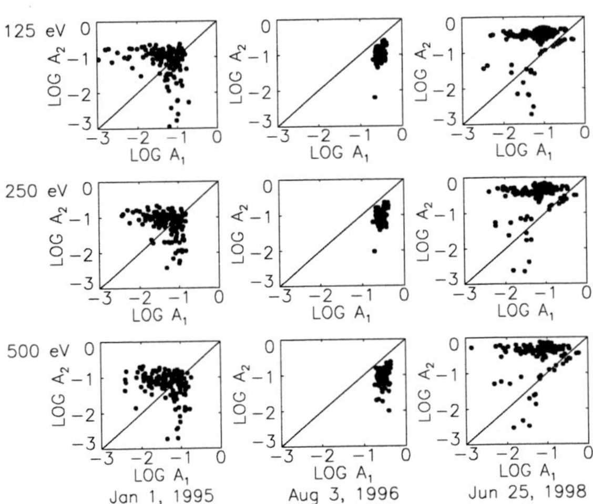

FIGURE 5. Plot of log($|A_2|$) against log($|A_1|$) for the three different electron energies and for the three different PAD distributions shown in Figure 3. The right column contains points from the day of obvious BDE PADs. Above the diagonal lines $|A_2| > |A_1|$.

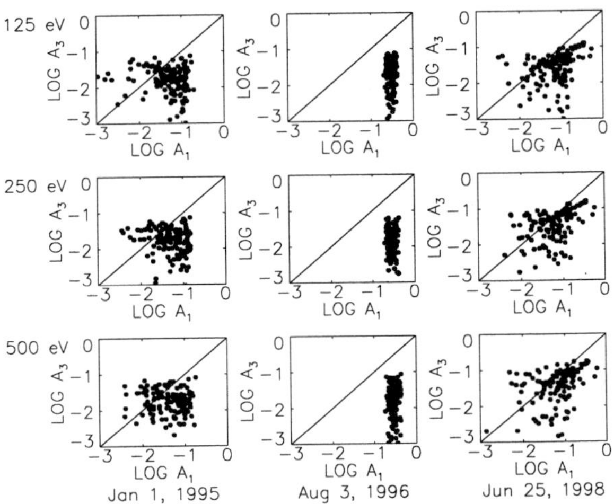

FIGURE 6. Plots of log($|A_3|$) against log($|A_1|$) with the same format as in Figure 5.

PAD points, but is rarely that large in the flat and directed PADs. In addition, while $A_2 > A_1$ in the BDE PADs and $A_2 < A_1$ in the directed PADs, perhaps as expected, we also find that usually $A_2 > A_1$ in the flat PADs. As a possible strategy for selecting criteria for BDEs, we could plot A_2 versus A_1 data points for the whole 1995-1999 period, looking for a separate population of points in the approximate ranges defined by $A_2 \geq A_1$ and $A_2 > 0.2$.

2.2. A_3 and the HF Flow Diagnostic

The most important goal in this analysis is to determine when we have a sufficiently clearly defined HF flow direction to decide the IMF polarity. From Figure 2 we see that when A_3 and A_1 have the same sign and $A_3 > A_1$, the flow direction is reversed from that indicated by A_1 alone. In Figure 6 we show plots of log($|A_3|$) against log($|A_1|$). Points above the diagonal line indicate cases in which A_3 is large enough to result in a net HF flow reversed from that of A_1 if both terms have the same sign. A substantial number of points in the flat and BDE PADs suggest such reverse flows. However, a complication here is that about 20-30% of the A_3 points have signs opposite those of the corresponding A_1 values and therefore enhance the net HF flow. In addition, large values of A_2 could act to diminish the net HF flows determined from the A_1 and A_3 values alone.

The ambiguities involved in trying to sort out a set of relationships among A_1, A_2, and A_3 suggest that a better approach may be simply to calculate the net HF Q_e over all electron energies. We have done this and show the plots for the three days in Figure 7. The red trace is Q_e parallel to the magnetic field direction, and the purple and green traces show the Q_e in directions orthogonal to the IMF. The magnitudes of the purple plots, calculated along the axis perpendicular to the plane of the IMF direction and the Sun, are comparable to the uncertainties of the calculated Q_e along the IMF and can be used as filters for valid HF flow directions. In particular, we see that for much of the 1 Jan 1995 flat PAD plot the magnitudes of many of the perpendicular HF points are comparable to the matching points of the parallel points, suggesting poorly defined HF directions. On the other hand, the parallel HF is much greater than the perpendicular HFs for all of the directed PAD plot of 3 August 1996. Note that the HF calculation used here does not produce the systematic 20° deviations between the Q_e and B directions found by Salem et al. [10].

2.3. A_4 and Depletions

The inclusion of the A_4 harmonic shown in Figure 2 enables us to look for the cases of depletions [6, 7], the significant decreases symmetrically centered on 90° PA. Figure 8 shows plots of log(A_4) against log(A_2) with the positive and negative values of log(A_4) plotted on separate panels. Recall that A_2 is nearly always positive in our selected data sets, producing a broad PAD depression at 90°. A negative A_4 term is consistent with the further narrow, symmetric depression at 90° that characterizes depletions. In most cases of Figure 8 $|A_4| < |A_2|$, and A_4 is negative (bottom panels). While these limited

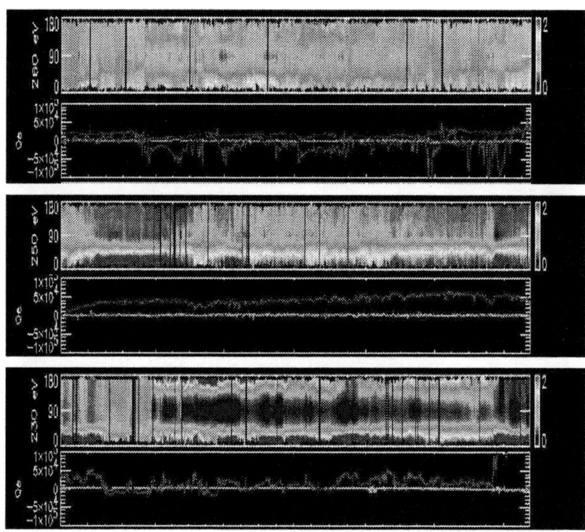

FIGURE 7. Pairs of 3DP 260 eV PADs (top) and Q_e (bottom) for the same three days of Figure 3. In the Q_e plots the red trace shows Q_e along the IMF direction, the purple along a perpendicular to the plane of the IMF direction and the Sun and the green along the third orthogonal axis.

results suggest that depletions may be defined by the criteria that $A_4 < 0$ and $|A_4| > 0.5 \times |A_2|$, a much larger survey of the 3DP data parameters will be required to justify those criteria.

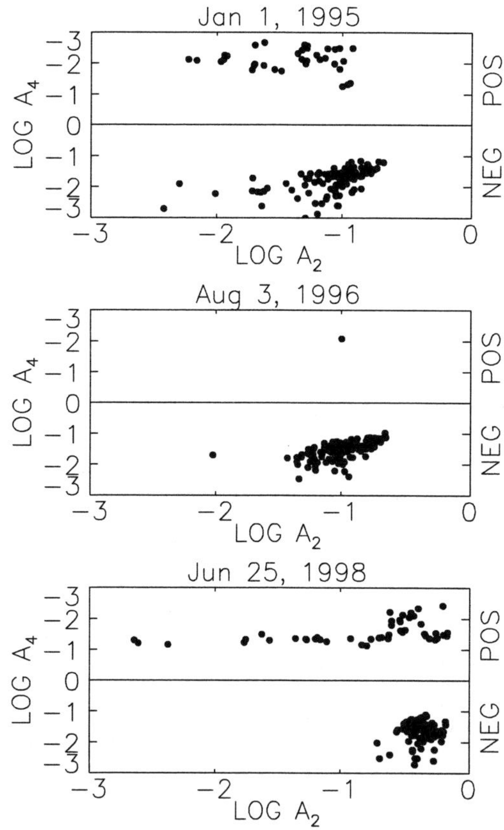

FIGURE 8. Plots of log(A_4) versus log(A_2) for the three days of the study. Here we distinguish between the positive (upper panels) and negative (lower panels) signs of A_4. The predominately negative values of A_4 are expected when PAD depletions [6, 7] occur.

3. DISCUSSION

The five-year data base of the 3DP HF electron data has been analyzed in terms of the first four Fourier cosine coefficients over 10-min averages. We intend to develop quantitative criteria for the A_n coefficients that will enable us to determine IMF solar polarities and to find periods of BDEs and depletions. In this initial work we have examined characteristic parametric tradeoffs among the A_n for three selected days with very different kinds of PADs.

ACKNOWLEDGMENTS

This work was supported by NASA under grants NAG5-10856 at Boston University and DPR W-19,926 at the Air Force Research Laboratory. We thank A. Ling for generating the computer plots.

REFERENCES

1. Kahler, S., and Lin, R.P., *Geophys. Res. Let.*, **21**, 1575, 1994.
2. Kahler, S.W., in *Coronal Mass Ejections*, edited by N. Crooker et al., Geophys. Mon. 99, AGU, p.197, 1997.
3. Gosling, J.T., et al., *J.Geophys. Res.*, **92**, 8519, 1987.
4. Kahler, S.W., Crooker, N.U., and Gosling, J.T., *J. Geophys. Res.*, **101**, 24373, 1996.
5. Feldman, W.C., et al., *Geophys. Res. Let.*, **26**, 2613, 1999.
6. Gosling, J.T., Skoug, R.M., and Feldman, W.C., *Geophys. Res. Let.*, **28**, 4155, 2001.
7. Gosling, J.T., Skoug, R.M., Feldman, W.C., and McComas, D.J., *Geophys. Res. Let.*, in press, 2002.
8. Lin, R.P., et al., *Space Sci. Rev.*, **71**, 125, 1995.
9. Richardson, I.G., and Reames, D.V., *Ap. J. Suppl. Ser.*, **85**, 411, 1993.
10. Salem, C., et al., *J. Geophys. Res.*, **106**, 21701, 2001.

Meandering Path to Solar Activity Forecast for Cycle 23

H.S. Ahluwalia

Department of Physics and Astronomy, The University of New Mexico, Albuquerque, NM 87131, USA

Abstract. Solar activity affects the shape of the corona as well as the size of the heliosphere. Also, it has a bearing on the quality of life issues on earth. So it is important that its level be forecast in a reliable manner, well ahead of time, to allow for proper planning. At Solar Wind Nine, I reported on the discovery of a three-cycle quasi-periodicity in the magnetized solar wind from the polar coronal holes late in the descending phase of a cycle. This led us to develop a procedure for forecasting the amplitude and the rise time of solar cycle 23. We predicted a moderate cycle 23 with a smoothed sunspot number (SSN) at its maximum, in early 2000: 131.5+33/-20. The solar astronomers criticized our prediction as being overly on the low side and unlikely to come true. So far, our forecast is right on the mark. We review the solar and geophysical data as of the end of May 2002 and describe the present state of cycle 23 in relation to those observed over a 400-year period and the lessons learned by us from this exercise.

INTRODUCTION

At Solar Wind Eight conference (Dana Point, CA), we questioned an early forecast for the solar cycle 23 that it would be more active than cycle 22 [Wilson, 1992]. Our study of the planetary index Ap data (1932-1994) had led us to conclude that the annual mean minimum of Ap in 1997 was likely to be like those observed in 1934 and 1965 [Ahluwalia, 1995]; the inferred three-cycle quasi-periodicity in the magnetized solar wind from the coronal holes, seemed to be similar to that observed in the frequency of occurrence of auroras for a period of about four and a half centuries and in the SSNs for 1868-1990 period reviewed by Silverman [1992]. The trend in the galactic cosmic ray variations appeared to be similar [Ahluwalia, 1997a]. We speculated that one may observe ~ 100 SSNs at the cycle 23 maximum. We reviewed the available data for Ap (1932-1997) and aa (1868-1997); these indices are commonly used as precursors for forecasting the size of the new cycle [Ahluwalia, 1998]. A new procedure was devised for computing the annual mean SSNs at the maximum (Rmax) for cycle 23 (119.3 +/-30); we predicted that it would be a moderate cycle (a la cycle 17). Our prediction was criticized as being overly on the low side and unlikely to come true [Wilson and Hathaway, 1999]. We defended our forecast [Ahluwalia, 1999a]. At the Solar Wind Nine conference (Nantucket, MA), we updated our forecast using the smoothed SSN data (to suppress the transients), taking account of the fact that the onset of cycle 23 occurred in May rather than in October 1996, assumed in our earlier work [Ahluwalia, 1996]. We reaffirmed that cycle 23 will be moderate with a smoothed SSN at its maximum (in early 2000): 131.5+33/-20.

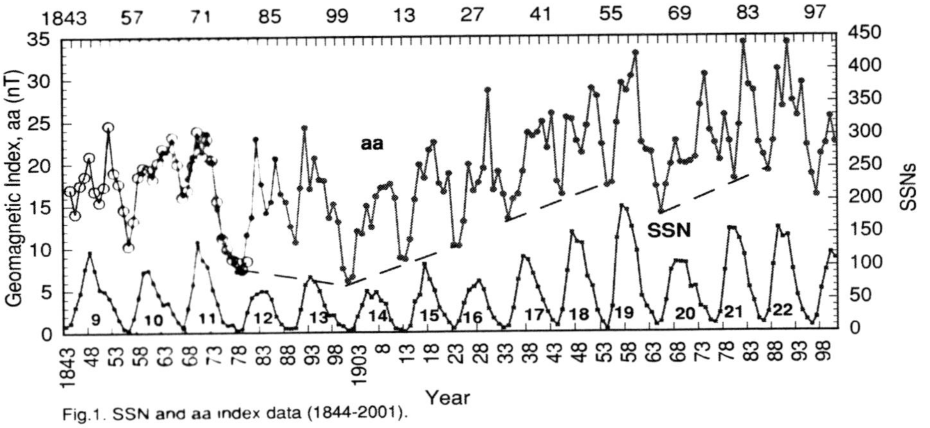

Fig.1. SSN and aa index data (1844-2001).

Cycle 23 Rise Time

We present a new method for computing the rise time (Tr) of a solar cycle, using aa index data (1844-2001) depicted in the upper half of the Figure 1; also shown are the corresponding SSN data for cycle 9 onwards. The aa index was devised by Mayaud [1972] to study

CP679, *Solar Wind Ten: Proceedings of the Tenth International Solar Wind Conference*, edited by M. Velli, R. Bruno, and F. Malara

the long term changes in the intensity of magnetic activity on planet earth. He combined the data from the two antipodal magnetic observatories to cancel the observed daily and seasonal variations in the record, leading to the variations of a planetary character (shown by points in the figure) from 1868 onwards. Nevanlinna and Kataja [1993] extended Mayaud's series back in time for two additional solar cycles (1844 onwards) using the magnetic declination observations made at the Helsinki magnetic observatory in Finland, indicated by open circles in Figure 1. The two data strings overlap nicely for the common period. The following characteristic features may be noted in the time series.

1. After 1901, the aa index seems to be riding on a line of a positive slope. A question arises whether this rise will continue indefinitely. Feynman [1982] ascribed this trend to the increasing phase of the long solar cycle with an average period of 87 years (Gleissberg cycle). This interpretation implies that aa indices should have decreased rapidly after the 1950s. This did not happen, indicating that the situation may be more complex and longer periods may be present in the time series. For example, Silverman [1992] notes that recurring minima near the turn of a new century are typical of the auroral occurrence data from 1500 onwards.

2. The three-cycle quasi-periodicity in the annual mean aa index (near solar cycle minima) are highlighted by the dashed lines; no such trend is present in SSN time series. Note that the slope of the dashed line is negative prior to 1901. A question arises whether the quasi-periodicity will continue after the solar minimum circa 2006 and in what direction. Our forecast for cycle 24 will be greatly influenced by what happens then.

3. Since the slope of the line marking the longer-term trend must turn negative eventually, the data suggest the presence of a cyclic variation of greater than 100 years in aa index.

4. Beginning with cycle 10, one observes a pattern where even cycles of the even-odd pairing are less active; it disappeared after cycle 21. The physical cause for this pattern is unknown. One wonders whether this pattern will manifest itself again in the future.

Figure 2 shows a plot of the annual mean aa index at the minimum (aamin) of the smoothed SSN cycle versus the rise time (Tr) of the cycle (defined as the months between the smoothed SSNs at the minimum and maximum), beginning with cycle 9 minimum (1855). The distribution of points indicates a linear trend. The fit parameters are given by,

$$\text{Tr (months)} = 61.6 — \text{aamin (nT)} \quad (1)$$

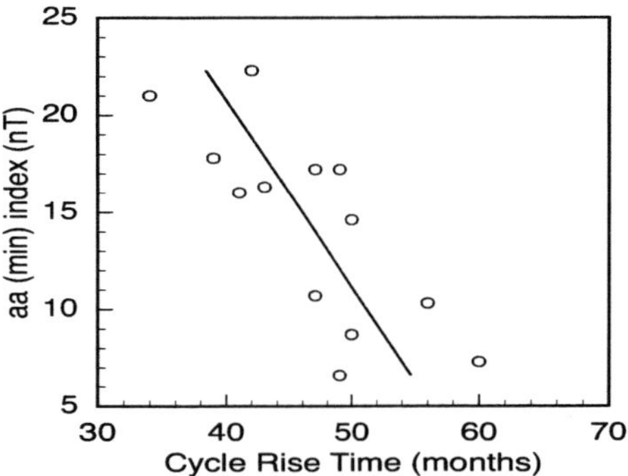

Fig.2. A linear correlation between aa and Tr.

with a correlation coefficient (cc) = - 0.76 at a confidence level (cl) > 99% [see Appendix C in Taylor, 1985]. For 1996, aamin = 18.6 nT, so Tr = 43 mos; with the observed scatter of the points within +/-6 mos off of the fit, giving a forecast for cycle 23 rise time Tr = 43 +/-6 mos. Cycle 23 reached maximum in April 2000 (47 months after onset), well within the limits set by (1).

Cycle 23 Timeline

In Figure 3, the timeline for cycle 23 is compared to those for 6 prior cycles (17 to 22) for 66 months after onset; cycles are normalized at the origin by subtracting SSN at onset from those for the subsequent months. The

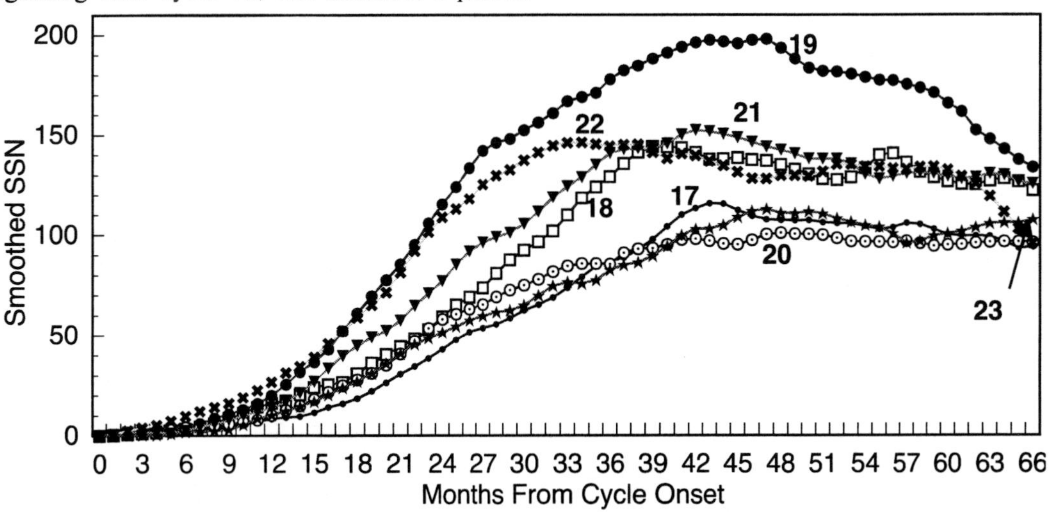

Fig. 3. timeline of cycle 23 is compared to those of 6 prior cycles for 66 months.

following features may be noted.

1. The timelines may be classified into 2 groups. Cycles 18, 19, 21, 22 exhibit above average activity; 19 was the most active cycle ever observed in nearly 400 years of SSN observing period. Cycles 17 and 20 exhibit moderate activity. A clear separation occurs between the two groups 30 months after their onsets; cycle 17 starts out less active than 20 but its timeline settles at a higher level after 39 months. On the other hand, cycle 22 starts out more active than 19 but became less so after 18 months.

2. Cycle 23 starts out mimicking cycle 20 for 21 months, drifting closer to cycle 17 timeline settling below it after 33 months, rising above cycle 20 timeline after 42 months and that for cycle 17 after 45 months and lingering at the higher level afterwards, reaching a broad maximum 47 months (April 2000) after onset. It is at its second maximum as of the end of May 2002; all solar and geophysical phenomena exhibit two maxima, known as the Gnevyshev gap [Ahluwalia, 2000b].

3. The only suspense left for cycle 23 is whether it would again mimic cycle 20 in its declining phase and end up being a long cycle with a length of 12 years or be a short cycle (~10 years) like others in recent times.

Waldmeier Effect

According to Waldmeier [1955], the rise time of a cycle is determined by a single parameter, namely the SSN at its maximum (Rmax). He showed that active cycles have a steeper rise and moderate ones rise more slowly.

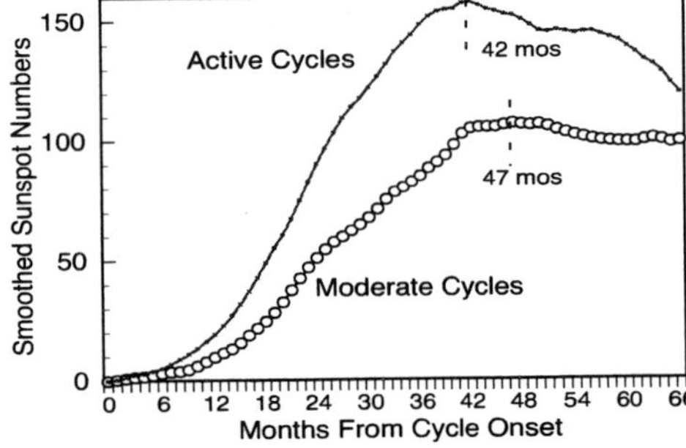

Fig. 4 Average active and moderate cycles

Figure 4 shows an average of the monthly smoothed SSNs for the recent active (18, 19, 21, 22) and moderate (17, 20, 23) cycles, plotted for 66 months after onset; all cycles are normalized at the origin, as in Fig.3. So the Waldmeier effect is alive and well but its physical cause is unknown. The moderate cycles attain their maximum an average of 5 months later and have a tendency to linger near the maximum. We have to wait another 6 years or so to learn how the fall time (Tf) compares to the rise time (Tr) for the two groups.

Historic Record

Figure 5 shows a 300-year record (1700-2000) of the annual mean SSNs. The following points may be noted.

1. There is a tendency for the solar cycles to be less active at the start of a new century. About six cycles in each century are moderate (Rmax ~ 100 or less). Four most active cycles (18, 19, 21, 22) all occurred towards the later half of the 20th century. One wonders whether this fact has a bearing on the cycles to come in the 21st century. No clear answer is available yet.

2. For the 18th and 19th centuries the average cycle length is about 11 years, since there are nine cycles per century present. However, for the 20th century more than nine cycles are present, including the rising phase of the cycle 23. This may be indicative of the presence of a cyclic variation of greater than 100 years in SSNs .

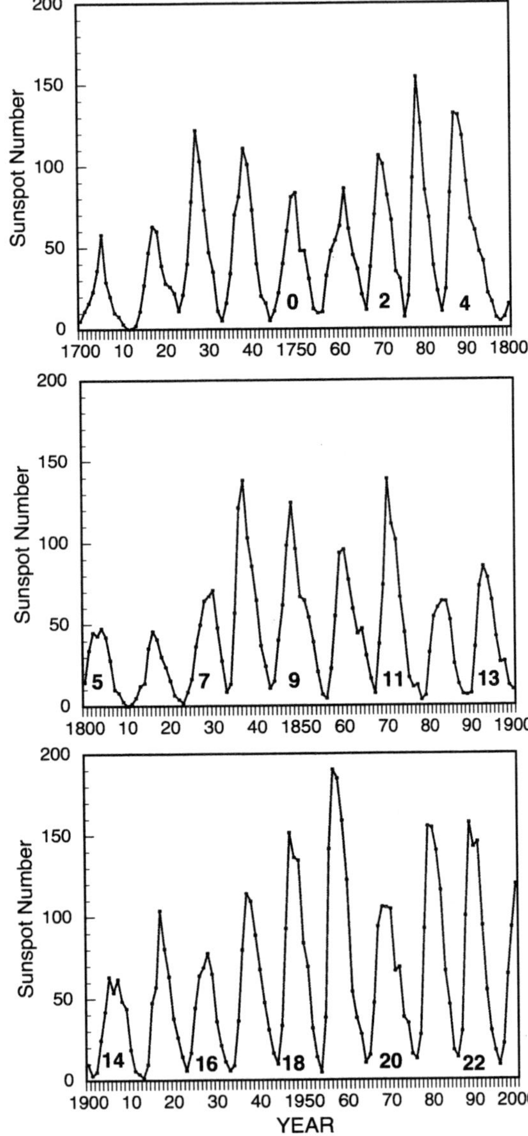

Fig. 5 Annual mean SSNs (1700-2000)

3. As noted earlier, one observes a pattern where an even cycle of the even-odd pairing is less active, beginning with cycle 10 and ending with cycle 21. It is not clear how this is linked to the workings of the solar dynamo.

DISCUSSION

Ohl [1971] was the first to realize that sun advertises in advance what to expect by way of its activity (SSNs) in a new cycle. He posited that sun's subliminal message is conveyed to earth via geomagnetic indices; he used sum Kp data. However, he could not explain how this incretion of sun s message actually occurs. At first sight it sounds incredible that planetary indices (aa, Ap, Kp) should be such good solar activity development indicators. We suspected that solar wind must be the carrier of the weak solar signal. So, we carried out a detailed analysis of the available solar wind data (1963-1998) to understand the sun-magnetosphere relationships and Ap's role in it [Ahluwalia, 2000b]. We discovered that the three-cycle quasi-periodicity in the annual mean Ap minima in a solar cycle may be ascribed to the corresponding time variations of the flux of the open field lines of the solar magnetic field carried by the high speed solar wind streams (HSSWS) from the coronal holes (measured in situ at earth's orbit), late in the declining phase of a cycle. We suggested that the pertinent information from the sun is transferred to the magnetosphere via temporal fluctuations of the induced interplanetary electric field, leading to the appropriate temporal variations of the planetary indices. These results appear to be in qualitative agreement with Babcock's Solar Dynamo model [1961] which outlines a scheme whereby high latitude solar poloidal fields near solar minimum appear as toroidal fields on the opposite sides of the solar equator in the new cycle. We speculate that the precursor solar poloidal fields are entrained in HSSWS and brought to earth's orbit. It is not clear whether the three-cycle quasi-periodicity is generated on the surface of the sun at high latitudes where Babcock's dynamo operates or whether it has to do with an intrinsic property of the circulation pattern in the convection zone under the solar surface. Some light may be shed on these issues if the solar modelers attempt to solve the MHD equations, applied self-consistently to the sun.

CONCLUSIONS

1. Ohl's hunch is correct; the sun advertises in advance what to expect by way of activity in a new cycle. The suggestion by Schatten et al [1978] that Ohl's hunch may be understood on the basis of Babcock's model of a solar dynamo appears to be plausible. While we have made considerable progress in understanding how the solar message may get embedded on the planetary indices, several details need to be worked out before we understand the physical link between the observations and the workings of the solar dynamo. Even so, our simplified operational procedure for forecasting the rise time and the amplitude for a new cycle appears to have been vindicated for cycle 23. For the first time, in the long history of observations of the sunspots, a forecast was made for a solar cycle; it was defended against peer criticism and has turned out to be right on the mark.

2. The only mystery left is whether cycle 23 will follow the timeline for cycle 20 in its declining phase and be a long cycle or end up with a duration ~ 10 years like other more recent cycles.

3. Waldmeier effect is alive and well but its physical cause is not known; it may be used as a confirming tool for a forecast ~ 30 months after the new cycle onset.

4. No explanation is available at present for the disappearance of the pattern observed from cycle 10 to 21 where even cycles of the even-odd pairings were observed to be less active. It is not clear whether the pattern will reappear in the future.

5. It is too early to make a forecast for cycle 24 in view of the uncertainties discussed in this paper. We have to wait for about 4 years more for cycle 23 to have run its course.

ACKNOWLEDGEMENTS

This research is supported by NSF Grant: ATM-9904989. Smoothed SSN data are from the Sunspot Index Data Center (SDIC), at Brussels, Belgium.

REFERENCES

Ahluwalia, H.S., *Solar Wind Eight Conf. Paper,* p. 469. AIP Conf. Proc. Eds. D. Winterhalter, J.T. Gosling, W.S. Kurth, and M. Neugebauer, Woodbury, NY, (1995).

Ahluwalia, H.S., *Int. Cosmic Ray Conf. 28th*, **2**, 109 (1997a).

Ahluwalia, H.S., *J. Geophys. Res.,* **102**, 24229 (1997b).

Ahluwalia, H.S., *J. Geophys. Res.,* **103**, 12103 (1998).

Ahluwalia, H.S., *J. Geophys. Res.,* **104**, 2559 (1999a).

Ahluwalia, H.S., *Solar Wind Nine Conf. Paper,* p. 415. AIP Conf Proc. Eds. S. Habbal, R. Risser, J. Hollweg, and P. Isenberg, Woodbury, NY, (1999b)

Ahluwalia, H.S., *Adv. Space Res.,* **26**, 187 (2000a)

Ahluwalia, H.S., *J. Geophys. Res.,* **105**, 27481 (2000b).

Babcock, H.W., *Astrophys. J.,* **133**, 572 (1961).

Feynman, J., *J. Geophys. Res.,* **87**, 6153 (1982).

Mayaud, P.N., *J. Geophys. Res.,* **77**, 6870 (1972).

Nevanlinna, H., and E. Kataja, *Geophys. Res. Lett.,* **20**, 2703 (1993).

Ohl, A.I., *Geomag. Aeron.,* **11**, 549 (1971).

Schatten, K.H., P. Scherrer, L. Svalgaard, and J.M. Wilcox, *Geophys. Res. Lett.,* **5**, 411 (1978).

Silverman, S.M., *Rev. Geophys.,* **30**, 333 (1992).

Taylor, J.P., *An Introduction to Error Analysis,* Univ. Sci. Books, Mill Valley, Calif., (1985).

Waldmeier, M., *Ergebnisse und Probleme der Sonnenforschung,* p. 154, 2nd Ed., Leipzig, (1955).

Wilson, R.M., *Solar Phys.,* **140**, 181 (1992).

Wilson, R.M., and D. Hathaway, *J. Geophys. Res.,* **105**, 2555 (2000).

Propagation of a Toroidal Magnetic Cloud through the Inner Heliosphere

Eugene Romashets[*] and Marek Vandas[⁎]

[*]Institute of Terrestrial Magnetism, Ionosphere, and Radio Wave Propagation of Academy of Sciences, Troitsk, Moscow Region, 142190, Russia, Email: romash@izmiran.rssi.ru
[⁎]Astronomical Institute, Academy of Sciences, Bocni II 1401, 14131 Praha 4, Czech Republic, Email:vandas@ig.cas.cz

Abstract. An analytical solution for a potential magnetic field with arbitrary intensity around a toroidal magnetic cloud has been found. The background external field may have a gradient. The solution is used for calculation of magnetic cloud propagation. Obtained velocity profiles show a good agreement with in situ observations near the Earth's orbit.

INTRODUCTION

The idea that magnetically isolated bodies, also called "magnetic clouds", may exist in inerplanetary space has been considered before[1-5]. There are many reasons for these structures to retain common field lines with the Sun [6,7]. On the other hand, the possibility for such objects to be isolated from the solar magnetic field may not excluded. In order to verify if this hypothesis is consistent with observations of magnetic clouds, including time delays of their launches and arrivals to the Earth's orbit, their velocities observed by coronographs close to the Sun and at 1 AU, we calculated dynamics of isolated bodies of toroidal shape and found their velocity profiles for $5R_s < R < 220R_s$, where R_S is solar radius.

CALCULATION

It is assumed that the main (homogeneous) magnetic field $\vec{B}_0$ is directed along the x axis. A small magnetic field gradient B_1/L along the x axis is added to the main field. The scalar magnetic potential of the resulting field is

$$\Phi_0 = B_0 x - \frac{B_1}{2L}\left(x^2 - y^2\right), \qquad (1)$$

which yields the magnetic field components, as follows:

$$B_x = B_0 - \frac{B_1}{L}x, \quad B_y = \frac{B_1}{L}y, \quad B_z = 0 \qquad (2)$$

Φ_0 is a harmonic function, i.e., $\Delta\Phi_0 = 0$, which ensures the condition $divB = 0$ is fulfilled. A toroid with the major radius R_0 and the minor radius r_0 is inserted into this field. The intrinsic system of the toroid will be described in the toroidal coordinates μ, η, and φ:

$$x = \frac{a \sinh\mu \cos\varphi}{\cosh\mu - \cos\eta}, \quad y = \frac{a \sinh\mu \sin\varphi}{\cosh\mu - \cos\eta}, \quad z = \frac{a \sin\eta}{\cosh\mu - \cos\eta}, \qquad (3)$$

The parameter a is defined as $a = \sqrt{R_0^2 - r_0^2}$, and the surface of the toroid is given by $\mu = \mu_0$, where $\cosh\mu_0 = R_0/r_0$. The expression for the scalar potential Φ_0 is arranged as a sum of the harmonic functions x, and $x^2 - y^2$. These functions can be expressed as a sum of toroidal harmonics :

CP679, *Solar Wind Ten: Proceedings of the Tenth International Solar Wind Conference*, edited by M. Velli, R. Bruno, and F. Malara

$$x^2 - y^2 = \sqrt{\cosh\mu - \cos\eta}\,\frac{2^{5/2}a^2}{3\pi}\sum_{n=-\infty}^{\infty} Q^2_{n-1/2}(\cosh\mu)\cos n\eta\cos 2\varphi.$$

$$x = \sqrt{\cosh\mu - \cos\eta}\,\frac{2^{3/2}a}{\pi}\sum_{n=-\infty}^{\infty} Q^1_{n-1/2}(\cosh\mu)\cos n\eta\cos\varphi \qquad (4)$$

$Q^m_{n-1/2}$ are the Legendre functions of the second kind. A way to obtain these relationships is described in [8]. Now we shall look for the potential magnetic field which (i) has normal components to the surface of the toroid equal to zero and which (ii) tends to the undisturbed field (2) at larger distances from the toroid. That is, we add to the scalar potential (1), expressed in the toroidal harmonics, additional toroidal harmonic functions, for which the additional magnetic field becomes negligible when $\mu \to 0$, and for which the total magnetic field component $B_\mu = 0$ for $\mu = \mu_0$. The scalar magnetic potential Φ_1 of the total field is

$$\Phi_1 = \sqrt{\cosh\mu - \cos\eta}\,\frac{2^{3/2}a}{\pi}\times \qquad (5)$$
$$\times\Big\{B_0\sum_{n=0}^{\infty}\varepsilon_n\Big[Q^1_{n-1/2}(\cosh\mu) - a^1_n P^1_{n-1/2}(\cosh\mu)\Big]\cos n\eta\cos\varphi -$$
$$-\frac{B_1 a}{3L}\sum_{n=0}^{\infty}\varepsilon_n\Big[Q^2_{n-1/2}(\cosh\mu) - a^2_n P^2_{n-1/2}(\cosh\mu)\Big]\cos n\eta\cos 2\varphi\Big\},$$

where the coefficients a^1_n and a^2_n were formally added and they will be selected in order for the two above mentioned conditions, (i) and (ii), to be satisfied. $P^m_{n-1/2}$ are the Legendre functions of the first kind, $\varepsilon_n = 1$ for $n = 0$ and $\varepsilon_n = 2$ for n>0. Condition (ii) is fulfilled by the selection of Legendre functions of the first kind [8]. Condition (ii) implies that $B_\mu = 0$ at $\mu = \mu_0$ for all η and φ.

In Figures 1-3 magnetic field magnitude contours and field lines for the case $R_0 / r_0 = 5$ are shown.

Once the magnetic field around the toroid is found, the diamagnetic force acting on it along the x axis can be calculated:

$$F_x = -\frac{1}{8\pi}\oint\left(B_\eta^2 + B_\varphi^2\right)e_x\cdot dS = F_{x\eta} + F_{x\varphi} = \qquad (6)$$
$$= -\frac{a^2\sinh\mu_0}{8\pi}\int_{-\pi}^{\pi}\int_0^{2\pi}\left(B_\eta^2 + B_\varphi^2\right)\frac{(\cosh\mu_0\cos\eta - 1)\cos\varphi}{(\cosh\mu_0 - \cos\eta)^3}\,d\varphi d\eta$$

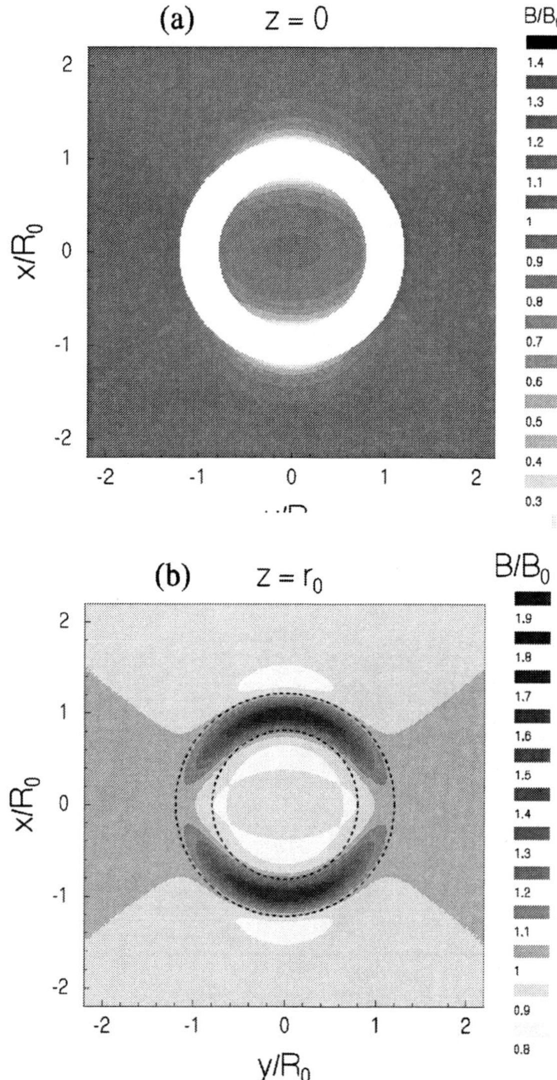

FIGURE 1. Contours of modified by the insertion of a toroid magnetic field's magnitude in xy plane, $z = 0$ (a), and $z = r_0$ (b).

In Figure 4 one can see radial velocities of toroidal magnetic cloud as a function of the radial distance from the Sun. The diamagnetic force given by (6) was used for calculation. Also gravity and the drag force of solar wind plasma were taken into account, in a way similar to that used in [8].

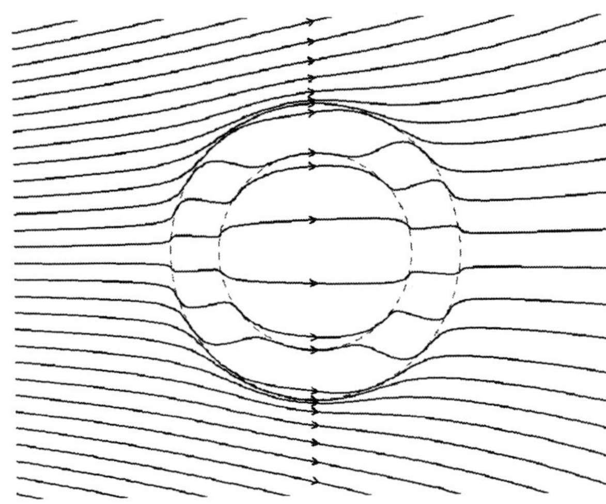

FIGURE 2. Field lines for the case of initially non-uniform field. $B_1 = B_0/4$, $L = 2\,R_0$.

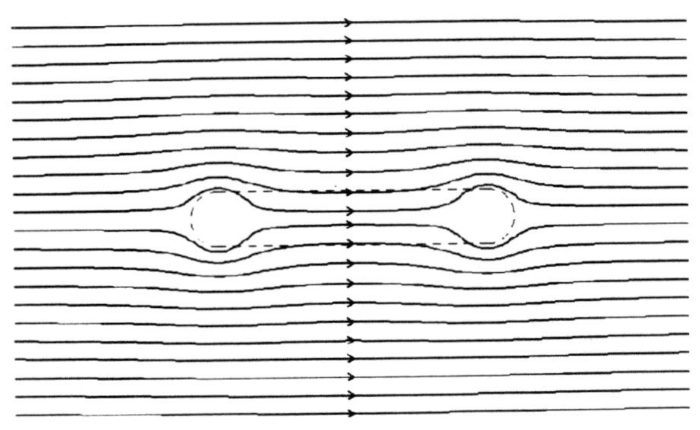

FIGURE 3. Field lines for the case of initially non-uniform field. Side view.

CONCLUSIONS

Magnetic field modification caused by the insertion of toroidal body with super-conductive walls into current free medium was calculated. This result can be applied for space and laboratory plasmas.

It is shown that the maximum field increase in the case of a sub-sonic cloud is of the order of 2 times in magnitude.

The obtained formulas for the diamagnetic force can be used for calculation of velocity profiles of toroidal clouds from the Sun to the Earth's orbit.

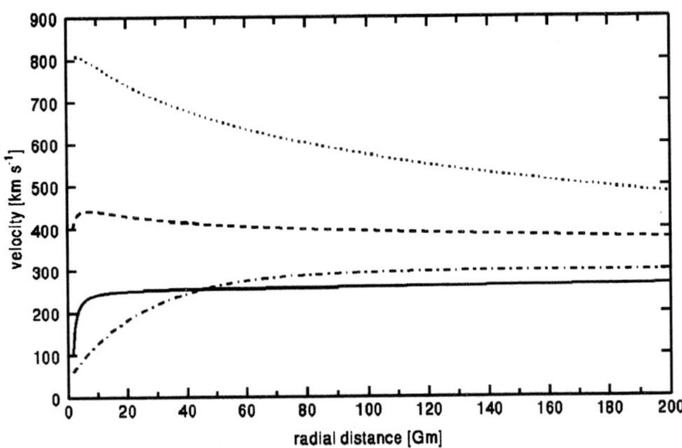

FIGURE 4. Velocity profiles of a toroidal magnetic cloud for the following initial velocities close to the Sun:100 km s^{-1} (solid line), 400 km s^{-1} (dashed line), and 800 km s^{-1} (dotted line). The dashed-dotted line shows the ambient solar wind velocity.

ACKNOWLEDGEMENTS

This work was supported by EU/INTAS/ESA grant 99-00727 and by AV CR project S1003006.

REFERENCES

1. Klein, L. W., and Burlaga, L. F., *J. Geophys. Res.*, ***87***, 613-624 (1982).
2. Ivanov, K. G., and Harshiladze, A. F., *Solar. Phys.*, ***98***, *379*-386 (1985).
3. Ivanov, K. G., Harshiladze, A. F., and Romashets, E. P., *Solar. Phys.*, ***143***, *365*-372 (1993).
4. Vandas, M., et al., *J. Geophys. Res.*, ***98***, *11467*-11475 (1993).
5. Vandas, M., et al., *J. Geophys. Res.*, ***98***, *21061*-21069 (1993).
6. Kahler, S. W., and Reames, D. W., *J. Geophys. Res.*, ***96***, 9419-9424 (1991).
7. Chen, J., and Garren, D. A., *Geophys. Res. Lett.*, ***20***, 2319-2322 (1993).
8. Romashets, E. P., and Vandas, M., *J. Geophys. Res.*, ***106***, 10615-10624 (2001).

Composition Measurements over the Solar Poles Close to Solar Maximum - Ulysses COSPIN/LET Observations.

M.Y. Hofer*, R.G. Marsden*, T.R. Sanderson* and C. Tranquille*

*Research and Scientific Support Dept. of ESA, ESTEC, Keplerlaan 1, 2201 AZ Noordwijk, The Netherlands.

Abstract. We present energetic particle composition measurements acquired on board the Ulysses spacecraft in 2000 and 2001 during its recent south and north polar passages. In an earlier study using data only from the south polar pass (Hofer et al., 2002), we found that the high-latitude composition data reflected the generally high level of solar activity present during that period. The observed particle populations during the south and north polar passages comprised predominantly of solar energetic particles (SEP) accelerated in association with CMEs, rather than particles related to SIR or CIRs. In this work, we compare the energetic particle composition signatures in the two helio-hemispheres, and find that the latest data from the north polar pass show the same transient-dominated signature as in the south.

INTRODUCTION

Following aphelion passage in 1998 the Ulysses spacecraft began the second climb to high southern heliographic latitudes, reaching its maximum latitude of 80.2° at a solar distance of ~2.27 AU on November 27, 2000 (DOY 332). On October 13, 2001 (DOY 286) the spacecraft was at the maximum northern heliographic latitude of 80.2°. The south polar pass lasted from September 6 in 2000 (DOY 250) to January 16 in 2001 (DOY 16) and the north polar pass from August 31 (DOY 243) to December 10 in 2001 (DOY 344), i.e. during which the spacecraft was above 70° heliolatitude. The second orbit of the spacecraft Ulysses during the solar maximum mission is shown in Figure 1.

A key question to be addressed by Ulysses during the polar passes at solar maximum is the origin of energetic particle populations observed at high heliolatitudes. Composition analysis can provide useful clues in this regard, allowing a clear distinction between particles accelerated at transients associated with coronal mass ejections (CMEs) and at corotating interaction regions (CIRs) or stream interaction regions (SIRs). Hofer et al. [2002], in a study of particles increases recorded at high latitudes by the COSPIN/LET instrument on board Ulysses during the south polar pass in 2000, found that the elemental particle composition in these events was consistent with coronal abundances. This suggests a transient-related, solar energetic particle (SEP) origin for the bulk of the particles, perhaps not an unexpected result given the near-maximum solar activity conditions.

In the present work, we extend this analysis done for the data recorded above the southern polar region to the subsequent time period between the two polar passes and the entire northern polar passage in 2001 as shown in Figure 1, and compare the characteristics of the particle populations in the two helio-hemispheres.

DESCRIPTION OF THE DATA

The particle data used in this study are from the Low Energy Telescope (LET) on board the Ulysses spacecraft. LET is one of the five telescopes in the Cosmic Ray and Solar Particle Investigation (COSPIN) recording the fluxes and the composition of solar energetic particles and of low energy cosmic ray nuclei from hydrogen up to iron over a range of energies from ~1 MeV/n to 50 MeV/n using solid state detectors [Simpson et al., 1992]. For the current particle composition analysis three-day-averaged values in the low MeV/n energy range are used.

The arrival times of interplanetary shocks at the Ulysses spacecraft are provided by R.J. Forsyth (private communication).

PROTON AND ALPHA INTENSITY

In order to provide context for the energetic particle composition measurements, we show in Figure 2, the hourly proton(1.2 – 3.0 MeV)/alpha(1.0 – 5.0 MeV/n) ratios and the same alpha intensity for the years 2000 and 2001. The time periods of the south and the north polar pass are marked with black horizontal lines in Figure 2 and are also shown in Figure 1.

CP679, *Solar Wind Ten: Proceedings of the Tenth International Solar Wind Conference*, edited by M. Velli, R. Bruno, and F. Malara

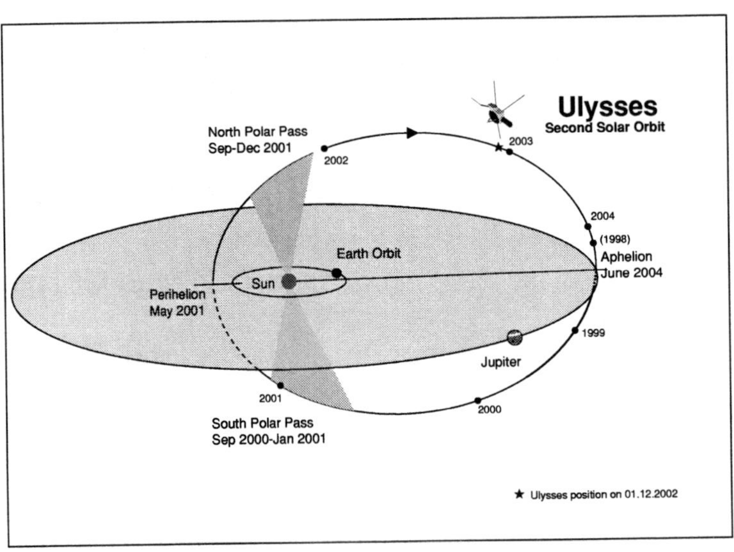

FIGURE 1. Second orbit of the spacecraft Ulysses during the solar maximum mission. The south and north polar passes are indicated in the figure.

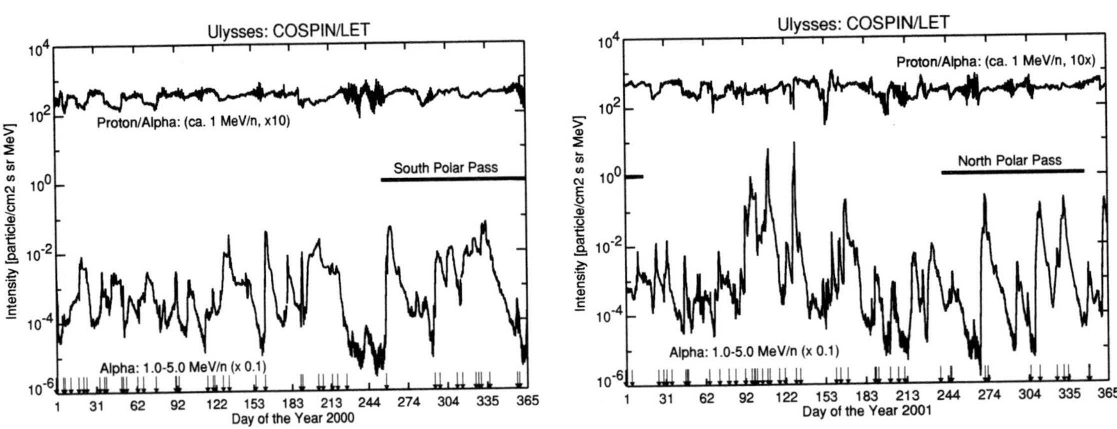

FIGURE 2. The hourly proton(1.2 – 3.0 MeV)/alpha(1.0 – 5.0 MeV/n) ratio and the alpha intensity (1.0 – 5.0 MeV/n) during the years 2000 and 2001 recorded by COSPIN/LET on the Ulysses spacecraft. The south (250/2000-16/2001) and the north polar pass (243/2001-344/2001) are marked with horizontal lines. The arrows at the bottom of the panels indicate the times of the shock arrival at the spacecraft.

During the south polar pass, transient events occurred in rapid succession, resulting in overlapping particle events. Intensities remained above background levels throughout the period.

The data from the north polar pass are similar, although the frequency of large transient MeV increases is somewhat lower, and occasionally quiet-time intervals are present. The decrease in frequency could be a direct result of the evolution of the solar cycle. The proton/alpha ratio is clearly more variable above the north solar polar region.

COMPOSITION ANALYSIS

With the energetic particle composition data it is possible to distinguish between the particle population accelerated in transients associated with CMEs from those energized by interaction regions, i.e. SIR or CIR. These two structures pick up and accelerate the material in different ways and at different locations in the heliosphere.

In Table 1, nominal CIR and SEP elemental abundances with respect to oxygen for the selected elements are listed as reported from Mason and Sanderson [1999].

The last column contains the ratios CIR/SEP and their errors. High He/O at around 2.9 and low Fe/O at around 0.6 could be used as indicators for a non-SEP situation, i.e. SIR or CIR dominated regimes.

TABLE 1. Reference CIR and SEP elemental abundances with respect to Oxygen [Mason and Sanderson, 1999] and their ratios and errors for selected elements.

	CIR	SEP	CIR/SEP
He/O	159 ± 1	55.2 ± 3	$\mathbf{2.9 \pm 0.2}$
C/O	0.89 ± 0.04	0.48 ± 0.02	1.9 ± 0.1
N/O	0.14 ± 0.01	0.13 ± 0.01	1.1 ± 0.1
Ne/O	0.17 ± 0.02	0.15 ± 0.01	1.1 ± 0.2
Fe/O	0.097 ± 0.01	0.16 ± 0.02	$\mathbf{0.6 \pm 0.1}$

In Figure 3, the three-day-averaged elemental abundance ratios of helium, carbon, nitrogen, neon and iron with respect to oxygen at approximately 5 MeV/n recorded in the years 2000 and 2001 divided by the corresponding reference SEP values are shown as a function of time. The corresponding error bars take the uncertainty of the measured element and the error of the oxygen value into account. The horizontal line is drawn at the the level of 1.0 for reference of a nominal SEP value of all elements. The 30 minutes proton intensity (1.2-3.0 MeV, 10x) multiplied by a factor of 10 is plotted for comparison in green.

In Figure 4, the three-day-averaged elemental abundance ratios of helium and iron with respect to oxygen at approximately 5 MeV/n recorded in the years 2000 and 2001 divided by the corresponding reference SEP values are shown. The horizontal lines are drawn at the the level of 2.9 and 0.6 for comparison.

Regarding the maxima of the intensity peaks measured during the north polar pass they are almost one order of magnitude higher then those measured above the south polar pass. Furthermore, the decay rates are slightly different to the southern polar ones.

In Figures 3 and 4, three enhanced He/O values close to 2.9 and Fe/O close to 0.6 mark SIR or CIR dominated regions. Values close to unity for all selected elements identify particles having an SEP signature. As noted in Hofer et al. [2002], the majority of the particle events during the south polar pass have an SEP signature, i.e. in Figure 3 the majority of values during the second solar polar pass lay within the error bar close to a measured-to-SEP ratio of unity. Inspection of the right-hand panel show that the same statement can be made with respect to the north polar pass in 2001. During the northern passage the He/O ratio never reaches 2.9. All the abundances look very similar to those recorded above the southern polar region.

During the time interval between the south polar pass and the north polar pass at least three enhanced He/O, i.e. according to the values in Table 1 SIR-like, values are found. The three enhanced He/O values do not occur with the characteristic time difference of 26 days which would be expected for a corotating structure. Therefore, a CIR dominated situation can be excluded. The first and the last high He/O ratios are accompanied by rather low Fe/O values which would also be expected for an SIR. But consulting the corresponding error bars of these values in Figure 2 it is only a slight indication for such a compression region.

The intensity profile is a result of overlaping intensity events. Comparing the three-day-average SIR-like values with the proton intensity time profile no striking time coincidence with a large increase in the intensity can be found.

DISCUSSION AND CONCLUSIONS

The recent high-latitude observations correspond to near-maximum activity conditions with large transient phenomena, e.g. Hofer et al. [2001], Marsden et al. [2001], McKibben et al. [2001] and Sanderson et al. [2001].

The data acquired during the north polar pass show a slightly reduced frequency in the number of ~5 MeV/n particle events, and occasional returns to near-background levels, indicating a small change in the level of activity with respect to the south solar pass. This is supported by the fact that continuous high-speed solar wind flow was measured at Ulysses during the north polar pass, presumably originating in the newly-formed northern polar coronal hole. Furthermore, the maxima and the decay rates of the events were found to be slightly different over the north polar region as over the south pole. Based on other observations, e.g. magnetic field measurements, the overall situation in the inner heliosphere turns out to be slightly different during the north polar pass in 2001 than during the south polar pass beginning at the end of 2000.

During the time period between the south and north polar passes, the majority of the MeV particles had an SEP signature. There were a small number of instances of three-day-averaged SIR-like composition with enhanced He/O ratios and lowered Fe/O ratios. The corresponding intensity time profiles reflect overlaping of several events. We suggest that, during the same time period a few single compression regions causing the characteristic change in the composition data were embedded in a large number of SEP populations.

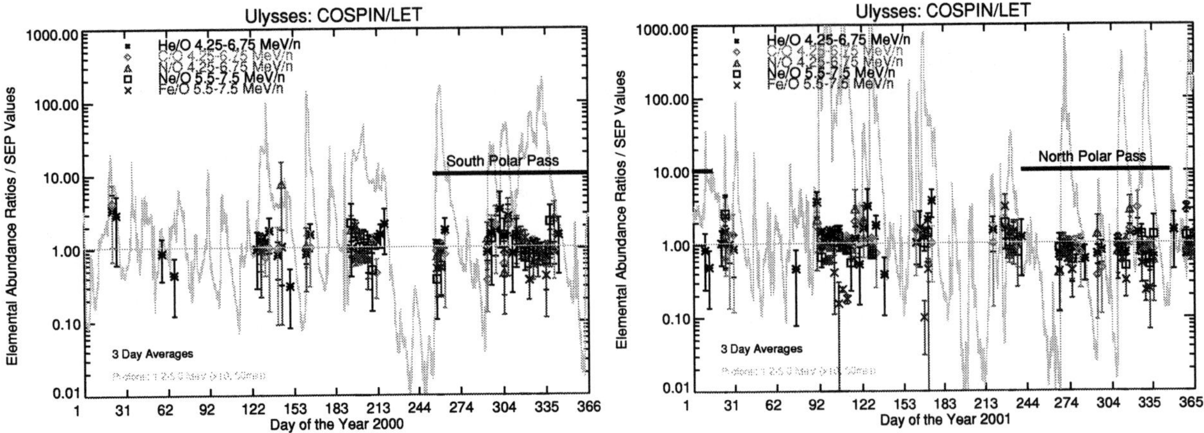

FIGURE 3. The three-day-averaged elemental abundance ratios for the energy ranges as indicated of helium, carbon, nitrogen, neon, and iron with respect to oxygen divided by the corresponding reference SEP values for the years 2000 and 2001. The solid horizontal line marks a nominal SEP value at unity. The proton intensity (1.2-3.0 MeV) is plotted for comparison.

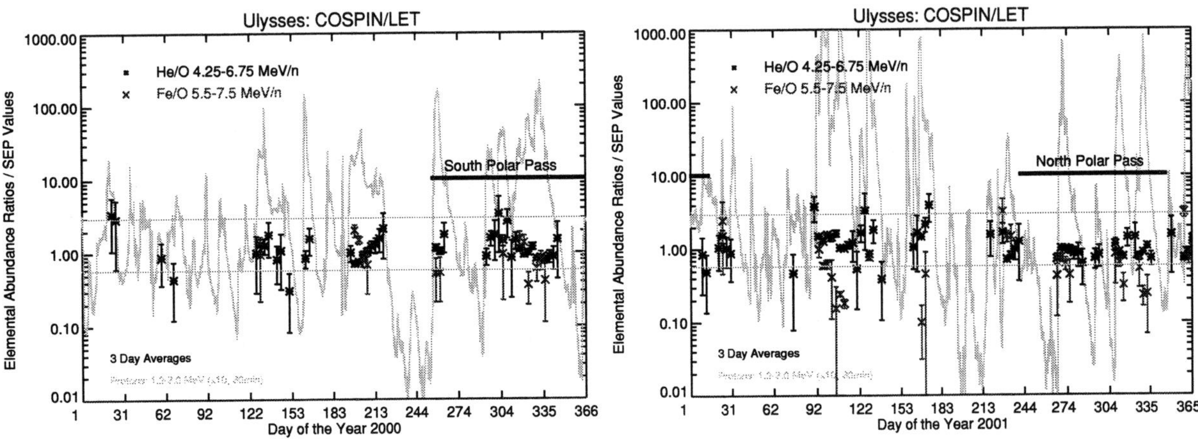

FIGURE 4. The three-day-averaged elemental abundance ratios for the energy ranges as indicated of helium, and iron with respect to oxygen divided by the corresponding reference SEP values for the years 2000 and 2001. The solid horizontal lines mark the ratios of the He/O and Fe/O, SIR-like value at 2.9 and at 0.6. The proton intensity (1.2-3.0 MeV) is plotted for comparison.

We also find that the populations of energetic particles at $\sim$5 MeV/n measured during the north polar passage, are predominantly of SEP origin, confirming the results obtained earlier during the south polar pass.

We conclude that the second south and north solar polar passage and the intermediate time interval are dominated by SEP events which are most probably associated with the coronal mass ejection (CME) shock acceleration phenomenon.

ACKNOWLEDGMENTS

We acknowledge the use of the Ulysses Data System in the preparation of this paper. MYH thanks ESA for the current research fellowship.

REFERENCES

1. Hofer, M.Y., Marsden, R.G., Sanderson, T.R. and Tranquille, C., *Int. Cosmic Ray Conf.*, **8**, 3116 (2001).
2. Hofer, M.Y., Marsden, R.G., Sanderson, T.R., and Tranquille, C., *Geophys. Res. Lett.*, **29(16)**, 10.1029/2002GL014944, (2002).
3. Marsden, R.G., Sanderson, T.R., Tranquille, C. and Hofer, M.Y., *Int. Cosmic Ray Conf.*, **8**, 3310 (2001).
4. Mason, G.M., and Sanderson, T.R., *Space Science Rev.*, **89**, 77 (1999).
5. McKibben, R.B., Connell, J.J., Lopate, C. et al., *Int. Cosmic Ray Conf.*, **8**, 3281 (2001).
6. Sanderson, T.R., Marsden, R.G., Tranquille, C., and Balogh, A., *Geophys. Res. Lett.*, **28(24)**, 4525 (2001).
7. Simpson, J.A., Anglin, J.D., Balogh, A. et al., *Astronomy&Astrophysics*, **92(2)**, 365 (1992).

Solar Wind Streams and Their Interactions

A. Lazarus [1], J. Kasper [1], A. Szabo [2], K. Ogilvie [2]

[1]*MIT Center for Space Research, Cambridge, MA;* [2]*NASA/GSFC, Greenbelt, MD*

Abstract. We present observations of solar wind parameters and magnetic fields from the Wind spacecraft during 1995. We concentrate on periods of interaction between fast and slow streams. We look at speeds, flow angles, alpha particle concentrations, and specific entropy to identify stream interfaces.

OBSERVATIONS

Figure 1 shows 62 days of observations from the Wind spacecraft which include at least three stream interfaces (indicated by "SI"). The time periods around each SI (A, B, and C) are expanded in Figures 2, 3, and 4. The three interfaces were initially chosen on the basis of the return of the flow angles to zero degrees.

Note the following well-known features related to the interaction between two streams of solar wind:

- An increase in density before the arrival of the higher-speed wind.
- The increase in thermal speed after the interface.
- The flow deflections: (slower wind coming first from the East and then from the West) as would be expected from a faster stream colliding with slower wind; the N/S flow angle change is less well understood.
- A sudden increase in entropy per proton ($T/N^{1/2}$) after the interface [1].

The change in alpha particle content and relation of the SI to magnetic field direction relative to the Parker spiral are clearly visible in the detailed plots.

SUMMARY AND QUESTIONS

Identification of interfaces is accomplished using well-known parameters, but some puzzles remain:

1. Though the East/West deflection of the streams is expected from co-rotating structures, what causes the North/South deflection? Is it due to streams at different latitude?

2. Although an increase in entropy is expected from the interaction between streams, our definition of entropy may be too limited in that it does not take into account the obvious compression of the magnetic field and associated fluctuations in the region of the interface. The sharp increase in entropy, as defined, is consistent with the typical lower density and higher temperature for higher speed wind. Neugebauer et al. [2] report a decrease in specific entropy before the interface; we do not observe such a decrease.

CP679, *Solar Wind Ten: Proceedings of the Tenth International Solar Wind Conference,*
edited by M. Velli, R. Bruno, and F. Malara

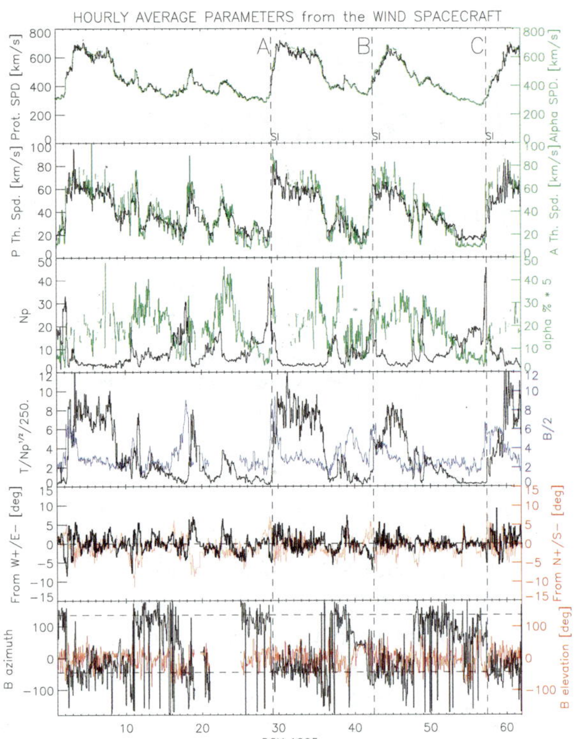

FIGURE 1. Solar wind parameters for 61 days in 1995. Stream interfaces are indicated by "SI".

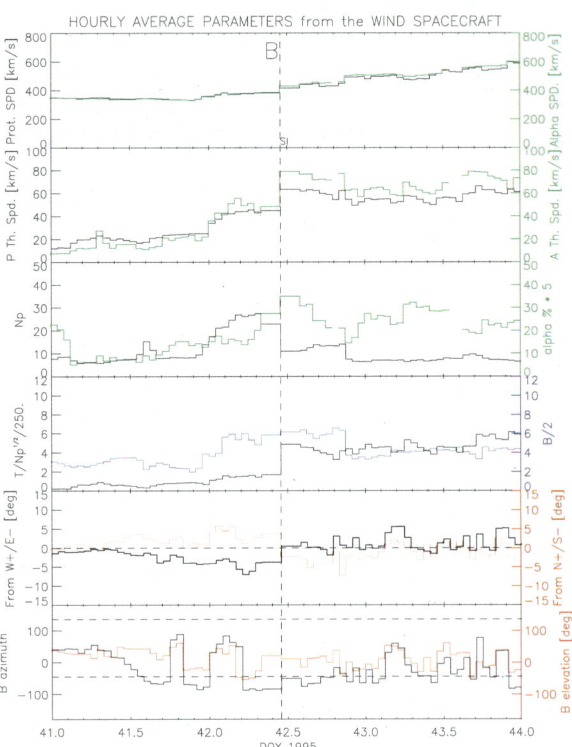

FIGURE 3. A detailed view of interface "B". In this case, there is no magnetic sector boundary.

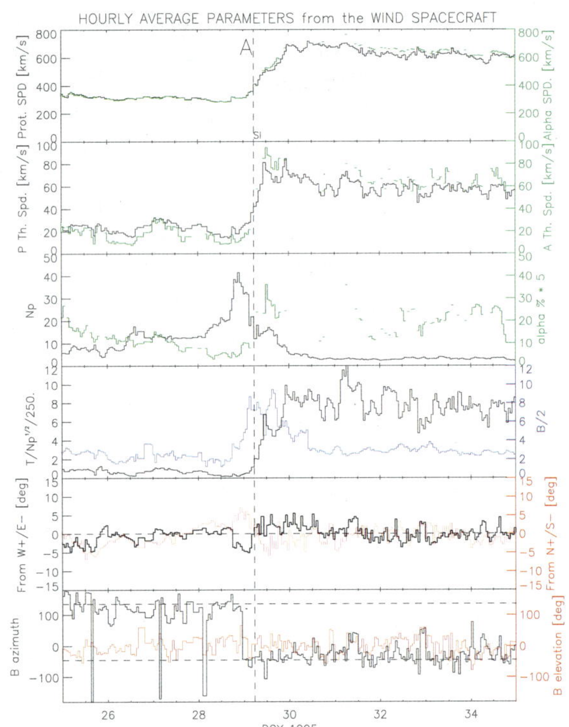

FIGURE 2. A detailed view of interface "A". Note the increase in the magnitude of the field surrounding the "SI".

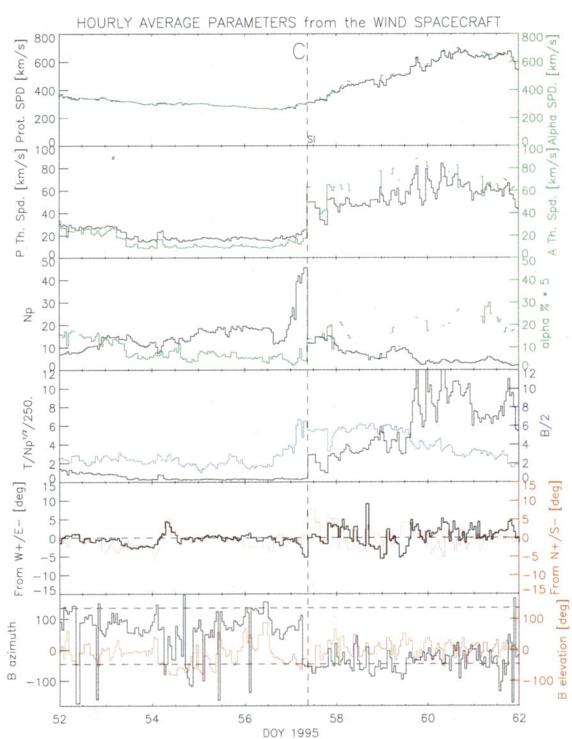

FIGURE 4. A detailed view of interface "C". The major change in flow direction is in the North/South direction.

ACKNOWLEDGEMENTS

This work was supported in part by NASA/Goddard under Grant NAG-10915. We appreciate the hard work of the organizers of the Solar Wind 10 Conference.

REFERENCES

1. Siscoe, G and D. Intriligator, *Geophys. Res. Lett.*, *20*, 2267-2270, 1993.
2. Neugebauer, M., P.C. Liewer, E.J. Smith, R.M. Skoug, and T.H. Zurbuchen, *J. Geophys. Res.*, in press, 2002.

Improved Method for Specifying Solar Wind Speed Near the Sun

Charles N. Arge[*], Dusan Odstrcil[*], Victor J. Pizzo[†], and Leslie R. Mayer[*]

[*]Cooperative Institute for Research in Environmental Sciences, University of Colorado, Boulder, CO 80309, USA
[†]Space Environment Center, National Oceanic and Atmospheric Administration, Boulder, CO 80305, USA

Abstract. We have found an improved technique for empirically specifying solar wind flow speed near the Sun (~0.1 AU) using a set of three simple inter-linked coronal/solar wind models. In addition to magnetic field expansion factor, solar wind speed also appears to be influenced by the minimum angular distance that an open field footpoint lies from a coronal hole boundary. We conduct our study using polar field corrected Mount Wilson Solar Observatory Carrington maps from 1995. During this period, the Sun was in the declining phase of the solar cycle and the solar wind had relatively simple global structure.

INTRODUCTION

Theoretical understanding of solar wind acceleration remains elusive. To date, the most common methods for predicting solar wind speed near Earth are based on empirical methods or interplanetary scintillation (IPS) techniques [1]. The focus of this paper is on the first approach.

In 1990, Wang and Sheeley (WS) [2] found a correlation between solar wind speed at L1 and magnetic field expansion factor (f_s) of open coronal field lines, where f_s is defined as the rate that a flux tube expands (compared to purely radial expansion) from 1 $R_\odot$ to a "source surface" positioned ~2-3 $R_\odot$ from Sun center. WS used a potential field source surface (PFSS) model [3,4,5] to identify open field lines and to calculate their expansion factors. Arge et al. [6] have recently applied a modification of the WS method for space weather forecasting purposes and now routinely predict solar wind speed and interplanetary field polarity (IMF) polarity at Earth (http://www.sec.noaa.gov/ws/). The modification consists of deriving an empirical relationship between f_s and solar wind speed (v_{sw}) at 2.5 $R_\odot$ (i.e., at the source surface) rather than at L1. This empirical function was found by iteratively testing various mathematical relationships between f_s and v_{sw}, using it to assign the solar wind speed at the source surface, propagating the wind out to L1, and then comparing the results with observations. The procedure was repeated until a best fit was found. A simple 1-D modified kinematic (1-DMK) model, which included an ad hoc method to account for stream interactions, was used to transport the wind to L1.

We report here on our effort to find an improved empirical relationship between solar wind speed near the Sun (~0.1 AU) and various photospheric and coronal field parameters. We have found a new relationship for specifying solar wind speed, which is a function of both f_s and the minimum angular separation (at the photosphere) between an open field footpoint and its nearest coronal hole boundary (θ_b). The new relationship works much better than one derived previously, which is a function of f_s only. We focus here on the solar wind during 1995, as we have had past difficulty successfully predicting the stream structure for this period.

APPROACH

Solar wind observations made at L1 must be mapped back to the Sun to compare with photospheric and coronal field parameters. We mapped WIND spacecraft observations from 1995 back to 0.1 AU (i.e., 21.5 $R_\odot$) assuming constant flow speed, neglecting stream interactions, but accounting for solar rotation. While simple, this approach has been shown to work reasonably well [7]. To relate the mapped-back solar wind with observed photospheric and derived coronal field parameters, a PFSS model (representing the field configuration of the inner corona from 1 to 2.5 $R_\odot$) was used in combination with the Schatten current sheet (SCS) model [8] (representing the field configuration of the outer corona between 2.5 and 21.5 $R_\odot$). Together, these models permit open field lines to be identified and then traced down to the photosphere. Daily updated photospheric field maps from Mount Wilson Solar Observatory (MWO) (grid resolution 5°×5°) were used as input to the models. The PFSS+SCS model combination was run for each daily updated map available from 1995.

Field lines located on the outer boundary of the SCS model and positioned at the appropriate sub-earth point for a given date and time are the relevant ones for this study. The (mapped-back) solar wind speeds associated

CP679, *Solar Wind Ten: Proceedings of the Tenth International Solar Wind Conference*, edited by M. Velli, R. Bruno, and F. Malara

with these particular field lines are compared with the following five photospheric and coronal parameters: expansion factor (f_s) evaluated at 2.5 $R_{\odot}$, photospheric (or footpoint) field strength (B_{ph}), and the minimum (spherical) angular separation between an open field footpoint and 1) sub-earth point (θ_{se}), 2) current sheet (θ_{cs}), and 3) nearest coronal hole boundary (θ_b). These parameters are plotted in Figure 1 in five separate panels with solar wind speed at 21.5 $R_{\odot}$ over-plotted in each.

Interpreting the results in Figure 1 is complicated. For instance, the procedure used to map the solar wind from L1 back to 21.5 $R_{\odot}$ is, without doubt, overly simple and certainly can result in large errors, potentially masking statistical correlations between the parameters and v_{sw}. Yet, it is surprising how clear the inverse correlation between f_s and v_{sw} appears in the top panel of Figure 1. In fact, we obtained a correlation coefficient of –0.55 between these two quantities, which is very similar to the one found originally by WS. Nevertheless, one can see that a given expansion factor does not necessarily correspond to the same solar wind speed. For instance, note the values of f_s and v_{sw} for the high-speed stream located between Carrington rotations (CR) 1893 and 1894 (i.e., Panel 1 of Figure 1), and compare the same quantities for the stream located just after the start of CR 1900. This clearly suggests that solar wind speed is a function of more than just f_s.

Because of all the uncertainties with the mapped-back solar wind and the possibility that v_{sw} is a complicated function of the above-mentioned parameters, we take the following approach. First, search for patterns between solar wind speed and the above five photospheric/coronal parameters. Second, deduce an empirical relationship for v_{sw} involving one or more these parameters. Third, assign v_{sw} at the outer boundary of the SCS model, and then, for each daily updated map from 1995, propagate the solar wind out to Earth using a 1-D modified kinematic (1-DMK) code. Fourth, directly compare predictions with observations. For this particular study, the 1-DMK code is preferred over a more advanced 3-D MHD code, as it significantly reduces the time required for each new trial run.

While tedious, the main advantage of the approach just described is that it relies only on the mapped-back solar wind for deducing the empirical velocity function. Propagating the solar wind from the Sun to L1 using our 3-step approach (i.e., PFSS+SCS+1-DMK model) is much more reliable (e.g., stream interactions are taken into account) and permits direct comparison of predictions with observations.

RESULTS

A large number of test functions were tried using various combinations of the five photospheric/coronal parameters. The most interesting and/or promising were f_s, B_{ph}, and θ_b. The B_{ph} factor is especially interesting since flux tubes can have the same expansion factor but different values of B_{ph}. Do more intense magnetic flux tubes yield the same solar wind speed as WS would suggest? We were unable to find a velocity relationship dependent on B_{ph} (plus other parameters) that matched the observed wind speed at L1 any better than one of our best (though unpublished) empirical relationships

$$v_{sw}(f_s) = 285+650/(f_s)^{5/9} \text{ km s}^{-1}, \qquad (1)$$

which is a function only of f_s.

Close inspection of the bottom panel of Figure 1 reveals the general pattern that the farther away an open field footpoint is from a coronal hole boundary (i.e., θ_b), the faster the corresponding solar wind speed. There are exceptions to this trend, but in most of those cases the tendency is for the expansion factors to be moderately large (e.g., just before the start of CR 1895 – many other case can be found). Note, that there is a broad data gap (with the values at the beginning and end of the gap connected by straight lines) between CR 1891 and 1892. The values there are not to be trusted. One will also note periods of rapid oscillation in θ_b (e.g., CR1897), we believe this primarily due to the coarseness of the grid used (5°×5°). The pattern described above and the clear correlation between v_{sw} and f_s, suggests using θ_b and f_s in combination to predict solar wind speed. We found that the function

$$v_{sw}(f_s,\theta_b) = 265 + \frac{25}{f_s^{\,2/7}}\left\{5 - 1.1e^{\left(1-\left(\theta_b/4\right)^2\right)}\right\}^2 \text{ km s}^{-1} \qquad (2)$$

matched the solar wind speed observations at L1 rather well for the entire sequence of 15 Carrington rotations (CR1890-1904), which corresponds to the time interval from December 1994 to the end of 1995. (There were a number of large gaps in the photospheric field data in the early part of 1995 that appear to have degraded the results there.) For the entire 15 Carrington rotation sequence, the predictions obtained with the use of equation 2 agreed (qualitatively) with the WIND satellite observations generally better than those obtained using equation 1. CR 1896 is an especially nice example. For this particular rotation, Figure 2 shows a comparison of the solar wind speed predictions at L1, using both equations 1 and 2, with WIND satellite observations.

We were also interested in comparing model predictions with and without the SCS model's inclusion. We therefore re-ran the entire sequence of daily updated MWO maps using just the PFSS and 1-DMK model in combination with the later now transporting the solar wind from 2.5 $R_{\odot}$ to L1. Once again the solar wind speed predictions obtained using equation 2 show better overall agreement with observations than those obtained using equation 1 (see Figure 3). While

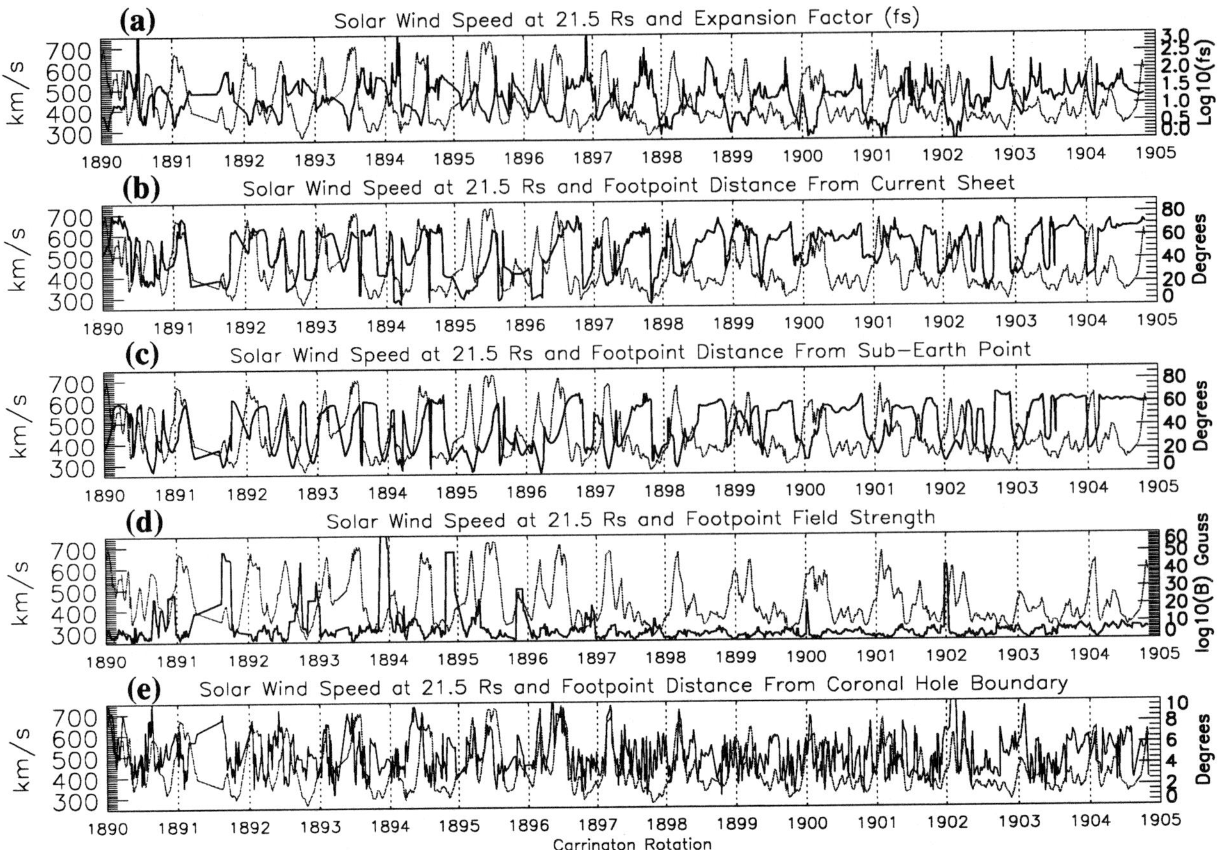

Figure 1. Five photospheric and coronal parameters (black lines) compared to solar wind speed (light gray lines). **(a)** Expansion factor (f_s), **(b)** footpoint distance from current sheet (θ_{cs}), **(c)** footpoint distance from sub-earth point (θ_{se}), **(d)** photospheric field strength (B_{ph}), and **(e)** footpoint distance from coronal hole boundary (θ_b).

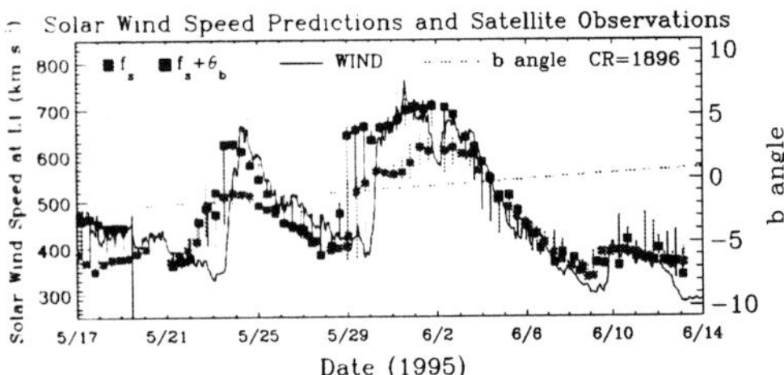

Figure 2. Comparison of solar wind speed observations (thin solid line) for CR 1896 with predictions using the PFSS+SCS+1-D kinematic model combination and equations 1 (black squares) and 2 (light gray squares). The vertical bars are uncertainty estimates determined by calculating the solar wind speeds for values located 2.5° above and below the sub-earth points.

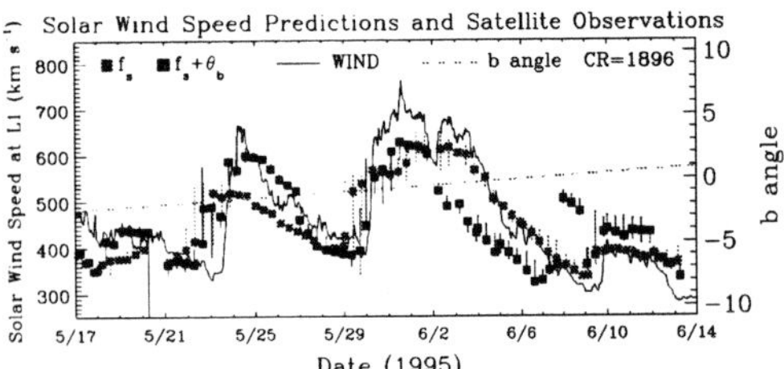

Figure 3. Same as Figure 2 expect now for the PFSS+1-D kinematic model combination.

equation 2 provides a clear improvement over equation 1, it is unclear whether the SCS model provides general improvement to the overall model prediction scheme. We find that when the solar wind source region (i.e., open field footpoints) is located at higher latitudes, the SCS+PFSS+1-DMK model combination produces results that typically agree better with observations than those made using the PFSS+1-DMK model combination. However, when the open field footpoints are located near the equator the reverse often appears to be true. The problem near the equator may be an artifact of the field line tracing routine used in the SCS model. It occasionally seems to get lost when the sub-earth point lies very close to the current sheet. Further investigation on this matter is required.

CR 1896 is a case where the inclusion of the SCS model improves model predictions at both high and low latitudes. Figure 4a shows the coronal holes as determined by the PFSS+SCS model combination, while Figure 5 shows them using just the PFSS model. The field polarity at the photosphere is indicated by the light (positive polarity) and dark (negative polarity) gray contours, while the light gray dots (the brighter the dot the smaller f_s and vice versa) identify the footpoints of the open field lines (i.e., the coronal holes) at the photosphere. The white crosses near the equator mark the daily positions of the sub-earth point. The black straight lines identify the connectivity between the outer (open) boundary (i.e., 2.5 $R_\odot$ for the PFSS model or 21.5 $R_\odot$ for the PFSS+SCS model combination) and the source regions of the solar wind at the photosphere. For the large coronal hole located in the southern hemisphere (i.e., longitudes 176-236°), the PFSS+SCS model combination indicates that the solar wind (i.e., the stream observed at L1 beginning on May 30, 1995 in Figure 2 and 3) emerged from deeper within the hole than that implied by the PFSS model. The PFSS+SCS model also appears to explain better the properties of the stream (i.e., its short duration and the sharp velocity spike) that arrived 6 days before the larger one just mentioned (see Figure 2). The source regions of this stream cut directly across the equatorial hole, unlike the latitudinal meandering suggested by the PFSS model. Figure 4b is a plot of the field polarity at the outer boundary of the SCS model for CR 1896. In this figure, note that those portions of the sub-Earth track connected to the two coronal holes just discussed lie far from the current sheet. The field line tracing routine is more reliable in such cases, which helps explain why the PFSS+SCS model performed so well for CR1896.

SUMMARY

In addition to magnetic field expansion factor (f_s), solar wind speed also appears to be influenced by the minimum (spherical) angular distance (θ_b) that an open field footpoint lies from a coronal hole boundary. We have found a new empirical relationship for specifying solar wind speed near the Sun that is a function both of f_s and θ_b and which generally works much better (for the year 1995) than a relationship which is a function of f_s only. It is unclear whether the Schatten current sheet model produces a general improvement to the overall model prediction scheme. Our results suggest that it may (e.g., result from CR 1896), if the field line tracing routine used in it were improved. To establish the robustness of this new empirical relationship, we will test it for different periods of the solar cycle and perform a rigorous statistical analysis. We also plan to test the new empirical relationship using a 3D MHD solar wind model.

ACKNOWLEDGMENTS

We thank K. Ogilive, J. Steinberg, and A. Lazarus for allowing us access to the SWE key parameter data. We thank R. Ulrich for providing us access to daily updated MWO photospheric field synoptic maps. This work is supported by grants NSF ATM-0001851, ONR N00014-01-F-0026, and AFOSR/MURI.

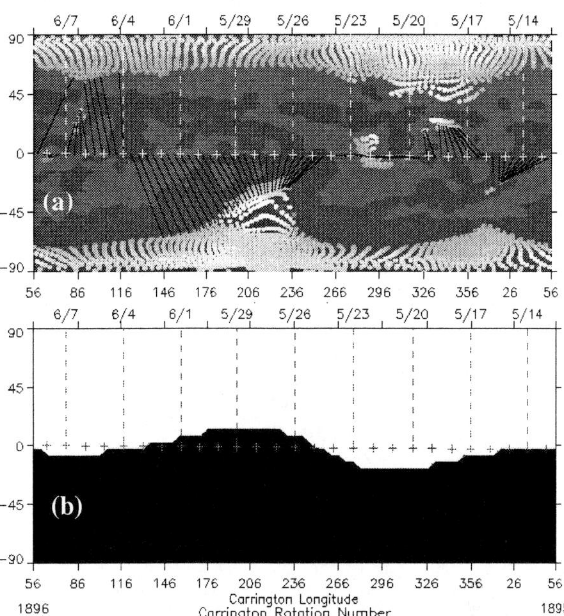

Figure 4. **(a)** Derived coronal holes for CR 1896 using the PFSS+SCS model combination. **(b)** IMF polarity at 21.5 $R_\odot$. Positive (negative) polarity is shaded white (black). See text for details.

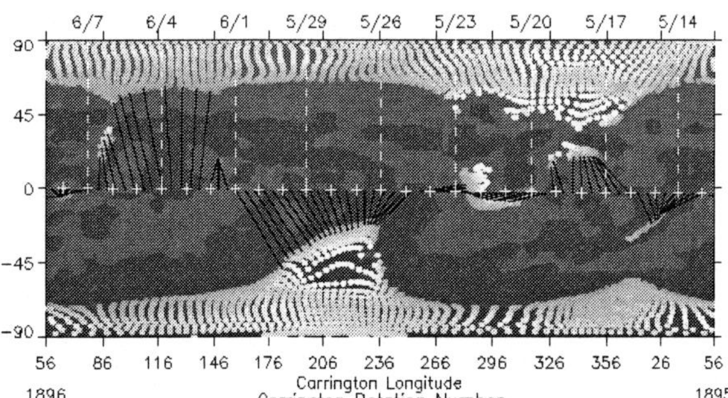

Figure 5. Same as Figure 4a except now using only the PFSS model.

REFERENCES

1. Hakamada, K., M. Kojima, Tokumaru, M., Ohmi, T., Yokobe, A., and Fujiki, K., *Sol. Phys.* in press (2002).
2. Wang, Y.-M., and Sheeley Jr., N. R., *Astrophys. J.* **355**, 726-732 (1990).
3. Schatten, K. H., Wilcox, J. M., and Ness, N. F., *Sol. Phys.* **9**, 442-455 (1969).
4. Altschuler, M. A., and Newkirk Jr., G., *Sol. Phys.* **9**, 131-149 (1969).
5. Wang, Y.-M., and Sheeley Jr., N. R., *Astrophys. J.* **392**, 310-319 (1992).
6. Arge, C. N., and Pizzo, V. J., *J. Geophys. Res.* **105**, 10,465-10,479 (2000).
7. Neugebauer, M., Liewer, P. C., Smith, E. J., Skoug, R. M., and Zurbuchen, T. H., *J. Geophys. Res.* in press (2002).
8. Schatten, K. H., *Cosmic Electrodyn.* **2**, 232-245 (1971).

Fluid Modeling of the VLISM/Solar Wind Interaction With the 13-Moment Formalism

Ralph L. McNutt, Jr.

The Johns Hopkins University Applied Physics Laboratory
Laurel, MD, U.S.A.

Abstract. The interaction of the solar wind with the Very Local Interstellar Medium is mediated largely by collisional charge exchange of interstellar neutral atomic hydrogen and protons in the heliosheath. This interaction provides an additional momentum to the interstellar population over that due to the charged population alone that leads to a decrease in the expected size of the heliospheric cavity. The interaction is complicated by time variations in the internal solar wind flow, the presence of both the interstellar and interplanetary magnetic fields, and the large mean free paths for charge exchange. Proper treatment of the problem calls for a fully six-dimensional, time-dependent kinetic interaction model, yet computational complexities inherent in such a model have precluded its full implementation. Although the applicability of fluid models to this problem has been questioned, one can expect them to provide fairly good estimates of the various boundary locations provided that all salient moments are included. While the ion population can be approximated to first order by a convected Maxwellian, given the relatively small ion gyroradii, the neutral population acquires significant non-Maxwellian features due to the large mean free paths for collisions. In this case the lowest-order moment description is the thirteen-moment description of Grad. The thirteen quantities are the density, temperature, velocity vector, heat flux vector, and five deviator components of the pressure tensor. Unlike the case of the Navier-Stokes equations, there is no a priori assumption about collisions; the only assumption is how the hierarchy of fluid equations is to be truncated. To connect the neutral and ion components, moments of the Boltzmann collision operator must be evaluated for representative distribution functions that give rise to such moments, using the appropriate collision cross section. New results are reported for the collision operator moments corresponding to the full set of pressure-tensor components and the heat flux vector components. All of these can be expressed in closed form in terms of sums of confluent hypergeometric functions of the first kind (Kummer's function). This derivation completes the formalism required for the implementation of a time-dependent, magnetized fluid model of the interaction using the thirteen-moment formalism to describe the neutral component.

INTRODUCTION

Charge exchange of interstellar atomic hydrogen and protons mediates interaction of the solar wind with the Very Local Interstellar Medium (VLISM)[1-10]. The problem is complicated by time variations in the internal solar wind flow, interstellar and interplanetary magnetic fields, and large mean free paths for charge exchange. There is a need for a fully six-dimensional, time-dependent kinetic interaction model to "do it right". However, such an approach is computationally intensive, requiring simplifying assumptions[3-5,10], and time-dependence introduces additional complexities[2].

FLUID APPROACH

Fluid models should provide good estimates of many details[1,6-9], e.g., boundary locations, and the ion population can be approximated by a convected Maxwellian distribution, given the relatively small ion gyroradii. The neutral population acquires significant non-Maxwellian features due to the large mean free paths for collisions. To capture these in a fluid context, we consider the lowest-order moment description that is the thirteen-moment description of Grad[11-13]. In this approach, the thirteen moments are the density, temperature, velocity vector, heat flux vector, and five deviator components of the pressure tensor.

Closure of the Moment Equations

Following Grad, consider distribution functions that contain non-trivial moments through the fourth order. These functions will then non-trivially have 15 fourth-order moments and 20 moments of orders 0, 1, 2, and 3 that must satisfy 20 moment equations. We can then reduce the 10 third-order moments to the

CP679, *Solar Wind Ten: Proceedings of the Tenth International Solar Wind Conference*,
edited by M. Velli, R. Bruno, and F. Malara

components of the heat flow vector, leaving 1 zeroth-order moment, 3 first-order moments, 6 second-order moments, and 3 non-trivial components of the 10 third-order moments. This is the "thirteen-moment approximation."

Boltzmann Collision Operator

Moments of the Boltzmann collision operator connect the neutral and ion components in the VLISM problem, so these components must be evaluated up through deviator components of pressure tensor and heat flux vector. We assume a power-law cross section, which provides a good fit to the charge-exchange cross section. In general, the moments of the collision operator can then be expressed in closed form in terms of sums of confluent hypergeometric functions of the first kind (Kummer's function). As a check on the derivations, we can then verify the general results against results for Maxwell molecules (molecules that obey a repulsive inverse-fifth-power force law).

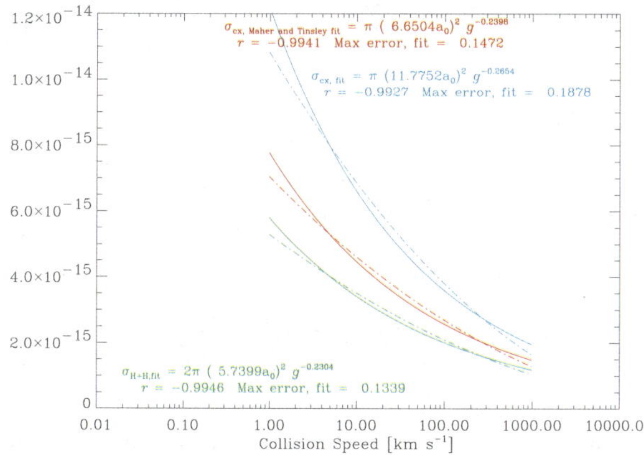

FIGURE 1. Exact (solid lines) and fit (broken lines) charge exchange cross sections for protons on atomic hydrogen and elastic scattering cross section and fit for atomic hydrogen scattering.

Model Equations

For the neutral distribution we assume a Maxwellian modified to have a non-zero set of deviator stress components and non-zero heat flux.

$$f_1 = f_{M,1}(1+\phi_1) = \frac{n_1}{\pi^{3/2} w_1^3} e^{-|\mathbf{v}_1 - \mathbf{u}_1|^2 / w_1^2} (1+\phi_1) \quad (1)$$

where

$$\phi_1 \equiv \frac{\overleftrightarrow{\mathbf{p}}_1 - p_1 \overleftrightarrow{\mathbf{I}}}{w_1^2 p_1} : (\mathbf{v}_1 - \mathbf{u}_1)(\mathbf{v}_1 - \mathbf{u}_1) + \tfrac{8}{5} \frac{1}{n_1 m_1 w_1^6} \mathbf{q}_1 \cdot (\mathbf{v}_1 - \mathbf{u}_1)(|\mathbf{v}_1 - \mathbf{u}_1|^2 - \tfrac{5}{2} w_1^2) \quad (2)$$

and the assumed distribution function for the neutral population is entirely equivalent to the 13-moment equation set.

For the ions, we assume an isotropic Maxwellian distribution

$$f_2 = f_{M,2} = \frac{n_2}{\pi^{3/2} w_2^3} e^{-|\mathbf{v}_2 - \mathbf{u}_2|^2 / w_2^2} \quad (3)$$

and we introduce the auxiliary quantities w_i via $\frac{1}{2} m_i w_i^2 \equiv kT_i$ as well as $\Delta \mathbf{U} \equiv \mathbf{u}_2 - \mathbf{u}_1$, $\kappa \equiv \frac{1}{w_1^2 + w_2^2}$, and $\alpha \equiv \sqrt{\kappa} \Delta \mathbf{U} = \frac{|\Delta \mathbf{U}|}{\sqrt{w_1^2 + w_2^2}}$.

Collision Operator

The collision operator for the moments of any dynamical quantity ϕ can be written as

$$\left(\frac{\delta \varphi_1}{\partial t} \right)_{collision} = \iiint (f_1' f_2' - f_1 f_2) \varphi_1 \frac{d\sigma}{d\Omega} d\Omega d^3 v_1 d^3 v_2$$
$$= \iiint f_1 f_2 (\varphi_1' - \varphi_1) \frac{d\sigma}{d\Omega} d\Omega d^3 v_1 d^3 v_2 \quad (4)$$

The cross section integrals are

$$\int d\Omega (1 - \cos^2\theta) \frac{d\sigma}{d\Omega} \equiv \sigma_P \quad (5)$$

and

$$\int d\Omega (1 - \cos\theta) \frac{d\sigma}{d\Omega} \equiv \sigma_M \quad (6)$$

where

$$\sigma_{M,P} = \sigma_{M0,P0} \, | \mathbf{v}_2 - \mathbf{v}_1 |^{-\nu} \quad (7)$$

Variation of the ratio of σ_M/σ_P[14] with ν is approximated by

$$\frac{\sigma_P}{\sigma_M} = \frac{\sigma_{P0}}{\sigma_{M0}} = \frac{A_2(\nu)}{A_1(\nu)} \approx 0.4108\nu + 0.6600$$
$$\approx 0.0326\nu^2 + 0.3425\nu + 0.6797 \quad (8)$$

Where the correlation coefficients for the linear and quadratic cases are 0.9955 and 0.9978, respectively.

Pressure Tensor and Heat Flux Vector

The pressure tensor and heat flux vector are given by $\vec{\mathbf{p}}_1 \equiv \int m_1(\mathbf{v}_1 - \mathbf{u}_1)(\mathbf{v}_1 - \mathbf{u}_1) f_1 d^3 v_1$ and $\mathbf{q}_1 \equiv \int \frac{1}{2} m_1(\mathbf{v}_1 - \mathbf{u}_1) |\mathbf{v}_1 - \mathbf{u}_1|^2 f_1 d^3 v_1$

Change to velocities in the center of mass frame and with respect to the center of mass. Define $\mu \equiv \frac{m_1 m_2}{m_1 + m_2}$, $\mathbf{g} \equiv \mathbf{v}_2 - \mathbf{v}_1$, $\mathbf{G} \equiv \frac{m_1\mathbf{v}_1 + m_2\mathbf{v}_2}{m_1 + m_2}$, and $\hat{\mathbf{G}} \equiv \mathbf{G} - \mathbf{u}_1$ to obtain

$$\left(\frac{\partial \vec{\mathbf{p}}_1}{\partial t}\right)_{collision} = \mu \iint f_1 f_2 g \sigma_M \left(\mathbf{g}\hat{\mathbf{G}} + \hat{\mathbf{G}}\mathbf{g}\right) d^3\hat{G} d^3 g \\ +\mu \frac{m_2}{m_1 + m_2} \iint f_1 f_2 g \sigma_P \left(\tfrac{1}{2} g^2 \vec{\mathbf{I}} - \tfrac{3}{2}\mathbf{g}\mathbf{g}\right) d^3\hat{G} d^3 g \tag{9}$$

$$\left(\frac{\partial \mathbf{q}_1}{\partial t}\right)_{collision} = \tfrac{1}{2}\mu \iint f_1 f_2 g \sigma_M \mathbf{g}\left(\hat{G}^2 + g^2\left[\tfrac{m_2}{m_1+m_2}\right]^2\right) d^3\hat{G} d^3 g \\ +\mu \iint f_1 f_2 g \sigma_M \hat{\mathbf{G}}(\hat{\mathbf{G}}\cdot\mathbf{g}) d^3\hat{G} d^3 g \\ +\mu \frac{m_2}{m_1+m_2} \iint f_1 f_2 g \sigma_P \left(\tfrac{1}{2} g^2 \hat{\mathbf{G}} - \tfrac{3}{2}\left[\hat{\mathbf{G}}\cdot\mathbf{g}\right]\mathbf{g}\right) d^3\hat{G} d^3 g \tag{10}$$

To evaluate, shift the variables once more using $\mathbf{v}_1 - \mathbf{u}_1 = \mathbf{G}^* - \kappa w_1^2(\mathbf{g} - \Delta\mathbf{U})$ and $\hat{\mathbf{G}} \equiv \mathbf{G} - \mathbf{u}_1 = \mathbf{G}^* + \kappa w_1^2 \Delta\mathbf{U} + \frac{2\kappa}{m_1+m_2}(kT_2 - kT_1)\mathbf{g}$, so that the operators now contain forms like

$$\frac{1}{\pi^{3/2}\kappa^{3/2} w_1^3 w_2^3} \int d^3 G^* e^{-\frac{G^{*2}}{\kappa w_1^2 w_2^2}} (G^*)^{2n} \tag{11}$$

which can be integrated immediately. The remaining integrands are then of the form

$$\int d^3 g e^{-\kappa |\mathbf{g} - \Delta\mathbf{U}|^2} g^{\nu+1} g_{i_1} g_{i_2} g_{i_3} \cdots g_{i_n} \tag{12}$$

that can be integrated analytically in terms of

$$y_{n,\nu}(\alpha) \equiv \frac{\Gamma(2 + n + \frac{1}{2}\nu)}{\Gamma(n + \frac{3}{2})} M(-\tfrac{1}{2} - \tfrac{1}{2}\nu, n + \tfrac{3}{2}, -\alpha^2) \\ \approx \left[\left(\tfrac{5}{4} + n + \tfrac{1}{4}\nu\right) + \alpha^2\right]^{\frac{1}{2}(1+\nu)} \tag{13}$$

To write out the results, define the unit vector $\mathbf{e}_1 \equiv \Delta\hat{\mathbf{U}} \equiv \Delta\mathbf{U}/|\Delta\mathbf{U}|$ and use it and the Kronecker delta to define the fully symmetric tensors

$$I_{hkji} \equiv \delta_{hk}\delta_{ij} + \delta_{hi}\delta_{kj} + \delta_{ki}\delta_{hj} \tag{14}$$

$$E_{hkji} \equiv e_{1h}e_{1k}\delta_{ij} + e_{1j}e_{1i}\delta_{hk} + e_{1h}e_{1i}\delta_{kj} \\ + e_{1k}e_{1j}\delta_{hi} + e_{1h}e_{1j}\delta_{ki} + e_{1k}e_{1i}\delta_{hj} \tag{15}$$

$$E_{hkj} \equiv \delta_{hk}e_{1j} + \delta_{hj}e_{1k} + \delta_{kj}e_{1h} \tag{16}$$

The full pressure-tensor collision operator is then given by

$$\left(\frac{\partial p^M_{1,ij}}{\partial t}\right)_{collision} = \mu \frac{n_1 n_2 \sigma_{M0}}{g_0^\nu \kappa^{1/2+\nu/2}} \\ \times \left\{ \begin{array}{l} \delta_{ij} y_{1,\nu} \frac{2(kT_2 - kT_1)}{m_1+m_2} + 2\alpha^2 e_{1i}e_{1j}\left(\frac{m_2}{m_1+m_2} y_{2,\nu} + \kappa w_1^2 [y_{1,\nu} - y_{2,\nu}]\right) \\ + \left(\frac{p_{1,hk}}{p_1} - \delta_{hk}\right) \left\{ \begin{array}{l} \frac{1}{2} I_{ijkh} \kappa w_1^2 \left(\frac{m_2 w_2^2}{m_1+m_2}[y_{2,\nu} - y_{1,\nu}] - \frac{m_1}{m_1+m_2}[w_1^2 y_{2,\nu} + w_2^2 y_{1,\nu}]\right) + \\ E_{ijkh}\alpha^2 \left([y_{3,\nu} - y_{2,\nu}] \frac{2\kappa w_1^2 (kT_2 - kT_1)}{m_1+m_2} + \frac{1}{2}\kappa w_1^2 [w_1^2 - w_2^2]\right) + \\ e_{1h}e_{1k}\delta_{ij}\alpha^2 [y_{2,\nu} - y_{1,\nu}] \frac{1}{2} w_1^2 \left[\frac{m_1 - m_2}{m_1+m_2}\right] + \\ e_{1i}e_{1j}e_{1h}e_{1k}\alpha^4 \left([y_{4,\nu} - 2y_{3,\nu} + y_{2,\nu}]\left[\frac{2m_2 w_1^2}{m_1+m_2} - \kappa w_1^4\right] + [y_{3,\nu} - 2y_{2,\nu} + y_{1,\nu}]\kappa w_1^4\right) \end{array} \right\} \\ + \frac{8}{5} \frac{\kappa^{1/2} q_{1,h} \alpha}{n_1 m_1} \left\{ \begin{array}{l} E_{ijh} \left(\begin{array}{l} [y_{2,\nu} - y_{1,\nu}]\left[\kappa w_2^2 - \frac{2(3+\frac{\nu}{2})\kappa(kT_2 - kT_1)}{m_1+m_2}\right] \\ + \frac{1}{2}[2 + \frac{\nu}{2}][y_{1,\nu} - y_{0,\nu}][\kappa w_2^2 - \kappa w_1^2] \end{array} \right) + \\ e_{1h}\delta_{ij} \frac{1}{2}[2 + \frac{\nu}{2}][y_{1,\nu} - y_{0,\nu}]\left[\frac{m_2 - m_1}{m_2 + m_1}\right] + \\ e_{1i}e_{1j}e_{1h} 2\alpha^2 \left(\begin{array}{l} [2 + \frac{\nu}{2}][-y_{2,\nu} + 2y_{1,\nu} - y_{0,\nu}]\kappa w_1^2 \\ + [y_{3,\nu} - 2y_{2,\nu} + y_{1,\nu}]\left[\kappa w_2^2 - \frac{2(3+\frac{\nu}{2})\kappa(kT_2 - kT_1)}{m_1+m_2}\right] \end{array} \right) \end{array} \right\} \end{array} \right\} \tag{17}$$

$$\left(\frac{\partial p^P_{1,ij}}{\partial t}\right)_{collision} = \mu \frac{n_1 n_2 \sigma_{P0}}{g_0^\nu \kappa^{1/2+\nu/2}} \frac{m_2}{m_1+m_2} \\ \times \left\{ \begin{array}{l} \frac{\alpha^2}{\kappa} y_{2,\nu} + \left(\frac{1}{2}\delta_{ij} - \frac{3}{2}e_{1i}e_{1j}\right) + \\ \frac{1}{2} w_1^2 \left(\frac{p_{1,hk}}{p_1} - \delta_{hk}\right) \left\{ \begin{array}{l} -\frac{3}{4} y_{2,\nu} I_{ijkh} + E_{ijkh} \frac{3}{2}\alpha^2 [y_{2,\nu} - y_{1,\nu}] \\ + e_{1h}e_{1k}\delta_{ij}\alpha^2 [3 + \frac{\nu}{2}][y_{2,\nu} - y_{1,\nu}] + \\ e_{1i}e_{1j}e_{1h}e_{1k}\alpha^2 \left(\begin{array}{l} -3[3 + \frac{\nu}{2}][y_{2,\nu} - y_{1,\nu}] \\ + \frac{21}{2}[y_{3,\nu} - y_{2,\nu}] \end{array} \right) \end{array} \right\} \\ + \frac{8}{5} \frac{\kappa^{1/2} q_{1,h}}{n_1 m_1} \alpha \{3 + \frac{\nu}{2}\} \left\{ \begin{array}{l} [2 + \frac{\nu}{2}][y_{0,\nu} - y_{1,\nu}][\frac{1}{2}\delta_{ij} - \frac{3}{2}e_{1i}e_{1j}]e_{1h} + \\ \frac{3}{2}[y_{2,\nu} - y_{1,\nu}][\frac{1}{2}E_{ijh} - \frac{5}{2}e_{1i}e_{1j}e_{1h}] \end{array} \right\} \end{array} \right\} \tag{18}$$

The expressions for the heat flux vector are even more complex and are not given here.

MAXWELL-MOLECULE LIMIT

These general expressions are complex. We can reduce the complexity and provide for a cross-check in the derivation, by going to the limit of Maxwell molecules for which ν=-1. Going back to the original definitions, for the moments of the collision operator, we obtain for the pressure tensor operator

$$\left(\frac{\partial \vec{\mathbf{p}}_1^P}{\partial t}\right)_{collision, \nu=-1} = \mu(g_0\sigma_{P0}) \frac{m_2}{m_1+m_2} \left[\begin{array}{l} \left(\frac{3}{2}\frac{n_1 p_2}{m_2} + \frac{3}{2}\frac{n_2 p_1}{m_1} + \frac{1}{2} n_1 n_2 \Delta U^2\right)\vec{\mathbf{I}} \\ -\left(\frac{3}{2}\frac{n_1 \vec{\mathbf{p}}_2}{m_2} + \frac{3}{2}\frac{n_2 \vec{\mathbf{p}}_1}{m_1} + \frac{3}{2} n_1 n_2 \Delta\mathbf{U}\Delta\mathbf{U}\right) \end{array} \right] \tag{19}$$

$$\left(\frac{\partial \bar{\mathbf{p}}_1^M}{\partial t}\right)_{collision,\nu=-1} = \mu(g_0\sigma_{M0})\left[\frac{2}{m_1+m_2}\left(n_1\bar{\mathbf{p}}_2 - n_2\bar{\mathbf{p}}_1\right) + \frac{2m_2}{m_1+m_2}n_1n_2\Delta\mathbf{U}\Delta\mathbf{U}\right] \tag{20}$$

Corresponding results for the heat flux vector are

$$\left(\frac{\partial \mathbf{q}_1^M}{\partial t}\right)_{collision,\nu=-1} = \tfrac{1}{2}\mu(g_0\sigma_{M0})\left[\begin{array}{l} 8\left(\frac{m_2}{m_1+m_2}\right)^2\frac{n_1\mathbf{q}_2}{m_2} - \frac{6m_1^2+2m_2^2}{(m_1+m_2)^2}\frac{n_2\mathbf{q}_1}{m_1} \\ +\Delta\mathbf{U}\cdot\left\{8\left(\frac{m_2}{m_1+m_2}\right)^2\frac{n_1\bar{\mathbf{p}}_2}{m_2} + 2\left(\frac{m_2-m_1}{m_1+m_2}\right)^2\frac{n_2\bar{\mathbf{p}}_1}{m_1}\right\} \\ +\Delta\mathbf{U}\left\{12\left(\frac{m_2}{m_1+m_2}\right)^2\frac{n_1p_2}{m_2} + 3\left(\frac{m_2-m_1}{m_1+m_2}\right)^2\frac{n_2p_1}{m_1} + 4n_1n_2\left(\frac{m_2}{m_1+m_2}\right)^2\Delta U^2\right\} \end{array}\right] \tag{21}$$

$$\left(\frac{\partial \mathbf{q}_1^P}{\partial t}\right)_{collision,\nu=-1} = \tfrac{1}{2}\mu(g_0\sigma_{P0})\frac{m_2}{m_1+m_2}\left[\begin{array}{l} -\frac{4}{m_1+m_2}(n_2\mathbf{q}_1 + n_1\mathbf{q}_2) \\ +\frac{m_1-3m_2}{m_1+m_2}\Delta\mathbf{U}\cdot\frac{\bar{\mathbf{p}}_1}{m_1}n_2 - \frac{4m_2}{m_1+m_2}\Delta\mathbf{U}\cdot\frac{\bar{\mathbf{p}}_2}{m_2}n_1 \\ +\Delta U\left(\frac{3m_1+m_2}{m_1+m_2}\frac{3p_1}{m_1}n_2 - \frac{2m_2}{m_1+m_2}\frac{3p_2}{m_2}n_1 - \frac{2m_2}{m_1+m_2}n_1n_2\Delta U^2\right) \end{array}\right] \tag{22}$$

These results can finally be compared with the limiting forms with ν=-1 by taking $\mathbf{q}_2$=0, $\mathbf{p}_2$ to be diagonal, and m_1=m_2=m. The results are

$$\left(\frac{\partial \bar{\mathbf{p}}_1^M}{\partial t}\right)_{collision,\nu=-1} = \tfrac{1}{2}(g_0\sigma_{M0})\left[\left(n_1p_2\bar{\mathbf{I}} - n_2\bar{\mathbf{p}}_1\right) + n_1n_2m\Delta\mathbf{U}\Delta\mathbf{U}\right] \tag{23}$$

$$\left(\frac{\partial \bar{\mathbf{p}}_1^P}{\partial t}\right)_{collision,\nu=-1} = \tfrac{1}{4}m(g_0\sigma_{P0})n_1n_2\left(-\tfrac{3}{4}w_1^2\left[\frac{\bar{\mathbf{p}}_1}{p_1} - \bar{\mathbf{I}}\right] + \left[\tfrac{1}{2}\Delta U^2\bar{\mathbf{I}} - \tfrac{3}{2}\Delta\mathbf{U}\Delta\mathbf{U}\right]\right) \tag{24}$$

$$\left(\frac{\partial \mathbf{q}_1^M}{\partial t}\right)_{collision,\nu=-1} = \tfrac{1}{4}(g_0\sigma_{M0})\left[-2n_2\mathbf{q}_1 + \Delta\mathbf{U}\left\{5n_1p_2 + n_1n_2m\Delta U^2\right\}\right] \tag{25}$$

$$\left(\frac{\partial \mathbf{q}_1^P}{\partial t}\right)_{collision,\nu=-1} = \tfrac{1}{8}(g_0\sigma_{P0})\left[\begin{array}{l} -2n_2\mathbf{q}_1 - \Delta\mathbf{U}\cdot\bar{\mathbf{p}}_1n_2 \\ +\Delta\mathbf{U}\left(6p_1n_2 - 5p_2n_1 - n_1n_2m\Delta U^2\right) \end{array}\right] \tag{26}$$

SUMMARY AND CONCLUSIONS

The complete formalism still requires the general formulation for the heat-flux vector. This approach does provide an internally consistent fluid model applicable to Knudsen numbers ~1 or greater, while voiding the inherent assumptions of Chapman-Enskog approach, viz. (1) the collisionality assumption is replaced with assumed distribution function form, (2) the lowest-order kinetic results should still be captured, and (3) the approach trades integration over velocity space for assumed closure in moments equations. Even the thirteen-moment approach is still inherently complex except for Maxwell molecules. In addition, a numerical model is still required to provide cross checks with kinetic results and explore further the parameter space for the interaction.

ACKNOWLEDGMENTS

This work was supported in part under NASA Grant NAG5-11060.

REFERENCES

1. Pauls, H. L., Zank, G. P., and Williams, L. L., *J. Geophys. Res.*, **100**, 21595 (1995).
2. Liewer, P. C., Karmesin, S. R., and Brackbill, J. U., *J Geophys. Res.*, **101**, 17199-17128 (1996).
3. Baranov, V. B., Izmodenov, V. V., and Malama, Y. G., *J. Geophys. Res.*, **103**, 9575 (1998).
4. Linde, T. J., Gombosi, T. I., Roc, P. L., Powell, K. G., and DeZeeuw, D. L., *J. Geophys. Res.*, **103**, 1889-1904 (1998).
5. Lipatov, A. S., Zank, G. P., and Pauls, H. L., *J. Geophys. Res.*, **103**, 20631 (1998).
6. McNutt, R. L., Jr., Lyon, J., and Goodrich, C. C., *J. Geophys. Res.*, **103**, 1905-1912 (1998).
7. McNutt, R. L., Jr., Lyon, J., and Goodrich, C. C., *J. Geophys. Res.*, **104**, 14803-14809 (1999).
8. McNutt, R. L., Jr., Lyon, J., Goodrich, C. C., and Wiltberger, M., "3D MHD Simulations of the Heliosphere-VLISM Interaction," in *Solar Wind Nine*, edited by S. R. Habbal et al., AIP Conference Proceedings 471, New York: American Institute of Physics, 1999, pp. 823-826.
9. McNutt, R. L., Jr., Wiltberger, M., Lyon, J., and Goodrich, C. C., "A Fluid Approach to the Heliosphere/VLISM Problem," in *The Outer Heliosphere: The Next Frontiers*, edited by K. Scherer et al., COSPAR Colloquia Series, Vol. 11, New York: Pergamon, 2001, pp. 89-98.
10. Müller, H.-R., Zank, G. P., and Lipatov, A. S., *J. Geophys. Res.*, **105**, 27419-27438 (2000).
11. Grad, H., *Commun. Pure and Appl. Math.*, **2**, 331-407 (1949).
12. Grad, H., "Principles in the Kinetic Theory of Gases" in *Encyclopedia of Physics*, Vol. XII Thermodynamics of Gases, edited by S. Flügge, Springer-Verlag, Berlin, 1958, pp. 205-294.
13. Gombosi, T. I., and Rasmussen, C. E., *J Geophys. Res.*, **96**, 7759-7778 (1991).
14. Chapman, S., *Manchester Mem.*, **66**, 1-8 (1922).

Interstellar Pioneer 10 EUV Data: Possible Constraints on the Local Interstellar Parameters

Vlad Izmodenov*, Pradip Gangopadhyay†, Mike Gruntman† and Darrell Judge†

*Lomonosov Moscow State University, Department of Aeromechanics, Faculty of mechanics and mathematics, Vorob'evy gory, Glavnoe Zdanie MGU, 119899, Moscow, Russia
†University of Southern California, Los Angeles, USA

Abstract. The neutral hydrogen and proton densities of the local interstellar cloud are still not well known even after many decades of space research. There is, however, a lot of diagnostic data available to investigate these parameters. We have used one such data set, the Pioneer 10 (P10) Lyman Alpha data obtained between the heliocentric distances 20 to 45 AU, to estimate the local interstellar hydrogen and proton densities. We have used state of the art neutral-plasma and radiative transfer models for the interpretation of the P10 EUV data. The results are presented and possible constraints are discussed.

INTRODUCTION

The heliospheric interface, formed due to the interaction between the solar wind and the local interstellar cloud (LIC), is a very complicated phenomenon where the solar wind and interstellar plasmas, interstellar neutrals, magnetic field, and cosmic rays play prominent roles. The heliosphere provides a unique opportunity to study in detail the only accessible example of a commonplace but fundamental astrophysical phenomenon - the formation of an astrosphere. The heliospheric interface is a natural "environment" of our star and knowledge of its characteristics is important for the interpretation and planning of space experiments.

Remote sensing of the heliospheric interface through the study of the interstellar hydrogen atoms is possible since H atoms play a very important role in the formation of the heliospheric interface. Interstellar H atoms are strongly coupled with plasma protons by charge exchange. Distribution of H atoms inside the heliosphere has imprints of the heliospheric interface. Thus, interstellar hydrogen atoms provide excellent remote diagnostics on the structure of the heliospheric interface. The study of the neutral hydrogen atoms in the outer heliosphere has been made possible by the presence of four deep space spacecraft, Pioneers 10 and 11 (P10 and P11), and Voyagers 1 and 2 (V1 and V2). The USC photometers on-board P10 and P11 and the ultraviolet spectrometers (UVS) on-board V1 and V2 have measured the interplanetary Ly α background radiation for more than twenty years. Various studies of P10/11 and V1/2 Ly α data have been published. Yet, the estimation of the interstellar H atom density varies greatly from study to study, ranging between 0.03 and 0.3 cm^{-3} [1].

In this paper we present the first results of our reanalysis of the P10 Ly α data to improve our knowledge of the very local interstellar neutral hydrogen and proton densities. This reanalysis uses the latest state of the art neutral hydrogen-plasma and radiative transfer models outlined in the later sections.

HELIOSPHERIC INTERFACE MODEL

The interaction of the solar wind with the interstellar medium influences the distribution of interstellar atoms inside the heliosphere. Further, it is now clear that the Local Interstellar Cloud is partly ionized and that the plasma component of the LIC interacts with the solar wind plasma to form the heliospheric interface (Figure 1). Interstellar H atoms interact with the plasma component through charge exchange. This interaction strongly influences both the plasma and neutral components. The main difficulty in the modeling of the H atom flow in the heliospheric interface is its kinetic character due to the large, i.e. comparable to the size of the interface, mean free path of H atoms with respect to the mean free path for charge exchange process. In this paper to get the H atom distribution in the heliosphere and heliospheric interface structure, we use the self-consistent model developed by Baranov and Malama in [2]. The kinetic equation for the neutral component and the hydrodynamic Euler equations were solved self-consistently by the method of global interactions. To solve the kinetic equation for H atoms, an advanced Monte Carlo method with splitting of trajectories [3] was used. Basic results of the model were

CP679, *Solar Wind Ten: Proceedings of the Tenth International Solar Wind Conference*, edited by M. Velli, R. Bruno, and F. Malara

reported in [4-7].

RADIATIVE TRANSFER MODEL

The LISM neutral hydrogen gas is an optically thick medium for solar Lyman Alpha photons. The scattering path length for neutral hydrogen density of 0.1 cm $^{-3}$ will be of the order of 10 to 15 AU. This implies that the radiative transfer calculation of Lyman Alpha photons at heliocentric distances greater than 15 AU must necessarily take into account multiple scattering. In fact, a full treatment of the solar Lyman Alpha radiative transfer problem must include the actual self-reversed solar line shape, multiple scattering, full angular and frequency redistribution function, Doppler and aberration effects, heliosphere-wide hydrogen temperature and velocity changes and Voigt Lyman α absorption profile. Details of our radiative transfer model are given in [8].

RESULTS

Monte Carlo radiative transfer calculations were carried out for seven models of neutral hydrogen density (Table 1). The calculated results, I_{model}, were then compared with P10 EUV data. An example of this is shown in figure 2. In order to properly compare the data with the calculated results, it was necessary to calculate the optimum P10 instrument calibration factor (CF) for each of the density models. This step is necessary since it is known that the P10 and V1/2 instrumental calibrations differ by a factor of 4.4 at Lyman α. The difference between the P10 photometer and V2 spectrometer calibration factors forces one to reproduce the distance dependence of the data rather than rely on the absolute value of the measured intensity. It should be stated here that the P10 photometer calibration did not drift with time during the period 1979-1988 considered here. The degradation of the P10 Bendix channel multipliers has been studied in the laboratory. It has been found that the electron multipliers can deliver about 16 coulombs of charge without any sign of fatigue. The P10 electron multiplier for the hydrogen channel is estimated to have delivered at most 4 coulombs of charge by 1988. The early degradation observed in the hydrogen channel of the P11 instrument is attributed to damage due to the hostile environment encountered by P11 during its flyby past Saturn. The optimum calibration factor for a density model is calculated by minimizing the least squares sum, LSS, where LSS is calculated by the following equation

$$LSS = \sum (I_{model} + bg - CF * I_{P10data})^2, \qquad (1)$$

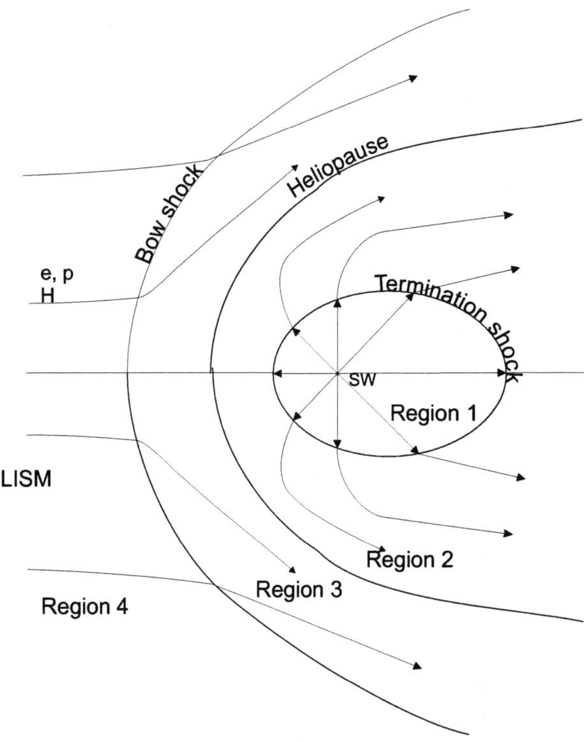

FIGURE 1. The heliospheric interface is the region of the solar wind interaction with LIC. The heliopause is a contact discontinuity, which separates the plasma wind from interstellar plasmas. The termination shock decelerates the supersonic solar wind. The bow shock may also exist in the interstellar medium. The heliospheric interface can be divided into four regions with significantly different plasma properties: 1) supersonic solar wind; 2) subsonic solar wind in the region between the heliopause and termination shock; 3) disturbed interstellar plasma region (or "pile-up" region) around the heliopause; 4) undisturbed interstellar medium.

where summation is over the P10 data points and bg is the Lyman α galactic background. Both CF and bg were varied to obtain the minimum LSS. Once the optimum *CF* and bg are found then P10 data are multiplied by *CF* and compared with the calculated intensity. Both *CF* and *LSS* for each of the 7 density models are given in Table 1.

It is clear from Table 1 that the model with the VLISM neutral hydrogen density of 0.15 cm $^{-3}$ and proton density of 0.07 cm $^{-3}$ yields the lowest *LSS* and so best reproduces the P10 data. The next best fit occurs for the model with neutral hydrogen density of 0.2 cm $^{-3}$ and proton density of 0.2 cm $^{-3}$. It is not at present possible to choose between the two neutral densities used in these two models as it is necessary to calculate model results for other ionization ratios for both of these cases and compare with P10 data. The Lyman α background, bg, was determined to be negligibly small for all the density models. In fact the best fit was obtained for bg equal

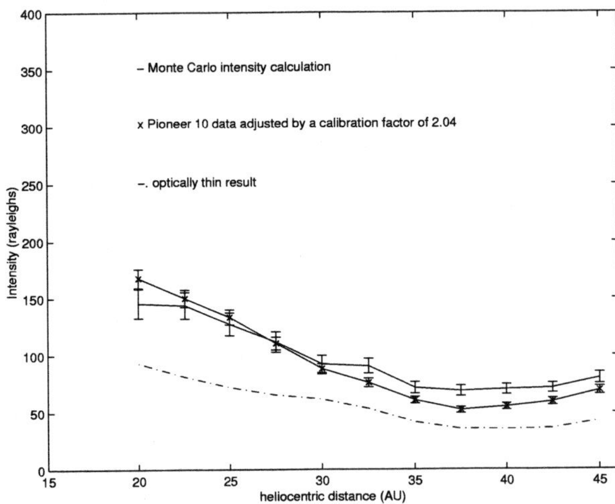

FIGURE 2. Comparison of Monte Carlo calculation using heliospheric model with neutral hydrogen density of 0.15 cm^{-3} and proton density of 0.07 cm^{-3} and Pioneer 10 Lyman α glow data adjusted by the constant calibration factor as function of heliocentric distance. Comparison with optically thin model is also shown.

TABLE 1. Sets of model parameters and results

$n_{H,LIC}$	$n_{p,LIC}$	**sqrt(Least squares sum)**	**Calibration factor**
0.15	0.07	40.9	2.04
0.15	0.10	52.2	1.96
0.18	0.15	59.4	1.84
0.2	0.05	71.8	3.16
0.2	0.10	50.7	2.26
0.2	0.20	50.4	1.80
0.25	0.10	56.9	3.08

to zero although the deviation of the data from the best fit curve was from + 22 to - 16 Rayleighs. The background glow is assumed to be approximately the same for all data points since the look directions are approximately the same with respect to the galactic plane. A look at Table 1 shows that the largest deviation from the P10 data occurs for the model in which neutral density is 0.2 cm^{-3} and proton density=0.05 cm^{-3}. The ionization ratio ($n_p/(n_H+n_p)$) for this model is 0.2.

The value 0.2 might well be the lower limit of the LISM ionization ratio. This is because the LISM neutral hydrogen density can not be too high because of the growing evidence that the neutral hydrogen density inside the termination shock is of the order of 0.1 cm^{-3} or less. For example, Wang and Richardson suggest in [9] that the Voyager 2 solar wind observations are best fitted with an interstellar neutral hydrogen density of 0.08 cm^{-3} at the termination shock. Such a low neutral hydrogen density at the termination shock would imply a neutral hydrogen density of 0.16 cm^{-3} at "infinity" even assuming that the neutral hydrogen suffers a 50 % depletion at the interface. Similarly, *Gloeckler and Geiss* in [10] found that the neutral hydrogen density at the termination shock is 0.115 cm^{-3} and obtained a value of 0.18 cm^{-3} for the interstellar hydrogen density assuming a 58 % filtration effect. We did not use a very high interstellar neutral hydrogen density as that will imply a large neutral density inside the heliosphere which would contradict observational evidence [9, 10].

It is possible to estimate the upper limit to the ionization ratio from the fact that the model with neutral and proton densities equal to 0.2 cm^{-3} (ionization ratio =0.5) gives better fit to the Pioneer 10 data than the models discussed previously except for the first model in Table 1. However, such a high proton density is ruled out as it would imply a solar wind shock too close to the sun contrary to observations. Thus, an ionization ratio as high as 0.5 would imply a low neutral density and a ratio higher than 0.5 is extremely unlikely. The relatively high value of the ionization ratio (0.2 to 0.5) estimated in this work clearly shows that the VLISM neutral hydrogen density can not be as high as 0.25 cm^{-3}. This is because it is clear from figure 8 in [8] and from the better fit obtained for the model with both proton and neutral densities equal to 0.2 cm^{-3} that an ionization ratio substantially higher than 0.3 would be necessary to better fit the neutral density, 0.25 cm^{-3}, model with the P10 data. However, even an ionization ratio of 0.4 for a neutral hydrogen density of 0.25 cm^{-3} would imply a proton density of about 0.18 cm^{-3} which would move the solar wind termination shock too close to the sun.

Another issue that needs to be discussed is the possible reasons for the deviation of the model calculation from the P10 data. The obvious reason is, of course, the difference of the model neutral hydrogen density from the actual heliospheric neutral density. Another reason is the possible variation of the solar Lyman α line center flux with respect to the integrated line [11]. We have plotted the ratio of the model intensity to the P10 data against solar Lyman α flux in order to see if the deviation is due to the variation in line center flux. There is a trend for the ratio to decline from greater than 1 to less than 1 as the solar flux increases. This trend might be due to the line center flux variation.

Figure 2 also shows a comparison between optically thick and optically thin radiative transfer calculations. The difference is significant. This result suggests that the interstellar gas is optically thick.

SUMMARY

The comparison of predicted Lyman Alpha glow using state of the art heliosphere model and Monte Carlo radiative transfer calculations with P10 data has yielded several constraints on the VLISM parameters. The ionization ratio is found to vary between 0.2 and 0.5. The upper limit to the neutral hydrogen density is found to be less than 0.25 cm^{-3}. It is found that the Lyman Alpha galactic background is negligibly small (less than a Rayleigh). The optimum calibration factor was found to vary from 1.8 to 3.2 for the seven models used in this work. The calibration factor is found to be 2 for the model ($n_{H,LIC}$ =0.15 cm^{-3}; $n_{p,LIC}$ = 0.07 cm^{-3}) that best fit the data. This work suggests that P10 UV photometer flux data values need to be increased by a factor of 2. Thus the difference in calibration factor between P10 photometer and V2 spectrometer is reduced from 4.4 to about 2.

ACKNOWLEDGMENTS

This work was supported by NASA grant NAG5-10989. V.I. was also supported in part by INTAS Award 2001-0270, RFBR grants 01-02-17551, 02-02-06011, 01-01-00759, and International Space Science Institute in Bern.

REFERENCES

1. Quémerais, E., Bertaux, J.-L., Sandel, B., Lallement, R., *Astronomy and Astrophysics* **290**, 941-955 (1994).
2. Baranov V. B., and Malama, Yu. G., *J. Geophys. Res.* **98**, 15157, 1993.
3. Malama, Y. G., *Astrophys. Space Sci.* **176**, 21, (1991).
4. Baranov, V. B., and Malama, Yu. G., *J. Geophys. Res.* **100**, 14755 (1995).
5. Izmodenov, V. V., Lallement, R., Malama, Yu. G., *Astron. Astrophys.* **342** L13-L16 (1999).
6. Izmodenov, V., *Astrophys. Space Science* **274**, 55-69 (2000).
7. Izmodenov, V., Gruntman, M., and Malama, Yu. G., *J. Geophys. Res.* **106**, 10,681-10,690 (2001).
8. Gangopadhyay P., Izmodenov, V., Gruntman, M., Judge, D., *J. Geophys. Res.*, submitted (2002).
9. Wang, C., and Richardson, J.D., *J. Geophys.Res.* **106** 29401, 2001.
10. Gloeckler, G. and Geiss, J., *Space Science Reviews* **97**, 169 (2001).
11. Lemaire, P., Charra, J., Jouchoux, A., Vidal-Madjar, A., Artzner, G.E., Vial, J.C., Bonnet, R.M., Skumanich, A., *Astrophys. J.* **223**, L55 (1978).

Narrow coronal holes in *Yohkoh* soft X-ray images and the slow solar wind

C.N. Arge*, K.L. Harvey, H.S. Hudson† and S.W. Kahler**

**CIRES, University of Colorado & NOAA, SEC*
†Space Sciences Laboratory, UC Berkeley
***Space Vehicles Directorate, Air Force Research Laboratory*

Abstract. Soft X-ray images of the solar corona sometimes show narrow dark features not obviously present in HE I 10830Å images. We term these "narrow coronal holes" (NCHs). A prototype for this type of structure crossed solar central meridian on October 29, 2001. Standard source-surface models showed open magnetic field lines in this feature, tending to confirm its identification as a coronal hole. The magnetic field in this example is relatively strong (above 100 G in the low-resolution Kitt Peak magnetograms), and the boundaries of the open-field domain fall within the unipolar area as expected. We have surveyed the *Yohkoh* SXT data for other examples of this phenomenon, and have found several candidates. From observations of the associated solar wind, and from modeling, we find these regions to be sources of slow solar wind.

INTRODUCTION

The solar wind arguably consists of two kinds of long-lived flows, the slow and fast solar winds. The fast wind originates in coronal holes (CHs), but the source(s) of the slow wind have been controversial. Perhaps the first idea was that the slow solar wind originates on open field lines within CHs but near the boundaries where the magnetic flux tube expansion is largest ([1], [2], [3]) This idea is complicated by the suggestion that at solar maximum the slow wind can apparently ([4], [5]) originate in small isolated CHs whose field lines may be strongly diverging, and by the observed association of active regions and solar-wind flow ([6], [7]). The appearance of white-light blobs in LASCO observations suggested to Wang et al. [8] that the slowest and densest solar wind originates in helmet-streamer loops but with a major component still originating from inside CHs near the boundaries.

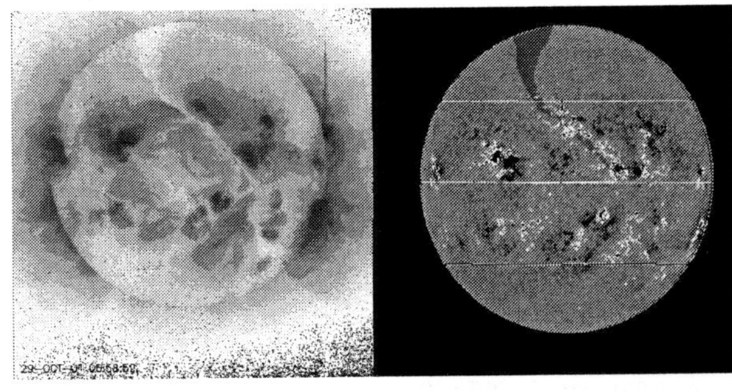

FIGURE 1. Representative narrow coronal hole (NCH) observed by *Yohkoh*, 29 October 2001. The north polar hole can be seen to extend to the southwest (lower right) into a strong-field region towards NOAA region 9678 (N07, W21), as shown in the rough overlay over the KPNO magnetogram on the right.

The boundaries between the slow and fast solar winds during the first southern polar pass of the *Ulysses* spacecraft were distinguished by relatively sharp spatial boundaries of the values of O^{+7}/O^{+6}, Mg/O, and Fe/O in the SWICS instrument [10]. The ratio O^{+7}/O^{+6} in particular reflected freeze-in temperatures of either ~1.6 MK in slow solar wind or ~1.3 MK in fast solar wind. The corresponding solar wind speed profiles at the boundaries were much more gradual, having been washed out by stream-stream dynamics. Longer time averages of the *Ulysses* O^{+7}/O^{+6} observations showed a simple low-temperature freezing-in process for the fast solar wind but a more complicated range of high charge states for the slow wind [11]. A more structured slow solar wind was observed in the ecliptic plane with the *ACE* SWICS O^{+7}/O^{+6} observations [12].

If the Earth traverses solar wind originating in an NCH, we can examine both the speed and O^{+7}/O^{+6} profiles at 1 AU to determine whether or how much of the solar wind from the NCH appears to be fast wind. Such a traversal would intersect solar wind flowing from both the NCH boundaries and the middle of the NCH.

CP679, *Solar Wind Ten: Proceedings of the Tenth International Solar Wind Conference*, edited by M. Velli, R. Bruno, and F. Malara

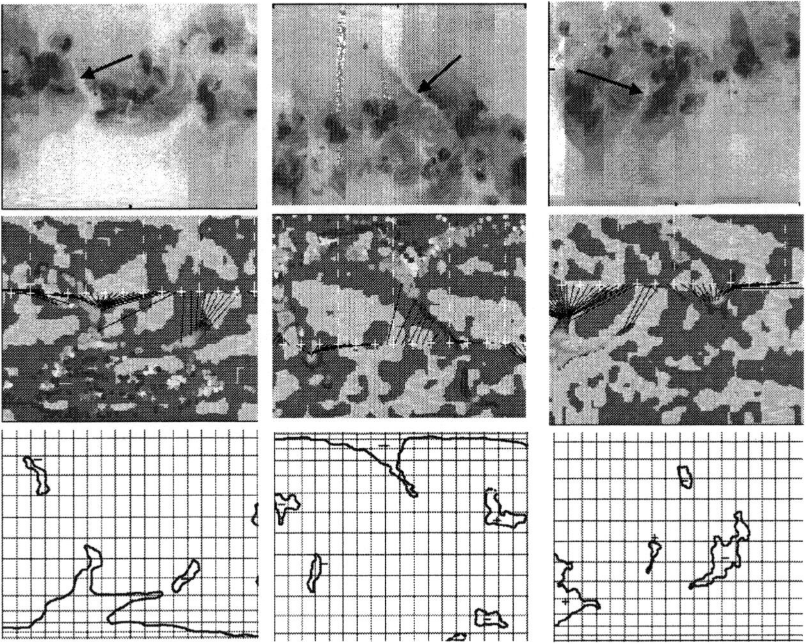

FIGURE 2. Synoptic maps (top, *Yohkoh* soft X-rays; middle, model magnetic field; and bottom, HE I 10830Å CH boundary map) for each of the three NCH episodes studied (left, CR 1851, the "YCH1" of [9]; middle, CR 1982; right, CR 1983). The arrows in the X-ray images point to the NCHs discussed in this paper. The colored dots represent photospheric footpoints of open field lines, with different colors used to indicate the expansion factors (or solar wind speed) associated with the flux tubes. Red corresponds to small expansion factors or fast wind, while blue corresponds to large expansion factors or slow wind. Areas shaded light gray (dark gray) are closed field lines with $B_r > 0$ ($B_r < 0$).

The lack of any signature of fast wind could suggest that in some cases an NCH could be the source of only slow wind, contrary to the concept of CHs as sources of fast wind, at least in their centers.

The *Yohkoh* images from the Soft X-ray Telescope (SXT) give clear views of coronal holes of all sorts. For the purposes of this paper, we use the term "coronal hole" in the sense of "open field region" and assume that the darkest regions of a soft X-ray image during active periods can be identified with open fields (note that Figure 2 shows negative images, such that the CHs are lighter in color). In particular the SXT data often show the existence of NCHs as sharply-defined narrow features especially clearly seen near disk center. We examine the relationship between these narrow CHs and the source of the slow solar wind, which may depend upon rapidity of field-line expansion in the lower corona [13]. This work continues a series of studies using the *Yohkoh* SXT data to characterize CH boundaries in the lower corona ([14], [9]).

DATA ANALYSIS

We have selected three prominent cases of NCHs in the *Yohkoh* SXT data, each of which is a low-latitude extension of a polar CH. The approximate central meridian passage dates are: 28 January 1992 (CR 1851); 29 October 2001 (CR 1982, also shown in Figure 1); and 6 December 2001 (CR 1983). In Figure 2 we show synoptic X-ray, calculated coronal magnetic field, and HE I 10830Å CH boundary maps. Each image was cropped at 45° latitude in the hemisphere opposite the NCH and to a longitude range of 180°. Note that the top two images are linear in latitude, while the He image is linear in cosine latitude. The color coding of the magnetic field maps indicates the magnetic expansion factors of the open field regions, and each NCH was found to be a field region open at the 2.5 $R_{\odot}$ source surface with large (shown as green or blue) expansion factors, although the agreement of the boundaries is not exact. The agreement between the X-ray and coronal field maps supports the interpretation of the X-ray NCHs as true open-field CHs. On the other hand, the He CH boundaries omit significant regions of the NCHs, as well as showing other large disagreements with both the X-ray and coronal magnetic field open areas. This suggests that at least in NCHs the

He images may fail to indicate the presence of open field regions.

The lines extending from the + signs in the middle panels of Figure 2 show the calculated source regions of the solar wind at Earth for each date. In each case we find that the NCH was a solar wind source. Since the general direction of the NCHs was north-south, the Earth traversed the NCH boundaries and central regions. For the two most recent events of the study we can examine the corresponding speed and O^{+7}/O^{+6} profiles measured by the SWICS instrument [15] on the ACE satellite. These are shown in Figure 3 with the approximate times of the calculated passages of the solar wind from the NCHs. In the October 2001 period the O^{+7}/O^{+6} ratio is about 0.5 and in the December 2001 period about 0.8. These values are generally in the range expected for slow solar wind [12]. In addition, the solar wind flow is about 350 km/s and declining through both periods. Note that shocks were tentatively identified in the ACE data at 1300 UT on 31 October and at 0155 UT on 5 November, before and after the NCH solar wind in the October period. In addition, two shocks were identified at 2210 UT on 6 December and at 2300 UT on 11 December, again before and after the period of the NCH solar wind flow in the December period. These transient features do not appear to compromise the general association of the slow wind flows with the NCHs.

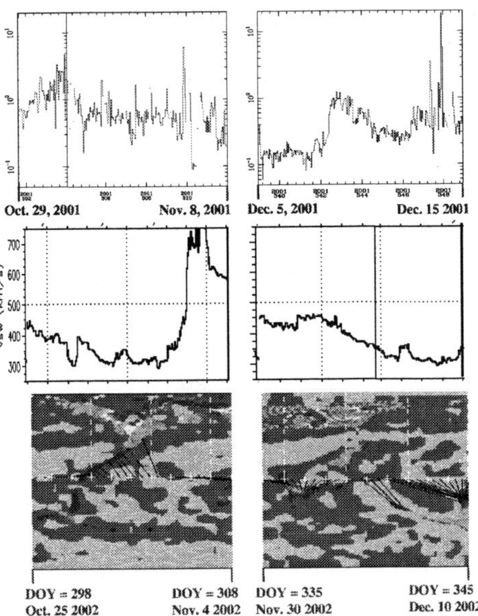

FIGURE 3. SW speed (middle panels) and O^{+7}/O^{+6} abundance ratios (upper panels) for the two most recent of the NCH examples. The panels on the left show the NCH of October, 2001, as imaged directly in Figure 1. The right column shows the NCH of December 2001. The lower panels show source-surface modeling; note that these images are reversed in time to match the SW data.

NARROW CORONAL HOLES

The initial motivation for this paper was the recognition that many of the small CH observed by *Yohkoh* SXT at low latitudes were in fact quite elongated. As Figure 1 illustrates (beware the negative representation), the apparent darkness of these features resembles that of larger-scale CHs. However the SXT data cannot easily be used to determine the physical parameters because of dynamic-range limitations, so the identification of these dark regions with "true" CH, defined as open-field regions, must remain an assumption.

In all respects that we have studied, these NCHs resemble ordinary CHs. In particular they do not exhibit obvious evidence for magnetic reconnection at their boundaries in the form of unusual heating or motions [9]. The NCH pattern is prominent even at SXT resolution (typically using 5″ pixels), and we believe that more systematic and detailed study at higher resolution may be fruitful as discussed below. The HE I signature has limited resolution because of its excitation by coronal EUV emissions, which illuminate relatively wide areas of the chromosphere. As pointed out, studies using HE I 10830Å may have systematically ignored these regions, which may contain relatively large magnetic fluxes (see Figure 1) [16]. The NCHs frequently extend into active regions, where 10830Å observations become confused, and we identify such cases with the open fields of active regions noted by Levine [17].

CONCLUSIONS

We have found NCHs to be sources of slow solar wind originating on open field lines. These may have escaped detection by HE I 10830Å imaging. The source-surface models of these regions show large expansion factors, consistent with the Wang-Sheeley explanation [13], but do not implicate streamers directly. If reconnection takes place freely at the boundaries of these NCHs, and if their boundaries do map to the streamer locations, our results would be consistent with the idea of associating some of the slow wind with streamers. Thus far searches in the X-ray observations [9] have not shown evidence for reconnection in the form of heating or sudden motions, but more definitive data analyses could be carried out. The immediate problem to resolve is the apparent association of slow wind with open field lines (our result) and the earlier indications, from stream properties and from the "streamer blob" assocation [8], that closed field lines play an important role in the formation of the slow wind.

NCHs may escape detection via the standard He I 10830Å observations, although the source-surface models for coronal fields may reveal them. Such features may have smaller expansion factors in the low corona, since they are long and thin – intuitively, the one-dimensional expansion from an NCH might not be so rapid as a two-dimensional one from a more symmetric CH. On the Wang-Sheeley hypothesis, this would suggest a weaker slow-wind flow pattern. However the NCHs may occur in strong-field regions (for example, the one shown in Figure 1), which would be consistent with larger amounts of solar-wind flow originating there. We do find from source-surface model calculations that the NCHs correspond to relatively large expansion factors. This preliminary look at the NCHs therefore suggests the need for a fuller survey that could confirm or deny the NCH/slow-wind relationship, and illuminate the role played by the expansion factor in these CHs.

In a recent work Neugebauer et al. [7] have confirmed the presence of open field lines in active regions (ARs). What we term NCHs may have even higher O^{+7}/O^{+6} ratios than the ones they cite for ARs, and for similarly active periods. Our NCHs often penetrate too close to active regions for good fidelity in the He I signature, and we plan to study a more extensive sample NCHs to clarify their relationship to these AR sources.

ACKNOWLEDGMENTS

The work of CNA was supported by grants NSF ATM-0001851 and ONR N00014-01-F-0026. NASA supported the work of KAH and HSH under contract NAS8-40801. NSO/Kitt Peak data used here are produced cooperatively by NSF/NOAO, NASA/GSFC, and NOAA/SEL. We note with sorrow the loss of our friend and co-author, Karen Harvey, as this work was being prepared.

REFERENCES

1. Wang, Y.-M., Hawley, S. H., and Sheeley Jr., N. R., *Science*, **271**, 464–469 (1996).
2. Bravo, S., and Stewart, G. A., *ApJ*, **489**, 992 (1997).
3. Neugebauer, M., Forsyth, R. J., Galvin, A. B., Harvey, K. L., Hoeksema, J. T., Lazarus, A. J., Lepping, R. P., Linker, J. A., Mikic, Z., Steinberg, J. T., von Steiger, R., Wang, Y.-M., and Wimmer-Schweingruber, R. F., *JGR*, **103**, 14587–14600 (1998).
4. Wang, Y.-M., *ApJ*, **437**, L67–L70 (1994).
5. Kojima, M., Fujiki, K., Ohmi, T., Tokumaru, M., Yokobe, A., and Hakamada, K., *JGR*, **104**, 16993–17004 (1999).
6. Hick, P., Jackson, B. V., Rappoport, S., Woan, G., Slater, G., Strong, K., and Uchida, Y., *GRL*, **22**, 643–646 (1995).
7. Neugebauer, M., Liewer, P. C., Smith, E. J., Skoug, R. M., and Zurbuchen, T. H., *JGR*, **in the press** (2002).
8. Wang, Y.-M., Sheeley, N. R., Walters, J. H., Brueckner, G. E., Howard, R. A., Michels, D. J., Lamy, P. L., Schwenn, R., and Simnett, G. M., *ApJ*, **498**, L165 (1998).
9. Kahler, S. W., and Hudson, H. S., *ApJ*, **in the press** (2002).
10. Geiss, J., Gloeckler, G., von Steiger, R., Balsiger, H., Fisk, L. A., Galvin, A. B., Ipavich, F. M., Livi, S., McKenzie, J. F., Ogilvie, K. W., and Wilken, B., *Science*, **268**, 1033 (1995).
11. von Steiger, R., Schwadron, N. A., Fisk, L. A., Geiss, J., Gloeckler, G., Hefti, S., Wilken, B., Wimmer-Schweingruber, R. F., and Zurbuchen, T. H., *JGR*, **105**, 27217–27238 (2000).
12. Zurbuchen, T. H., Hefti, S., Fisk, L. A., Gloeckler, G., and Schwadron, N. A., *JGR*, **105**, 18327–18336 (2000).
13. Wang, Y.-M., and Sheeley, N. R., *ApJl*, **372**, L45–L48 (1991).
14. Kahler, S. W., and Hudson, H. S., *JGR*, **106**, 29239–29248 (2001).
15. Gloeckler, G., Cain, J., Ipavich, F. M., Tums, E. O., Bedini, P., Fisk, L. A., Zurbuchen, T. H., Bochsler, P., Fischer, J., Wimmer-Schweingruber, R. F., Geiss, J., and Kallenbach, R., *Space Science Reviews*, **86**, 497–539 (1998).
16. Harvey, K. L., Harvey, J. W., and Sheeley, N. R., *Solar Phys.*, **79**, 149–160 (1982).
17. Levine, R. H., *ApJ*, **218**, 291–305 (1977).

A Numerical Study on the Evolution of CMEs and Shocks in the Interplanetary Medium

J. A. González-Esparza*, A. Lara*, A. Santillán† and N. Gopalswamy**

*Instituto de Geofísica, UNAM, México
†DGSCA-Cómputo Aplicado, UNAM, México
**The Catholic University of America, Washington DC

Abstract. We studied the evolution in the solar wind of four CMEs detected by SOHO-LASCO which were associated with ICMEs and interplanetary (IP) shocks detected afterward by Wind at 1 AU. The study is based on a 1-D hydrodynamic single fluid model using the ZEUS code. These simple numerical simulations of CME like pulses illuminate several aspects of the heliocentric evolution of the ICME front and its associated IP shock and we were able to reproduce some characteristics of the IP shocks and ICMEs inferred from the two-point measurements from spacecraft. The simulation shows that ICMEs and IP shocks follow different evolutions in the interplanetary medium both having phases of about constant speed propagation followed by an exponential deceleration with heliocentric distance. IP shocks always propagate faster than their associated ICME drivers and the former began to decelerate well before the IP shock. The results indicate that, in general, although an IP shock is driven by its ICME in the inner heliosphere in most of the cases this is not true any more when they approach to 1 AU.

INTRODUCTION

Coronal mass ejections (CME) and interplanetary (IP) shocks play a predominant role in solar-terrestrial relations. We study the dynamics between the Interplanetary counterpart of a CME (ICME) and its associated IP shock to answer the following question: how far from the Sun does an IP shock is driven by its ICME? We simulated four events which were observed by a coronograph near the Sun and afterward in-situ by an spacecraft at 1 AU, employing the numerical code ZEUS-3D (version 4.2). This code solves the system of ideal MHD equations (non-resistive, non-viscous) by finite differences on an Eulerian mesh [1]. In order to simplify the calculations, we neglected all magnetic effects and assume spherical symmetry. These simplified 1-D hydrodynamic numerical simulations of interplanetary disturbances have proved to be very useful in understanding the basic physical aspects of the injection and heliospheric evolution of solar disturbances [2, 3, 4, 5, 6, 7, 8]. Since the dynamics of an IP shock and its ICME driver have important 2-D and 3-D effects (e.g., the shock strength varies along the shock front, the shock-ICME separation is greater at the shock wings than at the shock nose, the IP shock has a larger angular extent that its driver, etc.), our 1-D results are limited to the direction of the nose of the IP shock and its ICME driver.

TABLE 1. Four CMEs detected by LASCO and their associated transient shocks and ICME detected posteriorly by Wind at 1 AU.

	LASCO CME time*	Wind shock time†	Wind ICME time
A	1/06/97 15:10	1/10/97 00:52	1/10/97 05:00
B	2/07/97 00:30	2/09/97 12:50	2/10/97 03:00
C	4/07/97 14:27	4/10/97 12:55	4/11/97 06:00
D	5/12/97 06:30	5/15/97 01:15	5/15/97 10:00

* taken from [9]
† taken from [10]

NUMERIC SIMULATIONS

The study is based on four halo CMEs (referred as A, B, C and D) observed by SOHO-LASCO, which were related to ICME signatures and IP shocks detected afterward by the magnetic and plasma experiments on board Wind spacecraft. Table 1 shows the times of the four CMEs near the Sun and their interplanetary counterparts at 1 AU.

Following a technique similar to that of Gosling and Riley [1996], we produced the ambient solar wind by specifying the fluid speed, density and temperature at an inner boundary located beyond the critical point (R_o=0.08 AU), and then allowing the code to evolve and reach an equilibrium state that mimics the observed val-

CP679, *Solar Wind Ten: Proceedings of the Tenth International Solar Wind Conference*, edited by M. Velli, R. Bruno, and F. Malara

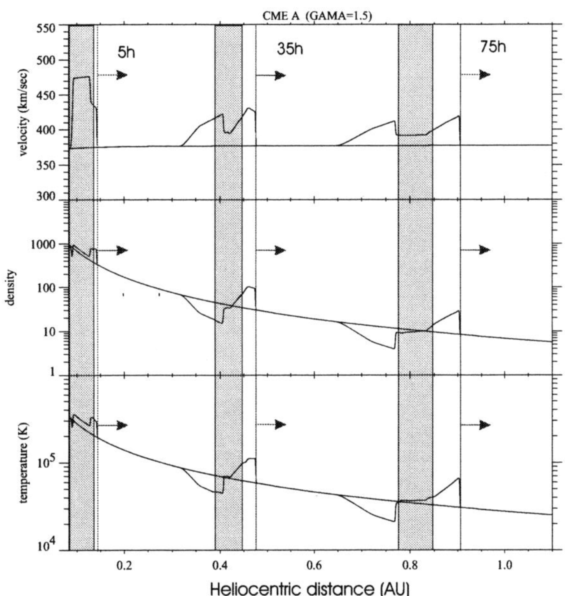

FIGURE 1. Propagation of ICME A (solid case) and its associated IP shock (arrow) at three different times after the CME injection. The continuous line in the background is the undistorted ambient wind.

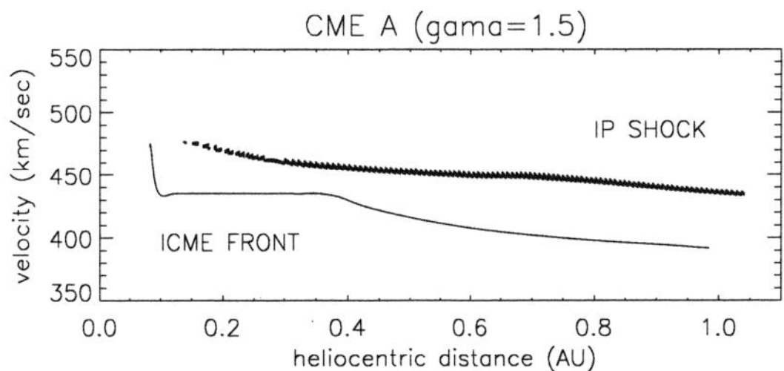

FIGURE 2. Speed evolution of the front of ICME A and its associated IP shock with heliocentric distance.

ues of the solar wind at 1 AU. After the ambient solar wind has been established we injected a perturbation from the inner boundary to simulate the propagation of a CME into the interplanetary medium. These CME-like perturbations were square pulses with a given initial speed characterized by small increments in density and temperature, over a finite extent of time [8]. We included the solar gravity and assumed that the solar wind is an ideal fluid with a ratio of specific heats, $\gamma = 1.5$. All numerical runs have a resolution of .0005 AU/zone with an in-flow condition at the inner-boundary and an out-flow condition at the outer-boundary.

Table 2 shows the initial conditions for the four simulations. At the left side there are the input parameters to produce the numeric ambient wind and the ambient wind (upstream of the IP shock) observed by Wind. Note that by defining arbitrary conditions at the inner boundary we are able to reproduce all the parameters of the ambient wind measured by Wind for each event. In Table 2, the events C and D presented an ambient wind with unusuallly very high densities (Nwind > 15 cm^{-3}), whereas in event B the solar wind speed (Vwind > 518 km/s) and temperature (Twind $\sim 10^5$ K) were relatively high compared to mean slow wind values. At the right side of Table 2 we have given the characteristics of the four CME pulses. The values of the CME velocities and the time durations were inferred from analyzing the SOHO-LASCO movies whereas the other CME values were assumed arbitrarily as small jumps with respect to the ambient wind ahead. We should keep in mind that there is a large uncertainty in estimating the physical characteristics of a halo CME based on coronograph observations [11].

RESULTS

The four events correspond to CMEs propagating with speeds faster than their ambient winds ahead. Figure 1 shows the evolution of ICME A and its associated IP shock as they propagate in the interplanetary medium. This event corresponds to the CME with the lowest increase of speed with respect to the ambient wind and shortest injection time (Table 2). The figure shows the plots of velocity, density and temperature at three different times after the CME initiation. At each time step the code tracks the leading and trailing edges of the ICME as well as the position of the IP shock and estimates the shock parameters by solving the Rankine-Hugoniot equations. The ICME expands radially as propagates outward from the Sun, while the IP shock separates from the ICME.

Figure 2 shows the speed evolution of the ICME front and its associated IP shock against heliocentric distance for event A. The ICME front speed and the IP shock speed have different evolution. The CME front suffers a clear deceleration at 0.37 AU, at about 27 hours after the CME injection, however, the IP shock propagates at a constant speed until 0.7 AU (47 hours after the CME injection) after which begins to decelerate. When the ICME front and the IP shock propagate at about constant speeds, the shock-ICME radial separation increases linearly with heliocentric distance; however, when the ICME reaches the critical distance and begins to decelerate this radial separation increases exponentially with heliocentric distance.

Figure 3 shows a similar plot that the previous figure but for event C. The ICME front has a clear deceleration from about 0.4 AU, 21 hours after the CME injection, whereas its associated IP shock began to decelerate be-

TABLE 2. Initial conditions for the four simulations. (*Left*) Physical characteristics (velocity, density and temperature) of the numeric ambient wind at the inner boundary (Ro=0.08 AU), the numeric wind at 1 AU and the ambient wind measured by Wind. (*Right*) Physical characteristics of the CME-like pulses: initial speed, jumps in density and temperature with respect to the ambient wind and CME temporal duration.

	Ambient solar wind									CME like pulse			
	Vo [km/s]	$V1_{AU}$ [km/s]	Vwind [km/s]	No [cm^{-3}]	$N1_{AU}$ [cm^{-3}]	Nwind [cm^{-3}]	To [K]	$T1_{AU}$ [K]	Twind [K]	V_{cme} [km/s]	ΔN	ΔT	dt_{cme} [h]
A	373	377	378	990	6.8	6.5	2.5e5	2.8e4	2.6e4	473	1.2	1.2	4.4
B	380	521	518	880	4.4	4.4	1.6e6	1.5e5	1.6e5	480	1.2	1.2	7.5
C	270	307	306	2280	13.9	15.2	4.5e5	4.6e4	4.6e4	880	1.2	1.2	5.6
D	308	322	321	2800	18.6	19.5	3.1e5	3.4e4	3.2e4	700	1.2	1.2	10.4

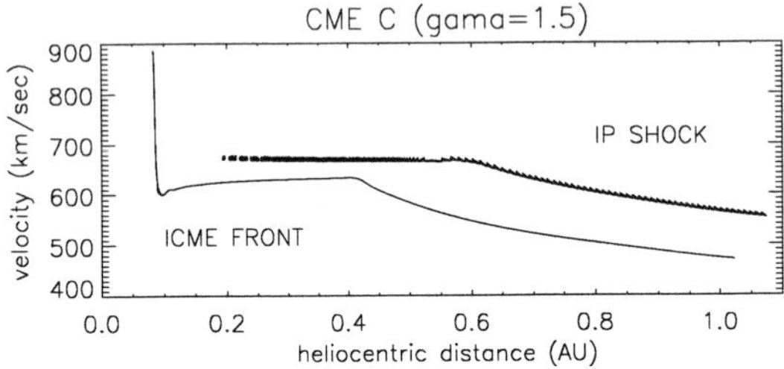

FIGURE 3. Speed evolution of the front of ICME C and its associated IP shock with heliocentric distance.

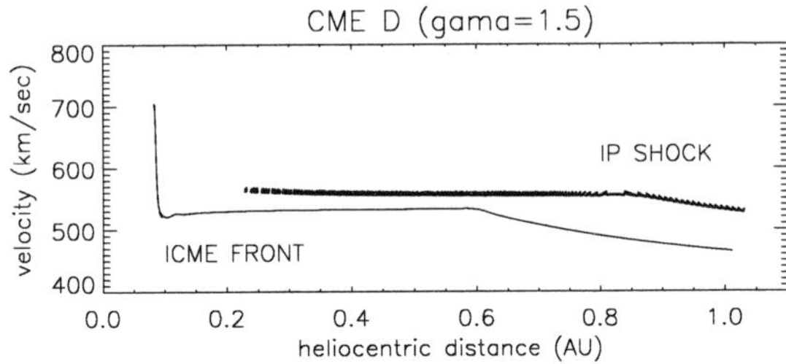

FIGURE 4. Speed evolution of the front of ICME D and its associated IP shock with heliocentric distance.

yond 0.6 AU at about 31 hours after the CME injection. Figure 4 shows the plot of the velocity versus heliocentric distance for event D. In this case the ICME front begins to decelerate at about 0.58 AU, 40 hours after the CME injection, while the IP shock begins a clear deceleration at about 0.85 AU, 61 hours after the CME injection. Note that this longer distance for which the front propagates at a constant speed is related to the long injection duration of this event. The results show that an ICME begins to decelerate well before its associated IP shock, indicating that, in general, most IP shocks are not driven any more by the time they reach to 1 AU.

Table 3 compares the numerical results with the Wind observations at 1 AU. The first three columns show the ICME front speed observed by Wind, the corresponding simulated results and their difference. In the case of event A we obtained a poor agreement between the ICME speed observation and the simulation with a difference of about -78 km/s. However, for the other three events the agreement is better and the difference is less than ten percent. The next three columns in Table 3 show the IP shock transit time. In this case there is an excellent agreement between observation and simulation for all four events, with the maximum difference being less than two hours. The next three columns show the ICME transit time. In this case there is a good agreement in 3 of the four events, but in event C the ICME arrival time was predicted about 17 hours before the Wind observation. This might be a geometrical effect (e.g., the shock wing passed by Wind instead of the shock nose), which cannot be studied with this 1-D model. Finally the last three columns show the temporal lag between the IP shock and the ICME passing at 1 AU. In this case there is a good agreement in event A and D, however in events B and C the model predicts that the shock and the ICME were a few hours closer than their corresponding Wind observation. Again this may be a geometrical effect.

In summary these simple simulations reproduced most of the features inferred from the observations.

DISCUSSION AND CONCLUSIONS

The ambient solar wind has a strong influence on the propagation of ICMEs in the interplanetary medium: Fast ICMEs decelerate as a consequence of interchange of momentum with the solar wind ahead. CME C was initiated about two times faster than CME B (Table 2), however both events resulted in the same transit times to 1 AU (Table 3). In the simulations the evolution of the front of the ICME shows three different phases with heliocentric distance [8]. After the initial injection, the ICME suffers a strong deceleration during a couple of hours reaching an intermediate speed with respect to the

TABLE 3. Comparison between Wind in-situ observations at 1 AU and the simulation results: ICME front speed at 1 AU, interplanetary shock transit time, ICME transit time and shock-ICME temporal lag.

	ICME speed at 1AU			IP shock transit time			ICME transit time			shock-ICME time lag		
	wind [km/s]	sim. [km/s]	dif. [km/s]	wind [h]	sim. [h]	dif. [h]	wind [h]	sim. [h]	dif. [h]	wind [h]	sim. [h]	dif. [h]
A	470	392	-78	82	84	+2	86	90	+4	4	6	+2
B	515	551	+36	60	60	-0	74	68	-6	14	8	-6
C	470	453	-17	60	61	+1	87	70	-17	17	9	-8
D	440	466	+26	67	68	+1	75	74	+1	8	6	-2

ambient wind[1], then it propagates at this approximately constant speed until at certain point (0.35-0.6 AU), at which it exponentially decelerates. This heliocentric evolution is consistent with results based on interplanetary scintillation (IPS) observations of ICMEs [12, 13]. On the other hand, the IP shock associated with a fast ICME has a different evolution with two main phases, one having a small and quasi-constant deceleration until at certain point (0.5 - 0.84 AU) at which its deceleration increases exponentially with increasing heliocentric distance. An IP shock always propagates faster than its ICME driver and the lateer began to decelerate well before the shock. In general, the simulations show that most transient shocks are not driven any more by their ICMEs when they reach 1 AU.

Finally, it is remarkable that the 1-D simulation is able to capture most of the dynamic behavior of CMEs in the interplanetary medium, consistent with the observations. Problems such as the neglect of magnetic fields, assumptions of spherical symmetry and homogeneity of the ambient solar wind, and the inner computational boundary conditions being too far from the Sun need to be considered in future simulations.

ACKNOWLEDGMENTS

We are grateful to Pete Riley for providing the trace subroutine and many useful discussions. This project was partially supported by CONACyT project J33127-E.

REFERENCES

1. Stone, J. M., and M. Norman, *Astrophys. J.*, **80**, 753 (1992).
2. Hundhausen, A. J., and R. A. Gentry, *J. Geophys. Res.*, **74**, 2908 (1969).
3. Dryer, M., *Space Sci. Rev.*, **7**, 363 (1994).
4. Gosling, J. T., and P. Riley, *Geophys. Res. Lett.*, **23**, 2867 (1996).
5. Riley, P., and J. T. Gosling, *Geophys. Res. Lett.*, **25**, 1529 (1998).
6. Riley, P., "CME dynamics in a structured solar wind," in *Solar Wind Nine*, edited by S. R. Habbal, R. Esser, J. V. Hollweg, and P. A. Isenberg, AIP Conference Proceedings 471, American Institute of Physics, New York, 1999, p. 131.
7. Riley, P., J. T. Gosling, and V. J. Pizzo, *J. Geophys. Res.*, **106**, 8291 (2001).
8. González-Esparza, J. A., A. Lara, E. Pérez-Tijerina, A. Santillán and N. Gopalswamy, *J. Geophys. Res.*, **108** (2003).
9. Gopalswamy, N., A. Lara, R.P. Lepping, M. L. Kaiser, D. Berdichevsky, and O. C. St. Cyr, *Geophys. Res. Lett.*, **27**, 145 (2000).
10. Berdichevsky, D. B., A. Szabo, R. P. Lepping and A. Vinñas, *J. Geophys. Res.*, **105**, 289–314 (2000).
11. Lara, A., J. A. González-Esparza and N. Gopalswamy, *Geofisica Internacional*, **41**, in press (2002).
12. Manoharan P. K., M. Tokumaru, M. Pick, P. Subramanian, F. M. Ipavich, K. Schenk, M. L. Kaiser, R. P. Lepping, and A. Vourlidas, *Astrophys. J.*, **559**, 1180–1189 (2001).
13. Yamashita, M., M. Tokumaru, M. Kojima, "Radial dependence of propagation speed of solar wind disturbance," in *this issue*, 2002.

[1] This impulsive deceleration might be overestimated by the model [8].

CMEs at High Northern Latitudes During Solar Maximum: Ulysses and SOHO Correlated Observations

Daniel B. Reisenfeld*, John T. Gosling*, John T. Steinberg*, Pete Riley†, Robert J. Forsyth¶, and O. Chris St. Cyr‡

*Space and Atmospheric Sciences, MS-D466, Los Alamos National Laboratory, Los Alamos, NM 87544
†SAIC, 10260 Campus Point Drive, San Diego, CA 92121
¶The Blacklett Laboratory, Imperial College, London, UK
‡Catholic University of America, Code 682, NASA-Goddard Space Flight Center, Greenbelt, MD 20771

Abstract. From September through November 2001, Ulysses was almost continuously immersed in polar coronal hole (CH) flow during its northern polar pass of the Sun. For much of this time, the flow was fast (> 700 km/s) and steady, quite similar to the steady unstructured flow observed during Ulysses' first polar orbit near solar minimum. During the three months Ulysses transited the northern polar CH it observed 5 coronal mass ejections (CMEs). Of these, two were clearly over-expanding and two were at least partially driven by overexpansion. The phenomenon of over-expansion was frequently observed at high latitudes during Ulysses' first orbit. The recurrence of over-expanding CMEs during the second orbit at high latitudes indicates that this is a phenomenon apparently unique to and typical of CMEs embedded in polar CH flow. Ulysses was nearly above the solar limb during this three-month interval, providing an opportunity to use LASCO/SOHO observations to study the initial velocity profiles of the CMEs observed further out by Ulysses. These initial conditions were used as inputs into a hydrodynamic code, the results of which are reported here.

INTRODUCTION

From September through November 2001, Ulysses was almost continuously immersed in polar coronal hole (CH) flow as it transited the northern polar regions of the Sun. This polar pass coincided with the phase of the solar cycle immediately following solar maximum when a new polar CH had formed over the North Pole [1]. This CH was much smaller in extent than that observed during the first orbit: during the first orbit, near solar minimum, Ulysses observed continuous CH flow at all latitudes above ~35°, whereas Ulysses observed the recent CH only above latitudes of 70° N. Although covering a much narrower range of latitudes, the properties of this flow were nearly identical to those of the first orbit: a fast, steady wind speed, a low density, and a high temperature. The one notable difference is that whereas during the first pair of polar passes, there were very few coronal mass ejections (CMEs), this was not the case here. During the first orbit, Ulysses encountered only 6 CMEs over the course of ~3 years of immersion in polar CH flow (1-1/2 years each in the southern and northern CHs, respectively). However, in the brief three-month period considered here, it intercepted 5 CMEs. Of these, 2 were clearly over-expanding (expansion driven by an internal over-pressure [2]) and 2 were at least partially driven by over-expansion. Because these 5 CMEs were embedded in otherwise steady flow, we were afforded an excellent opportunity to investigate CME evolution through interplanetary space without interference from other solar wind perturbations. This was an ideal situation for comparison to the hydrodynamic (HD) simulation of CME evolution. Additionally, because Ulysses was positioned almost directly above the Sun's North limb for this time period, the Ulysses observations could be correlated with remote sensing observations of CMEs made by instruments on board the Solar and Heliographic Observatory (SOHO).

We were able to identify the solar source of 3 of the 5 CMEs observed by Ulysses using LASCO and EIT images. We have used the LASCO observations to determine the initial speed profiles of these CMEs,

CP679, *Solar Wind Ten: Proceedings of the Tenth International Solar Wind Conference*, edited by M. Velli, R. Bruno, and F. Malara

and then used these as inputs into a 1-D HD model of CME evolution. Here, we describe the SOHO and Ulysses observations of 2 of these CMEs and compare the Ulysses observations to the simulation results. We then discuss the successes and pitfalls of such comparisons. This analysis is particularly relevant in view of the upcoming STEREO mission, which has the specific goal of investigating CME evolution via multi-spacecraft remote imaging and *in situ* observations.

SOHO AND ULYSSES OBSERVATIONS

Of the 5 CMEs, we present here the observational details of 2. Plasma observations were made by the Ulysses ion and electron spectrometers [3]. Solar coronagraphic images were obtained from the Large Angle Spectrometric Coronagraph (LASCO) experiment on SOHO [4]. Where possible, we make use of full-disk images obtained from the extreme ultraviolet imaging telescope (EIT) on SOHO [5] to locate the activity source leading to the CME.

The October 24, 2001 Event

The first event occurred at the Sun on October 24, 2001 at 23:06 UT, appearing in LASCO as a partial halo at the North Pole. The event was clearly accelerating. When it first appeared in the C3 white light optical system, the projected velocity toward Ulysses was only about 100 kms^{-1}, but by the time it left the LASCO field-of-view (FOV) (at 28 solar radii, or 0.13 AU) it was moving at an almost constant velocity of 700 km/s. That this event is directed toward Ulysses is supported by EIT 284 Å images that show the disappearance of an arcade having a northern boundary coincident with that of the North Pole CH.

We have associated these SOHO observations with an event observed by Ulysses on 29 October – 1 November. The CME, shown in Figure 1, exhibited nearly classic over-expansion [2]. We identify this as a CME by the presence of counter-streaming electrons, a descending speed profile, an enhanced magnetic field, and smooth field rotation (not shown), which indicates a magnetic flux rope. The central portion of the CME showed a pressure minimum, sandwiched between two pressure pulses traveling away from the center. At this distance from the Sun, only the reverse compression had steepened into a shock. The CME was immersed in very steady flow, with a central speed of 750 km s^{-1}, comparable to the speeds before and after the disturbance. Its dynamical evolution was therefore driven solely by the CME's internal over-pressure. We also note that the central speed of 750 km s^{-1} was very close to the leading edge speed of 700 km s^{-1} observed at the limit of the LASCO FOV.

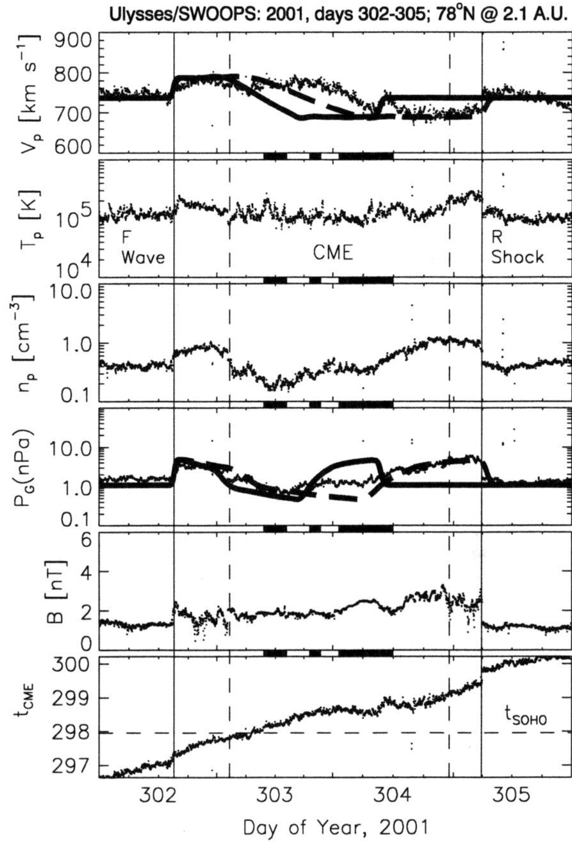

FIGURE 1. Ulysses plasma and magnetic field data for the CME associated with the 24 October event. The CME (between the dashed vertical lines) is bounded by a forward wave and a reverse shock. The panels show the solar wind proton flow speed v_p, proton temperature T_p, proton density n_p, total gas pressure (electrons + ions) P_G, magnetic field strength B, and the predicted time when the plasma departed the Sun t_{CME} (see text). The bottom panel also shows the time t_{SOHO} when the CME is observed to depart the Sun by SOHO. The dark bands between panels show times of electron counterstreaming. In the panels for v_p and P_G, the solid lines show the results from the 1-D HD simulation of the event, and the dashed lines show the simulation results scaled in time to match the duration of the observed CME.

The bottom panel of Figure 1 shows the time t_{CME} that the CME plasma is predicted to have left the Sun: $t_{CME} = t_U - r_U / v_U$, where t_U, r_U, and v_U are the time of observation, heliocentric distance, and flow speed at Ulysses [6]. The time t_{CME} can be compared to the time t_{SOHO} when the leading edge of the associated CME was first observed at the Sun by LASCO. The equation for t_{CME} assumes ballistic flow from the Sun, that is, no net acceleration or deceleration of the CME throughout its transit. The bottom panel of Figure 1 shows t_{CME}, and the horizontal dashed line represents the time t_{SOHO} (23:06 UT on 24 October) of the associated LASCO event. The time t_{SOHO} preceded t_{CME} for most of the CME, indicating that the plasma must have accelerated after initiation, which in fact is just what was observed in the LASCO images.

The November 22, 2001 Event

A large halo CME was observed by LASCO on 22 November associated with a pair of M-class flares, at 20:58 and 23:30 UT, respectively. The disk activity was observed by EIT in the SW quadrant of the Sun; thus it is not obvious that this event headed toward Ulysses. Nevertheless, by tracking the leading edge, we find that this event was moving extremely fast, with a projected speed of 1390 km s^{-1} toward Ulysses.

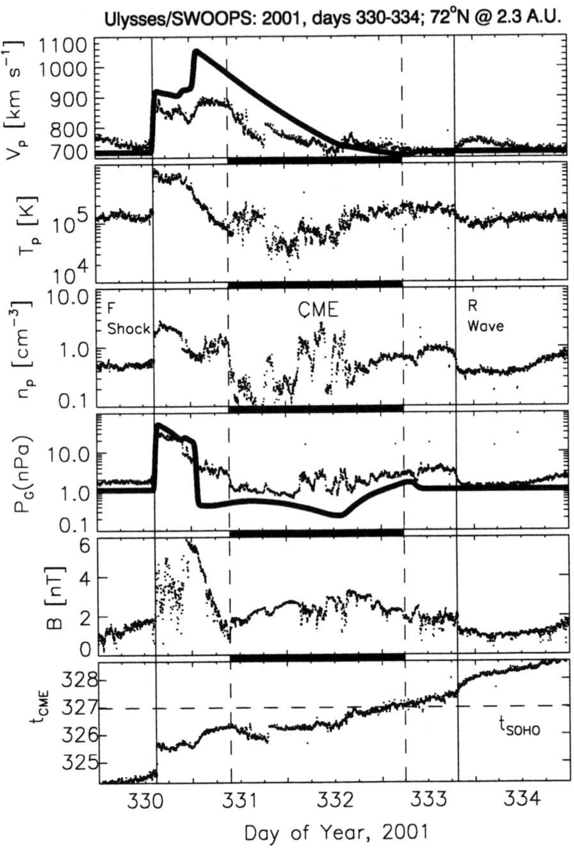

FIGURE 2. Ulysses plasma and field data for the CME associated with the LASCO 22 November event. The CME (between the dashed vertical lines) is bounded by a forward shock and a reverse wave. See Fig. 1 for explanation.

Figure 2 shows Ulysses plasma and field data at 2.3 AU for 26-29 November, which we have associated with the 22 November solar event. The event was remarkably strong, exhibiting counter-streaming; a forward shock at 890 km s^{-1} having a strength of 4.0, and a central temperature depression. The CME had a strong magnetic field enhancement and a low field variance. The center showed a weak pressure minimum, and the event ended with a reverse wave propagating into the wind behind it. Thus, whereas the front half of the event was typical of a fast CME plowing into slower wind ahead, the rear half of the event had the shape of an over-expanding CME. The reverse wave was weak since it was propagating into receding plasma. The bottom panel of Figure 2 shows that t_{CME} preceded t_{SOHO} throughout the event, implying there must have been significant deceleration in the heliosphere.

We believe that the Ulysses and SOHO observations correspond to the same event because (1) the Ulysses event was very energetic and no LASCO events which are likely to give rise to such a powerful event at Ulysses were observed at a reasonable time before or after the described event, and (2) an event at ACE is observed on 24-26 November (not shown) which has a very similar magnetic profile to the Ulysses event, and for which the calculated values of t_{CME} lead to an obvious association with the LASCO event. Furthermore, 2-D and 3-D HD simulations by Riley et al. [7] and Odstrcil & Pizzo [8], show that CME disturbances can undergo significant lateral expansion in the heliosphere; thus it is reasonable to expect that an event initially spanning a relatively localized angular extent near the Sun may span an angular region many times greater by the time it reaches heliocentric distances of order 1 AU.

SIMULATIONS

We have performed 1-D HD simulations to understand better the evolution of the above events through the heliosphere. Simulations of this sort have been performed in the past to model the evolution of the solar wind in general and CMEs in particular as they propagate through the heliosphere, but here we use SOHO observations to constrain partially the simulations by providing initial start times and speeds.

We use a simple 1-D HD simulation that reproduces the basic physics behind these events. The code predicts too strong an interaction between the CME and the ambient solar wind because it does not incorporate shear flows, which relieve pressure stresses. The code also neglects magnetic forces. Despite these limitations, our simulations have produced event profiles that are qualitatively, and sometimes quantitatively, similar to the disturbances observed far from the Sun.

All simulations were performed using a versatile numerical code (ZEUS) [9] that we have used extensively for solar wind studies. The calculations we present were initiated at an inner boundary of 0.14 AU, which lies well outside the critical point where the solar wind becomes supersonic. Speed, density, and temperature were first held steady at the inner boundary until a stationary, supersonic flow with a speed of 760 km s^{-1} at 5 AU and a density of 2.5 cm^{-3} at 1 AU, matching average high latitude conditions observed by Ulysses, filled the computational mesh. The computed speed and pressure at the appropriate distance were then compared to the observed plasma profiles. For each event, a number of different initiations were tried. Here we report only the results of the trials having best agreement.

The October 24 event was initiated as a square-wave density perturbation of 8 times the initial ambient density, and having a duration of 8 hours. As discussed in [2], modeling it as a density pulse creates the initial internal overpressure required to generate an over-expanding CME, which is exactly what results (Fig. 1). The duration of 8 hours was selected based on the observed duration of the event at the Sun. This is probably an overestimate, necessitated because we have found it difficult otherwise to reproduce the disturbance duration of 63 hours observed at Ulysses. Even at 8 hours, we still do not reproduce the Ulysses duration. Qualitatively, however, we do get good agreement with the overall shape of the disturbance at Ulysses. The agreement improves in Figure 1 when we stretch out the temporal scale of the simulation results (dashed line).

The November 22 event has a more complex initiation: a saw-tooth velocity pulse up to 1400 km s^{-1} is first introduced turning up suddenly and tailing off over 20 hours, and a bell-shaped density pulse of 8 times the ambient density is initiated at the midpoint of the velocity pulse, lasting 10 hours. We chose this initialization to generate the observed strong forward shock followed at the rear of the CME by a reverse compression wave (Fig. 2). The reverse wave does not appear if we introduce a velocity pulse alone, nor does it appear if the velocity pulse completely overlaps the density pulse and does not lead it. The physical rational for such an initial profile might be a filament eruption in which an overlying eruption drags out denser filament material below it.

We find that within the already mentioned limitations of a 1-D HD simulation, we do get reasonable qualitative agreement with the observations, suggesting that we have a basic understanding of the dynamical evolution of these CMEs. However, to reproduce the observations in any detail, we require not only more sophisticated simulations (multi-dimensional and magneto-hydrodynamic), but also better observational constraints. The primary limitations are a lack of measurements of the initial density, temperature or field, as well as the line-of-sight confusion resulting from single-viewpoint observations. It would be extremely useful to coordinate *in situ* measurements with CME density and temperature measurements like those obtained from the Ultraviolet Coronagraph Spectrometer on board SOHO [10]. The line-of-sight issue will be addressed by the STEREO mission, which consists of two spacecraft orbiting the Sun such that they produce stereoscopic images of the corona.

ACKNOWLEDGMENTS

Work at Los Alamos was performed under the auspices of the U. S. Department of Energy with support from NASA's Ulysses program. OCS acknowledges partial support from National Space Weather Program grant ATM-0196112 and from NASA contract S-8670-E. PR gratefully acknowledges the support of NASA and the NSF.

REFERENCES

1. D. J. McComas, H. A. Elliot, J. T. Gosling et al., *Geophys. Res. Lett.* **29** (2002).
2. J. T. Gosling, D. J. McComas, J. L. Phillips et al., *Geophys. Res. Lett.* **21**, 2271 (1994).
3. S. J. Bame, D. J. McComas, B. L. Barraclough et al., *Astron. Astrophys. Suppl. Ser.* **92**, 237 (1992).
4. G. E. Brueckner, R. A. Howard, M. J. Koomen et al., *Sol. Phys.* **162**, 357 (1995).
5. J. P. Delaboudiniere, G. E. Artzner, J. Brunaud et al., *Sol. Phys.* **162**, 291 (1995).
6. H. O. Funsten, J. T. Gosling, P. Riley et al., *J. Geophys. Res.* **104**, 6679 (1999).
7. P. Riley, J. T. Gosling, and V. J. Pizzo, *J. Geophys. Res.* **102**, 14677 (1997).
8. D. Odstrcil and V. J. Pizzo, *J. Geophys. Res.* **104**, 493 (1999).
9. J. M. Stone, D. Mihalas, and M. L. Norman, *Astrophys. J. Suppl. Ser.* **80**, 819 (1992).
10. A. Ciaravella, J. C. Raymond, F. Reale et al., *Astrophysical Journal* **557**, 351 (2001).

Sensitivity of Cosmic Ray Modulation to an Outer Scale of Turbulence

S. Parhi*, J. W. Bieber*, W. H. Matthaeus* and R. A. Burger†

*Bartol Research Institute, University of Delaware, Newark, DE 19716, USA
†School of Physics, Potchefstroom University for CHE, Potchefstroom, South Africa

Abstract. We develop an *ab initio* modulation model, in which the diffusion tensor is specified by particle transport theory based on observed properties of the turbulence, and the spatial variation is determined by turbulence transport models that specify how turbulence properties vary throughout the heliosphere. Using a recent perpendicular diffusion treatment based upon the Taylor-Green-Kubo equation we study the sensitivity of solar wind modulation to an outer scale of turbulence, the "ultrascale", which is very poorly understood. For larger values of the ultrascale (approximately hundreds of correlation lengths) we find that the radial profile of modulated cosmic ray protons compares well with the observational data from Voyager and IMP for both polarities.

INTRODUCTION

In modeling the solar modulation of cosmic rays, the diffusion tensor and its spatial variation are usually treated as free parameters that can be adjusted to fit observational data. This has been recognised not to be a satisfactory procedure, since the diffusion tensor in fact is not free, but rather is governed by the interplanetary turbulent magnetic field. Thus we develop an *ab initio* theory [1] using a recent treatment of perpendicular diffusion based on Taylor-Green-kubo formalism [2], which is one of several available theoretical or numerical formulations of perpendicular diffusion [3, 4, 5]. The perpendicular diffusion in two-component slab + 2D turbulence depends critically upon an outer scale of turbulence called the "ultrascale" [6, 7] which has been studied very little.

Another important factor which is essential for *ab initio* theory is the radial variation of both the parallel and perpendicular diffusion coefficient which are strongly dependent on radial variation of the ordinary correlation length. However, the latter is very poorly understood which can be attributed to the uncertain impact of pickup ion-driven turbulence [8, 9] and to the difficulties in measuring the parallel correlation length in the outer heliosphere.

Here, we explore the sensitivity of modulation to ultrascale by direct numerical solution of transport equations by introducing a new perpendicular diffusion formalism as described earlier.

DESCRIPTION OF THE ULTRASCALE

Recently, a nonperturbative approach has been developed for a two-component model (slab plus 2D turbulence) to better understand the diffusion of magnetic field lines [6, 9]. In the 2D plus slab model of magnetic turbulence we assume $\mathbf{B} = \mathbf{B_0} + \mathbf{b}\ (x,y,z)$, where $\mathbf{B_0} = B_0\hat{z}$, $\mathbf{b} \perp \hat{z}$, and $\mathbf{b} = \mathbf{b^{2D}}(x,y) + \mathbf{b}^{\text{slab}}(z)$. The magnetic field line diffusion coefficients are found to be a nonlinear combination of magnetic field line wandering for slab turbulence with l_{slab} as the parallel correlation length and the same associated with 2D fluctuations with ultrascale $\tilde{l}$ as an outer scale weighted by the 2D magnetic fluctuations. The field line diffusion coefficient in composite turbulence, $D_\perp$, is related to the separate slab and 2D diffusion coefficients, $D_\perp^{\text{slab}}$ and $D_\perp^{\text{2D}}$, as follows [6, 10]: $D_\perp = D_\perp^{\text{slab}} + (D_\perp^{2D})^2/D_\perp$, where $D_\perp^{\text{slab}} = b_{\text{slab}}^2 l_{\text{slab}}/2B_0^2$, $D_\perp^{2D} = b_{2D}\tilde{l}/B_0$. Here, variance of slab magnetic fluctuation $b_{\text{slab}}^2 = \langle |\ \mathbf{b}^{\text{slab}}(z)\ |^2 \rangle$, variance of 2D magnetic fluctuation $b_{2\text{D}}^2 = \langle |\ \mathbf{b}^{2\text{D}}(x,y)\ |^2 \rangle$, $k_\perp^2 = k_x^2 + k_y^2$, and $\mathbf{k} = (k_x, k_y, k_z)$ denotes wavevector.

Physically the ultrascale is governed by large scale solar wind fluctuations. It can be considered as the length obtained from the ratio of mean square magnetic flux to mean magnetic fluctuation energy and is, therefore, a measure of the mean size of poloidal (2D) flux structures which could be as large as the order of the system size [7]. On the other hand it is not clear as to how much fluctuation energy resides in ultrascale fluctuations though it is known that substantial amount of energy resides in spatial scales between ion inertial scale

CP679, *Solar Wind Ten: Proceedings of the Tenth International Solar Wind Conference*,
edited by M. Velli, R. Bruno, and F. Malara

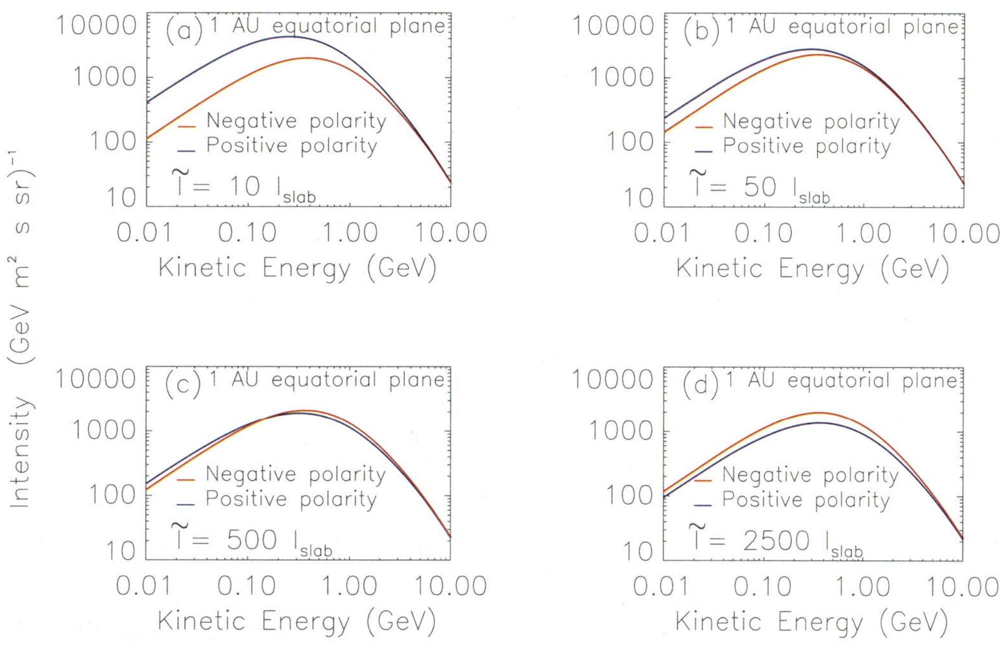

FIGURE 1. Results from an *ab initio* modulation model for $l_{slab} \propto r^{-0.3}$, and various ultrascales $\tilde{l}$ = 10 l_{slab}, 50 l_{slab}, 500 l_{slab} and 2500 l_{slab}. Panels display model predictions for 1 AU spectrum at equatorial plane. Red (blue) is used for results for negative (positive) solar polarity.

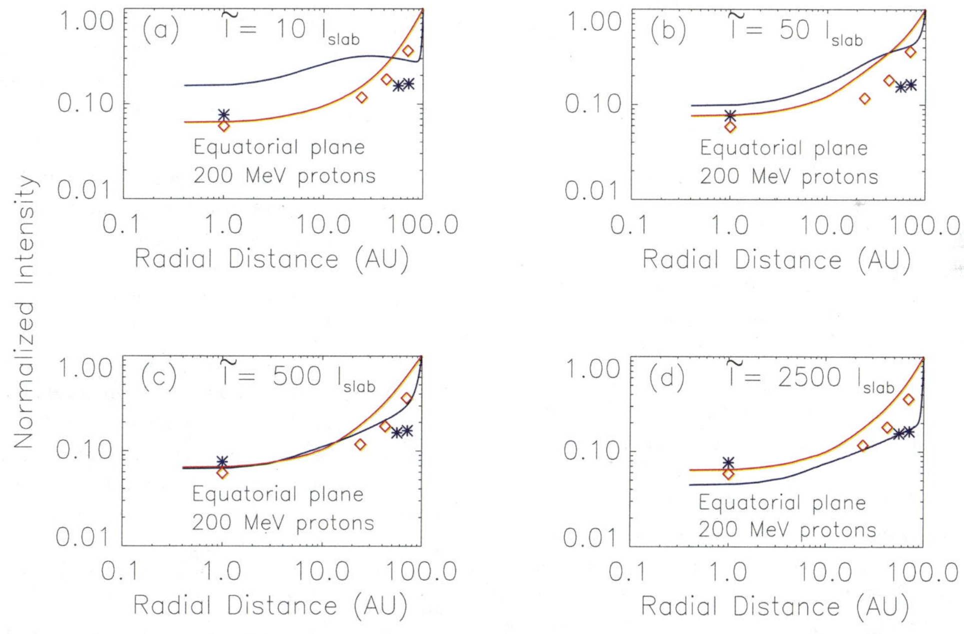

FIGURE 2. Panels display model predictions for radial profile for 200 MeV particles at equatorial plane for the same four values of $\tilde{l}$ as in Fig. 1. Red / diamond (blue / star) used for both model and observational results for negative (positive) solar polarity. Radial profile data [11] come from Voyager and IMP.

and the observed correlation scale. It seems very little has been done to describe these fluctuations intuitively, or in terms of mathematics of field line random walk. Nevertheless one can see from the definition of $D_\perp^{2D}$ and $D_\perp^{slab}$ that $D_\perp^{2D} \propto (b_{2D}/B_0)$ whereas $D_\perp^{slab} \propto (b_{slab}/B_0)^2$, and hence for small amplitude magnetic field fluctuations which means $b/B_0 \to 0$, the correct model is to include the ultrascale for 2D diffusion coefficient, and not to consider simply the slab or quasi linear approach which has smaller correlation length.

MODELS AND STRATEGY

The modulation of galactic cosmic rays is described by Parker's transport equation [12]. For our two dimensional model we describe the anisotropic diffusion by the radial diffusion coefficient $\kappa_{rr} = \kappa_\parallel \cos^2\psi + \kappa_\perp^{r\phi}\sin^2\psi$ and by $\kappa_\perp^{\theta\theta}$, the diffusion coefficient perpendicular to the mean magnetic field in the polar direction. $\kappa_\parallel$ and $\kappa_\perp^{r\phi}$ denote respectively the diffusion coefficients parallel and perpendicular to the mean background magnetic field in the $r-\phi$ plane. Here r and θ are heliocentric radial distance and colatitude (polar angle) respectively, and ψ is the spiral angle. In our two dimensional model $\kappa_\perp^{r\phi}$ acts only in radial direction, and the model simulates the effect of a wavy current sheet [13] by using an averaged drift field with only an r and θ component. We define $\kappa_\perp^{r\phi}$ as [2]

$$\kappa_\perp^{r\phi} = \frac{VR_L}{3}\frac{1}{\Omega\tau}\left(\frac{\Omega^2\tau^2}{1+\Omega^2\tau^2}\right), \qquad (1)$$

where $\Omega\tau = 2R_L/3D_\perp$. Here R_L is the particle Larmor radius, V is the particle speed, and $\Omega = V/R_L$ is the angular gyrofrequency. We also have introduced the timescale τ which is the effective timescale for the decorrelation of the particle trajectories. In this report we set $\kappa_\perp^{\theta\theta}$ as 5 times $\kappa_\perp^{r\phi}$ at high latitudes, in accord with Ulysses observations of the magnetic field variance and anisotropy at high latitudes [14]. Because the particle drifts are sensitive to the sign of the magnetic field, inclusion of drift terms leads to different predictions for different solar magnetic polarities. For drift we use the form $\kappa_A = VR_L/3$. The motivation for this weak scattering form is given in [13]. The tilt angle α of the wavy current sheet is set at 20°, a value appropriate for moderate solar modulation conditions. The solar wind velocity is radial. The solar wind speed is 400 km/sec in the equatorial plane and increases to 800 km/sec in the polar regions.

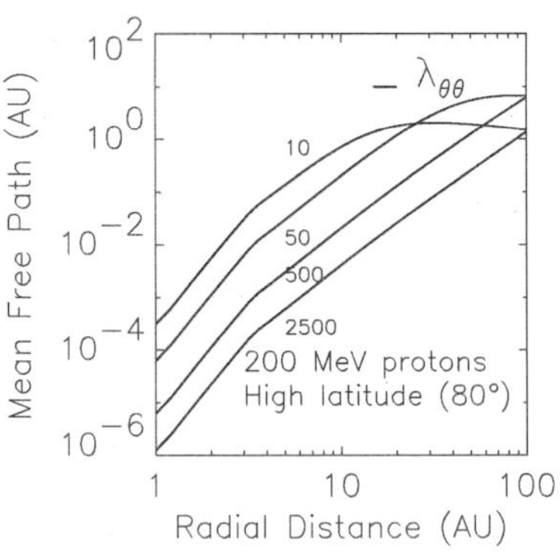

FIGURE 3. Latitudinal mean free path $\lambda_{\theta\theta} = 3\kappa_\perp^{\theta\theta}/V$ versus radial distance is displayed for the same four values of $\tilde{l}$ as in Fig. 1 at high latitude (80°) for 200 MeV protons.

NUMERICAL RESULTS

We employ Direct Numerical Simulation of Parker's transport equation by varying only $\tilde{l}/l_{slab}$. We study its impact on particle intensity, mean free path, and radial gradient in our attempt to build an ab initio model of modulation. In Fig. 1 the panels show the predicted energy spectrum for positive and negative polarity at 1 AU in the equatorial plane for various ultrascales. For the same ultrascales Fig. 2 shows radial profiles in the equatorial plane for 200 MeV protons. In all cases l_{slab} (and hence $\tilde{l}$) is taken to scale as $r^{-0.3}$, in accord with theoretical predictions for pickup ion driven turbulence [15].

The 1 AU spectrum (Fig. 1) indicates modulation becomes stronger for larger ultrascale in case of positive polarity. Similar trend does not exist for negative polarity. The spectra for the two polarities sometimes cross, as in Fig. 1 (cf [16]). For larger ultrascale as in Fig. 1 (d), the intensity for negative polarity is larger than that in positive polarity which does not happen for smaller ultrascale scans. Fig. 2 clearly indicates the radial profile in negative polarity does not vary much in comparison with the observed data when different ultrascales are considered, whereas the gradient in positive polarity is very sensitive to ultrascale. As a result the radial profile is a good approximation to the observed data for moderately large ultrascale scans.

The results presented in Figures 1 and 2 can be understood with reference to the effect of the ultrascale upon the latitudinal diffusion coefficient at high latitude. As shown in Figure 3, $\lambda_{\theta\theta}$ becomes smaller for larger ultrascale, leading to increased modulation. The effect on modulation is especially prominent during positive solar polarity, because positive charges enter the heliosphere at high latitudes and drift out along the current sheet.

Their access to the equatorial plane is thus impeded by a smaller diffusion coefficient at high latitudes. During negative solar polarity, positive charges drift in along the current sheet and are comparatively insensitive to conditions at high latitudes, consistent with our simulation.

CONCLUSIONS

We describe here an *ab initio* approach to modulation modeling, in which parameters governing diffusion tensor are computed from available theories of solar wind turbulence based upon particle transport, and the variation of turbulence properties through the heliosphere is determined by turbulence transport models. We have focused on one important but poorly understood aspect of *ab initio* modeling: the size of the turbulence ultrascale which determines the rate of field line random walk for 2D turbulence modes.

We conclude that modulation is quite sensitive to the ultrascale, particularly during positive solar magnetic polarity (for positive charges). A larger ultrascale leads to reduced latitudinal diffusion at high latitudes, and thus greater modulation observed in the equatorial region. For moderately large ultrascale (hundreds of correlation lengths), we obtain qualitatively reasonable agreement with observed radial gradients, and a spectrum crossover similar to that reported in observations [16].

ACKNOWLEDGMENTS

Supported by NASA grant NAG5–8134 (SECTP) and by NSF grants ATM–0000315 and ATM–0105254.

REFERENCES

1. Parhi, S., R. A. Burger, J. W. Bieber, and W. H. Matthaeus, Sensitivity of cosmic ray modulation to the correlation length, *Geophys. Res. Lett.*, **29(8)**, 991, 2002.
2. Bieber, J. W., and W. H. Matthaeus, Perpendicular diffusion and drift at intermediate cosmic ray energies, *Astrophys. J.*, **485**, 655, 1997.
3. Jokipii, J. R., and E. N. Parker, Stochastic aspects of magnetic lines of force with application to cosmic ray propagation, *Astrophys. J.*, **155**, 777, 1969.
4. Giacalone, J. and J. R. Jokipii, The transport of cosmic rays across a turbulent magnetic field, *Astrophys. J.*, **520**, 204-214, 1999.
5. Mace, R. L., W. H. Matthaeus, and J. W. Bieber, Numerical investigation of perpendicular diffusion of charged test particles in weak magnetostatic slab turbulence, *Astrophys. J.*, **538**, 192, 2000.
6. Matthaeus, W. H., P. C. Gray, D. H. Pontius, Jr., and J. W. Bieber, Spatial structure and field-line diffusion in transverse turbulence, *Phys. Rev. Lett.*, **75**, 2136, 1995.
7. Matthaeus, W. H., C. W. Smith, and J. W. Bieber, Correlation lengths, the ultrascale, and the spatial structure of interplanetary turbulence, in Proceedings of Solar Wind 9, AIP proceedings 471, edited by S. R. Habbal, R. Esser, J. V. Hollweg and P. A. Isenberg, 511, 1999.
8. Smith, C. W., W. H. Matthaeus, G. P. Zank, N. F. Ness, S. Oughton, J. D. Richardson, Heating of the low-latitude solar wind by dissipation of turbulent magnetic fluctuations, *J. Geophys. Res.*, **106**, 8253, 2001.
9. Zank, G. P., W. H. Matthaeus, J. W. Bieber, and H. Moraal, The radial and latitudinal dependence of the cosmic ray diffusion tensor in the heliosphere, *J. Geophys. Res.*, **103**, 2085, 1998.
10. Ruffolo, D., and W. H. Matthaeus, Field line separation in two-component magnetic turbulence, Proc. 27th Int. Cosmic Ray Conf., 3298-3301, 2001.
11. Webber, W. R. and J. A. Lockwood, Voyager and Pioneer spacecraft measurements of cosmic ray intensities in the outer heliosphere: Toward a new paradigm for understanding the global solar modulation process 1. Minimum solar modulation (1987 and 1997), *J. Geophys. Res.*, **106**, 1, 2001.
12. Parker, E. N., The passage of energetic charged particles through interplanetary space, *Planet. Space Sci.*, **13**, 949, 1965.
13. Burger, R. A., M. S. Potgieter, and B. Heber, Rigidity dependence of cosmic-ray proton latitudinal gradients measured by the Ulysses spacecraft: Implications for the diffusion tensor, *J. Geophys. Res.*, **105**, 27,477, 2000.
14. Balogh, A., R. J. Forsyth, T. S. Horbury, and E. J. Smith, Variances and fluctuations of the heliospheric magnetic field in solar polar flows, in *Solar Wind Eight*, edited by D. Wintehalter, J. T. Gosling, S. R. Habbal, W. S. Kurth and M. Neugebauer, 221-224, AIP Press, New York, 1996.
15. Zank, G. P., W. H. Matthaeus, and C. W. Smith, Evolution of turbulent magnetic fluctuation power with heliospheric distance, *J. Geophys. Res.*, **101**, 17093, 1996.
16. Lockwood, J. A. and W. R. Webber, Comparison of the rigidity dependence of the 11-year cosmic ray variation at the Earth in two solar cycles of opposite magnetic polarity, *J. Geophys. Res.*, **101** , 21,573, 1996.

Modeling of the tail region of the heliospheric interface

Dmitry Alexashov* and Vlad Izmodenov†

*Institute for Problems in Mechanis, Russian Academy of Sciences, Moscow, Russia
†Lomonosov Moscow State University, Department of Aeromechanics, Faculty of mechanics and mathematics, Vorob'evy gory, Glavnoe Zdanie MGU, 119899, Moscow, Russia

Abstract. The processes in the tail region of the interaction of the solar wind with the partially ionized local interstellar medium are investigated in a framework of the self-consistent kinetic gas dynamic model. It is shown that charge exchange of the hydrogen atoms with the plasma protons leads to suppression of the gas dynamic instabilities and disappearance the contact discontinuity at sufficiently large distances from the Sun. It is shown that the solar wind plasma temperature decreases and, ultimately, the parameters of the plasma and hydrogen atoms approach to the corresponding parameters of the pristine interstellar medium at large heliocentric distances.

INTRODUCTION

The solar wind interacts with partly ionized local interstellar cloud (LIC). The structure of the Solar Wind - LIC interaction region is shown in Figure 1. Contact discontinuity, or *the heliopause* (HP), separates solar wind and interstellar plasma. The heliopause is an obstacle to flow around by the supersonic (the Max number is about 10) solar wind, and also by the supersonic (the Max number is about 2) interstellar gas. A shock has to be formed in the case of supersonic flow. The supersonic solar wind passed through *the termination shock* (TS) to become subsonic. *The bow shock* is formed in the local interstellar cloud. The whole region of the solar wind interaction is called as *the heliospheric interface*. The interstellar H atoms interact with plasma component by charge exchange and strongly influence on the location of the shocks and the heliopause. To describe interstellar H atom flow in the interface it is necessary to solve kinetic equation. Self-consistent model of the heliospheric interface was proposed in [1] and realized in [2]. Last paper also present first calculations in the region of the heliospheric interface. Figure 1 shows a comparison of locations of the two shocks and the heliopause for two cases with H atoms and without them. The discontinuities are significantly closer to the Sun in the case with atoms. In the tail region the structure of the flow changes qualitatively. The termination shock becomes more spherical and Mach Disk (MD), reflected shock (RS) and tangential discontinuity (TD) disappear (Figure 1).

The model of the heliospheric interface allows answering two fundamental questions: 1. Where is the edge of the solar system? 2. How far is the influence of the solar system on the surrounding interstellar medium?

To give an answer on the first question we need to give definition of the solar system boundary. It is naturally to assume the boundary is the heliopause, which separate solar wind and interstellar plasmas. Note, that the influence of the solar system on the interstellar medium is significantly far than the heliopause. Secondary interstellar atoms, which are result of the charge exchange of original interstellar atoms and solar wind protons, disturbed the interstellar gas upwind the bow shock (e.g. [3,4] and reference therein). However, the most of papers have studied the upwind region effects mainly. At the same time the study of the heliotail region has also significant interest. For the heliotail we cannot say that the heliopause is the heliospheric boundary. It is seen in Figure 1, the heliopause is not closed surface and the solar wind fills the whole space into the downwind direction. The goal of the present work is in the study the effect of charge exchange on the tail region of the heliospheric interface.

MODEL

To study the effect of charge exchange on the structure of the heliotail we used kinetic-gasdynamic model by Baranov and Malama [1-4]. To describe the charged component (electrons and protons) we solve hydrodynamic Euler equations, where the effect of charge exchange is taken into account in the right parts of these equations. To calculate the flow of interstellar H atoms in the heliospheric interface we solve kinetic equation:

$$\mathbf{w}_{\mathrm{H}} \cdot \frac{\partial f_{\mathrm{H}}(\mathbf{r},\mathbf{w}_{\mathrm{H}})}{\partial \mathbf{r}} + \frac{\mathbf{F}}{m_{\mathrm{H}}} \cdot \frac{\partial f_{\mathrm{H}}(\mathbf{r},\mathbf{w}_{\mathrm{H}})}{\partial \mathbf{w}_{\mathrm{H}}} \qquad (1)$$

CP679, *Solar Wind Ten: Proceedings of the Tenth International Solar Wind Conference*, edited by M. Velli, R. Bruno, and F. Malara

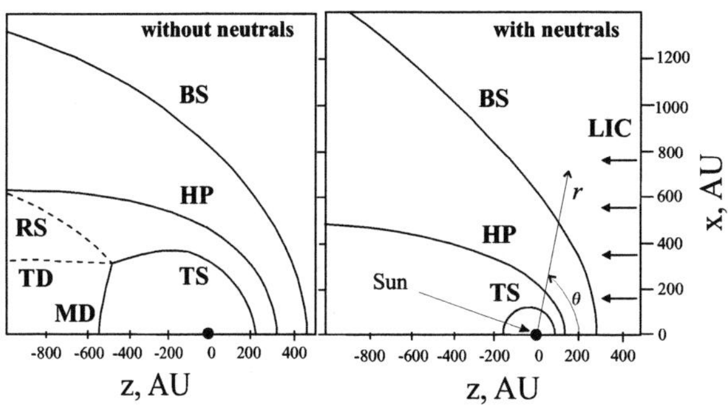

FIGURE 1. The structure of the heliospheric interface is the region of the solar wind interaction with the interstellar medium.

$$= -f_{\mathrm{H}}(\mathbf{r},\mathbf{w}_{\mathrm{H}}) \int |\mathbf{w}_{\mathrm{H}} - \mathbf{w}_p| \sigma_{ex}^{\mathrm{HP}} f_p(\mathbf{r},\mathbf{w}_p) d\mathbf{w}_p$$
$$+ f_p(\mathbf{r},\mathbf{w}_{\mathrm{H}}) \int |\mathbf{w}_{\mathrm{H}}^* - \mathbf{w}_{\mathrm{H}}| \sigma_{ex}^{\mathrm{HP}} f_{\mathrm{H}}(\mathbf{r},\mathbf{w}_{\mathrm{H}}^*) d\mathbf{w}_{\mathrm{H}}^*$$
$$-(\beta_i + \beta_{\mathrm{impact}}) f_{\mathrm{H}}(\mathbf{r},\mathbf{w}_{\mathrm{H}}).$$

Here $f_{\mathrm{H}}(\mathbf{r},\mathbf{w}_{\mathrm{H}})$ is velocity distribution function of H atoms; $f_p(\mathbf{r},\mathbf{w}_p)$ is locally Maxwellian velocity distribution of protons; $\mathbf{w}_p$ and $\mathbf{w}_{\mathrm{H}}$ is individual velocities of protons and H atoms, respectively; $\sigma_{ex}^{\mathrm{HP}}$ is the cross section of the charge exchange of H atoms and protons; β_i is the photoionization rate; m_{H} is the mass of H atom; β_{impact} is the electron impact ionization;and $\mathbf{F}$ is the sum of the solar gravitation and radiation pressure forces.

The main process of the plasma-neutral coupling is charge exchange. Photoionization and electron impact ionization are also taken into account in equation (1). The interaction of the charged and neutral components results in exchange of mass, momentum and energy between the components. The expressions for the source terms into the right parts of Euler equations are given in [2].

For boundary conditions in the unperturbed LIC we assume that velocity $V_\infty = 25$ km/s, interstellar H atom and proton number densities are 0.2 cm^{-3} and 0.07 cm^{-3}. The temperature of the LIC was assumed 6000 K. Velocity, number density and Max number at the Earth orbit are 450 km/s, 7 cm^{-3} and 10, respectively. Velocity distribution function of H atoms is assumed to be Maxwellian.

Euler equations with the source terms were solved self-consistently with the kinetic equation for H atoms. Unlike the previously published papers based on Baranov-Malama model, we performed the calculations in extended computation region toward the heliotail. We performed computations up to 50000 AU along the axis of symmetry and up to 5000 AU in the perpendicular to the symmetry axis direction. To estimate convergence of chosen numerical scheme we used different nonuniform non-stationary grids adapted to the flow structure. We found that for relatively small computational region ($|z_{max}| <$ 10000 AU) in the tail region there is strong dependence of the numerical results on the size of computational domain. However, for extended computational domains ($|z_{max}| >$ 10000 AU) dependence of the numerical solution on the computational domain size disappears. Our numerical tests show that the numerical solution does not depend strongly on grid resolution. For extended computational domains we found small difference of the numerical solution, when we change computational grid from 100 to 200 grid points in that direction.

QUALITATIVE ANALYSIS

In present work we consider effect of the charge exchange process ($H + H^+ \rightarrow H^+ + H$) on the plasma flow in the tail region of the heliospheric interface. The supersonic solar wind passes through the heliospheric termination shock, where its kinetic energy transfers to the thermal energy. Let us assume now that the heliopause surface is parallel to the direction of interstellar flow. In this case the solar wind can be considered as a flow in the nozzle with constant cross-section. Our computations show that the solar wind pressure downstream the termination shock is several times smaller than the interstellar pressure. Under these conditions the solar wind flow decelerates and has some minimal value at infinity. The minimal value is determined by the parameters of the solar wind downstream the termination shock and the interstellar pressure. Neither the interstellar proton number density nor the relative Sun-LISM velocity does not determine the minimal velocity. Therefore, in the case with no atoms in the frame of hydrodynamic approach it is possible to find solution where the solar wind (and, therefore, the solar system) extended in the heliotail up to infinity. Such a qualitative consideration can be easily generalized, when the heliopause is not parallel to the axis of symmetry. Then the solar wind flow can be considered as a flow in convergent or expanding nozzle.

Qualitatively other situation is realized in the case with interstellar atoms. Our calculations show that in the case the pressure is larger than the interstellar pressure. The solar wind should be accelerated in this case by the pressure gradient. However, interstellar atoms play significant role due to charge exchange. Due to large mean free path the interstellar atoms fulfill the heliotail. Part of original (or primary) interstellar atoms increases with the heliocentric distance. The temperature (7000 K) and velocity (25 km/s) of interstellar atoms are smaller than the velocity (100 km/s) and temperature (100000 K) of the post shocked solar wind. New protons, which are born from interstellar H atom, have smaller average and ther-

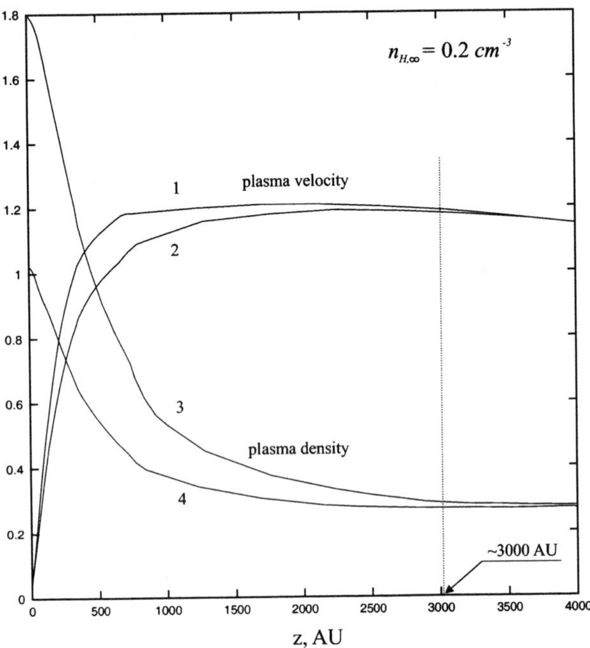

FIGURE 2. Velocities (curves 1 and 2) and densities (curves 3 and 4) from both sides of the heliopause as a function of the heliocentric distance along the heliopause. Curves 2 and 3 correspond to interstellar side; curves 1 and 4 correspond to solar wind side. The velocities and temperatures are normalized to their interstellar values.

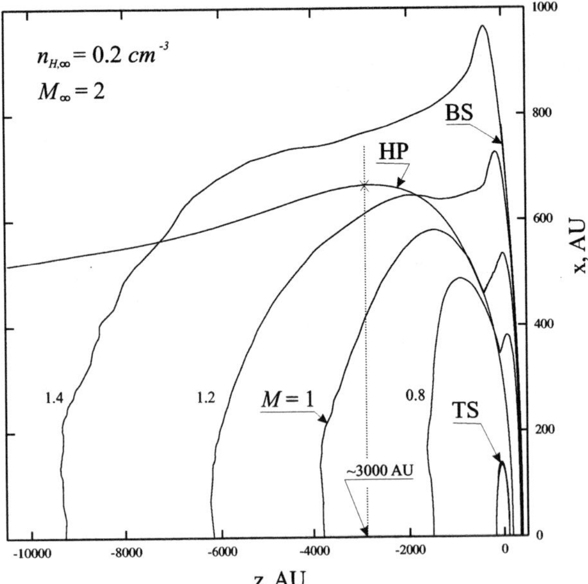

FIGURE 3. Isolines of Mach number M. It is seen that on the distances more than 4000 AU into the heliotail direction the solar wind flow is supersonic. The Mach number increases with increase of the heliocentric distance and approaches its interstellar number.

mal velocities than original solar protons. Therefore, the charge exchange process leads to effective cooling and deceleration of the solar wind. Since the part of primary interstellar atoms increases with the increasing of the heliocentric distance, it is naturally to expect approaching of the solar wind velocity, density and temperature to their interstellar values.

Despite a number of assumptions, given qualitative analysis is confirmed by our numerical calculations.

RESULTS AND DISCUSSION

Distributions of plasma parameters in the heliospheric tail region are presented in figures 2 and 3. In figure 2 we present distribution of density and velocity of plasma from both sides of the heliopause. In the classical hydrodynamics there are two conditions for tangential discontinuity. The conditions are 1) no mass transport through the discontinuity, 2) balance of pressures on the both sides of the discontinuity. These conditions permit a jump of density and tangential velocity through the heliopause. In the case with H atoms the jump of density and pressure will become weaker with increasing the distance along the heliopause due to mass transport caused by charge exchange. For $z \approx -3000$ AU, where z is the distance along the axis of symmetry and sign "-" means the direction along the interstellar flow direction, the jump of density and tangential velocity disappears (Figure 2). The velocity of the solar wind is about 100 km/s downstream the termination shock. Then the velocity becomes smaller due to new protons injected by charge exchange and approaches the value of interstellar velocity. The solar wind becomes also cooler due to charge exchanges. It is interesting to see whether the solar wind becomes supersonic due to effective cooling. Figure 3 shows isolines of Mach numbers in the heliospheric interface. The solar wind passes through the sound velocity at about 4000 AU, then the Mach number increases approaching its interstellar value. The heliopause is also shown in Figure 3. The line $z = -3000$ AU shows the region where there is no jump of density and velocity through the heliopause.

Figure 4 represents the densities, velocities and temperatures of the interstellar hydrogen along the different downwind directions. The angle θ in Figure 4 is the angle between line-of-sight and upwind directions (Figure 1). Parameters of H atoms are approaching their interstellar values on distances less than 20000 AU for all line-of-sights. The approaching is faster for smaller θ. It is interesting to note, that the hydrogen wall, the increase of H atom number density in the region between the heliopause and the bow shock [1-3] is visible even for large $\theta \approx 150 - 170^o$.

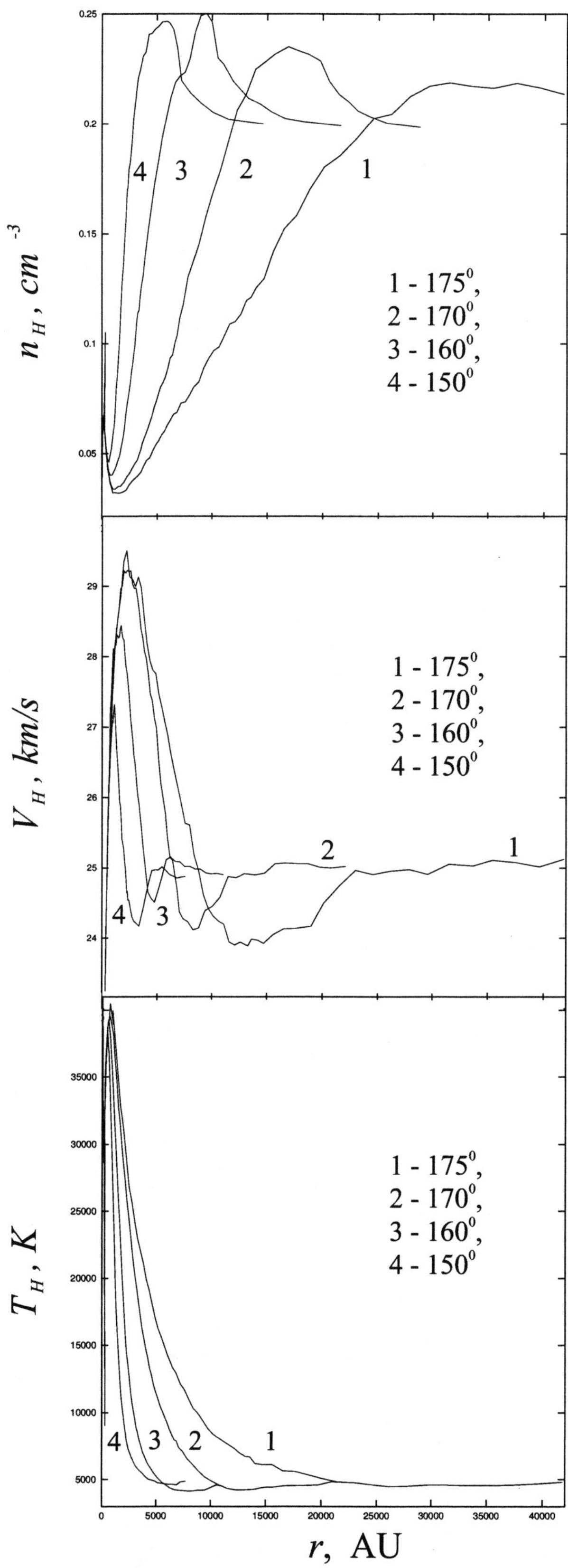

FIGURE 4. Number density, velocity and temperature of the interstellar H atom along downwind lines of sights θ = 150, 160, 170, 175 degrees.

To get numerical solution of the heliotail plasma becomes possible due to charge exchange. It is important that the solar wind is supersonic at the outer boundary. This allows fulfilling correct boundary conditions.

In this work we considered influence of the charge exchange process only. In future, influences of different hydrodynamic and plasma instabilities, latitude dependence of the solar wind, interstellar and heliospheric magnetic fields, which required 3D model, on the heliotail structure must to be considered.

SUMMARY

In this paper we consider effects of charge exchange on the structure of the heliotail region. In particular, it was shown that

1. The charge exchange process change the solar wind - interstellar interaction flow qualitatively in the tail region. The termination shock becomes more spherical and Mach disk, reflected shock and tangential discontinuity disappear (Figure 1). The jumps of density and tangential velocity through the heliopause becomes smaller into the tail and disappears at about 3000 AU.
2. Parameters of solar wind plasma and interstellar H atoms are approaching their interstellar values. It allows to estimates of influence of the solar wind, and, therefore, the solar system size into the downwind direction as about 20000- 40000 AU. Unlike the upwind direction the solar system boundary has diffusive nature in downwind.
3. The supersonic character of the solar wind flow in the heliotail allows us to perform correct numerical calculations. This is not possible in the case without H atoms.

ACKNOWLEDGMENTS

This work was done in the frame of ISSI team "Physics of the heliotail". The authors thank International Space Science Institute (ISSI) in Bern for excellent organization of the first team-meeting and for partial support. This work was supported in part by INTAS-ISSI Cooperation Project 2001-0270, RFBR grants 01-02-17551, 02-02-06011, 02-02-06012, 01-01-00759.

REFERENCES

1. Baranov V.B., Lebedev M.G., Malama Yu.G., Astrophys. J., 1991, V. 375, P. 347.
2. Baranov V.B., Malama Yu.G., J. Geophys. Res., 1993, V. 98, No A9, P. 15,157.
3. Izmodenov, 2000, Astronomy and Space Science, 274, 55-69, 2000.

Three-Dimensional Magnetohydrodynamics of the Solar Corona and of the Solar Wind with Improved Energy Transport

Roberto Lionello*, Jon A. Linker* and Zoran Mikić*

*SAIC, 10260 Campus Point Dr., San Diego, CA 92121-1578, USA

Abstract. We have developed a three-dimensional magnetohydrodynamic (MHD) model of the solar corona and of the solar wind. We specify a magnetic flux distribution on the solar surface and integrate the time dependent MHD equations to steady state. The model originally employed a polytropic energy equation. In order to improve the physics in our algorithm, we have incorporated thermal conduction along the magnetic field, radiation losses, and heating into the energy equation. The 2D version of the model is able to reproduce the contrast in density between the open and closed magnetic structures in the corona and the fast and slow streams of the solar wind. We now present preliminary results of 3D MHD simulations with improved thermodynamics. The results can be tested against observations by spacecraft and Earth based observatories, in situ solar wind and magnetic field measurements, heliospheric current sheet crossings.

INTRODUCTION

The solar corona is well described as a quasi-neutral, magnetized fluid, or plasma [1, p. 32], that responds dynamically in response to slowly evolving boundary conditions. The appropriate mathematical model for describing the low-frequency motions of such a fluid is MHD. Several multidimensional MHD algorithms have been developed to study the global corona [2, 3, 4, 5, 6, 7, 8, 9, 10, 11]. The 3D MHD, polytropic model of Linker et al. [12] could reproduce many features of the solar corona during Whole Sun Month, but the plasma density and temperature were not in quantitative agreement with the observations. In order to have a better match between the model and the observations, a more realistic energy equation is needed [13]. Suess et al. [14] and Wang et al. [15] used a 2D MHD model of the corona and of the solar wind with heating and thermal conduction. Suess et al. [16] included multifluid effects. Mikić et al. [17] developed a 3D MHD model that includes thermal conduction along the magnetic field, radiation losses, and coronal heating. The model has been used to study the 2D structure of the corona including the transition region and the upper chromosphere [18]. Here we present preliminary results of a study of the 3D structure of the solar corona during the Whole Sun Month. Using the data produced by the model we calculate emission images that can be compared with the observations.

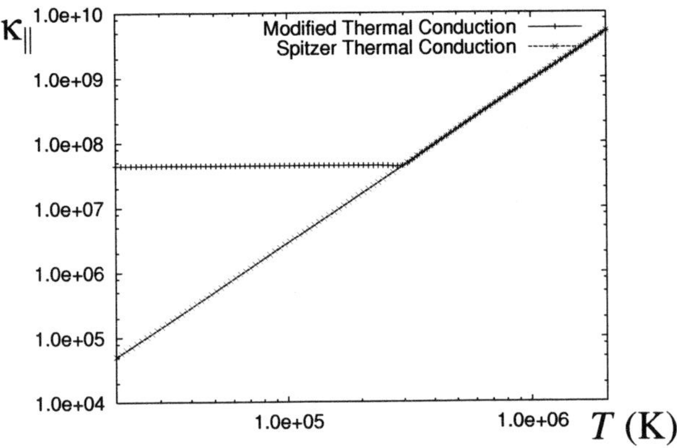

FIGURE 1. Parallel thermal conduction coefficient according to Spitzer and modified coefficient used in the present calculation to broaden the temperature gradient of the transition region.

THE MHD THERMODYNAMIC MODEL

In our time-dependent three-dimensional resistive and viscous MHD model in spherical coordinates, we solve the following set of equations here written in CGS units [17]:

$$\nabla \times \mathbf{A} = \mathbf{B}, \quad (1)$$

$$\frac{\partial \mathbf{A}}{\partial t} = \mathbf{v} \times \mathbf{B} - \frac{c^2\eta}{4\pi}\nabla \times \mathbf{B}, \quad (2)$$

CP679, *Solar Wind Ten: Proceedings of the Tenth International Solar Wind Conference*, edited by M. Velli, R. Bruno, and F. Malara

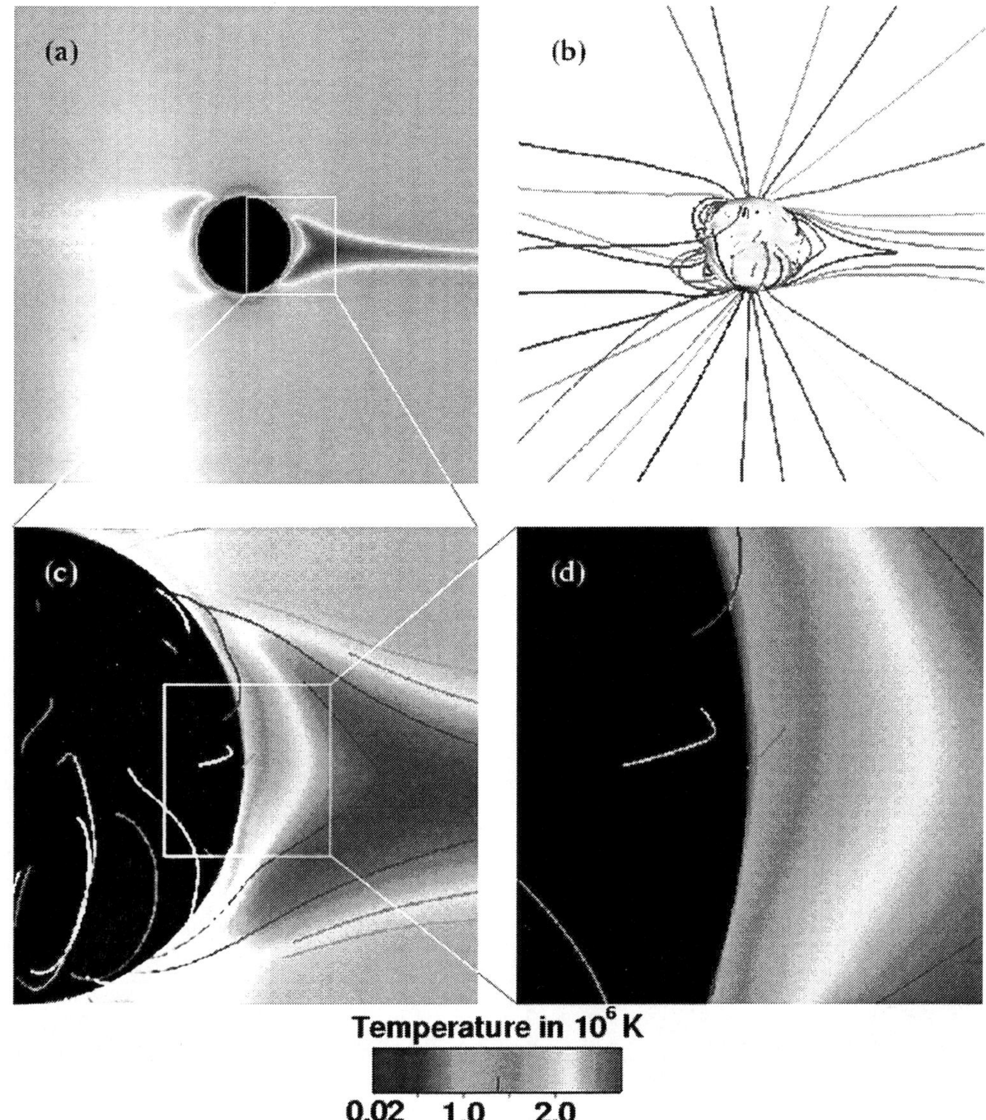

FIGURE 2. Results from the 3D MHD model for the Sun on August 15, 1996: (a,c,d) temperature at increasingly smaller scales; (b) magnetic field lines. Closed field regions, where the plasma is trapped, appear hotter.

$$\frac{\partial \rho}{\partial t} + \nabla \cdot (\rho \mathbf{v}) = 0, \tag{3}$$

$$\frac{1}{\gamma - 1}\left(\frac{\partial T}{\partial t} + \mathbf{v} \cdot \nabla T\right) = -T\nabla \cdot \mathbf{v} - \frac{m}{k\rho}(\nabla \cdot \mathbf{q} + n_e n_p Q(T) - H_{\mathrm{ch}}), \tag{4}$$

$$\rho\left(\frac{\partial \mathbf{v}}{\partial t} + \mathbf{v} \cdot \nabla \mathbf{v}\right) = \frac{\nabla \times \mathbf{B} \times \mathbf{B}}{4\pi} - \nabla p + \rho \mathbf{g} + \nabla \cdot (\nu \rho \nabla \mathbf{v}), \tag{5}$$

where $\mathbf{A}$ and $\mathbf{B}$ are respectively the magnetic vector potential and field, c is the speed of light, η the resistivity, and $\mathbf{v}$ the velocity. ρ is the plasma density, and T the temperature. The energy equation, (4), contains the heat flux $\mathbf{q}$, a radiation losses function Q as in Athay [19], and a coronal heating source H_{ch}. $\gamma = 5/3$ is the specific heats ratio, m the hydrogen atom mass, k is Boltzmann's constant, n_e (n_p) is the number density of electrons (protons). p is the plasma pressure, $\mathbf{g}$ is the acceleration of gravity, ν is the viscosity.

THE SIMULATION

We have used our improved model to investigate the aspect of the corona during the Whole Sun Month. A magnetic field distribution obtained from the synoptic charts of the National Solar Observatory at Kitt Peak is

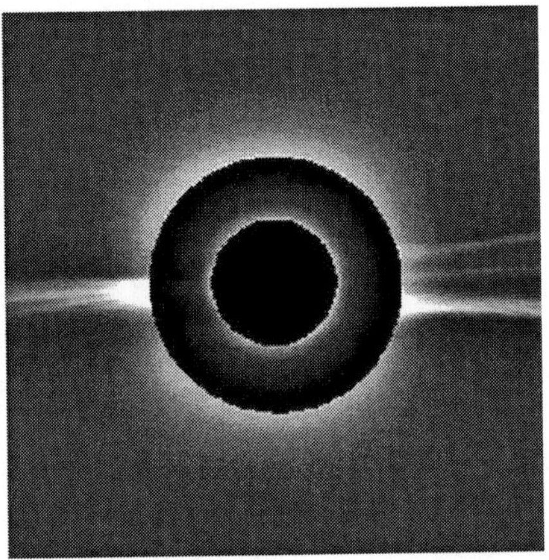

FIGURE 3. On the left: superimposed Lasco C1 (FeXIV) and C2 (polarized brightness) images; the solar surface is colored according to the magnetic flux. On the right: emission calculated using the MHD model.

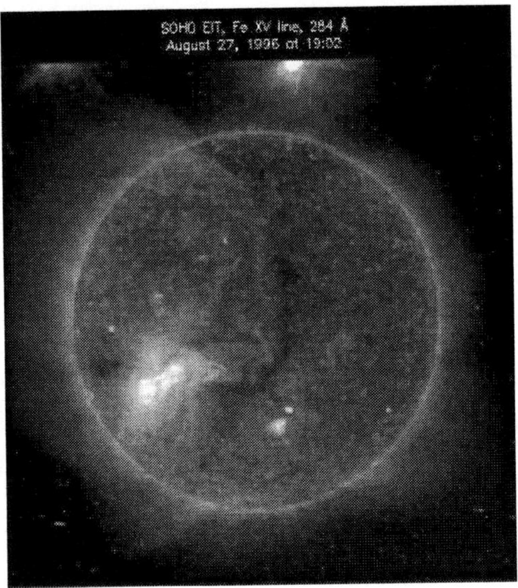

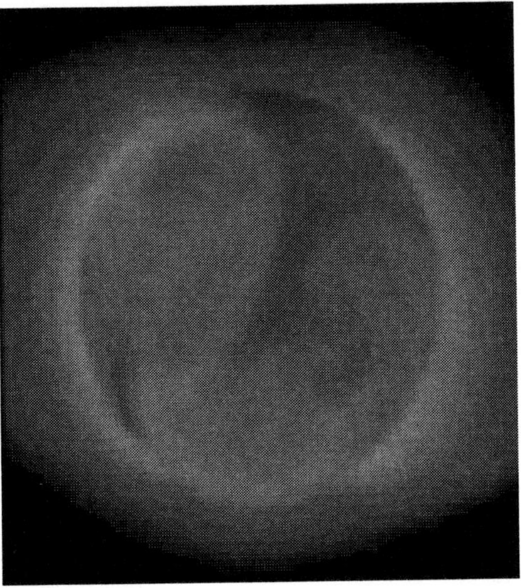

FIGURE 4. Comparison of 284 Å(FeXV) image taken by SOHO/EIT with the emission calculated using the MHD model.

specified at $r = R_\odot$. A uniform density ($\rho_0 = 10^{11}$ cm^{-3}) and temperature ($T_0 = 20,000$ K) are assumed at the base of the domain. Below $r \sim 10\,R_\odot$, thermal conduction is collisional, in the Spitzer form, and directed along $\mathbf{B}$:

$$\mathbf{q} = -\kappa_\parallel \hat{\mathbf{b}}\hat{\mathbf{b}} \cdot \nabla T, \tag{6}$$

where $\hat{\mathbf{b}}$ is the unit vector along the magnetic field and $\kappa_0 = 9 \times 10^{-7}$ in CGS units. In order to broaden the gradient in the transition region without altering qualitatively the solution, we use a special form of $\kappa_\parallel$ [20]:

$$\kappa_\parallel = \kappa_0[sT^{5/2} + (1-s)T^\alpha T_{\text{mod}}^{5/2-\alpha}], \tag{7}$$

where

$$s(T) = \frac{1}{2}\left(1 + \tanh\left(\frac{T - T_{\text{mod}}}{\Delta T_{\text{mod}}}\right)\right). \tag{8}$$

The values of the parameters are $T_{\text{mod}} = 3 \times 10^5$ K, $\alpha = 0$, and $\Delta T_{\text{mod}} = 2.5 \times 10^4$ K. A plot of $\kappa_\parallel$ is presented in Fig. 1. It shows that for $T < T_{\text{mod}}$ we have a constant $\kappa_\parallel$, and for $T > T_{\text{mod}}$ we have $\kappa_\parallel = \kappa_0 T^{5/2}$. Heating ($H_{\text{ch}}$) is non uniform in latitude and follows an exponentially decaying law with the length scale ranging from $\lambda = 0.7\,R_\odot$ at the poles to $\lambda = 0.2\,R_\odot$ at the equator. H_{ch} is chosen such that the heat flux at the base ranges from $q_0 = 1 \times 10^5$ to $q_0 = 2 \times 10^5$ erg cm^{-2} s^{-1}.

As initial condition for $\mathbf{A}$ we normally prescribe the potential field corresponding to the surface magnetic flux distribution. We then integrate Eqs. (2-5) to steady state. This process takes more than 2 days of real time. If high resolution is required to resolve the transition region temperature gradient, and the improved treatment of the energy equation is used, this translates into many days of computational time. In order to shorten the relaxation time, we have used as initial condition of the magnetic field a relaxed MHD state obtained in the past with the polytropic model. The vector potential ($\mathbf{A}_{\mathrm{Int}}$) and the current density ($\mathbf{J}_{\mathrm{Int}}$) from the past simulation have been interpolated on the present, finer mesh. In order to eliminate the noise introduced by the interpolation, we have relaxed $\mathbf{A}$ by advancing the following equation to steady state:

$$\frac{\partial \mathbf{A}}{\partial \tau} = \frac{4\pi}{c}\mathbf{J}_{\mathrm{Int}} - \nabla \times \nabla \times \mathbf{A}, \quad (9)$$

$$\mathbf{A}(\tau = 0) = \mathbf{A}_{\mathrm{Int}}. \quad (10)$$

The solution, $\mathbf{A}$, is used as initial condition of the magnetic field. For the plasma velocity, temperature, and density we use as initial condition a steady state wind solution obtained with a 1D version of our code.

RESULTS

We here show the results for a 3D simulation of the solar corona, transition region, and upper chromosphere during Whole Sun Month. In Fig. 2 we present the aspect of the solar corona on August 15, 1996. Figures 2a, 2c, and 2d show the plasma temperature at increasingly smaller scales. Figures 2b has a large scale view of selected magnetic field lines; the solar surface is colored according to the magnetic flux. Open field lines are rooted in the polar caps, whence the fast wind flows, while in the equatorial regions the field loops back to the solar surface. In Figures 2c and 2d some field lines are superimposed onto the temperature plot. The temperature appears higher in the closed field regions under the helmet streamers. Notice how it rises sharply from 20,000 at the base of the domain to more than two million degrees in the corona.

Using the data obtained from the simulation, we have calculated the emission in several lines and compared the result with observations. In Fig. 3a we show a composite image obtained on Sep 3, 1996 from the C1 (FeXIV) and C2 (polarized brightness) Lasco coronographs. The calculated emission is in Fig. 3b. The use of a synoptic magnetogram implies that the magnetic flux distribution does not correspond exactly to that of the days in which the observations were taken. That may account for some differences. In Fig. 4a we have an image of the corona taken by SOHO/EIT in the FeXV line. It can be compared with our result in Fig. 4b, which was obtained using data from a slightly different model that uses radiative balance conditions at the base [18]. Although the model reproduces some aspect of the "elephant trunk" coronal hole, the equatorial active region is not visible. This may be due to the use of a simplistic coronal heating function. Comparisons with solar-wind measurements are possible and will be part of future more comprehensive studies.

REFERENCES

1. Priest, E. R., *Solar Magnetohydrodynamics*, Dordrecht, Boston, Lancaster, D. Reidel Publishing Company, 1981
2. Endler, F. Ph.D. thesis, Gottingen Univ., 1971.
3. Pneuman, G. W., and Kopp, R. A., *Solar Phys.*, **18**, 258 (1971).
4. Steinolfson, R. S., Suess, S. T., and Wu, S. T., *Astrophys. J.*, **255**, 730 (1982)
5. Washimi, H., Yoshino, Y., and Ogino, T., *Geoph. Res. Lett.*, **14**, 487 (1987)
6. Linker, J. A., Van Hoven, G., and Schnack, D. D., *Geophys. Res. Lett.*, **17**, 2281 (1990).
7. DeVore, C. R., *J. Comput. Phys.*, **92**, 142 (1991).
8. Mikić, Z., and Linker, J. A., in *Solar Wind Eight: Proceedings of the Eight International Solar Wind Conference*, ed. by D. Winterhalter et al., AIP Conference Proceedings, New York, 1996, p. 104.
9. Usmanov, A. V., in *Solar Wind Eight: Proceedings of the Eight International Solar Wind Conference*, ed. by D. Winterhalter et al., AIP Conference Proceedings, New York, 1996, p. 141.
10. Groth, C. P. T., de Zeeuw, D. L., Gombosi, T. I., Powell, K. G., *Space Science Reviews*, **87**, 193 (1999)
11. Keppens, R., and Goedbloed, J. P., *Space Science Reviews*, **87**, 223 (1999)
12. Linker, J. A., Mikić, Z., Biesecker, D. A., Forsyth, R. J., Gibson, S. E., Lazarus, A. J., Riley, P., Szabo, A., and Thompson, B. J., *J. Geophys. Res.*, **104**, 9809 (1999)
13. Withbroe, G. L., *Astrophys. J.*, **325**, 442 (1988)
14. Suess, S. T., Wang, A. H., Wu, S. T., *J. Geophys. Res.*, **101**, 19957 (1996)
15. Wang, A. H., Wu, S. T., Suess, S. T., and Poletto, G., *J. Geophys. Res.*, **103**, 1913 (1998)
16. Suess, S. T., Wang, A. H., Wu, S. T., Poletto, G., McComas, D. J., *J. Geophys. Res.*, **104**, 4697 (1999)
17. Mikić, Z., Linker, J. A., Schnack, D. D. , Lionello, R., and Tarditi, A., *Phys. Plasmas*. **6**, 2217 (1999).
18. Lionello, R., Mikić, Z., and Linker, J. A., *Astrophys. J.*, **546**, 542 (2001).
19. Athay, R. G., *Astrophys. J.*, **308**, 975 (1986).
20. Linker, J. A., Lionello, R., Mikić Z., and Amari, T., *J. Geophys. Res.*, **106**, 25165 (2001).

Solar wind velocity structure around the solar maximum observed by interplanetary scintillation

K. Fujiki*, M. Kojima*, M. Tokumaru*, T. Ohmi*, A. Yokobe*, K. Hayashi*, D. J. McComas† and H. A. Elliott†

*Solar-Terrestrial Environment Laboratory, Nagoya University, 3-13 Honohara, Toyokawa, Aichi Japan
†Southwest Research Institute, San Antonio, TX, USA

Abstract. Ulysses observed a latitude structure of solar wind in its second fast latitude scan and found that the global structure of solar wind near the solar maximum is significantly different from that in the solar minimum. Also soon after the solar maximum, Ulysses measured that the fast solar wind which has magnetic polarity of the new solar cycle appeared at high latitude in northern hemisphere. This fast wind appeared and disappeared a few times. We introduced a new tomographic algorithm, time-series tomography, to reconstruct IPS velocity map using all data observed in the year from 1998 to 2001 and analyzed the variation of the solar wind structure through these four year. Especially in 2001, we compared the Ulysses' fast latitude scan data. As results, it is found that disappearance and recovery of the fast solar wind around the north pole precedes that around the south pole for several months. And also found that the IPS observation shows high level agreement to the Ulysses observation especially for high latitudes.

INTRODUCTION

It is well known that the solar wind is consisted by two velocity components (so called bimodal) in the solar minimum. The slow solar wind is confined around the Sun's streamer belt and the fast solar wind is onserved in other regions. When the solar activity increases the low speed solar wind region extends to higher latitude.

Ulysses observed a latitude structure of the solar wind in its second fast latitude scan (Nov., 2000 - Oct., 2001) and show us a remarkably different solar wind structure in the solar maximum from that in the solar minimum (McComas et al.[1]). The fast solar wind observed by Ulysses above $\sim$ N40°, soon after the solar maximum, disappeared and appeared a few times.

We study the variation of the solar wind velocity structure during the years from 1998 (rising phase) to 2001 (after the solar maximum) using interplanetary scintillation (IPS) tomographic analysis. The IPS has an advantage for studies of the solar wind velocity structure because it is possible to obtain the velocity structure for each Carrington rotation. In this study, we introduce a modified tomographic technique which is mentioned in the next section.

In this paper, we will briefly report the results of the solar wind velocity structure derived by the new tomography technique and the comparison between the IPS and Ulysses second fast latitude scan observations.

OBSERVATION

We employ the interplanetary scintillation (IPS) of natural radio sources at the Solar-Terrestrial Environment Laboratory, Nagoya University, Japan. Two-dimensional solar wind velocity distribution map (V-map) at 2.5 R_s is available for each carrington rotation. IPS signals of a natural radio source is affected by a bias which is caused by a weighted-integration along the line of sight, though tomographic analysis is introduced by Kojima et al.[2] to remove the bias and to improve spatial resolution. The tomographic analysis requires stable solar wind velocity structure over one solar rotation because all IPS data in a given Carrington rotation are used to reproduce a V-map. Around the solar maximum, however, the solar wind structure changes rapidly, which disturbs tomographic V-map.

A new tomographic technique called time-series tomography (TST) here is introduced to improve the reliability of the V-map around the solar maximum, which use all IPS data observed in a year and reproduces one V-map for the year. This technique allow us to analyze the continuously-changing velocity structure of the solar wind such as that around the solar maximum. We derived the velocity structure using the TST technique in the years from 1998 to 2001 which correspond to the period from rising phase to soon after the solar maximum.

Ulysses/SWOOPS daily averaged data during the second fast latitude scan in 2001 are compared to the TST

CP679, *Solar Wind Ten: Proceedings of the Tenth International Solar Wind Conference*, edited by M. Velli, R. Bruno, and F. Malara

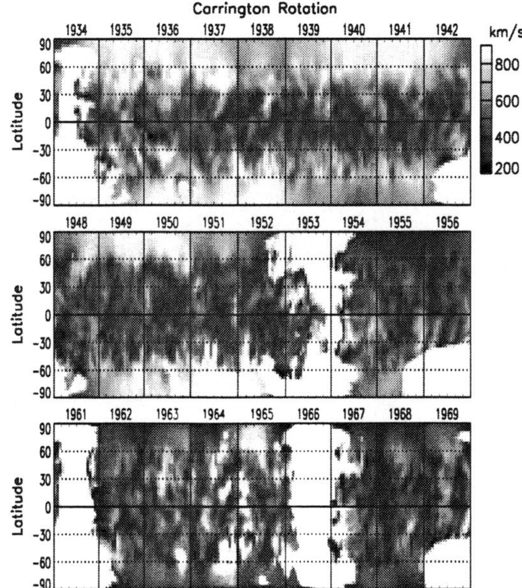

FIGURE 1. Solar wind velocity structure around the solar maximum. Top to bottom panels shows continuous variation of solar wind structure in the years from 1998 to 2001. White regions are data gap.

V-map. Although the second fast latitude scan started in November, 2000, we did not compare the data observed in 2000 because the data coverage in November and December in this year are not sufficient to reproduce the solar wind structure at high latitudes in southern hemisphere in which Ulysses passed during this period.

RESULTS

Figure 1 shows results of TST analysis for the year from 1998 (top panel) to 2000 (bottom panel). The velocity structure in 1998 is still bimodal, and the fast solar wind around the north pole disappeared in CR 1951 (1999). However the fast solar wind around the south pole still exists in this year. Then the fast solar wind at high latitudes disappeared completely in 2000. Note that there are about four month gap between these maps because IPS observations are stopped in winter season due to heavy snow.

Upper panel of figure 2 shows a v-map in 2001 (Carrington rotations from 1974 (end-March) to 1981 (end-October). In end of July, powerful lightening struck our observation system at the foot of Mt. Fuji unfortunately. The system was severely damaged and the observation stopped about one month (white region in CR 1979). According to this figure, the fast solar wind already appeared in CR 1974 at the north pole but longitudinally localized. The fast solar wind disappeared in CR 1975 and CR 1976 and appeared again in CR 1977, and finally the fast solar wind surrounds the north pole in 1980. At the south pole, on the other hand, the fast solar wind appeared in CR 1977 and disappeared in CR 1980. Especially for CR from 1977 to 1978 (also 1979), the fast winds at the both poles expand into lower latitudes, about 40 degree.

Circle plots show the Ulysses observation with same gray scale which are ballistically mapped back to 2.5 R_S using the observed solar wind speed. In this simple mapping method, the estimated longitudes at the source surface are slightly shifted with respect to the longitudes of true stream sources because the mapped surface is inside the Alfven surface. Gaps of the Ulysses plots, seen in the boundary between CR 1977 and 1978 for an example, attributes large difference of observed solar wind speed because a traveling time of slow solar wind observed by Ulysses has about two times longer than that of fast solar wind. Thus the gaps are always seen when Ulysses passes a boundary of fast and slow solar wind regions.

Lower panel in figure 2 are the comparison between IPS and Ulysses data. Error bar of the tomographic analysis is also plotted. Rapid increase of the solar wind velocity appeared in the low latitude around DOY 110 is coronal mass ejection. Ulysses observed fast solar wind around DOY 170, 200, 220 and then reaches fast solar wind region near pole (CR 1980 and 1981). According to the Ulysses trajectory in upper panel, it is clear that the Ulysses passed near the boundary of fast and slow solar wind regions. On the other hand, the IPS and Ulysses data in low latitude where slow solar wind are observed do not correlate well each other.

SUMMARY AND DISCUSSION

We employed time-series tomography for the analysis of the solar wind velocity structure around the solar maximum. As results, It is found that the fast solar wind at the north pole disappeared in mid-1999 and that at south pole disappeared in 2000. Slow solar wind region occupied almost area on the source surface ($2.5R_s$) in this year. Recovery of the fast solar wind at the north pole precedes that at the south pole for a few solar rotations.

Ulysses also observed recovery of the fast solar wind at from mid to high latitudes. The fast solar wind appeared and disappeared in a few times. According to IPS observation, it is clear that the appearance and disappearance of the fast solar wind attributes to that the fast solar wind region localized in limited longitudes and Ulysses passed near the boundary of the fast and slow solar wind regions. Although we have no data for CR 1979 the localized fast solar wind region existed for three Carring-

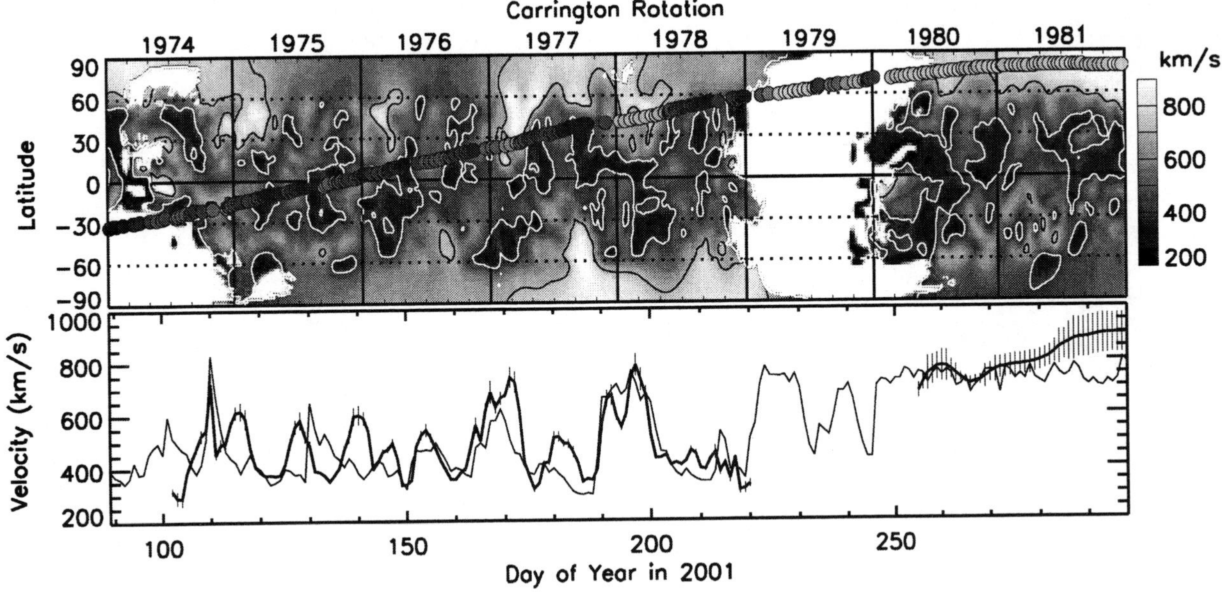

FIGURE 2. Comparison between IPS and Ulysses observations. Velocity map in Carrington rotations from 1974 to 1981 are estimated from IPS observation. Black and white contours are 700km/s and 400km/s, respectively. Circle plots show Ulysses data which is traced back to $2.5R_S$. The gray scale of the circle plots are same as the background image. (upper figure). Velocity observed by IPS along the Ulysses trajectory is compared in lower panel. Thick and thin lines shows IPS and Ulysses observations.

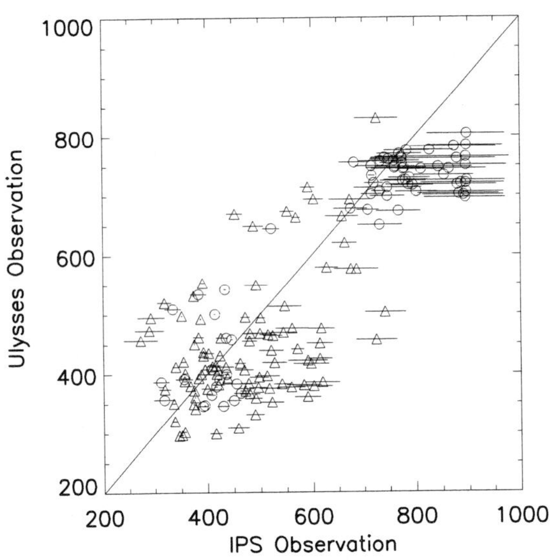

FIGURE 3. Correlation between IPS and Ulysses observation. Triangle and circle plots show the data at latitudes lower and higher than 45°, respectively.

ton rotations (CR 1978-1979) because Ulysses observed fast solar wind in almost same Carrington longitude in CR 1979. Then the fast solar wind region extend around the north pole in CR 1980. Difference between IPS and Ulysses observations becomes large in CR 1981. This difference comes from insufficient data coverage of the IPS observation around the north pole because error bars of tomographic process in this region are also large.

Figure 3 shows correlation between IPS and Ulysses observations. If we do not include the data observed in CR 1981 when the IPS data contain large error, latitudes higher than 45° (circle) are concentrate around $y = x$ line. On the other hand, the data at the latitude lower than 45° (triangle) are rather scattered. It is said that the slow wind is closely related to the active region (*e.g.*Uchida et al.[3] have found that the corona above an active region expands almost continually). The photospheric magnetic structure and shape of coronal streamer change rapidly in comparison with the period which is required to be stable structure by the TST analysis (about 2 weeks in this case). Though the velocity structure at low latitude reconstructed by the TST is temporary averaged and the difference from Ulysses observations become large.

ACKNOWLEDGMENTS

We would like to thank to the Japan Science Society which supported financially for this presentation in Pisa, Italy.

REFERENCES

1. D. J. McComas and H. A. Elliott, J. T. Gosling, D. B. Reisenfeld, and R. M. Skoug, B. E. Goldstein and M. Neugebauer, A. Balogh, Ulysses' second fast-latitude scan: Complexity near solar maximum and the reformation of polar coronal holes, *Geophys. Res. Lett.*, **29(9)**, 10.1029/2001GL014164, 2002.
2. Kojima, M., M. Tokumaru, H. Watanabe, A. Yokobe, K. Asai, B. V. Jackson, and P. L. Hick, Heliospheric tomography using interplanetary scintillation observations, 2, Latitude and heliospheric distance dependence of solar wind structure at 0.1-1 AU, *J. Geophys Res.*, **103**, 1981-1989, 1998
3. Uchida, U., A. McAsllister, K. T. Strong, Y. Ogawa, T. Shimizu, R. Matsumoto, and H. S. Hudson, Continual expansion of the active region corona observed by the Yohkoh soft X-ray telescope, *Publ. Astron. Soc. Jpn.*, L155-L160, **44**, 1992

A Technique For Comparing Solar Wind Structures Observed By ACE And Ulysses

H. A. Elliott and D. J. McComas

Southwest Research Institute, San Antonio, TX

P. Riley

Science Applications International Corporation, San Diego, CA

Abstract. Comparison of solar wind observations from the ACE spacecraft, in the ecliptic plane at ~1 AU, and the Ulysses spacecraft as it orbits over the Sun's poles, may provide valuable information about the latitudinal extent and variation of solar wind structures in the heliosphere. While qualitative comparisons can be made using average properties observed at these two locations, comparison of specific, individual structures requires a procedure to determine if a given structure has been observed by both spacecraft. Before addressing the challenge of comparing such structures at different latitudes, we seek to benckmark our procedure by comparing solar wind structures observed when both spacecraft were near the ecliptic plane. In this paper we assess the use of a 1-D hydrodynamic code to propagate ACE plasma measurements out to the distance of Ulysses and develop a procedure for compensating for the differing longitudes of the ACE and Ulysses spacecraft. In addition to comparing the plasma parameters and their characteristic profiles, we examine superthermal electron measurements and magnetic field polarity to determine if the same features are encountered at both ACE and Ulysses. We find that when both spacecraft are near the ecliptic plane; they frequently observe the same features.

INTRODUCTION

Coronagraph measurements have provided information about the size and three-dimensional shape of coronal structures within ~30 R_s of the Sun, but further out in the solar wind, much less is known about how such large-scale structures evolve. Over the past forty years, most of what has been learned about the heliosphere is based on in-ecliptic measurements. However, the unique orbit of Ulysses, which passes over the Sun's poles, provides in situ measurements as a function of latitude. Ulysses has already greatly enhanced our knowledge of the 3-D structure of the inner heliosphere during solar minimum when the large-scale structure varies slowly [e.g., *McComas et al.*, 2000, *and references therein*]. For example, Ulysses solar minimum measurements showed a remarkably simple solar wind structure with high-speed streams filling the high latitude regions when there are large, persistent polar coronal holes. In contrast, the near-maximum solar wind is much more complex with highly variable flows arising from a variety of coronal sources observed at all heliolatitudes [*McComas et al.*, 2002].

The goal of this paper is to develop an analysis technique that will provide more information about the three-dimensional configuration of solar wind structures in the heliosphere, particularly around solar maximum. We focus on the in-ecliptic measurements to determine if the same features can be observed at two spacecraft with small latitude separations. The results of this study demonstrate that such features can be reliably identified, enabling us to extend this work to larger latitude separations in future studies.

DATA AND MODEL

We use primarily ion and electron measurements from the Advanced Composition Explorer (ACE) and Ulysses spacecraft. These observations are taken from the ACE Solar Wind Electron Proton Alpha Monitor (SWEPAM) [*McComas et al.*, 1998] and the Ulysses Solar Wind Observations Over the Poles of the Sun (SWOOPS) instruments [*Bame et al.*, 1992]. We also use supporting observations from the magnetometers on Ulysses [*Balogh et al.*, 1992] and ACE [*Smith et al.*, 1998].

To propagate data from 1 AU (ACE) out to Ulysses, we use the Zeus Astrophysical single-fluid magnetohydrodynamic model [*Stone and Norman*, 1992]. The model uses a Eulerian finite difference

CP679, *Solar Wind Ten: Proceedings of the Tenth International Solar Wind Conference*, edited by M. Velli, R. Bruno, and F. Malara

scheme. Although the Zeus model can include magnetic fields and radiation transport, in this study they are neglected and only hydrodynamic effects are examined. We propagate magnetic sector by defining boundaries of given regions (such as a CME or high-speed stream) in the ACE data and then track those boundaries by advancing them at each time step in the simulation using the velocity and acceleration at that grid point. The model solves a system of continuity, momentum, and internal energy equations. We use a polytropic index of 3/2 since most solar wind measurements are generally consistent with this value [*Riley et al.*, 2001]. Although Zeus is a 2-D model, only one dimension is used so that we can use single point in situ measurements to drive the model. Using the 1-D model allows us to focus on the dynamic evolution of structures between 1 and 5.4 AU without

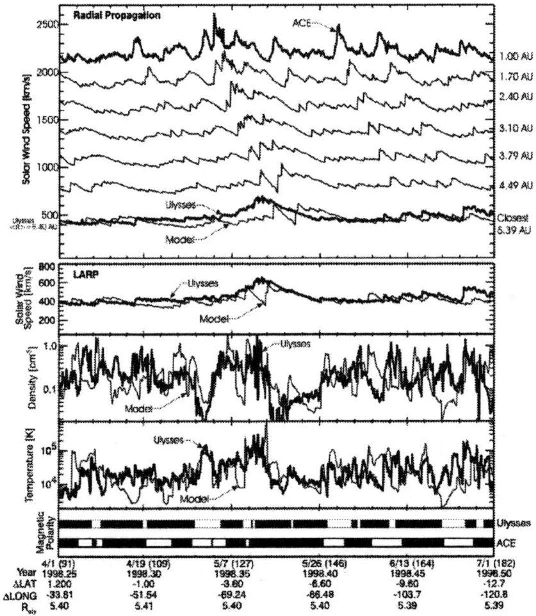

Figure 1. ACE solar wind speed data at 1 AU as a thick line, and model speeds at given propagation distances as thin lines (top panel). Subsequent panels compare Ulysses data and model results for the LARP method (velocity second panel, density third panel, temperature the forth panel and the polarity of the magnetic field last panel).

making assumptions about the solar wind properties at locations away from our measurements. *Gosling et al.* [1995; 1998] showed that a 1-D model predicts many key features of CMEs, which are also found in more complex 2-D models of CMEs [*Riley et al.,* 1997]. When Ulysses and WIND were in radial alignment, *De Keyser et al.* [2000] found that using a 1-D model they could often identify sector boundaries at both spacecraft, and *Crooker et al.* [2001] found that the current sheet is coherent on a global scale, but has a highly variable local structure. By comparing solar wind measurements from multiple spacecraft, *Richardson et al.* [2001] found that correlation lengths perpendicular to the flow are 70 R_E for the solar wind speed and 100 R_E for the density.

We use ACE measurements as the inner boundary conditions for each time step of the simulation. We then compare the propagated ACE measurements with the Ulysses measurements. Since the Ulysses spacecraft position varies slowly, the data are split into 4 segments per year to calculate an average radial position. For each segment we compare Ulysses SWOOPS measurements to ACE mapped data at the radial grid point that most closely matches the average radial distance of Ulysses. We use the magnetic polarity, magnitudes and profiles of the velocity, density, and temperature to identify specific features. The polarity is determined by comparing the magnetic field direction to the Parker spiral direction determined using the solar wind speed [*Forsyth et al.,* 1996]. Since we are interested in the large-scale structure, we use a 12-hour running mean to calculate the sector structure. Electron distributions are used to identify coronal mass ejections.

RESULTS

The top panel of Figure 1 displays a time series of ACE SWEPAM solar wind speed measurements at 1 AU (top curve), the radial propagation (RP) of the ACE data at intermediate grid steps (5 middle curves), and a direct comparison between Ulysses SWOOPS measurements and mapped ACE data (bottom curve). High-speed streams become steeper with distance as faster plasma overtakes slower plasma and shocks form. Many shocks form between 1.5 and 2.5 AU, and can easily be tracked out to 5 AU, and many high-speed structures are worn down as momentum and energy are transferred to lower speed plasma. For example, from 1998.24 to 1998.34 and from 1998.4 to 1998.47 the model predicts that the high-speed features are worn down, consistent with Ulysses measurements of a more uniform low speed solar wind speed. The peak in the velocity between 1998.34 and 1998.4, occurs at a later time in the radial propagation model curve than in the Ulysses data. For this large feature between 1998.34-1998.4, ACE and Ulysses electron measurements both show the presence of counterstreaming electrons indicating a CME. *Gosling* [1996] wrote a review paper on signatures of CMEs in the heliosphere and finds that the presence of counterstreaming electrons is perhaps the most reliable signature of a CME although no single signature works all the time [*Gosling*, 1996; *Neugebauer and Goldstein*, 1997].

The RP method does not take into account the longitudinal separation of the two spacecraft, which varies between 0 and 360 each year, as the Earth and L1 revolve around the Sun. To compensate for this effect (at least for co-rotating structures), we have "de-rotated" one set of observations with respect to the other. We will refer to the combination of longitude adjustment and radial propagation as (LARP). The propagation time for the LARP method can be thought of as the time required for a spiral at ACE to propagate to Ulysses. To examine Ulysses observations at higher latitudes, the LARP method may need to be refined to include the latitude dependence of the spiral winding. The second panel in Figure 1 overlays the speed profiles using the LARP method. This was accomplished by subtracting the inertial heliospheric longitudes for the two spacecraft and converting this difference into an effective time shift using the solar rotation rate. To minimize the effects of temporal evolution of the solar source, the data are rotated the shortest way around (forward or backward in time) so that comparisons are always made between data that is separated by less than two weeks. Using the LARP method causes the large feature in the Ulysses data at 1998.34 to align with the double peaked feature in the mapped data.

It appears that the series of peaks observed at ACE either do not merge properly in the model, or parts of a CME could evolve differently as some CME model results show [*Riley et al,*.1997]. The LARP method works particularly well for the CME at 1998.34. In total we have examined 15 counterstreaming intervals from February 1, 1998 to October 31, 1998, and find that the LARP method improves the sector alignment for 10 of those intervals. We plan to investigate more CMEs to determine if the LARP method is generally a better method to use for comparing the ACE Ulysses data than the RP method. If a CME stays magnetically connected to the Sun for a long time then it becomes aligned along a spiral as depicted by *McComas et al.* [1992] such that the westward flank becomes elongated more than the eastward flank. Also, the spiral angle of a CME will depend on the speed of the CME.

In the lower panels of Figure 1 we compare Ulysses density (third panel), temperature (forth panel), and magnetic polarity (bottom panel) measurements to the LARP ACE data. The LARP ACE speeds, average density, and temperature agree fairly well with the Ulysses data; however, the high frequency fluctuations in density and temperature do not. We would not expect high frequency fluctuations to agree well because while the two spacecraft observe the same large-scale solar wind structures, they never observe precisely the same parcels of plasma. ACE and Ulysses show similar sector structure throughout this entire interval. From 1998.31 to 1998.42, the mean mapped density and density variations agree well with the data. The mapped mean temperature and temperature profile agrees well with the measured temperature over a longer time period from 1998.31 to 1998.5, although from 1998.42 to 1998.5 the difference between the model and data average temperature is greater than earlier. It appears that the model mapped density and temperatures do not agree as well with the data as the velocity and sector do because the temperature and density are highly variable parameters.

In order to compare the average variations in the mapped density and temperature with the Ulysses observations, we calculated radial profiles. In Figure 2 the model results from February 5, 1998 through December 31, 1999 are binned in radial distance and averaged over time. The ACE and Ulysses measurements are analyzed similarly. The radial variation of density and temperature observed in the Ulysses data is consistent with the density being proportional to R^{-2} and temperature proportional to R^{-1} as predicted by the Zeus model.

CONCLUSIONS

We find that the mapped ACE data agrees well with average plasma properties at Ulysses when the latitude and longitude separations of the spacecraft are small. In addition, we find that adjusting the source longitude clearly improved the mapping of the large-scale structure. Thus, at small latitude separations, the sector and plasma profiles can be adjusted for source longitude and mapped from 1 to 5.4 AU such that good agreement is reached between these mapped plasma profiles and the Ulysses observations.

We plan to extend our analysis to examine the latitudinal structure of these solar wind structures by comparing the mapped ecliptic observations with Ulysses observations at times when Ulysses is out of the ecliptic plane. The ability to compare such observations will elucidate the large-scale solar wind structures. We will test our ability to extend this work to larger latitude separations (which occurs during solar maximum) by comparing ACE and Ulysses measurements with coronal hole maps [*Elliott et al.*, 2003]. Such comparisons will help determine when ACE and Ulysses are observing the same or different coronal holes. Our techniques can be applied to learn more about CMEs as well. For example, our two methods of propagation will allow us to probe the shape of CMEs, and will provide information about how long a CME stays magnetically connected to the Sun.

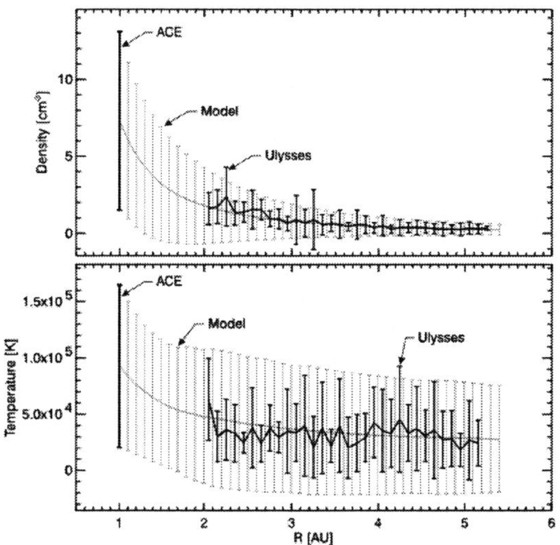

Figure 2. Density and temperature data divided into radial bins and averaged. Simulation results are shown in grey, ACE and Ulysses data are shown in black.

ACKNOWLEDGMENTS

We would like to thank N. Ness, C. Smith, and A. Balogh for use of the ACE and Ulysses magnetometer data. Our work is supported by NASA's ACE and Ulysses programs as a part of the SWEPAM and SWOOPS data analysis efforts. Pete Riley greatfuly acknowledges the support of NASA (SEC-GI program).

REFERENCES

Balogh A., et al., The magnetic field investigation on the Ulysses Mission: Instrumentation and preliminary scientific results, *Astron. and Astrophys., Suppl. Ser.*, **92**, 221, (1992.)

Bame S. J. et al., The Ulysses solar wind plasma experiment, *Astron. and Astrophys., Suppl. Ser.*, **92**, 237, (1992).

De Keyser, J., M. Roth, R. Forsyth, and D. Reisenfeld, Ulysses observations of sector boundaries at aphelion, *J. Geophys. Res.*, **105**, 15,689, (2000).

Elliott, H. A., D. J. McComas, and P. Riley, Latitudinal extent of large-scale structures in the solar wind, *Ann. Geophys.*, "Ulysses and Beyond", accepted (2003).

Forsyth, R. J., A. Balogh, E. J. Smith, G. Erdös, and D. J. McComas, The underlying Parker spiral structure in the Ulysses magnetic field observations 1990-1994., *J. Geophys. Res.*, **101**, 395, (1996).

Gosling, J. T., et al., A CME-driven solar wind disturbance observed at both low and high heliographic latitudes *Geophys. Res. Lett.*, **22**, 1753, (1995).

Gosling, J. T., Corotating and transient solar wind flows in three dimensions, *Annu. Rev. Astron. Astrophys,* **34**, 35-73, (1996).

Gosling, J. T., P. Riley, D. J. McComas, and V. J. Pizzo, Overexpanding coronal mass ejections at high heliographic latitudes: Observations and simulations, *J. Geophys. Res.*, **103**, 1941, (1998).

McComas, D. J., J. T. Gosling, and J. L. Phillips, Interplanetary Magnetic Flux: Measurement and Balance *J. Geophys. Res.*, **97**, 171-177, (1992).

McComas, D. J., et al. Solar wind electron proton alpha monitor (SWEPAM) for the Advanced Composition Expolorer, *Space Sci. Rev.*, **86**, 563, (1998).

McComas, D. J., et al., Solar wind observations over Ulysses' first full polar orbit, *J. Geophys. Res.*, **105**, 10419, (2000).

McComas, D. J., et al., Complexity near solar maximum and the reformation of polar coronal holes, *Geophys. Res. Lett.*, *29(9)*, 10.1029/2001GL014164, (2002).

Neugebauer, M., and R. Goldstein, *Coronal Mass Ejections Geophysical Monograph 99*, edited by N. Crooker, et al., Washington D. C., AGU, pp. 245, (1997).

Riley, P., J. T. Gosling, and V. J. Pizzo, A two-dimensional simulation of the radial and latitudinal evolution of a solar wind disturbance driven by fast, high-pressure CME ,J. Geophys. Res., **102**, 14,677, (1997).

Riley, P., J. T. Gosling, V. J. Pizzo, Investigation of the polytropic relationship between density and temperature within interplanetary coronal mass ejections using numerical simulations, *J. Geophys. Res.*, **106**, 8291, (2001.)

Smith, C. W., et al., The ACE magnetic fields experiment *Space Science Reviews*, **86**, 613, (1998).

Stone, J. M., and M. L. Norman, ZEUS-2-D: A radiation magnetohydrodyanmics code for astrophysical flows in two dimensions, I, The hydrodynamic algorithms, and tests, *Astrophys. J.*, **80**, 753. (1992).

Richardson, J. D., and K. I. Paularena, Plasma and magnetic field correlations in the solar wind, *J. Geophys. Res.*, **106**, 239, (2001).

Crooker et al., Scales of heliospheric current sheet coherence between 1 and 5 AU, *J. Geophys. Res.* **106**, 15963, (2001).

Repeated Structures Found After the Solar Maximum in the Butterfly Diagrams of Coronal Holes

M.Y. Hofer* and M. Storini†

*Research and Scientific Support Dept. of ESA, ESTEC, Keplerlaan 1, 2201 AZ Noordwijk, The Netherlands.
†IFSI/CNR, Via del Fosso del Cavaliere 100, 00133 Roma, Italy.

Abstract. The influence of the solar cycle evolution on the coronal hole space-time distribution is well known, for polar as well as for equatorial isolated sources of high speed solar wind. Among them the long-lived coronal holes occurrence from the sunspot cycle 21 on is investigated, using the coronal hole catalogue based on HeI (1083 nm) observations (Sanchez-Ibarra and Barraza-Paredes). In at least these two solar cycles (n. 21 and n. 22) a similar structure in the latitude-time diagram of coronal holes is found. The area occurs shortly after the solar maximum at around $\sim$35° heliolatitude and consists of over several Carrington Rotations stable coronal holes (>5 Carr. Rot.s). The diagonal disappears 2-3 years later at the helioequator. Furthermore, the analysis results in a close relation between long-lived isolated coronal holes and the soft X-class flares.

INTRODUCTION

A good knowledge of coronal hole (CH) evolution in time and in the different heliographic latitudinal belts is relevant for the understanding of the decay and/or the stability of large-scale solar magnetic fields (MFs) and their extensions in the heliosphere. The first investigations of these uni-polar and low density areas in the solar atmosphere were most probably made by Waldmeier [1956]. Coronal holes are the source of the high speed solar wind. Moreover, the coronal temperature and density are low in such regions.

The largest areas with uni-polar fields are the southern and northern polar CHs. The polar CHs change their shape during the solar activity cycle. They expand towards the heliographic equator during the decreasing phase of the solar activity cycle, shrink back to the poles in the ascending one, and disappear during the maximum activity phase for a certain time. The new polar CHs appear with inverted magnetic polarity. In addition, uni-polar field regions, isolated from the polar CHs, are observed at any heliographic belt below ±60° latitude.

In a previous analysis peculiar features in CH occurrence were underlined [Hofer and Storini, 2002a]. Indications for a north/south asymmetry in the number distribution of the polar coronal holes and a 22-year periodicity in the longitudinal width of the extensions of the polar CHs, i.e. MF regions observed below ±60° latitude being well connected to the polar coronal holes, were found.

Using the CH catalogue, compiled mainly from the HeI absorption line (1083 nm) measurements ([Sanchez-Ibarra and Barraza-Paredes, 1992], and updated from the NOAA-Boulder Web Pages), we here investigate the latitudinal distributions of long-lived CH occurrence (ages >5 Carr.Rot.s) mainly during the sunspot cycles 21 and 22.

USED DATA CATALOGUE

Data for the CH occurrence [Sanchez-Ibarra and Barraza-Paredes, 1992] were taken from the NOAA (Boulder) Web pages to analyze the variation of the CH heliographic coordinates from $\sim$1970.1 to $\sim$1995.4. These data were mainly obtained by observing the HeI absorption line (1083 nm) by the National Solar Observatory/Kitt Peak. The catalogue provides the central location of the CHs in heliographic coordinates and the longitudinal and latitudinal extensions (widths) of the CH area. The CHs were even identified over several Carr. Rot.s. using other data sources, such as the CH contours from the H_α synoptic charts.

The catalogue mainly lists two types of CHs:

i) The first class: EP-CHs (Extended Polar Coronal Holes), consists of uni-polar MF regions observed below the ±60° heliographic latitude; they are well connected with the polar CHs (i.e. they are extended polar CHs).

ii) The second class: I-CHs (Isolated Coronal Holes), are found at any heliographic latitudinal belt below the ±60° heliographic latitude and are therefore clearly isolated from the polar coronal holes.

CP679, *Solar Wind Ten: Proceedings of the Tenth International Solar Wind Conference*,
edited by M. Velli, R. Bruno, and F. Malara

BUTTERFLY DIAGRAM OF THE CORONAL HOLES

The butterfly diagrams of extensions of the polar (EP-CHs) and isolated (I-CHs) coronal holes are shown in Figure 1. The dashed long vertical lines mark the solar minima occurrence. The solar equator is shown by a horizontal dashed line.

In the left panel of Figure 1, the butterfly diagramm of the extension of the polar coronal holes (EP-CHs) is shown. The horizontal width of the strings corresponds to the observation time of the EP-CHs. The labeled short dashed verticals mark the CHs that extend to the solar equator in 1973, in 1984 and in 1994. The EP-CH reach the equatorial regions about four years after the maximum, two to three years before the minimum. There are no EP-CH during the maximum activity phase in 1980 and in 1990. There is a tendency that long-term EP-CHs are observed below 50° heliographic latitude.

In the right panel of Figure 1, the strings report the central heliographic latitude of the I-CHs, as a function of time (the horizontal length of the strings gives the total observation time of the CHs). The dotted curve on top of the distribution points out the external contour of the I-CH distribution. The ellipses show two repeated large regions without I-CHs. The dashed diagonals mark:

a) a selected region with long-lived I-CHs, appearing after 1979 (sunspot cycle 21);

b) an example for an area without any I-CHs (sunspot cycle 22).

From the external contour we see that it tends to follow the solar activity level. Shortly after the solar maxima, the widths are maximum as well. Furthermore, the solar minima are in vicinity of the minimum widths of the latitude-time distribution of the I-CHs. In other words, an 11-year cycle characterizes such distribution.

In Figure 2, the latitudinal distribution of EP-CHs are shown using a grey color. The I-CHs are represented by dark strings (left: long-term I-CHs > 3 Carr. Rot.s; right: long-term I-CHs >5 Carr.Rot.). The vertical dashed lines mark the times of the solar minima occurrence. During the time intervals ~1980-1983 (south) and ~1990-1992 (north) two diagonals of long-living I-CHs, starting at around 35° heliolatitude, shortly after the solar maximum, and ending close to the equator, two to three years later, can be found in Figure 2. Each line consists of more than six I-CHs with ages of more than 5 Carr.Rot.s.

In Figure 3, the latitudinal distribution of the solar optical flares associated with soft X-ray flares of X-class are shown. The slopes of long-lived isolated coronal holes on the selected diagonal (as shown in the panels of Figure 2) and the soft X-class flares are both steeper in the second time interval. They are therefore somehow related, whereas the equatorward boundaries of the distribution of the EP-CHs do not seem to change so dramatically from one to the next solar cycle as it can be seen in the left panel of Figure 1.

DISCUSSION AND SUMMARY

We investigated the coronal holes distribution from the sunspot cycle 21 on, using the coronal hole catalogue based on HeI (1083 nm) observations (Sanchez-Ibarra and Barraza-Paredes, 1992).

The influence of the solar cycle evolution on the envelop of the coronal hole space-time distribution is clearly visible, for the polar as well as for the low-latitude isolated sources of high speed solar wind.

The disappearance of the extensions of the polar coronal holes for a time period of about one year is expected because even the corresponding polar coronal hole is not formed, and consists for while of several regions with different polarities (see, for instance Sanderson et al. [2001] and Fox et al. [1998]). During the same time period the region around ±55° does not seem to be governed by quiet MF as is can be seen in the left panel of Figure 1. Furthermore, few isolated CHs are observed at high heliolatitude around solar activity maximum as it can be seen in right panel of Figure 1. Regarding Figure 3, active X-ray flares are found to occur up to ~ 45° heliographic latitude. Therefore, the region between 45° and 60° heliographic latitude during the maximum active phase does not contain large quiet MF regions and not even active areas. It looks like a belt between the two extremes. It would be interesting to analyse this region with respect to the well-known Gnevyshev gap (e.g. Feminella and Storini [1997] and references therein), during which a reduction of the solar activity effects is found also in the heliosphere (e.g. Storini and Felici [1994], Storini [1995], Storini and Pase [1995], for early works; Storini and Hofer [2001], Storini et al. [2002] for reviews).

Concluding, we remark that in at least two solar cycles (n. 21 and n. 22), we identified a similar structure in the latitude-time diagram of the isolated long-lived coronal holes. The edge of the area starts at around 35° heliolatitude, shortly after the solar maximum, and ends close to the equator, two to three years later. The structure occurs after and below the above mentioned latitudinal belt separating the two extreme areas. It is a diagonal region consisting of over several Carrington Rotations stable isolated coronal holes. More precisely, during the time intervals 1980-1984 (south) and 1990-1992 (north) two diagonals of long-living I-CHs emerged, being each one characterized by more than six I-CHs (each with an age of more than five Carr.Rot.s).

The found structure does not evolve symmetrically on the northern and the southern solar hemispheres. The

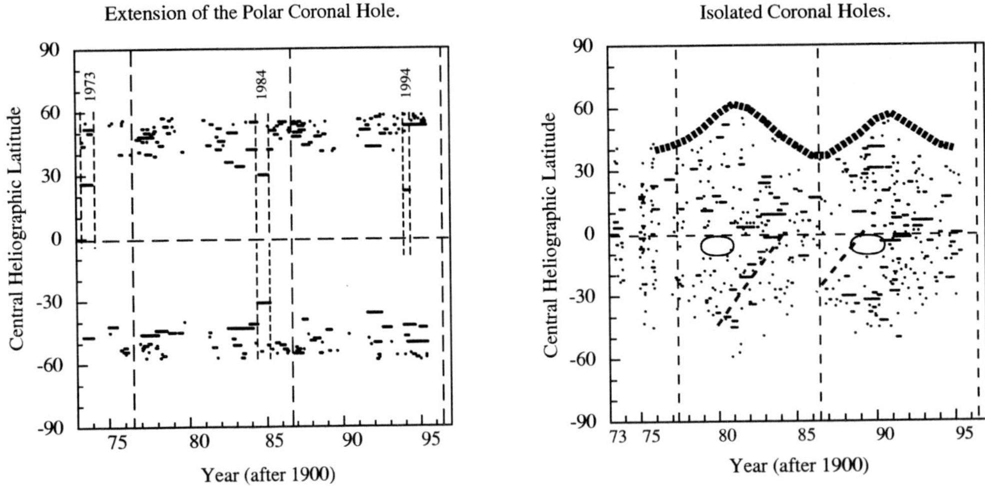

FIGURE 1. Butterfly diagrams of extensions of the polar coronal holes (left) and isolated coronal holes (right). The dashed vertical lines in both panels mark the times of the solar minima occurrences. The short labeled line represent the CHs that extend into the equatorial regions in 1973, 1984, 1994. The diagonals and the ellipses point out selected areas in the distributions. Both figures were adapted from Hofer and Storini [2000].

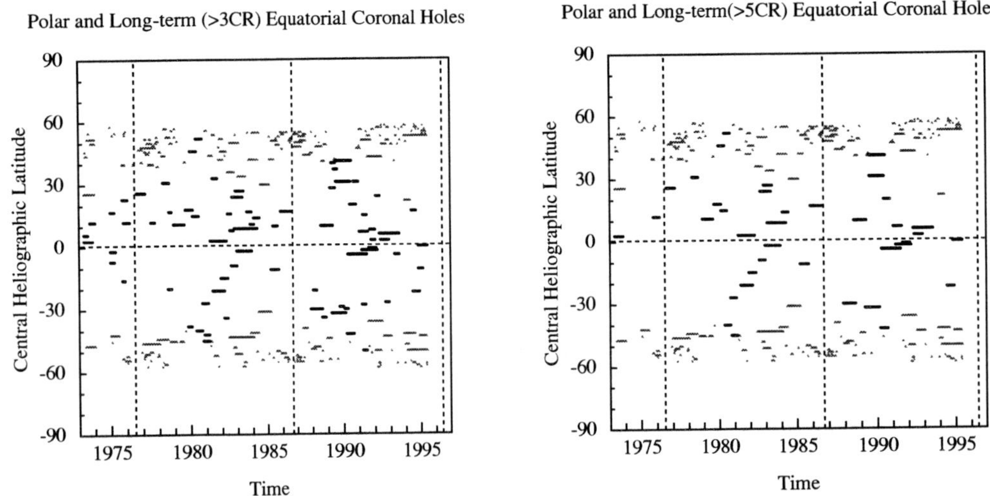

FIGURE 2. Extensions of the polar coronal holes and long-term isolated coronal holes. a) left: long-term I-CHs > 3 Carr. Rot.; b) right: long-term I-CHs >5 Carr.Rot. The I-CHs are represented by dark horizontal strings. The central latitude of the EP-CHs is shown with filled grey small triangles. The dashed vertical lines mark the times of the solar minima occurrences.

fact that the structure is found first in the South than in the North motivates two interpretations:

i) a reasonable 22-year periodicity (Hale cycle) in the coronal hole occurrence;

ii) the found structure is closely related to the time evolution of the nearby magnetic fields.

A north-south asymmetry is also found for the latitudinal extension of the I-CHs as reported by Hofer and Storini [2002a], Storini and Hofer [1999]. The second interpretation deserves special attention because, from our study, it results in a close relationship between long-lived isolated coronal holes and soft X-class flares. In a parallel work we show that the extensions of the polar coronal hole and the sunspot areas are related in a similar way, when their latitude-time distributions are consid-

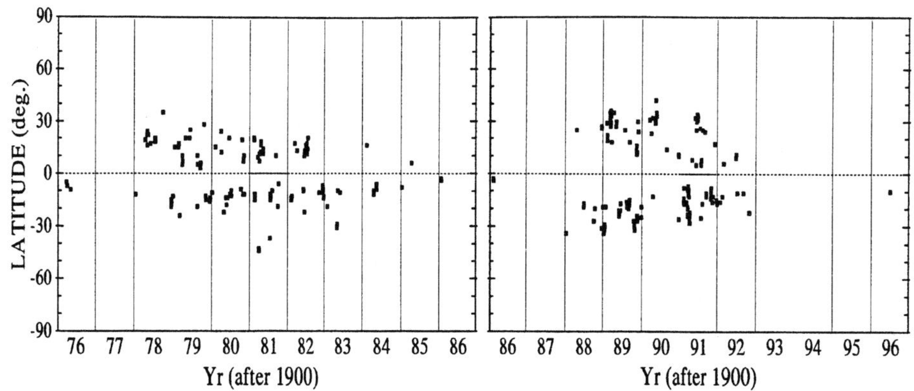

FIGURE 3. Latitudinal distribution of the solar optical flares associated with soft X-ray flares of X-class. The figure is adapted from Storini and Hofer [1999].

ered [Hofer and Storini, 2002b].

We conclude that the latitudinal distribution of the equatorial and polar coronal holes should be regarded more frequently for the understanding of the solar and heliospheric evolution in time.

ACKNOWLEDGMENTS

Part of this work was supported by the Swiss National Science Foundation (fellow grant 81BE-57318) and the National Antarctic Research Program (PNRA) of Italy in the frame of Science for Solar-Terrestrial Relations. MYH also thanks ESA for the present research fellowship.

REFERENCES

1. Feminella, F. and Storini, M., *Astron. Astrophys.*, **322**, 311-319 (1997).
2. Fox, P., McIntosh, P.S., Wilson, P.R., *Solar Phys.*, **177**, 193 (2000).
3. Hofer, M.Y., and Storini, M., *Report CNR/IFSI-2000-9*, May (2000).
4. Hofer, M.Y., and Storini, M., *Solar Phys.*, **207**, 1-10 (2002a).
5. Hofer, M.Y., and Storini, M., in preparation (2002b).
6. Sanchez, A., and Barraza-Paredes, M., WDCA, Boulder, UAG-102 (1992).
7. Sanderson, T.R. et al., *Geophys.Res.Lett.*, **28 (24)**, 4525 (2001).
8. Storini, M., and Felice, A., *Nuovo Cim.*, **17C**, 697-700 (1994).
9. Storini, M. and Hofer, M.Y., *Proc ESA*, **SP–448**, p.889 (1999).
10. Storini, M. and Hofer, M.Y., *Mem. S.A.It*, **72 (3)** , 637-640 (2001).
11. Storini, M., *Adv. Space Res.*, **16 (9)**, 51 (1995).
13. Storini, M. and Pase, S., *STEP GBRSC News*, **5 - Special Issue**, 267-274 (1995).
13. Storini, M., *Adv. Space Res.*, **16 (9)**, 51 (1995).
14. Storini, M., et al. *Adv. Space Res.*, **in press**, (2002).
15. Waldmeier, M., *Z. Astrophys.*, **39**, 219 (1956).

Investigation of the sources of the slow solar wind

Lucia Abbo*, Ester Antonucci*, Maria Adele Dodero† and Carlo Benna*

*Istituto Nazionale di Astrofisica (INAF), Osservatorio Astronomico di Torino, Pino Torinese, Italy
†Università degli Studi di Torino, Torino, Italy

Abstract. Aim of this analysis is to study the variation of the physical conditions of the coronal plasma across the streamer boundary in order to identify the coronal sources of the slow solar wind during the minimum of solar activity. The analysis is based on the observations of equatorial streamers, obtained in the outer corona during the years 1996 and 1997 with the Ultraviolet Coronagraph Spectrometer (UVCS) onboard SOHO. The outflow velocity, the electron density and the oxygen abundance relative to hydrogen of the coronal plasma have been determined, in the range between 1.6 and 3.5 solar radii ($R_\odot$), by means of a spectroscopic analysis of the OVI 1032, 1037 Å and the HI Lyα 1216 Å lines. Coronal expansion at low velocity, in the range 80–100 km/s, is observed along regions 15°–20° wide, surrounding the streamer boundary. Evidence for coronal plasma outflows at low velocity is also found further out in the region along the streamer axis. In this case the outflows become significant beyond 2.7 $R_\odot$. Hence, the slow solar wind during solar minimum flows just outside the denser and brighter zone of a streamer, characterized by closed magnetic field lines and in a lane around the heliospheric current sheet, forming just above the closed field line region.

INTRODUCTION

According to the Ulysses observations, during solar minimum the slow solar wind tends to be confined in the equatorial streamer belt, whereas the fast wind fills most of the heliosphere. This study is an attempt to identify the coronal regions where the heliospheric slow wind originates, by analyzing the observations obtained with the Ultraviolet Coronagraph Spectrometer (UVCS) onboard SOHO. The most probable sources of the slow wind are the regions of open magnetic field lines at the border of the large quiescent equatorial streamers observed during solar minimum. Therefore we focus on the comparison of the dynamical conditions observed in the central brightest streamer region, corresponding to plasma predominantly confined by closed field lines, and in the dim lanes running along the borders of these bright regions. The UVCS observations of the outer corona, extending from 1.5 $R_\odot$ to a few solar radii, provide a set of spectroscopic data apt to identify plasma outflows by means of the Doppler dimming analysis of the resonantly scattered coronal emission.

DIAGNOSIS OF THE CORONAL ULTRAVIOLET EMISSION

The spectroscopic diagnostic technique used in this study (Antonucci et al., 2002) allows us to determine the electron density and the outflow velocity of the OVI ion component of the coronal plasma uniquely on the basis of the OVI 1032, 1037 doublet, and to compute the abundance of oxygen relative to hydrogen by including also the HI Lyα 1216 line data. The technique fully accounts for the effect of Doppler dimming in an expanding corona.

The ultraviolet lines emitted in the extended corona are formed both via collisional and radiative excitation. The radiative contribution becomes predominant in low density conditions. The collisional and radiative components are separated by considering a pair of lines emitted from the same ion, such as the OVI 1032 and 1037 Å lines. The electron density is then derived from the ratio of the collisional to radiative component of a spectral line, being proportional to this ratio multiplied by a factor depending on the outflow velocity. Therefore in order to derive electron density and outflow velocity, at the same time, we need a further constraint given by the mass flux conservation along the flow tube connecting the corona and the heliosphere: $n_e\, w\, A$=const, where n_e is the electron density, w is the outflow speed, $A = F(r)\, r^2$ is the cross section of the flux tube and the quantity $F(r)$ takes into account the deviation from radial expansion.

The abundance of oxygen relative to hydrogen is computed from the ratio of the radiative components of the OVI 1032 line and the HI Lyα 1216 line, both functions of the outflow velocity (Antonucci and Giordano, 2001), by assuming conditions close to ionization equilibrium.

CP679, *Solar Wind Ten: Proceedings of the Tenth International Solar Wind Conference*, edited by M. Velli, R. Bruno, and F. Malara

TABLE 1. Coronal streamers used in the analysis.

Date			Streamer axis (PA, ^{o})	Altitude range ($R_\odot$)
19	Aug	1996	120	1.6 – 3.5
22	Aug	1996	60	1.6 – 3.5
30	Aug	1996	240	1.6 – 3.5
1	Sep	1996	280	1.6 – 3.5
30	Apr	1997	280	1.7 – 3.3
5	May	1997	80	1.8 – 3.3

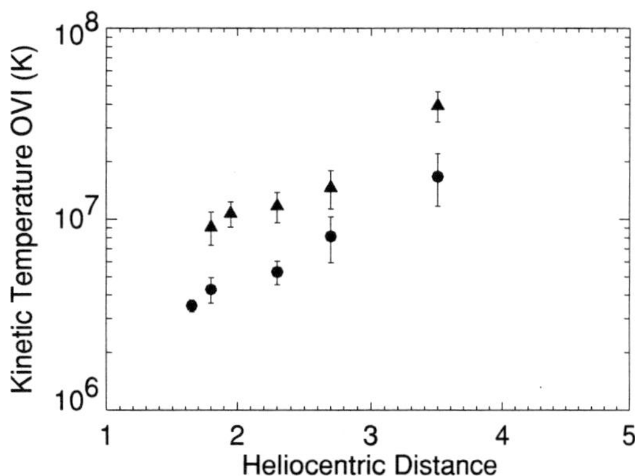

FIGURE 1. Kinetic temperature of OVI ions as a function of heliocentric distance (in solar radii) relative to both the streamers (full dots) and the regions adjacent to the streamer boundary (full triangles).

OBSERVATIONS AND DATA ANALYSIS

The present analysis is based on a set of six streamers, observed with UVCS at high spectral resolution. They are detected at mid and low latitudes during the solar minimum period 1996–1997. Table I reports the following quantities: date of observation, position angle (PA, in degrees, counterclockwise from the North Pole) of the streamer axis, range of the UVCS fields of view, in solar radii ($R_\odot$).

For each streamer we define the streamer boundary as the contour level corresponding to the 1/e value of the OVI line intensity maximum (different definitions of streamer boundary are discussed by Abbo and Antonucci, 2001). The intensities of the OVI 1032, 1037 and HI Lyα 1216 lines are then computed by integrating the line emission in two regions: within the streamer boundaries and in the external region, approximately 15°-20° wide, adjacent to the boundary. The integrated emissions are then fitted with a gaussian curve to obtain the line intensity and width, that can be expressed in terms of the kinetic temperature, T_k, quantity that measures the velocity distribution of the emitting particles along the line-of-sight (l.o.s.).

The kinetic temperatures derived from the OVI 1032 line profiles are plotted in Figure 1 as a function of heliocentric distance (the values found for streamers are reported as full dots and those relative to the regions surrounding streamers as full triangles). This quantity is clearly increasing with height and the values found outside are significantly higher than inside streamers (Antonucci et al., 1997).

The diagnostic technique to be applied to the UVCS data is based on the computation of resonance absorption along the direction of the incident radiation coming from the disk, within a solid angle subtended by the solar disk itself. Therefore the analysis has to include the distribution of the oxygen ion velocity in three dimensions, whereas only the distribution along the l.o.s. is observed. Within the streamer, that is, in higher density conditions, we assume an isotropic maxwellian velocity distribution with the width defined by the observed T_k. Outside the streamer, in the thinner regions adjacent to the streamer boundary, we assume a bi–maxwellian velocity distribution of the ions, with a kinetic temperature equal to the observed one in the plane perpendicular to the radial direction, and equal to the electron temperature along the radial direction. This assumption is dictated by the fact that line broadening in the regions close to streamers is about twice larger than in the streamer itself, thus representing an intermediate condition between closed field regions and the core region of coronal holes, where the ion velocity distributions are found to be highly anisotropic (e.g., Kohl et al. 1998, Cranmer at al. 1999, Antonucci et al., 2000).

The coronal electron temperature, T_e, assumed in the analysis of streamers is that derived by Gibson, et al. 1999. In the open field line regions it is inferred from the coronal hole measurements by David, et al. 1998 (see Antonucci et al. 2000).

In order to define the constraint imposed by mass flux conservation in the hypothesis of coronal expansion, the flux tube geometry has been modeled by assuming the expansion factor of the magnetic field lines as derived by Wang and Sheeley (1990) for the regions immediately adjacent to a streamer. In this model the magnetic field is taken to be potential everywhere except in an equatorial current sheet (with the inner boundary of the current sheet located at 2.5 $R_\odot$), where the magnetic field radial component changes sign discontinuously and the latitudinal component vanishes. In this model the open magnetic field lines adjacent to the streamer are characterized by a maximum expansion factor, $F(r)_{max} = 13.5$, much larger than that expected in the core of coronal holes, where $F(r)_{max} = 4$. In a previous analysis (Abbo and

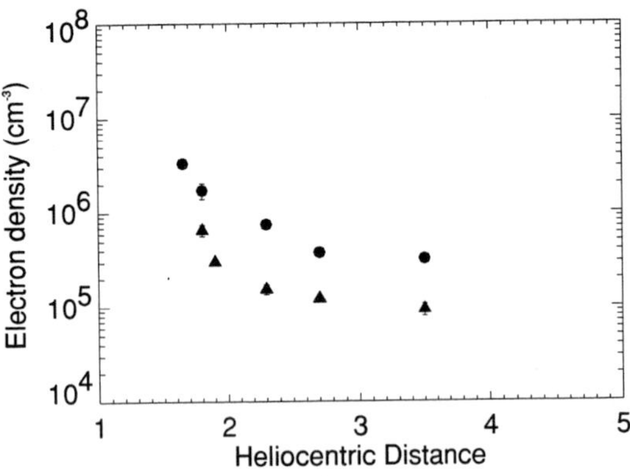

FIGURE 2. Electron density as a function of heliocentric distance (in solar radii) derived for streamers (full dots) and for the regions surrounding the streamer boundary (full triangles).

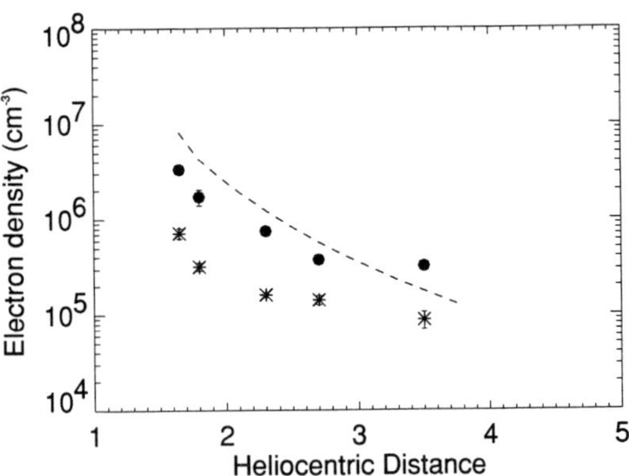

FIGURE 3. Streamer electron density as a function of heliocentric distance (in solar radii) calculated by assuming static conditions (full dots) and dynamic conditions (asterisks), compared with that derived by Gibson et al, 1999 (dashed line) from visible light observations.

Antonucci, 2002) we presented preliminary results of the same data set, obtained with a factor equal to the total expansion factor of a coronal hole, as measured by Munro and Jackson, 1977.

RESULTS

The electron densities averaged over an individual streamer and over its surrounding regions have been derived in both cases in the assumption of a dynamically expanding atmosphere and of a static plasma. The values have been further averaged over the set of streamers considered in the analysis. The results are plotted versus heliocentric distance in Figure 2.

A comparison of the streamer electron density with that derived from the visible light coronal observations by Gibson et al. 1999 is shown in Figure 3. It is clear that only the values obtained for a static atmosphere (full dots) are compatible with the white light results (dashed line), except for the density at 3.5 $R_\odot$, where both values, obtained in static and dynamic conditions (asterisks), are compatible with the visible light result. Hence, we conclude that the plasma in the bright emission region, delimited by the 1/e intensity contour line, is predominantly static out to 2.7 $R_\odot$. At higher altitudes data are also consistent with outward expansion.

In the regions adjacent to streamers, on the other hand, the results obtained for a static plasma are unrealistic, whereas those obtained for a dynamic expanding corona (triangles in Figure 2) are acceptable, implying that the narrow region, 15°–20^0 wide, along the streamer boundary, is dominated by open magnetic field lines.

It is interesting to note that all along the streamer border (that in the present set of data starts at 1.8 $R_\odot$, height corresponding to a portion of the spectrometer slit, set to 1.6 $R_\odot$, far from the center) and around the streamer axis above 2.7 $R_\odot$, the oxygen ions kinetic temperatures are higher than within the inner part of the streamer (Figure 1). If this large increase in kinetic temperature is interpreted as energy deposition as suggested by the coronal hole observations (Antonucci et al. 1997b, Kohl et al. 1997, Noci et al. 1997), these regions where the plasma expands are regions of energy dissipation. Such an effect, however, occurs only beyond 2.7 $R_\odot$ along the streamer axis, presumably in coincidence with the transition from closed to open magnetic field lines, that is, close to the inner boundary of the heliopsheric current sheet.

In the regions where coronal expansion is found to be compatible with the OVI emission data, we find an outflow velocity ranging from 90 km/s to 100 km/s along the streamer border and a value about 80–90 km/s above 2.7 $R_\odot$ along the streamer axis. These values are compatible with a regime of slow wind. In fact in coronal holes between 1.8 $R_\odot$ and 3.5 $R_\odot$ the fast wind expands much more rapidly with a speed that varies typically from about 100 km/s to 400 km/s.

A distinctive feature of the heliospheric slow and fast wind, besides the wind speed, is the plasma composition. Therefore we also derive the oxygen abundance relative to hydrogen, although, being a high FIP element, oxygen does not show a marked variation between slow and fast wind. The oxygen abundance is plotted in Figures 4, inside (full dots) and outside streamers (full triangles), as a function of heliocentric distance. The abundance decreases with increasing heliodistance, as found in quiescent streamers (Marocchi et al. 2001). In streamers it ranges from values close to the photospheric abundance,

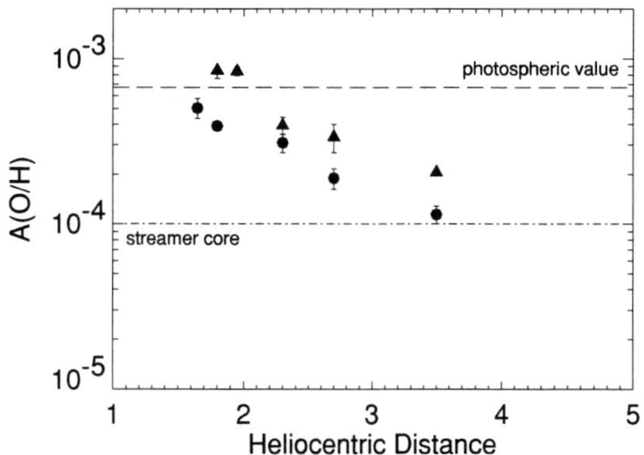

FIGURE 4. Abundance of oxygen relative to hydrogen as a function of heliocentric distance (in solar radii) derived for streamers (full dots) and for the regions surrounding the streamer boundary (full triangles). The abundance value for the streamer core, 1×10^{-4}, and the photospheric value, 6.7×10^{-4}, are plotted as dashed-dotted, dashed lines, respectively.

6.7×10^{-4}, to 1×10^{-4}, abundance found in the core of a streamer (e.g. Raymond et al., 1997). This supports the interpretation given by Marocchi et al. that the abundance of the wind close to the heliospheric current sheet might derive by open field lines existing within the streamer, and separating substreamers not always resolved in the outer corona (following the hypothesis of slow wind formed between substreamers formulated by Noci et al. 1997). In the regions surrounding the streamer border the decrease of abundance tends to be less marked. Since also the slow wind oxygen abundance tends to be lower than the fast wind abundance, the trend found in Figure 4 for the regions adjacent to the streamer border (full triangles) is supporting the conclusion that these are sources of slow wind.

This study, based on the OVI coronal emission, leads to the conclusion that the regions running along the streamer boundaries are site of coronal outflows at low velocities and therefore these are the likely sources of the slow wind. In addition, coronal expansion at low velocities is also likely to occur beyond 2.7 $R_\odot$ along the streamer axis, presumably where the transition between closed and open magnetic field lines takes place and the interplanetary current sheet forms. The values of the outflow velocity found, in the range between 80 km/s and 100 km/s, are consistent with the results obtained by Sheeley et al., 1997 and Wang et al., 1998 for the outward motions of the density inhomogeneities detected with LASCO (SOHO), which originate at the cusp of streamers. They are also consistent with the results found by Strachan et al., 2002, obtained by analyzing, with the traditional Doppler dimming technique, the coronal visible and ultraviolet emission of an equatorial streamer observed with UVCS. At last, considerations on the oxygen abundance in the regions of plasma outflows do support the interpretation of these regions as sources of the slow wind. The resulting scenario is compatible with the model proposed by Wang et al., 2000, of a two–component slow wind: one component flowing along the rapidly diverging open magnetic field lines adjacent to the streamer boundary, and the second one confined to the region of the denser equatorial plasma sheet.

REFERENCES

1. Abbo, L., and Antonucci, E., ESA Publication Division, Noordwijk, The Netherlands, ESA SP–**477**, 323-326 (2001).
2. Abbo, L. and Antonucci, E., ESA Publication Division, Noordwijk, The Netherlands, ESA SP–**508**, 477–480 (2002).
3. Antonucci, E., Noci, G., Kohl, J.L., et al., ASP Conference Series **118**, 273-277 (1997a).
4. Antonucci, E., Giordano, S., Benna, C., et al., ESA Publication Division, Noordwijk, The Netherlands, ESA SP–**404**, 175-182 (1997b).
5. Antonucci, E., Dodero, M.A., Giordano, S., *Sol. Phys.*, **197**, 115-134 (2000).
6. Antonucci, E. and Giordano, S., AIP Conference Proceedings, New York, **598**, 77-81 (2001).
7. Antonucci, E., Dodero, M.A., Giordano, S., *A&A, submitted* (2002).
8. Cranmer, S.R., Field, G.B. and Kohl, J.L., *ApJ*, **518**, 937-947 (1999).
9. David, C., Gabriel, A. H., Bely-Dubau, F., et al., *A&A*, **336**, L90-94 (1998).
10. Gibson, S. E., Fludra, A., Bagenal, F., et al., *JGR*, **104**, 9691-9700 (1999).
11. Kohl, J.L., Noci, G., Antonucci, E., et al., *Sol. Phys.*, **175**, 613-644 (1997).
12. Kohl, J.L., Noci, G., Antonucci, E., et al., *ApJ*, **501**, L127-131 (1998).
13. Marocchi, D., Antonucci, E. and Giordano, S., *Ann. Geophys.*, **19**, 135-145 (2001).
14. Munro, R.H. and Jackson, B.V., *ApJ*, **213**, 874-886 (1977).
15. Noci, G., Kohl, J.L., Antonucci, E., et al., and slow solar wind", in *Fifth SOHO Workshop*, ESA Publication Division, Noordwijk, The Netherlands, ESA SP–**404**, 75-84 (1997).
16. Raymond, J.C., Kohl, J.L., Noci, G., et al., *Sol. Phys.*, **175**, 645-665 (1997).
17. Sheeley, N.R., Jr., Wang, Y.-M., Hawley, S.H., et al., *ApJ*, **484**, 472-478 (1997).
18. Strachan, L., Suleiman, R., Panasyuk, A.V., et al., *ApJ*, **571**, 1008-1014 (2002).
19. Wang, Y.H. and Sheeley, N.R., Jr., *ApJ*, **355**, 726-732 (1990).
20. Wang, Y.H., Sheeley, N.R., Jr., Walters, J.H., et al, *ApJ*, **498**, L165-168 (1998).
21. Wang, Y.-M., Sheeley, N.R., Jr., Socker, D.G., et al, *JGR*, **105**, 25133-25142 (2000).

Topology and dynamics of the Sun's magnetic field

E. Gavryuseva*† and N. Kroussanova**†

*Florence University, Largo E. Fermi 5, 50125 Florence, Italy
†OAC, via Moiariello 19, Naples, Italy
**Sternberg Astronomical Institute, Moscow, Russia

Abstract.
The distribution of the magnetic field on the solar surface from 1975 up to 2002 is analyzed using the observations taken on the WSO observatory, revealing remarkable latitudinal zonal structure in both the northern and southern hemisphere. A preliminary model combining the dipole, quadrupole, octupole and other components for the longitude distribution is discussed. The active longitude problem has been investigated. The new active regions seem to appear with the sideral period of about 25.0 ± .1 days, corresponding to the solar rotation rate at 10 degree latitude at the 0.85±.03 solar radii.

1. DATA

We used the Wilcox Solar Observatory data about the Sun's magnetic field (MF) in the form of the Global Magnetic Field (GMF) and Local Magnetic Field (LMF) from the Synoptic Charts of the whole Sun [1, 2]. GMF data confirm the results obtained from the LMF analysis. WSO's Babcock solar magnetograph measures the line-of-sight component of the photospheric magnetic field using the Zeeman splitting of the 5250.2 A Fe I spectral line since May 1976. The noise level of each measurement is less than 10 $mcTl$. The resolution of the LMF in longitude is 5 degrees. In the latitude grid there are 30 data points of arcsine from +14.5/15 to -14.5/15. The magnetic field for the Carrington Rotations since 1642 to 1984 have been used. Carrington Rotations (CR) are a convenient coordinate system for locating positions on the Sun, defined a fixed solar coordinate system that rotates in a sideral frame exactly once every 25.38 days. This period was determined by watching low-latitude sunspots. CR 1642 begins at 1976:05:27.

The synodic rotation rate varies a little during the year because of the eccentricity of the Earth's orbit and its mean value is about 27.2753 days. The differential rotation of the Sun has to be taken into account in the further analysis. The problem of the rotation rate of the coordinate system is very important for the study of the longitude MF distribution, and it is less relevant for the latitude structure investigation until the corotation between the MF distribution and solar rotation is assumed.

2. LATITUDE DISTRIBUTION OF THE MAGNETIC FIELD THROUGH SOLAR CYCLES

There is a common view on the MF distribution on the Sun. A global dipole field describes the North–South polarity which is flipping each solar activity (SA) maximum. The total period of the dipole field is about 22-year. Additionally there are famous 11-years solar activity cycles known as an increase of the sunspots number on the middle latitudes.

We revealed the magnetic field temporal change on the latitudes from about + 75 degrees ($arcsin 14.5/15.$) North to - 75 degrees South with 30 steps in between and around the Sun with a 5 degree longitude step. Then we studied how the MF which is averaged along the longitude and along the latitude behaves in time. In this section we discuss the properties of the latitude distribution over the last three SA cycles.

A very well formed zonal structure in the MF latitude distribution has been revealed, and its temporal evolution was investigated. In the solar activity minimum the latitude distribution is almost linear, then as the activity increases three zones of alternative polarity appear in each hemisphere. They are mirror-symmetric to the equatorial plain. The polar and pre-equatorial zones have the polarity of the previous cycle with the corresponding mean MF 120 and 50-70 $McTl$. The intermediate zones occupy the latitudes from about 50 to 25 degrees in the North and in the South hemispheres during the solar activity increase. During the SA maximum the middle zone boundaries are shifted very little from the 25 degrees, instead the high latitude boundaries are moving to the poles, the

CP679, *Solar Wind Ten: Proceedings of the Tenth International Solar Wind Conference*,
edited by M. Velli, R. Bruno, and F. Malara

polar field is changing the sign, and its values continue to increase up to 120 $mcTl$ while the activity is decreasing. At the same time the amplitude of the mean field in the pre-equatorial zones increases during the maximum of the activity up to 120 $mcTl$, and then decreases slowly until it becomes almost zero inside the zone from 40 N to 40 S degrees. The latitude distribution of the magnetic field is turned to be of the opposite polarity in the SAC 22 minimum but with the same dependence on the latitude. It turns back in the next cycle, so the cycle 23 is similar to SAC 21. The most North latitude magnetic field changes from the positive value of 100-150 $mcTl$ (depends on the longitude) at the beginning of the Carrington Rotation 1642, through zero fifty rotations later (about CR 1692, during the solar activity maximum) to the negative value of -(100-200) $mcTl$ at the end of 21 cycle in the CR 1777. The high South latitude field is changing from the -100 $mcTl$ in CR 1642 through zero in CR 1696 to 150 $mcTl$ in CR 1984. The mean MF on the - 75 S degree is equal to about + 30 $mcTl$, on the + 75 N is equal to -(20-30) $mcTl$, and almost indifferent to the longitude. The yearly MF variations due to the Earth orbital motion take place on the high latitudes. Yearly MF variability on the high South latitudes are less visible than on the North.

In 22 solar activity cycle the situation is similar, while some small differences take place. For example, the MF yearly variation in the high latitude regions on the North and on the South are smaller than in the SAC 21. The 23th solar activity cycle is not yet finished, it is decreasing now. The change of the polarity took place during about CR 1966 about, 60 Carrington rotations since the beginning of the SAC 23 in CR 1906.

The latitude MF distribution is almost symmetric (with the polarity change) from the South to the North and from the increase to the decrease of the activity. In this sense we can talk about remarkable mirror-symmetry of the latitude MF distribution relatively to the equatorial plain.

Such a dynamic of the latitude distribution of the solar magnetic field is described by the following model: F=a1*(x-a5)*cos(a2*t)+a3*sin(a4*x)*sin(a2*t), where $a1$ =-8.2758 $mcTl$, $a2 = 2\pi/270$ =0.02327, $a3$ =-120 $mcTl$, $a4 = 2\pi/14.5$ =0.43332, $a5$ =14.5, x is an argument along the latitudes (the number of steps in the $arcsin(14.5-n)/15.$), and t is the time expressed in the CR rotation numbers.

The described properties of the latitude MF distribution are illustrated by the Fig. 1 with the LMF averaged over the periods of minimum, increasing, maximum and decreasing of SA in 21 and 22 cycles. The dotted lines correspond to the model described above related to the latitude distribution over the periods of SA maxima.

3. LONGITUDE DISTRIBUTION

The longitude structures are mainly formed in zone below the latitude of ± 40 degrees. Magnetic field polarity distribution along different mid-latitudes is reproduced in each rotation as it is expected due to the long living active regions. But it is interesting to mention that there are global large scale structure (dipole, quadruple) as well as small scale structure (composed of the 5, 6, 7, 8-th spherical harmonics) presented in the solar surface MF distribution over the cycle, slowly changing the polarity through the cycles. Large scale structures are better visible during the periods of low and high solar activity. An addition of higher harmonics is more significant during the intermediate periods of SA increasing and decreasing. The set of the main harmonics depends on the interval of a cycle.

The small scale longitude magnetic structures are well formed at the mid latitudes with the maximal amplitude deviations from the mean level at the latitude of 17.5 degrees about where they reach about $\pm 100 mcTl$. Latitude zone boundaries stay about at the +25 N, 0 and -25 S degree.

It is important to stress that the position of the longitude structures are relatively stable over the cycle. The polarity in longitude cells is changing to the opposite one from the minimum to the maximum of solar activity in the SAC 21. There is a correlation between the longitude deviations from the mean level from 35 to 0 degree in the North and in the South hemispheres around the Sun. The averaging over the total 21-th cycle composed of the 135 Carrington rotations shows well correlated polarity distribution along the longitudes on the different latitudes. They are so stable that even the averaging over 11 years doesn't remove them.

There is an obvious relation between the behavior of the MF in the North and in the South hemispheres during all periods of solar activity cycles. Fig. 2 illustrates 3 examples of such relation for different intervals SAC 21. There is a significant correlation during the period of decreasing activity (K_{corr}=0.8) and an anti-correlation during the maximum of activity (K_{corr}=-0.8). The intermediate picture takes place during the period of increasing activity, when for one half of the Sun there is a correlation between the North and the South hemispheres, while there is an anti-correlation for other half of the Sun. For different intervals of SAC 22 and SAC 23 the relationship between the MF in the North and the South is similar but not identical. It needs an additional investigation of the physical origin of this difference and its dependence on the rotation rate.

The models to describe the longitude and latitude distribution of the solar magnetic field and its variation in time are in progress. The best results probably could be achieved by fitting the longitude distribution to the sum

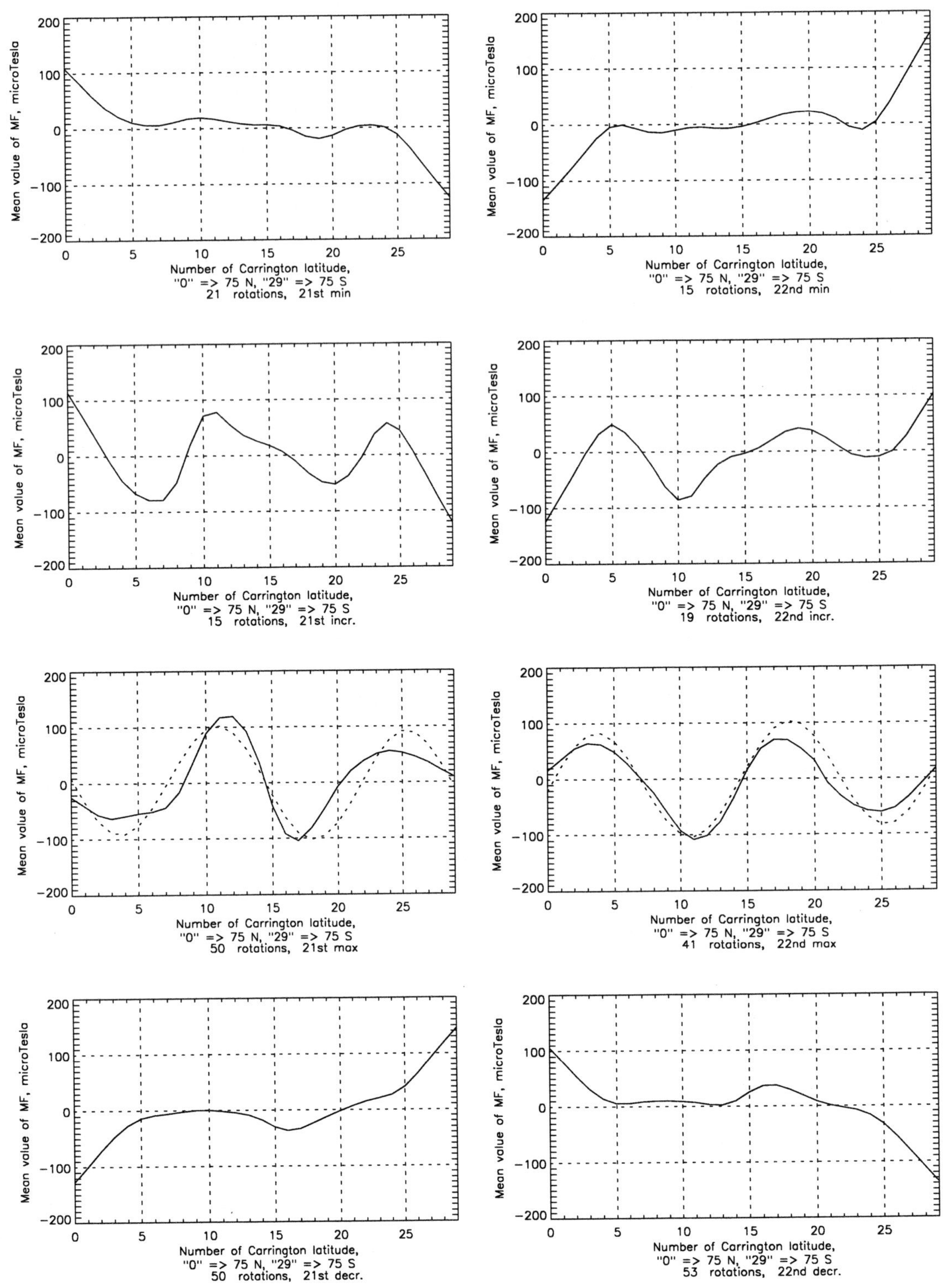

FIGURE 1. Latitude distribution of the solar magnetic field in the 21th (on the left) and in the 22th (on the right) solar cycles over the intervals of low activity, increasing activity, maximum and decreasing activity.

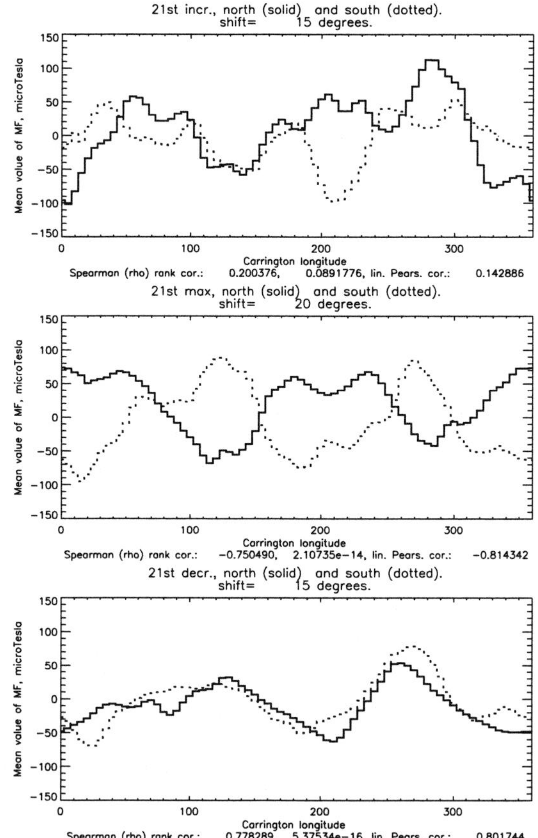

FIGURE 2. The correlations between the behavior of MF in the North and in the South hemispheres for the periods of intermediate and high activity of SAC 21.

of the three harmonics with the periods related to each other as about 1:2:4.

4. DIFFERENTIAL ROTATION

For the first stage of our analysis we assumed a rigid rotation for the MF data plotting on the solar surface coordinate system. The longitude variability is strongest over the SA cycle at the latitudes of 17-26 degrees. The longitude variability maxima lays in the North and in the South hemispheres at the 17.5 degree latitude. The MF at this latitude rotates with the Carrington rotation rate (CPR). The rotation of the MF at the higher latitudes is slower (up to the 8.5-10% at the 56 degree), and 2-3% faster at the 1.9 degree on the North and on the South.

There is the remarkable property of the appearance of the new active regions (NewARs) or new sectors of the activity. At the latitudes from 40 to 20 degrees the NewARs appear with a rotational rate 1.4±.1% faster than the CRR (the sideral period of about 25.0 ± .1 days). This NewARs period corresponds to the solar rotation at the 10 degree latitude at the 0.85±.03 solar radii [3] and very close to the main rotation rate shorter than the CRR found by many authors [4] .

The position of the longitude variability maxima doesn't coincide with the latitude variability maxima at the 10 degree latitude as it can be seen in Fig. 1. The appearance of sunspots at about 25 degrees coincides well with the position of the boundary between the zones of the opposite polarity in the latitude distribution of the MF averaged around the Sun through all the longitudes.

There is some North-South asymmetry, better visible during maxima of the solar activity in the cycles 21 and 22. This is probably attributed to the deviation of the rotation axis from the axis of the magnetic field distribution.

5. CONCLUSIONS

From our analysis of the MF distribution we can reach an important conclusion.

1. The magnetic field distribution has a very clear latitude zonal structure which is changing during the 22-year period.

The next two statements have a preliminary character.

2. Large scale (dipole and quadruple) and small scale (8,5,6,7th harmonics) structures have been revealed on the Sun during 21, 22, 23 SAC. Quadruple, dipole and octupole structures play an important role in the modulation and in the variation of the MF distribution during solar activity cycles.

3. New active regions appear with the sidereal period of about 25.0 ± .1 days. They are originated at the internal layers rotating faster than the CRR, and they have an "exit" from the solar interior on the boundary between the latitude zones of the opposite polarity on the 20-25 degrees.

REFERENCES

1. Sherrer, P. H. e. a., *Solar Phys.*, **54**, 353–361 (1977).
2. Hoeksema, J. T., and Sherrer, P. H., *Solar Phys.*, **105**, 205–211 (1986).
3. Schou, J. e. a., *ApJ*, **505**, 390–417 (1998).
4. Bumba, V., and Hejna, L., *Bull. Astron. Inst. Chechosl.*, **42**, 76–85 (1991).

Session II

Coronal Heating and the Acceleration of Solar and Solar-type Winds

Observational and theoretical constraints on the heating and acceleration of the fast solar wind

Ruth Esser*, Øystein Lie-Svendsen†, Richard Edgar* and Yao Chen*

*Harvard-Smithsonian Center for Astrophysics, 60 Garden Street, Cambridge MA 02138, USA
†Norwegian Defence Research Establishment, div. for electronics, P.O. Box 25, NO–2027 Kjeller, Norway

Abstract. Constraints on coronal plasma parameters derived from remote and in situ observations are reviewed. The coronal observations include measurements of polarized white light and the widths and intensities of spectral lines. Emphasis is placed on electron temperatures derived from these measurements. In situ observations include mass flux, velocity and ion composition. Some of these observational results are in contradiction to each other. It is discussed how these contradictions could be overcome and what it means for the physical properties of the coronal plasma. The observations will are placed in context with different theoretical models of ion formation and solar wind expansion. Emphasis is placed on the fast solar wind originating in the polar coronal holes.

CORONAL PLASMA PARAMETERS

Since the launch of the Ulysses and SOHO spacecraft a wealth of information on the conditions in the corona and in situ solar wind has become available. Of particular interest for solar wind heating and acceleration are the electron temperatures and densities derived from the Solar Ultraviolet Measurement of Emitted Radiation (SUMER) instrument, and the effective ion temperatures which includes thermal and other randomly appearing motions, as well as flow speeds derived from the Ultraviolet Coronagraph Spectrometer (UVCS) on board SOHO. The SUMER observations show that the electron temperature in the region below 1.5 R_S is less than 10^6 K (Fig. 1) [1], which is in agreement with electron temperatures previously derived from less reliable observations [e.g. 2, and references therein], and more recent results derived from emission measure distribution [3, and references therein]. UVCS observations reveal that the proton effective temperature is twice as high as the electron temperature at 1.5 R_S and keeps increasing to at least 3×10^6 K at 3 R_S (Fig. 1) [4].

The heavy ions are much hotter than the protons above 1.5 R_S [6]. The ion temperature diagnostic is based on the measurements of the line widths of selected spectral lines such as H I Ly-α 1215.67 Å, and the line pair O VI 1032 and 1037 Å. It is in principle possible to determine the electron temperature in the same way from the width of the Ly-α 1215 Å Thompson scattered component [7]. However, for most regions in the corona, in particular for the source regions of the fast solar wind, these measurements are below the sensitivity of current instruments. Even though an instrument with sufficient sensitivity has been designed (a detailed description can be found at: http://cfa-www.harvard.edu/asce/), at present the only means we have to determine the electron temperature is via line ratio diagnostic. The spectral lines used by Wilhelm et al. [1] are the two Mg IX 708 and 750 Å lines. Using two lines of the same ion has the advantage that the ratio is not sensitive to outflow speeds. The ratio might depend on non-Maxwellian tails on the

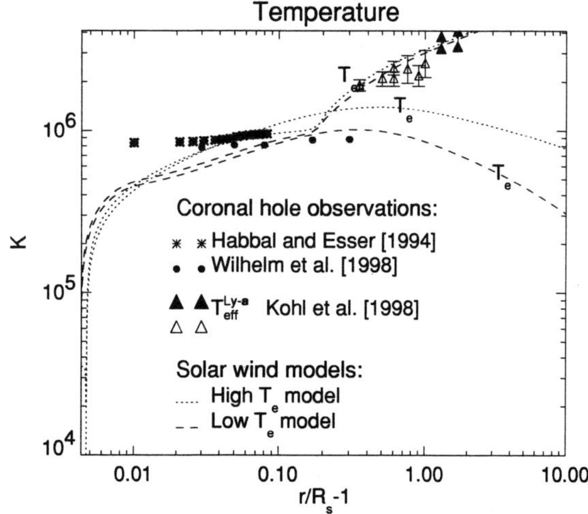

FIGURE 1. Observed electron temperatures (dots and stars), proton effective temperature derived from UVCS observations (triangles) [5, and references therein] and corresponding temperatures derived from a solar wind model for two different heat inputs (see text).

CP679, *Solar Wind Ten: Proceedings of the Tenth International Solar Wind Conference*, edited by M. Velli, R. Bruno, and F. Malara

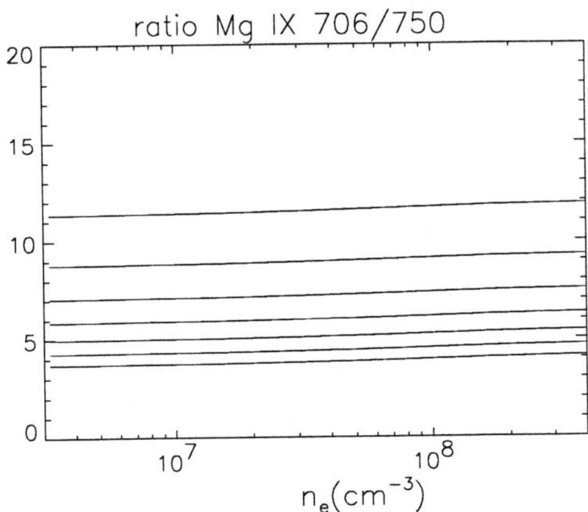

FIGURE 2. Ion ratio as a function of electron temperature when the electron density is varied from 10^6 to 4×10^8 cm^{-3}. The curves for the different densities fall almost on top of each other. The atomic data are from the CHIANTI data base [9].

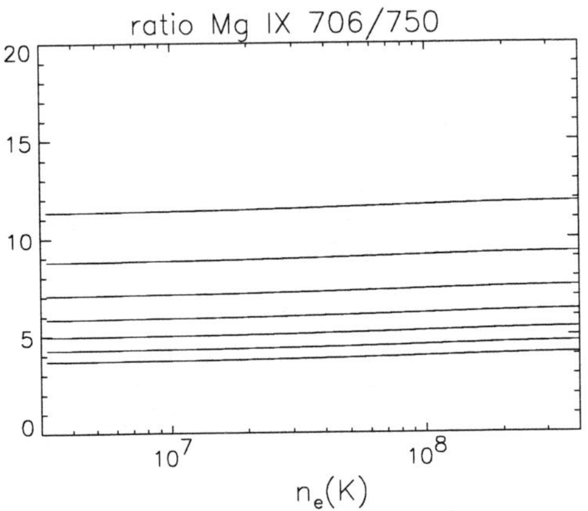

FIGURE 3. Ratio as a function of density when the electron temperature varies from 2×10^5 K (upper curve) to 3×10^6 K (lower curve).

distribution if they are present, but this sensitivity is expected to be small [8]. This Mg line ratio is relatively sensitive to electron temperature as can be seen in Fig. 2, which shows the ratio as a function of electron temperature for a range of densities that can be expected close to the coronal base, spanning from 10^6 to 4×10^8 cm^{-3}. Due to the small density dependence of this line ratio, the curves for the different densities fall almost on top of each other.

The small density dependence can also be seen in Fig. 3 which shows the Mg line ratio as a function of electron density for a number of different electron temperatures. The atomic data used in these calculations are from the CHIANTI data base [9].

The spectral line ratios observed by SUMER in a

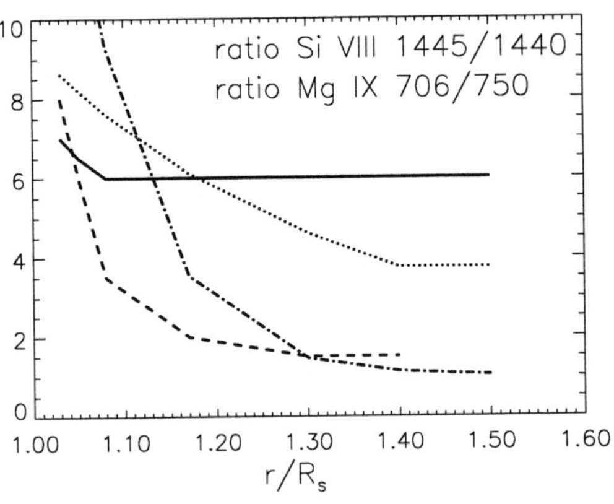

FIGURE 4. Mg (solid line) and Si (dashed line) line ratios observed with SUMER [1]. The Mg (dotted line) and Si (dash-dotted line) ratios calculated for the "freezing-in" temperature shown in Fig. 8 are also shown.

coronal dark lane which are assumed to be the source region of the high speed solar wind, are shown in Fig. 4 (solid line) [1]. The figure shows that the Mg line ratio is relatively constant over the observed distance range. A Si line ratio that is commonly used as density diagnostic is also shown (dashed line) [1].

Electron temperatures derived from models of the high speed solar wind are in agreement with temperatures derived from the SUMER observations. It has been known for quite some time that models of the high speed solar wind are in much better agreement with observed solar wind parameters such as in situ mass flux, flow speed and coronal electron densities if the coronal electron temperature in the models is low, $\leq 10^6$ K. In Fig. 1 we show two electron temperature profiles derived from a 16-moment fluid solar wind model [10]. This model has the lower boundary in the upper chromosphere which is unaffected by the energy deposition in the corona. The density and temperature structure in the transition region and corona are uniquely determined by the type and location of the energy deposition. If only the protons are heated in the corona, the electron temperature that results from such a model is of order 5×10^6 K since at coronal densities and flow speeds, the coupling between protons and electrons is not strong enough to transfer enough energy from the protons to the electrons. To increase the electron temperature to 10^6 K, a significant fraction of the energy has to be deposited directly into the electrons in the corona. To drive a high speed solar wind a total of about 4×10^5 erg cm^{-2}s^{-1} has to be deposited into the plasma above the chromosphere. In the low T_e case (Fig. 1, dashed line) 20% of that energy is deposited into the electrons in the corona. In the high T_e case (Fig. 1, dotted line) we have tried to heat the electrons as much

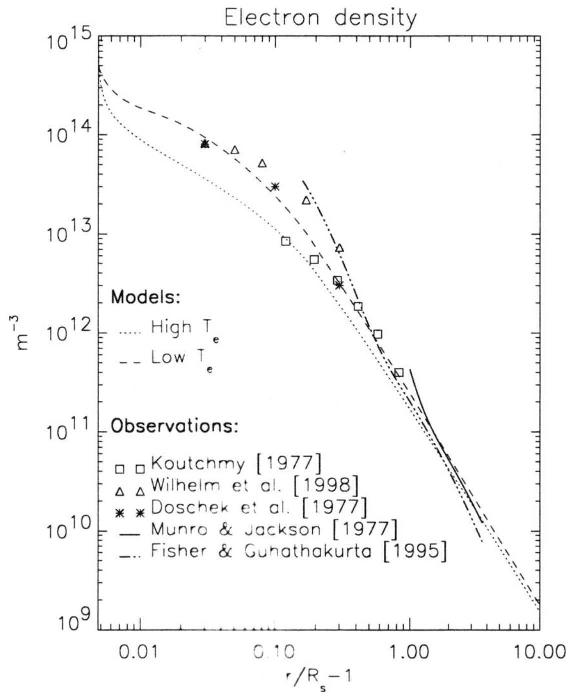

FIGURE 5. Coronal electron densities derived from observations and models. Observed densities are derived from measurements of polarized brightness of Thompson scattered white light, except densities by Wilhelm et al. [1] and Doschek et al. [11] which are derived using intensity ratios of selected spectral lines.

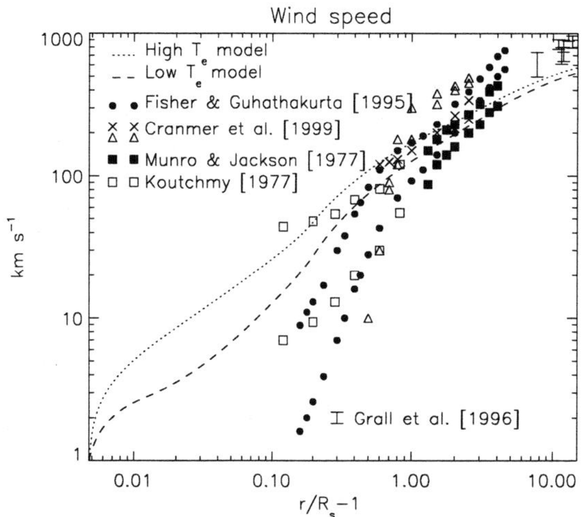

FIGURE 6. Flow speeds derived from the two theoretical models and different observational techniques (see text for details).

as possible without violating the other constraints placed by observations (e.g., mass flux, coronal density and proton temperature). To achieve an elctron temperature of 1.4×10^6 K more than 30% of the energy has to be deposited directly into the electrons. The conduction downward is significantly increased in this case and additional heating in the transition region has to decrease accordingly to lower the density and maintain a reasonable mass flux (of order $2\times\,10^8\ \mathrm{cm}^{-2}\mathrm{s}^{-1}$). The electron density in the corona is decreased to values slightly lower than observed which can be seen in Fig. 5 (dotted line). This figure shows the radial profiles of the electron density for the two theoretical models and the values derived from different coronal observations. We find that with reasonable mass fluxes, it is not possible to increase T_e even further, without violating other observed parameters (primarily density and flow speed).

The proton flow speed derived from the models is shown in Fig. 6 together with flow speeds derived from different observational constraints. Flow speeds are derived from the electron density [12] (n_e = n_p, where *p* stands for proton) assuming constant mass flux and a flow tube expansion spanning from radial (lower solid dots) to seven times faster than radially (upper solid dots). Flow speeds derived from the densities by Koutchmy [13] (open squares) and Munro and Jackson [14] (filled squares) using the same assumptions are also shown in the figure. Different Doppler dimming observations are also shown in the figure, [e.g. 6], where open triangles are flow speeds of O VI ions and crosses represent proton flow speeds. Some flow speeds derived from Interplanetary Scintillation observations are also plotted [15].

IONIZATION BALANCE CALCULATIONS

The above observational and theoretical constraints on the electron temperature in the inner corona are in contradiction to the ion charge states observed in situ in the solar wind. The in situ ion fractions are connected to the coronal plasma parameters through the ionization balance equation:

$$\frac{1}{A}\frac{\partial}{\partial r}(n_i v_i A) = n_e(n_{i-1}C_{i-1} - n_i(C_i + R_i) + n_{i+1}R_{i+1}). \tag{1}$$

Here A is the expansion of the flux tube, r the radial distance from sun center, n_i and n_e are the ion and electron number density, respectively, and v_i is the flow speed of the ion species. The ionization rate is:

$$C = 4\pi \int \sigma(v_e) v_e^3 G(v_e) dv_e \tag{2}$$

where v_e is the velocity of the electrons and G is their velocity distribution function. The recombination rates,

R, are of the same form but with a different cross section σ. If the electrons are Maxwellian distributed, then the C and R depend only on the electron temperature T_e. On the other hand if the distribution function is non-Maxwellian with a high energy tail, for example, then the exact shape of the distribution also plays a role.

This shows that the rate coefficients depend on the electron temperature as well as the details of the electron distribution function and the atomic physics. The formation of the ions, therefore, depends on the electron density, temperature, the detailed character of the distribution function, as well as the flow speeds of the individual ion species in the entire region where the ions form.

If it is assumed that the ions are static and that there are only two dominant charge states, as in the case of O^{+6} and O^{+7} in the high speed solar wind, then the ionization balance can be written in the very simplified form [e.g. 16]:

$$\frac{n_{i+1}}{n_i} = \frac{C_i + R_i}{R_{i+1}} = F(T_e). \tag{3}$$

If on the other hand v_i increases and n_e becomes very small, then for $v_i = v_{i+1}$, the ratio in Eq. (3) is constant. It is often assumed that the ions are coupled initially and then decouple, or freeze-in, instantly, such that the ion ratio measured in situ reflects the electron temperature at the point where the ions decouple.

If such simplified assumptions are not made then the set of coupled equations, Eqs. (1), has to be solved for a given $n_e(r)$ making assumptions about the ion flow speeds, $v_i(r)$, and the electron distribution function. Time dependent variations of plasma parameters that might be present in the corona, could also affect the ion formation. Since the in situ ion fractions are averaged over long time periods (see for example Ko et al. [17]), these time variations would not necessarily show up in the in situ observations. However, they should show up in the coronal observations, unless they have very peculiar time periods relative to the spectroscopic observing times.

If the ion fractions are interpreted in the traditional way, namely assuming a Maxwellian distribution of the electrons and that all ions of the same element flow at the same speed, then a coronal electron temperature can be derived. It is the temperature maximum that is the interesting physical parameter, as this determines how much of the total energy has to be deposited into the electrons in the corona (see above). It is, therefore, not particularly interesting to take just one, more or less accidental, ion ratio and determine that somewhere in the corona there is a temperature of 10^6 K. All the ions that are measured in situ should be taken into account and modeled to derive the temperature maximum [e.g. 17]. The results that are derived from such an approach should be compared to coronal plasma conditions derived from other observations (e.g., Figs. 1, 5 and 6).

Figure 7 (solid lines) shows the ion fractions measured in situ by the SWICS instrument on board ULYSSES. The ranges shown are the actual measured values plus/minus an estimated error (upper/lower solid lines). These measurements were carried out during the first polar passage of Ulysses and are described in Ko et al. [17].

SOLUTION FOR HIGH CORONAL ELECTRON TEMPERATURES

If it is assumed that the electron distribution function in the corona is Maxwellian and that all ions of the same element flow at the same speed, then solving the set of Eqs. (1) leads to a temperature of the form shown in Fig. 8 (best fit model in Ko et al. [17]). The ratios of C ions (as well as O ions, not shown in the figure) are formed at temperatures below 10^6 K, the Mg ion ratios need temperatures of order 1.4×10^6 K, and the higher charge states of Si form at a temperature of about 1.6×10^6 K. For these ratios to form in the same plasma it is necessary that the temperature has a profile with a marked and relatively narrow temperature peak. Whether a ratio freezes-in before or after the temperature peak is determined by the corresponding ionization and recombination rates. The rates of the Si ions shown in Fig. 7 are faster than O rates, for example. The Si ions couple, therefore, to larger distance where the electron density is lower and the flow speeds are higher. The difference in the rates is, however, not enough to separate the freezing-in distances of the ion ratios sufficiently. For the calculated ratios to fit the observations it is also generally necessary that the elements flow with very different flow speeds, with O and C flowing more slowly and Si and Fe flowing faster (e.g. Ko et al. [17]). In addition, in a situation where the gradients are large as in Fig. 8, the ions might pass through the plasma volume without ever coming to ionization equilibrium at the local temperature if the time an ion spends in that volume is short relative to the time it takes to ionize. In order to still ionize the same number of ions as in the equilibrium case, the electron temperature has to continue to increase above the equilibrium ionization temperature to compensate for the decrease in electron density. For example, the peak temperature has to be closer to 2×10^6 K if the Si flow speeds are close to the observed proton flow speeds.

The temperature profile shown in Fig. 8 leads to line ratios that are much steeper than the ones observed by SUMER as can be seen in Fig. 4 (dotted line). It is difficult to explain that disagreement with the uncertainty in the atomic physics as such an uncertainty is more likely to shift the curve up or down, rather than changing its gradient. For the temperature (density) diagnostic this

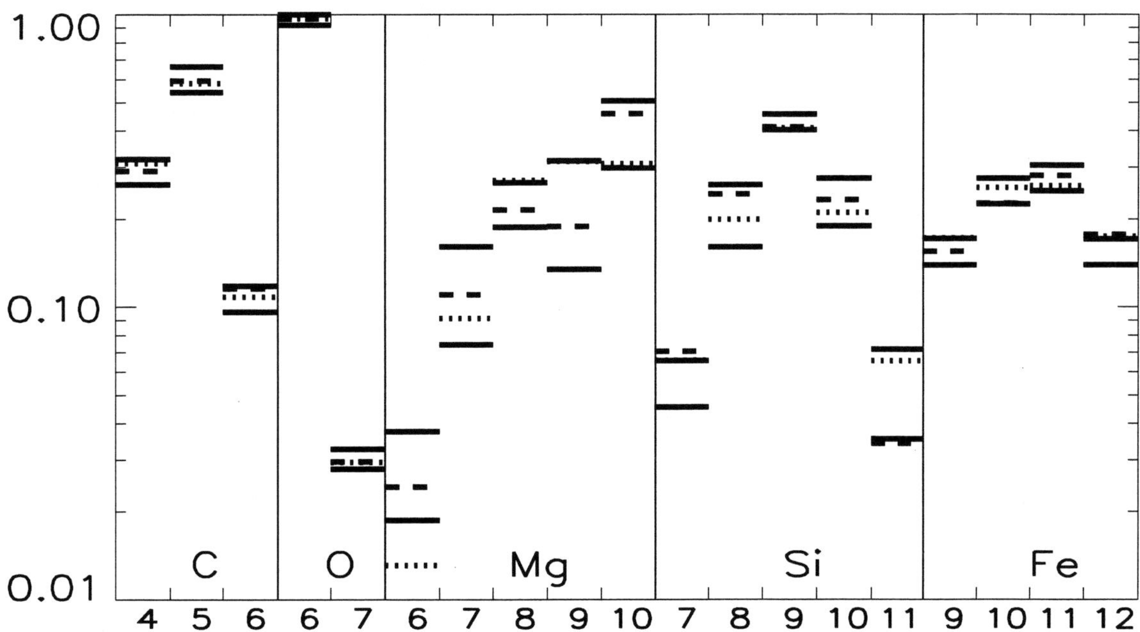

FIGURE 7. Observed ion fractions plus/minus error estimates (upper/lower solid lines) from Ko et al. [17]. Ion fractions calculated for observed coronal temperatures fit the measured values only if it is assumed that the electrons have a high energy tail already low in the corona (dotted lines) or that ions of the same element have large differential flow speeds (dashed lines). (In cases when not all four lines are shown for an ion, some of the lines fall on top of each other. For the Mg^{6+} ion, the calculated value for the non Maxwellian case is smaller than observed. However, the observed values of that ion are the least reliable of all the observations. We have, therefore, not paid much attention to that value.

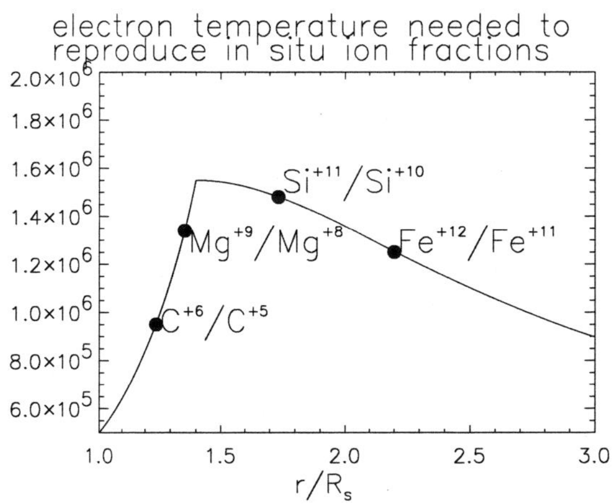

FIGURE 8. Electron temperature as derived from in situ ion charge states if the electrons are Maxwellian distributed, all ions of the same element flow with the same speed, and ion flow speeds are small compared to the proton flow speed in the coronal region where the ions form [17].

could only be the case if the ratio was very sensitive to density (temperature) and/or flow speed. Since only one ion is involved, the ratios are obviously not dependent on the flow speed. The density dependence of the Mg ratio is small as can be seen in Fig. 3. Likewise the temperature dependence of the Si ratio is also small [e.g. 1]. Due to the discrepancy between the coronal electron temperatures deduced from coronal observations and model calculations, and in situ charge state observations, it seems highly questionable at best that the in situ charge states can be the result of an electron temperature profile that is both narrowly peaked and has a high maximum value. Such a temperature profile is also hard to explain from a theoretical point of view since heat conduction, which is close to classical in that region, [18, e.g.], will be very efficient at such high temperatures and will therefore smear out large temperature gradients.

ORIGIN OF DISCREPANCY: RESULTS AND DISCUSSION

Since the parameters that are involved in determining the formation of the ion fractions in the corona are complex as was described above, there might be more than one factor giving rise to the discrepancy between coronal and in situ observations. Two possible explanations have been discussed in the literature: 1. Non-Maxwellian electron distribution functions, similar to the ones observed in situ, exist already in the ion forming region [8]; 2. Large differential flow speeds between ions of the same element are present in the corona [19]. Both, distribution functions and sets of flow speeds can be found that with a set of coronal electron core temperatures and densities, in agreement with the ones shown in Figs. 1 and 5, lead to frozen-in ion fractions inside the observational limits shown in Fig. 7 (dotted and dashed lines, respectively).

In order for the 5 elements to encounter the electron distribution function that is needed to reproduce the in situ ion fractions, the differential flow speed between elements has to be significant, as in the case where the in situ ion fractions are explained by a high, narrowly peaked electron temperature profile (see above). Since the non-Maxwellian character of the distribution function most likely increases with radial distance from the Sun (rather than decreases), the O ions have to flow fastest since they only need a small tail on the distribution, the Fe ions, on the other hand, need a larger tail, and have to flow more slowly so that they can decouple higher up in the corona where the tail is more developed. To calculate the values given in the figure, it was assumed that the halo on the distribution is fully developed at 1.4 R_S, contains 5% of the electrons, and has a temperature of 5 times the core, as is observed in situ in the solar wind (see Esser and Edgar [8] for details). In case of the differential flow speeds between ions of the same element, the ratio of adjacent flow speeds has values ranging from 2 (for O ions) to several orders of magnitude (for Si ions) (see Esser and Edgar [19] for details). However, using a multi-fluid solar wind code where continuity, momentum and energy equations were solved simultaneously (e.g. 15 sets of equations for Si), it was recently shown by Chen et al. [20] that it is impossible to produce such high differential flow speeds between ions in the regions where the ion formation takes place. As long as the ions form they couple to each other which leads to a coupling of the flow speeds as well, even if the ions are strongly preferentially heated. Large differential flow speeds can only be produced outside of the ion forming region. In case of the non-Maxwellian electron distribution, it was previously shown by Lie-Svendsen et al. [21] that halos and tails do not naturally develop in the corona. If they exist they must, therefore, be actively created by for example heating processes. The electrons must be heated in the corona in order to reach a temperature of 10^6 K. It is presently not known what that heating mechanism might be. It remains, therefore, to be seen whether this mechanism could be the origin of a halo in the electron distribution function sufficient to produce the ion fractions.

Given the complexity of the coronal plasma conditions and ion formation, additional coronal observations, particularly of the electron temperature and differential ion flow speeds, seem to be necessary. Such observations can only be carried out with spectroscopic instruments that have at least an order of magnitude increased sensitivity, such as The Advanced Spectroscopic and Coronagraphic Explorer (ASCE) mission recently worked out by Kohl et al. (http://cfa-www.harvard.edu/asce/). This type of observational approach will improve the possibility to interpret the ion fractions observed in situ. Presently we only know that they do not reflect the coronal electron temperature correctly. Progressing from there to a unique and correct prediction of coronal conditions is presently not possible.

ACKNOWLEDGMENTS

This work was supported by NASA grants NAG5-7055 and NAG5-9564, and by the Research Council of Norway under grant 136030/431. R. Esser thanks the Norwegian Defence Research Establishment (FFI) for their hospitality.

REFERENCES

1. Wilhelm, K., et al., *Astrophys. J.*, **500**, 1023 (1998).
2. Habbal, S. R., Esser, R., and Arndt, M., *Astrophys. J.*, **413**, 435 (1993).
3. Doschek, G. A., Feldman, U., Laming, J. M., Schühle, U., and Wilhelm, K., *Astrophys. J.*, **546**, 55 (2001).
4. Kohl, J. L., et al., *Astrophys. Lett.*, **501**, L133 (1998).
5. Esser, R., et al., *Astrophys. J.*, **510**, L63 (1999).
6. Cranmer, S. R., et al., *Astrophys. J.*, **511**, 481 (1999).
7. Fineschi, S., et al., in *Fifth SOHO Workshop: The Corona and Solar Wind Near Minimum Activity*, ESA, 1997, vol. SP-404.
8. Esser, R., and Edgar, R. J., *Astrophys. J.*, **532**, L71 (2000).
9. Dere, K. P., Landi, E., Mason, H. E., Monsignori-Fossi, B. M., and Young, P. R., *Astron. Astrophys.*, **125**, 149 (1997).
10. Lie-Svendsen, Ø., Leer, E., and Hansteen, V. H., *J. Geophys. Res.*, **106**, 8217 (2001).
11. Doschek, G. A., Warren, H. P., Laming, J. M., Mariska, J. T., Wilhelm, K., Lemaire, P., Schühle, U., and Moran, T. G., *Astrophys. J.*, **482**, L109 (1997).
12. Fischer, R. R., and Guhathakurta, M., *Astrophys. J.*, p. L139 (1995).
13. Koutchmy, S., *Sol. Phys.*, **51**, 399 (1977).
14. Munro, R. H., and Jackson, B. V., *Astrophys. J.*, **213**, 874 (1977).
15. Grall, R. R., et al., *Nature*, **379**, 429 (1996).
16. Hundhausen, A. J., *Coronal Expansion and Solar Wind*, Springer Verlag, New York, Heidelberg and Berlin, 1972.
17. Ko, Y.-K., Fisk, L. A., Geiss, J., Gloeckler, A., and Guhathakurta, M., *Sol. Phys.*, **171**, 345 (1997).
18. Lie-Svendsen, Ø., Holzer, T. E., and Leer, E., *Astrophys. J.*, **525**, 1056 (1999).
19. Esser, R., and Edgar, R. J., *Adv. Space Res.* (2001), accepted.
20. Chen, Y., Esser, R., and Hu, Y. Q., *Astrophys. J.* (2002), in press.
21. Lie-Svendsen, Ø., Hansteen, V. H., and Leer, E., *J. Geophys. Res.*, **102**, 4701 (1997).

Relation Between Polar Plumes and Fine Structure in the Solar Wind from Ulysses High-Latitude Observations

Yohei Yamauchi*, Steven T. Suess* and Takashi Sakurai†

*NASA/Marshall Space Flight Center, SD 50, Huntsville, AL 35812, U.S.A.
†National Astronomical Observatory of Japan, Osawa 2-21-1, Mitaka, Tokyo 181-8588, Japan

Abstract. Ulysses observations showed that pressure balance structures (PBSs) are a common feature in the high-latitude and high-speed solar winds near the solar minimum. PBSs have been hypothesized to be remnants of coronal plumes and to be related to network activity such as magnetic reconnection in the photosphere. This suggests that information on the magnetic structure of PBSs would help to study the relation between PBSs and polar plumes. We have investigated the magnetic structures of the 104 PBSs by applying a minimum variance analysis to Ulysses/Magnetometer data and by examining the pitch-angle distribution of energetic electrons measured with Ulysses/SWOOPS. We found that PBSs have relatively more tangential discontinuities rather than rotational from the minimum variance analysis and there is no difference between PBSs observed in north and south polar regions. From the analysis of energetic electron data, most PBSs also show local bi-directional electron flux or isotropic pitch-angle distribution expected in plasmoids or, less often, the distribution expected in association with current-sheet structures. This suggests the hypothesis that PBSs are generated due by network activity such as magnetic reconnection at the base of polar plumes.

INTRODUCTION

Pressure balance structures (PBSs) are intervals in which changes in the plasma and magnetic pressures balance one another while total pressure remains constant, and they permeate the high-speed solar wind [1]. Many observations and theoretical studies have suggested that PBSs are the interplanetary remnant of coronal plumes, since they have similar properties and plumes are common features in the solar corona, in particular at solar activity minimum [1, 2, 3, 4, 5, 6].

Given the inherent magnetic structure of plumes [7], it is logical to suppose that information on the magnetic structure of PBSs might be helpful in investigating the relation between PBSs and plumes in more detail. Previous studies imply that the formation of plumes is related to network activity such as magnetic reconnection, thus PBSs could have magnetic features associated with that activity (*e.g.*, plasmoids or current sheets). We investigated what kind of magnetic structure PBSs have by analyzing magnetic field data from the Ulysses/Magnetometer in north and south polar regions with a minimum variance analysis (MVA) [8]. We also examined whether bi-directional electron flux exists near magnetic discontinuities in PBSs using energetic electron data from Ulysses/SWOOPS measurements. We report results and discuss the possible relation between PBSs and polar plumes.

OBSERVATIONS AND ANALYSIS

Ulysses Observations

We identify PBSs using 1-hour averaged plasma data from Ulysses/SWOOPS [9], and 1-hour averaged magnetic data from Ulysses/Magnetometer [10] with the criteria of PBSs shown in [11]. We use 2-sec magnetic data for the MVA analysis. These data were taken in the north and south polar regions above 50° latitude. The details of the observations are shown in Table 1.

For the heat flux carrying energetic electrons, we use data from the electrostatic analyzer of Ulysses/SWOOPS. The electrostatic analyzer has 20 energy channels which are centered at a given energy from 1.69 to 814 eV with a width of about 12% of its energy. In this paper, we focus on the pitch-angle data in two energy channels at 84 and 116 eV, because the electrons in this energy range are directly associated with those of the million-degree solar corona. It should be noted that the electron data is of lower temporal resolution and precision than the magnetometer data so we regard our results here as supporting the magnetic field analysis rather than superseding or supplanting that analysis.

CP679, *Solar Wind Ten: Proceedings of the Tenth International Solar Wind Conference*, edited by M. Velli, R. Bruno, and F. Malara

TABLE 1. Parameters of Ulysses high-latitude observations

	Period	Latitude (deg)	Distance (AU)
North Pole	May 11, 1995 – Jan 23, 1996	50.0 – 80.5	1.50 – 3.19
South Pole	Jan 19, 1994 – Dec 21, 1994	-50.0 – -80.5	3.74 – 1.62

Analysis of magnetic data with MVA

We have investigated the magnetic structures inside of PBSs using 2-sec magnetic field data with an MVA, a method to derive a normal vector, n, to the plane of a magnetic directional discontinuity by computing the minimum value of the standard deviation, σ_{MVA}, of the magnetic field component in that direction [8]

$$\sigma_{MVA}^2 = \frac{1}{N}\sum_{i=1}^{N}[B_i \cdot n - \langle B\rangle \cdot n]^2, \tag{1}$$

where B_i is i-th vector field measurement and $\langle B\rangle$ is the average vector, $(1/N)\sum_{i=1}^{N} B_i$. Minimizing the value of the standard deviation is equivalent to finding the smallest eigenvalue, λ_{min}, of the covariant matrix although one derives three eigenvalues (λ_{min}, λ_{int}, λ_{max}) from Eq.(1), where $\lambda_{min} < \lambda_{int} < \lambda_{max}$. In this paper, we use a cut-off ratio of the intermediate to the minimum eigenvalues $\lambda_{int}/\lambda_{min} < 2$ to define a discontinuity.

The parameters $B \cdot n/|B|$ and $\Delta|B|/|B|$ are used to determine whether each event is a rotational or tangential discontinuity with the following event identification criteria [*e.g.*, [12]]:

Rotational (RDs): $B \cdot n/|B| \geq 0.4$, $\Delta|B|/|B| < 0.2$
Tangential (TDs): $B \cdot n/|B| < 0.4$, $\Delta|B|/|B| \geq 0.2$
Either (EDs): $B \cdot n/|B| < 0.4$, $\Delta|B|/|B| < 0.2$
Neither (NDs): $B \cdot n/|B| \geq 0.4$, $\Delta|B|/|B| \geq 0.2$

Here, NDs are inconsistent with RDs or TDs, while EDs could be either in principle.

Electron Pitch-Angle Distribution

We use energetic electron data to determine whether PBSs contain magnetic structures like plasmoids or current sheets, because the electron pitch-angle distribution depends on the configuration of the interplanetary magnetic field directly. The limitation with this is the low temporal resolution of SWOOPS electron data (~ 10 min or more).

The pitch angle θ is defined as

$$\theta = cos^{-1}\frac{B \cdot v}{|B||v|} = tan^{-1}\frac{v_\perp}{v_\parallel}, \tag{2}$$

where B is the magnetic field vector, and $v_\perp$ and $v_\parallel$ are components of solar wind velocity v perpendicular and parallel to B, respectively. Figure 1 shows three cases of magnetic structures which PBSs might have and their ideal influence on the electron pitch angle distributions; (a) is the case of a plasmoid, (b) a current sheet, and (c) a magnetic field bent by Alfvénic fluctuations. The figure assumes that the Sun is downward and the magnetic field at polar regions is positive so that the electron heat flux goes upward. In the first case, electron pitch angles show a bi-directional distribution at $\theta = 0°$ and $180°$ when Ulysses goes across the plasmoid. Or, the pitch angles may be nearly isotropic, since Ulysses observes the solar wind at far from the Sun (~ 2 – 3AU) where the scattering, for examples, due to the collisions among circulating electrons or wave-particle interactions could make the pitch-angle distribution isotropic rather than bi-directional. In the second case, the pitch angle θ changes from 0° (180°) to 180° (0°), and then it goes back to 0° (180°) as Ulysses crosses a pair of current sheets extending back to the Sun. In the last case, the pitch angle does not change even though Ulysses observes magnetic polarity reversals. This is because the direction of electron flux against local magnetic field never changes, *i.e.*, the pitch angle is a constant 0° or 180°.

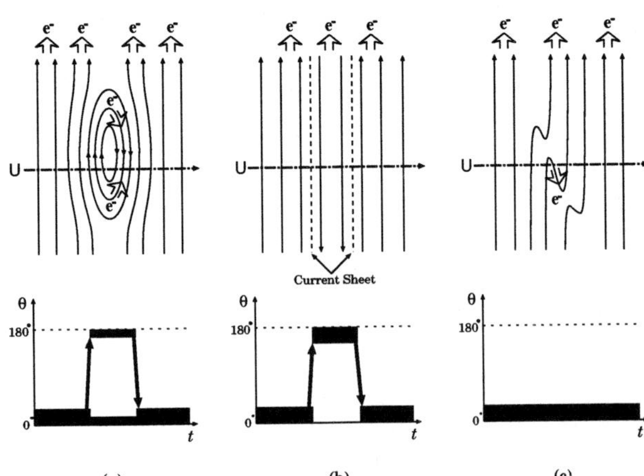

FIGURE 1. Relation between the magnetic structure and electron pitch-angle distribution: Top panels show magnetic structures which PBSs would have in three cases (a) plasmoid, (b) current sheets, and (c) Alfvénic fluctuations. The Sun is downward in this case. Bottom panels show the time variation of electron pitch-angle distribution in PBSs which Ulysses ("U") would measure if the spacecraft passed through each structure along the dash-dotted line.

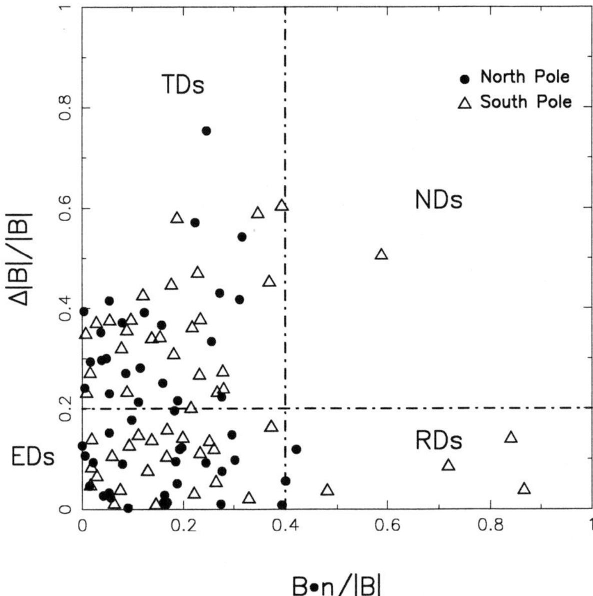

FIGURE 2. Scatter plot of the normal field component and relative field magnitude across PBSs from Ulysses observations in the north (•) and south (△) polar regions.

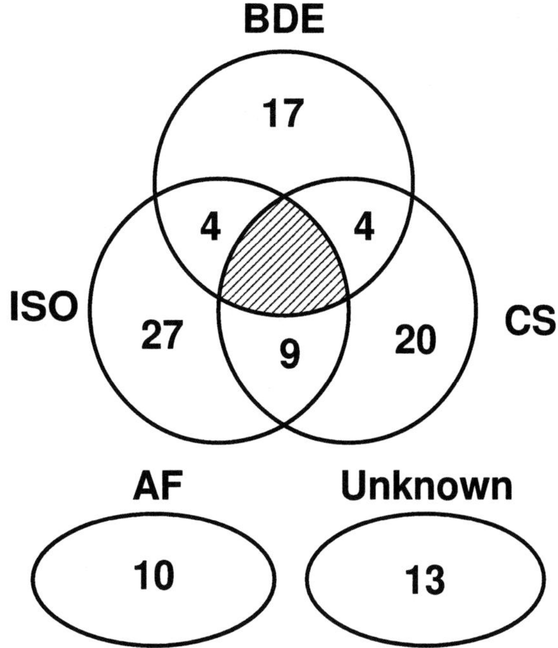

FIGURE 3. Distribution of events identified as bi-directional electron (BDE) flow, isotropic distribution (ISO), current sheet (CS), Alfvénic fluctuations (AF), or unknown case in the case of PBSs.

RESULTS

We have identified 51 PBSs from Ulysses north-polar observations and 53 PBSs from south observations, respectively. Figure 2 shows the results of the MVA analysis. The figure is a scatter plot of the normal field component and relative field magnitude across discontinuities in PBSs. The averaged thickness of the discontinuities is 52.7 ± 44.0 secs. *Tsurutani and Smith*[13] report that the width of discontinuities at 5 AU is $5-10$ times as thick as that at 1AU which is typically $\sim 2-3$ sec, and the radial distance range of Ulysses observations in our analysis period is $1.50-3.74$ AU, Therefore, the averaged thickness of discontinuities we obtained is reasonable. In Figure 2, there is no difference between PBSs observed in the north and south polar regions. The percentages that we find for RDs:TDs:EDs:NDs in PBSs is 6%:46%:47%:1%. Thus, there is a preponderance of TDs relative to RDs. This indicates that the structures of discontinuities contained in PBSs appear to be like current sheets or plasmoids.

Next, we have investigated energetic electron pitch-angle distributions in 104 PBSs. Figure 3 gives the overall distribution of the number of events identified as each case. The figure shows that 17 events were bi-directional electron (BDE) flow, 27 were isotropic distribution (ISO), 20 were current sheet (CS), and 10 were Alfvénic fluctuations (AF). We also have thirteen unknown events due to the limited time resolution of the pitch-angle data; there is only one pitch angle data point in 10 to 40 minutes. The data is therefore subject to aliasing and some of our examples may thus be misleading. This is the main reason why we have tried to examine a large number of examples. Crossing areas of BDE and ISO, BDE and CS, and ISO and CS in Figure 3 indicate that it is sometimes hard to identify the events in the areas to one of two cases and this is probably the reason. However, since 81 events are identified as BDE or ISO or CS, we conclude that PBSs usually contain the magnetic structure of plasmoids or current sheets. The results support the idea that PBSs are the remnants of polar plumes and are created by magnetic reconnection due to network activity in the photosphere of the Sun.

DISCUSSION

We have examined magnetic structures of 104 PBS with a minimum variance analysis using magnetic field data from Ulysses/MAG in the south polar regions in 1994 and north in 1995. We find that PBSs have tangential discontinuities, in preference to rotational. Further, we have examined energetic electron pitch-angle data taken by Ulysses/SWOOPS and found that most PBSs show the bi-directional or isotropic flux of plasmoids or flux associated with current sheets.

Thieme et al. [2, 3] found from Helios observations that the angular size of PBSs is ~ 2 degree (≈ 30000km as projected onto the Sun), which is the same angular size as network cells. They therefore suggested that PBSs are associated with open magnetic flux tubes originating from the network. Their definition of PBSs was when the gas and magnetic pressures anticorrelate while total pressure is almost constant. *Reisenfeld et al.* [6] suggested that the solar origin for PBSs appear to be polar plumes, since they found that the plasma beta and the helium abundance inside of PBSs correlate with each other and since solar wind abundances are fixed in the chromosphere and transition region. Therefore, our PBS criteria require the enhancement of the plasma beta and the helium abundance in PBSs as well, thus they are more restrictive and the PBSs we identified are a subset of those by *Thieme et al.* [2, 3]. Since polar plumes are the part of the structures on the network in the coronal holes and since most PBSs apparently show electron flux associated with plasmoids or current sheets, this supports the suggestion that PBSs are the remnants of the network activity (*e.g.*, magnetic reconnection) at the base of polar plumes. However, since plumes have a very small filling factor (~ 1-3%) in coronal holes, the possibility remains PBSs also originate elsewhere from the network activity.

In the model of *Yamauchi et al.* [11], which was based on the analysis of Ulysses magnetic data and previous studies, a local loop created by magnetic reconnection in a polar plume is expected to expand outward to the corona and form a pair of current sheets. The current sheets may reconnect with the surrounding flux and create plasmoids. The fraction of BDE and ISO is greater than CS as shown in Figure 3. This indicates that plasmoids are formed preferentially. Therefore our results suggest that the expanding loop is subject to magnetic reconnection before forming a pair of current sheets and a plasmoid is created, although only some loops expand to interplanetary space without the reconnection and form the current sheets. This process could occur anywhere in the network but plumes are known to have a favorable magnetic field for this to happen [14].

ACKNOWLEDGMENTS

This work was performed while Y. Yamauchi held a National Research Council NASA/MSFC Research Associateship. We are grateful to A. Balogh and G. H. Jones for preparation of 2-sec magnetometer data. We also thank J. T. Gosling and R. Skoug for preparing electron pitch-angle data. STS received support from the Ulysses/SWOOPS experiment.

REFERENCES

1. McComas, D. J., G. W. Hoogeveen, J. T. Gosling, J. L. Phillips, M. Neugebauer, A. Balogh, and R. Forsyth, Ulysses observations of pressure-balance structures in the polar solar wind, *Astron. and Astrophys.*, *316*, 368-373, 1996.
2. Thieme, K. M., E. Marsch, and R. Schwenn, Relationship between structures in the solar wind and their source regions in the corona, *in Proceedings Sixth International Solar Wind Conference*, edited by V. J. Pizzo, T. E. Holzer, and D. G. Sime, pp.317-321, *Tech. Note NCAR/TN-306 + Proc.*, Natl. Center for Atoms. Res., Boulder, Colo., 1988.
3. Thieme,K. M., E. Marsch, and R. Schwenn, Special structures in high speed streams as signatures of fine structures in coronal holes, *Ann. Geophysi.*, *8*, 713-724, 1990.
4. Velli, M., S. R. Habbal, and R. Esser, Coronal plumes and fine scale structure in high speed solar wind streams, *Space Sco. Rev.*, *70*, 391, 1994.
5. Casalbuoni, S., L. Del Zanna, S. R. Habbal, and M. Velli, Coronal plumes and the expansion of pressure-balanced structures in the fast solar wind, *J. Geophys. Res.*, *104*, 9947-9961, 1999.
6. Reisenfeld, D. B., D. J. McComas, and J. T. Steinberg, Evidence of a solar origin for pressure balance structures in the high-latitude solar wind, *Geophys. Res. Lett.*, *26*, 1805-1808, 1999.
7. Suess, S. T., A.-H. Wang, I. Cnseru, G. Poletto, and S. T. Wu, The geometric spreading of coronal plumes and coronal holes, *Solar Phys.*, *180*, 231-246, 1998.
8. Sonnerup, B. U. Ö., and L. J. Cahill, Magnetopause structure and attitude from Explorer 12 observations, *J. Geophys. Res.*, *72*, 171-183, 1967.
9. Bame, S. J., D. J. McComas, B. L. Barraclough, J. L. Phillips, K. J. Sofaly, J. C. Chavez, B. E. Goldstein, and R. K. Sakurai, The Ulysses solar wind plasma experiment, *Astron. and Astrophys., Suppl. Ser.*, *92*, 237-265, 1992.
10. Balogh, A., T. J. Beek, R. J. Forsyth, P. C. Hedgecock, R. J. Marquedant, E. J. Smith, D. J. Southwood, and B. T. Tsurutani, The magnetic field investigation on the Ulysses mission: Instrumentation and preliminary scientific results, *Astron. and Astrophys., Suppl. Ser.*, *92*, 221-236, 1992.
11. Yamauchi, Y., S. T. Suess, and T. Sakurai, Relation between pressure balance structures and polar plumes from Ulysses high latitude observations, *Geophys. Res. Lett.*, in press, 2002.
12. Neugebauer, M., D. R. Clay, B. E. Goldstein, B. T. Tsurutani, and D. Zwickl, A reexamination of rotational and tangential discontinuities in the solar wind, *J. Geophys. Res.*, *89*, 5395-5408, 1984.
13. Tsurutani, B. T., and E. J. Smith, Interplanetary discontinuities: temporal variations and the radial gradient from 1 to 8.5 AU, *J. Geophys. Res.*, *84*, 2773-2787, 1979.
14. Wang, Y.-M., Network activity and the evaporative formation of polar plumes, *Astrophys J.*, *501*, L145-L150, 1998.

Comparison of solar wind driving mechanisms: ion cyclotron resonance versus kinetic suprathermal electron effects

Sunny W. Y. Tam* and Tom Chang*

*Center for Space Research, Massachusetts Institute of Technology, Cambridge, MA 02139, USA

Abstract. The combined kinetic effects of two possible solar wind driving mechanisms, ion cyclotron resonance and suprathermal electrons, have been studied in the literature [1]. However, the individual contribution by these two mechanisms was unclear. We compare the two effects in the fast solar wind. Our basic model follows the global kinetic evolution of the solar wind under the influence of ion cyclotron resonance, while taking into account Coulomb collisions, and the ambipolar electric field that is consistent with the particle distributions themselves. The kinetic effects associated with the suprathermal electrons can be included in the model as an option. By comparing our results with and without this option, we conclude that, without considering any wave-particle interactions involving the electrons, the kinetic effects of the suprathermal electrons are relative insignificant in the presence of ion cyclotron resonance in terms of driving the solar wind.

INTRODUCTION

Kinetic ion cyclotron resonance has been shown to accelerate the solar wind ions to the observed high-speed range [1, 2]. Qualitative features associated with the resonance are consistent with observations. For example, the resonance involving sunward propagating electromagnetic waves was shown to produce double-peaked proton velocity distributions [1], which had occasionally been observed [3]. In addition, the cyclotron resonance based on observed power spectra [4] was shown to preferentially accelerate the alpha particles over the protons [1], in agreement with the observations by the Helios, Ulysses, and WIND spacecraft [5, 6, 7]. These observed power spectra may be produced by the small-scale reconnections near the coronal region. Recently, Chang [8] has suggested that such type of intermittent turbulence may be associated with the "complexity" generated by the sporadic localized mergings and interactions of the coherent magnetic structures near the coronal holes.

Another mechanism that may accelerate the solar wind is due to kinetic suprathermal electron effects (KSEE). Due to the velocity-dependence of the Coulomb collisional depth and the global kinetic nature of the solar wind flow, suprathermal tails may form in the electron distributions, giving rise to an anomalous outward electron heat flux [9]. The idea of this "velocity filtration effect" (VFE) was further pursued by Olbert [10], who suggested that the heat flux contribution by the suprathermal electrons may drive the solar wind. Such an idea has been applied to the ionospheric polar wind with photoelectrons playing the role of the suprathermal population [11, 12], and has successfully addressed various satellite observations. It has been shown that in the absence of wave-particle interactions, the suprathermal electron population can increase the ambipolar electric field, leading to higher ion velocities in the polar wind. Because the ionospheric polar wind and the solar wind are both outflows along open magnetic field lines, it is worthwhile to study KSEE in the solar wind.

The combined kinetic effects of ion cyclotron resonance and the suprathermal electron population have been considered by Tam and Chang [1]. The study associated the ion resonance with some of the solar wind observations, as discussed earlier. The bulk acceleration of the solar wind demonstrated in the study, however, consisted of contributions by both the ion resonance and the suprathermal electron effects. It was therefore unclear to what extent each of the two acceleration mechanisms contributes to the driving of the solar wind. Our goal in this study is to compare the relative importance of the kinetic effects due to suprathermal electrons and ion cyclotron resonance, and to determine which of the mechanisms is mostly responsible for driving the solar wind.

MODEL

Our model is adapted from a self-consistent hybrid model for the ionospheric polar wind [11, 12]. It is based on an iterative scheme between fluid and kinetic calculations. A set of fluid equations determines the ambipolar electric field and the properties of the bulk thermal electrons, whose distributions are assumed to be in a drifting

CP679, *Solar Wind Ten: Proceedings of the Tenth International Solar Wind Conference*, edited by M. Velli, R. Bruno, and F. Malara

Maxwellian. To solve the fluid equations, we impose the quasi-neutrality and current-free constraints for the solar wind, and make use of the results from the kinetic calculations in the model. The kinetic part of the model consists of multiple components, each describing the global evolution of a particle component along the solar wind. In the basic version of the model, the kinetic approach is applied only to the protons and alpha particles. The distributions of these ions evolve under the influence of Coulomb collisions, an ambipolar electric field, and cyclotron resonance. The results of the kinetic calculations are coupled with the fluid equations. This ensures the consistency of the ambipolar electric field with the particle distributions when an iteration between the fluid and kinetic parts of the model converges. In addition, a convergence of the results enables our kinetic calculations to correctly take into account the Coulomb collisions for the ions, including those among the same ion species.

To take into account the effects of cyclotron resonance for a given ion species, we incorporate quasilinear diffusion operators with coefficients $D_{\parallel}$ and $D_{\perp}$ into the steady-state collisional kinetic equations for the species:

$$\left[v_{\parallel}\frac{\partial}{\partial s}-\left(g-\frac{q}{m}E_{\parallel}\right)\frac{\partial}{\partial v_{\parallel}}-v_{\perp}{}^{2}\frac{B'}{2B}\left(\frac{\partial}{\partial v_{\parallel}}-\frac{v_{\parallel}}{v_{\perp}}\frac{\partial}{\partial v_{\perp}}\right)\right]f$$
$$=Cf+\left[\frac{\partial}{\partial v_{\parallel}}D_{\parallel}\frac{\partial}{\partial v_{\parallel}}+\frac{1}{v_{\perp}}\frac{\partial}{\partial v_{\perp}}\left(v_{\perp}D_{\perp}\frac{\partial}{\partial v_{\perp}}\right)\right]f, \quad (1)$$

where s is the distance along the radial magnetic field line, $f(s,v_{\parallel},v_{\perp})$ is the distribution function for the species, q and m are the electric charge and mass respectively, $E_{\parallel}$ is the field-aligned ambipolar electric field, g is the gravitational acceleration, B is the magnetic field, $B' \equiv dB/ds$, and C is a Coulomb collisional operator. The expressions for the diffusion coefficients are:

$$D_{\parallel}=\eta\frac{q^2}{4m^2}\int d\omega\, v_{\perp}{}^{2}P_C(\omega/2\pi)\,\delta(\omega-kv_{\parallel}-\Omega), \quad (2)$$

$$D_{\perp}=\eta\frac{q^2}{4m^2}\int d\omega\,(\Omega/k)^2P_C(\omega/2\pi)\,\delta(\omega-kv_{\parallel}-\Omega), \quad (3)$$

where Ω is the gyrofrequency of the species, ω and k are the frequency and wavevector for the resonance with the individual ion, and are related by the cold plasma dispersion relation for left-hand polarized waves. The argument of the δ-function corresponds to the resonance condition. P_C is the magnetic field wave power, based on an interpolation/extrapolation scheme of the Helios measurements [4] and taking into account both inward and outward propagation. The assumptions on P_C are discussed in more detail in Tam and Chang [1]. Lastly,

$$\eta=\eta_0\exp[(1-r)/0.5] \quad (4)$$

is an efficiency factor that adjusts the available wave power, and depends on the heliocentric distance r (in the unit of $R_{\odot}$). The reasons for these adjustments have been discussed in [1], one of those being consistent with a recent study of wave dissipation near the corona [13].

Equation (1) is solved with a Monte Carlo technique. This technique for describing ion resonant heating was originally applied to the auroral region by Retterer et al. [14]. With this technique, our model is able to incorporate the resonant effect of a given wave spectrum into an otherwise self-consistent solar wind description.

Up to this point, we have described the basic constituents of our model. This model enables us to follow the global evolution of the ion distributions along the solar wind flow, including the effect of cyclotron resonance. However, it does not take into account KSEE. To do so, we simply add another kinetic component into the model to describe the suprathermal electron population. The kinetic equation for the suprathermal electrons is similar to Eq. (1), but without the terms that represent wave-particle interactions. The results of this kinetic component, including the heat flux contribution, are coupled into the fluid part of the model. The suprathermal electrons in our model comprises the tail portion of the electron Maxwellian distribution at 1 $R_{\odot}$, the lower boundary of our model. They satisfy the criteria at 1 $R_{\odot}$: $(1/2)m_e v^2 > 8T_{e0}$ and $v_{\parallel} > 0$, where m_e is the electron mass, and T_{e0} is the thermal electron temperature. Because the suprathermal electrons originally constitute the high-energy portion of a Maxwellian, rather than a distribution with a significantly enhanced tail, and because we do not consider any interactions between the electrons and the waves, the KSEE in our calculations is essentially the VFE proposed by Scudder and Olbert [9]. By comparing solutions generated with and without the optional suprathermal electron component being included, we can identify the contribution by the VFE in driving the solar wind.

COMPARISON OF SOLAR WIND DRIVING MECHANISMS

We have generated two solar wind solutions for the range between 1 $R_{\odot}$ and 1 AU with identical parameters and boundary conditions, one with KSEE, and the other without. Boundary and initial conditions are imposed at 1 $R_{\odot}$: the boundary thermal electron temperature is 100 eV; initial distributions for the ion species are the upper halves of Maxwellian distributions, whose temperatures are also 100 eV; the densities for the protons and alpha particles are respectively 4.2×10^7 and 4.0×10^6 cm^{-3}. The parameter η_0 that characterizes the strength of the available wave power is 0.016.

A noticeable difference between the two solutions is the shape of the total electron distributions. Their reduced distributions in the parallel direction are shown in Fig. 1. We see that the total electron distribution deviates from a Maxwellian when KSEE is included. The formation of a significant tail in the outward portion of the distribution verifies the idea of VFE [9].

The suprathermal electron population influences the ion species mainly through the effect of its heat flux on the ambipolar electric field. With a Maxwellian approximation for the entire electron population, the overall electric potential drop from 1 $R_\odot$ to 1 AU is about 700 V. The potential difference increases by about 70 V when KSEE is taken into account. Such an increase suggests that VFE can accelerate the ions in the solar wind.

To evaluate the significance of the VFE as a solar wind driving mechanism, we should consider the energy input to the ions. Other physical processes that may deposit energy to the solar wind ions are cyclotron resonance and Coulomb collisions. In fact, like VFE, these other two processes also indirectly affect the ion energy through their influence on the ambipolar electric field. Whether an ion can reach large radial distances in the solar wind depends whether the overall kinetic energy it gains is high enough to overcome the gravitational potential. Therefore, we evaluate the significance of a physical mechanism in driving the solar wind by comparing its contribution to the energy input for the ions with the contributions from all the physical processes combined. Because we are interested in comparing only the processes that may drive the solar wind, we shall exclude the gravitational force in our energy consideration. After all, the gravitational potential profile is invariant under all solar wind conditions, and is independent of the presence of other physical processes. Hence, we define the following

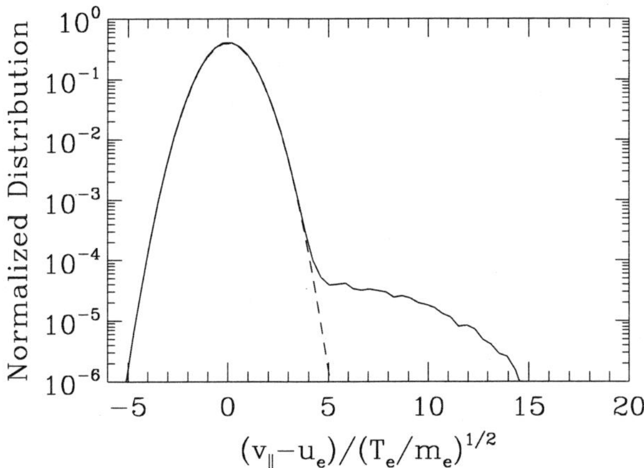

FIGURE 1. Reduced electron distributions in the parallel direction at 0.1 AU. Solid: with kinetic suprathermal electron effects; dashed: based on a Maxwellian approximation.

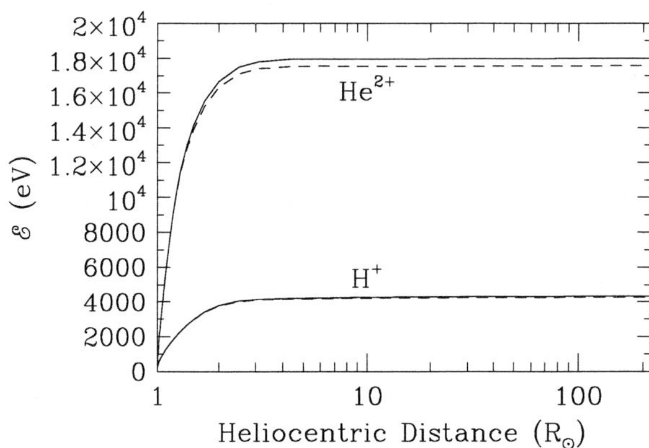

FIGURE 2. $\mathcal{E}$ for protons and alpha particles. Solid: with kinetic suprathermal electron effects; dashed: based on a Maxwellian approximation.

quantity for each ion species for our purpose of energy comparison:

$\mathcal{E} \equiv$ average kinetic energy +
gravitation potential energy per ion.

For a given ion species in our model, its increase in $\mathcal{E}$ along the solar wind flow is due to a combination of energy input from the ambipolar electric field, ion cyclotron resonance, and Coulomb collisions with other species. In particular, the ambipolar electric field also reflects the influence due to ion cyclotron resonance, and Coulomb collisions, and, in the case where KSEE is included, the VFE as well. The profiles of $\mathcal{E}$ for the protons and alpha particles in our two solar wind solutions are shown in Fig. 2. It appears from the figure that the difference in $\mathcal{E}$ is very small between the two solutions. Indeed, we find that with KSEE taken into account, the increase in $\mathcal{E}$ for the protons from 1 $R_\odot$ to 1 AU is only 1.5% larger than in the case of a Maxwellian approximation. For the alpha particles, the corresponding difference is 2.4%.

The influence by the KSEE on the ion outflow velocities can be seen in Fig. 3. The difference due to VFE is minimal, considering that the kinetic effect only increases the proton speed by 1.5% at 1 AU, and the alpha particle speed by 2.3%. Note that even under a Maxwellian approximation for the overall electron distributions, the ion speeds at 1 AU are as high as 650 km/s in the solution. Thus, the available wave power in these calculations seems to be able to drive the solar wind to the high-speed range. For such a strong wave-driven solar wind, it is clear from our results that VFE plays an insignificant role in the driving and acceleration of the solar wind.

Recently, Vocks and Marsch [15] have introduced a semi-kinetic solar wind model to describe the effects of strong ion cyclotron resonant heating. Like this study,

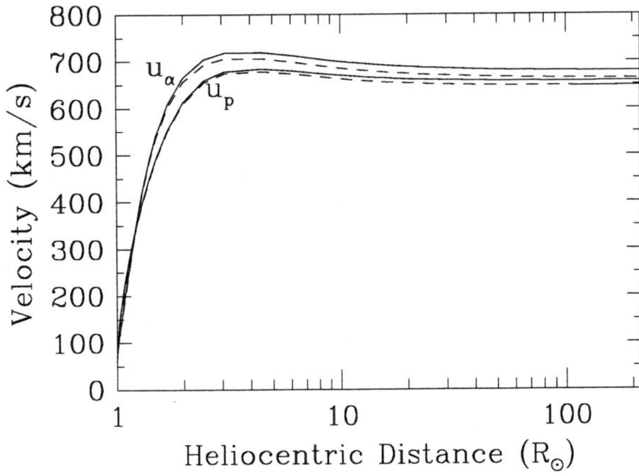

FIGURE 3. Profiles of the proton (u_p) and alpha particle (u_α) outflow velocities. Solid: with kinetic suprathermal electron effects; dashed: based on a Maxwellian approximation.

their model describes the wave-particle interaction in the framework of quasilinear theory. However, the treatment of the particles is different in the two models. Their model is based on reduced ion distributions, massless electron fluid, and an assumed electron temperature profile. With the approximations on the electrons, the model does not take into account KSEE. However, as we have shown that the VFE associated with the suprathermal electrons is negligible when strong ion resonant heating is present, the assumptions in [15] regarding the electrons seem to be reasonable approximations.

KSEE was proposed to be a solar wind acceleration mechanism under the assumption that no wave-particle interactions were present [10]. Therefore, it is worthwhile for us to investigate the VFE in the scenario of a weak wave-driven solar wind, in which case, the available wave power for ion resonance is not sufficient to drive the outflow to the high-speed velocity range. We reduce η_0 to 0.0075, less than half of the strong wave-driven case. With the reduced wave power, and without KSEE, the velocities of both ion species plateau at about 390 km/s. The inclusion of VFE increases the proton speed by only 2.5%, and the alpha particle speed by only 2.8%. As for the contribution in terms of driving the solar wind, VFE leads to an increase of less than 2% in the total energy input to the ions. Therefore, even in a weak wave-driven scenario, VFE is negligible in terms of driving and accelerating the solar wind.

We should point out, however, that KSEE can be more significant if the tail portion of the solar wind electron distributions is more enhanced (see e.g. [16]). A physical mechanism that can enhance this non-thermal feature is wave-particle interaction. One of our previous studies [17] showed that electron cyclotron resonant heating may increase the proton velocity by 16% in a strong wave-driven solar wind. Thus, even though ion cyclotron resonance appears to be the dominant solar wind driving mechanism, it is possible that KSEE can make a considerable contribution to the acceleration of the solar wind.

Epilogue. After submission of the manuscript, the authors realized that the modeling technique discussed here seems to satisfy the criteria suggested by Meyer-Vernet et al. [18] for a proper treatment of the overall electron population.

ACKNOWLEDGMENTS

This work is partially supported by AFOSR, NASA, and NSF.

REFERENCES

1. Tam, S. W. Y., and Chang, T., *Geophys. Res. Lett.*, **26**, 3189–3192 (1999).
2. Isenberg, P. A., Lee, M. A., and Hollweg, J. V., *Solar Phys.*, **193**, 247–257 (2000).
3. Marsch, E., Mühlhäuser, K.-H., Schwenn, R., Rosenbauer, H., Pilipp, W., and Neubauer, F. M., *J. Geophys. Res.*, **87**, 52–72 (1982).
4. Bavassano, B., Dobrowolny, M., Mariani, F., and Ness, N. F., *J. Geophys. Res.*, **87**, 3617–3622 (1982).
5. Marsch, E., Mühlhäuser, K.-H., Rosenbauer, H., Schwenn, R., and Neubauer, F. M., *J. Geophys. Res.*, **87**, 35–51 (1982).
6. Feldman, W. C., Barraclough, B. L., Phillips, J. L., and Wang, Y.-M., *Astron. Astrophys.*, **316**, 355–367 (1996).
7. Steinberg, J. T., Lazarus, A. J., Ogilvie, K. W., Lepping, R., and Byrnes, J., *Geophys. Res. Lett.*, **23**, 1183–1186 (1996).
8. Chang, T., "'Complexity' induced plasma turbulence in coronal holes and solar wind" (2002), in the proceedings.
9. Scudder, J. D., and Olbert, S., *J. Geophys. Res.*, **84**, 2755–2772 (1979).
10. Olbert, S., *NASA Conf. Publ.*, **2280**, 149–159 (1982).
11. Tam, S. W. Y., Yasseen, F., Chang, T., and Ganguli, S. B., *Geophys. Res. Lett.*, **22**, 2107–2110 (1995).
12. Tam, S. W. Y., Yasseen, F., and Chang, T., *Ann. Geophys.*, **16**, 948–968 (1998).
13. Cranmer, S. R., Field, G. B., and Kohl, J. L., *Astrophys. J.*, **518**, 937–947 (1999).
14. Retterer, J. M., Chang, T., and Jasperse, J. R., *Geophys. Res. Lett.*, **10**, 583–586 (1983).
15. Vocks, C., and Marsch, E., *Geophys. Res. Lett.*, **28**, 1917–1920 (2001).
16. Maksimovic, M., Pierrard, V., and Lemaire, J. F., *Astron. and Astrophys.*, **324**, 725–734 (1997).
17. Tam, S. W. Y., and Chang, T., *Geophys. Res. Lett.*, **28**, 1351–1354 (2001).
18. Meyer-Vernet, N., Mangeney, A., Maksimovic, M., Pantellini, F., and Issautier, K., "Basic aspects of solar wind acceleration" (2002), in the proceedings.

Some Basic Aspects of Solar Wind Acceleration

Nicole Meyer-Vernet*, Andre Mangeney*, Milan Maksimovic*, Filippo Pantellini* and Karine Issautier*

*LESIA, Observatoire de Paris, CNRS UMR8109, 92195 Meudon, France

Abstract. We discuss some effects related to particle coherent orbits in the acceleration region and in the wind. (1) In the distant wind, when collisions are negligible, the temperature of escaping electrons follows adiabatic anisotropic (CGL) relations, whereas those reflected by the electrostatic and/or mirror forces behave as an adiabatic isotropic fluid; hence, contrary to a widespread view, electrons do not follow a single adiabatic law in absence of collisions. (2) In the corona, if one superimposes a minute hot maxwellian tail to a maxwellian velocity distribution, the relative importance of the tail increases rapidly with height; this is fundamentally different from a Kappa distribution, whose non maxwellian character remains constant with height. Suprathermal electrons also produce a large heat flux because many of them escape from the electric potential. (3) Fluid models, exospheric models, and a numerical simulation including particle orbits and collisions agree on finding that when an accelerated transonic wind is produced, the potential energy of protons has a maximum, so that some protons can be reflected; the production of a transonic wind also requires the trapped electron orbits to be populated, which requires some collisions. Finally, since particles on different kinds of orbits behave very differently, fluid models should not consider each particle species as a single fluid; each species should be modelled instead as a superposition of several fluids having different transport properties.

INTRODUCTION

Since the solar wind acceleration region is weakly collisional and the particle free paths increase quickly with energy, suprathermal particles are virtually collisionless, making the acceleration problem non local, and letting the velocity distributions develop high energy tails. Thus, fluid models using local closure relations are not justified. However, kinetic models including a few collisions present major difficulties since with high energy tails, the Fokker-Planck and other usual approximations of the Boltzmann-Landau-Balescu formulation are questionable.

As a result, despite the avalanche of data and sophisticated models flowing unremittingly, there is no agreement on how the corona is heated and the solar wind is accelerated. Since the physics is not fully understood and the observations are inaccurate, detailed models should be handled with care, since virtually any mixture put into the computer kitchen may reproduce observation if the saucepan contains enough free parameters. A complementary approach is to explore the basic physics with simple models, which are not expected to reproduce observation, but to suggest instead which physical ingredients should be introduced into the recipe. The present paper is written in this spirit.

SOME BASIC OBSERVATIONS

The most recent analyses of observations in polar coronal holes yield a proton temperature T_p of $(1-3)\times 10^6$ K[3]. The electron temperature T_e is about twice as cold: in situ charge state observations from Ulysses find a maximum of 1.5×10^6 K at $1.4r_\odot$[9], whereas Soho observations yield values smaller than 10^6 K[24],[2]. The conflict on T_e might be resolved if the electron velocity distribution develops a significant suprathermal tail within a short distance from the coronal base[4].

With these parameters, thermal energy alone with classical conduction cannot drive the wind[1]. Consider the approximate energy equation between the base r_0 of the wind and large distances:

$$\frac{m_pV^2}{2} \approx \frac{Q_0}{n_0V_0}+\frac{5}{2}k(T_{e0}+T_{p0})-\frac{m_pM_\odot G}{r_0} \quad (1)$$

With the classical conductive heat flux $Q\approx -10^{-11}T_e^{5/2}dT_e/dr$ (in S.I. units), $T_{e0}\approx 10^6$ K and $T_e\propto r^{-\beta}$ with $\beta<1$, one finds $Q_0<10$ J/m^2/s at $r_0\approx r_\odot$. To estimate the flux n_0V_0 from the value of about $2.\times 10^{12}$ protons/m^2/s observed at 1A.U. in the high speed wind, we consider the simple case where the flux tubes expand as r^{-2}, which may hold at least in the central part of coronal holes[25]. This yields: $Q_0/n_0V_0<10^{-16}$ J/proton. With these parameters, the heat flux and enthalpy terms cannot even lift the protons

CP679, *Solar Wind Ten: Proceedings of the Tenth International Solar Wind Conference*, edited by M. Velli, R. Bruno, and F. Malara

out of the sun's gravitational well. The problem is made worse if the flux tubes expand faster than radially.

This result, however, should not be taken too seriously for several reasons. First of all, the conductive heat flux varies as $T_e^{7/2}$ and T_e is not well known (as is its gradient). Putting in Eq.(1): $T_{e0} = 1.5 \times 10^6$ K at $r_0 = 1.4 r_\odot$ as measured from Ulysses[9] would yield instead a wind speed of $900 \times \sqrt{\beta}$ km/s! Secondly, there are not enough collisions in the corona for the classical heat flux formula to hold securely because, even though thermal electrons are somewhat collisional, faster electrons - which contribute most to the heat flux - are not[21]. Thirdly, observations of minor ions suggest heating and acceleration through dissipation of high-frequency waves[6]; low frequency waves might also play a role. Finally, if the electron velocity distribution in the corona has a suprathermal energy tail, this should increase the heat flux[11] and contribute to the heating[21] and acceleration[14], as first suggested by Olbert [17].

The latter effects are related to the coherent motion of particles. To evaluate their importance, let us estimate the dynamic time scale. The electric force on electrons, which acts to ensure charge neutrality, can be estimated by noting that it approximately balances their pressure force, i.e.:

$$eE \approx -\frac{1}{n}\frac{\partial (nkT_e)}{\partial r} \tag{2}$$

Since T_e varies generally less rapidly than n, we have $eE \sim -kT_e \times \frac{1}{n}\frac{\partial n}{\partial r}$. The last term is the inverse of the density scale height H, so that the work done by the electric force along a mean free path l is: $eEl \sim kT_e \times l/H \sim kT_e$ since $l \sim H$ in the corona and in the wind. This means that the electric field is roughly equal to the Dreicer field in these regions (see also [22]).

The relation $eEl \sim kT_e$ may be rewritten as $(eE/m_e v_{the})^{-1} \sim l/v_{the}$, where v_{the} is the electron thermal velocity. Hence, the dynamic time scale of the particles is roughly equal to their collision time, so that coherent dynamics and collisions are of similar importance. We discuss below some consequences of coherent dynamics, and how collisions may change the picture.

ORBITS OF ELECTRONS

The electric field pulls electrons towards the sun, so that only those whose kinetic energy at distance r satisfies $m_e v^2/2 > e\phi$, where ϕ is the potential at r, may escape to infinity in absence of collisions. Electrons of lower energy are reflected and return inwards. There are two kinds of such electrons: the ones (labelled "ballistic" in Fig.1) whose inclination to **B** is small enough that they are not reflected by the magnetic mirror force have their orbits connected to the base r_0; the other ones are trapped due to reflection by the mirror force on the sun ward side and by the electric field on the outward side (Fig. 1, [8],[13]).

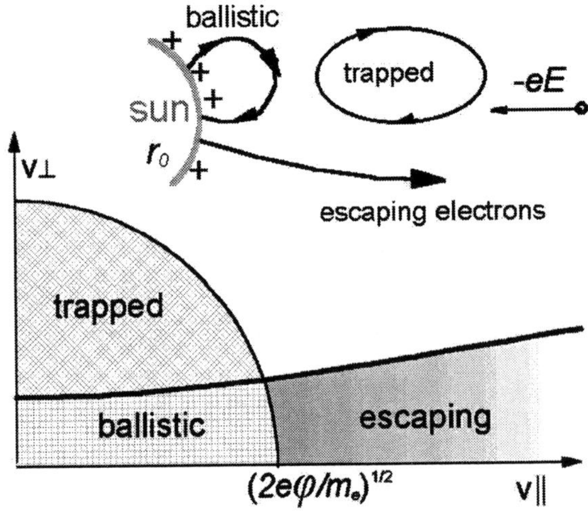

FIGURE 1. Sketch of electron orbits and of the corresponding regions in velocity space (the symbols || and ⊥ refer to the direction of **B**).

The angular limit for ballistic and escaping particles (heavy line in Fig.1) is set by energy and magnetic moment conservation between r_0 and r. At large distances, this simplifies to:

$$\frac{v_\perp^2}{v^2} < \theta_M^2 = \frac{B}{B_0}\left[1 + 2e(\phi_0 - \phi)/m_e v^2\right] \ll 1. \tag{3}$$

The diagram is symmetrical with respect to the $v_\parallel = 0$ axis except that for $v_\parallel < 0$, the region labelled "escaping" corresponds instead to particles coming to r_0 from infinity and is not populated in absence of collisions (unless there is a reservoir of particles at infinity); the white region corresponds to particles coming from and returning to infinite distance, and is not populated either. The population of trapped electrons, which is arbitrary in absence of collisions, is set to be in quasi-equilibrium with the ballistic population.

The velocity distribution at r is deduced from the one at r_0 from Liouville's theorem as $f(v) = f_0\left[v^2 + 2e(\phi_0 - \phi)/m_e\right]^{1/2}$. Let us deduce the radial temperature profile at large distances[16].

Consider first the non escaping particles. Their total density n_{ne} is the velocity volume integral of $f(v)$ in the sphere of radius $(2e\phi/m_e)^{1/2}$ (see Fig.1). Since at large distances the potential satisfies $\phi \ll [kT_{e0}/e, \phi_0]$, $f(v)$ is roughly constant (independent of v and r) within the integration volume, so that $n_{ne} \propto \phi^{3/2}$. Now, it can be shown that the non escaping electrons represent the main contribution to the total density, even at large distances, so that $n_{ne} \propto r^{-2}$; thus $\phi \propto r^{-4/3}$. The pressure P_{ne} of these particles is the volume integral of $v^2 f(v) \approx [v^2 \times \text{constant}]$ in the same sphere, i.e., $P_{ne} \propto \phi^{5/2}$. Hence, their temperature varies as $P_{ne}/n_{ne} \propto \phi^{5/2}/r^{-2} \propto r^{-4/3}$. This means that the temperature of non escaping electrons varies as for an adiabatic isotropic fluid.

Consider now the escaping electrons. We have seen that at distance r they populate the domain of velocity space [$v > (2e\phi/m_e)^{1/2}$, $\theta < \theta_M$]. To estimate their moments at large distances, the inequality $\phi \ll [kT_{e0}/e, \phi_0]$ enables us to make two approximations: (1) $f(v)$ is independent of ϕ thus of r; (2) the lower limit of the integral over v is equal to 0. With these approximations, their density and pressure have the same radial variation, as they are both proportional to θ_M^2. Hence the temperature of escaping particles is roughly independent of r at large distances. This is consistent with an adiabatic anisotropic fluid behaviour of this population since the CGL relations yield at large distances: $T \approx T_{\parallel} \approx$ constant.

Finally, since the densities of escaping and non escaping particles vary similarly (as r^{-2}) if $\mathbf{B}$ is radial, the total electron temperature is the sum of a term $\propto r^{-4/3}$ and a constant, so that the temperature profile flattens with distance, as observed[20]. One can show that both terms have similar orders of magnitude at 1 A.U.[16], and that the temperature profile agrees with Ulysses observations[7]. These results can be generalized to a spiral magnetic field[19].

How do collisions change these results? Because of the small electron mass, collisions with other particles do not change the electron energy, and collisions between electrons do not change their total temperature. Hence, collisions are not expected to modify locally the *total* electron temperature, even though they do affect the shape of the velocity distribution.

SUPRATHERMAL ELECTRONS

Some observations suggest that the electron velocity distribution in the corona might have a small suprathermal tail[4]. We make below an order-of-magnitude estimate to show that adding a minute hot maxwellian tail to a maxwellian distribution at the coronal base has two consequences: firstly, the relative importance of the tail increases rapidly with altitude[15], which might explain recent observations[4]; secondly, the heat flux and the wind speed are significantly enhanced, as was previously shown with different kinds of distributions presumably much farther from equilibrium[17], [14].

For illustrative purposes, let us assume that the distribution at the base $r_0 \approx r_\odot$ is a sum of two maxwellians: a cold one of temperature $T_c \approx 0.8 \times 10^6$ K, and a hot one at $T_h \approx 5T_c$, representing a minute proportion, say, 0.2 % of the total density (so that the total temperature is close to T_c). For an order-of-magnitude estimate, we neglect collisions and waves; we approximate the electric potential close to the sun by the Pannekoek-Rosseland one: $e\phi \approx m_p MG/2r$, which is generally an underestimate[8],[13]. We also calculate the distribution just above r_0 as in a bound atmosphere, i.e. we neglect the anisotropy due to the escaping particles; this approximation might hold sufficiently close to the base.

With these simplifications, Liouville's theorem tells us that the temperatures T_c and T_h remain constant with altitude, whereas the density of each maxwellian decreases with altitude as $e^{-e\phi/kT_c}$ and $e^{-e\phi/kT_h}$ respectively. Since $e\phi/kT_h < e\phi/kT_c$, the density of the hot maxwellian decreases more slowly than the cold one, so that the proportion of the hot population increases with altitude. At the distance where the potential has decreased by $\Delta\phi$, the ratio of hot to cold densities has increased by the factor

$$\exp\left[\frac{e\Delta\phi}{k}\left(\frac{1}{T_c}-\frac{1}{T_h}\right)\right] \qquad (4)$$

Let us calculate the distribution at, say, $r = 1.5 \times r_\odot$ (farther out, our simplifications may be too drastic). We have $e\Delta\phi \approx 0.17 \times m_p MG/r_\odot$, so that the factor in Eq.4 amounts to 49. Hence the proportion of hot electrons increases from 0.2 % to 10 % over half a solar radius. This makes the total electron temperature increase from about T_c to: $T_c \times (1 + 0.1T_h/T_c) \approx 1.5 \times T_c$.

Although this estimate is merely illustrative, it highlights the essence of the physics, that is velocity filtration by the attractive potential, first calculated by[21] with a kappa velocity distribution. With a sum of maxwellians - which is a well-behaved function, we find a fundamental difference with the kappa case: the hot tail contributes more and more to the density as altitude increases, so that the increase in total temperature stems from the increase in the non thermal character of the distribution. In contrast, with a kappa distribution, the value of κ, which is a measure of the non thermal character, remains constant with altitude, although the total temperature increases[21].

The minute hot maxwellian contributes negligibly to the density and temperature at r_0. This is not so, however, for the heat flux, because a significant part of the hot electron population can escape from the potential well. Let us make a simple estimate, assuming as previously a potential of about $e\phi_0 \approx m_p M_\odot G/2r_0$. The heat flux is produced by electrons of speed $v > \sqrt{2e\phi_0/m_e}$ at r_0, which are all escaping in absence of collisions. For a maxwellian of density n and temperature T:

$$Q_0 \approx n\sqrt{KT/m_e}kT\left(1 + e\phi_0/kT\right)^2 e^{-e\phi_0/kT}. \qquad (5)$$

Since $e\phi_0/kT_h \approx 2.8$, this yields $Q_0 \approx 4 \times 10^{-10} n_{h0}$, where n_{ho} is the density of the hot population at r_0. With a relative concentration of 2×10^{-3}, and a total density of $2 \times 10^{14}\mathrm{m}^{-3}$[5], one finds $Q_0 \approx 160\ \mathrm{Jm^{-2}s^{-1}}$, which is much larger than the conductive flux, and yields a wind speed of the order of 10^3km/s. This is an overestimate, for at least two reasons. Firstly, the actual electric potential should be different from the Pannekoek-Rosseland value in order to make the flux of escaping electrons balance the proton one. Secondly, collisions should reduce the heat flux[11], in particular by populating the orbits of incoming particles having $v > \sqrt{2e\phi_0/m_e}$ at r_0.

ORBITS OF IONS

The electron potential energy $-e\phi$ increases monotonously with distance; this is not so, however, for the total potential energy of protons $\psi = e\phi - m_p M_\odot G/r$, which is expected to have a maximum. Such a maximum occurs in simple fluid models because the temperature should decrease slower than r^{-1} in order to produce an accelerated transonic wind[18]. From Eq.2, the electric force is thus expected to decrease slower than r^{-2}, so that this force, which pushes the protons outwards, should dominate gravity at large distances. If the gravitational attraction dominates at the base of the wind, the total potential energy of protons has a maximum[8]. The existence of this maximum is of fundamental importance since it governs particle orbits. Indeed, ballistic and/or trapped ions are present out to the position of this maximum. In contrast, all ions present farther out are escaping, thus they all have $v_{\parallel} > 0$, so that the mean parallel velocity is expected to be of the order of magnitude of the thermal speed. Hence this maximum is expected not to be very far from the sonic radius[23]. Let us examine this point in more detail.

Consider the simplest fluid model, with equal electron and ion densities, temperatures and speeds, and assume $n \propto r^{-\alpha}$, $T \propto r^{-\beta}$ near the sonic radius r_S as in [18]. At r_S, the temperature T_S satisfies: $m_p M_\odot G/r_S = 4kT_S(1+\beta/2)$. The slope of the velocity profile at r_S: $dV/dr\,|_{r_S}$ calculated by Parker, together with nVr^2 =constant yields the relation between α and β: $\alpha = 2 + \left[-\beta + \sqrt{\beta^2 + 8(\beta+2)(1-\beta)}\right]/4$. This enables us to calculate the slope of the proton potential energy: $d\psi/dr = -eE + m_p M_\odot G/r^2$ at r_S. Substituting E from Eq.2 and the above relation between T_S and r_S yields: $d\psi/dr\,|_{r_S} = (4-\alpha+\beta)kT_S/r_S$. From the above value of α, elementary arithmetics shows that $d\psi/dr\,|_{r_S} \geq 0$ if and only if: $-3/4 \leq \beta < 1$.

Since $\psi \to 0^+$ at large distances, the positive slope at r_S means that ψ has a maximum located above r_S if and only if the temperature decreases slower than r^{-1} and does not increase faster than $r^{3/4}$ near r_S, which generally holds. (In the very special case $\beta = -3/4$ we retrieve the result found by[23]). If T increases even faster (but slower than r^2), there is still a maximum, but it shifts below r_S. As first noted by [23], the conditions for the ion potential to have a maximum with this simple model are similar to those yielding a transonic accelerated wind (i.e., $-2 < \beta < 1$).

A similar result is obtained with a numerical simulation taking into account particle orbits and model collisions[12]. With that model, a transonic wind is produced only with enough collisions near the sonic point, which populate trapped electron orbits and produce a maximum in proton potential energy located above the sonic point.

These results are consistent with exospheric models, which assume the orbits of trapped electrons to be populated in equilibrium with those emerging from the base, which requires some collisions, and yields a transonic wind with a maximum in the ion potential energy[10],[26].

Note, finally, that with a sonic point located below the maximum in potential energy, the wind acceleration region contains protons moving on reflected orbits - a point which is ignored in fluid models.

REFERENCES

1. Barnes A. et al., *Geophys. Res. Lett.* **23**, 3309 (1995)
2. David C. et al., *A & A* **336**, L90 (1998)
3. Esser R. et al., *Ap. J.* **510**, L63-L67 (1999)
4. Esser R., Edgar R.J. et al., *Ap. J.* **532**, L71 (2000)
5. Fludra A. et al., *J. Geophys. Res.* **104**, 9709 (1999)
6. Hollweg J. V., *Solar Wind Nine*, ed. by S. R. Habbal et al., The American Institute of Physics (1999)
7. Issautier K. et al., *Astrophys. Space Sci.* **277** (2001)
8. Jockers K., *A & A* **6**, 219 (1970)
9. Ko Y.- K. et al., *Sol. Phys.* **171**, 345,(1997)
10. Lamy et al, *Solar Wind 10*,(2002)
11. Landi S., Pantellini F.G., *A & A* **372**, 686 (2001)
12. Landi S., Pantellini F.G., *A & A*, in press (2003)
13. Lemaire J., Scherer M., *J. Geophys. Res.* **76,** 7479 (1971)
14. Maksimovic M. et al., *A & A* **324**, 725 (1997)
15. Meyer-Vernet N., *Planet. Space Sci.* **49**, 247 (2001)
16. Meyer-Vernet N., Issautier K., *J. Geophys. Res.* **103**, 29,705 (1998)
17. Olbert S., ESA SP161, 135 (1981)
18. Parker E.N., *Ap. J.* **139**, 72 (1962)
19. Pierrard V. et al., *Geophys. Res. Lett.* **28**, 223 (2001)
20. Pilipp W.G. et al., *J. Geophys. Res.* **95**, 6305 (1990)
21. Scudder J.D., *Ap. J.* **398**, 299-318, 1992
22. Scudder J.D., *J. Geophys. Res.* **101**, 13,461 (1996)
23. Scudder J. D., *J. Geophys. Res.* **101**, 11,039 (1996)
24. Wilhelm K. et al., *Ap. J.* **500**, 1023 (1998)
25. Woo R., Habbal S.R., *J. Geophys. Res.* **105**, 12,667 (2000)
26. Zouganelis et al., *Solar Wind 10* (2002)

Ion-Cyclotron Generation of the Fast Solar Wind: The Kinetic Shell Model

Philip A. Isenberg

Department of Physics and Institute for the Study of Earth, Oceans and Space
University of New Hampshire, Durham, NH 03824 USA

Abstract. We present a model of the resonant cyclotron dissipation of parallel-propagating ion-cyclotron waves in a coronal hole under the kinetic shell approximation. This approximation takes the resonant quasilinear wave-particle interaction to be much faster than the non-resonant processes affecting the proton distribution, essentially maintaining the resonant protons in a state of marginal stability with respect to this wave mode. Thus, the kinetic shell model represents the case of maximum possible dissipation by the resonant cyclotron mechanism. When the additional simplification of dispersionless waves is made, this model easily yields fast solar wind flows from plausible conditions at the model inner boundary. However, when ion-cyclotron dispersion is included the model fails, resulting in cooling of the proton population and weak acceleration to speeds on the order of 300 km s^{-1}. We conclude that resonant dissipation of parallel-propagating waves cannot be responsible for the fast solar wind.

INTRODUCTION

The generation of the fast solar wind, along with the required heating of the ions in coronal holes, is most often attributed to the cyclotron-resonant dissipation of intense fluxes of parallel-propagating ion-cyclotron waves (see [1, 2] for recent reviews in this area). Substantial evidence has accumulated over the years, particularly including the strong perpendicular ion heating in the heliocentric range 1.5 $R_S < r \leq 4$ R_S observed by SOHO/UVCS (e.g. [3]), which is consistent with expectations of this mechanism. In the context of such a proposed mechanism, the theoretical task is to test it by constructing models of the physical system which include all the relevant details, and compare the results with the observations.

However most such investigations treat the coronal hole ions as fluids, despite the fact that they are essentially collisionless beyond $r \sim 1.75$ R_S. This assumption is especially serious since the resonant interaction couples particular waves to particular particles, and does not energize the bulk plasma, leading to potentially important distortions of the ion distributions away from the Maxwellian or bi-Maxwellian shapes implied by a fluid picture.

Additionally, I contend that the fluid models are not well constrained. It is clear from the many fluid models in the literature that any model which provides sufficient perpendicular ion heating close to the Sun, whether by an arbitrary heating function or through an elaborate scenario of the detailed wave-particle interaction, finds that the mirror force acceleration will yield a plausible fast solar wind. Thus, a fluid analysis does not lead to a useful test of the resonant cyclotron mechanism, and it is necessary to investigate the kinetics of the interaction.

KINETIC RESONANT CYCLOTRON INTERACTION

There are essentially two aspects of the this kinetic process. First, the ion and the wave must satisfy the resonance condition, which for parallel-propagating ion-cyclotron waves is

$$\omega(k) - k v_{\|} = \Omega, \qquad (1)$$

where ω and k are the plasma frame frequency and wavenumber, respectively, of the wave, $v_{\|}$ is the speed of the ion in the plasma frame along the magnetic field, and Ω is the gyrofrequency of the ion. When this condition is satisfied, the Doppler-shifted wave frequency seen by the ion matches the ion gyrofrequency and the particle and wave can efficiently exchange energy. In a low-β plasma such as that in a

CP679, *Solar Wind Ten: Proceedings of the Tenth International Solar Wind Conference*,
edited by M. Velli, R. Bruno, and F. Malara

coronal hole, ion-cyclotron wave frequencies are confined to the range below the proton gyrofrequency. As a consequence, resonance between a proton and an ion-cyclotron wave requires $k\, v_{||} < 0$. Thus, we find that outward-propagating waves ($k > 0$) can only resonate with the sunward half of the proton distribution ($v_{||} < 0$), while a resonant interaction for the anti-sunward protons ($v_{||} > 0$) requires the presence of inward-propagating waves ($k < 0$).

The second aspect of the kinetic interaction comes from the quasilinear description of the process. When a continuous spectrum of waves over some range of k is resonant with an ion distribution over some range of $v_{||}$, the ions will diffuse in velocity space in a manner so as to conserve their energy as measured in the reference frame moving with the phase speed of the wave [4-6]. This constant-energy condition defines a set of surfaces in velocity space, and the resonant diffusion will act to reduce any ion density gradients along these surfaces. The quasilinear interaction conserves the combined energy of the waves and particles in the plasma reference frame, so if the diffusion results in an energy gain in this frame this must correspond to dissipation of the resonant waves. Conversely, if the diffusion results in a loss of ion energy, this indicates that waves in the resonant range of k are being generated.

THE KINETIC SHELL MODEL

This behavior suggests a useful simplification of the full kinetic problem. Let us assume that the wave-induced diffusion of resonant protons along the constant-energy surfaces takes place much faster than their non-resonant evolution in a coronal hole. If we then consider the system on the slower, non-resonant time scale, we can treat the proton density gradients along the constant-energy surfaces as being maintained at zero by the efficient resonant interaction. In this case, the resonant portions of the proton distribution will be composed of nested layers of shells in velocity space, each with a uniform density, corresponding to the set of non-intersecting constant-energy surfaces. These shells then evolve slowly under the influence of the non-resonant forces alone as the plasma flows away from the Sun.

The absence of gradients on the constant-energy surfaces is sometimes called a "quasilinear plateau", and is equivalent to the "marginally stable state" where the linear growth/damping rate of ion-cyclotron waves vanishes. By assuming that the rapid resonant interaction maintains the proton distribution in this state, we are modeling the effects of the *maximum possible dissipation* of ion-cyclotron waves by the resonant cyclotron mechanism. Once the distribution attains the marginally stable state at a given radial position, any further dissipation would cause a transport of protons along the constant-energy surfaces in a manner so as to increase the proton energy. But this transport would then correspond to the formation of unstable gradients in the distribution, so this cannot happen spontaneously. As the shells slowly evolve with radial position under the action of the non-resonant forces, the wave dissipation provides the energy needed to maintain the uniform density along the shells. However, the marginally stable proton distribution cannot take on any more wave energy through cyclotron resonance, even if more resonant waves are present. The plasma has become locally transparent to these waves.

This formalism also contains the implicit assumption that sufficient resonant waves are present to accomplish this rapid response of the ions. If the wave intensities are smaller, or if the resonant diffusion is somehow slower than that necessary to fill the shells on the fast time scale, this will correspond to less wave dissipation and reduced energization of the resonant protons.

This approximation leads to what I call the "kinetic shell" model for the radial evolution of the collisionless coronal hole proton distribution in response to the resonant cyclotron interaction with parallel-propagating ion-cyclotron waves [7-12]. We have derived a set of steady-state first-order equations for the density of resonant protons $f(r, \eta)$, where η is the shell label. We propagate the solution of these equations outward from a supersonic, collisionless proton distribution at $r = 2\, R_S$, assuming a given radial magnetic field $B(r)$ and electron temperature $T_e(r)$. With these input conditions, the model obtains as output: the self-consistent acceleration of the plasma flowing at $U(r)$, the collisionless proton distribution $f(r, v_{||}, v_{\perp})$ beyond the inner boundary, and the rates of wave dissipation and generation calculated from the radial changes in resonant proton energy flux.

It is noteworthy that the wave energy changes are outputs of this model, rather than inputs. Since we are modeling the effects of the maximum dissipation rate, which is a property of the proton distribution, we do not need to assume any details of the wave intensities or spectra. In particular, we can avoid the use of arbitrary heating functions or guesses as to the unobservable wave spectra in the low corona.

The radial evolution of the proton distribution under the kinetic shell assumption is governed by the shell motion as the plasma flows away from the Sun. This motion, in turn, results from the average of the non-resonant forces on the shell. The significant forces in a

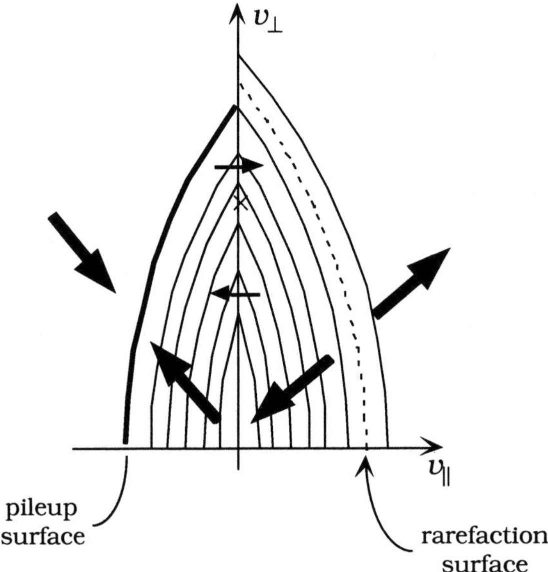

Figure 1. Velocity space diagram showing the kinetic shell response of the proton distribution to the non-resonant forces as the plasma flows away from the Sun. The thick arrows indicate the shell motion, converging on the sunward side and diverging on the anti-sunward side. The thin arrows indicate the non-resonant proton transport across the $v_\parallel = 0$ boundary, changing sign at the ×. The pileup and rarefaction surfaces are shown as a thick line and a dashed line, respectively.

coronal hole are gravity, the charge-separation electric field, and the mirror force due to the decreasing magnetic field. In this model, we also must include an inertial force since the wave phase speeds which determine the shell shapes are defined with respect to the accelerating plasma reference frame.

The interplay of these forces determines the evolution of the proton distribution under a strong cyclotron resonant interaction, and a qualitative discussion of their relative behavior provides considerable insight. The gravity and electric field are body forces, in that their strength is independent of the particle velocity. For any valid solar wind solution the gravitational effect is the stronger, so the net body force on a proton is sunward. The inertial force is proportional to $-(U + v_\parallel)\, dU/dr$, so this force is also sunward for a plasma accelerating away from the Sun. The only outward force is the mirror force, proportional to $v_\perp^2$. Thus shells close to the origin of velocity space, where $v_\perp$ is small, will feel an average sunward force in the plasma frame. Conversely, shells with sufficient portions of their surface at high $v_\perp$ will feel a net outward force. The rapid resonant diffusion will cause the shells to move without distortion, so the resulting motions will be with respect to the velocity space origin, as indicated by the thick arrows in Figure 1.

In the sunward half of the proton distribution, the shells close to the origin will expand as the plasma flows away from the Sun, while the larger shells contract. This configuration causes a convergence of shells toward a pileup surface, where the sunward protons become concentrated. On the anti-sunward side, the same forces lead to a divergence of the shells away from a rarefaction surface. These surfaces mark the positions where the shell-averaged inward and outward forces balance at some value of r. The dependence of the inertial force on $v_\parallel$ causes the rarefaction surface to always be further from the origin than the pileup surface, as shown.

The sunward and anti-sunward populations, resonating with independent sets of waves, are matched across the $v_\parallel = 0$ boundary by recognizing that the waves resonant with protons at the boundary have $k \rightarrow \infty$ and therefore have no power. In this case, protons will be transported across the boundary by the non-resonant forces, toward positive $v_\parallel$ when $v_\perp$ is high and the mirror force dominates the inward forces, and the reverse when $v_\perp$ is low. This transport is indicated by the thin arrows in Figure 1, where the value of $v_\perp$ at which the forces balance is denoted by the ×.

Although we hypothesize a flux of resonant outward-propagating waves which only has to match or exceed the level needed to keep the sunward protons at the marginally stable state, the appropriate intensity of the inward-propagating waves is not as well determined. We have shown that, if no inward waves are allowed and the anti-sunward protons are treated as free particles which conserve their energy and magnetic moment in the background fields, the anti-sunward proton distribution quickly becomes unstable [8]. More generally, the flux of protons which crosses the $v_\parallel = 0$ boundary from the concentrated pileup region at high $v_\perp$ will diffuse down the constant-energy surfaces, losing energy and therefore generating inward waves. Additional resonant inward waves may also be present due to turbulent wave generation or propagation from larger radial positions. To some extent, the cyclotron resonant interaction with these inward waves would cause anti-sunward shells to be filled as shown in Figure 1, but these shells cannot be maintained indefinitely. In particular, the shells which touch the $v_\parallel = 0$ axis above the sunward pileup region are no longer fed by the unstable proton flux across the boundary. Furthermore, the continued expansion of these shells eventually carries their protons to implausible energies [10]. So, the kinetic shell approximation must be

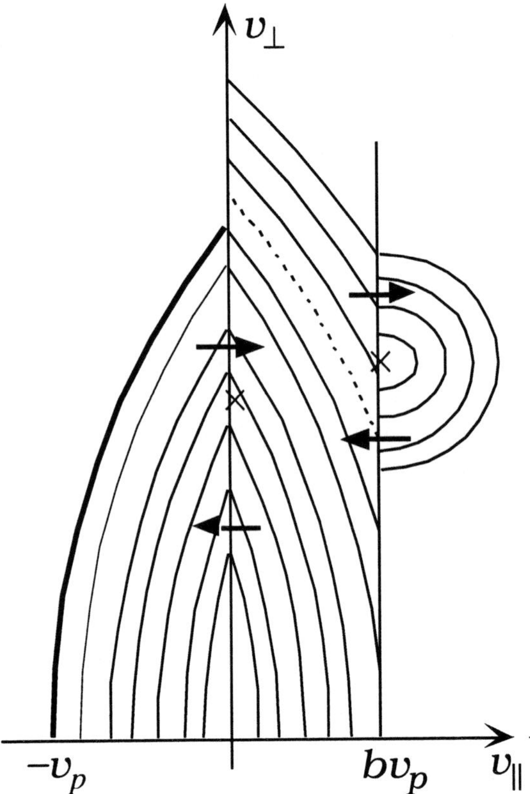

Figure 2. Velocity-space diagram showing the modification of the shell distribution for the anti-sunward protons to allow the effects of moderate instability. Rapid resonant diffusion is assumed for $0 < v_\parallel < b\, v_p$, and no resonant behavior for $v_\parallel > b\, v_p$. The pileup and rarefaction surfaces are shown as a thick line and a dashed line, respectively.

moderated when applied to the anti-sunward population.

In order to treat this problem within the context of the kinetic shell formalism, we take the efficient resonant diffusion of anti-sunward protons to be confined to a range of parallel speeds, $0 < v_\parallel < bv_p$, where v_p gives the parallel extent of the largest populated sunward shell and b is a constant, as illustrated in Figure 2. This construction considers the fact that the resonance condition (1) depends only on $v_\parallel$, so all protons in this range will be strongly scattered. As in [8], we model the protons with $v_\parallel > bv_p$ as free particles, conserving their magnetic moment and total energy in the reference frame of the Sun. Although this population will generally exhibit unstable density gradients as well, its density is substantially less than in [8] where the input flux of free particles came directly from the pileup region. We suggest that the overall behavior of the anti-sunward population will be reasonably approximated by ignoring the resonant effects where $v_\parallel > bv_p$ and taking them to be maximally efficient for $v_\parallel < bv_p$. We now have an additional boundary at $v_\parallel = bv_p$, and apply the same non-resonant boundary condition there as at $v_\parallel = 0$.

Of course, we also now have an additional free parameter in the model. Setting $b = 1$ gives a model where the instability due to the proton flux across $v_\parallel = 0$ is intense enough to fill all the shells which can be influenced by those particles. The case of $b < 1$ represents a weaker instability, while $b > 1$ would require additional inward waves from another source. In principle, one could allow b to vary in some manner dependent on the level of the unstable flux, but we will see that such an additional complication is unnecessary as our conclusions will be essentially independent of b.

Our initial explorations of the kinetic shell approximation used "dispersionless" waves, $\omega/k = V_A(r)$, to simplify the numerical effort. A wave phase speed independent of k is also independent of $v_\parallel$, so the constant-energy surfaces are sections of spheres centered on the Alfvén speeds, $v_\parallel = \pm V_A$, with respect to the plasma. For the very low β in coronal holes the populated proton shells all have $v_\parallel << V_A$, so the shells are very narrow in the parallel direction and result in highly perpendicular distributions. Apart from this artifact, the results of the dispersionless models have been very encouraging. We have easily obtained plasma flow speeds of 700 - 800 km s^{-1} within 8 R_s, along with perpendicular proton temperatures and derived heating rates consistent with the behavior of many of the fluid models of the fast solar wind. However, a physically valid model obviously requires the inclusion of dispersive waves, and we recently expanded our model to incorporate these effects.

THE DISPERSIVE MODEL

The only way that wave dispersion enters into the kinetic shell model is by changing the shape of the constant-energy surfaces. The phase speed of the resonant wave now depends in a systematic way on the parallel speed of the proton through the resonance condition (1) and the assumed dispersion relation. The constant-energy surfaces are now constructed by taking them to be only locally spherical and centered on the smoothly varying resonant phase speed. The procedure is described by [13] and [14].

In a coronal hole, we choose the dispersion relation for a cold electron-proton plasma, $\omega/k = \pm V_A (1 - \omega/\Omega_p)^{1/2}$. This choice gives shells which contact the $v_\parallel = 0$ axis perpendicularly, since the waves which resonate with these protons have zero phase speed. They are also considerably broader in $v_\parallel$ than the previous dispersionless shells, so will yield distributions with much more realistic anisotropies. (See the contributed paper in this volume [15] for the actual equations defining the dispersive shells.) The model equations which describe the shell evolution in r are now more complicated, but they are still first-order and can be readily solved numerically.

However, we have found to our dismay that the dispersive model does not yield a fast solar wind. Although the detailed behavior naturally varies for different conditions at the inner boundary, we obtain flow speeds $U \lesssim 300$ km s^{-1} at 6 - 8 R_s for apparently any plausible parameter values set at $r = 2\ R_s$ and for any value of the inward wave parameter b. Even more damning is the result that the non-resonant background forces invariably cause the pileup surface to contract towards the origin, yielding a perpendicular cooling of the proton distribution, directly contrary to observation. We note that dissipation of the outward-propagating waves still takes place in these models, at least for the first few solar radii beyond the inner boundary. The proton cooling comes from the adiabatic expansion of the flow, which dominates since the maximum allowable dissipation is not strong enough to compensate.

When compared to the earlier dispersionless results, the difference can be traced to the action of the inertial force on the wider dispersive shells. The dispersionless models all had extremely narrow shells, so the inertial force per unit mass, $-(U + v_{\|})\ dU/dr$, was only weakly dependent on $v_{\|}$. In these models, the evolution was such that the inward forces on the sunward shells initially increased with respect to the outward mirror force, leading to expansion of the pileup surface, strong heating of the distribution, and a rapid acceleration to fast wind speeds. Because the dispersive shells are broader in $v_{\|}$, the inward inertial force is substantially weaker on the sunward side ($v_{\|} < 0$). Furthermore, since $v_{\|}$ is no longer small compared to U, the expansion by the inward forces of a large dispersive shell becomes progressively more difficult since the expansion itself further weakens the inward force. The result in all our many attempts was a contraction of the pileup surface at all r, perpendicular cooling of the protons, and a mild acceleration of the flow limited to moderate speeds.

As a test of this effect (and of the dispersive code), we can force the dispersive shells to be as narrow as the previous dispersionless surfaces by greatly increasing the value of the Alfvén speed in the model. To construct dispersive shells which match the parallel width of the dispersionless distributions at the inner boundary, we need to set $V_A = 7 \times 10^4$ km s^{-1} at 2 R_s. When this is done, the model produces a fast wind similar to that obtained in the earlier dispersionless calculations. However, such narrow distributions required by the resonant cyclotron mechanism are definitively ruled out by the observations.

CONCLUSIONS

The kinetic shell model presented here poses a minimal test of the resonant cyclotron dissipation of parallel-propagating ion-cyclotron waves as the responsible mechanism for the generation of the fast solar wind in coronal holes. We have shown that the maximum allowable dissipation of outward-propagating waves, when treated kinetically, fails to yield perpendicular heating of protons or acceleration to fast solar wind speeds. We conclude that this dissipation cannot be the cause of the fast solar wind.

We emphasize that relaxing the extreme kinetic shell assumption will not remove the difficulty. Under a more realistic assumption the resonant shells will not be completely filled, but then the interaction will provide even less proton heating than in our model.

This conclusion is in contradiction to the results of Tam and Chang [16-18], who have also constructed a kinetic model of this same mechanism. However, their model assumes specific wave intensities and spectra, which are arbitrarily chosen and are not allowed to react to the ion evolution. In their model, the ions continually extract energy from the waves, even when this means their distribution would become unstable. Such a model would not exhibit the limitations inherent in the marginally stable state which are automatically included in the kinetic shell model.

The possibility remains that resonant cyclotron dissipation of obliquely-propagating waves could still produce the necessary heating and acceleration, but the kinetic shell formalism cannot treat this interaction due to the presence of multiple resonances. Otherwise, an entirely different mechanism must be found to generate the fast solar wind.

ACKNOWLEDGMENTS

The author is grateful for valuable conversations with T. Chang, T. G. Forbes, S. R. Habbal, J. V. Hollweg, M. A. Lee, and S. A. Markovskii. This work was supported in part by the NASA Sun-Earth Connections Theory Program under grant NAG5-8228 and by the NASA Living With a Star Program under grant NAG5-10835.

REFERENCES

1. Cranmer, S. R., S. R. Spangler, R. Esser and J. L. Kohl, *Space Sci. Rev.*, submitted (2002).
2. Hollweg, J. V. and P. A. Isenberg, *J. Geophys. Res.*, in press (2002).
3. Kohl, J. L., et al., *Astrophys. J.*, **501**, L127 (1998).
4. Kennel, C. F. and F. Engelmann, *Phys. Fluids*, **9**, 2377 (1966).
5. Rowlands, J., V. D. Shapiro and V. I. Shevchenko, *Sov. Phys. JETP*, **23**, 651 (1966).
6. Dusenbery, P. B. and J. V. Hollweg, *J. Geophys. Res.*, **86**, 153 (1981).
7. Isenberg, P. A., M. A. Lee and J. V. Hollweg, *Sol. Phys.*, **193**, 247 (2000).
8. Isenberg, P. A., M. A. Lee and J. V. Hollweg, *J. Geophys. Res.*, **106**, 5649 (2001).
9. Isenberg, P. A., *Space Sci. Rev.*, **95**, 119 (2001).
10. Isenberg, P. A., *J. Geophys. Res.*, **106**, 29,249 (2001).
11. Isenberg, P. A., *J. Geophys. Res.*, in press (2002).
12. Galinsky, V. L. and V. I. Shevchenko, *Phys. Rev. Lett.*, **85**, 90 (2000).
13. Gendrin, R., *J. Atmosph. Terr. Phys.*, **30**, 1313 (1968).
14. Isenberg, P. A. and M. A. Lee, *J. Geophys. Res.*, **101**, 11,055 (1996).
15. Isenberg, P. A., in *Solar Wind 10*, this volume (2003).
16. Tam, S. W. Y. and T. Chang, *Geophys. Res. Lett.*, **26**, 3189 (1999).
17. Tam, S. W. Y. and T. Chang, *Geophys. Res. Lett.*, **28**, 1351 (2001).
18. Tam, S. W. Y. and T. Chang, in *Solar Wind 10*, this volume (2003).

Collisional filtration model in the solar transition region

A. Greco* and P. Veltri*

* *Dipartimento di fisica, Università della Calabria (Italy)*

Abstract. In this work we will try to show the physical reasons for the steep temperature increase in the solar transition region, trying to look for a solution of the Boltzmann equation, which seems to be the most suitable tool for describing the plasma of the solar transition region. To solve the Boltzmann equation, we have used a Monte Carlo method; as a solution of this equation we have obtained a temperature enhancement and a density drop; besides, we have computed an electron heat flux towards the innersphere, showing a departure from the classical transport theory.

INTRODUCTION

A challenging problem in solar physics is represented by the attempt to understand the reasons for the steep temperature increase and density drop over a very short (with respect to solar scale lengths) distance (some thousand kilometers) in the transition region (TR). The electron mean free path increases from some hundred centimeters in chromosphere to several thousand kilometers in corona [2] and this means that in the transition region a matching is realized between a collisional medium, where a fluid description is adequate, and a medium where collisionless transport processes are expected to be dominant, so the plasma should be described in term of the kinetic Vlasov equation. For the above mentioned reasons two opposite approaches to this problem, have been pursued: the classical hydrodynamic approach [7, 8] yielding to the fluid models and the collisionless one yielding to the exospheric models [5]. The former is based on the assumption that the plasma is collision-dominated and on the assumption of weak external gradients. In this case there are only small departures from Maxwellian distribution functions and the Fourier's law for the heat flux holds. Collisional fluid descriptions for modelling the solar transition region are limited because of the presence of huge temperature and density gradients and because the thermal Knudsen numbers K_T are large enough for the classical transport coefficients to become modified [12]. On the other hand, the collisionless approach assumes that the effects of collisions are neglegible, so the Boltzmann equation for the particle flow is reduced to the Vlasov equation and any function of the total energy is a solution. If we start with a Maxwellian distribution function (DF) at the lower boundary, at each height a Maxwellian is found with the same temperature and the density is that of an isothermal atmosphere while heat flux is zero. To overcome the difficulty to reproduce the temperature jump, in the collisionless approach, Scudder [3, 4] proposed a new mechanism called *the velocity filtration effect*: if the lower boundary is overpopulated with high-velocity particles, this excess is recovered in the higher regions, giving rise to a temperature enhancement (see figure (3) and (4) of [3]) and this result is obtained analitically when using a generalized Lorentian (or Kappa) distribution for particles at the lower boundary. Scudder strategy was to baypass the need for a collisional calculation since the suprathermal tails which overcome the potential barrier and enter in corona, are essentially non collisional (the free path scales as $\lambda \sim v^4$). Collisionless exospheric models are limeted because collisions are not negligible and because they require non-thermal distribution functions (like Kappa-functions) to reproduce a temperature gradient.

For the above mentioned limitations, a complete description of the solar TR requires a kinetic model taking collisions into account; so, it is natural to consider the Boltzmann equation. Due to complexity of the physical problem and of the mentioned mathematical equation, we have made some simplifying assumptions, described in the next section.

NUMERICAL SIMULATION

Let us assume a stationary case, a planar one-dimensional geometry in the physical space and a cylindrical one around the vertical direction in the velocity space. In this case the DF depends on three variables, $f(x, u_\parallel, u_\perp)$, where x is the vertical coordinate (the distance from the solar surface), $u_\parallel = u_x$ is the velocity parallel to the vertical direction and $u_\perp$ is the

CP679, *Solar Wind Ten: Proceedings of the Tenth International Solar Wind Conference,* edited by M. Velli, R. Bruno, and F. Malara

modulus of the velocity perpendicular to this direction; then, the stationary Boltzmann equation for the electrons is written

$$u_x \frac{df(x,\mathbf{u})}{dx} + \frac{eE}{m_e}\frac{df(x,\mathbf{u})}{du_x} = \tag{1}$$

$$\int d\mathbf{v} \int_0^{b_{max}} bdb$$

$$* \int_0^{2\pi} |\mathbf{u}-\mathbf{v}| d\varepsilon (f(x,\mathbf{u}')f(x,\mathbf{v}') - f(x,\mathbf{u})f(x,\mathbf{v})),$$

where e and m_e are the electron charge and mass, respectively.

The right hand-side term of (1) is the collisional integral, in which two populations of particles are considered: the target ones with velocity $\mathbf{v}$ and the incident ones with velocity $\mathbf{u}$; $\mathbf{v}'$ and $\mathbf{u}'$ are their velocities after a collision, respectively; b is the impact parameter and, since we describe Coulomb collisions between electrons, the maximum impact parameter b_{max} is the Debye radius; ε is the angle made by the collision plane and an arbitrary plane, measured in a plane normal to the relative velocity vector $|\mathbf{u}-\mathbf{v}|$ (see figure 2 of [1]).

To avoid solving the Boltzmann equation for the protons, we have made the hypotheses of charge neutrality, equal bulk velocity V and temperature T in the momentum equations of the two kinds of particles; from these, we can obtain the expression of the electric field, where the electrons move $E = \frac{m_i - m_e}{2e}(V\frac{dV}{dx} + \frac{GM_S}{x^2})$, obtained in 1972 by Lemaire and Scherer.

We studied the solution of (1) in a region which starts at 2000 Km above the photosphere and extends up to about 7×10^4 Km in the corona. At the lower boundary we supposed that distribution function for the incoming electrons is Maxwellian; at the upper boundary, we assumed that the motion outside the region under study is ballistic, so that electrons which arrive with a velocity less than the escape one, come back to this boundary with a reversed sign of velocity (the escape velocity is considered as a parameter of the problem). Solving (1) with the above reported boundary conditions, we obtain electron distribution function at each level of the simulation box, which in turn allows us to determine the density, temperature, bulk velocity and heat flux profiles.

To solve the Boltzmann equation we use a Montecarlo method widely described in [1]; it is defined as a test particle method based on an iterative process, where the "random walk" of a particle (an electron) is studied while it interacts (through Coulomb collisions) with the field (target electrons). This tecnique shows the distinct advantage that is not necessary to introduce simplifying assumption to evaluate the collisional integral. The same method has been developed to study the Boltzmann equation in inhomogeneous situations, like the evolution of the distribution function through a shock wave profile. We will briefly recall here the main features of this technique. This method is based on five main points: (i) about 10^4 electrons are injected in the simulation box at the lower boundary with the assumed DF for positive parallel velocities; (ii) electrons propagate in the simulation box under the effect of the above electric field and are subjected to Coulomb collisions whose rate is determined by the DF of the target electrons; (iii) when particles arrive at the upper boundary we reverse their radial velocity if it is lower than the escape velocity; (iv) we stop to follow the particles trajectory when they arrive either at the lower boundary or to the upper one, but with a radial velocity higher than the escape one; (v) to compute the DF we have divided the phase space in cells and the DF for the incident electrons is computed by accumulating the time spent by the particles in each cell of the phase space.

Clearly, the knowledge of the target DF is needed, so we use an iterative procedure, in which the DF for the incident electrons obtained in the previous iteration is used as the DF of the target electrons in the following iteration. When the distribution functions of two following iterations are almost the same, the convergence has been obtained and we have a solution to the Boltzmann equation and the two populations (incident and target ones) can be considered identical. At the first iteration step, we started with a guess target DF and we have chosen for this a Maxwellian one; for the density and temperature profiles of the zero-order distribution, we have used those obtained as a solution of the Fourier's law with constant heat flux q_o [6]: $T(x) = (T_o^{7/2} - 7q_o(x-x_o)/2k)^{(2/7)}$ and from the guess assumption of constant pressure p_o, the density profile is given by $n(x) = p_o/k_B T(x)$. T_o, q_o and p_o are guess parameters, referred to their value at the lower boundary of the simulation box.

NUMERICAL RESULTS

Each length in the code is normalized to the Debye radius at the lower boundary r_{Do}, velocities to the elctron thermal velocity at the lower boundary v_{tho} and times to the inverse of the electron plasma frequency at the lower boundary $\omega_{peo} = v_{tho}/r_{Do}$. With this choice, density profiles are normalized to $n^* = r_{Do}^{-3}$, temperatures to $T^* = m_e v_{tho}^2/2K_b$ (T^* and T_o have the same value) and the heat fluxes to $q^* = n^* m_e v_{tho}^3$.

The figure 1 on the left shows the 6 iterations needed to reach the convergence of the moments of the distribution function; the plot on the top left and on the top right are the density and the average temperature, respectively and the plots on the bottom left and on the bottom right are the electron bulk velocity and the electron heat flux, respectively. The guess parameters for this simulation are $T_o = 10^4$K, $p_o = 6.9 \times 10^{-3}$ erg cm^{-3} ($n_o = 5 \times 10^9$ cm^{-3}) and $q_o \sim 5 \times 10^5$ erg s^{-1} cm^{-2} (see [6]). Besides, the density normalization n^* is equal to 10^6cm^{-3} and heat flux normalization q^* is about 150 erg s^{-1} cm^{-2}. From the two plots on the top, we can note that density decreases by a factor of about 100 and temperature increases by the same factor, reaching values of 1 million of degrees in corona. The electron temperature is an average between the parallel and perpendicular one and we have found that the perpendicular heating is much more effective than the parallel heating, but this result is not shown. As previously said, the escape velocity is an input parameter which determines the bulk velocity of the simulation; we have used a value of $10v_{tho}$ to obtain these results; the reader can note that the value of the obtained electron bulk velocity is very high and unrealistic, but it is a consequence of the very low value of the escape velocity, therefore many electrons have a parallel velocity greater than the escape one and these are the particles which provide such a high bulk velocity in corona. We would stress the fact that using an higher and more realstic value of escape velocity, the bulk velocity would be determined by the more energetic particles in the tails of the distribution function, where the statistic is poorer, so much more particles are needed in order to find some electron which has a parallel velocity geater than the escape one. Our efforts are going in this direction, but a very high computing power is required. As we can see from the fourth panel of the figure 1 on the left, we obtain an heat flux peak value of the order of $2000q^* = 3 \times 10^5$ erg/cm^2 sec. If we compare with the free-streaming value of the heat flux coming from corona with an electron thermal velocity v_{th1}, $q_{fs} = n_1 m_e v_{th1}^3 \sim 7 \times 10^6$ erg/cm^2 sec (n_1 is electron density at the upper boundary), we find that the electron heat flux is limeted by the value for streaming particles in a neutral gas [13]. Another upper limit fort the electron transported heat flux is the non collisional value $q_{NC} = Kn_L(T_o T_L^{1/2} - T_L T_o^{1/2})$ (K is a constant), where T_o and T_L are the temperatures at the boundaries of the simulation box and n_L is the density at the upper side [14]; if we suppose that the plasma is non collisional and that there are two Maxwellian distribution functions at the boundaries, the above expression gives $\sim -4.3 \times 10^5$ erg/cm^2 sec in each point, which is greater than the obtained value. So, the general behaviour of the system is not that of a collisionless plasma, because our heat flux is spatially variable and smaller than the collisionless value. Finally, the profile of the classical Spitzer-Harm heat flux [9] given by the expression $q_{SH} = -k_e T^{5/2} \frac{dT}{dx}$ (calculated with the obtained temperature and density profiles) is very close to the obtained one in corona, where temperature and density gradients are very small, but near the transition region, where the hypoteses for the classical heat conduction don't hold, the profile shows a significant departure from the obtained one.

The same figure 1 on the right shows the evolution of the computed electron DF, of the last iteration for the case $n_o = 5 \times 10^9$ cm^{-3}, versus parallel velocity normalized to v_{tho}, at four distances from the solar surface integrated on perpendicular velocities: at 7×10^4 Km (the upper boundary) the DF is not symmetric, as we can see from the comparison with a Maxwellian distribution function with the same temperature and density (dotted line). It is probably due to the lost particles having a parallel velocity greater than the escape one. As the altitude decreases (10^4 Km, 2.1×10^3 Km, 2×10^3 Km) the electron DF becomes more non-Maxwellian, in fact it exhibits an anisotropic, high-velocity tail going downward from corona to chromosphere, carring a heat flux, as it is possible to note in the fourth panel of the figure 1 on the left. In the same time an enthalpic flux $\frac{3}{2}nVK_bT + pV$ appears in order to balance the heat flux. Future works are going in the direction to check the energy balance for all the forms of energy in the system. At the lower boundary (2×10^3 Km) the distribution for positive velocities is consistent with the assumed Maxwellian conditions at the injection layer, whereas for negative velocities the distribution deviates from a Maxwellian at $2 - 3v_{tho}$.

DISCUSSION AND CONCLUSIONS

Concerning the broadening of the electron DF as the altitude increases, we think that a filtration process occurs in our system and this effect is mainly carried out by the collisions (besides the attractive potential), therefore in our case we do not need to inject any energy density in the tails at the base of TR: an electron which is going towards the corona has a great probability to collide with a hotter target electron, so it could gain energy and its free path could became very large (the free path $\lambda \sim v^4$). In this way it will suffer few collisions during its path and can have a quasi-ballistic motion up to corona, where the rate of collisions is decreased; so, if this electron has enough energy to overcome the potential well, it arrives to the upper boundary with large velocities. Particles which do not gain energy with collisions display a diffusive motion which keeps them for longer times in the lower regions. As result, in higher regions we mainly find high energy particles which in turn gives rise an enhancement of temperature.

The figure 2 illustrates very clearly this concept of filter due to collisions: the two temperature profiles depicted on the left are obtained from the simulation with two different density values at the lower boundary $n_o = 10^{10}$ cm^{-3} (solid line), and $n_o = 5 \times 10^9$ cm^{-3} (dashed line); the temperature jump is less for $n_o = 5 \times 10^9$ cm^{-3}; in-

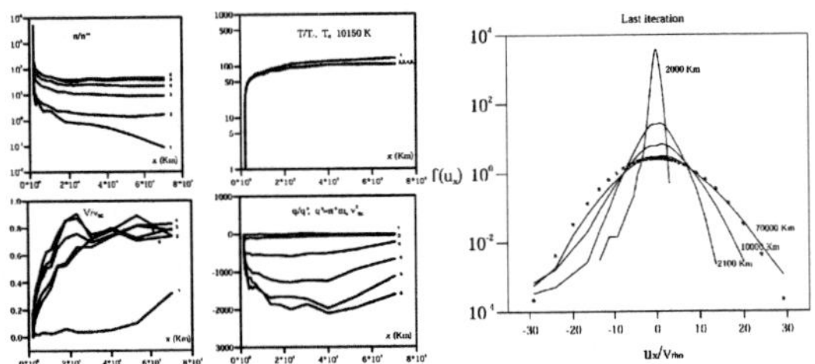

FIGURE 1. On the left: iterations for the moments of the distribution function, normalized at their value at the lower boundary of the simulation box. On the right: the distribution function of the last itaration, displayed for 4 distances above the solar surface. The dot line is explained in the text.

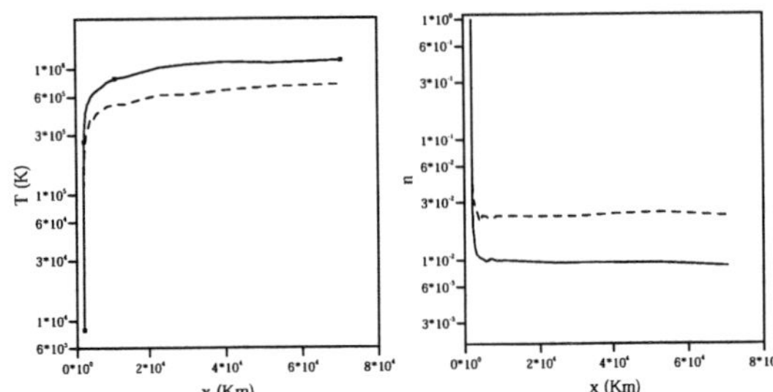

FIGURE 2. On the left: two temperature profiles for two values of density n_o. On the right: two density profiles normalized to their value n_o. For both panels solid lines refers to $n_o = 10^{10}$ cm^{-3} and dashed ones to $n_o = 5 \times 10^9$ cm^{-3}

deed in this case the filter due to collisions is less effective because density is lower (so the medium is less collisional), then the "cold"electrons from the bottom have not many possibilities to gain energy, so they reach the corona with relatively low velocities, and because of "cold" electrons are more numerous (they represent the core of the distribution), their density is higher in corona, as it is displayed in figure 2 on the right.

Clearly, we have made several simplyfing assumptions; on of these is to not consider electron-proton collisions. Because of the small mass ratio between electrons and protons, these collisions are not very effective for trasferring energy between the two species; however, even if protons do not matter for the energy distribution, they could be important for the angular distribution and for the isotropy of the electrons [10]. We have already said that the presence of protons is taken into account through the electric field in which electrons move; however, computing the proton distribution function in such a simulation where particle move in fixed external fields, requires that the charge neutrality has to be checked at each iteration step otherwise very huge electric fields could appear in the system.

ACKNOWLEDGMENTS

We gratefully acknowledge the lectures and notes of André Mangney, the helpfull discussions with Francesco Malara and the help of Fedele Stabile given for the parallelization of the Monte Carlo code on the AlphaServerSC cluster.

This work is part of a research programme which is financially supported by the Ministero dell'Università e della Ricerca Scientifica e Tecnologica (MURST), the Consiglio Nazionale delle Ricerche (CNR), contract no. 98.00148.CT02 and no. 98.00129.CT02, the Agenzia Spaziale Italiana (ASI), contract no. ARS98-82.

REFERENCES

1. Haviland J. K., The Solution of Two Molecular flow Problems by Monte Carlo Method in Methods in computational Physics, Academic Press, 4 (1965)
2. Mariska J. T., The Solar Transition Region, Cambridge Astrophysics Series (1992)
3. Scudder J. D., On the causes of temperature change in inhomogeneus low-density astrophysical plasmas, ApJ, 398 (1992)
4. Scudder J. D., Why all stars should posses circumstellar temperature inversion, ApJ, 398 (1992)
5. Maksimovic M., Pierrard, V., Lemaire, J. F., A kinetic model of the solar wind with Kappa distribution functions in the corona, A&A, 324 (1997)
6. Owocki S. P., Canfield, R. C., The role of non-classical electron transport in the lower transition region, ApJ, 300 (1986)
7. Spitzer L., Ḧarm, R., Phys. Rev, 89 (1953)
8. Braginskii, S., Transport Processes in a plasma in Reviews in Plasma Physics, 1 (1965)
9. Lie-Svendsen O., Holzer, E., Leer, E., Electron heat flux in the solar transition region: validity of the classical description, ApJ, 525 (1999)
10. Lie-Svendsen O., Leer, E., The electron velocity distribution in high-speed solar wind: Modelling the effects of protons, JGR, 105 (2000)
11. Shoub E. C., Invalidity of local thermodynamic equilibrium for electrons in the solar transition region. I. Fokker-Planck results, ApJ, 266 (1983)
12. Gray D. R., Kilkenny J. D., Plasma Phys., 22 (1980)
13. Campbell P. M., Transport phenomena in a completely ionized gas with large temperature gradients, Physical Review A, 30 (1984)
14. Landi S., Pantellini F. G. E., On the temperature profile and heat flux in the solar corona: Kinetic simulations, A&A, 372 (2001)

Large Amplitude Alfvén Waves In Open And Closed Coronal Structures

R. Grappin*, J. Léorat*, S. R. Habbal¶

*LUTH, CNRS, Observatoire de Paris-Meudon
¶University of Wales at Aberystwyth

Abstract. The time-dependent response of the corona in a spherical shell between 1.8 and 16 R_s to injection of low-frequency Alfvén waves at the inner boundary is considered in the MHD, isothermal and axisymmetric framework, without approximation for the wave-wind coupling. The magnetic field is the sum of an external dipole field assumed to be produced by the sun and of the field induced by the plasma motion in the spherical shell. Due to Alfvén wave injection, the wind and magnetic structure change, leading to an increased overexpansion of the high-latitudes flows and fields. Some of the factors which affect these changes: dissipation, and latitudinal distribution of the waves are explored, and the quantitative relation between wind speed and wave amplitude are discussed. We conclude that Alfvén waves alone lead ultimately to the disappearance of the slow wind, and that other factors, such as transverse structures and compressive waves, are necessary to explain the observed structure of the solar wind

INTRODUCTION

It is well-known that one-fluid models have limitations when dealing with the heating problem (see the review by Hollweg in the present proceedings). Nevertheless, we believe that the dynamics of the corona, and in particular its response to perturbations, can be adequately studied in the framework of MHD, as soon as we assume that it is approximately (or fully as done here) isothermal.

In fluid models of the solar wind, an additional simplifying assumption, namely that waves are pure Alfvén waves following the WKB prediction is frequently made, together with a damping length, to be fixed ad hoc. Such an assumption allows to represent the effect of the waves by a time-average pressure term (the wave pressure). However, in doing so, one misses the effects of other wave modes which are often strongly coupled with the Alfvén mode, as we will see here. Why are other coupled modes important? The basic reason is that the mean flow velocity is not solution of the primitive equations with time-averaged coefficients, because the auto-correlation of the fluctuations is not zero in general: this is the origin of the Alfvén wave pressure [1]. Other types of fluctuations also contribute, as found for instance in coronal hole simulations [2], and may lead to surprising results, as in the particular case of upward propagating acoustic fluctuations, where the wave pressure term appears with a negative sign, so that the wind gets decelerated [3].

These preliminary remarks may sound irrelevant, as in fact models based on averaged Alfvén wave-pressure terms are able to predict the important properties of the wind like the factor two contrast between fast and slow wind [4]. However, in recent time-dependent simulations which do not rely on the pure Alfvén wave WKB approximation [5], we find that the equatorial wind is also strongly accelerated by the waves, so that the contrast between fast and slow wind is not large enough to meet the factor two, when extrapolated to large distances. In this previous work, we proposed two possible explanations of the discrepancy: either the latitudinal distribution of wave injection (which includes the closed field region in our work, not in Usmanov et al's work) was too wide, or the Reynolds number is too low in our simulations.

In this previous work [5], we observed other long-time phenomena related to flows in the closed magnetic structures, but we choose here to concentrate on the issue of the overall wind structure due to Alfvén

CP679, *Solar Wind Ten: Proceedings of the Tenth International Solar Wind Conference*,
edited by M. Velli, R. Bruno, and F. Malara

waves, and in particular on the overexpansion of high-latitudes. We find that to cope with this question, compressive fluctuations cannot be neglected, so that it is difficult to trust pure Alfvén wave simulations. We conjecture that compressive fluctuations could disappear in presence of transverse quasi-equilibrium structures, but these also could not be described in the WKB limit.

METHOD

We consider the corona between 1.8 and 16 R_s. The parameters are as follows [5]: temperature is uniformly at 1.6 MK; the unperturbed wind has β at the inner radius equal to 0.017 at the poles, 0.086 at the equator. The magnetic field is the sum of a fixed dipole field and of the field induced by the flow. Boundaries are open, only upward propagating waves are specified, not fields. The wave amplitude is 140 km/s, compatible with the upper limit by Esser et al. [6]; the wave period is 18 minutes; eventually also 32 min and 1 hour 24 min.

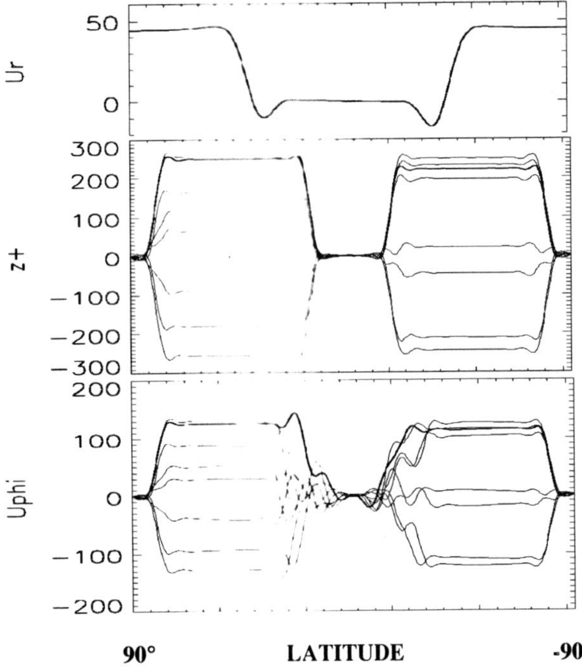

FIGURE 1. Wave injection conditions versus latitude at the inner boundary (1.8 R_s). Top: radial velocity profile before injection. Mid panel: $z_+=u_\phi-\text{sign}(B_r)b_\phi/\sqrt{\rho}$; bottom: u_ϕ; from start up to 16.8 hrs.

The injection of Alfvén waves in the simplest case is illustrated in Fig.1. We start from a relaxed wind with the usual properties of the Pneuman and Kopp [7] solution: at the inner boundary, there is a large region around the equator with a quasi-stagnation region (with albeit some small inflows). We excite outward propagating Alfvén waves via the incoming Alfvén characteristics, while there is no perturbation introduced via the incoming slow and fast characteristics [8]. This is done in a latitudinal range including a part of the stagnant region (mid panel). The poles are not excited, because this would be incompatible with the axisymmetric hypothesis, and a small region around the equator is excluded too, as this would also introduce singularities. In our basic run, waves are in phase in each hemisphere. The resulting azimuthal velocity (bottom panel) has an approximately uniform amplitude outside of the stagnant region, but is strongly varying within the stagnant region, due to interference there between waves propagating upward and downward along loops of closed field lines.

Note that the code has explicit dissipation terms: viscosity ν, as well as artificial dissipation σ for density, and filters. The overall dissipation is mainly due to σ and ν [5].

ALFVEN WAVES INCREASE OVEREXPANSION

Fig.2 shows, in the case of a non-monochromatic wave (three frequencies) with modulation of the phase with latitude, how waves injected in a large portion of the meridian propagate into the corona. The run starts with the unperturbed quasi-stationary wind, which shows a (mildly) slow equatorial wind with a current sheet where the density is slightly larger, and a (mildly) faster wind at high latitudes. Once injected, waves first propagate along the magnetic field lines without modifying them too much (note that the magnetic field lines are about parallel to the flow lines). The situation after 8.4 hrs is shown in the left panel. In a second phase (after 33.5 hrs injection, right panel), the flow lines are strongly bent towards the equator, so that the waves, which were previously excluded from this region, invade now almost completely the current sheet. As a consequence, the wind speed, which increased progressively at first well outside the equatorial region, is at the end substantially accelerated, even in the ecliptic.

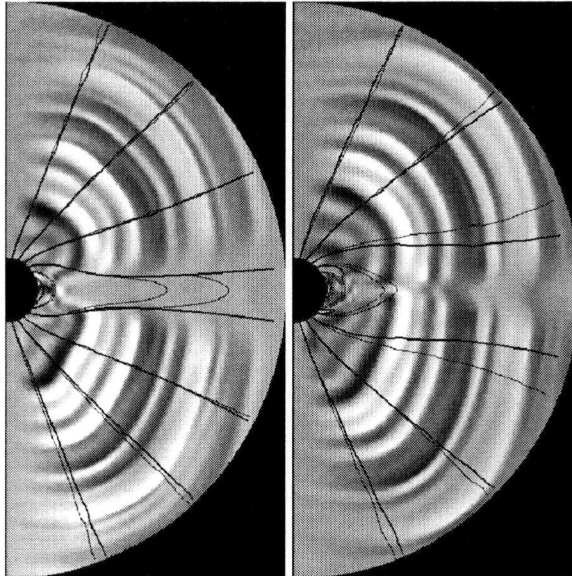

FIGURE 2. Wave field (azimuthal component of the velocity) propagating in the accelerating region between 1.8 and 16 R_s. Low dissipation, three frequencies, with modulation of the phase. Left panel: after 8.4 hrs injection; right panel: after 33.5 hrs injection. Lines are selected lines starting at a given set of latitudes (same for left and right panels). Bold lines are flowlines, thin lines are magnetic field lines.

Fig.3 shows, in the case of a monochromatic wave with uniform phase as in Fig.1, the wind speed at three stages: unperturbed wind, after 8.4 hours injection and after 33.5 hrs injection. The progressive closure of the near-equatorial field lines due to the overexpansion of the high latitudes is again observed, together with the progressive acceleration of the global flow.

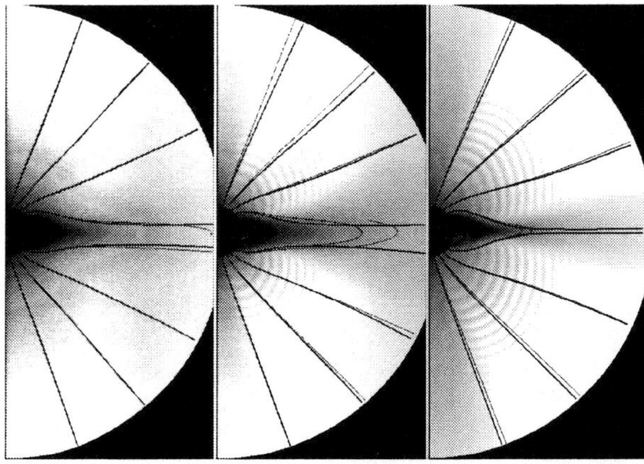

FIGURE 3. Monochromatic waves, uniform phases. Radial velocity field at three stages: unperturbed wind, transient and developed (8.4 hrs and 33.5 hrs). Lines: same as in Fig.2.

Due to the progressive closure of the near-equatorial field lines, and to the associated invasion of the ecliptic by the wave trains (Fig.2), the wind speed is not increased only at high latitudes, but also at the equator. As a result, the contrast between the high-latitude and the equatorial wind is not much increased, compared with the stationary Pneuman and Kopp wind [7], and finally it is far, at the outlet, from the factor two observed by Ulysses. We examine now whether this situation is generic or not, by varying some of the parameters of the numerical simulations.

SAVE THE SLOW WIND?

Diffusion leads to radial momentum transfer between fast and equatorial wind, and hence could be responsible for a large part for the acceleration of the slow wind [5]. To check this idea, we compare in Fig.4 the outlet radial velocity (16 R_s), obtained with two viscosities differing by a factor three. Profiles are drawn between 16.8 and 33.5 injection time, illustrating the evolution.

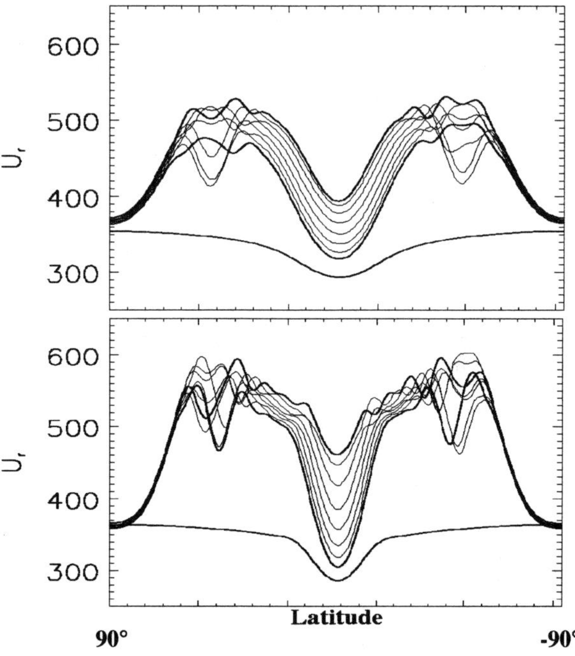

FIGURE 4. Radial velocity profiles versus latitude at outlet. Low Reynolds (top) and large Reynolds (bottom). Low profiles (below 400 km/s): unperturbed wind. High profiles: from 16.8 hrs up to 33.5 hrs after injection start. Note velocity is always increasing with time at the equator.

We find that decreasing the viscosity has two effects. First, the fast wind speed increases substantially. Second, the slow wind region becomes thinner, and its speed increases even more than that of the fast wind. Finally, the fast/slow wind contrast is not increased, but decreased, and it seems plausible that in the limit of infinite Reynolds number, the slow wind would completely disappear.

A direct way to prevent accelerating the equatorial wind is to restrict waves to the open regions. This is illustrated in Fig.5. The top and mid panels show the azimuthal velocity fluctuations, with respectively the previous, wide injection range, and the smaller domain, restricted to open regions. The bottom panel shows the radial velocity profile for reference, the lower curve showing the unperturbed velocity profile, the other curves showing the perturbed profiles generated by the waves when injected in the small domain.

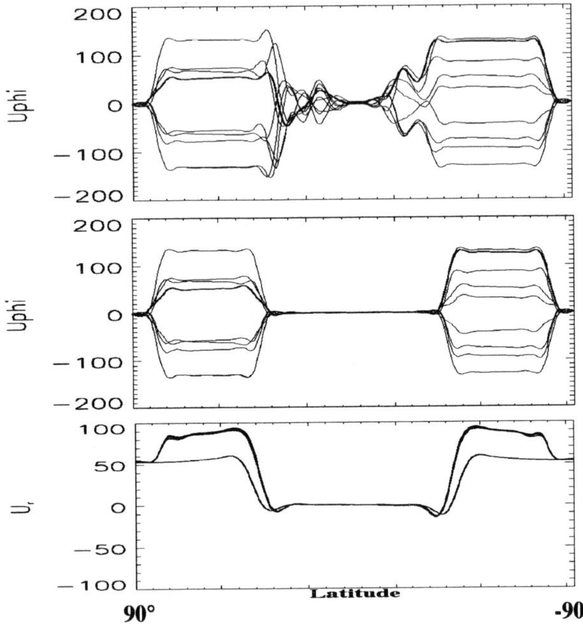

FIGURE 5. The two injection domains at inner boundary. Successive azimuthal velocity injection as in previous runs (top panel) and in a smaller, restricted domain (mid panel). Bottom: radial velocity profiles; unperturbed and perturbed with the restricted injection.

Fig.6 allows to compare propagation in the two cases, after 33.5 hrs injection. One sees immediately that the second injection mode (bottom) preserves, as expected, a large equatorial region from waves, as the overexpansion remains limited. Let us examine again the propagation with the previous injection mode where waves travel on the boundary of the closed region (top figure): one sees that, although wavetrains propagate at very different speeds and angle inside and outside the stagnant region, the wave packets outside remain connected with the wavepackets inside, so that wavetrains outside the closed region remain close to it, which ultimately lead to the overexpansion observed.

This description is of course valid only after long times. On the contrary, when the injection zone is restricted to the open region, the waves do not follow closely the stagnation region, and the unperturbed region remains large.

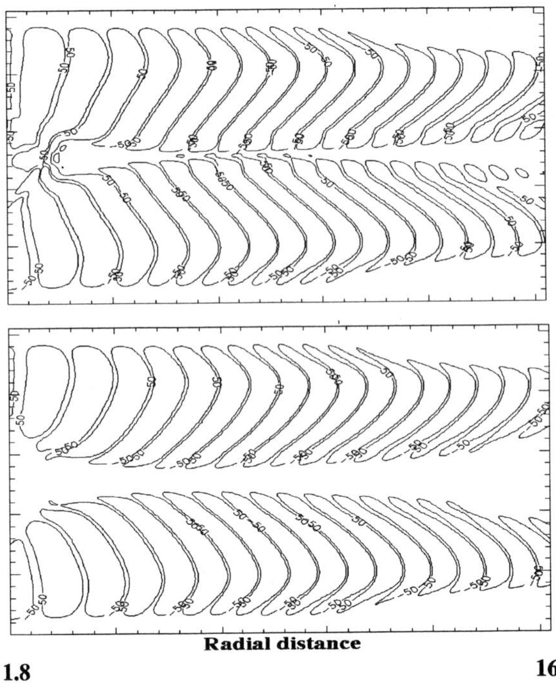

FIGURE 6. Distance-latitude maps of azimuthal velocity (±50km/s isolines) after 33.5 hrs injection. Top: large injection zone. Bottom: restricted injection zone.

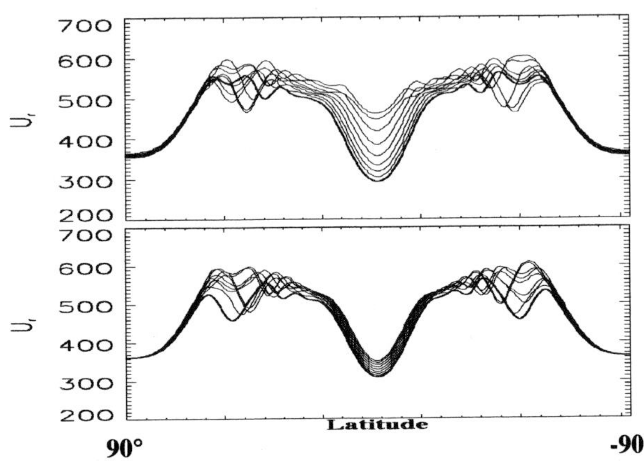

FIGURE 7. Radial velocity profiles at the outlet (16 Rs). Evolution between 16.8 hrs and 33.5 hrs injection Top: large injection. Bottom: restricted injection.

Fig.7 shows several radial velocity profiles at outlet, during the period between 16.8 and 33.5 hrs injection. One sees that the equatorial speed is increasing at a much slower rate when the injection region is limited to open lines. However, there is no indication that the process relaxes, and that the overexpansion stops increasing. In fact, with limited injection, the process is slowed down, but does not seem to be suppressed.

FAST WIND PROPERTIES

We now examine whether we can predict the wind speed from the wave amplitude. We consider for that a particular case, namely a simulation with three frequencies and a large injection region.

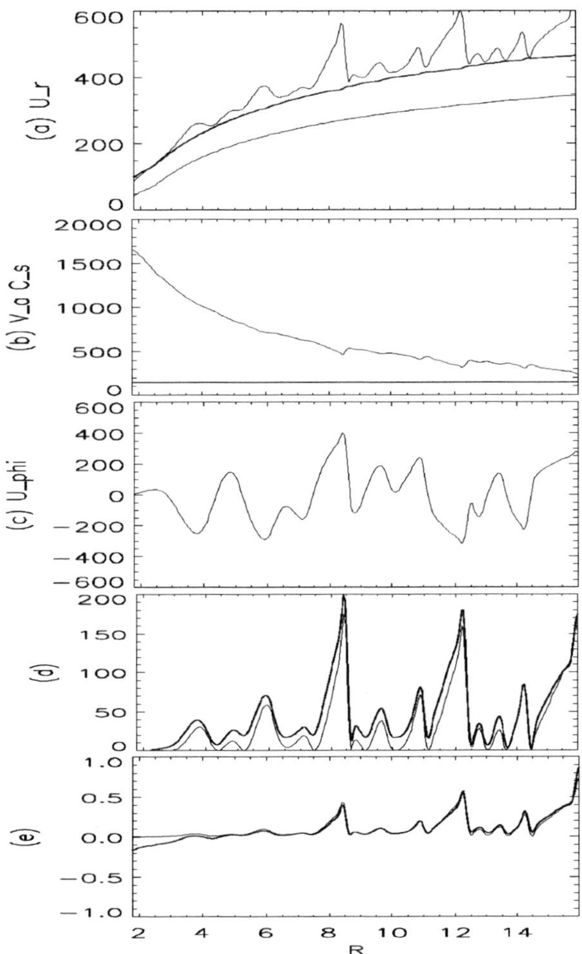

FIGURE 8. Radial cut after 33.5 hrs injection, at 39° from south pole (three frequencies). (a) radial velocity of unperturbed flow, of instantaneous perturbed flow, and estimated background flow U^1_r (bold, see text). (b) Alfvén and sound speeds. (c) azimuthal velocity (d) δu_r and $0.5\ u_f^2 / v_a$; (e) $\delta\rho/\rho$ and $\delta u_r/v_a$ (see text).

Let us examine how quantities vary with distance, at a given time (33.5 hours injection), and at a given latitude, 39° from south pole. Fig.8 shows several profiles. First one (a) shows three radial velocities: the unperturbed one, the instantaneous one (after 33.5 hrs injection), and a background profile U^1_r, which is empirically computed from the unperturbed profile $U°_r$ by the following formula:

$$U^1_r = U°_r + A \langle u_\phi^2 \rangle / (v_a c_s)^{1/2} \qquad (1)$$

where v_a is the Alfvén speed, c_s the sound speed, A is equal to 1 and $\langle u_\phi^2 \rangle$ is the variance of the azimuthal velocity along the radial line (panel c). As shown in the top panel, the predicted background profile U^1_r fits well the real background profile, that is, the profile obtained by erasing the peaks from the instantaneous wind. Note that the Alfvén speed varies by a large factor in the distance range, so that relation (1) should have some physical significance.

Panel (d) shows another fit, which relates the instantaneous amplitude of the fluctuating wind, $\delta u_r = u_r - U^1_r$, to an expression involving the instantaneous wave amplitude u_ϕ:

$$\delta u_r \approx 0.5\ u_\phi^2 / v_a \qquad (2)$$

Inspection by eye shows that errors in the fit come mainly from errors in the estimation of the background velocity U^1_r which in turn lead to errors in δu_r. The relation (2) is almost the one expected for sound waves driven by Alfvén waves propagating parallel to the mean field (except for a factor $1/(1-\beta)$, see panel (b)). One can check indeed in panel (e) that the radial velocity fluctuations are related to the density fluctuations by the expected linear relation:

$$\delta\rho/\rho \approx \delta u_r/v_a \qquad (3)$$

Note that the density fluctuation $\delta\rho$ is again computed as for δu_r by taking the difference between the instantaneous value and a background density. The three relations (1) to (3) are found to be quite general: they are valid at all latitudes not too close to the equator, with only some variations around unity for the value of the coefficient A in (1).

DISCUSSION

We studied in this work the propagation of monochromatic (or three frequencies) Alfvén waves in an isothermal, axisymmetric solar wind with external dipole field, and the resulting effect on the wind structure. The simulations show large compressive fluctuations which are difficult to reproduce via a usual WKB-type model, hence confirming that fully time-dependent simulations are necessary.

In summary, the injection of Alfvén waves increases much the overexpansion of high latitudes. The overexpansion is a slow process, which needs first that the overall radial speed be substantially larger than the unperturbed bulk speed. When this is achieved, the fast streams reach even the ecliptic, so that the slow wind practically disappears. The rate at which this is achieved depends on the Reynolds number and the latitudinal distribution of the wave injection. The process is faster when the Reynolds number increases, but slower when wave injection is restricted to initially open regions. Nevertheless, the slow wind seems also to ultimately disappear at long times, even in the latter case.

Although it is not excluded that by playing with injection (intermittency?) one could stop the overexpansion towards equator, it thus seems likely that some additional ingredient is necessary to recover the slow wind. The solution could come from the equator being a direct source of mass flux, which might be possible by considering a multipolar external field, and/or a direct injection of compressive waves.

Another result concerns the quantitative relations between the background fast wind, the instantaneous fast wind, the wave amplitude, and the compressive fluctuations. Relation (1) which predicts the background wind from the mean Alfvén amplitude, the Alfvén and sound speeds remains to be explained theoretically. On the other hand, relations (2) and (3) between Alfvén and radial (compressive) fluctuations are easily interpreted: they are a clear signature of quasi-parallel propagation of Alfvén waves. This holds also when injecting an oblique distribution of wavevectors via a fast phase modulation with latitude (see Fig.2). This is because, in spite of large transverse Alfvén speed gradients which in principle trigger deviations from parallel propagation, expansion restores rapidly quasi-radial wavectors by stretching wavefronts transversally [9]; this effect is visible in Fig.2 by comparing wavefronts angle fluctuations at small and large radial distance.

In the present simulations, in the 10 to 15 R_s range, both the relative rms wave amplitude $\delta B/B$ and the relative density fluctuation are around 0.5 (Fig. 8b, c and e): this is compatible with estimations based on Faraday rotation [10] and radio scintillation [11]. Note however that Fig. 8 indicates a rapid growth of the relative density fluctuation with distance. A less compressible regime has been obtained [8] by considering small amplitude Alfvén waves and temperature structures (channels), in which the waves adopted an oblique, almost incompressible propagation mode.

It would thus be interesting to investigate at what conditions one could recover the oblique quasi-incompressible regime but with large amplitude waves. This is relevant if we recall that sound waves decrease the average wind speed, contrary to Alfvén waves which increase it. Hence, in the regions where Alfvén waves would adopt an incompressible regime, the wind would be faster, while in other regions with more mixed conditions, the wind would be slower. This offers an alternative way to control the fast/slow wind speed ratio, resembling, superficially at least, the situation in the real solar wind [12].

REFERENCES

1. Alazraki, G., and Couturier, P., *Astron. Astrophys.* **13**, 380 (1971)

2. Ofman, L., and Davila, J.M., *J. Geophys. Res.* **103**, 23677 (1998)

3. Grappin, R., Cavillier, E., and Velli, M., *Astron. Astrophys.* **322**, 659 (1997)

4. Usmanov, A. V., Godstein, M. L., Besser, B. P., and Fritzer, J. M., *J. Geophys. Res.* **105**, 12675 (2000)

5. Grappin, R., Léorat, J., and Habbal, R. F., *J. Geophys. Res.* in press (2002)

6. Esser, R., Fineschi, S., Dobrzycka, D., Habbal, S. R., Edgar, R. J., Raymond, J. C., and Kohl, K.L., *Astrophys. J.* **510**, L63-67 (1999)

7. Pneuman, G. W., and Kopp, R.A., *Solar Phys.* **18**, 258 (1971)

8. Grappin, R., Léorat, J., and Buttighoffer, A., *Astron. Astrophys.* **362**, 342 (2000)

9. Völk, H. J., and Alpers, W., *Astrophys. and Sp. Sc.* **20**, 267 (1973)

10. Sakurai, T., and Spangler, S. R., *Astrophys. J.* **434**, 773 (1994)

11. Spangler, S. R., *Astrophys. J.* **576**, 997 (2002)

12. Grappin, R., Mangeney, A., and Velli, M., *Ann. Geophys.* **9**, 416 (1991)

Heating in coronal funnels by ion cyclotron waves

Xing Li

Department of Physics, University of Wales, Aberystwyth, UK

Abstract. Plasma heating by ion cyclotron waves in rapidly expanding flow tubes in the transition region, referred as coronal funnels, is investigated in a three-fluid plasma consisting of protons, electrons and alpha particles (α's). Ion cyclotron waves are able to produce a transition region and a hot corona over a distance range of 10^4 km by directly heating alpha particles. Although only alpha particles dissipate the waves, protons and electrons can also be heated to about 10^6 K due to Coulomb coupling. It is found that alpha particles can be much hotter and faster than protons. Beyond $1.02R_s$, the particles return to thermal equilibrium when the electrons reach about 10^6 K which is canonically defined as the base of the corona. These results lead to the following implications: (1) A transition region and corona may be energized by depositing energy to minor ions only. (2) If spectral lines formed at $T_e < 10^6$ K are observed at different heights, the inferred outflow velocities may vary by a factor of 5 to 6. (3) If minor ions are indeed much faster than protons and electrons at $T_e < 10^6$ K, one cannot reliably determine the bulk outflow velocity of the solar wind in that region by using minor ion outflow velocities. However, when the wave dissipation in the corona occurs much further away from the transition region, the loss of thermal equilibrium between plasma species is much less pronounced, or a transition region and a hot corona cannot be energized by the waves at all.

INTRODUCTION

Recent observations from the Ultraviolet Coronagraph Spectrometer (UVCS) and Solar Ultraviolet Measurements of Emitted Radiation (SUMER) on board Solar and Heliospheric Observatory (SOHO) have found that (1): the temperature of minor ions decreases with increasing mass per charge at the base of the corona in a southern coronal hole [22]; (2) ions not only have greater than mass-proportional temperatures in the inner corona, they are much hotter than electrons and are highly anisotropic as well [10, 15, 3]. These recent progresses strongly indicate that ion cyclotron resonance may play an important role in the coronal heating.

Transition region (TR) models often do not treat electrons and protons as two different fluids [16]. More recent TR models treat electrons and protons as two separate fluids [17, 6]. Ion cyclotron resonance as a mechanism to heat the corona is adopted [17, 6]. In a two fluid approach, a power law spectrum of ion cyclotron waves originating from below the TR was adopted as a mechanism to heat plasma in coronal funnels [17] and the waves are able to generate a hot corona. Electrons and protons start to lose thermal equilibrium when their temperature reaches 10^5K. The electron thermal pressure in [17] is low. As a result, the plasma can reach a bulk flow speed of 140 km/s at the top of the funnels.

Ion cyclotron waves propagating in the TR have important implications for minor ions: since minor ions have lower cyclotron frequency, the species with the lowest cyclotron frequency will be heated first. Since the Coulomb coupling is likely very strong in coronal funnels and the plasma beta value is small, those minor ions and protons are not expected to be far away from equilibrium. However protons and electrons may be not in thermal equilibrium in coronal funnels as shown in [17, 6]. A more relevant question is: is Coulomb coupling strong enough to transport energy from minor ions to the major species protons and electrons? More recently it was found that some TR spectral lines can be fitted by two Gaussians [18, 19]. One possible explanation is that the broad lines may be formed by ions in coronal funnels along open magnetic field lines [19] and the ions are preferentially heated by ion cyclotron waves.

In this paper, the heating of minor ions in coronal funnels along open magnetic field lines by high frequency Alfvén waves is investigated in a fluid approach. These waves are assumed to originate below the TR. It is shown that it is possible for ions in the transition region not to be in thermal equilibrium: they may have different temperature and outflow speed. Readers may also reference recent semi-kinetic calculations of cyclotron heating in coronal holes, oxygen ions are found much hotter (but not faster) than protons and alpha particles at the base of the corona [23, 24, 25].

CP679, *Solar Wind Ten: Proceedings of the Tenth International Solar Wind Conference*, edited by M. Velli, R. Bruno, and F. Malara

MODEL ASSUMPTIONS

Standard three-fluid plasma transport equations in an isotropic, charge-free plasma including electrons, protons and α's are considered. The time dependent mass, momentum and energy equations describing a one-dimensional, three-fluid conductive plasma flow in coronal funnels are the same as those used in [14, 8]. To save space, they will not be repeated here. In the following, various assumptions are summarized.

1. It is assumed that plasmas in coronal funnels are collision dominated and classical thermal conduction parallel to the magnetic field is included for the electrons, protons and α's [1, 12].

2. The radiative energy loss in the electron energy equation is assumed to have the form given by [20] for an optically thin medium. 3. A power law spectrum of Alfvén waves originating below the TR is assumed to be the only energy source of coronal heating. When the high frequency Alfvén waves propagate along open coronal magnetic field lines, ion cyclotron resonance may produce an upper limit on the frequency of Alfvén waves, f_H. This frequency is below the lowest gyrofrequency at a given height. Following [21, 8], f_H is taken as $f_H = \alpha_f f_{\alpha c} v_{\rm ph}/v_A$, where $f_{\alpha c}$ is the gyrofrequency of α's, $v_{\rm ph}$ is the phase speed of dispersionless Alfvén waves, v_A is the Alfvén speed, and α_f is a constant less than unity, taken to be 0.7 in this study. Previous studies have assumed $\alpha_f \leq 0.2$ [8, 12]. However, at such a low frequency, ion cyclotron waves are not in resonance with alpha particles in a low beta plasma such as the one treated here. At $\alpha_f = 0.7$, linear Vlasov theory shows that ion cyclotron waves already show some weak dispersion. This weak dispersion has been neglected. Ignoring nonlinear cascade processes, the evolution of Alfvén wave spectrum in a multi-ion fluid can be written as [9]

$$\frac{\partial P}{\partial t} + \frac{v_A^2}{v_{\rm ph}(v_{\rm ph} - v_{\rm cm})a}\frac{\partial}{\partial r}\left[\frac{v_{\rm ph}^2(v_{\rm ph} - v_{\rm cm})Pa}{v_A^2}\right] = 0, \quad (1)$$

where $v_{\rm cm}$ is the center of mass speed, $P(f,r)$ is the power spectrum density, related to the magnetic field variance $\langle\delta B^2\rangle$ and the Alfvén wave pressure p_w by$\langle\delta B^2\rangle = 8\pi p_w = \int_{f_L}^{f_H} P\,{\rm d}f$. It is also assumed that the Alfvén wave frequencies have a lower limit f_L. Integrating (1) with respect to f over the range (f_L, f_H), and ignoring nonlinear cascade processes leads to

$$\frac{\partial(2p_w)}{\partial t} + \frac{v_A^2}{v_{\rm ph}(v_{\rm ph} - v_{\rm cm})a}\frac{\partial}{\partial r}\left[\frac{2v_{\rm ph}^2(v_{\rm ph} - v_{\rm cm})p_w a}{v_A^2}\right]$$

$$= -Q, \; Q = -\frac{v_{\rm ph}P(f_H,r)}{4\pi}\frac{{\rm d}f_H}{{\rm d}r}, \quad (2)$$

where Q is a measure of energy transfer from waves to plasmas, a is the flow tube cross section area.

If the maximum frequency, f_d, of the Alfvén waves originating below the TR is still lower than the frequency f_H, ion cyclotron resonance does not occur. Hence, $Q = 0$, and the waves will not heat the plasma. Then p_w will be related to $P(f,r)$ as $8\pi p_w = \int_{f_L}^{f_d} P\,{\rm d}f$. Following [17], $P(f) \propto f^{-1}$ is assumed.

It was found that while Q is determined by (2), the total wave dissipation rate by resonant cyclotron interaction is $v_{\rm ph}(v_{\rm ph} - v_{\rm cm})Q/v_A^2$ [8]. The total dissipation rate is divided between the acceleration and heating for different ion species, and is determined by the microphysics of the resonant interaction. The method in [8] is used to determine this energy distribution, cold plasma dispersion relation and $P(k) \propto k^{-5}$ are assumed in the dissipation range, and only α's are heated by the waves.

NUMERICAL RESULTS

The same numerical method in [9, 14] is used in the paper. Model calculations start from a point in the lower TR where the temperature is 6×10^4K and extend to 15000km above it. The radial step dr gradually increases from 35m at the lower boundary, to about 300km at the top of the computational domain. The lower limit on the temperature is chosen such that both hydrogen and helium are almost fully ionized. The corresponding electron density at the boundary is 3.33×10^9cm^{-3}. Thus yielding an electron thermal pressure of $n_e T_e = 2 \times 10^{14}$ K cm^{-3} SUMER observations showed that thermal pressure is roughly a constant (log $n_e T_e = 14.2$) in the transition region when the electron temperature varies from 2×10^4K to 10^6K in coronal holes [26]. The total Alfvén wave amplitude at the base of the funnel is $\delta V = 27$km/s. The magnetic field geometry and a are taken from [11] and parameters are as follows: $f_m = 13$ is the expansion factor of the cross-section of a flux tube, $r_1 = 1.004R_S$ is the heliocentric distance at which the major expansion takes place. The parameter $\sigma = 0.005R_S$ represents the length over which the expansion occurs. The magnetic field at the bottom of the funnel is 130 G. This means that we assume that the coronal funnels considered here may originate from deep in the photosphere [5]. The low- and high-frequency ends of the wave spectrum are fixed at $f_L = 10^{-3}$ Hz and $f_d = 30000$ Hz.

Figure 1c shows the gyrofrequency of α's $f_{\alpha c}$ and $0.7f_{\alpha c}$ as a function of height h, which is defined by $h = r - r_0$. The r_0 is the heliospheric distance at the lower boundary (assumed to be R_S). In this study, flow velocities are much smaller than the Alfvén speed and $v_{\rm ph}$ is nearly identical to v_A. As a result, $f_H \sim 0.7f_{\alpha c}$, $0.7f_{\alpha c}$ is roughly the frequency at which cyclotron resonance dissipates the waves. If $f_d < f_H$, no wave heating will occur. Since the goal of the study is to investigate coronal fun-

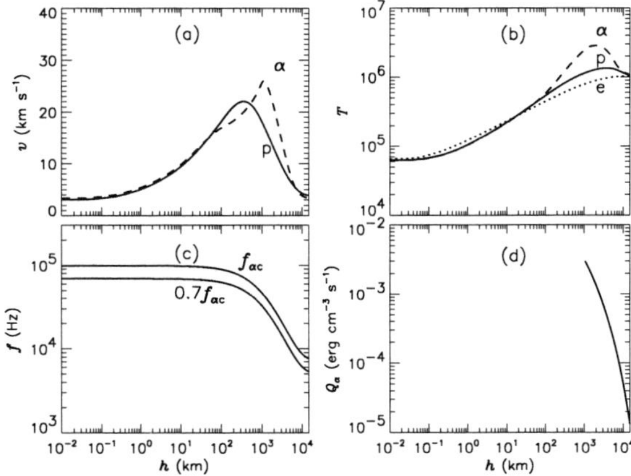

FIGURE 1. A coronal funnel model in which the plasma is heated by ion cyclotron waves. (a) speed of protons (solid line) and alpha particles (dashed line); (b) temperatures of electrons (dotted line), protons (solid line) and alpha particles (dashed line); (c) Gyrofrequency of alpha particles in a coronal funnel; (d) alpha particle heating rate. Here $f_d = 30000$ Hz.

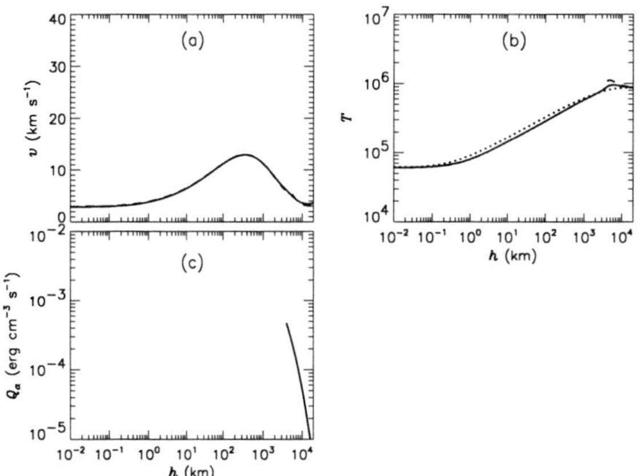

FIGURE 2. The same as Figure 1 but here $f_d = 12000$ Hz.

nels in coronal holes where fast solar wind is believed to originate, the abundance of α's is chosen at 6% [14, 8].

A steady state solution of the model calculations is shown in Figures 1a and 1b. Ion cyclotron waves indeed produce a very rapid temperature increase and a hot corona. When temperatures are still low, α's and protons are strongly coupled, so $T_p = T_\alpha$. At a height of 600km, α's start to become much faster and hotter than protons. The proton velocity at the lower boundary is $v_p = 2.91$km/s, a normalized proton flux density at 1 AU would be $n_p v_p/(215^2 \times f_m \times 6.5) = 2.2 \times 10^8$ cm^{-2} s^{-1}, a typical observed value in the fast wind. The α's reach a maximum velocity of 26 km/s at $h = 1080$km, totally due to the introduction of ion cyclotron resonance. Below this point, the wave frequency at the high end of the spectrum is smaller than the frequency where the α's can resonate with the waves, no wave dissipation occurs. Figure 1 shows that at the top of the TR or at the base of the corona, α's and protons (and electrons) can be significantly away from thermal equilibrium. This is the first time shown in model calculations that minor ions flow much faster and are much hotter than protons in the TR. In the nearby region below 600km, α's are significantly slower than protons. This is due to the higher temperature gradient of α's, and a smaller pressure gradient force for α's. Since ion cyclotron waves only heat alpha particles, the electrons are heated by Coulomb collisional coupling with protons and the α's, and by the electron heat flux.

Observations of TR spectral lines often find various ion velocities. These velocities can vary from a few to about twenty kilometers per second. For instance, it was found that in network boundaries an average outflow velocity is several kilometers per second when $T > 6.3 \times 10^5$K[7], but the outflow velocity of Ne^{7+} was 14km/s along open magnetic field lines from SUMER data [27]. To assume that other minor ions behave similarly to the α's, their velocity will vary in a similar manner. Hence, if spectral lines formed at $T_e < 10^6$ K are observed at different heights, the inferred outflow velocities may vary accordingly.

The fact that minor ions can flow much faster than the bulk plasma flow in the upper TR may have implications for the analysis of TR spectral line observations. Traditional atomic physics calculations on which the diagnostics of both temperatures and densities depend often assume equal ion and electron temperatures. From Doppler shift measurements of TR spectral lines, the mass flux in the transition region was found more than enough to account for in situ measured solar wind mass flux [4]. If minor ions are indeed much faster than protons and electrons at $T_e < 10^6$ K, one cannot reliably determine the bulk outflow velocity of plasmas in that region by using minor ion outflow velocities.

Figure 1d shows the alpha particle heating rate Q_α in the computation domain. This plasma heating rate is similar to those given by [17, 6]. In those studies, heat is directly deposited to protons. At the lower boundary of the computational domain, $f_H > f_d$, no wave heating is possible. At 1080km, the maximum frequency f_d finally reaches f_H and a strong heating is found. In Figure 1a when $h < 10$km, waves do not heat plasmas. The α's are slightly faster than protons due to a larger pressure gradient force. The electrons are little hotter than ions due to a much a greater heat flux.

When the maximum wave frequency f_d is smaller, wave dissipation will occur at a higher height. Shown in Figure 2 is a model in which the same parameters as those in Figure 1 except $f_d = 12000$ Hz are used. In this case, the initial plasma heating rate in Figure 2c is much smaller than that in Figure 1d. This is because the rapid decline of the magnetic field with height leads to a rapidly declining Alfvén speed and F_H. Hence Q will decline very rapidly with height as well. The plasma species are roughly in thermal equilibrium (Figures 2a and 2b), plasma species have the same flow speed. The α's are only slightly hotter than protons at the location where wave dissipation starts. SUMER observations have shown multi-million kinetic temperatures of minor ions at the base of the corona [22]. It seems that the result shown in Figure 1 may fit SUMER observations better than that in Figure 2. Note both models produce almost identical mass fluxes, the proton velocity is 2.85 km/s at the lower boundary in Figure 2 (compared to 2.91 km/s in Figure 1).

DISCUSSION

The examples given above have some implications for the coronal heating by a spectrum of ion cyclotron waves. If the high frequency limit of the spectrum is much higher than 30000Hz (Figure 1), more energy will be released closer to the lower boundary, the temperatures of species will increase more rapidly. It is expected that alpha particles will become much hotter and faster than protons in the region of initial ion cyclotron resonance. However, a steeper temperature gradient will also lead to a larger electron heat flux and a hotter electron temperature in the region close to the low boundary (as shown in the Figure 1). This will violate our assumption of equal species temperatures at the lower boundary. In such a case, it is necessary to develop a model to include chromosphere. A full treatment of hydrogen and helium ionization processes will have to be included.

On the other hand, if the high frequency limit is lower than 12000 Hz, it will lead to a smaller electron temperature gradient in the transition region. Eventually, a smaller electron heat flux will not be able to balance the radiation loss, and a hot corona cannot be created.

This study is limited to the case that the frequency of ion cyclotron waves is smaller than the gyrofrequency of α's. If waves with frequency between α and proton gyrofrequency also exist, protons will certainly become resonant with them. However these waves are difficult to excite according to linear Vlasov theory.

In summary, plasma heating in coronal funnels along open magnetic field lines by parallel propagating ion cyclotron waves is investigated. By heating alpha particles alone, the Coulomb coupling between α's and protons/electrons is strong enough to produce a TR and a hot corona. Minor ions and protons may lose thermal equilibrium in the transition region: they have different temperature and outflow speed. This study suggests that the corona and the solar wind may be energized solely through minor ions. In reality, the coronal heating physics may be far more complicated. Since there are many ions heavier than the alpha particles in the TR and their gyrofrequencies are smaller than that of α's, those heavy ions will be heated first if the waves considered here are indeed responsible for the coronal heating [2]. Due to their small abundances, it remains an open question whether the Coulomb coupling is able to transfer the energy from these minor ions to the major species.

ACKNOWLEDGMENTS

This work was supported by a PPARC rolling grant to UWA. Part of the work was supported by grant NAG5-10873 to Smithsonian Astrophysical Observatory (SAO).

REFERENCES

1. Braginskii, S.I., *Reviews of Plasma Physics*, **1**, 204, 1965.
2. Cranmer, S.R., *ApJ* **532**, 1197, 2000.
3. Cranmer, S.R., Kohl, J.L., et al., *ApJ* **511**, 481, 1999.
4. Dere, K.P., Bartoe, J.-D.F., Brueckner, G.E., and Recely, F., *ApJ* **345**, L95, 1989.
5. Dowdy, J.F., Robin, D., Moore, R.L., *Solar Phys.* **105**, 35, 1986.
6. Hackenberg, P., Mann, P., and Marsch, E., *A & A* **360**, 1139, 2000.
7. Hassler, D.M., Dammasch, I.E., Lemaire, P., et al., *Science* **28** 3, 810, 1999.
8. Hu, Y.Q., and Habbal, S.R., *J. Geophys. Res.* **104**, 1999.
9. Hu, Y.Q., Esser, R., and Habbal, S.R., *J. Geophys. Res.* **102**, 14,661, 1997.
10. Kohl, J.L., Noci, G., et al., *ApJ* **501**, L127, 1998.
11. Kopp, R. A., and Holzer, T.E., *Solar Phys.* **49**, 43, 1976.
12. Li, X., *ApJ*, 571, L67, 2002.
13. Li, X., and Habbal, S.R., *Solar Phys.* **190**, 485, 1999.
14. Li, X., Habbal. S.R., and Hu, Y.Q., *J. Geophys. Res.* **102**, 17,419, 1997.
15. Li, X., Habbal, S.R., Kohl, J., and Noci, G., *ApJ* **401**, 133, 1998.
16. Mariska, J.T., *The Solar Transition Region*, Cambridge University Press, Cambridge, 1992.
17. Marsch, E., and Tu, C.Y., *Solar Phys.* **176**, 87, 1997.
18. Peter, H., *A & A* **360**, 761, 2000.
19. Peter, H., *A & A* **374**, 1108, 2001.
20. Rosner, R., Tucker, W.H., and Vaiana, G.S., *ApJ* **503**, 475, 1978.
21. Tu, C.-Y., *Solar Phys.* **109**, 149, 1987.
22. Tu, C.-Y., Marsch, E., Wilhelm, K., and Curdt, W., *ApJ* **503**, 475, 1998.
23. Vocks, C., and Marsch, E., *Geophys. Res. Lett* **28**, 1917, 2001.
24. Vocks, C., *ApJ* **568**, 1017, 2002.
25. Vocks, C., and Marsch, E., *ApJ* **568**, 1030, 2002
26. Warren, H.P. and Hassler, D.M., *J. Geophys. Res.* **104**, 9781, 1999.
27. Wilhelm, K., Dammasch, I.E., Marsch, E., and Hassler, D.M., *A & A* **353**, 749, 2000.

Acceleration of the Solar Wind as a Result of the Reconnection of Open Magnetic Flux with Coronal Loops

L. A. Fisk[1], G. Gloeckler[1,2], T. H. Zurbuchen[1], J. Geiss[3], and N. A. Schwadron[4]

[1]*Department of Atmospheric, Oceanic, and Space Sciences, University of Michigan 2455 Hayward St., Ann Arbor, MI 48109-2143, USA*
[2]*Department of Physics and IPST, University of Maryland, College Park, MD 20742-4111, USA*
[3]*International Space Science Institute, Hallerstrasse 6, CH-3012 Bern, Switzerland*
[4]*Southwest Research Institute, San Antonio, TX 78228, USA*

Abstract. There are compelling observations of a clear anti-correlation between solar wind flow speed and coronal electron temperature, as determined from solar wind ionic charge states. A simple theory is presented which can account for these observations, including the functional form of the correlation: Solar wind flow speed squared varies essentially linearly as the inverse of the coronal electron temperature. In this theory, magnetic field lines in the corona that open into the heliosphere reconnect with coronal loops near their base. This process displaces the open field line, and disturbs and imparts energy into the overlying corona, thereby determining the Poynting vector into the corona. This process releases mass from the loop into the corona, and determines the mass flux of the solar wind. The Poynting and mass flux into the corona determine the final speed of the solar wind, and yield a relationship that provides an excellent fit to observations.

INTRODUCTION

We present a simple theory to explain the compelling observations of *Gloeckler et al.* [1] of a clear anti-correlation between the solar wind flow speed and the coronal electron temperature, as determined from solar wind ionic charge states. The anti-correlation is consistent with a specific curve, motivated by the theory that is presented in this paper: Solar wind flow speed squared varies essentially linearly as the inverse of the coronal electron temperature.

The theory is an outgrowth of our work on the transport of open magnetic flux on the Sun [2,3,4], and on the current understanding of the behavior of the coronal magnetic field and the development of coronal loops [5,6,7]. There is considerable evidence that open magnetic field lines on the Sun (those that open into the heliosphere) readily reconnect with closed magnetic loops. This results in a diffusive transport of the open flux that can account for the configuration of the heliospheric magnetic field, and offers an explanation for the formation of coronal holes and the apparent rotation of the large-scale current sheet that separates opposite polarities of the heliospheric magnetic field, during the solar cycle [4]. Processes similar to this are invoked to explain the evolution of the polar magnetic field of the Sun [8], and to explain the apparent ease with which the magnetic fields in coronal mass ejections (CMEs) become detached from the Sun, and do not result in a buildup of the magnetic field in the heliosphere [9].

CP679, *Solar Wind Ten: Proceedings of the Tenth International Solar Wind Conference*, edited by M. Velli, R. Bruno, and F. Malara

The theory for the solar wind presented here is based on this simple process. An open field line reconnects with a closed magnetic loop and is displaced in its location. This displacement disturbs the overlying coronal magnetic field and deposits energy into the corona. The reconnection permits the mass originally stored on the coronal loop to be released onto the open field line. As we shall demonstrate with a set of remarkably simple assumptions, it is possible to derive a formula that exactly accounts for the observations of *Gloeckler et al.* [1].

We begin by reviewing the observations of *Gloeckler et al.* [1]. We then summarize the theory that can explain these observations. Full details of the theory can be found in *Fisk* [10]. Finally, we discuss some of the implications of this theory for other theories of the acceleration of the solar wind.

THE OBSERVATIONS

In Fig. 1, from *Gloeckler et al.* [1], time variations of the inverse of the coronal electron temperature ($1/T$) in units of 10^6 K (open circles) and of the solar wind proton bulk speed (V_{sw}) in units of km/s (dotted curves) are plotted during a 166-day time period (August 27, 1996 – February 9, 1997) observed with SWICS on Ulysses. The coronal electron temperature is determined from the ratio of O^{7+}/O^{6+}, which freezes-in in the low corona. The data are 3-point running averages of the basic 12-hour averages. The tracking of the $1/T$ and V_{sw} curves is almost perfect except during the two time periods indicated by the shaded regions. Each of these two time periods coincides with a Coronal Mass Ejection (CME) event identified using bi-directional electron signatures in the Ulysses SWOOPS data (J. Gosling, private communication).

The relationship between solar wind speed and coronal electron temperature observed by *Gloeckler et al.* [1] is a surprise. There is no expectation that electrons in the corona have a major, direct role in the acceleration of the solar wind, particularly the fast solar wind. The temperatures and densities of the electrons, and the resulting pressure, are insufficient to accelerate the solar wind to the observed speeds of up to ~800 km/s. This has led to numerous models for the acceleration of the solar wind in which the protons must obtain the required large pressures [e.g.,

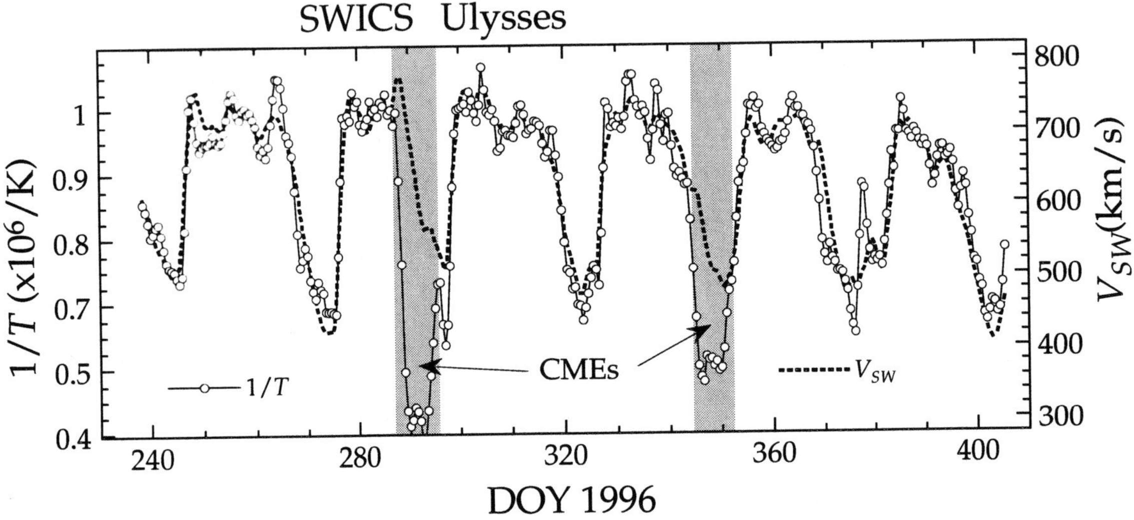

Figure 1. Time variations of the inverse of the electron temperature ($1/T$) in units of 10^6 K (open circles) and of the solar wind proton bulk speed (V_{sw}) in units of km/s (dotted curves) during a 166-day time period (August 27, 1996 – February 9, 1997) observed with SWICS on Ulysses.

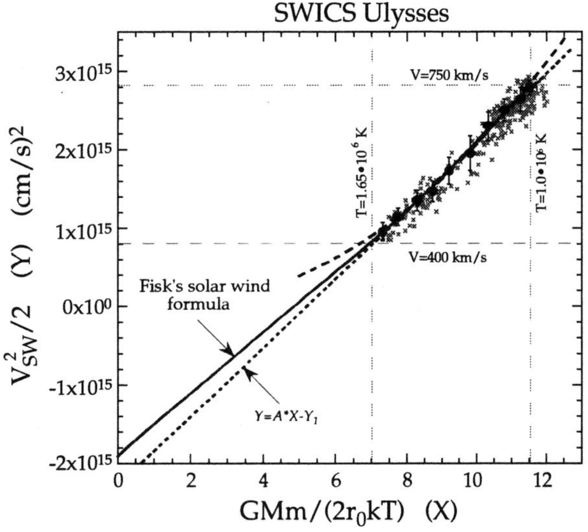

Figure 2. Scatter plot (gray crosses) of $Y = (V_{sw}^2)/2$ vs. $X = \mathrm{GMm}/(2\mathrm{r_0k}T)$ using data of Figure 1 except for the two CME periods.

[11]]. It is perhaps equally surprising that the solar wind speed is anti-correlated with the coronal electron temperature. In models where there is both proton and electron heating, and yet the electrons remain cooler due to heat conduction into the chromosphere [11], we might expect that higher proton temperatures, and thus high flow speeds, were directly correlated with the coronal electron temperature. Rather, the observations of *Gloeckler et al.* [1], that flow speed and coronal electron temperature are anti-correlated, is providing us with unique information on conditions and processes occurring in the corona, which are in turn responsible for the final speed of the solar wind.

In Fig. 2, from *Gloeckler et al.* [1], a scatter plot (gray crosses) of $Y = (V_{sw}^2)/2$ vs. $X = \mathrm{GMm}/(2\mathrm{r_0k}T)$ is created using data of Fig. 1 except for the two CME periods. The ten points indicated by solid circles are averages, and their error bars are standard deviations of (X_i, Y_i) pairs binned in ten equal X intervals.

There is, of course, scatter in the points, due to variations on the Sun and stream-stream interactions in the solar wind. Moreover, as would be expected, the specific curve, solar wind speed squared versus the inverse of the electron temperature, is most readily discernible when a broad range of solar wind speeds and coronal electron temperatures are considered. Such conditions are most prevalent at solar minimum, when both high and low speed flows occur. When a simple average is performed in these conditions, the specific curve holds for both fast solar wind from coronal holes and slower wind from elsewhere on the Sun. The only exception is solar wind plasma associated with CMEs, and even here, the relationship can be argued to hold with different choices for solar parameters [1].

THE THEORY

The theory for the solar wind model presented here is described in detail in *Fisk* [10]. The essential features of this model are based on a set of simple principles:

Magnetic loops are observed to occur everywhere on the Sun; they are believed to result from small bi-polar magnetic flux emerging through the solar surface and coalescing with each other by reconnection to form bigger loops (e.g. [6]), as is illustrated in Fig. 3. In coronal holes the loops are relatively small (heights <15,000 km) and cool (<800,000 K); outside of coronal holes the loops on the quiet Sun are larger (heights of 40,000-400,000 km) and hotter (~$1.5 \cdot 10^6$ K) [12].

Open field lines (magnetic field lines that open into the heliosphere) are present among the loops, in strong concentrations in coronal holes, but distributed also throughout the quiet Sun in lesser strength.

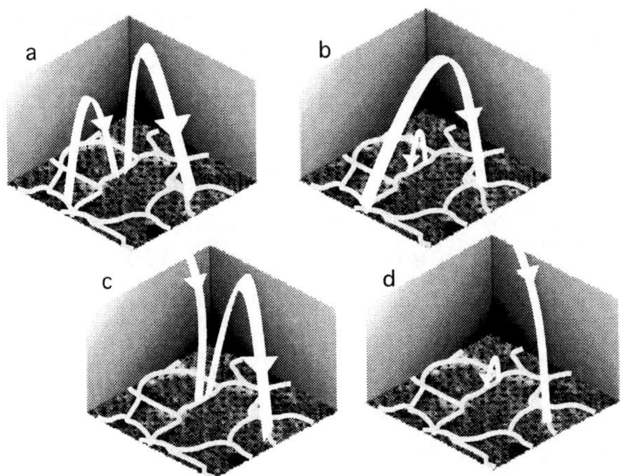

Figure 3. An illustration of the reconnection of loops and open field lines. In panel (a), the footpoints of two loops move with convective velocities along the lanes separating the granular and supergranular cells on the solar surface. In panel (b), two of the footpoints of the loops have reconnected to form a new larger loop and a small secondary loop that will subduct back into the photosphere. In panel (c), the footpoints of a loop and an open field line move along the lanes. In panel (d), a footpoint of the loop and the open field line have reconnected, the open field is displaced to lie over the location of another footpoint of the loop, and a small secondary loop is again formed that should subduct back into the photosphere.

As is illustrated in Fig. 3, an open field line can reconnect with the end of a loop with opposite magnetic polarity, presumably near the loop base, with three consequences: (a) The loop is effectively eliminated (a small secondary loop is created that appears to subduct back into the photosphere). (b) The open field line is displaced to lie over the location of the side of the original loop with the same polarity. (c) Mass is released from the loop onto the open field line.

The displacement of the open field line will disturb the overlying corona. Magnetic pressure variations will be introduced. When the coronal magnetic field relaxes back to equilibrium, work is done, and energy is assumed to be deposited in the corona in the form of heat. The amount of energy that is deposited can be readily calculated (see *Fisk* [10]).

The mass that is released through the reconnection process depends on the mass available in the loop. If the loops are isothermal, the available mass depends mainly on the scale height, which in turn depends on the isothermal temperature. For loop heights that are comparable to the scale height a small correction factor is required.

The energy that is supplied can be represented as a Poynting vector into the corona. Loops emerge through the solar surface, and thus represent an upward Poynting vector, for which there is no comparable downward Poynting vector since the loops are in large part eliminated by the reconnection with open field lines. The mass supplied through the reconnection process will determine the mass flux of the solar wind.

A simple energy balance equation can be used to specify the final solar wind speed squared, V_{sw}^2, in terms of the Poynting vector and mass flux [13]. It yields a unique formula (equation 11 of *Fisk* [10]) that predicts that the speed squared varies essentially linearly as the inverse of the loop temperature T,

$$\frac{V_{SW}^2}{2}=\left(\frac{B_{loop}}{\rho_{loop}}\right)\left(\frac{\int \mathbf{B}_{open}\cdot d\mathbf{h}}{4\pi r_0}\right)\left(\frac{GM_0 m_p}{2r_0 kT}\right)\beta\left(h_{loop},T\right)-\frac{GM_0}{r_0}$$

$$\beta\left(h_{loop},T\right)=\left\{1-\exp\left[-\left(1.75h_{loop}GM_0 m_p\right)/\left(2r_0^2 kT\right)\right]\right\}^{-1}. \quad (1)$$

Here, G is the gravitational constant, M_0 is the mass and r_0 the radius of the Sun, m is the proton mass, k the Boltzmann constant, and h_{loop} is the height of the loop above the point of reconnection. $(B/\rho)_{loop}$ is the ratio of the magnetic field strength to the density of the loop at the base, where the reconnection with the open field line occurs and $\int \boldsymbol{B}_{open}\bullet d\boldsymbol{h}$ is an integral along the open magnetic field from the surface of the Sun to large distances where B_{open} becomes negligible. *Fisk* [10] points out that if the open magnetic field in the corona can be described as a potential field, this integral will be approximately constant for all open field lines, regardless of whether they undergo a radial or a super-radial expansion, and has a value of $9.6\times10^{10}G$.

The factor $\beta(h_{loop},T)$ is significant only if the scale height and the loop height are comparable. For loops on the quiet Sun, *Feldman et al.* [12] find that the height of loops increases with increasing temperature of the material in the loops; hotter loops overlie cooler ones. Therefore, $\beta(h_{loop},T)$ should be primarily a function of temperature but not a strong function of temperature since the temperature dependence enters as h_{loop}/T.

Thus, provided that the quantity $(B/\rho)_{\text{loop}}$ is relatively constant on the Sun and, $\beta(h_{loop},T)$depends only weakly on T, we predict that the acceleration of the solar wind depends on only one parameter, the temperature of the material in the originating loops, and that the final speed squared varies essentially linearly as $1/T$. The quantity $(B/\rho)_{\text{loop}}$ would in fact be relatively constant, if loops expand such that the density and magnetic field strength stay proportional to one another.

The dependence on loop temperature in equation (1) arises simply because the mass available, and thus the mass flux, is proportional to the scale height, which in turn is proportional to temperature. The final speed squared of the solar wind varies inversely with the mass flux.

Equation (1) should hold in all forms of the solar wind: fast solar wind from coronal holes, where the loops involved are smaller and cooler; and slow solar wind from elsewhere on the Sun, where the loops are larger and hotter.

Equation (1) requires knowledge of the actual loop temperature. *Gloeckler et al.* [1] observe the solar wind ionic charge states and from that determine the electron temperature at the point in the corona where the charge states freeze-in. It is not unreasonable that these two temperatures will be nearly identical. First of all, the loops that are used are observed to have temperatures comparable to those inferred from solar wind charge states. A typical, relatively large coronal loop on the quiet Sun, which should be responsible for the slow solar wind, is observed to have temperatures $\sim1.5\bullet10^6$ K [12], whereas coronal electron temperature inferred from charge states is $\sim1.7\bullet10^6$ K [14]. Similarly, in the fast solar wind, the loops have temperatures ~800,000 K, whereas the temperature inferred from charge states is $\sim1.0\bullet10^6$ K. These apparent systematic differences could result from the reconnection process itself, when the open field line reconnects with the loop. A small amount of heat could be imparted to the loop. The actual release process itself, in which there is a sudden drop in density, could facilitate the freeze-in of ionic charge states at the point of release. Conversely, the free flow of electrons along the open field lines could preserve the electron temperature near the loop value, and the freeze-in occurs at the more traditional distances of a few solar radii [e.g. 15,16]. In the detailed numerical model of *Schwadron* [17], the calculated solar wind charge states are found to be representative of the electron temperatures in the loops. Protons, in contrast, need to be heated in the corona by the dissipation of the energy imparted by the displaced open field lines, in order to form the solar wind.

All that is required to test equation (1) using SWICS observations is that there is a one-to-one relationship between the coronal electron temperature inferred from charge state measurements and the actual temperature of electrons in the loop. For example, if these two temperatures differ by a constant amount, there is no change in the use of equation (1) to relate observed solar speed to freezing-in temperature from observed charge states, other than a small adjustment to the inferred dependence of loop heights on electron temperature in that loop.

COMPARISON WITH OBSERVATIONS

The dotted curve in Fig. 2 is a linear least-squares fit to the data, the dashed curve is a linear function, speed vs. $1/T$, and the solid curve is a fit using equation (1). Clearly, (1) does provide an excellent fit to the data. It is necessary in obtaining the detailed fit to have loop height vary directly with loop temperature. Note that the intercept of the curve, GM_o/r_o, the gravitational potential per unit mass of the Sun, is not an adjustable parameter.

CONCLUDING REMARKS

There are, of course, many solar wind theories more complicated than the one presented here [18, 19 and references therein; 20 and references therein; 21, 22, 23, 24]. No doubt, in time, some of these complexities will have to be added to our simple theory, and more completeness achieved. Nonetheless, the simple points made here do account for the observations, and, indeed, the challenge perhaps is to other theories. The clear anti-correlation between solar wind speed and coronal electron temperature of *Gloeckler et al.* [1], and the apparent simple relationship between these two quantities, represent a critical test against which all solar wind theories should be judged.

ACKNOWLEDGMENTS

This work was supported, in part, by NASA grant NAG5-10975, NSF grant NSF0096664, NASA/ JPL contract 955460, and NASA/CalTech contract NACG5-6912.

REFERENCES

1. Gloeckler, G., Zurbuchen, T. H., and Geiss, J., *J. Geophys. Res.*, in press (2002).
2. Fisk, L. A., *J. Geophys. Res.* **101**, 15,549 (1996).
3. Fisk, L. A., Zurbuchen, T. H., and Schwadron, N. A., *Astrophys. J.* **521**, 86 (1999).
4. Fisk, L. A., and Schwadron, N. A., *Astrophys. J.* **560**, 425 (2001).
5. Schrijver, C. J., Title, A. M., van Ballegooijen, A. A. H., Hagenaar, J., and Shine, R. A., *Astrophys. J.* **487**, 424 (1997).
6. Handy, B. N., and Schrijver, C. J., *Astrophys. J.* **547**, 1100 (2001).
7. Simon, G. W., Title, A. M., and Weiss, N. O., *Astrophys. J.* **561**, 427 (2001).
8. Schrijver, C. J., DeRosa, M. L., and Title, A. M., *Astrophys. J.*, in press (2002).
9. Crooker, N., Gosling, J. T., and Kahler, S. W., *J. Geophys. Res.* **107**, 148 (2002).
10. Fisk, L. A., *J. Geophys. Res.*, in press (2002).
11. Hansteen, V. H., Leer, E., and Holzer, T. E., in *Solar Wind Nine*, edited by S. R. Habbal et al., AIP Conference Proc. 471, Melville, New York, 1999, p. 17.
12. Feldman, U., Widing, K. G., and Warren, H. P., *Astrophys. J.* **522**, 1133 (1999).
13. Fisk, L. A., Schwadron, N. A., and Zurbuchen, T. H., *J. Geophys. Res.* **104**, 19,765 (1999).
14. Von Steiger, R., Schwadron, N. A., Fisk, L. A., Geiss, J., Gloeckler, G., Hefti, S., Wilken, B., Wimmer-Schweingruber, R. F., and Zurbuchen, T. H., *J. Geophys. Res.* **105**, 27,217 (2000).
15. Burgi A., and Geiss, J., *Solar Physics* **103**, 447 (1986).
16. Geiss, J., Gloeckler, G., von Steiger, R., Balsiger, H., Fisk, L. A., Galvin, A. B., Ipavich, F. M., Livi, S., McKenzie, J. F., Ogilvie, K. W., and Wilken, B., *Science* **268**, 1033 (1995).
17. Schwadron, N. A., *J. Geophys. Res.*, submitted (2002).
18. Parker, E. N., *Astrophys. J.* **128**, 664 (1958).
19. Isenberg, P. A., The solar wind, in *Geomagnetism*, vol. 4, edited by J. A. Jacobs, Academic Press, Orlando, 1991, p. 1.
20. Marsch, E., *Adv. Space Res.* 14, 103 (1994).
21. Hansteen, V. H., and E. Leer, *J. Geophys. Res.*, 11, 21,577 (1995).
22. Axford, W. I., and McKenzie, J. F., "The solar wind," in *Cosmic Winds in the Heliosphere*, edited by J. R. Jokipii, C. P. Sonett, and M. S. Giampapa, University of Arizona Press, Tucson, 1997, p. 31.
23. Hollweg, J. V., *J. Geophys. Res.* **105**, 15,699 (2000).
24. McKenzie, J. F., Axford, W. I., and Banaszkiewicz, M., *Geophys. Res. Lett.* **24**, 2877 (1997).

Quiet Sun Magnetic Fields

J. Sánchez Almeida

Instituto de Astrofísica de Canarias, E-38200 La Laguna, Tenerife, Spain

Abstract. The seemingly un-magnetized part of the solar surface is not really un-magnetized. It is occupied by magnetic structures producing low polarization which, therefore, escape detection in traditional measurements. Since most of the solar surface belongs to this category, the quiet Sun magnetic fields can easily carry most of the magnetic flux and energy existing in the photosphere at any given time. Consequently, they are a potentially important ingredient of the solar magnetism.

Most of the physical properties of the quiet Sun are still uncertain (distribution of field strengths, area coverage, influence on higher atmospheric layers, etc.).It is clear, however, that the topology of the field is complex, with field lines of very different properties coexisting in each resolution element. This fact hampers the detection of the quiet Sun magnetic fields. I argue that the best present measurements detect, at most, 30 % of the existing magnetic flux. Then the quiet Sun contains at least as much magnetic flux as all active regions and the network during the solar maximum.

INTEREST TO STUDY THE MAGNETISM OF THE QUIET SUN

Even during the maximum of the solar cycle, most of the solar surface appears as *non-magnetic* in traditional magnetic field determinations (e.g., the gray background in the magnetogram shown in Figure 1). This so-called *quiet Sun* does not produce enough polarization to show up in such measurements, however, one cannot infer from this fact that the magnetism of the quiet Sun regions is non-existing or un-important. Rather, the limited sensitivity of the standard measurements, together with the large surface coverage, indicate that the quiet regions may be very important in terms of the global magnetic properties of the Sun. A simple order-of-magnitude estimate illustrates the point. Magnetographs show a total unsigned magnetic flux across the solar surface of some 7×10^{23} Mx at solar maximum (e.g., Schrijver and Harvey [1]). If one divides this flux by the area of the solar surface, the flux density turns out to be of the order of 12 G. This figure is close to the noise level of the standard measurements (some 7 G for the magnetogram in Figure 1; see Jones et al. [2]). Consequently, signals below the usual sensitivity and covering most of the solar surface may contain as much magnetic flux as sunspots, active regions and the network all together.

If the quiet Sun carries a substantial fraction of the solar magnetic flux, weak polarization signals should appear upon improvement of sensitivity of the magnetometers. Such weak signals are actually observed. When the noise is in the few G level and the angular resolution about 1", then most of the solar surface becomes magnetic. Such sensitivity and angular resolution is frequently achieved by the new generation of solar spectropolarimeters which consistently show polarization signals almost everywhere (e.g., Grossmann-Doerth et al. [3]; Lin and Rimmele [4]; Lites [5]).

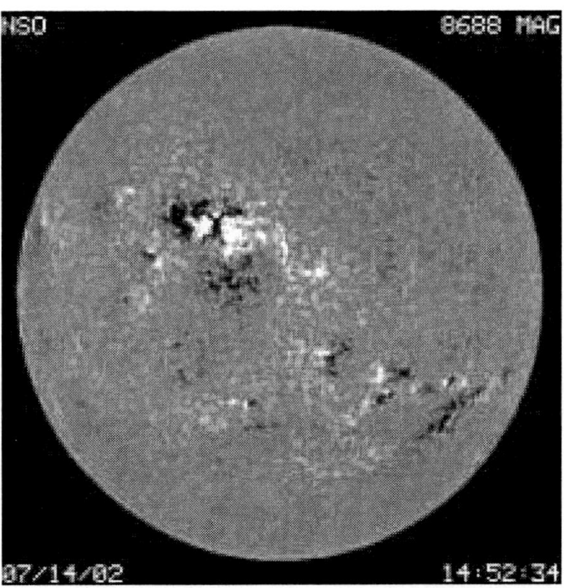

FIGURE 1. Typical magnetogram. Black and white represent magnetic signals with two different polarities. The gray background shows no signal above the sensitivity and corresponds to the part of the solar surface denoted along the text as *quiet Sun*. For details of this magnetogram, see Jones et al. [2]

Apart from the mere existence of this type of magnetism, little is known about the properties of the quiet Sun magnetic fields. They constitute a potentially impor-

CP679, *Solar Wind Ten: Proceedings of the Tenth International Solar Wind Conference*,
edited by M. Velli, R. Bruno, and F. Malara

tant ingredient of the solar magnetism that deserves careful study. One would like to know what is the amount of magnetic flux and energy that the quiet Sun contains. In particular, it would be important to find out whether and to what extent the quiet Sun magnetic fields are related to the other manifestations of the solar magnetism (active regions or network). Does it result from the decay of active regions? Does it emerge *as is* from sub-photospheric layers? Is it created in-situ by a dynamo driven by the granulation? Does it follow the solar cycle? Is it connected to the magnetic fields in the corona, from where the solar wind emanates? The importance of these and similar questions provides a clear rationale to study the magnetism of the quiet Sun.

A conspicuous observational feature characterizes the line polarization produced by the quiet Sun magnetic fields. The weak line polarization signals emerging from quiet Sun regions turn out to be highly asymmetric (Sánchez Almeida et al. [6]; Grossmann-Doerth et al. [3]; Sigwarth et al. [7]). They do not show the characteristic line shape to be expected if the magnetic and velocity fields were constant in the resolution element. Figures 2*b*-*f* include a set of observed Stokes *V* profiles (degree of circular polarization versus wavelength within the range of wavelengths of a spectral line). These profiles have to be compared with the perfectly antisymmetric signal that arise when magnetic and velocity fields are spatially resolved (Figure 2*a*). The existence of these so-called asymmetries of the Stokes profiles proves that the magnetic field in the quiet Sun varies within the typical 1" ($\equiv$ 725 km) resolution elements of the present observational setups. Moreover, the fact that the individual spectral lines generate net circular polarization indicates that part of this variation has to occur along the line-of-sight (LOS), within a fraction of the vertical extent of the photosphere (see the discussion in Sánchez Almeida [8], Sect. 2.2). The extreme character of some of the observed asymmetries suggests that the variation within the resolution element is not mild. For example, very often two different polarities seem to coexist in each resolution element (Sánchez Almeida et al. [6]; Sánchez Almeida and Lites [9]; Socas-Navarro and Sánchez Almeida [10]; Lites [5]). All magnetic field measurements are based on the correct interpretation of the observed polarization. The complexity the magnetic field in the quiet Sun warns against simplistic interpretations of this polarization. Oversimplifications often lead to the omission of magnetic structures and the underestimation of the magnetic flux existing in the quiet Sun. This work analyzes several observational biases that arise (and may be potentially important) when the interpretation assumes that a uniform magnetic field occupies each resolution element. I set lower limits to the missing flux due to (a) the coupling between magnetic field strength and density, (b) the presence of a wide range of field strengths, and (c) the existence of both polarities in each resolution element.

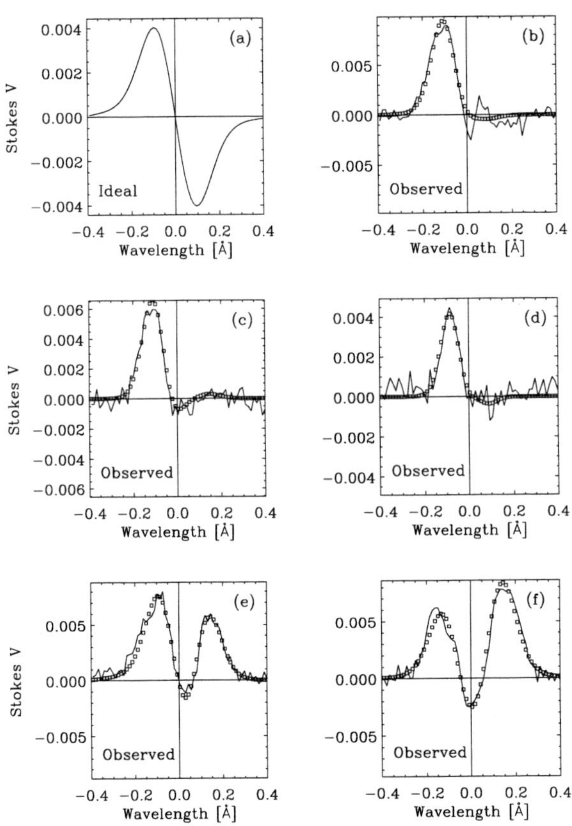

FIGURE 2. Ideal versus observed Stokes *V* profiles. If the magnetic field and the velocity were constant within the resolution element, then the observed Stokes *V* profiles *b*-*f* (the solid lines) should be like the profile in *a*, i.e., perfectly antisymmetric. The dots in figures *b*-*f* correspond to synthetic spectra able to reproduce the observed line profiles (Sánchez Almeida and Lites [9]).

BASIC ASSUMPTIONS OF THE STANDARD MEASUREMENTS

As it was pointed out above, the measurements of magnetic field direction, strength, etc., rely on the correct interpretation of the observed line polarization. In practice, this interpretation is based on a number of assumptions on the structure of the magnetic field that are at variance with the complications of the line polarization observed in the quiet Sun. A prototypical example of magnetic measurement is the magnetogram. Magnetograms employ the so-called magnetograph equation,

$$B_m = C_{cal} V, \tag{1}$$

which relates the observable *V* (Stokes *V* signal at a fixed wavelength within a spectral line) and an estimate of the magnetic flux density B_m. The symbol C_{cal} stands for a calibration constant. Under several hypotheses, in particular,

1. the magnetic field vector **B** is constant within the resolution element (more precisely, it is either constant or zero), and
2. the temperature and density that characterize the thermodynamic of the atmosphere are not modified by the presence of magnetic fields,

the magnetograph signal B_m is equal to the magnetic flux density $< B >$,

$$B_m = < B > \equiv \int_\Sigma \mathbf{B} \cdot \mathbf{n}\, ds / \int_\Sigma ds. \quad (2)$$

The integrals extend to the surface of the resolution element Σ, which is perpendicular to the unit vector along the LOS **n**. Note that $< B >$ represents the magnetic flux in the plane perpendicular the line-of-sight divided by the area of the resolution element.

Since the hypotheses (1) and (2) above are generally not satisfied in the quiet Sun,

$$< B > / B_m \gg 1. \quad (3)$$

How much larger? We do not know it yet, since it depends on (still unknown) details of the structure of the quiet Sun magnetic fields. However, one can estimate the deficit of B_m with respect to $< B >$ in specific cases. Bias arising form the breakdown of conditions (1) and (2) are analyzed in the forthcoming sections.

Consider a magnetic atmosphere whose physical properties are not uniform in the resolution element. The polarized spectrum emerging from such irregular atmosphere reflects some sort of ill-defined volume average of the local properties of the atmosphere. The nature of this average is easy to work out in two extreme cases corresponding to irregularities whose spatial scales are either much smaller or much larger than the typical photon-mean-free-path (see Figure 3). When the irregularities are optically thin then the average proceeds by first averaging the local absorption and emission, and then producing the polarized spectra corresponding to the mean atmosphere (this is the MISMA approximation put forward by Sánchez Almeida et al. [6]). When the irregularities are optically thick (but still spatially unresolved with the present instrumentation), the spectrum of each irregularity is first produced independently, and then the mean among these spectra renders the observable spectrum. (For irregularities of intermediate scales the synthesis is far more complicated since the two sorts of average are no longer uncoupled.) These ideas on the nature of the volume average corresponding to micro and macro irregularities will be used below.

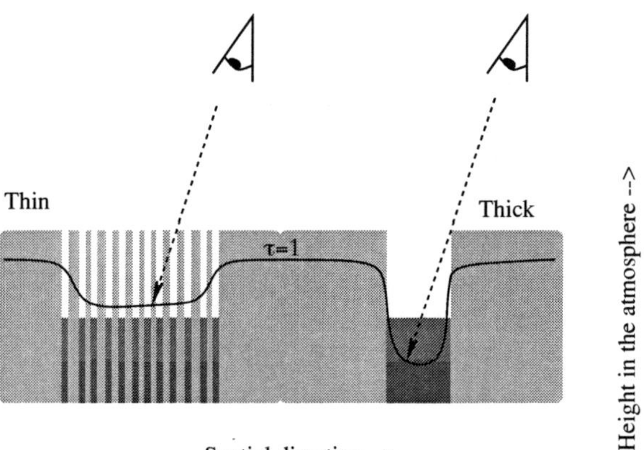

FIGURE 3. Cartoon of two irregular magnetic concentrations embedded in a non-magnetic background. The spatial scale of the irregularities is either smaller (left) or larger (right) than the photon-mean-free-path. The solid line points out the geometrical depth from where the observed photons escape. The dashed lines correspond to LOS.

COUPLING BETWEEN DENSITY AND MAGNETIC FIELD STRENGTH

Any magnetic structure that lasts long enough tends to evolve to a mechanical equilibrium configuration. The characteristic time of the process is of the order of size of the structure divided by the propagation speed of those perturbations that relieve the lack of equilibrium. This time scale is extremely short in the photosphere: using the sound speed for the propagation speed (some 10 km s^{-1}), the time scale is of the order of 100 s for a structure with the size of a granule, and 0.1 s for a 1 km wide magnetic concentration. Consequently, one should expect that most magnetic concentrations satisfy physical constrains characteristic of structures in equilibrium. In particular, consider an atmosphere where the magnetic field strength is not uniform. In order to keep horizontal mechanical balance, the variations of magnetic pressure associated with the variations of magnetic field have to be compensated by gas pressure variations. In other words, the total pressure P_t depends only on the height in the atmosphere z,

$$P_t(z) = \frac{B^2(z)}{8\pi} + P_g(z), \quad (4)$$

so that the magnetic field strength B $(= |\mathbf{B}|)$ and the gas pressure P_g have to be anti-correlated. Assume, for the sake of simplicity, that the temperature does not vary with B. Then Equation (4) renders a simple relationship between the local density ρ and the local magnetic field strength,

$$\rho/\rho_0 = 1 - (B/B_{\max})^2, \quad (5)$$

where $B_{max}(=\sqrt{8\pi P_t})$ is the maximum possible magnetic field strength, which corresponds to fully evacuated plasma ($\rho = 0$). The symbol ρ_0 in Equation (5) stands for the density of unmagnetized plasma ($B = 0$).

Now consider an irregular atmosphere where the magnetic field strength is not constant in planes perpendicular to the LOS. The emission and absorption of photons depend on the local density of the plasma that emits and absorbs. According to Equation (5), the more magnetic the plasma is the less dense and, consequently, the less polarization is expected to emit. This suggest that very strong magnetic fields ($B \sim B_{max}$) produce negligible polarized light and therefore may be missed in the magnetic field determinations. Such simplistic argument holds only if the light emitted by plasmas with different magnetic fields comes from the same atmospheric layers. This is the case when dealing with optically thin magnetic irregularities (Figure 3, left), since the mean opacity determines a single optically depth independent of the field strength (see last paragraph of the previous section). If the magnetic field fluctuations are optically thick, then the reduction of opacity associated with the decrease of density allows to see deeper in the more magnetic irregularities, and the global increase of density with depth may easily compensate the deficit of polarized emission[1].

Sánchez Almeida [11] studies this coupling between field strength and mass density. It is found to be responsible for the observed decrease of magnetic signals having $B > 0.9\,B_{max}$. By correcting for the low sensitivity to large field strengths, Sánchez Almeida [11] sets a lower limit for the missing flux density,

$$<B>/B_m \geq 2. \tag{6}$$

BIAS DEPENDING ON THE MAGNETIC FIELD STRENGTH

The magnetic field strength is not constant within the typical resolution elements, which biases the magnetic flux density determinations in yet another way. If a wide range of field strengths is present then, depending on the magnetic sensitivity of the spectral line used for measuring, one tends to select a particular part of the distribution of field strengths. In particular, if the whole range of magnetic field strengths from zero to the maximum possible value is present ($0 < B < B_{max} \sim 2$ kG), highly split Infra-Red (IR) lines tend to choose the sub-kG part of the distribution whereas visible lines are more sensitive to the kG fields. The mechanism is illustrated in Figure 4, which shows the synthetic Stokes V profiles of two lines as a function of the magnetic field strength (top row). The two synthetic lines, whose atomic parameters have been chosen to represent typical lines used in magnetic studies[2], differ only because of their Zeeman splittings. The visible line fully splits for $B > 1.5$ kG whereas 0.5 kG suffices in the IR (see the gray gap between the white and black bands in Figures 4*a* and 4*b*). In order to compute the spectra produced by the full distribution of field strengths, one would need to average the images 4*a* and 4*b* along the vertical direction[3]. In general, for an arbitrary distribution of field strengths, the average has to be weighted with the probability of finding a given field strength. For example, Figures 4*c* and 4*d* show Stokes V profiles resulting from two different probabilities having only sub-kG fields and mostly kG fields. They correspond to averages of the lower and upper parts of the images, respectively ($B < 0.5$ kG and $B > 0.5$ kG). The Stokes V profiles in Figures 4*c* and 4*d* are normalized to the flux density in the resolution element. First, note how the IR signal in Figure 4*d* decreases with increasing mean magnetic field strength. This is due to the spread in wavelength of the IR signals for kG fields. Such spread ceases for sub-kG fields where the signals accumulate at a constant wavelength (see Figure 4*b*). Should the quiet Sun contains both kG regions and sub-kG regions, one would preferentially detect those producing the largest signals, which in the IR correspond to the sub-kG fields. On the other hand, the sensitivity of the visible line is more balanced for weak and strong fields (Figure 4*c*). Note, however, that the visible signals are weaker, skewing the determinations to larger flux densities.

The measurement of the physical properties of the quiet Sun fields is still in a primitive phase. Nevertheless we already know that IR lines and visible lines render very different magnetic field strengths (see Socas-Navarro and Sánchez Almeida [10][12], where you can also find how to determine B independently of B_m). The difference goes in the sense of the bias described above, being the IR measurements those yielding the lower field strengths. Since the ranges of observed field strengths in the visible and in the IR have almost no overlap, one can assume that the two observations detect different magnetic structures. The observed IR and visible flux densities are similar (see the data collected by Sánchez Almeida et al. [13]). This fact suggests that approximately half of the flux is missing in estimates based on only IR lines or only visible lines. Using the notation of the previous section, this detection of only half of the ex-

[1] Sunspots correspond to this second case. They are strongly evacuated but produce polarized light.

[2] Fe I 6302.5 Å for the visible line and Fe I 15648 Å for the IR line.

[3] This approximation considers spatially unresolved structures that are optically thick (Figure 3, right), but the bias described in this section affects optically thin structures as well (Sánchez Almeida and Lites [9]).

isting flux can be summarized as,

$$< B > /B_m \sim 2. \tag{7}$$

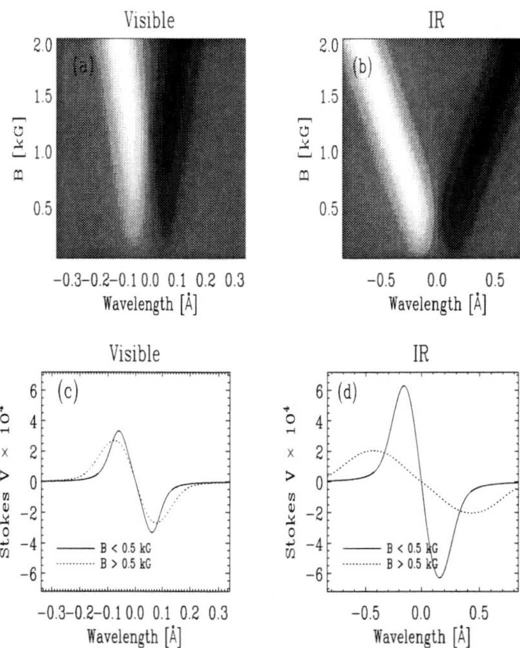

FIGURE 4. Top: Stokes V profiles (Stokes V versus wavelength) for a range of magnetic field strengths (B, in kG units). Image a shows the behavior of a magnetically sensitive visible line (e.g., Fe I 6302.5 Å). Image b represents the same line if it were in the near IR (say, at 1.5 μm). Bottom: average Stokes V profiles if the magnetic field strengths existing in the resolution element were mostly weak (the solid lines) or strong (the dotted lines). They have been computed by averaging the top images for $B < 0.5$ kG (the solid lines) and $B > 0.5$ kG (the dotted lines). The Stokes V profiles in c and d have been normalized to the same flux density in the resolution element: 1 G.

MIXED POLARITIES IN THE RESOLUTION ELEMENT

Stokes V profiles like those in Figures 2*f-g* point out the existence of two different polarities in the resolution element. The sign of the circular polarization reverses upon change of polarity, therefore, the existence of unresolved mixed polarities reduces the Stokes V signal and, via Equation (1), hides part of the existing magnetic flux. Since the presence of mixed polarities seems to be very common, this effect may severely bias the magnetic flux density determinations. Socas-Navarro and Sánchez Almeida [10] find that some 35% of the weak Stokes V profiles produced by the quiet Sun require mixed polarities to be reproduced. Assuming that each one of these mixed polarities cancels a fraction of the observed flux (say 50%), then

$$< B > /B_m \geq 0.35 \times 0.5 + 1 \simeq 1.2, \tag{8}$$

where $< B >$ represents the unsigned flux density to be observed if one could resolve the opposite polarities[4]. Again, the equation represents a lower limit to the ratio $< B > /B_m$ since only those magnetic structures where the annihilation between opposite polarities is not perfect leave a residual Stokes V able to reveal the presence of two polarities. (The imperfect cancellation is due to the existence of a Doppler shift associated with the change of polarity.)

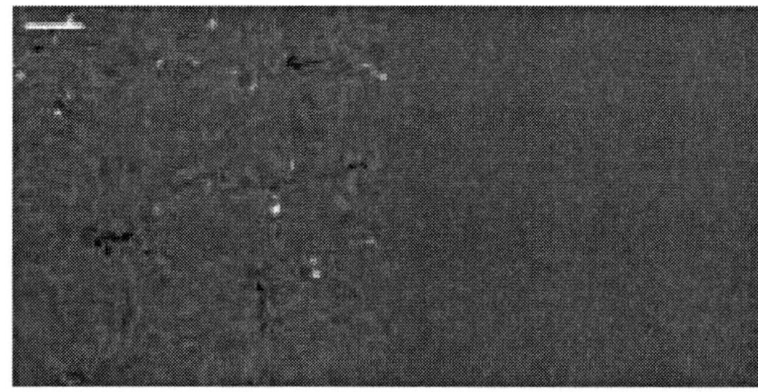

FIGURE 5. Synthetic magnetograms of magnetic fields produced by the turbulent dynamo numerical simulations of Cattaneo [14] and Emonet and Cattaneo [15]. Left, snapshot of the original simulation. Right, same magnetogram but observed with 1"angular resolution. 90% of the original signals go away after the spatial smearing. The gauge, shown for reference, corresponds to 1".

The degree of cancellation due to mixed polarities may be far more severe than the conservative limit set by the inequality (8). Figure 5 illustrates the huge decrease of polarization signals that can be induced by the presence of mixed polarities. It shows a synthetic magnetogram emerging from the numerical simulations of turbulent dynamo by Cattaneo [14] and Emonet and Cattaneo [15]. In this simulation the magnetic field grows out of the kinetic energy of the granular motions. It disappears by Ohmic diffusion when field lines of two polarities intertwine in the whirls of the granulation downdrafts. This turbulent dynamo mechanism may explain the origin of the quiet Sun magnetism and, in addition, it does not seem to contradict any observational constraint (Sánchez

---

[4] If $< B >_+$ and $< B >_-$ are the flux densities of the two opposite polarities, then $< B > \equiv < B >_+ + | < B >_- |$ whereas $B_m \sim | < B >_+ + < B >_- |$. We consider that at least 35% of the resolution elements have $< B > - B_m \simeq 0.5 B_m$, which leads to $< B > - B_m \geq 0.35 \times 0.5\, B_m$ and, consequently, to Equation (8).

Almeida et al. [13]). Figure 5 shows, side-by-side, both the original magnetogram at full resolution (some 15 km or 0."02) plus the magnetogram smeared to 1" resolution, typical of real observations. Most of the signals are gone. (The two images are shown in a common scale for direct comparison.) Only 10% of the original signals survive the spatial smearing so that, for this particular numerical simulation, $B_m \simeq 0.1 < B >$ or

$$< B > /B_m \simeq 10. \tag{9}$$

CONCLUSIONS

The quiet Sun is magnetic. Contrarily to the implicit assumptions of routine magnetic field determinations, the magnetic field of the quiet Sun is not uniform within the typical 1" resolution elements. This fact biases the measurements so that a fraction of the existing magnetic flux eludes detection. I have considered three among the possible biases associated with the lack of enough resolution, namely, the drop of polarized emission associated with local magnetic field strength enhancements, the sensitivity of spectral lines to specific ranges of fields, and the existence of mixed polarities in the resolution element. Lower limits to the missing flux due to these effects are given in Equations (6), (7) and (8). Since the three of them are independent, their contributions have to be added up to estimate the total effect, i.e.,

$$< B > -B_m = \sum_{i=1}^{3} (< B > -B_m)_i, \tag{10}$$

where $< B >$ is the true flux density and $(< B > -B_m)_i$ represents the deficit of observed flux density produced by the *i-th* bias. Then,

$$< B > /B_m = \sum_{i=1}^{3} (< B > /B_m)_i - 2 \geq 3.2, \tag{11}$$

implying that we detect less than (1/3.2 $\sim$) 30 % of the existing magnetic flux. This large relative error, together with the argument on the large surface coverage of the quiet Sun, points out that the flux carried by the quiet Sun is large in absolute terms. For $B_m \sim 5$ G (see, e.g., Sect. 4.1 in Sánchez Almeida et al. [13]), Equation (11) yields $< B > \; \geq 15$ G. This figure is larger than the magnetic flux in the form of active regions at solar maximum. The arguments for the potential importance of the quiet Sun magnetism put forward in the introductory section seems to be well-founded.

We still have not addressed any of the important questions on the role of the quiet Sun magnetism posed in the introduction. In particular, we do not know whether it bears any direct relationship to the coronal fields and the solar wind. However, there is a moral to be extracted from the difficulties to detect photospheric quiet Sun magnetic fields. It is a caveat of application to all atmospheric layers, including the corona. We know that

- an important structuring of the fields possibly exist at scales that cannot be observed due to technical limitations.
- Because of the insufficient spatial resolution, many magnetic structures (not necessarily irrelevant or secondary) elude detection.
- A rather complete knowledge of the topology and structure of the magnetic field is needed for a proper interpretation of the observations. Simplistic interpretations are bound to severe bias.

ACKNOWLEDGMENTS

ISO/Kitt Peak data used here (Figure 1) are produced cooperatively by NSF/NOAA, NASA/GSFC, and NOAA/SEL. Thanks are due to F. Cattaneo, T. Emonet, R. Grappin, S. Habbal, H. Socas-Navarro, and R. Woo for clarifying discussions. This work has been partly funded by the Spanish MCT, project AYA2001-1649.

REFERENCES

1. Schrijver, C. J., and Harvey, K. L., *Solar Phys.*, **150**, 1–2 (1994).
2. Jones, H. P., Duvall, T. L., Harvey, J. W., Mahaffey, C. T., Schwitters, J. D., and Simmons, J. E., *Solar Phys.*, **139**, 211–232 (1992).
3. Grossmann-Doerth, U., Keller, C. U., and Schüssler, M., *A&A*, **315**, 610 – 617 (1996).
4. Lin, H., and Rimmele, T., *ApJ*, **514**, 448 – 455 (1999).
5. Lites, B. W., *ApJ*, **573**, 431–444 (2002).
6. Sánchez Almeida, J., Landi Degl'Innocenti, E., Martínez Pillet, V., and Lites, B. W., *ApJ*, **466**, 537–548 (1996).
7. Sigwarth, M., Balasubramaniam, K. S., Knölker, M., and Schmidt, W., *A&A*, **349**, 941 – 955 (1999).
8. Sánchez Almeida, J., in *Three-Dimensional Structure of Solar Active Regions*, edited by C. E. Alissandrakis and B. Schmieder, ASP, San Francisco, 1998, vol. 155 of *ASP Conf. Ser.*, p. 54.
9. Sánchez Almeida, J., and Lites, B. W., *ApJ*, **532**, 1215 – 1229 (2000).
10. Socas-Navarro, H., and Sánchez Almeida, J., *ApJ*, **565**, 1323 – 1334 (2002).
11. Sánchez Almeida, J., *ApJ*, **544**, 1135 – 1140 (2000).
12. Socas-Navarro, H., and Sánchez Almeida, J., *in preparation* (2002).
13. Sánchez Almeida, J., Emonet, T., and Cattaneo, F., *ApJ*, **585**, in press (2003).
14. Cattaneo, F., *ApJ*, **515**, L39 – L42 (1999).
15. Emonet, T., and Cattaneo, F., *ApJ*, **560**, L197–L200 (2001).

The effect of time-dependent coronal heating on the solar wind from coronal holes

Øystein Lie-Svendsen*, Viggo H. Hansteen† and Egil Leer†

*Norwegian Defence Research Establishment, div. for electronics, P.O. Box 25, NO–2027 Kjeller, Norway
†Institute of Theoretical Astrophysics, Univ. of Oslo, P.O. Box 1029, Blindern, NO–0315 Oslo, Norway

Abstract. We have modelled the solar wind response to a time-dependent energy input in the corona. The model, which extends from the upper chromosphere to 1 AU, solves the time-dependent transport equations based on the gyrotropic approximation to the 16-moment set of transport equations, which allow for temperature anisotropies. Protons are heated perpendicularly to the magnetic field, assuming a coronal heating function that varies sinusoidally in time. We find that heating with periods less than about 3 hours does not leave visible manifestations in the solar wind (the oscillations are efficiently damped near the Sun); heating with periods of order 10 hours leads to perturbations comparable to Ulysses observations; while heating with periods of order 100 hours results in a series of steady-state solutions. Mass flux perturbations tend to be larger than perturbations in wind speed. Heating in coronal holes with periods of order 30 hours leads to large mass flux perturbations near Earth, even when the amplitude of the change in heating rate in the corona is small.

INTRODUCTION

Does the fairly steady-state high speed wind that, e.g., Ulysses observes from polar coronal holes [1] indicate a steady-state coronal heating process, or merely that the solar wind has filtered out the high-frequency variations before they reached Ulysses' orbit? Several of the proposed coronal heating mechanisms, e.g., by "nanoflares" [e.g. 2] or by "jets" [e.g. 3, 4], are inherently time dependent. The level of, or lack of, variability in the solar wind may put severe constraints on such mechanisms, or even rule out some of them.

To study the solar wind response to a time-dependent energy input in the corona, we have employed a fluid solar wind model based on the 16-moment approximation, a model that extends from the chromosphere to 1 AU [5]. In addition to yielding constraints on the coronal heating mechanism, modelling a time-dependent solar wind can also provide information about the evolution of shocks in the solar wind.

Only a few previous studies [6, 7, 8] have considered the effect of time-dependent energy input in the corona. The main improvement of the model used here is that the coupling between the chromosphere and corona is included, and that it provides a better description of energy transport between the chromosphere, transition region and corona.

THE MODEL

The model includes neutral hydrogen, protons (which are produced dynamically through ionization of HI), and electrons. For each species s the coupled equations for density n_s, drift speed u_s, temperature parallel and perpendicular to the magnetic field ($T_{s\|}$ and $T_{s\perp}$), and radial heat flux densities of parallel and perpendicular thermal motion ($q_{s\|}$ and $q_{s\perp}$), are solved. These equations are described elsewhere [5, 9]. We choose a rapidly expanding flow geometry, in which the flux tube area $A(r)$ increases by a factor 5 relative to radial (r^2) expansion near the Sun, and increases radially beyond a few solar radii.

Except for a small energy input in the transition region in order to maintain a minimum pressure in the corona, the main heating is contained in the proton perpendicular heating term, which is specified as the divergence of a "mechanical" energy flux,

$$Q_{pm\perp}(r,t) = -\frac{1}{A}\frac{\partial F_m(r,t)}{\partial r}. \tag{1}$$

We choose a simple analytical form for the applied energy flux,

$$F_m(r,t) = F_{m0}(t)\exp\left(-\frac{r-R_S}{H_m}\right), \tag{2}$$

where t denotes time and we choose $H_m = 0.5\,R_S$ where R_S is the solar radius. With the chosen parameters most of the energy will be deposited within a solar radius

CP679, *Solar Wind Ten: Proceedings of the Tenth International Solar Wind Conference*, edited by M. Velli, R. Bruno, and F. Malara

above the solar surface. For the time-dependence we choose a purely sinusoidal variation in time,

$$F_{m0}(t) = F_0 \left[1 + a_p \sin\left(\frac{2\pi t}{t_p}\right)\right]. \quad (3)$$

We choose $F_0 = 400$ W m^{-2}, which in the case of a constant heating rate ($a_p = 0$) leads to a reasonable high-speed wind with a speed of about 700 km/s and a mass flux scaled to Earth $(nu)_E \simeq 2.4 \times 10^{12}$ m^{-2} s^{-1}. For the amplitude of the change in heating rate we choose $a_p = 0.75$, so that $F_{m0}(t)$ varies between 100 and 700 W m^{-2} over a period t_p.

RESULTS

Figure 1 shows a snapshot of the solution after the model has been run for so long that the perturbations have had time to reach 1 AU. As can be seen, periods of order 10^3 s are filtered out close to the Sun and are therefore not visible in the solar wind. This is also the case with shorter periods; choosing, e.g., $t_p = 300$ s to simulate the so-called 5-minute oscillations, the solution is essentially steady state for $r > 2\ R_S$. Heating with a period of 10^4 s leads to an essentially steady-state wind beyond about 1/2 AU. For longer periods the solar wind is no longer able to smear out the perturbations by 1 AU, and for $t_p = 10^5$ s the figure shows that the mass flux perturbations even *increase* with distance. We attribute this to a "snowplow" effect in which faster parcels of plasma overtake slower parcels so that matter piles up behind the forward shock while behind the reversed shock the mass flux becomes very low. This process only works as long as there is sufficient difference in speed between the different regions. When this is no longer the case, in the outer solar wind, the enhanced pressure in the high-density regions will cause this plasma to expand. Eventually this expansion will damp the oscillations. Note also that, in agreement with e.g., Ulysses observations, the perturbations in mass flux are larger than the perturbations in wind speed. The temperature panels show that the perturbations in electron temperature are smaller than the proton temperature perturbation, indicating that electron heat conduction can smear out the small-scale electron perturbations. For even longer heating periods than 10^5 s the solution essentially turns into a series of steady-state solutions corresponding to the different values for the coronal heating rate.

Figure 2 shows how the mass flux evolves as a function of time and distance for heating periods $t_p = 10^4$ s and 10^5 s. We note again that the 10^4 s perturbations are gradually damped and disappear around 0.5 AU, while 10^5 s perturbations even grow in amplitude and remain clearly visible at 1 AU. Note also that the perturbations behave like "linear" waves, despite that the magnitude of the perturbations can be very large: If we launch perturbations in the corona with a period of 10^5 s, say, we shall find exactly the same period in the solar wind (as long as the perturbations are detectable). In that sense the waves maintain their identity, and we find no tendency for the solar wind to generate "higher harmonics."

A time-dependent coronal heating will have a similar effect in the slower solar wind generated in a radially expanding flow geometry. The main difference is that shocks tend to be somewhat less pronounced in a radially expanding flow, and that the lower wind speed allows the high-pressure interaction regions more time to expand before they reach Earth orbit.

What mechanism damps the oscillations in the solar wind? As alluded to above, the perturbations are damped mainly by expansion of the high-pressure regions (and compression of the low pressure regions). Such pressure differences are generated directly by changing the amount of coronal heating; for the shortest periods the gas does not have time to expand within one period of the heating. Pressure perturbations are also created indirectly in the solar wind when faster parcels of plasma overtake slower parcels. As the fluid moves outwards these high-pressure regions expand and cool, hence damping the oscillations.

In addition, electron heat conduction can in principle smear out temperature differences in the electron gas. To study the role of electron heat conduction, we have rerun the model with $t_p = 10^4$ s, but effectively switching off electron heat conduction beyond $r = 5\ R_S$. As expected, the electron temperature perturbations then become much larger (than they do in the $t_p = 10^4$ s panel of Figure 1), with the relative perturbations in T_e being comparable to the relative perturbations in T_p seen in the figure. The perturbations in T_e are still damped however, even in the absence of heat conduction, and disappear beyond about 0.5 AU, as do the perturbations in T_p. But most importantly, switching off electron heat conduction has essentially no effect whatsoever on the perturbations in wind speed u_p and mass flux $(nu)_E$. Hence electron heat conduction is of little importance in damping the solar wind oscillations. This is to be expected since the electron temperatures in the models are much lower than the proton temperatures, so that the dynamics will be dominated by the proton pressure gradients.

In order to damp the oscillations significantly the high-pressure regions must have time to expand a distance of order the distance between the high- and low-pressure regions. Hence for rapid changes in the coronal heating rate, and hence small distances between the high- and low-pressure regions, the required time will be shorter than for the long periods of the heating. Since this damping takes place as the solar wind propagates outwards, a more rapid damping for small values of t_p means that the

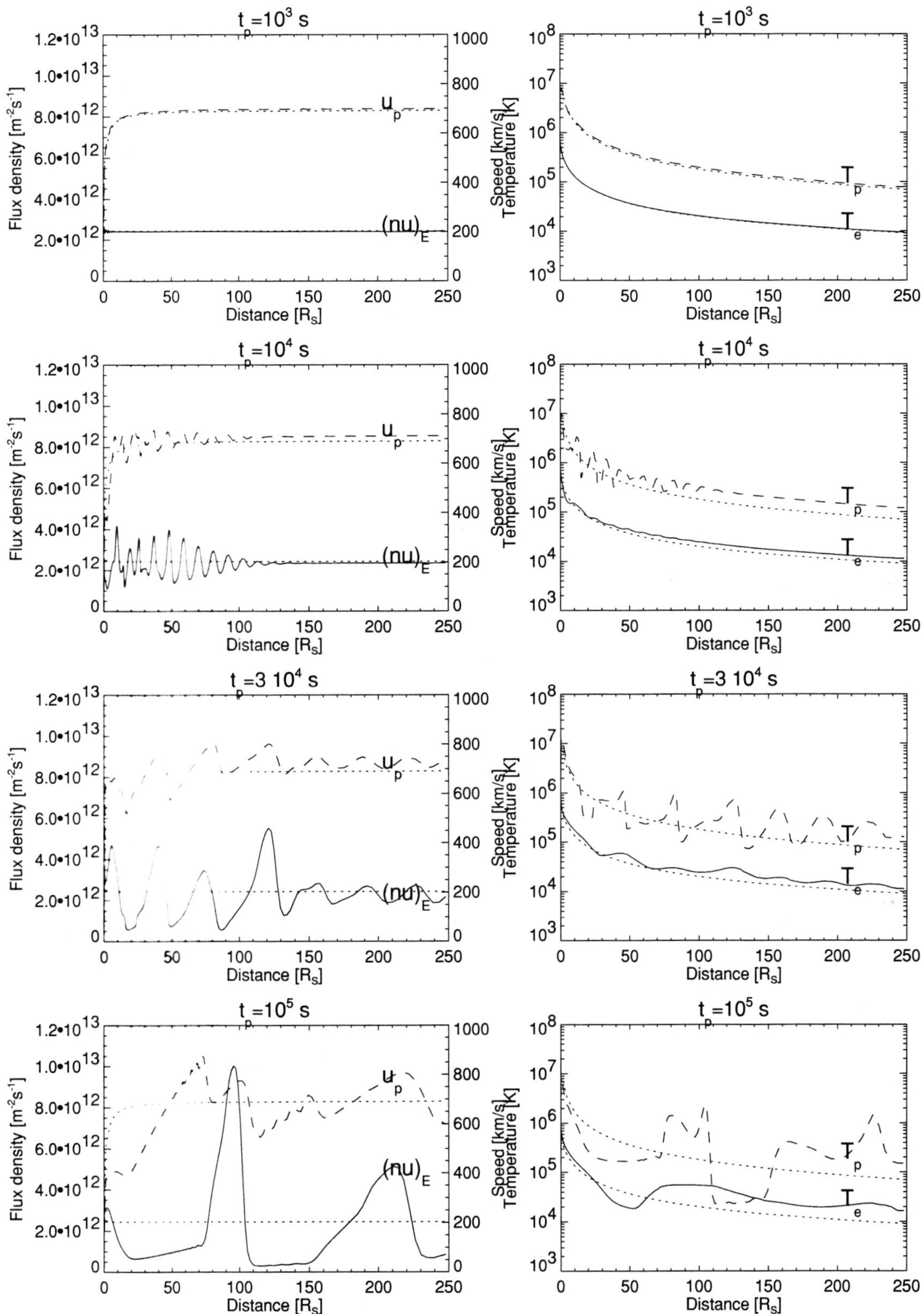

FIGURE 1. Snapshot of the solution for $t_p = 10^3$ s, 10^4 s, 3×10^4 s, and 10^5 s. Here $(nu)_E$ (solid curves) denote the proton particle flux scaled to 1 AU; u_p (dashed curves) is the proton speed, T_e (solid curves) is the mean electron temperature, and T_p (dashed curves) is the mean proton speed ($T_p \equiv (T_{p\|} + 2T_{p\perp})/3$). The dotted curves denote the corresponding steady-state values ($a_p = 0$).

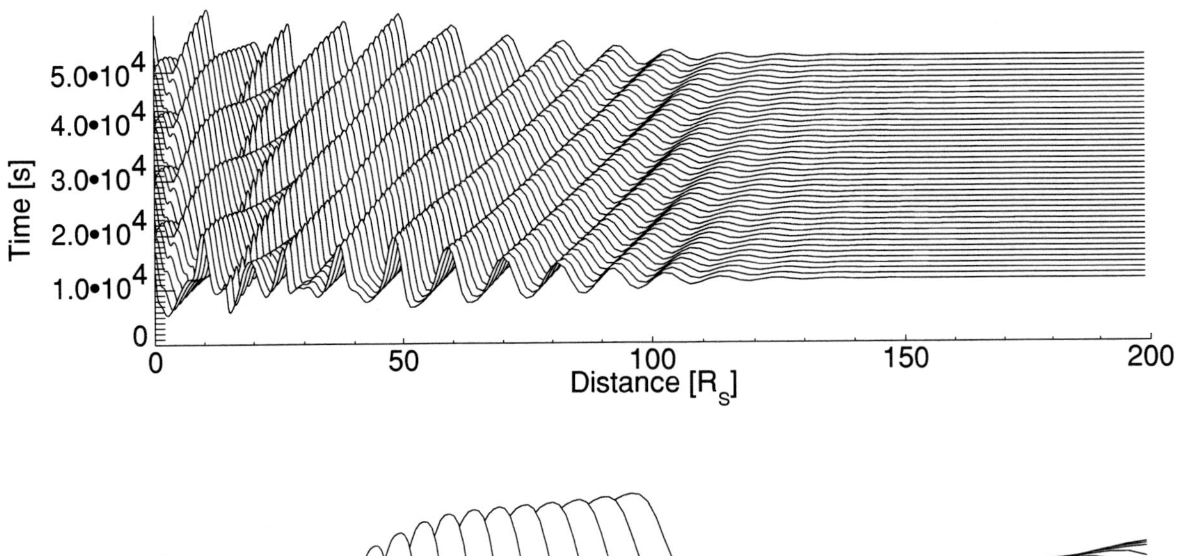

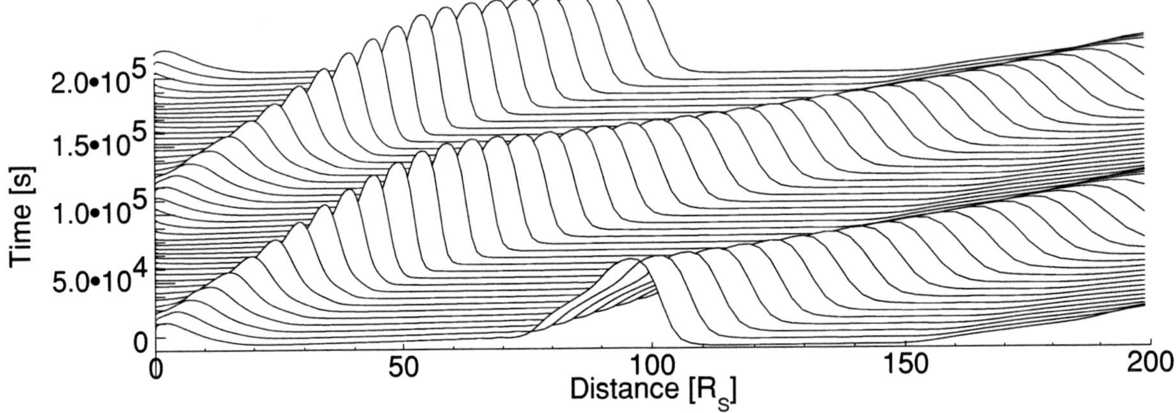

FIGURE 2. Proton particle flux scaled to 1 AU for $t_p = 10^4$ s (top panel) and $t_p = 10^5$ s (bottom panel) as a function of time and distance.

damping occurs closer to the Sun. For the longest period shown in Figure 1, $t_p = 10^5$ s, the plasma has not come to approximate pressure equilibrium even at 1 AU. For very short periods, $t_p \leq 10^3$ s, the coronal plasma hardly has time to respond to the changes and the initial perturbations in the corona will therefore be smaller, too.

A more detailed presentation of this study is given by Lie-Svendsen et al. [9].

ACKNOWLEDGMENTS

This work was supported in part by the Research Council of Norway under grants 121076/420 and 136030/431.

REFERENCES

1. McComas, D. J., et al., *J. Geophys. Res.*, **105**, 10419–10433 (2000).
2. Parker, E. N., *Astrophys. J.*, **330**, 474–479 (1988).
3. Feldman, W. C., Habbal, S. R., Hoogeveen, G., and Wang, Y.-M., *J. Geophys. Res.*, **102**, 26905–26918 (1997).
4. Wang, Y.-M., N. R. Sheeley, J., Socker, D. G., Howard, R. A., Brueckner, G. E., Michels, D. J., Moses, D., Cyr, O. C. S., Llebaria, A., and Delaboudinière, J.-P., *Astrophys. J.*, **508**, 899–907 (1998).
5. Lie-Svendsen, Ø., Leer, E., and Hansteen, V. H., *J. Geophys. Res.*, **106**, 8217–8232 (2001).
6. Grappin, R., Mangeney, A., Schwartz, S. J., and Feldman, W. C., *J. Geophys. Res.*, **104**, 17033–17043 (1999).
7. Hansteen, V. H., Leer, E., and Holzer, T. E., "The Origin of the High-Speed Solar Wind," in *Solar Wind Nine*, edited by S. R. Habbal, R. Esser, J. V. Hollweg, and P. A. Isenberg, The American Institute of Physics, 1999, vol. CP471, pp. 17–21.
8. Hansteen, V. H., Leer, E., and Lie-Svendsen, Ø., "Advances in modelling the fast solar wind," in *Proc. 9th European Meeting on Solar Physics, 'Magnetic Fields and Solar Processes,' Florence, Italy, 12–18 September 1999*, ESA, 1999, vol. SP-448, pp. 1091–1100.
9. Lie-Svendsen, Ø., Hansteen, V. H., and Leer, E., *J. Geophys. Res.*, **107** (2002), doi:10.1029/2001JA009144.

On the acceleration and wave heating of the solar wind: implications of the mean free path of solar energetic particles

T. Laitinen*, R. Vainio [1]* and H. Fichtner†

*Space Research Laboratory, VISPA and Department of Physics, University of Turku, Finland
†Institut für Theoretische Physik, Lehrstuhl IV: Weltraum- und Astrophysik,Ruhr-Universität Bochum, Germany

Abstract. Wave damping and cascading processes have been found to be important for the heating and acceleration of the solar wind. However, it remains a difficult task to extract details of these processes from observations of the thermal plasma only. The wave power required for efficient heating and acceleration of the solar wind also affects the acceleration and transport of solar energetic particles. Thus, their observation could provide valuable clues for the actual evolution of the wave power close to the coronal base and, in turn, give constraints for solar wind modeling. Pursuing this idea, we have developed a steady-state two-fluid model for the wave heating and acceleration of the solar wind. The dissipation frequency determining the heating is obtained from a cyclotron damping rate that depends on the plasma beta and, thus, differs from the usual assumption, a fixed fraction of the ion cyclotron frequency. We present first results obtained with the two-fluid code and, in particular, discuss the implications of the corresponding mean free path of energetic particles.

INTRODUCTION

Wave related phenomena are considered a strong candidate for heating solar wind on open field lines. Waves transfer energy efficiently from the sun to corona, and exert pressure to drive the plasma. Cyclotron damping has been suggested to strongly contribute to heating [1], and a global model using dissipation at cyclotron resonance for wave heating was presented by Tu and Marsch [2]. Subsequently the model has been extended to include cascading that efficiently transfers wave power to dissipation [3]. These models compare well with observations of coronal velocity, density and temperature profiles in the fast solar wind.

However, information on the available wave power is more difficult to obtain. Recently, Vainio and Laitinen [4] suggested that solar energetic particles could provide an important observational constraint for the waves. While the transient energetic particle events, related to solar flares or coronal mass ejections (CMEs), do not contribute significantly to the solar wind dynamics, the waves isotropize the energetic particle pitch angle distributions very efficiently [e.g., 5]. Vainio and Laitinen [4] estimated the energetic particle mean free path, an essential parameter for both particle acceleration and transport, resulting from the requirements for the model of Tu and Marsch [2]. According to their results, the wave power needed for heating was orders of magnitude too high to explain the properties of energetic particles observations in equatorial plane.

The study made in [4] however lacked solar wind modeling and was in this sense not consistent. Thus, in order to improve the study, we have created a steady-state two-fluid solar wind model using the self-consistent dissipation frequency, and re-iterated the energetic particle study. The model itself is presented in more detail in Laitinen et al. [6], this paper concentrates on the consequences for energetic particles.

SOLAR WIND MODEL

The basic solar wind model is based on a set of hydrodynamical two-fluid equations for mass, momentum and energy continuity. We include Coulomb collisions and heat conduction for both electrons and protons, but neglect radiative losses. The proton heating is related to the wave power by Tu and Marsch [2] as

$$Q_p = -\left(u + v_A \cos\psi\right) \frac{P_L(f_H, r)}{4\pi} \frac{df_H}{dr}, \qquad (1)$$

where $P_L(f,r)$ is the wave power of left-hand polarized waves. The wave power spectrum evolves radially as

$$P(f,r) = \theta\{f_T(r) - f\} P_0(f) \frac{M_0(M_0+1)^2}{M(r)[M(r)+1]^2}, \qquad (2)$$

[1] Present address: Department of Physical Sciences, University of Helsinki, Finland

CP679, *Solar Wind Ten: Proceedings of the Tenth International Solar Wind Conference*, edited by M. Velli, R. Bruno, and F. Malara

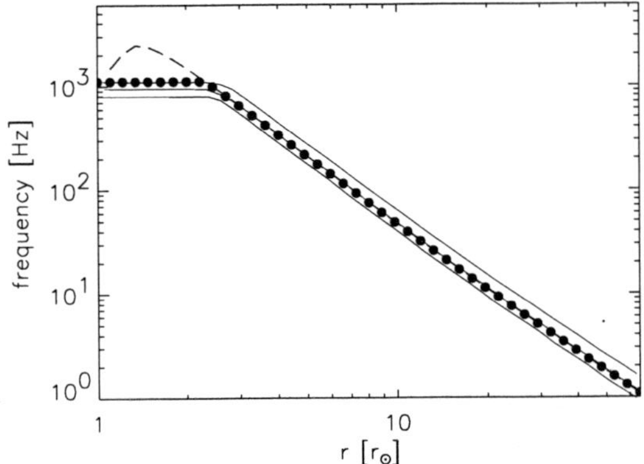

FIGURE 1. Examples of the dissipation and the spectral termination frequencies (dashed curve and filled circles, respectively) and contours of the relative damping decrement (solid curves) at levels 0.01, 1, and 100.

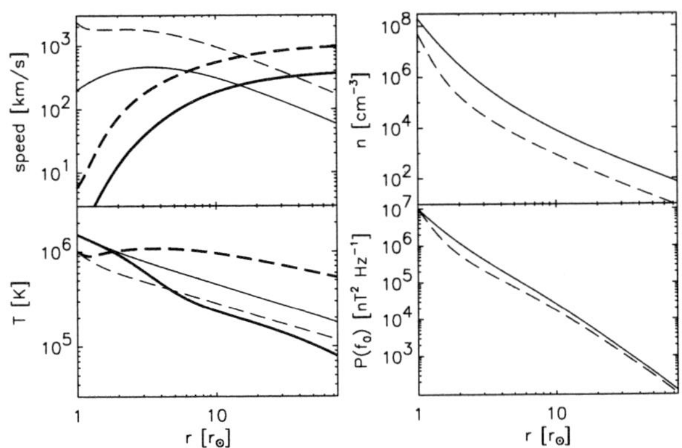

FIGURE 2. Solar wind solution for typical slow solar wind (solid lines) and coronal hole (dashed lines). On the top left panel thick lines represent the wind velocity, thin lines Alfvén velocity. On bottom left panel thick lines represent proton temperature, thin lines electron temperature. The bottom right panel shows the spectral power at $f_0 = 1$ Hz.

where $M = u/v_A \cos\psi$ is the Alfvén Mach number and $v_A = B/\sqrt{4\pi m_p n}$, and $f_T(r)$ is the frequency at which the spectrum terminates. The spectrum at the solar surface is a single power law $P_0(f) = P_\odot f^{-1}$. The magnetic field is described by a parametrization suggested by Kopp and Holzer [7], which allows faster than radial expansion close to the sun, with radial behaviour in the interplanetary space, and Archimedean spiral with spiral angle ψ. We also assume vanishing net magnetic helicity, i.e. $P_L = P_R$, which results in approximately two times stronger wave pressure gradient for momentum equation, compared to [6] and earlier solar wind heating models.

In earlier approaches the dissipation in cyclotron heating models has been taken to occur at a fixed fraction of the local ion cyclotron frequency. We employ a more self-consistent treatment, suggested by Vainio and Laitinen [4], which includes the thermal effects and results in dissipation frequency

$$f_H(r) = \frac{\Omega_p}{2\pi}\left(\frac{u}{v_A\cos\psi}+1\right)\left[\beta_p \ln\left(\frac{Y}{\sqrt{\ln Y}}\right)\right]^{-1/2}$$

$$Y(r) = \frac{\sqrt{\pi}\,\Omega_p(r-r_\odot)}{u+v_A\cos\psi}$$

where $\Omega_p = eB/(m_p c)$ is the proton cyclotron frequency and $\beta_p = 8\pi nT/B^2$ the proton plasma beta.

As shown in Fig. 1 (dashed line), this form of dissipation frequency is not necessarily a monotonously decreasing function of heliocentric distance. If the spectral termination frequency $f_T < f_H$, there are no waves available for heating. Thus, if the waves are generated only at solar surface, we may write f_T as

$$f_T(r) = \min\{f_H(r')\,|\,r' < r\}. \tag{3}$$

ENERGETIC PARTICLE TRANSPORT

The standard quasilinear theory (SQLT) relates Alfvén waves with energetic particle transport through the Fokker-Planck pitch angle diffusion coefficient

$$D_{\mu\mu} = \frac{\pi}{2}\Omega(1-\mu^2)\frac{|k_{\rm res}|I(k_{\rm res})}{B^2}, \tag{4}$$

where the resonant waves are described by the wavenumber spectrum $I(k)$, and the particles scatter at a single resonant wavenumber, $k_{\rm res} = -\Omega/(v\mu)$.

In the case of a spectral cut-off, the used form of the resonance condition will cause a resonance gap on pitch angles $|\mu| < \Omega/(vk_T)$, where there are no waves to scatter particles. The problem of such resonance gap has been addressed by several authors, with modeling of thermal damping [8], turbulent spectral evolution [9] or medium scale fluctuations [10], which act in widening the resonance, or with a power law dissipation region spectrum with waves propagating in both directions [11]. Here we assume that these effects close the resonance gap efficiently and use the SQLT model of the pitch angle diffusion coefficient.

If we further assume vanishing net magnetic helicity, we may derive the wave number spectrum from the frequency power spectrum by $|k|I(k) = (2fP_L(f))/2$, where $f = (u+v_A)|k|/(2\pi)$. The mean free path λ is then, for power law spectrum with spectral index of 1, given by

$$\lambda = \frac{3v}{8}\int_{-1}^{1} d\mu \frac{(1-\mu^2)^2}{D_{\mu\mu}} = \frac{2}{\pi}\frac{v}{\Omega}\frac{B^2}{2P_{\odot,L}R(r)}, \tag{5}$$

where $R(r)$ is the radial dependence in Eq. (2).

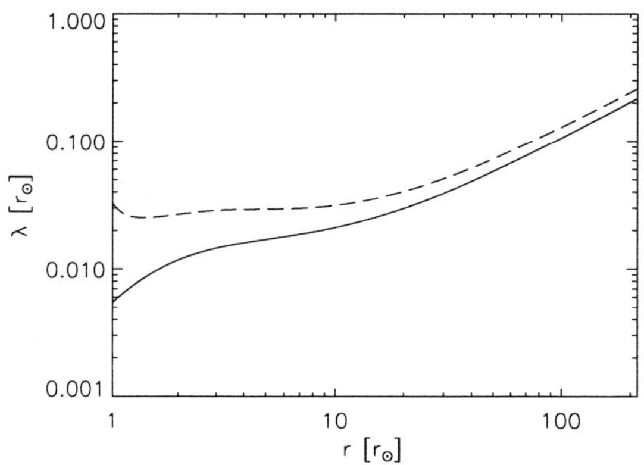

FIGURE 3. Energetic particle SQLT mean free path for 10 MeV protons, for slow wind (solid curve) and a coronal hole model (dashed curve).

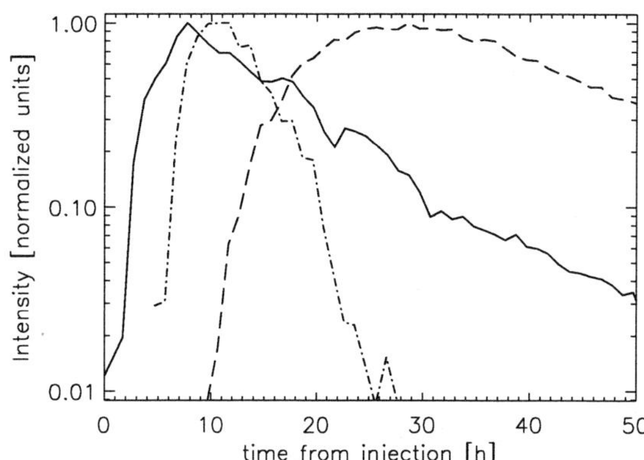

FIGURE 4. An energetic particle event (solid curve) compared with transport simulation in slow (dashed curve) and fast (dash-dotted curve) solar wind solution for energy 12-20 MeV, with wave power reduced to 10 %.

RESULTS AND DISCUSSION

We have used the solar wind model to generate two distinct cases of solar wind to represent a typical slow solar wind and a coronal hole with fast wind. The velocity, density, temperature and spectral power profiles are shown in Fig. 2. The magnetic field geometry for the coronal hole wind is super-radial, with the magnetic field 6 times higher than for the slow wind model in the low corona. Both models require coronal base wave power of $P_{\odot,L} = 5 \cdot 10^{-4} G^2$.

The SQLT energetic particle mean free paths derived for these winds, for 10 MeV protons, are presented in Fig. 3. In the corona, the stronger magnetic field of the fast wind model results in a longer mean free path, as suggested by Eq. (5). The interplanetary mean free path for both models remains very short, and the resulting energetic particle transport is dominated by convection, as the diffusion time scale, $t_d = r^2/(2v\lambda)$, is an order of magnitude longer than the convection time scale.

We compare the resulting transport with an energetic particle event observed by ERNE instrument onboard SOHO spacecraft on 28 December 1999, analysed in more detail by Torsti et al. [12]. We use Monte Carlo simulations [see, e.g., 13] of 12 – 52 MeV protons with a spectrum $\propto p^{-2}$, where p denotes particle momentum. We increase the mean free path by an order of magnitude from the SQLT values, in order to ensure that it isn't underestimated due to the assumptions of the SQLT. Mean free paths obtained from the energetic particle observations can be more than an order of magnitude longer than what the SQLT suggests, as phenomena such as less scatter-effective wave modes are ignored [14].

As shown in Fig. 4, the observed event onset, t_o=1.6 h, is approximately twice the the scatter free arrival time ($t_f \approx 0.7$ h for 20 MeV protons to 1 AU), with injection fixed to an X-ray flare maximum. The slow wind model, representative to equatorial conditions, on the other hand has $t_o/t_f \approx 15$, clearly contradicting the observations.

The coronal hole fast wind, on the other hand, provides very fast convection and, thus, the onset occurs earlier, with $t_o/t_f \approx 7$. The three events during ULYSSES's first south polar pass reported by Bothmer et al. [15] give the ratios of around 10 during quiet sun conditions. Dalla et al. [16] analyzed the events during the second pass, during active sun conditions, and found the transport to be much faster, with $t_o/t_f \sim 1 - 3$ in fast wind events. This discrepancy would suggest that fast wind streams are significantly different during quiet and active sun. However, the low statistics make more accurate analysis difficult, as e.g. bad magnetic connection to acceleration site may cause large delays. In addition, as the spacecraft was over 3 AU from the sun for the first pass events and around 2 AU for the second, onset time comparisons between the two and with 1 AU simulations may be unreliable and difficult to interpret.

Waves contribute also to particle acceleration in corona and interplanetary space, as the estimated mean free path provides favorable conditions for particle acceleration by CME driven shocks. To demonstrate this, we simulate shock acceleration with a Monte Carlo code [13], using the slow wind profiles and a shock with speed $u_s = 1000$ km/s. Particles of spectrum $\propto p^{-5}$, from 50 keV to 50 MeV, are injected as an outward directed beam in front of the shock as the shock is launched at the sun. The intensities observed at 0.2 AU are shown in Fig. 5, as a function of time and energy.

The acceleration is limited by adiabatic focusing in the diverging magnetic field. Particles that are focused more efficiently than diffused back towards the shock, escape the shock and are not accelerated further. The characteristic velocity v_c of the escaping particles can be obtained from $v_c\lambda(v_c,r)/(3u_1) \sim r/2$ where $u_1 = u_s - (u + v_A)$ is the scattering center velocity in shock frame [13]. This is shown in Fig. 5 as a function of $t = (r - r_\odot)/u_s$.

As can be seen, the high-energy particles escape the shock immediately, without re-acceleration. Particles with lower energies, on the other hand, are either trapped to the acceleration region or escape downstream. As a result, lower energy particles are not observed until the shock gets closer to the observer, contradicting observations.

Thus, both particle transport and acceleration observations suggest that the scattering wave power must decrease in the interplanetary space. This can be achieved by including spectral transfer, as in [3]. Also the effect of minor ions on the wave power should be considered, as noted by Tu and Marsch [17], as well as the effect of 2D fluctuations on heating.

Energetic particle observations may provide valuable information also for wave power reduction mechanisms. An increase in the mean free path may result in a gradually decreasing characteristic energy, which would subsequently be observed as a delayed release of lower energy particles. The observed intensities will thus contain information of the radial evolution of the wave power. This information may also be obtained from analysing interplanetary transport of flare accelerated particles. Both studies would benefit from observations made closer to the sun, as long interplanetary transport brings uncertainty to the analysis.

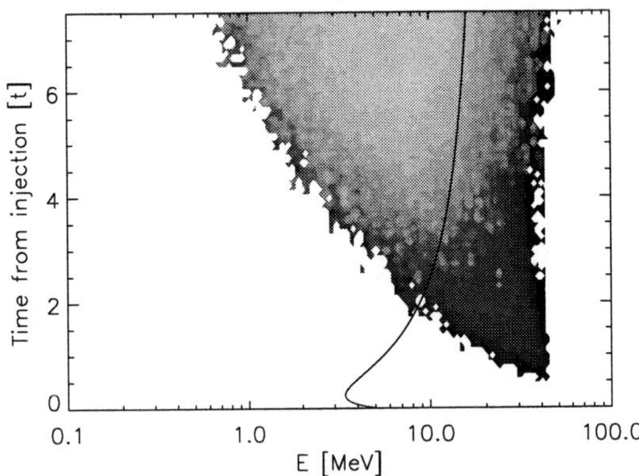

FIGURE 5. Energetic particles accelerated by a shock with $u_s = 1000$ km/s as observed at 0.2 AU. Contours are set for each half order of magnitude. The solid line describes the characteristic energy for the shock acceleration

CONCLUSION

We have studied an important constraint for solar wind modeling provided by energetic particle observations. We generated a typical slow and fast solar wind using a new, self-consistent treatment for the wave dissipation frequency, suggested by [4], which includes the thermal dependence of the cyclotron damping process.

We find that the slow wind produces interplanetary energetic particle transport conditions inconsistent with energetic particle observations. However, in the corona, the short mean free path provides favorable conditions for particle acceleration by e.g. CME driven shocks. The fast wind transport may fit better to the model results, but as the observations are sparse and made at larger distances, no definite conclusions can be made.

The phenomena reducing wave power must thus be included in solar wind modeling, to the extent that energetic particle observations may be explained. Energetic particle observations made closer to the sun with the planned Solar Orbiter spacecraft will provide an important contribution to understanding solar wind heating.

REFERENCES

1. Marsch, E., Goertz, C. K., and Richter, K., *J. Geophys. Res.*, **87**, 5030 (1982).
2. Tu, C.-Y., and Marsch, E., *Sol. Phys.*, **171**, 363 (1997).
3. Hu, Y. Q., Habbal, S. R., and Li, X., *J. Geophys. Res.*, **104**, 24819 (1999).
4. Vainio, R., and Laitinen, T., *A&A*, **371**, 738 (2001).
5. Kallenrode, M., *J. Geophys. Res.*, **98**, 19037 (1993).
6. Laitinen, T., Fichtner, H., and Vainio, R., *J. Geophys. Res.* (2002), manuscript no. 2002JA009479. Accepted.
7. Kopp, R. A., and Holzer, T. E., *Sol. Phys.*, **49**, 43 (1976).
8. Schlickeiser, R., and Achatz, U., *J. Plasma Phys.*, **49**, 63 (1993).
9. Bieber, J. W., Matthaeus, W. H., Smith, C. W., et al., *ApJ*, **420**, 294 (1994).
10. Ng, C. K., and Reames, D. V., *ApJ*, **453**, 890 (1995).
11. Vainio, R., *ApJS*, **131**, 519 (2000).
12. Torsti, J., Kocharov, L., Laivola, J., Pohjolainen, S., Plunkett, S. P., Thompson, B. J., Kaiser, M. L., and Reiner, M. J., *Sol. Phys.*, **205**, 123–147 (2002).
13. Vainio, R., Kocharov, L., and Laitinen, T., *ApJ*, **528**, 1015 (2000).
14. Bieber, J. W., Wanner, W., and Matthaeus, W. H., *J. Geophys. Res.*, **101**, 2511 (1996).
15. Bothmer, V., Marsden, R. G., Sanderson, T. R., Trattner, K. J., Wenzel, K.-P., Balogh, A., Forsyth, R. J., and Goldstein, B. E., *Geophys. Res. Lett.*, **22**, 3369 (1995).
16. Dalla, S., Balogh, A., Krucker, S., Posner, A., Müller-Mellin, R., Anglin, J. D., Hofer, M. Y., Marsden, R. G., Sanderson, T. R., Heber, B., Zhang, M., and McKibben, R. B., *Annales Geophysicae* (2002), Submitted.
17. Tu, C.-Y., and Marsch, E., *J. Geophys. Res.*, **106**, 8233 (2001).

Ion Heating Due to Plasma Microinstabilities in Coronal Holes and the Fast Solar Wind

S. A. Markovskii* and Joseph V. Hollweg*

*Space Science Center, University of New Hampshire, Durham NH 03824, USA

Abstract. There is growing evidence that the heating of ions in coronal holes and the fast solar wind is due to cyclotron resonant damping of ion cyclotron waves. At the same time, the origin of these waves is much less understood. We suggest that the source of the waves in the coronal holes is a heat flux coming from the Sun. The heat flux generates ion cyclotron waves through plasma microinstability, and then the waves heat the ions. We use a new view according to which the heat flux is launched intermittently by small-scale reconnection events (nanoflares) at the coronal base. This allows the heat flux to be sporadically large enough to drive the instabilities, while at the same time to satisfy the time-averaged energy requirements of the solar wind. Depending on the plasma parameters, the heat flux can excite shear Alfvén and electrostatic ion cyclotron waves. We show that, for reasonable parameters, the heat flux is sufficient to drive the instability that results in significant heating of protons and heavy ions in the inner corona.

INTRODUCTION

It is now widely believed that the heating of the ions in solar coronal holes and the resulting generation of the fast solar wind is due to cyclotron-resonant damping of ion cyclotron waves; see, e.g., recent reviews by Hollweg and Isenberg [1] and Cranmer [2]. Much less understood is the origin of these waves in coronal holes. The mechanisms of the wave generation proposed so far are associated with a number of difficulties; see, e.g., Ref. [3] for a review. We suggest that the waves in the proton cyclotron frequency range are generated by a plasma microinstability due to a heat flux coming from the Sun. The electron distribution functions observed in the solar wind often indicate the presence of a nonzero heat flux in the solar wind frame (e.g., [4–6]). Therefore, it is natural to assume that the heat flux also exists in the inner corona. The heat flux can then excite microinstabilities that heat ions. Similar ideas are widely used in ionospheric physics and they were also discussed in connection with the solar corona by Forslund [7], Toichi [8], and Coppi [9].

It is usually thought that the heat flux arises simply from the gradients in a steady corona and solar wind. We explore the consequences of a new view. We assume that the heat flux is launched intermittently by small-scale reconnection events (nanoflares) at the coronal base. This allows the heat flux to be sporadically large enough to drive the instabilities, while at the same time to satisfy the time-averaged energy requirements of the solar wind.

In the reconnection process, a significant part of the reconnected magnetic field energy is released into the thermal energy of electrons. The thermal energy density of the electrons nT can be estimated as the energy density $B^2/8\pi$ of the reconnecting magnetic field. Then, for reasonable parameters at the base of a coronal hole, $B = 12$ G and $n = 4\cdot 10^8$ cm^{-3}, the electron temperature $T \sim 10^8$ K. In a collisionless plasma, the electrons escaping from the site of local heating can form a beam-like distribution function, through velocity dispersion, at least at the initial stage of the evolution (e.g., [10]). The situation at the coronal base is certainly more complicated. Nevertheless, it appears to be reasonable to use a two-component model of the electron distribution function consisting of a Maxwellian core and a heat-flux carrying component that moves with respect to it (Figure 1).

LINEAR ANALYSIS

We investigate an electron heat-flux ion cyclotron instability. The instability is due to the fact that the maximum of the electron distribution function is displaced with respect to that of the ion distribution function (Figure 1), so that the destabilization results from the electron Landau resonance. In some sense, it is similar to the instability driven by a parallel current considered, for instance, by Forslund et al. [11], except that we assume a zero net current condition to be valid.

At large plasma beta, $\beta \gtrsim 0.1$, the primary growing heat flux instability is a whistler mode (e.g., [12]). If the plasma beta is between 0.1 and 0.01, then a shear Alfvén heat flux instability is the most important one [13]. In

CP679, *Solar Wind Ten: Proceedings of the Tenth International Solar Wind Conference*, edited by M. Velli, R. Bruno, and F. Malara

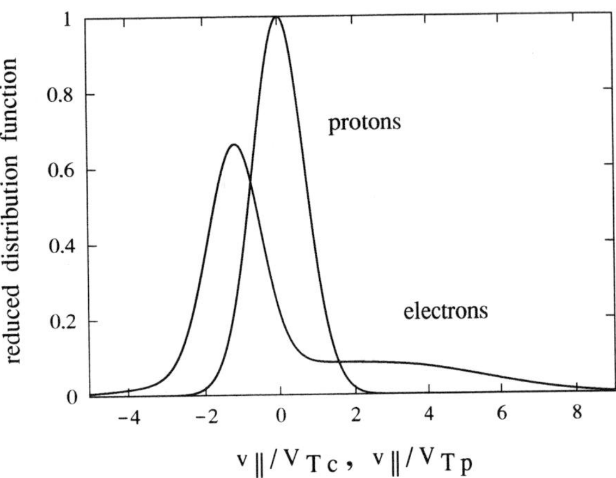

FIGURE 1. Schematic of the reduced distribution function, i.e., integrated over velocities perpendicular to the background magnetic field (arbitrary units), for protons and electrons.

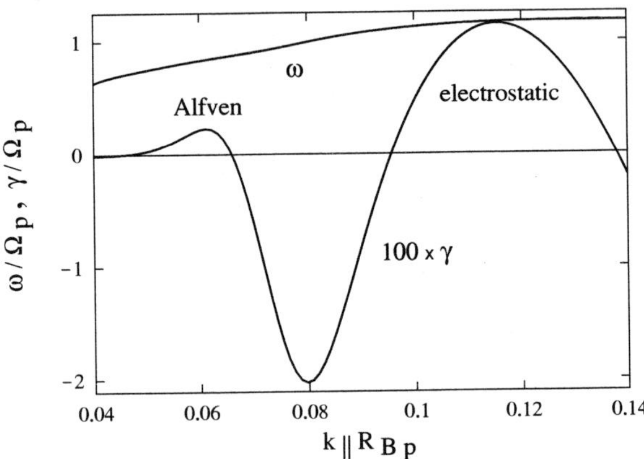

FIGURE 2. Frequency and growth rate of shear Alfvén and electrostatic instabilities as functions of the parallel wave number between the growth rate maxima at $\beta_p = \beta_c = 0.002$ and $u_c = 3.7V_{Tc}$. The plot is a cut in the $k_\perp$–$k_\|$ plane along a line going though the points where the growth rate is maximum.

coronal holes, the plasma beta is presumably between 0.01 and 0.001. In this case, an electrostatic heat flux instability is competitive with the shear Alfvén instability [14]. The reason is that the growth rate of the electrostatic instability increases much faster than that of the shear Alfvén instability with increasing heat flux above the threshold. Therefore, the electrostatic instability can dominate even though its threshold is somewhat greater.

We solve the full electromagnetic linear dispersion equation for a quasi-neutral electron-proton plasma. For simplicity we use an isotropic Maxwellian distribution function for the protons. The electron distribution function consists of two isotropic Maxwellian components, a cold and dense core denoted by the subscript "c" and a hot and tenuous heat-flux carrying halo ("h"). The two electron components move with respect to each other along the background magnetic field, so that a zero-current condition is satisfied.

The results of our calculations are displayed in Figures 2–6. Both instabilities are proton cyclotron resonant. As can be seen from Figure 2, at $\beta = 0.002$, the resonance factor $|\omega - \Omega_p|/k_\| V_{Tp}$ for the fastest growing electrostatic waves is ≈ 1.36. Therefore, the electrostatic waves are strongly proton-resonant. The proton resonance with Alfvén waves is weaker but still not negligible: the resonance factor is ≈ 2.56.

Because the instabilities are driven by the Landau resonance with the core electrons, the waves propagate upstream in the solar wind. Similar instabilities on a qualitative level were considered for the first time by Forslund [7]. It should be emphasized that if the waves are in a region of the solar wind where they can actually propagate toward the Sun in the inertial frame, they will be able to reach locations where they are resonant with He^{++} and other heavy ions, even though they were only proton resonant at the generation site.

At greater plasma beta, the Alfvén instability takes over. Figures 2-4 show the growth rate and frequency at $u_c/V_{Tc} = 0.37$ for different values of beta. At $\beta = 0.01$, the electrostatic instability merges into the Alfvén instability. It is interesting to note that the Alfvén waves can then have significant growth rate at a frequency equal to the proton gyrofrequency. As a result, they become strongly proton-resonant. The fact the ion cyclotron waves can be excited at $\omega = \Omega_p$ is also suggested by the solution obtained by Voitenko and Goossens [15] for an ion beam driven instability. However, their approximate solution of the dispersion equation is not valid at $\omega = \Omega_p$, in contrast with our exact solution. Furthermore, we show in Figures 2-4 that the physical reason why the waves are excited at $\omega = \Omega_p$ is coupling between shear Alfvén and electrostatic ion cyclotron waves.

ION HEATING

Let us now compare the heat flux q that excites the instability with the energy flux at the coronal base q_0 needed to drive the fast solar wind. Taking a typical density and temperature in the vicinity of the coronal base $n_e = 4 \cdot 10^8$ cm^{-3} and $T_c = 10^6$ K and using the parameters $n_h = 0.1n_p$, $T_h = 10T_c$, and $u_c = 0.3V_{Tc}$, we obtain $q \sim 2 \cdot 10^8$ erg cm^{-2} s^{-1}. The energy flux needed to drive the fast solar wind can be estimated assuming that the expansion factor is 10 and taking into account the fact that at 1 AU the kinetic energy dominates over other forms of energy. Then, it can be shown that $q_0 \sim 10^6$ erg cm^{-2} s^{-1} for $n_p =$

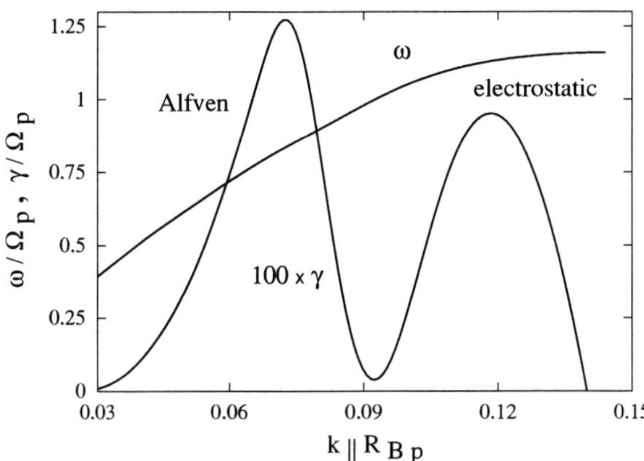

FIGURE 3. Same as Figure 3 at $\beta_p = \beta_c = 0.005$.

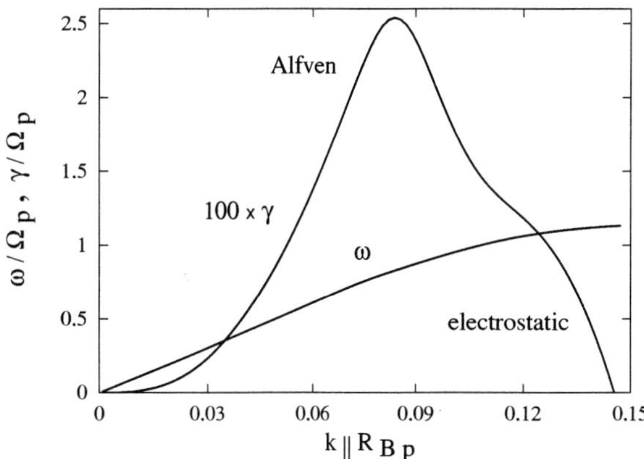

FIGURE 4. Same as Figure 3 at $\beta_p = \beta_c = 0.01$. Electrostatic wave merges into the Alfvén wave.

4 cm^{-3} and $V_{SW} = 750$ km/s at 1 AU. Because $q \gg q_0$, the heat flux must be launched intermittently. This means that a sporadic heat flux, high enough to excite the instability, exists for a short period of time followed by a long period of a much lower flux, so that the time-averaged flux gives the right energy of the solar wind.

It is important that the waves driven by the time-averaged heat flux can provide effective proton heating in coronal holes, even though the time-averaged flux is much smaller than the sporadic one. To show this, we calculate the proton heating rate by waves from quasilinear theory as

$$n(dT/dt) \sim \gamma W, \tag{1}$$

where W is the energy density of the turbulent fluctuations excited by the instability. The heating rate in the vicinity of the coronal base can be estimated by order of magnitude as 10^{-6} erg cm^{-3} s^{-1} from kinetic [16] and fluid [17] models that give the right parameters of the solar wind.

Taking the growth rate of the instability $\gamma = 10^{-2}\Omega_p$ with $\Omega_p = 10^5$ rad/s, we obtain the following relative wave energy density

$$W/nT \sim 10^{-7}. \tag{2}$$

In (1) and (2) we imply that the heating rate and the energy density correspond to a time-averaged heat flux. The sporadic wave energy density is thus a factor of $q/q_0 \approx 200$ greater. Nonetheless, this wave energy density is consistent with the one produced by a current-driven instability at a moderately supercritical drift in the quasilinear limit [18]. Note that the Alfvén waves at large propagation angles are almost electrostatic, because their electric field perturbation is almost parallel to the wavevector. Therefore, the quasilinear theory well-developed for the electrostatic instability gives qualitatively right results for the Alfvén instability.

The threshold of the instabilities depends on the plasma beta and the ratio of the proton to core temperature. The threshold of the Alfvén instability u_c^* is plotted in Figure 6 as a function of the heliocentric distance for typical coronal hole parameters. We used the simple approximate formula

$$u_c^* = 0.18 V_{Tc} (\beta_p/0.01)^{-1/3} \sqrt{T_p/T_c}. \tag{3}$$

to plot the curve. The beta-dependence of u_c^* has the same origin as that in the case of a parallel-current instability in a low-beta plasma [11] and the numerical factor is calculated at $\beta_p = 0.01$ and $T_p = T_c$. This approximate formula gives $u_c^* \approx 0.16 V_{Tc}$ at $\beta = 0.04$ and $T_p/T_c = 2$, while the exact value of u_c^* is $0.15 V_{Tc}$.

In the region of increased threshold the heating is less effective or even stops, if the actual core–proton velocity $u_c = -(n_h/n_c)u_h$ is constant. However, this is not necessarily the case. If the velocity of the halo u_h is initially almost constant, the halo density decreases with distance faster than the density of the background electron population, which plays the role of the current-neutralizing core. Therefore, the ratio n_h/n_c can increase with the distance together with u_c, so that u_c is always well above u_c^*. At later times, the intermittently launched halo expands in the radial direction. As a result, its density decreases faster and the ratio n_h/n_c does not become too high. In any case, at large enough distances, the threshold of the instability decreases and nearly all of the launched heat flux is transformed to the ion kinetic energy flux.

ACKNOWLEDGMENTS

We are grateful to Terry Forbes, Philip Isenberg, Martin Lee, and Bernard Vasquez for useful discussions.

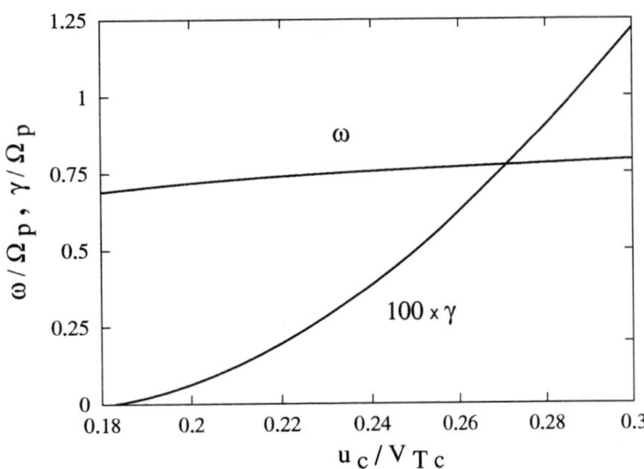

FIGURE 5. Growth rate of the Alfvén waves at $\beta_p = \beta_c = 0.01$ maximized over the wavenumbers and propagation directions and the corresponding frequency as functions of the core–proton velocity u_c.

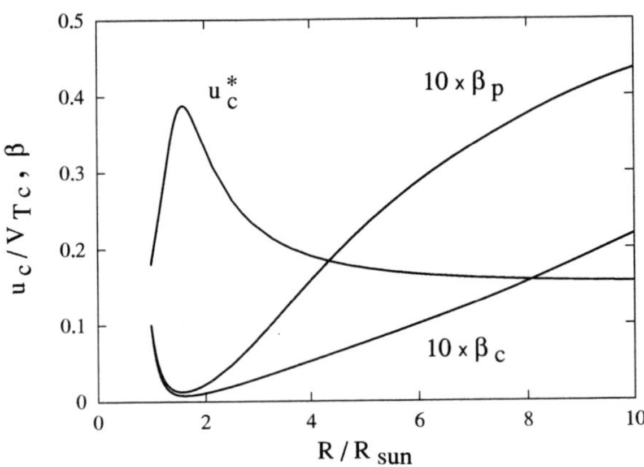

FIGURE 6. Threshold core–proton drift velocity and core and proton betas for typical coronal hole parameters as functions of the heliocentric distance.

This work is supported by NASA Sun-Earth Connection Theory program under grants NAG5-8228 and NAG5-11797, Solar and Heliospheric Physics SR&T program under grant NAG5-10988, and Living With a Star program under grant NAG5-10835 to the University of New Hampshire.

REFERENCES

1. Hollweg, J. V., and Isenberg, P. A., *J. Geophys. Res.*, 10.1029/2001JA000270 (2002).
2. Cranmer, S. R.,"Coronal holes and the high-speed solar wind," *Space Sci. Rev.*, submitted (2002).
3. Markovskii, S. A., and Hollweg, J. V., "Parametric cross-field current instability in solar coronal holes," *J. Geophys. Res.*, in press (2002).
4. Feldman, W. C., Asbridge, J. R., Bame, S. J., Montgomery, M. D., and Gary, S. P., *J. Geophys. Res.*, **80,** 4181 (1975).
5. Rosenbauer, H., Schwenn, R., Marsch, E., Meyer, B., Meggenrieder, H., Montgomery, M. D., Mühlhäuser, K.-H., Pilipp, W., Voges, W., and Zink, S. M., *J. Geophys.*, **42,** 561 (1977).
6. Marsch, E., in *Physics of the Inner Heliosphere, 2, Particles, Waves and Turbulence,* edited by R. Schwenn, and E. Marsch, Springer-Verlag, Berlin, 1991, p. 45.
7. Forslund, D. W., *J. Geophys. Res.*, **75,** 17 (1970).
8. Toichi, T., *Solar Phys.*, **18,** 150 (1971).
9. Coppi, B., "Collisionless collective modes including transverse ion heating," unpublished talk presented at Solar Wind 9 Conference, Nantucket Island, Massachusetts, USA, October 5–9, 1998.
10. Ledenev, V. G., and Starygin, A. P., *Plasma Phys. Rep.*, **27,** 652 (2001).
11. Forslund, D. W., Kindel, J. M., and Stroscio, M. A., *J. Plasma Phys.*, **21,** 127 (1979).
12. Gary, S. P., *Theory of Space Plasma Microinstabilities,* Cambridge Univ. Press, Cambridge, UK, 1993, p. 156.
13. Gary, S. P., Newbury, J. A., and Goldstein, B. E., *J. Geophys. Res.*, **103,** 14559 (1998).
14. Markovskii, S. A., and Hollweg, J. V., *Geophys. Res. Lett.*, 10.1029/2002GL015189 (2002).
15. Voitenko, Y., and Goossens, M., *Solar Phys.*, **206,** 285 (2002).
16. Isenberg, P. A., Lee, M. A., and Hollweg, J. V., *J. Geophys. Res.*, **106,** 5649 (2001).
17. Esser, R., Habbal, S. R., Coles, W. A., and Hollweg, J. V., *J. Geophys. Res.*, **102,** 7063 (1997).
18. Muschietti, L., and Dum, C. T., *J. Geophys. Res.*, **95,** 173 (1990).

Role of driving scales in a model of coronal heating

O. Podladchikova*†, B. Lefebvre* and V. Krasnoselskikh*

*LPCE/CNRS & Université d'Orléans UMR 6115
3A av. de la Recherche Scientifique, 45071 Orléans, FRANCE
†Kiev Polytechnic Institute, Kiev, UKRAINE

Abstract. Models of coronal heating based on the dissipation of small-scale current sheets generally assume energy injection at large scales by photospheric motions. However due to the nature of these motions, excitation mechanisms may occur on a wide range of scales. We study the role of the dominant scales of energy injection in the framework of the lattice model introduced in [1]. In particular it is shown that as the weight of large scales increases, the probability densities of dissipated energy and of waiting times between heating events develops fatter tails tending toward power-laws.

INTRODUCTION

The dissipation of many small scale current sheets resulting in the so-called nano-flares was conjectured by Parker to be a mechanism of coronal heating [2]. Various formation mechanisms of these small scale currents have been proposed, whose origin ultimately lies in photospheric convection inducing complex motions of coronal magnetic field lines footpoints.

Observed statistics of flare, microflares, and smaller events have suggested a certain scale-invariance between these phenomena ([3, 4] and references therein). Extrapolation of a power-law distribution for the dissipated energy to smaller energies indicate that heating by nanoflares is a viable mechanism if its index is smaller than -2 [5]. Different interpretations of recent observations in EUV of small scale events in terms of dissipated energy provide indices scattered around -2 [6, 7, 8] thereby leaving the question quite open.

These particular statistics have stimulated a number of simple (lattice) models whose statistics can be extensively studied. For instance, flare models based on Self-Organized-Criticality (SOC) [9, 10] have met some success in reproducing certain statistics of observed flare and micro-flare associated emissions [4]. Especially for applications to coronal heating, it is essential to provide lattice models allowing for a clear interpetation by physical processes rather than concentrating on SOC (e.g. [11, 12, 13, 1, 14]). The model we shall consider here is the one introduced in [1] whose two fundamental elements are on one hand the dissipative processes mimicking anomalous resistivity or magnetic reconnection and on the other hand the sources perturbing the magnetic field configuration and generating the currents.

The emphasis in this note shall be on the role played by the dominant scales in the source term of the model. Either Parker's model or models of coronal heating based on MHD turbulence explicitely make the classical assumption that energy is injected at large scales only from photospheric motions and cascades down to small scales [15]. However, photospheric convection involves a wide range of different motions, including shear-generated turbulent-like fluctuations between granules [16]. Sources perturbing the system on a wide range of scales are likely to affect its dynamics, having some effects on the energy dissipation and conversion into heat. This is what we are going to study in the framework of the lattice model of [1] using sources with a power-law spectrum. This form easily allows to change the relative weigths of the different characteristic scales of the source term.

MODEL

The model consists of magnetic field distributed into cells of a 2D grid, perpendicular to the grid, and subject to driving and dissipation mechanisms. Such a simplification allows to easily select between different sources and dissipation mechanisms. Periodic boundary conditions on a square grid of size $N \times N$ are assumed.

The source spectrum is a power-law spectrum in a certain range of wavenumbers. This allows to easily control the relative weights of different wavenumbers, thereby controlling the dominant scales of energy injection. Therefore the source provides a magnetic field in-

CP679, *Solar Wind Ten: Proceedings of the Tenth International Solar Wind Conference*,
edited by M. Velli, R. Bruno, and F. Malara

crement of the form

$$\delta B(x,y,t) = C_{\alpha,k_{\min},k_{\max}} \sum_{k_x,k_y} |\mathbf{k}|^{-\alpha} e^{i(k_x x + k_y y + \phi(t))}, \quad (1)$$

where C is such that the total intensity is normalized to 1. The phases ϕ are independant random variables uniformly distributed in $[0,2\pi)$. The dominant scales in δB can be controlled by changing both α and the range of wavenumbers in the summation.

The dissipation mechanisms considered in our previous studies are of two types, namely reconnection or anomalous resistivity. Both depend on currents, which propagate on the border of the cells and are computed as $\mathbf{j} = \nabla \times \mathbf{B}$. Anomalous resistivity arises when a current exceeds a given threshold as the result of micro-instabilities. Reconnection in this simple framework requires an additional condition mimicking an X-point. In both cases, the energy dissipated into heat is supposed to be proportional to j^2. See [1] for more details. In the following the grid size is 256×256, dissipation is provided by anomalous resistivity, and the dissipation threshold is $j_{\max} = 3$.

MULTISCALE DRIVING

In this section the source acts at all available scales, but different exponents in eq. (1) are used. In particular, the case $\alpha = 0$ corresponds to white (in space) noise [1, 17].

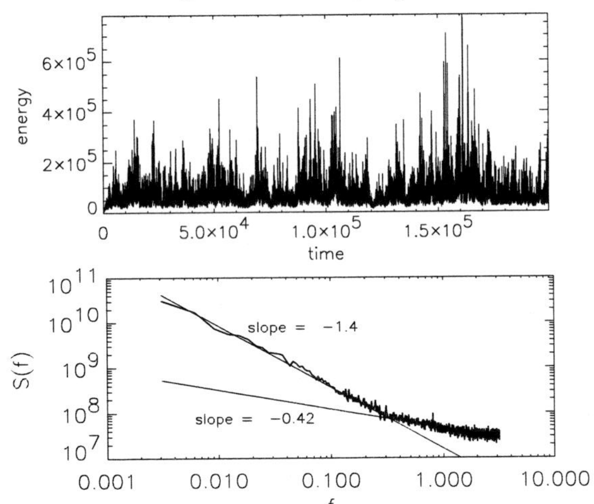

FIGURE 1. Time series of the total dissipated energy over the grid and its power spectrum for $\alpha = -2$.

As α increases, the time series of the total dissipated energy displays more intense and sporadic bursts (Fig. 1). At the same time, the power-spectrum tends toward a more pronounced power-law shape (Fig. 1, lower panel). This power-law breaks down at high-frequencies, where the spectrum flattens. This breaking point drifts toward higher-frequencies as α increases. Interestingly on each side of the breaking point spectra weakly depend on α, at least in the interval $[1,3]$.

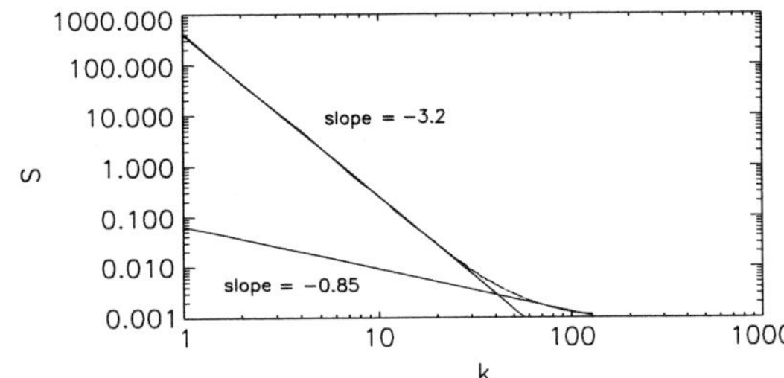

FIGURE 2. Power spectrum in k of the magnetic field for $\alpha = -2$.

The spatial power spectrum of B shows a similar tendency to a power-law at small k which breaks down to a flatter spectrum at small scales (Fig. 2). The energy contained in the steepest part and at the largest scales increases relatively to the energy contained in the small scale part when α increases (96% for $\alpha = 1$, 99% for $\alpha = 1$ and 99.8% for $\alpha = 3$). At large scales the spectrum decreases faster than the source, which means that the development of the "cascade" is somehow slowed down.

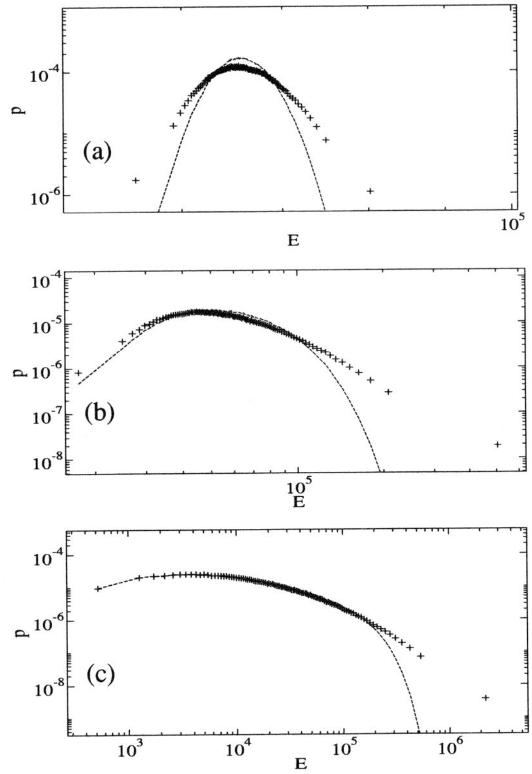

FIGURE 3. Log-log plot of the dissipated energy PDF for different values of α. (a) $\alpha = 1$, (b) $\alpha = 2$, (c) $\alpha = 3$. The dashed line represents the Gaussian with same mean and variance.

White noise sources combined with anomalous resistivity dissipation provide dissipated energies whose Probability Density Functions (PDF) weakly depart from the Gaussian [1]. When α increases to positive values, this departure is more and more marked by a heavy tail corresponding to large deviations toward high energy events (Fig. 3). However the tail does not tend toward a power-law, since on the log-log plot it exhibits clear concavity. A more precise characterization could be provided using Pearson curves [18]. Another statistics of solar flares extensively discussed is the waiting time distribution between two solar flares (here between two peaks of dissipated energy above a given threshold). For $\alpha = 3$ it follows a power-law (Fig. 4), in agreement with some observations (see the discussion concluding the paper).

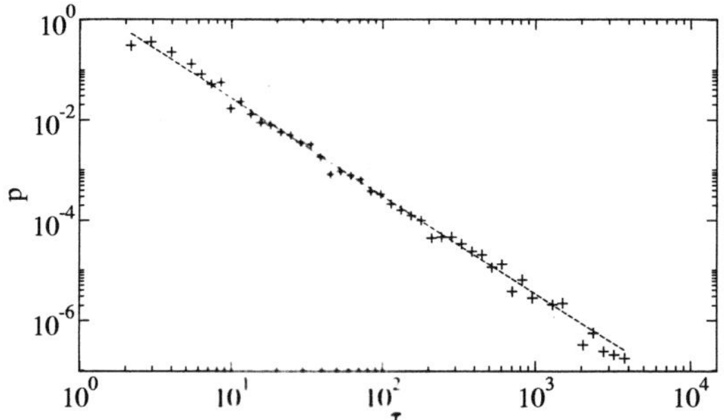

FIGURE 4. PDF of waiting times for $\alpha = -3$. A least-squares power-law fit (dashed line) yields an exponent -1.95 ± 0.10.

The source also has an effect on the spatial characteristics of the magnetic field. In [17] these characteristics where studied by Singular Value Decomposition (SVD). From the singular values μ_i it is possible to obtain an entropy which provides a quantitative estimate of spatial complexity, defined as [19, 17]

$$H = -\lim_{N \to \infty} \frac{1}{\ln N} \sum_{k=1}^{N} E_k \ln E_k,$$

with $E_k = \mu_k^2 / \sum_i \mu_i^2$. H is defined in such a way that $H = 1$ when energy is equidistributed over all singular modes (maximal disorder) and $H = 0$ when all the variance is contained in a single mode (minimal disorder).

TABLE 1. Entropy as a function of the exponent α defined in eq. (1).

α	0	1	2	3
H	0.75	0.55	0.22	0.12

As clearly seen in Table 1, the entropy significantly decreases when α increases. Thus when large scales are given more and more weight in the sources, the magnetic field has a natural tendency to have a more coherent structure at large scales.

DRIVING AT LARGE SCALES

Here we consider a quasi-periodic source involving only the two largest modes.

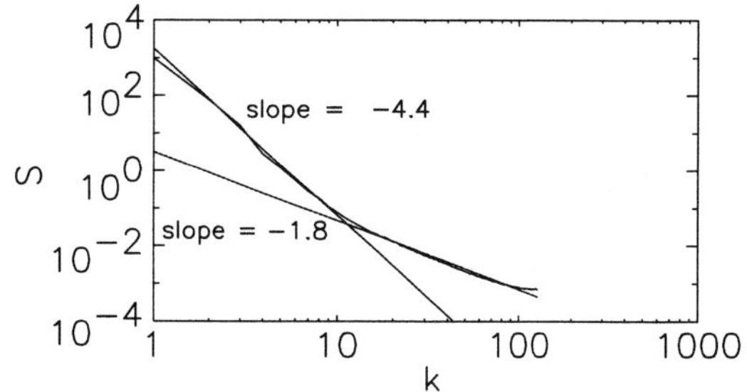

FIGURE 5. Power spectrum in k of the magnetic field, when the system is driven by the two largest modes.

Although the system is driven only at large scales, the spectrum develops toward smaller scales due to the local coupling between cells. In a statistically steady state, it may be approximated by a power-law which flattens at small scales (Fig. 5). The dissipated energy PDF displays a fat high-energy tail which can be roughly approximated by a power-law, with rather small exponent -1.19 ± 0.08. The waiting time distribution also exhibits a slowly decaying tail (Fig. 7), with a power-law shape with exponent -2.28 ± 0.12.

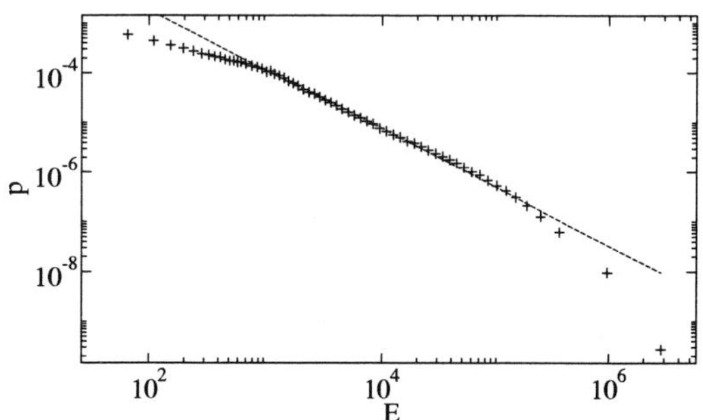

FIGURE 6. PDF of the dissipated energy. A power-law fit to its tail (dashed line) has exponent -1.19 ± 0.08.

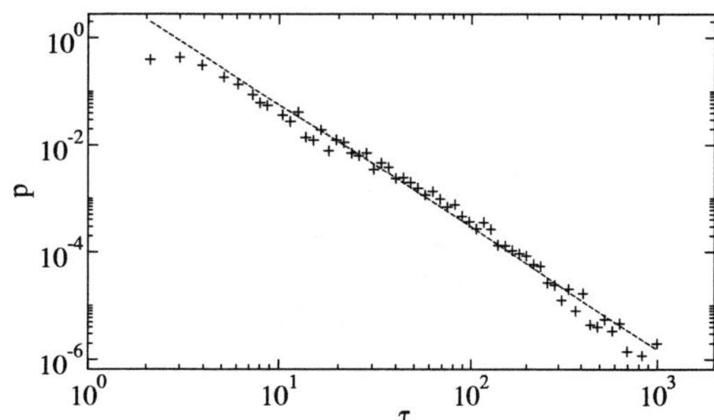

FIGURE 7. PDF of waiting times. A power-law fit to its tail (dashed line) has exponent -2.28 ± 0.12.

DISCUSSION

The role of the scales in the source term of a lattice model for coronal heating was studied. For this purpose a power-law spectrum was used, increasing the relative role of large scales with respect to the limit case of white-noise-in-space sources considered in [1, 17].

The magnetic field fluctuations appear to have a power-law spectrum which is at large scales generally steeper than the one of the source, showing that the system does not exhibit a trivial response to the source but rather is ruled by its own dynamics. Particularly at small scales it is rather insensitive to the form of the source's spectrum. Furthermore, a similar power-law power spectrum appears when the system is driven only at large scales, showing that it is an intrinsic feature of such a model when large scales dominate the energy injection.

As the weight of large scales plays a more significant role in energy injection, a long tail is formed in the PDF of dissipated energy as well as in the waiting time distribution between energetic events. Indeed, in this case large scale gradients form and their dissipation produces rare but energetic dissipative events resulting from many small-scale local dissipations.

The power-laws found in the PDF of dissipated energy are in poor agreement with those found in experiment: the exponents are too small or the tail does not convincingly follow a power-law, and the PDF steepens toward high energies although flares seem to follow a flatter distribution than microflares and nanoflares. This situation may improve by considering sources with potentially large amplitude as found in [20], or using a different and more sophisticated dissipation process such as the reconnection used in [1, 17] and which was shown to produce significant changes in the statistics of dissipated energy and fatter tails in its PDF.

However, we have found heavy tails in waiting time distributions, whose power-law decay seems to agree with most observations [21, 22, 23]. Attention on waiting time distribution was drawn in particular by the remark that pure SOC models exhibit only a Poisson waiting time distribution [22]. Attempts to correct this have used "non-conventional" properties of the source, whereas they are non-stationary [24] or correlated [25] in time, or acting continuously in time [26]. The present study suggests that sources acting at large scales or representing a wideband spectrum dominated by large scales (as can be expected on physical grounds) also help to recover a more correct waiting time distribution.

ACKNOWLEDGMENTS

The authors are thankful the Referee for his help to improve the presentation of this paper. BL acknowledges financial support from the CNES.

REFERENCES

1. Krasnoselskikh, V., Podladchikova, O., Lefebvre, B., and Vilmer, N., *A&A*, **382**, 699–712 (2002).
2. Parker, E. N., *ApJ*, **330**, 474–479 (1988).
3. Aschwanden, M. J., Poland, A., and Rabin, D., *Annu. Rev. Astron. Astrophys.*, **39**, 175–210 (2001).
4. Georgoulis, M. K., Vilmer, N., and Crosby, N. B., *A&A*, **367**, 326–338 (2001).
5. Hudson, H. S., *Solar Physics*, **133**, 357–369 (1991).
6. Benz, A., and Krucker, S., *Sol. Phys.*, **182**, 349–363 (1998).
7. Mitra-Kraev, U., and Benz, A. O., *A&A*, **373**, 318–328 (2001).
8. Aschwanden, M. J., Tarbell, T. D., Nightingale, R. W., Schrijver, C. J., Title, A., Kankelborg, C. C., Martens, P., and Warren, H. P., *ApJ*, **535**, 1047–1065 (2000).
9. Lu, E. T., and Hamilton, R. J., *ApJ*, **380**, L89–L92 (1991).
10. Charbonneau, P., McIntosh, S., Han-Li, L., and Bogdan, T., *Solar Phys.*, **203**, 321 (2001).
11. Isliker, H., Anastasiadis, A., Vassiliadis, D., and Vlahos, L., *A&A*, **335**, 1085–1092 (1998).
12. Einaudi, G., and Velli, M., *Phys. Plasmas*, **6**, 4146–4153 (1999).
13. Longcope, D. W., and Noonan, E. J., *ApJ*, **542**, 1088–1099 (2000).
14. Buchlin, E., Aletti, V., Galtier, S., Velli, M., and Vial, J., *these proceedings* (2002).
15. Gómez, D. O., Dmitruk, P. A., and Milano, L. J., *Solar Phys.*, **195**, 299–318 (2000).
16. Nesis, A., Hammer, R., Kiefer, M., Schleicher, H., Sigwarth, M., and Staiger, J., *A&A*, **345**, 265–275 (1999).
17. Podladchikova, O., Dudok de Wit, T., Krasnoselskikh, V., and Lefebvre, B., *A&A*, **382**, 713–721 (2002).
18. Podladchikova, O., PhD thesis, Université d'Orléans, 2002.
19. Aubry, N., Guyonnet, R., and Lima, R., *J. Stat. Phys.*, **64**, 683–739 (1991).
20. Georgoulis, M., and Vlahos, L., *ApJ*, **469**, L135 (1996).
21. Wheatland, M. S., Sturrock, P. A., and McTiernan, J. M., *ApJ*, **509**, 448–455 (1998).
22. Boffetta, G., Carbone, V., Giuliani, P., Veltri, P., and Vulpiani, A., *Phys. Rev. Lett.*, **83**, 4662–4665 (1999).
23. Lepreti, F., Carbone, V., and Veltri, P., *ApJ*, **555**, L133–L136 (2001).
24. Norman, J. P., Charbonneau, P., McIntosh, S. W., and Liu, H., *ApJ*, **557**, 891–896 (2001).
25. Sanchez, R., Newman, D., and Carreras, B., *Phys. Rev. Lett.*, **88**, 068302 (2002).
26. Hamon, D., Nicodemi, M., and Jensen, H., *A&A*, **387**, 326–334 (2002).

A new exospheric model of the solar wind acceleration: the transsonic solutions

I. Zouganelis*, M. Maksimovic*, N. Meyer-Vernet*, H. Lamy† and V. Pierrard†

*LESIA, Observatoire de Paris, CNRS FRE2461, 92195 Meudon, France
†Institut d'Aéronomie Spatiale de Belgique, Brussels, Belgium

Abstract. This paper presents basic issues for the solar wind acceleration with a collisionless model when the base of the wind is sufficiently low for the potential energy of the protons to have a maximum, thereby producing a transsonic wind. Using a formulation in terms of the particle invariants of motion, we study the existence of different categories of ion orbits and the consequences on the wind acceleration. We also study how a suprathermal tail in the electrons velocity distribution enhances the wind acceleration and makes the electron temperature increase within a few solar radii.

INTRODUCTION

The exospheric approach of the solar wind acceleration is based on the assumption that the Coulomb collisional mean free path of the particles in the corona is much greater than the density scale height above a given altitude called the exobase. The exobase is therefore usually defined as the altitude where these two characteristic quantities become equal. In these collisionless models, the interplanetary electrostatic field, which pulls electrons inward and accelerates protons outward in order to preserve zero electric charge and current, is calculated self-consistenly [2], [5].

In previous models [6], the altitude of the exobase ($> 5R_\odot$) was so high that the gravitational attraction on protons was everywhere smaller than the outward electric force, so that the total potential (electric+gravitational) of the protons was decreasing monotonically. This high exobase altitude may be adequate to model the slow solar wind emanating from equatorial regions where the density is high enough. This is not so, however, for the fast solar wind, which emanates from the coronal holes where the exobase is located deeper in the corona. In that case, the gravitational force is stronger than the electric one at the base of the wind so that the total potential energy of the protons is attractive out to some distance where the two forces balance each other. Farther out, the outward electric force dominates. That means that the total potential for the protons is not monotonic, presenting a maximum at a certain distance from the exobase [2]. The existence of this maximum has important consequences on the physics of the wind acceleration (see [8] in this issue) in governing the ion orbits, and is fundamentally associated to the production of a transsonic wind.

Since the electrostatic potential is determined by the balance of electric charge and current, it is rather sensitive to the velocity distribution function (VDF) of the electrons. Following Scudder's pioneer work [10], recent exospheric models have modelled the electron VDFs with generalized Lorentzian ("kappa") functions in order to allow for the possibility of electron suprathemal tails. Such non-thermal VDFs enable us to obtain increased terminal bulk speeds [6].

In the present paper, we present some basic issues emerging from an exospheric model allowing both a non-monotonic potential and non-maxwellian electron VDFs (see [4] in this issue). We examine in particular how the wind properties vary with the non thermal character of the electrons VDF and set the basis for an improved exospheric description of the solar wind acceleration, which maps particle orbits in terms of the invariants of motion.

NON-MONOTONIC PROTON POTENTIAL ENERGY

As the plasma is assumed to be collisionless, the total energy and the magnetic moment of the particles are conserved, i.e.

$$E = \frac{mv^2}{2} + m\phi_g + Ze\phi_E = cst \tag{1}$$

$$\mu = \frac{mv_\perp^2}{2B} = cst \tag{2}$$

where v is the velocity of the particle of mass m, Ze its charge, $\phi_g(r) = -M_\odot G/r$ the gravitational potential

CP679, *Solar Wind Ten: Proceedings of the Tenth International Solar Wind Conference*, edited by M. Velli, R. Bruno, and F. Malara

and $\phi_E(r)$ the interplanetary electrostatic potential. The total potential energy for the protons is then $\Psi(r) = m\phi_g(r) + e\phi_E(r)$, which reaches a maximum at some radial distance r_{max}. Below this altitude, the gravitational force is larger than the electrostatic one and the total potential is attractive. The opposite is true above r_{max}, forcing all the protons present at these altitudes to escape from the system. Whereas in previous models [6] all protons were escaping, the existence of a maximum allows other kinds of proton trajectories to exist below r_{max}. There are now protons (called "ballistic") which do not have enough energy to escape from the gravitational well of the Sun and are therefore returning towards it. Another kind of non-escaping protons are the "trapped" ones which do not have enough energy to escape, but whose inclination to the magnetic field lines is large enough that they are reflected by the magnetic mirror force before returning back to the exobase r_0. Note that there are no particles coming from infinity.

Although previous exospheric models with non-maxwellian electron VDFs at the exobase [6] assumed a monotonic potential energy for the protons, the method of treating non-monotonic potentials for the solar wind had already been developed by Jockers [2] who did not allow for suprathermal electrons.

When dealing with the Vlasov equation, it is convenient to use the total energy E and the magnetic moment μ as primary coordinates. The velocity distribution function can then be written as $f(E,\mu)$ as well as its moments. This choice is very convenient because it removes the spatial dependence of the VDF and only the region of integration (over which the moments are calculated) in $E-\mu$ space changes. The basic problem is the accessibility of the different particle populations in the $E-\mu$ space, described in [3] for an arbitrary potential energy structure.

The conservation laws (1) and (2) determine the region where the function f is defined as:

$$v_{\parallel}^2 \geq 0 \Rightarrow E \geq \mu B(r) + \Psi(r) \tag{3}$$

where $B(r)$ is the magnetic field which is assumed to be radial. The relation (3) defines the line $v_{\parallel} = 0$ for each altitude r; the distribution function f is defined only above this line. Note that the slope of this line is just the amplitude of the magnetic field B. Figure 1 shows the regions of integration for the different particle populations. The function f is defined at the exobase r_0 only above the corresponding line (0). Escaping protons are present at all altitudes including r_{max}. This means that they are defined above *all* $v_{\parallel} = 0$ lines, a region which corresponds to the upper part of the figure. The filled region corresponds to ballistic protons which cannot escape from r_{max} or lower altitudes. Trapped protons can exist in the region filled of circles as they are not present at r_0 nor at r_{max}.

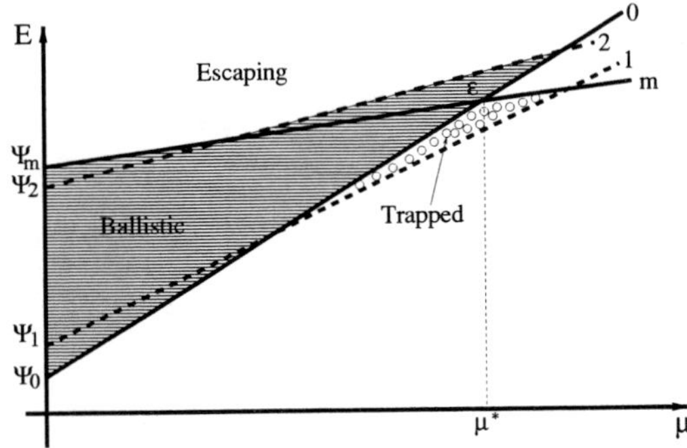

FIGURE 1. Accessibility of the different particle populations in $E-\mu$ space.

Note that if all $v_{\parallel} = 0$ lines are located above the intersection point (E^*, μ^*) between the lines (0) and (m) (with $E^* = \mu^* B_0 + \Psi_0$), there are no trapped protons. We can therefore show that there are no trapped protons at altitudes where:

$$\Psi(r) \geq \left(\Psi_{max} - \Psi_0\right) \frac{r_{max}^2}{r^2} \frac{r_0^2 - r^2}{r_0^2 - r_{max}^2} + \Psi_0 \tag{4}$$

As we will see below, there are some cases in which the inequality (4) holds for all r so that no trapped protons exist.

The situation presented in figure 1 is not complete, since only the influence of two altitudes (r_1 and r_2) is shown. Actually we have to take into account the influence of *all* altitudes between the exobase r_0 and r_{max}. In order to simplify the calculations, we make the following assumption. All particles which can be present at both r_0 and r_{max} are considered as escaping. i.e. we consider that the region labelled ε in figure 1 corresponds to escaping particles at r_{max}, instead of ballistic which do not overcome r_2. This important approximation is used by Jockers as the resulting error in the particle flux is in general less than 1% [2]. However, the errors due to this approximation may be much greater with suprathermal electrons. The relation (4) is valid only with this approximation.

In this paper, this assumption has been made for the protons for two reasons: the induced error is low enough with the chosen parameters and the simplicity of the method enables us to deduce basic aspects of the acceleration physics. The expressions given by Jockers [2] are therefore used for the protons, assuming a maxwellian VDF. An equivalent approach is to perform the integrations of the VDF in the velocity space [5]. This is leading to the same results as our approximative model (see [4] in this issue).

SUPRATHERMAL ELECTRONS

For the electrons, the situation is much simpler since their total potential energy ($\approx -e\phi(r)$) is monotonically increasing with distance. The attractive potential allows the three kinds of particle orbits (escaping, ballistic and trapped) to be present at all altitudes (see [8] in this issue).

In the solar wind, the electron distributions are observed to have important high velocity tails and the most convenient way to fit them is with kappa functions [7], which play an important role in accelerating the wind by a strong filtration mechanism [10].

The electrons VDFs in the corona might also be non-maxwellian. The fundamental reason is that fast electrons collide much less frequently than slow ones because of their greater free path, so that they cannot relax to the equilibrium maxwellian distribution. This is why electron VDFs could have suprathermal tails, although the core distribution should be closer to a maxwellian. Recent exospheric models have therefore used kappa VDFs at the exobase [6] given by the expression

$$f_{\kappa e} = \frac{n_{e0}}{(\pi \kappa v_{th}^2)^{3/2}} \frac{\Gamma(\kappa+1)}{\Gamma(\kappa-1/2)} \left(1+\frac{v^2}{\kappa v_{th}^2}\right)^{-(\kappa+1)} \quad (5)$$

where the thermal speed is defined by $v_{th} = \left(\frac{2\kappa-3}{\kappa}\frac{k_B T_e}{m_e}\right)^{1/2}$. This function tends to a maxwellian when $\kappa \rightarrow \infty$.

In our model we use this electron distribution, with the expressions of the moments given by [9]. Note, however, that the value of κ in eq.(5), which represents the non thermal character, remains constant with altitude, which does not seem to agree with recent observations [1]. We use this distribution as a first step to represent suprathermal tails, having in mind that a more realistic VDF could be a sum of two maxwellians [8] (which, however, needs one more free parameter) or a superposition of a maxwellian and a truncated power law. We will return to this point below.

RESULTS AND DISCUSSION

As in previous exospheric models, the electrostatic potential is calculated by imposing the quasi-neutrality and by equalizing the electron and proton fluxes at all altitudes [2],[6]. In order to study the basic physics and since the exobase location is defined from parameters which are not accurately known, we put the exobase at $r_0 = 1R_{\odot}$. This stands as an approximation for a more realistic value which should be slightly higher and has no important qualitative impact on the results. The initial temperature values at the exobase are taken as $T_{e0} = 10^6 K$ and

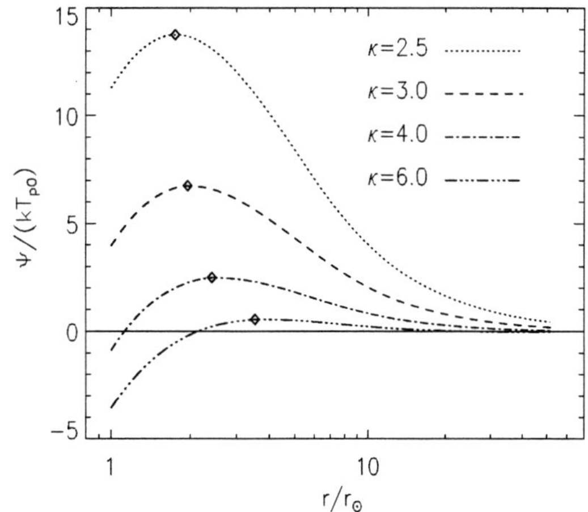

FIGURE 2. Total proton potential energy for different values of $\kappa = 2.5, 3.0, 4.0$ and 6.0. The energy is normalized in $k_b T_{p0}$. The maximum in each case is designated by the diamont.

$T_{p0} = 2T_{e0}$. The total potential energy for the protons is shown in figure 2 for different values of κ ranging from $\kappa = 6$ to $\kappa = 2.5$, a case with a conspicuous suprathermal tail. We can see that the value of the maximum of potential increases as κ decreases. This is because with more suprathermal electrons, a stronger electric potential is needed to preserve quasi-neutrality. Note also that r_{max} increases with increasing κ. In any case the potential tends to zero at large distances.

When the electric potential is known, the bulk speed can be calculated at all distances [6]. This is shown in figure 3 for the same values of κ as in fig.2. It can be seen that a high terminal bulk speed (>700km/s) is obtained when the suprathermal electron tail is conspicuous ($\kappa = 2.5$). This is due to the large value of the maximum in ion potential energy ($\approx 14 k_b T_{p0}$), which is transformed into kinetic energy of the escaping protons as they are accelerated above r_{max}. An important remark is that the major part of this high terminal bulk speed is obtained within small distances ($\approx 10R_{\odot}$) which is due to the high acceleration represented by the important slope of the potential after r_{max}.

The large suprathermal tail assumed has another important consequence. It makes the electron temperatures increase considerably up to a maximum ($\approx 7\times 10^6 K$) within a few solar radii. This maximum in electron temperature is smaller for larger values of κ and disappears as $\kappa \rightarrow \infty$ as shown in fig.4. This heating is a direct consequence of filtration of the non maxwellian VDF by the attracting potential [10]. A problem with the ε region approximation is a discontinuity in the calculated proton temperatures obtained just above r_{max}. The same

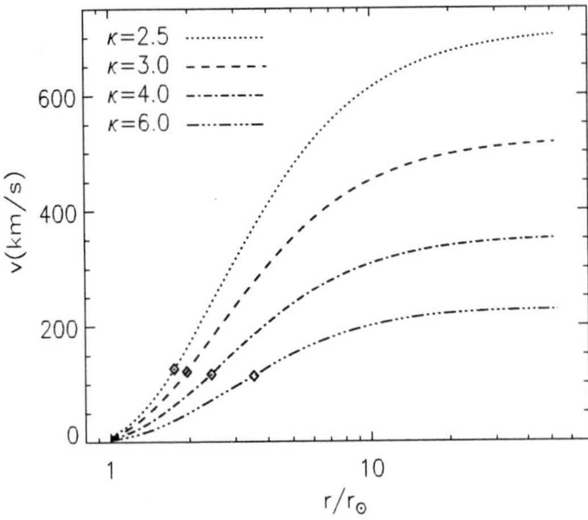

FIGURE 3. Solar wind bulk speed for different values of $\kappa = 2.5$, 3.0, 4.0 and 6.0. The potential maximum location in each case is designated by the diamont.

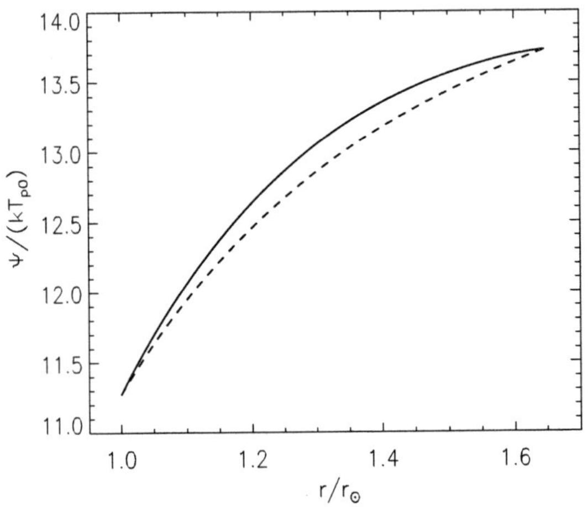

FIGURE 5. Calculated values (full line) of the total proton potential energy between r_0 and r_{max} for $\kappa = 2.5$. The dashed line corresponds to the right hand side of the expression (4).

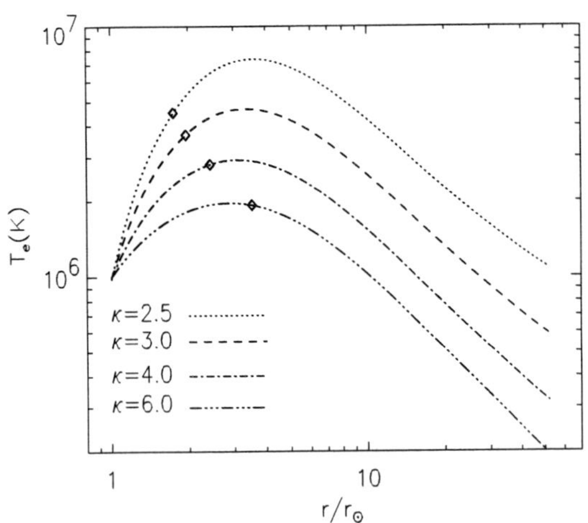

FIGURE 4. Electron temperatures for different values of $\kappa = 2.5$, 3.0, 4.0 and 6.0. The potential maximum location in each case is designated by the diamont.

problem was present in Jockers' models [2] and could be probably eliminated with the implementation of a generalized model with no assumption regarding the ε region ($fig.1$).

The $E - \mu$ formalism enables us to examine whether trapped protons are present. In figure 5 the dashed line represents the right hand side of eq.(4) between r_0 and r_{max} for $\kappa = 2.5$, whereas the full line corresponds to the calculated $\Psi(r)$ values. We can see that the inequality (4) holds for all r, which means that there are no trapped protons. This result is very important as it shows that trapped protons are not necessary to produce a wind, when the electron VDFs are highly non-thermal. This no longer holds when the distribution is close to a maxwellian as the two lines intersect in the space (r_0, r_{max}) for the case $\kappa = 6$. Trapped protons are present up to the intersection point and there are only ballistic (and escaping) ones beyond it. The presence or not of trapped particles is an important issue as they are ad hoc populated in fully collisionless models.

Note, finally, that our results suggest that the Kappa velocity distribution may not be adequate to model VDF having suprathermal tails in the corona. Indeed, producing large enough speeds requires irrealistically small values of κ, which, furthermore, yield an electron temperature increase much larger than observed. Alternatively, more realistic values of κ produce an electron temperature increase close to the observed values, but a too small terminal velocity. A future study will include more realistic non thermal distributions.

REFERENCES

1. Esser R., Edgar R.J. et al., *Ap. J.*, **532**, L71 (2000)
2. Jockers K., *Astron. Astrophys.*, **6**, 219 (1970)
3. Khazanov G. et al. *J. Geophys. Res.* **103,** 6871(1998)
4. Lamy et al, *Solar Wind 10*,(2002)
5. Lemaire J., Scherer M., *J. Geophys. Res.* **76,** 7479 (1971)
6. Maksimovic M. et al., *Astron. Astrophys.* **324**, 725 (1997)
7. Maksimovic M. et al. *Geophys. Res. Let.* **24,** 1151(1997)
8. Meyer-Vernet et al, *Solar Wind 10*,(2002)
9. Pierrard V., Lemaire J., *J. Geophys. Res.* **101,** 7923 (1996)
10. Scudder J.D., *Ap. J.* **398**, 299-318, 1992

Equatorial Coronal Holes and Their Relation to High-Speed Solar Wind Streams

Lidong Xia and Eckart Marsch

Max-Planck-Institut für Aeronomie, D-37191 Katlenburg-Lindau, Germany

Abstract. Using together SUMER, EIT and MDI onboard SOHO, we examine plasma properties and magnetic fields at the base of three equatorial coronal holes (ECHs) occurring during August and October 1996 near solar minimum. We estimate the electron density, flow speed deduced from UV/EUV lines as well as the average magnetic field of the photosphere. These ECHs produced distinct high-speed streams in the ecliptic plane with average flow speeds of larger than 500 kms^{-1}. With SWE and MFI both onboard the WIND satellite, we also determine the parameter values of the plasma and magnetic field of these high-speed streams at 1 AU in the Earth's orbit. We discuss the relationships between observations of high-speed streams at 1 AU and coronal holes at the coronal base.

INTRODUCTION

The nascent fast solar wind may first be accelerated in magnetic funnels [1]. This idea is supported by measurements of Doppler shifts of the Ne VIII line with SUMER (Solar Ultraviolet Measurements of Emitted Radiation), which show that large blue shifts occur mainly along the chromospheric network [2]. Another interesting result is that regions of large blue shifts spatially coincide with very dark regions, if we compare the Dopplergram of the hole with its intensity image in the same line [3]. However, the cited studies were carried out in polar coronal holes (PCHs). In that case there are several observational disadvantages due to the line-of-sight geometry. In this contribution, we study three ECHs observed during the solar minimum in the second half of 1996 and investigate their relation to interplanetary high-speed streams.

OBSERVATIONS AND DATA ANALYSIS

We selected three ECHs that have been observed by SUMER (Table 1). The instrument has been described in detail elsewhere [4]. The detector B and slit 2 with the size of 1"×300" were used. In this study, simultaneous data obtained by five instruments onboard SOHO and WIND were analysed (Table 2).

TABLE 1. ECHs observed by SUMER

Item	Date	Wavelengths
ECH1	August 27, 1996	Ref. Spec.: 660-1500Å
ECH2	October 12/13, 1996	Ref. Spec.: 660-1500Å
ECH3	October 19/20, 1996	Raster: 750-790Å

TABLE 2. Instruments

Instruments	Measurements
SUMER	intensity, Doppler shift, line width and electron density (Mg IX 694/706)
EIT	coronal hole boundaries (Fe XII 195Å)
MDI	photospheric magnetic field
SWE	solar wind speed and density at 1AU
MFI	magnetic field at 1AU

The three ECHs and their related high-speed streams are shown in Figs. 1 and 2. ECHs appear as dark regions in the EIT (Extreme ultraviolet Imaging Telescope) meridian maps obtained in the Fe XII line (195Å).

The ECH1 and ECH2 including the positions of the SUMER slit are shown in Figs. 3 and 4. Because the coronal lines are quite weak in these ECHs, we have binned the data along the slit in the Y direction selecting areas obtained from homogeneous dark parts of the holes, from 190" to 335" for ECH1 (Fig. 3) and from -60" to 50" for ECH2 (Fig. 4).

We deduce the outflow speed by measuring Doppler shifts of the Mg X line (624.968Å, log T_e=6.04) observed in 2nd order. For wavelength calibration we used cool lines of C I. Chae *et al.* [5] have estimated that C I lines have only a small average red shift of about 1.5 kms^{-1}. This value has been subtracted from the average Doppler shift of the Mg X line. The Ne VIII line (770.428Å, log T_e=5.8) observed in ECH3 is strong enough, so that line center positions can be determined for every pixel.

After getting the intensity ratio of two Mg IX lines 694/706 (log T_e=5.98), the CHIANTI atomic database [6] has been used to calculate the electron density in

CP679, *Solar Wind Ten: Proceedings of the Tenth International Solar Wind Conference*, edited by M. Velli, R. Bruno, and F. Malara

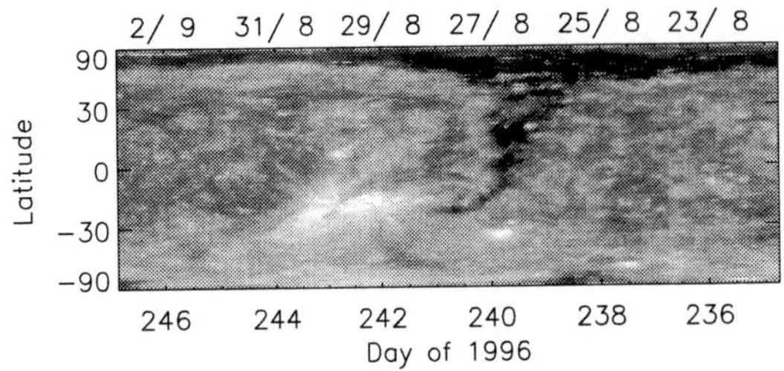

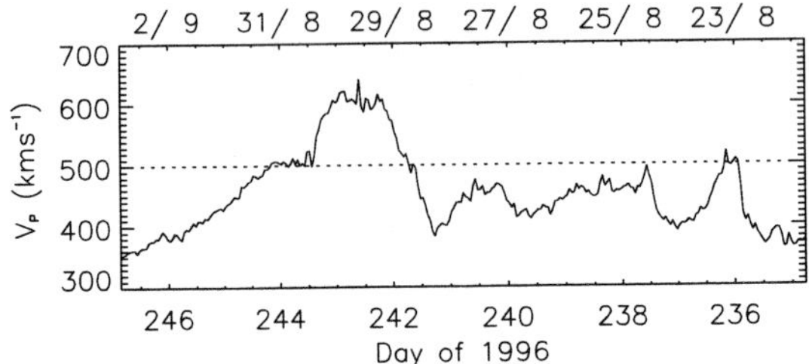

FIGURE 1. Left: EIT meridian map during August 22 and September 3, 1996, showing the recurrent ECH "Elephant's Trunk". Right: Solar wind speed observed by SWE at 1AU during the same period showing a related high-speed stream. Note the displacement in time, which is due to the 3-days travel time of the stream from the corona to 1AU.

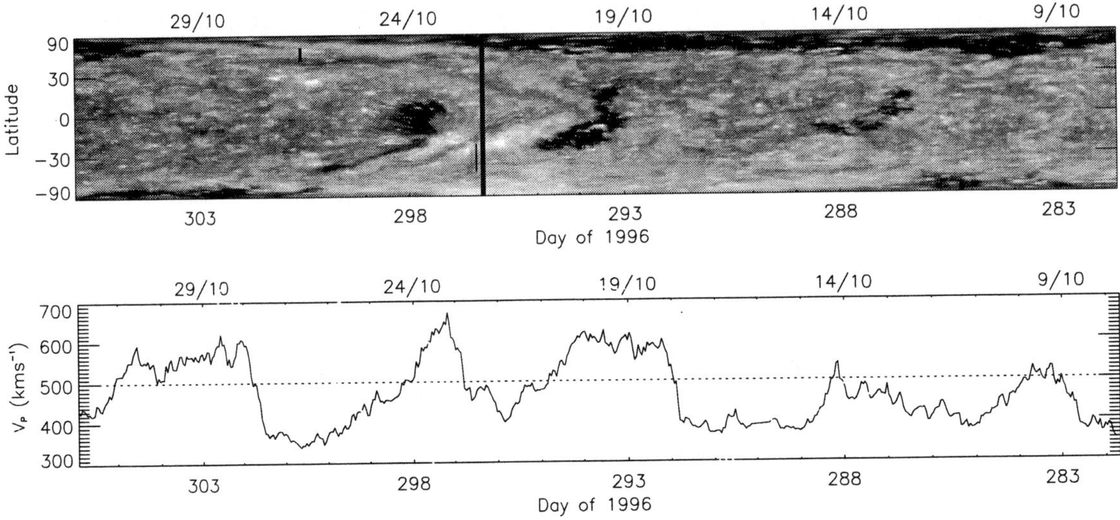

FIGURE 2. Top: EIT meridian map during October 8 and November 1, 1996, showing three ECHs. Bottom: Solar wind speed observed by SWE at 1AU during the same period showing three related high-speed streams.

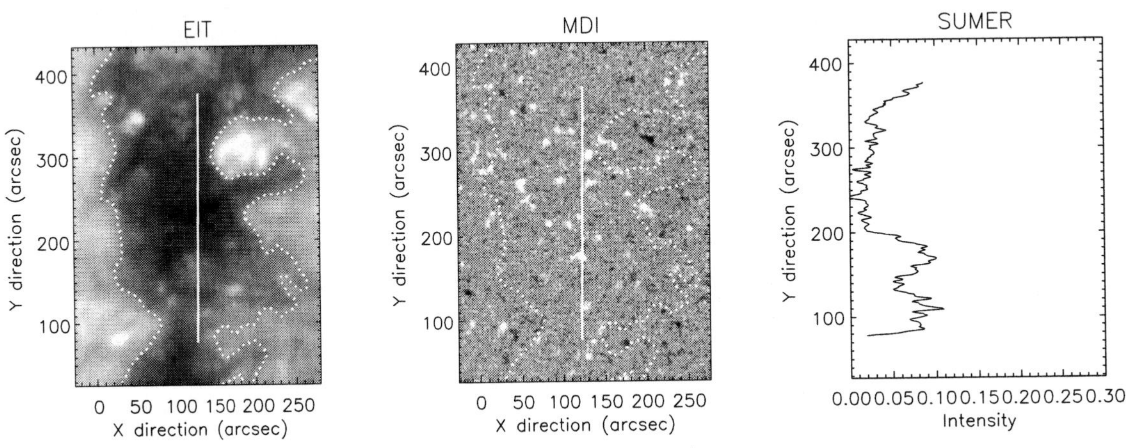

FIGURE 3. Left: EIT map on August 27, 1996, showing part of the ECH "Elephant's Trunk". Middle: MDI magnetoheliogram. Right: Intensity of the Mg IX 706Å line along the slit of SUMER. Note the coronal hole boundary indicated by a dotted line and the SUMER slit position indicated by a solid line in the EIT and MDI images.

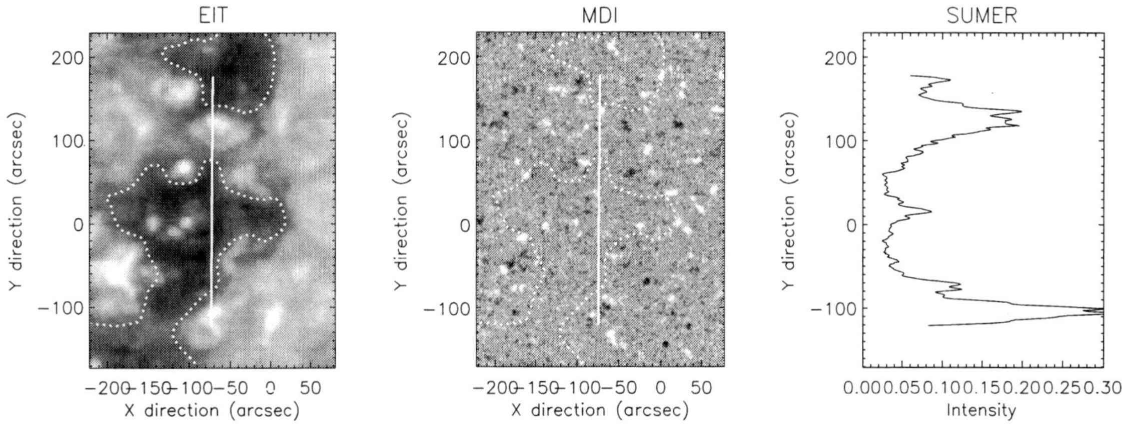

FIGURE 4. The same as Fig. 3 but for the coronal hole observed on October 12, 1996.

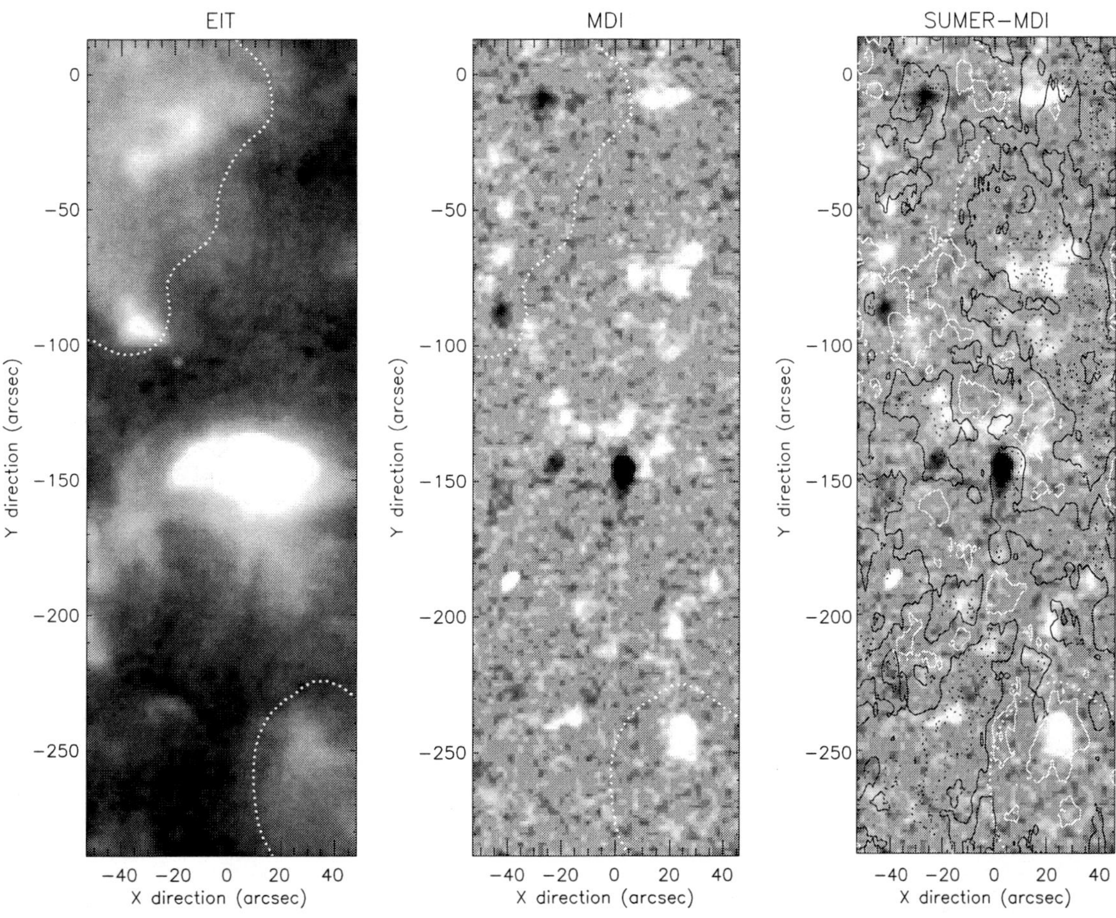

FIGURE 5. Left: EIT map on October 20, 1996, showing part of the ECH3. Middle: MDI magnetoheliogram (white: positive polarity; black: negative polarity). Right: Contour plots of Doppler-shifts of the Ne VIII line overlaid on the MDI magnetoheliogram. We take the average speed outside this ECH as 0 kms^{-1}, since we have no absolute wavelength calibration in this spectral window (black dotted: -6, black solid: -2, white solid: 2, and white dotted: 6 kms^{-1}; negative speed for blue shifts and positive for red shifts). Note the coronal hole boundary indicated by a dotted line in the EIT and MDI images.

ECH1 and ECH2. The average electron density along the slit below 335" in ECH1 has been measured to be about 2×10^8cm^{-3}, which is consistent with the result obtained by Del Zanna and Bromage [7] in the same hole.

MDI (Michelson Doppler Imager) full-disk magnetograms (sampled at a rate of 15/day) are used to estimate the average photospheric magnetic field across areas that spatially coincide with the ones selected to deduce the outflow speed and electron density by using SUMER data (Figs. 1 and 2).

A simple computation is performed to determine the parameter values of the plasma and magnetic field of the high-speed streams at 1AU, by averaging the measured values across the least variable part of the high-speed streams between the leading and trailing edges.

TABLE 3. Coronal parameters deduced from SUMER and MDI

	N_e (cm^{-3})	V (kms^{-1})	f_e=N_e*V (cm^{-2}s^{-1})	B_0 (G)
ECH1	0.8×10^8	7.9	6.3×10^{13}	3.1
ECH2	1.8×10^8	1.4	2.5×10^{13}	1.0

TABLE 4. In-situ parameters deduced from SWE and MFI

	N_p (cm^{-3})	V_p (kms^{-1})	f_p=N_p*V_p (cm^{-2}s^{-1})	B_r (nT)
ECH1	4.3	606	2.6×10^8	2.8
ECH2	3.6	594	2.2×10^8	2.7

RESULTS

By inspection of Table 3, containing the data obtained by SUMER and MDI, and Table 4, giving the parameters obtained from in-situ observations of SWE (Solar Wind Experiment) and MFI (Magnetic Fields Investigation), we can infer that the two high-speed streams produced by ECH1 and ECH2 have similar properties. If we trace back the proton flux and magnetic field under the assumption of a radial expansion of the streams, they should have an average proton flux of about 10^{13}cm^{-2}s^{-1} and a magnetic field of about 1 G at the Sun's surface. Comparison with the photospheric magnetic field measured by MDI indicates that ECH1 has an expansion factor of about 3, and ECH2 of about 1. This may be interpreted as ECH1 having a stronger photospheric magnetic field. If we determine the expansion factor by means of the mass flux measured by SUMER in the holes and by SWE at 1AU, this expansion factor should be about 5.5 for ECH1 and about 2.5 for ECH2. The lower expansion factors determined by the magnetic field in both holes may mainly result from our underestimation of the net field strength caused by the noise level of the field data measured by MDI. In addition, Measurements of the line widths show that both the Mg IX and Mg X lines have larger Doppler widths for ECH1 than for ECH2.

Finally, we give 2-D images of ECH3 in Fig. 5 to show the close relationship between the Doppler shift and the magnetic network. Except for the area occupied by a bright point, where a mixed-polarity magnetic structure is present and the velocity field becomes very complicated, the Ne VIII line is more blue shifted inside than outside the hole. Moreover, the contour plots show that the larger blue shifts with speeds above 6 kms^{-1} are associated mainly with those photospheric regions where large magnetic flux with a single polarity is concentrated.

SUMMARY AND CONCLUSION

We have correlated coronal with in-situ measurements. We found that the larger blue shifts, as deduced from coronal lines, are associated mainly with those photospheric regions where large magnetic flux with a single polarity is concentrated. This observation agrees with the model prediction that the fast solar wind is initially accelerated in the coronal funnels rooted there. Coronal holes with a larger photospheric magnetic flux may result in a larger expansion factor of the solar wind stream tube, and by mass and magnetic flux conservation, have a larger initial flow speed at the coronal base.

ACKNOWLEDGMENTS

The SUMER project is financially supported by DLR, CNES, NASA and the ESA PRODEX program (Swiss contribution). We thank the MDI, EIT, SWE and MFI teams for use of their data.

REFERENCES

1. Marsch E. and C. Y. Tu, *Sol. Phys.*, **176**, 87-106, 1997.
2. Hassler D. M., I. E. Dammasch, P. Lemaire *et al.*, *Sci.*, **283**, 810-813, 1999.
3. Wilhelm K., I. E. Dammasch, E. Marsch, *et al.*, *Astron. Astrophys.*, **535**, 749-756, 2000.
4. Wilhelm K., W. Curdt, E. Marsch, *et al.*, *Sol. Phys.*, **162**, 189-231, 1995.
5. Chae J., H. S. Yun and A. I. Poland, *Astrophys. Suppl.*, **114**, 151-164, 1998.
6. Dere K. P., E. Landi, H. E. Mason *et al.*, *Astro. Astrophys. Suppl. Ser.*, **125**, 149-173, 1997.
7. Del Zanna G. and B. J. I. Bromage, *J. Geophys. Res.*, **104**, 9753-9766, 1999.

A Magnetohydrodynamic Test of the Wang-Sheeley Model

S. A. Ledvina*, J. G. Luhmann* and W. P. Abbett*

*Space Sciences Laboratory, University of California, Berkeley, USA, 94720-7450

Abstract. The Wang-Sheeley relationship relates the solar wind speed at the Earth to the divergence rate of open magnetic flux tubes in the solar corona. This relationship is based on a statistically significant correlation between the flux tube divergence parameter "f_s" derived from a photospheric field-based potential field source surface model, and satellite observations of the solar wind speed. The fast solar wind emanates from regions of small magnetic divergence, while slow solar wind comes from regions of high magnetic divergence. Arge and Pizzo [2] improved the reliability of the method by relating the coronal flux tube expansion factor to the solar wind speed at the source surface instead of the satellite. We use a three-dimensional MHD model of the solar corona to further investigate the implications of the Wang-Sheeley relationship for solar wind acceleration. The results suggest what additional heating and momentum inputs may be necessary in an MHD model to obtain the observed relationship between flux tube divergence and solar wind speed.

INTRODUCTION

The Wang-Sheeley relationship relates the solar wind speed observed at the Earth to the amount an open coronal magnetic flux tube expands in a potential field source surface model. A coronal expansion factor, f_s, is defined as:

$$f_s = \left(\frac{r_s}{r_{ss}}\right)^2 \frac{B_s}{B_{ss}} \tag{1}$$

where r_s and r_{ss} are the photospheric and source surface radii (typically $r_{ss} = 2.5r_s$), and B_s and B_{ss} are the magnetic field strengths at these locations along the flux tube. The source surface is a sphere at which the field is presumed to be everywhere radial, mimicking the effect of a solar wind outflow on the coronal field.

Wang & Sheeley [13] found that the fast solar wind emanates from regions of small magnetic divergence, while slow solar wind comes from regions of high magnetic divergence. Wang and Sheeley [14] and Wang [15] argued that higher expansion factors are associated with regions of large mass flux. Thus for a uniform energy flux at the base of the corona, the high mass flux regions will have less energy per particle then the lower mass flux regions. This results in a decrease in the solar wind speed associated with increasing flux tube divergence.

Arge and Pizzo [2] improved the performance of the Wang-Sheeley relationship by relating the expansion factor to the solar wind speed at the source surface instead of the Earth's orbit. Their new empirical relationship:

$$v(f_s) = 267.5 + \frac{410}{f_s^{2/5}}, \tag{2}$$

applies strictly near the solar equator. The subsequent kinematic extrapolation to 1 AU, including an approximate treatment of stream interaction effects, is not of interest here. Our objective is to use their empirical relationship to better characterize the heating and momentum deposition in a global MHD coronal model.

Although it is established that the slow solar wind has a more complicated origin than the high speed wind, including a probable transient component and a contribution from low latitude coronal holes, we assume, as in Neugebauer et al. [7] that for low solar activity periods, a significant fraction comes from the edges of polar coronal holes. Admittedly, the details of the diverse nature of the slow solar wind are averaged-over in the Wang-Sheeley empirical relationship. We nonetheless use the coronal magnetic field geometry from an MHD model to calculate solar wind speeds at the effective source surface with equation 2, and compare the results to solar wind speeds obtained directly from the MHD model. By doing this, we hope to learn more about both the physics behind the empirical relationship, and the acceleration terms needed in MHD simulations to model the solar wind.

THE SOLAR CORONA MODEL

We adopt a simple polytropic energy relation with $\gamma = 1.05$ (in order to achieve a trans-sonic solar wind). Following Mikic et al [6], it is known that polytropic models cannot account for the observed solar wind speeds using this low value of γ, without increasing the temperature at the base of the corona to an unphysically high value,

CP679, *Solar Wind Ten: Proceedings of the Tenth International Solar Wind Conference*,
edited by M. Velli, R. Bruno, and F. Malara

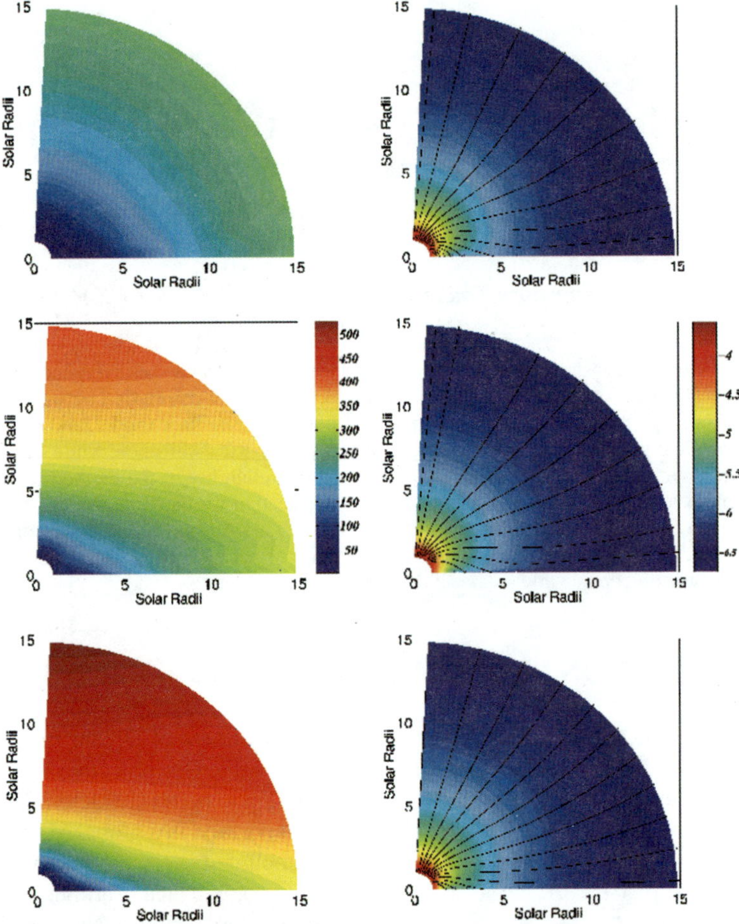

FIGURE 1. Solar wind speed (right column) and Log of the magnetic field strength (left column) for each MHD case: polytropic (top), Alfvén waves $\langle \delta V^2 \rangle = 35$ km/s (middle), and Alfvén waves $\langle \delta V^2 \rangle = 70$ km/s (bottom).

and/or incorporating additional sources of heating and momentum. Since Alfvén waves may provide a potentially significant momentum source (Alazraki and Couturier [1]). We incorporate them into our MHD model by using the WKB approximation to evolve an effective Alfvén wave pressure. Here we assume that the scale size for variations in the corona are small over a typical Alfvén wavelength. The use of the WKB approximation has been validated by the analyses of Roberts [8] and Smith et al. [9] (see Usmanov et al., [12], for further discussion).

Our model is a three dimensional single fluid MHD model with a dipolar field geometry. We solve the following MHD equations (Mikic et al [6], Usmanov et al. [12]):

$$\frac{\partial \rho}{\partial t} + \nabla \cdot (\rho \mathbf{v}) = 0 \tag{3}$$

$$\frac{\partial \rho \mathbf{v}}{\partial t} + \nabla \cdot (\rho \mathbf{v}\mathbf{v}) = -\nabla P - \nabla p_w - \mathbf{J} \times \mathbf{B} - \rho \mathbf{g} \tag{4}$$

$$\frac{\partial e}{\partial t} + \nabla \cdot (e\mathbf{v}) = -P\nabla \cdot \mathbf{v} \tag{5}$$

$$\frac{\partial \mathbf{B}}{\partial t} = \nabla \times (\mathbf{v} \times \mathbf{B}) \tag{6}$$

where ρ is the plasma density, $\mathbf{v}$ is the plasma flow velocity, P is the total thermal pressure (both electrons and ions), $\mathbf{J}$ is the current density, $\mathbf{B}$ is the magnetic field and e is the energy density. The Alfvén wave pressure p_w, is governed by:

$$\frac{\partial \varepsilon}{\partial t} + \nabla \cdot \left(\frac{3}{2}\mathbf{v} + \mathbf{v_A}\right) \varepsilon = \mathbf{v} \cdot \nabla p_w \tag{7}$$

where ε is the Alfvén wave energy density ($\varepsilon = 2p_w$) and $\mathbf{v}_A$ is the Alfvén velocity.

The equations are solved in three-dimensions in spherical coordinates (r, θ, ϕ), using the Zeus-3D (Stone and Norman, [10] and [11]) code developed at the Laboratory for Computational Astrophysics at NCSA. Each case is axially symmetric (e.g. independent of ϕ) with a resolution of 75×30 in the r and θ directions. The radial

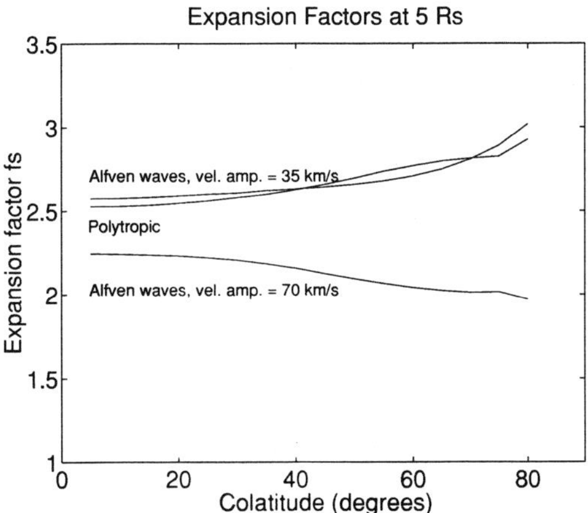

FIGURE 2. The calculated expansion factors "f_s" at 5 R_s for each of the MHD cases

points are non-uniformly spaced; resolution ranges from 3.2×10^4 km near 1 R_s, to 3.1×10^5 km at 15 R_s. The azimuthal points are uniformly spaced from 0 to π with an angular resolution of 6°.

The initial values of the hydrodynamic variables at 1 R_s are set to $\rho_0 = 1.67 \times 10^{13}$ kg/m^3 and $T_0 = T_i = T_e = 1.6 \times 10^6$ K. We assume a dipole field strength of 1 G at the equator and hold B_r constant at the inner boundary (are parameters are similar to those used in earlier simulations by Linker and Mikic [5]). The values of v_r, v_θ and B_θ are calculated at this boundary. In addition, the Alfvén wave energy density at the lower boundary is assumed to be given by:

$$\varepsilon_0 = \rho_0 \langle \delta V^2 \rangle (1 - \sin^4 \theta) \tag{8}$$

where $\langle \delta V^2 \rangle$ is the mean square amplitude of the velocity fluctuations. The angular dependence, which is somewhat arbitrary, is adopted so as to maintain a closed streamer belt. The radial dependence is initialized with the one-dimensional solution to equations 3-7, found by Usmanov et al. [12].

We consider three heating/acceleration scenarios:

1. A simple polytropic model $\gamma = 1.05$ and $\varepsilon_0 = 0$ (no Alfvén waves).
2. A polytropic model, with Alfvén waves having mean square amplitude velocity fluctuations of 35 km/s at the lower coronal boundary.
3. A polytropic model, with Alfvén waves having mean square amplitude velocity fluctuations of 70 km/s at the lower coronal boundary.

In case 2 the 35 km/s mean square amplitude velocity fluctuations are the same value used by Usmanov et al. [12] and near the upper limit of the non-thermal velocity fluctuation amplitudes inferred from observations by Hassler et al. [3].

RESULTS

The model was initiated with a radially symmetric hydrodynamic Parker solar wind solution combined with an initial dipole magnetic field configuration. The system was allowed to relax for 110 hours of solar time. Based on the resulting field configuration, the source surface used for the Wang-Sheeley wind speed calculations was taken to be at 5 R_s, where the last closed field line occurs in the MHD coronal model.

Figure 1, shows contours of the solar wind speed and magnetic field strength for each of the three cases. The flow speeds are highest in the polar regions, decreasing toward the equator above the closed field line region. A stagnation region exists inside of the closed field. The presence of the Alfvén waves significantly increases the plasma flow speed near the poles over the simple polytropic case. The flow speed near the ecliptic is also increased over the polytropic case, though not to the same extent. The current model does not have sufficient spatial resolution to fully resolve the flow details above the closed field region. Thus while it is sufficient for the purposes of this analysis over most of the 5 R_s sphere, we cannot make conclusions concerning the first open flux tubes. The resolution will be improved in future runs.

The magnetic field configurations are similar for cases 1 and 2 despite the different flow speeds. The last closed field line in each case occurs just inside 5 R_s, beyond this distance the field lines are open. These results imply that the magnetic field configuration is not very sensitive to our addition of Alfvén waves. This assertion is further supported by examining the calculated expansion factor "f_s" (figure 2) at the source surface for these cases. The magnetic field divergence in case 3 is slightly different than the other cases but the magnitude of the field has not significantly changed. A few more field lines are open in this case due to the higher flow speed, and the effective source surface has moved out slightly. These differences are further reflected in the "f_s" calculation. This difference however, does not significantly change the speed predicted by the Wang-Sheeley relationship.

The flow speeds calculated at the source surface by the improved Wang-Sheeley relationship are compared to the flow speed results from each MHD case in figure 3. The Arge and Pizzo [2] Wang-Sheeley relationship results, which were essentially the same in the 3 cases, yield higher predicted wind speed values at the source surface than any of the MHD models. The gradual drop off in the flow speed with colatitude for the Alfvén wave cases (compared to the Wang-Sheeley and polytropic

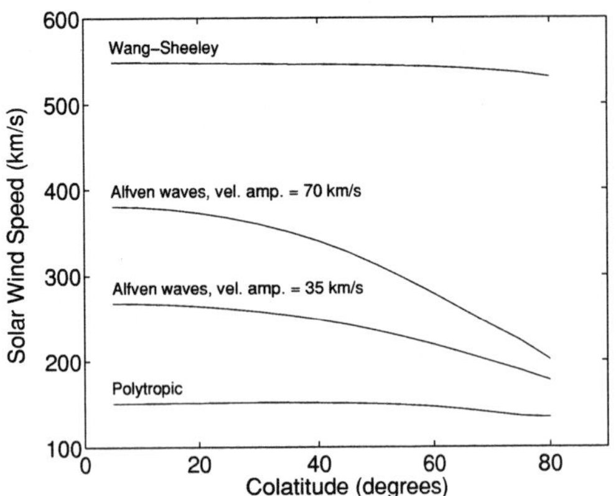

FIGURE 3. Comparison of the calculated solar wind speeds (at $r = 5R_s$) from the improved Wang-Sheeley relationship and the output from each MHD cases.

case) is a result of our assumed angular dependence in ε. This implies that Alfvén waves alone cannot be the only acceleration process acting on the solar wind. Another source is needed that acts within the region between the lower corona and the source surface.

CONCLUSIONS

We have used an MHD model of the solar corona to compare solar wind speeds from such a model to those predicted by the Wang-Sheeley empirical relation. The implications of this study are:

- The magnetic field configuration is not very sensitive to the details of the plasma flow speed. This implies that a simple polytropic model can be used to understand many effects related to the magnetic field configuration.
- Alfvén waves of reasonable amplitude introduced at the base cannot produce wind speeds in agreement with the Wang-Sheeley predictions. Therefore, another acceleration process that acts between the lower corona and the source surface seems to be needed.

Here we only examine the effects of Alfvén waves within the WKB approximation on the solar wind speed. Other heat sources and sinks (such as Spitzer conductivity, or optically thin radiative losses may also dramatically affect the calculation. If the Alfvén waves were to undergo a turbulent cascade and transfer energy to the high frequencies, then ion cyclotron resonances may provide additional acceleration (see Hollweg, [4] for further discussion). However, exact expressions for these processes, especially within an MHD model context, are not well known. The Wang-Sheeley relationship provides an additional constraint that can be used to tune representations of physical processes within an MHD model.

ACKNOWLEDGMENTS

This work was supported by the DoD/AFOSR Muri grant 'Understanding Magnetic Eruptions and their Interplanetary Consequences', and Living with a Star NASA grant NAG 510925. We also wish to thank the Institute for Theoretical Physics at UCSB for support through their program on Solar Magnetism and their hospitality during our stay in 2002.

REFERENCES

1. Alazraki, G., & Couturier, P., *A&A*, **13**, 380-389, (1971).
2. Arge, C. N., & Pizzo, V. J., *J. Geophys. Res.*, **105**, 10465-10479, (2000).
3. Hassler, D. M., Rottman, G. J., Shoub, E. C., & Holzer, T. E., *ApJ*, **348**, L77-L80., (1990).
4. Hollweg, J. V., *J. Geophys. Res.*, **105**, 7573-7581, (2000).
5. Linker, J. A., & Mikic, Z., *ApJ*, **438**, L45-L48, (1995).
6. Mikic, Z., Linker, J. A., Schnack, D. D., Lionello, R. & Tarditi, A., *Phys. Plasmas*, **6**, 2217-2224, (1999).
7. Neugebauer, M. et al., *J. Geophys. Res.*, **103**, 14,587-14600, (1998).
8. Roberts, D. A., *J. Geophys. Res.*, **94**, 6899-6905, (1989).
9. Smith, E. J., Balogh, A., Neugebauer, M., & McComas, D., *Geophys. Res. Lett.*, **22**, 3381-3384, (1995).
10. Stone, J. M., & Norman, M. L., *ApJ Sup.*, **80**, 753-790, (1992).
11. Stone, J. M., & Norman, M. L., *ApJ Sup.*, **80**, 791-818, (1992).
12. Usmanov, A. V., M. L. Goldstein, B. P. Besser, & J. M. Fritzer, *J. Geophys. Res.*, **105**, 12675-12695, (2000).
13. Wang, Y.-M., *ApJ*, **410**, L123-L126, (1993).
14. Wang, Y.-M., and Sheeley, N. R. Jr., *ApJ*, **355**, 726-732 (1990).
15. Wang, Y.-M., and Sheeley, N. R. Jr., *ApJ*, **372**, L45-L48, (1991).

Solar wind acceleration in low density regions

L. Teriaca*, G. Poletto*, M. Romoli† and D. Biesecker**

*Osservatorio Astrofisico di Arcetri, Largo E. Fermi 5, 50125 Firenze, Italy
†Dipartimento di Astronomia e Scienza dello Spazio, Università di Firenze, Largo Fermi 5, 50125 Firenze, Italy
**Emergent I. T., Inc., NASA/GSFC, Greenbelt, USA

Abstract.
High speed solar wind is known to originate in polar coronal holes which, however, are made up of two components: bright, high density regions known as *plumes*, and dark, weakly emitting low density regions known as *interplumes*. Recent space observations have shown that the width of UV lines is larger in interplume regions [see e. g. 1, 2] while observations of the ratio of the O VI doublet lines at 1032 and 1037 Å, at the altitude of 1.7 solar radii, suggest higher outflows in interplume regions than in plumes [3]. These results seem to locate the source of the fast solar wind in the interplume regions.

The present work aims at identifying the outflow speed vs. altitude profile of the O VI ions, at heights up to 2 solar radii, both in plumes and interplume regions. To this end, we examined SUMER and UVCS data taken in the North polar coronal hole on June 3, 1996 over the altitude range between 1 and 2 solar radii.

A Doppler dimming analysis applied to our data allows us to determine the outflow speed in interplume regions throughout the range covered by the observations. Our results favor interplumes as sources of fast wind. However, models mimicking observations in plume regions will also be discussed.

INTRODUCTION

Polar coronal holes have long been recognized to be the sources of the high speed solar wind [4], but only recently some light has been shed on where, within coronal holes, the solar wind originates. UVCS and SUMER measurements of line widths and of the ratio of the O VI doublet lines at 1032 and 1037 Å in the low corona (at heliocentric distances lower than 2.5 $R_\odot$) seem to indicate that the low density background plasma, rather than the high density plume plasma, is the site where high speed wind originates. Recent analyses of SUMER and UVCS data [see e. g. 3, 2, 5] have shown that the width of UV lines is larger in interplume than in plume regions, hinting to interplumes as the site where energy is preferentially deposited and, possibly, fast wind emanates. Moreover, the analysis of fast polar wind performed by [6] seems to favor the low density regions as sources of the fast wind streams, while no evidence of outflow motions in bright points and plumes observed within a coronal hole in the Ne VIII 770 Å line has been found by [7].

In this paper we present strong evidence supporting the idea that interplume regions are sources of the fast solar wind. The outflow velocity profile for O VI ions in interplume areas will be given together with a model reproducing plume observations.

OBSERVATIONS

The observations discussed here were acquired using several instruments aboard SoHO on 3 June 1996 and are shown in Fig. 1. SUMER observations comprise two rasters of the North Polar Coronal Hole (NPCH) obtained using the $4'' \times 300''$ slit and covering a total area of $\sim 280'' \times 470''$ (for further details [see 2]). These rasters were aimed to obtain high signal-to-noise O VI 1032 and 1037 Å line profiles in the plume and interplume regions above the limb out to 1.5 $R_\odot$. Seven UVCS spectra comprising the O VI doublet and the Hydrogen Ly *alpha* were acquired with the slit tangent to the solar limb at altitudes up to 2.1 solar radii. A CDS raster scan of the solar limb in the Mg IX 368 Å line provides informations on the roots of the structures onto the solar disk.

Polar plumes are linear structures that are apparent over the solar poles in visible light, in extreme ultraviolet and in soft X-rays [see 8, and references therein]. After selecting a plume and an interplume region, binning was applied to SUMER and UVCS data to increase adequately the S:N ratio. In such a way, line profiles were determined at 19 different locations above the solar limb for both plume and interplume areas.

CP679, *Solar Wind Ten: Proceedings of the Tenth International Solar Wind Conference*,
edited by M. Velli, R. Bruno, and F. Malara

DATA ANALYSIS

SUMER spectral profiles were carefully fitted in order to obtain C II 1037, O VI 1032 and O VI 1037 Å line intensities. Despite the very high quality of the telescope mirror, the level of stray light is not negligible when observations of lines that are bright on disk are carried out above the limb [9]. The level of stray light in SUMER spectra was determined using the C II 1037 Å line as described in [2]. Stray light in UVCS spectra was also estimated and removed.

Errors on line intensities were calculated through Poissonian statistics and, finally, their propagation in the O VI 1032/1037 line ratio was evaluated.

RESULTS AND DISCUSSION

O VI 1032 and 1037 Å spectral lines arise from transitions well described by a two level atomic model where, at coronal conditions, the upper level is populated by electron impact excitation (collisional component) and by resonant absorption of the O VI radiation from the transition region (radiative component),

$$I_{obs} = I_{Coll.} + I_{Rad} = 0.85 \frac{\Delta E}{4\pi} A_{O/H} \times \\ \times \left[\langle q(T_e)R(T_e)N_e^2\rangle + BI_{\odot}\langle D(T_O,W)R(T_e)N_e\rangle\right] \quad (1)$$

where ΔE is the energy of the transition, $A_{O/H}$ is the oxygen abundance relative to hydrogen, $q(T_e)$ the collisional excitation rate coefficient, B the Einstein absorption coefficient, $I_{\odot}$ the O VI disk-averaged line intensity, $D(T_O,W)$ accounts for Doppler dimming and geometrical dilution factors and is function of the oxygen temperature T_O (consisting of the components parallel $T_o^{\|}$ and perpendicular $T_o^{\perp}$ to the direction of the magnetic field lines) and the wind velocity W. N_e is the electron number density, $R(T_e)$ is the oxygen ionic fraction calculated in ionization equilibrium and 0.85 is the value of the hydrogen to electron number density ratio for a fully ionized plasma with composition given by [10]. The quantities in brackets $\langle ... \rangle$ are integrated along the line of sight. We are interested in obtaining the O VI outflow profile through the comparison of observed O VI line intensities and ratio with values computed using Eq. 1. Both radiative and collisional components are strongly dependent on the electron density (N_e), electron temperature (T_e), and oxygen elemental abundance ($A_{O/H}$), while the radiative component depends also on the adopted values of $I_{\odot}$, $T_o^{\|}$, $T_o^{\perp}$ and W.

$I_{\odot}$ was evaluated assuming a $1/cos(\theta)$ center-to-limb line intensity variation [11] and adopting an intensity at disk centre of 280 mW m^{-2} St^{-1} for the O VI 1032 Å line. The above value was obtained from a disk centre spectrum obtained on June 4th 1996.

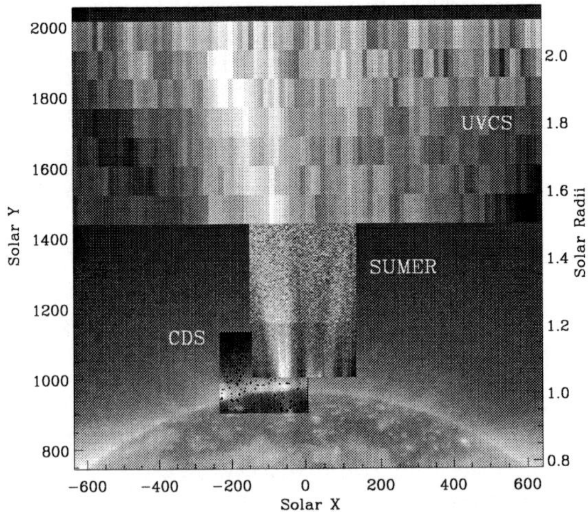

FIGURE 1. June 3rd 1996 map of the north polar coronal hole as obtained using four of the instruments aboard SoHO. Strong plumes and inter plume lanes can be identified in this diagram out to 2.0 $R_{\odot}$ above the limb. An EIT image in the Fe XII 195 Å line provides the disk image showing a well developed coronal hole. A CDS raster in the Mg IX 368 Å line shows the root of a large plume structure. A SUMER scan in the O VI 1032 Å line shows how the plume develops with altitude up to 1.5 solar radii. Above this altitude the structure of the coronal hole is obtained throughout seven spectra acquired in the O VI 1032 Å line with the UVCS slit normal to the solar radius. The contrast of the SUMER and UVCS maps was enhanced by dividing each map for its average intensity profile in the *y*-direction.

In the case of the observations here discussed, we can assume the magnetic field lines to be perpendicular to the line of sight. This allow us to identify $T_o^{\perp}$ with the effective temperature T_{eff} associated with the Doppler width (in km s^{-1}) of the observed spectral line. Doppler width in both plume and interplume regions were obtained for the SUMER data of this dataset by [2]. UVCS line widths were obtained from our data.

I_{Rad} is not a strong function of $T_o^{\perp}$ and, due to the small difference between measured line widths in plume and interplume, a unique Doppler width profile was assumed as representative of both regions. Both cases of $T_o^{\|}$ equal to T_e (anisotropic case) and $T_o^{\|}$ equal to $T_o^{\perp}$ (isotropic case) have been explored.

Spectroscopically derived T_e values never exceed 10^6 K in the range 1.05–1.3 R$_{\odot}$. [5] provide the only measurements of T_e in plume and interplume regions below 1.2 solar radii while, at higher altitudes, an upper

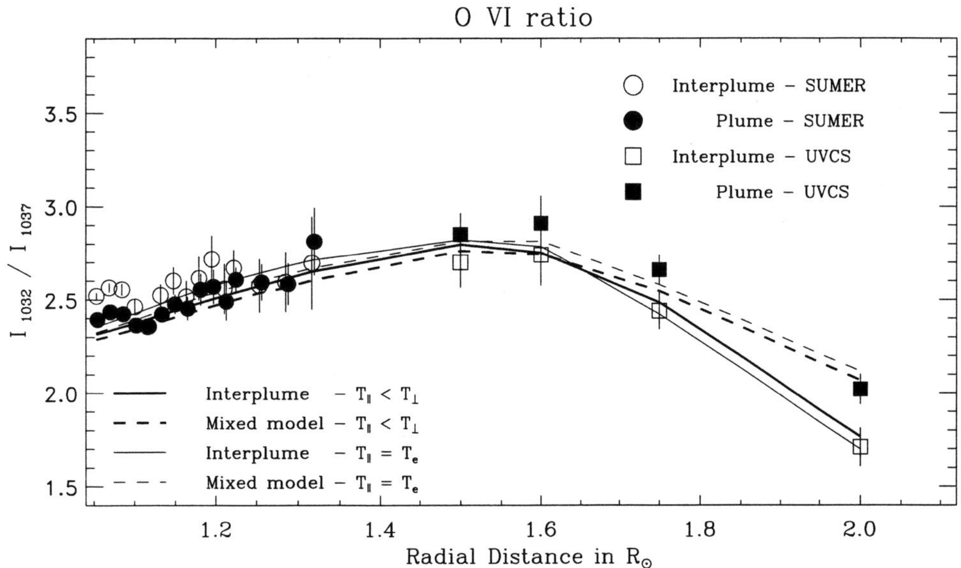

FIGURE 2. O VI 1032 to O VI 1037 intensity line ratio as a function of altitude in polar coronal holes. Values obtained in interplume regions: SUMER data (*open circles*), UVCS data (*open squares*). Values obtained in plumes: SUMER data *filled circles*, UVCS data (*filled squares*). Solid lines represent models for interplume conditions in the anisotropic (thin lines) and partially isotropic case (thick lines), respectively. Dashed lines represent a mixed model comprising plume and interplume lanes.

limit to the electron temperature is given by the value of 1.075 MK given by [12] from *in situ* measurements of the O VII/O VI ion ratio in fast wind streams. For this work we adopted the T_e values given by [5] below 1.2 solar radii and we assumed constant temperature above this altitude.

The O VI 1032/1037 line ratio is insensitive to T_e [14] and does not depend on the elemental abundance. It is, however, strongly dependent on the electron density N_e as well as on the wind speed W. In order to reproduce the observed line ratio, the N_e profile as a function of height needs to be known, together with the exciting transition region radiation. Above 1.5 solar radii average electron densities were obtained from the UVCS WLC and the LASCO C2 coronograph for the day of our observations. Densities in coronal holes have been evaluated by several authors [e. g. 15, 16]. However, in the present analysis only data obtained by [17, 5, 18, 19, 15] in 1996 (*i. e.* closer in time to our data) were used to integrate our measurements. In particular, the behaviour at lower altitudes was evaluated from data published by [5] for both plume and interplume regions.

Using those electron density and temperature profiles the O VI 1032 and 1037 observed line intensity ratios below 1.2 solar radii were reproduced throughout Eq. 1 assuming no outflow speed. In these conditions ($W = 0$), O VI line intensities depend on the elemental abundance only (once the density and temperature profiles have been fixed). We found that observed line intensities below 1.2 solar radii could be reproduced provided an oxygen abundance of 8.5 is adopted.

Using the adopted interplume T_e and N_e profiles and an oxygen abundance of 8.5, we found the outflow profiles able of reproducing our interplume data (see Fig. 2) for both isotropic and anisotropic conditions. Above 1.3 solar radii, unless wind is accelerated and lines are Doppler dimmed, we would be unable to reproduce UVCS data (see Fig. 3).

A realistic attempt to model our plumes observations requires a combination of plume and interplume emissivities along the line of sight. A mixed plume-interplume model has been, hence, created assuming a single plume with no outflowing plasma embedded in the outflowing interplume plasma. The line of sight fraction occupied by the plume was estimated from Fig. 1. This model is able to reproduce our plume observations (see Fig. 2) supporting the hypothesis that interplumes are the regions from where fast wind streams emanate.

CONCLUSIONS

Our interpretation of the measurements of the O VI 1032/1037 line intensity ratio shows that negligible outflow velocities are consistent with the values measured below 1.3 $R_\odot$, while the wind starts being accelerated above this altitude. This is the first time that the O VI outflow profile is provided below 1.5 solar radii: results

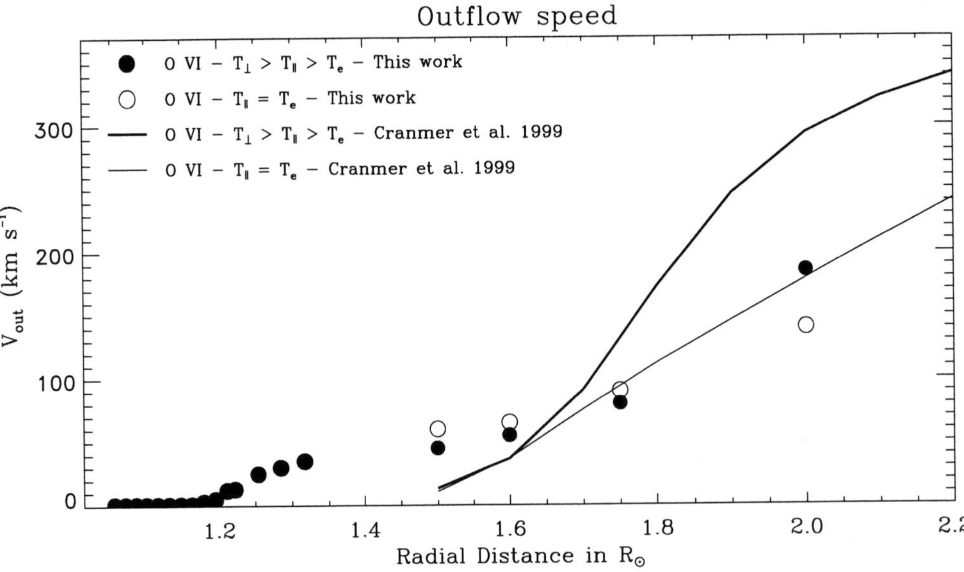

FIGURE 3. O VI ions outflow speed as a function of altitude in polar coronal holes. Values obtained in interplume for the anisotropic case are shown as *open circles* while values obtained in the partially isotropic case are represented using *filled circles*. As a reference, the outflow profiles given by [13] for an average coronal hole are also shown.

presented here show that the wind acceleration initiates at lower altitudes than usually assumed ([see e.g. 13]).

ACKNOWLEDGMENTS

L. T. and G. P. are partially supported by MURST and ASI. SOHO is a mission of international cooperation between ESA and NASA.

REFERENCES

1. Giordano, S., Antonucci, E., Benna, C., and *others*, "Plume and interplume regions and solar wind acceleration in polar coronal holes between 1.5 and 3.5 $R_\odot$", in *Fifth SOHO Workshop: The Corona and Solar Wind Near Minimum Activity*, edited by A. Wilson, ESA SP 404, ESA Publications Division, ESTEC, Noordwijk, The Netherlands, 1997, pp. 413–416.
2. Banerjee, D., Teriaca, L., Doyle, J. G., and Lemaire, P., *Solar Physics*, **194**, 43–58 (2000).
3. Giordano, S., Antonucci, E., Noci, G., and *others*, *Astrophysical Journal*, **531**, L79–L82 (2000).
4. Krieger, A. S., Timothy, A. F., and Roelof, E. C., *Solar Physics*, **29**, 505–525 (1973).
5. Wilhelm, K., Marsch, E., Dwivedi, B. N., and *others*, *Astrophysical Journal*, **500**, 1023–1038 (1998).
6. Antonucci, E., Dodero, M. A., and Giordano, S., *Solar Physics*, **197**, 115–134 (2000).
7. Wilhelm, K., Dammasch, I. E., Marsch, E., and Hassler, D. M., *Astronomy and Astrophysics*, **353**, 749–756 (2000).
8. DeForest, C. E., Hoeksema, J. T., Gurman, J. B., and *others*, *Solar Physics*, **175**, 393–410 (1997).
9. Lemaire, P., Wilhelm, K., Curdt, W., and *others*, *Solar Physics*, **170**, 105–122 (1997).
10. Feldman, U., Mandelbaum, P., Seely, J. F., and *others*, *Astrophysical Journal Supplement*, **81**, 387–408 (1992).
11. Wilhelm, K., Lemaire, P., Dammasch, I. E., and *others*, *Astronomy and Astrophysics*, **334**, 685–702 (1998).
12. Wimmer-Schweingruber, R., Von-Steiger, R., Geiss, J., and *others*, "O^{5+} in high speed solar wind streams: SWICS/Ulysses results", in *Solar Composition and its Evolution - From Core to Corona*, edited by C. Frohlich, M. C. E. Huber, and S. K. Solanki, Kluwer Academic Press, 1998, pp. 387–396.
13. Cranmer, S. R., Kohl, J. L., Noci, G., and *others*, *Astrophysical Journal*, **511**, 481–501 (1999).
14. Li, X., Habbal, S. R., Kohl, J. L., and Noci, G., *Astrophysical Journal*, **501**, L133–L137 (1998).
15. Kohl, J. L., Noci, G., Antonucci, E., and *others*, *Astrophysical Journal*, **501**, L127–L131 (1998).
16. Fisher, R., and Guhathakurta, M., *Astrophysical Journal*, **447**, L139–L142 (1995).
17. Doyle, J. G., Teriaca, L., and Banerjee, D., *Astronomy and Astrophysics*, **349**, 956–960 (1999).
18. Zangrilli, L., Poletto, G., Nicolosi, P., and *others*, *Astrophysical Journal*, **574**, in press (2002).
19. Lamy, P., Quemerais, E., Liebaria, A., and *others*, "Electronic Densities in Coronal Holes from LASCO-C2 Images", in *Fifth SOHO Workshop: The Corona and Solar Wind Near Minimum Activity*, edited by A. Wilson, ESA SP 404, ESA Publications Division, ESTEC, Noordwijk, The Netherlands, 1997, pp. 491–494.

2D MHD MODELS OF THE LARGE SCALE SOLAR CORONA

Eirik Endeve*†, Thomas E. Holzer† and Egil Leer*†

*Institute of Theoretical Astrophysics, P.O. Box 1029 Blindern, N-0315 Oslo Norway
†High Altitude Observatory, NCAR, P.O. Box 3000, Boulder Colorado 80307 USA

Abstract. By solving the equations of ideal MHD the interaction of an isothermal coronal plasma with a dipole-like magnetic field is studied. We vary the coronal temperature and the magnetic field strength to investigate how the plasma and the magnetic field interact to determine the structure of the large scale solar corona. When our numerical calculations are initiated with an isothermal solar wind in a dipole magnetic field, the equations may be integrated to a steady state. Open and closed regions are formed. In the open regions the atmosphere expands into a super-sonic wind, and in the closed regions the plasma is in hydrostatic equilibrium. We find that the magnetic field configuration in the outer corona is largely determined by the equatorial current sheet.

INTRODUCTION

Observations from the Ulysses space craft indicate that the solar wind exhibits large variations with heliographic latitude [1]. At solar minimum the solar magnetic field is close to that of a tilted dipole. Fast wind emanates from higher latitudes, whereas the slow wind is observed at lower latitudes. This variation is believed to be closely related to the interaction between the solar wind and the coronal magnetic field. The inertial, gravitational and pressure gradient forces in the stratified solar atmosphere together with magnetic forces determine the large scale structure of the solar corona.

In order to study the force balance in an axially symmetric magnetized solar corona, and the coupling between the outflowing plasma and the coronal magnetic field, we use the equations of ideal MHD to model an isothermal flow from the base of the corona ($r = 1R_s$) to 15 solar radii. Here R_s is the radius of the Sun.

This is essentially the problem solved by Parker in 1958 [2], but here we have included a solar dipole field, and the problem is solved self consistently in a MHD formulation. Problems of this type were first addressed by Pneuman and Kopp in 1971 [3]. Since the paper by Pneuman and Kopp 1971 many have worked on problems on gas-magnetic field interactions in the solar wind: See Bravo and Stewart 1997 [4] and the references therein.

EQUATIONS

To study the interaction between the ionised plasma of the outer solar atmosphere and the coronal magnetic field we apply the time dependent equations of ideal MHD: Continuity equation:

$$\frac{\partial \rho}{\partial t} + \nabla \cdot (\rho \mathbf{u}) = 0 \qquad (1)$$

Momentum equation:

$$\frac{\partial (\rho \mathbf{u})}{\partial t} + \nabla \cdot (\rho \mathbf{u}\mathbf{u}) = -\nabla P - \frac{GM_s}{r^2}\rho \mathbf{e}_r + \mathbf{J} \times \mathbf{B} \qquad (2)$$

Ampère's law:

$$\mathbf{J} = \frac{1}{\mu_0}\nabla \times \mathbf{B} \qquad (3)$$

Ohm's law:

$$\mathbf{E} = -(\mathbf{u} \times \mathbf{B}) \qquad (4)$$

Induction equation:

$$\frac{\partial \mathbf{B}}{\partial t} = -\nabla \times \mathbf{E} \qquad (5)$$

Here ρ $(= m_p n), \mathbf{u}, P, \mathbf{J}, \mathbf{B}$ and $\mathbf{E}$ are mass density, flow velocity, gas pressure, electric current density, magnetic field and electric field, respectively. G, M_s, m_p and μ_0 are the gravitational constant, the solar mass, the proton mass and the magnetic permeability. The gas pressure is given by the ideal gas law, $P = 2nkT$.

In order to study the acceleration of high- and low-speed solar wind one must include a realistic energy

CP679, *Solar Wind Ten: Proceedings of the Tenth International Solar Wind Conference*, edited by M. Velli, R. Bruno, and F. Malara

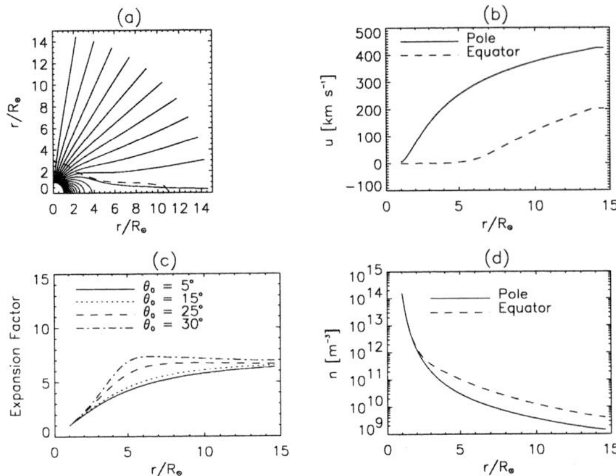

FIGURE 1. Solution to the MHD equations for a temperature $T = 1.25 \times 10^6$ K, and a coronal base pressure $P_0 = 5.5 \times 10^{-3}$ N m^{-2}. At the coronal base the magnetic field strength is $B_0 = 7 \times 10^{-4}$ T at the pole. Panel (a): Magnetic field configuration. The dashed line is where the flow becomes super-sonic. Panel (b): Radial flow velocity versus heliocentric distance at the pole (Solid line) and the equator (dashed line). Panel (c): Expansion factor, f_{exp}, versus heliocentric distance for selected field lines emanating from the solar surface at a colatitude θ_0. Panel (d): Electron density versus radial distance for the pole (solid line) and the equator (dashed line).

equation including an appropriate heating mechanism. However, this is not the focus of the present paper. We are interested in studying the force balance in the system. In order to make the problem as simple as possible we take the temperature to be a constant throughout the solar atmosphere.

RESULTS

The equations are solved in spherical coordinates (r, θ, ϕ), assuming axial symmetry. We have discretized the equations on a 129×65 nonuniform (r, θ) grid, where our computational domain extends from $r = 1R_s$ to $15R_s$ in r, and from 0 to π in θ. We use a MHD code developed by Nordlund and Galsgaard 1995 (only available on the web http://www.astro.ku.dk/~kg) modified to handle spherical coordinates. Our boundary conditions at the coronal base, $r = 1R_s$, are uniform in θ. At the outer boundary the flow is super-sonic. There we use simple extrapolations to calculate our boundary conditions.

Figure 1 is a plot of a solution to Eqs. (1)-(5) initiated with a dipole field and a spherically symmetric isothermal solar wind with a temperature of 1.25×10^6 K and a base pressure of $P_0 = 5.5 \times 10^{-3}$ N m^{-2}. The magnetic field strength at the coronal base is $B_0 = 7 \times 10^{-4}$ T at the pole. The solar wind causes the dipole to open and form flow tubes down to a colatitude of about 30 degrees. In the closed regions the atmosphere is in hydrostatic equilibrium. In this model the closed region extends out to about 5 solar radii at the equator. In the open regions a field aligned wind is allowed to expand into interplanetary space. The open regions and the hydrostatic closed regions, frequently referred to as a helmet streamer, are separated by a thin current sheet. Above the helmet streamer, in the equatorial plane, there is a current sheet separating the northern and southern hemispheres where the magnetic field has opposite polarity. This equatorial current sheet determines the structure of the large scale magnetic field, and the geometry of the individual flow tubes.

A flow tube with a cross section A_0 at the coronal base expands with heliocentric distance to a cross section $A(r)$ at a radial distance r. To study the expansion of an individual flow tube we define the expansion factor [5]:

$$f_{exp}(r) = (\frac{B_0}{B(r)})(\frac{R_s}{r})^2, \tag{6}$$

where B_0 is the magnetic field strength in the flow tube at the coronal base, and $B(r)$ is the field strength in the flow tube at a radial distance r.

In Figure 1 (c) we plot the expansion factor versus radial distance for flow tubes emanating at a colatitude θ_0 at the solar surface. Notice that the individual flow tubes expand super-radially in the inner corona approaching a constant value of about 7 at the outer boundary. Beyond the outer boundary the magnetic field falls off as r^{-2}. At the outer boundary there is a difference of more than 200 km s^{-1} in the radial flow speed from the equator to the pole (see Figure 1 b). At the equator the flow is sub sonic inside $r = 11R_s$, whereas the flow becomes super-sonic at about 3 solar radii in the polar region. In panel (d) we see that the density scale height is larger at the equator than near the pole. The high pressure in the equator region is localized in θ; it extends over some five degrees. This high pressure in the equator produces an axial current. The $\mathbf{J} \times \mathbf{B}$ force balances the pressure gradient force.

Varying coronal temperature

In Figure 2 and Figure 3 we plot characteristics of the solutions to the MHD equations where the coronal temperature has been varied from 1.0×10^6 K to 2.0×10^6 K. The electron density at the inner boundary is adjusted in order to fix the coronal base pressure to $P_0 = 5.5 \times 10^{-3}$ N m^{-2}. The magnetic field strength at the coronal base is 7×10^{-4} T at the pole. This corresponds to a plasma β $(= \frac{P}{B^2/2\mu_0})$ of about 3×10^{-2}.

The increase in temperature causes the proton flux density at 1 AU to increase from 4.6×10^{11} (5.7×10^{11})

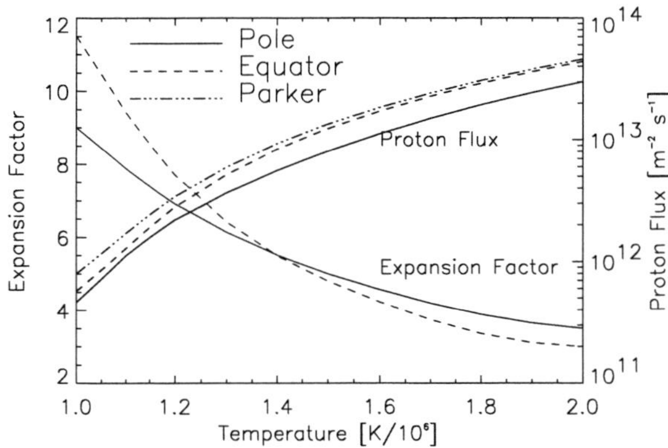

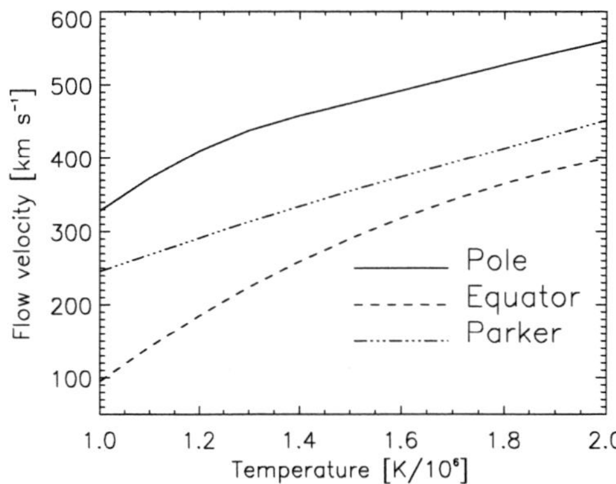

FIGURE 2. Solutions to the MHD equations where the temperature in the corona is varied from 1×10^6 K to 2×10^6 K. The electron density at the inner boundary is adjusted in order to fix the coronal base pressure to $P_0 = 5.5 \times 10^{-3}$ N m^{-2}. The magnetic field strength at the coronal base is $B_0 = 7 \times 10^{-4}$ T. The ascending lines are plots of the proton flux, scaled to 1 AU, for the polar (solid line) and equatorial (dashed line) regions. We also plot the proton flux for a spherically symmetric isothermal flow (dash-dot-dot-dot line). The descending lines are plots of the expansion factor at the outer boundary, $r = 15R_s$, for field lines emanating from latitudes near the pole (solid line) and for open field lines emanating from latitudes near the closed field regions (dashed line).

FIGURE 3. Same solutions as in Figure 2, but here we plot the radial flow velocity at the outer boundary, $u_r(15R_s)$, versus coronal temperature for the polar (solid line) and the equatorial (dashed line) regions. We also plot the flow speed at $r = 15R_s$ for a spherically symmetric isothermal Parker wind (dash-dot-dot-dot).

m^{-2} s^{-1} to 3.0×10^{13} (4.4×10^{13}) m^{-2} s^{-1} for the polar (equatorial) regions. For a more massive wind the magnetic flux at the outer boundary increases. The expansion factor at $15R_s$ decreases from 9 (11.5) to 4.0 (3.5). The increase in magnetic flux in the wind is consistent with an increased current in the equatorial region. The integrated current in the equatorial sheet increases by more than a factor five when the temperature is increased by a factor two. The flow velocity at the outer boundary increases from 328 (90) km s^{-1} to 560 (400) km s^{-1}. For comparison we also plot solutions for a spherically symmetric isothermal corona. The dash-dot-dot-dot lines (termed Parker in Figures 2 and 3) are solutions to the exact same problem where the magnetic field is set to zero (or is purely radial). Notice that the super-radial expansion of flow tubes at high latitudes (produced by the equatorial current sheet) leads to a faster wind than what was found by Parker [6].

Varying magnetic field strength

In Figures 4 and 5 we plot results for models where the magnetic field strength at the coronal base has been varied from 5×10^{-4} T to 1×10^{-3} T at the pole. The temperature is set to 1.25×10^6 K, and the coronal base pressure is fixed to $P_0 = 5.5 \times 10^{-3}$ N m^{-2}. This corresponds to a plasma β changing from 1.4×10^{-2} to 5.5×10^{-2}.

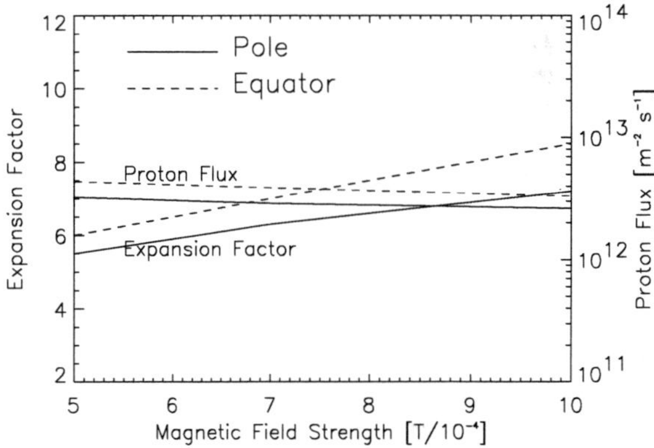

FIGURE 4. Solutions to the MHD equations where the magnetic field strength at the inner boundary is varied. The field strength at the pole, B_0, is varied from 5×10^{-4} T to 1×10^{-3} T. The coronal temperature is 1.25×10^6 K, and the coronal base pressure is $P_0 = 5.5 \times 10^{-3}$ N m^{-2}. The plot is similar to Figure 2. Ascending lines are the expansion factor at 15 solar radii. Descending lines are the proton flux scaled to 1 AU.

For an increasing magnetic field strength the proton flux at the orbit of Earth decreases from 3.3×10^{12} (4.4×10^{12}) m^{-2} s^{-1} to 2.6×10^{12} (3.3×10^{12}) m^{-2} s^{-1} for the polar (equatorial) regions. The expansion factor at the outer boundary, $r = 15R_s$, increases from 5.5 (6.0) to 7.2 (8.5). The flow velocity at the outer boundary increases from 405 km s^{-1} to 450 km s^{-1} in the polar region, whereas at the equator it decreases from 240 km s^{-1} to

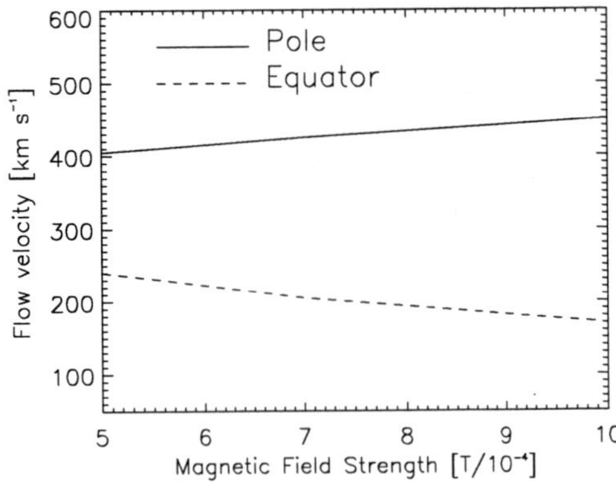

FIGURE 5. Same solutions as in Figure 4. Here we plot the radial flow velocity at the outer boundary, $r = 15R_s$, for the pole (solid line) and the equator (dashed line), versus magnetic field strength at the coronal base.

170 km s^{-1}. Notice that a non-magnetic, isothermal solar wind, with $T = 1.25 \times 10^6$ K and $P_0 = 5.5 \times 10^{-3}$ N m^{-2} yields a proton flux at 1 AU of $(nu)_E = 4.6 \times 10^{12}$ m^{-2} s^{-1}, and a flow speed at $r = 15R_s$ of 303 km s^{-1} (see Figures 2 and 3).

For a stronger magnetic field at the coronal base the magnetic forces in the inner corona becomes more dominant. For $B_0 = 5.0 \times 10^{-4}$ T the atmosphere is in hydrostatic equilibrium out to $4R_s$ at the equator, while for $B_0 = 1.0 \times 10^{-3}$ T the closed field region extends out to $6R_s$ at the equator. A stronger magnetic field does not produce the significant changes in the corona as the change in temperature. The proton flux at 1AU, the expansion factor and the flow speed at the outer boundary change by less than 50 percent. The current sheet in the equatorial region stays relatively unchanged. When the magnetic field strength in the inner corona increases the closed field region extends to a higher latitude. The magnetic flux at the outer boundary relative to the flux at the coronal base decreases. This causes the expansion factor at $r = 15R_s$ to increase. The change in proton flux is determined by two competing factors. For a stronger field the critical point moves inward causing an increase in proton flux, but the super-radial expansion beyond the critical point leads to a decrease in the flux. In our models the latter effect seems to dominate.

DISCUSSION

Our numerical experiments show that for an isothermal solar coronal we are able to construct steady state solutions to Eqs. (1)-(5). The solutions show the same exponential dependence of proton flux with temperature as in spherically symmetric models where the magnetic field is purely radial [7]. We also find that the expansion factor of individual flow tubes decreases rapidly with increasing temperature. This is not the case when we vary the magnetic field strength at the base of the corona.

The isothermal models are able to maintain a fairly high pressure in the equatorial current sheet whereas at higher latitudes, where the flow becomes super-sonic close to the Sun, the pressure falls off more rapidly with heliocentric distance. This results in a pressure gradient force away from equator which is balanced by magnetic forces. Our models demonstrate that an increasing density and temperature in the corona produces a larger solar wind mass flux, a higher pressure in the equator, a larger magnetic flux in the outer corona, and a stronger equatorial current sheet.

In more realistic models, where the temperature is allowed to decrease with heliocentric distance, the equatorial pressure may fall off more rapidly and steady state solutions may not even exists yielding a strongly time dependent reconnection process at equatorial latitudes.

ACKNOWLEDGMENTS

This work was supported by the Norwegian Research Council (NFR) under contract 145519/432. Eirik Endeve thanks the High Altitude Observatory for their hospitality during his visit.

REFERENCES

1. McComas, D. J., Phillips, J. L., Bame, S. J., Gosling, J. T., Goldstein, B. E., and Neugebauer, M., *Space Sci. Rev.*, **72**, 93–98 (1995).
2. Parker, E. N., *Astrophys. J.*, **128**, 664–676 (1958).
3. Pneuman, G. W., and Kopp, R. A., *Sol. Phys.*, **18**, 258–270 (1971).
4. Bravo, S., and Stewart, G., *Astrophys. J.*, **489**, 992–999 (1997).
5. Kopp, R. A., and Holzer, T. E., *Sol. Phys.*, **49**, 43–56 (1976).
6. Holzer, T. E., *J. Geophys. Res.*, **82**, 23–35 (1977).
7. Holzer, T. E., and Leer, E., *J. Geophys. Res.*, **85**, 4665–4679 (1980).

A solar cellular automata model issued from reduced MHD

E. Buchlin[*†], V. Aletti[*], S. Galtier[*], M. Velli[†**] and J.-C. Vial[*]

[*]I.A.S., CNRS – Université Paris-Sud, bât. 121, 91405 Orsay Cedex, France
[†]Dipartimento di Astronomia e Scienza dello Spazio, Università di Firenze, 50125 Firenze, Italy
[**]Istituto Nazionale Fisica della Materia, Sezione A, Università di Pisa, 56100 Pisa, Italy

Abstract. A three-dimensional cellular automata (CA) model inspired by the reduced magnetohydrodynamic equations is presented to describe impulsive events generated along a coronal magnetic loop. It consists of a set of planes, distributed along the loop, between which the information propagates through Alfvén waves. Statistical properties in terms of power laws are obtained in agreement with SoHO observations of X-ray bright points of the quiet Sun. Physical meaning and limits of the model are discussed.

INTRODUCTION

It is now commonly accepted that the ultimate source of energy for coronal heating lies in the photosphere, and that the questions to address are the transfer, the storage, and the release of this energy. These processes involve MHD and structures extending over a large range of scales. This is consistent with observational statistics of impulsive events, whose luminosity, peak luminosity and duration distributions are power-laws (Dennis 1985 [1], Crosby et al. 1993 [2]).

This CA model attempts to describe the statistics of such events. It represents a coronal loop whose foot-points are anchored in the photosphere and are randomly moved.

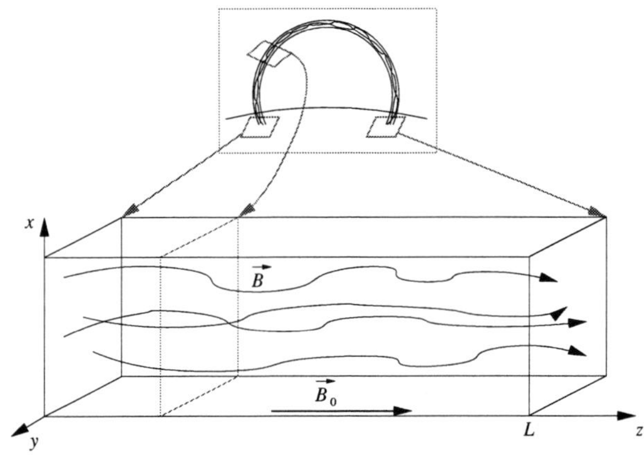

FIGURE 1. Coronal loop above the photosphere (top) and correspondence with the geometry of the loop model (bottom).

DESCRIPTION OF THE MODEL

In our CA model, a 3D regular grid is made up of a set of planes distributed along the loop and orthogonal to its axis, as shown on Fig. 1. Both boundary planes represent the photospheric footpoints, while the intermediate planes represent the loop itself, as if it were unbent.

The presence of a strong axial magnetic field in the loop leads to essentially 2D dynamics, *i.e.* perpendicular to the mean magnetic field. A quite good model in this case is given by the reduced magnetohydrodynamics (RMHD) equations, which describe the evolution of the magnetic and velocity fields, and include Alfvén wave propagation, energy dissipation and non-linear dynamics:

$$\begin{aligned}
\partial_t \vec{v}_\perp + (\vec{v}_\perp \cdot \vec{\nabla}_\perp)\vec{v}_\perp &= b_0 \partial_z \vec{b}_\perp + \nu \Delta_\perp \vec{v}_\perp \\
&\quad + (\vec{b}_\perp \cdot \vec{\nabla}_\perp)\vec{b}_\perp - \vec{\nabla}(b_\perp^2/2) \\
\partial_t \vec{b}_\perp + (\vec{v}_\perp \cdot \vec{\nabla}_\perp)\vec{b}_\perp &= b_0 \partial_z \vec{v}_\perp + \eta \Delta_\perp \vec{b}_\perp \\
&\quad + (\vec{b}_\perp \cdot \vec{\nabla}_\perp)\vec{v}_\perp
\end{aligned}$$

The non-linear dynamics are modeled through an *on-off* mechanism, triggered by a *dissipation criterion*. For our model, we choose this dissipation criterion to be a current density threshold.

Energy is input as random magnetic and velocity fields on both photospheric planes, with a *2D spatial power-law spectrum* of index α, and it is *transported* by Alfvén waves to all loop planes. As a result, energy grows until

CP679, *Solar Wind Ten: Proceedings of the Tenth International Solar Wind Conference*, edited by M. Velli, R. Bruno, and F. Malara

a quasi-stationary state is obtained, when dissipation is sufficient to compensate for the energy input.

RESULTS

Dissipations. Energies of dissipations span a wide range of values (four orders of magnitude). They grow until the quasi-stationary phase is reached, as indicated on Fig. 2 (c) and (d).

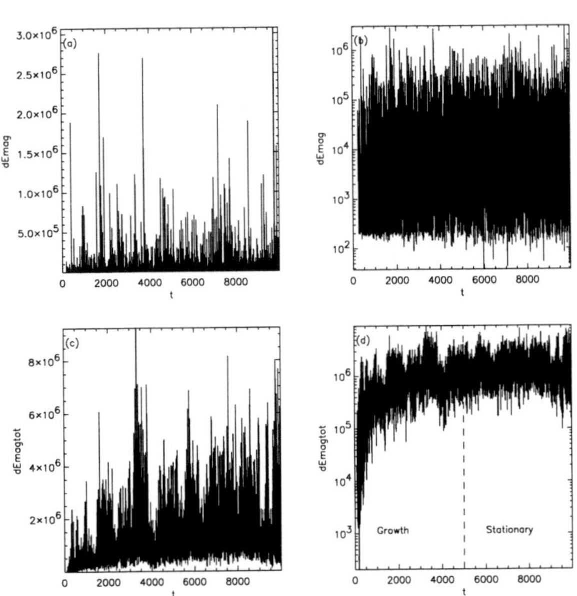

FIGURE 2. Time series of energy dissipations E_i (a: lin; b: log) and $dE = \sum dE_i$ (c: lin; d: log).

Energy histograms power-laws. Histograms of elementary energy dissipations dE_i in a plane i fit to power-laws over 2 to 3 decades, as seen on Fig. 3. The indices ζ of these power-law depend on the parameters, and their absolute values are between 1 and 2. However, power-laws of histograms of energies dissipated in the whole simulation box (*i.e.* seen at a lower spatial resolution) are narrower and present steeper slopes (Fig. 4).

The signal $dE_i(t)$ "seems intermittent" (Fig. 2) and its histograms are consistent with what is expected. However, this is not sufficient: intermittence in MHD turbulence needs also to be characterized by wide wings of PDFs at small scales, or, equivalently, some properties of structure functions or flatness.

Location of dissipations. Dissipations do not mainly occur in current sheets, and these structures are not predominant structures as in MHD simulations (see Fig. 5). This is a limit due to the dissipation criterion chosen to model the non-linear dynamics: it is indeed well established that the terms of the (R)MHD equations which

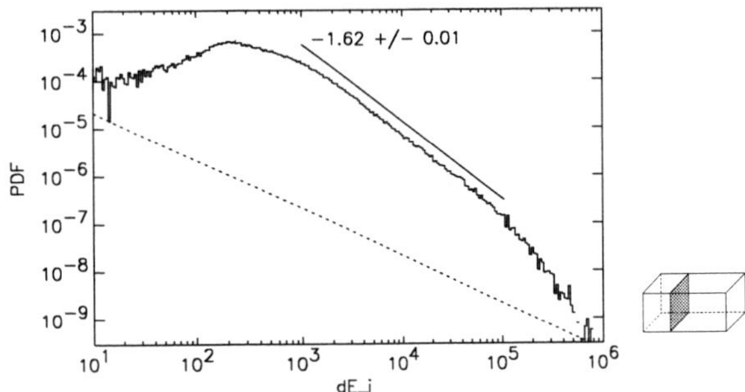

FIGURE 3. Example of histogram of magnetic energy dissipation dE_i in a plane i of the simulation box.

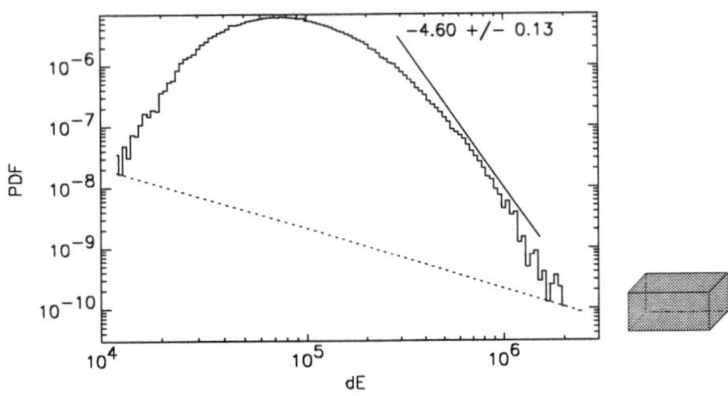

FIGURE 4. Example of histogram of magnetic energy dissipation $dE = \sum dE_i$ in the whole simulation box.

lead to structures like current sheets, where reconnection occurs, are the non-linear terms.

It could be interesting to use a more realistic dissipation criterion. However, a CA model is only supposed to produce realistic statistics, not realistic fields.

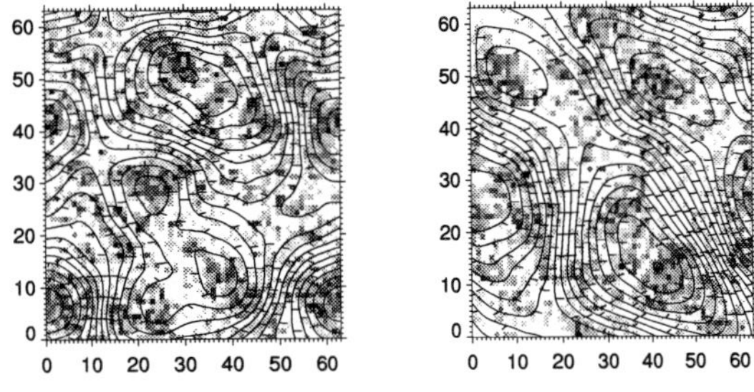

FIGURE 5. Two examples of magnetic field (lines) and current density (background) in a plane.

Durations of events. Durations of elementary events extend over two decades. Histograms can be obtained, but these two decades of duration span are not enough to perform relevant power-law fitting (Fig. 6).

The duration of events is correlated with their energy, like $dE_i \propto dt_i^{1.75}$ (Fig. 7), in agreement with observations (see discussion). This correlation can also be seen by plotting event energy histograms for different ranges of event durations, as on Fig. 8.

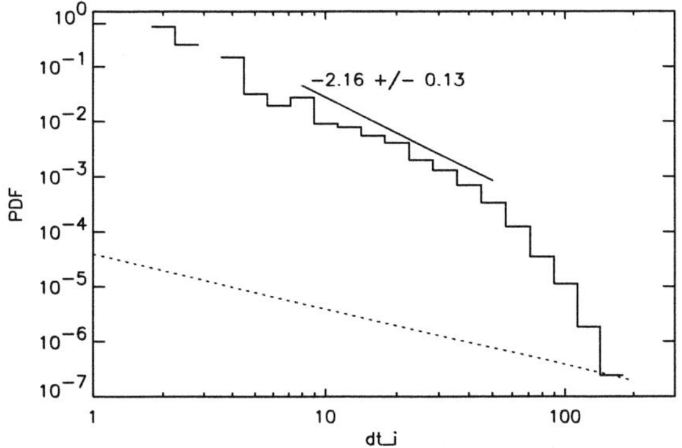

FIGURE 6. Typical histogram of events durations in a plane of the simulation box.

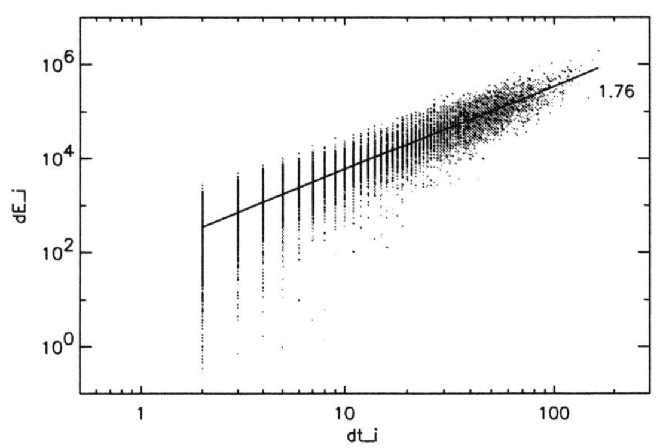

FIGURE 7. Correlation between events duration and energy.

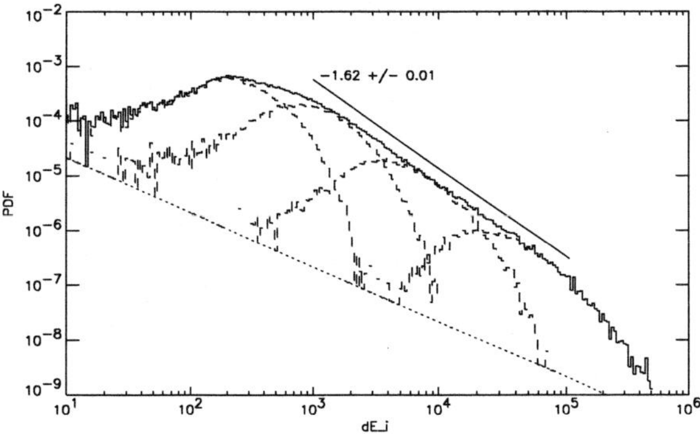

FIGURE 8. Events durations separate populations, from low duration (left, dashed) to long duration (right, dashed), which have different energy distributions. The solid line is the sum of other histograms; it is the same curve as in Fig. 3.

Parametric study. An extensive parametric study was performed, with at least 200000 time steps computed for more than 20 parameters sets. It shows the variability of the event energy histogram slope ζ as a function of the loading index α (Fig. 9). For low α's, the characteristic power-law shape of histograms is more difficult to obtain, thus the ζ's are perhaps not very relevant power-law indices. On the contrary, for high α's, the power-laws are wide and robust, and $\zeta \approx -1.6$, appears to be a "universal" power-law index for the histograms of dE_i. This "universal" behavior is similar to the behavior of SOC (self-organized criticality) systems, from basic sandpile models (Bak et al. 1988 [3]) to more elaborate solar-like SOC models like Vlahos et al. 1995 [4] or Isliker et al. 2000 [5].

Changing the other parameters (resistivity, dissipation efficiency) does not really affect the histograms, confirming the "universal" behavior of the slope.

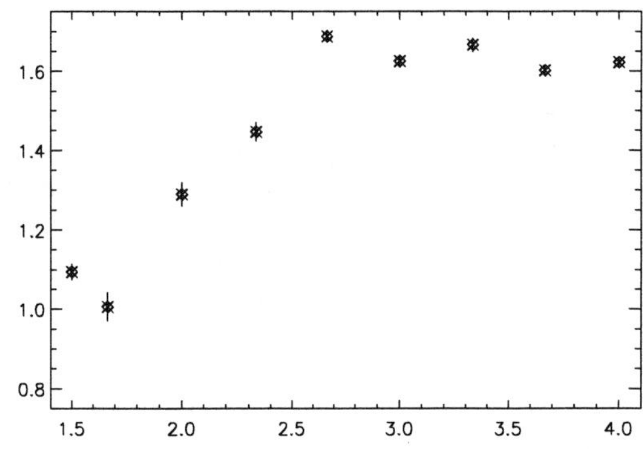

FIGURE 9. Variability of event energy histogram slope $|\zeta|$ *vs.* loading spectrum index α.

DISCUSSION

Comparison with observations

Statistical observations of bright points luminosities lead to power-law histograms of index -1.6 to -2.6. When taking into account an observational bias due to temperature, Aschwanden and Charbonneau [6] show that these observations lead to a "universal" index which could be even less steep than -1.6. These values are compatible with the power-law slopes of event energy distributions produced by our model.

Furthermore, the existence of such a bias emphasizes the importance for new models to produce *observables*, *i.e.* variables whose statistics can be directly compared to the statistics of observational data. An event energy is indeed not the same variable as an event luminosity and could have different statistics.

We also pointed out in Fig. 3 and 4 a possible effect of spatial resolution on the statistics, which is discussed on SoHO/EIT observational data in Aletti et al. [7]. Indeed, the interpretation of several subresolution events in a single pixel as one single event, which is unavoidable when analyzing observations, changes the slope and the shape of histograms compared to histograms of really elementary events.

At last, observations from Berghmans et al. [8] show that event durations scale like their radiative loss at the power 0.5, which is in quite good agreement with our prediction $1/1.76 = 0.57$ (see Fig. 7).

Conclusion

This CA model, which tries to stay close enough to the MHD equations and to the physics of coronal magnetic loops, succeeds in reproducing some of the characteristics of the statistics of events observed on the Sun. However, progress still remains to be done to model more accurately the non-linear terms of the MHD equations leading to the magnetic reconnection process, either by a better dissipation criterion in the frame of threshold dynamics, or even better, by incorporating these non-linear terms in a more realistic way in the dynamics of a new model. More accurate event distribution power-law indices could then be compared to Hudson's [9] critical index of -2 and thus give a clue about Parker's hypothesis of nanoflares; these works could help interpreting the statistical observations of flares.

ACKNOWLEDGMENTS

The authors acknowledge partial financial support from PNST (Programme National Soleil–Terre). E. Buchlin thanks the Scuola Normale Superiore of Pisa for support and accommodation.

REFERENCES

1. Dennis, B. R., *Sol. Phys.*, **100**, 465–490 (1985).
2. Crosby, N. B., Aschwanden, M. J., and Dennis, B. R., *Sol. Phys.*, **143**, 275–299 (1993).
3. Bak, P., Tang, C., and Wiesenfeld, K., *Phys. Rev. A*, **38**, 364–374 (1988).
4. Vlahos, L., Georgoulis, M., Kluiving, R., and Paschos, P., *Astron. Astrophys.*, **299**, 897+ (1995).
5. Isliker, H., Anastasiadis, A., and Vlahos, L., *Astron. Astrophys.*, **363**, 1134–1144 (2000).
6. Aschwanden, M. J., and Charbonneau, P., *ApJ*, **566**, L59–L62 (2002).
7. Aletti, V., Velli, M., Bocchialini, K., Einaudi, G., Georgoulis, M., and Vial, J.-C., *ApJ*, **544**, 550–557 (2000).
8. Berghmans, D., Clette, F., and Moses, D., *Astron. Astrophys.*, **336**, 1039–1055 (1998).
9. Hudson, H. S., *Sol. Phys.*, **133**, 357–369 (1991).

A three-fluid, 16-moment gyrotropic bi-Maxwellian fast solar wind: the effect of He^{++}

Lorraine Allen and Xing Li

Department of Physics, University of Wales, Aberystwyth, UK

Abstract. We present a 16-moment, three-fluid description of the solar wind consisting of electrons, protons, and alpha particles. We assume gyrotropic flow (transport across the magnetic field is neglected) which reduces the 16-moment set of transport equations to a six-moment set yielding the density, velocity, temperatures parallel and perpendicular to the magnetic field, and parallel heat conductive fluxes from the parallel and perpendicular directions for each particles species. The model incorporates the effects of Coulomb collisions. It allows for non-radial divergence of the magnetic field and heating and momentum addition to the particles. We investigate the influence of the heat conductive flux in shaping the temperature anisotropy.

INTRODUCTION

SUMER/ and UVCS/SoHO observations of minor ions in the inner corona imply that particles with greater mass-to-charge ratios are rapidly accelerated and preferentially heated in the direction perpendicular to the magnetic field in the fast solar wind [7, 19, 18]. To adequately model the thermal behavior of minor ions, it is necessary to know how to treat their heat conductive flux which is usually assumed negligible but may be significant in shaping the temperature profile.

In the collision dominated case, the classical value for the heat flux [17] can be recovered from the dominant terms in the heat conductive flux equation which is derived by taking velocity moments of the Boltzmann equation. However, since the solar wind is collision dominated only a short distance above the solar base, the classical Spitzer-Harm values of the heat conductive flux are not valid descriptions in the solar wind. Helios observations show that the heat conductive flux of the protons in interplanetary space is well below that obtained from the classical expression [16, 11]. Gyrotropic 16-moment fluid models, incorporating the heat conductive flux equations for protons and electrons, yield a proton heat flux which is lower than the classical value and which influences the behavior of the proton temperature anisotropy [14, 9, 10].

Little attention has been paid to minor ions and specifically to alpha particles. On account of the empirical evidence that minor ions are interacting with waves that may heat and accelerate the solar wind, more exploration of their behavior is needed. With an empirical abundance of 3-5 % that of the protons in interplanetary space, helium nuclei account for $\sim 20\%$ of the solar wind mass density and contribute a non-negligible portion of the flow's momentum and energy flux at 1 AU [16]. If ion cyclotron resonance heats and drives the flow in the solar wind acceleration region, the alphas are in queue immediately before the protons to be heated, if at all, by the waves and their behavior can serve as a probe of the waves. Because of their lack of emission, alpha particles are not observable in the inner corona, but Helios observations in interplanetary space show that the core distribution of the alphas is anisotropic with $T_{\|} > T_{\perp}$ at 0.3 AU while the protons, with an implied ratio of $T_{\perp}/T_{\|} \sim 1-4$ in the inner corona, are anisotropic with $T_{\perp} > T_{\|}$ until close to 1 AU [11]. Our goal is to determine if the heat conductive flux of the alpha particles plays a significant role in governing the temperature anisotropy of the alphas and how it can be treated in models.

THE MODEL

The 16-moment fluid equations for protons, electrons, and alphas are obtained by expanding the distribution function of each of the fluids about a bi-Maxwellian and taking velocity moments of the Boltzmann equation. We assume the gyrotropic limit in which the off-diagonal terms of the pressure tensor are neglected and thermal transport occurs only in the direction parallel to the magnetic field. This is valid in the strong magnetic field region close to the Sun and may also be justified in interplanetary space where observations have shown the presence of field-aligned distribution functions and that the

CP679, *Solar Wind Ten: Proceedings of the Tenth International Solar Wind Conference*, edited by M. Velli, R. Bruno, and F. Malara

solar wind heat flux is parallel to the magnetic field [11]. We derive our equations for one-dimensional flow from the formulation in [1] as

$$\frac{\partial n_i}{\partial t} + \frac{1}{a}\frac{\partial (n_i v_i a)}{\partial r} = 0 \tag{1}$$

$$\frac{\partial v_i}{\partial t} + v_i \frac{\partial v_i}{\partial r} + \frac{1}{n_i m_i}\frac{\partial (n_i k T_{\parallel i})}{\partial r} + \frac{k(T_{\parallel i} - T_{\perp i})}{m_i}\frac{1}{a}\frac{da}{dr}$$

$$+\frac{Z_i}{m_i}\left[\frac{1}{n_e}\frac{\partial (n_e k T_{\parallel e})}{\partial r} + k(T_{\parallel e} - T_{\perp e})\frac{1}{a}\frac{da}{dr} - m_e\left(\frac{\delta v_e}{\delta t}\right)_c\right]$$

$$+\frac{GM_s}{r^2} - \frac{D_i}{n_i m_i} = \left(\frac{\delta v_i}{\delta t}\right)_c \tag{2}$$

$$\frac{\partial T_{\parallel j}}{\partial t} + v_j \frac{\partial T_{\parallel j}}{\partial r} + 2T_{\parallel j}\frac{\partial v_j}{\partial r} + \frac{1}{n_j k a}\frac{\partial (a q_{\parallel j})}{\partial r}$$

$$-\frac{2q_{\perp j}}{n_j k a}\frac{da}{dr} - \frac{2}{n_j k}Q_{\parallel j} = \left(\frac{\delta T_{\parallel j}}{\delta t}\right)_c \tag{3}$$

$$\frac{\partial T_{\perp j}}{\partial t} + v_j \frac{\partial T_{\perp j}}{\partial r} + \frac{v_j T_{\perp j}}{a}\frac{da}{dr} + \frac{1}{n_j k a}\frac{\partial (a q_{\perp j})}{\partial r}$$

$$+\frac{q_{\perp j}}{n_j k a}\frac{da}{dr} - \frac{1}{n_j k}Q_{\perp j} = \left(\frac{\delta T_{\perp j}}{\delta t}\right)_c \tag{4}$$

$$\frac{\partial q_{\parallel j}}{\partial t} + v_j \frac{\partial q_{\parallel j}}{\partial r} + 4q_{\parallel j}\frac{\partial v_j}{\partial r} + \frac{v_j q_{\parallel j}}{a}\frac{da}{dr}$$

$$+\frac{3 n_j k^2 T_{\parallel j}}{m_j}\frac{\partial T_{\parallel j}}{\partial r} = \left(\frac{\delta q_{\parallel j}}{\delta t}\right)'_c \tag{5}$$

$$\frac{\partial q_{\perp j}}{\partial t} + v_j \frac{\partial q_{\perp j}}{\partial r} + 2q_{\perp j}\frac{\partial v_j}{\partial r} + \frac{2 v_j q_{\perp j}}{a}\frac{da}{dr} + \frac{n_j k^2 T_{\parallel j}}{m_j}\frac{\partial T_{\perp j}}{\partial r}$$

$$+\frac{n_j k^2 T_{\perp j}}{m_j}\frac{(T_{\parallel j} - T_{\perp j})}{a}\frac{da}{da} = \left(\frac{\delta q_{\perp j}}{\delta t}\right)'_c \tag{6}$$

The subscript i denotes alphas or protons and subscript j denotes alphas, protons, or electrons. We assume overall charge neutrality and negligible current so that $n_e \approx \Sigma Z_i n_i$ and $n_e v_e \approx \Sigma Z_i n_i v_i$ where Z_i is the charge of ion i. The partial pressures are given by $p_{*j} = n_j k T_{*j}$ where $*$ denotes the parallel or perpendicular direction and k is Boltzmann's constant. The mean temperature and total heat conductive flux of species j are $T_j = T_{\parallel j}/3 + 2T_{\perp j}/3$ and $q_j = q_{\parallel j}/2 + q_{\perp j}$, respectively, where $q_{\parallel j}$ and $q_{\perp j}$ are the conductive fluxes from the parallel and perpendicular direction, respectively. We use collision terms, denoted by subscript c, derived in [2] using the Bhatnagar-Gross-Krook approximation and written out explicitly in [3]. We do not include the ionization and recombination terms used in their model.

We assume radial flow through a diverging flux tube with cross-sectional area $a = r^2 f(r)$ where $f(r)$ is from [8]. We allow for arbitrary momentum and heating to the species of the form

$$D_i = D_{i0}e^{(R_s - r)/\sigma_i} + D_i^{(ext)}(1 - e^{(R_s - r)/(10R_s)})\frac{R_s^4}{r^4}, \tag{7}$$

$$Q_{*j} = Q_{*j0}e^{(R_s - r)/\sigma_j}, \tag{8}$$

where $*$ denotes the parallel or perpendicular direction for species j and $D_{i0}, \sigma_i, D_i^{(ext)}, Q_{*j0}$, and σ_j are constants. In the absence of knowing what heats and accelerates the solar wind, the functional forms above allow for a wide range of mechanisms. Eq.8 originates from Q been written as the negative divergence of an energy flux density that is dissipated over length scale σ [5, 20, 15]. It is reasonable to assume that a similar form applies to the momentum addition. The second term in Eq.7 allows for extended momentum addition to the particles. It was originally added to counteract the large inertia of the helium ions by preventing the excessive dip in the alpha particle velocity in the inner corona and by aiding the helium velocity to exceed the proton velocity in interplanetary space. Since observations indicate and models have incorporated extended proton heating [12, 13, 6], such as by the dissipation of waves, there is likely also extended momentum addition to the particles.

The equations are simultaneously solved using a fully implicit, time dependent code described in [6].

RESULTS

We present in Figure 1 a model solution with the following parameters:

Geometry: $f_m = 5, r = 1.31\mathrm{R}_s$, $\sigma = 0.51\ \mathrm{R}_s$

Momentum Addition:

$D_{p0} = 5 \times 10^{-15}$ g cm^{-2} s^{-2}, $\sigma_p = 0.6\ \mathrm{R}_s$

$D_p^{(ext)} = 5 \times 10^{-12}$ g cm^{-2} s^{-2}

$D_{\alpha 0} = 5 \times 10^{-16}$ g cm^{-2} s^{-2}, $\sigma_\alpha = 0.6\ \mathrm{R}_s$

$D_\alpha^{(ext)} = 5 \times 10^{-12}$ g cm^{-2} s^{-2}

Heating:

$Q_{\perp p0} = 6 \times 10^{-8}$ erg cm^{-3} s^{-1}, $\sigma_p = 0.5\ \mathrm{R}_s$

$Q_{\perp \alpha 0} = 1 \times 10^{-8}$ erg cm^{-3} s^{-1}, $\sigma_\alpha = 0.5\ \mathrm{R}_s$

$Q_{\parallel e0} = 2 \times 10^{-7}$ erg cm^{-3} s^{-1}, $\sigma_e = 0.2\ \mathrm{R}_s$

$Q_{\perp e0} = 2 \times 10^{-7}$ erg cm^{-3} s^{-1}, $\sigma_e = 0.2\ \mathrm{R}_s$

At the coronal base the electron density is 5×10^8 cm^{-3} and the temperatures are equal to 6×10^5 K. The solution matches empirical electron density and inferred temperature profiles in the inner corona and is consistent with proton flux ($n_p v_p = 3.3 \times 10^8$ cm^{-2} s^{-1}), velocity ($v = 700$ km s^{-1}), and Helium abundance ($n_\alpha / n_p = .048$) at 1 AU. The steep acceleration of the velocity profile in

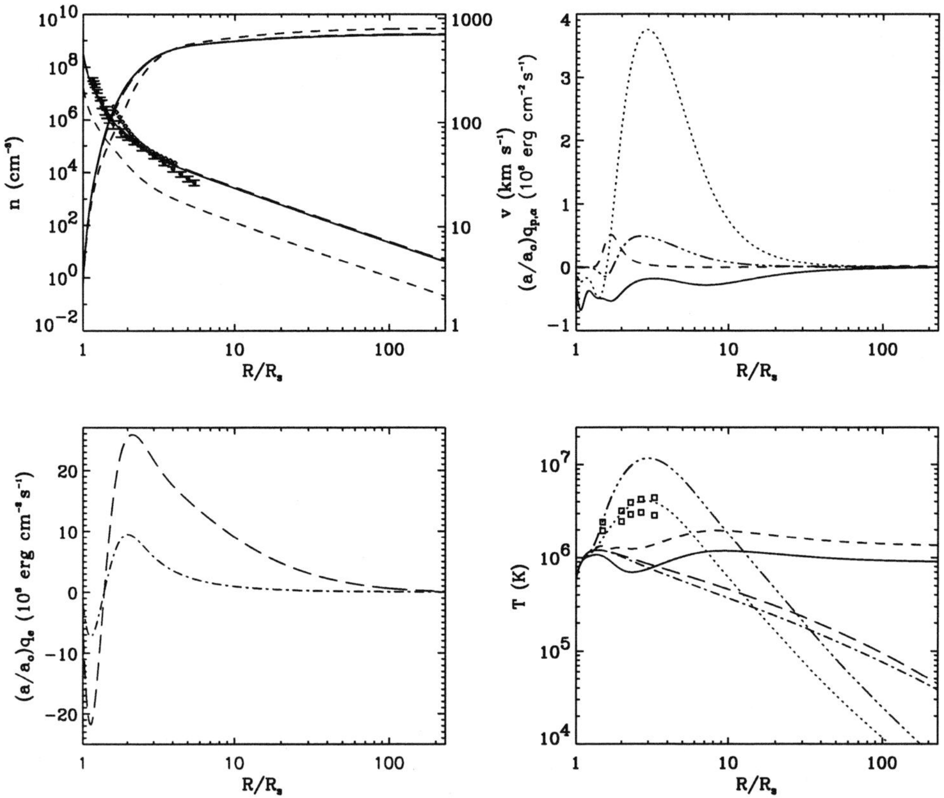

FIGURE 1. Model solution for parallel proton parameters (solid line), perpendicular proton parameters (dotted line), parallel alpha parameters (short-dash line), perpendicular alpha parameters (dash-dot-dot-dot line), parallel electron parameters (long-dash line), and perpendicular electron parameters (dash-dot line). The heat flux density is multiplied by the cross-sectional area a of the flux tube at distance R divided by the cross-sectional area a_o at the solar base to account for the expansion of the flux tube. The electron density is fitted to empirical data [4] (solid dots) and [7] (diamonds). The proton temperature is also fitted to empirical data [7].

the inner corona results in a noticeable dip in $T_{\parallel p}$ and $T_{\parallel \alpha}$ below 3R$_s$. The alphas initially lag behind the protons because of their large inertia. The extended heating to the alphas is mainly responsible for the dominance of the alpha velocity over the proton velocity in the outer corona. Since $n_\alpha << n_p$, the momentum addition per particle is significantly higher for the alphas.

Clearly the temperatures at 1 AU are not in agreement with observations. $T_{\perp p}$ and $T_{\perp \alpha}$, initially dominated by the heating terms $Q_{\perp p}$ and $Q_{\perp \alpha}$, respectively, peak in the inner corona, then plummet due to adiabatic cooling (the $\frac{vT_\perp}{a}\frac{da}{dr}$ term). Extended heating is necessary to increase $T_{\perp p,\alpha}$ in the outer corona to match observations in interplanetary space. Because thermal transport is only in the parallel direction, the decrease in $T_\perp$ results in a large rise in $T_\parallel$ for both the protons and the alphas. As noted previously in the literature, this is a serious problem for the protons since empirically $T_{\parallel p} \sim 3 \times 10^5$K at 1 AU, although ion cyclotron resonance may serve to decrease the parallel temperature in the outer corona. Interestingly, a high parallel temperature is less of a concern regarding the alphas since $T_{\parallel \alpha} \sim 0.9 - 2 \times 10^6$K at 1 AU. For this solution $T_{\parallel \alpha}$ is in agreement with observations, although $T_{\perp \alpha}$ is much farther below its empirical value ($\sim 0.7 - 1 \times 10^6$ K) at 1 AU than is $T_{\perp p}$.

Within the first 10 R$_s$, the heat flux profiles are in general proportional to the temperature and not to the gradient of the temperature. The alpha parallel and perpendicular heat conductive flux is several orders of magnitude below that of the protons and especially of the electrons, as expected, but its role is not insignificant. The creation of $q_{\parallel \alpha}$ (Eq.5) is a delicate balance between competing terms in each region of the corona. The flux $q_{\parallel \alpha}$ peaks slightly below 2R$_s$ where there is a local maximum in $T_{\parallel \alpha}$ due to the $\frac{\partial v}{\partial r}$ term in equation 3. The parallel heat flux $q_{\parallel \alpha}$ does not occur in the perpendicular energy equation (4) and does not dominate in any of the terms in the parallel energy equation (3). It is not particularly significant in affecting either temperature profile. The perpendicular heat flux, $q_{\perp\alpha}$, is not influential in any of the terms in the perpendicular energy equation. Instead, $T_\perp$ is controlled by the heating and collision terms in the inner corona and by the $\frac{vT_\perp}{a}\frac{da}{dr}$ and $v\frac{\partial T_\perp}{\partial r}$ terms farther out. However, the $\frac{2q_\perp}{nka}\frac{da}{dr}$ term in the parallel energy equation is dominant from $\sim 3-9$ R$_s$, transporting energy from the perpendicular into the parallel direction. This causes the noticeable rise in $T_{\|\alpha}$ after $T_{\|\alpha}$ decreases from 2-3 R$_s$ due to the sharp rise in v_α. $T_{\|\alpha}$ remains high into interplanetary space. Similar behavior exists for the protons.

Thus $q_{\perp\alpha}$ affects $T_{\|\alpha}$. The flux $q_{\perp\alpha}$ is negligible in the first $0.5R_s$ due to almost perfect balance between the collision and $nT_\|\frac{\partial T_\perp}{\partial r}$ terms in equation (6). These two terms decrease and the temperature anisotropy $[T_{\perp\alpha}(T_{\|\alpha}-T_{\perp\alpha})]$ term becomes dominate at ~ 2 R$_s$. This generates a large $q_{\perp\alpha}$ value which peaks ~ 2 R$_s$, near to the peak of $T_{\perp\alpha}$. Two other terms in Eq.6 are also prominent in this region: $\frac{2vq_\perp}{a}\frac{da}{dr}$ which serves to decrease $q_{\perp\alpha}$, is slightly larger than $v\partial q_\perp/\partial r$, but both are much less than the anisotropic term. Beyond ~ 4 R$_s$, $\frac{2vq_\perp}{a}\frac{da}{dr}$ dominates over the anisotropic term, which falls rapidly, and $\frac{2vq_\perp}{a}\frac{da}{dr}$ and $v\partial q_\perp/\partial r$ nearly balance to create a small $q_{\perp\alpha}$ term in interplanetary space.

CONCLUSION

In conclusion we find that, because of the large inertia of the alpha particles, extended momentum addition is very beneficial in creating an alpha velocity comparable to or greater than the proton velocity in interplanetary space. Extended heating in the perpendicular direction is also necessary for both the protons and the alphas if $T_\perp$ at 1 AU is to match observed values. As has been noted in the literature, cooling is needed in the parallel direction for the protons because $T_{\|p}$ is too large in interplanetary space. However, because of the high empirical value of $T_{\|\alpha}$ at 1 AU, the large interplanetary values of $T_{\|\alpha}$ in this model are not excessive. Additional heating directly to the electrons is needed to match the empirical electron temperature in interplanetary space.

The parallel and perpendicular heat conductive flux are several orders of magnitude lower than those of the protons and electrons, but $q_{\perp\alpha}$ is responsible for increasing $T_{\|\alpha}$ in the region several solar radii above the coronal base. A high positive $q_{\perp\alpha}$ value is generated in this region predominantly in response to the large temperature anisotropy that develops due to the increase in $T_{\perp\alpha}$ from the perpendicular heating. The perpendicular heat flux $q_{\perp\alpha}$ itself does not significantly influence $T_{\perp\alpha}$ and the parallel heat flux $q_{\|\alpha}$ does not appear to play an important role in governing the behavior of $T_{\|\alpha}$ or $T_{\perp\alpha}$.

Future work will include removal of the ad hoc functions in the Eqs. (7) and (8) and the addition of waves to heat and accelerate the solar wind. These will provide a more realistic approach to determining the amount and location of energy addition and the resulting model parameters. This should allow a better comparison of model solutions with observations, with the hope that the model will help explain the different thermal behavior of the alphas and protons in interplanetary space and at 1 AU.

REFERENCES

1. Barakat, A. R., and Schunk, R. W., *Plasma Physics*, **24**, 389, 1982.
2. Burgers, J. M., *Flow equations for composite gases*, Academic, San Diego, Calif., 1969.
3. Gombosi, T. I., and Rasmussen, C. E., *J. Geophys. Res.*, **96**, 7759, 1991.
4. Habbal, S. R., Esser, E., Guhathakurta, M., and Fisher, R. R., *Geophys. Res. Lett.*, **22**, 1465, 1995.
5. Hammer, R., *Astrophys. J.*, **259**, 767, 1982.
6. Hu, Y.-Q., Esser, R., and Habbal, S. R., *J. Geophys. Res.*, **102**, 14661, 1997.
7. Kohl, J. L., et al., *Astrophys. J.*, **501**, L127, 1998.
8. Kopp, R. A., and Holzer, T. E., *Solar Phys.*, **49**, 43, 1976.
9. Li, X., *J. Geophysics. Res.*, **104**, 19773, 1999.
10. Lie-Svendsen, O., Leer, E., and Hansteen, V. H., *J. Geophys. Res.*, **106**, 8217, 2001.
11. Marsch, E., Kinetic Physics of the Solar Wind Plasma, in *Physics of the Inner Heliosphere*, vol. 2, Schwenn, R. and Marsch, E. (Eds.), Springer-Verlag, Berlin, 1991.
12. Marsh, E., and Tu, C.-Y., *Astron. Astrophys.*, **319**, L17, 1997.
13. McKenzie, J. F., Axford, W. I., and Banaszkiewicz, M., *Geophys. Res. Lett.* , **24**, 2877, 1997.
14. Olsen, E. L. and Leer, E., *J. Geophys. Res.*, **104**, 9963, 1999.
15. Olsen, E. L., Leer, E., and Lie-Svendsen, O., *Astron. Astrophys.*, **338**, 747,1998.
16. Schwenn, R., Large-Scale Structure of the Interplanetary Medium, in *Physics of the Inner Heliosphere*, vol 1, Schwenn, R. and Marsch, E. (Eds.), Springer-Verlag, Berlin, 1990.
17. Spitzer, L., Jr., and Harm, R., *Phys. Rev.*, **89**, 977, 1953.
18. Tu, C.-Y., Marsch, E., and Wilhelm, K. *Space Sci. Rev.*, **87**, 331, 1999.
19. Wilhelm, K., et al., *Astrophys. J.*, **500**, 1023, 1998.
20. Withbroe, G. L., *Astrophys. J.*, **325**, 442, 1988.

Coronal MHD transport theory and phenomenology

L.J. Milano*, W.H. Matthaeus*, P. Dmitruk* and S. Oughton†

*Bartol Research Institute, University of Delaware
†Department of Mathematics, University of Waikato

Abstract. In the presence of a weakly inhomogeneous background, magnetohydrodynamic fluctuations are transported, reflected and at small scales, dissipated. In contrast to orderings appropriate to outer solar wind conditions, here we explore transport in a regime relevant for solar coronal heating and solar wind acceleration, in which effects of the order of the Alfvén speed are retained while disregarding the solar wind velocity. We consider the general properties of the transport equations as well as some solutions of interest.

INTRODUCTION

MHD fluctuations in a nonuniform plasma such as the solar corona or solar wind are often studied by means of a two-scale expansion method. The best known approach, so-called "WKB theory" [1, 2], is valid for relatively high-frequency, non-interacting Alfvén waves. However, if the fluctuations are low frequency, or if strong mode-mode coupling (e.g., turbulence) is present, then the WKB orderings break down and a more general form of transport theory is required. Fairly general approaches for carrying out such two-scale expansion have been presented [3, 4] including effects associated with the large-scale (wind) flow $\mathbf{U}_0$, and the large-scale magnetic field $\mathbf{B}_0$ and density ρ. Transport formulations of this type, adapted for outer heliospheric solar wind conditions in the super-Alfvénic wind ($U_0 \gg V_A$ where the Alfvén speed is $V_A = B_0/\sqrt{4\pi\rho}$), have proven useful [5]. For these applications one typically simplifies the transport equations by assuming that $V_A/U_0 \ll 1$ and dropping terms of order of this ratio. In the present paper we consider another interesting regime, one that is appropriate to the corona. Here we assume $U_0/V_A \ll 1$ and drop terms accordingly in the full transport formalism. This leads to relatively simple equations that can provide insights that complement direct calculations based upon the full primitive equations.

We begin by decomposing the magnetohydrodynamic (MHD) fields into mean and fluctuating components as follows:

$$\frac{\mathbf{B}}{\sqrt{4\pi\rho}} = \mathbf{V}_A(\mathbf{s},t) + \mathbf{b}(\mathbf{s},\mathbf{x},t),$$
$$\mathbf{v} = \mathbf{U}_0(\mathbf{s},t) + \mathbf{v}(\mathbf{s},\mathbf{x},t), \qquad (1)$$

where ρ is the plasma density, and $\mathbf{s}$ and $\mathbf{x}$ are formally independent variables, representing respectively slow and fast variations. The slow coordinate $\mathbf{s}$ spans large-scale distances, of the order of $1\,R_\odot$, while the fast coordinate is associated to local turbulent dynamics. The capitalized fields are obtained via an ensemble average over the total fields ($\mathbf{U}_0 \equiv \langle \mathbf{v} \rangle$, $\mathbf{V}_A \equiv \langle \mathbf{B}/\sqrt{4\pi\rho} \rangle$), and represent slowly varying background fields (respectively the Wind and Alfvén velocities). The ensemble average is assumed to filter out fast variations, and can be thought of as a volumetric average over the fast variable. Inserting Eqs. (1) in the MHD equations, one obtains [4]:

$$\partial_t \mathbf{z}^\pm + \mathbf{Z}_0^\mp \cdot \nabla \mathbf{z}^\pm + \mathbf{z}^\mp \cdot \nabla \mathbf{Z}_0^\pm \pm R^\pm \mathbf{b} \pm r^\pm \mathbf{V}_A = -\frac{1}{\rho}\nabla p + N_0^\pm + D^\pm, \qquad (2)$$

where

$$\mathbf{z}^\pm \equiv \mathbf{u} \pm \mathbf{b}, \qquad r^\pm = \nabla \cdot (\frac{1}{2}\mathbf{u} \pm \mathbf{b}),$$
$$\mathbf{Z}_0^\pm \equiv \mathbf{U}_0 \pm \mathbf{V}_A, \qquad R^\pm = \nabla \cdot (\frac{1}{2}\mathbf{U}_0 \pm \mathbf{V}_A), \qquad (3)$$

$\nabla = \nabla_s + \nabla_x$ is the total gradient operator, p is the total (kinetic plus magnetic) pressure, $D^\pm$ represents the dissipation terms, and $N_0^\pm$ the nonlinear terms. The last term in the LHS of Eq. (2) was not present in Eq. (21) of Ref. [4], where terms of order V_A were neglected with respect to terms of order U_0, which is a valid approximation at distances $\geq 1\,$AU. Here we retain terms of order V_A, and we assume local incompressibility, as well as homogeneity and isotropy in the fast variable. Fine points of the general derivation are discussed in Ref. [4].

CP679, *Solar Wind Ten: Proceedings of the Tenth International Solar Wind Conference*,
edited by M. Velli, R. Bruno, and F. Malara

TRANSPORT IN RMHD

We further simplify the treatment by postulating a rectangular geometry in which $\mathbf{s}$ is orthogonal to $\mathbf{x}$:

$$\mathbf{x}=(x,y,0),\ \ \nabla_x=(\partial_x,\partial_y,0),\ \ \mathbf{z}^{\pm}=(z_x^{\pm},z_y^{\pm},0),$$
$$\mathbf{s}=(0,0,s),\ \ \nabla_s=(0,0,\partial_s),\ \ \mathbf{Z}_0^{\pm}=(0,0,U_0\pm V_A). \quad (4)$$

Under this approximation, expansion and curvature effects are neglected, thus reducing mathematical complexity. We postpone investigation of these higher-order corrections to future efforts.

In rectangular geometry, the scale-separated MHD equations Eq. (2) take the reduced (hence the name Reduced MHD, or RMHD) form:

$$\partial_t\mathbf{z}^{\pm}+(U_0\mp V_A)\partial_s\mathbf{z}^{\pm}\pm\frac{1}{2}R^{\pm}(\mathbf{z}^{+}-\mathbf{z}^{-})=N_0^{\pm}+D^{\pm}, \quad (5)$$

where we explicitly wrote $\mathbf{b}=\frac{1}{2}(\mathbf{z}^{+}-\mathbf{z}^{-})$. We define the usual correlation tensors [4]:

$$H_{ij}^{\pm}\equiv\langle z_i^{\pm}z_j^{\pm'}\rangle, \quad (6)$$

$$\Lambda_{ij}\equiv\langle z_i^{+}z_j^{-'}\rangle, \quad (7)$$

where the average is over the fast variable $\mathbf{x}$, and the symbol $'$ denotes evaluation at the displaced position $\mathbf{x}+\mathbf{r}$; i.e., if $\mathbf{z}^{\pm}=\mathbf{z}^{\pm}(x,y,s,t)$ then $\mathbf{z}^{\pm'}=\mathbf{z}^{\pm}(x+r_x,y+r_y,s,t)$.

The evolution equations for the tensors $H_{ij}^{\pm}$ and Λ_{ij} are obtained by multiplying Eqs. (5) by $z_{i,j}^{\pm'}$ and then averaging. While the LHS of Eqs. (5) give a fairly straightforward contribution, which describes the linear transport of these quantities, the RHS needs to be modeled phenomenologically, as we discuss in the following sections.

LINEAR ALFVÉNIC TRANSPORT

Proceeding as explained above, one obtains:

$$\partial_t H_{ij}^{\pm}+L_{WKB}^{\pm}H_{ij}^{\pm}-R^{\pm}\Lambda_{ij}=0, \quad (8)$$

$$\partial_t\Lambda_{ij}+\left(U_0\partial_s+\frac{\partial_s U_0}{2}\right)\Lambda_{ij}-\frac{R^{-}H_{ij}^{+}-R^{+}H_{ij}^{-}}{2}+V_A M_{ij}=0, \quad (9)$$

where

$$L_{WKB}^{\pm}\equiv(U_0\mp V_A)\partial_s+R^{\pm}, \quad (10)$$

$$M_{ij}\equiv\langle z_i^{+}\partial_s z_j^{-'}-z_i^{-'}\partial_s z_j^{+}\rangle. \quad (11)$$

The equation for Λ_{ij} involves a new tensor M_{ij}, whose evolution should be computed independently in a complete treatment. However, the equation for $\partial_t M_{ij}$ involves not just D_{ij} and Λ_{ij}, but additional tensors, with higher-order derivatives in s. This is a familiar type of closure problem. Our approach herein will involve modeling of M_{ij}.

By Alfvénic transport we mean that terms of $O(U_0/V_A)$ are neglected. In this particular case the equations reduce to:

$$\partial_t H_{ij}^{\pm}\mp L_{WKB}H_{ij}^{\pm}\mp R\Lambda_{ij}=0, \quad (12)$$

$$\partial_t\Lambda_{ij}+\frac{1}{2}R(H_{ij}^{+}-H_{ij}^{-})+V_A M_{ij}=0, \quad (13)$$

where

$$L_{WKB}\equiv-V_A\partial_s+R \quad\text{and}\quad R\equiv\partial_s V_A. \quad (14)$$

Defining the energies $E^{\pm}\equiv\langle|\mathbf{z}^{\pm}|^2\rangle=H_{ii}^{\pm}|_{r=0}$, and energy difference $D^{\pm}\equiv\langle\mathbf{z}^{+}\cdot\mathbf{z}^{-}\rangle=\langle|\mathbf{u}|^2-|\mathbf{b}|^2\rangle=\Lambda_{ii}|_{r=0}$, one readily obtains,

$$\partial_t E^{\pm}\mp L_{WKB}E^{\pm}\mp RD=0, \quad (15)$$

$$\partial_t D+\frac{1}{2}R(E^{+}-E^{-})+V_A M=0, \quad (16)$$

where $M\equiv M_{ii}|_{r=0}=\langle\mathbf{z}^{+}\cdot\partial_s\mathbf{z}^{-}-\mathbf{z}^{-}\cdot\partial_s\mathbf{z}^{+}\rangle$. Note that a conveniently defined cross helicity is $H_c\equiv 4\langle\mathbf{u}\cdot\mathbf{b}\rangle=E^{+}-E^{-}$.

As a first case, we look for steady solutions of Eqs. (15–16). For this case Eq. (16) shows that $V_A M=-\frac{1}{2}RH_c$. Subtracting Eqs. (15) we obtain:

$$L_{WKB}H_c=0 \quad\Rightarrow\quad H_c(s)=aV_A(s), \quad (17)$$

where a is a constant to be determined by the boundary conditions. It follows that the steady state expression for M is given by:

$$M(s)=-\frac{a}{2}R(s). \quad (18)$$

These steady linear solutions are exact. However, we still cannot integrate Eqs. (15) to obtain E^{+} and E^{-}, since in steady state Eq. (16) does not provide any information on D. Note that M is proportional to the reflection coefficient, and thus tends to zero when $\partial_s V_A\to 0$. In fact, in the absence of reflections the steady solution is trivial: both E^{+} and E^{-} are constant in s, and the energy difference does not play any significant role in the (linear) dynamics.

As a second example, we can gain insight into the role of the energy difference D by looking at the early evolution starting from the initial condition: $E^{-}=E_0^{-}=const$, $E^{+}=0\Rightarrow D=0$. For short times

($t \ll 1/R_{max}$), we assume that $\partial_s \mathbf{z}^{\pm} = 0$ and drop all ∂_s terms in Eqs. (15–16), including M. We obtain:

$$E^+(t) = E_0^- \frac{1}{4} R^2 t^2,$$
$$E^-(t) = E_0^- (1 - \frac{1}{4} R^2 t^2),$$
$$D(t) = E_0^- \frac{1}{2} Rt. \quad (19)$$

Note that D grows linearly in time and helps transfer energy from the dominant field (z^- in this example) to the weaker field (z^+) This linear growth will saturate when M balances the cross helicity contribution in Eq. (16).

We now consider a case in which the system of equations (15)–(16) can be closed. This is the case in which only one of the fields, say z^-, is injected at a low frequency ($\omega_0 \ll \omega_A$) from the boundaries, and the reflections are weak ($R \ll \omega_A$), where ω_A is the Alfvén frequency $\omega_A \equiv 2\pi/t_A$, with the Alfvén transit time $t_A \equiv L_s/\langle V_A \rangle_s$ defined in terms of the s-averaged Alfvén velocity $\langle V_A \rangle_s$ and the length of the system in the slow direction L_s, which is typically about $1\,R_\odot$. In this case, we expect the system to be close to the zero-order solution (the solution for $R = 0$): $\mathbf{z}^+_{(0)} = 0$, and $\mathbf{z}^-_{(0)}$ a solution to the wave equation $\partial_t \mathbf{z}^-_{(0)} = -V_A \partial_s \mathbf{z}^-_{(0)}$. That is,

$$\mathbf{z}^+ = \mathbf{z}^+_{(1)},$$
$$\mathbf{z}^- = \mathbf{z}^-_{(0)} + \mathbf{z}^-_{(1)},$$
$$z^+_{(1)} \sim z^-_{(1)} \ll z^-_{(0)}, \quad (20)$$

where $z^{\pm} \equiv \langle |\mathbf{z}^{\pm}|^2 \rangle^{1/2}$. We will show that under these conditions we can write

$$M = -\partial_s D + \delta M \quad (21)$$

where $\delta M \equiv M + \partial_s D$ is a small correction that tends to zero when both ω_0 and R tend to zero. This relationship (evaluated at $\delta M = 0$) allows a closed system of equations (15)–(16) for E^+, E^- and D.

We first note that M can be rewritten as:

$$M = -\partial_s D + 2\langle \mathbf{z}^+ \cdot \partial_s \mathbf{z}^- \rangle, \quad (22)$$

where we used the product rule on $\partial_s(\mathbf{z}^+ \cdot \mathbf{z}^-)$ and the definition of $D = \langle \mathbf{z}^+ \cdot \mathbf{z}^- \rangle$. Under the assumed conditions, the second term on the RHS of Eq. (22) is negligible. To show this, we estimate the s derivatives of the fields. While the zero-order solution is driven by the boundary condition at frequency ω_0, the perturbed component oscillates at the much faster Alfvén frequency ω_A:

$$V_A \partial_s z^-_{(0)} \sim \omega_0 z^-_{(0)},$$
$$V_A \partial_s z^{\pm}_{(1)} \sim \omega_A z^{\pm}_{(1)}. \quad (23)$$

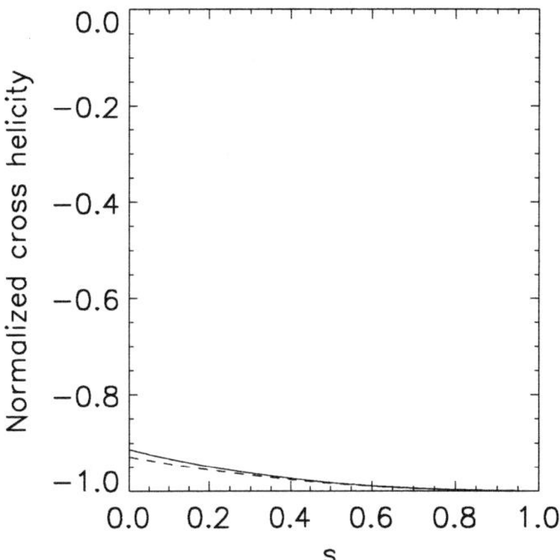

FIGURE 1. Time averaged linear solution of the MHD equations (dashed line), compared with the (linearized) closure equations (3-4,10). The Alfvén profile is linear, with $R\ L_s/V_A = 0.5$ and $w_0/w_A = 0.1$.

These estimates imply $V_A \partial_s D \sim \omega_A z^-_{(0)} z^+_{(1)}$, and thus,

$$\frac{\langle \mathbf{z}^+ \cdot \partial_s \mathbf{z}^- \rangle}{\partial_s D} \sim \frac{\omega_0}{\omega_A} + \frac{z^-_{(1)}}{z^-_{(0)}} \ll 1\ , \quad (24)$$

as we anticipated above.

We test this approximate closure and show a diagnostic plot in Fig. 1. The figure compares the normalized cross helicity [$\sigma = (E^+ - E^-)/(E^+ + E^-)$] as computed: (i) exactly, from the integration of the linearized MHD equations for a linear Alfvén profile, and (ii) from our closure linear equations. We use a linear Alfvén profile, featuring moderate reflections ($R = 1/2\omega_A$). The boundary conditions are: no incoming waves [$z^+(s = L_s) = 0$], and low-frequency ($\omega_0 = \omega_A/10$) outgoing (z^+) waves injected from the photosphere (at $s = 0$). The exact solution is obtained from the RMHD equations for a single perpendicular mode, as in [6] (linearized Eqs. (10–12) from Ref. [6]). Since the periodicity in the boundary condition introduces a time periodicity in the solutions, we time-average over one period ($2\pi/\omega_0$) and make use of the steady state solutions. The comparison shows fairly good agreement. Flatter Alfvén profiles give better agreement, as expected.

NON-LINEAR TRANSPORT

Anticipating that the solar corona may be in a turbulent state, we briefly discuss a strategy to handle MHD non-linearities in the transport equations.

So far we have neglected the nonlinear and dissipative terms (respectively $N_0^{\pm}$ and $D^{\pm}$) in the RHS of Eq. (5). It is well known that the non-linear terms transport directly cascading ideal invariants (such as $E^{\pm}$) through intermediate (*inertial range*) scales, toward small scales where they are dissipated. Even though the problem seems conceptually simple, the mathematical description is extremely difficult, with no complete rigorous theory of the process being known. For a macroscopic description of the problem where the focus is in the transport and dissipation of the fluctuations, an effective way to handle it consists of two main steps: (1) neglect dissipative terms, an excellent approximation at large scales; (2) replace the exact non-linear terms by phenomenologically modeled non-linear terms. The second step is the more difficult. One example of such a phenomenology is given by the following model [7]:

$$\partial_t E_N^{\pm} = -\alpha^{\pm}\frac{\sqrt{E^{\mp}}}{\lambda^{\pm}}E^{\pm}, \tag{25}$$

where $\partial_t E_N^{\pm}$ stands for *the non-linear contribution* to $\partial_t E^{\pm}$, which has to be added to the RHS of Eqs. (15). Here $\alpha^{\pm} \sim 1$ and $\lambda^{\pm}$ are the correlation lengths for $E^{\pm}$, that is,

$$\lambda^{\pm} \equiv \frac{\int_0^{\infty} H_{ii}^{\pm} dr}{H_{ii}|_{r=0}} \equiv \frac{L^{\pm}}{E^{\pm}}. \tag{26}$$

It turns out that the time evolution and space distribution of $\lambda^{\pm}(s,t)$ is not necessarily trivial. In order to show the complexity of the problem, we work with the linearized Eqs. (8) to obtain:

$$\partial_t L^{\pm} \mp L_{WKB} L^{\pm} \mp R L^D = 0 \ , \tag{27}$$

and assuming that $E^+ > 0$ and $E^- > 0$,

$$\partial_t \lambda^{\pm} \mp V_A \partial_s \lambda^{\pm} \mp R\frac{D}{E^{\pm}}(\lambda^D - \lambda^{\pm}) = 0 \ , \tag{28}$$

where L^D and λ^D follow from definitions analogous to those for $L^{\pm}$ and $\lambda^{\pm}$. Note that the linear equation (28) admits a steady solution $\lambda^+ = \lambda^- = \lambda^D = constant \equiv \lambda_0$. The use of this simple solution, i.e. the choice of one single similarity length characterizing all the turbulent correlations, together with Eqs. (15), (16), (21) and (25), gives a complete set of equations for the energies and dissipation of the outgoing and incoming fluctuations.

CONCLUSIONS

We have proposed a model for transport of MHD fluctuations in the solar corona, through the study of two-point correlation functions. The model extends previous work [4] to include a coronal scenario, for which the Alfvén velocity is much greater than the wind speed. Reflections, derived from the inhomogeneity of the large-scale Alfvén velocity, play an essential role in redistributing energy between incoming and outgoing fluctuations. Thus, an initial configuration with only outgoing waves would, after a transient period of time of about $1/R$ (where R is the reflection coefficient), reaches a state in which both outgoing and incoming fluctuations coexist, allowing for non-linear interactions and turbulence.

The system of equations for the time evolution of two-point correlations suffers a closure problem. However, this problem can be overcome in the asymptotic limit in which reflections are smooth, and only one type of fluctuation (either outgoing or incoming) is injected at a low frequency. In this case, an approximate closure can be used and a closed system of equations obtained.

We also briefly discussed a strategy to deal with non-linearities in the model equations. However, a better understanding of the phenomenology of MHD turbulence is required for a more accurate modeling of the non-linear terms in the transport equations.

ACKNOWLEDGMENTS

We acknowledge support by the NSF (ATM 0105254), NASA (NAG5-8134), and UK PPARC (PPA/G/S/1999/00059). This research has made use of NASA's Astrophysics Data System.

REFERENCES

1. Parker, E. N., ApJ, **143**, 32–+ (1966).
2. Hollweg, J. V., J. Geophys. Res., **78**, 3643–3652 (1973).
3. Marsch, E., and Tu, C.-Y., *Journal of Plasma Physics*, **41**, 479–491 (1989).
4. Zhou, Y., and Matthaeus, W. H., J. Geophys. Res., **95**, 10291–10311 (1990).
5. Smith, C. W., Matthaeus, W. H., Zank, G. P., Ness, N. F., Oughton, S., and Richardson, J. D., J. Geophys. Res., **106**, 8253–8272 (2001).
6. Dmitruk, P., Milano, L. J., and Matthaeus, W. H., ApJ, **548**, 482–491 (2001).
7. Hossain, M., Gray, P. C., Pontius, D. H., Matthaeus, W. H., and Oughton, S., *Physics of Fluids*, **7**, 2886–2904 (1995).

The timescales and heating efficiency of MHD wave-driven turbulence in an open magnetic region

Pablo Dmitruk and William H. Matthaeus

Bartol Research Institute, University of Delaware, Newark, DE 19716, USA

Abstract. Incompressible MHD waves propagating in a non-homogeneous background, with Alfven speed parallel gradients, can interact with their reflections producing a strong turbulent energy cascade to small perpendicular length scales. In an open magnetic region, with a strong background magnetic field, this nonlinear process can be described by the reduced MHD equations, in which incompressible or weakly compressible fluctuations are transverse to the mean field and their gradients in the parallel direction are much weaker than in the perpendicular direction. The system is forced by Alfven waves injected through the bottom boundary, with characteristic perpendicular length scale, amplitude and frequency. We identify the relevant timescales of the system and propose an ordering which favors the turbulent dissipation efficiency. We have applied this mechanism for the heating of the lower corona required in models of the origin of the solar wind.

The presence of magnetohydrodynamic (MHD) turbulence in coronal holes, as regions of the origin of the fast solar wind, is an attractive idea, since it can provide a mechanism by which the energy found in low-frequency motions and large scales can be *cascaded* (i.e., transferred) to small scales and higher frequencies where it can be more efficiently dissipated. Provided a kinetic mechanism can be identified for the dissipation of the cascaded energy, the proper amount of heating can be obtained for the requirements imposed in fluid models of the acceleration of the solar wind. From the point of view of MHD turbulence theory however, a number of important considerations have to be taken into account to properly address this turbulent cascade in an open magnetic region. First, the presence of a strong radially directed magnetic field $\mathbf{B_0}$ impose an anisotropy of the turbulent cascade: the energy transfer in the direction of the magnetic field is suppressed in favor of a transfer of energy to small perpendicular (to $\mathbf{B_0}$) length scales [1, 2, 3] This is important if a dissipation mechanism is invoked, since it has to take into account that anisotropy in the fluctuations scale dependence. Second, the same presence of the strong magnetic field also induces the fast propagation of almost non-dissipative Alfven waves along $\mathbf{B_0}$. As a consequence, the injected energy would leave the open region considered if no other process is identified to allow the nonlinear transfer to more dissipative small scales. Both considerations have been addressed in [4] by doing numerical simulations of the reduced MHD equations in a basic model of an open magnetic region. The reduced MHD approximation [5, 6, 7] describes the anisotropy found in the turbulent cascade. The presence of Alfven speed parallel gradients produce reflections of the injected waves [8, 9, 10, 11] and the coexistence of counter-propagating waves is a necessary condition to allow nonlinearities to activate and sustain the turbulence [12, 13, 14, 15, 16]. A more subtle issue is that non-propagating structures favor turbulence sustainment [1, 2, 17] while the presence of such structures is controlled by the nature of the boundary conditions [4] We further analyze this system here, by identifying the relevant timescales and proposing an ordering of those timescales that favors the presence of turbulence and hence the enhancement of dissipation. The present report offers a brief discussion of the basic physical issues and associated timescales that affect the sustainment and efficiency of MHD turbulence in an open region. We have addressed elsewhere [18] the application of this model to more realistic density and magnetic field radial profiles and their effect on the heating in a coronal open magnetic region.

The reduced MHD approximation [5, 6, 7] describes an incompressible or weakly compressible plasma in the presence of a strong background magnetic field $\mathbf{B_0}$. A condition in RMHD is that gradients of fluctuations in the direction parallel to $\mathbf{B_0}$ are much weaker than gradients in the perpendicular direction, thus, the anisotropy of the turbulent cascade is inherent to the description. Transverse fluctuations of the magnetic $\mathbf{b}$ and velocity fields $\mathbf{v}$ are conveniently described by the upward/downward propagating fluctuations $\mathbf{z}_{\mp} = \mathbf{v} \mp \mathbf{b}$. In an inhomogeneous background, with an Alfven speed varying in the parallel direction $V_A(s) = B_0/\sqrt{4\pi\rho(s)}$, where $\rho(s)$ is the smoothly varying background density,

CP679, *Solar Wind Ten: Proceedings of the Tenth International Solar Wind Conference*, edited by M. Velli, R. Bruno, and F. Malara

the RMHD equations are:

$$\frac{\partial \mathbf{z}_\pm}{\partial t} \mp V_A \frac{\partial \mathbf{z}_\pm}{\partial s} = \mp \frac{1}{2}\frac{dV_A}{ds}\mathbf{z}_\pm \pm \frac{1}{2}\frac{dV_A}{ds}\mathbf{z}_\mp$$
$$-\nabla_\perp p' - \mathbf{z}_\mp \cdot \nabla_\perp \mathbf{z}_\pm + \eta \nabla_\perp^2 \mathbf{z}_\pm$$

where η is the resistivity (assumed equal to the viscosity) and p' is the (magnetic plus thermal) pressure. The injection of waves from the bottom, which allows non-propagating structures to be present in the system (see [4]), is described by the boundary condition:

$$\frac{\partial \mathbf{z}_-}{\partial s}^{\text{bot}}(\mathbf{k}_\perp) = A(k_\perp)\cos(2\pi f\, t) \quad ,\text{if } k_1 \le k_\perp \le k_2 \tag{1}$$

where f is the frequency of the forcing and A controls the amplitude of the injected waves. The forcing is narrow in Fourier space, with a k-band given by k_1 and k_2. At the top boundary, no waves are injected, so, the condition imposed is:

$$\frac{\partial \mathbf{z}_+}{\partial s}^{\text{top}}(\mathbf{k}_\perp) = 0 \quad ,\forall\, \mathbf{k}_\perp \tag{2}$$

The dynamical RMHD equations and the boundary conditions, contain several timescales and it is of relevance for the dissipation efficiency of the turbulent cascade to identify them. We will proceed now to describe those timescales.

Ignoring the right hand side Eq. (1) leaves us with a description of the linear propagation of waves in the system. For constant V_A, the solutions $\mathbf{z}_\pm \sim \exp[ik_\parallel(s \pm V_A t)]$ correspond to downward (+) and upward (-) propagating Alfven waves, with speed V_A. The timescale in which a linear fluctuation would propagate out of the box of height L is $t_A = L/V_A$, which can be computed using the average V_A when there are gradients.

The wave period (given by the forcing frequency of the boundary conditions at the bottom) and the wave crossing time are the fundamental scales of the problem until the additional effects contained in the r.h.s. of Eq. (1) are considered, namely, reflection, nonlinear couplings and dissipation. The first two terms on the r.h.s. of Eq. (1) describe the reflection of the waves due to the presence of parallel gradients of the Alfven speed $V_A(s)$. Those reflections introduce another timescale, $t_R = L/\Delta V_A$, due to the variation of the Alfven speed within the box. If $t_R \sim t_A$ a moderate amount of downward fluctuations can be generated from a population of only upward injected fluctuations. In that case, the nonlinear term in the r.h.s. of Eq (1) $\mathbf{z}_\mp \cdot \nabla_\perp \mathbf{z}_\pm$ is not trivial and the system can develop turbulence. If, on the other hand, there are no Alfven speed gradients, $t_R = \infty$, and it can be seen [4] that the turbulence can not be sustained. Even a very small amount of reflected waves (less than 5%) can be enough for sustaining the turbulence. The nonlinear term also introduce the timescales $t_{NL}(k) \sim l_k/z_k$ for fluctuations z_k at different length scales $1/k$. On the bottom boundary, the amplitude of the waves is given by a typical imposed fluctuation amplitude, δz_0 and at a typical perpendicular length scale $l_0 = 1/k_0$, where $k_1 < k_0 < k_2$ is a wavenumber included in the narrow band forcing. The timescale introduced by this forcing is $t_0 = l_0/\delta z_0$. If there is a turbulent cascade, the ordering is $t_{NL}(k) < t_0$ and it is actually t_0 which determines the typical timescale of nonlinearities.

Both the RMHD approximation and the efficiency of the turbulence on transferring the energy to smaller scales is favored if the ordering $t_0 < t_A$ is satisfied. If Alfven waves travel too fast through the box (i.e, if $t_A << t_0$) the energy of the fluctuations is mostly transmitted and lost through the top boundary. The condition $t_0 < t_A$ also means that the correlation scale of the forcing fluctuations (given by the forcing length scale l_0) is small enough to favor nonlinearities to develop, $l_0 < L(\delta z_0/B_0)$. The "competition" between the emptying behavior given by the Alfven wave crossing time t_A and the turbulent cascade development given by the nonlinear timescale t_0 is key to the system and determines the efficiency on dissipating the energy injected through the bottom.

Another important timescale is given by the frequency of the forcing, $t_f = 1/f$, since it is well known from linear studies of waves in inhomogeneous media [8, 9, 10, 11] that high frequency waves can travel almost unaffected by reflections and hence nonlinearities can not survive in that case, as stated before. So, the condition which favors the dissipation efficiency regarding that issue is $t_R < t_f$. Finally, the dissipative term in Eq. (1) contains the timescale $t_\eta = l_0^2/\eta$ for the dissipation of the large scale forcing structures. The actual dissipation mechanism of the system is in general more complicated, and it should be described considering kinetic theory. For the macroscopic MHD description considered here, it suffices to say that the dissipation timescale associated with the large scale energy containing structures is very large. This is to say that the Reynolds number of the system (comparison of the nonlinear term to the dissipative term) is very large.

From the previous considerations, we can conjecture that the timescale ordering which would favor the development of a strong turbulent cascade and sustained heating efficiency is given by:

$$t_{NL}(k) < t_0 < t_A \sim t_R < t_f < t_\eta \tag{3}$$

This relationship embodies a balance among numerous effects – wave propagation speed, the spectral transfer rate of turbulence, the strength of driving, the large scale Reynolds number, the self-similarity of the cascade, the degree of inhomogeneity and strength of reflections, the ratio of the system size to the turbulence energy contain-

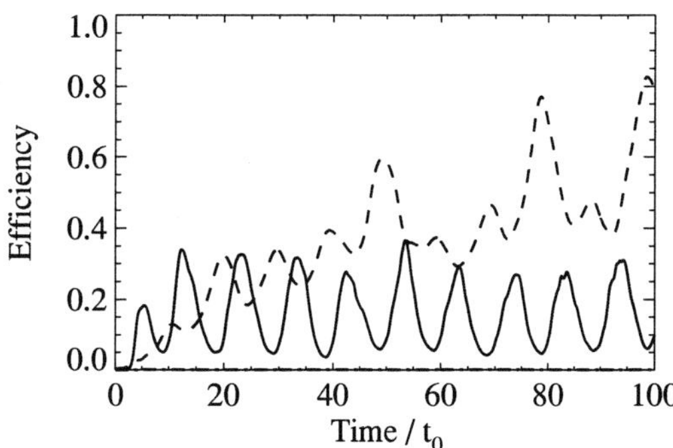

FIGURE 1. Turbulent dissipation efficiency for different timescale ratios, $t_0/t_A = 0.1$ (dashed), $t_0/t_A = 1$ (continuous) and $t_0/t_A = 10$ (dotted, but of such low value it is almost invisible)

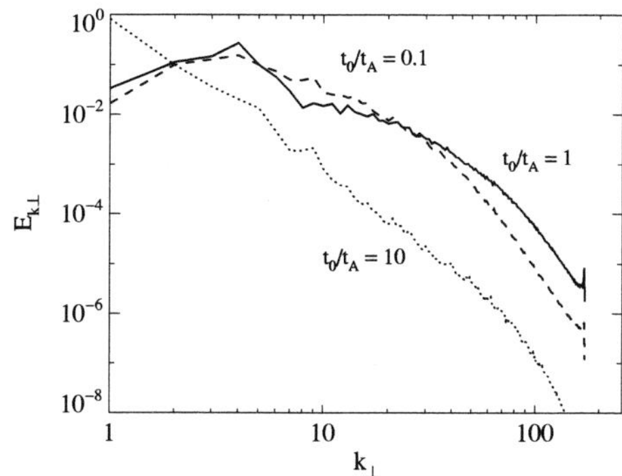

FIGURE 2. Perpendicular spectra for different timescale ratios, $t_0/t_A = 0.1$ (dashed), $t_0/t_A = 1$ (continuous) and $t_0/t_A = 10$ (dotted)

ing scale – each of which enters into the overall balance that determines whether a strong cascade driven by waves can persist in an inhomogeneous medium with open field line boundary conditions. Extensive study using numerical simulations for a variety of conditions and parameters will be required to fully investigate the validity and consequences of the orderings implied by this inequality. This substantial effort will be deferred for the present. However to explore some simple consequences of the above discussion, we present a comparison of numerical simulations of Eqs (1)-(3), using a pseudo-spectral type code (see [4] for a previous use of this code), for three cases corresponding to different timescale ratios $t_0/t_A = 0.1, 1, 10$ and fixed forcing frequency, $t_0/t_f = 0.05$, dissipative timescale $t_0/t_\eta = 0.001$ and $t_R \sim t_A$ in each case. The dissipation efficiency, defined as the turbulent energy transfer rate divided by the energy injection rate is shown in Fig. 1. Clearly, the $t_0/t_A << 1$ case is the most efficient, while in the case with $t_0/t_A >> 1$ the dissipation efficiency is very low. The ratio of downward propagating fluctuations to the upward propagating fluctuations is in all these cases less than 5% which indicates that (for the $t_0/t_A = 0.1, 1$ cases) turbulence is still efficient even for low reflected waves population.

In Fig. 2 we show the perpendicular energy spectra for the same cases. For the cases $t_0/t_A = 0.1, 1$ the spectrum is broad band and consistent with a Kolmogorov cascade description. The spectrum for the $t_0/t_A = 10$ is much steeper, indicating the fact that the emptying property of fast Alfven waves is dominating over the nonlinear turbulent transfer. The spectral properties of the system under different timescale conditions deserves further study.

In conclusion, we have identified the relevant timescales in a wave-driven turbulent system in an open magnetic region forced through the boundaries. The conditions which favor the development of turbulence and the enhancement of heating efficiency have been described. In general terms, high dissipation efficiency can be obtained when the nonlinear timescale is less than the Alfven wave crossing time, even if the population of the necessary reflected waves is low. The considered problem is of relevance for the heating in coronal holes and the consequent acceleration of the solar wind. We have discussed such heating application of the present model in [18].

ACKNOWLEDGMENTS

This research is supported in part by National Aeronautics and Space Administration Sun-Earth Connection Theory Program grant NAG5-8134, NASA grant NAG5-7164 and National Science Foundation grant ATM-0105254 to the University of Delaware Bartol Research Institute.

REFERENCES

1. J. V. Shebalin, W. H. Matthaeus and D. C. Montgomery, J. Plasma Phys. **29**, 525 (1983).
2. S. Oughton, E. R. Priest and W. H. Matthaeus, J. Fluid Mech. **280**, 95 (1994)
3. W. H. Matthaeus, S. Ghosh, S. Oughton and D. A. Roberts, J. Geophys. Res. **101**, 7619 (1996)
4. P. Dmitruk, W. H. Matthaeus, L. J. Milano and S. Oughton, Phys. Plasmas **8**, 2377 (2001).
5. H. R. Strauss, Phys. Fluids **19**, 134 (1976)
6. D.C. Montgomery, Physica Scripta **T2/1**, 83 (1982)
7. G. P. Zank and W. H. Matthaeus, J. Plasma Phys. **48**, 85 (1992)

8. M. Heinemann and S. Olbert, J. Geophys. Res. **85**, 1311 (1980)
9. J. V. Hollweg, Solar Phys. **70**, 25 (1981)
10. C.-H. An, Z. E. Musielak, R. L. Moore and S. T. Suess, Astrophys. J. **345**, 597 (1989)
11. M. Velli, Astron. Astrophys. **270**, 304 (1993)
12. R. H. Kraichnan, Phys. Fluids **8**, 138 (1965)
13. M. Dobrowolny, A. Mangeney and P. Veltri, Phys. Rev. Lett. **45**, 144 (1980)
14. R. Grappin, U. Frisch, J. Léorat and A. Pouquet, Astron. Astrophys. **105**, 6 (1982)
15. A. Pouquet, U. Frisch and M. Meneguzzi, Phys. Rev. A **33** 4266 (1986)
16. S. Ghosh, W. H. Matthaeus and D.C. Montgomery, Phys. Fluids **31**, 2171 (1988)
17. D. C. Montgomery and W. H. Matthaeus, Astrophs. J. **447**, 706 (1995)
18. P. Dmitruk, W. H. Matthaeus, L. J. Milano, S. Oughton, G. P. Zank and D. J. Mullan, Astrophys. J., in press (2002).

A search for turbulent wave heating and acceleration signatures with SOHO/SUMER observations : Measurements of the widths of off-limb Iron lines

L. Dolla*, P. Lemaire*, J. Solomon* and J.-C. Vial*

*Institut d'Astrophysique Spatiale, Unité Mixte CNRS-Université Paris XI, Bât. 121, 91405 Orsay, France

Abstract. The widths of coronal ions lines may contain important information about the energetics of the solar wind and corona. We present a method to measure these widths, taking into account the problems of instrumental stray light inherent to SoHO/SUMER. The Iron lines are interesting to set an upper limit on the "unresolved" velocity, that may be a signature of turbulent or wave motion in the corona.

INTRODUCTION

For many years, and more particularly since Ulysses observations of a fast solar wind at high heliolatitudes, it has appeared that the heating of the solar corona and the acceleration of the solar wind require additionnal (non-thermal) energy input. One possible energy input could result from wave-particle interactions (Alfvén waves, ion cyclotron resonance, turbulence, *e.g.* [1], [2]). A possible signature of such turbulent or Alfvénic motion could be the apparent non-thermal broadening of coronal ions lines as discussed in [3], [4], for instance.

The non-thermal broadening of spectral lines of coronal ions

The width of an optically thin line is a signature of the particle velocity distribution of the plasma which emitted the light (unfortunately modified by the integration along the line of sight). Coronal (and transition region) spectral lines show a width larger than the thermal width corresponding to the ion formation temperature.To account for this excess broadening, we can add a non-thermal or "unresolved" velocity ξ, so that the Gaussian width is given by :

$$\sigma^2 = \frac{\lambda^2}{2c^2}(\frac{2kT}{M} + \xi^2) + \sigma_I^2.$$

(where σ_I is the instrumental width)

As the thermal contribution is inversely proportional to the ion mass, it is smaller for heavy ions, like iron. Therefore, the width of iron lines can be used to set an upper limit on this unresolved velocity.

THE DATA

The SUMER instrument is described in [5]. We just remind here some characteristics of the data. The data that we get from the detector are made of images of 1024×360 pixels (wavelength versus spatial dimension). Each pixel dimension corresponds to $\approx$ 43 mÅ and $\approx$ 1 arcsec.

We use data acquired during the MEDOC Campaign #7 (May 2001), in the corona above a "Quiet Sun" region. The position of the slit was (860",-600") (in solar disc coordinates). The altitudes covered are from 13 to 180 arcsec above the solar limb. Two Iron lines are visible in the same spectral range : Fe X and Fe XI, at 1463.50 Å and 1467.06 Å, respectively.

A 1" wide slit is used for a better accuracy in measuring the width.

To reduce the noise, several files can be summed, thus increasing the total exposure time. For the same purpose, we can average the data over several pixels in the spatial dimension.

The accuracy of the width given by a gaussian fit can be estimated by simulating fits of noisy lines (lines with known width, plus some noise proportional to the square root of the total number of counts in the lines). In this simulation, we reproduce the main characteristics of the studied lines (average width, discrete sampling on the detector pixels, etc...). With sufficient counts in the line, the width uncertainty is a fraction of a pixel.

Using the results of the simulation, we can perform temporal and/or spatial sums to get the counts necessary for a given accuracy.

CP679, *Solar Wind Ten: Proceedings of the Tenth International Solar Wind Conference*, edited by M. Velli, R. Bruno, and F. Malara

DEALING WITH THE INSTRUMENTAL STRAY LIGHT

As the SUMER instrument has been designed to observe the solar disc, there is no occulter. Therefore, when observing off-limb corona, the SUMER slit is "dazzled" by the light coming from the solar disc. Thus, in the spectra acquired in off-limb observations, one must take into account the stray light, especially for radial analysis of the linewidth [6].

Yet, SUMER remains the unique spectrometer able to provide information about the corona below 1.5 solar radii, complementing SoHO/UVCS field of view.

Choice of a "stray light spectrum"

We first have to establish a stray light spectrum, *i.e.* the spectral distribution that contributes to the off-limb spectrum as stray light. We can use a reference spectrum directly acquired on the disc, but this solution is not satisfactory. Indeed, such data were not available at the time of our observations. Moreover, an on-disc spectrum is local, while the off-limb stray light is made of contributions from all the points of the solar disc, including structures such as active regions or sunspots, which are known to exhibit different kinds of spectra (see [8]).

For these reasons, we preferred to use data obtained the same day at a higher altitude z_0 ($\approx$ 350 arcsec above limb), which can be considered as due to stray light only. These data provide a reference spectrum $I(\lambda)_{\text{ref}}$.

Estimation of the stray light intensity

A method to estimate the stray light intensity, using cold lines, is described in [7]. It is based on the assumption that all counts observed in off-limb cold lines are due to stray light (very low emission of cold ions in the hot corona).

Here we extend this assumption to the continuum: we suppose that there is no emission of continuum at the altitude above limb where we are observing: then, all the continuum observed off-limb is due to stray light.

By using $I(\lambda)_{\text{ref}}$, we have at any altitude $z < z_0$, and for any wavelength λ:

$$I(\lambda,z)_{\text{stray light}} = \frac{I(z)^{(\textit{off-limb})}_{\text{continuum}}}{I^{(\textit{ref})}_{\text{continuum}}} \cdot I(\lambda)_{\text{ref}}$$

During all these intensity measurements, we take into account the average noise from the detector itself.

We can, next, compare $I(\lambda,z)_{\text{stray light}}$ to the observed spectrum $I(\lambda,z)$ (Fig. 1): it appears that the two Iron lines are blended by stray light from chromospheric lines.

For the sake of comparison, we tried to use an on-disc spectrum as an alternative stray light reference, instead of the high altitude spectrum. Without a doubt, this latter better matched the intensities of the known cold lines appearing as stray light in the off limb spectra. An accurate determination of the stray light spectrum is very important for width measurements, which are more sensitive to the line profile than intensity measurements.

Two kinds of blending by stray light are possible:

1. by the same resonance line, if this one is very intense on the disc
2. by a cold ion line, which should not exist in the hot corona.

Presently, we are in the second case: both Iron lines are blended by C I lines (possibly with some contribution from Ni II). It is noticeable that the stray light intensity decreases less rapidly than the intensity of the emission lines (Fig. 1: comparison at two different altitudes).

MEASURING THE WIDTH

To get the width of the effective coronal emission line, we fit the observed spectrum with two gaussians, one of them fitting the stray light contribution. Nevertheless, this method is very sensitive to the wavelength of the blending line. Thus we had to adjust the wavelength scales between the data and the reference spectrum for every radial position, to take into account some remnants of the geometric distorsion of the detector (this distorsion is not completely corrected by applying the SUMER data reduction software).

Performing such gaussian fits also produces some errors, as the blending lines do not look like pure gaussians, mainly because they are blended themselves. These errors become critical when the blending is strong, and will be taken into account in a future work. Here we present the results yielded by using gaussian fits.

"Radial" variation

Figures 2 and 3 show the radial evolution of the widths of Fe X and Fe XI lines, with a 12600 s total exposure time. We perform spatial averaging over several pixels in the slit height, in such a way as to provide a quite constant size of the errors bars (using at least 10 pixels).

We have plotted both linewidths before and after the correction from the stray light contribution (removing

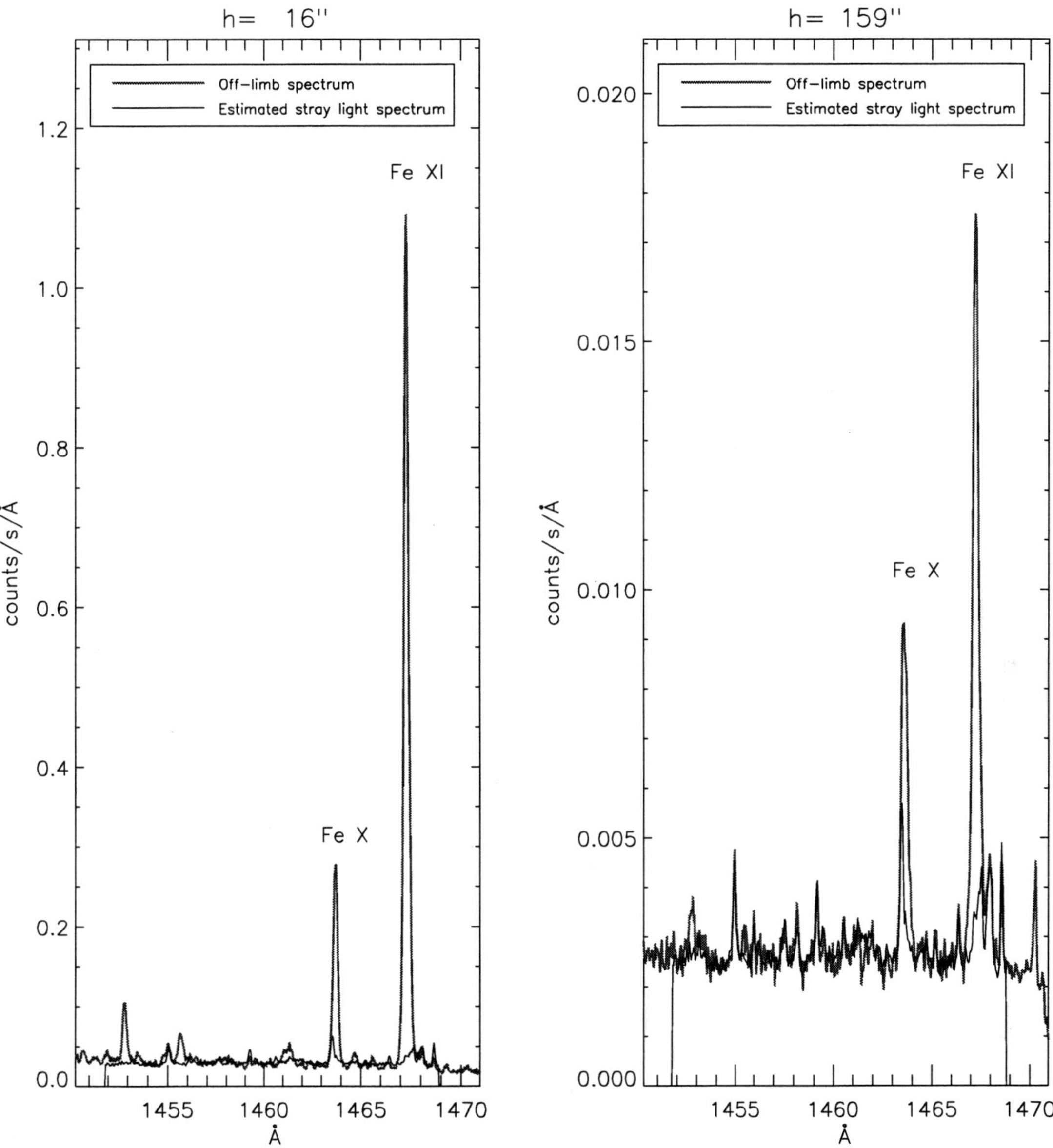

FIGURE 1. Examples of spectra at two altitudes (16 and 159 arcsec above the solar limb) : the observed spectrum (in grey) is the sum of stray light contribution (in black) and effective off-limb emission.

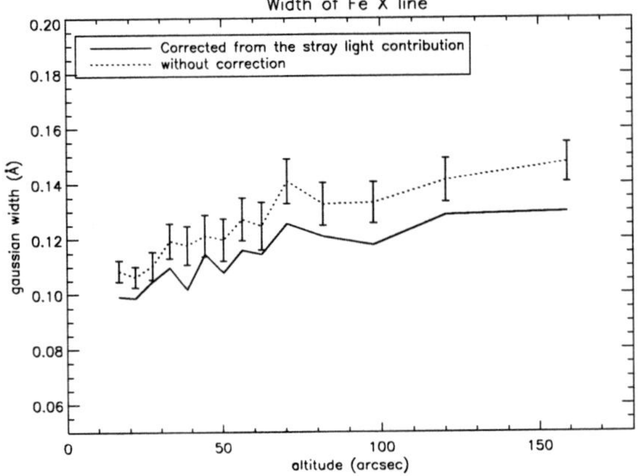

FIGURE 2. Radial variation of the gaussian width of Fe X: the large correction necessary to account for stray light make this line dubious to use for width measurements.

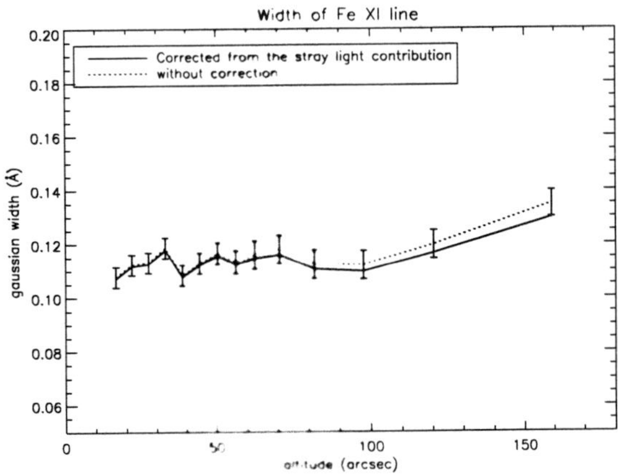

FIGURE 3. Radial variation of the gaussian width of Fe XI (error bars are only drawn for the values before correction from the stray light).

the instrumental width contribution). The error bars are given for the non-corrected results. It is obvious that the stray light correction is not important for the Fe XI line (the corrected curve is within the error bars of the non-corrected one), at least for the considered altitude range. On the contrary, there is no doubt about the critical effect of the stray light on the width of the Fe X line.

Radial or latitudinal variation ?

One should notice that when moving along the SUMER slit, both the radial and latitudinal positions change (13 to 180 arcsec above limb, and -27° to -41° in latitude). This way, the field of view may intersect different structures (a check with EIT image shows that it is the case for the present observations). This may explain some of the sharp variations displayed by the Fe XI line.

CONCLUSION

Width measurements of relatively hot Iron lines in the low corona have been discussed in some details. Particularly, we have stressed the importance of the stray light spectrum which can increase dramatically the error on width measurements, if not taken into account. In this context, analysis of recent observations of various coronal ions lines is in progress.

ACKNOWLEDGEMENTS

SoHO is a mission of international cooperation between ESA and NASA. The SUMER project is financially supported by DLR, CNES, NASA and ESA PRODEX Programme (Swiss contribution). The observations were performed during the MEDOC Campaign #7 (May 2001).

REFERENCES

1. Cranmer, S. R., Field, G. B., and Kohl, J. L., *Ap. J*, **518**, 937–947 (1999).
2. Marsch, E., and Tu, C.-Y., *J. Geophys. Res.*, **106**, 227–238 (2001).
3. Seely, J. F., Feldman, U., Schuehle, U., Wilhelm, K., Curdt, W., and Lemaire, P., *Ap. J. Let.*, **484**, L87–+ (1997).
4. Tu, C.-Y., Marsch, E., Wilhelm, K., and Curdt, W., *Ap. J*, **503**, 475+ (1998).
5. Wilhelm et al., K., *Solar Phys.*, **162**, 189–231 (1995).
6. Doschek, G. A., Feldman, U., Laming, J. M., Schühle, U., and Wilhelm, K., *Ap. J*, **546**, 559–568 (2001).
7. Feldman, U., Doschek, G. A., Schühle, U., and Wilhelm, K., *Ap. J*, **518**, 500–507 (1999).
8. Curdt, W., Brekke, P., Feldman, U., Wilhelm, K., Dwivedi, B. N., Schühle, U., and Lemaire, P., *Astron. Astrophys.*, **375**, 591–613 (2001).

MHD resonant flow instability in the magnetotail

R. Erdélyi and Y. Taroyan

Space and Atmosphere Research Center (SPARC), Department of Applied Mathematics, University of Sheffield, The Hicks Building, Hounsfield Road, Sheffield S3 7RH, UK. (e-mail: Robertus@sheffield.ac.uk)

Abstract. Resonant flow instability (RFI) and Kelvin-Helmholtz instability (KHI) are investigated as possible wave generating mechanisms in the mantle-like boundary layer of the Earth's magnetotail where all equilibrium quantities transition continuously from magnetosheath values to values more characteristic of the tail lobe. It is shown that as in the case of a sharp interface the KHI requiring high flow speeds in the magnetosheath is unlikely to be operative under typical conditions. RFI which is physically distinct from KHI may appear at lower flow speeds due to the inhomogeneity of the mantle-like boundary layer. It is shown that RFI can be important when the variation length-scale of the flow velocity is smaller than the variation length-scales of other equilibrium quantities such as density and magnetic field strength. Interpretation in terms of the wave energy flux is presented and the applicability to the magnetotail is discussed. The obtained results could explain the observed low power of ULF waves in the tail lobes compared with other parts of the magnetosphere.

INTRODUCTION

The Kelvin-Helmholtz instability (KHI) on the magnetopause may be important for coupling solar wind momentum into the magnetosphere and is often invoked as a source mechanism for various magnetospheric phenomena.

Most of the studies mainly concentrate on KHIs at a tangential discontinuity representing the magnetopause. However, observations (see, e.g., [1, 2, 3]) have identified nonuniform boundary layers inside the magnetopause extending from the subsolar point to the distant tail. Its thickness tends to increase with distance from the subsolar point. The inclusion of such boundaries leads to resonances which complicate the analysis. Few works ([4, 5, 6]) have investigated the effects introduced by nonuniform boundary layers on the KHI at the flanks of the magnetosphere. However, in all these works resonances were avoided by taking sufficiently high flow speeds or by imposing additional simplifying constraints on plasma parameters (e.g., no magnetic field in the magnetosheath and in the boundary layer).

The inclusion of a boundary layer, where one or several equilibrium quantities undergo continuous variation across the inhomogeneity, not only could modify the KHI but it could also lead to the appearance of a new type of instability known as resonant flow instability (RFI) which is physically distinct from the KHI. In [7, 8] the authors investigated the effect of velocity shear on the rate of resonant absorption in a steady plasma and found negative absorption rate above a certain velocity threshold. Negative absorption rate in a driven wave-plasma interaction means that instead of wave energy dissipation the wave subtracts energy from the system (where the source of energy is the reservoir of plasma flow) resulting in amplified wave amplitudes. Later this phenomenon was explained in terms of negative energy waves (see, e.g., [9, 10, 11]. All these works considered discontinuous flow profiles which facilitates the study significantly but makes them less applicable to the magnetotail where the flow speed is known to vary continuously.

The aim of the present paper is to find and characterize the main conditions under which RFI could be an important wave generating mechanism in a tail-like geometry and to check if this mechanism is operative in a realistic magnetotail model where all equilibrium quantities undergo continuous variation across a mantle-like boundary layer. Effects introduced by these continuous variations on the KHI are also examined.

MODEL AND EIGENVALUE PROBLEM

The model shown in Fig. 1 represents the tail region where the plasma flow in the magnetosheath is approximately parallel to the open field lines in the magnetosphere. All equilibrium quantities change only in the direction of the x-axis and divide the plasma into two semi-infinite homogeneous regions separated by a nonuniform layer of thickness L. The equilibrium quantities in the homogeneous region $x < 0$ ($x > L$) are denoted by subscript 1 (2), respectively. The magnetic field is directed along the z-axis, i.e., $\mathbf{B_0} = (0, 0, B_0(x))$. For the sake of simplicity we assume that the plasma flow

CP679, *Solar Wind Ten: Proceedings of the Tenth International Solar Wind Conference*, edited by M. Velli, R. Bruno, and F. Malara

$\mathbf{u_0} = (0, 0, u_0(x))$, the Alfvén speed $c_A \equiv B_0/\sqrt{\mu_0 \rho_0}$ and the cusp speed $c_T \equiv c_s c_A/\sqrt{c_s^2 + c_A^2}$ are linear in the inhomogeneous layer and constant elsewhere:

$$u_0(x) = \begin{cases} V, & x < 0, \\ V\dfrac{h-x}{h}, & 0 < x < h, \\ 0, & x \geq h, \end{cases} \quad (1)$$

$$c_{A,T}(x) = \begin{cases} c_{A1,T1}, & x \leq 0, \\ c_{A1,T1} + (c_{A2,T2} - c_{A1,T1})\dfrac{x}{L}, & 0 < x < L, \\ c_{A2,T2}, & x \geq L. \end{cases} \quad (2)$$

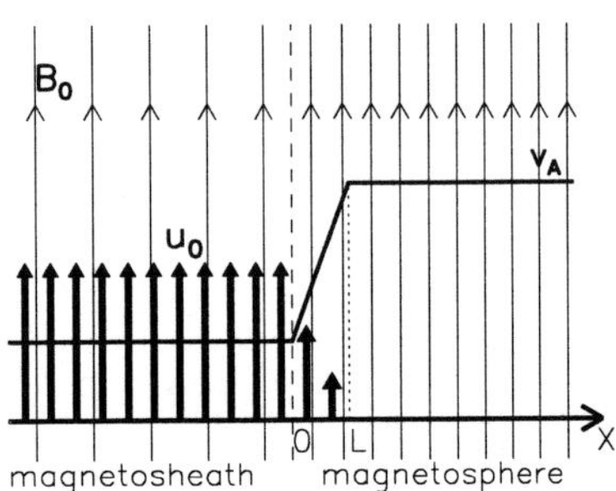

FIGURE 1. The model cartoon .

Since the equilibrium quantities depend on the x-coordinate only, the perturbed quantities can be Fourier analyzed with respect to y, z and t. We thus seek solutions of the form $f(x, k_y, k_z, \omega) \exp i(k_y y + k_z z - \omega t)$. The linearized ideal MHD equations are then reduced to a set of two coupled ordinary differential equations for the normal component of the Lagrangian displacement (ξ_x) and the Eulerian perturbation of total pressure (P). In the homogeneous regions 1 and 2 the solutions can be represented analytically. Inside the nonuniform boundary layer the set of equations has regular singularities at the positions $x = x_A$ and $x = x_T$ where

$$\omega = k_z[u_0(x) \pm c_A(x)] \quad \text{or} \quad \omega = k_z[u_0(x) \pm c_T(x)], \quad (3)$$

respectively. Connection formulae crossing the regular singularities were originally derived in [12, 13, 14, 15] for plasmas where the only dissipative process is resistivity.

The procedure for solving the eigenvalue problem for MHD surface modes is a shooting method from $x = 0$ to $x = L$. Starting with the analytical solutions for $x < 0$, we numerically integrate the ideal MHD equations. If a resonance is encountered during the calculations, then connection formulae are applied. After having passed through the resonant layer the integration of the ideal equations is resumed. The same procedure is repeated whenever any other resonance is encountered until the final point $x = L$ is reached. The application of the continuity conditions for ξ_x and P at $x = L$ yields the dispersion relation.

RESULTS

The magnetosheath is usually taken to be a high β plasma, whereas the plasma in the tail lobes has a very small β (of the order of 10^{-4}) and, therefore, it can be considered as cold. For representative numerical evaluations we set $\gamma = 5/3$, $\beta_1 = 2.6$ in the magnetosheath and (a) $\beta_2 = 0.6$ or (b) $\beta_2 = 0$ (i.e., cold plasma approximation) in the tail lobes. Case (a) gives a more general picture of existing wave modes and their interactions, whereas case (b) is more realistic representing the cold plasma in the tail lobes. The equation of magnetohydrostatic balance yields $B_{02}/B_{01} = 1.5$ or $B_{02}/B_{01} = 1.89$, respectively. The ratio of the Alfvén speeds on both sides is taken to be $v_{A2}/v_{A1} = 4.5$ and, according to Eq. (2), it is an increasing function in the boundary layer. One can show that in case (a) the cusp speed is an increasing function in the inhomogeneous layer, whereas in case (b) it is a decreasing function. A propagation angle of $\theta = \arctan k_y/k_z = 50°$ with respect to the magnetic field is chosen. Here $k_y = k \sin\theta$, $k_z = k \cos\theta$ and k is the wavenumber. The length, speed and magnetic field strength are normalized with respect to $1/k$, v_{A1} and B_{01}, respectively.

Narrow shear flow layer ($h/L = 0.1$)

Let us now consider an inhomogeneous boundary layer in the tail region across which all equilibrium quantities vary continuously from their characteristic values in the magnetosheath to their counterparts in the tail lobes. In general the inhomogeneity length-scales can be different for the various physical quantities (e.g., velocity, density, etc.). First we consider the case when the variation length-scale of the boundary layer plasma flow (h) is small compared to the variation length-scales of other equilibrium quantities (L), i.e., $h/L \ll 1$. Further we assume that $kL = 0.1$ which corresponds to the long wavelength approximation. When the magnetosheath flow, V, is high enough a given mode at a given frequency can be in resonance with backward and forward continua simultaneously. The real and imaginary

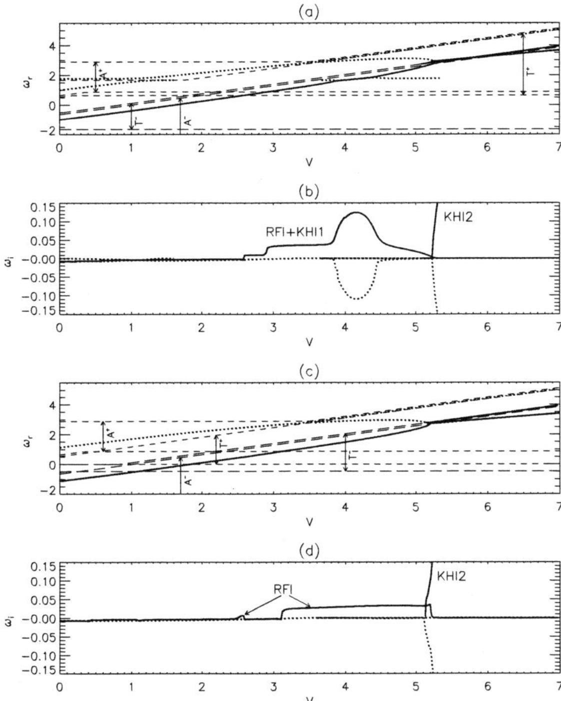

FIGURE 2. The real and imaginary parts of oscillation frequencies as function of the magnetosheath flow, V, for a narrow shear flow layer in the long wavelength approximation ($h/L = 0.1$, $kL = 0.1$), (a,b) for $\beta_2 = 0.6$ and (c,d) for $\beta_2 = 0$.

parts of the mode frequencies as function of the flow strength are plotted in Fig. 2. The first two plots (a, b) show the eigenfrequencies when there is a non-zero kinetic pressure in the magnetosphere ($\beta_2 = 0.6$) and the remaining two plots (c, d) correspond to $\beta_2 = 0$ (cold plasma in the magnetosphere). Dashed lines (long dashed lines) indicate the lower and upper boundaries of forward (backward) Doppler shifted cusp and Alfvén continua. Forward propagating modes are denoted by dotted lines and backward propagating modes are denoted by solid lines. The modes discussed in the previous subsection (primary and β modes) are still present with the only difference that the β modes on side 2 are absent in Figs. 2c, d, since $\beta_2 = 0$. However, now both the forward and backward propagating primary modes (p_f and p_b) are damped in the absence of flow due to the mechanism of resonant absorption (e.g., Fig. 2b). The damping rate changes due to the Doppler shift. At a certain value of magnetosheath flow ($V = 1.3$) the backward propagating primary mode p_b reverses its direction of propagation. It starts to propagate in the Sun-Earth direction in the rest frame of the magnetosphere remaining a backward propagating mode in the rest frame of the magnetosheath. The p_b mode becomes resonantly unstable as its frequency consecutively enters the forward Doppler shifted cusp and Alfvén continua (e.g., at $V = 2.5$ and $V = 2.9$, correspondingly, in Fig. 2b). Note that for the case shown in Figs. 2c, d the cusp RFI has very narrow velocity ranges. Unlike the RFI the KHI is present when the inhomogeneous boundary layer is reduced to a true tangential discontinuity (L=0). When $\beta_2 = 0$ the KHI1 disappears because of vanishing β modes on side 2. Note also that in Figs. 2b and d the magnetosheath flow thresholds for RFI overlap the KHI thresholds.

DISCUSSION

The existence of an extended geomagnetic tail has been known for many decades. It was suggested relatively early that the large size of this cavity might allow the existence of MHD eigenmodes with frequencies of the order of mHz and below, as an explanation for observations of such frequencies in magnetometer data. The KHI at the magnetopause is a possible external mechanism for the generation of waves. However, as shown by several authors, the flow speeds required for the KHI are too high compared with the flow speeds usually observed in the magnetosheath. The introduction of the mantle-like boundary layer, where equilibrium quantities undergo continuous variation, leads to the appearance of RFI. The assumption of a discontinuous flow profile adopted in previous studies [10, 11] of RFI is not valid for realistic magnetotail. According to [2] near the tail boundary beyond about 100 R_E GEOTAIL often encounters a plasma mantle-like boundary layer in which the plasma flowing tailward transitions smoothly from magnetosheath values of speed and density to much smaller values, more characteristic of the tail lobe. We have studied the possible resonant amplification of the surface waves at this boundary layer for different representative variation profiles of the equilibrium quantities. The surface waves can have several resonances with local cusp or Alfvén waves throughout the entire inhomogeneous boundary layer. When due to an increased Doppler shift the backward propagating primary mode p_b changes its direction of propagation, i.e., starts to propagate tailward in the rest frame of the magnetosphere, and its frequency enters the forward cusp and/or Alfvén continua it can become resonantly unstable. However, this is not necessarily true and numerical investigation is required, since at the same time the p_b mode is subject to backward resonances.

Let us finally investigate how the threshold of RFI and KHI depends on the thickness of the shear flow layer. The dependence of critical velocities of KH and RF instabilities on the parameter h/L is plotted in Fig. 3. When $\beta_2 = 0$ (in the limit of a cold plasma in the magneto-

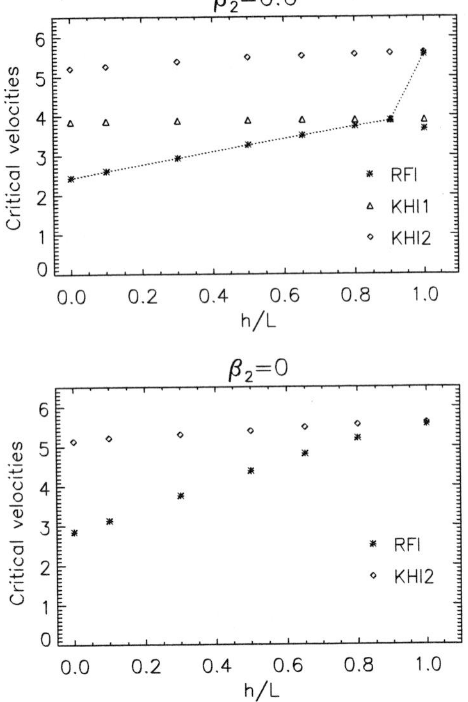

FIGURE 3. Critical velocities for the RFI and KHI as a function of h/L, where $L = 0.1$.

sphere) the KHI1 is absent since there are no β modes on side 2. The critical threshold velocities for both KHI1 and KHI2 are not affected very much by the introduction of the inhomogeneous boundary layer and the variation of the dimensionless parameter h/L, where $L = 0.1$. The critical velocity for the KHI2 increases slightly with increasing h. The plasma β in the tail lobes is usually of the order of 10^{-4}. In that case, as shown in Fig. 3b, KHI1 is absent and the critical velocity required for the KHI2 is approximately equal to $c_{A1} + c_{A2}$ which is too high for realistic flow seeds. The critical velocity for the RFI is lower, however, it becomes higher when the parameter h/L is increased. In our study we have taken $h/L \leq 1$. Based on this study, it is not difficult to conclude that the critical velocities for RFIs become less important and may become even higher than those for the KHIs when $h/L \geq 1$. It is also important to note the following: in the present study we have taken a flow parallel with the magnetic field. The antiparallel case could be included if negative velocities were considered. However, taking into account symmetry considerations we can easily show that in both cases the resonantly unstable waves will propagate tailward away from the Earth and that the same critical velocities will be present with positive or negative signs.

In conclusion the KHI threshold velocities are almost unaffected by the change in continuous variation profiles of equilibrium quantities within the boundary layer. If MHD waves are generated by the mechanism of RFI, then this should indicate a flow velocity variation length-scale being smaller than the variation length-scales of other equilibrium quantities, such as the density and the magnetic field strength. Conversely, if the variation length-scale of the flow velocity is small then the outer boundary of the magnetotail should be subject to RFIs.

ACKNOWLEDGMENTS

RE acknowledges M. Kéray for patient encouragement. RE also acknowledges NSF, Hungary (OTKA, Ref. No. TO32462) and The Nuffield Foundation (Ref. No. NAL/99-00) for financial support. YT is grateful to the White Rose Consortium for financial support.

REFERENCES

1. Hones, E.W., Jr., Asbridge, J.R., Bame, S.J., Montgomery, M.D., Singer, S. and Akasofu, S.-I., *J. Geophys. Res.77*, 5503, 1972.
2. Siscoe, G.L., Frank, L.A., Ackerson, K.L., Paterson, W.R., *Geophys. Res. Lett.21*, 2975, 1994.
3. Eastman T.E. et al., *J. Geophys. Res.103*, 23,503, 1998.
4. Walker, A.D.M., *Planet. Space Sci. 29*, 1119, 1981.
5. Miura, A., *J. Geophys. Res.97*, 10,655, 1992.
6. Mills, K.J., Longbottom, A.W., Wright, A.N. and Ruderman, M.S., *J. Geophys. Res.105*, 27,685, 2000.
7. Hollweg, J.V., Yang, G., Cadez, V.M., Gakovich B., *Astrophys. J.349*, 335, 1990.
8. Erdélyi, R. and Goossens, M., *Astron. Astrophys.313*, 664, 1996.
9. Tirry, W.J., Cadez, V.M., Erdélyi, R., Goossens, M., *Astron. Astrophys.332*, 786, 1998.
10. Ruderman, M.S. and Wright, A.N., *J. Geophys. Res.103*, 26,573, 1998.
11. Taroyan, Y. and Erdélyi, R., *Phys. Plasmas 9*, 3121, 2002.
12. Sakurai, T., Goossens, M., Hollweg, J.V., *Solar Phys.133*, 227, 1991.
13. Goossens, M. Hollweg, J.V. and Sakurai, T., *Solar Phys.138*, 233, 1992.
14. Erdélyi, R. Goossens, M. and Ruderman, M.S., *Solar Phys.161*, 123, 1995.
15. Erdélyi, R., *Solar Phys.171*, 49, 1997.
16. Pu, Z.-Y. and Kivelson M.G., *J. Geophys. Res.88*, 841, 1983.
17. Andries, J. and Goossens, M., *Astron. Astrophys.375*, 1100, 2001.

Parametric decay of non-linear circularly polarised Alfvén waves

Rim Turkmani* and Ulf Torkelsson*

*Chalmers University of Technology/Göteborg University, Department of Astronomy & Astrophysics, S-412 96 Gothenburg, Sweden

Abstract. We study the evolution of non-linear monochromatic circularly polarised Alfvén waves by solving numerically the time-dependent equations of magnetohydrodynamics in one dimension. We find that in a low β plasma the waves may undergo a parametric decay. This is because the wave excites a density enhancement that travels slower than the wave itself and thus interacts with the wave. When $\beta \sim 1$ the density enhancement does not interact with the wave and no decay takes place.

INTRODUCTION

Large amplitude, low frequency, Alfvén waves have been observed in the solar corona for over 30 years (e.g. Belcher & Davis [1]) and are thought to play a role in heating the corona and accelerating the solar wind. During the last decade Ulysses has provided plasma and magnetic field measurements that have allowed extensive investigations on the behaviour of Alfvénic turbulence in the high-latitude solar wind. The data shows a strong correlation between the fluctuations in velocity and magnetic fields [3], which reveals the presence of both inward and outward propagating Alfvén waves (e.g Bavassano et al. [2]).

An infinitely long circularly polarised Alfvén wave is an exact solution to the equations of magnetohydrodynamics (MHD). Therefore circularly polarised Alfvén waves are more likely to reach high altitude, whereas the linearly polarised Alfvén wave can form current sheets at the nodes of the fluctuating magnetic field and dissipate at a lower altitude (Boynton & Torkelsson [6]).

Different mechanisms, in which the wave may lose its linearity have been studied, like phase mixing due to a transverse gradient in the phase velocity (e.g. Heyvaerts & Priest [4]) or the nonlinear coupling of the Alfvén wave with other modes (e.g. Wentzel [5]). In the high latitude solar wind with a smooth density profile the parametric decay of circularly polarised Alfvén waves has been proposed to play a role in generating turbulence and inward propagating Alfvén waves beyond the critical point where the Alfvén speed becomes lower than the solar wind speed.

A forward propagating Alfvén wave can generate a forward propagating acoustic wave and a backward propagating Alfvén wave through a parametric instability in the presence of a density fluctuation. The parametric decay of circularly polarised Alfvén waves has been studied both analytically (e.g. [7] and [8]) and numerically by several groups (e.g. [9], [10], [11] and [12]). Most of these studies were restricted to periodic boundary conditions (except [9] and [11]), and external noise in the form of density fluctuations were added to excite the decay.

In this paper we study the evolution of non-linear circularly polarised Alfvén waves by solving numerically the time-dependent MHD-equations in one dimension assuming plane parallel geometry. We examine the behaviour of the waves for different values of beta without adding any external perturbations, which makes our model an almost ideal case. We therefore expect our model to underestimate the importance of the parametric decay compared to a real medium with density inhomogeneities. In all our models we extend the box so that the right boundary does not affect the solution. The plan of the paper is the following; in Sec. II we describe the basic MHD equations that we are solving numerically and some of the properties of the parametric decay. In Sec. III we describe the parameters of our models and discuss the results, and the concluding remarks are in Sec. IV.

MATHEMATICAL FORMULATION

Fundamental properties of Alfvén waves

The equations of ideal isothermal MHD can be written as

CP679, *Solar Wind Ten: Proceedings of the Tenth International Solar Wind Conference*, edited by M. Velli, R. Bruno, and F. Malara

$$\frac{\partial \rho}{\partial t} + \nabla \cdot (\rho \mathbf{v}) = 0, \tag{1}$$

$$\frac{\partial (\rho \mathbf{v})}{\partial t} + \nabla \cdot (\mathbf{v} \rho \mathbf{v}) = -\nabla p + \mathbf{J} \times \mathbf{B}, \tag{2}$$

$$\frac{\partial \mathbf{B}}{\partial t} = \nabla \times (\mathbf{v} \times \mathbf{B}), \tag{3}$$

$$\nabla \cdot \mathbf{B} = 0, \tag{4}$$

where ρ is the density, $\mathbf{v}$ the velocity, p the pressure, $\mathbf{B}$ the magnetic field, and $\mathbf{J} = \nabla \times \mathbf{B}/\mu_0$ the current density. The constraint $\nabla \cdot \mathbf{B} = 0$ is fulfilled by Eq. (3) if it is imposed as an initial condition.

In a homogeneous medium with a density ρ_0 and a background magnetic field $\mathbf{B_0} = B_0\hat{\mathbf{z}}$ a circularly polarised forward propagating Alfvén wave is described by the transverse magnetic field

$$\mathbf{B}_{\perp} = B_{\perp}[\cos(kz - \omega t)\hat{\mathbf{x}} + \sin(kz - \omega t)\hat{\mathbf{y}}], \tag{5}$$

and the velocity

$$\mathbf{v}_{\perp} = -\frac{\mathbf{B}_{\perp}}{\sqrt{\mu_0 \rho_0}}. \tag{6}$$

The wave obeys the dispersion relation

$$\omega = v_{\mathrm{A}} k \tag{7}$$

with the Alfvén velocity

$$v_A = \frac{B_0}{\sqrt{\mu_0 \rho_0}}. \tag{8}$$

For this wave the magnetic pressure ${B_{\perp}}^2/(2\mu_0)$ is the same everywhere inside the wave. This is the physical reason why an infinitely extended circularly polarised Alfvén wave is incompressible and an exact solution of the nonlinear MHD equations. In the one-dimensional problem that we study, there is only one additional wave mode, an acoustic wave obeying the dispersion relation

$$\omega = c_{\mathrm{S}} k, \tag{9}$$

where the isothermal speed of sound is

$$c_{\mathrm{S}} = \sqrt{\frac{p_0}{\rho_0}}. \tag{10}$$

The amplitude of the density oscillation $\Delta\rho$ is related to that of the longitudinal velocity, v_z, through

$$v_z = \frac{\Delta \rho}{\rho} c_{\mathrm{S}}. \tag{11}$$

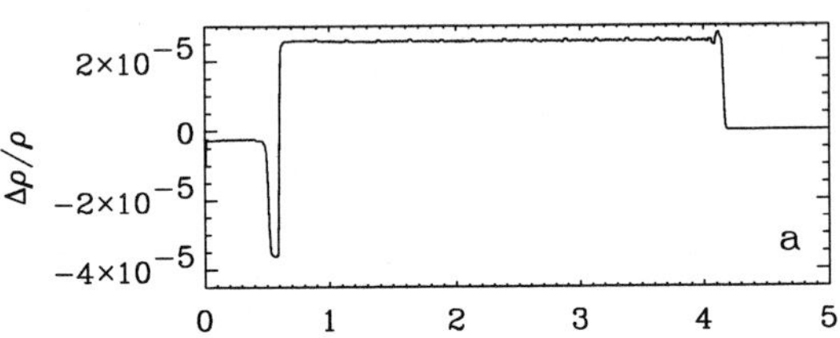

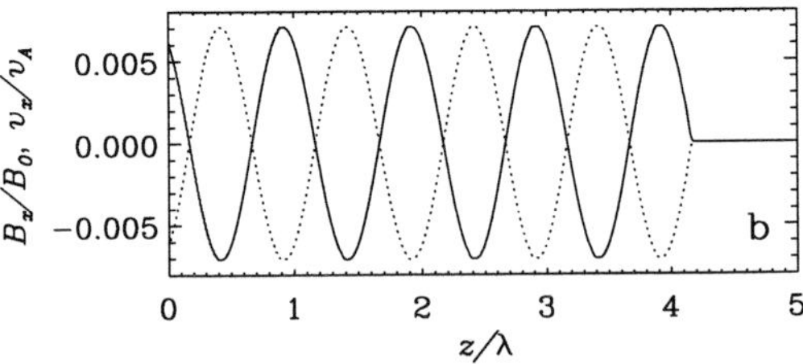

FIGURE 1. MHD waves propagating through a homogeneous medium of low β (Run 1a). **a** $\Delta\rho/\rho = \frac{\rho - \rho_0}{\rho_0}$ versus z at $t = 4.2P$ s. **b** B_x/B_0 (solid line) and v_x/v_{A} (dashed line) versus z/λ at the same time.

TABLE 1. Simulations of Alfvén waves in a homogeneous medium. The wave is characterised by the two quantities η, the amplitude of the imposed Alfvén wave in terms of the vertical magnetic field, and $\beta = 2\mu_0 p/B_0^2$ the plasma beta. For the different Runs we further specify the length of the time step Δt in terms of the period of the wave, P, the length of the computational domain, L in terms of the wave length, λ, and the number of grid points, N.

Run	$\Delta t/P$	N	L/λ	η	β
1a	0.003 3	3 600	78	0.007	0.042
1b	0.003 3	3 600	78	0.07	0.042
1c	0.003 3	3 600	78	0.7	0.042
2a	0.003 3	13 500	193	0.007	0.96
2b	0.003 3	13 500	193	0.07	0.96
2c	0.003 3	13 500	193	0.7	0.96

The modulational instability

Galeev & Oraevskii [13] showed that an Alfvén wave with a frequency $\omega_0 = v_A k_0$ and a wave number k_0 can decay into a backward propagating Alfvén wave with a frequency ω_- and a wave number k_- and a forward propagating acoustic wave with a frequency ω and a wave number k that fulfill the resonance conditions

$$\omega_0 = \omega + \mid \omega_- \mid \tag{12}$$

and

$$k_0 = k - k_- \tag{13}$$

In the limit of low β and high wave amplitude, the growth rate of the reflected wave is (Galeev & Oraevskii

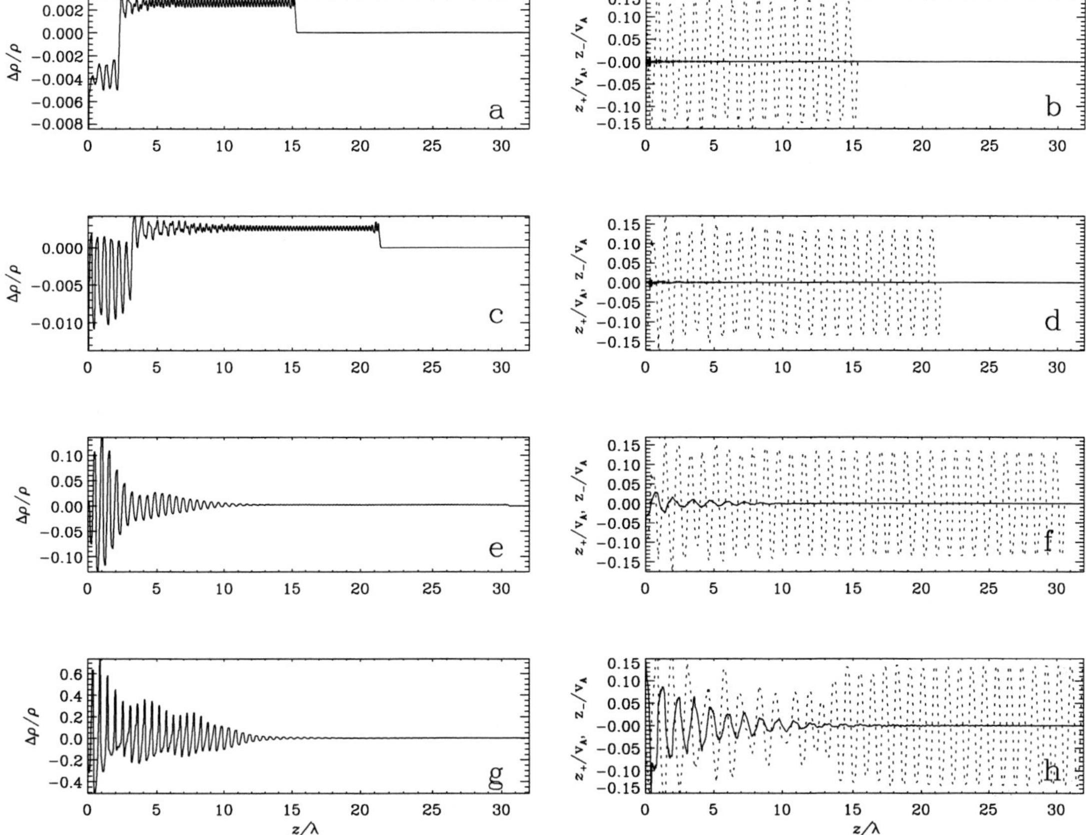

FIGURE 2. MHD waves propagating through a homogeneous medium of low β (Run 1b). **a, c, e, g** $\Delta\rho/\rho = \frac{\rho-\rho_0}{\rho_0}$ as a function of distance z at $t/P = 16.7, 23.3, 33.3, 46.7$. Note the changes in the $\Delta\rho/\rho$-scale between the frames. **b, d, f, h** z_-/v_A (solid line) and z_+/v_A (dashed line) as a function of distance z at the same times.

1963)

$$\gamma \simeq \frac{\omega_0}{2^{3/4}} \frac{\eta^{1/2}}{\beta^{1/4}}, \tag{14}$$

where $\beta = 2\mu_0 p_0/B_0^2$, and $\eta = B_\perp/B_0$.

Since we want to be able to separate the forward and backward propagating Alfvén waves in our simulations, it is useful to introduce the Elsässer variables

$$z_\pm = \left| \mathbf{v}_\perp \mp \frac{\mathbf{B}_\perp}{\sqrt{\mu_0 \rho}} \right|, \tag{15}$$

which describe the forward and backward propagating Alfvén waves, respectively.

RESULTS

We modify the numerical code of Boynton & Torkelsson [6] to simulate circularly polarised Alfvén waves. The waves are driven on the left boundary. In all the runs the grid is sufficiently extended that the wave does not hit the right boundary. The different models are described in Tab. 1.

The propagation of a low amplitude wave in a low β medium (model 1a) is shown in Fig. 1b. The Alfvén wave excites a density enhancement (Fig. 1a), whose right edge coincides with the front of the Alfvén wave, while the left edge propagates with the lower speed c_s. The density enhancement is a second order effect, and its amplitude is consequently proportional to the square of the amplitude of the Alfvén wave. The Alfvén wave is propagating through the medium at a speed equal to v_A without decaying.

The density discontinuity at the front of the Alfvén wave excites a secondary compressional wave, which can be seen as a weak modulation of the density in Fig. 1a, and also of the magnetic pressure $|\mathbf{B}_\perp|^2/(2\mu_0)$. This density fluctuation serves as the necessary seed for the parametric instability, but due to the low amplitude of the

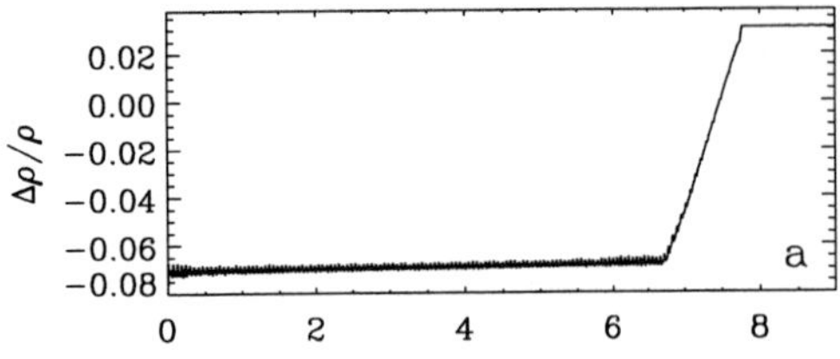

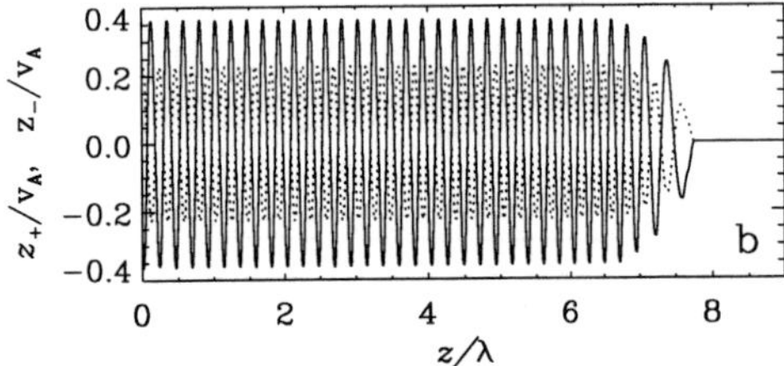

FIGURE 3. MHD waves propagating through a homogeneous medium of $\beta \sim 1$ (run 2b) . **a** $\Delta\rho/\rho = \frac{\rho-\rho_0}{\rho_0}$ as a function of distance z at $t = 33.3P$. **b** z_+/v_A (solid line) and z_-/v_A (dashed line) as a function of distance z at the same times.

fluctuation the instability grows slowly. The evolution of the backward-propagating Alfvén wave that is generated by the instability can be followed in Figures 2b, d, f and h. The backward-propagating wave grows in amplitude away from the wave front, which enhances the growth rate of the parametric instability upstream. Eventually the instability is so strong that it becomes an efficient source of a forward-propagating sound wave (Figs 2a, c, e and g). In agreement with the theoretical prediction the sound wave has a wave number $k = 1.8k_0$. Figure 2h shows that there is a phase shift of $\frac{\pi}{2}$ between the backward- and the forward-propagating Alfvén waves as we expect if the backward-propagating wave is generated through the parametric instability.

The sound waves stimulate the parametric decay of the forward-propagating Alfvén wave, but since the Alfvén speed is larger than the sound speed, the first section of the Alfvén wave will remain unaffected. On the other hand we see a region with $z < 13\lambda$ in Fig. 2g and h, in which the waves are strongly interacting. In this region there is a strong damping of the forward-propagating Alfvén wave, and the acoustic wave is amplified until it becomes so nonlinear that it steepens into shocks.

When $\beta \sim 1$ the density enhancement remains ahead of the Alfvén wave, and the parametric decay does not take place regardless of the amplitude of the wave. Instead the Alfvén wave is reflected off the density gradient at the back of the density enhancement (Fig. 3). This process will be studied in more details in a future paper.

CONCLUSION

Circularly polarised Alfvén waves are subject to a parametric instability that generates a backward propagating Alfven wave and a forward propagating sound wave. We find that the compression of the background medium that takes place at the wave front of an Alfvén wave in a low β plasma is sufficient to trigger this instability. However when $\beta \sim 1$ the Alfvén wave generates an acoustic precursor that remains ahead of the Alfvén wave, which therefore is not subject to the parametric decay.

The aim of our model is to demonstrate some of the basic properties of a propagating circularly polarised Alfvén waves. These results can then guide us in future investigations of the dynamics of Alfvén waves in more realistic configurations.

REFERENCES

1. Belcher, J. W. & Davis, L., JGR, 79, 4174 (1971).
2. Bavassano, B., Pietropaolo, E., & Bruno, R., JGR, 105, 15959 (2000).
3. Smith, E. J., Balogh, A., Neugebauer, M., McComas, D., GRL, 22, 3381 (1995).
4. Heyvaerts, J., & Priest, E. R., A&A, 117, 220 (1983).
5. Wentzel, D. G., Solar Phys., 39, 129 (1974).
6. Boynton, G. C., Torkelsson, U., A&A, 308, 299 (1996).
7. Cohen, R. H., Kulsrud, R. M., Phys. Fluids, 17, 2215 (1974).
8. Jayanti, V., Hollweg, J. V., JGR, 98, 19049 (1993).
9. Del Zanna, L., Velli, M., Londrillo, P., A&A, 367,705 (2001).
10. Malara, F., Primavera, L., Veltri, P., Phys. Plasma, 7, 2866 (2000).
11. Pruneti, F., Velli, M., In Proceedings of the Fifth SOHO Workshop: The Corona and Solar Wind Near Minimum Activity, ESA SP-404, 623 (1997).
12. Ghosh, S., Goldstein, M. L., JGR, 99, 13351 (1994).
13. Galeev, A. A., & Oraevskii, V. N., Sov. Phys. Dokl., 7, 988 (1963).

Observations of Anisotropic Compressive Turbulence Near The Sun

W. A. Coles[1], A. P. Rao[2] and S. Ananthakrishnan[2]

[1]*Electrical and Computer Engineering, University of California at San Diego*
[2]*National Center for Radio Astrophysics, Tata Institute of Fundamental Research, Pune, India*

Abstract. Small-scale density-fluctuations cause angular-broadening of distant radio sources observed through the solar wind. Observations of angular broadening near the Sun show that the compressive microstructure is highly field aligned. Recent work strongly suggests that the density fluctuations are caused by obliquely propagating MHD waves. With this in mind we have made a new series of angular scattering observations using the Giant Meter Wavelength Radio Telescope (GMRT) located near Pune. We selected radio sources passing over the solar poles at distances near 10 Rs. By comparing these observations with simultaneous white light scattering from the LASCO C3 instrument, we hoped to learn how the waves are related to the large scale coronal structures, i.e. coronal holes, polar plumes, and streamers.

INTRODUCTION

Observations of angular broadening have three distinctive characteristics, which suggest the presence of obliquely propagating MHD waves. The scales of the density fluctuations range upwards several decades from the proton inertial scale. The spatial spectrum of the density fluctuations in this range is flatter than "Kolmogorov", and most of the energy is transverse to the local magnetic field[1]. Observations also indicate that the velocity of the fluctuations is faster than the flow speed by approximately the local Alfven speed. Such waves may carry enough energy to contribute significantly to the acceleration of the solar wind, so their distribution with respect to coronal structures is of interest.

PREVIOUS OBSERVATIONS

1. The Spatial Spectrum of N_e

Spacecraft observations provide a measure of the low frequency (large scale) structure. The spectrum of Ulysses observations over the north pole shows a clear break at an "outer scale" which is related to the transit time. This spectrum agrees well with estimates made from Helios observations by Tu and Marsch[2]. The space-craft spectra show a Kolmogorov power-law form at frequencies above the inverse of the transit time.

Radio observations of phase scintillation made using dual-frequency transponders on spacecraft provide a measure of the mid-scale structure overlapping with the direct observations[3,4]. These observations match the space-craft spectra at low frequencies and have approximately Kolmogorov behavior. However they tend to flatten somewhat at higher frequencies. Spectral broadening measurements can be made using the same space craft transponder[5]. These extend the spectral estimates to smaller scales, past the ion-inertial scale in fact. The spectra become considerably flatter than Kolmogorov as the ion-inertial scale is approached.

2. Anisotropy of spatial spectrum

Two-dimensional measurements of the electric field correlation can be made at scales of: (1) 1000-10,000 km using the VLBA; (2) at scales of 1-30 km by the VLA or GMRT. The VLBA measurements are nearly isotropic, whereas the VLA/GMRT observations are very anisotropic[7].

The observed axial ratio of the microstructure decreases with increasing solar distance. However it is

CP679, *Solar Wind Ten: Proceedings of the Tenth International Solar Wind Conference*,
edited by M. Velli, R. Bruno, and F. Malara

significantly reduced by the line of sight integration, so the intrinsic axial ratio AR must be recovered by model fitting. An empirical model has the form[8]

$$(AR-1) \approx 160/R^{1.5} \quad \text{where R is in } R_s .$$

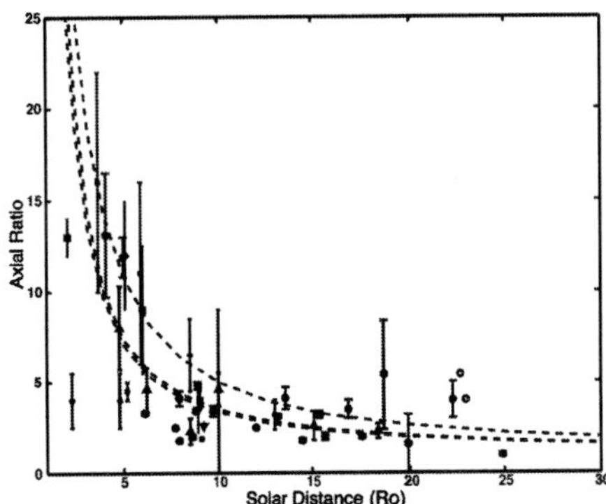

FIGURE 1. The observed axial ratio vs solar distance. The upper dashed line is a model of the intrinsic axial ratio of the medium. The lower dashed line shows the effect of the line of sight averaging.

3. The spectral cutoff

The spectra show a distinct cutoff at high frequencies, which we interpret as a dissipation or "inner" scale. This is visible in VLA/GMRT angular broadening, spectral broadening, and intensity scintillation measurements, which are sensitive to the smallest scales. It changes with radial distance roughly as S_i (km) $\approx$ R (R_s). This is very close to the behavior of the ion inertial scale[5].

4. The velocity distribution

The apparent velocity distribution of the density fluctuations is characterized by a high mean and a very wide spread in radial velocity[8,9]. Other measurements and theoretical values tend to cluster around the lower envelope of the radio scattering observations.

The radio observations can be explained under the assumption that the scattering is caused by waves. which are traveling out-wards with respect to the flow. Very good model fits can be obtained assuming that the waves are obliquely propagating Alfven waves or if they include a 50% admixture of the slow magneto-acoustic mode. A wide range of apparent speeds is observed because the line of sight includes a very wide range of wave speeds.

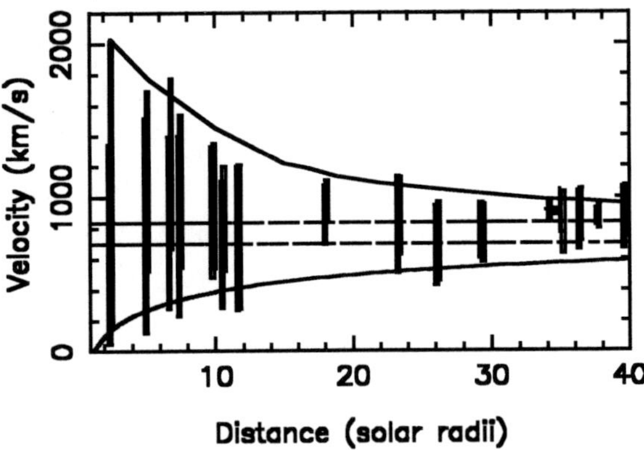

FIGURE 2. The observed distribution of parallel velocity in the polar wind. A model is fit to the data in which it is assumed that the velocity transverse to the line of sight is distributed over a range. The vertical bars show this range.

THE NEW OBSERVATIONS

The objective of these observations was to learn how the characteristics of these MHD waves which are presumed to cause the radio scattering are related to plasma conditions and large scale structures in the corona. This requires correlating some measure of the MHD waves with another measure of the corona. The most obvious measures of coronal structure are the LASCO white light coronagraphs on SOHO. They have the sensitivity to detect fine structure and they have a field of view that includes the regions accessible to radio scattering.

It is difficult to make continuous radio measurements of the velocity because the measurements require long radial-aligned baselines, which only persist for a short time because of the Earth's rotation. However angular scattering measured with arrays such as the VLA or GMRT can be made anytime a suitable source is present. As a result angular scattering measurements can be compared with other solar measurements more readily. The level of turbulence and the axial ratio of the angular scattering, which indicates the "obliqueness" of the wave spectrum, are particularly robust.

The spectral exponent and the inner scale can also be determined but they are more dependent on accurate calibration.

The GMRT operates at 150, 233, 327, 610, and 1420 MHz. It has a range of baselines up to 35 km. The system is under development and has just been opened to outside use. At this time the two higher frequency bands are better developed. We found that good observations could be made around 10 Rs at 600 MHz and most of our observations are at that frequency.

The GMRT gain calibration is not as precise as that of the VLA, because the instrument is not as mature. At present the GMRT data requires an additional "self-calibration" step. This has little effect on the determination of the anisotropy, but it can slightly bias the shape of the spatial spectrum. Thus the spectral exponent and inner scale estimates are not reliable but the axial ratio and its orientation are well determined.

RESULTS

The location of the scattered source was registered on the LASCO C3 images. A cut through the C3 image is taken for comparison with the axial ratio and turbulent intensity. Since we are searching for correlation in the small-scale structure we used the standard C3 daily average from which a long-term lower-bound had been subtracted. This procedure emphasizes deviations from spherical symmetry which is what we are searching for.

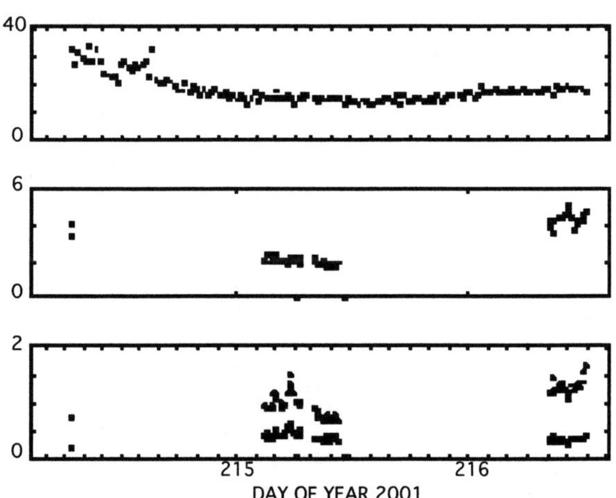

FIGURE 3. Observations over the north solar pole on Aug. 2, 3, and 4 of 2001 at a distance of 10 Rs. The top panel is the white light brightness, the middle panel is the axial ratio, and the lower panel is the level of turbulence in the parallel and perpendicular directions (parameterized by the phase structure function at 10 km).

In the first three observations the orientation of the microstructure was radial within the statistical errors. In the fourth case the orientation was significantly non-radial. In all cases significant variations were seen in the radio scattering and also in the C3 brightness. However these variations did not appear to be correlated in any simple way.

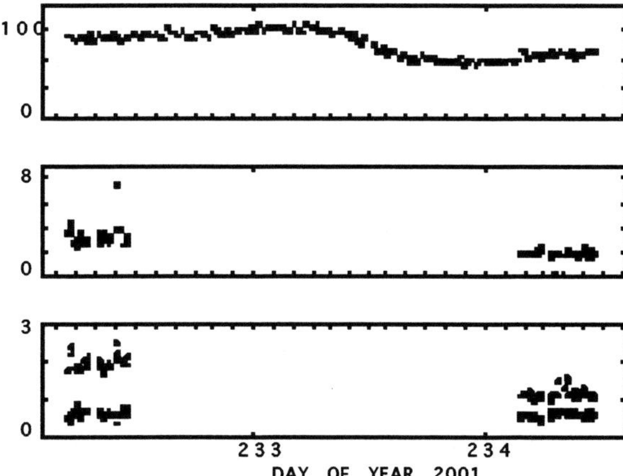

FIGURE 4. Observations over the north solar pole on Aug. 20, 21, and 22 of 2001 at a distance of 10 Rs.

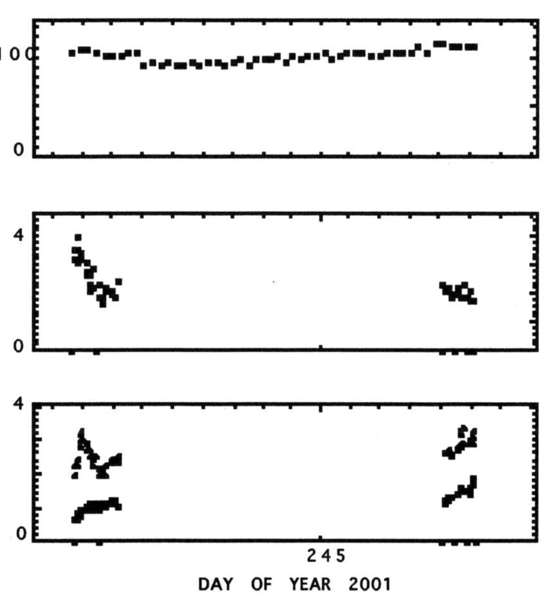

FIGURE 5. Observations under the south solar pole on Sept. 1 and 2 of 2001 at a distance of 10 Rs. The tics on the time axis mark intervals of 2 hrs.

CONCLUSIONS

We undertook these observations with the assumption that anisotropic scattering is caused by obliquely propagating MHD waves. We searched for variations in the behavior of these MHD waves in different coronal structures using the white-light brightness observed by the LASCO C3 instrument as a measure of the large scale coronal structure. During the observations the scattering region passed through different coronal structures and there were significant variations in the axial ratio and in the level of the scattering. However these variations were not correlated in an obvious way with the coronal structure. This was a surprise to us and we do not have an explanation.

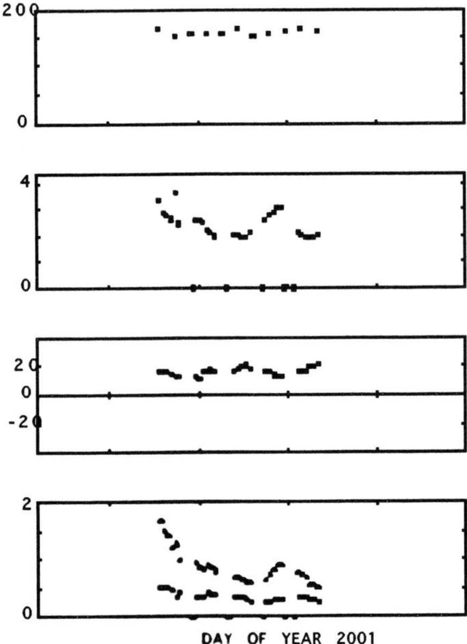

FIGURE 6. Observations off the west limb on Sept. 23, 2001 at a distance of 10 Rs. The third panel is the orientation of the structure in degrees. The tics mark 2 hr, or 2 10^5 km, intervals.

It is clear that anisotropic scattering is observed in all phases of the quasi-static solar wind: the fast polar holes; the quiet slow wind; and in the coronal streamers. We did not observe any transient events, but other observations[10] suggest that anisotropic scattering is observed in CME's aligned with the local magnetic field. The angular scattering can vary significantly in regions which appear to be homogeneous in white-light. We also observed a case where the orientation of the scattering (which we assume is the direction of the magnetic field) remained significantly non-radial over an extended region as shown in Figure 6.

We have to conclude: (1) MHD waves are present in all phases of the solar wind and (2) the waves can vary significantly within large scale structures. It is possible to have large variations in scattering with no obvious change in white-light brightness because, although both observations are line-of-sight integrals, the weighting factors are quite different. The radio scattering is weighted by the square of the electron density, whereas the white-light brightness is weighted linearly by the electron density. Thus the radio scattering is particularly sensitive to thin dense filaments whereas the white-light image is sensitive to smaller variations in large-scale structures.

The observations should be repeated closer to the Sun where the contrast, both in large-scale structures and in radio scattering, is known to be larger.

ACKNOWLEDGMENTS

We are indebted to the GMRT telescope operators who were very helpful in completing the observations. One of us (WAC) was partially supported by a Sarojini Damodaran International Fellowship during this work.

REFERENCES

1. Harmon, J. K., *J. Geophys. Res.* **94**, 15399-15405, (1989).
2. Tu, C.-Y., and Marsch, E., *Space Sc. Rev.* **73**, 1-210, (1995).
3. Woo, R., and Armstrong, J. W., *J. Geophys. Res.* **84**, 7288-7296, (2000).
4. Paetzold, M., Karl, J., and Bird, M. K., *Astron. Astrophys.* **316**, 449-456, (1996).
5 Coles, W. A., Liu, W., Harmon, J. K., and Martin, C. L., *J. Geophys. Res.* **96**, 1745-1755, (1991).
6. Armstrong, J. W., Coles, W. A., Kojima, M., and Rickett, B. J. , *Astrophys. J.*, **358**, 685, (1990).
7. Grall, R. R., W. A. Coles, S. R. Spangler, T. Sakurai, and J. K. Harmon, *J. Geophys. Res.* **102**, 263-273, (1997).
8. Massey, W., *Measuring intensity scintillations at the very long baseline array (VLBA) to probe the solar wind near the Sun*, M.S. Thesis, University of California, San Diego, (1998).
9. Klinglesmith, M., *The polar wind from 2.5 to 40 solar radii: results of intensity scintillation measurements*, Ph.D. Thesis, University of California, San Diego, (1997).
10. Lynch. B. J., Coles, W. A., and Sheeley, N.R., *GRL*, in press, October (2002).

Solar wind kinetic exospheric models with typical coronal holes exobase conditions

H. Lamy*, M. Maksimovic†, V. Pierrard* and J. Lemaire*

*BIRA-IASB, Avenue Circulaire 3, 1180 Bruxelles, Belgium
†LESIA, Observatoire de Paris, CNRS FRE2461, 92195 Meudon, France

Abstract. In coronal holes where densities are lower than in equatorial streamers, the exobase, which is the radial distance where collisions between solar wind particles become negligible, is located closer to the basis of the corona where the downward gravitationnal force exceeds the upward electrostatic force acting on the protons. Therefore, the total potential of the protons is a non-monotonic function of the radial distance with a maximum value at a distance r_{max}. A new kinetic model with an exobase located at low radial distances is presented and compared to previous kinetic models where the exobase was located at larger radial distance. With such a low exobase, the electric field that ensures the quasi-neutrality of the plasma is strongly enhanced which has for consequence to accelerate the solar wind to values comparable to the velocities observed in the fast solar wind. We show that the lower the radial distance of the exobase, the larger the bulk velocity of the solar wind at large distance.

INTRODUCTION

The mechanisms accelerating the solar wind to supersonic speeds have been intensively studied either by hydrodynamic/fluid models assuming a collision-dominated plasma, or by kinetic theories in a collisionless plasma. Early in situ observations by space probes revealed that the bulk velocities are drastically different in the slow solar wind ($\sim 300 - 400\,\mathrm{km\,s^{-1}}$) and in the fast solar wind ($\sim 700 - 800\,\mathrm{km\,s^{-1}}$). The fast solar wind originates from the high latitude coronal holes where at least the electron temperature is lower than in the equatorial streamers. Since in current models the velocities of the particles are related to their temperatures in the corona, these observations represent a severe constraint that yet no solar wind models, either hydrodynamic or kinetic, have been able to explain without postulating too large coronal temperatures or an additional ad-hoc deposition of heat/momentum. In this paper, we review the basic concepts of the kinetic models and show their evolution since the early models of [6]. We will emphasize on our most recent kinetic model taking into account a non-monotonic potential for the protons and accelerating the solar wind to velocities similar to those observed in the fast solar wind [5].

THE BASICS OF KINETIC EXOSPHERIC MODELS

Exospheric kinetic models assume that above a sharp level called the exobase, the particles are collisionless and move freely under the influence of the Sun's gravitational field and of the interplanetary electrostatic and magnetic fields. Their trajectories solely depend on the conservation of their energy and of their magnetic moment, assuming that the guiding center approximation is valid. Depending on their energies and their pitch angles, four classes of particles can populate the region above the exobase : the escaping particles,which have sufficiently large kinetic energy to escape from the Sun's gravitationnal potentiel well, the ballistic particles that do not have enough energy to escape and fall back into the corona, the trapped particles which are continuously bouncing up and down the magnetic field lines between a magnetic mirror point and a gravitational mirror point, and the incoming particles arriving from the interplanetary regions. In kinetic models these latter particles are usually assumed to be negligible [6]. For simplicity, the interplanetary magnetic field is assumed to be radial since a spiral structure does not significantly change the bulk speed of the solar wind in these kinetic models [10]

Since electrons are more lighter than protons, their gravitational binding is smaller and so they tend to slightly separate from the protons. A polarisation electric field parallel to the magnetic field lines appears and adjusts itself in order to maintain the electrical neutrality

CP679, *Solar Wind Ten: Proceedings of the Tenth International Solar Wind Conference*, edited by M. Velli, R. Bruno, and F. Malara

of the plasma. This is the Pannekoek- Rosseland electric field which appears in a plasma of electrons and protons in hydrostatic equilibrium. However, the solar wind is an expanding medium with no particles coming from infinity. The fluxes of electrons and protons escaping from the corona are proportional to their thermal velocities which vary as the inverse square root of their masses. Therefore, to prevent the accumulation of negative charges at large distance from the Sun and of positive charges at the basis of the corona, a larger electrostatic potential difference between the exobase and infinity must establish in order to decelerate the electrons and accelerate the protons outwards. This larger electrostatic field accelerates the solar wind particles to supersonic velocities.

EXOSPHERIC MODELS WITH MONOTONIC ENERGY POTENTIAL FOR THE PROTONS

The exobase level, r_0, is defined as the radial distance where the mean free path of the particles is equal to the density scale height and is usually located at a distance of $\sim 6-10$ solar radii (R_s) according to typical coronal parameters in the equatorial regions [6]. In this case, the total potential for the electrons, essentially due to the electrostatic potential, is attractive from the exobase to infinity and increases monotonically. Only the electrons with enough kinetic energy can overcome the potential barrier and escape. The situation is different for the protons since their total potential energy, given by the sum of their gravitational and electrostatic potentials, is repulsive and monotonically decreasing. All protons are moving outwards and can escape, in contrast with the electrons for which only high speed particles contribute to the escape flux. The first exospheric models with an exobase located at these radial distances predicted values for the solar wind bulk velocities around $300\,\mathrm{km\,s^{-1}}$ at Earth's orbit in satisfactory agreement with the average observations of the slow solar wind but were unable to reproduce the velocities observed in the fast solar wind.

Later on, these models were modified in order to use electrons velocity distribution functions (VDF) with an excess of high energy particles compared to a Maxwellian distribution [9], as is actually observed in the fast solar wind at large distance [7]. These suprathermal electrons increase the number of particles with a kinetic energy sufficient to overcome the potential well and escape to infinity. The electric field that warrants the electrical neutrality of the plasma increases in order to accelerate the solar wind protons to larger values. Therefore, the solar wind bulk speed increases as well.

These suprathermal electrons are well fitted by Lorentzian (or Kappa) distributions characterized by the value of a κ index

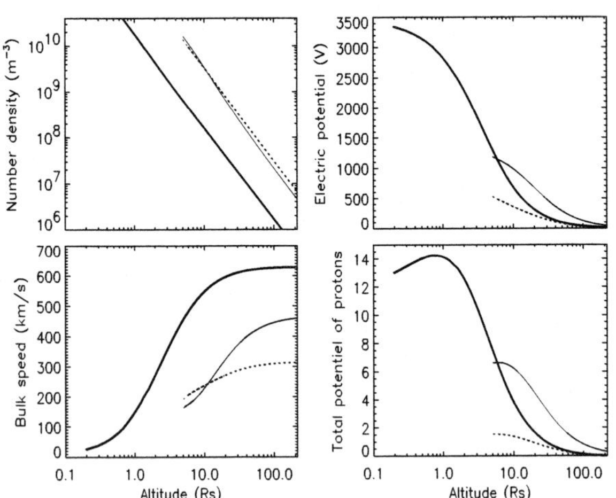

FIGURE 1. Comparison of the various exospheric kinetic models of the solar wind. For each model, we present the number density, the electrostatic potential and the bulk speed of the solar wind as well as the total (gravitational+electrostatic) potential of the protons. The dotted line corresponds to a model with a monotonically decreasing potential for the protons (r_0 located at 6R$_s$) and a Maxwellian VDF for the electrons. The solid line represents a model with a monotically decreasing potential for the protons (r_0 located at 6 R$_s$) and a Kappa distribution for the electrons ($\kappa = 3$). The thick line is our generalized model with r_0 located at 1.1 R$_s$ and $\kappa = 3$. The temperature at the exobase is $1,5\,10^6$ K and is assumed identical for electrons and protons for simplicity.

$$f\kappa = \frac{n_0}{(\pi\kappa v_{th}^2)^{3/2}}\frac{\Gamma(\kappa+1)}{\Gamma(\kappa-1/2)}\left(1+\frac{v^2}{\kappa v_{th}^2}\right)^{-(\kappa+1)} \quad (1)$$

where $v_{th} = \left(\frac{2\kappa-3}{\kappa}\frac{k_B T_0}{m_e}\right)^{1/2}$ is the thermal speed of electrons, n_0 and T_0 are respectively the electronic density and temperature at the exobase. These VDF decrease as a power law of the electrons velocity instead of exponentially for a Maxwellian. As $\kappa \to \infty$, the kappa function approaches a Maxwellian distribution.

Although the kinetic models based on these electrons VDF were able to accelerate the solar wind to larger bulk velocities [8] (see Figure 1), they were still unable to reach the velocities observed in the fast solar wind without assuming exobase temperatures of the order of 2×10^6 K, in disagreement with current observations of the coronal holes (electron temperatures $\sim 10^6$ K) [7].

CORONAL HOLES EXOBASE CONDITIONS AND NON-MONOTONIC ENERGY POTENTIAL FOR THE PROTONS

In coronal holes, the densities are lower than in equatorial streamers. As a consequence, the mean free path of the particles is larger and the altitude of the exobase is located deeper in the corona, at distances of only $1.1-3\,\mathrm{R_s}$. At these low radial distances, the gravitational force acting on the protons becomes larger than the outward repulsive electrostatic force. Above the exobase, the gravitational force decreases as $1/\mathrm{r}^2$ and the repulsive electric force decreases as $\sim 1/\mathrm{r}$, so that there is a radial distance, $\mathrm{r_{max}}$, for which the two forces balance each other. Therefore, the protons are first located in an attractive potential, from the exobase to $\mathrm{r_{max}}$ and then, in a repulsive potential, from $\mathrm{r_{max}}$ to infinity.

Below $\mathrm{r_{max}}$, the protons now can also be ballistic or trapped, so that the flux of escaping protons is reduced. The situation is not changed for electrons since their attractive electrostatic potential is still much larger than their gravitational potential. Therefore, in order to guarantee equal fluxes of escaping protons and electrons, a larger electric field than in the case with the exobase located above r_{max} is needed to increase the number of protons with an energy allowing them to overcome the potential barrier. This mechanism simply explains how the solar wind emerging from the coronal holes can be accelerated to larger velocities even with lower electron temperatures at the exobase.

In Figure 1, the different kinetic models are compared. The electrostatic potential difference between the exobase and infinity strongly increases either when we add suprathermal electrons or when the radial distance of the exobase is decreased. As a consequence, the solar wind is accelerated to large values.

The details of this generalized exospheric kinetic model are given elsewhere [5]. A description of this model in the formalism developed by [4] is also given in [12]. Here, we briefly describe the characteristics of this new model. The main difficulty of all kinetic models is to determine the radial distribution of the electrostatic potential, $V(\mathrm{r})$ in order to compute densities, fluxes and temperatures for the different classes of electrons and protons. For that purpose, we solve at every radial distance above the exobase a quasi-neutrality equation,

$$n_p(\mathrm{r}) = n_e(\mathrm{r}) \tag{2}$$

where $n_p(\mathrm{r})$ and $n_e(\mathrm{r})$ are respectively the protons and electrons densities at the radial distance r above the exobase. In previous exospheric models with monotonically decreasing potential for the protons, the only additional unknown in this equation was V_0, the electrostatic potential at the exobase which determines the potential barrier the electrons have to overcome in order to escape to infinity. In this new model with a non-monotonic potential for the protons, the situation is a bit more complicated : in addition to $V(\mathrm{r})$ and V_0, the densities n_p and n_e also depend on two additional parameters : r_{max}, the altitude of the maximum of the total potential of the protons and V_{max}, the electrostatic potential at r_{max} which determines the potential barrier for the protons. All these parameters depend on the values chosen for the model which are essentially the radial distance of the exobase, r_0, the kappa index of the electrons VDF, κ, and the protons and electrons temperatures at the exobase, T_0.

In order to determine these unknowns, we followed a method developed by Jockers [3] for Maxwellian electrons VDF and generalized it to the case of a Kappa VDF. Practically, the value of r_{max} is fixed and V_0 and V_{max} are found by simultaneously resolving a zero-current equation and a quasi-neutrality equation at $r = r_{\mathrm{max}}$,

$$F_p(r_{\mathrm{max}}) = F_e(r_{\mathrm{max}}) \tag{3}$$

$$n_p(r_{\mathrm{max}}) = n_e(r_{\mathrm{max}}) \tag{4}$$

where F_e and F_p are respectively the electrons and protons fluxes. Once these parameters are known, the radial distribution of $V(\mathrm{r})$ can be calculated from the exobase to infinity. There is only one value of r_{max} for which this distribution is continuous [5] [3]. The bulk velocity of the solar wind is then simply the ratio between the fluxes and the densities.

In Figure 2, we illustrate the influence of the radial distance of the exobase on the bulk velocity of the solar wind at large distance with realistic temperatures ($T_0 = 10^6\,\mathrm{K}$ for electrons and $T_0 = 2.10^6\,\mathrm{K}$ for protons) and kappa value of the electrons VDF at the exobase. In that particular case where the exobase is located very low ($r_0 = 1.1\,\mathrm{R_S}$), the bulk speed at 1 AU is $\sim 700\,\mathrm{km\,s^{-1}}$, of the order of the velocities observed in the fast solar wind. The densities required in order to have the exobase located at $1.1\,\mathrm{R_s}$ can be derived by equalizing the mean free path of the particles to the scale height. For that purpose, we use equations (11) and (12) of [6] and find that, for a proton temperature $T_0 = 2.10^6\,\mathrm{K}$ at the exobase, the densities are $n_0 \sim 2,5.10^6\,\mathrm{cm^{-3}}$, in good agreement with current observations [1] [2].

CONCLUSION

We have presented a new kinetic model of the solar wind with an exobase located very close to the basis of the corona, a situation which adequately describes the coronal holes where the densities are lower. With the use of suprathermal tails in the electrons VDF, these new kinetic models are able to reproduce the large velocities

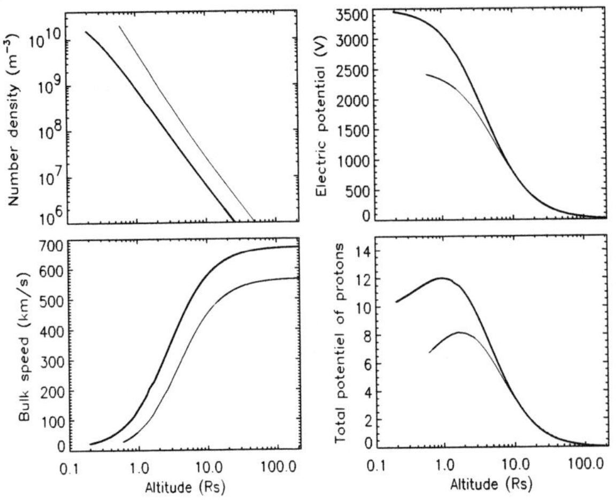

FIGURE 2. Influence of a change of the radial distance of the exobase on the number density, the electrostatic potential and the bulk speed of the solar wind for the case of $\kappa = 3$ and of electron and proton temperatures at the exobase respectively of 10^6 K and 2.10^6 K. The total potential of the protons is given in the lower right panel. The thicker line corresponds to the case of an exobase located at a radial distance $r_0 = 1.1\,\mathrm{R_s}$ while the other line corresponds to $r_0 = 1.5\mathrm{R_s}$.

observed in the fast solar wind without the need of additional ad-hoc heat/momentum deposition and even if the electron temperatures are lower than in other parts of the corona. The problem with models using Kappa distributions for the electrons is that they predict an increase of electron temperatures (as a consequence of the velocity filtration effect described by Scudder [11]) within a few solar radii, in disagreement with observations [12]. Although this problem has to be more carefully studied, it may suggest the use of a more appropriate method to describe the electrons suprathermal tails observed in the corona.

ACKNOWLEDGMENTS

H. Lamy and V. Pierrard thank the Belgian FNRS and the SSTC for their support.

REFERENCES

1. Esser R., et al., *Ap. J.*, *510*, L63, 1999.
2. Habbal S.R., Esser R., Guhathakurta M., Fisher R.R., *Geophys. Res. Let.*, *22*, 1465, 1995.
3. Jockers K., *Astron. Astrophys.*, *6*, 219, 1970.
4. Khazanov G., et al., *J. Geophys. Res.*, *103*, 6871, 1998.
5. Lamy H., Pierrard V., Maksimovic M., Lemaire J., *submitted to J. Geophys. Res.*, 2002
6. Lemaire J., Scherer M., *J. Geophys. Res.*, *76*, 31, 1971a.
7. Maksimovic M., Pierrard V., Riley P., *Geophys. Res. Let.*, *24*, 1151, 1997a.
8. Maksimovic M., Pierrard V., Lemaire J., *Astron. Astrophys.*, *324*, 725, 1997b.
9. Pierrard V., Lemaire J., *J. Geophys. Res.*, *101*, 7923, 1996.
10. Pierrard V., Issautier K., Meyer-Vernet N., Lemaire J., *Geophys. Res. Let.*,*28*, 2, 223, 2001.
11. Scudder J.D., *Ap. J.*, *398*, 299-318, 1992.
12. Zouganelis I., Maksimovic M., Meyer-Vernet N., Lamy H., Pierrard V., *Solar Wind 10 Proceedings*, 2002.

Evolution of Wake Instabilities and the Acceleration of the Slow Solar Wind: Melon Seed and Expansion Effects

A. F. Rappazzo*, M. Velli†**, G. Einaudi*** and R. B. Dahlburg‡

*Dipartimento di Fisica, Università di Pisa, Pisa
†Dipartimento di Astronomia e Scienza dello Spazio, Università di Firenze, Firenze
**Istituto Nazionale di Fisica della Materia, Sez. A, Università di Pisa, Pisa
‡Laboratory for Computational Physics & Fluid Dynamics, Naval Research Laboratory, Washington, D. C.

Abstract. We extend previous 2D simulation studies of slow solar wind acceleration due to the nonlinear evolution of the instability of the plasma/current sheet above streamers. We include the effects of the melon-seed force due to the overall magnetic field radial gradients on the plasmoid formed by the instability, as well as the subsequent expansion effects using the Expanding Box Model.

INTRODUCTION

Explaining how the slow component of the solar wind is accelerated is one of the outstanding problem in solar physics. Although the association between the slow solar wind and the streamer belt is broadly recognized, the mechanism which leads to such an acceleration is still a matter of debate.

Einaudi et al. [1] developed a magnetohydrodynamic model that accounts for many of the typical features observed in the slow component of the solar wind: the region above the cusp of a helmet streamer is modelled as a current sheet embedded in a broader wake flow. Previous incompressible studies (Dahlburg et al. [2]; Einaudi et al. [1]) show that reconnection of the magnetic field occurs at the current sheet and that in the non-linear regime, when the equilibrium magnetic field is substantially modified, a Kelvin-Helmoltz instability is allowed to develop, leading to the acceleration of the central part of the wake (Dahlburg [3]).

In this work we extend the previous model by including compressibility, and then the expansion and the melon seed force, both of them due to the spheric geometry of the problem.

GOVERNING EQUATIONS

In our simulations we used the compressible, dissipative MHD equations, that we write here in dimensionless form:

$$\frac{\partial \rho}{\partial t} + \nabla \cdot (\rho \mathbf{u}) = 0, \quad (1)$$

$$\rho\left(\frac{\partial \mathbf{u}}{\partial t} + (\mathbf{u}\cdot\nabla)\mathbf{u}\right) = -\nabla\left(P + \frac{\mathbf{B}^2}{2}\right) + (\mathbf{B}\cdot\nabla)\mathbf{B} + \frac{1}{\mathcal{R}}\nabla\cdot\xi, \quad (2)$$

$$\frac{\partial \mathbf{B}}{\partial t} = \nabla\times(\mathbf{u}\times\mathbf{B}) + \frac{1}{\mathcal{R}_m}\nabla^2\mathbf{B}, \quad (3)$$

$$\text{where} \quad \xi_{ij} = \frac{\partial u_i}{\partial x_j} + \frac{\partial u_j}{\partial x_i} - \frac{2}{3}\delta_{ij}\nabla\cdot\mathbf{u}, \quad P = \rho T, \quad (4)$$

and symbols have the usual meaning. For simplicity we consider an isothermal equation of state.

To render non-dimensional the equations we used the quantities L^*, u^*, ρ^*, t^*, T^*, B^*, where

$$t^* = \frac{L^*}{u^*}, \quad T^* = (u^*)^2, \quad (B^*)^2 = 4\pi\rho^*(u^*)^2. \quad (5)$$

The Reynolds numbers $\mathcal{R}$ and $\mathcal{R}_m$ are defined as:

$$\frac{1}{\mathcal{R}} = \frac{\mu}{\rho^* u^* L^*}, \qquad \frac{1}{\mathcal{R}_m} = \frac{\eta c^2}{4\pi u^* L^*}, \quad (6)$$

μ and η are respectively the shear viscosity and the resistivity that we suppose constant and uniform.

In the following we refer to the spatial coordinate aligned with the mean flow as the streamwise direction (y), the spatial coordinate along which the mean flow varies as the cross-stream direction (x), and to the remaining (spanwise) direction as z. The system has periodic boundary conditions in the streamwise direction, along which a pseudospectral collocation method is used

CP679, *Solar Wind Ten: Proceedings of the Tenth International Solar Wind Conference*,
edited by M. Velli, R. Bruno, and F. Malara

in the numerical computation, and nonreflecting boundary conditions in the cross-stream direction, where a finite difference technique is used. The dimension of the numerical box are $L_y = 2\pi/a$, where a is the streamwise wavenumber, and $L_x = \pm 10.63$; furthermore we used 361 collocation points in the cross-stream direction and 128 points in the streamwise direction.

INITIAL CONDITIONS

We model a planar section of the solar streamer belt as a current sheet of thickness a_B embedded in a wake of thickness a_V, flowing at the speed of the fast solar wind at the edges and at a much lower velocity at the sheet. Indicating with the index "∞" the values of the physical quantities at the edges, we choose $L^* = a_V$, $\rho^* = \rho_\infty$, $u^* = u_\infty$, and the basic fields are then given by:

$$u_{0y}(x) = 1 - \operatorname{sech}(x), \tag{7}$$

$$B_{0y}(x) = A \tanh(\delta x), \qquad B_{0z}(x) = A \operatorname{sech}(\delta x), \tag{8}$$

$$\rho = 1, \qquad T = \frac{1}{M^2}, \tag{9}$$

where $\delta = \frac{a_V}{a_B}$ is the ratio of the two widths. A and M are respectively the Alfvén number and the sonic Mach number of the system at the edges:

$$A = \frac{B_\infty}{\sqrt{4\pi\rho_\infty} u_\infty} = \frac{c_{A\infty}}{u_\infty}, \qquad M = \frac{u_\infty}{\sqrt{T_\infty}} = \frac{u_\infty}{c_{s\infty}}. \tag{10}$$

Observational data suggest us that near the cusp of a helmet streamer $\delta \gg 1$, $A \sim 2.5$ and $M \sim 6$. Here we use the more computationally accessible values $\delta = 5$, $A = 1.5$, $M = 1$.

AVOIDING DIFFUSION OF EQUILIBRIUM FIELDS

Solar corona Reynolds numbers are so high that fields (7)-(9) are substantially an equilibrium solution of equations (1)-(4) over the time scales of our simulations. But in our computational domain Reynolds numbers are limited by resolution to values around $\mathcal{R} = \mathcal{R}_m = 200$, values which are so small that unphysical diffusion of equilibrium fields takes place.

The pattern of the diffusion terms in the foregoing equations is

$$\frac{\partial f}{\partial t} = \ldots + \frac{1}{\mathcal{R}} \frac{\partial^2 f}{\partial x^2}, \tag{11}$$

and we can avoid such an unphysical diffusion using instead of eq. (11) the following one:

$$\frac{\partial f}{\partial t} = \ldots + \frac{1}{\mathcal{R}} \frac{\partial^2}{\partial x^2}\left(f - \hat{f}_0\right). \tag{12}$$

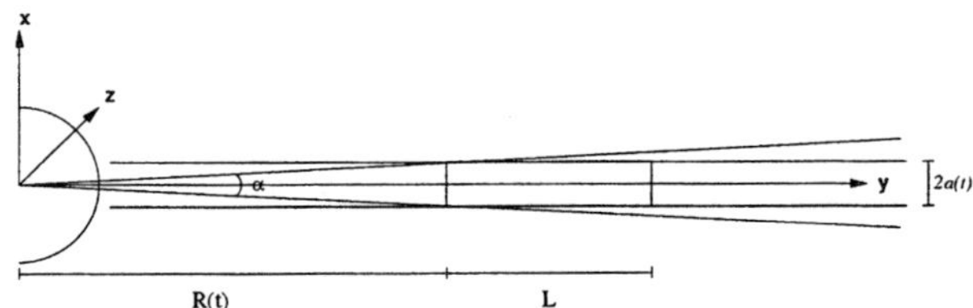

FIGURE 1. Expanding Box Model: coordinate system and notations.

In this way the modes $k = 0$ of the Fourier series in the y direction, that we indicate with $\hat{f}_0$, do not diffuse; this could have potentially given rise to development of excessively small scales in x on the $k = 0$ harmonic which empirically does not occur.

EXPANDING BOX MODEL

In the region in which takes place the acceleration of the solar wind the equilibrium fields are not homogeneous but are spherically simmetric. Looking at figure 1, we approximate the spheric sector of angular extent α in the cross-stream direction and β in the spanwise direction with a parallelepiped whose cross lengths are given by:

$$a(t) = \frac{\alpha}{2} R(t) = \frac{a_0}{R_0} R(t), \qquad b(t) = \frac{\beta}{2} R(t) = \frac{b_0}{R_0} R(t), \tag{13}$$

where, as usual, a_0, R_0 are the values of these lengths at $t = 0$. In the same way, for a particle which moves radially outward holds:

$$x(t) = \frac{x_0}{R_0} R(t), \qquad z(t) = \frac{z_0}{R_0} R(t), \tag{14}$$

and its cross component of the velocity is given by

$$\mathbf{u}_\perp = \frac{U(t)}{R(t)}\Big(x(t), 0, z(t)\Big), \quad \text{where} \quad U(t) = \dot{R}(t). \tag{15}$$

In the fluid case we replace this velocity with the field $\mathbf{U}(x,t) = \frac{U(t)}{R(t)}(x, 0, z)$, which inserted in the momentum equation gives the force for unity of mass

$$\mathbf{F}(\mathbf{x},t) = \frac{\partial \mathbf{U}}{\partial t} + (\mathbf{U}\cdot\nabla)\mathbf{U} = \frac{\dot{U}}{R}\Big(x, 0, z\Big). \tag{16}$$

Hence this is the expression of the force that we insert "by hand" in the momentum equation, where the velocity of the box and its distance from the center of the sun are defined as:

$$U(t) = \frac{1}{L_y}\int_0^{L_y} u_y(0, y, 0, t)\,\mathrm{d}y, \tag{17}$$

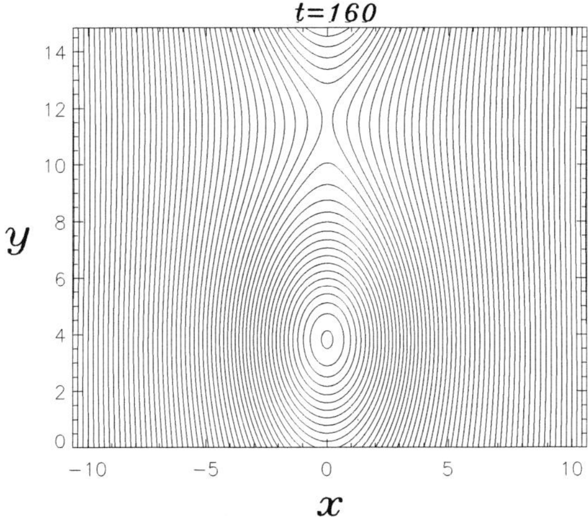

FIGURE 2. Run A: magnetic field lines at $t = 160$.

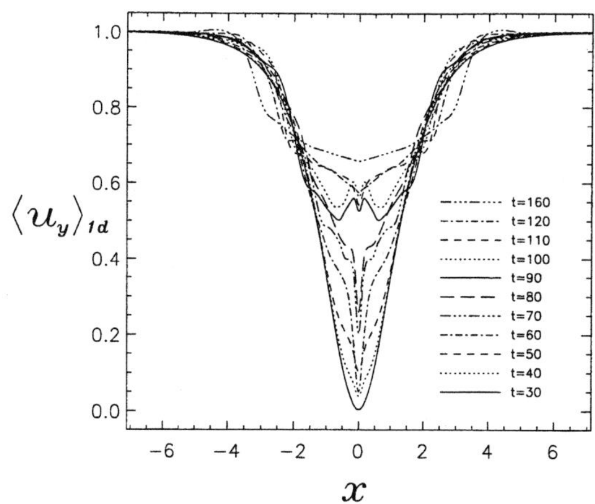

FIGURE 3. Run A: streamwise component of the velocity (u_y) averaged in the y direction as a funtion of x at selected times.

$$R(t) = R_0 + \int_0^t U(t')\mathrm{d}t'. \tag{18}$$

In this paper we consider another force, which acts on a magnetic island in a radial magnetic field, the *melon-seed force* (Schmidt and Cargill [4]) . The total force acting on the magnetic island is of the form

$$\mathbf{F}_m = -V\frac{\mathrm{d}}{\mathrm{d}R}\left(\frac{B_e^2}{8\pi}\right)\hat{\mathbf{e}}_y, \tag{19}$$

where $B_e = B_{e0}\left(\frac{R_0}{R}\right)^2$ is the solar magnetic field and V the volume of the plasmoid, B_{e0} being costant. Hence we introduce the melon-seed force in the form of a force for unitary volume whose expression is

$$\mathbf{F}_m = \Theta(x,y,t)\frac{B_{e0}^2}{2\pi R_0}\left(\frac{R_0}{R}\right)^5\hat{\mathbf{e}}_y, \tag{20}$$

in which $\Theta = 1$ inside magnetic island and $\Theta = 0$ outside.

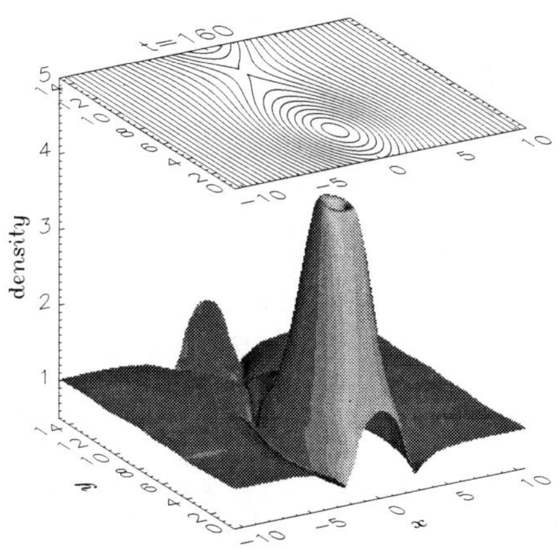

FIGURE 4. Run A: density of matter (*bottom*) and magnetic field lines (*top*) at $t = 160$ showing density enhancement of magnetic island.

RESULTS

This section presents the results of two numerical simulations. In the first one the melon seed and expansion effects are neglected, while in the second we consider them both. In all calculations a small amplitude ($\epsilon = 10^{-4}$) perturbation of the resistive-varicose type ([2]), at the wavenumber $a = 0.42$ was added to the initial state; except for some crucial differences in their details, magnetic reconnection and the formation of moving plasmoids are the characteristic phenomena for all our simulations. We made also a test run without perturbation and verified that over the time scales of our simulations the equilibrium fields (7)-(9) are not modified within numerical errors.

Run A

As expected for a classic tearing instability, magnetic islands (figure 2) form with streamwise length equal to the perturbation wavelength, while their cross-stream width grows from infinitesimal in the linear regime to the order of a_B at the beginning of the non-linear phase, when a striking feature takes place. In fact until time $t = 30$, the beginning of the non-linear phase, the wake profile is substantially unmodified (figure 3), while after this time, when the perturbed magnetic field attains finite amplitude, modifying the equilibrium field, the velocity average grows up to the value ~ 0.7 at $t = 160$.

Another important feature is the density enhancement

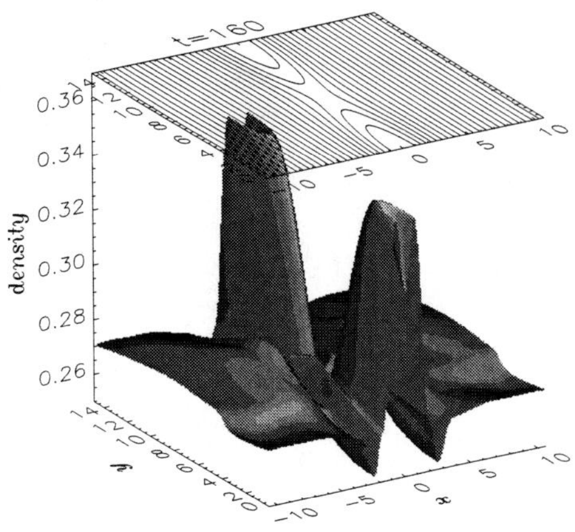

FIGURE 5. Run B: density of matter (*bottom*) and magnetic field lines (*top*) at $t = 160$ showing density enhancement of magnetic island.

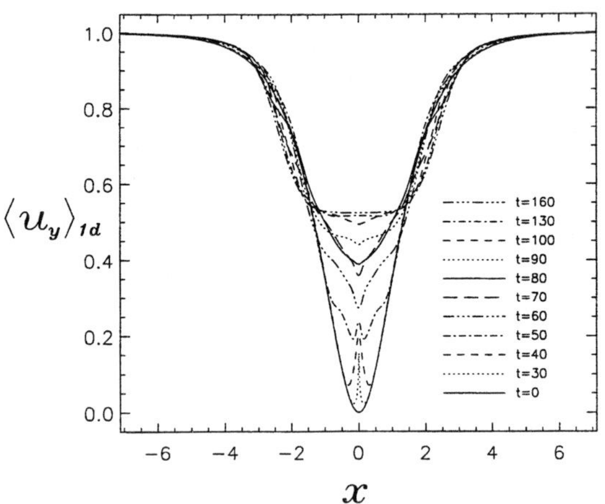

FIGURE 6. Run B: streamwise component of the velocity $\langle u_y \rangle$ averaged in the y direction as a funtion of x at selected times.

of the magnetic island shown in figure 4. This feature is qualitatively similar to the density enhancements, called "blobs", observed with the LASCO instrument onboard SOHO (Sheeley *et al.* [5], Wang *et al.* [6]).

Run B

This run provides us with information on the effects of melon-seed force and expansion. In order to keep our computational grid fixed we used the coordinate scaling

$$x_F = \frac{R(t)}{R_0}\, x, \tag{21}$$

where x_F represents the physical grid (*which expands*), and x the numerical grid (*fixed*). As $L^* = a_V \sim 0.1 R_\odot$, we used as starting radius $R_0 = 60$ which in dimensional units is $\sim 6R_\odot$, approximately the radius of the region located near the cusp of a helmet streamer.

Again we have the formation of a density enhanced magnetic island (figure 5), but as now expansion is occurring the average density of matter decreases as $\left(\frac{R_0}{R}\right)^2$. In the same way, as the magnetic field components decrease while expansion proceeds, the acceleration of the central part of the wake is lower (figure 6).

In figure 7 we plot the average streamwise velocity at selected x values (the fixed numerical grid) as a function of the distance between the box and the center of the sun, which reaches the value $R \sim 115$, i.e. $R \sim 12R_\odot$ in conventional units, at $t = 160$. The melon seed force results in the initial rapid acceleration, while interaction with the fluid instability shapes the global acceleration.

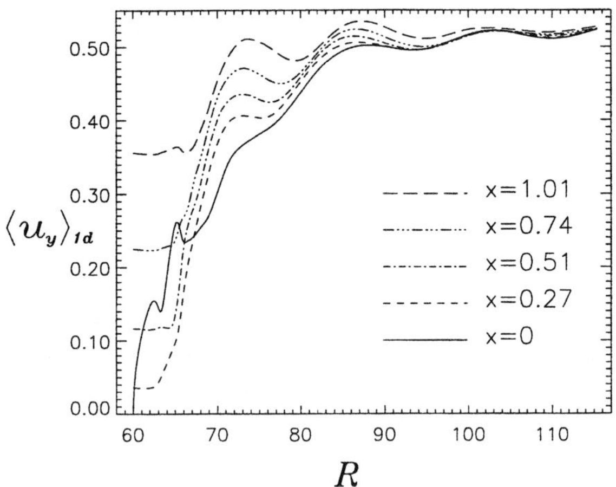

FIGURE 7. Run B: average streamwise component of the velocity ($< u_y >$) at selected x values (x representing the fixed numerical grid) as a function of *R*, the distance of the expanding box from the center of the sun.

REFERENCES

1. Einaudi, G., Boncinelli, P., Dahlburg, R. B., and Karpen, J. T., *J. Geophys. Res.*, **104**, 521 (1999).
2. Dahlburg, R. B., Boncinelli, P., and Einaudi, G., *Phys. Plasmas*, **5**, 79 (1998).
3. Dahlburg, R. B., *Phys. Plasmas*, **5**, 133 (1998).
4. Schmidt, J. M., and Cargill, P. J., *J. Geophys. Res.*, **105**, 10455 (2000).
5. Sheeley, N. R. et al., *Astrophys. J.*, **484**, 472 (1997).
6. Wang, Y.-M. et al., *Astrophys. J.*, **498**, L165 (1998).

Session III

Waves, Turbulence, and Kinetic Physics

Observations of Alfvénic turbulence evolution in the 3-D heliosphere

Bruno Bavassano

Istituto di Fisica dello Spazio Interplanetario (CNR), Roma, Italy

Abstract. The heliospheric plasma is an excellent laboratory to study the behaviour of collisionless Alfvénic turbulence. This is a fundamental topic in both plasma physics and astrophysics. The impressive amount of observations at different solar distances and latitudes collected in last three decades allows to get a good view of the turbulence behaviour in the heliosphere inside 5 AU. In present review the focus will be on the evolution of the Alfvénic turbulence in both low- and high-latitude solar wind, as derived from old observations on the ecliptic and from recent observations by Ulysses at polar latitudes.

INTRODUCTION

The solar wind, a high Reynolds' numbers collisionless plasma, is the most accessible medium in which to study Alfvénic turbulence. This is a topic of fundamental importance for both plasma physics and astrophysics. Alfvénic turbulence in solar wind 1) contributes to plasma heating and acceleration, 2) accelerates particles to high energies, and 3) affects cosmic ray propagation. In the seventies and the eighties impressive advances have been made in the knowledge of Alfvénic turbulent phenomena in solar wind. In those days spacecraft observations were confined to a small latitudinal belt around the solar equator. In the nineties, with the launch of the Ulysses spacecraft, investigations have been extended to the high-latitude regions of the heliosphere. This has allowed to study how Alfvénic turbulence evolves in polar wind, a plasma flow in which the effects of large-scale inhomogeneities are considerably less important than in low-latitude wind. With this new laboratory relevant advances have been made. In present review the Ulysses observations of turbulence evolution in polar wind will be discussed and compared to those typical of near-equatorial solar wind.

A short overview of the parameters that are generally used to describe Alfvénic fluctuations will be preliminarily given. Basic quantities are the Elsässer's variables ($\mathbf{z}_\pm$). They are defined as $\mathbf{z}_\pm = \mathbf{v} \pm \mathbf{b}$, where $\mathbf{v}$ and $\mathbf{b}$ are the velocity and magnetic field vectors, respectively, and $\mathbf{b}$ is scaled to Alfvén units (i.e., divided by $\sqrt{4\pi\rho}$, with ρ the mass density). Taking into account how the sign of the Alfvénic correlation depends on the propagation direction with respect to the background magnetic field, it has become common to use the above definition in the case of a background magnetic field pointing to the Sun, while the equation $\mathbf{z}_\pm = \mathbf{v} \mp \mathbf{b}$ is taken for the opposite polarity. With this choice we have that, whatever the polarity is, $\mathbf{z}_+$ ($\mathbf{z}_-$) fluctuations always correspond to modes with an outward (inward) direction of propagation, with respect to Sun, in the plasma frame. An exhaustive discussion on the use of Elsässer variables in solar wind turbulence studies has been performed by *Tu and Marsch* [1995]. The relative weight of the energies (per unit mass) e_+ and e_- associated to $\mathbf{z}_+$ and $\mathbf{z}_-$ fluctuations, respectively, is measured by the Elsässer ratio $r_E = e_-/e_+$. An analogous measure for the energies e_V and e_B of the $\mathbf{v}$ and $\mathbf{b}$ fluctuations is given by the Alfvén ratio $r_A = e_V/e_B$. Other related parameters are the normalized cross-helicity $\sigma_C = (e_+ - e_-)/(e_+ + e_-) = (1 - r_E)/(1 + r_E)$ and the normalized residual energy $\sigma_R = (e_V - e_B)/(e_V + e_B) = (r_A - 1)/(r_A + 1)$.

Both outward and inward propagating fluctuations are observed in the solar wind. Outward fluctuations mainly have a solar origin (or, more precisely, inside the Alfvén critical point). Conversely, inward propagating fluctuations can only be generated outside such critical distance. As well known, the presence of both kinds of fluctuation leads to the development of nonlinear interactions.

The main features of the Alfvénic turbulence evo-

CP679, *Solar Wind Ten: Proceedings of the Tenth International Solar Wind Conference*, edited by M. Velli, R. Bruno, and F. Malara

lution versus radial distance (i.e., versus transit time to the observer) will be now discussed for ecliptic and polar wind. Ecliptic results are mainly based on Helios and Voyagers observations, while polar results obviously rely on Ulysses measurements.

ECLIPTIC TURBULENCE

Since the first studies on Alfvénic fluctuations in ecliptic wind it was clear that fast streams (or, more precisely, their trailing edge) are the best places to observe them. Helios spacecraft, with a systematic covering of the inner heliosphere from 0.3 to 1 AU, gave the first opportunity to study the turbulence evolution with solar distance.

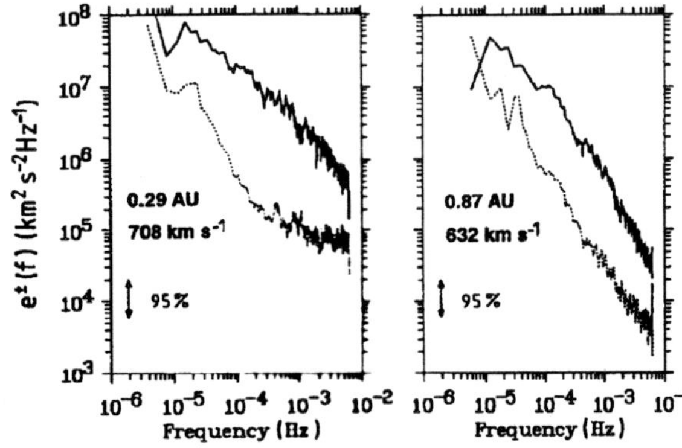

FIGURE 1. Power spectra of e_+ and e_- (solid and dotted line, respectively) in the trailing edge of fast streams at 0.29 and 0.87 AU (adapted from *Marsch and Tu* [1990], copyright 1990 American Geophysical Union, modified by permission of American Geophysical Union).

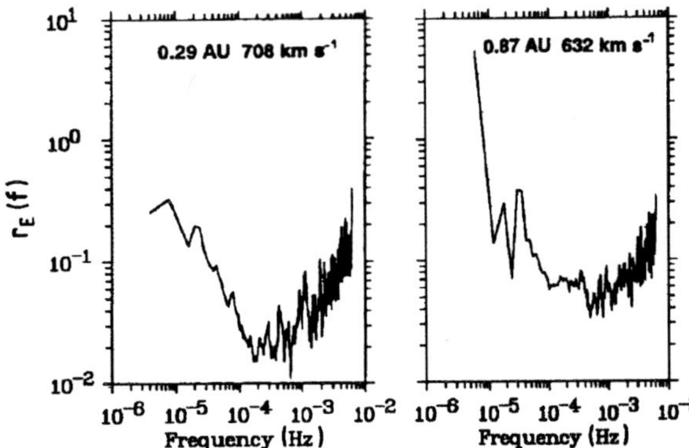

FIGURE 2. Power spectrum of the Elsässer ratio r_E ($= e_-/e_+$) in the trailing edge of fast streams at 0.29 and 0.87 AU (adapted from *Marsch and Tu* [1990], copyright 1990 American Geophysical Union, modified by permission of American Geophysical Union).

Figure 1 (from *Marsch and Tu* [1990]) shows how e_+ and e_- power spectra (solid and dotted line, respectively) vary when solar distance increases from 0.3 to 0.9 AU. The e_+ spectrum declines faster than that of e_-, with the result that the two spectra approach each other. At the same time the spectral slopes evolve in such a way that an extended inertial regime expands to low frequencies.

The corresponding variation in the relative weight of outward and inward fluctuation energies is shown by Figure 2 (from *Marsch and Tu* [1990]). The small values of r_E observed at 0.3 AU in the core of the Alfvénic regime (frequencies around $10^{-4} - 10^{-3}$ Hz) have disappeared at 0.9 AU. However, in spite of this r_E increase, the predominance of $\mathbf{z}_+$ fluctuations remains a clear feature of the Alfvénic turbulence observed by Helios inside 1 AU. Bavassano et al. [2001], using Ulysses data from the ecliptic phase of the mission, have shown that this situation persists at least up to 5 AU.

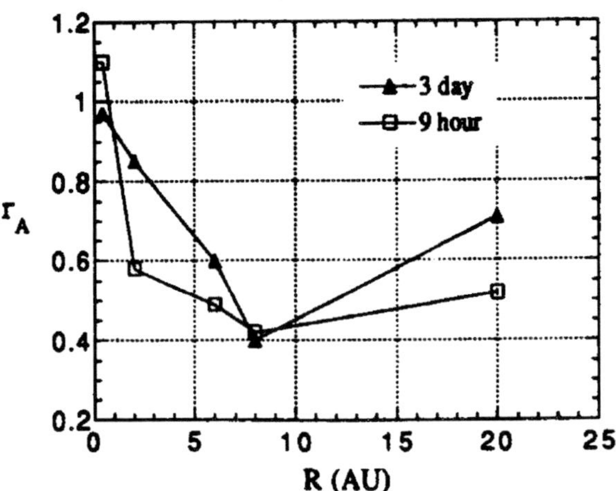

FIGURE 3. Radial variation of the Alfvén ratio r_A ($=e_V/e_B$) as seen by Helios and Voyagers between 0.3 and 20 AU (from *Roberts et al.* [1990], copyright 1990 American Geophysical Union). The 9-hr curve (see squares) shows that r_A, after the fast decrease at small distance, remains nearly unchanged.

Another relevant feature of the solar wind Alfvénic fluctuations is the emergence, for increasing solar distance, of a predominance of the magnetic fluctuation energy. This decreasing trend for the Alfvén ratio r_A ($=e_V/e_B$) clearly appears in Helios data [*Bruno et al.*, 1985; *Marsch and Tu*, 1990]. It has been shown by *Bavassano and Bruno* [2000] that this change in the relative amplitude of $\mathbf{v}$ and $\mathbf{b}$ fluctuations occurs without any relevant variation of their correlation. The decreasing trend of r_A, however, is not without limit, in other words a seemingly final stage is reached in which the value of r_A remains nearly unchanged. This is seen in Figure 3 (from *Roberts et al.* [1990]), combining Helios and Voyagers observations to give an overall view of the radial variation of r_A between 0.3 and 20 AU. The 9-hr curve, that of the two reported curves best describes the typical MHD scales, shows that r_A, after a fast and pronounced decrease, remains nearly unchanged.

POLAR TURBULENCE

Observations by Ulysses during its first out-of-ecliptic orbit have shown that, at low solar activity, the high-latitude solar wind is a fast and relatively steady flow. A relevant feature of the polar wind is the ubiquitous presence of an intense flow of Alfvénic fluctuations (e.g., *Goldstein et al.* [1995]; *Horbury et al.* [1995]; *Smith et al.* [1995]; *Bavassano et al.* [1998]). Similarly to previous ecliptic observations in fast streams, a largely dominant fraction of these fluctuations is outward propagating, with respect to the Sun, in the solar wind frame. The relevance of the Ulysses observations, besides their exploratory character, is that the relative absence of structure in polar flow offers the opportunity of studying the evolution of Alfvénic turbulence under almost undisturbed conditions.

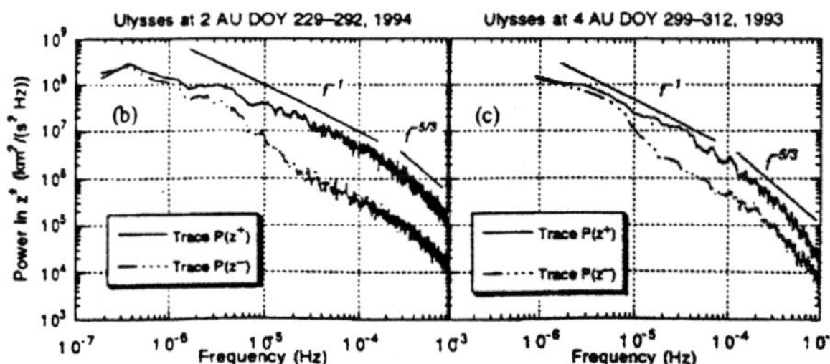

FIGURE 4. The plot shows power spectra of $\mathbf{z}_+$ and $\mathbf{z}_-$ (upper and lower curve, respectively) in polar wind at (left panel) $\sim$ 2 AU and (right panel) $\sim$ 4 AU (adapted from *Goldstein et al.* [1995], copyright 1995 American Geophysical Union, modified by permission of American Geophysical Union).

In Figure 4 power spectra of $\mathbf{z}_+$ and $\mathbf{z}_-$ at about 2 and 4 AU in polar wind as obtained by *Goldstein et al.* [1995]) are shown. The spectral evolution appears qualitatively similar to that observed in ecliptic wind, with the development of a turbulent cascade with increasing distance that moves to lower frequencies the breakpoint between the f^{-1} and $f^{-5/3}$ regimes. However, in polar wind the breakpoint is at higher frequency than at similar distances in ecliptic wind (see *Horbury et al.* [1996]). Thus, spectral evolution in polar wind is slower than in ecliptic wind.

Further evidence on this is given in Figure 5. The plot shows the power spectra of magnetic field components (solid circles) and magnitude (squares) as estimated at 1 AU from trends in power scalings measured by Ulysses in polar wind between 1.4 and 4.1 AU (solid lines) and by Helios in ecliptic fast wind inside 1 AU (dashed lines) [*Horbury and Balogh*, 2001]. At time scales smaller than few hours (see scale on top) the general agreement appears good, in both shape and magnitude. However, it is clearly seen that the spectrum of magnetic components is steeper in ecliptic wind, as expected for a faster evolution.

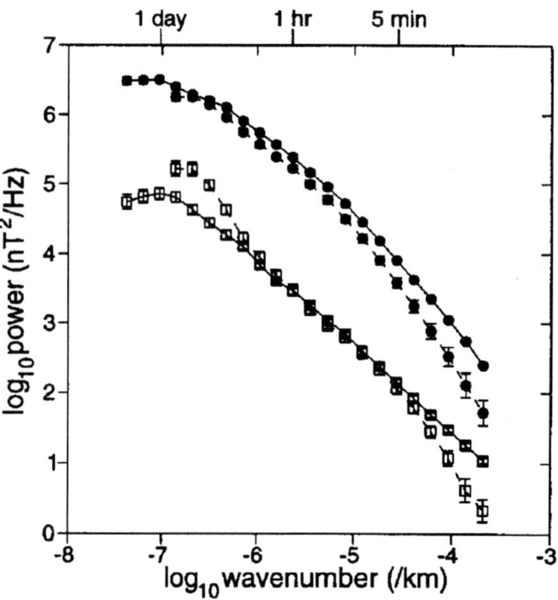

FIGURE 5. The plot shows the power spectra of magnetic field components (solid circles) and magnitude (squares) as estimated at 1 AU from trends in power scalings at Ulysses in polar wind between 1.4 and 4.1 AU (solid lines) and Helios in ecliptic fast wind inside 1 AU (dashed lines), from *Horbury and Balogh* [2001] (copyright 2001 American Geophysical Union).

A general agreement exists on the fact that the slower evolution for polar turbulence has to be ascribed to the lack of a large-scale stream structure. The role of such a structure in accelerating turbulence evolution has been recently stressed by *Bavassano et al.* [1998] using Ulysses data at mid latitudes (in the first out-of-ecliptic orbit), where strong gradients were dominant in the velocity pattern. It should also be mentioned that the turbulence appears younger in polar flow since fluctuations at a given distance have had less time to evolve due to the higher wind speed (e.g., see *Matthaeus et al.* [1999]).

The radial variation of the Elsässer ratio r_E and the Alfvén ratio r_A in polar wind is shown in Figure 6 (see also *Bavassano et al.* [2000a]). Both the decline of the e_+ predominance and the increase of the e_B predominance do not go beyond some limits, in agreement with ecliptic observations.

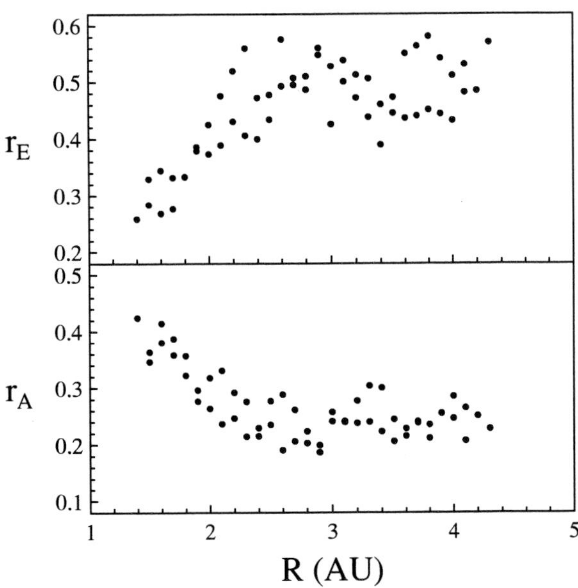

FIGURE 6. Radial variation of (top) Elsässer ratio r_E (=e_-/e_+) and (bottom) Alfvén ratio r_A (=e_V/e_B) in polar wind, as obtained from hourly total variances of the fluctuating vectors.

Agreement between polar and ecliptic observations is found also for the radial variation of magnetic field fluctuations. Figure 7 shows the decline with solar distance for hourly variances of magnetic field components and magnitude in polar wind [*Forsyth et al.* 1996]. Best fit power laws (R^{-n}) for data in the range from 1.5 to 3 AU (data at larger distances have not been included to avoid effects related to compressive features) indicate a radial slope n of about 3.4 for the magnetic components. This value is in good agreement with that found in ecliptic wind, at the same scale and for a similar range of distances, by *Bavassano and Smith* [1986] with Pioneer 10 and 11 data.

Alfvénic turbulence evolution, however, is better described by the variation of $\mathbf{z}_+$ and $\mathbf{z}_-$, rather than by magnetic field. Figure 8 is a composite plot combining the Ulysses observations in polar wind with those by Helios in the trailing edge of fast streams on the ecliptic. As regards polar wind, the e_+ values (hourly variances) exhibit the same radial gradient over all the investigated range of distances. In contrast, for e_- a change of slope around 2.5 AU is clearly apparent. Another remarkable feature is the good agreement of the Ulysses gradients with Helios data.

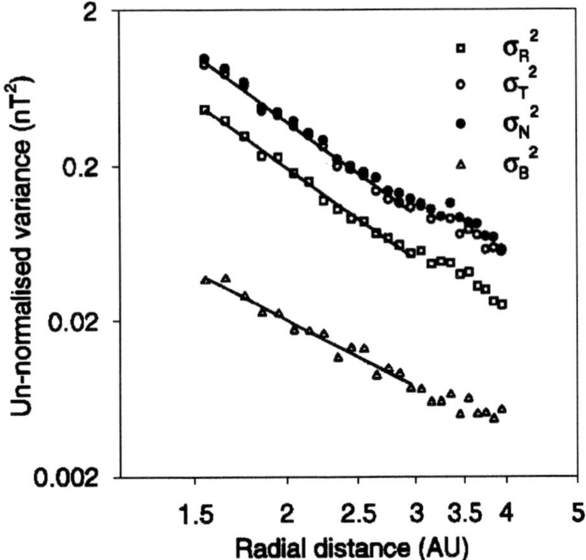

FIGURE 7. Radial variation of hourly variances of magnetic field components and (lower curve) magnitude (from *Forsyth et al.* [1996], copyright 1996 American Geophysical Union).

The e_- behaviour highlighted by Figure 8 is probably related to the presence, inside ~2.5 AU, of local generation effects. Turbulence generation in polar wind is a very relevant point. Velocity shear probably is an important factor to generate turbulence and drive its evolution in ecliptic solar wind, but certainly this does not hold for polar wind, where large-scale velocity gradients are almost absent. It has been proposed (e.g., see *Malara et al.* [2000] and *Del Zanna* [2001]) that in polar wind a role might be played by the parametric decay. Simulations of the non-linear development of the parametric decay for large-amplitude non-monochromatic Alfvénic fluctuations [*Malara et al.*, 2000] have shown that the final state strongly depends on the value of β (thermal to magnetic pressure ratio). For $\beta < 1$ the normalized cross-helicity σ_C decreases, from an initial value of 1, to values close to 0. Thus, the instability appears able to completely destroy the initial Alfvénic correlation. In sharp contrast, for $\beta = 1$ (a value closer to real solar wind conditions) σ_C is around 0.5 in the final state. In this case the parametric instability is not able to go beyond some limit in the disruption of the initial correlation between velocity and magnetic field fluctuations. This solution surely is qualitatively reminiscent of the Ulysses data behaviour shown in Figure 8 (see also the r_E variation in Figure 6). Analogous results are reached by *Del Zanna* [2001] for arc-polarized Alfvén waves. It should be stressed, however, that these models have many limitations (e.g., see discussion by *Malara et al.* [2001]). Turbulence generation in a flow with a low level of large-scale velocity shear is still an open issue.

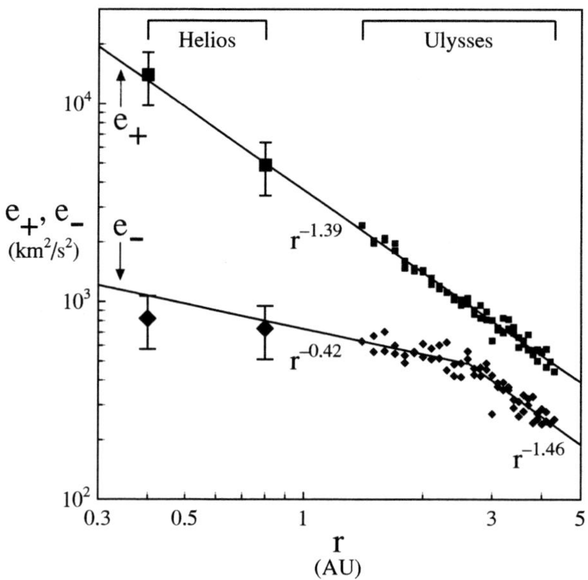

FIGURE 8. A composite plot combining Ulysses observations in polar wind with those by Helios 1 and 2 inside 1 AU on the ecliptic plane (adapted from *Bavassano et al.* [2000b], copyright 2000 American Geophysical Union, modified by permission of American Geophysical Union). The values of e_+ (e_-) are shown as squares (diamonds), small for Ulysses and large for Helios. Best fit lines and radial power laws for Ulysses data are given.

A last comment is about the nature of the Alfvénic turbulence variations observed in polar wind (i.e., radial, or latitudinal, or both). Both distance and latitude concurrently vary along the Ulysses trajectory, thus this point needs to be carefully examined. Several analyses of Ulysses data have indicated that the variation of turbulence properties in high-latitude wind is essentially radial, rather than latitudinal, in nature (e.g., *Goldstein et al.* [1995]; *Horbury et al.* [1995]; *Forsyth et al.* [1996]). A further robust argument in favor of the radial character of the turbulence variation comes from the agreement between gradients observed in high-latitude wind and in fast streams on the ecliptic. This has been seen to hold both for magnetic field fluctuations (see above) and for outward and inward Alfvénic fluctuations (as shown by *Bavassano et al.* [2000b] and [2001]). However, recent results of *Horbury and Balogh* [2001] have indicated that fluctuations in magnetic field components at time scales shorter than about 1 day (in the spacecraft frame) exhibit a non negligible dependence on latitude. Bavassano et al. [2002] have reexamined this point applying the same method of *Horbury and Balogh* [2001] to the Elsässer variables and carefully removing data intervals with compressive features (as those seen in Figure 7). Their conclusion is that, at least for the core of the Alfvénic regime, latitude does not appear to have an appreciable influence on the Alfvénic turbulence evolution in polar wind.

CONCLUDING REMARKS

Ulysses observations, combined with previous results in the ecliptic, allow to get a quite complete view of the Alfvénic turbulence evolution in the heliosphere from 1 to 5 AU.

Polar wind observations, when compared to previous ecliptic results, do not appear as a dramatic break. Polar evolution is similar to that in the ecliptic, although slower. This well agrees with the view proposed by *Bruno* [1992], a middle course between a non-relaxing turbulence (due to the lack of velocity shear, *Grappin et al.* [1991]) and a quick evolution (due to the large amplitude of fluctuations relative to the mean magnetic field, *Roberts* [1990]).

A still open question is about the driving mechanism(s) for polar turbulence evolution, given the low level of large-scale velocity shear in the polar flow. Parametric decay could be a good candidate, but models still need relevant improvements.

It has been seen that in all the examined region the Alfvénic fluctuations appear characterized by 1) a predominance of outward fluctuations (i.e., a positive cross-helicity) and 2) a predominance of magnetic fluctuations (i.e., a negative residual energy).

As regards outward fluctuations, in high-latitude wind their dominant character probably extends well far away from the Sun. At low solar activity the polar wind fills a large fraction of the heliospheric cavity. For such conditions the outward fluctuations may play a leading role in acceleration and diffusion of high-energy particles. In other words, models based on the assumption of a negligible cross-helicity should not be able to give a good description of these phenomena.

As regards the imbalance in favour of magnetic energy, it does not appear to go beyond a given limit. Several ways to get a different from zero residual energy have been proposed, for instance, 2-D processes (e.g., *Oughton et al.* [1994]; *Matthaeus et al.* [1996]) or propagation in a non uniform medium (e.g., *Hollweg and Lee* [1989]). However, convincing arguments to account for the existence of such limit have not been given yet.

In conclusion, though non negligible advances have still to be made, a satisfactory understanding of the solar wind Alfvénic turbulence does not seem too far away.

ACKNOWLEDGMENTS

Useful discussions with R. Bruno are gratefully acknowledged. The present work has been supported by the Italian Space Agency (ASI).

REFERENCES

1. Bavassano, B., and Bruno, R., *J. Geophys. Res.*, **105**, 5113-5118 (2000).
2. Bavassano, B., and Smith, E. J., *J. Geophys. Res.*, **91**, 1706-1710 (1986).
3. Bavassano, B., Pietropaolo, E., and Bruno, R., *J. Geophys. Res.*, **103**, 6521-6529 (1998).
4. Bavassano, B., Pietropaolo, and Bruno, R., *J. Geophys. Res.*, **105**, 12,697-12,704 (2000a).
5. Bavassano, B., Pietropaolo, E., and Bruno, R., *J. Geophys. Res.*, **105**, 15,959-15,964 (2000b).
6. Bavassano, B., Pietropaolo, E., and Bruno, R., *J. Geophys. Res.*, **106**, 10,659-10,668 (2001).
7. Bavassano, B., Pietropaolo, E., and Bruno, R., *this issue*, (2002).
8. Bruno R., in *Solar Wind Seven*, eds. E. Marsch and R. Schwenn, COSPAR Colloquia Series vol. 3, pp. 423-428, Pergamon Press, Tarrytown, N.Y., (1992).
9. Bruno R., Bavassano, B., and Villante, U., *J. Geophys. Res.*, **90**, 4373-4377 (1985).
10. Del Zanna, L., *Geophys. Res. Lett.*, **28**, 2585-2588 (2001).
11. Forsyth, R. J., Horbury, T. S., Balogh, A., and Smith, E. J., *Geophys. Res. Lett.*, **23**, 595-598 (1996).
12. Goldstein, B. E., Smith, E. J., Balogh, A., Horbury, T. S., Goldstein, M. L., and Roberts, D. A., *Geophys. Res. Lett.*, **22**, 3393-3396 (1995).
13. Grappin, R., Velli, M., and Mangeney, A., *Ann. Geophys.*, **9**, 416-426 (1991).
14. Hollweg, J. V., and Lee, M. A., *Geophys. Res. Lett.*, **16**, 919-922 (1989).
15. Horbury, T. S., Balogh, A., Forsyth, R. J., and Smith, E. J., *Geophys. Res. Lett.*, **22**, 3401-3404 (1995).
16. Horbury, T. S., Balogh, A., Forsyth, R. J., and Smith, E. J., *Astron. Astrophys.*, **316**, 333-341 (1996).
17. Horbury, T. S., and Balogh, A., *J. Geophys. Res.*, **106**, 15,929-15,940 (2001).
18. Malara, F., Primavera, L., and Veltri, P., *Phys. Plasmas*, **7**, 2866-2877 (2000).
19. Malara, F., Primavera, L., and Veltri, P., *Nonlinear Proc. Geoph.*, **8**, 159-166 (2001).
20. Marsch, E., and Tu, C.-Y., *J. Geophys. Res.*, **95**, 8211-8229 (1990).
21. Matthaeus, W. H., Ghosh, S., Oughton, S., and Roberts, D. A., *J. Geophys. Res.*, **101**, 7619-7629 (1996).
22. Matthaeus, W. H., Zank, G. P., Smith, C. W., and Oughton, S., *Phys. Rev. Lett.*, **82**, 3444-3447 (1999).
23. Oughton, S., Priest, E. R., and Matthaeus, W. H., *J. Fluid Mech.*, **280**, 95-117 (1994).
24. Roberts, D. A., *Geophys. Res. Lett.*, **17**, 567-570 (1990).
25. Roberts, D. A., Goldstein, M. L., and Klein, L. W., *J. Geophys. Res.*, **95**, 4203-4216 (1990).
26. Smith, E. J., Balogh, A., Neugebauer, M., and McComas, D., *Geophys. Res. Lett.*, **22**, 3381-3384 (1995).
27. Tu, C.-Y., and Marsch, E., *Space Sci. Rev.*, **73**, 1-210 (1995).

Electric Fluctuations and Ion Isotropy

P. J. Kellogg,* M.K. Dougherty,[¶] R.J. Forsyth[¶], D.A. Gurnett,[§] G.B. Hospodarsky,[§] and W.S. Kurth[§]

*School of Physics and Astronomy, University of Minnesota, 116 Church St. S.E., Mpls., MN 55455, USA
[§]Department of Physics and Astronomy, University of Iowa, Iowa City, IA 52240, USA
[¶]Physics Department, Imperial College, London, SW72AZ, UK

Abstract. In a recent paper observations of electric field fluctuations in the range .5 to 25 Hz were reported, observations by the RPWS experiment on Cassini. We have found that the data presented in that paper are affected by broadband interference which appears to be generated in the spacecraft wake. Evidence for a wake instability will be presented. We present a new spectrum for electric fluctuations in the solar wind taken when the antennas are outside the wake. In earlier papers we have tried to understand why the solar wind behaves as a collisional plasma although collisions are very rare. Whether these electric fluctuations might replace the effects of collisions will be discussed.

INTRODUCTION

In an earlier paper, observations were reported of electrostatic fluctuations in the frequency range below 25 Hz, from the RPWS experiment on the Cassini spacecraft. We now have found that the interpretation was wrong, and that the fluctuations are due to an instability on the wake of the spacecraft, specifically on the wake of the 4 m diameter high gain antenna. The evidence for this wake turbulence is presented, and then a new spectrum, taken with antennas outside the wake, is presented, and evidence is given that this represents waves in the free solar wind

The solar wind behaves as a collisional plasma although collisions are very rare. MHD seems to work, and the ions are relatively isotropic [1,2,3]. The ions, through conservation of magnetic moment, ought to have $T_{\|} / T_{\perp}$ of a few hundred, whereas it is observed to be within a factor of 2 of unity. It is probable that fluctuations of the fields replace collisions. To be most effective, electric fluctuations should be nearly resonant with the ions, and with the Doppler shift of reasonable candidate wave modes, would, at 1 AU, then appear in the range around and below 1 Hz. This range of electric fields has not been well explored experimentally. Because of photoelectric variations of the potential of a cylindrical antenna, this frequency range cannot be measured on a spacecraft spinning in the usual direction, i.e., around an axis nearly perpendicular to the sun direction Cassini is the first 3 axis stabilized spacecraft with a measurement channel devoted to this frequency range [4]. Kellogg et al., [3], reported such measurements while Cassini was in the range of 1 to 1.2 AU from the sun. However, on 1 Oct 2000 when a series of maneuvers was begun to allow various of the instruments to observe Jupiter, it was seen that the signal level in the frequency range .2 to 25 Hz depended strongly on spacecraft attitude in a way that did not seem consistent with natural signals.

EXPERIMENT DESCRIPTION

The measurements reported here are from the RPWS experiment on Cassini. [4]. Cassini is a large spacecraft, 6 m long, carrying a fixed 4 m diameter telemetry antenna at one end. RPWS measures electric fields using 3 orthogonal monopole antennas 10 m long and 2.86 cm diameter, made of beryllium-copper and magnetic fields using three mutually orthogonal search coils. Normally two of the electric monopoles are connected as a dipole, Ex, whose electric axis is approximately in the spacecraft X direction, and the third monopole, called Ew, is operated alone. Cassini must be oriented with the parabolic (high gain) antenna pointing toward the earth when downlink is required from distant positions. A crude sketch of the aspect of Cassini with respect to the sun in this case is shown in the lower two panels of Fig. 1. In this only the main structure of the spacecraft and the high gain antenna are shown together with the electric field monopoles. The bases of the antennas actually issue from a mechanism

CP679, *Solar Wind Ten: Proceedings of the Tenth International Solar Wind Conference*, edited by M. Velli, R. Bruno, and F. Malara

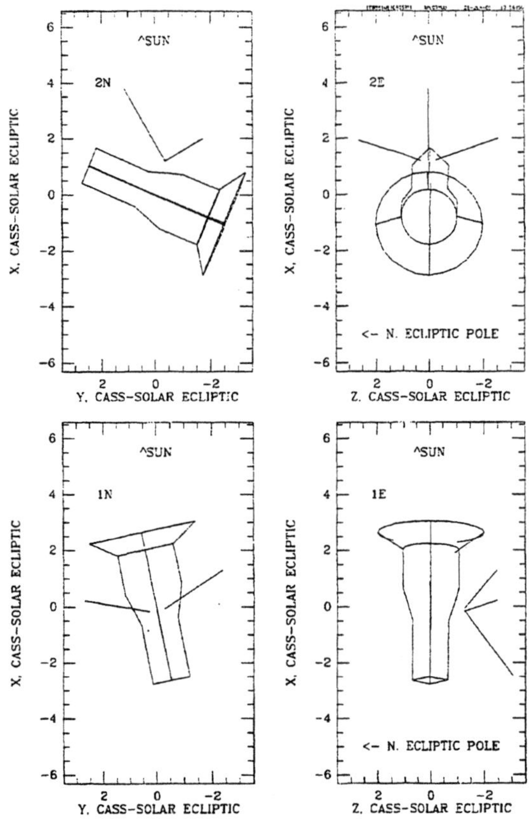

FIGURE 1. A rough sketch of the Cassini spacecraft, showing the positions of the high gain antenna and the RPWS antennas (heavy lines) in earth pointing attitude (lower two panels) and out of the wake (upper panels). N and E refer to a view from the North ecliptic pole and from the East, and 1 and 2 refer to different periods of Fig. 2.

which is not shown. One important appendage which is not shown is the magnetometer boom, which nearly bisects the Ex monopoles but is perpendicular to the spacecraft axis.

It will be seen, assuming that the solar wind is flowing in a radial direction, that the bases and lower parts of the RPWS antennas are in the plasma wake of the high gain antenna, and that they protrude through the wake surface. As Cassini traveled toward Jupiter, the sun-earth angle was never large enough to bring the RPWS antennas out of the wake until 1 Oct 2000.

WAKE INTERFERENCE

The first indications that something did not fit the interpretation of waves in the freely streaming solar wind came from comparing the signal intensities in the Ex and Ew antennas. The signal on the Ew monopole was several times more intense than the signal on the

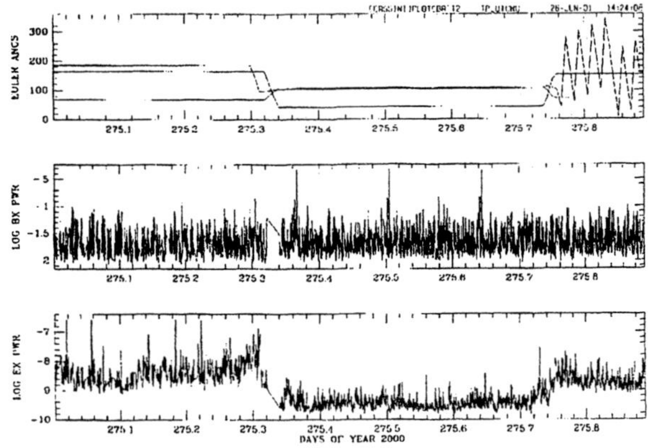

FIGURE 2. RPWS observations on 1 Oct, 2000 showing electric and magnetic fields and how they change as the orientation of Cassini is changed. The upper panel shows the Euler angles of the attitude without designation to show when the attitude changed.

Ex dipole. An explanation came on 1 Oct 2000 when Cassini was rotated so that some experiments could view Jupiter for the first time. The relative power observed in the Ex and Bx channels is plotted for the day in the lower two panels of Fig. 2, and the Euler angles of the spacecraft attitude are plotted without identification in the topmost panel, just to show times when the attitude changes. Relative power is power from the spectrum multiplied by freq $^{1.67}$, to make all frequencies equally important—otherwise the power is mainly just the power at the lowest frequency.

The attitude of Cassini at day 275.11 is shown in the lower panels of Fig. 1, while the attitude at day 275.55, (when the signal is Ex is considerable reduced) is shown in the upper panels of the figure. The changes were too large to be explained as response to an anisotropic signal, and furthermore show complete correlation with antenna position on following days.

The Ew-Ex difference was found to decrease as Cassini traveled outward. We now interpret this Ew-Ex difference as mainly due to an instability which is larger on the Ew antenna as it is farther downstream in the wake, and the decrease as due to a slowing of the growth rate as the plasma becomes less dense. Some Ew-Ex difference may be due to the different response of a monopole antenna to density fluctuations however, which are largely cancelled out by a dipole.

In summary, then, the observations lead to the belief that the signals are contaminated by an instability which is most intense on the surface of the wake of the high gain antenna, and which grows in the downstream direction. To investigate an instability on the surface of the wake, plots of signal vs. angle of the X antenna which was closest to the antisolar direction, were made.

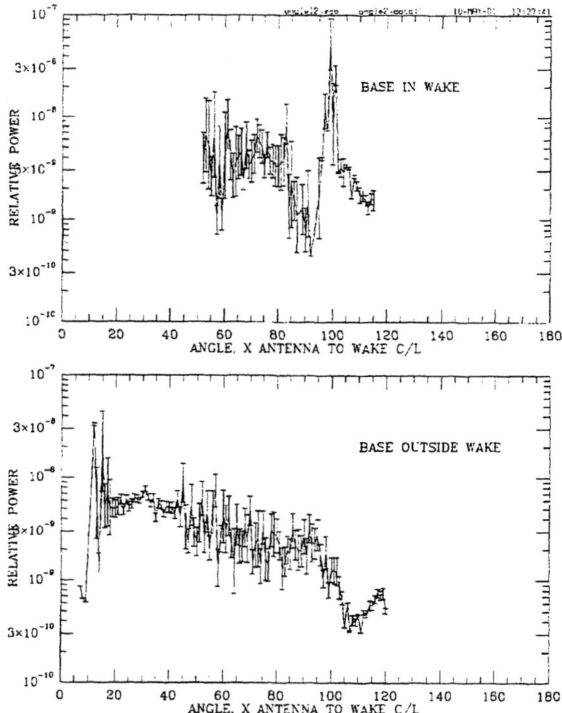

FIGURE 3. Relative wave power as a function of RPWS antenna position with respect to the anti-sun direction, taken as being the wake center line.

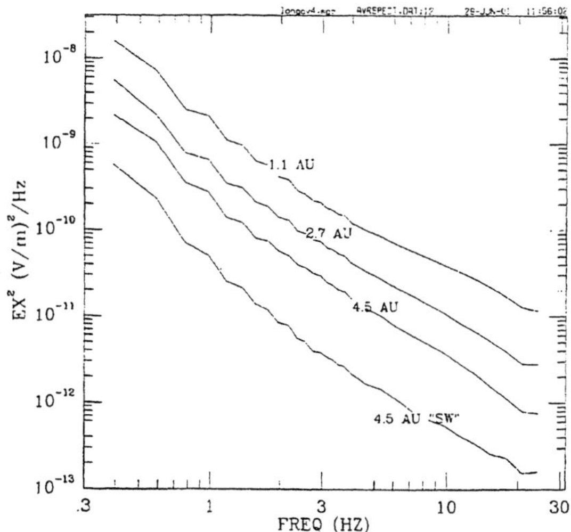

FIGURE 4. Observed electric field power spectrum in the spacecraft frame at various distances from the sun. The upper three curves are probably from wake turbulence. The lowest curve may be a true representation of the spectrum in the undisturbed solar wind at 4.5 AU.

In Fig. 3 plots of relative power in the .5-25 Hz band vs. the angle between the anti-sun direction (called the wake), and whichever X monopole is closest to the anti-sun direction, are shown.

The signals have been averaged in 3° bins, and the error bars shown are the expected standard deviation of the means. It will be seen that there is a sharp drop in the signal amplitude when the nearest X antenna is at an angle of more than 103° with the antisolar direction, which we interpret as meaning that the wake turbulence is radiated outward at a large angle. For base out of wake, there is a peak at about 10°, which we suppose means that the antenna is lying right in the turbulent layer around the wake, and the relative power also drops substantially at 100-105°.

Fluctuations in spacecraft wakes in the magnetosphere have been reported by several investigators and some theoretical work has been done [5,6,7,8]. However, the plasma regime here is quite different from the Earth's magnetosphere, both ions and electrons being unmagnetized. The Debye length is much larger than the spacecraft dimensions so that electrons can freely enter the wake even though ions are absent, and so only small electron density gradients are expected, conditions which argue against lower hybrid waves for this case. The electrons in the wake should be streaming toward the spacecraft and antenna, whereas the ions of the solar wind are streaming away and this suggests a Kelvin-Helmholtz instability on the wake-solar wind boundary due to this velocity difference. Note that the configuration of the wake of the Cassini high-gain antenna is quite different from most spacecraft wakes, in that the body of the spacecraft, presumed to be electrically positive with respect to the plasma, occupies much of the wake, while in most situations the wake is negatively charged.

Fig. 4 shows electric field spectra averaged over a solar rotation of the signals measured at three different solar distances. Three of the spectra are those when Cassini's attitude was like that shown in the lower panels of Fig. 1 so that they are spectra of the wake turbulence. At 3 Hz, the turbulence power varies as $r^{-1.3}$, more slowly than the plasma density. The lowest spectrum, marked 4.5 AU "SW", is from data when the antennas are upstream from the spacecraft body. We believe that this spectrum represents waves in the free solar wind. Not only is this spectrum about 10 dB weaker than the corresponding wake spectrum, but also its slope is different.

As a working hypothesis, it is assumed that the "SW" data represent electric fields in the free solar wind. To try to justify this assumption, in Fig. 5 Cassini data are compared to STO experiment data from Ulysses at about the same distance from the sun and for a heliographic latitude near 0, but at a higher frequency.

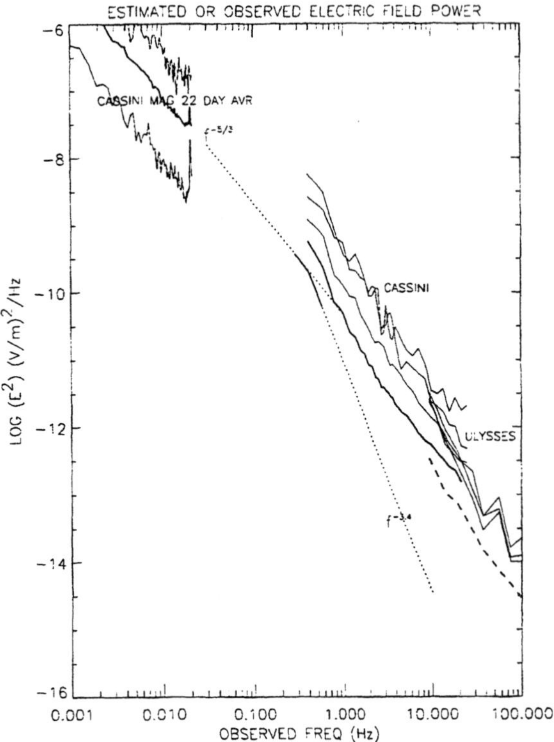

FIGURE 5. Estimation and observation of electric fields at 4.5 AU plotted against observed (Doppler-shifted) frequency. In the upper left are shown expected electric fields from the Lorentz transformation of magnetic fluctuations. The heavy line running between .4 z and 25 Hz is the averaged observed "SW" spectrum. See text.

In the region near 4.5 AU, the noise threshold of the STO experiment in the 9-448 Hz range is near the average signals from Cassini RPWS so that comparison of averages is not meaningful. Rather, we have selected a few periods of strongest signals. These are shown as lighter lines in Fig. 5. The dashed line near the ULYSSES data is the noise threshold for the instrument. Now they agree perfectly, though in our earlier paper [3] the Cassini spectrum was out of line with Ulysses. Hence, these direct measurements of electric fields give some confirmation that the "SW" electric field is actually the electric field in the undisturbed solar wind. Unfortunately, except for a very short period, the high gain antenna was pointed either at Earth or at the sun until Cassini had reached 4.5 AU on 1 Oct. 2000, aspects which put the antennas in the wake, so that no uncontaminated data are available closer to the sun.

ELECTRIC FIELDS IN THE SOLAR WIND

A contribution to observed electric field fluctuations in the solar wind comes from the Lorentz transform of the magnetic fluctuations. It is essential to determine whether the observed fields are from this Lorentz transform or from either electrostatic waves or the longitudinal component of the wave modes of the magnetic fluctuations, because the ions are not affected by Lorentz transform fields. The Lorentz transform of magnetic fluctuations gives a field of the order of $E(f) = V_{sw}B(f)$. Since, in the region considered a typical Alfven speed V_A is only about 15 km/s while the typical solar wind speed, V_{sw}, is about 500 km/s, the intrinsic electric field which the ions see might be much smaller than the field measured here if the observed fields are the Lorentz transform of the magnetic fluctuations.

However, at 4.5 AU, the Cassini search coil is not sufficiently sensitive to be used to measure the magnetic component of these waves. Therefore, we have to use some indirect arguments which, however, will not lead to a definite conclusion.

It is well known that magnetic power spectra follow the Kolmogorov power law, -5/3, for the inertial range. Magnetic fluctuations are well known to be intermittent, with power which varies from one period to another by orders of magnitude. However, during any interval of 10 minutes or more, the spectrum seems always to be roughly a power spectrum with index near -5/3. We therefore consider the spectra at lower frequencies obtained from the Cassini MAG experiment of Imperial College and extrapolate them to the desired frequency range. Some magnetic fluctuations are shown in Fig. 5 in the upper left corner.

What is shown is an average spectrum of the transverse (to the antisolar direction) magnetic field, averaged over the same period as the electric field spectrum multiplied by V_{sw} to give the electric field of the expected Lorentz transform (heavy line). Also shown are two spectra (light lines) representing large and small daily averages, to give an idea of the variation encountered. A dashed line has been drawn from the high frequency limit of the magnetic power spectrum at the –5/3 power law. It is also well known [9,10] that this power law ceases to be valid (beginning of the "dissipation" range) at a frequency of the order of the ion cyclotron frequency. At the frequency given by Neugebauer, i.e, at a Doppler shifted frequency whose wavenumber is equal to the inverse proton cyclotron radius we have continued it at a typical power law found by Beinroth and Neubauer [11], namely $f^{-3.4}$. This is our attempt to use what is known of magnetic field fluctuations to extrapolate the magnetic measurements to a higher frequency range and to calculate the electric field which would result from the Lorentz transformation.

It will be seen that the averaged Cassini electric field spectrum (heavy line) lies slightly above the estimates from these extrapolations, especially at frequencies above about 1 Hz. However, in the region expected to be resonant with the ions, namely around .4 Hz marked with a heavier line, the electric field is larger than the extrapolation only by an amount not large compared to the variation of the magnetic spectra (as shown by the spectra in light lines in Fig. 5), and so we consider that it is uncertain whether the averaged Cassini electric field is larger than the Lorentz transform of the magnetic field or not. Note that if electrostatic waves were actually to be measurable above the Lorentz transform of B, then their electric fields would have to be quite large in the plasma frame, larger by a factor of $V_{sw}/V_A \approx 30$ than the electric fields of order $V_A B$ due to electromagnetic modes. Measurement of density fluctuations [12] provides the best way to measure electrostatic waves in the relevant frequency range.

If, as is true at frequencies below one cycle per minute, the spectra remain $f^{-5/3}$ power laws even though their amplitudes vary, then it might be expected that there is a correlation between the amplitudes of the magnetic spectra from MAG at frequencies below .02 Hz and the amplitudes of electric field fluctuations in the Cassini range above .4 Hz. No such correlation was found and, in fact, the calculated correlation is slightly negative, even though both quantities are positive. This provides some evidence in favor of electrostatic waves.

DISCUSSION

Since the evidence for electrostatic waves is inconclusive, the question of whether the observed fluctuations are sufficient to isotropize the ions will be discussed according to each of the possibilities. In earlier papers [1] (but there is a typographical error) and [2], a rough estimate of the diffusion of the velocity perpendicular (to the magnetic field) in time τ in a fluctuating electric field was obtained and compared to the decrease of the perpendicular component due to the conservation of magnetic moment, $v_\perp^2/B$. The order of magnitude calculations in those papers will not be repeated here for lack of space. The spectrum used in [2] was erroneously calculated from density fluctuations, and was relatively flat compared to what we now think is correct. Hence the diffusion, proportional to the electric power, is very sensitive to the lower frequency limit. We integrate the electric power under the heavy line in the octave around .4 Hz in Fig 5. If it is assumed that the observed electric fields are actually those of electrostatic waves, then the electric fields are more than sufficient to maintain isotropy.

At 4.5 AU, there are qualitative differences with the situation at 1 AU and nearer the Sun. As discussed by [2], the dominant contribution to diffusion of ions inside 1 AU is electric fluctuations, even if these are only from electromagnetic fluctuations (ion cyclotron waves and whistlers), since the ratio of E to B in these fluctuations is of order Alfven speed, v_A, while the ratio of forces is the thermal speed. At 4.5 AU however, the thermal speed is expected to be larger than the Alfven speed, and so magnetic fluctuations provide the dominant diffusive force (unless there are electrostatic fluctuations). In this case, the fluctuations appear only marginally able to maintain isotropy. However, the calculation is only correct to an order of magnitude so that the fluctuations might be sufficient. Unfortunately, the parameters of the plasma and the solar wind are such that it is difficult to decide this question at 4.5 AU.

SUMMARY AND CONCLUSIONS

Our main purpose is to identify fluctuating fields which might replace collisions to validate MHD and to maintain the isotropy of the ion distributions. It turns out that it is difficult to do this at 4.5 AU, given the parameters of the solar wind and the thresholds of the Cassini RPWS instrument complement. Significant electric fields have been observed but it is difficult to know for sure that these are electrostatic waves since the magnetic fluctuations in the same frequency range are too small to be observed. If these electric fields are electrostatic, then they are plenty large enough to account for the isotropy of the ions. If they are the Lorentz transform of the magnetic fluctuations, then they are perhaps marginally large enough.

The evidence leans toward electrostatic waves in that (1) the electric fields are slightly larger than extrapolations of Lorentz transformed magnetic fields, and (2) there is no correlation with the amplitude of magnetic field spectra. However, if the waves are not electrostatic and the magnetic fluctuations are dominant, then the picture is that isotropy is just due to the heating of the perpendicular component of the ion velocity by absorption of the magnetic fluctuations, a picture which was suggested long ago [13].

ACKNOWLEDGMENTS

The work at the University of Iowa was carried out under Contract 961152 with the National Aeronautics and Space Administration. The work at Imperial College was carried out under a PPARC research grant.

All of the authors would like to thank the many members of the Cassini project and the experiment teams who made possible the success of this mission, particularly D.L. Kirchner and T.F. Averkamp.

REFERENCES

1. Kellogg, P. J., and Lin N., "Ion Isotropy and Fluctuations in the Solar Wind, "*in SP-415, The 31st ESLAB Symposium, Correlated Phenomena at the Sun, the Heliosphere and in Geospace*, A. Wilson, Ed., ESTEC, Noordwijk, the Netherlands, 1997.
2. Kellogg, P. J., *Ap. J.*, **528**, 480-485 (2000).
3. Kellogg, P. J., Gurnett, D.A., Hospodarsky, G.B., and Kurth, W.S., *Geophys.Res.Lett* **28**, 87-90 (2001).
4. Gurnett, D.A., et al., The Cassini Plasma Wave Investigation, in press, *Space Sci. Rev.* (2002).
5. Gurnett, D.A., Kurth, W.S., and Steinberg, J.T., *Geophys.Res.Lett* **15**, 760-763 (1988).
6. Murphy, G. B., Reasoner, D.L., Tribble, A., D'Angelo, N., Pickett, J.S., and Kurth, W.S., *JGR* **94**, 6866-6872 (1989).
7. Keller, A. E., Gurnett, D.A., Kurth, W. S., *Planet. Space Sci.*, **45**, 201-219 (1997).
8. Samir, U., Comfort, R.H., Singh, N., Hwang, K.S., Stone, N.H., *Planet. Space.Sci.* **37**, 873 (1989).
9. Neugebauer, M., *JGR* **80**, 998-1002 (1975).
10. Leamon, R. J., Smith, C.W., Ness, N.F., Mattheus, W.H., and Wong, H.K., *JGR* **103**, 4475 (1998).
11. Beinroth, H. J., and Neubauer. F.M., *JGR* **86**, 7755 (1981).
12. Kellogg, P. J., Goetz, K., Monson, S.J., and Wygant, J.R., *JGR* **104**, 12627 (1999).
13. Tu, C.-Y., *JGR.* **93**, 7 (1988).

Cyclotron-resonant diffusion regulating the core and beam of solar wind proton distributions

C.-Y. Tu*, E. Marsch† and L.-H. Wang*

*Department of Geophysics, Peking University, 100871, Beijing, China
†Max-Planck-Institut für Aeronomie, 37191 Katlenburg-Lindau, Germany

Abstract. Ion diffusion as predicted by quasi-linear theory has been compared with in-situ solar wind proton measurements. It is found that the observed phase-space-density contours match very well those corresponding to the time-asymptotic plateau generated by proton diffusion in cyclotron-wave resonance. Observations show that the perpendicular temperature of the beam distribution is of the same order as its parallel one. A perpendicular heating mechanism is needed to balance the radial tendency for adiabatic cooling. Outward and inward propagating cyclotron waves may together be able to control the thermal anisotropy of the core distribution. However, there are hardly any cyclotron waves, which could resonate with a proton beam having a drift velocity equal to or greater than the Alfvén speed. Therefore, we consider also outward-propagating waves, with both left and right hand polarization, on a second dispersion branch existing in a cold plasma with electrons, protons and alpha particles. These waves can resonate with the beam protons. The resulting diffusion can indeed explain the shape of the beam distribution. A time-dependent kinetic code, in two-dimensional velocity space, has been developed to integrate the quasi-linear diffusion equation. An initial shuttle-like distribution function is shown to develop into a distribution having a core and a beam. The beam is found to drift at the Alfvén speed and be less anisotropic than the core. The radial evolution of the beam density in the model is found to be consistent with the observations.

INTRODUCTION

The velocity distribution of solar wind protons is known to have usually two components, an anisotropic core and a drifting beam. Such non-thermal distributions were found first near 1 AU in the early days of space observations [5] and later on by Helios near 0.3 AU [16]. Recently, these non-thermal features were also identified in the Ulysses [8] and Wind [1], [13] spacecraft observations.

Many theoretical papers were published to explain these phenomena, which have nevertheless not been understood fully. Several papers, (such as [2], [9], [10], [17], [24], [25], and [29]) tried to explain the perpendicular heating by means of the heating rates as calculated in the quasilinear theory of cyclotron resonance. It was further assumed that the velocity distribution function (VDF) was a fixed drifting bi-Maxwellian. These theoretical works yielded proton anisotropies somewhat lower than the observed ones. However, especially when considering the damping of the waves at frequencies near the ion gyro-frequencies [24], such kind of models failed in describing the large anisotropy of the O^{5+} oxygen ions in the solar corona [14]. A common weakness of the models is that resonant diffusion is not properly described. In order to consider these wave-particle processes the rigid bi-Maxwellian VDF must be given up.

RECENT DEVELOPMENTS

Here we will briefly review some recent work. The thermal anisotropy instability was invoked ([7], [8]) to explain the limits of the observed value of the proton thermal anisotropy, $A = T_{p\perp}/T_{p\|} - 1$. Recently, the anisotropy was analysed again, using data obtained in low-speed wind by the WIND plasma instrument [13], and found to have an upper bound close to the theoretical linear instability threshold. However, the majority of the data points are far from this threshold and cannot be explained as being limited by the instability. The theoretical instability threshold, $A = 0.6\beta^{0.40}$ [7], is somewhat lower than the observed values of A, in particular when the plasma beta, β, is less than 0.1 as in high-speed wind near 0.3 AU [26]. This failure to predict a proper upper limit may be considered as a clear indication that the core VDF in high-speed wind is not described adequately by a bi-Maxwellian VDF, which was used in the instability calculation. Moreover, a theory involving a shape-invariant VDF can also not explain how the thermal anisotropy is formed in the first place.

Concerning wave dissipation and generation, it should be pointed out that the energy source required for the primary wave heating of the protons is entirely different from the energy source of the secondary waves resulting from marginal stability. With regard to primor-

CP679, *Solar Wind Ten: Proceedings of the Tenth International Solar Wind Conference*,
edited by M. Velli, R. Bruno, and F. Malara

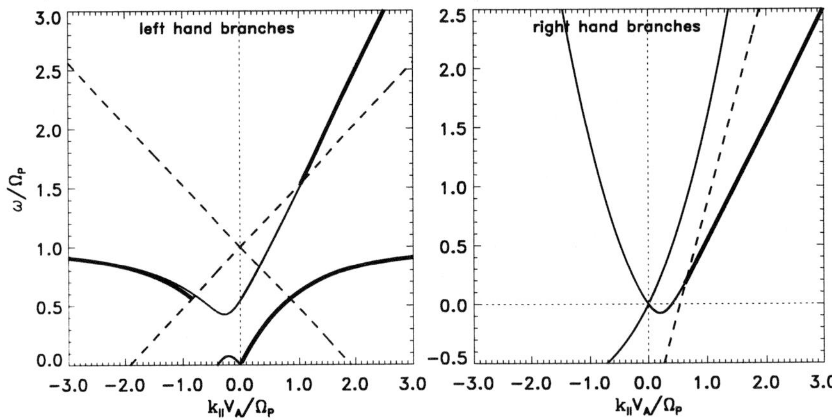

FIGURE 1. The dispersion relation of a cold plasma with alpha-particle abundance, $\eta_\alpha = 0.05$, and normalized drift, $U_\alpha = 1$. The left panel shows the dispersion relation of LHP waves. The dashed lines delineate the cyclotron-resonance condition with $v_{\|}/V_A = -0.52$ and $v_{\|}/V_A = 0.52$, respectively. The thick curves show the dispersion relation used in this numerical simulation. The right panel shows the dispersion relation for RHP waves. The dashed line shows the resonance condition with $v_{\|}/V_A = 1.86$.

dial proton heating, pre-existing high-frequency waves are damped while being in resonance with the particles. The energy may be replenished from waves in the low-frequency range, perhaps by cascading [2], [20], [22] or the sweeping mechanism [21], [23]. In contrast, in the case of an instability the wave energy comes from the kinetic thermal energy of the ions, and the threshold is determined by assuming a small growth rate. It is clear that these mechanisms cannot work simultaneously at the same frequency, although they both predict asymptotic states with small absolute values of the growth rate.

The proton VDF is expected to be isotropic near the base of the corona as the result of collisions. Due to the mirror force, the proton thermal anisotropy, A, will initially decrease with the expanding solar wind and thus tend to have negative values. Consequently, instability theory does not explain how the proton core anisotropy can attain values between 3 and 4, as observed near 0.3 AU in high-speed streams.

There are only few papers in which one tried to explain the original formation of the proton beam distribution. Tam and Chang [19] presented a hybrid solar wind model showing the formation of a proton beam. However, the differential speed of the beam is about three times the Alfvén velocity (see their Figure 3). This value is too large when campared with observations. This large differential speed may come from the resonance of the protons with RHP waves. Clack et al. [1] compared the proton beam differential velocity observed by WIND with the limit set by the Alfvén instability [4] and found that most data points lie considerably below the instability threshold, especially in cases where the core plasma beta is low. They concluded that how exactly the proton-proton relative velocity is constrained remains unclear.

It is obvious that non-thermal features of solar wind protons should be explained by resonant diffusion. Recently, quasilinear diffusion theory has been used to develop kinetic solar wind models [3], [6], [9], [11], [12], [28]. Marsch and Tu [18] provided from Helios observations the first direct evidence of proton velocity-space diffusion driven by cyclotron resonance. Tu and Marsch [26] showed that the observed proton core distribution (and its anisotropy) could be explained by quasilinear diffusion driven by resonance of protons with LHP waves, having an inward as well as outward sense of propagation. Marsch [15] presented a brief review of this topic in these proceedings. Tu et al. [27] showed that the proton beam might be understood as the result of the resonance of protons with outward propagating waves (both LHP and RHP), which follow the dispersion of the second branch that exists in a plasma with alpha particles drifting along the field at about the Alfvén speed.

FORMATION OF THE PROTON BEAM DISTRIBUTION

For a given wave power spectrum density, the quasilinear diffusion equation [2] is solved by means of a two-dimensional code based on the Lax-Wendroff integration scheme [27]. In Figure 1 the used dispersion relation is shown, which pertains to a plasma consisting of protons, alpha particles and electrons. The relative abundance of the alpha particles is taken to be 0.05, and their drift

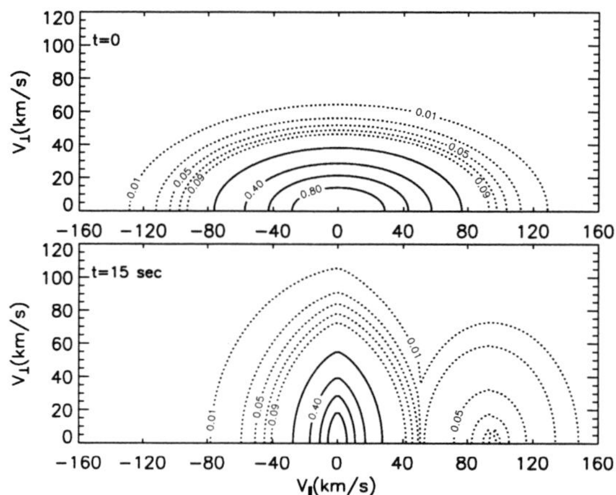

FIGURE 2. Formation of a proton beam VDF. The upper panel shows the initially assumed bi-Maxwellian distribution with parameters: $v_\parallel = 60$ km/s, $v_\perp = 30$ km/s, and $u_\parallel = 0$ km/s. The bottom panel shows the simulation results after 50000 time steps in the numerical run. The magnetic field is $B = 40$ nT and the density, $n_p = 82.5$ cm^{-3}

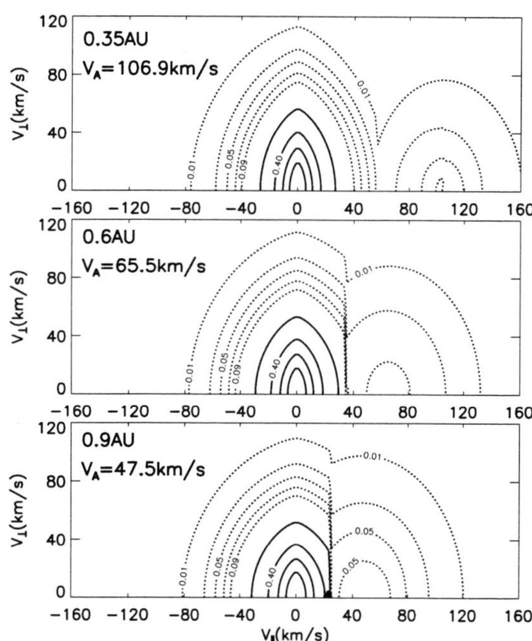

FIGURE 3. Radial evolution of a proton beam VDF. The upper panel shows the final stable VDF resulting at 0.3 AU, which is simulated from an initial distribution function that is the same as the lower distribution in Figure 2. The middle panel shows the final stable distribution at 0.6 AU. The bottom panel shows the final stable VDF at 0.9 AU.

velocity relative to the protons is V_A. The solid lines show the dispersion relation and the dashed lines show the resonance conditions.

The LHP first-branch waves with outward sense of propagation are in resonance with inward moving protons. The LHP first branch waves with inward propagation sense (see the thick solid line) are in resonance with outward moving protons, having a parallel velocity less than a critical velocity, V_C, say of 50 km/s. The protons with velocity greater than 50 km/s, but less than $V_A = 96$ km/s are in resonance with the second-branch LHP waves with outward sense of propagation. Protons with velocities greater than V_A are in resonance with second-branch RHP waves with outward sense of propagation.

The wave power spectrum, P_B, is assumed to have a power-law form: $P_B(k_\parallel) = P_0 \mid k_\parallel/k_0 \mid^{-\gamma}$, where $\gamma = 5/3$, $k_0 = 10^{-10}$ cm^{-1}, and $P_0 = 7180$ G^2 cm, which is consistent with a power density of 1 nT^2/Hz near the proton gyrofrequency according to the Helios observations at 0.3 AU. The power densities of the four wave modes involved are all given. The initial velocity distribution is assumed to be elongated along the magnetic field direction and shown in the upper panel of Figure 2. After integrating the diffusion equation for 50000 steps in time, corresponding to 15 seconds ($9\times2/\Omega_p$, where $\Omega_p = 3.83$ rad/s), we obtain a rather stable distribution, which is shown in the lower panel of Figure 2 and reveals two components, core and beam. The resulting anisotropies of the two components are 1.8 and 1.1, respectively. The velocity difference between beam and core is 96 km/s and about equal to V_A. In conclusion, the numerical results describe the observations well.

EVOLUTION OF THE DENSITY RATIO BETWEEN CORE AND BEAM PROTONS

With this simulation code we can also calculate the radial evolution of the ratio of the beam density to the core density. Assuming an initial distribution function at $r = 0.35$ AU and typical plasma parameters there, i.e. $V_A = 106.9$ km/s, we ran the numerical program and obtained finally a stable distribution (still for $r = 0.35$ AU). Then we used this final distribution function as the initial input distribution at $r = 0.6$ AU and typical plasma parameters at that distance, where $V_A = 65.5$ km/s, and found another final distribution. We further applied the same procedure at $r = 0.9$ AU and thus got finally the stable VDF at $r = 0.9$ AU. We also assumed that the ratio, V_C/V_A is constant. The radial evolution of the distribution function is shown in Figure 3.

The corresponding radial evolution of n_b/n_p is shown in Figure 4, where n_b is the beam density and n_p the total proton density. The crosses show the Helios observations

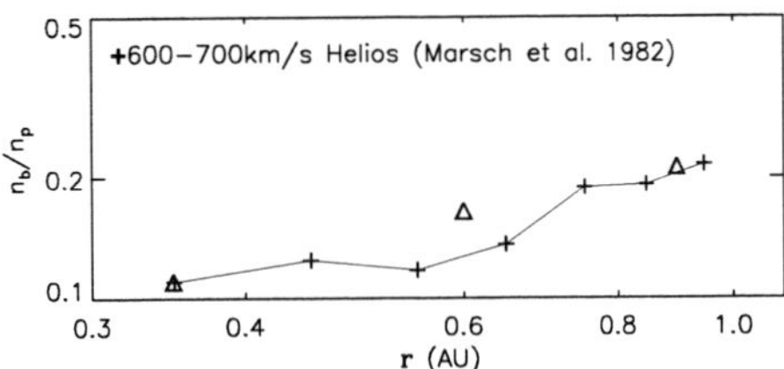

FIGURE 4. Radial evolution of the relative beam density (n_b/n_p). The crosses are the Helios observations (taken from [16]), and the triangles are our simulation results with the model density and magnetic field from [22].

in high-speed solar wind [17]. Both model results and observations exhibit the same radial trend.

However, the possible weakness of this comparison should be pointed out. The observational data shown do not have error bars. The discrimination of the beam from the core protons in not easy and may lead to a somewhat ambiguous separation in the measured data. Moreover, in the model calculations we did not consider the self-consistent evolution of the wave dispersion relation. Also, the effects of the magnetic field direction changing between 0.3 AU and 0.9 AU were not considered.

CONCLUSION AND DISCUSSION

The proton core temperature anisotropy and the differential speed of the proton beam have been explained by invoking cyclotron resonance diffusion. This scenario is supported by observations as well as numerical simulations. The core protons can be in resonance with first-branch LHP waves with outward sense and first-branch LHP waves with inward sense of propagation. The beam protons can be in resonance with second-branch LHP and RHP waves having an outward propagation direction.

So far, it has been impossible to identify the second-branch waves from magnetic field observations made on single spacecraft. To detect and observe the second-branch waves and to identify the possible different wave modes remains a task for future solar wind missions. One should also try to find a physical mechanism to create the second-branch waves in the solar wind plasma. For example, a highly anisotropic alpha-particle beam may be able to generate the second-branch waves.

The dispersion relation in our model calculations was simply assumed. This dispersion relation may not be consistent with the evolution of the model distribution in the numerical run. A self-consistent treatment of both, the velocity distribution function and the dispersion relation, is required and should be considered within the rationale of this model as the next reasonable step.

ACKNOWLEDGMENTS

Tu's and Wang's work was supported by the National Natural Science Foundation of China under projects with contract numbers 40174045 and 49990452, and by the Foundation of Major Projects of National Basic Research under contract number G-2000078405.

REFERENCES

1. Clack, D., A.J. Lazarus, J.C. Kasper, *SW10*, this volume, (2002)
2. Cranmer, S.R., G.B. Field, and J.L. Kohl, *Astrophys. J.*, 518, 937 (1999)
3. Cranmer, S.R., *J. Geophys. Res.*, 106, 24937 (2001)
4. Daughton, W. and S.P. Gary, *J. Geophys. Res.*, 103, 20613 (1998)
5. Feldman, W. C., J.R. Asbridge, S.J. Bame, and M.D. Montgomery, *Geophys. Space Phys.*, 4, 715 (1974)
6. Galinsky, V.L., and V.I. Shevchenko, *Phys. Res. Lett.*, 85(1), 90 (2000)
7. Gary, S.P., R.M. Skoug, J.T. Steinberg, and C.W. Smith, *Geophys. Res. Lett.*, 28, 2579 (2001)
8. Goldstein, B., M. Neugebauer, L.D. Zhang and S.P. Gary, *Geophys. Res. Lett.*, 27, 53, (2000)
9. Hu, Y.-Q., R. Esser, and S.R. Habbal, *J. Geophys. Res.*, 102, 14661 (1997)
10. Isenberg, P.A., *J. Geophys. Res.*, 89, 6613 (1984)
11. Isenberg, P.A., M.A. Lee, and J.V. Hollweg, *J. Geophys. Res.*, 106, 5649 (2000)
12. Isenberg, P.A., *J. Geophys. Res.*, 106, 29249 (2001)
13. Kasper, J.C., A.J. Lazarus, S.P. Gary, *SW10* (2002)
14. Kohl, J.L., et al., *Astrophys. J.*, 501, L127 (1998)
15. Marsch, E., *SW10*, this volume, (2002)
16. Marsch, E., K.-H. Mühlhauser, R. Schwenn, H. Rosenbauer, W. Pilipp, and F.M. Neubauer, *J. Geophys. Res.*, 87, 52 (1982a)
17. Marsch, E., K.-H. Mühlhauser, H. Rosenbauer, R. Schwenn, and F.M. Neubauer, *J. Geophys. Res.*, 87, 35 (1982b)
18. Marsch, E. and Tu, C.-Y., *J. Geophys. Res.*, 106, 8357 (2001)
19. Tam, S.W.Y and T. Chang, *Geophys. Res. Lett.*, 26, 3189 (1999)
20. Tu C.-Y.,Z.-Y. Pu, and F.-S. Wei., *J. Geophys. Res.*, 89, 9695 (1984)
21. Tu, C.-Y., *Solar Phys.*, 109, 149 (1987)
22. Tu, C.-Y., *J. Geophys. Res.*, 98, 7 (1988)
23. Tu C.-Y. and E. Marsch, *Solar Phys.*, 171, 363 (1997)
24. Tu C.-Y. and E. Marsch, *J. Geophys. Res.*, 106, 8233 (2001)
25. Tu C.-Y. and E. Marsch, *Astron. Astrophys.*, 368, 1071 (2001)
26. Tu C.-Y. and E. Marsch, *J. Geophys. Res.*, in press (2002)
27. Tu, C-Y., Wang, L-H., and E. Marsch, *J. Geophys. Res.*, in press (2002)
28. Vocks, C., and E. Marsch, *Astrophys, J.*, 568, 1030 (2002)
29. Li, X., and S.R. Habbal, Solar Phys., 190, 485 (1999)

Three-dimensional MHD modeling of the solar corona and solar wind

Arcadi V. Usmanov*† and Melvyn L. Goldstein*

*Code 692, NASA Goddard Space Flight Center, Greenbelt, MD 20771, USA
†Institute of Physics, University of St.-Petersburg, St.-Petersburg 198504, Russia

Abstract. A global MHD model is developed to reproduce Ulysses observations during its fast latitude transition in 1994-1995. The governing polytropic single-fluid MHD equations are solved for a steady coronal outflow. The model includes Alfvén wave momentum and energy addition into open field regions. We combine a solution for a tilted dipole magnetic field in the inner computational region (1-20 $R_\odot$) with a three-dimensional solution in the outer region which extends to 1 AU. The inner region solution is essentially the same as in [1], but obtained with a different numerical algorithm and rotated to match the inclination inferred for the solar dipole from observations during the Ulysses transversal. The steady solution in the outer region is constructed by a marching-along-radius method and accounts for solar rotation. We show that the simulated variations of plasma and magnetic field parameters and in particular the extension of slow wind belt agree fairly well with the Ulysses observations.

INTRODUCTION

Ulysses observations during its first fast latitude scan in 1994-1995 revealed a bimodal solar wind with a sharp transition from a uniform fast wind at high latitudes to a relatively slow wind around solar equator [2]. The magnetic field was dominantly outward (inward) in the northern (southern) hemisphere and in the fast wind showed no prominent dependence on latitude [3]. Those observations were taken just prior solar activity minimum when the dominant component of the solar source magnetic field was a dipole inclined by about 10° to the solar rotation axis [4]. The lack of a significant latitudinal gradient in the radial magnetic field implies that magnetic field is transported to lower latitudes by non-radial coronal expansion.

The large-scale structure of the expanding solar corona is determined largely by the pattern of magnetic fields on the solar photosphere. A number of simulation studies have tried to match the Ulysses observations by solving the equations of magnetohydrodynamics (MHD) with the photospheric field as boundary condition [5, 6, 7, 8, 9, 10]. To produce a fast wind and to get a reasonable agreement with typical coronal data, however, an additional source of momentum must be incorporated into the models. There are a number of candidates for this role, one of which is Alfvén waves. The ability of the waves to bring models into agreement with observations both near the Sun and at large distances was recognized three decades ago [11, 12] and was extensively exploited in one-dimensional models with the flow tube geometry prescribed more or less arbitrarily *ab initio*, e.g., [13, 14]. It appears to be very attractive and natural to combine the wave acceleration mechanism with two- and three-dimensional approaches to solar corona and solar wind MHD modeling in which the flow geometry is determined self-consistently.

Usmanov et al. [1] developed an axisymmetric model in which a steady coronal outflow was simulated in a dipolar magnetic field. In that work, the WKB Alfvén waves were explicitly invoked as a means of heating and accelerating the solar wind flow and a steady-state two-dimensional solution was obtained. The dipole field strength and the amplitude of Alfvén waves at 1 $R_\odot$ (solar radius) were chosen to obtain a good fit to Ulysses data. A self-consistent solution was constructed by applying a time-relaxation technique in the region near the Sun. In the outer computational region, a marching-along-radius numerical algorithm was used. The solution formed a bimodal structure of fast and slow wind, as observed, and the computed parameters were generally consistent with Ulysses data and with typical parameters at the coronal base.

Although the bimodality is already present in the models without waves — faster and more tenuous wind from polar regions and slower and denser wind above the streamer in the heliospheric plasma sheet, the contrast is relatively small to match the pronounced latitudinal variation observed by Ulysses [1]. The addition of waves increases the velocity of the faster wind and lowers its den-

CP679, *Solar Wind Ten: Proceedings of the Tenth International Solar Wind Conference*, edited by M. Velli, R. Bruno, and F. Malara

sity to observed values, but has a relatively small effect on the slower wind belt, where the divergence of flow tubes is much higher, plasma is denser, and the plasma beta is much higher than unity. Note also that in [1] the Alfvén wave energy drops to zero (i.e., has no acceleration effect) at the neutral sheet, which is embedded in the plasma sheet where the magnetic field is essentially zero.

The two-dimensional study [1] neglected all gradients in the azimuthal direction and the north-south symmetry was enforced by assuming the solar magnetic field to be a dipole perpendicular to the solar equatorial plane. The dipole assumption led to a heliospheric current sheet (HCS) that was aligned with the equatorial plane, while the actual HCS deviated slightly from that plane [15]. To account for that effect the model curves in [1] were shifted artificially by 15° north and south, emulating a relatively wide belt of slow wind near solar equator.

In the present study, we relax the assumptions of axial and north-south symmetry to account for that HCS warping more consistently. The basic idea is to use essentially the same solution for the inner region as in [1], but to transform it to match the observed position of the solar dipole, and than extend that tilted-dipole solution to the Earth's orbit through the outer region II using a three-dimensional model. By taking into account the solar rotation in the outer region, we incorporate into our model the interaction between faster and slower solar wind streams and the azimuthal component of magnetic field. To solve the problem in the inner region, we applied a newer TVD (Total Variation Diminishing) Lax-Friedrichs algorithm with the Woodward limiter [16]. The field-interpolated central difference approach suggested in [17] is used to maintain the $\nabla\cdot\mathbf{B}=0$ constraint.

MODEL FORMULATION

The governing MHD equations for a single-fluid polytropic flow driven by thermal and Alfvén wave pressure gradients, including solar rotation are

$$\frac{\partial\rho}{\partial t}+\nabla\cdot\rho\mathbf{v}=0, \tag{1}$$

$$\frac{\partial(\rho\mathbf{v})}{\partial t}+\nabla\cdot\left[\rho\mathbf{v}\mathbf{v}+\left(P+\frac{\mathcal{E}}{2}+\frac{B^2}{8\pi}\right)\mathbf{I}-\frac{1}{4\pi}\mathbf{B}\mathbf{B}\right]+\rho\left[\frac{GM_{\odot}}{r^2}\hat{r}+2\boldsymbol{\Omega}\times\mathbf{v}+\boldsymbol{\Omega}\times(\boldsymbol{\Omega}\times\mathbf{r})\right]=0, \tag{2}$$

$$\frac{\partial\mathbf{B}}{\partial t}=\nabla\times(\mathbf{v}\times\mathbf{B}), \tag{3}$$

$$\frac{\partial}{\partial t}\left[\frac{\rho}{2}(v^2-|\boldsymbol{\Omega}\times\mathbf{r}|^2)+\frac{P}{\gamma-1}+\frac{B^2}{8\pi}-\frac{\rho GM_{\odot}}{r}+\mathcal{E}\right]+\nabla\cdot\left\{\left[\frac{\rho}{2}(v^2-|\boldsymbol{\Omega}\times\mathbf{r}|^2)+\frac{\gamma P}{\gamma-1}-\frac{\rho GM_{\odot}}{r}\right]\mathbf{v}+\frac{\mathbf{B}}{4\pi}\times(\mathbf{v}\times\mathbf{B})+\left(\frac{3}{2}\mathbf{v}+\mathbf{V}_A\right)\mathcal{E}\right\}=0, \tag{4}$$

$$\frac{\partial\mathcal{E}}{\partial t}+\nabla\cdot[(\mathbf{v}+\mathbf{V}_A)\mathcal{E}]=-\frac{\mathcal{E}}{2}\nabla\cdot\mathbf{v}-|\mathbf{v}+\mathbf{V}_A|\frac{\mathcal{E}}{L}, \tag{5}$$

where the dependent variables ρ, $\mathbf{v}$, $\mathbf{B}$, P, and $\mathcal{E}$ are the plasma density, the flow velocity in the frame rotating with the Sun, the magnetic field, thermal pressure, and the Alfvén wave pressure, respectively. $M_{\odot}$ is the solar mass, $\boldsymbol{\Omega}$ the solar angular velocity vector, γ the polytropic index, t the time, r the heliospheric distance, G the gravitational constant, $\hat{r}$ a unit vector in the radial direction, $\mathbf{V}_A=\mathbf{B}/(4\pi\rho)^{1/2}$ the velocity of outward propagating Alfvén waves, and $\mathbf{I}$ the unit matrix. The Alfvén wave effects are incorporated into the governing equations in the WKB limit and it is assumed that the waves are damped by a mechanism that may be characterized by a dissipation length L.

As in [1], we separate the computational domain into two regions so that the flow in the outer region would be super-sonic and super-Alfvénic: the inner region I (1-20 $R_{\odot}$), where the equations (1–5) are solved by the time-relaxation method, i.e., the governing equations are integrated in time up to a steady state, and the outer region II (20 $R_{\odot}$- 1 AU) where the solution is constructed by forward integration along the hyperbolic radial coordinate [18].

The governing equations: region I

In this region, we solve a two-dimensional problem of axisymmetric flow in the dipole field and neglect the solar rotation. The equations (1–5) can be then rewritten in component form as

$$\frac{\partial}{\partial t}(r^2\rho)=-\frac{\partial}{\partial r}(r^2\rho u_r)-\frac{r}{\sin\theta}\frac{\partial}{\partial\theta}(\sin\theta\rho u_\theta), \tag{6}$$

$$\frac{\partial}{\partial t}(r^2\rho u_r)=-\frac{\partial}{\partial r}\left[r^2\rho\left(u_r^2+\frac{P}{\rho}+\frac{\mathcal{E}}{2\rho}+\frac{B_\theta^2-B_r^2}{8\pi\rho}\right)\right]-\frac{r}{\sin\theta}\frac{\partial}{\partial\theta}\left[\rho\sin\theta\left(u_r u_\theta-\frac{B_r B_\theta}{4\pi\rho}\right)\right]+\rho r\left(u_\theta^2-\frac{GM_{\odot}}{r}+\frac{2P}{\rho}+\frac{\mathcal{E}}{\rho}+\frac{B_r^2}{4\pi\rho}\right), \tag{7}$$

$$\frac{\partial}{\partial t}(r^2\rho u_\theta)=-\frac{\partial}{\partial r}\left[r^2\rho\left(u_r u_\theta-\frac{B_r B_\theta}{4\pi\rho}\right)\right]$$

$$-r\frac{\partial}{\partial\theta}\left[\rho\left(u_\theta^2+\frac{P}{\rho}+\frac{\mathcal{E}}{2\rho}+\frac{B_r^2-B_\theta^2}{8\pi\rho}\right)\right]$$
$$-r\rho\left[u_r u_\theta-\frac{B_r B_\theta}{4\pi\rho}+\cot\theta\left(u_\theta^2-\frac{B_\theta^2}{4\pi\rho}\right)\right], \quad (8)$$

$$\frac{\partial}{\partial t}(rB_r)=\frac{1}{\sin\theta}\frac{\partial}{\partial\theta}[\sin\theta(u_r B_\theta-u_\theta B_r)], \quad (9)$$

$$\frac{\partial}{\partial t}(rB_\theta)=-\frac{\partial}{\partial r}[r(u_r B_\theta-u_\theta B_r)], \quad (10)$$

$$\frac{\partial}{\partial t}\left[r^2\left(\frac{\rho u^2}{2}+\frac{P}{\gamma-1}+\frac{B^2}{8\pi}+\mathcal{E}\right)\right]$$
$$=-\frac{\partial}{\partial r}\left\{r^2\left[u_r\left(\frac{\rho u^2}{2}+\frac{\gamma P}{\gamma-1}\right)+\frac{B_\theta}{4\pi}(u_r B_\theta-u_\theta B_r)+\left(\frac{3}{2}u_r+V_{Ar}\right)\mathcal{E}\right]\right\}$$
$$-\frac{r}{\sin\theta}\frac{\partial}{\partial\theta}\left\{\sin\theta\left[u_\theta\left(\frac{\rho u^2}{2}+\frac{\gamma P}{\gamma-1}\right)+\frac{B_r}{4\pi}(u_\theta B_r-u_r B_\theta)+\left(\frac{3}{2}u_\theta+V_{A\theta}\right)\mathcal{E}\right]\right\}$$
$$-\rho u_r GM_\odot, \quad (11)$$

$$\frac{\partial\mathcal{E}}{\partial t}=-\frac{1}{r^2}\frac{\partial}{\partial r}[r^2(u_r+V_{Ar})\mathcal{E}]-\frac{1}{r\sin\theta}\frac{\partial}{\partial\theta}[\sin\theta(u_\theta+V_{A\theta})\mathcal{E}]-\frac{\mathcal{E}}{2}\nabla\cdot\mathbf{u}-|\mathbf{u}+\mathbf{V}_A|\frac{\mathcal{E}}{L}, \quad (12)$$

where $\mathbf{u}=(u_r,u_\theta)$ is the velocity vector in the inertial frame of reference, B_r and B_θ are the magnetic field components, $u^2=u_r^2+u_\theta^2$ and $B^2=B_r^2+B_\theta^2$.

The governing equations: region II

In region II, we assume the absence of a steady electric field [1, 18]. Consequently, plasma is flowing along the magnetic field in the rotating frame and the equations for tangential magnetic field components may be excluded from the governing set by using the relations: $B_\theta=u_\theta B_r/u_r$ and $B_\phi=v_\phi B_r/u_r$, where $v_\phi=u_\phi-w$ is the azimuthal velocity in the rotating frame and $w=\Omega r\sin\theta$. Using the divergence-free condition instead of the induction equation (3) for the magnetic field we have

$$\frac{\partial}{\partial r}(r^2\rho u_r)=-\frac{r}{\sin\theta}\frac{\partial}{\partial\theta}(\sin\theta\rho u_\theta)-\frac{r}{\sin\theta}\frac{\partial}{\partial\phi}(\rho v_\phi), \quad (13)$$

$$\frac{\partial}{\partial r}\left[r^2\rho\left(u_r^2+\frac{P}{\rho}+\frac{\mathcal{E}}{2\rho}+\frac{B_\theta^2+B_\phi^2-B_r^2}{8\pi\rho}\right)\right]$$
$$=-\frac{r}{\sin\theta}\frac{\partial}{\partial\theta}\left[\rho\sin\theta\left(u_r u_\theta-\frac{B_r B_\theta}{4\pi\rho}\right)\right]-\frac{r}{\sin\theta}\frac{\partial}{\partial\phi}\left[\rho\left(u_r v_\phi-\frac{B_r B_\phi}{4\pi\rho}\right)\right]$$
$$+r\rho\left(u_\theta^2+u_\phi^2-\frac{GM_\odot}{r}+\frac{2P}{\rho}+\frac{\mathcal{E}}{\rho}+\frac{B_r^2}{4\pi\rho}\right), \quad (14)$$

$$\frac{\partial}{\partial r}\left[r^3\rho\left(u_r u_\theta-\frac{B_r B_\theta}{4\pi\rho}\right)\right]$$
$$=-r^2\frac{\partial}{\partial\theta}\left[\rho\left(u_\theta^2+\frac{P}{\rho}+\frac{\mathcal{E}}{2\rho}+\frac{B_r^2+B_\phi^2-B_\theta^2}{8\pi\rho}\right)\right]-\frac{r^2}{\sin\theta}\frac{\partial}{\partial\phi}\left[\rho\left(u_\theta v_\phi-\frac{B_\theta B_\phi}{4\pi\rho}\right)\right]$$
$$-r^2\rho\cot\theta\left(u_\theta^2-u_\phi^2-\frac{B_\theta^2-B_\phi^2}{4\pi\rho}\right), \quad (15)$$

$$\frac{\partial}{\partial r}\left[r^3\rho\left(u_r u_\phi-\frac{B_r B_\phi}{4\pi\rho}\right)\right]$$
$$=-r^2\frac{\partial}{\partial\theta}\left[\rho\left(u_\theta u_\phi-\frac{B_\theta B_\phi}{4\pi\rho}\right)\right]-\frac{r^2}{\sin\theta}\frac{\partial}{\partial\phi}\left[\rho\left(u_\phi v_\phi+\frac{P}{\rho}+\frac{\mathcal{E}}{2\rho}+\frac{B_r^2+B_\theta^2-B_\phi^2}{8\pi\rho}\right)\right]$$
$$-2r^2\rho\cot\theta\left(u_\theta u_\phi-\frac{B_\theta B_\phi}{4\pi\rho}\right), \quad (16)$$

$$\frac{\partial}{\partial r}(r^2 B_r)=-\frac{r}{\sin\theta}\frac{\partial}{\partial\theta}(\sin\theta B_\theta)-\frac{r}{\sin\theta}\frac{\partial B_\phi}{\partial\phi}, \quad (17)$$

$$\frac{\partial}{\partial r}\left\{r^2\left[u_r\left(\frac{\rho u^2}{2}+\frac{\gamma P}{\gamma-1}\right)-\frac{wB_r B_\phi}{4\pi}+\left(\frac{3}{2}u_r+V_{Ar}\right)\mathcal{E}\right]\right\}$$
$$=-\frac{r}{\sin\theta}\frac{\partial}{\partial\theta}\left\{\sin\theta\left[u_\theta\left(\frac{\rho u^2}{2}+\frac{\gamma P}{\gamma-1}\right)-\frac{wB_\theta B_\phi}{4\pi}+\left(\frac{3}{2}u_\theta+V_{A\theta}\right)\mathcal{E}\right]\right\}$$
$$-\frac{r}{\sin\theta}\frac{\partial}{\partial\phi}\left\{v_\phi\left(\frac{\rho u^2}{2}+\frac{\gamma P}{\gamma-1}\right)\right.$$

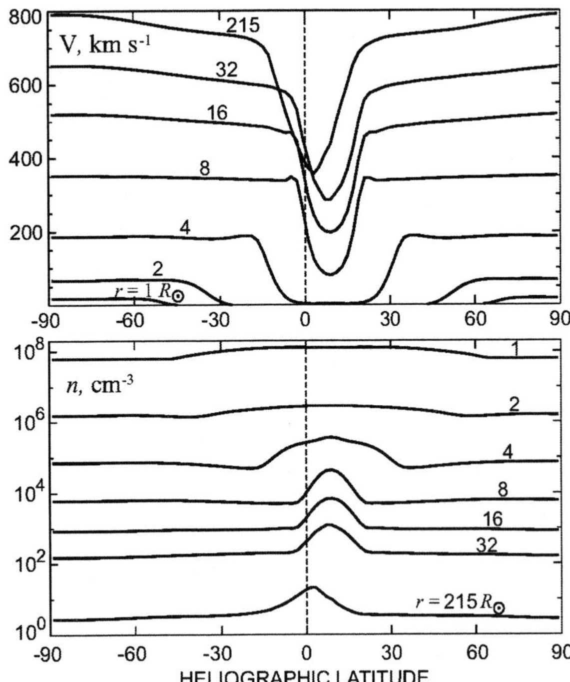

FIGURE 1. The radial velocity u_r and the number density n versus heliolatitude at the indicated radial distances in the meridional plane $\phi = 0$.

$$+w\left(P+\frac{\mathcal{E}}{2}+\frac{B_r^2+B_\theta^2-B_\phi^2}{8\pi}\right)$$

$$+\left(\frac{3}{2}v_\phi+V_{A\phi}\right)\mathcal{E}\Big\} - \rho u_r GM_\odot, \qquad (18)$$

$$\frac{\partial}{\partial r}\left[\frac{r^2 u_r (v+V_A)^2 \mathcal{E}}{vV_A}\right]$$

$$= -\frac{r}{\sin\theta}\frac{\partial}{\partial\theta}\left[\frac{\sin\theta u_\theta (v+V_A)^2\mathcal{E}}{vV_A}\right]$$

$$-\frac{r}{\sin\theta}\frac{\partial}{\partial\phi}\left[\frac{v_\phi (v+V_A)^2\mathcal{E}}{vV_A}\right] - \frac{r^2(v+V_A)^2}{V_A}\frac{\mathcal{E}}{L}, \qquad (19)$$

where now $u^2 = u_r^2 + u_\theta^2 + u_\phi^2$, $v = (u_r^2 + u_\theta^2 + v_\phi^2)^{1/2}$, $V_A = (V_{Ar}^2 + V_{A\theta}^2 + V_{A\phi}^2)^{1/2}$. The set of equations (13-19) is similar to that in [18] except that Alfvén wave effects are included in the WKB approximation.

Model parameters

As in [1], the polytropic index is chosen to be different from the adiabatic value to account implicitly for thermal conduction: $\gamma = 1.12$ in region I, and 1.46 [20] in region II. The driven Alfvén wave velocity amplitude at $1\ R_\odot$ is assumed to be 35 km s^{-1}, close the upper limit value inferred by Hassler et al. [19]. The strength of the dipole field at the coronal base was chosen to be 12 G. The plasma temperature and density in the initial state at $1\ R_\odot$ are 1.8×10^6 K and 7.5×10^7 particles cm^{-3}, respectively, and the dissipation length for Alfvén waves is $L = 80\ R_\odot$. Note that the values above were selected to optimize the fit to Ulysses data.

SIMULATION RESULTS

We start from an initial state with radial flow in a dipolar magnetic field [1] and integrate the region I equations (6-12) in time until a steady state is achieved. The computations are performed on a grid with the angular spacing of 2° and the radial step which increases linearly from $0.02\ R_\odot$ at $1\ R_\odot$ to $0.4\ R_\odot$ at $20\ R_\odot$. The boundary conditions are similar to those in [1]. Once the axisymmetric solution is obtained, it is transformed into a three-dimensional distribution that matches approximately the orientation of solar dipole during the first fast latitude scan of Ulysses (September 1994 - July 1995). The dipole orientation is computed from the expansion coefficients for the photospheric field inferred at the Wilcox Solar Observatory (WSO) from the line-of-sight boundary condition and the source surface at $2.5\ R_\odot$ [4]. Athough the dipole orientation changed slightly during the Ulysses transition, for the entire interval we used the dipole parameters from solar rotation 1887 (at the beginning of the transition, September-October 1994), when the dipole axis deviated by 9.7° from the rotation axis and its azimuth in the northern hemisphere was 330°.

The boundary between regions I and II is placed in the supersonic and super-Alfvénic flow, so a solution in region II depends only on the flow characteristics on the boundary. The transformed solution at the upper radial level in region I is used to initialize integration of equations (13-19) along radius through region II to 1 AU.

Figure 1 shows variations of velocity and density with latitude at $\phi = 0$ for various heliocentric distances from $1\ R_\odot$ to 1 AU ($\sim 215\ R_\odot$). A slower wind around the equator turns into a uniform fast wind at higher latitudes. The flow structure is not symmetric about the equator in the meridional plane and the slow wind pattern at 1 AU is shifted towards the equator due to solar rotation. The slow wind speed at the Earth's orbit is less than 400 km s^{-1} at minimum, jumps to $\sim$700 km s^{-1} by $\sim$15°, and then slowly increases towards the pole. The number density is $\sim$3 cm^{-3} in the fast wind and reaches $\sim$20 cm^{-3} near the equator.

Contour maps of the flow parameters in the heliographic coordinates are presented in Figure 2. The left three panels show the radial magnetic field B_r, radial velocity u_r, and number density n at the coronal base

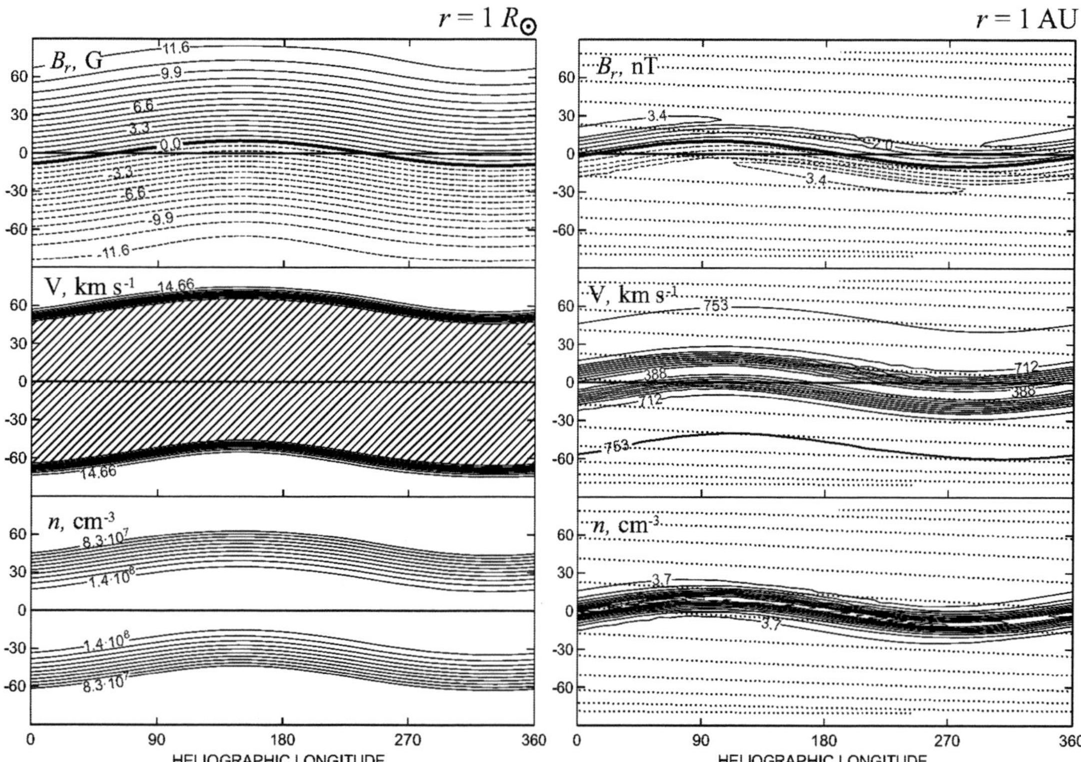

FIGURE 2. Contour plots of the radial magnetic field, the radial velocity and the number density in the heliographic coordinates at 1 $R_\odot$ (left panels) and at 1 AU (right panels). Ulysses' trajectory is shown on the right panels by dotted lines. The stagnation region (of no mass outflow) is hatched on the left middle plot. Negative levels of B_r are shown by dashed lines.

(1 $R_\odot$). The B_r distribution at 1 $R_\odot$ is dipolar and is kept as boundary condition during the relaxation process. The map of u_r at 1 $R_\odot$, where the hatching highlights the region without outflow ($u_r = 0$), provides a view of solar wind sources at the coronal base: the wind is streaming out of the polar regions extending down by $\sim 30°$ from the dipole axis. In these regions which can be regarded as "coronal holes," the outflow speed is ~ 16 km s^{-1} (see also Figure 1), while $n \sim 7.5 \times 10^7$ cm^{-3} which is markedly lower than in the stagnation belt nearby where n sharply increases to $\sim 1.5 \times 10^8$ cm^{-3}. The distribution of plasma temperature (not shown) is similar to that of density; the temperature changes from 1.8×10^8 K in polar regions to 2.0×10^8 K near the equator.

The computed solar wind parameters at 1 AU are presented on the right plots in Figure 2. The computed wind is clearly bimodal: no prominent variations both in plasma and magnetic field parameters except for a narrow equatorial belt where the flow is relatively dense and slow and B_r changes its sign. The trajectory of Ulysses in the heliographic coordinates during solar rotations 1887–1898 is superimposed on the plots and consists of slightly inclined lines depicting its travel from south to north polar regions.

A direct comparison of Ulysses observations with the model results is shown in Figure 3. The daily Ulysses data were scaled to 1 AU assuming that B_r and n to fall off with radial distance as r^2, the azimuthal magnetic field B_ϕ as r^{-1}, and the temperature T as $r^{-2(\gamma-1)}$ with $\gamma = 1.46$. It is seen from Figure 3 that our attempt to account for the HCS warping by inclining the solar dipole with respect to the solar rotation axis has provided fairly good agreement between the model and the Ulysses observations, including the latitudinal extension of the slower wind belt. Note that the normalization of Ulysses data to 1 AU is not a necessary condition for comparison with the model because the latter can be easily extended to ~ 2.3 AU to cover the range of heliospheric distances scanned by Ulysses. Our decision in favor of the normalization was made mainly to eliminate relatively easy-to-follow radial variations and emphasize the latitudinal structure of the solar wind flow. The computed temperature is somewhat higher than that observed, but could be adjusted length L: larger L would provide less wave dissipation and ultimately lower temperatures. The electron temperature in the distant solar wind is larger than the proton temperature, so that the single-fluid temperature used in our simulation can be higher than the proton temperature observed by Ulysses and presented in Figure 3.

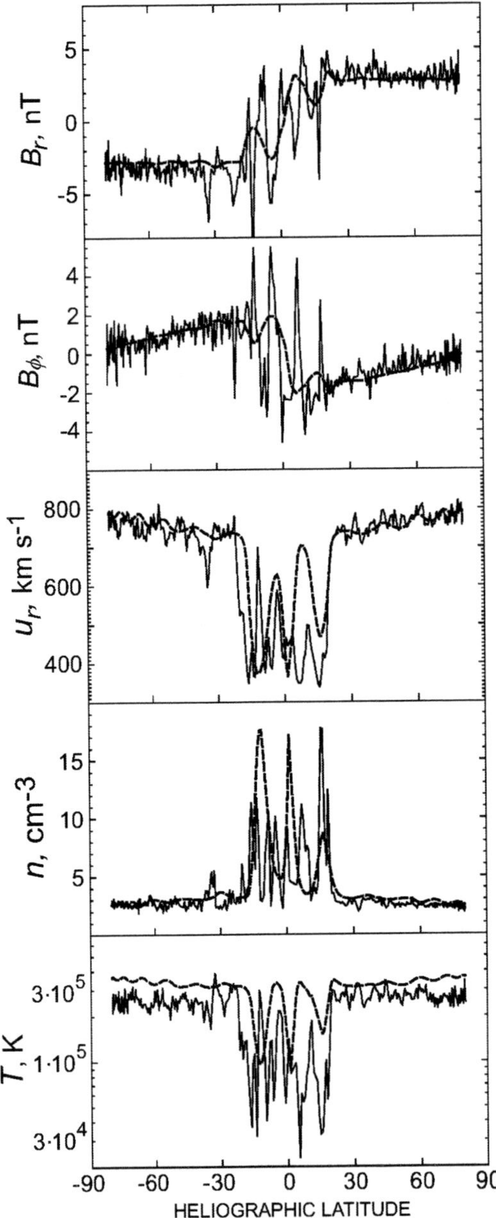

FIGURE 3. The model output (dashed lines) versus Ulysses data from the first fast latitude scan in 1994-1995 (solid lines). The Ulysses data are normalized to conditions at 1 AU.

SUMMARY

We have produced a simulation of the inner heliosphere during solar activity minimum as determined by boundary conditions at the coronal base and compared output from the model with Ulysses observations in 1994-1995. The bimodality of solar wind with a rapid change in flow parameters with latitude is reproduced along with the observed extension of the slower wind belt in latitude. In the present tilted-dipole model, this extension results from latitudinal oscillations of a relatively narrow region of slow velocities (see Figure 1) due to warping of the heliospheric current sheet and the solar rotation. To reproduce the slow wind observations in more detail, we plan to implement a fully three-dimensional model which would incorporate higher-than-dipolar harmonics of solar magnetic field into simulation.

ACKNOWLEDGMENTS

This work was performed while one of the authors (AVU) held a National Research Council Senior Research Associateship at the NASA/Goddard Space Flight Center. Ulysses plasma and magnetic field data were obtained from the NSSDC COHOWeb. Wilcox Solar Observatory data were obtained from `http://quake.stanford.edu/~wso`, courtesy of J. T. Hoeksema.

REFERENCES

1. Usmanov, A. V., Goldstein, M. L., Besser, B. P., and Fritzer, J. M., *J. Geophys. Res.* **105**, 12,675-12,695 (2000).
2. Phillips, J. L., et al., *Geophys. Res. Lett.* **22**, 3301-3304 (1995).
3. Forsyth, R. J., Balogh, A., Horbury, T. S., Erdös, G., Smith, E. J., and Burton, M. E., *Astron. Astrophys.* **316**, 287-295 (1996).
4. `http://quake.stanford.edu/~wso`.
5. Mikić, Z., and Linker, J. A., "The large-scale structure of the solar corona and inner heliosphere," in *Solar Wind Eight*, edited by D. Winterhalter et al., AIP Press, Woodbury, New York, 1996, pp. 104-107.
6. Stewart, G. A., and Bravo, S., *J. Geophys. Res.* **102**, 11,263-11,272 (1997).
7. Sittler, E. C., and Guhathakurta, M., *Astrophys. J.* **523**, 812-826 (1999).
8. Mikić, Z., Linker, J. A., Schnack, D. D., Lionello, R., and Tarditi, A., *Phys. Plasmas* **6**, 2217-2224 (1999).
9. Groth, C. P. T., De Zeeuw, D. L., Gombosi, T. I., and Powell, K. G., *J. Geophys. Res.* **105**, 25,053-25,078 (2000).
10. Riley, P., Linker, J. A., and Mikić, Z., *J. Geophys. Res.* **106**, 15,889-15,901 (2001).
11. Belcher, J. W., *Astrophys. J.* **168**, 509-524 (1971).
12. Alazraki, G., and Couturier, P., *Astron. Astrophys.* **13**, 380-389 (1971).
13. Hollweg, J. V., *Rev. Geophys.* **16**, 689-720 (1978).
14. Jacques, S. A., *Astrophys. J.* **226**, 632-649 (1978).
15. Smith, E. J., Balogh, A., Burton, M. E., Erdös, G., and Forsyth, R. J., *Geophys. Res. Lett.* **22**, 3325-3328 (1995).
16. Tóth, G., and Odstrčil, D., *J. Comput. Phys.* **128**, 82-100 (1996).
17. Tóth, G., *J. Comput. Phys.* **161**, 605-652 (2000).
18. Pizzo, V. J., *J. Geophys. Res.* **87**, 4374-4394 (1982).
19. Hassler, D. M., Rottman, G. J., Shoub, E. C., and Holzer, T. E., *Astrophys. J.* **348**, L77-L80 (1990).
20. Totten, T. L., Freeman, J. W. and Arya, S., *J. Geophys. Res.* **100**, 13-17 (1995).

The Microscopic State of the Solar Wind

Eckart Marsch

Max-Planck-Institut für Aeronomie, 37191 Katlenburg-Lindau, Germany

Abstract. The microscopic state of the solar wind is reviewed, in particular the measurements and models of proton and electron velocity distributions and kinetic features of heavy ions in the fast solar wind and coronal holes. Apparently, electron distributions are largely determined by Coulomb collisions. Concerning the ions, there is mounting evidence that pitch-angle diffusion in resonance with ion-cyclotron waves is the main process forming the shape of ion velocity distributions. Moreover, the absorption of high-frequency waves seems to play a major role in the heating of the corona and solar wind. Dispersive plasma waves and associated wave-particle interactions are the key to this problem. Plasma stability analyses and model calculations, as well as observations adressing these subjects are briefly reviewed, while focussing on the critical issues.

INTRODUCTION

Recent observational and theoretical results on the microscopic state of the solar wind are reviewed, in particular the proton velocity distributions (VDFs) and kinetic features of heavy ions in the fast solar wind and coronal holes. We summarize some of the well established observations of electrons, and discuss especially the Coulomb collision effects on their VDFs. Whereas for the electrons collisions are sufficient to explain the main kinetic features, for the ions wave-particle interactions are the key processes shaping their VDFs.

There is mounting evidence that pitch-angle diffusion (PAD) of thermal protons, in resonance with dispersive ion-cyclotron waves, is essential. Plateau formation plays a crucial role in the wave absorption or opacity for kinetic waves near the ion cyclotron resonances [18]. PAD has been included in recent kinetic models of the evolution of proton VDFs in the solar corona and wind, in addition to collisions. Finally, regulation mechanisms through microinstabilities, e.g for the ion temperature anisotropy, are briefly discussed.

VELOCITY DISTRIBUTIONS

This paper reviews selected results, returned from various mission, on particle VDFs in the solar wind. Comprehensive surveys are available in the reviews [15], [3] and [20]. The solar corona and wind are inhomogeneous, and their plasma parameters show strong gradients with respect to distance from the sun. Concerning the *macrostructure* or fluid properties we have as main scales: heliocentric distance, $r = 1$ AU ($= 150$ Gm$=$ $215\ R_S$); solar radius, $R_S = 696000$ km; wavelength of an Alfvén wave, $\lambda = 30 - 100$ Mm. Concerning the *microstructure* or kinetic properties we have at 1 AU as the important scales: Coulomb free path, $l \approx 0.1 - 10$ AU; ion inertial length, $V_A/\Omega_p \approx 100$ km; ion gyroradius, $r_L \approx 50$ km; and Debye length, $\lambda_D \approx 10$ m. For comparison, the spacecraft diameter $d \approx 3 - 5$ m. Here V_A is the Alfvén speed and Ω_p the proton gyrofrequency.

Kinetic processes in the solar corona and solar wind are important, because the plasma is dilute, multi-component and nonuniform. As a result of this and macroscopic forces, significant deviations from local thermal equilibrium arise, and complexity is caused in phase space through strong distortions of the VDFs in the thermal regime and by the occurence of suprathermal particles, e.g., the electron strahl or nonthermal ion beams and ion differential streaming. Owing to the weak collisionality there is a remaining influence of the global boundary conditions (in the corona), which are still reflected locally in the microscopic state of the solar wind.

Ions

From [3] we reproduce here Figure 1, for reminding the reader of the proton VDFs. The most prominent features are: Proton beams (B, E, H), temperature anisotropy in the core, with $T_\perp > T_\parallel$ (F, J, H), and ion differential motion with $\Delta V_i \leq V_A$ (see, e.g. [22]) not shown here, whereby all heavies appear to drift together with the alpha particles ahead of the protons in fast streams and reveal the temperature scaling, $T_i/T_p \geq m_i/m_p$, indicating preferential heating by waves more than in mass proportion. The free energy contained in these nonthermal features can be converted by microinstabilities into plasma waves of vari-

CP679, *Solar Wind Ten: Proceedings of the Tenth International Solar Wind Conference*, edited by M. Velli, R. Bruno, and F. Malara

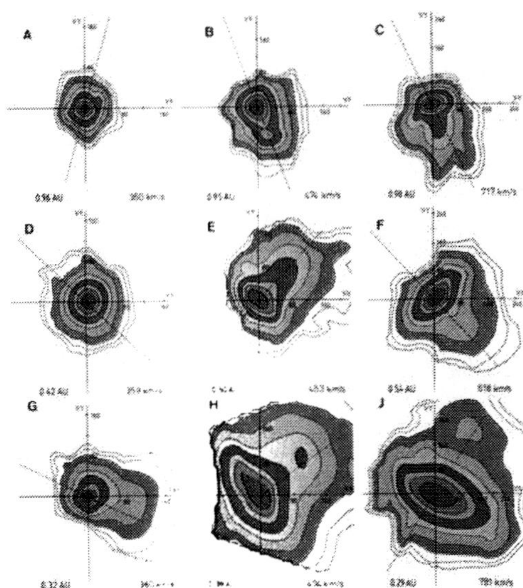

FIGURE 1. Solar wind proton velocity distribution functions as measured by Helios at different locations and for various flow speeds. Contours correspond to 80, 60, 40, 20, 10, 1, 0.1 percent of the maximum.

ous kinds, most prominently ion cyclotron waves propagating quasi-parallel to the background magnetic field. However, sometimes collisions are strong enough in slow wind to establish local Maxwellians (A, D).

Electrons

Solar wind electrons (see, e.g., Figures 1-3 in [3]) are characterized by two main populations: The thermal core electrons (typically 96%), which are bound to the Sun by being trapped in the electrostatic potential well associated with the interplanetary electric field, $e\mathbf{E} \approx -\nabla p_e$, and the suprathermal halo electrons (about 4%), which can escape from the Sun and surmount the potential barrier. In simplified terms, the core population is local, collisional, bound by the electrostatic potential ($\Phi \approx 500 - 1000$ eV between Sun and Earth) and reflected by the magnetic mirror in the corona, whereas the halo population is more global, almost collisionless, and free to escape to the outer heliosphere. Yet, halo electrons may scatter by magnetic inhomogeneities and wave-particle intercations.

The electrons in slow wind (from the closed streamer belt in the corona) are comparatively hot, but the electrons from coronal holes that are open magnetically are much cooler [2]. The halo electrons seem to carry most of the heat [14] and are hotter, with $T_H \approx 7T_C$. Heat conduction in the solar wind does not obey Fourier's law, resulting from a small skewness associated with a gentle temperature gradient, i.e. $Q_e \neq -\nabla T_e$. The electron energy spectra obtained from Ulysses have often been fitted [13] by kappa functions (generalized Lorentzians with $\kappa \approx 3-4$), which have much stronger tails than a Maxwellian for the hot halo but the wrong curvature of the spectrum at suprathermal energies. Moreover, the electron pitch-angle distributions are non-isotropic and highly focused (strahl) in fast wind at higher energies (beyond about 100 eV) according to the still best-resolved measurements made by Helios [21]).

Therefore, that suprathermal electrons could drive solar wind solely through the electric field is not compatible with the new coronal and old in-situ observations. The expanding corona is not an electron exosphere, but a weakly collisional medium for which Liouville's theorem cannot be exploited straightforwardly. Electrons do not matter dynamically in the solar wind acceleration, and velocity filtration is a negligable effect [9]. As a result, the fast solar wind ions have certainly to be driven by their own pressure-gradient force (see the reviews [19] and by Hollweg, in this volume).

KINETIC TRANSPORT EQUATIONS

General remarks

The most challenging but complete description of the expanding corona and the solar wind is by means of the Boltzmann-Vlasov kinetic equations for the ions, i.e. protons, alpha-particles (with typically 4% in abundance), minor heavier ions, as well as for the electrons (core, halo, and strahl). Such a kinetic description of the microscopic state of the corona and wind has, in the past and still today, been a numerically almost intractable problem, because the solar wind is multi-component and the plasma is nonuniform, rather dilute and turbulent on the fluid as well as microscopic scales. Yet, serious attempts have been made to overcome these difficulties.

Coulomb collisions of ions and electrons

What is theoretically well understood is collisional transport in terms of the Fokker-Planck operator. The importance of Coulomb collisions varies strongly with distance from the solar surface. Table 1 gives typical numbers for the electron density, temperature and collision-free path. Note the wide range of parameters involved.

Concerning collisional effects on solar wind ions, in [11] a kinetic equation with a simple relaxation term was integrated. Since in most types of wind $N \leq 1$, with N being the number of collisions of a proton per transit time through the scale height r, Coulomb collisions require a

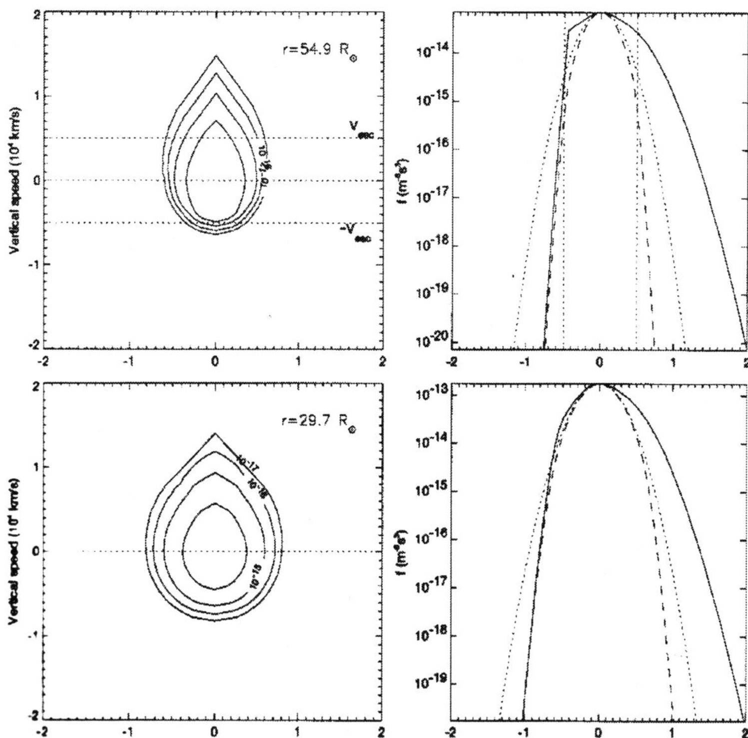

FIGURE 2. Solar wind electron velocity distribution functions as modeled by the kinetic Boltzmann equation including Coulomb collisions via the Landau collision integral. Left: isocontours at 54.9 R_S and 29.7 R_S (bottom) with pitch-angle focussing. Right: corresponding cuts along the field illustrating the distinct skewness on the anti-sunward and clear cutoff on the sunward side.

TABLE 1. Collisional free path and plasma parameters

Parameter	Chromosphere (1.003 R_S)	Corona (1.2 R_S)	Solar wind (1 AU)
n_e/cm^{-3}	10^{10}	10^7	10
T_e/K	10^4	(1-2)10^6	10^5
λ/km	10	10^3	10^7

kinetic treatment. It can be shown, though, that even very few collisions suffice to remove the otherwise extreme exospheric anisotropies. In slow wind one has $N > 5$ for about 10% of the time, and $N > 1$ for about (30-40)%. Therefore, here collisions do certainly matter, a result confirmed also by the Ulysses measurements at lower heliographic latitudes (see the review [20]).

The most advanced model for the collisional evolution of solar wind electrons has been presented in [9], in which the Fokker-Planck equation for electrons was integrated in a realistic background model solar wind. Some of the results are illustrated in Figure 2. Coulomb collisions were found to maintain a fairly isotropic core to large heliocentric distances. The VDF is determined primarily by the electric field and the expanding geometry, whereby velocity filtration is a rather weak effect. The model VDFs compare well with the observed ones, qualitatively in the pitch-angle distributions and even quantitatively in the energy spectra. In [24] it has recently been shown that the contribution of the halo to the total electric field (in terms of the halo partial pressure gradient) is dynamically negligable.

In conclusion, the behaviour of the majority electrons is essentially understood on the basis of collisions alone, and with the exception of the strahl electrons any wave-related process is hardly needed. Electrons seem to play a passive role [19] in the acceleration of the solar wind (which is to say of protons and other ions).

Quasilinear diffusion of ions

However, in order to explain the detailed kinetics of ion VDFs, wave-particle interactions are indispensible, and the Boltzmann equation complemented by wave terms must be used. If the wave amplitudes are small and the spectra broad in Fourier space, which is the case in the solar wind kinetic regime, quasilinear theory (QLT) should be adequate to describe the wave-particle couplings. The diffusion equation for any type of waves, propagating obliquely to the field in a magne-

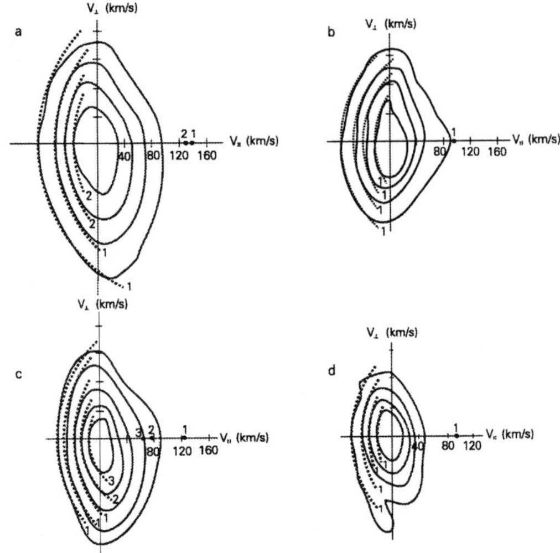

FIGURE 3. Comparison of some measured proton velocity distributions obtained by Helios 2 with the theoretical cyclotron-resonance plateaus as predicted by quasilinear theory. The thick dotted lines on the left-hand sides are the circular arcs delineating the plateau, whereby the related centers of the circles are marked on the $v_{\parallel}$ axis by full dark dots (with the same numbers attached to the contours). The dots indicate the locations of the effective phase speed of the waves.

tized plasma, has originally been derived by [8].

The quasilinear diffusion equation describes the evolution of the velocity distribution function, $f_j(v_{\parallel}, v_{\perp}, t)$, of any particle species j, e.g., in the solar inertial frame of reference, in which the particles and waves are supposed to propagate. With the nomenclature used in [16], the diffusion equation can be generally written as

$$\frac{\partial}{\partial t} f_j(v_{\parallel}, v_{\perp}, t) = \sum_M \sum_{s=-\infty}^{+\infty} \frac{1}{(2\pi)^3} \int_{-\infty}^{+\infty} d^3k\, \hat{\mathcal{B}}_M(\mathbf{k}) \times \frac{1}{v_{\perp}} \frac{\partial}{\partial \alpha} \left(v_{\perp} \nu_j(\mathbf{k}, s; v_{\parallel}, v_{\perp}) \frac{\partial}{\partial \alpha} f_j(v_{\parallel}, v_{\perp}, t) \right), \quad (1)$$

where the pitch-angle gradient in the wave frame was introduced. It is given by the velocity derivative

$$\frac{\partial}{\partial \alpha} = v_{\perp} \frac{\partial}{\partial v_{\parallel}} - \left(v_{\parallel} - \frac{\omega_M(\mathbf{k})}{k_{\parallel}} \right) \frac{\partial}{\partial v_{\perp}}. \quad (2)$$

The sums extend over the Bessel function index, s, and wave mode number, M. The magnetic field fluctuation spectrum is $\hat{\mathcal{B}}_M(\mathbf{k})$, which is normalized to the background-field energy density. It turns out to be physically meaningful to introduce the ion-wave relaxation or collision rate denoted by $\nu_j(\mathbf{k}, s; v_{\parallel}, v_{\perp})$. For its definition see [16]. The quantities $v_{\parallel}$ and $v_{\perp}$ are the the plasma-frame velocity components parallel and perpendicular to the magnetic field, $\mathbf{B}$. Other symbols used are: $\omega_M(\mathbf{k})$, the frequency of a linear wave mode M in the plasma frame, and $\mathbf{k}$ the wave vector. Strong wave-particle intercation, and thus diffusion in the wave frame, occurs whenever an ion at speed $v_{\parallel}$ fulfils the resonance condition:

$$\omega_M(\mathbf{k}) - s\Omega_j - k_{\parallel} v_{\parallel} = 0, \quad (3)$$

with the gyrofrequency $\Omega_j = q_j B/(m_j c)$, speed of light, c, and q_j and m_j, the charge and mass of the ion.

Note that the well-known quasilinear plateau in the VDF implies a vanishing pitch-angle gradient, i.e. $\partial f_j / \partial \alpha = 0$. The wave growth rate or absorption coefficient is essentially proportional to it. If the growth rate remains small, the slowly varying part of the ion VDF is controlled by diffusion. Then the time evolution of (1) will, if the wave power is large enough, lead to a time-asymptotic state of the VDF given by:

$$f_j(v_{\perp}, v_{\parallel}) = f_j \left(\frac{v_{\perp}^2 + v_{\parallel}^2}{2} - \int_{v_{\parallel 0}}^{v_{\parallel}} dv'_{\parallel} \frac{\omega_M(\mathbf{k})}{k_{\parallel}}(v'_{\parallel}) \right), \quad (4)$$

where $v_{\parallel 0}$ is the initial value of the parallel speed $v_{\parallel}$, which satisfies the resonance condition (3). In the case of plateau formation, the particles conserve their energy in the frame of reference propagating at the effective phase speed, $V_{ph}(\mathbf{k}) = \omega(\mathbf{k})/k_{\parallel}$.

Plateau formation in proton VDF

Observational evidence from Helios plasma data has been obtained for the occurrence of pitch-angle diffusion of solar wind protons [17]. Their VDFs show plateaus defined by vanishing pitch-angle gradients (implying marginal plasma stability). Parts of the isodensity contours in velocity space shown in Figure 3 are outlined well by a sequence of segments of circles centered at the phase speed V_{ph} (dots indicate its location), which is assumed to vary slightly and to be due to dispersion smaller than the local Alfvén speed. For the contours between 0.2 and 0.4 of the maximum density, the plateau can be as wide as 70 degrees in pitch angle.

KINETIC MODELS FOR THE IONS

The diffusion in velocity space of plasma ions being in resonance with waves is an old plasma physics subject [8] and has been studied in the literature in very much detail. Only recently [5], [7], [1], [29], [30], [23] has QLT been applied also to the solar corona and wind. The principle feature QLT predicts is that ions in resonance with waves undergo merely PAD, while conserving their kinetic energy in the frame moving at speed V_{ph}.

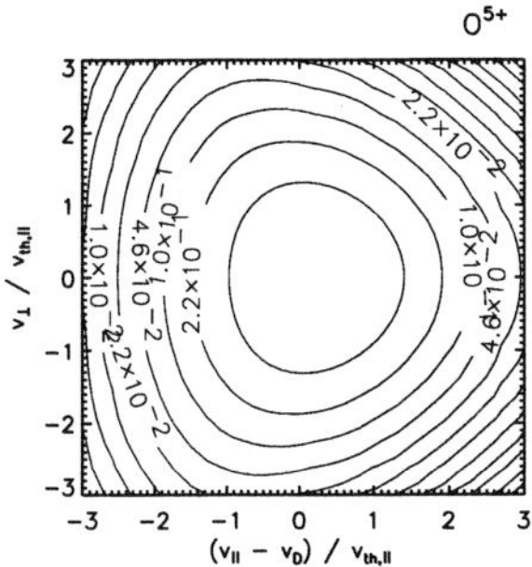

FIGURE 4. Two-dimensional gyrotropic VDF of the heavy coronal ion O^{5+}. Note the contours with a perpendicular temperature anisotropy and skewness along the magnetic field.

Semi-kinetic model for coronal ions

A semi-kinetic model has been developed by [29], [30] for the plasma dynamics of ions in the lower solar corona. This model consists of a closed set of reduced (with respect to the perpendicular velocity component) quasilinear diffusion equations. They involve one-dimensional "reduced VDFs", as they occur also in the wave dispersion relations. This numerical model includes wave-particle interactions within the framework of QLT and Coulomb collisions calculated by using the Landau collision integral. Coupled Vlasov/Boltzmann equations for these reduced VDFs were derived. The semi-kinetic diffusion equations were solved for a coronal funnel and hole [30].

The results obtained for heavy ions in a coronal funnel show good agreement with SOHO observations and yield preferential heating of the heavy ions, such as oxygen. This is illustrated in Figure 4. It was also found that sizable temperature anisotropies and heat fluxes develop. The reduced VDFs of the heavy ions exhibit pronounced deviations from a Maxwellian, which cannot be described adequately by higher-order polynomial expansions of the type that have been used [12], [10] to model fast wind VDFs. The non-Maxwellian characteristics tend to increase further with height due to the ever decreasing efficiency of Coulomb collisions. The wave damping/growth rate shows that the VDFs can reach marginal stability over a wide range of resonance speeds, where wave absorption ceases. Such effects could never be obtained if rigid VDFs were assumed at the outset.

Kinetic models for solar wind ions

The kinetic effects of wave-particle interactions have been investigated [23] using a global hybrid model, which takes the wave spectra as given, includes the QLT wave-diffusion operator and Coulomb collisions, and then follows the proton VDF in the expanding wind. The model can account for the bulk acceleration, produces preferential resonant heating of alpha particles and occasionally a double proton beam. However, the shape of the proton VDF does not correspond in detail to the observed one, and the model lacks selfconsistency concerning wave absorption or instability and the transport of the wave energy.

In another recent hybrid model [25], [26] the VDFs were fixed as bi-Maxwellians, but the wave spectrum evolution was allowed to evolve selfconsistently. From the results the following conclusions were drawn: It is problematic to use a spectrum with a fixed spectral slope near the cyclotron resonance, when one calculates the partition of wave energy among the different ionic species. This assumption neglects the important effects of wave absorption in the dissipation domain, and thus renders energy supply possible at extremely low amplitudes of the waves. But if the spectrum is allowed to evolve through wave damping, the high perpendicular temperature anisotropy as observed by SOHO does not occur in the model. In conclusion, serious efforts must be made to calculate the wave opacity selfconsistently without making restrictive assumptions about the VDF.

The hybrid model in [5] considered outward waves only, but the cyclotron-resonant wave-particle interactions were calculated selfconsistently on the basis of a scale separation between the small turbulent wave field and large nonuniform background magnetic field. In contrast, the recently proposed kinetic-shell model [6], [7] assumes again a rigid shape of the VDF, which is arranged in terms of shells centered according to equation (4) at $V_{ph} \approx V_A$. Yet, fixing the VDF is not selfconsistent, and the kinetic-shell approach applies only to a subset of the observed VDFs (see Figure 1). However, the inclusion of outward waves allows energy transfer across the $v_{\|} = 0$ boundary, and thus enables dissipation of outward waves and generation of inward waves. In [6] these processes were not calculated locally, since by the shell assumption the absorption is implied to be zero, but globally by means of the overall energy conservation laws.

Temperature anisotropy regulation

The observed velocity distributions such as in Figure 1 are often at the margin of stability. Wave-particle interactions are a key to understand ion kinetics in the solar corona and wind, however the non-linear evolution has not been fully investigated. In Figure 5 we show the proton temperature anisotropy as observed [27] and obtained from various regulation models [4]. Apparently, the core anisotropy is constrained and tuned by plateau formation.

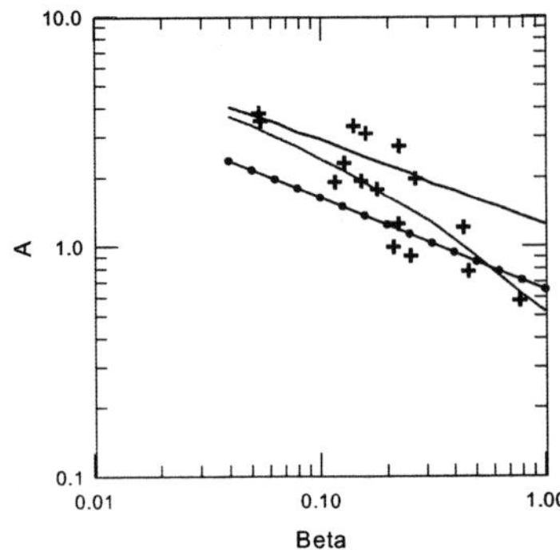

FIGURE 5. Proton core-temperature anisotropy as a function of the core-temperature plasma beta. The crosses show the proton data taken from various high-speed VDFs. The solid line shows the anisotropies as determined by the diffusion plateau, calculated by means of the cold plasma dispersion relation. The thin middle line is determined empirically by the pitch-angle plateau. The line with full dots shows a numerical threshold yielding: $A = 0.6\beta^{0.40}$ according to [4].

Similar investigations of the proton beam (see Figure 1) indicate that the beam speed is also regulated by pitch-angle diffusion [28] at the margin of resonantly driven proton beam instabilities, with a modified dispersion in the presence of alpha particles. Limited space does not permit to fully discuss here such kinetic instabilies.

CONCLUSIONS

This short review of the microscopic state of the solar wind emphasised recent work on modelling the evolution of electrons by collisons and ions by cyclotron-wave-induced diffusion. It seems clear that pure fluid and hybrid-kinetic models cannot grasp the effects of wave-particle interactions. To describe them adequately, especially in the context of high-frequency plasma waves heating the corona, requires kinetic physics. Theoretical models using reduced or shape-invariant particle VDFs and fixed wave energy spectral densities (ESDs) have provided valuable first insights into the kinetics of the solar wind but appear insufficient. Qualitatively new numerical results, which seem to approximate to the observations, were obtained in the past years.

However, the model assumptions of fixed VDFs or rigid ESDs have to be given up. QLT when applied selfconsistently in the case of weak turbulence seems to capture the wave-particle kinetic processes in the solar corona and wind, also for oblique wave propagation. The problem of wave-energy generation, transport and cascading (dissipation) in the dispersive domain remains to be solved. As the next meaningful and feasible step, selfconsistent kinetic calculations of the particle VDFs together with the wave ESDs and opacities are suggested.

REFERENCES

1. Cranmer, S., *J. Geophys Res.* **106**, 24937 (2001).
2. David, C., A.H. Gabriel, F. Bely-Dubau, A. Fludra, P. Lemaire, and K. Wilhelm, *Astron. Astrophys.* **336**, L90 (1998).
3. Feldman, W.C., and E. Marsch, in *Cosmic Winds and The Heliosphere*, J. R. Jokipii, C. P. Sonett, and M. S. Giampapa (Eds.), pp. 617, University of Arizona. Press, Tucson (1997).
4. Gary, S.P., R.M. Skoug, J.T. Steinberg, and C.W. Smith, *Geophys. Res. Lett.* **28** , 2759 (2001).
5. Galinsky, V.L. and V.I. Shevchenko, *Phys. Rev. Lett.* **85(1)**, 90 (2000).
6. Isenberg, P.A., M.A. Lee and J.V. Hollweg, *J. Geophys. Res.* **106**, 5649 (2001).
7. Isenberg, P.A., *J. Geophys. Res.* **106**, 29249 (2001).
8. Kennel, C. F., and F. Engelmann, *Phys. Fluids* **9**, 2377 (1966).
9. Lie-Svendsen, O., V.H. Hansteen, and E. Leer, *J. Geophys. Res.* **102**, 4701 (1997).
10. Li, X., *J. Geophys. Res.* **104**, 19773 (1999).
11. Livi, S., and E. Marsch, *J. Geophys. Res.* **92**, 7255 (1987).
12. Leblanc, F. and D. Hubert, *Astrophys. J.* **501**, 375 (1998).
13. Maksimovic, M., V. Pierrard, J.F. Lemaire, *Astron. Astrophys.* **324**, 725 (1997).
14. McComas, D.J., S.J. Bame, W.C. Feldman, J.T. Gosling, and J.L. Phillips, *Geophys. Res. Lett.* **19**, 1291 (1992).
15. Marsch, E., in *Physics of the Inner Heliosphere*, R. Schwenn and E. Marsch (Eds.), Springer-Verlag, Heidelberg, Germany, Vol. II, 45 (1991).
16. Marsch, E., and C.-Y. Tu, *J. Geophys. Res.* **106**, 227 (2001a).
17. Marsch, E., and C.-Y. Tu, *J. Geophys. Res.* **106**, 8357 (2001b).
18. Marsch, E., C. Vocks, and C.-Y. Tu, *Nonlinear Processes Geophys.*, in press (2003).
19. Marsch E., W.I. Axford, and J.F. McKenzie, in *The Dynamic Sun*, B. Dwivedi (Ed.), Cambridge University Press, in press (2002).
20. Neugebauer, M., in *The Heliosphere near Solar Minimum*, The Ulysses Perspective, A. Balogh, R.G. Marsden, and E.J. Smith (Eds.), Springer-Verlag, Heidelberg, 43 (2001).
21. Pilipp, W.G., H. Miggenrieder, M.D. Montgomery, K.-H. Mühlhäuser, H. Rosenbauer, and R. Schwenn, *J. Geophys. Res.* **92**, 1075 (1987).
22. von Steiger, R., J. Geiss, G. Gloeckler, and A.B. Galvin, *Space Sci. Rev.* **72**, 71 (1995).
23. Tam, S.W.Y., and T. Chang, *Geophys. Res. Lett.* **26**, 3189 (1999).
24. Tam, S.W.Y., and T. Chang, *Astron. Astrophys.*, submitted (2002).
25. Tu C.-Y., and E. Marsch, *J. Geophys. Res.* **106**, 8233 (2001a)
26. Tu C.-Y., and E. Marsch, *Astron. Astrophys.* **368**, 1071 (2001b)
27. Tu C.-Y., and E. Marsch, *J. Geophys. Res.*, in press (2002a)
28. Tu C.-Y., Wang, L.-H., and E. Marsch, *J. Geophys. Res.*, in press (2002b)
29. Vocks, C., *Astrophys. J.* **568**, 1017 (2002a).
30. Vocks, C., and E. Marsch, *Astrophys. J.* **568**, 1030 (2002b).

The Effect of Microstreams on Alfvénic Fluctuations in the Solar Wind

M. L. Goldstein*, D. A. Roberts* and A. Deane†

*Code 692, NASA Goddard Space Flight Center, Greenbelt, MD 20771, USA
†Institute of Physical Science and Technology, University of Maryland, College Park, MD, USA

Abstract. Even in nominally uniform solar wind flows, such as over the solar poles near solar minimum, the wind velocity exhibits fluctuations of order 40 km/s. We have shown previously that such variations will shear planar parallel propagating magnetic fluctuations leading to the generation of transverse wave vectors. Here we extend our previous two-dimensional magnetohydrodynamic (MHD) simulations to three dimensions and describe how, starting from an initial spectrum of circularly polarized Alfvén waves with radial wave vectors, such "microstreams" might produce fluctuations that could be described as "quasi-two-dimensional". Our goal is to elucidate the origin of the "two-component" nature of the correlation function of magnetic fluctuations.

1. BACKGROUND

The earliest observations of plasma and magnetic fields in the solar wind revealed that fluctuations resembling outward propagating Alfvén waves were ubiquitous [1, 2, 3] (also see [4, 5]). There ensued a lengthy debate as to whether the highly Alfvénic nature of the fluctuations reflected a remnant of turbulence in the corona that was convected into the solar wind, or whether the solar wind was an evolving turbulent magnetofluid. The former interpretation was supported by the fact that pure Alfvén waves are exact solutions of the incompressible ideal equations of magnetohydrodynamics and, as such, do not undergo further evolution. However, the large velocity shears between fast and slow solar wind flows contain sufficient free energy to drive a turbulent cascade. Furthermore, log-log plots of the power spectra of magnetic fluctuations typically show an "inertial" range of nearly constant slope $\simeq -5/3$, which is characteristic of fully developed fluid turbulence. A resolution of the debate was offered [6] (see also the review [7]) where it was argued that the nearly pure outward propagating Alfvénic fluctuations did reflect coronal processes, but that the fluctuations were also stirred *in situ* by velocity shears that led to an evolution of the spectrum with heliocentric distance from a spectral index $\simeq -1$ near 0.3 AU to $\simeq -5/3$ at 1 AU, and beyond. The stirring by velocity shears also produced inward-propagating fluctuations which could then participate in a nonlinear cascade.

From single-spacecraft magnetic field data, one cannot easily characterize the symmetry properties of the solar wind fluctuations. A magnetofluid filled with parallel propagating transverse Alfvén waves would have slab symmetry, while fully developed fluid turbulence might be axisymmetric or, possibly, isotropic. That the solar wind was neither slab nor isotropic was suggested by [3, 8, 9]. Evidence for a two-component symmetry can be found in [10, 11, 12] (also see [13] and the review by Oughton in this volume). A common interpretation of those analyses is that solar wind fluctuations comprise two dominant populations: planar shear Alfvénic fluctuations propagating nearly parallel to the background magnetic field and a quasi-two-dimensional component with wave vectors predominately perpendicular to the background field. The nature and origin of this second component has yet to be determined. In the next section we review briefly some of its characteristics and discuss possible origins. Work presented in this volume by C. W. Smith suggests that the percentage of fluctuations with nearly perpendicular wave vectors, especially at high heliocentric latitudes, may be highly variable.

2. PROPERTIES OF THE NEARLY-PERPENDICULAR COMPONENT

Interest in the perpendicular wave number component of solar wind fluctuations is motivated by the possibility that they represent a true quasi-two-dimensional population. Quasi-two-dimensional fluctuations do not resonantly scatter energetic particles because $k_{\|} \simeq 0$. For the same reason, in a quasi-two-dimensional magnetofluid, the direction of minimum variance of magnetic fluctua-

CP679, *Solar Wind Ten: Proceedings of the Tenth International Solar Wind Conference*, edited by M. Velli, R. Bruno, and F. Malara

tions should lie along $\mathbf{B}_\circ$, the local mean magnetic field. In addition, as shown in [14], two-dimensional fluctuations produce a stochastic diffusion of flux tubes that for particles magnetically tied to the local magnetic field will result in significant perpendicular diffusion.

In fact, cosmic rays appear to have scattering mean free paths that are longer than is expected from quasi-linear diffusion in slab-symmetry Alfvénic turbulence (see, *e.g.*, [15]). Furthermore, cosmic rays exhibit enhanced diffusion perpendicular to the background magnetic field (see, *e.g.*, [16]). Finally, the direction of minimum variance of interplanetary fluctuations tends to lie along $\mathbf{B}_\circ$ [3, 17]. Thus, proof that quasi-two-dimensional fluctuations are the dominant mode of interplanetary fluctuations could provide a simple solution to three heretofore vexing problems in solar wind research.

2.1. Possible origins of the nearly-perpendicular wave number component

True quasi-two-dimensional turbulence arises in the limit of a strong DC magnetic field and incompressible flow [18, 19]. The solar wind, however, is not incompressible, which has motivated several generalizations of the Strauss and Montgomery theories. A *nearly incompressible* theory has been developed [20] and generalized ([21], also see [22], and Zank *et al.* and Bhattacharjee and Ng, these proceedings). That magnetic fluctuations become anisotropic when a strong magnetic field is present has been confirmed in a series of numerical simulations [23, 24, 25, 26]. The general lack of density and magnetic field magnitude fluctuations suggest that the $k_\perp$ component does not arise from fast mode waves or from pressure-balance structures, which have $\delta\mathbf{b}$ parallel to $\mathbf{B}_\circ$. Other processes can generate perpendicular wave numbers, most notably the fact that parallel propagating planar Alfvén waves in sheared velocity fields experience phase mixing [25, 27, 28] that refracts the wave vectors toward the direction perpendicular to $\mathbf{B}_\circ$. Alternatively, the $k_\perp$ component might arise in the solar atmosphere in regions of strong magnetic field, perhaps as a result of magnetic reconnection (see Chang, these proceedings, and [29]).

The two most likely origins of the $k_\perp$ component, therefore, are a coronal source of true quasi-two-dimensional turbulence and/or phase mixing of planar Alfvénic fluctuations by velocity shears. Independent of any coronal source (*e.g.*, magnetic reconnection), fluctuations in solar wind velocity will shear the phase fronts of all planar waves. To explore how small velocity shears transform the two-dimensional correlation function of an initially nearly-planar Alfvénic wave packet, we designed three-dimensional simulations that included small velocity-sheared flux tubes. The results are described in the next section.

3. SIMULATION OF THE EFFECT OF MICROSHEARS ON AN ALFVÉNIC WAVE PACKET

3.1. The MHD equations

The algorithm used to solve the compressible ideal equations of MHD is described in [30]. The code incorporates the rotation and tilt of the solar corona and allows for an influx of Alfvén waves, two-dimensional turbulence, and/or pressure balance structures. Only Alfvénic fluctuations are germane to this paper. Neither the heliospheric current sheet (HCS) nor gravity were included for this application.

The general MHD equations describe conservation of mass,

$$\frac{\partial \rho}{\partial t} + \nabla \cdot \rho \mathbf{v} = 0, \tag{1}$$

conservation of momentum,

$$\frac{\partial \rho \mathbf{v}}{\partial t} + \nabla \cdot \left[\left(p + \frac{B^2}{8\pi} \right) \mathbf{I} + \rho \mathbf{v}\mathbf{v} - \frac{1}{4\pi} \mathbf{B}\mathbf{B} \right] = 0, \tag{2}$$

conservation of energy,

$$\frac{\partial}{\partial t} \left(\rho e + \frac{1}{2}\rho \mathrm{v}^2 + \frac{B^2}{8\pi} \right) + \nabla \cdot \left[\left(\rho e + \frac{1}{2}\rho \mathrm{v}^2 + p \right) \mathbf{v} - \frac{1}{4\pi} (\mathbf{v} \times \mathbf{B}) \times \mathbf{B} \right] = 0, \tag{3}$$

and Faraday's Law,

$$\frac{d\mathbf{B}}{dt} = -c \nabla \times \mathbf{E}, \tag{4}$$

where

$$\mathbf{E} = -\frac{1}{c} \mathbf{v} \times \mathbf{B}. \tag{5}$$

In addition, the internal energy, e, is related to the pressure, p, through the relation

$$p = (\gamma - 1)\rho e. \tag{6}$$

These equations were transformed to dimensionless units and solved. Here $\mathbf{I}$ is the unit tensor with components δ_{ij} and γ is the ratio of specific heats, here taken to be $5/3$ (see [31, 32]). The equations were solved on either a $304 \times 154 \times 154$ or 154^3 grid in r, θ, and φ spherical coordinates.

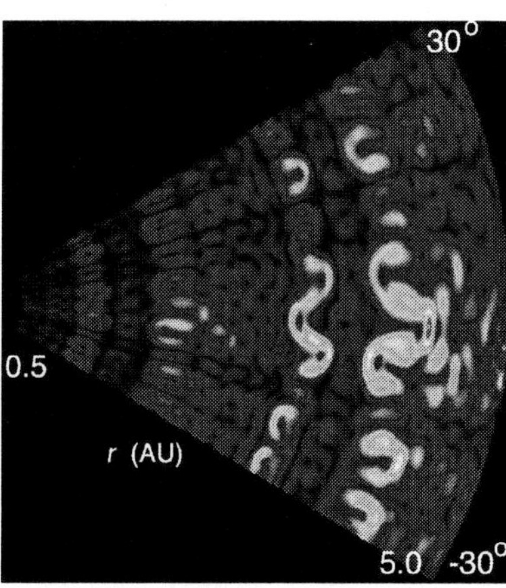

FIGURE 1. Magnitude of the vorticity, ω in the $r-\theta$-plane from $r = 0.5-5.0$ AU from the $304 \times 154 \times 154$ run. The color scale is rainbow with minimum ω black and maximum ω white.

FIGURE 2. A three-dimensional view of the first simulation from the $304 \times 154 \times 154$ point run. r extends from $0.5-5.0$ AU. The color shading at the in-flow boundary shows the initial distribution of the velocity fluctuations (again using a rainbow color bar); the shading in the $r-\theta$-plane is the fluctuation in $\delta B_{\perp}$; and, in the $r-\varphi$-plane, magnetic field lines are shown in white along with a color shading of $|\mathbf{v}|$. The data are normalized so that the minimum value of the plotted quantity is black and the maximum value is white.

3.2. Microshears in high speed wind

We varied an otherwise steady fast wind by adding a superposition of 8×8 small amplitude modes ($\delta v/v_{\circ} \sim 8\%$). We ran two cases; the first extended in radius from $0.5-5$ AU, while in the second case, the radial domain was restricted to $0.1-1$ AU. In both cases latitudinal and longitudinal range was $\pm 30°$. A Parker spiral and associated changes in the flows were produced by moving the field across the input surface at a rate consistent with the 28-day solar rotation period. A packet of plane polarized Alfvén waves was introduced at the in-flow boundary, either 0.1 or 0.5 AU. The magnitude of the vorticity, $\omega = \nabla \times \mathbf{v}$, in the $r-\theta$-plane at the end of the run, which extended to 5 AU, is shown in Figure 1, illustrating that significant evolution of the flow occurs by ~ 1 AU.

The microstreams in three dimensions are illustrated in Figure 2. The color contour at the in-flow boundary) at 0.5 AU is the initial distribution of the velocity fluctuations; the contour in the $r-\theta$-plane is the fluctuation in $\delta B_{\perp} = \sqrt{(\delta B_{\theta}^2 + \delta B_{\varphi}^2)}$; and in the $r-\varphi$-plane, magnetic field lines are plotted in white along with a color contour of $|\mathbf{v}|$. The microstreams produce fluctuations in v whose structure follows approximately that of the Parker spiral field. Not shown is that similar contour maps of $\delta B_{\perp}$ also conform to the spiral field structure.

In Figure 3 illustrates the evolution of the magnetic fluctuations from plane-polarized waves to something reminiscent of the "Maltese cross" pattern [10]. Both panels are a collage of three color contours: The smallest in the lower lefthand corner is a color image of $\delta B_{\perp}$; the alignment of these contours approximately follows that of the spiral field structure, *i.e.*, $\mathbf{B}_{\circ}$ is nearly parallel to r on the left and is closer to the φ-direction on the right. The abscissa of the radial ranges was $r = 0.1-0.43$ AU and $r = 4-5$ AU for the left and right panels, respectively. The ordinate spans the φ domain of $\pm 30°$. The next larger color image is the two-dimensional correlation function of δB_{θ} and δB_{φ}, also shown in the same rectangular $r-\varphi$ representation and computed over the same radial ranges. The largest contour is the sum of the two-dimensional power spectra of δB_{θ} and δB_{φ}. The abscissa and ordinate are k_r and k_{φ}, respectively.

The two-dimensional correlation function at small radial distance is aligned nearly perpendicular to the wave vector of the fluctuations, as expected of radial wave vectors. Close to $r = 5$ AU the contours are nearly parallel to fluctuating magnetic field structures; furthermore, the shape of the correlation function (particularly, the *green* level), is reminiscent of that from ISEE-3 [10]. Although the two-dimensional power spectrum taken between $\sim 4-5$ AU still contains some power in wave vectors parallel to the spiral field direction, the power in the perpendicular direction dominates. Thus, the small velocity shears deform the distribution of initially parallel-propagating plane-polarized Alfvén waves so that by several AU the correlation function contains substantial power nearly perpendicular to the direction of the background magnetic field. The interval analyzed by [10] was a mixture of fast and slow wind from near the ecliptic plane, while the simulation parameters chosen above are closer to those encountered by Ulysses at high latitudes. However, C. W. Smith (these proceedings) has presented evidence that the two-component model of solar wind fluctuations also fits periods of Ulysses data.

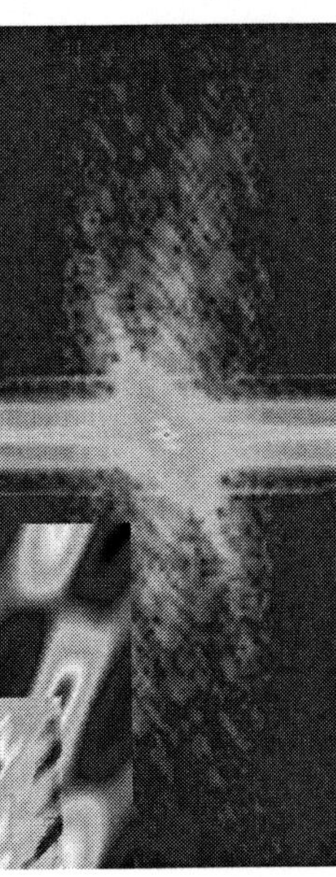

FIGURE 3. $B_\perp$, two-dimensional correlation function, and two-dimensional power spectrum (see text). The lefthand panel covers $r = 0.1 - 0.43$ AU with $\varphi \pm 30°$. The righthand panel covers $r = 4 - 5$ AU and the same range in φ. The data are normalized and a rainbow color bar is used.

4. SUMMARY

The shearing of Alfvénic fluctuations by even small velocity shears similar to those observed by Ulysses at high heliographic latitudes can yield nearly perpendicular wave vectors. Such distributions of wave vectors, while similar to that deduced from ISEE-3 data, are not truly quasi-two-dimensional in that $\delta\mathbf{B}, \delta\mathbf{k}$, and $\mathbf{B}_\circ$ are not mutually perpendicular. (Note that $\mathbf{B}_\circ \| \hat{\varphi}$ and much of the spectral power in $\delta B_\perp$ comes from B_φ.) We conclude that the effect of microstreams must be included in any explanation of the observed anisotropies in solar wind fluctuations and that the solar wind may not contain a dominant, truly quasi-two-dimensional, component.

ACKNOWLEDGMENTS

We would like to acknowledge the support of James Fischer, Manager for High Performance Computing at the Goddard Space Flight Center, for helping to arrange use of computing resources at the NASA Advanced Supercomputing Division.

REFERENCES

1. Unti, T. W., and Neugebauer, M., *Phys. Fluids*, **11**, 563 (1968).
2. Coleman, P. J., *Phys. Rev. Lett.*, **17**, 207 (1966).
3. Belcher, J. W., and Davis, L., *J. Geophys. Res.*, **76**, 3534 (1971).
4. Matthaeus, W. H., and Goldstein, M. L., *J. Geophys. Res.*, **87**, 6011 (1982).
5. Goldstein, M. L., *Astrophys. and Space Sci.*, **277**, 349–369 (2001).
6. Roberts, D. A., Goldstein, M. L., Klein, L. W., and Matthaeus, W. H., *J. Geophys. Res.*, **92**, 12,023 (1987).
7. Goldstein, M. L., Roberts, D. A., and Matthaeus, W. H., *Ann. Rev. Astron. and Astrophys.*, **33**, 283 (1995).
8. Sari, J. W., and Valley, G. C., *J. Geophys. Res.*, **81**, 5489 (1976).
9. Matthaeus, W. H., Goldstein, M. L., and King, J. H., *J. Geophys. Res.*, **91**, 59 (1986).
10. Matthaeus, W. H., Goldstein, M. L., and Roberts, D. A., *J. Geophys. Res.*, **95**, 20,673 (1990).
11. Bieber, J. W., Wanner, W., and Matthaeus, W. H., *J. Geophys. Res.*, **101**, 2511–2522 (1996).
12. Carbone, V., Malara, F., and Veltri, P., *J. Geophys. Res.*, **100**, 1763 (1995).
13. Richardson, J. D., and Paularena, K. I., *Geophys. Res. Lett.*, **25**, 2097–2100 (1998).
14. Matthaeus, W. H., Gray, P. C., D. H. Pontius, J., and Bieber, J. W., *Phys. Rev. Lett.*, **75**, 2136 (1995).
15. Bieber, J., Matthaeus, W., C. Smith, W. W., Kallenrode, M.-B., and Wibberenz, G., *Astrophys. J.*, **420**, 294 (1994).
16. Giacalone, J., Jokipii, J. R., and Mazur, J. E., *Astrophys. J.*, **532**, L75–L78 (2000).
17. Völk, H. J., and Alpers, W., *Astrophys. Space Sci.*, **20**, 267 (1973).
18. Strauss, H. R., *Phys. Fluids*, **19**, 134 (1976).
19. Montgomery, D., *Physica Scripta*, **T2/1**, 83 (1982).
20. Zank, G. P., and Matthaeus, W. H., *Phys. Fluids A*, **3**, 69 (1991).
21. Bhattacharjee, A., Ng, C. S., Ghosh, S., and Goldstein, M. L., *J. Geophys. Res.*, **104**, 24,835–24,844 (1999).
22. Bavassano, B., Bruno, R., and Klein, L. W., *J. Geophys. Res.*, **100**, 5871–5876 (1995).
23. Shebalin, J. V., Matthaeus, W. H., and Montgomery, D., *J. Plasma Phys.*, **29**, 525 (1983).
24. Matthaeus, W. H., and Lamkin, S. L., *Phys. Fluids*, **28**, 303 (1985).
25. Roberts, D. A., Goldstein, M. L., Matthaeus, W. H., and Ghosh, S., *J. Geophys. Res.*, **97**, 17,115 (1992).
26. Oughton, S., Priest, E., and Matthaeus, W. H., *J. Fluid Mech.*, **280**, 95 (1994).
27. Roberts, D. A., and Ghosh, S., *J. Geophys. Res.*, **104**, 22,395–22,399 (1999).
28. Ruderman, M. S., Goldstein, M. L., Roberts, D. A., Deane, A., and Ofman, L., *J. Geophys. Res.*, **104**, 17,057–17,068 (1999).
29. Matthaeus, W., and Goldstein, M. L., *Phys. Rev. Lett.*, **57**, 495 (1986).
30. Goldstein, M. L., Roberts, D. A., Burlaga, L. F., Siregar, E., and Deane, A. E., *J. Geophys. Res.*, **106**, 15,973 (2001).
31. Siscoe, G., and Intriligator, D., *Geophys. Res. Lett.*, **20**, 2267 (1993).
32. Totten, T. L., Freeman, J. W., and Arya, S., *J. Geophys. Res.*, **100**, 13–17 (1995).

Magnetic Turbulence, Fast Magnetic Field line Diffusion and Small Magnetic Structures in the Solar Wind

G. Zimbardo*, P. Pommois†* and P. Veltri*

* *Dipartimento di fisica, Università della Calabria (Italy)*
†*Center for High Performance Computing, Università della Calabria (Italy).*

Abstract. The influence of magnetic turbulence on magnetic field line diffusion has been known since the early days of space and plasma physics. However, the importance of "stochastic diffusion" for energetic particles has been challenged on the basis of the fact that sharp gradients of either energetic particles or ion composition are often observed in the solar wind. Here we show that fast transverse field line and particle diffusion can coexist with small magnetic structures, sharp gradients, and with long lived magnetic flux tubes. We show, by means of a numerical realization of three dimensional magnetic turbulence and by use of the concepts of deterministic chaos and turbulent transport, that turbulent diffusion is different from Gaussian diffusion, and that transport can be inhomogeneous even if turbulence homogeneously fills the heliosphere. Several diagnostics of field line transport and flux tube evolution are shown, and the size of small magnetic structures in the solar wind, like gradient scales and flux tube thickness, are estimated and compared to the observations.

INTRODUCTION

The magnetic turbulence found in many plasmas causes a magnetic field line random walk which "destroys" the magnetic surfaces [1, 2, 3, 4, 5, 6], and causes a fast plasma transport across the magnetic field structure, a phenomenon which is sometimes called stochastic diffusion. Recently it was shown that magnetic field line diffusion gives a natural explanation for the Ulysses observations of energetic particles at high southern heliographic latitudes [7]. Nevertheless, the relevance of stochastic diffusion in space plasmas has been challenged on the basis of the fact that sharp gradients of particular ion composition are often observed in space (see e.g. Ref. [8]). Also, sharp intensity variations in impulsive energetic particle events seen by the SWICS instrument on ACE may suggest that turbulent diffusion is not smoothing out the gradients [9]. Yet, it was shown by Giacalone et al. [10], by means of a numerical simulation, that magnetic field line random walk is consistent with both fast diffusion and small-scale gradients in energetic particle intensity.

In this paper we argue that this is a general feature of the transport of a passive tracer by a turbulent field, that turbulent diffusion is different from Gaussian diffusion, and show by a numerical simulation that fast magnetic field line and plasma transport can be found simultaneously to small magnetic flux tube structure and field line mixing. Moreover, we argue on the type of anisotropy present in the solar wind turbulence.

TURBULENT VERSUS GAUSSIAN DIFFUSION

In a normal, Gaussian diffusion process, the diffusing particles move randomly in all directions, because of either molecular diffusivity, in an ordinary fluid, or collisions, in a plasma. Such random motion tends to smooth out a density gradient, and a point like initial concentration of, say, blue dye, evolves with the well known Gaussian smooth profile. Also, a Gaussian random walk is characterised by a finite mean square step length, and by random, uncorrelated directions of motion. In such a case normal diffusion results, $\langle \Delta x_i^2 \rangle = 2D_i t$, where D_i is the diffusion coefficient and t the time, and, in particular, the diffusing particles moves in all direction with equal probability. This leads to the smoothing of density or concentration gradients.

We consider here turbulent diffusion as the result of the advection of a passive tracer by a turbulent velocity (or magnetic) field. The transport of the passive tracer (or of the test particles in a turbulent magnetic field) is due to the chaotic evolution of the non linear dynamical system. Such evolution can be very complex and "unpredictable", but still deterministic. This means that each passive tracer does not move randomly, but follows a flow (or field) line for long distances regardless of the density or concentration gradients. Basically, this is what happens when you are mixing a blue dye with a white paint: at least for some time (until molecular diffusion becomes important) blue stripes stand out on a white

CP679, *Solar Wind Ten: Proceedings of the Tenth International Solar Wind Conference*,
edited by M. Velli, R. Bruno, and F. Malara

background. In a turbulent medium, long range correlations among the fluctuation occurr, and, indeed, anomalous transport regime can be found, $\langle \Delta x_i^2 \rangle = 2D_i t^{\alpha_i}$ ($i = x, y$), with $\alpha_i > 1$ (see Pommois et al. [5, 6]). The main features of such chaotic (not random) transport can be grasped by considering the evolution of a magnetic flux tube in the presence of magnetic turbulence. In a uniform background field, a flux tube with an initially circular cross section will be elongated in one direction and squeezed in the other (because of $\nabla \cdot \mathbf{B} = 0$, this is the typical behaviour). Then the flux tube cross section will be distorted and further on it will develop branches (see figures by Rechester and Rosenbluth [3]; and Isichenko [11]). A very complex and ramified structure can be obtained, with the outer size of the flux tube growing fast, but with many "voids" in the cross section. A numerical simulation of the flux tube evolution is reported here.

The magnetic field model is

$$\mathbf{B}(\mathbf{r}) = B_0 \hat{e}_z + \delta\mathbf{B}(\mathbf{r})$$

that is an uniform background field plus zero-average fluctuations. Here the magnetic fluctuations are constructed in a parallelepipedal box with sides $L_i \simeq 4l_i$, where l_i is the correlation length in the i-th direction ($i = x, y, z$). The magnetic fluctuations are represented as

$$\delta\mathbf{B}(\mathbf{r}) = \sum_{\mathbf{k},\sigma} \delta B(\mathbf{k}) \mathbf{e}^{(\sigma)}(\mathbf{k}) \exp i[\mathbf{k}\cdot\mathbf{r} + \phi_{\mathbf{k}}^{(\sigma)}]$$

Magnetic turbulence appears to be anisotropic both in the solar wind [12, 13] and in the interstellar medium [14]. Therefore, the Fourier intensity is given by

$$\delta B(\mathbf{k}) = \frac{C}{({k_x}^2{l_x}^2 + {k_y}^2{l_y}^2 + {k_z}^2{l_z}^2)^{\gamma/4+1/2}}$$

where C is a normalization constant. Note that with this choice of $\delta B(\mathbf{k})$, the Fourier amplitude depends on all of l_x, l_y, l_z, and that the isolevels of $\delta B(\mathbf{k})$ are ellipsoids in $\mathbf{k}$ space (see, Pommois et al. [5, 6], for more details). Once the magnetic field model is set up, we can trace the magnetic field structure by integrating the magnetic field line equations $d\mathbf{r}/ds = \mathbf{B}(\mathbf{r})/|\mathbf{B}(\mathbf{r})|$. It is usually found that the level of stochasticity increases with the turbulence level $\delta B/B_0$. Here, δB is the rms value of fluctuations. In a similar way the level of stochasticity increase with the ratio of correlation lengths l_z/l_x [6, 15, 16]. A suitable parameter to caracterize the level of stochasticity and the transport regime was shown to be the Kubo number $R = \delta B l_z / B_0 l_x$ [15, 16]. This expression of R was found to be valid even when $l_x \neq l_y$ [16]. Besides, the study of magnetic field line transport carried out in [15, 16] shows that anomalous diffusion is found for $R \ll 1$, quasilinear diffusion, i.e. $D_i \propto \delta B^2$, is found for $0.2 \lesssim R \lesssim 1$, and the percolation scaling of the diffusion coefficient, $D_i \propto \delta B^{0.7}$, is found for $R \gtrsim 10$.

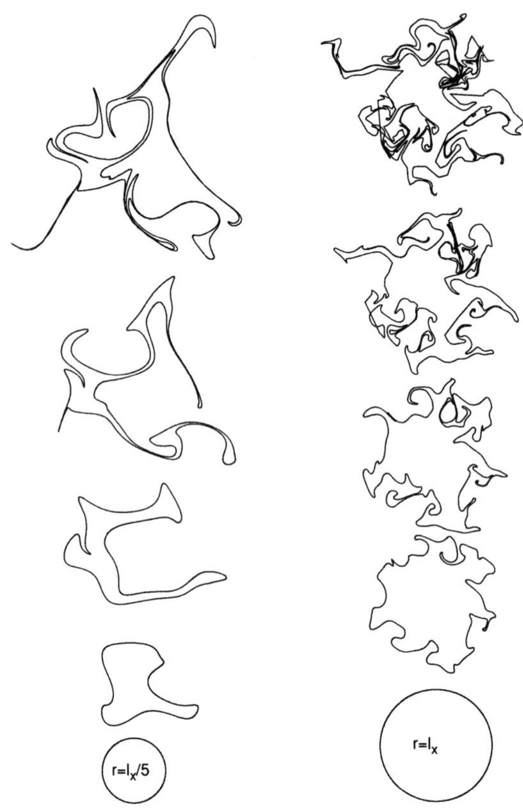

FIGURE 1. Evolution of a magnetic flux tube cross section in the case of quasi 2-D spectrum with $l_x/l_z = l_y/l_z = 0.1$ (from bottom to top). The magnetic flux tube is starting in the circle of radius r at bottom of the figure and is traced with 5000 points. Each adjacent line is integrated, and we plotted here the evolution for interval $Z = 0.25L_z \sim l_z$ (where L_z is the box simulation dimension, and l_z is the correlation length in z direction). The turbulence level is $\delta B/B_0 = 0.4$. Flux tube on the left, $r = 0.05L_z \sim 0.2l_z$; flux tube on the right, $r = 0.25L_z \sim l_z$.

NUMERICAL RESULTS

Figures 1, 2, and 3 show the evolution of a magnetic flux tube for different realizations of the turbulent magnetic field. All of them have the same fluctuation level $\delta B/B_0 = 0.4$, and different degrees of anisotropy. In each case, we choose the starting positions of the field lines on a path representing the boundary of a circular flux tube cross section. Then, integrating up to 5000 adjacent field lines, we can devise the evolution of the shape of a flux tube cross section [3, 11, 17]. We argue that the spatial distribution of not too energetic particles injected in a magnetic flux tube will follow the shape of the flux tube, except for finite Larmor radius effects.

In Figure 1 a quasi-2D anisotropy is considered, with $l_z = 10l_x = 10l_y$. In such a case we have relatively large

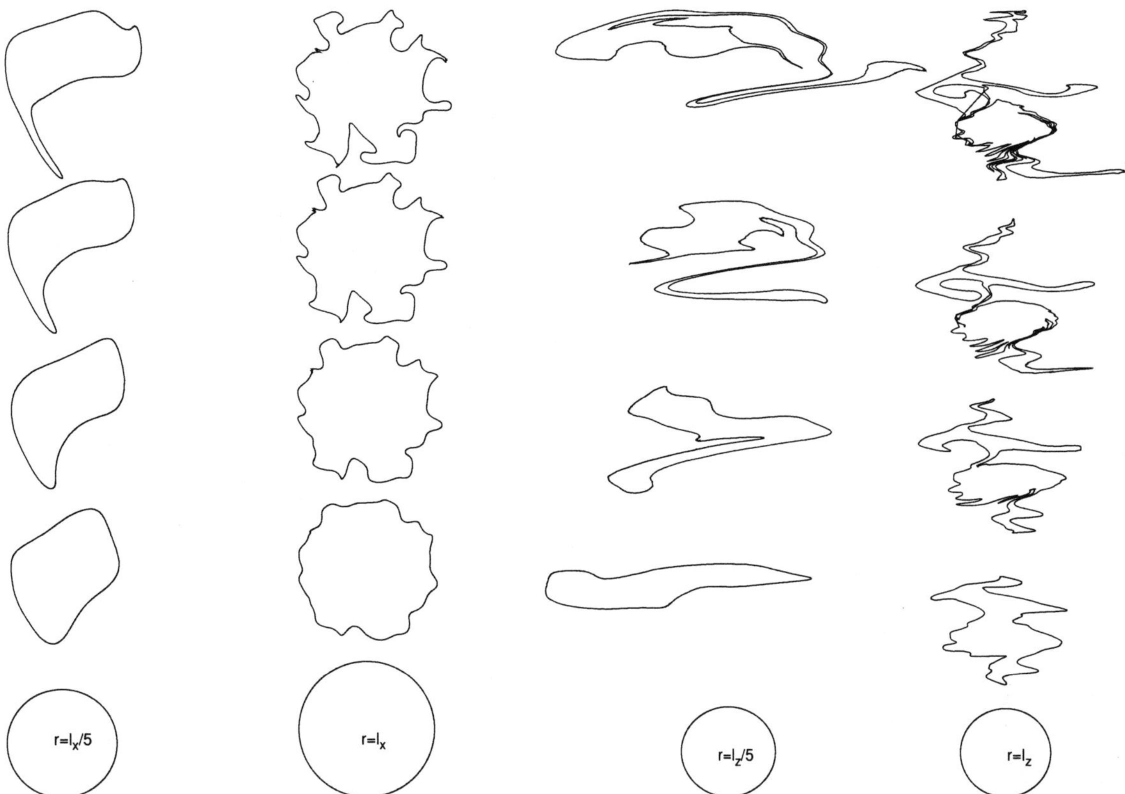

FIGURE 2. Same Figure as Figure 1. We use here a Slab-like spectrum with $l_x/l_z = l_y/l_z = 10$ for the magnetic turbulence, and the flux tube cross section from bottom to top are plotted at intervals $Z = 2L_z$ (which correspond to $\sim 8L_z$). Flux tube on the left, $r = 0.05L_x \sim l_x/5$; flux tube on the right, $r = 0.25L_x \sim l_x$.

FIGURE 3. Same Figure as Figure 1. We use here an anisotropic spectrum with $l_x/l_z = 8$ and $l_y/l_z = 1$ for the magnetic turbulence, and the flux tube cross section from bottom to top are plotted at intervals $Z = 2L_z$ (which correspond to $\sim 8l_z$). Flux tube on the left, $r = 0.05L_z \sim l_z/5$; flux tube on the right, $r = 0.25L_z \sim l_z$.

Kubo number, $R = (\delta B/B_0)(l_z/l_x) = 4$, and, indeed, a very fast evolution of the magnetic flux tube is obtained. In the left "column", from bottom top, the initial size of the flux tube, that is the radius r, is much smaller than the turbulence correlation length, $r = 0.05L_z \sim 0.2l_z \ll l_z$. On the right side of Figure 1, the initial size of the flux tube is comparable to the turbulence correlation length, $r = 0.25L_z \sim l_z$. It can be seen that very elongated and ramified structures quickly form. This is in agreement with the magnetic field line evolution reported by Zank et al. [18]. In Figure 2 a quasi-slab anisotropy is considered, with $l_z = l_x/10 = l_y/10$. In such a case we have a relatively small Kubo number, $R = 0.04$, and the evolution of the flux tube is slower. Indeed, it can be seen that the flux tube is distorted, but this process is much slower than in the previous case. It appears that the Kubo number also quantifies the speed of ramification of the flux tubes.

In Figure 3, an anisotropic spectrum with $l_x = 8$, $l_y = 1$, and $l_z = 1$ was used, as appropriate to the solar wind case (see [13, 7]). In this case the Kubo number is $R = 0.05$. In Figure 3, the flux tube has been followed for larger distances, up to $32l_z$, in order to make a direct comparison with the observations in the solar wind. Indeed, considering that at 1 AU from the Sun, the correlation length is about 0.03 AU [7], we can argue that to travel for about 1 AU in the heliosphere (the typical distance where most of the spacescraft take measures) we need to cover about 30 correlation lengths l_z Each plot, from bottom to top, is spaced by a distance in the z direction which corresponds to $8l_z$. At the top of the flux tube cross section of left panel of Figure 3, we can observe filamentation of the curves, which corresponds to small structures in the magnetic field. If we consider that a particular ion composition or energetic particles are injected at the base of the magnetic flux tube, we can understand how sharp gradients can be found in the solar wind. Here two filaments are separed by 2–$3l_z$, which are the largest structures observable in the plot, even though smaller structures are present. Converted in the solar wind dimension this corresponds to structures of about 0.06–0.09 AU and is consistent with the findings of Giacalone *et al.* [10]. Clearly the smaller structures will give rise to shorter time variations.

On the other hand, the flux tube evolution represented in Figures 1, 2, and 3 shows that this is very sensitive to the anisotropy of the turbulence spectrum. Different kinds of anisotropy give rise to rather different evolution (see also [4, 18]). In particular, the details of the flux tube fine structure can be different for different kind of anisotropies. Hence, we argue that an accurate study of the morphology of impulsive energetic particle events when compared to the simulation results, can give information on the anisotropy of solar wind magnetic turbulence.

Indeed, when the Kubo number is large (like for quasi 2-D anisotropy) the evolution of the flux tube is very fast, with the formation of fine scale structures. Consequently, the energetic particle intensity time profile will exhibit many short time peaks, possibly with a hierarchy of durations. (Indeed, a fractal structure is often the result of the evolution of a strongly nonlinear system). On the other hand, when the Kubo number is small (like for slab turbulence, or when $l_x \gg l_z$), the evolution of the flux tube is slow, and only a few, well defined periods of high energetic particle fluxes will be found in the intensity profile.

CONCLUSIONS

In this paper we have pointed out the different features of normal diffusion due to collisions and of turbulent diffusion due to the convection of a passive tracer by turbulence. We have shown by means of a numerical simulation of the evolution of a magnetic flux tube that fast magnetic field line transport and the formation of small structures in the flux tube cross section go together. We also have found that the rate of small structure formation depends on the anisotropy of turbulence, and in particular on the Kubo number.

If we assume that either energetic particles accelerated by a flare, or particular ion compositions are injected from the corona in the solar wind in a given flux tube, we can imagine that the charged particles in their propagation will follow to a good extend the magnetic field lines. Hence, the spatial distribution of particles will be organized as the cross sections of the magnetic flux tubes shown in Figures 1 to 3. As a consequence, when such structure is convected over the spacecraft from the solar wind flow, sudden variations of energetic particles or ion concentration will be seen, corresponding to sharp gradients. At the same time, fast spreading out, that is fast perpendicular transport, is obtained with respect to the average magnetic field. Further study may help to use the observation of the fine structure in the ion composition or energetic particle fluxes to infer information on the anisotropy of turbulence.

ACKNOWLEDGMENTS

This work is part of a research programme which is financially supported by the Ministero dell'Università e della Ricerca Scientifica e Tecnologica (MURST), the Agenzia Spaziale Italiana (ASI), contract no. I/R122/01, and the High Performance computing center at the University of Calabria.

REFERENCES

1. Rosenbluth, M. N., Sagdeev, R. Z., Taylor, G. B., and Zaslavsky, G. M., *Nucl. Fusion*, **6**, 297 (1966).
2. Jokipii, J. R., and Parker, E. N., *Phys. Rev. Lett.*, **21**, 44 (1968).
3. Rechester, A. B., and Rosenbluth, M. N., *Phys. Rev. Lett.*, **40**, 38 (1978).
4. Matthaeus, W. H., Gray, P. C., Pontius, Jr., D. H., and Bieber J. W., *Phys. Rev. Lett.*, **75**, 2136 (1995).
5. Pommois, P., Zimbardo, G., Veltri, P., *Phys. Plasmas*, **5**, 1288 (1998).
6. Pommois, P., Veltri, P., Zimbardo, G., *Phys. Rev.* E, **59**, 2244 (1999).
7. Pommois, P., Veltri, P., Zimbardo, G., *J. Geophys. Res.*, **106** (A11), 24,965 (2001).
8. Zurbuchen, T. H., Hefti, S., Fisk, L A., Gloecker, G., and Schwadron, N. A., *J. Geophys. Res.*, **105**, 18,327 (2000).
9. Mazur, J. E., Mason, G. M., Dwyer, J. R., Giacalone, J., Jokipii, J. R., Stone, E. C., *Astrophys. J.*, **532**, L79 (2000).
10. Giacalone, J., Jokipii, J., Mazur, J. E., *Astrophys. J.*, **532**, L75 (2000).
11. Isichenko, M. B., *Rev. Modern Phys.*, **64**, 961 (1992).
12. Klein, L., Bruno, R., Bavassano, B., Rosenbauer, H., *J. Geophys. Res.*, **98** (A5), 7837 (1993).
13. Carbone, V., Malara, F., Veltri, P., *J. Geophys. Res.*, **100** (A2), 1763 (1995).
14. Bhattacharjee, A., and Ng, C. S., *Astrophys. J.*, **548**, 318 (2001).
15. Zimbardo, G., Veltri, P., Pommois, P., *Phys. Rev.* E, **61**, 1940 (2000).
16. Pommois, P., Veltri, P., Zimbardo, G., *Phys. Rev.* E, **63**, 066405 (2001).
17. Isichenko, M. B., *Plasma Phys. Control. Fusion*, **33**, 809 (1991).
18. Zank, G., Matthaeus, W., and Zhou, Y., Nearly Incompressible Hydrodynamics in the Inhomogeneous Solar Wind, *this volume*.

The Geometry of Turbulent Magnetic Fluctuations at High Heliograpahic Latitudes

Charles W. Smith

Bartol Research Institute, University of Delaware, Newark, Delaware

Abstract. Although the orientation of wave vectors for interplanetary magnetic fluctuations is often discussed or assumed when modelling the solar wind turbulence, there were few if any direct measurements of the wave vector orientation until very recently. Indirect inferences abound. For instance, the transverse nature of magnetic fluctuations has often been used to infer a wave vector parallel to the mean magnetic field, but transverse fluctuations do not necessarily lead to parallel wave vectors – the fluctuations of two-dimensional turbulence are fully tranverse, but their wave vectors are also transverse to the mean field. *Bieber et al.* [3] offer the first single-point algorithm for separating field-aligned wave vectors from perpendicular wave vectors using the component spectra. This method has since been used by others [5, 6, 8, 9] and is here applied to Ulysses measurements in a comparison of high latitude turbulence with the results of previous studies of near-ecliptic observations.

INTRODUCTION

Although the anisotropy of interplanetary magnetic field (IMF) fluctuations has been recognized for some time [1], little is known of the orientation of the associated wave vectors. Under limited conditions and assumptions the minimum variance direction corresponds to the direction of the wave vector (nonlinear or eliptically polarized Alfvén and fast mode waves). However, under many more general assumptions (multiple waves projecting onto the same spacecraft frame frequency and 2-dimensional (2-D) turbulence where the wave vector is normal to the mean field direction B_0, but the minumum variance direction is parallel to B_0) the wave vector and minimum variance direction are unrelated.

Two clear conclusions of the observed correlation function for IMF fluctuations at 1 AU [7] are: (1) the angular width of the correlation function is sufficient to permit multiple wave vectors to project onto the same spacecraft frequency and (2) wave vectors at large angles to B_0 with $k_{\|}/k_{\perp} \approx 0$ are present. Therefore, it is suspect that minimum variance directions are an inadequate prescription by which to determine the orientation of the wave vector.

The relative importance of the large $k_{\perp}$ oscillations during energetic solar particle events was made clear by *Bieber et al.* [1996] who demonstrated that ~85% of the energy in the observed IMF fluctuations at 1 AU resided in the 2-D component. Previously, *Bieber et al.* [2] demonstrated that nearly this same 20/80 ratio of slab (1-D) to 2-D energy could account for the observed mean free path of solar energetic particles.

Until now, this method has not been applied to the Ulysses observations at high heliographic latitude.

TECHNIQUE

We adopt the technique described by *Bieber et al.* [3]. The observed spectrum of IMF fluctuations is assumed to be an admixture of transverse fluctuations, some of which have wavevectors parallel to the mean magnetic field ($\mathbf{k} \parallel \mathbf{B_0}$) and some having $\mathbf{k} \perp \mathbf{B_0}$. This is a simplification of the "Maltese Cross" correlation function for IMF fluctuations at 1 AU [7]. Figure 1 illustrates the two conditions.

The IMF data is rotated into the same mean field coordinate system used by Belcher and Davis [1]. In this coordinate system one component perpendicular to $\mathbf{B_0}$ is also perpendicular to the wind velocity $\mathbf{V_{SW}}$ and one is not. We denote the "Y" component as that component which is perpendicular to both $\mathbf{B_0}$ and $\mathbf{V_{SW}}$ while the

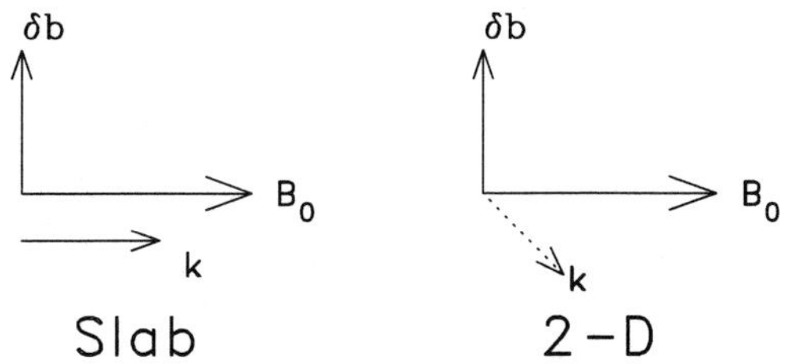

FIGURE 1. Illustration of the two **k** orientation for slab (1-D) and 2-D geometries. Both have IMF fluctuations $\delta\mathbf{b} \perp \mathbf{B_0}$.

CP679, *Solar Wind Ten: Proceedings of the Tenth International Solar Wind Conference*, edited by M. Velli, R. Bruno, and F. Malara

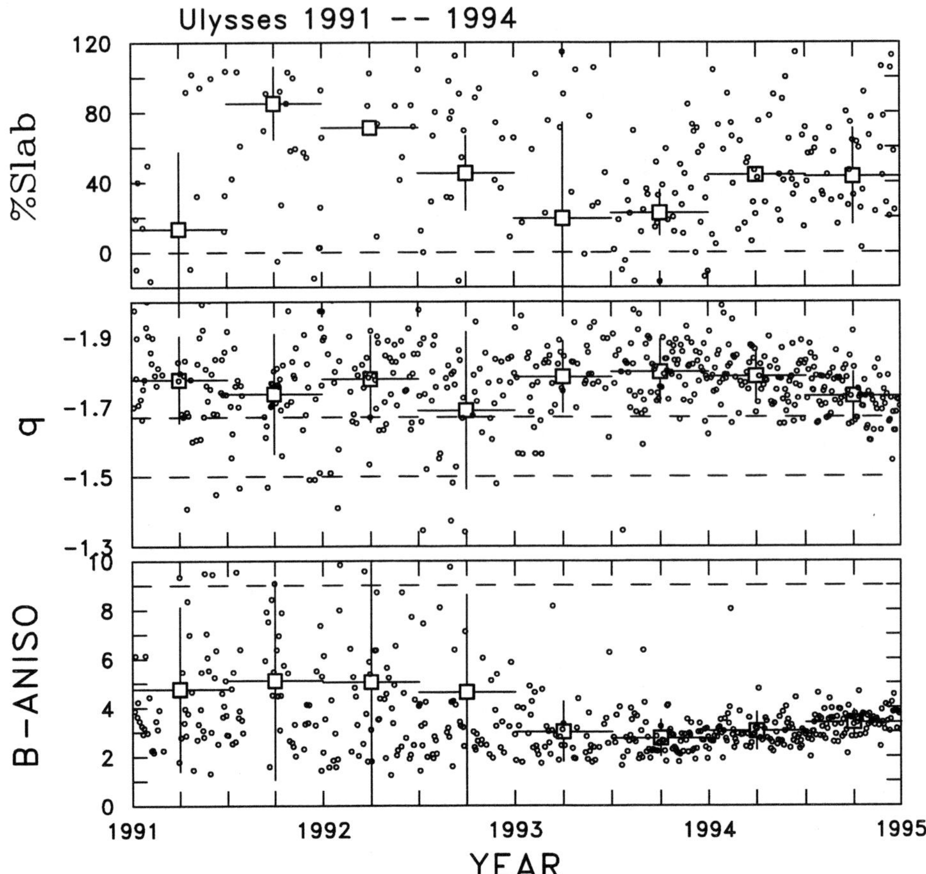

FIGURE 2. The %Slab, power law index $-q$, and IMF fluctuation anisotropy are plotted for the analyzed intervals. Square symbols represent averages over the time intervals represented by horizontal lines. Vertical lines give the variance of the underlying subset.

"X" component is $\perp \mathbf{B_0}$ but has a nonzero projection onto $\mathbf{V_{SW}}$. It can then be shown [3]:

$$\frac{P_Y}{P_X} = \frac{k_s^{1-q} + r' \left(\frac{2q}{1+q}\right) k_2^{1-q}}{k_s^{1-q} + r' \left(\frac{2}{1+q}\right) k_2^{1-q}} \tag{1}$$

where

$$k_s = 2\pi\nu/V_{SW}cos(\Theta_{BV}) \tag{2}$$

and

$$k_2 = 2\pi\nu/V_{SW}sin(\Theta_{BV}). \tag{3}$$

Θ_{BV} is the angle between the $\mathbf{B_0}$ and $\mathbf{V_{SW}}$, where the wind velocity is assumed to be in the Radial **R** direction, $-q$ is the power law spectral index, and ν is the spacecraft frame frequency. From this we can obtain:

$$r' = [tan(\Theta_{BV})]^{1-q} \left(\frac{P_Y/P_X - 1}{q - P_Y/P_X}\right) \left(\frac{1+q}{2}\right) \tag{4}$$

where

$$r = \frac{1}{1+r'} \tag{5}$$

is the fraction of energy contained in the slab (1-D) component.

From this it follows that $P_Y/P_X = 1$ if $r = 1$ and the spectrum is entire composed of $\mathbf{k} \parallel \mathbf{B_0}$ wave vectors. If $P_Y/P_X = q$, then $r = 0$ and the spectrum is entirely composed of 2-D fluctuations. Also, $1 < P_Y/P_X < q$ for all Θ_{BV}. *Belcher and Davis* [1] found that $P_Y/P_X \sim 5:4$. This ratio implies that $r = 0.57$ for $\Theta_{BV} = 45°$ and $q = 5/3$ indicating that the interval studied by Belcher and Davis contains an approximately equal admixture of slab (1-D) waves and 2-D turbulence.

The geometry analysis assumes that the spectrum in the rest frame of the plasma is axisymmetric and invariant to rotation about the mean magnetic field. One test of this assumption is that both transverse components should have the same power law index. Sometimes they do and often they do not. Intervals when the two power law indices differ substantially are excluded from this analysis.

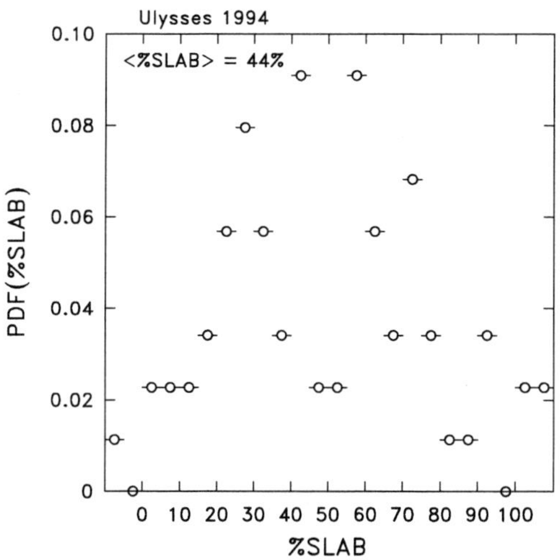

FIGURE 3. Probability distribution function for %Slab as computed from Ulysses data intervals during 1994.

ANALYSIS

We have examined 530 intervals of Ulysses data, each 12 hours to 2 days in duration with 1 minute data resolution obtained from the National Space Science Data Center (NSSDC). The analysis begins with day 298 of 1990 and extends over the southern solar pole at the end of 1994. The analysis applies an automated badpoint removal algorithm before rotating to mean field coordinates. A first-order difference filter is applied followed by a Blackman-Tukey correlation function and spectral analysis. The resulting spectra are post-darkened to correct for the first-order differencing filter and the geometry analysis is applied.

Figure 2 shows the results of the analysis. The geometry analysis neglects the parallel fluctuation component and assumes the fluctuations are all transverse to the mean field. The bottom panel addresses this assumption by plotting the ratio of the perpendicular and parallel components of the fluctuation $E^B_\perp/E^B_\parallel$. Although highly variable at low latitudes, the magnetic power ratio settles down to a consistent value of ~ 3 beyond $-30°$ latitude and slowly increases as the spacecraft approaches the southern solar pole.

The power law index $-q$ is also critical to the analysis. While again variable at low latitudes, it also settles into a consistent value of -1.75 at southern latitudes below $-30°$ and approaches $-5/3$ over the solar pole. The relatively high values of q are troubling and demonstrate the leading reason why the results shown here are considered preliminary. Low latitude values of q are normally found to be $-5/3 < -q < -3/2$.

The results of the %Slab= $r \times 100$ analysis is shown in Figure 2. It must be noted that this analysis can yield a slab percentage less than zero or greater than 100%. This reflects statistical fluctuations within the ensemble that can push the ratio P_Y/P_X beyond the bounds of the geometry analysis. Only the statistics of the distribution are truly meaninful. The low-latitude %Slab results are unusually high compared with the low-latitude, 1 AU results of *Bieber et al.* [3] and *Leamon et al.* [5, 6], but settle down to the expected value by mid-1993. As the spacecraft climbs in latitude toward the south solar pole the slab fraction of the spectrum climbs to a value of $\sim$ 40%, a value that is in excess of the nominal low latitude results at 1 AU. Temporarily setting aside the low latitude results of this analysis, this means that the high latitude wind at solar minimum is more wave-like than the low latitude wind, but is not exclusively composed of parallel propagating waves. Both waves oof the slab (1-D) component and turbulence as represented by the 2-D component are present in approximately equal proportion.

Figure 3 shows the probability distribution function of %Slab values computed by this analysis for the year 1994 when Ulysses was closest to the south solar pole. The poorness of the PDF is in part due to the excessive application of the axisymmetry assumption which reduced the number of usable intervals. Nevertheless, a peak near 50% is evident as opposed to a bi-modal distribution peaked at 0% and 100% that would be the case if the two populations of slab (1-D) and 2-D geometries were spatially isolated. This distribution, whose mean is 44%, indicates that the two populations of slab (1-D) and 2-D wave vectors coexist in the same plasma elements.

Figure 4 compares the observed variation of P_Y/P_X with the prediction based on the geometry analysis for varying %Slab using the observed average value of the power law index during this time $q = 1.75$ for the year 1994. Although the fit is not as good as seen by *Bieber et al.* [3] at low latitudes, it is generally consistent with an equal admixture of slab (1-D) and 2-D geometries.

CONCLUSIONS

We have applied the technique and assumptions of *Bieber et al.* [3] to Ulysses IMF measurements in an attempt to obtain some insight into the relative orientation of wave vectors at high latitudes during solar minimum. This method assumes that the wave vectors are either parellel or orthogonal to $\mathbf{B_0}$ in keeping with a simplification of the observed correlation function for 1 AU near-ecliptic observations. We find the IMF fluctuation energy to be almost equally divided between the two assumed extremes.

It is sometimes assumed but never proven that the $k_\parallel$ population is waves while the $k_\perp$ fluctuations constitute

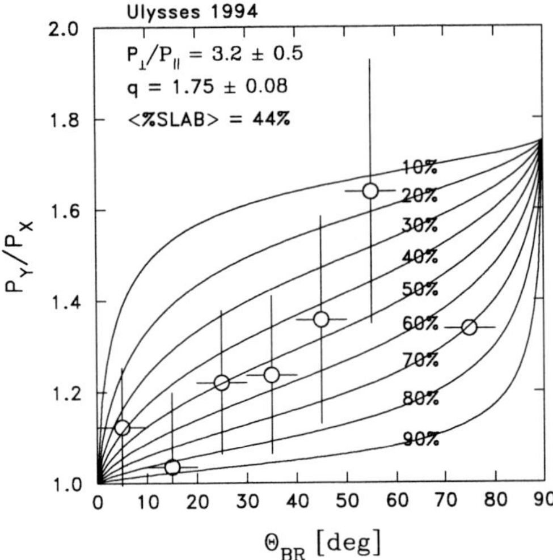

FIGURE 4. Ratio of the observed average P_Y/P_X as a function of Θ_{BR} compared with the prediction from the geometry analysis for varying fractions of the slab (1-D) component. Horizontal lines on the observed means represent the averaging interval and vertical lines are the variance of the underlying population. Note that the observed value for $60° < \Theta_{BR} < 70°$ contains only 1 observation and so it has not variance but a great deal of uncertainty.

MHD turbulence and, in fact, this language has been adopted here. Neither assumption is necessarily true and may be extreme simplifications. It is true that the $k_\perp$ fluctuations have been shown to arise in simulations from initial conditions of $k_\parallel$ waves and low-level background noise [4], but it may also be the case that $k_\parallel$ fluctuations arise from an abundance of preexisting $k_\perp$ modes in keeping with the predictions of Nearly Incompressible MagnetoHydroDynamcis (NIMHD) [10]. The $k_\perp$ modes may have wave-like components and the $k_\parallel$ modes must have a turbulent dynamic in order to drive the $k_\perp$ modes in simulations. All this analysis seeks to demonstrate is the distribution of energy between the two populations of slab (1-D) and 2-D wave vectors which is roughly equally distributed at high latitudes during solar minimum.

ACKNOWLEDGMENTS

This work was supported by NASA Sun Earth Connection Guest Investigator program grant NAG5-10911 to the Bartol Research Institute.

REFERENCES

1. Belcher, J. W., and L. Davis Jr., Large-amplitude Alfvén waves in the inerplanetary medium, 2, *J. Geophys. Res.*, *76*, 3534–3563, 1971.
2. Bieber, J. W., W. H. Matthaeus, C. W. Smith, W. Wanner, M.-B. Kallenrode, and G. Wibberenz, Proton and electron mean free paths: The Palmer consensus revisited, *Astrophys. J.*, *420*, 294–306, 1994.
3. Bieber, J. W., W. Wanner, and W. H. Matthaeus, Dominant two-dimensional solar wind turbulence with implications for cosmic ray transport, *J. Geophys. Res.*, *101*, 2511–2522, 1996.
4. Ghosh, S., W. H. Matthaeus, D. A. Roberts, and M. L. Goldstein, Waves, structures, and the appearance of two-component turbulence in the solar wind, *J. Geophys. Res.*, *103*, 23,705–23,715, 1998.
5. Leamon, R. J., C. W. Smith, N. F. Ness, W. H. Matthaeus, and H. K. Wong, Observational constraints on the dynamics of the interplanetary magnetic field dissipation range, *J. Geophys. Res.*, *103*, 4775–4787, 1998a.
6. Leamon, R. J., C. W. Smith, and N. F. Ness, Characterisitcs of magnetic fluctuations within coronal mass ejections: The January 1997 event, *Geophys. Res. Lett.*, *25*, 2505–2509, 1998b.
7. Matthaeus, W. H., M. L. Goldstein, and D. A. Roberts, Evidence for the presence of quasi-two-dimensional nearly incompressible fluctuations in the solar wind, *J. Geophys. Res.*, *95*, 20,673–20,683, 1990.
8. Smith, C. W., D. J. Mullan, N. F. Ness, R. M. Skoug, and J. Steinberg, Day the solar wind almost disappeared: Magnetic field fluctuations, wave refraction and dissipation, *J. Geophys. Res.*, *106*, 18,625–18,634, 2001.
9. Smith, C. W., D. J. Mullan, N. F. Ness, R. M. Skoug, and J. Steinberg, Day the solar wind almost disappeared: Magnetic field fluctuations, wave refraction and dissipation, this volume, 2002.
10. Zank, G. P., and W. H. Matthaeus, The equations of reduced magnetohydrodynamics, *J. Plasma Phys.*, *48*, 85–100, 1992.

The interaction of turbulence with shock waves

G.P. Zank*, Ye Zhou†, W.H. Matthaeus** and W.K.M. Rice‡

*Institute of Geophysics and Planetary Physics, University of California, Riverside
†Lawrence Livermore National Laboratory, University of California, Livermore
**Institute of Geophysics and Planetary Physics, University of California, Riverside, and // Bartol Research Institute, The University of Delaware, Newark
‡The University of St Andrews, St Andrews, Fife

Abstract. The interaction of turbulence and shock waves is considered self-consistently so that the back-reaction of the turbulence and its associated reaction on the turbulence is addressed. Upstream turbulence interacting with a shock wave is found to mediate the shock by 1) increasing the mean shock speed, and 2) decreasing the efficiency of turbulence amplification at the shock as the upstream turbulence energy density is increased. The implication of these results is that the energy in upstream turbulent fluctuations, while being amplified at the shock, is also being converted into mean flow energy downstream. The variance in both the shock speed and position is computed, leading to the suggestion that, in an ensemble-averaged sense, the turbulence-mediated shock will acquire a characteristic thickness given by the standard deviation of the shock position. Lax's geometric entropy condition is used to show that as the upstream turbulent energy density increases, the shock is eventually destabilized, and may emit one or more shocks to produce a system of multiple shock waves. Finally, turbulence downstream of the shock is shown to decay in time t according to $t^{-2/3}$.

INTRODUCTION

The solar wind is intrinsically turbulent and shock waves, either propagating or formed upstream of obstacles such as planets and comets, must interact with low-frequency turbulent fluctuations. Many shocks, especially quasi-parallel shocks, generate turbulence in their very extended foreshocks, and this interacts eventually with the shock ramp. In this paper, we take the first steps towards addressing the interaction of turbulence and shock waves at a self-consistent level. In this, we adopt a perspective rather different from previous studies in that we employ a statistical description for the fluctuations from the outset, rather than attempting to use a linearized perturbation analysis of individual waves and fluctuations. A schematic of the problem that we address is given in Fig. 1. Figure 1 illustrates a shock distorted by its interaction with upstream turbulence, and the subsequent transmission of the turbulence into the downstream subsonic region. We wish to compute the mean location and speed of the shock as a function of the upstream energy density of turbulence and of course the variance of the shock position and speed. As a consequence, the amplification of the turbulence at the shock is also addressed, together with its subsequent dissipation downstream.

MATHEMATICAL FORMULATION

For hypersonic flows in which ram pressure dominates thermal quantities, the gas equations reduce to the inviscid Burgers' equation,

$$\frac{\partial u}{\partial t}+\frac{\partial}{\partial x}\left(\frac{1}{2}u^2\right)=0 \Longleftrightarrow \frac{\partial u}{\partial t}+u\frac{\partial u}{\partial x}=0. \quad (1)$$

Solutions to equation (1) require the insertion of a shock to counteract wave steepening and breaking. If we let $x=\phi(t)$ describe the shock location, then the shock normal is given by $\mathbf{n}=(-\phi_t,1)$, and the R-H condition is given classically by

$$-\phi_t\,[u]+\left[\frac{1}{2}u^2\right]=0. \quad (2)$$

To solve (2), we may prescribe the initial data

$$u(t=0)=\left\{\begin{array}{l} u_1 \\ u_2 \end{array}\right., \quad (3)$$

and to ensure that the geometric entropy condition [1] holds, we must have $u_2<u_1$. One obtains immediately from (2) and (3) the well-known classical result for the shock speed

$$\phi_t=\frac{\left[\frac{1}{2}u^2\right]}{[u]}=\frac{1}{2}(u_1+u_2), \quad (4)$$

CP679, *Solar Wind Ten: Proceedings of the Tenth International Solar Wind Conference*, edited by M. Velli, R. Bruno, and F. Malara

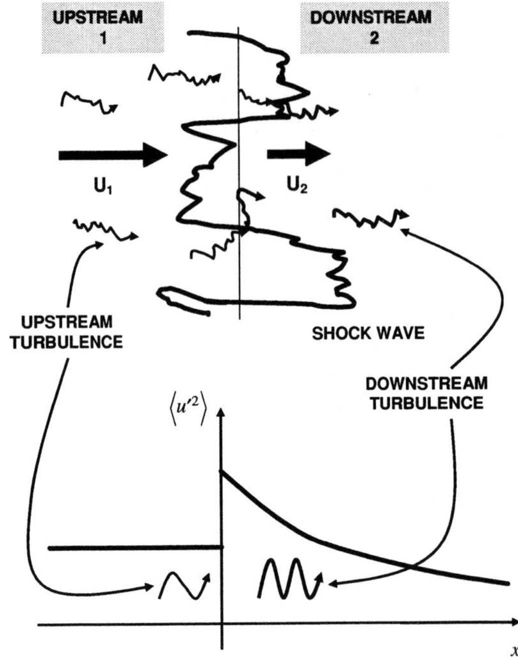

FIGURE 1. Schematic of the problem in the stationary shock frame. The top figure depicts a shock wave that is highly distorted by the repeated random interaction of upstream turbulence with the shock. Quantities upstream of the shock, where the flow is supersonic, carry the subscript 1 and downstream quantities have the subscript 2. The flow velocity is denoted by **U**. The problem is to compute the mean shock velocity, position, and variance of these quantities in the presence of upstream turbulence. In addition, as illustrated in both the top and bottom figures, we wish to compute the transmission characteristics of the upstream turbulence (the amplification it may experience) and its subsequent decay as it is dissipated into the downstream flow. The bottom figure illustrates schematically the spatial evolution of the turbulent fluctuation energy from upstream of the shock to downstream.

which is constant and yields $x = \phi(t) = \frac{1}{2}(u_1 + u_2)t$ for the shock position.

In the following section, we reformulate the classical problem (1)-(3) to include turbulent fluctuations upstream of the shock. Subject to the restrictions imposed by our assumption of a hypersonic flow, the basic questions that we address are 1) what is the propagation speed of a shock in a turbulent medium? 2) Can upstream turbulence destabilize an apparently stable shock wave? 3) What is the effective or "averaged" thickness of the shock as a result of the presence of turbulent upstream fluctuations? 4) How strongly is turbulence amplified by a shock? And 5) What is the decay law for turbulence downstream of a shock?

Consider again the classical inviscid Burgers' equation problem (1)-(3), but now assume that the upstream flow is turbulent. We may express the flow incident on the shock wave by a mean and fluctuating part, so that $u = \bar{u} + u'$ and $\langle u \rangle = \bar{u}$ after ensemble averaging to eliminate fast time scales. Of course, the R-H condition (2) continues to hold exactly for u, and $x = \phi(t)$ as before. However, we shall assume that a detailed solution is either inaccessible or undesirable and instead seek a statistical formulation of the problem. For the 1D problem (1) and (2), the shock position fluctuates in response to variation in u, i.e., with respect to u', so that $\phi(t) = \Phi(t) + \phi'(t)$, $\langle \phi \rangle = \Phi(t)$. The mean field form of the R-H condition is thus given by

$$-\Phi_t[\bar{u}] - [\langle \phi_t' u' \rangle] + \frac{1}{2}[\bar{u}^2 + \langle u'u' \rangle] = 0. \qquad (5)$$

The boundary condition associated with the fluctuating component is

$$\begin{aligned} -\phi_t[\bar{u}] - \Phi_t[u'] &- [\phi_t' u' - \langle \phi_t' u' \rangle] + [\bar{u}u'] \\ &+ \frac{1}{2}[u'u' - \langle u'u' \rangle] = 0. \qquad (6) \end{aligned}$$

By comparing the time scale for spectral transfer in the fluctuating quantities to the transmission time scale for fluctuations across the shock, we reach the important conclusion [2] that spectral transfer is unimportant across narrow shocks for most fluctuations in the turbulence spectrum, and derive a lower bound on the scale size of upstream turbulent fluctuations that can interact "linearly" with the shock wave. Since spectral transfer is unimportant across the shock, the quantities $u'u'$ and $\phi_t' u'$ are independent of the fast nonlinear coupling time scale in the transition from upstream to downstream of the shock. We therefore have the remarkable result that, across the thin shock, the boundary condition for the fluctuating components in a mean-field decomposition of the variables satisfies the *linear* equation

$$-\phi_t[\bar{u}] - \Phi_t[u'] + [\bar{u}u'] = 0. \qquad (7)$$

Equation (7) possesses the formal structure of a linearized equation by virtue only of the cancellation of certain terms. We emphasize that (7) does not result from the linearization of the exact boundary condition (6) and we have not assumed that the fluctuations are of small amplitude. The argument leading to (7) is a form of rapid distortion theory.

By employing (7), we can determine various correlations across the shock. Details can be found in [2]. The Rankine-Hugoniot conditions for hypersonic gas dynamic shocks in a turbulent flow number four and are conveniently collected together as

$$\Phi_t = \left(1 - \frac{\langle [u']^2 \rangle}{[\bar{u}]^2}\right)\left(\frac{1}{2}\frac{[\bar{u}^2 + \langle u'u' \rangle]}{[\bar{u}]} - \frac{[u'\bar{u}][\langle u' \rangle]}{[\bar{u}]^2}\right); \qquad (8)$$

$$-\Phi_t[\langle u'^2 \rangle] + [\bar{u}\langle u'^2 \rangle] = [\bar{u}]\langle u'^2 \rangle \exp\left(\frac{[\bar{u}]\tau}{\bar{\nu}}\right); \qquad (9)$$

$$\left[\langle u'^2\rangle\ell\right] = -\left[\bar{u}\right]\langle u'_1 u'_2\rangle\tau; \quad (10)$$

$$[\ell] = [\bar{u}]\frac{\ell_1}{\bar{u}_1}. \quad (11)$$

In the above, $2\bar{\nu}$ is the viscous scale length for the shock, ℓ denotes the correlation length, and τ is a characteristic interaction time of fluctuations with the shock. The Rankine-Hugoniot conditions relate the upstream and downstream states across a shock, but now generalized to include the turbulent energy density $\langle u'^2\rangle$, the correlation length ℓ, and the cross-correlation across the shock $\langle u'_1 u'_2\rangle$. In this regard, equations (8) - (11) describe statistically the interaction of turbulence with shocks, with the back-reaction of the turbulence accounted for self-consistently. This distinguishes our approach from the linearized approach of [3].

Before solving the R-H conditions explicitly, let us consider the free decay of turbulence downstream of the shock (see Fig. 1, bottom). Immediately downstream of the shock, the turbulence may be characterized by $\langle u_2'^2\rangle$ and ℓ_2. By means of an energy-containing model, using 1-point correlations, we can derive the relation $\langle u'^2\rangle \sim t^{-2/3}$ (see [4] for a related approach).

RESULTS AND CONCLUSIONS

The Rankine-Hugoniot conditions yield a quadratic equation in the mean shock speed Φ_t, and the remaining downstream quantities and correlations can be derived [2]. Two other quantities of interest are the variance in the shock speed and is the variance in shock position. The latter quantity may be interpreted as the "turbulent shock thickness."

In Figures 2 - 4, we plot various solutions of the R-H conditions (8) - (11). A solution to the quadratic equation for the mean shock speed Φ_t is plotted in Fig. 2 as a function of normalized upstream energy density $\langle u_1'^2\rangle$. From the quadratic equation for Φ_t, two solutions exist for each $\langle u_1'^2$, and, the classical result is recovered in the absence of turbulent fluctuations. The second solution (not shown) begins from the classical solution and decreases with increasing $\langle u_1'^2\rangle$. However, not all solutions are admissible. For a shock to exist, Lax's geometric entropy condition [1] must be satisfied. This requires that the forward and backward characteristics which intersect at the shock wave can both be traced back to the initial data. The curve plotted in Fig. 2 can be shown to be admissible [2]. By contrast, the decreasing solution is never intermediate to the forward and backward characteristics. In fact, the positive root solutions illustrated in Fig. 2 continue to increase with increasing $\langle u_1'^2\rangle$ to above $\Phi_t = 1$, and these larger solutions are also inadmissible. Thus, unique stable shock solutions exist for only a limited range of upstream turbulence levels. Outside these solutions,the shock becomes unstable and we cannot assume that either a mean speed or position for the shock is possible. Thus, upstream turbulence, at least for the simple hypersonic 1D case considered here, leads to an increase in shock speed as the turbulent fluctuations merge with and transmit through the shock, but as the intensity increases, upstream turbulence renders the shock unsteady.

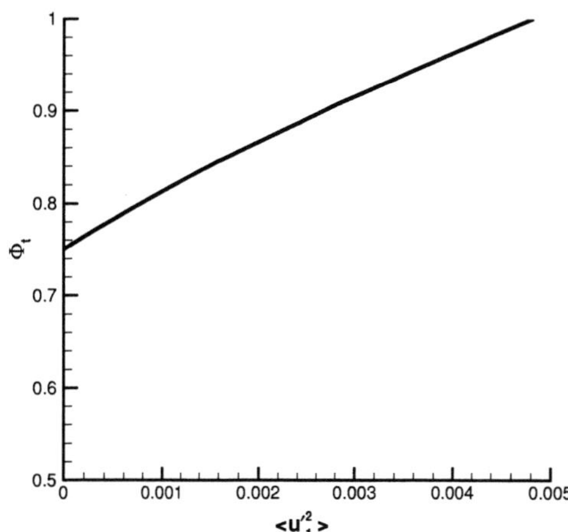

FIGURE 2. A plot of the mean shock speed Φ_t as a function of the upstream turbulence energy density $\langle u_1'^2\rangle$ normalized to the square of the upstream mean flow speed. In the absence of upstream turbulence, the classical result is recovered for $\langle u_1'^2\rangle = 0$.

Figure 3 plots the corresponding amplifications of the upstream turbulence by the shock, and Figure 4 shows the variance of the shock speed $\langle \phi_t'^2\rangle$ (essentially proportional to the variance in shock position, i.e., "turbulent shock thickness") and the cross-correlation $\langle u'_1 u'_2\rangle$.

Our results may be summarized as follows.

1. Although based on a simple energy-containing model for an idealized hypersonic fluid, our particular results emphasized the importance of a self-consistent coupling of turbulence and the mean shock variables. Specifically, the mean shock speed was found to increase with increasing levels of upstream turbulence. Correspondingly, the efficiency of upstream turbulence amplification by the shock decreased i.e., $\langle u_2'^2\rangle/\langle u_1'^2\rangle$ was a decreasing function of increasing $\langle u_1'^2\rangle$. The implication of this result is that the energy in upstream turbulent fluctuations, while being amplified at the shock, is also being converted into mean flow energy downstream. Thus, models which consider the amplification of

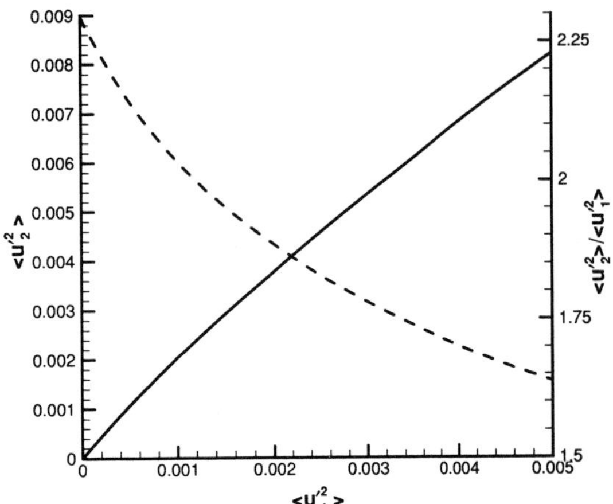

FIGURE 3. Two plots of the energy density in downstream turbulent fluctuations $\langle u_2'^2 \rangle$. The solid line (left axis) plots the downstream energy density normalized to the upstream ram energy and the dashed line (right axis) is the ratio of the downstream to upstream turbulent energy densities and is therefore a measure of the efficiency with which upstream fluctuations are amplified by the shock wave. Thus, although the amplitude of the energy density in transmitted fluctuations increases with an increasing upstream turbulent energy density, the corresponding efficiency decreases.

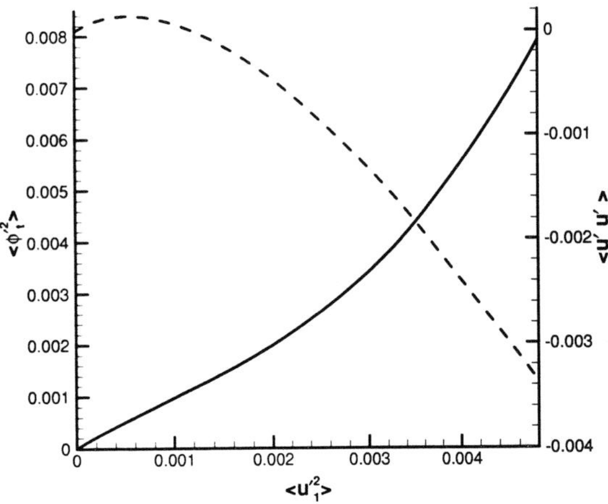

FIGURE 4. Plot of the shock speed variance $\langle \phi_t'^2 \rangle$ (solid curve, left axis) as a function of the upstream turbulent energy density. The shock speed variance is proportional to the variance in shock position. The dashed curve (right axis) is a plot of the cross-correlation of the upstream and downstream turbulent fluctuations $\langle u_1' u_2' \rangle$ as a function of the upstream turbulent energy density. Note that $\langle u_1' u_2' \rangle$ increases from zero and then decreases becoming negative with larger values of $\langle u_1'^2 \rangle$.

turbulence at a shock wave without considering the subsequent "turbulent mediation" of the shock will tend to over-estimate the levels of downstream turbulence.

2. The variance in the shock speed increases with increasing values of the upstream turbulent energy density. The variance in the shock position can be used as a measure of the "shock thickness", and we find that the "ensemble averaged" shock is no longer infinitesimally thin but instead has a shock thickness length scale given by $L_{shock} \sim \sqrt{\langle \phi'^2 \rangle}$. This suggests the possibility of introducing a "turbulent viscosity" to model shock structure in the presence of upstream turbulence.
3. It was found that surprisingly low levels of upstream turbulence energy density could destabilize the shock. As the mean shock speed increased with increasing $\langle u_1'^2 \rangle$, Lax's geometric entropy condition [1] was eventually violated and an upstream state could no longer be traced back to the initial data. In this case, a steady shock wave is no longer possible and instead a combination of shocks is needed to satisfy the Riemann problem. Thus, high levels of upstream turbulence would drive the original shock to eventually emit one or more shock waves, so producing a system of multiple shocks, the detailed study of which is beyond the scope of the present approach.
4. Finally, we found that in the case of a steady shock mediated self-consistently by turbulence, the emitted amplified turbulence decayed with a $t^{-2/3}$ dependence in the downstream region.

ACKNOWLEDGMENTS

This work was supported in part by an NSF grant ATM-0296113.

REFERENCES

1. Lax, P.D., *Hyperbolic Systems of Conservation and the Mathematical Theory of Shock Waves*, SIAM, Philadelphia (1990).
2. Zank, G.P., Zhou, Ye, Matthaeus, W.H., and Rice, W.K.M., *Phys. Fluids*, in press (2002).
3. McKenzie, J.F., and Westphal, K.O., *Phys. Fluids*, **11**, 2350 (1968).
4. Zank, G.P., Matthaeus, W.H., and Smith, C.W., *J. Geophys. Res.*, **101**, 17,093 (1996).

Solar Wind Fluctuations: Waves and Turbulence

Sean Oughton

Department of Mathematics, University of Waikato, Hamilton, Private Bag 3105, New Zealand

Abstract. We present a brief review of observations and theory regarding the nature and radial evolution of MHD-scale solar wind fluctuations. Emphasis is placed on the fact that the fluctuations consist of both waves and turbulence, and on their dual dynamical roles.

1. INTRODUCTION

The realization that the interplanetary medium contains fluctuations in the velocity, magnetic field, and density followed hard on the heels of the presentation of the first theoretical model of the solar wind [1]. Spacecraft observations soon confirmed the existence of such fluctuations, *e.g.*, [2]. Parker's original model [1] predicted flow which was radial with a speed of $\sim 400\,\mathrm{km\,s^{-1}}$ and an Archimedean spiral magnetic field; over forty years later, the *mean* wind is still adequately described and understood using essentially this model. However, understanding of how the associated fluctuations evolve, as they are advected outwards by the mean wind, is much less advanced. Here we review some of the advances in understanding which have occured over these four decades regarding the *origin, nature*, and *radial evolution/transport* of the fluctuations. Attention is focused on the role of waves and turbulence in regions close to the ecliptic plane. We denote the fluctuating fields as $\mathbf{v}$, $\mathbf{b}$, and $\delta\rho$, and the mean fields as $\mathbf{U}_0$, $\mathbf{B}_0$, and ρ_0. (Magnetic fields are measured in Alfvén speed units, *e.g.*, $\mathbf{b}$ has been redefined as $\mathbf{b}/\sqrt{4\pi\rho_0}$.)

Table 1 highlights a few important milestones regarding investigations of MHD-scale solar wind fluctuations. Early observations provided evidence for the existence of both Alfvén waves (via highly correlated time series for $\mathbf{v}$ and $\mathbf{b}$, *e.g.*, [3, 4]) and also for strongly nonlinear processes such as turbulence (*e.g.*, power-law energy spectra [5]). However, initial theoretical efforts predominantly assumed that the fluctuations were (Alfvén) waves, and often non-interacting waves, *e.g.*, [3, 6–9], although see [5]. See Table 2 for some important distinctions between the behaviour of waves and turbulence.

Prior to the 1980s the associated transport and evolution models typically employed (leading-order) WKB theory, wherein the wavelength of the fluctuations is assumed much smaller than the lengthscale on which the mean fields vary. While this assumption is well-grounded, it is now recognised that it it is a necessary—but not sufficient—condition for the validity of WKB theory as applied to MHD-scale fluctuations in a supersonic and super-Alfvénic flow [10–12]. Moreover, in standard leading-order WKB theory there is no interaction between inward and outward type fluctuations, where these terms refer to the sense of mode propagation along $\mathbf{B}_0$ (however, see [10–13]).

TABLE 1. A rough (decadal) history of solar wind fluctuation studies

1950s	Birkeland deduces existence of solar wind. Parker proposes a theoretical model for it [1].
1960s	First satellite data; Power-law spectra and evidence for waves. First application of WKB theory to evolution of fluctuations [8]. Suggestion that fluctuations are turbulent [5].
1970s	Plentiful wave observations in ecliptic plane. Many wave transport models (mostly WKB).
1980s	First turbulent transport models and analysis [14–22].
1990s	Newer data (Ulysses, WIND, ACE, ...). Improvements in both transport models and understanding of the nature of fluctuations.

In fact, many of the predictions of WKB theory disagree with observational results (see below). These shortcomings prompted consideration of various other approaches, including turbulence-based transport models.

Before proceeding to the observational results, it is instructive to discuss the major factors which influence the evolution of solar wind fluctuations (Table 3). Although expansion effects dominate the radial evolution behaviour, it is departures from this spherical expansion which comprise the "interesting" physics, since the former holds little mystery. The remaining factors in the table can lead to such departures. All fluctuations are carried along by the mean wind, and since at small enough lengthscales the flow is essentially uniform, fluctuations

CP679, *Solar Wind Ten: Proceedings of the Tenth International Solar Wind Conference*, edited by M. Velli, R. Bruno, and F. Malara

TABLE 2. Some distinctions between turbulence and waves

TURBULENCE	WAVES
Inherently nonlinear $\Rightarrow$ *spectral* transfer.	(Small-ampl.) theory = linear $\Rightarrow$ no spectral transfer.
Advection ("self"-distortion).	Propagation $\rightarrow$ *spatial* transport of energy.
No dispersion relation; all dynamical length/ time-scales coupled.	Dispersion relation, *i.e.*, each timescale depends on only a few lengthscales *e.g.*, $\omega = \pm \mathbf{k} \cdot \mathbf{V}_A$

TABLE 3. Some factors influencing fluctuation evolution

Expansion	Dominant factor. Controls overall "decay" with R.
Advection	By mean flow.
Preferred directions	Radial and (local) mean magnetic field. At least partly responsible for various observed anisotropies.
Nonlinear interactions	*Spectral* transfer of energy (from large scales to small ones).
Wave effects	Propagation. *Spatial* transport of energy.
Large-scale gradients	So-called MECS effects (Mixing, Expansion, Compression, Shear): Gradients in mean fields mediate couplings amongst fluctuations.
Pickup ions	Inject energy into fluctuations. Important for $R \gtrsim 8\,\mathrm{AU}$.

at these scales are unaltered by the advection. On the other hand, larger scale fluctuations notice the spatial dependence of the mean wind and are influenced by the associated gradients. This is one contribution to the MECS effects also listed in the table. Note that these are typically non-WKB effects [10–12, 22–25], which bleed energy from the mean fields into fluctuations, as in the Kelvin–Helmholtz instability. It is straightforward to estimate that the lengthscales on which these large-scale gradients vary is of order the heliocentric distance R [22, 26]. For example, in spherical geometry $(\nabla \cdot \mathbf{U}_0)/U_0 = 2/R$ for $\mathbf{U}_0 = U_0\hat{\mathbf{R}}$. Nonlinear interactions tend to cause *spectral* redistribution of energy, primarily transfer from large scales to small ones, whereas wave effects are associated with *spatial* transport of energy.

The influence of the two preferred directions (radial and $\mathbf{B}_0$) is also significant. The simplest models of the solar wind are often one-dimensional, with that direction being the radial since this is parallel to the direction of the mean flow. More sophisticated 3D models further emphasize the importance of the radial direction to the fluctuation dynamics, due to both expansion and shear effects; *e.g.*, [27–30]. A strong mean $\mathbf{B}$ tends to induce the dynamical development of spectral (and sometimes also variance) anisotropies, in both incompressible and compressible plasmas, *e.g.*, [31]. The effect is to make the flow quasi-two-dimensional (quasi-2D) in the sense that correlation lengths perpendicular to $\mathbf{B}_0$ become much shorter than those parallel to it, *e.g.*, [32–34].

2. OBSERVATIONAL RESULTS

These have been well-reviewed elsewhere (*e.g.*, [35–37]) so that here we present only a few pertinent points regarding radial evolution and the geometry of the fluctuations.

2.1. Radial Evolution

Observational studies of *in situ* data collected near the ecliptic in both the inner and outer heliosphere suggest that certain features of the radial evolution of fluctuations are rather robust [15–18, 38–43].

The magnetic energy (per unit *mass*) of the fluctuations, denoted $\langle \mathbf{b}^2 \rangle$ with $\langle \cdots \rangle$ thought of as a spatial average (and $\mathbf{b}$ in Alfvén speed units), typically decreases with heliocentric distance and indeed inside $\sim 8\,\mathrm{AU}$ follows an approximate powerlaw of $\sim R^{-1}$. Recall that (leading-order) WKB theory predicts an R^{-1} dependence [8]. However, one should also note that the observed dependence is not a strict powerlaw, with scatter in the data probably being due to multiple effects including variations in wind speed, solar cycle phase, and distance from the current sheet. For distances $\gtrsim 8\,\mathrm{AU}$, the magnetic energy is significantly above the WKB powerlaw level and the suggestion has been made that this "excess" energy is provided by pickup ions [24, 44–46].

Kinetic energy per unit mass, $\langle \mathbf{v}^2 \rangle$, often evolves similarly to the magnetic energy, and it is convenient to consider the evolution of their ratio, known as the the Alfvén ratio $r_\mathrm{A} = \langle \mathbf{v}^2 \rangle / \langle \mathbf{b}^2 \rangle$. At around 0.3 AU this takes values near to or slightly larger than unity. Thereafter, it tends to decrease with distance, to a value of $\approx \frac{1}{2}$ by 5 AU and remain at roughly this level out to at least 20 AU [47].

Note that r_A can be used as a diagnostic of the prevalence of fluctuations which are Alfvén waves. This follows since the energy in an individual Alfvén wave is equipartitioned between its kinetic and magnetic components (when averaged over a wave period) and thus should have $r_\mathrm{A} = 1$. Departures from this value suggest the presence of fluctuations which are not Alfvén waves, such as turbulence. In particular, 2D and 3D MHD simulations with an energetically weak $\mathbf{B}_0$ tend to have $r_\mathrm{A} < 1$, indicating an excess of fluctuation magnetic energy, *e.g.*, [33, 48–50]. This may be associated with the current sheets and vorticity quadrupoles characteristic of magnetic reconnection sites [50]. Alternatively, $r_\mathrm{A} < 1$ may be indicative of local (in Fourier space) dynamo action in the turbulent magnetofluid [51]. These two explanations for $r_A < 1$ are based on nonlinear processes. In addition, modeling results indicate that the expansion itself, in the form of "mixing" effects—which are linear (see §3.1)—can also lead to $r_A < 1$ [52, 53].

Another quantity which can provide information regarding the abundance of waves is the normalised cross helicity: $\sigma_c = 2\langle \mathbf{v} \cdot \mathbf{b} \rangle / [\langle \mathbf{v}^2 \rangle + \langle \mathbf{b}^2 \rangle]$. This is defined such that $|\sigma_c| < 1$, with purely outward propagating Alfvén waves having $\sigma_c = 1$. Helios and Voyager observations found that σ_c decreased from values of ≈ 1 near 0.3 AU, to values scattered around zero beyond about 5 AU [42]. Note that this is very different from the (leading-order) WKB prediction of $\sigma_c \approx 1$ [7]. (Systematic behaviour of σ_c at *high* latitudes has also been studied [40].)

Density fluctuations are typically observed to be $\sim$ 10% at small (1 hour) scales, in both compression and rarefaction regions [43]. Such results provide support for the use of incompressible and nearly incompressible models for the fluctuations [54–56] (however cf. [57]).

Although it is difficult to give a unique interpretation to the behaviour summarised above, one can nonetheless conclude that (i) fluctuations are not just (outward) propagating Alfvén waves, and (ii) inward-type modes tend to become relatively more abundant with distance.

2.2. Fluctuation Geometry

Determining the nature of the geometry (*e.g.*, quasi-2D *vs.* slab waves) of the fluctuations using data from a single spacecraft is problematic for at least two reasons. First, even full (vector) amplitude information is often insufficient to distinguish between different types of fluctuation. For example, both (quasi-)parallel-propagating Alfvén waves and (quasi-)2D turbulence have amplitudes which are (almost) transverse to $\mathbf{B}_0$. Consequently, they both have a minimum variance direction (MVD) which is $\approx \hat{\mathbf{B}}_0$, although their Fourier wavevector orientations are very different (respectively $\|$ and $\perp$ to $\mathbf{B}_0$.)

Second, single spacecraft data typically consists of time series at approximately the same spatial position which can be used to calculate power spectra as a function of (Fourier) frequency, $P(f)$, say. Invoking the Taylor frozen-in flow hypothesis (justified because of the supersonic flow speed of the wind) enables these frequency spectra to be converted into *reduced* wavenumber spectra, $S^{\mathrm{red}}(k_{\mathrm{red}}) = \int S(\mathbf{k}) d\mathbf{k}_\perp$, where $S(\mathbf{k})$ is the full wave*vector* spectrum and integration is over coordinates perpendicular to the sampling one. Unfortunately, except for particularly strong symmetries (*e.g.*, isotropic) determination of fluctuation geometry usually requires knowledge of the full wave*vector* spectrum [58, 59].

Thus, most spacecraft datasets do not support full determination of fluctuation geometry. Nonetheless, using various strategies it is sometimes still possible to extract additional information regarding the nature of the fluctuations. For example, one can analyse data intervals which have $\mathbf{B}_0$ oriented at many different angles to the sampling (radial) direction. Such an analysis has been performed [60], and under the assumption of axisymmetry about $\mathbf{B}_0$, a magnetic correlation function was obtained as a function of coordinates parallel and perpendicular to $\mathbf{B}_0$ (the so-called "Maltese cross"). The results indicate that the fluctuations could consist of two (or more) distinct populations, the first characterised by little variation in the perpendicular directions (*e.g.*, slab waves), and the second by little variation in the parallel direction (*e.g.*, quasi-2D turbulence). Carbone et al. [61] performed a related study with the dataset restricted so that only Alfvénic intervals were used and an assumed decomposition into linear wave modes. They also found evidence for a two-component nature of the fluctuations, although of a different kind to the Maltese cross study, presumably because of the differences in both the data selection policies and in the underlying assumptions in the data analysis. In any case, "two-component" descriptions of solar wind fluctuations have subsequently been widely employed (cf. §3).

More quantitative results have also been presented. Bieber et al. [62] used data for cosmic ray mean free path lengths to infer that the energy partitioning for a quasi-2D/slab model for solar wind fluctuations would be $\approx$ 80%-20%. In a direct test based on observed (inertial range) power spectra, it was also found that this 80-20 partitioning produced the best fit to the data [63]. Still a third test derives from Mach number scalings associated with nearly incompressible theory [54], and for typical observed Mach numbers is also consistent with the 80-20 split. Thus there is abundant—and consistent—evidence for a slab/quasi-2D two-component description, with the quasi-2D component being energetically dominant.

Efforts have also been made to explain the dynamical origin of the two-component model(s) using simulation studies [64, 65], and have met with some limited success. The suggestion is that the two components can appear at different stages of the evolution, or alternatively that the "2D" component is associated with non-zero (but still small) $k_\|$ values, and is therefore actually *quasi*-2D.

Returning to the MVD data, Voyager observations inside of 10 AU indicate that the MVD for $\mathbf{b}$ is centered around the local $\mathbf{B}_0$ direction, and similarly for the $\mathbf{v}$ fluctuations, although their MVD moves towards the radial with increasing R [66]. Compressible (polytropic) 3D MHD simulations can reproduce similar MVD data, both in terms of direction and power ratios between components, although there is quite a strong plasma beta dependence [32]. Also contentious is the issue of whether the MVD is aligned with the fluctuations wavevector, as would be the case for Alfvén waves but not for quasi-2D fluctuations [60, 66].

3. TURBULENCE-BASED MODELS

The first published suggestion that solar wind fluctuations could be turbulent appears to be due to Coleman [5]. It took another 16 years, however, for the first turbulence-based transport model for fluctuations to appear [21]. Prior to this, transport models were predominantly based on the application of WKB theory to short-wavelength Alfvén waves [6–9, 67, 68], although there were some studies which explored non-WKB effects, *e.g.*, [69].

Results from WKB theory are, in general, *not* in agreement with observations,[1] with some of the more well-known failures listed in Table 4 [10, 22, 37, 66]. Consequently there was a need to develop non-WKB models.

Below we briefly review a particular class of turbulence-based models which have had some success in matching a wider range of observational data. Note that various other interesting and important non-WKB models have also been investigated, such as the *expanding box model* of Grappin and Velli [27, 28], but they are not our focus here (see also [71–73]).

3.1. Scale-separation models

The underlying idea here is similar to that of the Reynolds decomposition in classical hydrodynamic turbulence theory *e.g.*, [74]. More specifically, one assumes that $\mathbf{V}$, $\mathbf{B}$, and ρ can each be meaningfully decomposed into a mean component and a fluctuating component using a multiple-scales approach. For example, that $\mathbf{V} = \mathbf{U}_0(\mathbf{R}) + \mathbf{v}(\mathbf{x},t;\mathbf{R})$, where $\mathbf{R}$ is the heliocentric position vector, $\mathbf{x}$ is a small spatial displacement at each given $\mathbf{R}$, and $\mathbf{U}_0(\mathbf{R}) = \langle \mathbf{V} \rangle$ can be thought of as the average of the velocity with respect to the small scales, $\mathbf{x}$.

An important point is that whereas in the WKB approach fluctuations are assumed to be Alfvén waves from the outset, the present approach requires no such assumption: the fluctuations can be waves and/or turbulence. Also, the amplitude of the fluctuations need not be small.

Equations for the evolution of the fluctuations can be obtained by substituting these decompositions into the

[1] The major exception is the radial evolution of $\langle \mathbf{b}^2 \rangle$ inside $\sim 10\,\mathrm{AU}$, which is indeed close to the WKB prediction of R^{-1}. However, it turns out that many other approaches also yield this form. See *e.g.*, [24, 47, 70] and discussion below.

TABLE 4. Some failures of (leading-order) WKB theory

WKB PREDICTION	OBSERVATION ($1 \lesssim R \lesssim 10\,\mathrm{AU}$)
No inward modes above Alfvén critical radius.	Significant amounts
$r_A \approx 1$; $\sigma_c \approx 1$	$r_A \approx 0.5$; $\sigma_c(R) \to \approx 0$
MVD for $\mathbf{v}$, $\mathbf{b}$: radial	$\approx$ parallel to $\hat{\mathbf{B}}_0$ (but that for $\mathbf{v}$ moves closer to $\hat{\mathbf{R}}$ with incr R)

MHD equations, averaging, and then subtracting the results from the unaveraged equations. Switching to Elsässer variables, $\mathbf{z}^{\pm} = \mathbf{v} \pm \mathbf{b}$, the equations can be written

$$\left(\frac{\partial}{\partial t} + L_{\pm}^{WKB}\right)\mathbf{z}^{\pm} + M^{\pm}\mathbf{z}^{\mp} = NL^{\pm}, \qquad (1)$$

where the (tensor) operators $L_{\pm}^{WKB}$ represent WKB effects and the $M^{\pm}$ "mixing" effects, which depend solely on the gradients of the mean fields. Nonlinear terms are collected on the RHS and denoted only symbolically. Note that the mixing effects are linear and appear at the same formal order as the WKB effects [11, 12, 22, 23].

While it is sometimes advantageous to work directly with these equations [11, 24, 25], it is often more convenient to form the equations for the evolution of the associated spectra (or, equivalently, correlation functions). To make further progress, one can assume that the fluctuations have particular symmetry properties, based on observational [60–63] and theoretical results [75, 76]. In the most general case this would yield 16 transport equations, one for each of the 16 scalar functions (such as energies and helicities) which collectively characterise the fluctuations [22, 76]. In some of the simpler cases the equations reduce to

$$\frac{\partial E^{\pm}(\mathbf{k})}{\partial t} + L_{WKB}^{\pm} E^{\pm}(\mathbf{k}) + M^{\pm} F(\mathbf{k}) = NL^{\pm}(\mathbf{k}) + S^{\pm}, \quad (2)$$

where $E^{\pm}(\mathbf{k}) = \langle \mathbf{z}^{\pm}(-\mathbf{k}) \cdot \mathbf{z}^{\pm}(\mathbf{k}) \rangle$ are the spectra for the Elsässer energies, $F(\mathbf{k}) \sim \langle \mathbf{z}^{+}(-\mathbf{k}) \cdot \mathbf{z}^{-}(\mathbf{k}) \rangle \propto \langle \mathbf{v}^2 - \mathbf{b}^2 \rangle$, and at least two more equations similar to (2) are required to close the system [13, 22, 23, 25, 77]. Also included are possible source terms, $S^{\pm}$. Although the associated models by no means reproduce all the observational trends, the degree of agreement is encouraging [22, 24, 37, 52, 70, 77–80].

Recently, this approach was used to model the radial evolution of $\langle \mathbf{b}^2 \rangle$, a correlation length for the magnetic energy, and the proton temperature T_p, from 1–40 AU [70, 78]. Three major assumptions were adopted, namely that (i) fluctuations are primarily quasi-2D turbulence (with $\sigma_c \approx 0$), (ii) energy is injected into the fluctuations by large-scale shear and compressions, and also, at large R, by pickup ions, and (iii) energy cascades to small *perpendicular* scales where it is converted into heat.

Using the scale-decomposition approach described above, one obtains a system of three (ordinary) differential equations, assuming steady-state. The linear structure is of the form (2)—although further simplifications are made—while the nonlinear terms are modeled using homogeneous turbulence theory (*e.g.,* [81] and references therein). Source terms modeling the injection of pickup ion energy are also included and these start to become effective for $R \gtrsim 8\,\mathrm{AU}$. Numerical solutions of the equations for realistic values of the variables at the inner (1 AU) boundary, and the wind speed, etc. show excellent agreement with observational data from the Voyager and Pioneer missions in the case of $\langle \mathbf{b}^2 \rangle$ and T_p (the agreement with the correlation scale is less persuasive).

Richardson has recently improved this model by using the observed correlation between wind speed and T_p at 1 AU. This produces a model wherein the radial temperature dependence fits the observational peaks and troughs quite strikingly (see his paper in this volume).

Finally, we wish to comment on the apparent success of WKB theory in predicting $\langle \mathbf{b}^2 \rangle \sim R^{-1}$ [24, 47, 53, 70]. For reasonable models of the mean fields the mixing terms (*e.g.,* $M^{\pm} F$) are only important for $R \lesssim 2\,\mathrm{AU}$ [22, 52]. Thus, in the outer heliosphere departures from WKB evolution of the energies would appear to be due to nonlinear effects and/or source terms [see Eq. (2)]. However, for inertial range scales in strong turbulence $NL^{\pm}(\mathbf{k}) \approx 0$ [*e.g.,* 22]. Thus, in the absence of sources (2) reduces to the WKB energy transport equation and the WKB prediction is recovered (coincidentally), *despite its lack of applicability* in these circumstances.

4. CONCLUSIONS AND SUMMARY

It is now well-accepted that solar wind fluctuations include both turbulence and waves, with each playing important roles. Models based purely on wave effects generally give inadequate agreement with observations, whereas turbulence-based models are giving encouraging agreement over an enormous range of distances [13, 22–25, 37, 70, 77, 78, 80].

Returning to the three aspects of solar wind fluctuations mentioned in the introduction (origin, evolution, and nature), we note a few points about each one.

Origin. Outward propagating waves probably predominantly originate in the corona, with the bulk of the inward type modes being generated by (spatially local) *in situ* dynamics. Generation of the inward type modes could be due to either linear effects (*e.g.,* mixing) or nonlinear ones, such as turbulence, or a combination.

Evolution. Despite some apparent successes, the WKB approximation is usually inappropriate and inadequate, whereas observations, theory, and modeling all suggest that turbulence is likely to play a crucial role. In particular, driving by shear appears to be essential in accounting for the observed super-adiabatic $T_p(R)$ profile.

Nature. The fluctuations are definitely not pure Alfvén waves (either outward type or a linear superposition of inward and outward), although such intervals may exist. There is multiple support—from observations, theory, and simulations—for the fluctuations being comprised of (at least) two components, namely quasi-parallel-propagating Alfvén waves and quasi-2D turbulence, with the latter energetically dominant.

Finally, we note that there are many other issues pertaining to solar wind fluctuations which we have not discussed here. These include the details of the dissipation mechanism, the difficulty of achieving a parallel cascade of energy in MHD when $\langle \mathbf{b}^2 \rangle / B_0^2 < 1$, the situation out of the ecliptic, the radial evolution of spectra, and variance and polarization anisotropies. Many of these topics are discussed elsewhere in this volume.

Acknowledgments. Valuable discussions with W.H. Matthaeus, C.W. Smith, and G.P. Zank are happily acknowledged. This research was supported by grants from the NSF, NASA, and the UK PPARC.

REFERENCES

1. Parker, E. N., *Astrophys. J.*, **123**, 644 (1958).
2. Neugebauer, M., and Snyder, C. W., *J. Geophys. Res.*, **71**, 4469 (1966).
3. Belcher, J. W., and Davis Jr., L., *J. Geophys. Res.*, **76**, 3534 (1971).
4. Coleman, P. J., *Phys. Rev. Lett.*, **17**, 207 (1966).
5. Coleman, P. J., *Astrophys. J.*, **153**, 371 (1968).
6. Hollweg, J. V., *J. Geophys. Res.*, **78**, 3643 (1973).
7. Hollweg, J. V., *J. Geophys. Res.*, **79**, 1539 (1974).
8. Parker, E. N., *Space Sci. Rev.*, **4**, 666 (1965).
9. Whang, Y. C., *J. Geophys. Res.*, **78**, 7221 (1973).
10. Hollweg, J. V., *J. Geophys. Res.*, **95**, 14 873 (1990).
11. Matthaeus, W. H., Zhou, Y., Zank, G. P., and Oughton, S., *J. Geophys. Res.*, **99**, 23 421–23 430 (1994).
12. Zhou, Y., and Matthaeus, W. H., *J. Geophys. Res.*, **95**, 14 863 (1990).
13. Velli, M., Grappin, R., and Mangeney, A., *Phys. Rev. Lett.*, **63**, 1807 (1989).
14. Dobrowolny, M., Mangeney, A., and Veltri, P., *Phys. Rev. Lett.*, **45**, 144 (1980).
15. Bavassano, B., Dobrowolny, M., Fanfoni, G., Mariani, F., and Ness, N. F., *Solar Phys.*, **78**, 373 (1982).
16. Bavassano, B., Dobrowolny, M., Mariani, F., and Ness, N. F., *J. Geophys. Res.*, **87**, 3617 (1982).
17. Matthaeus, W. H., and Goldstein, M. L., *J. Geophys. Res.*, **87**, 6011 (1982a).
18. Matthaeus, W. H., and Goldstein, M. L., *J. Geophys. Res.*, **87**, 10 347 (1982b).
19. Tu, C., *Solar Phys.*, **109**, 149 (1987).
20. Tu, C.-Y., *J. Geophys. Res.*, **93**, 7 (1988).
21. Tu, C.-Y., Pu, Z.-Y., and Wei, F.-S., *J. Geophys. Res.*, **89**, 9695 (1984).

22. Zhou, Y., and Matthaeus, W. H., *J. Geophys. Res.*, **95**, 10 291 (1990).
23. Tu, C.-Y., and Marsch, E., *J. Plasma Phys.*, **44**, 103 (1990).
24. Zank, G. P., Matthaeus, W. H., and Smith, C. W., *J. Geophys. Res.*, **101**, 17 093 (1996).
25. Velli, M., Grappin, R., and Mangeney, A., "Alfvén wave propagation in the solar atmosphere and models of MHD turbulence in the solar wind," in *Proceedings of Solar Wind 7, COSPAR Colloq. Ser.*, edited by E. Marsch and R. Schwenn, Pergamon, Oxford, UK, 1992, vol. 3, p. 569.
26. Hundhausen, A. J., *Coronal Expansion and the Solar Wind*, Springer-Verlag, New York, 1972.
27. Grappin, R., and Velli, M., *J. Geophys. Res.*, **101**, 425 (1996).
28. Grappin, R., Velli, M., and Mangeney, A., *Phys. Rev. Lett.*, **70**, 2190 (1993).
29. Roberts, D. A., Ghosh, S., and Goldstein, M., *Geophys. Rev. Lett.*, **23**, 591 (1996).
30. Roberts, D. A., Goldstein, M. L., Matthaeus, W. H., and Ghosh, S., *J. Geophys. Res.*, **97**, 17 115 (1992).
31. Montgomery, D. C., *Physica Scripta*, **T2/1**, 83 (1982).
32. Matthaeus, W. H., Ghosh, S., Oughton, S., and Roberts, D. A., *J. Geophys. Res.*, **101**, 7619–7629 (1996).
33. Oughton, S., Priest, E. R., and Matthaeus, W. H., *J. Fluid Mech.*, **280**, 95–117 (1994).
34. Shebalin, J. V., Matthaeus, W. H., and Montgomery, D., *J. Plasma Phys.*, **29**, 525 (1983).
35. Goldstein, M. L., Roberts, D. A., and Matthaeus, W. H., *Ann. Rev. Astron. Astrophys.*, **33**, 283 (1995).
36. Matthaeus, W. H., Bieber, J. W., and Zank, G. P., *Rev. Geophys. Supp.*, **33**, 609 (1995).
37. Tu, C.-Y., and Marsch, E., *Space Sci. Rev.*, **73**, 1 (1995).
38. Bavassano, B., and Bruno, R., *J. Geophys. Res.*, **94**, 11 977 (1989).
39. Bavassano, B., and Bruno, R., *J. Geophys. Res.*, **97**, 19 129 (1992).
40. Bavassano, B., Pietropaolo, E., and Bruno, R., *J. Geophys. Res.*, **106**, 10 659 (2001).
41. Bavassano, B., and Smith, E., *J. Geophys. Res.*, **91**, 1706 (1986).
42. Roberts, D. A., Goldstein, M. L., Klein, L. W., and Matthaeus, W. H., *J. Geophys. Res.*, **92**, 12 023 (1987b).
43. Roberts, D. A., Klein, L. W., Goldstein, M. L., and Matthaeus, W. H., *J. Geophys. Res.*, **92**, 11 021 (1987a).
44. Lee, M., and Ip, W.-H., *J. Geophys. Res.*, **92**, 11041 (1987).
45. Williams, L. L., Zank, G. P., and Matthaeus, W. H., *J. Geophys. Res.*, **100**, 17 059 (1995).
46. Zank, G. P., *Space Sci. Rev.*, **89**, 413–688 (1999).
47. Roberts, D. A., Goldstein, M. L., and Klein, L. W., *J. Geophys. Res.*, **95**, 4203 (1990).
48. Biskamp, D., and Welter, H., *Phys. Fluids B*, **1**, 1964 (1989).
49. Fyfe, D., Montgomery, D., and Joyce, G., *J. Plasma Phys.*, **17**, 369 (1977).
50. Matthaeus, W. H., and Lamkin, S. L., *Phys. Fluids*, **29**, 2513 (1986).
51. Pouquet, A., Frisch, U., and Léorat, J., *J. Fluid Mech.*, **77**, 321 (1976).
52. Oughton, S., and Matthaeus, W. H., *J. Geophys. Res.*, **100**, 14 783–14 799 (1995).
53. Zhou, Y., and Matthaeus, W. H., *Geophys. Rev. Lett.*, **16**, 755 (1989).
54. Zank, G. P., and Matthaeus, W. H., *J. Geophys. Res.*, **97**, 17 189 (1992).
55. Zank, G. P., and Matthaeus, W. H., *Phys. Fluids A*, **5**, 257 (1993).
56. Zank, G. P., Matthaeus, W. H., and Klein, L. W., *Geophys. Rev. Lett.*, **17**, 1239 (1990).
57. Leamon, R. J., Smith, C. W., Ness, N. F., Matthaeus, W. H., and Wong, H. K., *J. Geophys. Res.*, **103**, 4775 (1998).
58. Batchelor, G. K., *The Theory of Homogeneous Turbulence*, CUP, Cambridge, 1970.
59. Fredricks, R. W., and Coroniti, F. V., *J. Geophys. Res.*, **81**, 5591 (1976).
60. Matthaeus, W. H., Goldstein, M. L., and Roberts, D. A., *J. Geophys. Res.*, **95**, 20 673 (1990).
61. Carbone, V., Malara, F., and Veltri, P., *J. Geophys. Res.*, **100**, 1763 (1995).
62. Bieber, J. W., Matthaeus, W. H., Smith, C. W., Wanner, W., Kallenrode, M., and Wibberenz, G., *Astrophys. J.*, **420**, 294 (1994).
63. Bieber, J. W., Wanner, W., and Matthaeus, W. H., *J. Geophys. Res.*, **101**, 2511 (1996).
64. Ghosh, S., Matthaeus, W. H., Roberts, D. A., and Goldstein, M. L., *J. Geophys. Res.*, **103**, 23 691 (1998).
65. Ghosh, S., Matthaeus, W. H., Roberts, D. A., and Goldstein, M. L., *J. Geophys. Res.*, **103**, 23 705 (1998).
66. Klein, L. W., Roberts, D. A., and Goldstein, M. L., *J. Geophys. Res.*, **96**, 3779 (1991).
67. Hollweg, J. V., *Rev. Geophys. Space Phys.*, **13**, 263–289 (1975).
68. Weinberg, S., *Phys. Rev.*, **126**, 1899 (1962).
69. Heinemann, M., and Olbert, S., *J. Geophys. Res.*, **85**, 1311 (1980).
70. Smith, C. W., Matthaeus, W. H., Zank, G. P., Ness, N. F., Oughton, S., and Richardson, J. D., *J. Geophys. Res.*, **106**, 8253–8272 (2001).
71. Malara, F., Veltri, P., Chiuderi, C., and Einaudi, G., *Astrophys. J.*, **396**, 297 (1992).
72. Veltri, P., and Malara, F., *Il Nuovo Cimento*, **20**, 859 (1997).
73. Lou, Y.-Q., *J. Geophys. Res.*, **98**, 3563 (1993).
74. Tennekes, H., and Lumley, J. L., *A First Course in Turbulence*, MIT Press, Cambridge, MA, 1972.
75. Matthaeus, W. H., and Smith, C., *Phys. Rev. A*, **24**, 2135 (1981).
76. Oughton, S., Rädler, K.-H., and Matthaeus, W. H., *Phys. Rev. E*, **56**, 2875–2888 (1997).
77. Mangeney, A., Grappin, R., and Velli, M., "MHD Turbulence in the Solar Wind," in *Advances in Solar System Magnetohydrodynamics*, edited by E. R. Priest and A. W. Hood, CUP, Cambridge, 1991, p. 327.
78. Matthaeus, W. H., Zank, G. P., Smith, C. W., and Oughton, S., *Phys. Rev. Lett.*, **82**, 3444–3447 (1999).
79. Oughton, S., *Transport of Solar Wind Fluctuations: A Turbulence Approach*, Ph.D. thesis, University of Delaware, Newark, Delaware, 19716 (1993).
80. Tu, C.-Y., and Marsch, E., *J. Geophys. Res.*, **98**, 1257 (1993).
81. Matthaeus, W. H., Zank, G. P., and Oughton, S., *J. Plasma Phys.*, **56**, 659–675 (1996).

Turbulent dissipation in the solar wind and corona

W. H. Matthaeus*, P. Dmitruk*, S. Oughton† and D. Mullan*

*Bartol Research Institute, University of Delaware, Newark, DE 19716 USA
†Department of Mathematics, University of Waikato, Hamilton NZ

Abstract. Models based upon anisotropic magnetohydrodynamic (MHD) cascade offer promising explanations for observations of both interplanetary and coronal turbulence and heating, which are reviewed here. In the standard picture the cascade proceeds by driving at the energy-containing scales, transfers through the inertial range, and into small scales where it drives small-scale random turbulent reconnection events. In order to understand more fully the heating and dissipation processes, one also needs to understand how small-scale MHD-driven reconnection — involving current sheets and filaments — induces kinetic plasma processes that thermalize the fluid energy. Here we suggest that in these reconnection sites MHD electric fields drive ion beam instabilities and nonlinear electron dynamics involving electron solitary wave structures, in analogy with the kinetic physics observed near parallel electric field auroral regions by the FAST spacecraft.

INTRODUCTION

A familiar signature of turbulence, whether it be hydrodynamic [1] or magnetohydrodynamic [2], heliospheric [3], solar [4] or astrophysical [5], is a power-law wavenumber fluctuation spectrum that extends over a wide range of spatial scales. Powerlaw spectra are indicative of scale invariant processes that distribute the fluctuation energy across a wide range of scales, but which on balance transfer energy towards smaller scales The cascade leads to small scale dissipation, which in a collisionless plasma may be anisotropic and dependent upon detailed plasma physics processes. An example is the role of the Hall effect in collisionless reconnection [6, 7]. For some purposes, the inertial range itself can "drop out" of the picture, acting as a lossless pipeline that connects the large-scale energy sources to the small-scale dissipation. In the solar wind this pipeline spans four orders of magnitude, from the correlation scale at $\sim 5\times 10^{11}$ cm down to around the ion inertial scale at $\sim 2\times 10^{7}$ cm. Important physics lies in the identification of the energy sources for the cascade, as well as identification of dissipation mechanisms. Here we will focus on three heliospheric problems; namely to explain why (1) the solar wind is "too hot" at 1 AU; (2) the solar wind is "too hot" at 30 AU; and (3) the corona is "too hot" at $2\,R_s$. Each of these puzzles can be addressed using simple energy-containing range treatments of MHD turbulence in which large scale dynamics govern both the cascade rate and the rate of dissipation.

ANISOTROPIC MHD CASCADE

MHD turbulence in regimes of incompressibility or near-incompressibility gives rise to anisotropic spectral transfer relative to an externally imposed dc magnetic field direction [8, 9]. There is an equivalent type of anisotropy relative to a local mean magnetic field direction [10, 11]. The preferred spectral transfer generates strong gradients across the mean magnetic field, and greater excitation in perpendicular wavevectors than in parallel wavevectors. Frequently, "perpendicular transfer" leads to low-frequency, quasi-2D turbulence for which Reduced MHD (RMHD) [48] is a good approximation.

The perpendicular cascade ends when it reaches the dissipation range of wavenumbers. In hydrodynamics the large scale nonlinear time is λ/u for turbulent velocity fluctuation amplitude u, and energy-containing scale λ. The estimated dissipation scale is $\lambda_{diss} \approx (\nu^3/\epsilon)^{1/4}$, ν the kinematic viscosity and ϵ the dissipation rate of energy per unit mass. A key approximate connection between turbulence energy per unit mass u^2, the energy-containing scale, and the dissipation rate is $\epsilon \approx u^3/\lambda$ [12]. Simple MHD models include both a viscosity and a resistivity μ. Provided that these are not too dissimilar in value, it is well established that the anisotropic MHD cascade will terminate in randomly distributed reconnection sites. This is illustrated in Fig. 1. When there is a high degree of anisotropy caused by a (locally) uniform mean magnetic field, the reconnection sites tend to be highly oblique, with elec-

CP679, *Solar Wind Ten: Proceedings of the Tenth International Solar Wind Conference*, edited by M. Velli, R. Bruno, and F. Malara

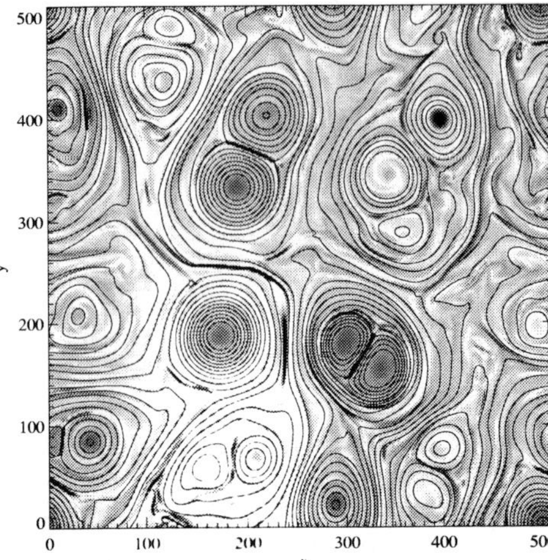

FIGURE 1. Cross section of a spectral method simulation of decaying 2D MHD turbulence showing magnetic field-lines and electric current density (gray scale) in the out-of-page direction. Current is concentrated in sheets or filaments between interacting magnetic flux structures. The cascade process is characterized by random driven reconnection events, involving magnetic island merger and generation of strong current channels.

tric current density tending to be aligned with the mean field. Because of anisotropy, the extension of the decay phenomenology [13] for low cross helicity is hydrodynamic-like with $\epsilon \approx Z^3/\lambda_\perp$, except that the lengthscale that appears, $\lambda_\perp$, is a *perpendicular* energy-containing scale. Here $Z = (\langle \mathbf{u}^2 + \mathbf{b}^2 \rangle)^{1/2}$, $\mathbf{u}$ the turbulent velocity, $\mathbf{b}$ the turbulent magnetic field in Alfvén speed units, and $\langle \ldots \rangle$ denotes an ensemble or volume average.

SOLAR WIND AT 1 AU

From the days of early spacecraft exploration of the solar wind it has been known that the proton temperature at 1 AU is much higher than what would be expected based upon reasonable coronal base temperatures and an adiabatic expansion [14]. Some local heating appears to be required. The first suggestion that a strong hydrodynamic-like cascade [12, 15] might be responsible for this heating was due to Coleman [16]. Influential papers by Tu et al. [17] and Hollweg [18] formulated spectral and energy-containing range theories (respectively) that would quantitatively account for this turbulent heating of the wind. There are complications in the inner heliosphere, such as nonzero cross helicity effects, and proximity to the initial data at distances $<$ 1 AU [19–22]; however, apart from details, it seems fairly certain that a nonlinear, turbulent cascade contributes to the heating of the solar wind at 1 AU.

Recent studies by *Leamon et al.* [23–26] examine turbulent dissipation at 1 AU in WIND data by examining the spectral breakpoints in magnetic spectra near 1 Hz. Some main conclusions are: (1) simple parallel cyclotron resonance does not provide a consistent explanation; (2) association of the breakpoint with oblique structures of size equal to the ion inertial scale provides a better fit than either the cyclotron frequency itself, or the parallel resonant wavenumber; (3) there is evidence for about equal mixture of processes with cyclotron signatures and other processes that lack cyclotron signatures (e.g., Landau damping). These results are consistent with involvement of current sheets having wavevectors with average orientation of about 70–80° to the mean magnetic field.

SOLAR WIND AT 30 AU

Interpretation of solar wind observations beyond 1 AU have also presented difficulties. On the one hand [27] a transport theory ("WKB") for noninteracting waves seems to give approximately the correct radial variation of fluctuation amplitude. However, persistence of powerlaw spectra [3] and the highly non-adiabatic radial proton temperature profile [28, 29] point to an ongoing deposition of energy in the extended solar wind plasma. A plausible explanation is that the turbulence cascade is maintained at near steady conditions, so that driving balances dissipation [27], although development of an appropriate transport theory [30] was needed to demonstrate that a near-WKB radial energy density variation of $\sim r^{-3}$ would be maintained to 10 AU or so. A source of driving for the turbulence is needed, and simple estimates of shear-driving due to instability of interfaces between high and low speed solar wind appears to work well [30]. Finally, turbulence theory [31, 32] accounts well for the observed radial variation of proton temperature. (See paper by J. Richardson, these proceedings.) The structure of this theory includes a one-point closure model of the turbulence decay rate, which in accordance with the discussion of the previous section, is chosen to vary as $Z^3/\lambda_\perp$ where $\lambda_\perp$ is the transverse energy-containing lengthscale, which controls the cascade rate. The assumption of an anisotropic cascade is central in the turbulence theory explanation of heating of the outer heliospheric solar wind. (See paper by Oughton, these Proceedings.)

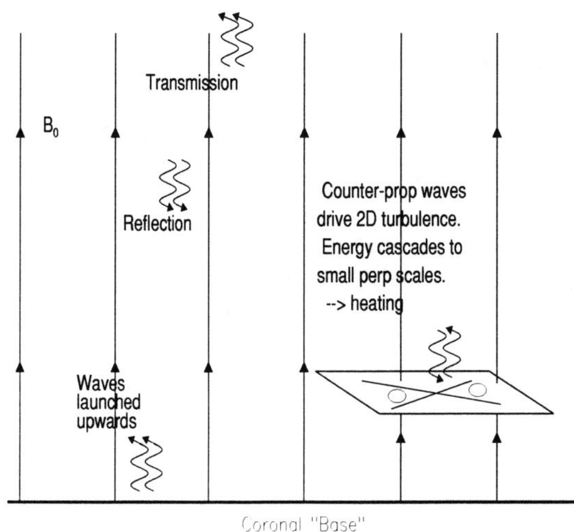

FIGURE 2. Diagram of an open field-line coronal heating model that is powered by low-frequency turbulence. Incident waves periods of $\sim 1000\,\mathrm{s}$ and transverse scales of $\sim 30,000\,\mathrm{km}$, enter through the coronal base. Inhomogeneity of the Alfvén speed induces some reflection. Counterpropagating wave trains interact through nonlinear couplings, driving a quasi-2D low-frequency cascade. Some wave energy exits through the "top" and is lost to the system. If the cascade is sustained, then some fraction of the total energy flux supplied is converted into heat by turbulent dissipation.

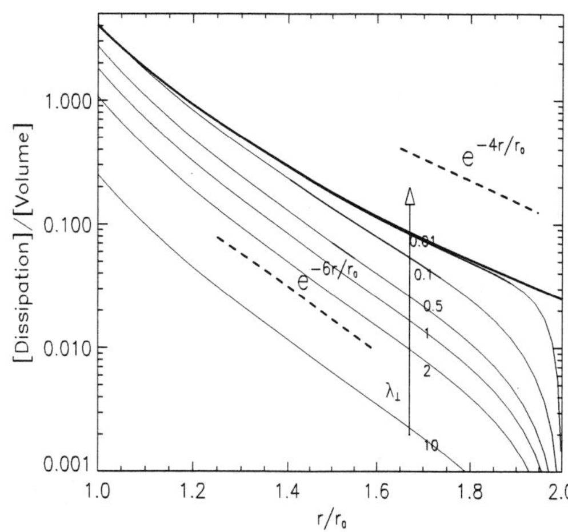

FIGURE 3. Profiles of heating (per unit volume) from an analytical phenomenology that describes coronal heating by waves, reflection, and cascade. The various curves correspond to different values of the perpendicular energy-containing scale. All resemble an exponential, with e-folding length a fraction of a solar radius, similar to the *ad hoc* heat functions used in models of the acceleration of the solar wind.

OPEN FIELD CORONA AT 2 R_S

Magnetic fluctuations in some form are a likely conduit that transports the required $6\times10^5\,\mathrm{erg/cm^2}$-sec [33] into the corona. The search for a heating process is constrained by Spartan [36] and UVCS [37] observations of proton temperatures that reach $T > 10^6\,\mathrm{K}$ within $2\,R_s$of the photosphere. Accordingly observations by UVCS indicate a solar wind speed $> 200\,\mathrm{km/s}$ at $r \approx R_s$. The wind speed reaches $500\,\mathrm{km/s}$ or more by $r \approx 6\,R_s$ [38]. Theories of solar wind acceleration [39, 40] confirm that deposition of heat close to the coronal base can accelerate this fast polar wind.

Low compressibility upwards propagating Alfvén waves are candidates for transporting energy into the corona. These might be high-frequency waves generated in a "furnace" region in the chromospheric network [33] which has a typical transverse scale of $\approx 30,000\,\mathrm{km}$. High-frequency waves near the cyclotron frequency, perhaps 1 kHz, may be readily transmitted from the network to the corona and dissipated by cyclotron absorption [33–35, 40]. Alternatively, in view of the fact that the photospheric motions have convective time scales of 100–1000 s, the waves might be of much lower frequency.

There are several reasons to consider driving by low-frequency waves, apart from the better match with the convective photospheric timescales. One reason is that cyclotron resonance may be, in a sense, too efficient [41–43]. The cumulative effect of many minor ion resonances [41] may be to leave insufficient wave energy to heat the protons. This would then require that the wave spectrum be regenerated within the corona. However to resupply the coronal high-frequency wave spectrum through an MHD cascade requires cascade to $k_{\parallel} \sim \Omega_{ci}/B_0$, i.e., to high parallel wavenumbers. But MHD cascade [8, 9] mainly generates high perpendicular wavenumber excitations, and so it may be very difficult to use a nearly incompressible [44] cascade to resupply the high-frequency spectrum.

The above reasons (See paper by Isenberg, these proceedings.) provide ample motivation to examine an alternative scenario for coronal heating, illustrated in Fig. 2. Low-frequency (period $\approx 10^3\,\mathrm{s}$) Alfvénic fluctuations enter the corona and propagate upwards. These represent excitations of only one (say, Z^-) of the two ($Z^\pm$) Elsässer amplitudes. However, to excite strong incompressible wave-wave couplings and turbulence, both Elsässer amplitudes are needed, as upwards-type waves interact nonlinearly only with downwards-type waves, and vice versa. Consequently there needs to be a source of downwards waves, and in this regard the inhomogeneity of the medium enters in a crucial way. Density changes, or more precisely Alfvén speed gradients [45–47], produce refraction and reflection of Alfvén waves. Downward waves produced in this way engage in nonlinear interactions with the upwards waves. With a strong vertical magnetic field, and low plasma beta, this results in a quasi-2D nearly incompressible cascade that can be appropriately described by the equations of RMHD [48]

The efficacy of the above described model can be evaluated using a variety of analytical and numerical modeling approaches. Numerical studies have established some of the general conditions for maintaining a cascade driven by waves [49], confirming the general features of the above picture. Analytical phenomenologies can be developed that include transport, reflection, and simple cascade models [50], and these have suggested that reasonable efficiencies (10% to 30%) can be maintained. Numerical RMHD simulations [51] confirm that these are attainable efficiencies. Most recently, numerical RMHD simulations have been extended to employ reasonable models of coronal density and magnetic field profiles [52]. Results show that the deposition of heat by this mechanism is, on a per-unit-volume basis, concentrated near the coronal base, while the per-unit-mass heating is extended to at least several solar radii (see Fig. 3). The model has the interesting property that the deposition of heat by the cascade follows a radial profile that is very similar in shape to the assumed coronal density profile. For a simple isothermal gravitationally stratified coronal model [45] the density behaves similarly to an exponential in altitude. In general, in the wave reflection-driven turbulence model the macroscopic density profile can control the cascade [52].

RECONNECTION & DISSIPATION

The physical picture remains incomplete until we understand how the kinetic plasma responds to the cascade and generates heat. As discussed above, most of the energy in a nearly incompressible MHD cascade is expected to dissipate through small-scale reconnection processes. This points to a suggestion: *the key to understand the dissipation and heating associated with MHD turbulence in collisionless plasmas is to be found in the kinetic plasma response to turbulent reconnection.* In particular, turbulence generates patches of strong electric field surrounding the reconnection sites; see Fig. (4). Furthermore, when there is a large-scale magnetic field, anisotropy of the cascade implies that magnetic reconnection sites and their associated kinetic responses will be distributed anisotropically. What kind of kinetic response do we expect from driven, turbulent reconnection? This is a difficult problem that appears not to have been completely solved as yet. There are several areas in which we can seek some guidance however.

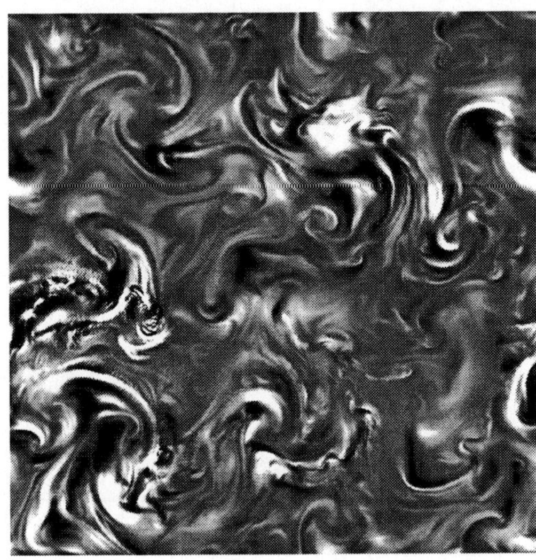

FIGURE 4. Cross-section of the vertical MHD electric field from an RMHD simulation with coronal density profile, and wave driving at the bottom boundary. Large scale Reynolds number and magnetic Reynolds number are ≈ 2000. The snapshot is $\approx 1/10\,R_s$ above the base, after a statistically steady state has been achieved. Regions of intense electric field (intense white or black shading) are concentrated around sites of magnetic reconnection. Kinetic response of ions and electrons will be strongly influenced by these patches of electric field, which last for several Alfvén times (10^2–10^3 s in the corona).

Test particle response can be viewed as an upper limit to the impact that reconnection fields can have on the kinetic plasma. Pending newer results, we anticipate some of the physics by looking back at the test particle results [53] from a simulation of decaying 2D reconnection in the presence of a finite-amplitude broadband turbulence. The reconnection dynamics used in these simulations admits strong fluctuations in space and time. There are multiple and highly dynamic reconnection zones, and a growth of reconnected magnetic islands faster than either tearing mode theory or laminar Sweet–Parker theory (see, e.g., [54]). Accordingly, the test particle response is complex. The highest energy particles tend to be those that are temporarily trapped in the turbulent reconnection zones, in which typically there are several highly dynamic X-point regions. Fig. 5 illustrates this feature.

What we can conclude based upon test particle results is fairly limited. However, they confirm the expectation that the electric fields appearing in and near turbulent reconnection sites excite beam-like distributions of particles. Being much lighter, electrons would absorb energy faster than protons. Very rapid electron dynamical timescales imply that self-consistent modifications of the distributions would be manifest quickly. Vlasov and PIC simulations (e.g., [55]) provide some insight into electron response to low-frequency electric fields. Electron plasma oscillations and low-frequency ion acoustic waves are generated, followed by damping of the high-frequency waves and formation of suprathermal tails. Implications for proton distribution are not yet clear from these studies, presumably due to computational limitations. Another view of the response of kinetic plasma to reconnection is obtained in the simulations of the smooth, non-driven, non-turbulent reconnection simulations that have been carried out in the GEM collaborations [6]. For example, it has been observed [7] that whistler turbulence is excited by kinetic effects in laminar reconnection, and electron Hall currents give rise to distinctive magnetic field signatures near the boundaries of the ion reconnection layer.

Numerical modeling of reconnection and turbulence is moving rapidly, and we expect additional findings will pave the way to advancement in our understanding of dissipation processes in a variety of space plasma environments in which reconnection and turbulence occur. On the observational side, further breakthroughs in understanding solar microphysics will await results of proposed missions, including SDO, RAM, ASCE, and Solar Probe. In the interplanetary environment, important results about microphysics and dissipation will come from an operational multi-spacecraft cluster, such as a Multi-spacecraft Heliospheric Mission, or an L1 Cluster of existing missions such as ACE, WIND, Genesis, SOHO, and Triana. Geospace missions, especially in the auroral acceleration regions, are providing important insights now.

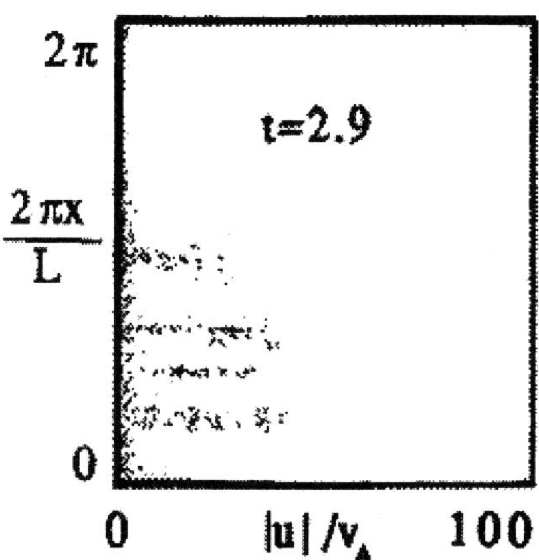

FIGURE 5. Test particles speeds vs. position along the nominal current sheet in a turbulent reconnection simulation, 2.9 eddy turnover times from the initial state, a 2D current sheet extended in the x-direction. Turbulence triggers highly dynamic reconnection. Higher energy particles are produced in small regions along the current sheet, the positions corresponding to small transient X-point regions. At this time particles are collimated in the z-direction (not shown); later the particles isotropize. Adapted from Ambrosiano et al [1988].

CLUES IN AURORAL PHYSICS

The accuracy and time resolution of current spacecraft experiments makes it possible to explore previously unobservable kinetic phenomena. An excellent example is the FAST mission [56] study of the auroral acceleration region in which there is now clear evidence for electric field parallel to the local magnetic field [57]. FAST shows distinct signatures depending upon the electric field direction. Upwards electric field drives electrons downwards, generating the aurora. In this case protons stream upwards, forming beams and associated cyclotron instabilities. For downwards electric field, there are upwards electron beams, and associated solitary structures (highly nonlinear phase space "holes") that propagate upwards. In the latter case protons are observed to have high perpendicular temperature, presumably due to very fast time scale encounters with propagating solitary structures [58, 59]. Moreover, in this case there also appears to be a *deficit* of power at harmonics of the proton cyclotron frequency.

There are great differences between the origins of the parallel electric field in the aurora and in the corona. However from auroral studies and analogy with the corona, we may be able to learn something about the favored mode of response of a collisionless plasma to a parallel electric field. Production of nonlinear electron beams, copious solitary structures and associated broadband noise, and perpendicular heating of protons, may well be the favored response when electrons are the preferred charge carriers. When proton beams are formed, cyclotron beam instabilities and production of cyclotron waves may be preferred. Since the corona, too, is gravitationally stratified, and the vertical electric fields associated with random reconnections are alternately upwards or downwards directed, both types of phenomena may occur in the corona, in about equal proportions. This may be a promising way to think about nonlinear kinetic heating in the corona, and perhaps in other space environments as well. If so, it would be quite gratifying that plasma physics discovered in one space venue might elucidate fundamental physics in other applications.

ACKNOWLEDGMENTS

Supported by NASA (NAG5-8134), NSF (ATM-0105254), and the UK PPARC.

REFERENCES

1. Grant, H. L., Stewart, R. W., and Moilliet, A., *J. Fluid Mech.*, **12**, 241 (1962).
2. Biskamp, D., and Müller, W.-C., *Phys. Plasmas*, **7**, 4889 (2000).
3. Matthaeus, W. H., and Goldstein, M. L., *J. Geophys. Res.*, **87**, 6011 (1982a).
4. Harmon, J. K., and Coles, W. A., *EOS* (2001).
5. Armstrong, J., Cordes, J., and Rickett, B., *Nature*, **291**, 561 (1981).
6. Birn, J. *et al*, *J. Geophys. Res.*, 3715 (2001).
7. Shay, M. A., *et al.*, *J. Geophys. Res.*, **103**, 9165 (1998).
8. Shebalin, J. V., Matthaeus, W. H., Montgomery, D., *J. Plasma Phys.*, **29**, 525 (1983).
9. Oughton, S., Priest, E. R., and Matthaeus, W. H., *J. Fluid Mech.*, **280**, 95–117 (1994).
10. Cho, J., and Vishniac, E., *Astrophys. J.*, 273 (2000).
11. Milano, L. J., *et al.*, *Phys. Plasmas*, **8**, 2673 (2001).
12. von Kármán, T., and Howarth, L., *Proc. Roy. Soc. London Ser. A*, **164**, 192 (1938).
13. Hossain, M., *et al.*, *Phys. Fluids*, **7**, 2886(1995).
14. Hundhausen, A. J., *Coronal Expansion and the Solar Wind*, Springer-Verlag, New York, 1972.
15. Batchelor, G. K., *The Theory of Homogeneous Turbulence*, CUP, Cambridge, 1970.
16. Coleman, P. J., *Astrophys. J.*, **153**, 371 (1968).
17. Tu, C.-Y., Pu, Z.-Y., and Wei, F.-S., *J. Geophys. Res.*, **89**, 9695 (1984).
18. Hollweg, J. V., *J. Geophys. Res.*, **91**, 4111 (1986).
19. Roberts, D. A., *et al*, *J. Geophys. Res.*, **97**, 17115 (1992).
20. Zhou, Y., and Matthaeus, W. H., *Geophys. Rev. Lett.*, **16**, 755 (1989).
21. Bavassano, R., Bruno, R., and Klein, L. W., *J. Geophys. Res.*, **100**, 5871 (1995).
22. Bavassano, B., Pietropaolo, E., and Bruno, R., *J. Geophys. Res.*, **106**, 10659 (2001).
23. Leamon, R. J., *et al. J. Geophys. Res.*, **103**, 4775 (1998).
24. Leamon, R. J., *et al.*, *Astrophys. J.*, **507**, L181 (1998).
25. Leamon, R. L., *et al.*, *J. Geophys. Res.*, **104**, 22331 (1999).
26. Leamon, R. L., *et al.*, *Astrophys. J.*, **537**, 1054 (2000).
27. Roberts, D. A., Goldstein, M. L., and Klein, L. W., *J. Geophys. Res.*, **95**, 4203 (1990).
28. Richardson, J. D., *et al.*, *Geophys. Rev. Lett.*, **22**, 325 (1995).
29. Gazis, P. R., *et al.*, *J. Geophys. Res.*, **99**, 6561 (1994).
30. Zank, G. P., Matthaeus, W. H., and Smith, C. W., *J. Geophys. Res.* , **101**, 17093 (1996).
31. Matthaeus, W. H., *et al.*, *Phys. Rev. Lett.*, **82**, 3444 (1999).
32. Smith, C. W., *et al.*, *J. Geophys. Res.*, **106**, 8253 (2001).
33. Axford, W. I., and McKenzie, J. F., in *Cosmic Winds and the Heliosphere*, Ariz. U. Press, 1997, p. 31.
34. Marsch, E., and Tu, C. Y., *Solar Phys.*, **176**, 87 (1997)
35. Tu, C. Y., and Marsch, E., *Solar Phys.*, **171**, 363 (1997)
36. Kohl, J. L. *et al.*, *Space Sci. Rev.*, **72**, 29 (1995).
37. Kohl, J. L. *et al.*, *Solar Phys.*, **175**, 613 (1997).
38. Grall, R. R., *et al.*, *Nature*, **379**, 429 (1996).
39. Habbal, S. R., *et al.*, *Geophys. Rev. Lett.*, **22**, 1465 (1995).
40. McKenzie, J., Banaszkiewicz, M., and Axford, W. I., *Astron. Astrophys.*, **303**, L45 (1995).
41. Cranmer, S. R., *Astrophys. J.*, **532**, 1197 (2000).
42. Tu, C.Y., and Marsch, E., *J. Geophys. Res.*, **106**, 8233 (2001a)
43. Tu, C.Y., and Marsch, E., *Astron. Astrophys.*, **368**, 1071 (2001b)
44. Zank, G. P., and Matthaeus, W. H., *J. Plasma Phys.*, **48**, 85 (1992).
45. An, C.-H., *et al.*, *Astrophys. J.*, **345**, 597 (1989).
46. Velli, M., *Astron. Astrophys.*, **270**, 304 (1993).
47. Hollweg, J. V., *Solar Phys.*, **91**, 269 (1984).
48. Montgomery, D. C., *Physica Scripta*, **T2/1**, 83 (1982).
49. Dmitruk, P., *et al.*, *Phys. Plasmas*, **8**, 2377–2384 (2001).
50. Dmitruk, P., Milano, L. J., and Matthaeus, W. H., *Astrophys. J.*, **548**, 482 (2001).
51. Oughton, S., *et al. Astrophys. J.*, **551**, 565–575 (2001).
52. Dmitruk, P., *et al. Astrophys. J. 575*, 571 (2002).
53. Ambrosiano, J., *et al.*, *J. Geophys. Res.*, **93**, 14383 (1988).
54. Matthaeus, W. H., and Lamkin, S. L., *Phys. Fluids*, **29**, 2513 (1986).
55. Vinas, A. F., Wong, H. K., and Klimas, A. J., *Astrophys. J.*, 509–523 (2000).
56. Carlson, C. W., Pfaff, R. F., and Watzin, J. G., *Geophys. Rev. Lett.*, 2013–2016 (1998).
57. Ergun, R. E. *et al.*, *Phys. Rev. Lett.*, 045003–1 (2001).
58. Carlson, C. W. *et al.*, *Geophys. Rev. Lett.*, 2017–2020 (1998).
59. Ergun, R. E. *et al.*, *Geophys. Rev. Lett.*, 2025 (1998).

Anisotropic MHD Turbulence in the Interstellar Medium and Solar Wind

A. Bhattacharjee and C. S. Ng

Center for Magnetic Reconnection Studies,
Department of Physics and Astronomy, The University of Iowa, Iowa City, IA 52242

Abstract. A theoretical model is given of anisotropic magnetohydrodynamic turbulence in the interstellar medium and the solar wind. The model is motivated by observations that show significant deviations from the Kolmogorov power-law. Dimensional and heuristic arguments are given and critically assessed. On the basis of the weak turbulence approximation in which three-wave interactions dominate, analytical and numerical results are obtained for the anisotropic energy spectrum produced by the random scattering of shear Alfvén waves propagating parallel to a large-scale magnetic field. The energy spectrum is shown to be proportional to $k_\perp^{-2}$, qualitatively consistent with some observations and wave kinetic theory.

1. INTRODUCTION

Our observational knowledge of the small-scale density fluctuations in the ionized interstellar medium (ISM) is primarily due to interstellar scintillations [1,2]. These observations show two qualitatively important features of density fluctuation spectra: they obey power laws and are anisotropic. However, some delicate observational and theoretical issues complicate precise quantitative results on the power-law exponent(s) and the degree of anisotropy.

If we write the power spectrum in the form

$$P_N(\mathbf{k}) = C_N{}^2 k^{-\alpha},\ k_{out} \le k \le k_{in}, \tag{1}$$

where $k = |\mathbf{k}|$ is the magnitude of the wave number and $C_N{}^2$ is a positive constant, a large number of observations report that the exponent α is approximately equal to $11/3$ over many decades in k [3-7]. Equivalently, if the power spectrum is expressed as a one-dimensional form in k-space, the exponent is given by $\beta = \alpha - 2 \approx 5/3$ which is identical to that predicted by Kolmogorov's well-known inertial range spectrum for turbulent fluids [8]. Although this is suggestive [3,9], it is far from obvious why Kolmogorov's spectral law for an incompressible and isotropic neutral fluid should apply at all to the ISM which is a compressible ionized medium permeated by a large-scale and directed magnetic field.

A definitive theoretical interpretation of the observed spectra is difficult for at least two reasons. First, the exponent α (or β) depends sensitively on the mechanism that produces the turbulence and needs to be determined with a high level of precision in order to discriminate between different theoretical models. A number of observations show that $\alpha(\beta)$ is less than 4(2), but this is not precise enough to discriminate between different theoretical models. Second, the claim that the power-law should be attributed to an inertial range spectrum carries with it not only the challenge of establishing the power law over several decades in k, but also the identification of an outer wave number k_{out} where energy is injected and an inner wave number k_{in} above which energy is dissipated. The realization of both of these objectives simultaneously by independent measurements is difficult. There has been a general tendency among observers to settle for the Kolmogorov exponent $\beta = 5/3$ [10-14]. However, a significant number of observations show deviations from Kolmogorov scaling [13,15-18]. The error bars in these observations are sufficiently small that they raise questions regarding the universal validity of a Kolmogorov scaling for ISM turbulence. In fact, these examples suggest that β lies in the range between 1.9 and 2. Lambert and Rickett [18] have shown recently that many features of diffractive measurements can be accounted for by a non-turbulent $\beta = 2$ model with abrupt (or discontinuous) changes in the density profile of the ISM [19], but due to the presence of several discrepant features in the data they rule out both the $\beta = 2$ model as well as the Kolmogorov model as universal models for ISM fluctuations.

CP679, *Solar Wind Ten: Proceedings of the Tenth International Solar Wind Conference*,
edited by M. Velli, R. Bruno, and F. Malara

They, therefore, suggest the development of different spectral models for different lines of sight.

As mentioned above, there is significant observational evidence to suggest that the interstellar scintillation spectrum is anisotropic [6,13,14,20] . The density irregularities have a cigar-like structure, with long spatial scales parallel and short spatial scales perpendicular to the background field [21]. The degree of anisotropy is different for different sources (that is, different lines of sight through the ISM). Averaging along the line of sight can cause a reduction in the measured degree of anisotropy.

Non-Kolmogorov power laws for velocity and magnetic field fluctuation spectra have also been observed in the turbulent solar wind. Observations from Voyager 1 and 2 spacecrafts between 13 AU and 25 AU show k^{-2} spectra at low frequencies at low heliographic latitudes [22,23]. A possible explanation of this spectrum is that it is mainly due to the presence of shocks and discontinuities [24,25]. However, it is also possible that the spectra may have a turbulent origin, with turbulence eventually steepening to produce shocks. As in the ISM, anisotropy is a persistent feature of solar wind turbulence and manifests itself in several in situ observations as more power perpendicular than parallel to the local magnetic field [26-28].

In this paper, we present a new calculation of the anisotropic energy spectrum in a plasma permeated by a uniform background magnetic field. The closure employed in this calculation is weak turbulence [29,30] which has been shown to be dominated by three-wave interactions [31-33]. We show by means of a novel simulation of the random scattering of shear Alfvén waves that the inertial range anisotropic energy spectrum is proportional to $k_\perp^{-2}$, obtained earlier by heuristic analysis [34,35] and wave kinetic theory [33]. Although the geometry of our model is simple and weak turbulence closure is restrictive, the calculation provides qualitative support for some of the observations on non-Kolmogorov power-laws.

The following is a layout of this paper. In §2, we review the dimensional and heuristic arguments for the Kolmogorov, Iroshnikov-Kraichnan (IK) [36,37] and anisotropic MHD energy spectra. We do so because although such arguments have been successful and are widely used, they can be problematic, reinforcing the need for careful dynamical calculations. In §3, we present analytical and numerical calculations that test and verify the heuristic arguments for anisotropic weak MHD turbulence. We conclude in §4 with a summary of our results, a discussion of the limitations of our theoretical model, and implications for observations of turbulence in the ISM and the solar wind.

2. DIMENSIONAL AND HEURISTIC ANALYSIS

Kolmogorov derived his celebrated energy spectrum for hydrodynamics (HD) essentially by dimensional analysis [8]. He made two crucial assumptions: the turbulence is isotropic and the dominant interactions between eddies are local in k-space.

If the turbulence is isotropic in k-space, the energy can be written $\int E(k)dk$ where $E(k)$ is the energy spectrum. Assume, following Kolmogorov, that there exists an inertial range such that the energy transfer rate $\varepsilon(k)$ is a constant independent of k and furthermore, that the energy transfer process is local in k-space. Since the dimension of ε is L^2T^{-3}, k is L^{-1} and $E(k)$ is L^3T^{-2} (where L is the dimension of length and T is of time), dimensional homogeneity of the relation $\varepsilon \sim k^\alpha E_k^\beta$ yields the inertial-range energy spectrum $E(k) \propto \varepsilon^{2/3}k^{-5/3}$.

IK extend Kolmogorov's analysis to incompressible magnetohydrodynamic (MHD) turbulence. As discussed by Kraichnan [37], the small wave number components act like a background magnetic field which cannot be removed by a Galilean transformation and support Alfvén wave packets propagating in both directions with the Alfvén speed V_A. An Alfvén wave packet can interact with another wave packet only if the two collide, with the interaction time given typically by $\tau_k \sim (kV_A)^{-1}$. Note that τ_k has an inherently nonlocal character in k-space because it depends not only on the typical spatial dimension (k^{-1}) of a wave (a local property) but also on the magnitude of the large-scale (small k) magnetic perturbation that determines V_A (a nonlocal property). In many cases of physical interest $\tau_k \sim (kV_A)^{-1}$ is much shorter than the eddy turnover time $(kv_k)^{-1}$, with the consequence that the energy cascade is more inhibited in MHD than it is in HD. By treating the $k=0$ component of the magnetic field at any spatial location as the background uniform field, and assuming that the energy cascade is isotropic and local in k-space, Kolmogorov's dimensional analysis arguments can be repeated, now with ε depending on k, $E(k)$ and V_A (with dimension LT^{-1}). Writing $\varepsilon \sim k^{\alpha} E_k^{\beta} V_A^{\gamma}$, we can deduce the spectral index ν of the inertial-range energy spectrum:

$$\nu = \frac{\alpha}{\beta} = \frac{5-\gamma}{3-\gamma}. \tag{2}$$

(Note that the Kolmogorov spectrum $\nu = 5/3$ for HD is obtained as a special case of (2) if we set $\gamma = 0$.) To find ν for MHD, we must determine γ. This can be done in the limit of weak turbulence when the lowest-order interaction involves 3-wave interactions during which two wave packets collide for a typical time scale $\tau_k \sim (kV_A)^{-1}$ and produce a third wave with typical velocity magnitude $\delta v_k \sim \dot{v}_k \tau \sim v_k^2 / V_A$. By the relation $\varepsilon \propto (\delta V_k)^2 / \tau_K \propto V_A^{-1}$, we obtain $\gamma = -1$ and thus $\nu = 3/2$, which yields the IK spectrum $E(k) \propto \varepsilon^{1/2} k^{-3/2}$.

The scaling results obtained above by dimensional analysis for isotropic MHD turbulence can also be obtained by an alternate heuristic physical argument. Let each of the two colliding wave packets have amplitude of the order v_k and spatial scale k^{-1}. We assume again that the energy transfer is local in k-space, which means that a wave will interact dominantly with another wave characterized by the same length scale but moving in the opposite direction. From the MHD equations, we estimate that $\dot{v}_k \sim k v_k^2$, where an overdot denotes time derivative. If 3-wave interactions dominate, we can write $\delta v_k \sim \dot{v}_k \tau \sim v_k^2 / V_A$. In the weak limit, it will take a large number of random collisions, $N \sim (v_k / \delta v_k)^2 >> 1$ to change a wave packet amplitude by a factor of unity. Noting that $E(k) \sim k^{-1} v_k^2$, we obtain $\varepsilon \sim v_k^2 / N\tau \sim k^2 E(k) V_A / N \sim k^3 E^2(k) V_A^{-1}$ which implies again that $E(k) \propto \varepsilon^{1/2} k^{-3/2}$.

The IK theory, which has provided the physical underpinnings of much subsequent work on MHD turbulence, neglects anisotropy. Subsequently, numerous analytical and computational studies have attempted to address different aspects of anisotropic turbulence [32-34,38-43]. In the presence of a uniform magnetic field, the spectrum is anisotropic. Dimensional analysis, by itself, cannot then provide a definite result since it cannot discriminate between the two length scales perpendicular ($k_\perp^{-1}$) and parallel ($k_\parallel^{-1}$) to the uniform magnetic field. Let us assume that the energy cascade occurs entirely in the direction perpendicular to the uniform field so that the total energy can be written $\int E(k_\perp) dk_\perp dk_\parallel$. (This assumption is shown to be true in §3.) If we now repeat the scaling argument given in the last paragraph with $\tau_k \sim (k_\parallel V_A)^{-1}$ and $E(k_\perp) \sim (k_\parallel k_\perp)^{-1} v_k^2$, we obtain

$$\delta v_k \sim k_\perp v_k^2 / (k_\parallel V_A) \sim k_\perp^2 E(k_\perp) / V_A. \tag{3}$$

Thus, $\varepsilon \sim k_\perp^4 k_\parallel E^2(k_\perp)/V_A$ which implies that the anisotropic spectrum is $E(k_\perp) \propto \varepsilon^{1/2} k_\perp^{-2}$ for weak MHD turbulence dominated by 3-wave interactions. In what follows, we test this scaling by a vigorous analytical calculation and numerical simulations.

3. RANDOM THREE-WAVE INTERACTIONS AND THE ANISOTROPIC SPECTRUM

We assume for simplicity that the plasma fluid is permeated by a spatially uniform magnetic field $\mathbf{B} = \hat{\mathbf{z}}$. Then the nonlinear MHD equations can be reduced rigorously to the so-called reduced MHD (RMHD) [44-46] equations

$$\frac{\partial \Omega}{\partial t} - \frac{\partial J}{\partial z} = [A, J] - [\phi, \Omega], \tag{4}$$

$$\frac{\partial A}{\partial t} - \frac{\partial \phi}{\partial z} = -[\phi, A], \tag{5}$$

where the magnetic field is given by $\mathbf{B} = \hat{\mathbf{z}} + \nabla_\perp A \times \hat{\mathbf{z}}$ with A as the magnetic flux function, the flow velocity is given by $\mathbf{v} = \nabla_\perp \phi \times \hat{\mathbf{z}}$ with ϕ as the stream function, and $[\phi, A] \equiv \phi_y A_x - \phi_x A_y$. Here $\Omega = -\nabla_\perp^2 \phi$ is the parallel vorticity and $J = -\nabla_\perp^2 A$ is the parallel current density. Note that we have normalized the background uniform magnetic field in the $\hat{\mathbf{z}}$-direction to have unit magnitude, and the density has been chosen so that the Alfvén speed $V_A = 1$.

Recent work has demonstrated conclusively that weak MHD turbulence in the presence of a uniform magnetic field is dominated by three-wave interactions that mediate the collisions of shear-Alfvén wave packets [31-33]. Using the ideal RMHD equations, Ng and Bhattacharjee (NB) calculate in closed form the three-wave and four-wave interaction terms, and show the former to be asymptotically dominant if the wave packets have non-zero $k_{\parallel}=0$ components. These three-wave interaction terms provide the basis for our Monte-Carlo simulation of the random scattering of Alfvén waves, discussed below.

For weak interactions between two colliding shear-Alfvén wave packets $f^{\pm}$ traveling in the $\pm\hat{\mathbf{z}}$ directions, we write perturbative solutions of the form

$$\phi = f^-(\mathbf{x}_\perp, z^-) + f^+(\mathbf{x}_\perp, z^+) + \phi_1 + \phi_2 + \cdots, \tag{6}$$

$$A = f^-(\mathbf{x}_\perp, z^-) - f^+(\mathbf{x}_\perp, z^+) + A_1 + A_2 + \cdots, \tag{7}$$

where $\mathbf{x}_\perp = (x, y)$ is perpendicular to $\hat{\mathbf{z}}$ and $z^{\pm} = z \mp t$. Here $f^{\pm}(\mathbf{x}_\perp, z^{\pm})$ represents Alfvén wave packets that propagate non-dispersively with the Alfvén speed $V_A = 1$. For given zero-order fields $f^{\pm}$, we can then calculate the first-order fields from the equations

$$\frac{\partial \Omega_1}{\partial t} - \frac{\partial J_1}{\partial z} = 2\{[f^+, \nabla_\perp^2 f^-] + [f^-, \nabla_\perp^2 f^+]\} \equiv F, \tag{8}$$

$$\frac{\partial A_1}{\partial t} - \frac{\partial \phi_1}{\partial z} = 2[f^-, f^+] \equiv G. \tag{9}$$

Equations (8) and (9) are radiation equations for the first-order fields, with the source term determined by the overlap of the given zero-order fields f^+ and f^-. The asymptotic expressions of ϕ_1, A_1 can be written,

$$\phi_1(\mathbf{x}_\perp, t \to \infty) \to f_1^-(\mathbf{x}_\perp, z^-) + f_1^+(\mathbf{x}_\perp, z^+), \tag{10}$$

$$A_1(\mathbf{x}_\perp, t \to \infty) \to f_1^-(\mathbf{x}_\perp, z^-) - f_1^+(\mathbf{x}_\perp, z^+), \tag{11}$$

where

$$f_1^{\pm}(\mathbf{x}_\perp, z) = \pi \int [\tilde{F}'(\mathbf{k}, \pm k_z) \mp \tilde{G}(\mathbf{k}, \pm k_z)] e^{i\mathbf{k}\cdot\mathbf{x}} d\mathbf{k} \tag{12}$$

Here $\tilde{F}'(\mathbf{k}, \omega) \equiv \tilde{F}(\mathbf{k}, \omega)/k_\perp^2$ and $\tilde{F}(\mathbf{k}, \omega)$ is the Fourier transform of $F(\mathbf{x}, t)$, defined by

$$F(\mathbf{x}, t) = \int \tilde{F}(\mathbf{k}, \omega) e^{i(\mathbf{k}\cdot\mathbf{x} - \omega t)} d\mathbf{k} d\omega.$$

The Fourier transform $\tilde{G}(\mathbf{k}, \omega)$ is similarly defined. For simplicity, we consider the case when the functions $f^{\pm}(\mathbf{x}_\perp, z)$ are separable, i.e., $f^{\pm}(\mathbf{x}_\perp, z) = f_\perp^{\pm}(\mathbf{x}_\perp) f^{\pm}(z)$. NB show that

$$\tilde{F}(\mathbf{k}, \omega) = \frac{1}{2} \tilde{F}_\perp(\mathbf{k}_\perp) \tilde{f}^+(\kappa^+) \tilde{f}^-(\kappa^-),$$

$$\tilde{G}(\mathbf{k}, \omega) = \frac{1}{2} \tilde{G}_\perp(\mathbf{k}_\perp) \tilde{f}^+(\kappa^+) \tilde{f}^-(\kappa^-)$$

where $\kappa^{\pm} \equiv (k_z \pm \omega)/2$, $\tilde{f}(\kappa^{\pm})$ is the one-dimensional Fourier transforms of $f^{\pm}(z^{\pm})$, and $\tilde{F}_\perp$ and $\tilde{G}_\perp$ are the two-dimensional Fourier transforms of $F_\perp$ and $G_\perp$. It follows that

$$f_1^{\pm}(\mathbf{x}_\perp, z^{\pm}) = \pi u_\perp^{\pm}(\mathbf{x}_\perp) \tilde{f}^{\mp}(0) f^{\pm}(z^{\pm})/2, \tag{13}$$

where

$$u_\perp^{\pm}(\mathbf{x}_\perp) = \int [\tilde{F}_\perp' \mp \tilde{G}_\perp] e^{i\mathbf{k}_\perp \cdot \mathbf{x}_\perp} d\mathbf{k}_\perp, \tag{14}$$

and $\tilde{f}^{\pm}(0)$ is the $k_z = 0$ Fourier component of $f^{\pm}(z)$, with $\tilde{F}'(\mathbf{k}_\perp) \equiv \tilde{F}(\mathbf{k}_\perp)/k_\perp^2$. We note that the expression (14) for three-wave interactions preserves the z-dependence of the zero-order fields. This implies that there is no energy transfer parallel to the magnetic field when three-wave interactions dominate [32,33,39,41,47]. (Note that this conclusion holds independent of the assumption of separability of the functions $f^{\pm}(\mathbf{x}_\perp, z)$.)

Imposing periodic boundary condition in $\mathbf{x}_\perp$, we write

$$f_\perp^{\pm}(\mathbf{x}_\perp) = \sum_{mn} f_{mn}^{\pm} e^{2\pi i(mx+ny)} \tag{15}$$

where $f_{mn}^{\pm}$ are constants. We define the energy $\int E_{\pm}(k_\perp) dk_\perp$ with the spectral functions

$$E_{\pm}(k_\perp) \propto k_\perp^{-\mu_\pm} \text{ or } \left|f_{mn}^{\pm}\right| \propto (m^2 + n^2)^{-(3+\mu_\pm)/4}, \tag{16}$$

where $\mu_\pm$ are the spectral indices. Assuming that the energy is randomly distributed in the zeroth-order fields, we can calculate the spectra of the first-order fields using (13). Our main objective is to determine how the spectrum of an Alfvén wave packet changes in time after many collisions with wave packets coming from the opposite direction. To be specific, let us consider the evolution of a f^+ field interacting with a sequence of random f^- fields. We write

$$\frac{\partial \Psi^+}{\partial t} = -[f^-, \Psi^+] + [f_x^-, f_x^+] + [f_y^-, f_y^+] \ , \tag{17}$$

where $\Psi^+ = -\nabla_\perp^2 f^+$. Numerically, the Fourier amplitudes f_{mn}^- are randomly chosen for a given spectral index μ_- in every time step τ_A. The time-step is chosen small enough so as to satisfy the weak turbulence assumption and to keep each wave packet in the sequence of f^- uncorrelated with any other in the sequence. Also, in order to realize a well-resolved inertial range for the f^+ spectrum, a hyper-dissipation term of the form $\eta\nabla_\perp^6\Psi^+$ is added to the right of equation (17). We optimize the simulation so that the inertial range index is insensitive to the value of η. Equation (17) is solved by a pseudo-spectral method for different values of μ_- and for different levels of resolution (up to 1024^2) until the f^+ spectrum reaches a quasi-steady state.

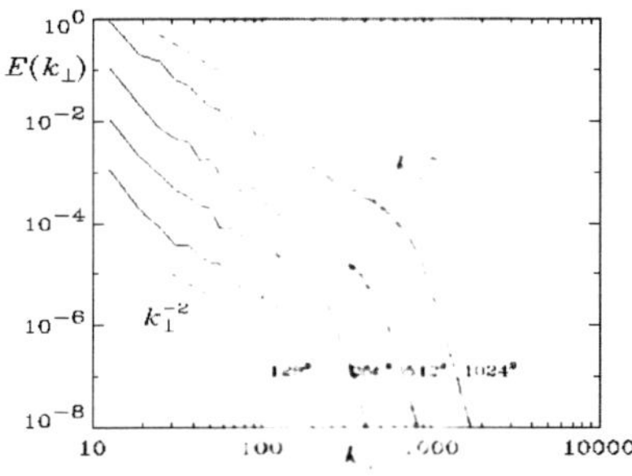

FIGURE 1. The spectra of the f^+ field with $\mu_- = 2$ for different levels of resolution. A vertical separation has been added in between each pair of curves for clarity.

Figure 1 shows the f^+ spectra for the case with $\mu_- = 2$ for different levels of resolution. We see that the inertial range for all runs have roughly the same index, $\mu_+ \approx 2$. We have checked numerically the relation $\mu_+ + \mu_- \approx 4$ for a range of values of μ^+ and μ^-. For $\mu_+ = \mu_- = 2$, which corresponds to the case considered in §2, we obtain the anisotropic energy spectrum $k_\perp^{-2}$, consistent with the heuristic result.

4. CONCLUSION

Prompted by observations of non-Kolmogorov and anisotropic turbulent spectra in the ISM and the solar wind, we have discussed a theoretical model of weak MHD turbulence, which produces an anisotropic energy spectrum proportional to $k_\perp^{-2}$. The anisotropic energy cascade in our model is due to the random scattering of shear-Alfvén waves dominated by three-wave interactions. We have reviewed critically the assumptions underlying the heuristic derivation of scaling laws in HD and MHD turbulence and underscored the need to verify these scaling laws by dynamical calculations. Our dynamical calculation provides independent confirmation of the $k_\perp^{-2}$-spectrum derived earlier from heuristic arguments [34,35] and wave kinetic theory [33].

Since measurements of fluctuations in the ISM are line-integrated, an interesting question is how the anisotropic spectrum obtained above for a uniform magnetic field might show up in observations of the ISM, which is generally permeated by a spatially varying magnetic field. If we make the drastic but simplifying assumption that the background magnetic field $\mathbf{B_0}$ takes all possible directions with equal probability, it is easy to show by averaging over three-dimensional wave vector space that the spectrum will be proportional to k^{-2} [33]. However, because all possible directions for $\mathbf{B_0}$ are not equally probable, one might expect deviations from the k^{-2} scaling. This remark is also applicable to observations of anisotropic turbulence in the solar wind which are based so far entirely on single spacecraft observations [48]. Despite this limitation, by analyzing Helios 2 data, Carbone *et al* [27] have obtained significant quantitative information on the three-dimensional structure of anisotropic turbulence in the solar wind.

We conclude with a few cautionary remarks on the limitations of our model. Although one of the strengths of weak turbulence theory is that it provides rigorous closure, it is far from clear that weak turbulence is a valid approximation for ISM or solar wind turbulence, which is often strong and compressive. Furthermore, we have calculated energy spectra, not density fluctuation spectra. It is often assumed that density fluctuations are enslaved to energy fluctuations, but this is not necessarily so [46,49]. In future work, we will attempt to remedy some of these limitations.

ACKNOWLEDGMENTS

We thank R. Bruno, S. Galtier, R. Mutel, A. Newell, A. Pouquet and S. Spangler for useful discussions. This research is supported by the NSF Grant No. ATM-0001317 and the DOE Cooperative Agreement No. DE-FC02-01ER54651 under the auspices of the program on Scientific Discovery through Advanced Computing. Numerical computations were carried out on the National Partnership for Advanced Computational Infrastructure.

REFERENCES

1. Rickett, B. J. 1990, ARA&A, 28, 561
2. Narayan, R. 1992, Phil. Trans. R. Soc. Lond. A, 341,151
3. Armstrong, J. W., Cordes, J. M., & Rickett, B. J. 1981, Nature, 291, 561
4. Cordes, J. M., Weisberg, J. M., & Boriako+ff, V. 1985, ApJ, 288, 221
5. Gwinn, C. R., Moran, J. M., Reid, M. J., & Schneps, M. H. 1988, ApJ, 330, 817
6. Spangler, S. R., & Cordes, J. M. 1988, ApJ, 332, 346
7. Mutel, R. M., & Lestrade, J. F. 1990, ApJL, 349, L47
8. Kolmogorov, A. N. 1941, C. R. Akad. Sci. SSSR, 30, 301
9. Lee, L. C., & Jokipii, J. R. 1975, ApJ, 196, 695
10. Rickett, B. J., & Lyne A. G. 1990, MNRAS, 244, 68
11. Spangler, S. R., & Gwinn, C. R., 1990, ApJ, 353, L29
12. Gupta, Y., Rickett, B. J., & Coles, W. A. 1993, ApJ, 403, 183
13. Wilkinson, P. N., Narayan, R., & Spencer, R. E. 1994, MNRAS, 269, 67
14. Molnar, L. A., Mutel, R. L., Reid, M. J., & Johnston, K. J. 1995, ApJ, 438, 708
15. Cordes, J. M., & Wolszcan, A. 1986, ApJL, 307, L27
16. Moran, J. M., Rodriguez, L. F., Greene, B., & Backer, D. C. 1990, ApJ, 348, 147
17. Mutel, R. M., Molnar, L. A., & Spangler, S. R. 1999, private communication
18. Lambert, H. C., & Rickett, B. J. 2000, ApJ, 531, 883
19. Blandford, R. & Narayan, R., 1985, MNRAS, 213, 591
20. Frail, D. A., Diamond, P. J., Cordes, J. M., & Van Langevelde, H. J. 1994, ApJL, 427, L43
21. Higdon, J. C. 1984, ApJ, 285, 109
22. Burlaga, L. F., & Goldstein, M. L. 1984, J. Geophys. Res., 89, 6813
23. Burlaga, L. F., Ness, N. F., & McDonald, F. B. 1987, J. Geophys. Res., 92, 13647
24. Burlaga, L. F., & Mish, W. H. 1987, J. Geophys. Res., 92, 1261
25. Roberts, D. A., & Goldstein, M. L. 1987, 92, 10105
26. Klein, L., Bruno, R., Bavassano, B., & Rosenbauer, H. 1993, J. Geophys. Res., 98, 17461
27. Carbone, V., Malara F., & Veltri P., 1995, J. Geophys. Res., 100, 1763.
28. Bieber, J. W., Wanner, W., & Matthaeus, W. H. 1996, J. Geophys. Res., 103, 6521
29. Sagdeev, R. Z., & Galeev, A. 1969, Nonlinear Plasma Theory (Benjamin: New York)
30. Zakharov, V. E., L'vov, V. S., Falkovich, G. 1992, Kolmogorov Spectra of Turbulence I (Berlin: Springer)
31. Montgomery, D. & Matthaeus, W. H. 1995, ApJ, 447, 706
32. Ng, C. S. & Bhattacharjee, A. 1996, ApJ, 465, 845
33. Galtier, S., Nazarenko, S. V., Newell, A. C., and Pouquet, A. 2000, J. Plasma Phys., 63, 447
34. Ng, C. S. & Bhattacharjee, A. 1997, Phys. Plasmas, 4, 605
35. Goldreich, P., & Sridhar, S. 1997, ApJ, 485, 680
36. Iroshnikov, P. S. 1963, AZh, 49, 742
37. Kraichnan, R. H. 1965, Phys. Fluids, 8, 1385
38. Montgomery, D. & Turner, L. 1981, Phys. Fluids, 24, 825
39. Shebalin, J. V., Matthaeus, W. H., & Montgomery, D. 1983, J. Plasma Phys., 29, 525
40. Sridhar, S., & Goldreich, P. 1994, ApJ, 432, 612
41. Oughton, S., Priest, E. R. & Matthaeus, W. H. 1994, J. Fluid Mech., 280, 95.
42. Matthaeus, W. H., Ghosh, S., Oughton, S., & Roberts D. A. 1996, J. Geophys. Res., 101, 7619
43. Chen, S., & Kraichnan, R. H. 1997, J. Plasma Phys., 57, 187
44. Strauss, H. R. 1976, Phys. Fluids, 19, 134
45. Zank, G. P., & Matthaeus, W. H. 1992, J. Plasma Phys., 48, 8
46. Bhattacharjee, A., Ng, C. S., & Spangler, S. R. 1998, ApJ, 494, 409
47. Kinney, R. M., & McWilliams, J. C. 1998, Phys. Rev. E, 57, 7111
48. Horbury, T. S., in Plasma Turbulence and Energetic Particles in Astrophysics, ed. M. Ostrokowski & R. Schlickeiser (Krakow: Uniwersytet Jagiellonski), 1999
49. Terry, P. W., Fernanadez, E., & Ware, A. S. 1998, ApJ, 504, 821

Intermittency of turbulence in the solar wind

V. Carbone*, L. Sorriso–Valvo*, F. Lepreti*, P. Veltri* and R. Bruno†

*Dipartimento di Fisica, Universitá della Calabria and Istituto Nazionale di Fisica della Materia, sezione di Cosenza, Italy.
†Istituto di Fisica dello Spazio Interplanetario/CNR, Rome, Italy.

Abstract.
We review some of the work done to investigate statistical features of turbulence in the inner solar wind, within slow–speed streams. We present results related to anomalous scaling laws of fluctuations and to the scaling behaviour of Probability Density Function (pdf). The results are compared with turbulence in other systems, that is neutral fluid flows and laboratory plasma. Some of the statistical properties are shared with low–resolution cascade models which describe the gross features of turbulence.

INTRODUCTION

Turbulence is a phenomenon in which chaotic dynamics and power law statistics coexist, and is characterised by randomness in both space and time, unpredictability and instability to every small perturbation. Flows in turbulent conditions are often characterised by the presence of structures on all scales, energetically dominant with respect to the remaining part of the flow. In these conditions low–resolution dynamical models can describe quite realistically the flow [1]. Here we briefly summarize the work done by studying low–frequency plasma turbulence using the solar wind as a "wind tunnel" and spacecrafts as probes. We will compare the results with other turbulent systems.

SCALING LAWS FOR HYDROMAGNETIC TURBULENCE

Low–frequency turbulence in plasmas is the result of nonlinear dynamics of incompressible Magnetohydrodynamic (MHD) equations, which can be written as

$$\frac{\partial \mathbf{z}^{\pm}}{\partial t} + \left(\mathbf{z}^{\mp} \cdot \nabla\right) \mathbf{z}^{\pm} = -\nabla(P/\rho) + \nu \nabla^2 \mathbf{z}^{\pm} + \mathbf{f}^{\pm} \qquad (1)$$

with the condition $\nabla \cdot \mathbf{z}^{\pm} = 0$. We use the Elsässer variables $\mathbf{z}^{\pm} = \mathbf{u} \pm \mathbf{B}/\sqrt{4\pi\rho}$, being $\mathbf{u}$ and $\mathbf{B}$ respectively the velocity and magnetic field, P is the total pressure, ρ the constant mass density, ν represents the kinematic viscosity (assumed to be equal to the magnetic diffusivity) and $\mathbf{f}^{\pm}$ represent some external forcing terms.

Let us introduce the differences of fields, along the direction of r, between two points separated by a distance r, that is $\delta z_r^{\pm} = [\mathbf{z}^{\pm}(\mathbf{x}+\mathbf{r}) - \mathbf{z}^{\pm}(\mathbf{x})] \cdot \mathbf{e_r}$ which represent characteristic fluctuations across eddies at the scale r ($\mathbf{e}_r$ being the direction of $\mathbf{r}$). These are the main quantities involved in the study of statistical properties of turbulence. In fact, under suitable hypothesis (homogeneity, stationarity and isotropy, see Ref. [2]), an exact relation for the inertial range can be obtained from ideal MHD equations

$$< (\delta z_r^{\pm})^2 \delta z_r^{\mp} >= -\frac{4}{3}\varepsilon^{\pm} r \qquad (2)$$

being $\varepsilon^{\pm}$ the energy transfer rates in the stationary state, for both Alfvénic fluctuations (brackets represent ensemble averages). This relation is the MHD analogous of the $4/5$–law Kolmogorov's law [3]. Since the 3–th order moment is different from zero the energy cascade generates fluctuations with some phase correlations. Moreover ideal incompressible MHD equations (1) are invariant providing lengths and velocities are scaled respectively as $r \to \lambda r$ and $\mathbf{z}^{\pm} \to \mathbf{z}^{\pm}\lambda^h$ for each value of h (or both $\mathbf{v} \to \mathbf{v}\lambda^h$ and $\mathbf{B} \to \mathbf{B}\lambda^h$). Then we expect scaling laws where $\delta z_r^{\pm} \sim \delta \mathbf{v} \sim \delta \mathbf{B} \sim r^h$. Since equations are invariants for each value of h, this can be fixed only with suitable phenomenlogical considerations. The compressible case is a little bit different. Since scaling laws are expected also for the density fluctuations, velocity and magnetic fluctuations must have scaling laws with different exponents.

In the fluid–like case, by introducing the pseudo–energies dissipation rate $\varepsilon^{\pm} \sim (\delta z_r^{\pm})^2/T_r^{\pm}$, and using the eddy–turnover times $T_r^{\pm} \sim \tau_{NL}^{\pm} \sim r/\delta z_r^{\mp}$, we obtain the scaling relations $(\delta z_r^{\pm})^2 \delta z_r^{\mp} \sim \varepsilon^{\pm} r$, which lead to the

CP679, *Solar Wind Ten: Proceedings of the Tenth International Solar Wind Conference*,
edited by M. Velli, R. Bruno, and F. Malara

Kolmogorov's scaling $h = 1/3$ when $\varepsilon^{\pm}$ are constant. When the charged fluid is magnetically dominated, the Alfvén time $\tau_A \sim r/C_A$ ($\mathbf{C}_A = \mathbf{B}_0/\sqrt{4\pi\rho}$ is the Alfvén velocity related to the average magnetic field) can become smaller than the eddy–turnover times, so that the nonlinear interactions between opposite travelling eddies, are lowered. This means that the energy cascade is realised in a time $T_r^{\pm} \sim \tau_{NL}^{\pm}(\tau_{NL}^{\pm}/\tau_A)$ [4], and the scaling relation becomes $(\delta z_r^{\pm})^2(\delta z_r^{\mp})^2 \sim \varepsilon^{\pm} r$, which leads to the Kraichnan scaling law $h = 1/4$ if the pseudo–energies dissipation rate are constant.

Homogeneous and isotropic turbulence in MHD has been described by the pseudo–energies density spectra which are related to the 2–th order moment of fluctuations $kE^{\pm}(k) \sim <(\delta z_r^{\pm})^2> (r \sim 1/k)$. Using the scaling laws with $h = 1/3$, a Kolmogorov's spectrum $E^{\pm}(k) \sim k^{-5/3}$ is obtained, while in the other case ($h = 1/4$) the Iroshnikov–Kraichnan spectrum $E^{\pm} \sim k^{-3/2}$ is recovered. Satellite observations do not give definite answer as regarding to the phenomenology preferred by the solar wind [5]. Of course in the usual fluid flows the Kolmogorov's spectrum is observed in all cases, and this indicates that the predictability lost at dynamical level, can be reintroduced at a statistical level [3].

To get some insigth of turbulence we have to look at higher order moments of fluctuations. In fact, since $\delta z_r^{\pm}$ are stochastic variables, the Probability Density Functions $pdf(\delta z_r^{\pm})$ are uniquely determined when the entire set of moments of fluctuations are known. For a gaussian process the 2–th order moment *suffices* to fully determine pdf and then to characterize turbulence. It can be immediately seen what kind of prediction can be made for the high–order moments. In the fluid–like case we have $\delta z_r^{\pm} \sim (\varepsilon^{\pm} r)^{1/3}$, so that, by defining the p–th order moment $S_p^{\pm}(r) = <(\delta z_r)^p>$, we obtain $S_p(r) \sim (\varepsilon^{\pm})^{p/3} r^{p/3}$. In the magnetically dominating case we have $\delta z_r^{\pm} \sim (\varepsilon^{\pm} r)^{1/4}$, so that $S_p(r) \sim (C_A \varepsilon^{\pm})^{p/4} r^{p/4}$. In both cases the scaling exponent ζ_p, defined through $S_p^{\pm}(r) \sim r^{\zeta_p}$, are linear $\zeta_p = ph$.

SOLAR WIND LOW–FREQUENCY FLUCTUATIONS

The satellite observations of both velocity and magnetic field in the interplanetary space, offer us an almost unique possibility to gain information on the turbulent MHD state in a very large scale range, say from 1 AU (Astronomical Units) up to 10^3 km. Here we report some analyses of plasma measurements of the bulk velocity $V(t)$ and magnetic field intensity $B(t)$. These analysis are mainly based on plasma measurements as recorded by the instruments on board Helios 2 during its primary mission in the inner heliosphere. The original data were collected in 81 s. bins and we choose a set of subintervals of 2 days each. The subintervals were selected separately within low speed regions and high speed regions selected in a standard way according to a threshold velocity [6]. For each subinterval we calculated the velocity and magnetic increments at a given time scale τ through $\delta V_\tau = V(t+\tau) - V(t)$ and $\delta B_\tau = B(t+\tau) - B(t)$. Of course in the supersonic solar wind moving at speed V_{SW}, the usual Taylor's hypothesis is verified, and we can get information on spatial scale r through $\tau = r/V_{SW}$. In the following we report only results relative to the slow periods. The incompressibility assertion is perhaps correct in the Alfvén waveband, in fast streams, but it is very approximative in the slow streams, even in the same waveband. Also, at still lower wavelengths, it is certainly false.

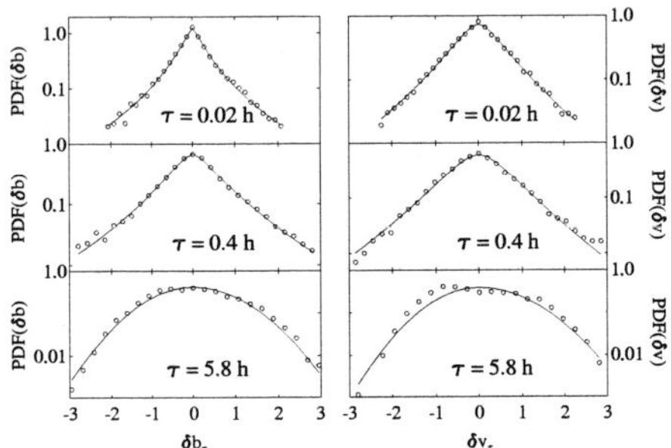

FIGURE 1. We report pdfs of fluctuations for velocity (right hand panels) and magnetic field (left hand panels) at different scales τ for solar wind data.

Let us consider the pdfs of the normalized quantities $\delta u_\tau = \delta V_\tau / <\delta V_\tau^2>^{1/2}$ and $\delta b_\tau = \delta B_\tau / <\delta B_\tau^2>^{1/2}$ (using the ergodic theorem brackets are now time averages). The interest of this normalization is the fact that pdfs of these fluctuations can be compared as far as the scaling properties are concerned. In particular if pdfs at two different scales become identical, the phenomenon is self–similar [7]. A plot of pdfs calculated for these quantities (see fig. 1) as a function of the scale r shows that: 1) the pdfs are not gaussian, at least at small scales; 2) the shape of pdfs depends on the scale τ. In particular, while for large scales the differences are gaussian distributed, as the scale becomes smaller the wings of pdfs becomes more and more important [8, 9]. This accumulation of high–probability intense fluctuations to small scales is one of manifestation of intermittency in turbulence. Since the process is not gaussian the energy density does not play any privileged role. Turbulence must be characterized not only by the second–order moment, rather by the whole set of moments or by the scaling behaviour of pdfs.

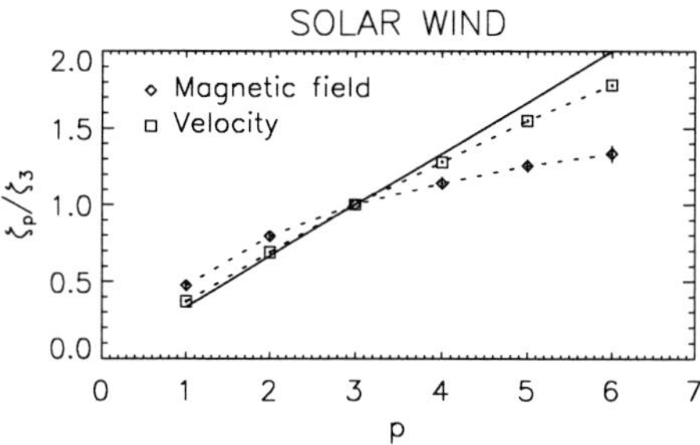

FIGURE 2. Normalized scaling exponents ζ_p/ζ_3 of p–th order moment for both velocity and magnetic field for solar wind data.

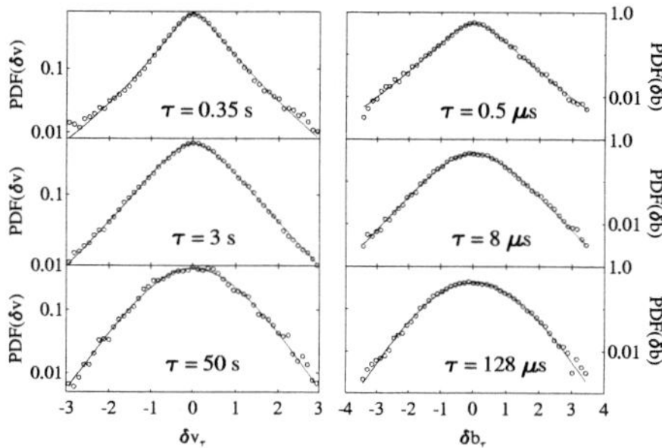

FIGURE 3. Pdfs of fluctuations for velocity field (fluid flows, left hand panel), and magnetic field (laboratory plasma, right hand panel) at different scales τ.

Since the pdfs are scale–independent, turbulence appears to be not globally self–similar, so that moments of fluctuations can have anomalous scaling relations. From the scaling properties of p–th moments of fluctuations $< \delta V_\tau^p >$ and $< \delta B_\tau^p >$ [6] The relative scaling exponents ζ_p/ζ_3 allow us to compare them with the scaling exponents obtained in usual fluid flows [10, 6]. In figure 2 the scaling exponents as a function of the order p are reported. It is evident that the behaviour of ζ_p/ζ_3 is different from the linear shape of the classical phenomenologies (see also Ref. [11]): the shape of ζ_p/ζ_3 turns out to be a nonlinear function of p.

The above behaviours of pdfs and moments are quite universal. We analysed a sample of fluid turbulence collected in the earth's boundary layer [12] where the longitudinal velocity field $v(t)$ and the temperature field $T(t)$ are recorded. Furthermore some samples of magnetic turbulence collected at the edge of RFX have been examined. RFX is a device in PAdova (Italy) where the plasma is confined in a reversed field pinch configuration [13] designed for thermonuclear fusion. In figures 3 the pdfs of the normalized velocity $\delta w_\tau = \delta v_\tau / < \delta v_\tau^2 >^{1/2}$ (for the fluid flow), and normalized magnetic fluctuations $\delta b_\tau = \delta B_\tau / < \delta B_\tau^2 >^{1/2}$ for RFX, have been reported. The scaling behaviour of pdfs looks to be similar to what has been observed in the solar wind. The pdfs are gaussian at large scales, and develop fat tails at small scales. Furthermore we examined the normalized scaling exponents for velocity and temperature fields in the fluid flows. The values of the normalised exponents for the velocity turn out to be the same for the solar wind and for the fluid sample. It is worthwhile to realize that in this experiment the temperature field acts like a passive scalar, and is transported by the velocity field. Using the differences $\Delta_p = p/3 - (\zeta_p/\zeta_3)$ as a measure of the intensity of intermittency, it can be seen that the passive scalar is more intermittent than the velocity field (fig. 4). In the solar wind the same behaviour is visible for the magnetic field. Actually magnetic field, which behaves like a "passive vector", results to be more intermittent than the velocity field, and this seems to be a characteristic of MHD flows because this is visible also in numerical simulations [14]. As regards laboratory plasma, intermittency of magnetic fluctuations in RFX [15] depends on the distance from the external wall where measurements have been performed (fig. 5). We found that intermittency increases going towards the external wall.

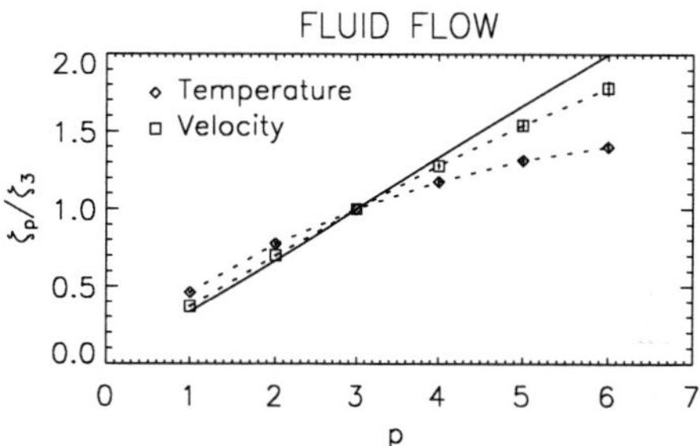

FIGURE 4. Normalized scaling exponents ζ_p/ζ_3 of p–th order moment for both velocity and temperature fields, for fluid flows data.

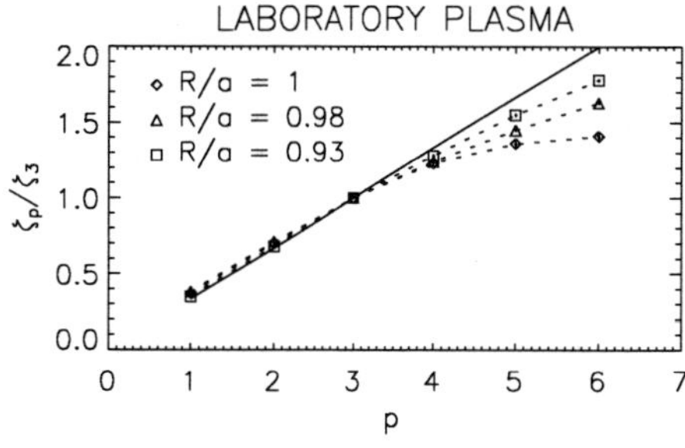

FIGURE 5. Normalized scaling exponents ζ_p/ζ_3 for RFX, calculated from turbulent samples taken at different distances R from the external wall. Distances are normalized with the minor radius $a = 0.457$ m of the torus.

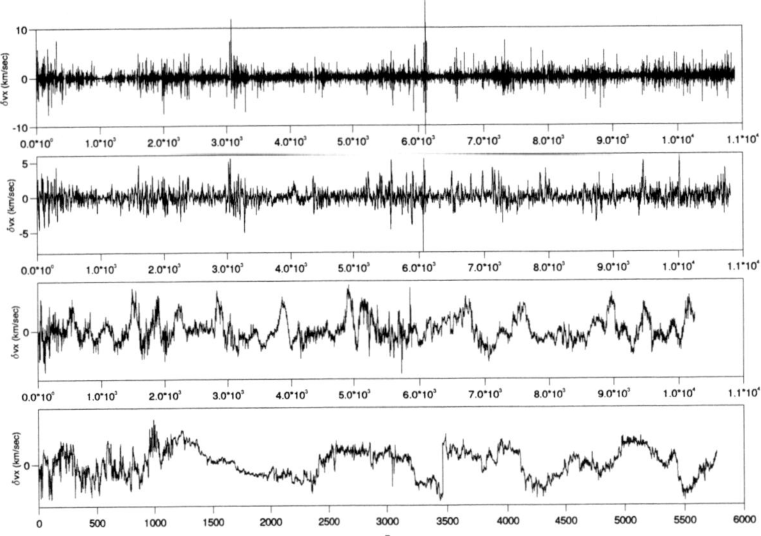

FIGURE 6. We report the time evolution of velocity fluctuations $\delta V_\tau(t)$ at four different scales for solar wind data. Scales increases from the top to the bottom.

WHAT IS INTERMITTENCY? A MULTIFRACTAL MODEL

Let us look at figure 6, where we reported the time evolution of δV_τ for four different values of τ for the solar wind data. Fluctuations at large scales appear to be smooth, while as the scale becomes smaller, intense fluctuations are visible. In fact these intense fluctuations are not distributed in a continuous way, instead they are relatively rare, and we see that there are periods with relative quiet activity alternating to small periods where the turbulent activity is very high. This is precisely the meaning of intermittency in fully developed turbulence. Starting from this point, it is natural to conjecture that, even if the fluid cannot be globally self–similar, self–similarity can be reintroduced as a local property. This is the basis of the multifractal model of intermittency, cf. e.g. [3], in which it is conjectured that turbulent flows can be made of an infinite set of points $S_h(\mathbf{x})$, each characterised by a scaling law $\delta z^{\pm}_{\tau} \sim \tau^h$ and a local scaling exponent $h(\mathbf{x})$. The dimension of the set is variable $D(h)$. With this in mind it can be shown that the high–order moments can be described by $\zeta_p = min_h[ph+3-D(h)]$. In this way the departure of ζ_p from a linear scaling, and then intermittency, can be characterized through the changing of generalized dimensions $D(h)$, as h is varied. That is as p increses, we are probing regions of fluids where even more rare and intense events exists. These regions are characterised by a smaller value of h, and by stronger singularities of the gradient of the field.

Because of the idea of self–similarity underlying the energy cascade process in turbulence [1], a different point of view can be introduced [16, 9]. That is a model which tries to characterize the behaviour of the pdfs through the scaling law of a parameter describing how the shape of the pdf changes in going towards small scales. In its simplest form the model can be introduced by saying that the pdf of the increments $\delta\psi_r$ (representing here both velocity and magnetic fluctuations) at a given scale r, is made by a convolution of the typical Gaussian distribution of widths $\sigma = <\delta\psi^2>^{1/2}$, whose distribution is given by a function $G_\lambda(\sigma)$

$$P(\delta\psi_r) = \frac{1}{\sqrt{2\pi}}\int_0^\infty G_\lambda(\sigma)\exp\left(-\delta\psi_r^2/2\sigma^2\right)\frac{d\sigma}{\sigma} \quad (3)$$

In a purely self–similar situation, where the energy cascade generates only a trivial variation of σ with the scale, a Gaussian distribution for $P(\delta\psi_r)$ is recovered. When the cascade is not strictly self–similar, the width of the distribution G_λ is different from zero, and the scaling behaviour of the width of this distribution, namely λ^2, can be used to characterize intermittency.

In order to make a quantitative analysis of the energy cascade leading to the process just described, the distributions have been fitted by using the log–normal ansatz [16]

$$G_\lambda(\sigma) = \frac{1}{\sqrt{2\pi}\lambda} \exp\left(-\frac{\ln^2 \sigma/\sigma_0}{2\lambda^2} \right) \quad (4)$$

The width of the log–normal distribution of σ is given by $\lambda(r) = <(\Delta \ln \sigma)^2>^{1/2}$.

The expression (3) have been fitted on the experimental pdfs for both velocity and magnetic intensity, and the corresponding values of the parameter λ can be recovered. In figures 1 we plotted, as full lines, the curves relative to the fit, showing that the scaling behaviour of pdfs in all cases is very well described by the function (3). At each scale τ, we get a value for the parameter $\lambda^2(\tau)$, which for $\tau \leq 1$ hour, can be fitted with a power law $\lambda^2(\tau) = \mu\tau^{-\gamma}$. The values of μ and γ obtained in the fitting procedure are $\mu \simeq 0.75 \pm 0.03$ and $\gamma = 0.18 \pm 0.03$ for the magnetic field, while $\mu \simeq 0.38 \pm 0.02$ and $\gamma = 0.20 \pm 0.04$ for the velocity field, in the range of scales $\tau \leq 0.72$ hours.

TURBULENT STRUCTURES AND NON–POISSONIAN EVENTS

The nonlinear energy cascade towards smaller scales accumulates fluctuations only in relatively small regions of space, where gradients become singular. These regions can be viewed as localized zones of fluid where some phase correlation exists (coherent structures). Structures continuously appear and disappear apparently in a random fashion, in some random location of fluid, and they carry the great quantity of energy of flows. The turbulent flow can be viewed as a superposition of non–gaussian structures, within the sea of gaussian background.

The presence of structures can be evidenced by using for example a Wavelets transform. Unlike the Fourier basis, Wavelets allow a decomposition both in time and frequency (or space and scale) (see for example [17] and references therein). That is a function $f(t)$ can be projected on a wavelet basis with coefficients $w(\tau,t)$. Since a Parceval's theorem exists, the square modulus $|w(\tau,t)|^2$ represents the energy content of fluctuations $f(t+\tau) - f(t) \sim w(\tau,t)$ at the scale τ at time t. It is useful to introduce a measure of local intermittency [17], as for example $lim = |w(\tau,t)|^2 / < |w(\tau,t)|^2 >$ (averages are made over all times at a given scale τ). This represent the energy content of fluctuations at a given scale, with respect to the standard deviation of fluctuations at that scale. The whole set of wavelets coefficients can then be splitted in two set: a "gaussian" set $w_g(\tau,t)$ and a "structure" set $w_s(\tau,t)$. Then $w(\tau,t) = w_g(\tau,t) \oplus w_s(\tau,t)$, according to wheter *lim* is respectively lesser or greater of a given threshold (the symbol $\oplus$ stands here for union of disjoint sets). An inverse wavelet transform, performed separately on w_g and w_s, gives two separate time series: a "file" for the gaussian background, and another for structures. In figure 7 we give an example of that behaviour.

Apart from recognizing the typical structures in the space [18, 19], some statistics can be made. The interesting statistics is about the time separation of structures. Let us call Δt the waiting time between two consecutive structures, that is between $w_g(\tau,t)$ and $w_g(\tau,t+\Delta t)$ at a scale τ, and let us consider the pdf $P(\Delta t)$. In figures 8 we report the pdf for magnetic structures calculated for solar wind and RFX, and the pdf obtained for velocity structures in fluid flows. As it can be seen the waiting times are distributed according to a well defined power law $P(\Delta t) \sim \Delta t^{-\beta}$ with some values for β, extended over at least two decades. A similar investigation in the usual fluid flows and in the laboratory plasma, shows the same phenomenon (see fig. 8). This property is very interesting, because this means that the underlying process of cascade is non–poissonian [20, 21]. In fact waiting times occurring between isolated poissonian events, must be distributed according to an exponential function [22]. The power law for $P(\Delta t)$ represents the asymptotic behaviour of a Lévy function with characteristic exponent $\alpha = \beta - 1$ [23]. This function describes self–affine processes and are obtained from the central limit theorem by relaxing the hypothesis that the variance of variables is finite. The power law for waiting times we found is a clear evidence that long–range correlation (or in some sense "memory") exists in the underlying cascade process [23].

CONCLUSIONS

We reported some of the work done on solar wind turbulence, and we compared the behaviour with other turbulent systems. Intermittency manifests itself through a breakdown of pure self–similarity of fluctuations, leading to anomalous scaling laws, and non–gaussian tails for pdfs at small scales. At small scales the statistics of waiting times between intermittent isolated events, results to be non–poissonian. This indicates that the underlying cascade process which generates these events conserves memory. In some sense, the cascade continuously transmit to small scales a phase–correlated excitation into varying subsets of the fluid. To what extent dynamical models can describe all behaviours is an interesting task [21]. A simplified shell model describing the gross features of MHD turbulence [24] is able to reproduce all statistics observed. Time intermittency, that is the occurrence of bursts of chaoticity concentrated on the dissipative shells, generates a break of the global scaling invariance in the shell model, which is responsible for the observed departure from self–similarity.

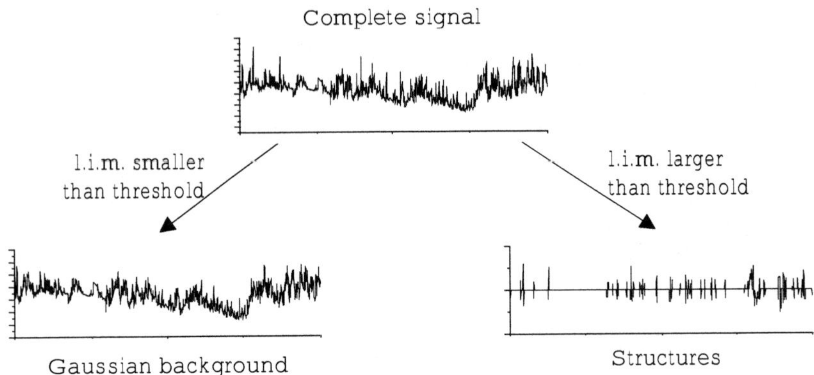

FIGURE 7. An example of the procedure used to recognize structures on a given scale. In the top panel of the figure we report the original time serie (here a sample of velocity field in the solar wind). Then we operate with the *lim* procedure (see text), and we obtain the two series at the bottom of the figure.

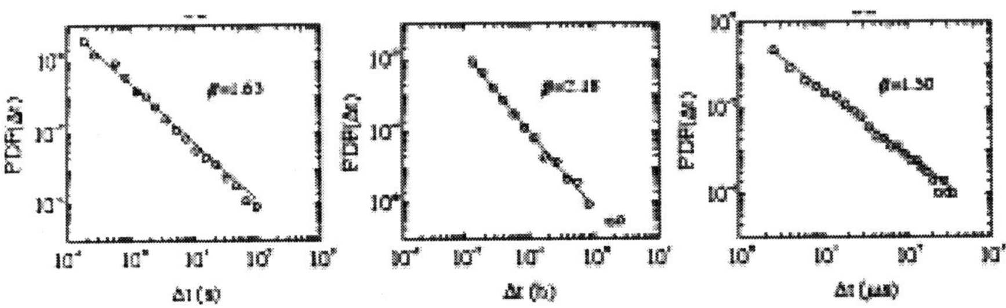

FIGURE 8. The distribution of waiting times between structures at the smallest scale for the velocity (fluids, left panel) and magnetic fluctuations (solar wind, central panel and RFX, left panel).

We are grateful to Roland Grappin for the critical reading of the manuscript.

REFERENCES

1. Bohr, T., Jensen, M.H., Paladin, G., and Vulpiani, A., *Dynamical System Approach to Turbulence* (Cambridge Univ. Press., Cambridge) 1998.
2. Politano, H, and Pouquet, A., *Phys. Rev. E 57*, R21, 1998.
3. Frisch, U., *Turbulence: the legacy of A.N. Kolmogorov* (Cambridge Univ. Press., Cambridge) 1995.
4. Dobrowolny, M., Mangeney, A., and Veltri, P., *Phys. Rev. Lett. 45*, 144, 1980; Carbone, V., *Phys. Rev. Lett. 71*, 1546, 1993.
5. Carbone, V., *Ann. Geophys. 12*, 585, 1994.
6. Carbone, V., Veltri, P., and Bruno, R., *Phys. Rev. Lett. 75*, 3110, 1995.
7. Van Atta, C.W., and Park, J., *Lect. Notes in Phys. 12*, 402, 1975.
8. Marsch, E., and Tu, C.Y., *Ann. Geophys. 12*, 1127, 1994
9. Sorriso–Valvo, L., et al., *Geophys. Res. Lett. 23*, 121, 1996
10. Benzi, R., et al., *Phys. Rev. E 48*, R29, 1993.
11. Burlaga, L.F., *J. Geophys. Res. 96*, 5847, 1991; Marsch, E., and Tu, C.Y., *Ann. Geophys. 11*, 227, 1993
12. Albertson, P., et al., *Phys. Fluids 10*, 1725, 1998.
13. Carbone, V. et al., *Phys. Rev. E 62*, R49, 2000.
14. Politano, H., Pouquet, A., and Carbone, V., *Europhys. Lett. 43*, 516, 1998; Sorriso-Valvo, L., Carbone, V., Veltri, P., Politano, H., and Pouquet, A., *Europhys. Lett. 51*, 520, 2000
15. Carbone, V. et al., *Phys. Rev. E 62*, R49, 2000
16. Castaing, B., Gagne, Y., and Hopfinger, E.J., *Physica D 46*, 177, 1990.
17. Farge, M., *Ann. Rev. Fluid Mech. 24*, 395, 1992.
18. Veltri, P., and Mangeney, A., in *Proceeedings of Solar Wind Nine*, Ed. S.R. Habbal, R. Esser, J.V. Hollweg, and P.A. Isenberg, AIP Conf. Proc. No 471 (AIP, Woodbury, N.Y.), 1999.
19. Bruno, R., V. Carbone, P. Veltri, E. Pietropaolo and B. Bavassano, *Planetary Space Sci., 49*, 1201, 2001.
20. Boffetta, G., Carbone, V., Giuliani, P., Veltri, P., and Vulpiani, A., *Phys. Rev. Lett., 83*, 4662, 1999
21. Carbone, V. et al., *Europhys. Lett. 58*, 349, 2002.
22. Feller, W., *An introduction to probability theory and its applications, Vol 1*, 2d Ed. Wiley New York, 1968.
23. Lepreti, F., Carbone, V., and Veltri, P., *Astrophys. J. 555*, L133, 2001.
24. Giuliani, P., and Carbone, V., *Europhys. Lett. 43*, 527, 1998.

On the Outer Scale of Turbulence in the Solar Wind

I.V. Chashei*, M.K. Bird† and A.I. Efimov**

*Lebedev Physical Institute, Russian Academy of Science, Moscow, 117924, Russia
†Radioastronomisches Institut, Universität Bonn, 53121 Bonn, Germany
**Inst. for Radio Engineering & Electronics, Russian Academy of Science, Moscow, 101999, Russia

Abstract.
The outer scale of turbulence in the solar wind has been estimated from frequency fluctuation data recorded during radio sounding experiments using the Galileo and Ulysses spacecraft carrier signals at 2295 MHz. The outer scale was observed to increase approximately linearly with increasing heliocentric distance in the range between 7 and 80 $R_{\odot}$. A model is presented here for the formation and evolution of the turbulence outer scale. The model is based on the assumption that the density fluctuations are caused by nonlinear interactions of Alfvén waves and that the rates of linear and nonlinear processes are nearly equal in the spectral range near the inverse outer scale. The dependence of the turbulence outer scale on heliocentric distance and local plasma parameters is investigated for three possible nonlinear wave coupling mechanisms: strong interactions (Kolmogorov turbulence), three-wave interactions (Kraichnan turbulence), and four-wave interactions. Comparisons of the model with the observations indicate that the three-wave decay processes with participation of both Alfvén and magnetosonic waves are the main type of nonlinear interactions in the inertial range of the turbulence power spectra.

INTRODUCTION

The properties of the solar wind's outer scale of turbulence, particularly its dependence on heliocentric distance and local plasma parameters, have not been studied sufficiently, especially for the inner solar wind. Moreover, the outer scale of turbulence, in addition to its significance for the problem of solar wind formation and evolution, is a crucially important parameter for the physics of turbulence itself. A model of the formation and evolution of the density turbulence outer scale in the inner solar wind is developed here and compared with observational data obtained during radio occultation experiments with the Galileo and Ulysses spacecraft.

OBSERVATIONAL RESULTS

The density turbulence outer scale makes itself apparent, for example, in long, uninterrupted intervals of frequency fluctuation measurements. The specific data discussed here were obtained in 1995/1996 using the Galileo and Ulysses spacecraft carrier signals at the frequency 2295 MHz. Temporal power spectra were calculated for frequency fluctuations recorded by ground-based tracking stations during solar conjunction for solar ray path proximate points in the range of heliocentric distances between 7 and 80 $R_{\odot}$ ($R_{\odot}$ = solar radius). Frequency fluctuation power spectra, an example of which is shown in Fig. 1, generally display a clearly defined peak at temporal frequencies $\nu_{max} \simeq 0.1$ mHz and fall off beyond this frequency with an index α_f (inertial range). These peaks have been interpreted as a manifestation of the density turbulence outer scale in the outward-flowing solar wind plasma. The peak associated with the turbulence outer scale is much more pronounced in the frequency fluctuation spectra than, for example, in quasi 2D-spectra for the phase fluctuations or the quasi 1D-spectra for *in situ* measurements, because the instantaneous frequency is equal to the time derivative of the radio wave phase (Armand et al., 1987).

Values of the density turbulence outer scale L_o were estimated from the relation $L_o = \sqrt{2/\alpha_f}\, v_c/\nu_{max}$, with the solar wind convection velocity v_c determined from a cross-correlation analysis of frequency fluctuations measured simultaneously at two widely-spaced ground stations (Bird et al., 2002). As illustrated in Fig. 2, it was found that the turbulence outer scale $L_o(R)$ increases with increasing heliocentric distance R of the radio ray path proximate point.

Values with error bars in Fig. 2 were derived using simultaneous two-station correlation determinations of solar wind convection velocity. Values without error bars use v_c from a theoretical model. Solid (open) data points are for occultation ingress (egress). A linear regression best fit to the entire data set, shown by the thick solid line in Fig. 2, yields a radial dependence $L_o(R) = a[R/R_{\odot}]^m$, with $a \simeq 0.23$ $R_{\odot}$, and $m \simeq 0.82$. Preliminary analysis

CP679, *Solar Wind Ten: Proceedings of the Tenth International Solar Wind Conference*,
edited by M. Velli, R. Bruno, and F. Malara

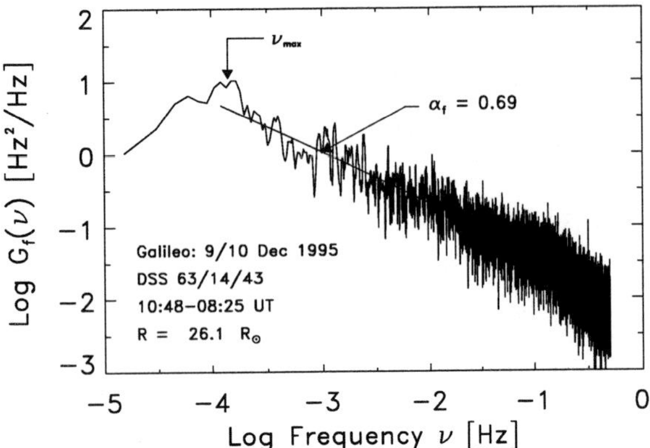

FIGURE 1. *Example of a frequency fluctuation power spectrum. The peak at the fluctuating frequency $\nu_{max} \simeq$ 0.1 mHz is associated with the density turbulence outer scale. The spectrum decreases as a power-law $\nu^{-\alpha_f}$ for $\nu > \nu_{max}$ (adapted from Bird et al., 2002).*

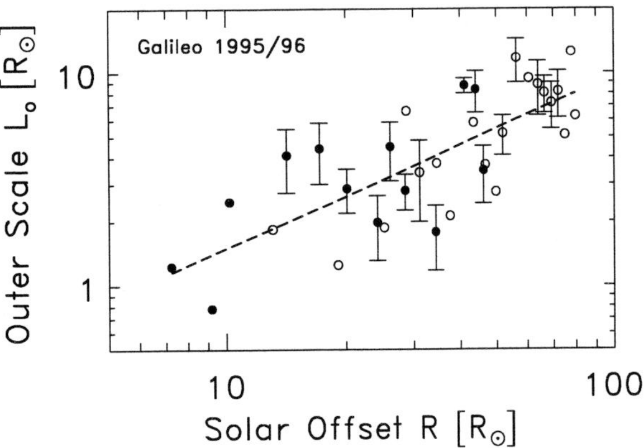

FIGURE 2. *Radial dependence of the density turbulence outer scale $L_o(R)$ deduced from Galileo measurements during the solar conjunction in 1995/96 (adapted from Bird et al., 2002).*

of the Ulysses radio occultation data yields very similar results with $a \simeq 0.25$ $R_\odot$, and $m \simeq 0.88$ (Wohlmuth et al., 2001).

FORMATION MECHANISM OF THE TURBULENCE OUTER SCALE

We now consider a possible formation mechanism for the turbulence outer scale in the outward flowing nonuniform solar wind plasma. We assume that: (a) the main energy-containing disturbances are Alfvén waves, generated initially at the coronal base and propagating away from the Sun, and (b) density fluctuations are associated with magnetosonic waves excited locally by nonlinear Alfvén wave interactions. The propagation of Alfvén waves can be described by the following equation (Tu and Marsch, 1995):

$$\nabla \cdot \left[(\frac{3}{2}\vec{v}_c + \vec{v}_a) W_k \right] - \frac{1}{2}\vec{v}_c \cdot \nabla W_k = -\frac{\partial F_k}{\partial k} \tag{1}$$

where W_k is the spectral density of the 1D Alfvén wave power spectrum, k is the wavenumber, and $\vec{v}_c$ and $\vec{v}_a$ are the solar wind convection speed and Alfvén speed, respectively. The left-hand side of Eq. (1) basically corresponds to a WKB approach, and the right-hand side describes the various nonlinear cascading processes, with F_k being a cascading function. Following the work of Chashei and Shishov (1981), we define the inverse turbulence outer scale $k_o = 2\pi/L_o$ as that wavenumber at which the typical time scales of *linear* $t_l = R/(v_c + v_a)$ and *nonlinear* $t_{nl} = F_k/k$ processes are approximately equal. Those waves with $k < k_o$ propagate in the linear regime, and the wave energy is pumped to higher k by cascading processes to form an inertial scaling spectrum with $F_k = const$ in the range $k > k_o$, dominated by nonlinear processes. Magnetosonic waves (and consequently density fluctuations) are generated locally by nonlinear wave interactions in the spectral range $k > k_o$. The power spectra of the Alfvén waves (magnetic field turbulence) and the magnetosonic waves (density turbulence) have a similar shape at $k > k_o$ (Chashei and Shishov, 1985). In the model considered here, we represent the turbulence power spectrum by a broken power law:

$$W_k = \begin{cases} Ck^{-\gamma} & \text{for} \quad k < k_o \\ Ck_o^{\alpha-\gamma}k^{-\alpha} & \text{for} \quad k > k_o \end{cases} \tag{2}$$

where C is the structure constant, γ is an initial (shallow) power exponent, and the power exponent α $(\alpha > \gamma)$ is defined by the k-dependence of the cascading function F_k (see Fig. 3). The exponent α is related to the index α_f of the frequency fluctuation spectrum (previous section) by $\alpha = \alpha_f + 1$.

Because the spectral range $k \approx k_o$ serves as an energy source for the spectral energy flux and for the turbulence at $k > k_o$, the value of k_o shifts to lower k with increasing distance from the Sun. The cascading function F_k can be specified in the following form:

$$F_k = \varepsilon k^2 v_a \left[\frac{4\pi k W_k}{B^2} \right]^n W_k \tag{3}$$

where $\varepsilon = const \approx 0.1$ (Tu and Marsch, 1995), and B is the local ambient magnetic field strength. The values of the power exponent n in Eq. (3) and the corresponding values of the inertial spectral range power exponents α in Eq. (2), as obtained from dimensional analysis, are equal to:

$$\begin{array}{ll} n = n_1 = 1/2, & \alpha = \alpha_1 = 5/3 \\ n = n_2 = 1, & \alpha = \alpha_2 = 3/2 \\ n = n_3 = 2, & \alpha = \alpha_3 = 4/3 \end{array} \tag{4}$$

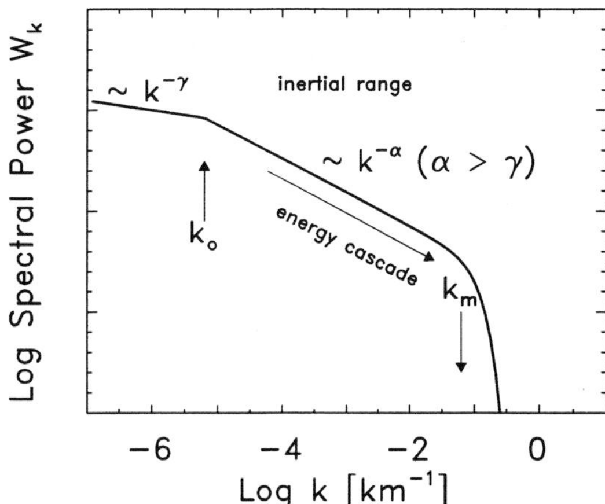

FIGURE 3. *Turbulence power spectrum given by Eq. (2). The intrinsic spectrum for $k < k_o$ is characterized by a spectral index γ. The spectrum steepens in the inertial range ($k > k_o$), starting from the wavenumber corresponding to the outer scale and extending up to a dissipation scale indicated here by k_m.*

The index $n_1 = 1/2$ corresponds to the Kolmogorov spectrum with strong turbulence (Tu and Marsch, 1995). The cases $n_2 = 1$ and $n_3 = 2$ apply to the weak three-wave (Kraichnan spectrum) and weak four-wave interactions, respectively (Chashei and Shishov, 1977).

OUTER SCALE DEPENDENCE ON HELIOCENTRIC DISTANCE AND LOCAL PLASMA PARAMETERS

Assuming that the typical rates of linear and nonlinear processes are equal, we find the following radial dependence for the turbulence outer scale in the developed solar wind (where $v_c = const$, $B \sim R^{-2}$, $v_a \sim R^{-1}$, and, as follows from the WKB approach, $C \sim R^{-3}$):

$$L_o = \frac{2\pi}{k_o} \sim R^m \tag{5}$$

with

$$\begin{aligned} m_1 &= 1/(3-\gamma) \\ m_2 &= 1/(2-\gamma) \\ m_3 &= 1/(3-2\gamma) \end{aligned} \tag{6}$$

where the indicies 1,2,3 are related to the corresponding cascading functions given by Eqs. (3) and (4). Similarly, we can find the dependence of L_o on local plasma parameters. This can be reduced to the dependence of L_o on the solar wind speed v_c if we assume that the solar wind acceleration is associated mainly with the Alfvén waves and the solar wind mass flux density is approximately constant at a fixed distance R. We then have:

$$L_o = \frac{2\pi}{k_o} \sim v_c^{\mu} \tag{7}$$

with

$$\begin{aligned} \mu_1 &= (1-2\gamma)/2(3-\gamma) \\ \mu_2 &= 2(1-\gamma)/(2-\gamma) \\ \mu_3 &= (5-4\gamma)/2(3-2\gamma) \end{aligned} \tag{8}$$

COMPARISON WITH OBSERVATIONS

We now compare the radial dependence of the turbulence outer scale predicted by the models with the Galileo observations presented in Fig. 2. It is found that $\gamma_1 = 7/4$, $\gamma_2 = 3/4$, and $\gamma_3 = 1/4$. As known from *in situ* measurements of magnetic field fluctuations near $R = 1$ AU (Matthaeus and Goldstein, 1986), as well as from Faraday rotation fluctuations near the Sun (Chashei et al., 2000), that the low frequency exponent of the power spectra is $\gamma_{obs} \approx 1$ (so-called "flicker" spectrum).

A comparison of the above estimates $\gamma_{1,2,3}$ with $\gamma_{obs} = 1$ shows convincingly that the model with $n = n_2 = 1$ agrees best with the observed radial dependence of the turbulence outer scale. The estimate $\gamma_1 = 7/4$, in fact, would mean that the turbulence spectrum does not even have an outer scale, because $\gamma_1 = 7/4 > \alpha_1 = 5/3$. Finally, it is noted that the value $\gamma_3 = 1/4$ is reasonable, in principle, but differs rather strongly from the observational value $\gamma_{obs} = 1$.

Within the framework of our model with $n = n_2 = 1$, we can use the numerical values of the turbulence outer scale L_o to estimate the fractional turbulence level and its dependence on heliocentric distance:

$$\frac{4\pi C}{B^2} \approx 0.3 \left[\frac{R}{50R_\odot}\right] \tag{9}$$

where the right-hand side of Eq. (9) is in reasonably good agreement with the *in situ* measurements of the Helios spacecraft (Tu and Marsch, 1995) over the radial distances 0.3 AU $< R <$ 1.0 AU, as well with the radial dependence of C expected from the WKB approach (Belcher and Davis, 1971).

CONCLUSIONS

The density turbulence outer scale deduced from the Galileo and Ulysses radio occultation data increases approximately linearly with increasing heliocentric distance.

Comparison of the observed radial dependence of the turbulence outer scale with different versions of the theoretical model for turbulence evolution shows that the

Kraichnan type model with n = 1 appears to yield better agreement than the models with n = 1/2 or n = 2. The physical consequences of this fact, as well as comparisons with other observational data, should be considered in more detail in future work.

As follows from the estimate given in Eq. (9) for the Galileo data, the fractional level of turbulence is moderate for the developed low-latitude solar wind in the inner heliosphere, at least for the period of solar activity minimum.

Further studies of the turbulent cascading mechanism could include a more detailed analysis of the parametric dependence of the turbulence outer scale on the solar wind speed. Indeed, the dependence of the outer scale on solar wind speed, $L_o(v_c)$ in Eqs. (7) and (8), differs for various cascading mechanisms. For example, L_o decreases with increasing solar wind speed v_c for the Kolmogorov turbulence with $n = 1/2$, is almost independent of v_c for the Kraichnan turbulence with $n = 1$, and increases with increasing v_c for the four-wave interactions with $n = 2$ (or for higher-order interactions with $n > 2$).

ACKNOWLEDGMENTS

This work presents results of a bi-national Research Project partially funded by the Deutsche Forschungsgemeinschaft (DFG) and the Russian Foundation for Basic Research (RFBR), Grant 00-02-04022. Additional support from the RFBR, Grant 00-02-17845, is acknowledged.

REFERENCES

1. Armand, N.A., Efimov, A.I., Yakovlev, O.I., 1987, A model of the solar wind turbulence from radio occultation experiments, *Astron. Astrophys. 183*, 135-141.
2. Belcher, J.W., Davis, Jr., L., 1971, Large amplitude Alfvén waves in the interplanetary medium, 2, *J. Geophys. Res. 76*, 3534-3563.
3. Bird, M.K., Efimov, A.I., Samoznaev, L.N., Chashei, I.V., Edenhofer, P., Plettemeier, D., Wohlmuth, R., 2002, Outer scale of coronal turbulence near the Sun, *Adv. Space Res. 30(3)*, 447-452.
4. Chashei, I.V., Shishov, V.I., 1977, On the solar wind turbulence, *Geomagn. Aeron. 17*, 984-993.
5. Chashei, I.V., Shishov, V.I., 1981, A mechanism for generating the turbulence spectrum of interplanetary plasma, *Astron. Zh. Lett. 7*, 500-506.
6. Chashei, I.V., Shishov, V.I., 1985, MHD turbulence spectra of the interplanetary plasma with allowance for nonlinear absorption, *Geomagn. Aeron. 25*, 1-6.
7. Chashei, I.V., Efimov, A.I., Samoznaev, L.N., Bird, M.K., Pätzold, M., 2000, The spectrum of magnetic field irregularities in the solar corona and in interplanetary space, *Adv. Space Res. 25(9)*, 1973-1978.
8. Matthaeus, W.H., Goldstein, M.L., 1986, Low frequency 1/f noise in the interplanetary magnetic field, *Phys. Rev. Lett. 57*, 495-498.
9. Tu, C.Y., Marsch, E., 1995, MHD structures, waves and turbulence in the solar wind: observations and theories, *Space Sci. Rev. 73*, 1-210.
10. Wohlmuth, R., Plettemeier, D., Edenhofer, P, Bird, M.K., Efimov, A.I, Andreev, V.E., Samoznaev, L.N., Chashei, I.V., 2001, Radio frequency fluctuation spectra during the solar conjunctions of the *Ulysses* and *Galileo* spacecraft, *Space Sci. Rev. 97*, 9-12.

Alfvénic turbulence in high-latitude solar wind: Is latitude a relevant parameter?

Bruno Bavassano[1], Ermanno Pietropaolo[2], and Roberto Bruno[1]

[1]*Istituto di Fisica dello Spazio Interplanetario (CNR), Roma, Italy*
[2]*Dipartimento di Fisica, Università dell'Aquila, L'Aquila, Italy*

Abstract. Plasma and magnetic field measurements by Ulysses during its first out-of-ecliptic orbit have allowed extensive investigations on the behavior of Alfvénic turbulence in high-latitude solar wind. Most analyses have shown that the turbulence evolution in high-latitude wind is radial, rather than latitudinal, in nature. However, a recent study based on magnetic field fluctuations has suggested that latitude might play a non negligible role. Here we further examine this possibility by using Elsässer's variables, that directly are related to the Alfvénic content of solar wind fluctuations. Our conclusion, supported by a comparison between polar and ecliptic observations, is that latitude does not appear to have an appreciable influence on the turbulence evolution in high-latitude solar wind.

INTRODUCTION

Observations by Ulysses during its first out-of-ecliptic orbit have shown that, at low solar activity, the high-latitude (or polar) solar wind is a fast and relatively steady flow. A relevant feature of the polar wind is the ubiquitous presence of an intense flow of Alfvénic fluctuations (e.g., *Goldstein et al.* [1995]; *Horbury et al.* [1995]; *Smith et al.* [1995]). Similarly to previous ecliptic observations in fast streams (e.g., *Tu and Marsch* [1995]), a largely dominant fraction of these fluctuations is outward propagating, with respect to the Sun, in the solar wind frame. These outward fluctuations mainly have a solar origin (or, more precisely, inside the Alfvén critical point). Conversely, inward propagating fluctuations observed in the interplanetary space can only be generated outside such critical distance.

A relevant point to be discussed is the nature of the Alfvénic turbulence variations observed in high-latitude solar wind (i.e., radial, or latitudinal, or both), given that both distance and latitude change along the Ulysses trajectory. Several analyses of Ulysses data have indicated that the variation of turbulence properties in high-latitude wind is essentially radial, rather than latitudinal, in nature (e.g., *Goldstein et al.* [1995]; *Horbury et al.* [1995]; *Forsyth et al.* [1996]). A further robust argument in favor of the radial character of the turbulence variation comes from the agreement between gradients observed in high-latitude wind and in fast streams on the ecliptic. This has been seen to hold both for magnetic field fluctuations (as observed by *Forsyth et al.* [1996] in high-latitude wind and by *Bavassano and Smith* [1986] in ecliptic wind) and for outward and inward Alfvénic fluctuations (as shown by *Bavassano et al.* [2000] and [2001]).

In spite of all these pieces of evidence, we have decided of re-examining the role of latitude on Alfvénic turbulence variations in high-latitude solar wind. We have been motivated to this by the recent results of *Horbury and Balogh* [2001], showing that fluctuations in magnetic field components at time scales shorter than about 1 day (in the spacecraft frame) exhibit a non negligible dependence on latitude. In present analysis the same method of *Horbury and Balogh* [2001], based on a multiple regression, will be used, but we will directly look at the Alfvénic component of solar wind fluctuations, rather than at the magnetic fluctuations. This is a remarkable difference. In fact, magnetic fluctuations, though obviously related to the Alfvénic turbulence, also include non negligible contributions from other disturbances and structures convected past the spacecraft by the plasma flow. A problem with the regression analysis is that it is based on variables (distance and latitude of Ulysses) that are not mutually independent. For this reason, when discussing the regression results, we will take advantage of comparisons between polar and ecliptic observations.

CP679, *Solar Wind Ten: Proceedings of the Tenth International Solar Wind Conference*,
edited by M. Velli, R. Bruno, and F. Malara

DATA ANALYSIS

The analyzed intervals are highlighted in Figure 1, where the solar wind velocity V, the proton number density N' normalized to 1 AU (assuming an inverse square scaling with distance), the spacecraft heliocentric distance r, and the spacecraft heliographic latitude λ are shown for years 1993 to 1996. The two thick lines at the top, labeled s and n, indicate the two intervals (southern and northern, respectively) of full immersion of Ulysses in polar wind. The two thin horizontal lines with label f highlight two polar wind intervals selected in the phase of 'fast latitudinal scan' around the perihelion. The first interval is from the maximum southern latitude (dashed vertical line on the left) to the exit from southern polar wind, the second one from the entry into the northern polar wind to the maximum northern latitude (dashed vertical line on the right). These two intervals allow to get a quick, but complete, latitudinal survey of the polar wind for a reduced range of distances (see the r and λ variations in the two lower plots). Finally, the horizontal line (of intermediate thickness) labeled n_{HB} indicates the period investigated by *Horbury and Balogh* [2001].

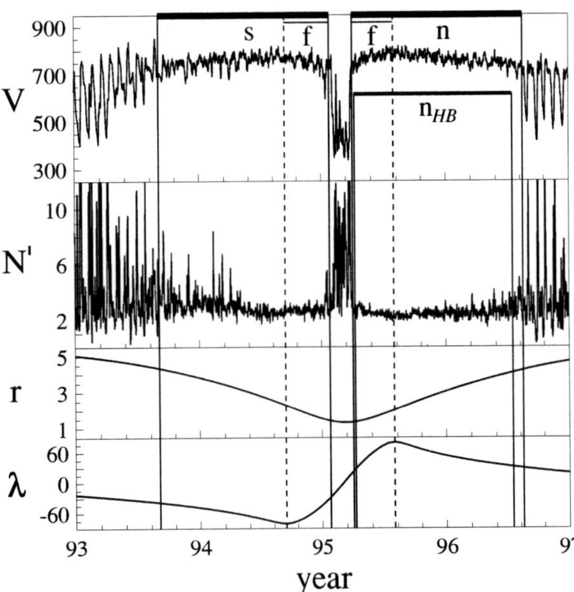

FIGURE 1. Solar wind data and spacecraft coordinates for years 1993 to 1996. Ulysses data have been made available, through NASA/GSFC World Data Center, by D. J. McComas and A. Balogh.

A point to be stressed in Figure 1 is the presence of spike-like variations of the normalized density N' also during polar wind phases. These spikes are weaker than those observed at lower latitudes, nevertheless they indicate that also in polar wind non negligible compressive effects are developed by interacting flows (see also the velocity profile in the top plot). They mainly affect polar wind observations towards the aphelion phase, when the spacecraft moves slowly in latitude and spends a lot of time close to the low-latitude boundary of the polar wind region. Here stream interactions, typical of the ecliptic wind, still persist, though with weaker effects. All this indicates that in order to get a clean evaluation of the polar turbulence variation a data selection is needed, especially when observations at large distance are involved. This point has already been stressed by *Bavassano et al.* [2000] (hereafter BPB), who derived, starting from the s and n polar phases, a 'selected' data sample where all intervals with large changes in plasma velocity, and/or plasma density, and/or magnetic field magnitude were rejected. In the present study we will use this 'selected' sample, exceptions will be explicitly indicated.

Our analysis is based on the use of the Elsässer's variables ($\mathbf{z}_\pm$), a well known tool to identify Alfvénic fluctuations. They are defined as $\mathbf{z}_\pm=\mathbf{v}\pm\mathbf{b}$, where $\mathbf{v}$ and $\mathbf{b}$ are the velocity and magnetic field vectors, respectively, and $\mathbf{b}$ is scaled to Alfvén units (i.e., divided by $\sqrt{4\pi\rho}$, with ρ the mass density). Taking into account how the sign of the Alfvénic correlation depends on the propagation direction with respect to the background magnetic field, we will use the above definition for the case of a background magnetic field pointing to the Sun, while the equation $\mathbf{z}_\pm=\mathbf{v}\mp\mathbf{b}$ will hold for the opposite polarity. In this way $\mathbf{z}_+$ ($\mathbf{z}_-$) fluctuations will always correspond to modes with an outward (inward) propagation, with respect to Sun, in the plasma frame.

Analogously to the analysis of *Bavassano et al.* [2000] (hereafter BPB), the results discussed here refer to hourly variances of $\mathbf{z}_+$ and $\mathbf{z}_-$. These variances have been radially averaged on bins of 0.05 AU and the resulting values (in the following e_+ and e_-) have been used, similarly to *Horbury and Balogh* [2001], for a multiple regression study based on the equation $log(e_\pm) = A_\pm + B_\pm\ log(r) + C_\pm\ sin(\theta)$, where θ is the absolute value of λ.

A relevant point to be stressed is that r and θ, being coupled each other by the equation of the Ulysses orbit in a very specific manner, do not represent a set of independent variables. Violating the assumption that the variables in the regression are independent is risky and renders the results of doubtful validity, so that they can hardly be trusted. Though obvious, this caveat has to be kept well in mind in the following. Our approach is that of performing the multiple regression analysis mainly to highlight, with respect to that of *Horbury and Balogh* [2001], the effect of using clean data samples and of looking at the behavior of $\mathbf{z}_+$ and $\mathbf{z}_-$, instead of the magnetic field. Our conclusions about the role of latitude on the polar Alfvénic turbulence evolution will mosty come from comparisons with other observations on the ecliptic, where latitudinal effects are absent.

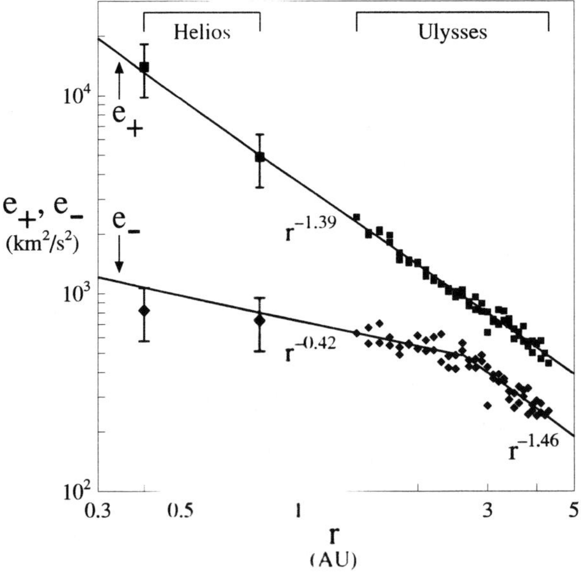

FIGURE 2. This plot combines Ulysses observations in polar wind with those by Helios inside 1 AU on the ecliptic plane. The values of e_+ (e_-) are shown as squares (diamonds), small for Ulysses and large for Helios. Best fit lines and radial power laws for Ulysses data are given. The figure is adapted from *Bavassano et al.* [2000] (copyright 2000 American Geophysical Union, modified by permission of American Geophysical Union).

Since our analysis is based on the same data set of *Bavassano et al.* [2000], with the only difference that they used radial bins of 0.1 AU (instead of 0.05 AU), it is useful to briefly recall their results. To this end we show in Figure 2 a slightly modified version of one of their figures. This is a composite plot combining Ulysses observations in polar wind with those by Helios 1 and 2 in the trailing edge of fast streams on the ecliptic (as obtained from average spectra around 0.4 and 0.8 AU, see *Tu and Marsch* [1990]). It easily seen that the e_+ values observed by Ulysses exhibit the same radial gradient over all the investigated range of distances. In contrast, for e_- a change of slope around 2.5 AU is clearly apparent. Another remarkable feature is the good agreement of the Ulysses gradients with Helios data.

OUTWARD FLUCTUATIONS

The values of the B_+ and C_+ coefficients for the investigated samples are shown in Table 1. The first row is for the *Horbury and Balogh* [2001] interval (n_{HB}). Then, in the other rows, we give the results for the northern hemisphere (n), the southern hemisphere (s), the northern and southern hemispheres altogether (n+s), and finally the fast latitudinal scan f (with and without data selection).

A first comment is about the n_{HB} interval, the same used by *Horbury and Balogh* [2001]. Here we find, without any data selection, a value of C_+ of -0.11 ± 0.04, while at the same scale they observed a value around -0.3. Thus, just by using $\mathbf{z}_+$ instead of magnetic field, we are led to a remarkable reduction of the latitudinal effect. When selected data are used, C_+ becomes even smaller. For instance, for both northern and southern polar phases (n+s sample) we obtain a C_+ value of -0.03 ± 0.03. A latitudinal effect, if any, surely is far to be significant.

We would like to stress the results obtained for the sample f. In this case, with a latitudinally fast moving spacecraft, the effect of disturbances close to the low-latitude boundary of polar wind is greatly reduced and data selection becomes less important. As a matter of fact, using all available data in the sample a value for C_+ of -0.02 ± 0.07 is obtained. This does not leave doubts about the absence of a significant role of latitude in the e_+ variation.

The above conclusion is definitely confirmed by the good agreement (see Figure 2) between the e_+ radial gradient observed by Ulysses in polar wind and the Helios observations on the ecliptic (i.e., done in the absence of any significant change of latitude).

Table 1. Outward fluctuations regression: Radial (B_+) and latitudinal (C_+) coefficients

sample	B_+	C_+
n_{HB}*	-1.44 ± 0.05	-0.11 ± 0.04
n	-1.44 ± 0.04	-0.06 ± 0.04
s	-1.41 ± 0.04	-0.01 ± 0.04
n+s	-1.42 ± 0.03	-0.03 ± 0.03
f	-1.50 ± 0.17	-0.01 ± 0.07
f *	-1.48 ± 0.17	-0.02 ± 0.07

*without data selection

INWARD FLUCTUATIONS

As seen in Figure 2, the variation of e_- cannot be explained by a simple radial power law as for e_+. Rather, two different radial regimes seem to characterize its behavior, with a break at a distance of $\sim$ 2.5 AU. The dual regime is probably related to local generation effects that, important in the inner region, become negligible at larger distance (e.g., see *Malara et al.* [2001] and *Del Zanna* [2001]).

Applying the multiple regression separately to distances below and above 2.6 AU (see BPB), it has been found (Table 2) that C_-, negligible in the inner region, becomes significantly different from zero in the outer region. This inconsistency (why should latitude be important at large distance only?) confirms that it is hard to trust on a multiple regression analysis based on variables that are not independent. This is especially true in the region outside 2.6 AU, where the e_- signal has become quite noisy, due to both ambient (i.e., related to local plasma structures) and instrumental disturbances. Thus, to get a reliable view about the role of latitude on the e_- variation we have to look for other arguments, in particular we have to compare the polar trends with observations done on the ecliptic plane.

Under this approach, the lack of a latitudinal effect for e_- in polar wind inside 2.6 AU comes from the agreement of the Ulysses gradient in this region with Helios observations on the ecliptic (Figure 2). As regards the region outside 2.6 AU, Figures 2 and 3 show that there e_- declines at approximately the same rate as e_+. Having established that e_+ is not appreciably affected by latitude, we may infer that this holds for e_- too. A second, unequivocal, argument in favor of this conclusion comes from the agreement, within errors, with the e_- gradient observed at these distances in the ecliptic leg of the Ulysses trajectory (Figure 3 and *Bavassano et al.* [2001]).

All these pieces of evidence offer a robust conclusion in favor of a radial nature for the inward Alfvénic turbulence evolution in polar wind.

Table 2. Inward fluctuations regression: Radial (B_-) and latitudinal (C_-) coefficients

r (AU)	B_-	C_-
1.4 - 2.6	-0.54 ± 0.16	0.07 ± 0.10
2.6 - 4.3	-1.01 ± 0.23	0.28 ± 0.12

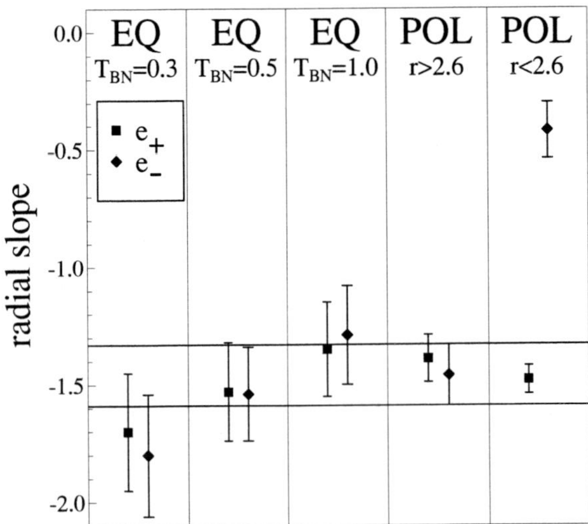

FIGURE 3. Radial slopes of e_+ (squares) and e_- (diamonds) observed by Ulysses in ecliptic and polar winds. The first three columns (label EQ) indicate ecliptic results obtained with different upper limits T_{BN} for density and magnetic intensity fluctuations. Last two columns (label POL) refer to polar wind results outside and inside 2.6 AU, respectively. This figure is from *Bavassano et al.* [2001] (copyright 2001 American Geophysical Union, modified by permission of American Geophysical Union).

REFERENCES

1. Bavassano, B., and Smith, E. J., *J. Geophys. Res.*, **91**, 1706-1710 (1986).
2. Bavassano, B., Pietropaolo, E., and Bruno, R., *J. Geophys. Res.*, **105**, 15,959-15,964 (2000).
3. Bavassano, B., Pietropaolo, E., and Bruno, R., *J. Geophys. Res.*, **106**, 10,659-10,668 (2001).
4. Del Zanna, L., *Geophys. Res. Lett.*, **28**, 2585-2588 (2001).
5. Forsyth, R. J., Horbury, T. S., Balogh, A., and Smith, E. J., *Geophys. Res. Lett.*, **23**, 595-598 (1996).
6. Goldstein, B. E., Smith, E. J., Balogh, A., Horbury, T. S., Goldstein, M. L., and Roberts, D. A., *Geophys. Res. Lett.*, **22**, 3393-3396 (1995).
7. Horbury, T. S., Balogh, A., Forsyth, R. J., and Smith, E. J., *Geophys. Res. Lett.*, **22**, 3401-3404 (1995).
8. Horbury, T. S., and Balogh, A., *J. Geophys. Res.*, **106**, 15,929-15,940 (2001).
9. Malara, F., Primavera, L., and Veltri, P., *Nonlinear Proc. Geoph.*, **8**, 159-166 (2001).
10. Smith, E. J., Balogh, A., Neugebauer, M., and McComas, D., *Geophys. Res. Lett.*, **22**, 3381-3384 (1995).
11. Tu, C.-Y., and Marsch, E., *Space Sci. Rev.*, **73**, 1-210 (1995).

On the role of coherent and stochastic fluctuations in the evolving solar wind MHD turbulence: Intermittency

R. Bruno*, V. Carbone†, L. Sorriso–Valvo† and B. Bavassano*

*Istituto Fisica Spazio Interplanetario del CNR, 00133 Roma, Italy
†Dipartimento di Fisica Università della Calabria, 87036 Rende (Cs), Italy

Abstract. One of the most interesting findings about solar wind turbulence is the intermittent character of its fluctuations. So far, intermittency has been studied, within solar wind context, only for scalar quantities neglecting to include in the study also vector fluctuations. In this paper we compare magnetic field and wind velocity intermittency computed for both scalar and vector fluctuations. We confirm that intermittency generally increases with radial distance from the sun only for fast wind and we suggest that at the basis of this behavior is a competing action between coherent structures imbedded in the wind and Alfvénic stochastic fluctuations propagating within the wind. The radial evolution experienced by these two components is rather different and consistent with the observed radial increase of intermittency in the inner heliosphere.

INTRODUCTION

Solar wind turbulence is strongly anisotropic and poorly single scale–invariant (see [13] and references therein for an extensive review on this and on related subjects), two fundamental hypotheses at the base of Kolmogorov's theory [8]. Moreover, the study of these fluctuations by means of conventional spectral analysis is strongly limited by the non-Gaussianity of their probability distribution function (PDF), which makes the second order moment of the distribution no longer the limiting order. As a consequence, it becomes necessary to look at higher order moments of the PDFs, which are built directly from the differences of the fluctuating field over all the possible spatial scales. This is usually done in the framework of the so-called multifractal approach [7]. The first time this new approach was employed within the solar wind context [4], it showed that interplanetary magnetic field and velocity fluctuations are highly intermittent. This feature implies that small turbulent eddies are less and less space filling if turbulence is looked at in the framework of a classical Richardson's cascade. Thus, the global scale invariance required in the Kolmogorov theory would become a local scale invariance where different fractal sets are characterized by different scaling exponents [5]. These first results obtained for the solar wind [4] also showed an unexpected similarity to those obtained for laboratory turbulence, unraveling consistency between observations on scales of 1 AU and laboratory observations on scales of meters. Further studies [9] addressed the radial evolution of intermittency within the inner heliosphere concluding that the Alfvénic turbulence observed in fast streams starts from the Sun as self–similar but then, during the expansion, decorrelates becoming multifractal. On the contrary, it was found that slow wind did not suffer a similar radial evolution [12].

Sorriso et al., [11] studied quantitatively the effect of intermittency on the PDFs of the increments over different scales adopting Castaing's model [6]. This model is based on the idea of a log–normal energy cascade and the non-Gaussian behavior of the PDF's at small scales can be represented by a convolution of Gaussians whose variances are distributed according to a log-normal distribution. As a matter of fact, large scales PDFs closely resemble Gaussian distributions but, at smaller scales the tails of the distributions become more and more stretched.

Moreover, techniques recently adopted in the context of solar wind turbulence and based on the wavelet decomposition of time series [14] allowed to show that some of the events causing intermittency are either compressive phenoma like shocks or planar sheets like tangential discontinuities separating adjacent flux–tubes [3].

Intermittency can be estimated looking at the behavior of the flatness factor of the PDFs built at different scales. If this parameter increases as the scale of interest becomes smaller our time series can be considered intermittent [7]. Having adopted this definition of intermittency, we studied the radial behavior of the flatness factor at different scales as a function of the solar wind velocity regime within the inner heliosphere. The study was performed for magnetic field and velocity fluctuations as reported in the following sections.

CP679, *Solar Wind Ten: Proceedings of the Tenth International Solar Wind Conference*,
edited by M. Velli, R. Bruno, and F. Malara

TABLE 1. From left to right: time interval in dd:hh, heliocentric distance in AU, average wind velocity in km/sec

time interval	radial distance	< V >
46:00–48:00	0.90	433
49:12–51:12	0.88	643
72:00–74:00	0.69	412
75:12–77:12	0.65	630
99:12–101:12	0.34	405
105:12–107:12	0.29	729

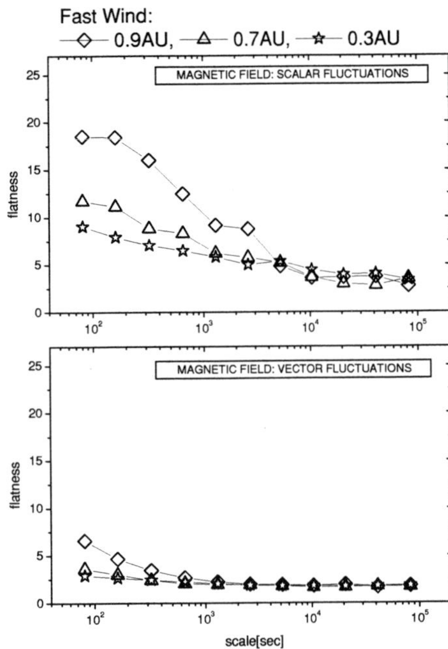

FIGURE 1. Flatness factor F for scalar and vector magnetic field differences as a function of time scale expressed in seconds are shown at the top and bottom panels, respectively.The three different symbols refer to three different heliocentric distances as illustrated at the top of the Figure. Time intervals refer to fast wind.

DATA ANALYSIS

For this analysis we used 81 sec averages of magnetic field and plasma data recorded by Helios 2 s/c during its first solar mission in 1976. Helios orbit was such that allowed us to observe the same corotating fast stream at different heliocentric distances during consecutive solar rotations. Thus, in the hypothesis of stationary solar wind source conditions, we can estimate possible radial gradients of physical parameters. Together with these fast streams we also included in our analysis the slow wind interval preceding each stream having care of avoiding any stream interface. Time intervals, heliocentric distances and solar wind speed are reported in Table 1. One of the characteristics of these fast streams is that they are notorious for being dominated by strong Alfvénic fluctuations and offer a unique opportunity to observe the radial evolution of MHD turbulence within the inner heliosphere [13].

We computed, scale by scale, the following flatness estimator

$$F(\tau) = \frac{<\xi^4(\tau)>}{<\xi^2(\tau)>^2} \tag{1}$$

where τ is the scale of interest and $\xi^p(\tau) = <|V(t+\tau) - V(t)|^p>$ is the Structure Function of order p of the generic function $V(t)$.

The flatness factor can be studied for both scalar and vector fluctuations (also called compressive and directional fluctuations, respectively) of velocity and magnetic field vectors. As a matter of fact, a fluctuating vector field encompasses two distinct contributions, a compressive one that modifies the strength, or intensity, of the fluctuations and can be expressed as $\delta|\vec{B}(t,\tau)| = |\vec{B}(t+\tau)| - |\vec{B}(t)|$ and a directional contribution that acts on the vector orientation and can be expressed as $\delta\vec{B}(t,\tau) = \sqrt{\sum_{i=x,y,z}(B_i(t+\tau)-B_i(t))^2}$. Obviously, $\delta\vec{B}(t,\tau)$ takes into account also compressive contributions and the expression $\delta\vec{B}(t,\tau) \geq |\delta|\vec{B}(t,\tau)||$ is always true. In the section that follows we will describe the radial behavior of the flatness factor F for both scalar and vector fluctuations, for both velocity and magnetic field and for both fast and slow wind. From the comparison of the radial behavior of these two quantities we can infer useful hints that can help us to better interpret the radial evolution of the intermittency observed in the solar wind MHD turbulence.

Radial dependence of magnetic field and velocity intermittency

Flatness factor F for both scalar and vector differences are shown in Figure 1 for magnetic field as a function of time scale expressed in seconds. The three different curves in each panel, which refer to fast wind observed at the three heliocentric distances we chose, show that flatness increases with distance for both kinds of fluctuations. In particular, the smallest scales are the least Gaussian while, the largest scales are rather Gaussian (flatness close to 3) and not influenced by the radial excursion. In addition, since F increases at small scales more slowly for directional fluctuations than for scalar fluctuations, these last ones can be considered more intermittent. It is interesting to compare this behavior to that shown within slow wind as illustrated in Figure 2. The most striking

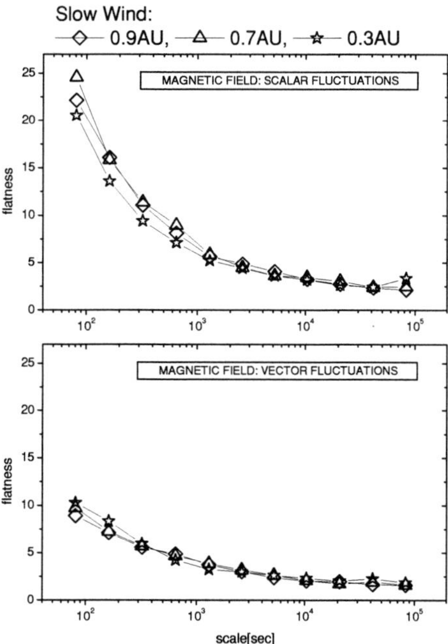

FIGURE 2. Flatness factor F for scalar and vector magnetic field differences as a function of time scale are shown in the top and bottom panels in the same format used for Figure 1.Time intervals refer to slow wind.

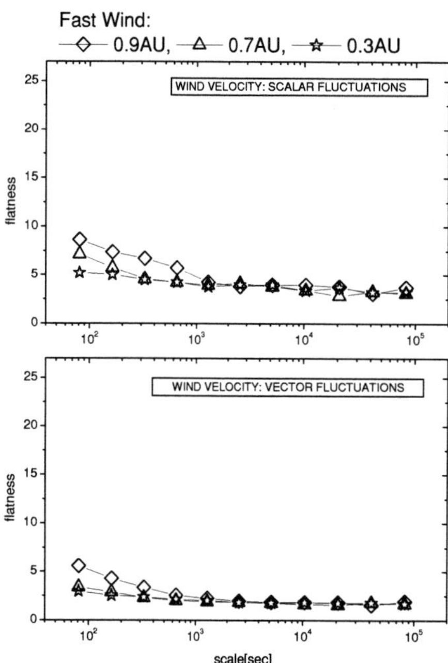

FIGURE 3. Flatness factor F for scalar and vector velocity fluctuations as a function of time scale expressed in seconds are shown at the top and bottom panels, respectively. The three different symbols refer to three different heliocentric distances as illustrated at the top of the Figure. Time intervals refer to fast wind.

feature is that intermittency does not evolve with distance either for scalar or for vector fluctuations. Moreover, at small scales, F is generally higher than in fast wind, suggesting a higher intermittency level. In particular, F for vector fluctuations starts to increse at much larger scales than in the fast wind, suggesting that intermittency is already present at hourly scales. Results relative to fast wind velocity fluctuations are shown in Figure 3 in the same format as of the previous Figures. Although the general trend shown in both panels confirms that intermittency for scalar and vector fluctuations increases with distance also for fast solar wind velocity, we easily recognize that scalar fluctuations (upper panel) are generally less intermittent than the corresponding magnetic field fluctuations (Figure 1, upper panel). On the contrary, the behavior of F for velocity vector fluctuations (lower panel) suggests that intermittency of these fluctuations is quite similar to that of magnetic field vector fluctuations (Figure 1, lower panel). This last result, as it will be discussed later on, clearly derives from the strong contribution due to the Alfvénic component of these fluctuations.

Results similar to those obtained for magnetic field observations within slow wind, are shown for velocity fluctuations in Figure 4. The three curves, for both panels, show that F increases from the largest to the smallest scales but its behavior doesn't depend on heliocentric distance given that these curves intermingle with one another especially at the smallest scales where the existence of a radial trend is expected to be more clear. However, this complicated behavior might also be due to a poor stationarity at the source regions of the wind itself. However, also in this case, especially for scalar fluctuations, wind velocity confirms to be less intermittent than magnetic field.

SUMMARY AND DISCUSSION

We have studied the radial evolution of intermittency associated with magnetic field and wind velocity fluctuations between 0.3 and 1 AU on the ecliptic plane for time scales ranging between 81 sec and 24 hours . Since one of the main effects of intermittency is that of increasing the value of the flatness factor of the PDFs of the fluctuations as we look at smaller and smaller scales [6, 7, 11], we concentrated our attention on the behavior of the flatness factor at different scales, analyzing both scalar and vector fluctuations. This last point is important since intermittency of vector fluctuations contains also contributions due to stochastic fluctuations like the Alfvénic ones.

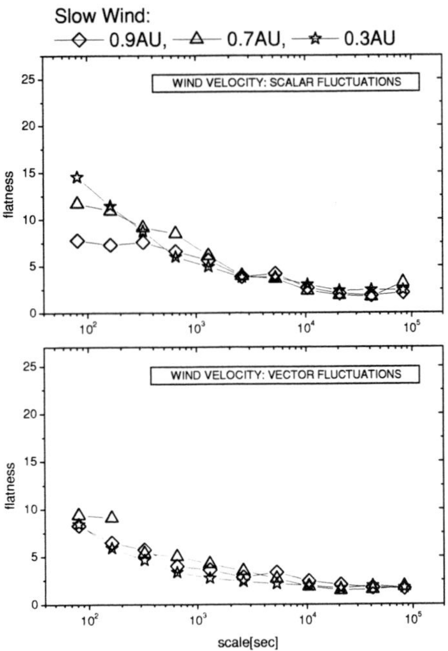

FIGURE 4. Flatness factor F for scalar and vector velocity fluctuations as a function of time scale are shown at the top and bottom panels in the same format used for Figure 3. Time intervals refer to slow wind.

Our first result shows that magnetic field scalar fluctuations are always more intermittent than the corresponding velocity fluctuations as it was already noticed [2]. This is probably due to the fact that, differently from wind velocity, magnetic field has to obey to $\nabla \cdot B = 0$ and it has to readjust its topology continuously to satisfy this relation during the wind's expansion.

We have also found that: a) generally, scalar (i.e. compressive) fluctuations are more intermittent than vector (i.e. directional) fluctuations, b) that slow wind does not show any radial evolution while c) fast wind intermittency, for both magnetic field and velocity, increases with distance. The interpretation we give of our observations derives from a view of the MHD turbulence that has already been given [1, 12] but that finds new evidence in our analysis. In other words, we can imagine interplanetary fluctuations made of two distinct components: one due to coherent, non propagating structures convected by the wind and, another one made of propagating, stochastic fluctuations, namely Alfvénic modes. While the first component tends to increase the intermittency level because of its coherent nature, the second one tends to decrease it because made of stochastic fluctuations. At 0.3 AU power associated to directional fluctuations largely exceeds that associated to compressive fluctuations and thus the corresponding intermittency of vector fluctuations is very low. However, as the wind expands, the Alfvénic contribution is depleted because of turbulent evolution [13] and, consequently, the underlying coherent structures convected by the wind, strengthen further on by stream–stream dynamical interaction, assume a more important role. Obviously, slow wind doesn't show a similar behavior because Alfvénic fluctuations have a less dominant role than within fast wind and their turbulent evolution is much slower. An alternative view [10] is that these coherent structures, contributing to increase intermittency, are locally created by parametric decay instability of large amplitude Alfvén waves. Probably both mechanisms are present at the same time and confirm that the radial evolution of the intermittency level of interplanetary fluctuations is strongly related in any case to the turbulent evolution of the spectrum of these fluctuations.

ACKNOWLEDGMENTS

We thank F. Mariani and N. F. Ness, PI's of the magnetic experiment and, H. Rosenbauer and R. Schwenn, PI's of the plasma experiment onboard Helios 1 and 2, for allowing us to use their data.

REFERENCES

1. Bruno, R. and Bavassano, B., 1993, Planet. Space Sci., 41, 677–685.
2. Bruno, R., B. Bavassano, E. Pietropaolo, V. Carbone and P. Veltri, 1999, Geophys. Res. Lett., 26, 3185–3188
3. Bruno, R., V. Carbone, P. Veltri, E. Pietropaolo and B. Bavassano, 2001, Planetary Space Sci., 49, 1201–1210.
4. Burlaga, L., 1991b, Geophys. Res. Lett. 18, 69–72.
5. Carbone, V., R. Bruno and P. Veltri, 1995, Scaling laws in the solar wind turbulence, in Lecture Notes in Physics, Ed. Springer–Verlag, 462, 153
6. Castaing, B., Gagne, Y., and Hopfinger, 1990, Physica D 46, 177–200.
7. Frisch, U., 1995, Turbulence: the legacy of A. N. Kolmogorov, Cambridge University Press
8. Kolmogorov, A. N., 1941, C. R. Akad. Sci. SSSR 30, 301.
9. Marsch, E., and Liu, S., 1993, Ann. Geophysicae 11, 227–238.
10. Primavera L., Malara F and P. Veltri, Parametric instability in the solar wind: numerical study of the nonlinear evolution., paper SIII-28, these Proceedings.
11. Sorriso–Valvo, L. , Carbone, V., Veltri, P., Consolini, G., Bruno, R., 1999, Geophys. Res. Lett. 26, 1801–1804.
12. Tu, C.-Y and Marsch, E., 1993, J. Geophys. Res. 98, 1257–1276.
13. Tu, C.-Y, Marsch, *Space Sci. Rev.*, **73**, 1-210 (1995)
14. Veltri, P., and Mangeney, A., 1999, in Solar Wind IX, edited by S. Habbal, AIP Conf. Publ., 543–546.

Numerical MHD Simulation of Flux-Rope Formed Ejecta Interaction With Bimodal Solar Wind

Wang, A. H.,* Wu, S. T.*and Tan, A.**

*Center for Space Plasma and Aeronomic Research
Department of Mechanical and Aerospace Engineering
University of Alabama in Huntsville
Huntsville, AL 35899
**Department of Physics, Alabama A & M University
Normal, AL 35762

Abstract. A theoretic numerical simulation of interaction between CME ejecta and bimodal solar wind from solar surface to 30 solar radii has been presented. A comparison with an interaction between CME ejecta with homogeneous and bimodal solar wind is given. The results show that the bimodal solar wind changes the topology of CME ejecta with faster propagation speed away from the equator, and fast wind will energize the CME. Also the results of steady-state bimodal solar wind characteristics that extend to 1AU are presented.

INTRODUCTION

Observation, especially Ulysses, showed that the solar wind consists of both fast and slow components in the solar minimum [1,2]. In this bimodal characteristics of solar wind there are an almost homogeneous high-speed flow over coronal hole and the low-speed flow over coronal streamer with the division between them at about 70° from the pole. The bimodal solar wind property is very important to study the propagation of Coronal Mass Ejection (CME) which is major large scale solar eruption. The fast solar wind at the coronal hole will affect the propagation speed and the topology of CME.

We combined our bimodal solar wind model [3] and the flux-rope model [4] to investigate the interaction between CME and bimodal solar wind. The purpose of this study is to reveal the physical mechanism that causes the changes of the speed and topology of the CME during the propagation. CMEs are also the major disturb to the Earth environment. In order to study the disturbed solar wind parameters at 1AU due to the CME propagation in a bimodal solar wind environment, we have build a steady-state bimodal solar wind model to present here. Their effects due to propagation of CME will be given elsewhere [5].

MHD MODELS AND RESULTS

In our simple solar wind model we assume an axisymmetric, time-dependent, MHD flow of a single-fluid, polytropic, and fully ionized plasma. In order to obtain bimodal solar wind model, we added volumetric heating, momentum addition, and thermal conduction. The flow is calculated in a meridional plane defined by the axis of the magnetic field. The detailed description of the governing equations can be found in [3].

In simple solar wind model the polytropic index γ=1.05. Temperature and density at lower boundary (solar surface) are uniform from equator to pole. The plasma β=1.0 at equator and 0.2 at pole. In bimodal solar wind model the polytropic index γ=5/3. The temperature and density are not uniform from equator to pole. The plasma β=1.0 at equator and 0.02 at pole. The volumetric heat source is given by

$$Q = Q_0 \frac{\rho}{\rho_0} e^{-(r-Rs)/(LRs)} \quad (1)$$

where Q_0 is 5x 10^{-8} erg $cm^{-3}s^{-1}$ and ρ_0 is the base density. A heating term similar to this was used by Hartle and Barnes [6]. The thermal conductive fluxes for a Lorentz gas is given by

$$q = \kappa_{\|} T^{5/2} (\vec{B} \cdot \nabla T) \frac{\vec{B}}{B^2} \quad (2)$$

where κ is the collisional thermal conductivity along the magnetic field lines as given by Spitzer [7]. The momentum addition is given by

$$D = \frac{D_0 a^2}{(r-a)^2 + a^2} \{1 - \arctan[5(\frac{\theta}{\Delta\theta} - 14.5)]/4\} \quad (3)$$

CP679, *Solar Wind Ten: Proceedings of the Tenth International Solar Wind Conference*,
edited by M. Velli, R. Bruno, and F. Malara

where D_0 with the value of $5x10^3$ dyn/g, D reaches its maximum value at $\boldsymbol{a}$, and θ is latitudinal angle.

The computational domain is in meridional plane from the pole to the equator in the θ direction and from the solar surface to 30Rs for the interaction of flux rope and solar wind and to 1AU for quasi-steady state Sun-Earth bimodal solar wind in the radial direction, respectively.

The symmetric boundary conditions are used at two side boundaries. The lower boundary is a physical boundary and the non-reflected characteristic boundary conditions are used. The upper boundary is a computational boundary, and since the flow at this boundary is supersonic and superAlfvenic the linear extrapolation is used. The detailed description of the boundary conditions are given by [4].

Fig. 1 shows the distributions of the magnetic field, density and velocity of the initial steady state for the simulation of the interaction with a simple and a bimodal solar wind, respectively. The contrast of the density and velocity from the pole to the equator is larger in the bimodal solar wind than that in the simple solar wind. Also the current sheet is thinner in the bimodal solar wind than in the simple solar wind.

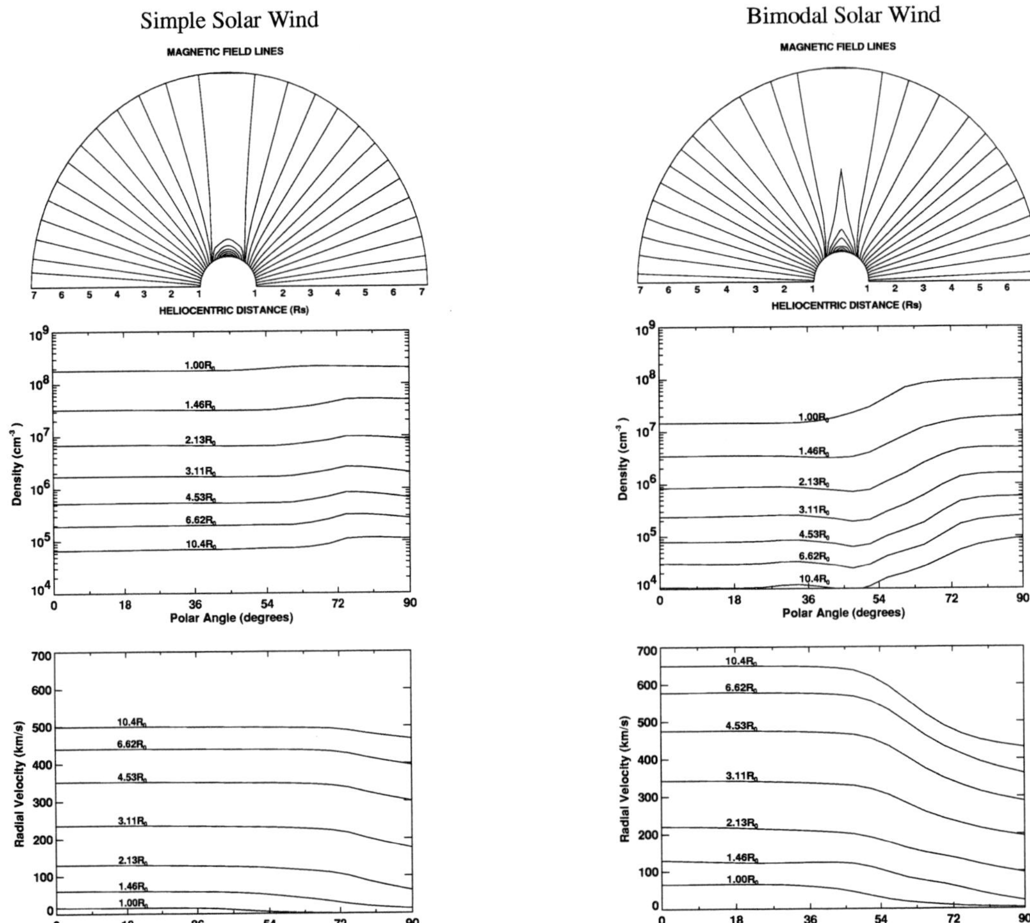

FIGURE 1. Initial field lines, density and velocity distributions for simple and bimodal solar wind are showed.

CME-SOLAR WIND INTERACTION

To show CME and solar wind interaction a flux-rope emerged from the solar surface centered at the equator. The flux rope's radius is 0.5Rs. The center thermal pressure is 50 times larger than its edge. The plasma beta at the edge is 0.1 and its emerging speed from the solar surface is 50 km/s. Fig. 2 shows the simulation results at 12 hours. In this figure the evolutionary magnetic field lines, velocity and the density enhancement for both cases, i.e. simple solar wind and bimodal solar wind, are depicted. From this figure the large distortion of the flux rope after interacting with fast speed flow in the bimodal solar wind at open field line region could be seen. The side edges of the flux rope move faster than its center because the fast speed solar wind in open field region which carries the edge of the flux rope with them.

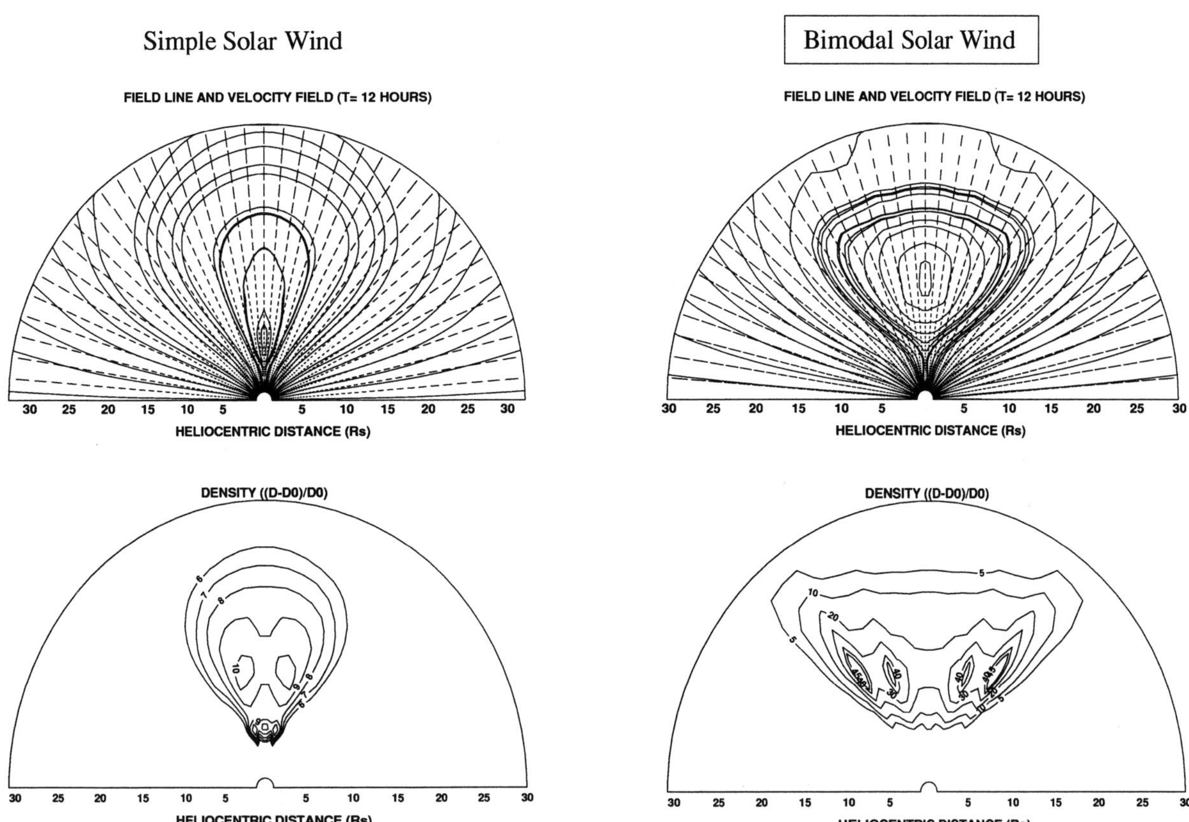

FIGURE 2. Upper panels are magnetic field and velocity at 12 hours, and lower panels are density enhancement.

R-T Diagram of Single Flux Rope's Front and Center

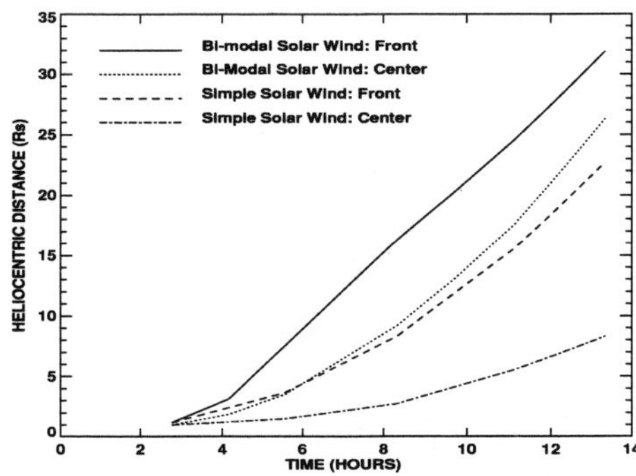

FIGURE 3. Distance vs time of flux rope front and center

Further, Fig. 3 show the distance-time curve for the edge of the flux-rope. It is clearly indicated that the solar wind has energized the propagation of the flux-rope. As expected this effect will modify the CME induced shock strength which related to particle acceleration.

BIMODAL SOLAR WIND FROM SOLAR SURFACE TO 1AU

To extent the steady-state bimodal solar wind to 1AU we have to choose the following parameters of a and D_0 in momentum addition of equation (3). These two parameters are not constants and are function of latitude. The specific equations for them are given as follows:

$$a(\theta) = a_0 - 0.36\arctan(5(\frac{\theta}{\Delta\theta} - 14.5)) \qquad (4)$$

and

$$D_0(\theta) = D_0'(1 - \arctan(5(\frac{\theta}{\Delta\theta} - 14.5))/3.5) \qquad (5)$$

where D_0' has the same value of D_0 in the equation (3).

The heating length L in equation (1) is redefined as follows:

$$L = 10 \qquad r \le 20Rs$$
$$L = 12.5 \qquad r > 20Rs$$

Some steady-state results are shown in Fig. 4 and Fig. 5.

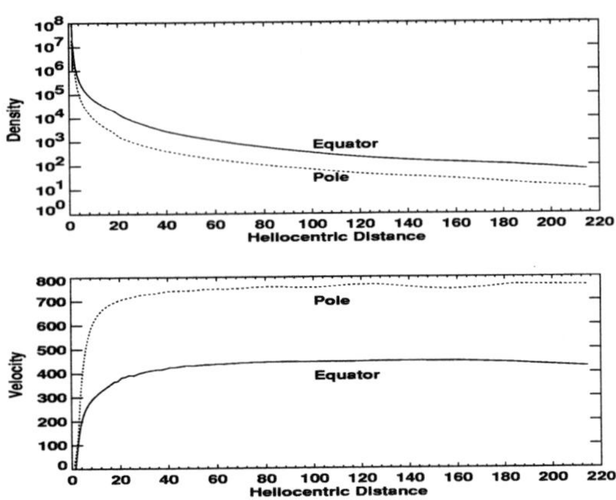

FIGURE 4. Densities and velocities vs heliocentric distance at the pole and the equator are showed.

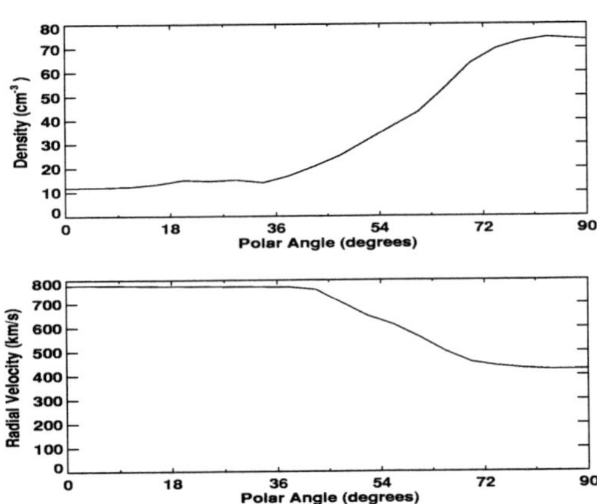

FIGURE 5. Densities and velocities vs latitude at 1AU are showed.

By examining Figs. 4 and 5, the solar wind characteristics of this study are : at the solar surface n (at equator) = $10^8/cm^3$ and n (at pole) = $2x10^7/cm^3$; v_r (at equator) ≅ 0 and v_r (at pole) = 25km/s. At 1AU, n (at equator) = $73/cm^3$ and n (at pole) = $12/cm^3$; v_r (at equator) = 420km/s and v_r (at pole) = 790km/s. Comparing with the average observational data the number density at equator and pole in our bimodal solar wind model are 2-3 times higher, but the velocity is matched well.

SUMMARIES AND CONCLUSIONS

Theoretic numerical simulations for interaction of CME with simple solar wind or bimodal solar wind have been presented up to inner heliosphere in this study. Also the steady-state bimodal solar wind results are showed up to 1AU. The purpose is to lay the ground to study the disturbed solar wind properties due to the CME propagating to the Earth's environment. We have the following conclusions:

1. Fast speed solar wind will energize the CME.
2. The shape and speed of the flux rope will change during the propagation because the interaction with bimodal solar wind.
3. The forms of heating and momentum addition functions are important to obtain realistic Sun-Earth bimodal solar wind.

ACKNOWLEDGMENTS

This research work performed by AHW and STW is supported by NSF grant (ATM0070385) and AFOSR grant (F49620-00-0-0304). AHW and AT also are supported by NASA grant (NAG5-10202).

REFERENCES

1. Phillips, J. L., et al., *Geophys. Res. Lett.,* **22,** 3301 (1995).
2. Grall, R. R., Cole, M. T. Klinglesmith, A. R. Breen, P. J. S. Williams, J. Markkanen, and R. Esser, Nature, 379, 429 (1996).
3. Wang, A. H., Wu, S. T., Suess, S. T., Poletto, G., J. Geophys. Res., 103, 1913 (1998).
4. Wu, S. T. and Guo, W. P., Coronal Mass Ejections, in Geophysical Monograph 99 (AGU Press, Washington DC), edited by N. Crooker, J. Joslyn, and J. Feynman, 1997, pp. 83-89.
5. Wang, A. H. and Wu, S. T., Solar Physics (to be submitted).
6. Hartle, R. E. and Barnes, A., J. Geophys. Res., 75, 6915 (1970).
7. Spitzer, L., Physics of Fully Ionized Gases. 2nd rev., edited by John Wiley, New York, 1962.

Kinetics of Electrons in the Corona and Solar Wind

C. Vocks* and G. Mann*

*Astrophysikalisches Institut Potsdam, An der Sternwarte 16, 14482 Potsdam, Germany

Abstract. The velocity distribution functions (VDFs) of electrons as measured in the solar wind show pronounced deviations from a Maxwellian. They seem to be composed of a thermal core and energetic tails, called halo. These VDFs can be fitted very well by kappa distributions. The formation of the energetic tails in the corona or in the solar wind is investigated. The relaxation of a kappa distribution under the influence of Coulomb collisions in the coronal plasma is calculated. This allows an estimation if the energetic tails of the VDFs can be formed in the corona. Resonant interaction between the electrons and electron cyclotron waves is suggested as a mechanism for the generation of the energetic tails. A kinetic model for electrons is presented. Coulomb collisions and wave-particle interactions are considered. With this model, electron VDFs can be calculated from the transition region up into the solar wind.

INTRODUCTION

Observations of solar wind electron velocity distribution functions (VDFs) reveal non-Maxwellian high-energy tails [1]. The VDFs are often described as core, halo and superhalo populations. They can also be fitted by κ - distributions, especially at higher energies of several keV [2]. Thus, at higher energies much more electrons are present than expected from a simple Maxwellian setup.

The question arises how these high-energy tails of the electron VDFs are formed. In [3] exospheric model results are presented that show a core and halo population, but no tails at higher energies. However, these models do not include an energy source for the formation of high-energy tails.

In this paper, resonant interaction between electrons and plasma waves is discussed as a possible candidate for the generation of the high-energy tails.

CORONAL ORIGIN OF HIGH-ENERGY TAILS?

The high-energy tails of the electron VDFs are measured in-situ in the solar wind. The question arises whether a coronal origin of these energetic electrons is possible, or if they have to be accelerated in the interplanetary space.

A coronal origin of the high-energy tails requires that these electrons have relaxation times due to Coulomb collisions in the corona, that are larger than the time for the passage through a characteristic length like the coronal pressure scale height of $5 \cdot 10^4$ km.

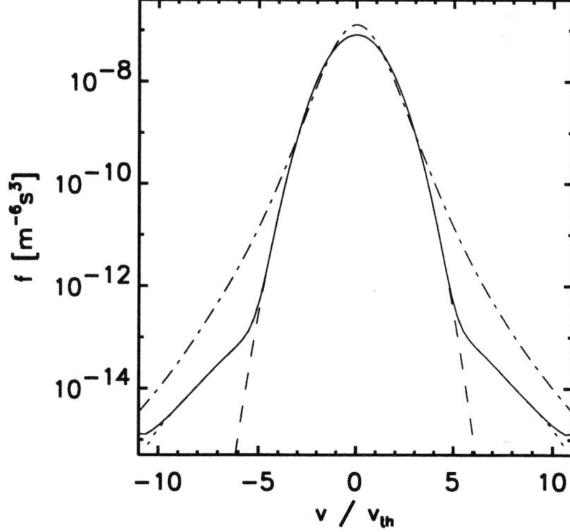

FIGURE 1. Electron VDF in $v_{\|}$ (solid line) and $v_{\perp}$ (dotted line) after a simulation time of 5 s. Shown are also a Maxwellian (dashed line) and the initial κ distribution (dash-dotted line) with $\kappa = 10$.

Numerical simulation of relaxation process

To assess the relaxation process due to Coulomb collisions in the corona, a numerical simulation of a homogenous, force-free proton-electron plasma is performed. With a particle density $N = 10^{14}\,\mathrm{m}^{-3}$ and a temperature $T = 10^6$ K, typical coronal conditions are chosen. The model includes no heating process. The proton distribution is set to a Maxwellian one. The initial condition for the electrons is an isotropic κ - distribution with $\kappa = 10$.

CP679, *Solar Wind Ten: Proceedings of the Tenth International Solar Wind Conference*,
edited by M. Velli, R. Bruno, and F. Malara

Figure 1 shows the model electron VDF after a simulation time of 5 s. For low speeds, $v < 5v_{th}$, the distribution function has already relaxed to a Maxwellian. But at higher speeds, $v > 5v_{th}$, the high-energy tails of the original κ - distribution are still present.

This result is due to the v^{-3} dependence of the Coulomb collision frequency. Consequently, for low speeds the relaxation process is much faster than for higher speeds.

At a coronal temperature of $T = 10^6$K, the electron thermal velocity is $v_{th} = 3893$km/s, thus for $v = 5v_{th}$ the transit time through one pressure scale height ($5 \cdot 10^4$km) is 2.6 s. This is just half of the simulation time needed for the results displayed in figure 1.

Conclusions from the simulation

Since electrons with higher speeds, $v > 5v_{th}$, can leave the corona on a time scale that is smaller than the time needed to deform their VDF significantly due to Coulomb collisions, it is concluded that they can preserve non-Maxwellian VDFs while they are propagating from the corona towards the solar wind.

Thus, it is possible that high-energy tails of the electron VDFs as they are measured in the interplanetary space can be formed already in the solar corona.

QUASILINEAR WAVE-ELECTRON INTERACTION

In this paper, resonant interaction between electrons and electron cyclotron waves is discussed as a possible mechanism for the generation of high-energy tails of the electron VDF in the corona.

The wave-particle interaction is described within the framework of quasilinear theory [4]. Only waves propagating anti-sunward and parallel to the background magnetic field are considered.

Since electron cyclotron waves are restricted to frequencies ω less than the electron cyclotron frequency, Ω_e, if follows from the resonance condition

$$\omega - kv_{\parallel} - \Omega_e = 0 \qquad (1)$$

that only sunward moving electrons, $v_{\parallel} < 0$, can interact resonantly with the outward propagating waves.

Maximum effect of the waves

The resonant interaction of the electrons with the waves leads to pitch-angle diffusion in the reference frame of the waves. If the wave-particle interaction is the dominant process in the plasma, this leads to the formation of "shells" with no pitch angle gradients of the VDF [5].

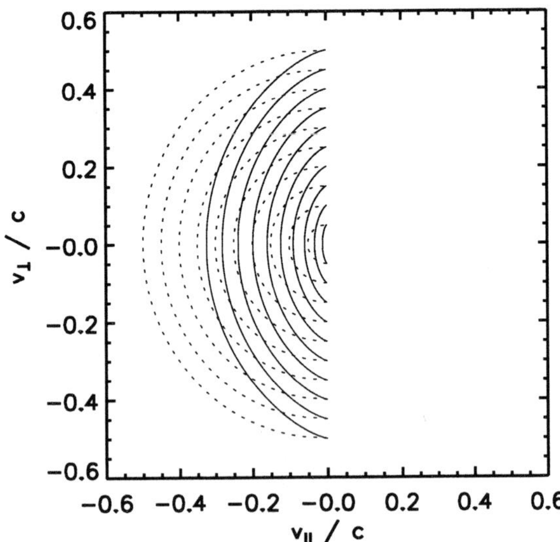

FIGURE 2. Kinetic shells for electron-wave interaction in a plasma with $\omega_p/\Omega_e = 2$ (solid lines) and isocontours of a Maxwellian distribution (dotted lines)

Figure 2 displays the shells calculated under consideration of the dispersional relation of the electron cyclotron waves [6]. The ratio between the plasma frequency, ω_p, and the electron cyclotron frequency, Ω_e, is set to a value of 2, that is typical for the lower corona.

These shells indicate a strong temperature anisotropy with $T_{\perp} \gg T_{\parallel}$. This is characteristic for a heating process due to resonant interaction with cyclotron waves.

Generation of high energetic electrons

As an example, we consider an electron with $v_{\perp} = 0$ and $v_{\parallel} = 0.08\,c$, i.e. $6.16\,v_{th}$ at $T = 10^6$K. According to figure 2, it can reach the position ($v_{\parallel} = 0, v_{\perp} = 0.2\,c$) due to quasilinear diffusion along its resonance shell. This corresponds to an energy of 10 keV.

Of course, the production of high-energy electrons due to quasilinear wave-particle interaction can work only if a sufficient supply of wave energy is present.

Resonant interaction between ions and ion cyclotron waves is discussed as a heating mechanism for the solar corona ([7], [8], [9]), and ion distributions measured in the solar wind show signatures of interaction with cyclotron waves ([10]).

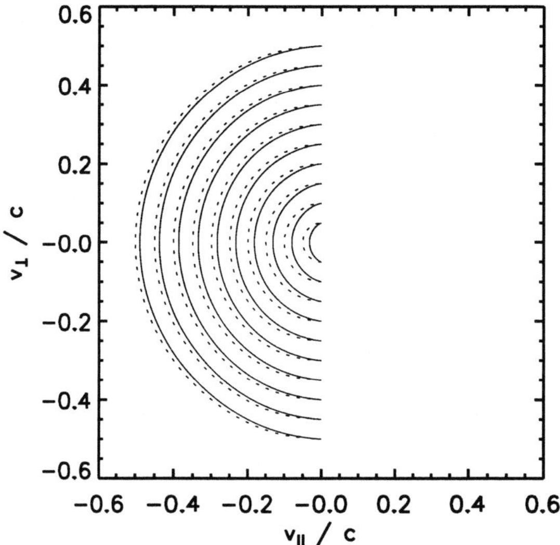

FIGURE 3. Kinetic shells for electron-wave interaction in a plasma with $\omega_p/\Omega_e = 20$ (solid lines) and isocontours of a Maxwellian distribution (dotted lines)

So wave energy is likely to be present at ion cyclotron frequencies. Since the electron cyclotron frequency is considerably higher, much less spectral wave energy is expect to be found there. For a power law spectrum with ω^{-1}, the ratio of spectral wave energies at the electron and proton cyclotron frequency would correspond to the mass ratio, m_e/m_p.

However, the bulk of the electrons is not to be heated as the ions are in the corona. From the resonance condition, eq. (1), it follows that electrons with higher energies interact with waves with lower frequencies. As the waves propagate away from the sun, their frequencies increase in units of the local electron cyclotron frequency. Thus, at a certain height a wave will interact with the most energetic electrons. At larger heights, it can interact with less energetic electrons, if it hasn't been absorbed at lower heights. Thus, the most energetic electrons experience a preferred acceleration due to electron cyclotron resonance.

The above example deals with electrons with an initial speed of 6 thermal speeds. In comparision to the total number density of both ions and electrons, only a minor number of the electrons has such an initial speed. Thus, it is reasonable to assume that the wave energy is fully sufficient to have a significant effect on them.

Figure 3 shows the resonance shells for a plasma with $\omega_p/\Omega_e = 20$. This higher ratio is typical for the solar wind. The shells hardly differ from the isocontours of a Maxwellian distribution, thus indicating only a weak energetization of electrons.

From the results in this section it is concluded that resonant wave-electron interaction as described by quasilinear theory is a possible mechanism of accelerating electrons to energies of the order 10keV. This mechanism is restricted to the corona, since the parameter ω_p/Ω_e increases towards the solar wind, thus leading to a less efficient acceleration of the electrons. The numerical simulation in the previous section shows that the energetic electrons produced in the corona can leave the corona towards the interplanetary space without being relaxed by Coulomb collisions with ions or other electrons.

KINETIC MODEL OF ELECTRON-WAVE INTERACTION

To study the evolution of the electron VDF from the corona into the solar wind a kinetic model has to be developed. The model is based on the kinetic model for ions in the solar corona of [11], but here two velocity coordinates $(v_{\|}, v_{\perp})$ are considered. Gyrotropy is still assumed. This reduces the number of spatial coordinates to one, s, along the background magnetic field.

The Coulomb collisions are calculated using the Landau collision integral (e.g. [12]). For the wave-particle interaction, the quasilinear theory is employed. The Boltzmann-Vlasov equation for the VDF $f(s, v_{\|}, v_{\perp})$ reads:

$$\frac{\partial f}{\partial t} + v_{\|} \cdot \frac{\partial f}{\partial s} + \left(g_{\|} - \frac{e}{m_e} E_{\|}\right) \cdot \frac{\partial f}{\partial v_{\|}} +$$

$$\frac{v_{\perp}}{2A} \frac{\partial A}{\partial s} \left(v_{\perp} \cdot \frac{\partial f}{\partial v_{\|}} - v_{\|} \cdot \frac{\partial f}{\partial v_{\perp}}\right) = \left(\frac{\delta f}{\delta t}\right)_{Coul.} + \left(\frac{\delta f}{\delta t}\right)_{w.-p.} \quad (2)$$

Here, $g_{\|}$ and $E_{\|}$ are the components of the graviational and electric field parallel to the background magnetic field, and $A(s)$ is the cross sectional area of the magnetic flux tube under consideration.

A simple fluid model yields densities, drift velocities and temperatures for a proton-electron plasma as background conditions. This background is constant in time, and the electron VDF represents a test particle population. The background is also used to provide Maxwellians as initial condition for the electron VDFs. From this initial condition, the temporal evolution of the VDF is calculated using the Boltzmann-Vlasov equation (2).

The simulation box extends from the low corona up to 0.67 AU into the interplanetary space. The velocity coordinates cover electron energies up to 100keV. In the corona, the magnetic field decreases with height according to the geometry of a coronal funnel as modelled by [13]. At larger heights, it decreases radially.

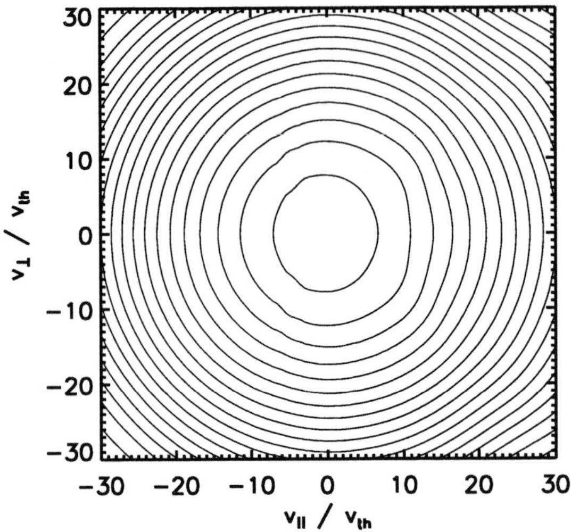

FIGURE 4. Isolines of the electron VDF at a height of $s = 0.68R_\odot$

The plasma waves are assumed to be emitted from the sun, i.e. only outward propagating waves are considered. They enter the simulation box at the lower boundary with a given power law spectrum proportional to ω^{-1}. The evolution of the spectrum within the simulation box is within the scope of the model, and energy conservation between waves and electrons is guaranteed.

Kinetic results

Figure 4 shows the electron VDF after a simulation time of 100s. The initial Maxwellian distribution has developed an asymmetry between $v_\parallel < 0$ and $v_\parallel > 0$ and a temperature anisotropy with $T_\perp > T_\parallel$.

It follows from the resonance condition, eq. (1), that only electrons with $v_\parallel < 0$ can interact with the waves propagating away from the sun. Indeed, the isolines of the electron VDF show the same deformation and anisotropy as in figure 2. Thus, they are in coincidence with the theoretical expectations for a plasma in which quasilinear wave-particle interaction plays a major role.

The mirror force in the opening magnetic field geometry of the coronal funnel and in the solar wind has the tendency to bring particles with high $v_\perp$ towards high positive $v_\parallel$. Since it scales with $v_\perp^2$, see eq. (2), electrons with higher energies are pushed more strongly towards the anti-sunward side.

This explains the deformation of the electron VDF on the anti-sunward side in figure 4. Electrons accelerated on the sunward side due to wave-particle interaction cross the line $v_\parallel = 0$ and are just being focused towards high $v_\parallel$ and low $v_\perp$ by the mirror force.

CONCLUSIONS

In this paper, a mechanism for generating high-energetic electrons in the solar corona with resonant wave-particle interaction is presented. It is demonstrated that the waves are capable of accelerating electrons from a few thermal speeds up to energies of the order 10keV. The demand on wave energy is not too high, since it is not necessary to heat the bulk of the electron VDF. So sufficient wave energy is likely to be present in the corona.

This process is restricted to the corona due to the increase of ω_p/Ω_e from the corona towards interplanetary space. However, the assessment in the first section shows that electrons with only 5 thermal speeds can leave the corona towards the solar wind without significant changes of their VDFs. Electrons with higher energies will even more easily propagate from the corona into the interplanetary space.

The numerical results presented in the last section show that the electron VDF is influenced by the waves in the way expected from quasilinear theory.

Thus, it is concluded that interaction between electrons and plasma waves in the solar corona is a promising candidate for the generation of the high-energy tails of the electron VDFs that are observed in the solar wind.

REFERENCES

1. Lin, R. P., *Space Science Rev.*, **86**, 61–78 (1998).
2. Maksimovic, M., Pierrard, V., and Lemaire, J. F., *Astron. Astrophys.*, **324**, 725–734 (1997).
3. Lie-Svendsen, Ø., Hansteen, V. H., and Leer, E., *J. Geophys. Res.*, **102**, 4701–4718 (1997).
4. Kennel, C. F., and Engelmann, F., *Phys. Fluids*, **9**, 2377 (1966).
5. Isenberg, P. A., Lee, M. A., and Hollweg, J. V., *J. Geophys. Res.*, **106**, 5649–5660 (2001).
6. Isenberg, P. A., and Lee, M. A., *J. Geophys. Res.*, **101**, 11055–11066 (1996).
7. Marsch, E., and Tu, C.-Y., *J. Geophys. Res.*, **106**, 227–238 (2001).
8. Vocks, C., and Marsch, E., *Geophys. Res. Lett.*, **28**, 1917–1920 (2001).
9. Vocks, C., and Marsch, E., *Astrophys. J.*, **568**, 1030–1042 (2002).
10. Marsch, E., and Tu, C.-Y., *J. Geophys. Res.*, **106**, 8357–8361 (2001).
11. Vocks, C., *Astrophys. J.*, **568**, 1017–1029 (2002).
12. Ljepojevic, N. N., and Burgess, A., *Proc. R. Soc. Lond. A*, **428**, 71–111 (1990).
13. Hackenberg, P., Mann, G., and Marsch, E., *Space Science Rev.*, **87**, 207–210 (1999).

Fine Structure of the Solar Wind Turbulence Inferred from Simultaneous Radio Occultation Observations at Widely-Spaced Ground Stations

M.K. Bird*, P. Janardhan†, A.I. Efimov, L.N. Samoznaev, V.E. Andreev**, I.V. Chashei‡ and P. Edenhofer, D. Plettemeier, R. Wohlmuth§

**Radioastronomisches Institut, Universität Bonn, 53121 Bonn, Germany*
†*Astronomy & Astrophysics Division, Physical Research Laboratory, Ahmedabad, 380 009, India*
***Inst. for Radio Engineering & Electronics, Russian Academy of Science, Moscow, 101999, Russia*
‡*Lebedev Physical Institute, Russian Academy of Science, Moscow, 117924, Russia*
§*Institut für HF-Technik, Universität Bochum, 44780 Bochum, Germany*

Abstract. Coronal radio sounding experiments with the Ulysses spacecraft at superior conjunction provided numerous opportunities for simultaneous observations of the downlink signals at two widely-spaced ground stations. In some instances the duration of these observations extended for up to four hours, thereby allowing studies of solar wind turbulence dynamics at spatial scales comparable with the corona-projected distance between ground stations (a few thousand km). The frequency and phase fluctuations produced by electron density inhomogeneities are normally quite well correlated on these scales. The spectral index of the temporal frequency fluctuation spectra varied over a wide range during the observations. The cross-correlation coefficient reached maximal values (≈ 0.5) when the spectral index was high (≈ 1), but no correlation could be detected when the spectral index became small (< 0.4). Similar behavior in many of the data sets implies that this is a common, if not permanent, feature of the solar wind. Possible reasons for the fluctuation decorrelation are analysed. The decorrelation at heliocentric distances $\approx 10\ R_\odot$ most likely results from continual deformation of the solar wind density irregularities during their motion across the radio ray paths.

INTRODUCTION

Coronal sounding observations of occulted radio signal parameters such as fluctuations of amplitude, phase, frequency or Faraday rotation at two or more separated ground stations is an effective method to study solar wind motion, especially in those regions inaccessible to *in situ* measurements. A general assumption underlying these observations is that the modulating irregularities are convected with a speed approximately equal to the bulk velocity of the solar wind. Strictly speaking, this assumption is valid only at large heliocentric distances where the solar wind is superalfvenic and supersonic. For sufficiently accurate measurements of correlation time delay, the spacing between the ray paths to each ground station should not be much less than the typical spatial scale of the measured fluctuations. This is the diffraction (Fresnel) scale ($\simeq 100$ km) in the case of amplitude (intensity) fluctuations (Armstrong and Coles, 1972; Coles and Kaufman, 1978), and the dominant energy containing scale in the electron density spatial power spectrum in the case of the other fluctuating parameters. On the other hand, the spacing between antennas should not be too much larger than the typical scale of the irregularities in order to avoid degradation of the cross correlation.

In contrast to phase or Faraday rotation fluctuations, frequency fluctuation observations produce a flatter temporal power spectrum (Armand et al., 1987; Wohlmuth et al., 2001) that enables use of smaller spacings between ray paths and, consequently, between ground-based antennas. Another important aspect of spaced observations is the length of the fluctuation record. Whereas a long observation time is needed to reduce statistical errors, the actually attainable interval length is always limited under real experimental conditions. Some typical features of spaced frequency fluctuation observations are studied in this paper using measurements obtained as part of the Ulysses Solar Corona Experiment (SCE) during that spacecraft's solar conjunctions at the large tracking antennas of the NASA Deep Space Network (DSN). Results similar to those presented below were found in other two-station observations with the Galileo spacecraft. Emphasis is placed here on the dependence of the cross correlation on the spatial spectrum of density irregulatities at heliocentric (solar offset) distances near $10\ R_\odot$. Earlier indications for the existence of such a de-

CP679, *Solar Wind Ten: Proceedings of the Tenth International Solar Wind Conference*, edited by M. Velli, R. Bruno, and F. Malara

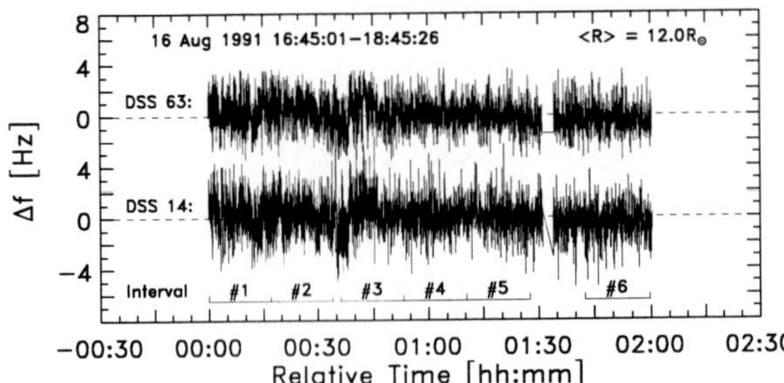

FIGURE 1. *Ulysses frequency residual records on 16 August 1991 at the ground stations Madrid (DSS 63) and Goldstone (DSS 14), divided into 6 subintervals.*

pendence were reported by Armand and Efimov (1984).

OBSERVATIONS AND DATA PROCESSING

Records of the Ulysses (Doppler) frequency residuals (carrier frequency: 2.295 GHz) at the two DSN ground stations Madrid (DSS 63) and Goldstone (DSS 14) are presented in Fig. 1. The frequency residuals were recorded at a sampling rate of 1 s^{-1}. The mean heliocentric distance and heliographic latitude of the solar proximate points along the ray paths were $\langle R \rangle = 12.0\ R_\odot$ and $\langle \phi \rangle = 18.7°$, respectively. The ground baseline projection in the corona for the combination Madrid/Goldstone is closer to the radial direction than the projections of the other possible ground station pairs.

The records for the two ground stations in Fig. 1 are qualitatively very similar. The two-hour records shown in Fig. 1 were divided into 6 subintervals, each of 1024 s duration. Temporal power spectra and cross-correlation functions were calculated for each interval and the results are summarized in Table 1.

The following are listed in Table 1 from top to bottom:

- UT start and end time for the intervals 1-6
- heliocentric distance of ray path proximate point: R
- coronal separation of ray path proximate points: ΔS
- the radial projection of $\Delta\vec{S}$: ΔR
- RMS frequency fluctuations at DSS 14: σ_{f14}
- RMS frequency fluctuations at DSS 63: σ_{f63}
- spectral power exponent at DSS 14: α_{f14}
- spectral power exponent at DSS 63: α_{f63}
- $\langle\alpha_f\rangle = (\alpha_{f14} + \alpha_{f63})/2$

The last rows of Table 1 contain the following quantities for the raw data (effective filtration time T_1 = 1 s) and filtered data (T_2 = 13 s):

- maximum cross-correlation time lag: $\tau_{\max}$
- maximum value of the temporal cross-correlation function: $K_{\max} = K(\tau_{\max})$
- estimated solar wind speed: $v_c = \Delta R/\tau_{\max}$.

A comparison of the data measured separately at DSS 14 and DSS 63 demonstrates their close similarity for both stations. The values of σ_f differ by no more than 15%, and the values of α_f differ by no more than 25% for all intervals listed in Table 1. These variations are only marginally higher than the formal errors (~10%) associated with the determination of these quantities. At the same time, the spectral parameters vary considerably from one interval to another, i.e., the typical scale of temporal variations is of the order of 20 minutes.

The strength of the cross correlation was found to be particularly variable. The maximum cross-correlation coefficient $K_{\max}$ for the raw data (T_1 = 1 s) varies from the rather high value of 0.42 for Interval 3 down to indistinguishable values for the Intervals 4-6. In order to increase the accuracy of the time lag derived from the cross-correlation functions (see Table 1), the records of Fig. 1 were passed through a high-frequency (low-pass) filter. The cross-correlation data before (effective filtration time $T_1 = 1s$) and after the filtration procedure ($T_2 = 13s$) can be compared in Table 1. Removal of the high-frequency fluctuations produces a considerable increase in the cross-correlation level as well as the appearance of a detectable time lag near the expected numerical value.

CROSS CORRELATIONS BETWEEN SPACED FREQUENCY FLUCTUATIONS

The increase in the cross-correlation level following the filtration procedure is shown graphically in Figs. 2-4.

The data of Table 1 also indicate that the maximum cross correlation of the frequency fluctuations $K_{\max}$ increases with the index of the fluctuation power spectrum $\langle\alpha_f\rangle$. This approximately linear dependence is shown in Fig. 5 for both filtered and unfiltered data.

In order to explain the behavior in Fig. 5 qualitatively, we briefly consider possible causes of decorrelation between temporal fluctuations measured simultaneously at spaced sites. The principle candidates relevant to our observations are the following:

- Motion of the irregularities transverse to the coronal projection of the ground baseline
- Solar wind velocity spread connected with the presence of streams with different speeds in the modulated propagation medium (Chashei et al., 2000);

TABLE 1. Results of frequency fluctuation measurements at Goldstone (DSS 14) and Madrid (DSS 63)

Interval	#1	#2	#3	#4	#5	#6
UT	16:45-17:02	17:02-17:19	17:21-17:38	17:38-17:55	17:55-18:12	18:27-18:44
R [$R_\odot$]	12.10	12.07	12.04	12.01	11.98	11.92
ΔS [km]	6555	6447	6295	6143	5981	5471
ΔR [km]	6533	6415	6249	6083	5905	5344
σ_{f14} [Hz]	1.28	1.06	1.02	0.90	0.96	1.02
σ_{f63} [Hz]	1.19	1.02	1.13	0.92	0.81	0.93
α_{f14}	0.58	0.57	1.06	0.37	0.30	0.25
α_{f63}	0.66	0.43	0.97	0.36	0.39	0.28
$\langle\alpha_f\rangle$	0.62±0.07	0.50±0.06	1.02±0.08	0.37	0.35	0.26
unfiltered data [T_1 = 1 s]						
$\tau_{\max 1}$ [s]	-18.8±2.2	-14.1±1.4	-14.3±1.3	-	-	-
$K_{\max 1}$	0.21±0.02	0.15±0.09	0.42±0.09	-	0.06	-
v_{c1} [km s^{-1}]	348±41	455±45	437±40	-	-	-
filtered data [T_2 = 13 s]						
$\tau_{\max 2}$ [s]	-15.5±1.1	-15.2±0.8	-16.1±0.7	-14.3	-12.4	-
$K_{\max 2}$	0.63±0.03	0.60±0.09	0.84±0.10	0.33	0.41	-
v_{c2} [km s^{-1}]	421±30	422±22	388±17	425	476	-

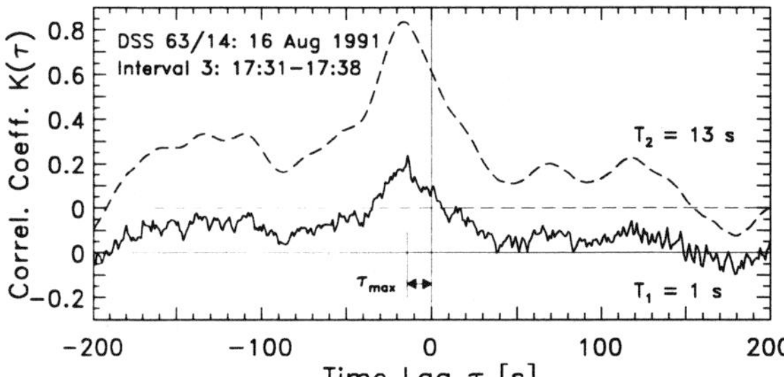

FIGURE 2. *Frequency fluctuation cross-correlation functions for Interval 3 with $<\alpha_f>$ = 1.015. Lower curve: raw data before filtration; upper curve: after filtration (curve and scale displaced upward by 0.2).*

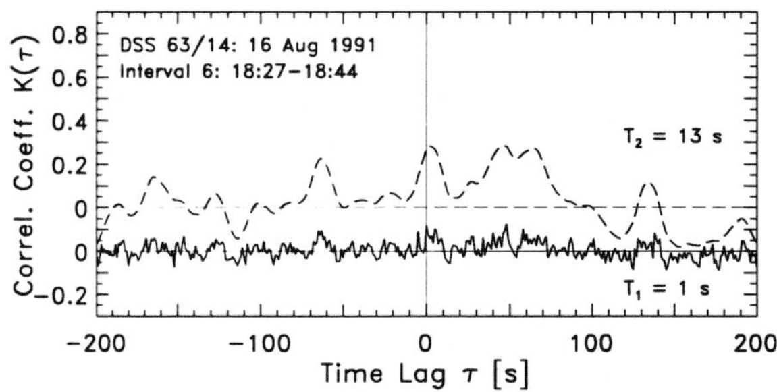

FIGURE 3. *Frequency fluctuation cross-correlation functions for Interval 6 with $<\alpha_f>$ = 0.264 (curves as in Fig. 2).*

- Growth of irregularities during their motion between the radio ray paths, the so-called "bubbling pattern" (Little and Ekers, 1971).

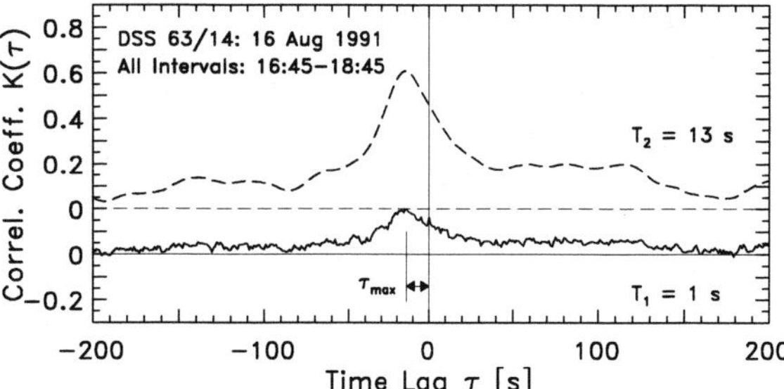

FIGURE 4. *Frequency fluctuation cross-correlation functions for the entire record in Fig. 1 (curves as in Fig. 2).*

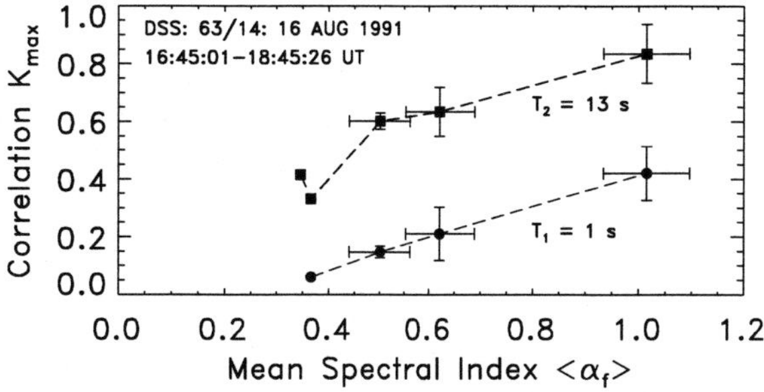

FIGURE 5. *Cross-correlation coefficient $K_{\max}$ versus power exponent $\langle\alpha_f\rangle$ before (circles) and after (squares) filtration.*

For all three above cases, it can be shown that the maximum cross correlation of fluctuations registered at spaced sites will increase upon decreasing the ratio "site spacing / spatial correlation size". The first possible cause above is most probably not responsible for the dependence of Fig. 5, because the angle between the coronal projection of the baseline and the radial direction is quite small (see Table 1). Furthermore, the expected value of the ratio between the instantaneous convection velocity spread and the mean solar wind speed at small heliocentric distances should not be greater than at large distances where this ratio is much less than unity (Chashei et al., 2000). For the last possible cause, random changes of moving irregularities can be characterized by a chaotic velocity w_{ch}. Using spaced interplanetary scintillation observations, Ekers and Little (1971) found that the chaotic velocity is comparable to the solar wind speed ($w_{ch} \gtrsim v_c$) at heliocentric distances less than 10 $R_\odot$, i.e., in that region appropriate to the observations presented in this paper. The increase in the cross correlation with increasing spectral power exponent is thus explained naturally by a corresponding increase in the spatial turbulence correlation scale.

Moreover, weak cross correlation for the flat temporal fluctuation spectra indicates that the typical "lifetime" of the irregularities at a given scale L is comparable to the convection time L/v_c. Consequently, since $w_{ch} \gtrsim v_c$, the cross correlation is dominated by irregularities with scales $L \gtrsim \Delta R$, where ΔR is the radial coronal projection of the radio ray path separation. The increase in cross correlation resulting from high-frequency filtration evidently has the same explanation.

CONCLUSIONS

A statistical analysis of frequency fluctuations recorded during coronal radio sounding experiments with the Ulysses and Galileo spacecraft shows a high degree of similarity between the power spectral parameters measured simultaneously at widely-spaced ground stations with short-time averaging. However, considerable temporal variations of the spectral parameters, particularly the power exponent, were detected on typical time scales of about 20 min.

The cross-correlation coefficient of frequency fluctuations at spaced ground stations was also found to be quite variable from one 20-minute interval to the next. The cross-correlation level $K_{\max}$ was found to increase roughly linearly with increasing spectral power exponent. Whereas no cross-correlation time lag could be determined at high sampling rate for the case of flat temporal spectra ($\alpha_f < 0.4$), $K_{\max}$ reached sufficiently high levels $\simeq 0.5$ for the case of steep spectra ($\alpha_f \geq 1$).

High-frequency filtration of the initial records results in an even steeper increase of $K_{\max}$. The dependence of $K_{\max}$ on α_f and the filtration effect show that those solar wind density irregularities with scales less than the radial projection of the baseline spacing do not produce correlated frequency fluctuations. The decorrelation of fast frequency fluctuations is best explained for the range of heliocentric distances near 10 $R_\odot$ by the temporal changes of density irregularities during their convection between the separated radio ray paths to the ground stations. The typical time scale for these changes in the irregularities is greater than, but comparable with the convection time. The conclusions of Ekers and Little (1971) on fast changes of irregularities with scales $\lesssim$ 100 km is thus extended here to the range of irregularities of size $10^3 - 10^4$ km. The possible cause of this frequency fluctuation bubbling is most likely associated with the propagation and damping of wave-like density irregularities (Chashei et al., 2000).

ACKNOWLEDGMENTS

This work presents results of a bi-national research project partially funded by the Deutsche Forschungsgemeinschaft (DFG) and by the Russian Foundation for Basic Research (RFBR), Grant 00-02-04022. Additional support from the RFBR, Grant 00-02-17845, and from the Russian Ministry of Industry, Technology and Science, is acknowledged.

REFERENCES

1. Armand, N.A., Efimov, A.I., 1984, Correlation of frequency fluctuations of waves propagating in turbulent media, *Radiotechn. Electron. 29*, 1649-1657.
2. Armand, N.A., Efimov, A.I., Yakovlev O.I., 1987, A model of solar wind turbulence from radio occultation experiments, *Astron. Astrophys. 183*, 135-141.
3. Armstrong, J.W., Coles, W.A., 1972, Analysis of three-station interplanetary scintillation, *J. Geophys. Res. 77*, 4602-4610.
4. Chashei, I.V., Shishov, V.I., Kojima, M., Misawa, H., 2000, Velocity fluctuations in the interplanetary scintillation pattern, *J. Geophys. Res. 105*, 27409-27417.
5. Coles, W.A., Kaufman, J.J., 1978, Solar wind velocity estimations from multi-station interplanetary scintillation, *J. Geophys. Res. 83*, 1413-1420.
6. Ekers, R.D., Little, L.T., 1971, The motion of the solar wind close to the Sun, *Astron. Astrophys. 10*, 310-316.
7. Little, L.T., Ekers, R.D., 1971, A method for analyzing drifting random patterns in astronomy and geophysics, *Astron. Astrophys. 10*, 306-309.
8. Wohlmuth, R., Plettemeier, D., Edenhofer, P, Bird, M.K., Efimov, A.I, Andreev, V.E., Samoznaev, L.N., Chashei, I.V., 2001, Radio frequency fluctuation spectra during the solar conjunctions of the Ulysses and Galileo spacecraft, *Space Sci. Rev. 97*, 9-12.

Characteristics of the Near-Sun Solar Wind Turbulence from Spacecraft Radio Frequency Fluctuations

A.I. Efimov, N.A. Armand, L.N. Samoznaev*, M.K. Bird†, I.V. Chashei** and P. Edenhofer, D. Plettemeier, R. Wohlmuth‡

*Inst. for Radio Engineering & Electronics, Russian Academy of Science, Moscow, 101999, Russia
†Radioastronomisches Institut, Universität Bonn, 53121 Bonn, Germany
**Lebedev Physical Institute, Russian Academy of Science, Moscow, 117924, Russia
‡Institut für HF-Technik, Universität Bochum, 44780 Bochum, Germany

Abstract.
Frequency fluctuation temporal spectra measured during spacecraft radio occultation experiments are shown to have a maximum at a well defined temporal frequency if the density turbulence power spectrum of the solar wind has a power law form in the wavenumber range between the outer and inner turbulence scales. The frequency of this maximum and the associated maximal spectral power of the frequency fluctuations both depend on the density variance, the solar wind convection speed, the turbulence outer scale, and the power index of the density turbulence spectrum in the region of the propagation medium near the line-of-sight proximate point. If the solar wind speed can be estimated from simultaneous frequency fluctuation measurements at widely-spaced ground stations, then estimates can be derived for both the density turbulence outer scale and density variance from the measured values of the power exponent of the frequency fluctuation temporal spectra. Results of coronal radio sounding experiments in 1997 with the Galileo spacecraft were analysed using the above method. Distinct differences were found between the temporal frequency fluctuation spectra observed at large and at small heliocentric distances. Values of the fractional density variance in the solar wind at low heliolatitude during a period of low solar activity are presented for the range of heliocentric distances between 7 $R_{\odot}$ and 31 $R_{\odot}$.

INTRODUCTION

Investigations of the spatial and temporal (including solar cyclic) variations of the circumsolar plasma turbulence are of great importance for the physics of the solar wind. Coronal radio sounding experiments (Yakovlev et al., 1980; Bird, 1982) are valuable tools for studying the solar wind turbulence. One modification of the radio occultation method is based on measurements of the radio frequency fluctuations. This observable has some advantages for investigation of density variance compared with others (for instance, amplitude fluctuation measurements), because the large-scale regime of the spatial density turbulence spectrum, where the energy density is greatest, produces most of the fluctuations of the radio signal frequency (Wohlmuth et al., 2001).

In order to determine density turbulence spectra and obtain estimates of the solar wind density fluctuation variance we study here frequency fluctuation measurements of the Galileo radio signals at the carrier frequency f = 2.295 GHz during the spacecraft's solar occultation in January-February 1997. Our results are thus applicable to a period of very low solar activity level and to low heliolatitudes. Based on these estimates and published data on the average electron density, we derive the radial dependence of the fractional density variance over the specific range of heliocentric distances which includes the region of solar wind acceleration. We concentrate here particularly on the fractional density fluctuations, because this parameter is of crucial importance for understanding the turbulence regimes and evolution in the solar wind. We also compare frequency fluctuation spectra observed in the developed solar wind with those applicable to the acceleration region.

THEORETICAL RELATIONS

In this section we consider the relation of the frequency fluctuation temporal spectrum measured in radio occultation experiments to the basic spatial spectrum of solar wind turbulence and to the solar wind speed. We assume that the spatial density fluctuation spectrum in the solar wind is a power law, nearly isotropic, and can be represented by the following form (Armand et al., 1987):

$$\Phi_N(q,r) = C^2(r)\frac{\exp(-q^2/q_m^2)}{(q^2+q_o^2)^{-p/2}} \qquad (1)$$

CP679, *Solar Wind Ten: Proceedings of the Tenth International Solar Wind Conference*,
edited by M. Velli, R. Bruno, and F. Malara

where r is the heliocentric distance, q is the wavenumber, $L_o = 2\pi/q_o$ and $L_m = 2\pi/q_m$ ($L_o >> L_m$) are the turbulence outer and inner scales, respectively, and p is the 3D power exponent. If the spectrum of Eq. (1) is normalized to the density fluctuation variance $\sigma_N^2(r)$ then the temporal power spectrum of frequency fluctuations $G_f(\nu)$ is defined by the relation:

$$G_f(\nu) = B\pi^{-1}(\lambda r_e)^2\sigma_N^2(R)L_e v_c(R)\nu_o^\alpha \nu^2 \times(\nu_o^2+\nu^2)^{-(\alpha+2)/2}\exp(-\nu^2/\nu_m^2) \quad (2)$$

where $\alpha = \alpha_f = p-3$ is the power exponent of the temporal frequency fluctuations spectrum, λ is the radio wavelength, $r_e = 2.82\times10^{-13}$ cm is the classical electron radius, R is the solar offset distance of the radio ray path, $L_e \sim R$ is the effective thickness of the slab near the solar proximate point along the ray path which contains the radio frequency modulating turbulence, v_c is the solar wind convection speed, $\nu_o = v_c/L_o$ and $\nu_m = v_c/L_m$ are the frequencies corresponding to the outer and inner scales, and the constant B is equal to

$$B = \begin{cases} \alpha & \text{for } 0<\alpha\leq 1 \\ [\ln(2q_m/q_0)]^{-1} & \text{for } \alpha = 0 \end{cases} \quad (3)$$

In contrast to phase or amplitude fluctuation spectra, the frequency fluctuation spectrum in Eq. (2) exhibits a maximum at the fluctuation frequency $\nu_{\max}$ given by

$$\nu_{\max} = \begin{cases} (2/\alpha)^{1/2}\nu_o & \text{for } 0<\alpha\leq 1 \\ (\nu_o\nu_m)^{1/2} & \text{for } \alpha = 0 \end{cases} \quad (4)$$

Numerical values of $\nu_{\max}$ from Eq. (4) are expected to be of order of 0.1 mHz in the regions of the developed solar wind $r > 20\ R_\odot$ with $\alpha \approx 2/3$ and $v_c = const$ (Wohlmuth et al., 2001). In the solar wind acceleration region ($r < 10\ R_\odot$), however, for which spectra with $\alpha \approx 0$ were reported (Woo and Armstrong, 1979), the values of $\nu_{\max}$ may be considerably higher, up to $\nu_{\max} > 10$ mHz. In such cases the spectral maximum is weakly pronounced and it is difficult to use the lower expression of Eq. (4) for estimates of ν_o or ν_m.

Substituting $\nu_{\max}$ from Eq. (4) into Eq. (2) leads to the following relation for the maximum spectral density

$$G_f(\nu_{\max}) = \frac{B_1}{\pi}(\lambda r_e)^2\sigma_N^2(R)L_e v_c \quad (5)$$

with

$$B_1 = \begin{cases} 2\left[\frac{\alpha}{\alpha+2}\right]^{(\alpha+2)/2} & \text{for } 0<\alpha\leq 1 \\ [\ln(2q_m/q_o)]^{-1} & \text{for } \alpha = 0 \end{cases} \quad (6)$$

Based on frequency fluctuation observations of power index α, solar wind speed v_c and maximum spectral power $G_f(\nu_{\max})$, Eqs. (5) and (6) are used below for estimating the density variance $\sigma_N^2(R)$.

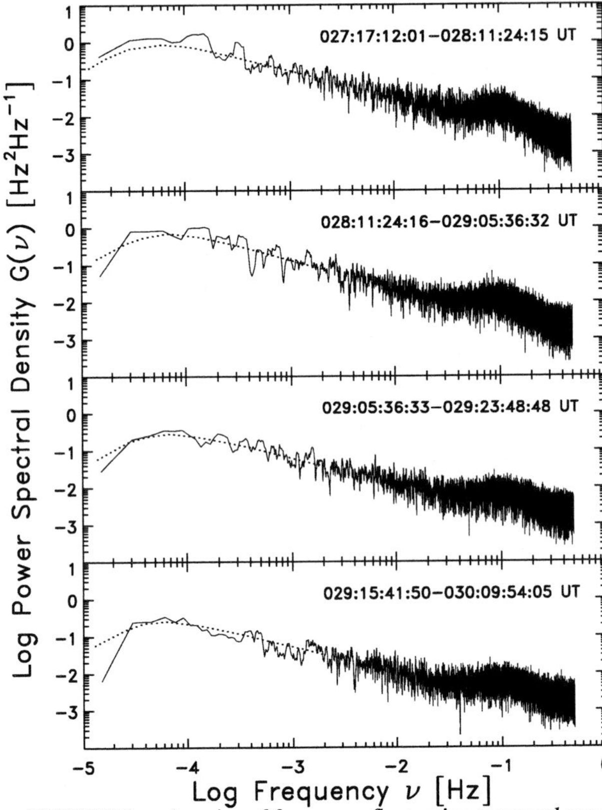

FIGURE 1. *A series of frequency fluctuation temporal spectra. The start and stop times are labeled for each spectrum. Dotted lines show fitted theoretical spectra from Eq. (2) normalized to the maximum spectral density given by Eq. (5).*

OBSERVATIONAL DATA

Long observation records of frequency fluctuations at high sampling rate (1 s^{-1}) were processed according to the above presented method. The Galileo S-band radiometric signal parameters were recorded with the three large antennas of the global NASA Deep Space Network (DSN): Goldstone (DSS 14), Canberra (DSS 43) and Madrid (DSS 63). The Doppler residual time series was calculated after subtracting a slowly varying component from the raw data to compensate for the spacecraft motion relative to the observer known from navigation data. Two continuous frequency fluctuation records are described in this section. The first one is typical for comparatively large heliocentric distances ($\simeq$ 30 $R_\odot$); the other for the region closer to the Sun $\simeq 7\ R_\odot$.

A series of successive temporal frequency fluctuation power spectra from the first record is presented in Fig. 1. The spectral parameters α and $\nu_{\max}$ were estimated from the normalized spectra (dotted lines in Fig. 1).

Simultaneous observations, and a corresponding capability for estimating the solar wind speed, occur during the overlap time intervals between DSN stations. In these

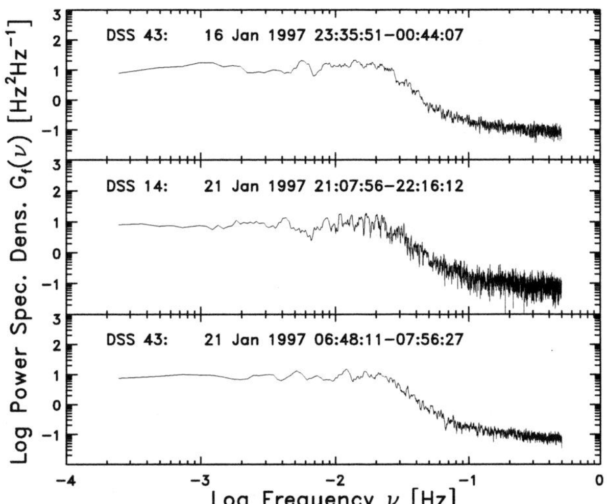

FIGURE 2. *Temporal frequency fluctuation spectra for data recorded at smaller solar offset distances.*

cases the solar wind speed was found as the ratio of the calculated 2-station ray path radial separation to the time lag of maximum frequency fluctuation cross correlation between the two stations.

Spectral parameters for the Galileo 1997 observations are summarized in Table 1. The Spectra #1–4 were obtained at a larger solar offset distance and have generally different characteristics from those recorded at smaller solar distances (Spectra #5–8). Estimates of outer scale were obtained using Eq. (4) for observed values of spectral index α and solar wind velocity v_c. The RMS values of density fluctuations in the solar wind σ_N were found from Eq. (5). Results from Viking ranging measurements (Muhleman and Anderson, 1981), which are applicable to the solar minimum period, were used for computing the mean electron density N_e in Table 1.

Three temporal frequency fluctuation spectra corresponding to the Doppler residuals recorded at solar ray path offsets much closer to the Sun are shown in Fig. 2. These spectra, in strong contrast to those shown in Fig. 1, display a very flat low frequency part (power exponent $\alpha \approx 0$) and a very sharp spectral break (power exponent > 2) at frequencies $\nu \geq 0.02$ Hz.

Because the spectra of Fig. 2 have a flat low frequency part and a sharp high frequency break, we cannot determine the frequency ν_{max} and corresponding turbulence outer scale L_o. Furthermore, the frequency fluctuation variance for spectra of this type is dependent only on the nearly constant spectral density at low frequencies G_{flf} and the break frequency ν_b.

The values of G_{flf} and ν_b needed for the estimates of σ_f and σ_N, as well as the minimal frequency in the temporal spectra ν_{min} (defined by the length of the data record), are presented in Table 1 (Spectra #5–8). The same method described in the previous section was used

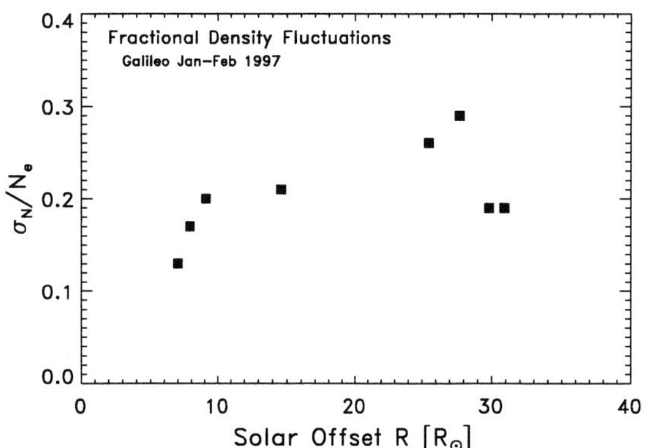

FIGURE 3. *Fractional density fluctuations σ_N/N_e versus R.*

to calculate σ_N from the frequency fluctuation spectra. The only difference is that the ratio ν_b/ν_{min} was substituted in Eqs. (5),(6) instead of q_m/q_0. It should be noted that the variances σ_f and σ_N for spectra #5–8 of Table 1 are defined mainly by the spectral density in the spectral range near the break frequency ν_b.

σ_N/N_E IN THE INNER SOLAR WIND

We use now the frequency fluctuation data to derive the fractional density fluctuations in the range of heliocentric distances between 7 $R_\odot$ and 31 $R_\odot$. The values of fractional density fluctuations σ_N/N_e are given in Table 1 and shown in Fig. 3.

The fractional density fluctuations over the investigated heliocentric distance range, as shown in Fig. 3, are typically of the order of 0.1–0.3 at low heliolatitudes and low solar activity levels. These estimates are (on average) in good agreement with the results of Woo et al. (1995) found from investigation of Ulysses dual-frequency ranging data, applicable to temporal frequencies in the range 6×10^{-6} Hz $< \nu < 8 \times 10^{-4}$ Hz.

The data of Fig. 3 display a slight tendency for an increase in fractional density fluctuations σ_N/N_e with increasing heliocentric distance. Such behavior would be quite reasonable from a physical point of view. Indeed, the typical temporal spectra observed at low solar activity close to the Sun (see, for instance, Fig. 2) correspond to spatial density fluctuations with an initial power spectrum that slowly evolves as it is carried outward by the solar wind flow. No cascading is expected for such spectra. Switching on the cascading processes, with a corresponding transition to developed Kolmogorov turbulence, would be possible at larger distances from the Sun if the fractional turbulence level increases with increasing heliocentric distance as a result of smooth radial gradients in the ambient plasma parameters. It should be noted that Woo et al. (1995) found a radial increase of

TABLE 1. Summary of derived spectral parameters: Galileo 1997

Spectrum	#1	#2	#3	#4	#5	#6	#7	#8
Date, Jan 1997	27/28	28/29	29	29/30	16-17	21	22	24
Start time [UT]	17:12:01	11:24:16	05:36:33	15:41:50	23:35:51	21:07:56	06:48:11	03:38:14
Stop time [UT]	11:24:15	05:36:32	23:48:48	09:54:05	00:44:07	22:16:12	07:56:27	10:14:21
R [$R_\odot$]	25.5	27.7	29.8	30.9	7.0	7.9	9.1	14.6
σ_f [mHz]	73.4	63.0	51.6	46.9	699	548	579	149
α	0.677	0.775	0.667	0.640	0.0	0.0	0.0	0.446
v_c [km/s]	249±60	208±29	259±73	262±89	145	110	120	160
$\nu_{\max}$ [10^{-4} Hz]	0.6	0.7	0.7	0.65	-	-	-	-
$G_f(\nu_{\max})$, [Hz^2/Hz]	0.9	0.72	0.28	0.27	-	-	-	-
L_o, [$R_\odot$]	10.2±2.5	6.8±1.0	9.2±2.6	10.2±3.5	-	-	-	-
$\nu_{\min}$ [10^{-4} Hz]	-	-	-	-	1.0	1.0	1.0	3.0
G_{flf} [Hz^2/Hz]	-	-	-	-	10.0	8.0	8.0	1.4
ν_b [10^{-2} Hz]	-	-	-	-	2.0	2.0	2.0	8.0
σ_N [10^2 cm^{-3}]	1.2	1.1	0.62	0.56	1.4	1.3	1.2	0.44
N_e [10^2 cm^{-3}]	4.5	3.8	3.2	3.0	10.5	8.0	6.1	2.1
σ_N/N_e	0.26	0.29	0.19	0.19	0.13	0.17	0.20	0.21

σ_N/N_e for the fast solar wind streams. Although more data would be needed to confirm this preliminary conclusion, we find evidence in this work that the same tendency may well hold for the slow solar wind.

CONCLUSIONS

Continuous Galileo spacecraft signal frequency fluctuation records in the period of low solar activity at low heliolatitudes reveal a strong contrast between density turbulence spectra observed at large ($R > 20\ R_\odot$) and small ($R < 10\ R_\odot$) solar offset distances.

Typical density turbulence power spectra have a pronounced outer scale and power-law decrease at high frequencies with a 3D power exponent $p \approx 3.6 - 3.7$ far from the Sun. Such spectra correspond to developed turbulence of the Kolmogorov type. In contrast, the power exponent close to the Sun is typically $p \approx 3$, or even less, in the low frequency spectral range, and becomes very large ($p > 4$) at frequencies above about 0.02 Hz (break frequency). Spectra of this type are possible only in the absence of nonlinear cascading processes.

The fractional density fluctuation level lies within the range 0.1–0.3 for heliocentric distances between 7 $R_\odot$ and 31 $R_\odot$. The slight increasing trend toward larger R requires further investigations for confirmation.

ACKNOWLEDGMENTS

This work presents results of a bi-national research project partially funded by the Deutsche Forschungsgemeinschaft (DFG) and by the Russian Foundation for Basic Research (RFBR), Grant 00-02-04022. Additional support from the RFBR, Grant 00-02-17845, and from the Russian Ministry of Industry, Technology and Science, is acknowledged.

REFERENCES

1. Armand, N.A., Efimov, A.I., Yakovlev O.I., 1987, A model of solar wind turbulence from radio occultation experiments, *Astron. Astrophys. 183*, 135-141.
2. Bird, M.K., 1982, Coronal investigations with occulted spacecraft signals, *Space Sci. Rev. 33*, 99-126.
3. Muhleman, D.O., Anderson J.D., 1981, Solar wind elerctron densities from Viking dual frequency radio measurements, *Astrophys. J. 247*, 1093-1101.
4. Wohlmuth, R., Plettemeier, D., Edenhofer, P, Bird, M.K., Efimov, A.I, Andreev, V.E., Samoznaev, L.N., Chashei, I.V., 2001, Radio frequency fluctuation spectra during the solar conjunctions of the Ulysses and Galileo spacecraft, *Space Sci. Rev. 97*, 9-12.
5. Woo, R., Armstrong, J.W., 1979, Spacecraft radio scattering observations of the power spectrum of electron density fluctuations in the solar wind, *J. Geophys. Res. 84*, 7288-7296.
6. Woo, R., Armstrong, J.W., Bird, M.K, Pätzold, M, 1995, Variation of fractional electron density fluctuations inside 40 $R_\odot$ observed by Ulysses ranging measurements, *Geophys. Res. Lett. 22*, 329-332.
7. Yakovlev, O.I., Efimov, A.I., Razmanov, V.M., Shtrykov, V.K., 1980, Inhomogeneous structure and velocity of the circumsolar plasma based on data of the Venera-10 station, *Astron. Zh. 57*, 790-798.

Turbulence Regimes of the Solar Wind in the Region of its Acceleration and Initial Stage of Supersonic Motion

L.N. Samoznaev, A.I. Efimov, V.E. Andreev*, M.K. Bird†, I.V. Chashei** and P. Edenhofer, D. Plettemeier, R. Wohlmuth‡

*Inst. for Radio Engineering & Electronics, Russian Academy of Science, Moscow, 101999, Russia
†Radioastronomisches Institut, Universität Bonn, 53121 Bonn, Germany
**Lebedev Physical Institute, Russian Academy of Science, Moscow, 117924, Russia
‡Institut für HF-Technik, Universität Bochum, 44790 Bochum, Germany

Abstract.
Coronal radio sounding experiments were carried out during the solar conjunctions of the spacecraft Ulysses and Galileo, providing information on the solar wind plasma over a wide range of heliocentric distances and heliolatitudes on both East and West limbs of the Sun. An important component of these investigations is to identify the turbulence regimes of the solar wind in its acceleration and initial supersonic regions. This work concentrates on the variation of the spectral index of the temporal frequency fluctuation spectrum α_f. The analysis leads to the following preliminary conclusions: (1) At low heliolatitudes the turbulence becomes 'developed', with α_f reaching the Kolmogorov value of 2/3, at distances beyond 20 $R_\odot$; (2) At high heliolatitudes (poleward of 65°) the solar wind turbulence remains undeveloped out to distances of at least 30 $R_\odot$; (3) At distances close to the Sun (less than 7 $R_\odot$) the spectrum sometimes becomes a double power-law with small spectral index $\alpha_f \simeq$ 0.03-0.11 at low fluctuation frequencies ($\nu < 0.02$ Hz), but a sharp decrease ($\alpha_f > 1.2$) in the fluctuation regime beyond the break frequency.

INTRODUCTION

Many years of solar wind plasma investigations by radio occultation (Yakovlev et al., 1980; Bird, 1982; Bird and Edenhofer, 1990; Wohlmuth et al., 2001) and *in situ* methods have shown that turbulence is a permanent property of the interplanetary plasma at all heliocentric distances and heliolatitudes. Fluctuations of the electron density, magnetic field, velocity, etc. are characterized by spatial and temporal spectra covering many decades in the wavenumber or temporal frequency domains, respectively. Turbulence evolution and its interaction with the background plasma is of great importance for solar wind physics, especially near the Sun where the plasma flow accelerates to supersonic and superalfvenic velocities.

This paper presents some results of coronal sounding experiments carried out during four extended intervals with the spacecraft Ulysses (1991 DOY 218-248; 1995 DOY 053-073) and Galileo (1995/96 DOY 336-014; 1996/97 DOY 360-045). Temporal frequency fluctuations of the spacecraft's downlink signals at the carrier frequency 2295 MHz were measured over a wide range of heliocentric distances, between 4.3 and 79.8 $R_\odot$, heliolatitudes up to 89°, and solar activity levels with sunspot numbers from 0 to 300. Analysis of the frequency fluctuation spectra shows that the shape of the electron density spatial spectrum is strongly dependent on heliocentric distance, heliolatitude and solar activity. Double power-law spectra with sharp breaks in spectral density at temporal frequencies near 0.01–0.03 Hz were detected for the first time.

OBSERVATIONS

Coronal sounding experiments were conducted using radio signals of the spacecraft Galileo and Ulysses during their active missions in the years 1991-1997. The signal frequencies were recorded at a 1 s^{-1} sampling rate during the spacecraft solar conjunctions at the large ground stations of the NASA Deep Space Network (DSN). The radio subsystem configuration on the Ulysses spacecraft for radio science investigations consists of an S-band uplink and dual-frequency S/X-band downlinks, both downlink frequencies being phase coherent with the uplink. The plasma-induced frequency variations (differential Doppler residuals) were calculated by subtracting out the nondispersive effects imposed onto the two coherent downlinks. In the case of Galileo, the single S-band downlink was generated by an ultrastable oscilla-

CP679, *Solar Wind Ten: Proceedings of the Tenth International Solar Wind Conference*,
edited by M. Velli, R. Bruno, and F. Malara

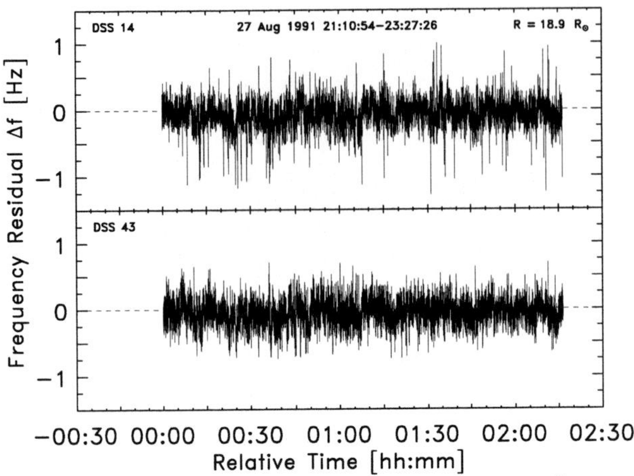

FIGURE 1. *Differential Doppler residuals Δf of the Ulysses signals during simultaneous observations at the DSN ground stations Goldstone (DSS 14) and Canberra (DSS 43); time interval: 21:10:54 to 23:27:26 UT on 27 Aug 1991 (occultation egress phase). The heliodistance of the ray path proximate point was $R = 18.9\ R_\odot$, the heliolatitude was $\varphi \approx 19°$.*

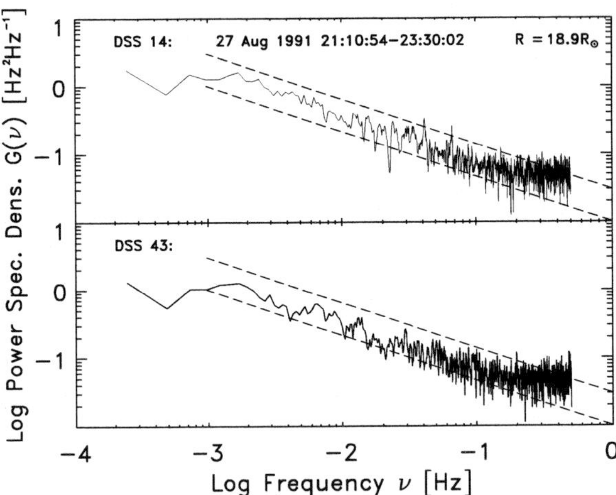

FIGURE 2. *Average FFT spectra (2 × 4096 data points) of Ulysses frequency fluctuations shown in Fig. 1. For comparison, the dashed lines are theoretical Kolmogorov spectra ($\alpha_f = 2/3$) for $\nu > 1$ mHz.*

tor (USO) and the Doppler residuals were determined as a difference between the measured frequencies and the predicted nondispersive values calculated from navigational data.

Descriptive information about the four coronal sounding experiments discussed in this work is presented in Table 1. The heliolatitude of the ray path proximate point is φ and the mean sunspot number for the interval is $\overline{W}$.

As seen in Table 1, the solar activity levels differed strongly between the 1991 and the 1996/1997 observation sessions. Whereas the solar activity was high with W reaching 290-300 on some days of 1991, W was essentially zero during 25 days of the 1996/97 Galileo occultation.

An example of Ulysses differential Doppler residuals measured simultaneously at two DSN ground stations is presented in Fig. 1.

Temporal power spectra of the frequency fluctuations were calculated using an FFT algorithm for all observational intervals with a duration of more than 4×4096 samples within the periods shown in Table 1. Spectral indices α_f were determined for the spectral ranges where the power density can be represented as a power law. The spectral index α_f is related to the power exponent of the spatial 3D density turbulence power spectrum α_d by $\alpha_d = \alpha_f + 3$. We also investigated the possibility of a break in an individual temporal power spectrum that could be described by a distinct change in α_f. The frequency fluctuations are better for detecting such breaks because the power spectra are flatter than the temporal spectra of phase (index p) or amplitude (index a) fluctuations, $\alpha_p = \alpha_a = \alpha_f + 2$ (Yakovlev et al., 1980).

THE POWER EXPONENT OF FREQUENCY FLUCTUATION SPECTRA

Frequency fluctuation spectra for the data of Fig. 1, which are typical for a comparatively large heliocentric distance, are illustrated in Fig. 2.

The two spectra of Fig. 2 are similar and have a power law shape in the frequency range 10^{-3} Hz $< \nu < 10^{-1}$ Hz. The spectral index is found to be very close to the Kolmogorov value $\alpha_f \approx 2/3$ (see dashed lines).

The dependence of the frequency fluctuation spectral indices on heliodistance is summarized in Fig. 3 for all Galileo and Ulysses observations at low heliolatitudes ($\varphi < 10°$). The measurements indicate that the spectral index is roughly constant for $R > 20\ R_\odot$ and corresponds to developed turbulence with $\alpha_f \approx 2/3$. The value of α_f increases with R from about 1/3 to about 2/3 over the range $5\ R_\odot < R < 20\ R_\odot$, in agreement with the phase scintillation analysis of Woo and Armstrong (1979).

The Ulysses 1995 data are well suited for investigating the heliolatitude dependence of the spectral index α_f. As shown in Fig. 4, $\alpha_f \approx 2/3$ in the range of heliolatitude -50° $< \varphi <$ 0°, but has a tendency to decrease when the line of sight approaches the south polar region. These results agree with the spectral changes found in phase fluctuation spectra by Pätzold et al. (1996).

The data of Fig. 4 most likely reflect a real heliolatitude dependence, because the variations in solar offset were rather small ($22\ R_\odot < R < 32\ R_\odot$) during this measurement interval. Furthermore, no changes in α_f were detected within this distance range at low latitudes (see Fig. 3). It may thus be concluded that the solar wind turbulence remains undeveloped in the high latitude regions, at least for heliocentric distances out to 30 $R_\odot$.

TABLE 1. Coronal sounding experiments 1991-1997

Observation period	Spacecraft	Solar offset [$R_\odot$]	φ[°]	$\overline{W}$
1991 Aug 06 – 1991 Sep 06	Ulysses	4.3 – 40.8	1.4 – 89	183
1995 Feb 22 – 1995 Mar 14	Ulysses	21.5 – 32.2	–89 – +1	34
1996 Dec 02 – 1996 Jan 14	Galileo	4.3 – 79.8	1 – 14	12
1996 Dec 25 – 1997 Feb 14	Galileo	6.9 – 73.9	2 – 20	6

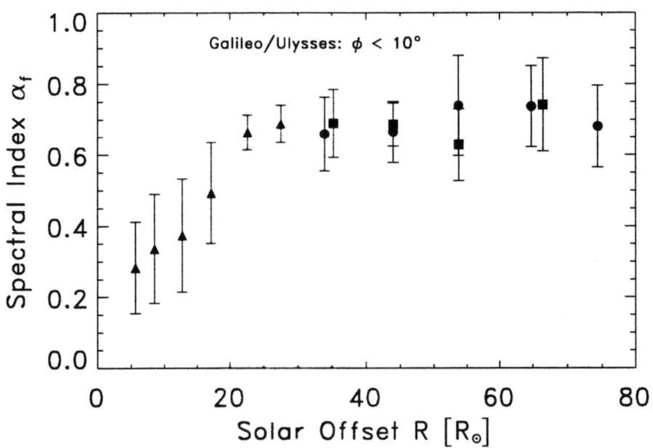

FIGURE 3. *Dependence of spectral index α_f on heliocentric distance. Average values of α_f are presented for 10 $R_\odot$ bins in the range 30–80 $R_\odot$, 5 $R_\odot$ bins in the range 10–30 $R_\odot$, and 2.5 $R_\odot$ bins in the range 5–10 $R_\odot$. The solid circles (squares) denote measurements on the West (East) solar limb. Triangles represent data averages from both solar limbs.*

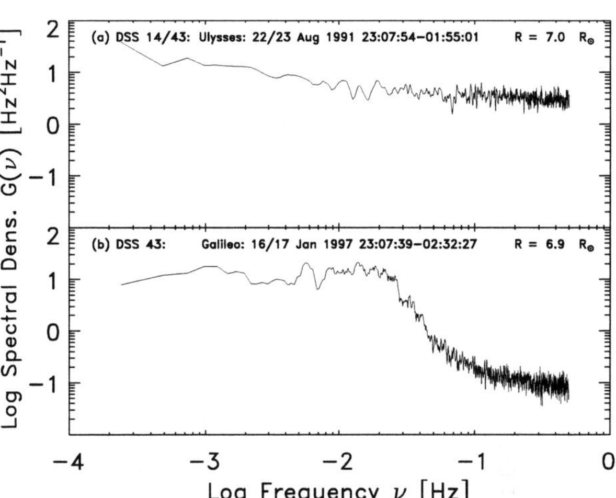

FIGURE 5. *Average FFT power spectra (3 × 4096 points): (a) single power law, high solar activity; (b) double power law with break frequency, low solar activity.*

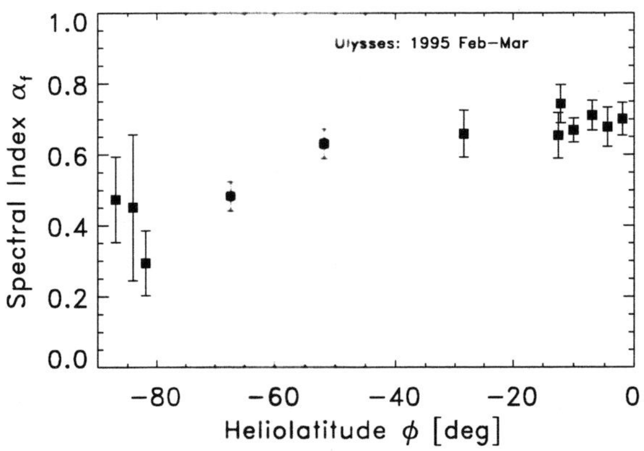

FIGURE 4. *Dependence of α_f on heliolatitude φ (Ulysses 1995).*

SPECTRAL FEATURES OBSERVED AT SMALL HELIOCENTRIC DISTANCES

Many Galileo frequency fluctuation spectra recorded at small heliocentric distances were found to have a sharp spectral break. As a rule, the spectral breaks are observed inside 10 $R_\odot$ at relatively low levels of RMS frequency fluctuations. Contrasting examples of single power law and double power-law (broken) spectra are presented in Fig. 5.

Descriptive data for the spectra in Fig. 5 are as follows: (a) Ulysses, very high solar activity level $W = 212$ (22/23 Aug 1991), heliolatitude $\simeq 40°$, and (b) Galileo, extremely low solar activity $W < 9$ (16/17 Jan 1997), heliolatitude $\simeq 20°$. Whereas the heliocentric distances were nearly identical ($R \simeq 7.0$ $R_\odot$), the levels of the RMS frequency fluctuations were very different: (a) $\sigma_f = 2.552$ Hz; (b) $\sigma_f = 0.662$ Hz. The power exponents for spectrum (a) and for the low-frequency part of spectrum (b) are approximately equal, and are close to the values presented in Fig. 3 for the corresponding heliocentric distance.

The break in spectrum (b) of Fig. 5 takes place at the frequency $\nu_b \approx 0.024$ Hz. Spectra with obvious break frequencies are not observed at high fluctuation levels or in regions outside 10 $R_\odot$. More data would be required to determine just what environmental conditions (heliolatitude, solar activity level, others?) are necessary for producing the unusual spectra with the break frequency.

Fig. 6 presents all observed values of the spectral break frequency ν_b measured at different heliocentric distances. One can see from Fig. 6 that the measured values of ν_b range between the limits 0.01 Hz and 0.03 Hz.

There appears to be a slight tendency for an increase of ν_b with increasing solar offset distance R, but the

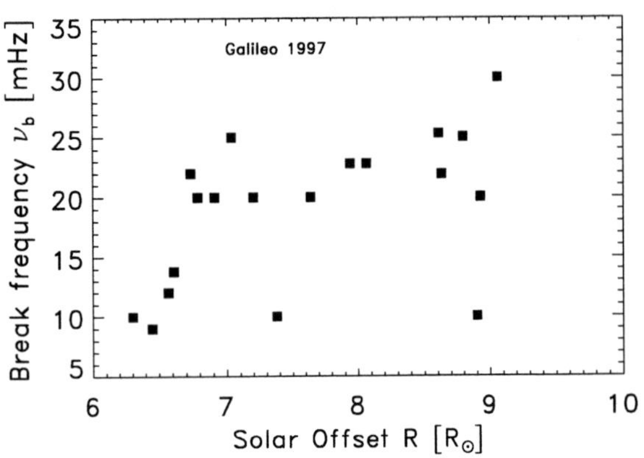

FIGURE 6. *Dependence of the break frequency on solar offset distance R (Galileo 1997).*

data statistics are insufficient for drawing a more definite conclusion about this dependence.

CONCLUSIONS

The power exponent of the 3D density turbulence spectrum in the low-latitude solar wind increases with increasing heliocentric distance for $R < 20\ R_\odot$, reaching a roughly constant value $\alpha_d \approx 3.6$–3.7 for $R > 20\ R_\odot$, in agreement with the previous findings of Woo and Armstrong (1979). The high-latitude solar wind turbulence near solar activity minimum, on the other hand, definitely remains undeveloped out to 30 $R_\odot$.

The turbulence spectra are sometimes found to display a spectral break at frequencies 0.01–0.03 Hz in the range $R < 10\ R_\odot$ for low solar activity levels. Breaks were not observed for $R > 20\ R_\odot$, or for $R < 10\ R_\odot$ under high solar activity conditions. The origin and properties of these spectral breaks are a topic for future theoretical and experimental investigations.

The turbulence regimes may well be closely connected to the regimes of solar wind flow. While the flat 3D turbulence spectra with $\alpha_d \simeq 3$ are typical for the region of plasma acceleration and transition to supersonic and superalfvenic flow, the steep spectra with $\alpha_d \simeq 3.6 - 3.7$ occur in the developed flow with constant speed. The rearrangement of turbulence spectra is apparently yet another manifestation of the mechanism that supplies the energy needed for the additional acceleration of the solar wind.

ACKNOWLEDGMENTS

This work presents results of a bi-national research project partially funded by the Deutsche Forschungsgemeinschaft (DFG) and by the Russian Foundation for Basic Research (RFBR), Grant 00-02-04022. Additional support from the RFBR, Grant 00-02-17845, and from the Russian Ministry of Industry, Technology and Science, is acknowledged.

REFERENCES

1. Bird, M.K., 1982, Coronal investigations with occulted spacecraft signals, *Space Sci. Rev. 33*, 99-126
2. Bird, M.K., Edenhofer, P., 1990, Remote sensing observations of the solar corona, In: R. Schwenn and E. Marsch (Eds.), *Physics of the Inner Heliosphere 1*, Springer-Verlag, Heidelberg, 13-90
3. Pätzold, M., Karl, J., Bird, M.K., 1996, Coronal radio sounding with Ulysses: dual-frequency phase scintillation spectra in coronal holes and streamers, *Astron. Astrophys. 316*, 449-456
4. Wohlmuth, R., Plettemeier, D., Edenhofer, P, Bird, M.K., Efimov, A.I, Andreev, V.E., Samoznaev, L.N., Chashei, I.V., 2001, Radio frequency fluctuation spectra during the solar conjunctions of the Ulysses and Galileo spacecraft, *Space Sci. Rev. 97*, 9-12
5. Woo, R., Armstrong, J.W., 1979, Spacecraft radio scattering observations of the power spectrum of electron density fluctuations in the solar wind, *J. Geophys. Res. 84*, 7288-7296
6. Yakovlev, O.I., Efimov, A.I., Razmanov, V.M., Shtrykov, V.K., 1980, Inhomogeneous structure and velocity of the circumsolar plasma based on data of the Venera-10 station, *Astron. Zh. 57*, 790-798

Particle Transport in the Solar Wind Magnetic Turbulence: a Numerical Investigation

P. Pommois*†, P. Veltri† and G. Zimbardo†

*Center for High Performance Computing, Università della Calabria (Italy).
† Dipartimento di fisica, Università della Calabria (Italy)

Abstract. In order to interpret the fluxes of particles accelerated at corotating interaction regions and to forecast the solar energetic particle events, the propagation of energetic particles is investigated numerically. Particle transport is influenced in a strong way by the magnetohydrodynamic turbulence in the solar wind. We carried out a simulation in which ions are injected in a numerical realization of magnetic turbulence superimposed on a background uniform magnetic field. Particle transport is studied as a function of particle energy, pitch angle distribution, turbulence level, and turbulence anisotropy with parameter values appropriate to the solar wind. The ratio of perpendicular to parallel diffusion coefficients is estimated. For strong enough turbulence $\delta B/B_0 \sim 1$, as typical of the solar wind, we find that the particle pitch-angle distribution is quickly isotropized, and that perpendicular transport grows with the particle energy. Varying the turbulence anisotropy, that is the turbulence correlation lengths $l_{\parallel}$ and $l_{\perp}$, we find that in the cases with $l_{\parallel} \gg l_{\perp}$ (i.e., in the limit of 2D turbulence) the particle perpendicular transport is faster than in the cases with $l_{\parallel} \ll l_{\perp}$ (i.e., for the so called slab turbulence).

INTRODUCTION

The study of particle transport in the presence of magnetic turbulence is of great interest for several physical problems, like the transport of cosmic rays, the energetic particles propagation in the solar wind, and the plasma confinement in laboratory experiments. In strong turbulence analytic treatments are not easily feasible, and it is necessary to perform numerical simulations. In the solar wind, these require large computational resources, in order to describe a long turbulence spectrum, which, ideally, should include wave lengths comparable with the Larmor radius of the considered particles.

In this paper we show the results obtained by a numerical simulation where the particles trajectory are integrated in a three-dimensional (3D) magnetic turbulence. This work is an extension of the study of magnetic field line transport which we performed previously [1, 2, 3, 4, 5, 6]. Here we present the preliminary results, for which we kept the ratio between Larmor radius r_L and the shortest turbulence wavelength $\lambda_{\min}$ small, in order to check whether the findings obtained for field line diffusion actually hold for particle having small r_L.

NUMERICAL MODEL

The kinetic energies of accelerated particles in the solar wind vary over wide ranges. Having in mind the forecast of solar energetic particle events, we consider in this study proton energies varying from 100 keV to 1000 MeV. In the solar wind rest frame we can neglect the electric field, and the particle relativistic equations of motion can be written as:

$$\frac{d\mathbf{r}}{dt} = \mathbf{v} \quad , \quad \frac{d\mathbf{v}}{dt} = \frac{q}{mc\gamma}\mathbf{v} \wedge \mathbf{B}$$

where $\mathbf{r}$ is the position of the particle having mass m and charge q, $\mathbf{v}$ is the particle velocity, and $\gamma = (1 - v^2/c^2)^{-1}$ is the Lorentz factor which is constant because the only force is the magnetic force. A 3D magnetic field in a periodic simulation box is set up (see Pommois et al. [2, 3]), $\mathbf{B}(\mathbf{r}) = \mathbf{B}_0 + \delta\mathbf{B}(\mathbf{r})$ where $\delta\mathbf{B}(\mathbf{r})$ is the sum of static magnetic perturbations

$$\delta\mathbf{B}(\mathbf{r}) = \sum_{\mathbf{k},\sigma} \delta B(\mathbf{k}) \mathbf{e}^{(\sigma)}(\mathbf{k}) \exp i[\mathbf{k}\cdot\mathbf{r} + \phi_{\mathbf{k}}^{(\sigma)}]$$

Considering that the magnetic perturbations propagate with the Alfvén velocity, the assumption of static perturbations is well satisfied for energetic particles in the solar wind. The wave vectors are chosen as

$$\mathbf{k} = \frac{2\pi}{N_{\min}} \left(\frac{n_x}{l_x}, \frac{n_y}{l_y}, \frac{n_z}{l_z} \right)$$

in which l_x, l_y and l_z are the correlation lengths (note that in the axi-symmetric cases, $l_{\parallel} = l_z$ and $l_{\perp} = l_x = l_y$). The spectrum has a cut-off for both the short and the

CP679, *Solar Wind Ten: Proceedings of the Tenth International Solar Wind Conference*, edited by M. Velli, R. Bruno, and F. Malara

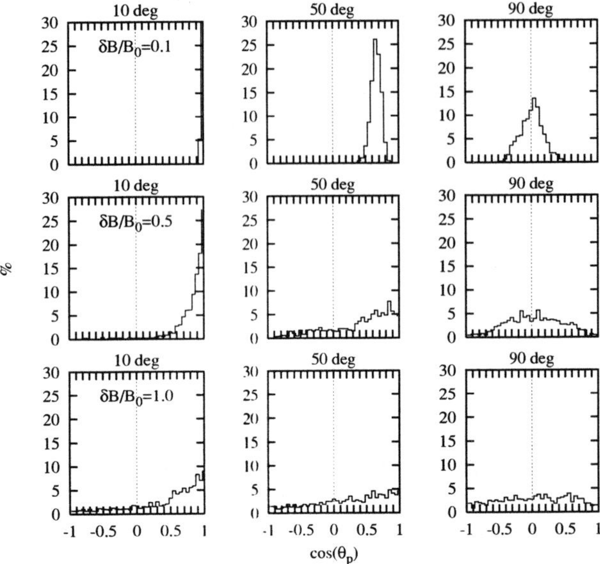

FIGURE 1. Distribution of the cosine of the pitch-angle θ_p for 1000 particle. Here we show the influence of the intensity of the fluctuation level $\delta B/B_0$ and of the initial pitch angle on the change of the velocity distribution after an integration time of $10^4/\omega_{bi}$, of particle with 1MeV energy in an isotropic turbulence spectrum. Left panels: $\theta_p = 10°$; middle panels: $\theta_p = 50°$; right panels: $\theta_p = 90°$. Upper panels: $\delta B/B_0 = 0.1$; center panels: $\delta B/B_0 = 0.5$; lower panels: $\delta B/B_0 = 1.0$.

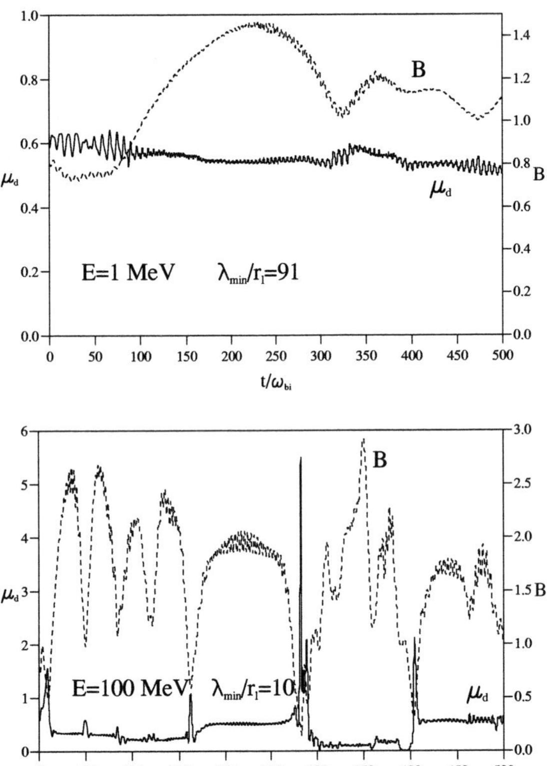

FIGURE 2. Evolution of the dimensionless magnetic moment μ_d (full line) and of the local magnetic field modulus (dashed line), for particle having 1 MeV energy (upper panel) and 100 MeV energy (lower panel), in a turbulent magnetic field with $\delta B/B_0 = 1.0$. Here $\mu_d = \mu/\mu_0$ where $\mu_0 = v_0^2/2mB_0$.

long wavelengths (band spectrum) with: $N_{\min}^2 \leq n_x^2 + n_y^2 + n_z^2 \leq N_{\max}^2$. We have $\mathbf{e}^{(\sigma)}(\mathbf{k}) \cdot \mathbf{k} = 0$, where $\mathbf{e}^{(\sigma)}(\mathbf{k})$ are the polarization unit vectors with $\sigma = 1,2$, and $\phi_{\mathbf{k}}^{(\sigma)}$ are random phases. The perturbation amplitude $\delta B(\mathbf{k})$ represents a self-similar spectrum:

$$\delta B(\mathbf{k}) = \frac{C}{(k_x{}^2 l_x{}^2 + k_y{}^2 l_y{}^2 + k_z{}^2 l_z{}^2)^{\beta/4+1/2}}$$

where $\beta = 3/2$ is the Kraichnan spectral index. More details on the numerical simulation can be found in Refs. [1, 2, 3, 4, 5, 6]. We report in the following on the main features of particle transport in such a magnetic configuration. Indeed, the above equations of motion are integrated in the presence of various levels of magnetic turbulence and various kinds of turbulence anisotropy. A background field $B_0 = 10$ nT is assumed, and time is measured in units of the inverse of the nonrelativistic proton gyrofrequency, $\omega_{bi} = qB_0/mc$.

EVOLUTION OF THE PARTICLE VELOCITY DISTRIBUTION

In the runs presented here the injection energy was constant (in other words, the distribution function corresponds to an energy shell rather than to a Maxwellian), in order to better discriminate the effects of different energies. Two types of initial particle distribution were used in the simulation: conic distribution (with fixed injection pitch angle θ_p between the initial velocity $\mathbf{v}_0$ and magnetic field $\mathbf{B}_0$) and isotropic distribution (where the direction of $\mathbf{v}_0$ is random).

During the integration the conic distribution is evolving and is progressively isotropised, see Fig. 1. The factors that speed up pitch angle "diffusion" are found to be: The intensity of the magnetic fluctuation $\delta B/B_0$; and the energy of the particles.

From Fig. 1, it can be seen that there is no particular difficulty for pitch angle diffusion across $\theta_p = 90°$ (e.g. [7]), and it is shown that stronger magnetic fluctuation give a flatter distribution. Also shown is the way in which a different initial pitch angle can influence the final distribution, after the same integration time of $10^4\ \omega_{bi}^{-1}$.

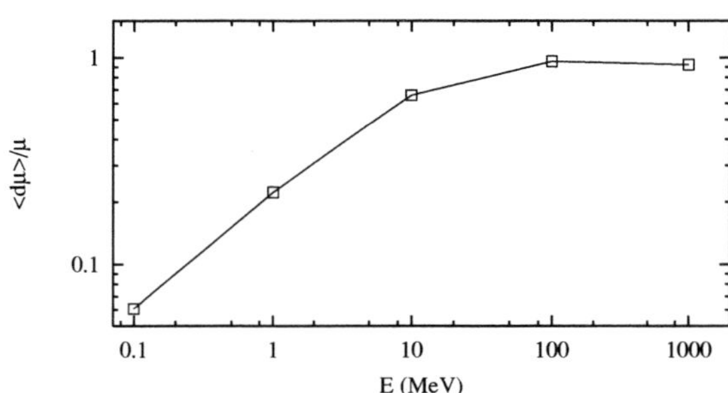

FIGURE 3. Relative change of the magnetic moment in function of the particle energy. Here for each points, the simulation are done for 100 particles integrating up to $500/\omega_{bi}$, and with fluctuation level $\delta B/B_0 = 1.0$.

CONSERVATION OF THE MAGNETIC MOMENT μ

We find two types of changes for the magnetic moment (the first adiabatic invariant) $\mu = mv_\perp^2/2B(\mathbf{r})$ during the integration: first, a periodic variation associated with cyclotron motion in a non uniform magnetic field; this variation is more or less the same for the different energies (see Fig. 2). Here $B(\mathbf{r})$ is the modulus of the local total magnetic field. Second, a non adiabatic variation associated with large magnetic field changes (strong magnetic gradients). This variation increases with energy (that is with the Larmor radius r_L) as it should. Indeed, it can be seen that the change in μ can be large even when $r_L \sim \lambda_{\min}/10$, and that the change in μ is particularly strong when B is weak.

In Fig. 3, we plot the relative variation of the magnetic moment $< d\mu > /\mu$ (which correspond to the non adiabatic variation) as a function of particle energy. We note that for the 1 MeV particles, $< d\mu > /\mu \sim 0.2$ which is quite high considering that the smallest length in the spectrum $\lambda_{\min}$ is about $100 r_L$. This shows that turbulence has a strong influence on particle motion and that guiding center approximation is not suitable for particles having energies equal or larger than 1 MeV with the parameter used in these simulations: $B_0 \sim 10$ nT and a correlation length $l_c \sim 0.03$ AU, which is tipical in solar wind at 1 AU from the Sun. In particular, we find that the first adiabatic invariant μ is not conserved. This shows that the many small variations of μ that a particle undergoes when moving in the presence of turbulence can give rise to the non conservation of the first adiabatic invariant even if $r_L \ll \lambda_{\min}$. This effect should be even more important in the actual solar wind turbulence because of the presence of much smaller wavelengths than those found in our numerical model.

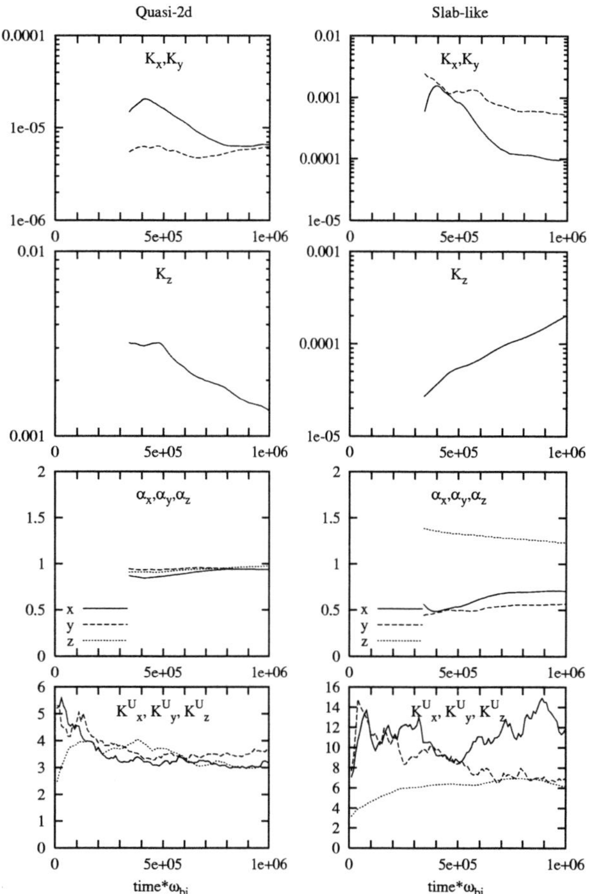

FIGURE 4. Evolution of $\mathcal{K}_i$, α_i, and K_i^U in function of time. The 1 MeV particles are integrated in a spectrum with quasi-2D turbulence ($l_{\|} \gg l_\perp$, panel on the left) and Slab-like turbulence ($l_{\|} \ll l_\perp$, panel on the right) with a fluctuation level $\delta B/B_0 = 1.0$.

TRANSPORT PROPERTIES

In order to determine quantitatively the transport properties, we compute the variances $\langle \Delta x_i^2 \rangle$, where $\Delta x_i = x_i - x_i^{(0)}$ $(i = x, y, z)$, as a function of time t. Then we make a fit of $\langle \Delta x_i^2 \rangle$, with the anomalous transport law $\langle \Delta x_i^2 \rangle = 2\mathcal{K}_i\, t^{\alpha_i}$ and determine α_i and $\mathcal{K}_i$ when t is large enough to attain asymptotic values (see [1, 2, 3, 5]. The results presented here were obtained with $t = 10^6 \omega_{bi}^{-1}$; in physical units, this corresponds to more than 10 days for $B = 10$ nT. Here, the exponent α_i characterizes the transport law: $\alpha_i = 1$ in the diffusive regime (Gaussian random walk); $\alpha_i = 2$ in the ballistic regime; $\alpha_i < 1$ in the case of trapping (subdiffusive regime), and $1 < \alpha_i < 2$ in the case of Lévy random walk (superdiffusive regime). The results for $\mathcal{K}_i$, α_i and for the kurtosis $K_i^U = \langle \Delta x_i{}^4 \rangle / \langle \Delta x_i{}^2 \rangle^2$ are shown in Fig. 4. The Gaussian value of K_i^U is 3.

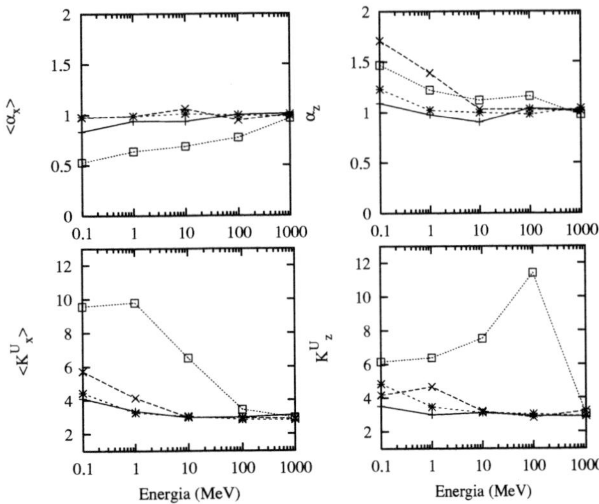

FIGURE 5. Exponent α and Kurtosis for different particle energy and different magnetic turbulence. Full line and pluses: quasi-2D with $\delta B/B_0 = 1.0$; long-dashed line and crosses: isotropic spectrum with $\delta B/B_0 = 0.5$; short-dashed line and stars: isotropic spectrum with $\delta B/B_0 = 1.0$; dotted line and squares: Slab-like with $\delta B/B_0 = 1.0$. The left panels show this coefficient averaged in the x and y directions which are perpendicular to the mean magnetic field direction.

In general we obtain Gaussian diffusion for the spectrum having larger $l_{\|}/l_{\perp}$ (that is quasi 2D turbulence, see Fig. 4 left panels). In such cases the coefficient $\mathcal{K}_i$ are the diffusion coefficients. We found also non Gaussian cases for lower energy, an example is shown in Fig. 4, right panels, with superdiffusion along the direction z, and subdiffusion along x and y. The values of the Kurtosis, reported in the bottom panels of Fig. 4, confirm that we are in Gaussian regime for quasi-2D turbulence, and in anomalous, non Gaussian regimes for Slab-like turbulence. The superdiffusion along z indicates that the particle motion along the magnetic field is intermediate between diffusive and scatter free.

Figure 5 shows the values of α_i and K_i^U as a function of particle energy. It can be seen that the larger the energy, the closer we are to a Gaussian regime.

The present simulation allows to directly evaluate the ratio $\mathcal{K}_{\perp}/\mathcal{K}_{\|}$ at least in the cases of Gaussian diffusion. This ratio is often required in astrophysical studies of particle transport (e.g., [8]). We note that $\mathcal{K}_{\perp}/\mathcal{K}_{\|}$ increases with the particle energy, with the turbulence fluctuation level, and passing from Slab turbulence to 2D turbulence spectrum, that is with the ratio $l_{\|}/l_{\perp}$. More complete results will be reported elsewhere.

CONCLUSIONS

We studied particle transport for 3 different cases of magnetic turbulence (isotropic, 2D and Slab spectrum), and for different particle energies. The main results are: (1) fast pitch angle diffusion leads to a rapid isotropization of the velocity distribution function; (2) the first adiabatic invariant μ is not well conserved if $r_L/\lambda_{\min}$ exceeds 10^{-2}; (3) motion along the average field is not ballistic but diffusive, even in the small Larmor radius cases considered up to now; (4) anomalous transport are found for $l_{\|} \ll l_{\perp}$; (5) perpendicular transport increase with particle energy, and with the ratio $l_{\|}/l_{\perp}$; (6) the ratio $\mathcal{K}_{\perp}/\mathcal{K}_{\|}$ varies in the range reported in the literature, between $\sim 10^{-4}$ and $\sim 10^{-1}$ [8].

ACKNOWLEDGMENTS

This work is part of a research programme which is financially supported by the Ministero dell'Università e della Ricerca Scientifica e Tecnologica (MURST), the Agenzia Spaziale Italiana (ASI), contract no. I/R122/01, and the High Performance computing center at the University of Calabria.

REFERENCES

1. Zimbardo, G., Veltri, P., Basile, G., Principato, S., Anomalous diffusion and Lévy random walk of magnetic field lines in three dimensional turbulence, *Phys. Plasmas*, **2**, 2653 –2663 (1995).
2. Pommois, P., Zimbardo, G., Veltri, P., Magnetic field line transport in three dimensional turbulence: Lévy random walk and spectrum models, *Phys. Plasmas*, **5**, 1288 –1297 (1998).
3. Pommois, P., Veltri, P., Zimbardo, G., Anomalous and Gaussian transport regimes in anisotropic 3-D magnetic turbulence, *Phys. Rev.* E, **59**, 2244 –2252 (1999).
4. Zimbardo, G., Veltri, P., Pommois, P., Anomalous, quasilinear, and percolative regimes for magnetic-field-line transport in axially symmetric turbulence, *Phys. Rev.* E, **61**, 1940 –1948 (2000).
5. Pommois, P., Veltri, P., Zimbardo, G., Kubo number and magnetic field line diffusion coefficient for anisotropic magnetic turbulence, *Phys. Rev.* E, **63**, 066405 (2001a).
6. Pommois, P., Veltri, P., Zimbardo, G., Field line diffusion in the solar wind magnetic turbulence and energetic particle propagation across heliographic latitudes, *J. Geophys. Res.*, **106** (A11), 24,965 –24,979 (2001b).
7. Bieber, J. W., Matthaeus, W. H., Smith, C. W., Proton and electron mean free paths: The palmer consensus revisited, *Astrophys. J.*, **420**, 294 –306 (1994).
8. Giacalone, J., Jokipii, J., Mazur, J. E., Small-scale gradients and large-scale diffusion of charged particles in the heliospheric magnetic field, *Astrophys. J.*, **532**, L75 –L78 (2000).

"Complexity" Induced Plasma Turbulence in Coronal Holes and the Solar Wind

Tom Chang

Center for Space research, Center for Theoretical Geo/Cosmo Plasma Physics
Massachusetts Institute of Technology, Cambridge, MA 02139 USA

Abstract. Global kinetic wave-particle interaction theories [1-4] have demonstrated that plasma turbulence emanating from coronal holes may efficiently accelerate the solar wind to observed characteristics and speeds [5,6]. The origin of such turbulence is not known although it is sometimes attributed to the small-scale reconnection processes. We demonstrate that sporadic, localized creation of magnetic coherent structures that arose from Alfvénic resonances can produce the type of turbulence as generally expected. When conditions are favorable, the coherent structures can merge, interact, bifurcate, convect and evolve into a "complex" state of forced and/or self-organized criticality (FSOC) leading to a broad-band power-law spectrum of plasma fluctuations of all scales [7,8]. Phenomenological models are constructed to represent the dynamic fluctuations near the coronal hole base. Dynamic renormalization-group calculations yield values of the scaling exponents that seem to agree with previously conjectured estimates [9]. This result is equally relevant to the basic understanding of the fraction of the non-propagating pseudo-2D plasma fluctuations that are prevalent in the solar wind [10]. A brief discussion on resonant heating of solar wind ions by non-propagating turbulent fluctuations that do not satisfy wave dispersion relations is also included [11,12].

INTRODUCTION

"Complexity" has become a hot topic in nearly every field of modern physics. In this short communication, we demonstrate that the sporadic and localized interactions of magnetic coherent structures are the origin of "complexity" in coronal hole and non-propagating pseudo-2D solar wind turbulence. The intermittent localized interactions, which generate the anomalous diffusion, transport and evolution of the macroscopic state variables of the overall dynamical system, may be modeled [13] by a triggered localized chaotic growth equation of a set of relevant order parameters. Such processes would generally pave the way for the global system to evolve into a "complex" state of long-ranged interactions of non-propagating fluctuations, displaying the phenomenon of forced and (or) self-organized criticality (FSOC). Dynamic renormalization-group analyses (DRG) for simple phenomenological models mimicking coronal hole turbulence yield reasonable ω-scaling exponents for the correlation of the transverse flux function and the trace of the transverse magnetic correlation tensor. These results and their generalizations should also have relevance in addressing the fraction of the non-propagating pseudo-2D turbulence that are generally detected in the solar wind [10]. We also address the concept of resonant energization of solar wind ions by non-propagating turbulent fluctuations that do not satisfy wave dispersion relations [11,12].

PLASMA RESONANCES AND COHERENT STRUCTURES

Most field theoretical discussions begin with the concept of propagation of waves. For example, in the MHD formulation, one can combine the basic equations and express them in the following propagation forms:

$$\rho d\mathbf{V} / dt = \mathbf{B} \cdot \nabla \mathbf{B} + \cdots \tag{1}$$

$$d\mathbf{B} / dt = \mathbf{B} \cdot \nabla \mathbf{V} + \cdots \tag{2}$$

where the ellipses represent the effects of the anisotropic pressure tensor, the compressible and dissipative effects, and all notations are standard. Equations (1,2) admit the well-known Alfvén waves. For such waves to propagate, the propagation vector **k** must contain a field-aligned component, i.e., $\mathbf{B} \cdot \nabla \rightarrow i\mathbf{k} \cdot \mathbf{B} \neq 0$. However, at sites where the parallel component of the propagation vector vanishes (i.e., at the resonance sites), the fluctuations are localized. Around these resonance sites (usually in the form of curves), it may be shown that the fluctuations are held back by the background magnetic field, forming Alfvénic coherent structures in the form of flux tubes or current filaments [8,11].

CP679, *Solar Wind Ten: Proceedings of the Tenth International Solar Wind Conference*,
edited by M. Velli, R. Bruno, and F. Malara

Generally, there exist various types of propagating modes (whistler modes, lower hybrid waves, etc.) in a continuum plasma. Thus, we envision a corresponding number of different types of plasma resonances and associated coherent structures that typically characterize the dynamics of the plasma medium under the influence of a background magnetic field. These coherent structures will wiggle, migrate, deform and undergo different types of motions under the influence of the local plasma and magnetic topology. In the next section, we will consider how the coherent structures can interact and produce the type of intermittency generally observed in dynamical plasmas.

INTERACTIONS OF COHERENT STRUCTURES

When filamentary current structures of the same polarity migrate toward each other, strong current sheets are generated [8,14,15]. As the electrons travel across the magnetic field lines, they would excite whistler fluctuations. Now, in analogy to the Alfvén resonances, singularities of $k_{\|} = \mathbf{k} \bullet \mathbf{B} = 0$ generally can develop at which whistler fluctuations cannot propagate. These "whistler resonances" can provide the nuclear sites for the emergence of coherent whistler structures, which is the analog of the coherent Alfvénic structures but with much smaller scales. The intermittent turbulence resulting from the intermixing and interactions of the coherent whistler and small scale Alfvénic structures in the intense current sheet region can then provide the coarse-grain averaged dissipation that allows the larger Alfvénic coherent structures to merge, interact, or breakup [8].

CORONAL HOLE TURBULENCE

The above intermittency description for non-propagating plasma turbulence may be modeled by the combination of a localized chaotic functional growth equation for a set of relevant order parameters and a functional transport equation for the control parameters [13]. The resulting transport processes will generally be sporadic and anomalous [16]. Due to the limitation of space in this short communication, we shall consider instead below two simple phenomenological models, which may have some relevance to coronal hole turbulence.

To evaluate the scaling properties of these kinetic models near complex states of long-ranged correlations (FSOC), we shall employ the functional method of classical path integrals and the dynamic renormalization-group (DRG) to be outlined below [17,18].

Model I

Assuming that the parallel mean magnetic field B_0 is sufficiently strong and the magnetic fluctuations dominate in the transverse directions, we introduce the flux function ψ for the non-propagating transverse fluctuations as follows,

$$\mathbf{B} = \mathbf{e}_z \times \nabla \psi + B_0 \mathbf{e}_z \tag{3}$$

This insures the magnetic field to be divergence free.

The coherent structures for such a system are generally flux tubes approximately aligned in the mean parallel direction [11]. Conservation of helicity indicates that the integral of ψ over a flux tube is approximately constant. Instead of invoking the standard reduced MHD formalism, here we simply consider ψ as a dynamic order parameter. As the flux tubes merge and interact, they may correlate over long distances, which, in turn, will induce long relaxation times near FSOC [7]. Let us assume that the base of the coronal hole is sufficiently broad compared to the cross sections of the coherent structures (or flux tubes), such that we may invoke homogeneity and assume the dynamics to be independent of boundary effects. We may then model the dynamics of flux tube mergings and interactions, in the crudest approximation, in terms of the following order-disorder intermittency equation:

$$\partial \psi_k / \partial t = -\Gamma_k \partial F / \partial \psi_{-k} + f_k \tag{4}$$

where ψ_k are the Fourier components of the flux function, Γ_k an analytic function of k^2, f_k a random noise which includes all the other effects that had been neglected in this crude model, and $F(\psi_k, k)$ the state function.

Model II

In the above model, we have neglected both the effects of diffusion and convection. We next construct a phenomenological model that includes the transport of cross-field diffusion. We now assume the state function to depend on the flux function ψ and the local "pseudo-energy" measure ξ. Thus, in addition to the dynamic equation (4), we now also include a diffusion equation for ξ. In Fourier space, we have

$$\partial \xi_k / \partial t = -Dk^2 \partial F / \partial \xi_{-k} + h_k \tag{5}$$

where ξ_k are the Fourier components of ξ, $D(k)$ is the diffusion coefficient, the state function is now $F(\psi_k, \xi_k, k)$, and h_k is a random noise. By doing so, we separate the slow transport due to diffusion of the local "pseudo-energy" measure ξ from the noise term of (4).

We shall now proceed to study the complexity and FSOC that can arise from the critical dynamics of these phenomenological models.

DYNAMIC RENORMALIZATION-GROUP ANALYSIS

For nonlinear stochastic systems near criticality, the correlations among the fluctuations of the random dynamical fields are extremely long-ranged and there exist many correlation scales. The dynamics of such systems are notoriously difficult to handle either analytically or numerically. On the other hand, since the correlations are extremely long-ranged, it is reasonable to expect that the system will exhibit some sort of invariance under scale transformations. A powerful technique that utilizes this invariance property is the technique of the dynamic renormalization-group (17,18, and references contained therein). As it is described in these references, based on the path integral formalism, the behavior of a nonlinear stochastic system far from equilibrium may be expressed in terms of a "stochastic Lagrangian L". Then, the renormalization-group (coarse-graining) transformation may be formally expressed as:

$$\partial L / \partial \ell = RL \qquad (6)$$

where R is the renormalization-group transformation (coarse-graining) operator and ℓ *is* the coarse-graining parameter for the continuous group of transformations. It will be convenient to consider the state of the stochastic Lagrangian in terms of its parameters $\{P_n\}$. Equation (6), then, specifies how the Lagrangian, L, flows (changes) with ℓ in the affine space spanned by $\{P_n\}$.

Generally, there exists a number of fixed points (singular points) in the flow field, where $dL / d\ell = 0$. At a fixed point, the correlation length should not be changing. However, the renormalization-group transformation requires that all length scales must change under the coarse-graining procedure. Therefore, to satisfy both requirements, the correlation length must be either infinite or zero. When it is at infinity, the system is by definition at criticality. The alternative trivial case of zero correlation length will not be considered here.

To study the stochastic behavior of a nonlinear dynamical system near a particular criticality, we can linearize the renormalization-group operator R about it. The mathematical consequence of this approximation is that, close to criticality, certain linear combinations of the parameters that characterize the stochastic Lagrangian L will correlate with each other in the form of power laws. This includes, in particular, the (k, ω), i.e. mode number and frequency, spectra of the correlations of the various fluctuations of the dynamic field variables. In addition, it can be demonstrated from such a linearized analysis that generally only a small number of (relevant) parameters are needed to characterize the stochastic state of the system near criticality (i.e., low-dimensional behavior; see [7]).

We have performed dynamic renormalization-group (DRG) analyses for the two kinetic models described in the previous section. We note that under the DRG transformation, the correlation function C of ψ_k should scale as:

$$e^{a_c \ell} C(k,\omega) = C(k e^{\ell}, \omega e^{a_\omega \ell}) \qquad (7)$$

where ω is the Fourier transform of the time t, ℓ the renormalization parameter as defined in the previous section, and (a_c, a_ω) the correlation and dynamic exponents. Thus, $C / \omega^{a_c / a_\omega}$ is an absolute invariant under the DRG, or $C \sim \omega^{-\lambda}$, where $\lambda = -a_c / a_\omega$. DRG analysis for Model I with Gaussian noise yields the value of λ approximately equal to 2.0.

DRG analyses performed for Model II for Gaussian noises for several approximations yield the value for λ to be approximately equal to 1.88 - 1.66.

Interestingly, for both models, DRG calculations give a value -1.0 for the ω-exponent for the trace of the transverse magnetic correlation tensor. Matthaeus and Goldstein [9] had suggested that such an exponent might represent the superposition of discrete structures emerge from the solar convection zone. We now turn our attention to the solar wind.

PSEUDO-2D NON-PROPAGATING TURBULENCE IN THE SOLAR WIND

Although the above discussions have been addressed to the turbulence to be expected in the coronal hole base, these results and their generalizations should also have direct bearing on the observed pseudo-2D non-propagating turbulence [10] and flux tubes [19] in the solar wind. It is expected that as the turbulent fluctuations emanate from the coronal hole base, some of the fluctuations will be mode-converted either linearly or nonlinearly into propagating Alfvén wave turbulence. Eventually, as the flux tubes in the open field lines interact, some of the wave turbulence would again be re-converted to non-propagating pseudo-2D fluctuations. The resulting fraction of the non-propagating turbulence can again be addressed by the ideas introduced in this short communication. Generally the magnetic fluctuations are composed of both the propagating and non-propagating components as originally pointed out by Belcher and Davis [20]. These ideas explain why generally the Alfvén ratio in the solar wind is less than one [21]. We demonstrate in the next section that non-propagating turbulent fluctuations can be an important contribution to the resonant energization of solar wind ions.

ENERGIZATION OF IONS

The non-propagating fluctuations discussed above entail a broadband power-law spectrum $\varepsilon(\mathbf{k},\omega)$ of the transverse electric field fluctuations. Because of the broad band nature of the fluctuations, they can provide continuous resonant energization to the solar wind ions. If the process is due to resonant phase-space diffusion, the perpendicular diffusion coefficient to the lowest order would then simply be [12]

$$D_{\perp} = (q/m)^2 \int\int d\omega d\mathbf{k}\, \varepsilon(\mathbf{k},\omega) R(\mathbf{k},\omega,\mathbf{v}) \qquad (8)$$

where (q, m, **v**) are the ion charge, mass, and velocity, respectively, and $R(\mathbf{k},\omega,\mathbf{v})$ is the resonance function that accommodates the required Doppler shifts for resonance interactions, coherence conditions, and stochastic broadening of interactions among the ions and fluctuations. We note that, for resonant energization of the ions, the turbulent $(\mathbf{k},\omega)$-spectra do not need to satisfy any dispersion relations as the fluctuations are generally non-propagating.

CONCLUSION

We have provided a theory of plasma turbulence in the coronal holes and solar wind in terms of complexity generated non-propagating fluctuations. Such non-propagating fluctuations are characterized by the mergings and interactions of coherent structures (generally in the form of flux tubes aligned in the parallel direction of the mean magnetic field) that arose from Alfvénic plasma resonances. Simple phenomenological kinetic models are introduced to characterize such dynamic interactions. Dynamic renormalization-group (DRG) analyses near forced and/or self-organized criticality (FSOC) for these models yield the scaling law $C \sim \omega^{-\lambda}$, where the exponent λ is estimated to be 2.0 and 1.88 - 1.66, and an ω-exponent for the trace of the transverse magnetic correlation tensor approximately equal to -1.0.

These ideas are equally applicable to the solar wind turbulence where non-propagating pseudo-2D fluctuations are commonly observed. The resulting broadband spectrum of the electric field fluctuations can conveniently energize solar wind ions without the fluctuations satisfying any wave dispersion relations.

ACKNOWLEDGMENTS

The author thanks Drs. Giuseppe Consolini, W.H. Matthaeus, Sunny W.Y. Tam and Cheng-chin Wu for useful discussions. The author also wishes to thank Drs. M.L. Goldstein for specifically mentioning this work during his invited presentation at the SW10. This research is supported in part by the AFOSR, NASA, and NSF.

REFERENCES

1. Tam, S.W.Y., and Chang, T., *Geophys. Res. Lett.*, **26**, 3189-3192 (1999).
2. Tam, S.W.Y., and Chang, T., *Geophys. Res. Lett.*, **28**, 1351-1354 (2001).
3. Vocks, C., and Marsch, E., *Geophys. Res. Lett.*, **28,** 1917-1920 (2001).
4. Isenberg, P.A., Lee, M.A., Hollweg, J.V., *J. Geophys. Res.*, **106**, 5649-5660 (2001).
5. Kohl, J.L., et al., *Astrophys. J*, **501**, L127-L131 (1998).
6. Cranmer, S.R., Field, G.B., and Kohl, J.L., *Astrophys. J.*, **518**, 937-947 (1999).
7. Chang, T., *IEEE Trans. Plasma Sci.*, **20**, 691-694 (1992).
8. Chang, T., *Phys. Plasmas*, **6,** 4137-4145 (1999).
9. Matthaeus, W.H., and Goldstein, M.L., *Phys. Rev. Lett.*, **57**, 495-498 (1986).
10. Matthaeus, W.H., Goldstein, M.L., and Roberts, D.A., *J. Geophys. Res.*, **95**, 20673-20683 (1990).
11. Chang, T., *Physica Scripta*, **T89**, 80-83 (2001).
12. Chang, T., *Nonlinear Processes in Geophysics*, **8**, 175-179 (2001)
13. Chang, T., and Wu, C.C., *Phys. Plasmas*, **9**, 3679-3684, 2002.
14. Wu, C.C., and Chang, T., *Geophys. Res. Lett.*, **27**, 863-866 (2000).
15. Wu, C.C., and Chang, T., *J. Atm. Solar-Terres. Phys.*, **63**, 1447-1453 (2001).
16. Chang, T., Wu, C.C., and V. Angelopoulos, *Physica Scripta*, **T98**, 48-51 (2002).
17. Chang, T., Nicoll, J.F., and Young, J.E., *Phys. Lett.*, **67A**, 287-290 (1978).
18. Chang, T., Vvedensky, D.D., and Nicoll, J.F., *Phys. Reports*, **217**, 279-362 (1992).
19. Bruno, R., Carbone, V., Veltri, P., Pietropaolo, E., and Bavassano, B., *Planet. Space Sci.*, **49**, 1201-1210 (2001).
20 Belcher, J.W., and Davis, L., *J. Geophys. Res.*, **76**, 3534-3563 (1971).
21. Tu, C.Y., and E. Marsch, *Space Sci. Rev.*, **73**, 1-210 (1995); and references contained therein.

A global three dimensional hybrid simulation of the interaction between a weakly magnetized obstacle and the solar wind

P. Trávníček*, P. Hellinger* and D. Schriver†

*Institute of Atmospheric Physics, AS CR, Boční II. čp. 1401, 14131 Prague 4, Czech Republic.
†Institute of Geophysics and Planetary Physics, UCLA, Los Angeles, 90095-1567, U.S.A.

Abstract. We study an interaction between the solar wind flow and a conductive obstacle with a weak dipole magnetic field using a three dimensional implementation of the hybrid code. We show that the hybrid approach is capable of describing most of the structures formed due to the interaction between the solar wind and a magnetized planet like the bow shock, proton foreshock, magnetopause, magnetosheath, northern and southern cusps and the current sheath.

INTRODUCTION

The interaction between a collisionless solar wind flow and an obstacle with a typical scale comparable to ion scales is beyond the applicability of standard magnetohydrodynamic (MHD) models. To include ion kinetic effects one can use a hybrid scheme, where electrons are considered as a massless fluid while ions are treated kinetically by the second order particle in cell scheme. Hybrid codes were applied successfully in studies of the interaction between the solar wind and unmagnetized obstacles such as Venus, Mars, and comets [1, 2, 3, 4, 5].

In this paper we are interested in the interaction between the solar wind plasma and a magnetized obstacle. In a hybrid model the interaction between the solar wind flow and a magnetized obstacle is determined by three parameters: the radius of the obstacle R, the velocity of the solar wind flow v_{sw}, and the dipole moment M. Table 1 gives an overview of different parameters for different planets of the Solar system. We present results of the three-dimensional (3-D) global simulation of a small planet with radius $R = 8.5\ L_{\mathrm{in}}$ (where L_{in} is proton inertial length) with a weak dipole magnetic field given by dipole moment $M = 1.5 \cdot 10^4\ B_{\mathrm{sw}} L_{\mathrm{in}}^3/\mu_0$ (where B_{sw} is the mean value of the solar wind magnetic field) embedded into a solar wind plasma flow with the velocity $v_{\mathrm{sw}} = 3.0\ v_A$ (v_A is the Alfvén speed). Table 1 shows that appart from unmagnetized planets Venus and Mars, Mercury is the most likely candidate for the description by global numerical experiments based on a kinetic model, however, the scales in our numerical experiment do not match any of the real parameters yet, since we are limited by available computing resources. In this paper we mainly describe qualitative results of the overall features of the interaction between a weakly magnetized obstacle and the solar wind.

We concentrate mainly on transition regions. One of these is a collisionless shock wave that results from the interaction between the "supersonic" flow and the planet.

The shock wave resulting from the interaction with a small obstacle usually has a multiple, "shocklet" structure. Observations of this shocklet structure in front of Mars have been discussed by Dubinin et al. [6] (see also [7]), who referred to the detailed numerical analysis by Omidi and Winske [8] (see also [9]). Dubinin et al. [6] studied the role of O^+ heavy ions in the process of formation of this structure. Shimazu [5] reported the formation of the shocklet structure in front of an unmagnetized obstacle in a 3-D global hybrid simulations for a pure electron-proton plasma. Shimazu [5] proposed that the formation of the shocklet structure is related to the finite Larmor radius of protons and appears for sufficently small obstacles. We compare our simulation results with these results and predictions.

THE MODEL

We consider a 3-D numerical model of a small planet with a weak dipole magnetic field in a solar wind flow with a bulk speed $v_{\mathrm{sw}} = 3.0\ v_A$ along the x_+ axis. The initial ambient magnetic field $\vec{B}\,(x,y,z) = \vec{B}\,(\cos\varphi\,\cos\psi, -\sin\varphi\,\cos\psi, \sin\psi)$ has been defined by setting $\varphi = 30^o$, $\psi = 0$ (no north-south component). The dipole field of

CP679, *Solar Wind Ten: Proceedings of the Tenth International Solar Wind Conference*, edited by M. Velli, R. Bruno, and F. Malara

the planet is given by the standard formula

$$\vec{B}_{\rm dip} = \frac{\mu_0 M}{4\pi r^3}\left[-2\sin\lambda\,\vec{e}_r + \cos\lambda\,\vec{e}_\lambda\right]$$
$$= \frac{M'}{r'^3} B_{\rm sw}\left[-2\sin\lambda\,\vec{e}_r + \cos\lambda\,\vec{e}_\lambda\right]$$

where M stands for the magnetic dipole moment of the planet, and r is the distance measured from the center of the planet. M and r are in SI units, while M' and r' are in the dimensionless units used in the simulation ($B_{\rm sw} L_{\rm in}^3/\mu_0$ and $L_{\rm in}$ respectively). Here also $\vec{e}_r$ and $\vec{e}_\lambda$ are unit vectors in the r and λ directions, with r, λ being radius and magnetic latitude respectively in a spherical coordinate system which has its origin in the center of the obstacle. No gravitational terms are included in the model. The numerical scheme describing the dynamics of the plasma is based on a 3-D version of the CAM algorithm developed by Matthews [10].

The spatial dimensions of the simulation box are $(L_x, L_y, L_z) = (104, 92, 92)\ L_{\rm in}$ with $dx = dy = dz = 0.8\ L_{\rm in}$. The radius of the planet is $R = 8.5\ L_{\rm in}$, which is roughly ten times smaller than the radius of Mercury. We use the magnetic dipole strenght $B_{\rm dip}$ given by magnetic field moment M (see Table 1), $M' = 1.5\cdot 10^4\, B_{\rm sw}\, L_{\rm in}^3\ /\ \mu_0$.

Initially the simulation box is filled with a homogeneous plasma (32 protons per cell) with $\beta_e = \beta_p = 0.5$. The box has open boundaries along the x axis. Solar wind protons, with a Maxwellian distribution function, are injected at the boundary plane $x = -0.33\ L_x$ with the bulk speed $v_{\rm sw} = 3\ v_A$. The particles can freely escape the box at the opposite ($x = 0.66\ L_x$) side. The volume has periodic boundary conditions in both directions perpendicular to the solar wind plasma flow. The other boundary in the model consists of the surface of the planet. In this particular case we use a conductive obstacle with all fields defined in the interior. For the particles reaching the surface of the planet we use reflective boundary condition. The sphere representing the planet is placed at the origin of the Cartesian coordinate system so that $x_{\rm min} = -0.33\ L_x$ and $x_{\rm max} = 0.66\ L_x$. The x_- axis points toward the Sun, z_+ points toward the magnetic north pole. The length scales in figures 1-3 are expressed in units of R.

The duration time of the simulation $20\ \Omega_{pi}^{-1}$ (Ω_{pi}: the proton gyro frequency in the solar wind) is comparable to the characteristic time $R/v_A = 8.5\Omega_{pi}^{-1}$. Due to the short time duration we do not consider the rotation of the planet.

RESULTS

In this section we summarize the results of the numerical experiment. The initial conditions lead to the formation of a complicated structure of transition regions that are similar to those observed around magnetized planets. The nine panels of Figures 1–3 show the main results of the simulation at time $t = 10\ \Omega_{pi}^{-1}$. The overall structure of the interaction is already formed, but the simulation has not yet reached its stationary state. Later on, at $t \sim 20\ \Omega_{pi}^{-1}$, boundary effects start to affect the flow.

Figure 1 shows the number density n as a grey scale plot. The left panel displays the equatorial-plane cut $n = n(x, y, z = 0)$, the middle panel displays the first meridian-plane cut $n = n(x, y = 0, z)$ and the right panel displays the second meridian-plane cut $n = n(x = 0, y, z)$. Arrows give the magnetic field vector $\vec{B} = \vec{B}\ (x, y, z)$ projected onto the given plane. One can clearly recognize the *quasi-perpendicular* and *quasi-parallel shock* regions on the left panel. The left panel also shows the anti-clockwise injection of protons from the front behind the planet around the planetary surface in the equatorial plane. The middle panel shows the formation of the *current sheath* where the density n is higher, while the vector $\vec{B}$ of the magnetic field changes its direction.

Figure 2 shows the magnitude of the magnetic field $|\vec{B}|$ in the same format as Figure 1. Arrows represent the magnetic field vector $\vec{B} = \vec{B}\ (x, y, z)$ projected onto the given plane. The magnetopause is clearly visible in the numerical experiment on the middle panel together with the formation of a *current sheath*. The position of the *magnetopause* can be approximately predicted using the expression for the balance between dynamic pressure of the solar wind flow and the magnetic pressure of the model dipole magnetic field $R_{MP}/R = [(B_{\rm eq}/B_{\rm sw})/(\sqrt{2}\ v_{\rm sw}/v_A)]^{1/3}$ where R_{MP} stands for the distance of the nose of the magnetosphere from the center of the planet and $B_{\rm eq}$ denotes the magnitude of the magnetic field at the planet's equator $B_{\rm eq} = 1.5\cdot 10^4/8.5^3\ B_{\rm sw}$. This expression provides a rough theoretical estimate for the position of the nose of the magnetosphere $R_{MP,\,MHD} \sim 1.7\ R$. Closer evaluation of the B^2 in the x_- direction suggests, that the magnetopause is formed around $R_{MP,\,model} \sim 1.4\ R$, however the shock front is not yet well separated from the magnetopause as we have not reached stationary state of the simulation. The structure of a *magnetosphere* begins to appear as seen by the formation of northern and southern *cusp regions*.

The last set of panels shown on Figure 3 describes flows of protons in the simulation box. The density n is shown in the the same format as Figure 1. Arrows represent the proton flow velocity $\vec{u}_p = \vec{u}_p\ (x, y, z)$ projected onto the given plane. Figures 1–3 reveal an asymmetry of the magnetosphere (see all right panels). The shock observed in the simulation has the shocklet structure (apparent especially in the area of the perpendicular shock, seen in all of the left panels).

TABLE 1. Some parameters of the plasma environment around planets of the Solar system obtained from different sources determining scales and units used in our numerical model. The first column gives the distance of the planet from Sun in $[AU]$. The next two columns list parameters of the solar wind environment of the planet: the proton particle density n_p and the strength of the interplanetary magnetic field B_{sw}. Values of the solar wind velocity v_{sw} are equal to the upstream Alfvén Mach number M_A of the solar wind flow surrounding the given planet. The Earth magnetic dipole moment is $M_{Earth} = 8 \cdot 10^{22}\ Am^2$.

Name	distance [AU]	n_p [cm^{-3}]	B_{sw} [nT]	L_{in} [km]	v_A [km s^{-1}]	radius R_M [km]	[L_{in}]	$v_{sw} = 300km/s$ [v_A]	dipole moment M [M_{Earth}]	[$L_{in}^3\, B_{sw}/\mu_0$]
Mercury	0.31	73	46	26.7	118	2 439	**91**	**2.5**	$4.7 \cdot 10^{-4}$	**$5.4 \cdot 10^{7}$**
	0.47	32	21	40	82	2 439	**60**	**3.6**	$4.7 \cdot 10^{-4}$	**$3.5 \cdot 10^{8}$**
Venus	0.72	14	10	61	59	6 052	**99**	**5**	-	-
Earth	1.0	7	6	86	50	6 378	**74**	**6**	1.0	**$2.6 \cdot 10^{9}$**
Mars	1.5	3.1	3.4	129	42	3 397	**26**	**7**	-	-

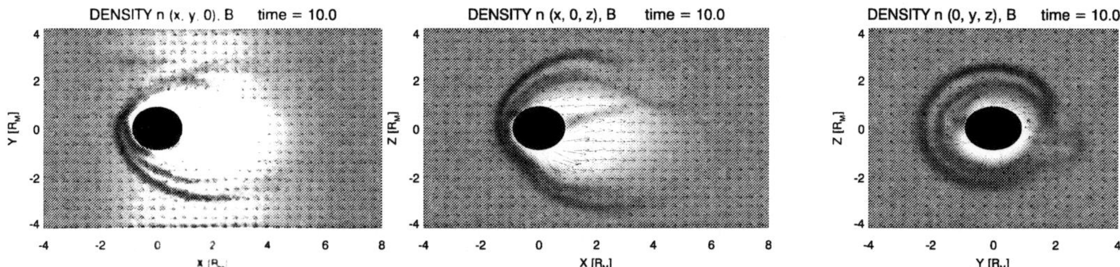

FIGURE 1. Gray scale plots of the density $n = n(x,y,z)$ in three perpendicular planes (left) $z = 0$, (middle) $y = 0$, (right) $x = 0$ through the center of the planet. Regions with higher density of protons are darker. Arrows correspond to the magnetic field vector $\vec{B} = \vec{B}\ (x,y,z)$ projected onto the given plane.

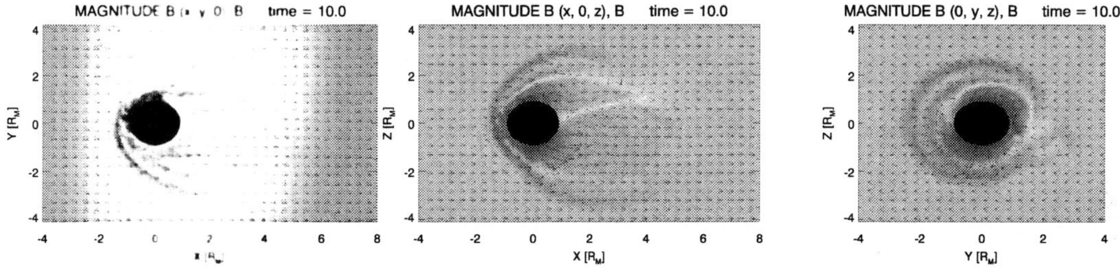

FIGURE 2. Gray scale plots of the magnitude of magnetic field $|\vec{B}| = B(x,y,z)$ in three perpendicular planes (left) $z = 0$, (middle) $y = 0$, (right) $x = 0$. Arrows represent the magnetic field vector $\vec{B} = \vec{B}\ (x,y,z)$ projected onto the given plane.

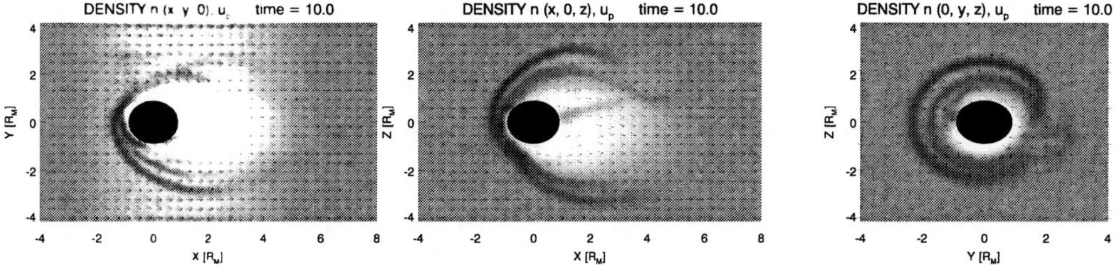

FIGURE 3. Gray scale plot of the density $n = n(x,y,z)$ in three perpendicular planes (left) $z = 0$, (middle) $y = 0$, (right) $x = 0$. Arrows correspond to the proton flow velocity $\vec{u}_p = \vec{u}_p\ (x,y,z)$ projected onto the given plane.

CONCLUDING REMARKS

In this paper we discuss results of a 3-D numerical model of a small planet ($R = 8.5\ L_{in}$) with a weak dipole magnetic field ($M = 1.5 \cdot 10^4\ B_{sw} L_{in}^3 / \mu_0$) embeded in a solar wind plasma flow with bulk speed $v_{sw} = 3.0\ v_A$. These values can be compared with the corresponding parameters describing the plasma environment of the solar system planets in Table 1.

The main results are shown on Figures 1–3: The simulation clearly shows important kinetic effects like the differences between a turbulent *quasi-parallel* part and a more stable *quasi-perpendicular* part of the bow-shock region (Figures 1–3, left panels).

The quasi-perpendicular part of the bow-shock has a *shocklet* structure. Shimazu [5] suggested that this structure forms due to finite Larmor radius effects of protons. Shimazu [5] used a 3-D hybrid code and showed that the shocklet bow-shock structure exists for an (unmagnetized) obstacle with a radius $R = 1.6\ r_L$ (where $r_L = v_{sw}\ \Omega_{pi}^{-1}$), but does not exist for an obstacle with a radius $R = 6.4\ r_L$. In our case the effective radius of the magnetized obstacle is about $R \sim 3\ r_L$ and the result is consistent with the predictions and results of Shimazu [5]. The formation of the shocklet structure deserves further study. There is still an open question on whether this effect is a transient [8], caused by nonstationarity of the numerical experiment, or whether it is a consequence of the kinetic processes around a relatively small obstacle. One may also ask whether the spatial resolution $dx = 0.8\ L_{in}$ (and worse, cf. [5, 4]) is sufficient to resolve the shock structure with a typical shock thickness of about L_{in}.

The spatial size of the *magnetosphere* with respect to the radius of the obstacle R roughly corresponds to the proportions of the Hermean magnetosphere. Hence the magnetic dipole field is hidden well inside the obstacle forming a very thin magnetosphere. The *magnetopause* is more apparent in the northern hemisphere on the middle panels and its position $R_{MP,\,model} \sim 1.4\,R$ is in a good agreement with predicted theoretical value $R_{MP,\,MHD} \sim 1.7\ R$. We have noticed the early stage of the formation of northern and southern polar *cusps*. The *magnetotail* is not yet formed due to the limited time span of the simulation. However we have seen regions with bi-directional proton flows together with bi-directional magnetic field in the simulation. This is a sign of the existence of *reconnection sites*.

The initial bulk speed v_{sw} of the solar wind plasma is set to $3.0\ v_A$ in entire volume of the simulation box. This flushes out particles initially located behind the planet and produces a wake with very low particle density. This wake begins to fill by the anti-clockwise injection of protons in the equatorial plane thus forming a temporary belt around the planetary surface. The ring continues to form until the end of the simulation at $t = 20\,\Omega_{pi}^{-1}$. The *current sheath* is initially formed in northern hemisphere slowly moving to the equatorial plane. The current sheath makes an angle of roughly 10° with respect to the equatorial plane. This changes in the latter stage of the simulation and the current sheath moves down towards the equator. We have also noticed formation of sites with relatively negligible particle flow in the magnetotail (vanishing $\vec{u}_p$ on left panel of Figure 3).

Small planets with small magnetospheres relative to the scales given by the plasma parameters (like Mercury) provide an opportunity to study the interaction between the solar wind and a real magnetosphere using a fully kinetic model. The size of the simulation box could easily be doubled in all three dimensions and further doubled in two chosen dimensions with respect to sizes we have used. This makes global three dimensional kinetic simulations of a magnetosphere of small planets possible such that we can study the influence of all parameters involved with the formation of the magnetosphere.

Due to limited computing resources we have not reached a stationary state in our simulation (cf. [9, 5]). The stationarity, the use of more realistic parameters, and formation of the shocklet structure will be the subject of future work.

ACKNOWLEDGMENTS

Authors acknowledge support of the grant ESA PRODEX 14529/00/NL/SFe and NSF International Grant INT-0010111. Authors thanks to N. Meyer and F. Pantellini who had useful suggestions to the manuscript.

REFERENCES

1. Brecht, S. H., *Geophys. Res. Lett.*, **17**, 1243–1246 (1990).
2. Brecht, S. H., and Ferrante, J. R., *J. Geophys. Res.*, **96**, 11209–11220 (1991).
3. Brecht, S. H., Ferrante, J. R., and Luhmann, J. G., *J. Geophys. Res.*, **98**, 1345–1357 (1993).
4. Brecht, S. H., *J. Geophys. Res.*, **102**, 4743–4750 (1997).
5. Shimazu, H., *J. Geophys. Res.*, **106**, 8333–8342 (2001).
6. Dubinin, E. M., Sauer, K., Baumgärtel, K., and Srivastava, K., *Earth Planets Space*, **50**, 279–287 (1998).
7. Dubinin, E. M., Sauer, K., Lundin, R., Baumgärtel, K., and Bogdanov, A., *Geophys. Res. Lett.*, **23**, 785–788 (1996).
8. Omidi, N., and Winske, D., *J. Geophys. Res.*, **92**, 13,409–13,426 (1987).
9. Omidi, N., and Winske, D., *J. Geophys. Res.*, **95**, 2281–2300 (1990).
10. Matthews, A., *J. Comput. Phys.*, **112**, 102–116 (1994).

Solar Wind Particle Distribution Function Fitted via the Generalized Kappa Distribution Function: Cluster Observations

M. N. S. Qureshi*, G. Pallocchia*, R. Bruno*, M. B. Cattaneo*, V. Formisano*, H. Reme†, J. M. Bosqued†, I. Dandouras†, J.A. Sauvaud†, L. M. Kistler**, E. Möbius **, B. Klecker‡, C. W. Carlson§, J. P. McFadden§, G. K. Parks§, M. McCarthy¶, A. Korth||, R. Lundin††, A. Balogh‡‡ and H. A. Shah§§

Istituto di Fisica dello Spazio Interplanetario (CNR) - Via Fosso del Cavaliere, 100 - 00133 Rome - Italy
†*CESR, Toulose, France*
***University of NewHampshire, Durham, NH, USA*
‡*MPE, Garching, Germany*
§*University of California, Berkeley, CA, USA*
¶*University of Washington, Seattle, WA, USA*
||*MPAe, Lindau, Germany*
††*SISP, Kiruna, Sweden*
‡‡*Imperial College, London, England*
§§*Government College University, Lahore. Pakistan*

Abstract. One of the major issues in space plasma physics is that in spite of the inhomogeneity of interplanetary plasma and the complicated magnetic field topology we do not find strong deviations from Maxwellian distributions as it would be expected for a quasi-collisionless plasma. However, the presence of high energy tail and shoulders in the profile of distribution function stimulate to look for a better analytic representation of the observed distributions. Therefore, here we adopt a non-Maxwellian distribution function such as the Ellipsian distribution function, which is the generalized form of the Kappa distribution function. In this paper we have analysed the solar wind data recorded by Cluster s/c during early 2001 and 2002 when the s/c were repeatedly immersed in the solar wind, ahead of the Earth's bow shock. Data were modeled with the help of the Ellipsian distribution function and values of the best fit parameters were successively used to characterize the solar wind kinetics at different locations of one of the four Cluster s/c.

INTRODUCTION

In the natural space environment, e.g. planetary magnetospheres, the solar wind and astrophysical plasmas are generally observed to possess a particle distribution function with a non-Maxwellian tail (Chun-yu and Summers, 1998; Summers et. al., 1994) and heat flux shoulders (Marsch et. al., 1982). An appropriate particle distribution function for modelling such plasmas would be a generalized kappa distribution function such as "Ellipsian" distribution function defined in equation-1. This distribution function is characterized by spectral indices k and m, where k is the index of inverse power-law tail and m represents the shoulders of the distribution function, and can be reduced to Maxwellian for m=1 and $k \to \infty$. In this paper we use "Ellipsian" distribution function to model the solar wind ion data, because the fit can be best achieved in both the high energy tail as well as shoulders in the distribution function.In the present study we used plasma measurements recorded by the CLUSTER mission.

CLUSTER MISSION AND CLUSTER ION SPECTROMETRY (CIS)

The ESA Cluster mission consists of four identical spacecraft which allow to study in three dimensions the small-scale plasma structures in the near Earth environ ment. The orbital parameters of the four spacecraft are slightly different to obtain a tetrahedral configuration in the regions of scientific interest. The size of this tetrahedron will be varied from 100 *km* to 18000 *km* during the course of the mission. The spacecraft cross the various near-Earth plasma regions through the year, for example

CP679, *Solar Wind Ten: Proceedings of the Tenth International Solar Wind Conference*, edited by M. Velli, R. Bruno, and F. Malara

the Earth magnetotail in the summer and the polar cusp and solar wind six months later. (Escoubet, C. P.; et al., 1997).

The Cluster Ion Spectrometry (CIS), one of the 11 experiments on board the cluster spacecraft, is capable of measuring both the cold and hot ions from the solar wind, the magnetosheath, and the magnetosphere (including the ionosphere) with good angular, energy and mass resolution (Rème, H.; et al., 1997).

CIS experiment employs two sensors to obtain the full three-dimensional ion distribution of the major species. One sensor, ion Composition and Distribution Function analyzer (CODIF) is a top-hat electrostatic analyzer followed by a time of flight section which measures the distribution of the major ion species (H^+, He^+, He^{++}, and O^+) from 0 to 40 keV q^{-1} with an angular resolution of 22.5^o x 11.2^o and two different sensitivities. The high sensitivity side has a larger geometric factor (used for low fluxes), the low sensitivity side has a smaller geometric factor (used for high fluxes). The other sensor, the Hot Ion Analyzer (HIA), is also a top-hat electrostatic analyzer which measures the distribution of the ions without distinction of mass from 5 eV q^{-1} to 32 keV q^{-1} with a maximum angular resolution of 5.6^o x 5.6^o and two different sensitivities (Réme, H.; et al., 1997).

ELLIPSIAN DISTRIBUTION FUNCTION

The distribution function which we shall use to model this data is called an "Ellipsian" distribution function, and has the form

$$f_E = A\left[1+\left\{\left(\frac{v_{\|}}{kv_{T\|}}\right)^2+\left(\frac{v_{\perp}}{kv_{T\perp}}\right)^2\right\}^m\right]^{-k} \qquad (1)$$

where

$$A = \frac{1}{\pi^{3/2}\, v_{T\perp}^2\, v_{T\|}}$$

This distribution function is a generalized version of the Lorentzian (kappa) distribution function, where m=1 always (Summers et.al. 1991, 1992, 1994; Thorne et. al. 1991). Data fits of the magnetosheath electrons (Qureshi, M. N. S.; M.Phil thesis) showed that fits for the distribution functions are better achieved by our model distribution function than the Lorentzian (kappa) distribution function and bi-Maxwellian.

We transform this model distribution function into a form, which we find more convenient, from the point of view of numerical modeling. For this purpose, we define different parameters as:

$$b = \frac{v_{T\perp}^2}{v_{T\|}^2} \quad and \quad c = \frac{1}{v_{T\perp}^2} \qquad (2)$$

And expressing $v_{\|}$ and $v_{\perp}$ through angle θ with respect to the ambient magnetic field B_o, as

$$v_{\|} = v\cos\theta \quad ; \quad v_{\perp} = v\sin\theta$$

and by taking the log on both sides of equation-1, we obtain

$$log f_E = a - k\,log\left[1+\left\{\frac{cv^2}{k}(b\cos^2\theta+\sin^2\theta)\right\}^m\right] \qquad (3)$$

where

$$a = log\,A$$

Thus equation-3 is in the form, which we use to model the distribution function from the data given. We note that each experimental point is characterized by a value of v (the velocity of the particles) and θ (from the $v_{\|}$, $v_{\perp}$ planes) and we determine (numerically) the values of a, b, c, k and m and hence the values of log f_E.

In order to achieve the best possible fit of data with the model distribution function given by equation-1, we carry out the following steps.

1. For each measured point which is characterized by θ and v, we take the square of the differences between the observed data and model, in the following form:

$$dif_i = [log f_{oi} - log f_{mi}]^2 \qquad (4)$$

Where f_o and f_m are the observed data and theoretical values of the distribution functions.

2. Secondly we take the sum of the differences dif_i.

$$sum = \sum dif_i$$

3. Now we minimize this sum to obtain the best possible values of a, b, c, m and k.

DATA SELECTION AND TREATMENT

In order to fit the data with our model distribution function we need the data in the form of log of distribution function, corresponding magnitude of the velocities of the particle and θ values on the $v_{\|}$ and $v_{\perp}$ plane in the plasma reference frame.

In order to select only protons, we used CODIF data. Unfortunately, we cannot use HIA data which would have a better angular resolution in phase space but cannot separate masses. Moreover, as fluxes are high in the solar wind, we used data only from the low sensitivity side.

Initially we have the data in the form of distribution function with thirty one energies and eighty eight angles measured every sixteen or eight seconds and correspondingly the number density, bulk speed and magnetic field data in the GSE coordinate system which are then transformed in the instrumen coordinate system.

In order to improve the statistics we average 30 successive distribution functions (sometimes 35) and normalized the averaged distribution function by number density. Then we transform the measured velocities from the instrument to the plasma frame of reference.

Each measured velocity (in the plasma reference frame) is then represented in the $v_{\parallel}$ and $v_{\perp}$ plane and is identified by v and θ.

To select the data, we chose time intervals in which the plasma parameters had little fluctuations and computed the time averages over such intervals, as described above. One such interval of 21st Feb. 2001 can be seen in the Fig.-2.The other time intervals are similar to this one in this respect but not in position. In that plot the selected time interval is from 77482 to 77947 seconds just before the shock. For this day we can also see the positions of the spacecraft in the solar wind in Fig.-1.

All the selected intervals which are considered in this paper are from the s/c4, because this came out to be the only spacecraft available in the solar wind mode in low sensitivity as explained above.

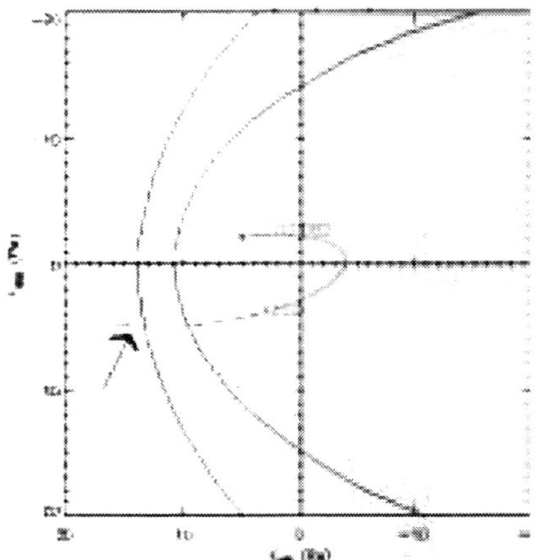

FIGURE 1. Position of the s/c4 on 21st Feb. 2001. The selected interval is indicated by the arrow in the solar wind.

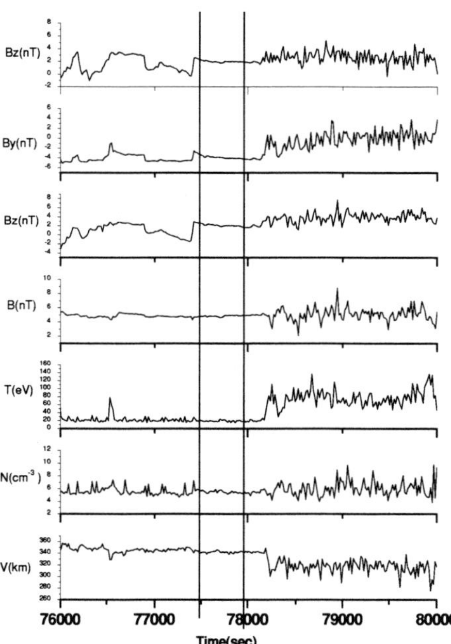

FIGURE 2. 16 sec. averages of magnetic field and Plasma parameters on 21st Feb. 2001 and the selected interval in between the vertical lines.

FITTING RESULTS

We now present some fitting results and the corresponding values of the different parameters. In this procedure we fit all the points but here we are showing the results only in certain directions in the $v_{\parallel}$ and $v_{\perp}$ plane. The solar wind ion distributions are focused in a very narrow angular range and few energy steps in phase space, therefore we never have a large number of points in certain directions.

In Fig.-3 we can see the fits in certain directions for the selected intervals from 21st Feb. 2001, 28th March 2001 and 11th Feb. 2002 and the corresponding parameters in the Table-1. For sake of brevity, we are not showing fit for 22nd Feb. 2001, although the parameters are given for this day in the Table-1, because of the similar type of fit in the similar conditions of slow solar wind as day 21st Feb. 2001.

SUMMARY AND DISCUSSION

In this paper we fitted the solar wind ion data with "Ellipsian" distribution function. All the selected intervals are from the slow solar wind except the interval of 28TH Mar. 2001 in which solar wind is intermediate. This interval is also different from the other intervals in the respect that it is in a magnetic cloud. Data fits in Fig.-3 show that the Ellipsian distribution function fits quite well for high energy tails as well as shoulders in the distribution function.

The parameters m and k in the Table-1 are the spectral indices and in general represent the flat part and the high energy tail of the distribution function respectively.

TABLE 1. Distribution function's parameters for the Solar Wind.

Day	a	b	c	k	m
21st Feb. 2001	-15.1582	0.6925	0.000628	0.8456	2.8958
22nd Feb. 2001	-15.0088	0.7511	0.001186	0.9798	2.2758
28th Mar. 2001	-15.8503	0.9936	0.000268	0.8348	2.6785
11th Feb. 2002	-14.9241	0.6278	0.0004619	2.8642	1.2156

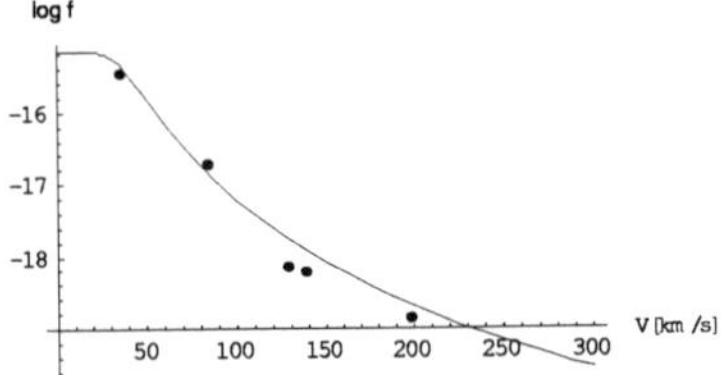

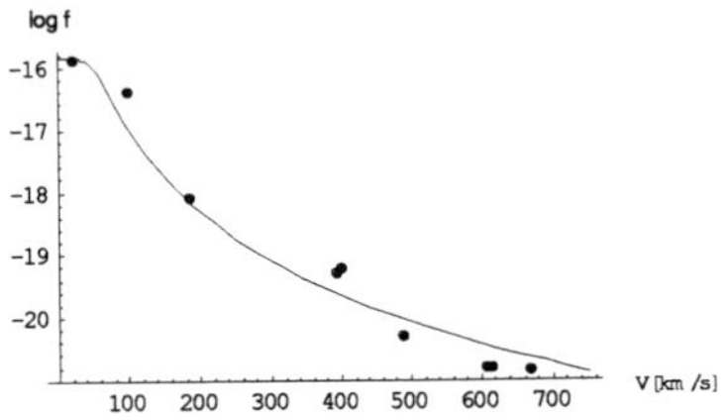

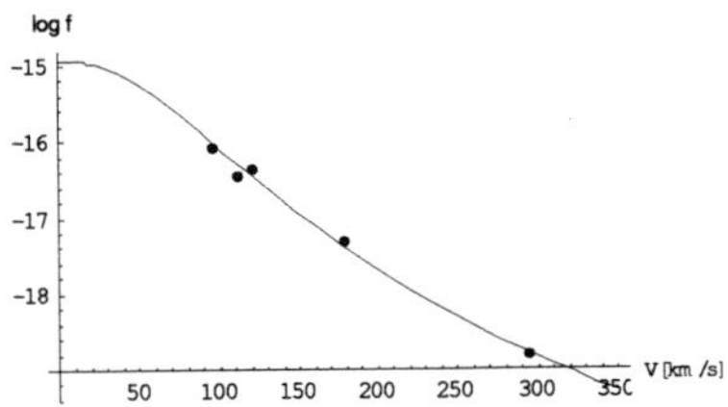

FIGURE 3. Data fits for 21st Feb. 2001, 28th March 2001 and 11th Feb. 2002 are shown in the upper, middle, and lower panels, respectively. The dots indicate the observed data points and lines indicate the model distribution functions. Each plot refers to only one direction in $v_{\|}$ and $v_{\perp}$ plane.

If value of k becomes higher it tends to shift towards Maxwellian (Leubner and Schupfer 2000; Summers et al. 1991). Typical values of k in space plasmas lies in the range $1 < k \leq 5$ (Summers and Thorne, 1992, and references therein whereas in our case it lies in the range $0 < k \leq 5$). The other parameters like a which contains log of number density and thermal speeds is almost the same for all the intervals and b, which is the anisotropy ratio (see equation.-2), is not much different in all three intervals except 28th March, where the value of b is almost one. This particular time interval is in the magnetic cloud and due to the compression and thermal heating of the plasma, anisotropy is much reduced. In the last interval of 11th Feb. 2002, the value of m is lowest and $k$is highest than the other intervals indicating the presence of less high energy tail in the distribution function. Also the value of b is less than from the other intervals, indicating high anisotropy with parallel temperature greater than the perpendicular temperature.

We conclude that data fits of the magnetosheath electrons (Qureshi M. N. S.; M.Phi Thesis) and solar wind ions distribution functions can be best achieved by our "Ellipsian" distribution function (with $m > 1$) than by Generalized Lorentzian (Kappa) distribution function where m=1 always.

This is a preliminary study and only one CLUSTER spacecraft is available up to now for this type of data analysis. But in the future, we want to carry out the same analysis with the data from at least two spacecraft at the same time, located at different positions to see the spatial variation in plasma.

REFERENCES

1. Chun-yu, Ma., Summers, D., "Formation of Power-Law Energy Spectra in Space Plasmas by Stochastic Acceleration due to Whistler-Mode Waves ", Geophysical Research Letters,Vol.25,No.79: 4099-4102, 1998.
2. Escoubet, C. P., Schmidt, R., M. L. Goldstein, "CLUSTER-Science and Mission Overview", Space Science Reviews 79: 11-32, 1997.
3. Leubner, M. P., Schupfer, N., "A General Kinetic Mirror Instability Criterion for Space Applications", Journal of Geophysical Research, 106, A7, 12993-12998, July 1, 2001.
4. Marsch, E., et al., "Solar Wind Protons: Three-Dimensional Velocity Distributions and Derived Plasma Parameters Measured Betweeen 0.3 ans 1 AU", Journal of Geophysical Research, 87, A1, 52-72, Jan. 1, 1982.
5. Qureshi, M. N. S., "Parallel Propagating Waves In A Plasma With An Ellipsian Distribution Function" M. Phil Thesis 1999-2001.
6. Rème, H., et al., "The Cluster Ion Spectrometry (CIS) Experiment", Space Science Reviews 79: 303- 350, 1997.
7. Summers, D., Thorne, R. M., "The Modified Plasma Dispersion Function", Physics Fluids B 3 (8), August 1991.
8. Summers, D., Thorne, R. M., "A New Tool for Analyzing Microinstabilities in Space Plasmas Modeled by A Generalized Lorentzian (Kappa) Distribution", Journal of Geophysical Research, 97, A11, 16827-16832, November 1, 1992.
9. Summers, D., Xue, S., Thorne R. M., "Calculation of The Dielectric Tensor for A Generalized Lorentzian (Kappa) Distribution Function", Physics Plasmas 1 (6), June 1994.
10. Thorne, R. M., Summers, D., "Landau Damping in Space Plasmas", Physics Fluids B 3 (8), August 1991.

A Kinetic Shell Description of the Ion Cyclotron Anisotropy Instability

Philip A. Isenberg

Department of Physics and Institute for the Study of Earth, Oceans and Space
University of New Hampshire, Durham, NH 03824 USA

Abstract. We consider the simulation investigations of the electromagnetic ion cyclotron anisotropy instability which have been undertaken by Gary and co-workers, in particular their claim that an unstable bi-Maxwellian ion distribution will retain its bi-Maxwellian character when it reaches the asymptotic stable state. Quasilinear theory predicts that the time-asymptotic ion distribution will be given by a form of the kinetic shell distribution, with no density gradients along surfaces of constant energy in the wave frame. These constant energy surfaces are not ellipses, as would be required for contours of a bi-Maxwellian distribution. We derive an expression for this asymptotic distribution predicted by quasilinear theory. We then present an example, and compare it with a published simulation result by Gary et al. We find some aspects of our results in agreement with the simulation, but discrepancies appear at small $v_{\|}$. Further comparative studies could provide useful information on the applicability of quasilinear theory and/or hybrid simulations to wave-particle interactions in space plasmas.

INTRODUCTION

Highly anisotropic ion distributions with $T_{\perp} >> T_{\|}$ are linearly unstable and will spontaneously reduce their anisotropy while generating ion cyclotron waves. This anisotropy instability has been investigated in detail by Gary and co-workers, primarily through the use of hybrid simulations[1-6]. These simulations start with highly perpendicular bi-Maxwellian distributions, and proceed through the generation of ion cyclotron waves and the reduction of the anisotropy until a stable state is reached. Qualitatively this evolution is consistent with theoretical expectations, but the authors also claim that the final stable distribution has remained a bi-Maxwellian. However, such a conclusion is at odds with our understanding of the quasilinear resonant cyclotron interaction which is thought to be responsible for the instability.

In our recent investigations of this interaction in the context of coronal heating and solar wind acceleration [7-9], we have developed some tools for constructing distributions with a stable "quasilinear plateau". Here, we will use these tools to derive the stable distribution produced by an unstable bi-Maxwellian and compare it with the simulation results.

RESONANT CYCLOTRON INTERACTION

The quasilinear description of the resonant cyclotron interaction invokes a diffusion of the ions in pitch angle along surfaces of constant ion energy in the reference frame of the resonant wave. As with any diffusive process, the interaction brings about a reduction in the density gradients along these surfaces. An ion cyclotron instability results when the gradient reduction causes the ions to lose energy in the plasma frame. This energy goes into generating the ion cyclotron waves.

The stable asymptotic state then has no density gradients on these constant-energy surfaces. We have termed this state of marginal stability a "kinetic shell distribution".

KINETIC SHELL DESCRIPTION

If we know the dispersion relation of the resonant waves and can assume that it does not change significantly during the interaction, we can calculate the shape of these constant-energy surfaces.

Then, by redistributing the initially bi-Maxwellian ions on these surfaces, we can obtain the final stable state predicted by quasilinear theory. In principle, we

CP679, *Solar Wind Ten: Proceedings of the Tenth International Solar Wind Conference*, edited by M. Velli, R. Bruno, and F. Malara

can also determine the intensity and spectrum of the generated waves by tabulating the energy losses of the ions.

The shell shapes in $(v_{\|}, v_{\perp})$ space are given by the expression [10-12]

$$v_{\|}^2 + v_{\perp}^2 - 2\int_0^{v_{\|}} V_{ph}(v_{\|}')\,dv_{\|}' = \text{const} = \eta^2 \quad (1)$$

where V_{ph} is the phase speed of the wave resonating with an ion of streaming speed $v_{\|}$, and η is a constant coordinate label for each shell.

These shells are symmetric in $v_{\|}$, since the sign of V_{ph} for the resonant wave changes when the sign of $v_{\|}$ of the resonant ion changes. Throughout this paper, we will take the evolution of the system to be symmetric in $v_{\|}$: The initial bi-Maxwellian generates equal wave intensities in both directions along the magnetic field, and approaches the symmetric kinetic shell distribution.

The redistribution of the bi-Maxwellian ions is simply effected by rewriting the initial distribution as a function of the new variables $(\eta, v_{\|})$

$$F\left(v_{\|}, v_{\perp}^2\right)$$

$$= F\left[v_{\|}, \eta^2 - v_{\|}^2 + 2\int_0^{v_{\|}} V_{ph}(v_{\|}')\,dv_{\|}'\right] \quad (2)$$

then averaging this function over $v_{\|}$, holding η fixed, to obtain the total density on each shell.

So, for an initial bi-Maxwellian (parallel thermal speed v_{th}, anisotropy $T_{\perp}/T_{\|}$)

$$F\left(v_{\|}, v_{\perp}\right) = A\,\exp\left[-\frac{v_{\|}^2}{2v_{th}^2}\right]\exp\left[-\frac{v_{\perp}^2}{2v_{th}^2}\frac{T_{\|}}{T_{\perp}}\right], \quad (3)$$

the asymptotic stable distribution is

$$f(\eta) = \frac{A}{v_o(\eta)}\int_0^{v_o(\eta)} dv_{\|}\,\exp\left[-\frac{v_{\|}^2}{2v_{th}^2}\right]\exp\left[-\frac{T_{\|}}{2v_{th}^2 T_{\perp}}\left(\eta^2 - v_{\|}^2 + 2\int_0^{v_{\|}} V_{ph}\,dv_{\|}'\right)\right] \quad (4)$$

where $v_o(\eta)$ is the maximum $v_{\|}$ of the ηth shell.

The shell shapes given by (1) will not be ellipses centered on the origin in $(v_{\|}, v_{\perp})$ space, as would be required for the asymptotic distribution to be truly bi-Maxwellian. However, it is possible that the asymptotic distribution could exhibit bi-Maxwellian features under some circumstances. This will be shown in the example below.

EXAMPLE

As an example, we consider the first case presented by Gary et al. [4], which simulated the evolution of an electron-proton plasma with an initial proton bi-Maxwellian distribution of $\beta_{\|p} = 0.5$ and $T_{\perp p}/T_{\|p} = 2.94$.

Such a plasma is not really cold, but we will take the dispersion relation for parallel-propagating ion cyclotron waves in a cold electron-proton plasma,

$$V_{ph} = \pm V_A\sqrt{1 - \frac{\omega}{\Omega_p}} \quad (5)$$

which results in shells given in parametric form as [12]

$$v_{\perp}^2(y) = \eta^2 - V_A^2\left[y^{-2} - \ln\left(\frac{|y| + \sqrt{y^2+4}}{|2y|}\right)\right] \quad (6)$$

$$v_{\|}(y) = V_A\left(\frac{\pm\sqrt{y^2+4} - y}{2} - y^{-1}\right) \quad (7)$$

We obtain the kinetic shell distribution which diffuses along these shells from the initial bi-Maxwellian by a straightforward numerical integration. The resulting asymptotic distribution is shown in Figure 1, as a function of the normalized shell parameter η/V_A, where it is understood that the instability operates on

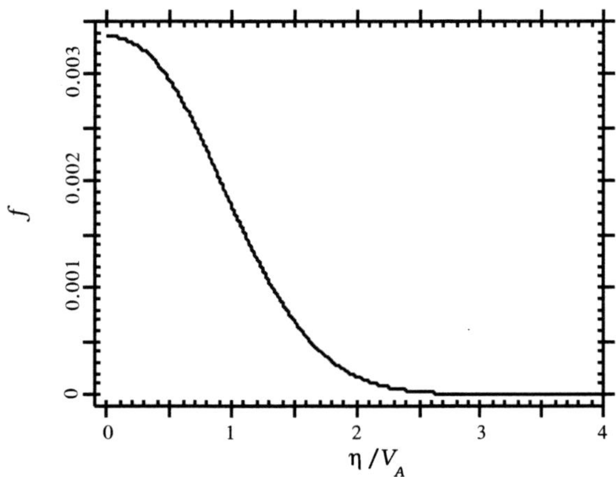

FIGURE 1. Asymptotic shell distribution obtained from (4) using the cold plasma dispersion relation (5). We have set $A\upsilon_{th}^2$ = 3.356 x 10^{-3} to match the bi-Maxwellian (3) to the initial distribution of Gary et al.

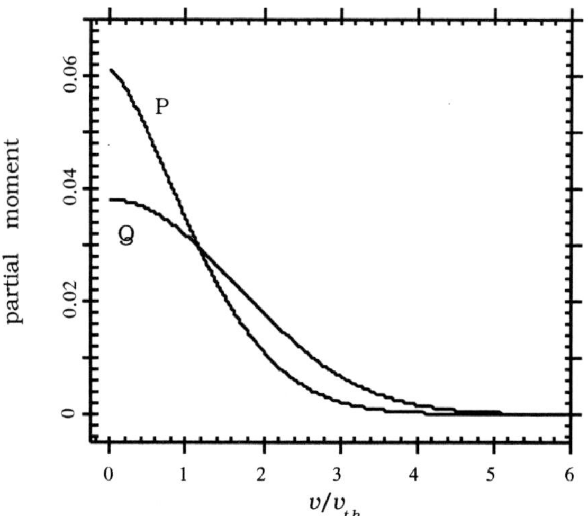

FIGURE 2. Partial moments (8) and (9), of the distribution in Figure 1.

the gradients on both sides of $\upsilon_{||} = 0$ and the distribution remains symmetric in $\upsilon_{||}$.

Gary et al. characterize the distributions in their simulations by calculating partial moments, equivalent to the ion density on slices through the distribution along planes oriented normal or parallel to the magnetic field:

$$P(\upsilon_{||}) = 2\pi \int f \, \upsilon_{\perp} \, d\upsilon_{\perp} \tag{8}$$

$$Q(\upsilon_z) = \int f \, d\upsilon_y \, d\upsilon_{||} \tag{9}$$

where y and z are directions perpendicular to the magnetic field.

These partial moments would have Maxwellian shapes for a bi-Maxwellian distribution. The various partial moments of the initial and final distributions in Gary et al. are shown in their Figure 1. The authors point out that the partial moments of their final state are well fit by Maxwellians, with linear correlation coefficients of R = 0.997 or 0.998. They do not give values for the thermal widths of these moments, but we estimate from their figure that the width of P is ~ 1.2 times the initial thermal speed, and the width of Q is ~ 1.6 times this speed. They base their conclusion that a bi-Maxwellian shape is maintained by the instability on this result.

We show in Figure 2 the equivalent partial moments of the distribution in our Figure 1, as functions of the initial parallel thermal speed.

We see that the transverse moment, Q, agrees well with the equivalent moments in Gary et al. The function Q is extremely well fit by a Maxwellian function of thermal width 1.635 υ_{th}, with correlation coefficient R = 0.99998. However, the parallel moment, P, is taller and narrower than the curve given by the simulation results, best fit by a Maxwellian of thermal width 0.9929 υ_{th}, with correlation coefficient R = 0.998.

These results indicate that our shell calculation leaves more protons at small $\upsilon_{||}$ than in the simulation of Gary et al. There are several points to consider when addressing this discrepancy:

a) The cold plasma dispersion relation (5) may not give an accurate description of the shell shape in this thermal plasma. If this is so, we would expect the differences to show up at small $\upsilon_{||}$.

b) The simulation cannot treat the effect of waves with inverse wavenumber smaller than the spatial resolution used by Gary et al. Again, this would affect the protons at small $\upsilon_{||}$, which are resonant with large wavenumber waves.

c) The shell calculation given here assumes an unstable interaction with ion cyclotron waves up to infinite wavenumber. That is, we do not consider the existence of a "resonance gap" which may also lead to noticeable differences at small $\upsilon_{||}$.

These considerations are not independent, and it is also not clear what the true physical resolution should be. Additionally, we expect that non-linear processes are influencing the simulation results, and these processes may be either physically valid or artificial numerical effects.

CONCLUSIONS

It would be particularly interesting to explore numerical simulations at low plasma beta with high spatial resolution. Further comparative investigations could provide useful information on the applicability of quasilinear theory and/or hybrid simulations to wave-particle interactions in space plasmas.

We note as well that the question of the asymptotic stable state has been addressed with an entropy analysis by Nakamura [13] under the assumption of dispersionless waves. It will be worthwhile to investigate the results of this procedure when wave dispersion is included.

ACKNOWLEDGMENTS

The author is grateful for valuable conversations with S. P. Gary, J. V. Hollweg, and B. J. Vasquez. This work was supported in part by the NASA Sun-Earth Connections Theory Program under grants NAG5-8228 and NAG5-11797 and by the NASA Living With a Star Program under grant NAG5-10835.

REFERENCES

1. Gary, S. P. and M. A. Lee, *J. Geophys. Res.*, **99**, 11,297 (1994).
2. Gary, S. P., M. B. Moldwin, M. F. Thomsen and D. Winske, *J. Geophys. Res.*, **99**, 23,603 (1994).
3. Gary, S. P., M. F. Thomsen, L. Yin and D. Winske, *J. Geophys. Res.*, **100**, 21,961 (1995).
4. Gary, S. P., V. M. Vazquez and D. Winske, *J. Geophys. Res.*, **101**, 13,327 (1996).
5. McKean, M. E., D. Winske and S. P. Gary, *J. Geophys. Res.*, **99**, 11,141 (1994).
6. Ofman, L., A. Viñas and S. P. Gary, *Astrophys. J.*, **547**, L175 (2001).
7. Isenberg, P. A., M. A. Lee and J. V. Hollweg, *J. Geophys. Res.*, **106**, 5649 (2001).
8. Isenberg, P. A., *J. Geophys. Res.*, **106**, 29,249 (2001).
9. Isenberg, P. A., *J. Geophys. Res.*, in press (2002).
10. Rowlands, J., V. D. Shapiro and V. I. Shevchenko, *Sov. Phys. JETP*, **23**, 651 (1966).
11. Gendrin, R., *J. Atmosph. Terr. Phys.*, **30**, 1313 (1968).
12. Isenberg, P. A. and M. A. Lee, *J. Geophys. Res.*, **101**, 11,055 (1996).
13. Nakamura, T. K., *Phys. Plasmas*, **7**, 4812 (2000).

A fluid description of kinetic effects for Alfvén wave trains

T. Passot and P.L. Sulem

CNRS, Observatoire de la Côte d'Azur, B.P. 4229, 06304 Nice Cedex 4, France

Abstract. A generalized Kinetic Derivative Nonlinear Schrödinger equation for the multidimensional dynamics of Alfvén wave trains is derived from the Vlasov-Maxwell equations. It includes the coupling to the mean effect of the longitudinal fields averaged along the direction of propagation, that plays a significant role in the transverse dynamics of extended wave packets, in particular in the phenomenon of transverse collapse and Alfvén wave filamentation. This equation can be viewed as an asymptotically exact fluid model including the Landau damping, that can be used as a benchmark for more general magnetohydrodynamic type descriptions of collisionless plasmas.

INTRODUCTION

Astrophysical and geophysical plasmas are usually permeated by an ambient magnetic field, leading to a strong anisotropy of the observed fluctuations. In the case of incompressible fluids, the dynamics is, to leading order in an expansion in terms of the magnitude of the external field, two-dimensional in planes perpendicular to the ambient field with a coupling between these planes that reduces to linear Alfvén waves [1, 2]. The longitudinal transfer is subdominant and the energy spectrum displays a rapid decay in this direction. Furthermore, the nature of the transverse turbulence is prescribed by the ratio of the local Alfvén and eddy-turnover times. In the range of scales for which the Alfvén time is the shortest, a weak turbulence regime is expected, characterized by a k^{-2} energy spectrum [3, 4]. In contrast, in the spectral domain where the opposite ordering holds, the turbulence is strong with a spectrum whose behavior is still debated between $k^{-5/3}$ and $k^{-3/2}$ [5, 6, 7, 8]. In the $(k_\perp, k_\|)$-plane, the two regimes are separated by a "critical curve" of the form $k_\| = k_\perp^{2/3}$, according to the Goldreich-Shridhar theory [9].

The presence of compressibility can alter this picture. As long as the Alfvénic Mach number is small, the system is governed by the reduced MHD equations, where the transverse dynamics is to leading order incompressible [10, 11]. This dynamics is purely 2D when the β of the plasma, defined as the square ratio of the sound and Alfvén speeds, is far from unity, while it becomes 2.5D as $\beta = 1$ is approached. The reduced MHD regimes do not necessarily require that the longitudinal characteristic scale be much larger than the transverse one. Indeed, as shown in [12], the derivatives of the longitudinal fields along the ambient magnetic field are comparable to the transverse gradients of the perpendicular components. This situation results from the presence of nonlinear Alfvén waves that, for parallel propagation in a specific direction, are governed by the Derivative Nonlinear Schrödinger equation (DNLS) and can coexist with the transverse turbulence without affecting it, at least when β is far from unity. When Alfvén waves are propagating in both directions, they contribute to feed the transverse dynamics by the usual coupling between contra-propagating Alfvén pulses. In this case, even if not initially present, backward propagating waves will be created by the decay instability when $\beta < 1$ [13, 14, 15].

The Alfvén wave dynamics results from the competition between nonlinear, dispersive and dissipative processes. The first two effects were extensively studied in the past few years. As an example, a possibly important process that can affect the weak turbulence dynamics results from the filamentation instability [16] that leads to transverse collapse of a weakly nonlinear quasi-monochromatic dispersive Alfvén wave train and formation of intense magnetic filaments [17]. This transverse dynamics is essentially governed by the two-dimensional nonlinear Schrödinger equation for the amplitude of the wave. This envelope equation admits solutions that blow up in a finite time [18, 19]. As shown in [20], this asymptotic description provides a quantitatively accurate description of the early stages of the collapse. The influence of the oblique instabilities is also captured in this formalism, by retaining the coupling to magnetosonic waves. When such instabilities can develop, they tend to weaken the process of wave energy concentration, while enhancing the development of density gradients. Note

CP679, *Solar Wind Ten: Proceedings of the Tenth International Solar Wind Conference*, edited by M. Velli, R. Bruno, and F. Malara

that a transverse concentration of wave energy is still observed in the numerical simulations, for Alfvén waves of moderate amplitude (that are not amenable to an envelope description), leading to the formation of helicoidal magnetic filaments [21]. The possibility for such structures to form in a context of wave turbulence and their effect on the spectral transfer of energy is an important question that has not yet been addressed.

Concerning dissipation in a collisionless medium, the proper mechanisms to be considered originate from the Landau damping of the ion acoustic waves and from the ion cyclotron resonance (when considering small enough scales). A quantitative description of such effects is delicate since a fluid formalism neglecting kinetic effects is not appropriate, while the use of the full Vlasov-Maxwell equations is much beyond the capacities of the present day computers. When considering waves whose wave length is not too large compared to the ion gyro-radius, hybrid simulations which treat ions as particles and electrons as a massless fluid, are feasible and it was shown that ion kinetics play a crucial role, not only by reducing instability growth rates but also by destabilizing ranges of wavenumbers that are stable in a fluid description, especially for β of order one or larger [22]. This approach nevertheless fails when considering much longer waves or smaller amplitudes. The question thus arises of developing asymptotic approaches that incorporate the kinetic effects that are the most important for the dynamics.

THE KINETIC DNLS EQUATION

The dynamics of Alfvén waves propagating along a strong ambient field are amenable to an asymptotic expansion, directly from the Vlasov-Maxwell equation, when involving scales that are large compared to the ion Larmor radius and amplitudes small enough to keep linear dispersive effects relevant [23]. For this purpose, one writes the Vlasov-Maxwell equations in the form

$$\partial_t F_r + v\cdot\nabla F_r + \frac{q_r}{m_r}(E + \frac{1}{c} v\times B)\nabla_v F_r = 0 \quad (1)$$

$$\frac{1}{c}\partial_t B = -\nabla\times E \quad (2)$$

$$\nabla\times B = \frac{4\pi}{c}\Sigma_r q_r n_r \int v F_r d^3v \quad (3)$$

$$4\pi\Sigma_r q_r n_r \int F_r d^3v = 0, \quad (4)$$

where the subscript r refers to the particle species and where the displacement current has been neglected.

For an ambient field of strength B_0 pointing in the x-direction, the approach consists in rescaling the longitudinal variable $\xi = \varepsilon^2(x-\lambda t)$, the transverse ones as $\eta = \varepsilon^3 y$ and $\zeta = \varepsilon^3 z$, and in introducing a time scale $\tau = \varepsilon^4 t$. One also expands the distribution function and the electric and magnetic fields in the form

$$F_r = F_r^{(0)} + \varepsilon(f_r^{(0)} + \varepsilon f_r^{(0)} + \ldots) \quad (5)$$

$$B = B_0 + \varepsilon(b^{(0)} + \varepsilon b^{(1)} + \ldots) \quad (6)$$

$$E = \varepsilon(e^{(0)} + \varepsilon e^{(1)} + \ldots). \quad (7)$$

Inserting this long-wave expansion into the Vlasov-Maxwell equations and averaging over the velocity, one selects the Alfvén waves, characterized by the correlation $b_\perp^{(0)} = -\frac{B_0}{\lambda}u_\perp$ between the transverse components of the magnetic field and of the hydrodynamic velocity $u_\perp = \frac{1}{\rho^{(0)}}\Sigma_r m_r n_r \int v_\perp f_r^{(0)} d^3v$, where $\rho^{(0)} = \Sigma_r m_r n_r \int F_r^{(0)} d^3v$. At this order, the longitudinal magnetic field is $b_\parallel^{(0)} = 0$. The propagation velocity λ is given by

$$\lambda^2\rho^{(0)} = \frac{1}{4\pi}|B_0|^2 + p_\perp^{(0)} - p_\parallel^{(0)}, \quad (8)$$

where the usual expression for the Alfvén velocity is affected by the anisotropy of the equilibrium pressure tensor, characterized by its longitudinal and transverse components

$$p_\parallel^{(0)} = \Sigma_r m_r n_r \int v_\parallel^2 F_r^{(0)} d^3v \quad (9)$$

$$p_\perp^{(0)} = \Sigma_r m_r n_r \int \frac{v_\perp^2}{2} F_r^{(0)} d^3v. \quad (10)$$

The time evolution of the transverse magnetic field is obtained at a higher order of the expansion in the form

$$\partial_\tau b_\perp^{(0)} + \frac{\delta}{2\Omega_i}\partial_{\xi\xi}(\widehat{x}\times b_\perp^{(0)}) - \frac{B_0}{2\lambda\rho^{(0)}}\nabla_\perp\widetilde{P} + \frac{1}{2\lambda\rho^{(0)}}\partial_\xi(\widetilde{P} b_\perp^{(0)}) + \partial_\xi(\overline{\mathscr{U}} b_\perp^{(0)}) = 0 \quad (11)$$

$$\partial_\xi b_\parallel^{(1)} + \nabla_\perp b_\perp^{(0)} = 0, \quad (12)$$

with a dispersion coefficient $\delta = \frac{1}{\rho^{(0)}}(\frac{B_0^2}{4\pi} + 2p_{\parallel i}^{(0)} - p_{\perp i}^{(0)})$ and fluctuations of total pressure in the direction transverse to the local magnetic field

$$\widetilde{P} = \frac{B_0^2}{4\pi}\widetilde{A} + \widetilde{p}_\perp^{(1)}. \quad (13)$$

The magnetic contribution involves the magnetic field strength perturbation $A = \frac{1}{2B_0^2}|b_\perp^{(0)}|^2 + \frac{1}{B_0}b_\parallel^{(1)}$ defined by expressing the amplitude of the total magnetic field in the form $|B| = B_0(1+\varepsilon A + O(\varepsilon^2))$. Separating the mean value in the longitudinal direction (denoted by $\langle\cdot\rangle_\xi$) from the fluctuations, one writes $A = \langle A\rangle_\xi + \widetilde{A}$. The fluctuations of transverse thermodynamical pressure

$$\widetilde{p}_\perp^{(1)} = (2p_\perp^{(0)} + N - M^2L^{-1})\widetilde{A} \quad (14)$$

are sensitive to the distortion of the local magnetic field produced by the wave and also includes the Landau damping effect through the operators

$$L = 2\pi\Sigma_r \frac{q_r^2}{m_r} n_r \int_0^\infty \mathcal{G}_r d(\frac{v_\perp^2}{2}) \tag{15}$$

$$M = 2\pi\Sigma_r q_r n_r \int_0^\infty \frac{v_\perp^2}{2} \mathcal{G}_r d(\frac{v_\perp^2}{2}) \tag{16}$$

$$N = 2\pi\Sigma_r m_r n_r \int_0^\infty \frac{v_\perp^4}{4} \mathcal{G}_r d(\frac{v_\perp^2}{2}) \tag{17}$$

where, denoting by $\mathcal{H}$ the Hilbert transform with respect to the ξ variable,

$$\mathcal{G}_r = p.v. \int \frac{1}{v_\parallel - \lambda} \frac{\partial F_r^{(0)}}{\partial v_\parallel} dv_\parallel + \pi \frac{\partial F_r^{(0)}}{\partial v_\parallel}\bigg|_{v_\parallel = \lambda} \mathcal{H}. \tag{18}$$

Note that this operator has a symbol of order zero. As a result, all the scales are affected by the Landau damping. In particular, the criteria for modulational instability in the direction of propagation are strongly modified by this effect [24, 25] and new dissipative structures were also reported [26]. Note that the action of the Landau damping on the Alfvén waves is mediated by the coupling with ion-acoustic waves that are directly affected.

The above system is closed in the case of localized pulses that vanish as $\xi \to \infty$ or, in the physical variables, at distances large compared with ε^{-2}, since in this case, $\overline{\mathcal{U}} = 0$. These equations were derived by Rogister [23] using a Fourier space formalism. Mjølhus and Wyller [27] showed that the same equations can be obtained in a fluid formalism starting from an extension of the Hall-MHD equations including the electron pressure and finite Larmor radius effects [28]. They included the kinetic effects by estimating the transverse pressure fluctuation within the guiding center approximation.

WAVE-TRAIN DYNAMICS AND FILAMENTATION

As noted in [29] and recently illustrated numerically on direct numerical simulations of the Hall-MHD equations [30], the filamentation instability, associated with the transverse collapse of the Alfvén wave, can only occur in the case of wave packets whose extension significantly exceeds the wave length of the carrier. In such a regime, the contributions of fields that only depend on the transverse coordinates are to be retained. They correspond to the $\overline{\mathcal{U}}$ term in eq. (11) that is computed by pushing the expansion to the next order, in the form

$$\overline{\mathcal{U}} = \langle u_\parallel \rangle_\xi + \frac{\lambda}{2B_0} \langle b_\parallel^{(1)} \rangle_\xi + \frac{1}{\lambda \rho^{(0)}} (p_\parallel^{(0)} - p_\perp^{(0)}) \langle A \rangle_\xi + \frac{1}{2\lambda\rho^{(0)}} (\langle p_\perp^{(1)} \rangle_\xi - \langle p_\parallel^{(1)} \rangle_\xi). \tag{19}$$

The parallel hydrodynamic velocity averaged over the longitudinal coordinate, given to leading order by

$$\langle u_\parallel \rangle_\xi = \frac{1}{\rho^{(0)}} \Sigma_r m_r n_r \int v_\parallel \langle f^{(1)} \rangle_\xi d^3v, \tag{20}$$

obey

$$\partial_\tau \langle u_\parallel \rangle_\xi = \frac{1}{\rho^{(0)} B_0} \nabla_\perp \cdot \langle \widetilde{P} b_\perp^{(0)} \rangle_\xi, \tag{21}$$

while the mean longitudinal magnetic field $\langle b_\parallel^{(1)} \rangle_\xi$ is related to the mean density fluctuations (also constructed from $\langle f^{(1)} \rangle_\xi$), by

$$\frac{1}{\rho^{(0)}} \langle \rho^{(1)} \rangle_\xi = \frac{1}{B_0} \langle b_\parallel^{(1)} \rangle_\xi. \tag{22}$$

One clearly sees that the mean fields are driven by the transverse gradients. Furthermore,

$$\frac{B_0^2}{4\pi} \langle A \rangle_\xi + \langle p_\perp^{(1)} \rangle_\xi = \text{const}. \tag{23}$$

The system is completed by writing dynamical equations for the mean value of the pressure perturbations parallel and transverse to the local magnetic field, that to leading order read

$$\langle p_\perp^{(1)} \rangle_\xi = \Sigma_r m_r n_r \int \frac{|v_\perp|^2}{2} \langle f^{(1)} \rangle_\xi d^3v - \frac{B_0^2}{4\pi} \langle \frac{|b_\perp^{(0)}|^2}{2B_0^2} \rangle_\xi \tag{24}$$

$$\langle p_\parallel^{(1)} \rangle_\xi = \Sigma_r m_r n_r \int |v_\parallel|^2 \langle f^{(1)} \rangle_\xi d^3v + 2(p_\parallel^{(0)} - p_\perp^{(0)}) \langle \frac{|b_\perp^{(0)}|^2}{2B_0^2} \rangle_\xi. \tag{25}$$

The magnetic contributions in the above expressions reflects the distortion of the magnetic field lines. Dynamical equations governing the time evolution of these quantities are also obtained. They however seem too complicated to be explicitly written here and will be published elsewhere. The important point is that these equations complete the system that is then closed. It provides an asymptotically exact description of the nonlinear dynamics of Alfvén wave trains that retain the effect of the Landau damping mediated by the interaction with the ion-acoustic waves.

It is of interest to note the similarity (up to the Landau damping) between the above system and the equations derived from the Hall MHD equations for a polytropic gas where the pressure is assumed to be isotropic and to adiabatically follow the density variations. These equations read

$$\widetilde{P} = \frac{1}{1-\beta} \frac{B_0^2}{4\pi} \widetilde{A} \tag{26}$$

$$\frac{1}{\rho^{(0)}}\langle\rho^{(1)}\rangle_\xi = \frac{1}{B_0}\langle b_\parallel^{(1)}\rangle_\xi \tag{27}$$

$$\langle A\rangle_\xi + \frac{\beta}{1+\beta}\frac{\langle|b_\perp^{(0)}|^2\rangle_\xi}{2B_0^2} = \text{const.} \tag{28}$$

$$\overline{\mathscr{U}} = \langle u_\parallel\rangle_\xi + \frac{\lambda}{2B_0}\langle b_\parallel^{(1)}\rangle_\xi, \tag{29}$$

where the mean longitudinal hydrodynamic velocity $\langle u_\parallel\rangle_\xi$ obeys the same equation as in the kinetic theory. The identity between the kinetic and MHD systems in the limit of a cold plasma is conspicuous.

Several problems can be addressed using the above kinetic version of the DNLS equation, valid for wave trains. One of them concerns the influence of the Landau damping on the filamentation phenomenon. In the Hall-MHD context, the envelope formalism predicts that filamentation of long Alfvén waves requires the condition $\beta > 1$. Direct evidence of this effect was demonstrated by numerical integration of three-dimensional DNLS equation including the mean fields [31], and also by comparison with direct numerical simulations of the Hall-MHD equations [20]. The extension of the envelope analysis to the kinetic DNLS system (with the mean fields) is in project and should enable one to characterize the influence of Landau damping on Alfvén wave filamentation.

Furthermore, the above system that, as already mentioned, provides an asymptotically exact fluid description for the evolution of long Alfvén waves, can also be used as a benchmark for Landau fluid models recently proposed to describe the large scale dynamics of a collisionless plasma permeated by a strong ambient field [32]. Such systems are constructed as fluid moment equations from the guiding center distribution function and can be viewed as a generalized magnetohydrodynamic description that retains ion kinetic Landau damping. In this approach, longitudinal and transverse pressure fluctuations are not assumed to adiabatically follow the density variations but obey dynamical equations that involve the heat flux and are consequently unclosed. The system is then completed by expressing the heat fluxes in terms of the lower moments and of the magnetic field perturbation. These relations are determined by matching with the linear kinetic density and perpendicular pressure response [32]. Several models of different degrees of complexity were proposed. Their respective accuracy can be evaluated by comparing their predictions for the dynamics of long Alfvén waves with the results of the exact asymptotic theory.

ACKNOWLEDGMENTS

This work benefitted of support from CNRS program "Soleil-Terre" and from INTAS contract 00-292.

REFERENCES

1. Strauss, H.R., *Phys. Fluids*, **19**, 134-140 (1976).
2. Montgomery, D.C., *Physica Scripta*, **T2**, 83-88 (1982).
3. Ng, C.S., and Bhattacharjee, A., *Astrophys. J.*, **465**, 845-854 (1996).
4. Galtier, S., Nazarenko, S.V., Newell, A.C., and Pouquet, A., *J. Plasma Phys.* **63**, 447-488 (2000).
5. Matthaeus, W.H. and Zhou, Y., *Phys. Fluids B* **1**, 1929-1931 (1989).
6. Maron, J., and and Goldreich, P., *Astrophys. J.* **554**, 1175-1196 (2001).
7. Cho, J., Lazarian, A., and Vishniac, E.T., *Astrophys. J.* **564**, 291-301 (2002).
8. Bhattacharjee, A., oral comunication at Solar Wind 10 (2002).
9. Goldreich, P., and Sridhar, S., *Astrophys. J.* **438**, 763-(1995); **485**, 680-688 (1997).
10. Zank, G.P., and W.H. Matthaeus, W.H., *J. Plasma Phys.* **48**,85-100 (1992).
11. Gazol, A., Passot, T., and Sulem, P.L., *Phys. Plasmas* **6**, 3114-3122 (1999).
12. Gazol, A., Passot, T., and Sulem, P.L., *Rev. Mex. de Astro. y Astrofís (Serie de Conferencias)* **9**, 80-82 (2000).
13. Wong, H.K. and Golstein, M.L., *J. Geophys. Res.*, **91**, 5617-5628 (1986).
14. Champeaux, S., Laveder, D., Passot, T., and Sulem, P.L., *Nonlinear Processes in Geophysics* **6**, 169-176 (1999).
15. Del Zanna, L., Velli, M., and Londrillo, P., *Astron. Astrophys.* **367**, 705-718 (2001).
16. Shukla, P.K. and Stenflo, L., *Astro. Space Sci.*, **155**, 145-147 (1989).
17. Champeaux, S., Passot, T., and Sulem, P.L., *J. Plasma Phys.* **58**, 665-690 (1997).
18. Bergé, L., *Phys. Reports* **303**, 259-370 (1998).
19. Sulem, C., and Sulem, P.L., *The nonlinear Schrödinger equation: self-focusing and wave-collapse*, Applied Mathematical Sciences, Vol. 139, Springer (1999).
20. Laveder, D., Passot, T., and Sulem, P.L., *Phys. Plasmas* **9**, 293-304 (2002).
21. Laveder, D., Passot, T., and Sulem, P.L., *Phys. Plasmas* **9**, 305-314 (2002).
22. Vasquez, B.J., *J. Geophys. Res.*, **100**, 1779-1792 (1995).
23. Rogister, A., *Phys. FLuids* **12**, 2733-2739 (1971).
24. Spangler, S.R., *Phys. Fluids B* **1**, 1738-1746 (1989); **2**, 407-418 (1989);
25. Medvedev, M.V., and Diamond, P.H., *Phys. Plasmas* **3**, 863-873 (1996).
26. Medvedev, M.V., Shevchenko, V.I., Diamond, P.H., and Galinsky, V.L., *Phys. Plasmas* **4**, 1257-1285 (1997).
27. Mjølhus, E., and Wyller, J., *J. Plasma Phys.* **40**, 299-318 (1988).
28. Yajima, N., *Prog. Theor. Phys.* **36**, 1-16 (1966).
29. Passot, T., and Sulem, P.L., *Phys. Rev. E* **48**, 2966-2974 (1993).
30. Laveder, D., Passot, T., Sulem, C., Sulem, P.L., Wang, D., and Wang, X.P., *Wave collapse in dispersive magnetohydrodynamics: direct simulations and envelope modeling*, Submitted to Physica D.
31. Laveder, D., Passot, T., and Sulem, P.L., *Physica D* **152-153**, 694-704 (2001).
32. Snyder, P.B., Hammett, G.W., and Dorland, W., *Phys. Plasmas* **4**, 3974-3985 (1997).

Three Second Waves Observed Upstream Of The Earth's Bow Shock

X. Blanco-Cano[1], C. T. Russell[2], J. Ramírez[1], and G. Le[3]

[1]*Instituto de Geofísica, UNAM, México D.F.;* [2]*IGPP, UCLA, Los Angeles;* [3]*Godard Space Flight Center, Maryland.*

Abstract. A new class of ULF waves were discovered in the foreshock from the ISEE magnetometer data by Le et al. [1]. These unusual type of waves differ greatly from the more commonly observed 30 s waves, shocklets and SLAMS. The new waves have periods near 3 s and always show a very narrow spectrum that is in contrast with the lower frequency waves observed in the foreshock, which usually have broader spectra. Three second waves have large amplitudes and are observed in the upstream region only when the interplanetary magnetic field intersects the bow shock and when the plasma beta is high. Three second waves can be divided in three types. *Isolated* waves are associated with reflected cold beams and are observed in regions where the magnetic field is very quiet. *Superposed* waves are associated with more intermediate ion distributions, and are superposed on non steepened lower frequency waves. *Irregular* waves are associated with diffuse ions and are observed as trains surrounded by an irregular magnetic field. Three second waves are generated by the right-hand non resonant instability, being right-handed and propagating downstream in the plasma frame. This instability grows due to the interaction of the solar wind distribution with a reflected ion beam. In this work we map the location of the different types of three second waves with respect to the bow shock, and study their evolution in the foreshock. We also study which instability is generating the lower frequency waves observed with the superposed 3 s waves.

INTRODUCTION

This paper continues the study of three second waves discovered in the foreshock from the ISEE magnetometer data by Le et al. [1]. These unusual type of waves differ greatly from other ultra low frequency (ULF) waves, such as 30 second waves, shocklets, and SLAMS [2, 3]. They have periods ~3 s and always show a very narrow spectrum that is in contrast with the broader spectrum of other ULF waves. They are always right-hand nearly circularly polarized in the spacecraft frame and are convected downstream by the solar wind. Three second waves are observed in the upstream region only when the interplanetary magnetic field intersects the bow shock and when the plasma beta is high. ULF waves are generated in the foreshock by kinetic ion instabilities produced by the interaction of the solar wind with suprathermal ion beams. Their study is important because they play an active role in the interaction of the solar wind with the bow shock; they participate in wave-particle interactions, and in the transmission of wave energy from one region to another.

WAVE CHARACTERISTICS

Three second waves can be found in three different environments [4]. Figure 1 shows an example of each type of wave, and of the suprathermal ions (bottom panels) that are observed with the waves. Data are from ISEE magnetometer [5] and ISEE fast plasma experiment [6]. *Isolated* waves are observed in regions where the magnetic field is very quiet (Fig. 1a). *Superposed* waves (Fig. 1b) are observed on top of lower frequency (10^{-2} Hz) quasi-sinusoidal nonsteepened waves. *Irregular* waves (Fig. 1c) are observed as trains surrounded by a perturbed magnetic field. Three second waves are associated with different types of suprathermal ions. While isolated waves are associated with reflected field-aligned cold fast beams, irregular waves are accompanied by more isotropic diffuse hot ions, with slow drift speed. Superposed waves are accompanied by ions whose characteristics are *intermediate*. Three second waves propagate at small angles ≤30° with respect to the background magnetic field and have amplitudes $\delta B \approx$ 2-7 nT.

CP679, *Solar Wind Ten: Proceedings of the Tenth International Solar Wind Conference*, edited by M. Velli, R. Bruno, and F. Malara

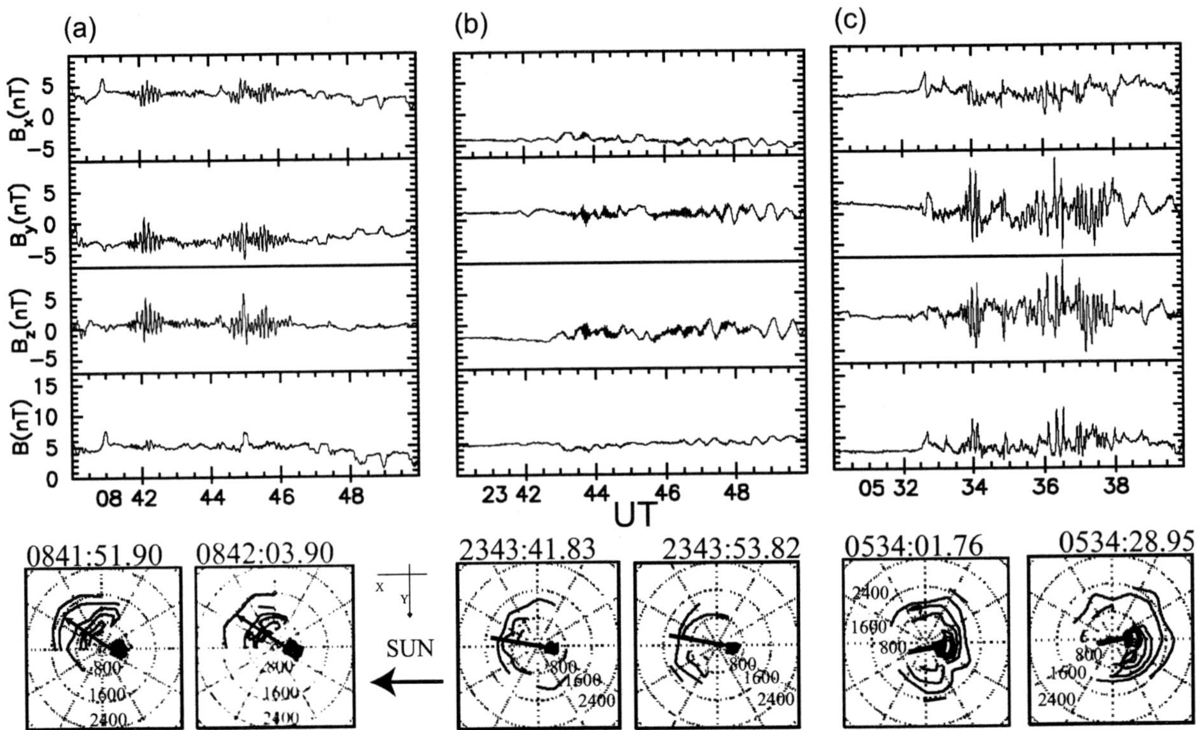

FIGURE 1. Three second waves observed by ISEE 1: (a) isolated waves (0841:56-0842:44, and 0844:30-0846:00 on November 3, 1978), (b) superposed waves (2343:33-2343:56, 2344:09-2344:4, and 2345:52-2346:49 on September 7, 1979), and (c) irregular trains (0533:45-0534:44, and 0536:45-0537:45 on October 2, 1978). Bottom panels show suprathermal ions associated with the waves. Distributions are given as contours of constant phase-space density separated logarithmically (two contours per decade). The solar wind appears as tighten contours at positive v_x. Concentric circles indicate speeds of 800, 1600 and 2400 km/s. The arrow through the solar wind distribution is the projection of the average magnetic field.

We compared three second wave properties (frequency, wavenumber, polarization, magnetic compression, and noncoplanar ratio) with the characteristics of kinetic ion instabilities and showed that three second waves can be identified as the right-hand non resonant instability, being right-handed and propagating downstream in the plasma frame [4]. We found that isolated and superposed waves can be generated locally by the right-hand nonresonant instability that grows due to the solar wind interaction with the reflected beams associated with these waves. In contrast, the diffuse ions observed with irregular waves can not generate the right-hand nonresonant mode, and we suggested that irregular waves are generated upstream of where they are observed by the right-hand nonresonant instability.

SPATIAL VARIATION OF WAVES

The properties of waves and plasma are position depending within the foreshock. Three second waves are observed immediately upstream from the bow shock, as well as near ISEE orbit apogee, far from the nominal bow shock position [1]. In order to investigate if isolated, superposed and irregular waves permeate the same regions or not, we find the position of the waves with respect to the bow shock. The position of the shock was found using the model of Farris and Russell [7]. Figure 2 shows dx, the wave position with respect to the bow shock nose, and $\delta B/B_o$ the normalized amplitude of three second waves observed by ISEE 1. dx is the distance along the X direction, and as in all cases the solar wind velocity was mainly along X, the distance from the shock along the solar wind flow direction can be approximated as dx. It is possible to see that superposed and irregular waves are found at distances $\leq 3\ R_E$ ($R_E \equiv$ earth radius). In contrast, most isolated waves are found at much larger distances, reaching $\approx 7\ R_E$. The amplitudes of superposed and isolated waves are in between 0.1-0.8 B_o, while irregular waves have amplitudes ≈ 0.5 up to 1.2 B_o.

The fact that different types of three second waves are found at different distances from the shock is in

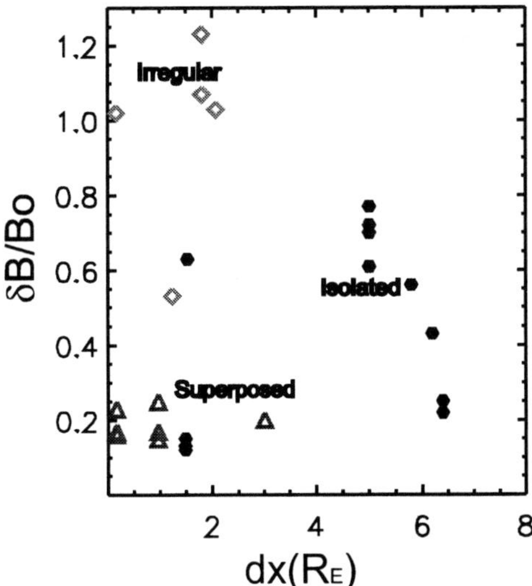

FIGURE 2. Wave amplitude ($\delta B/B_o$) and distance (dx) from the bow shock along the X-direction (≈solar wind flow direction). Solid dots correspond to isolated waves, triangles to superposed and diamonds to irregular waves.

agreement with the association of three second waves with suprathermal ions, and with the magnetic environment surrounding the three types of waves. Isolated waves are observed with reflected cold beams and are found in regions where the magnetic field is very quiet. Reflected cold beams are produced in the bow shock and can reach large distances upstream. The fastest beams can reach the upstream edge of the foreshock. Far from the bow shock there are few waves and this is reflected in the cold non-disturbed ion distributions observed with isolated waves, which have not been heated by wave scattering. The interaction of these beams with the solar wind makes the plasma unstable and waves start to grow. These waves are convected by the solar wind downstream, where via wave-particle interactions they start to heat the ion distributions. Closer to the shock, wave activity increases and wave-particle interactions result in ion scattering leading to the wider distributions associated with superposed three second waves. As we discuss below, in these regions the plasma can be unstable to more than one mode leading to the generation of the lower frequency waves observed with superposed three second waves. Even closer to the shock (dx ≤ 2 R_E), where irregular waves are observed, ion distributions are hotter and isotropic as a result of continuos ion scattering due to the waves that permeate the deep foreshock. The fact that irregular waves have larger amplitudes than superposed and isolated waves is consistent with previous findings [8] that show that closer to the shock ULF waves have larger amplitudes. This can be explained as a result of wave growing which can take place as the waves convect with the solar wind.

Comparison of observed three second wave properties with ion kinetic instabilities allowed us [4] to identify the right-hand non resonant instability as the mode that is generating three second waves. In our previous work we concluded that while isolated and superposed waves can be generated locally by their associated suprathermal ions, irregular waves must be generated upstream from where they are observed. The fact that irregular waves are observed closer to the shock and have larger amplitudes supports our prediction. The growth of waves in an unstable region is not instantaneous, and in fact, waves can grow as they convect towards the bow shock, providing that the free energy is available from the suprathermal ions. Thus, irregular waves are cases of "well-developed" isolated waves that were generated upstream and grow as they approached the shock. In the parcels of plasma where these waves are observed other perturbations grow while approaching the shock creating the very perturbed environment that is observed beside irregular three second waves. This scenario is sketched in figure 3.

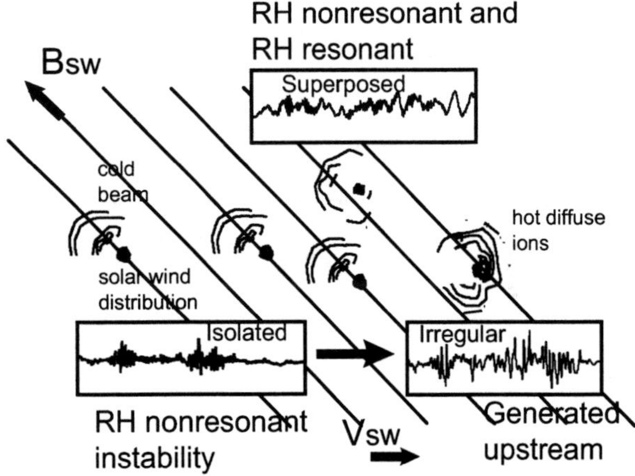

FIGURE 3. Drawing sketch of position, generation, and evolution of three second waves.

LOWER FREQUENCY WAVES

The ULF waves that are observed near three second superposed waves have frequencies f~0.02-0.03 Hz. In contrast to three second waves, they are left-handed and elliptically polarized in the spacecraft frame. As stated above, in regions with superposed waves the suprathermal ions have intermediate

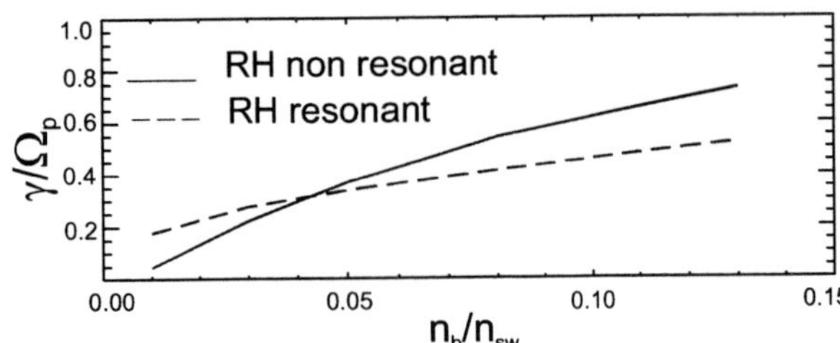

FIGURE 4. Growth rate of the right-hand resonant and non resonant instabilities as a function of beam density.

characteristics, i.e. they are field-aligned and have suffered some heating by wave scattering.

The interaction of the solar wind with these beams can generate the right-hand resonant and the right-hand non resonant instabilities. Figure 4 shows growth rate as a function of beam density for the right-hand resonant and non resonant instabilities. The beam characteristics resemble the suprathermal ions observed with superposed waves (T_b=30T_{sw}, v_{ob}=20v_A, more details are given in [4]). It is possible to see that when the beam is sufficiently dense the right hand nonresonant mode dominates, and when the beam density is small (<0.04) the right-hand resonant mode has the largest growth. Therefore, we believe that the lower frequency waves observed with superposed three second waves can be identified as the right-hand resonant mode, which suffers a reversal in polarization and is left-handed in the spacecraft frame. In regions with superposed waves, beams with different densities interact with the solar wind, leading to regions where the nonresonant mode is dominant and regions where the resonant mode has the higher growth.

CONCLUSIONS:

Three second waves can be divided in three types: isolated, superposed and irregular. The waves are associated with different suprathermal ions: Isolated waves are accompanied by reflected ion beams, superposed are observed with intermediate distributions, and irregular waves are associated with diffuse ions.

We found that irregular waves are observed closer to the shock than isolated waves. This supports an scenario that we proposed in the past [4] in which isolated and superposed waves are generated locally by the right-hand nonresonant instability, while irregular waves are generated upstream of where they are observed by the same instability.

The lower frequency (f~0.02-0.03 Hz) waves observed near superposed three second waves are identified as the right-hand resonant mode.

To our knowledge three second waves have only been studied using ISEE data. It would be very interesting to study in more detail the properties of these waves with present missions such as CLUSTER.

ACKNOWLEDGMENTS

We thank J. T. Gosling of LANL for kindly providing the suprathermal ion data. We are grateful to J. Newbury for helping us to process suprathermal ion distributions.

REFERENCES

1. Le, G., C. T. Russell, M. F. Thomsen, and J. T. Gosling, Observation of a new class of upstream waves with periods near 3 sec, *J. Geophys Res.*, 97, 2917-2925, 1992.

2. Greenstadt, E. W., G. Le, and R. J. Strangeway, ULF waves in the foreshock, *Adv. Space Res.*, 15 (8/9) 71-84, 1995.

3. Burgess, D., What do we really know about upstream waves?, *Adv. Space Res.*, 20 (4/4), 673-682, 1997.

4. Blanco-Cano, X., G. Le, and C. T. Russell, Identification of foreshock waves with 3-s periods, *J. Geophys. Res.*, 104, 4643-4656, 1999.

5. Russell, C. T., The ISEE 1 and 2 fluxgate magnetometers, *IEEE Trans. Geosci. Electron.*, GE-16, 239-242, 1978.

6. Bame, S. J., J. R. Asbridge, H. E. Felthauser, J. P. Glore, G. Paschmann, P. Hemmerich, K. Lehmann, and H. Rosenbauer, ISEE-1 and ISEE-2 fast plasma experiment, *IEEE Trans. Geosci. Electron.*, GE-16, 216, 1978.

7. Farris M. H., and C. T. Russell, Determining the standoff distance of the bowshock: Mach number dependence, *J. Geophys. Res.* 99, 17681, 1994.

8. Le, G, and C. T. Russell, The morphology of ULF waves in the Earth's Foreshock, in *Solar wind sources of magnetospheric Ultra Low Frequency waves*, Ed. M. J. Engebretson, K. Takahashi, and M. Scholer, Geophys. Monograph 81, 81-98, AGU, 1994.

Parametric instability in the solar wind: numerical study of the nonlinear evolution

Leonardo Primavera[*†], Francesco Malara[*†] and Pierluigi Veltri[*†]

[*]*Dipartimento di Fisica, Università della Calabria, 87036 Rende (CS), Italy*
[†]*Istituto Nazionale per la Fisica della Materia, Unità di Cosenza, via P. Bucci, 87036 Rende (CS), Italy*

Abstract. A possible mechanism to explain the decrease of the Alfvenic correlation with the distance from the sun, observed in fast speed streams in the solar wind, is the parametric instability. Starting from a circularly polarized Alfven wave, this instability naturally produces backward propagating waves and compressive fluctuations, by destroying the initial correlation of the wave. However, the applicability of this mechanism to the solar wind is debatable, since it works better when the plasma beta is much lower than 1 and the initial wave is monochromatic. To elucidate better this phenomenon, we numerically simulated the propagation of a turbulent spectrum of Alfven waves on a uniform background magnetic field by using a pseudo-spectral, one dimensional, MHD code. We used values of the plasma beta about one and very high values of the physical diffusivities, by using "artificial" numerical diffusivities to ensure the stability of the numerical scheme. We found that, even under such conditions, the initial alfvenic correlation of the waves is progressively destroyed. At the saturation of the instability, the forward and inward propagating waves have comparable energies in the spectra, at the larger scales, while the forward propagating fluctuations dominate the spectrum at smaller scales. The two spectra tend to approach each to the other at subsequent times. A tentative comparison of these results with the observations of the alfven waves present in the solar wind was carried out, with good qualitative agreement.

INTRODUCTION

The turbulence present in the high latitude solar wind shows very striking Alfvénic characteristics, with high correlation between velocity and magnetic field fluctuations and relatively low level of density perturbations. However, several observations ([1, 2, 3, 4]) also showed that the Alfvénicity of the fluctuations depends on the distance from the sun. By using the Elsässer notation: $\mathbf{Z}^{\pm} = \mathbf{v} \pm \mathbf{b}/\sqrt{\rho}$, the outward propagating waves can be identified with $\mathbf{Z}^{+}$ modes, and the ones travelling in the opposite direction with $\mathbf{Z}^{-}$. Then we define the *pseudo-energies* associated with the Elsässer variables: $e^{\pm} = \frac{1}{2} < |\mathbf{Z}^{\pm}|^2 >$, and the *normalized cross helicity of the fluctuations*:

$$\sigma_c = \frac{e^+ - e^-}{e^+ + e^-}$$

where the symbol: $<>$ identifies running averages on a given time scale. Under these assumptions, the radial evolution of the Alfvénicity in the solar wind can be summarized as follows: *a*) close to the sun, at a distance of about 0.3 AU, the hourly average cross helicity σ_c has a value close to 1, meaning that the fluctuations are almost pure $\mathbf{Z}^{+}$ (namely outward propagating) fluctuations; *b*) with increasing distance from the sun, σ_c decreases, because the pseudo-energies $e^{\pm}$ both decrease; *c*) however, the decrease rate for e^+ is faster than that for e^-, up to a distance of ~ 2.5AU; *d*) after that distance, the two quantities lower at the same rate. This behaviour is clearly displayed in fig. 1, where the evolution of the quantity $E = e^-/e^+$, i.e. the ratio between the hourly averaged energies of $\mathbf{Z}^{-}$ over $\mathbf{Z}^{+}$, is shown. Since e^+ decreases faster than e^-, according to the point *c*) above, the ratio increases up to ~ 1.5AU, then it stays approximately on the same level, since e^+ and e^- decrease in the same way.

Moreover, the fluctuations show a strong intermittent character, as shown from the analysis of the flatness of the fluctuations carried out by Bruno et al.(2002), in these proceedings ([5]). They showed that in the fast polar wind, the flatness of both velocity and magnetic field intensities increases, so they become more and more intermittent with decreasing the length scale, and in general, the intermittency attains higher and higher values as the distance from the sun increases.

It was conjectured by several authors that in the high latitude wind, where the medium is more homogeneous, the radial evolution of the quantities can be influenced by the parametric instability ([6, 7, 8, 11]). In fact, the parametric decay process of a circularly polarized Alfvén-wave (for example a $\mathbf{Z}^{+}$ mode) leads to the growth of a forward propagating sound wave, and a backscattered Alfvén wave with a correlation opposite to that of the mother wave (a $\mathbf{Z}^{-}$ mode), thus producing a decrease of

CP679, *Solar Wind Ten: Proceedings of the Tenth International Solar Wind Conference*,
edited by M. Velli, R. Bruno, and F. Malara

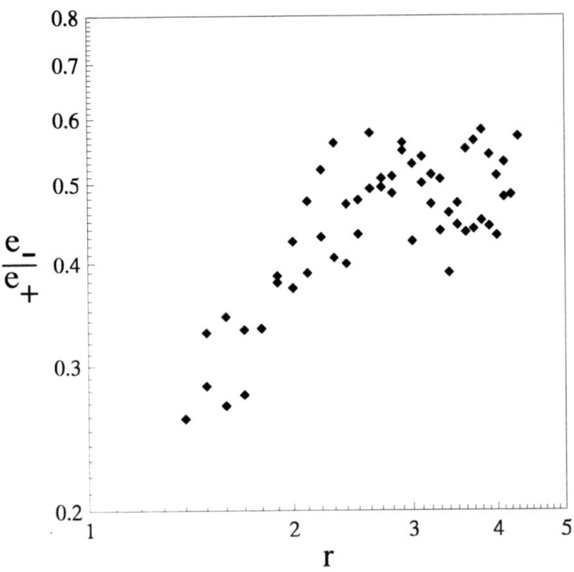

FIGURE 1. Radial evolution of the hourly average for the Alfvénratio e^-/e^+ in the solar wind.

the initial correlation. The parametric instability has been studied both numerically and analitically in several situations. Recently, Del Zanna et al. ([9, 10]) studied numerically the evolution of a *monochromatic* Alfvénwave in a fully three dimensional, compressible situation, subject to the parametric instability, by studying its fully nonlinear evolution and saturated state. One important result of them is that the instability is a robust process, namely it works practically in the same way independently of the dimensionality of the problem.

However, in order to understand the importance of such a mechanism for the solar wind, one has to look at what happens when the initial wave is non monochromatic, but it has an extended spectrum, as it happens in the solar wind. Malara et al. ([11]) studied the evolution of the parametric instability when the initial wave has a broad band spectrum. They found that, when the plasma $\beta \sim 1$, the time evolution of the spectra of $\mathbf{Z}^+$ and $\mathbf{Z}^-$ has several common characteristics with the evolution of the spectra with distance, observed in the solar wind ([4, 12, 13]).

In this paper we want to try to understand whether the parametric instability maight account for the observed behaviour of the e^-/e^+ ratio with the radial distance and for the cited high intermittent behaviour found for the fluctuations in the solar wind. This possibility has been pointed out in these proceedings by Bavassano ([14]) and Del Zanna et al. ([15]).

NUMERICAL MODEL

We solved numerically the fully compressible, nonlinear, MHD equations in a one-dimensional confi guration. We used a pseudospectral numerical code, with periodicity boundary conditions in the spatial domain $x \in [0,2\pi]$. Further details about the numerical technique can be found elsewhere [7]. The code calculates the time evolution of the density, the velocity, the magnetic fi eld and the temperature, starting from a given initial condition. The above quantities are normalized to characteristic values and the unit time is the Alfvén time. We neglect viscosity and resistivity in the code, although we use hyperviscosity and hyper-resistivity in order to ensure numerical stability.

The initial condition consists of a broad-band, large amplitude Alfvén wave, polarized in the yz plane, propagating on a uniform background magnetic fi eld $\mathbf{b}_0$. The total fi eld is chosen to have uniform intensity everywhere:

$$\mathbf{b}(x,0) = b_0\hat{\mathbf{e}}_x + b_1\{\cos[\phi(x)]\hat{\mathbf{e}}_y + \sin[\phi(x)]\hat{\mathbf{e}}_z\} \qquad (1)$$

Here $b_1 = 0.5$ is the ratio between the amplitude of the mother wave and the background magnetic fi eld $b_0 = 1$. The function $\phi(x)$ represents the phase of the pump wave. Its form determines the spectrum of the initial wave. Measures by Helios spacecraft in fast streams at distances $r \simeq 0.3$ AU (the closest available distance) show that fluctuations are dominated by outward-propagating Alfvén waves, with a double-slope spectrum: at frequencies $f < f_0$, with $f_0 \sim 3\text{-}5 \times 10^{-4}$ Hz, the spectrum is flatter (slope ~ -1), while at frequencies $f > f_0$ the spectrum is steeper (slope $\sim (-1.5)$-(-2)) [12]). In our model the phase $\phi(x)$ has been chosen so that the initial fluctuations have also an approximately double-slope spectrum, though the slopes are steeper than the measured values. The frequency f_0 corresponds in the model to the wavenumber $k_0 \simeq 60$.

The initial wave being Alfvénic, the velocity fi eld is given by:

$$\mathbf{v}(x,0) = \frac{c_{A0}}{b_0}\delta\mathbf{b}(x,0) \qquad (2)$$

where $c_{A0} = 1$ is the Alfvén speed and the δ operator indicates the fluctuating part of quantity f: $\delta f = f - \langle f \rangle$, $\langle f \rangle$ being the spatial average of f. The initial density $\rho(x,0) = 1$ and temperature $T(x,0) = T_0$ are uniform. The value T_0 is used to fi x the initial plasma β, defi ned by $\beta = \gamma_a T_0/c_{A0}$, with $\gamma_a = 5/3$ the adiabatic index. In the simulations we used $\beta = 1$, which is of the order of the value measured in the high-latitude solar wind.

The above-described confi guration is an exact solution of the ideal MHD equations, which would propagate undistorted in the limit of a vanishing dissipation. An initial noise of amplitude 10^{-3} is added to the density to trigger the instability.

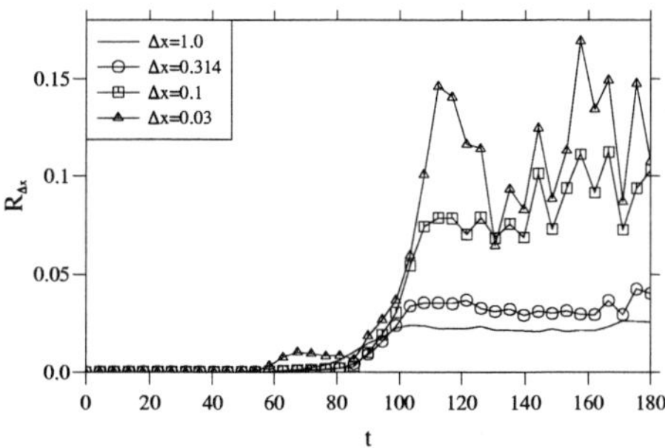

FIGURE 2. $R_{\Delta x}$ as a function of t, for various values of Δx.

COMPARISON WITH THE OBSERVATIONS

Our aim is to compare the results of our numerical simulations with the solar wind data. The plot in fig. 1 represents the spatial evolution of the ratio $E = e^-/e^+$ of the hourly averaged pseudo-energies of the inward and outward propagating fluctuations. In our simulations, the evolution with time emulates the change of the quantities with distance from the sun, whilst the initial broad band spectrum of the waves mimics the fluctuations present in the quantities measured in the solar wind during its transit in front of the spacecraft.

In fig. 2 we plotted the ratio $R_{\Delta x}(t) = e^-_{\Delta x}(t)/e^+_{\Delta x}(t)$ between pseudoenergies, calculated at a given spatial scale Δx, corresponding to a characteristic wavenumeber $k = 2\pi/\Delta x$. We show the behaviour of the ratio $R_{\Delta x}$ as a function of t, for different values of $\Delta x = 1.0, 0.314, 0.1, 0.03$, corresponding to the wavenumbers $k = 6, 20, 60, 100$. $R_{\Delta x}(t)$ is initially very small, the initial velocity and magnetic field being highly correlated. With increasing time, the instability tends to reduce the Alfvénic correlation and $R_{\Delta x}(t)$ increases at all scales Δx. This growth stops at time $t_{sat} = 90$-100, corresponding to the instability saturation. For subsequent times, the ratio $R_{\Delta x}(t)$ remains almost constant or it oscillates around a final average value that increases as the averaging length scale decreases.

In order to try a comparison between these results and measures in the high-latitude solar wind, we consider the ratio e^-/e^+, which has been calculated by [8], shown in fig. 1, for fluctuations averaged on a time basis of 1 hour, obtained from Ulysses measures. In order to establish a corrispondence between frequencies measured in the solar wind and wavevectors in the numerical model, we assume that the frequency $f_0 \sim 5 \times 10^{-4}$ Hz ~ 1 h^{-1}, where the slope of the e^+ spectrum changes, at $r = 0.29$ AU, corresponds to the wavenumber $k_0 \simeq 60$; the behavior of $R_{\Delta x}(t)$ for $k \simeq 60$ is among those plotted in Figure 2. However, the saturation value of the ratio e^-/e^+ shown in fig. 1 is slightly higher ($e^-/e^+ \simeq 0.5$) with respect to the one obtained in our simulations for the corresponding length scale ($e^-/e^+ \simeq 0.1$ for $\Delta x = 0.1$). Nevertheless, we stress that, in spite of the crudeness of the extimations made for evaluating the length scale in the initial spectrum of the waves, the magnitude order of the values are in fairly good agreement. We conclude that, even from a quantitative point of view, the behavior of e^-/e^+ found in the solar wind is partially reproduced by the numerical model.

Finally we remark that the evolution of the instability produces shock waves that, in turn, give rise to an intermittent behaviour for the intensity and direction of the fluctuating fields, due to the presence of strong gradients in those quantities. Bruno et al. (2002), analyzed the behaviour of the flatness of the velocity and magnetic field fluctuations observed in the solar wind at several distances from the sun. The flatness at a length scale l, for a generic quantity f (the intensity of the velocity or magnetic field fluctuations), is defined as:

$$F_f(l) = \frac{< [f(x+l) - f(x)]^4 >}{< [f(x+l) - f(x)]^2 >^2} \tag{3}$$

They showed that:

1. the flatness for both the velocity and magnetic field has a value of 3 at large scales l (that indicates a Gaussian distribution of the strongest gradients of the fluctuations), while it increases for decreasing the length scale l, pointing out that intermittency dominates the turbulence at small scales;
2. the flatness for the magnetic field intensity is higher than the corresponding values at the same scale l for the velocity field, demonstrating that the magnetic field is in general "more intermittent" than velocity;
3. the flatness increases when the distance from the sun increases, meaning that the fluctuations become more and more intermittent during the travel in the heliosphere.

In fig.s 3 and 4 we show the plots of the flatness, defined as in (3), for several length scales l, ranging from the highest (the one of the numerical box), down to the smallest length scales available in the simulation, at several times. We recall that time evolution in the simulation is analogous to evolution with distance in the observational data. A look at the result shows a striking analogy between the results of the simulations and the observation by Bruno et al. (2002) [5]. In particular, we point out that: *a*) the flatness increases with decreasing the length scale (see point 1 above), *b*) the flatness increases with increasing time, likewise it does with distance from the sun in the data. However, we point out that the final values of the flatness are much higher in our simulation than in the data. Moreover, the magnetic field in our simulations does not appear to be more intermittent than the velocity fi eld, as found in the data (see point 2 above). We conclude that the agreement of the values found in the simulations with the observational data is only qualitative, then our simplifi ed model does not account for all the characteristic features of the polar wind. Some ingredient is missing, such as the expansion of the wind and the three-dimensionality of the problem. Further investigation in these directions are necessary to draw a conclusion.

ACKNOWLEDGMENTS

The authors are grateful to B. Bavassano, P. Pietropaolo, R. Bruno and R. Grappin for clarifying details about the comparison with the solar wind data and interesting discussions.

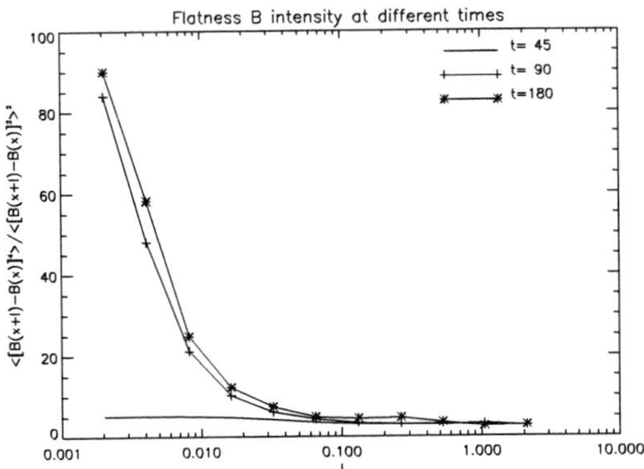

FIGURE 3. Flatness (as defined in (3)) of the magnetic field intensity fluctuations at several length scales l from the simulations.

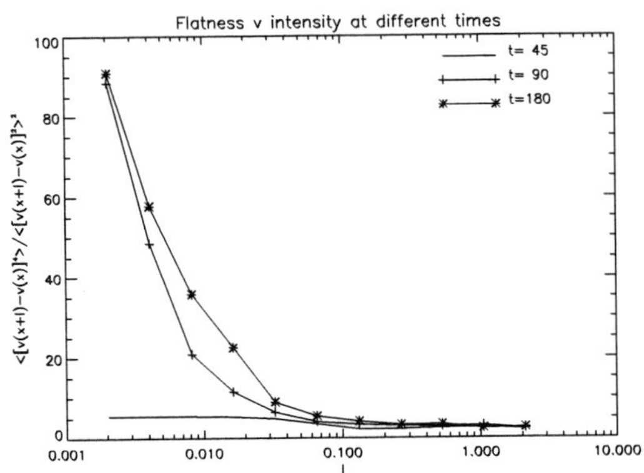

FIGURE 4. Flatness (as defined in (3)) of the velocity field intensity fluctuations at several length scales l from the simulations.

REFERENCES

1. Roberts, D. A., Goldstein, M. L., Klein, L. W., Matthaeus, W. H., Origin and evolution of fluctuations in the solar wind: Helios observations and Helios-Voyager comparisons, *J. Geophys. Res.*, **92**, 12023-12035, 1987.
2. Roberts, D. A., Goldstein, M. L., Klein, L. W., The amplitude of interplanetary fluctuations: stream structure, heliocentric distance, and frequency dependence, *J. Geophys. Res.*, **95**,4203-4216, 1987.
3. Bavassano, B., Bruno, R., Evidence of local generation of Alfvénenic turbulence in the solar wind, *J. Geophys. Res.*, **94**, 11977-11982, 1989.
4. Grappin, R., Mangeney, A., Marsch, E., On the origin of solar wind turbulence: Helios data revisited, *J. Geophys. Res.*, **95**, 8197-8210, 1990.
5. Bruno, R., Carbone, V., Bavassano, B., Pietropaolo, E., On the role of coherent and stochastic fluctuations in the evolving solar wind MHD turbulence, this issue.
6. Tu, C.-Y., Marsch, E., MHD structures, waves and turbulence in the solar wind: Observations and theories, *Space Sci. Rev.*,**73**, 1-210, 1995.
7. Malara, F., Primavera, L., Veltri, P., Compressive fluctuations generated by time evolution of Alfvénenic perturbations in the solar wind current sheet, *J. Geophys. Res.*,**101**, 21597-21617, 1996.
8. Bavassano, B., Pietropaolo, E., Bruno, R., On the evolution of outward and inward Alfvénenic fluctuations in the polar wind, *J. Geophys. Res.*, **105**, 15959-15964, 2000.
9. Del Zanna, L., Velli, M., Londrillo, P., Parametric decay of circularly polarized Alfvénwaves: Multidimensional simulations in periodic and open domains, *Astron. Astrophys.*, **367**, 705-718, 2001.
10. Del Zanna, L., Parametric decay of oblique arc-polarized Alfvén waves, *Geophys. Res. Lett.*, **28**, 2585-2588, 2001.
11. Malara, F., Primavera, L., Veltri, P., Nonlinear evolution of parametric instability of a large amplitude nonmonochromatic Alfvénwave, *Phys. Plasmas*,**7**, 2866-2877, 2000.
12. Marsch, E., Tu, C.-Y., On the radial evolution of MHD turbulence in the inner heliosphere, *J. Geophys. Res.*, **95**, 8211-8229, 1990.
13. Goldstein, B. E., Smith, e. J., Balogh, A., Horbury, T. S., Goldstein, M. L., Roberts, D. A., Properties of magnetohydrodynamic turbulence in the solar wind as observed by Ulysses at high heliographic latitudes, *Geophys. Res. Lett.*, **22**, 3393-3396, 1995.
14. Bavassano, B., Observations of Alfvénic turbulence evolution in the 3D heliosphere, this issue.
15. Del Zanna, L., Velli, M., Londrillo, P., Nonlinear evolution of large-amplitude Alfvénwaves in parallel and oblique propagation, this issue.

Day the Solar Wind Almost Disappeared: Magnetic Field Fluctuations and Wave Refraction

Charles W. Smith*, Dermott J. Mullan*, Norman F. Ness*, Ruth M. Skoug† and John Steinberg†

*Bartol Research Institute, University of Delaware, Newark, Delaware
†Los Alamos National Laboratory, Los Alamos, New Mexico

Abstract. On May 11, 1999 the Advanced Composition Explorer (ACE) spacecraft observed a rarefied parcel of solar wind that has come to be known as "The Day the Solar Wind Disappeared." Little if any change is seen in the large-scale interplanetary magnetic field during this time, but the magnetic field fluctuations are depressed and significantly more transverse to the mean field. The high Alfvén speed resulting from the constant field intensity and low ion density enhances wave refraction, and we examine this as a possible explanation for the fluctuation properties.

INTRODUCTION

Midday on May 10, 1999 (day 130), the ACE spacecraft observed the beginning of a period of depleted solar wind ion density that continued until midday on May 12 with a density minimum that lasted ~6 hours at the end of May 11. During this extended 48-hour period the density dropped from a fairly typical value of $\sim 5\,\mathrm{cm}^{-3}$ to $\sim 0.1\,\mathrm{cm}^{-3}$. We hereinafter refer to this entire 48-hour period as the "rarefaction interval" without intending to attribute any particular source explanation [3, 5, 8]. During this interval the Alfvén speed was exceptionally high, while the ion temperature and wind speeds were low. The exceptional nature of this parcel of solar wind provides an opportunity to examine in some detail two distinct physical processes that operate in the solar wind. One involves macrophysics (refraction) and the other involves microphysics (dissipation).

Although our presentation at the meeting discussed our analysis of magnetic dissipation and demonstrated that the onset of the dissipation range for interplanetary magnetic fluctuations occurs at the ion inertial scale $L_{ii} \sim V_A/\Omega_{ic}$ where $V_A = B/\sqrt{4\pi N_P m_P}$ is the Alfvén speed, B is the interplanetary magnetic field (IMF) intensity, N_P is the proton density, m_P is the proton mass, and $\Omega_{ic} = eB/m_P c$ is the proton cyclotron frequency, space prevents us from discussing this result here [6]. We therefore limit this discussion to refraction effects only.

ANALYSIS

Figure 1 shows an abreviated overview of the plasma and field measurements during May 10–12, 1999 (days 130–132). The IMF magnitude B is shown. Not shown is the field latitude δ, and longitude λ which are equally unremarkable. The IMF magnitude and direction on day 131 are more steady than on the days before and after, but not unusually so. There is no depletion of the field magnitude during the time in question.

The root-mean-square fluctuation level of the IMF, B_{RMS}, computed from 3 vectors s^{-1} data using a 16-s mean is shown. We draw attention to the fact that B_{RMS} is unusually small (0.1-0.2 nT) during the rarefaction interval. Before and after this interval, B_{RMS} is close to 1 nT. During the rarefaction interval, B_{RMS} undergoes a factor of 5-10 reduction compared to the neighboring plasma.

The proton density N_P is shown to decrease through ~2 orders of magnitude with a minimum at ~1800 UT on day 131. The wind speed V_{SW} (not shown) is seen to decrease monotonically at the same time. The proton temperature T_P (not shown) is unusually low for typical solar wind observations during the last 6 hours of day 131, but *Richardson et al.* [5] demonstrate that T_P and V_{SW} display the nominally expected correlation for solar wind observations.

The Alfvén speed V_A is shown. The low N_P, in combination with the nearly constant B, leads to an elevated V_A during this time. The simultaneously decreasing V_{SW} leads to nearly sub-Alfvénic wind flow during the last 6 hours of day 131.

The apparent depletion of magnetic fluctuations on

CP679, *Solar Wind Ten: Proceedings of the Tenth International Solar Wind Conference*, edited by M. Velli, R. Bruno, and F. Malara

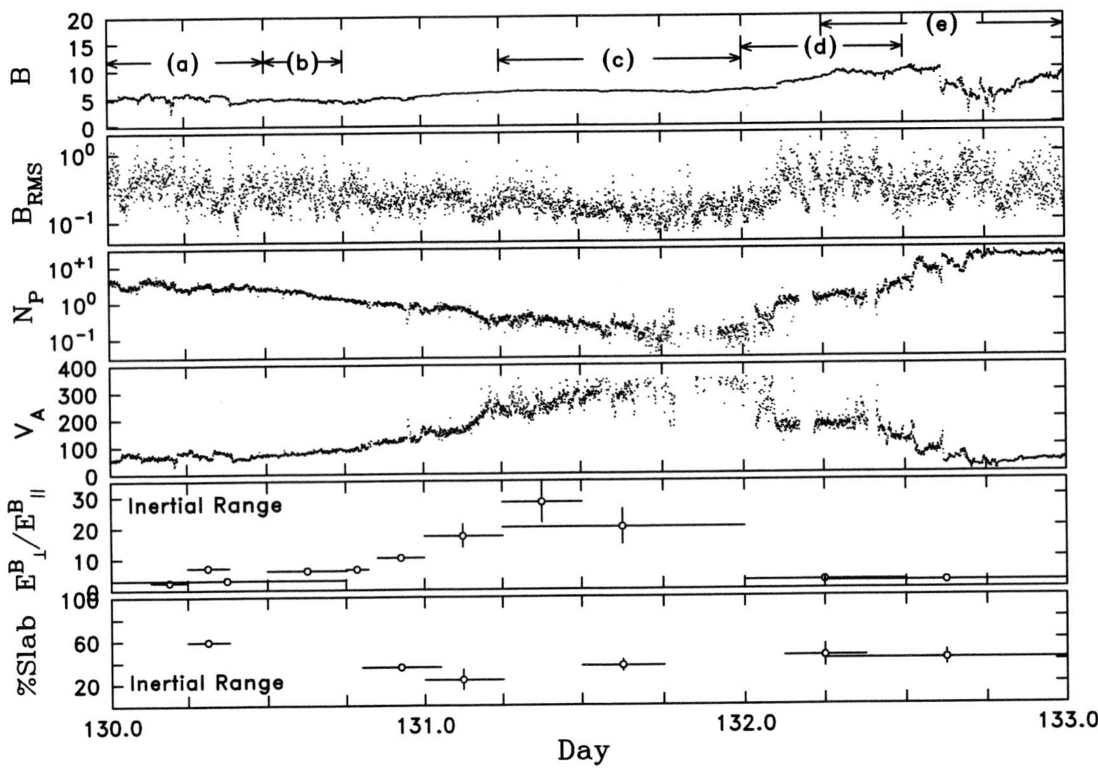

FIGURE 1. Days 130 through 132, 1999, when the solar wind density is observed to drop to 0.1 p cm^{-3}. IMF intensity B (nanoTeslas) as well as RMS level of the IMF fluctuations B_{RMS} (nanoTeslas) are provided by the MAG instrument. The proton density N_P (cm^{-3}) is provided by the SWEPAM instrument. The Alfvén speed V_A (km s^{-1}) is computed from data supplied by both instruments. The anisotropy of the IMF fluctuation spectra $E^B_\perp/E^B_\parallel$ in the inertial range is shown along with the anisotropy of the wave vector (expressed in terms of percent slab component also in the inertial range).

day 131 begs an explanation. We draw our inspiration from solar physics. MHD waves propagate through the magnetically structured corona in a predictable manner. Specifically, fast-mode waves are refracted away from regions of high Alfvén speed [7]. Conversely, fast-mode waves are focused toward regions of low Alfvén speed. If this mechanism is to explain the observations here, then several predictions must be satisfied: (1) Evidence must be provided that the fluctuations are indeed waves, (2) The fluctuations must be less compressive within the high-V_A region from which the compressive fast-mode waves are expelled, and (3) The fluctuation energy outside the refraction region should (hopefully) demonstrate an excess wave energy due to the build up of waves in the low-V_A region.

Figure 1 shows the results of two additional analyses. The computed anisotropy of the magnetic fluctuation energy for components parallel $E^B_\parallel$ and perpendicular $E^B_\perp$ to the mean field is shown for frequencies 2×10^{-3} to $(1-2) \times 10^{-1}$ Hz representing the high frequency end of the inertial range. Horizontal bars represent the data interval used in the analysis while vertical bars give the variance of the computed ratio. The key feature of the anisotropy analysis is that during the period of low solar wind density the IMF fluctuations in the inertial range are distinctly more transverse to the mean field than elsewhere. *Belcher and Davis* [2] state that the typical IMF fluctuations in the inertial range possess an anisotropy $E^B_\perp/E^B_\parallel$ that is 9:1 on average. We find that on days 130-133 the anisotropy $E^B_\perp/E^B_\parallel$ is significantly smaller than this before and after the interval in question, but more than twice this value during the period of lowest solar wind density. On day 131 the inertial range fluctuations are exceptionally transverse and demonstrate high anisotropy ($E^B_\perp/E^B_\parallel > 20{:}1$ and briefly exceeding 28:1). Transverse fluctuations are frequently cited as an indication for the presence of noncompressive Alfvén waves, implying that the usual compressive component is absent from the high-V_A region but present in excess of typical levels immediately outside the high-V_A region.

Bieber et al. [1] and *Leamon et al.* [4] demonstrate that on average only 20% of the fluctuation energy at 1 AU is associated with wave vectors that are parallel to the mean field (one-dimensional (1-D) or slab geometry

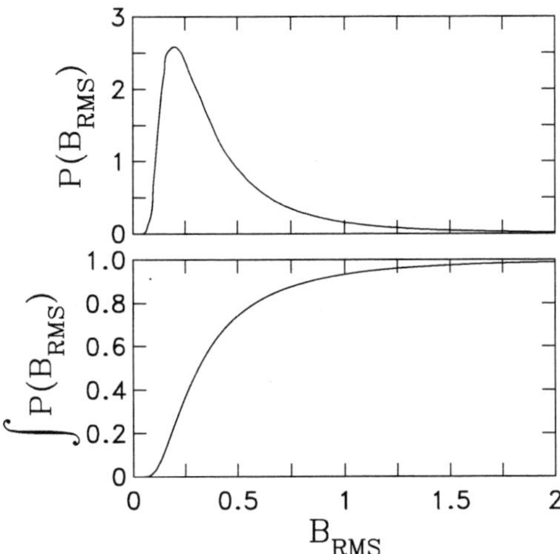

FIGURE 2. (top) The unusually low RMS level (nanoTeslas) of the IMF fluctuations during the interval in question is demonstrated by computing the distribution of RMS levels over the period day 1, 1998, through day 169, 2000. (bottom) The integral over the distribution from 0 to the prescribed value emphasizes this point.

generally regarded as parallel-propagating waves). The remaining 80% of the fluctuation energy is associated with wave vectors that are normal to the mean field (2-D turbulence). We employ this same analysis in the bottom panel in Figure 1. It is interesting to note that the percent slab geometry in the inertial range early in the rarefaction interval (when $V_A \leq 50\,\mathrm{km\,s^{-1}}$) is consistently higher than previously seen by Bieber et al. and Leamon et al. However, when ACE lies deep inside the rarefaction interval, the slab fraction is consistent with the earlier analyses and less than the values immediately outside the high-V_A region. There appears to be an enhancement of the slab wave component early and late in the 3-day period when V_A is more nearly typical of solar wind values at 1 AU, but a depletion of the slab component deep within the rarefaction interval where V_A is high. This is consistent with the refraction argument.

From the apparent fact that our 3-day interval is more wavelike than typical IMF observations and less compressive while at the same time it demonstrates a reduced overall fluctuation level, we argue that wave refraction is (or has been) effective at attenuating the wave component within the region of high V_A leading to reduced RMS levels there. An apparent build-up of compressive wave energy is implied in the surrounding region.

STATISTICS

Figure 2 shows the distribution of B_{RMS} levels for ACE/MAG measurements from day 1, 1998, through day 169, 2000. The distribution peaks at $\sim$0.2 nT but exhibits a long tail extending to 1.5 nT and beyond. In order to demonstrate the relative rarity of RMS levels < 0.2 nT, we integrate the distribution function in the lower panel. The computed value for $\int_0^{0.1} Pd(B_{\mathrm{RMS}}) = 0.02$ while $\int_0^{0.2} Pd(B_{\mathrm{RMS}}) = 0.24$, $\int_0^{0.5} Pd(B_{\mathrm{RMS}}) = 0.74$ and $\int_0^{1.0} Pd(B_{\mathrm{RMS}}) = 0.93$. This indicates that while B_{RMS} levels which range from 0.5 to 1.0 both before and after day 131 are larger than the median value, the levels during day 131 tend to the lower limits of the distribution.

CONCLUSIONS

We have demonstrated that an extended region of reduced solar wind density resulting in an elevated local Alfvén speed displays reduced magnetic fluctuation levels consistent with wave refraction. A combination of data analysis methods including spectral analysis techniques [6] supports this interpretation.

The interval studied here is not the sole event that demonstrates the behavior described here. Figure 3 shows a 6-day interval 14 days prior to the period studied here that is centered on a 24-hour period when B was constant, N_P was low, V_A was high, and B_{RMS} was reduced. Other similar intervals can be found in the ACE dataset. Supression of the IMF fluctuation level is evident at the same time that the Alfvén speed is elevated. Preliminary analysis of this interval shows properties similar to that found for the interval studied here and suggests that refraction may explain this interval as well. Moreover, refraction may be one of several competing processes throughout the solar wind that determines the local level of the wave energy and the degree of compressive fluctuations.

Wave refraction is ray optics and requires that the refracting (high-V_A) region be large in comparison with the wavelength of the wave. *Smith et al.* [6] demonstrated that such intervals exist in the solar wind in addition to the one shown here and that large, high-V_A regions have reduced B_{RMS} on average. However, resolution of the refraction signatures is not always possible owing to competing dynamics including the presence of local sources of turbulence within the wind. With this in mind, one must wonder if refraction in the early stages of ICME formation is not at least in part responsible for the low fluctuation levels generally seen in these high-V_A structures.

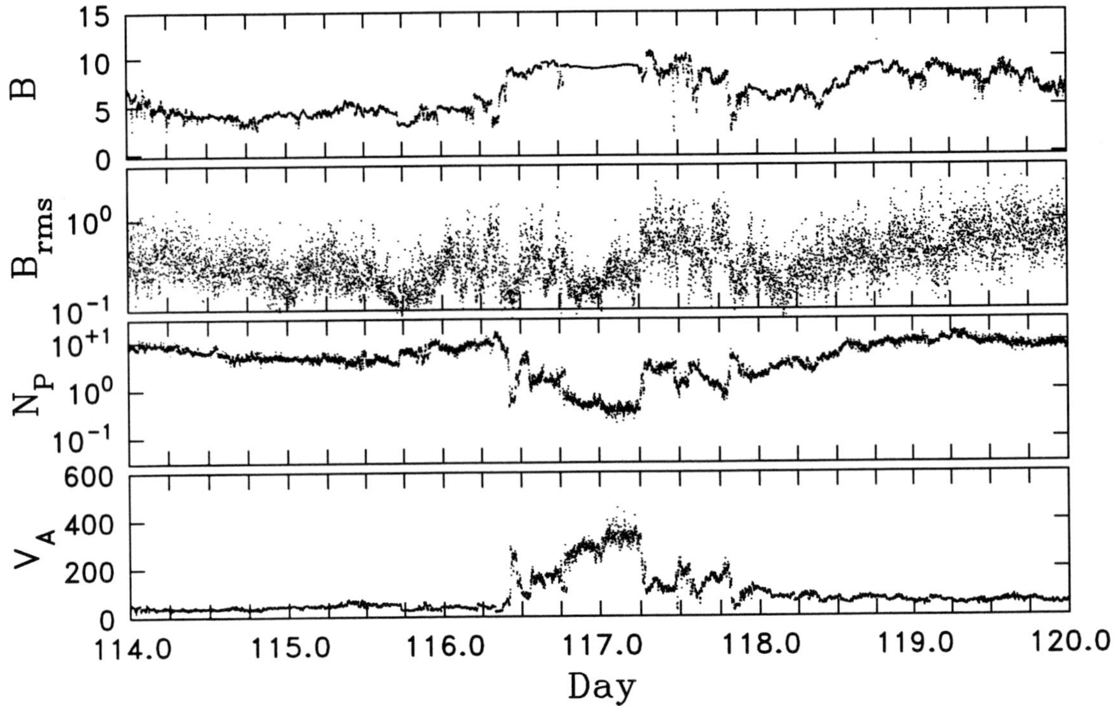

FIGURE 3. B, B_{RMS}, N_P, and V_A observed by ACE for days 114 through 119 of 1999. Note another period of low N_P and high V_A on days 116 and 117 when B was constant and B_{RMS} was low.

ACKNOWLEDGMENTS

Efforts at the Bartol Research Institute were supported by CIT subcontract PC251439 under NASA grant NAG5-6912 for support of the ACE magnetic field experiment and by the NASA Delaware Space College Grant. Work at Los Alamos was performed under the auspices of the U.S. Department of Energy with financial support from the NASA ACE program.

REFERENCES

1. Bieber, J. W., W. Wanner, and W. H. Matthaeus, Dominant two-dimensional solar wind turbulence with implications for cosmic ray transport, *J. Geophys. Res.*, *101*, 2511–2522, 1996.
2. Belcher, J. W., and L. Davis Jr., Large-amplitude Alfvén waves in the interplanetary medium, 2, *J. Geophys. Res.*, *76*, 3534–3563, 1971.
3. Crooker, N. U., S. Shodham, J. T. Gosling, J. Simmerer, R. P. Lepping, J. T. Steinberg, and S. W. Kahler, Density extremes in the solar wind, *Geophys. Res. Lett.*, *27*, 3769–3772, 2000.
4. Leamon, R. J., C. W. Smith, N. F. Ness, W. H. Matthaeus, and H. K. Wong, Observational constraints on the dynamics of the interplanetary magnetic field dissipation range, *J. Geophys. Res.*, *103*, 4775–4787, 1998.
5. Richardson, I. G., D. Berdichevsky, M. D. Desch, and C. J. Farrugia, Solar-cycle variation of low density solar wind during more than three solar cycles, *Geophys. Res. Lett.*, *27*, 3761–3764, 2000.
6. Smith, C. W., D. J. Mullan, N. F. Ness, R. M. Skoug, and J. Steinberg, Day the solar wind almost disappeared: Magnetic field fluctuations, wave refraction and dissipation, *J. Geophys. Res.*, *106*, 18,625–18,634, 2001.
7. Uchida, Y., Flare-induced MHD disturbances in the corona: Moreton waves and Type II shocks, in *High Energy Phenomena in the Sun*, edited by R. Ramaty and R. G. Stone, NASA Spec. Publ., *NASA SP-342*, 577–588, 1973.
8. Usmanov, A. V., M. L. Goldstein, and W. M. Farrell, A view of the inner heliosphere during the May 10-11, 1999 low density anomaly, *Geophys. Res. Lett*, *27*, 3765–3768, 2000.

Weak Double Layers in the Solar Wind and their Relation to the Interplanetary Electric Field

C. Salem*, C. Lacombe†, A. Mangeney†, P. J. Kellogg** and J.-L. Bougeret†

*Space Sciences Laboratory, University of California, Berkeley, USA
†LESIA, Observatoire de Paris-Meudon, Meudon, France
**School of Physics and Astronomy, University of Minnesota, Minneapolis, USA

Abstract. In the solar wind at 1 AU, coherent electrostatic waveforms in the ion acoustic frequency range (between the ion and electron plasma frequencies) have been recently observed by the WAVES/TDS instrument on WIND. Many of these structures have been interpreted in terms of Weak Double Layers (WDL), since they sustain a net potential drop of roughly 1 mV directed towards the Earth. The TDS data are compared to the continuous measurements of thermal and non thermal electric spectra above 4 kHz obtained by the WAVES/TNR instrument : this allows us to determine the frequency of occurrence of the WDL at the L1 Lagrange point. Extrapolating this result provides a total potential drop of about 300 to 1000 Volts on the Sun-Earth distance, compatible with the potential needed to maintain the global charge neutrality in the solar wind. This suggests that the interplanetary electrostatic potential is not continuous but results from a succession of WDL, distributed intermittently between the Sun and the Earth. We also find that the energy of the non thermal fluctuations on TNR between 4 and 6 kHz is correlated to the interplanetary electrostatic field, parallel to the spiral magnetic field, calculated with a two-fluid model, thus providing further evidence of a relation between the interplanetary electrostatic field and the electrostatic fluctuations in the ion acoustic range.

1. INTRODUCTION

The solar wind is the outward extension of the million-degree hot solar corona. It is a weakly collisional, strongly turbulent plasma in a supersonic and super-Alfvénic spherical expansion. Since the electrons are less gravitationally bounded by the Sun than the protons, they tend to be displaced outward with respect to the protons. To maintain the global charge neutrality of the solar wind plasma, an interplanetary electrostatic potential difference $\Delta\Phi_{IP}$ sets in between the solar corona and infinity. The corresponding electric field E_{IP} is directed antisunward and plays a key role in the solar wind expansion. Values of $\Delta\Phi_{IP}$ can be obtained from different models for solar wind expansion, for example, in a two-fluid model (where E is related to the electron pressure) or in an exospheric model [1, 2] (where E is such that the flux of the escaping electrons is equal to the proton flux). These models predict a potential difference $\Delta\Phi_{IP}$ of the order of 400 to 1000 Volts between the solar corona and the Earth orbit. Such large-scale potentials can of course not be measured directly in-situ.

Since the solar wind is a weakly collisional plasma, it is usually argued that wave-particle interactions replace binary collisions in order to restore the fluid character of the flow by regulating the energy transport and dissipation [3, 4]. Among the waves that can play a role in this respect, electrostatic waves in the Doppler-shifted ion acoustic frequency range, *i.e.* with frequencies f between the proton and the electron plasma frequencies ($f_{pi} \leq f < f_{pe}$), have been observed by several spacecraft in the solar wind. This broadband ion acoustic activity is an intermittent but almost permanent feature of the solar wind [5, 6, 7]. Neither the wave mode nor the source of these waves have yet been unambiguously identified [5].

Recently, high-time resolution data from the WAVES experiment on WIND have led to a major contribution to our understanding of this ion-acoustic-like wave activity in the solar wind, by revealing for the first time its highly coherent nature [7, 8]. Indeed, Coherent Electrostatic Waves (CEW hereafter) have been observed, as a mixture of quasi-sinusoidal wave trains and solitary like structures with scales of tens of Debye lengths [7]. The latter appear to be Weak Double Layers (WDL hereafter) with a net potential difference $\Delta\phi \geq 1$ mV across the structure. The observed potentials usually drop towards the Earth, in the same sense as the interplanetary electrostatic potential Φ_{IP}. It is then tempting to speculate that Φ_{IP} is actually the result of a succession of small potential drops in WDL, due to small charge separations between the protons and the escaping electrons [8]. In this paper, we propose to check this hypothesis by estimating the rate of occurrence of the WDL in the solar wind. We also check whether the waves play a role in the solar

CP679, *Solar Wind Ten: Proceedings of the Tenth International Solar Wind Conference*,
edited by M. Velli, R. Bruno, and F. Malara

wind energy transport by looking for a relation between the energy of the waves and the solar wind properties.

2. THE WAVE MEASUREMENTS

The WAVES experiment on WIND measures electric and magnetic plasma waves over a large range of frequencies [9]. In the present study, we consider the electric field fluctuation measurements provided by two instruments, the Thermal Noise Receiver (TNR) and the Time Domain Sampler (TDS), with the x antenna, a wire dipole of physical length $2L_x$ tip-to-tip ($L_x = 50$ m), spinning in the Ecliptic plane.

The TDS is a "snapshot" waveform sampler. It detects all the electric signals above a programmable threshold of 50 μV/m, and generates 2048 point events. Due to telemetry constraints, only a few waveforms are transmitted to the ground. During the period analyzed here in 1995, the transmitted event is not the most intense but the most recently recorded, roughly every 10 min. In this study, we consider only high bit rate events sampled at 120,000 samples per second, so that an event duration is 17.07 ms. The electric field E is obtained by dividing the measured potential difference at the antenna terminals by the length L_x [7].

The TNR is a very sensitive digital spectrum analyzer designed to do thermal noise spectroscopy in the ambient solar wind plasma [4, 10]. In its lowest frequency band (4-16 kHz), it measures continuously electric power spectra V^2 in V^2/Hz every 4.5 s with an integration time of 1.472 s. The square electric field E^2 (V^{2}m^{-2}Hz^{-1}) is obtained by dividing V^2 by L_x^2 [11].

We also use here hourly averages of the magnetic field components (MFI experiment [12]), the solar wind speed V_{sw} and the electron and proton temperature, T_e and T_p (3D-Plasma experiment [13]), the electron density N_e from electron thermal noise [10]. Detailed electron distribution functions [13] have been integrated to give hourly averages of the components of the heat flux vector $\mathbf{Q}_e$ and of the parallel to perpendicular temperature ratio $T_{e\parallel}/T_{e\perp}$ [10].

3. SUMMARY OF OBSERVATIONS

Our observations were taken in the ambient solar wind, at the Lagrange point L1, from May 20 to June 26, 1995. This interval is typical of the solar wind close to the last minimum of solar activity, because Wind has explored high-speed as well as low-speed streams [7, 14]. During this interval, the TDS detected about 2160 Coherent Electrostatic Waves (CEW hereafter). These CEW display two main typical shapes : sinusoidal wave packets and non-sinusoidal isolated spikes lasting less than 1 ms. These solitary like structures are found to be Weak Double Layers (WDL hereafter), with a net potential difference across the structure implying a non-zero average electric field almost always directed towards the Earth [7, 8]. Figure 1 shows an example of WDL observed in the solar wind. Figure 1a displays the electric field $E_\parallel$ parallel to $\mathbf{B}$ (in mV/m, and smoothed over 10 points in order to eliminate the high frequency noise) during a time interval of 4 ms. The corresponding electric potential, normalized to the local electron temperature T_e is shown in Figure 1b. About 30% of the CEW in our sample are WDL. A statistical study indicates that a typical spatial size of the WDL is $\sim 25\lambda_D$ and a typical value for the potential drop across the WDL is $\Delta\phi \geq 1$ mV, or $e\Delta\phi/k_BT_e \approx 10^{-4}$–$10^{-3}$ [7].

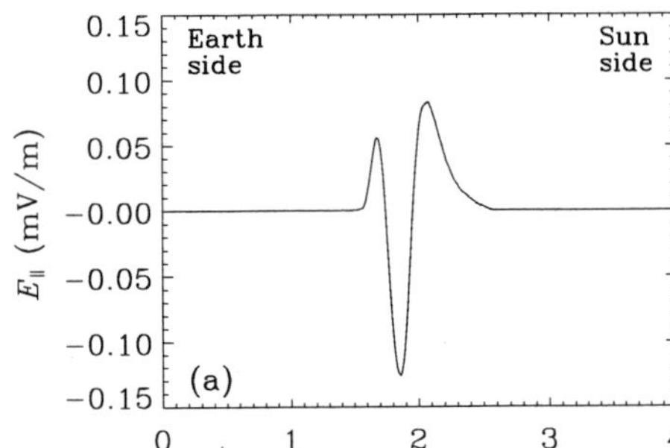

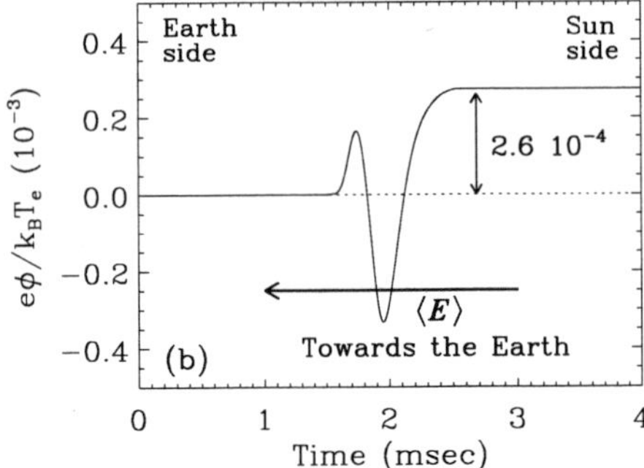

FIGURE 1. Typical weak double layer detected in the solar wind : (a) the measured electric field smoothed over 10 points (positive if directed towards the Earth) ; (b) the corresponding electric potential profile, normalized to k_BT_e.

TDS observations strongly suggest that these structures are one-dimensional, varying only along the magnetic field $\mathbf{B}$, and that they are convected by the solar wind since their velocity in the plasma frame is much smaller than the solar wind speed [7]. These WDL probably manifest small-scale charge separation due to a partial decoupling between electrons and protons on distances comparable to a Debye length scale.

TDS observations are not continuous so that the rate

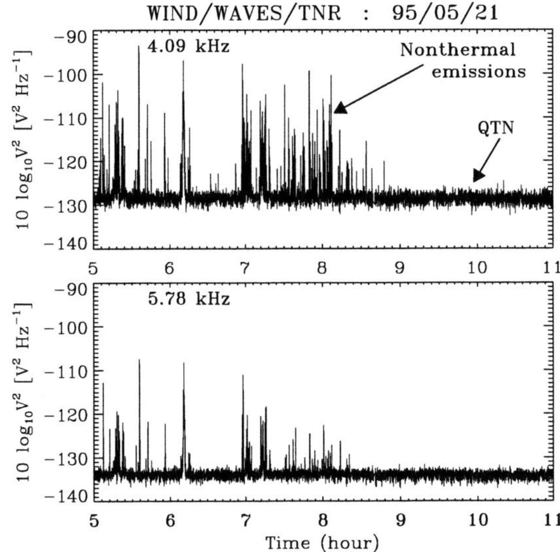

FIGURE 2. Time profile of the spectral power of the electric potential at two frequencies on TNR during 6 hours (sampling time : 4.5 s).

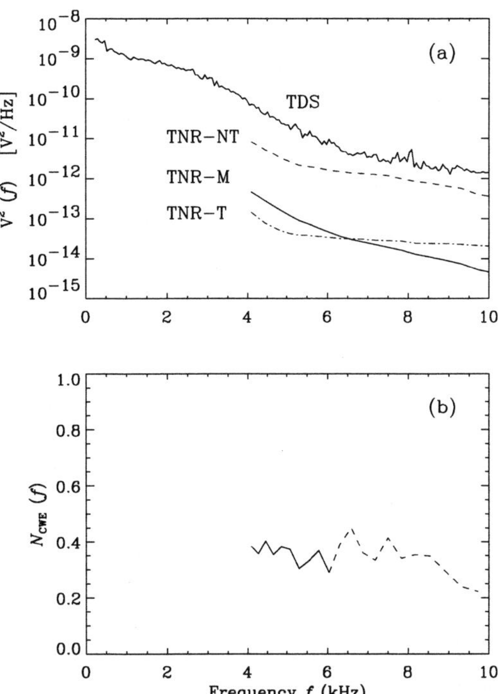

FIGURE 3. (a) The different average TDS and TNR spectra (see text below) ; (b) the rate of occurrence in s^{-1} of CEW in the solar wind at 1 AU as a function of the frequency (Eq. 1).

of occurrence of WDL in the solar wind can only be estimated if we combine wave observations from both TDS and TNR instruments. The TNR instrument is sensitive enough to measure both the thermal and the non-thermal fluctuations. This can be seen on Figure 2 which displays the temporal profile of the electrostatic fluctuations (in V^2/Hz) measured on TNR at 4.09 kHz (upper panel) and 5.78 kHz ($f_{pe} \simeq 23-25$ kHz). The thermal fluctuations, or Quasi-Thermal Noise (QTN) [15], depend on N_e, T_e, V_{sw}, and T_p [16] and their intensity distribution is Gaussian. The non-thermal high intensity fluctuations are very sporadic, with a more or less power law intensity distribution [11]. Their intensity and rate of occurrence decrease when the frequency increases. These nonthermal fluctuations are the spectral counterparts of the waveforms seen on TDS[11].

4. RATE OF OCCURRENCE OF WDL

We estimate the rate of occurrence of the CEW and of the WDL in the solar wind by comparing the average spectral densities on TDS and on TNR in their common frequency range, *i.e.* between 4 and 6 kHz. The upper line of Figure 3a gives the spectral density $V^2_{TDS}(f)$ of the CEW averaged over the 2160 waveform spectra. Most of these spectra have a peak power around 2 kHz [7, 11]. However, the measured frequencies over the whole sample vary between 0.2 and 8 kHz, and about 11% of the CEW have significant power above 4 kHz [11]. The dashed line $TNR-T$ is the spectrum of the purely thermal fluctuations averaged over the sample of thermal spectra, and the dashed line $TNR-NT$ the spectrum of the non-thermal fluctuations averaged over the sample of non-thermal spectra during 38 days. The solid line spectrum $TNR-M$ is the average of the non-thermal TNR spectral energy over the total number of spectra, thermal plus non-thermal ; it is thus a time average of the non-thermal energy in the ambient solar wind. So the average number of CEW in the solar wind per second will be given by a comparison of V^2_{TDS} with V^2_{TNR-M}. If there is only one TDS event during the TNR integration time τ, then its TNR intensity at a given frequency will be $V^2_{TNR} = V^2_{TDS}/\beta$, $\beta = 1.472/0.01707 = 86$ being the ratio between the TNR and the TDS integration times. Thus, the number of CEW observed in the solar wind during 1 s above 4 kHz is

$$N_{CEW}(f) = \beta V^2_{TNR-M}/(\tau V^2_{TDS}). \quad (1)$$

$N_{CEW}(f)$ is plotted in Figure 3b. Above 4 kHz, the average number of CEW per second in the solar wind is $N_{CEW} \sim 0.36$.

Since only 11% of the observed CEW contribute to the frequency range above 4 kHz, and 30% are WDL, we conclude that an estimate of the *number of WDL* drifting past the Wind spacecraft per second is

$$N_{WDL} \sim 1 \text{ s}^{-1} \quad (2)$$

Assuming (i) an average travel time of 3 10^5 s for a solar wind plasma element between the solar corona and the Wind orbit, (ii) that N_{WDL} and the average potential difference $\Delta\phi$ across a WDL remain both constant from the solar corona to the Earth, one may estimate the *total potential difference* at 1 AU [11] :

$$300 \leq \Delta\Phi_{1\,AU} \leq 1000 \text{ Volts.} \tag{3}$$

This range of values for $\Delta\Phi_{1\,AU}$ is the one needed to maintain charge neutrality in the solar wind [17]. So these results suggest that *the interplanetary electric potential is not continuous but is actually established through a succession of WDL*, distributed intermittently along the radial direction.

5. ENERGY OF THE WAVES

We look here for correlations between the electric field of the *ion acoustic like* waves and some properties of the solar wind plasma, as those found in Helios data [18]. For that, we consider hourly averages of the square electric field E^2 of the TNR non-thermal fluctuations between 4 and 6 kHz [11]. We find no correlation of $\log E^2$ with T_e/T_p, nor with V_{sw} or Q_e, in contrast to the Helios results. However, weak correlations (0.31 or 0.32) are found between $\log E^2$ and T_e, $T_{e\parallel} - T_{e\perp}$, and $\cos\chi = \cos(\mathbf{V_{sw}}, \mathbf{B})$. One possible interpretation is that these parameters to which $\log E^2$ seems to be related play a role in the interplanetary electrostatic potential. According to *Pilipp et al.* [19] the interplanetary electrostatic potential Φ_{PG} due to the electron pressure gradient, in a two-fluid model, can be written as a function of T_e, $T_{e\parallel} - T_{e\perp}$ and $\cos^2\chi$ for a spiral magnetic field in the ecliptic plane

$$eN_e \frac{d\Phi_{PG}}{dr} = \frac{d}{dr}\left[p_{e\perp} + (p_{e\parallel} - p_{e\perp})\cos^2\chi\right] + \frac{1}{r}(2\cos^2\chi - \sin^2\chi)(p_{e\parallel} - p_{e\perp}) \tag{4}$$

where $p_{e\parallel,\perp} = N_e k_B T_{e\parallel,\perp}$ is the electron pressure parallel and perpendicular to the B field. Using the relation

$$\cos^2\chi = V_{sw}^2/(V_{sw}^2 + \Omega^2 r^2) \tag{5}$$

where Ω is the angular frequency of the Sun rotation, and assuming that $T_e \propto r^{-\alpha}$, the radial component of the interplanetary electric field at 1 AU is

$$E_r(V/m) = -\frac{d\Phi_{PG}}{dr} = \left[(2+\alpha)T_{e\perp} + (T_{e\parallel} - T_{e\perp})\right.$$
$$\left.[1 + (1+\alpha)\cos^2\chi - 2\cos^4\chi]\right] / 1.5\,10^{11} \tag{6}$$

where the temperatures are in eV ; and the component of this electric field along the magnetic field is

$$E_{IP\parallel}(V/m) = |\cos\chi|\, E_r\,. \tag{7}$$

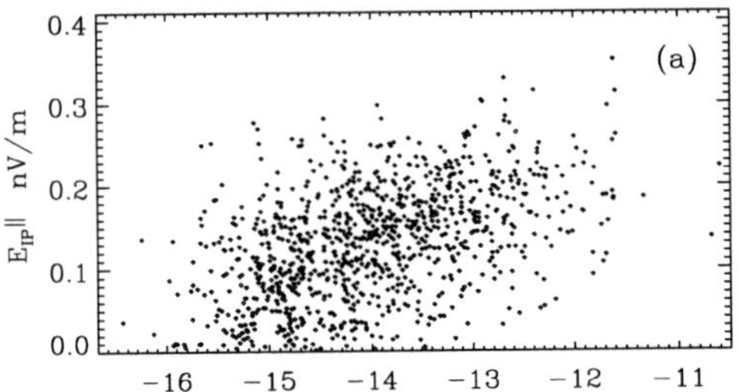

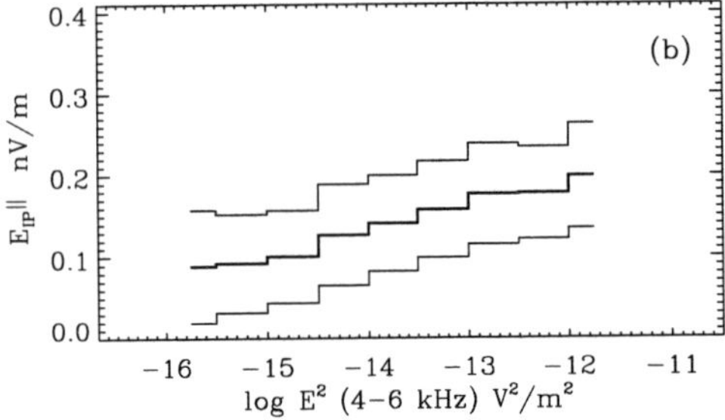

FIGURE 4. The interplanetary electric field parallel to the spiral magnetic field in a two-fluid model (Eqs 4 to 7) as a function of the hourly energy E^2 of the nonthermal emissions between 4 and 6 kHz : (a) scatter plot, (b) average and standard deviation in equal bins of $\log E^2$.

$E_{IP\parallel}$ is displayed in Figure 4 as a function of $\log E^2$. Note that $E_{IP\parallel}$ is expressed in nV/m, not in mV/m. $E_{IP\parallel}$ refers to the large-scale interplanetary electric field, calculated in the framework of a two-fluid model (Eqs. 6 and 7), while the electric field E in Figure 1 or in the abscissae of Figure 4 refers to the local electric field fluctuations measured by TDS and/or TNR with an antenna (see section 2). The correlation between $\log E^2$ and $E_{IP\parallel}$, shown in Figure 4 is better (~ 0.45) than the individual correlations (0.31 or 0.32) between $\log E^2$ and $T_{e\parallel} - T_{e\perp}$, T_e, or $\cos\chi$, providing *further evidence* of a relation between the small-scale coherent electrostatic waves and the large-scale interplanetary electric field.

6. CONCLUSION

Observations made by the Time Domain Sampler, an electric waveform analyzer onboard WIND, have shed a new light on the nature of the "ion acoustic" electrostatic turbulence in the solar wind. We showed that this turbulence consists of small amplitude coherent waves and solitary like structures, many of which are very Weak Double Layers (WDL) with small potential drops of roughly 1mV over a few tens of Debye lengths, directed towards the Earth [7]. This coherent electrostatic wave activity seems to be a common feature of collisionless space plasmas. Indeed, electrostatic solitary structures have been observed almost everywhere in the Earth's environment. Most of the observations available, in the auroral terrestrial regions, in the Earth's magnetotail as well as in the solar wind, have been rewiewed by *Salem et al.* [8]. The properties of these waves depend on the region of observation which determines the plasma regime.

WIND observations have allowed for the determination of the rate of occurrence of WDL in the solar wind, $N_{WDL} \sim 1\mathrm{s}^{-1}$. We show that extrapolating this result leads to a total potential difference of 300 to 1000 Volts between the solar corona and 1 AU, which is in the range of values needed to maintain charge neutrality in the solar wind plasma [17]. This gives the first observational indication of the existence of the large-scale electric field in the interplanetary medium, which plays a fundamental role in the expansion of the solar wind [20]. Furthermore, a correlation is found between the energy of the coherent ion acoustic waves and the amplitude of the interplanetary electric field expected in a two-fluid model, with a spiral magnetic field.

These results suggest that the observation of weak double layers in the solar wind is related to the existence of the interplanetary electric field. The corresponding electric potential difference between the solar corona and the Earth orbit would actually be established through a succession of small potential drops across a multitude of WDL, distributed intermittently along the radial direction.

ACKNOWLEDGMENTS

Work at UC Berkeley is supported by NASA grant FDNAG5-11804 to the University of California. The french contribution is supported by the Centre National d'Etudes Spatiales and the Centre National de la Recherche Scientifique.

REFERENCES

1. Lemaire, J., and Scherer, M., *J. Geophys. Res.*, **76**, 7479 (1971).
2. Pierrard, V., *et al..*, *Geophys. Res. Let.*, **28**, 233 (2001).
3. Kellogg, P. J., *Astrophys. J.*, **528**, 480 (2000).
4. Salem, C., *Ondes, turbulence et phénomènes dissipatifs dans le vent solaire à partir des observations de la sonde WIND*, Ph.D. thesis, Université de Paris VII/Observatoire de Paris, France (2000).
5. Gurnett, D. A., "Waves and instabilities," in *Physics of the Inner Heliosphere II - Particles, Waves and Turbulence*, edited by R. Schwenn and E. Marsch, Springer-Verlag, 1991, p. 135.
6. MacDowall, R. J., *et al..*, *Astron. & Astrophys.*, **316**, 396 (1996).
7. Mangeney, A., *et al..*, *Ann. Geophysicae*, **17**, 307 (1999).
8. Salem, C., *et al..*, "Coherent Electrostatic Nonlinear Waves in Collisionless Space Plasmas," in *LNP Vol. 536: Nonlinear MHD Waves and Turbulence*, edited by P.-L. S. T. Passot, Springer, 1999, p. 251.
9. Bougeret, J.-L., *et al..*, *Space Sci. Rev.*, **71**, 231 (1995).
10. Salem, C., *et al..*, *J. Geophys. Res.*, **106**, 21701 (2001).
11. Lacombe, C., *et al..*, *Ann. Geophysicae*, **20**, 609 (2002).
12. Lepping, R.P., *et al..*, *Space Sci. Rev.*, **71**, 207 (1995).
13. Lin, R.P., *et al..*, *Space Sci. Rev.*, **71**, 125 (1995).
14. Lacombe, C., *et al..*, *Ann. Geophysicae*, **18**, 852 (2000).
15. Meyer-Vernet, N., and Perche, C., *J. Geophys. Res.*, **94**, 2405 (1989).
16. Issautier, K., *et al..*, *J. Geophys. Res.*, **104**, 6691 (1999).
17. Scudder, J. D., and Olbert, S., *J. Geophys. Res.*, **84**, 6603 (1979).
18. Gurnett, D. A., *et al..*, *J. Geophys. Res.*, **84**, 2029 (1979).
19. Pilipp, W. G., *et al..*, *J. Geophys. Res.*, **95**, 6305 (1990).
20. Meyer-Vernet, N., *Eur. J. Phys.*, **20**, 167 (1999).

Weak turbulence of anisotropic shear-Alfvén waves

S. Galtier*, S.V. Nazarenko†, A.C. Newell** and A. Pouquet‡

*I.A.S., Université Paris-Sud, bâtiment 121, 91405 Orsay Cedex, France
†Mathematics Institute, University of Warwick, Coventry, CV4 7AL, UK
**Department of Mathematics, University of Arizona, P.O. Box 210089, 617 N. S. Rita, Tucson, AZ 85721, USA
‡ASP/NCAR, P.O. Box 3000, Boulder, CO 80307-3000, USA

Abstract. We review recent results obtained in [14] on weak turbulence of shear-Alfvén waves in the limit of strongly anisotropic pulsations that are elongated along a strong external magnetic field. The kinetic equation for energy is presented at the level of three-wave interactions. It is in agreement with the Galtier et al. [13] formulation of the full three–dimensional helical case when taking the proper limit. The simplified formalism induced by the new approach renders the applicability conditions for weak turbulence more transparent. It provides an attractive theoretical framework for describing anisotropic MHD turbulence in astrophysical contexts where a strong magnetic field is present and for which shear-Alfvén waves are important.

INTRODUCTION

For many turbulent plasmas encountered in astrophysics [45, 20, 30, 43, 49] as well as in laboratory devices such as tokamaks [51, 17], magnetohydrodynamics (MHD) offers a first satisfactory description. MHD turbulence differs significantly from hydrodynamic turbulence in the fact that a strong magnetic field has a non-trivial effect on the dynamics. It was Iroshnikov [21] and Kraichnan [24] (hereafter IK) who first recognised that the presence of Alfvén waves travelling in opposite directions along a local large magnetic field leads to the weakening of energy transfer to small scales and therefore to a modification of the scaling of the (Kolmogorov) energy spectrum from $k^{-5/3}$, for neutral fluids, to $k^{-3/2}$. In the IK phenomenology MHD turbulence is assumed to be 3D isotropic. However in many realistic situations the presence of strong magnetic fields is observed which makes MHD turbulence strongly anisotropic. Similarly to rotating flows, anisotropy is manifested in a two dimensionalisation of the turbulence spectrum in a plane transverse to the locally dominant magnetic field and in inhibiting spectral energy transfer along the direction parallel to the field [34, 35, 32, 22]. Replacing the 3D isotropy assumption by a 2D one, and retaining the rest of the IK dimensional analysis, gives a $k_\perp^{-2}$ spectrum ($\mathbf{b_0} = b_0 \hat{\mathbf{e}}_\parallel$, the applied magnetic field, $k_\parallel = \mathbf{k} \cdot \hat{\mathbf{e}}_\parallel$, $\mathbf{k}_\perp = \mathbf{k} - k_\parallel \hat{\mathbf{e}}_\parallel$, $k_\perp = |\mathbf{k}_\perp|$) [40].

Density variations are observed in the solar wind and the interstellar medium [1, 3, 19, 28, 29] which means in particular that any theory based on incompressible MHD has to be seen as a first attempt to describe such turbulent plasmas. It is a hard task to include compressibility effects in a rigorous theory like weak wave turbulence. Note that a first attempt has been realized by Kuznetsov [25] in the low β limit. Weak MHD turbulence has to be seen as a useful theoretical framework to understand media like the solar wind or the interstellar medium. It cannot be seen *stricto sensu* as a model for such media, since we observe only a moderate anisotropy (moderate background magnetic field). However this asymptotic model (limit of strong background magnetic field), by revealing properties which are quite different to the isotropic case (strong turbulence), tells us towards what the dynamics must tend when the degree of anisotropy increases and thus it gives important physical information about the medium itself. For example several observations reveal spectra with a power-law index between -1.9 and -2 [9, 10, 48], which are definitively steeper than the (Kolmogorov or IK) phenomenological predictions for isotropic turbulence. As indicated in Bhattacharjee and Ng [5] such a steepening of the spectrum can be due to shocks and discontinuities but it could also be due to turbulence.

IK proposed independently a phenomenology for incompressible, homogeneous, isotropic MHD turbulence based on three-wave processes of interactions of counter-propagating Alfvén waves embedded in a local mean magnetic field. The main differences with the neutral hydrodynamic turbulence case is a slowing down of the energy transfer to small scales and a $k^{-3/2}$ scaling for the energy spectrum when correlations between the velocity and magnetic field are negligible. In hydrodynamic

CP679, *Solar Wind Ten: Proceedings of the Tenth International Solar Wind Conference*,
edited by M. Velli, R. Bruno, and F. Malara

turbulence, the Kolmogorov $k^{-5/3}$ energy spectrum prediction is well supported by many experimental and numerical observations (see for a review Frisch [12]). On the other hand, in magnetized flows there is still a debate about the predicted scaling in the absence of a uniform magnetic field $\mathbf{b}_0$ [7, 44, 8] or in the anisotropic case [11, 27]. Indeed, the interpretation of the Alfvén wave effect on the dynamics is still a subject of discussion [6], including in the case of the Solar Wind and the interstellar medium [2, 18]. However it is well known that the presence of a large scale applied magnetic field changes drastically the physics by imposing a bi-dimensionalisation of the turbulence with a slower spectral energy transfer along $\mathbf{b}_0$. This can be seen for example from numerical simulations of incompressible [34, 47, 41, 39, 33], reduced [22] and compressible MHD [42, 26]. It is interesting to note that in the context of the same regime of weak turbulence a tendency towards bidimensionalisation is being analytically determined for other type of waves like whistler waves described by electron MHD equations [15] or inertial waves for incompressible fluid under rapid rotation [16].

In the present paper a weak wave turbulence formalism is reviewed [14] where only shear-Alfvén waves are taken into account. To simplify further the derivation, the problem will be considered in the limit of strongly anisotropic pulsations elongated along a strong external magnetic field. The result is a much simpler formalism than the one derived in Galtier et al. [13] where the complexity in the algebra of deriving the kinetic equations for the eight correlators involved in the full case of weak MHD turbulence may hinder our understanding of the physical mechanisms at play in weak MHD turbulence. Part of the complexity stems from the inclusion of helicity Such a complexity of the weak turbulence in the general case makes it harder to use, to analyze and to establish the conditions of its applicability. On the other hand, the kinetic equations become much easier to analyze in the limit of nearly bi-dimensional turbulence, $k_\parallel \ll k_\perp$ [13]. In other words, we are able to show that making an assumption about strong anisotropy ($k_\parallel \ll k_\perp$) in the original MHD equations allows, in particular, for a reliable control of the assumptions made during the derivation and results in transparency of the applicability conditions. The latter point is especially important because of the active debate around the role of three-wave *vs.* four-wave processes [50, 39, 13, 37]. The resulting kinetic equation is simple and it provides an attractive theoretical framework for applying anisotropic MHD turbulence theory to astrophysics.

DERIVATION OF THE KINETIC EQUATION FOR SHEAR-ALFVÉN WAVES

We introduce the "perturbed" Elsässer variables $\varepsilon \mathbf{z}^s = \mathbf{v} + s(\mathbf{b} - \mathbf{b}_0)$ where $\mathbf{b_0} = b_0 \hat{\mathbf{e}}_\parallel$ is a strong external magnetic field ($\hat{\mathbf{e}}_\parallel$ is a constant unit vector), $s = \pm 1$ is a polarization index indicating the wave propagation direction and ε is a small parameter measuring the intensity of the wave turbulence. Here, we choose units such that the velocity $\mathbf{v}$ and the magnetic field $\mathbf{b}$ have the same physical dimension. The inviscid 3D incompressible MHD equations written in terms of the "perturbed" Elsässer variables are

$$(\partial_t - s b_0 \partial_\parallel) z_j^s = -\varepsilon \partial_{x_m} z_m^{-s} z_j^s - \partial_{x_j} P_* , \qquad (1)$$

where P_* is the total pressure and $\partial_\parallel$ is the derivative along $\hat{\mathbf{e}}_\parallel$. After Fourier transforming equation (1) and separating the fast and slow time dependencies, we have

$$\partial_t a_j^s(\mathbf{k}) = \qquad (2)$$

$$-i\varepsilon k_m P_{jn} \int a_m^{-s}(\boldsymbol{\kappa})\, a_n^s(\mathbf{L})\, e^{i(-s\omega_k - s\omega_\kappa + s\omega_L)t}\, \delta_{\mathbf{k},\boldsymbol{\kappa}\mathbf{L}}\, d_{\boldsymbol{\kappa}\mathbf{L}}$$

where $d_{\boldsymbol{\kappa}\mathbf{L}} = d\boldsymbol{\kappa} d\mathbf{L}$, $\delta_{\mathbf{k},\boldsymbol{\kappa}\mathbf{L}} = \delta(\mathbf{k} - \boldsymbol{\kappa} - \mathbf{L})$, $z_j^s(\mathbf{x},t) = \int a_j^s(\mathbf{k},t)\, e^{i(\mathbf{k}\cdot\mathbf{x} + s\omega_k t)}\, d\mathbf{k}$ and $P_{jn}(k) = \delta_{jn} - k_j k_n / k^2$ is the projection operator which ensures the incompressibility condition. The small parameter ε measures the strength of the nonlinear coupling, and the Alfvén waves frequency is $\omega(\mathbf{k}) = \mathbf{b_0} \cdot \mathbf{k} = b_0 k_\parallel$.

The first simplification consists to assume in equation (2) that the turbulence is strongly anisotropic, *i.e.* $k_\parallel \ll k_\perp$. Here we use explicitly the property of anisotropy already observed in the presence of a strong $\mathbf{b}_0$ [34]. The second assumption concerns the type of waves that is going to be considered. Only shear-Alfvén waves are taken into account; they are described by the transverse part of $\mathbf{a^s}$, namely $\mathbf{a^s}_\perp = (a_1^s, a_2^s)$. The divergence free condition allows one to express a_2^s in terms of a_1^s as $a_2^s = -(k_1/k_2) a_1^s - (k_\parallel/k_2) a_\parallel^s \approx -(k_1/k_2) a_1^s$. Introducing the notation $a^s = a_1^s$ and neglecting terms that contain factors $k_\parallel$ on the right hand side (RHS), we finally obtain

$$\partial_t a^s(\mathbf{k}) = \qquad (3)$$

$$-i\varepsilon \int \frac{k_2}{\kappa_2 L_2 k_\perp^2} (\mathbf{k}_\perp \cdot \mathbf{L}_\perp)\, (\mathbf{k} \times \boldsymbol{\kappa})_\parallel\, a^{-s}(\boldsymbol{\kappa})\, a^s(\mathbf{L})\, e^{-2isb_0\kappa_\parallel t}$$

$$\delta_{\mathbf{k},\boldsymbol{\kappa}\mathbf{L}}\, d_{\boldsymbol{\kappa}\mathbf{L}} ,$$

where the solution of the resonance conditions has been used. Note that the integration in this equation is still over the three-dimensional vectors $\boldsymbol{\kappa}$ and $\mathbf{L}$ and that a^{-s}

depends on all three of the wavenumber components. We note that the fundamental equation (3) is much simpler than the one obtained in the most general case [13].

The next step consists in introducing the wave turbulence spectra $q^s(\mathbf{k})$ as follows,

$$\langle a^s(\mathbf{k})\, a^{s'}(\mathbf{k}')\rangle = q^s(\mathbf{k})\,\delta(\mathbf{k}+\mathbf{k}')\,\delta(s-s'), \qquad (4)$$

where the averaging is taken over an initial ensemble. Spatial homogeneity means space averaging is equivalent. Further, because of the linear dynamics, the initial ensemble evolves to a state for which the random phase approximation holds. Here, as usual, $\delta(\mathbf{k}+\mathbf{k}')$ appears due to the turbulence homogeneity and $\delta(s-s')$ is due to the fast decorrelation of the oppositely propagating waves. Following standard weak turbulence approach [4, 53] it is possible to derive the following kinetic equation (for more details, see Galtier et al. [14]),

$$\partial_t q^s(\mathbf{k}) = \qquad (5)$$

$$\frac{\pi\varepsilon^2}{b_0}\int \frac{k_2\,(\mathbf{k}_\perp\cdot\mathbf{L}_\perp)^2\,(\mathbf{k}\times\boldsymbol{\kappa})_\parallel^2}{k_\perp^2 L_2 \kappa_2^2}\, q^{-s}(\boldsymbol{\kappa})$$

$$\left[\frac{k_2}{L_2 k_\perp^2}\, q^s(\mathbf{L}) - \frac{L_2}{k_2 L_\perp^2}\, q^s(\mathbf{k})\right]\,\delta(\kappa_\parallel)\,\delta_{\mathbf{k},\boldsymbol{\kappa}\mathbf{L}}\, d_{\boldsymbol{\kappa}\mathbf{L}}\,.$$

The energy spectrum $e^s(\mathbf{k})$ of the shear-Alfvén waves is the sum of the perpendicular components of the energy tensor or, using the divergence free condition, $e^s(\mathbf{k}) = (k_\perp^2/k_2^2)q^s(\mathbf{k})$; therefore it obeys to the following kinetic equation,

$$\partial_t e^s(\mathbf{k}) = \qquad (6)$$

$$\frac{\pi\varepsilon^2}{b_0}\int \frac{(\mathbf{k}_\perp\cdot\mathbf{L}_\perp)^2\,(\mathbf{k}\times\boldsymbol{\kappa})_\parallel^2}{k_\perp^2 L_\perp^2\,\kappa_\perp^2}\, e^{-s}(\boldsymbol{\kappa})$$

$$[e^s(\mathbf{L}) - e^s(\mathbf{k})]\,\delta(\kappa_\parallel)\,\delta_{\mathbf{k},\boldsymbol{\kappa}\mathbf{L}}\, d_{\boldsymbol{\kappa}\mathbf{L}}\,.$$

Equation (6) is our main result and it describes a three-wave process which dominates the turbulence dynamics at low turbulence intensities. This equation coincides with equation (46) of Galtier et al. [13] which was obtained from the general kinetic equations of weak Alfvénic turbulence in the limit $k_\parallel \ll k_\perp$. This shows in particular that the taking of the two limits involved does commute (anisotropic limit before or after having apply the weak turbulence formalism to the MHD equations).

The conditions of validity of the kinetic equation are already given in Galtier et al. [14]. The first condition is

$$\frac{k_\parallel}{k_\perp} \gg \varepsilon^2, \qquad (7)$$

which can be satisfied at any finite wavenumber for sufficiently weak turbulence. Note however that, for any turbulence intensity ε, there always exists a region of small $k_\parallel$ where the condition is violated; this corresponds to the *non-uniform* validity of the kinetic equation. Such behavior any three and four wave interactions is discussed in Newell, Nazarenko and Biven [38].

Another applicability condition is that the spectrum must change slowly when crossing the wavenumber cone (7), that is that the spectrum $q^s(\mathbf{k})$ stays approximately constant as a function of $k_\parallel$ in the range $-\varepsilon^2 k_\perp \ll k_\parallel \ll \varepsilon^2 k_\perp$ at fixed $k_\perp$, which means that long spatial correlations along the external magnetic field are assumed to be absent.

One of the well-known consequences of the three-wave process for turbulence of Alfvén waves in the case $k_\parallel \ll k_\perp$ is the $k_\perp^{-2}$ energy spectrum. Recently, it was shown to be an exact constant-flux solution of equation (6) [13]. Perhaps, a less-known fact is that a stationary constant-flux solution of equation (6) does not have to be necessarily $k_\perp^{-2}$: it may have its exponents anywhere in the range from -1 to -3 depending on the degree of asymmetry of the forcing of the $s=1$ and $s=-1$ waves [13]. Such asymmetric wave pumping is very common in astrophysics (*e.g.*, there are more Alfvén waves traveling away from the Sun than toward it in the Solar Wind (see *e.g.* Matthaeus, Goldstein and Roberts [31]).

DISCUSSION

The weak turbulence kinetic equation can be derived in a much shorter way if the limit $k_\parallel \ll k_\perp$ is taken before the statistical averaging. The simplicity of our derivation also allows for an easier understanding of the two applicability conditions. The first one can always be checked based on the solution of the kinetic equation. On the other hand, the second condition cannot be checked based on the weak turbulence theory itself because this theory is invalid near $k_\parallel = 0$ and, therefore, cannot be used to see if any spikes are present in this region or not. However, at present it seems unlikely that any strong turbulence mechanism could lead to long parallel correlations and to a strong condensation of turbulence near $k_\parallel = 0$ that would preclude the validity of the weak turbulence theory in incompressible MHD.

It is shown in Galtier et al. [13] that the $k_\perp^{-2}$ energy spectrum is an exact finite flux Kolmogorov solution of the weak wave turbulence kinetic equation at the level of three–wave interactions. It is interesting to note that in the context of kinetic Alfvén waves the same scaling can also be predicted [52]. Note also that when the four–wave interactions are dominant, the predicted energy spectrum is $k_\perp^{-7/3}$ [50]. For strong anisotropic MHD turbulence, there is still a debate in the absence of a rigorous theory; either $k_\perp^{-5/3}$ [50] or $k_\perp^{-3/2}$ [36]

are predicted, using the *ad hoc* EDQNM closure or the Lagrangian DIA approximation.

The limit $k_{\parallel} \ll k_{\perp}$ presented here appears to be important in the astrophysical context (*e.g.* like for coronal magnetic loop) for which further applications have also to be developed. Note also that a recent work [46] analyzing data from the Galileo spacecraft observing the Jovian magnetosphere [23] may be providing a new observational support for the present theory.

ACKNOWLEDGMENTS

Grants from CNRS (PNST and PCMI) and from EC (FMRX-CT98-0175) are gratefully acknowledged.

REFERENCES

1. Armstrong, J. W., Cordes, J. M., and Rickett, B. J., *Nature* **291**, 561–564 (1981).
2. Amstrong, J. W., Rickett, B. J., and Spangler, S. R., *ApJ* **443**, 209–221 (1995).
3. Bavassano, B., Dobrowolny, M., Fanfoni, G., Mariani, F., and Ness, N.F., *Solar Phys.* **78**, 373–384 (1982).
4. Benney, D. J., and Newell, A. C., *Stud. Appl. Math.* **48**, 29–53 (1969).
5. Bhattacharjee, A., and Ng, C. S., *ApJ* **548**, 318–322 (2001).
6. Biskamp, D., *Magnetic Reconnection in Plasmas* Cambridge : Cambridge Univ. Press, 2000.
7. Biskamp, D., and Welter, H., *Phys. Fluids B* **1**, 1964–1979 (1989).
8. Biskamp, D., and Müller, W.-C., *Phys. Plasmas* **7**, 4889–4900 (2000).
9. Burlaga, L. F., and Goldstein, M. L., *J. Geophys. Res.* **89**, 6813–6817 (1984).
10. Burlaga, L. F., Ness, N. F., and McDonald, F. B., *J. Geophys. Res.* **92**, 13647–13652 (1987).
11. Cho, J., and Vishniac, E.T., *ApJ* **539**, 273 (2000).
12. Frisch, U., *Turbulence*, Cambridge : Cambridge Univ. Press, 1995.
13. Galtier, S., Nazarenko, S. V., Newell, A. C., and Pouquet, A., *J. Plasma Phys.* **63**, 447–488 (2000).
14. Galtier, S., Nazarenko, S. V., Newell, A. C., and Pouquet, A., *ApJ* **564**, L49-L52 (2002).
15. Galtier, S., and Bhattacharjee, A., *A whistler wave turbulence theory for electron MHD*, in preparation.
16. Galtier, S., *An inertial wave turbulence theory for incompressible fluid under rapid rotation*, in preparation.
17. Gekelman, W., and Pfister, H., *Phys. Fluids* **31**, 2017 (1988).
18. Gomez, T., Politano, H., and Pouquet, A., *Phys. Fluids* **11**, 2298–2306 (1999).
19. Grappin, R., Mangeney, A., and Marsch, E., *J. Geophys. Res.* **95**, 8197–8209 (1990).
20. Heiles, C., Goodman, A.A., Mc Kee, C.F., and Zweibel, E.G., "Magnetic fields in star-forming regions: observations", *in Protostars and Planets* **IV**, Tucson : Univ. Arizona Press, 279, 1993.
21. Iroshnikov, P., *Sov. Astron.* **7**, 566-571 (1963).
22. Kinney, R. M., and McWilliams, J. C., *Phys. Rev. E* **57**, 7111–7121 (1998).
23. Kivelson, M. G., Khurana, K. K., Russell, C. T., and Walker, R. J., *Geophys. Res. Lett.* **24**, 2127 (1997).
24. Kraichnan, R., *Phys. Fluids* **8**, 1385–1387 (1965).
25. Kuznetsov, E.A., *JETP*, in press (2002).
26. Mac Low, M. M., *ApJ* **524**, 169 (1999).
27. Maron, J., and Goldreich, P., *ApJ* **554**, 1175–1196 (2001).
28. Marsch, E., and Tu, C. Y., *J. Geophys. Res.* **95**, 8211–8229 (1990).
29. Marsch, E., and Tu, C. Y., *Ann. Geophys.* **11**, 659–677 (1993).
30. Marsch, E., and Tu, C.Y., *Ann. Geophys.* **12**, 1127–1138 (1994).
31. Matthaeus, W. H., Goldstein, M. L., and Roberts, D. A., *J. Geophys. Res.* **95**, 20673–20683 (1990).
32. Matthaeus, W.H., Ghosh, S., Oughton, S., and Roberts, D.A., *J. Geophys. Res.* **101**, 7619–7629 (1996).
33. Milano, L. J., Matthaeus, W. H., Dmitruk, P., and Montgomery, D. C., *Phys. Plasmas* **8**, 2673–2681 (2001).
34. Montgomery, D., and Turner, L., *Phys. Fluids* **24**, 825–831 (1981).
35. Montgomery, D.C., and Matthaeus, W.H., *ApJ* **447**, 706–707 (1995).
36. Nakayama, K., *ApJ* **556**, 1027–1037 (2001).
37. Nazarenko, S. V., Newell, A. C., and Galtier, S., *Physica D* **152-153**, 646–652 (2001).
38. Newell, A. C., Nazarenko, S. V., and Biven, L., *Physica D* **152-153**, 520–550 (2001).
39. Ng, C. S., and Bhattacharjee, A. *ApJ* **465**, 845–854 (1996).
40. Ng, C. S., and Bhattacharjee, A., *Phys. Plasmas* **4**, 605–610 (1997).
41. Oughton, S., Priest, E. R., and Matthaeus, W. H., *J. Fluid Mech.* **280**, 95–117 (1994).
42. Oughton, S., Matthaeus, W. H., and Ghosh, S., *Phys. Plasmas* **5**, 4235 (1998).
43. Parker, E.N., *Spontaneous current sheets in magnetic fields with applications to stellar X-rays*, Oxford : Oxford University Press, 1994.
44. Politano, H., Pouquet, A., and Sulem, P.-L., *Phys. Fluids B* **1**, 2330–2339 (1989).
45. Priest, E.R., *Solar Magnetohydrodynamics*, D. Reidel Pub. Comp., 1982.
46. Saur, J., Politano, H., Pouquet, A., and Matthaeus, W.H., *A&A* **386**, 699–708 (2002).
47. Shebalin, J. V., Matthaeus, W. H., and Montgomery, D., *J. Plasma Phys.* **29**, 525–547 (1983).
48. Spangler, S. R., and Gwinn, C. R., *ApJ* **353**, L29–L32 (1990).
49. Spangler, S.R., *ApJ* **522**, 879–896 (1999).
50. Sridhar, S., and Goldreich, P., *ApJ* **432**, 612–621 (1994).
51. Taylor, J.B., *Rev. Mod. Phys.* **58**, 741–763 (1986).
52. Voitenko, Y.M., *J. Plasma Phys.* **60(3)**, 497–514 (1998).
53. Zakharov, V.E., L'vov, V., and Falkovich, G.E., *Kolmogorov Spectra of Turbulence I : Wave Turbulence*, Berlin : Springer, 1992.

Electromagnetic Ion/Beam Instabilities In The Fast Solar Wind: Proton Core Temperature Anisotropy Effects On The Relative Drift Speed And Ion Heating

Jaime A. Araneda[1], Adolfo F.-Viñas[2], and Hernán F. Astudillo[1]

[1]*Departamento de Física, Facultad de Ciencias Físicas y Matemáticas, Universidad de Concepción, Casilla 160-C, Chile*
[2]*Laboratory for Extraterrestrial Physics, Mail Code 692, NASA Goddard Space Flight Center, Greenbelt, MD 20771, USA*

Abstract. Ion velocity distributions in the fast solar wind are usually characterized by nonthermal features such as multicomponent ions, temperature anisotropies and average relative drifts. These nonthermal features lead to the growth of several electromagnetic instabilities. Here, two-dimensional hybrid simulations are used to study these instabilities in a homogeneous, magnetized collisionless plasma model. We show that for conditions typical of the fast solar wind the proton core temperature anisotropy plays a significant role in modifying the wave-particle scattering of each proton component as compared to the isotropic cases. As a consequence, the scattering reduces both the heating and anisotropy enhancement of the proton beam and decreases the relative proton/ion flow speed below the corresponding isotropic instability thresholds in agreement with recent observations and with our previous one-dimensional study. Our results are consistent with recent satellite observations and provide support to the physical scenario in which core temperature anisotropies play a regulating effect on the instability thresholds.

INTRODUCTION

Helios and Ulysses *in situ* plasma measurements in the fast solar wind ($U_{sw} \gtrsim 600$ km/s) have revealed a variety of nonthermal features in the ion velocity distributions [See e.g., Refs.1-3]. Proton distributions exhibit two ubiquitous characteristics: double-peaked streams with a typical relative drift speed of $U_{pp}/V_A \lesssim 3$, and core temperature anisotropies with temperatures ratios of $T_\perp/T_\parallel \sim 1$-4. Each ion component, (core and beam) is usually modeled by a bi-Maxwellian distribution with a density n_j, a flow speed U_j, and $T_{\parallel j}$ and $T_{\perp j}$, temperatures parallel and perpendicular to $\mathbf{B}_0$, the background magnetic field. In addition, minor ions such as alpha particle are observed to flow faster than protons with drift speeds which, on average, follow the local Alfvén speed ($U_{\alpha p} \lesssim 1.0\ V_A$). Furthermore, the alpha particles are also typically observed to be hotter ($T_\alpha \approx 2$-4 T_p). Each of these nonthermal characteristics can trigger a number of electromagnetic ion/ion instabilities such as the magnetosonic and Alfvén/cyclotron instabilities.

Although electromagnetic ion/ion instabilities have been studied extensively in the context of several plasma environments [e.g., Ref. 4], fewer investigations have been concerned with instabilities derived from typical fast solar wind parameters. Based upon model distribution functions of plasma observations near 1 AU and linear theory, Montgomery et al., [5-6] found that if the proton/proton relative drift speed approaches the Alfvén speed two modes become unstable: A magnetosonic and several oblique Alfvén modes. The proton/proton magnetosonic instability has the larger growth rate in the direction parallel to the background magnetic field with a weak beam density, whereas the oblique Alfvén modes are more unstable at sufficiently large beam density and relatively small β core [7].

Linear Vlasov theory predicts that threshold conditions for the magnetosonic instability correspond to $U_{pp}/V_A \lesssim 2$ [6,7]. Nevertheless, Helios data analysis of Marsch and Livi [8] showed a number of observations matching $2 \lesssim U_{pp}/V_A \lesssim 4$, with a large

CP679, *Solar Wind Ten: Proceedings of the Tenth International Solar Wind Conference*, edited by M. Velli, R. Bruno, and F. Malara

number corresponding to apparently unstable cases. In contrast, Ulysses observations of the proton/proton relative drift speed reported by Goldstein et al. [3] lie below the predicted threshold with mean values considerable smaller. Because the two data sets were obtained at different distances from de Sun, there must be some regulating mechanism for the differential streaming and probably ion heating. Recently, the effect of a proton core anisotropy on the relative drift speed and beam heating was investigated by Araneda et al., [9]. Using linear Vlasov theory and one-dimensional hybrid simulations, these authors found that a proton core temperature anisotropy $T_{\perp c}/T_{\parallel c} > 1$ play a key regulator for the relative streaming and preferential heating of the more tenuous proton component. The wave-particle scattering by electromagnetic fluctuations from the proton cyclotron instability not only reduces the relative streaming speed but also constraint the proton beam anisotropy to $T_{\perp b}/T_{\parallel b} \lesssim 1$ values. Moreover, although a $T_{\perp \alpha}/T_{\parallel \alpha} < 1$ anisotropy can significantly increase the growth rate of the alpha/proton magnetosonic instability [10], they shown that the same mechanism reduces the alpha/proton relative flow and constrains also the enhancements of the temperature ratio of the alphas $T_{\perp \alpha}/T_{\parallel \alpha}$. Simulations with initial isotropic components reduce also the core/beam relative drift but yields to a strong heating of the more tenuous beam component temperature perpendicular to $\mathbf{B}_0$ [11-13].

The purpose of this paper is to extend the nonlinear study of the proton/proton instabilities of Ref. [9] by using two-dimensional simulations. We examine the effect of an initial proton core temperature anisotropy on the magnetosonic instability under conditions in which this instability has the largest growth rate in the 1-D case.

COMPUTER SIMULATIONS

We have used a two-dimensional hybrid code that treats the ions as discrete particles and the electrons as a massless fluid. In order to verify our calculations we have employed two codes with completely different schemas obtaining similar results. The simulations are performed in the center of mass frame. At t = 0, the core and proton beam components, each assumed to be a drifting bi-Maxwellian distribution, stream parallel to $\mathbf{B}_0$ (the x direction) and with relative velocity $\boldsymbol{U}_j$. Each ion distribution is represented by a fixed number of superparticles. On average, about 256 particles per cell are used for each component. For each simulation we used a 128 × 64 computational grid with the simulation system length chosen to be about 8 times the wavelength of the fastest growing mode and an integration time step of $\Omega_p \Delta t = 0.05$. Charge neutrality and the zero current condition were imposed at $t = 0$. We use the subscripts c to denote the more dense proton component and b for the less dense beam.

We first describe results from run I for which we assume that $\beta_{\parallel c} = 8\pi n_e k_B T_{\parallel c} / B_0^2 = \beta_{\parallel b} = 1.0$ and $T_{c\perp} / T_{\parallel c} = T_{b\perp} / T_{b\parallel} = 1.0$. Other initial parameters are $n_b/n_e = 0.05$, and $U_{bc}/V_A = 2.63$. These parameters are similar to those of the simulation of the magnetosonic instability illustrated in Figure 1 of Ref. 11 and the results are analogous. The time histories of the beam and core anisotropies, the beam/core relative drift speed, and the beam/core temperature ratio are shown in Figure 1. The less dense beam component develops a relative strong temperature anisotropy $A_b \equiv T_{b\perp} / T_{\parallel b}$, whereas $T_{\parallel b} / T_{\parallel c}$ decreases, and $A_c \equiv T_{c\perp} / T_{\parallel c}$ remain approximately without variations. Note that the relative drift speed exhibit also a moderate reduction indicating that the magnetosonic instability alone is inefficient to produce significant changes in U_{bc}.

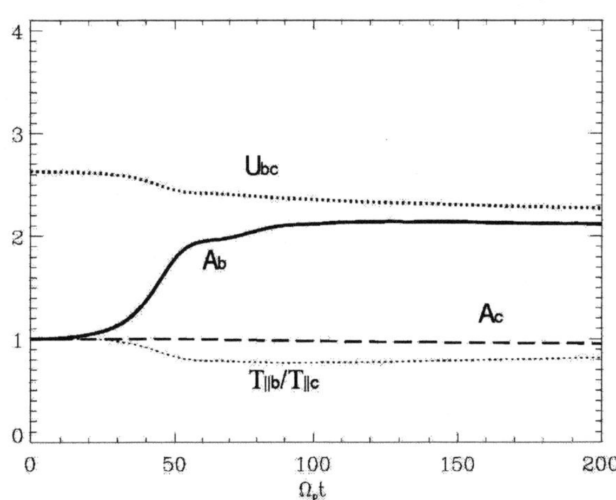

FIGURE 1. Time histories for run I: The relative drift speed indicated by heavy dotted lines. The solid lines represent the beam proton anisotropy, the thin dotted lines the beam/proton temperature ratio, and the dashed lines the core temperature anisotropy.

We now examine the consequences of a initial core temperature anisotropy on the magnetosonic instability. Specifically, our run II used the same parameters as in run I, except that $T_{c\perp} / T_{\parallel c} = 3.0$. Under these parameters the magnetosonic instability quench significantly whereas the proton cyclotron modes (both forward and backward) dominate [9].

The time histories for this case are shown in Figure 2 in the same format as Figure 1. Between $\Omega t = 0$ and 40, we observe the same characteristic increase of the beam temperature anisotropy, similar to the isotropic case of run I. However, afterward a substantial slowdown of both the relative drift speed and beam temperature anisotropy takes place. Following saturation, the normalized values for these parameters drop around 1 in agreement with Ulysses observations [3]. The tendency toward an isotropic proton beam distribution is due to the scattering from the backward-propagating ion cyclotron waves resulting from the relaxation of the anisotropic proton core. In fact, observing the individual particles in the phase-space (not shown) it can be seen that beam particles localized on the negative tail of the distribution (as seen in the beam frame) experience a force $\delta \mathbf{v}_\perp \times \delta \mathbf{B}_\perp$ in the parallel direction. This causes an increase in the proton beam parallel temperature and also contributes to reduce the proton/proton relative drift speed. Because the Doppler effect, the only waves able to resonate with the beam proton in this range of velocities are the backward propagating ion cyclotron waves. Since the magnetosonic waves will tend to heat the beam proton distribution in the perpendicular direction, the net effect is an almost isotropic heating of the beam.

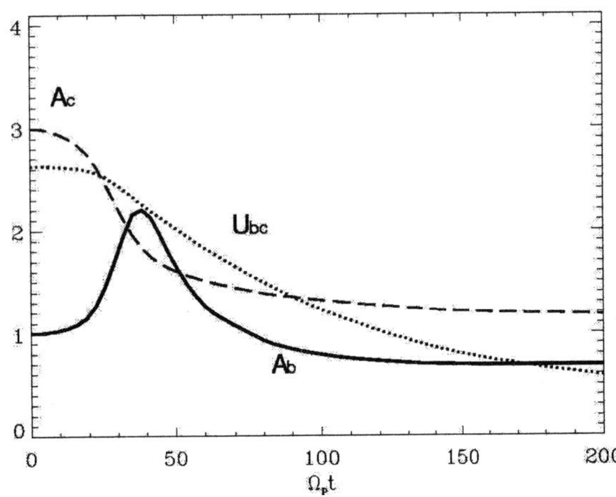

FIGURE 2. Time history for run II: The dotted lines represent the relative drift speed. The solid lines represent the beam proton anisotropy, and the dashed lines the core temperature anisotropy.

The last simulation that we examine (run III) is one in which the initial parameter are again the same, but now with $T_{c\perp} / T_{\parallel c} = 4.0$. This value corresponds to the maximum proton anisotropy observed in the solar wind from the Helios spacecraft [8].

Figure 3 shows the time histories for run III in the same format as Figure 2. As in run II, we note the initial perpendicular heating of the beam followed by a decrease, continuing almost constant below $A_b = 1$. For this extreme case, we note a strong reduction of the relative drift speed and the evolution of the plasma toward a stable configuration with $A_c \gtrsim 1$.

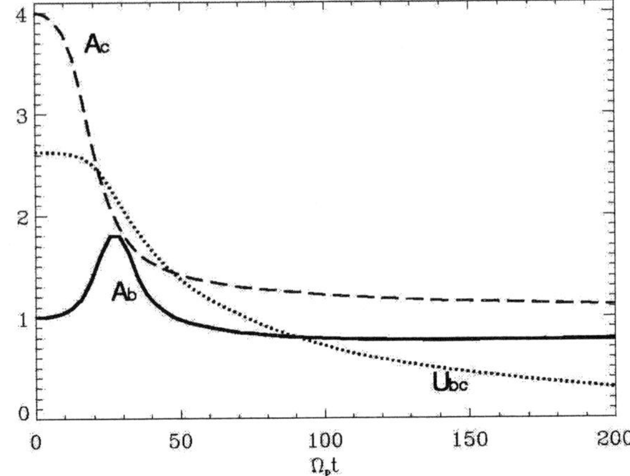

FIGURE 3. Time history for run III: The dotted lines represent the relative drift speed. The solid lines represent the beam proton anisotropy, and the dashed lines the core temperature anisotropy.

CONCLUSIONS

We used two-dimensional hybrid simulations of the proton/proton and proton core driven Alfvén/cyclotron instabilities using initial conditions similar to those observed in the fast solar wind near 0.3 AU. The 2D simulations confirm previous results from one-dimensional hybrid simulations that the proton core temperature anisotropy play a key regulator of both the perpendicular heating of streaming protons and the relative drift speed. This implies that a statistical study on correlations of observed core/beam distributions with theoretical threshold conditions must include the effects of the proton core anisotropy and probably also the beam temperature anisotropy.

ACKNOWLEDGMENTS

This work was partially supported by Fondecyt grant 1000/7000320 and by the University of Concepción. A. F. Viñas thanks the NASA Theory Program at GSFC for his support.

REFERENCES

1. Feldman, W. C., J. R. Asbridge, S. J. Bame, and M. D. Montgomery, *J. Geophys. Res.*, **78**, 2017, (1973).

2. Marsch, E., K.-H. Mühlhäuser, H. Rosenbauer, R. Schween, and F. M. Neubauer, *J. Geophys. Res.*, **87**, 35, (1982).

3. Goldstein B. E., M. Neugebauer, L. D. Zhang, and S. P. Gary, *Geophys. Res. Lett*, **27**, 53, (2000).

4. Gary, S. P., *Theory of Space Plasma Microinstabilities*, Univ. Press, Ney York, 1993.

5. Montgomery, M. D., S. P. Gary, D. W. Forslund, and W. C. Feldman, *Phys. Rev. Lett.* **35**, 667 (1975).

6. Montgomery, M. D., S. P. Gary, W. C. Feldman, and D. W. Forslund, *J. Geophys. Res.*, **81**, 2743 (1976).

7. Daughton, W., and S. P. Gary, *J. Geophys. Res.,* **103**, 20,613 (1998).

8. Marsch, E., and S. Livi, *J. Geophys. Res.*, **92**, 7263, (1987).

9. Araneda, J. A., A.-F. Viñas, and H. F. Astudillo, *J. to be published in J. Geophys. Res.* (2002).

10. Li, X., and S. R. Habbal, *J. Geophys.Res.,* **105**, 7483 (2000).

11. Daughton, W., S. P. Gary, and D. Winske, *J. Geophys. Res.,* **104**, 4657 (1999).

12. S. P. Gary, L. Yin, D. Winske, and D. B. Reisenfeld *Geophys. Res. Lett.*, **27**, 1355 (2000).

13. S. P. Gary, L. Yin, D. Winske, and D. B. Reisenfeld *J. Geophys. Res.*, **105**, 20,989 (2000).

Temperature Anisotropies of Heavy Solar Wind Ions from Ulysses-SWICS

R. von Steiger* and T. H. Zurbuchen†

*International Space Science Institute, Hallerstrasse 6, CH-3012 Bern, Switzerland
†Dept. of AOSS, Univ. of Michigan, Ann Arbor, MI 48109, USA

Abstract. We report the first in-situ measurements of temperature anisotropies of heavy ions in the solar wind, obtained with the Solar Wind Ion Composition Spectrometer (SWICS) on the Ulysses spacecraft. Since SWICS measures only 1-dimensional cuts through the full, 3-d velocity distribution functions we resort to a statistical approach, separating the particle data according to the instantaneous magnetic field angle. We apply this analysis to the ions of He^{++} and O^{6+} during extended time periods in the fast streams from both the south and the north polar coronal holes that Ulysses traversed in 1993–96. In both cases we find anisotropies of the order of $T_\perp/T_{||} = 0.8$. The results of this study are discussed in relation to the observations made on Helios for He^{++} in the 1970s, and to recent observations made on SOHO-UVCS, which show extreme temperature anisotropies of O VI, or O^{5+}, at a few solar radii.

MOTIVATION

The kinetic properties of heavy ions in the solar wind are indicative of processes affecting their distribution functions in the solar corona and in interplanetary space. Earlier observations at 1 AU have established that all heavy ion species flow approximately at the same bulk speed and have approximately equal thermal speeds (i.e., mass-proportional temperatures), with exceptions at times when the solar wind density was unusually high. At 5 AU such exceptions no longer occur and the basic picture applies with very high accuracy [1, 2, 3]. This was interpreted as evidence for the growing dominance of wave-particle interactions over Coulomb collisions with increasing heliocentric distance.

Even though the solar wind is often modelled by fluid or MHD equations, it is important to understand the underlying physics and processes that determine its expansion and acceleration. These processes seem to occur on a kinetic scale, leading to deviations from a thermal Maxwell distribution.

Recent results from SOHO-UVCS [4, 5] suggest that kinetic processes may even dominate very close to the Sun. Of particular interest is the temperature anisotropy in O VI, or O^{5+}, which is observed to be very large, $T_\perp/T_{||} \gg 1$. All these UVCS remote solar wind measurements of the outer corona are done using emissions from heavy ions, such as from O and Fe.

In this paper we address these topics by investigating the distribution functions of heavy ions observed on SWICS/Ulysses at several AU, with emphasis on suprathermal tails and anisotropies. The principal question is whether or not we can find any residual signature in interplanetary space of the anisotropy in heavy ions found with SOHO-UVCS in the corona.

PREVIOUS RESULTS

Anisotropies were found both in the H^+ and He^{++} distribution functions with the Helios plasma experiment in the 1970s [cf. 6, and references therein]. The core of the distribution functions often showed an anisotropy perpendicular to the field direction, i.e., in the same sense as now seen with SOHO-UVCS. At the distances closest to the Sun, $\lesssim 0.5$ AU, it sometimes was as strong as $T_\perp/T_{||} \simeq 2$, but decayed with increasing distance out to 1 AU. The decay was more pronounced in the slow solar wind, most likely owing to the longer expansion time [6, Fig. 8.1]. On the other hand, the full distribution functions were often associated with a component drifting outwards along the magnetic field direction, i.e., with an opposite anisotropy of $T_\perp/T_{||} \simeq 0.5$.

A similar, yet weaker, anisotropy was found with Ulysses-SWOOPS in the distribution functions of He^{++} at larger heliospheric distances of 1.5–4.2 AU: Reisenfeld et al. [7] report $T_\perp/T_{||} = 0.87$, which was essentially constant with distance.

Observations of heavy ions with Ulysses-SWICS [8] at ~5 AU show no differential streaming between species, nor any significant deviation from mass-proportional temperatures. This demonstrates the

CP679, *Solar Wind Ten: Proceedings of the Tenth International Solar Wind Conference*,
edited by M. Velli, R. Bruno, and F. Malara

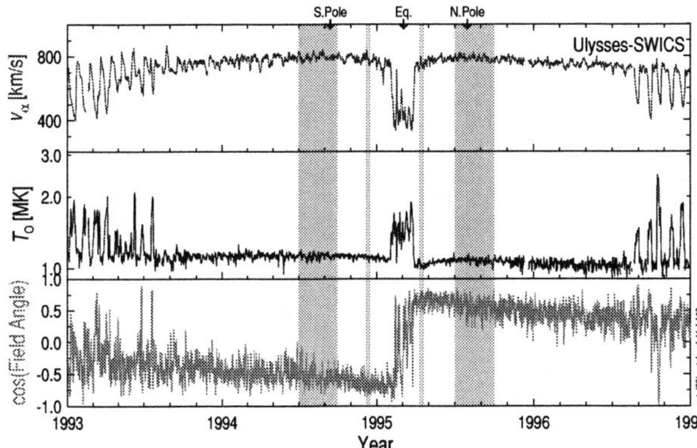

FIGURE 1. Overview of the time periods selected for this analysis. Red shade: He^{++} at full, 13-minute time resolution; blue shade: O^{6+} at 3-hour time resolution.

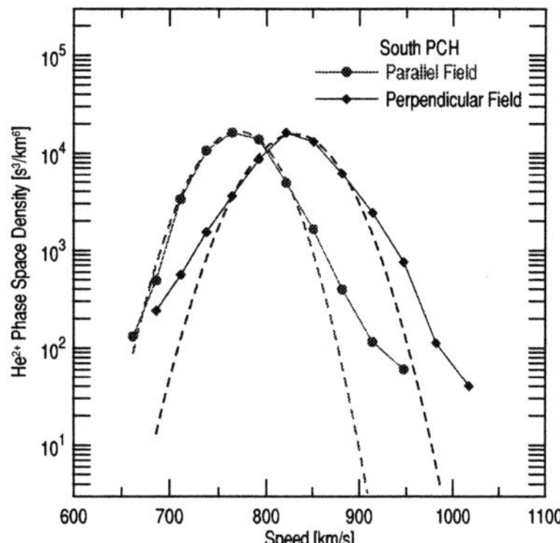

FIGURE 2. Sample distribution functions of He^{++} obtained in the southern fast stream. Red: magnetic field parallel to the radial direction; blue: perpendicular field. Note the presence of suprathermal tails except in the Sunward direction.

increasing importance of wave-particle interactions, which are heating and accelerating the heavy species, with heliocentric distance [3]. Extending these observations we now search for possible anisotropies in these heavy ion distribution functions.

OBSERVATIONS

We analyse distribution functions of He^{++} and O^{6+} obtained in polar high-speed streams at $\sim 2\,\mathrm{AU}$ with Ulysses-SWICS, with particular aim at the kinetic temperatures and their anisotropies. We picked two time periods each in the high-speed streams from the south and the north polar coronal holes during solar minimum (see Fig. 1). In the case of He^{++} the flux was sufficient so spectra could be taken every 13 minutes, which is the the full time resolution of SWICS. We therefore only analysed two short time periods at high latitudes for this ion, shaded in red in Fig. 1. Since the flux of O^{6+} ions is much lower, we had to accumulate the distribution functions over 3 hours (corresponding to about 14 spectra) in order to obtain sufficient statistics. Consequently, a much longer time period was analysed for this ion, shaded in blue in Fig. 1.

Strictly speaking, it is not possible with SWICS to observe a temperature anisotropy, as the sensor measures only 1-d cuts along the radial direction through the full, 3-d distribution functions. We therefore have to rely on a statistical approach in order to determine thermal anisotropies (a similar approach was used for alpha particles from Ulysses-SWOOPS by Reisenfeld et al. [7]). The spectra are sorted according to the average magnetic field angle during their accumulation. This angle between the radial direction and the magnetic field, ϕ, is measured on Ulysses with the VHM/FGM experiment [9] and obtained from the Ulysses Data System at a time resolution of 1 minute, which is then averaged over the accumulation period. A plot of the thermal speed as a function of the field angle finally reveals the average anisotropy.

Two sample distribution functions of He^{++} are shown in Fig. 2, one for parallel field (red) and another, taken at a different time, for perpendicular field (blue). The cores of both samples are very closely approximated by Maxwellian fits. However, at $\lesssim 1\,\%$ below the maximum significant suprathermal tails can be seen *except* on the low-speed side of the parallel spectrum, i.e., in the direction back towards the Sun.

The thermal speed is obtained as the second moment of the full distribution functions observed by SWICS including the tails. The thermal speed determined from a 1-d cut at a field angle ϕ through the 3-d velocity distribution function is related to its field-parallel and field-perpendicular components by

$$\frac{1}{v_{\mathrm{th}}^2(\phi)} = \frac{\cos^2\phi}{v_{\mathrm{th},\|}^2} + \frac{\sin^2\phi}{v_{\mathrm{th},\perp}^2}. \qquad (1)$$

In the limit of small anisotropy, i. e. $v_{\mathrm{th},\perp} \simeq v_{\mathrm{th},\|}$, this can be shown to yield a linear relation in $\cos^2\phi$,

$$v_{\mathrm{th}}(\phi) \simeq v_{\mathrm{th},\perp} + (v_{\mathrm{th},\|} - v_{\mathrm{th},\perp})\cos^2\phi. \qquad (2)$$

Plotting v_{th} from a large number of spectra obtained at all field angles versus $\cos^2\phi$ thus may reveal an average anisotropy if it shows a systematic trend. This is done in Fig. 3 for He^{++} and in Fig. 4 for O^{6+}. The data from the southern stream (left panels) are plotted versus $-\cos^2\phi$ simply to illustrate that the field was pointing inward there. Thick green data points indicate periods when the average magnetic field direction was well-defined over the accumulation time of the spectrum, i.e. the variability of ϕ remained below a certain threshold,

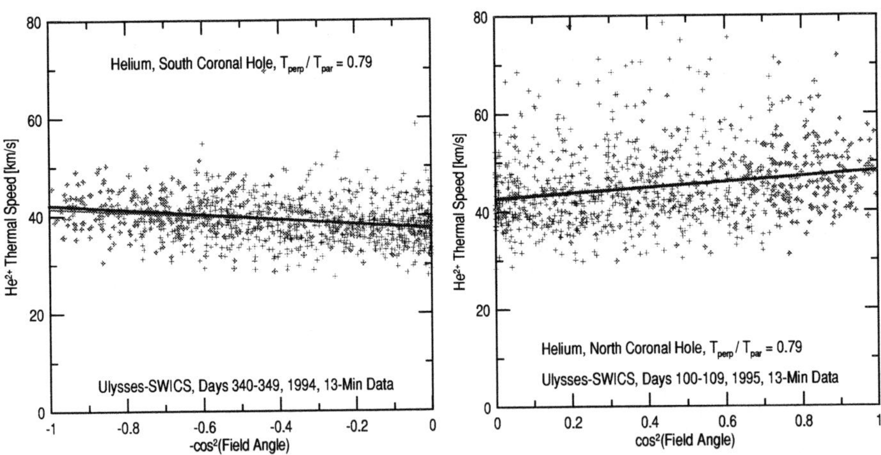

FIGURE 3. He^{++} thermal speed from ~1000 13-minute spectra each in the southern (left) and northern (right) fast stream, plotted vs. $\cos^2$ of the field angle. The thick, green data points indicate spectra taken when the magnetic field variability was small, while the others are marked with thin blue symbols. The anisotropy is found from the parameters of the fit, which is marked as a thick red line for the green data points only and a thin orange line for all data points. Both plots show a consistent trend, which is interpreted as an average anisotropy of $T_\perp/T_\parallel(\mathrm{He}^{++}) = 0.79$.

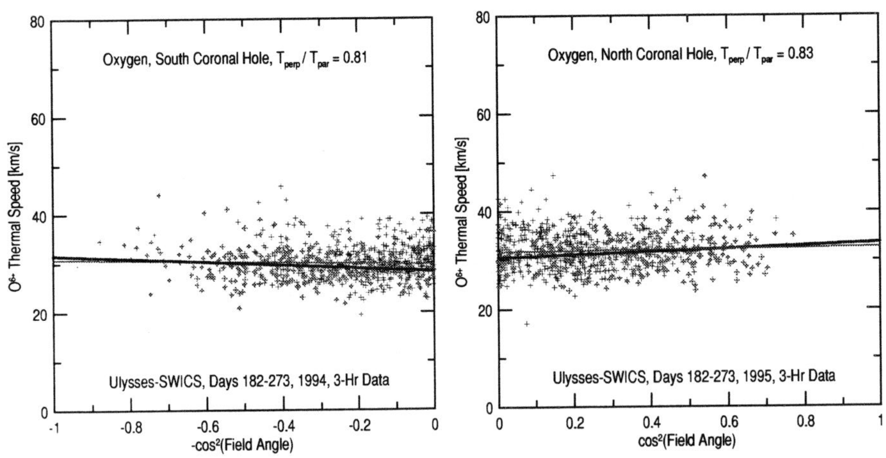

FIGURE 4. Same as Fig. 3, but for ~750 3-hour spectra of O^{6+}. Again, an average anisotropy of $T_\perp/T_\parallel(\mathrm{O}^{6+}) = 0.82$ is found.

whereas the thin blue data points refer to periods with more variability. A fit is calculated according to Eq. 2 to all data points, indicated as a thin orange line, and to the subset of the green points, indicated as a thick red line. From the fit parameters we finally obtain the anisotropy, $T_\perp/T_\parallel = v_{\mathrm{th}}^2(\pi/2)/v_{\mathrm{th}}^2(0)$.

It is evident from these figures that an anisotropy exists in all cases. For helium (Fig. 3) we find $T_\perp/T_\parallel(\mathrm{He}^{++}) = 0.79$ both in the southern and the northern polar stream, although with larger scatter there. This is smaller (i.e., the anisotropy is larger) than the result from SWOOPS [7], but considering the ~10 % accuracy of both results they are consistent. In both panels of Fig. 3 it is irrelevant whether all data points are included in the fit or only those with small field variability, i.e. on the time scale of 13 minutes the variability has no effect on the result.

For oxygen (Fig. 4) we first note that there are significantly fewer data points near $\cos\phi = \pm 1$. This was to be expected since these spectra were accumulated over 3 hours, and it is less likely for the field to remain parallel on this longer time scale. When we restrict ourselves to the spectra with little field variability we obtain an anisotropy of similar magnitude as for He^{++} both in the southern and the northern polar stream. If we had taken all data points we would still have found an anisotropy, albeit a weaker one. Disregarding the points with large field variability we obtain an average anisotropy of $T_\perp/T_\parallel(\mathrm{O}^{6+}) = 0.82$. Once again, we stress that this is not the anisotropy of individual spectra, but a gross average over the two three-month periods indicated in Fig. 1.

DISCUSSION

Many theories in the corona and heliosphere predict or assume $T_\perp/T_{||} > 1$ [10, 11, 12]. The interaction of particles with the ambient turbulence occurs through cyclotron interactions, and the energy is dissipated from resonant waves through cyclotron resonance, i.e., $\omega(k) - \mathbf{k}\cdot\mathbf{v} = \Omega(m/q)$. These theories therefore dissipate only high-frequency waves, whereas most of the energy generally resides in low frequency waves. Also, there should thus be a mass-per-charge dependency in the interaction term that should be observable.

On the other hand, $T_\perp/T_{||} < 1$ is predicted for a statistical acceleration process [13] that assumes the energy being dissipated by Landau resonance, i.e., $\omega(k) - \mathbf{k}\cdot\mathbf{v} = 0$. The energy is directly dissipated from magnetic field compressions. This heating can happen at all wavelengths as outwards-propagating waves interacting with outwards-propagating particles (relative to the plasma frame). Statistical theories also predict that there should be tails on the distribution functions, as is observed.

While the first, cyclotron resonant picture probably applies in the corona and innermost heliosphere our results, taken at 2–3 AU, appear to favour the second, statistical picture: We observe $T_\perp/T_{||} < 1$ in both species investigated, equal to within a few percent although their mass-per-charge differs by 30 %, and we find outward-directed suprathermal tails. This is consistent with the Helios results mentioned above that saw the opposite anisotropy decay with increasing distance, while suprathermal tails were building up.

CONCLUSIONS

We have reported the first in-situ measurements of a temperature anisotropy of oxygen ions at a heliocentric distance of 2–3 AU:

$$T_\perp/T_{||}(O^{6+}) = 0.82$$

These are the first measurements of the full dynamic properties of O, particularly of its temperature anisotropy. The velocity distribution functions of He^{++} and O^{6+} are found to be very similar, and both show a very similar anisotropy. The distribution functions of these (and most likely all other) heavy ions are clearly non-Gaussian, with extended suprathermal tails, except in the Sunward direction.

These observations rise questions about the approximation of bi-Maxwellians commonly used in coronal physics to be applicable in the heliosphere beyond 1 AU. They show that there are significant deviations in both the parallel and perpendicular directions, affecting the internal pressure and the nature of the distribution function. Ion cyclotron models seem to predict temperature anisotropies > 1, slow energy diffusion (and therefore very weak tails), and mass-per-charge dependencies between different heavy ion species, neither of which is observed in the outer heliosphere.

A statistical model has qualitative advantages because it implies a temperature anisotropy < 1, the development of tails that are broadly directed away from the Sun (since most waves propagate outwards in coronal hole-associated wind [14]), and the similarities of He and O (and other heavy ions). However, no quantitative theory is currently available that can be used to compare with our data.

REFERENCES

1. von Steiger, R., Geiss, J., Gloeckler, G., and Galvin, A. B., *Space Sci. Rev.*, **72**, 71–76 (1995).
2. Zurbuchen, T. H., Fisk, L. A., Schwadron, N. A., and Gloeckler, G., "Observations of non-thermal properties of heavy ions in the solar wind," in [15], pp. 215–220.
3. von Steiger, R., and Zurbuchen, T. H., *Adv. Space Res.* **30**, 73–78 (2002).
4. Kohl, J. L., Noci, G., Antonucci, E., Tondello, G., Huber, M. C. E., Cranmer, S. R., Strachan, L., Panasyuk, A. V., Gardner, L. D., et al., *Astrophys. J.*, **501**, L127–L131 (1998).
5. Cranmer, S. R., Kohl, J. L., Noci, G., Antonucci, E., Tondello, G., Huber, M. C. E., Strachan, L., Panasyuk, A. V., Gardner, L. D., et al., *Astrophys. J.*, **511**, 481–501 (1999).
6. Marsch, E., "Kinetic Physics of the Solar Wind Plasma," in *Physics of the Inner Heliosphere*, edited by R. Schwenn and E. Marsch, Springer-Verlag, Berlin, 1991, vol. 2, chap. 8, pp. 45–133.
7. Reisenfeld, D. B., Gary, S. P., Gosling, J. T., Steinberg, J. T., McComas, D. J., Goldstein, B. E., and Neugebauer, M., *J. Geophys. Res.*, **106**, 5693–5708 (2001).
8. Gloeckler, G., Geiss, J., Balsiger, H., Bedini, P., Cain, J. C., Fischer, J., Fisk, L. A., Galvin, A. B., Gliem, F., et al., *Astron. Astrophys. Suppl.*, **92**, 267–289 (1992).
9. Balogh, A., Beek, T. J., Forsyth, R. J., Hedgecock, P. C., Marquedant, R. J., Smith, E. J., Southwood, D. J., and Tsurutani, B. T., *Astron. Astrophys. Suppl.*, **92**, 221–236 (1992).
10. Marsch, E., Goertz, C. K., and Richter, K., *J. Geophys. Res.*, **87**, 5030–5044 (1982).
11. Cranmer, S. R., *Astrophys. J.*, **532**, 1197–1208 (2000).
12. Isenberg, P. A., and Hollweg, J. V., *J. Geophys. Res.*, **88**, 3923–3935 (1983).
13. Fisk, L. A., Gloeckler, G., Zurbuchen, T. H., and Schwadron, N. A., "Ubiquitous statistical acceleration in the solar wind," in [15], pp. 229–233.
14. Horbury, T. S., and Balogh, A., *J. Geophys. Res.*, **106**, 15929–15940 (2001).
15. Mewaldt, R. A., Jokipii, J. R., Lee, M. A., Möbius, E., and Zurbuchen, T. H., editors, *Acceleration and Transport of Energetic Particles Observed in the Heliosphere*, vol. 528 of *AIP Conference Proceedings*, AIP Press, Woodbury, NY, 2000.

The multifractal spectrum for the solar wind flow

Wiesław M. Macek

Faculty of Mathematics and Science, Cardinal Stefan Wyszyński University in Warsaw, Dewajtis 5, 01-815 Warszawa; Space Research Centre, Polish Academy of Sciences, Bartycka 18 A, 00-716 Warszawa, Poland

Abstract. We analyze time series of velocities of the low-speed stream of the solar wind plasma including Alfvénic fluctuations measured in situ by the Helios spacecraft in the inner heliosphere. We demonstrate that the influence of noise in the data can be efficiently reduced by a moving average filter. We calculate the multifractal spectrum for the solar wind flow directly from the cleaned experimental signal. We also show that due to nonlinear noise reduction we get with much reliability estimates of the Kolmogorov entropy and the largest Lyapunov exponent. For intermediate length scales the Lyapunov exponent and the entropy are plausibly positive locally, which exhibits sensitivity to initial conditions. This shows that the slow solar wind in the inner heliosphere is most likely a deterministic chaotic system, where noise is not dominant.

The generalized dimensions of attractors are important characteristics of *complex* dynamical systems [1]. Since these dimensions are related to frequencies with which typical orbits in phase space visit different regions of the attractors, they provide information about dynamics of the systems [2]. If the measure has different fractal dimensions on different parts of the support, the measure is multifractal [3].

Following space physics applications, e.g., [4, 5], we consider the inner heliosphere. The solar wind plasma flowing supersonically outward from the Sun is quite well modeled within the framework of the hydromagnetic theory. This continuous flow has two forms: slow ($\approx 300\,\mathrm{km\,s^{-1}}$) and fast ($\approx 900\,\mathrm{km\,s^{-1}}$) [6]. The fast wind is associated with coronal holes and is relatively uniform and stable, while the slow wind is quite variable in terms of velocities. We limit our study to the low-speed stream. Indication for a chaotic attractor in the slow solar wind has been given in [7, 8, 9, 10]. In particular, Macek [7] has calculated the correlation dimension of the reconstructed attractor and has provided tests for *nonlinearity* in the solar wind data, including a powerful method of statistical surrogate data tests [11]. Further, Macek and Redaelli [9] have shown that the Kolmogorov entropy of the attractor is *positive* and finite, as it holds for a *chaotic* system. The entropy is plausibly constrained by a *positive* local Lyapunov exponent that would exhibit sensitive dependence on initial conditions of the system.

Recently, we have extended our previous results on the dimensional time series analysis [7]. Namely, we have applied the technique that allows a realistic calculation of the generalized dimensions of the solar wind flow, directly from the cleaned experimental signal by using the Grassberger and Procaccia method [12]. The resulting spectrum of dimensions shows multifractal structure of the solar wind in the inner heliosphere [10]. The obtained multifractal spectrum is consistent with that for the multifractal measure on the self-similar weighted Cantor set. In this paper we demonstrate the influence of noise on these results and show that noise can efficiently be reduced by a simple moving average filter.

We analyze the Helios data using plasma parameters measured *in situ* in the heliosphere near the Sun, at 0.3 AU, Ref. [6]. The radial velocity component of the plasma flow, v, has been investigated in [7, 9]. In this paper we also analyze one of the so-called Elssäser variables $x = v + v_A$, where $v_A = B/(\mu_o \rho)^{1/2}$ is the Alfvénic velocity calculated from the experimental data: the radial component of the magnetic field of the plasma B and the mass density ρ (μ_o is the permeability of free space). These raw data of $N = 4{,}513$ points, with sampling time of $\Delta t = 40.5$ s, are shown in Figure 1 (a).

As in [10], slow trends ($420.06 - 15.4\,t - 82.31\,t^2$, with t being a fraction of total sample) were subtracted from the original data $x_i = v(t_i) + v_A(t_i)$, in km $\mathrm{s^{-1}}$, $i = 1, \ldots, N$ and the data with the initial several percent noise level, were (8-fold) smoothed (replacing each data point with the average of itself and its two nearest neighbors). The detrended and smoothed data are shown in Figure 1 (b) of Ref. [10]. Certainly, this moving average filter removes considerable amount of noise, leaving only about 1%. The nonlinear filtering, which allows calculation of the entropy, has been discussed in [9]. It has been shown that after the nonlinear Schreiber filter the calculated dimension has been somewhat reduced. In this paper, we focus on the calculations of dimensions.

CP679, *Solar Wind Ten: Proceedings of the Tenth International Solar Wind Conference*, edited by M. Velli, R. Bruno, and F. Malara

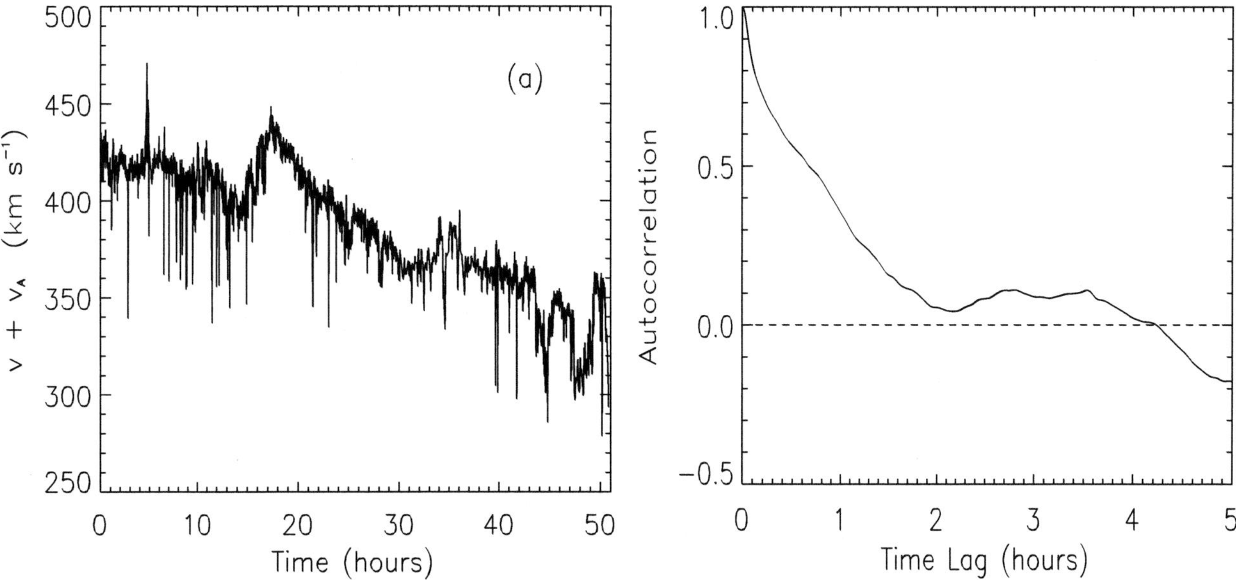

FIGURE 1. (a) The flow velocity with included Alfvénic velocity, $v+v_A$ (Elssäser variable) observed by the Helios 1 spacecraft in 1975 from 67:08:20.5 to 69:11:07 (day:h:min) at distances 0.32 AU from the Sun, for the raw data, (b) the normalized autocorrelation function as a function of the time lag for the detrended and smoothed data.

Therefore, we use moving average filtering, cf. [7].

Table 1, taken from Ref. [10], summarizes selected calculated characteristics of the detrended data cleaned by using the moving average filter. The probability distributions are clearly non-Gaussian. We have a large skewness of ~ 0.26 (as compared with its normal standard deviation 0.06) and a very large kurtosis of 0.88 (the latter was small for the analysis with no magnetic field), cf. Ref [7]. We have also estimated Lempel-Ziv measure of *complexity*, relative to white noise [13]. The calculated value ~ 0.17 is even smaller than in [7] (≈ 0.20); maximal complexity, or randomness, would have a value of 1.0, while a value of zero denotes perfect deterministic nonlinear predictability.

As shown in Figure 1 (b), the normalized autocorrelation function first fells steeply by a factor of $1/e$ in one-third of hour then decreases nearly linearly (reaching a value of 1/2 at 2/3 h) to $1/e$ at $t_a \approx 1$ h, cf. [8, a lower inset to Fig. 1] (see also Table 1). Obviously, for a periodic system the optimum time delay for attractor reconstruction would be one-quarter of the natural orbital period, i. e., the first zero of the autocorrelation function. Therefore, we choose a time delay $\tau = 150\,\Delta t$, where the autocorrelation function has the first minimum and its value is very small. This value is smaller than $t_0 = 376\,\Delta t$, the first zero of the autocorrelation function, $(\langle x(t)x(t+t_0)\rangle - \langle x(t)\rangle^2)/\sigma^2 = 0$ with average velocity $\langle x\rangle = -0.36$ km s^{-1} and standard deviation $\sigma = 8.54$ km s^{-1}, cf. also t_a in Table 1 when the autocorrelation function decreases to $1/e$.

Using our time series of equally spaced, detrended and cleaned data, we construct a large number of vectors $\mathbf{X}(t_i) = [x(t_i), x(t_i+\tau), \ldots, x(t_i+(m-1)\tau)]$ in the embedding phase space of dimension m, where $i = 1, \ldots, n$ with $n = N-(m-1)\tau$. Then, we divide this space into a large number $M(r)$ of equal hypercubes of size r which cover the presumed attractor. If p_j is the probability measure that a point from a time series falls in a typical j-th hypercube, using the q-order function $I_q(r) = \sum(p_j)^q, j = 1, \ldots, M$, the q-order generalized dimension is given by [2]

$$D_q = \frac{1}{q-1} \lim_{r \to 0} \frac{\ln I_q(r)}{\ln r}, \qquad (1)$$

We see from Eq. (1) that the larger q is, the more strongly are the higher probability cubes (visited more frequently by a trajectory) weighted in the sum for $I_q(r)$. Only if $q = 0$, all the cubes are counted equally, $I_o = M$, and we recover the box-counting dimension, D_0.

Writing $I_q(r) = \sum p_j (p_j)^{q-1}$ as a weighted average $\langle (p_j)^{q-1} \rangle$, one can associate bulk with the generalized average probability per hypercube $\mu = \sqrt[q-1]{\langle (p_j)^{q-1} \rangle}$, and identify D_q as a scaling of bulk with size, $\mu \propto r^{D_q}$. Since the data cannot constrain well the capacity dimension D_0, we look for higher order dimensions, which quantify the multifractality of the probability measure on the attractor. For example, the limit $q \to 1$ leads to a geometrical average (the information dimension). For $q = 2$ the generalized average is the ordinary arithmetic average (the standard correlation dimension), and for $q = 3$ it

TABLE 1. Characteristics of the solar wind, $v+v_A$, filtered data.

	Smoothed	Shuffled data	Shuffled phases
Skewness, κ_3	0.26	0.26	-0.26
Kurtosis, κ_4	0.88	0.88	0.26
Relative complexity	0.17	1.0	0.23
Autocorrelation time, t_a	3.5×10^3 s	40 s	3.3×10^3 s
Capacity dimension, D_0	4.4	5.8	4.6
Information dimension, D_1	4.0	5.0	4.5
Correlation dimension, D_2*	3.5	5.2	4.6

* The average slope for $6 \leq m \leq 8$ is taken as D_2.

is a root-mean-square average. In practice, for a given m and r,

$$p_j \simeq \frac{1}{n-2n_c-1} \sum_{i=n_c+1}^{n} \theta(r-\mid \mathbf{X}(t_i)-\mathbf{X}(t_j)\mid) \quad (2)$$

with $\theta(x)$ being the unit step function, and $n_c = 2$ is the Theiler's correction [14]. Finally, $I_q(r)$ is taken to be equal to the generalized q-point correlation sum [12]

$$C_q(m,r) = \frac{1}{n_{\text{ref}}} \sum_{j=1}^{n_{\text{ref}}} (p_j)^{q-1}, \quad (3)$$

where $n_{\text{ref}} = 500$ is the number of reference vectors. For large dimensions m and small distances r in the scaling region it can be argued that $C_q(m,r) \propto r^{(q-1)D_q}$, where D_q is an approximation of the ideal limit $r \to 0$ in Eq. (1) for a given q, Ref. [12].

First, we calculate the natural logarithm of the standard ($q=2$) correlation sum $C_m(r) = C_2(m,r)$ versus $\ln r$ (normalized) for various embedding dimensions: $m = 4$ (dotted curve), $m = 5$ (diamonds), 6 (triangles), 7 (squares), and 8 (crosses) signs. The slopes $D_{2,m}(r) = \mathrm{d}[\ln C_m(r)]/\mathrm{d}(\ln r)$ in the scaling region of r should provide the correlated dimension. The results obtained using the moving average filter are presented in Figure 2, while those obtained using the singular-value decomposition and nonlinear Schreiber filters have been discussed in [7, 9]. Since the correlation sum is simply an arithmetic average over the numbers of neighbors, this can yield meaningful results for the dimension even when the number of neighbors available for some reference points is limited in most real dynamical systems. If the D-dimensional attractor exists, we expect a plateau of the slopes for $m \geq D$ and in the worst case for $m > 2D$. For m large enough an average slope in the scaling region indicates a proper correlation dimension D_2. We have a clear plateau which appears already for $m = 4$ (dotted curve) and $m = 5$. For higher dimensions, $m \geq 8$, the plateau is still present but more smeared out by the statistical fluctuations at small r. In our case the slope of the calculated correlation sum saturates for $m > 5$, with an average for $6 \leq m \leq 8$ of $D_2 = 3.5 \pm 0.1$, cf. Ref. [7]; this is consistent with the attractor of the low-dimension.

Second, the generalized dimensions D_q in Eq. (1) as a function of q are shown in Figure 3. The spectrum of dimensions shows multifractal structure of the solar wind in the inner heliosphere. For comparison, an extremely simple example of the multifractal system is the weighted Cantor set, where the probability of visiting one segment is p (say $p \leq 1/2$) and the probability of visiting the other segment is $1-p$. In this case the results can be obtained analytically; for any q in Eq. (1) one has

$$(1-q)D_q = log_3[p^q + (1-p)^q]. \quad (4)$$

The difference of the maximum and minimum dimension, associated with the least dense and most dense points on the attractor, correspondingly, is $D_{-\infty} - D_{\infty} = log_3(1/p - 1)$ and in the limit $p \to 0$ this difference rises to infinity. Hence, the parameter p can be regarded as a degree of multifractality. The results of $D_q + 3$ calculated for $p = 0.1$ in Eq. (4) are also shown in Figure 3 by a dash-dotted line. We see that for $q > 1$ the multifractal spectrum of the solar wind is roughly consistent with that for the multifractal measure on the self-similar weighted Cantor set, with a single weighting parameter p. The obtained value of this parameter demonstrates that some cubes that cover the attractor of this dynamical system are visited one order of magnitudes more frequently than some other cubes, as is illustrated in our previous paper, see Figure 5 of Ref. [7].

We also estimate the Kolmogorov correlation entropy, K_2, and the largest positive Lyapunov exponent, λ_{max}. The vertical spacings between the parallel lines in Figure 2 of Ref. [10] averaged in the saturation region $8 \leq m \leq 10$ are taken as K_2 yielding the value of ≈ 0.1 (per delay time τ). Using the algorithm of Ref. [15] and nonlinear noise reduction, one obtains the magnitude of $\lambda_{\text{max}} \sim 0.1$ in the same units as for K_2 (base e). In general, the entropy K_q is at most the sum of the positive Lyapunov exponents $\sum \lambda_i$, e.g., [2]. The value of the Lyapunov exponent is consistent with the Kolmogorov ($q=2$) entropy, which should be its lower

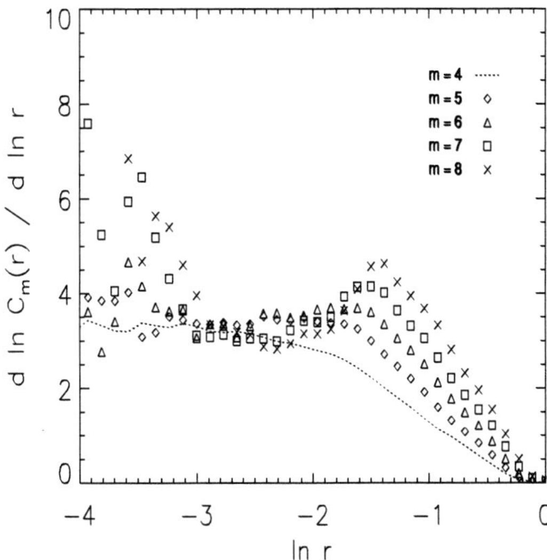

FIGURE 2. The slopes $D_{2,m}(r) = \mathrm{d}(\ln C_m(r))/\mathrm{d}(\ln r)$ for the correlation sum $C_m(r)$ versus $\ln r$ (normalized) obtained for the cleaned experimental signal for various embedding dimensions: $m = 4$ (dotted curve), $m = 5$ (diamonds), 6 (triangles), 7 (squares), and 8 (crosses).

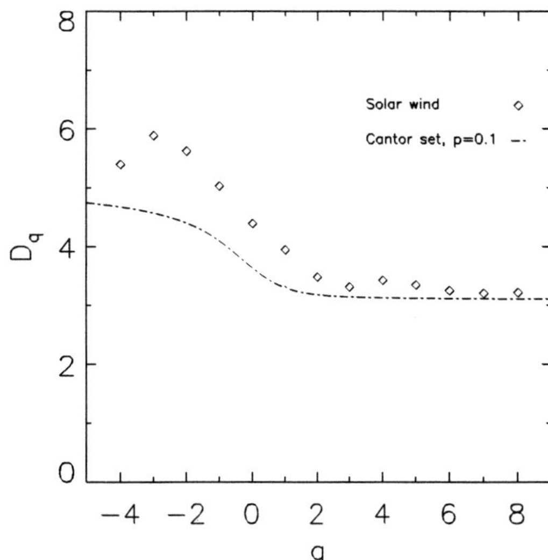

FIGURE 3. The generalized dimensions D_q in Eq. (1) as a function of q. The correlation dimension is $D_2 = 3.5 \pm 0.1$, see Table 1. The values of $D_q + 3$ calculated analytically for the weighted Cantor set with $p = 0.1$ (dash-dotted).

bound: $K_2 \leq \sum \lambda_i$ (positive). The time over which the meaningful prediction of the behavior of the system is possible is roughly $\sim 1/\lambda_{\max}$, e.g., [2]. Hence the predictability of the system is limited to hours.

The obtained measures of the attractor have been subjected to the surrogate data test [11]. As has been demonstrated in Figure 8 of Ref. [7], if the original data are indeed deterministic, analysis of these surrogate data will provide values that are statistically distinct from those derived for the original data, see also Table 1. In particular, the slope of the correlation sum increases with m (no saturation), and Lempel-Ziv complexity calculated for shuffled-data becomes clearly 1.0, as it should be for a purely stochastic system. Again, we have found that the solar wind data are sensitive to this test.

In conclusion, we have shown that the moving average filter removes some amount of noise, which is sufficient to calculate the generalized dimensions of the solar wind attractor. The obtained multifractal spectrum is consistent with that for the multifractal measure on the self-similar weighted Cantor set with a degree of multifractality of $p \sim 0.1$. The obtained characteristics of the attractor are significantly different from those of the surrogate data. Thus these results show multifractal structure of the solar wind in the inner heliosphere. Hence we suggest that there exists an inertial manifold for the solar wind, in which the system has *multifractal* structure, and where noise is certainly not dominant.

This work has been done in the framework of the European Commission Research Training Network Grant No. HPRN-CT-2001-00314.

REFERENCES

1. Hentschel, H. G. E., and Procaccia, I., Physica D, **8**, 435–444 (1983); Grassberger, P., Phys. Lett. A, **97**, 227–230 (1983); Halsey, T. C., Jensen, M. H., Kadanoff, L. P., Procaccia, I., and Shraiman, B. I., Phys. Rev. A, **33**, 1141–1151 (1986).
2. Ott, E., *Chaos in Dynamical Systems*, Cambridge University Press, Cambridge, 1993.
3. Mandelbrot, B. B., Multifractal measures, especially for the geophysicist, in *Pure and Applied Geophysics*, **131**, 5–42, Birkhäuser Verlag, Basel (1989).
4. Kurths, J., and Herzel, H., Physica D, **25**, 165–172 (1987).
5. Burlaga, L. F., Geophys. Res. Lett., **18**, 69–72 (1991).
6. Schwenn, R., Large-scale structure of the interplanetary medium, in *Physics of the Inner Heliosphere*, edited by R. Schwenn and E. Marsch, Springer-Verlag, Berlin, 1990, Vol. 20, pp. 99–182.
7. Macek, W. M., Physica D, **122**, 254–264 (1998).
8. Macek W. M., and Obojska, L., Chaos, Solitons & Fractals, **8**, 1601–1607 (1997); **9**, 221–229 (1998).
9. Macek W. M., and Redaelli, S., Phys. Rev. E, **62**, 6496–6501 (2000).
10. Macek, W. M., Multifractality and chaos in the solar wind, in *Experimental Chaos*, edited by S. Boccaletti, B. J. Gluckman, J. Kurths, L. M. Pecora, and M. L. Spano, American Institute of Physics, New York, 2002, Vol. 622, pp. 74–79.
11. Theiler, J., Eubank, S., Longtin, A., Galdrikian, B., and Farmer, J. D., Physica D, **58**, 77–94 (1992).
12. Grassberger P., and Procaccia, I. , Physica D, **9**, 189–208 (1983).
13. Kaspar, F., and Schuster, H. G., Phys. Rev. A, **36**, 842–848 (1987).
14. Theiler, J., Phys. Rev. A, **34**, 2427–2432 (1986).
15. Kantz, H., Phys. Lett. A, **185**, 77–87 (1994).

Comparison of VLF Wave Activity in the Solar Wind During Solar Maximum and Minimum: Ulysses Observations

Naiguo Lin, P. J. Kellogg,
University of Minnesota, Minneapolis, Minnesota, USA

R. J. MacDowall,
NASA Goddard Space Flight Center, Greenbelt, Maryland, USA

D. J. McComas,
Southwest Research Institute, San Antonio, Texas, USA

A. Balogh, and R. J. Forsyth
The Blackett Laboratory, Imperial College, London, UK

Abstract. We have compared observations of VLF waves (0.2 to 448 Hz) made by Ulysses during its second fast latitude scan (near the solar maximum) with the wave observations during the first fast latitude scan in 1995, when the solar activity was approaching a minimum. The occurrences and properties of the waves are found to be similar during the solar maximum and solar minimum periods for slow and intermediate speed solar wind. The maximum intensity of the electromagnetic waves for the two solar cycle periods is comparable. These similarities suggest that the plasma conditions for the waves' excitation are similar for the slow and intermediate solar wind in both solar maximum and minimum phases. It is also found that the electric field noise detected in the low band channels, which are measuring less than 9 Hz signals, is contaminated by the spin modulation of the electric field due to photoelectrons around the spacecraft, especially when the ambient plasma density is low. This noise increases with increasing solar aspect angle.

INTRODUCTION

Observations of VLF waves made by the Unified Radio and Plasma Wave Experiments (URAP) of Ulysses during its first orbit, which occurred when the solar activity was approaching minimum, have found that there are different patterns and properties of wave activity occurring in the slow to intermediate (~300 to 600 km/s) solar wind and in the fast solar wind (1, 2). During the solar minimum period, the slow to intermediate solar wind was observed at low and mid-latitudes, i.e. within the heliospheric current sheet region. In this region enhanced electromagnetic waves in the whistler mode frequency range were observed near interplanetary (IP) shocks, during heliospheric current sheet crossings, and in other solar wind turbulence. These waves were most intense in compression regions of high-speed stream interfaces, which were observed in periods of increasing solar wind velocity. Electromagnetic whistler mode waves in the solar wind have been suggested to be excited by the whistler heat-flux instability and to be responsible for electron heat flux regulation in the solar wind [cf. (3) and references therein]. Comparison of the electron heat flux data with wave activity, however, is not conclusive: high levels of electron heat flux are often absent during periods of intense, low frequency electric wave activity (4).

Lin et al. (1) have also reported that enhanced electric fluctuations with little or no magnetic fluctuations are often observed at low band channels

CP679, *Solar Wind Ten: Proceedings of the Tenth International Solar Wind Conference*, edited by M. Velli, R. Bruno, and F. Malara

of the wave form analyzer (WFA) of the URAP instrument. The central frequencies of these channels range from 0.2 to 5.3 Hz. These electric enhancements often occur in the expanding solar wind, i.e., in intervals when the measured solar-wind velocity is decreasing. The intensity of magnetic waves decreased dramatically in these intervals, indicating that these intervals did not favor the excitation of the whistler mode. As will be addressed below, the electric field enhancements seen in the low band channels contain electric field variations caused by photo-electron clouds around the spacecraft.

Outside the heliospheric current sheet region, in relatively stable fast solar wind streams, Ulysses observed nearly continuous electric wave activity with peak power near the local electron gyrofrequency, f_{ce}, (a few tens Hz). These waves may also have a magnetic component but their levels have decreased below the search coil background at distances beyond ~2.5 AU.

The Ulysses mission has extended the observations to a full solar cycle. As Ulysses started the second orbit, the Sun was approaching maximum activity conditions. In this period, the spacecraft observed an irregularly structured mixture of slow and intermediate-speed solar wind and no entry into the fast solar wind (5). Strong and frequent wave activity has been observed, not only near the ecliptic plane and at mid-latitudes, but also at high latitudes. The plasma conditions of slow and intermediate solar wind existing at all latitudes during the second orbit, are responsible for the lack of latitudinal variations of the wave activity. The features of VLF wave activity described above, which were seen in low and mid-latitudes during the solar minimum mission, were also observed at high latitudes during solar maximum. This paper presents the observations of VLF waves made during the second fast latitude scan, and compares the observations with those obtained during the period when Ulysses was in the slow to intermediate solar wind in the first fast latitude scan in 1995.

OBSERVATIONS AND COMPARISON

The second fast latitude scan occurred throughout the year 2001. Figure 1 displays the observations for days 100–170 of 2001, at heliographic latitudes ~ –20° to 27°. The top and bottom panels show the electric field and magnetic field spectra using 15-min averaged data. The wave power is expressed by the relative intensity, which is the measured signal minus the background noise of the instrument, divided by the background noise (1). A local f_{ce} line is overplotted in each panel of the spectra.

The figure shows similar features of wave activity as those observed in the heliospheric current sheet region during the first orbit. The peak power of the wave activity is below f_{ce}, which indicates the waves are in the whistler mode frequency range. Electromagnetic waves, which are observed in both electric and magnetic field spectra, occur near interplanetary shocks (marked by the vertical dashed lines in the middle two panels) and where there is an increase in the solar wind velocity. These waves are seen most clearly in the high band channels (>9Hz). In expanding solar wind stream regions, i.e. during the intervals when the solar wind velocity is decreasing, magnetic noise reduces significantly. The above features have been observed during the second orbit (starting at the beginning of the year 1998) at all latitudes. Electric waves are frequently observed in lower band data (<9 Hz). We notice that these are periods with low plasma density, as seen in the third panel where the electron density (calculated as N_p+2N_α, where N_p and N_α are the proton and helium ion density, respectively) is plotted. Some clear examples of these periods are marked with the yellow stripes in the middle two panels. The anti-correlation between the electric field enhancements and the electron density has led us to believe that the enhancements of the electric field are caused by photoelectron clouds around the spacecraft. This effect becomes stronger when the solar aspect angle (the angle of Sun-Ulysses-Earth) becomes larger in the second latitude scan than in the first.

In order to compare the intensity of the waves occurring during solar maximum and minimum periods, we compare the observations of VLF waves taken in the first and second latitude scans for a heliographic latitude range of –20° to 20°, where Ulysses was traveling in the slow to intermediate solar wind, at similar distances from the Sun for both scan periods. Observations for the first latitude scan in 1995 have been reported in paper (1). Figure 2 shows the comparison of spin plane magnetic and electric signals measured at the channel with a central frequency of 14 Hz, which is typical for whistler mode waves observed in the solar wind at these distances. The first panel displays the power density of magnetic waves vs. heliographic latitude. In general, the maximum intensity of the waves in the second latitude scan in 2001 (red line) is comparable to that in the first scan (green line) in 1995, except for the strong enhancement of the wave on day 130, 2001, at a latitude of ~ –5°, which is associated with a strong IP

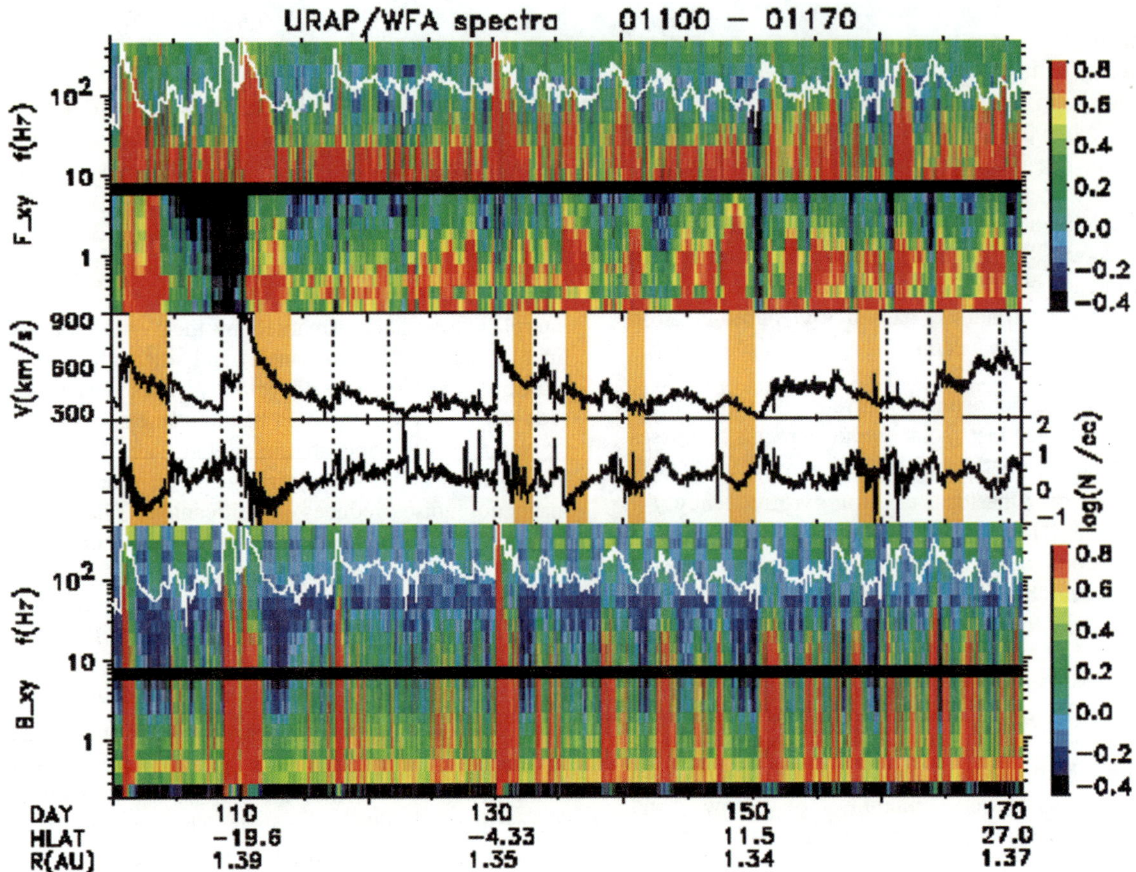

FIGURE 1. From top to bottom: Spectra of 15-min averaged relative intensity of spin plane electric wave power for the period from day 100 to day 170 of 2001 (the local f_{ce} is overplotted as a white line); the solar wind velocity in km/s; the electron density; and the magnetic wave power in the same format as that of the electric field spectra. The vertical dotted lines mark IP shock crossings. The vertical yellow stripes in the two middle panels indicate periods of low-density level.

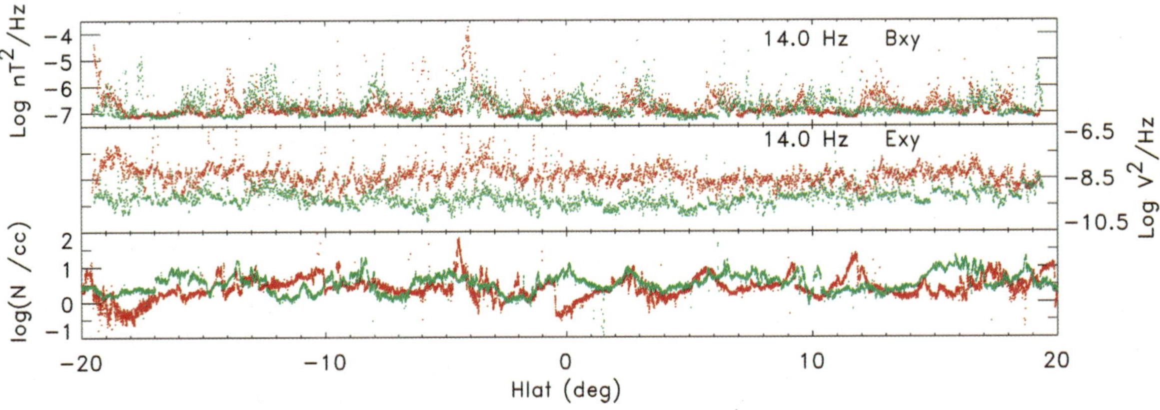

FIGURE 2. From top to bottom: The power density of spin plane magnetic field waves (in the log scale) at 14.0 Hz for the first fast latitude scan in 1995 (green line) and for the second latitude scan in 2001 (red line), plotted vs. heliographic latitude; the same as the first panel but for the electric field power density; the electron density calculated as N_p+2N_α.

shock. The second panel displays the comparison of electric wave power (expressed in V^2/Hz), which shows that the electric waves in the second scan (during the solar maximum period) are apparently slightly stronger than those in the first scan (during the solar minimum period). The electric measurements of Ulysses are affected by the variation of the solar aspect angle, with larger angles producing higher instrument background noise (1). The effect mainly exists in the low band channels (<9 Hz), while the high band channels are less affected. During the second latitude scan, the solar aspect angle for the period displayed was higher (~25°–40°) than that during the first latitude scan (<13°), which may contribute to the difference in the intensity for both periods. Latitude variations of the electron density for the two fast scan periods are plotted in the third panel, which shows the density varies in about the same range for both periods, indicating similar plasma conditions for the regions examined during solar maximum and minimum. In summary, our observations have shown that during the ascending phase of the solar cycle, when Ulysses was embedded in the slow and intermediate speed solar wind at all latitudes, and frequently encountered enhanced solar wind turbulence, interplanetary shocks, and current sheet crossings, the spacecraft observed VLF wave activity with an occurrence pattern similar to that observed in the low and mid-latitudes during the declining phase of the solar cycle (1). The intensity of the electromagnetic whistler mode waves, which are very commonly observed in the solar wind, is comparable for the two periods of different solar cycle phases. These similarities suggest that the plasma conditions for the waves' excitation are similar for the slow and intermediate solar wind in both solar cycle phases.

THE PHOTOELECTRON EFFECT

Electric field measurements in low band channels (< 9Hz) of the WFA are contaminated by the spin-modulated signal produced by the asymmetric spacecraft photoelectron cloud. This one cycle per spin signal leaks into all of the low band channels, because the 16-point Walsh transform for the low band data does not have sufficient resolution. This effect is stronger when the ambient electron density decreases, and thus produces an apparent anti-correlation between the electric field noise and the density. In the expanding solar wind where the density becomes low, the leaking of the spin period signal produces enhanced electric noise in low band channels. This effect is more significant for larger solar aspect angles (as in the second latitude scan, compared to the first scan), which produce stronger asymmetry of photoelectron cloud. This effect is also seen in the high band channels (>9Hz) but is less significant. For example, if we express the relation between the electric wave power P_E (in V^2/Hz) and the density N(cm^{-3}) as $\log P_E = A + B \log N$, for day 111-170 of year 2001 (the second latitude scan), for the low band channel at 5.33 Hz, we have B=-0.33, and the correlation coefficient, C, between $\log P_E$ and logN is as high as –0.74; while for the high band channel of 14 Hz, C is only –0.35. For the first scan with lower aspect angle, taking data from day 40 to 75 of 1995, we have C~-0.35 for 5.33 Hz channel, and C~0.05 for 14 Hz channel. Because of this photoelectron effect, the low band data is only useable for intense events. This problem was not addressed in paper (1). We will present a more detailed analysis in a future paper.

ACKNOWLEDGMENTS

The URAP experiment is a collaboration of NASA Goddard Space Flight Center, the Observatoire de Paris-Meudon, the University of Minnesota, and the Centre d'etude des Environnements Terrestres et Planetaires, Velizy, France. We thank the SWOOPS team and the Magnetometer team of Ulysses for the support of this work. N.L. thanks K. Goetz and S. J. Monson of the University of Minnesota for support in conducting this study.

REFERENCES

1. Lin, N., P. J. Kellogg, R. J. MacDowall, E. E. Scime, A. Balogh, R. J. Forsyth, D. J. McComas, and J. L. Philips, Very low frequency waves in the heliosphere: Ulysses observations, *J. Geophys. Res.*, **103**, 12,023 (1998).

2. MacDowall, R. J., and P. J. Kellogg, "Waves and instabilities in the 3-D heliosphere", in *The Heliosphere Near Solar Minimum: The Ulysses Perspective*, edited by A. Balogh, R. G. Marsden, E. J. Smith, London, Springer-Praxis, 2001, pp. 229-258.

3. Gary, S. P., E. E. Scime, J. L. Phillips, and W. C. Feldman, The whistler heat flux instability: Threshold conditions in the solar wind, *J. Geophys. Res.*, **99**, 23,391 (1994).

4. Scime, E. E., J. E. Littleton, S. P. Gary, and N. Lin, Solar cycle variations in the electron heat flux: Ulysses observations, *Geophys. Res. Lett.* **28**, 2,169 (2001).

5. McComas, D. J., B. Goldstein, J. T. Gosling and R. M. Skoug, Ulysses' second orbit: Remarkably different solar wind, *Space Sci. Rev.* **97**, 99 (2001).

Solar Wind Temperature Anisotropies

J. C. Kasper[1], A. J. Lazarus[1], S. P. Gary[2], A. Szabo[3]

[1]*MIT/CSR, 77 Massachusetts Avenue, Cambridge, USA,* [2]*LANL/NIS-1, M.S. D466, Los Alamos, USA,* [3]*NASA/GSFC, Code 696, Greenbelt, USA*

Abstract. Solar wind proton and alpha spectra from the Faraday Cup portion of the SWE experiment on the Wind spacecraft have been analyzed to determine the temperature anisotropy of each species, under the assumption of convected, bi-Maxwellian distributions. From the start of the mission in late 1994 to date we have collected over 2 million measurements of the anisotropies in the solar wind. This dataset is sufficiently large to conduct a statistical study, comparing the observed temperature anisotropies to various limits imposed by instabilities. Specifically we will discuss the effects of the firehose, mirror, and cyclotron instabilities. With a limit to the proton temperature anisotropy established, we examine several cases where this limit is approached or exceeded and comment on magnetic field activity and alpha parameters during these intervals. In the large plasma beta regime we illustrate evidence of a transition from the cyclotron to the mirror instability as the dominant limit.

INTRODUCTION

The first-order departure from a simple Maxwellian velocity distribution function (VDF) of a particle species in the solar wind is due to the existence of a temperature anisotropy. Generally this anisotropy is well described by a convected, field-aligned, bi-Maxwellian VDF, with two temperatures, $T_\perp$ and $T_\parallel$ perpendicular and parallel to the ambient magnetic field $\mathbf{B}_o$. The solar wind is a collisionless plasma, and one might expect it to satisfy the double adiabatic equations of state,

$$\frac{d}{dt}\left(\frac{T_\perp}{B}\right)=0;\frac{d}{dt}\left(\frac{T_\parallel B^2}{n^2}\right)=0 \qquad (1)$$

In this case an initially isotropic parcel of solar wind would be expected to develop a temperature anisotropy, R,

$$R=T_\perp/T_\parallel-1 \qquad (2)$$

which might vary by orders of magnitude. While a radial evolution of the proton anisotropy has been observed [1], generally $|R| \leq$ unity.

Consider the solar wind observations on April 30, 1997 by the Wind spacecraft which are summarized in Figure 1. There was relatively little activity in the first 15 hours of the day; in general the density was increasing while the magnetic field strength decreased. The observed parallel and perpendicular temperatures

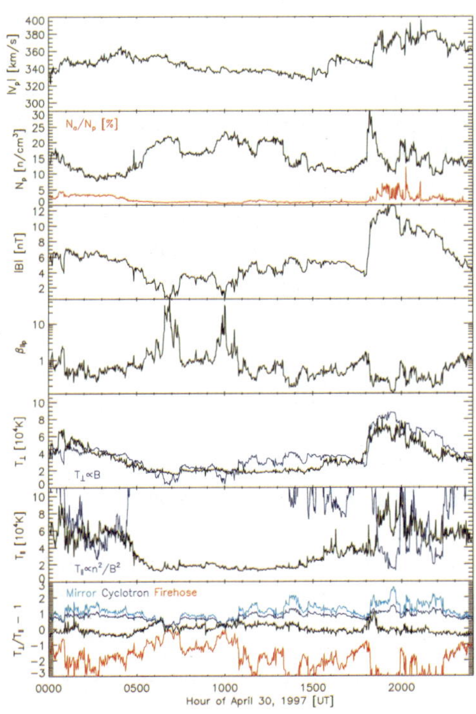

FIGURE 1. Solar wind observations by the Wind spacecraft on April 30, 1997. The alpha particle abundance is very low and may be neglected. Note that the predicted value of $T_\parallel$ is very high and off the scale of the plot for 0500-1500 UT.

CP679, *Solar Wind Ten: Proceedings of the Tenth International Solar Wind Conference,* edited by M. Velli, R. Bruno, and F. Malara

are shown along with predictions using (1). As the solar wind evolves, $T_\perp$ agrees with the adiabatic predictions, while $T_\parallel$ decreases and dramatically disagrees with that calculation.

The permissible range of R is constrained by the effects of kinetic micro-instabilities driven by the temperature anisotropy. In a collisionless magnetized plasma with R>0 mirror [2] and cyclotron [3] instabilities may arise. These growing modes have real frequency ω_r with $\omega_r \sim \Omega_p$ for the cyclotron instability and $0<\omega_r<\Omega_p$ for the cyclotron instability, where Ω_p is the proton cyclotron frequency. Electrons are non-resonant with these modes and to first order the alpha particles may be neglected, so we may study them with proton and magnetic field observations. For a given fixed value of the dimensionless growth rate γ/Ω_p of these instabilities, linear theory and hybrid simulations [4] have shown that the maximum values R_m allowed by the mirror, and R_c by the cyclotron instability take the form,

$$R_{m,c} = S_p \big/ (\beta_{\parallel p})^{\alpha_p} \tag{3}$$

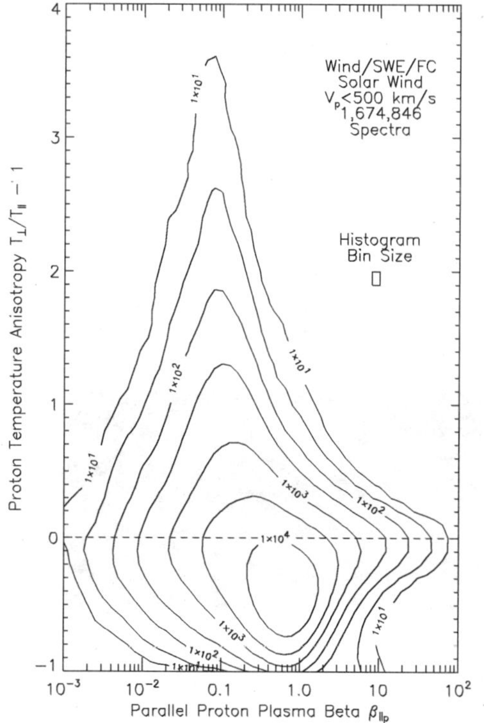

FIGURE 2. Two-dimensional histogram of all solar wind spectra with speeds less than 500 km/s as a function of anisotropy and plasma beta.

where $\beta_{\parallel p}$ is the parallel proton plasma beta, $\beta_{\parallel p}=8\pi n_p k_B T_{\parallel p}/B_o^2$, S_p is of order unity and determined by the choice of γ/Ω_p, and α_p is roughly constant over a range of values. For R<0 the firehose instability [5] provides the lower limit, with

$$R_f = -S_p \big/ (\beta_{\parallel p})^{\alpha_p} \tag{4}$$

The resulting electromagnetic fluctuations scatter the protons and drive the proton VDF to isotropy. The final panel in Figure 1 shows the measured value of R along with the theoretical upper bounds due to the cyclotron (purple) and mirror (blue) instabilities using (3), and the lower bound due to the firehose (red) instability calculated with (4). It is clearly seen that these instabilities limit the anisotropy accessible to the solar wind protons, especially in the region with $\beta_{\parallel p}>20$. The effects of the mirror and cyclotron instabilities have been studied extensively in the magnetosheath [6], but the only previous study of the firehose instability is based on several hours of Vela 4 data [7]. We have performed a bi-Maxwellian analysis of solar wind ions observed by the Solar Wind Experiment (SWE) Faraday Cup (FC) instruments on the Wind spacecraft [8]. A typical FC spectrum contains 300 measurements of the proton VDF. By comparing these observations with a model distribution function using a non-linear fitting routine we are able to extract the best-fit parameters, their uncertainties, and χ^2 per degree of freedom (dof). These data, with the addition of magnetic field measurements by the MFI experiment, have been used to produce the first statistical study of the firehose instability in the solar wind. The results of that study and a complete description of the instrument analysis techniques may be found in [9]. In this paper we outline two separate approaches for understanding the effects of these instabilities on the solar wind: by statistical studies of a large sample of ion spectra over a range of solar wind parameters and by examining the detailed properties of specific events.

STATISTICAL METHOD

For the statistical study we look at the values of R which the protons have access to as a function of plasma beta. Based on (3) and (4) we expect to see signatures of a bounds on R as a function of beta. Figure 2 is a two-dimensional histogram of the 1.6 million observations with V_p<400 km/s, χ^2/dof < 10, and an uncertainty σ_R<0.3 collected by Wind from 1994-2001. The average uncertainty in R was 0.18, or roughly the size of each bin in the histogram. The most probable state of the protons is R=-0.3, $\beta_{\parallel p}$=0.8. It is clear that for $\beta_{\parallel p}$>1 the protons are more strongly

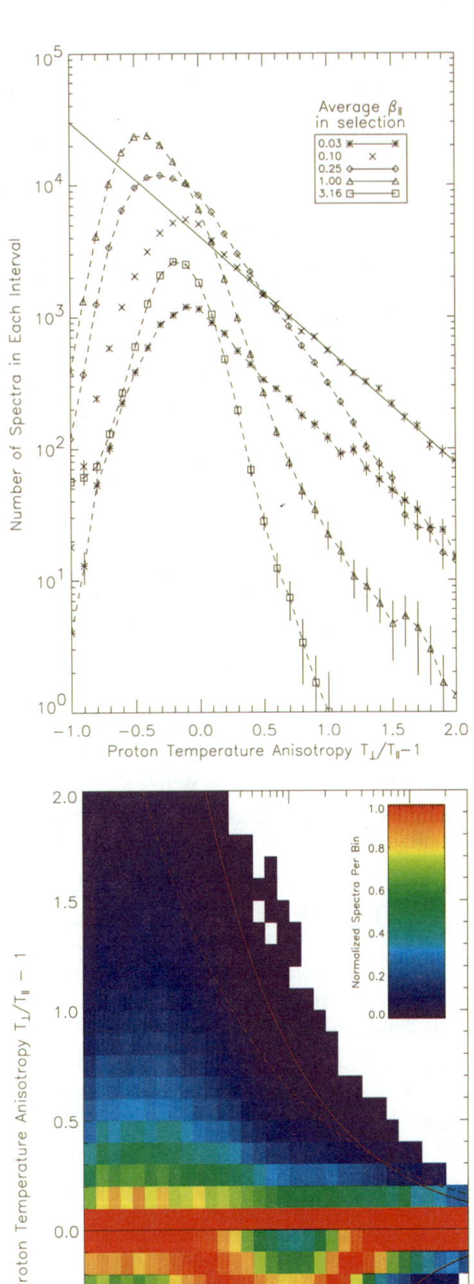

FIGURE 3. Two views of the histogram in Figure 2. Upper: Number of spectra vs R for several intervals in $\beta_{\|p}$, fit to $\beta_{\|p}$=0.1; Lower: The original histogram normalized to illuminate the effect of the instabilities: Each column in $\beta_{\|p}$ is separated into the parts with R>0 and R<0, and each of those parts is normalized to unity. Limits from instabilities R_m(solid,R>0), R_c (dashed), R_f(solid,R<0).

constrained to isotropy as $\beta_{\|p}$ increases, in agreement with the predictions of theory. To compare the observed distribution of spectra with the predictions of theory we employ two methods, both shown in Figure 3. In the top panel the number of spectra seen as a function of R are plotted for several values of $\beta_{\|p}$. The nature of the curves for R>0 are determined by the competing effects of the mirror and cyclotron instabilities, and observationally they are well described by exponential curves. A fit to the $\beta_{\|p}$ =0.1 interval is shown. In [9] the same analysis was applied to the R<0 data to quantify the effect of the firehose instability, in which it was assumed that for each interval in $\beta_{\|p}$ the observed limiting anisotropy $R_f(\beta_{\|p})$ is given by the value of R at which the number of spectra falls to 10% of its maximum value,. The best fit of (4) to the measured values of $R_f(\beta_{\|p})$ was given by (S_p, α_p)=(1.21±0.26, 0.76±0.14), compared to hybrid simulation results [5] of (1, 0.74) . The bottom panel shows a version of the histogram in Figure 2 which has been normalized to remove the distribution of solar wind spectra in $\beta_{\|p}$. The theoretical limits to R are indicated. New features may be seen, such as the clear competition between the expansion of the solar wind and the effect of the firehose instability for R<0.

CLOSE STUDY OF SINGLE EVENTS

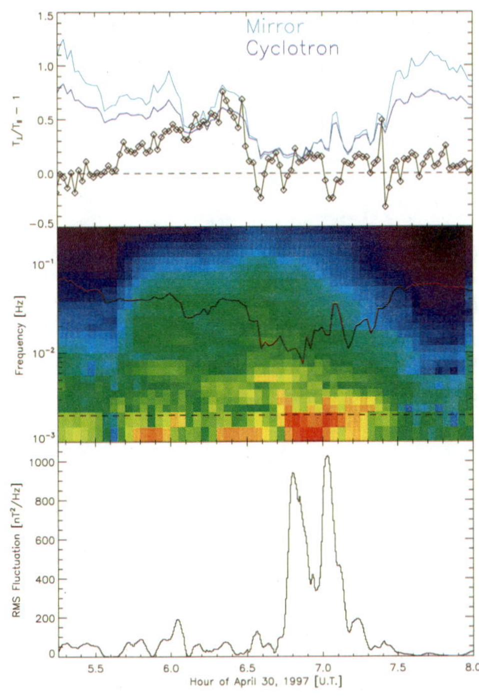

FIGURE 4. Close inspection of three hours of observations comparing anisotropy, limits, and magnetic fluctuations.

We have shown that the properties of the cyclotron and firehose instabilities may be probed with large statistical studies of the allowed values of R as a function of $\beta_{\|p}$, but there are limitations to this method. It is clear from Figure 3 that the mirror instability is not expected to contribute to a great extent until $\beta_{\|p}$ >5, so there are not enough measurements to perform a similar statistical study; and instead we should focus on individual periods with high $\beta_{\|p}$. Additionally, while the variation of the slope of the curves in the top panel of Figure 3 is likely related to the growth rate of the instabilties as a function of $\beta_{\|p}$, a more direct measurement of that growth may come from a study of the electromagnetic fluctuations seen during individual events [10].

We can sort through this large dataset of plasma conditions to identify ideal periods for close examination. The interval shown in Figure 1 is such an example: The alpha abundance was extremely low, the anisotropy was right at the theoretical limit, and $\beta_{\|p}$ was so high that the mirror instability contributed equally to the enhanced field fluctuations which scattered the proton VDF and prevented it from becoming even more anisotropic. Figure 4 shows a period in which R is near the limiting values R_c and R_m. The second panel shows the RMS fluctuation of the magnetic field about its ambient value as a function of frequency and time. The red curve is Ω_p. The bottom panel shows the power in the frequency range marked in the middle panel by the dotted line. Note that for the duration of the interval there is an enhancement of power near Ω_p, but that between 0630 and 0730 large fluctuations are seen at $\omega_r<\Omega_p$, exactly in agreement with the theoretical signiatures of those instabilities discussed above.

CONCLUSIONS

We have used a large set of measurements by the SWE experiment on the Wind spacecraft of the proton temperature anisotropy in the solar wind to constrain theoretical models of the mirror, cyclotron, and firehose instabilities. Our results may be summarized as follows:

1. Statistically, the observed bounds on R are in agreement with the limits imposed by the cyclotron and firehose instabilities.
2. Solar wind with R<0 is pinned between the competing effects of adiabatic expansion and the firehose instability.
3. An unanticipated lower limit to R at $\beta_{\|p}<1, R<0$, is indicative of the existence of another instability.
4. Frequency and temporal dependence of magnetic fluctuations during a high-$\beta_{\|p}$ interval provides evidence of the mirror instability.

More case studies of single events will permit a quantitative probe of the mirror instability. We are studying the effect of alpha particles. Work should be done to relate the observed exponential fall-off shown in Figure 3 to the growth rate of the instabilities and possibly to the rate of cyclotron heating for R>0.

REFERENCES

1. Marsch, E., K.-H. Muhlauser, R. Schwen, H. Rosenbauer, W. Pilipp and F. M. Neubauer, Solar wind protons: Three dimensional velocity distributions and derived plasma parameters measured between 0.3 and 1 A.U., *J. Geophys. Res.*, **87**, A1, 35, 1982.

2. Chandrasekhar, S., A. N. Kaufman and K. M. Watson, The stability of the pinch, *Proc. Roy. Soc. London, Ser. A*, **245**, 435, 1958.

3. Kennel, C. F. and H. E. Petschek, Limit on stably trapped particle fluxes, *J. Geophys. Res.*, **71**, 1, 1966.

4. Gary, S. P., L. Yin, and D. Winske, Electromagnetic proton cyclotron instability: Wave-particle scattering rate, *Geophys. Res. Lett.*, **27**, 2457, 2000.

5. Gary, S. P., H. Li, S. O'Rourke, and D. Winske, Proton resonant firehose instability: Temperature anisotropy and fluctuating field constraints, *J. Geophys. Res.*, **103**, 14567, 1998.

6. Phan, T.-D., G. Paschmann, W. Baumjohan, W. Sckope, and H. Lühr, The magnetosheath region adjacent to the dayside magnetopause: AMPTE/IRM observations, *J. Geophys. Res.*, **99**, 121, 1994.

7. Eviatar, A. and M. Schultz, Ion-temperature anisotropies and the structure of the solar wind, *Planetary Space Sci.*, **18**, 321, 1970.

8. Ogilvie, K. W., et. al., SWE: A comprehensive plasma instrument for the Wind spacecraft, *Space Sci. Rev.*, **71**, 55, 1995.

9. Kasper, J. C., A. J. Lazarus, and S. P. Gary, Wind/SWE observations of firehose constraint on solar wind proton temperature anisotropy, *Geo. Res. Lett.*, *in press*, 10.1029/2002GL015128.

10. Anderson, B. J., S. A. Fuselier, S. P. Gary, and R. E. Denton, Magnetic spectral signatures in the Earth's magnetosheath and plasma depletion layer, *J. Geophys. Res.*, **99**, 5877, 1994.

Electric Field Statistics in the Solar Wind

B. Breech*, L. J. Milano*, W. H. Matthaeus* and C. W. Smith*

*Bartol Research Institute, University of Delaware, Newark, Delaware

Abstract. The induced electric field, a fundamental quantity in magnetohydrodynamics (MHD), having important possible implications in turbulence and in particle scattering, is examined both theoretically and observationally. Using the assumption of normally distributed velocity and magnetic field fluctuations, we find that the induced electric field component distributions are exponential or exponential-like, in accordance with the degree and type of correlations present. Observations at 1 AU using four day samples of one-hour data confirm that this distribution is a very good approximation for low frequency MHD scales.

INTRODUCTION

Studies and theories involving interplanetary magnetohydrodynamic (MHD) fluctuations typically characterize the magnetic and velocity field fluctuations while ignoring the induced electric field (relevant references are numerous, e.g., [1, 2, 3, 4, 5, 6, 7]; however for a discussion of the electric field spectrum, see [8]). Yet, the evolution of the MHD fluctuation spectrum, the heating of the solar wind, and the scattering of energetic particles all involve to some degree the generation of, and response to, electric field fluctuations.

The dynamical importance of the electric field is embodied in the magnetic field $\mathbf{B}$ induction equation,

$$\frac{\partial \mathbf{B}}{\partial t} = -\nabla \times \mathbf{E}. \tag{1}$$

In MHD models of plasma dynamics, the electric field is usually expressed by a constitutive relation, a generalized Ohm's Law,

$$\mathbf{E} = -\mathbf{V} \times \mathbf{B} + \eta \mathbf{j} + \mathbf{E_c} \tag{2}$$

where η is the resistivity, $\mathbf{j}$ is the electric current, and $\mathbf{V}$ is the velocity field. Kinetic contributions to the electric field, such as Hall and electron pressure effects, are included in $\mathbf{E}_c$. For space plasmas, such as the solar wind, η is typically small and can be ignored. Furthermore, we will neglect the collisionless term, $\mathbf{E}_c$. At very small time scales, additional terms in equation (2) would appear. However, as will be mentioned later, we will concern ourselves with one hour data so these terms will also be ignored.

We subject $\mathbf{V}$ and $\mathbf{B}$ to the usual decomposition,

$$\mathbf{V} = \mathbf{V}_0 + \mathbf{v}, \quad \mathbf{B} = \mathbf{B}_0 + \mathbf{b}, \tag{3}$$

where $\mathbf{V}_0 = \langle \mathbf{V} \rangle$, $\mathbf{B}_0 = \langle \mathbf{B} \rangle$ and the $\langle\ \rangle$ operator denotes an appropriate ensemble average. Note that, by definition,

$$\langle \mathbf{v} \rangle = 0 = \langle \mathbf{b} \rangle. \tag{4}$$

We are concerned with electric fields produced by the fluctuating velocity and magnetic fields, which we define as,

$$\delta \mathbf{e} \equiv -\mathbf{v} \times \mathbf{b} + \langle \mathbf{v} \times \mathbf{b} \rangle. \tag{5}$$

We also define,

$$\mathbf{e} \equiv -\mathbf{v} \times \mathbf{b} \tag{6}$$

Obviously, $\mathbf{e} = \delta \mathbf{e}$ if $\langle \mathbf{e} \rangle = \langle \mathbf{v} \times \mathbf{b} \rangle = 0$.

In our previous work (Milano et al. [9]), we derived the probability distribution function, PDF, for $\mathbf{e}$, assuming that $\mathbf{v}$ and $\mathbf{b}$ had Gaussian distributions. In particular, we examined several cases involving varying degrees of correlation between the component of $\mathbf{e}$. If the components are totally uncorrelated, such that $\langle \mathbf{v} \cdot \mathbf{b} \rangle = 0 = \langle \mathbf{e} \rangle$, then the PDF, $f(e_z)$ is,

$$f(e_z) = \frac{1}{\sqrt{2}\sigma_{e_z}} exp\left(-\frac{\sqrt{2}}{\sigma_{e_z}}|e_z|\right). \tag{7}$$

In the more complicated case of correlated cross helicity where $\langle \mathbf{v} \cdot \mathbf{b} \rangle \neq 0$, we found that

$$f(e_z) = \frac{1}{\sqrt{2}\sigma_{e_z}} gexp\left(\rho_1, \rho_2, -\frac{\sqrt{2}}{\sigma_{e_z}}|e_z|\right), \tag{8}$$

CP679, *Solar Wind Ten: Proceedings of the Tenth International Solar Wind Conference*, edited by M. Velli, R. Bruno, and F. Malara

where the ρ values provide a measure of the cross correlation. For the particular case of e_z, we have

$$\rho_1 \equiv \frac{\langle v_x b_x \rangle}{\sigma_{v_x}\sigma_{b_x}}, \quad \rho_2 \equiv \frac{\langle v_y b_y \rangle}{\sigma_{v_y}\sigma_{b_y}} \tag{9}$$

Here $gexp$ is a "generalized exponential" defined by

$$gexp(\rho_1,\rho_2,z) \equiv$$
$$\frac{\sqrt{1-\rho_1\rho_2}}{2\pi}\int_0^{2\pi} exp\left(\frac{\Theta_1}{\Theta_2}\sqrt{\frac{1-\rho_1\rho_2}{1-\rho_1^2}}\, z\right)\frac{d\theta}{\Theta_1\Theta_2}, \tag{10}$$

where

$$\Theta_i \equiv \sqrt{1-\rho_i sin(2\theta)}. \tag{11}$$

Equation (8) reduces correctly to equation (7) if $\mathbf{v}$ and $\mathbf{b}$ are uncorrelated.

An important characterization of a distribution is the kurtosis ($\equiv \langle x^4\rangle/\langle x^2\rangle^2$, where $\langle x\rangle = 0$). The inverse of the kurtosis provides a sense of how much the distribution "fills" the space. In particular, a Gaussian distribution has a kurtosis of 3. The $gexp$ function above has a kurtosis between 6 and 9, inclusive depending on the values of ρ_i, with 6 being obtained in the limit of uncorrelated velocity and magnetic fields.

We now wish to test our theory by comparing it to direct observations in the solar wind.

OBSERVATIONS

Before we begin with the observations, it should be noted that, to this point, we have assumed that the $\langle\rangle$ is an "appropriate" ensemble averaging operator with no further restrictions on it. We now define two averaging operators that we will use throughout the rest of the paper. The first is

$$\langle x\rangle_I = \frac{1}{N(I)}\sum_{i=1}^{N(I)} x_i \tag{12}$$

which is the typical averaging of data for a single interval with $N(I)$ points in it. The second averaging operator is

$$\langle f\rangle = \sum_{I=1}^{N_{int}} \frac{N(I)}{N_{tot}} < f >_I \tag{13}$$

which is a weighted average of the quantity f calculated across N_{int} intervals. N_{tot} is the total number of data points available.

We examine data from the NSSDC Omnitape dataset (King and Papitashvili [10]) which consists of 1 hour magnetic and velocity field averages from over 30 years at 1 AU. The data is broken down into 4 day intervals, which is much longer than a typical correlation time in the solar wind, and much shorter than the solar rotation period (about 28 days). Furthermore, the mean fields can be determined reasonably well over 4 days and there is a good chance that the interval will not have sections consisting of different statistical properties. Earlier studies [7, 11] have shown that averaging intervals on the order of four days give reasonable and stable results for many solar wind fluctuation properties of interest.

Omnitape has gaps in its data. Furthermore, in some data points, there may be a measure of $\mathbf{B}$, but not the corresponding measure of $\mathbf{V}$ (or vice-versa). These points must be rejected. Finally, we avoid satellite sector crossings by performing sector rectification and allowing only data from the away sector. It should be noted that the toward sector could also be used. There are no significant differences in our results from using the toward sector.

Now we convert $\mathbf{B}$ into Alfvén speed units, $\mathbf{B} \rightarrow \mathbf{B}/\sqrt{4\pi\rho}$ where ρ is the four-day measured mean proton number density. Throughout the remainder of the paper, both velocity and magnetic field will have units of km/sec and the electric field will be in km^2/sec^2. The means of $\mathbf{B}$ and $\mathbf{V}$ are calculated for each interval. Following that, coordinates are transformed into mean field coordinates [12]. For the rest of the paper, the "z" direction corresponds to the direction of the mean magnetic field, with "x" and "y" directions being perpendicular to "z" in a right handed coordinate system.

We are now able to calculate $\delta\mathbf{e}$ according to equation (5). We find that a typical value for $\delta\mathbf{e}$ is 1930 km^2/sec^2. Furthermore, we find that the mean of the components of $\langle\mathbf{v}\times\mathbf{b}\rangle$ are 11, -3.8, and 18.8 (km^2/sec^2), and the r.m.s. values for these components are 195, 386 and 288 (km^2/sec^2). Since the means are small we have, as a consequence, $\delta\mathbf{e} \approx \mathbf{e}$. Thus, we can compare our theory, which was only for $\mathbf{e}$, to the observations, $\delta\mathbf{e}$, with negligible error. For the rest of this paper, we will use $\mathbf{e}$, rather than $\delta\mathbf{e}$.

In order to calculate the PDF of the observed data, we first normalize each of the fields so that each component has a variance of 1. This is a rather drastic step that requires justification. During our investigation, we found that the magnetic field on each interval was roughly Gaussian, but with varying widths. This is consistent with other work (see, for instance, [11] and references therein). We found similar behavior for the velocity field. This presents us with a problem as we would like to compare the theory with the entire 30 year data set and not just an interval's worth. Yet, if all the data were combined together, then a skewed distribution would result since adding two Gaussian distributions together can yield a non-Gaussian distribution. Thus, we choose to normalize the fields to unit variance. Others, such as Sorriso-Valvo et al. [13] have also used this normalization. As will be seen later, the importance of this normalization is that it reveals the underlying Gaussian nature of the velocity and magnetic fields. Finally, it should be noted that this normalization is done to each of the fields separately. In particular, **e** is first calculated from the raw data for **v** and **b** and then normalized, rather than calculated from normalized **v** and **b**.

Figures 1 and 2 show the PDFs for the z component of the velocity and magnetic fields, respectively. For illustration purposes, the dashed line shows a Gaussian distribution using the same half-width calculated from the data. The components of **v** and **b** typically are not very far from Gaussian distributions. The kurtosis values are 2.6 and 3.4 for v_z and b_z, respectively. A true Gaussian distribution would have a kurtosis of 3.0.

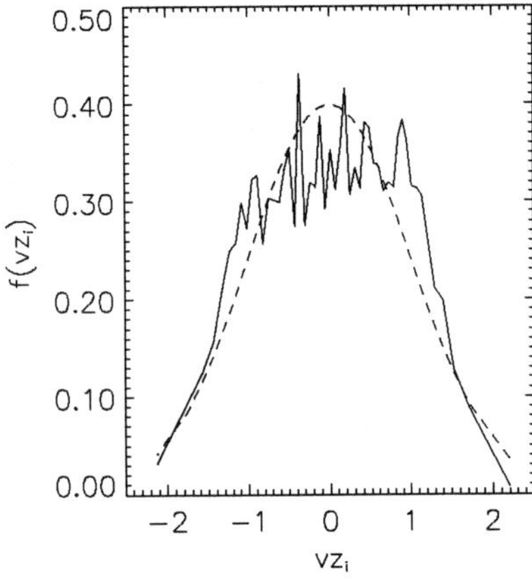

FIGURE 1. PDF for z-component of the velocity field. The dashed curve is a Gaussian distribution with the same σ value.

Finally, figures 3, 4, and 5 show the PDF for the x, y and z components of **e**, respectively. The dashed curve is eq. (8). Note the remarkable agreement between theoretical and observed distributions. The kurtosis values for these distributions are 7.9, 5.9 and 7.2, respectively. The theory predicts kurtosis values between 6 and 9, inclusive, where 6 is obtained in the limit that *gexp* tends to an exponential.

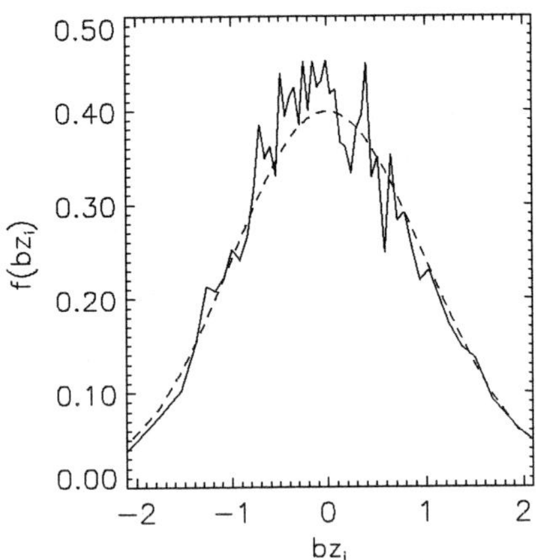

FIGURE 2. PDF for z-component of the magnetic field. The dashed curve is a Gaussian distribution with the same σ value.

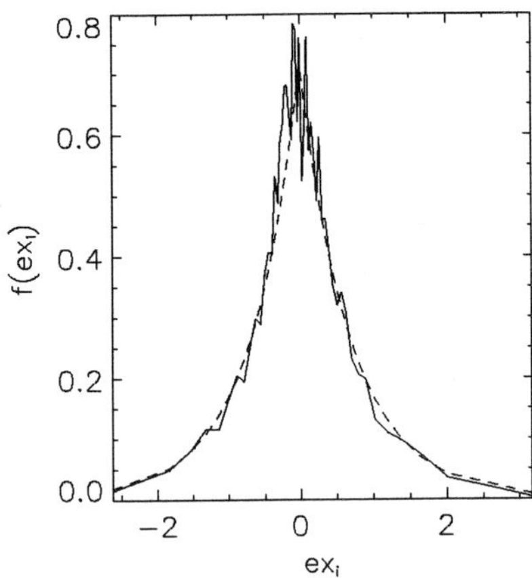

FIGURE 3. Distribution function of the x-component of induced electric field computed from normalized data. The dashed curve is eq. (8).

CONCLUSIONS

We have presented analytical expressions for the PDF of the turbulent electric field in MHD turbulence. In particular, assuming the underlying velocity and magnetic fields are given by Gaussian distributions, the resulting electric field distribution is given by *gexp*, a function with exponential like behavior.

Excellent agreement is found between the theoretical expressions and observed data at 1 AU, using four day intervals to compute the means. There is of course the possibility that the statistical prop-

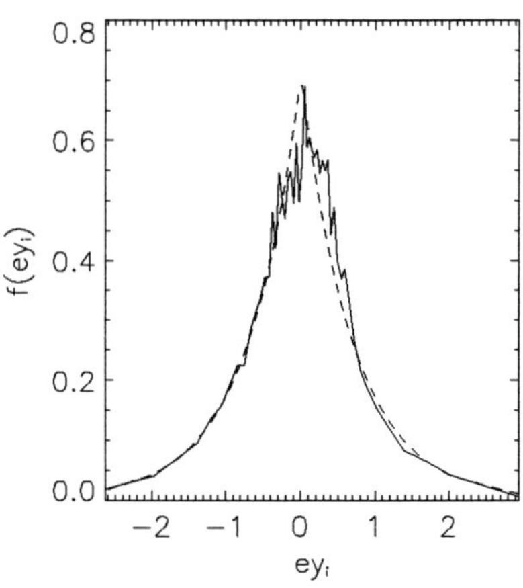

FIGURE 4. Distribution function of the y-component of induced electric field computed from normalized data. The dashed curve is eq. (8).

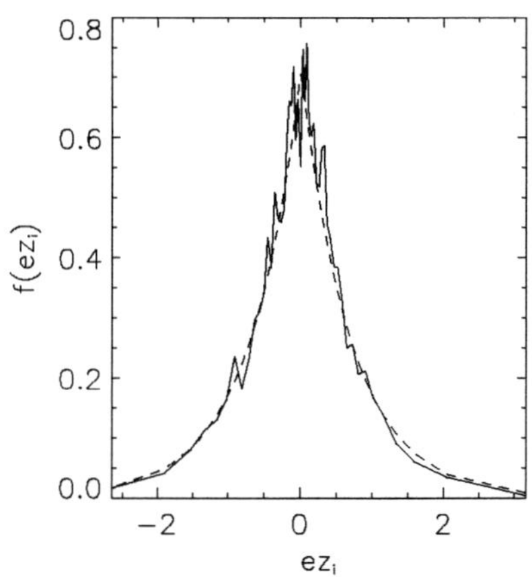

FIGURE 5. Distribution function of the z-component of induced electric field computed from normalized data. The dashed curve is eq. (8).

erties of the electric field will be different in other ranges of space and time scales, and we defer such studies to future efforts along these lines. Induced electric field properties may also vary with heliocentric distance. As the Alfvénic correlations decrease with increasing heliocentric distance Roberts et al. [14] the relative strength of induced electric field is expected to increase.

ACKNOWLEDGMENTS

This work was supported by NASA Sun Earth Connection Guest Investigator program grant NAG5-10911 and NAG5-8134, NAG5-6570 and by NSF grant ATM-0105254. to the Bartol Research Institute.

REFERENCES

1. Jokipii, J. R., *Rev. Geophys. Space Phys.*, **9**, 27 (1971).
2. Goldstein, M. L., Roberts, D. A., and Matthaeus, W. H., *Ann. Rev. Astron. Astrophys.*, **33**, 283 (1995).
3. Tu, C.-Y., and Marsch, E., *Space Sci. Rev.*, **73**, 1 (1995).
4. Matthaeus, W. H., Zank, G. P., Smith, C. W., and Oughton, S., *Phys. Rev. Lett.*, **82**, 3444–3447 (1999).
5. Bieber, J. W., Matthaeus, W. H., Smith, C. W., Wanner, W., Kallenrode, M., and Wibberenz, G., *Astrophys. J.*, **420**, 294 (1994).
6. Droege, W., *Space Sci. Rev.*, **93**, 121 (2000).
7. Matthaeus, W. H., and Goldstein, M. L., *J. Geophys. Res.*, **87**, 6011 (1982a).
8. Marsch, E., and Tu, C. Y., "Electric Field Fluctuations and Possible Dynamo Effects in the Solar Wind," in *Proceedings of Solar Wind 7, COSPAR Colloq. Ser.*, edited by E. Marsch and R. Schwenn, Pergamon, Oxford, UK, 1992, vol. 3, p. 505.
9. Milano, L., Matthaeus, W. H., Breech, B., and Smith, C. W., *Phys. Rev. E*, **65** (2002).
10. King, J., and Papitashvili, N., *Interplanetary Medium Data Boook - Supplement 5, 1988-1993*, (Rep. NSSDC/WDC-A-R&S 94-08, NASA, Greenbelt, MD, 1994.
11. Padhye, N., Smith, C. W., and Matthaeus, W. H., *J. Geophys. Res.*, **A106**, 18635 (2001).
12. Belcher, J. W., and Davis Jr., L., *J. Geophys. Res.*, **76**, 3534 (1971).
13. Sorriso-Valvo, L., Carbone, V., Veltri, P., Consolini, G., and Bruno, R., *Geophysical Research Letters*, **26**, 1801–1804 (1999).
14. Roberts, D. A., Goldstein, M. L., Klein, L. W., and Matthaeus, W. H., *J. Geophys. Res.*, **92**, 12 023 (1987b).

Cross Helicity Correlations in the Solar Wind

S. Dasso*, L.J. Milano†, W.H. Matthaeus† and C.W. Smith†

*Instituto de Astronomía y Física del Espacio (IAFE), Buenos Aires, Argentina.
Instituto de Ciencias, Universidad de Gral. Sarmiento, Los Polvorines, Buenos Aires, Argentina.
On leave from Instituto de Física del Plasma (INFIP), Buenos Aires, Argentina.
† Bartol Research Institute, University of Delaware, USA

Abstract. Over the last decade, several magnetohydrodynamic models of the solar wind proposed a two component structure for the fluctuations: a "slab" (Alfvénic) component with wavenumbers parallel to the ambient dc magnetic field and a quasi two-dimensional (turbulent) component with wavenumbers mostly perpendicular to the magnetic field. Initial support and motivation for these models was given in part from the study of three dimensional correlation functions for the magnetic field from solar wind data (W.H Matthaeus, M.L. Goldstein and D.A. Roberts 1990, JGR 95, 20673). We extend here this study to the analysis of the cross-correlation between the velocity and the magnetic field. The cross-correlation function is simply related to the cross helicity power spectrum, a quantity of great interest for solar wind models. This quantity provides, on one hand, a measure of the relative importance of outgoing and incoming Alfvénic fluctuations. On the other hand, the turbulent properties of the system are greatly influenced by the amount of cross helicity present in it. We analyze ACE data and present preliminary results for the three dimensional cross-correlation function. Special emphasis is given to the implications for solar wind models.

INTRODUCTION

The solar wind is a privileged scenario where in-situ observations unveil many aspects concerning magnetohydrodynamic (MHD) turbulence in a magnetized plasma. Over the last two decades, much progress has been made by several authors by means of spacecraft observations of solar wind turbulence (see for instance [1] and references therein).

As opposed to the hydrodynamic (HD) case, two dynamic fields (velocity and magnetic fields) interplay to determine the evolution of an incompressible MHD turbulent system. A new rugged invariant, namely the cross helicity (or cross correlation between these two fields), gets into the scene. A useful dimensionless expression for this quantity is the normalized cross helicity (σ_c), defined as the ratio of the cross helicity to the total energy, and ranging from -1 to 1. Monopropagating Alfvén waves have maximum $|\sigma_c|$ ($\sigma_c = \pm 1$, depending on the sense of propagation). The cross helicity can be obtained as the difference of the energy of outgoing to incoming Alfvén waves, thus giving a measure of the imbalance between the two. The structure of the incompressible MHD equations is such that the non-linear terms vanish when the cross helicity is maximum. To develop turbulence it is necessary to have counterpropagating fluctuations along the mean magnetic field. These fluctuations are thought to interact non-linearly to produce an energy cascade in perpendicular wavenumbers (see Dmitruk et al. [2] and references therein). High levels of turbulence in the solar wind are usually accompanied by a value of σ_c close to zero (e.g., see [3] and references therein).

The normalized cross helicity and its spectrum have been determined from single-point measurements, and show the dominance of outgoing Alfvénic fluctuations [4]. However, there is to present no observational study of possible anisotropy in the solar wind cross helicity. The solar wind magnetic fluctuations are not isotropic, and strong evidence of the presence of two populations have been provide in [5] and [6]. The former shown the presence of: (a) Alfvénic (slab) fluctuations with correlation lengths stretched in the direction transverse to the background field (B_0) and (b) quasi-two dimensional (turbulent) fluctuations with elongated correlation lengths parallel to B_0.

We present hereafter our preliminary efforts to analyze possible anisotropy of the normalized cross helicity (σ_c), in the same spirit of the analysis of anisotropy in magnetic field self correlations performed in Ref.[4]. The second section describes the technique we use to process the interplanetary data in order to calculate two-dimensional correlation functions. The results are summarized in third section. Finally, we conclude in the last section.

CP679, *Solar Wind Ten: Proceedings of the Tenth International Solar Wind Conference*,
edited by M. Velli, R. Bruno, and F. Malara

DATA PROCESSING

We analyze magnetic and bulk velocity fields measured by the Advanced Composition Explorer (ACE) spacecraft, from January 23, 1998 to March 3, 1999. The data have been analyzed with a cadence of one minute. The solar wind observations we analyze here correspond to a distance of $\sim$ 1 AU from the Sun, and essentially on the ecliptic plane.

We group our whole set of data in 4-day intervals, thus obtaining N sub-series (or intervals). For every interval I ($I = 1,...,N$) and from the observed magnetic ($\mathbf{B}^I_j$) and velocity ($\mathbf{V}^I_j$) fields, we define the fluctuation fields ($\mathbf{b}^I_j$, $\mathbf{v}^I_j$, and the Elsässer variables $\mathbf{z}^{I,\pm}_j$), where the index j labels the time t^I_j from the beginning of every sub-series I (i.e., $t^I_0 = 0$), as follows:

$$\mathbf{v}^I_j = \mathbf{V}^I_j - \mathbf{U_0}^I$$
$$\mathbf{b}^I_j = \frac{\mathbf{B}^I_j}{\sqrt{4\pi\rho^I}} - \mathbf{V_A}^I$$
$$\mathbf{z}^{I,\pm}_j = \mathbf{v}^I_j \pm \mathbf{b}^I_j \qquad (1)$$

Here, $\mathbf{U_0}^I = \langle \mathbf{V}^I_j \rangle$ and $\mathbf{V_A}^I = \langle \mathbf{B}^I_j \rangle / \sqrt{4\pi\rho^I}$ are respectively the time average for the plasma velocity and for the Alfvén velocity, within the interval I; ρ^I is the mean density of mass for the interval I.

In order to compute statistics from the observed data, we need to normalize the fluctuating fields so that the amplitude of the fluctuations in the different intervals be comparable. This is a rather drastic step that requires justification. It has been established that different time intervals in the solar wind have similar statistics, but when comparing one interval to any other there is usually a scaling factor relating the amplitude of the fluctuation in one with respect to the other. For example, the magnetic field on each interval usually is roughly a Gaussian variable, but with varying widths (see, for instance, [7] and references therein). Yet, if all the data were combined together, then a skewed distribution would result since adding two Gaussian distributions together can yield a non-Gaussian distribution. Thus, we choose to normalize the fields so that the energy in each interval is the same, and it is equal to the mean energy (kinetic plus magnetic) of the whole (raw) dataset E. That is, if E^I is the mean energy of the (raw) data in the interval I, we rescale the fluctuating fields ($\mathbf{v}$ and $\mathbf{b}$) with a similarity factor $\lambda^I = \sqrt{(E/E^I)}$. Others, such as Sorriso-Valvo et al. [8] have also used similar normalization schemes.

The two-point velocity correlation function is defined as

$$R_{vv}(\mathbf{r}) = \langle \mathbf{v}(\mathbf{x}) \cdot \mathbf{v}(\mathbf{x}+\mathbf{r}) \rangle \qquad (2)$$

Analogous definitions hold for R_{bb}, R_{vb}, and for the correlations in the Elsässer variables: R_{++} and R_{--}.

The ACE spacecraft provides time series of velocity and magnetic field, thus the correlation functions constructed from these data are essentially two-time single-point. However, due to the fact that the mean speed ($\mathbf{V_{sw}}$) of the solar wind is super-Alfvénic, it is possible calculate the spatial correlation functions from the measured temporal fluctuations using the relationship $R(0,t) = R(-\mathbf{V_{sw}}t, 0)$ [4]. These approximations are the MHD analogues of the Taylor 'frozen-in-flow' hypothesis [9].

For a given interval I, the mean speed of the solar wind $\mathbf{U_0}^I$ gives the direction of the lag $\mathbf{r}$, which is almost along the radial direction as measured from the Sun. So, we calculate $R^I(r)$, where r is a distance along $\mathbf{U_0}^I$. An in-house numerical code, which employs the 'Blackman-Tukey' technique [10] was used to calculate the different correlation functions $R^I(r)$. The maximum lag taken, when $R^I(r)$ is calculated, corresponds to two days.

In order to analyze the anisotropy of the fluctuations, we label each interval according to the value of the angle (θ^I) between the direction of the mean field ($\mathbf{V_A}^I$) and $\mathbf{U_0}^I$, and study variations in several statistical quantities as a function of θ, as shown below.

As mentioned before, the Elsässer variables give information on the level of the activity of waves traveling either parallel or anti-parallel to the background magnetic field. To give physical meaning to our analysis, we have grouped the fluctuations according to whether they are traveling outwards from the Sun ("out"), or towards the Sun ("in"), and consistently re-labeled the Elsässer variables as $\mathbf{z}_{out}$ and $\mathbf{z}_{in}$ in each interval. The reduced energy spectra for kinetic and magnetic energy ($E_v(k)$ and $E_b(k)$) , and for the Alfvén waves activity 'outward' ('inward') propagating, $E_{out}(k)$ ($E_{in}(k)$), are obtained by means of a Fast Fourier Transform of the corresponding correlation functions [3]. The spectra are normalized to give the total energies as usual:

$$E_{\{v,b\}} = \frac{1}{2}\langle \{v,b\}^2 \rangle = \int E_{\{v,b\}}(k)dk \qquad (3)$$

$$E_{\{in,out\}} = \frac{1}{4}\langle \{z_{in}, z_{out}\}^2 \rangle = \int E_{\{in,out\}}(k)dk \qquad (4)$$

From these energy spectra, the reduced cross helicity spectrum (in the direction of the mean wind velocity) can be obtained from

$$H_c(k) = (E_{out}(k) - E_{in}(k))/2, \qquad (5)$$

and the normalized cross helicity from

$$\sigma_c(k) = \frac{E_{out}(k) - E_{in}(k)}{E_{out}(k) + E_{in}(k)}. \qquad (6)$$

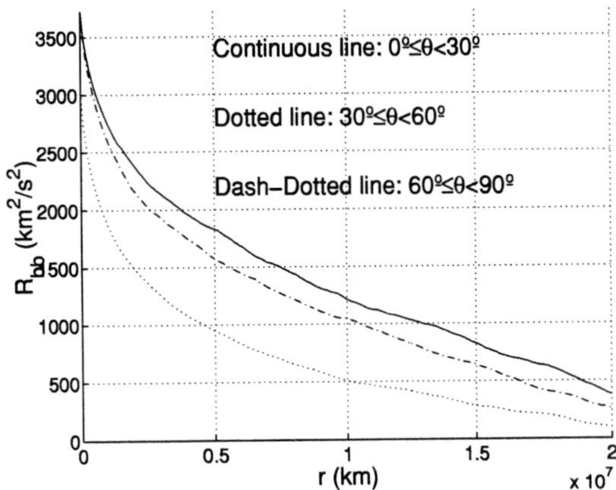

FIGURE 1. Conditioned average of the magnetic field self correlation function $R_{bb}(r)$, in Alfvén units. The averages have been conditioned to that intervals of the whole dataset where the value of the angle θ between the direction of the mean field $\mathbf{V}_A$ and mean wind velocity $\mathbf{U}_0$ is in a selected range. Continuous, dotted, and dash-dotted lines correspond to $0 < \theta < 30, 30 < \theta < 60$, and $60 < \theta < 90$, respectively. The larger scale shown corresponds to ~ 0.1 AU.

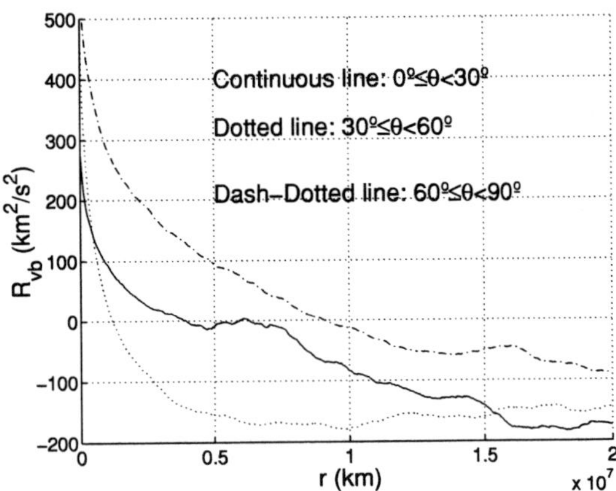

FIGURE 2. Cross helicity correlation function $R_{vb} = (R_{out} - R_{in})/4$. Different curves correspond to different ranges of θ as in Figure 1. Correlation decay more slowly in the direction perpendicular to the mean field.

RESULTS

In order to study spectral anisotropy (or alternatively spatial anisotropy for the correlation functions) we define three ranges for θ. The chosen ranges for θ and the number of intervals that correspond to every range are:

- $0 \leq \theta < 30$ (10 intervals),
- $30 \leq \theta < 60$ (53 intervals),
- $60 \leq \theta < 90$ (24 intervals).

Thus, from the correlation functions of every interval, $R^I(r)$, we carry out conditional averages considering only those intervals which correspond to a given range of θ values obtaining $R(r)$.

Figure 1 shows the conditional average of the magnetic field self correlation function according to θ. It is evident that the curve with steeper correlation corresponds to the oblique direction ($30 \leq \theta < 60$), in full consistency with Figure 3 of [4]. The results shown in this figure support the two component (slab + 2D) model.

The global degree of correlation between **v** and **b** is measured by the cross helicity $H_c = R_{vb}(r = 0)/2$. Considering the whole dataset, a value of $H_c \sim 230\text{km}^2/\text{sec}^2$ is obtained. The total (kinetic plus magnetic) mean energy resulted $\sim 3\text{x}10^3\text{km}^2/\text{sec}^2$. These numbers yield a normalized global cross helicity value of $\sigma_c = 2H_c/E \sim 0.15$. The two-point cross helicity correlation $R_{vb}(r)$ is shown in Figure 2 for the three different ranges of the angle θ. The figure seems to indicate that the correlations along the intermediate direction decay fastest than the other two components do. An alternative view of the angular distribution of cross correlations is given in figure 3. The figure shows the (reduced) normalized cross helicity power spectrum $\sigma_c(k)$ along different directions (the angle ranges defined above), what allows the analysis of different scales at different angles. Note that at intermediate angles ($\theta \sim 45$), for wave numbers larger than $5\text{x}10^{-7}$ km^{-1}, $\sigma_c(k)$ is larger than for the other two directions (perpendicular and along the mean field), what seems in contradiction with the picture of the two populations: the Alfvenic-slab and the turbulent quasi-2D. However we note that our statistics are still too low at extreme angles as to make any definite conclusions as yet. The opposite occurs at large scales ($k < 5\text{x}10^{-7}$ km^{-1}). It is important to stress the preliminary character of these results, what forces us to be cautious and avoid any physical interpretation until these results are either confirmed or corrected with a more complete analysis (see next section). Finally, figure 4 shows the Alfvén ratio spectrum $r_A(k)$. The three curves for different θs are very close to each other, and overall they seem consistent with the values reported in Table III of Ref.[3].

SUMMARY AND CONCLUSIONS

We present preliminary results from a study of anisotropy in the velocity, magnetic, and cross helicity correlation functions (also power-spectra) by considering spatial lags (wave-vectors) at different angles θ with respect to the background magnetic field $\mathbf{B}_0$.

The magnetic self correlations are consistent with previously published results, supporting the two component model of the solar wind. That is, the presence of two pop-

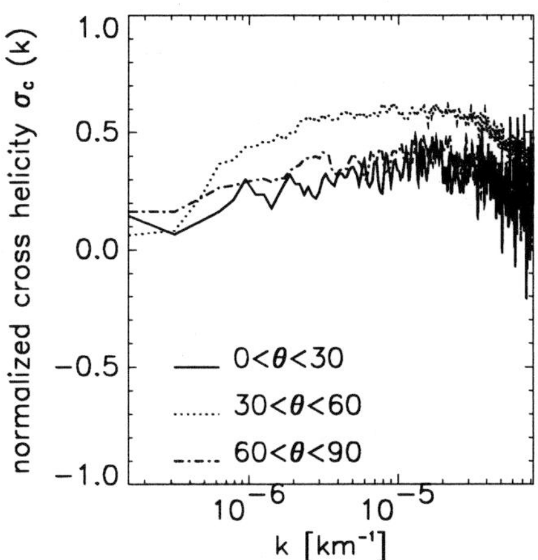

FIGURE 3. Plot of the conditioned average of the normalized cross helicity ($\sigma_c(k) = 2H_c(k)/E(k)$) power spectrum. The average has been conditioned for intervals with different values of θ, as in Figure 1.

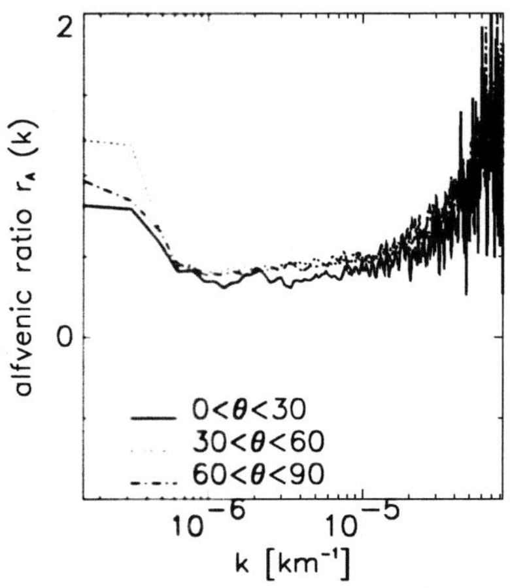

FIGURE 4. Figure showing the reduced spectrum of the Alfvén ratio, $r_A(k) = E_v(k)/E_b(k)$. Different curves correspond to different θ angle ranges, as in Figure 1.

ulations: a "slab" (or Afvénic) population aligned with the main magnetic field and with wavenumbers parallel to it, and a "quasi-2D" (or turbulent) population with almost perpendicular wavenumbers, as it is typical of anisotropic turbulence in the presence of a mean magnetic field.

We have also presented the methodology, techniques and some preliminary results of the study of angular dependence of the power spectrum (or alternatively its self correlation function) for very relevant quantities such as the dimensional and normalized cross helicity and the Alfvén ratio. The progress of our research at this point is still that of an early stage. To achieve more dependable results we need to: (a) extend our temporal base to a larger dataset, where more intervals with extreme angles $\theta \sim 0$ and $\theta \sim 90$ can be found; (b) consequently include more bins in our angular discretization (currently we can only use 3 because of lack of statistics and for simplicity in the analysis); (c) implement a stationary test and exclude data intervals with sector crossings; (d) study the noise we see in our spectra at very high k (close to the Nyquist frequency).

It is for all of these reasons that we resist the temptation of drawing any major conclusions out of the present preliminary analysis. We plan to extend our research in a timely fashion and publish our final results elsewhere. In the meantime, we believe that the physics involved in this research project are worth the effort, and will help achieve a better of the nature of the solar wind and MHD turbulence.

ACKNOWLEDGMENTS

SD acknowledges partial support by the Argentinian UBA grant UBACYT X059. SD is a fellow of CONICET. WHM and LJM acknowledge support by the NSF (ATM 0105254) and NASA (NAG5-8134). ACE data was provided by the ACE Science Center. CWS is supported by JPL contract PC251459 under NASA grant NAG5-6912 for support of the ACE/MAG instrument. This research has made use of NASA's Astrophysics Data System.

REFERENCES

1. Goldstein, M. L., Roberts, D. A., and Matthaeus, W. H., ARA&A, **33**, 283–326 (1995).
2. Dmitruk, P., Milano, L. J., and Matthaeus, W. H., ApJ, **548**, 482–491 (2001).
3. Tu, C.-Y., and Marsch, E., *MHD structures, waves and turbulence in the solar wind: observations and theories*, Dordrecht: Kluwer, |c1995, 1995.
4. Matthaeus, W. H., and Goldstein, M. L., J. Geophys. Res., **87**, 6011–6028 (1982).
5. Matthaeus, W. H., Goldstein, M. L., and Roberts, D. A., J. Geophys. Res., **95**, 20673–20683 (1990).
6. Carbone, V., Malara, F., and Veltri, P., J. Geophys. Res., **100**, 1763–1778 (1995).
7. Padhye, N. S., Smith, C. W., and Matthaeus, W. H., J. Geophys. Res., **106**, 18635–18650 (2001).
8. Sorriso-Valvo, L., Carbone, V., Veltri, P., Consolini, G., and Bruno, R., Geophys. Res. Lett., **26**, 1801–+ (1999).
9. Taylor, G., "The Spectrum of the turbulence," in *Proc. R. Soc. London Ser. A, 164*, 1938, pp. 476+.
10. Blackman, R., and Tukey, J., *Measurements of power spectra*, Dover, Mineola, NY, |c1958, 1958.

Castaing Scaling and Kernels for Theories of the Turbulent Cascade

Miriam A. Forman

Department of Physics and Astronomy
State University of New York at Stony Brook NY11794-3800 USA

Abstract. The Castaing cascade kernel function G(u;L,L') relates the pdf of velocity fluctuations on scale L' to those on scale L. The shape of G(u;L,L') and how it depends on scale is a measure of intermittency alternative to measures such as the exponent of the structure function or multifractal spectra. This paper provides analytical forms and graphic illustrations of the Castaing kernels corresponding to six models of the turbulent cascade. The popular P-model and Poisson models of turbulence have discrete Castaing kernels, which seems non-physical.

INTRODUCTION

Castaing scaling [1,2,3,4] describes the cascade of turbulent energy in a fluid from scale L to smaller scale L' by relating the probability distribution functions (pdfs) of the velocity increments δV at the different scales with a cascade kernel G(u:L',L) such that

$$P_{L'}(\delta V) = \int_{-\infty}^{+\infty} G(u;L',L) e^{-u} P_L\left(e^{-u}\delta V\right) du \cdot \tag{1}$$

The kernel G(u;L',L) represents the probability that turbulence of amplitude $e^{-u}\delta V$ on scale L, cascades to turbulence of amplitude δV on scale L'. Castaing scaling is linear in δV, in the sense that the same G(u) relates $e^{-u}\delta V$ at the larger scale to δV at the smaller scale, for every $-\infty < \delta V < \infty$. Since Castaing kernels directly describe the distribution of turbulent amplitudes at larger scales that contribute to a given amplitude at small scales, they provide a graphic description of the cascade. Multifractality and intermittency appear in the cascade when the kernel G(u) is anything other than a delta-function. The details depend on the shape of G(u; L',L) versus u, and its dependence on L'.

If both G(u) and $P_L(\delta V)$ are Gaussian functions of their arguments, $P_{L'}(u)$ will be a winged "Castaing distribution" characterized by the mean and width of G(u). Such distributions can be conceptually useful particularly for looking at the dependence of the parameters on scale [5, 11, 13]. However, neither G(u) nor $P_L(u)$ can be precisely Gaussian; if they were, the scaling would be log-normal, and the third moment of the parallel velocity increment would be zero, both of which are refuted fundamentally by Frisch [6]. The Castaing pdf is only approximate and should be used with care. This paper deals only with Castaing *scaling* which is much more general.

RELATION OF CASTAING KERNEL G, TO THE STRUCTURE FUNCTION ζ

Following equation 1, the scaling properties of all the moments of the pdf arise from the scaling of G(u;L,L). In particular, since the moment of order q at scale L' is

$$S(q,L') = \int_{-\infty}^{+\infty} d(\delta V)(\delta V)^q \int_{-\infty}^{\infty} G(u;L',L) e^{-u} P_L\left(e^{-u}\delta V\right) du \tag{2a}$$

$$= \left[\int_{-\infty}^{\infty} G(u;L',L) e^{qu} du\right] S(q,L) \tag{2b}$$

all of the dependence on L' (in other words, the

CP679, *Solar Wind Ten: Proceedings of the Tenth International Solar Wind Conference,*
edited by M. Velli, R. Bruno, and F. Malara

scaling) is contained in G(u:L',L). Thus when the scaling $\zeta(q) = \dfrac{d\,ln\,S(q,L')}{d\,ln\,L'}$ occurs on scales from L' to L, (2) implies that the Fourier transform of G(u;L',L) is equal to [4]

$$\hat{G}(k;L',L) = exp[-ln(L/L')\zeta(-ik)] \quad . \tag{3}$$

Scaling in actual data sets is usually described with ζ(q), and turbulence theories with a mathematical form for ζ(q). In practice ζ(q) can be calculated from finite and imperfect real data sets only over a limited range of q, because very large and very small δV will be missing. If (1) applies, the *scaling is completely subsumed into the kernel G, and does not involve the pdf at large scale*. Also, the scaling of the moments of the absolute value of δV is the same as the scaling of the moments of δV for integer q ≥ 0. The odd moments of the absolute values are not the same as the odd moments of δV itself, *but they scale the same way* if equation (1) is valid. The linear form of (1) assures that moments of the absolute value of δV at odd, non-integer and negative q provide additional useful information about G(u;L',L).

Although the idea of Castaing scaling underlies turbulence theory already, it is instructive to examine the actual kernels G(u;L',L) for popular and reasonable models of the turbulent cascade, to better appreciate the differences between the models, and to encourage further turbulence theories to describe themselves with their kernels.

COMPARISON OF CASTAING KERNELS FOR TURBULENCE MODELS

Table 1 lists in historical order seven models of inertial turbulence, with their ζ(q) and G(u;L',L), minimum scaling exponent, and the dimension on the minimum scaling exponent. All models have ζ(3) =1 in accord with the Kologorov 4/5 law [6] for inertial turbulence. G(u;L',L) was calculated by inverse Fourier transform of eq. (3) [4]. The minimum scaling exponent and dimension on the minimum scaling exponent were calculated from ζ(q) by the method given in [6].

Intermittent turbulence implies a Castaing kernel of finite width; equivalently, a ζ(q) with negative curvature. The Kolmogorov-Obkuhov 1962 (KO62) form is the simplest extension of the K41 delta function, to a normal distribution in u. Its ζ(q) is the unique parabola which has ζ(3) = 1, and ζ(q)⇒q/3 when μ⇒0. Some experts [e.g., 6] dismiss this model in incompressible fluid turbulence because the ζ(q) should not have a peak and its slope should be non-negative at all q. A peculiarity of the KO62 kernel is that at positive u it is non-zero (although small) and increases with scale. This allows a non-zero probability for arbitrarily large fluctuations at smaller scales to arise from smaller-amplitude fluctuations at larger scales. These dominate the moments of large positive order, and confuse their scaling. Whether this is also strictly impossible in compressible and/or magnetized fluids such as the solar wind is unclear. Solar wind velocity increments do have ζ(q) with negative slopes in some scale ranges at which the intermittency is very large [7].

The binomial and Poisson models [8,9,10] avoid the negative slopes in ζ(q) and, since their kernels are zero at u>0, never let small δV at large scale lead to larger δV at smaller scales. Of these, DeBrulle's [10] 2-parameter generalized Poisson model is the most flexible, and can have $D(h_{min}) = 2$ (Δ = 0.4, β = 0.6 works well). However, the binomial and Poisson kernels are uncomfortably discontinuous sums of delta functions. In the binomial and Poisson models, δV at small scale can arise only from certain discrete values of δV at larger scale, which is implausible.

The Universal Multifractal model [13] is a generalization of the KO62 model to a kernel that is a continuous Levy function of u, near the most probable values of u. Crucial truncations at extreme values of u arise from plausible self-organized criticality and limited data sets, preventing the slope of ζ(q) from becoming negative. Such modifications could be applied to the KO 1962 model as well.

The intermittent models in Table 1 produce similar-looking ζ(q) for |q-3|<5 where ζ(q) might be reliably estimated from data.

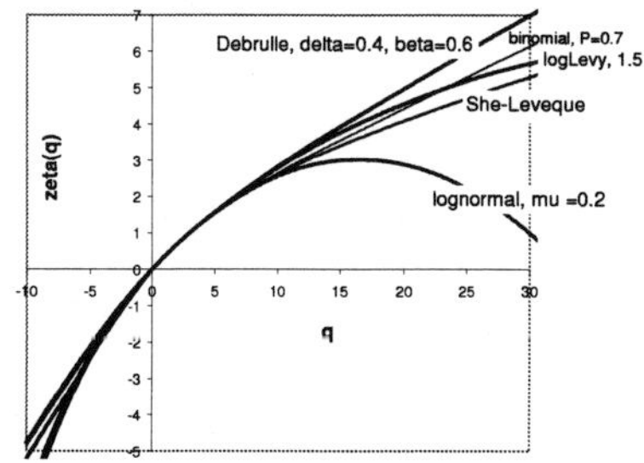

FIGURE 1. Exponents of the structure function for six models of the intermittent turbulent cascade. Parameters for all models were chosen such that ζ(4)/4 -ζ(2)/2 = 0.022.

TABLE 1. $\zeta(q)$ and $G(u)$ for models of the turbulent cascade

Turbulence model	$\zeta(q)$ Exponent of structure function $\langle V_L^{q}\rangle \propto L^{\zeta(q)}$	$G(u,\eta)$ Castaing kernel for $L_0 > L$ $u \equiv \ln(\delta V_L/\delta V_{L0})$ $\eta \equiv \ln(L_0/L)$	h_{min}	and $D(h_{min})$
Kolmogorov 1941	$\frac{q}{3}$	Delta function: $\delta(u + \eta/3)$	$\frac{1}{3}$	3
Kolmogorov-Obukhov 1962 (log-normal)	$\frac{q}{3}+\frac{\mu}{2}\left[\frac{q}{3}-\left(\frac{q}{3}\right)^2\right]$ $\mu \approx 0.2$	Gaussian $\frac{3}{\sqrt{2\pi\mu\eta}}\, exp\left\{-\frac{9}{2\mu\eta}\left[u+\frac{\eta}{3}\left(1+\frac{\mu}{2}\right)\right]^2\right\}$	$-\infty$	NA
Meneveau and Sreenivasan's P-model 1991 (binomial)	$1-log_2\left[P^{\frac{q}{3}}+(1-P)^{\frac{q}{3}}\right]$ $.5<P<1$	binomial $\frac{1}{2^n}\sum_{0}^{n}\delta\left(u-\frac{m}{3}ln(P)-\frac{n-m}{3}ln(1-P)\right)\frac{n!}{(n-m)!\,m!}$ for $(L_0/L) = 2^n$	$\frac{1}{3}log_2\left(\frac{1}{P}\right)$	2
She and Leveque 1994 (Poisson)	$\frac{q}{9}+2\left[1-\left(\frac{2}{3}\right)^{\frac{q}{3}}\right]$	Poisson $\sum_{m=0}^{\infty}\delta\left(u+\frac{\eta}{9}+m\frac{ln1.5}{3}\right)\frac{(2\eta)^m}{m!}e^{-2\eta}$	$\frac{1}{9}$	1
De Brulle 1994 (Poisson)	$\frac{1-\Delta}{3}q+\frac{\Delta}{1-\beta}\left(1-\beta^{\frac{q}{3}}\right)$ $0<\Delta,\beta<1$	Poisson $\sum_{m=0}^{\infty}\delta(u+a\eta+mW)\frac{(b\eta)^m}{m!}e^{-b\eta}$ $a=\frac{1-\Delta}{3}$ $W=\frac{ln\frac{1}{\beta}}{3}$ $b=\frac{\Delta}{1-\beta}$	$\frac{1-\Delta}{3}$	$3-\frac{\Delta}{1-\beta}$
Shertzer and Lovejoy's (1997) Universal Multifractal (log-Levy)	$\frac{q}{3}+\frac{\mu}{2(\alpha-1)}\left[\frac{q}{3}-\left\lvert\frac{q}{3}\right\rvert^{\alpha}\right]$ $1<\alpha<2$ (most likely 1.5)	Levy function of order α, Scale factor $\gamma=\frac{3^{-\alpha}\mu}{2(\alpha-1)}\eta$ Mean $= -\left[\frac{1}{3}+\frac{\mu}{6(\alpha-1)}\right]\eta$	$-\infty$	NA
S +L's UM + self-organized criticality and finite data set	As UM, except straight line for $q > q_D$	Truncated Levy $h_{min}=\frac{1}{3}+\frac{\mu}{6(\alpha-1)}\left[1-\alpha\left(\frac{q_D}{3}\right)^{\alpha-1}\right]$		$D(h_{min}) =$ $3-\frac{\mu}{2}\left[\frac{q_D}{3}\right]^{\alpha}$

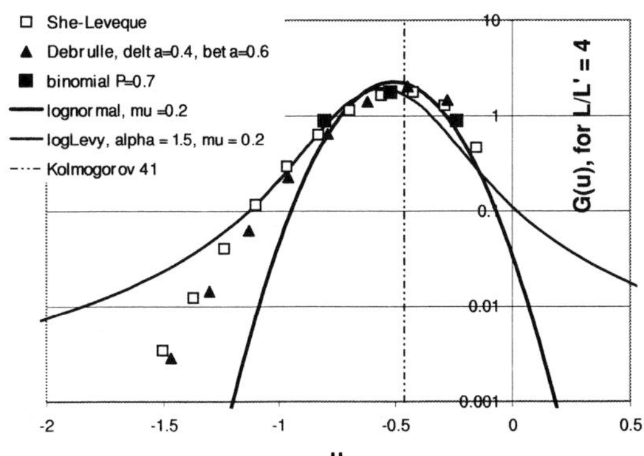

FIGURE 2. Castaing kernels (Eq. 1) for the models and parameters in Figure 1, for a scale L/L' = 4. The formulas in Table 1 were used. The discrete models are shown as isolated values scaled by the Δu between points, to compare with the continuous models. The integral under all curves = 1. The vertical line shows the location of the delta function implied by the original Kolmogorov 1941 theory.

Agreement is good in the central parts of the G(u) for all models, corresponding to $|q-3|<5$, except for the strange discreteness in the binomial and Poisson models. Both (non-truncated) continuous models have wings at u>0, but the logLevy has much bigger wings than the lognormal. Shertzer et al.'s [13] assertion that self-organized criticality will modify $\zeta(q)$ at its extremes, will also modify the wings of the G(u) for the lognormal and logLevy models.

CONCLUSIONS

The Castaing kernels describe the distribution of amplitudes of turbulence on larger scales that give rise to turbulence of a given amplitude on a smaller scale, as a function of the ratio of amplitudes (ln(u)) and scale. The Kolmogorov 1941 theory has a single delta function as a kernel, but theories of intermittent turbulence have more spread-out kernels. The popular binomial (P) model and the Poisson models which have well-behaved structure functions, have discrete kernels that are sums of delta functions, implying that small-scale turbulence of a given amplitude can arise only from certain discrete amplitudes on larger scale. This discreteness seems unphysical. On the other hand, the full Kolmogorov 1962 lognormal theory and its extension to logLevy, have continuous Castaing kernels, but problematic wings and structure functions. The lognormal or logLevy theories, modified by self-organized criticality and effect of finiteness of data sets as proposed by Shertzer, et al. [13], avoids these problems and seems the most acceptable among the classes considered here.

The non-linear processes causing self-organized criticality of extreme fluctuations may make Eq. (1) invalid at very extreme u anyway.

ACKNOWLEDGMENTS

It is a pleasure to acknowledge discussions with Leeonard F. Burlaga and colleagues at NASA/Goddard Spaceflight Center, USA, and the kindness of the Conference organizers. This work was supported by NASA grant NAG 510995.

REFERENCES

1. B. Castaing, Y. Gagne and E. Hopfinger, *Physica D*, *46*, 177 (1990)
2. Chaubaud, A. Naert, J. Peinke, F. Chilla, B. Castaing, and B. Hebral, *Phys. Rev. Lett.*, ***73***, 3277 (1994)
3. Arneodo, S. Roux, and J.F. Muzy, *J. Phys. II France*, ***7***, 363 (1997)
4. J.F. Muzy, J. Delour, and E. Bacry, arXiv:cond-mat/0005400 24 May 2000
5. L. Sorriso-Valvo, V. Carbone, P. Veltri, G. Consolini, and R. Bruno, *Geophys. Rev. Lett.*, ***26***, 1801 (1999)
6. U. Frisch, Turbulence, *The Legacy of A. N. Kolmogorov*, Cambridge: Cambridge University press, 1995
7. P. Veltri and A. Mangeney, in *Solar Wind Nine*, edited by Habbal et al., AIP Conference Proceedings 471, Woodbury, NY: American Institute of Physics, 1999, pp. 543-546
8. C. Meneveau and K.R. Sreenivasan, *Phys. Rev. Lett.*, 59, 1424 (1987)
9. Z-S. She and E. Leveque, *Phys. Rev. Lett*, 72, 336 (1994)
10. B. Dubrulle, *Phys. Rev. Lett*, 73, 959 (1994)
11. C. Pagels and A. Balogh, *Nonl. Proc. in Geophys.*, 8, 313 (2001)
12. Shertzer, D., S. Lovejoy, F. Schmitt, Y. Chigirinskya, and D. Marsan, *Fractals*, 5, 427 (1997)
13. Forman, M.A., and L.F. Burlaga, in *Solar Wind Ten*, edited by M. Velli, et al., AIP Conference Proceedings, Woodbury, NY: American Institute of Physics (this volume)

Exploring the Castaing Distribution Function to Study Intermittence in the Solar Wind at L1 in June 2000

Miriam A. Forman[1] and Leonard F. Burlaga[2]

[1]*Department of Physics and Astronomy, SUNY/Stony Brook, NY 11794-3800 USA*
[2]*Code 692, NASA/Goddard Space Flight Center, Greenbelt MD 20771, USA*

Abstract. We considered 31,561 consecutive 64-second values of radial solar wind speed reported by the SWEPAM instrument (D. McComas, Los Alamos, Principal Investigator) on the ACE spacecraft at L1 upstream of the earth's bow shock beginning day 157 of the year 2000. Running values, moments and probability density functions (pdfs) were calculated for the speed differences over a range of lags from 64 seconds to several days. Running values show local intermittency in their amplitudes, and correlate with local solar wind speed. Moments of order greater than 6 are dominated by the largest values, which increase slowly with lag in the inertial range causing the exponent of the structure function at large q to be a straight line. The pdfs are compared to "Castaing distributions" which are superpositions of Gaussians whose standard deviations are log-normally distributed. Although the Castaing distribution does not and in principle cannot precisely fit actual pdfs of velocity increments in the solar wind, it looks good and provides a basis for a handy two-parameter description of the pdfs. The run of those two parameters with scale provides a further handy description of the intermittency in the cascade that is independent of any particular model of the cascade. Proving the disability of the Castaing distribution, the third moment of the longitudinal velocity increments does exist and it scales consistent with Kolmorogov's 4/5 law that is a requirement for all theories of the inertial cascade.

INTRODUCTION

The solar wind is turbulent on a wide range of scales. Furthermore, its turbulence is intermittent: the local amplitude of the fluctuations fluctuates irregularly from time to time and place to place. Figure 1 illustrates this fluctuation in the amplitude of fluctuations. Some, but not all, is related to the local average solar wind speed. Power spectra do not reveal intermittence. Apparently, intermittence occurs because the cascade of turbulent amplitudes from large to small scales distributes the energy non-uniformly in space at smaller scales (see all of the references). Intermittence is usually studied with statistics of the increments $\Delta V_L(t) \equiv V(t+L)-V(t)$ such as those shown in Figure 1. The set of moments $M(q,L) \equiv \langle (\Delta V_L(t))^q \rangle$ is called the structure function. It measures the shape of the probability distribution functions (pdf), such as shown in Figure 2, of $\Delta V_L(t)$ at different L Its exponent $\zeta(q) \equiv \partial \log(M(q,L))/\partial \log(L)$ which is independent of L where scaling occurs [1,2,3, and references therein], measures the evolution of the pdf with scale rather than the pdf at any particular scale. See Fig. 6.

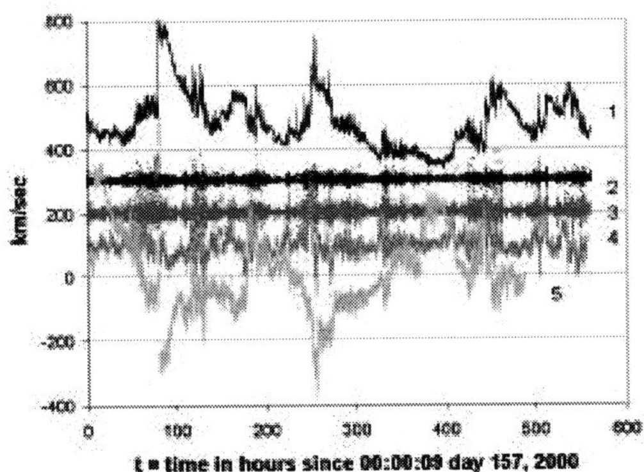

FIGURE 1. Data used in this study. (1) Values of solar wind radial component measured by the SWEPAM instrument on ACE at L1 every 64 seconds. (2) $\Delta V_L \equiv V(t + L)-V(t)$ for lag, L = 64 seconds, shifted up by 300 km/sec for clarity. (3) as (2), for lag = 1024 seconds, shifted 200 km/sec. (4) as (2) for lag = 16384 seconds, shifted 100 km/sec. (5) as (2) for lag = 262144 seconds (72.8 hours), unshifted.

CP679, *Solar Wind Ten: Proceedings of the Tenth International Solar Wind Conference*, edited by M. Velli, R. Bruno, and F. Malara

Various conceptions of the uneven (intermittent) nature of the cascade of turbulent energy yield theoretical predictions for the detailed shape of $\zeta(q)$ [4,5,6,7,8,9,10,11]. However, the finiteness of data sets makes measured moments at large q depend mostly on the single largest data point (e.g. Figure 4), so $\zeta(q)$ generally show a straight line for large enough q (e.g., Figure 6). In some sense this scaling relates to the turbulence physics [10], but not in the same way as theories of $\zeta(q)$ for a continuous, infinite data set. Figure 4 shows that $\zeta(q)$ at $q>6$ should not be used to compare this data set of 31,561 values with theories for infinite data sets, that is, with any theory except possibly [10].

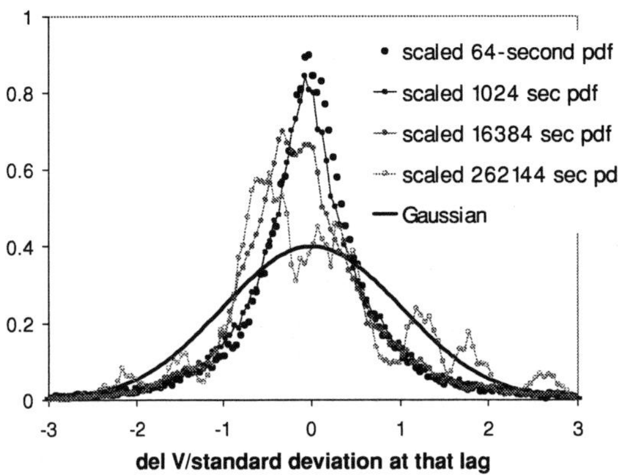

FIGURE 2. Probability density functions (pdfs) of the velocity increments in figure 1.

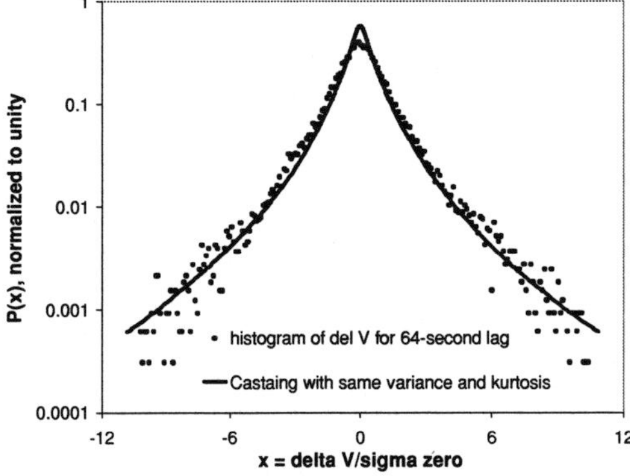

FIGURE 3. Fit of a "Castaing pdf" to the 64-second increments, curve 2 in figure 1 and the most peaked pdf in figure 2. Although the fit looks good to the eye, note that the data falls short of the Castaing pdf at very small and very large velocity increments, as if the smallest and largest values of σ are missing from Equation (2). A closer look, using the cumulative distribution, shows the pdf is slightly skewed to positive increments (as it must be in the inertial range to satisfy Kolmogorov's 4/5 law [3]) and is a power-law at large increments.

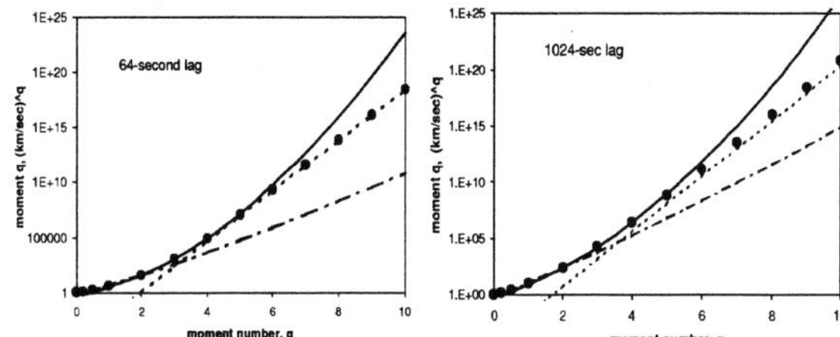

FIGURE 4. Moments of the distribution of velocity differences. Large filled circles: moments calculated from data in Figure 1. Full line: Castaing model moments, fitted to second and fourth moment of this data. Fits are very good from q=1 to q = 5. Dotted line: moments calculated using only the single largest value (straight line). Dot-dashed line: moments of a Gaussian with the same standard deviation.

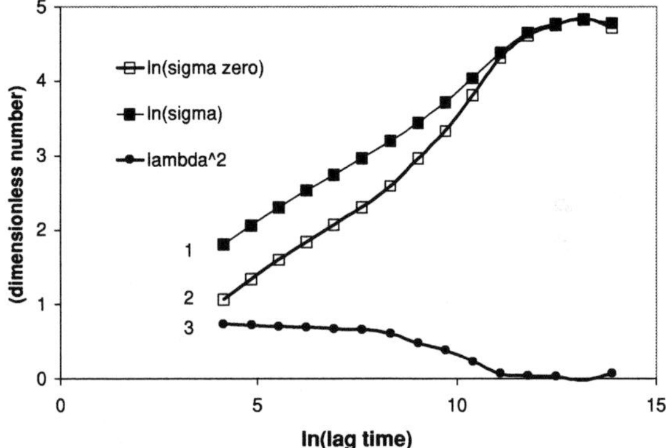

FIGURE 5. Castaing parameters calculated from the second and fourth moments of velocity differences in the entire data set in fig. 1, using equations 2 and 3. The increase of λ^2 towards smaller scales describes the systematic deviation from a Gaussian shown in figure 2. When portions of these curves are straight lines, the Castaing model implies that the scaling of the structure function is lognormal, with quadratic coefficient given by $0.5d\lambda^2/d\ln(\text{scale}) \approx 0.011$ in the inertial range. Strictly speaking, if the fit to Castaing pdf is very good, and plots in this figure are straight lines, the scaling is log-normal in that range of q and of scales.

DATA SET AND PRELIMINARY ANALYSIS

Level 2 data from the SWEPAM instrument on the ACE spacecraft are provided on the ACE website at <http://www.srl.caltech.edu/ACE/ASC/level2/lvl2DATA_SWEPAM.html>. We took as our data set, 64-second values of the radial component of the ion speed beginning at 00:00:09 on day 157 of the year 2000, through day 180 at 08:59:49. We ended the data series then because of a data gap of over 10 minutes. We filled the few gaps of mostly one or two data points by linear interpolation. We had then a set of 31,561 consecutive values of radial solar wind speed. (Fig. 1)

In this analysis, we make no distinction between episodes of fast and slow solar wind, or any other macro or meso-scale attributes of the data, but treat the whole set and look for its statistical properties at different scales. We formed sets of running differences at time t, and different lags, L: $\Delta V_L(t,L) = V(t+L)-V(t)$. Defined this way, $-\Delta V_L$ is the longitudinal velocity increment which is the subject of Kolmogorov's 4/5 law [3], in the inertial range. We took lags from one data point (31,560 64-second increments) to 2^{14} data points. Figure 1 also shows the running values of four of the 15 data sets so formed.

CASTAING PDF BASICS

Sorriso-Valvo, et al [12] first used Castaing distributions [13,14] for the solar wind. Pdfs of ΔV_L in the inertial range are highly kurtotic, and look like the Castaing distribution (Figs. 2 and 3). The Castaing pdf (Cpdf) is a convolution of a parent Gaussian of width σ, with a lognormal distribution of σ, whose width is λ. The parent Gaussian may be thought of as pertaining to very small sub-sets of the data, or as the distribution of ΔV at large scale. The Cpdf is given by

$$P(\Delta V;\sigma_0,\lambda) = \int \frac{e^{-\frac{(\Delta V)^2}{2\sigma^2}}}{\sqrt{2\pi}\sigma} \frac{e^{-\left[\frac{(ln\sigma - ln\sigma_0)^2}{2\lambda^2}\right]}}{\sqrt{2\pi}\lambda} d\,ln\sigma \cdot \quad (1)$$

The odd moments of the Cpdf are identically zero because the Cpdf is symmetric. The moments of the *absolute values* are

$$M(q,L) \equiv \left\langle |\Delta V|^q \right\rangle = a(q) e^{\frac{q^2\lambda^2}{2}} (\sigma_0)^q \qquad (2)$$

where a(q) is the qth moment of a Gaussian of unit width. We evaluated the Castaing parameters σ_0 and λ^2 from Eq. 2, using the second and the fourth moment of the velocity increments at each scale, without going through a tedious curve-fitting process. $\lambda^2 = 0.25\,ln\left(\frac{M(4,L)}{3[M(2,L)]^2}\right)$. Once λ^2 and σ_0 are calculated, we use Equation (1) to evaluate the whole Cpdf.

RESULTS

Figure 3 shows a pdf and Cpdf for our data. Figure 4 shows the extent of the fit of equation (3) to data in the inertial range, and figure 5 shows how the Castaing parameters vary with scale. In figure 5, λ^2 describes the deviation of the pdf at each scale from Gaussian; it is very large in the inertial range below an hour or so, but still increasing slowly with decreasing scale. The large *value* of λ^2 denotes how intermittent the fluctuation amplitudes in Figure 1 are, and how big the wings on the pdfs in figures 2 and 3 are. The steady *change* in λ^2 with scale indicates that the turbulence *cascades* intermittently in that range. That slope, = $-d(\lambda^2)/d(\ln L) = \zeta(2)/2 - \zeta(4)/4$, is a simple robust measure of local intermittency in the cascade process independent of any detailed model, or the size of a (reasonably large) data set and even independent of the assumption that the Cpdf is a reasonable representation of observations.

Kolmogorov's "4/5" law [3] states that the third moment of the *signed* values of the longitudinal velocity increment is *not* zero, but = -0.8*E*L, where E in the energy dissipation rate per unit mass, and L is the scale, in the inertial range. This is why all theories of inertial turbulence predict $\zeta(3)=1$. Figure 7 shows our data set fits this relation, with E about 25,000 Joules per second per kilogram. Such results are impossible if the pdf were an intrinsically symmetric Cpdf. In addition, the actual pdf lack both small and large values of σ compared to the lognormal distribution in the Cpdf that fits most of the pdf (Figs. 3 and 4). While this appears mostly due to the finite resolution and length of the data set, self-organized criticality for extreme increments may play a role [10].

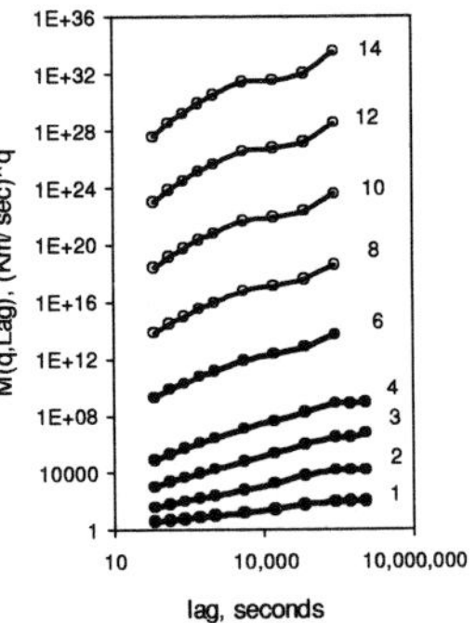

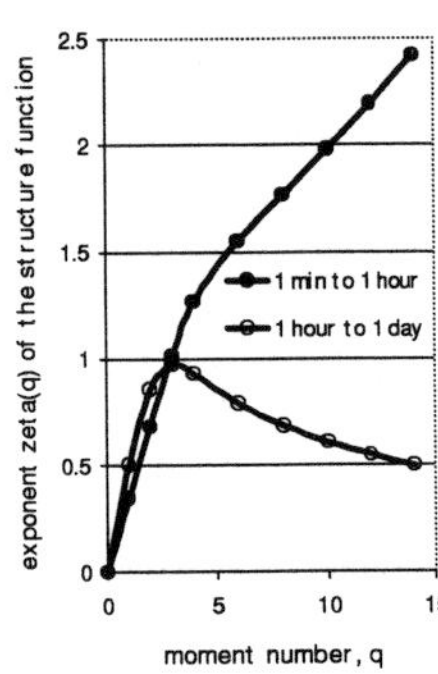

FIGURE 6. Left panel: Structure function of the radial component of the solar wind (this data set). Right panel: Exponent of this structure function, for the inertial range (1 min to 1 hour) and for the interaction range.

We looked at the fastest and lowest-speed epochs, each about 36 hours long, in our data. σ and σ_0 were about twice as large in the fast epoch, consistent with Fig. 7. The λ^2 in the inertial range is about 0.3 in each, also consistent with Figure 7. The high-speed episodes apparent in Figures 1 and 7 are much less intermittent than the recurrent "fast wind" analyzed by Liu and Marsch [15]. This is probably because June 2000 is practically solar maximum, when classic fast wind is rare.

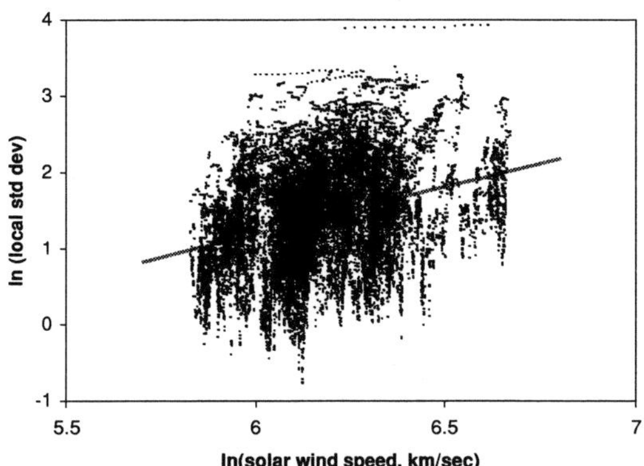

Figure 7. The *running* local standard deviation, σ, of speed increments (in this case 16 64-second increments, 16 points on curve 2 in figure 1) versus the solar wind speed averaged over the same interval of 1024 seconds. Although σ tends to increase with solar wind speed, the (fairly constant) intrinsic spread in $\ln\sigma$ at each speed due to local intermittency is about as large as the variation between the slowest and fastest solar wind in this data set. Both effects contribute to the apparent intermittency in the 24-day data set. Overall, the pdf of the $\ln\sigma$ in this figure is very like a truncated Gaussian, consistent with the deviations from a true Castaing pdf apparent in Figures 3 and 4. Horizontal trails of points at large σ in this figure are due to shocks.

SUMMARY AND CONCLUSIONS

On the basis of a statistical study of 24 days of solar wind radial velocities at a cadence of 64 seconds at L1 in June 2000, we find:

Although the Castaing distribution does not and in principle cannot precisely fit actual pdfs of velocity increments in the solar wind, it looks good and provides a basis for a handy two-parameter description of the pdfs. The run of those two parameters with scale provides a further handy description of the intermittency in the cascade that is independent of any particular model of the cascade. Proving the disability of the Castaing distribution, the third moment of the longitudinal velocity increments does exist and it scales consistent with Kolmorogov's 4/5 law that is a requirement for all theories of the inertial cascade [3].

ACKNOWLEDGMENTS

We are very grateful to NASA, to the ACE project and SWEPAM experimenters for the beautiful data of the SWEPAM instrument available on the ACE website. We are also grateful to a careful referee. MAF is supported by NASA grant NAG58106.

REFERENCES

1. Burlaga, L.F., *J. Geophys. Res., 96*, 5847 (1991)
2. Burlaga, L.F., *Interplanetary Magnetohydrodynamics*, New York: Oxford University Press, 1995
3. Frisch, U., Turbulence, *The Legacy of A. N. Kolmogorov*, Cambridge: Cambridge University Press, 1995
4. Kolmogorov, A. N., *Dokl. Akad. Nauk, SSSR, 30,* 299 (1941)
5. Kolmogorov, A.N., *J. Fluid Mech., 13,* 82 (1962)
6. Obukhov, A.M., *J. Fluid Mech., 13*, 77 (1962)
7. C. Meneveau and K.R. Sreenivasan, *Phys. Rev. Lett.*, 59, 1424 (1987)
8. Z-S. She and E. Leveque, *Phys. Rev. Lett*, 72, 336 (1994)
9. B. Dubrulle, *Phys. Rev. Lett*, 73, 959 (1994)
10. Shertzer, D., S. Lovejoy, F. Schmitt, Y. Chigirinskya, and D. Marsan, *Fractals*, 5, 427 (1997)
11. Forman, M.A., Castaing Scaling and Kernels for Theories of the Turbulent Cascade, in *Solar Wind Ten*, edited by M. Velli et al., AIP Conference Proceedings, Woodbury, NY: American Institute of Physics (this volume)
12. Sorriso-Valvo, L., V.Carbone, P.Veltri, G.Consolini, and R. Bruno, *Geophys. Res. Lett. 26*, 1801 (1999)
13. Castaing, B., Y. Gagne, and E.J. Hopfinger, *Physica D 46*, 177-200 (1990)
14. Chabaud, B., A. Naert, J. Pienke, F. Chilla, B. Castaing, and B. Hebral, *Phys. Rev. Lett, 73*, 3227-3230 (1994)
15. Marsch, E., and S. Liu, *Annales Geophysicae, 11*, 227-238 (1993)

Alfvén Turbulence Driven by High-Dimensional Interior Crisis in the Solar Wind

A. C.-L. Chian*,†, E. L. Rempel*,†, E. E. N. Macau†, R. R. Rosa† and F. Christiansen**

*World Institute for Space Environment Research-WISER, NITP, University of Adelaide, SA 5005, Australia
†National Institute for Space Research-INPE, P.O. Box 515, 12227-010 São José dos Campos, SP, Brazil
**Solar-Terrestrial Physics Division, Danish Meteorological Institute, Lyngbyvej 100, DK-2100 Copenhagen, Denmark

Abstract. Alfvén intermittent turbulence has been observed in the solar wind. It has been previously shown that the interplanetary Alfvén intermittent turbulence can appear due to a low-dimensional temporal chaos [1]. In this paper, we study the nonlinear spatiotemporal dynamics of Alfvén waves governed by the Kuramoto-Sivashinsky equation which describes the phase evolution of a large-amplitude Alfvén wave. We investigate the Alfvén turbulence driven by a high-dimensional interior crisis, which is a global bifurcation caused by the collision of a chaotic attractor with an unstable periodic orbit. This nonlinear phenomenon is analyzed using the numerical solutions of the model equation. The identification of the unstable periodic orbits and their invariant manifolds is fundamental for understanding the instability, chaos and turbulence in complex systems such as the solar wind plasma. The high-dimensional dynamical system approach to space environment turbulence developed in this paper can improve our interpretation of the origin and the nature of Alfvén turbulence observed in the solar wind.

INTRODUCTION

In a recent paper by Chian, Borotto and Gonzalez [1], Alfvén turbulence driven by a low-dimensional chaos in the solar wind was studied. Through a numerical analysis of the stationary solutions of the driven-dissipative nonlinear Schrödinger equation, two types of interplanetary Alfvén intermittency were identified: type-I Pomeau-Manneville intermittency and interior crisis-induced intermittency. The nonlinear Schrödinger equation and the derivative nonlinear Schrödinger equation have also been used to model the propagation of finite-amplitude Alfvén waves [2, 3] and the formation of magnetic holes [4] in the solar wind, as well as the solar corona heating [5]. The spatiotemporal chaos of Alfvén waves appears in a system of MHD coupled wave equations that are the generalization of the nonlinear Schrödinger equation [6].

It was demonstrated by Lefebvre and Hada [7] that under the assumption of weak instability and wave-packet limit the derivative nonlinear Schrödinger equation, that describes the modulational instability of quasi-parallel Alfvén waves of moderate amplitudes in a finite β plasma, reduces to a complex Ginzburg-Landau equation. In this paper, we study the spatiotemporal dynamics of a nonlinear Alfvén wave in the solar wind governed by the Kuramoto-Sivashinsky equation, which under certain approximations describes the phase evolution of the complex amplitude of the Ginzburg-Landau equation.

THE KURAMOTO-SIVASHINSKY EQUATION

The one-dimensional Kuramoto-Sivashinsky equation can be written as [8]

$$\partial_t u = -\partial_x^2 u - \nu \partial_x^4 u - \partial_x u^2, \tag{1}$$

where ν is a 'viscosity' damping parameter. We assume that $u(x,t)$ is subject to periodic boundary conditions $u(x,t) = u(x+2\pi,t)$ and expand the solutions in a discrete spatial Fourier series

$$u(x,t) = \sum_{k=-\infty}^{\infty} b_k(t) e^{ikx}. \tag{2}$$

Substituting Eq. (2) into Eq. (1) yields an infinite set of ordinary differential equations for the complex Fourier coefficients $b_k(t)$

$$\dot{b}_k(t) = (k^2 - \nu k^4) b_k(t) - ik \sum_{m=-\infty}^{\infty} b_m(t) b_{k-m}(t), \tag{3}$$

CP679, *Solar Wind Ten: Proceedings of the Tenth International Solar Wind Conference*, edited by M. Velli, R. Bruno, and F. Malara

where the dot denotes derivative with respect to t. Reality of $u(x,t)$ implies that $b_{-k} = b_k^*$. We restrict our investigation to the subspace of odd functions $u(x,t) = -u(-x,t)$ assuming that $b_k(t)$ are purely imaginary by setting $b_k(t) = -ia_k(t)/2$, where $a_k(t)$ are real. Equation (3) then becomes

$$\dot{a}_k(t) = (k^2 - \nu k^4)a_k(t) - \frac{k}{2}\sum_{m=-\infty}^{\infty} a_m(t)a_{k-m}(t), \quad (4)$$

where $a_0 = 0$, $1 \leq k \leq N$, N is the truncation order. We integrate the high-dimensional dynamical system given by Eq. (4) using a fourth-order variable step Runge-Kutta integration routine. We choose $N = 16$, since numerical tests indicate that for the range of the control parameter ν used in this paper the solution dynamics remains essentially unaltered for $N > 16$. In all the computational results presented in this paper, higher order truncations yield the same numerical conclusions as the 16-mode truncation. We adopt a Poincaré map with the $(N-1)$ dimensional hyperplane defined by $a_1 = 0$, with $\dot{a}_1 > 0$.

Nonlinear Dynamical Analysis

A bifurcation diagram can be obtained from the numerical solutions of the 16-mode truncation of Eq. (4) by varying the control parameter ν and plotting the Poincaré points of one Fourier mode after discarding the initial transient. Figure 1a shows a period-3 (p-3) window where we plot the Poincaré points of the Fourier mode a_6 as a function of ν. The corresponding behavior of the maximum Lyapunov exponent is shown in Fig. 1b. Evidently, the high-dimensional temporal dynamics of the K-S equation preserves the typical dynamical features of a low-dimensional dynamical system. The dotted lines in Fig. 1a denote the Poincaré points of the p-3 unstable periodic orbit (UPO) which emerges via a saddle-node bifurcation at $\nu = 0.02992498$, marked SN in Fig. 1a. In this paper, we will analyze the role played by this p-3 UPO in the onset of interior crisis at $\nu_{IC} = 0.02992021$, marked IC in Fig. 1.

The interior crisis at ν_{IC} occurs when the p-3 UPO collides head on with the 3-band weak strange attractor evolved from a cascade of period-doubling bifurcations, as seen in Fig. 1a.

The interior crisis leads to a sudden expansion of a strange attractor, turning the weak strange attractor (WSA) into a strong strange attractor (SSA), as seen in Fig. 2. Figure 2 is a 3-dimensional projection (a_1, a_{10}, a_{16}) of the strong strange attractor (light line) defined in the 15-dimensional Poincaré hyperplane right after the crisis ($\nu = 0.02992020$), superimposed by the 3-band weak strange attractor (dark line) at crisis

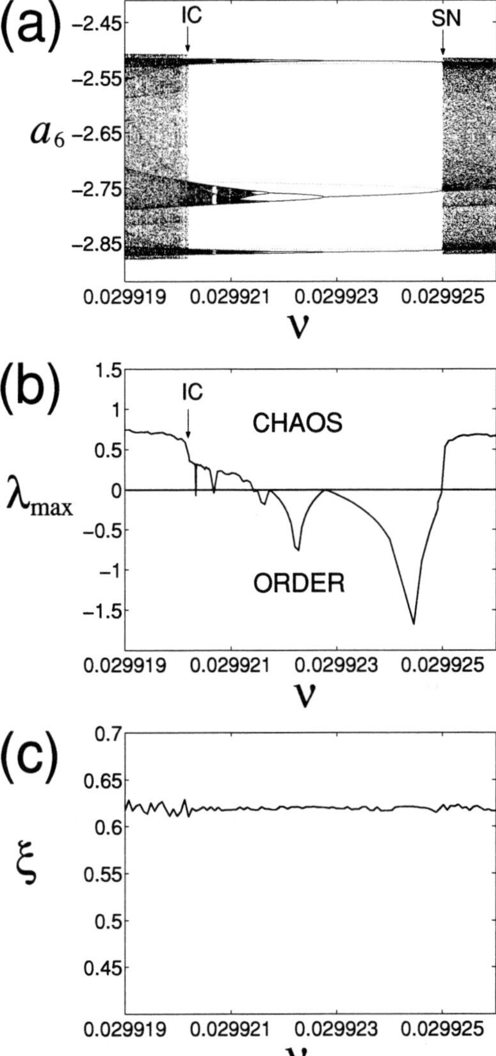

FIGURE 1. (a) Bifurcation diagram of a_6 as a function of ν. IC denotes interior crisis and SN denotes saddle-node bifurcation. The dotted lines represent a period-3 unstable periodic orbit. (b) Variation of the maximum Lyapunov exponent λ_{max} with ν. (c) Variation of the correlation length σ with ν.

($\nu = 0.02992021$). The abrupt increase in the system's chaoticity after the interior crisis can be characterized by the value of the maximum Lyapunov exponent (λ_{max}), plotted in Fig. 1(b). At crisis ($\nu_{IC} = 0.02992021$), $\lambda_{max} = 0.35$, and after the crisis at $\nu = 0.02992006$, $\lambda_{max} = 0.62$. In Fig. 1(c) we plotted the spatial correlation length σ, which remains basically unaltered throughout the whole range of ν used in Fig. 1. This means that there is little variance in the spatial disorder of the pattern. An inspection of the spatiotemporal pattern $u(x,t)$ after the crisis ($\nu = 0.02992006$), shown in Fig. 3, reveals that for the

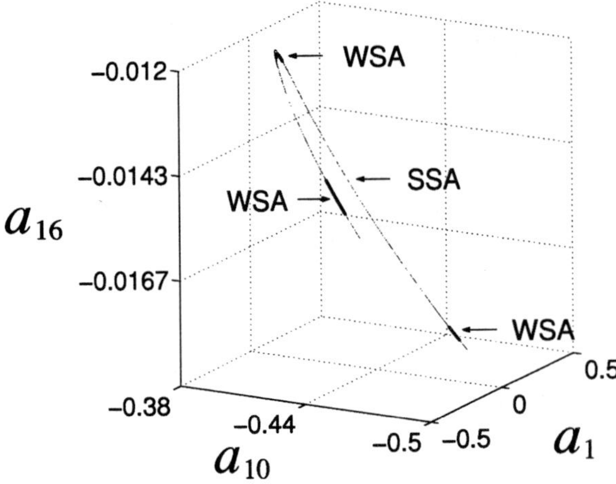

FIGURE 2. Three-dimensional projection (a_1, a_{10}, a_{16}) of the strong strange attractor SSA (light line) defined in the 15-dimensional Poincaré hyperplane right after crisis at $\nu = 0.02992020$, superimposed by the 3-band weak strange attractor WSA (dark line) at crisis $(\nu = 0.02992021)$.

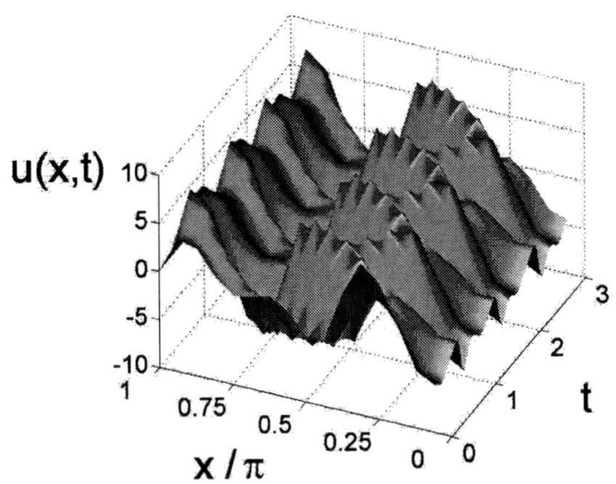

FIGURE 3. The spatiotemporal pattern of $u(x,t)$ after the crisis at $\nu = 0.02992006$. The system dynamics is chaotic in time but coherent in space.

chosen values of ν and the spatial system size $L = 2\pi$, the dynamics of the Kuramoto-Sivashinsky equation is coherent in space.

On the Poincaré hyperplane an unstable periodic orbit turns into a saddle fixed point, with its associated invariant stable and unstable manifolds. At crisis, only one of the 16 stability eigenvalues of the p-3 UPO illustrated in Fig. 1a has an absolute value greater than 1. This implies that its invariant unstable manifold is a one-dimensional curve embedded in the 15-dimensional Poincaré space. Of the remaining eigenvalues, one has absolute value equal to unity and all the other fourteen have absolute values less than one, implying that the invariant stable

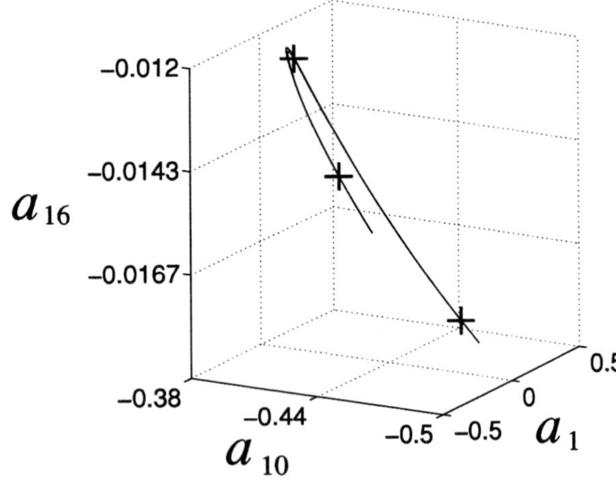

FIGURE 4. Three-dimensional projection (a_1, a_{10}, a_{16}) of the invariant unstable manifolds of the period-3 saddle (crosses) right after the crisis at $\nu = 0.02992020$.

manifolds have dimension fourteen. Although we have adopted a 16-mode truncated system in our analysis, all the calculations performed can be extended to an arbitrary high number $(N < \infty)$ of modes for an appropriate choice of ν and L. Figure 4 is a plot of the projection of the invariant unstable manifold of the p-3 UPO onto three axes (a_1, a_{10}, a_{16}) at $\nu = 0.02992020$, right after the interior crisis. The crosses represent the p-3 UPO. The invariant unstable manifolds consist of infinitely many distinct, discrete Poincaré points whose backward orbits converge to the saddle.

We proceed next with the characterization of the high-dimensional crisis at ν_{IC} by showing in Fig. 5 the collision of the weak strange attractor with the p-3 UPO in the reduced 2-dimensional Poincaré plane (a_5 vs. a_6), in the vicinity of the upper cross in Fig. 4. The dark line denotes the strange attractor, and the light line denotes the numerically computed invariant unstable manifold of the saddle periodic orbit. Figures 5a,b,c display the dynamics before, at, and after the crisis, respectively. Note that the strange attractor always "overlaps" the invariant unstable manifold. At the crisis point $\nu_{IC} = 0.02992021$, the chaotic attractor is the closure of one branch of the unstable manifold of the p-3 UPO, as seen in Fig. 5b. The "head-on" collision of the weak strange attractor with the p-3 UPO at ν_{IC}, shown in Fig. 5b, proves the occurrence of an interior crisis. Although we are showing only two of the 16 modes, the collision can be seen in any choice of Fourier modes. This collision leads to an abrupt expansion of the strange attractor, as seen in Fig. 5c. A comparision of Figs. 2 and 4 confirms that, after the crisis, the strong strange attractor and the invariant unstable manifold "overlap" with each other.

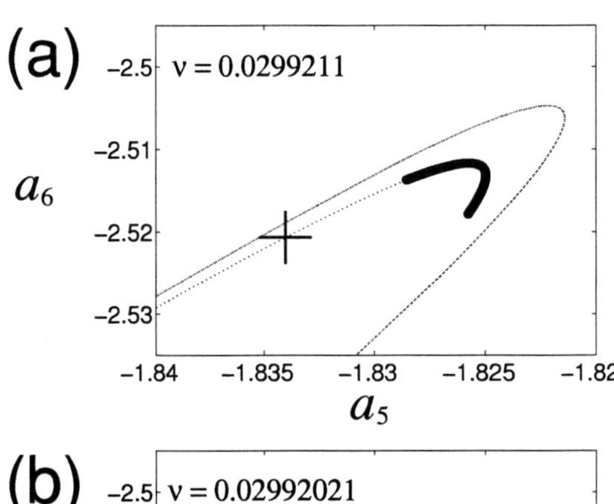

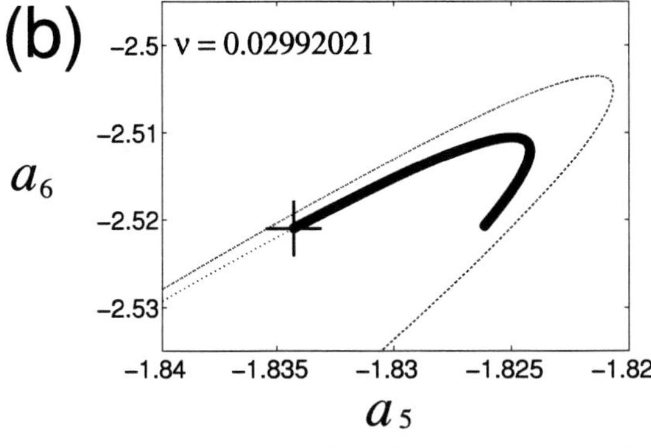

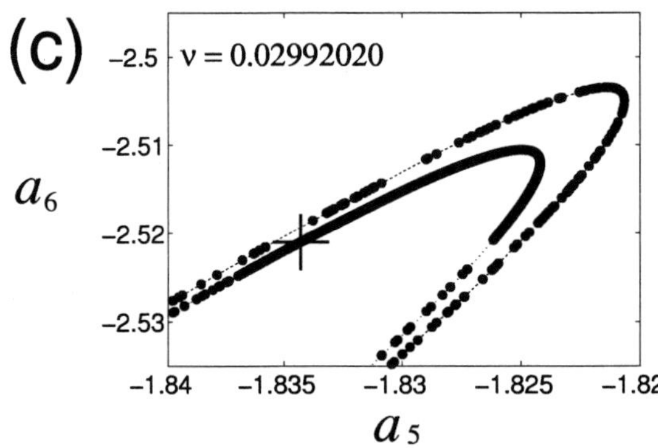

FIGURE 5. The plots of the strange attractor (dark line) and invariant unstable manifolds (light lines) of the saddle before (a), at (b) and after (c) crisis. The cross denotes one of the saddle points.

DISCUSSION

The high-dimensional interior crisis discussed in this paper can improve our understanding of Alfvén intermittent turbulence in the Solar wind. Observations of Alfvénic intermittent turbulence in the solar wind were reported by Marsch and Liu [9] and Tu and Marsch [10]. Using the *Helios 2* data in the inner solar wind between 0.3 and 1.0 AU, they identified the multifractal nature of interplanetary Alfvénic fluctuations and the dependence of Alfvénic intermittent turbulence on stream speed and radial distance from the Sun. It was shown by Chian, Borotto and Gonzalez [1] that an interior crisis can induce a temporal Alfvén intermittency, yielding a power-law spectrum similar to the typical power spectra of Alfvénic turbulence observed in the interplanetary medium. The present paper suggests that the spatiotemporal intermittency of interplanetary Alfvén waves can be driven by an interior crisis. There is observational evidence of Alfvén chaos in the solar wind. A nonlinear time series analysis of Alfvén waves in the low-speed streams of the solar wind detected by the Helios spacecraft in the inner heliosphere shows that the Lyapunov exponent and the entropy are positive, which indicates that the solar wind plasma is in a chaotic state [11].

ACKNOWLEDGMENTS

This work is supported by CNPq, FAPESP and AFOSR. A. C.-L. Chian and E. L. Rempel wish to thank Professors Tony Thomas and Tony Williams of the University of Adelaide for their kind hospitality.

REFERENCES

1. Chian, A. C.-L., Borotto, F. A., and Gonzalez, W. D., *Astrophys. J.*, **505**, 993-998 (1998).
2. Ghosh, S., and Papadopoulos, K., *Phys. Fluids*, **30**, 1371-1387 (1987).
3. Hada, T., Kennel, C. F., Buti, B., and Mjϕlhus, E., *Phys. Fluids B*, **2**, 2581-2590 (1990).
4. Baumgartel, K., *J. Geophys. Res.*, **104**, 28295-28308 (1999).
5. Champeaux, S. et al., *Astrophys. J.*, **486**, 477-483 (1997).
6. Oliveira, L. P. L., Rizzato, F. B., and Chian, A. C.-L., *J. Plasma Phys.*, **58**, 441-453 (1997).
7. Lefebvre, B., and Hada, T. "Spatiotemporal Behavior of a Driven System of MHD Waves", AGU Fall Meeting, San Francisco (2000).
8. Chian, A. C.-L., Rempel, E. L., Macau, E. E. N. Rosa, R. R., Christiansen, F., *Phys. Rev. E* **65**, 035203(R) (2002).
9. Marsch, E., Liu, S. *Ann. Geophys.*, **11**, 227 (1993).
10. Tu, C.-Y., Marsch, E., *Space Sci. Rev.*, **33**, 1 (1995).
11. Macek, W. M., and Redaelli, S., *Phys. Rev. E*, **62**, 6496-6504 (2000).

Solitary Waves
Observed By Cluster In the Solar Wind

M.Fränz*, T.S.Horbury†, V.Génot**, O.Moullard‡, H.Rème, I.Dandouras§, A.N.Fazakerley¶, A.Korth* and F.Frutos-Alfaro*

**MPI für Aeronomie,D-37191 Katlenburg-Lindau, D; Email: fraenz@linmpi.mpg.de*
†Imperial College,London SW7 2BW,UK
***Queen Mary,Univ.London,London E1 4NS,UK*
‡ESTEC, 2200AG Noordwijk,NL
§CESR,F-31028 Toulouse cedex 4,F
¶MSSL, Dorking RH5 6NT, UK

Abstract. Short dropouts of the magnetic field intensity have been frequently observed in the solar wind on interplanetary spacecraft. But so far it could not be established whether these are caused by kinetic instabilities or whether they can be described as solitary MHD waves. The multi-satellite observations of the Cluster-mission allow for the first time to measure proton and electron distributions with a sufficient temporal and spatial resolution to tackle this question. We use measurements by the FGM magnetometer, the CIS ion spectrometer, the PEACE electron instrument and the Whisper plasma wave instrument to investigate the role of protons, heavy ions and electrons for the stability of the structures. We also use the 4-satellite observations of the Cluster magnetic field instrument to determine the proper motion of these structures relative to the solar wind. The presence of foreshock waves close to the Earth bowshock strongly limits the event selection. In the current paper we discuss a 10 s linear wave without sufficient particle data resolution and a 4 min wave for which particle distributions are available. The larger wave shows that the stability of the structure might be caused by changes in the thermal electron distributions while proton and α distributions are unaffected.

MAGNETIC HOLES - SOLITONS?

Magnetic field dropouts are commonly observed in planetary magnetosheaths in high plasma β conditions. These observations are usually explained by mirror mode instabilities characterized by anisotropic proton distributions [1, 2]. Ulysses data have shown that field dropouts are also common in the solar wind [3]. But it could not be established that these favorably occur in high plasma β conditions and are correlated with proton temperature anisotropies [4]. Recently it was suggested that these dropouts can be explained as solitary MHD waves in MHD models with additional friction [5] or bi-ion plasmas [6]. The frequency of field dropouts in the interplanetary field suggests that they can significantly affect energetic ion diffusion [7]. The existence of MHD solitary waves in astrophysical plasmas would be of high interest for MHD theory and would have impact on the occurrence of plasma instabilities.

CLUSTER OBSERVATIONS

4-spacecraft observations in the solar wind allow for the first time to determine proper speed and spatial structure of magnetic holes in the solar wind. For that purpose a working group has been established involving the Cluster ion and electron instruments (CIS, PEACE) and the magnetic and electric field investigations (FGM, EFW, STAFF, WHI)[1]. This paper concentrates on preliminary observations of the CIS ion spectrometer[8], FGM magnetometer[9], the PEACE electron spectrometer [10] and the WHI wave spectrometer[11]. The Cluster spacecraft encountered the solar wind for several hours on each orbit for the periods January to June 2001 and again from January 2002. We first determined the solar wind periods by eye using the CIS proton energy spectra. Then we did an automatic search through those periods looking for field drops of more than 3 σ in 4 min windows using a 12 s sliding average of the 4 s field data similar to the method used previously for Ulysses

[1] See: http://www.linmpi.mpg.de/~fraenz/magholes

CP679, *Solar Wind Ten: Proceedings of the Tenth International Solar Wind Conference,* edited by M. Velli, R. Bruno, and F. Malara

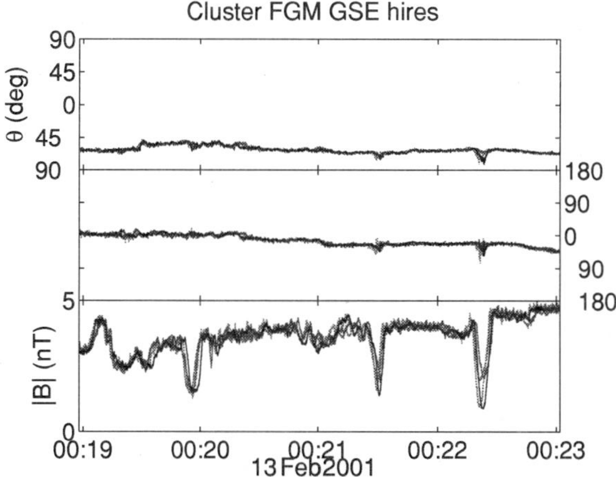

FIGURE 1. First set of solitary linear waves observed by Cluster upstream of the Earth bowshock at 19.5 Re 12.6 MLT on 13 Feb 2001. Shown are magnetic field GSE polar and azimuth angles and total magnitude at 22 vectors/s for all 4 S/C (gray tones). The extension of the structures in time is about 10 s corresponding to a size of 5000 km at a solar wind speed of V_p=450 km/s.

data[4]. This gave us a list of 14 events for data until 15 Feb 2002. The main problem was then to exclude events related to the proximity of the Earth bowshock, specifically fore-shock cavities as discussed by [12]. It turned out that the only safe method to do this is to compare with data from an upstream satellite. From the 14 events there was one left which was also visible in ACE key parameters and not associated with a field reversal. This is the second event discussed below.

SELECTED EVENTS

The first set of solitary waves (Fig. 1) observed by Cluster on 13 Feb 2001 is shown here because of its obvious wave form. It was not discovered by an automatic search. Unfortunately the size of only 10 s (5000 km) is too small to get any reliable particle measurements from within the waves. Also the wave is not visible in the public (16s) ACE data sets. This makes it difficult to exclude the proximity of the Earth bow shock as a possible source of these waves. Plasma conditions in the surrounding plasma were: $n_p = 12$ cm^{-3}, $n_\alpha = 0.9$ cm^{-3}, $T_p = 0.2$ MK, $v_{P_{GSE}}$ = (-450, 20, -10) km/s. This corresponds to a ion thermal pressure of $P_i = 0.04$ nPa or β_i = 4.4 and an Alfv́en speed $v_A = 27$ km/s for a 5 nT field.

The motion of 3 structure boundaries in the plasma rest frame has been determined from the 4-point measurement to be between 7-14 km/s. This is still in agreement with convection with the solar wind - taking into account the error in calculation.

The large wave (Fig. 2, bottom) observed on 22 Feb 2001 has a temporal extent of 4 min -indicated by the black frame - corresponding to a size of about 100000 km. Data in Fig. 2 have a time resolution of 4 s if not stated otherwise. The event is also seen in 16 s ACE magnetic field data (bottom panel) 210 RE upstream of Cluster. Thus it is clearly of solar wind origin. Also the time delay of 50 min between ACE and Cluster proves the stability of the structure. Though the holes are not linear (panel 8,9), there is no indication that the event forms a magnetic cloud, nor that it is associated with a current sheet crossing since there is no field reversal. The large size allows detailed particle measurements within the structure. There is no strong signature in thermal proton moments (panel 5,7), also no change in α densities(panel 6). The proton and α moments with 4 s time resolution are onboard moments . while moments with 16 s resolution (red traces and panel 7) are calculated on ground from the 3D distributions. The 16 s proton velocity distribution (Fig. 3) shows no distortion associated with the field dropouts. The only strong signal we can observe so far are in the thermal electron data (panel 1,2): Electron temperature decreases are observed associated with the field dropouts on SC2 (panel 1). A similar behavior is not visible on C1 and C4 since PEACE data cannot be corrected for spacecraft potential - the data in panel 1 can only be used qualitatively. Also it cannot be determined whether the parallel temperature dropouts correspond to a change in anisotropy. Electron densities from WHI in panel 2 may be used quantitatively and show that electron density variations are much stronger than proton density variations but correlation with the field dropouts are complex. We have not yet analyzed electric field data for this event. Plasma conditions in the surrounding plasma where: $n_p = 6$ cm^{-3}, $n_\alpha = 0.5$ cm^{-3}, $T_p = 0.1$ MK, $v_{P_{GSE}}$ = (-340, 20, -30) km/s, which corresponds to $P_i = 0.01$ nPa, $\beta_i = 1.1$ and $v_A = 38$ km/s. Thus this event does not occur in high β conditions.

4-point measurements of the motion of 4 structure boundaries results in a speed of 5 km/s for two boundaries and 30 km/s for the other two. While the second value is closer to the Alfvén speed we must emphasize that the precision of the determination is currently not good enough to draw strong conclusions from that. We hope to improve on these measurements by considering the timing measurements relative to the ACE spacecraft.

The observations suggest that the electrons play a crucial role in the description of these waves. In contrast to magnetospheric electron holes [13] there are no clear density drops, but probably temperature anisotropies. An explanation by electron mirror modes might be possible (*R.A. Treumann, personal comm.*).

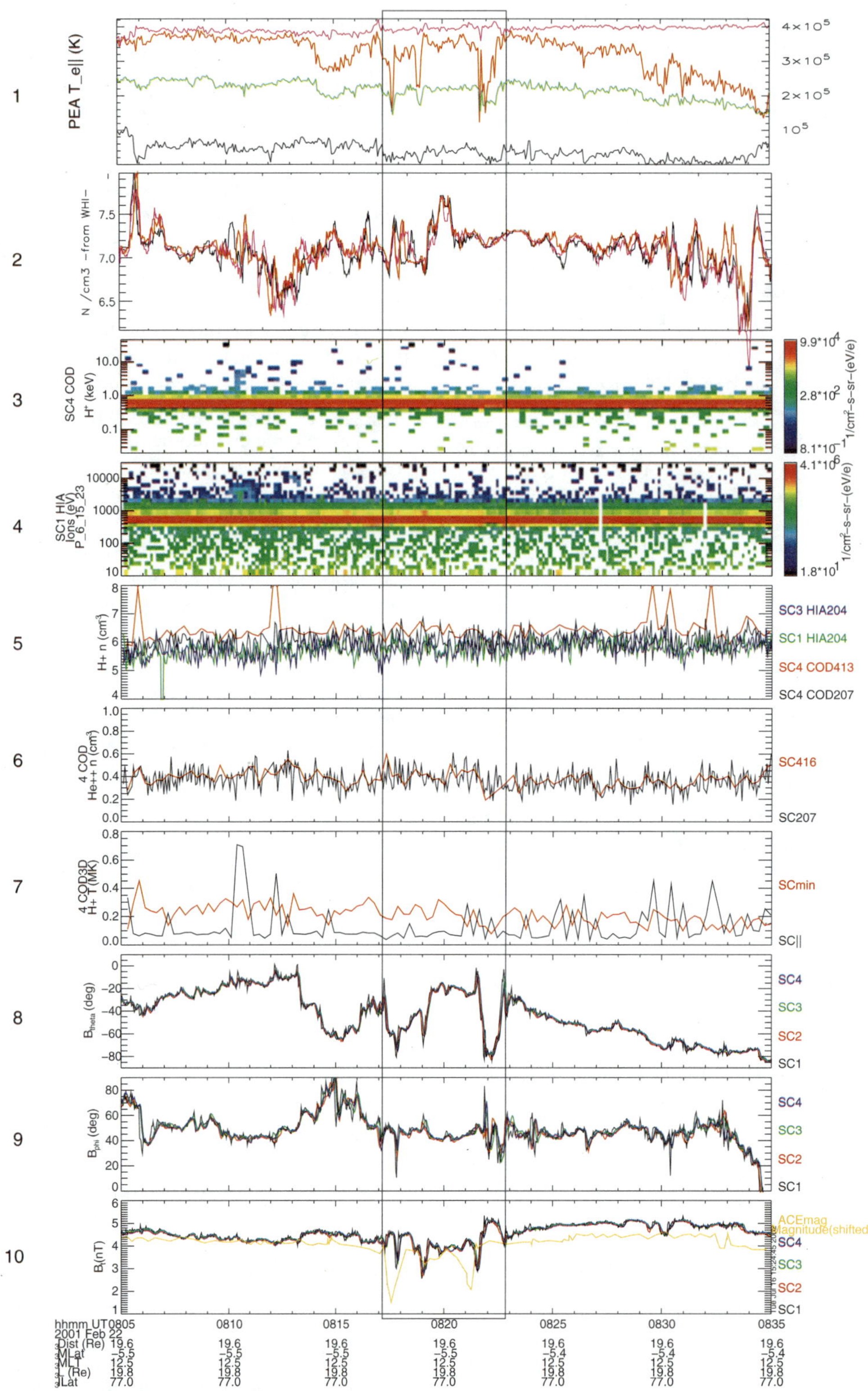

FIGURE 2. Large wave observed by ACE and Cluster on 22 Feb 2001 at 19.6 RE and MLT=12.5 h. The panels show from top to bottom: (1)Parallel electron temperature (PEACE) for C1(black), C2(red), C3(green), C4(magenta), not corrected for SC-potential; (2)Electron density (WHI) for C1,C2,C4; 16s proton spectra of the CIS C1 CODIF(3) and HIA(4) sensors, respective onboard proton densities(5), the ground(16 s) and onboard(4 s) α density on C4(6), ground(16s) parallel and perpendicular temperature(7); magnetic field GSE theta(8), phi(9) angles and magnitude(10) for all 4 SC. The bottom panel shows also 16 s magnetic field magnitude at the ACE spacecraft shifted in time by about 50 min, using the ACE distance and solar wind velocity (orange).

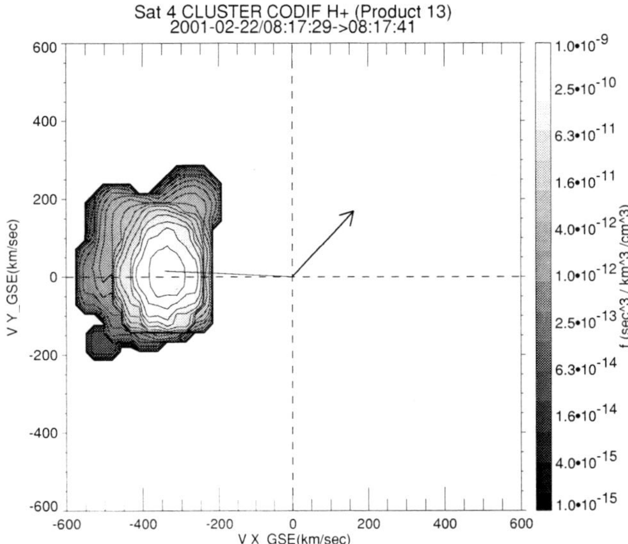

FIGURE 3. CIS Proton Velocity Distribution passing through a magnetic depression at 20010222T08:17, taken by the CODIF sensor on C4 projected onto the GSE XY-plane. The plot also contains the magnetic field vector projection (at center).

CONCLUSIONS

For the detection of solar wind magnetic holes with Cluster it is very advisable to correlate data with an upstream spacecraft. This sets a lower limit of about 30 s on the size of detectable structures. Also proton velocity distributions on Cluster are only obtained with 16 s time resolution at 11° angular resolution in normal telemetry mode. The comparison with ACE data allows to clearly identify holes of solar wind origin. Nevertheless we show one event which has a smooth solitary and linear wave form and might be described as a MHD soliton. It is the only event of this type discovered in Cluster data for 2001. But the size of only 8 s does not allow a more detailed particle analysis. The 4 s onboard moments do not show significant changes.

A systematic search for events larger than 30 s visible at ACE and Cluster resulted in only one event without field reversal for which particle distributions are available. This event does neither show significant proton temperature anisotropies indicative of proton mirror modes, nor density increases or velocity vector sweeps expected for Hall-MHD solitons. The signature in electrons - specifically temperature dropouts and density variations are much stronger. This agrees with the observation that magnetic holes observed by Ulysses are often associated with Langmuir waves[14]. Determination of proper speed is difficult at Cluster separations of $\sim$ 600 km. Both events discussed show proper speeds in agreement with convection with solar wind. For the second event we hope to improve the analysis using higher resolution data from ACE. Another large hole observed by Ampte and ISEE1 upstream of the Earth bowshock [15] was also convected with the solar wind and has some features in common with the event discussed here.

We are confident that the combination of Cluster ion, field and wave data at larger spacecraft separations in the later parts of the mission will elucidate the physics of magnetic holes in the solar wind.

ACKNOWLEDGMENTS

We are very grateful for the support from the Cluster community and specifically the CIS team and the working group on solitary waves. We acknowledge the provision of the ACE MAG (N. Ness at Bartol) and SWEPAM (D. McComas at LANL) data through the CDAWeb system and the support by K-H. Fornaçon and K-H. Glaßmeier (Univ. Braunschweig). T. S. Horbury and V. Génot are supported by PPARC (UK) fellowships.

REFERENCES

1. Tsurutani, B., Smith, E., Andersen, R., Ogilvie, K., Scudder, J., Baker, D., and Bame, S., *J. Geophys. Res.*, **87**, 6060–6072 (1982).
2. Baumgärtel, K., *Planet. Space Sci.*, **49**, 1239–1247 (2001).
3. Winterhalter, D., Neugebauer, M., Goldstein, B., and Smith, E., *J. Geophys. Res.*, **99**, 23,371–23,381 (1994).
4. Fränz, M., Burgess, D., and Horbury, T., *J. Geophys. Res.*, **105**, 12725–12732 (2000).
5. Baumgärtel, K., *J. Geophys. Res.*, **104**, 28295–28308 (1999).
6. McKenzie, J., *J. Plasma Phys.*, **65**, 181–195 (2001).
7. Tsurutani, B., Lakhina, G., Winterhalter, D., Arballo, J., Galvan, C., and Sakurai, R., *Nonlinear Process Geophys.*, **6**, 235–242 (1999).
8. Rème, H., Aoustin, C., Bosqued, M., Dandouras, I., and the CIS-team, *Ann. Geophys.*, **19**, 1303–1354 (2001).
9. Balogh, A., Carr, C., Acuña, M., Dunlop, M., and the FGM-team, *Ann. Geophys.*, **19**, 1207–1217 (2001).
10. Johnstone, A., Alsop, C., Burge, S., Carter, P., and the PEACE-team, *Space Sci. Rev.*, **79**, 351–398 (1997).
11. Decreau, P., Fergeau, P., Krasnoselskikh, V., and the Whisper-team, *Ann. Geophys.*, **19**, 1241–1258 (2001).
12. Sibeck, D., Decker, R., Mitchell, D., Lazarus, A., Lepping, R., and Szabo, A., *J. Geophys. Res.*, **106**, 21675–21688 (2001).
13. Chen, J., and Parks, G., *Nonlinear Process Geophys.*, **9**, 111–119 (2002).
14. Lin, N., Kellogg, P. J., Macdowall, R. J., Balogh, A., Forsyth, R. J., Phillips, J. L., Buttighofer, A., and Pick, M., *Geophys. Res. Lett.*, **22**, 3417–3420 (1995).
15. Chisham, G., Schwartz, S. J., Burgess, D., Bale, S. D., Dunlop, M. W., and Russel, C. T., *J. Geophys. Res.*, **105**, 2325–2335 (2000).

Nonlinear evolution of large-amplitude Alfvén waves in parallel and oblique propagation

Luca Del Zanna*, Marco Velli* and Pasquale Londrillo†

*Dipartimento di Astronomia e Scienza dello Spazio, Largo E. Fermi 2, I-50125 Firenze, Italy
†INAF, Osservatorio Astronomico di Bologna, Via C. Ranzani 1, 40127, Bologna, Italy

Abstract. The stability of monochromatic large-amplitude Alfvén waves is investigated via MHD numerical simulations. In a compressible medium, such as the heliospheric environment, these waves are subject to the parametric decay instability. The mother wave decays in a compressive mode, that soon steepens and dissipates thermal energy, and in a backscattered Alfvénic mode with lower amplitude and frequency, thus starting an inverse cascade. This well known process is shown here to be very robust, since it occurs basically unchanged regardless of the dimensionality of the spatial domain and, above all, even linear or arc-polarized waves in oblique propagation, most often found in solar wind data, appear to behave in the same way. This physical process could help to explain the observed radial decrease of cross helicity in the fast polar wind, as measured by Ulysses.

INTRODUCTION

The low-frequency part of fluctuations spectra in high-speed and polar regions of the solar wind is known to be dominated by large-amplitude Alfvén waves propagating outwards (Belcher & Davis [3]; Goldstein et al. [7]), probably originated in coronal holes at the Sun, where the magnetic field is open and mainly unipolar. Fluctuations seen in slow solar wind streams (on the equatorial plane) evolve rapidly towards a classical Kolmogorov turbulent cascade, driven by velocity and magnetic shears ubiquitous there (e.g. Malara et al. [10]), while in polar regions, where these features are rare (at least on a large scale) the turbulence evolution is much slower: the spectrum has initially a smaller index, reaching values close to -5/3 only at large distances, and at the same time the prevalence of outward propagating modes decreases monotonically (Roberts et al. [13]; Bavassano et al. [2]).

The origin of this residual turbulence evolution is still a matter of debate: incompressible fluctuations with an initial imbalance between outward and inward propagating modes should evolve in a way as to increase this imbalance, rather than to decrease it (*dynamical alignment*, Dobrowolny el al. [6]). The observed process was first supposed to be due just to overall solar wind expansion effects (e.g. Grappin & Velli [8]), but a promising alternative is provided by compressible wave-wave nonlinear interactions, namely parametric decay.

When an Alfvén wave propagates by keeping an overall constant magnetic field strength, regardless of its amplitude, the coupling to compressive perturbations leads to a decay of the mother wave in favour of a backscattered Alfvénic mode and of a steepening magnetosonic wave. Therefore the process *reduces* the initial Alfvénic imbalance, as needed. When compressive waves start to dissipate into shock fronts the resonance is lost and the decay saturates. This process is rather robust, since it occurs for a variety of wave amplitudes and plasma betas; in general the evolution in a low-beta plasma is faster but even in equipartition conditions, like in the solar wind, growth rates are reasonably high for sufficient wave amplitudes (see Del Zanna et al. [4] for a broad overview of results and references).

Another clue that parametric decay might be indeed responsible for the (slow) evolution of the fast solar wind turbulence is the observational result that the so-called Alfvénic ratio $r_A = E^-/E^+$, where $E^\pm$ are the fluctuations energies in the outward and inward modes, respectively, stops to increase out to a certain distance (around 2.5 AU, from Ulysses data, Bavassano et al. [2]). This saturation reminds strongly the saturation of parametric decay, where r_A ceases to increase in a very similar way, and we will see that also the predicted distance for the saturation is in the right range. The same interpretation and scenario are also supported by other works, both observational (the review paper by Bavassano in these proceedings) and theoretical (Malara et al. [11]; Primavera et al., again in these proceedings).

In this paper we will briefly summarize the main results found in DZ1, that dealed with waves in parallel propagation alone, whereas we will discuss in a broader

CP679, *Solar Wind Ten: Proceedings of the Tenth International Solar Wind Conference*,
edited by M. Velli, R. Bruno, and F. Malara

TABLE 1. Simulation parameters for the three runs, labeled A, B and C (see text).

run	η	β	k_0	k_c	k_t^-	γ
A	0.2	0.1	4	6	2	0.41
B	0.5	0.5	4	5	1	0.39
C	1.0	1.2	4	5	1	0.41

way the case of oblique propagation, that is certainly more common in realistic situations (the results are taken from Del Zanna [5], DZ2 from now on). Both cases demonstrate the robustness of the parametric decay process and naturally lead to the conclusion that parametric decay really seems to be the best candidate to explain the observed fast wind turbulence evolution.

PARALLEL PROPAGATION

When Alfvén waves propagating along a main uniform field $\mathbf{B}_0$ are circularly polarized, the total field intensity is constant and these waves are exact solutions of the full MHD equations regardless of their amplitude. However, it is also well known that these solutions are actually unstable in the presence of arbitrarily small compressive fluctuations. After a transient period, the initial pump wave (with wave number k_0) decays into a sound-like compressive wave (with greater wave number k_c) and into a backward propagating Alfvén-like transverse mode (with wave number $k_t = -(k_c - k_0)$).

In the small $\eta = B_\perp/B_0$ and $\beta = c_s^2/v_A^2$ limit, the linear dispersion relations for sound and Alfvén waves can be employed in the following resonance conditions

$$k_0 = k_c + k_t, \quad \omega_A(k_0) = \omega_s(k_c) + \omega_A(k_t), \qquad (1)$$

yielding the unstable wave numbers and the linear growth rate γ as (Sagdeev & Galeev [14]):

$$k_c \simeq 2k_0, \quad k_t \simeq -k_0; \quad \gamma \simeq \omega_0 \eta \beta^{-1/4}. \qquad (2)$$

Here, like in DZ1 and DZ2, three sets of parameters are considered (see Table 1), with values appropriate for solar wind conditions near the Sun (A), intermediate (B), and at the Earth's orbit (C). Notice that the growth rates are in all cases very similar.

In DZ1 the development of the parametric decay was studied via numerical simulations (using a high-order shock-capturing 3D-MHD code, Londrillo & Del Zanna [9]), for the three cases, in a periodic numerical box where the exact solution Alfvén wave with $k = k_0$ was initially set, together with a small (10^{-4}) density white noise to trigger the instability. Both the linear phase and the nonlinear saturation were followed by plotting the time histories of the density fluctuations, of the Elsässer energies $E^\pm$ and of the normalized cross helicity, defined by

$$\sigma = \frac{E^+ - E^-}{E^+ + E^-}, \quad E^\pm = < \frac{1}{2}|\mathbf{z}^\pm|^2 >, \qquad (3)$$

where $\mathbf{z}^\pm = \delta\mathbf{v} \mp \delta\mathbf{B}$ ($\mathbf{z}^+$ is here the outgoing wave) and brackets indicate spatial averaging.

The results show that after the exponential growth of the daughter waves and of density fluctuations during the linear phase, there is a saturation when nonlinear steepening and heat dissipation occurs. Correspondingly r_A increases ($\sigma = (1-r_A)/(1+r_A)$ decreases) until it reaches a final constant value that appears to depend strongly on the plasma beta. In high-beta conditions an asymptotic balance between outward and inward modes is found, as observed with increasing heliocentric distance in the solar wind, while for low-beta conditions the cross helicity usually reverses its sign (the backscattered wave becomes the dominant mode) and sometimes even multiple decays are observed. Similar results are also found in 2-D and 3-D and even in non-periodic domains, confirming the robustness and universality of the parametric decay process.

OBLIQUE PROPAGATION

All solar wind *in situ* data show clearly that Alfvénic magnetic field oscillations are never circularly polarized, as would be natural for parallel propagating Alfvén waves preserving $B \sim$ const, but rather show a spherical arc-type polarization, with the tip of $\delta\mathbf{B}$ describing circular arcs (rotations are usually less than 180°) on the surface of an imaginary sphere with radius $B = |\mathbf{B}|$. According to Riley et al. [12], who performed a systematic analysis by studying *Ulysses* fast-stream data in the ecliptic plane, up to 10% of Alfvénic modes are pure arc-polarized waves, meaning that the plane containing the maximum and intermediate variance directions remains fixed in time. The majority of the events was best explained by planar (1-D) arc-polarized Alfvén waves propagating obliquely to the background magnetic field and by embedded phase-steepened structures (rotational discontinuities: RDs).

Arc polarization for large-amplitude oblique Alfvénic modes was first proposed by Barnes and Hollweg [1] who showed, by performing a second-order expansion of the 1-D MHD nonlinear equations, that a monochromatic Alfvén wave which is initially linearly polarized develops another oscillatory component achieving arc polarization and $B^2 =$ const. Magnetoacoustic modes should also form in the process, but Landau damping was proposed to be efficient in the almost collisionless solar wind plasma so as to leave the exact Alfvénic $B^2 =$ const mode

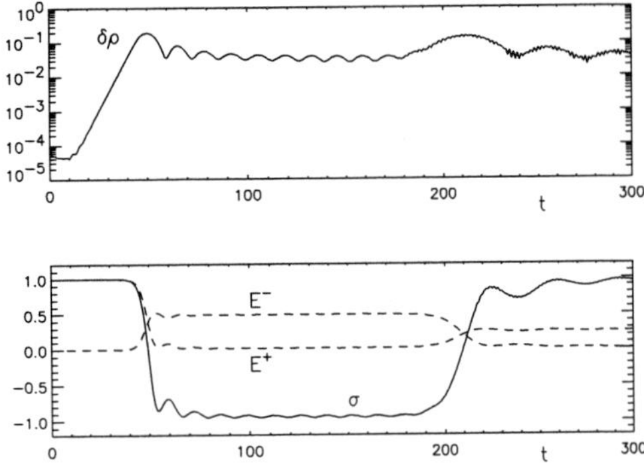

FIGURE 1. Time histories of *rms* density fluctuations (top), normalized cross helicity σ and normalized Elsässer energies $E^{\pm}$. Parametric decay occurs twice, at $t = 50$ and $t = 210$. Here run A parameters with $\theta = 30°$ are employed.

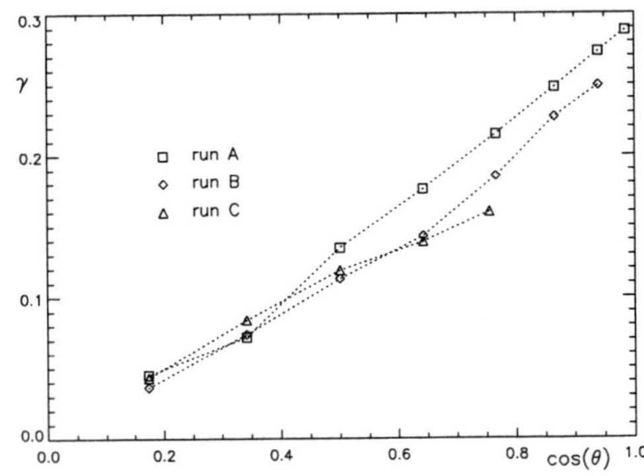

FIGURE 2. The measured growth rates γ as a function of $\cos\theta$, with θ ranging from 10° to 80°.

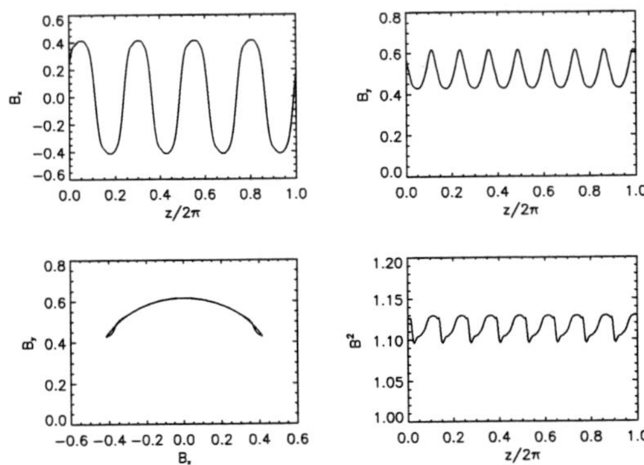

FIGURE 3. Magnetic field components (B_z is constant and it is not shown), hodogram in the $B_x - B_y$ plane and B^2 for $t = 3\pi$. Here $\eta = \beta = 0.5$ and $\theta = 30°$.

dominant, as observed. This was confirmed by Vasquez and Hollweg [15] by using hybrid simulations, who indeed found asymptotic Alfvénic states of approximate arc polarization and constant B (both waves and RDs for sufficiently high amplitudes).

The first result that we show here is the total independence of the parametric decay process by the actual polarization of the mother wave. We take as initial condition an Alfvén wave in oblique propagation (k_0 is along z and B_0 makes an angle θ with it, with another component along y), given by

$$B_x = \eta B_0 \cos(k_0 z), \quad B_y = \sqrt{C^2 - B_x^2}, \tag{4}$$

where $C^2 = B_x^2 + B_y^2 = B^2 - B_{0z}^2$ is the constant squared module of the transverse magnetic field and is obtained by imposing $< \delta B_y >= 0$, see DZ2 for details.

In Fig. 1 we plot the *rms* density fluctuations, the normalized cross helicity σ and the Elsässer energies $E^{\pm}$ (normalized to $E^{+}(0)$) for run A parameters and for $\theta = 30°$. Like in DZ1 multiple decays may be seen, at least in this low-beta case, where density fluctuations first increase and then saturate and where the dominant wave looses about half of its energy at every instability saturation. Correspondingly, the cross helicity reverses its sign flipping between the ± 1 states for pure Alfvénic modes. The plots for run B and C are also very similar to those in DZ1 and are not reported here. The result that for $\beta \sim 1$ the asymptotic value of σ approaches zero after the first occurrence of the instability still holds (B: $\sigma = -0.3$, C: $\sigma = 0.2$).

In spite of the striking similarities between the behavior of circular and arc-polarized waves, the evolution of the instability is slower in the latter case. This is mainly due to the fact that the phase velocity is smaller by a factor $\cos\theta$, since the dispersion relation for the oblique mother wave is $\omega_0 = k_0 v_A \cos\theta$ and both times and growth rates should be normalized correspondingly. In Fig. 2 the derived growth rates for a set of different angles are shown for the three runs. The linear dependence on $\cos\theta$ is especially apparent for small amplitudes, as expected.

Consider now the evolution of an oblique Alfvénic mode linearly polarized along the x direction but without any y component. This time B^2 is not constant and both a driven wave and magnetoacoustic waves, all with $k = 2k_0 = 8$, are expected to form (density noise at $t = 0$ is not needed, due to initial non-equilibrium). The driven wave, that travels at the same speed as the mother wave, tends to restore a $B^2 \simeq$ const condition by creating a B_y component in arc polarization. The situation may be seen in Fig. 3, where an almost exact arc with nearly constant total magnetic strength is achieved. The parameters are those of run B, $\theta = 30°$ and $t = 3\pi$. Note that the B_x profile has almost phase-steepened to a RD, although this is much more apparent for larger amplitudes.

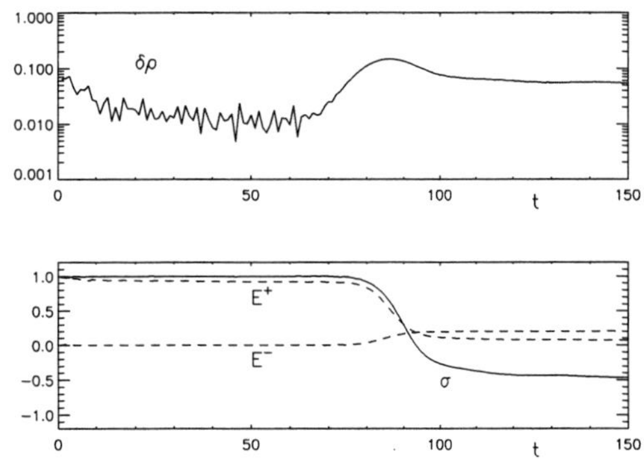

FIGURE 4. Same quantities as in Figure 1 but for the run of Fig. 3, with a linearly polarized wave as initial condition. After an initial dissipation phase, parametric decay sets in and saturates around $t = 85$.

Density fluctuations first increase very rapidly due to the ponderomotive force, but soon steepen and dissipate through shock heating, causing the *rms* value to decrease. In general, for smaller β and larger η density fluctuations are larger, but also dissipate more rapidly. When this value reaches a noise level, the situation is very similar to that of the previous section because we have a traveling arc-polarized wave with an almost constant field strength: parametric decay finally sets in and the mother wave decays as usual (see Fig. 4). The linear growth rates are about the same as those of the corresponding case of initial arc polarization, just around 10% less because of a stronger wave amplitude decay: for example $\gamma = 0.120$ and $\gamma = 0.093$ are found for the cases corresponding to runs A and B in Table 1. No decay is found for run C parameters, because the wave amplitude damping is too strong for $\eta = 1$ and the instability interval does not contain integer wave numbers any longer.

CONCLUSIONS

In this paper we have shown, by reporting selected results from two previously published papers (DZ1 and DZ2), that the parametric decay process invariably occurs whenever large-amplitude Alfvén waves are involved. This process is shown to be present under a wide variety of conditions: for parallel propagating waves with circular polarization and for oblique waves with arc-polarization (both are exact solutions of the MHD equations), with almost identical statistical properties. Moreover, even linearly polarized waves in oblique propagation are seen to first reach the state of arc-polarization, as predicted by Barnes & Hollweg [1], and later to decay parametrically as usual.

Parametric decay is certainly a slow process in solar wind conditions, and both partial reflection due to Alfvén speed gradients or interaction with velocity shear layers are able to shape the turbulent spectrum in much shorter time-scales. However, this process seems to be able to explain rather naturally the evolution of the turbulent spectrum in the smooth polar wind, as described in the introduction. By taking our parameter sets A, B and C, that simulate different conditions from the outer corona to the Earth and for which the ratio $\gamma/\omega_0 \sim 5\%$ is constant (say it does not depend on the distance from the Sun), a wave with $P_0 \sim 10^4$ s will decay with a characteristic e-folding distance $d \sim (1/2\pi)(\gamma/\omega_0)^{-1} v_{SW} P_0 \sim$ 0.2 AU for $v_{SW} \sim 800$ km/s, that has the correct order (it may be slightly larger when considering non-monochromatic waves or dispersive effects). Moreover, the *Ulysses* polar wind data by Bavassano et al. [2] show that the Alfvénic ratio ceases to increase at 2.5 AU, its value being $r_A \sim 0.5 \Rightarrow \sigma = (1-r_A)/(1+r_A) \sim 0.3$, to be compared with our asymptotic value of $\sigma \sim 0.2$ for solar wind conditions. However, for more realistic and quantitative modeling, non-monochromatic wave trains, dispersion and radial expansion should be all included in future simulations.

REFERENCES

1. Barnes, A., and J. V. Hollweg, *J. Geophys. Res.* **79**, 2302 (1974).
2. Bavassano, B., E. Pietropaolo, and R. Bruno, *J. Geophys. Res.* **105**, 15,959 (2000).
3. Belcher, J. W., and L. Davis, *J. Geophys. Res.* **76**, 3534 (1971).
4. Del Zanna, L., M. Velli, and P. Londrillo, *Astron. Astrophys.* **367**, 705 (2001, DZ1).
5. Del Zanna, L., *Geophys. Res. Lett.* **28**, 2585 (2001, DZ2).
6. Dobrowolny, M., A. Mangeney, and P. Veltri, *Phys. Rev. Lett.* **45**, 144 (1980).
7. Goldstein, B. E., M. Neugebauer, and E. J. Smith, *Geophys. Res. Lett.* **22**, 3389 (1995).
8. Grappin, R., and M. Velli, *J. Geophys. Res.* **101**, 425 (1996).
9. Londrillo, P., and L. Del Zanna, *Astrophys. J.* **530**, 508 (2000).
10. Malara, F., P. Veltri, C. Chiuderi, and G. Einaudi, *Astrophys. J.* **396**, 297 (1992).
11. Malara, F., L. Primavera, and P. Veltri, *Phys. Plasmas* **7**, 2866 (2000).
12. Riley, P., C. P. Sonnet, B. T. Tsurutani, A. Balogh, R. J. Forsyth, and G. W. Hoogeveen, *J. Geophys. Res.* **101**, 19,987 (1996).
13. Roberts, D. A., M. L. Goldstein, L. W. Klein, and W. H. Matthaeus, *J. Geophys. Res.* **92**, 12,023 (1987).
14. Sagdeev, L. Z., and A. A. Galeev, *Nonlinear Plasma Theory*, Benjamin, New York (1969).
15. Vasquez, B. J., and J. V. Hollweg, *J. Geophys. Res.* **101**, 13,527 (1996).

A Three Dimensional Magnetohydrodynamic Pulse in a Transversely Inhomogeneous Medium

D. Tsiklauri and V.M. Nakariakov

Physics Department, University of Warwick, Coventry, CV4 7AL, U.K. tsikd@astro.warwick.ac.uk

Abstract. Interaction of impulsively generated MHD waves with a one-dimensional plasma inhomogeneity, transverse to the magnetic field, is considered in the three-dimensional regime. Because of the transverse inhomogeneity, MHD fluctuations, even if they do not include initially any density perturbation, evolve toward states where the compressible components tend to become predominant. The propagating MHD pulse asymptotically reaches a quasi-steady state with the final levels of density perturbation weakly depending on the degree of non-planeness of the pulse in the homogeneous transverse direction and somewhat stronger depending on plasma β. Our study demonstrates the necessity of incorporation of compressible and 3D effects in theory of Alfvén wave phase mixing. However, as far as the dynamics of weakly non-plane Alfvén waves is concerned it can still be qualitatively understood in terms of the previous 2.5D models.

INTRODUCTION

Problems of heating of the open corona of the Sun and acceleration of the solar wind are closely related with the interaction of MHD waves with plasma inhomogeneities and, in particular, with coupling of compressible and incompressible components of the waves. The original idea of the phase mixing of *incompressible* Alfvén waves ([2]) was based on the following argument: when plasma has a density gradient perpendicular to the magnetic field, local Alfvén speed is a function of the transverse coordinate. Thus, when an Alfvén wave propagates along the field its perturbations on the adjacent field lines become out of phase. This stretching of Alfvén wave front creates progressively smaller spatial scales across the field. In turn, because the dissipation is proportional to the wave number squared, phase mixing leads to enhanced dissipation of the Alfvén wave directly. In the *compressible* plasma, as demonstrated by [3, 4, 5, 1, 7, 8], phase mixing of linearly polarized plane Alfvén waves leads to the enhanced nonlinear generation of fast magnetoacoustic waves, which dissipate more efficiently than incompressible Alfvén waves. Also, the compressive waves, as opposed to the Alfvén waves, can transport energy across the magnetic field, due to the fact that their propagation in space is not constrained by the field lines. Here we consider fully three dimensional geometry and study interaction of linear (with the non-linear effects totally ignored) MHD waves with a one-dimensional inhomogeneity of the plasma, taking into account *compressibility* of the plasma and the *localization* of the MHD pulse in the direction perpendicular to both the magnetic field and the inhomogeneity gradient. In particular, we shall study how the phenomenon of phase mixing is affected by these factors. In this work we study the propagation part of the problem, i.e. we consider an ideal plasma limit. The work is in progress to include finite plasma resistivity in order to investigate quantitatively the dependence of the decay of Alfvénic part of the MHD pulse upon the coupling to the existing compressive waves.

THE MODEL

In our model we use equations ideal MHD

$$\rho\frac{\partial\vec{V}}{\partial t}+\rho(\vec{V}\cdot\nabla)\vec{V}=-\nabla p-\frac{1}{4\pi}\vec{B}\times\mathrm{curl}\vec{B}, \quad (1)$$

$$\frac{\partial\vec{B}}{\partial t}=\mathrm{curl}(\vec{V}\times\vec{B}), \quad (2)$$

$$\frac{\partial p}{\partial t}+\vec{V}\cdot\nabla p+\gamma p\nabla\cdot\vec{V}=0, \quad (3)$$

where $\vec{B}$ is the magnetic field, $\vec{V}$ is plasma velocity, ρ is plasma mass density, and p is plasma thermal pressure. In what follows we use $5/3$ for the value of γ. We solve equations (1)-(3) in Cartesian coordinates (x,y,z). Note that as we solve a fully 3D problem we retain variation in the y-direction, i.e. $(\partial/\partial y\neq 0)$. The equilibrium state is taken to be an inhomogeneous plasma of density $\rho_0(x)$ and a uniform magnetic field B_0 in the z-direction. We consider a plasma configuration similar to the one

CP679, *Solar Wind Ten: Proceedings of the Tenth International Solar Wind Conference*, edited by M. Velli, R. Bruno, and F. Malara

investigated in [3, 4, 5, 1, 7], i.e. the plasma has a one-dimensional inhomogeneity in the equilibrium density $\rho_0(x)$ and temperature $T_0(x)$. The unperturbed thermal pressure, p_0, is taken to be constant everywhere.

Next, we do usual linearization of the Eqs. (1)-(3) and write them in component form as following

$$\rho_0(x)\frac{\partial V_x}{\partial t}+\frac{\partial p}{\partial x}-\frac{B_0}{4\pi}\left(\frac{\partial B_x}{\partial z}-\frac{\partial B_z}{\partial x}\right)=0, \quad (4)$$

$$\rho_0(x)\frac{\partial V_y}{\partial t}+\frac{\partial p}{\partial y}+\frac{B_0}{4\pi}\left(\frac{\partial B_z}{\partial y}-\frac{\partial B_y}{\partial z}\right)=0, \quad (5)$$

$$\rho_0(x)\frac{\partial V_z}{\partial t}+\frac{\partial p}{\partial z}=0, \quad (6)$$

$$\frac{\partial B_x}{\partial t}-B_0\frac{\partial V_x}{\partial z}=0, \quad (7)$$

$$\frac{\partial B_y}{\partial t}-B_0\frac{\partial V_y}{\partial z}=0, \quad (8)$$

$$\frac{\partial B_z}{\partial t}+B_0\left(\frac{\partial V_x}{\partial x}+\frac{\partial V_y}{\partial y}\right)=0, \quad (9)$$

$$\frac{\partial p}{\partial t}+\gamma p_0\left(\frac{\partial V_x}{\partial x}+\frac{\partial V_y}{\partial y}+\frac{\partial V_z}{\partial z}\right)=0. \quad (10)$$

It is useful to re-write Eqs. (4)-(10) in a form of three coupled wave equations as following

$$\left[\partial^2_{tt}-\left(c_s^2(x)+c_A^2(x)\right)\partial^2_{xx}-c_A^2(x)\partial^2_{zz}\right]V_x \quad (11)$$
$$-\left[\left(c_s^2(x)+c_A^2(x)\right)\partial^2_{xy}\right]V_y-\left[c_s^2(x)\partial^2_{xz}\right]V_z=0,$$
$$\left[\partial^2_{tt}-\left(c_s^2(x)+c_A^2(x)\right)\partial^2_{yy}-c_A^2(x)\partial^2_{zz}\right]V_y \quad (12)$$
$$-\left[\left(c_s^2(x)+c_A^2(x)\right)\partial^2_{xy}\right]V_x-\left[c_s^2(x)\partial^2_{yz}\right]V_z=0,$$
$$\left[\partial^2_{tt}-c_s^2(x)\partial^2_{zz}\right]V_z-\left[c_s^2(x)\partial^2_{xz}\right]V_x \quad (13)$$
$$-\left[c_s^2(x)\partial^2_{yz}\right]V_y=0,$$

where $c_A(x) = B_0/\sqrt{4\pi\rho_0(x)}$ and $c_s(x) = \sqrt{\gamma p_0/\rho_0(x)}$ denote local Alfvén and sound speeds respectively. We solve Eqs.(4)-(10) numerically after re-writing them in a dimensionless form using following normalization: $B_{x,y,z} = B_0\bar{B}_{x,y,z}$, $(x,y,z) = a_*(\bar{x},\bar{y},\bar{z})$, $c_A(x) = B_0/\sqrt{4\pi\rho_0(x)} = B_0/\sqrt{4\pi\rho_*}/\sqrt{3-2\tanh(\lambda x)} = c_A^*/\sqrt{3-2\tanh(\lambda x)}$, $t = (a_*/c_A^*)\bar{t}$, $V_{x,y,z} = c_A^*\bar{V}_{x,y,z}$. Note, that $c_s(x) = \sqrt{\gamma\beta/2}c_A(x)$, where β stands for the ratio of thermal and magnetic pressures $\beta = p_0/(B_0^2/8\pi)$. Here, λ is a free parameter which controls the steepness of the density profile gradient. In our simulations we use $\lambda = 0.5$. Our choice of density variation mimics an edge of a coronal plume, across which density variation is typically a factor of 4. In what follows we omit bars on top of the physical quantities.

NUMERICAL RESULTS

In order to solve Eqs.(4)-(10) numerically we have written a new numerical code *dt4dx10*, which uses a high order finite difference scheme. Namely, it evaluates 10-th order centered spatial derivatives, and advances solution in time using 4-th order Runge-Kutta algorithm. The simulation cube size is set by the limits $-25.0 \le x \le 25.0$, $-25.0 \le y \le 25.0$ and $-25.0 \le z \le 25.0$. Boundary conditions used in all our simulations are zero-gradient in all three spatial dimensions. We have performed calculation on various resolutions in attempt to achieve convergence of the results. The graphical results presented here are for the spatial resolution 128^3, which refers to number of grid points in x, y and z directions respectively. We have also performed calculation on the spatial resolution 256^3 and we found that the results converge perfectly. In the numerical simulations the MHD perturbation is initially a plane (with respect to x-coordinate) pulse, which has a Gaussian structure in y and z-coordinates

$$V_y(x,y,z,t=0) = c_A(x)\exp\left(-(\alpha_y y)^2-(\alpha_z z)^2\right). \quad (14)$$

Here, α_y and α_z are free parameters which control the strength of gradients in y and z direction of the initial perturbation. As the problem considered is linear, the wave amplitude can be taken to be normalized to unity. In the geometry considered, when the initial perturbation depends on y it cannot be regarded as pure Alfvénic one. In fact, in such a pulse, all three waves – Alfvén, fast and slow magnetosonic waves – are inter-coupled so that there is no use of their separation *per se*. The pulse is set to be initially plane in the x-direction. This allows us to emphasize the effect of the inhomogeneity on the pulse evolution. From the point of view of applications, this simply means that the initial characteristic size of the pulse in that direction is greater than the scale of the inhomogeneity. The particular choice of the initial condition Eq. (14) is motivated by the argument that when we set B_y initially to zero we automatically guarantee fulfillment of div$\vec{B} = 0$ when $\alpha_y \neq 0$, and as the system evolves it adjusts itself which mode to excite (depending whether α_y is zero or not). For instance, if we choose our initial conditions as V_y given by Eq. (14), $B_y(x,y,z,t=0) = \exp\left(-(\alpha_y y)^2-(\alpha_z z)^2\right)$, with $\alpha_y = 0$ and the rest of physical quantities set to zero, we would excite a pure Alfvénic pulse traveling in the negative direction along z-axis. However, as long as $\alpha_y \neq 0$ we have to set $B_y(x,y,z,t=0) = 0$ in order to fulfill div$\vec{B} = 0$. Besides, our choice of the localized "kinematic" (the velocity is initially perturbed only, while the magnetic field is constant) perturbation Eq. (14) is well motivated by the fact that such perturbations can arise, e.g., during coronal mass ejections, solar flares, or other violent events (see, e.g., [6]), etc.

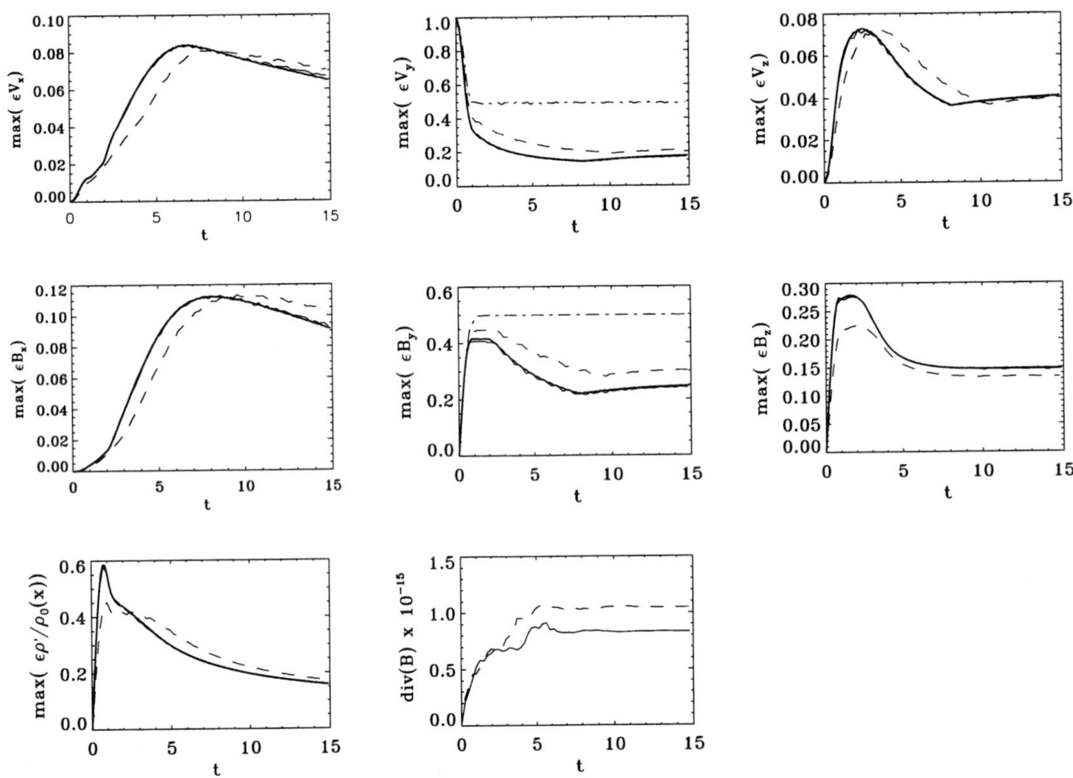

FIGURE 1. Maximum of the absolute values over the whole 3D simulation box of all physical quantities as a function of time. Here, $\beta = 0.5$ and $\alpha_z = 1.0$. Thick solid lines correspond to the resolution 256^3, while thin solid lines to that of 128^3, both for $\alpha_y = 0.6$. Note that on all seven panels these two curves practically overlap, which serves as a proof of convergence of our simulation results. Dashed lines represent $\alpha_y = 0.4$, while dash-dotted lines show $\alpha_y = 0$. The last panel shows div$\vec{B}$ versus time.

A fairly good quantity describing the relation between different components of the pulse is the maximum of absolute value over the whole simulation domain. We plot this quantity for all physical variables in Figure 1. In this figure, thick solid lines correspond to the resolution 256^3, while thin solid lines to that of 128^3, both for $\alpha_y = 0.6$. It is remarkable that on all panels these two curves practically do overlap, which serves as a proof of convergence of our simulation results. The last figure in the bottom row presents the maximum of absolute value over the whole simulation domain of div $\vec{B}$, and we indeed observe almost perfect fulfillment of the fundamental law, div $\vec{B} = 0$, which comes as a bonus of high-order, centered, finite difference numerical scheme. As expected, Fig. 1 shows that when $\alpha_y = 0$, the pulse is perfectly Alfvénic (perturbing V_y and B_y only). This conclusion can be deducted either from analyzing Eqs. (11)-(13) or resorting to a classic mechanical analogy of coupled pendulums. In effect, Eqs. (11)-(13) also describe three inter-coupled mathematical pendulums which are located in xOz plane. Thus, as long as we do not perturb these pendulums such that $\alpha_y \neq 0$, they will always oscillate in the xOz plane. Thus, what we see in Fig. 1 when $\alpha_y = 0$ (dash-dotted lines) is that initial kinematic Alfvén perturbation (V_y) is split in half-amplitude D'Alambert's solutions (both for V_y and B_y) and no other physical quantity is generated. However, when we switch on the coupling, $\alpha_y \neq 0$, compressible perturbations are generated.

In Figure 2 we investigate the energetics of our numerical simulation. In fact, we observe nearly perfect ($\pm 0.2\%$ error) conservation of total energy by *dt4dx10* numerical code. The major conclusion which can be drawn from this graph is that when $\alpha_y = 0$ there is no internal (compressive) energy generation, while with the increase of α_y its final, asymptotic levels increase progressively. The parametric space of the problem is studied in Figure 3 by plotting the final levels of relative density perturbation, approached by the MHD pulse, as a function of the initial localization of the pulse in the y-direction α_y, and plasma β. We gather from Fig. 3 that the achieved levels of density fluctuations depend weakly on α_y, while there is somewhat stronger dependence on β.

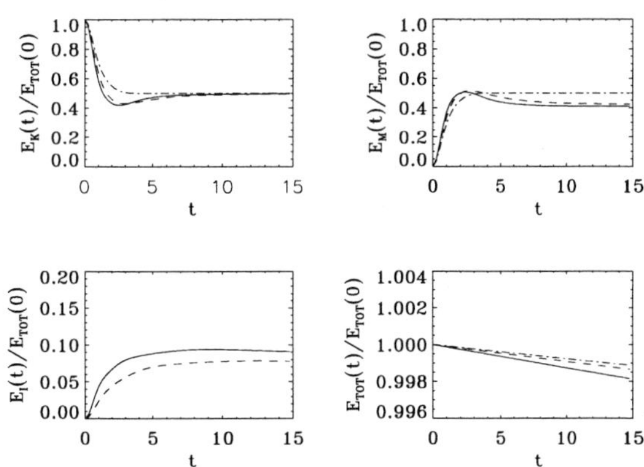

FIGURE 2. Normalized kinetic, magnetic, internal and total energies (all of them integrated over the volume) as a function of time. Here, $\beta = 0.5$ and $\alpha_z = 1.0$. Solid lines correspond to $\alpha_y = 0.6$, while dashed and dash-dotted lines show cases of $\alpha_y = 0.4$ and $\alpha_y = 0$ respectively.

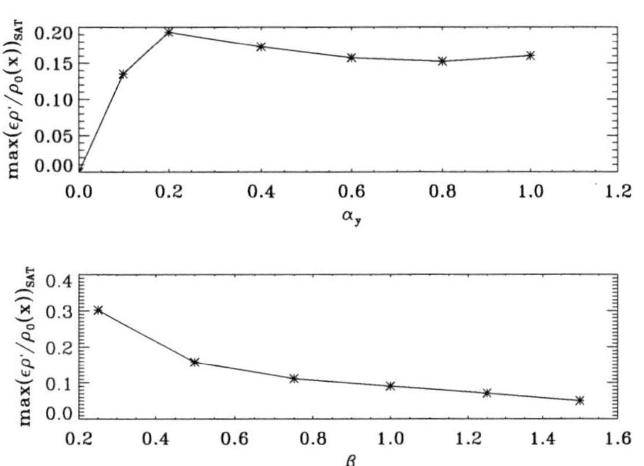

FIGURE 3. Top panel: the final levels of relative density perturbations as a function of α_y (for $\beta = 0.5$ and $\alpha_z = 1.0$). Bottom panel: the same but as a function of β (for $\alpha_y = 0.6$ and $\alpha_z = 1.0$).

CONCLUSIONS

In this paper we present results of numerical modeling of interaction of an initially localized MHD pulse with a transverse inhomogeneity of the plasma. The pulse is non-plane in all three spatial directions and is subject to the effect of phase mixing. An impulsively generated perturbation of the plasma velocity develops to an anisotropically propagating pulse, which has a significant compressible component. More specifically: (1) The non-uniformness of the pulse in the homogeneous transverse (or, in other words, in the third, perpendicular to both the magnetic field and the inhomogeneity gradient) direction (non-zero α_y) leads to appearance of compressive perturbations in the pulse. (2) In the presence of plasma inhomogeneity, a non-uniform MHD pulse is essentially compressible. More specifically, when the pulse is initially plane in the homogeneous transverse direction, $\alpha_y = 0$, there is no internal (compressive) energy generation, while with the increase of α_y, the levels of the compressible energy increase progressively. (3) A propagating MHD pulse asymptotically reaches a quasi-steady state. The final levels of density perturbation, which can be considered as a measure of the compressibility in the pulse, depend weakly on α_y, while there is somewhat stronger dependence on β. (4) A smooth 1D transverse inhomogeneity of the plasma supports propagation of *compressible* MHD pulses. This mechanism of wave guiding is different from the well-understood phenomenon of refraction of fast magnetosonic waves, as the profile of the Alfvén speed in the inhomogeneity does not have a minimum. (5) Weakly non-plane MHD waves are subject to phase mixing and can qualitatively be considered in terms of the 2.5D theory of Alfvén wave phase mixing. Consequently, the 2.5D theory of Alfvén wave phase mixing remains relevant to the 3D case too. However, quantitative theories of the interaction of MHD waves with plasma inhomogeneities should include the compressibility of the plasma as a necessary ingredient.

Our main conclusion is that the effects of *three-dimensionality, compressibility and inhomogeneity* should be all together taken into account in the wave-based theories of coronal heating and solar wind acceleration, and as well as in the theories of MHD turbulence.

acknowledgements The authors are grateful to Tony Arber for valuable advise when *dt4dx10* code was written and for a number of valuable discussions. DT acknowledges financial support from PPARC. Numerical calculations of this work were performed using the PPARC funded Compaq MHD Cluster at St Andrews and Astro-Sun cluster at Warwick.

REFERENCES

1. Botha, G. J. J., Arber, T. D., Nakariakov, V. M., & Keenan, F.P. 2000, A&A, 363, 1186
2. Heyvaerts, J., & Priest, E. R. 1983, A&A, 117, 220
3. Malara, F., Primavera, L., & Veltri, P. 1996, ApJ, 459, 347
4. Nakariakov, V. M., Roberts, B., & Murawski, K. 1997, Solar Phys., 175, 93
5. Nakariakov, V. M., Roberts, B., & Murawski, K. 1998, A&A, 332, 795
6. Roussev, I., Galsgaard, K., Erdelyi, R., & Doyle, J.G. 2001, A&A 370, 298
7. Tsiklauri, D., Arber, T. D., & Nakariakov, V. M. 2001, A&A, 379, 1098
8. Tsiklauri, D., Nakariakov, V. M., & Arber, T. D., 2002, A&A, 395, 285

Session IV

Internal State, Composition, High Energy Particles

Solar Wind Composition

Robert F. Wimmer-Schweingruber[1]

Physikalisches Institut, Universität Bern, Sidlerstrasse 5, CH-3012 Bern, Switzerland

Abstract.
Accurate knowledge of the composition of the solar wind and of solar system abundances allows us to improve our understanding of various processes affecting the elemental, isotopic, and charge-state abundances of the solar wind. During the evolution of the Sun, isotopic and elemental abundances have been affected by migration processes at the interface between the well mixed outer convective zone and the radiative zone. In the solar atmosphere, the isotopic and elemental composition can be affected by Coulomb drag and wave-particle interactions. In addition, elemental abundances are clearly fractionated by some process that is controlled by the first ionization potential (FIP) of the elements. The abundances of the charge states are determined by processes in the corona, and, in interplanetary space, serve as valuable tracers for the coronal origin of the solar wind. The abundances of certain key elements and their isotopes can safely and accurately be determined from their meteoritic values. Their abundances in the solar wind can be used to obtain estimates for the degree of fractionation undergone by other elements and their isotopes, for which the meteoritic abundances do not necessarily reflect photospheric values, or for which it is not clear which meteoritic component corresponds to solar values. In short, the composition of the solar wind can yield valuable information about processes in the early solar system and its evolution, the evolution of the Sun, processes in the solar interior, atmosphere, and corona, as well as in interplanetary space. In addition, solar (wind) composition has implications for for astronomy and astrophysics, for stellar models, and for models of galactic chemical evolution.

INTRODUCTION

Knowledge of the composition of the Sun and the solar wind is important for other fields than just solar wind studies. Apparent changes in solar metallicity with improving measurements have implications for models of the chemical evolution of the galaxy and for the formation of planet-bearing stellar systems, to name two important examples. Moreover, because the Sun is the star we can study best of all, it is a unique test bed for theories of stars and their winds. In addition, detailed studies of solar system and solar abundances may tell us how to interpret abundances measured in other stars. Can they be taken straightforwardly, 1:1 from spectra? What corrections need to be made? Studies of solar abundances are of immense help in interpreting abundances throughout the galaxy and probably beyond.

Inspite of the fact that we know solar abundances to a remarkable degree of accuracy (that we will discuss below), there are several puzzles in solar and galactic composition that currently elude our understanding. Why does the Sun appear to be chemically equally evolved as the local interstellar medium [1]? Why does the isotopic composition of non-volatile elements in the galactic cosmic rays agree so well with their solar values [2]? We would expect the galaxy and hence the interstellar medium to have evolved chemically since the formation of the solar system some 4.57×10^9 years ago.

Nevertheless, the badly understood, but good, agreement of solar and galactic abundances are an improvement to what appeared as an even greater mystery just about 10 years ago, the "old Sun problem". Compared to the local neighborhood, the Sun was metalrich and hence presumed to be older. Galactic gradients of element abundances are expected to be negative, i.e. the abundances of heavy elements generally decrease with increasing distance from the galactic center. Part of the confusion (but not all) in the past has been due to badly known solar abundances. While the old Sun problem has been partly resolved by a finally available volume-limited sample of solar-like stars, recent corrections in solar metallicity (to which the oxygen abundance is the prime contributor) also contributed to its resolution [3]. This underlines the importance of determining true abundances (i. e. of some element X relative to H, X/H) as opposed to the more traditional abundance ratios with

[1] now at: Institut für experimentelle und angewandte Physik, Christian-Albrechts-Universität zu Kiel, D-24098 Kiel, Germany

CP679, *Solar Wind Ten: Proceedings of the Tenth International Solar Wind Conference*, edited by M. Velli, R. Bruno, and F. Malara

respect to some favorite element, e. g. O or Si. Intriguingly, the old Sun problem still appears to be live and kicking when one considers isotopic galactic gradients. There the Sun still lies off the trend expected from galactic chemical evolution models based on measurements of galactic isotope gradients. Will this again be solved by an apparent "change" in solar abundances, this time isotopic abundances?

Whatever the answers to the mentioned questions may turn out to be, they have certainly motivated why the study of solar and solar wind composition is important. We will now proceed to discuss just how well we know solar and solar wind abundances.

SOLAR SYSTEM ABUNDANCES

As the solar system formed some 4.57×10^9 years ago, it closed itself off from the interstellar medium as it existed at that time and at that location. While the bulk of the matter was concentrated in the Sun itself, only a small fraction of it is available to us for composition studies. From characteristic differences in the composition of various phases or separates of different solar system bodies we have been able to acquire a good understanding of the origin of the solar system. For instance, among the nearly 23'000 cataloged meteorites, there is one class, the CI chondrites, that exhibit a composition that is nearly identical to photospheric. Therefore, that class is considered the most primitive class and often used to determine solar abundances. However, there are only five CI chondrites, of which there is an abundant amount of sample material from only one, Orgeuil [4]. This should illustrate that we should be cautious when equating solar and solar system abundances. This is often done because meteorites can be studied in the laboratory and hence to a high degree of accuracy. However, it is often forgotten that this accuracy is limited by sample-to-sample variations of 3 - 10% [H. Palme, pers. comm.,2002] [see e. g. 5, for an illustrative figure], while photospheric abundance determinations are limited in accuracy to 25 - 30% by unknowns in the processes on the solar surface and in atomic parameters [3, 6]. Despite the small number of CI chondrite samples, let us not forget that one gram of such material contains many more atoms than have been analyzed by in-situ composition experiments during the space age.

When the accuracy of abundance determinations will improve sometime in the future (e. g. with the sample return from the Genesis mission), there will be a new set of problems that will need to be investigated. For instance, the interplay of radiative forces, collisions, and gravitation in the (time-evolving) outer convective zone of the Sun leads to a depletion of the heavy element abundances by nearly 10% [see e. g. 5, for a detailed comparison with meteoritic abundance ratios]. So just how representative are photospheric abundances of those of the bulk Sun and the solar system?

In the early phases of the formation of the solar system, the Sun most likely underwent a violent T-Tauri phase. Just how did the protoplanetary disk and the nascent Sun interact? In this phase, a few 10^{-6} solar masses of disk material are processed per year. Some material may have been expelled from the solar system by an X wind [7], some was funneled into the Sun, non-volatile material may have been ejected back into the disk along ballistic trajectories, allowing the formation of chondrules. Given this violent youth, just how representative of solar abundances is the bulk disk material?

Let me illustrate one more question that may be solved with more precise abundance measurements. Butler *et al.* [8] found that planet-bearing stars had a higher metallicity than the Sun, see Figure 1. So far, only stars with

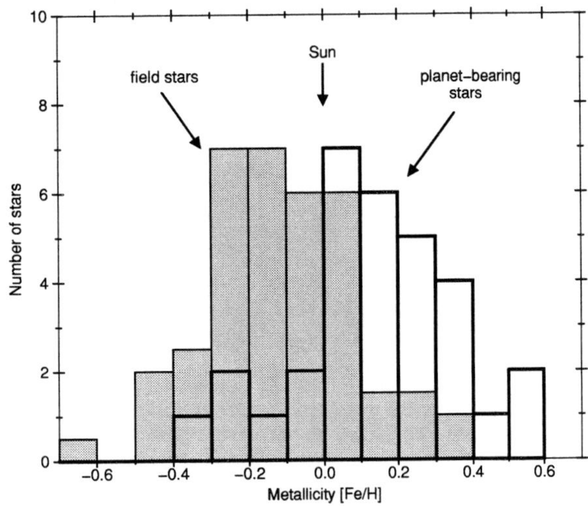

FIGURE 1. Comparison of the metallicity of planet-bearing and field stars. Data are from Butler *et al.* [8]. Plotted are the number of field stars (shaded, thin line) and planet-bearing stars (empty, thick line) in a given metallicity bin. Metallicities are normalized to solar and given on a logarithmic scale (dex).

at least Saturn-sized companions have been detected [8]. Smaller planets cannot presently be detected because their influence on the central star is too small. Because giant gas planets only form beyond the ice line, their detection closer to their star means that they have migrated inward. And any Earth-like, rocky planet must have migrated too far... Planet-bearing stars also show a striking surface enhancement of Li [9], an element that is largely destroyed by nucleosynthetic processes in the early phases of the evolution of Sun-like stars. That Li is depleted in the solar photosphere would then imply that the Sun did not "eat up" a substantial amount of planetary matter after the formation of an outer convective zone. This implies that the disk cleared faster than the migration time for terrestrial planets, possibly due to intense radiation of neighboring giant stars which were born in the same star-forming region as the Sun and which may have contributed to or altered the composition of disk material.

SOLAR WIND COMPOSITION

The composition of the solar wind is one step removed from that of the photosphere. It is altered by a multitude of fractionation mechanisms which may not all be active at the same time and locations. The most striking effect is the fractionation according to first ionization potential (FIP) which appears to enhance the abundance of elements with low FIP (below about 10 eV) with respect to elements with a higher FIP. However, the previously clear cut FIP step between low- and high-FIP elements does not appear to be as clear as it used to be. Plotting the familiar FIP plot with logarithmic axes results in the plot shown in Figure 2. A dashed line ($\propto \mathrm{FIP}^{-1}$) has been drawn to guide the eye and appears to organize the plot just as well as a step-like curve.

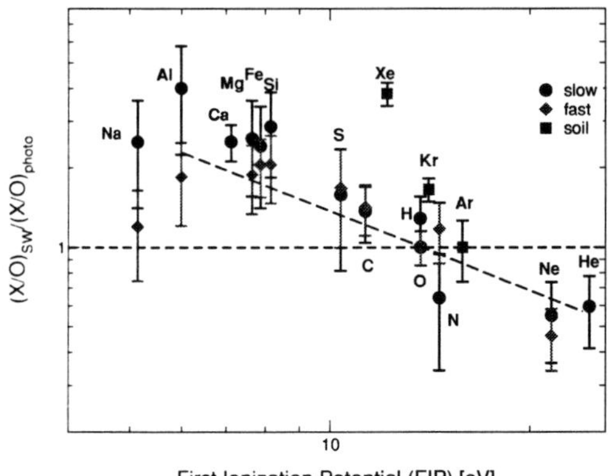

FIGURE 2. Comparison of solar wind abundance ratios normalized with O with their photospheric values. Elements with low first ionization potential (FIP) are enhanced with respect to high-FIP elements. The dashed horizontal line indicates photospheric values. The diagonal dashed line corresponds to an abundance enhancement ordered by 1/FIP and forced to be unity for O. It is only drawn to guide the eye. Solar wind abundances were taken from the review of [10] and normalized to the photospheric abundances of [11]

The origin of the FIP effect seems to lie in some mechanism that separates atoms and ions [12, 13, 14]. Measurements of noble gas abundances in lunar soils [15] may help to better understand this mechanism because they do not appear to follow the same trend as the other elements in Figure 2. Assuming solar matter to be ionized by UV and EUV photons, one may derive a first ionization time (FIT) which appears to order all elements, including Ar, Kr, and Xe, better than FIP [16]. The FITs of other ionizing scenarios, e. g. electron impact ionization, have not been computed to my knowledge, and we do not know how they would organize a FIT plot.

Other fractionation mechanisms such as inefficient Coulomb drag or wave-particle interactions further modify the abundances of the nascent solar wind from chromospheric to what we measure in interplanetary space [17, 18]. A good understanding of these effects is necessary for an accurate determination of solar abundances from solar wind measurements. With the sample-return mission Genesis now collecting solar wind (see the paper by Reisenfeld et al. presented at this conference) we need to address fractionation mechanisms to a much higher degree of accuracy [See e. g. 19, for a review.].

SOLAR WIND COMPOSITION AS A TRACER

So far, we have considered the use of solar (wind) composition for what might be termed "studies of origins" - the origins of the Sun and its planets, the solar system, or the evolution of the galaxy. Of course, certain models need to be developed to make corrections for fractionation mechanisms or other processes that could modify the abundances of the elements. One of the beauties of composition studies is the possibility to use composition measurements to test these models, to infer the importance of the various processes, but also to to use the many possible combinations of elements, isotopes, and ions for studies of the underlying plasma physics which determines the properties of the solar atmosphere, the heliosphere, and beyond. Let us consider just a few examples of how changes or differences in solar wind composition can be used to infer the importance of various processes active in the chromosphere, the corona, and interplanetary space.

The Chromosphere

The chromosphere appears to be the most likely site of the FIP effect. Relating abundance enhancements in a FIP plot to FIT allows us to determine timescales for the atom-ion separation process, typically a few to a few 10s of seconds. The process gradually depletes the high-FIP elements until the material is released to the solar wind. Widing and Feldman [20] have investigated the FIP enhancement in active regions over time. They found that the material in the loops is unfractionated right after appearance of the loop and that the ratio of Mg/Ne (their proxy for low FIP to high FIP elements) increases approximately linearly with time. Hence solar wind abundance measurements should give information about the lifetime of loops prior to disruption.

The Corona

On its way from the photosphere to interplanetary space, the solar wind is further modified in the corona e. g. by collisions with coronal electrons. The charge-state composition of an element is determined by by a competition of timescales. As the solar wind is accelerated through the corona it experiences a diminishing electron number density and hence an increase in the ionization timescale, but a decrease in the expansion timescale because the solar wind is being accelerated. Where the two timescales are equal, we traditionally say that the charge states freeze in [21] at that location and electron temperature. Unfortunately, this picture is overly simplistic and one cannot determine a coronal electron temperature profile from charge-state measurements in any straightforward way (See the paper by Esser et al., presented at this conference.). Not all ions experience the same acceleration and hence differential streaming between different ions tries to establish itself. This is counteracted by the change of identity of ions when their charge state is modified by ionization or recombination. Depending on where in the corona the ions form, differential streaming is an important factor in determining the charge-state composition of the solar wind [22].

Recent Developments

The recently proposed model for the heliospheric magnetic field around solar activity minimum [23] not only is a possible explanation for the transport of energetic particles to high heliographic latitudes, but also for the quasi-rigid rotation of coronal holes and possibly even for the origin of the solar wind. This model differs from similar earlier work of Wang and Sheeley [24] by the introduction of a systematic and self consistent transport of field lines across the corona. An interesting consequence of this transport is the necessity of reconnection and migration of open field lines in the coronal hole and in the streamer belt. The reconnecting field lines open up previously closed loops, setting free the enclosed plasma, thus giving rise to the solar wind. In this picture the older and larger loops in the streamer belt have had more time to fractionate (see subsection chromosphere) and, by virtue of being higher than in coronal holes, should exhibit higher electron temperatures than the smaller and younger loops in coronal holes. The well known anti correlation of freeze-in temperatures with solar wind speed [25, 26, 27]is then interpreted as due to different loop heights [28] and is experimentally found to result from physically sensible parameters [29]. It is not clear, however, how the systematic ordering of and large differences in freeze-in temperatures arises in this scenario. If they are established higher up in the corona, as is generally assumed, then how should the plasma retain its memory of the loop temperature or loop height? If the temperature is set in the loops, why do charge-state ratios of C, O, Mg, Si, S, Fe, etc. all show different temperatures?

Interplanetary Science

In-situ measurements of solar wind heavy ions can give information about kinetic processes in the heliosphere. Early measurements of the speeds and kinetic temperatures of protons and alpha particles showed that the latter often flow faster then the former and that the temperatures are mostly mass proportional [see e. g. 30, 31, for a review]. Early measurements of heavy ions showed that they generally flow at a speed in between the proton and alpha-particle speed [32, 33, 34, 35, 36], a finding that has been confirmed recently [37]. The mass proportionality of kinetic temperatures also holds for heavy ions at high temperatures, however, the cold solar wind appears to be isothermal [37]. These results must be compared with the observation of von Steiger *et al.*, who found that, at 5 AU heliocentric distance, all heavy ions have kinetic temperatures which are mass proportional. In other words, the transition from collision dominated (isothermal) cold solar wind to wave heated (mass proportional temperatures) solar wind takes place between 1 and 5 AU. At 1 AU slightly warmer wind lies in a regime between collision and wave-particle-interaction dominated. In other words, 1 AU is an interesting place for in-situ instruments because that is where this transition takes place. Because wave-particle interaction is sensitive to the mass and charge of the interacting ions, and heavy ions contribute a large number of mass numbers and charge states, accurate measurement of their 3-d velocity distribution functions in various solar wind regimes for instance by the PLASTIC instrument on STEREO will be very interesting.

Coronal Mass Ejections

Coronal mass ejections (CMEs) are often but not always associated with compositional peculiarities. Large abundances of alpha particles and sometimes of singly charged helium have been reported [38], sometimes associated with high charge states of iron [38], even Fe^{20+} has been reported [39]. Occasionally, large overabundances of the very rare isotope ^{3}He are observed [40], some spectacular CMEs have very unusual composition, such as the May 1998 event [41, 42, 43]. A class of CMEs appears to exhibit evidence for mass-proportional fractionation, possibly due to leakage of material out of loop structures before the eruption [44, 45, and the article by Wurz *et al.* in these proceedings.]. In a study of 42 CMEs observed with SWICS on Ulysses, Neukomm [46] found that CMEs can be divided into three classes, one similar to the slow and the fast wind, respectively, and one distinct from any type of solar wind. Such surveys are important, because they drive home the point that not all CMEs are unusual, in fact, typical CMEs are quite unspectacular composition-wise. We often tend to concentrate on some "interesting cases" because they are spectacular - and then to forget that they are spectacular, not typical. Zurbuchen *et al.* give a summary of CME composition in these proceedings.

The composition of CMEs is probably determined in two phases with quite different time scales. The elemental (and isotopic) composition is largely determined in the time preceding the eruption. During this phase, various processes can fractionate elements according to atomic properties such as FIP, mass, or mass per charge. Recalling the results of Widing and Feldman [20], we should be able to determine how long the CME material was contained in an active region loop prior to eruption (it would also be interesting to compare with the Wurz model [45]). After the CME has begun to rise through the solar atmosphere, the charge states of the many different elements are affected by the rapidly changing electron energy distribution function.

SUMMARY, DISCUSSION, AND CONCLUSIONS

The current uncertainties of solar abundance determinations are large compared to the small signatures left in various components of solar system matter during and after the formation of the solar system. However, the sample-return mission Genesis will improve the accuracy of solar wind abundance ratios by about an order of magnitude, which in turn implies that many small effects which have so far been neglected need to be studied in more detail. Solar wind abundance measurements have the potential to improve in accuracy beyond that possible with meteoritic samples because possible sampling biases can be much better understood and corrected for.

With the increasing time (and ion species) resolution of in-situ mass spectrometers solar wind composition measurements are increasingly being used as a new tool for a wide variety of problems. Solar wind composition is used as a tracer for its origin in complicated situations such as co-rotating interaction regions (CIRs) [47, 48]. The structure of CMEs is also (at least partially) resolved in composition measurements [e.g. 44, 42]. High time resolution measurements of velocity distributions of heavy ions will give new information about plasma processes active in the inner heliosphere.

Accurate knowledge of solar abundances are important for a variety of applications, not just in our field, but also in astronomy and astrophysics. Most important, the Sun is the only star which we can study in the detail to which we have studied the Sun. For instance, here we can test model predictions for abundance alterations by comparing solar system with solar abundances. This cannot be done for other stars. Thus the solar and heliospheric community is in a unique position and communication between the solar and stellar composition communities needs to be increased.

ACKNOWLEDGMENTS

It is a pleasure to acknowledge stimulating discussions with P. Bochsler, P. Wurz, R. v. Steiger, H. Holweger, G. Gloeckler, T. H. Zurbuchen, and the many participants of the Joint SOHO/ACE workshop on "Solar and Galactic Composition". Furthermore, I wish to thank the organizers of SW 10 for their flawless organization of a very inspiring conference. This work was supported in parts by the Swiss National Science Foundation and the Canton of Bern.

REFERENCES

1. Gloeckler, G., and Geiss, J., "Composition of the local interstellar cloud from observations of interstellar pickup ions", in *Solar and Galactic Composition*, edited by R. F. Wimmer-Schweingruber, AIP conference proceedings, Melville, NY, 2001, pp. 281 – 289.
2. Wiedenbeck, M. E., Binns, W. R., Christian, E. R., Cummings, A. C., Davis, A. J., George, J. S., Hink, P. L., Israel, M. H., Leske, R. A., Mealdt, R. A., Stone, E. C., von Rosenvinge, T. T., and Yanasak, N. E., "Constraints on the Nucleosynthesis of Refractory Nuclides in Galactic Cosmic Rays", in *Solar and Galactic Composition*, edited by R. F. Wimmer-Schweingruber, AIP conference proceedings, Melville, NY, 2001, pp. 269 – 274.
3. Holweger, H., "Photospheric Abundances: Problems, Updates, Implications", in *Solar and Galactic Composition*, edited by R. F. Wimmer-Schweingruber, AIP conference proceedings, Melville, NY, 2001, pp. 23 – 30.
4. Grady, M. M., *Catalogue of Meteorites*, Cambridge University Press, 2000.
5. Turcotte, S., and Wimmer-Schweingruber, R. F., *J. Geophys. Res.* (2002), accepted for publication.
6. Del Zanna, G., Bromage, B. J. I., and Mason, H. E., "Elemental abundances of the low corona as derived

from SOHO/CDS observations", in *Solar and Galactic Composition*, edited by R. F. Wimmer-Schweingruber, AIP conference proceedings, Melville, NY, 2001, pp. 59 – 64.

7. Shu, F. H., Adams, F. C., and Lizano, S., *Annu. Rev. Astron. Astrophys.*, **25**, 23 – 81 (1987).
8. Butler, R. P., Vogt, S. S., Marcy, G. W., Fischer, D. A., Henry, G. W., and Apps, K., *Astrophys. J.*, **545**, 5–4 – 511 (2000).
9. Israelian, G., Santos, N. C., Mayor, M., and Rebolo, R., *Nature*, **411**, 163 – 166 (2001).
10. Wimmer-Schweingruber, R. F., *Adv. Space Sci.* (2002), invited review for COSPAR 2000, in press.
11. Grevesse, N., and Sauval, A. J., *Space Sci. Rev.*, **85**, 161 – 174 (1998).
12. Meyer, J. P., "A Tentative Ordering af all Available Solar Energetic Particle Abundance Observations: I- The Mass Unbiased Baseline", in *17th International Conference on Cosmic Rays, Paris*, CEN Saclay, Gif-sur-Yvette, France, 1981, vol. 3, p. 145.
13. Meyer, J. P., "A Tentative Ordering af all Available Solar Energetic Particle Abundance Observations: II- Discussion and Comparison with Coronal Abundances", in *17th International Conference on Cosmic Rays, Paris*, CEN Saclay, Gif-sur-Yvette, France, 1981, vol. 3, p. 149.
14. Geiss, J., *Space Sci. Rev.*, **33**, 201 – 217 (1982).
15. Wieler, R., *Space Sci. Rev.*, **85**, 303 – 314 (1998).
16. v. Steiger, R., *Space Sci. Rev.*, **85**, 407 – 418 (1998).
17. Bodmer, R., and Bochsler, P., *Astron. Astrophys.*, **337**, 921 – 927 (1998).
18. Bodmer, R., and Bochsler, P., *J. Geophys. Res.*, **105**, 47 – 60 (2000).
19. Bochsler, P., *Rev. Geophys.*, **38**, 247 – 266 (2000).
20. Widing, K. G., and Feldman, U., *Astrophys. J.*, **555**, 426 – 434 (2001).
21. Hundhausen, A., Gilbert, H., and Bame, S., *J. Geophys. Res.*, **73**, 5485 – 5493 (1968).
22. Chen, Y., Esser, R., and Hu, Y. Q., *J. Geophys. Res.* (2002), accepted for publication.
23. Fisk, L. A., *J. Geophys. Res.*, **101**, 547 – 553 (1996).
24. Wang, Y. M., and Sheeley, Jr., N. R., *Astrophys. J.*, **414**, 916 – 927 (1993).
25. von Steiger, R., Wimmer-Schweingruber, R. F., Geiss, J., and Gloeckler, G., *Adv. Space Res.*, **15**, 3 – 12 (1995).
26. Hefti, S., *Solar Wind Freeze-in Temperatures and Fluxes Measured with SOHO/CELIAS/CTOF and Calibration of the CELIAS Sensors*, Ph.D. thesis, University of Bern, Switzerland (1997).
27. Aellig, M. R., *Freeze-in temperatures and relative abundances of iron ions determined with SOHO/CELIAS/CTOF*, Ph.D. thesis, University of Bern, Switzerland (1998).
28. Fisk, L. A., *J. Geophys. Res.* (2002), accepted for publication.
29. Gloeckler, G., Geiss, J., and Zurbuchen, T. H., *J. Geophys. Res.* (2002), accepted for publication.
30. Marsch, E., Mühlhäuser, K. H., Rosenbauer, H., Schwenn, R., and Neubauer, F. M., *J. Geophys. Res.*, **87**, 35 – 51 (1982).
31. Neugebauer, M., *Fundam. Cosmic Phys.*, **7**, 131 – 199 (1981).
32. Ogilvie, K. W., Coplan, M. A., and Zwickl, R. D., *J. Geophys. Res.*, **87**, 1763 – 7369 (1982).
33. Bochsler, P., Helium and oxygen in the solar wind: Dynamic properties and abundances of elements and helium isotopes as observed with the ISEE-3 plasma composition experiment, Habilitationsschrift (1984), Physikalisches Institut, Universität Bern, Switzerland.
34. Bochsler, P., *J. Geophys. Res.*, **94**, 2365 – 2373 (1989).
35. Schmid, J., Bochsler, P., and Geiss, J., *Astrophys. J.*, **329**, 956 – 966 (1988).
36. Bochsler, P., Geiss, J., and Joos, R., *J. Geophys. Res.*, **90**, 10779 – 10789 (1985).
37. Hefti, S., Grünwaldt, H., Ipavich, F. M., Bochsler, P., Hovestadt, D., Aellig, M. R., Hilchenbach, M., Galvin, A. B., Geiss, J., Gliem, F., Gloeckler, G., Kallenbach, R., Klecker, B., Marsch, E., Möbius, E., Neugebauer, M., and Wurz, P., *J. Geophys. Res.*, **103**, 29697 – 29704 (1998).
38. Schwenn, R., Rosenbauer, H., and Mühlhäuser, K.-H., *Geophys. Res. Lett.*, **7**, 201 – 204 (1980).
39. Wimmer-Schweingruber, R. F., Kern, O., and Hamilton, D. C., *Geophys. Res. Lett.*, **26**, 3541 – 3544 (1999).
40. Ho, G. C., Hamilton, D. C., Gloeckler, G., and Bochsler, P., *Geophys. Res. Lett.*, **27**, 309 – 312 (2000).
41. Skoug, R. M., Bame, S. J., Feldman, W. C., Gosling, J. T., McComas, D. J., T.Steinberg, J., Tokar, R. L., Riley, P., Burlaga, L. F., Ness, N. F., and Smith, C. W., *Geophys. Res. Lett.*, **26**, 161 – 164 (1999).
42. Gloeckler, G., Fisk, L. A., Hefti, S., Schwadron, N., Zurbuchen, T., Ipavich, F. M., Geiss, J., Bochsler, P., and Wimmer, R., *Geophys. Res. Lett*, **26**, 157 – 160 (1999).
43. Wimmer-Schweingruber, R. F., Bochsler, P., Gloeckler, G., Ipavich, F. M., Geiss, J., Kallenbach, R., Fisk, L. A., Hefti, S., and Zurbuchen, T. H., *Geophys. Res. Lett.*, **26**, 165–168 (1999).
44. Wurz, P., Ipavich, F. M., Galvin, A. B., Bochsler, P., Aellig, M. R., Kallenbach, R., Hovestadt, D., Grünwaldt, H., Hilchenbach, M., Axford, W. I., Balsiger, H., Bürgi, A., Coplan, M. A., Geiss, J., Gliem, F., Gloeckler, G., Hefti, S., Hsieh, K. C., Klecker, B., Lee, M. A., Managadze, G. G., Marsch, E., Möbius, E., Neugebauer, M., Reiche, K. U., Scholer, M., Verigin, M. I., and Wilken, B., *Geophys. Res. Lett.*, **25**, 2557 – 2560 (1998).
45. Wurz, P., Bochsler, P., and Lee, M. A., *J. Geophys. Res.*, **105**, 27239 – 27249 (2000).
46. Neukomm, R. O., *Composition of CMEs*, Ph.D. thesis, University of Bern, Switzerland (1998).
47. Wimmer-Schweingruber, R. F., von Steiger, R., and Paerli, R., *J. Geophys. Res.*, **102**, 17407 – 17417 (1997).
48. Wimmer-Schweingruber, R. F., v. Steiger, R., and Paerli, R., *J. Geophys. Res.*, **104**, 9933 – 9945 (1999).

Ubiquitous Suprathermal Tails on the Solar Wind and Pickup Ion Distributions

George Gloeckler

Department of Physics and IPST, University of Maryland, College Park, Maryland 20742, USA

Abstract. Using time-of flight mass spectrometers on Ulysses and ACE we measure the velocity distributions of particles in the poorly explored suprathermal range that extends from about twice to fifty times the solar wind speed. We find that H^+, He^{++} and He^+ are always present in this energy (or speed) range at variable intensity levels that depend on solar wind conditions. Contrary to most common expectations, even during most quiet times, in the absence of shocks and other interplanetary disturbances, suprathermal power-law tails on solar wind distributions are observed that extend to the highest energies measured. Suprathermal tails are observed in distributions of solar wind and interstellar pickup ions, not just for protons but also in distributions of all heavy ions that can be measured. Tails are seen in the quiet and disturbed slow wind, as well as in the super-quiet fast wind from polar coronal holes. Instead of diminishing rapidly with heliocentric distance, tails are found to persist to at least 5.4 AU and become harder. We suggest that pre-accelerated particles contained in these tails may well be the population that is injected for further acceleration by shocks. It remains unknown what processes produce these tails and whether these tails persist to very large distances.

INTRODUCTION

There is little dispute that shocks are among the prime accelerators of particles in and beyond our solar system. A prime example is Anomalous Cosmic Rays that are most likely accelerated to their highest energies (hundreds of MeV) by the heliospheric termination shock. Yet a major difficulty with the shock acceleration mechanism is the injection problem. Shocks alone may not be able to efficiently accelerate particles below some threshold energy of about 100 keV/nuc. Observations of heliospheric suprathermal ions begin to provide some answers. We find that in the quiet solar wind, in the absence of all kinds of interplanetary shocks and strong turbulence associated with such shocks, in the absence of waves and compression regions, etc., solar wind and interstellar pickup ion velocity distributions have pronounced and persistent high-energy tails. The distribution functions of these ions are non-maxwellian with strong power-law tails to the highest energies observed. These tails are not generated in the process of solar wind acceleration because they also appear in distributions of interstellar pickup ions that are absent near the Sun. They are observed not just for protons but also in distributions of all heavy ions that were measured. Pre-accelerated particles contained in these tails may well be the population that is injected for further acceleration by shocks. We review observations from Ulysses and ACE of energy spectra of solar wind and pickup ions observed in the fast and slow quiet solar wind, and contrast these with those found in association with heliospheric shocks. Explaining the origin of these ubiquitous tails remains a challenge.

OBSERVATIONS OF SUPRATHERMAL TAILS

Measurements presented here were made using the Solar Wind Ion Composition Spectrometer (SWICS) instruments on Ulysses (1) and ACE (2). SWICS is an ion mass vs. mass/charge spectrometer in which energy/charge analysis, followed by post-acceleration, is combined with a time-of-flight and energy measurement to determine the mass/charge, mass and energy of ions from ~0.6 to 60 (Ulysses) and ~100 (ACE) keV/charge. The Ulysses orbit spans heliocentric distances between 1.4 and 5.4 AU and all latitudes from +80° to –80°. ACE orbits around L-1 at ~1 AU in the ecliptic plane.

CP679, *Solar Wind Ten: Proceedings of the Tenth International Solar Wind Conference*, edited by M. Velli, R. Bruno, and F. Malara

In order to establish a baseline against which other spectra will be compared, we show in Figure 1 the average velocity distributions of H^+, He^+ and He^{++} that were obtained during a two-year time period when Ulysses was in the slow wind, near its aphelion at ~5 AU, and close to the ecliptic plane. The spectra are complicated, being a mixture of several distinct components: (a) the solar wind which peaks at $W \approx 1$ and has a non-maxwellian distribution, (b) the relatively flat spectra of the interstellar pickup ions with a cutoff at $W \approx 2$, and (c) well developed suprathermal tails above $W \approx 2$ that are reasonably good power laws in the solar wind frame with indices of 5.1 for H^+, He^{++} and 5.6 for He^+. In these tails there is more He^+ than He^{++} and the H^+/He^{++} ratio of 60 indicates acceleration occurs at $W \sim 1.5$ (where this ratio is also 60) and not the at the bulk SW, at $W = 1$, where the H^+/He^{++} ratio is considerably larger.

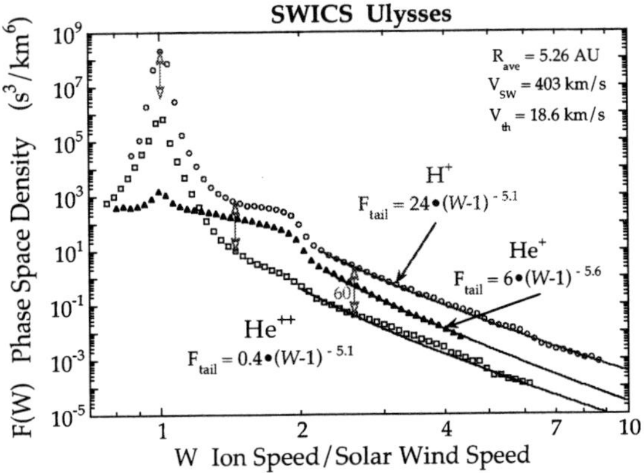

FIGURE 1. Phase space density of H^+, He^{++} and He^+ vs. W, the normalized ion speed. These baseline distributions, measured with SWICS on Ulysses, were averaged over a two year period (1997.108-1999.108). No time periods were excluded. These represent the average typical spectra at the bow shock of Jupiter during a 2-year period.

The distributions of every species measured show power law tails extending to 50 times the solar wind speed. This is illustrated in Figure 2 (from reference 3) which shows the fluence of He, O and Fe measured with mass spectrometers on ACE. In these long-term averaged spectra we see the slow and fast solar wind, interstellar pickup He (the small bump at about $2 \bullet 10^4$ eV), particles accelerated in CIRs and by traveling shocks, and solar energetic particles. Suprathermal tails extend to energies beyond tens of MeV and are power laws between ~20 to 2,000 keV/nuc. The power law index is ~-1.75 which is equivalent to 5.5 for phase space density vs. ion speed. The power law index is about the same for He, O and Fe. These distributions were averaged over long time periods and thus include contributions from particles of various sources,

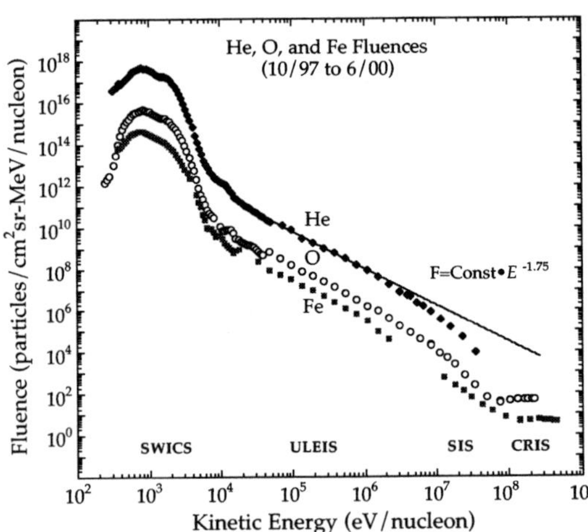

FIGURE 2. Fluence of He, O, and Fe during a 33-month long time period at 1 AU during nearly the same phase in the solar cycle as in Figure 1. (From reference 3). These are the average spectra seen by the bow shock of Earth. No time periods were excluded from these averages.

such as shocks and compression regions of CIRs that are known to produce distributions with strong suprathermal tails (4, 5) and to accelerate particles (see e.g. reference 6). These accelerated particles, however, would rapidly lose energy by adiabatic cooling in the expanding solar wind and the suprathermal tails of their distributions would decay in a matter of days.

In order to establish the time history of particle densities in the suprathermal tail regions we show in Figure 3 the count rate of all ions (mostly protons) with speeds above 2.3 times the ambient solar wind speed (i.e. $W > 2.3$). The top panel is for a time period during the declining phase from solar maximum when forward-reverse shocks and compression regions of CIRs were common. Acceleration of suprathermal ions by shocks (indicated by solid and dotted vertical lines) is clearly visible especially for the strongest shocks and in CIRs between forward-reverse shock pairs, such as the one identified by the shaded area in the panel. Sometimes there is modest acceleration when no shocks are observed, for example, between DOY91 ~110 to ~115, ~175 to 185 and 192 to 200. However, there are few, if any, 6-hour intervals when no suprathermal ions are seen.

In the middle panel we show a 110 day time period in 1998 when Ulysses was at low latitudes, at ~5 AU in the slow wind during the ascending phase from solar minimum. CMEs were predominant during that phase of the solar cycle. The maximum rate observed, indicated by the arrow head on the left, is almost a factor of 10 lower than the maximum rate in 1991. During counterstreaming electron events, most likely CMEs, the rates are low. The highest rates are

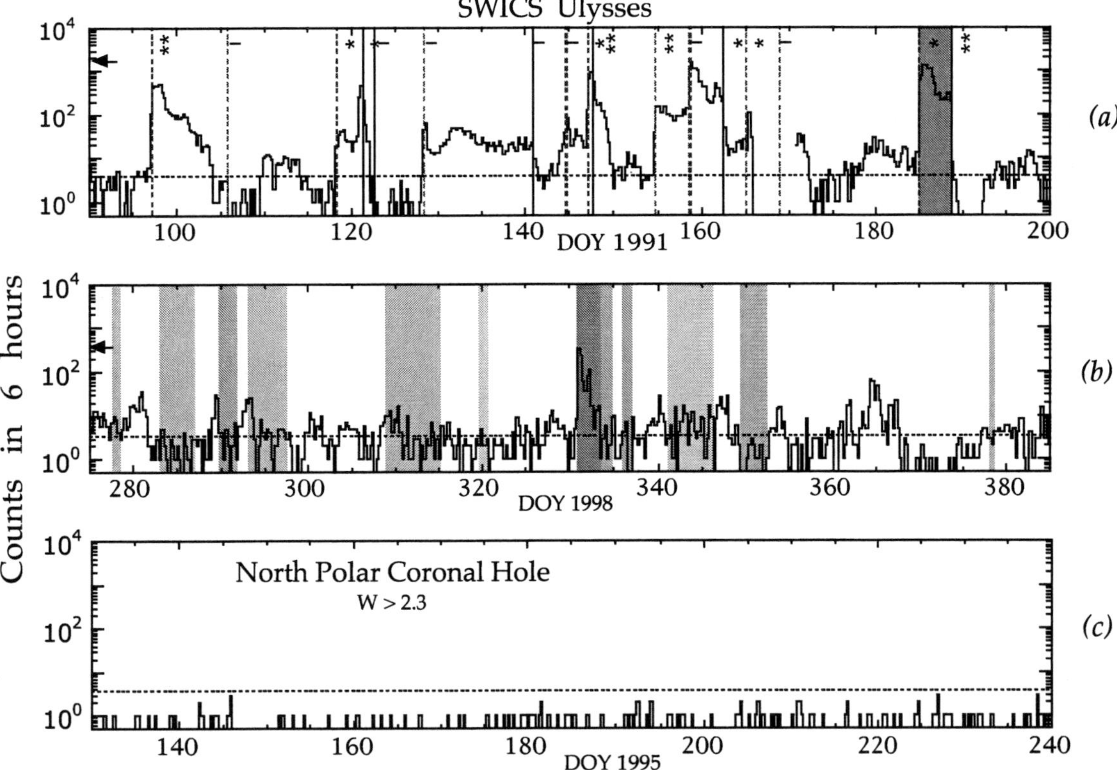

FIGURE 3. Counts in 6 hours of ions (mostly protons) in the suprathermal regions ($W > 2.3$) of the velocity distributions in three 110-day long periods: (a) during the declining phase from solar maximum in 1991 at low latitudes between ~3 and ~4 AU; (b) during the ascending phase from solar minimum in 1998 at low latitudes and ~5 AU; and (c) at solar minimum in 1995 at high latitudes in the north polar coronal hole. In panel (a) forward shock times are indicated by dotted vertical lines and reverse shocks by solid lines (from a list in reference 7). Approximate shock strengths are indicated by '**' for M_S (sonic Mach number) ≥ 2, '*' for $1.5 \leq M_S < 2$ and '–' for $M_S < 1.5$. In panel (b) counterstreaming electron events (J. Gosling, private communication) are indicated by the hatched vertical bars.

observed behind shocks (e.g. around DOY98.331). There are hardly any 6-hour intervals when no suprathermal ions ($W > 2.3$) are seen.

The bottom panel of Figure 3 shows the counting rate of suprathermal ions during a 110-day period in 1995 when Ulysses was at high latitudes in the north polar coronal hole at solar minimum. No shocks were recorded during the entire time period. The rates are very low but steady. Detailed mass vs. mass/charge analysis of the triple coincidence pulse-height data clearly shows that the counts are real (H^+, He^{++} and He^+) and not due to some instrumental background. Clearly suprathermal particles are present in the super-quiet solar wind, far removed from any shocks.

Velocity Distribution Behind Shocks

Shocks are powerful particle accelerators. They heat the solar wind and produce strong suprathermal tails as is illustrated in Figure 4 (see also references 4 and 5). For all three species the distribution functions between the forward–reverse shock pairs of a CIR are quite similar and the suprathermal tails are power laws (in the solar wind frame), each having the same index of 4.5. The spectra directly behind the forward shock of November 27, 1998 (most likely associated with the CME that follows it (see panel (b) of Figure 3) are also similar to one another. The suprathermal tails (above $W \approx 2.3$) are again power laws. The indices, while close to 4.5, are somewhat different among the three ion species. The tail spectra of all the species in both the CIR and CME associated shocks are harder (less steep) than the tails in the 1997-1999 baseline distributions shown in Figure 1.

In Figure 5 we compare the baseline proton distribution from Figure 1 and that measured from 1991.088 to 1993.108 during the declining phase from solar maximum, to the CIR shock-associated H^+ spectrum shown in the top panel of Figure 4. As expected, the

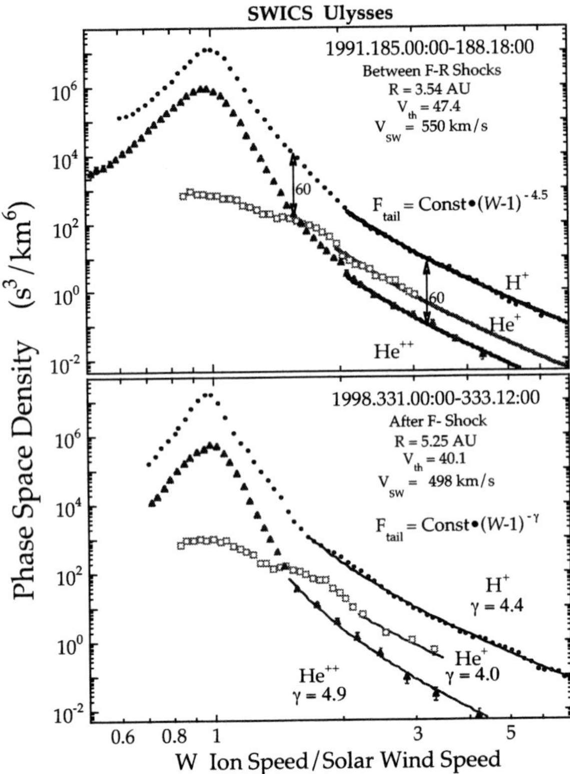

FIGURE 4. Averaged velocity distribution of H^+, He^+ and He^{++} associated with CIR forward and reverse shocks (top panel) and a CME driven shock (bottom panel). The averaging time periods are indicated in each panel and correspond to the dark shaded regions in Figure 3. The location of Ulysses and average solar wind bulk and thermal speeds are also listed in each panel.

most intense (strongest) tails are found in the shock associated proton distribution which is also characterized by the highest thermal speed. The 1991-93 baseline tail has the same power law index as the CIR tail but is about a factor of 10 weaker. The 1997-99 baseline distribution has the steepest and weakest tail, and the lowest solar wind thermal speed compared to the two other populations. This implies that the CIR-dominated solar wind of 1991 to 1993 is more effective in heating the solar wind and producing suprathermal tails than the CME-dominated solar wind of 1997 to 1999 following solar minimum.

Suprathermal Tails During Quiet Times

In the super-quiet period in the polar coronal hole wind during solar minimum the count rate of ions above $W = 2.3$ never exceeded 3 counts in a six-hour interval as was shown in Figure 3(c). We use this maximum rate to select quiet times in the turbulent slow solar wind (see top two panels of Figure 3) to be only those times which have less than 4 counts in any six-hour period. This selection threshold is indicated by the dotted horizontal lines in all panels of Figure 3. It is evident from Figure 3 that most of such time periods are far removed from shocks.

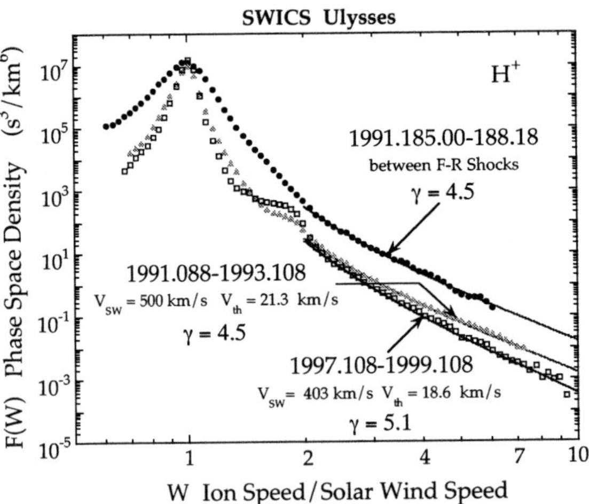

FIGURE 5. Two-year averaged baseline distributions of H^+ (1991.088-1993.108 and 1997.108-1999.108) and the proton spectrum in the CIR compression region from Figure 4.

The proton velocity distributions during quiet times in the CME-dominated slow wind (1997-99) and in the super-quiet wind of the two polar coronal holes are shown in Figure 6. The intensity in the tail of the quiet-time slow wind is down by about a factor of 5 to 6 from the baseline (dashed curve) tail intensity, but the spectral shape is the same for both. In the polar coronal holes a power-law suprathermal tail is also observed. However, the tail found in polar coronal holes is very weak and soft (steep) with an index of about 8. It seems extremely unlikely that these quiet-time tails, especially in the polar coronal holes, are remnants of particles accelerated by remote shocks.

Quiet-Time Tails at 1 AU and 5.3 AU

We examine changes in the quiet-time suprathermal tails with heliocentric distance by comparing in Figure 7 the spectra of protons measured at 1 AU by SWICS on ACE in 1999, with that measured at 5.26 AU by SWICS on Ulysses during 1997.108-1999.108, the CME-dominated phase of the solar cycle. The velocity distributions differ in several respects. The solar wind at 5.26 AU is cooler than it is at 1 AU. Pickup H^+, invisible at ACE, is clearly seen at 5.26 AU for $1.4 < W < 2.2$. But the most interesting result is that the quiet-time proton tail spectrum above $W \approx 2.3$ at 5.26 AU is stronger and harder than at 1 AU. Evidently, suprathermal proton tails, rather than diminishing due to adiabatic cooling, continue to be regenerated in the expanding solar wind, even during quiet times in the absence of shocks.

DISCUSSION

We have shown that everywhere in the heliosphere where measurements could be made, suprathermal tails, extending to the highest measurable energies, are always present. They exist not only in the baseline solar wind, but also during quiet times far removed from shocks that are well known to accelerate particles. Suprathermal tails are seen even in the super-quiet solar wind of polar coronal holes. They are observed in solar wind and pickup ion distributions,

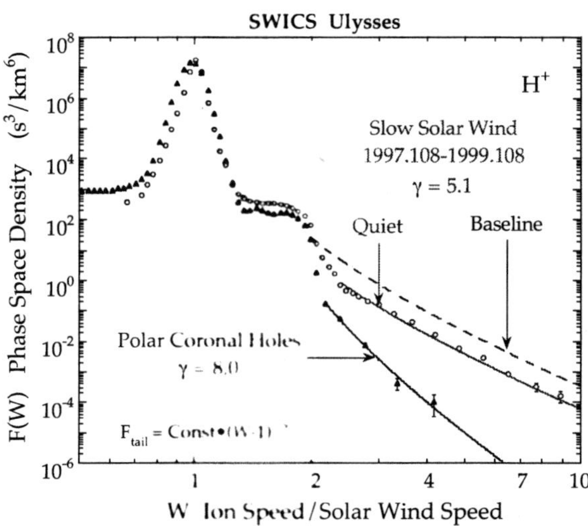

FIGURE 6. Quiet-time proton spectra in the CME-dominated slow solar wind (1997.108-1999.108) and in polar coronal holes (1994.001-365 and 1995.090-1996.210). The 1997-99 baseline proton suprathermal tail (from Figure 1) is indicated by the dashed curve.

and for light and heavy ions. These tails are power-laws in speed (or energy/nucleon) with similar indices for different species. The hardest spectra and the strongest tails are found in shock-associated distributions, the softest and weakest tails in the super-quiet wind of solar-minimum polar coronal holes.

The ions contained in the quiet-time tails are most likely to be the particles accelerated most efficiently by CIR and CME shocks, as well as bow shocks of planets such as Earth and Jupiter. Since the tails extend to beyond 10s of keV, they provide a pre-accelerated population, thus solving the shock injection problem. If tails continue to be generated in the quiet wind beyond the ~5 AU of present observations, all

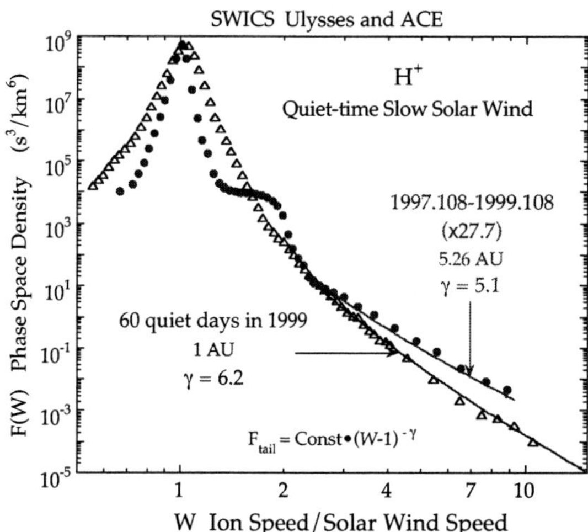

FIGURE 7. Quiet-time spectra of protons at 1 AU and 5.26 AU in 1999. The Ulysses spectrum is that from Figure 6 multiplied by $R^2 = 27.7$. During the CME-dominated phase the tail at Ulysses is harder (less steep).

the way to the termination shock, then ions in these tails would be the seed populations for the Anomalous Cosmic Rays (ACRs) accelerated by the termination shock. In Figure 8 we show that this may indeed be the case. The quiet-time He^+ spectrum measured at ~5 AU was projected to the termination shock assumed to be at 100 AU. A least- squares power-law fit to the suprathermal tail between 5 and 19 keV/nuc, extrapolated to 1 MeV/nuc connects to the lowest two He points measured at 77 AU by Voyager 1 (9) which are most likely the high-energy end of the He^+ tail at 77 AU rather than the He ACRs. Presence of suprathermal tails at the termination shock would solve the injection problem for termination shock acceleration of ACRs, especially in the strong- shock case.

SUMMARY AND CONCLUSIONS

Long term averages of velocity distributions (baseline distributions) of H^+, He^+, and He^{++} in the slow solar wind show well developed tails that are approximately power-laws in speed W (in the solar wind frame) with indices between ~5 and ~5.5. These baseline distributions include particles accelerated in a variety of ways. In the tails, the intensity of He^+ is higher than that of He^{++} implying that ions in the tails come predominantly from outside the thermal part of the solar wind distribution ($W \approx 1$), e.g. from portions of the distribution where $W > \sim 1.5$. Power-law tails of baseline He, O and Fe distributions at 1 AU have the same indices, ~5.5, and extend to energies of several MeV/nuc (speeds up to 50 times the solar wind speed).

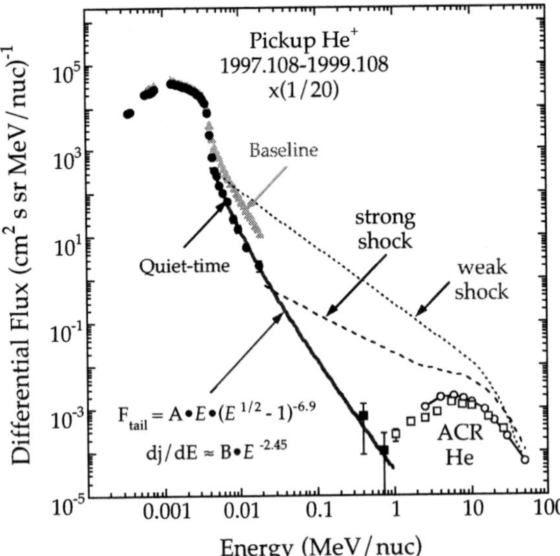

FIGURE 8. Differential intensity vs. energy per nucleon of baseline He^+ (from Figure 1) and quiet-time He^+ during 1997 to 1999, and ACR He measured by Voyager 1 at ~75 AU in 1998-1999.182 (8), open circles, and in 2000 (9), open squares. Dashed and dotted curves are respectively strong and weak shock model fits to the 1998-99 ACR He (9). The He+ distributions were reduced by a factor of 20 to account for the $1/R$ decrease in density of interstellar pickup helium at the termination shock at ~100 AU. The solid curve is the fit of the form indicated in the figure to the quiet-time He^+ spectrum from 5 to 19 keV/nuc, extended to 1 MeV/nuc.

Tails are observed during quiet time periods in the slow solar wind. These periods are far removed from shocks, waves, CIR compression and high turbulence regions which are known to accelerate particles. The intensity of quiet-time tails is lower by a factor of 5 to 6 from baseline but their power-law index is the same as baseline. Power-law tails are present in polar coronal holes, where no shocks, waves or CIRs are observed. Compared to baseline distributions, these super-quiet tails are weaker by factors of 100 or more and steeper (index ~8). Quiet-time tails persist undiminished to distances of at least 5 AU (as sampled by Ulysses) and become harder (index decreases). Because quiet-time tails persist to large distances and are seen also for pickup He, we can rule out that they are formed during solar wind acceleration close to the Sun. What mechanism produces the quiet-time suprathermal tails still remains an open question.

Quiet-time suprathermal tails are most likely the seed population for particles accelerated by CIR and CME shocks and bow shock of planets. Should these tails continue to the outer regions of the heliosphere, as appears to be the case, then they could well be the source of ACRs accelerated by the termination shock.

ACKNOWLEDGMENTS

The essential contributions of the many individuals (see reference 1 and 2) at the Universities of Maryland and Bern, the Max-Planck-Institut fur Aeronomie, and the Technical University of Braunschweig who contributed to the success of the SWICS experiments on Ulysses and ACE are most gratefully acknowledged. I thank L. A. Fisk, N. A. Schwadron and T. H. Zurbuchen for stimulating discussions, G. M. Mason for kindly provided the time periods of the 60 quiet days in 1999, and Christine Gloeckler for her essential help with data reduction. This work was supported in part by NASA/Caltech grant NAG5-6912 and NASA/JPL contract 955460.

REFERENCES

1. Gloeckler, G., et al., *Astron. Astrophys. Suppl. Ser.* **92**, 267-289 (1992).
2. Gloeckler, G., et al., *Space Sci. Rev.* **71**, 79-124 (1995).
3. Mewaldt, R. A., et al., "Long-Term Fluences of Energetic Particles in the Heliosphere,, in *Solar and Galactic Composition*, edited by R. F. Wimmer-Schweingruber, AIP Conference Proceedings 598, New York: American Institute of Physics, 2001, pp. 165-169.
4. Gloeckler, G., *Space Sci. Rev.* **89**, 91-104 (1999).
5. Gloeckler, G., Fisk, L. A., Zurbuchen, T. H., Schwadron, N. A., "Sources, Injection and Acceleration of Heliospheric Ion Populations,, in *Acceleration and Transport of Energetic Particles Observed in the Heliosphere*, edited by R. A. Mewaldt et al., AIP Conference Proceedings 528, New York: American Institute of Physics, 2000, pp. 221-228.
6. Mason, G. M., and Sanderson, T. R., *Space Sci. Rev.* **89**, 77-90 (1999).
7. Balogh, A., Gonzales-Esparza, J., A., Forsyth, R. J., Burton, M. E., Goldstein, B. E., and Bame, S. J., et al., *Space Sci. Rev.* **72**, 171-180 (1995).
8. Cummings, A. C., Stone, E. C., and Steenberg, C. D., *Astrophys. J.* **578**, 194-210 (2002).
9. Krimigis, S. M., Decker, R. B., Hamilton, D.C., Hill, M. E., and Gloeckler G., "Survey of Energetic Particles Observed at Voyager 1 and 2 During 1999-2001,, in *Proceedings of the 27th ICRC*, pp. 3607-3610.

Solar Flares, Type III Radio Bursts, CMEs, and Energetic Particles

H. V. Cane[†]

Laboratory for High Energy Astrophysics, NASA/GSFC, Greenbelt, MD USA
Bruny Island Radio Spectrometer, Tasmania, Australia

Abstract. Despite the fact that it has been well known since the earliest observations that solar energetic particle events are well associated with solar flares it is often considered that the association is not physically significant. Instead, in large events, the particles are considered to be only accelerated at a shock driven by the coronal mass ejection (CME) that is also always present. If particles are accelerated in the associated flare, it is claimed that such particles do not find access to open field lines and therefore do not escape from the low corona. However recent work has established that long lasting type III radio bursts extending to low frequencies are associated with all prompt solar particle events. Such bursts establish the presence of open field lines. Furthermore, tracing the radio bursts to the lowest frequencies, generated near the observer, shows that the radio producing electrons gain access to a region of large angular extent. It is likely that the electrons undergo cross field transport and it seems reasonable that ions do also. Such observations indicate that particle propagation in the inner heliosphere is not yet fully understood. They also imply that the contribution of flare particles in major particle events needs to be properly addressed.

INTRODUCTION

Solar energetic particle (SEP) events are often categorized into two classes based on various properties [1]. The most significant difference is the duration of the particle emission. The less energetic events of short duration, are called "impulsive". Such events have heavy ion abundances about a factor of 10 above those determined for the corona and solar wind. A significant number of these events are associated with CMEs, but the CMEs tend to be rather small in angular extent. In contrast, major, long lasting solar energetic particle events, are associated with large, fast CMEs [2]. As a consequence they are also associated with strong interplanetary shocks which can accelerate coronal/solar wind material. Thus it is difficult to differentiate and determine the composition of the material accelerated at the Sun. However data returned from the SIS instrument on ACE [3] suggest that this material is also enhanced in heavy ions [4].

Impulsive particle events have charge states that indicate a source region that is heated to some 10 MK. Such temperatures are typical of the source regions for soft Xray flares, so it is widely accepted that such particle events have their origins in the magnetic reconnection regions associated with flares. The charge state measurements of Luhn et al. [5] indicating much lower temperatures argue against a flare contribution in major particle events [1]. However the Luhn et al. [5] study made observations at about 1 MeV/nuc. At these energies, large particle events are dominated by shock acceleration in the interplanetary medium. Thus the mean Fe charge state of ~15 that they found, (typical for a plasma of T_e ~2 MK), does not represent material accelerated at the Sun. More recent results from the MAST experiment on SAMPEX, at much higher energy of ~28-65 MeV/nuc., indicate Fe charge states in the range ~14-21 [6]. The lower values were obtained for events with strong interplanetary shocks (Cane 2002, in preparation). For other events, where the measurements were dominated by particles accelerated at the Sun, the charge states are consistent with "flare–heated material".

Clearly the question of the contribution from flares associated with major proton events needs to be addressed. One problem in ongoing discussions is one of semantics. Some people use the term flare to describe localized brightenings in Hα. This is clearly very restrictive. Cliver [7] discusses eruptive flares as being "formed by reconnection of field lines opened by a CME" but does not give a definition of the term. The definition provided by Hudson et al. [8] "A flare is a sudden energy release in the solar atmosphere" is not unreasonable and will be

CP679, *Solar Wind Ten: Proceedings of the Tenth International Solar Wind Conference,*
edited by M. Velli, R. Bruno, and F. Malara

used in this study. Clearly flare phenomena include particle acceleration; nonthermal particles produce the classic Hα signatures. Another classic flare signature is that of a type III radio burst caused by streaming electrons. Since the emission is plasma radiation, the frequency of emission is proportional to the square root of the plasma density. Thus the bursts drift rapidly to lower frequencies as the electron streams encounter decreasing densities further from the Sun. Below about 20 MHz, the emission must originate more than 1 solar radius above the solar surface. Bursts that extend to lower frequencies indicate the escape of flare particles along open field lines into the interplanetary medium. Thus the presence or absence of type III bursts must indicate whether or not flare particles escape in a particular solar event. However, even if particles do escape there is another question that needs to be addressed: What regions of the inner heliosphere are accessible to these particles? Most researchers attempting to understand solar energetic particle increases consider the interplanetary magnetic field to generally follow a smooth Parker spiral. In such a picture, flare particles are released onto and remain confined to a relatively small cone of field lines. Thus flare particles are presumed to be detectable at Earth only from events at about W60° on the Sun. However the interplanetary medium is complex with numerous structures that can affect the flow of energetic particles. For example, an Earth–based observer can be "well–connected" to an eastern hemisphere flare if Earth is inside an interplanetary CME. Richardson et al. [9] showed how particles arrived from field lines to the east of the Earth–Sun line during the ground–level enhancement of October 1981 associated with a flare at E31°. Another important aspect is the suggestion based on recent observations that particles are transported across field lines (e.g. McKibben et al. [10]). Thus if flare particles can both escape to the interplanetary medium in large CME events, and can cross field lines, they could make a significant contribution to major solar particle increases. This paper briefly presents the results of two papers which show that both these conditions are fulfilled.

THE OBSERVATIONS

Cane et al. [11] have determined the solar associations for 123 events comprising essentially all >20 MeV proton events detected by the Goddard experiment on IMP 8 for the period 1997 to mid–2001. They found that all but two were preceded by type III radio bursts. The two exceptions were weaker events for which particles were not measured until several hours after the flare. The radio bursts occurred in conjunction with Hα flares for all events determined to originate on the front side of the Sun. The flares occurred at the times of large, fast CMEs. For the large proton events (i.e. excluding the >20 MeV proton events that would be described as 'impulsive') the radio bursts were not normal type III bursts. Such bursts occur at the onset of flares, typically last about 5–10 minutes, and commence at frequencies above ~200 MHz. Instead the radio bursts associated with major proton events were longer–lasting, started at frequencies below about 100 MHz and were much more fragmented than a group of normal type IIIs. At frequencies observable from the ground (typically above 15 MHz) they were not particularly intense and usually superimposed on slow drifting features. Slow drift (type II) bursts are attributed to shocks but these bursts are generally considered to occur behind CMEs (e.g. Leblanc et al. [12]). Independent of the nature of these type II bursts, the type III electron streams appear to originate lower in the corona and therefore are unlikely to be shock accelerated. Below 1 MHz, as observed from the WAVES experiment on Wind, the type III events associated with major proton events are very intense and generally much more prominent that any type II emission. Cane et al. [11] called these events "type III-l". Examples (above 18 MHz) may be seen on the solar/Culgoora/historical data/SEP related events, section of the following website: `http://www.ips.gov.au`. The low frequency counterparts may be seen at `http://lep694.gsfc.nasa.gov/waves/waves.html`. Cane et al. [11] show composite spectra covering the entire frequency range from 2 GHz to 20 kHz. The coverage is completed by including data from the Bruny Island Radio Spectrometer (`http://fourier.phys.utas.edu.au/birs/`) which bridges the gap between the Culgoora and WAVES frequency ranges.

Cane et al. [11] also performed a counter study looking for type III-l bursts that were not associated with SEP events. This was undertaken in two ways. First a random study was made of long–lasting radio bursts seen in the WAVES data. Most were found to be either type III-l bursts with associated SEP events or long lasting groups of type III bursts, as reported by ground based observers. Second, a study was performed to look at the radio bursts associated with all large (angular extent >140°) CMEs in 1997-2000 which were frontside events as determined by the SOHO observers. Again, most type III-l bursts found were already identified as being associated with an SEP event. It was also found that large

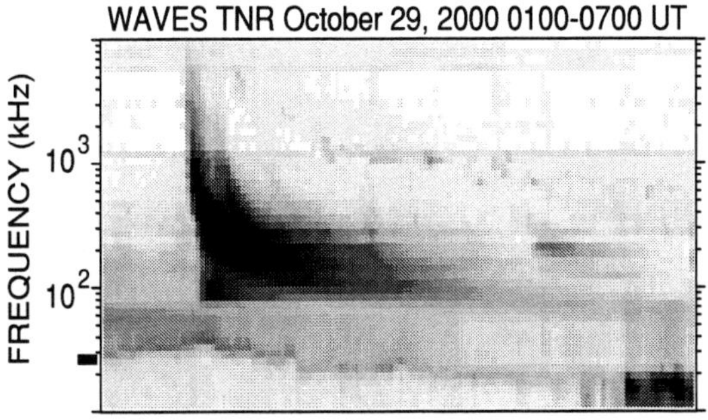

FIGURE 1. A type III-*l* burst associated with a solar event from E35°. Note that the drift rate of the burst becomes almost zero and the event does not extend below 50 kHz implying that the associated electron beam did not intercept the spacecraft. The associated slow particle increase is shown in Figure 3.

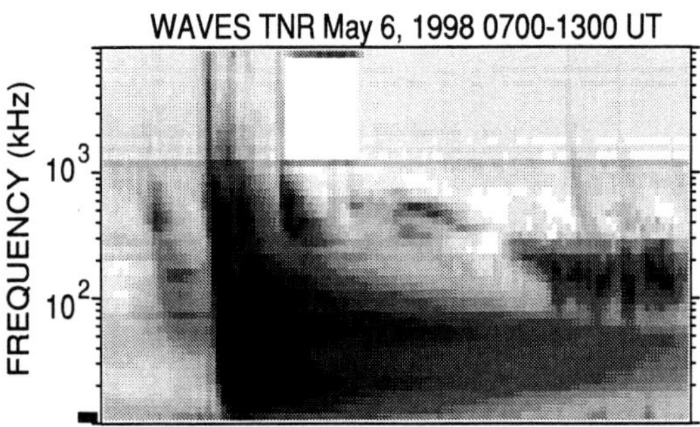

FIGURE 2. A type III-*l* burst associated with a solar event from W65°. Note that the burst drifts very rapidly. The associated particle increase rose very rapidly as may be seen in Figure 3.

CMEs without associated radio bursts were in general slow. It was found that type III-*l* bursts without associated SEP events were from the eastern hemisphere of the Sun. In these cases the radio emission did not extend to the lowest frequencies observed by WAVES indicating that the associated electrons did not reach the vicinity of the spacecraft consistent with the absence of an SEP event. A number of type III-*l* bursts that were associated with energetic particle increases also did not extend to the lowest frequencies and all but one these came from the eastern hemisphere. An example is shown in Figure 1 (the solar event occurred on October 29, 2000) which shows data from the TNR receiver on the WAVES experiment. In this case, and others like it, the particle event had a very slow rise. For the event illustrated the near–Earth ~25 MeV intensity did not rise above background until about 10 hours after the solar event (see Figure 3). However at Ulysses, located at 3 AU, at extreme southern latitudes, and east of the Earth–Sun line the event began about 6 hours earlier. Consistent with the more rapid particle increase, the type III-*l* burst drifted more rapidly to low frequencies in the Ulysses radio data. (The radio and particle data are shown in Cane and Erickson [13]).

Figure 2 shows data for a radio event on May 6, 1998 that drifted very rapidly to the lowest frequencies consistent with the implied good magnetic connection to the flare region; the associated flare was located at W63°. Figure 3 shows the proton profile for the associated particle event and also for the increase associated with the event of Figure 1. The particle increases have rise times consistent with the low frequency drift rates of the type III-*l* bursts.

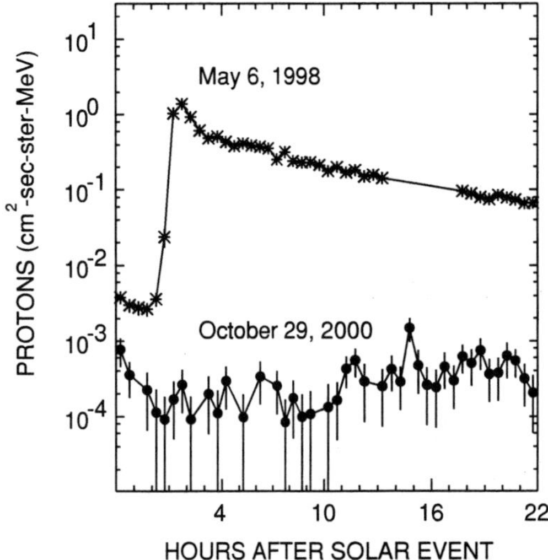

FIGURE 3. Proton intensities in the range 24-29 MeV associated with the two radio events illustrated in Figure 1.

Similarly from a statistical study of the type III-*l* bursts associated with >20 MeV proton events it can be shown [13] that the drift rates of the radio emission organize the proton data. Figure 4 shows how the "radio delay" is related to the magnetic connection of the observer. The radio delay is the time interval between the start of the burst at the Sun and when its leading edge reached the local plasma frequency. The local plasma frequency is determined from noise in the antenna which appears as a dark intensification across the bottom of the TNR data, indicated by a bar along the left edges of Figures 1 and 2. The magnetic connection is the difference between the flare longitude and the presumed footpoint of the field line of the observer

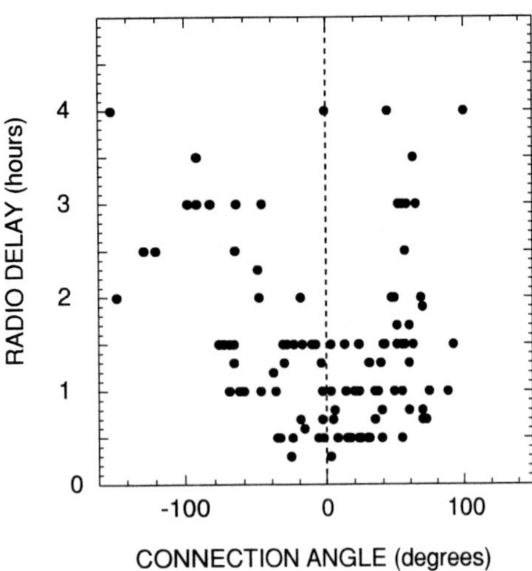

FIGURE 4. Radio burst delays as a function of magnetic connection to the flare longitude.

taking into consideration the solar wind speed.

Note that some of the radio bursts take several hours to drift to the local plasma frequency whereas others take less than an hour. This does not mean that the responsible electrons are faster in some events than in others - the radio emission is generated by those electrons which form a positive slope in the parallel distribution function which at 1 AU is typically in the 2-20 keV range. Rather the delays reflect the time it takes the responsible electrons to propagate to 1 AU. Delays in the arrival of solar energetic particles as a function of connection angle of the associated flare have been known for years [14] but attributed to delays occurring at the Sun. In recent years delays were attributed to the time required for CME–driven shocks to deposit particles on the observer's field line. If this were the case for the radio generating electrons, one would then expect a separate signature at the Sun of their escape, in some cases several hours after the flare. This is not what is observed. Instead the radio producing electrons actually may take up to several hours to propagate from the Sun to near Earth. Since some events originate at locations well removed from the region of good magnetic connection, there must be cross field transport for the radio generating electrons. Since the bursts also organize the proton characteristics it seems reasonable that the ions undergo cross field transport as well.

CONCLUSIONS

1. Type III radio bursts show that there are open field lines from flaring regions accompanying all solar particle events.
2. The lowest frequency emissions of the radio bursts show that the responsible electrons can gain access to a large region of the inner heliosphere. The radio data organize proton onsets suggesting that flares could contribute to all solar particle events.
3. Interplanetary transport of energetic particles, in particular transport across the mean field, is not yet fully understood.

ACKNOWLEDGMENTS

The use of the data made available via the NSSDC CDAWeb is acknowledged. This work was partially funded by a NASA contract with USRA.

REFERENCES

1. Reames, D. V., *Space Sci. Rev.* **90**, 413–(1999).
2. Cane, H. V., *Proc. 27th Internat Cosmic Ray Conf. (Hamburg)* **8**, 3231 (2001).
3. Stone, E. C., et al., *Space Sci. Rev.* **86**, 357 (1998).
4. von Rosenvinge, T. T., et al., in *Solar and Galactic Composition*, AIP Conference Proceedings 598, New York: American Institute of Physics, 2001, pp. 343–348.
5. Luhn, A., et al., *Proc. 19th Internat. Cosmic Ray Conf. (La Jolla)* **4**, 241–244 (1985).
6. Leske, R. A., et al., in *Solar and Galactic Composition*, AIP Conference Proceedings 598, New York: American Institute of Physics, 2001, pp. 171–176.
7. Cliver, E. W., *Solar Physics* **157**, 285 (1995).
8. Hudson, H., *J. Geophys. Res.* **100**, 3473 (1995).
9. Richardson et al., *J. Geophys. Res.* **96**, 7853 (1991).
10. McKibben, R. B. et al. *Proc. 27th Internat Cosmic Ray Conf. (Hamburg)* **8**, 3281–3284 (2001).
11. Cane, H. V., et al. *J. Geophys. Res.* **107**, 1315 (2002).
12. Leblanc et al. *J. Geophys. Res.* **105**, 18,215 (2000).
13. Cane, H. V. and W. C. Erickson, *J. Geophys. Res.* in press 2003.
14. Reinhard, R. and Wibberenz, G., *Solar Phys.* **36**, 473 (1974).

Pickup Ion Acceleration in the Heliosphere: Consequences of Organized Footpoint Motion on the Sun

N. A. Schwadron

Southwest Research Institute, P.O. Drawer 28510, San Antonio, TX, 78228-0510

Abstract. Suprathermal and pickup ion tails exist ubiquitously in slow solar wind. What is the intrinsic property of slow solar wind that causes ubiquitous particle acceleration? Observations from Ulysses have shown that statistical acceleration through transit time damping of fluctuations in field strength is the vehicle by which pickup ion tails are produced. But the source of compressive fluctuations in slow solar wind has not been established. We show that shearing of magnetic fields by small-scale fluctuations in the solar wind speed provides a strong and continuous source of compressive fluctuations. This effect occurs naturally if slow solar wind is generated from material stored in coronal loops and released due to organized drifts of open magnetic field footpoints on the Sun. Another consequence of these organized footpoint drifts is the creation of large-scale magnetic field structures that connect fast and slow solar wind – located for example in the outer heliosphere at the latitudinal boundaries between fast and slow solar wind. These structures are sites for particle acceleration throughout the heliosphere. In summary, we argue that the organized drifts of open magnetic field footpoints significantly alter the structure of the heliospheric magnetic field in regions where there are strong changes in the solar wind speed, on small and large spatial scales, creating a previously unknown role for solar wind shearing, and providing a new and needed vehicle for the particle acceleration throughout the heliosphere.

OBSERVATIONAL CONSTRAINTS

It has been well established that particles with energies of order 1 MeV per nucleon can be accelerated at the forward and reverse shocks which surround Co-rotating Interaction Regions (CIR's), where high and low speed streams interact in the solar wind [e.g., 1, 2, 3]. [4] and [5] have suggested a two step acceleration process is called for. In the first step, pickup ions born with random speeds up to twice that of the solar wind are accelerated by transit-time damping of fluctuations in the magnetic field magnitude within CIRs, achieving random speeds between four and six times that of the solar wind. In the second step, these already pre-accelerated ions may efficiently move along magnetic field lines and are thereby injected into diffusive shock acceleration at the shocks (particle the reverse shock) that bound CIRs.

Observational evidence for this two-step acceleration process has not been hard to find. Figure 1 shows the results of an observational test conducted by [5] in which the energy in initially accelerated pickup ions was not found to to be strongly correlated with the presence shocks. A strong correlation was however found with the presence of fluctuations in the magnetic field magnitude that may be transit-time damped by pickup ions.

[7] and [8] show the remarkable ubiquity of acceleration of pickup ions and solar wind suprathermal ions within slow solar wind as demonstrated through univer-

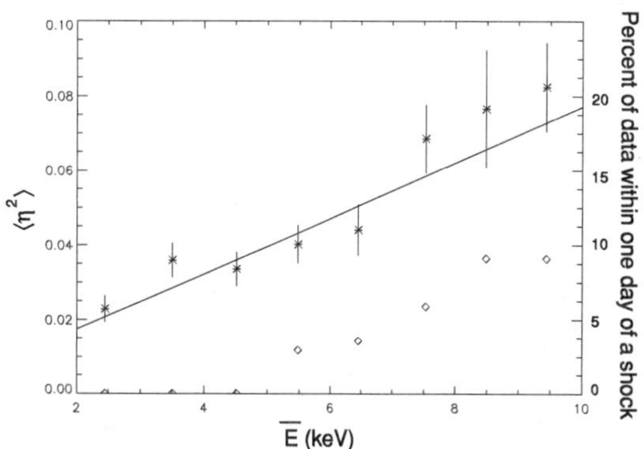

FIGURE 1. Evidence that the initial acceleration of pickup ions in slow solar wind is caused by transit-time damping, not acceleration near CIR shocks [5]. The x-axis is average energy, $\bar{E}$, in pickup ion tails (for random speeds from twice to six times the solar wind speed). The y-axis is the average intensity of fluctuations in the magnetic field magnitude, $\eta^2 = \delta|B|^2/|B|^2$, which may be transit time damped by pickup ions. The data here were taken from Ulysses observations for 150 days during 1992 when Ulysses was at low-latitudes and encountered 23 shocks associated with CIRs.*Plotted as diamonds* are the percentage of data within each energy bin that fall within one day of a shock. *Plotted as stars* are the average intensity of fluctuations in field magnitude versus the average energy in pickup ion tails. Data were binned according to the energy. *The solid line* represents the results of an analytic model presented by [5] that describes transit-time damping by pickup ions.

CP679, *Solar Wind Ten: Proceedings of the Tenth International Solar Wind Conference*, edited by M. Velli, R. Bruno, and F. Malara

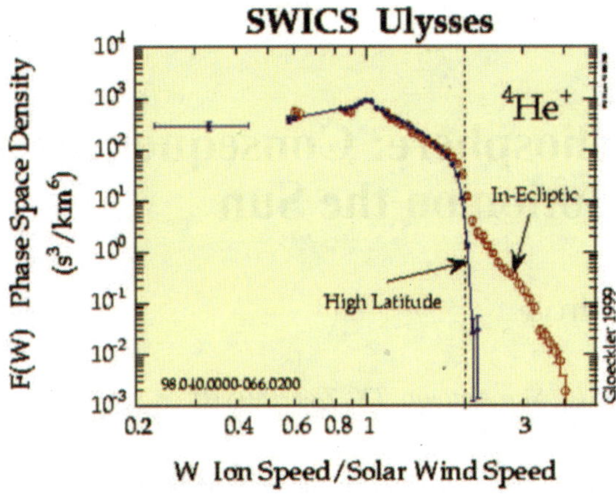

FIGURE 2. Distributions of interstellar pickup He^+ in high-latitude fast solar wind (blue) and in-ecliptic slower solar wind (red) observed by Ulysses/SWICS [6].

sal presence of pickup ion and solar wind tails (see Figure 2). Comparisons between fast and slow solar wind are also illuminating, as demonstrated by Figure 2. It appears that the suprathermal and pickup ion tails are significantly stronger in slow solar wind than in fast solar wind [4, 9, 7, 10, 8]. The comparison begs the question: *what are the properties intrinsic to slow solar wind that cause the ubiquity of suprathermal ion and pickup ion tails?*

STATISTICAL ACCELERATION THROUGH TRANSIT TIME DAMPING

Statistical acceleration through transit time damping was overviewed recently by [11]. The close parallel was made to a magnetic pump which can be treated in a very straightforward manner. Consider an ion that propagates along a field line in which the field magnitude varies. In the frame of reference of a compressive fluctuation (moving along the mean magnetic field) the ion's first adiabatic invariant is preserved. A fluctuation of field magnitude, with amplitude $(\delta|B|)/|B|$, results in a change of the perpendicular ion speed, $v_\perp$, given by $(\delta v_\perp)/v_\perp = (\delta|B|)/(2|B|)$. Since we are dealing here with compressive magnetosonic waves which propagate at an angle with respect to the mean magnetic field, the component of wave propagation parallel to the mean magnetic field with speed is $v_{ph}\lambda_\parallel/\lambda_\perp$, where v_{ph} is the phase speed of the wave and $\lambda_\parallel$ and $\lambda_\perp$ are the parallel and perpendicular correlation lengths, respectively. Energy is conserved in the frame moving along the mean field with the fluctuation so that $v_\perp \delta v_\perp = -(v_\parallel - v_{ph}\lambda_\parallel/\lambda_\perp)\delta v_\parallel$, where $v_\parallel$ is the ion speed parallel to the mean magnetic field in the plasma frame. This also yields an energy change in the plasma frame given by $\delta E = p\delta p/m = m\delta v_\parallel v_{ph}\lambda_\parallel/\lambda_\perp$. Given also that the interaction time is $\delta t = \lambda_\parallel/|v_\parallel - v_{ph}\lambda_\parallel/\lambda_\perp|$, the diffusion coefficient in momentum (or equivalently, speed) is

$$\frac{D_{pp}}{p^2} \approx \left\langle \frac{(\delta p)^2}{p^2 \delta t} \right\rangle = \left(\frac{v_\perp}{v}\right)^4 \frac{\lambda_\parallel}{\lambda_\perp^2} \frac{v_{ph}^2}{|v_\parallel - v_{ph}\lambda_\parallel/\lambda_\perp|} \frac{\langle \eta^2 \rangle}{4} \quad (1)$$

Detailed quasi-linear calculations of this effect were performed by, e.g., [12] and [5], and [13]. The form for the coefficient of momentum diffusion depends critically on the ion gyroradius and on the nature of the compressive turbulence. [5] found that they could account for observed pickup ion tails in slow solar wind if they assumed the turbulence was highly elongated (with parallel correlation lengths 30 times those of perpendicular correlation lengths). Correlation lengths perpendicular to the mean magnetic field of 0.01 AU were used and are typical of interplanetary turbulence.

ENERGY SOURCES FOR TRANSIT TIME DAMPING

Transit time damping is great for producing acceleration, but acts so quickly that it readily destroys fluctuations in field strength. It is for precisely this reason that fluctuations in field strength are observed to be small in the solar wind. Nonetheless, if fluctuations in field strength are observed, even at a low level, there must be a continuous source that produces them.

We point out here that shearing of the magnetic field by the solar wind is inherent to the boundaries between faster and slower streams if there is an organized motion of open magnetic field footpoints on the Sun through these boundaries. More generally, we show that shearing of the magnetic field may be an inherent property of slow solar wind, particularly on small spatial scales. The effect continuously transfers energy from the solar wind flow into compression and rarefaction regions. On small spatial scales, the effect helps to explain the general presence of compressive waves in slow solar wind. On large spatial scales, it suggests a new and different structure to the boundaries between fast and slow solar wind; structures that are quite favorable for particle acceleration, even in the distant heliosphere.

It is clear from solar wind observations that transitions in speed are ubiquitous. Traditionally it has been thought the these speed transitions do not shear the magnetic field since the footpoints of open field lines and solar wind transition regions rotate bodily with the Sun. It was realized recently by, e.g., [14], [15] and [16] that the footpoints of open magnetic field lines are not fixed on a bodily rotating Sun; instead, field line footpoints move through coronal hole boundaries, as illustrated in Figure

5. The motion of field line footpoints across a region with an abrupt change in solar wind speed (for instance, a coronal hole boundary) leads to strong shearing of the magnetic field.

[17] considered this shearing effect in co-rotating rarefaction regions (CRRs) which are mapped out from the trailing edges of coronal holes. In CRRs, the heliospheric magnetic field is strongly underwound if drifts of open magnetic field footpoints through the coronal hole boundary are present. The underwinding is caused by solar wind shearing since the faster portions of the stream draw out the magnetic field more quickly than the slower portions. The effect may be quantified simply. The tangent of the angle ψ, the field direction relative to the radial direction in the $\phi - r$ plane, is given by

$$\tan\psi = \frac{B_\phi}{B_r} = -\frac{(\Omega_\odot + \omega_\phi) r \sin\theta}{V - (r/V)\mathbf{u}_\perp \cdot \nabla_S V}. \tag{2}$$

Here, V is the solar wind speed, r is the heliocentric radius, $\nabla_S V$ is the gradient in solar wind speed along source surface, $\mathbf{u}_\perp$ is the velocity of footpoints along the source surface (in the frame of reference that co-rotates with the Sun), $\Omega_\odot$ is the equatorial rotation rate, and ω_ϕ is the footpoint rotation rate in the azimuth. The resulting orientation of the underwound heliospheric magnetic field is shown in Figure 4. *At large distances the field orientation in CRRs becomes fixed, whereas a Parker spiral becomes increasingly azimuthal.* The effect has been confirmed by observations of the heliospheric magnetic field [18, 19].

In the case of rarefaction regions, this shearing effect causes a strong reduction in the field magnitude [17]. In compression regions, shearing will cause compression of the magnetic field. These magnitude changes are due to more than the mere compression of the plasma. This is exemplified by considering the magnetic field generated in heliospheric space when footpoints move across a *latitudinal* transition in the wind speed, such that the wind speed, $V = V(\theta)$, is given as a function of co-latitude, θ. In this case, the magnetic field can be solved exactly

$$\mathbf{B} = B_{SS}\frac{R_{SS}^2}{r^2}\left[\left(1 - \frac{r\omega_\theta}{V^2}\frac{\partial V}{\partial\theta}\right)\hat{\mathbf{e}}_r - \frac{r\omega_\theta}{V}\hat{\mathbf{e}}_\theta - \frac{\Omega_\odot r \sin\theta}{V}\hat{\mathbf{e}}_\phi\right] \tag{3}$$

where $\hat{\mathbf{e}}_r$, $\hat{\mathbf{e}}_\theta$, and $\hat{\mathbf{e}}_\phi$ are the unit vectors in the radial, co-latitude and azimuthal directions, respectively. The radial position of the source surface is R_{SS}. The term caused by shearing, $-[B_{SS}R_{SS}^2\omega_\theta/(rV^2)](\partial V/\partial\theta)\hat{\mathbf{e}}_r$, leads to an additional component of the field in the radial direction which may either add to or subtract from the nominal radial component (depending on the direction of the speed gradient relative to the direction of footpoint motions). In this case, the plasma itself experiences no compression or rarefaction since the speed transitions occur across latitude. The fluctuations in magnitude arise because of the additional radial shearing term.

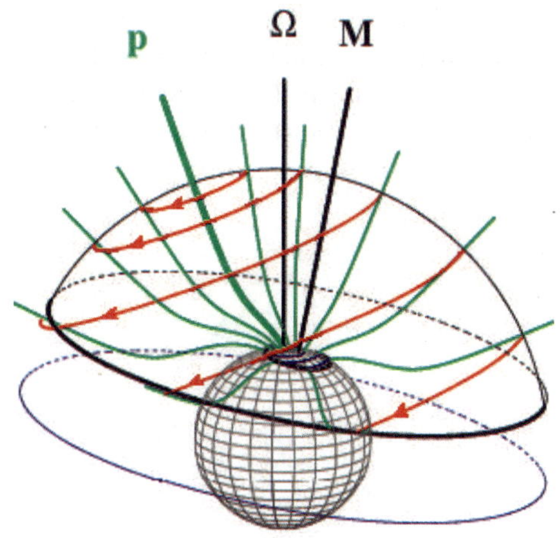

FIGURE 3. An illustration of the motions of the open magnetic field in the solar corona, in the polar corona [20, 14, 15]. The figure is drawn in the frame co-rotating with the equatorial rotation rate. The axis of expansion symmetry is "M", the solar rotation axis is "Ω. Open field lines are shown in green, with "p" the field line that connects to the heliographic pole. The red curves are the trajectories of the field lines, the motion of which is driven by differential rotation of the photosphere. The thick black curves show the coronal hole boundaries.

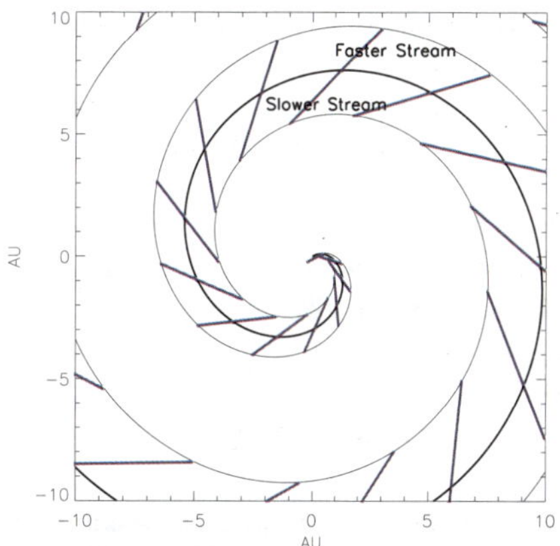

FIGURE 4. The orientation of the magnetic field in rarefaction regions under the influence of shearing and footpoint motion. Streamlines are shown by the solid spiral lines (black), and segments (blue) show the field orientation in the rarefaction regions. The speed gradient is given by a 300 km/s speed change over a 5 degree transition region.

In the outer heliosphere, at boundaries between fast and slow solar wind, the shearing induces large-scale regions where the field is strongly perturbed (and more radial). The fact that these transition regions are magnetically connected implies that, along a given field line, there is a speed gradient from slow to fast wind speeds. As ions propagate and scatter along the field lines, they diffuse in energy (or equivalently, momentum) by interacting with the speed gradients. Hence, these are important sites for acceleration.

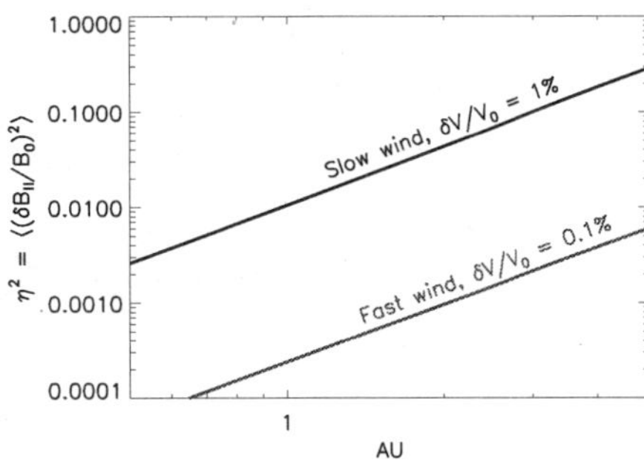

FIGURE 5. Fluctuations in the magnetic field strength produced by shearing.

The effects of shearing have been applied above to large-scale regions, but they apply on small spatial scales as well. As an example, we may consider a solar wind speed that varies on small spatial scales as a function of latitude in the presence of a large-scale footpoint drift in the latitudinal direction. Equation (3) may be used to solve for mean energy in compressive magnetic field fluctuations. The results are shown in Figure 5. This is a purely kinematic description and we have neglected damping of these compressive fluctuations. Hence, the amplitude of these fluctuations are overestimated. However, the calculation illustrates that shearing would play an important role by continuously feeding energy into these small-scale compressive waves.

DISCUSSION

Small-scale speed fluctuations are observed to be larger in slow than in fast solar wind. This may be a natural consequence of forming slow solar wind from material stored and subsequently released by large coronal loops with highly variable properties [e.g., 15]. [16] and [21] argue that this source for slow solar wind follows from the sporadic reconnection between open field lines and closed loops which is inherently tied to the dynamic organization and continuous reconfiguration of open magnetic flux on the Sun. We have shown here that the formation of slow solar wind from large loops and the global drift of open field footpoints naturally causes a state of slow solar wind in which shearing of the magnetic field (on small spatial scales) continuously feeds energy into fluctuations in the field strength – the source of energy for transit-time damping and statistical acceleration of pickup ions.

ACKNOWLEDGMENTS

This work was supported by NSF grant ATM 0100659. I wish to thank Dave McComas, Heather Elliott, Len Fisk and George Gloeckler for useful discussions.

REFERENCES

1. Barnes, C. W., and Simpson, J. A., *Astrophys. J.*, **210**, L91 (1976).
2. McDonald, F. B., Teegarten, B. J., Trainor, J. H., von Rosenvinge, T. T., and Webber, W. R., *Astrophys. J. Lett.*, **203**, L149 (1976).
3. Fisk, L. A., and Lee, M., *Astrophys. J.*, **237**, 620 (1980).
4. Gloeckler, G., Geiss, J., Roelof, E. C., Fisk, L. A., Ipavich, F. M., Ogilvie, K. W., Lanzerotti, L. J., von Steiger, R., and Wilken, B., *J. Geophys. Res.*, **99**, 17637 (1994).
5. Schwadron, N. A., Fisk, L. A., and Gloeckler, G., *Geophys. Res. Lett.*, **23**, 2871–2874 (1996).
6. Gloeckler, G., *Space Science Reviews*, **89**, 91–104 (1999).
7. Gloeckler, G., and Geiss, J., *Space Sci. Rev.*, **86**, 127 (1998).
8. Gloeckler, G., *This Proceedings* (2002).
9. Collier, M. R., Hamilton, D. C., Gloeckler, G., Bochsler, P., and Sheldon, R. B., *Geophys. Res. Lett.*, **23**, 1191–1194 (1996).
10. Chotoo, K., Schwadron, N. A., Mason, G. M., Zurbuchen, T. H., Gloeckler, G., Posner, A., Fisk, L. A., Galvin, A. B., Hamilton, D. C., and Collier, M. R., *J. Geophys. Res.*, **105**, 23107–23122 (2000).
11. Fisk, L. A., Gloeckler, G., Zurbuchen, T. H., and Schwadron, N. A., "Ubiquitous Statstical Acceleration in Solar Wind," in *Acceleration and Transport of Energetic Particles Observed in the Heliosphere: ACE 2000 Symposium*, edited by R. A. Mewaldt et al., AIP, 2000, pp. 229 – 233.
12. Fisk, L. A., *J. Geophys. Res.*, **81**, 4633–4640 (1976).
13. Schwadron, N. A., *Theoretical and Observation Studies of Ion Transport in the Heliosphre*, Ph.D. thesis, U. Michigan (1996).
14. Fisk, L. A., Zurbuchen, T. H., and Schwadron, N. A., *Astrophys. J.*, **521**, 868 (1999).
15. Schwadron, N. A., Fisk, L. A., and Zurbuchen, T. H., *Astrophys. J.*, **521**, 859 (1999).
16. Fisk, L. A., and Schwadron, N. A., *Astrophys. J.*, **560**, 425 (2001).
17. Schwadron, N. A., *Geophys. Res. Lett.*, **In Press** (2002).
18. Murphy, N., Smith, E. J., and Schwadron, N. A., *Geophys. Res. Lett.*, **In Press** (2002).
19. Murphy, N., Smith, E., and Schwadron, N., "Strongly Underwound Magnetic Fields in Co-rotating Rarefaction Regions: Observations and Implications," in *Solar Wind X*, 2002, vol. This Proceedings.
20. Zurbuchen, T. H., Schwadron, N. A., and Fisk, L. A., *J. Geophys. Res.*, **102**, 24175 (1997).
21. Schwadron, N. A., Fisk, L. A., and Schulz, M., *J. Geophys. Res.*, **Submitted** (2002).

CME-driven Coronal Shock Acceleration Of Energetic Electrons

G.M. Simnett* and E.C. Roelof†

*School of Physics and Astronomy, University of Birmingham, Birmingham, B15 2TT, UK.
†Johns Hopkins University/Applied Physics Laboratory, Laurel, MD 20723-6099, USA

Abstract.
53 impulsive (38-315 keV) near-relativistic solar electron events with beam-like pitch-angle distributions were observed by the ACE/EPAM experiment while the SOHO/LASCO coronographs were observing coronal mass ejections (CME) between 2.5 and 30 $R_\odot$. Simnett, Roelof and Haggerty [in companion papers to be published in Ap. J., 2002] report a close association among the impulsive electron beams, solar electromagnetic emissions, and western hemisphere CMEs, jets, etc. They find that the electron injections are delayed $\sim$10 minutes after the electromagnetic emissions and $\sim$20 minutes after the CME launches, so that the electron release occurs when the CME has travelled 1-2 $R_\odot$ beyond the CME launch altitude. The median exciter speed of the associated solar type III radio bursts (deduced from WIND/WAVES decametric spectrograms) is 0.08c, implying that the characteristic electron energies in the exciter front are only a few keV. Since no prompt near-relativistic electrons are injected until $\sim$10 minutes after the type III burst, the energy spectrum of the type III associated electrons must be steep at these energies. Therefore the near-relativistic electrons that must be present to produce the microwave and hard X-ray bursts also do not escape promptly with intensities measurable by ACE/EPAM. Inverse correlation between the finite delays of near-relativistic electrons after the CME launch confirms that the electrons are injected when the CMEs are $\sim$1-2 $R_\odot$ above the photosphere. The positive correlation between CME speed and electron intensity (as well as spectral hardness) is consistent with the process of shock acceleration. Therefore we conclude that the simplest explanation of the observational associations is that the electrons are accelerated by CME-driven shocks in the corona at altitudes $\sim$1-2 $R_\odot$ above the photosphere. We see no reason why ions should not also be accelerated concurrently in the corona by this same process, although the final velocity of the ions may be less than that of the electrons.

INTRODUCTION

The Sun is a prolific source of energetic electrons, but the processes governing their acceleration and release have remained elusive. Recently it has been recognised (Haggerty and Roelof, 2002) that the best way of identifying within a few minutes when such electrons are released from the Sun is only to analyse events whose pitch angle distribution forms a highly anisotropic outward flowing beam. A spacecraft, say at 1 AU, is connected magnetically to the Sun on one side, and to some region beyond 1 AU on the other side. Electrons released from the Sun become field-aligned within a few solar radii, and thus will first cross the 1 AU radius as a beam. However, somewhere, normally beyond 1 AU, significant scattering takes place, probably at compression regions where fast and slow solar wind streams interact. This means that for electrons released onto distant field lines which do not immediately pass through the observer, back-scattering from the interaction region can produce an onset which is relatively isotropic. Thus an impulsive intensity increase is not in itself indicative of a recent release unless it is highly anisotropic. Beam-like events have for many years been referred to as "scatter-free". A related criterion for identifying scatter-free propagation of the earliest-arriving electrons is their velocity dispersion (Krucker *et al*, 1999). The difference between injection and onset times must be equal to the distance travelled divided by the electron velocity.

Haggerty and Roelof (2002) have identified scatter-free electron events in the $\sim$40-300 keV energy region using the EPAM instrument on the ACE spacecraft, which is near the L_1 Lagrangian point on the Earth-Sun line. They showed (their Fig. 1) that is was possible from the onset times in different electron energy bands to derive an injection time at the Sun. When they examined radio, X-ray and optical emission from the Sun, they found that the electron release time was delayed by at least 10 minutes from the peaks of the microwave and hard X-ray emission (when present) and that the events were frequently associated with chromospheric activity in the western hemisphere. The latter is not surprising, as the nominal magnetic connection from ACE back to the Sun is to a longitude around W60°. This will be discussed in more detail in section IV. They also found that the scatter-free electron events were accompanied by de-

CP679, *Solar Wind Ten: Proceedings of the Tenth International Solar Wind Conference*, edited by M. Velli, R. Bruno, and F. Malara

cametric type III radio bursts in the high corona. These were observed by the WAVES instrument (Bougeret *et al.*, 1995) on the WIND spacecraft, and had drift rates such that the characteristic energies of the electrons in the exciter front were only a few keV (Haggerty *et al.*, 2001). This, plus the delayed injection of the near-relativistic electrons, led Haggerty and Roelof to conclude that the electrons that produced the microwave and hard X-ray emission in the chromosphere do not escape promptly into space at intensities measurable by ACE/EPAM; nor do near-relativistic electrons escape at the time of the high coronal type III bursts.

Simnett *et al.* (2002) searched for CMEs in association with the beam events, and found that the majority of the latter occurred in association with a west-hemisphere CME. In this paper we review these results and examine what inferences may be made from the data to improve our understanding of electron acceleration and release from the solar environment.

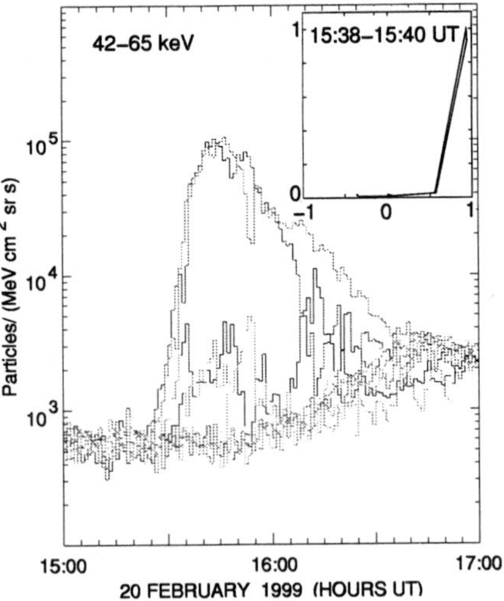

FIGURE 1. The 42-65 keV electron sectored intensities from LEFS60 for the 20 February 1999 beam event. Four of the eight sectors are around background level until after 16:00 UT. Inset: The normalised pitch cosine distribution from 15:38-15:40 UT; the abscissa is the cosine of the pitch angle.

THE OBSERVATIONS

The electron observations we use were made with the EPAM instrument (Gold *et al.*, 1998) on the ACE spacecraft from 25 August 1997 to 9 September, 2000. Solar electrons were measured in either the LEFS60 or the LEMS30 system; for the latter the electrons were magnetically deflected into a separate detector, thus eliminating the possibility of ion contamination which exists in the LEFS60 telescope. However, at the onset ion contamination is not a problem because they arrive much later. The LEFS60 and LEMS30 telescopes are oriented on the spinning ACE spacecraft at 60° and 30° respectively to the spin axis; and the spin axis is generally within a few degrees of the Sun-Earth line. Thus these telescopes are well suited for studying prompt solar electron events.

Fig. 1 shows an example of the beamed electron event on 20 February, 1999. Other examples are given in Simnett *et al.*, (2002). Four of the eight sectors of the LEFS60 telescope are at the background level until after 16:00 UT, which is some 15 minutes after the intensity maximum. Inset is the pitch cosine distribution, normalised to the intensity in the peak sector, from 15:38-15:40 UT. The measured angular width is comparable to the opening angle of the telescope collimator, so the actual beam width is probably unresolved. The strong beam is maintained for over 30 minutes. This event is typical of the electron events we have used in this study.

The CME observations were made with the LASCO (Brueckner *et al.*, 1995) C2 and C3 coronagraphs on the SOHO spacecraft, which is also near the L_1 point. Simnett *et al.* (2002) examined running difference images for 52 out of 79 scatter-free electron events analysed by Haggerty and Roelof (2002), and were able to determine the onset times at a nominal 1 $R_{\odot}$. The other events occurred when LASCO was not observing. We have since found another event, so the statistics in this paper are from 80 events, of which 53 are observed by LASCO.

We also made use of data from the WAVES instrument on the WIND spacecraft; and from Solar Geophysical Data, (US Dept. of Commerce, Boulder, CO).

ASSOCIATIONS AND CORRELATIONS

Haggerty and Roelof (2002) showed that 46 out of the original 79 electron events occurred in association with an Hα flare. However, when they allowed for the fact that the distribution of flares extended beyond the west limb, a conservative estimate of the true fraction of Hα flare associations was ~85%. Thus most of the beam events were associated with chromospheric flare activity and the location of such activity was consistent with good magnetic connectivity to ACE. The majority of the beamed events were observed in 3 or 4 of the EPAM electron channels. Using the onsets at the highest measured energies, Haggerty and Roelof (2002) were able to determine the injection time at the Sun. This time is, of course, only the beginning of an extended injection phase that may last tens of minutes. The injection must last for at least as long as the duration of the beam-like anisotropy. In the example in Fig. 1 the duration is 40 - 60 minutes, typical of most of the beams we have detected.

The majority of the beamed electron events are also

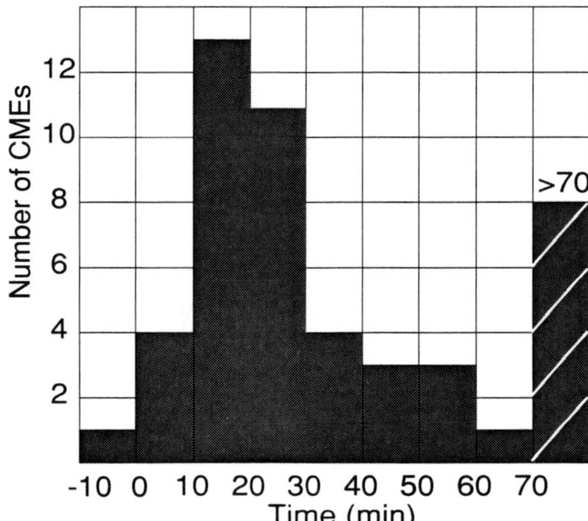

FIGURE 2. The electron delay from the CME launch time. The final column represents eight events with a delay of more than 70 minutes. (After Simnett *et al.*, 2002)

associated with CMEs observed (in projection onto the plane of the sky) by LASCO off the solar west limb. Of the 53 events occurring during LASCO observations, 48 were distinct enough that the height-time profile could be mapped back to the Sun. We define, for consistency, a CME onset to be when the height-time profile maps back to 1 $R_\odot$. This radius is somewhat artificial as the erupting structure must form above the photosphere; however this serves as a useful reference point from which to construct a statistical analysis. Of the 48 CMEs, 35 we considered as "classical" CMEs (including 6 halo events) which are large, loop-like structures, as distinct from small blobs.

From the two data sets we can construct a histogram of the time between the CME onset and the start of the electron injection, and this is shown in Fig. 2 (from Simnett *et al.*, 2002). It is clear that the distribution is peaked at around 19 minutes, with the electron injection delayed from the CME launch. It is useful to examine in more detail the accuracy or confidence we have in this result, which is based on extrapolations of two entirely independent data-sets.

Krucker *et al.* (1999) discussed the origin of impulsive electron events below 300 keV and concluded that there were two types. Low energy events, below 25 keV, were released in good temporal association with metric type III radio bursts, although timing of the injection to within a few minutes becomes more difficult at lower energies: a 5 keV electron takes 70 minutes to travel along the interplanetary magnetic field line to Earth. The higher energy events, comparable in energy to those we discuss here, were delayed by up to half an hour from such bursts. Haggerty and Roelof (2002) have also analysed the timing of the beam events (lower energy threshold 38 keV) with respect to type III emission, and reach a conclusion consistent with Krucker *et al.*, for the higher energy electrons. They estimated that the uncertainty in the electron release time is of the order of three minutes. Recall that it is only the low energy electrons that are released at the same time as the type III radio emission is observed; the higher energies (>38 keV) are delayed.

Although many of the CME height-time profiles show a constant speed, there is inevitably a finite error associated with the estimated onset time. This is difficult to establish, and an attempt to do this for each CME is not fruitful, as the notional starting altitude is somewhat arbitrarily chosen at 1 $R_\odot$ simply for consistency. We estimate that the uncertainty in CME onset time is also a few minutes. However, we have a different way to verify this, namely by examining the shape of the distribution in Fig. 2. If the two times were physically unrelated, then the expectation value of the delay would be zero. Any uncertainties in the estimates of either the CME launch time or the electron injection time would broaden the distribution; systematic errors could shift the median. If we interpret the distribution in Fig. 2 as a Gaussian plus a tail that includes the events with long delay times, the half width gives a value of the standard deviation, $\sigma \sim 10$ minutes. In order to assess the statistical significance of the median delay (19 $\pm$10 minutes), suppose that the half-width were entirely due to measurement uncertainties in the estimates of the CME launch time (σ_{CME}) and the electron injection (σ_{elec})

If we suppose that:

$$\sigma = (\sigma_{CME}^2 + \sigma_{elec}^2)^{1/2}$$

this gives $\sigma_{CME} \sim \sigma_{elec} \sim 0.7\sigma = 7$ minutes, assuming that the uncertainties are comparable. However, we seriously doubt that the timing uncertainties could be this large. That is because there is only one event with a negative delay, and we would consider it a remarkable coincidence that a putative systemmatic error could combine with random measurement errors to give delays that are essentially always positive. Since the actual delay times must have some variation from event to event, any random measurement error must be <<7 minutes. Therefore our earlier estimates of a few minutes for the uncertainty of both the electron injection and CME onset are justified by the data in Fig. 2.

For each event we may determine the height of the CME at the time of the electron injection, and the distribution of these distances is shown in Fig. 3 (from Simnett *et al.*, 2002). The CME was below a projected altitude of 5 $R_\odot$ for the majority (44) of the injections. The median value of this distribution, ignoring the 4 events beyond 5 $R_\odot$, is 2.3 $R_\odot$. There is a further consideration, namely that the CMEs are seen projected onto the plane of the sky. This will underestimate some CME speeds, but not

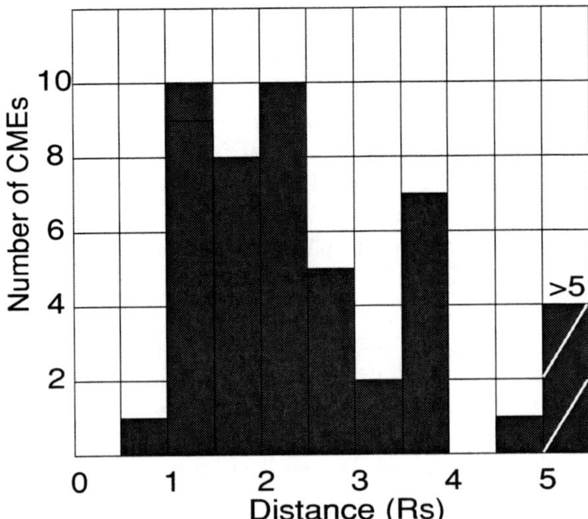

FIGURE 3. The height of the CME above Sun-center at the time of the electron release in the corona. (After Simnett *et al.*, 2002)

their onset times. However, the effect is relatively small, as it varies as the cosine of the angle out of the plane of the sky and our CMEs were almost all visible off the west limb. In view of this effect we regard 3.0 $R_{\odot}$ as a more conservative estimate of the mean radius for electron injection.

We have examined the CME speeds corresponding to the events. Fig. 4 (from Simnett *et al.*, 2002) shows the distribution of CME speeds versus the electron release delay time from the CME onset time. There is a trend towards anticorrelation. Note that the 8 events with long delays follow the general trend because they correspond to the slowest CMEs. The continuity of this ordering with respect to CME velocity shows that there is no separable sub-class of near-relativistic electron injections. This anticorrelation is what produces the injections at a relatively small range of coronal altitudes, as indicated by Fig. 3. We would not expect a perfect anticorrelation as the magnetic topology of the high corona is not expected to be homogeneous, nor to have a simple radial dependence.

There are two other features of the electron events which are worthy of note, namely the peak intensity and the energy spectrum. Fig. 5 (from Simnett *et al.*, 2002) shows the peak intensity for the four electron channels in the LEFS60 detector. We have used this as it has a larger geometrical factor than the deflected electron detector, and is therefore statistically more accurate. As mentioned above, ion contamination is negligible as any ions associated with the events which could satisfy the detector logic requirements for electrons take hours to reach Earth, and cannot contribute to the peak intensity in these electron events, which typically have intensity-time profiles like Fig. 1. The lines drawn through the points are least-squares fits to the data. The positive slope to each line increases with energy. This means that the faster (slower) the CME the more (less) efficient it is at accelerating electrons, and moreover that the effect is strongest for the highest energy electrons that we measure.

Fig. 6 (from Simnett *et al.*, 2002) shows the power law index of the spectrum, defined from the intensities measured in adjacent electron energy channels, as a function of CME speed. The straight lines represent least-squares fits to the points. The decreasing slopes mean that the faster (slower) CMEs produce harder (softer) electron spectra. This trend is slightly stronger at higher energies. There might be a sensitivity effect, in that small events with steep spectra will not appear; however, if we accept the trend evident in Fig. 4, these should correspond to lower speed CMEs. Therefore they would have increased the anticorrelation had they been measured.

Physically the trends in Figs. 5 and 6 show that faster CMEs not only accelerate more electrons, but they also produce harder electron spectra. Both trends are consistent with the general theory of shock acceleration; it is likely that most of the CMEs drive shocks in the corona near the Sun. Both trends can be made even more visible by constructing composite spectra derived from the least-squares fits in Fig. 5. These linear fits of the log-mean intensity (at each of the four channel energies) to the associated CME velocity can be converted to four-point spectra of the form $j(E,V)=j0(E)\exp[a(E)V]$, where $j0(E)$ is taken from the ordinate intercepts in each panel of Fig. 5 and $a(E)$ is the slope of each fit (indicated in the panel). These composite spectra are plotted in Fig. 7 and each is labeled with the associated CME velocity $200<V<1200$ km/s. It is clear that (1) the faster the CME, the more intense the 40 - 300 keV electron intensity tends to be, and (2) the faster the CME, the harder (flatter) is the electron spectrum. Trends (1) and (2) are consistent with shock acceleration theory.

DISCUSSION

It was the comprehensive study of electron events observed near solar minimum by Krucker et al. (1999) that called out the common occurrence of significant delays ($\sim$10s of minutes) of near-relativistic electron injection after prompt solar flare electromagnetic emission. However, Krucker et al. interpreted their results in terms of two classes of scatter-free electron events, namely those with negligible delays and those with significant delays with respect to metric type III bursts. As pointed out in our discussion of Fig. 4, we find only one inseparable class of events, with delays ranging from the small to

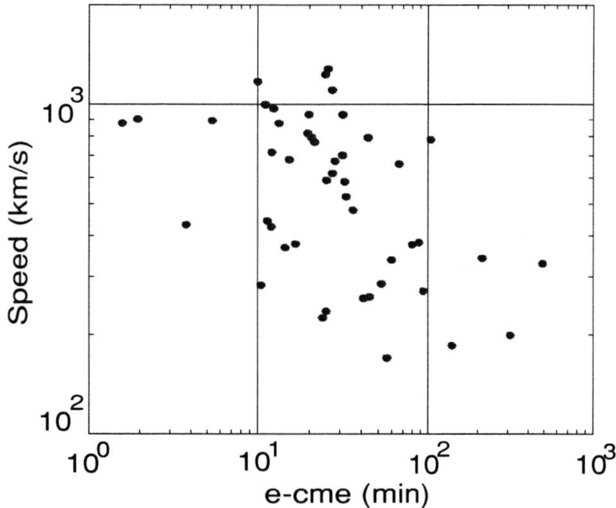

FIGURE 4. The electron injection delay from the CME launch time versus the CME speed. (After Simnett *et al.*, 2002)

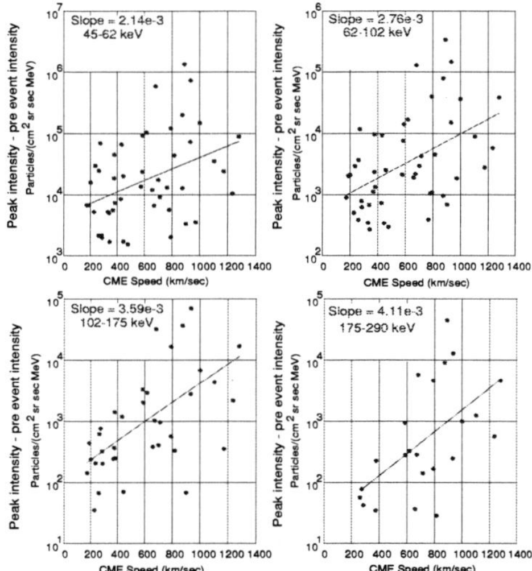

FIGURE 5. The electron peak intensity in the four electron channels of LEFS60, as a function of the CME speed. (After Simnett *et al.*, 2002)

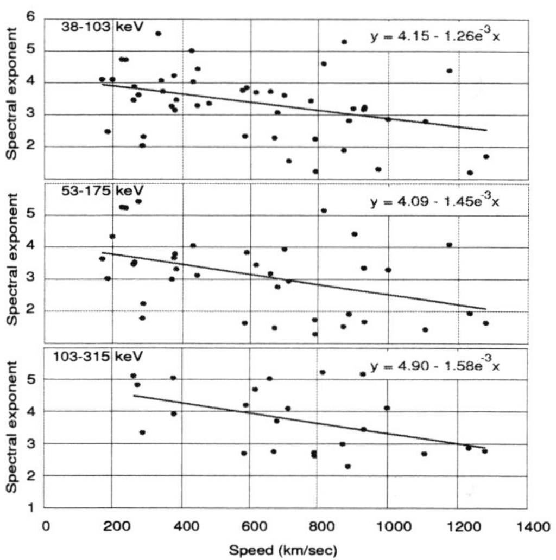

FIGURE 6. The two point electron intensity power-law spectral exponent derived from adjacent pairs of the four deflected electron channels as a function of CME speed. The upper plot is from DE1 and DE2; the middle plot is from DE2 and DE3; and the lower plot is from DE3 and DE4. (After Simnett *et al.*, 2002)

the large because the small delays are associated with fast CMEs and the large delays with slow CMEs. We believe the results of Krucker et al. are actually consistent with this latter interpretation to which we were led by the close associations we found with CMEs. This consistency can be seen in Fig. 8 which shows the distribution of the 58 3DP scatter-free electron injection delays (Krucker et al., 1999) with that of our own 48 EPAM events associated with metric type III bursts (taken from the list of Haggerty and Roelof, 2002). Allowing for a possible solar cycle effect (see below), we think that both distributions represent a single class of events.

Solar electron acceleration and emission is clearly complex and Krucker *et al.*, (1999) have suggested that it could be solar cycle dependent. Low energy electron events without metric type III associations such as those reported by Potter *et al.*, (1980) from ISEE-3 during solar maximum had around 40% of total events that were measurable only below 15 keV. Krucker *et al.* point out that from the WIND data around solar minimum, events seen only at low energies are rare. This suggests that more than one physical process must be responsible for the acceleration. It seems clear that the low energy events

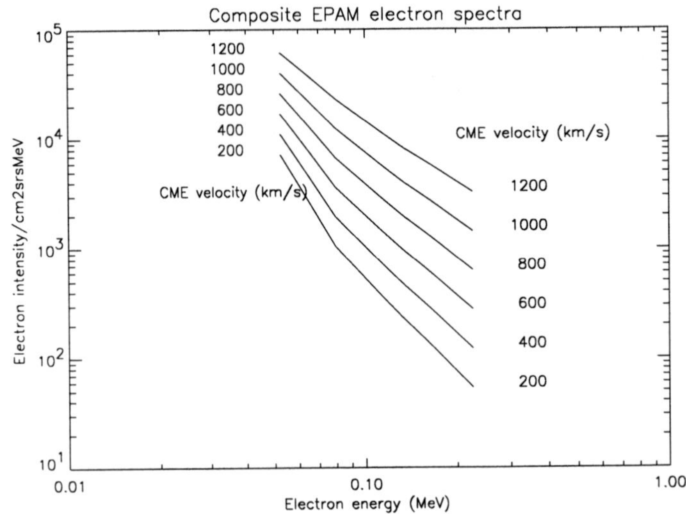

FIGURE 7. "Composite" electron spectra constructed from the least-squares fits of intensity vs. CME velocity from Fig. 5. See text for details. The velocity dependence is consistent with that generally predicted by shock acceleration theory.

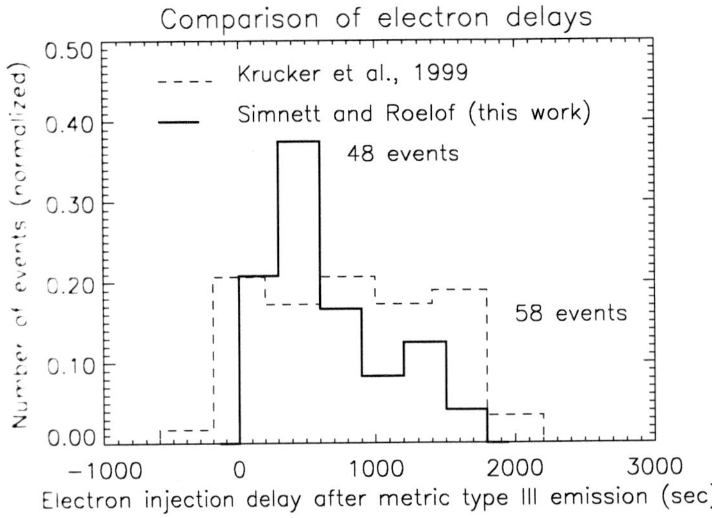

FIGURE 8. Distributions of near-relativistic electron delays after solar type III radio bursts: 58 events from WIND/3DP (Krucker et al., 1999) and 48 ACE/EPAM events taken from the list of Haggerty and Roelof (2002).

are formed from a process taking place high in the corona (Lin, 1985), whereas a typical high energy event comes from a CME-driven shock. The events are only seen at a given spacecraft under favorable magnetic connections.

However, the situation is more complicated than suggested by this simple analysis. LASCO observations show that there are many CMEs which are fast enough to drive shocks through the corona. Simnett et al (2002) estimated conservatively that if all such CMEs accelerated near-relativistic electrons, there should be at least five times as many electron beams seen at ACE than we have detected. The clue to the puzzle lies in the near perfect association of strong decametric type III bursts seen by the WIND spacecraft in assocation with the electron beams. Haggerty et al., (2001) showed that the exciter speed was equivalent to electrons of only a few keV. Therefore we believe the CME-driven shocks will not accelerate near-relativistic electrons out of the ambient coronal plasma, but that a seed population of non-thermal electrons (*e.g.* a few keV) needs to be present before significant electrons are accelerated by the CME. The more energetic/intense the seed population, the more energetic/intense the subsequent event. Some very intense electron events, such as 20 February 1999, 18 February, 2000, and 12 July, 2000, occurred when the associated CME was the second of a pair, separated by only a few hours. It is possible that electrons trapped in the magnetic structure of the first CME cannot escape into the interplanetary medium, but populate the near-solar environment behind the first event and thus become available as a seed population for the second CME, and thereby produce an exceptionally intense event.

There are other puzzling features of the events which are worth exploring in more detail. (1) Why do the beams not themselves produce radio emission in the corona? A possible answer to this is that there is a finite risetime (>10 minutes) for the acceleration (see Fig. 1) such that the energy spectrum is always in quasi-equilibrium, and it never develops a significant "bump on the tail" which is necessary for the production of a type III burst. (2) Why are some beams associated with CMEs of low speeds whereas other CMEs with much higher speeds do not appear to produce them? The answer may lie in the physical conditions in the medium through which the CME propagates. If the local Alfvén velocity is low, either because of a low magnetic field or a high density, then the Mach number may be enough to form a shock even though the absolute CME speed is relatively low.

CONCLUSIONS

The new results from our recent study (Simnett *et al.*, 2002) are the following:

(1) There is a temporal delay between the electron injection and the CME launch time, which suggests that the CME drives the acceleration mechanism.

(2) The correlations we have found between the CME velocity and the electron intensity and spectral index are consistent with the signatures of shock acceleration.

(3) The conclusion is that CMEs drive the shocks that accelerate the electrons.

There is one further implication of these findings. If the acceleration is predominantly a velocity-dependent process there is no reason why ions should not be accelerated also. However, observationally ions are not typically seen to as high a velocity as electrons.

ACKNOWLEDGMENTS

We thank Dr D.K. Haggerty for his help in updating some of the figures used in this paper; and our colleagues in the LASCO and EPAM consortia for their support and contribution to the data analysis. We acknowledge the support of Starlink for computing facilities at the University of Birmingham. ECR was supported in part by NASA Contract NAS5-97271(009).

REFERENCES

. Bougeret, J.-L. et al, 1995, Sp. Sci. Rev., 71, 231
. Brueckner, G.E., et al, 1995, Solar Physics, 162, 357
. Gold, R.E. et al, 1998, Sp. Sci. Rev., 86, 541
. Haggerty, D.K., Roelof, E.C. and Kaiser, M.L., 2001, Proc. PRE-V, (Graz, Austria), Austrian Academy of Sciences, 437
. Haggerty, D.K.and Roelof, E.C, 2002, Ap. J., 579, 841
. Krucker, S., Larson, D.E., Lin, R.P. and Thompson, B.J. 1999, Ap. J., 519, 864
. Lin. R.P. 1985, Solar Physics, 100, 537
. Potter, D.W., Lin, R.P. and Anderson, K.A., 1980, Ap. J., 236, L97
. Simnett, G.M., Roelof, E.C and Haggerty, D.K., 2002, Ap. J., 579, 854

The Composition of Interplanetary Coronal Mass Ejections

Thomas H. Zurbuchen[*], L. A. Fisk[*], S. T. Lepri[*], and R. von Steiger[†]

[*] *Department of Atmospheric, Oceanic and Space Sciences, University of Michigan, 2455 Hayward St., Ann Arbor, MI 48109-2143, USA*
[†] *International Space Science Institute, Hallerstrasse 6, CH-3012 Bern, Switzerland*

Abstract. Interplanetary coronal mass ejection (ICME) associated plasma can exhibit signatures in elemental, ionic and isotopic composition. These signatures occur in less than 50% of all ICMEs, but are very indicative of ICME plasma. We review these compositional anomalies and briefly discuss a physical scenario that could be responsible for these anomalies.

INTRODUCTION

The interplanetary consequences of coronal mass ejections (CMEs) have been a central topic of heliospheric physics since their discovery in the 1970s. Research on Interplanetary Coronal Mass Ejections (ICMEs) has focused on two major aspects: the identification of and the physical drivers for ICME material.

First, the identification of ICME material has been the focus of many publications, e.g., by Richardson et al. [1, 2]. Common plasma signatures of ICMEs in the solar wind include counterstreaming electrons, anomalously low proton and electron temperatures, strong magnetic fields, low plasma beta, and smooth field rotations [3]. These studies have stressed the diverse nature of ICME-associated plasmas and the resulting lack of a coherent set of signatures that can be identified as "necessary and sufficient" for the presence of ICME plasma.

Second, progress is being made in relating CME signatures to the underlying physical processes that govern CME initiation in the low corona, and propagation of CMEs from the corona into the heliosphere. In situ measurements provide crucial constraints for these CME models [4] and can serve as "ground-truth" measurements for various CME models. Even though substantial progress has been made, CME models are still not mature enough for some of these detailed tests [5].

Composition measurements of ICMEs provide important contributions for both research aspects discussed above. Before we summarize recent results, we would like to note that some valuable solar wind composition data of ICMEs preceded modern dedicated composition experiments. A comprehensive description of such data can be found in a paper by Bame [6] summarizing his 1982 Solar Wind conference presentation.

In this paper, two major issues will be addressed using Ulysses and ACE composition data measured by each of their Solar Wind Ion Composition Spectrometers (SWICS) [7]. First, we will focus on the compositional signatures of average ICMEs. We specifically concentrate on the elemental composition of "driver gas" identified by an unusually high alpha-to-proton (He^{2+}/H^{+}) ratio. We will then propose a physical scenario that could produce these compositional measurements.

COMPOSITIONAL SIGNATURES

This section discusses the observed compositional anomalies of ICMEs. The diversity and geometrical complexity of ICMEs, based on in situ plasma and field signatures [3], become evident when looking at these compositional signatures. It is therefore very difficult to define average properties of this highly diverse plasma population. We now have access to 11 years of Ulysses data and 4 years of ACE data. From these data we know that there are two classes of relatively rare ICMEs that are fundamentally different than the vast majority of ejections observed by these instruments. First, there are unusually "cold events" that are characterized by a substantial fraction of all ele-

CP679, *Solar Wind Ten: Proceedings of the Tenth International Solar Wind Conference*, edited by M. Velli, R. Bruno, and F. Malara

ments with low charge states. One prominent example is discussed by Gloeckler et al. [8] based on ACE-SWICS data. These events show unusual fractionation patterns [9] uncharacteristic of the vast majority of ICMEs. Second, there are high-latitude events described by Gosling [10], which have a composition very similar to coronal-hole-associated wind. These two classes of ICMEs are of considerable interest but will not be discussed further in this paper because of their apparently insignificant relative contribution to ICMEs during the solar cycle.

The following subsections will summarize our present knowledge of compositional anomalies in the majority of the ICMEs (those that fall outside the two categories described above) observed by ACE and Ulysses.

Unusual Ionic Composition

It has been pointed out that a strong preference exists for ICMEs to have charge states indicative of unusually hot electron temperatures [11, 12, 13]. Figure 1 shows plasma and composition data of an ICME observed by the combined ACE plasma and field sensors. The figure shows a time period of 14 days during 1999 that contains ICME ejecta, the trailing edge of coronal-hole-associated wind, and variable slow wind. The ionic composition during large parts of this time period is unusual. Both O and Fe show charge states that are clearly above the average solar wind composition. The compositional signatures are highly time-dependent, indicating hotter and cooler parts in the ICME. In a comprehensive study by Lepri et al. [12], roughly 50% of all ICMEs, as identified by a large set of plasma and field signatures, had long Fe charge state enhancements. Conversely, over 95% of all Fe charge state enhancements that extended over 20 hours were found to be ICMEs. Even though there are no similar studies on O charge states, similar behavior might be expected when studying O^{7+}/O^{6+} [13].

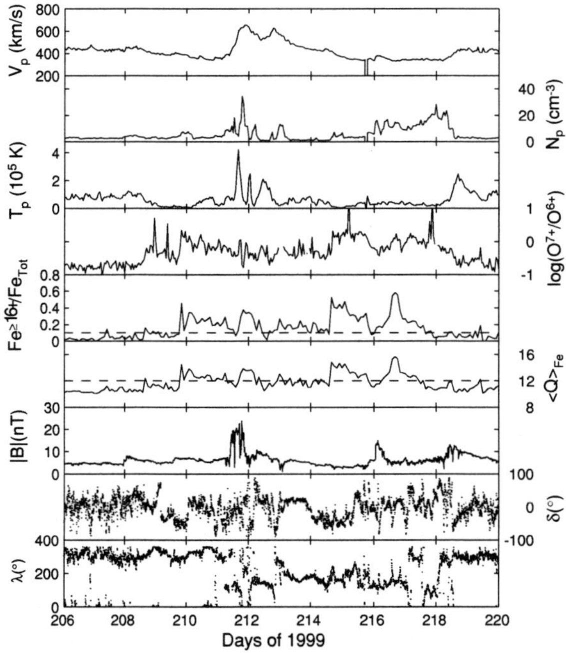

FIGURE 1. ICME event observed by ACE. The time period shows an elongated complex structure that exhibits compositional anomalies that are typical of ~50% of all ICMEs.

The observation of predominantly hot charge states in ICMEs might be surprising. Remotely observed CMEs often have a three-part structure with a bright rim, indicative of the interaction region of CME plasma with the ambient solar wind, a dark cavity, and a bright high-density core that is presumably associated with an erupting prominence. Out of over 200 ICMEs observed at ACE, only <2 have very low Fe charge states. If cool prominence plasma ever makes it into space, it either has to undergo rapid ionization during the CME release, or it needs to be confined to a distinguishably small volume within the ICME. In the absence of any of these explanations, the observed small fraction of "cold" ICMEs is not consistent with any prominence plasma making it into the heliosphere.

Unusual Elemental Composition

One of the most prominent ICME signatures is the He^{2+}/H^{+} ratio [3]. Typically, $He^{2+}/H^{+}>0.08$ is considered unusually high and can be used for the identification of ICMEs from the Sun into the outer heliosphere [14] (does this mean you can observe high alpha to proton ratios at the sun and out to Ulysses? This isn't quite clear here). The He^{2+}/H^{+} ratio has been measured over many decades using electrostatic systems and Faraday cup systems [15, 16]. Traditionally, periods of unusually large He^{2+}/H^{+} ratio have been identified as "driver gas" even before these data were put into context with CME observations at the Sun [6]. It is now clear that these signatures exist in only a fraction of the plasma associated with an ICME. The geometrical association of these periods with parts of the traditional three-part structure is unclear.

Using SWICS data, ICMEs can be studied using a more comprehensive set of data. Reinard et al. [13] examined elemental composition data from over 20 ICMEs in 1998-1999 identified using bi-directional electron signatures. She compared them with the composition of slow solar wind offset by 1-2 days from the ejecta. The result was that the average elemental composition of Fe/O of ICMEs was indistinguishable from the slow solar wind composition. Several of the events in question had substantial He/H enhancements accompanied by similar He/O enhancements. The enhancements in He therefore seemed to occur without

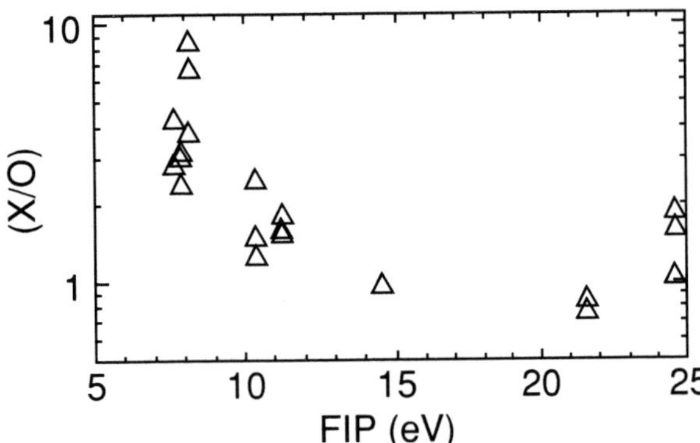

FIGURE 2. Elemental composition of driver gas during the Ulysses mission. The average composition is very similar to the composition of the solar wind with the exception of He/O. Si/O is artificially enhanced due to "blending" with Fe of high charge state.

significantly affecting the O and Fe abundances relative to H.

A more comprehensive test can be carried out using Ulysses-SWICS data. A search has been performed, over the entire Ulysses data set, for full days that have $He^{2+}/H^{+}>0.8$ in order to ensure sufficient statistical accuracy for all rare elements in question. Due to the intermittent nature of these enhancements, it is difficult to find those that span the entire day, except in a few cases. Figure 2 shows the elemental abundance of Fe, Mg, Si, S, Ne, and He of this "driver gas" ordered by First Ionization Potential (FIP). Si is found to be contaminated by Fe of a very high charge state (perhaps 16+), and therefore has an unusually high elemental abundance. All other elements have abundances that are very similar to the slow solar wind, fractionated according to the FIP. We therefore conclude that, even though ICMEs can exhibit order-of-magnitude enhancement of He^{2+}/H^{+}, the elemental abundances show little or no fractionation beyond the common FIP fractionation of the slow solar wind.

Unusual Isotopic Composition

Unusual enhancements of $^3He/^4He$ in ICMEs have been reported based on early plasma measurements [6]. These enhancements are very interesting because of the similar fractionation effect observed in Solar Energetic Particles (SEP) [16]. Figure 3 compares $^3He/^4He$ of slow, fast, and ICME-associated solar wind. Slow solar wind values are from Geiss [17] and Ogilvie et al. [18], fast solar wind values are from Gloeckler and Geiss [19] and the ICME values came from Gloeckler et al. [8] and Ho et al. [20]. We do not currently know how often these enhancements occur or how these signatures correlate with other ICME plasma and compositional signatures. There are no reports of isotopic fractionation in ICMEs of elements other than He that are above the detection limit.

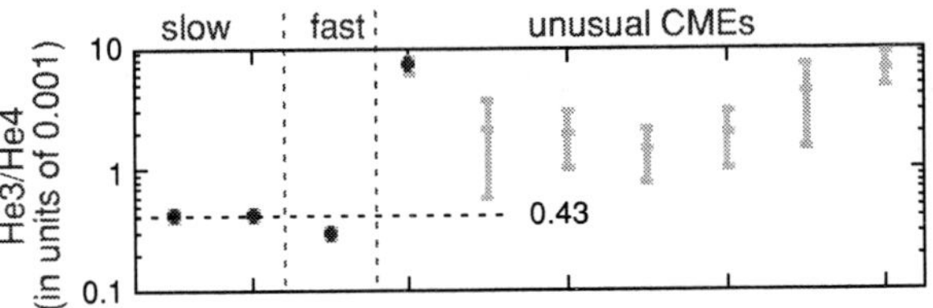

FIGURE 3. The He isotopic ratio in various solar wind regimes.

PHYSICAL SCENARIO

This section discusses an interpretation of the compositional signatures of ICMEs in the context of the physical processes governing CMEs in the solar corona. The compositional signatures to be explained are fourfold:

a) FIP fractionation,
b) He/H enhancements,
c) Unusually high charge states of heavy ions, and
d) $^3He/^4He$ enhancements.

The overall chemical composition of ICMEs is very similar to the composition of the slow solar wind, which is closely associated with coronal loops [21]. These loops are presumed to undergo successive reconnections in the vicinity of their footpoints, as sketched in Fig. 4a. These reconnection events lead to successive FIP fractionation, perhaps by a process not unlike the one described by Von Steiger and Geiss [22]. During this process, gravitational settling presumably also plays a role, most severely affecting He/H. Therefore, the depletion of He in these loops must be accompanied by an enhancement of He/H in the vicinity of the loop footpoints.

When the magnetic structure of these loops becomes unstable, the erupting field will reconnect with the overlaying magnetic field-lines. This process will drive reconnection currents in these erupting loops, as sketched in Fig. 4b. These currents are mostly carried by supra-thermal electrons. The electrons have two primary effects: (1) they first ionize the ambient plasma, leading to heating and rapid ionization of heavy ions; and (2) they can also interact with the chromosphere, leading to chromospheric evaporation at the footpoints of the loops where He/H is enhanced. The latter effect then gives rise to an excess of He/H in the erupting loops. In a plasma that carries a current and has elevated He/H, wave-particle interactions with ^3He will lead to heating of ^3He [16]. As a result, this heating will enhance the $^3He/^4He$ ratio. The eruption will therefore tend to produce all three compositional

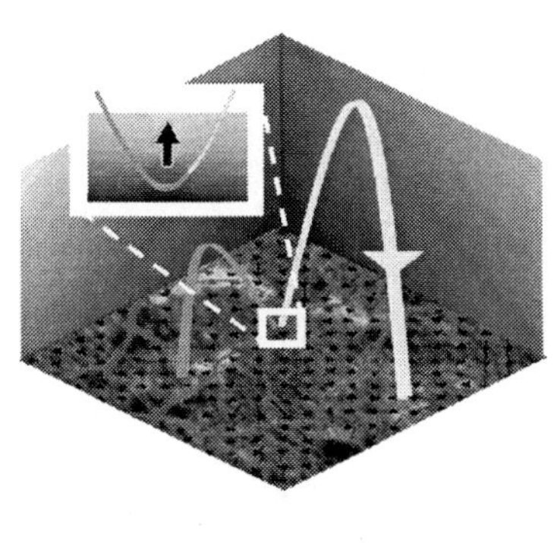

(a)

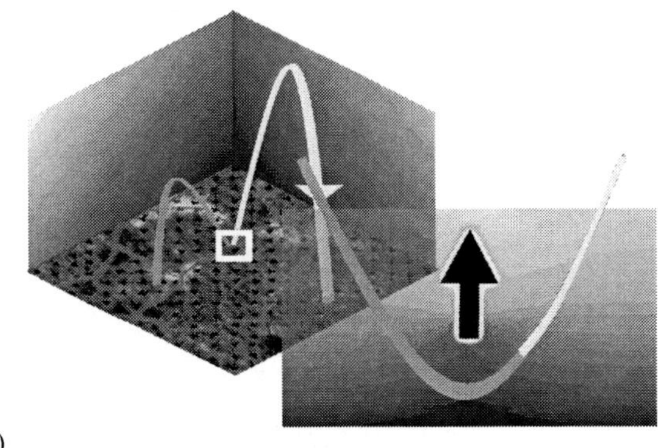

(b)

FIGURE 4. A physical scenario to explain the observed compositional signatures. Plasma is transported into loops through successive reconnection events, He/H is depleted in these loops, enhancing He/H below these loops to large values. During CME eruption, reconnection-driven currents lead to compositional changes in the loop plasma, and lead to chromospheric evaporation.

signatures that distinguish solar wind from ICME plasma.

CONCLUSIONS

Compositional signatures of ICMEs have been described, covering elemental, ionic, and isotopic composition anomalies. Even though the study of these signatures in situ has matured [2], these data have not been sufficiently analyzed in the context of the physical processes that govern CME eruptions. Physical scenarios such as described above are possible and need to be incorporated into a next-generation CME model therefore providing important constraints on the generation and eruption of CMEs and their propagation to Earth.

REFERENCES

1. Richardson, I. G., Cliver, E. W., and Cane, H. V., *J. Geophys. Res.* **28**, 2569 (2000).
2. Richardson, I. G. et al., these proceedings, 2002.
3. Gosling, J. T., "Coronal Mass Ejections: An Overview," in *Coronal Mass Ejections*, edited by N. Crooker, J. A. Joselyn, and J. Feynman, Geophys. Monograph 99, 1997, p. 9.
4. Riley, P., Linker, J. A., and Mikic, Z., *J. Geophys. Res.* **106**, 15889 (2001).
5. Manchester, W., et al., these proceedings, 2002.
6. Bame, S. J., "Solar Wind Minor Ions – Recent Observations," in *Solar Wind 5 Proceedings*, edited by M. Neugebauer, NASA Conference Proceedings 2280, 1983.
7. Gloeckler, G., et al., *Astron. Astrophys. Suppl. Ser.* **92**, 267, 1992.
8. Gloeckler, G., et al., *Geophys. Res. Lett.* **26**, 157 (1999).
9. Wurz, P., et al., these proceedings.
10. Gosling, J. T., Riley, P., McComas, D. J., and Pizzo, V. J., *J. Geophys. Res.* **103**, 1941 (1998).
11. Henke, T., et al., *Geophys. Res. Lett.* **25**, 3465 (1998).
12. Lepri, S. T., et al., *J. Geophys. Res.* **106**, 21,231 (2001).
13. Reinard, A.A., et al., "Comparison between Average Charge States and Abundances of Ions in CMEs and the Slow Solar Wind," in *SOHO/ACE workshop Solar and Galactic Composition*, edited by R. F. Wimmer-Schweingruber, AIP Conference Proceedings 598, Melville, New York, 2001, p. 139.
14. Paularena, K. I., Wang, C., von Steiger, R., and Heber, B., *Geophys. Res. Lett.* **28**, 2755 (2001).
15. Aellig, M. R., Lazarus, A. J., and Steinberg, J. T., *Geophys. Res. Lett.* **28**, 2767 (2001).
15. Borrini, G., Gosling, J. T., Bame, S. J., and Feldman, W. D., *Solar Physics* **83**, 367 (1983).
16. Fisk, L. A., *Astrophys. J.* **224**, 1045 (1978).
17. Geiss, J., Eberhardt, P., Buehler, F., Meister, J., and Signer, P., *J. Geophys. Res.*,**75**, 5972 (1970).
18. Ogilvie, K. W., Coplan, M. A., Bochsler, P., and Geiss, J., *J. Geophys. Res.*, **85**, 6021 (1980).
19. Gloeckler, G., and Geiss, J., *Space Sci. Rev.* **84**, 275 (1998).
20. Ho, G., Hamilton, D. C., Gloeckler, G., and Bochsler, P., *Geophys. Res. Lett.* **27**, 309 (2000).
21. Zurbuchen, T. H., Fisk, L. A., Gloeckler, G., and Schwadron, N. A., *Space Sci. Rev.* **85**, 397 (1998).
22. Von Steiger, R., and Geiss, J., *Aston. Astrophys.* **225**, 222 (1989).

Effect of CME Interactions on the Production of Solar Energetic Particles

N. Gopalswamy*, S. Yashiro†*, G. Michalek†*, M. L. Kaiser*, R. A. Howard**, R. Leske‡, T. von Rosenvinge* and D. V. Reames*

**NASA Goddard Space Flight Center, Greenbelt, MD 20771.*
†Catholic University of America, Washington, DC 20064
***Naval Research Laboratory, Washington, DC 20375.*
‡California Institute of Technology, Pasadena, CA 91125.

Abstract.
We analyzed a set of 52 fast and wide, frontside western hemispheric (FWFW) CMEs in conjunction with solar energetic particle (SEP) and radio burst data and found that 42 of these CMEs were associated with SEPs. All but two of the 42 SEP-associated FWFW CMEs (95%) were interacting with preceding CMEs or dense streamers. Most of the remaining 10 SEP-poor FWFW CMEs had either insignificant or no interaction with preceding CMEs or streamers, and were ejected into a tenuous corona. There is also a close association between type II radio bursts in the near-Sun interplanetary medium and SEP-associated FWFW CMEs suggesting that electron accelerators are also good proton accelerators.

INTRODUCTION

Copious production of nonthermal electrons and ions is an important aspect of solar eruptive events such as flares and coronal mass ejections (CMEs). Nonthermal electrons are inferred from the radio signatures they produce or detected by in situ observations. On the other hand, energetic ions need to be observed only in situ. Spectacular radio signatures indicating the production of nonthermal electrons at several solar radii from the Sun have been recently identified and found to be the result of colliding CMEs [1, 2]. A natural question would be whether such CME interactions affect the production of nonthermal ions, commonly referred to as solar energetic particles (SEPs). A preliminary statistical study by Gopalswamy et al. [3, hereinafter Paper 1] suggests that a vast majority of the large SEP events are associated with fast CMEs that interacted with one or more preceding CMEs within about 20 solar radii. Within the current paradigm that CME-driven shocks [see, e.g., 4, 5] accelerate the SEPs from the upstream solar wind material, one would expect that the shocks pass through the material of the preceding CMEs. CMEs contain plasma structures with temperatures ranging from a few thousand K to several million K, so the nature of the accelerated particles will depend on the physical properties of the source material that enters the shock. There are good indications that the composition of the SEPs significantly differs from that of the solar wind [see, e.g. 6]. Therefore, the statistical association between CME interaction and large SEP events may be important in understanding the real situation in the acceleration sites of SEPs. One aspect of the statistical analysis in Paper 1 was to perform an inverse study of all the fast and wide, frontside western hemispheric (FWFW) CMEs to check their level of interaction with other CMEs and their association with SEPs. The inverse study confirmed the importance of CME interaction for SEP production. However, we did not examine the SEP-poor FWFW CMEs and the SEP-associated FWFW CMEs with no interactions in detail. In this paper, we report on our further analysis of the FWFW CMEs.

DATA

The FWFW CMEs observed by the Solar and Heliospheric Observatory (SOHO) mission's Large Angle and Spectrometric Coronagraph (LASCO) were selected from the SOHO/LASCO CME catalog maintained on line (`http://cdaw.gsfc.nasa.gov/`) by requiring that: 1. the CME speed (V) is > 900 km/s, 2. the width (W) > 60°, 3. the CME span includes position angle 270°, and 4. the solar source of the CME has a longitude between W0 and W90. During the period January 1996 to December 2001, we identified 52 CMEs, whose basic properties are listed in Table 1 (Date, UT, speed (km/s), W, and solar source in columns 1-5). We

CP679, *Solar Wind Ten: Proceedings of the Tenth International Solar Wind Conference*, edited by M. Velli, R. Bruno, and F. Malara

TABLE 1. Characteristics of the FWFW events

	CME					Type II	SEP	
Date	Time	Speed	Width	Location	Int		I	Time
96/07/12	15:37	1085	68	S10W80	N	nN	N	———
96/11/28	16:50	984	101	N05W90	F1	nN	e	21:00
97/11/06	12:10	1556	H	S18W63	F1	yY	M	12:30
98/04/20	10:07	1863	165	S43W90	F1	yY	M	11:30
98/05/02	14:06	938	H	S15W15	NH	yY	M	14:00
98/05/06	08:29	1099	190	S11W65	F1*	yY	M	08:30
98/05/09	03:35	2331	178	N26W90	F1*	yY	M	05:00
98/06/16	18:27	1484	281	S17W90	F1*	yY	m	21:30
98/11/05	20:44	1118	H	N22W18	F1	nY	m	$03:00^n$
99/06/04	07:26	2230	150	N17W69	F1?	yY	M	08:30
99/06/24	13:31	975	H	N29W13	F1	nN	e	$04:30^n$
99/06/28	21:30	1083	H	N22W44	P1*	yN	N	———
99/07/25	13:31	1389	H	N38W81	F1?	yN	N	———
99/08/28	01:26	1147	98	S26W16	P1	nN	N	———
99/09/21	03:30	1402	125	N19W90	P1*	yN	N	———
99/09/23	15:54	1150	77	S14W47	P1?	nN	N	———
00/02/09	19:54	910	H	S17W40	F1*	nN	N	———
00/02/12	04:31	1107	H	N26W23	F1*	yY	m	06:00
00/04/04	16:32	1188	H	N16W66	F2	yY	M	17:00
00/04/23	12:54	1187	H	N12W90	N	nN	m	15:00
00/04/27	14:30	1110	138	N32W90	F1*	nY	e	17:00
00/05/04	11:26	1404	170	S17W90	F2	yY	e	16:30
00/05/15	16:26	1212	165	S24W67	F1*	nY	m	19:00
00/06/10	17:08	1108	H	N22W38	F2*	yY	M	17:30
00/06/15	20:06	1081	116	N20W65	NS	yY	e	$01:00^n$
00/06/25	07:54	1617	165	N16W55	P1	yY	m	11:30
00/06/28	19:31	1198	134	N20W90	F1?	yN	e	20:00
00/07/14	10:54	1674	H	N22W07	F1*	yY	M	10:30
00/07/22	11:54	1230	105	N14W56	F1*	yY	M	11:30
00/08/11	07:31	1071	70	N27W90	P1	nY	M	12:00
00/09/12	11:54	1550	H	S17W09	F1*	yY	M	13:00
00/10/16	07:27	1336	H	N05W90	F2*	yY	M	08:00
00/11/08	23:06	1345	H	N10W77	F1*	nY	M	23:30
00/11/24	05:30	994	H	N20W05	F2*	yY	m	06:00
00/11/24	15:30	1245	H	N22W07	F2*	yY	M	15:30
01/01/28	15:54	916	250	S04W59	NS	nY	M	16:30
01/02/10	05:54	956	H	N30W07	N	nN	N	———
01/02/11	01:31	1183	H	N24W57	F2	yY	m	02:30
01/03/29	10:26	942	H	N20W19	F1	yY	M	12:00
01/04/02	11:26	992	80	N17W60	F1	yY	m	12:00
01/04/02	22:06	2505	244	S19W72	F1*	yY	M	23:00
01/04/09	15:54	1192	H	S21W04	N	yY	m	16:00
01/04/10	05:30	2411	H	S23W09	F1	yY	M	07:30
01/04/12	10:31	1184	H	S19W43	NS	yY	M	11:30
01/04/15	14:06	1199	167	S20W85	F3	yY	M	14:00
01/04/26	12:30	1006	H	N17W31	F1?	yY	M	14:00
01/05/07	12:06	1223	205	N25W35	F2	nY	M	13:30
01/06/20	19:54	1407	H	N08W17	N	nN	N	———
01/07/19	10:30	1668	166	S08W62	N	nN	N	———
01/10/01	05:30	1405	H	S20W90	F4	nY	M	13:30
01/10/19	16:50	901	H	N15W29	P1*	yY	M	17:30
01/11/04	16:35	1810	H	N06W18	NS	yY	M	16:30

TABLE 2. FWFW CMEs, Interactions, and SEP Association

	With SEPs	without SEPs
No interaction	7 (2)	4
Interaction	35 (40)	6

examined the GOES proton data for possible association of the FWFW CMEs with SEP events. If an SEP event was associated, we have given the intensity level (I) in the > 10 MeV energy range and the onset times in columns 8 and 9, respectively (A superscript n denotes the time corresponds to the next day). On the basis of the observed peak intensity (I), each event is classified as a major (M, with I > 10 pfu), minor (m, with 1 pfu < I < 10 pfu) or marginal (e, with I < 1 pfu) event. The marginal events are those, which clearly stand above the noise level, but had an intensity < 1 pfu. We also examined the association of the CMEs with metric type II bursts from the online Solar Geophysical Data (SGD) and interplanetary type II bursts from the Wind/WAVES catalog (`http://lep694.gsfc.nasa.gov/waves/waves.html`). These data will tell us whether the CMEs were responsible for electron acceleration. The presence (y,Y) or absence (n,N) of type II radio bursts are indicated in column 7, with small (capital) letters referring to metric (DH) type II bursts. Using movies of SOHO/LASCO images and the CME catalog, we identified potential instances of interaction between the FWFW CMEs and other preceding CMEs. This information is given in column 6. If the extent of position angle (PA) overlap between FWFW CMEs and the preceding ones is > 30°, the interaction is characterized as full (F), and partial (P) otherwise. The number of preceding CMEs is also indicated as a subscript to F and P. An * in column 7 means there were additional interactions, and a ? means the interaction was beyond 30 solar radii, but within 50 solar radii. NL and NS denote that there was no intersection of trajectories, but there was interaction with the leg of a preceding CME (NL) or with a dense streamer (NS). One event (1998 may 02, marked NH in column 6) with no apparent interaction was preceded by halo CMEs and hence might have interacted with them along the line of sight.

ANALYSIS AND RESULTS

The first thing we notice in Table 1 is that 42/52 (81 %) of the FWFW CMEs were associated with SEP events identified from GOES data. For the remaining 10 events (19%), we did not find an SEP event above the noise level in the GOES proton plots. Table 2 gives the extent of CME interaction and the SEP association for the 42 events. For simplicity, we have combined the full and partial interactions together as simply "interactions". Similarly, we did not distinguish an SEP event whether it is a major, minor, or marginal event. This is roughly the level of association found in the study of SEPs in Paper 1. It is clear that the lower left cell in Table 2 is the dominant one, suggesting that 83% of all FWFW CMEs that had SEPs were preceded by CME interaction, similar to the level of interaction found starting from SEP events in Paper 1.

TABLE 3. SEP-associated CMEs without interaction

Date	V (km/s)	W (deg)	Location	Type
98/05/02	938	360	S15W15	Major, NH
00/04/23	1187	360	N12W90	minor
00/06/15	1081	116	N20W65	marginal, NS
01/01/28	916	250	S04W59	Major, NS
01/04/09	1192	360	S21W04	minor
01/04/12	1184	360	S19W43	Major, NS
01/11/04	1810	360	N06W18	Major, NS

TABLE 4. SEP-poor CMEs with Interaction

Date	V (km/s)	W (deg)	Location	Type
99/06/28	1083	360(28)	N22W44	N/P TB
99/07/25	1389	360(97)	N38W81	N/F TB
99/08/28	1147	98(12)	S26W16	N/P TB
99/09/21	1402	125(14)	N19W90	N/P TB
99/09/23	1150	77(9)	S14W47	N/P TB
00/02/09	910	360(37)	S17W40	N/F TB

SEP-associated CMEs without Interaction

We now look at the 17% of the FWFW CMEs that were associated with SEPs, but were apparently not preceded by CME interaction (see Table 3). Four of the seven FWFW CMEs without interaction had major SEPs. This is somewhat embarrassing because we have claimed that CME interaction is an important aspect of SEP production. However, LASCO movies showed that the primary CMEs in these events interacted with dense streamers, completely destroying them. Most streamers represent the pre-eruption manifestations of CMEs and have the same three-part structures as the CMEs. Thus, the interaction with the streamers should be similar to the interaction with other CMEs. Out of the seven FWFW CMEs in Table 3, four interacted with streamers (denoted by NS in the last column of Table 3). In the case of 1998 May 2 event, the FWFW CME in question was preceded by two halo CMEs from the same active region (noted as "NH" in Table 3). The height-time plots that describe the interaction correspond to the sky plane measurements, which generally underestimate CME speeds of halo CMEs. Therefore one cannot rule out interaction with preceding CMEs. Thus only two of the seven FWFW CMEs can be regarded to have no interactions with any confidence. Thus, allowing for interaction with dense streamers, we see that 40/42 (95%) FWFW CMEs with SEPs interacted with one or more preceding CMEs, again consistent with Paper 1. In the remaining 2 cases (5%) without interaction, the SEP events were minor.

SEP-Poor CMEs with Interaction

If CME interaction is important for SEP production, why were the six FWFW CMEs in Table 2 (also listed in Table 4) that were preceded by CME interaction were not associated with any SEPs? In the last column of Table 4, N/P (N/F) means no SEP but there was partial (full) interaction. To answer this question, we examined the properties of the preceding CMEs. The most striking aspect of the preceding CMEs is that all but one were very narrow (see the number within parentheses in the W column of Table 4), there by reducing the extent of overlap to the width of the preceding CMEs (see Fig 1).

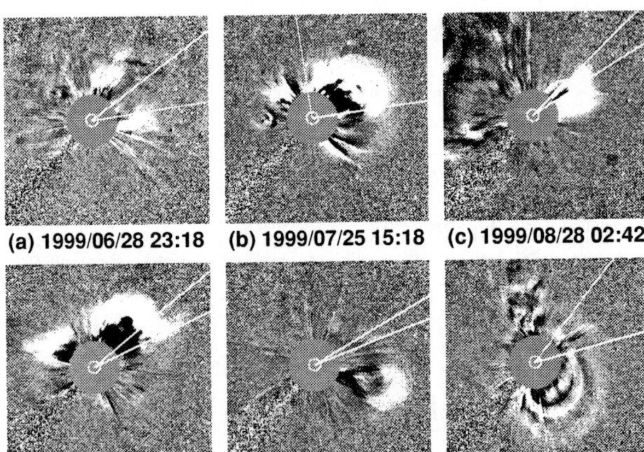

FIGURE 1. One SOHO/LASCO/C3 running difference image for each of the SEP-poor CMEs in Table 4. The width of the preceding CME is indicated by the two solid lines.

In Paper 1, the average extent of overlap between preceding and primary CMEs was found to be $\sim 50°$. Thus, the interaction is not expected to be severe in all but one case. The small overlap between the primary and preceding CMEs also reduces the chance of the interaction region having a magnetic connection to the Earth. In the one case (1999 July 25), the extent of overlap is quite extensive, but there was another problem: even though the trajectories showed intersection, the preceding CME had departed the Sun some 17 hours earlier and faded to the background level even before the arrival of the primary CME. Thus in all the cases the severity of interactions is rather low and for practical purposes, the events in Table 4 must be considered as FWFW CMEs with no interaction.

SEP-poor CMEs with No Interaction

Finally, we examine the four FWFW CMEs without SEPs and without interaction (see Table 5). In the last column of Table 5, NN denotes no interaction at all and NE means interaction above eastern hemisphere, which is not relevant to SEPs. This is the crucial sample that supports our suggestion that CME interaction is an important aspect of SEP production. Unfortunately, the sample size is rather small. We need to collect more events of this kind to arrive at firmer conclusions. Nevertheless, it is instructive to examine the circumstances of these eruptions to gain some insight as to why these FWFW CMEs were SEP-poor. Figure 2 shows a LASCO/C3 snapshot of each of the four CMEs. The width of the first event was only slightly above average thus barely satisfying the width requirement to be classified as FWFW events. Two events are halos but there

TABLE 5. SEP-poor CMEs with No interaction

Date	V (km/s)	W (deg)	Location	Type
96/07/12	1085	68	S10W80	NN
01/02/10	956	360	N30W07	NE TB
01/06/20	1407	360	N08W17	NE TB
01/07/19	1668	166	S08W62	NN TB

is no way of knowing their true widths. The first event was close to the solar minimum and hence was confined to the extent of the equatorial streamer belt. In the other three events, the CMEs were ejected into a tenuous background (noted as TB in the last column of Table 5) compared to the opposite hemisphere. This is also the case with all the events in Table 4. The lower density means higher Alfven speed in the medium and weaker shocks. The situation is opposite to that of running into a dense structure such as a CME or a streamer.

(a) 1996/07/12 20:56

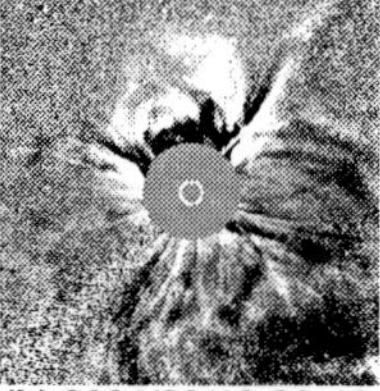
(b) 2001/02/10 08:42

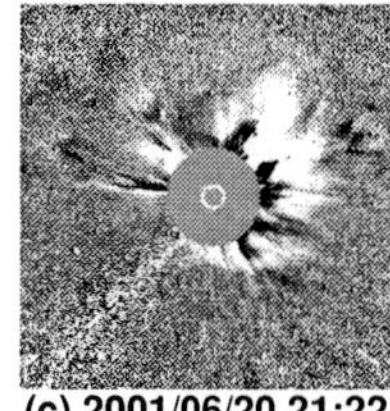
(c) 2001/06/20 21:22

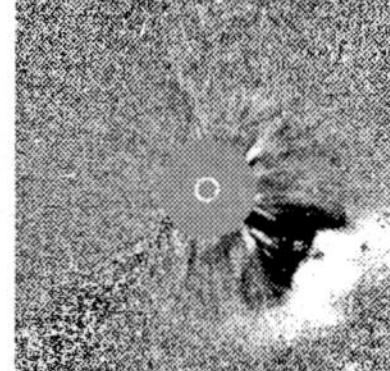
(d) 2001/07/19 12:18

FIGURE 2. One SOHO/LASCO/C3 running difference image for each of the SEP-poor CMEs in Table 5.

Electron and Proton Accelerations

Metric and IP type II bursts are indicative of nonthermal electrons accelerated by MHD shocks. The same shocks may also accelerate protons and heavier ions. From Table 1, we found that 38/52 (73%) of FWFW CMEs had DH type II bursts while 35/52 (67%) had metric type II bursts. For the 42 SEP-associated FWFW CMEs, the association was better: 90% and 79% for DH and metric type II bursts, respectively. As was pointed out in Paper 1, SEP-poor FWFW CMEs were also electron-poor (no DH type II). However, three of the 10 SEP-poor FWFW CMEs were associated with the metric type II bursts. Three marginal and one minor SEP events were not associated with DH type II bursts (the 2000 June 25 event had a weak DH type II, but was listed as a no DH event in Paper 1), although the minor and one marginal SEP events had metric type II burst association. The difference between metric and DH type II bursts has been attributed to different radial profiles of the Alfven speed in the corresponding regions of the solar corona [7].

DISCUSSION AND CONCLUSIONS

The analysis presented in this paper confirms the importance of CME interaction in the production of SEPs. A detailed analysis of the SEP-poor FWFW CMEs with interaction reveals that the interactions are insignificant. Similarly, FWFW CMEs with SEPs but no apparent interaction with preceding CMEs had significant interaction with large-scale streamers. The extreme cases of FWFW CMEs with no interaction and no SEPs were ejected into very tenuous regions of the corona with no apparent structures to interact with. The close association between DH type II bursts and fast and wide CMEs reported elsewhere [8] is consistent with the results of this paper that electron accelerators are also proton accelerators. As we had pointed out before, the efficiency of particle acceleration is boosted when a fast shock runs into preceding CMEs. This might result in shock strengthening due to the enhanced density in the preceding CMEs as compared to the solar wind (provided the magnetic field is not enhanced significantly). Particles may also be trapped in the closed field lines of the preceding CMEs or the associated turbulence so they are subject to repeated acceleration by the shock for the time the shock takes to transit through the preceding CME. The preceding CMEs may also drive weak shocks that may preaccelerate suprathermal ions even if they are too weak to accelerate > 1 MeV ions. What happens after the shock passage is an interesting problem: the main body of the fast CME comes in contact with the slower CME and might form a merged resultant CME. We refer to this merger as CME cannibalism and needs further investigation.

REFERENCES

1. Gopalswamy, N. et al., *ApJ*, 548, L91, 2001a
2. Gopalswamy, N. et al., *GRL*, 29(8), 10.1029/2001GL013606, 2002a
3. Gopalswamy, N. et al., *ApJ*, 572, L103, 2002b
4. Reames, D. V., *Space Sci. Rev.*, 90, 413, 1999
5. Mason, G. et al., *ApJ*, 525, L133, 1999.
6. Mewaldt, R. et al. 2002, Adv. Space Res., in press, 2002
7. Gopalswamy, N. et al., *JGR*, 106, 25261, 2001b
8. Gopalswamy, N. et al., *JGR*, 106, 29219, 2001c

Development of shocks waves in the solar corona and the interplanetary space

G. Mann*, A. Klassen*, H. Aurass* and H. T. Classen*

*Astrophysikalisches Institut Potsdam, An der Sternwarte 16, D-14482 Potsdam, Germany

Abstract. At the Sun shock waves are produced either by flares or by coronal mass ejections and are regarded to be the source of solar energetic particle events. The propagation of a disturbance away from an active region through the corona into the interplanetary space is considered by evaluating the radial behaviour of the Alfvén speed. The magnetic field of an active region is modelled by a magnetic dipole superimposed on that of the quiet Sun. Such a magnetic field structure leads to a local mimimum of the Alfvén speed in the middle of the corona and a maximum of 740 km/s at a radial distance of 6 solar radii from the center of the Sun. The occurrence of these extrema of the Alfvén speed has consequences for the formation and development of shock waves in the corona and interplanetary space.

INTRODUCTION

In the solar corona shock waves can occur as blast waves due to the flare process [1, 2] or shocks driven by coronal mass ejections (CMEs) [3]. They can continue as travelling shocks in the interplanetary space. Coronal and interplanetary shocks can be the source of type II radio radiation [4]. Cane [5] and Gopalswamy et al. [6] suggested the appearance of two different kinds of shocks, the flare produced blast wave associated with the solar type II radio bursts in the inner corona and the CME driven interplanetary shocks, which are regarded to be the source of solar energetic particle events (SEPs) [7, 8].

Yohkoh and *SOHO* images of the Sun and complementary spatial radio-heliographic observations revealed that sources of solar type II radio bursts are mainly propagating non-radially away from active regions [9, 10, 11, 12]. In order to study the formation and development of shock waves in the solar corona and the interplanetary space (see Section 3) it is necassary to evaluate the radial behaviour of the Alfvén speed (see Section 2) as the characteristic speed in a magnetoplasma.

THE ALFVÉN SPEED MODEL

The Alfvén speed is depending on the magnitude B of the magnetic field and the full particle number density N. Therefore, a model of the magnetic field of an active region superimposed on that of the quiet Sun and a density model is necessary to know, in order to study the radial behaviour of the Aflvén speed in the solar corona and the interplanetary space.

Since coronal shock waves are established near but out of active regions [9, 10, 11, 12], a one-fold Newkirk model [13]

$$N_e = N_0 \cdot 10^{4.32 R_S/R} \tag{1}$$

($N_0 = 4.3 \cdot 10^4 cm^{-3}$; R_S, solar radius; R, radial distance from the Sun) is assumed as an appropriate model of the electron number density N_e. This model is well fitting the conditions above quiet equatorial regions as white-light scattering observations [14] showed. Mann et al. [15] presented a heliospheric density model as a special solution of Parker's [16] wind equation. It agrees well with the observations from the corona up to the interplanetary space at 5 AU. Since a density model above quiet equatorial regions is required for this study, the one-fold Newkirk [13] model is adopted in the region $R \leq 1.8R_S$, whereas that by Mann et al. [15] has to be taken at regions $R \geq 1.8R_S$. That results in a radial behaviour of the electron number density as presented in Figure 1.

In the solar corona the magnetic field $\vec{B}$ is composed of that of an active region $\vec{B}_{ar}$ and of the quiet Sun $\vec{B}_{qS}$, i. e. $\vec{B} = \vec{B}_{ar} + \vec{B}_{qS}$. Here, an active region is modelled by a magnetic dipole with the moment $\vec{M}$ and the length λ (see Figure 2). It is located in a depth $\lambda/2$ under the photosphere. Therefore, $\lambda = 0.1R_S$ is taken because of the assumed width of the hydrogen convection zone of the Sun [17]. The field of a magnetic dipole [18] is given by

$$\vec{B}_{ar} = \frac{3(\vec{M} \cdot \vec{r})\vec{r}}{r^5} - \frac{\vec{M}}{r^3} \tag{2}$$

where r denotes the distance of the point P from the center of the magnetic dipole (see Figure 2). The axis

CP679, *Solar Wind Ten: Proceedings of the Tenth International Solar Wind Conference*, edited by M. Velli, R. Bruno, and F. Malara

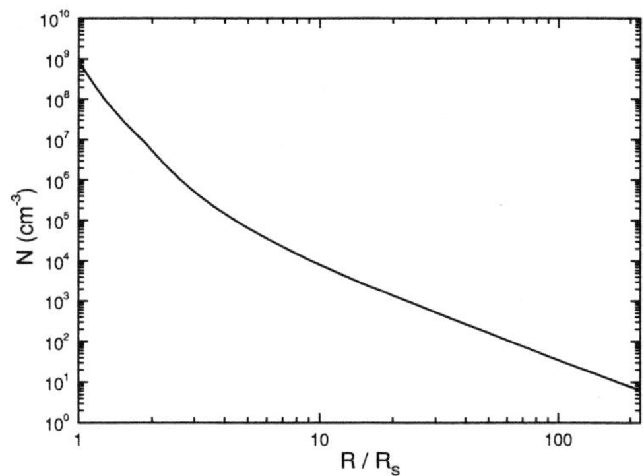

FIGURE 1. Radial behaviour of electron number density model from the low corona up to 1 AU

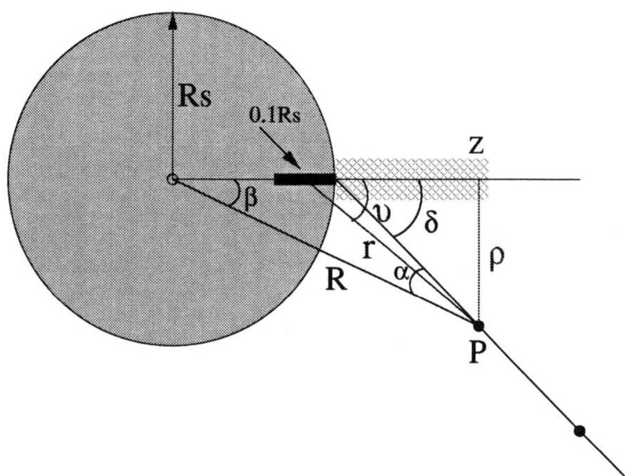

FIGURE 2. Scheme of reference frame

of the dipole is directed along the z-axis, i. e. $\vec{M} = M \cdot \vec{e}_z$ with $\vec{e}_z$ as the unit vector along the z-axis, as indroduced in Figure 2. Then, the magnetic field (see Eq. (2)) is expressed by

$$B_{ar,z}(r) = B_0 \cdot \frac{\lambda^3}{16} \cdot \left(\frac{R_S}{r}\right)^3 \cdot (3\cos\vartheta^2 - 1) \qquad (3)$$

$$B_{ar,\rho}(r) = B_0 \cdot \frac{\lambda^3}{16} \cdot \left(\frac{R_S}{r}\right)^3 \cdot (3\cos\vartheta \sin\vartheta) \qquad (4)$$

in cylindrical coordinates (see Figure 2). B_0 is defined as the magnitude of the magnetic field at the z-axis on the photospheric level, i. e. at $r = \lambda/2$ and $\vartheta = 0^o$. Then, M can be calculated by $M = B_0\lambda^3/16$. Here, $B_0 = 0.8\ kG$ is taken as a typical value of the magnetic field in the center of an active region [17].

The magnetic field of the quiet Sun is assumed to be radially directed according to $\vec{B}_{qS} = B_S \cdot (R_S/R)^2 \cdot \vec{e}_R$ with $\vec{e}_R$ as the unit vector along the radial direction. $B_S = 2.2\ G$ as the magnitude of the quiet Sun at the photospheric level has been deduced from the analysis of coronal transient (or EIT) waves [19]. Then, the magnetic field of the quiet Sun is given by

$$B_{qs,z}(R) = B_{qS} \cdot \cos\beta \qquad (5)$$

$$B_{qS,\rho}(R) = B_{qS} \cdot \sin\beta \qquad (6)$$

in the same frame of reference (see Figure 2).

The magnetic fields of the dipole (active region) and the quiet Sun can be superimposed by two different ways. Along the z-axis both fields can be directed either parallely ("+" sign) or anti-parallely ("-" sign). Henceforth, these cases are called "parallel" or "anti-parallel". The composed magnetic field is obtained after vector addition. Its magnitude is finally entering into the calculation of the Alfvén speed.

Since solar type II radio burst sources are mainly travelling non-radially away from an active region [9, 10, 11, 12], the behaviour of the Alfvén speed along a straight line away from the center of the magnetic dipole should be studied. For a quantitative discussion this line is chosen to be inclined by an angle $\delta = 45^o$ with respect to the z-axis (see Figure 2). Then, each point P on this line is unambiguously determined either by ρ or R, which are related by

$$\frac{\rho}{R_S} = 0.707 \cdot \sqrt{\frac{R^2}{R_S^2} - 0.5} - 0.5 \qquad (7)$$

to each other. Then, the angle ϑ can be calculated by

$$\vartheta = \arctan\left[\frac{1}{1 + \frac{\lambda}{2} \cdot \frac{R_S}{R}}\right] \qquad (8)$$

leading to the determination of the radial distance r

$$\frac{r}{R_S} = \frac{\rho}{R_S} \cdot \frac{1}{\sin\vartheta} \qquad (9)$$

of the point P from the center of the magnetic dipole (see Figure 2). Now, Alfvén speed can be calculated at each point P along the straight line according to $v_A = B/(4\pi\mu m_p N)^{1/2}$ (m_p, proton mass) via the determination of the magnitude of the magnetic field B and the electron number density N_e. The full particle number density N is related to N_e by $N = 1.92 N_e$ for a mean molecular weight $\mu = 0.6$ [17]. The radial behaviour of the so found Alfvén speed is presented for the parallel and anti-parallel case in Figure 3. As Figure 3 shows, the local Afvén speed has a local minumum in the middle of the corona and a local maximum of 740 km/s at a distance of $3.8 R_S$ from the center of the Sun. The mixture of the local magnetic field of the active region and the gobal magnetic field of the quiet Sun causes the appearance

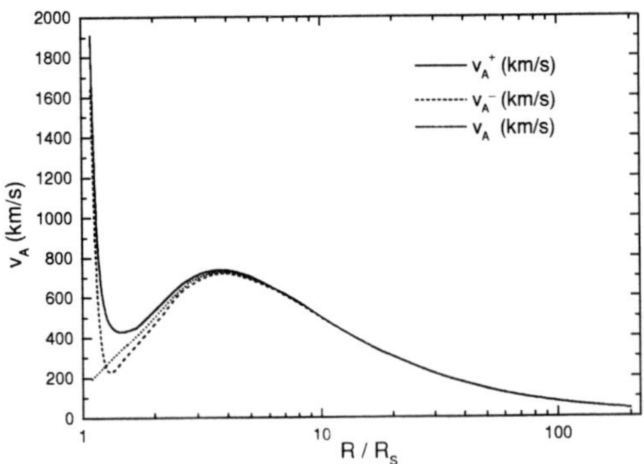

FIGURE 3. Alfvén speed along a straight line with $\delta = 45^0$ as a function of R. The full and dashed lines indicate the parallel and anti-parallel case, respectively. The dotted line represents the behaviour of the Alfvén speed only due to the magnetic field of the quiet Sun.

of the local minimum of the Alfven speed. The influence of the active region on the complete magnetic field is disappearing in the upper corona beyond a radial distance of $2R_S$. The occurrence of a local maximum of the Alfvén speed in the near-Sun interplanetary space was already deduced by Mann et al. [19]. Gopalswamy et al. [9] already reported on such local extrema of the Alfvén speed. They only considered the purely radial behaviour of the Alfvén speed above active region using the magnetic field model by Dulk and McLean [20] and the density model by Saito et al. [21]. In contrast to Gopalswamy et al. [9] the presented approach also allows to study the non-radial propagation of coronal disturbances. That is much more appropriate because of the observations of a non-radial movement of type II radio burst sources [9, 10, 11, 12].

DISCUSSION

The occurrence of such local extrema of the Alfvén speed has consequences for the formation and development of shock waves in the solar corona and the interplanetary space. In order to discuss that, a disturbance is considered to be propagating with a velocity V along a straight line away from an active region. This straight line should be inclined with respect to the radial direction. Such a disturbance has initially been generated by a flare or the lift off a CME. Shock waves can be established at regions, where the travelling disturbance becomes super-Alfvénic, i. e. $V > v_A$. The disturbances associated with solar type II radio bursts have typical radial velocities in the range $270 - 1220 km/s$ with a mean value at $730 km/s$ [23]. Then, they can become super-Alfvénic in the middle of the corona, i. e. $1.2 < R/R_S < 2$ and beyond $6R_S$. That fact might be an explanation of the occurrence of two kinds of shocks as claimed by several authors [5, 6, 22].

The local Alfvén speed has a local minimum of $428 km/s$ and $227 km/s$ at a radial distance of $1.45R_S$ and $1.32R_S$ in the parallel and anti-parallel case (see Figure 3), respectively. Note that the sound speed has a value of $180\ km/s$ for a temperature of $1.4 \cdot 10^6\ K$, which is typical in the corona. Thus, a shock wave is propagating into region of decreasing Alfvén speed after its formation in the low corona, i. e. the Alfvén-Mach number is increasing. Especially, it becomes a high Alfvén-Mach number shock near regions of the Alfvén speed minimum. It is well-known that high Alfvén-Mach number shocks are preferably accelerate particles [24]. This fact could be the root of the time delay between the flare on-set and the production of highly energetic electrons as recently reported by several authors [25, 26, 27]. These authors [25, 26, 27] claim an additional generator of highly energtic electrons in the upper corona. According to the presented approach a shock wave would need about $370s$ from the initial energy release (flare) up to reaching the place of the Alfvén speed mimimum, where the shock waves become efficient particle accelerators because of their high Alfvén-Mach number.

Furthermore, interplanetary shocks are regarded to be driven by CMEs. Since CMEs have velocities with a mean value in the range $400 - 500\ km/s$ [28], they become super-Alfvénic at radial distances beyond $6R_S$ from the center of the Sun, i. e. in the near-Sun interplanetary space . They are formed after about 4000 s after the lift off of the CME and subsequently generate highly energetic protons, which are observed as SEPs [7, 8].

ACKNOWLEDGMENTS

This work was supported by the German *Deutsche Forschungsgemeinschaft, DFG* under project number MA 1376/14-1.

REFERENCES

1. Uchida, Y. Altschuler, D., and Newkirk, G., *Sol. Phys.*, **28**, 495, 1973.
2. Vršnak, B. Ruždjak, V., and Aurass, H., *Sol. Phys.*, **158**, 1995.
3. Stewart, R. T., Howard, R. a., Hansen, F., Gergely, T., and Kundu, M. R., *Sol. Phys.*, **36** 219, 1974.
4. Nelson, G. S. and Melrose, D., "Type II bursts", in *Solar Radio Physics*, edited by D. J. McLean and

N. R. Labrum, Cambridge Univ. Press, Cambridge, 1985, pp. 333–360.

5. Cane, H., *Solar Wind Five*, NASA Conf. Publ., CP-2280, 703, 1983.
6. Gopalswamy, N., Kaiser, M. L., Lepping, R. P., Kahler, S., W., Ogilvie, K., Berdichevski, D., Kono, T., Isobe, T., and Akioka, M., *J. Geophys. Res.*, **103**, 307, 1998.
7. Kahler, S. W., *Astrophys. J.*, **428**, 837, 1994
8. Reames, D. V., Barbier, L. M., and Ng, C. K., *Astrophys. J.*, **466**, 473, 1996
9. Aurass, H., Hofmann, A., and Urbarz, H., *Astron. Astrophys.*, **334**, 289, 1998.
10. Klassen, A., Aurass, H., Klein, K.-L., Hofmann, A., and Mann, G., *Astron. Astrophys*, **343**, 287, 1999.
11. Klein, K.-L., Khan, J. I., Vilmer, N., Delouis, J.-M., and Aurass, H., *Astron. Astrophys.*, **346**, L53, 1999.
12. Gopalswamy, N., Lara, A., Kaiser, M. L., and Bougeret, J.-L., *J. Geophys. Res.*, **106**, 25,261, 2001.
13. Newkirk, G. A., *Astrophys. J.*, **133**, 983, 1961.
14. Koutchmy, S., *Adv. Space Res.*, **14(4)**, 29, 1994.
15. Mann, G., Jansen, F., MacDowall, R. J., Kaiser, M. L., and Stone, R. G., *Astron. Astrophys.*, **348**, 614, 1999.
16. Parker, E. N., *Astrophys. J.*, **128**, 664, 1958.
17. Priest, E. R., *Solar Magnetohydrodynamcis,* Reidel, Dordrecht, 1982, pp. 82–86.
18. Landau, L. D. and Lifschitz, E. M., *The classical field theory*, Pergamon Press, Oxford, 1975, pp. 85–92.
19. Mann, G., Aurass, H., Klassen, A., Estel, C., and Thompson, B. J., in *Proc. 8th SOHO Workshop*, ESA **SP-446**, 129, 1999.
20. Dulk, G. A. and McLean, D. J., *Astron. Astrophys.*, **66**, 315, 1978
21. Saito, K., Poland, A. I., and Munro, R. H., *Sol. Phys.*, **55**, 121, 1977.
22. Classen, H. T. and Aurass, H., *Astron. Astropys.*, **384**, 1098, 2002.
23. Klassen, A., Aurass, H., Mann, G., and Thompson, B. J., *Astron. Astrophys.*, **141**, 357, 2000.
24. Kennel, C. F., Edmiston, J. P., and Hada, T., "A Quarter Century of Collisionless Shock Research", in *Collisionless Shocks in the Heliosphere: A Tutorial Review*, edited by R. G. Stone and B. T. Tsurutani, Geophysical Monograph **34**, AGU, Washington D.C., 1985, pp. 1–36.
25. Krucker, S. Larson, D. E., Lin, R. P., and Thompson, B. J., *Astrophys. J.*, **519**, 864, 1999.
26. Haggerty, D. K. and Roelof, E. C., in *Proc. Internat. Cosmic Ray Conf. XXV*, Copernicus Society, 3228, 2001.
27. Klassen, A., Bothmer, V., Mann, G., Reiner, M. J., Krucker, S., Vourlidas, A., and Kunow, H., *Astron. Astrophys.*, **385**, 1078, 2002.
28. St. Cyr, O. C. and 13 co-authors, *J. Geophys. Res.*, **105**, 18,169, 2000

The Coronal Isotopic Composition as Determined Using Solar Energetic Particles

R. A. Leske*, R. A. Mewaldt*, C. M. S. Cohen*, E. R. Christian†, A. C. Cummings*, P. L. Slocum**‡, E. C. Stone*, T. T. von Rosenvinge† and M. E. Wiedenbeck**

**California Institute of Technology, Pasadena, CA 91125 USA*
†NASA/Goddard Space Flight Center, Greenbelt, MD 20771 USA
***Jet Propulsion Laboratory, Pasadena, CA 91109 USA*
‡Presently at The Aerospace Corporation, El Segundo, CA 90245 USA

Abstract. Solar energetic particles (SEPs), like the solar wind, provide a direct sample of the Sun. Although SEP abundances show a variable amount of mass fractionation, it is possible to develop methods of correcting for it in order to deduce the composition of the corona. Using high-resolution measurements from the Solar Isotope Spectrometer on the Advanced Composition Explorer, we have studied the isotopic composition of 10 abundant elements from C to Ni in 32 large SEP events from late 1997 to the end of 2001 at energies >15 MeV/nucleon. We show that various isotopic and elemental enhancements are correlated with each other, discuss the first order corrections used to account for the variability, and obtain estimated coronal abundances. We compare the coronal values and their uncertainties inferred from SEPs with those that are available from solar wind and meteoritic measurements and find generally good agreement. We include C and Ni isotopic abundances, for which no solar wind measurements have yet been reported.

INTRODUCTION

Studies of solar material in both the solar wind and solar energetic particles (SEPs) may be used to determine the Sun's composition, and each approach has its own challenges. For example, solar wind composition measurements typically require detailed knowledge of instrument response functions, effi ciencies, and backgrounds, whereas SEP abundances are often highly skewed by fractionation processes during particle acceleration or transport. It is therefore valuable to compare solar abundances obtained from both types of studies.

The seed material for gradual SEP events is thought to be solar wind or coronal material accelerated by large shocks driven by fast coronal mass ejections [1]. Elemental abundances vary greatly from event to event but are correlated with the ionic charge to mass ratio, Q/M [2]. After correcting for this fractionation [2, 3] or averaging over many events [4], SEP abundances reveal the coronal elemental composition. In principle, the coronal isotopic composition can be similarly determined [5, 6].

Recent studies using the Solar Isotope Spectrometer (SIS) on the Advanced Composition Explorer (ACE) have found large enhancements and event-to-event variability in SEP isotopic abundance ratios [see, e.g. 7, and references therein]. In the present work, we present ACE/SIS isotopic abundance measurements for C, O, Ne, Mg, Si, S, Ar, Ca, Fe, and Ni in as many as 32 individual SEP events. Although the isotopic composition is highly variable, we can use the observed correlations to empirically correct for the variations to fi rst order and obtain coronal isotopic abundances from SEPs, which we compare with solar wind and terrestrial abundances.

OBSERVATIONS AND ANALYSIS

The SIS instrument uses the dE/dx versus residual energy technique in a pair of silicon solid-state detector telescopes to obtain the nuclear charge, Z, mass, M, and total kinetic energy, E, for particles with energies of $\sim$10 to $\sim$100 MeV/nucleon [8]. For this study, we examined all SEP events with high-energy heavy ion fluxes large enough to yield statistically meaningful isotope abundances for at least a few isotope ratios. This selection resulted in 32 large SEP events, including the 18 discussed in our earlier work [7] and an additional 14 in 2001. During the very highest rate periods, mass resolution is degraded by chance coincidences between heavy ions and low energy protons. Therefore, time periods near the peaks of the 14 July 2000, 9 November 2000, 24 September 2001, 4 November 2001, and 22 November 2001 events were not used for the isotopic analysis. Mass resolution depends on Z and E; for the species

CP679, *Solar Wind Ten: Proceedings of the Tenth International Solar Wind Conference*, edited by M. Velli, R. Bruno, and F. Malara

and energies studied here it ranges from ~0.15 to ~0.3 amu. Obtaining event-integrated isotope abundance ratios from these high-resolution data is straightforward; further details are given elsewhere [9, 10].

Deriving coronal abundances from SEP data is complicated by the fact that SEP isotopic abundances may vary significantly from event to event [10], as shown for the ^{22}Ne/^{20}Ne ratio in Figure 1. There is some indication that this variability may depend on the phase of the solar cycle [11]; further investigation may help determine if this is indeed the case or if the apparent changes in variability are merely statistical aberrations.

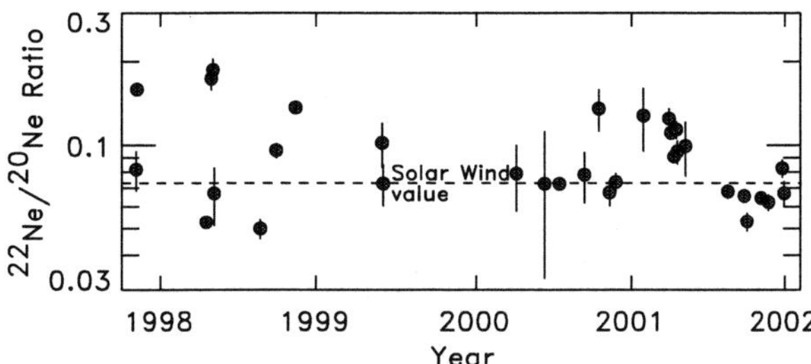

FIGURE 1. The SEP ^{22}Ne/^{20}Ne ratio measured by ACE/SIS at $E > 15$ MeV/nucleon plotted versus the date of the event, compared with the average solar wind value [12] (dashed line).

As mentioned above, earlier studies [e.g. 2] have shown that SEP heavy ion elemental abundances are also quite variable from event to event, with abundance ratios that scale approximately as a power law in Q/M with a different power law index for each SEP event. Any process which fractionates elements based on Q/M will also affect isotopes with different M, and there should be a predictable correlation between elemental and isotopic abundances. Following [5], if we base the power law fractionation index on the abundance ratio of any two reference species, such as Fe/O, Na/Mg, or, in general terms, R_1/R_2, power-law fractionation in Q/M implies that the enhancement or depletion of any other SEP abundance ratio X_1/X_2 should be:

$$\frac{(X_1/X_2)_{SEP}}{(X_1/X_2)_{corona}} = \left(\frac{(R_1/R_2)_{SEP}}{(R_1/R_2)_{corona}}\right)^{\frac{\ln[(Q/M)_{X_1}/(Q/M)_{X_2}]}{\ln[(Q/M)_{R_1}/(Q/M)_{R_2}]}} . \quad (1)$$

We find that isotopic abundances are indeed correlated with elemental abundances, as illustrated in Figure 2, which reinforces the trends seen with many fewer events in earlier work [10]. To compare the correlations with those expected from equation (1), the ionic charge states Q of the species involved must be known. It is reasonable to assume that $Q(^{22}\mathrm{Ne})=Q(^{20}\mathrm{Ne})$, so the values of $Q(X_1)$ and $Q(X_2)$ in equation (1) factor out, but this will not be true for the charge states of Fe/O or Na/Mg used in Figure 2. For most SEP events Q is not measured at energies of tens of MeV/nucleon. Lower energy measurements may not apply as several events have been found

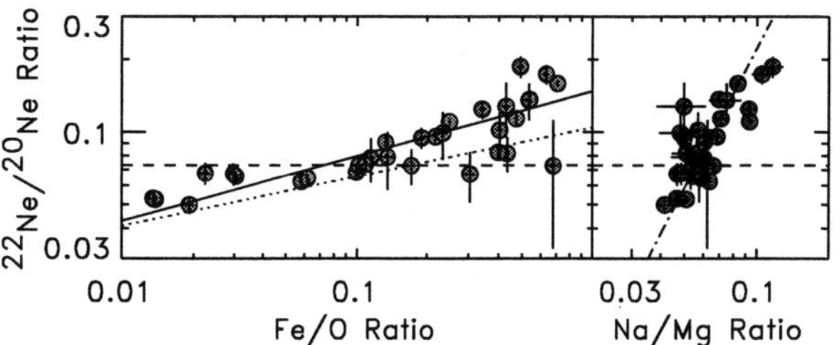

FIGURE 2. The SEP ^{22}Ne/^{20}Ne isotopic ratio plotted versus the Fe/O (left) and Na/Mg (right) elemental abundance ratios. Diagonal lines show the correlations expected from equation (1), assuming parameters discussed in the text. The horizontal dashed line marks the solar wind ^{22}Ne/^{20}Ne value.

with energy-dependent charge states for heavy elements [e.g. 13], while those higher energy charge state measurements that do exist show considerable event-to-event variability [see, e.g. 14, and references therein]. Also, note that in the derivation of equation (1) it was implicitly assumed that any elemental fractionation associated with the first ionization potential (FIP) effect is the same magnitude in the SEP event as it is in the corona. However, the size of the FIP effect in SEP events is variable [15, 16], which can affect certain elemental but not isotopic ratios and may blur the expected correlations.

If we evaluate equation (1) with Q(Fe)=15 [17] (and a FIP step of 4), we obtain the dotted line in Figure 2. If instead we use Q(Fe)=18 and a reduced FIP step of 1.7, both of which were found for the 6 November 1997 event [18], we obtain the steeper solid line. Mean iron charge states both lower than 15 and higher than 18 have been observed [14], so the illustrated lines indicate only some of the range of variability that might be expected.

The ^{22}Ne/^{20}Ne ratio correlates much better with Na/Mg, as shown in the right panel of Figure 2. Note that those events with ^{22}Ne/^{20}Ne values near that of the solar wind exhibit Fe/O ratios which span nearly 2 orders of magnitude, while all their Na/Mg ratios are tightly clustered around the coronal value. Na and Mg have similar FIP values, and their charge states are much less variable than those of Fe. As noted in [19], both Na and Mg ions should have ~2 electrons attached over a broad range of coronal temperatures [20], and ^{23}Na is neutron-rich compared to ^{24}Mg, so they differ significantly in Q/M. The dot-dashed line in Figure 2 shows the expected correlation if Q(Na)=9 and Q(Mg)=10. This very simple model provides a good first order fit to the data, but the actual correlation line is a bit shallower than predicted.

The predicted correlations are very sensitive to Q for reference ratios involving similar values of Q/M, and changing Q(Na)/Q(Mg) by only 1% changes the expected slope considerably [7]. Even if Q could be measured to this accuracy at SIS energies, it is Q at the time

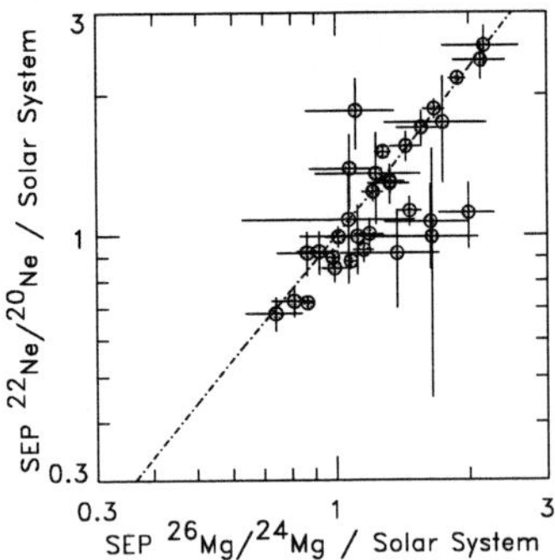

FIGURE 3. The ^{22}Ne/^{20}Ne versus ^{26}Mg/^{24}Mg isotopic ratios in each of the 32 SEP events, normalized to standard solar system values [21]. The diagonal line shows the correlation expected using equation (1).

of fractionation that is relevant, which may differ from that at 1 AU if fractionation happens early and if stripping occurs in acceleration through the corona [see, e.g. 22]. The sensitivity of the correlations to variability in the Na and Mg charge states may account for the scatter still present in the right panel of Figure 2.

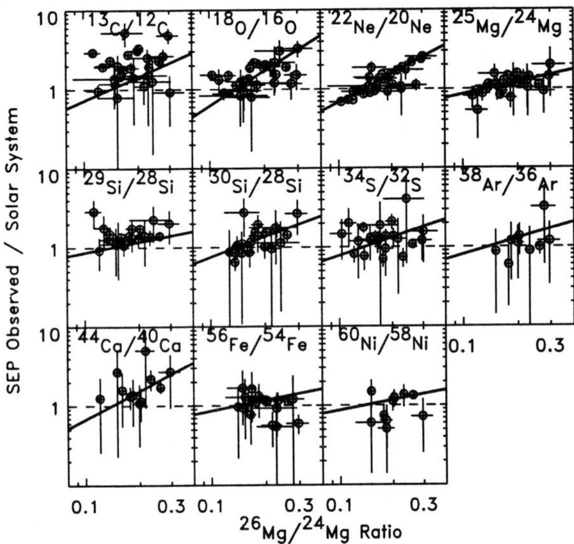

FIGURE 4. Eleven SEP isotope abundance ratios (normalized to standard solar system abundances [21]) in up to 32 SEP events plotted versus the ^{26}Mg/^{24}Mg ratio. Diagonal lines show the correlations expected using equation (1).

Since Q should be the same for isotopes of the same element, if we use an isotope ratio as the reference value in equation (1) the charge states factor out and we might expect a tighter correlation. This is illustrated by the plot of ^{22}Ne/^{20}Ne versus ^{26}Mg/^{24}Mg in Figure 3. Most of the events with small uncertainties agree very well with the expected line, while the outliers tend to have the largest uncertainties. Across the range of observed values no systematic deviation from the expected trend is evident.

RESULTS

The SEP abundance values for 11 isotope ratios for elements from C to Ni are shown plotted versus ^{26}Mg/^{24}Mg in Figure 4, with the correlations expected from equation (1) indicated by diagonal lines. The data roughly agree with the expected trends for O, Ne, Mg, and Si. Elements heavier than Si tend to have fewer data points and larger uncertainties, and it is unclear whether there is any fractionation at all, much less whether it follows the expected trends. The correlations may break down for species with Q/M far from that of the Mg reference ratio if the actual dependence on Q/M is not a simple power law as we assumed. At lower Z, ^{13}C is often enhanced or ^{12}C is depleted in SEP events relative to terrestrial abundances, but the reason for this is not understood. Higher ^{13}C values do not appear to be due to spillover from ^{12}C, as the two mass peaks are generally well separated [9].

Using ^{26}Mg/^{24}Mg as the reference ratio, taking its coronal value to be the terrestrial value [21], and assuming Q is the same for different isotopes of the same element, we solve equation (1) for the coronal isotope ratios for each SEP event. Averaging over all the SIS measurements for each isotope ratio results in the SEP-derived coronal values shown in Figure 5. We also show the weighted average without correcting for fractionation, which may be more appropriate for cases where the data may not follow the expected fractionation correlations. Also, the uncorrected number gives the average SEP composition arriving at 1 AU, and for ^{22}Ne/^{20}Ne we find it to be consistent with the value of ~0.09 of the so-called SEP component found implanted in lunar soils [23] (but see also [24]). Our SEP-derived values are also compared with standard solar system values [21] and existing solar wind values [12, 25, 26, 27] in Figure 5.

Although our fractionation correction is simplistic, the resulting coronal abundances mostly appear to be quite reasonable. All but 2 values are within 2σ of the standard "solar system" values [21] (which are actually terrestrial values except for Ne and Ar, for which solar wind values were used). Neither C nor Ni isotope abundances have yet been reported from solar wind data. Our ^{13}C measurements suggest significant additional fractionation of SEP C isotopes, but ^{60}Ni/^{58}Ni appears to be much less variable and the average is more likely representative of the corona. For ratios such as ^{18}O/^{16}O, ^{34}S/^{32}S, and ^{54}Fe/^{56}Fe, uncertainties on the SEP-derived coronal abundances are comparable to those obtained from the solar wind. Additional SEP events may reduce the uncertainties for the heaviest species where there are still only a few measurements, but a better theoretical understanding of the mass fractionation process would allow much further progress. Continued study of puzzles such as apparent changes in fractionation with the solar cycle and the frequent enhancement of ^{13}C may provide clues

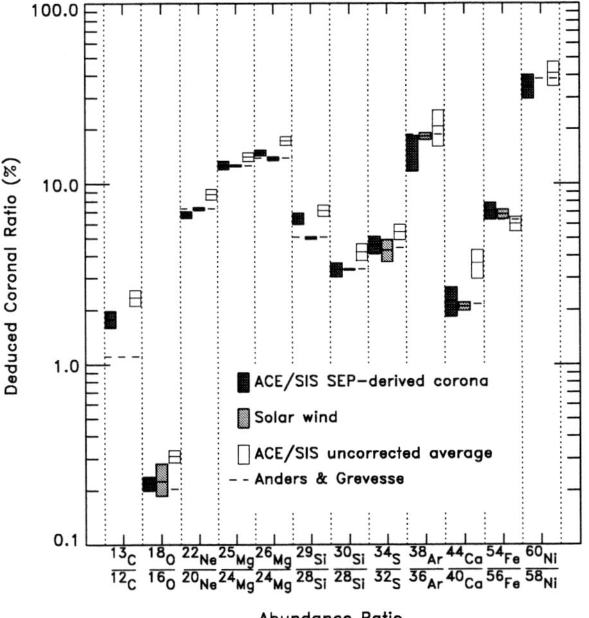

FIGURE 5. Average coronal isotopic abundance ratios from SIS SEP measurements after correcting for fractionation (dark grey boxes). The $^{26}Mg/^{24}Mg$ ratio served as the reference value in equation (1) for everything other than the $^{26}Mg/^{24}Mg$ ratio, for which $^{22}Ne/^{20}Ne$ was used. For comparison, averages without fractionation corrections (open boxes), standard solar system values (dashed lines; [21]) and measured solar wind values (light grey boxes; [12, 25, 26, 27]) are shown.

to the nature of the fractionation process.

ACKNOWLEDGMENTS

This work was supported by NASA at the California Institute of Technology (grant NAG5-6912), the Jet Propulsion Laboratory, and the Goddard Space Flight Center.

REFERENCES

1. Reames, D. V., *Revs. Geophys.*, **33**, 585–589 (1995).
2. Breneman, H. H., and Stone, E. C., *Astrophys. J. Lett.*, **299**, L57–L61 (1985).
3. Garrard, T. L., and Stone, E. C., *Proc. 23rd Internat. Cosmic Ray Conf. (Calgary)*, **3**, 384–387 (1993).
4. Reames, D. V., *Adv. Space Res.*, **15**, (7)41–(7)51 (1995).
5. Mewaldt, R. A., and Stone, E. C., *Astrophys. J.*, **337**, 959–963 (1989).
6. Williams, D. L., Leske, R. A., Mewaldt, R. A., and Stone, E. C., *Space Sci. Rev.*, **85**, 379–386 (1998).
7. Leske, R. A., et al., "Isotopic Abundances in the Solar Corona as Inferred from ACE Measurements of Solar Energetic Particles," in *Solar and Galactic Composition*, edited by R. F. Wimmer-Schweingruber, AIP Conf. Proc. 598, AIP, New York, 2001, pp. 127–132.
8. Stone, E. C., et al., *Space Sci. Rev.*, **86**, 357–408 (1998).
9. Leske, R. A., et al., *Geophys. Res. Lett.*, **26**, 153–156 (1999).
10. Leske, R. A., et al., *Geophys. Res. Lett.*, **26**, 2693–2696 (1999).
11. Leske, R. A., et al., *Eos, Trans. AGU*, **83(19)**, abstract SH42B–04 (2002).
12. Kallenbach, R., "Isotopic Composition Measured In-Situ in Different Solar Wind Regimes by CELIAS/MTOF on Board SOHO," in *Solar and Galactic Composition*, edited by R. F. Wimmer-Schweingruber, AIP Conf. Proc. 598, AIP, New York, 2001, pp. 113–119.
13. Mazur, J. E., Mason, G. M., Looper, M. D., Leske, R. A., and Mewaldt, R. A., *Geophys. Res. Lett.*, **26**, 173–176 (1999).
14. Leske, R. A., et al., "The Ionic Charge State Composition at High Energies in Large Solar Energetic Particle Events in Solar Cycle 23," in *Solar and Galactic Composition*, edited by R. F. Wimmer-Schweingruber, AIP Conf. Proc. 598, AIP, New York, 2001, pp. 171–176.
15. Garrard, T. L., and Stone, E. C., *Adv. Space Res.*, **14**, (10)589–(10)598 (1994).
16. Mewaldt, R. A., et al., "Variable Fractionation of Solar Energetic Particles According to First Ionization Potential," in *Acceleration and Transport of Energetic Particles Observed in the Heliosphere*, edited by R. A. Mewaldt et al., AIP Conf. Proc. 528, AIP, New York, 2000, pp. 123–126.
17. Luhn, A., et al., *Proc. 19th Internat. Cosmic Ray Conf. (La Jolla)*, **4**, 241–244 (1985).
18. Cohen, C. M. S., et al., *Geophys. Res. Lett.*, **26**, 149–152 (1999).
19. Cohen, C. M. S., et al., *Geophys. Res. Lett.*, **26**, 2697–2700 (1999).
20. Arnaud, M., and Rothenflug, R., *Astron. Astrophys. Suppl.*, **60**, 425–457 (1985).
21. Anders, E., and Grevesse, N., *Geochim. Cosmochim. Acta*, **53**, 197–214 (1989).
22. Barghouty, A. F., and Mewaldt, R. A., "Simulation of Charge-Equilibration and Acceleration of Solar Energetic Ions," in *Acceleration and Transport of Energetic Particles Observed in the Heliosphere*, edited by R. A. Mewaldt et al., AIP Conf. Proc. 528, AIP, New York, 2000, pp. 71–78.
23. Wieler, R., *Space Sci. Rev.*, **85**, 303–314 (1998).
24. Mewaldt, R. A., Ogliore, R. C., Gloeckler, G., and Mason, G. M., "A New Look at Neon-C and SEP-Neon," in *Solar and Galactic Composition*, edited by R. F. Wimmer-Schweingruber, AIP Conf. Proc. 598, AIP, New York, 2001, pp. 393–398.
25. Wimmer-Schweingruber, R. F., Bochsler, P., and Gloeckler, G., *Geophys. Res. Lett.*, **28**, 2763–2766 (2001).
26. Ipavich, F. M., Paquette, J. A., Bochsler, P., Lasley, S. E., and Wurz, P., "Solar Wind Iron Isotopic Abundances: Results from SOHO/CELIAS/MTOF," in *Solar and Galactic Composition*, edited by R. F. Wimmer-Schweingruber, AIP Conf. Proc. 598, AIP, New York, 2001, pp. 121–126.
27. Wimmer-Schweingruber, R. F., *Adv. Space Res.* (2002), in press.

Thermal forces and the coronal helium abundance

V.H. Hansteen*, Ø. Lie-Svendsen*† and E. Leer*

*Institute of Theoretical Astrophysics, P.O. Box 1029, Blindern, NO–0315 Oslo, Norway
†Norwegian Defence Research Establishment, P.O. Box 25, NO–2027 Kjeller, Norway

Abstract. The interaction between protons and minor ions in the chromosphere-corona transition region produces an upward force on the minor ions and an enhanced coronal abundance. In this presentation we compare a "classical" hydrodynamical model of a hydrogen – helium solar wind and a model based on a 16-moment fluid description where the heat flux is treated in a self-consistent manner.

BACKGROUND

The observed abundance variations of Helium, as well as other elements, between the photosphere and corona/solar wind have raised considerable interest. A study of the separation mechanisms could possibly reveal insight into processes controlling the presumably magnetically open fast solar wind streams as well as the partially closed source regions of the slow solar wind. Helium is a special case in that it alone may not be regarded as a minority speicies. Indeed previous studies [1, 2] showed that a Helium rich corona could play an important role in the solar wind acceleration mechanism through the altered electric field set up the α-particles as well as by the frictional coupling between protons and α-particles. Later studies by [3], showed that the corona could in certain cases become Helium rich. These studies invoked the large thermal force between Helium and Hydrogen in the transition region to transport Helium into the corona.

In a follow up study the entire chromosphere, transition region, corona, solar wind system was considered consistently ([4]). From these models it was concluded that, indeed, it was often the case that the thermal force produced coronae with a high Helium abundance. However, it was also found that the models which reproduced the observations best were those that were characterized by a low electron temperature and hence an insignificant electric field. These "best fit" models derived their acceleration from the large pressure gradients that the direct heating of ions results in. As such these models predicted the high ion temperatures that were later observed by the UVCS instrument on board SOHO but also that the role Helium plays in the accelaration of the largely proton wind was insignificant, even when Helium abundances greater than 50% were found in the corona. This because, as noted earlier, the electric field is unimportant as long as the electron temperature is on the order of 1MK and because the frictional coupling between protons and α-particles is very small when the ion temperatures are as high.

Even so a Helium abundances on the order of 50% would challenge our ideas of the solar corona and revisiting the problem is certainly justified. The Helium abundace increase in the corona in the models mentioned is due the thermal force which in turn is a consequence of the heat flux in the ion fluids. The classical description of the ion heat flux has recently been shown to be incorrect for conditions in the lower corona (e.g. [5]). In addition it seems likely that the deposition of energy may be in the moment perpendicular to the magnetic field, through for example cyclotron resonance. This will decrease the amount of ion heat flux that is brought down into the transition region. Both of these phenomena may have consequences for the thermal force between Hydrogen and Helium in the transition region. This paper is a first attempt at a study of the behaviour of Helium in models in light of a "better than classical" description of the ion heat flux.

Another finding in the [4] paper was that the frictional coupling between neutral Hydrogen and Helium in the chromsophere had to be greater than nominal in order to be able to pull any Helium into the corona/solar wind at all. We will also revisit this problem.

NEW SIMULATIONS

In these new simulations we use the gyrotropic approximation to solve for the particle density, velocity, temperature and heat flux moments parallell and perpendicular to the magnetic field. The gyrotropic approximation

CP679, *Solar Wind Ten: Proceedings of the Tenth International Solar Wind Conference*,
edited by M. Velli, R. Bruno, and F. Malara

to the 16-moment transport equations has been adopted from [6]. We solve the equations for a plasma containing eletrons, protons, singly ionized helium and α-particles as well as neutral hydrogen and helium. We require charge (quasi) neutrality and no current. The electron-ion plasma is created dynamically, in part through photoionization of the neutral hydrogen and helium gas, in part through collisional ionization in the transition region. For each particle species s we get six conservation equations for density, momentum, parallel and perpendicular temperature, and heat flux densities along the magnetic field of parallel and perpendicular thermal motion:

Particle conservation:

$$\frac{\partial n_s}{\partial t} = -\frac{1}{A}\frac{\partial}{\partial r}(n_s u_s A) + \frac{\delta n_s}{\delta t} \tag{1}$$

Particle momentum:

$$\begin{aligned}\frac{\partial u_s}{\partial t} &= -u_s\frac{\partial u_s}{\partial r} - \frac{k}{m_s}\frac{\partial T_{s\|}}{\partial r} - \frac{kT_{s\|}}{n_s m_s}\frac{\partial n_s}{\partial r} \\ &\quad -\frac{1}{A}\frac{dA}{dr}\frac{k}{m_s}(T_{s\|}-T_{s\perp}) \\ &\quad +\frac{e_s}{m_s}E - \frac{GM_S}{r^2} + \frac{1}{n_s m_s}\frac{\delta M_s}{\delta t}\end{aligned} \tag{2}$$

Particle energy:

$$\begin{aligned}\frac{\partial T_{s\|}}{\partial t} &= -u_s\frac{\partial T_{s\|}}{\partial r} - 2T_{s\|}\frac{\partial u_s}{\partial r} - \frac{1}{n_s k}\frac{\partial q_{s\|}}{\partial r} \\ &\quad -\frac{1}{A}\frac{dA}{dr}\frac{q_{s\|}}{n_s k} + \frac{2}{A}\frac{dA}{dr}\frac{q_{s\perp}}{n_s k} + \frac{1}{n_s k}Q_{sm\|} \\ &\quad +\frac{1}{n_s k}\frac{\delta E_{s\|}}{\delta t}\end{aligned} \tag{3}$$

$$\begin{aligned}\frac{\partial T_{s\perp}}{\partial t} &= -u_s\frac{\partial T_{s\perp}}{\partial r} - \frac{1}{A}\frac{dA}{dr}u_s T_{s\perp} - \frac{1}{n_s k}\frac{\partial q_{s\perp}}{\partial r} \\ &\quad -\frac{2}{A}\frac{dA}{dr}\frac{q_{s\perp}}{n_s k} + \frac{1}{n_s k}Q_{sm\perp} \\ &\quad +\frac{1}{n_s k}\frac{\delta E_{s\perp}}{\delta t}\end{aligned} \tag{4}$$

Particle heat flux:

$$\begin{aligned}\frac{\partial q_{s\|}}{\partial t} &= -u_s\frac{\partial q_{s\|}}{\partial r} - 4q_{s\|}\frac{\partial u_s}{\partial r} - u_s q_{s\|}\frac{1}{A}\frac{dA}{dr} \\ &\quad -3\frac{k^2 n_s T_{s\|}}{m_s}\frac{\partial T_{s\|}}{\partial r} + \frac{\delta q_{s\|}'}{\delta t}\end{aligned} \tag{5}$$

$$\begin{aligned}\frac{\partial q_{s\perp}}{\partial t} &= -u_s\frac{\partial q_{s\perp}}{\partial r} - 2q_{s\perp}\frac{\partial u_s}{\partial r} - 2u_s q_{s\perp}\frac{1}{A}\frac{dA}{dr} \\ &\quad -\frac{k^2 n_s T_{s\|}}{m_s}\frac{\partial T_{s\perp}}{\partial r} \\ &\quad -\frac{1}{A}\frac{dA}{dr}\frac{k^2 n_s T_{s\perp}}{m_s}(T_{s\|}-T_{s\perp}) + \frac{\delta q_{s\perp}'}{\delta t}.\end{aligned} \tag{6}$$

Here k is Boltzmann's constant, G is Newton's gravitational constant, M_S is the solar mass, m_s and e_s are the mass and charge of the particle (thus for electrons $e_e = -e$ and for protons $e_p = e$, where e is the elementary charge), and r is the heliocentric distance. A is the area of the radial flow tube ($A \propto 1/B$ where B is the strength of the radial magnetic field) and E is the radial electric field.

The parallel and perpendicular temperatures and heat fluxes are related to the average temperature T_s and total heat flux q_s through

$$T_s = \frac{1}{3}(T_{s\|} + 2T_{s\perp}) \tag{7}$$

$$q_s = \frac{1}{2}(q_{s\|} + 2q_{s\perp}). \tag{8}$$

The term $\delta n_s/\delta t$ in (1) is the rate of production (or loss) of particles due to ionization and recombination. The following particle destruction and creation reactions have been included: radiative recombinations from [7], collisional ionization and three-body recombination from [8], and photoionization which is roughly modelled by simply assuming constant rates taken from [9].

The collision terms $\delta M_s/\delta t$, $\delta E_{s\|(\perp)}/\delta t$, and $\delta q'_{s\|(\perp)}/\delta t$ contain both elastic and inelastic (ionization, recombination and charge exchange) collisions. Charge exchange between neutral hydrogen and protons is included using the coefficients found by [10]. The expressions for elastic collisions, as given by [6], are very extensive. We therefore use much simplified expressions for these.

RESULTS

In order to illustrate the type of models we expect to arrive at with this set of equations we have computed three sets of models, all with roughly the same energy input (coronal heating) per particle. Energy is inserted into the temperature moment perpendicular to the magnetic field and it is apportioned between protons and α-particles such that protons receive 85% while α-particles receive 15% of the deposited energy. (Electrons are also heated with 10 Wm^{-2}in the lower corona in all simulations).

- Models with radial (r^2) geometry. In these models we insert 110 Wm^{-2}; 10 Wm^{-2}in the inner corona with a scale height of 14 Mm, the remaining 100 Wm^{-2}with a scale height of 0.5R_S. The models differ in the amount of frictional coupling between neutral hydrogen and neutral helium in the chromosphere; the three models presented have 10× nominal, 5× nominal and nominal (e.g. [11]) coupling. The radial model with only nominal coupling in the

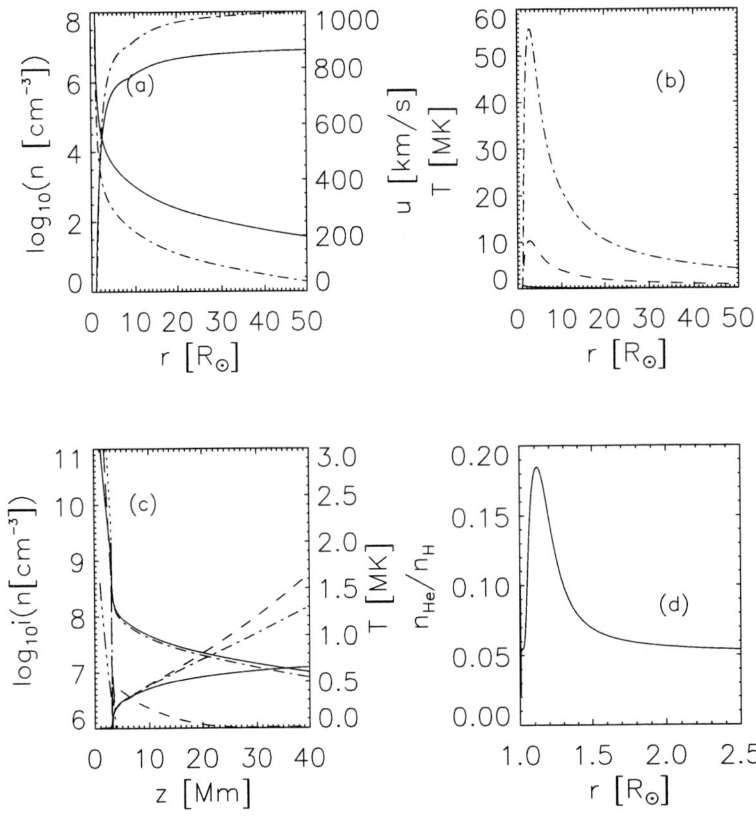

FIGURE 1. Rapidly expanding ($f_{max} = 5$) geometry and 5$\times$ nominal coupling between neutral hydrogen and neutral helium. All of the energy deposited in the corona (500 Wm^{-2}) is dissipated with a large scale height ($0.5R_S$), in addition 25% of this energy is put into the perpendicular moment of the α-particle temperature as opposed to 15% for all other models. The proton flux $n_p u_p = 1.75 \times 10^{12}\mathrm{m}^{-2}\mathrm{s}^{-1}$, while the α-particle flux $n_\alpha u_\alpha = 1.03 \times 10^{10}\mathrm{m}^{-2}\mathrm{s}^{-1}$at 1 AU which is 5.9% of the proton flux.

chromosphere is not shown, as in that model helium settled gravitationally and almost no helium was pulled into the corona or solar wind.
(For reference we find a proton flux $n_p u_p = 4.78 \times 10^{12}\mathrm{m}^{-2}\mathrm{s}^{-1}$, while the α-particle flux is $n_\alpha u_\alpha = 1.3 \times 10^{8}\mathrm{m}^{-2}\mathrm{s}^{-1}$at 1 AU; 2.7×10^{-3}% of the proton flux in that model.)

- Models with rapidly expanding geometry (expansion factor 5 as defined by [12]) and with coronal heat inputs of 500 Wm^{-2}; 50 Wm^{-2}in the inner corona with a scale height of 14 Mm, the remaining 450 $\mathrm{Wm}^{-2}$with a scale height of $0.5R_S$. A small amount, 10 Wm^{-2}, is inserted into the electron fluid. Again the difference between the models lies in the coupling factor between neutral hydrogen and neutral helium in the chromosphere, using the same values as above.
- Models with rapidly expanding geometry where we do not deposit energy in the inner corona (except a small amount, 10 Wm^{-2}, in the electron fluid), but rather deposit all 500 $\mathrm{Wm}^{-2}$with a scale height of $0.5R_S$. Here, again, we vary the frictional coupling between neutral hydrogen and neutral helium in the chromosphere using 10$\times$ nominal, 5$\times$ nominal coupling. But we also vary the percentage of energy deposited in protons and α-particles. A model where α-particles receive 25% of the deposited energy is presented.

The various models are summarized in Table 1 while a specific example with properties close to those observed in high speed streams is plotted in Figure 1.

CONCLUSIONS

We can vary the behaviour of the models by changing four parameters: The ratio of the energy dissipated to the expansion factor, the amount of energy dissipation in

TABLE 1. Summary of the runs shown above. Below f_{exm} is the geometrical expansion factor of the flow tube (1 is radial), F_{m0} is the total energy (including that in the inner corona) deposited in the corona, $F_{m,inner}$ the energy deposited in the inner corona. The two next columns signify the collisional coupling between neutral hydrogen and neutral helium in the chromosphere and the percentage of deposited energy put into the α-particles. For comparison we have included the equivalent data for the old 'standard' model in the last row of the table.

f_{exm}	F_{m0} Wm^{-2}	$F_{m,inner}$ Wm^{-2}	H i – He ii coupling	heating % α-paricles	$n_p u_p$ $m^{-2}s^{-1}$	$n_\alpha u_\alpha$ $m^{-2}s^{-1}$
1	110	10	10×	15	2.99×10^{12}	1.31×10^{11}
1	110	10	5×	15	3.95×10^{12}	1.05×10^{11}
1	110	10	nominal	15	4.78×10^{12}	1.30×10^{8}
5	500	50	10×	15	2.32×10^{12}	1.72×10^{11}
5	500	50	5×	15	2.43×10^{12}	1.63×10^{11}
5	500	50	nominal	15	2.85×10^{12}	3.10×10^{10}
5	500	0	10×	15	1.57×10^{12}	1.11×10^{11}
5	500	0	5×	15	1.72×10^{12}	1.00×10^{11}
5	500	0	5×	25	1.75×10^{12}	1.03×10^{11}
1	100	0	15×	15	2.02×10^{12}	1.07×10^{11}

the inner corona, the frictional coupling between helium and hydrogen in the chromosphere, and the percentage of energy going into α-particles.

The first number more or less sets the total mass flux from the model. This is modified by the amount of heating that occurs in the inner corona: A greater percentage of heating there leads to a greater heat flux in the transition region and thus a denser corona and a slower, denser wind. Likewise, if heating is concentrated in the outer corona or if the expansion factor is large, heat flow back to the transtion region is inhibited and a tenuous, faster wind arises.

Due gravitational settling helium drains out of the upper chromosphere on a timescale of hours unless the frictional coupling between hydrogen and helium is (artificially) increased. On the sun some process (such as turbulent motions?) must hinder this settling. Note that the helium flux in the models above is set by this number more or less alone. Note also that helium is not a minority species in these models and the proton flow is modified by the prescence of helium; i.e. helium 'holds' the protons back, through frictional coupling in the lower corona. This effect is less pronounced in the rapidly expanding models with no lower corona heating and indetecable in the old standard model where coronal heating took place quite far out in the corona. Note also that the coronal abundance of Helium found in these models varies greatly, from roughly 20% found in the model presented in Figure 1 up to greater than 80% in the radial model with 10× frictional coupling in the chromosphere and with 15% of the alloted energy deposited in the coronal α-particle fluid.

The α-particle speed, but little else, is set by the percentage of energy dissipated into said α-particles.

ACKNOWLEDGMENTS

This work was supported in part by teh Research Council of Norway under grants 121076/420 and 136030/431.

REFERENCES

1. Leer, E., and Holzer, T. E., *Ann. Geophysicae*, **9**, 196–201 (1991).
2. Bürgi, A., *J. Geophys. Res.*, **97**, 3137–3150 (1992).
3. Hansteen, V. H., Leer, E., and Holzer, T. E., *Astrophys. J.*, **428**, 843–853 (1994).
4. Hansteen, V. H., Leer, E., and Holzer, T. E., *Astrophys. J.*, **482**, 498–509 (1997).
5. Olsen, E. L., and Leer, E., *J. Geophys. Res.*, **104**, 9963–9973 (1999).
6. Demars, H., and Schunk, R., *J. Phys. D. Appl. Phys.*, **12**, 1051–1077 (1979).
7. Allen, C. W., *Astrophysical Quantities*, Athlone Press, Univ. London, 1976.
8. Arnaud, M., and Rothenflug, R., *A&AS*, **60**, 425 (1985).
9. Vernazza, J. E., Avrett, E. H., and Loeser, R., *Astrophys Supp.*, **45**, 635–725 (1981).
10. Janev, R. K., Langer, W. D., Evans, K., and Post, D. E., *Elementary Processes in Hydrogen-Helium Plasmas*, Springer, 1987.
11. Banks, P. M., and Kockarts, G., *Aeronomy*, Academic Press Inc., New York, 1973.
12. Kopp, R. A., and Holzer, T. E., *Sol. Phys.*, **49**, 43–56 (1976).

Composition Variations during Large Solar Energetic Particle Events

G.C. Ho[1], E.C. Roelof[1], G.M. Mason[2,3], D. Lario[1], R.E. Gold[1], J.E. Mazur[4], J.R. Dwyer[5]

[1]*Applied Physics Laboratory, Johns Hopkins University, Laurel, MD 20723*
[2]*Department of Physics, University of Maryland, College Park, MD 20742*
[3]*Institute for Physical Science and Technology, University of Maryland, MD 20742*
[4]*Department of Physics and Space Sciences, Florida Institute of Technology, FL 32901*
[5]*Aerospace Corporation, El Segundo, CA 90245*

Abstract. Recent theoretical work has posited that proton-generated Alfvén waves scatter ions according to their rigidity during propagation from the acceleration region to the observed position. In particular, predictions have been made for the initial behavior of the high energy (> 2 MeV/nucleon) He/H and Fe/O ratios during large solar energetic particle (SEP) events based on this theory. In this paper, we have examined the temporal variations of low energy (~0.3 MeV/nucleon) hydrogen, helium and iron ions throughout thirty large SEP events using the ULEIS instrument on the ACE spacecraft. These events originated from different solar longitudes; hence we compare events that have similar magnetic connection to minimize the transport effect. All events exhibited an eventual decrease in the He/H ratio as is suggested by the wave-particle model, but the degree of variation of He/H varies from event to event. However, the time variation of the Fe/O ratio does not correlate at all with that of the He/H in most of the events that we studied.

INTRODUCTION

Large solar energetic particle (SEP) events are believed to be associated with Coronal Mass Ejection driven shocks. Mason *et al.* [1] first reported that there are large variations of the He/H, O/He and Fe/O ratios within SEP events. Recently, Ng, Reames & Tylka [2] suggested that self-generated Alfvén waves at the traveling shock and throughout the transport region can modify the abundance ratio of species that have different rigidity. In particular, they predict that any abundance ratio of high-rigidity/low rigidity ions at the same velocity (example; He/H, Fe/O) will begin at high value and decline with time. According to their model, the Alfvén waves will scatter ions with lower *A/Q* compared with those with high *A/Q* (since rigidity $\propto A/Q$). Hence high-rigidity ions should arrive earlier than low-rigidity ions, even when they share the same velocity. However, in some events He/H ratios may also rise with time, which is a signature of events that have hard spectra according to [3]. Reames, Ng & Tylka [3] using *Wind, ICE,* and *IMP8* data have examined four large SEP events and claim a qualitative agreement with Ng *et al.* theory [2]. In this paper, using composition data from the Ultra-Low Energy Ion Spectrometer (ULEIS) on the *Advanced Composition Explorer* (ACE) spacecraft, we examine the low energy (230-450 keV/nucleon) He/H and Fe/O ratios in thirty large SEP events (including those examined by [3]).

INSTRUMENT DESCRIPTION

The ULEIS instrument is a high-resolution mass spectrometer that measures both elemental and isotopic ion compositions from 50 keV/nucleon to a few MeV/nucleon. The overall instrument geometric factor is about 1.3 cm^2-sr. This large geometric factor enables us to examine low energy ion elemental ratio variations at hourly resolution or better. Mason *et al.* [4] describe the instrument in full detail.

CP679, *Solar Wind Ten: Proceedings of the Tenth International Solar Wind Conference,*
edited by M. Velli, R. Bruno, and F. Malara

OBSERVATIONS

We selected gradual SEP events in the ACE/ULEIS data set from November 1997 to December 2002. The criteria for our selection are as follows:

(1) the event onset at 0.23-2.0 MeV/nucleon is isolated from other increases (necessary for onset ratio study);

(2) there are SOHO/LASCO observations to provide us an estimate launch location and time of the associated Coronal Mass Ejection back at the Sun (except for the event in September 1998);

(3) positive identification on the associated Hα flare location;

(4) no compound events (necessary for fluences study)

With these criteria we should minimize effects that could obscure the transport mechanism. The CME observations are taken from the LASCO web site at http://cdaw.gsfc.nasa.gov/CME_list/. For most of the events, the Hα associations have been analyzed in [5].

For this paper, we study both the time history and event averaged He/H and Fe/O variations in large SEP events as a function of their associated solar longitudes, and event maximum intensities.

Event Averaged Ratio

Figure 1 shows the event averaged He/H, C/O and Fe/O for all selected events as a function of their Hα flare longitudes. The event with the highest Fe/O ratio in our survey is associated with a Hα flare located at W69, a magnetically well-connected region. However, the remainder of the events show no clear longitude signature. If we average the Fe/O ratios for those events at W61-W70, the mean Fe/O = 0.63 is more than three times the SEP based coronal value (0.17). However, the spread of those ratios is also large for the four events in that longitude (Fe/O from 0.1 to 1.3). In addition, the C/O ratios for all events have a relatively constant value as compared to the variations we observed for the Fe/O ratios.

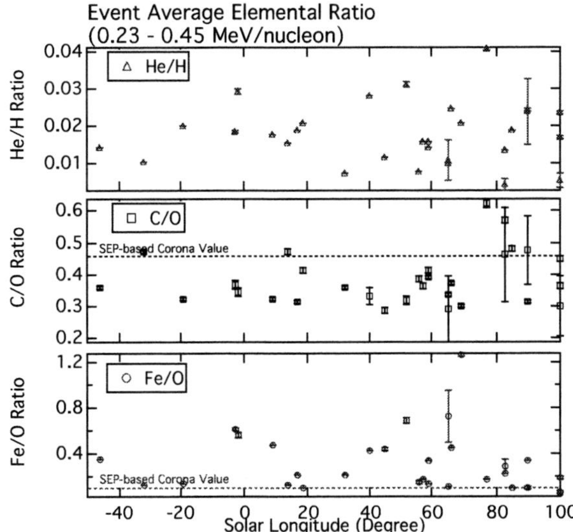

FIGURE 1. Event averaged elemental ratios for all selected thirty large SEP events from September 1997 to December 2002. Horizontal dashed lines indicate the coronal values from [7].

von Rosenvinge *et al.* [6] have found at higher energy (12-60 MeV/nucleon) that Fe/O ratios are higher for magnetically well-connected events. In Figure 2, we plot the Fe/O ratio as a function of both the solar longitude and indicate the maximum event proton intensity at 230-450 keV (by the size of the circular marker). Only the prompt impulsive peak intensity was included regardless of the delayed peak of the shock associated ESP event. The highest intensity event in our survey had maximum proton intensity of 52500 particles/[cm^2-sec-str-MeV] at ~300 keV which was associated with an Hα flare at W83, while the lowest intensity event had maximum proton intensity of only 98 particles/[cm^2-sec-str-MeV] that was associated with an Hα brightening at E02.

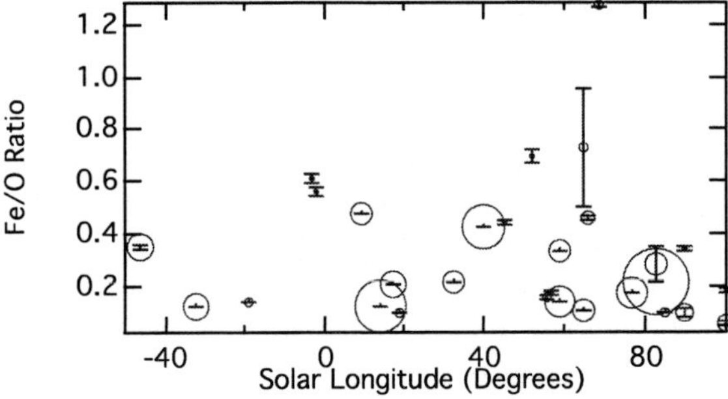

FIGURE 2. Fe/O ratios for events with different maximum intensities (excluding the ESP portion of the SEP event) as a function solar longitude. Larger circles denote higher intensity events.

Temporal Variations

Prior to the launch of ACE, limited data sets were available to study the temporal variation of low energy elemental ratios within large SEP events [1]. However, from [1] we know that the He/H, O/He and Fe/O can have large fluctuations about their means within a large SEP event. The temporal variations of the elemental ratio within an event can be attributed to many factors. The model by [2] suggests these variations are caused by transport effects through self-generated Alfvén waves.

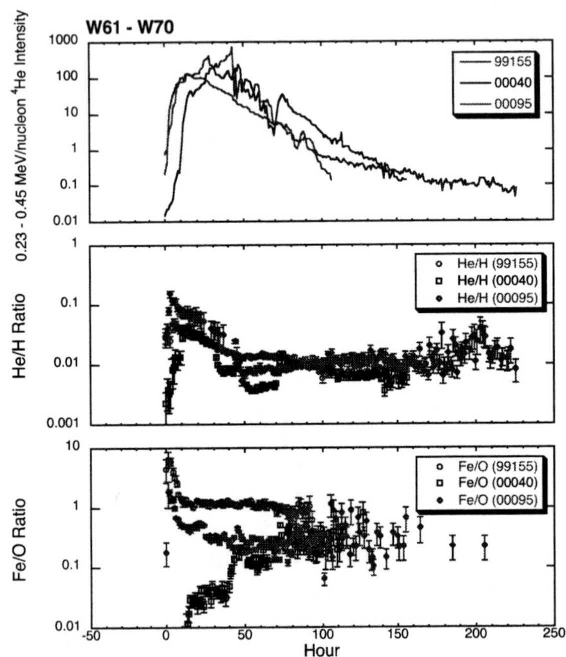

FIGURE 3. Figure 3 shows three well-connected events (W61-W90) having similar duration and intensity. The hour zero is defined as the time when intensity rises above the pre-event level. Top: ^{4}He hourly intensity. Middle and Bottom: hourly He/H and Fe/O ratios respectively.

Given the possible close relationship between the temporal elemental ratio variations and transport effects [2], we first attempt to minimize effects of magnetic connection by organizing our events according to their respective longitudes. Figure 3 shows three selected events that have similar intensity and duration between W61 to W70 in solar longitude. The hour t = 0 is defined as the time when the proton intensity rises above the pre-event level. The first panel in Figure 3 shows the background-subtracted hourly ^{4}He intensity. Both He/H and Fe/O ratios are shown in the second and third panels respectively. All three events have a similar He/H pattern (i.e. a rapid increase and then decreasing He/H). However, the Fe/O ratio for February 9, 2000 (Day of Year 40, blue) event increases throughout the event, which is opposite to the other two despite similar event connections.

Figure 4 shows three events that originated from western longitudes (W81-W90) and that have similar particle intensities (same format as Figure 3). Again, all three events have a similar He/H pattern. However, the Fe/O ratio for the April 1998 (DOY 110, red) event, that was modeled by [8] at higher energies (>2 MeV/nucleon), shows drastically different temporal variations compared the other two similar events. At higher energy (>2.5 MeV/nucleon), [8] shows a rapid increase and then an exponential decrease of the Fe/O for the same event. However, at 0.23-0.45 MeV/nucleon, the Fe/O ratio shows only a gradual increase and then stay relatively constant throughout the event.

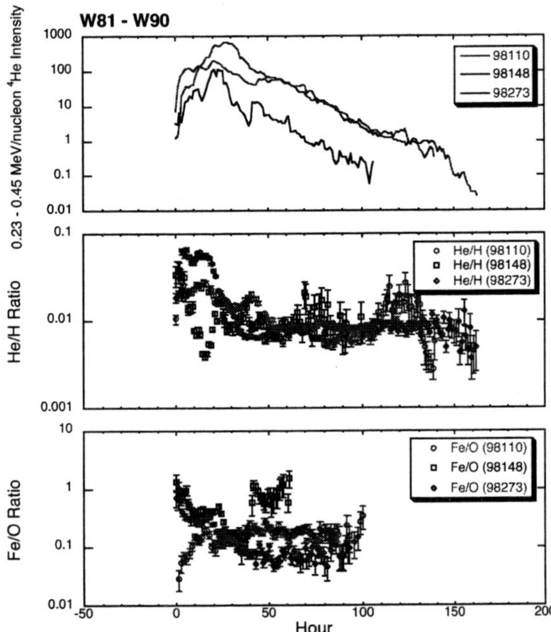

FIGURE 4. Three west limb events (W81-W90) with similar intensity and duration. The April 1998 (red) event has been modeled by [8] at higher energy (>2 MeV/nucleon), but it is also the only event of the three that shows an increase in the Fe/O ratio during the onset phase of the event.

Discussion

Some years ago Mason *et al.* [1] reported that during the onset of large, well-connected SEP events, the Fe/O, O/He, and He/H ratios showed large temporal changes that in most, but not all, cases could be understood as an interplanetary propagation effect caused by the scattering mean free path increasing with particle rigidity. Recently, Ng, Reames and Tylka [2] explored another approach to these onset phase abundance variations by modeling the effects that proton-generated Alfvén waves would have on the scattering of energetic ions during propagation from the acceleration region to the observer. This rigidity dependent scattering can also modify the abundance ratio of species even when they have the same velocity. In particular, the latter theory predicts in events with soft energetic particle spectra an initial decrease in both the He/H and Fe/O ratios at energies >2 MeV/nucleon during large SEP events.

Recently, investigators on both ACE and Wind spacecraft have shown that ion composition variations observed in a few large SEP events at the higher energies (>2 MeV/nucleon) qualitatively agreed with these rigidity dependent transport models [3,8]. However, the same type of variation has not been studied systematically at lower energies, and we present here initial results of such a study.

We examined the event averaged low energy (~300 keV/nucleon) abundance ratio of He/H and Fe/O ratios as a function of longitude and event maximum intensity. No obvious correlation is evident. However, the mean Fe/O ratio for those magnetically well-connected events (W61-W60) has a larger value (0.63) than those from other longitudes. But the spread of those ratios is also large for the four events in that longitude (Fe/O from 0.1 to 1.3). A similar result at higher energy (12-60 MeV/nucleon) was also reported by [6].

In Figures 3 and 4, we showed the temporal variations of He/H and Fe/O for selected events originating from well-connected and west limb longitudes. We thus attempt to minimize effects of magnetic connection by comparing events that have similar longitudes. The majority of the events that we studied exhibited a rapaid increase, and then an eventual decrease in the He/H ratio as suggested by the wave-particle model. However, unlike the well eventual decrease of He/H ratio, we observed large fluctuations in the Fe/O ratio in most of our events. These large fluctuations happened not just at the onset of an event (February, 2000 event in Fig. 3). In the three events that we show in Figure 4, the Fe/O ratio in the April 1998 event (DOY 110) even increases within the event, which is completely different from the expected pattern at higher energy.

SUMMARY AND CONCLUSIONS

Thirty large SEP events were selected for this study from September 1997 to December 2002. We only selected clear and isolated events to study both the event averaged and within event elemental ratios variations. We have grouped our events according to their associated Hα longitudes; hence we minimize any connection effect for these large SEP events. We have found the following:

(1) The event averaged Fe/O ratios show a large range of values for those events between W61 and W70. The largest value is more than three times the coronal value.

(2) No correlation is observed for the event averaged Fe/O as a function of event maximum intensity.

(3) When we examined the temporal elemental ratios within an event, all exhibited an eventual decrease in the He/H ratio as suggested by the wave-particle model, but the degree of variation of He/H varies from event to event.

(4) We observed large temporal fluctuations in the Fe/O ratio in most of our events; and these large fluctuations were not restricted just to the onset of the events.

(5) Even in events that have similar characteristics (magnetic connection, and intensity), the Fe/O ratios can vary drastically from one to another.

REFERENCES

1. Mason, G.M., Gloeckler, G., Hovestadt, D., *ApJ,* **267**, 844-862, (1983).
2. Ng, C.K., Reames, D.V., Tylka, A.J., *Geophys. Res. Lett.,* **27**, 309-312, (1999).
3. Reames, D.V., Ng, C.K., Tylka, A.K., *ApJ,* **531**, L83-L86, (2000).
4. Mason, G.M. *et al., Space Sci. Rev.,* **86**, 409-448, (1998).
5. Cane, H.V. *et al., JGR,* **in press**, (2002).
6. von Rosenvinge, T.T. *et al., AIP Conf. Proc.,* **598**, 343, (2001).
7. Reames, D.V., *Space Science Revs.,* **90**, 413, (1999).
8. Tylka, A.K., Reames, D.V., Ng, C.K., *Geophys. Res. Lett.,* **26**, 2141-2144, (1999).

Cosmic Ray Spectra and the Solar Magnetic Polarity: Preliminary Results from 1994-2002

J. W. Bieber[1], J. Clem[1], M. L. Duldig[2], P. Evenson[1], J. E. Humble[3], and R. Pyle[1]

1 Bartol Research Institute, University of Delaware, Newark, DE 19716, U.S.A.
2 Australian Antarctic Division, Kingston, Tasmania 7050, Australia.
3 School of Mathematics and Physics, University of Tasmania, GPO Box 252-21, Hobart, Tasmania 7001, Australia.

Abstract. Each year, beginning in 1994, a U.S./Australia collaboration has conducted a neutron monitor latitude survey from the United States to McMurdo, Antarctica and back over a ~6-month period. This experiment permits us to observe the high-energy cosmic ray spectrum as it evolves from the last solar activity minimum, through the recent maximum and magnetic polarity reversal, and into the early stages of the new polarity epoch. In this work, we focus on the controversial issue of whether a high-energy (~8 GeV) "cross-over" exists in spectra observed during opposite magnetic polarities. We report on a preliminary analysis of the data and discuss our findings. In particular, we investigate the sensitivity of the modulated cosmic ray spectrum to the sign of the Sun's magnetic field.

1. INTRODUCTION.

Over the past eight years we have conducted an annual latitude survey which traversed the Pacific Ocean from Seattle, USA to McMurdo, Antarctica and return during a ~6-month interval each year. The monitor, a standard 3-NM64 design, is carried aboard one of two U.S. Coast Guard icebreakers, the *Polar Sea* or the *Polar Star*. The data from the surveys cover a wide range of cutoff rigidities, from ~0 GV at McMurdo to over 14 GV in the mid-Pacific. The survey technique ([1, 2] and references therein) has been used for many years to improve our knowledge of the neutron monitor response function and to test geomagnetic cutoff models. Differentiating the curve relating counting rate and cutoff rigidity produces the neutron monitor "differential response," which is a measure of the cosmic ray spectrum. There have been recent reports of one or more 'crossovers' in the spectral forms from two opposite magnetic polarity epochs (e.g. [1, 3, 4]. One of the major goals of this series of surveys is to try to observe such a crossover by deriving a series of cosmic ray spectra through a solar magnetic polarity change (as occurred in 1999/2000). We believe that this is the first set of observations which span the entire period from solar minimum to solar maximum at semi-regular intervals. Data reported in this paper extend through the first half of the 2001/2002 survey.

2. DATA AND METHODS.

Data were taken on eight separate trips from Seattle to McMurdo and return. These voyages are plotted in Fig. 1.; also shown are selected 1980 vertical geomagnetic cutoff contours. Counts from the three counter tubes are recorded once a second, together with data from pitch and roll inclinometers. Once a minute, pressure data and the GPS-derived latitude, longitude and time are recorded. In this preliminary study, we have not yet corrected for any possible effects resulting from non-level operation; however, the data we are utilizing in this paper are from regions where the geomagnetic cutoff is greater than 2 GV, which eliminates most periods of rough seas, especially near Antarctica.

We have calculated the effective geomagnetic cutoff for each hour of the surveys, at the exact time and location of the measurement, taking into account the applicable DGRF magnetic field model, geomagnetic activity level and tilt angle of the geomagnetic dipole (using the Tsyganenko model magnetosphere); [5, 6, 7, 2, 8].

During each survey, the monitor spent several weeks in the harbor at McMurdo, near the McMurdo neutron monitor. We used this period to normalize the total counting rate to the McMurdo monitor. This compensates for any instrumental changes which may have occurred from year to year. In December 2001 we installed the monitor in a new shipping container

CP679, *Solar Wind Ten: Proceedings of the Tenth International Solar Wind Conference*,
edited by M. Velli, R. Bruno, and F. Malara

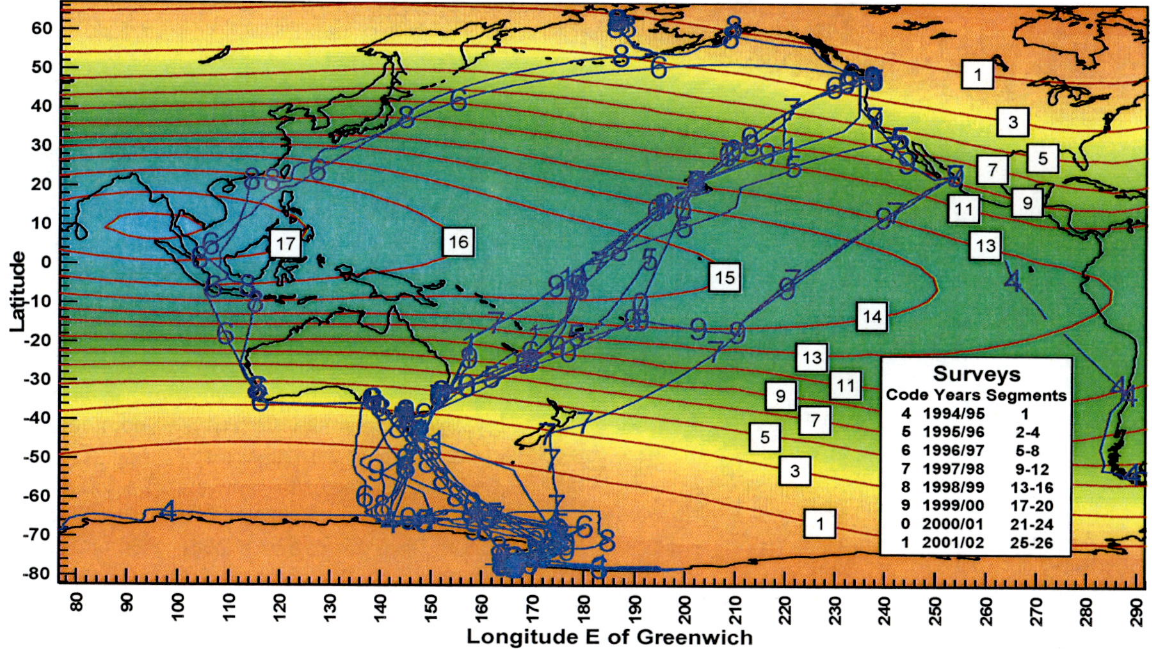

FIGURE 1. Course plots for the 8 surveys used in this paper. Each is labeled at 5-day intervals by the start year of the survey (e.g. 7 for 1997/98). "Segment" codes are given in the inset. Selected vertical cutoff contours are labeled.

during a port call in Hobart, Australia, and used the nearby Kingston neutron monitor to renormalize the data. Otherwise, during each survey year (approximately November-May), care was taken not to make any instrumental changes which might have affected the normalization.

In order to remove various noise problems encountered during the trips, the counting rate data were corrected on a minute-by-minute basis, time-corrected using onboard GPS clock data, and then pressure-corrected to 760 mm_{Hg} using a pressure coefficient varying with cutoff rigidity as follows: β = 0.983515 - 0.00698286*Pc, where β is in percent per mm_{Hg} and Pc is in GV [8]. Since this series of observations was conducted during a period of frequent and often extreme changes in modulation level, we have organized the data to yield the highest time resolution possible, consistent with a significant sweep over a large range of cutoff rigidities. Therefore, we have divided the 8 surveys into 26 segments, with each traverse to and from the Magnetic Equator (or highest vertical cutoff value) treated separately. Some segments were adjusted to avoid the inclusion of major Forbush decreases. Each voyage, therefore, would be expected to yield 4 data segments, but equipment failures, etc, result in four of the possible segments not being available. For each segment, the hourly data points were plotted against the effective vertical cutoff rigidity at the center of the hour. A least-square fit to a three-parameter Dorman function, (e.g. [1]) was performed for all data above 2 GV. The resulting fit was then differentiated to give the differen-

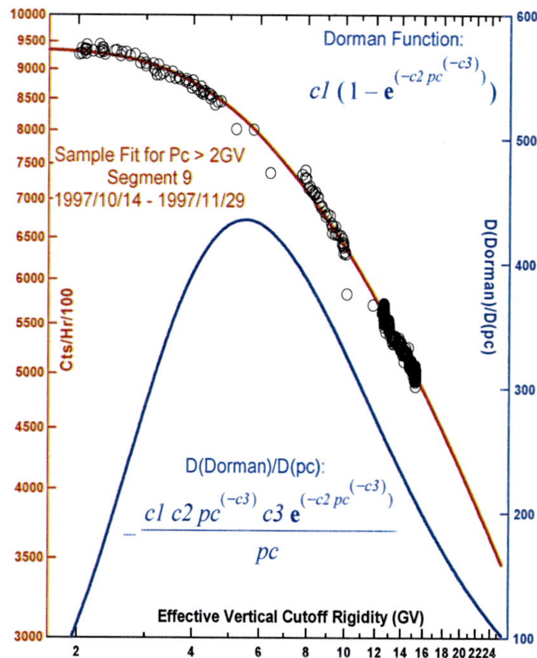

FIGURE 2. Sample fit of a segment's data to a Dorman function, along with the corresponding derivative.

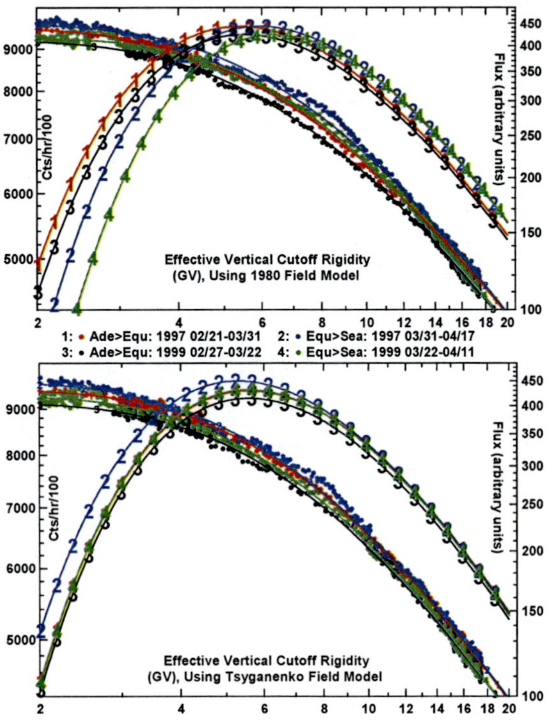

FIGURE 3. Western Pacific Segments: using 1980 Cutoffs (top) *vs* Tsyganenko Cutoffs (bottom).

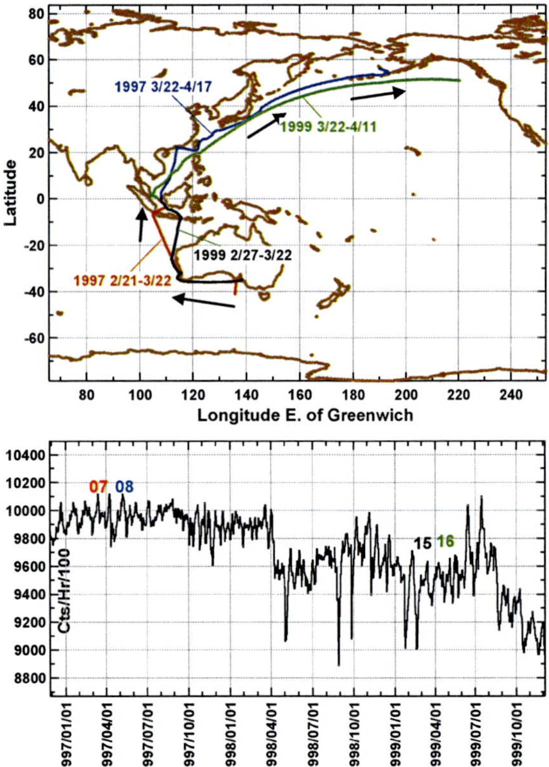

FIGURE 4. Course plots for the 4 western Pacific segments (top). Times of the segments on a plot of the McMurdo counting rate (bottom).

tial response. For one segment, a sample set of results is shown in Fig. 2.

3. RESULTS.

In our previous analysis of these data [9], we used constant 1980 vertical cutoff values. We found that all 4 segments which went west of Australia and through the western Pacific (segments 7, 8, 15, 16 from the 1996/7 and 1998/9 surveys - see Fig. 1) formed a separate and distinct set from those segments east of Australia and in the mid-Pacific. These four spectra did not agree among themselves, even at high cutoff rigidities (>10 GV). We attributed this difference to the secular drift of the earth's geomagnetic field and/or changes in the overall level of geomagnetic activity, in combination with our use of a simple vertical cutoff.

With the use of the new cutoff calculations, this discrepancy has been resolved. In Figure 3 we depict the results from our earlier paper (based on the 1980 effective cutoffs with a static field model) as well as the same spectra analyzed using the Tsyganenko model magnetosphere. Figure 4 displays the two surveys (segments 7, 8, 15, and 16) together with a plot of the McMurdo counting rate with the times of the segments indicated.

The key point is that the differential response curves should converge at high rigidity, where solar modulation becomes weak. This convergence was not obtained with the 1980 cutoffs (top panel of Figure 3), but it does occur with the new Tsyganenko cutoffs (bottom panel).

In Figure 5, the rigidity of the response function

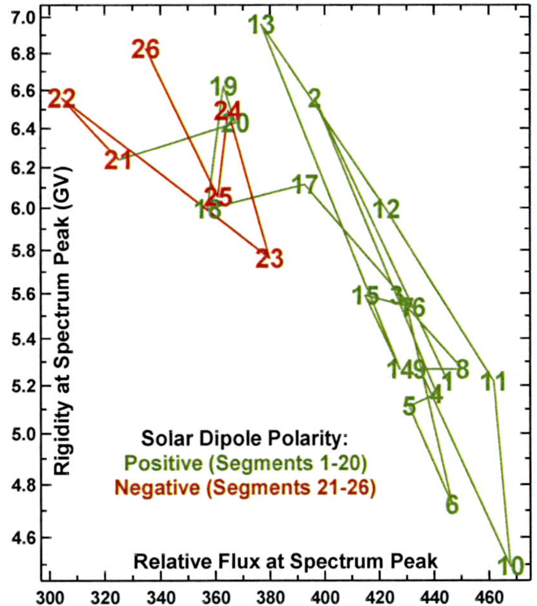

FIGURE 5. Rigidity at the spectrum maximum is plotted versus relative flux at the maximum for each of the 26 segments.

maximum is plotted versus the relative flux at the maximum for each of the 26 segments. As solar activity increased from 1994 to 2000, the points move towards upper left. With increased modulation, the spectrum peak is lower, but it occurs at a higher rigidity. The more recent surveys occurred during negative solar magnetic polarity (shown red in the plot), and they appear to have moved off the trend line established during positive magnetic polarity (shown green). The results are suggestive of a nascent "hysteresis loop." If confirmed by further analysis, this hysteresis effect would be evidence for effects of drift and/or magnetic helicity in solar modulation.

Moraal et al. [1] previously reported evidence of an 8 GeV spectrum cross-over between positive and negative solar polarity, with harder spectra in negative polarity. If the negative (red) polarity points in Figure 5 continue to stay to the left of the positive (green) polarity points as solar activity decreases, this would however be consistent with harder spectra in positive polarity (i.e., for a fixed peak flux, the rigidity of the peak is higher in positive polarity), which is in the opposite direction of the effect reported by Moraal et al.

4. DISCUSSION.

The existence of a 22-year hysteresis loop or of a spectrum cross-over at 8 GeV would indicate the influence of transport effects sensitive to magnetic polarity. Only two such transport mechanisms are known: drift and scattering by turbulence containing nonzero magnetic helicity. Drift models of solar modulation can rather easily produce a spectrum cross-over near 400 MeV, but have difficulty producing one as high as 8 GeV [4]. On the other hand, the low-frequency turbulence responsible for scattering > 10 GeV cosmic rays has been shown to possess finite magnetic helicity with a definite dominant sign [10]. However quantitative modeling of this effect has yet to be performed.

In the near future we will be incorporating additional (small) corrections in the analysis by taking into account 1) the tilt of the monitor, especially important during periods of rough seas; 2) corrections due to anisotropy (by fitting the Spaceship Earth data to measure the first-order anisotropy [11]) and the effects of spectral variations (by using a network of stations at various cutoff rigidities); 3) examination of the contribution of off-vertical directions to the calculated cosmic ray cutoffs .

We plan to continue these yearly surveys at least until the next solar minimum, so that a complete 11-year modulation cycle can be studied in detail.

5. ACKNOWLEDGMENTS.

This work was supported in part by NSF grant ATM- 0000315 (JB, JC, PE and RP). L. Shulman, J. Roth assisted with the onboard equipment. We thank the officers and crew of the United States Coast Guard Cutters *Polar Star* and *Polar Sea* for their assistance in these surveys.

6. REFERENCES.

1. Moraal, H., Potgieter, M.S., Stoker, P.H., and van der Walt, A.J., "NM Latitude Survey of the Cosmic Ray Intensity During the 1986/87 Solar Minimum", *J. Geophys. Res.* **94**, 1,459–1,464, 1989.
2. Bieber, J.W., Evenson, P.E., Humble J.E., and Duldig, M.., 1997, "Cosmic Ray Spectra Deduced from Neutron Monitor Surveys", *Proc. 25th Intl. Cosmic Ray Conf. (Durban)* **2**, 45– 48, 1997.
3. Lockwood, J.A. and Webber, W.R., "Comparison of the rigidity dependence of the 11-year cosmic ray variation at the Earth in two solar cycles of opposite magnetic polarity", *J. Geophys. Res.*, **101**, 21573-21580, 1996.
4. Reinecke, J.P.L., Moraal, H., Potgieter, M.S., McDonald, F.B., and Webber, W.R., "Different Cross-overs", *Proc. 25th Int. Cosmic Ray Conf. (Durban)* **2**, 49–52, 1997.
5. Lin, Z., J.W. Bieber and P. Evenson, "Electron trajectories in a model magnetosphere: Simulation and observation under active conditions", *J. Geophys. Res.*, **100**, 23,543-23,549, 1995.
6. Flückiger, E.O. and Kobel, E., "Aspects of Combining Models of the Earth's Internal and External Magnetic Fields", *J. Geomag. GeoElec.*, **42**, 1123-1128, 1990.
7. Cooke, D.J., Humble, J.E., Shea, M. A., Smart, D.F., Lund, N., Rasmussen, I.L., Byrnak, P., Goret, P. and Petrou, N., "On Cosmic Ray Cut-off Terminology", *Nuovo Cimento Soc. Ital. Fis. C*, **14**, 213, 1991.
8. Clem, J.M., Bieber, J.W., Evenson, P., Hall, D., Humble, J.E., and Duldig, M., "Contribution of Obliquely Incident Particles to Neutron Monitor Counting Rate", *J. Geophys. Res.*, **102**, 26919, 1997.
9. Bieber, J.W., Clem, J., Duldig, M.L., Evenson, P., Humble, J.E. and Pyle, R., "A continuing yearly neutron monitor yearly survey: Preliminary results from 1994-2001", *Proc. 27th Intl. Cosmic Ray Conf. (Hamburg)* **10**, 4087-4090, 2001.
10. Smith, C. W., and Bieber, J. W., "Detection of Steady Magnetic Helicity in Low-Frequency IMF Turbulence", *Proc. 23rd Internat. Cosmic Ray Conf. (Calgary)*, **3**, 493-496, 1993.
11. Bieber, J.W. and Evenson, P., "Spaceship Earth — an Optimized Network of Neutron Monitors", *Proc. 24th Intl. Cosmic Ray Conf. (Rome)* **4**, 1078-1081, 1995.

Comparison Of The Genesis Solar Wind Regime Algorithm Results With Solar Wind Composition Observed By ACE

Daniel B. Reisenfeld[*], John T. Steinberg[*], Bruce L. Barraclough[*], Eric E. Dors[*], Roger C. Wiens[*], Marcia Neugebauer[†], Alysha Reinard[‡], and Thomas Zurbuchen[‡]

[*]*Space and Atmospheric Sciences, MS-D466, Los Alamos National Laboratory, Los Alamos, NM 87544*
[†]*Lunar and Planetary Laboratory, Bldg. 92, University of Arizona, Tucson, AZ 85721*
[‡]*Department of Atmospheric, Oceanic and Space Sciences, University of Michigan, Ann Arbor, MI 48109*

Abstract. Launched on 8 August 2001, the NASA Genesis mission is now collecting samples of the solar wind in various materials, and will return those samples to Earth in 2004 for analysis. A primary science goal of Genesis is the determination of the isotopic and elemental composition of the solar atmosphere from the solar wind material returned. In particular, Genesis will provide measurements of those species that are not provided by solar and in situ observations. We know from in situ measurements that the solar wind exhibits compositional variations across different types of solar wind flows. Therefore, Genesis exposes different collectors to solar wind originating from three flow types: coronal hole, coronal mass ejection (CME), and interstream flows. Flow types are identified using in situ measurements of solar wind protons, alphas, and electrons from electrostatic analyzers carried by Genesis. The flow regime selection algorithm and subsequent collector deployment on Genesis act autonomously. We present an assessment of composition variations of O, He, and Mg ions observed by ACE/SWICS concurrent with Genesis observations, and compare these to the Genesis algorithm decisions. Not only does this serve as a test of the algorithm, the compilation of composition vs. regime will be important for comparison to the abundances determined from sample analysis at the end of the mission.

INTRODUCTION

The Genesis spacecraft, launched on 8 August 2001, will be the first spacecraft ever to return from interplanetary space. Its goal is to collect samples of solar wind and return them to Earth for isotopic and elemental analysis [1]. Two instruments are used to collect samples. One is a set of panels arrayed with wafers that passively collect solar wind material. The second is the solar wind ion concentrator, which uses a set of large electrostatic elements to focus solar wind ions onto diamond and silicon carbide substrates and thereby increase the flux of ions implanted [2]. The resulting concentrated sample will be important for determining the solar wind oxygen isotope ratios to the desired accuracy.

The principal scientific rational of Genesis is to understand the composition of the solar nebula from which our solar system formed, and to use this as a baseline for comparison with present-day planetary compositions. The outer layers of the Sun are thought to be relatively unchanged since the formation of the solar nebula. The solar wind is known to be elementally fractionated relative to the solar photosphere [3], though fractionation of isotopes is likely to be relatively minor. In addition, we know from *in situ* measurements that the solar wind exhibits compositional variations across different types of solar wind flows [4]. Significant elemental variations are seen between fast and slow wind types. Coronal mass ejections (CMEs) exhibit yet another set of abundances. Therefore, Genesis exposes different arrays of collectors to solar wind originating from these three flow types: fast wind, slow wind, and CMEs. Flow types are identified using *in situ* measurements of solar wind protons, alphas, and electrons from electrostatic analyzers carried by Genesis [5]. An on-board regime selection algorithm uses these data to determine in which of the three regimes the spacecraft is currently immersed, and the appropriate collection panel is deployed [6]. The algorithm takes into account the proton speed, proton temperature, alpha particle abundance, and the

CP679, *Solar Wind Ten: Proceedings of the Tenth International Solar Wind Conference,*
edited by M. Velli, R. Bruno, and F. Malara

presence of counter-streaming electrons as determined onboard.

The solar wind flow regimes serve as proxies for the solar wind abundance regimes, which are ultimately what we desire to isolate on different panels; thus, we would like to know how successfully the algorithm discriminates among abundance regimes. Furthermore, the concentrator focuses oxygen ions of different charge states differently, so we need to know the relative amounts of the different oxygen charge states collected over the mission in order to model the oxygen implantation distribution across the target. Genesis does not have instruments capable of performing time-resolved heavy-ion composition measurements; thus, to address these issues, we turn to the Solar Wind Ion Composition Spectrometer (SWICS) on board the Advanced Composition Explorer (ACE) [7] to provide composition information that we can then correlate with the Genesis data.

An equally important goal of this study is to make use of the Genesis regime selection algorithm to investigate the scientific question: "How well does speed correlate with composition?" This question has been investigated by many [8], but here we extend the analysis by addressing the following issue: Because of hydrodynamic processing of the solar wind as it propagates outward from the Sun, the solar wind speed measured at 1 AU may be different from the flow speed as it left the inner corona, especially for plasma at flow interfaces. Originally fast wind may be slowed and vice-versa. Thus, the speed measured at 1 AU at these interfaces may not correlate with the composition as expected. However if we consider the speed history, we can assess whether the wind has been accelerated or decelerated and then use SWICS observations to evaluate whether taking this into account improves the regime separation.

REGIME SELECTION ALGORITHM

The Genesis regime selection algorithm processes data from the Genesis electron and ion spectrometers, determining the solar wind regime in real time on board the spacecraft. The algorithm is fully described in Neugebauer et al. [6]. Here we focus on those aspects of the algorithm relevant to composition. Figure 1 shows the decision flow chart used to determine one of three cases: whether the observed solar wind is (1) fast wind originating from a coronal hole (CH), (2) slow wind from an interstream source (IS), or (3) CME plasma. Based on this determination, collection array changes are made within minutes of determination of a regime change.

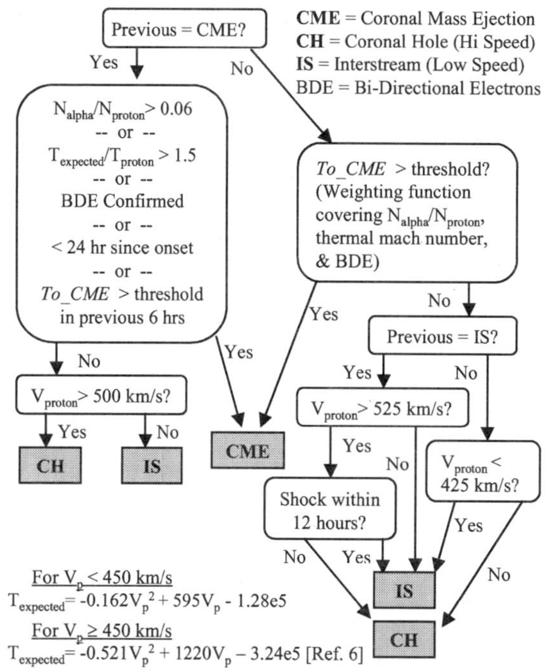

FIGURE 1. Flow chart showing the Genesis on-board logic for determining the solar wind regime in real time.

The following assumptions underlie the algorithm:

1) CH and IS flow (collectively referred to as *quasi-stationary* flow) originate from solar sources that exhibit distinct abundance signatures. Such flows are characterized by speed. We therefore identify CH and IS material based on solar wind speed and use the speed to determine which collector array to deploy.

2) Elemental abundances in CMEs are highly variable and are not necessarily related to CH or IS abundances. We therefore identify and collect CME material on a separate collector.

3) Although solar wind speed is related to the outflow source (CH or IS), the solar wind undergoes hydrodynamic evolution which can obscure the source signature. The recent history of the solar wind can be used to partially recover some of this information. In particular: (a) on the leading edge of a stream, initially slow wind has been accelerated; thus, although at the location of observation the wind is fast, it will have an abundance expected of slower wind. Thus when a slow-to-fast transition occurs, the speed set point for retracting the IS collector and deploying the CH collector is set relatively high: 525 km s^{-1}. (b) On the trailing side of a fast stream, originally fast wind has been decelerated due to the rarefaction caused by fast

wind outrunning slow wind; thus, although at the location of the observation the trailing wind is slow, it will have the abundance of faster flow. Thus when a fast-to-slow transition is observed, the speed set point for retracting the CH collector and deploying the IS collector is set relatively low: 425 km s^{-1}.

SOLAR WIND COMPOSITION MEASUREMENTS

To test the performance of the regime algorithm, we considered a combination of elemental abundance ratios and charge-state ratios derived from ACE/SWICS observations coincident with Genesis operation from 20 September 2001 to 6 May 2002. For this time, the on-board algorithm determined the following fractions of time spent in each regime: fast wind (11%), slow wind (57%), and CME (32%).

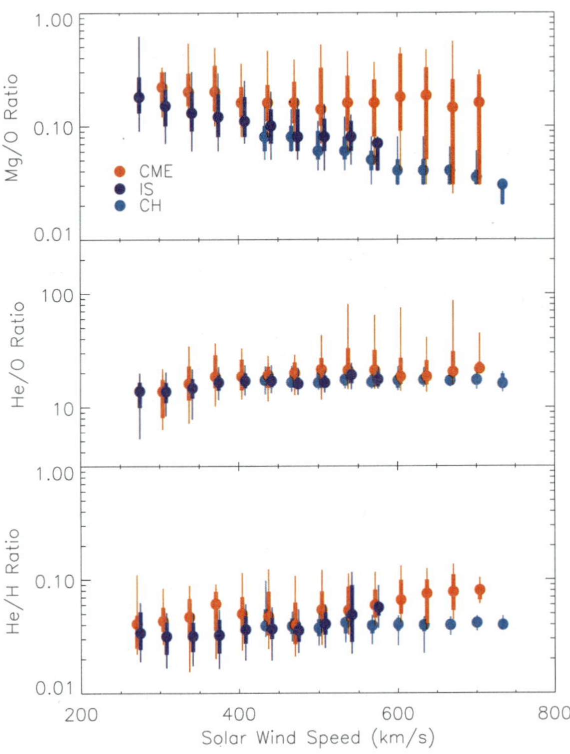

FIGURE 2. Elemental abundance ratios for Mg/O, He/O and He/H as a function of solar wind proton speed and Genesis solar wind regime. The points indicate the median values for data within bins 33 km/s wide. The thin lines span the 10- to 90-percentile ranges, and the heavy lines span the 25- to 75-percentile ranges of the data within each bin.

Elemental fractionation of the solar wind depends on the time it takes for an element to become ionized in the chromosphere, which in turn, depends strongly on the first ionization potential (FIP). Low-FIP elements (FIP < 10 eV) are enhanced over their photospheric abundance relative to high-FIP elements [9]. This effect is most strongly observed in IS flow, where the enhancement factor is 4-5, whereas the factor is ≤ 2 for CH flow [4]. In Figure 2 we show selected abundance ratios measured by SWICS, separated according to the three solar wind regimes as determined by Genesis, and further sorted by solar wind proton speed.

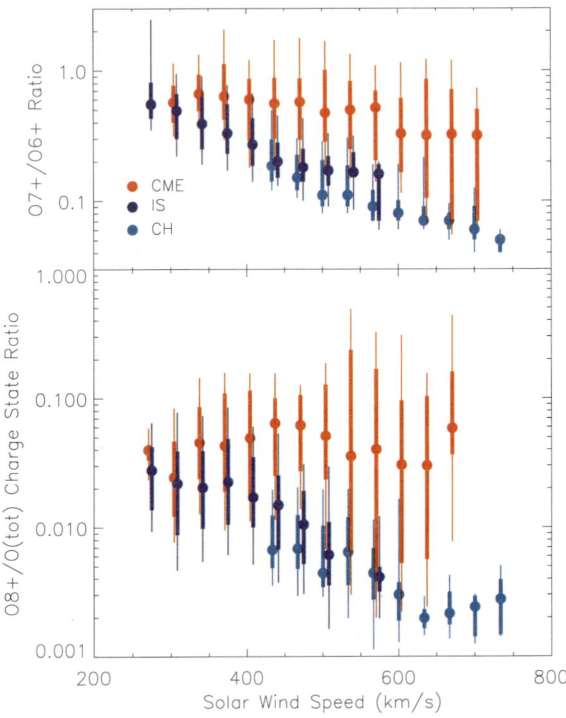

FIGURE 3. Charge state ratios for O^{7+}/O^{6+} and O^{8+}/O_{total} as a function of proton speed and Genesis solar wind regime. O_{total} is the total number of oxygen ions of all charge states.

The top panel of Figure 2 shows the ratio of magnesium (a low-FIP element) to oxygen (a high-FIP element) [10]. For the quasi-stationary wind, we see a steady monotonic decrease in the Mg/O ratio with speed, varying across a factor of 6. The smooth decrease of Mg/O with speed shows us that, although it is common to refer to different FIP fractionation values for IS and CH flows (as we have done in the paragraph above), there is not a sharp abundance boundary between high and low speed. Rather, at least to first order, fractionation is a continuous function of proton speed. Importantly, where the IS and CH regimes overlap, we do see a clear separation in the Mg/O ratio. We attribute this to the interplanetary processing at flow boundaries describe above. We also note that the Mg/O ratio shows significantly less variance for the CH regime than for the IS regime.

In contrast, the Mg/O ratio for CME flow is remarkably steady at an average value of 0.20, equal to the ratio seen in IS flow at ~300 km s^{-1}. Although the Mg/O average value remains constant, the distribution skews to smaller values with increasing proton speed. We attribute this to the conservative design of the algorithm: to minimize the risk that the CH and IS arrays are contaminated by CME material, the algorithm maintains the CME regime state for at least 18 hours, plus an additional 6 hours after CME signatures are last detected. In this way, because CME signatures are often intermittent, enough persistence is designed into the algorithm to keep it from leaving the CME decision state prematurely. As a consequence, the CME array is unavoidably contaminated with quasi-stationary flow, and this is reflected in the skewness of the Mg/O distribution.

The second panel of Figure 2 shows the He/O abundance ratio. As these are both high-FIP elements, little variation with speed or across regimes is expected or seen. A slight increase occurs between 250 and 425 km s^{-1}, which may be attributed to the very high FIP of helium compared to oxygen [9]. The third panel shows He/H abundance ratio [He] measured by Genesis. The values for the quasi-stationary wind range from about 3.4% to 4.1% with speed, slightly lower than previous surveys of [He] [11]. For the CME regime, the ratio is consistently higher, which is expected since an enhanced [He] value (> 7%) is one of the algorithm's possible criteria for CME selection. The [He] value in CMEs ranges from 4.4% to 8.3%, increasing with speed, indicating that larger helium enhancements are generally found in faster CMEs.

We have also determined the oxygen charge state ratios as a function of velocity and for the different regimes. The oxygen charge state distribution in the solar wind is a measure of the electron temperature at the point in the corona where the distribution "freezes in". It is well established that the ratios of charge states, particularly for O^{7+}/O^{6+}, can be used to identify different coronal sources, in the same manner as elemental abundance ratios [4]. We see from the top panel of Figure 3 that the O^{7+}/O^{6+} ratio distribution closely matches the Mg/O ratio, even to the extent that there is a clear separation at intermediate speeds between the IS and CH flow regimes. This is a further indication that the regime algorithm is correctly differentiating between solar sources. In the bottom panel of Figure 3 we show the O^{8+}/O_{total} ratio, where O_{total} is the total number of oxygen ions of all charge states. As we can see, the O^{8+} content is only about ~2% in IS flow, and < 1% in CH flow, thus most often it is a negligible constituent. However, in CMEs, the O^{8+} content can extend beyond 10%, and can contribute significantly to the concentrator collection sample. Thus, it is important to track all three charge states of oxygen in order to understand the distribution of oxygen on the concentrator targets.

CONCLUSIONS

By applying the Genesis algorithm results to ACE/SWICS abundance and charge-state data, we show that the solar wind speed history can be used to test the discrimination between the IS and CH flow types. By using a lower speed threshold for fast-to-slow than for slow-to-fast regime transitions, the Genesis algorithm effectively compensates for effects that are due to transit to 1 AU. Furthermore, we show that on average, CME composition is independent of speed, having a composition signature most typical of the slow wind < 400 km s^{-1}. However, differences between CMEs and the slow wind exist (see He/H and O^{8+}). The algorithm is successfully isolating the CME population, and thus protecting the CH and IS samples from contamination by CME material.

REFERENCES

1. D. S. Burnett, B. L. Barraclough, R. Bennett et al., *Space Science Reviews*, in press (2002).
2. J. E. Nordholt, R. C. Wiens, R. A. Abeyta et al., in press (2002).
3. J. Geiss, G. Gloeckler, and R. Von Steiger, *Space Science Reviews* **72**, 49 (1995).
4. R. Von Steiger, N. A. Schwadron, L. A. Fisk et al., *Journal of Geophysical Research* **105**, 27217 (2000).
5. B. L. Barraclough, E. E. Dors, R. A. Abeyta et al., *Space Science Reviews*, in press (2002).
6. M. Neugebauer, J. T. Steinberg, R. L. Tokar et al., *Space Science Reviews*, in press (2002).
7. G. Gloeckler, J. Cain, F. M. Ipavich et al., *Space Science Reviews* **86**, 497 (1998).
8. R. von Steiger, R. F. Wimmer Schweingruber, J. Geiss et al., *Advances in Space Research* **15**, 3 (1995).
9. J. Geiss, *Space Science Reviews* **33**, 201 (1982).
10. R. von Steiger and J. Geiss, *Advances in Space Research* **13**, 63 (1993).
11. W. C. Feldman, J. R. Asbridge, S. J. Bame et al., in *The solar output and its variation*, edited by O. R. White (Colorado Associated University Press, Boulder, CO, 1977).

Particle transport at CME-driven shocks

Gang Li*, Gary P. Zank* and W.K.M. Rice†

*IGPP, University of California, Riverside, CA 92521 USA.
†School of Physics and Astronomy, University of St. Andrews, St. Andrews, Fife KY169SS, Scotland.

Abstract. It has been commonly accepted that in large solar energetic particle (SEP) events, particles are often accelerated to MeV energies (and perhaps up to GeV energies) at shock waves driven by coronal mass ejections (CMEs). As a CME-driven shock propagates, expands and weakens, particles accelerated diffusively at the shock can escape upstream and downstream into the interplanetary medium. These escaping energized particles then propagate along the interplanetary magnetic field, experiencing only weak scattering from fluctuations in the interplanetary magnetic field (IMF). In this paper, we concentrate on the transport of energetic particles escaping from a CME-driven shock using a Monte-Carlo approach. This work, along with our previous work on particle acceleration at shocks, allows us to investigate the characteristics (intensity profiles, spectra, angular distribution, particle anisotropies) of high-energy particles arriving at various distances from the sun and form an excellent basis with which to interpret observations of high-energy particles made at 1 AU by ACE and WIND.

INTRODUCTION

It has been well established that gradual Solar Energetic Particles (SEPs) events result from particle acceleration at CME-driven coronal and interplanetary shocks (e.g. [1], [2], [3], and [4]). Here the measured abundances of ion charge states indicate that gradual events are associated with a temperature of T $\approx 2\times10^6$ K, in good agreement with coronal material (see [5, 6, 7, 8, 9]).

The commonly accepted mechanism for particle acceleration at a shock is first-order Fermi acceleration, also known as diffusive shock acceleration (see e.g. [10]). In the context of CME-driven shocks, explicit model calculations [11] suggest that particles can be accelerated up to GeV energies. Subsequent investigations for a shock with arbitrary strength verifies the conclusion [12]. The transport of high energy particles accelerated at a shock front based on the work of [11] and [12] is also underway [13].

As particles are accelerated at the shock wave, they are able to escape from shock front and propagate ahead of the shock wave. The detection of these high energy particles thus provides a way of predicting subsequent disruptive events by the shock itself. Upon leaving the shock, these energetic particles gyrate along interplanetary magnetic field lines, described by the Parker field, with occasional pitch angle scatterings. In this work, we consider the transport of these energetic particles using a Monte-Carlo approach. We are interested particularly in the intensity, the spectrum and the momentum distribution asymmetry observed at 1 AU.

MODEL DESCRIPTION

The Boltzmann-Vlasov equation describes the evolution of the distribution function $f(r,p,t)$ for energetic particles escaping from the shock front into the interplanetary medium,

$$\frac{\partial f}{\partial t} + \frac{\mathbf{p}}{m}\cdot\nabla f + \mathbf{F}\cdot\nabla_p f = \left.\frac{\partial f}{\partial t}\right|_{coll}. \qquad (1)$$

The right hand side of (1) describes particle collisions. In our context, this term describes the scattering of charged particles by fluctuations in the interplanetary magnetic field (IMF) and is replaced by $\frac{\partial}{\partial\mu}(D_{\mu\mu}\frac{\partial f}{\partial\mu})$, where $D_{\mu\mu}$ is the corresponding Fokker-Plank coefficient.

Between two consecutive pitch angle scatterings, charged particles move in the solar wind by gyrating along a field line. As the magnetic field expands radially, particles will experience focusing effects as a consequence of the adiabatic invariants of the motion. In the context of solar wind, the **B** field is given by the usual Parker spiral,

$$B = B_0(\frac{R_0}{r})^2[1+(\frac{\Omega_0 R_0}{u})(\frac{r}{R_0}-1)^2 sin^2\theta]^{1/2}, \qquad (2)$$

where θ is the colatitude of the solar wind with respect to the solar rotation axis. Ω_0 is the solar rotation rate, u is the radial solar wind speed, and B_0 is the interplanetary magnetic field (IMF) at the co-rotation radius R_0 (typically, $R_0 = 10R_\odot$, $B_0 = 1.83\times10^{-6}T$, $u = 400$ km/s, and $\Omega_0 = 2\pi/25.4$ days). The components of **B** along $\hat{\mathbf{r}}$

CP679, *Solar Wind Ten: Proceedings of the Tenth International Solar Wind Conference*,
edited by M. Velli, R. Bruno, and F. Malara

and $\hat{\phi}$, B_r and B_ϕ respectively, satisfy

$$\frac{dr}{B_r} = \frac{r\sin\theta d\phi}{B_\phi}. \quad (3)$$

On taking $\theta = \pi/2$ (i.e. we consider the equatorial plane), the path length ds of the particle along the B field line is then,

$$ds = \sqrt{dr^2 + (rd\phi)^2} = \sqrt{1 + (B_\phi/B_r)^2}dr. \quad (4)$$

Equation (4) describes the "free" motion of charged particles between pitch angle scatterings. The length of this "free" motion, which is the mean free path $\lambda_{//}$, is given in the formalism of the quasi-linear theory [14] by,

$$\lambda_{//} = \lambda_0 (\frac{p}{1\mathrm{GeV}})^{\alpha} (\frac{r}{1\mathrm{AU}})^{\beta} \quad (5)$$

where λ_0, an input parameter, is taken to be 0.8AU here. Parameters α and β describe momentum and heliocentric distance dependence, with α between 0 and $1/3$ and β between 0 and $2/3$. In our simulation, we consider three cases of α and β, being, 1) $\alpha = 1/3$, $\beta = 0$, 2) $\alpha = 0$, $\beta = 0$ and 3) $\alpha = 0$, $\beta = 2/3$. In the context of a Monte-Carlo simulation, the motion of charged particles is followed individually when they escape from shock front. If we denote l as the distance between two scatterings, then the probability density $p(l)$ for scattering satisfies,

$$p(l) = e^{-l/\lambda_{//}}. \quad (6)$$

We can then characterize the motion between two pitch angle scatterings using l by

$$l = \int_{r_i}^{r_f} \frac{v\mu(t)dt}{dr} dr, \quad (7)$$

where r_i is the initial position and r_f is the final position of the particle, v is the particle velocity and $\mu(t)$, the pitch angle, is a function of time (thus r). If $\mu(t_i)$ is less than zero, then the particle is moving back toward the sun and it may experience mirroring (when μ becomes 0 and changes sign). On the other hand, if $\mu(t_i)$ is greater than zero, the particle will experience focusing as it moves away from the sun.

To facilitate the simulation, we need to know the position of the shock as function of time and the phase space distribution of charged particle at the shock front. This is done in our previous work [11, 12]. There, the accelerated particle distribution function $f(r(p,t),p,t)$ is obtained for a series of times t_1, t_2, ... , t_n. Integrating $f(r(t),p,t)$ over p^2dp, we get

$$N(t) = \int f(r(t),p,t)p^2dp, \quad (8)$$

and $N(t)$ is the function from which we sample the escape time (and thus location) of our initial particles. For a given random number ξ, the following equation

$$\int_{t_s}^{t_e} N(t')dt'N(t') = \xi \quad (9)$$

determines t, which is the time for a test particle leaving the shock. In the above, t_s and t_e denote the time at which the shock starts and the time when the shock reaches 1 AU. Once t is determined, the momentum of the particle can be determined in a similar way through

$$\int_{p_{inj}}^{p_{max}} f(r(p',t),p',t)dp' = \xi', \quad (10)$$

where ξ' is another independent random variable. Once t and p are decided, the corresponding location can then be inferred from $r(p,t) = r_{sh}(t) + l_{esc}(p)$, where $r_{sh}(t)$ represent the location of the shock complex at time t and $l_{esc}(p)$ is the escape length for particle with momentum p.

After the initial condition (r, t, p) of a test particle is decided, its subsequent motion is simply an interplay between "free" motion and pitch angle scatterings. The scatterings themselves are assumed to be isotropic and Markovian, thus the new pitch angle after a scattering does not have any memory of the previous pitch angle.

RESULTS

We consider simulations of a weak shock. The CME driven shock is introduced at 0.1AU by temporarily increasing the number density and solar wind velocity at 0.1AU by a factor of 3 for half an hour. This corresponds to an energy injection of the order 10^{32} erg, a typical value for a coronal mass ejection. We assume $\gamma = 5/3$ to model the solar wind. The CME driven shock is initiated at 0.1 AU where the solar wind is already supersonic. The weak shock has an initial velocity of 900 km/s and drops to about 600 km/s at 1 AU, taking 2.25 days to reach 1 AU.

Figure 1 plots the relative intensity profile for energetic particles observed at 1 AU from 2 MeV to 50 MeV. The *y*-axes show the relative numbers of particles crossing over 1 AU at different times with a total of 100 time intervals. The three panels in each of the figures represent different choices of α and β as in equation (5). The curves in each of the panel represent, from top to bottom, $K = 2.5$, 5.5, 11, 21 and 51 MeV respectively. We have set λ_0 in equation (5) to be 0.8AU.

From Figure 1 we observe that the weak dependence of particle mean free path on momentum and heliocentric distance has little affect on the simulation. For particles with energy $K < 11$ MeV, the arrival of the shocks

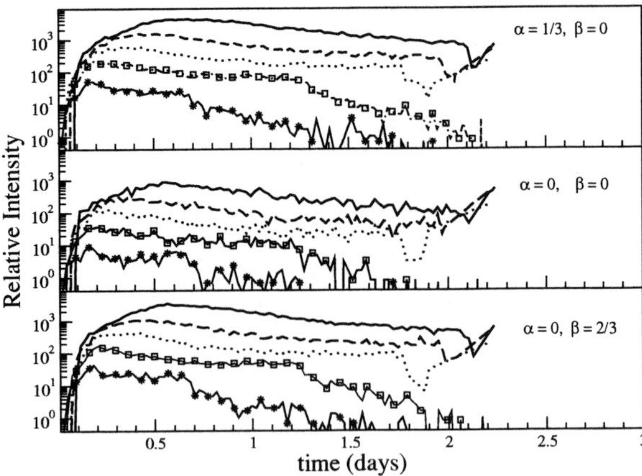

FIGURE 1. Relative intensity for a weak shock, plotted as a function of time for three possible choices of mean free path as in equation (5). Each curve represents a distinct particle energy with the solid curve for $K = 2.5$ MeV, the dashed curve for $K = 5.5$ MeV, the dotted curve for $K = 11$ MeV, the curve with square for $K = 21$ MeV and the curve with star for $K = 51$ MeV.

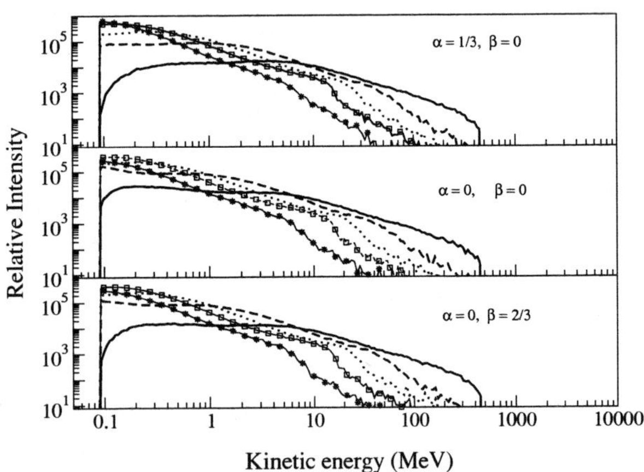

FIGURE 2. The particle spectrum observed at 1 AU for three possible choices of mean free path as in equation (5). Each curve represents a distinct time period with the solid curve for $0 - 1/5$ T, the dashed curve for $1/5 - 2/5$ T, the dotted curve for $2/5 - 3/5$ T, the curve with square for $3/5 - 4/5$ T and the curve with star for $4/5 - 1$ T, where T = 2.25 days is the arrival time for the shock wave.

at 1 AU corresponds to a peak in the intensity profile. However, for particles with energy $K > 20$ MeV, no such peaks exist. The absence of a peak at the shock in the particle intensity was noted earlier in [11]. The present work confirms that interplanetary shocks, especially weak shocks, can only accelerate particles up to high energies during the early stage of their propagation. As the shock weakens, and the IMF strength becomes smaller away from the sun, the maximum possible accelerated particle energy decreases. The simple assumption, that longer-lived shocks can accelerate particles to higher energies in the solar wind, is incorrect. For the weak shock example of Figure 1, the shock stopped accelerating particles up to energies $K > 20$MeV when it was still close to the sun. As can also be seen from Figure 1, even though the shock has stopped accelerating particles above a certain energy by 1 AU, higher energy particles can remain trapped in the post-shock complex ($K \sim 20$ MeV). At even higher energies, particles accelerated earlier have completely escaped and are no longer trapped in the complex region downstream of the shock (e.g. $K \sim 50$ MeV). The intensity profiles of the $K \geq 20$ MeV particles resemble those that are seen in impulsive events.

Figure 2 plots particle spectra observed at 1 AU at different times. Again, the three panels in the figure represent three choices of α and β. The five curves in each panel correspond to spectra at 1 AU for different time periods after the shock was initiated. If we denote by T the time of shock arrival, which is 2.25 days for our simulation, then the curves, from left (curve with stars) to right (solid curve), correspond to $4/5 - 1$ T, $3/5 - 4/5$ T, $2/5 - 3/5$ T, $1/5 - 2/5$ T, $0 - 1/5$ T respectively. In Figure 2, the spectra correspond to $0 - 1/5$ T (the solid curve) is quite different from those at later times. The more plateau-like shape for the $0 - 1/5$ T spectra is due to the "free-streaming" limit, that particles reaching 1 AU at early times are those freely stream out without pitch angle scatterings, thus lower energy particles have smaller speeds and it takes longer for them to reach 1 AU than for high energy particles. The spectra further softens with time, indicating that fewer very energetic particles are accelerated by the shock at later times. The spectra at later times are approximately power laws which assume a broken form. Comparison of the three panels reveals again, that the choice of α and β does not have a pronounced affect on the simulation result.

Finally, we discuss the time evolution of the particle distribution function $f(\mathbf{p}, r = 1\text{AU})$. In particular, we are interested in the angular distribution of f with $\hat{B} \cdot \hat{p}$. Figure 3 plots $f(\mathbf{p}, r = 1\text{AU})$ at four different times for the case of $\lambda_0 = 0.8$ AU and $\alpha = \beta = 0$. In the figure, Z_x and Z_y are the measures of particle velocity and related to particle momentum $\mathbf{p}$ and the Parker field $\mathbf{B}$ through,

$$Z_x \equiv cos(\theta_{\hat{\mathbf{B}},\hat{\mathbf{p}}})(log(p/\text{MeV}) - 4.25), \quad (11)$$

$$Z_y \equiv sin(\theta_{\hat{\mathbf{B}},\hat{\mathbf{p}}})(log(p/\text{MeV}) - 4.25). \quad (12)$$

The $+\mathbf{Z_x}$ axis coincides with the Parker field. The plot is obtained for a total of eight energy intervals and forty pitch angle intervals. Note that it is an even function about the Z_y axis, reflecting the gyro-symmetry of pitch angle scattering. From the innermost circle to the outermost circles, the corresponding particle energies

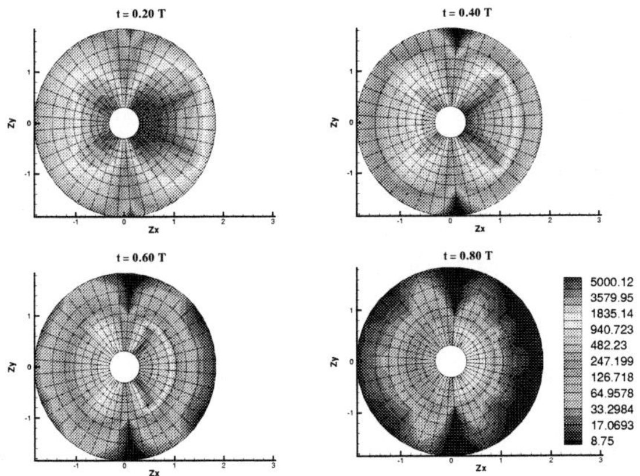

FIGURE 3. Phase space distribution of the energetic particles observed at 1 AU at 4 different times. The coordinates Z_x, Z_y correspond to $v_{//}$ and $v_{\perp}$ (see text for details). Different colors represent different relative number density. The legends are the same for all 4 figures ranging from 8.75 to 5000.

are $K = 4.88$, 8.12, 10.47, 15.35, 21.06, 30.75, 50.80, 100.13 MeV respectively. It is clear from the figure that, at early times, there are more high energy particles and most of them cross 1 AU along $+\hat{\mathbf{B}}$ direction. At later times, however, the density peak moves toward low energies and more low energy particles cross at 1 AU since their speeds are smaller and it takes longer for them to reach 1 AU. Also apparent in this figure is the interesting feature of the "reverse propagation" of high energy particles at later times. Indeed, at $t = 0.80\,\mathrm{T}$, for energies $K > 15.35$ MeV, there are more particles propagating inward against the Parker field than those propagating along the Parker field. If we define the asymmetry $\mathscr{A}$ with respect to the direction of $\mathbf{B}$ observed at 1 AU by

$$\mathscr{A} \equiv \frac{\sum_i^N \mu_i}{N}, \tag{13}$$

then the panel corresponding to $t = 0.8\mathrm{T}$ implies a negative $\mathscr{A}$ at later times for high energies. Finally, the gap at $\mu = \pm 1$ in the figure shows that particles must have a component along the $\pm\hat{\mathbf{B}}$ direction to be observed.

CONCLUSION

We have investigated the transport of energetic particles subject to pitch angle scattering after they are accelerated at CME-driven shocks. Particle intensity and spectra are obtained for a weak shock, using several models for the mean free path of pitch angle scattering. Such weak shocks occur quite frequently. We have also studied the particle's angular distribution with respect to particle pitch angle and find a negative asymmetry $\mathscr{A}$ at later times. Our simulation, together with our earlier theoretical work on particle acceleration at CME-driven shocks, provides a sophisticated theoretical framework for interpreting corresponding observations.

ACKNOWLEDGMENTS

This work has been supported in part by a NASA grant NAG5-10932 and an NSF grant ATM-0296113.

REFERENCES

1. Reames, D.V., Particle acceleration at the Sun and in the heliosphere, Space Sci. Rev., 90, 413-491, 1999.
2. Cane, H.V., The structure and evolution of interplanetary shocks and the relevance for particle-acceleration, Nucl. Phys. B.,39A, 35-44, 1995.
   Ann. Rev. Astron. Astrophys., 30, 113-141, 1992.
3. Gosling, J.T., The solar-flare myth, J. Geophys. Res., 98, 18937-18949, 1993.
4. Cliver, E.W., Cane, H.V., Gradual and Impulsive Solar Energetic Particle Events, EOS Vol. 83, number 7, 2002.
5. Luhn, A, B. Klecker, D. Hovestadt, and E. Mobius, The mean ionic charge state of silicon in He-3-rich solar-flares, Astrophys. J., 317, 951-955, 1987.
6. Leske, R.A., J.R. Cummings, R.A. Mewaldt, E.C. Stone, and T.T. von Rosenvinge, Measurements of the ionic charge states of solar energetic particles using the geomagnetic-field, Astrophys. J., 452, L149-L152, 1995.
7. Mason, G.M., J.E. Mazur, M.D. Looper, and R.A. Mewaldt, Charge-state measurements of solar energetic particles observed with SAMPEX, Astrophys. J., 452, 901-911, 1995.
8. Tylka, A.J., P.R. Boberg, J.H. Adams, L.P. Beahm, W.F. Dietrich, and T. Kleis, The mean ionic charge state of solar energetic Fe ions above 200 MeV per nucleon, Astrophys. J., 444, L109-L113, 1995.
9. Oetliker, M., B. Klecker, D. Hovestadt, G.M. Mason, J.E. Mazur, R.A. Leske, R.A. Mewaldt, J.B. Blake, and M.D. Looper, The ionic charge of solar energetic particles with energies of 0.3-70 MeV per nucleon, Astrophys. J., 477, 495-501, 1997.
10. Axford, W.I., E. Leer, and G. Skadron, Proc. 15th Int. *Cosmic Ray Conf. (Plovdiv), 11*, 132, 1977.
11. Zank, G.P., W.K.M. Rice, and C.C. Wu, Particle acceleration and coronal mass ejection drive shocks: A theoretical model, J. Geophys. Res. (Space), 105, 25079-25095, 2000.
12. Rice, W.K.M. , Zank, G.P. and Li, G., Particle acceleration at coronal mass ejection drive shocks: for arbitrary shock strength. In preparation.
13. Li, G., Zank, G.P. and Rice, W.K.M., Particle transport at coronal mass ejection drive shocks. In preparation.
14. Zank, G.P., Matthaeus, W. H., Bieber, J. W., Moraal, H. The radial and latitudinal dependence of the cosmic ray diffusion tensor in the heliosphere, J. Geophys. Res. (Space), 103, 2085-2097, 1998.

ACE Observations of Energetic Particles Associated with Transient Interplanetary Shocks

D. Lario*, G.C. Ho*, R.B. Decker*, E.C. Roelof*, M.I. Desai† and C.W. Smith**

*The Johns Hopkins University, Applied Physics Laboratory, Laurel, MD 20723
†University of Maryland, College Park, MD 20742
**Bartol Research Institute, University of Delaware, Newark, DE 19716

Abstract. We present preliminary results from a survey of the effects of interplanetary shocks on energetic >47 keV ions and >38 keV electrons as observed by the field (MAG), plasma (SWEPAM) and energetic particle (EPAM) experiments on the ACE spacecraft. From September 1997 to December 2001 ACE observed over 270 shocks, from which we have selected a total of 168 forward transient interplanetary shocks. Particle events associated with these shocks range from large particle intensity enhancements lasting several hours to small shock spikes lasting only a few minutes, as well as events where no intensity increases are observable. Our survey of the proton and electron intensity-time profiles has revealed the following: (1) intensity increases associated with a shock passage are more frequently observed in the low-energy ion fluxes; (2) electron shock events, which occur less frequently than ion shock events, are mainly spikes or step-like post-shock increases; and, (3) peak intensities are usually observed within ~2 minutes of the shock passage with a clear trend towards occurrence in the downstream region of the shock.

INTRODUCTION

Increases of energetic charged particle intensities observed in association with the passage of transient interplanetary shocks are known as energetic storm particle (ESP) events. Phenomenological classification of interplanetary shocks and associated ESP events has revealed a richness of different shock structures and a wide variety of different types of ESP events [1, 2]. In this paper we present preliminary results of a statistical study aimed at characterizing the time-intensity profiles of ions (>47 keV) and electrons (>38 keV) during the passage of forward transient interplanetary shocks. Such studies help one to assess what conditions are necessary for shock acceleration to occur and what role irregularities in the ambient solar wind and/or shock surface may play in producing and modifying shock-accelerated particle populations.

EVENT CLASSIFICATION

We use observations of magnetic fields, solar wind plasma, and energetic ions and electrons from the ACE spacecraft. We include interplanetary shocks observed during the period September 1997 (shortly after ACE launch) through December 2001, i.e., the rise to maximum of the solar cycle 23. The solar wind plasma [3] and magnetic field [4] instruments on ACE, at the L1 Lagrangian point ~230 R_E upstream of the earth, have detected over 270 shocks during this time interval. A preliminary list of the shocks is available at http://www.bartol.udel.edu:80/~chuck/ace/ACElists/obs_list.html. In order to analyze only those energetic particle events associated with forward transient interplanetary shocks, we have excluded from this study reverse shocks and shocks associated with corotating interaction regions or stream interaction regions. In addition, shocks associated with the most intense solar energetic particle (SEP) events, such as the Bastille Day 2000 and November 2001 events, were also excluded because of possible instrument anomalies during these time intervals. A total of 168 forward transient interplanetary shocks were left for our analysis.

We use energetic particle data from the EPAM instrument [5] on ACE, in particular from the Low Energy Magnetic Spectrometer (LEMS) telescopes that measure 47-4800 keV ions and 38-315 keV electrons. We have analyzed high-time resolution data (12 seconds) as well as longer-time averages (1 min, 5

CP679, *Solar Wind Ten: Proceedings of the Tenth International Solar Wind Conference*, edited by M. Velli, R. Bruno, and F. Malara

min and 1 hour) of spin-averaged ion and electron intensities during several time intervals (1 hour, 10 hours and 1 day) around the shock passage. We find that the time-intensity profiles in association with the shock passages are best described by dividing them into 6 different categories:

- Type 0 → No obvious intensity variation above the pre-existing intensity level (Nothing).
- Type 1→ Slow rise of the particle intensity beginning several hours before the shock (Classic ESP event).
- Type 2→ Intensity spike of a few (~10) minutes duration at or near the shock (Spike).
- Type 3→ A classic ESP event with a spike at or near the shock superimposed on it (ESP+spike).
- Type 4→ Step-like post-shock increase (Step-like).
- Type 5→ Irregular time-intensity profile with flux variations not coincident with the shock passage and not fitting into the types described above (Irregular).

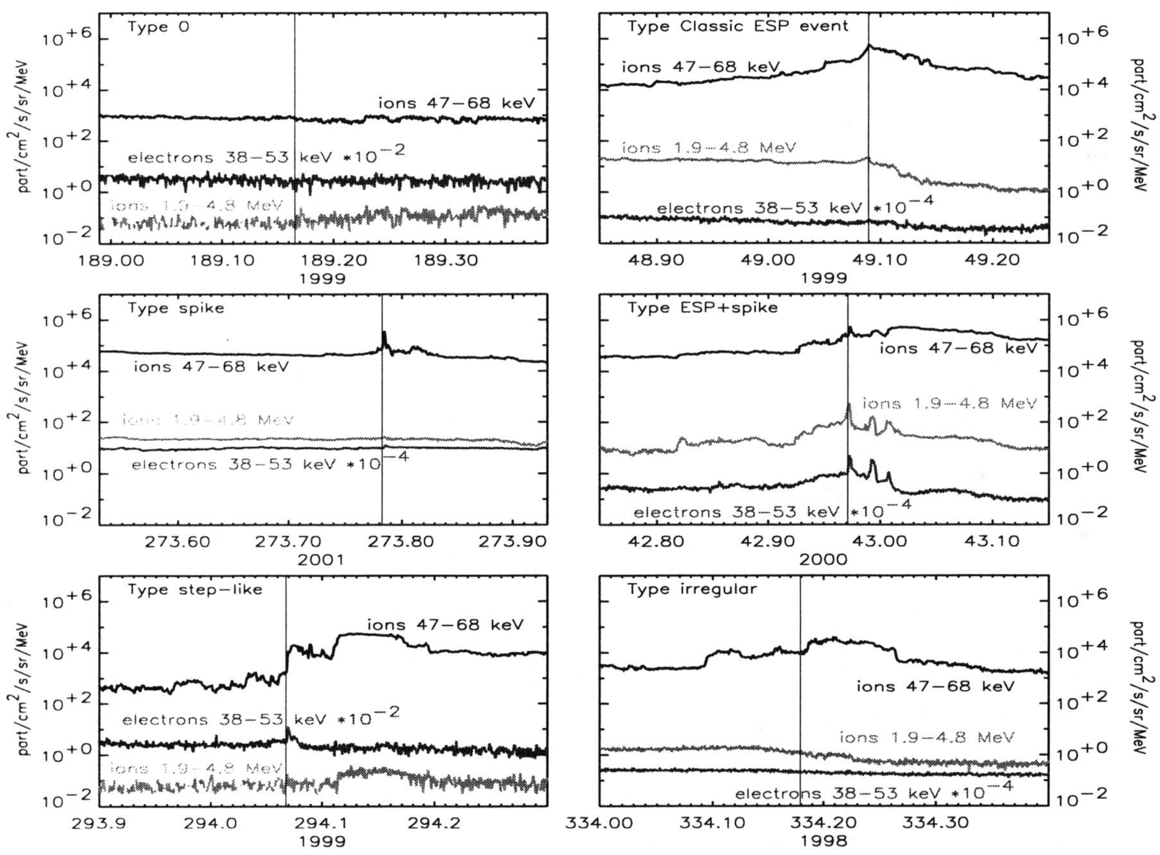

FIGURE 1. Six examples of shock-associated energetic particle events in energy channels 47-68 keV and 1.9-4.8 MeV for ions and 38-53 keV for electrons. The event classification is based on the 47-68 keV ion channel. Note the same time-scale was used in our event classification. Plotted fluxes are 1-min spin-averages as measured by LEMS120 for ions and LEMS30 for electrons.

Figure 1 shows 6 examples of shock-associated particle events, corresponding to each of the 6 types of events. It is important to note that a single event may fall into different categories for the different energy ranges we analyzed. For example the event in the right bottom panel would be classified as Type 5 (irregular) for the 47-68 keV ion channel, but only as Type 0 (nothing) for the 1.9-4.8 MeV ion channel. Thus the classifications given in Figure 1 are assigned only on the basis the 47-68 keV ion channel.

Figure 2 shows the number distribution of events for each different type. Solid bars are for 47-68 keV ions, gray bars for 1.9-4.8 MeV ions and hatched bars for 38-53 keV electrons. It is clear that shock associated changes in ion intensities are more prominent at lower energies. A large fraction of the shock events (83%) show no measurable change in the flux of >38 keV electrons. Irregularities in ion intensities due to either confinement by plasma structures (other than the shock) or to passage by the spacecraft of field lines bearing non-locally shock-accelerated particles, are abundant and more prominent at lower ion energies. Spike events, either isolated or superimposed on ESP events, constitute a significant fraction of the events with measurable particle flux enhancements for both ions (33%) and electrons (43%). ESP events (either with or without spikes) constitute a large fraction of events with particle flux enhancements for both low-energy (46%) and high-energy (58%) ions, but a smaller fraction for electrons (28%).

In comparison with ion events, step-like post-shock increases are much more common in the electron fluxes. As suggested by Tsurutani and Lin [2], energetic electrons may have gyroradii comparable to the shock thickness. Plasma waves at the shock transition may be able to scatter the energetic electrons in pitch-angle as they pass through the shock, thus providing some coupling of these electrons to the solar wind plasma as it is compressed in the shock transition. On the other hand, shock drift acceleration might also provide a mechanism for producing these post-shock enhancements [2].

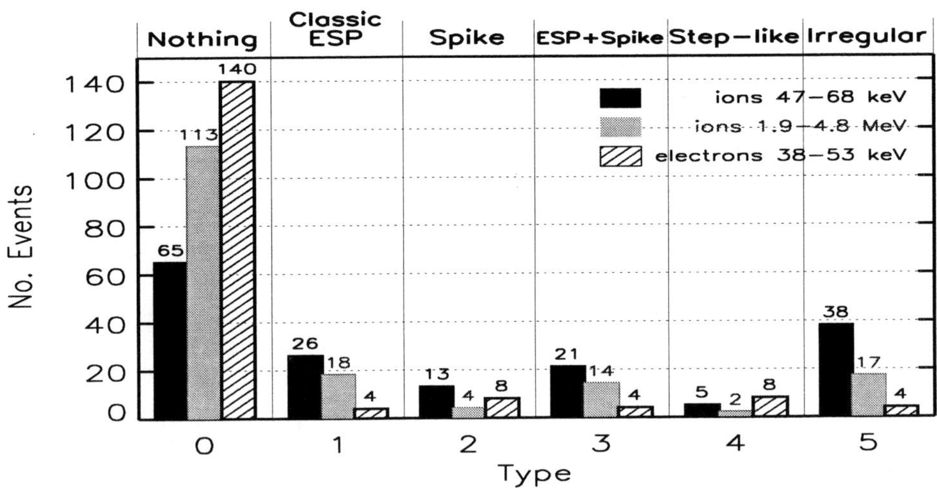

FIGURE 2. Number distribution of shock-associated energetic particle events for the energy channels 47-68 keV and 1.9-4.8 MeV for ions and 38-53 keV for electrons. Different energies or species in the same event may be assigned to different types.

For those events with energetic particle flux enhancements (Type > 0), we computed the time delay between the shock passage and the time when the local peak intensity is observed. We used 12-second averaged magnetic field (**B**) and energetic particle data to locate the shock arrival and the peak intensity observed around the shock passage. Figure 3 shows the number distribution of these time delays for the 47-68 keV and 1.9-4.8 MeV ion channels and for the 38-53 keV electron channel. The solid line marks the time of the shock passage, which we identified as the time when |**B**| starts to increase (i.e., the foot of the shock). Times to the left of the line mean that the peak intensity is observed prior to the shock, while those to the right mean that the peak intensity is observed after the shock passage. Peak intensities typically tend to occur about ~0.5-3.5 minutes after the shock passage. For instance note that all the events in Figure 1 (except the two events in the two top panels) showed peak intensities after the shock passage. Events with long time delays (positive or negative) are irregular events. The tendency for peak intensities to lie downstream of the shock, which is most prevalent in the low-energy ion and electron fluxes, probably arises from several effects, including: (1) our identification of the shock passage with the foot of the shock, which may introduce a systematic offset (i.e., our criterion assigns too early a time to passage of the shock transition); (2) the presence of fluctuations in **B** in the downstream region of the shocks; and, (3) variations in the pre-shock plasma, field, and seed particle levels that introduce random variations in intensities of shock-accelerated, post-shock particles. Fluctuations in **B**, associated with the overshoot of the shock or formed in the compressed plasma left behind the shock, may be very efficient at trapping energetic particles with small gyroradii and therefore produce the peak intensity several minutes after the shock passage. Particles with larger gyroradii are less affected by these downstream fluctuations, and the peak intensity tends to occur closer to the shock passage. Also, with regard to (3), random variations in the angle θ_{Bn} between the local **B** and shock normal can produce random "bursts" of downstream intensity increases. For example, ions incident on the shock within a flux tube where, locally, $\theta_{Bn}\approx 90°$, will rapidly pass downstream, having their energy component transverse to **B** enhanced by a factor equal to the shock compression ratio. Once downstream, this transient intensity peak will decay, with a dissipation rate increasing with ion energy, i.e., as faster ions leave the localized enhancement more rapidly. Thus, such intensity bursts will survive longer for lower energies, as observed in the top two panels of Figure 3. Other singular cases may be those associated with shocks with slow transitions (>1 min) between upstream and downstream magnetic fields and those cases when non-local magnetic connection between the spacecraft and a remote section of the shock efficient in particle acceleration is established after the shock passage, thus producing peak intensities well behind the passage of the shock.

SUMMARY

Solar cycle 23 has provided us a rich and diverse set of data with several types of shock-associated energetic particle events. Classification of these events is the first step to further comparison with shock-acceleration theories and hence assess the relative merits of shock-acceleration models. Correlation between energetic particle intensities and shock characteristics, such as θ_{Bn}, local shock speeds, magnetic field and density compression ratios, as well as plasma and magnetic field peculiarities of the shocks, will allow us to determine the origins of the great variety of shock-associated particle events.

ACKNOWLEDGMENTS

We acknowledge the use of ACE Level 2 data and thank the ACE Science Center for providing these data. DL, GCH and CWS were partially supported by NASA under grants NAG5-10787, NAG5-10836, and NAG5-6912, respectively.

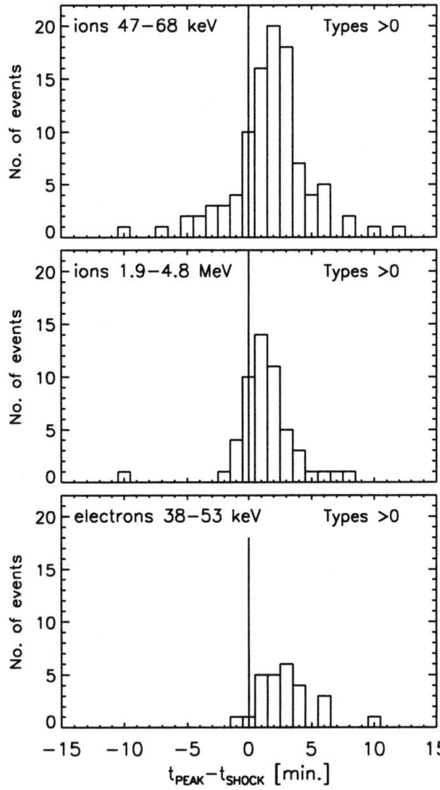

FIGURE 3. Number distribution of the time delays between peak intensities and shock passage for three different energy channels. Three irregular events are out range in the top panel, two events in the middle panel and two events in the bottom panel.

REFERENCES

1. van Nes, P., et al., *J. Geophys. Res.* **89**, 2122-2132, (1984).
2. Tsurutani, B.T., and Lin, R.P., *J. Geophys. Res.* **90**, 1-11, (1985).
3. McComas, D.J., et al., *Space Sci. Rev.* **86**, 563-612, (1998).
4. Smith, C.W., et al., *Space Sci. Rev.* **86**, 613-632, (1998).
5. Gold, R.E., et al., *Space Sci. Rev.* **86**, 541-562, (1998).

Galactic cosmic rays in the global heliosphere: an axisymmetric model

V. Florinski*, G. P. Zank* and N. V. Pogorelov†

*Institute of Geophysics and Planetary Physics, University of California, Riverside
†Institute for Problems in Mechanics, Russian Academy of Sciences, Moscow

Abstract. We present a new axisymmetric model of the heliosphere that includes the three principal particle species (plasma with magnetic field, interstellar neutral atoms and galactic cosmic rays). An important improvement over the previous models is the kinetic description for the cosmic rays instead of the usual fluid approach. We compute the cosmic-ray diffusion coefficients from the quasi-linear theory assuming a constant power spectrum of the fluctuations in both the solar wind and the interstellar medium. Our model predicts small cosmic rays gradients implying little modification of the plasma flow, except, possibly, in the heliosheath region.

INTRODUCTION

It is widely recognized that galactic cosmic rays (GCR) are capable of changing the flow pattern of the solar wind and the surrounding local interstellar medium (LISM) provided the particles' coupling to the plasma is sufficiently strong. While the interstellar cosmic-ray population spectra are reasonably well constrained [1], large uncertainties still exist in our knowledge of diffusion coefficients that ultimately determine the cosmic-ray pressure gradients. Currently, a fluid approach is most widely used to describe the GCR in the context of a global heliospheric model (GHM) [2, 3]. The deficiency of this description is the necessity to impose *apriori* the momentum-averaged diffusion coefficient as well as the average cosmic-ray "adiabatic index". Clearly this approach suffers from the fact that the average quantities are themselves dependent on the cosmic-ray gradients.

In this paper we overcome the problems inherent in fluid models by introducing a kinetic cosmic-ray GHM. Our model can be seen as an extension of the model [4] to include the region beyond the heliopause (HP). The model, in a sense, bridges the gap between self-consistent GHMs and the cosmic-ray modulation models.

A major feature of our model is the introduction of the 3-dimensional heliospheric magnetic field in a kinematic approximation. This gives a more accurate representation of the field and the diffusion coefficients in the heliosheath and the heliotail than by using a Parker's spiral field as was done in [2]. The current model is axisymmetric with respect to the interstellar wind (IW) direction. Neutral atoms and charge exchange are also included (region 1 population only) and effects of the presence of the neutrals on the cosmic rays discussed. Since the model is two-dimensional, we are constrained to a plane through the IW axis at an angle ϕ to the equator. We compare our results with the fluid models and provide data to validate their choice of momentum-averaged parameters.

MODEL DESCRIPTION

We use the MHD model of the plasma described by the usual set of conservation laws with charge exchange terms and CR pressure gradients included. Due to axial symmetry of the model, only the interstellar field is included in the MHD equations (see, e.g., [5]). However, we include the heliospheric field kinematically by assuming Parker's spiral field at the inner boundary and following its evolution in time and space under assumption that the solar wind velocity is axially symmetric, i.e., the velocity vector $(u,0,w)$ satisfies $\partial u/\partial\phi = \partial w/\partial\phi = 0$. In the cylindrical coordinate system the equations are

$$\frac{\partial B_r}{\partial t} + \frac{\partial(wB_r - uB_z)}{\partial z} = -u\left[\frac{1}{r}\frac{\partial(rB_r)}{\partial r} + \frac{\partial B_z}{\partial z}\right] \tag{1}$$

$$\frac{\partial B_z}{\partial t} - \frac{\partial(wB_r - uB_z)}{\partial r} = -w\left[\frac{1}{r}\frac{\partial(rB_r)}{\partial r} + \frac{\partial B_z}{\partial z}\right] + \frac{wB_r - uB_z}{r} \tag{2}$$

$$\frac{\partial B_\phi}{\partial t} + \frac{\partial(uB_\phi)}{\partial r} + \frac{\partial(wB_\phi)}{\partial z} = 0 \tag{3}$$

CP679, *Solar Wind Ten: Proceedings of the Tenth International Solar Wind Conference*, edited by M. Velli, R. Bruno, and F. Malara

TABLE 1. Diffusion parameters

	heliospheric	interstellar
A^2	0.01 [7]	$0.01-0.1$ [8]
l_c	4.5×10^{11} cm [7]	$10^{17}-10^{18}$ cm [8]
$\kappa_\perp/\kappa_\parallel$	0.002 [9, 7]	1.0

This description is valid for any ϕ, however the field reversal at the neutral sheet cannot be taken into account properly. We ignore the $\mathbf{j}\times\mathbf{B}$ force produced by this field because the latter will necessarily destroy the axial symmetry.

The model includes the neutral particles and charge exchange using the Pauls–Zank–Williams formalism [6]. Only interstellar neutrals are considered, while energetic neutral atoms produced in the solar wind are currently ignored. Both MHD and neutral equations are solved numerically on a polar grid lying in the half-plane $\phi =$ const using the TVD Lax–Friedrichs scheme [5].

We describe the galactic cosmic-ray population with the usual Parker–Skilling transport equation. Drift velocities are not included because they contain a component perpendicular to the half-plane. We also note that diffusion in the direction perpendicular to the plane is necessarily ignored, which is a serious limitation of the axial symmetry. To get a better idea about the amount of CR modulation, one needs to consider the angle-averaged cosmic-ray distribution computed for a range of ϕ angles. We plan to include such analysis in a future publication.

We compute the diffusion coefficients following [7] by assuming that a fully developed energy-inertial turbulent spectrum exists in both the solar wind and the interstellar medium. The diffusion coefficients are

$$\kappa_\parallel = \frac{27}{35}\frac{r_g^{1/3} l_c^{2/3} v}{A^2}\left[\frac{7}{27}\left(\frac{r_g}{l_c}\right)^{5/3}+1\right], \quad (4)$$

where l_c is the correlation length (turbulence outer scale) and $A^2 = \langle\delta B_\perp^2\rangle/B^2$ for slab turbulence, and

$$\kappa_\perp = \alpha A^2 \kappa_\parallel, \quad (5)$$

where α is a constant. The last expression corresponds to the second ("modified QLT") model of [7] and predicts a relatively large perpendicular diffusion coefficient. The various parameters used in computing the diffusion coefficients are summarized in Table 1. The interstellar diffusion coefficient computed according to (5) is several orders of magnitude larger than the heliospheric. This is consistent with the notion that the outer scale of the turbulence in the LISM is considerably larger than in the Solar system. Because the direction of the LISM magnetic field is poorly known, we use isotropic diffusion in this region.

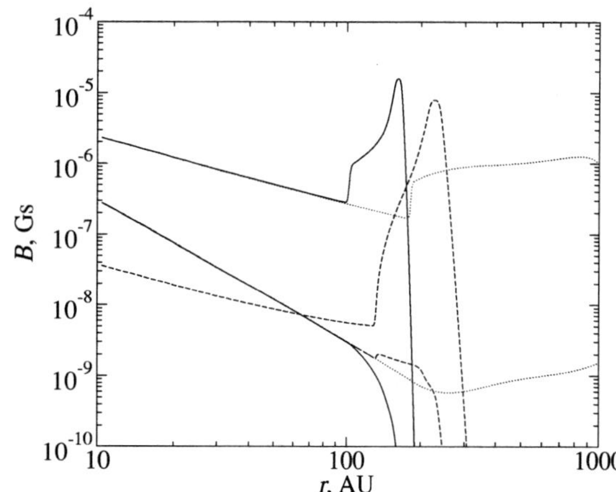

FIGURE 1. Heliospheric fields B_r and $|B_\phi|$ for $\theta = 0^0$ (solid), 90^0 (dashed) and 180^0 (dotted lines). B_ϕ is the larger of the two at low latitudes, but smaller at 90^0.

Our current model uses the results obtained in [10] showing that the turbulence level decays slower than in the WKB approximation because pickup ions and stream interaction create additional amounts of turbulence. We therefore approximate both A^2 and l_c by their respective values at the inner boundary, located at 10 AU. While this approximation is rather crude in the heliosheath region, we must defer model refinement until sufficient improvements are made to the turbulence evolution theory.

Finally, we use the Jokipii–Kota modified field [11] to suppress diffusion in the regions where the field is very small. We specify a modified field B_m in the ϕ direction described by an equation similar to (3) with $B_m/B_r = 0.2$ at 1 AU.

RESULTS

In this paper we present results for the single value of the azimuthal angle $\phi = 0$, i.e., the meridional plane. The simulation domain was $400\times100\times170$ grid points in the r, θ and p directions, respectively. The momentum domain corresponds to proton kinetic energies between 10 MeV and 10 GeV. For the typical interstellar spectrum [1] this includes about 80% of the total pressure.

Figure 1 shows all three components of the heliospheric magnetic field at different angles relative to the IW axis (the LISM and outer heliosheath field is not shown). One important feature of this plot is the existence of a region with a very large magnetic field in the vicinity of the heliopause. This field is found to considerably reduce the ability of the lower energy cosmic rays to penetrate into the heliosphere creating a "modulation barrier".

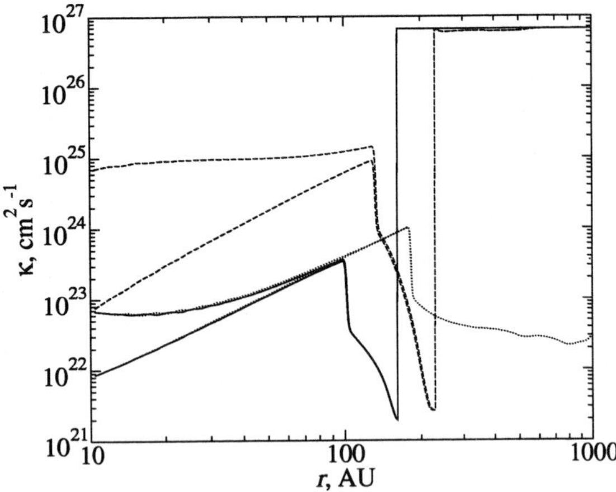

FIGURE 2. 1 GeV proton diffusion coefficients for $\theta = 0^0$ (solid), 90^0 (dashed) and 180^0 (dotted lines). κ_{rr} and $\kappa_{\theta\theta}$ are shown, κ_{rr} is larger at small solar distances.

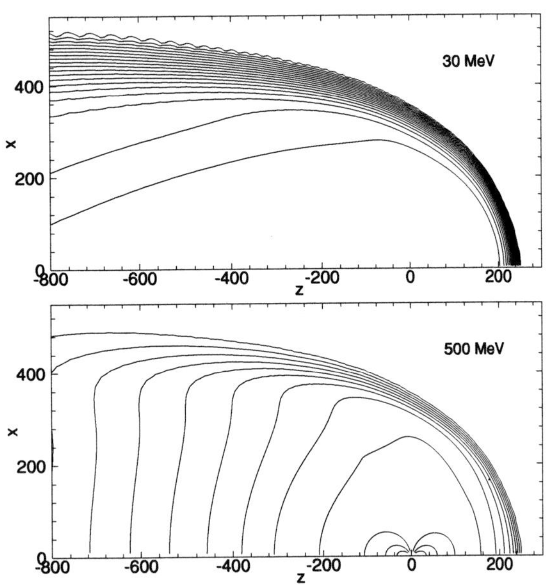

FIGURE 3. GCR phase space density contours at 30 MeV (top) and 500 MeV (bottom panel). Contour lines are drawn with 5% intensity decrements.

We show 1 GeV proton diffusion coefficients in Figure 2. Note that κ is very large in the LISM and causes no modulation even at the lowest energy on the scales of the problem. However, the barrier is evident in the inner heliosheath.

Figure 3 shows the phase space density contours of cosmic ray protons at two different energies for the no-neutral case, while Figure 4 plots the same quantity for the case when neutral atoms were included. It can be seen that low energy particles are mostly filtered by the small diffusion region in the inner heliosheath. Particle distribution inside the termination shock shows the typical lobe structure characteristic of the modulation models [12]. There is an enhancement in 500 MeV intensity in the heliotail region which is caused by re-acceleration due to slowing down of the heliotail flow and the contraction of the tail caused by charge exchange.

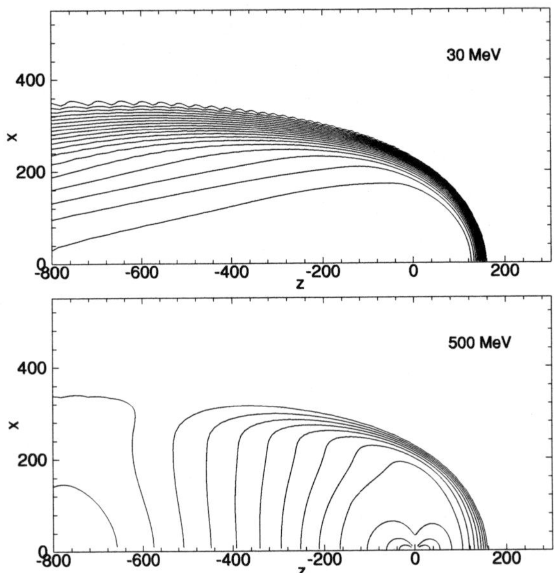

FIGURE 4. Same as Figure 3, but with the neutrals included.

Modulated proton spectra are shown in Figure 5 together with the available solar-minimum observational data. The computed spectra are seen to be in agreement with the observations. The crossover is visible in the heliotail region indicating GCR re-acceleration. Despite significant (up to 50%) reduction in the size of the modulation cavity, GCR intensity in the presence of neutral atoms differs by no more than 5% at 10 AU.

Finally, we would like to compare our momentum-dependent diffusion coefficients with the $\overline{\kappa_{ij}}$ used by fluid models. The latter can be computed according to

$$\overline{\kappa_{ij}} = \frac{4\pi}{3}\left(\frac{\partial P_c}{\partial x_j}\right)^{-1} \int \kappa_{ij} \frac{\partial f}{\partial x_j} p^3 v dp, \tag{6}$$

where P_c is the cosmic-ray pressure. The average diffusion coefficient was found to correspond approximately to $\kappa(1\,GeV)$, which is considerably larger than that used in fluid models [3], when large GCR effects were obtained. We therefore expect that galactic cosmic rays would not change the structure of the termination shock, but merely produce a wide and shallow precursor.

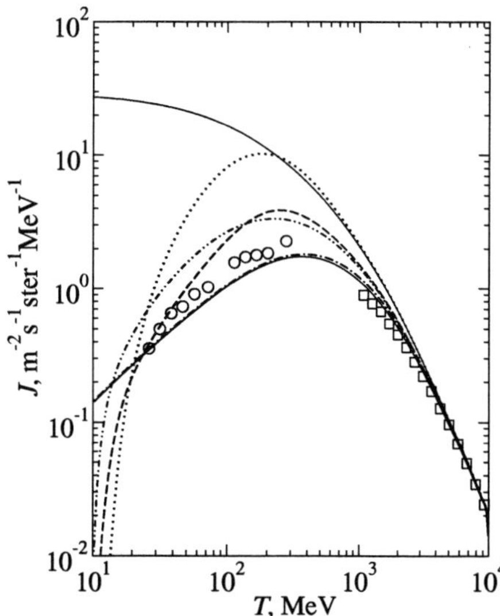

FIGURE 5. Modulated proton spectra in the heliosheath at $1.1r_s$, $\theta = 0$ (dashed line), in the crosswind direction at $1.1r_s$, $\theta = 90°$ (dash-double dotted), in the heliotail at 800 AU, $\theta = 180°$ (dotted) for the case with H atoms. The 10 AU spectrum is plotted with a solid line and the no-neutral 10 AU spectrum is shown with a dash-dotted line for comparison. The input (galactic) spectrum at 1000 AU is also shown with a solid line. Experimental data from BESS (squares) and IMP8 (circles) are shown for comparison.

DISCUSSION AND FUTURE WORK

We believe the new model of the CR-modified heliosphere to be a useful extension of the current fluid GHMs. Our results indicate that cosmic rays are not coupled strongly to the plasma in the heliosphere and the surrounding interstellar medium, except in the inner heliosheath, where the average diffusion coefficient is 2-3 orders of magnitude below the heliospheric values. We find that while high-energy (1 GeV and higher) particles can penetrate into the heliosphere quite easily, lower-energy cosmic rays are stopped by the modulation barrier.

Due to space constraints we do not show P_c, however the latter was found to vary little throughout the heliosphere. We find only 0.07 eV/cm^3 decrease in P_c between the outer and the inner boundaries, which is insignificant when compared with the wind dynamic pressure of 0.65 eV/cm^3 at the termination shock. Galactic cosmic rays are therefore not likely to modify the shock significantly.

We also note that neutral atoms may have an important effect on the cosmic-ray propagation in the heliotail region. When neutrals are present, the heliosphere contracts due to reduction in the momentum density of the solar wind as well as cooling and decelerating caused by charge exchange in the heliotail [6]. In this case, GCR experience additional acceleration in the convergent flow ($\nabla \cdot \mathbf{u} < 0$).

To summarize, the most important result of our model is that modulation in the heliosheath cannot be ignored. In fact, the amount of attenuation experienced by particles with $T < 1\,\mathrm{GeV}$ is greater in the heliosheath than in the solar wind. One implication of this is the existence of large gradients in P_c in this region, which may result in the solar wind being strongly mediated by the cosmic rays in the upwind heliosheath region. We also found that, despite a large reduction in the size of the heliosphere when the neutral hydrogen is included in the model, GCR intensities differ only by 5% in the inner heliosphere for these two cases. We have also found that GCR upwind-downwind asymmetries are too small at small heliocentric distances to provide a useful tool for detecting the outer heliospheric structures.

For the future, we plan to improve the model by including a better turbulence description. In particular, the model [10] needs to be extended to 3 dimensions and the restriction on the Alfvén speed being small removed. One should also consider including turbulence production from pickup ions, whose density can be directly computed from the charge exchange rate. This extra source of turbulence may reduce diffusion in the outer heliosheath causing additional attenuation of GCR intensity.

This work was supported, in part, by the NASA grant NAG5-11621, the NSF grant ATM-0296114 and the RFBR grant 02-01-0948.

REFERENCES

1. Ip, W.-H., and Axford, W. I., *Astrophys. J.*, **149**, 7–10 (1985).
2. Fahr, H. J., Kausch, T., and Scherer, H., *Astron. Astrophys.*, **357**, 268–282 (2000).
3. Myasnikov, A. V., Alexashov, D. B., Izmodenov, V. V., and Chalov, S. V., *J. Geophys. Res.*, **105**, 5167–5177 (2000).
4. Florinski, V., and Jokipii, J. R., *Astrophys. J.*, **523**, L185–L188 (1999).
5. Pogorelov, N. V., and Semenov, A. Y., *Astron. Astrophys.*, **321**, 330–337 (1997).
6. Pauls, H. L., Zank, G. P., and Williams, L. L., *J. Geophys. Res.*, **100**, 21,595–21,604 (1995).
7. le Roux, J. A., Zank, G. P., and Ptuskin, V. S., *J. Geophys. Res.*, **104**, 24,845–24,862 (1999).
8. Spangler, S. R., *Astrophys. J.*, **376**, 540–555 (1991).
9. Giacalone, J., and Jokipii, J. R., *Astrophys. J.*, **520**, 204–214 (1999).
10. Zank, G. P., Matthaeus, W. H., and Smith, C. W., *J. Geophys. Res.*, **101**, 17,093–17,107 (1996).
11. Jokipii, J. R., Kota, J., Giacalone, J., Horbury, T. S., and Smith, E. J., *Geophys. Res. Lett.*, **22**, 3385–3388 (1995).
12. Potgieter, M. S., *Space Sci. Rev.*, **83**, 147–158 (1998).

Relative Abundance Variations of Energetic He^{+}/He^{2+} in CME Related SEP Events

H. Kucharek, E. Möbius, W. Li, C. Farrugia, M. Popecki, A. Galvin,*, B. Klecker†, M. Hilchenbach** and P. Bochsler‡

* *University of New Hampshire, Space Science Center, Durham, 03824 United States*
† *Max-Planck-Institut für extraterrestrische Physik, P.O. Box 1603, Garching, 85740 Germany*
***Max-Planck-Institut für Aeronomie, Max-Planck-Str. 2, Katlenburg-Lindau, 37819 Germany*
‡*University of Bern, Sidler Str. 5, Bern, 3012 Switzerland*

Abstract. We have investigated several CME-related SEP events with unusually high abundance of He^{+} relative to He^{2+} in the energetic particle population which have been observed between 1998 and 2000 with ACE/SEPICA and SOHO/CELIAS. Usually the abundance of He^{+} is below a few percent whereas at these times the He^{+}/He^{2+} ratio can be closer to one. Possible sources for He^{+} are interstellar pickup ions or cold plasma in CME's. We have investigated in detail the temporal evolution and the energy spectra of these events. We find that the maximum of the He^{+}/He^{2+} ratio usually coincides with the arrival of the shock or a solar wind structure. This is a strong indication for local acceleration of these ions. The He^{+} enhancement does not seem to be associated with cold plasma within CME's itself. Therefore, most probably interstellar pickup ions are the source for the He^{+} enhancement. Furthermore, the He^{+}/He^{2+} ratio appears to be consistently lower at higher energies. Also, the observed temporal variability decreases with increasing energy. These two results seem to indicate two different populations for He^{+} and He^{2+} with different energy spectra.

INTRODUCTION

One of the first charge state measurements of energetic He^{+} during solar energetic particle events have been reported by Hovestadt et al. [1984]. Analyzing data from the ULEZEQ instrument on board the ISEE 3 spacecraft they observed a large variability of the He^{+}/He^{2+} ratio in the energy range of 0.41 - 1.05 MeV/nuc ranging between 0.1 and 1.0. They concluded that a possible source for these ions could be an admixture of cold solar material. However, in solar energetic particle SEP events with a high He^{+} abundance the charge state distributions for heavy ions did not show an indication of lower charge states. This seemed to speak against an admixture of cold solar material during these events. The detection of interstellar He^{+} pickup ions by Möbius et al. [1985] suggested an other possible source for He^{+}. These particles enter the heliosphere as interstellar neutral particles and are then ionized by UV radiation and picked-up by the interplanetary field. These ions had been suggested already earlier by Fisk et al. [1974] as the source for the anomalous component of cosmic rays. In a study by Gloecker et al. [1994] these ions have been observed at corotating interaction regions (CIR) at a radial distance of 4.5 AU. They identified interstellar pickup He^{+} as the major contributor of the suprathermal He in CIRs with Ulysses SWICS for energies up to 60 keV. More recently, He^{+} was also found as part of the suprathermal CIR population at 1 AU up to 200 keV/Q with SOHO STOF (Hilchenbach et al. 1999) and with Wind STICS (Chotoo et al. 2000). These observations led to the suggestion that pickup ions may constitute a generally important source of suprathermal ions for further acceleration at interplanetary shocks (Gloeckler, 1999).

For a recent series of unusual CIR's close to solar maximum in 1999 and 2000, Möbius et al. [2001a] have reported a substantial fraction (10 - 30%) of He^{+} in the energy range 0.25 - 0.8 MeV/nuc, which they attributed to interstellar pickup ions. For the same CIR events, Morris et al. [2001] have shown that the He^{+}/He^{2+} ratio increases as time elapses during successive passage of the CIR. On the other hand in a recent publication, Skoug et al. [1999] pointed out a prolonged He^{+} enhancement within a coronal mass ejection in the solar wind. They indicated that the source for the He^{+} could be prominence material. Klecker et al. [2001] and Kucharek et al. [2001] have reported substantial variation of He^{+}/He^{2+} abundance ratio during the passage of a coronal mass ejection. They investigated twelve hour averages of the He^{+}/He^{2+} abundance ratio of SOHO/CELIAS/STOF and ACE/SEPICA data during CME events and they found substantial variations ranging from 0.01 up to 0.8.

CP679, *Solar Wind Ten: Proceedings of the Tenth International Solar Wind Conference*, edited by M. Velli, R. Bruno, and F. Malara

Furthermore Klecker et al. [2001] correlated daily averages of the He^+ abundance with the solar wind parameters and found an general anti-correlation of the He^+ abundance with the solar wind velocity and the solar thermal velocity. In this paper we provide a survey of the He^+/He^{2+} abundance ratios in energetic particle events during 1998 - 2000. Another major topic addressed in this paper is the dependence of the temporal variability of the energetic He^+ on the energy. We extend the work by Klecker et al. [2001] to higher energies to test for the He^+/He^{2+} correlation with the solar wind velocity. In addition we test for suggested sources by studying carefully the time variation of the He^+/He^{2+} abundance ratio during one series of CME events.

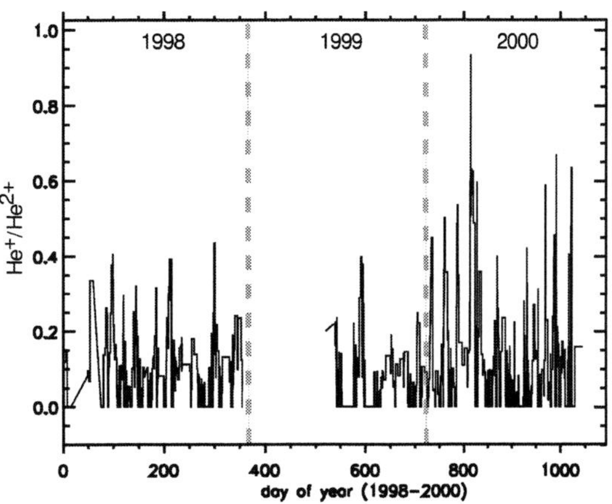

FIGURE 1. shows the He^+/He^{2+} ratio of ACE/SEPICA for the years 1998 - 2000 (separated by dashed lines) in the energy range of 250 - 800 keV/nuc.

INSTRUMENTS AND SPACECRAFT

For this investigation we used data from two different sensors on board two different spacecraft, the Advanced Composition Explorer (ACE) spacecraft and the SOlar and Heliospheric Observatory (SOHO) spacecraft have halo orbits around L1. The Suprathermal Time Of Flight (STOF) sensor is part of the Charge ELement and Isotopic Analyzing System (CELIAS) instrument and is designed to study the composition of the solar wind and of solar and interplanetary energetic particles on the SOHO spacecraft. This sensor covers the energy range from 35 to 630 keV/Q by employing an electrostatic deflection system. As a second sensor we used the Solar Energetic Particle Ionic Charge Analyzer (SEPICA). This sensor is the main instrument on board the ACE spacecraft to determine the ionic charge states of solar and interplanetary energetic particles in the energy range from 0.2 MeV/nuc to 5 MeV/charge.

OBSERVATIONS

We have determined the He^+/He^{2+} abundance ratio from the pulse height data of ACE/SEPICA for the entire time period 1998 - 2000, by employing a double gaussian fit curve to the actual charge state distribution. The integration times for the charge state distribution were variable and chosen according the minimal statistics requirements of 3000 counts within a certain time interval. Figure 1 shows the result of the survey for the energetic He^+/He^{2+} abundance within the energy range of 250 - 800 keV/nuc ratio during 1998 - 2000. As one can see, the ratio shows a significant variation between 0. and 0.8. During this time period many shocks associated with CME's, flares, and CIR's at 1 AU have been reported. For the same time period SOHO/CELIAS/STOF data are available. In Figure 2 we show twelve hour averages of the He^+/He^{2+} abundance ratio for both data sets as a function of the solar wind bulk speed for 1999 according to Klecker et al. [2001]. This data set indicate a dependence on solar wind speed. Higher solar wind bulk speed show a lower He^+/He^{2+} ratio. However, when we determine the He^+/He^{2+} ratios from twelve hour averages by using the ACE/SEPICA data in the same time interval, we found no obvious correlation between solar wind bulk speed and abundance ratio in the energy range of 0.25 MeV/nuc - 0.8 MeV/nuc. The He^+/He^{2+} ratio seems to increase towards the suprathermal energies. In addition, the ratio is highly variable over the course of individual events, with a stronger variation at the lower energies. This strong variability is under current investigation. The key questions at this point are, what are the possible sources for He^+ and what causes the observed variation. In the rest of the paper we will address to the determination of the source of the He^+. In order to identify the structures which are associated with He^+ enhancements. We have investigated a CME event in high time resolution which has a substantial enhancement of He^+ in the solar wind. The time span chosen contains a sequence of events corresponding to the first very active burst in the rising phase of solar cycle 23. Some of these have been also investigated by other authors, such as Skoug et al. [1999] and Farrugia et al. [2002]. Figure 3 show the He^+/He^{2+} ratio of a ten day time period of 1998. This is a very interesting episode. On DOY 120 (April, 30) a shock passes the spacecraft followed two days later (May, 2) by a CME event, which contains a magnetic cloud. Skoug et al. [1999] found in the ACE/SWEPAM a prolonged He^+ enhancement in the solar wind population at the same time that Farrugia et al. [2002] noted a complete dropout of electron halo population. Skoug et al. [1999] suggested that the source for the He^+ could be

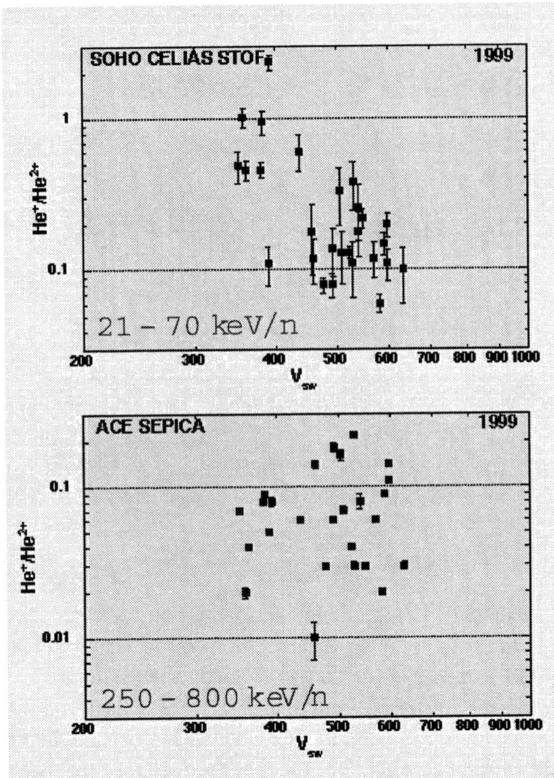

FIGURE 2. shows the He^+/He^{2+} abundance ratio versus solar wind speed for SOHO/CELIAS/STOF (top panel) and for ACE/SEPICA (bottom panel)

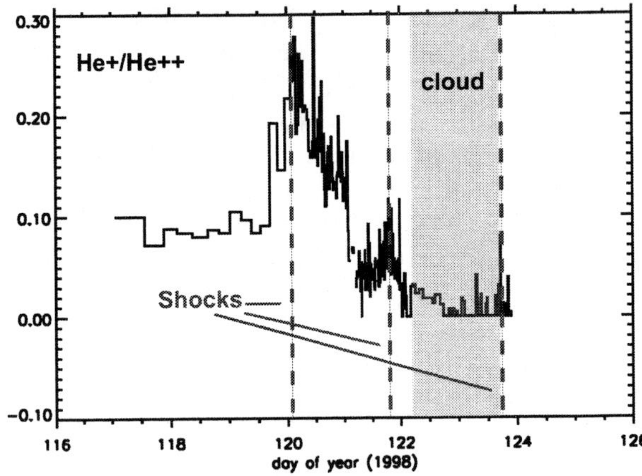

FIGURE 3. This figure shows a time period in the year 1998 which contains shocks (dashed lines) and a well investigated CME/cloud event with a high He^+ abundance on the solar wind (shaded band).

prominence material. As we see from Figure 3 there are three marked shocks outside the cloud. A significant enhancement of the energetic He^+/He^{2+} ratio is seen at the shock arrival at DOY 120. At the second shock, driven by the CME, we also see an enhancement of He^+, although less pronounced than at the previous shock. It should be pointed out that at the cloud associated with the CME has not arrived yet. In the cloud itself, where a large amount of He^+ is found in the solar wind, no significantly enhanced values of He^+/He^{2+} ratio in the high energetic population are observed. In fact the only, very moderate, enhancement occurs in the later part of DOY 123 at the shock presumably overtaking the cloud (Farrugia et al. 2002).

DISCUSSION

One of the key observation of our CME the case study is that the enhancement of the He^+/He^{2+} ratio peaks at shocks. This association clearly indicates that the acceleration occurs locally at these interplanetary disturbances. It should be noted here that many of the enhancements in the survey (Fig. 1) which are associated with shocks.

The two suggested sources for He^+, i.e. cold solar atmospheric material, which is occasionally found in the solar wind (Gosling et al. 1974, Schwenn et al. 1987, Skoug et al. 2001), and interstellar pickup ions (Möbius et al. 1985; Gloeckler, 1999), are indeed present locally. Our case study also provides clues to distinguish between the two sources.

At the energies in our study, we see the largest enhancement at the shock at DOY 120, which is well separated from the cloud. The shocks closer to the cloud shows only a very moderate He^+/He^{2+} enhancement. In fact the smallest enhancement occurs at the shock on DOY 123 which is presumably overtaking the cloud. If cold solar material were the major source we would expect the highest enhancement to take place at this shock. Although, some contribution from cold solar material cannot be excluded, for the adjacent shocks, it is more compelling that the majority comes from pickup ions. This conclusion is also supported by the fact, that He^+ enhancements are ubiquitous in our survey, while He^+ rich clouds are very rare. These observations are also consistent with theoretical considerations that compared to the solar wind alpha particle pickup ions seems to be preferentially injected into an acceleration process. Scholer [1999] and Scholer and Kucharek [1999] have demonstrated that pickup ions are indeed essentially suprathermal ions in the frame of the solar wind, while the solar wind itself is rather cold.

The observed substantial increase of the He^+/He^{2+} ratio towards lower energies, and the variability during individual events and the correlation with solar wind speed (Klecker et al. 2001), may provide further evidence for the source. The strong dependence on solar wind speed at lower energies could reflect a strong variation of the pickup densities with solar wind speed or, more likely, is related to the variability of the thermal distribution of solar wind He^{2+} or pickup ions, injected into the acceleration process. At higher energies other processes may play an important role. Due to dispersion and transport effects the dominance of the locally injected ions may be washed out, and energetic ions from other sources, i.e. flare material and ions accelerated closer to the Sun may be added. This would reduce the ratios and time variability and it would destroy any correlation with solar wind parameters.

SUMMARY AND CONCLUSIONS

In this paper we have presented a survey of energetic He^{+}/He^{2+} abundance ratio for the 3 year period 1998-2000. For this investigation we used data from ACE/SEPICA in the energy range of 250 - 800 keV/nuc and SOHO/CELIAS/STOF data in the lower energy range of 85 - 280 keV/nuc. The ratio in these two energy ranges is very variable. Ratios starting from 0.01 up to 1 have been observed. The He^{+}/He^{2+} ratio appears to be consistently lower at higher energies. In addition, the temporal variability decreases with increasing energy. We found significant enhancements at interplanetary shocks, shocks associated with CME's. A detailed investigation of the temporal evolution and the energy spectra of one of these events have shown that the maximum of the He^{+}/He^{2+} ratio usually coincides with the arrival of the shock. This is a clear indication of local acceleration of these ions. Also, the He^{+} enhancement does not seem to be associated with cold plasma in the CME. Therefore, the most likely source of the ions are interstellar pickup ions. Apart from CME events, as discussed in detail here, where we found a substantial enhancement of the He^{+}/He^{2+} ratio there are different interplanetary configurations within this survey which are able to produce a substantial enhancement of this ratio. However, a detailed discussion of these events is deferred to a future study.

ACKNOWLEDGMENTS

The work on the SEPICA instruments was supported by NASA under Contract NAS5-32626 and the data analysis under Grant NAG 5-6912 where as the work on STOF/CELIAS was supported under DLR contracts 50OC89056 and 50OC96059, NASA Grant NAG 5-2754.

REFERENCES

1. Chotoo, K., et al.,*The suprathermal seed population for corotating interaction region ions at 1 AU deduced from composition and spectra of H^{+}, He^{2+}, and He^{+} observed on Wind*, J. Geophys. Res., 105, 23107 - 32122, 2000.
2. Farrugia, C. et al., *Wind and ACE observations during the great flow of May 1-4 1998: Relation to solar activity and implications for the magnetosphere*. J. Goephys. Res., 107 (A9), 1240, doi:10.1029/2001JA000188,2002.
3. Fisk, L.A., B. Kozlowsky, and R. Ramaty, *An interpretation of the observed oxygen and nitrogen enhancements in low-energy cosmic rays*, Astrophys. J., 190, L35 - L38, 1974
4. Gloeckler, G., *Observation of injection and pre-acceleration processes in the slow solar wind*, Space Sci. Rev., 89, 91 - 104, 1999.
5. Gosling, J.T., *Mass ejections from the Sun: A view from Skylab*, J. Geophys. Res., 79, 4581, 1974.
6. Hilchenbach, M., et al., *Observation of suprathermal helium at 1 AU: charge states in CIRs*, in: Solar Wind Nine, S.R. Habbal, R. Esser, J.V. Hollweg and P.A. Isenberg, Editor, 1999.
7. Hovestadt, D., B. Klecker, G. Gloeckler, F. M. Ipavich, and M. Scholer, *Survey of He^{+}/He^{2+} abundance ratios in energetic particle events*, ApJ., 282, L39 - L42, 1984.
8. Klecker, B., et al., *On the variability of suprathermal He^{+} at 1 AU.*, Proc. of the 27th. International cosmic ray conference, Hamburg, 2001.
9. Kucharek, H., et al., *Variable abundance of energetic He^{+} in CME related SEP events*, Proc. of the 27th. International cosmic ray conference, Hamburg, 2001.
10. Möbius, E., et al., *Survey of Ionic Charge States of Solar Energetic Particle Events During the First Year of ACE*, in: Acceleration and Transport of Energetic Particles Observed in the Heliosphere, R. A. Mewaldt et al. ed., AIP Conf. Proc., 528, 131 - 134, 2000.
11. Möbius, E., et al., *Direct observation of He^{+} pick-up ions of interstellar origin in the solar wind*, Nature, 318, 426 - 429, 1995.
12. Möbius, E., et al., *Charge states of energetic ions obtained from a series of CIRs in 1999 & 2000 and implications on source populations*, Geophys. Res. Lett., subm., 2001a.
13. Möbius, E., et al., *Variation of Energetic He^{+}, He^{2+} and Heavy Ions Across Co-Rotating Interaction Regions*, Proc. of the 27th. ICRC, Hamburg, 2001b.
14. Morris, D., et al., *Implications for source populations of energetic ions in co-rotating interaction regions from ionic charge states*, in: ACE-SOHO Workshop on Solar and Galactic Composition, ed. R. Wimmer-Schweingruber et al., AIP Conf. Proc., p. 201, 2001.
15. Scholer, M., *Injection and acceleration processes in corotating interaction regions: theoretical concepts*, Space, Sci. Rev., 89, 105, 1999.
16. Scholer, M., and H. Kucharek, *Interaction of pickup ions with quasi-parallel shocks*, Geophys. Res. Lett., 26, 29, 1999.
17. Schwenn, R., *The solar wind*, In ESA Space Astronomie sna Solar System Exploration, p. 131 - 141, 1987.
18. Skoug R., *A prolonged He^{+} enhancement within a coronal mass ejection in the solar wind*, Geophys. Res. Lett., 26, 2613, 1999.

How Common is Energetic ^{3}He in the Inner Heliosphere?

M. E. Wiedenbeck*, G. M. Mason†, E. R. Christian**, C. M. S. Cohen‡, A. C. Cummings‡, J. R. Dwyer§, R. E. Gold¶, S. M. Krimigis¶, R. A. Leske‡, J. E. Mazur||, R. A. Mewaldt‡, P. L. Slocum||, E. C. Stone‡ and T. T. von Rosenvinge**

*Jet Propulsion Laboratory, California Institute of Technology, Pasadena, CA 91109 USA
†University of Maryland, College Park, MD 20742 USA
**NASA/Goddard Space Flight Center, Greenbelt, MD 20771 USA
‡California Institute of Technology, Pasadena, CA 91125 USA
§Florida Institute of Technology, Melbourne, FL 32901 USA
¶Applied Physics Laboratory, Johns Hopkins University, Laurel, MD 20723 USA
||The Aerospace Corporation, El Segundo, CA 90009 USA

Abstract. Using data from the SIS and ULEIS instruments on the Advanced Composition Explorer (ACE) we have identified periods during which energetic ^{3}He is present in near-Earth interplanetary space between November 1997 and May 2002. The data, which cover the energy intervals 0.2–1 MeV/nuc (ULEIS) and 4.5–16.3 MeV/nuc (SIS), show that ^{3}He is present a significant fraction of the time, as would be required if these suprathermal particles were the major source of the ^{3}He being accelerated by shocks in the interplanetary medium. Specifically, we find that energetic ^{3}He is present at least $\sim$ 60% of the time, and perhaps significantly more often.

INTRODUCTION

The existence in some solar energetic particle (SEP) events of large enhancements of the ^{3}He/^{4}He abundance ratio, sometimes by factors $> 10^4$ over the solar-wind ratio of 4×10^{-4}, has been a subject of great interest since the discovery of these "^{3}He-rich events" in the 1970s (see, e.g., [1] and references therein). The extreme isotopic fractionation of helium in these events is thought to occur as the result of some resonant form of wave heating and/or acceleration in solar flares (see [1] for references). Adopting the commonly-used terminology, we will call these events "impulsive"[1] to distinguish them from "gradual" events (see below).

Measurements made by new instruments with improved mass resolution and large collecting power on the Advanced Composition Explorer (ACE) [2, 3] have shown that previous empirical definitions of ^{3}He-rich events need to be broadened. Specifically, it has been found that ^{3}He/^{4}He enhancement factors range from < 10 to $> 10^4$. The threshold previously used to define these events, ^{3}He/^{4}He> 0.1, was simply related to the inability of earlier instruments to identify lower fractional abundances of ^{3}He.

In addition, ACE measurements [5, 4, 6, 7, 8] have shown that sizeable enhancements of the ^{3}He/^{4}He ratio also occur in a significant fraction of large gradual SEP events, in which particles are believed to be accelerated not as a direct result of reconnection in solar flares but rather by shocks driven by coronal mass ejections (CMEs) as they propagate through the solar corona and interplanetary space. In some gradual events ^{3}He/^{4}He ratios > 0.01 have been measured.

Mason et al. [6] addressed the question of the origin of the ^{3}He excesses in these gradual events, proposing that a background of energetic ^{3}He particles from numerous small impulsive events is commonly present in the inner heliosphere and that a coronal/interplanetary shock encountering this remnant impulsive material will accelerate these suprathermal particles more efficiently than particles from the solar-wind thermal distribution. Thus one could obtain excesses, relative to the solar wind, of ^{3}He and of other ions (e.g., Fe) that tend to be enhanced in impulsive events. In this model, the enhancement of ^{3}He depends on the occurrence of appropriately-located impulsive SEP events in the hours or days preceding the passage of the shock.

Desai et al. [9] found enhanced abundances of energetic ^{3}He in more than 40% of particle events associ-

[1] In this paper we treat "impulsive" as a synonym for "^{3}He-rich".

CP679, *Solar Wind Ten: Proceedings of the Tenth International Solar Wind Conference*, edited by M. Velli, R. Bruno, and F. Malara

Table 1
Energy Intervals

ULEIS-L	0.2-0.4 MeV/nuc
ULEIS-H	0.4-1.0 MeV/nuc
SIS-L	4.5-7.6 MeV/nuc
SIS-H	7.6-16.3 MeV/nuc

ated with the passage of interplanetary shocks near Earth. Mewaldt (private communication, 2002; see also [4, 8]) found a similar fraction of large, gradual SEP events having significant ^{3}He excesses. In the events studied by Mewaldt and collaborators, acceleration is attributed to CME-driven shocks relatively close to the Sun. One possible explanation of these two results is that energetic ^{3}He is present and available for further acceleration in the interplanetary medium a sizeable fraction of the time and over a sizeable range of heliocentric radii.

In this paper we examine the Mason et al. model [6] for the origin of shock-accelerated ^{3}He by directly measuring how frequently energetic ^{3}He is present in the interplanetary medium near Earth.

OBSERVATIONS

The present study uses data from two instruments on ACE that measure helium isotopes in complementary energy intervals, the Ultra-Low-Energy Isotope Spectrometer (ULEIS) [3] and the Solar Isotope Spectrometer (SIS) [2]. Each data set was divided into a low-energy (L) and a high-energy (H) interval as indicated in Table 1. For each instrument the lower-energy interval provides the best statistical accuracy while the higher-energy interval offers improved mass resolution and background rejection.

In order to identify time periods with ^{3}He present, plots of measured particle mass vs. time were made for each of the four energy intervals. An example is shown in Figure 1 for one solar rotation in late 1999 (Bartels rotation 2271). In regions of the plot having low densities of particles, each detected particle is indicated by an individual dot. Where densities are higher (e.g., along the mass 4 line near the maxima of SEP events) intensities are indicated by a grey scale averaged over rectangles encompassing 1 hr $\times$ 0.05 amu. Using similar plots, we partitioned the data for Bartels rotations 2241 through 2302 (September 1997 to April 2002) into time intervals with durations ranging from 0.25 to 10.5 days during which the characteristics of the He isotope fluxes remained relatively constant. In each interval and in each of the four energy ranges the observations were subjectively assigned one of 7 classifications, as defined in Table 2. The classifications assigned for periods in rotation 2271 are indicated along the top of Figure 1.

Table 2
Time Interval Categorization Scheme

ID	Definition
0	no clear ^{3}He
1	low-rate ^{3}He, SEP events not distinguishable
2	^{3}He-rich SEP event
3	^{3}He event, possibly continuation
4	^{3}He event, possibly overlapping events
5	probable ^{3}He, contaminated by spillover
6	too much spillover to distinguish real ^{3}He

Although the ^{3}He/^{4}He ratio is of interest in individual impulsive events, this ratio is not particularly useful for determining the presence of ^{3}He during a general time interval, since one very commonly encounters ^{3}He-rich events during which the ^{4}He flux is dominated by particles from preceding events. An example is seen on days 336–337 in Figure 1. Thus we based our classifications only on the ^{3}He observations. Clearly our ability to detect the presence of small ^{3}He fluxes is limited by the sensitivity of the instruments and the duration of the averaging interval. For reference, the ^{3}He intensities corresponding to one event recorded per hour in SIS and ULEIS are $\sim 2 \times 10^{-6}$ and $\sim 5 \times 10^{-3}$ particles/cm^2-sr-s-MeV/nuc, respectively. There are periods (e.g., days 351–355 in SIS) where no more than a few ^{3}He particles are detected per day. As shown by the upper set of mass histograms in Figure 2 (and also evident in Fig. 1) genuine ^{3}He signals can be observed during such periods, but the count rates are too low to distinguish individual ^{3}He-rich SEP events. In fact, near solar minimum (1997–98) the ^{3}He observed at SIS energies may be dominated by galactic cosmic-ray ^{3}He that has undergone adiabatic deceleration entering the heliosphere. The lower histograms in Figure 2 are from such a solar-minimum period.

The ^{3}He observations have a complex time structure and it is frequently not possible to determine whether a period of enhanced ^{3}He intensity (e.g., days 343–349 in Fig. 1) is attributable to multiple ^{3}He injections at the Sun or propagation effects. These complexities have been discussed by Mazur et al. [10]. In fact, it is possible that different sources of ^{3}He are sometimes being observed at the different energies, particularly since transit times from the Sun to Earth differ by about 0.3 days over the energy range we are considering, even in the absence of scattering.

In both SIS and ULEIS, instrumental effects can cause misidentification of a small fraction of the detected ^{4}He as ^{3}He. During some periods (e.g., ULEIS-H on day 338–339.5) there is a reasonable indication of a real ^{3}He contribution, together with some spillover. During other times (e.g., ULEIS-L and SIS-L during this same period)

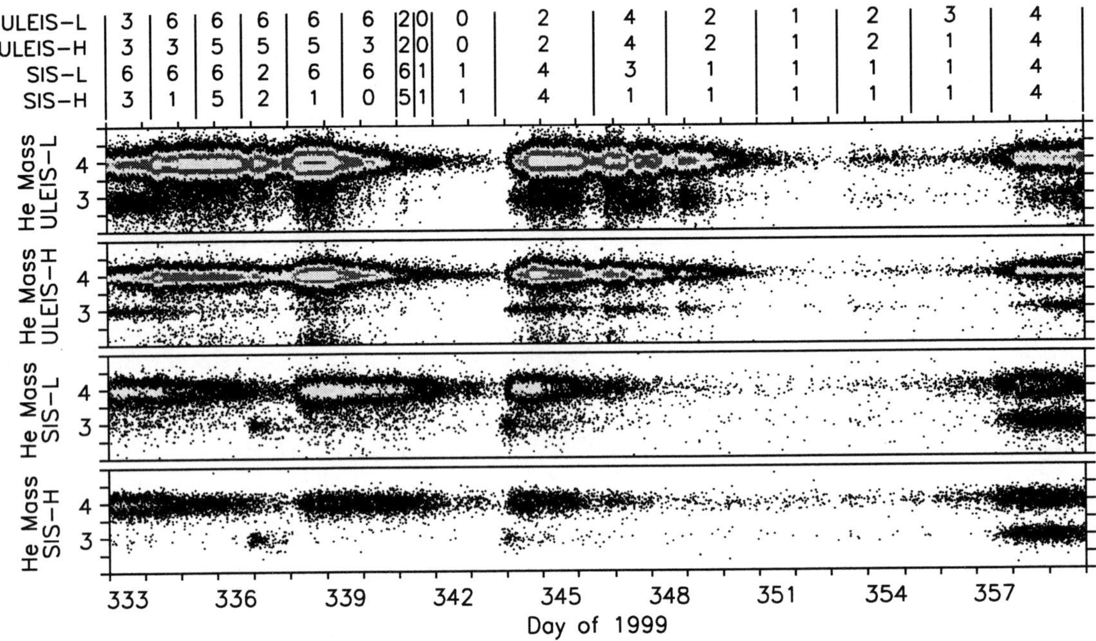

FIGURE 1. Measured He mass vs. time for detected particles in 4 energy intervals. Numbers above the plots indicate how various time intervals were classified (see text). Grey shading is used in portions of the plots where event densities are highest.

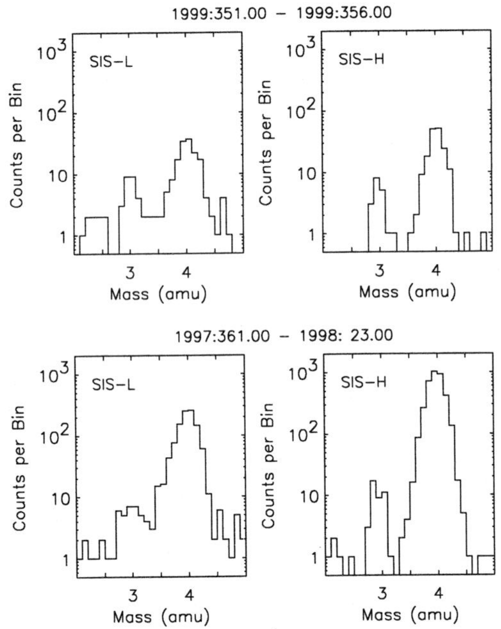

FIGURE 2. SIS He mass histograms from low-intensity periods at solar maximum (upper panels) and solar minimum (lower panels).

the spillover obscures any ^{3}He that may be present.

Using the assigned classifications, we derived limits on the fraction of the time that energetic ^{3}He is present. A very conservative lower limit is obtained by including only times which fall in category 2, 3, or 4 for at least one of the four energy intervals. In these categories there is an unambiguous ^{3}He enhancement. This limit is plotted for each Bartels rotation as unfilled circles in Figure 3. The dotted curve shows the running average over 6 Bartels rotations.

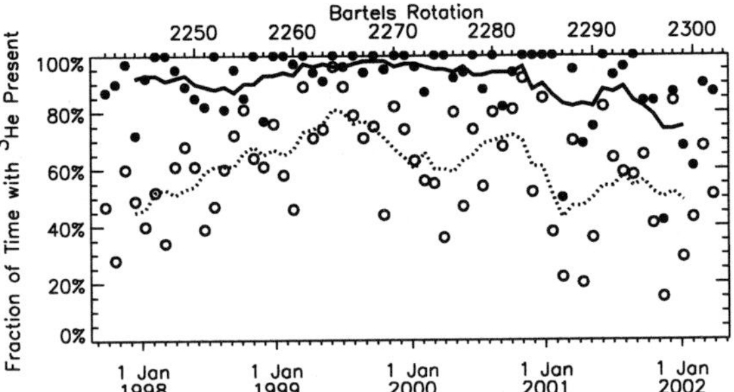

FIGURE 3. Fraction of time with ^{3}He present. Points are shown for individual Bartels rotations; lines show running average over 6 rotations. Unfilled circles and dotted line: unambiguous ^{3}He (categories 2–4); filled circles and solid line: probable ^{3}He (categories 1 and 5) also included.

A less-conservative limit is obtained by also including times in categories 1 or 5. This limit is indicated by the filled circles and solid line in Figure 3. We emphasize that these values are also best thought of as lower limits. Intervals in category 5 (as opposed to category 6) have low-enough levels of spillover that the presence of actual ^{3}He is probable. Category-1 intervals have clear ^{3}He, but at levels too low to distinguish distinct SEP events.

DISCUSSION

The more-conservative limit (dotted line) plotted in Figure 3 indicates that ^{3}He in the energy interval studied was present $\sim (60 \pm 20)\%$ of the time over the period from November 1997 to May 2002. This is similar to the fraction of interplanetary shock events in which Desai et al. [9] report ^{3}He acceleration. However, we note that the ^{3}He observed by Desai et al. is accelerated primarily from suprathermal particles less energetic than those investigated here. We also point out that when we use our looser criterion for identifying periods with ^{3}He, we find that in a sizeable fraction of the solar rotations studied there is ^{3}He present virtually all the time (filled circles plotted along the 100% line in Fig. 3).

The distribution in time, space, and energy of ^{3}He originating in impulsive SEP events is probably quite complex. As the particles propagate out from the Sun their spatial distribution broadens due to geometry, velocity dispersion, and scattering. Scattering from magnetic structures in the expanding solar wind also causes adiabatic energy loss, while shocks can increase particle energies. Thus it is not straightforward to infer from our observations the fraction of time that CME-driven shocks would encounter energetic ^{3}He at a few solar radii. Furthermore, particles could be detected from acceleration occurring anywhere that the field line connected to the observer is crossed by the shock front.

The fraction of time with ^{3}He present changed by at most a factor of ~ 2 over the 4.5 years included in our study. This relatively small change may be due, in part, to the contribution of galactic ^{3}He at MeV energies near solar minimum when there is a lower rate of SEP events. This galactic component, which should be much less significant at lower energies and at smaller heliocentric radii, probably is not making an major contribution to the source population being accelerated by interplanetary shocks.

A striking feature seen in the particle mass vs. time plots such as shown Figure 1 is the common occurrence of intervals in which ^{3}He is being detected by the SIS and/or ULEIS instruments at a rate of a only few particles per day. Previous studies [11, 12] have shown that ^{3}He energy spectra accumulated over many days of solar quite time exhibit low-energy turn-ups, indicating a solar (not galactic) origin of these particles. We suggest that the ^{3}He we are observing during such quite periods could be the result of a large number of impulsive SEP events with peak intensities too low to be detected as such with present instrument sensitivity. One gets a visual impression (Fig. 1) that the distribution of waiting times between ^{3}He particles is not the exponential expected if the flux were truly constant. A quantitative study of this distribution, which may help constrain the source of these particles, is beyond the scope of the present work.

While the present study shows that energetic ^{3}He is present in the interplanetary medium frequently enough to provide the seed material for acceleration by interplanetary shocks, further work will also be needed to establish whether the *quantity* of ^{3}He is also sufficient.

ACKNOWLEDGMENTS

This work was supported by the National Aeronautics and Space Administration at the California Institute of Technology and the Jet Propulsion Laboratory under grant NAG5-6912 and at the University of Maryland under grant PC 251429.

REFERENCES

1. Reames, D.V. 1999, Space Sci. Rev. 90, 413–491.
2. Stone, E.C. et al. 1998, Space Sci. Rev. 86, 357–408.
3. Mason, G.M. et al. 1998, Space Sci. Rev. 86, 409–448.
4. Cohen, C.M.S. et al. 1999, Geophys. Res. Lett. 26, 2697–2700.
5. Mason, G.M. et al. 1998b, Geophys. Res. Lett. 26, 141–144.
6. Mason, G.M., Mazur, J.E., and Dwyer, J.R. 1999, Astroph. J. Lett. 525, L133–L316.
7. Wiedenbeck, M.E. et al. in *Acceleration and Transport of Energetic Particles Observed in the Heliosphere*, edited by R.A. Mewaldt et al., AIP Conf. Proc. 528, AIP, New York, 2000, pp. 107–110.
8. Leske, R.A. et al. 2000, in *High Energy Solar Physics: Anticipating HESSI*, edited by R. Ramaty and N. Mandzhavidze, ASP Conf. Series, vol. 206, pp. 118–123.
9. Desai, M.I. et al. 2001, Astroph. J. Lett. 553, L89–L92.
10. Mazur, J.E., et al. 2000, Astroph. J. Lett. 532, L79–L82.
11. Richardson, I.G. et al. 1990, Astroph. J. Lett. 363, L9–L12.
12. Slocum, P.L. et al. 2002, Adv. Space Res. 30, 97–104.

Characterization of SEP events at high heliographic latitudes

S. Dalla*, A. Balogh*, S. Krucker†, A. Posner**, R. Müller-Mellin**, J.D. Anglin‡, M.Y. Hofer§, R.G. Marsden§, T.R. Sanderson§, B. Heber¶, M. Zhang|| and R.B. McKibben††

*Blackett Laboratory, Imperial College, London, UK
†Space Sciences Laboratory, University of California, Berkeley, USA
**University of Kiel, Germany
‡National Research Council of Canada, Ottawa, Canada
§Research and Scientific Support Dept. of ESA, ESTEC, The Netherlands
¶University of Osnabrück, Germany
||Florida Institute of Technology, USA
††University of Chicago, USA

Abstract. Between February 2000 and May 2002, the Ulysses spacecraft made the first ever measurements of solar energetic particles (SEPs) at high heliographic latitudes. Nine large gradual SEP events were detected at latitudes greater than 45°, their signatures being clearest at high particle energies, i.e. protons >30 MeV and electrons >0.1 MeV. In this paper we measure the onset times of Ulysses high latitude events in several energy channels, and plot them versus inverse particle speed. We repeat the procedure for near Earth observations by Wind and SOHO. Velocity dispersion is observed in all the events near Earth and in most of them at Ulysses. The plots of onset times versus inverse speed allow to derive an experimental path length and time of release from the solar atmosphere. We find that the derived path lengths at Ulysses are longer than the length of a Parker spiral magnetic field line connecting it to the Sun, by a factor between 1.2–2.7. The time of particle release from the Sun is typically between 100 and 200 mins later than the relase time derived from in-ecliptic measurements. Unlike near Earth observations, Ulysses measurements are therefore not compatible with scatter-free propagation from the Sun to the spacecraft.

INTRODUCTION

Ulysses measurements have shown that solar energetic particles (SEPs) can easily reach high latitudes [1]. Three possible explanations for these observations are as follows: (a) the CME shocks accelerating the particles extended to high latitudes and crossed the interplanetary magnetic field lines connecting to Ulysses; (b) the CME shocks did not extend to high latitudes but significant particle cross-field diffusion took place [2]; and (c) magnetic field lines connecting high latitudes with low latitude active regions existed in the solar corona, allowing particles to reach high latitudes close to the Sun [3].

Onset time analysis can provide a measurement of the distance travelled by particles prior to their arrival at a spacecraft (path length), and of their release time from the solar atmosphere. Experimental values of the path length for high latitude SEP events could be used to constrain models of particle and cosmic ray propagation.

In this paper we analyse onset times for 9 high latitude events using data from the Ulysses COSPIN experiment. We measure the onset times in several energy channels, and plot them versus c/v where v is the particle speed and c the speed of light. This allows us to verify whether dispersion is observed, and to calculate the apparent path length travelled by the particles. We compare Ulysses measurements with those by near Earth spacecraft.

DATA ANALYSIS

Figure 1 shows an overview plot of SEP measurements by the Ulysses COSPIN/KET instrument between January 2000 and May 2002. During this time Ulysses travelled over the South Pole of the Sun, then moved northwards crossing the ecliptic and passing over the North Pole. We considered times when the spacecraft latitude was >45° (the non shaded parts of Figure 1). We selected SEP events producing large count rates in the KET 38-125 MeV proton channel and obtained a list of 9 events. Details of the events, and of their associated Soft-X-Ray flares, are given in Table 1. During the events, Ulysses' distance from the Sun was between 1.63 and 3.28 AU, and for most its latitude was between 70° and 80°.

All 9 events in our list were observed near Earth. In fact, a comparison of Ulysses with IMP8 profiles shows

CP679, *Solar Wind Ten: Proceedings of the Tenth International Solar Wind Conference*,
edited by M. Velli, R. Bruno, and F. Malara

TABLE 1. SEP events at high latitudes

event n	year	date	doy	SXR onset	flare loc.	SXR class	$R_{Ulysses}$(AU)	Ulysses heliolat
1	2000	14 Jul	196	10:03	N22 W07	X 5.7	3.17	-62.1
2	2000	12 Sep	256	12:13	S17 W09	M 1.0	2.80	-70.9
3	2000	8 Nov	313	22:42	N10 W77	M 7.4	2.41	-79.3
4	2001	15 Aug	227	23:54*	-	-	1.63	+63.1
5	2001	24 Sep	267	09:36	S16 E23	X 2.6	1.90	+78.2
6	2001	4 Nov	308	16:03	N06 W18	X 1.8	2.18	+77.6
7	2001	22 Nov	326	20:22	S25 W67	M 3.8	2.31	+73.7
8	2001	26 Dec	360	04:32	N08 W54	M 7.1	2.54	+66.7
9	2002	21 Apr	111	00:59	S14 W84	X 1.5	3.28	+47.9

* onset of associated CME (no SXR flare)

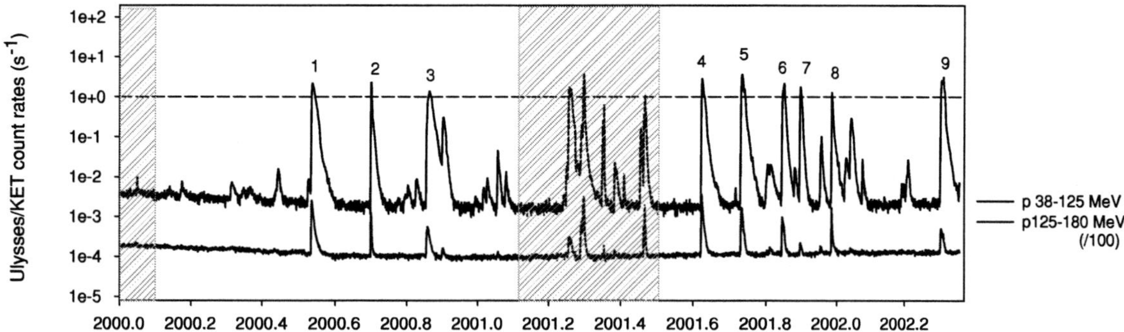

FIGURE 1. Overview of Ulysses SEP high latitude measurements, for two COSPIN/KET proton channels. Times when the spacecraft was at latitudes <45° are shaded in grey and were not considered in this study.

that at energies above ~10 MeV almost all large events near Earth produce significant increases at Ulysses [1].

Onset time analysis

For a particle of energy E, the time $t_{s/c}$ of arrival at the detecting spacecraft is given by:

$$t_{s/c}(E) = t_{Sun} + L/v(E) \qquad (1)$$

where t_{Sun} is the time of particle release at the Sun, L is the path length and v is the particle speed. The quantities t_{Sun} and L are assumed to be independent of energy in eq.(1). If this is the case, onset times plotted versus $1/v$ lie on a straigth line, whose slope gives the path length and intercept with the y-axis the time of particle release.

In their analysis of Wind/3DP data Krucker et al. measured L and t_{Sun} for several medium-sized SEP events, and found that they could be divided into 2 classes [4]. The first class consisted of events with both electron and proton dispersion compatible with path length of ~1.2 AU, but proton release delayed by about 1 hour from electron release. The second class had events with simultaneous release of protons and electrons (within uncertainties of 20 mins), but proton dispersion indicating a path length of ~2 AU.

Instruments and onset time measurement

We used energetic particle data from the Ulysses COSPIN experiment and from Wind/3DP and Soho/COSTEP near Earth. The COSPIN experiment consists of several separate energetic particle detectors. We measured electron onset times from the HET instrument (electrons 3–5 MeV). For proton measurements we have used channels from the HET, ATs and HFT. Most of the HET proton channels are contaminated by electrons during the onset of SEP events. For some of these channels we were able to subtract the electron contributions. For Ulysses, we used spin-averaged data. The

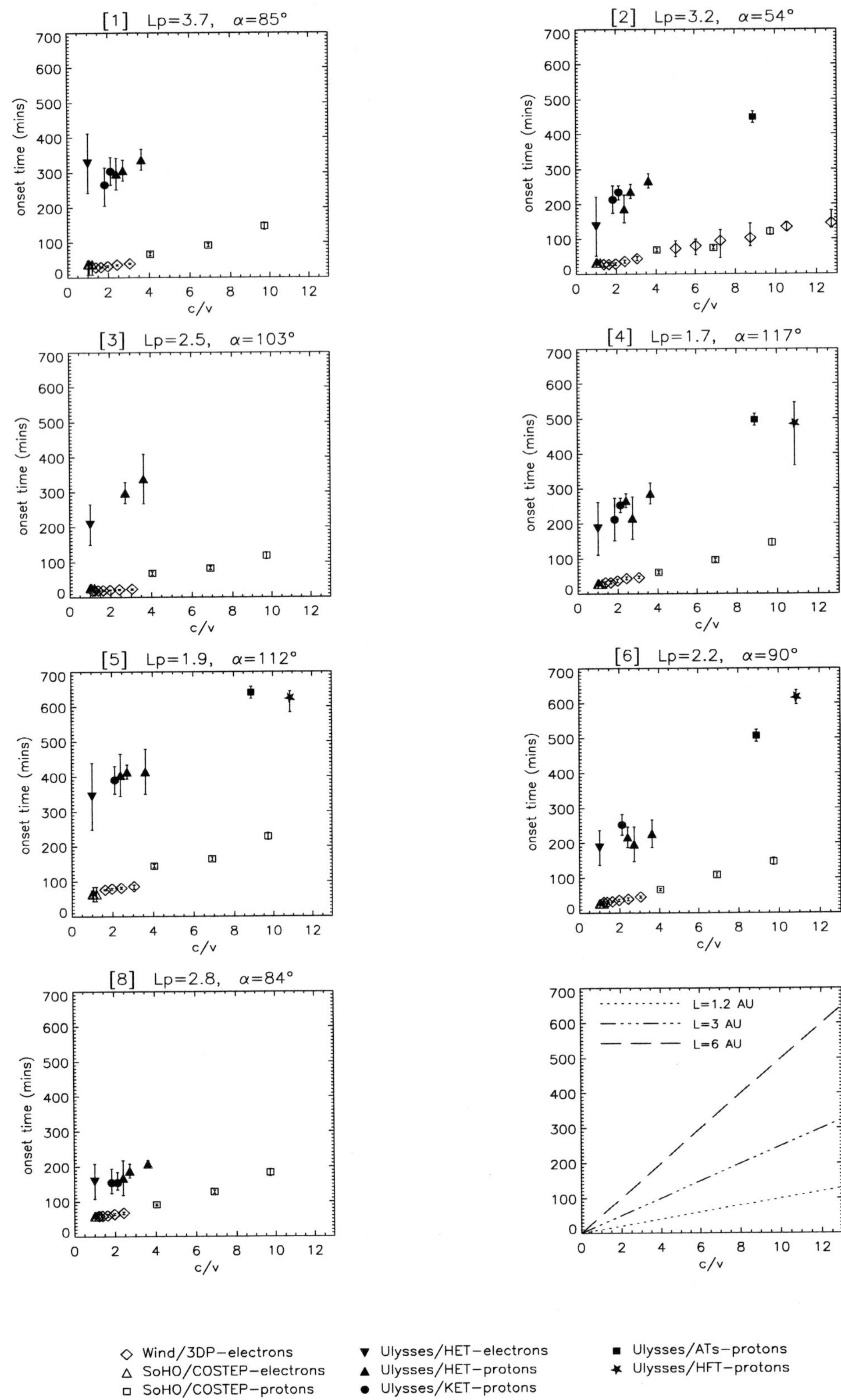

FIGURE 2. Plots of onset time versus c/v near Earth (empty symbols) and at Ulysses (filled symbols). The theoretical length L_p (in AU) of the Parker spiral to Ulysses and the angle α between the flare and Ulysses' footpoint are given. The last plot gives an indication of the dispersion curves that would be expected for 3 different path lengths and injection at t=0.

pitch angle of the particles responsible for the measured onset might not be zero, as is implied by eq.(1). This would result in the experimental path length being larger than the actual path length by a factor $1/\cos\theta$, with θ the pitch angle of the first arriving particles. A non zero pitch angle would lead to underestimating the release time. Wind/3DP electron data at zero pitch angle and Soho/COSTEP electron and proton data were used.

We determined onset times by following the procedure outlined by Krucker et al. [5]. This involves subtraction of the pre-event background and calculation of its standard deviation σ. The upper limit t_{upper} to the onset time is the time at which the particle count rate reaches 4σ. The onset time t_{onset} is the first time prior to t_{upper} for which particle counts are above zero. The error bar on each value of the onset time is given by 2 (t_{upper}-t_{onset}).

Onset times were determined from 10 or 20 min averaged data for Ulysses and from 64 s averages for Wind data. For a channel measuring particles of energy in the range between E_{min} and E_{max}, the measured onset time has been assigned to the speed $v(E_{max})$, within the assumption that the fastest particles arrive first. For the few channels for which a more accurate estimate of the most probable speed existed, its value has been used.

RESULTS AND DISCUSSION

Figure 2 shows plots of onset times versus c/v for 7 high latitude events. Empty symbols are for near Earth channels and filled symbols for Ulysses ones. In the plots, t=0 is the time of onset of the associated SXR flare.

For each event we have calculated the nominal length L_p of the Parker spiral magnetic field line that would connect Ulysses to the Sun, using the measured solar wind speed. We have also calculated the angle α between the location of the Ulysses footpoint and the solar flare associated to the event.

Both at Ulysses and near Earth, reasonable agreement in the onset times measured by different instruments is seen. For events 1,3 and 8, it was not possible to determine onset times at low energies at Ulysses. For event 1, this is due to intensities not being at background level prior to the flare, for 3 to data gaps and for 8 to very likely contamination by higher energy protons in the low energy channels. Plots of onset times are not reproduced for events 7 and 9, the former having large pre-event fluxes which make onsets unreliable, and the latter because some of the data was not yet available.

We observe that Wind/3DP electron onsets are consistent with scatter-free propagation along a field line of length ~1.2 AU. For event 3, instrumental problems are hiding the velocity dispersion. A first correction for the instrumental effect shows that also this event is consistent with scatter-free propagation.

The release time deduced from Ulysses onset times is much later than the one found from near Earth observations. For most events, the Ulysses release time is approximately 100 to 200 mins after the Wind release time. Exceptions are event 8, for which the delay in release is ~60 mins, and event 3 for which it is ~260 mins.

Path lengths deduced from the Ulysses plots are larger than the nominal length of the Parker spiral to the spacecraft. Approximate fits to the plots in Figure 2 give a ratio between L and L_{Parker} between 1.2 and 2.7.

SUMMARY

- In 6/7 of the events, the release time deduced from high latitude onsets is delayed by more than 100 minutes with respect to the release time deduced from near Earth measurements.
- Path lengths derived from Ulysses measurements are larger than the length of a Parker spiral through the spacecraft, by factors between 1.2 and 2.7.
- We conclude that onset times at high latitudes are not compatible with direct scatter-free propagation along a magnetic field line.
- The large path lengths and late release times suggest that propagation to high latitudes requires scattering.

ACKNOWLEDGMENTS

We acknowledge use of the Ulysses Data System. S.D. acknowledges support from PPARC through a Post-Doctoral Fellowship. M.Y.H. thanks ESA for the current research Fellowship.

REFERENCES

1. McKibben R.B., et al., *Ulysses COSPIN observations of the energy and charge dependence of the propagation of solar energetic particles to the Sun's south polar regions*, Proc. 27th ICRC, Hamburg, 3281 (2001); McKibben et al, *Ulysses COSPIN observations of cosmic rays and SEPs from the South Pole to the North Pole of the Sun during solar maximum*, submitted to Annales Geophys. (2002)
2. Zhang M., et al., *Ulysses observations of solar energetic particles from the July 14, 2000 event at high heliographic latitudes*, Proc. 27th ICRC, Hamburg (2001)
3. Neugebauer M., et al, *Sources of the solar wind at solar activity maximum*, J. Geophys. Res., in press (2002)
4. Krucker S., et al., *Two classes of solar proton events derived from onset time analysis*, Ap. J. 542, L61 (2000)
5. Krucker S., et al., *On the origin of impulsive electron events observed at 1 AU*, Ap. J. 519, 864 (1999)

Multi-spacecraft observations of decay phases of SEP events

S. Dalla

Blackett Laboratory, Imperial College, London, UK

Abstract. A multi-spacecraft analysis of the decay phase of 26 SEP events is presented, based upon Helios 1 and 2 and IMP8 data. The Helios spacecraft were magnetically connected to the far side of the Sun for part of their lifetime, and detected SEP events at large longitudinal separation from the location of the associated flares. In this study, 26 SEP events are considered, 19 of which observed by three spacecraft and 7 by two. For each event, the total event duration at 1 electron and 2 proton energies is measured. A plot of event duration versus the longitudinal distance $\Delta\phi$ between the associated flare location and the footpoint of the magnetic field line through the spacecraft reveals asymmetries in the detection and duration of SEP events. First, SEP events associated with flares far to the east of the spacecraft footpoint are 5 times more likely than events associated with flares far to the west. Second, the event duration shows a tendency to decrease as the location of the associated flare changes from east to west. We show that the first asymmetry is not a result of the trajectory of the spacecraft.

INTRODUCTION

Gradual SEP events have typically durations of a few days at Earth orbit. After reaching their peak, particle intensities generally show a long duration, quasi exponential decay.

For many years the decay phase of SEP events was interpreted as resulting from particle scattering by the turbulence of the interplanetary (IP) magnetic field. Particle intensity profiles were fitted by means of scattering models and values of the IP diffusion coefficient were derived from the fit. Once the CME shock acceleration paradigm for gradual SEP events became established, however, decay phase energetic particles were attributed to continuous acceleration by the CME shock [1]. By continuous acceleration it is meant that particles are still being accelerated by the shock as it travels through IP space, many days after the flare and CME took place on the Sun. There are two problems with this model: the first is that the fractionation patterns observed in gradual SEP events are not those typical of the solar wind [2] and the second is that their decay phases appear very similar at spacecraft widely separated in longitude [3]. If shock acceleration produces very different profiles at different longitudes during the onset phase, how can it give rise to nearly equal intensities in the decay phase? To explain the latter observation the concept of magnetic bottle was introduced: particles accelerated by the shock are also trapped by it [4]. Consequently, after the shock has passed by, a spacecraft starts detecting the magnetically trapped particles, a spatially nearly uniform population. An explanation along similar lines had been put forward earlier to explain the observation of nearly zero gradients in particle intensities within the inner heliosphere: these would result from trapping by magnetic barriers created by earlier solar events, resulting in particle reservoirs [5].

In this paper we investigate decay phases of SEP events using data from the Helios spacecraft. We measure the duration of 26 events at several energies and spacecraft, and plot it versus the angle between the location of the associated flare and the footpoint of the IP field line through the observing spacecraft. This reveals asymmetries in the detection and duration of SEP events.

DATA ANALYSIS

The Helios 1 and 2 spacecraft (from hereon H1 and H2) were launched in 1974 and 1976 respectively. They orbited the Sun in highly eccentric trajectories, at radial distances between 0.3 and 1 AU. For part of their lifetime the spacecraft were magnetically connected to the back of the Sun, providing in a few cases measurements of SEPs at large longitudinal separation from the sites of the associated flares, as seen from Earth.

The starting point for our study was the list of 77 Helios SEP events compiled by Kallenrode et al. [6] (from hereon K92) using data from the Cosmic Ray Particles experiment. K92 selected the events were by requiring an increase in intensity of a factor 20 above background in the 0.3–0.8 MeV electron channel, and features of velocity dispersion in the onset. A flare association was also required. Of the 77 events, 52 were classified as gradual and 25 as impulsive.

In this study we considered Helios energetic particle data for the 52 gradual events in the K92 list. Data from

CP679, *Solar Wind Ten: Proceedings of the Tenth International Solar Wind Conference*, edited by M. Velli, R. Bruno, and F. Malara

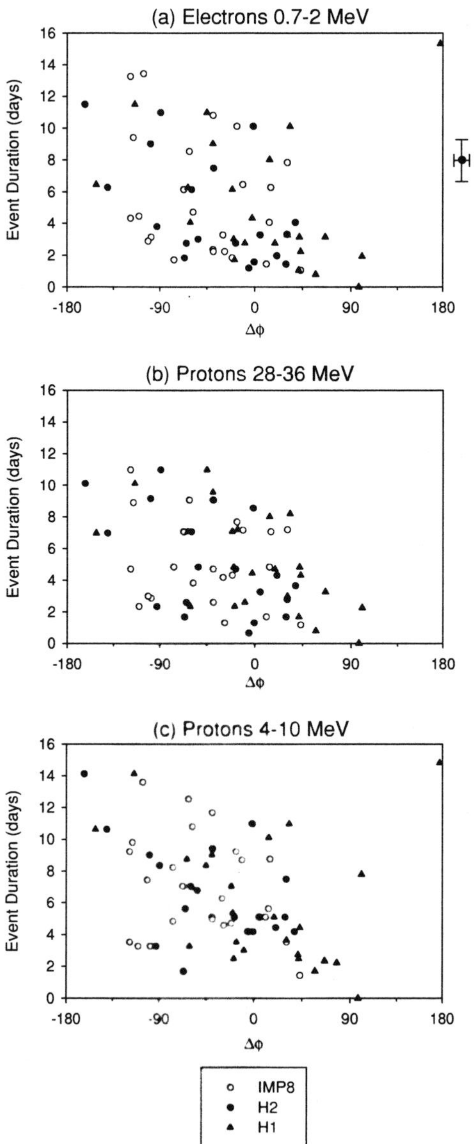

FIGURE 1. Duration of SEP events versus $\Delta\phi = \phi_{\text{flare}} - \phi_{\text{footpt}}$. Filled triangles = H1, filled circles = H2 and empty circles = IMP8. Positive values of $\Delta\phi$ mean that the flare is western (with respect to the spacecraft connection point), negative values that it is eastern. An estimate of the maximum error bar associated to a data point is shown to the right of panel (a).

H1 and H2 were complemented with IMP8 data from the GME and CRNC instruments [7, 8]. A multi-spacecraft plot of intensity profiles was produced for each event. For protons, two energy ranges were considered: low energies (Helios 4–13 MeV; IMP8/GME 4–6 MeV) and high energies (Helios 27–37 MeV; IMP8/GME 29–35 MeV). For electrons, one energy range was considered (Helios 0.8–2 MeV; IMP8/CRNC 0.7–2 MeV). After looking at the plots of the 52 events, a subset was excluded from further analysis. These were events with no flux in the selected channels, or those for which the event duration was an ambiguous quantity. For example, events with decay phase interrupted by the start of a new event were not considered. The final list for analysis comprised 26 SEP events. Of these, 19 were observed by three spacecraft (H1, H2 and IMP8) and 7 by two (H1 and IMP8).

The duration of the SEP event in the three energy channels was measured. Here the event duration is defined as the time between the onset of the SEP event and the time when intensity goes back to the pre-event level. The longitude with respect to Central Meridian of the flare associated with an event, as given in K92, is indicated as ϕ_{flare}. All but one of the 26 SEP events considered are reported by K92 to have a confident flare association. From spacecraft trajectory data, the longitude ϕ_{footpt} with respect to Central Meridian of the coronal footpoint of the nominal IP magnetic field line through the spacecraft was calculated, by using the Parker model and the actual measured solar wind speed. For the very few events for which solar wind speed measurements were not available, a speed of 430 km/s is assumed. The longitudinal separation $\Delta\phi$ between the flare location and position of magnetic footpoint is defined as $\Delta\phi = \phi_{\text{flare}} - \phi_{\text{footpt}}$. Positive values of $\Delta\phi$ indicate western flares and negative values eastern ones, where these terms are intended with respect to spacecraft footpoint rather than Central Meridian. The degree symbol in angles will be omitted from now on.

When we plotted the event duration versus $\Delta\phi$, for the electron and two proton channels, we obtained the plots shown in Figure 1. All three plots display asymmetry in the parameter $\Delta\phi$, in two ways. First, there appear to be many more SEP events in the far left of each plot than in the far right. We will refer to this in the following as the detection asymmetry. Second, the plots show a trend for the event duration to be a decreasing function of $\Delta\phi$. We will call this second asymmetry the duration asymmetry.

DISCUSSION

Detection asymmetry

The first thought that crosses one's mind when presented with the detection asymmetry is that it might result from the particular trajectories of the observing spacecraft. The value of $\Delta\phi$ for each point in Figure 1 is the result of where the flare took place on the visible disk and where the spacecraft magnetic footpoint was located (depending on the position of the spacecraft and the solar wind speed). The footpoints of H1 and H2 spanned the

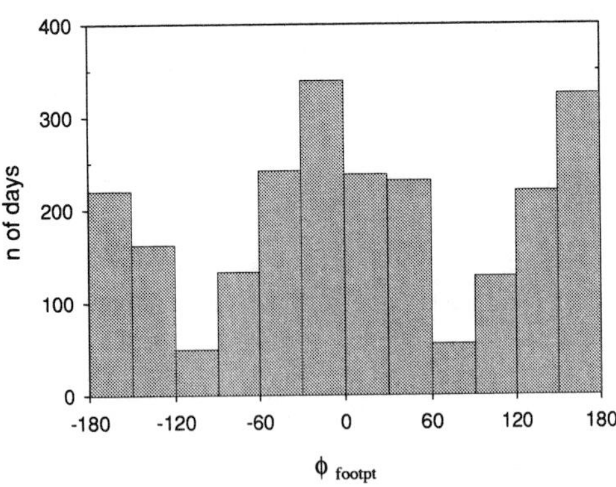

FIGURE 2. Histogram of the number of days spent by the H1 and H2 footpoints in 30° longitude bins. Central Meridian is at 0.

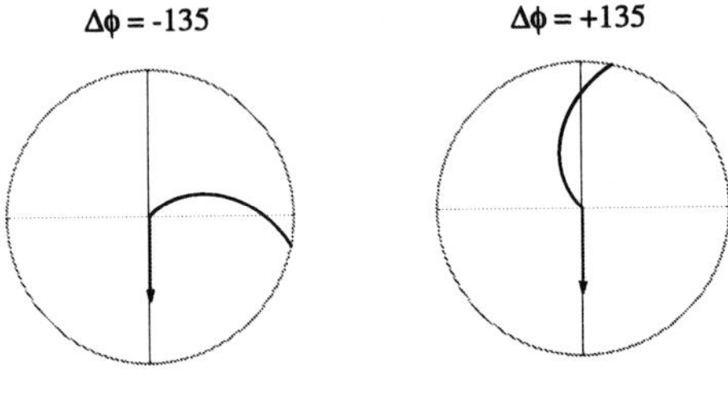

FIGURE 3. Diagram showing the geometry for the cases: $\Delta\phi$=-135 and $\Delta\phi$=+135. The arrow indicates the position of the nose of the shock with respect to the spacecraft footpoint.

entire interval of longitudes [-180,+180]. Consequently for SEP events observed by H1 and H2 any value of $\Delta\phi$ could in principle occur. For IMP8, ϕ_{footpt} takes values in the range between 30 and 70. Taking a representative value of $\phi^{IMP}_{\text{footpt}} \approx 55$, we get that $\Delta\phi$ for SEP events observed by IMP8 could only be in the range [-145,+35], i.e. IMP8 (and any other near Earth spacecraft) cannot span the far western range. We excluded IMP8 data points from Figure 1 and found that the detection asymmetry was still present: there are only 2 H1 or H2 particle events with $\Delta\phi$ in the range [+90,+180], while there are 11 events in [-180,-90] (the data point with zero duration at $\Delta\phi = 100$ corresponds to an event for which SEPs were detected at other spacecraft but not at H1).

It should be noted that the detection asymmetry was already visible in Figure 4 of K92, where the authors commented that they did not think it had a physical reason, but rather was related to the fact that data transmission from H1 and H2 was poor at times when the spacecraft footpoint was in the eastern hemisphere. This will be commented upon below. It should be pointed out that Figure 4 of K92 contains a different set of data points from our Figure 1, because in K92 for each event only data from the spacecraft closest to the flare were included. The data points of Figure 1 correspond to a subset of the K92 events, where however for each event data from another one or two spacecraft further away from the flare were added. Even with this addition the lack of SEP events associated with far western flares is still present.

To verify whether there is a bias in the Helios dataset towards observing far eastern SEP events rather than far western ones, we estimated the fraction of total observing time spent by the footpoint of H1 and H2 in several longitude bins. We calculated ϕ_{footpt} for the spacecraft from daily averaged trajectory data. To exclude solar minimum times, we restricted the time interval of the analysis to times when the smoothed monthly sunspot number was greater than 60. Furthermore, we excluded data gaps of duration >20 days in energetic particle data. The gaps were identified by yearly plots of the 0.3-0.8 MeV electron channel, which was used in K92 to select the events for study. The data gaps were mostly in H1 data, 50% of the days with no data being in 1983. SEP events at times of smaller data gaps appear to have been included in the K92 analysis.

Figure 2 shows a histogram of the number of days spent by the spacecraft connection point ϕ_{footpt} in 30 degree longitude bins at times of high solar activity. Would this distribution of ϕ_{footpt} give rise to a smaller probability for SEP events associated with far western flares? To have $\Delta\phi$ in [+90,+180], one would need a flare located to the west of the spacecraft footpoint, at longitudinal distance greater than 90. Now, flares are seen from the Earth only in the interval [-90,+90], where we recall that 0 represents Central Meridian. From Figure 2 we can see that while the Helios connection point did not spend a long time in the longitude bin [-180,-90], it did spend a considerable fraction of the total time in [+90,+180], making eastern flares during this time good candidates for a $\Delta\phi$ in [+90,+180]. This shows that far western SEP events would not require the footpoint to be in the eastern hemisphere as stated by K92, and that poor data transmission cannot be the reason for the detection asymmetry.

We conclude that there is no evidence from trajectory consideration that the Helios spacecraft would be less likely to detect an SEP event with $\Delta\phi$ in the far western regions. It cannot completely be ruled out that a particular pattern of spatial distribution of flares on the solar disk might have resulted in the low number of far western events observed. All indications are however that the detection asymmetry is the result of a physical process.

According to the current paradigm for gradual events,

SEPs originate from direct connection of the spacecraft to a CME shock front. The lack of detection of far western events can be explained by this model as resulting from the curvature of the interplanetary magnetic field lines. This can be seen in Figure 3 which shows the location of footpoint and magnetic field line to 1 AU for the cases $\Delta\phi$=-135 and $\Delta\phi$=+135. The latter case would require a shock of wider longitudinal extent than the former one, and is therefore less likely.

It should be noted however, that while the geometry for the $\Delta\phi$=+135 case appears unfavourable for detection of SEPs by the spacecraft at times when the shock is within 1 AU, eventually a connection between shock and spacecraft will be established. Enhancements starting several days after the flare would be expected from this model, but are not seen. The detection asymmetry therefore supports the idea that most of the acceleration in SEP events takes place very close to the Sun.

Duration asymmetry

The panels in Figure 1 suggest a trend for the duration of SEP events to decrease with the angle $\Delta\phi$. The longest duration events are associated with far eastern flares, with the point on the top right corner of the plots being very close to 180 and compatible with a definition of far eastern. Very few events of short duration are seen in the far eastern region of the plot, and there are no long duration events west of $\Delta\phi$=45 (apart from a single data point in panel (c)).

It is also true that there is a lot of scattering in the plots, with event durations at a fixed value of $\Delta\phi$ spanning a very wide range. On the other hand, the lack of events in the far western region of the plot (detection asymmetry) seems to be the obvious continuation of the average event duration going to zero as one moves from east to west.

The duration asymmetry is suggestive of corotation playing a very important role in the decay phase of SEP events. Therefore if the long duration of SEP events were due to a magnetic bottle effect, this would have to still be effective at times when the shock has travelled several AU into interplanetary space.

As far the interplanetary scattering model of SEP decays is concerned, it follows from Figure 1 that fitting SEP profiles with scattering models would give a larger scattering coefficient for eastern flares than for western ones, an unphysical result.

CONCLUSIONS

The main results of the analysis are as follows:

1. Based on Helios data, a spacecraft is 5 times more likely to detect SEP events from flares far to the East of its magnetic footpoint, than from flares far to the West (detection asymmetry). Only 2 far western SEP events were detected by the Helios spacecraft. These observations can be explained by the CME shock acceleration mechanism, when the curvature of the IP magnetic field lines is taken into account. They would however require a large number of shocks of very wide longitudinal extent, which are not observed at 1 AU [7].
2. The Helios data show a possible trend for the total duration of a particle event to decrease as the location of the associated flare changes from eastern to western longitudes with respect to the magnetic footpoint of the detecting spacecraft, within the range of values of $\Delta\phi$ in [-180,+90].

ACKNOWLEDGMENTS

The author acknowledges use of the NSSDC data archive. She thanks the teams of the Helios Cosmic Ray Particles, IMP8 GME and IMP8 CRNC instruments for their work in providing the data used in this paper. She thanks E. Roelof, H.V. Cane, T. Horbury and G.H. Jones for interesting discussions. Support from PPARC through a Post-Doctoral Fellowship is acknowleged.

REFERENCES

1. Reames, et al., *The spatial distribution of particles accelerated by coronal mass ejection-driven shocks*, Astrophys. J., 466, 473–486 (1996)
2. Mewaldt R.A., et al., *Are solar energetic particles an accelerated sample of solar wind?* Proc. 27th Int. Cosmic Ray Conf. Hamburg, 3132–3134 (2001)
3. McKibben R.B., *Azimuthal propagation of low-energy solar flare protons as observed by spacecraft very widely separated in azimuth*, J. Geophys. Res., 77, 3957–3984 (1972)
4. Reames, D.V., et al., *Spatial and temporal invariance in the spectra of energetic particles in gradual solar events*, Astrophys. J., 491, 414–420 (1997)
5. Roelof, E.C., et al. *Low energy solar electrons and ions observed at Ulysses February-April 1991: the inner heliosphere as a particle reservoir*, Geophys. Res. Lett., 19, 1243–1246 (1992)
6. Kallenrode, M.B., et al., *Composition and azimuthal spread of solar energetic particles from impulsive and gradual flares*, Astrophys. J. 391, 370–379 (1992)
7. Cane H.V., *The structure and evolution of interplanetary shocks and the relevance for particle acceleration*, Nucl. Phys B (Proc. Suppl.) 39A, 35–44 (1995)
8. Daibog, E.I., et al., *Decay phases in gradual and impulsive solar energetic particle events*, Proc. 27th Int. Cosmic Ray Conf. Hamburg, 3631–3634 (2001)

Transport in random magnetic fields: diffusion, subdiffusion and nonlinear second diffusion

G Qin*, W H Matthaeus† and J W Bieber†

*Department of Physics and Space Sciences, Florida Institute of Technology, Melbourne, FL 32901 USA

†Bartol Research Institute, University of Delaware, Newark, DE 19716 USA

Abstract. We present numerical results that show, first, conditions under which parallel scattering reduces the effectiveness of perpendicular scattering, leading to the phenomenon of subdiffusion ("compound diffusion"), and second, that when sufficiently strong three dimensional effects are present, true diffusion is restored, with a suppressed perpendicular diffusion coefficient that depends upon the parallel mean free path.

INTRODUCTION

Standard quasilinear scattering theory [5] provides a framework for understanding transport of charged particles in directions both parallel to, and perpendicular to, a mean magnetic field. In this approximation the two effects are necessarily distinct from one another. In reality there should be connections between parallel and perpendicular transport, and there have been suggestions concerning its nature [6, 16, 14]; however as yet these effects are incompletely understood, and the realm of different theoretical formulations has not been well established. Here we describe two effects, each of which represents departures from quasilinear theory, and which involves interplay between parallel and perpendicular transport. First, recent numerical results [13] have demonstrated that scattering of charged particles in the direction parallel to the mean magnetic field can suppress diffusive transport perpendicular to the magnetic field. In such cases transport across the mean field is subdiffusive. Second, we show, again using numerical simulations, that a regime of second diffusion can be recovered, provided that the transverse structure of the turbulence is of sufficient complexity.

Diffusion of charged particles in directions perpendicular to the large scale average magnetic field remains incompletely understood [3, 11]. Heliospheric observations relating to perpendicular transport are puzzling, as on the one hand persistence of sharp boundaries [12] suggests a diminished role of transverse diffusion, while other evidence indicates enhanced transport of charged particles to widely separated latitudes [9],requiring robust transport across the mean interplanetary spiral magnetic field. A physically appealing picture is based upon the tendency of charged particles (or, more precisely, particle gyrocenters) to follow magnetic field lines. Field lines follow a random walk, and therefore so do the particles. However, fundamental questions can be raised regarding the applicability of this Field Line Random Walk (FLRW) limit to particle transport in certain geometries, and especially when there are one or more ignorable coordinates [7]. It is also troubling that numerical computations of low energy particle transport have so far failed to confirm FLRW behavior, or for that matter, any articulated theory of perpendicular diffusive transport [3, 11].

One possible complication is that perpendicular transport might not be a diffusion process. There are at least two ways in which this can occur. First, charged particles can be trapped and therefore have bounded displacements. A variation on this is the idea that in some circumstances the field lines themselves are non-diffusive [e.g., 4, 17]. Particles trying to follow such trapped or bounded field lines would then be restricted to non-diffusive transport. A second major possibility is that parallel scattering causes charged particles to scatter back along the same or a similar field line - by "retracing their steps" the particles experience a reduction in the rate of increase of perpendicular displacements [16]. In this "compound" transport scenario, particle transport is relegated to a subdiffusive rate even if the field lines are globally diffusive. Below we review an example of the type of numerical result [13] that verifies the compound subdif-

CP679, *Solar Wind Ten: Proceedings of the Tenth International Solar Wind Conference*, edited by M. Velli, R. Bruno, and F. Malara

fusive phenomenon, which occurs for model magnetic fluctuations that are only weakly dependent on the transverse coordinates. In this regard, we can examine the discussions of compound subdiffusion [16, 8, 13] which in essence hold that the accumulation of mean square perpendicular displacement is suppressed when parallel scattering reverses particle guiding center trajectories relative to a field line. On the other hand, the particles actually sample a bundle of field lines, and these may not be identical. As simulations verify, subdiffusion ensues when the field lines sampled within this field line bundle are almost identical. However under very similar circumstances the particles might well become randomized if the field lines sampled by the particles' gyro orbits have substantial dissimilarities. Thus it is a reasonable hypothesis that perpendicular diffusion is recovered if the magnetic field lines have sufficient transverse structure. We now proceed to demonstrate both subdiffusion and the latter effect – the recovery of diffusion due to nonlinear effects.

NUMERICAL SIMULATIONS

We compute test particle trajectories in magnetic turbulence with an adaptive step fourth order Runge Kutta method with a normalized per-step-accuracy of one part in 10^9. The particles with mass m and velocity $\vec{v}$ obey

$$m\frac{d\mathbf{v}(t)}{dt} = \frac{q\mathbf{v}(t)}{c} \times \mathbf{B}(\mathbf{X}), \qquad (1)$$

where the laboratory frame electric field is neglected. The model magnetic field $\mathbf{B} = \mathbf{B}_0 + \mathbf{b}$ consists of a uniform mean magnetic field $\mathbf{B}_0$ (in the cartesian z-direction) to which is added a broad band spectrum of magnetic fluctuations. We choose a composite transverse magnetic fluctuation model $\mathbf{b} = (b_x(x,y,z), b_y(x,y,z), 0)$ consisting of a two dimensional (2D) part $\mathbf{b}^{2D}(x,y)$ and a one dimensional "slab" part $\mathbf{b}^{slab}(z)$. We control the amount of transverse structure by varying the ratio $E^{slab} : E^{2D}$. For more details see Mace et al. [11] and Qin et al. [13].

We calculate simultaneously the perpendicular and parallel diffusion coefficients employing the computed trajectories of charged particles. The calculations extend for time scales up to one thousand vt/λ_c (particle speed $\times$ time/correlation length). Both slab and 2D spectra become flat at wavenumbers k much less than the correlation scale and goes over to a $k^{-5/3}$ form at high k. We compute the running diffusion coefficients $\tilde{\kappa}_{xx} = \mathrm{d}\langle(\Delta \mathrm{x})^2\rangle/2\mathrm{dt}$ and $\tilde{\kappa}_{zz} = \mathrm{d}\langle(\Delta \mathrm{z})^2\rangle/2\mathrm{dt}$, where the time derivative is

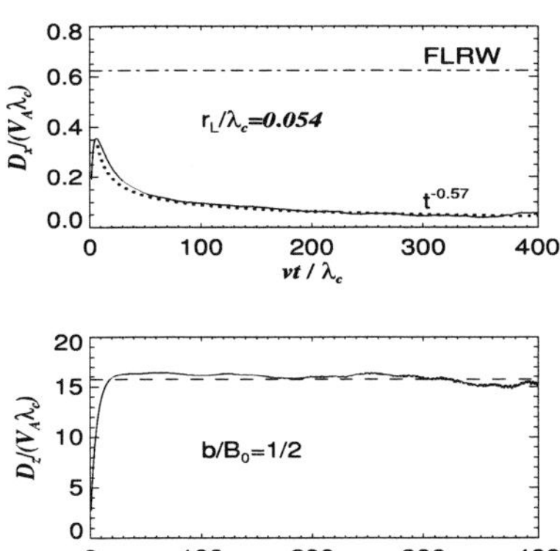

FIGURE 1. Perpendicular diffusion (upper panel) and parallel diffusion (lower panel) in nearly pure slab turbulent magnetic field ($E^{slab} : E^{2D} = 9999 : 1$) with low particle energy, $r_L/\lambda_c = 0.054$ and strong turbulence level, $b/B_0 = 0.5$. (Bottom panel) Solid line is running parallel diffusion coefficient vs. time. Dashed line is quasilinear theory result. (Top panel) Solid line is running perpendicular diffusion coefficient. Dotted line corresponds to subdiffusion. FLRW theoretical result is also shown. The running perpendicular diffusion coefficient $\sim t^{-0.57}$ in range vt/λ_c:[20,400].

computed using a first order finite difference. When the mean square displacements are diffusive ($\propto t$), the running diffusion coefficient is identical to the usual one.

NUMERICAL RESULTS

First we study test charged particle transport in nearly pure slab turbulent magnetic field, $E^{slab} : E^{2D} = 9999 : 1$ with large fluctuation amplitudes $b/B_0 = 1/2$ and small perpendicular correlation scale $\lambda_x = 0.1\lambda$, periodic box dimension, $L_z = 10000\lambda$, and the grids $N_z = 2^{22} = 4194304$.

The results are shown in Figure 1. The solid lines show perpendicular running diffusion (top panel) and parallel running diffusion (bottom panel). We can see at very short time scale running parallel diffusion performs free streaming while running perpendicular diffusion reaches its maximum value and then begins decreasing. This maximum value can be called first standard perpendicular diffusion. Then the running parallel diffusion settles into a constant value at a time scale when most particles have traveled a parallel mean free path. After parallel diffusion sets in the running perpendicular diffusion coefficient behaves very nearly as $\kappa_{xx} \propto 1/t^{1/2}$ which agrees with subdiffusion [16, 8].

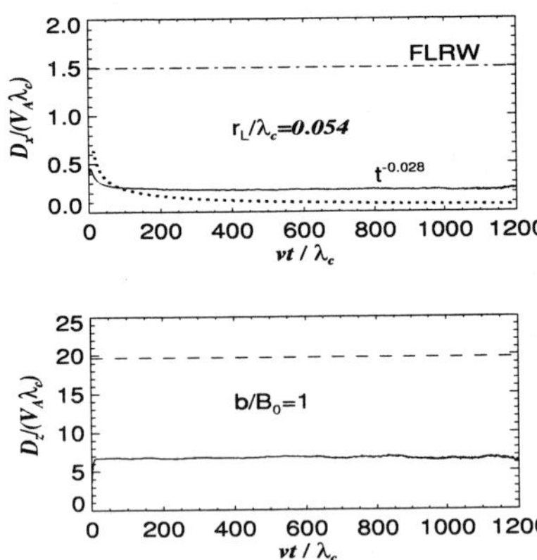

FIGURE 2. Diffusions in composite turbulent magnetic field ($E^{slab}:E^{2D}=20:80$) with low particle energy, $r_L/\lambda_c=0.054$ and strong turbulence level, $b/B_0=1$. The running perpendicular diffusion coefficient $\sim t^{-0.028}$ in range vt/λ_c:[200,1200] so the second regime of perpendicular diffusion is recovered.

The above example is typical of cases in which the turbulence is mostly of the slab type, so that the fluctuations vary weakly in the transverse directions. Nevertheless the turbulence field has no strictly ignorable coordinate [7]. The weak 2D fluctuations have a relatively small perpendicular correlation scale, and this does not favor the appearance of subdiffusion (see below). However, the result shows that on balance the particles in this example have been able to retrace their steps, thus suppressing perpendicular transport.

Now we turn to test particle diffusion in composite turbulent magnetic field with much stronger transverse structure. The spectrum of the slab and 2D model components now is chosen to have much more energy in the 2D component. In particular, we chose $E_{slab}:E_{2D}=20:80$, large fluctuation amplitudes $b^2/B_0^2=1$, and small perpendicular correlation scale $\lambda_x=0.1\lambda$. The periodic box has dimension $L_z=10000\lambda$ and $L_x=L_y=100\lambda$. The grids are $N_z=2^{22}=4194304$ and $N_x=N_y=4096$. The results are shown in Figure 2 with similar format as that of Figure 1. We can see, in composite model turbulence, when parallel diffusion sets in, the running perpendicular diffusion coefficient decreases, qualitatively as it did in the subdiffusive regime. However now it tends towards another constant having a lower level than either FLRW or the first standard perpendicular diffusion coefficient. True diffusion is recovered. We refer to this as "second diffusion." Note that the parallel diffusion is less than the standard quasilinear theory [5] result, presumably due to nonlinear effects.

DISCUSSION: SUBDIFFUSION AND SECOND DIFFUSION

Inherent in the QLT approach [5] is the assumption that the various types of transport, such as parallel and perpendicular scattering, can be computed independently. This assumption enters in QLT when the random forces are integrated along unperturbed trajectories. Without parallel scattering, if the field lines themselves randomize diffusively, the particles will also experience perpendicular diffusion. This can be explained that if the spread of field lines is at a rate of $D_\perp=\langle(\Delta x)^2\rangle/2\Delta z$ and particle speed is v, the (FLRW) diffusion of particles can be written as

$$\begin{aligned}\kappa_{xx} &= \frac{\langle(\Delta x)^2\rangle}{2\Delta t}\\ &= \frac{\Delta z}{\Delta t}\frac{\langle(\Delta x)^2\rangle}{2\Delta z}\\ &\sim vD_\perp. \end{aligned} \tag{2}$$

As long as the particles are free-streaming along field lines, the question as to whether their perpendicular transport becomes diffusive falls back on the issue of the complexity of the magnetic field viewed along field lines. Early on *Jokipii* [6] recognized that the structure of the magnetic field fluctuations perpendicular to the mean field might influence perpendicular transport, because the field lines sampled by a gyrating particle would tend to separate. In this way, *Jokipii* reasoned that FLRW perpendicular transport of particles would be accurate for low energy particles, which have gyroradii too small to sample very much transverse structure.

Skilling et al [15] carried out a calculation of cosmic ray scattering in the galaxy, noting apparently for the first time that divergence of neighboring field lines is crucial in situations in which particles' parallel scattering brings them back into the system from which they originate (in their case, scattering back into the galaxy). They noted that in order to scatter particles more than a gyroradius from their original field line, one needed to invoke field line separation. They used this idea to estimate, in effect, the distance a particle would scatter in the perpendicular direction for each unit of length transported along the magnetic field. The calculation of *Skilling et al.* is rather specialized in that it applies to a bounded inhomogeneous system (the galaxy) with parallel scattering centers located externally to the system; however, it is noteworthy that this is perhaps the first calculation in which it transpires that collisionless perpendicular transport depends explicitly upon the parallel transport.

It is not until somewhat later that it was recognized that the logical conclusion of the above reasoning is that in the limit of weak transverse structure, perpendicular transport is not diffusive. Urch [16] pointed out that particles' motions are not free streaming, but rather represent a parallel random walk, so that when particles back scatter through 90°, the perpendicular displacement is decreased. Accordingly, the estimate $\Delta z/\Delta t \sim v$ is incorrect, instead of, $\Delta z/\Delta t \approx (2\kappa_{zz}t)^{1/2}$, one finds instead that

$$\tilde{\kappa}_{xx} = D_{\perp}\sqrt{\frac{\kappa_{xx}}{\pi t}}. \tag{3}$$

(Kóta and Jokipii [8] revisited the issue of compound subdiffusion in the presence of strong parallel diffusion, and reached the same conclusion.) From this perspective, starting from weak transverse structure, what is needed is some effect to *restore* perpendicular diffusion.

The year following *Urch*, the same issue was again discussed by *Rechester and Rosenbluth* [14], but from a different perspective, namely that when particles' gyrocenters follow magnetic field lines, then the phase space density structure in directions perpendicular to the magnetic field evolves as an area preserving map. Field line wandering can make the phase space structure wrap up and fold, but without some additional effect, these surfaces cannot merge or break, so that in the presence of parallel scattering the entire process can be undone. The assertion (given essentially without proof) is that a small amount of scattering (in their case [14], presumably due to collisions) can cause the complex transverse phase space structure to blur slightly, so that parallel scattering can no longer cause the full restoration of the initial state. The conclusion [14] is that this slight additional collisional effect restores perpendicular transport to the FLRW rate, Eq. (2). These arguments are physically appealing but quantitatively inconsistent with our simulation results. There have been various calculations (e.g., [2]) that have adopted this assertion as proven, and, moreover, applicable to the collisionless limit. However to our knowledge our current line of research is the first to examine the issue of loss and restoration of perpendicular diffusion directly and quantitatively using accurate numerically determined particle orbits.

From our simulation research, examples of which are shown in the present paper, we conclude that for magnetic turbulence with little transverse structure subdiffusion is a long-lived state and mostly likely permanent state. But for magnetic turbulence with strong transverse structure, a regime of second diffusion is recovered. We view the "first" diffusion to be the evolution towards the FLRW limit. If there is no parallel scattering, this first diffusion limit is achieved (we will show evidence for this in a subsequent publication). Parallel scattering suppresses this tendency, decreasing the mean square transverse separation relative to the FLRW expectation. Subsequently one either gets subdiffusion, or if there is sufficient transverse structure, a new regime of second diffusions appears. This conclusion is qualitatively in accord with the physical reasoning of *Jokipii*, *Skilling et al*, and *Rechester and Rosenbluth*. However, none of these seems to have recognized that the (Coulomb) collisionless limit can result in a stable diffusion regime at a rate lower than FLRW, which is nevertheless manifestly nonlinear in a manner at least partially anticipated by these authors. We will present a more complete treatment of these effects in subsequent reports.

This research supported in part by NSF grants ATM-9977692, ATM-0000315, and ATM-0105254.

REFERENCES

1. Bieber, J. W., Wanner, W. and Matthaeus, W. H. 1996, *J. Geophys. Res.*, 101, 2511
2. Chandran, B. D. G., and Cowley, S. C., *Phys. Rev. Lett.*, **80** 3077 (1998)
3. Giacalone, J., and Jokipii, J. R. 1999, *Astrophys. J.*, 520, 204
4. Isichenko, M. B. 1991, *Plasma Phys. Contr. Fusion*, 33, 809
5. Jokipii, J. R. 1966, *Astrophys. J.*, 146, 480
6. Jokipii, J. R., *Astrophys. J.* **183** 1029 (1973)
7. Jokipii, J. R., Kóta, J. and Giacalone, J. 1993, *Geophys. Res. Lett.*, 20, 1759
8. Kóta, J., and Jokipii, J. R. 2000, *Astrophys.J.*, 531, 1067
9. Lario, D., Roelof, E. and Reisenfeld, D. B. 2002, AGU Spring Meeting 2002, paper SH51A-04
10. Matthaeus, W. H., Goldstein, M. L. and Roberts, D. 1990, *J. Geophys. Res.*, 95, 20, 673
11. Mace, R. L., Matthaeus, W. H. and Bieber, J. W. 2000, *Astrophys. J.*, 538, 192
12. Mazur, J. E., Mason, G. M., Dwyer, J. R., Giacalone, J., Jokipii, J. R. and Stone, E. C. 2000, *Astrophys. J.*, 532, L79
13. Qin, G., Matthaeus, W. H. and Bieber, J. W. 2002, *Geophys. Res. Lett.*, 29, 7-1
14. Rechester, A. B., and Rosenbluth, M. N., *Phys. Rev. Lett.*, **40**, 38 (1978)
15. Skilling, J., McIvor, I, and Holmes, J. A., *Mon. Not. R. Astr. Soc.*, **167**, 87P (1974)
16. Urch, I. H. 1977, *Astrophys. Space Sci.*, 46, 389
17. Zimbardo, G. and Veltri, P. 1995, *Phys. Rev. E*, 51, 1412

Charge-to-mass fractionation during injection and acceleration of suprathermal particles associated with the Bastille Day event: SOHO/CELIAS/HSTOF data

K. Bamert*, R. F. Wimmer-Schweingruber*, R. Kallenbach†, M. Hilchenbach** and B. Klecker‡

*Physikalisches Institut, University of Bern, Sidlerstrasse 5, CH-3012 Bern, Switzerland
†International Space Science Institute, Hallerstrasse 6, CH-3012 Bern, Switzerland
**Max-Planck-Institut für Aeronomie, D-37189 Katlenburg-Lindau, Germany
‡Max-Planck-Institut für Extraterrestrische Physik, D-85740 Garching, Germany

Abstract. We present SOHO/CELIAS/HSTOF data on suprathermal H, He, CNO, and Fe ions in the energy range 0.035–2 MeV/amu associated with the Bastille Day coronal mass ejection event, July 14–16, 2000. We observe a complicated evolution of the spectra in the plasma upstream from the strong interplanetary shock on July 15, in the downstream turbulent compression region, and in the magnetic cloud following the compression region. The spectra of suprathermal H, CNO, and Fe ions from the solar wind source fit a scheme of ordering by their charge-to-mass (Q/A) ratio. This scheme suggests that the ions are stochastically accelerated in the turbulence region downstream from the shock and then injected into first-order Fermi acceleration. The suprathermal He flux consists of solar wind $^4He^{2+}$ ions matching the Q/A-ordering scheme, and an additional component.

INTRODUCTION

Suprathermal particles in the solar wind plasma are the source population for particles accelerated to high energies by CME-driven shocks in gradual solar energetic events. The STOF sensor offers one of the first opportunities to verify the evolution of ions in the energy range between 35 keV/amu and a few MeV/amu. In this energy range, the injection into first-order Fermi acceleration is believed to occur.

Here, we evaluate spectra of suprathermal ions associated with the strongest interplanetary shock of the Bastille Day event. The spatial evolution of the energy spectra in the plasma upstream from the shock, in the strong downstream turbulence, and in the following magnetic cloud is analyzed.

DATA ANALYSIS

The CELIAS/HSTOF (Highly Suprathermal Time Of Flight) mass spectrometer on board SOHO measures the elemental and charge composition of ions with suprathermal energies in the range 35–2000 keV/amu. The sensor applies electrostatic deflection, time-of-flight measurement, and the detection of the residual energy in a solid state device to determine the mass, charge, and energy of an ion. A detailed description of the CELIAS sensors is given in Ref. [1].

The data analysis is based on two sets of data: (1) The Pulse Height Analysis (PHA) words indicate the time of flight, the energy stored in the solid-state stop detector, the position on the stop detector and on the microchannel plates creating the start signal, the gain flag of the solid-state detector, and the energy per charge of an ion when entering the instrument. If the particle fluxes are high (as they are for July 14–16, 2000) only a subset of PHA words is transmitted. (2) The rates are accumulated on board after a classification and selection scheme. The rates have better statistics, whereas the PHA words contain more information about a single event. We have verified that both data sets deliver consistent results.

OBSERVATIONS

On July 14, 2000 (DOY 196, Bastille Day) at 1024 UT an X-class flare, one of the brightest ever seen by SOHO, was detected by SOHO/EIT. The flare was accompagnied by a very strong energetic proton storm. At 1054 UT

CP679, *Solar Wind Ten: Proceedings of the Tenth International Solar Wind Conference*,
edited by M. Velli, R. Bruno, and F. Malara

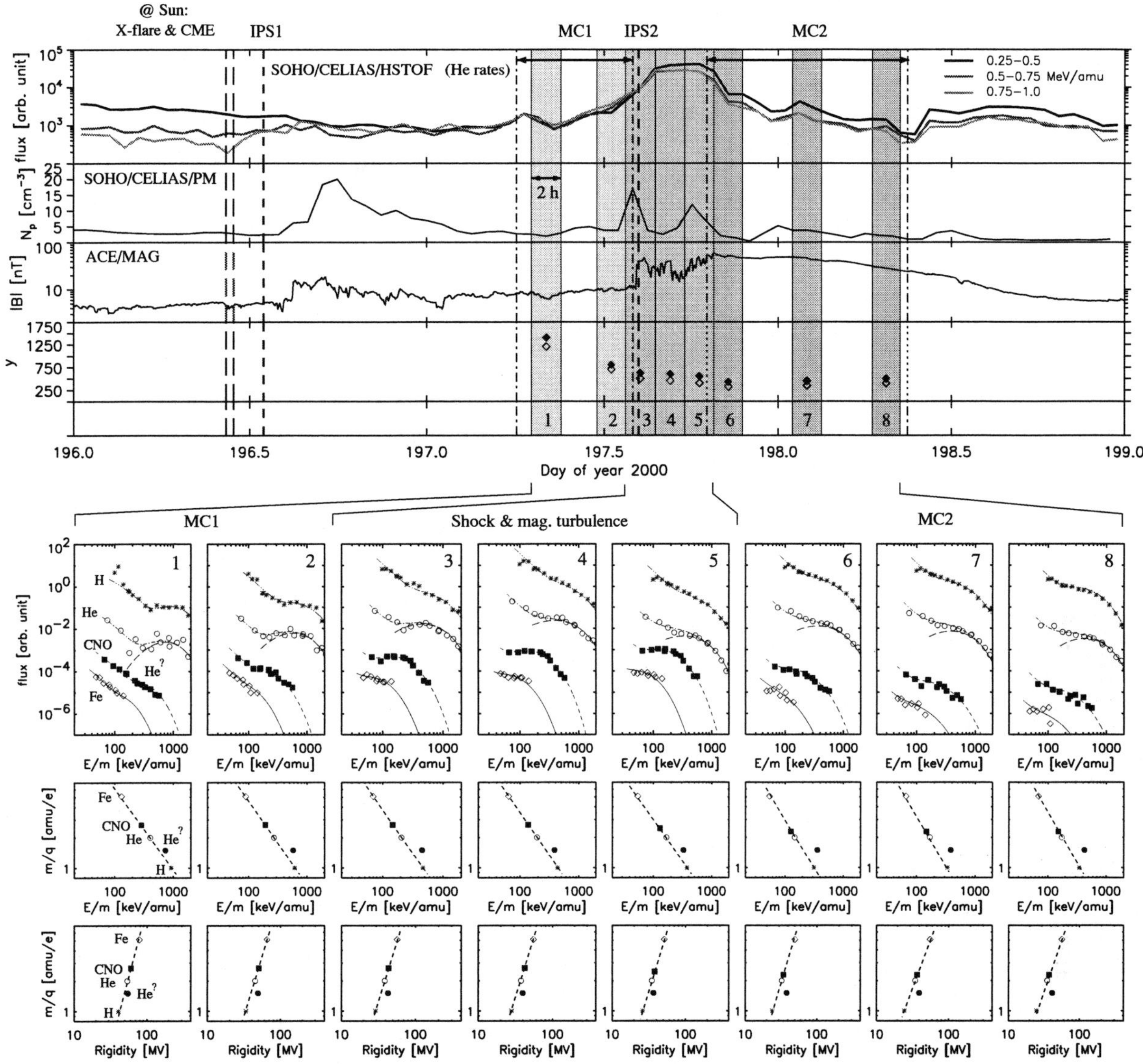

FIGURE 1. Overview of the Bastille Day event. We show from top to bottom: The He intensities in three energy ranges derived from rates data; the proton density from proton monitor (PM) data of SOHO/CELIAS (Carrington Rotation web site `http://umtof.umd.edu/pm/crn/`); the magnetic field strength from Level 2 data of the MAG experiment on board ACE (`http://www.srl.caltech.edu/ACE/ASC/level2/lvl2DATA_MAG.html`); the values for y (see Eq. 2) used to fit the spectra (open symbols: H, He^{++}, CNO, and Fe, closed symbols: $He^{?}$); the spectra of H, He, CNO and Fe (2-hour averages) during the passage of the first magnetic cloud (MC1, 1+2), during the passage of the strong interplanetary shock (IPS2), during the following strong magnetic turbulence (3–5), and during the passage of the second magnetic cloud (MC2, 6–8); and the maxima of the function f_2 for H, He^{++}, CNO and Fe, and in addition for $He^{?}$, vs. E/m and rigidity, respectively. The two interplanetary shocks are marked by dashed lines. The boundaries of the magnetic clouds are indicated by dash-dotted lines. In addition, the onsets of the X-flare and the CME as observed by SOHO/EIT and SOHO/LASCO, respectively, are marked by bold dashed lines.

the onset of an Earth-directed halo coronal mass ejection (CME) was observed by SOHO/LASCO. The CME left the Sun at a speed of ≈ 1775 km/sec. The associated solar energetic particle event was one of the strongest in this cycle. The interplanetary shock driven by the CME arrived at SOHO, which is in a halo orbit around the first Lagrangian point L1, about 28 hours later. The events during the three-day period from DOY 196 until DOY 199 are marked in Figure 1, which presents the data of HSTOF and other sensors on board SOHO and ACE.

The intensity of suprathermal He is lower before the shock passage, and higher in the magnetic turbulence after the shock. This indicates a strong locally generated suprathermal particle population. In the second magnetic cloud the fluxes drop successively to rise again after its passage. The spatial variations of the suprathermal He fluxes in the different energy ranges look qualitatively similar. In the first magnetic cloud and during the shock passage, particles with higher energies, i. e. in the ranges 0.5–0.75 and 0.75–1.0 MeV/amu, are as abundant as particles in the lowest energy range (0.25–0.5 MeV/amu). However, after the passage of the shock, in the magnetic turbulence and during the passage of the second magnetic cloud, the flux of low-energy particles is higher.

The spectra derived from the HSTOF data are fitted by a sum of two semi-empirical distribution functions, $f = f_1 + f_2$. The function f_1 describes the ions resulting from first-order Fermi acceleration with spectral index γ_1 [2]:

$$f_1 = n_1 \left(\frac{E}{m}\right)^{-\gamma_1} \exp\left(-\frac{E}{m}\frac{1}{c_1}\left(\frac{A}{Q}\right)^{\delta_1} - \frac{m}{E} d_1 \frac{Q}{A}\right), \quad (1)$$

where E/m is the energy of the particle in keV/amu. The spectral index γ_1 is about 2.4. The variables c_1 and δ_1 may be related to the rigidity-dependent escape from first-order Fermi acceleration. The second argument in the exponential term denotes the propagation of ions in the upstream magnetic cloud. The variable d_1 depends on the distance to the shock. It presumably results from a spatial integral over the plasma bulk speed divided by the diffusion coefficient [2].

The function f_2 describes the suprathermal population generated in the strong magnetic turbulence:

$$f_2 = n_2 \cdot \left(\frac{E}{m}\right)^{-\gamma_2} \cdot \exp\left(-\frac{E}{m} \cdot \frac{1}{c_2} \cdot \left(\frac{A}{Q}\right)^{\delta_2}\right) \cdot$$

$$\cdot \exp\left(-\left(\frac{E}{m}\right)^{-\beta_2} \cdot y \cdot \left(\frac{Q}{A}\right)^{\alpha_2}\right). \quad (2)$$

The parameters c_2, δ_2, and γ_2 describe spectra resulting from stochastic acceleration and are presumably related to the plasma conditions in the downstream turbulence and in the downstream magnetic cloud, e.g. the relative amplitude of the magnetic fluctuations and their spectral distribution [3]. The same applies for the parameters β_2 and α_2. The parameter y changes with distance from the turbulence region. It may in part represent the plasma conditions in the turbulence region, but also describes diffusion away from the turbulence region. The exponents β_2 and α_2 also seem to be valid to describe the spatial diffusion in the plasma.

The suprathermal ion spectra (differential flux) described by Eq. 2 have a maximum between 60 keV/amu and 1 MeV/amu for H, He^{++}, CNO and Fe (Fig. 1). The position of this maximum on the energy-per-mass scale and on the rigidity scale varies spatially and for ion species with different mass-per-charge ratio A/Q. For ions with small A/Q the maxima are at higher energies per mass than for ions with large A/Q. The energy per mass of maximum differential flux is higher in the downstream turbulence region than in the downstream magnetic cloud. Before the shock passage the dispersion in A/Q becomes larger with increasing distance from the shock. Presumably, the ion species are fractionated due to propagation effects which depend on the charge-to-mass ratio of the ions. We also observe this behavior during the April 4–7, 2000 CME event which served as a reference. For future purpose, we note that the parameter $\beta_2 \approx 1$, and c_2 and y have been determined to be ≈ 500 keV/amu and ≈ 250 keV/amu, respectively, in the downstream turbulence region.

The component $He^?$ in the spectra represents a special species, which is not identified unambiguously, yet. The distribution functions of $He^?$ measured by HSTOF do not fit the A/Q ordering scheme derived from the other species, neither under the assumption that these ions are $^4He^{++}$, nor under the assumption that they are $^4He^+$ or $^3He^{++}$ ions. The mass spectra and studies on other CME events suggest that $He^?$ is $^3He^{++}$ from an impulsive flare.

MODEL

The empirical functions used to fit the spectra match a model [3] applying theoretical considerations of Refs. [2, 4, 5]. It combines the mechanisms of stochastic acceleration in the turbulence near interplanetary shocks, first-order Fermi acceleration, and stationary diffusion in the field irregularities of magnetic clouds. The data suggest that the model may be valid, if the spectral index of the magnetic field irregularities is about $n \approx 1$.

DISCUSSION

The observations on ions in the energy range 0.035 – 2 MeV/amu associated with the Bastille Day event and theoretical considerations suggest that the dominant suprathermal particle population is the one generated by stochastic acceleration in the turbulent compression region between the strong interplanetary shock and the magnetic cloud downstream from the shock. The evolution of spectra during the event indicates that 1) most ions of H, He, CNO, and Fe have solar wind charge states and are stochastically accelerated in the downstream turbulence region. If this turbulence is one-dimensional Alfvénic turbulence, its spectral index is $n \approx 1$. 2) This seed population is injected into first-order Fermi acceleration at the shock.

Upstream from the shock, the ions resulting from the first-order Fermi process dominate. They propagate through the upstream magnetic cloud with magnetic field irregularities that may have a spectral index $n_{\mathrm{MC1}} \approx 1$, pending an analysis of magnetic field data and pending an evaluation of other models on spatial and momentum diffusion of suprathermal ions in the solar wind plasma. Downstream from the shock, mainly the suprathermal population resulting from stochastic acceleration is observed. These ions diffuse from the turbulence region into the downstream magnetic cloud. The spectra near this cloud indicate that the suprathermal ions of CNO and Fe have somewhat higher charge than usually observed in the solar wind. This may be due to contributions from the source of bulk ions confined in the downstream magnetic cloud [6].

The He spectra show a signature in addition to the signatures of solar wind ions. Possibly, this signature is due to suprathermal $^3He^{++}$ ions from an impulsive flare.

ACKNOWLEDGMENTS

The STOF sensor onboard the SOlar and Heliospheric Observatory (SOHO), a joint project between ESA and NASA, is part of the Charge, ELement, and Isotope Analysis System (CELIAS), which is a joint effort of five hardware institutions under the direction of the Max-Planck-Institut für extraterrestrische Physik (pre-launch) and the University of Bern (UoB; post-launch). The Max-Planck-Institut für extraterrestrische Physik is the prime hardware institution for (H)STOF. The University of Bern provided the entrance system. The data processing unit (DPU) was provided by the Technical University of Braunschweig.

This work was supported by the Swiss National Science Foundation.

REFERENCES

1. Hovestadt, D., Hilchenbach, M., Bürgi, A., Klecker, B., Laeverenz, P., Scholer, M., Grünwaldt, H., Axford, W. I., Livi, S., Marsch, E., Wilken, B., Winterhoff, H. P., Ipavich, F. M., Bedini, P., Coplan, M. A., Galvin, A. B., Gloeckler, G., Bochsler, P., Balsiger, H., Fischer, J., Geiss, J., Kallenbach, R., Wurz, P., Reiche, K.-U., Gliem, F., Judge, D. L., Ogawa, H. S., Hsieh, K. H., Möbius, E., Lee, M. A., Managadze, G. G., Verigin, M. I., and Neugebauer, M., *Sol. Phys.*, **162**, 441–481 (1995).
2. Forman, M. A., and Webb, G. M., *Washington DC American Geophysical Union Geophysical Monograph Series*, **35**, 91–114 (1985).
3. Kallenbach, R., Bamert, K., and Wimmer-Schweingruber, R. F., "Charge-to-mass Fractionation of Suprathermal Ions Associated with Interplanetary CMEs", in *these proceedings*, 2002.
4. Hasselmann, K., and Wibberenz, G., *Zeitschrift für Geophysik*, **34**, 353 (1968).
5. Möbius, E., Scholer, M., Hovestadt, D., Klecker, B., and Gloeckler, G., *Astrophys. J.*, **259**, 397–410 (1982).
6. Neukomm, R. O., and Bochsler, P., *Astrophys. J.*, **465**, 462–472 (1996).

Charge-to-mass Fractionation of Suprathermal Ions Associated with Interplanetary CMEs

R. Kallenbach*, K. Bamert† and R.F. Wimmer-Schweingruber†

*International Space Science Institute, Hallerstrasse 6, CH-3012 Bern, Switzerland
†Physikalisches Institut, University of Bern, Sidlerstrasse 5, CH-3012 Bern, Switzerland

Abstract. A model for the acceleration and transport of suprathermal ions associated with interplanetary coronal mass ejections (CMEs) is presented. The combined mechanisms of stochastic acceleration in the turbulence near interplanetary shocks, first-order Fermi acceleration, stationary spatial diffusion in field irregularities of magnetic clouds, and time-dependent propagation in the upstream solar wind are described. Ions from the bulk solar wind are considered as source populations of suprathermal ions. The results of the model are compared to the spectra of suprathermal ions observed with SOHO/CELIAS/HSTOF during the Bastille Day CME of 14–16 July 2000.

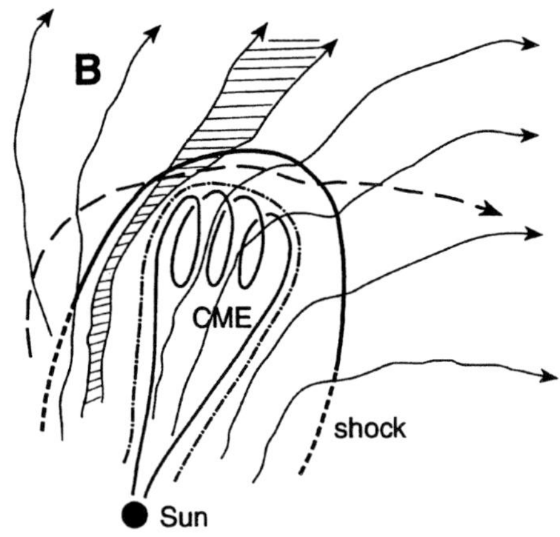

FIGURE 1. Configuration of a CME [1].

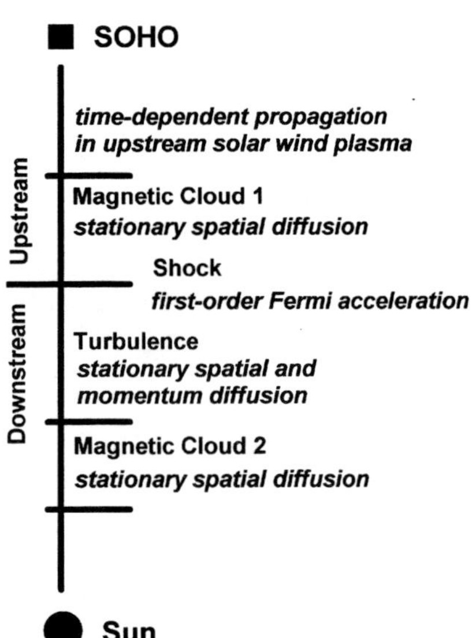

FIGURE 2. Plasma regions associated with the Bastille Day CME and the dominant processes effecting the spatial and momentum distribution of suprathermal ions (not to scale).

INTRODUCTION

Solar energetic particle (SEP) events are commonly classified as gradual or impulsive. Energetic ions associated with gradual events are believed to be accelerated near interplanetary shocks driven by CMEs. Very often, the ejecta of the CME propagate into the solar wind in form of magnetic clouds with spiral field configurations (Figure 1). The compression region between the shock and the cloud contains magnetic turbulence. The Bastille Day event is a CME with multiple shocks and magnetic clouds. We evaluate the suprathermal ion distributions near the main shock, which is preceded by another magnetic cloud upstream (MC1) in addition to the magnetic cloud (MC2) following the downstream turbulence.

The processes of stochastic acceleration of solar wind ions i.e. momentum diffusion in magnetic field irregularities, stationary spatial diffusion in the turbulence region and inside the magnetic clouds, time-dependent propagation in the upstream solar wind plasma, and first-order Fermi acceleration are considered (Figure 2). The ion energy range between 35 keV/amu and 2 MeV/amu is studied, the energy range in which the CELIAS/HSTOF instrument is sensitive.

CP679, *Solar Wind Ten: Proceedings of the Tenth International Solar Wind Conference*, edited by M. Velli, R. Bruno, and F. Malara

THE TRANSPORT EQUATION

The evolution of suprathermal ion populations near a parallel shock between upstream and downstream plasma with bulk speeds $|\mathbf{V}| \approx V_1$ and $|\mathbf{V}| \approx V_2$, respectively, is described by the standard transport equation:

$$\begin{aligned} \frac{\partial f}{\partial t} + \mathbf{V}\cdot\nabla f &= \nabla\cdot(\kappa\nabla f) + I - S + \frac{p}{3}\frac{\partial f}{\partial p}\nabla\cdot\mathbf{V} \\ &+ \frac{1}{p^2}\frac{\partial}{\partial p}\left(p^2 D_{pp}\frac{\partial f}{\partial p}\right), \end{aligned} \quad (1)$$

where $p = v/v_0$ with v_0 the speed of 1 MeV protons. The left hand side denotes the explicit time dependence and convection of suprathermal ions, whereas the terms on the right hand side describe spatial diffusion, sources (I), sinks (S), adiabatic deceleration, and momentum diffusion. For symmetrically counter-propagating Alfvén waves and isotropic distribution functions f, spatial and momentum diffusion [2] are related by

$$D_{pp}\kappa = \frac{1}{9}V_A^2 p^2,\ V_A^2 = \frac{B^2}{\mu_0 m_p n_p},\ D_{pp} = p^{-2}D_{TT}, \quad (2)$$

with V_A the Alfvén speed, B the ambient magnetic field, and n_p the proton density. In the hard-sphere approximation [3], the energy diffusion parameter is

$$\begin{aligned} D_{TT} &= \frac{4}{9}\frac{V_A^2}{\kappa_0}\left(\frac{Q}{A}\right)^{2-\alpha} T^{(\alpha+1)/2}, \\ \kappa_0 &= 6\times 10^5 \frac{B}{\delta B^2}\,\mathrm{Tm^2 s^{-1}}, \end{aligned} \quad (3)$$

where δB is the amplitude of the magnetic field irregularities with power spectral density $P = P_0 f^{-\alpha}$ resonating with the ion of atomic mass A and charge Q. This yields

$$\begin{aligned} D_{pp} &= \frac{4}{9}\frac{V_A^2}{\kappa_0}\left(\frac{Q}{A}\right)^{2-\alpha} p^{\alpha-1} \text{ and} \\ \kappa &= \frac{1}{4}\kappa_0\left(\frac{Q}{A}\right)^{\alpha-2} p^{3-\alpha}. \end{aligned} \quad (4)$$

Stationary Spatial and Energy Diffusion

If the time scales for acceleration and diffusion of suprathermal ions in the downstream turbulence near the main shock are shorter than the dynamical time scales of the CME structure, the explicite time derivative, the convection term for an observer in the downstream plasma frame, and the adiabatic deceleration term of the transport equation may be neglected. It is further assumed that sources and sinks of the populations in the suprathermal energy range studied here can be neglected, i.e. injection into the process of spatial and energy diffusion occurs at lower energies, the solar wind bulk energies. Finally, it is assumed that one spatial coordinate x is sufficient to describe the situation of a parallel shock. We obtain a partial differential equation solvable by separating the variables i.e. $f(x,p) = g(x)h(p)$:

$$\frac{\partial}{\partial x}\left(\kappa\frac{\partial f}{\partial x}\right) + \frac{1}{p^2}\frac{\partial}{\partial p}\left(p^2 D_{pp}\frac{\partial f}{\partial p}\right) = 0, \quad (5)$$

All Q/A- and p-dependent terms are incorporated into the momentum part of the equation, whereas all x-dependent parts are absorbed in the spatial differential equation. Both parts must equal a constant $\pm C$ because x and p are independent variables. Therefore,

$$\frac{\partial^2 h}{\partial p^2} + \frac{\alpha+1}{p}\frac{\partial h}{\partial p} - C\left(\frac{A}{Q}\right)^{4-2\alpha} p^{4-2\alpha}h = 0. \quad (6)$$

Neglecting terms with factors of order p^{-2F-2} and p^{-4F-2}, which fall off more rapidly than others for $\alpha < 3$ and for the parameters of our experimental situation (which needs to be checked if applied to another problem), we find

$$\begin{aligned} h(p) &= h_0 p^{-2\beta}\exp\left(-e_{ro}p^{2F} - Gp^{2H}\right), \\ \beta &= \frac{3}{4},\ e_{ro} = \frac{\sqrt{C}}{2F}\left(\frac{A}{Q}\right)^{2-\alpha}, \\ F &= \frac{3-\alpha}{2},\ G = \frac{9-6\alpha}{32F^2 e_{ro}},\ H = -F. \end{aligned} \quad (7)$$

The spatial differential equation becomes

$$\frac{9\kappa_0}{4V_A^2}\frac{\partial}{\partial x}\left(\frac{\kappa_0}{4}\frac{\partial g(x)}{\partial x}\right) = -Cg(x). \quad (8)$$

With the new dimensionless variable

$$\xi := \int_0^x \frac{4V_A}{3\kappa_0}dx' \quad (9)$$

the differential equation is transformed to

$$\frac{\partial^2 g(\xi)}{\partial \xi^2} + \frac{1}{V_A}\frac{\partial V_A}{\partial \xi}\frac{\partial g(\xi)}{\partial \xi} + Cg(\xi) = 0. \quad (10)$$

If, for example, $V_A = V_{A,0}\exp\left[-\xi^2/(2\xi_A^2)\right]$, then the intensity profile of the suprathermal ion flux, while passing the CME's turbulence region, is

$$g(\xi) = g_0\exp\left(-\frac{\xi^2}{2\xi_0^2}\right) \quad (11)$$

with $\xi_0 = \xi_A$. The constant C is identified as $1/\xi_0^2$, and, hence, is related to the scale length L (in meters) of the turbulence region by

$$C \approx \frac{1}{\xi_0^2} \approx \frac{9\kappa_0^2}{16L^2V_A^2}. \quad (12)$$

Stationary Spatial Diffusion

The magnetic clouds of the Bastille Day event also contain magnetic field irregularities, although with smaller relative amplitude $|\delta B/B|$ than in the turbulence region near the shock. Thus, suprathermal ions are produced locally in the clouds to less extent, but they diffuse from the turbulence region into the clouds. This boundary condition prohibits removing the p- and Q/A-dependence in the spatial differential equation

$$V_{\mathrm{MC1(2)}}\frac{\partial f}{\partial x}=\frac{\partial}{\partial x}\left[\frac{1}{4}\kappa_0\left(\frac{Q}{A}\right)^{\alpha-2}p^{3-\alpha}\frac{\partial f}{\partial x}\right], \qquad (13)$$

where $V_{\mathrm{MC1(2)}}$ is the bulk speed of the magnetic cloud in the plasma frame of the turbulence region, which has, within about one Alfvén speed, the shock speed.

In steady state, the bulk speeds $V_{\mathrm{MC1(2)}}$ are important parameters in the spatial diffusion profile [4]. With $x_{1(2)}$ as the location of the boundary of the magnetic cloud towards the shock or turbulence region, we find

$$\begin{aligned} f(x,p) &= f(x_{1(2)},p)\exp(-\eta), \\ \eta &:= \int_{x_{1(2)}}^{x}\frac{4p^{\alpha-3}V_{\mathrm{MC1(2)}}}{\kappa_0}\left(\frac{Q}{A}\right)^{2-\alpha}dx', \end{aligned} \qquad (14)$$

where $f(x_{1(2)},p)$ is the spectrum evaluated e.g. in the previous subsection. The parameter η describes the stationary dispersion in speed and in Q/A of the ions, while diffusing away from the source i.e. the turbulence region. The speed V_{MC2} of the downstream magnetic cloud in the shock frame is probably not much larger than the Alfvén speed, whereas V_{MC1} is quite large, on the order of the upstream plasma speed V_1.

Time-Dependent Solution

The propagation of ions in the upstream solar wind plasma is explicitely time-dependent as the distance between the CME event and the SOHO spacecraft changes. The process is described in the SOHO spacecraft frame, in which the solar wind bulk plasma speed is V_{SW}. Only the propagation along the direction x of SOHO's magnetic connection to the CME is considered:

$$\frac{\partial f}{\partial t}+V_{\mathrm{SW}}\frac{\partial f}{\partial x}=\frac{\partial}{\partial x}\left[\kappa\frac{\partial f}{\partial x}\right]. \qquad (15)$$

For a spatially constant diffusion parameter κ in the upstream solar wind, the solution's kernel is

$$K(x,t)=K_0\frac{1}{\sqrt{\kappa t}}\exp\left[-\frac{(x-V_{\mathrm{SW}}t)^2}{4\kappa t}\right]. \qquad (16)$$

The suprathermal ions are emitted from the CME as a moving source. The earliest arrival time for the ions from the CME spaced by $x=x_{\mathrm{MC1}}-x_{\mathrm{SOHO}}$, with x_{MC1} the boundary of the upstream magnetic cloud towards SOHO, is $t=x/v$. We integrate the source term over time from x/v to large t, when the CME was closer to the Sun ($t\approx\infty$). We neglect variations in κ and the Δt of the arrival times due to diffusive effects. This yields

$$f(\zeta)=f_{\mathrm{MC1}}\left[1-\mathrm{erf}(\zeta)\right], \quad \zeta=\sqrt{\frac{vx}{4\kappa}}. \qquad (17)$$

The population f_{MC1} is the one at x_{MC1}.

First-order Fermi Acceleration

The injection into first-order Fermi acceleration [4] of the suprathermal ions that are stochastically accelerated in the downstream turbulence is given by

$$f_{\mathrm{F}}(0,p)=\frac{h_0}{p^{\gamma}}\int_{p_{\mathrm{inj}}}^{p}dp'\frac{\exp\left(-e_{\mathrm{ro}}p'^{2F}-Gp'^{2H}\right)}{p'^{1+2\beta-\gamma}}. \qquad (18)$$

This expression yields a power law $f_{\mathrm{F}}(0,p)=f_{\mathrm{F},0}p^{-\gamma}$ with spectral index $\gamma=3V_1/|V_1-V_2|$, if $p\gg p_{\mathrm{inj}}$. If the latter is not the case, the expression becomes a more complicated function of momentum p.

A problem may arise because the ions that participate in the first-order Fermi process change momentum during one excursion in the turbulence region downstream. The momentum diffusion is $\Delta p\approx\sqrt{D_{pp}t}$, whereas the spatial diffusion is $\Delta x\approx\sqrt{\kappa t}$. The expectation value of the time for an ion to return to the shock from the turbulence region is $\tau_{\mathrm{r}}\approx\Delta x/V_2\approx\sqrt{\kappa\tau_{\mathrm{r}}}/V_2$, where V_2 is the downstream plasma speed in the shock frame. In a first-order Fermi process, the momentum change $\Delta p(\tau_{\mathrm{r}})$ by diffusion should be smaller than the momentum gain $\Delta P\approx 2|V_1-V_2|/v_0$. Therefore,

$$\Delta p(\tau_r)\approx\frac{\sqrt{D_{pp}\kappa}}{V_2}=\frac{V_{\mathrm{A}}}{3V_2}p<\Delta P. \qquad (19)$$

The ratio V_{A}/V_2 is of order 1, and ΔP is of order 0.1, so that the shock acceleration is a pure first-order Fermi process up to energies of ≈ 100 keV/amu. This estimate is only valid for an exactly parallel shock, where perpendicular diffusion and gyro-motion of the ions do not contribute to the ion return to the shock. Therefore, in reality the problem should be less severe, i.e. pure first-order Fermi acceleration can be expected up to higher energies.

Further parameters are relevant for the processes at the shock and in the turbulence region, such as the mean free path λ of the ions, the shock width L_{sh}, the return time scale τ_r, the shock width L_{sh}, and the injection threshold p_{inj} (the momentum for which the ion mean free path equals the shock width). The shock width L_{sh} typically has a value between the electron inertial length $L_e = c/\omega_{pe}$ and the proton inertial length $L_p = c/\omega_{pp}$, where $\omega^2_{pe(p)} = n_{e(p)}e^2/(4\pi\varepsilon_0 m_{e(p)})$. For $B \approx 10\,\mathrm{nT}$, $B^2/\delta B^2 \approx 100$, $\alpha = 1$, $L \approx 10^{10}\,\mathrm{m}$, $n_p \approx n_e \approx 10^7\,\mathrm{m}^{-3}$, and $V_2 \approx V_A \approx 7 \times 10^4\,\mathrm{m/s}$ the values for these parameters are:

$$\begin{aligned}
\tau_r &\approx \frac{\kappa}{V_2^2} \approx \frac{\kappa_0}{4V_2^2}\left(\frac{Q}{A}\right)^{\alpha-2} p^{3-\alpha} \approx 3 \times 10^5\,\mathrm{s} \times \frac{A}{Q}p^2; \\
\lambda &\approx \frac{3\kappa}{v_0 p} \approx 4 \times 10^8\,\mathrm{m} \times \frac{A}{Q}p\,; \\
p_{inj} &\approx \frac{4cv_0}{3\omega_{pp}\kappa_0}\frac{Q}{A}\sqrt{\frac{L_{sh}}{L_p}} \approx 10^{-3}\frac{Q}{A}\sqrt{\frac{L_{sh}}{L_p}}\,; \\
\tau_L &\approx \frac{L}{v} \approx \frac{L}{v_0 p} \approx \frac{10^3\,\mathrm{s}}{p}\,.
\end{aligned} \tag{20}$$

The estimate for the injection threshold certainly is too low because the model does neither describe the shock potential nor the ion motion of low-energy particles on the shock length scale. The momentum for downstream escape, however, can be derived from $\tau_r \approx \tau_L$ as

$$p_{esc} \approx 0.14\left(\frac{Q}{A}\right)^{1/3}. \tag{21}$$

This estimate shows that first-order Fermi accelerated ions should be observable upstream and downstream from the shock.

SOHO/CELIAS/HSTOF DATA

The suprathermal ions of the Bastille Day event have been observed at 1 AU with SOHO/CELIAS/HSTOF [5]. The spectra are fitted by a sum of two distribution functions. One of them represents the ions stochastically accelerated in the turbulence downstream of the shock. They diffuse into the two magnetic clouds upstream and downstream and farther out. Near the shock, they are also injected into first-order Fermi acceleration. This results in a second population which also diffuses into the magnetic clouds and farther out.

Within the experimental uncertainties of about 20%, the model is consistent with the spectra of suprathermal H, CNO, and Fe ions assuming that 1) the ions have solar wind charge states, 2) the turbulence in the compression region downstream from the shock is Alfvénic with spectral index $\alpha \approx 1$, and 3) the field irregularities in the magnetic clouds have a spectral index $\alpha \approx 1$ as well. The model needs to be verified by an evaluation of ACE/MAG magnetic field data. Solar wind turbulence very often has a Kolmogorov type power spectral index $\alpha \approx 5/3$, but proton-beam generated waves near the shock may have a different spectral index. Other theories for the diffusion coefficients also need to be compared to the data.

For the case of He, the spectra match the model at lower energies, if one assumes that the He ions are solar wind $^4He^{++}$. However, at higher energies, there appears to be an additional component, which may be $^3He^{++}$ from impulsive flares [5].

CONCLUSIONS

The presented model and data from the HSTOF sensor on board SOHO suggest that solar wind ions are pre-accelerated by momentum diffusion in the turbulence region downstream from the shock and subsequently injected into first-order Fermi acceleration. The efficiency for the momentum diffusion scales with the charge-to-mass ratio of the ions. The propagation of the suprathermal ions through the magnetic field irregularities shows the signatures of dispersion in speed and in the charge-to-mass ratios of the ions.

ACKNOWLEDGMENTS

This work was supported by the Swiss National Science Foundation and by the INTAS grant WP 270.

REFERENCES

1. Lee, M.A., Particle acceleration and transport at CME-driven shocks, *in:* N.U. Crooker, J.A. Joselyn, and J. Feynman (eds.), *Coronal Mass Ejections*, AGU Press, p. 227, 1997.
2. Hasselmann, K., and Wibberenz, G., Scattering of charged particles by random electromagnetic fields, *Zeitschrift für Geophysik, 34*, 353, 1968.
3. Möbius, E., Scholer, M., Hovestadt, D., Klecker, B., and Gloeckler, G., Comparison of helium and heavy ion spectra in He-3-rich solar flares with Model calculations based on stochastic Fermi acceleration in Alfvén turbulence, *Astrophys. J., 259*, 397, 1982.
4. Forman, M.A., and G.M. Webb, Acceleration of Energetic Particles, *in:* R.G. Stone and B.T. Tsurutani (eds.), *Collisionless Shocks in the Heliosphere: A Tutorial Review*, AGU, Washington, D.C., 1985, p. 91.
5. Bamert, K., Wimmer-Schweingruber, R.F., Kallenbach, R., Hilchenbach, M., and Klecker, B., Charge-to-mass fractionation during injection and acceleration of suprathermal particles associated with the Bastille Day event: SOHO/CELIAS/HSTOF data, these proceedings.

Solar Wind High-Speeds Observed Near the Earth

V.M.Silbergleit [+,*]

[+]Departamento de Física, FIUBA, Av.Paseo Colón 850, Piso 2 – C1063ACV Capital Federal – Argentina
[*]CONICET of Argentina

Abstract. To predict the occurrence of major solar wind velocities near the Earth, hourly solar wind speed magnitudes from November 1963 to May 2000 are considered by applying Gumbel's first distribution. According to the present study a maximum value equal to (1017 ± 50) km/sec is expected to observe during the current solar cycle as a consequence of this result, we infer the possibility to detect intense geomagnetic storms on the Earth.

INTRODUCTION

During active periods, fast streams of energetic particles are ejected from the place of a flare, a coronal hole, or other variable solar-surface CME region. Studies of the statistical properties of the solar wind were understanding using available spacecraft data., most from spacecraft near Earth orbit. Interesting results were obtained focalizing the correlations between plasma parameters and the solar cycle variations. Solar cycle variations were observed in the solar wind speed and pressure [3].

Dynamic processes on the Sun release a plasma charged particles (mainly protons and electrons) and associated fields to the Earth's environment, being an important cause of geomagnetic disturbances at the Earth's surface. Large nonrecurrent geomagnetic storms, shock wave perturbations in the solar wind, and energetic particle events in interplanetary medium frequently appear in close connection with coronal mass ejections. Geomagnetic activity is vitalized by high solar wind speed and by a southward direction of the interplanetary magnetic field, in solar- coordinates.

Sudden commencements (SCs) are the sign of one of the most important facts that are a consequence of the interaction of the solar wind with the Earth's magnetosphere.

Siscoe [4], studied Gumbel's first distribution to the first, second and third largest geomagnetic storms in nine solar cycles. He used the average half-daily of the aa indices.

To study the prediction of the major solar wind speeds Gumbel's [2] asymptotic distribution extreme values is considered.

SOLAR WIND SPEED PEAKS

The temporal distribution of hourly solar wind speed values (V) from November 2, 1963 to May 31, 2000 as obtained by NASA home page is considered. For each event, the solar wind speed peaks (Vi, here i =1, ...n is the number if each solar cycle) is used. The data set, is constructed by:

i- The compilation of hourly data from November 2, 1963 to May 31, 2000 as they are presented via homepage of NASA at http://www. NASA.com/

CP679, *Solar Wind Ten: Proceedings of the Tenth International Solar Wind Conference*,
edited by M. Velli, R. Bruno, and F. Malara

ii- The data processing is applied to digital hourly V values (to obtain the higher Vi values as selected in part i).

iii- The selection procedure to obtain the maximum Vi peaks related to each solar cycle (using the data recorded in ii) is considered.

EXTREME VALUE DISTRIBUTION

Gumbel's first distribution includes most credible distributions, such as the normal, lognormal and exponential distributions. The distribution function is given by Gumbel [2]:

$$G_1(x) = \exp\{-\exp[-(M_1 x - M_2)]\} = \exp\{-\exp[-\alpha(x-m)]\} \quad (1)$$

where m is the mode and A, B and α are constants. The best fitting adjustment obtained for solar wind speed peaks considering the Gumbel's asymptotic distribution give us the parameter values $M_1 = (156 \pm 3)10^{-4}$ sec/km and $M_2 = 15.4 \pm 0.3$.

Extreme values of the absolute magnitude of the Vi (with i = 1,2 and 3) are obtained by using the Vi values detected for each solar cycle and considering these data in ascending order. Each observed value was assigned a probability according to the mean ranking method as published by Gringorten [1]. A frequently used approximation is:

$$P_i = (i-0.44)(N+0.12)^{-1} \quad (2)$$
with $i = 1,2...N$

where i is the ordinal number related to the observed value.

The return intervals T(x) between such extreme values are calculated by using the expression:

$$T(x) = [1-\exp\{-\exp[-(M_1 x - M_2)]\}]^{-1} \quad (3)$$

T(x) is the expected time required to have one event with the extreme equal to or exceeding x.

The median value is estimated by plotting the observations (Fig. 1 shows the ascending curveT(x) and the descending one $T(x)[T(x)-1]^{-1}$).
When the ordinate is equal to 2, the abscissa is equal to the median value (mv).

The standard statistical parameters of the variable $(M_1 x - M_2)$ are found from the

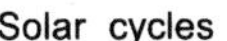
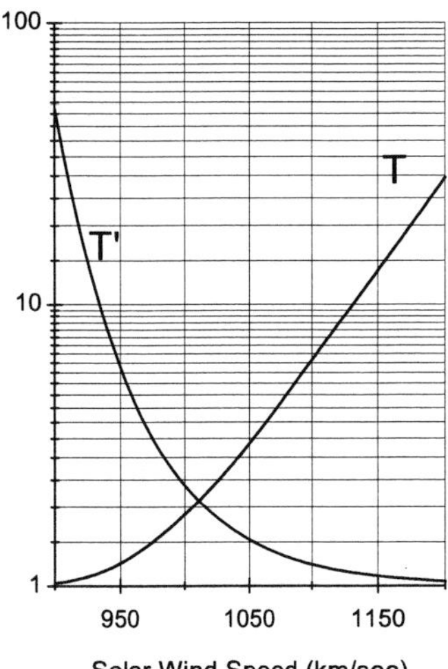

FIGURE 1. The descending /ascending branches show the waited number of periods required to detect one with extremes less than /equal to or exceeding x.

distribution function [2]. The arithmetic mean (am, which is the average of all possible results), is related to m by the expression:

$$am = m + 0.57722\, A^{-1} \quad (4)$$

The relative dispersion (rd) acording to Siscoe [4] is :

$$rd = \pi (A\sqrt{6}\, m)^{-1} \quad (5)$$

The statistical characteristics of the largest Vi peaks per solar cycle obtained are: m = 987 km/sec, mv = 1012 km/sec, am = 1024 km/sec and rd = 0.083 km/sec.

The hourly maximum Vi values for the last three solar cycles are: 951 km/sec, 1021 km/sec and 1090 km/sec. Although it is interesting to note that the level of mean hourly Vi peaks is now much higher that some years ago.

CONCLUSION

The results obtained to study the Vi magnitudes are shown in Fig. 1.

For the three cycles studied, the solar wind velocity peaks considered are equal to 951 km/sec, 1021 km/sec and 1090 km/sec respectively, showing that the level of magnitude from cycle to cycle was increasing. According to the present study we conclude that during the current solar cycle (23) it is predicted an observed maximum value of the hourly solar wind speed between 967 and 1067 km/sec, this interval indicates a predicted value less that the maximum observed during the prior solar cycle but sufficiently strong to cause important geomagnetic disturbances.

The predicted values obtained for the solar wind speed peaks indicate that the solar wind particles are expected to be numerous enough and fast to modify the Earth´s magnetic field and cause effects on Earth.

To increase the confident results more extended data intervals are demanded.

ACKNOWLEDGMENTS

This work was partially supported by agreements of Facultad de Ingenieria (Universidad de Buenos Aires) and CONICET of Argentina.

REFERENCES

1. Gringorten, ,I. I., A Plotting Rule for Extreme Probability Paper. *J. Geophys. Res.* *68*, *813, (1963).*
2. Gumbel, E. J., Statistical Theory of Extreme Values and Some Practical Applications. *Nat. Bur. Standards Appl. Math. Ser.* *33*, *Washington D.C., 51,.(1954).*
3. Neugebauer, M., Large scale solar cycle variations of the solar wind. *Space Sci. Rev.* *17*, *221, (1975).*
4. Siscoe, G. L., On the Statistics of the Largest Geomagnetic Storms per Solar Cycle. *J. Geophys. Res.*, *81*, *4782, (1976).*

V.M.Silbergleit, Departamento de Física de la Facultad de Ingeniería (UBA) and CONICET. Av.
Paseo Colon 850 - C1063ACV-Buenos Aires - Argentina
Fax: 54 11 49 63 20 62.
email: vsilber@fi.uba.ar

Session V

Dynamic Activity of the Corona and Heliosphere

Spatial Relationship of Signatures of Interplanetary Coronal Mass Ejections

I. G. Richardson[1], H. V. Cane[1], S. T. Lepri[2], T. H. Zurbuchen[2], and J. T. Gosling[3]

[1]*Laboratory for High Energy Astrophysics, NASA/Goddard Space Flight Center, Greenbelt, MD 20771, USA;*
[2]*Dept. of Atmospheric, Oceanic and Space Sciences, University of Michigan, Ann Arbor, MI 48109, USA;*
[3]*Los Alamos National Laboratory, Los Alamos, NM 87545, USA*

Abstract. Interplanetary coronal mass ejections (ICMEs) are characterized by a number of signatures. In particular, we examine the relationship between Fe charge states and other signatures during ICMEs in solar cycle 23. Though enhanced Fe charge states characterize many ICMEs, average charge states vary from event to event, are more likely to be enhanced in faster or flare-related ICMEs, and do not appear to depend on whether the ICME is a magnetic cloud.

INTRODUCTION

Interplanetary coronal mass ejections (ICMEs), the counterparts in the solar wind of coronal mass ejections (CMEs) at the Sun, are characterized by various signatures (e.g., 1, 2, and references therein). These include: abnormally low solar wind proton temperatures (T_p), bi-directional suprathermal electron strahls (BDEs), low plasma β, helium abundance enhancements, low magnetic field variances, non-Parker-spiral field directions and, in a subset of events, the enhanced, smoothly-rotating magnetic fields and low β plasma defining "magnetic clouds". Energetic particle signatures include intensity depressions (cosmic ray Forbush decreases) and bi-directional flows at ~ MeV to cosmic ray energies (e.g., 3). Each ICME is unique to some extent and most do not exhibit all of the characteristic signatures. Thus, there is no obvious "necessary and sufficient" condition for the identification of an ICME. In addition, as has been well known for many years (e.g., 4), the boundaries of the various signatures may not be co-located. Hence, a number of signatures should be compared in order to arrive at a consensus. From considering these various ICME signatures, we have identified ~200 probable ICMEs in the near-Earth solar wind since beginning of the current solar cycle in 1996 until the end of 2001. Detailed observations of plasma composition anomalies in many of these ICMEs are available from the SWICS instrument on ACE. Recently, Lepri et al. (5) have reported that intervals of enhanced iron charge states with durations ≥ 20 hours observed in 1998-mid 2000 were associated with ~50% of the ICMEs that we identified independently during this period. The high Fe charge states, which remain unchanged ("frozen-in") as the solar wind propagates through the heliosphere, are consistent with coronal temperatures often exceeding ~2 MK and rapid plasma expansion during CME initiation. The emphasis in this paper is on examining further the relationship between enhanced Fe charge states and ICMEs, including their relative boundaries. Since a significant fraction of ICMEs do not appear to be associated with Fe events as defined by (5), we will consider whether high Fe charge states are present in other ICMEs but for intervals < 20 hours, whether they are absent in some ICMEs, and whether the characteristics of the ICME or the related solar event (such as the occurrence of a flare) may have any bearing on the Fe charge states inside an ICME.

OBSERVATIONS

The ICME of September 25-26, 1998 (Figure 1) is an example of a "classic", well-studied, event. It is not particularly typical however in that it exhibits a wide range of signatures that are essentially co-located. The main panel shows 5-minute averaged ACE observations of the magnetic field intensity and polar

CP679, *Solar Wind Ten: Proceedings of the Tenth International Solar Wind Conference*, edited by M. Velli, R. Bruno, and F. Malara

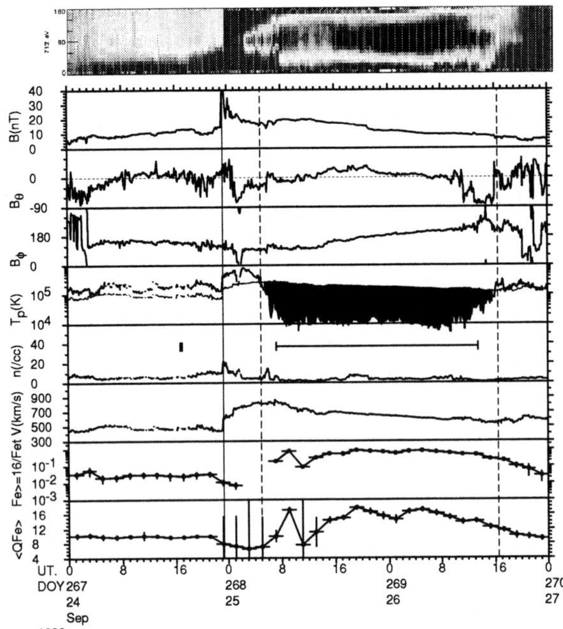

FIGURE 1. An ICME on 25-26 September, 1998 which drove a shock (S) ahead of it (solid vertical line). BIF = bi-directional ~1 MeV ions observed by IMP 8 (6). The top panel shows pitch-angle distributions for 712 eV solar wind electrons.

and azimuthal angles; the plasma proton temperature, density and speed; and two-hour averages of the ratio $Fe\geq16+/Fe_{tot}$, and the mean Fe charge state. In the T_p panel, the second trace indicates the "expected temperature" for normally-expanding solar wind (T_{ex}), which is correlated with the solar wind speed (2). Shading indicates when $T_p < 0.5T_{ex}$, frequently a signature of ICMEs (2). This fast (~700 km/s) ICME generated a strong shock ahead of it and originated in association with an M7/3B flare at N18°E09° on September 23. The vertical dashed lines are the ICME boundaries suggested by the low T_p region. A discontinuity terminating an interval of southward magnetic field is evident at the trailing edge. The magnetic field in the ICME is relatively smooth and rotates slowly in azimuth. The enhanced Fe charge states coincide with this region, to within the 2-hour resolution of the measurements, and reach $Fe\geq16+/Fe_{tot}$ ~1 and $\langle Q_{Fe}\rangle$~16, indicating freeze-in temperatures T_{Fe} ~7 MK (5). The top panel in Figure 1 shows pitch-angle distributions of 712 eV solar wind electrons from the ACE/SWEPAM experiment. Bands at the top and bottom of the plot (clearer in the colored original) indicate electrons flowing in both directions along the magnetic field, i.e., a BDE. This starts ~3 hours ahead of the ICME leading edge based on T_p and is consistent with an abrupt decrease in the > 60 MeV cosmic ray intensity observed at IMP 8 (6), suggesting that entry into a closed magnetic structure, and the true

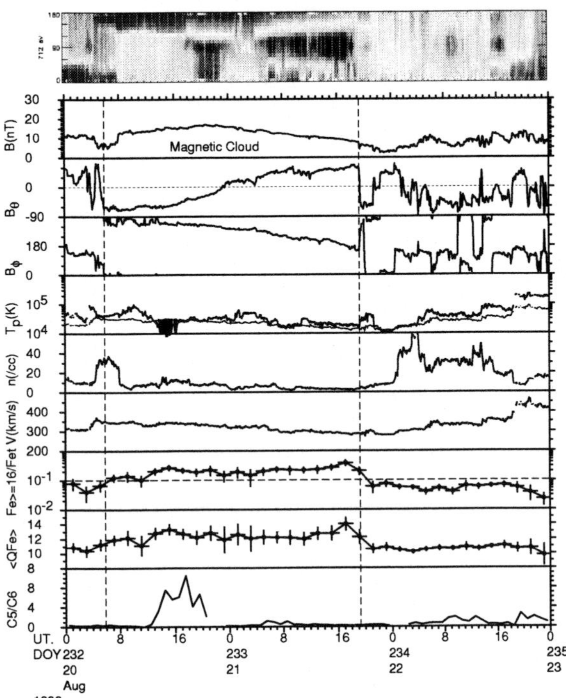

FIGURE 2. An ICME on 20-21 August, 1998 with a "magnetic cloud" configuration.

ICME leading edge, most likely occurred at this time. The interval of bi-directional ~1 MeV ion flows observed by IMP 8 in this ICME (6) is indicated by "BIF" in Figure 1.

An ICME on August 20-21, 1998 with the enhanced, smoothly rotating magnetic field characteristic of a magnetic cloud (MC) is shown in Figure 2. The MC interval has $Fe\geq16+/Fe_{tot}\geq 0.1$. However, this ratio (~0.2-0.3) and $\langle Q_{Fe}\rangle$ (~12-13; T_{Fe} ~1.5 MK) are significantly less than during the ICME in Figure 1. During most of the MC, T_p~T_{ex}. The brief period of low T_p (shaded) clearly does not denote the true extent of the ICME. The 712 eV electron distribution (top panel) indicates BDEs during a high density/depressed field region near the MC leading edge and intermittently during the trailing half of the MC. Elsewhere in the MC, the absence of BDEs suggests that magnetic fields were open (7). An interesting feature is the peak in the C^5/C^6 ratio associated with the low T_p region. Similar features are seen in some other ICMEs (see Figure 3) and suggest that occasionally, regions of low T_p at 1 AU may reflect low coronal temperatures during ICME initiation, though this needs further investigation.

Another ICME (on August 26-27, 1998) is shown in Figure 3. This relatively fast ICME (associated with an X1/3B flare at N35°E09° on August 24) also generated a shock ahead of it. The ICME interval is

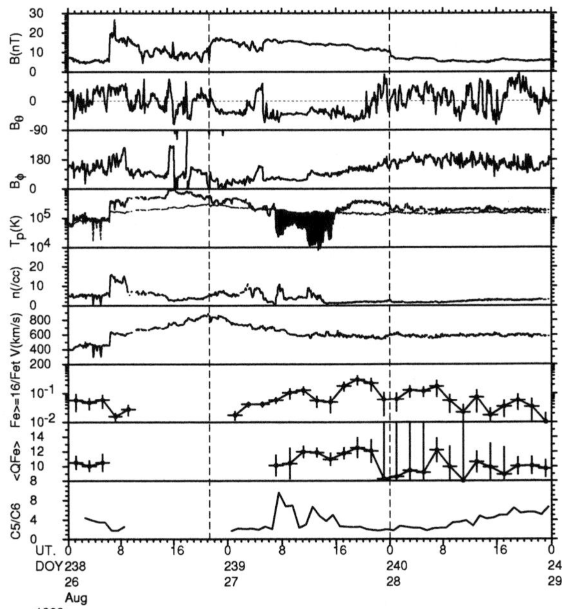

FIGURE 3. An ICME on 26-27 August, 1998 that generated a shock ahead of it. Note the lack of high Fe charge states compared to the ICME in Figure 1.

principally inferred from the enhanced, relatively low-variance magnetic field commencing ~15 hours following the shock. Abnormally-low *Tp* only occupies the center third of this interval and, as in Figure 2, is associated with enhanced C^5/C^6. Only a ~6 hour period inside the ICME trailing edge has $Fe{\geq}16+/Fe_{tot} \geq 0.1$, and average charge states within

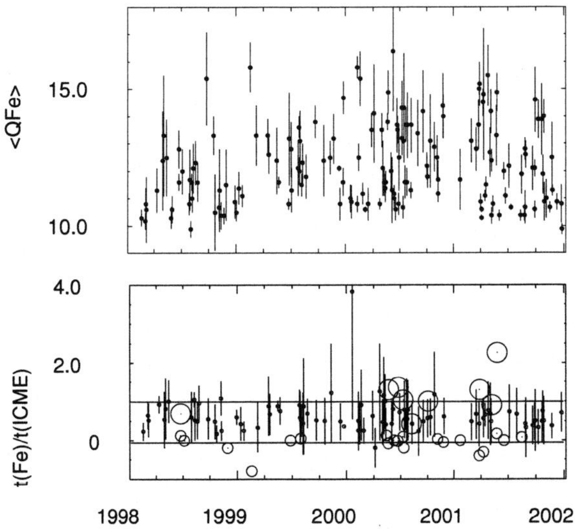

FIGURE 4. (a) Mean Fe charge states (average of 2-hr averages) in 157 probable ICMEs in 1998-2001. (b) Times of intervals with $Fe{\geq}16+/Fe_{tot} \geq 0.1$ relative to the ICME interval (0-1). Small (large) circles = start (end) time only available.

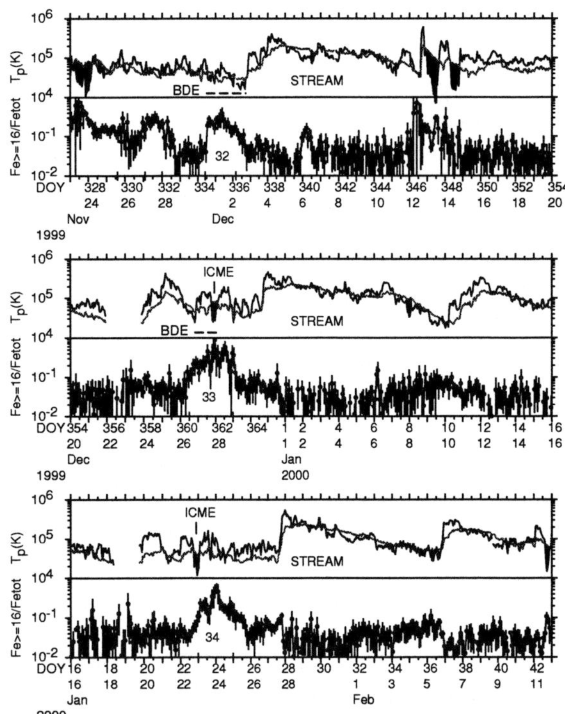

FIGURE 5. The middle panel shows an ICME embedded in an extended high Fe charge-state interval which appears to recur at the 27-day solar-rotation period (top and bottom panels). T_p and T_{exp} are shown in addition to $Fe{\geq}16+/Fe_{tot}$. Fe event numbers are from (4). BDEs are also indicated.

the ICME are 10-12 ($T_{Fe} \leq$ 1.5 MK). Though not shown here, a BDE was also present in this ICME.

The above observations suggest that Fe charge states vary from event to event, and enhanced charge states do not characterize every ICME. Figure 4(a) shows $<Q_{Fe}>$ (mean of 2-hr averages) during 157 probable ICMEs in 1998-2001. The all-event average is 12.2±1.5, only slightly above the average for all solar wind during this period (11.1±1). In 22 events (14%), $<Q_{Fe}> \geq 14$ ($T_{Fe}>$ 2 MK). A preliminary examination suggests that 10 of these high charge state ICMEs have associations with solar flares (C7-X6). A further 8 have probable associations, one was associated with a LASCO halo CME with no reported flare, and the associations are unclear for 2 events. Finally, one ICME was embedded in an extended high charge state event (December 26-28, 1999; middle panel of Figure 5) that appears to be part of a sequence of such events, nos. 28, and 32-34 of (5), that recur at approximately the solar rotation period and precede a corotating high-speed stream. Events 32-34 are shown in Figure 5. Event 32 was coincident with an extended BDE. A BDE was also present during the first part of Event 33, extending to the end of the low T_p signature.

To examine the spatial relationship between ICME boundaries (defined by a consensus of signatures) and enhanced Fe charge states, Figure 4(b) summarizes the locations of intervals with $Fe{\geq}16+/Fe_{tot} \geq 0.1$ relative to the ICME boundaries, normalized to the ICME duration (i.e., 0 corresponds to the ICME leading edge, and 1 to the trailing edge). The enhanced Fe charge states generally commence close to or inside the leading edge and finish inside the trailing edge. They do not favor specific locations in the ICME (e.g, inside the leading or trailing edge, or center). They tend to extend some distance beyond the ICME trailing edge more frequently than they do at the leading edge. This may be because it is generally more difficult to locate ICME trailing edges, often the Fe charge state ratio declines gradually to the ~0.1 threshold level, and the statistical accuracy of the Fe data will be reduced in the lower density plasma that is often present in ICME trailing edges. Nonetheless, an interesting possibility, which requires further study, is that a region of high Fe charge states follows some ICMEs. In a few cases, abrupt changes in Fe charge states occur within the ICME, suggestive of the presence of substructures, possibly multiple ICMEs. The most notable differences between the ICME and high Fe charge state intervals are for those ICMEs embedded in "recurrent" events as discussed above.

We have also examined whether there is any relationship between the mean Fe charge state and other ICME parameters. Around 30% of ICMEs have $<Q_{Fe}> \geq 13$, irrespective of whether they are "reported" magnetic clouds, have some evidence of a magnetic field rotation but have not been reported as magnetic clouds, or have no cloud-like magnetic signature. Thus, enhanced Fe charge states do not appear to depend on the presence or absence of a magnetic cloud structure. There is some evidence of a speed-dependence. Only 21% of ICMEs in our sample with in-situ speeds < 500 km/s have $<Q_{Fe}> \geq 13$, compared with 56% of those with speeds >500 km/s. There is also evidence of a solar-activity dependence. In 1998, 10% of the ICMEs have $<Q_{Fe}> \geq 13$ compared with 32% in 1999, 43% in 2000 (year of sunspot maximum), and 29% in 2001.

SUMMARY AND DISCUSSION

The finding (5) that only ~50% of the ICMEs identified independently using other signatures had high Fe charge states (as defined by (5)) arises because such charge states are not present in all ICMEs, or, if present, are detected for less than the 20 hour criterion required by (5). The charge-states (and other observed ICME characteristics) will also be influenced by the spacecraft trajectory relative to the ICME. In general, high Fe charge states lie within the ICME, though they occasionally continue beyond the ICME trailing edge. ICMEs with high Fe charge states appear to be predominantly related to solar flares, although the event in Figure 3 indicates that not all flare-related events have high charge states. Faster ICMEs are also more likely to be associated with high charge states. The charge states in ICMEs also increase with solar activity levels. Other charge state variations (e.g., C^5/C^6) in some ICMEs are spatially related to variations in T_p/T_{ex}, suggesting that the T_p signatures at 1 AU may occasionally contain vestiges of conditions during CME formation. An interesting subset of high Fe charge state events appear to form a corotating sequence, and further investigation of the associated solar wind structures is required.

ACKNOWLEDGEMENTS

We acknowledge the use of ACE MAG, SWEPAM and SWICS data from the ACE Science Center.

REFERENCES

1. Gosling, J. T., "Coronal mass ejections and magnetic flux ropes in interplanetary space," in *Physics of Magnetic Flux Ropes*, Geophys. Monogr. Ser., vol. 58, edited by C. T. Russell, E. R. Priest, and L. C. Lee, Washington, D.C., AGU, 1990, p. 343.

2. Richardson, I. G., and Cane, H. V., *J. Geophys. Res.*, **100**, 23,397 (1995).

3. Richardson, I. G., Dvornikov, V. M., Sdobnov, V. E., and Cane, H. V., *J. Geophys. Res.*, **105**, 12,597 (2000).

4. Zwickl, R. D., J. R. Asbridge, S. J. Bame, W. C. Feldman, J. T. Gosling, and E. J. Smith, "Plasma properties of driver gas following interplanetary shocks observed by ISEE 3", in *Solar Wind Five*, NASA Conf. Pub., CP-2280, 1983, p. 711.

5. Lepri, S. T., Zurbuchen, T. H., Fisk, L. A., Richardson, I. G., Cane, H. V., and Gloeckler, G., *J. Geophys. Res.*, **106**, 29,231 (2001).

6. Richardson, I. G., Dvornikov, V. M., Sdobnov, V. E., and Cane, H. V., *Proc. 27th Int. Cosmic Ray Conf.*, **9**, 3498 (2001).

7. Gosling, J. T., Birn, J., and Hesse, M., *Geophys. Res. Lett.*, **22**, 869 (1995).

Composition of magnetic cloud plasmas during 1997 and 1998

P. Wurz*, R. F. Wimmer-Schweingruber*, P. Bochsler*, A. B. Galvin†, J. A. Paquette**, F. M. Ipavich** and G. Gloeckler**

*Physics Institute, University of Bern, Sidlerstrasse 5, CH-3012 Bern, Switzerland
†EOS Space Sciences, University of New Hampshire, Morse Hall, Durham, NH 03824, USA
**Dept. Physics and Astronomy, University of Maryland, College Park, MD 20742, USA

Abstract. We present a study of the elemental composition of a sub-set of coronal mass ejections, namely events which have been identified of being of the magnetic cloud type (MC). We used plasma density data from the MTOF sensor of the CELIAS instrument of the SOHO mission and plasma ionization data from the SWICS instrument of the ACE mission. So far we have investigated MCs of 1997 and 1998. The study covers the proton and heavy ion elemental abundances. Considerable variations from event to event exist with regard to the density of the individual species with respect to regular "slow" solar wind preceding the MC plasma. However, two general features are observed. First, we observe for the heavy elements (carbon through iron), which can be regarded as tracers in the solar wind plasma, a mass-dependent enrichment of ions monotonically increasing with mass. The enrichment can be explained by a previously published theoretical model assuming coronal plasma loops on the solar surface being the precursor structure of the MC. Second, when comparing the MC plasma to regular solar wind composition preceding the event, a net depletion of the lighter ions is always observed. Proton and alpha particle abundances have to be regarded separately since they represent the main plasma.

INTRODUCTION

In a recent study it was reported that the plasma of the coronal mass ejection (CME) of 6 January 1997 was strongly mass-fractionated favoring heavier elements with respect to lighter ones [1, 2]. From the magnetic field measurements on WIND it has been concluded that the 6 January 1997 CME falls into the group of magnetic cloud (MC) events [3]. An overview of the 6 January 1997 CME event has been given by Fox et al. [4], which covers the launch of the CME, its propagation through interplanetary space, and its effect on the Earth's magnetosphere. Following this observation, the measured mass fractionation was successfully modelled by assuming large coronal loops, being the precursors of the magnetic cloud (MC) plasma, where the mass fractionation is established by diffusion across magnetic field lines [5].

A review covering the present understanding of CMEs has been given recently by Gosling [6], for the compositional aspects of CMEs see the review by Galvin [7]. CMEs with magnetic cloud topology generally exhibit somewhat higher freeze-in temperatures, i.e., the charge-state distribution of a particular element is shifted toward higher charge states than for the ambient undisturbed solar wind [8, 9]. Observational signatures of MCs consist of an enhanced magnetic field strength, a smooth rotation of the magnetic field direction as the cloud passes the spacecraft, and a low proton temperature. It has been found earlier that near 1 AU about one third of all CMEs in the ecliptic plane are magnetic cloud events [10].

In this paper we present an improved analysis of the magnetic cloud events during 1997 and 1998 we reported on earlier [11]. The present analysis combines the plasma density data from the CELIAS/SOHO instrument [12] with the charge state distribution data from the SWICS/ACE instrument [13]. Aside from minor improvements in the MTOF instrument function, including the charge-state distributions is the major improvement compared to the previous study. There were two MCs during 1997 and five MCs during 1998, which passed the SOHO and ACE spacecraft located close to the Earth at first Lagrangian point, L1. These events are listed in Table 1. We present the analysis for six of these MCs. Most of these events have charge-state distributions of the heavy ions that are similar to what is observed in regular solar wind with a shift to higher average charge-states in MC plasmas mentioned before. The sequence of three CME events from 2–3 May 1998, of which the second one is of MC nature, is the exception because it had

CP679, *Solar Wind Ten: Proceedings of the Tenth International Solar Wind Conference*, edited by M. Velli, R. Bruno, and F. Malara

TABLE 1. List of magnetic clouds during 1997 and 1998 that could be observed with solar wind particle instrumentation near Earth. Exact times of analyzed intervals for the reference periods and the MCs cloud are given as day-of-year (DOY).

MC Event*	Ref. start time [DOY]	Ref. end time [DOY]	MC start time [DOY]	MC end time [DOY]	Solar wind speed† [km/s]	Remarks
10–11 Jan 1997	8.00	9.00	10.27	10.98	440	MC followed by filament
7 Nov 1997	301.00	302.20	311.30	312.50	420	CME preceding MC
2–3 May 1998	118.00	119.30	121.88	122.96	510	Multiple CMEs
2 June 1998	151.72	153.00	153.43	153.65	430	
24 June 1998	173.00	174.00	175.42	176.00	390	
25 Sep 1998	—	—	—	—	—	SOHO not operational
8 Nov 1998	310.00	311.25	312.18	313.73	520	

* Date the event was observed at 1 ~ AU
† Average speed during MC duration

very unusual charge-state distributions [14]. For the 25 September 1998 MC event the SOHO spacecraft was not operational, thus no data can be presented. The present analysis gives further experimental evidence for the elemental fractionation of heavy ions in MC plasmas.

DATA ANALYSIS

In this study we evaluated the densities of the elements C, N, O, Ne, Na, Mg, Al, Si, S, Ar, Ca, and Fe in the solar wind and the MC plasma. From the ions recorded with the MTOF sensor, the CELIAS data processing unit accumulates time-of-flight (TOF) spectra for 5 minutes, which then are transmitted to ground. The raw counts for each mass peak of the different elements were extracted from each of the transmitted TOF spectra by fitting a model function of the peak shape and the background [15]. Subsequently, the overall efficiency of the MTOF sensor was calculated for each element and for each accumulation interval. To obtain particle fluxes for the chosen elements, the instrument response of the MTOF sensor comprising the transmission of the entrance system and the response of the isochronous TOF mass spectrometer, was taken into account in great detail [15]. The application of the instrument function to the measured count rates yielded densities for the different elements. We extensively checked if the instrument function introduces a mass bias, but so far we did not find such an effect in the data analysis.

The actual solar wind plasma parameters, which were measured by the Proton Monitor (PM), a sub-sensor of the MTOF sensor, are needed as input parameters for the instrument response function of the MTOF sensor. The quality of the determination of the solar wind plasma parameters with the PM is quite good [16] and better than required for the determination of densities with the MTOF sensor.

Another input parameter needed for the determination of the MTOF instrument response, in particular for the determination of the transmission of the entrance system, is the charge-state distribution of each element for each accumulation interval. The MTOF sensor determines only the mass of the incoming ion (with high resolution, however), but not its charge. The charge-state distribution is measured by the SWICS/ACE instrument for C, O, Ne, Mg, Si, S, and Fe with a time resolution of one hour. These data are interpolated to the MTOF time resolution and adapted for the elements not covered by SWICS/ACE. Since ACE science operations started early 1998, we had to resort to a model for the charge-state distributions of the elements for the two MC events during 1997. We derive the so-called freeze-in temperature from a semi-empirical model using the solar wind velocity as input parameter [15]. This model also accounts for ionization resulting from non-maxwellian electron distributions. From the freeze-in temperature we obtained charge distributions for each element by assuming an ionization equilibrium in the corona and by applying ionization and recombination rates for electronic collisions from Arnaud and co-workers [17, 18].

The MTOF sensor settings are cycled in a sequence consisting of four to six steps, which were optimized to cover a broad range of solar wind conditions. The stepping sequence includes two voltage settings for the entrance system and up to three values for the potential difference between the entrance system and the TOF mass spectrometer (negative, zero, and positive potential difference). In principle a time resolution of five minutes, the dwell time for each step, can be obtained if the sensitivity of the MTOF sensor is high enough for the particular element considered. For typical solar wind conditions and for the more abundant heavy elements in the solar wind, it is indeed possible to derive densities with such a high time resolution, as has been demonstrated earlier [2]. However, in the present analysis we report on the average densities of the MC plasmas for the whole event.

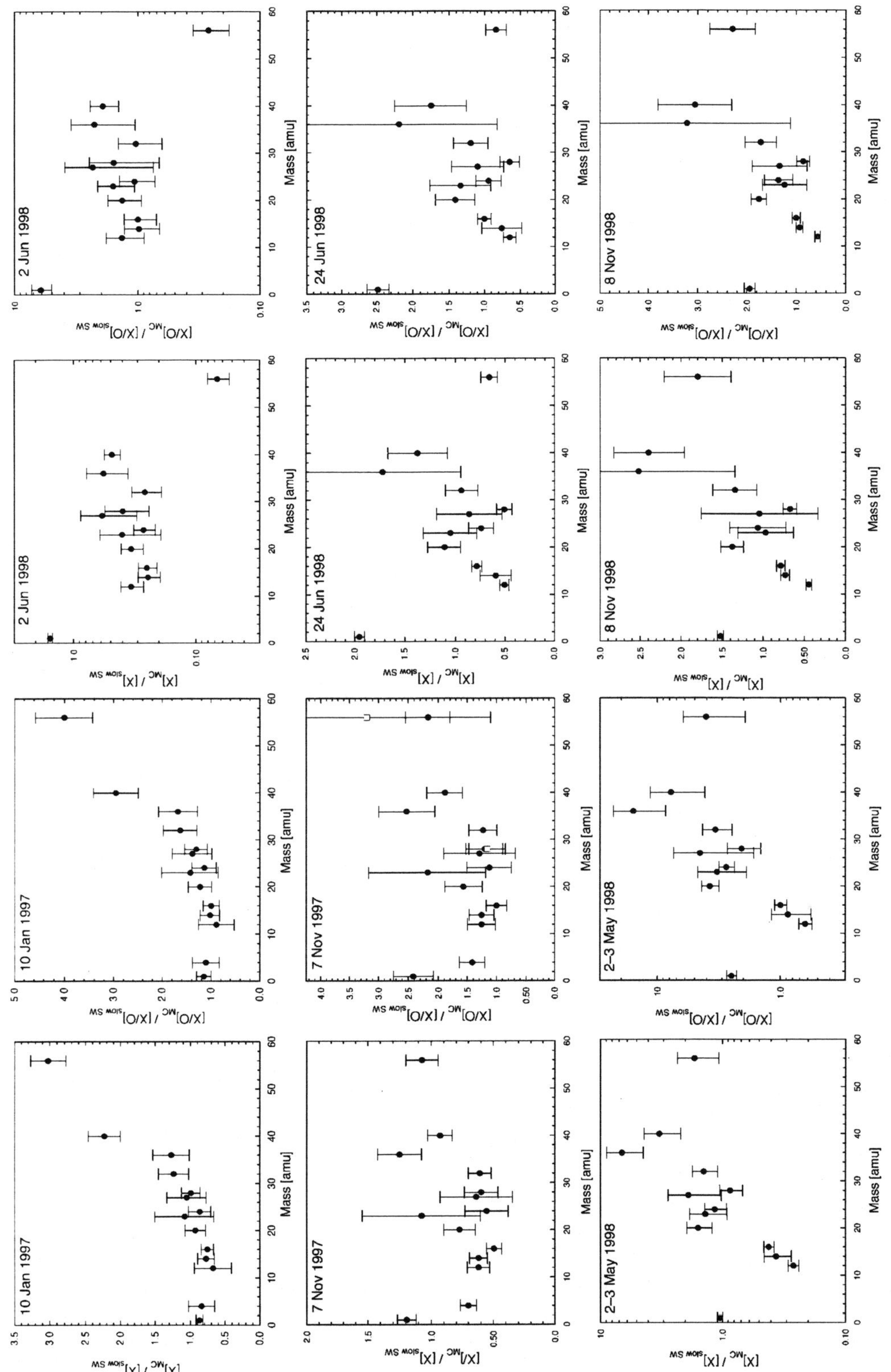

FIGURE 1. Results of the analysis of six MC events. The densities in the MC are compared to the respective densities in the preceding reference period of slow solar wind (see Table 1). Data are shown twice: one panel gives the density ratios of MC plasma and preceding SW plasma, the adjacent panel at the right shows the abundance ratios with respect to oxygen. The open symbols in the panel for the 7 November 1997 event were taken from [19].

RESULTS

We compare the densities for the different elements during the MC with the corresponding densities during a reference solar wind period preceeding the event. To account for the variability of the solar wind with time, or with location of origin or with solar activity, we used for reference a day of slow solar wind preceding the MC by about a day. Since we chose reference periods preceding the MC events we can safely assume that these reference plasmas are unaffected by the disruption that caused the CME release. Note that the slow solar wind, e.g. solar wind associated with the streamer belt, is already fractionated by mechanisms governed by the first ionization potential (the FIP effect) [20, 21]. The exact time periods for the MC and the reference solar wind used in the analysis are given in Table 1.

Figure 1 shows the results for all five MCs in two formats, one where data are given with reference to oxygen and one where direct comparison of MC and solar wind plasma is plotted. We derived the proton data from measurements with the PM. The heavy ion data were derived from mass spectra recorded with the MTOF sensor using the charge-state distributions recorded by SWICS/ACE.

The second and forth column in the Figure 1 shows the ratio of abundances with reference to oxygen in the MC cloud versus the solar wind reference period (that is $[X/O]_{MC}/[X/O]_{ref.\ SW}$ is plotted), which is a commonly used format to display variations in heavy ion abundances. The top panel shows a re-evaluation of the 10–11 January 1997 MC plasma considering more elements than in the original paper [2]. Within the error bars the initial results [2] have been reproduced. For the 7 November 1997 event there are two earlier measurements from WIND/MASS available [19], $[Si/O]_{MC}/[Si/O]_{slowSW}$ and $[Fe/O]_{MC}/[Fe/O]_{slowSW}$, which have been added to the plot. The reported Si and Fe abundance ratios agree with the present analysis within the error bars. In addition, for the 2–3 May 1998 event an Fe/O ratio of 0.28 ± 0.10 was derived from ACE/SWICS measurements [14], which is an increase of the Fe/O ratio by more than a factor of 2 compared to regular solar wind. This result is in good agreement with our analysis.

For all six events we find that the composition of the MC is markedly different from the preceding solar wind plasma. In all six events we find a more or less organized mass fractionation. i.e., monotonic with mass, for the heavy elements with heavier ions being enriched more than the lighter ones. There is of course some event-to-event variability in the abundance of the ions and in the magnitude of the enrichment of the heavy elements. For example, in the 2 June 1998 event the mass fractionation is rather small. Iron is in four cases the exception to the monotonic behavoir of the mass fractionation. Although significantly enriched compared to its reference solar wind (in the range of 1.5 to 4), its density in the MC falls below the expectation for a more or less monotonic increase in the mass fractionation with mass. For the 2 June 1998 and the and 24 June 1998 events we find that the iron abundance is lower, by about a factor of five and two, respectively, than in the preceding solar wind.

To get a better understanding of what is actually going on in the MCs we have to consider the ratios of the densities in the MC versus the solar wind reference period (e.g. $[X]_{MC}/[X]_{ref.\ SW}$). These data are shown in the first and thrid column of Figure 1. At first glance the data looks qualitatively the same as the abundance data revealing again the mass fractionation. However, the striking difference is that for all events the lighter elements, with the exception of hydrogen, are actually depleted to about half to the density they have in the preceding solar wind plasma. Depending on the strength of the mass fractionation the densities of heavier elements reach solar wind values (events 7 November 1997 and 24 June 1998) or are even enriched compared to the solar wind (events 10 January 1997, 2–3 May 1998, and 8 November 1998). Only the 2 June 1998 event does not match this pattern and we find that the iron density is a factor 20 below its solar wind value.

DISCUSSION AND CONCLUSIONS

Two things appear to be in common for the MC events presented in this study: the mass-dependent fractionation and the substantial depletion of lighter elements. These two findings are stick out the data shown in Figure 1. The mass-dependent fractionation is observed for all minor ions. Protons, being the major constituent of the plasma, have their own life and have to be considered separately. Since the protons constitute the main plasma and the heavy elements are only tracer particles in the plasma, a different behavior of the protons is not surprising. The third observation is that Fe may play a special role, since its enrichment in the MC plasma is in four out of six events not as high as the monotonic behavoir of the mass fractionation would suggest.

The mass fractionation of the heavy ions can be explained well by a recent theoretical model that was developed to explain the observed mass fractionation in the plasma of MC of the 10–11 January 1997 event [5]. This model explains the mass fractionation by assuming large coronal loops as the precursor structure for the MC in which different elements are depleted as a result of diffusion across magnetic field lines. Since this diffusion is mass dependent a mass-dependent fractionation is established. This model can also reproduce the present data

TABLE 2. Model fits to the mass fractionation observed for the magnetic clouds during 1997 and 1998. The model has been described previously [5]. The fitted loop parameters are the electron temperature, T_e, the magnetic field in the loop, B, and the life time of the coronal loop, t_L. The other parameters are $a = 10^4 m$, $T_1 = T_2 = 10^5 K$, $n_p = 2.5 \cdot 10^{17} m^{-3}$.

MC Event	T_e [K]	B [mT]	t_L [s]
10–11 Jan 1997	$2.3 \cdot 10^6$	0.5	37'500
7 Nov 1997	$2.0 \cdot 10^6$	0.5	12'500
2–3 May 1998	$2.5 \cdot 10^6$	0.6	85'000
2 June 1998	$2.0 \cdot 10^6$	0.4	15'000
24 June 1998	$2.0 \cdot 10^6$	0.7	40'000
25 Sep 1998	—	—	—
8 Nov 1998	$2.6 \cdot 10^6$	0.6	42'500

quite well. The necessary model parameters explaining the mass fractionations are listed in Table 2. Note that the model is based on depletion of elements from the precursor structure of the MC, which is in good agreement with the present finding of a substantial depletion of the lighter elements.

The presented sample of CME events are all magnetic cloud events. In addition, five of the six events have charge state distributions that are similar to what is observed in regular solar wind. By analyzing 56 CME events recorded with Ulysses/SWICS it was observed that MCs generally exhibit somewhat higher freeze-in temperatures compared to the ambient undisturbed solar wind [8, 9], i.e., the charge-state distributions are shifted to higher charges states. In the ecliptic plane the increase in freeze-in temperature correlates with solar wind speed and is largest for solar wind speeds exceeding 700 km/s [8, 9]. The MC events we analyzed are at moderate solar wind speeds (see Table 1) and therefore the charge-state distributions are similar to regular solar wind, as seen in the ACE/SWICS charde-state data. The investigated sample of MC events might introduce a bias in the result, in the sense that MCs with charge-state distributions significantly different from regular solar wind might also show different mass fractionations, if at all. On the other hand, the 2–3 May 1998 event does show a pronouced mass fractionation despite its very unusual charge-state distribution.

We considered six out of a total of seven events from the 1997–1998 time period in the present analysis. In the future we will analyze also the MC events from 1999 until present including more events with unusual charge-state distribution, like the 2–3 May 1998 event we presented. We have to await these analyses to see if the common features we found for MCs so far, the mass-dependent fractionation and the substantial depletion of lighter elements, will be observed there as well.

ACKNOWLEDGMENTS

CELIAS is a joint effort of five hardware institutions under the direction of the Max-Planck Institut für Extraterrestrische Physik (pre-launch) and the University of Bern (post-launch). The University of Maryland was the prime hardware institution for MTOF, the University of Bern provided the entrance system, and the Technical University of Braunschweig provided the DPU. This work is supported by the Swiss National Science Foundation.

REFERENCES

1. Wurz, P., Ipavich, F. M., Galvin, A. B., Bochsler, P., Aellig, M. R., Kallenbach, R., Hovestadt, D., Grünwaldt, H., Hilchenbach, M., Axford, W. I., Balsiger, H., Bürgi, A., Coplan, M. A., Geiss, J., Gliem, F., Gloeckler, G., Hefti, S., Hsieh, H. C., Klecker, B., Lee, M. A., Livi, S., Managadze, G. G., Marsch, E., Möbius, E., Neugebauer, M., Reiche, K.-U., Scholer, M., Verigin, M. I., and Wilken, B., *ESA*, **SP-415**, 395–400 (1997).
2. Wurz, P., Ipavich, F., Galvin, A., Bochsler, P., Aellig, M., Kallenbach, R., Hovestadt, D., Grünwaldt, H., Hilchenbach, M., Axford, W., Balsiger, H., Bürgi, A., Coplan, M., Geiss, J., Gliem, F., Gloeckler, G., Hefti, S., Hsieh, H. C., Klecker, B., Lee, M. A., Managadze, G. G., Marsch, E., Möbius, E., Neugebauer, M., Reiche, K.-U., Scholer, M., Verigin, M. I., and Wilken, B., *Geophys. Res. Lett.*, **25(14)**, 2557–2560 (1998).
3. Burlaga, L., Fritzenreiter, R., Lepping, R., Ogilvie, K., Szabo, A., Lazarus, A., Steinberg, J., Gloeckler, G., Howard, R., Michels, D., Farrugia, C., Lin, R. P., and Larson, D. E., *J. Geophys. Res.*, **103**, 277–286 (1998).
4. Fox, N. J., Peredo, M., and Thompson, B. J., *Geophys. Res. Lett.*, **25(14)**, 2461–2464 (1998).
5. Wurz, P., Bochsler, P., and Lee, M. A., *J. Geophys. Res.*, **105(A12)**, 27239–27249 (2000).
6. Gosling, J. T., "Coronal mass ejections: An overview", in *Coronal Mass Ejections*, edited by N. Crooker, J. A. Joselyn, and J. Feynman, Geophys. Monograph 99, American Institute of Physics, 1997, pp. 9–16.
7. Galvin, A. B., "Minor ion composition in CME-related solar wind", in *Coronal Mass Ejections*, edited by N. Crooker, J. A. Joselyn, and J. Feynman, Geophys. Monograph 99, American Institute of Physics, 1997, pp. 253–260.
8. Henke, T., Woch, J., Mall, U., Livi, S., Wilken, B., Schwenn, R., Gloeckler, G., v. Steiger, R., Forsyth, R. J., and Balogh, A., *Geophys. Res. Lett.*, **25**, 3465–3468 (1998).
9. Neukomm, R., *Composition of Coronal Mass Ejections Derived with SWICS/Ulysses*, Phd thesis, University of Bern, Bern, Switzerland (1998).
10. Gosling, J. T., "Coronal mass ejections and magnetic flux ropes in interplanetary space", in *Physics of Magnetic Flux Ropes*, edited by C. T. Russell, E. R. Priest, and L. C. Lee, Geophys. Monograph 58, American Institute of Physics, 1990, pp. 343–364.
11. Wurz, P., Wimmer-Schweingruber, R. F., Issautier, K., Bochsler, P., Galvin, A. B., and Ipavich, F. M.,

"Composition of magnetic cloud plasmas during 1997 and 1998", in *Solar and Galactic Composition*, edited by R. Wimmer-Schweingruber, CP-598, American Institute Physics, 2001, pp. 145–151.

12. Hovestadt, D., Hilchenbach, M., Bürgi, A., Klecker, B., Laeverenz, P., Scholer, M., Grünwaldt, H., Axford, W. I., Livi, S., Marsch, E., Wilken, B., Winterhoff, P., Ipavich, F. M., Bedini, P., Coplan, M. A., Galvin, A. B., Gloeckler, G., Bochsler, P., Balsiger, H., Fischer, J., Geiss, J., Kallenbach, R., Wurz, P., Reiche, K.-U., Gliem, F., Judge, D. L., Hsieh, K. H., Möbius, E., Lee, M. A., Managadze, G. G., Verigin, M. I., and Neugebauer, M., *Solar Physic*, **162**, 441–481 (1995).
13. Gloeckler, G., Cain, J., Ipavich, F. M., Tums, E. O., Bedini, P., Fisk, L. A., Zurbuchen, T., , Bochsler, P., Fischer, J., Wimmer-Schweingruber, R. F., Geiss, J., and Kallenbach, R., *Space Sci. Rev.*, **86**, 497–539 (1998).
14. Gloeckler, G., Hefti, S., Zurbuchen, T. H., Schwadron, N. A., Fisk, L. A., Ipavich, F. M., Geiss, J., Bochsler, P., and Wimmer-Schweingruber, R. F., *Geophys. Res. Lett.*, **26(2)**, 157–160 (1999).
15. Wurz, P., *Heavy ions in the solar wind: Results from SOHO/CELIAS/MTOF*, Habilitation thesis, University of Bern, Bern, Switzerland (1999).
16. Ipavich, F. M., Galvin, A. B., Lasley, S. E., Paquette, J. A., Hefti, S., Reiche, K.-U., Coplan, M. A., Gloeckler, G., Bochsler, P., Hovestadt, D., Grünwaldt, H., Hilchenbach, M., Gliem, F., Axford, W. I., Balsiger, H., Bürgi, A., Geiss, J., Hsieh, K. C., Kallenbach, R., Klecker, B., Lee, M. A., Mangadze, G. G., Marsch, E., Möbius, E., Neugebauer, M., Scholer, M., Verigin, M. I., Wilken, B., and Wurz, P., *J. Geophys. Res.*, **103(A8)**, 17205–17214 (1997).
17. Arnaud, M., and Rothenflug, R., *Astron. Astrophys. Suppl. Ser.*, **60**, 425–457 (1985).
18. Arnaud, M., and Raymond, J., *Astrophys. J.*, **398**, 394–406 (1992).
19. Wimmer-Schweingruber, R. F., Kern, O., and Hamilton, D. C., *Geophys. Res. Lett.*, **26(23)**, 3541–3544 (1999).
20. von Steiger, R., "Solar wind composition and charge states", in *Solar Wind Eight*, edited by D. Winterhalter, J. T. Gosling, S. R. Habbal, W. S. Kurth, and M. Neugebauer, AIP Press, 1995, pp. 193–198.
21. Hénoux, J.-C., *Space Sci. Rev.*, *85*, 215–226 (1998).

Comparison of Simulated and Observed Interplanetary Flux Ropes

M. Vandas*, S. Watari† and A. Geranios**

*Astronomical Institute, Academy of Sciences, Boční II 1401, 14131 Praha 4, Czech Republic
†Communications Research Laboratory, 4-2-1 Nukuikita, Koganei, Tokyo 184-8795, Japan
**Physics Department, Nuclear and Particle Physics Section, Athens University, Panepistimioupoli-Kouponia, Athens 15771, Greece

Abstract. Three dimensional magnetohydrodynamic numerical simulations of propagating interplanetary flux ropes are presented and compared with "in-situ" spacecraft measurements. Flux ropes are injected near the Sun with various inclinations. Specific features followed from simulations, as double-peak magnetic field profiles or a possibility to observe one flux rope two times by the same spacecraft, are searched in observational data.

INTRODUCTION

Interplanetary flux ropes are a subject of many studies (e.g., [1, 2, 3, 4, 5, 6]). The most distinct interplanetary flux ropes are magnetic clouds, defined by [1] as regions in the solar wind with higher and smoothly rotating magnetic fields and with lower proton temperatures, with spatial extents of the order of 0.1 AU near 1 AU.

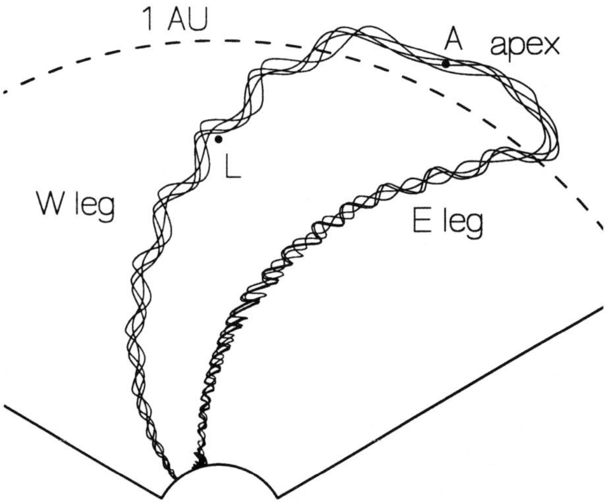

FIGURE 1. Flux rope in the ecliptic plane as a result of numerical simulations. The dashed line shows 1 AU distance.

Several years ago we have performed [7] MHD simulations of a propagating flux rope in the inner heliosphere in two dimensions. The flux rope was perpendicular to the computational domain and was introduced by perturbations of quantities at the inner boundary, which was supermagnetosonic. A model of the flux rope was a cylinder with a constant-α force-free field inside. For three dimensional simulations we used the same way, only the cylinder was replaced by a toroid; initially, a constant-α force-free field was inside the toroid and the external field was a potential field around a toroid, described in [8]. During injection of the toroid, only the magnetic field was perturbed at the inner boundary, other quantities remained unchanged. When a half of the toroid emerged, its feet were kept at the inner boundary and shifted westward to simulate the solar rotation. The magnetic field strength in the computational domain was one fifth of a typical value to suppress numerical diffusion (i.e., ≈ 1 nT at 1 AU).

More details on these simulations are given in [9]. The simulations were made for several inclinations of the flux rope to the ecliptic plane.

RESULTS OF SIMULATIONS

Figure 1 shows a shape of a flux rope when its leading edge (apex) crossed 1 AU. The flux rope was injected with 0° inclination to the ecliptic plane. Labels in Figure 1 identify parts of the flux rope, the apex and two legs (western and eastern). The points A and L (bullets) are located in the apex and the western leg, respectively.

CP679, *Solar Wind Ten: Proceedings of the Tenth International Solar Wind Conference*, edited by M. Velli, R. Bruno, and F. Malara

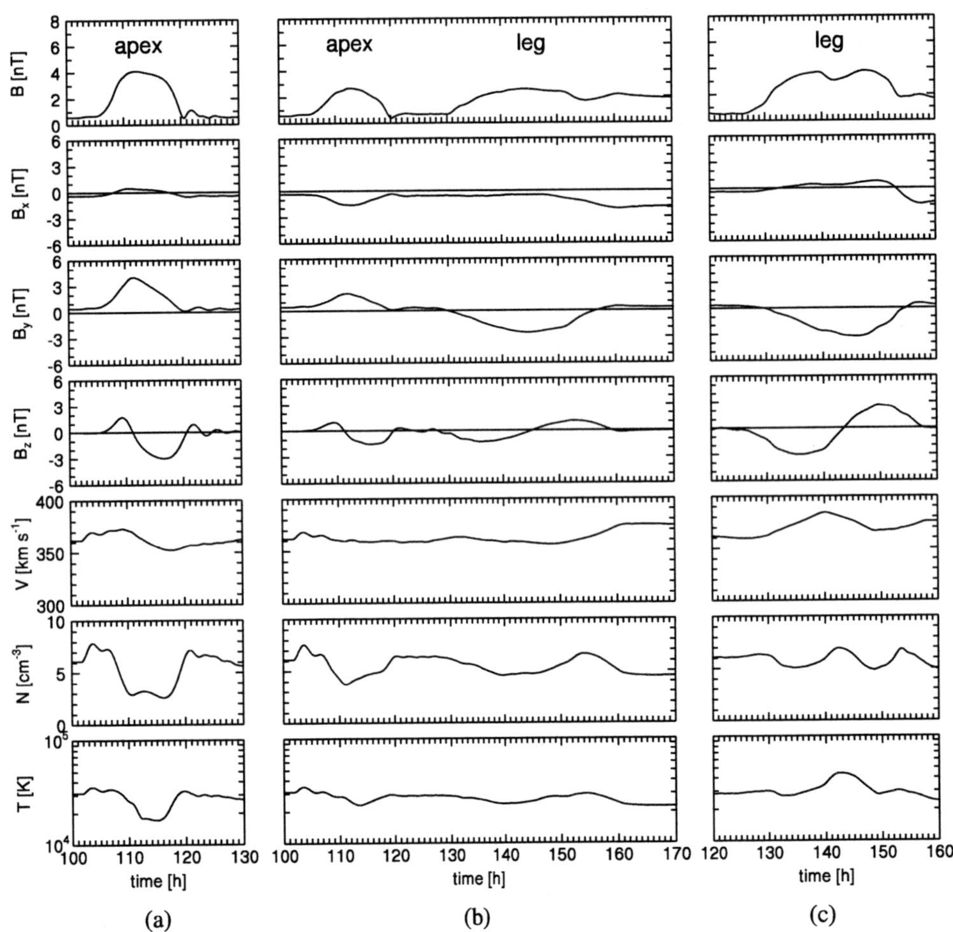

FIGURE 2. Simulated observations at 1 AU of a flux rope with an inclination 20°. Hypotetic spacecraft were located in places with the same longitude, the first one 4° below the ecliptic plane (a), the second one in the ecliptic plane (b), and the third one 4° above it (c). B is the magnetic field magnitude, B_x, B_y, and B_z are the solar ecliptic components of the magnetic field, V, N, and T are the velocity, density, and temperature of the solar wind, respectively.

Simulations give us temporal profiles of solar wind quantities at various locations. Inspection of these profiles reveals that they are highly variable and depend on which part of the flux rope was observed. A hypothetic spacecraft sitting in A (Figure 1) will observe the flux rope two times, its apex and its eastern leg (imagine a radial line connecting A with the Sun).

Figure 2 shows simulated observations of a flux rope with an inclination 20° by three hypothetic spacecraft at 1 AU, which are located quite close together. The first spacecraft observed only an apex, the second one both the apex and a leg, and the third one only the leg. Rotation of the magnetic field vector is pronounced in both the apex and the leg. But behaviors of plasma parameters are different in these parts. There are distinct decreases in the density and temperature in the apex, so the apex may be identified as a magnetic cloud. Contrary, in the leg, variations of the density and temperature are not so pronounced and they have even higher values near the leg's center. A saddle in B profile may develop at a leg (Figure 2c).

Figure 3 displays the different spatial behaviors of quantities at an apex and legs where distribution of B and N on planes perpendicular to the flux rope at points A and L (Figure 1) are shown. The black lines give projections of magnetic field lines, approximately indicating a boundary of the flux rope. The apex has an oblate shape, the shape of the leg is more circular. The leg has an increase in density (and temperature) near the center. We understand this fact by different external conditions for the apex and the leg. The leg moves more radially and its central part does not expand so much as the diverging

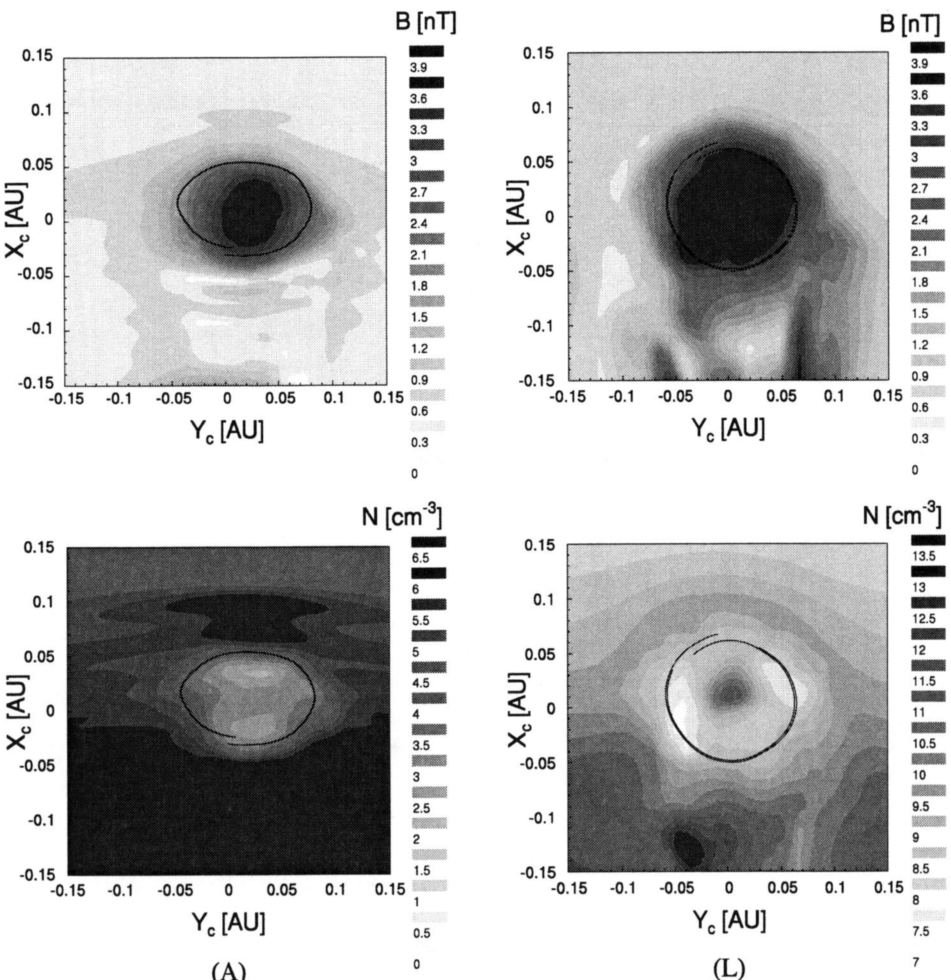

FIGURE 3. Spatial distribution of the magnetic field magnitude B and density N at the cross-section of the flux rope at the apex (A) and the leg (L). The temperature has a similar behavior as the density. A local X_c axis lies in the ecliptic plane.

ambient solar wind.

COMPARISON WITH OBSERVATIONS

Simulations show that candidates for a leg would have relatively lower inclinations ($\lesssim 30°$) and would be more radially aligned (the last item is not always the case for eastern legs, see Figure 1). Figure 4a displays a flux rope of June 8–9, 1997, which might be a candidate for a leg observation. The magnetic field and plasma behavior look similar to simulated profiles. An estimated inclination of the flux rope is $\vartheta_c = -4°$ and its azimuthal angle $\varphi_c = 201°$. The density and temperature have increases in the center of the rope. Another example could be the interplanetary flux rope of February 18, 1998, where the density and temperature are enhanced inside the rope.

The simulations show that when a double-peak in B (Figure 2c) develops, the saddle in the B profile is located where the B_z component changes its sign (for low inclined flux ropes, B_z has a profile like the $\pm\sin$ function, see again Figure 2c). And this is really the case which is met in observations. Figure 4b documents this fact. The flux rope of September 25–26, 1982, has been analyzed by [4], who determined $\vartheta_c = 10°$ and $\varphi_c = 111°$. Another example could be the case of May 25-26, 1975, analyzed by [1], who estimated $\vartheta_c = -31°$ and $\varphi_c = 206°$.

Any clear example of two subsequent crossings of one flux rope was not found. A flux rope should be low inclined and have two subsequent increases in B_z with the same polarity (see Figure 2b). One suspicious case is the event of June 2–4, 1998, with $\vartheta_c \approx 9°$. A real situation may disfavored this possibility: a larger B and

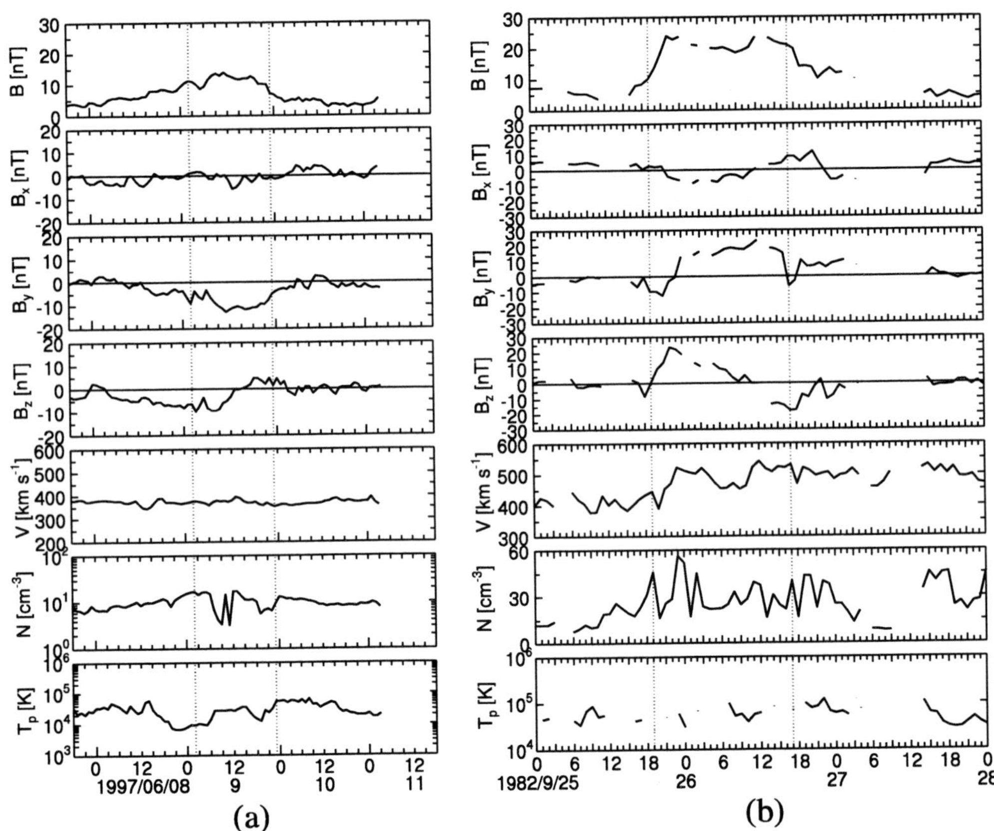

FIGURE 4. "In-situ" observations of interplanetary flux ropes by (a) the Wind and (b) IMP-8 spacecraft. The quantities have the same meaning as in Figure 2, T_p is the proton temperature.

faster motion than they are in the simulations, may cause a flux rope is not so bended; or magnetic field rotation is not very distinct in some parts of legs.

CONCLUSIONS

Simulations show a variety of profiles through interplanetary flux ropes and may help to understand what part of a rope is observed by spacecraft. In particular, they suggest a distinction in plasma parameters in an apex and legs.

ACKNOWLEDGMENTS

The observational data were provided by the computer services of the National Space Science Data Center (WDC-A-R&S); PIs are: R. P. Lepping for magnetic field data, and J. T. Gosling and K. W. Ogilvie for plasma data. This work was supported by grant A3003003 and projects S1003006 and K3012103 from the Academy of Sciences of the Czech Republic.

REFERENCES

1. Klein, L. W., and Burlaga, L. F., *J. Geophys. Res.*, **87**, 613–624 (1982).
2. Marubashi, K., *Adv. Space Res.*, **6**(6), 335–338 (1986).
3. Ivanov, K. G., Harshiladze, A. F., Eroshenko, E. G., and Styazhkin, V. A., *Sol. Phys.*, **120**, 407–419 (1989).
4. Lepping, R. P., Jones, J. A., and Burlaga, L. F., *J. Geophys. Res.*, **95**, 11,957–11,965 (1990).
5. Vandas, M., Fischer, S., Pelant, P., and Geranios, A., *J. Geophys. Res.*, **98**, 21,061–21,069 (1993).
6. Shimazu, H., and Marubashi, K., *J. Geophys. Res.*, **105**, 2365–2374 (2000).
7. Vandas, M., Fischer, S., Dryer, M., Smith, Z., and Detman, T., *J. Geophys. Res.*, **100**, 12,285–12,292 (1995).
8. Romashets, E., and Vandas, M., Propagation of a toroidal magnetic cloud through the inner heliosphere, *this issue*.
9. Vandas, M., Odstrčil, D., and Watari, S., Three dimensional MHD simulation of a loop-like magnetic cloud in the solar wind, *J. Geophys. Res.*, in press.

Cancellations and structures in the solar photosphere: signature of flares

L. Sorriso-Valvo*, V. Abramenko†, V. Carbone*, A. Noullez**, H. Politano**, A. Pouquet‡, P. Veltri* and V. Yurchyshyn§†

*Dipartimento di Fisica, Universitá della Calabria and Istituto Nazionale di Fisica della Materia, sezione di Cosenza, Italy.
†Crimean Astrophysical Observatory, Nauchny, Crimea, Ukraine
**Observatoire de la Côte d'Azur, Nice, France
‡ASP/NCAR, Boulder, CO
§Big Bear Solar Observatory, Big Bear City, CA

Abstract. The topological properties of the typical current structures in a turbulent magnetohydrodynamic flow can be measured using the cancellations analysis. In two-dimensional numerical simulations, this reveals current filaments being the most typical current structures. The observations of the topology of photospheric current structures within active regions shows that modifications occur correspondingly with strong flares.

INTRODUCTION

Solar flares are sudden, transient energy release above active regions of the Sun (Priest, 1982). The magnetic energy is released, and thus observed, in various form as thermal, soft and hard X-ray, accelerated particles etc. It seems natural to look for hints of flaring activity in the magnetic field in the photosphere, but the direct observation of the magnetic field itself gives no unambiguous results (*e.g.* Hagyard *et al.*, 1999 and references therein). Recently, unambiguous observations of changing have been reported by Yurchyshyn *et al.* (2000). The authors observed some typical changes of the scaling behavior of the current helicity calculated inside an active region of the photosphere, connected to the eruption of big flares above that active region. In the present paper we conjecture that the changes in the scaling behavior of the observed quantity is related to the occurrence of changes in the topology of the magnetic field at the footpoint of the loop.

SIGNED MEASURE, CANCELLATIONS AND STRUCTURES

Topological properties of scalar fields which oscillate in sign can be studied through the scaling of signed measures. First of all, given a meanless, scalar field $f(x)$, let us introduce the signed measure

$$\mu_i(r) = \int_{Q_i(r)} f(x)dx \qquad (1)$$

through a coarse-graining of non overlapping boxes $Q_i(r)$ of size r, covering the whole field defined on a region of size L. It has been observed (Ott *et al.*, 1992) that, for fields presenting self-similarity, this quantity displays well defined scaling laws. That is, in a range of scales r, the partition function $\chi(r)$, defined as

$$\chi(r) = \sum_{Q_i(r)} |\mu_i(r)| \sim r^{-\kappa} \qquad (2)$$

where the sum is extended over all boxes occurring at a given scale r, follows a power-law behavior

$$\chi(r) \sim r^{-\kappa}. \qquad (3)$$

The scaling exponent κ has been called cancellation exponent (Ott *et al.*, 1992) because it represents a quantitative measure of the scaling behavior of imbalance between negative and positive contributions in the measure. For example, a positive definite measure or a smooth field have $\kappa = 0$, while $\kappa = d/2$ for a completely stochastic field in a d-dimensional space. As the cancellations between negative and positive part of the measure decreases toward smaller scales, we get $\kappa > 0$, and this is the interesting situation. It is clear that the presence of structures, seen as smooth parts of the field, has an important effect on the cancellation exponent. For example, values of $\kappa < d/2$, where d is the dimension of the

CP679, *Solar Wind Ten: Proceedings of the Tenth International Solar Wind Conference*, edited by M. Velli, R. Bruno, and F. Malara

space (in the present paper $d = 2$), indicate the presence of sign-persistent (*i. e.* smooth) structures.

Within turbulent flows, the value of the cancellation exponent can be related to the characteristic fractal dimension D of turbulent structures on all scales using a simple geometrical argument (Sorriso-Valvo *et al.*, 2002). Let λ be the typical correlation length of that structures, of the order of the Taylor microscale (see for example Frisch, 1995), so that the field is smooth (correlated) in D dimensions with a cutoff scale λ, and uncorrelated in the remaining $d-D$ dimensions. If the field is homogeneous, the partition function (2) can be computed as $(L/r)^d$ times the integral over a generic box $Q(r)$ of size r. The scaling of the latter can be estimated integrating over regular domains of size λ^d and considering separately the number of contributions coming from the correlated dimensions of the field and those from the uncorrelated ones. The integration of the field over the smooth dimensions will bring a contribution proportional to their area $(r/\lambda)^D$, while the uncorrelated dimensions will contribute as the integral of an uncorrelated field, that is proportional to the square root of their area $(r/\lambda)^{(d-D)/2}$. Thus, when homogeneity is assumed, collecting all the contributions in (2) leads to scaling $\chi(r) \sim r^{-(d-D)/2}$ for the partition function, so that one can obtain the simple relation

$$\kappa = (d-D)/2\,. \tag{4}$$

NUMERICAL SIMULATIONS

Using high resolution numerical simulation of two-dimensional ($d = 2$) turbulent magnetohydrodynamic flows (Politano *et al.*, 1998; Sorriso-Valvo *et al.*, 2000; Sorriso-Valvo *et al.*, 2001), we can build up the signed measure for different fields. For example, since the geometry of the magnetic field $B(x,y) = (B_x, B_y, 0)$ is two-dimensional, the current $J(x,y) = \nabla \times B = (0,0,J_z)$ has only the z component, perpendicular to the 2-d simulation box, *i. e.* the plane (x,y). In Figure 1 we display the current field $J(x,y)$ for the numerical data, using ten snapshots in the statistically steady state, from $t = 168$ up to $t = 336$ in non-linear times units, τ_{NL}. As can be seen, the presence of positive and negative structures is evident. The signed measure of the current can be then computed as

$$\mu_i(r) = \int_{Q_i(r)} J_z(x,y)\,dxdy\,,$$

and the scaling properties of the time averaged partition function are reported in Figure 2. A power-law scaling (3) is clearly visible in a range extending from the large scales (near the integral scale of the flow $\ell_0 \sim 0.2L$, $L = 2\pi$ being the size of the simulation box) down to a correlation lenght r^* of the order of the Taylor microscale $\lambda \sim 0.02L$ of the flow (see for example Frish, 1995). In this region, we fit the partition function to obtain the cancellation exponent $\kappa = 0.43 \pm 0.06$. A saturation of the partition function is observed at a scale r_S which is found to be of the order of the dissipative scale of the flow. In fact, for scales smaller than r_S the dissipation stops the structures formation cascade, so that cancellations are stopped too. The fractal dimension of the current structures has been computed using the relation (4), which gives $D \simeq 1$, indicating current structures similar to filaments. The presence of filaments can be clearly observed by a direct inspection of the current field contour plot, confirming the reliability of the model (see Sorriso-Valvo et al., 2002).

SOLAR DATA

To get a quantitative measure of the change of the scaling of current helicity inside active regions, we used observations of the vector magnetic field obtained with the Solar Magnetic Field Telescope of the Beijing Astronomical Observatory (China). Measurements were recorded in the FeI 5324.19 Å spectral line. The field of view is about 218" x 314", corresponding to 512 x 512 pixels on CCD. The magnetic field vector at the photosphere has been obtained through the measurements of the four Stokes parameters, and the current density $J_z(x,y)$ has been calculated as a line integral of the transverse field vector over a closed contour of dimension 1.72" x 1.86" (cf. Yurchischin *et al.*, 2000, for details). The current helicity $H_c = \mathbf{B}\cdot\mathbf{J}$ (where $\mathbf{B}$ represents the magnetic field and $\mathbf{J} = \nabla \times \mathbf{B}$ the current density) is a measure of small scales activity in magnetic turbulence. It indicates the degree of clockwise or anti-clockwise knotness of the current density. Let us consider a magnetogram of size L taken on the solar photosphere of an active region, and let $\mathbf{B}_\perp(x,y)$ the observed magnetic field perpendicular to the line of sight ((x,y) are the coordinate on the surface of the sun). Through this field we can measure the surrogate of current helicity, that is $h_c(x,y) = B_z(x,y)J_z(x,y)$ being $J_z(x,y) = [\nabla \times \mathbf{B}_\perp] \cdot \hat{e}_z$. A signed measure can be defined from this quantity

$$\mu_i(r) = \int_{Q_i(r)} h_c(x,y)dxdy\,. \tag{5}$$

In Figure 3 we show, as example, the scaling behavior of $\chi(r)$ vs. r for a flaring active region (NOAA 7315) which started to flare on October 22, 1992. At larger scales we find $\chi(r) \sim$ *const.*, and this is due to the complete balance between positive and negative contributions. The same behavior does not appear at smaller scales, showing that the resolution of the images is not high enough

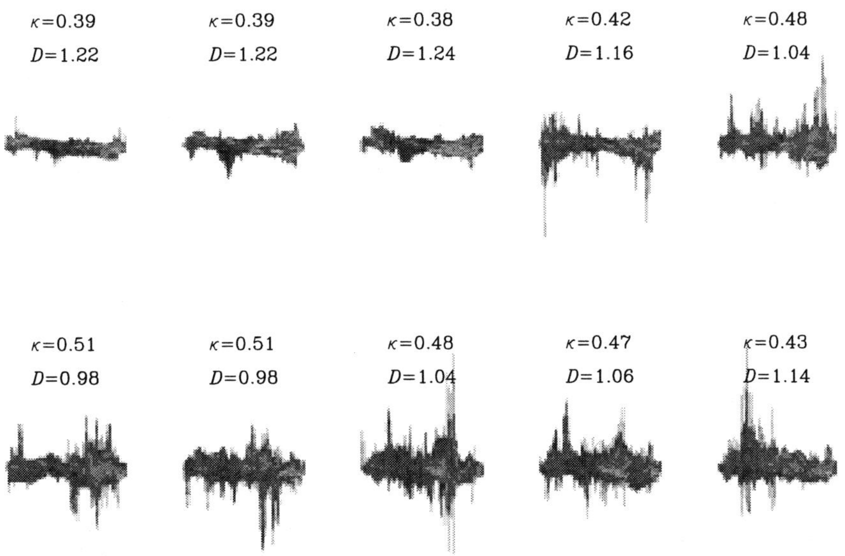

FIGURE 1. The current field J obtained from high resolution two-dimensional numerical simulation of MHD equations. The different plots are ten snapshots in the statistically steady state, from about $t = 168$ up to $t = 336$ in non-linear times units, τ_{NL}. As for the solar data, the presence of positive and negative strucutres on all scales is clear. We report, over each plot, the measured values of κ and D.

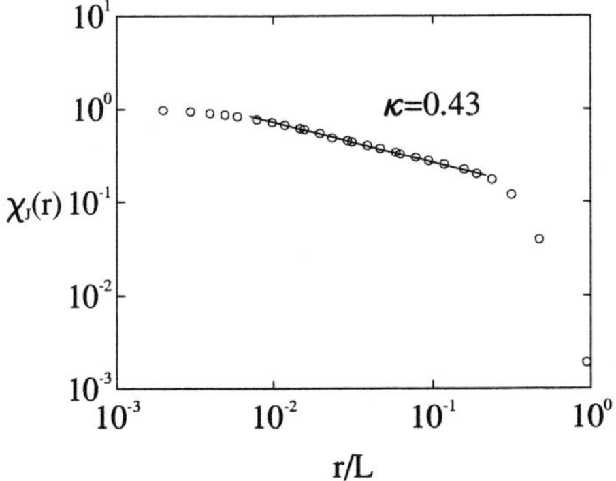

FIGURE 2. The scaling of the partition function for the current obtained from numerical data. This result is obtained by averaging the time evolution, in order to increase the statistics. The power-law fit is indicated as a straight line. The scales are normalized to the simulation box size $L = 2\pi$.

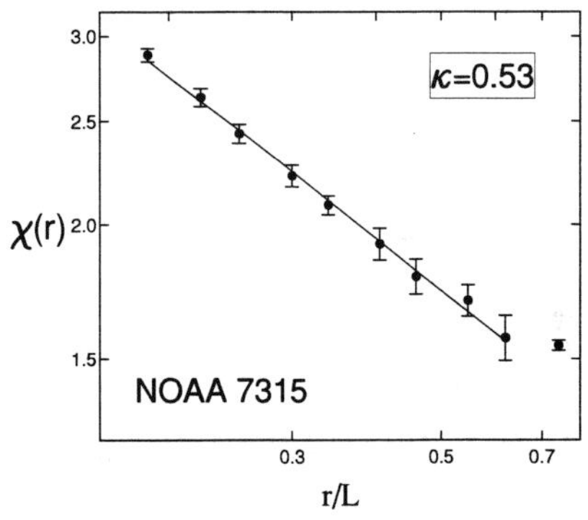

FIGURE 3. The scaling of the partition function for a flaring active regions (NOAA 7315), which started to flare on October 22, 1992. The power-law fit is indicated as a dotted line.

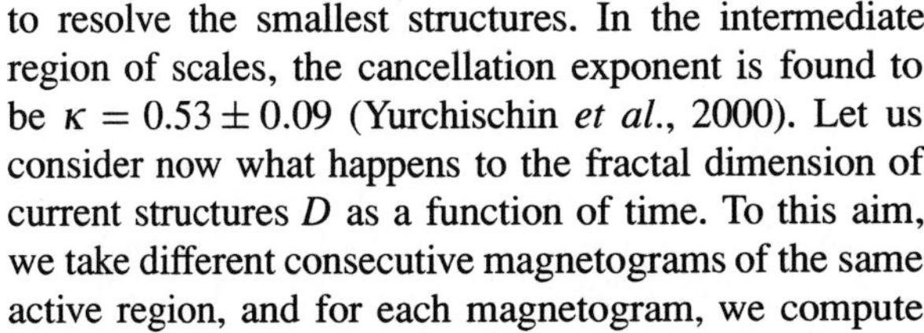
to resolve the smallest structures. In the intermediate region of scales, the cancellation exponent is found to be $\kappa = 0.53 \pm 0.09$ (Yurchischin *et al.*, 2000). Let us consider now what happens to the fractal dimension of current structures D as a function of time. To this aim, we take different consecutive magnetograms of the same active region, and for each magnetogram, we compute the value of κ and then of D through relation (4). Note that, since cancellations in the vertical photospheric magnetic field $B_z(x,y)$ itself have been found to be very small (Lawrence *et al.*, 1993; Abramenko *et al.*, 1998), with a cancellation exponent of the order of 10^{-2}, cancellations of the current helicity are entirely due to the current structures.

In Figure 4, we report the time evolution of D superimposed to the flares occurred in two active regions,

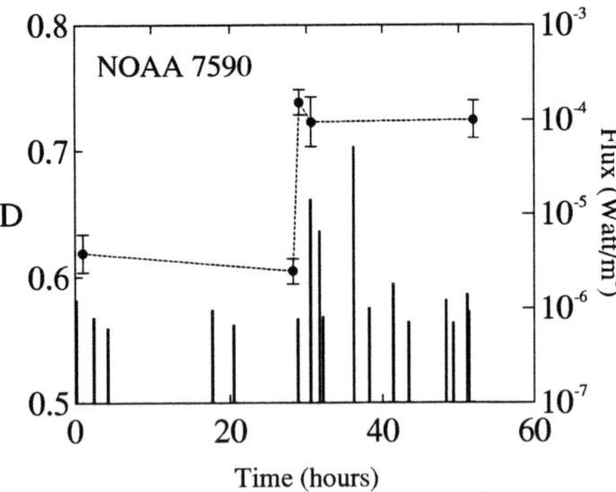

FIGURE 4. We present one flaring event observed in 1993 (see text for description) in the bottom part of the plot (line bars, in arbitrary units). The corresponding time variation of the fractal dimension D is reported (symbols).

namely NOAAs 7315, and 7590 (which flared on October 1, 1993). Quite surprisingly we observe that the fractal dimension D, starting from a given value ($D < 1$), becomes abruptly larger in correspondence with a sequence of big flares occurring at the top of the active region into the corona. The same behavior has been found for all calculations in all active regions we examined. The increase of the dimension of the structures may be the signature that dissipation has occured. In fact, annihilation is responsible for the smoothing of the small scales structures. The change towards larger D is then probabily due to the occurrence of dissipation of smaller and smaller flux tubes, that is magnetic energy is suddenly transferred towards small scales. This is the occurrence of an energy cascade towards smaller scales.

CONCLUSIONS

In this paper we point out that the changes in the scaling behavior of cancellations, measured through the cancellation exponent κ, are due to the topology changes of the structures present in the field, and are thus related to the importance of dissipative effects. The non-linear turbulent cascade, underlying the formation os such structures on all scales, can be considered as one important input mechanism for flares. The results obtained from the analysis of the numerical simulations can be considered as a test for our model for the fractal dimension of structures, thus supporting our interpretation of the observational results for the photospheric magnetic field in the active regions.

To conclude, it is evident that the behavior we found can be used as a signature of the occurrence of big flares. High energy solar flares become of great interest because they can produce severe damages on Earth. Power blackouts, break up of communications and mainly damage of satellites or space flights, can be ascribed to energy released during big solar flares. It is then evident that the possibility of forecasting, even if partially, high energy flares has a wide practical interest to prevent the effects of flares on Earth and its environment. We build up a model which allows us to recognize without ambiguity changing behavior of the photospheric magnetic field of active regions. These changes, pointed out through the variation of a scaling index for current helicity, can be seen mainly before the eruption of big flares. The change of scaling index is due to the turbulent and intermittent energy cascade towards smaller scales, a mechanism which could be identified as the input of flaring activity, where energy is dissipated. The method could allows us to forecast, in real time, the appearence of the strongest flaring activity above active regions.

REFERENCES

1. Abramenko, V.I., Yurchyshyn V.B., and Carbone, V., Does the photospheric current take parott t in the flaring process?, *Astron. Astrophys., 334*, L57, 1998.
2. Boffetta, G., Carbone, V., Giuliani, P., Veltri, P., and Vulpiani, A., Power laws in solar flares: self-organized Criticality or Turbulence?, *Phys. Rev. Lett., 83*, 4662, 1999.
3. Frisch, U., *Turbulence: the legacy of A. N. Kolmogorov*, Cambridge U. P., 1995.
4. Hagyard, M.J., Stark, B.A., and Venkatakrishnan, P., A search for vector magnetic field variations associated with the M-class flares of 10 June 1991 in AR 6659, *Sol. Phys., 184*, 133, 1999.
5. Lawrence, J.K., Ruzmaikin, A.A., and Cadavid, A.C., *Astrophys. J., 417*, 805, 1993
6. Lepreti, F., Carbone, V., and Veltri, P., Solar flare waiting time distribution: varying-rate Poisson or Levy function?, *Astrophys. J., 555*, L133, 2001.
7. Ott, E., Du, Y., Sreenivasan, K.R., Juneja, A., and Suri, A.K., *Phys. Rev. Lett., 69*, 2654, 1992.
8. Politano, H., Pouquet, A., and Carbone, V., Determination of anomalous exponents of structure functions in two-dimensional magnetohydrodynamic turbulence, *Europhys. Lett., 43*, 516, 1998.
9. Priest, E., *Solar Magnetohydrodynamic*, D. Reidel Publishing Company, Dordrecht, 1982.
10. Sorriso-Valvo, L., Carbone, V., Noullez, A., Politano, H., Pouquet, A., and Veltri, P., Analysis of cancellation in two-dimensional MHD turbulence, *Phys. Plasmas, 9*, 89, 2002.
11. Sorriso-Valvo, L., Carbone, V., Veltri, P., Politano, H., and Pouquet, A., Non-gaussian probability distribution functions in two-dimensional magnetohydrodynamic turbulence, *Europhys. Lett., 51*, 520, 2000.
12. Yurchyshyn, V.B., Abramenko, V.I., and Carbone, V., Flare-related changes of an active region magnetic field, *Astrophys. J., 538*, 968, 2000.

Numerical Simulation of Interacting Magnetic Flux Ropes

Dusan Odstrcil*, Marek Vandas†, Victor J. Pizzo** and Peter MacNeice‡

*University of Colorado, CIRES-NOAA/SEC, 325 Broadway, Boulder, Colorado 80305, USA
†Astronomical Institute, Academy of Sciences, Boční II 1401, 14131 Praha, Czech Republic
**NOAA/Space Environment Center, 325 Broadway, Boulder, Colorado 80305, USA
‡Drexel University, NASA/GSFC, Greenbelt, Maryland 20771, USA

Abstract. A $2\frac{1}{2}$-D MHD numerical model is used to investigate the dynamic interaction between two flux ropes (clouds) in a homogeneous magnetized plasma. One cloud is set into motion while the other is initially at rest. The moving cloud generates a shock which interacts with the second cloud. Two cases with different characteristic speeds within the second cloud are presented. The shock front is significantly distorted when it propagates faster (slower) in the cloud with larger (smaller) characteristic speed. Correspondingly, the density behind the shock front becomes smaller (larger). Later, the clouds approach each other and by a momentum exchange they come to a common speed. The oppositely directed magnetic fields are pushed together, a driven magnetic reconnection takes a place, and the two flux ropes gradually coalescence into a single flux rope.

INTRODUCTION

Coronal mass ejections (CMEs) represent a major transient release of mass and energy from the Sun. Recently, evidence of interacting CMEs was found in radio observations [1]. In this paper, we will investigate the dynamic interaction between two magnetic flux ropes using numerical magnetohydrodynamic (MHD) simulation. The aim is to provide some qualitative picture of the shock-cloud and cloud-cloud interactions.

NUMERICAL SIMULATIONS

The $2\frac{1}{2}$-D ideal MHD equations are solved in Cartesian coordinates using an explicit, multi-dimensional version of the TVDLF scheme [2]. The Paramesh adaptive mesh refinement (AMR) package [3] is used to obtain high resolution of fine structures.

We consider the case of a moderate ambient magnetic field with $\beta = 4.8$ (where β is the ratio of the thermal to magnetic pressure) with an adiabatic index $\gamma = 5/3$, initial temperature $T_0 = 3/5$, initial density $\rho_0 = 1$, and initial background magnetic field $B_0 = 0.5$. We use units where the gas constant and magnetic permeability are set to unity. Thus the sound velocity $C_0^S = (\gamma T_0)^{1/2} = 1$ and the Alfven velocity $C_0^A = |B_0|/\rho_0{}^{1/2} = 0.5$. Two magnetic flux ropes (clouds), left and right, are considered. The left cloud has the central field strength $B_L = 3 \times B_0$ and plasma density $\rho_L = 3 \times \rho_0$. The right cloud is specified with two different cases for its field strength and plasma density. Namely, $B_R = 4\times\ (2\times)\ B_0$ and $\rho_R = 2\times\ (4\times)\ \rho_0$ for Case 1 (2). The temperature is adjusted to provide constant thermal pressure everywhere. The magnetic pressure in the right cloud is a factor 16 (4) larger than in the external medium and the minimum β inside the cloud is 0.3 (1.2) in Case 1 (2).

Figure 1 shows the initial profile of various quantities for both cases. The density, temperature, and velocity are constant within the clouds. The magnetic field components follow Lundquist's force-free solution surrounded by a potential field. The clouds are initially cylindrical, the radius $R_L = R_R$ is set to unity, and this is chosen as the unit of length. The unit of time is set to the cloud sound crossing time $\tau = R_R/C_0^S$. Further, the flow velocity is expressed in units of the sound velocity. The rectangular domain of our simulation ($-12.5 \leq x \leq 12.5$ and $-12.5 \leq y \leq 12.5$) is chosen so that the boundaries are sufficiently far away from the clouds to avoid numerical artifacts. The center of the left cloud is located at $x = -7.5$ and $y = 0$. The center of the right cloud is located at $x = 0$ and $y = 0$. A maximum numerical resolution corresponds to a uniform grid with 512×512 computational cells. This gives a resolution of 20 zones per initial cloud radius, which is sufficient to capture basic cloud evolution.

RESULTS AND DISCUSSION

At time $t = 0$ the left cloud is set initially into motion with velocity $V_L = 1.5 \times C_0^S$ parallel to the ambient magnetic field. This leads to the immediate formation of a shock pair at the right (leading) edge of the cloud and rarefaction waves at the left edge of the cloud. A forward shock propagates ahead of the cloud to the right and a reverse shock propagates through the cloud to the left. We

CP679, *Solar Wind Ten: Proceedings of the Tenth International Solar Wind Conference*,
edited by M. Velli, R. Bruno, and F. Malara

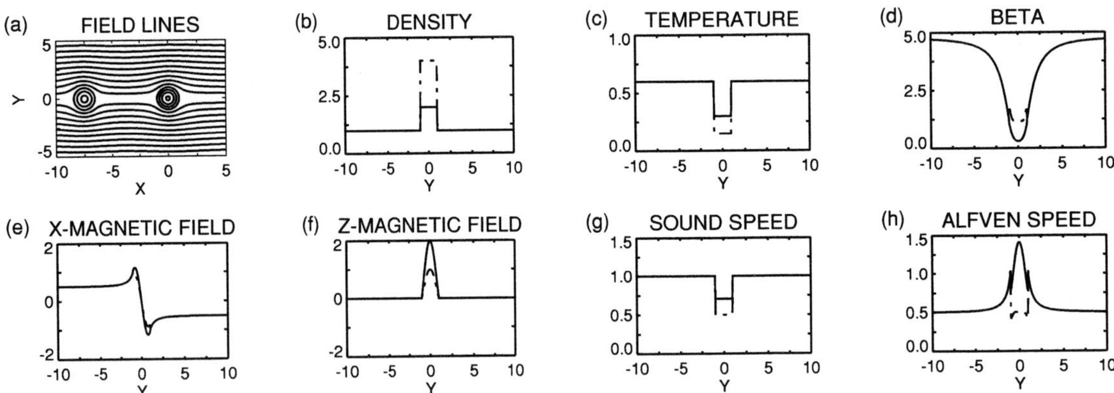

FIGURE 1. Initial conditions used for simulations of two interacting magnetic flux ropes. Two different cases with the same magnetic field topology (a) and parameters of the left cloud but differing parameters within the right cloud (b-h; profiles are through the cloud center, x=0) are considered in this paper. Case 1 (Case 2) has cloud parameters yielding larger (smaller) maximum characteristic speed than the background medium, as indicated by solid (dash-dot) line profiles.

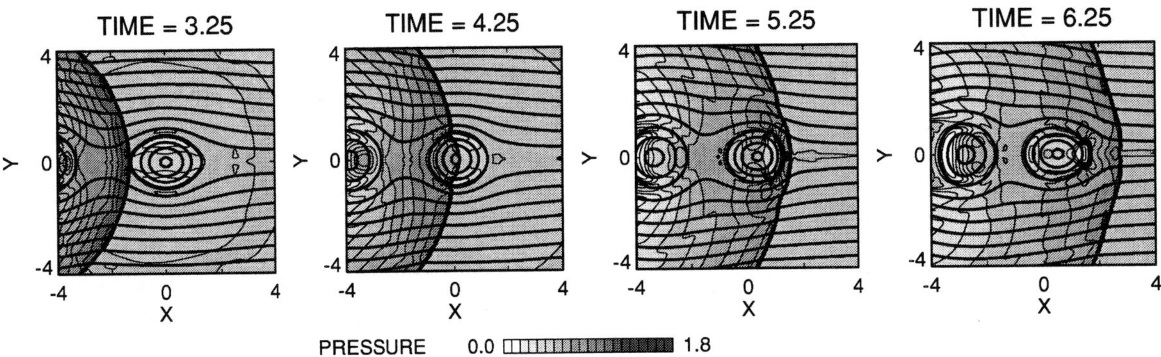

FIGURE 2. Interacting magnetic flux ropes for Case 1. Distribution of the thermal pressure (grey shading and light lines) and magnetic field lines (thick lines), is shown at four different times.

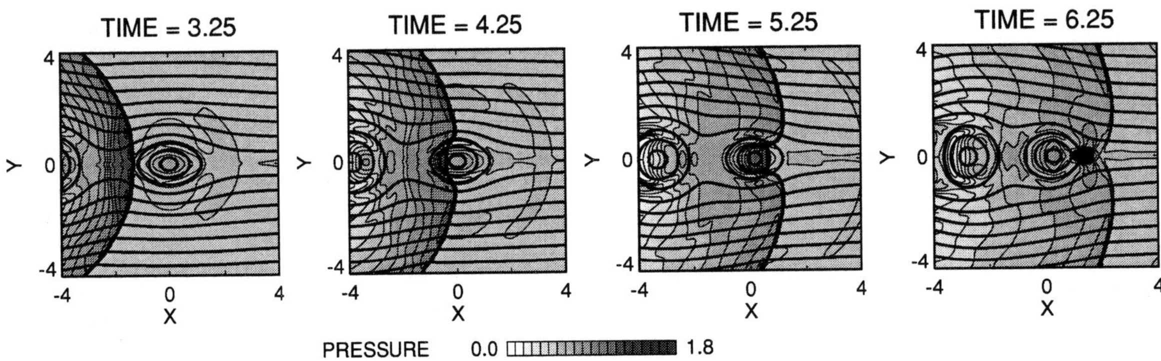

FIGURE 3. Interacting magnetic flux ropes for Case 2. Distribution of the thermal pressure (grey shading and light lines) and magnetic field lines (thick lines), is shown at four different times.

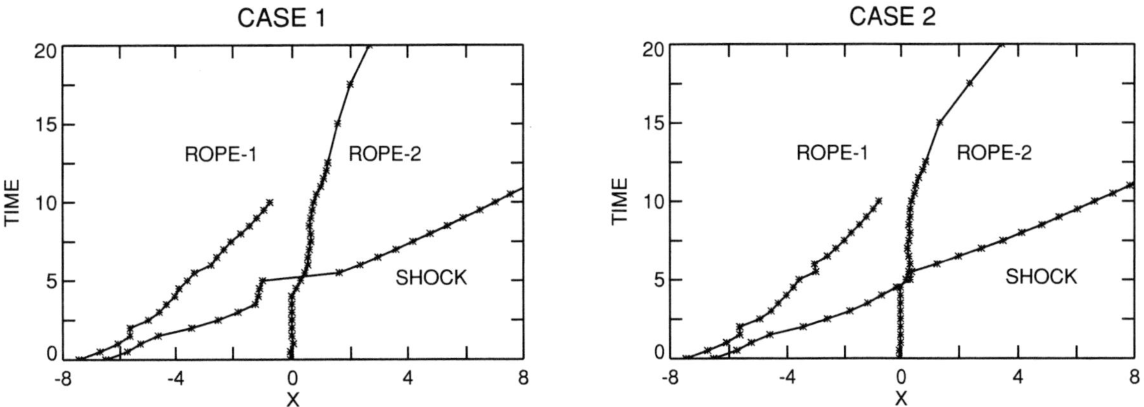

FIGURE 4. Trajectories of the magnetic flux rope centers (field strength maxima) and the shock (pressure maximum) for Case 1 (left panel) and Case 2 (right panel).

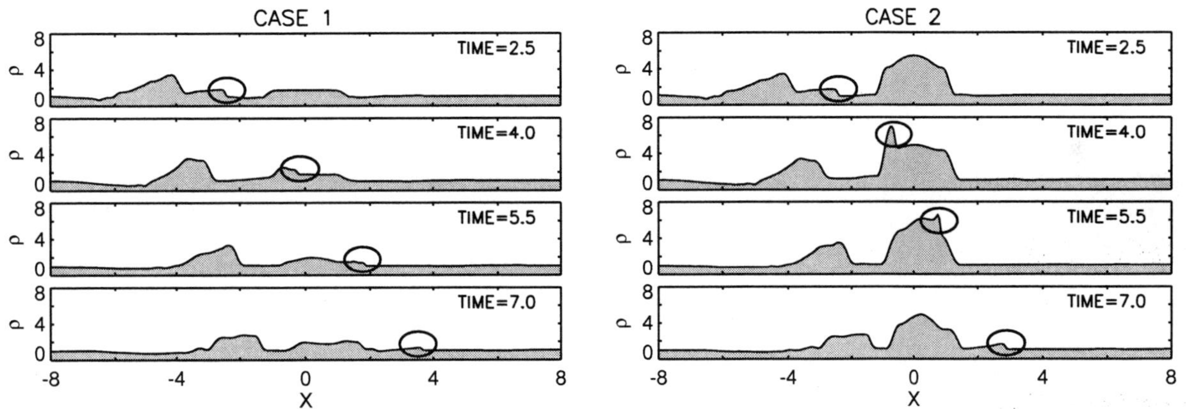

FIGURE 5. Density profiles through the magnetic flux rope centers (y=0) at four different times for Case 1 (left panel) and Case 2 (right panel). Positions of the shock are marked by a thick oval.

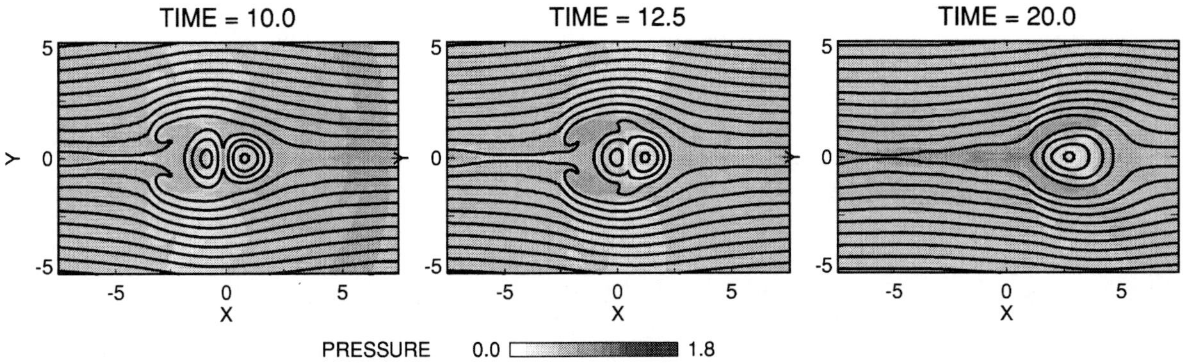

FIGURE 6. Interacting magnetic flux ropes for Case 1. Distribution of the thermal pressure (grey shading) and magnetic field lines (thick lines), is shown at three different times.

will discuss only the forward, "piston-driven", shock that forms in front of the left cloud and propagates toward the right cloud.

Figures 2 and 3 show the evolution of the dynamic interaction for Case 1 and Case 2, respectively. The left cloud generates a forward shock that starts to enter the right cloud at about t = 3.25. However, the differing parameters within the right cloud (see Fig. 1) cause differences during the shock-cloud interaction process. In Case 1, the larger characteristic speed causes faster shock propagation and concave-outward distortion of the shock front (Fig. 2). The shock leaves the cloud at about t = 5.0, visibly ahead of the shock portions that propagated through the surrounding medium. Since the shock is relatively weak, its effect on the cloud is weak as well. In Case 2, the lower characteristic speed causes slower shock propagation and concave-inward distortion of the shock front (Fig. 3). The shock leaves the cloud at about t = 5.75, significantly behind the shock portions that propagated through the surrounding medium. Effects on the cloud are larger due to the stronger shock. The magnetic structure is compressed and becomes oblate. Note that at larger times the shock fronts in both cases gradually acquire a uniform shape (after t = 8) and they "forget their history".

Figure 4 shows positions of the cloud centers and shocks as function of time for two different cases. When the shock enters the cloud, the momentum of the post-shock flow accelerates the cloud. This effect has been reported previously [4]. However, this acceleration is not permanent in our examples. Further, our results show that the shock can either accelerate or decelerate within the cloud depending on physical parameters within the cloud. Note that only shock acceleration was described in the interplanetary shock-cloud simulation [4].

Figure 5 shows the plasma density profiles through the cloud centers. The shock front is accelerated (decelerated) in the cloud with larger (smaller) characteristic speed. Correspondingly, the density behind the shock front becomes smaller (larger). This latter behavior is similar to ocean waves; as they approach the shallows near the coast, they slow down and their height increases.

Later, the clouds approach each other and in the subsequent momentum exchange they come to a common speed. Note that both flux ropes have the same rotation of the magnetic field, i.e., there are oppositely directed magnetic fields at the leading edge of the left cloud and at the trailing edge of the right cloud. These oppositely directed magnetic fields are pushed together and a driven magnetic reconnection takes a place. The reconnection process (caused by numerical diffusion) proceeds slowly and two flux ropes coalescence into a single flux rope gradually (Fig. 6).

CONCLUSIONS

The shock-cloud and cloud-cloud interactions are rather complex and different parameters can produce different outcomes. We present only two cases here, each having the same background state, cloud shape, and cloud velocity, but with different clouds characteristic speeds. We have found that: (1) the shock front is significantly distorted because it propagates faster (slower) in the cloud with larger (smaller) characteristic speed; and (2) correspondingly, the density behind the shock front becomes smaller (larger). Later on, the clouds approach each other, with the results that: (1) the clouds acquire the same speed; (2) the oppositely directed magnetic fields are pushed together and driven magnetic reconnection takes a place; and (3) the two flux ropes gradually coalescence into a single flux rope. The parameters of the clouds and background state we have considered are quite ideal, and they provide only a qualitative picture of the true interaction of CMEs and shocks.

During the interaction all characteristic parameters of the initial structures are modified, and this may lead to observable effects in remote observations of the coronal white-light and radio-emission, as well as for in-situ observations of energetic particles, plasma parameters, and magnetic field. Specifically, patchy enhancements of type II radio bursts and apparently irregular changes in frequency drift rate may be associated with localized shock strengthening and distortion, shock acceleration or deceleration, and the generation of complex shock patterns and reflections. Further, shock-cloud and cloud-cloud interactions may enhance electron and ion acceleration and/or magnetic field reconnection processes that will affect energetic particles. Finally, such interactions between the Sun and Earth can: (1) modify parameters of a single transient disturbance (shock strength, momentum, southward magnetic field); (2) reduce the number of shocks and magnetic clouds by "cannibalism"; and (3) lead to compound events with extended durations. All these effects complicate space weather forecasting in ways that are yet poorly understood.

ACKNOWLEDGMENTS

This work was supported by the DoD/AFOSR MURI Grant and by the Czech Academy of Sciences Grant A3003003 and Project S1003006. DO is on leave from Astronomical Institute, Ondřejov, Czech Republic.

REFERENCES

1. Gopalswamy, N., *Astrophys. J. Letts.* **548**, L91-L94 (2001).
2. Toth, G., and Odstrcil, D., *J. Comput. Phys.* **128**, 82-100 (1996).
3. MacNeice, P., Olson, K. M. Mobarry, C., Faichstein, de R., and Packer, C., *Comput. Phys. Commun.* **126**, 330-354 (2000).
4. Vandas, M., Fischer, S., Dryer, M., Smith, Z., and Detman, T., *J. Geophys. Res.* **102**, 22,295-22,300 (1997).

Models of Coronal Mass Ejections: A Review with A Look to The Future

Jon A. Linker*, Zoran Mikić*, Pete Riley*, Roberto Lionello* and Dusan Odstrcil†

*SAIC, 10260 Campus Point Dr., San Diego, CA 92121-1578, USA
†CIRES, University of Colarado, and NOAA/SEC, Boulder, CO 80305, USA

Abstract. Coronal mass ejections (CMEs) are a major transient input of mass and energy into the solar wind. We review some of the past and present concepts that influence the development of models of coronal mass ejections, both for CME initiation and CME evolution and propagation in the solar wind. We use the flux cancellation model to illustrate present research on CMEs. Primarily for convenience, modeling of CME propagation has usually been treated separately from the initiation problem. We suggest that future computational modeling of interplanetary CMEs is likely to emphasize the need to study coronal initiation and solar wind propagation together.

1. INTRODUCTION

Coronal mass ejections (CMEs) are dynamic, large-scale events in the solar corona that expel plasma and magnetic fields into the solar wind. CMEs typically appear as loop-like features that disrupt helmet streamers in the solar corona [1]. They were first observed with space-based coronagraphs in the early 1970's on OSO 7 [2] and Skylab [3]. Subsequent observations from the Solwind [4] and Solar Maximum Mission spacecraft [5] allowed identification of many of the properties of CMEs. The Solar and Heliospheric Observatory spacecraft has now extensively observed CME events from solar minimum in 1996 into the present maximum phase of the solar cycle. Halo events observed with the Large Angle Spectrometric Coronagraph (LASCO) [6, 7] now provide the most effective means of identifying earthward-directed CMEs, which are believed to be the primary cause of large, non-recurrent geomagnetic storms [8].

In situ signatures that are now recognized to be the interplanetary manifestations of CMEs have been measured for many years [9]. Perhaps the most frequently referred to example of an interplanetary CME is a magnetic cloud [10, 11] Magnetic clouds are identified in the solar wind as low beta plasmas associated with high field strength flux-rope structures; they are often (but not always) preceded by interplanetary shock waves [12]. While magnetic clouds are fairly common occurrences, many interplanetary structures that do not meet the criteria of magnetic clouds are also believed to be interplanetary CMEs. Other typical signatures of CMEs in the solar wind include counterstreaming suprathermal electrons [13] helium abundance enhancements [14] and low proton temperatures [15]. No single plasma or magnetic field characteristic is exhibited by all the structures identifed as interplanetary CMEs (see reviews by Gosling [16] and Neugebauer and Goldstein [17]).

Despite years of study, we still don't understand key aspects of CMEs; specifically, how are they initiated in the solar corona, and how they evolve to produce the signatures that are measured with interplanetary spacecraft. Clearly modeling must play a key role if we are to clarify these issues. There is a huge amount of literature on CMEs in general and CME modeling in particular, and we will not attempt to review the topic in detail. Rather, we show what some of the primary themes in CME modeling are today in the context of previous work, for both CME initiation and heliosphereic CME models. We briefly discuss the flux cancellation model as one of the present candidates for explaining CME initiation, and we discuss results obtained when the model is extended out into the interplanetary medium.

2. HOW ARE CMES INITIATED?

Models of CME initiation have been recently reviewed by Forbes [18], Klimchuk [19], and Low [20]; the reader is referred to these papers for a more comprehensive discussion. Our purpose here is to briefly outline the key issues regarding CME initiation, and what the likely direction of future research will be. CMEs can carry > 10^{32} ergs of kinetic energy, so the most obvious question in studying this phenomena is where does the energy come from? Indeed, models of CME initiation can be broadly classified by their postulated energy source: (1)

CP679, *Solar Wind Ten: Proceedings of the Tenth International Solar Wind Conference*,
edited by M. Velli, R. Bruno, and F. Malara

Energy storage models; (2) energy driven models; (3) thermal blast models. (Note that [19] has a significantly more detailed classification scheme.) We will discuss types (2) and (3) first.

Energy driven models hypothesize that magnetic energy can be injected at a sufficiently rapid rate to drive an eruption directly. They have been proposed since the 1970s [21] but are in contradiction with a number of observations [18]. Driving a large CME directly requires on the order of 10^{32}ergs to be dumped into the corona in a few thousand seconds; the required Poynting flux implies magnitudes for photospheric motions of magnetic fields that are not observed [22]. Chen [23, 24] has proposed a flux injection mechanism that has similar energy requirements to these previous models. The model appears to describe coronal observations of CMEs in terms of flux ropes reasonably well [25], and also has been used to model interplanetary magnetic clouds [24]. However, the proposed initiation mechanism suffers from the same observational difficulties as previous energy driven models [19]

The thermal blast models were the earliest explanation for CMEs, and hypothesize that they are initiated by a sudden release of thermal energy in the lower corona from a solar flare [26, 27]. These models are similar to energy-driven models in that they require an impulsive energy release, and this is the only part of the CME process that is modeled. However, thermal blast models also assume that the energy for the flare was stored in the coronal magnetic field prior to the flare occurence. The thermal blast model appeared to be a plausible explanation for CMEs based on the initial discoveries of CMEs. However, subsequent observations have revealed numerous problems with this model as an explanation for the vast majority of CMEs; for example, less than about 20% of CMEs are associated with a large flare [8], and many flare-associated CMEs are initiated prior to the flare occurrence [28, 29].

Energy storage models are generally considered to be the most likely candidates at the present time. These models assume that the energy that drives CMEs and other forms of solar activity is stored slowly in the magnetic field prior to eruption. Highly non-potential coronal magnetic fields in active regions have been observed frequently [30, 31, 32, 33], indicating that there is more than enough magnetic energy to drive coronal eruptions. This energy may be stored by photospheric motions shearing and twisting the coronal field, or the magnetic fields may already be twisted when they emerge from below the photosphere. Present estimates indicate that most of the twist in active region magnetic fields actually emerges from below the photosphere when the regions are born [34]. In any case, how the magnetic energy is stored is not a critical feature of energy storage models. The key question for these models is how is this energy released.

Another constraint on CME models is that CMEs open (i.e., drag out into the solar wind) at least a portion of the coronal magnetic field. In strong magnetic field regions low in the corona, the magnetic field pressure dominates both the plasma pressure and the gravitational force, so that fields that are in equilibrium are essentially force-free. Aly [35, 36] and Sturrock [37] have shown that the energy of the *open field* (for a given magnetic flux distribution, the magnetic field with all field lines extending to infinity) is the maximum energy for a force-free magnetic field. This appears to present a paradox: how can the magnetic field be opened while releasing energy? CME models have been devised that circumvent this constraint in a variety of ways; for example, magnetic reconnection (the limit only applies to ideal MHD solutions) or by only partially opening the magnetic field.

CME initiation models are now progressing to the point that they can provide at least zero-order explanations for the energy release seen in coronal mass ejections as well as some of the qualitative features observed in white light (e.g., the LASCO coronagraph). As an example, we describe the flux cancellation model [38], in which converging flows near the neutral line causes the formation of flux ropes, which are a candidate structure for supporting prominences. Flux cancellation has been defined as the mutual disappearance of magnetic fields of opposite polarity at the neutral line separating them [39]. Observations have shown this process to be active at filament sites [40]. Calculations by Forbes and Isenberg [41] and Lin et al. [42] have suggested that once a flux rope is formed, continuation of the flux cancellation process can result in a loss of equilibrium. The new, lower energy equilibrium contains a current sheet and a higher height for the flux rope. In resistive MHD simulations, this current sheet is the site of rapid magnetic reconnection [43, 44, 45]. Figure 1 shows how a simulated helmet streamer configuration is disrupted by this process. The helmet streamer configuration shown has highly sheared magnetic field lines near the neutral line (see Linker et al. [45] for details), as is commonly observed in filament channels [46]. Flux cancellation first forms a stable flux rope configuration within the helmet streamer (2nd frame). The high density in the flux rope (seen in the white light image) is reminiscent of a prominence but because of the simplified (polytropic) energy equation, the plasma does not have the correct thermodynamic properties (it is too hot). (Linker et al. [44] have shown that when a more detailed energy equation is used, cool prominence-like material is lifted into the corona.) When flux cancellation continues, the helmet streamer is destabilized at $t_0 + 16$ hours (not shown) and subsequently erupts outward into the corona ($t_0 + 18$ and $t_0 + 20$ hours). The white light images show, albeit in a very idealized way, the 3 part structure often seen in CMEs.

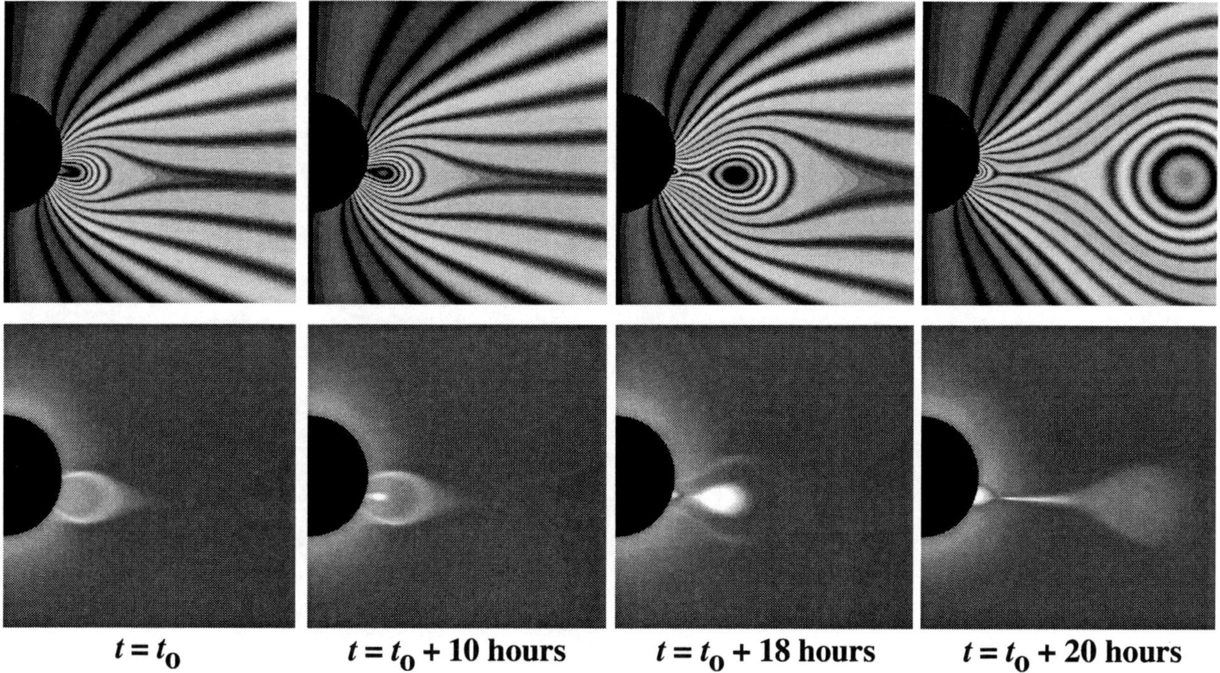

FIGURE 1. MHD Simulation of a helmet streamer eruption triggered by flux cancellation. The stripes in the top panels show projected field lines (there is also a component of the magnetic field, B_ϕ, out of the plane) and the bottom panels shows the polarization brightness that would be observed by a coronagraph if this were a real CME. At $t_0 + 10$ hours a high density flux rope has formed; this structure is stable if flux cancellation is halted. With continued flux cancellation, the configuration erupts.

The "breakout" model [47] is another viable initiation mechanism. It requires a multipolar flux distribution, and like the flux cancellation model, requires strongly sheared fields near the neutral line as is observed in filament channels. As shear in the central arcade is increased, slow reconnection transfers overlying flux in the central arcade to the neighboring arcades. As the restraining force of the overlying flux is decreased, the sheared field lines rises further, causing yet more overlying flux to be diverted. Eventually all of the overlying flux is removed and the sheared central arcade erupts explosively. While MHD simulations for the breakout model have not yet been performed on a more realistic helmet streamer configuration, they have demonstrated an explosive energy release.

Flux cancellation and breakout are two of the best developed models at the moment, but there are many others (including a recent model by Low and Zhang [48]). Why can't we tell which (if any) of these models is correct? In many cases the expected observational differences between the models are subtle. For example, flux cancellation requires a flux rope prior to eruption, while breakout does not. Distinguishing between field lines that wrap around each other ("flux rope") from a collection of strongly-sheared dipped field lines is quite difficult for realistic fields. We note that after eruption, the ejected material is expected to be embedded in a flux rope in both models. Indeed, both coronal and interplanetary observations indicate that some fraction of CMEs contain flux ropes, but unfortunately this information does not discriminate between any of the actively considered models. (Incidentally, this explains why the Chen [23] flux rope model, which appears to have untenable assumptions about CME energetics, can still model LASCO images of CMEs reasonably well.)

While breakout, flux cancellation, and other energy storage models can explain some aspects of the observations, so far they have only been studied for relatively idealized configurations. The challenging task for all CME models is to increase their sophistication to the point where calculations of specific events can be performed, and observable quantities, such as disk emission seen in Yohkoh soft X-ray images and SOHO Extreme ultraviolet Imaging Telescope (EIT) images, can be realistically predicted and tested against the observations. This is especially true for observations that will be available in the next few years (e.g., the Solar-B and STEREO missions).

Kopp and Pneuman (1976) - "Post-Flare" Loops Formed From Reconnection

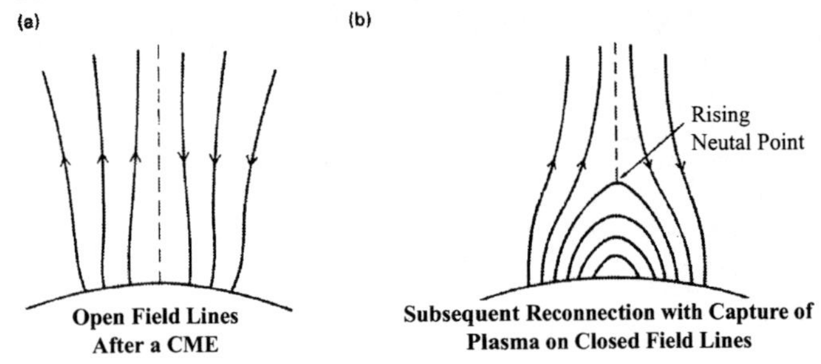

MHD Simulation of Streamer Disruption

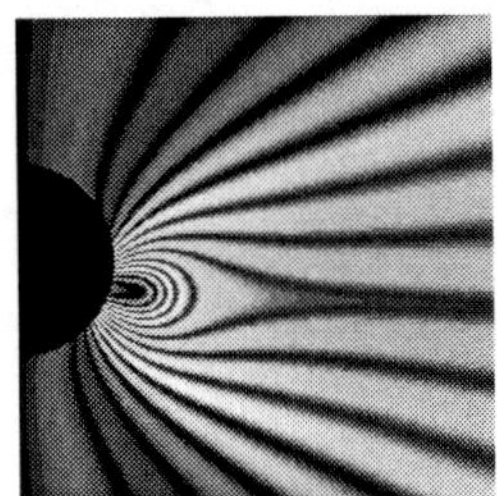

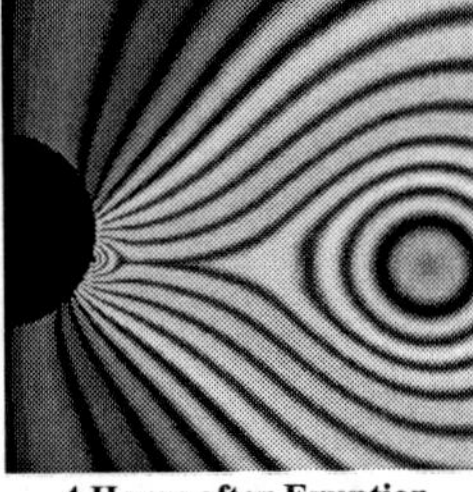

Energized Streamer | **4 Hours after Eruption** | **2 days After Eruption**

FIGURE 2. (a) (After Kopp and Pneuman [49].) Eruption opens previously closed magnetic field lines. (b) reconnection reforms the closed field, resulting in images of successively higher "postflare" loops. (c) Reformation of the helmet streamer in the aftermath of a CME (from the flux cancellation simulation of Figure 1).

3. WHAT WE THINK WE DO UNDERSTAND ABOUT CMES?

While the underlying cause of CMEs is unresolved, the eruption process in the corona has been documented by many observations. At least for the simplest mass ejections, a picture that began to emerge over 25 years ago reconciles a number of the observations. When CMEs disrupt helmet streamers, they open previously closed magnetic field regions (Figure 2a). Kopp and Pneuman [49] suggested that the reclosing of these magnetic fields led to the formation of the successivly higher "postflare" loops seen in the aftermath of eruptive events (Figure 2b). The phenomenon has since been associated with CMEs [50]. MHD simulations of helmet streamer disruptions via flux cancellation show the formation of new loops in the aftermath of the eruptive process (Figure 2c). This effect appears to be independent of the mechanism that disrupts the helmet streamer; for example, the post-eruptive loop formation is seen in simulations of CMEs initiated by photospheric shearing flows [51]. The well known observation of bands moving away from the neutral line in two-ribbon flares observed in Hα [52] is explained by this mechanism as the reconnection line at the top of the arcades moving upward. As the arcades reform and increase in size, the Hα ribbons (bounding the arcades) move outward.

Figure 3 shows an example of post-eruption loop formation for the CME event of September 12, 2000. In the pre-event state, a prominence is seen in both EUV and Hα. A halo CME was subsequently observed in the LASCO C2 and C3 coronagraphs, during which time the formation of the post-eruption loops were observed in EIT and Yohkoh.

4. CME EVOLUTION IN THE SOLAR WIND

Until recently, most multi-dimensional models of interplanetary CMEs in the solar wind eliminated the inner corona from consideration by starting at $20-40R_S$ (beyond the Alfvén and sonic points in the corona). This approach enormously simplifies the inner boundary condition; beyond these critical points the MHD characteristics all point into the computational domain and the upstream quantities can be arbitrarily specified. The earliest studies looked at the propagation of interplanetary shockwaves [53, 54] based on the "thermal blast" model of CMEs (see section 2). Later models recognized

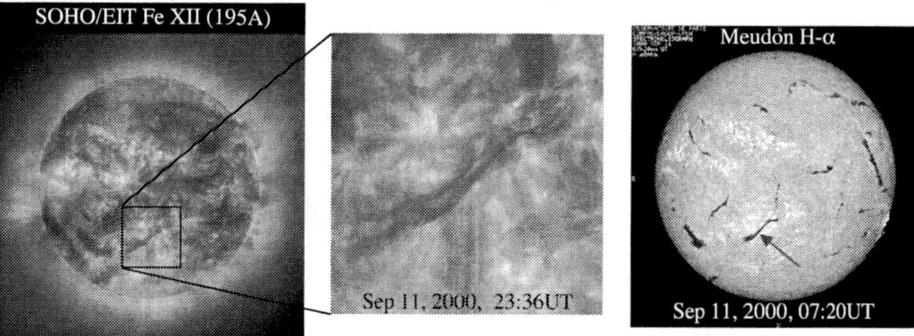

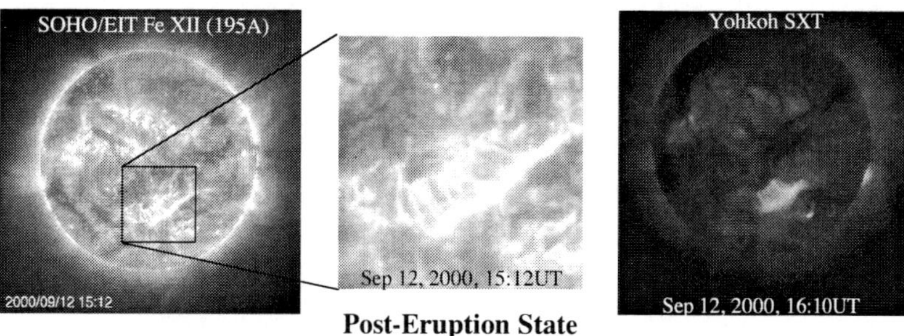

FIGURE 3. Pre- and post event images for the September 12,2000 CME. (a) EIT and Hα images of a filament prior to eruption. The red arrow in the Hα image indicates the position of the filament. (b) EIT and Yohkoh images of the post-eruptive state, showing the presence of post-eruption loops after the disappearance of the filament. During this time the CME was observed in LASCO images.

the importance of the magnetic topology of the CME and investigated the propagation of spheromak structures and cylindrical flux ropes [55, 56, 57], including models of the localized effects of propagation on the flux-rope structure [58]. Riley et al. [59] and Odstrcil and Pizzo [60, 61] investigated the effects of solar wind velocity structure on CME ejecta, using an idealized density pulsed injected into a two-state (fast and slow) solar wind. Odstrcil and Pizzo [62] also investigated how the background solar wind magnetic field is affected by the ejecta. The complicated density structures that emerged from these calculations show the richness of behavior that is possible.

While models that provide solutions beyond the critical points have instructed us about important aspects of CME propagation, the properties of the CME ejecta in these calculations are somewhat arbitrary. By definition, these models cannot directly connect the initiation of a CME with the subsequent interplanetary observations, this requires a calculation starting at the Sun. Such calculations have already been performed. The three-dimensional (3D) computation by Linker and Mikić [63] showed how differential rotation on the Sun can eventually disrupt a helmet streamer, ejecting a portion of the streamer belt out into the solar wind. Wu et al. [64] performed a 2D calculation that followed the evolution of a simulated helmet streamer eruption from the corona out to Earth orbit, and Groth et al. [65] performed a 3D computation that included the interaction of the simulated CME with the Earth's magnetosphere. All of these models were useful for demonstrating both CME initiation and propagation in a single calculation, but the initiation mechanisms themselves were not very realistic. The differential rotation studied by Linker and Mikić [63] is unlikely to be a major source of energization for coronal magnetic fields. The Wu et al. [64] model imposed an ad hoc increase of the magnetic field perpendicular to the plane of the computation (B_ϕ in spherical coordinates) for an arbitrarily specified portion of the grid; this created a flux rope that eventually disrupted the helmet streamer [66]. Groth et al. [65] relied on a pressure pulse similar to the earlier thermal blast models.

Ideally, we would like the calculations of CME initiation described in section 2 to be extended out into the solar wind; solar wind measurements might possibly then provide another test of the proposed initiation mechanisms. As a demonstation of this idea, Odstrcil et al. [67] have coupled the SAIC coronal model with the

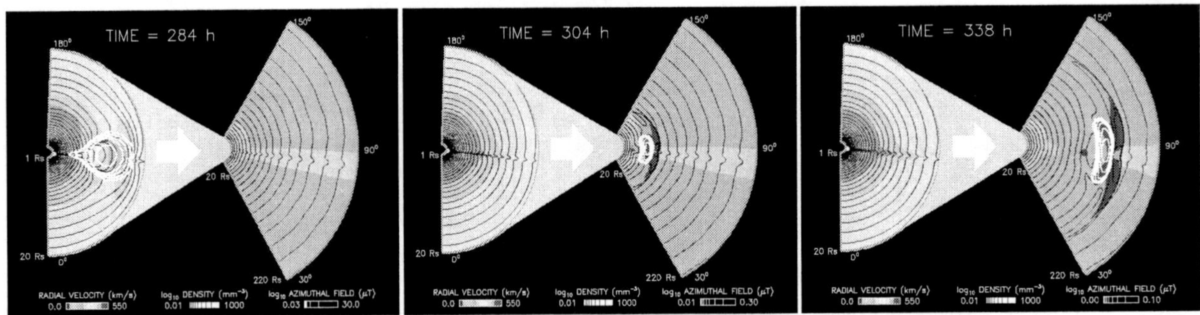

FIGURE 4. Computation of a CME initiated by flux cancellation out to 1 A.U., using the coupled SAIC and NOAA/SEC MHD models. In each frame, the leftmost hemisphere shows the domain of the coronal simulation, which feeds into the heliospheric simulation (wedge-shaped domain on the right). Black lines show contours of density. Colors show the solar wind speed. White lines show contours of B_ϕ and mark the approximate location of the flux rope. The region of highest velocity (red) and density pileup show the location of the shockwave.

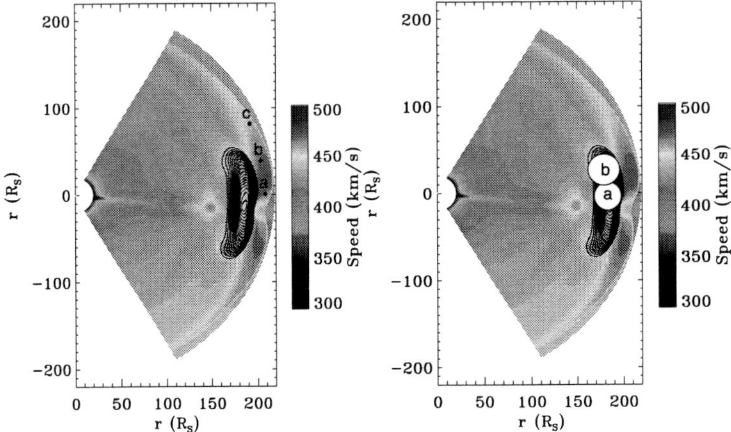

FIGURE 5. In the left frame, the simulated CME near 1 A.U. (from the same calculation of Figure 4) is shown together with a color map of the solar wind speed. Black lines are contours of B_ϕ and show the location of the flux rope. The positions of hypothetical spacecraft that encounter the simulated CME are marked a, b, and c. The right frame is the same as the left, but the white circles show the size and shape of the flux ropes inferred by fits to a linear force-free (LFF) model.

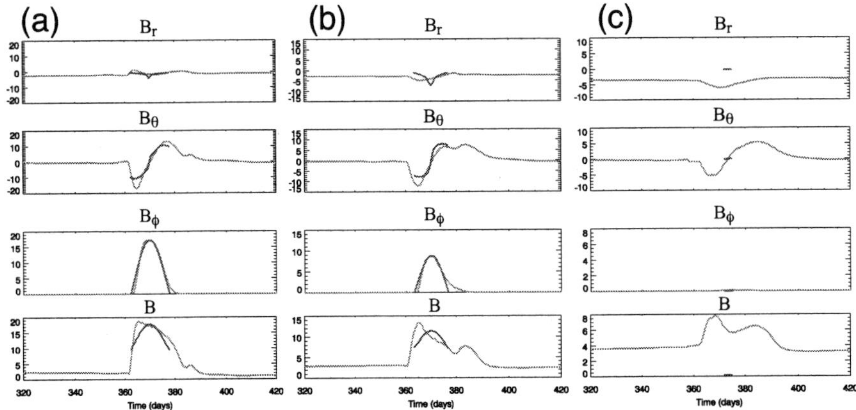

FIGURE 6. The time series of the magnetic field (blue lines) as "measured" by the hypothetical spacecraft a, b, and c in Figure 5. The field resulting for a best fit to a linear force free cylindrical flux rope is shown in red; the inferred flux ropes are the white circles shown in Figure 5.

NOAA/SEC heliospheric model to extend the flux cancellation simulation of Figure 1 out to 1 A.U. Figure 4 shows results from these coupled models. On the left of each frame, the domain of the coronal model is shown while on the right the domain of the heliospheric model is depicted. The simulation shows that a flux rope (identified by the white contours showing the concentration of B_ϕ) propagates outward. Close to the Sun the shape of the flux rope is nearly circular but it becomes wedge-shaped as it drives into the ambient slower solar wind; this is accompanied by the formation of a fast mode shockwave ahead of the ejecta.

The calculation of Figure 4 is highly idealized, nevertheless it shows that even the simple propagation of a flux rope can lead to the formation of a relatively complex structure. It is interesting to investigate how data for this simulated flux rope might be appear for spacecraft situated in different locations. The leftmost frame of Figure 5 shows the plasma speed as a color map and the position of the flux rope (black contours) when the CME is near Earth orbit. (The rightmost frame of Figure 5 is discussed below.) The position of three hypothetical spacecraft (a,b,c) is also shown. At this point in time, the shockwave has just passed over all of the spacecraft. Spacecraft (a) and (b) are about to encounter the simulated CME ejecta. The blue curves in Figure 6 shows the simulated time series of the magnetic field "measured" by the three spacecraft.

Linear (constant α) force-free field models are frequently found to give good fits to interplanetary flux ropes [68]. We derived a fit to a simple linear force-free (LFF) model to the time series of data observed by each hypothetical spacecraft. The red lines in Figure 6 show the results of this fitting procedure. The LFF model finds quite a good fit to the magnetic field for the time series from spacecraft (a), and the model also finds a reasonable fit for the data from spacecraft (b). Spacecraft (c) misses the flux rope entirely and the LFF model correctly finds no fit. This simple LFF model assumes a cylindrical flux rope shape, a frequent assumption of the models used to interpret interplanetary data. The white circles in the rightmost frame of Figure 5 show the position and shape of the flux ropes inferred by the LFF model. We see that the sizes and shapes of the inferred flux ropes are very misleading, despite the fact that reasonable fits were obtained. Our results indicate that one needs to be careful about assuming that flux ropes are actually cylindrical, even if a cylindrical model gives a good fit to the data. Non-force free [69] or noncylindrical [70] approaches might yield a more accurate description of the shape. Given the relative simplicity of this idealized MHD simulation, one can easily imagine why in the real solar wind, CME ejecta can develop far more complicated configurations. MHD simulations may help to give us insights into the best way to utilize measurements of these structures. Interplanetary observations in turn provide an important test of CME models.

5. SUMMARY

Modeling of CMEs has traditionally been separated into coronal calculations that primarily focus on CME initiation, and heliospheric models that focus on interplanetary evolution of CMEs. Solving the puzzle of CME genesis will likely require more sophisticated CME simulations that can be applied to specific events. As CME initation models are driven to more realism in order to confront the detailed coronal images we now obtain in a variety of wavelengths, the natural course is to also extend the initiation models into interplanetary space. We have shown here an example for flux cancellation, a candidate initiation mechanism. Coupled initiation and interplanetary propagation models of CMEs are likely to increase in sophistication in the near future.

Acknowledgments

The images in Figure 3 are from the EIT instrument aboard the SOHO spacecraft, the Yohkoh soft X-ray telescope, and the spectroheliograph at Meudon observatory. This work was supported by NASA's SR&T, LWS, and SECTP programs, and by NSF grants ATM0000950 and ATM0120950 (The Center for Integrated Space Weather Modeling). Computations were performed at the San Diego Supercomputer Center. We also wish to acknowledge beneficial discussions at the SHINE workshops.

REFERENCES

1. Hundhausen, A. J., *J. Geophys. Res.*, **98**, 13,177, 1993.
2. Tousey, R., *Bull. Amer. Astronom. Soc.*, **5**, 419, 1973.
3. MacQueen, R. M., Eddy, J. A., Gosling, J. T., Hildner, E., Munro, R. H., Newkirk, G. A., Jr., Poland, A. I., and Ross, C. L., *Astrophys. J.*,**187**, L85, 1974.
4. Michels, D. J., Howard, R. A., Koomen, M. J., and Sheeley Jr., N. R., in *Radio Physics of the Sun*, ed. by M. R. Kundu and T. Gergely, D. Reidel, Hingham, MA, 1980, p. 439.
5. MacQueen, R. M., Csoeke-Poeckh, A., Hildner, E., House, L., Reynolds, R., Stanger, A., Tepoel, H., and Wagner, W., *Solar Phys.*, **65**, 91, 1980
6. Brueckner, G. E., Howard, R. A., Koomen, M. J., Korendyke, C. M., Michels, D. J., Moses, J. D., Socker, D. G., Dere, K. P., Lamy, P. L., Llebaria, A., Bout, M. V., Schwenn, R., Simnett, G. M., Bedford, D. K., Eyles, C. J., *Solar Phys.*, **162**, 375, 1995.
7. Howard, R. A., et al., in Coronal Mass Ejections (N. Crooker, J. Joselyn and J. Feynman, eds.), *Geophys. Monogr.*, **99**, 17, 1997.

8. Gosling, J. T., *J. Geophys. Res.*, **98**, 18,937, 1993.
9. Gosling, J. T., Pizzo, V., Bame, S., *J. Geophys. Res.*, **78**, 2001, 1973.
10. Burlaga, L., Sittler, E., Mariani, F., Schwenn, R., *J. Geophys. Res.*, **86**, 6,673, 1981.
11. Klein, L. W., Burlaga, L. F., *J. Geophys. Res.*, **87**, 613, 1982.
12. Marubashi, K., in Coronal Mass Ejections (N. Crooker, J. Joselyn and J. Feynman, eds.), *Geophys. Monogr.*, **99**, 147, 1997.
13. Gosling, J. T., Thomsen, M. F., Bame, S. J., Zwickl, R. D., *J. Geophys. Res.*, **92**, 12,399, 1987.
14. Borini, G., Gosling, J. T., Bame, S. J., Feldman, W. C., *J. Geophys. Res.*, **87**, 7,370, 1981.
15. Richardson, I. G., and Cane, H. V., *J. Geophys. Res.*, **100**, 23,397, 1995.
16. Gosling, J. T., in Coronal Mass Ejections (N. Crooker, J. Joselyn and J. Feynman, eds.), *Geophys. Monogr.*, **99**, 9, 1997.
17. Neugebauer, M., Goldstein, R., in Coronal Mass Ejections (N. Crooker, J. Joselyn and J. Feynman, eds.), *Geophys. Monogr.*, **99**, 245, 1997.
18. Forbes, T. G., *J. Geophys. Res.*, **105**, 23,153, 2000.
19. Klimchuk, J. A., in *Space Weather*, ed. by P. Song, H. Singer, and G. Siscoe, *Geophys. Monogr.*, **125**, AGU, Washington, 2001, p. 143.
20. Low, B. C., *J. Geophys. Res.*, **106**, 25,141, 2001.
21. Heyvaerts, J., *Sol. Phys.*, **38**, 419, 1974.
22. McClymont, A. N., Fisher, G. H., in Solar System Plasma Physics, (J. H. Waite, J. L. Burch, and R. L. Moore, eds.), *Geophys. Monogr.*, **54**, 219, 1989.
23. Chen, J., *Astrophys. J.*, **338**, 453, 1989.
24. Chen, J., *J. Geophys. Res.*, **101**, 27,499, 1996.
25. Chen, J., Howard, R. A., Brueckner, G. E., Santoro, R., Krall, J., Paswaters, S. E., St. Cyr, O. C., Schwenn, R., Lamy, P., Simnett, G. M., *Astrophys. J.*, **490**, L191, 1997.
26. Dryer, M., *Space Sci. Rev.*, **33**, 233, 1982.
27. Wu, S. T., *Space Sci. Rev.*, **32**, 115. 1982.
28. Harrison, R. A., *Astron. and Astrophys.*, **162**, 283, 1986.
29. Hundhausen, A. J., in Proceedings of the Sixth International Solar Wind Conference (Solar Wind 6) (V. J. Pizzo, T. E. Holzer, and D. G. Sime, eds), NCAR/TN-306+Proc, Boulder, 181, 1988.
30. Gary, G. A., Moore, R. L., Hagyard, M. J., Haisch, B. M. , *Astrophys. J.*, **314**, 1987.
31. Hagyard, M. J., *Sol. Phys.*, **115**, 1988.
32. Canfield, Richard C., de La Beaujardiere, J.-F., Fan, Y., Leka, K. D., McClymont, A. N., Metcalf, T. R., Mickey, D. L., Wuelser, J-P., Lites, B. W., *Astrophys. J.*, **411**, 362, 1993.
33. Leka, K. D., Canfield, Richard C., McClymont, A. N., de La Beaujardiere, J.-F., Fan, Y., and Tang, F., *Astrophys. J.*, **411**, 370, 1993.
34. Démoulin, P., Mandrini, C. H., van Driel-Gesztelyi, L., Thompson, B. J., Plunkett, S., Kovári, Z., Aulanier, G., Young, A., *Astron. and Astrophys.*, **382**, 2002.
35. Aly, J. J., *Astrophys. J.*, **283**, 349, 1984.
36. Aly, J. J., *Astrophys. J.*, **375**, 61L, 1991.
37. Sturrock, P. A., *Astrophys. J.*, **380**, 655, 1991.
38. van Ballegooijen, A. A., and Martens, P. C. H., *Astrophys. J.*, **361**, 971, 1989.
39. Martin, S. F., S. H. B. Livi, and Wang, J., *Australian J. Phys.*, **38**, 929, 1985.
40. Litvinenko, Y. E., Martin, S. F., *Solar Phys.*, **190**, 45, 1999.
41. Forbes, T. G., Isenberg, P. A., *Astrophys. J.*, **373**, 294, 1991.
42. Lin, J., Forbes, T. G., Isenberg, P. A., Démoulin, P., *Astrophys. J.*, **504**, 1,006, 1998.
43. Amari, T., Luciani, J. F., Mikić, Z., and Linker, J., *Astrophys. J.*, **529**, 49L, 2000.
44. Linker, J. A., Lionello, R., Mikić Z., and Amari, T., *J. Geophys. Res.*, **106**, 25165, 2001.
45. Linker, J. A., Mikić Z., Lionello, R., Riley, P., Amari, T., and Odstrcil, D., *Phys. Plasmas*, to appear, 2003.
46. Martin, S. F., and C. R. Echols, in *Solar Surface Magnetism*, ed. by R. J. Rutten and C. J. Schrijver, Kluwer Academic, Dordrecht, Netherlands, 1994, p. 339.
47. Antiochos, S. K., MacNeice, P. J., Spicer, D. S., and Klimchuk, J. A., *Astrophys. J.*, **512**, 985, 1999.
48. Low, B. C., Zhang, M., *Astrophys. J.*, **564**, L53, 2002.
49. Kopp, R. A., Pneuman, G. W., *Solar Phys.*, **50**, 85, 1976.
50. Hiei, E., Hundhausen, A. J., Sime, D. G., *Geophys. Res. Lett.*, **20**, 2,785, 1993.
51. Linker, J. A., Mikić, Z., *Astrophys. J.*,**438**, L45, 1995.
52. Zirin, H., *Astrophysics of the Sun*, Cambridge, 1988.
53. Wu, S. T., Dryer, M., Han, S. M., *Solar Phys.*, **84**, 395, 1983.
54. Dryer, M., *Space Sci. Rev.*, **67**, 363, 1994.
55. Detman, T. R., Dryer, M., Yeh, T., Han, S. M., Wu, S. T., *J. Geophys. Res.*, **96**, 9,531, 1991.
56. Vandas, M., Fischer, S., Dryer, M., Smith, Z., and Detman, T., *J. Geophys. Res.*, **101**, 2505, 1996.
57. Vandas, M., Fischer, S., Dryer, M., Smith, Z., and Detman, T., *J. Geophys. Res.*, **101**, 15,645, 1996.
58. Cargill, P. J., Chen, J., Spicer, D. S., Zalesak, S. T., *J. Geophys. Res.*, **101**, 4,855, 1996.
59. Riley, P., Gosling, J. T., and Pizzo, V. J., *J. Geophys. Res.*, **102**, 14,677, 1997.
60. Odstrcil, D., Pizzo, V., *J. Geophys. Res.*, **104**, 483, 1999.
61. Odstrcil, D., Pizzo, V., *J. Geophys. Res.*, **104**, 493, 1999.
62. Odstrcil, D., Pizzo, V., *J. Geophys. Res.*, **104**, 28,225, 1999.
63. Linker, J. A., Mikić, Z., in *Coronal Mass Ejections*, ed. by N. Crooker, J. Joselyn and J. Feynman, *Geophys. Monogr.*, **99**, AGU, Washington, 1999, p.269.
64. Wu, S. T. Guo, W. P., Michels, D. J., Burlaga, L. F., *J. Geophys. Res.*, **104**, 14,789, 1999.
65. Groth, C. P. T., de Zeeuw, D. L., Gombosi, T. I., Powell, K. G., *J. Geophys. Res.*, **105**, 25,053, 2000.
66. Wu, S. T., Guo, W. P., in Coronal Mass Ejections (N. Crooker, J. Joselyn and J. Feynman, eds.), *Geophys. Monogr.*, **99**, 83, 1997.
67. Odstrcil, D., Linker, J. A., Lionello, R., Mikić, Z., Riley, P., Pizzo, V. J., and Luhmann, J., *J. Geophys. Res.*, **107**, 10.1029/2002JA009334, 2002.
68. Lepping, R. P., Burlaga, L. F., Jones, J. A., *J. Geophys. Res.*, **95**, 11,957, 1990.
69. Hidalgo, M. A., Cid, C., Medina, J., Viñas, A. F., *Solar Phys.*, **194**, 165, 2000.
70. Mulligan, T, Russell, C. T., *J. Geophys. Res.*, **106**, 10,581, 2001.

Solar Wind Disturbances and Their Sources in the EUV Solar Corona

A. N. Zhukov*†, I. S. Veselovsky†, F. Clette*, J.-F. Hochedez*, A. V. Dmitriev†, E. P. Romashets**, V. Bothmer‡ and P. Cargill§

*Royal Observatory of Belgium, Avenue Circulaire 3, B–1180 Brussels, Belgium
†Institute of Nuclear Physics, Moscow State University, Moscow 119992, Russia
**IZMIRAN, Troitsk, Moscow Region 142190, Russia
‡Max-Planck-Institut für Aeronomie, Max-Planck-Str. 2, D–37191 Katlenburg-Lindau, Germany
§Imperial College of Science, Technology and Medicine, London, SW7, 2BZ, UK

Abstract. We investigate possible links between the activity manifestations in the solar corona and conditions in the solar wind. For the reduction of this immense task we have selected 206 events in the solar wind in 1997 – 2000 corresponding to geomagnetic events with $A_p > 20$ (compiled into a database at `http://alpha.sinp.msu.ru/apev`). Up to now, 24 events during the epoch of low solar activity (January 1997 – January 1998) are investigated. The solar wind conditions monitored by ACE and WIND spacecraft were traced back to the solar corona observed by SOHO/EIT. The search for coronal signatures which are probably associated with the disturbed solar wind conditions was performed. The coronal sources of these 24 events are identified, namely: eruptions in active regions, filament eruptions and coronal holes. It is shown that halo and partial halo CMEs observed within the SOHO/LASCO sensitivity limits are not necessary indicators of Earth-directed eruptions, and coronal EUV dimmings can be used as a complementary indicator. We also found that a structure now conventionally called a "sigmoid" cannot be represented as a single S-shaped loop (flux tube), but exhibits an assembly of many smaller structures. It could be formed and destroyed via eruptions.

INTRODUCTION

Physical conditions on the Sun responsible for the production of appreciable geomagnetic perturbations are still under investigation. The biggest geomagnetic storms develop in association with the appearance of strong heliospheric perturbations seen in plasma and magnetic field parameters measured by satellites in the solar wind. They are commonly related to powerful nonstationary processes in the solar corona and deeper layers of the solar atmosphere visualized by different observations. Impulsive and long-duration solar flares, disappearing filaments, coronal mass ejections (CMEs), transient brightenings (dimmings) and coronal holes represent the most popular signatures to date used for diagnostic purposes.

The aim of this paper is to report several examples illustrating the preliminary results on the search of the geomagnetic disturbances sources in the SOHO/EIT data.

ANALYSIS AND RESULTS

We have selected 206 events in 1997 – 2000 corresponding to the days with daily A_p index more than 20 (an upper limit of the one standard deviation interval in 1997 – 2000). The statistical analysis of these events was done by Bothmer et al. [1]. The events were compiled into the database located at `http://alpha.sinp.msu.ru/apev`. To date 24 events occurred in January 1997 – January 1998 were investigated, thus the epoch of low solar activity is considered in this paper.

The procedure of analysis is as follows. First, using ACE and WIND spacecraft data, we identify the solar wind disturbance which produced increase of A_p index (shock, discontinuity of another type, region of strong negative B_z, increase in velocity, pressure etc). Average velocity of the disturbance is found and the approximate start time from the Sun is estimated assuming constant solar wind velocity en route from the Sun to the Earth. To establish more precise timing, halo (angular width 360°) and partial halo (angular width more than 120°) CMEs detected by SOHO/LASCO close to the estimated start time were identified using the CSPSW/NRL CME catalog (`http://cdaw.gsfc.nasa.gov/CME_list/index.html`). In most cases a possible source halo/partial halo CME is found to happen in one – two days interval around the estimated start time. The coronal activity observed by EIT close to the initiation

CP679, *Solar Wind Ten: Proceedings of the Tenth International Solar Wind Conference*,
edited by M. Velli, R. Bruno, and F. Malara

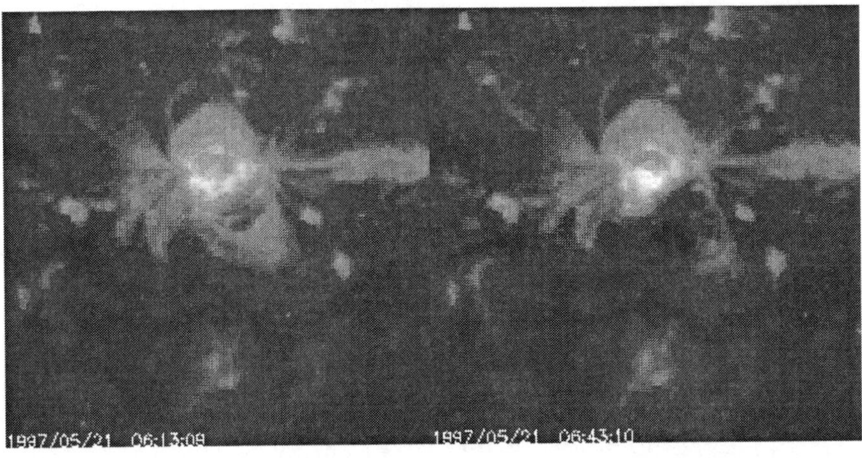

FIGURE 1. NOAA AR 8040 observed by SOHO/EIT in the Fe XII bandpass (195 Å) on May 21, 1997 at 06:13:09 UT (left panel) and 06:43:10 UT (right panel). Note the dimming in the SW part of the AR in the right panel.

of this CME was studied to identify the coronal structure and/or process which is the most probable candidate to be the source of the CME (and of the geomagnetic disturbance). Sometimes no halo/partial halo CMEs were observed during these days. In this case we investigate the EIT data in the 3-day window around the estimated start time.

The list of events with their identified sources is shown in Table 1. Three events (numbers 2, 3, 10) are produced by flows from coronal holes (CH), either equatorial ones (event 3) or extensions of polar CHs to the equator (events 2, 10). 7 events (9, 12, 13, 15, 16, 20, 21) are produced by eruptions in active regions (AR). Event 11 is a combined event: plasma erupted from the AR is pushed by a faster flow from the CH. 3 events (14, 23, 24) are produced by the eruptions of filaments inside ARs or connected to them. 4 events (4, 6, 17, 18) are produced by filament eruptions outside ARs. (A filament producing event 17 is situated in the remains of decayed AR 8076, and at the time of eruption no sunspots could be seen there.) 4 events (5, 7, 19, 22) have the same source as events of the previous day (4, 6, 18, 21 respectively). These events are either long-duration events or represent long (more than a day) relaxation of the A_p index to undisturbed values. Sources of 2 events (1 and 8) are not identified as EIT was baking out.

Following examples illustrate the obtained results. Consider the event 13. We note first that it cannot be produced by a flow from a CH because the average velocity of the geoeffective solar wind disturbance was very low (about 360 km s^{-1}). During previous two weeks LASCO did not observe a halo/partial halo CME. According to e. g. Thompson et al. [2], Hudson et al. [3], eruptions could be accompanied by coronal dimmings. Indeed, on May 21 EIT detected two eruptions in the NOAA AR 8040 associated with dimmings (the first eruption is shown in Figure 1). Hence, halo/partial halo CMEs (as observed by LASCO) are not necessary indicators of Earth-directed eruptions.

A most interesting problem linked with eruptions is their precursors. It was reported [3, 4, 5] that an eruption in AR is often accompanied by the sigmoid-to-arcade restructuring of this AR seen by Yohkoh/SXT. It is argued [5] that sigmoid represents an S-shaped loop (or a flux tube) stretched along the photospheric magnetic neutral line – an unstable configuration which is then erupted [6]. AR transforms into an unsheared arcade with its axis along the pre-eruption sigmoid.

Canfield et al. [4] found 23 sigmoidal ARs in 1997. Later, Glover et al. [7] reported that only 7 of them were truly sigmoidal, others being either projected sigmoidal (when many structures collectively form an S-shaped feature) or non-sigmoidal. Glover et al. [7] stress the need of observational definition of the term "sigmoid". In the recent work Glover et al. [8] state that a truly sigmoidal AR for most of the time exhibits two J-shaped loops which reconnect and form a true sigmoid just before the eruption.

Our analysis of the EIT data leads to a stronger conclusion: S-shaped loops along the neutral line are never observed (at least in our limited data set), see e. g. NOAA AR 8038 – truly sigmoidal according to a modified classification by Glover et al. [7] – in the right panel of Figure 3. In the SXT data "sigmoids" are either interrupted, or partly saturated, or too diffuse to reveal their magnetic structure. EIT images (Figure 3, right panel) show that the overall configuration of AR indeed could be S-shaped (it is even better seen in the 284 Å bandpass), but no single S-shaped loop (flux tube) is observed.

Another interesting feature of NOAA AR 8038 (which erupted on May 12, 1997 to produce the event 12, see Thompson et al. [2]) is the origin of its "sigmoidal" struc-

TABLE 1. List of geomagnetic events in January 1997 – January 1998 with $A_p > 20$ and their coronal sources observed by SOHO/EIT

Event number	Date dd.mm.yy	Coronal source seen by SOHO/EIT	Start from the Sun
1	10.01.97	(EIT bakeout)	–
2	28.01.97	extension of the southern polar CH to the equator	end of Jan. 23
3	08.02.97	CH near the disk center	Feb. 5
4	10.02.97	southern polar crown filament eruption	Feb. 7, 01:05 UT
5	11.02.97	same as event 4	–
6	27.02.97	filament eruption in the southern hemisphere	Feb. 22, 01:18 UT
7	28.02.97	same as event 6	–
8	28.03.97	(EIT bakeout)	–
9	11.04.97	two eruptions in NOAA AR 8027	Apr. 6, 23:10 UT and Apr. 7, 14:00 UT
10	17.04.97	extension of the southern polar CH to the equator?	Apr. 13
11	01.05.97	eruption in NOAA AR 8035 followed by the flow from the southern polar CH	Apr. 27, 03:08 UT
12	15.05.97	eruption in NOAA AR 8038	May 12, 04:50 UT
13	27.05.97	eruption in NOAA AR 8040	May 21, 06:13 UT and/or 20:42 UT
14	09.06.97	filament eruption from the west of NOAA AR 8048	Jun. 1, 22:35 UT
15	03.08.97	eruption in NOAA AR 8066	Jul. 31, 01:43 UT
16	03.09.97	eruption in decaying NOAA ARs 8076, 8078, 8079	Aug. 29, 23:24 UT
17	01.10.97	filament eruption in remains of NOAA AR 8076	Sep. 27, 17:58 UT
18	10.10.97	southern polar crown filament eruption	Oct. 6, 12:13 UT
19	11.10.97	same as event 19	–
20	07.11.97	eruption in NOAA AR 8100	Nov. 4, 05:58 UT
21	22.11.97	eruption in NOAA AR 8108	Nov. 17, 03:51 UT
22	23.11.97	same as event 21?	–
23	30.12.97	eruption of the filament channel extending out from NOAA AR 8124	Dec. 25, 21:49 UT
24	07.01.98	eruption of the filament-like structure to the east of NOAA AR 8130	Jan. 2, 22:34 UT

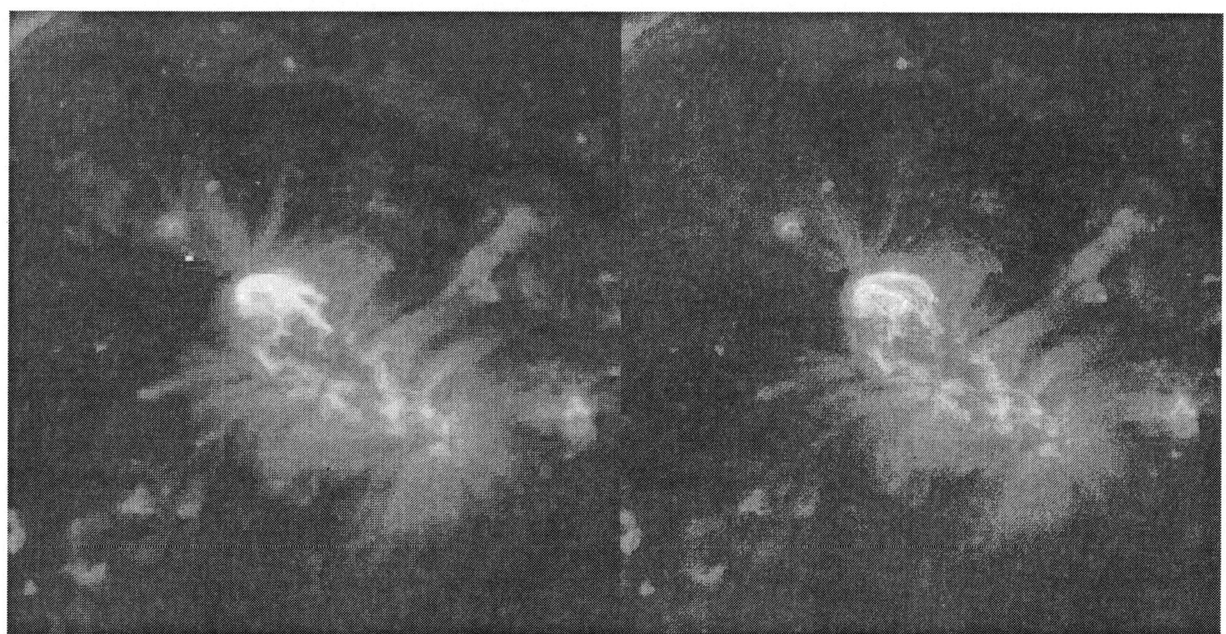

FIGURE 2. Remains of NOAA ARs 8076, 8078 and 8079 observed by SOHO/EIT in the Fe XII bandpass (195 Å) on August 30, 1997 at 00:49:26 UT (left panel; note the "sigmoidal" structure) and 01:12:31 UT (right panel; note rising post-eruption loops).

FIGURE 3. NOAA AR 8038 observed by SOHO/EIT in the Fe XII bandpass (195 Å) on May 10, 1997 at 11:06:10 UT (left panel), 14:34:10 UT (middle panel) and 23:55:17 UT (right panel). Note the dimming associated with the eruption to the south of the AR in the middle panel and an assembly of small-scale structures forming an overall S-shaped feature in the right panel.

ture on May 10. It appears that the "sigmoid" is formed as a result of another eruption as shown in Figure 3. This AR exhibits an example of the process when "sigmoidal" configuration (unstable?) is formed via an eruption and is destroyed via an eruption as well.

Finally, consider event 16 produced by an eruption in the decaying NOAA ARs 8076, 8078 and 8079 on August 29–30, 1997. In the left panel of Figure 2 one can see a "sigmoidal" structure (close to saturation). It is interesting to note that post-eruption loops have the same footpoints as the "sigmoid", and there is no signature of reconnection although the "sigmoidicity" of that part of the AR disappears.

CONCLUSIONS

We have investigated the coronal sources of geoeffective interplanetary disturbances during the low activity epoch (January 1997 – January 1998). Using the EIT observations of the EUV solar corona, we have preliminarily identified the sources of all 24 geomagnetic events with daily A_p index values more than 20. The identification of the source is mostly unambiguous; it is doubtful only in 2 cases (events 10 and 22).

It is found that halo/partial halo CMEs observed within LASCO sensitivity limits are not necessary indicators of Earth-directed eruptions (events 13, 14, 15). We propose to use the coronal dimmings observed by EIT as a complementary indicator. Indeed, transient dimmings are associated with all of the events except those produced by flows from CHs.

A structure now conventionally called a "sigmoid" cannot be represented as a single S-shaped loop (flux tube) but instead exhibits an assembly of many structures of smaller scale.

ACKNOWLEDGMENTS

SOHO/LASCO and SOHO/EIT consortiums are acknowledged for the data used during the preparation of this paper. SOHO is a joint ESA–NASA project. The CME catalog used is generated and maintained by the Center for Solar Physics and Space Weather, The Catholic University of America in cooperation with the Naval Research Laboratory and NASA. The authors are grateful to the ACE and WIND teams for open access via Internet to the solar wind data. This work is supported by the Belgian OSTC program for cooperation with Central and Eastern Europe, INTAS–ESA 99-00727 and INTAS 00-752 grants. The work in the Moscow State University is supported by the RFBR, State Program "Astronomy" and the Program "Universities of Russia" grants.

REFERENCES

1. Bothmer, V., Cargill, P., Dmitriev, A. V., Romashets, E. P., Veselovsky, I. S., Zhukov, A. N., and Yakovchouk, O. S., *Solar System Research*, **36**, 499–506 (2002).
2. Thompson, B. J., Plunkett, S. P., Gurman, J. B., Newmark, J. S., St. Cyr, O. C., and Michels, D. J., *GRL*, **25**, 2465–2468 (1998).
3. Hudson, H. S., Lemen, J. R., St. Cyr, O. C., Sterling, A. C., and Webb, D. F., *GRL*, **25**, 2481–2484 (1998).
4. Canfield, R. C., Hudson, H. S., and McKenzie, D. E., *GRL*, **26**, 627–630 (1999).
5. Sterling, A. C., Hudson, H. S., Thompson, B. J., and Zarro, D. M., *ApJ*, **532**, 628–647 (2000).
6. Rust, D. M., and Kumar, A., *ApJ*, **464**, L199–L202 (1996).
7. Glover, A., Ranns, N. D. R., Harra, L. K., and Culhane, J. L., *GRL*, **27**, 2161–2164 (2000).
8. Glover, A., Ranns, N. D. R., Brown, D. S., Harra, L. K., Matthews, S. A., and Culhane, J. L., *Journal of Atmospheric and Solar-Terrestrial Physics*, **64**, 497–504 (2002).

ICME Observations During the Ulysses Fast Latitude Scan

R. J. Forsyth*, A. Rees*, D. B. Reisenfeld†, S. T. Lepri¶, and T. H. Zurbuchen¶

*The Blackett Laboratory, Imperial College, London SW7 2BW, UK
†Los Alamos National Laboratory, Los Alamos, NM 87544, USA
¶University of Michigan, Ann Arbor, MI 48109, USA

Abstract. Between November 2000 and October 2001 the Ulysses spacecraft performed a fast traversal of the heliospheric latitudes between 80°S and 80°N, a period close to the activity maximum of the current solar cycle. This paper provides an overview of the Ulysses observations of the transient solar wind structures associated with coronal mass ejections (ICMEs) during this period. Compared to the previous Ulysses fast latitude scan near solar minimum in 1995, many more ICME related signatures were observed in the present data set. Events were encountered spread over the full latitude range between 80°S and 80°N. Those at high northern latitudes, where fast solar wind from a northern polar coronal hole had become re-established, were of the over-expanding type first identified in Ulysses data at mid-latitudes near solar minimum. The signatures of these events and their latitude dependence are discussed and some ongoing and possible future studies with this data set are described.

INTRODUCTION

Coronal mass ejections (CMEs) are explosive eruptions of plasma from regions of the solar atmosphere that were previously magnetically closed, seen to dramatic effect by space-borne coronagraph instruments. These CMEs propagate out into the heliosphere embedded in and interacting with the solar wind. Here, their markedly different properties from the ambient solar wind lead to their identification as they pass over in-situ interplanetary spacecraft. The aim of this paper is to provide a first report and overview of the Ulysses observations of the transient solar wind structures associated with coronal mass ejections (known as ICMEs) during the recent *fast latitude scan* in 2000–2001.

The *fast latitude scans* are defined as the perihelion phases of the Ulysses orbit during which the spacecraft traverses heliographic latitudes between 80.2°S and 80.2°N, with a perihelion of 1.3 AU, in a little over 10 months. Trajectory information for the recent fast scan, which took place between November 24 (day 329) 2000 and October 11 (day 284) 2001, is provided in Figure 1. This was the second such traversal of Ulysses through this latitude range. Whereas the first in 1995 took place during a period approaching solar activity minimum, the present fast latitude scan occurred close to the maximum of the present solar cycle (no. 23). Since CME occurrence rates have been shown to be a strong function of the solar cycle [1], as well as their tendency to occur at higher latitudes [2], we would expect this to be a prime period for their study. This data set thus provides unique opportunities for studying the latitude dependence of ICME occurrence and physical signatures. In this paper we first provide a catalogue of the well defined ICME events identified by the Ulysses instruments during the fast latitude scan and discuss latitude dependencies, including example events. Figure 1 also shows that for much of the fast latitude scan Ulysses was located in a direction off the west limb of the Sun as seen from the Earth. We explore the resulting opportunities for correlative studies linking Ulysses in-situ ICMEs with LASCO coronagraph [3] observations from SOHO.

ICME OBSERVATIONS

We have endeavored to identify and catalogue all the ICME events observed by Ulysses during the fast latitude scan. To do this we have employed data from three instruments, the magnetometer [4], the SWOOPS

CP679, *Solar Wind Ten: Proceedings of the Tenth International Solar Wind Conference*,
edited by M. Velli, R. Bruno, and F. Malara

ion and electron instruments [5] and the solar wind composition instrument (SWICS) [6]. The list of events is presented in Table 1 along with a tabulation of the ICME signatures shown by each. It is well documented in the literature that ICMEs often have remarkably different magnetic field and plasma signatures from the solar wind in which they are embedded [e.g. 7]. Those referred to in Table 1 are, by column, counter-streaming suprathermal electrons (CSE), smooth magnetic field rotations consistent with flux rope-like field structures (FR), low plasma beta (Beta), whether the event is driving a shock wave (Sh) as it propagates through the solar wind, low proton thermal temperature (Tp), enhanced helium (alpha/ proton) abundance (He), existence of high iron charge states (Fe), and enhanced O^{7+}/O^{6+} ratio (O7/O6) when compared to the surrounding solar wind. The events were mostly first selected by searching for at least two from among CSE signatures, magnetic field rotations, low beta, and enhanced He, although in one case the initial identification came from the charge state data. We then checked which of the other signatures were present. In general, ICME identification is made harder by the fact that not all ICMEs exhibit all the above signatures and also that the majority of these signatures can also arise from other causes. Thus the list presented here represents what we believe to be a complete set of the ICMEs identifiable from the Ulysses data under study, but cannot be absolutely guaranteed to be so.

For some of the signatures a few additional words of explanation are necessary. Of those events listed as having a flux rope signature (FR), those marked with 'x' (rather than '?') also exhibit the enhanced field magnitude and low proton temperature necessary to be classed as a magnetic cloud [8]. The driving of a shock wave is included as a signature since CIR shocks have typically not yet formed at the distances covered by Ulysses during this period. In addition, this paper represents the first time that both oxygen and iron charge state information have been combined together with both the field and plasma data in identifying ICMEs with Ulysses data. Studies of the systematic differences in the charge state distributions of heavier elements including oxygen [9] and iron [10] between ICMEs and the ambient solar wind are relatively recent. Their inclusion here is particularly interesting in view of the latitude range covered. The criteria for the charge state signatures were an average iron charge state greater than 12.0 and an O^{7+}/O^{6+} ratio greater than 0.7. The charge state distributions are accumulated in 3 hour bins. We only included events where at least two contiguous bins satisfied the identification criteria. It should be noted that not all the signatures were present for the full duration of each event. The *Start Time* and *End Time* columns list the earliest possible start time and latest possible stop time obtained from combining all the signatures. These times are accurate to the nearest half hour at best due to the ICME signatures often not having sharp boundaries or alternatively due to the presence of multiple discontinuities, any of which could be the actual ICME boundary. The latitude quoted is that corresponding to the start time.

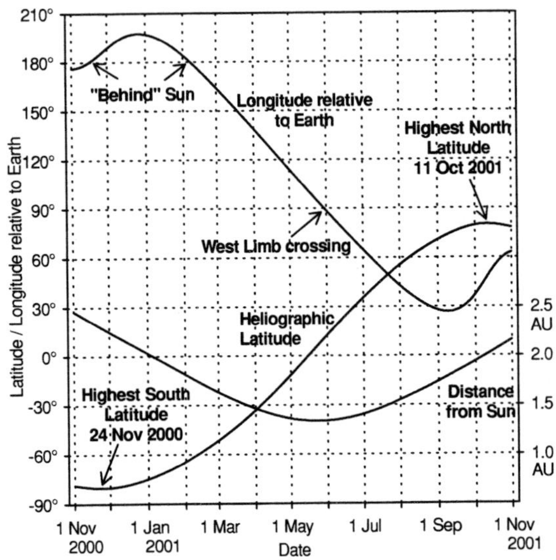

FIGURE 1. The trajectory of Ulysses through the fast latitude scan showing heliocentric distance, heliographic latitude and the longitude of the spacecraft relative to Earth.

Latitude Distribution of Events

Figure 2 shows the locations in heliocentric distance and heliographic latitude of the ICMEs identified from Ulysses data during the solar maximum fast latitude scan (filled circles) and during the previous solar minimum fast latitude scan (open circle). It can be seen that during the solar minimum epoch only one ICME event was encountered during the ~10 month period [11], at a relatively low latitude of 23°S consistent with the latitude range covered by the streamer belt at that time. In contrast, we have identified 19 events during the equivalent solar maximum period, consistent with the CME rate being a strong function of solar activity.

These events are distributed over the full latitude range, in particular with ICMEs identified both close to 80°S and 80°N. Although the figure does give an impression that there are fewer events at high latitudes, there are too few events to draw a firm conclusion

TABLE 1. ICMEs Identified During the Ulysses Fast Latitude Scan.

Start Time	End Time	Lat.	CSE	FR	Beta	Sh	Tp	He	Fe	O7/O6
2000 341 0300	345 0800	–79.6°	x	x	x		x	x	?	
2001 023 2300	024 2030	–67.4°	x	x	x		x			
032 1500	034 1900	–64.0°	x	?	x		x			
077 0800	078 2330	–41.7°	x	x	x	x	x			
080 0100	081 0600	–40.0°		x				x	x	x
090 1800	093 0800	–33.3°			x		x	x		
100 2100	103 2100	–26.4°	x	x	x	x	x	x	x	x
104 1700	106 0800	–23.5°	x			x		?	x	x
110 1000	114 0800	–19.6°	x	x	x	x	x	x	x	x
130 0430	133 0600	–4.5°	x	?	x	x	x	x	x	x
139 0000	142 0000	2.4°			x	x		x		
156 1100	160 0400	16.2°	x	x			x	x	x	x
161 0500	164 2330	20.0°	x	x	x		x	x	x	x
177 1330	179 1200	32.2°	x					x		
185 1000	188 1330	37.8°	x	x			x	x	x	x
204 1000	208 2200	50.3°	x	x	x	x	x			
224 1800	225 1600	61.8°	x		x			x		
236 1300	237 0100	67.5°	x	x	x		x	x		
271 1230	272 1630	79.1°	x	x	x		x	x		

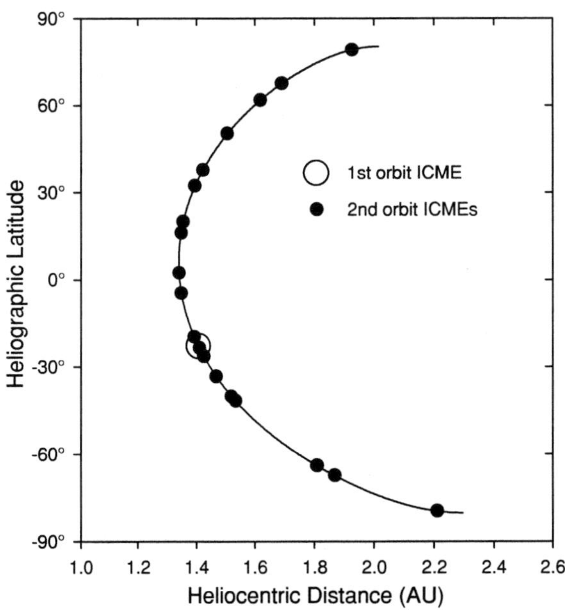

FIGURE 2. The locations of ICME events identified in Ulysses data during the fast latitude scans on both the first solar minimum orbit and the second solar maximum orbit.

from this data alone. However, when combined with events identified in 1999 and 2000 as Ulysses traversed high southerly latitudes prior to the fast latitude scan [12], it becomes clear that even at solar maximum, Ulysses has identified fewer events at high latitudes than at low to mid latitudes. This result is consistent with past results on the latitude distribution of CMEs obtained from coronagraph observations [2, 13] and with the present solar maximum observations by LASCO [O. C. St. Cyr, private communication].

Solar Wind Context of Events

During the solar maximum fast latitude scan the solar wind conditions were very different in the southern and northern polar regions [14]. In the south the solar wind was found to be a complex mixture of slow and intermediate speed streams whereas by the time Ulysses had reached the north polar regions fast solar wind from a newly reformed polar coronal hole had become re-established. The characteristics of the two highest latitude ICMEs observed in these different contexts were remarkably different. To illustrate this, a selection of the field and plasma parameters for the two events, the first and last from Table 1, are plotted in Figures 3 and 4. From top to bottom the six panels show the magnetic field magnitude ($|\mathbf{B}|$), the two angles describing the magnetic field direction (ϕ_B, θ_B), the solar wind speed (v_p), the proton temperature (T_p) and the alpha particle to proton ratio (α/p).

The highest southerly event (Figure 3) is similar in character to many other events observed by Ulysses at all latitudes at times when the solar wind is dominated by slow or intermediate speeds [e.g. 12]. The leading edge of the ICME can be seen to be propagating at a similar speed (~450 km/s) to the solar wind in front, thus the event is not driving a shock wave. The declining speed profile shows that the event is expanding as a result of the trailing solar wind being slower (~340 km/s). The magnetic cloud signature can be identified from the slow rotation seen in the two magnetic field angles coincident with a depression in the proton temperature. However, noting that the event

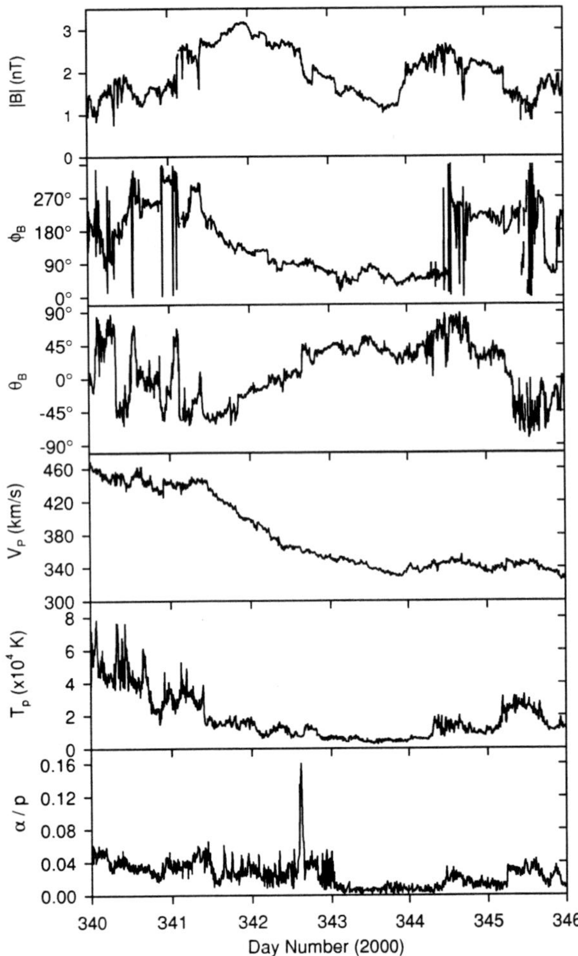

FIGURE 3. An example of an ICME observed by Ulysses at 79.6°S. The panels are explained in the text.

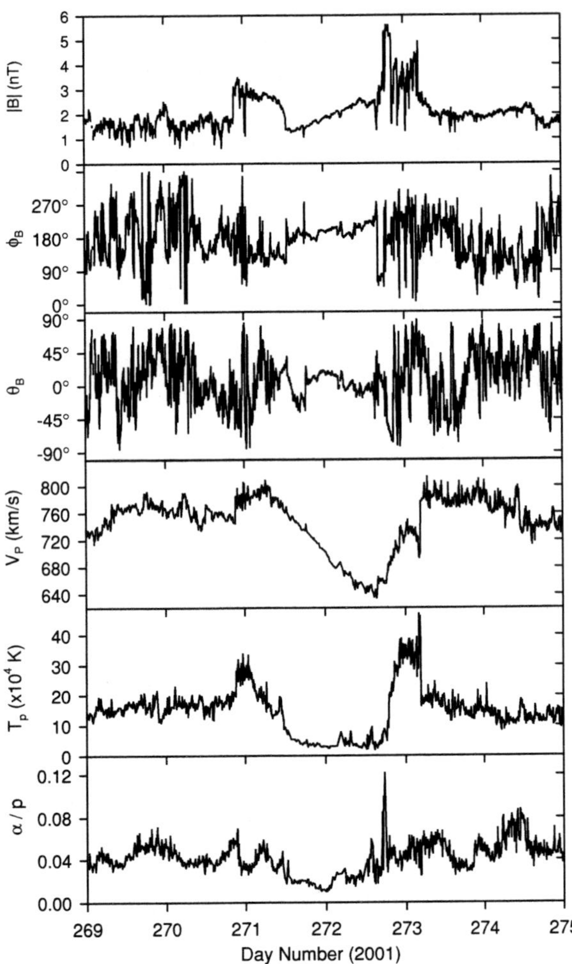

FIGURE 4. An example of an ICME observed by Ulysses at 79.1°N. The panels are explained in the text.

continues into day 345, it can be seen that this ICME has a second part distinct from the magnetic cloud. This two (or more) part structure is typical of many of the ICMEs observed by Ulysses during the solar maximum period. It may be that the second part is a result of part of the same solar eruption with different properties or origin or due to a separate ICME immediately behind the first.

In contrast, the highest northerly event (Figure 4) has quite different characteristics. Here the solar wind speed both before and after the ICME event is fast (~750 km/s), showing the typical characteristics of solar wind originating from a polar coronal hole. The speed profile again indicates expansion, but here the expansion is due to the internal pressure of the ICME itself, typical of the over-expanded ICMEs previously observed embedded in the fast solar wind during the Ulysses first orbit closer to solar minimum [15]. Thus this type of ICME dynamic signature appears to be unique to ICMEs propagating within high-speed solar wind streams [16]. This ICME also shows a magnetic cloud signature, indicating that magnetic clouds can be found at all latitudes at solar maximum. Previously at solar minimum, no magnetic clouds were observed at latitudes above ~40°, even though ICME events were observed up to ~60°.

ICME Signatures

In Table 1 we have provided an indication of the signatures exhibited by each of the fast latitude scan ICME events. In this section we give a first report on two specific results that are emerging from their study.

Firstly we comment on the latitude distribution and characteristics of those events showing the strongest iron and oxygen charge state signatures. It is notable that these events are confined within a latitude range within 40° of the equator and are absent from the higher latitude events. All the well defined magnetic clouds in this latitude range show a clear difference in charge state distribution from the surrounding solar wind although the converse does not hold. This is similar to the results found for oxygen at solar minimum [9]. At that time there was an almost one-to-one correspondence between magnetic clouds and anomalous composition signatures. These results suggest that this correlation does not hold for magnetic clouds observed at greater than 40° from the equator. The reason for this is not obvious. It may be indicative of a recently suggested close association of these signatures with solar active regions and flares [17].

A second result that is emerging from this data concerns the axis orientation of the magnetic cloud events. This can be determined to a good approximation by fitting a force-free flux rope model to the magnetic field data [18]. The results show that axes of southern hemisphere magnetic clouds are predominantly directed in a westwards direction while those in the northern hemisphere are predominantly directed eastwards [A. Rees and R. J. Forsyth, manuscript in preparation]. Given that the data is obtained in an odd numbered solar cycle, this behavior is consistent with the predictions of [19] derived from near-ecliptic observations, dependent on the preferred orientations of prominences in the low solar atmosphere.

CORRELATION WITH SOLAR OBSERVATIONS

Returning to Figure 1, we now focus on the solar longitude of Ulysses with respect to that of the Earth. This is an important parameter for identifying periods when there should be good opportunities to compare CME images obtained by SOHO with in-situ observations from Ulysses. In particular, when this angle is 90° Ulysses is in a prime position to intercept CMEs which erupt from the west limb of the Sun as seen from Earth, and Ulysses was within 30° of this condition for a significant fraction of the fast latitude scan. Even during the period marked *"Behind Sun"* on Figure 1 when the longitude relative to Earth is 180°, at 80°S and 2 AU from the Sun the spacecraft is still in a good position to intercept a subset of CMEs that appear in projection to erupt from the southern polar regions of the Sun. Using coronagraph observations to identify the particular solar event associated with an ICME event has been relatively successful close to solar minimum [e.g. 20], when the number of CMEs is relatively low. First attempts at a similar study with the recent Ulysses data have proved much harder due to confusion caused by the larger number of events. At high northern latitudes, immediately following the fast scan the solar sources of 3 out 5 ICMEs embedded in the fast solar wind were identifiable [16].

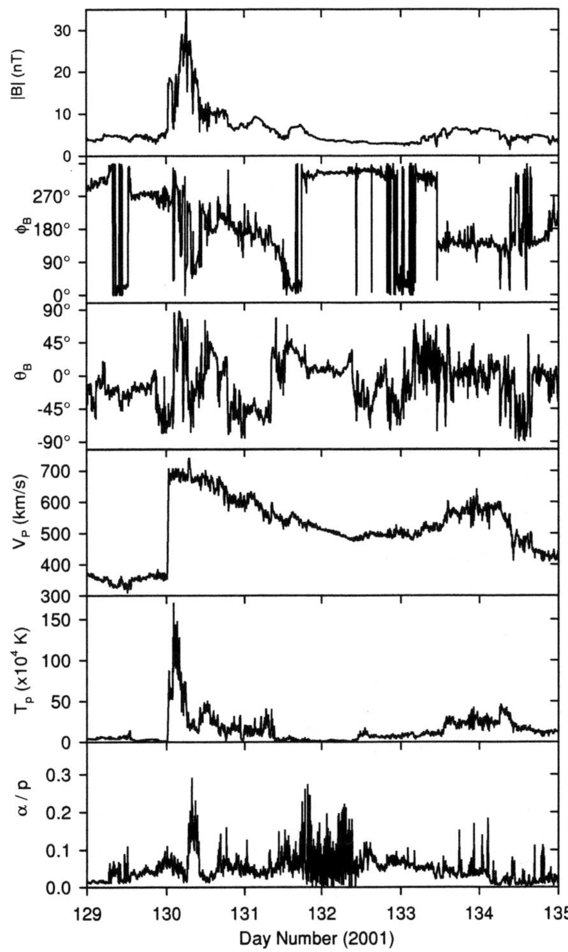

FIGURE 5. An example of an ICME observed by Ulysses near the equator associated with a west limb CME. The panels are the same as in Figures 3 and 4.

We conclude this paper by reporting on one dramatic west limb CME which occurred during the Ulysses fast scan where the solar to in-situ correlation was unmistakable. Figure 5 shows the in-situ signatures of this event, encountered by Ulysses on May 10 (day 130) 2001 at a latitude of 4.5°S. The associated CME could be identified unambiguously as an eruption seen by LASCO on 2001 May 7 at 1206 UT which passed through the LASCO field of view at a speed of ~1200 km/s [http://cdaw.gsfc.nasa.gov/CME_list]. The time of arrival and speed of the ICME at Ulysses indicate that the ejecta was being decelerated as it interacted with slower solar wind as it traveled out to Ulysses. The interaction produced an unusually strong shock wave leading the event as it passed Ulysses. The magnetic field strength of ~35 nT was the highest recorded by the Ulysses magnetometer in interplanetary space during the 11 year mission so far. The LASCO images show a complex multi-part eruption while the leading part of the event at Ulysses consisted of many field rotations and current sheets. It is likely that the complexity seen at Ulysses is due both to the structures of the original CME and the strong compression that has clearly taken place during the journey out from the Sun. Analysis of the planar structuring of the discontinuities in the sheath region ahead of the ICME suggests that the bulk of the ejecta was located at more positive longitudes relative to Ulysses [21], that is further behind the limb as seen from the Earth. It was not possible to identify a clear magnetic cloud in the leading part of the event but the low beta and high helium content clearly indicate the presence of magnetic cloud-like material.

SUMMARY

We have presented an overview of the ICME events observed by Ulysses during the solar maximum fast latitude scan, cataloguing their occurrence and signatures, as well as presenting three example events. Clearly there is much scope for further work with this data set, for example on the changes noted in the composition signatures with latitude, and for a more detailed correlative study of the Ulysses events with those observed by LASCO in the same time period.

ACKNOWLEDGMENTS

Ulysses research at Imperial College London is supported by the UK Particle Physics and Astronomy Research Council. Work at Los Alamos was performed under the auspices of the US Department of Energy with support from NASA.

REFERENCES

1. Webb, D. F., and Howard, R. A., *J. Geophys. Res.* **99**, 4201-4220 (1994).
2. Hundhausen, A. J., *J. Geophys. Res.* **98**, 13177-13200 (1993).
3. Brueckner, G. E., Howard, R. A., Koomen, M. J., et al., *Solar Phys.* **162**, 357-402 (1995).
4. Balogh, A., Beek, T. J., Forsyth, R. J., et al., *Astron. Astrophys. Suppl. Ser.* **92**, 221-236 (1992).
5. Bame, S. J., McComas, D. J., Barraclough, B. L., et al., *Astron. Astrophys. Suppl. Ser.* **92**, 237-266 (1992).
6. Gloeckler, G., Geiss, J., Balsiger, H et al., *Astron. Astrophys. Suppl. Ser.* **92**, 267-290 (1992).
7. Neugebauer, M., and Goldstein, R., "Particle and field signatures of coronal mass ejections in the heliosphere," in *Coronal Mass Ejections,* edited by N. Crooker, J. A. Joselyn, and J. Feynman, Washington D. C.: AGU, 1997, pp. 245-251.
8. Burlaga, L. F., "Magnetic clouds," in *Physics of the inner heliosphere II,* edited by R. Schwenn and E. Marsch, Berlin: Springer-Verlag, 1991, pp. 1-22.
9. Henke, T., Woch, J., Mall, U., et al., *Geophys. Res. Lett.* **25**, 3465-3468 (1998).
10. Lepri, S. T., Zurbuchen, T. H., Fisk, L. A., et al., *J. Geophys. Res.* **106**, 29231-29238 (2001).
11. Gosling, J. T., Bame, S. J., Feldman, W. C., et al., *Geophys. Res. Lett.* **22**, 3329-3332 (1995).
12. Gosling, J. T., and Forsyth, R. J., *Space Sci. Rev.* **97**, 87-98 (2001).
13. St. Cyr, O. C., Howard, R. A., Sheeley, N. R., et al., *J. Geophys. Res.* **105**, 18169-18185 (2000).
14. McComas, D. J., Elliott, H. A., Gosling, J. T., et al., *Geophys. Res. Lett.* **29**, 10.1029/2001GLO14164 (2002).
15. Gosling, J. T., Riley, P., McComas, D. J., and Pizzo, V. J., *J. Geophys. Res.* **103**, 1941-1954 (1998).
16. Reisenfeld, D. B., Gosling, J. T., Steinberg, J. T., et al., in *Proceedings of Solar Wind 10*, this volume.
17. Richardson, I. G., Cane, H. V., Lepri, S. T., et al., in *Proceedings of Solar Wind 10*, this volume.
18. Lepping, R. P., Jones, J. A., and Burlaga, L. F., *J. Geophys. Res.* **95**, 11957-11965 (1990).
19. Bothmer, V., and Schwenn, R., *Ann. Geophys.* **16**, 1-24 (1998).
20. Funsten, H. O., Gosling, J. T., Riley, P., et al. *J. Geophys. Res.* **104**, 6679-6689 (1999).
21. Jones, G. H., Rees, A., Balogh, A., and Forsyth, R. J., *Geophys. Res. Lett.* **29**, 10.1029/2001GL014110 (2002).

Emission of Doppler-shifted photons from excited energetic neutral atoms created in the solar wind

A. Czechowski*, M. Hilchenbach† and K.C. Hsieh**

*Space Research Centre, Polish Academy of Sciences, Bartycka 18A, PL 00-716 Warsaw, Poland
†Max-Planck-Institut für Aeronomie, D-37191 Katlenburg-Lindau, Germany
**Physics Department, University of Arizona, Tucson , AZ 85721, U.S.A.

Abstract. The fast-moving protons from the solar wind plasma convert into energetic hydrogen atoms by charge-exchange with the background hydrogen from the interstellar medium. If created in the excited state, the energetic atoms will emit photons, with frequency Doppler-shifted away from the resonance frequency range of the background hydrogen. We estimate the flux and the spectrum of photons from the de-excitation of the H(2p) state of the energetic hydrogen atoms deriving from the pick-up protons in the solar wind and from the thermal protons in the hot plasma region beyond the solar wind termination shock. We also discuss the possibility of detection of the photon flux from this source.

INTRODUCTION

One of potential methods to observe the distant heliosphere is by means of photons emitted by excited atoms in this region. This could supplement the other method, relying on direct observations of the flux of energetic neutral atoms (Hsieh et al. [1], Czechowski et al. [2]). Although expected photon flux intensity from this source is low, for energetic atoms the Doppler shift, by moving the emission away from the line centre, might increase the chance to make the signal observable. In addition, separating the signal from the background may in some cases be helped by the shape of the spectrum.

In the following we estimate the flux and spectrum of Ly-α photons originating in two populations of energetic protons in the solar wind: the pick-up protons in the supersonic solar wind upstream of the termnination shock and thermal protons in the shocked solar wind. We also consider the contribution from protons in the supersonic solar wind upstream of the shock.

Photons are emitted when those energetic protons are converted into neutral atoms in excited state (H(2p)) by capturing an electron from hydrogen and helium atoms of the background gas (the H(2s) state decays by two-photon emission). The momentum exchange during such reactions is small, so that the neutral atoms preserve approximately the momentum of the original ions. The Doppler-shifted spectrum of emitted photons reflects therefore the velocity distributions of energetic proton populations.

In our previous study (Hilchenbach et al., [3]) we have considered the analogous contribution from the anomalous cosmic ray protons (by oversight, the contribution from H(2s) state was calculated using one-photon mechanism).

Gruntman [4] considered the emission from the He^{+}(2p) state ions created from the solar wind α-particles picking up electrons from the background atoms. Most of the signal in this case comes from the inner heliosphere (within 10 A.U.) and would be useful for imaging the three-dimensional structure of the solar wind within this distance. The X-ray emission from heavier species was calculated by Cravens [5]. His calculation was restricted to the region inside the termination shock: he estimates, however, that the outer heliosphere could make a comparable contribution.

THE MODEL

The parameters of the plasma flow in the heliosphere are taken from Kausch five component gas-dynamical solution (Fahr et al. [6]), which also provides the pick-up ion density and the neutral hydrogen flow. We assume that the ion populations considered: protons in the supersonic solar wind, hot protons in the shocked solar wind downstream from the shock, and the pick-up protons in the solar wind upstream, are described by isotropic distributions in the plasma frame. For solar wind protons we use the Maxwellian with position-dependent temperature provided by the Kausch solution. For the pick-up proton velocity distribution we use two alternative forms. One corresponds to the assumption used in the Kausch model to derive the pick-up proton pressure: the distribution

CP679, *Solar Wind Ten: Proceedings of the Tenth International Solar Wind Conference*,
edited by M. Velli, R. Bruno, and F. Malara

function is assumed to be a velocity independent constant for the plasma frame pick-up proton speed $|v'| \leq V_{SW}$ and zero for $|v'| > V_{SW}$:

$$f_{PUI}(\mathbf{r}, v') = \frac{3}{4\pi V_{SW}^3} n_{PUI}(\mathbf{r}) \quad (|v'| \leq V_{SW}) \quad (1)$$

$$f_{PUI}(\mathbf{r}, v') = 0 \quad (|v'| > V_{SW}) \quad (2)$$

The other assumption, which is based on the distribution derived by Vasyliunas and Siscoe [7] (see also Gloeckler et al. [8]) for the isotropized (fast scattering in the pitch angle) case, is that the distribution for $|v'| \leq V_{SW}$ behaves as $|v'/V_{SW}|^{-3/2}$. In each case the normalization factor is chosen to reproduce the pick-up proton density of the Kausch model:

$$f_{PUI}(\mathbf{r}, v') = \frac{3}{8\pi V_{SW}^3} n_{PUI}(\mathbf{r}) w^{-3/2} \; (v_{min} \leq |v'| \leq V_{SW}) \quad (3)$$

$$f_{PUI}(\mathbf{r}, v') = 0 \quad (|v'| > V_{SW}, |v'| < v_{min}) \quad (4)$$

Here $w = |v'/V_{SW}|$. We have introduced the cutoff speed $v_{min} = V_{SW}(r_{min}/r)^{3/2}$ where $r_{min} = 5$ AU is the characteristic size of the region with low neutral hydrogen density close to the Sun. Our formula is a simplified version of that of Vasyliunas and Siscoe, corresponding to taking the neutral background density to be constant in space for $r > r_{min}$ and zero otherwise.

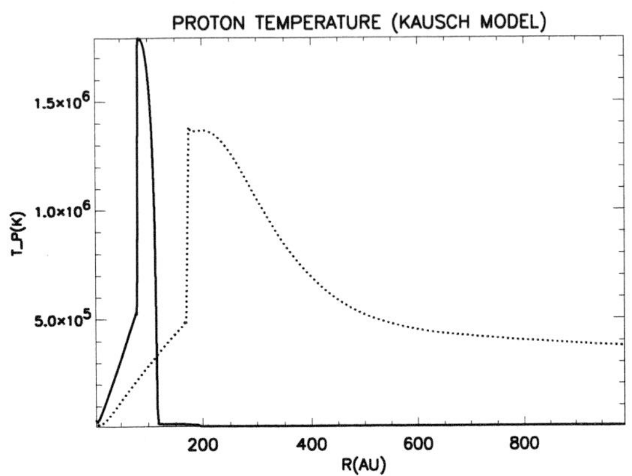

FIGURE 1. Proton temperature (Kausch model) plotted as a function of heliocentric distance for two directions: LISM apex (solid line) and the heliotail (dotted line).

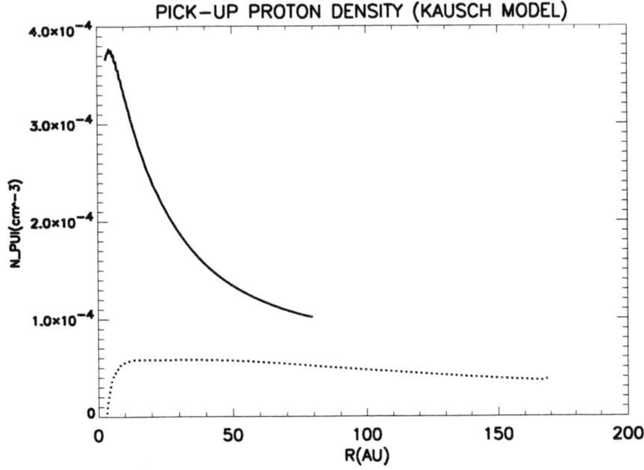

FIGURE 2. Pick up proton density upstream of the termination shock (Kausch model) as a function of heliocentric distance for two directions: LISM apex (solid line) and LISM anti-apex (the heliotail, dotted line).

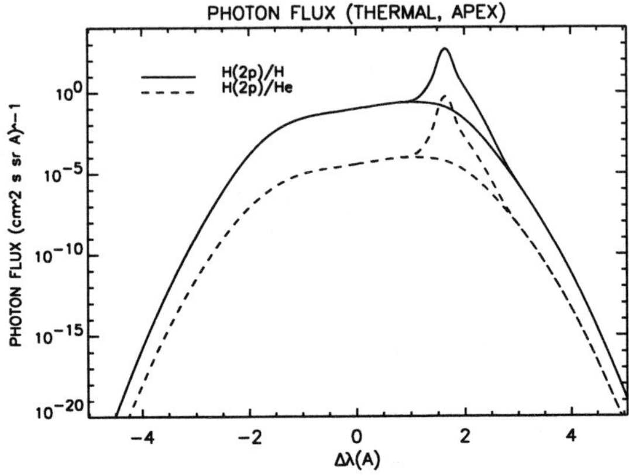

FIGURE 3. Flux of photons (from the LISM apex direction) from de-excitation of the H(2p) state of the energetic hydrogen atoms originating from hot shocked solar wind protons neutralized by charge-exchange with hydrogen (solid line) and helium (dashed line) atoms of the background gas. $\Delta\lambda = \lambda - \lambda_0$ where λ_0 is the Lyman-α wavelength. The superimposed peaks show the contribution from thermal protons upstream of the shock.

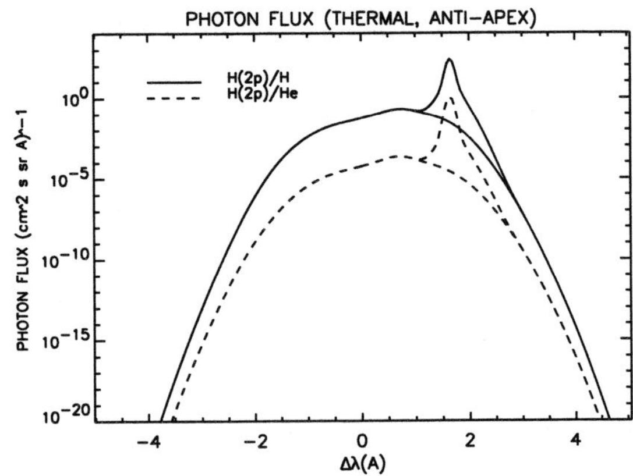

FIGURE 4. As Fig. 3, but for the flux from the anti-apex direction (heliotail).

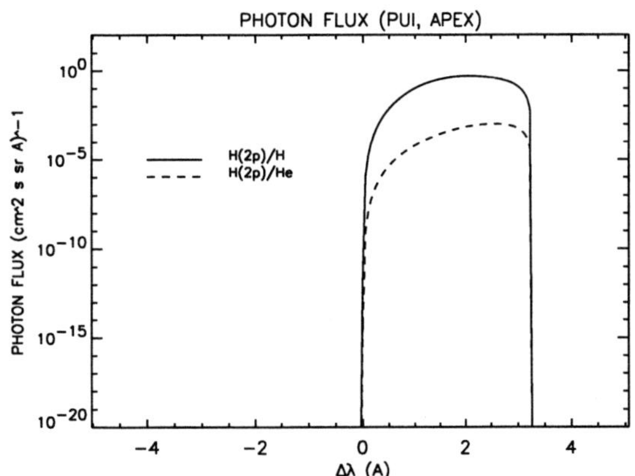

FIGURE 5. Flux of photons (from the LISM apex direction) from de-excitation of the H(2p) state of energetic hydrogen atoms originating from pick-up protons upstream of the shock (the distribution function given by Eqs. 1 & 2)

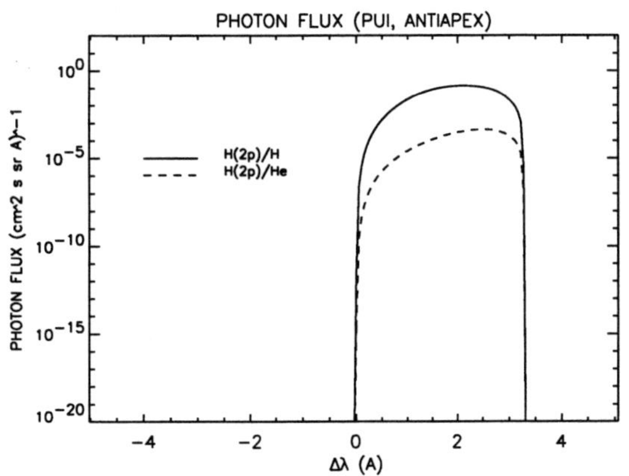

FIGURE 6. As Fig. 5, but for the anti-apex direction (heliotail).

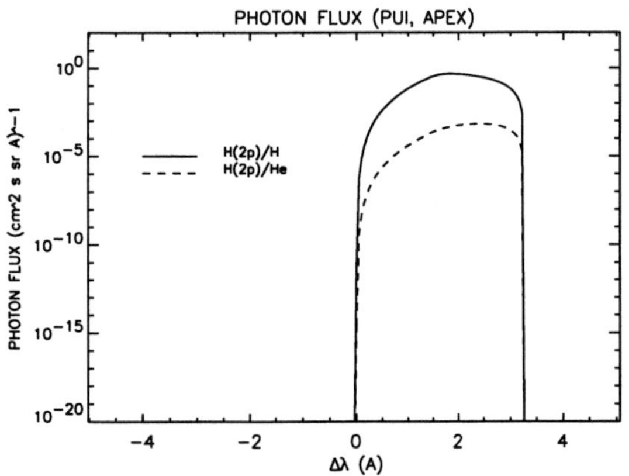

FIGURE 7. Flux of photons (from the LISM apex direction) from de-excitation of the H(2p) state of energetic hydrogen atoms originating from pick-up protons upstream of the shock (the distribution function given by Eqs. 3 & 4)

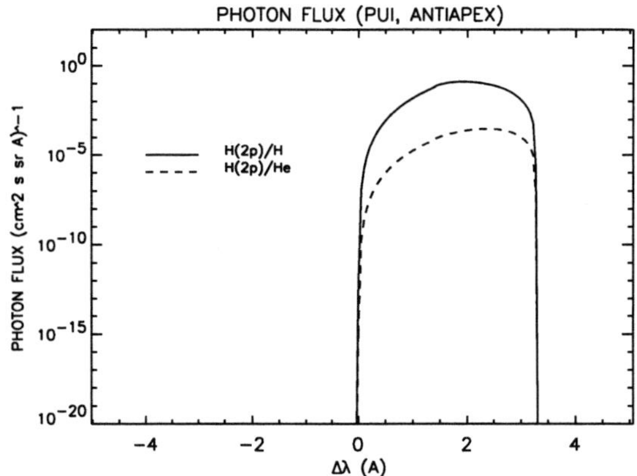

FIGURE 8. As Fig. 7, but for the anti-apex direction (heliotail)

We calculate the photon flux from direction $\hat{\mathbf{n}}$ as the integral along the line-of-sight:

$$\frac{dJ}{d\nu} = \frac{1}{4\pi}\int ds \int d^3v' f(\mathbf{r},v')v_{rel}(\sigma_H(v_{rel})n_H + \sigma_{He}(v_{rel})n_{He})\delta\left(\nu - \nu_0\left(1+\frac{\mathbf{v}\cdot\hat{\mathbf{n}}}{c}\right)\right) \quad (5)$$

where $dJ/d\nu$ is the differential flux of photons per unit area, time, solid angle and frequency, $\mathbf{v}$ and $\mathbf{v}'$ the proton velocities in the observer and the plasma frame, respectively, v_{rel} the relative speed between the proton and the background gas, σ_H and σ_{He} the cross sections (Barnett [9]) for electron capture by proton from H and He atoms into the H(2p) state ($H^+ + H \rightarrow H(2p) + H^+$, $H^+ + He \rightarrow H(2p) + He^+$), and finally, n_H, n_{He} the number densities of H and He atoms of the background gas. We have assumed that the emission of photons occurs at the frequency ν_0 in the atom rest frame and that it can be taken as approximately isotropic in the observer frame. The flux per unit wavelength is given by $dJ/d\lambda = (c/\lambda^2)dJ/d\nu$.

The model of the heliosphere which we consider is a simplification. The flow is assumed to be axially symmetric with respect to the LISM apex-antiapex line: consequently, the three-dimensional structure of the heliosphere, in particular the latitudinal variation of the solar wind, is not included. On the other hand, the apex-antiapex asymmetry of the heliosphere, including asymmetric shape of the termination shock and the extended heliotail, is described by the model. We concentrated on this asymmetry and calculated the photon fluxes from the apex and the anti-apex directions.

DISCUSSION

The results are shown in Figs. 3-8. For the photons originating in thermal protons outside the shock (Figs. 3 and 4) the spectra show a shift by about 0.5 angstrem to higher wavelength, corresponding to outward plasma flow at the speed of 100 km/s downstream from the shock. Despite larger size of the source region in the heliotail (antiapex) direction, the flux of photons from the heliotail direction is not larger than from the apex. This is partly due to higher proton temperature in the forward region (Fig. 1), which also affects the width of the spectrum, and partly to higher density of the neutral hydrogen. The sloping part in the center is caused by the variation in the relative speed between the protons and target atoms. As the cross sections for the capture of electron into the H(2p) excited state in the relevant energy range increase with energy, the emission rate from the (neutralized) protons moving outward is larger than from those moving inward. Note also that only the high energy tail of the proton distribution contributes to the emission (the cross section below ~600 eV is taken to be zero). The superimposed peaks show the contribution from thermal protons upstream of the shock.

The photons coming from the pick-up proton population have a spectrum with sharp cutoff at $\Delta\lambda = 0$ and at $\Delta\lambda$ corresponding to twice the solar wind speed. The results are approximately the same for both models of pick-up proton distribution function considered by us (Eqs. 1 or 3). The emission from the apex direction (Figs. 5, 7) is higher than from the heliotail (Figs. 6, 8): this is because both the pick-up proton (Fig. 2) and the neutral hydrogen density are higher in the forward region. We did not calculate the contribution from the pick-up ions downstream of the shock (the flux of energetic neutral atoms from this source is calculated in Czechowski et al. [10]).

In each case the flux intensity is very low (10^{-5} rayleigh maximum). The contribution from thermal protons inside the shock is much higher (10^{-3} R) because the source region in this case includes the region of high proton density close to the Sun. The spectra originating in the inner (narrow peak) and the outer heliosphere are sufficiently different in shape to make their separation possible in principle. However, except for the contribution from the inner region, the intensity of the signal is much below the interplanetary background.

ACKNOWLEDGMENTS

A.C. acknowledges support from the KBN grant 8 T12E 029 20. A.C. also wishes to thank Max-Planck-Institut für Aeronomie in Lindau for hospitality.

REFERENCES

1. Hsieh, K.C., Shih, K.L., Jokipii, J.R. and Grzedzielski, S., *ApJ* **393**, 756 (1992).
2. Czechowski, A., Fichtner, H., Grzedzielski, S., Hilchenbach, M., Hsieh, K.C., Jokipii, J.R., Kausch, T., Kota, J. and Shaw, A., *A&A* **368**, 622-634 (2001).
3. Hilchenbach, M., Hsieh, K.C. and Czechowski, A., in: *Proceedings of the COSPAR Colloquium on: The Outer Heliosphere: the New Frontiers*, Potsdam 24-28 July 2000, 2001, p.281
4. Gruntman, M., *J. Geophys. Res.* **106**, 8205-8216 (2001).
5. Cravens, T.E., *ApJ* **532**, L153-L156 (2000).
6. Fahr, H.J., Kausch, T. and Scherer, H., *A&A* **357**, 268-282 (2000).
7. Vasyliunas, V.M., Siscoe, G.L., *J. Geophys. Res.* **81**, 1247-1251 (1976).
8. Gloeckler, G., Schwadron, N.A., Fisk, L.A. and Geiss, J., *GRL* **22**, 2665-2668 (1995).
9. Barnett, C.F. (Ed.), *Atomic Data for Fusion. Collisions of H, H2, He and Li Atoms and Ions with Atoms and Molecules*, Oak Ridge Natl. Lab. Report, ORNL-6086-V1, Oak Ridge, Tenn., 1990.
10. Czechowski, A., Fahr, H.J., Lay, G. and Hilchenbach, M., *A&A* **379**, 601 (2001).

The interaction and evolution of interplanetary shocks from 1 to beyond 60 AU

Chi Wang[1] and John D. Richardson[2]

[1] *Center for Space Science and Applied Research, Chinese Academy of Sciences, Beijing, China*
[2] *Center for Space Research, Massachusetts Institute of Technology, Cambridge, MA 02139, USA*

Abstract. During the current solar maximum (Cycle 23), several major CMEs associated with solar flares produced large transient flows and shocks which were observed by widely-separated spacecraft such as Wind at Earth and Voyager 2 beyond 60 AU. Using data from these spacecraft and numerical models, we study shock propagation and interaction in the outer heliosphere. We demonstrate that a strong shock in the distant heliosphere could be an outer heliospheric remnant of a strong shock in the inner heliosphere ("one to one" relationship), or it could be an outcome of the successive interaction and merging of a series of interplanetary shocks ("one to many" relationship).

INTRODUCTION

While Voyagers continue to make solar wind measurements in the outer heliosphere (Voyager 2 is located at 67 AU as of June, 2002), several Earth-orbited spacecraft such as ACE, Wind are monitoring solar wind conditions at 1 AU. With the development of both observational and theoretical studies, our insight into the dynamical processes in the solar wind throughout the heliosphere has been much improved. Shocks are an important component of the solar wind structures. Generally speaking, there are two classes of shocks observed in the solar wind: co-rotating shocks and transient shocks. Co-rotating shocks result from the interaction of the fast and slow solar wind streams as a consequence of the solar rotation and the tilt of the solar dipole. They are more likely to occur near solar minimum when the solar wind has a relatively simple configuration with high-speed streams at high latitudes and slow-speed streams at low latitudes. On the other hand, transient shocks are in general produced near the Sun by fast ejecta from violent events on the Sun. They are most frequent near solar maximum when the Sun is most active. The observations made by a fleet of spacecraft during past three decades have inspired enormous interests in studying the evolution and interaction of shocks (see the review by *Whang*(1)). These studies, however, are limited to the distance within 10 AU or do not appreciate the importance of pickup ions in the distant heliosphere (beyond ~30 AU).

With the accumulation of Voyager observations in the outer heliosphere, we now have the opportunity to study the evolution of a transient flow system (2) and its shocks from the inner to distant heliosphere where pickup ions play an important role in the flow dynamics and shocks (3-5). Especially during the current solar maximum (Cycle 23), Voyager 2 observed serval relatively strong shocks beyond 60 AU, including the well-known Bastille Day 2000 CME-driven shock and the October 2001 shock, the strongest one have been recorded in the outer heliosphere since 1991. In order to find their inner heliospheric origins, we take advantage of the Wind observations at 1 AU and use numerical models to propagate the solar wind structures from 1 AU to the location of Voyager 2. We follow the evolution and interaction of shocks until they pass Voyager 2, and compare the model predictions with Voyager 2 observations. The solar wind in the outer heliosphere is fundamentally different from that in the inner heliosphere, with the influences from the local interstellar medium becoming profound. In this study, we employ an one-dimensional multi-fluid MHD model(6), which assumes a spherically symmetry of the heliosphere and takes into account the interaction of the solar wind protons with the interstellar neutral hydrogen. We follow the approach of *Isenberg* (7) to assume the solar wind consists of there co-moving there particle populations: protons, pickup ions and electrons. The solar wind protons are coupled with the neutral hydrogen via charge exchange. We also allow the energy transfer between the solar wind pro-

CP679, *Solar Wind Ten: Proceedings of the Tenth International Solar Wind Conference*, edited by M. Velli, R. Bruno, and F. Malara

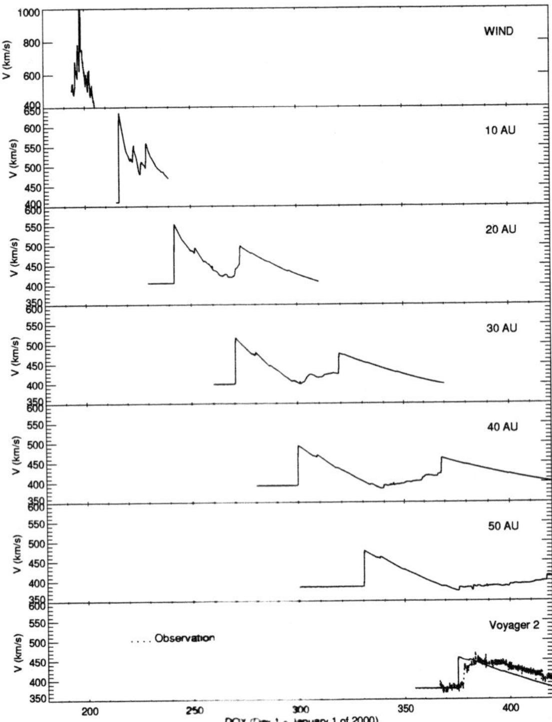

FIGURE 1. Speed profiles observed by Wind (top panel) and predicted by the model with the Wind data as input at various radial distance. The comparison of the model result with Voyager 2 (63 AU) is shown in the bottom panel. Note that the strong shock observed by Voyager 2 is the outer heliospheric remanet of the strong shock in the inner heliosphere.

tons and pickup ions. The energy partition ratio (6) is taken be 0.05, which means about 5% of the total energy from the pickup process goes to the solar wind wind protons, in order to reproduce the temperature profile observed by Voyager 2 in the outer heliosphere. Furthermore we use the hydrodynamical approach to calculate the distribution of the neutral hydrogen in an self-consistent manner. The interstellar neutral hydrogen density is chosen as 0.09 cm^{-3} at the termination shock to match the slowdown of the solar wind.

ONE TO ONE RELATIONSHIP

A large solar event took place on the Sun on July 14 (Bastille Day), 2000. Many aspects of this storm event can be found in the December 2001 topical issue of Solar Physics. The passage of the ejecta at Earth produced a very large high-speed stream on July 15 with a speed jump from ~600 to over 1050 km s^{-1} and a few small streams. The propagation and evolution of the Bastille Day CME-driven shock in the outer heliosphere and their interaction with the heliospheric boundaries have been studied by serval authors with different approaches (8-11). Figure 1 shows the speed profiles observed by Wind at 1 AU (top panel) and predicted by the model with the Wind data as input at different distances from 10 AU to Voyager 2. The interaction and evolution of the dominant large high-speed stream and serval small streams observed by Wind at Earth has evolved into a well-defined strong shock in the front by 10 AU, followed by a complicated solar wind structure. As they continue to propagate into the heliosphere, the speed profile becomes a relatively simple "jump-ramp" structure. The leading forward shock (we called it the "Bastille Day shock") decays significantly with distance, while other shocks/discontinuties almost disappear at Voyager 2. Figure 2 plots the shock strength (indicated by the density compression ratio) and propagation speed as functions of distance.

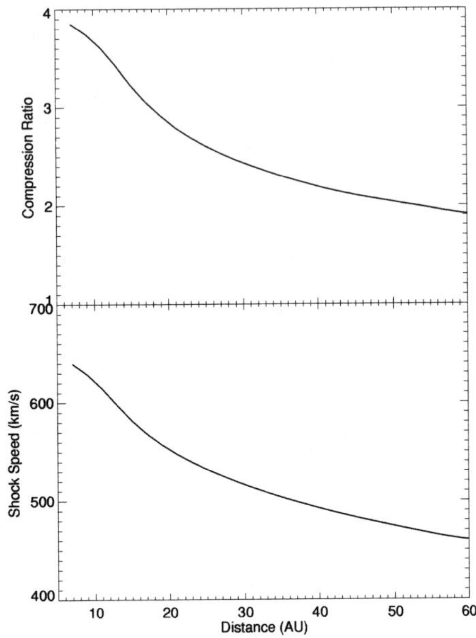

FIGURE 2. (a)The shock strength (represented by the density compression ratio) and (b) the shock propagation speed decay with distance of the leading forward shock.

The compression ratio decreases steady from ~4 in the inner heliosphere to ~1.8 at Voyager 2, and the shock propagation speed decreases from above 650 to 460 km s^{-1} at Voyager 2 (~63 AU). The model predicted the Bastille Day shock would arrive at Voyager 2 on January 9, 2001 with a speed jump of ~ 70 km s^{-1}. Within a few days of the predicted date, Voyager 2 saw a relatively strong shock on January 12, 2001 with a speed jump of ~ 65 km s^{-1}. The thermal pressure increases across the shock by a factor of 2.5.

The bottom panel in figure 1 shows the comparison of the model results (dotted line) with the Voyager 2 observations. The timing and speed profile are in reasonably good agreement with the observations. Therefore, we conclude the strong shock observed by Voyager 2 at ~63 AU on January 13, 2001 is naturally the outer heliospheric remanet of the strong Bastille Day shock in the inner heliosphere. Not surprisingly, there exists an one to one relationship between an strong shock in the outer heliosphere and a strong shock in the inner heliosphere driven by a big solar event. However, not all strong shocks in the outer heliosphere can find their counterparts in the inner heliosphere. There exists another type of relationship which we will discuss in the following section.

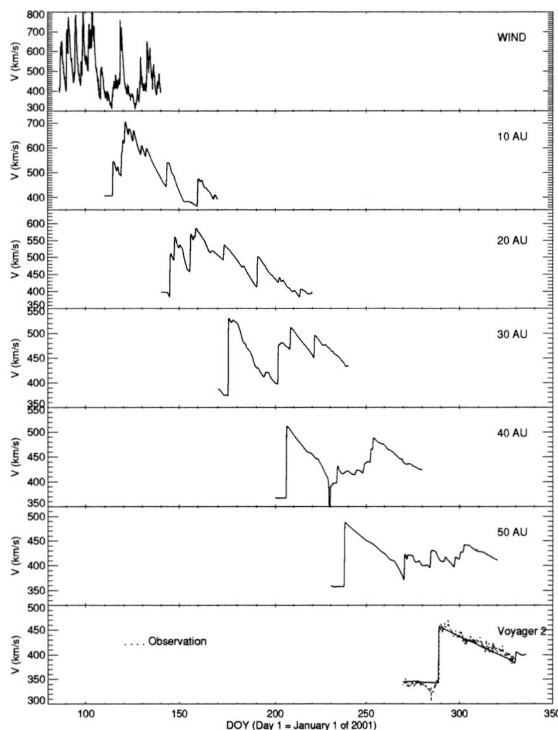

FIGURE 3. Speed profiles observed by Wind (top panel) and predicted by the model with the Wind data as input at various radial distance. The comparison of the model result with Voyager 2 (65 AU) is shown in the bottom panel. Note that the strong shock observed by Voyager 2 is the outcome of the interaction and merging of a series of interplanetary shocks.

ONE TO MANY RELATIONSHIP

On October 16, 2001, Voyager 2 recorded a strong shock at ~65 AU with a speed jump of 105 km s^{-1} across the shock, and the thermal pressure increases by a factor of 5.9. In contrast to the Bastille Day shock at Voyager 2 (with a speed jump of 65 km s^{-1} and a thermal pressure increases by a factor of 2.5), it is a much stronger shock. As a matter of fact, it is the strongest shock has been observed by Voyager 2 in the outer heliosphere since 1991. In an attempt to identify its inner heliospheric source, initially we search the Wind data for a big event similar to the Bastille Day 2000 event, but failed to pinpoint a single solar source which could have been responsible. Instead, we notice the active regions on the Sun produced a series of solar flares and CMEs in April, 2001. The consequences of these solar events at Earth are the group of high-speed streams observed by Wind during this time period, each separated by only a few days and last for almost one solar rotation. By comparing the solar wind plasma measurements and other aspects such as the geomagnetic impact and Forbush decrease at Earth, none of these events by itself is as large as the Bastille Day 2000 event. Therefore we hypothesize that as a group they can coalesce and evolve into a stronger shock in the outer heliosphere (12). As before, we insert the Wind data in April into the inner boundary at 1 AU.

Figure 3 shows the interaction and merging of shocks in the same format as that in figure 1. By the distance of 10 AU, the stream structures observed by Wind has evolved into a big triangular speed structure with a leading forward shock. Most obviously, the big triangular speed structure is then slitted into a forward and reverse shock pairs. (In fact, numerous interplanetary shocks have developed from the stream interactions and we will not follow the detail here). The forward shock overtakes the leading shock and forms a strong shock by 30 AU in the outer heliosphere. The reverse shock collides with the trailing shocks and discontinuities and produces lots of small scale solar wind structures. From 30 to 60 AU, the leading forward shock decays slightly but not significantly because the leading triangle structure continues to interact with the small trailing shocks . The bottom panel of the figure 3 shows the comparison of the model predictions with the Voyager 2 observations. Both the timing and overall character of the propagated Wind speed profile match the Voyager 2 observations quite well. Hence we believe that the strong shock observed by Voyager 2 in October 2001 is the result of the interaction and merging of a series of the interplanetary shocks evolved from the high-speed streams observed by Wind at Earth in April 2001. These exists the one to many relationship between a strong shock in the outer heliosphere and a series of interplanetary shocks as results of multiple solar events.

DISCUSSION AND SUMMARY

During the current solar maximum (Cycle 23), several major CMEs associated with solar flares produced large transient shocks which were observed by widely-separated spacecraft such as Wind at Earth and Voyager 2 beyond 60 AU. Using data from these spacecraft and numerical models which include the interaction between solar wind protons, pickup ions and interstellar neutrals, we study shock propagation and interaction in the outer heliosphere. The model we used is a one-dimensional multi-fluid model. Since Wind and Voyager 2 are not generally radially aligned, the radial projection of the speed profile at Wind to Voyager 2 is impossible. However, considering the latitudinal dependence of the solar wind is small near solar maximum according to the Ulysses observations (13), to first approximation, we only need to worry about the longitudinal separation between the two spacecraft near solar maximum . For the the Bastille Day event case, Wind was, fortunately, in the similar longitude as Voyager 2. As for the October shock case, considering the magnitude of the solar activity in April 2001, we believe that the spatial extent of those events is likely large enough to make our 1-D assumption feasible. Nevertheless, the model predictions at Voyager 2 are in good agreements with observations.

A strong shock at Earth undergoes a dramatic change while propagating outward. For example, the Bastille Day 2000 CME shock had a speed jump of over 400 km s^{-1} at Earth and was detected by Voyager at 63 AU with a speed jump of 65 km s^{-1} about 6 months later. However, a strong shock at Voyager 2 does not necessarily correspond to a strong shock at Earth. On October 16, 2001, Voyager 2 at 65 AU observed a strong shock with a speed jump over 100 km s^{-1}, the strongest shock recorded since 1991, we could not find a single solar event which is directly linked to this shock. Instead, a series of solar events in April 2001 is found to be responsible. The model results show that successive merging and interaction of relatively small interplanetary shocks could form a well-developed strong forward shock beyond 30 AU. In a world, a strong shock in the distant could be a outer heliospheric remnant of a strong shock in the inner heliosphere such as the Bastille Day shock ("one to one" relationship), or it could be a outcome of the interaction and evolution of a series of interplanetary shocks such as the October shock("one to many" relationship). Our demonstration that large shocks can and do form from the merging mechanism may have important consequences for the formation of merged interaction regions and the triggering of the heliospheric radio emission.

ACKNOWLEDGMENTS

This work was supported under NASA contract 959203 from JPL to MIT and NASA grant NAG5-11623. C. Wang is grateful to the one-hundred talent program of the Chinese Academy of Sciences.

REFERENCES

1. Whang, Y.C., *Space Sci. Rev.*, **57**, 339 (1991)
2. Burlaga, L.F., McDonald, F.B., and Ness, N.F., *J. Geophys. Res.*, **98**, 1 (1993)
3. Zank, G.P. and Pauls, H.L., *J. Geophys. Res.*, **102**, 7037, (1997)
4. Whang, Y.C. and Burlaga, L.F., *J. Geophys. Res.*, **104**, 6721, (1999)
5. Wang, C., Richardson, J.D., Gosling, J.T., *Geophys. Res. Lett.*, **27**, 2429, (2000)
6. Wang, C. and Richardson, J.D., *J. Geophys. Res.*, **106**, 29401, (2001)
7. Isenberg, P.A.,*J. Geophys. Res.*, **91**, 9965 (1986)
8. Wang, C., Richardson, J.D. and Paularena.,K.I., *J. Geophys. Res.*, **106**, 13,007, (2001)
9. Wang, C., Richardson, J.D. and Burlaga, L.F., *Solar Phys.*, **204**, 411, (2001)
10. Whang, Y.C., et al., *Solar Phys.*, **204**, 255, (2001)
11. Zank, G.P, et al., *J. Geophys. Res.*, **106**, 29363, (2001)
12. Wang, C. and Richardson, J.D., *Geophys. Res. Lett.*, in press, (2002)
13. McComas, D.J., J.T. Gosling, R.M. Skoug, *Geophys. Res. Lett.*, **27**, 2487 (2000)

Global Structure of Interplanetary Coronal Mass Ejections Retrieved from the Model Fitting Analysis of Radio Scintillation Observations

Munetoshi Tokumaru*, Masayoshi Kojima*, Ken'ichi Fujiki* and Masahiro Yamashita*

**Solar-Terrestrial Environment Laboratory, Nagoya University, Toyokawa 442-8507, Japan*

Abstract. Interplanetary scintillation (IPS) measurements made with the Solar-Terrestrial Environment Laboratory, Nagoya University, multi-station system at 327 MHz are used to study the global structure of the interplanetary counterpart of coronal mass ejections (so-called interplanetary CMEs or ICMEs). We have analyzed our IPS data of four ICME events which occurred successively in July 2000; those are 2000 July 10, 11, 12 and 14 events (the last one corresponds to "Bastille day event"). We have employed the model fitting method to obtain unbiased three-dimensional properties of ICME from IPS observations. The parameters determined here include the location of the ICME center, the e-folding radial thickness and angular span, anisotropy of the angular extent, the local enhancement factor, the (average) expansion speed, and its angular dependence. The results suggest that the global shape of July 12 and 14 events were loop-like; the latitude angular span of these events was much smaller than the longitude one. The July 10 and 11 events are found to be explained excellently by a shell-shape ICME model. The mass contained by the ICME has been estimated using information on the global structure obtained here.

INTRODUCTION

The global properties of the interplanetary counterpart of coronal mass ejections (interplanetary CMEs or ICMEs) are little understood owing to a lack of imaging observations for the solar wind plasma, although they provide crucial information for studying the dynamics of CMEs in the solar wind. Interplanetary scintillation (IPS) measurements with a high sensitivity system have an unrivalled potential to clarify the global properties of ICMEs, since they allow us to produce an plane-of-sky projection map of the solar wind plasma (e.g. [1], [2], [3]). The spatial resolution of the IPS solar wind map depends on the number of line-of-sight (los) employed for IPS observations in a day. Although it has proven from earlier studies that IPS solar wind maps act as an effective tool to detect and track ICMEs traveling from the sun to beyond the earth orbit, the three-dimensional structure of ICMEs is still unclear, since an apparent feature of ICMEs in the IPS solar wind map is influenced significantly by the los integration effect and the plane-of-sky projection effect. So that, we need to remove these effects in order to clarify three-dimensional properties of ICMEs from IPS observations.

Recently, we have investigated the global structure of ICME associated with the 2000 July 14 flare event (so-called the Bastille day event), using our IPS observations [4] (hereafter Paper I). In Paper I, we have performed the model fitting analysis to extract unbiased three-dimensional information from our IPS data. In this paper, we have analyzed another three ICME events, which occurred successively just prior to the Bastille Day event, by employing the model fitting method developed in the previous study.

IPS OBSERVATIONS OF ICME

IPS observations at 327 MHz have been carried out daily between April and December of every year with the four-station system of the Solar-Terrestrial Environment Laboratory (STEL), Nagoya University [5]. Our IPS observations covers the radial distance range between 0.2 AU and 1 AU from the sun, and number of line-of-sight (los) used in a day for our IPS observations is about 40. From our IPS data, we derive solar wind velocity and g-value, which represents the relative level of solar wind (density) turbulence, ΔN_e. In this study, the g-value data have been analyzed to study global properties of ICMEs.

The g-value is normalized to the level of the ambient solar wind. When a highly turbulent plasma associated with an ICME passes across the los for a given radio source, the g-value increases abruptly. By plotting the g-value data of an ICME event on the plane of sky, we can make the spatial distribution map of turbulent (high ΔN_e)

CP679, *Solar Wind Ten: Proceedings of the Tenth International Solar Wind Conference*, edited by M. Velli, R. Bruno, and F. Malara

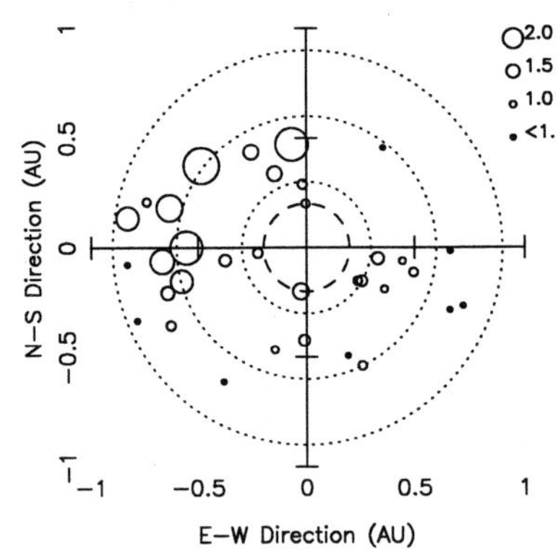

FIGURE 1. (right) All-sky g-map obtained from our IPS observations for July 11 22h UT – July 12 7h UT, 2000, and (left) SOHO/LASCO C3 image for the CME on July 10, 2000.

TABLE 1. Solar-terrestrial phenomena associated with 2000 July ICME events analyzed. Information on flare and ssc, which acts as a proxy of the IP shock occurrence at the earth, is from Solar Geophysical Data, and information on CME is from SOHO/LASCO observations.

Event Date	July 10, 2000	July 11, 2000	July 12, 2000	July 14, 2000
Solar Flare				
Peak Time (UT)	21:42	13:10	10:37	10:24
Class/Importance	M5.7/2B	X1/2N	X1.9/2B	X5.7/3B
Heliographic Location	N18E49	N18E27	N17E21	N22W07
CME				
Occurrence	Y	Y (Full-halo)	Y	Y (Fast-halo)
IP shock/ssc				
Occurrence Time (UT)	July 13 09:42		July 14 15:32	July 15 14:37
V_{shock} (km/s)	690		790	1470

plasma. Such a plane-of-sky projection map of the solar wind is called an all-sky g-map. In the right panel of Figure 1, an all-sky g-map obtained from our IPS observations between July 11 22h UT and July 12 7h UT, 2000, is indicated. The center of this map corresponds to the location of the sun, and dotted concentric circles are iso-R contours for 0.3, 0.6, and 0.9 AU. Here, $R = R_{SE}\sin\varepsilon$, and R_{SE}, ε are the Sun-Earth distance (i.e. 1 AU) and the solar offset angle, respectively. Circles in the figure show the location of los, and the size of circles denotes the strength of g-value. Enhancements of g-value found in the north-east quadrant of the map are considered to be an interplanetary consequence by the CME which was observed on July 10, 2000, with SOHO/LASCO measurements [7] (the left panel of Figure 1). Here, we note that the location of the g-enhanced region in the map shows an excellent agreement with the radial extension of brightness enhancements in the LASCO image. While the crescent-shape appearance of g-enhanced region may suggest that the ICME had a shell-shape structure with a thin radial thickness, we must draw attention to the fact that this feature includes effects of the los integration effect and the plane-of-sky projection.

MODEL FITTING ANALYSIS OF G-VALUE DATA

The g-value, g, is related to the integration of solar wind density fluctuations ΔN_e via the following formula,

$$g^2 \propto \int \Delta N_e{}^2 w(z)dz, \tag{1}$$

for weak scattering of a radio wave (i.e. for $R > 0.2$ AU in the case of 327 MHz wave, where R is the radial distance). Here, z is the distance along the line-of-sight, and $w(z)$ is the IPS weightening function [6]. Equation (1) allows us to calculate the g-value, provided that the three-dimensional ΔN_e distribution in the solar wind is given by a suitable model. Thus, we can determine the best-fit model of ΔN_e distribution by comparing the model calculations to observed g-value data. This analysis yields unbiased information of three-dimensional ΔN_e distribution in the solar wind [8]. In the present analysis, we have made the model fitting analysis for three ICME events which occurred just prior to the Bastille day event; those are July 10 (Figure 1), July 11, and July 12. The solar-terrestrial phenomena associated with these ICME events

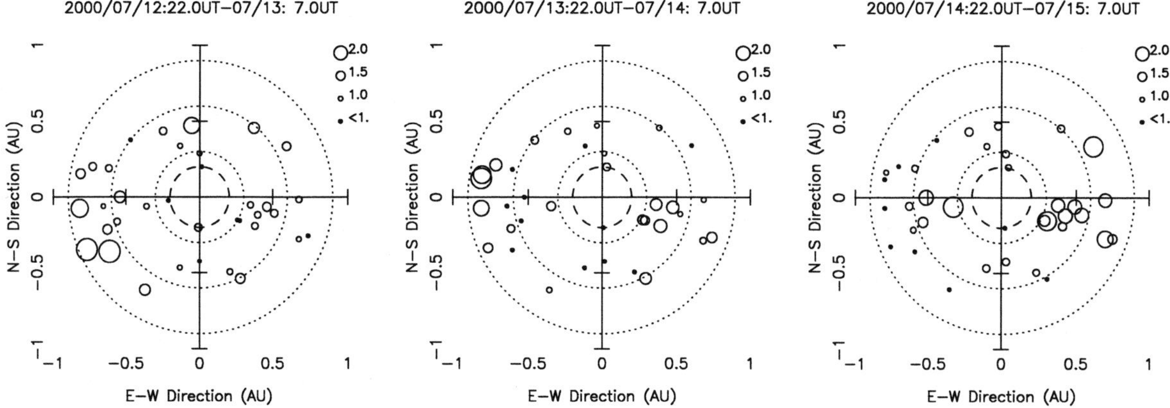

FIGURE 2. All-sky g-maps obtained from our IPS observations on (left) July 13, (middle) July 14, and (right) July 15, 2000. These events correspond to the July 11, 12 and 14 flare/CME events.

(including the Bastille day event) are summarized in Table 1. All-sky g-maps obtained from our IPS observations for July 11, 12 and 14 events are indicated in Figure 2.

For the fitting analysis, we have employed a simple ΔN_e model, which assumes a Gaussian-form enhancement by an ICME and R^{-2} radial fall of the background turbulence level. In this model, an anisotropic angular extent of ICME is taken into account, and an iso-ΔN_e contour at a given radial distance is assumed to be in an elliptic form which is defined by an e-folding major angular span, θ_0, the ratio of the minor angular span to the major one, AR, and the twist angle of the major angular span direction to the heliographic equator, β. The expansion speed V_S of ICME is assumed to depend on the separation angle θ to the ICME center axis, and a function of $V_S = V_{S0}\cos^{\alpha}(\theta/2)$ (where V_{S0}, α are constant) is used as the angular dependence of V_S. The effect due to deceleration or acceleration of ICME is neglected in this model. Besides θ_0, AR, β, V_{S0}, and α, free parameters of this model are the Carrington coordinate of the ICME center axis (λ_0, ϕ_0), the e-folding radial thickness D, and the local enhancement factor at the center C_1 (The ΔN_e model used here is the same as one used in Paper I, and more detailed description on it is presented therein).

As shown in the middle panel of Figure 2, g enhancements associated with the July 12 flare/CME event occur at two areas, which are located symmetrically with respect to the Sun-Earth line in the map. This fact suggests that the ICME had a loop- or rope-like structure which elongated in longitude, as the case of the Bastille day event (Paper I). On the other hand, g enhancements associated with July 10 and 11 events are found mostly in the eastern hemisphere (see the right panel of Figure 1 and the left panel of Figure 2). This implies that the central part of those events propagated in the eastward direction with a considerable offset angle to the Sun-Earth line. In the case of such broadside observations, anisotropy of the angular extent is hardly determined from the analysis of g-value data without a priori information, even if it exists (At least, we may conclude from our IPS data that the longitude extent of ICME is not so large as to arise g enhancements in the western hemisphere). Therefore, we assume in the analysis for these events that the ICME had a shell shape structure with an isotropic angular extent; i.e. $AR = 1$. This is equivalent to an assumption that the latitude extent of observed g-enhanced region is roughly equal to the longitude extent, and a similar assumption is often employed in the analysis of plane-of-projection data such as coronagraph data.

RESULTS AND DISCUSSIONS

Calculated g-values g_{cal} with the ΔN_e model have been compared with observed g-values g_{obs} for each ICME event, and free parameters of the model have been adjusted to minimize the rms deviation σ between g_{cal} and g_{obs}. Parameters determined by this analysis are listed in Table 2. The correlation coefficient ρ between g_{cal} and g_{obs} is also indicated in the table. The table includes the results of the model fitting analysis for the Bastille day event (Paper I). Provided that the global distribution of ΔN_e enhancements due to ICME is determined by the current analysis, the mass contained by the ICME, M_{ICME}, can be estimated by integrating the enhanced ΔN_e level over the volume occupied by the ICME [9]. Here, the solar wind density is assumed to be proportional to ΔN_e, and SOHO in situ measurements are used to determine the ambient level of the solar wind density, which is necessary to estimate M_{ICME}. Estimated values of M_{ICME} are shown in the bottom row of Table 2.

We summarize results of the present analysis as follows;

TABLE 2. Results of the model fitting analysis.

Event	July 10	July 11	July 12	July 14
Location	N13E19	S19E37	N02E41	N11W17
V_{S0} (km/s)	880	810	880	1540
α	2.9	0.	1.9	3.3
D (AU)	0.07	0.08	0.07	0.13
θ_0 (deg.)	51	18	202	130
C_1	4.2	6.9	4.8	6.4
AR	$\sim$1	$\sim$1	0.04	0.15
β (deg.)			0	28
σ	0.205	0.220	0.218	0.140
ρ	0.798	0.843	0.827	0.762
M_{ICME} (g)	2.6×10^{16}	8.7×10^{15}	6.5×10^{15}	5.5×10^{16}

1). Our IPS data suggests that the ICME event associated with the 2000 July 12 flare/CME had a loop-like (or toroidal) shape which stretched considerably in longitude. The minor angular span of this ICME event is found to be 0.04 times as small as the major one. A similar feature of the ICME has been reported from the analysis of the Bastille day event (Paper I). From the comparison with solar magnetograph measurements, it is demonstrated that the loop orientation of July 12 and 14 (Bastille day) events is approximately normal to the magnetic neutral line on the source surface. The loop-like structure revealed for July 12 and 14 events might be related to the magnetic structure of coronal ejecta, although the details are left to a future study.

2). For ICME events associated with July 10 and 11 flares/CMEs, it has been demonstrated that observed g enhancements are fitted nicely by a shell shape ICME model. Since the current analysis is based on a priori information of AR, we cannot conclude unambiguously that there are two types (loop or shell) of the ICME global structure. However, g enhancements with a halo-shape appearance have been observed for other halo CME events (e.g. the 2000 June 6 event), and this fact supports a shell shape model of ICME. Thus, we need to investigate this point from further analysis.

3). Estimated values of the radial thickness D for ICME events show good agreement with a typical thickness of the compression region driven by the IP shock at 1 AU [10]. The small angular span of ICME θ_0 for the July 11 event is consistent with the missing encounter of the IP shock to the earth. Locations of the ICME center determined here are found to be close to those of the flare site for all cases except for the July 11 event. Although a slight difference of locations between the flare site and the ICME center exists for July 10, 12 and 14 events, it is considered to be within the error inherent to our IPS observations. To examine whether this difference is intrinsic or not, IPS observations with a higher density are required.

4). The estimated values of ICME mass are in good agreement with a typical value of a CME mass. In comparing between ICME and CME mass estimates, it must be noted that an ICME mass is considered to increase significantly due to the sweeping effect, which is one of important processes governing the deceleration of an ICME in the interplanetary medium. Our results obtained here are useful to clarify the sweeping effect and the deceleration profile of ICME, and a study from this viewpoint is regarded as a promising future work (The deceleration of ICMEs between the corona and the earth orbit has been studied from our IPS observations [11]).

ACKNOWLEDGMENTS

This work was supported by the IPS project of the Solar-Terrestrial Environment Laboratory, Nagoya University.

REFERENCES

1. Gapper, G. R., A. Hewish, A. Purvis, and P. J. Duffet-Smith, *Nature*, **296**, 633–636 (1982).
2. Manoharan, P. K., M. Tokumaru, M. Pic, S. Subramanian, F. M. Ipavich, K. Schenk, M. L. Kaiser, R. P. Lepping, and A. Vourlidas, *Astrophys. J.*, **559**, 1180–1189 (2001).
3. Tokumaru, M., M. Kojima, K. Fujiki, and A. Yokobe, *J. Geophys. Res.*, **105**, 10435–10453 (2000).
4. Tokumaru, M., M. Kojima, K. Fujiki, M. Yamashita, and A. Yokobe, *Submitted to J. Geophys. Res.*, (2002).
5. Kojima, M., and T. Kakinuma, *Space Sci. Rev.*, **53**, 173–222 (1990).
6. Young, A. T., *Astrophys. J.*, **168**, 543–562 (1971).
7. Brueckner, G. E., et al., *Solar Phys.*, **162**, 357–402 (1995).
8. Tappin, S. J., *Planet. Space Sci.*, **35**, 271–283 (1987).
9. Gothoskar, P., and A. P. Rao, *Sol. Phys.*, **185**, 361–390 (1999).
10. Borrini, G., J. T. Gosling, S. J. Bame, and W. C. Feldman, *J. Geophys. Res.*, **87**, 4365–4373 (1982).
11. Yamashita, M., M. Tokumaru, and M. Kojima, *Proceedings of 10th International Conference on Solar Wind* (this issue), (2002).

Complexity Of The 18 October 1995 Magnetic Cloud Observed by Wind and the Multi-Tube Magnetic Cloud Model.

V.A. Osherovich[1], J. Fainberg[2], A. Vinas[2] and R. Fitzenreiter[2]

[1]*EER Systems Inc., NASA/GSFC, Greenbelt, MD 20771,*
[2]*NASA/GSFC Greenbelt, MD 20771, US*

Abstract. The notion that a magnetic cloud may consist not of one but of a few closely interacting magnetic tubes came from the analysis of the relation between the electron temperature T_e and plasma density N_e ([1]Fainberg et al. 1996). The log-log plot of T_e versus N_e of data in the 10-13 June 1993 Ulysses magnetic cloud revealed two polytropes (both with $\gamma_e < 1$) which have been attributed by [2]Osherovich, Fainberg and Stone (1999a) to two helices embedded in the same cylindrical flux rope. In contrast to configurations with cylindrical symmetry, magnetic configurations with helical symmetry allow a description of many helices which in cross section look like distorted flux ropes. We present elements of the magnetic and thermodynamic structure of the Oct 18-21, 1995 magnetic cloud observed by Wind. Our research is complementary to previous studies of this cloud by other authors ([3]Larson et al. 1997; [4]Lepping et al. 1997; [5]Janoo et al. 1998). The log-log plots of T_e vs N_e suggest that this cloud consists of eight magnetic tubes with polytropic indices below unity and a few non-coherent structures for which the polytropic relation is not valid. The solar wind quasi-invariant ([6]Osherovich et al., 1999b) for this cloud strongly anti-correlates with the Dst index for the resulting magnetic storm.

MAGNETIC STRUCTURE OF THE CLOUD AND SELF-SIMILAR (OFB MODEL) FOR A SINGLE FLUX ROPE

The October 18-21, 1995 cloud observed by Wind has all three magnetic cloud elements which according to Burlaga [[7]1981] are: a) a significant enhancement of the total magnetic field B (figure 1, black line), b) a smooth rotation of the total magnetic field $\vec{B}$ (figure 2, red line) and c) a low proton temperature T_p. The front sheath is marked by the shock at 11.111h (hours referenced to start of October 18). Starting from the shock, $\vec{B}$ rotates (blue line in figure 2), but the rotation in the z-y plane is not smooth. Only on entering the cloud at 19.1h, can one see the beginning of the smooth rotation during the period 19.104-48.917h. After this period, the $\vec{B}$ rotation continues in the back sheath (orange curve in figure 2), but the smoothness typical for inside the cloud is lost.

The magnetic structure of the October 18-21,1995 cannot be described by any force-free magnetic flux rope (neither constant nor non-constant alpha). The general solution for a force-free rope [Lust and Schluter, 1954] can be expressed through a generating function where

$$B_\phi^2 = -(R/2)\frac{df}{dR} \tag{1}$$

$$B_{z^*}^2 = f - B_\phi^2 \tag{2}$$

where (z^*, ϕ, R) are cylindrical coordinates with z^* along the axis of the flux rope. Thus, for a "flat top" cloud like the Oct 18-21,1995 cloud, $\frac{df}{dR} \approx 0$ and therefore B_ϕ should be close to 0. In fact, we see a large B_ϕ component (green curve in Figure 1).
The self-similar model of Osherovich, Farrugia and Burlaga (OFB 1993) has

$$\vec{B} = B_0(\eta)y^{-2}(t)\vec{e}_{z^*} + B_\phi(\eta)y^{-1}(t)\vec{e}_\phi \tag{3}$$

$$\vec{v} = \frac{y'}{y}R\vec{e}_R + v_0(\eta)\vec{e}_{z^*} \tag{4}$$

where $\eta = R/y(t)$ is a self-similar parameter. The evolution function y(t) obeys the equation

$$\frac{d^2y}{dt^2} = Sy^{-2} - Qy^{-1} + Ky^{(-2\gamma+1)} - \nu\frac{dy}{dt} \tag{5}$$

CP679, *Solar Wind Ten: Proceedings of the Tenth International Solar Wind Conference,*
edited by M. Velli, R. Bruno, and F. Malara

where S, Q, K and ν are constants; γ is a polytropic index. In order for the OFB flux rope to expand, γ

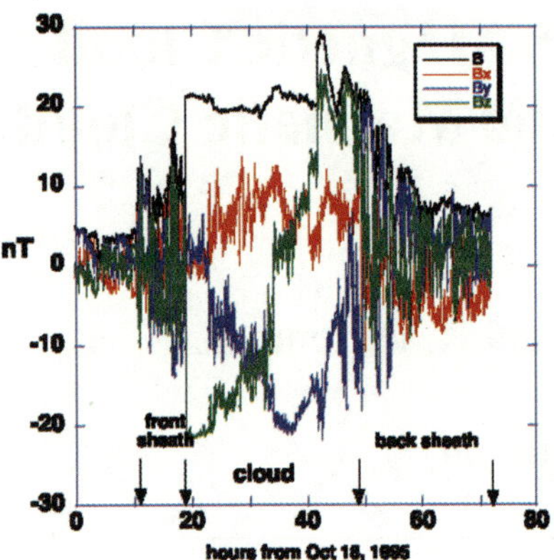

Figure 1. Total magnetic field components for October 18-21, 1995 magnetic cloud observed by Wind.

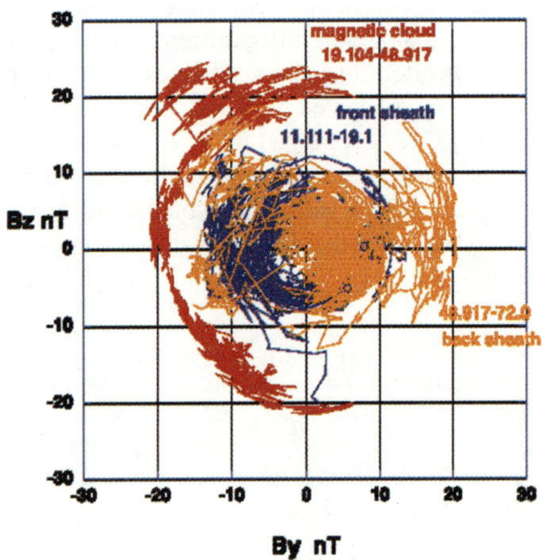

Figure 2. Rotation of Magnetic Field Vector in October 18-21,1995 Magnetic Cloud Observed by Wind.

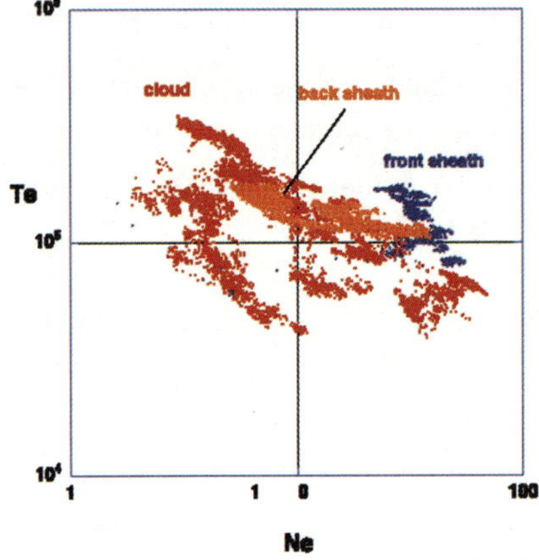

Figure 3. Total electron temperature vs. density in the magnetic cloud and the sheath

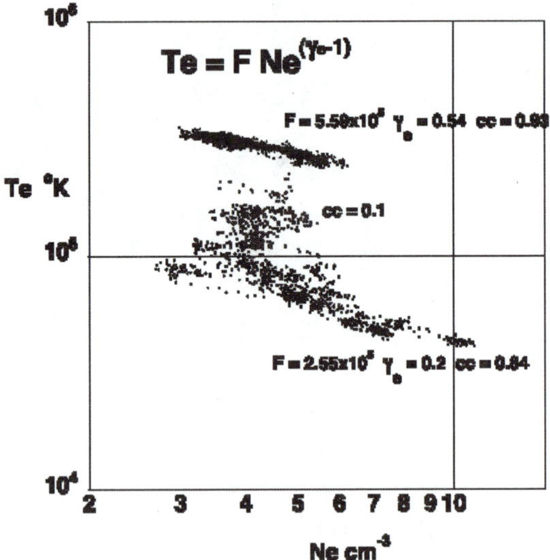

Figure 4. Three types of structures inside the magnetic cloud.

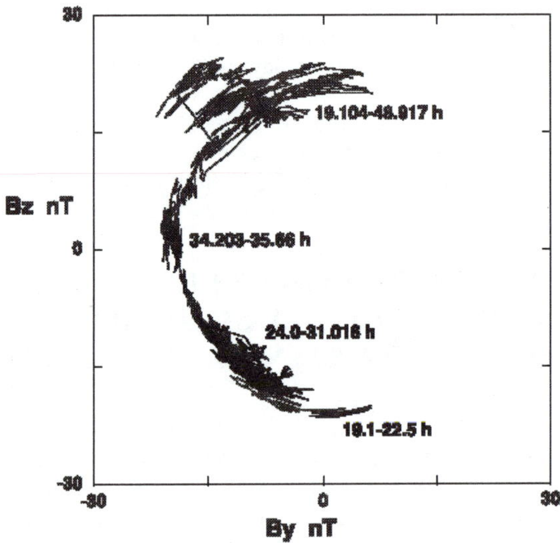

Figure 5. Position of three structures inside the magnetic cloud

must be below unity for either $v_0(\eta) = 0$ or $v_0(\eta)$ 0 which we include for purpose of generality. The OFB model does accommodate a flat top B profile for an old cloud, when y(t) is large enough and $B_{\phi\,max}$ is comparable to $B_{z^*\,max}$ (over-grown pinch).

ANTI-CORRLELATION OF Te AND Ne IN THE CLOUD AND THE SHEATH

The dependence of T_e on N_e in the cloud (red dots in figure 3) show eight extended periods with significant anti-correlation (cc > 0.7) between T_e and N_e, for which the polytropic relation

$$T_e = FN_e^{\gamma_e - 1} \qquad (6)$$

holds. Between those regions (which we interpret as magnetic tubes) there are non-coherent structures with cc < 0.7. For the whole cloud (including non-coherent structures), F=5.87x10^5 cgs units and γ_e = 0.39. For the sheath (orange and blue points in figure 3), F is lower and γ_e is somewhat higher than in the cloud, but is still less than unity. For the back sheath, for example, γ_e = 0.75 and F = 2.66x10^5. Three types of structures (highly coherent, coherent and non-coherent) inside the cloud are illustrated in figure 4. The corresponding correlation coefficients are cc =0.93 (blue color) cc = 0.84 (black) and cc = 0.1 (green). Figure 5 illustrates the position of the corresponding regions on the z-y plane of $\vec{B}$ rotation. The magnetic structure of the back sheath includes two regions: (strong field region – gray color in figures 6 and 7 and the weak field region – black color on the same figures). Thus, magnetically and thermodynamically, such regions can be separated from each other. Further division (fine structure) is possible.

ANOMALY OF SOLAR WIND QUASI-INVARIANT (QI) AND MAGNETIC STORM INDUCED BY THE OCTOBER 18-21, 1995 CLOUD

Recently, a new index of solar activity, the solar wind quasi-invariant

$$QI \equiv (B^2/8\pi)/(\rho v^2/2) \qquad (7)$$

has been suggested (Osherovich, Fainberg and Stone, 1999). QI is based only on parameters measured in the solar wind, namely magnetic field strength B, solar wind speed v and density ρ. While the yearly median values of QI follow yearly sunspot numbers with cc = 0.98, every magnetic cloud can be viewed as a violation of QI (increase by a factor of 20-100). Figure 7 illustrates this increase and also the strong anti-correlation with Dst drop during related magnetic storms.

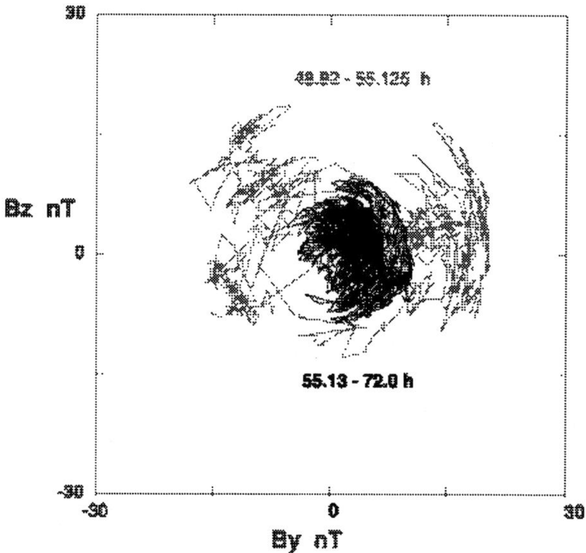

Figure 6. Rotation of magnetic field in back sheath

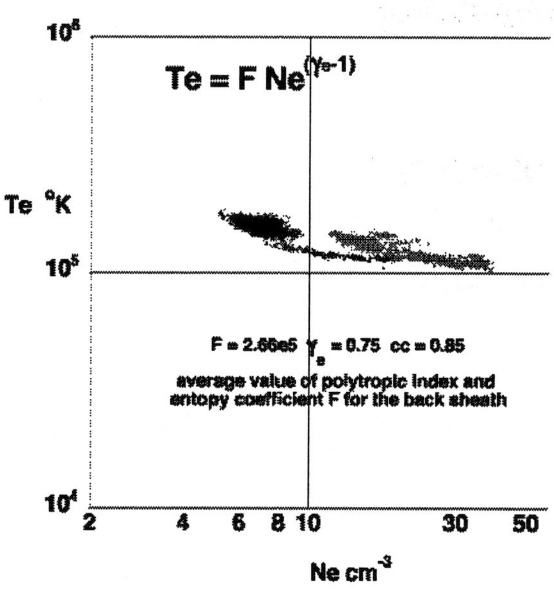

Figure 7. Structures inside the back sheath of the magnetic cloud.

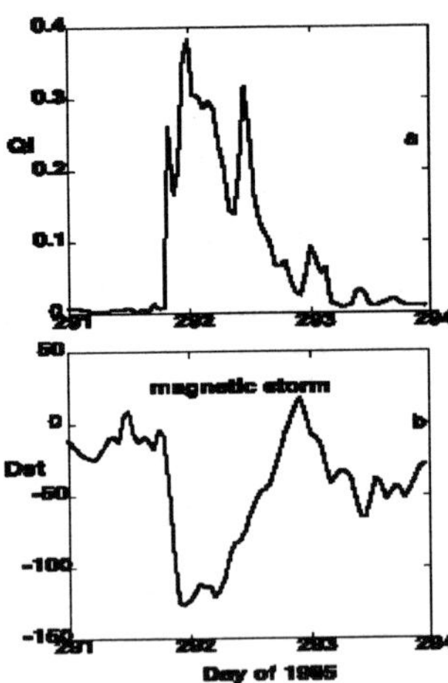

Figure 8. Solar wind quasi-invariant increase for the magnetic cloud and the related drop in Dst index.

DISCUSSION AND CONCLUSIONS

From the analysis of total T_e as a function of N_e, we infer that most of clouds consist of a few magnetic tubes wrapped up in a magnetic flux rope (container). Our exact bounded MHD solutions have depicted such configuration as multiple helices embedded in the cylindrical flux rope (Krat and Osherovich, 1976; Osherovich, Fainberg and Stone, 1999b). The topological complexity of such solutions depends on the magnetic flux function (ground-state solution, first excited state, etc.). The continuity of $\vec{B}$ everywhere and finite magnetic energy per unit of length of the main tube, makes such solutions, mathematically, similar to eigen-functions in quantum mechanics (Osherovich, 1975). The complex structure of magnetic clouds like the October 18-21, 1995 cloud, we believe has consequences for the interaction of such clouds with planetary magnetospheres and comets. The strongly non-Maxwellian distribution of electrons in magnetic clouds (discovered by Fainberg et al. 1996) leads to an abnormally high population of highly ionized heavy ions, which in turn can cause soft-x-rays during the collision with comets. This internal complexity could be a source of additional variability of such x-rays.

CONCLUSIONS

1. The October 18-21, 1995 cloud contains eight magnetic tubes, for which T_e anti-correlates with N_e with cc > 0.7 and a number of non-coherent structures between tubes with cc < 0.7.

2. For the whole cloud (including non-coherent regions), γ_e = 0.39 and F = $5.87\text{x}10^5$. For the back sheath, these values are γ_e = 0.75 and F = 2.66 $\text{x}10^5$. Thus, for both the cloud and the sheath $\gamma_e < 1$ but γ_e is significantly larger in the sheath as was previously observed in the June 10-13, 1993 Ulysses cloud [Fainberg et al. 1996; Osherovich and Burlaga 1997].

3. The cloud is old (expansion subsided) and as such is in agreement with the OFB model where it has an overgrown B_ϕ component which leads to a flat profile.

4. We interpret the complexity of this cloud (8 tubes) as a case of interacting helices embedded in one cylindrical flux rope (our multi-tube model).

5. The abnormally large QI for this cloud closely anti-correlates with the Dst index, which dropped during the associated magnetic storm.

REFERENCES

1. Fainberg, J. et al., Solar Wind 8, 554, 1996.

2. Osherovich, V.A., J. Fainberg and R.G. Stone GRL 26(3), 401, 1999a..

3. Larson, DE et al., Adv. Space Res. 20 (4-5), 655, 1997.

4. Lepping RP et al., J Geophys Res-Space 102 (A7): 14049, 1997.

5. Janoo, L et al., J Geophys Res-Space 103 (A8): 17249-17259, 1998.

6. Osherovich, V.A., J. Fainberg and R.G. Stone GRL 26(16), 2597, 1999b.

7. Burlaga, L.F., E. Sittler, F. Mariani and R. Schwenn, J. Geophys. Res. 86, 6673, 1981.

8. Lust, R. and A. Schluter, Z. Astrophys. 34, 263, 1954.

9. Osherovich, V.A. C. Farrugia and L.F. Burlaga, Adv. Space Res/. 13, 6(6), 57, 1993.

10. Krat, V.A. and V.A. Osherovich, Solar Phys., 59, 43, 1978.

11. Osherovich, V.A., Soln. Dann. 8, 1975.

12. Osherovich, V.A. and L.F. Burlaga, in Coronal Mass Ejections, eds N. Crooker, J. Joselyn and J. Feynman, 157, 1997.

A self-similar solution of expanding cylindrical flux ropes for any polytropic index value

Hironori Shimazu *† and Marek Vandas **

* *Applied Research and Standards Division, Communications Research Laboratory, Koganei, Tokyo 184-8795 Japan*
† *Department of Earth and Space Sciences, Box 351310, University of Washington, Seattle, WA 98195-1310 USA*
** *Astronomical Institute, Academy of Sciences, Boční II 1401, 141 31 Praha 4, Czech Republic*

Abstract. We found a new class of solutions for MHD equations that satisfies the condition that cylindrical flux ropes can expand self-similarly even when the polytropic index γ is larger than 1. We achieved this by including the effects of elongation along the symmetry axis as well as radial expansion and assuming that the radial expansion rate is the same as the elongation rate.

INTRODUCTION

Interplanetary magnetic flux ropes are structures in which magnetic field vectors rotate in one direction through a large angle. The sun often ejects flux ropes as coronal mass ejections (e.g., [1]). When flux ropes reach the earth, they often cause a geomagnetic storm (e.g., [2]). Observationally, they were first identified as "magnetic clouds" in interplanetary magnetic field data by [3].

While the flux rope propagates in interplanetary space, it expands because the ambient pressure decreases with its distance from the sun. The technique of converting time derivatives to self-similarity expansion parameters in MHD (magnetohydrodynamics) equations was originally applied by [4] and [5] to the explosion of a supernova. The self-similar approach simplifies time-dependent problems and makes them analytically tractable. A class of self-similar solutions of the expanding solar corona was found by [6] in the spherical coordinates when the polytropic index γ is exactly $4/3$. The same approach was used by [7], and a theoretical MHD model was presented describing the time-dependent expulsion of a three-dimensional coronal mass ejection.

Another class of self-similar solutions was found by [8] and [9]. The solution showed that cylindrical flux ropes expand self-similarly only when γ is less than 1 for this class of self-similar solutions. This formulation was applied to real interplanetary flux ropes [10]. Observations of solar wind electrons in interplanetary magnetic flux ropes were interpreted as having γ less than 1, and that is why flux ropes expand in interplanetary space [11].

However, coronal mass ejections, or flux ropes, expand in MHD simulations even when γ is larger than 1 [12] [13] [14] [15] [16]. Moreover, theoretical models of [17] and [18] showed expanding flux ropes with $\gamma > 1$. Recent comprehensive analysis by [19] has concluded that an observed negative correlation between the temperature and the density of electrons, on which the derivation of γ was based, cannot be a measure of γ. It was shown that single-point measurements cannot be used to determine γ value [20] [21]. To determine γ value observationally continues to be a matter for debate [22] [23].

This paper[1] describes a new class of solutions for MHD equations that satisfies the condition that cylindrical flux ropes can expand self-similarly even when γ is larger than 1. We achieved this by including effects of elongation along the symmetry axis, which is not included in the previous solutions. This new class of solutions puts new light to the very discussed problem on the relation between γ and the expansion. Mainly it emphasizes a role of flux rope elongation for its expansion. Recently a paper which deals with a similar topic to ours was submitted by [25].

SOLUTION

The MHD equations are solved as follows:

$$\frac{\partial \rho}{\partial t} + \nabla \cdot (\rho v) = 0, \tag{1}$$

[1] This paper is a summary of a theoretical part of our paper [24].

CP679, *Solar Wind Ten: Proceedings of the Tenth International Solar Wind Conference*,
edited by M. Velli, R. Bruno, and F. Malara

$$\rho\frac{\partial v}{\partial t}+\rho(v\cdot\nabla)v=-\nabla P+\frac{1}{\mu}(\nabla\times B)\times B, \quad (2)$$

$$\frac{\partial(P\rho^{-\gamma})}{\partial t}+(v\cdot\nabla)(P\rho^{-\gamma})=0, \quad (3)$$

and

$$\frac{\partial B}{\partial t}=\nabla\times(v\times B), \quad (4)$$

where t is time, ρ is mass density, P is pressure, μ is permeability, v is velocity, and B is the magnetic field. These MHD equations will be solved in cylindrical coordinates (r, θ, z) moving with a flux rope. The cylindrical flux rope has its axis along z and an axial symmetry is assumed, that is, the flux rope has a circular cross section and the quantities do not depend on θ.

Previous self-similar solution without elongation

Before showing the new solution, we will review the previous solution. This solution assumes self-similarity and no dependence on z (one-dimension and r-dependence only). Following the procedure described in [9], the solutions for (1), (3), and (4) were

$$v_r=\eta\dot{y}, \quad (5)$$

$$B_\theta=(-\eta f'/2)^{1/2}y^{-1}, \quad (6)$$

$$B_z=(2\mu SD)^{1/2}y^{-2}, \quad (7)$$

$$\rho=-D'\eta^{-1}y^{-2}, \quad (8)$$

and

$$P=KDy^{-2\gamma}, \quad (9)$$

where

$$D=\frac{f+\eta f'/2}{2\mu S\chi}, \quad (10)$$

and χ, S, and K are positive constants. The center dot means the derivative by the time t, and the prime is the derivative by the self-similar parameter η satisfying

$$\eta=ry^{-1}, \quad (11)$$

where y is the evolution function of time. The f is the generating function of η, satisfying

$$f'\leq 0, \quad (12)$$

$$f+\eta f'/2\geq 0, \quad (13)$$

and

$$(f+\eta f'/2)'\leq 0. \quad (14)$$

From the r component of the equation of motion (2), y satisfies

$$\ddot{y}=-\frac{dU}{dy}, \quad (15)$$

$$U=\begin{cases}\frac{K}{2\gamma-2}y^{2-2\gamma}+(1/2)Sy^{-2}+\chi S\ln y & (\gamma\neq 1)\\ (\chi S-K)\ln y+(1/2)Sy^{-2} & (\gamma=1).\end{cases} \quad (16)$$

Equation (15) can be regarded as an equation of motion under potential U. When γ is larger than 1, U takes a minimum value. Thus, flux ropes do not expand but oscillate when γ is larger than 1.

New solution with elongation

In this study we include effects of axial elongation (z-dependence) as well as radial expansion. The v_θ and B_r are assumed to be 0. An additional self-similar parameter ξ is introduced by

$$\xi=zy^{-1}. \quad (17)$$

This is consistent with the assumption that the radial expansion rate is the same as the elongation rate. We will solve (1) - (4) that satisfy

$$\frac{\partial f}{\partial\xi}=0 \quad (18)$$

for the simplest case. A new solution for (1), (3), and (4) is given by

$$v_r=\eta\dot{y}, \quad (19)$$

$$v_z=\xi\dot{y}, \quad (20)$$

$$B_\theta=(-\eta f'/2)^{1/2}y^{-2}, \quad (21)$$

$$B_z=(2\mu SD)^{1/2}y^{-2}, \quad (22)$$

$$\rho=-G'\eta^{-1}y^{-3}, \quad (23)$$

and

$$P=KGy^{-3\gamma}, \quad (24)$$

where G is a function of η and ξ.

From the r component of (2), U is expressed as

$$U=\begin{cases}\frac{K}{3\gamma-3}y^{3-3\gamma}+(D'/G')S(1-\chi)y^{-1} & (\gamma\neq 1)\\ -K\ln y+(D'/G')S(1-\chi)y^{-1} & (\gamma=1),\end{cases} \quad (25)$$

where D is also defined by (10) in this new solution. The θ component of (2) is zero under the condition (18). From the z component of (2), we get

$$U=\begin{cases}\frac{K}{3\gamma-3}y^{3-3\gamma}\frac{\eta(\partial G/\partial\xi)}{\xi(\partial G/\partial\eta)} & (\gamma\neq 1)\\ -K\ln y\frac{\eta(\partial G/\partial\xi)}{\xi(\partial G/\partial\eta)} & (\gamma=1).\end{cases} \quad (26)$$

For (25) to agree with (26), the conditions

$$\chi=1 \quad (27)$$

and

$$\frac{1}{\eta}\frac{\partial G}{\partial \eta} = \frac{1}{\xi}\frac{\partial G}{\partial \xi} \tag{28}$$

must be satisfied. From (28) we get

$$G = a\exp\left[-c(\eta^2+\xi^2)\right], \tag{29}$$

where a and c are constants. Thus, the pressure distribution is expressed as

$$P = Ka\exp\left[-c(r^2+z^2)/y^2\right]y^{-3\gamma}. \tag{30}$$

The ρ is given from (23) by

$$\rho = 2ac\exp\left[-c(r^2+z^2)/y^2\right]y^{-3}. \tag{31}$$

For the density to be positive anywhere and to be zero in infinitely distant regions, a and c must both be positive. In this new solution we do not need the condition (14), under which the density was positive or zero in the previous solution. The temperature T is expressed as

$$T = \frac{mK}{2kc}y^{-3\gamma+3}, \tag{32}$$

where k is Boltzmann's constant and m is the average mass of the solar wind particles.

Finally, we get

$$U = \begin{cases} \dfrac{K}{3\gamma-3}y^{3-3\gamma} & (\gamma \neq 1) \\ -K\ln y & (\gamma = 1). \end{cases} \tag{33}$$

This expression shows that U is a monotonically decreasing function of y. Thus, flux ropes obeying this solution expand for any γ value.

Equation (15) with the given potential (33) can be integrated yielding

$$dy/dt = \begin{cases} \pm\left[\dfrac{2K}{3-3\gamma}y^{3-3\gamma}+c_1\right]^{1/2} & (\gamma \neq 1) \\ \pm[2K\ln y + c_1]^{1/2} & (\gamma = 1), \end{cases} \tag{34}$$

where c_1 is a constant. In case $\gamma = 5/3$, we can integrate this equation further, and get

$$y = \left[c_1(t+t_0)^2 + K/c_1\right]^{1/2}, \tag{35}$$

where t_0 is a constant; $v_r = v_z = 0$ everywhere at $t = -t_0$. We took the plus sign assuming that y is a monotonically increasing function of t. In our solution y^{-1} is finite at $t = -t_0$, while in the previous solution y^{-1} is infinitely large. Thus, an infinitely large magnetic field strength is suppressed in our solution.

Features of the new solution

The most important effect of elongation of the axial length appears in the azimuthal component of the magnetic field:

$$B_\theta \sim y^{-1} \quad \text{(excluding elongation)} \tag{36}$$

is modified into

$$B_\theta \sim y^{-2} \quad \text{(including elongation)}. \tag{37}$$

The difference of B_θ dependence on y comes from the volume increase in the axial direction. This is the essential effect of the elongation. It was suggested the lack of a three-dimensional effect as the reason the previous solution does not expand when γ is greater than 1 [18]. This statement agrees with our solution.

We consider that a flux rope has a finite volume. The radius $r_1 = r_0 y$ and the axial length $l = l_0 y$ are determined by the initial radius r_0 and the initial length l_0 (at $t = -t_0$), respectively. r_1 is taken for B_z to be zero at the boundary.

A flux rope model for interplanetary magnetic clouds was constructed assuming that the total magnetic helicity is conserved by [26]. Our model agrees with their model in the dependence on l of the magnetic field strength, mass density, radius, and volume. Their results also showed that a flux rope evolves self-similarly, which agrees with our assumption.

The dynamics of flux ropes was investigated by [18]. His model starts with an equilibrium state of pressure between a flux rope and the ambient medium. He treated propagation in interplanetary space by the change of physical parameters of the ambient medium. In our model, we do not assume an initial equilibrium as other self-similar models do not assume. However, our model can trace a flux rope evolution during propagation in interplanetary space by the decreases of the ambient pressure and density with time.

In our solution the magnetic flux is conserved at all times. The magnetic energy decreases with increasing time. In the models of [26] and [18], the magnetic energy decreases as the flux rope expands. Our solution agrees with their results.

In our solution the magnetic flux is conserved at all times. The magnetic energy is expressed as

$$\begin{aligned} &\frac{1}{2\mu}\int_0^l\int_0^{r_1}(B_\theta^2+B_z^2)2\pi r dr dz \\ &= \frac{1}{y}\frac{\pi}{\mu}\int_0^{l_0}\int_0^{r_0}(2\mu SD - \eta f'/2)\eta d\eta d\xi. \end{aligned} \tag{38}$$

The magnetic energy decreases with increasing y and with increasing time. Our solution agrees with the models of [26] and [18], in which the magnetic energy decreases as the flux rope expands.

The r component of the Lorentz force is given by

$$(1/\mu)((\nabla \times B) \times B)_r = (\chi - 1)SD'y^{-5}. \quad (39)$$

The other components are 0. Thus, when $\chi = 1$ (27), a force-free state is maintained at all times. Expansion is caused by the pressure gradient force. The ambient and internal pressures control the expansion. This point is different from the solution of [9], in which the force-free state occurred only for one specific time.

Our solution is different from the models of [26] and [18] in maintaining the force-free state. In their models the force-free state was not maintained. The difference comes from the internal magnetic field configuration, which may be caused by the curvature of the cylindrical axis. If we consider a curved tube as a flux rope like [26] or [18], it may be difficult to maintain the force-free state.

As with some other self-similar solutions, velocity grows without limit as r or z increases. Although we consider finite volume around the origin as a flux rope, infinitely large velocity in infinitely distant regions (for large r as well as for large $|z|$) seems strange. However, in our opinion it is significant to show possible expansion in the medium where γ is greater than 1 in the same framework of self-similarity as shown in the previous solution.

The relation between T and ρ is considered. From (32) T decreases with time when γ is greater than 1. Thus T and ρ show a positive correlation in this case. If γ is less than 1, they show a negative correlation. If γ is equal to 1, T is constant. The relation between the correlation and γ is the same as the previous solution.

SUMMARY

We found a new class of solutions for two-dimensional MHD equations that satisfies the condition that a cylindrical flux rope can expand self-similarly for any γ value by considering the effects of axial elongation as well as radial expansion. This new class of solutions puts new light to the very discussed problem on the relation between γ and the expansion. Mainly it emphasizes a role of flux rope elongation for its expansion. This solution showed that the flux rope expands maintaining a force-free state.

ACKNOWLEDGMENTS

The authors thank Dr. Katsuhide Marubashi (Communications Research Laboratory, Japan), and Dr. Shin-ichi Watari (Communications Research Laboratory, Japan) for their stimulating and insightful comments and suggestions during the course of this work. H. S. was supported by the JSPS overseas research fellowship. M. V. was supported by project S1003006 from AV ČR.

REFERENCES

1. Marubashi, K., "Interplanetary magnetic flux ropes and solar filaments," in *Coronal mass ejections*, edited by N. C. et al., Geophysical Monograph 99, American Geophysical Union, Washington, D. C., 1997, pp. 147–156.
2. Tsurutani, B. T., Gonzalez, W. D., Tang, F., Akasofu, S. I., and Smith, E. J., *J. Geophys. Res.*, **93**, 8519–8531 (1988).
3. Burlaga, L., Sittler, E., Mariani, F., and Schwenn, R., *J. Geophys. Res.*, **86**, 6673–6684 (1981).
4. Bernstein, I. B., and Kulsrud, R. M., *Astrophys. J.*, **142**, 479–490 (1965).
5. Kulsrud, R. M., Bernstein, I. B., Kruskal, M., Fanucci, J., and Ness, N., *Astrophys. J.*, **142**, 491–506 (1965).
6. Low, B. C., *Astrophys. J.*, **254**, 796–805 (1982).
7. Gibson, S. E., and Low, B. C., *Astrophys. J.*, **493**, 460–473 (1998).
8. Osherovich, V. A., Farrugia, C. J., and Burlaga, L. F., *J. Geophys. Res.*, **98**, 13225–13231 (1993).
9. Osherovich, V. A., Farrugia, C. J., and Burlaga, L. F., *J. Geophys. Res.*, **100**, 12307–12318 (1995).
10. Farrugia, C. J., Burlaga, L. F., Osherovich, V. A., Richardson, I. G., Freeman, M. P., Lepping, R. P., and Lazarus, A. J., *J. Geophys. Res.*, **98**, 7621–7632 (1993).
11. Osherovich, V. A., Farrugia, C. J., Burlaga, L. F., Lepping, R. P., Fainberg, J., and Stone, R. G., *J. Geophys. Res.*, **98**, 15331–15342 (1993).
12. Vandas, M., Fischer, S., Dryer, M., Smith, Z., and Detman, T., *J. Geophys. Res.*, **100**, 12285–12292 (1995).
13. Vandas, M., Fischer, S., Dryer, M., Smith, Z., and Detman, T., *J. Geophys. Res.*, **101**, 15645–15652 (1996).
14. Wu, S. T., Guo, W. P., and Dryer, M., *Solar Phys.*, **170**, 265–282 (1997).
15. Odstrčil, D., and Pizzo, V. J., *J. Geophys. Res.*, **104**, 483–492 (1999).
16. Vandas, M., and Odstrčil, D., *J. Geophys. Res.*, **105**, 12605–12616 (2000).
17. Chen, J., and Garren, D. A., *Geophys. Res. Lett.*, **20**, 2319–2322 (1993).
18. Chen, J., *J. Geophys. Res.*, **101**, 27499–27519 (1996).
19. Gosling, J. T., *J. Geophys. Res.*, **104**, 19851–19857 (1999).
20. Skoug, R. M., Feldman, W. C., Gosling, J. T., McComas, D. J., and Smith, C. W., *J. Geophys. Res.*, **105**, 23069–23084 (2000).
21. Skoug, R. M., Feldman, W. C., Gosling, J. T., McComas, D. J., Reisenfeld, D. B., Smith, C. W., Lepping, R. P., and Balogh, A., *J. Geophys. Res.*, **105**, 27269–27275 (2000).
22. Osherovich, V., *J. Geophys. Res.*, **106**, 3703–3707 (2001).
23. Gosling, J. T., Riley, P., and Skoug, R. M., *J. Geophys. Res.*, **106**, 3709–3713 (2001).
24. Shimazu, H., and Vandas, M., *Earth Planets Space*, in press (2002).
25. Berdichevsky, D. B., Lepping, R. P., and Farrugia, C. J., *Phys. Rev. Lett.*, submitted (2002).
26. Kumar, A., and Rust, D. M., *J. Geophys. Res.*, **101**, 15667–15684 (1996).

The Geoeffectiveness of Magnetic Clouds as a Function of Their Orientation

C. Cid, T. Nieves-Chinchilla, M. A. Hidalgo, E. Sáiz and Y. Cerrato

Departamento de Física. Universidad de Alcalá. Alcalá de Henares, Madrid, Spain

Abstract. Trying to get light into the paradigm of forecasting geomagnetic activity, we have looked for a relationship between geoeffectiveness and the orientation and helicity of magnetic clouds. During the years 1995-2000, we have selected all the geomagnetic storms with Dst index less than -70 nT. Then, we have inspected WIND data looking for a possible magnetic cloud related to every storm event. When a magnetic cloud is encountered, we have fitted to experimental data a model that we have developed for the magnetic cloud topology in order to obtain the attitude of the magnetic cloud and its helicity. On the basis of the results obtained, a close relationship is observed between the orientation of the magnetic cloud and its helicity, and the geomagnetic activity.

INTRODUCTION

The defining feature of a geomagnetic storm is a global decrease in the horizontal component, H, of the geomagnetic field, and a gradual recovery to its average level ([1], [2]). Analyses of magnetic storm morphology have been made by Chapman [3], Vestine et al. [4], and many others (see [5]). These studies have shown that at equatorial and mid latitudes the decrease in H during a magnetic storm can approximately be represented by a uniform magnetic field parallel to the geomagnetic dipole axis and directed toward the South. The magnitude of the decrease in H represents the severity of disturbance, and the disturbance field is denoted as Dst index ([6],[7]). Thus, the Dst variation provides a quantitative measure of geomagnetic disturbance that can be correlated with other solar and geophysical parameters (e.g. [8]).

The variation in H during a geomagnetic storm is due to changes in the terrestrial ring current (see e.g.. [9]). This current ring consists in energetic charged particles flowing toroidally around the Earth, and creating a ring of westward electric current, centered at the equatorial plane and extended from geocentric distances of about 3 to 8 R_e. This ring is a critical element in understanding the onset and development of space weather disturbances in geospace.

In this paper we look for the response of the current ring as a magnetic cloud (MC) passes through it. Depending on the helicity of the MC and on the orientation of its axis, the induced current is calculated and related to the Dst index. We also compare the theoretical expectations that we obtain with experimental data.

THEORETICAL APPROACH

We consider a MC that passes through the ring current located around the Earth at a distance R_{rc}, with a thick δR_{rc}. We assume that the magnetic field of the MC is given by the model proposed by Hidalgo et al. [10]. In this model the magnetic field vector can be decomposed only in two components: an axial and a poloidal one (there is no radial component). These components of the magnetic field are obtained from Maxwell equations, assuming a circular cross section for the MC and a current density vector with no radial component, and with axial and poloidal components constants. The expressions obtained for the magnetic field components are the following

$$\begin{aligned} B_{poloidal}^{MC} &= \frac{\mu_0}{2} j_{axial} r \\ B_{axial}^{MC} &= \mu_0 j_{poloidal} \left(R_{MC} - r \right) \end{aligned} \quad (1)$$

CP679, *Solar Wind Ten: Proceedings of the Tenth International Solar Wind Conference*,
edited by M. Velli, R. Bruno, and F. Malara

where R_{MC} is the radius of the MC.

When the MC travels through the interplanetary medium, its axis makes an angle θ (latitude) with respect to the ecliptic plane and it presents a longitude angle, ϕ, relative to the Sun-Earth line. Then, the theoretical magnetic field of the MC has to be expressed in the new reference system. As we assume that the ring current lies on the ecliptic plane, we will consider the GSE reference system as a proper system for this analysis. The velocity of the MC is also refereed to GSE system, obtaining two components in the ecliptic plane: a parallel (v_x) and a perpendicular (v_y) to the Sun-Earth line (see Figure 1).

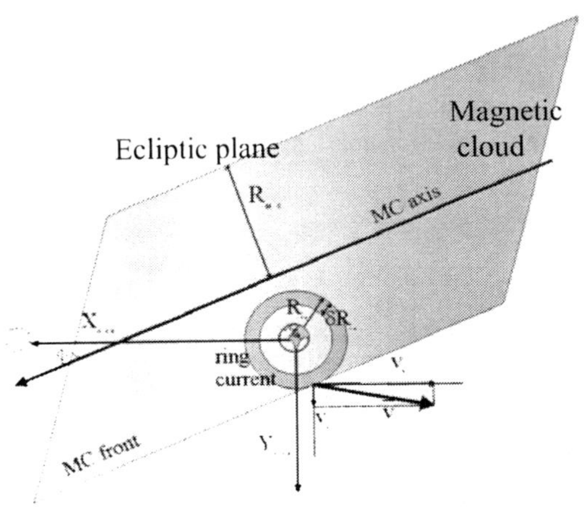

FIGURE 1. A schematic view of the ring current (in brown colour) and the projection of the MC (in blue) on the ecliptic plane. The orientation of the axis, the radius and the velocity of the MC are also indicated.

In experimental data we can see that v_x decreases linearly with time and v_y is almost constant. Then, we will consider that

$$v_x = v_x^0 + a_x t, \quad v_y = cte \tag{2}$$

Having into account the situation described above, the electric field induced in the ring current due to the passage of the MC is given by

$$\nabla \times \vec{E}^{GSE} = -\frac{\partial \vec{B}^{GSE}}{\partial t} \tag{3}$$

that we solve in cylindrical coordinates under the following considerations:

(1) We suppose that the electric field has only one component, E_φ, that is, it follows the angular direction in a cylindrical coordinate system, and

(2) We consider that B^{GSE} can be decomposed in two different contributions: the terrestrial magnetic field (that does not vary with time), and the MC magnetic field (that varies with time). In order to simplify the analysis, we only take into account the z-GSE component due to the MC magnetic field.

The electric field induced, E_φ, obtained from equation (3) produces a variation in the intensity of the ring current, δI, given by

$$\delta I = \sigma \int_{R_{rc}}^{R_{rc}+\delta R_{rc}} E_\varphi dr \tag{4}$$

where σ is the conductivity of the ring current, that we consider constant. This δI produces a magnetic field in the z direction, δB, whose effect in the center of the ring current can be approach as

$$\delta B = \frac{\mu_0 \delta I}{2R_{rc}} \tag{5}$$

This magnetic field, δB, is the responsible of the perturbation of the geomagnetic field, and can be compared with the Dst index. In this analysis we take into consideration the term of the theoretical Dst index that only depends on the longitude angle.

Then, from all the above theoretical considerations, we obtain

$$Dst \approx C_0 \left[sin\phi(2a_x t + v_x^0) - \cos\phi v_y \right] \tag{6}$$

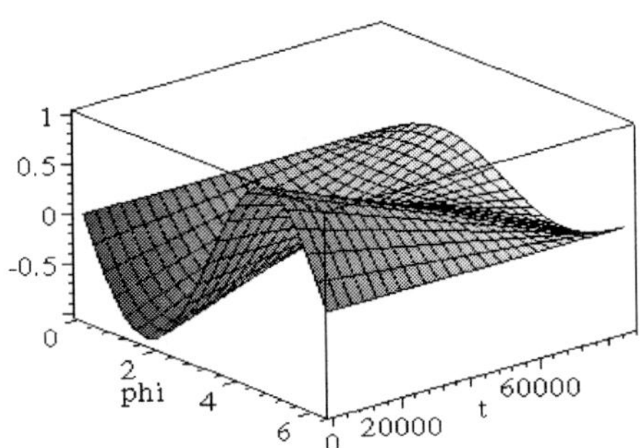

FIGURE 2. Simulation of the Dst index obtained from equation (6). For the simulation, we have taken a R_{rc}=4.5R_e, δR_{rc} =2R_e, | j_{axial} |=10^{-12} Cm^{-2}s^{-1}, v_y=0, v_x^0=-800 kms^{-1}, a_x=5.8 ms^{-2} and. As the conductivity has been assumed to be one, the Dst is expressed in arbitrary units.

where $C_0 = \mu_0^2 j_{axial} \sigma R_{rc} (\delta R_{rc})^2 / 16$. Figure 2 shows a simulation of the Dst index obtained.

Influence of the MC Helicity in the Geoeffectiveness

A MC will be geoeffective when a South magnetic field disturbs the Earth magnetic field as it passes, that is, when a negative value is obtained for the Dst index (at this stage, we do not consider the magnitude of Dst index, but just its sign). The influence of the MC helicity in equation (6) is included in the sign of the axial component of the current density. A left (right) handed MC corresponds to a positive (negative) j_{axial}. In order to explain clearly this influence, we will analize separatelly the case when the velocity of the MC as it goes away from the Sun presents only x-GSE component, and when the velocity present a y-GSE component not nule

The Magnetic Cloud Velocity Is Parallel To The X-GSE Direction

If we assume that the MC moves parallel to the Sun-Earth line (v_y=0), there are two possibilities to obtain we obtain from equation (6) a Dst negative value: (1) if it is left handed and its axis longitude angle is included in the interval between 0 and π rad or (2) if it is right handed and its axis longitude angle is between π and 2π rad. Then we have obtained that the helicity and orientation of a MC need to be considered in order to analyze its geoeffectiveness.

The Magnetic Cloud Velocity Is Not Parallel To The X-GSE Direction

In the last section, in order to obtain the geoeffective latitude intervals, we have considered that the MC travels far away from the Sun, directed right to Earth. Then, the function that let us obtain the theoretical Dst value (eq. 6) is the same that *sin* function. Anyway, if the cloud does not travel following the Sun-Earth line, that is, if $v_y \neq 0$, then a shift in the longitude angles that gives a negative Dst index is produced (Figure 3).

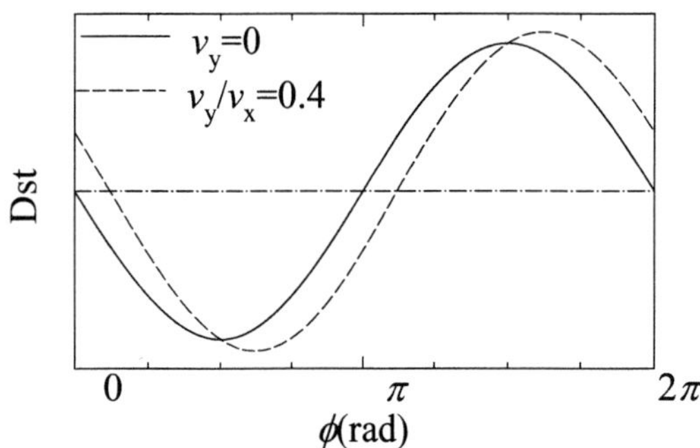

FIGURE 3. Theoretical Dst index obtained at t=0 for a left-handed MC with v_y=0 (solid line) and $v_y = v_x/4$ (dotted line).

DATA ANALYSIS

We have analyzed Dst index from Space Physics Interactive Data Resource (SPIDR) during the years 1995 to 2000. From this data, we have selected those events where the index was less than -70nT unless for one day, then, a sample of 61 events is considered for this work. Once a "storm event" has been selected, we examine magnetic field and plasma data from the Wind spacecraft looking for a MC event in the solar wind associated with the storm. Only for 11 storm events (a 16 per cent) there was no MC associated. For the rest of storms, we fit our model to magnetic field data in the MC interval obtaining the axial component of the current density and the longitude angle of the MC axis, as well as other parameters that are not related to this work.

Trying to check our theoretical results, we have plotted the sign of the axial component of the current density of all these MC events (as a measurement of the helicity) *versus* the longitude angle of its axis (Figure 4).

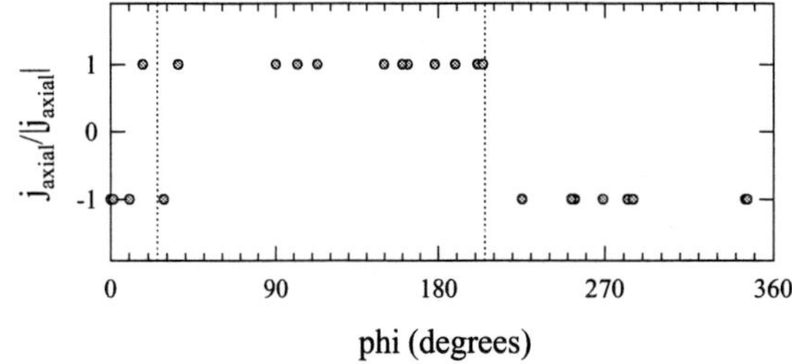

FIGURE 4. The figure represents the sign of the axial component of the current density *versus* the longitude angle of the MC axis. The reference lines indicate a shift of +25° from the Sun-Earth line.

The results obtained show a close relationship between the helicity and the longitude of the MC axis that is shifted from the Sun-Earth line direction (0°-180°). Following the theoretical results, this shift could be associated to the existence of a non-null component of the solar wind velocity in the perpendicular direction to the Sun-Earth line.

SUMMARY AND CONCLUSIONS

In this work, we have analyzed the influence of a MC in the magnetic field of the Earth. Using a model that we have developed previously for the magnetic field of a MC, we have calculated the electric field induced on the ring current around the Earth, and the influence of the appearance of this electric field on the Earth magnetic field. We have compared the disturbed magnetic field with the Dst index, trying to determine the characteristics of a MC to be geoeffective. We have focused our study on the influence of the longitude angle of the MC axis. A close relationship is obtained between this angle and the helicity of the clouds for the events analyzed. Although this relationship is clear, a shift from the expected values of the longitude angles is obtained. We have explain this shift as due to a component of the velocity perpendicular to the Sun-Earth line, although a detailed analysis of the velocity vector of these events would be desirable. We will try this analysis in a nearly future.

ACKNOWLEDGMENTS

We want to thank K. Ogilvie, R. Fitzenreiter and R. Lepping (Goddard Space Flight Center, Greenbelt, Mariland) for permission to use Wind data, and Space Physics Interactive Data Resource (SPIDR) for providing Dst index data. This work has been supported by the Comisión Interministerial de Ciencia y Tecnología (CICYT) of Spain, grant ESP97-1776.

Consuelo Cid wants also to acknowledge to the Solar Wind X organizing committee for their financial support.

REFERENCES

1. Broun, J. A., "On the horizontal force of the Earth's magnetism", Proceedings Roy. Soc. Edinburg 22, 1861, pp. 511.
2. Adams, W. G., *Phil. Trans. London (A)* **183**, 131 (1892).
3. Chapman, S., *Annali di Geofisica* **5**, 481 (1952).
4. Vestine, E.H., Laporte, L., Lange, I., and Scott, W.E., *The geomagnetic field, its description and analysis*, Carnegie Institution of Washington Publication 580, Washington D.C., 1947.
5. Akasofu, S.-I. and Chapman, *Solar Terrestrial Physics*, Oxford University Press, Oxford, 1972.
6. Mayaud, P.N., *Derivation, Meaning, and Use of Geomagnetic Indices*, Geophysical Monograph 22, American Geophysical Union, Washington D. C., 1980.
7. Rangarajan, G.K., "Indices of geomagnetic activity" in *Geomangetism*, p.323, ed. By J.A. Jacobs, Academic Press, London (1989).
8. Daglis, I. A., Kasotakis, G., Sarris, E.T., Kamide, Y., Livi, S., and Wilken, B., Influence of interplanetary disturbances on the terrestrial ionospheric outflow, *Physics and Chemistry of the Earth* **24**, 229-232, 1999).
9. Williams, D.J., *Planet. Space Sci.* **21**, 1195 (1981).
10. Hidalgo et al., *J. Geophys. Res.* **107**, pp1-pp7 (2002).

Interstellar magnetic field effects on the heliosphere

Romana Ratkiewicz*, Lotfi Ben-Jaffel†, James F. McKenzie**‡ and Gary M. Webb§

*Space Research Center PAS, Bartycka 18a, 00-716 Warsaw, Poland
†Institut d'Astrophysique de Paris, 98bis Blvd Arago, 75-014 Paris, France
**Max-Planck-Institut für Aeronomie, Max-Planck-Strasse 2, 37191 Katlenburg-Lindau, Germany
‡School of Pure and Applied Physics, University of Natal, Durban, South Africa
§Lunar and Planetary Laboratory, University of Arizona, Tucson, Arizona 85721, USA

Abstract. This paper summarizes the numerical results obtained by three-dimensional MHD simulations of the interaction between the solar wind and interstellar medium in *Ratkiewicz and Ben-Jaffel* [2002], *Ratkiewicz and McKenzie* [2002], and *Ratkiewicz and Webb* [2002]. We analyze the configuration in which Maxwell stresses lead to squeezing and/or pushing the heliospheric boundary. In particular, we explain the mechanism giving rise to a suction effect of the heliopause. Numerical results for the case of aligned interstellar MHD flow are compared with previous studies.

INTRODUCTION

The interaction between the solar wind (SW) and the interstellar gas (LISM) leads to the formation of the termination shock (TS) in the solar wind, the heliopause (HP) - the interface separating both media - and a bow shock (BS) in the interstellar plasma if the inflow is "supersonic". This complex interaction involves stationary or time-dependent plasma-plasma interactions, magnetic field stresses, effects of neutrals, creation of pick-up ions and their acceleration at shocks (in particular, at the termination shock, where they give rise to anomalous cosmic rays (ACRs)), energetic neutral atoms (ENA), modification of the supersonic solar wind by ACRs and galactic cosmic rays (GCRs), and the modulation of the structure of the termination shock. A thorough review of the subject has been given by *Zank* [1999]; and the reader is also referred to papers by *Ben-Jaffel et al.* [2000], *Fahr* [2000], *Izmodenov* [2000], *Ratkiewicz et al.* [2000], *Fichtner* [2001], and references therein.

A theoretical model taking into account all these factors in a self-consistent way does not yet exist. The most complete studies so far have been given by *Fahr et al.* [2000]; a 5-fluid hydrodynamic model of the interaction between the solar wind and interstellar medium includes protons, hydrogen atoms, pick-up ions, galactic and anomalous cosmic rays, which does not however include the effects of the interplanetary and interstellar magnetic fields. Magnetic field effects have been studied in pure MHD models or MHD models including the effect of neutrals [*Fujimoto and Matsuda,* 1991; *Baranov and Zaitsev,* 1995; *Washimi and Tanaka,* 1996; *Linde et al.,* 1998; *McNutt et al.,* 1998, 1999; *Pogorelov and Matsuda,* 1998, 2000; *Ratkiewicz et al.,* 1998, 2000, 2002; *Aleksashov et al.,* 2000; *Ratkiewicz and Ben-Jaffel,* 2002; *Ratkiewicz and Webb,* 2002; *Ratkiewicz and McKenzie,* 2002]. The results highlight the importance of magnetic field effects in causing asymmetries and distortions of the heliospheric boundary.

MAIN ASSUMPTIONS OF THE MODEL

The solar wind expands radially from the Sun and interacts with the uniformly flowing magnetized interstellar plasma and neutral hydrogen. In the zeroth order approximation we may neglect the change in the distribution of the neutral hydrogen, resulting from the interaction, and assume a constant flux of neutral hydrogen through out. The interaction with neutral hydrogen occurs due to charge exchange between protons and H atoms both inside and outside the termination shock. The charge exchange cross-section σ is taken to be $2 \times 10^{-15} cm^2$. All other processes such as creation of pick-up ions and their acceleration at shocks, modification of the supersonic solar wind by ACRs and galactic cosmic rays (GCRs), the interplanetary magnetic field, heliolatitudinal dependence of solar wind are neglected.

Following, for example, *Ratkiewicz and Ben-Jaffel* [2002], we use the same set of MHD equations with a source term **S** on the RHS describing charge exchange with the constant flux of hydrogen:

$$\frac{\partial \mathbf{U}}{\partial t} + \nabla \cdot \bar{\mathbf{F}} = \mathbf{Q} + \mathbf{S} \tag{1}$$

CP679, *Solar Wind Ten: Proceedings of the Tenth International Solar Wind Conference,*
edited by M. Velli, R. Bruno, and F. Malara

where $\mathbf{U}$, $\mathbf{Q}$, and $\mathbf{S}$ are column vectors, and $\bar{\mathbf{F}}$ is a flux tensor defined as:

$$\mathbf{U} = \begin{vmatrix} \rho \\ \rho\mathbf{u} \\ \mathbf{B} \\ \rho E \end{vmatrix}$$

$$\bar{\mathbf{F}} = \begin{vmatrix} \rho\mathbf{u} \\ \rho\mathbf{u}\mathbf{u} + \mathbf{I}(p + \frac{\mathbf{B}\cdot\mathbf{B}}{8\pi})\rho\mathbf{u}\mathbf{u} + \mathbf{I}(p + \frac{\mathbf{B}\cdot\mathbf{B}}{8\pi}) - \frac{\mathbf{B}\mathbf{B}}{4\pi} \\ \mathbf{u}\mathbf{B} - \mathbf{B}\mathbf{u} \\ \rho H\mathbf{u} - \frac{\mathbf{B}(\mathbf{u}\cdot\mathbf{B}}{4\pi}) \end{vmatrix}$$

$$\mathbf{Q} = -\begin{vmatrix} 0 \\ \frac{\mathbf{B}}{4\pi} \\ \mathbf{u} \\ \mathbf{u}\cdot\frac{\mathbf{B}}{4\pi} \end{vmatrix} \nabla\cdot\mathbf{B}$$

$$\mathbf{S} = \rho\nu_c \begin{vmatrix} 0 \\ \mathbf{V_H} - \mathbf{u} \\ \mathbf{0} \\ \frac{1}{2}{V_H}^2 + \frac{3k_BT_H}{2m_H} - \frac{1}{2}u^2 - \frac{k_BT}{(\gamma-1)m_H} \end{vmatrix}.$$

Here, ρ is the ion mass density, $p = 2nk_BT$ is the pressure, n is the ion number density, T and T_H (T_H = const) are ion and H atom temperatures, and $\mathbf{u}$ and $\mathbf{V_H}$ ($\mathbf{V_H}$ = **const**) are the ion and H atom velocity vectors, respectively; $\mathbf{B}$ is the magnetic field vector, $E = \frac{1}{\gamma-1}\frac{p}{\rho} + \frac{\mathbf{u}\cdot\mathbf{u}}{2} + \frac{\mathbf{B}\cdot\mathbf{B}}{8\pi\rho}$ is the total energy, and $H = \frac{\gamma}{\gamma-1}\frac{p}{\rho} + \frac{\mathbf{u}\cdot\mathbf{u}}{2} + \frac{\mathbf{B}\cdot\mathbf{B}}{4\pi\rho}$ is the total enthalpy per unit mass, γ is the ratio of specific heats. $\mathbf{I}$ is the 3 x 3 identity matrix. The charge exchange collision frequency is $\nu_c = n_H\sigma u_*$, where n_H (n_H = const) is H atom number density, σ is the charge exchange cross-section, and $u_* = ((\mathbf{u} - \mathbf{V_H})^2 + 128k_B(T + T_H)/(9\pi m_H))^{1/2}$ is the effective average relative speed of protons and H atoms, assuming a Maxwellian spread of velocities both for protons and H atoms. The flows are taken to be adiabatic with $\gamma = 5/3$. The additional constraint of a divergence-free magnetic field, $\nabla\cdot\mathbf{B} = 0$, in the numerical simulations is accomplished by adding the source term $\mathbf{Q}$ to the RHS of (1), which is proportional to the divergence of the magnetic field. By adding $\mathbf{Q}$ to the RHS of (1) assures that any numerically generated $\nabla\cdot\mathbf{B} \neq 0$ is advected with the flow, and allows one to limit the growth of $\nabla\cdot\mathbf{B} \neq 0$.

The coordinate system is Sun-centered, and the spherical (r,θ,ψ) grid is chosen with a Cartesian correspondence of $x = -r\cos\theta$, $y = r\sin\theta\cos\psi$, $z = r\sin\theta\sin\psi$. The LISM velocity and magnetic field vectors define the x-y plane. The LISM velocity vector is in the positive x direction. The interstellar inclination angle α is the angle between interstellar velocity $\vec{V}_{is}$ and magnetic field $\vec{B}_{is}$ vectors.

GENERAL BEHAVIOUR OF THE HELIOSPHERIC BOUNDARY

We have carried out a parametric study encompassing 90 cases of different boundary conditions of the interstellar medium in order to investigate how ionized and neutral interstellar hydrogen, and the interstellar magnetic field influence the interaction between the solar wind and LISM [*Ratkiewicz and Ben-Jaffel,* 2002]. For the solar wind we fix the boundary conditions. The inner boundary of the solar wind is taken at $r_0 = 30AU$, and corresponds to an unperturbed solar wind at $1AU$ with number density $n_E = 8cm^{-3}$, velocity $V_E = 400kms^{-1}$, and Mach number $M_E = 8$. The unperturbed interstellar magnetized plasma is taken to have the same velocity (magnitude and direction) and temperature as neutral hydrogen. At the outer boundary, located at $15000AU$, we take constant $V_{is} = 26kms^{-1}$ and $T_{is} = 7000°K$, and variable $n_{is} = 0.034, 0.043, 0.061cm^{-3}$, $B_{is} = 1.0, 1.4, 1.8, 2.2\mu G$, $\alpha = 30°, 35°, 40°, 45°, 50°$, and $n_H = 0.14, 0.24, 0.34cm^{-3}$ [*Ben-Jaffel et al.,* 2000].

The heliospheric boundary structures reveal that the size of the heliosphere inside the termination shock decreases with an increase of the neutral hydrogen number density as well as with an increase of the ionized LISM component number density. However, the effect of the ions is much more pronounced.

The inflow of LISM neutral atoms into the heliosphere creates the pick-up ions by charge exchange with protons, and causes heating of the supersonic solar wind and cooling of the plasma beyond the termination shock. The intensity of the effective heating (by introducing a new separate hot ion component into the solar wind) and cooling demonstrates the key role played by neutral hydrogen in heating or cooling the plasma of the solar wind. Simultanously very weak changes in temperature are observed with variations of interstellar plasma density.

The solar wind Mach number inside the termination shock decreases due to the interaction with the neutral hydrogen, while no changes are observed with variations of interstellar plasma density. Our results also show that the solar wind is decelerated by neutral hydrogen. (However, since filtration, which also slows the neutrals, is not taken into account, this introduces some errors in the deceleration of the solar wind).

For the assumed boundary conditions, the TS is located between 100 AU and 140 AU from the Sun; the HP is located between 170 AU and 220 AU, respectively. As concerns a BS its strength and location depend strongly on the interstellar magnetic field inclination angle and intensity.

The results show that the main features of asymmetries do not depend on the hydrogen number density (except for a shift of the whole structure toward the Sun).

In contrast, the dependence of the heliospheric boundary structure on the inclination angle is significant. Steps of 5° changes in the angle cause substantial changes in the structure of the flow, especially exterior to the heliopause. Also the importance of the role of the inclination angle particularly with regard to super/subsonic nature of the plasma flow is confirmed. For example in the case of 30° there is clearly a bow shock in the interstellar medium, whereas the case with 50° shows that the bow shock is very weak.

The strength of the magnetic field greatly affects the boundary structure: in a weak magnetic field the bow shock is well-pronounced (as in the axisymmetric case with the same physical parameters). However an increase of even a step of $\delta B_{is} = 0.4\mu G$ shows a significant difference in the heliospheric boundary, in particular, the bow shock weakens.

It follows that the heliosphere can be squeezed and/or pushed by the action of the interstellar magnetic field [*Ratkiewicz and McKenzie,* 2002]. In the axisymmetric field-aligned flow case the parallel magnetic field squeezes the heliosphere in directions perpendicular to the heliopause due to the combined action of the magnetic tension and total pressure forces. If the strength of the magnetic field is sufficiently large another feature appears, namely a sucking effect acting on the heliopause (see next sections). For the case of a perpendicular magnetic field, the effective total pressure reaches a maximum in the vicinity of the stagnation line and pushes the heliopause towards the Sun. In the oblique case the heliosphere is subject to both squeezing and pushing. Which effect dominates depends on some critical inclination angle. If we define δ as the angle between the tangent to the field line behind the bow shock and the unperturbed LISM flow direction, the following simple criterion predicts which effect is greater: (a) for $\alpha + \delta \leq 45°$, the outward squeezing effect exceeds the inward pushing effect and the heliopause moves outward, (b) for $\alpha + \delta > 45°$, the inward pushing effect exceeds the outward squeezing effect and the nose of the HP moves inward.

FIELD-ALIGNED MHD FLOW

In this section we examine numerical results for the case of aligned interstellar MHD flow with fixed boundary conditions for the solar wind and the interstellar plasma in which the magnetic field strength is varied between 1.0 to $3.0\mu G$ [*Ratkiewicz and McKenzie,* 2002]. The inner boundary conditions of the solar wind are the same as in previous section. At the outer boundary (at $15000AU$), we take $V_{is} = 26 km s^{-1}$, and $T_{is} = 7000°K$, $n_{is} = 0.07 cm^{-3}$, $n_H = 0.24 cm^{-3}$. The interstellar magnetic field $\vec{B}_{is}$, taken to be parallel to the velocity vector $\vec{V}_{is}$, varies in intensity in steps of $0.4\mu G$.

The location of the heliopause as displayed by thermal pressure isobars shows up more clearly for larger magnetic field strength.

The changes in the bow shock shape and its location, can be grouped into four classes:
I. $B_{is} = 1.0$ and $1.4\mu G$, II. $B_{is} = 1.8$, III. $B_{is} = 2.2$, and IV. $B_{is} = 2.6$ and $3.0\mu G$.
For $V_A < c$ the shocks are fast and evolutionary [*Landau, Lifshitz and Pitaevskii,* 1984] (class I).
For $V_A > c$ there are three types of shocks, for super-Alfvénic flows [*Landau, Lifshitz and Pitaevskii,* 1984]:
1) the flow speed ahead, $u_{n1} > \sqrt{4V_{A1}^2 - 3c_1^2}$. The shocks are fast and evolutionary (class II).
2) the flow speed ahead appears in the non-evolutionary range
$V_{A1} < u_{n1} < \sqrt{4V_{A1}^2 - 3c_1^2}$, with $\vec{B}_{t1} = \vec{B}_{t2} = 0$, (class III).
3) the flow speed ahead appears in the non-evolutionary range, but $\vec{B}_{t1} = 0$, $\vec{B}_{t2} \neq 0$. Then a *switch-on shock* occurs (M_{A1} lies in the range $1 \leq M_{A1} \leq 2\sqrt{1 - 3c_1^2/4V_{A1}^2}$) (class IV).

If the BS becomes a switch-on shock the flow behind splits, and the stagnation line is evacuated. The heliopause is therefore sucked out towards the bow shock. As shown by *Ratkiewicz et al.* [2000] in the case of a parallel interstellar magnetic field, the magnetic pressure goes to zero in the vicinity of the stagnation point. On the other hand, the combined magnetic tension and total pressure forces attain their maximum values near the flanks of the heliopause. Therefore, under the influence of compressive magnetic forces the heliopause is squeezed out. This is due to the Lorentz force $\vec{J} \times \vec{B}$, which compresses the plasma on the flanks of the heliopause, and pushes the plasma in the anti-sunward direction just upstream of the stagnation point, where $\vec{B} \simeq 0$. A "tooth-paste tube" or a "suction" effect appears [*Ratkiewicz and Webb,* 2002]. This effect is a feature of all possible aligned MHD flows with sufficiently high magnetic field regardless of whether the flows are in the hyperbolic or elliptic regimes [*Jeffrey and Taniuti,* 1964].

Recently *De Sterck and Poedts* [2001] have studied the disintegration and reformation of intermediate shock segments in 3D MHD bow shock flows. They note that intermediate shocks can be stable when dissipation is properly taken into account for a wide range of dissipation coefficients. However, the exact conditions for shock stability depends on the magnitude and type of the perturbations and the magnitude of the dissipation coefficients. In fact both dispersion and dissipation play an important role in the stability of the shock. The switch-on interstellar bow shock here is an intermediate shock which can be stabilized due to the dissipation provided by the interaction of the plasma with the interstellar neutrals.

COMPARISON WITH PREVIOUS RESULTS

In this section we discuss results obtained by our model [*Ratkiewicz and Webb,* 2002] in the context of these obtained by *Aleksashov et al.* [2000]. We consider the interaction of the solar wind with the interstellar MHD flow for the unperturbed SW and LISM exactly the same parameters as *Aleksashov et al.* [2000]. For the solar wind at $1AU$ the number density is $n_E = 7cm^{-3}$, the velocity is $V_E = 450kms^{-1}$, and the sonic Mach number is $M_E = 10$ (the corresponding values at the inner computational boundary at 30 AU are then determined from the standard spherically-symmetric solar wind model). For the unperturbed interstellar magnetized plasma at the outer boundary ($15000AU$) the velocity is taken $V_{is} = 25kms^{-1}$, the Mach number $M_{is} = 2$ (with the sound speed $c = 12.5kms^{-1}$), the interstellar ionized hydrogen number density $n_{is} = 0.07cm^{-3}$. We consider two cases in which the interstellar neutral hydrogen number density $n_H = 0.0$ and $0.2cm^{-3}$. The interstellar magnetic field $\vec{B}_{is}$ is parallel to the velocity vector $\vec{V}_{is}$ and varies from $B_{is} = 0.0, 2.6, 3.5\mu G$, which correspond to Alfvén Mach numbers of ∞, 1.18 and 0.9, respectively.

Our results show three different types of solutions.

1. For $M_A = \infty, M_A > M_{is}, V_A < c$, the shock is fast and evolutionary. The solution displays the pure gas shock limit behavior.

2. For $M_A = 1.18$, $M_A < M_{is}$, $V_A > c$, the flow speed ahead appears in the non-evolutionary range $V_A < V_{is} < \sqrt{4V_A^2 - 3c^2}$, a *switch-on shock* occurs, i.e., $\vec{B}_{t1} = 0$, $\vec{B}_{t2} \neq 0$. The solution is typical for a switch-on shock with a suction effect acting on the heliopause manifesting itself in the decrease of the number density in front of the heliopause along the stagnation line [*Webb et al.,* 1994]. The heliopause is therefore sucked out towards the bow shock.

3. For $M_A = 0.9, M_A < 1$ and $M_{is} > 1$, the unperturbed flow speed appears in the elliptic regime, therefore there is no shock [*Jeffrey and Taniuti,* 1964]. The solution shows how a large parallel interstellar magnetic field ($B_{is} = 3.5\mu G$) moves the heliopause outward.

The results are qualitatively similar to these of *Aleksashov et al.* [2000] for super-Alfvénic, supersonic, field-aligned interstellar flow. In this regime the steady MHD equations are hyperbolic. For example, the nose of the heliopause moves outward from the Sun and the nose of the bow shock moves inward as the interstellar magnetic field strength increases from $B_{is} = 0\mu G$ to $B_{is} = 2.6\mu G$ (i.e., the Alfvén Mach number decreases from $M_A = \infty$ to $M_A = 1.18$, and the corresponding sonic Mach number $M_{is} = 2$). But still our calculations indicate the appearance of a switch-on shock at an interstellar magnetic field strength $B_{is} = 2.6\mu G$.

The results are qualitatively different from *Aleksashov et al.* [2000] in the sub-Alfvénic ($M_A = 0.9, B_{is} = 3.5\mu G$) and supersonic ($M_{is} = 2$) interstellar flow regime, where the steady MHD characteristic are complex (i.e. in the elliptic flow regime). There is no bow shock in our model, whereas there is a bow shock in the corresponding results of *Aleksashov et al.* [2000].

To understand the differences in the results obtained by *Aleksashov et al.* [2000] and those presented here, one must note the different treatment of the problem in both models. The model presented here includes the neutral hydrogen as a constant flux, while in *Aleksashov et al.* [2000], the interstellar magnetic field is taken into account within the framework of an MHD model with source terms in which the trajectories of the hydrogen atoms are calculated by the Monte Carlo method. In the *Aleksashov et al.* [2000] model, the mutual interactions between the neutral component and plasma are included. This may cause a decrease in the influence of the magnetic field on the heliospheric boundary region.

SUMMARY

The interaction of a spherically symmetric solar wind with the magnetized interstellar plasma in the presence of a constant neutral hydrogen flux leads to the following phenomena: the supersonic solar wind is heated by the inclusion of pickup ions created through charge exchange with hot neutrals; the supersonic solar wind is decelerated; the heating and deceleration imply that the sound speed increases, and the hydrodynamic Mach number decreases with increasing heliocentric distance; changes in the shape and size of the termination shock, and reduction of heliocentric distances to the boundaries (TS, HP, and BS); the main features of asymmetries introduced by the interstellar magnetic field are the same as in the case without neutrals. However, this analysis does not take into account the magnetic field of the solar wind. The reconnection between the interpanetary and interstellar magnetic fields at the heliopause will modify this picture.

For the case of field-aligned flow, depending on the interstellar magnetic field intensity B_{is}: (a) the shock could be fast and evolutionary, or (b) a *switch-on shock*, or (c) a bow shock does not exist. In cases (b) and (c) the Lorentz force on the plasma results in a plasma evacuation in front of the heliopause and the suction effect occurs. The characteristic features of the solutions for such a flow configuration depends on the interplay between two effects:

I. the suction which moves the heliopause and bow shock outward for increasing B_{is}, and

II. interstellar neutrals which push these discontinuity surfaces (when they exist) inward as n_H increasing.

ACKNOWLEDGMENTS

RR and LBJ acknowledge support from the Centre National de la Recherche Scientifique of France (CNRS) and Polish Academy of Sciences (PAN) under Program 5037 (CNRS-PAN protocol) and program *Jumelage France-Pologne*. RR acknowledges support from the MPAE Katlenburg-Lindau, Germany, and in part from the KBN Grant No. 2P03C 005 19. GMW is supported in part by NASA grant NAG5-10974. JFM, RR and GMW acknowledge support from IGPP at the University of California, Riverside. The work of RR and GMW is also supported in part by the NSF Award ATM-0296114 of Prof. Gary Zank.

REFERENCES

1. Aleksashov, D.B., V.B. Baranov, E.V. Barsky, and A.V. Myasnikov, An Axisymmetric Magnetohydrodynamic Model for the Interaction of the Solar Wind with the Local Interstellar Medium, *Astronomy Letters*, *26*, No. 11, 743-749, 2000.
2. Baranov, V.B., and N.A. Zaitsev, On the problem of the solar wind interaction with magnetized interstellar plasma *Astron. Astrophys.*, *304*, 631-637, 1995.
3. Ben-Jaffel, L., O. Puyoo, and R. Ratkiewicz, Far-ultraviolet echoes from the frontier between the solar wind and the local interstellar cloud, *Astrophys. J.*, *533*, 924-930, 2000.
4. De Sterck, H., and S. Poedts, Disintegration and reformation of intermediate-shock segments in the three-dimensional MHD bow shock flows, *J. Geophys. Res.*, *106*, 30,023-30,037, 2001.
5. Fahr, H.J., The Multifluid Character of the 'Baranov' Interface, *Astrophys. Space Sci.*, *274*, No. 1-2, 35-54, 2000.
6. Fahr, H.J., T. Kausch, and H. Scherer, A 5-fluid hydrodynamic approach to model the solar system-interstellar medium interaction, *Astronom. Astrophys.*, *357*, 268-282, 2000.
7. Fichtner, H., Anomalous Cosmic Rays: Messengers from the Outer Heliosphere, *Space Sci. Rev.*, *95*, 639-754 , 2001.
8. Fujimoto, Y., and T. Matsuda, *Preprint No. KUGD91-2*, Kobe Univ., Japan, 1991.
9. Jeffrey, A., and T. Taniuti, *Non-linear wave propagation*, 369 pp., Academic Press, New York, London, 1964.
10. Izmodenov, V.V., Physics and Gasdynamics of the Heliospheric Interface, *Astrophys. Space Sci.*, *274*, No. 1-2, 55-69, 2000.
11. Landau, L.D., E.M. Lifshitz, and L.P. Pitaevskii, *Electrodynamics of Continuous Media*, 460 pp., Pergamon Press, Oxford, New York, Toronto, Sydney, Paris, Frankfurt, 1984.
12. Linde, T.J., T.I. Gombosi, P.L. Roe, K.G. Powell, and D.L. DeZeeuw, Heliosphere in the magnetized local interstellar medium: Results of a three-dimensional MHD simulation, *J. Geophys. Res.*, *103*, A2, 1889-1904, 1998.
13. McNutt Jr., R.L., J. Lyon, and C.C. Goodrich, Simulation of the heliosphere: Model, *J. Geophys. Res.*, *103*, A2, 1905-1912, 1998.
14. McNutt Jr., R.L., J. Lyon, and C.C. Goodrich, Simulation of the heliosphere: Generalized charge-exchange cross sections, *J. Geophys. Res.*, *104*, A7, 14,803-14,809, 1999.
15. Pogorelov, N.V., and T. Matsuda, Influence of the interstellar magnetic field direction on the shape of the global heliopause, *J. Geophys. Res.*, *103*, 237-245 , 1998.
16. Pogorelov, N.V., and T. Matsuda, Nonevolutionary MHD shocks in the solar wind and interstellar medium interaction, *Astronom. Astrophys.*, *354*, 697-702, 2000.
17. Ratkiewicz, R., A. Barnes, G.A. Molvik, J.R. Spreiter, S.S. Stahara, M. Vinokur, and S. Venkateswaran, Effect of varying strength and orientation of local interstellar magnetic field on configuration of exterior heliosphere: 3D MHD simulations, *Astronom. Astrophys.*, *335*, 363, 1998.
18. Ratkiewicz, R., A. Barnes, and J.R. Spreiter, Local interstellar medium and modeling the heliosphere, *J. Geophys. Res.*, *105*, 25,021-25,031, 2000.
19. Ratkiewicz, R., and L. Ben-Jaffel, Effects of Interstellar Magnetic Field $\vec{B}$ and Constant Flux of Neutral H on the Heliosphere *J. Geophys. Res.*, *107*, **SSH 2** - 1-13, 2002.
20. Ratkiewicz, R., A. Barnes, H.-R. Müller, G.P. Zank, and G.M. Webb, Modeling the heliosphere: Influence of the interstellar magnetic field in the presence of LISM neutral hydrogen, *Adv. Space Res.*, *29*, No 3, 433-438, 2002.
21. Ratkiewicz, R., and G.M. Webb, On the Interaction of the Solar Wind with the Interstellar Medium: Field Aligned MHD Flow *J. Geophys. Res.*, *in press*, 2002.
22. Ratkiewicz, R., and J.F. McKenzie, Interstellar magnetic field effects on the termination shock, heliopause and bow shock: aligned MHD flow *J. Geophys. Res.*, *submitted*, 2002.
23. Washimi, H., and T. Tanaka, 3-D Magnetic Field and Current System in the Heliosphere, *Space Sci. Reviews 78*, 85, 1996.
24. Webb, G.M., M. Brio, and G.P. Zank, Steady MHD flows with an ignorable co-ordinate and the potential transonic flow equation, *J. Plasma Physics*, *52*, Part 1, 141-188, 1994.
25. Zank, G.P., Interaction of the solar wind with the local interstellar medium: A theoretical perspective, *Space Sci. Rev.*, *89*, 413-687, 1999.

Flows in coronal loops driven by Alfvén waves: 1.5 MHD simulations with transparent boundary conditions

R. Grappin*, J. Léorat*, L. Ofman¶

*LUTH, CNRS, Observatoire de Paris-Meudon
¶The Catholic University of America and
NASA GSFC, Code 682, Greenbelt, MD 20771
Visiting Associate Professor at Tel Aviv University

Abstract. We investigate time-dependent siphon flows in coronal loops driven by Alfvén waves. We consider a 1.5 D isothermal, MHD model in which the coordinate is the abscissa along the loop, with an external gravity field reversing sign in the middle, and a uniform magnetic field parallel to the x-axis. We use transparent boundary conditions, meant to describe the upper part of the loop. The reaction of the loop to Alfvén waves depends entirely on whether we allow or not incoming parallel velocity fluctuations: only in the latter case do transonic flows arise, but the flow is in that case generated by a nonlinear coupling of the waves with the boundaries.

INTRODUCTION

We consider in this paper the interaction between Alfvén waves and the plasma flow along a closed coronal loop. The response of closed magnetic structures to Alfvén waves has been much studied in the MHD framework, in the context of coronal heating, but the generation of parallel flows by wave pressure has been studied in open field regions only. We consider a loop model with weak stratification, and transparent boundary conditions. In doing so, we basically model the coronal part of the loop, and exclude wave trapping, reflection and resonance. Our motivation is that, if Alfvén waves are to contribute to acceleration of the plasma in open regions to generate the fast solar wind, they should also be present in closed regions. An indication that Alfvén waves could have significant effects also in closed loops is given by recent observations [1], and by a numerical study of a global solar wind model [2] where Alfvén waves induce supersonic flows along closed loops in the equatorial region. In order to understand the conditions for the onset of such flows, we consider here the more restricted problem of a single closed loop, modeled as a one-dimensional domain with uniform magnetic field, and a stratification maintained by an artificial gravity changing sign in the middle of the loop.

The problem of interest is a time-dependent version of siphon flows, while in previous studies, only stationary flows were investigated, with fixed pressure boundary conditions [3,4]. Siphon flows show properties combining those of stellar winds for the upward flow and of accretion flows for the downflow. We investigate here whether Alfvén waves can trigger the transition from a static stratification to a supersonic flow. It is indeed possible that large pressure deviations result from small wave pressure imbalance, in particular when the plasma β is low enough, as the Alfvén wave becomes unstable [5]. For simplicity, we consider circularly polarized waves. As we will see, the result depends closely on boundary conditions.

MODEL AND BOUNDARY CONDITIONS

The initial hydrostatic equilibrium stratification has a density equal to 2.7 at the boundaries, and 0.4 in the rarefied middle of the loop. It is determined by an artificial gravity $g_x = -T_0/H_0 \sin(x)$ where T_0 is the temperature in convenient units $T=P/\rho= c^2$, where c is the sound speed), and H_0 is the scale height. The latter is about equal to the height of the loop: $H_0=1$. The

CP679, *Solar Wind Ten: Proceedings of the Tenth International Solar Wind Conference*,
edited by M. Velli, R. Bruno, and F. Malara

plasma is isothermal (γ=1), so that the temperature is uniform and constant. As a result, the initial density profile is $\rho = \exp(\cos x)$. The length of the loop is 2π. The parallel component B_x of the magnetic field is constant, since divB=0; it is fixed to be unity, so the initial magnetic field is $B_0 = (1,0,0)$. The initial Alfvén speed thus varies between 0.6 (at boundaries) and 1.65 (middle of the interval), so that the traversal time of an Alfvén wave is about 2π. The temperature is $T_0 = 0.1$ corresponding to a sound speed c=0.33, and an initial plasma βsuch that $0.073 < \beta < 0.54$, but we will mention runs with lower β, $0.007 < \beta < 0.054$ (T_0 = 0.01).

The injected wave is circularly polarized, with angular frequency ω=5 (which lead to about 5 wavelengths inside the domain). The wave amplitude is progressively increased in a quarter of a wave period, using a ramp function $A(t) = 1-\exp(-(t/\tau)^4)$, with $\tau=\pi/(2\omega)$. The amplitude of the wave is $\varepsilon = B_{perp}/B_x = 0.05$. The wave components injected are:

$$B^\circ_y = A(t)\varepsilon\cos(\omega t);\ B^\circ_z = A(t)\varepsilon \sin(\omega t) \quad (1)$$

The wave is injected both at x=0 and 2π, with a non zero delay δT at $x=2\pi$. This delay produces a small magnetic pressure difference between both boundaries. Let us show it in the non-stratified case. The addition of two waves propagating in opposite directions produces a magnetic pressure profile which is:

$$B_{perp}^2 = 2\varepsilon^2+2\varepsilon^2\cos(2kx+\phi) \quad (2)$$

where k is the wavenumber and ϕ is the phase difference between the two waves. The resulting pressure difference between the two boundaries is

$$\delta B_{perp}^2 = 2\varepsilon^2(\cos(\phi)-\cos(4,,k+\phi)) \quad (3)$$

As soon as 2k is not an integer, there is a (small) wave pressure difference between the two boundaries that scales as ε^2. If the frequencies of the two waves differ, the total magnetic pressure fluctuates with time, but this time dependence remains small if the frequencies are close to one another.

We use a finite difference compact spatial scheme, and a third order Runge-Kutta temporal scheme with a kinematic viscosity and a magnetic diffusivity, as well as a (small) artificial diffusion for the density.

There are six degrees of freedom in isothermal MHD. For subsonic and subalfvénic flows, we must specify the three incoming characteristics: Alfvén, slow and fast, denoted here as $L_{a,s,f}^+$ at x=0 and $L_{a,s,f}^-$ at $x=2\pi$ [2,5]. It sounds reasonable to inject Alfvén waves via the Alfvén characteristics L_a, as in global solar wind simulations [6,2], but in the present 1.5D case we found that this leads to a systematic drift of the magnetic pressure. To prevent the pressure drift, we control separately the injected magnetic field components, as done in [5]. (Note that the magnetic pressure of the injected wave is constant but the total magnetic pressure may vary with time, although with no systematic drift). Consider, to be specific, the x=0 boundary. The equations for the injected magnetic field components read, in term of the three rightward propagating characteristics:

$$(1/\sqrt{\rho})\partial b_y^+/\partial t=-\beta_z L_a^+-\beta_y/\rho\{\alpha_f L_s^+ + \alpha_s L_f^+\} \quad (4)$$

$$(1/\sqrt{\rho})\partial b_z^+/\partial t=-\beta_y L_a^+-\beta_z/\rho\{\alpha_f L_s^+ + \alpha_s L_f^+\} \quad (5)$$

where $\alpha_{f,s}$ are defined in terms of the Alfvén, fast and slow velocities [5]. In order to prescribe $\partial b_y^+/\partial t = dB^\circ_y/dt$ and $\partial b_z^+/\partial t = dB^\circ_z/dt$, where $B^\circ(t)$ is the function given in (1), we have to satisfy to (using $\beta_y^2 + \beta_z^2 = 1$):

$$L_a^+ = -\beta_y (1/\sqrt{\rho}) dB^\circ{}_z^+/dt + \beta_z (1/\sqrt{\rho})\, dB^\circ_y/dt \quad (6)$$

$$\alpha_f L_s^+ + \alpha_s L_f^+ = -\beta_y \sqrt{\rho}\, dB^\circ{}_z^+/dt - \beta_z \sqrt{\rho}\, dB^\circ_y/dt \quad (7)$$

A natural choice for the third condition is to set to zero the slow mode injection,

$$L_s^+ = 0 \quad (8)$$

The set (6-8), to be denoted C0, appears appropriate because in the linear limit it reduces to injecting the two polarizations of the Alfvén modes and no acoustic wave.

RESULTS

Fig.1 shows the plasma response to waves injected at both boundaries with a slight phase difference, in the moderate β case. The parallel (leftward) flow is seen to reach quasi-periodically a transonic regime. Correspondingly, there are large, low-frequency pressure fluctuations with amplitude more than an order of magnitude larger than the magnetic pressure difference between the two boundaries (Fig.2).

Fig.3 shows how the instability begins: the pressure first becomes larger at left boundary. It is finally not clear whether the long-term oscillation is due to the particular choice of boundary conditions, or if it is due to the variation of the conditions inside the loop induced by the wave.

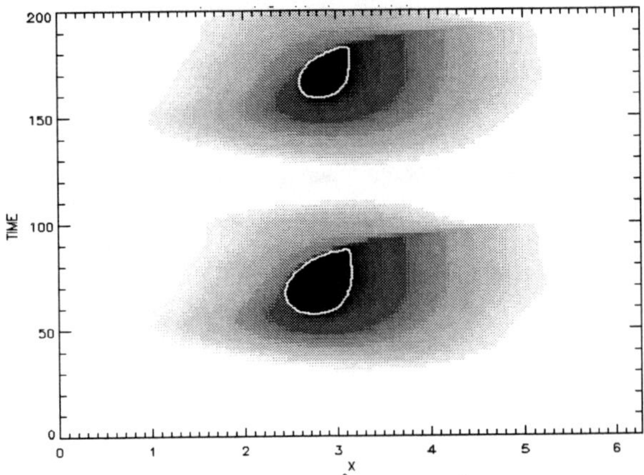

FIGURE 1. Parallel velocity of the plasma in space-time, in response to Alfvén wave injection, with boundary conditions C0.. Moderate β case. White lines: Mach=-1 isolines.

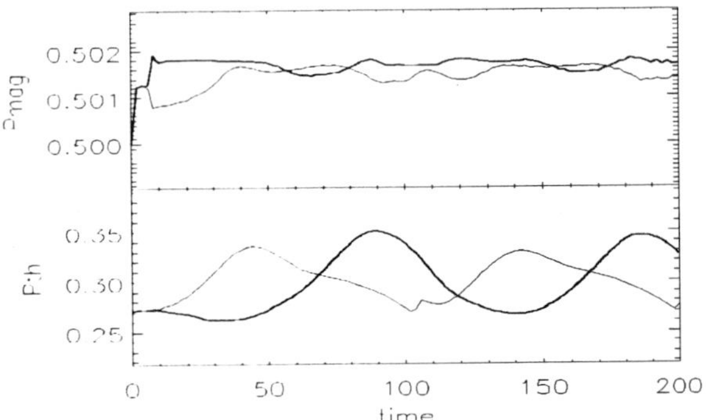

FIGURE 2. Pressure variations versus time at the boundaries (bold, x=0; plain, x=2π). Top: magnetic pressure; bottom: thermal pressure. Same run as Fig.1.

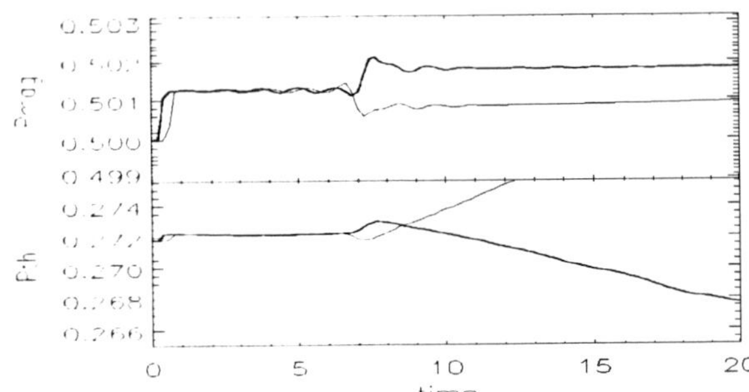

FIGURE 3. Same as Fig.2, short time evolution.

DISCUSSION

We have seen that the thermal pressure variation largely amplifies the magnetic pressure variation associated with the waves. To examine this process, we consider the simpler non-stratified case, and focus on the time period when the two wave trains reach the opposite boundaries.

Fig.4 shows the magnetic pressure, thermal pressure and parallel velocity at successive times. One sees that the magnetic pressure adopts the profile (2), but that the fluid pressure shows un abrupt variation at the boundaries when the waves reach the boundaries: the thermal pressure rises at x=0, and decreases at x=2π. Accordingly, there is a strong acceleration rightward at both boundaries. This corresponds to a (spurious) injection of u_x fluctuations, resulting from a nonlinear reflexion of the Alfvén wave on the boundary, associated with a change in wave mode..

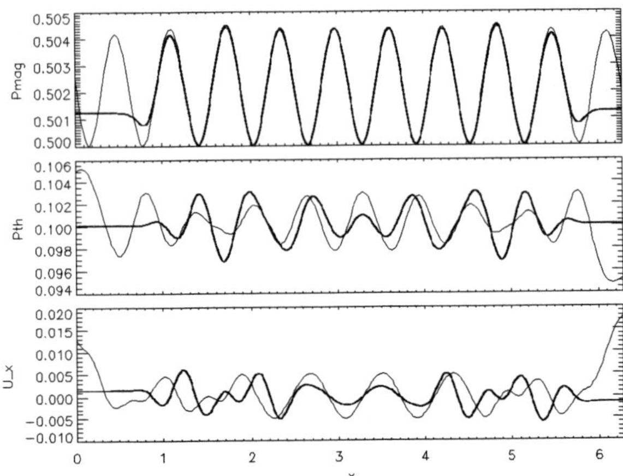

FIGURE 4. Profiles of the wave pressure (top), thermal pressure (mid panel) and parallel velocity (bottom), as the two waves reach the opposite boundaries, conditions C0. Times 6 (bold), and 8.

When one examines the equation for the parallel (incoming) velocity in terms the incoming characteristics, one sees indeed that, except in the linear limit, the condition of no slow injection is not sufficient to prevent injecting some parallel velocity fluctuation, as soon as the fast mode injection is non zero:

$$\partial u_x^+/\partial t = (1/c)(v_s \alpha_s L_s^+ - v_f \alpha_f L_f^+) \qquad (9)$$

(c is the sound speed and $v_{f,s}$ are the fast and slow speeds).

To examine whether we can suppress this coupling with the boundary, we consider the even simpler model of an isothermal non conducting gas, to which we apply from start an external force corresponding exactly to the gradient of the magnetic pressure of the previous situation, frozen to the state shown in Fig.3 at time t=8 when the waves have travelled the whole domain. We find in that case no spurious acceleration at the boundary, and the velocity fluctuations generated within the medium by the external force propagate away at the sound speed. There remains only a stationary pattern of thermal pressure fluctuations necessary to balance the external force mimicking the wave pressure, with a very slow systematic flow.

It is important to remark that, in this hydrodynamic version of the problem, there is only one characteristic to impose, namely $L_s^+ = 0$, the acoustic characteristics, which is here completely equivalent to $\partial u_x^+/\partial t = 0$. In the MHD case, we have seen that in the nonlinear case at least, no slow wave injection is not equivalent to no injection of parallel velocity. When looking at (9), we find that $L_s^+ = 0$ is equivalent to $\partial u_x^+/\partial t = 0$ only in either in the linear limit, or when dealing with a simple wave, that is, not in presence of two waves propagating in opposite directions.

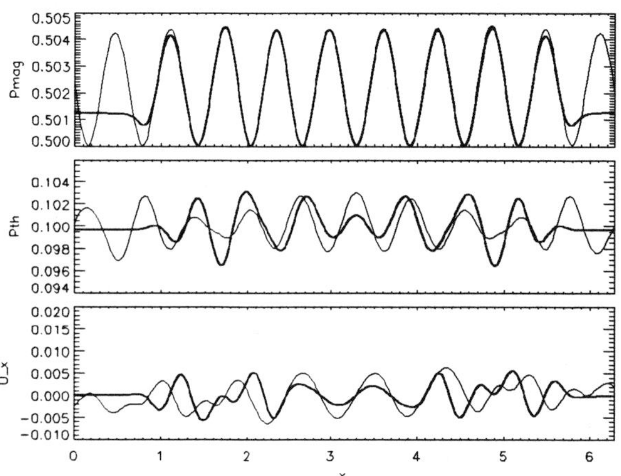

FIGURE 5. Same caption as Fig.4, boundary conditions C2.

We now come back to the MHD problem and replace the condition $L_s^+ = 0$ by $\partial u_x^+/\partial t = 0$, or:

$$v_s \alpha_s L_s^+ - v_f \alpha_f L_f^+ = 0 \quad (10)$$

We call C2 the set (6-7) and (10). When applying it first to the unstratified case, as shown in Fig.5, we see that the boundaries no longer suffer spurious acceleration, and that the pressure fluctuations remain at a small level, actually comparable to that obtained in the hydrodynamic case. Last, we apply conditions C2 to the stratified MHD case; we again find no large pressure imbalance (Fig.6), and, as a consequence, after transient fluctuations have escaped the loop, only a weak subsonic flow remains.

This remains true when we decrease the plasma β: although the wave decays into density fluctuations, the resulting flow does not reach a large Mach number. In conclusion, the reaction of the loop to Alfvén waves depends entirely on wether we allow or not incoming parallel velocity fluctuations: only in the latter case do transonic flows arise. On the other hand, we checked that purely reflecting conditions also do not lead to supersonic flows. True boundary conditions are more complex than just transparent or reflecting: in fact, the transparency of the chromospheric transition is frequency-dependent [7], and simulations including it are needed to decide whether Alfvén waves can trigger supersonic flows or not.

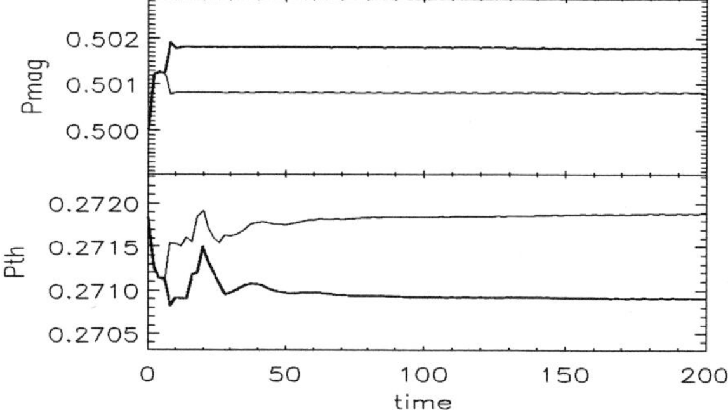

FIGURE 6. Pressure variations at the boundaries (bold, x=0; plain, x=2π). Top: magnetic pressure; bottom: thermal pressure (compare Fig. 2). Boundary conditions C2.

ACKNOWLEDGMENTS

RG thanks P. Londrillo and G. Belmont for fruitful discussions, the Centre International d'Ateliers Scientifiques for hospitality during July 2001. LO would like to aknowledge support by NASA Sun-Earth theory program, and NASA grant NAG5-11841

REFERENCES

1. Winebarger, A. R., Warren, H., van Ballegooijen, A., DeLuca, E.E., Golub, L., *Astrophys. J.* **567**, L89 (2002)
2. Grappin, R., Léorat, J., and Habbal, S.R., *J. Geophys. Res.* in press (2002)
3. Meyer, F., and Schmidt, H.U., *Z. Angew. Math. Mech.* **48**, 218 (1968)
4. Cargill P.J., and Priest, E.R., *Solar Physics* **65** 251 (1980)
5. Del Zanna, L., Velli, M., and Londrillo, P., *Astr. Astrophys.* **367**, 705-718 (2001)
6. Grappin, R., Léorat, J., and Buttighoffer, A., *Astr. Astrophys.* **362**, 342 (2000)
7. Hollweg, J.V., *Astrophys J.* **277**, 392 (1984)

Radial dependence of propagation speed of solar wind disturbance

Masahiro Yamashita*, Munetoshi Tokumaru* and Masayoshi Kojima*

*Solar-Terrestrial Environment Laboratory, Nagoya University, Japan.

Abstract. We studied the propagation of interplanetary counterpart of coronal mass ejections (ICMEs) by using interplanetary scintillation (IPS) measurements of the Solar-Terrestrial Environment Laboratory at 327MHz. In this study, we analyzed data of the solar wind disturbance factor, so-called g-value, derived from our IPS measurements. We analyzed two ICME events observed in 1997 and 2000. We assumed a shell-shape ICME model, defined by six parameters (location, propagation directions, radial thickness, angular width, and enhancement factor). We determined the most suitable set of those parameters by matching the model calculations of g-value to IPS observations. Then, we determined the radial variation of the propagation speed of CME or ICME by comparing our IPS data with coronagraph data and in-situ data. As result, it is found that the propagation speed of ICME decelerated as $V \propto r^{-1.05}$ for the 2000 event and $V \propto r^{-0.55}$ the 1997 event.

INTRODUCTION

Coronal mass ejections (CMEs) are important phenomena because of their large influence on the magnetosphere of the Earth. Therefore, a lot of studies of the dynamics of CMEs near the Sun have made by using coronagraph observations. The knowledge on dynamics of interplanetary counterpart of CMEs (ICMEs) is important, particularly from the viewpoint of the space weather prediction. However, dynamics of ICMEs are little understood, because ICME observations depend on a limited number of spacecraft. From the study using in-situ data near the earth, ICMEs are found to propagate at around the ambient solar wind speed which ranges from 400 to 800 km/s [1]. On the other hand, it has been reported from the coronagraph measurements that the propagation speeds of CMEs range from less than 50 to greater than 2000 km/s [2]. These facts suggest that the radial evolution of the ICME propagation speed takes place between the corona and 1AU through the interaction with the ambient solar wind.

Interplanetary scintillation (IPS) measurements conducted by the Solar-Terrestrial Environment Laboratory (STEL), Nagoya University, at 327 MHz [3] allows us to map the global structure of ICMEs at radial range between 0.2 and 1 AU [4, 5]. In the present paper, we investigated the three-dimensional (3-D) structure of ICMEs by using our IPS observations and studied radial dependence of the propagation speed of the ICME by combining our data with coronagraph and in-situ observations.

IPS G-VALUE

Radio waves from a cosmic radio source are scattered by the solar wind. As a result, intensity fluctuations are observed as IPS on the ground. An amplitude of the IPS depends on the solar wind density fluctuations along the line-of-sight (Los). The scintillation index m, which is defined by

$$m = \frac{< \Delta I >}{I} \tag{1}$$

represent the strength of IPS. Here, I is an observed source intensity, and ΔI is an intensity fluctuation. For weak scattering, the m-index is related to the solar wind fluctuations via the following formula

$$m^2 = \int \omega(z)[\Delta N_e]^2 dz \tag{2}$$

where $\omega(z)$ is the weighting function, ΔN_e is the level of solar wind density fluctuations, and z is the distance from the observer to the radio source along the Los. The m-index decreases with the radial distance from the Sun, because solar wind density fluctuations decrease approximately as r^{-2}. In addition, the m-index depends on the apparent size of radio source.

To correct these dependences, we have calculated a normalized factor, g-value [6], which is given by

$$g = \frac{m}{\overline{m}} \tag{3}$$

where $\overline{m}$ is the mean level of IPS for a given distance. If we observe the quiet solar wind, the g-value becomes around an unity. If density enhancements due to ICME

CP679, *Solar Wind Ten: Proceedings of the Tenth International Solar Wind Conference*,
edited by M. Velli, R. Bruno, and F. Malara

pass across the Los, the g-value enhances abruptly to $g > 1$.

ICME MODEL AND FITTING ANALYSIS

The g-value obtained form IPS observations corresponds to an integration of density fluctuations along the Los. (Equation (2)) Therefore, we have to remove the effect of the Los integration in order to extract 3-D information on ICMEs from the g-value data. Although we have removed the effect of the Los integration from IPS observations successfully by using the computer assisted tomography (CAT) method [7, 8, 9], the CAT method is only applicable to the analysis of a quasi-stationary structure in the solar wind, therefore we can't use the CAT method for a transient component such as ICMEs.

In this study, we have performed the model fitting analysis to obtain unbiased estimates of the ICME structure from g-value data. If the spatial distribution of ΔN_e is given by a appropriate model, the g-value can be calculated by using Equation (2). This means that we can determine the ΔN_e distribution by optimizing the model calculations to observed g-value data. We assume here that the disturbance has a shell-shape and an isotropic expansion speed in the radial direction. In this model, density fluctuations ΔN_e are given as follows.

$$\Delta N_e(r,\theta) = \frac{1}{r^2}(f_{enh}exp(-(\frac{r-r_0}{dr})^2)exp(-(\frac{\theta-\theta_0}{d\theta})^2)+1) \quad (4)$$

where f_{enh} is the enhancement factor of ΔN_e at the center, dr is the radial thickness, $d\theta$ is the angular width of the disturbance, r_0 is the distance from the Sun to the disturbance, r is the distance to a given point, and $\theta - \theta_0$ is the angle between the propagation direction and the direction of a given point. Figure 1 illustrates the ΔN_e model. In order to determine the best fit parameters, we calculated the rms deviation χ^2 between observed and calculated g-values. We searched a set of parameters that minimize χ^2 by adjusting model parameters ;

$$\chi^2[g_{obs,i}, g_{cal,i}(r_0,\theta,dr,d\theta,f_{enh}); (i=1,n)] \Rightarrow min \quad (5)$$

where n is the number of the Los.

ANALYZED EVENTS

In this study, we analyzed ICME events observed by IPS measurement on July 12, 2000 and November 8, 1997.

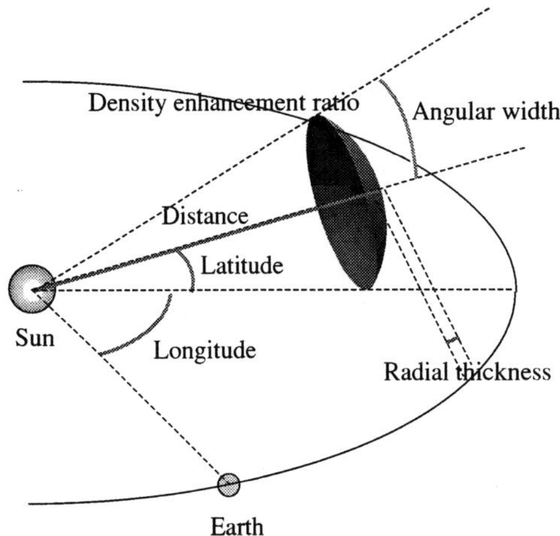

FIGURE 1. Shell-shape ICME model defined by six parameters (location, propagation directions, radial thickness, angular width, and enhancement factor). We made the g-value calculation using this model and determined the best fit parameters.

Event on July 12, 2000

A large active region NOAA 9077 generated some large flares accompanied by CMEs on July 2000. An ICME was observed by our IPS measurements on July 12, 2000, and this is considered to be associated with the M5.7/2B flare occurred at 21:05 UT on July 10, 2000. A CME was observed by the LASCO coronagraph in association with this flare. The associated shock arrival was reported from in-situ measurements of SOHO/Proton Monitor at 9 UT on July 13, 2000.

Event on November 8, 1997

An ICME observed by our IPS observation on Nov. 8 1997 is considered to be associated with the X9.4/2B flare occurred at 11:55 UT on Nov. 6, 1997 and this flare was accompanied with a CME. This CME had a high speed, and showed deceleration in the field of view of LASCO coronagraph. No shock arrival was observed for this ICME event.

CME SPEED MEASUREMENTS FROM LASCO DATA

We estimated the CME speed in the field of view of LASCO coronagraph by using C2 and C3 images for two events analyzed here. We calculate CME speeds from the location of the leading part of the CME along the propagation direction of the ICME determined by the model fitting analysis of IPS data.

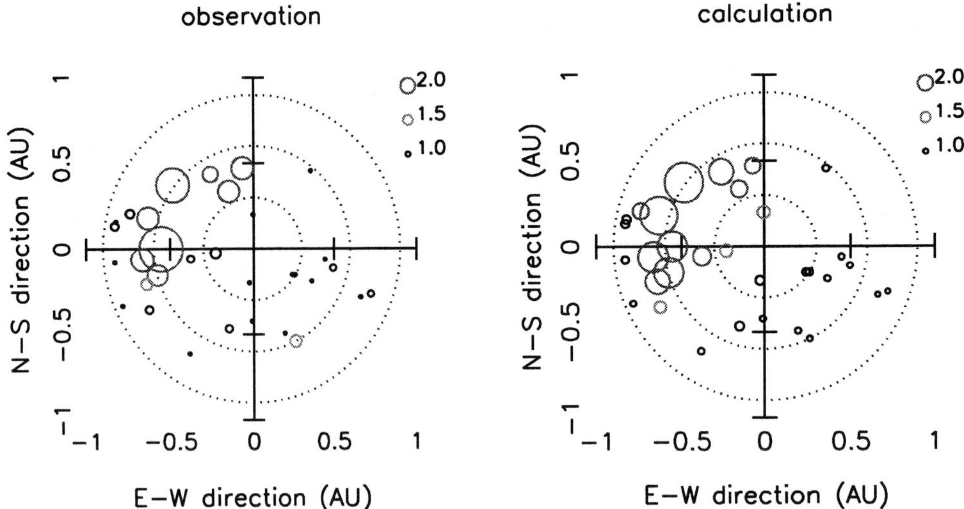

FIGURE 2. (left) The observed all-sky map of g-value on July 12, 2000 and (right) the result of the model calculations using the best fit parameters. The dotted circles are constant r contours drawn every 0.3 AU. A center of open circles corresponds to the location of radio source and a diameter of them represents a level of the g-value.

RESULTS

Event on July 12, 2000

Table 1 shows the best fit parameters obtained by the model fitting analysis. Figure 2 shows observed g-values (left) and the result of model calculations using the best fit parameters (right).

The propagation speeds of the CME derived from LASCO C2 and C3 images are plotted diamonds and stars respectively in Figure 3. Here, a mid point between two adjacent LASCO observations is used as a reference point for CME speed data, and horizontal bars denote the distance interval. By comparing coronagraph data taken at the largest distance with the result of IPS analysis, we calculated a mean propagation speed between the corona and the inter planetary (IP) space. This value is plotted by a square in the figure. By combining the IP shock data at 1AU with the result of IPS analysis, we also calculated a mean speed between the IP space and the earth orbit. This speed is shown by a circle in Figure 3. As shown in Figure 3, the CME accelerated rapidly within 10 Rs, and then decelerated with increasing distance. In this study, we focused on the deceleration of ICME. To evaluate a strength of the ICME deceleration, we fitted speed data by a function of $V = \alpha r^{-\beta}$, where V is an ICME speed relative to the ambient solar wind speed, and r is the distance from the Sun. For the ambient solar wind speed, we used the proton speed of 500 km/s observed by SOHO/Proton Monitor just before the shock arrival. As a result, we obtained $\beta = 1.05$.

TABLE 1. Obtained the best fit parameters for observation on Jul. 12, 2000

Distance	Long.	Lat.	Thickness	Angular width	Density enhancement
0.56	-36	15	0.12	33	7.2

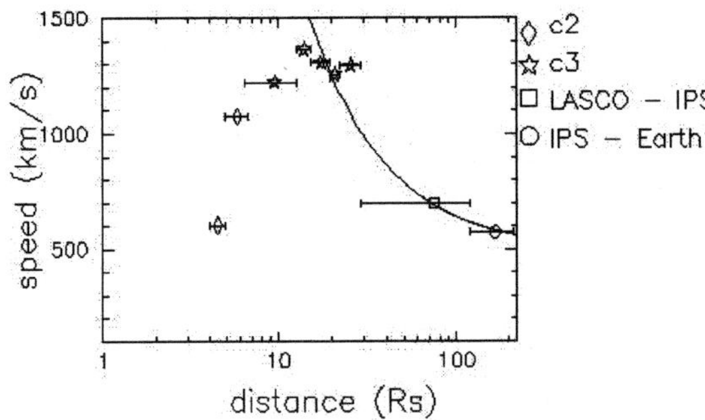

FIGURE 3. The propagation speed of CME and ICME for July 12, 2000 event. Speed data from C2 and C3 images, LASCO - IPS and IPS - in-situ are denoted by diamonds, stars, a square and a circle symbols respectively. The solid curve corresponds to the deceleration curve $V \propto r^{-\beta}$ with $\beta = 1.05$.

Event on November 8, 1997

The best fit parameters obtained from the model fitting analysis are shown in Table 2. The radial variation of propagation speeds of CME or ICME are shown in Figure 4. The figure suggests that the CME decelerated slowly in the corona, and also suggests that the ICME decelerated rapidly as moving from the Sun. A solid line corresponds to the deceleration curve $V \propto r^{-\beta}$ with the

TABLE 2. Obtained the best fit parameters for observation on Nov. 8, 1997

Distance	Long.	Lat.	Thickness	Angular width	Density enhancement
0.60	-50	11	0.13	30	7.4

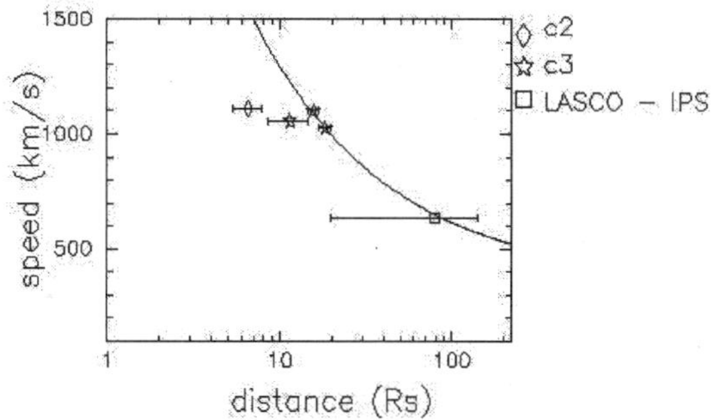

FIGURE 4. The propagation speed of CME and ICME for July 8, 1997 event. The solid curve corresponds to the deceleration curve $V \propto r^{-\beta}$ with $\beta = 0.55$.

best fit parameter of $\beta = 0.55$. Here, we used the speed of 400 km/s obtained from the CAT analysis of our IPS observations for the ambient solar wind speed.

DISCUSSION

Smart and Shea [10] reported that the blast wave shock front speed should vary as $r^{-0.5}$ when the solar wind density decrease as $1/r^2$. We obtained $r^{-0.55}$ speed dependence for the Nov. 8, 1997 event. This result agrees with the Smart and Shea model. On the other hand, the Jul. 12, 2000 event showed a radial deceleration of $r^{-1.05}$. This value is steeper than that in the result by Smart and Shea [10], i.e. $V_s \propto r^{-0.5}$. Thus, our result suggest that the deceleration profile of IP shock does not necessarily follow the $r^{-0.5}$ dependence, and that it depends on the condition of the ambient solar wind.

Vršnak and Gopalswamy [11] reported the aerodynamic drag force in the interplanetary space acts dominantly as a controlling factor of the ICME propagation. An amplitude of this force is determined by the density of ambient solar wind and the speed difference between ICME and ambient solar wind. In the solar maximum, the slow speed component of the solar wind becomes dominant in the entire space of the heliosphere, so that more intense drag force effect is expected to take place for the solar maximum ICME event, since the ICME has to penetrate into a dense slow wind. The fact that the event Jul. 12, 2000 event showed a steep deceleration is consistent with this expectation. We need further study to examine the effect of ambient solar wind on the ICME propagation in more detail.

SUMMARY

In this study, we made the g-value calculations using a shell-shape ICME model, and determined six parameters of the model by matching observed g-value with model calculations. By combining this result with coronagraph data and in-situ observations, we clarified the radial evolution of ICME speeds between the corona and the Earth orbit. We analyzed the ICME events occurred on Jul. 12, 2000 and Nov. 8, 1997. As a result, the ICMEs were found to decelerate rapidly as they propagated from the Sun. We compared the strength of deceleration between those events. As a result, it is found that the propagation speed of ICME decelerated as $V \propto r^{-1.05}$ for the 2000 event and $V \propto r^{-0.55}$ the 1997 event.

ACKNOWLEDGMENTS

We used CME images taken by the SOHO/LASCO coronagraph and in-situ data taken by the SOHO/Proton Monitor. We would like to thank engineering support from Y. Ishida and N. Yoshimi.

REFERENCES

1. Lindsay, G. M., Luhumann, J. G., Russell, C. T., and Gosling, J. T., *J. Geophys. Res.*, **104**, 12,515 (1999).
2. Howard, R., Sheeley, N. R., Koomen, M. J., and Michels, D. J., *J. Geophys. Res.*, **90**, 8173 (1985).
3. Kojima, M., and Kakinuma, T., *Space. Sci. Rev.*,**53**, 173 (1990).
4. Tokumaru, M., Kojima, M., Fujiki, K., and Yokobe, A., *J. Geophys. Res.*, **105**,10,435 (2000).
5. Tokumaru, M., Kojima, M., Fujiki, K., and Yamashita, M., this volume (2002).
6. Gapper, G. R., Hewish, A., Purvis, A., and Duffet-Smith, P. J., *Nature, 296*, 633 (1982).
7. Asai, K., Ishida, Y., Kojima, M., Maruyama, K., Misawa, H., and Yoshimi, M., *J. Geomag. Geoelectr.*,**47**, 1107 (1995).
8. Jackson, B. V., Hick, P. L., Kojima, M., and Yokobe, A., *J. Geophys. Res.*, **103**, 12,049, 1998.
9. Kojima, M., Tokumaru, M., Watanabe, H., Yokobe, A., Asai, K., Jackson, B. V., and Hick P. L., *J. Geophys. Res.*,**103**, 1981 (1998).
10. Smart, D. F. and Shea, M., A., *J. Geophys. Res.*, **90**, 183 (1985).
11. Vršnak, B. and Gopalswamy, N., *J. Geophys. Res.*, **107**, SSH 2 (2002).

Solar-Heliospheric-Magnetospheric Observations on March 23-April 26, 2001: Similarities to Observations in April 1979.

D. B. Berdichevsky[1,2], C. J. Farrugia[3], R. P. Lepping[2], I. G. Richardson[4,2], A. B. Galvin[3], R. Schwenn[5], D. V. Reames[2], K. W. Ogilvie[2], and M. L. Kaiser[2]

[1]*L-3 Communications Analytics Corporation, Largo, Maryland 20774, USA*

[2]*NASA Goddard Space Flight Center, Greenbelt, Maryland 20771, Mail Code 600, USA*

[3]*Space Science Center, University of New Hampshire, Durham, NH 03824, USA*

[4]*Department of Astronomy, University of Maryland, College Park, MD 20742, USA*

[5]*Max-Plank-Institut für Aeronomie, D 37191 Katlenburg-Lindau, Germany*

Abstract. We discuss the similarities and differences of two intervals of extreme interplanetary solar wind conditions, separated almost precisely by two solar cycles, in April 1979 and March-April 2001. The similarities extend to various data-sets: Energetic particles, solar wind plasma and interplanetary magnetic field. In April 1979 observations were made by three spacecraft covering a wide longitudinal range (~ 70°) in the heliosphere. Data are presented from Helios 2, located 28° East of the Sun-Earth line at ~ 2/3 AU, and from near the Earth. Observations of the 2001 interval are from Wind. We examine the geomagnetic activity during each interval.

1. INTRODUCTION

Interplanetary coronal mass ejections (ICME) also named ejecta cause the largest geomagnetic disturbances at Earth (e.g., Richardson et al., 2001). Their effect may be altered (enhanced or reduced) if instead of single ejecta, we have a sequence interacting with each other. We focus on the interplanetary causes of two geomagneticaly active intervals, one during the maximum of the current solar cycle (March-April 2001), the other at the maximum of cycle 21 (April 1979).

The 2001 interval was dominated by activity associated with the largest sun spot group in 10 years, consisting of three or more active regions (ARs) centered near AR 9393. For this period of unusually intense solar activity we discuss here energetic particle, radio, solar wind plasma and magnetic field parameters observed by the Wind spacecraft. An unusual sequence of fast ejecta accompanied by a comprehensive set of related signatures (long-lasting, intense radio emissions, gradual solar energetic particle (SEP) enhancements, unusually strong magnetic fields and extreme plasma conditions) were detected. In this paper we list the sequence of disturbances during the extended interval of March-April 2001. We also compare part of this interval with April 1 to 7, 1979. The geomagnetic response of the Dst index will be contrasted.

2. INTERPLANETARY OBSERVATIONS

Figure 1 shows the observations at Wind from March 26-April 26, 2001.

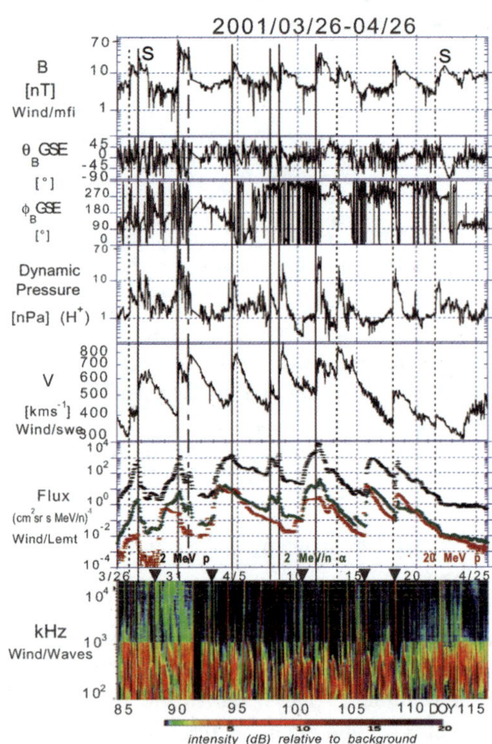

Figure 1. Presented IMF and plasma observations are 5-min averages. SEP: 2 MeV protons (black), 8 MeV α (green), and 20 MeV protons (red) are 1-hour averages. The radio frequency spectrogram shows 1-min average with color scale 0.1-20 decibels. For details see text.

CP679, *Solar Wind Ten: Proceedings of the Tenth International Solar Wind Conference*, edited by M. Velli, R. Bruno, and F. Malara

From top to bottom are plotted the total magnetic field and its components in GSE coordinates, the proton dynamic pressure ($m_pN_pV_{SW}$), bulk speed, SEP intensities at Wind in the MeV range, and radio signals in the range 10^2–1.210^4 kHz. These observations were made while Wind was executing a distant prograde orbit at (X1, Y1, Z1~) 0, –240, 0 R_E (GSE coordinates). Vertical lines in Figure 1 mark the passage by Wind of 11 shock candidates. Solid vertical lines indicate solar wind discontinuities at shocks with MeV energetic particle enhancement, other shock passages are marked with dashed-lines. Another less distinct shocks may be present. An example is indicated by the long-dashed vertical line late on March 31. The onsets of the SEP events –two to three days prior to the passage of the shock – align with type III radio bursts (vertical radio intensification lines which are in the bottom panel). Those associated with the lift-off of five of the CMEs that generated moderately strong SEP events are indicated with solid inverted triangles between the two bottom panels. During the interval in Figure 1, 11 halo CMEs were observed with LASCO/SOHO, 7 candidate ejecta intervals (suggested by the slow rotation of the IMF) may be identified, and there were at least 3 extended regions of very low, almost disappearing solar wind plasma (dynamic pressure was ≤ 1 nPa).

We will now focus on the period March 28-April 1, 2001. During this interval, SOHO LASCO and EIT observed signatures of two full halo CMEs directed toward Earth. One lifted off at ~ 1100-1200UT, from ~0° longitude on March 28, the other at ~1030UT from ~ N20W19 on March 29. The speed of the March 28 CME in the plane of the sky was estimated at 500 km/s over the south pole and there was negligible deceleration. This CME is possibly associated with a flare in AR 9397 [see e.g.,

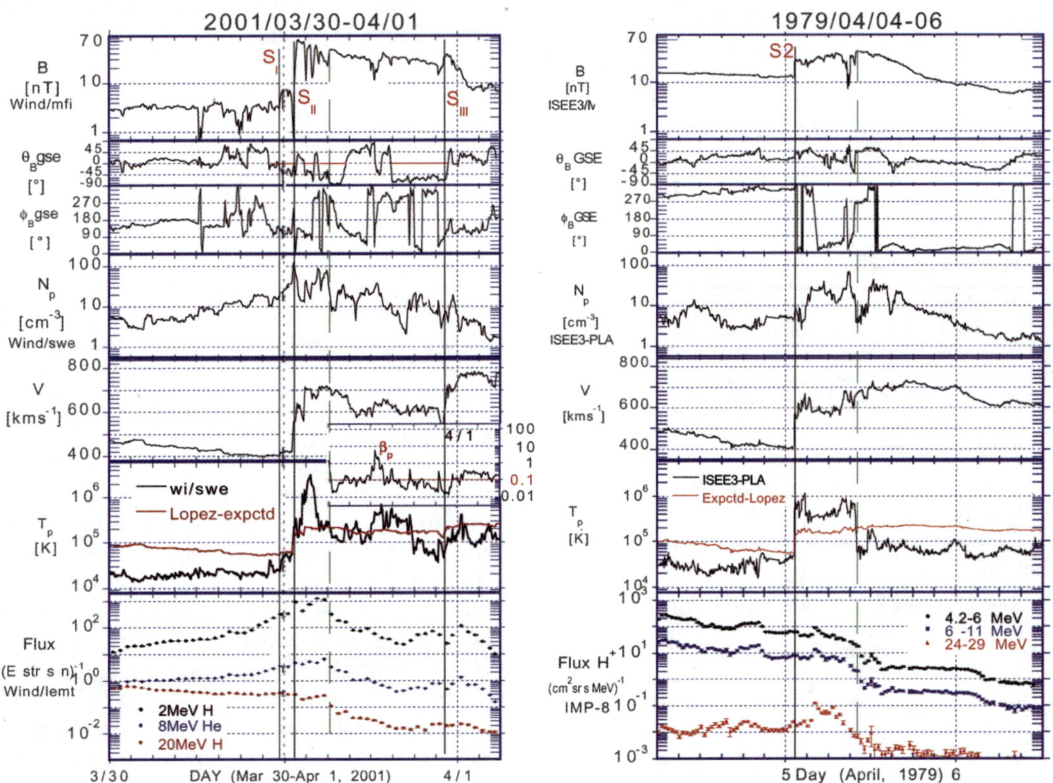

Figure 2a: Observations for the period 00UT March 30 to 06UT April 1, lines are 5-min average. Starting from the top: IMF strength B, latitudinal, longitudinal orientation of **B** in GSE coordinates, proton density N_p, solar wind speed V, and proton temperature T_p. The bottom panels contain the intensity of 2 MeV proton (black), 8 MeV α (blue) and 20 MeV proton (red).

Figure 2b: ISEE-3 observed for the period 06UT April 4–12UT April 6, 1979, same as in Fig. 2a except that bottom panel shows IMP-8 energetic proton chanels: 4.2-6 MeV (black, higher flux), 6-10 MeV (blue, medium), and 24-29 MeV (red, lower).

Sun et al., 2002]. The March 29 halo CME

had the much higher plane of the sky speed of ~1000 km/s with apparently significant deceleration. It was associated with an X1.7 flare in AR 9393. Figure 2a shows at Wind the total interplanetary magnetic field, and components in GSE coordinates, the proton density, bulk speed, temperature, and finally energetic particles in the MeV range. Figure 2a reveals two, almost coincident, shocks, S_I at 2330 UT on March 30 and S_{II} at 0111UT on March 31. These are more clearly identified from the twofold impulsive rise in T_p panel. A third shock S_{III} was seen by Wind 21 hours later. What is the relationship between these three shocks and the two halo CMEs? Both the arrival time and high speed of S_{III} at Wind exclude any association with the halo CMEs. The two first shocks are more likely related to the halo CMEs. Their proximity in time is consistent with the observation that the second halo CME twice as fast as the first one. Approximately six hours after the shocks S_I and S_{II} are two distinct regions of very strong magnetic fields (approximately ~18 hours of ≥30 nT) of low variance, each with low proton beta plasma ($\beta_p \leq 0.1$, illustrated in inset in Figure 2a). They are separated by a narrower region where the field is weaker (10 nT at ~12-13UT) and proton beta is high. These may be the ejecta corresponding to the two halo CMEs. The weakness of the first shock relative to the second is in our view an indication that the ejecta are in the process of coalescing, and the first shock in the process of disappearing. The complex nature of the interval is seen from the SEP profile with spike at the strong shock S_{II} and later at the start of the first extended low β_p region. A decrease in the proton temperature (T_p) and a substantial drop in the flux of MeV particles help to identify the start of the ejecta interval in Figure 2a, indicated with a vertical long-dashed line.

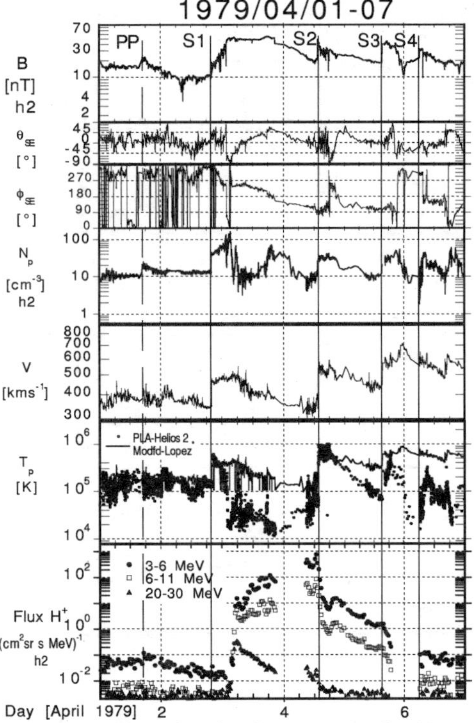

Figure 3: Helio 2 data for the time period 12UT April 1-12UT April 7, 1979, in the same format as in Figures 2a and 2b. The bottom panel shows Helios 2 energetic proton chanels: 3-6 MeV (full circles, higher flux), 6-11 MeV (open squares, medium), and 20-30 MeV (full triangles, lower).

Next we focus on a similar interval on April 1-7 1979 in which CMEs were observed in the process of overtaking each other (Burlaga et al., 1987), in observations from Helios 2 (see Figure 3, same parameters as in Figure 2a). The location of Helios 2 and 1 are shown in Figure 4.

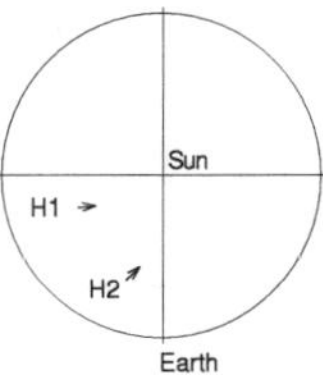

Figure 4. Shown are locations at approximately 68° of Helios 1, 28° of Helios 2 on the interval April 1-9, 1979. Length of line indicates the approximate displacement of the Helios spacecraft for the interval and arrow the direction of displacement.

Helios 1 observes different solar wind streams, placing in that way a limit to the longitudinal extent of the ejecta observed at Earth and Helios 2. There are at least four shocks at Helios 2, each preceding an ejecta. As during the March 27-April 1, 2001 period, a single SEP event (associated to a sun disk lift-off location W43° (W15°) from Helios 2 (Earth) at ~ 1100 UT on April 3, 1979) was present (Richardson and Cane, 1996). This was related to the ejecta driving shock S2. Did these ejecta observed in the inner heliosphere eventually coalesce? In fact those associated with S2 and S3 had done so by the time they arrived at Earth. Figure 2b shows data from ISEE-3 (situated near L1 point). Only one shock corresponding to S2 at Helios was still present at Earth. The disturbances in the B and V profiles are now much attenuated. The shock S3 and a possible shock S4 at Helios have dissipated. It is suggested that the ejecta passing Helios 2, and driving shocks S2, S3 and possibly S4 coalesced within about 0.3 AU downstream and 27° West of Helios 2, to form a "complex-ejecta", to use the terminology of Burlaga et al. 2002.

Both the March 30-April 1, 2001 ISTP era events and the April 4-6, 1979 Helios 2 events are related to solar CMEs that followed in quick succession (Sun et al., 2002; Burlaga et al., 1987). Similarities in Figures 2a and 2b are: i) the shocks S_{II} and S2 moving into low proton beta regions, ii) extraordinary heating and extreme densities in the shocked regions, iii) sharp decrease of the MeV SEP fluxes (~ 50%) at the beginning of the ejecta intervals, indicated by vertical long-dashed line in Figures 2a and 2b. There are also differences such as the latitudinal orientation of the IMF, the attenuated variations in IMF, V, T_p and SEP on April 5, 1979 at Earth, in Figure 2b when compared with those on March 31, 2001, in Figure 2a.

3. CONCLUSIONS

We have examined two intervals separated by 22 years, where solar wind conditions show some stricking similarities, in particular the merging of multiple ejecta.

How do the two events compare in their geoeffectiveness? In 2001, a great storm (minimum Dst < -380 nT, the largest of the current solar cycle) started on March 31 with a recovery phase lasting up to April 5. The Kp index exceeded 6 for the whole day and reached saturation at 9 UT. A second major storm was observed on April 11 (minimum Dst <-240 nT), with a recovery lasting to April 18. Two further, moderate storms (minimum Dst ~ -100 nT) followed.

By contrast during the 1979 period only one moderate-major storm was recorded (minimum Dst <-180 nT), on April 4, 1979, due to the passage of a magnetic cloud with a substantial southward excursion of the IMF (Burlaga et al., 1987). The following, putative coalesced ejecta discussed here barely affected the recovery phase of the earlier magnetospheric disturbances. Thus, there is a sharp contrast between the magnetic disturbances elicited by the respective magnetic configurations. This weaker activity in 1979 is due to the absence of a strong southward orientation of the IMF. This may be the result of a combination of factors: (1) the coalescence of the ICMEs, and/or (2) the longitudinal separation between Helios 2 and the Sun-Earth line. This is the subject of further study.

Acknowledgements. This work is supported by the Wind Grant NAG5-11803, and by NASA Living with a Star Grants NAG5-10883 and NASW-02025.

References

Burlaga, L.F., K.W. Behannon, and L.W. Klein, Compound streams, Magnetic Clouds, and Major Geomagnetic Storms, J. Geophys. Res., 92, 5725, 1987.

Burlaga, L.F., S.P. Plunkett, and O.-C. St Cyr, Successive CMEs and Complex Ejecta, J. Geophys. Res., in press, 2002.

Lopez R. E., and Freeman, J. W., The solar wind proton temperature-velocity relation, J. Geophys. Res., 91, 1701, 1986.

Richardson, I.G., and H.V. Cane, Particle flows observed in ejecta during solar event onsets and their implication for the magnetic field topology, J. Geophys. Res., 101, 27,521, 1996.

Richardson, I.G., E.W. Cliver, and H.V. Cane, Sources of geomagnetic storms for solar minimum and maximum conditions during 1972-2000, Geophys. Res. Lett., 28, 2569, 2001.

Sun, W., M. Dryer, C.D. Fry, C.S. Deehr, Z. Smith, S.-I. Akasofu, M.D. Kartalev, and K.G., Grigorov, Real time forecasting of ICME shock arrival at L1 during the "April fool's day" Epoch: 28 March – 21 April 2001, Ann. Geophys., in press, 2002.

Influence of the time-dependent heliosphere on global structure

G.P. Zank* and H.-R. Müller*†

*Institute of Geophysics and Planetary Physics, University of California, Riverside
†Bartol Research Institute, University of Delaware, Newark

Abstract. The solar cycle leads to important changes in the solar wind, which can have an important effect on the structure of the global heliosphere. In the ecliptic, the ram pressure can vary from one cycle to the other, being greater during periods of minimum activity. We investigate the response of the heliosphere to a temporally varying solar wind. Since neutral hydrogen is a key component in determining the global structure of the heliosphere, we employ a multi-fluid model in which the self-consistent charge-exchange interaction between neutral hydrogen and protons is included self-consistently. The variability of the termination shock location is described and the response of the hydrogen wall to the temporal solar wind is discussed.

INTRODUCTION

The solar wind varies on many spatial and temporal scales and some aspects of this variation have been considered in the context of global heliospheric structure. Most studies have concentrated on the response of the termination shock to relatively short ($\sim$ 180 day and less) variations in the solar wind ram pressure, often associated with interplanetary shocks (see [1] for references). These results, which are 1D, have been extended in a limited way to 2D by *Steinolfson* [2] and *Wang & Belcher* [3]. Less attention has been devoted to determining the global structure of the heliosphere as it responds to either long-term variation in ram pressure or very long-lived spatial variation.

As has been discussed extensively, the inclusion of interstellar neutral hydrogen (H) has a profound effect on global heliospheric structure. Of particular importance to time-dependent models of the global heliosphere is the role of pickup ions in the outer heliosphere. As discussed by numerous authors, pickup ions exert a profound influence on the dynamics of the steady solar wind, causing it to decelerate and thereby reduce the ram pressure of the wind (see [1] for an exhaustive review). Observations of the solar wind by the Voyager spacecraft [4] indicate that the ram pressure (ρu^2, where ρ denotes the solar wind density and u the radial solar wind speed) varies by a factor of 2 with solar cycle. This observation prompted *Karmesin et al.* [5] and *Wang & Belcher* [3] to consider the response of the global 2D heliosphere to variations in solar wind ram pressure with a 22 year period. *Karmesin et al.* [5] did not include interstellar neutrals in their simulation, and this was redressed by *Wang & Belcher* [3], who modeled neutrals as a second fluid following the approach developed by *Pauls et al.* [6]. Here, we consider more carefully the implications of the ram pressure varying with solar cycle on global heliospheric structure, using the somewhat more sophisticated description of the neutral component developed by *Zank et al.* [7]. This multi-fluid neutral description includes the neutral heating of the outer heliosheath regions, which is necessary to determine the bow shock stand-off distance. The use of the multi-fluid description allows us to consider the distribution of neutral hydrogen throughout the heliosphere, which was not discussed at all in [3].

SIMULATION MODEL

The heliospheric-LISM plasma environment is composed of three thermodynamically distinct regions: (i) the supersonic solar wind, with a relatively low temperature, large radial speeds, and low densities; (ii) the shock-heated subsonic solar wind with much higher temperatures and densities, and lower flow speeds, and (iii) the LISM, where the plasma flow speed and temperature is low. As discussed in detail in [7], each of the thermodynamically distinct regions contributes a distinct population of neutral atoms produced by charge exchange with the ambient plasma and neutrals entering the region. The self-consistent inclusion of neutral hydrogen in models of the solar wind-LISM interaction is fundamental to understanding the large-scale structure of the heliosphere.

Two basic classes of model have been developed to describe neutral H in and around the heliosphere: multi-fluid models of varying degrees of sophistication [3, 6–

CP679, *Solar Wind Ten: Proceedings of the Tenth International Solar Wind Conference*, edited by M. Velli, R. Bruno, and F. Malara

10] which treat the neutral atoms as a multi-fluid, and kinetic models which solve the neutral atom kinetic equation, either by a Monte-Carlo technique [11, 12], or by a particle-mesh method [13, 14]. The long charge exchange mean free path for neutral hydrogen may mean that the fluid description for neutrals within the heliosphere is not completely justifiable. Multi-fluid and Boltzmann models differ in the detailed predictions that each admits for the neutral atom distribution and this can lead to 10-15% differences in predicted neutral H densities and temperatures within the heliosphere. Nevertheless, the basic morphological predictions of both models remain the same.

The kinetic codes, when coupled self-consistently to the background solar wind and LISM plasma, are computationally intensive, and so both kinetic approaches are restricted to steady-state conditions so far. However, the solar wind properties vary on an 11-year scale. Since the time for a neutral atom to enter the heliosphere and reach 1 AU is 15-20 years, the local neutral atom distribution has experienced both a variable charge exchange and photoionization rate, as well as a supersonic solar wind whose extent, velocity and density is highly variable. Thus, the problem of neutral atom interaction with the solar wind is inherently time-dependent and non-stationary. Neutral atom characteristics will therefore depend on solar cycle, with the overall distribution being a mixture of atoms created in temporally different solar wind environments, since they cannot be lost to the system on time scales shorter than the solar cycle. This requires a time dependent approach to the modeling of the solar wind-LISM neutral atom interaction. Accordingly, since the multi-fluid models provide a computationally feasible approach to investigating time-dependent problems in the solar wind, we adopt this approach here.

In the simulations discussed below, the long-term variation in solar wind ram pressure is introduced at the inner boundary (1 AU) sinusoidally by varying the flow speed. We used a base of 400 km/s, an amplitude of 83 km/s and a period of 11 years. This yielded a ram pressure which varied by a factor of 2.

RESULTS

For reasons of space, we describe only our results from a 2-shock model. In this case, the incident interstellar plasma flows supersonically onto the heliospheric obstacle. The interstellar flow velocity is set to 26 km/s, at a temperature of 8000 K. An interstellar plasma density of 0.1 cm^{-3} and neutral density of 0.14 cm^{-3} is assumed for the model. Corresponding numbers for the solar wind at 1 AU are 5 cm^{-3} and 10^5 K, respectively. The steady-state 2-shock heliosphere is depicted in Figure 1. The figure is a 2D plot of the plasma temperature showing the labeled boundaries. The extent of the heliosphere in the nose direction (the "stagnation axis") is ~ 80 AU to the termination shock, ~ 110 AU to the heliopause, and ~ 230 AU to the bow shock.

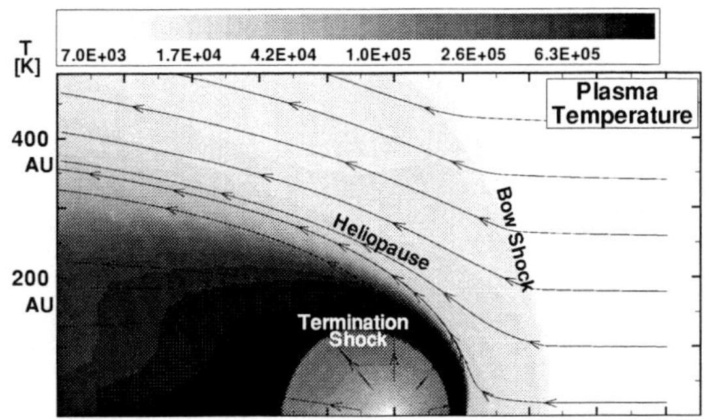

FIGURE 1. The 2D steady-state, 2-shock heliosphere showing the temperature distribution of the solar wind and interstellar plasma, together with velocity streamlines. The plasma boundaries, termination shock, heliopause, and bow shock are labeled. The plasma temperature is plotted logarithmically. The distances along the x and y axes are measured in astronomical units (AU).

Once the steady-state solution is computed, the inner boundary velocity at 1 AU is allowed to vary cyclically. Shortly after the initiation of the time-varying boundary condition, the heliosphere begins to exhibit an arrhythmic "breathing" as the termination shock begins to move in and out in response to the varying solar wind ram pressure. The arrhythmic motion of the termination shock is a consequence of the steady state heliosphere itself being asymmetric, and the local outward and inward motion of the TS being asymmetric. We note that, since interstellar neutrals reduce the asymmetry in the global structure compared to models that do not include neutral H [2, 5], the "breathing" of the heliosphere is less arrhythmic than in the corresponding plasma-only case. Also, the movement of the termination shock in the self-consistent case is somewhat smaller (~ 10 AU in the nose direction) than is exhibited in the plasma-only model. This is due to the mediation of the solar wind velocity, and hence ram pressure, by pickup ions.

The motion of the termination shock drives pressure waves into the inner heliosheath (the region between the termination shock and heliopause) which steepen as they propagate. Depending on the distance to the heliopause, the pressure waves may eventually steepen sufficiently to form weak shocks which then collide with the heliopause. However, the impact of the additional pressure on the heliopause leads to the transmission and/or emission of weak shocks into the outer heliosheath. This is illustrated in Figure 2, where differences in the plasma temperature are seen both in the heliotail and the in-

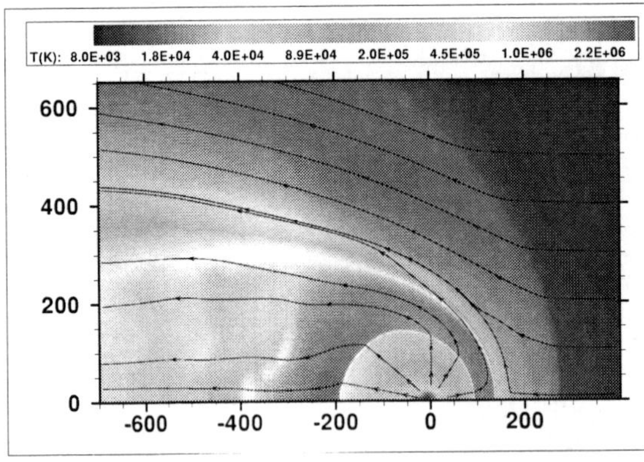

FIGURE 2. A snapshot during the time-dependent solar wind simulation. Note the presence of transmitted shocks beyond the heliopause and in the heliotail.

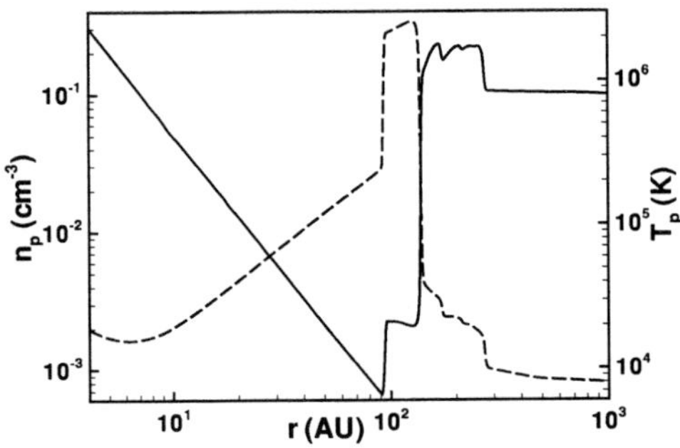

FIGURE 3. A 1D profile of the plasma density (solid) and temperature (dashed) as a function of radial heliocentric distance for Figure 2 along the stagnation axis in the upstream direction. Observe the train of shocks in the outer heliosheath (between the heliopause and bow shock).

ner and outer heliosheath (the lighter shading of the colors). The shocks transmitted across the heliopause in the upwind direction are more easily distinguished in a 1D cut along the stagnation axis, Figure 3, showing density and temperature, where the train of driven shocks is clearly apparent. The dynamical pressure associated with the train of shocks acts to communicate the varying solar wind ram pressure (although on a very much longer timescale) to the LISM and pushes the bow shock further out than in the steady-state case. This, of course, reflects the result that the solar wind ram pressure, suitably averaged, is larger than that of the steady-state wind. The steady bow shock is located at about 230 AU and the time-dependent bow shock at about 270 AU.

The time-dependent response of the heliosphere to dynamical variations in ram pressure is illustrated in Figure 4. Figure 4 is a color plot of the plasma temperature along the stagnation axis (upwind and downwind directions) as a function of time/phase. Since Figure 4 is a space-time diagram, the dynamical response of the boundaries is exhibited clearly and structures such as transmitted shocks or driven pressure waves which heat the plasma as they propagate can be discerned easily. The most striking feature of Figure 4 is the enormous excursions made by the termination shock in the downwind direction about its mean, steady-state location. The downwind termination shock can move by as much as $\sim$ 50 AU compared to $\sim$ 10AU in the upwind direction over a solar cycle. The larger downwind than upwind termination shock movement is a consequence of the additional interstellar ram pressure in the upwind direction. Within the orange band (hot plasma in the inner heliosheath) one can observe lighter colored streaks. The back of a light-colored streak corresponds to a steepening pressure pulse or shock front. The darker orange color corresponds to heliosheath material that has been compressed and heated by the driven shock wave. The light orange is therefore the ambient steady-state temperature of the heliosheath plasma in the absence of temporal solar wind driving. By comparing the temperatures (not the colors!) of Figures 1 and 2 or 3, the maximum temperature for the steady-state is $\sim 1.8 \times 10^6$ K, compared to $\sim 3.5 \times 10^6$ K for the temporal solar wind case. Clearly, both the upwind inner heliosheath and the heliotail within ~ -500 AU experience considerable additional heating by steepening pressure waves and shock waves driven by the solar cycle varying solar wind ram pressure. In the upwind direction, the bands continue beyond the inner heliosheath into the cooler outer heliosheath (shocked LISM plasma) but, as can be seen from the changed gradient of the bands, at a much slower speed (the light blue streaks now correspond to heated plasma). As illustrated in Figure 3, the slow structures are weak shock waves. The shock waves propagating into the heliosheath in the downwind direction also slow and weaken with increasing distance, as illustrated in Figure 4 by the slowly decaying exponential-like tracks. At any given time, multiple shocks can be present in the outer heliosheath and in the heliotail.

The solar cycle varying solar wind has some impact on the characteristics of neutral H entering the heliosphere. The increased separation distance between the BS and HP leads to a greater deceleration and heating of the hydrogen wall compared to its stationary counterpart. The wall amplitude is slightly larger (0.321 vs. 0.316 cm^{-3}) and the filtration of interstellar H is a little more pronounced. Both the neutral H density and temperature exhibit variability beyond 20 AU. The 1D stagnation axis profiles of Figure 5 show neutral density and temperature for two phases within a solar cycle. The change in neutral H temperature about half a period apart is most remarkable, and reflects the change in heliosheath and ISM plasma temperature due to the propagation of weak shocks. The front of the hydrogen wall, located approximately at the heliopause, exhibits some variation. Propagating

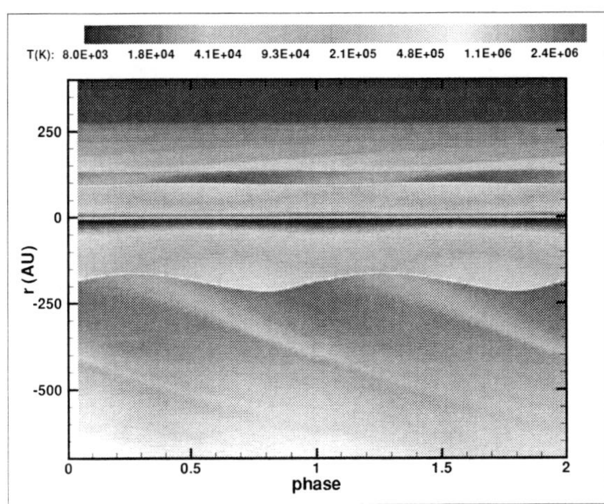

FIGURE 4. "Space-time" plot of the plasma temperature along the stagnation axis (upwind and downwind directions) over two solar cycles, illustrating the response of the termination shock, heliopause, and bow shock to variation in solar wind ram pressure. Note the much larger excursions of the termination shock in the downwind direction than in the upwind, the smaller response of the heliopause than the termination shock, and the formation of pressure waves and weak shocks driven by the motion of the termination shock.

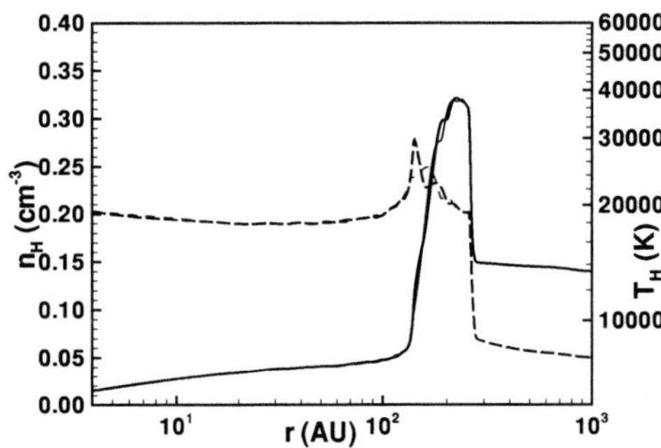

FIGURE 5. 1D plot of neutral hydrogen density (solid) and temperature (dashed) along the stagnation axis, for two representative phases within a solar cycle.

structures can also be seen, moving slowly outward toward the bow shock but decaying before then. These are the result of enhanced charge exchange due to the higher density and temperature of the plasma at and behind the weak transmitted shocks.

CONCLUSIONS

On the basis of a multi-fluid description of the solar wind-LISM interaction, we have investigated the structure of the global heliosphere when the observed solar cycle variation in ram pressure is included. We describe here only results obtained for a 2-shock model. The varying ram pressure moves the termination shock ~ 10 AU in the upwind direction and ~ 40 - 50 AU in the downwind direction, and the heliosphere itself exhibits a highly arrhythmic "breathing." Weak shocks are driven by the temporal solar wind, propagating into the outer heliosheath and heliotail, and acting to further heat these regions. The neutral hydrogen distribution is also found to vary beyond ~ 20 AU in its density and temperature.

ACKNOWLEDGMENTS

This work was supported in part by an NSF-DOE grant ATM-0296114 and a NASA grant NAG5-611621.

REFERENCES

1. Zank, G.P., *Space Sci. Rev.*, **89**, 413-688 (1999).
2. Steinolfson, R.S., *J. Geophys. Res.*, **99**, 13307 (1994).
3. Wang, C., and J.W. Belcher, *J. Geophys. Res.*, **103**, 247 (1998).
4. Lazarus, A.J. and R.L. McNutt, Jr., in *Physics of the Outer Heliosphere*,229, ed.'s S. Grzedzielski and D.E. Page, Pergamon (1990).
5. Karmesin, S.R., P.C. Liewer, and J.U. Brackbill, *Geophys. Res. Lett.*, **22**, 1153 (1995).
6. Pauls, H.L., G.P. Zank, and L.L. Williams, *J. Geophys. Res.*, **100**, 21595 (1995).
7. Zank, G.P, H.L. Pauls, L.L. Williams, and D.T. Hall, *J. Geophys. Res.*, **101**, 21,639 (1996).
8. Liewer, P.C., S.R. Karmesin, and J.U. Brackbill, *J. Geophys. Res.*, **101**, 17119 (1996).
9. Fahr, H. J. and Kausch, T. and Scherer, H., *Astron. Astrophys.*, **357**, 268-282 (2000).
10. Florinski, V., Zank, G.P., and Pogorelov, N.V., *Proc. Solar Wind 10*, this issue (2002).
11. Baranov, V.B., and Y.G. Malama, *J. Geophys. Res.*, **98**, 15157 (1993).
12. Izmodenov, V.V., J. Geiss, R. Lallement, G. Gloeckler, V.B. Baranov, and Y.G. Malama, *J. Geophys. Res.*, **104**, 4731-4741 (1999).
13. Lipatov, A.S., Zank, G.P. and Pauls, H.L., *J. Geophys. Res.*, **103**, 20,631, 1998.
14. Müller,H.R., Zank, G.P. and Lipatov, A.S., *J. Geophys. Res.*, **105**, 27,419-27,438 (2000).

Coherence Lengths of the Interplanetary Electric Field: Solar Cycle Maximum Conditions

Charles J. Farrugia, Hiroshi Matsui, Roy B. Torbert

Space Science Center, University of New Hampshire, Durham, NH.

Abstract.
It is increasingly being realized that by affecting geoeffective scale lengths the interplanetary electric field (IEF) is a key quantity in space weather discussions. In this work we derive and analyze statistically IEF coherence lengths in year 2000, i.e., near maximum of solar cycle 23, working in a much used formulation for the IEF. We focus on the frequency domain. We use magnetic field and plasma data sets acquired by Wind and ACE. During year 2000, ACE-Wind separations were very variable and, in particular, Wind's first dayside distant prograde orbit resulted in a Y-separation comparable to the X-separation ($\sim$220 R_E). We find IEF coherence lengths of 200-250 R_E (X) and 50-100 R_E (Y). The coherence is mainly carried by the low frequency components ($f < 0.01$ min^{-1}).

INTRODUCTION

The interplanetary electric field (IEF) plays an increasing important role in both theoretical and observational discussions of the geoeffectiveness of interplanetary configurations, i.e., the level of geomagnetic disturbances which they elicit inside the magnetosphere. For example, it is thought that the transpolar potential, but not the Dst, saturates for large IEF ($>$ 3 mV m^{-1}) (13). It follows that a knowledge of the coherence lengths of the IEF parallel and perpendicular to the Sun-Earth line is a crucial quantity in space weather considerations. Yet, no studies addressing these issues have been attempted to date. The need for such investigations was highlighted by (2) in their study of geoeffective interplanetary scale sizes.

In this paper we examine coordinated observations made by Wind and ACE in year 2000. This year is chosen because: (i) it is close to the maximum of solar cycle 23 when interplanetary configurations are expected to be strongly geoeffective on average; (ii) data coverage at both spacecraft is optimal; (iii) The orbit of Wind relative to ACE allows a study of the IEF as functions of both X and Y, the latter varying over $\sim$ 500 Earth radii, R_E. (By comparison, coordinate Z is small.)

As methodology we adopt the approach taken in a pilot study by (8). Rather than cross-correlating two time series using various time windows, they decomposed the signals into their Fourier components and could thus examine the level of coherence at the two observing sites also as a function of frequency. Spectral analysis has distinct advantages over the more common analysis in the time domain (8). We work in a formulation of the IEF derived from considerations of the maximum merging rate at the dayside magnetopause (15,6): IEF = $VB_T sin^2(\theta/2)$, where $B_T \equiv (Bygsm^2 + Bzgsm^2)^{1/2}$, V is the bulk speed, and θ is the IMF clock angle (i.e., the polar angle in the GSM YZ plane.)

We show a systematic decrease in the coherence level with the east-west separation ΔY during Wind's first dayside distant prograde orbit (DPO). From this we obtain the coherence length of the IEF in the east - west direction. We also examine the coherence level of the IEF with X, analysing data from the early part of the year when Wind made an excursion to the L1 Lagrangian point.

WIND AND ACE ORBITS IN YEAR 2000

Figure 1a shows the position of Wind during year 2000 in GSE coordinates. In January, Wind was executing orbits mostly in the geomagnetic tail. These data are not used further. During February-March, Wind made an excursion to the neighbourhood of L1. Then follow dayside and nightside, high latitude Petal orbits (April - August). From mid-August to the end of the year Wind starts its dayside DPOs. The months October - December contain one-half of such an orbit, taking Wind from -250 R_E to 250 R_E. The spacecraft spends relatively long times at the "apex" of the DPO. The DPOs afford a unique opportunity to study coherence lengths in the Y-direction over a long baseline. The resulting ACE-Wind separations are plotted in Figure 1b. The excursion by Wind to near L1 gives a range of 200 R_E in the Sun-Earth

CP679, *Solar Wind Ten: Proceedings of the Tenth International Solar Wind Conference,*
edited by M. Velli, R. Bruno, and F. Malara

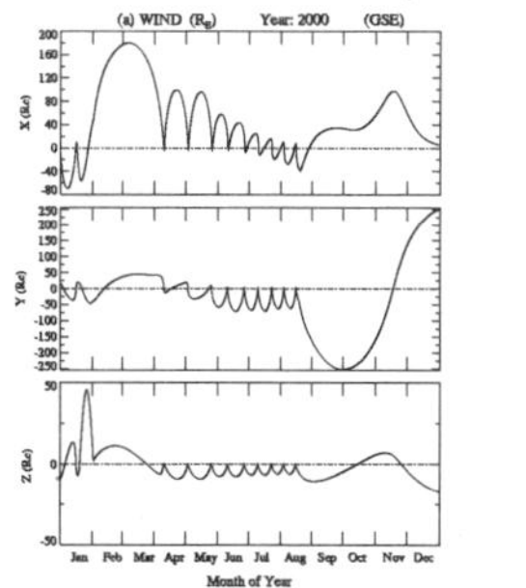
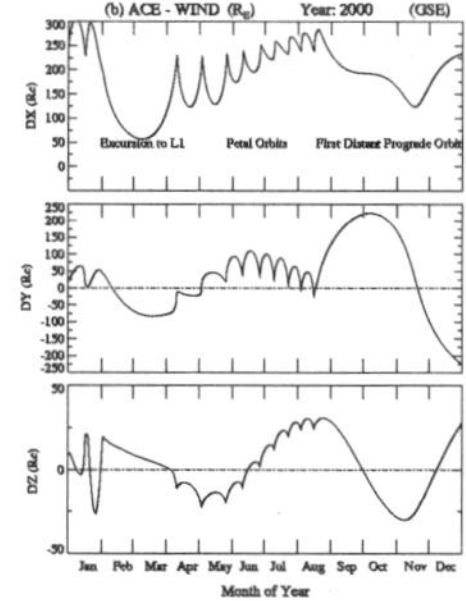

FIGURE 1. 1a: Position of Wind in year 2000. 1b: Wind-ACE separation in year 2000.

direction X. The first DPO gives a Wind-ACE separation of the same magnitude both to the east and west of Earth.

The magnetic field data were acquired by the Magnetic Field Investigation (MFI) on Wind (7) and the Magnetic Field Experiment (MAG) on ACE (14) with time resolutions of 90 s and 16 s, respectively. The plasma data are from the Solar Wind Experiment (SWE) on Wind (10) and the Solar Wind Electron Proton Alpha Monitor (SWEPAM) on ACE (9) with time resolutions of 94 s and 64 s, respectively. (These data are courtesy of the NASA CDA Web site.) The data resolution is equalized by linear interpolation. We employ a time resolution of 1 min. Data spikes have been removed before the analysis. For the magnetic field we remove those data points which are larger than neighboring points by 5 nT or more. Similar criteria are adopted for the velocity (a jump of 50 km s^{-1}). We interpolated only data gaps which are shorter than, or equal to, 10 data points.

A NOTE ON THE FOURIER TECHNIQUE

This is described in detail in (8). At each spacecraft we calculate the Fourier amplitude and phase of the IEF by FFT (11). From this we derive (i) the amplitude ratio, $R(f)$, (ii) the coherence, $C(f)$, and (iii) the phase lag, $\Delta\phi(f)$ between the two spacecraft as a function of frequency f. These three analysis quantities are defined as follows:

$$R(f) = \sqrt{\frac{\Sigma_{i=0}^{3}|Q_{i,ACE}(f)|^2}{\Sigma_{i=0}^{3}|Q_{i,Wind}(f)|^2}}, \quad (1)$$

where $Q_{i,Wind}(f)$ and $Q_{i,ACE}(f)$ correspond to the Fourier components of IEF, respectively. The summation is over four FFTs (see below). The quantities returned by the FFT routine are complex, calculated for each frequency and for each time.

$$C(f) = \frac{|\Sigma_{i=0}^{3}(Q_{i,ACE}(f)Q_{i,Wind}^{*}(f))|^2}{(\Sigma_{i=0}^{3}|Q_{i,ACE}(f)|^2)(\Sigma_{i=0}^{3}|Q_{i,Wind}(f)|^2)}, \quad (2)$$

where * denotes complex conjugation.

$$\Delta\phi(f) = \arg(\sum_{i=0}^{3}(Q_{i,ACE}(f)Q_{i,Wind}^{*}(f))). \quad (3)$$

We set the number, n, of samples per FFT as 512, corresponding to 512 min per FFT. Data have been weighted by a Hamming window. To obtain $C(f)$ we need results from multiple FFTs (e.g., 3, 5). Increasing the number of FFTs reduces random errors (1) but at the expense of time resolution. In addition, the assumption of time stationary conditions inherent in the method is more likely to be violated when many FFTs are summed. In this analysis we use 4 FFTs, the same number which we used successfully in an analysis of IMF and solar wind parameters (8). Neighboring FFTs are overlapped by one-half the number of samples per FFT. One cross-spectrum is thus based on 1280 samples per FFT.

The coherence is similar to the correlation of the phases between $Q_{i,Wind}(f)$ and $Q_{i,ACE}(f)$. When the phases of the fluctuations of the quantity are randomly distributed, the numerator of equation (2) is 0, which indicates lack of coherence. When the phase lag between $Q_{i,Wind}(f)$ and $Q_{i,ACE}(f)$ is constant, $C(f)$ is 1.

The phase lag $\Delta\phi(f)$ contains information about the propagation of the IEF between spacecraft. This value is significant only when the coherence is high or, in other words, when the correlation of the phases is high. Consider a front propagating with constant velocity, V. If the normal to the structure is parallel to the X axis, the relation between the phase lag and the frequency is

$$\Delta\phi(f) = 360 \cdot \frac{\Delta X}{V_X} f \text{ (deg.)}, \quad (4)$$

where ΔX is the separation between the two spacecraft in the X direction. Quantity $\Delta X/V_X$ corresponds to the convection delay time between the two spacecraft.

ILLUSTRATION OF THE METHOD

Figure 2 shows frequency-time spectra of IEF for October – December, 2000. The vertical axis gives the frequency in min^{-1}. The panels show from top to bottom the Fourier amplitudes at Wind and ACE; the amplitude ratio, $R(f)$, between Wind and ACE; the coherence, $C(f)$; and (e) the phase lag, $\Delta\phi(f)$. The scale of the amplitudes shown on the right side is in units of

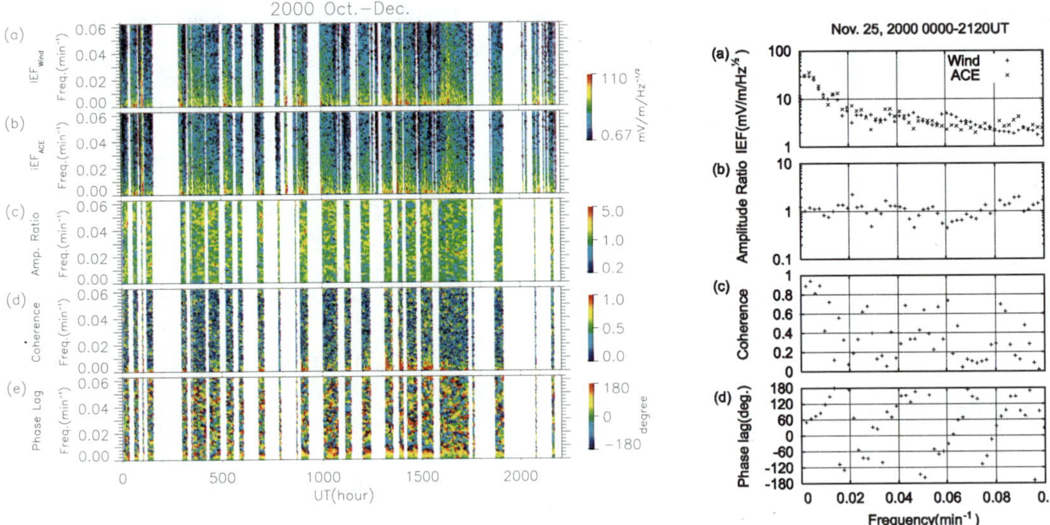

FIGURE 2. Frequency-time spectrograms for October-December, 2000. For further details, see text.

FIGURE 3. November 25, 2000: The IEF at Wind and ACE, the amplitude ratio, the coherence and the phase lag.

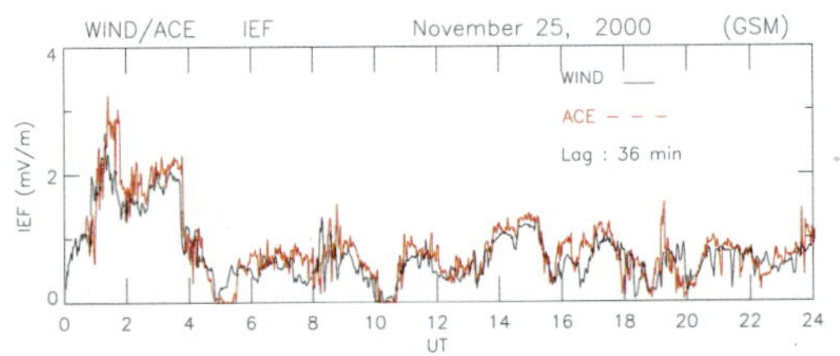

FIGURE 4. a superposition of the ACE (red) and Wind IEFs for November 25, 2000 with a 30 min propagation delay included.

mV/m/$\sqrt{\text{Hz}}$. One can make the following points. Typically, the coherence is high only at low frequencies, $f < 0.01$ min^{-1}. In November, with ΔY passing through 0, the coherence is generally high up to $f \leq 0.03$ min^{-1}. The phase lag panel shows a green band corresponding to no phase lag at the lowest frequencies. Then a yellow band at somewhat higher frequencies. Above f = 0.02 min^{-1}, the phase lag is random, indicating that it is difficult to predict the arrival time at Wind of high frequency components of the IEF observed at ACE.

Figure 3 shows an example of high coherence on November 25 (hours 1344-1368) and compares the spectral analysis technique with the time series analysis. The average positions of ACE and Wind were (224, 35, -17) R_E and (80, 82, 0.5) R_E, respectively, and the solar wind speed $\approx$ 400 km s^{-1}. The coherence > 0.6 for $f < 0.015$ min^{-1} . An approximately linear phase lag plot (bottom panel) below this frequency can be seen. The propagation velocity estimated from this phase lag is 30 min. Compare this with the convection delay X/V_x=38 min. Analyzing the two time series, we find that the lag at maximum correlation, R (= 0.88) = 36 min. In Figure 4 we overlap the two electric fields (ACE in light trace) for November 25, 2000 taking into account a 36 min delay. Thus the quantities are comparable and there is self-consistency between the analyses in the frequency and time domains.

By contrast poor coherence was achieved on October 6 when ACE and WIND were located at (224, -26, -10) R_E and (32, -250, -3) R_E, respectively, i.e. they had a an X- and Y-separation each $\sim$200 R_E. The low coherence is probably due to this large separation.

STATISTICAL RESULTS

We now summarize the main results of the statistical analysis on all year 2000 data acquired in the solar wind. Figure 5 shows the dependence of the coherence of IEF on period (frequency). As the period increases (the frequency decreases), coherence values increase. Values exceeding 0.6 are reached for periods of $\sim$150 min or longer (f $\sim$ 0.007 min^{-1} or lower).

Figure 6 gives the dependence of coherence on the X-separation between Wind and ACE. Calculations for 3 different times are shown as indicated. Highest correlation correspond to longer times (256 min). For this trace, the coherence decreases with X. (The range $\Delta X < 50 R_E$ is an exception because it is affected by the large ΔY, see Figures 1a, b.) Adopting a definition whereby the coherence length scale corresponds to the distance when the coherence decreases by 0.1 from its highest value (12,

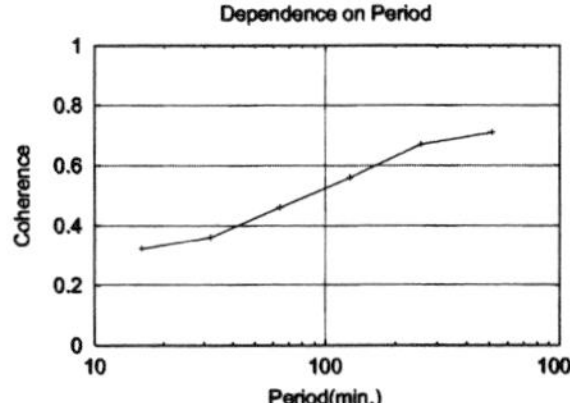

FIGURE 5. Dependence of the coherence of the IEF on period (frequency)

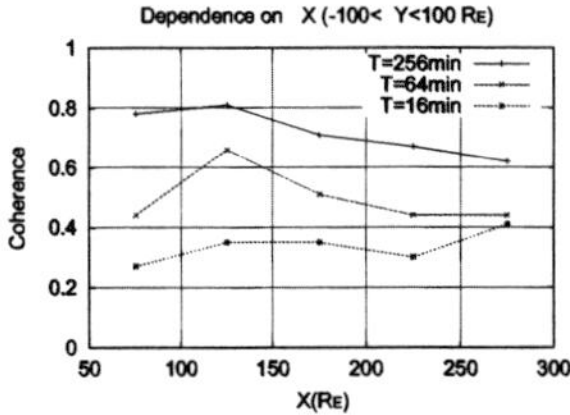

FIGURE 6. Dependence of the coherence of the IEF on X.

8), we obtain that the coherence length of IEF in the X-direction is of order 200-250 Re.

Figure 7 summarizes the statistical dependence of the coherence on the separation between Wind and ACE in the Y-direction. A coherence length of IEF in the Y-direction of 50-100 Re is indicated. Here typically the coherence peaks at a negative value of Y and then drops off, i.e., it is not symmetric about $\Delta Y = 0$ (for T = 256 min). The offset to negative GSE Y (west) may have several causes: (i) aberration effect due to the motion of the Earth; (ii) the Parker spiral orientation of the IMF; (iii) encounter with highly coherent structures preferentially on one side of the Earth-Sun line on this DPO.

CONCLUSIONS

Our statistical study over 1 year of joint ACE-Wind observations constitutes the first calculations of the coherence length of a quantity so central to discussions on the interaction of the solar wind with the magnetosphere: the interplanetary electric field, IEF. Utilizing Wind's distant

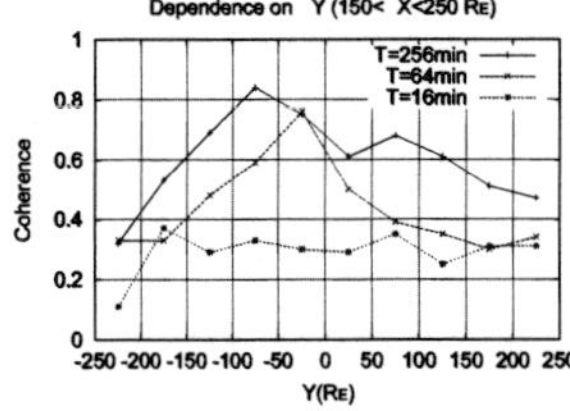

FIGURE 7. Dependence of the coherence of the IEF on Y.

prograde orbits, which yield an ACE-Wind Y-separation comparable to the X-separation ($\sim$ 220 R_E) we carried out a statistical study using all solar wind data returned by Wind and ACE in year 2000.

We find two different scale lengths for the IEF: 200 - 250 R_E in the X-, and 50-100 R_E in the Y-directions. In practical terms, this means that if a monitor orbiting L1, say, is displaced from the Sun-Earth line by less than 50 R_E, there is a high probability for the IEF it measures to be equal to that at the magnetopause nose. The coherence lengths of the IEF, in both X and Y, are similar to those of the IMF (8, and references therein).

As a function of frequency, the general behavior is for there to be good coherence at low frequencies. The coherence is invariably lost at the higher frequencies. Similar conclusions as to the frequency dependence of the coherence of the IMF and solar wind parameters were reached by (8). That work showed that the power spectral densities follow a -5/3 law in frequency, indicative of turbulence. This may also be the reason here.

ACKNOWLEDGMENTS

We thank the PIs of the magnetic field and plasma instruments on Wind and ACE for use of key parameter data from their instruments. This work is supported by the Wind Grant NAG5-11803, and NASA Living with a Star Grant NAG5 - 10883.

REFERENCES

1. Benignus, V. A., *IEEE Trans. Audio Electroacoust., 17,* 145–150, 1969.
2. Burke, W. J., et al., *J. Geophys. Res., 104,* 9989, 1999.
3. Eriksson, A. I., in *Analysis Methods for Multi-Spacecraft Data, ISSI Sci. Rep. SR-001,* pp. 5–42, ESA Publications, Noordwijk, Netherlands, 1998.
4. Farrugia, C. J., et al., *J. Geophys. Res., 107*(A9), 1240, doi:10.1029/2001JA000188, 2002.
5. Holmgren, G, and P. M. Kintner, *J. Geophys. Res., 95,* 6015, 1990.
6. Kan, J. R. and Lee, L. C., *Geophys. Res. Lett.,* 6, 577, 1979.
7. Lepping, R. P., et al., *Space Sci. Rev., 71,* 207, 1995.
8. Matsui, H., et al., *J. Geophys. Res., 107*(A11), 1355, doi:10.1029/2002JA009251, 2002.
9. McComas, D. J., et al., *Space Sci. Rev., 86,* 563–612, 1998.
10. Ogilvie, K. W., et al., *Space Sci. Rev., 71,* 55, 1995.
11. Press, W. H., et al., *Numerical recipes in C,* Cambridge Univ. Press, 2nd ed., Cambridge, 1992.
12. Richardson, J. D., and K. I. Paularena, *J. Geophys. Res., 106,* 239, 2001.
13. Russell, C. T., et al., *Space Sci. Rev., 71,* 563, 1995.
14. Smith, C. W., et al., *Space Sci. Rev., 86,* 613–632, 1998.
15. Sonnerup, B. U. Ö, *J. Geophys. Res., 79,* 1546, 1974.

Long-distance Correlations of Interplanetary Parameters: A Case Study with HELIOS

H. Matsui, C. J. Farrugia, H. Kucharek*, D. Berdichevsky†, R. B. Torbert, V. K. Jordanova*, I. G. Richardson†**, A. B. Galvin*, R. P. Lepping† and R. Schwenn‡

*Space Science Center, University of New Hampshire, Durham, NH 03824
†NASA Goddard Space Flight Center, Greenbelt, MD 20771
**Also at: Department of Astronomy, University of Maryland, MD
‡Max Planck Institut für Aeronomie, Katlenburg-Lindau, Germany

Abstract.
In recent work, promising agreement has been obtained between measured indices of geomagnetic activity (Dst, and cross-polar cap potential) and their predicted values using interplanetary input from probes in the inner heliosphere ($\sim$0.7 AU) when the probe was close to, (5), and even substantially displaced from, (4), the Earth-Sun line. Implicit in this agreement is a good correlation of, at least, the basic temporal profiles of the major interplanetary parameters at the two observing sites. In this work we discuss a case study using Helios 1 and 2 data when the spacecraft are lined - up and separated by an almost constant radial distance of 0.2 AU. In the period studied, the interplanetary medium consists of a fast stream being trailed by a magnetic cloud in a slower flow. Good correlation is found between the plasma and field observations at the two sites. Two lag times, reflecting the two types of major structures in the interval chosen, are determined. Evidence of evolutionary processes are briefly discussed. Spectral analysis confirms the results obtained from time series analysis.

INTRODUCTION AND AIMS OF THE STUDY

Together with accuracy, a long lead time is a desideratum of space weather predictions. In recent years good prediction of the temporal variation of the Dst index and cross-polar cap potential has been achieved using solar wind measurements in the inner heliosphere. Thus (5) could predict many features and long–term trends of the storm-time Dst index with (2)'s empirical formula, using data from monitors located at $\sim$0.7 AU and close to the Sun-Earth line at low heliographic latitudes. This would yield a lead time of $\sim$1 day. In a study on the effect of heliographic longitude on the quality of predictions, (4) examined a disturbed period in March 1979. They used a drift-loss model for the energization of the ring current (e.g., 3) with interplanetary inputs from 2 probes: Helios 2, at 0.7 AU and displaced east of the Sun-Earth line by 30°, and from ISEE 3, orbiting the L1 Lagrangian point. The temporal variation of the Dst measurements during three storms could be well reproduced in both cases. Further, the cross-polar cap potential calculated from (7)'s model showed a similar variation in both cases. Such results are good news for space weather efforts.

Implicit in this good agreement is a long coherence/correlation scale length of at least the gross features of major interplanetary parameters (such as V, B_z, B, etc.) In this work we wish to pursue this issue further. To this end, we investigate a 10-day period where the Helios 1 (H1) and Helios 2 (H2) probes were approximately lined-up and separated by $\sim$0.2 AU. The questions we pose are: (i) Are the basic features of the solar wind well correlated 0.2 AU downstream? (ii) Do we find systematic signatures of evolution?

The solar wind segment we study consisted of two different interplanetary structures moving at different speeds. After cross-correlating various field and plasma parameters, we identify two different lag times, each reflecting these two major structures. Good correlation is found between the measurements at the two probes, indicating that the integrity of the solar wind structures is preserved, at least in this instance. We further determine empirical factors by which parameters scale with radial distance. Elements of spectral analysis are used to confirm and extend the results of the analysis in the time domain.

CP679, *Solar Wind Ten: Proceedings of the Tenth International Solar Wind Conference*,
edited by M. Velli, R. Bruno, and F. Malara

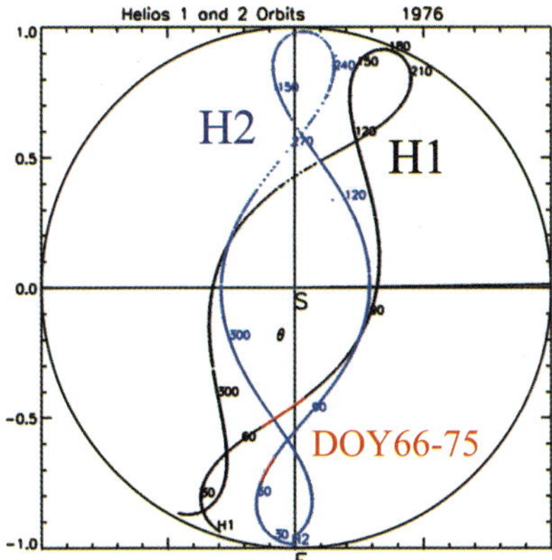

FIGURE 1. Ecliptic projections of the orbits of Helios 1 and 2 in year 1976. The Earth-Sun line is fixed. The red segments mark the 10-day interval analyzed here.

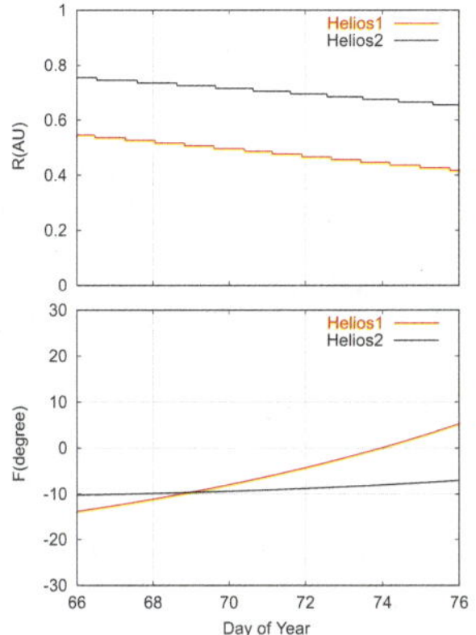

FIGURE 2. The heliographic radii of the spacecraft (top) and their longitudinal separation.

HELIOS 1 AND 2 OBSERVATIONS ON DOY 66-75, 1976

Figure 1 shows an ecliptic projection of the Helios orbits in 1976. The Earth (E) - Sun (S) line is fixed. The red segments indicate the selected 10-day study period during which the spacecraft are approximately lined up. The spacecraft separation varies between $R = 0.21$ and 0.24 AU, and their longitudinal separation, ΔF, from -3.6 to 12.3°, as can be seen in Figure 2 where we plot the heliospheric radii of the probes (top panel) and the difference in their longitudes.

Figure 3 shows the variation of select parameters, namely, the density, n_p, bulk speed, V_p, total field B, its z-component, B_z in solar ecliptic (SE) coordinates, and proton temperature T_p. Visually, it is clear that both spacecraft observe the same structures but shifted in time (see below). This solar wind segment consists of two main structures: (i) a high frequency oscillation lasting many days riding on a fast ($\sim$ 700 km s^{-1}) solar wind stream. These are most probably Alfvén waves. (ii) A

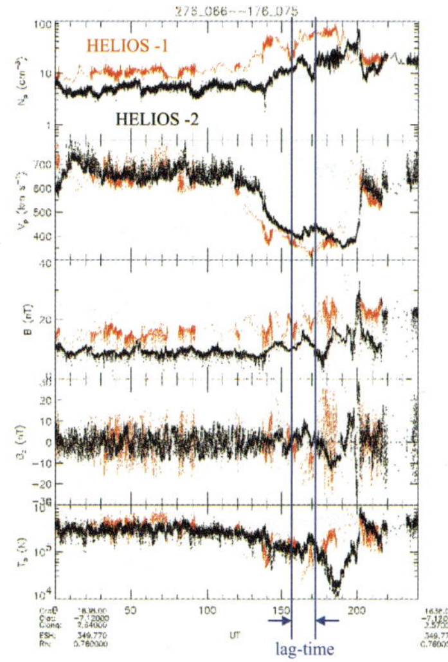

FIGURE 3. The variation of the density, n_p, bulk speed, V_p, total field B, z-component of the field, B_z in solar ecliptic (SE) coordinates, and proton temperature T_p at H1 (red trace) and H2 over a 10-day period.

low-temperature, enhanced-field region embedded in a depression of the bulk flow speed. In interval (ii) B_z and other magnetic field components (not shown) execute smooth rotations. Taken together, observations in (ii) are signatures of a magnetic cloud (1).

In Figure 4 we overlay the temporal profiles. A time shift of 17 hours aligns the magnetic signatures of the magnetic cloud well. The leading edge of the high speed stream trailing the cloud is overtaking the ejecta, as evidenced by the mismatch at the speed gradient. We may obtain how these parameters scale with heliospheric radius, R, treating the high-speed stream and the magnetic cloud separately. For the high speed stream between hours 17 and 140 in the figure we have: $n_p \sim R^{-2.1\pm0.2}$ $B, B_z \sim R^{-1.5\pm0.4}$, while V_p and T_p are approximately the same. For the magnetic cloud between hours 180 and 190: $n_p \sim R^{-2.6\pm0.4}$, $B \sim R^{0.9\pm0.5}$, and $T_p \sim R^{-2.2\pm1.8}$, the latter reflecting a cooling presumably due to the radial expansion of the magnetic cloud.

CROSS-CORRELATION ANALYSIS

We first have to equalize the time resolutions of the H1 and H2 data sets. We interpolate and/or average the data to a uniform 10 minute resolution. Data are analyzed in the time domain first and then in the frequency domain. The two approaches are complementary to each other. In the time domain, data are shifted in time in order to determine the lag corresponding to maximum correlation

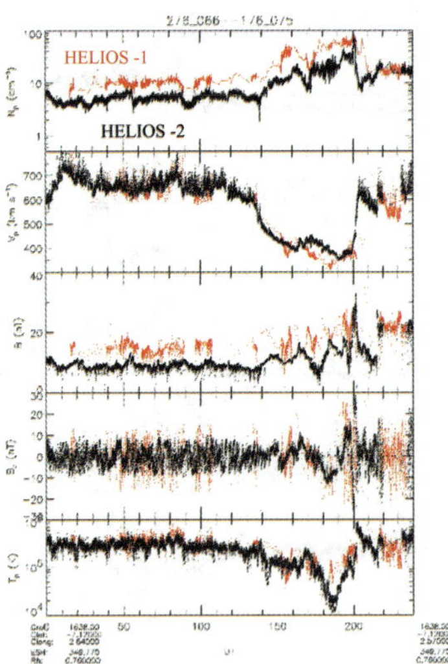

FIGURE 4. Same as Figure 3 but the universal time of H1 measurements have been delayed by 17 hours, this being the delay time we determine for the magnetic cloud signatures.

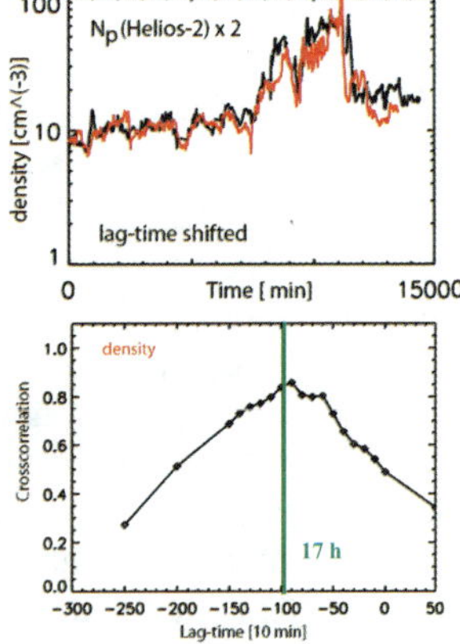

FIGURE 5. The correlation coefficient of n_p as a function of lag time (in units of 10 min) (bottom) and the superposition of the profiles.

between the spacecraft (see below).

As a first step we study the inter-spacecraft correlations of a plasma (n_p) and two magnetic field parameters (B, and B_z). The correlation coefficient of n_p is shown as a function of lag time (in units of 10 min) in the bottom panel of Figure 5. A peak cross-correlation of 0.85 is reached at a lag time of 17 hours. The resulting profiles are shown superposed in the top panel. Note that the density at H2 has been multiplied by a factor 2. The agreement is seen to be good. A second peak in the cross-correlation is present, corresponding to a lag of about 8 hours. We discuss this below when we consider correlations on B, where the two-peaked correlation plot is more pronounced.

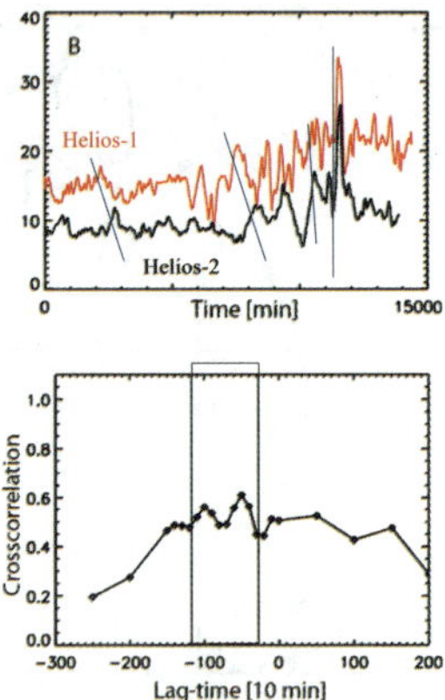

FIGURE 6. The correlation coefficient of B as a function of lag time. The correlation peaks at 2 different times, already evident in Figure 5. Top panel shows the superposition of the profiles with the longer delay time taken into account.

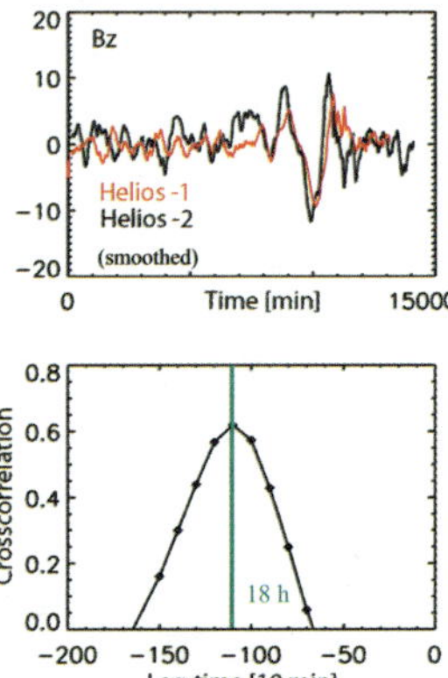

FIGURE 7. Similar to Figure 6 but for B_z.

Figure 6 shows the cross-correlation results for the total field, B. As the bottom panel shows, the routine picks out two peaks with the same lag times as those for the density. In the top panel we overlay the two time series using the longer time delay. This aligns the magnetic field in the magnetic cloud (indicated by the right vertical guideline). The slanted lines indicate correspondences of features in the high speed stream and confirms that these are subject to a different lag.

Figure 7 shows a similar calculation for the B_z component of the magnetic field. One peak, at a lag of $\sim$ 18 hours, is evident in the bottom panel. This corresponds to the large rotation of the field in the magnetic cloud. At this (low) resolution, the second, weaker peak corresponding to the oscillation in the high speed stream is not seen. Note that we use smoothed B_z data. This we believe is justified because, as we show below, the high frequency components lose coherence over very short distances and high coherence between two different points in the solar wind is carried by the low frequency components of the signals, i.e. their long-period variations.

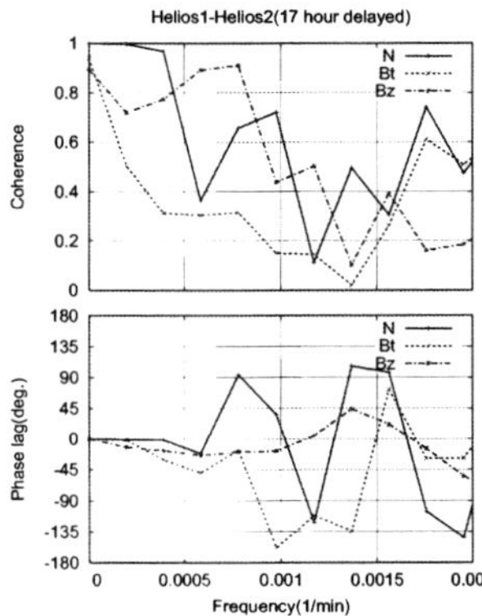

FIGURE 8. Coherence values for N_p, B_z and B as functions of frequency.

SPECTRAL ANALYSIS

We have also examined the data using a spectral analysis. The technique is described in detail in (6), to which we refer the reader. Typically one can obtain the coherence, the phase lag and amplitude ratio between the signals at the two spacecraft. In the interests of brevity, we discuss here the coherence only, reserving a fuller discussion to a future publication.

Figure 8 shows the coherence of N_p, B and B_z as functions of frequency. The data sampling interval is 10 min and the number of data points per FFT is chosen as 512, so that the time interval needed to obtain one spectrum is 5120 min (= 85.3 hours). In Figure 8, we have first shifted the H1 data forward by 17 hours, taking this result from the time series analysis reported above.

We may see that at low frequencies the coherence is relatively large for N_p and B_z, while it is small for B. The frequency at which the coherence drops below $\sim$0.5 is $f \sim 0.001$ min^{-1}. The coherence is being maintained by the low-frequency components of the signals, as shown in (6) and anticipated above.

When we examine the phase lag diagram (not shown) we find that there is a shallow linear gradient in the lowest frequency range ($\leq$ 0.0005 min $^{-1}$). The phase lag $\Delta\phi$ is related to the frequency by $\Delta\phi = 360 f \Delta t$ (deg), where the Δt is the propagation delay time. From the shallow gradient we infer an additional time shift of $\sim$1.5 hours to the lag estimated from time series. Thus the time- and frequency-analyses agree well with each other.

CONCLUSIONS

We have examined Helios 1 and 2 magnetic field and plasma data for a 10-day interval in 1976 when the spacecraft were approximately aligned and close to the Sun-Earth line. The solar wind segment we examined consisted of 2 structures: (i) large-amplitude waves in a high speed stream, and (ii) an interplanetary magnetic cloud embedded in a slower flow. Motivated by recent successes in predicting gross features of geomagnetic disturbances from interplanetary measurements made in the inner heliosphere, we concentrated on large separations (fraction of an AU). Here, Helios 2 was $\sim$0.2 AU downstream of Helios 1.

We pursued the questions: (i) Are basic features of the solar wind still correlated so far downstream? (ii) And do we find evidence of evolutionary signatures? The cross-correlation analysis gives an affirmative answer to the first question. It furthermore picks out two delay times which reflect the two main structures in the wind and their different evolution over 0.2 AU. We also determined empirical scaling factors of interplanetary parameters with heliospheric radius. We noted that these scaling factors are different in the fast stream and in the magnetic cloud. This latter point is a subject of future detailed study because of its relevance to space weather predictions from inner heliospheric probes.

Spectral analysis yielded lag times similar to those in the time series analysis. High signal coherence was reached for $f < 0.0005$ min $^{-1}$, which is the typical frequency of variation in the line plots. At higher frequencies the coherence is lost. As noted above, spectral analysis has increased frequency resolution but at the expense of time resolution.

ACKNOWLEDGMENTS

This work is supported by the Wind Grant NAG5-11803, and NASA Living with a Star Grant NAG5 - 10883.

REFERENCES

1. Burlaga, L. F., E. Sittler, Jr., F. Mariani, and R. Schwenn, *J. Geophys. Res.*, *86*, 6673, 1981.
2. Burton, R. K., R. L. McPherron, and C. T. Russell, *J. Geophys. Res.*, *80*, 4204, 1975.
3. Jordanova, V. K., C. J. Farrugia, J. M. Quinn, R. B. Torbert, J. E. Borovsky, R. B. Sheldon, and W. K. Peterson, *J. Geophys. Res.*, *104*, 429, 1999.
4. Lepping, R. P., C. J. Farrugia, V. K. Jordanova, D. B. Berdichevsky, I. G. Richardson, A. Galvin, and R. Schwenn, *Eos Trans. AGU*, *82*(47), Fall Meet. Suppl., Abstract SH31A-0703, 2001.
5. Lindsay, G. M., C. T. Russell, and J. G. Luhmann, *J. Geophys. Res.*, *104*, 10,335, 1999.
6. Matsui, H., C. J. Farrugia, and R. B. Torbert, *J. Geophys. Res.*, *107*(A11), 1355, doi:10.1029/2002JA009251, 2002.
7. Weimer, D. R., *J. Geophys. Res.*, *106*, 407, 2001.

Statistical properties of soft X-ray solar flares

F. Lepreti*†, V. Carbone*†, P. Veltri*† and P. Giuliani**

*Dipartimento di Fisica, Università della Calabria, I-87036 Rende (CS), Italy
†Istituto Nazionale per la Fisica della Materia, Unità di Cosenza, Italy
**Chemical Physics Department, Weizmann Institute of Science, 76100 Rehovot, Israel

Abstract. We investigate some statistical properties of soft X-ray bursts produced by solar flares. The Probability Density Functions (PDFS) of soft X-ray intensity fluctuations are shown to display wide, non-gaussian tails. The shape of the PDFs is nearly unchanged as the timelag, used to calculate fluctuations, varies. A very similar behavior is found for PDFs of energy dissipation fluctuations in a shell model of Magnetohydrodynamic (MHD) turbulence. Recalling also that both flare soft X-ray bursts and dissipative bursts in the MHD shell model are characterized by a power law distribution for waiting times between successive bursts, we suggest that the results shown in this paper support the idea that solar flares could represent bursty dissipative events of MHD turbulence.

INTRODUCTION

Bursts of X-ray emission are one of the main signatures of solar flares. Statistical properties of solar X-ray bursts have been extensively studied in order to investigate the physical mechanisms underlying flares. Several authors showed that probability distributions of flare peak flux, fluence and duration are well represented by power laws [1, 2, 3, 4, 5, 6]. An explanation of these results was proposed by Lu and Hamilton [7] by using "avalanche models" (also called "sandpile models") based on the idea that the coronal magnetic field is in a state of self-organized criticality (SOC) [8]. These models are able to reproduce the power law behavior of solar flare size distributions.

However, it has recently been pointed out that the statistics of time intervals Δt between two successive bursts (also called waiting times) is very important for solar flare modeling [9, 10]. In sandpile models, avalanches are independent on each other and follow a Poisson statistics. This gives rise to an exponential waiting time distribution (WTD) [9, 10]. On the other hand, recent analyses of hard X-ray (HXR) and soft X-ray (SXR) observations showed that the solar flare WTD clearly deviates from the exponential behavior expected from avalanche models [9, 10].

In particular, Boffetta et al. [10], by using SXR data acquired by the *Geostationary Operational Environmental Satellites* (*GOES*), found that the WTD follows a power law $P(\Delta t) \propto \Delta t^{-\beta}$, with $\beta \simeq 2.4$, for waiting times greater than a few hours. Lepreti et al. [11] extended this analysis, by showing that the sequence of SXR bursts is not consistent with a local Poisson process and that the observed WTD is well described by a Lévy function, which asimptotically displays a power law behavior. These results indicate that waiting times are statistically self-similar (see also Fig. 3 in ref. [11]) and suggest the presence of long-range correlations in the flaring process.

Boffetta et al. [10] showed that energy dissipation bursts in a shell model of magnetohydrodynamic (MHD) turbulence are characterized by power law distributions, including the WTD, as in the case of solar flare observations. On this basis, they suggested that the presence of long-range correlations could be related to the nonlinear dynamics occurring in fully developed MHD turbulence.

To the aim of investigating in more detail the physical origin of solar flare statistics, in this work we will analyze, besides the waiting time distribution, the scaling behavior of the Probability Density Functions (PDFs) of SXR intensity fluctuations. As a comparison, the same analysis will be performed on fluctuations of energy dissipation rate in a shell model of MHD turbulence.

ANALYSIS OF *GOES* SOFT X-RAY DATA

The SXR data used in this paper were acquired by the *GOES* 10 satellite in the 1-8 Å band, during the time interval between 1998 August 1 and 2000 July 29. The SXR flux $f(t)$ (shown in Fig. 1) was measured with a sampling time of 1 minute.

In order to calculate waiting times, bursts are defined as the time intervals during which the condition $f(t) \geq f_{th}$ is satisfied. The threshold is defined as $f_{th} = \langle f(t) \rangle + 3\sigma$, where the average and the standard deviation are calculated through an iterative procedure, excluding the

CP679, *Solar Wind Ten: Proceedings of the Tenth International Solar Wind Conference*,
edited by M. Velli, R. Bruno, and F. Malara

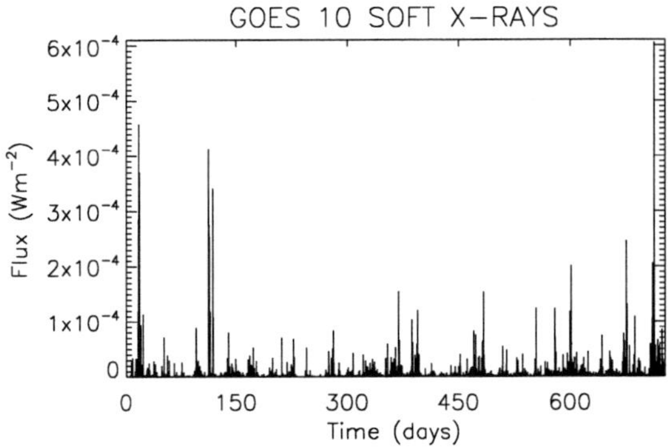

FIGURE 1. Soft X-ray flux measured by the *GOES* 10 satellite in the 1-8 Å band, in the interval between 1998 August 1 and 2000 July 29.

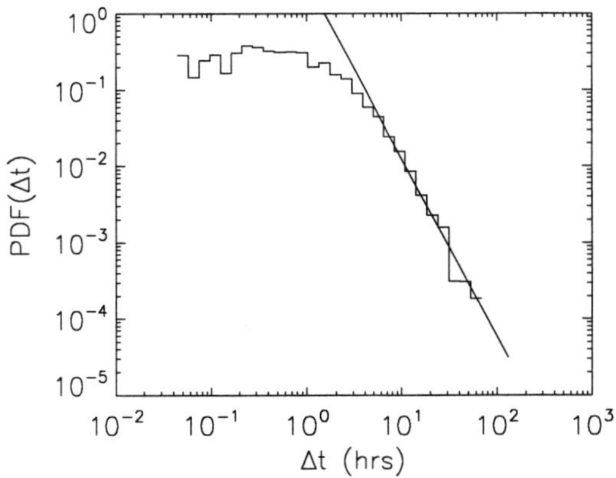

FIGURE 2. Distribution of waiting times between successive soft X-ray flares. The solid line represent a power law with an exponent $\beta = 2.33$.

bursts [10]. The waiting time distribution obtained from our analysis is similar to WTDs shown in previous works [10, 11], that is, it displays a power law tail, for $\Delta t \gtrsim 5$ hrs, with an exponent $\beta = 2.33 \pm 0.16$ (see Fig. 2), despite the fact we used a dataset covering a shorter period and we selected flares with a different criterion.

In addition to the WTD, we are interested in analyzing the scaling bahavior of SXR intensity fluctuations $\delta f_\tau = f(t+\tau) - f(t)$. This is done by calculating the PDFs of standardized fluctuations $\delta F_\tau = (\delta f_\tau - \langle \delta f_\tau \rangle)/\langle \delta f_\tau^2 \rangle^{1/2}$ (where brackets represent time averages) at different lagtimes τ. In Fig. 3, we report the PDFs of δF_τ for 0.5 hrs $\leq \tau \leq 8.7 \times 10^3$ hrs. It can be seen that the PDFs don't change their shape significantly, as τ varies, and display strong, non gaussian tails.

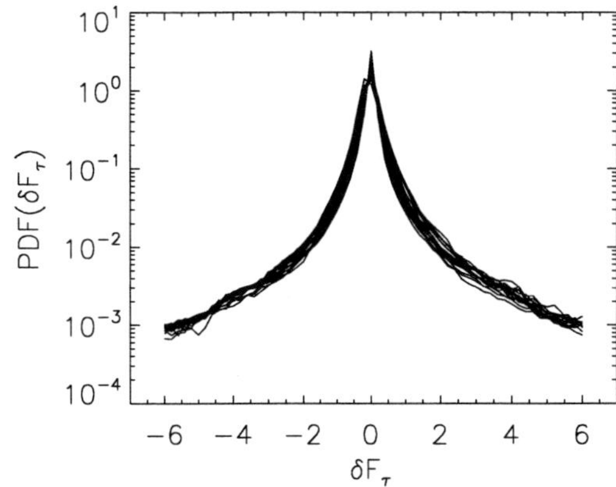

FIGURE 3. PDFs of soft X-ray intensity fluctuations at different timelags τ in the interval 0.03 hrs $\leq \tau \leq 8.7 \times 10^3$ hrs

ANALYSIS OF AN MHD SHELL MODEL

In this section, the same analysis tools described above will be applied to a shell model of magnetohydrodynamic turbulence. This model is a dynamical system which aims to reproduce the main features of nonlinear dynamics occurring in MHD turbulence [10, 12].

The wavevector space is divided in discrete shells of radius $k_n = 2^n k_0$. Two complex dynamical variables $u_n(t)$ and $b_n(t)$, representing, respectively, velocity and magnetic field increments on an eddy of scale $l \sim k_n^{-1}$, are assigned to each shell. The evolution equations for $u_n(t)$ and $b_n(t)$ are obtained by: a) introducing general quadratic couplings between neighbouring shells; b) imposing the conservation of MHD ideal invariants. In this way, the following set of nonlinear Ordinary Differential Equations can be obtained:

$$\frac{du_n}{dt} = -\nu k_n^2 u_n + f_n + T_n(u_n, b_n)\,, \tag{1}$$

$$\frac{db_n}{dt} = -\mu k_n^2 b_n + G_n(u_n, b_n)\,, \tag{2}$$

where where ν and μ are, respectively, the kinematic viscosity and the resistivity, f_n is an external forcing term, T_n and G_n are the nonlinear quadratic terms [10, 12].

In this work, we are interested in analyzing the statistical features of the energy dissipation rate $\varepsilon(t)$, given by

$$\varepsilon(t) = \nu \sum_n k_n^2 |u_n|^2 + \eta \sum_n k_n^2 |b_n|^2\,. \tag{3}$$

Time series $\varepsilon(t)$ (see Fig. 4) can be obtained from numerical simulatons of the model [10], and dissipation bursts can be found through the condition $\varepsilon(t) > \varepsilon_{th}$, where ε_{th} is a suitable threshold value for $\varepsilon(t)$.

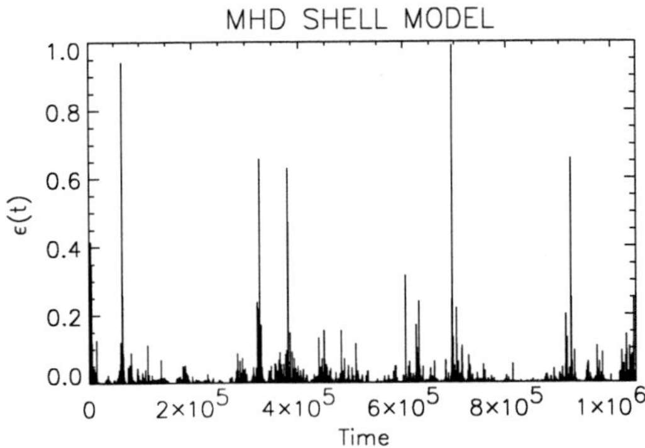

FIGURE 4. Time series of energy dissipation $\varepsilon(t)$ for the MHD shell model.

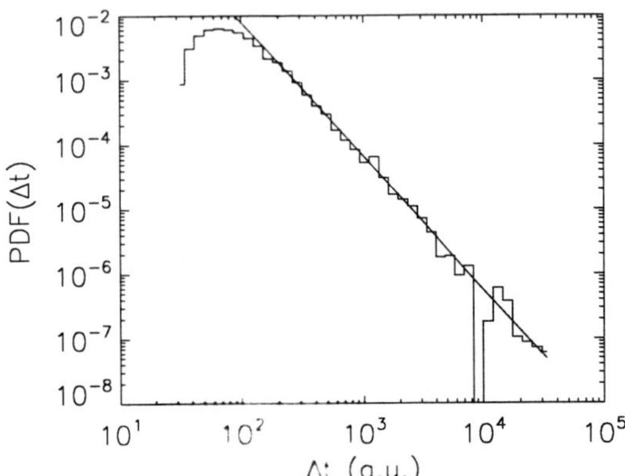

FIGURE 5. Distribution of waiting times between successive dissipative bursts in the MHD shell model. The solid line represent a power law with an exponent $\beta = 2.07$.

Boffetta et al. [10] already showed that dissipative bursts in the MHD shell model reproduce the statistical properties of X-ray solar flares, that is, the power law distributions for peak flux, total energy, durations and waiting times. The threshold chosen in the present paper to select dissipation bursts and calculate waiting times is given by $\varepsilon_{th} = \langle \varepsilon(t) \rangle + 3\sigma$ where $\langle \varepsilon(t) \rangle$ and σ are calculated in the same way as for *GOES* SXR data. The waiting time distribution (see Fig. 5) is characterized by the presence of a power law tail with a scaling exponent 2.07 ± 0.10.

As in the case of *GOES* SXR data, we also calculated the PDFs of standardized fluctuations of energy dissipation at different lagtimes τ, that is, $\delta E_\tau = (\delta \varepsilon_\tau - \langle \delta \varepsilon_\tau \rangle)/\langle \delta \varepsilon_\tau^2 \rangle^{1/2}$, where $\delta \varepsilon_\tau = \varepsilon(t+\tau) - \varepsilon(t)$. Fig. 6 shows that the PDFs of δE_τ have strong non-gaussian tails and don't change their shape significantly as τ varies, in good agreement with the behavior observed for

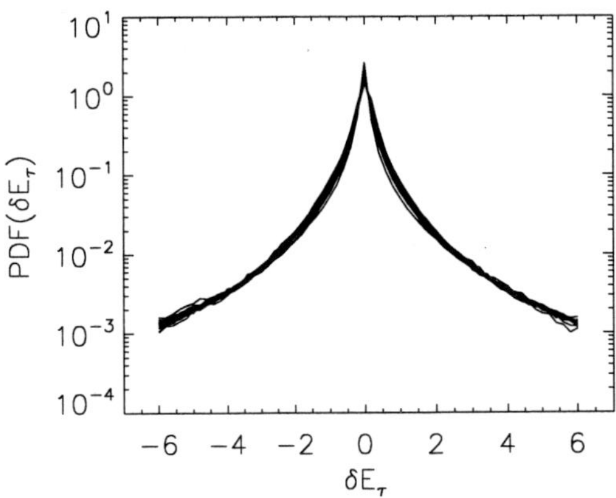

FIGURE 6. PDFs of energy dissipation fluctuations, for the MHD shell model at different timelags τ.

SXR data.

CONCLUSION

In this paper, we performed a statistical analysis of the soft X-ray bursts produced by solar flares and compared the results with a shell model of MHD turbulence. As already evidenced in previous papers[10, 11], we point out that the waiting time distribution displays a power law tail, both for flare SXR bursts and for energy dissipation bursts in the MHD shell model.

Besides the waiting time distribution, we investigated the scaling behavior of SXR intensity fluctuations. We showed that the PDFs of these fluctuations are characerized by the presence of wide, non-gaussian tails and have the same shape at different timescales. A very similar behavior was found for PDFs of energy dissipation fluctuations in the MHD shell model.

In our opinion, these results suggest that solar flares could represent bursty dissipative events of MHD turbulence, as proposed by Boffetta et al. [10]. Following this idea, the power law behavior of the WTD can be attributed to long-range correlations arising from the nonlinear dynamics acting in the system, while the non-gaussian PDFs can be interpreted as the result of strong events accumulating on the dissipative scale through the occurrence of nonlinear interactions.

ACKNOWLEDGMENTS

We thank Luca Sorriso-Valvo for useful discussions.

REFERENCES

1. Datlowe, D., Elcan, M. J., and Hudson, H. S., *Solar Phys.* **39**, 155-174 (1974).
2. Lin, R. P., Schwartz, R. A., Kane, S. R., Pelling, R. M., and Hurley, K. C., *Astrophys. Jour.* **283**, 421-425 (1984).
3. Dennis, B. R., *Solar Phys.* **100**, 465-490 (1985).
4. Crosby, N. B., Aschwanden, M. J., and Dennis, B. R., *Solar Phys.* **143**, 275-299 (1993).
5. Lee, T. T., Petrosian, V., and McTiernan, J. M., *Astrophys. Jour.* **412**, 401-409 (1993).
6. Bromund, K. R., McTiernan, J. M., and Kane, S. R., *Astrophys. Jour.* **455**, 733-745 (1995).
7. Lu, E. T., and Hamilton, R. J., *Astrophys. Jour.* **380**, L89-L92 (1991).
8. Bak, P., Tang, C., and Wiesenfeld, K., *Phys. Rev. Lett.* **59**, 381-384 (1987).
9. Wheatland, M. S., Sturrock, P. A., and McTiernan, J. M., *Astrophys. Jour.* **509**, 448-455 (1998).
10. Boffetta, G., Carbone, V., Giuliani, P., Veltri, P., and Vulpiani, A., *Phys. Rev. Lett. 83*, 4662-4665 (1999).
11. Lepreti, F., Carbone, V., and Veltri, P., *Astrophys. Jour.* **555**, L133-L136 (2001)
12. Giuliani, P., "Shell Models of MHD Turbulence", in *Nonlinear MHD Waves and Turbulence*, edited by T. Passot and P.L Sulem, Springer Lecture Notes in Physics 536, Berlin, 1999, pp.331-344

SOHO CTOF Observations of Interstellar He^+ Pickup Ion Enhancements in Solar Wind Compression Regions

L. Saul[1], E. Möbius[1], Y. Litvinenko[1], P. Isenberg[1], H. Kucharek[1], M. Lee[1], H. Grünwaldt[2], F. Ipavich[3], B. Klecker[4], P. Bochsler[5]

[1]*University of New Hampshire,* [2]*Max Planck Institute for Aeronomy,* [3]*University of Maryland,* [4]*Max Planck Institute for Astrophysics,* [5]*University of Bern*

Abstract. We present a recent analysis with 1996 SOHO CELIAS CTOF data, which reveals correlations of He^+ pickup ion fluxes and spectra with the magnetic field strength and solar wind density. The motivation is to better understand the ubiquitous large variations in both pickup ion fluxes and their velocity distributions found in interstellar pickup ion datasets. We concentrate on time periods of that can be associated with compression regions in the solar wind. Along with enhancements of the overall pickup ion fluxes, adiabatic heating and acceleration of the pickup ions are also observed in these regions. Transport processes that lead to the observed compressions and related heating or acceleration are discussed. A shift in velocity space associated with traveling interplanetary compression regions is observed, and a simple model presented to explain this phenomenon based on the conserved magnetic adiabatic moment.

INTRODUCTION

Although interstellar pickup ions (PUIs) stem from a supposedly stationary source of neutral gas, large variations of pickup He+ fluxes and their energy spectra in the solar wind have been observed on a variety of time scales. Some causes of these variations have been identified, including a depletion of pickup ions in the anti-sunward part of the distribution during time periods of radial interplanetary magnetic fields [3,7], variation of the ionization rate, and variation of the cutoff velocity of the He+ spectra due to the non-zero inflow speed of the neutral helium[1,8]. However, variations of the fluxes up to an order of magnitude, and substantial variations of the pickup ion spectra still remain unexplained. An understanding of these variations is necessary for using PUI measurements as probes of their source populations or for plasma transport parameters.

The CTOF instrument, in the CELIAS package on board SOHO, produced 150 days of data under relatively steady conditions in the upwind region of the interstellar flow. The relatively large geometric factor of the instrument, and it's location at the L1 point, make high time resolution observations of PUI spectra possible. This makes the dataset interesting for studying the short term variations in both total flux and velocity distribution.

The observations presented here were made during solar minimum, mostly in slow solar wind. The solar wind parameters used for this study are taken from the MTOF proton monitor on board SOHO. The magnetic field data are taken from WIND MFI, and convected back to the location of SOHO using the interplanetary magnetic field (IMF) orientation and measured solar wind speed. Our analysis of this dataset is in two parts. The first is a statistical analysis of the entire dataset, comparing PUI fluxes to solar wind parameters. The second analysis is of individual events in the dataset, for which enhancement of the PUIs in especially pronounced, either in the bulk or tail of the distribution. We look in detail at the evolution of the PUI distribution through these events.

STATISTICAL ANALYSIS

By combining PUI events from time periods with specified solar wind conditions, and binning them in normalized velocity V/V_{sw}, spectra are formed that are representative of the chosen solar wind parameters. This method of statistical analysis can tell us not only whether the PUI flux and the solar wind parameters are correlated, but also where in the velocity spectrum the correlation is the strongest. We carried out this analysis with bulk solar wind parameters, including

CP679, *Solar Wind Ten: Proceedings of the Tenth International Solar Wind Conference,* edited by M. Velli, R. Bruno, and F. Malara

proton density, thermal velocity, and solar wind speed. We also carried out the analysis with IMF magnitude and direction.

The observed fluxes of the He^+ PUIs were found to be strongly correlated with the proton density (Fig. 1). The energy flux was always higher during periods of higher proton density for this analysis. A peak near the injection speed ($2V_{sw}$) is clearly visible in the spectra for $30cm^{-3} < n_p < 40cm^{-3}$. Also note that the peak energy of the ions is shifted toward the cutoff in higher density solar wind

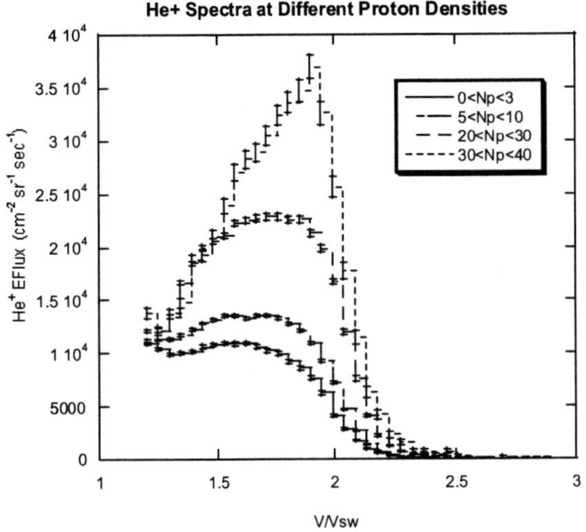

FIGURE 1. PUI differential energy flux spectra for four ranges of proton density in cm^{-3}. Error bars represent the statistical error.

Another strong correlation observed was with the IMF magnitude (Fig. 2). This complements previously observed correlations with the orientation of the IMF [3,7]. The PUI flux was found to not notably increase until |B| > 6nT, after which it increased monotonically with |B|. A peak near the injection speed was visible in the highest fields, similar to the peak in regions of highest proton density. We also performed the same analysis for restricted set of data based on IMF orientation. We found the correlation of PUI fluxes with |B| persisted for all orientations, becoming more pronounced as the field became more perpendicular to the solar wind velocity.

Because the PUI fluxes showed correlations with both IMF strength and field density, we then looked in more detail at times of solar wind compression, which have both enhanced IMF and proton density.

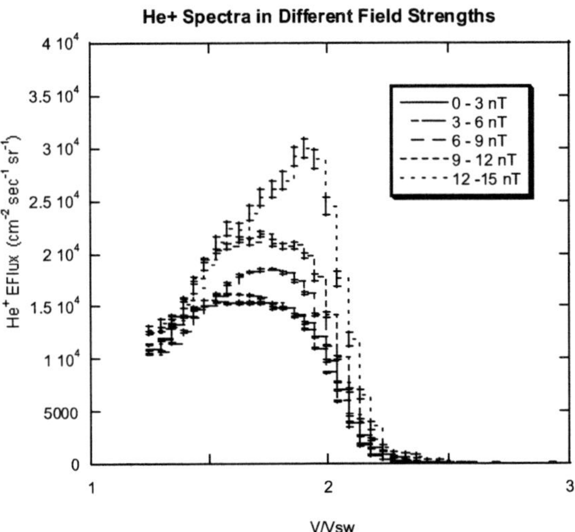

FIGURE 2. PUI differential energy flux spectra for five different ranges of field strength, integrated over the entire dataset.

COMPRESSION REGIONS

SOHO saw several solar wind compression events during 1996. Other compression events in high Heliospheric latitudes have been shown to affect PUI fluxes [10]. Similar effects were observed here in the solar wind at 1AU in the ecliptic. Accelerated particles are observed in conjunction with the compression, especially downstream of the regions, here and elsewhere [9].

Some CIRs and transient stream interaction regions have also been observed over this time period with other instruments on SOHO [5]. We also observed the strong transient stream interaction of DOY 170 (Fig. 3), and a magnetic cloud CME around DOY 149. This cloud displayed compression of the solar wind and He^+ before and after its passage, with stronger compression trailing the region. The details of the PUI spectra varied in these compression regions, though the major features showed many similarities. The compression of the PUIs was strongest in the velocity range near the injection speed, analogous to the statistical correlations discussed above. Accelerated PUIs (last panel Fig. 3) were enhanced downstream of the regions. Some times with what looks like high PUI tail flux (e.g. high energy range near the magnetic discontinuity in Fig. 3) are actually due to a shift in velocity space of the distribution.

Transient Interaction Region – DOY 170

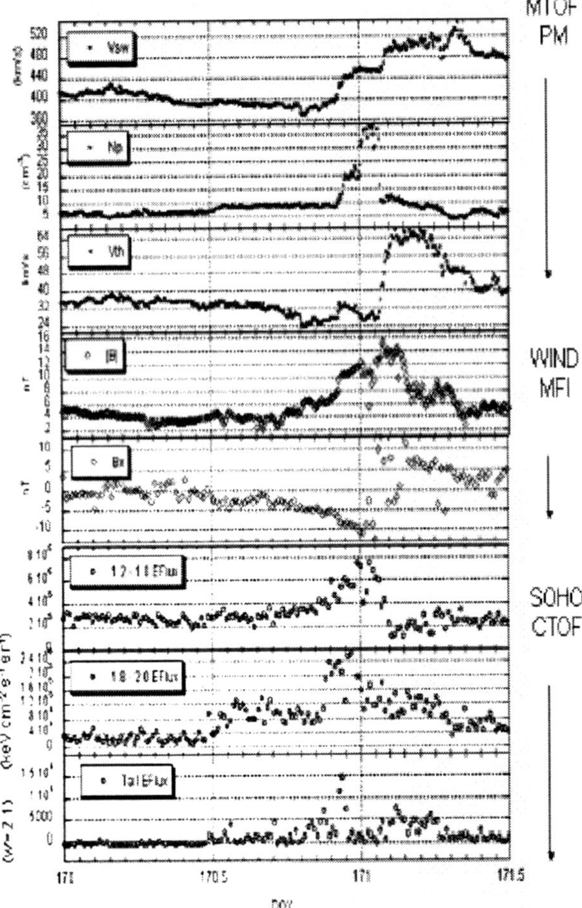

FIGURE 3. The stream interaction region above can be identified by the jumps in solar wind speed (1st panel), thermal speed (3rd panel), and the reversal of the radial component B_x (5th panel). The PUIs are divided into 3 energy ranges in w = V/V_{sw}.

Cutoff Shift at Magnetic Discontinuities

In some cases, the PUI spectra showed a shift in the cutoff velocity around the compression region (Figs. 3, 4). We hypothesized that this was due to conservation of $v_\perp^2/|B|$ as the distribution entered a compression region with enhanced IMF. Consistent with this explanation, the shift was only observed close to a jump in magnetic field magnitude. In addition, no shift was observed unless the discontinuity was traveling, i.e. fast wind pushing the discontinuity into slower wind.

This shift due to a conserved adiabatic moment will only affect perpendicular velocity. Therefore, the maximum PUI velocity shift is predicted for a ring distribution, for which the shift is: $v_2 = v_1\sqrt{B_2/B_1}$. To find the shift for an isotropic distribution we integrate over all pitch angles:

$$v_2 = v_1 \frac{\pi}{2} \int_0^{\pi} \sqrt{\frac{B_2}{B_1}\sin^2(\alpha) + \cos^2(\alpha)}\, \sin\alpha d\alpha \,,$$

which can be approximated as $v_2 \sim v_1\left(B_2/B_1\right)^{0.4}$. The initial velocity is taken (in the solar wind frame) to be the solar wind speed plus the radial component of the bulk neutral (Keplerian) velocity at the spacecraft's position in the Heliosphere [1,8].

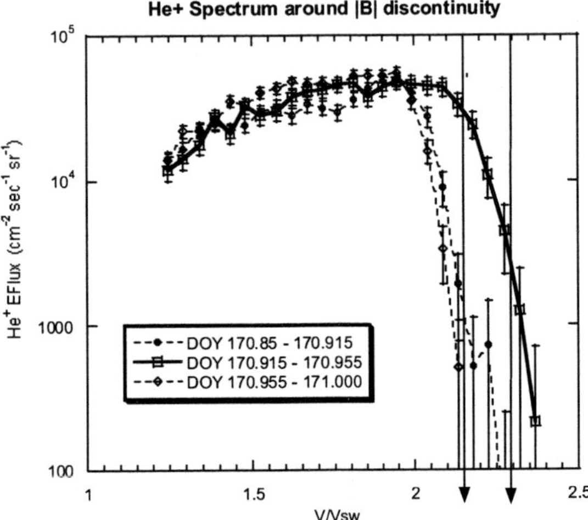

FIGURE 4. Logarithmic PUI differential energy flux spectra corresponding to three one hour time periods right at the magnetic discontinuity in Fig. 3.

The model described here for an isotropic distribution predicts the initial and compressed cutoff velocities shown in fig. 4 with downward arrows: v_1=2.15 and v_2=2.29 (for DOY 170, B2/B1 = 5/4, and assuming V_{lism} = 27km/s).

To further test this interpretation the shift ratio v_2/v_1 was estimated using the inflection points of polynomial fits to the spectra (in linear representation). The results are shown for four compression events (fig. 5), including the one shown here at DOY 170, the trailing edge of the event at DOY 149, and two smaller compressions on DOY 99. The results are found to be consistent with the models. The exact type of distribution depends on the pitch angle scattering rate as the PUI distribution evolves. A real model of the spectra (rather than a polynomial fit) would be needed to differentiate between the different types of distributions.

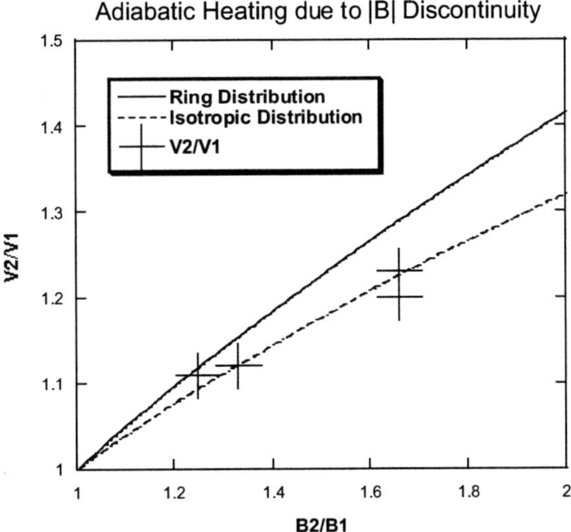

FIGURE 5. Predicted velocity shift ratios (lines) for isotropic and ring distributions are shown vs. the corresponding magnetic field compression ratio. The estimated shifts in 4 observed distributions are also shown.

DISCUSSION

The CTOF measurements show strong correlations with solar wind parameters, especially the proton density. This is perhaps not a surprise; one would expect conditions that compress the bulk solar wind to compress the PUIs contained in the wind. The observed correlation with the IMF magnitude was similar in character to that of the proton density, pointing to a similar cause for these two correlations. The observations are consistent with the hypothesis that compression regions are responsible for the many of the PUI flux variations.

The compressions not only enhance the overall flux, but change the shape of the distribution, most notably in an enhancement near the injection velocity in compressed regions. The observations are consistent with the compression compensating for some of the adiabatic cooling of the PUIs, thus keeping their speeds in the area around 2Vsw that would have cooled in the expanding wind. However, strong shocks are known to affect PUI distributions in other ways, including heated electrons increasing the ionization rate [6]. It could be that something similar takes place in compression regions, contributing to the enhancement near the cutoff.

The observed velocity shift due to magnetic discontinuities can be explained with adiabatic effects. Further work is needed to model it more fully and interpret its influence on different distributions or in different solar wind conditions.

ACKNOWLEDGMENTS

This work is partially supported by NASA grant NGT5-50381, NAG5-10890, and NST ATM-9800781. Thanks to the WIND/MFI team for the IMF data.

REFERENCES

1. Chalov, S.V., Fahr, H.J.; *Keplerian injection velocities reflected in helium pick-up ion spectra*; Astron. Astrophys 2000; ppL21-L24

2. Fisk, L.A., Schwadron, N.A., Gloecker, G.; *Implications of fluctuations in the distribution functions of interstellar pick-up ions for the scattering of low rigidity particles*; GRL 1997; pp93-96

3. Gloeckler, G., Schwadron, N.A., Fisk, L.A., Geiss, J.; *Weak pitch angle scattering of few MV rigidity ions from measurements of anisotropies in the distribution function of interstellar pickup H+*; GRL 1995; vol.22 pp.2665

4. Gloeckler, G. et al; *Acceleration of interstellar pickup ions in the disturbed solar wind observed on Ulysses*; JGR 1994; vol.99 pp17637

5. Hilchenbach, M. et al; Obervation of suprathermal helium at 1 AU: Charge states in CIRs; Solar Wind 8 proceedings, Volume 471, Issue 1, pp. 605-608

6. Isenberg, P.A., Feldman, W.C.; Electron-impact ionization of interstellar hydrogen and helium at interplanetary shocks; GRL 1995; vol.22 pp.873-875

7. Möbius, E., Rucinski, D., Lee, M.A., Isenberg, P.A.; *Decreases in the antisunward flux of interstellar pickup He+ associated with radial interplanetary magnetic field*; JGR 1998; vol.103 pp257-265

8. Möbius, E. et al; *Direct evidence of the interstellar gas flow velocity in the pickup ion cut-off as observed with SOHO CELIAS CTOF*; GRL 1999; vol.26 pp.3181-3184

9. Schwadron, N.A., Fisk, L.A., Gloeckler, G.; *Statistical acceleration of interstellar pick-up ions in co-rotating interaction regions*; GRL 1996; pp2871-2874

10. Schwadron, N.A., Zurbuchen, T.H., Fisk, L.A., Gloecler, G.; *Pronounced enhancements of pickup hydrogen and helium in high-latitude compressional regions*; JGR 1999; vol.104 pp.535-548

The Transition of Interplanetary Shocks through the Magnetosheath

A. Szabo*, C. W. Smith† and R. M. Skoug**

*Laboratory for Extraterrestrial Physics, NASA Goddard Space Flight Center, Greenbelt, MD 20771
†Bartol Research Institute, University of Delaware, Newark, DE 19716
**Los Alamos National Laboratory, Los Alamos, NM 87545

Abstract. WIND, ACE, IMP 8 and Geotail data shows that the magnetosheath signature of IP shocks is primarily a fast-mode shock or pressure pulse with a wide ramp. No model predicted secondary discontinuities could be identified above the magnetosheath background fluctuation level. Even though the interplanetary surface geometry of IP shocks could be significantly corrugated, this study suggests that there is no significant deceleration of the pressure front in the magnetosheath.

INTRODUCTION

A strong correlation of interplanetary (IP) shocks impinging on the magnetosphere and geomagnetic disturbances have been reported by many observers [1, 2, 3]. IP shocks, as all solar wind pressure events, tend to disrupt the magnetopause surface leading to magnetopause transient events [4] that, in turn, can initiate magnetic reconnection resulting in substorm onset [5]. IP shocks have also been connected to sudden commencements and auroral brightening [6]. However, before IP shocks reach the magnetopause, they have to cross the Earth's bow shock and traverse the magnetosheath introducing both geometrical and physical modifications. Following IP shocks through the magnetosheath with in-situ observations is the topic of this paper.

The interaction of IP shocks with the bow shock and its transmission through the magnetosheath to the magnetopause, has been studied mainly by gas dynamic modeling [7, 8, 9]. These models, by construction, allow the generation and propagation of only one type of wave in the magnetosheath. Therefore, it is not surprising that they find only a single, fast mode pressure pulse (or fast shock in the supersonic flanks) propagating through the magnetosheath. Interestingly the predicted disturbance geometry or shape in the magnetosheath remains nearly planar (the shape of the undisturbed IP shock) propagating with the same speed as the undisturbed IP shock [9]. This prediction is investigated in detail in this paper.

Attempts have been made to include the effect of the magnetic field on the interaction of the IP shock and bow shock by using various MHD and hybrid formulations. The one-dimensional model of *Whang* [10] was very successful at describing outer heliospheric observations of IP shocks. It allowed the merger of two IP shocks if they are both forward or reverse, and predicted the transmission of the two interacting shocks if one is forward and the other reverse with a tangential discontinuity (TD) forming between them. This model, however, is limited to the treatment of perpendicular shocks. *Cargill* [11] relaxed this requirement with the use of a one-dimensional hybrid code. He showed that while the collision between two perpendicular collisionless shocks gives rise to a TD located between the two transmitted shocks, as predicted by one-dimensional MHD theory, the collision between two oblique shocks produces a much more extensive and turbulent region between the two transmitted shocks, possibly a contact discontinuity (CD). The CD, like the TD, shows jumps in the plasma and magnetic field components and therefore should be identifiable in observational data. Moreover the transmitted shock is deflected from its original orientation that should also be observable. Aside for *Zhuang et al.* [12] reporting a case of ISEE 1 and 3 measurements where such a sequence of disturbances was possibly observed in the Earth's magnetosheath, observational evidence in the literature remains rather limited and is one of the topics of this paper.

Next the data sets and analytical techniques used in this study is discussed followed by the results

CP679, *Solar Wind Ten: Proceedings of the Tenth International Solar Wind Conference*,
edited by M. Velli, R. Bruno, and F. Malara
2003 American Institute of Physics 0-7354-0148-9/03/$20.00

of the interplanetary and magnetosheath shape and signatures of shocks. Finally, a short summary will be presented.

DATA AND ANALYSIS TECHNIQUES

During 1998–1999, twelve IP shocks have been identified that were observed by at least 2 solar wind monitors and by another spacecraft in the magnetosheath. This interval was selected because IP shocks were not very frequent and clearly observable (unlike the later solar maximum years), and there was nearly continuous coverage by WIND, ACE, IMP 8 and Geotail. Moreover, the relatively large separation between consequent IP shocks assured that there was only minimal possibility of interaction between them before reaching 1 AU keeping their surface geometry the simplest possible. In order to establish the undisturbed surface geometry of the incoming IP shocks the observed solar wind plasma and magnetic field data before and after the shocks were fitted by the non-linear least squares "Rankine-Hugoniot" technique originally developed by *Vinas and Scudder* [13] and further enhanced by *Szabo* [14]. This fitting technique provides the best possible local shock normal direction and speed determination by a single spacecraft with the associated uncertainties. Data from the WIND Magnetic Field Investigation (MFI) [15], and Solar Wind Experiment (SWE) [16], IMP 8 magnetic field and plasma experiments [17], and ACE magnetic field experiment (MAG) [18] and solar wind plasma instrument (SWEPAM) [19] were used for the analysis. Preliminary use was also made of the Geotail magnetic field (MGF) [20] and solar wind (CPI) [21] data.

IP SHOCK GEOMETRY IN INTERPLANETARY SPACE

Generally it is assumed that incoming IP shocks are planar on the scale size of the magnetosphere. Indeed, in a recent study *Russell et al.* [23] analyzed a single IP shock with four solar wind satellites and found that three of them were consistent with the planarity assumption. However, deviation from planarity has been reported before [24, 25]. Specifically, *Szabo et al.* [25] have found that IP shocks driven by small magnetic clouds have a highly corrugated surface geometry on the scale-size of the magnetosphere. To further illustrate that significant deviations from planarity is possible, 17 IP shocks observed by at least two solar wind monitors during the rising phase of the previous solar cycle (1997–1999)have been fitted and the shock normal directions compared. Figure 1 shows the angular deviation between the corresponding shock normal directions as a function of the inter-spacecraft separation perpendicular to the Sun-Earth line. Solid circles mark shocks observed by WIND and ACE, while open circles refer to WIND and IMP 8 observations (note the significantly larger associated uncertainties). Considerable deviation from planarity is apparent. Also the degree of curvature does not appear to be a simple function of the inter-spacecraft separation. In addition, those IP shocks that were clearly driven by magnetic clouds are marked by a solid cross indicating that some clouds drive shocks that are very nearly planar, on the other hand, some cloud driven shocks can be as corrugated as some of those for which the drivers are uncertain. This result complicates the analysis of the magnetosheath propagation of IP shocks as clearly planarity cannot be assumed (unlike for computer simulations) and some allowance for local curvature will have to be made.

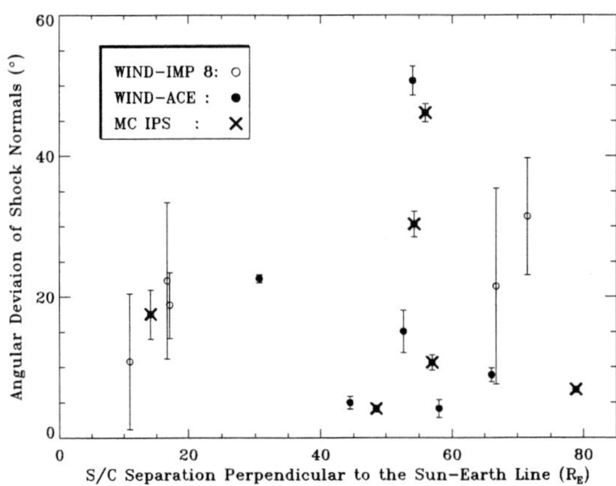

FIGURE 1. Angular deviation between shock normal directions for the same event as observed by WIND and ACE (solid circles) and WIND and IMP 8 (open circles). Crosses mark those shocks that were clearly driven by magnetic clouds.

IP SHOCKS IN THE MAGNETOSHEATH

During the time period of 1998–1999 twelve IP shocks have been identified that were observed by at least 2 solar wind monitors (WIND and ACE) to place some limit on the interplanetary curva-

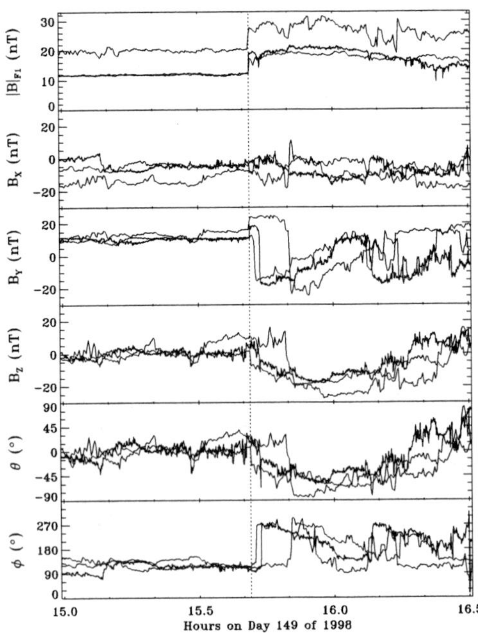

FIGURE 2. WIND, ACE and IMP 8 magnetic field magnitude and GSE Cartesian and spherical components on May 29, 1998. The WIND and ACE data is time shifted to line up the IP shock marked by the dashed line.

ture of the incoming shock, while IMP 8 provided magnetosheath observations. Some tentative Geotail sheath events were also identified. For some IP shocks a corresponding pressure pulse is clearly identifiable in the sheath observations. Figure 2 shows 90 minutes of observations of the magnetic field and its components by WIND, ACE and IMP 8 on May 29, 1998. The WIND and ACE data has been time shifted by 29 and 37 minutes, respectively to line up the shock observations with the sheath pressure pulse event (the higher overall field values correspond to the IMP 8 compressed sheath measurements). The sheath pressure pulse is very clear in both magnetic field and plasma observations. Also it should be noted that the nearby large field rotation corresponding to a TD has a markedly different advection time delay. This is consistent with the pressure pulse corresponding to the IP shock that travels faster than the strictly advecting TD. Such a clear sheath signature was not always apparent. Weaker and reverse shocks had significantly broader ramps, some reaching over 10 minutes. On the other hand, even for the clearest cases no other discontinuity nearby could be identified. This does not prove that the secondary discontinuities, predicted by MHD and hybrid codes, do not exist as the gen-

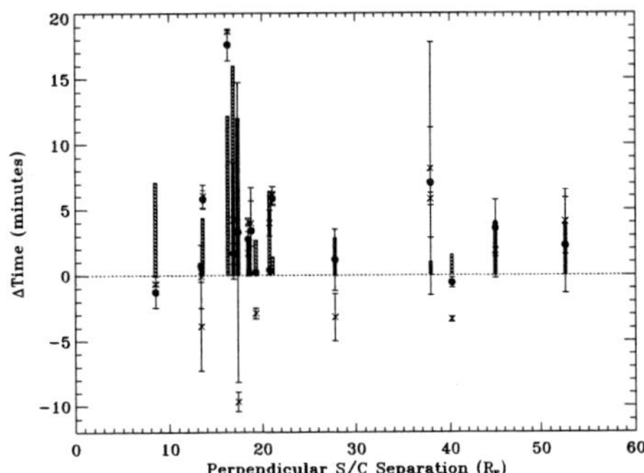

FIGURE 3. Difference in the predicted and observed arrival times in the magnetosheath as a function of the spacecraft separation. See the text for details.

eral magnetosheath background fluctuations could easily mask small variations. However, it does point out that for magnetospheric energy and momentum input the leading fast-mode pressure jump or shock is the most significant.

In order to make some assessment of the magnetosheath geometry of the transmitted disturbance, the IP shock surface normal directions and speeds, fitted in the solar wind data, were used to estimate a predicted arrival time at the magnetosheath monitor. This predicted arrival time was compared to the actually observed time delay. The thus obtained difference is plotted in Figure 3. Positive difference time refer to the actual magnetosheath observation being later than predicted based on the upstream shock fit results. This would be the expected case if the pressure front is decelerated in the sheath as is the case for TDs. Negative difference times correspond to earlier than expected arrival times. This could be due to an unlikely acceleration of the front or more likely to the intrinsic curvature of the IP shock. All time differences are calculated with respect to the beginning of the sheath pressure pulse. The length of the pressure ramp is indicated by the gray bars. A time difference within the gray bars would be consistent with an unaltered shock disturbance front. The same procedure is repeated for both solar wind monitors. The results corresponding to the solar wind monitor closest to the sheath monitor perpendicular to the Sun-Earth line is plotted as a solid circle (the other result is plotted as a cross) as a function of this cross-wind separation. Just as in the case of the undisturbed solar wind IP shocks (Figure 1), there is no clear dependence on the spacecraft separation.

The same data is plotted in a similar format in Figure 4 as a function of the distance of the sheath monitor behind the bow shock. The location of the bow shock was estimated using the *Peredo* model

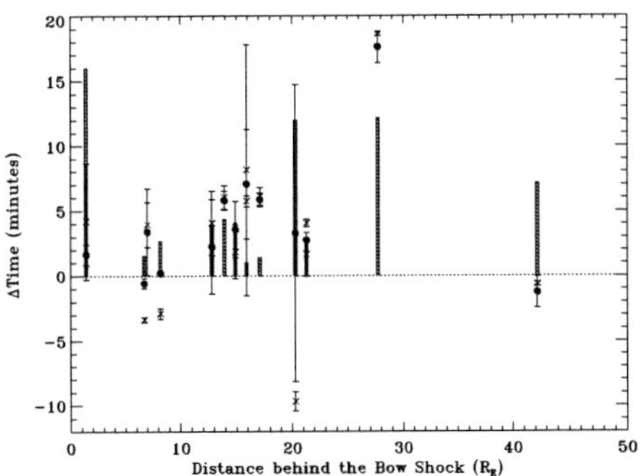

FIGURE 4. Difference in the predicted and observed arrival times in the magnetosheath as a function of the distance of the sheath monitor behind the bow shock. See the text for details.

[26] and the measured solar wind conditions. Clearly, better determination could be made of the sheath position of a spacecraft, but this plot serves to show that some of the better timing agreements happened for cases where the sheath monitor was deep inside the sheath, hence any systematic deceleration of the pressure front would be most noticeable.

SUMMARY

While the data set presented is very limited and the question of the transition of IP shocks through the magnetosheath complicated, at least a few preliminary assessments could be made. A clear fast-mode shock or pressure pulse with a wider ramp could be identified for most solar wind observed IP shocks. However, the model predicted secondary discontinuities were not apparent in the highly fluctuating sheath background indicating that for energetics the leading pressure pulse is the most relevant. Even though the intrinsic curvature or corrugation of IP shocks complicates the determination of the geometrical effects of the magnetosheath on the transmitted shocks, the data presented suggests that there is no systematic deceleration of the pressure front. That is, the uncertainty of the arrival time of an IP shock to the magnetopause based on upstream solar wind observations (a value that could be near 10 minutes) is not due to the effects of the magnetosheath but most likely to the unknown interplanetary geometry of the shock surface fronts.

ACKNOWLEDGMENTS

The authors are grateful to A. J. Lazarus, J. D. Richardson and the MIT IMP 8 and WIND team for providing plasma data, to S. Kokubun and L. A. Frank for the use of the Geotail key parameter magnetic field and plasma data.

REFERENCES

1. Tsurutani, B.T., et al., *J. Geophys. Res.* **100**, 21,717(1995).
2. Gonzales, W.D., Tsurutani, B.T., and Gonzales, C.de, *Space Sci. Rev.* **88**, 529–562 (1999).
3. Gonzales, W.D., and Tsurutani, B.T., "The interplanetary causes of magnetic storms: A Review," in *Magnetic Storms*, AGU Monograph 98, Washington, D.C., 1998.
4. Sibeck, D.G., and Newell, P.T., *J. Geophys. Res.* **100**, 21,773 (1995).
5. Wu, C.C., *J. Geophys. Res.* **105**, 7533 (2000).
6. Tsurutani, B.T., et al., *J. Atmos Sol.-Terr. Phys.* **63**, 513 (2001).
7. Shen, W.W., and M. Dryer, *J. Geophys. Res.* **77**, 4627 (1972).
8. Dryer, M., *Radio Sci.* **8**, 893 (1973).
9. Spriter, J.R., and Stahara, S.S., "Computer modeling of solar wind interaction with Venus and Mars," in *The Comperative Study of Venus and Mars: Atmospheres, Ionospheres and Solar Wind Interactions*, AGU Monograph 66, Washington, D.C., 1992, pp. 345–383.
10. Whang, Y.C., *Space Sci. Rev.* **57**, 339 (1991).
11. Cargill, P.J., *J. Geophys. Res.* **95**, 20,731 (1990).
12. Zhuang, H.C., Russell, C.T., Smith, E.J., and Gosling, J.T., *J. Geophys. Res.* **86**, 5590 (1981).
13. Vinas, A.F., and Scudder, J.D., *J. Geophys. Res.* **91**, 39 (1986).
14. Szabo, A., *J. Geophys. Res.* **99**, 14,737 (1994).
15. Lepping, R.P., et al., *Space Sci. Rev.* **71**, 207 (1995).
16. Ogilvie, K.W., et al., *Space Sci. Rev.* **71**, 55 (1995).
17. Lepping, R.P., et al., *NASA/GSFC LEP int. doc.*, (1992).
18. Smith, C.W., et al., *Space Sci. Rev.* **86**, 613 (1998).
19. McComas, D.J., et al., *Space Sci. Rev.* **86**, 563 (1998).
20. Kokubun, S., et al., *J. Geomag. Geoelectr.* **46**, 7 (1994).
21. Frank, L.A., et al., SES-TD-92-007SY, (1992).
23. Russell, C.T., et al., *J. Geophys. Res.* **105**, 25,143 (2000).
24. Russell, C.T., et al., *J. Geophys. Res.* **88**, 9941 (1983).
25. Szabo, A., et al., "The evolution of interplanetary shocks driven by magnetic clouds", in *Proceedings of "Solar Encounter: The First Solar Orbiter Workshop"*, ESA SP-493, The Netherlands, 2001, pp. 383–387.
26. Peredo, M., Slavin, J.A., Mazur, E., Curtis, S.A., *J, Geophys. Res.* **100**, 7907 (1995).

The Magnetic Helicity of an Interplanetary Hot Flux Rope

S. Dasso*, C.H. Mandrini† and P. Démoulin**

*Instituto de Astronomía y Física del Espacio (IAFE), Buenos Aires, Argentina.
Instituto de Ciencias, Universidad de Gral. Sarmiento, Los Polvorines, Buenos Aires, Argentina.
On leave from Instituto de Física del Plasma (INFIP), Buenos Aires, Argentina.
† Instituto de Astronomía y Física del Espacio, Buenos Aires, Argentina.
**Observatoire de Paris, DASOP, France.

Abstract. In the last years, interest in the study of the relationship between the magnetic helicity of solar active regions and the one contained in the interplanetary structures has grown. This has lead us to compute the helicity content of an interplanetary hot tube observed by Wind on October 24-25, 1995, applying three different approaches in cylindrical geometry: a linear force-free field, a constant twist angle, and a non force-free model with constant current. We have fitted the set of free parameters for each of the three models, finding that the determined magnetic helicity values are very similar when using the same orientation for the flux tube. From our point of view, these results imply that, whatever be the model used, magnetic helicity is a well-determined quantity and, thus, it is worth using it to understand the link between solar and interplanetary phenomena.

INTRODUCTION

Observations of helical magnetic structures in the solar atmosphere and solar wind have attracted considerable attention in the last years, with the consequent interest in magnetic helicity studies, both in the solar and interplanetary contexts. Magnetic helicity is one of the few global quantities which is preserved even in resistive MHD on time scales shorter than the global diffusion time scale [1]. Therefore, it can be used to link phenomena under very different physical conditions.

Interplanetary flux ropes, of which magnetic clouds are a subset, present in general a helicoidal structure that is supposed to be the trace of the torsion of the corresponding plasmoids ejected from the solar surface. These interplanetary phenomena can be modeled in cylindrical geometry using three different approaches: a linear force-free field model [2], a uniform twist model (or Gold-Hoyle model, see e.g. [3]) or a non force-free model with constant current [4]. These models are physically different, being not evident at all which of them give the best representation of interplanetary flux ropes.

In this work, we first derive the analytical expressions of the magnetic helicity for the three models mentioned above. Then, we apply them to a hot tube observed by Wind on October 24-25, 1995. We fit the set of free parameters for each of the three models and present our results. We find that the values of the magnetic helicity computed for each of the models are very similar for the same orientation of the tube. This orientation is computed using a minimum variance analysis [5]. However, if the orientation of the flux rope is fitted together with the model dependent physical parameters, what we have done for the constant current model [4], we obtain a value that is lower by a factor of $\sim$ 4-5. Even though, the three models give quite close values when we compute the magnetic helicity per unit of volume, $\sim$ 0.3-0.4 nT^2 AU. We suggest that our results imply that magnetic helicity is a well-determined quantity, and can be exploited to study the link between solar and interplanetary phenomena.

MAGNETIC HELICITY OF FLUX ROPES

The magnetic helicity of a field $\vec{\mathbf{B}}$ within a volume V is defined by $H = \int_V \vec{\mathbf{A}} \cdot \vec{\mathbf{B}}\, dV$, where the vector potential $\vec{\mathbf{A}}$ satisfies $\vec{\mathbf{B}} = \vec{\nabla} \times \vec{\mathbf{A}}$. However, the helicity defined as above is physically meaningful only when the magnetic field is fully contained inside the volume V (i.e., at any point of the surface S surrounding V, the normal component $B_n = \vec{\mathbf{B}} \cdot \hat{\mathbf{n}}$ vanishes). This is so because the vector potential is defined only up to a gauge transformation ($\vec{\mathbf{A}}' = \vec{\mathbf{A}} + \vec{\nabla}\Phi$), then H is gauge-invariant only when $B_n = 0$. For cases where $B_n \neq 0$ (as on both legs of the interplanetary flux tubes) it has been shown that a relative magnetic helicity (H_r) can be defined [6]. This relative helicity is obtained subtracting the helicity of a

CP679, *Solar Wind Ten: Proceedings of the Tenth International Solar Wind Conference*,
edited by M. Velli, R. Bruno, and F. Malara

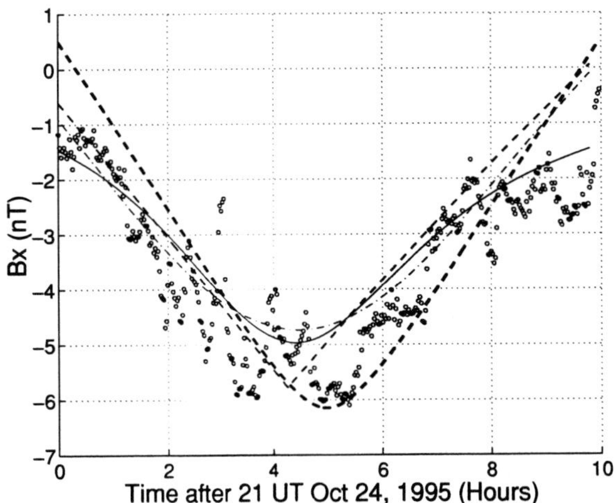

FIGURE 1. B_x component of the magnetic field (in GSE) for the flux rope observed on 24-25 October, 1995. Circles correspond to the observed field, solid line to Gold-Hoyle model, dash-dotted line to the linear force-free field model; while thin and thick dashed lines to the constant current model using a minimum variance method and a direct fit for the orientation of the tube, respectively.

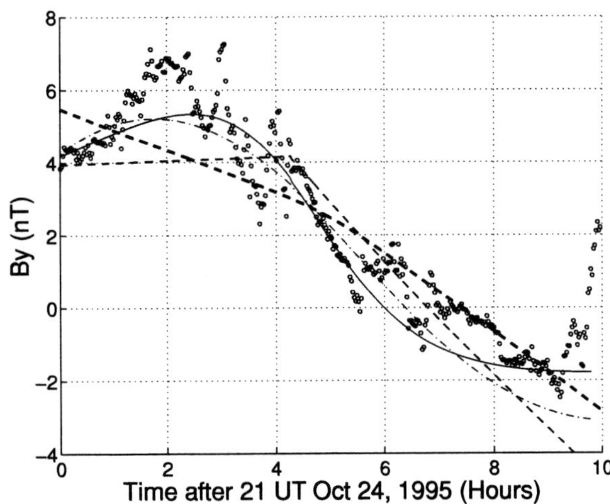

FIGURE 2. B_y component of the magnetic field (in GSE) for the flux rope observed by Wind on 24-25 October, 1995. The convention for the curves and the data is the same as in Fig. 1.

reference field $\vec{\mathbf{B}}_0$ having the same distribution of B_n on S:

$$H_r = H - \int_V \vec{\mathbf{A}}_0 \cdot \vec{\mathbf{B}}_0 \, dV \quad . \tag{1}$$

H_r is gauge−invariant and it does not depend on the common extension of $\vec{\mathbf{B}}$ and $\vec{\mathbf{B}}_0$ outside V, as was shown by [6] and [7]

The magnetic topology of an interplanetary flux rope can be modeled locally as a cylindrical structure with $\vec{\mathbf{B}}(\vec{\mathbf{r}}) = B_\phi(r)\widehat{\phi} + B_z(r)\widehat{\mathbf{z}}$. The reference field $\vec{\mathbf{B}}_0$ can be chosen as $\vec{\mathbf{B}}_0(r) = B_z(r)\widehat{\mathbf{z}}$, and the vector potential as $\vec{\mathbf{A}}_0 = A_{0,\phi}(r)\widehat{\phi} + A_{0,z}\widehat{\mathbf{z}}$, with $A_{0,z}$ a constant value and $A_{0,\phi}(r)$ satisfying $rB_{0,z}(r) = \frac{\partial}{\partial r}(rA_{0,\phi}(r))$, in order to satisfy $\vec{\mathbf{B}}_0 = \vec{\nabla} \times \vec{\mathbf{A}}_0$.

Thus, the relative magnetic helicity per unit length (L) along the tube can be expressed independently of A_0 and B_0 as,

$$H_r/L = 4\pi \int_0^R rdr \;\; A_\phi B_\phi \tag{2}$$

where R is the radius of the magnetic tube.

MAGNETIC HELICITY OF THE LINEAR FORCE-FREE FIELD

The general static, axially symmetric magnetic field of a linear force-free configuration ($\vec{\nabla} \times \vec{\mathbf{B}} = \alpha\vec{\mathbf{B}}$) was obtained by [2]. However, it has been shown that only one harmonic of this solution is enough to describe the main tendency of 'in situ' measurements for interplanetary magnetic flux ropes, as magnetic clouds (e.g., [8, 9, 10]). Thus, the field is well modeled by

$$\vec{\mathbf{B}} = B_0 J_0(\alpha r)\widehat{\mathbf{z}} + B_0 J_1(\alpha r)\widehat{\phi} \tag{3}$$

where J_n is the Bessel function of the first kind of order n, B_0 is the strength of the field and α is a constant.

Computing H_r from Eq. (2) and taking $\vec{\mathbf{A}} = \vec{\mathbf{B}}/\alpha$, it is possible to obtain the relative helicity for this force-free field as follows,

$$\frac{H_r}{4\pi L} = \frac{B_0^2}{\alpha} \int_0^R dr \; rJ_1^2(\alpha r) \tag{4}$$

A numerical integration of this equation gives [11]:

$$\frac{H_r}{L} \sim 0.70 B_0^2 R^3 \tag{5}$$

MAGNETIC HELICITY OF A UNIFORMLY TWISTED FIELD

The non-linear force-free field having a uniform twist has been used to model interplanetary flux ropes (e.g., [12]). The components of $\vec{\mathbf{B}}$ for this configuration are [3],

$$\vec{\mathbf{B}} = \frac{B_0}{1+b^2r^2}\widehat{\mathbf{z}} + \frac{B_0 br}{1+b^2r^2}\widehat{\phi} \tag{6}$$

In this magnetic configuration the amount by which a given line is twisted when going from one end of the tube to the other (b) is independent of the radius of the tube r.

From (2), and considering:

$$\vec{\mathbf{A}} = -\frac{B_0}{2b}\ln(1+b^2r^2)\widehat{\mathbf{z}} + \frac{B_0}{2b^2r}\ln(1+b^2r^2)\widehat{\phi}, \qquad (7)$$

the relative helicity results

$$\frac{H_r}{L} = \frac{\pi B_0^2}{2b^3}[\ln(1+b^2R^2)]^2 \qquad (8)$$

MAGNETIC HELICITY OF A CONSTANT CURRENT FLUX ROPE

A non force-free model has been recently proposed by [13] and [4] to describe interplanetary structures. This model assumes a constant current density such as $\vec{\mathbf{J}}(\vec{\mathbf{r}}) = J_\phi\widehat{\phi} + J_z\widehat{\mathbf{z}}$, where J_ϕ and J_z are constants. Thus, the magnetic field of this configuration is obtained as $B_z(r) = \frac{4\pi}{c}J_\phi(R-r)$ and $B_\phi(r) = \frac{2\pi}{c}J_z r$, where R is the radius of the interplanetary tube and c is the speed of light. Being the trajectory of the spacecraft $r(t)$, it can be seen that $B_z(r(t))$ and $B_\phi(r(t))$ behave as the function modulo when the impact parameter (p) of the satellite is $p = 0$.

From (2), and considering:

$$\vec{\mathbf{A}} = \frac{4\pi}{c}J_\phi r(R/2 - r/3)\widehat{\phi} - \frac{\pi}{c}J_z r^2\widehat{\mathbf{z}}, \qquad (9)$$

the relative helicity results,

$$\frac{H_r}{L} = \frac{28\pi^3}{15c^2}J_\phi J_z R^5 \qquad (10)$$

RELATIVE MAGNETIC HELICITY IN THE INTERPLANETARY HOT TUBE

We apply the analytical results derived in the previous section to the hot tube observed by Wind on October 24-25, 1995. The one minute cadence magnetic data have been downloaded from the public site: *http://cdaweb.gsfc.nasa.gov/cdaweb/istp-public/*.

Using a minimum variance analysis [5], [12] found a well-defined direction for the principal axis of the tube and $p \approx 0$ for this particular event. The variance coordinates have been used to obtain the physical parameters that best fit the observations of the flux tube for the three models discussed above. Furthermore, in order to test the validity of the minimum variance method, we have found the direction of the flux tube fitting directly its orientation (and consequently its radius) and physical parameters for the constant current model using GSE (Geocentric Solar Ecliptic) coordinates, as described in [13]. The fitting has been done in all cases using the Levenberg-Marquardt

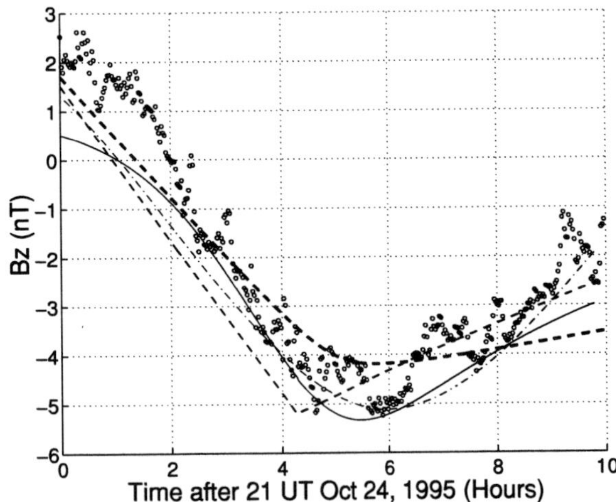

FIGURE 3. B_z component of the magnetic field (in GSE) for the flux rope observed by Wind on 24-25 October, 1995. The convention for the curves and the data is the same as in Fig. 1.

method [14]. Figures 1, 2, and 3 show the three components of the measured magnetic field, together with the curves obtained from every model.

Using the expressions derived for the relative magnetic helicity and the values of the parameters coming from the fitting, we compute the relative helicity per unit length (L) and per unit of volume (Vol). The value of a χ^2 test and the characteristics of the flux rope for every case are:

- Linear force-free field ($\chi^2_{Lfff} = 1.7$ nT):

$$|\vec{\mathbf{B}}(\vec{\mathbf{r}}=0)| = 7.2\text{ nT}, R = 0.035\text{ AU},$$
$$\alpha = 65.8\text{ AU}^{-1}$$
$$H_{r,Lfff} = 0.0016\text{ nT}^2\text{ AU}^3 L$$
$$H_{r,Lfff}/Vol = 0.40\text{ nT}^2\text{ AU}$$

- Uniform twist ($\chi^2_{Ut} = 1.5$ nT):

$$|\vec{\mathbf{B}}(\vec{\mathbf{r}}=0)| = 7.5\text{ nT}, R = 0.035\text{ AU}, b = 46.2\text{ AU}^{-1}$$
$$H_{r,Ut} = 0.0015\text{ nT}^2\text{ AU}^3 L$$
$$H_{r,Ut}/Vol = 0.39\text{ nT}^2\text{ AU}$$

- Constant current model: Results taking the orientation of the tube as computed with the minimum variance method (see [12]):

$$\chi^2_{ConstJ} = 2.1\text{ nT},$$
$$|\vec{\mathbf{B}}(\vec{\mathbf{r}}=0)| = 8.8\text{ nT}, R = 0.035\text{ AU},$$
$$J_z = 0.0435\text{ nT/s}, J_\phi = 0.0401\text{ nT/s},$$
$$H_{r,ConstJ} = 0.0013\text{ nT}^2\text{ AU}^3 L$$
$$H_{r,ConstJ}/Vol = 0.34\text{ nT}^2\text{ AU}$$

Results fitting simultaneously the orientation and the physical parameters of the tube:

$$\chi^2_{ConstJ} = 1.9\text{ nT},$$

$$|\vec{\mathbf{B}}(\vec{\mathbf{r}}=0)| = 8.7 \text{ nT}, R = 0.020 \text{ AU},$$
$$J_z = 0.12 \text{ nT/s}, J_\phi = 0.07 \text{ nT/s},$$
$$H_{r,ConstJ} = 0.00041 \text{ nT}^2 \text{ AU}^3 L$$
$$H_{r,ConstJ}/Vol = 0.31 \text{ nT}^2 \text{ AU}$$

In this latter case, the best fit is found when both the orientation and physical parameters are fitted together; this is evident from both the values of χ^2_{ConstJ} and Figures 1-3. This may imply that the minimum variance method does not give the best results for the flux tube orientation (and consequently radius). It is the aim of a future paper to explore the range of validity of this method. However, when the same orientation is used for the three models, the values of the relative magnetic helicity per unit length are in a very good agreement; being the same true for its value per unit volume.

DISCUSSION AND CONCLUSIONS

Coronal mass ejections are originated by an instability of the solar coronal field. The plasmoid ejected from the Sun will carry part of the magnetic helicity of its original field. This structure will consequently appear as an interplanetary magnetic flux rope. There is increasing evidence that this is the case from observations showing that the helicity sign in magnetic clouds matches that of their source regions (see e.g., [15, 16, 17]). Therefore, the computation of the magnetic helicity is an important tool to compare interplanetary and solar phenomena.

We have estimated the relative magnetic helicity values for a hot interplanetary flux tube observed by Wind on October 24-25, 1995. The measured magnetic field components of the structure have been fitted using three different approaches: a linear force-free field, a uniform twist and constant current model. The three models fit relatively well the observations being the quality of the fit, according to a χ^2 test, in ascending order: the uniform twist, the linear force-free field and the constant current model. The relative magnetic helicity per unit length derived for the three cases is very similar when the orientation of the tube and, therefore, its radius is the same. The latter is computed using a minimum variance method, as discussed in [12]. Only in the case when the orientation is computed independently, by fitting it together with the physical parameters depending on the model, we find a value which is a factor $\sim 4-5$ lower. We suggest that this is an indication of the limitation of the minimum variance method, what we intend to quantify in a following paper. However, when we compute the helicity per unit of volume, that is to say when the results are independent of the radius, we find again very similar results for the three models, $H_r/Vol \sim 0.3-0.4 \text{ nT}^2$ AU.

We conclude that magnetic helicity is a quantity that can be used to study the link between coronal and interplanetary phenomena, in the sense that, no matter the model used, its value is well-determined. The case studied here is an example. Our next step is to extend our analysis to a large dataset of interplanetary phenomena, in particular, magnetic clouds.

ACKNOWLEDGMENTS

This research has made use of NASA's Space Physics Data Facility (SPDF). This work was partially supported by the Argentinean UBA grant UBACYT X059. S.D. is a fellow of CONICET and C.H.M. is a member of the Carrera del Investigador Científico, CONICET. C.H.M. and P.D. thank ECOS (France) and SETCIP (Argentina) for their cooperative science program A01U04.

REFERENCES

1. Berger, M. A., *Geophysical and Astrophysical Fluid Dynamics*, **30**, 79–104 (1984).
2. Lundquist, S., *Ark. Fys.*, **2**, 361+ (1950).
3. Priest, E. R., *Solar magneto-hydrodynamics*, Dordrecht, Holland ; Boston : D. Reidel Pub. Co. ; Hingham,, 1982.
4. Hidalgo, M. A., Cid, C., Medina, J., and Viñas, A. F., Sol. Phys., **194**, 165–174 (2000).
5. Sonnerup, B. U., and Cahill, L. J., J. Geophys. Res., **72**, 171–183 (1967).
6. Berger, M. A., and Field, G. B., *J. Fluid. Mech.*, **147**, 133+ (1984).
7. Finn, J. H., and Antonsen, T. M., *Comments Plasma Phys. Contr. Fusion*, **9**, 111+ (1985).
8. Burlaga, L., Sittler, E., Mariani, F., and Schwenn, R., J. Geophys. Res., **86**, 6673–6684 (1981).
9. Burlaga, L. F., J. Geophys. Res., **93**, 7217–7224 (1988).
10. Lepping, R. P., Burlaga, L. F., and Jones, J. A., J. Geophys. Res., **95**, 11957–11965 (1990).
11. Démoulin, P., Mandrini, C. H., van Driel-Gesztelyi, L., Thompson, B. J., Plunkett, S., Kovári, Z., Aulanier, G., and Young, A., A&A, **382**, 650–665 (2002).
12. Farrugia, C. J., Janoo, L. A., Torbert, R. B., Quinn, J. M., Ogilvie, K. W., Lepping, R. P., Fitzenreiter, R. J., Steinberg, J. T., Lazarus, A. J., Lin, R. P., Larson, D., Dasso, S., Gratton, F. T., Lin, Y., and Berdichevsky, D., "A Uniform-Twist Magnetic Flux Rope in the Solar Wind," in *AIP Conf. Proc. 471: Solar Wind Nine*, 1999, pp. 745–748.
13. Hidalgo, M. A., Cid, C., Medina, J., Viñas, A. F., and Sequeiros, J., J. Geophys. Res., **107**, 10.1029+ (2002).
14. Press, W. H., Teukolsky, S. A., Vetterling, W. T., and Flannery, B. P., *Numerical Recipes*, Cambridge University Press, 1992.
15. Bothmer, V., and Schwenn, R., *Space Science Reviews*, **70**, 215–+ (1994).
16. Rust, D. M., Geophys. Res. Lett., **21**, 241–244 (1994).
17. Marubashi, K., "Interplanetary Magnetic Flux Ropes and Solar Filaments," in *Coronal Mass Ejection, Geophys. Monograph 99*, 1997, pp. 147–156.

Dust in the wind: The dust geometric cross section at 1 AU based on neutral solar wind observations

Michael R. Collier*, Thomas E. Moore*, K. Ogilvie*, D.J. Chornay*, J. Keller*, S. Fuselier†, J. Quinn**, P. Wurz‡, M. Wüest§ and K.C. Hsieh¶

*NASA/Goddard Space Flight Center, Code 692, Greenbelt, Maryland 20771
†Lockheed Martin Advanced Technology Center, Palo Alto, CA 94304
**University of New Hampshire, Durham 03824
‡University of Bern, CH-301, Switzerland
§Southwest Research Institute, 6220 Culebra Road, San Antonio, TX 78228-0510
¶Department of Physics, University of Arizona, Tucson, AZ 85721

Abstract. We report observations of the neutral component of the solar wind from the Low Energy Neutral Atom (LENA) imager on the NASA IMAGE spacecraft from year 2001. There is a pronounced annual modulation of the neutral solar wind, and the flux outside of the upstream region is used to place an upper limit on the dust geometric cross section in the sunward direction at 1 AU of $\Gamma^{1AU} < 6\text{x}10^{-19}$ cm^{-1}. This value agrees with inferences made from the zodiacal light.

INTRODUCTION

Because solar wind charge-state distributions are "frozen in" close to the Sun, they reflect high (ionization) temperatures ($> 10^6$ K) [1]. Consequently, a negligible fraction of the solar wind is expected to be neutral based on equilibrium charge state distributions. Yet, the solar wind is not completely ionized. Our observations at 1 AU indicate that a substantial ($\sim 10^{-5} - 10^{-3}$ fractionally) neutral component of the solar wind enters the Earth's magnetosphere at all times. This neutral solar wind forms between the Sun and the orbit of the Earth from solar wind ions exchanging charge with matter in this region.

There are three major sources of neutral solar wind: the Earth's geocorona, interstellar neutrals, and dust in the inner solar system. The Earth's geocorona extends far beyond the magnetopause and drops off rapidly with distance from the Earth [2]. So when the solar wind dynamic pressure increases dramatically, large fluxes of solar wind neutrals are created due to solar wind charge exchange with neutral hydrogen atoms surrounding the Earth but outside of the magnetopause. This mechanism for forming neutral solar wind was first discussed by *Dessler et al.* [3] in 1961 who considered it a potential source of background for observations of the energetic neutral atom flux resulting from proton ring current decay induced by charge exchange with hydrogen atoms.

Neutral solar wind may also form by solar wind protons exchanging charge with interstellar neutrals, a gas of mostly hydrogen and helium atoms streaming through the heliosphere at about 25 km/s coming from a point such that the Earth is upstream of the Sun in the neutral flow in early June each year (Specifically, the downstream flow direction is at a heliographic longitude of 74° so that the Earth is upstream on June 2 each year [4]). Because the interstellar neutral hydrogen becomes depleted due to photoionization and charge exchange with the solar wind, a one to three order of magnitude annual variation is expected in the neutral solar wind flux observed at the Earth as it moves from upstream, where the interstellar and solar wind neutral hydrogen fluxes are high, to downstream, where they are low [5]. *Fahr* [6] and *Hundhausen et al.* [7] both pointed out in 1968 that solar wind protons would charge exchange with neutral hydrogen near the Sun and produce a neutral solar wind.

Dust in the inner solar system is a source of neutrals for solar wind charge exchange close to the Sun [8]. The best-known effect of this dust is the zodiacal light, discovered and correctly identified by Cassini in 1683 as being due to sunlight scattered off small particles orbiting the Sun. The size range of these particles is from below one to more than 100 μm. The dust in the inner solar system, at least, is thought to originate primarily from comets and asteroids [9]. There are two possible mechanisms for dust to generate neutral solar wind shown in Figure 1. In the top panel, solar wind ions saturate the dust particles which later release neutral atoms that subsequently charge exchange with solar wind protons creating a neutral solar wind [10]. In the bottom panel, solar wind ions traverse small dust particles and

CP679, *Solar Wind Ten: Proceedings of the Tenth International Solar Wind Conference*, edited by M. Velli, R. Bruno, and F. Malara

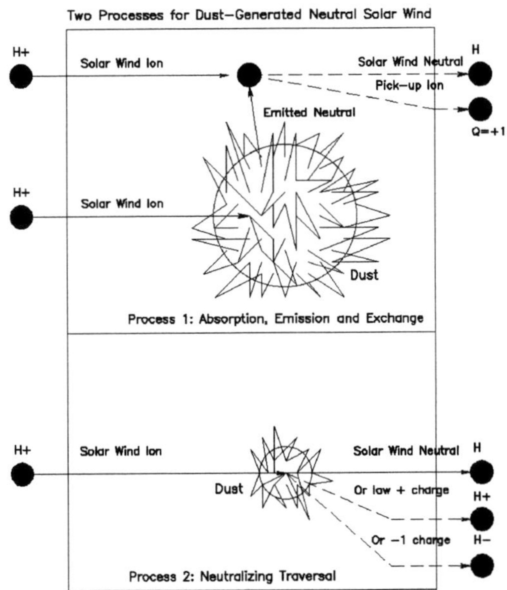

FIGURE 1. Dust creates solar wind neutrals either through a process of absorption, re-emission, and charge exchange or through a solar wind ion penetrating the dust grain and emerging on the other side in a neutral state.

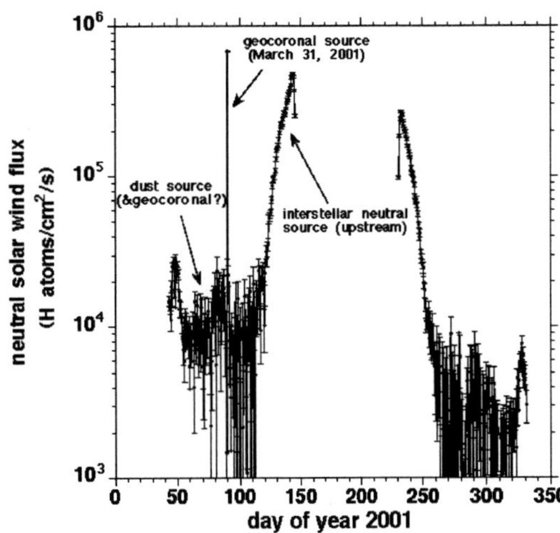

FIGURE 2. Neutral solar wind fluxes observed at IMAGE apogee during year 2001 illustrate the three major sources of neutral solar wind. Error bars indicate statistical uncertanties.

emerge neutral or in low charge states [11][1].

NSW OBSERVATIONS DURING 2001

The neutral solar wind (NSW) was observed for the first time by the Low Energy Neutral Atom (LENA) Imager on the polar orbiter, IMAGE [13, 14].

Figure 2 shows the neutral solar wind flux as observed by LENA during the year 2001 [15]. The observations were taken during the hour around apogee ($\sim$ 8 R_E) on each orbit (about every 14 hours) and cover slightly more than half the year (days 43-143 and 230-330) because LENA has the Sun in its $\sim$90° field-of-view only about half the time.

The data illustrate the three different sources of neutral solar wind: The spike on day 90 (31 March 2001) is due to solar wind charge exchange with the Earth's geocorona during a period when the solar wind ram pressure was unusually high and the terrestrial magnetopause was highly compressed [16]. The broad peak beginning about day 120 and lasting until about day 250 occurs when the Earth is upstream of the Sun in the interstellar neutral flow and a larger fraction of the interstellar neutrals penetrate to inside of 1 AU and charge exchange with the solar wind. It should be noted, however, that this peak may not be due to direct charge exchange with the $\sim$25 km/s interstellar neutrals but rather with a higher energy component that began as interstellars, e.g. neutralized pick up ions [17]. Finally the more uniform background flux between about $2\text{x}10^3$ and $2\text{x}10^4/\text{cm}^2/\text{s}$ is due primarily to solar wind charge exchange with dust or dust-generated neutral atoms with probably a geocoronal contribution, as well. However, note that the uniform background does appear to have systematic variations in it of about a factor of two near the center and edges which are consistent with the degree of variability observed during calibration in the polar angle response. In addition, the LENA efficiency may have trended downward in the latter part of 2001, perhaps the result of sputtering agents on the tungsten surface becoming depleted.

THE DUST GEOMETRIC CROSS SECTION AT 1 AU

We consider a quantity called the dust geometric cross section (Γ), which is the dust density times the effective cross section of a dust particle. This may be thought of as the total cross-sectional area per unit volume or, equivalently, the probability per unit length, for small distances, a solar wind proton will interact with the dust, either by being absorbed, as in process 1, or traversing the dust, as in process 2 (see Fig. 1). The rate, then, at which a solar wind proton will be absorbed by dust is the

[1] This is analogous to the behavior of ions penetrating the thin carbon foils used in time-of-flight instrumentation. More energetic ($\sim$50 keV) neutrals resulting from ions penetrating dust in Saturn's rings have been considered by Mauk et al. [12]

TABLE 1. Inferences of the dust geometric cross section at 1 AU based on various measurement techniques

Observation Method	Γ^{1AU} **(cm^{-1})**	**reference**
Zodiacal light observations	$10^{-21} - 2.0\text{x}10^{-19}$	[10]
Rocket and satellite data	$3.8\text{x}10^{-18}$	[10]
Thermal emission of the F corona	$8.0\text{x}10^{-18}$	[10]
He^+ in the solar wind	$6.5\text{x}10^{-17}$	[10]
Balmer emissions	$3.8\text{x}10^{-17}$	[10]
Micro-craters and in-situ	$4.6\text{x}10^{-21}$	[9]
Inner source pickup ions	$> 1.3\text{x}10^{-17}$	[18]
Neutral solar wind observations	$< 6\text{x}10^{-19}$	this work

solar wind speed times Γ. Multiplying by the solar wind number density gives the volumetric rate of solar wind interaction with the dust: $n_{sw}v_{sw}\Gamma = \Phi_{sw}\Gamma$, where Φ_{sw} is the solar wind flux $\sim 3\text{x}10^8/\text{cm}^2/\text{s}$.

We take the dust density to decrease with distance from the Sun as $1/r$ [19, 20] (which is a theoretically based slope although *Helios* observations place the fall-off close to this at $1/r^{1.3}$ [8]) and the solar wind density to decrease as $1/r^2$. We can then relate the volumetric rate of solar wind absorption to 1 AU quantities

$$\Phi_{sw}\Gamma = \Phi_{sw}^{1AU}\Gamma^{1AU}(r_0/r)^3, \tag{1}$$

where r_0 is 1 AU.

In either of the two processes shown in Fig. 1, the rate of neutral solar wind creation in steady state will be roughly the rate of solar wind absorption given by (1). In the case of process 1, absorption, emission, and charge exchange, this neglects the probability that an emitted neutral will become photoionized, which occurs at a much lower rate than charge exchange [21]. In the case of process 2, neutralizing traversal, this neglects both the probability that an emitted neutral will be photoionized as well as the probability that it will emerge in a low positive charge state or with a charge -1. However, the ions that emerge as ions after traversing the dust may proceed to interact with a second dust particle and those emerging with charge -1 may be neutralized soon thereafter.

To a good approximation, then, (1) represents the source term for the dust-generated neutral solar wind and continuity dictates

$$\frac{\partial \rho_{NSW}^{dust}}{\partial t} + \nabla \cdot (\rho_{NSW}^{dust}\mathbf{v}_{sw}) = \Phi_{sw}^{1AU}\Gamma^{1AU}\left(\frac{r_0}{r}\right)^3, \tag{2}$$

which in steady-state is

$$\frac{1}{r^2}\frac{\partial}{\partial r}\left(r^2\Phi_{NSW}^{dust}\right) = \Phi_{sw}^{1AU}\Gamma^{1AU}\left(\frac{r_0}{r}\right)^3. \tag{3}$$

Integrating from $10R_\odot$, taken because from this point on the assumptions made about the solar wind density drop and constant velocity are valid, to 1 AU, we get

$$\Phi_{NSW}^{dust}(1\text{ AU}) = \Phi_{sw}^{1AU}\Gamma^{1AU} \cdot 4.6\text{x}10^{13}\text{ cm}. \tag{4}$$

Taking a typical value for the dust-source neutral solar wind of 8300 /cm^2/s from the 2001 neutral solar wind data shown in Fig. 2, and applying (4) we get

$$\Gamma^{1AU} = 6\text{x}10^{-19}\text{ cm} \tag{5}$$

for the total dust geometric cross section at 1 AU. Since this assumes that the observed neutral solar wind flux is due solely to dust-generated neutrals, it should probably be viewed as an upper limit on Γ, although the assumptions discussed above in deriving (2) will tend to somewhat lower the estimate of Γ, mitigating the error introduced by the geocoronal contribution to the neutral solar wind flux.

DISCUSSION

This value, $\Gamma^{1AU} < 6\text{x}10^{-19}\text{ cm}^{-1}$, provides a useful upper limit because of the large uncertainty in measurements of dust in the inner solar system. As indicated in Table 1, measurements of the total geometric cross section at 1 AU vary from 10^{-21} through $6.5\text{x}10^{-17}\text{ cm}^{-1}$, about four to five orders of magnitude.

This upper limit is consistent with the zodiacal light observations, but is significantly lower than the other determinations and, interestingly, our upper limit is lower than the lower limit of *Schwadron et al.* [18]. Although it is unclear why this should be the case, the determinations depend critically on the measurement technique, perhaps because the various techniques are sensitive to dust in different size ranges.

This value is estimated at one phase in the solar cycle. The zodiacal light brightness does not appear to vary with solar cycle suggesting that the dust-generated neutral solar wind will not vary with solar cycle either [22]. However, even if the dust distribution is steady, the variability of the solar wind may cause some variation in the dust-generated neutral solar wind. Although this number contains a contribution from interstellar dust, this contribution is small in comparison to interplanetary dust, about 4% [23, 24], and exhibits a different spatial distribution, in particular, the presence of a downstream focussing cone [25].

One concern with this estimate involves the assumption that the contribution due to dust is close to azimuthally symmetric. The dust population may not be azimuthally symmetric or may deviate from a plane as was concluded by *Vrtilek and Hauser* [26]. However, there do not appear to be structures in zodiacal light measurements of the scale of $10^5 - 10^6$ km [20].

CONCLUSION

The properties of dust in the inner solar system and of the interaction of the solar wind with the interplanetary dust play a key role in understanding the origin of inner source pickup ions [27] and in determining the contribution of the interstellar neutral and geocoronal sources to the neutral solar wind. Solar wind-dust interaction was undoubtedly even more important during earlier, dusty periods in the solar system and may be of paramount importance near stars surrounded by very dense dust clouds [10].

Neutral solar wind observations provide a powerful and new probe of the cloud of interplanetary dust particles that pervades our solar system. This technique is not apparently sensitive to the size of interplanetary dust particles, requiring only that a solar wind ion interacts with a dust particle, which provides an advantage over other observational methods that are restricted to dust particles within certain size ranges [8].

ACKNOWLEDGMENTS

Special thanks to Melvyn Goldstein without whom this manuscript would not have materialized. Special thanks also to Ed Roelof, David McComas and Martin Hilchenbach for very helpful conversations during the meeting. This research is supported by the Explorer Program at NASA's GSFC under Mission Operations and Data Analysis UPN 370-28-20.

REFERENCES

1. Owocki, S., and Scudder, J., *Astrophys. J.*, **270**, 758–768 (1983).
2. Rairden, R., Frank, L., and Craven, J., *J. Geophys. Res.*, **91**, 13,613–13,630 (1986).
3. Dessler, A., Hanson, W., and Parker, E., *J. Geophys. Res.*, **66**, 3631–3637 (1961).
4. Geiss, J., and Witte, M., *Space Sci. Rev.*, **78**, 229–238 (1996).
5. Bzowski, M., Fahr, H. J., and Ruciński, D., *Icarus*, **124**, 209–219 (1996).
6. Fahr, H., *Astrophys. Space Sci.*, **62**, 496–503 (1968).
7. Hundhausen, A., Gilbert, H., and Bame, S., *J. Geophys. Res.*, **73**, 5485–5493 (1968).
8. Leinert, C., and Grün, E., "Interplanetary Dust," in *Physics and Chemistry in Space - Space and Solar Physics*, edited by R. Schwenn and E. Marsch, Physics of the Inner Heliopshere I 20, Springer-Verlag, Berlin, 1990, pp. 207–275.
9. Grün, E., "Dust in the Solar System," in *The Outer Heliosphere: Beyond the Planets*, edited by K. Scherer, H. Fichtner, and E. Marsch, Copernicus Gesellschaft e.V., Katlenburg-Lindau, 2000, pp. 289–304.
10. Banks, P., *J. Geophys. Res.*, **76**, 4341–4348 (1971).
11. Wimmer-Schweingruber, R. F., and Bochsler, P., *Geophys. Res. Lett.*, **in press** (2002).
12. Mauk, B.H. et al., *Planet. Space Sci.*, **46**, 1349–1362 (1998).
13. Moore, T.E. et al., *Geophys. Res. Lett.*, **28**, 1143–1146 (2001).
14. Collier, Michael R. et al., *J. Geophys. Res.*, **106**, 24,893–24,906 (2001).
15. Moore, T.E. et al., *Space Science Reviews*, **91**, 155–195 (2000).
16. Collier, Michael R. et al., "LENA observations on March 31, 2001: Magnetosheath remote sensing," in *EOS Transactions American Geophysical Union*, Fall Meeting Supplement 82(47), AGU, 2001, pp. F1071–1072 Abstract SM41C–05.
17. Gruntman, Mike et al., *J. Geophys. Res.*, **106**, 15,767–15,781 (2001).
18. Schwadron, N.A. et al., *J. Geophys. Res.*, **105**, 7473–7481 (2000).
19. Weinberg, J., *Ann. Astrophys.*, **27**, 718–738 (1964).
20. Richter, I., Leinert, C., and Planck, B., *Astron. Astrophys.*, **110**, 115–120 (1982).
21. Ruciński, D. et al., *Space Sci. Reviews*, **78**, 73–84 (1996).
22. Leinert, C., Richter, I., and Planck, B., *Astron. Astrophys.*, **110**, 111–114 (1982).
23. Mann, I., and Kimura, H., *J. Geophys. Res.*, **105**, 10,317–10,328 (2000).
24. Grün, Eberhard et al., *J. Geophys. Res.*, **105**, 10,403–10,410 (2000).
25. Landgraf, M., *J. Geophys. Res.*, **105**, 10,303–10,316 (2000).
26. Vrtilek, J. M., and Hauser, M. G., *Astrophys. J.*, **455**, 677–692 (1995).
27. Gloeckler, G., Fisk, L., Geiss, J., Schwadron, N., and Zurbuchen, T., *J. Geophys. Res.*, **105**, 7459–7463 (2000).

Interaction Of Magnetic Clouds In The Inner Heliosphere

E. Romashets[*], P. Cargill[‡], and J. Schmidt[‡]

[*]*Institute of Terrestrial Magnetism, Ionosphere, and Radio Wave Propagation of Russian Academy of Sciences (IZMIRAN), Troitsk, Moscow Region, Email: romash@izmiran.rssi.ru*
[‡]*Imperial College of Science, Technology, and Medicine, London, UK.*

Abstract. A method of potentials has been used in the past for the calculation of the force acting on isolated magnetic bodies in solar corona and inner heliosphere, where large gradients of magnetic pressure exist. Since recent observations showed that coronal mass ejections (CME) can leave the Sun more frequently than was expected before 1995, it is clear that interactions between CMEs can play important role in the formation of geo-effective structures near the Earth's orbit. We present here an evaluation of two interacting CMEs and the field distribution around them, using potential solution in bi-cylindrical coordinates.

INTRODUCTION

In attempts to find the value of a force acting on an isolated body in solar corona, Parker [1] proposed a potential magnetic field, and found an expression for the force: the so called "melon seed mechanism". The magnetic structure inside the cloud was assumed to be force-free [2]. Later another form of this distribution in cylindrical geometry [3] was used in the literature for interpretation of magnetic clouds observations. Romashets and Vandas [4] used the method for the study of more complicated, toroidal-shape clouds dynamics in solar wind. There were many studies of non-potential flow around such kinds of cylindrical and spherical bodies [5-7] using an MHD approach, both analytically and numerically. But no explicit expression for field and force were found.

CALCULATION: POTENTIAL FIELD

Assume two cylinders aligned along the Z axis are inserted into initially uniform X-directed field. The cloud magnetic fields have no connection with the ambient field. In bi-cylindrical coordinates the initial field potential is

$$\Psi_0 = B_0 x = B_0 a \sum_{n=0}^{\infty} \varepsilon_n e^{-n\mu} \cos n\eta \cdot \quad (1)$$

Here a is a parameter of the coordinate system, $\varepsilon_n = 1$ for $n = 1$, $\varepsilon_n = 2$ for $n \geq 2$. The following relations hold between cartesian coordinates and new coordinates μ, η and Z:

$$x = \frac{a \sinh \mu}{\cosh \mu - \cos \eta}, \quad (2)$$

$$y = \frac{a \sin \eta}{\cosh \mu - \cos \eta}, \quad (3)$$

$$z = Z, \quad (4)$$

and the radii of cylinders are equal to $r_0 = \frac{a}{\sinh \mu_0}$, the distance between their centers is: $D = 2 r_0 \cosh \mu_0 = 2a \coth \mu_0$. The equation $\mu = |\mu_0|$ defines the cylinder's surfaces in new coordinates. On insertion of the two clouds, the ambient field has been changed and distorted in such a way that

$$B_\mu \big|_{\mu=\mu_0} = 0 \quad (5)$$

CP679, *Solar Wind Ten: Proceedings of the Tenth International Solar Wind Conference,*
edited by M. Velli, R. Bruno, and F. Malara

The new potential is then

$$\Psi_1 = B_0 a \sum_{n=0}^{\infty} \varepsilon_n \left(e^{-n\mu} + e^{n\mu - 2n\mu_0}\right) \cos n\eta, \quad (6)$$

chosen among harmonic functions satisfying (5) and asymptotic to equ. (1) at infinity.

THE DISTURBED FIELD

The components of the disturbed field are:

$$B_x = 2B_0 \left\{ (1 - \cosh\mu\cos\eta) \sum_{n=1}^{\infty} n\left(e^{n\mu-2n\mu_0} - e^{-n\mu}\right)\cos n\eta + \right.$$
$$\left. + \sinh\mu\sin\eta \sum_{n=1}^{\infty} n\left(e^{n\mu-2n\mu_0} + e^{-n\mu}\right)\sin n\eta \right\} \quad (7)$$

$$B_y = 2B_0 \left\{ -\sinh\mu\sin\eta \sum_{n=1}^{\infty} n\left(e^{n\mu-2n\mu_0} - e^{-n\mu}\right)\cos n\eta + \right.$$
$$\left. + (\cosh\mu\cos n\eta - 1) \sum_{n=1}^{\infty} n\left(e^{n\mu-2n\mu_0} + e^{-n\mu}\right)\sin n\eta \right\} \quad (8)$$

where

$$\mu = \frac{1}{2} ln \frac{(x+a)^2 + y^2}{(x-a)^2 + y^2}, \quad (9)$$

$$\eta = \frac{i}{2} ln \frac{(y-ia)^2 + x^2}{(y+ia)^2 + x^2}. \quad (10)$$

THE FORCE ON THE FLUX ROPES AND EXAMPLE

The force acting on each cylinder is

$$F_a = \int_S \frac{B_\eta^2}{8\pi} e_x dS = -\frac{2B_0^2}{\pi} aL_z \cdot$$
$$\cdot \int_{-\pi}^{\pi} (\cosh\mu_0\cos\eta - 1) \left(\sum_{n=1}^{\infty} n e^{-n\mu_0} \sin n\eta \right)^2 d\eta \quad (11)$$

and is directed along X axis. This force moves the CMEs closer to each other. The following example can describe the situation near the Earth's orbit more adequately. The initial external field is

$$B_x = B_0 + B_1 \frac{x}{a} \quad (12)$$

$$B_y = -B_0 - B_1 \frac{y}{a} \quad (13)$$

where B_0 is the averaged value of x-component near the location being considered, for example near the 1 AU. $\frac{B_1}{a}$ measures the gradient of each component. In this case the disturbed potential is:

$$\Psi_1 = B_0 a \sum_{n=0}^{\infty} \varepsilon_n \left(e^{-n\mu} + e^{m\mu - 2n\mu_0}\right) \cos n\eta -$$
$$- B_0 a \sum_{n=0}^{\infty} 2\left(e^{-n\mu} + e^{m\mu - 2n\mu_0}\right) \sin n\eta +$$
$$+ B_1 a \sum_{n=0}^{\infty} n\left(e^{-n\mu} + e^{m\mu - 2n\mu_0}\right) \cos n\eta \quad (14)$$

In Figure 1 one can see contours of magnetic field magnitude.

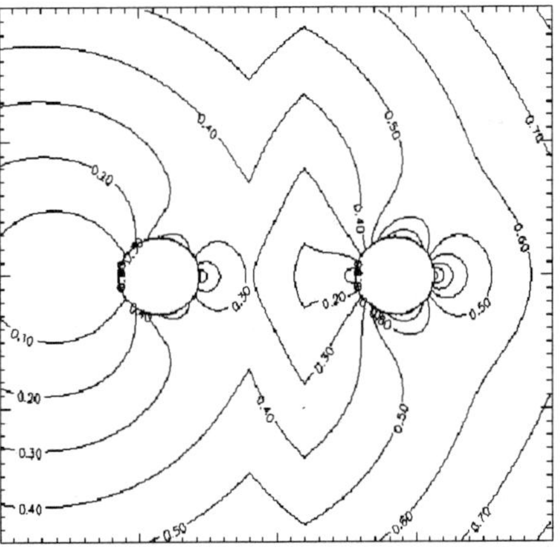

FIGURE 1. Sample contours of magnetic field magnitude disturbed by insertion of two cylindrical magnetic clouds into medium with initially non-uniform field.

The force acting on each of two cylinders has the following components:

$$F_x = -\frac{2aL_z}{\pi}\int_{-\pi}^{\pi}(\cosh\mu_0\cos\eta - 1)\left(B_p(\mu_0,\eta)\right)^2 d\eta \quad (15)$$

$$F_y = -\frac{2aL_z}{\pi}\int_{-\pi}^{\pi}\sinh\mu_0\sin\eta\left(B_p(\mu_0,\eta)\right)^2 d\eta \quad (16)$$

If the equations of motions are solved for both cylinders using (15) and (16), it will be seen that two bodies are rotating with respect to each other. There is only a stable orientation for $\varphi = 45^0$.

For evaluation of (15) and (16), we can use an approximation of interplanetary magnetic field (IMF):

$$B_r = \frac{B_e}{r^2}r_e^2,\ B_\varphi = \frac{B_e}{r}r_e, \quad (17)$$

where $B_e = 5\,nT$ is the averaged value of IMF components at 1 AU, $r_e = 1AU$, r is the distance from the center of the Sun. Now we can find gradient of magnetic field.

$$\frac{\partial B_r}{\partial r} = -\frac{2B_e}{r^3}r_e^2 \quad (18)$$

$$\frac{\partial B_\varphi}{\partial r} = -\frac{B_e}{r^2}r_e \quad (19)$$

From (12), (13) and (18), (19) it can be seen that we can take $B_0 = B_e$ and $B_1 = -\frac{3B_e a}{2r_e}$ for conditions at 1 AU. Using these values in (14) we have:

$$B_p(\mu,\eta) = 2B_e\sum_{n=1}^{\infty} n e^{-n\mu_0}\left(2+\frac{3na}{2r_e}\right)\cosh(n\mu-\mu_0)\cos\left(n\eta+\frac{\pi}{4}\right) \approx$$
$$\approx 2B_e e^{-\mu_0}\left(2+\frac{3na}{2r_e}\right)\cosh(\mu-\mu_0)\cos\left(n\eta+\frac{\pi}{4}\right) \quad (20)$$

and from (15), (16) we have the force:

$$F_x = 32aL_z B_e^2 e^{-2\mu_0}\left(1+\frac{3a}{2r_e}\right) \quad (21)$$

CONCLUSIONS

Modification of the IMF around two cylindrical CMEs was found using a potential field formulation in bi-cylindrical coordinates. The results can be used for a calculation of the force acting on both CMEs during their motion from the Sun. Plasma parameters distribution for this geometry were not found but can be if the entire system of MHD equations is solved for V, n, and T and will be the subject of future work. The maximum increase of B around both clouds is a factor of 2-3, and one can expect that the velocity will increase by a similar amount. Streamlines can be slightly different from field lines but resemble them in general.

ACKNOWLEDGEMENTS

This work was supported by EU/INTAS/ESA grant 99-00727.

REFERENCES

1. Parker, E. N., *Astrophys. J. Suppl. Ser.*, ***3***, 51-76, (1957).
2. Chandrasekhar, S., and Kendall, P. C., *Astrophys. J.*, ***126***, 457-460, (1957).
3. Lunquist, S., *Ark. Fys.*, ***2***, 361- 366, (1950).
4. Romashets, E. P., Vandas, M., *J. Geophys. Res.*, ***106***, 10615-10624, (2001).
5. Parbhakar, K. J. , and Uberoi, M. S. , *Phys. Fluids*, ***10***, 2083-2089, (1969).
6. Farrugia, C. J. et al., *Planet. Space Sci.*, ***35***, 227-240, (1987).
7. Detman, T. R. et al., *J. Geophys. Res.*, ***96***, 9531-9540, (1991).
8. Morse, P. M., and H Feshbach, *Methods of Theoretical Physics*, New York, McGraw-Hill, 1953, pp. 1300-1305.

Session VI

New Missions, Opportunities, and Techniques for Heliospheric Physics

Where do we go with Solar and Heliospheric Physics?

E. Möbius

Space Science Center and Department of Physics, University of New Hampshire, Durham, NH, USA

Abstract. After about 40 years of solar wind and heliospheric space research questions about the structure and the origin of the solar wind still await an answer. We still don't understand how the corona is heated and how the solar wind is accelerated, what are the sources for the fast and slow wind. Recent findings seem to indicate that the magnetic patterns below the sun's surface and the coronal structure are intimately connected, but quantitative connections are difficult to make. Modeling shows that the strength of the solar wind and coronal mass ejections controls the spatial and temporal response of the heliospheric boundary, while the physical state of the surrounding interstellar medium sets the boundary conditions. With a fleet of solar and heliospheric spacecraft throughout the heliosphere we have a unique situation that enables us to make substantial progress, and "Living with a Star" will add a comprehensive 3D view of the inner heliosphere. However, the fundamental questions cannot be solved without in-situ sampling to within a few solar radii, and the heliospheric boundary will not be fully understood without crossing it.

1. INTRODUCTION

Since the beginning of the space age in 1957 our satellites and probes have ventured the regions beyond the Earth's atmosphere. Our knowledge about the Earth's magnetosphere, other planets and their magnetospheres, interplanetary space and about the sun has advanced substantially. Spacecraft have visited almost every planet, they have flown by comets and asteroids. Probes have sampled the solar wind and even pickup ions, neutral atoms and dust particles of extrasolar origin, extensively in the ecliptic plane. The Ulysses spacecraft explores the high latitude regions beyond 1.5 AU, which has led to a better understanding of the 3-dimensional structure of the heliosphere. The Helios probes have extended our knowledge inward to 0.3 AU from the sun, and the Voyagers are expected to reach the first indication of the distant boundary of the heliosphere, the solar wind termination shock, within the next few years. However, we have not yet probed and understood the origin of the solar wind – the inner frontier, nor have we crossed and probed its boundary – the outer frontier, not to mention exploring the space beyond.

Concerning the origin of the solar wind, we still don't understand the physical processes that heat the corona to $1 - 2 \cdot 10^6$ K, whereas the photosphere is only at 5500 K. In layman's terms: why is the "cooking pot" hotter than the "stove"? There are still debates about the mechanisms that lead to the acceleration of the wind to supersonic speeds, about the sources of the fast and the slow wind, and where and how energetic particles are produced. Conversely, looking outward we need to understand how our galactic neighborhood interacts with the solar wind. This interaction determines the size of the heliosphere, how much of the galactic cosmic rays and interstellar dust is kept out, and how the interstellar neutral gas is filtered before it flows through our system. On a broader scheme, our heliosphere is the one example of an "astrosphere" - typical for many stars - that we can study with in-situ methods. Similarly, the local interstellar cloud (LIC) is the one sample of present day interstellar matter whose elemental and isotopic composition and physical state we can study in detail.

Why should we strive for such far-reaching goals, and will there be any tangible benefits for our society? In a nutshell there are three profound reasons:

• Firstly, it is a matter of comprehension; we need to understand our larger environment, in which the crucial Sun-Earth relationships are embedded. As we have learned over the past decade, humankind has made itself vulnerable to "space weather" [*1*], with continent-wide power grids that are sensitive to magnetic impulses, with application satellites whose electronics can be destroyed, and with humans in space who could be harmed by excessive radiation. Unraveling the source of the solar wind and its outer boundary is a fundamental step towards the basic understanding needed for successful "space weather" forecast, which is currently in the state of weather forecast about 100 years ago. Also, the heliosphere is the first

CP679, *Solar Wind Ten: Proceedings of the Tenth International Solar Wind Conference*,
edited by M. Velli, R. Bruno, and F. Malara

defensive shield of the terrestrial bio-system against high-energy galactic cosmic radiation. The inner two shields are the Earth's magnetosphere and atmosphere.
• Secondly, it is a matter of investigation, how our universe, the sun and its planets and we ourselves evolved. Taking two disparate samples of galactic material, i.e. from the solar system, which was born 4.5 billion years ago, and from the LIC, which represents present day galactic matter, we can learn about the evolution of matter from the Big Bang to date.
• Thirdly, sending a Solar Probe into the atmosphere of the sun, our star, and sending an Interstellar Probe into our local galactic environment, is exploration at its best. This is an indispensable aspect of cultural activity since the dawn of mankind. So far no exploratory mission has ever returned without new and unexpected findings that greatly expand our knowledge.

Only a few key ideas of possible future directions can be presented here. A more detailed account has been given elsewhere [*2*]. We will explore with a few examples how progress can be made taking a three-step approach. The first step is to make the most out of our existing space assets. With ACE, SOHO, Wind, Ulysses and the Voyagers a remarkable fleet of heliospheric spacecraft is scouting simultaneously the inner, outer and 3D heliosphere. The termination shock is within reach, while continuous observations of key regions and solar variability can be compiled, provided operations continue. It should be emphasized that the sun has a 22-year cycle! In addition, we will point at innovative uses of the existing spacecraft. The second step is to close obvious gaps in our knowledge by improving existing techniques and with modest missions between 0.2 and 5 AU. The third step is to implement frontier probes to the sun and into the interstellar medium, which require a serious commitment and in the latter case substantial technology development.

2. THE INNER FRONTIER

Figure 1 shows schematically the extent of our knowledge of the sun and the inner heliosphere together with the techniques that are used. Over the past two decades great strides have been made in the in-situ sampling of the solar wind from 1 to 5 AU with detailed composition measurements on Ulysses, SOHO and ACE [*3, 4*] and in the understanding of the 3D structure of the wind [*5*]. Observations have clearly demonstrated how strongly structured the solar wind is and how sharp boundaries between adjacent regions are [*6*]. Regions with starkly different conditions remain separate on their way from the sun to us [*7*]. Modeling and optical observations with SOHO UVCS have demonstrated that temperature differences between species and flux tubes are established very close to the sun [*8, 9*], but the responsible processes can only be inferred. To date no in-situ measurements have been made inside 0.3 AU.

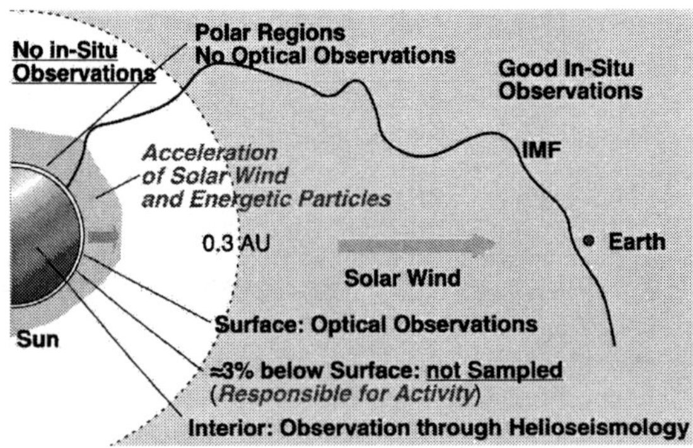

Fig. 1: Schematic view (not to scale) of the extent of our current knowledge of the sun and the inner heliosphere.

With SOHO and GONG helioseismology has provided a remarkable view of the sun's interior [*10*]. The radial structure and the large-scale convection pattern of the sun as well as the energy transport through its interior have been revealed. However, this technique provides no information about the last 3% of the interior below the sun's surface. Yet it is this critical subsurface layer, where solar activity with its complex magnetic fields is generated. Also the important high latitude regions of the sun have been barely visible to any optical observation, while it is here where the fast wind emerges during solar minimum.

2.1 Main Goals

To understand the generation and the variability of the heliosphere these are very critical gaps, which can only be closed with in-situ studies near the sun and pole-to-pole coverage. In addition, we will have to monitor the 3D heliosphere at strategic distances from the sun continuously over time scales substantially longer than the true solar cycle of 22 years. Turning this emerging knowledge into predictive power will require a continuous long term monitoring network comparable to the weather stations on Earth.

2.2 Use of Assets and Next Steps

3D Coverage and Monitoring: It can be anticipated and should be wholeheartedly supported that the substantial effort to monitor and understand the impact of solar variability on the inner heliosphere within NASA's Living with a Star Program [*11, 12*] will expand our detailed knowledge of solar wind structure and composition, electric and magnetic fields, and waves, as well as the energetic particle population. High-resolution imagers in a wide wavelength range will yield more detailed pictures of the sun's surface. However, true closure of the questions of how the solar wind is accelerated, how its distinct bi-modal structure is produced, and how the solar activity cycle is controlled, can only be reached by closing the two existing gaps in our coverage.

The Sun's Subsurface Layers: Penetrating the last 3% of the sun's interior beneath the surface will involve significant refinement of helioseismological observations in frequency and mode range together with a detailed simulation of the emerging structures and phenomena as observed on the surface. One attempt in this direction is high-resolution helioseismology of localized features, such as sunspots [*13*]. Such results can be tested and improved by comparing them with simulations of these upper layers of the sun's interior [*14*]. To achieve enough precision with the observations higher resolution and an extension to higher frequencies is needed. Because coronal holes and fast solar wind mostly emerge from high latitude regions, it will be extremely important to extend solar oscillation observations to the full latitude range on the sun. Missions, such as Solar Probe, Solar Orbiter and/or Solar Pole Sitter can achieve this goal.

Tracing Structures from the Sun to 1 AU: The lack of in-situ observations at <0.3 AU includes particle distributions and composition, with the need to extrapolate from a distance. It is even more glaring for magnetic fields, on which virtually no information is available outside the near sun corona. Before we look towards Solar Probe in this respect, let us present an example of what can be done with existing spacecraft and then is enhanced with a logical extension of current techniques - by taking steps 1 and 2 first. One important question is how to trace any magnetic field or particle information back to a specific location on the sun's surface. This would give us a better handle to infer processes on and below the sun's surface. It may also lead to better predictions of how solar events translate into effects at Earth.

For decades interplanetary space was thought to have a quasi-static pattern of interplanetary magnetic fields that rotate with sun, and particles generally were expected to disperse via diffusion. However, more recently a quite different picture has emerged. Motivated by Ulysses observations of structures extending to high latitudes in interplanetary space, Fisk [*15*] suggested that the differential rotation of the sun leads to large-scale latitudinal transport of field lines. A model emerged that can explain the reconfiguration of the solar magnetic field over the solar cycle through field line transport across the sun's surface via successive reconnection [*16*]. On the smaller scale it is consistent with SOHO observations that supergranules typically reconfigure through emerging magnetic flux in ≈1.5 days [*17*]. The model even offers an intriguing explanation for the energy source that may drive the solar wind. Interesting in connection with our discussion is a) that this model can be tested and b) that it can lead the way to more accurate tracing of phenomena from the sun's surface into space.

Contrary to the earlier naïve expectation that such a complex and small scale mixing of magnetic flux tubes on their way out from the sun might lead to enhanced mixing of particle populations equivalent to efficient diffusion, regions of solar wind with different composition [6] and different energetic particle populations [*18*] are found to remain separate on a very small scale. It is the latter that may be instrumental in probing the variable surface structures and their transport away from the sun. Mazur et al. [*18*] observed distinct dropouts in the clear time-dispersive structure of energy-time diagrams of energetic ions from compact impulsive solar events. Giacalone et al. [*19*] explained them in terms of a quickly changing flux tube topology through the combination of emerging flux, reconnection and convection with the solar wind.

In other words, particles from a localized source are spread (yet in discrete flux tubes) over an extended volume that expands with distance from the sun. The typical size of such a region can be estimated from the size of supergranules and radial expansion to reach a few million km at 1 AU. A mini-cluster of satellites, equipped with high collecting power energetic particle spectrometers, solar wind sensors and a magnetometer can be used to map individual supergranules with energetic particles and to determine the flux tube transport through interplanetary space. Existing spacecraft at and near L1, like ACE, SOHO and Wind, can serve as a pathfinder configuration.

2.3 The Challenge: Solar Probe

In spite of all progress, models of solar wind acceleration, coronal heating, particle acceleration in flares and at coronal shocks, as well as the role of magnetic fields, waves and instabilities in these processes can only be tested with in-situ observations. A Solar Probe with comprehensive particles and fields instrumentation is indispensable. In addition to its unique capability to close this glaring observational gap in solar and inner heliosphere physics, Solar Probe has the fabulous potential to be an inspiring mission of exploration, as it will be the first close-up visit of a star.

To unravel the particle-fields interaction that lead to heating, solar wind and energetic particle acceleration the payload must provide detailed particle velocity distributions, at least separate for key elements and isotopes, such as H, ^{3}He, ^{4}He, C, O, and Fe. In addition, magnetic and electric fields must be resolved in the key frequency range that covers important plasma waves, such as ion cyclotron, hybrid, whistler and Alfvén waves. The fast passage of the perihelion, as dictated by orbital dynamics, requires high time resolution to within seconds, for key wave features and the bulk of ions possibly also with sub-second resolution.

In order to relate the time series of in-situ observations to coronal structures simultaneous context observations are paramount. This is partly satisfied with a near quadrature of the probe orbit, which allows a complete view of the orbital plane with coronagraphs from Earth. For tomographic viewing and to locate the in-situ findings in the corona a coronagraph with forward viewing is highly desirable on Solar Probe.

With its anticipated polar pass Solar Probe may also provide a unique platform to gather a first pole-to-pole helioseismological map of the important high latitude regions. It could also take close-up images in the visible and X-ray band to study solar surface structures beneath its path. The proximity to the sun could provide advantages in spatial resolution for these observations. A description of such a Solar Probe spacecraft, payload and baseline mission may be found in the Solar Probe Definition Report [*20*].

At this point Solar Probe has been pushed back in its schedule due to the recent cancellation of the instrument selection. Substantial support from the scientific community is needed to move the process forward. As shown in the report Solar Probe is technically feasible and sound, yet the resources to accommodate the full payload are extremely tight. In this possibly over-constrained situation benefit may be gained from the simultaneous effort towards a Solar Orbiter by the European Space Agency, with a perihelion of 20 solar radii and the attempt to reach high latitudes during an extended mission [*21*]. With a full complement of in-situ and remote sensing instruments Solar Orbiter will be complementary to Solar Probe, cover different regions and longer periods. Remote sensing instruments could benefit even from being on Solar Orbiter and not on Solar Probe because of the slower relative motion and long cadence of observations. If resources are the ultimate issue towards implementation of Solar Probe, a decision to take advantage of the complementarity of the two missions and to concentrate remote sensing on Solar Orbiter may help to get Solar Probe on the way. "Ceterum Censeo", to close the in-situ observation gap close to the sun Solar Probe is absolutely indispensable and must be pursued.

3. THE OUTER FRONTIER

Let us now turn towards the edge of the heliosphere and into our galactic neighborhood. Here several fundamental questions may be summarized as follows:
1) What is the nature of the interstellar medium and the implications for the origin and evolution of matter?
2) How does the interstellar medium influence the heliosphere, its size, interior conditions and variations?
3) What is the impact of the solar system on the interstellar medium as an example of the interaction of a stellar system with its environment?

The broad nature of these questions gives them fundamental significance in several fields. In particular, the combination of in-situ observations and the astrophysical nature of the questions suggest a cross-disciplinary approach of space physics and astrophysics. They address objectives of all four NASA themes. "Sun-Earth-Connections", studying the influence of the sun and the surrounding space on Earth and other bodies in the solar system is addressed in all three questions. "Origins", which is geared towards formation and evolution of stars, planetary systems and life can relate to 1) and 2). "Structure and Evolution of the Universe" is addressed in 1), and "Planetary Exploration" topics are found in 2). Therefore, the outer frontier of "Sun-Earth Connections" can become a focus for all science disciplines.

3.1 Main Objectives

Cradle of the Stars: The interstellar medium is the cradle of the stars and provides the raw material for all bodies in stellar systems, including those of our own. This material has undergone continuous evolution from the Big Bang until today. The Big Bang produced only light nuclei, such as H, He, their isotopes ^{3}He and D, and some ^{7}Li [*22*]. Stars synthesize the heavier elements [*23*], and high-energy galactic cosmic rays contribute very rare elements, such as Be and B. Consequently, the abundance of elements and isotopes changes over time, and its knowledge for several points in time will provide the input to understand nucleosynthetic evolution.

Our current knowledge of the origin of the elements and their isotopes is mainly derived from composition measurements in the solar system. The relative abundance of nearly 300 nuclear species has been derived for the proto-solar nebula, which represents a sample of galactic matter from 4.5 billion years ago. Meteorites also provide some isotopic ratios in stellar grains, which represent very specific information about certain stellar sources, such as supernovae. Finally, spectroscopic data on elemental abundances (rarely on isotopes) are available for a variety of astrophysical objects. Missing is a sample of the present-day galaxy with reliable observations of a number of important elemental and isotopic abundance ratios. In-situ measurements of interstellar material inside and just outside the heliosphere, combined with remote absorption spectroscopy, will fill this gap.

Influence of the LIC on the Heliosphere: The state of the LIC controls the boundary of the heliosphere and, together with the solar wind, determines its size and inner state. With its (current) large size, its magnetic structure and its dynamics the heliosphere is the first shield of three that keep high-energy cosmic rays away from the Earth. With its journey through the galactic environment, which contains interstellar clouds with widely varying densities, the heliosphere must have varied substantially in size, most likely changing terrestrial conditions over time [*24*].

The interplay of the partially ionized LIC and the heliosphere leads to a complex boundary region that is controlled by the plasma flow around this obstacle and charge exchange with the neutral gas, which normally is not affected by the magnetic nature of the heliosphere. Depending on species this leads to stronger or weaker slowdown, accumulation at the boundary, heating and filtered penetration into the solar system. The main interstellar component, hydrogen, is among the species most affected and forms a so-called H-wall [25, 26]. To understand the physical processes and to quantitatively determine the elemental composition outside, these effects must be understood.

Further inside, at the solar wind termination shock, products of the LIC-heliosphere interaction, i.e. interstellar atoms, which are ionized in the inner heliosphere and subsequently transported outward with the solar wind, are accelerated to form the anomalous component of cosmic rays (ACRs) [e.g. *27*]. These energetic particles that fill the solar system with a unique composition of mostly singly charged ions with high ionization potential must also be messengers of our heliosphere into the galactic neighborhood. Although there is a rather good understanding of the main source [*28*], the key acceleration process, and the modulation inside the heliosphere [e.g. *29*], their injection into the acceleration process is still unclear [*30*], and the recent detection of minor contributions of species with low ionization potential [*31*] so far escapes an explanation.

Example for Astrospheres: The heliosphere cannot be unique. It is just an example – in fact, the only example that can be studied in detail with in-situ observations – of "astrospheres" around other stars. Any star with a magnetic field and a stellar wind must be surrounded by a similar sphere of influence and send messengers into its neighborhood, such as ACRs and fast neutral atoms that stem from charge exchange with the charged stellar wind.

The deceleration and accumulation of neutral hydrogen on the upwind side of the heliosphere and other astrospheres has already led to the identification of such H-walls in our own system and at several nearby star systems, through typical absorption features in the Lyα profile [*32, 33*]. In-situ studies of the outer reaches of our home system will sharpen these tools to understand the surroundings of many star systems and thus provide a direct link to astrophysics.

3.2 Future Pilot Missions

It is self-evident that an advance into the LIC proper requires extraordinary effort and probably is many years away. Yet, it should be emphasized that there is still much to do short of an Interstellar Probe.

First, it should be noted that two probes are on their way out of the heliosphere, Voyager 1 and 2, which have just surpassed their 25th anniversary. Although not optimized for the study of the interstellar environment, they will provide us with the first genuine in-situ information about the space beyond the termination shock and with a solid scale for the heliosphere. Because of this unique capability and because any variations happen on a scale of the solar cycle these probes, along with their inner heliosphere cousins, must be kept operational as long as technically feasible.

Second, energetic particles generated by the heliosphere, such as the ACRs and their sources can be studied in detail within the inner heliosphere, and it is the radial variation of composition and charge state that will yield new and valuable information on the sources and related acceleration processes. In addition, the suprathermal and energetic ion populations at the heliospheric boundary produce energetic neutral atoms (ENAs) through charge exchange with the interstellar neutral gas and can be used to provide a full sky image these regions [*34*].

Last, but not least, an interstellar wind blows through our system and carries neutral gas inside 3 AU. Hitching a ride on missions that are focused on other objectives, the sun or the Earth's magnetosphere, and often using instruments that are optimized for different problems, great strides have been made by investigating the UV glow [*35, 36*], pickup ions [*37, 3*], and neutral He atoms [*38*] of the interstellar gas. What is needed is a dedicated effort on a 1 by 3 AU orbit with instruments optimized for interstellar gas studies to effectively harness this information.

Particle Transport and Acceleration: As outlined above ACRs constitute a major source of energetic ions that emerges from the heliosphere and fills its interior. It is widely accepted that interstellar pickup ions with their highly suprathermal velocity distribution have a significant advantage for injection into shock acceleration [e.g. *39*]. This can explain the prevalence of interstellar pickup ions as the source of ACRs. Based on this paradigm it was suggested [*40, 41*] that the minor ions with low ionization potential in ACRs could be explained by pickup ions from the so-called inner source [*40, 42*]. However, recent findings in co-rotating interaction regions (CIR) suggest that inner source pickup ions are not effectively accelerated at 1 AU. In CIRs energetic heavy ions exhibit a charge state compatible with that of the solar wind with only a small contribution of Ne^+, while He^+ - clearly of interstellar origin - is effectively accelerated [*43*].

Because pickup ions are effectively cooled during radial transport with the solar wind and contrary to the interstellar source no ions of the inner source are added outside 1 AU, the ability to accelerate inner source ions should decrease rather than increase with distance from the sun. However, Gloeckler [*44*] has found strong suprathermal tail distributions in the solar wind and in pickup ion distributions, which even appear to be strengthened with distance from the sun. This observation suggests that an increasing pre-acceleration may be at work, which can propel particles into the acceleration at the termination shock.

To study this behavior quantitatively pickup ion, suprathermal and energetic particle populations need to be observed at 1 - 5 AU with elemental and ionic charge resolution and a collection power comparable to the instruments on ACE. By following the acceleration in CIRs and coronal mass ejections at increasing distances from the sun, it will be possible to delineate the injection and preacceleretion processes. These objectives can be achieved with an ACE-like mission on a 1 by 5 AU orbit using current instrumentation, within the scope of a Solar-Terrestrial Probe.

Probing the LIC inside the Heliosphere: Another mission that concentrates on studies of the LIC and its interaction with the heliosphere with accessible messengers has been coined Interstellar Pathfinder, with the following key scientific questions:
- What is the nucleosynthetic status of a present-day sample of the galaxy, and what are the implications for Big Bang cosmology, galactic evolution, and stellar nucleosynthesis?
- What is the physical state of the LIC and the nature of its interaction with the heliosphere?
- What are the characteristics and the 3D topology of the heliospheric termination shock?

Composition measurements require a pickup ion mass spectrometer with good resolution and a large geometric factor, which can be built, based on flight-proven time-of-flight instruments on Ulysses and ACE [*3*]. Mass resolution, energy range and geometric factor (up to a factor of 500 higher) must be tailored specifically to the pickup ion investigation, as past and current instruments were designed for other purposes. A neutral gas instrument that measures the interstellar He velocity distribution has been successfully flown on Ulysses [*38*]. Concepts to increase angular resolution and geometric factor have been developed [*45*]. A sensor with similar front end, surface conversion of neutrals into negative ions and subsequent time-of-flight analysis, similar to LENA on IMAGE, can extend the neutral gas observations to O and possibly H [e.g. *46*]. An energetic neutral atom imager that is optimized for the energy distribution of H neutrals from the termination shock with energies that range from a few 100 eV to several keV is derived from instruments on Cassini, IMAGE and TWINS [*34*].

In order to survey the interstellar gas and pickup ions effectively an orbit between 1 and 3 AU is needed. The gravitational deflection of the interstellar flow by the sun that is used by the neutral gas instrument to infer the flow velocity from the direction of the incoming neutrals is most pronounced very close to the sun. Conversely, interstellar pickup ions other than He can only be observed effectively at distances > 2 AU. For example, it should be pointed out that for O at 1 AU the inner source pickup ions become the most prominent component. Only beyond 1.4 AU can we distinguish unambiguously between the inner source and interstellar pickup ions. The farther away from the sun, the more of the interstellar pickup ion distribution becomes visible without interference from the inner source. To cover both, inner source and interstellar particles, an elliptical orbit in the ecliptic plane into the side-wind direction relative to the interstellar gas flow represents the ideal compromise (Fig. 2). This orbit will also cut through the gravitational focusing cone of He. With an aphelion of 3.16 AU the orbital period is exactly 3 years, thus providing for the opportunity of an Earth flyby after one orbit, which could be used to moderately change the orbital plane for a different cut through the gravitational focusing cone.

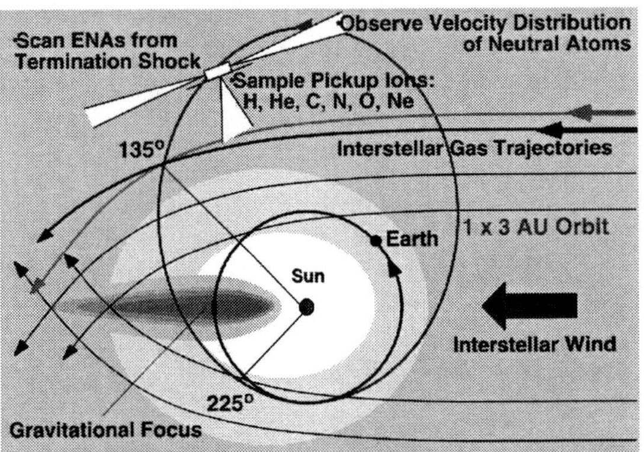

Fig. 2: Typical orbit, main objectives and viewing of an Interstellar Pathfinder with trajectories of the interstellar gas flow and a qualitative distribution of interstellar gas.

3.3 The Grand Challenge

Substantial new information on the LIC and its interaction with the heliosphere can be gathered with moderate missions in the inner heliosphere. However rich the harvest of such missions will be, they cannot supplant an ultimate mission into the LIC proper, yet they serve as extremely valuable precursor missions.

Objectives that must be studied outside: The LIC is a partially ionized environment with ionization fractions that are thus far only inferred [*47*]. Neutral gas can enter the heliosphere relatively unimpeded, but plasma is completely excluded and only the largest dust grains come through. This limits our study of isotopic and elemental abundances to H, N, O and the noble gases. Cosmic radiation with energies less than 100 MeV/nucleon is heavily shielded and modulated by the heliosphere, and probably the major fraction of this radiation has escaped our detection. Finally, the surrounding magnetic field is completely undetectable from inside and can only be inferred. Access to these components of our galactic neighborhood requires an Interstellar Probe. Direct observation of the interface layers in the heliospheric boundary region is much superior to inference from afar. In addition, infrared and radio observations of cosmic sources can be pushed to much lower intensities and frequencies. Close-up observations of Kuiper Belt objects may come into reach. Last but not least, the outer frontier is the final frontier of our heliosphere and thus awaits its exploration.

Interstellar Probe Implementation: Currently, two functioning spacecraft are on their way to leave the heliosphere, Voyager 1 and 2. These spacecraft were launched in 1977 and are now at ≈83.5 and ≈ 67.2 AU from the sun, respectively. They will probably pass the termination shock within the coming few years [*48*]. However, to reach, for example, the heliopause will need another 15-20 years, and may never happen while enough energy is available. Beyond this the payload was never designed for the interstellar medium, so useful measurements will not be possible.

Needed is a craft that can carry a full particles and fields payload, optimized for plasma, neutral gas, cosmic rays, magnetic and electric fields in the LIC. According to a recent study this will require a payload of at least 25 - 30 kg. It should reach a minimum distance of 200 AU within ≈ 15 - 20 years (together with the preparation the time span of a researcher's fruitful work) and should be able to return data from out to 400 AU [*49*]. This requires a final speed of 10 - 15 AU per year, which cannot be achieved with conventional rockets and swingby techniques. Two options have been studied, nuclear electric propulsion and a combination of a near sun passage with solar sailing. Under current circumstances the solar sail appears to be the most viable concept, based on the sizing of an interstellar probe, on the versatility of this technology, and on the acceptance in the wider community. While studies of the basic technologies are underway, solar sailing lacks the effort for a reasonably sized demonstration mission in the inner solar system. Clearly, before Interstellar Probe can be implemented, substantial effort has to go into propulsion. In addition, sensor and spacecraft system technology with emphasis on miniaturization and autonomy must be developed.

Solar sail technology not only has the potential to enable many long duration and low maintenance flights to a variety of targets within the solar system and just outside the heliosphere, it may serve as one of the few viable paths to sending probes deeper into interstellar space. Coupled with large-scale laser arrays as a fixed home-based energy source, light sails may be the only known way to propel a spacecraft without the need to carry all the necessary fuel, a requirement that hampers rocket technology. This has been described in a technically correct and brilliant way in the (truly) science fiction novel "Rocheworld" [*50*].

ACKNOWLEDGMENTS

Support by NASA under NAS5-32626 and NAG 5-6912 and valuable discussions with T. Forbes, P.A. Isenberg and M.A. Lee are gratefully acknowledged.

REFERENCES

1. Lanzerotti, L.J., *AGU Geophys. Monograph*, *125*, 11, 2001.
2. Möbius, E., G. Gloeckler, L.A., Fisk, R.A. Mewaldt, in: *The Heliosphere: Beyond the Planets*, K. Scherer et al. eds., *Copernicus Gesellschaft e.V.*, 357, 1999.
3. Gloeckler, G., J. Geiss, *Space Sci. Rev.*, *86*, 127, 1998.
4. Von Steiger, R., et al., *J. Geophys. Res.*, *105*, 27217, 2000.
5. McComas, D.J., et al., *J. Geophys. Res.*, *105*, 10419, 2000.
6. Aellig, M., et al., *J. Geophys. Res.*, *103*, 17215, 1998.
7. Bürgi, F., J. Geiss, *Solar Phys.*, *103*, 347, 1986.
8. Kohl, J. L., et al., *Solar Phys.*, *175*, 613, 1997.
9. Ko, Y.-K., et al., *Solar Phys.*, *171*, 345, 1997.
10. Hill, F., in: *Synoptic Solar Physics*, K.S. Balasubramaniam, J.W. Harvey, and D.M. Rabin eds., 33, 1998.
11. Withbroe, G.L., *AGU Geophys. Monograph*, *125*, 45, 2001.
12. Guhathakurta, M., *this volume*, 2002.
13. Kosovichev, A.G., et al., *Solar Phys.*, *192*, 159, 2000.
14. Fisher, G. H., Y. Fan, D.W. Longcope, M.G. Linton, A.A. Pevtsov, *Solar Phys.*, *192*, 119, 2000.
15. Fisk, L.A., *J. Geophys. Res.*, *101*, 15547, 1996.
16. Fisk, L.A., N.A. Schwadron, T. Zurbuchen, *J. Geophys. Res.*, *104*, 19765, 1999.
17. Shrijver, C.J., et al., *Nature*, *394*, 152, 1998.
18. Mazur, J.E., et al., *AIP Conf. Proc.*, *528*, 47, 2000.
19. Giacalone, J., et al., *AIP Conf. Proc.*, *528*, 157, 2000.
20. Gloeckler, G., et al., Report of the NASA Science Definition Team for the Solar Probe mission, *JHU APL*, 1999.
21. Marsden, R., *this volume*, 2002.
22. Schramm, D.N., *Space Sci. Rev.*, *84*, 3, 1998.
23. Prantzos, N., *Space Sci. Rev.*, *84*, 22, 1998.
24. Frisch, P.C., *Am. Scientist*, *86*, 52, 2000.
25. Baranov, V.B., *Space Sci. Rev.*, *52*, 89, 1990.
26. Zank, G.P., H.L. Pauls, *Space Sci. Rev.*, *78*, 95, 1996.
27. Klecker, B., *Space Sci. Rev.*, *72*, 419, 1995.
28. Fisk, L.A., B. Kozlowsky, R. Ramaty, *Astrophys. J.*, *190*, L35, 1974.
29. Jokipii, J.R., *Space Sci. Rev.*, *86*, 161 1998.
30. Lee, M.A., *AIP Conf. Proc.*, *528*, 3, 2000.
31. Reames, D.V., *Astrophys. J.*, *518*, 473, 1999.
32. Linsky, J.L., *Space Sci. Rev.*, *78*, 157, 1996.
33. Wood, B.E., J.L. Linsky, G.P. Zank, *Astrophys. J.*, *537*, 304, 2000.
34. Gruntman, M., *Rev. Sci. Instrum.*, *68*, 3617, 1997
35. Bertaux, J.L., J.E. Blamont, *Astron. Astrophys.*, *11*, 200, 1971.
36. Lallement, R., *Space Sci. Rev.*, *78*, 361, 1996.
37. Möbius, E., et al., *Nature*, *318*, 426, 1985.
38. Witte, M., M. Banaskiewicz, H. Rosenbauer, *Space Sci. Rev.*, *78*, 289, 1996.
39. Scholer, M., H. Kucharek, *Geophys. Res. Lett.*, *26*, 29, 1999.
40. Gloeckler, G. L. A. Fisk, J. Geiss, N. A. Schwadron, T. H. Zurbuchen, *J. Geophys. Res.*, *105*, 7459, 2000.
41. Cummings, A.C., et al., *Astrophys. J.*, *578*, 194, 2002.
42. Geiss, J., G. Gloeckler, L.A. Fisk, and R. von Steiger, *J. Geophys. Res.*, *100*, 23373, 1995.
43. Möbius, E., et al., *Geophys. Res. Lett.*, *29*, DOI 10.1029/2001GL013410, 2002.
44. Gloeckler, G., *Space Sci. Rev.*, *89*, 91 - 104, 1999.
45. Livi, S., et al., *this volume*, 2002.
46. Wurz, P., et al., *Opt. Eng.*, *34*, 2365, 1995.
47. Frisch, P.C., J.D. Slavin, *Space Sci. Rev.*, *78*, 223, 1996.
48. Stone, E.C., *this volume*, 2002.
49. Mewaldt, R.A., P.C. Liewer, *COSPAR Coll. Ser.*, *11*, 451, 2001.
50. Forward, R.L., *Rocheworld*, 1984.

Parallel, Adaptive-Mesh-Refinement MHD for Global Space-Weather Simulations

Kenneth G. Powell*, Tamas I. Gombosi*, Darren L. De Zeeuw*, Aaron J. Ridley*, Igor V. Sokolov*, Quentin F. Stout* and Gábor Tóth*†

*Center for Space Environment Modeling
University of Michigan
†Department of Atomic Physics, Loránd Eötvös University, Budapest, Hungary

Abstract. The first part of this paper reviews some issues representing major computational challenges for global MHD models of the space environment. These issues include mathematical formulation and discretization of the governing equations that ensure the proper jump conditions and propagation speeds, regions of relativistic Alfvén speed, and controlling the divergence of the magnetic field. The second part of the paper concentrates on modern solution methods that have been developed by the aerodynamics, applied mathematics and DoE communities. Such methods have recently begun to be implemented in space-physics codes, which solve the governing equations for a compressible magnetized plasma. These techniques include high-resolution upwind schemes, block-based solution-adaptive grids and domain decomposition for parallelization. We describe the space physics MHD code developed at the University of Michigan, based on the developments listed above.

INTRODUCTION

Global computational models based on first principles represent a very important component of efforts to understand the intricate processes coupling the Sun to the geospace environment. The hope for such models is that they will eventually fill the gaps left by measurements, extending the spatially and temporarily limited observational database into a self-consistent global understanding of our space environment.

Presently, and in the foreseeable future, magnetohydrodynamic (MHD) models are the only models that can span the enormous distances present in the magnetosphere. However, it should not be forgotten that even generalized MHD equations are only a relatively low-order approximation to more complete physics; they provide only a simplified description of natural phenomena in space plasmas.

NON-RELATIVISTIC MAGNETOHYDRODYNAMICS

The governing equations for an ideal, non-relativistic, compressible plasma may be written in a number of different forms. While the different forms of the MHD equations describe the same physics at the differential equation level, there are important practical differences when one solves discretized forms of the various formulations.

According to the Lax-Wendroff theorem [1] only conservative schemes can be expected to get the correct jump conditions and propagation speed for a discontinuous solution. This fact is much less emphasized in the global magnetosphere simulation literature than the more controversial divergence of B issue. In some test problems the non-conservative discretization of the MHD equations can lead to significant errors, which do not diminish with increased grid resolution.

Fully Conservative Form

The fully conservative form of the equations is

$$\frac{\partial \mathbf{U}}{\partial t} + (\nabla \cdot \mathbf{F})^{\mathrm{T}} = 0\,, \qquad (1)$$

where $\mathbf{U}$ is the vector of conserved quantities and $\mathbf{F}$ is a flux diad,

$$\mathbf{U} = \begin{pmatrix} \rho \\ \rho \mathbf{u} \\ B \\ E_{mhd} \end{pmatrix} \qquad (2)$$

CP679, *Solar Wind Ten: Proceedings of the Tenth International Solar Wind Conference*, edited by M. Velli, R. Bruno, and F. Malara

$$\mathbf{F} = \begin{pmatrix} \rho\mathbf{u} \\ \rho\mathbf{u}\mathbf{u} + \left(p + \frac{1}{2\mu_0}B^2\right)\mathbf{I} - \frac{1}{\mu_0}\mathbf{B}\mathbf{B} \\ \mathbf{u}\mathbf{B} - \mathbf{B}\mathbf{u} \\ \mathbf{u}\left(E_{mhd} + p + \frac{1}{2\mu_0}B^2\right) - \frac{1}{\mu_0}(\mathbf{u}\cdot\mathbf{B})\,\mathbf{B} \end{pmatrix}^{\mathrm{T}} \quad (3)$$

where E_{mhd} is the magnetohydrodynamic energy, given by

$$E_{mhd} = \frac{1}{2}\rho u^2 + \frac{1}{\gamma - 1}p + \frac{1}{2\mu_0}B^2 \quad (4)$$

Symmetrizable Formulation

Symmetrizable systems of conservation laws have been studied by Godunov [2] and Harten [3], among others. One property of the symmetrizable form of a system of conservation laws is that an added conservation law

$$\frac{\partial(\rho s)}{\partial t} + \frac{\partial(\rho s u_x)}{\partial x} + \frac{\partial(\rho s u_y)}{\partial y} + \frac{\partial(\rho s u_z)}{\partial z} = 0$$

for the entropy s can be derived by a linear combintion of the system of equations. For the ideal MHD equations, as for the gasdynamic equations, the entropy is $s = \log(p/\rho^\gamma)$. Another property is that the system is Galilean invariant; all waves in the system propagate at speeds $u \pm c_w$ (for MHD, the possible values of c_w are the Alfvén, magnetofast and magentoslow speeds). Neither of these properties holds for the fully conservative form of the MHD equations.

Godunov showed that the fully conservative form of the MHD equations (eq. 1) is not symmetrizable [2]. The symmetrizable form may be written as

$$\frac{\partial \mathbf{U}}{\partial t} + (\nabla\cdot\mathbf{F})^{\mathrm{T}} = \mathbf{Q}\,, \quad (5)$$

where

$$\mathbf{Q} = -\nabla\cdot\mathbf{B}\begin{pmatrix} 0 \\ \frac{1}{\mu_0}\mathbf{B} \\ \mathbf{u} \\ \frac{1}{\mu_0}\mathbf{u}\cdot\mathbf{B} \end{pmatrix} \quad (6)$$

Vinokur separately showed that eq. (5) can be derived starting from the primitive form, if no stipulation is made about $\nabla\cdot\mathbf{B}$ in the derivation. Powell showed that this symmetrizable form can be used to derive a Roe-type approximate Riemann solver for solving the MHD equations in multiple dimensions [4].

The MHD eigensystem arising from eq. (1) or eq. (5) leads to eight eigenvalue/eigenvector pairs. The eigenvalues and associated eigenvectors correspond to an entropy wave, two Alfvén waves, two magnetofast waves, two magnetoslow waves, and an eighth eigenvalue/eigenvector pair that depends on which form of the equations is being solved. This last wave (which describes the jump in the normal component of the magnetic field at discontinuities) has a zero eigenvalue in the fully conservative case, and an eigenvalue equal to the normal component of the velocity, u_n, in the symmetrizable case. The expressions for the eigenvectors, and the scaling of the eigenvectors, are more intricate than in gasdynamics [5].

We note that while eq.(1) is fully conservative, the symmetrizable formulation (given by eq. 5) is formally not fully conservative. Terms of order $\nabla\cdot\mathbf{B}$ are added to what would otherwise be a divergence form. The danger of this is that shock jump conditions may not be correctly met, unless the added terms are small, and/or they alternate in sign in such a way that the errors are local, and in a global sense cancel in some way with neighboring terms. This downside, however, has to be weighed against the alternative; a system (i.e., the one without the source term) that, while conservative, is not Gallilean invariant, has a zero eigenvalue in the Jacobian matrix, and is not symmetrizable.

SEMI-RELATIVISTIC PLASMAS

While the solar-wind speed remains non-relativistic in the solar system, the intrinsic magnetic fields of several planets in the solar system are high enough, and the density of the solar wind low enough, that the Alfvén speed,

$$V_{\mathrm{A}} = \sqrt{\frac{B^2}{\mu_0\rho}} \quad (7)$$

can reach appreciable fractions of the speed of light. In the case of Jupiter, the Alfvén speed in the vicinity of the poles is of order ten! Even Earth has a strong enough intrinsic magnetic field that the Alfvén speed reaches twice the speed of light in Earth's near-auroral regions.

Limiting the Alfvén Speed

For these regions, solving the non-relativistic ideal MHD equations does not make sense. Having waves in the system propagating faster than the speed of light, besides being non-physical, causes a number of numerical difficulties. However, solving the fully relativistic MHD equations is overkill. What is called for is a semi-relativistic form of the equations, in which the flow speed and acoustic speed are non-relativistic, but the Alfvén speed can be relativistic. A derivation of these semi-relativistic equations from the fully relativistic equations is given in [6]; the final result is presented here.

The semi-relativistic ideal MHD equations are of the form

$$\frac{\partial \mathbf{U}_{sr}}{\partial t} + (\nabla \cdot \mathbf{F}_{sr})^{\mathrm{T}} = 0 \qquad (8)$$

where the state vector, $\mathbf{U}_{sr}$, and the flux diad, $\mathbf{F}_{sr}$, are

$$\mathbf{U}_{sr} = \begin{pmatrix} \rho \\ \rho\mathbf{u} + \frac{1}{c^2}\mathbf{S}_{\mathrm{A}} \\ \mathbf{B} \\ \frac{1}{2}\rho u^2 + \frac{1}{\gamma-1}p + e_{\mathrm{A}} \end{pmatrix} \qquad (9)$$

$$\mathbf{F}_{sr} = \begin{pmatrix} \rho\mathbf{u} \\ \rho\mathbf{u}\mathbf{u} + p\mathbf{I} + \mathbf{P}_{\mathrm{A}} \\ \mathbf{u}\mathbf{B} - \mathbf{B}\mathbf{u} \\ \left(\frac{1}{2}\rho u^2 + \frac{\gamma}{\gamma-1}p\right)\mathbf{u} + \mathbf{S}_{\mathrm{A}} \end{pmatrix}^{\mathrm{T}} \qquad (10)$$

In the above,

$$\mathbf{S}_{\mathrm{A}} = \frac{1}{\mu_0}(\mathbf{E}\times\mathbf{B}) \qquad (11)$$

$$e_{\mathrm{A}} = \frac{1}{2\mu_0}\left(B^2 + \frac{1}{c^2}E^2\right) \qquad (12)$$

$$\mathbf{P}_{\mathrm{A}} = e_{\mathrm{A}}\mathbf{I} - \frac{1}{\mu_0}\mathbf{B}\mathbf{B} - \frac{1}{\mu_0 c^2}\mathbf{E}\mathbf{E} \qquad (13)$$

are the Poynting vector, the electromagnetic energy density, and the electromagnetic pressure tensor, respectively. The electric field $\mathbf{E}$ is related to the magnetic field $\mathbf{B}$ by Ohm's law.

Lowering the Speed of Light

This new system of equations has wave speeds that are limited by the speed of light; for strong magnetic fields, the modified Alfvén speed (and the modified magnetofast speed) asymptote to c. The modified magnetoslow speed asymptotes to a, the acoustic speed. This property offers the possibility of a rather tricky convergence-acceleration technique for explicit time-stepping schemes, first suggested by Boris [7]; the wave speeds can be lowered, and the stable time-step thereby raised, by artificially lowering the value taken for the speed of light. This method is known as the "Boris correction."

The equations in section are valid in physical situations in which $V_{\mathrm{A}} > c$. A slight modification yields a set of equations, the steady-state solutions of which are independent of the value taken for the speed of light. Defining the true value of the speed of light to be c_0, to distinguish it from the artificially lowered speed of light, c, the equations are:

$$\frac{\partial \mathbf{U}_{sr}}{\partial t} + (\nabla \cdot \mathbf{F}_{sr})^{\mathrm{T}} = \mathbf{Q}_{c_0} \qquad (14)$$

where the state vector, $\mathbf{U}_{sr}$, and the flux diad, $\mathbf{F}_{sr}$, are as defined above, and the new source term in the momentum equation is

$$\mathbf{Q}_{c_0} = \frac{1}{\mu_0}\left(\frac{1}{c_0^2} - \frac{1}{c^2}\right)\mathbf{E}\nabla\cdot\mathbf{E} \qquad (15)$$

An implementation of the semi-relativistic equations has been made in the BATSRUS code developed at the University of Michigan [8, 6].

SOLUTION TECHNIQUES

Finite-Volume Scheme

The MHD equations are well suited for finite volume methods when the governing equations are integrated over a computational cell i, yielding

$$\frac{d\mathbf{U}_i}{dt} = -\frac{1}{V_i}\sum_{\text{faces}}\mathbf{F}\cdot\hat{\mathbf{n}}A - \frac{\mathbf{Q}_i}{V_i}\sum_{\text{faces}}\mathbf{B}\cdot\hat{\mathbf{n}}A, \qquad (16)$$

where V_i is the volume of cell i, A is the surface area of the faces forming the computational cell, $\hat{\mathbf{n}}$ is the unit vector normal to the cell faces, $\mathbf{U}_i$ is the cell-averaged conserved solution vector, and $\mathbf{Q}_i$ is given by

$$\mathbf{Q}_i = -\begin{bmatrix} 0 \\ \frac{1}{\mu_0}\mathbf{B}_i \\ \mathbf{u}_i \\ \frac{1}{\mu_0}\mathbf{u}_i\cdot\mathbf{B}_i \end{bmatrix}. \qquad (17)$$

The numerical face fluxes, $\mathbf{F}\cdot\hat{\mathbf{n}}$, are defined in terms of the left and right interface solution states, $\mathbf{U}_{\mathrm{L}}$ and $\mathbf{U}_{\mathrm{R}}$, as follows

$$\mathbf{F}\cdot\hat{\mathbf{n}} = \mathscr{F}(\mathbf{U}_{\mathrm{L}}, \mathbf{U}_{\mathrm{R}}, \hat{\mathbf{n}}), \qquad (18)$$

where $\mathbf{U}_{\mathrm{L}}$ and $\mathbf{U}_{\mathrm{R}}$ are the state vectors at the left and right sides of the interface.

TVD-MUSCL

Because the MHD equations are a system of hyperbolic conservation laws, many of the techniques that have been developed for the Euler equations can be applied relatively straightforwardly. In particular, the high-resolution finite-volume approach of van Leer [9] (i.e. approximate Riemann solver + limited interpolation scheme + multi-stage time-stepping scheme) is perfectly valid. The Rusanov/Lax-Friedrichs approximate Riemann solver can be applied directly; no knowledge of the eigensystem of the MHD equations is required other than the fastest wave speed in the system. A Roe-type scheme [10] can be constructed for non-relativistic

MHD, but requires more work, because of the complexity of the eigensystem. In addition, an HLLE-type Riemann solver has been derived by Linde [11]; it is less dissipative than the Rusanov/Lax-Friedrichs scheme, but less computationally intensive than the Roe scheme. Whichever approximate Riemann solver is chosen to serve as the flux function, standard interpolation schemes and limiters can be used to construct a finite-volume scheme.

CONTROLLING $\nabla \cdot \mathbf{B}$

One way in which the numerical solution of the MHD equations differs from that of the gasdynamic equations is the constraint that $\nabla \cdot \mathbf{B} = 0$. Enforcing this constraint numerically, particularly in shock-capturing codes, can be done in a number of ways, but each way has its particular strengths and weaknesses. This issue is explained more fully in the references such as [12], [4, 8], [13], [14], [15]. Tóth has published a numerical comparison of many of the approaches for a suite of test cases [16].

BLOCK-BASED AMR ON CARTESIAN GRIDS

Adaptive mesh refinement techniques that automatically adapt the computational grid to the solution of the governing PDEs can be very effective in treating problems with disparate length scales. Methods of this type avoid underresolving the solution in regions deemed of interest (e.g., high-gradient regions) and, conversely, avoid overresolving the solution in other less interesting regions (low-gradient regions), thereby saving orders of magnitude in computing resources for many problems. For typical solar wind flows, length scales can range from tens of kilometers in the near Earth region to the Earth-Sun distance (1.5×10^{11} m), and timescales can range from a few seconds near the Sun to the expansion time of the solar wind from the Sun to the Earth ($\sim 10^5$ s). The use of AMR is extremely beneficial and almost a virtual necessity for solving problems with such disparate spatial and temporal scales.

Adaptive Blocks

Borrowing from previous work by Berger and coworkers [17, 18, 19, 20, 21] and Quirk [22, 23], and keeping in mind the desire for high performance on massively parallel computer architectures, a relatively simple yet effective block-based AMR technique has been developed and is used in conjunction with the finite-volume scheme described above. The method has some similarities with the block-based approaches described by Quirk and Hanebutte [23] and Berger and Saltzman [21]. Here the governing equations are integrated to obtain volume-averaged solution quantities within rectangular Cartesian computational cells. The computational cells are embedded in regular structured blocks of equal sized cells. The blocks are geometrically self-similar with dimensions $\tilde{\ell}_x \times \tilde{\ell}_y \times \tilde{\ell}_z$ and consist of $N_x \times N_y \times N_z$ cells, where $\tilde{\ell}_x$, $\tilde{\ell}_y$, and $\tilde{\ell}_z$ are the nondimensional lengths of the sides of the rectangular blocks and N_x, N_y, and N_z are even, but not necessarily all equal, integers. Typically, blocks consisting of anywhere between $4 \times 4 \times 4 = 64$ and $12 \times 12 \times 12 = 1728$ cells are used (see Figure 1). Solution data associated with each block are stored in standard indexed array data structures. It is therefore straightforward to obtain solution information from neighboring cells within a block.

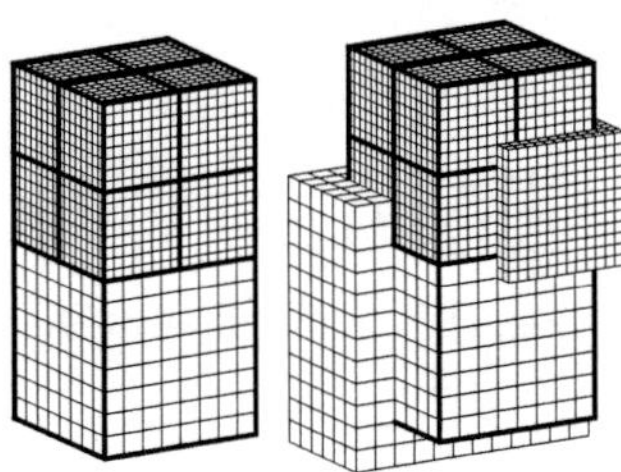

FIGURE 1. (left) Self-similar blocks used in parallel block-based AMR scheme. (right) Self-similar blocks illustrating the double layer of ghost cells for both coarse and fine blocks.

Computational grids are composed of many self-similar blocks. Although each block within a grid has the same data storage requirements, blocks may be of different sizes in terms of the volume of physical space that they occupy. Starting with an initial mesh consisting of blocks of equal size (i.e., equal resolution), adaptation is accomplished by the dividing and coarsening of appropriate solution blocks. In regions requiring increased cell resolution, a "parent" block is refined by dividing itself into eight "children" or "offspring." Each of the eight octants of a parent block becomes a new block having the same number of cells as the parent and thereby doubling the cell resolution in the region of interest. Conversely, in regions that are deemed overresolved, the refinement process is reversed, and eight children are coarsened and coalesced into a single parent block. In this way, the cell resolution is reduced by a factor of 2. Standard multigrid-type restriction and prolongation operators are used to evaluate the solution on all blocks created by the coarsening and division processes, respectively.

Two neighboring blocks, one of which has been refined and one of which has not, are shown in Figure 1. Any of the blocks shown in Figure 1 can in turn be refined, and so on, leading to successively finer blocks. In the present method, mesh refinement is constrained such that the cell resolution changes by only a factor of 2 between adjacent blocks and such that the minimum resolution is not less than that of the initial mesh.

In order that the update scheme for a given iteration or time step can be applied directly to all blocks in an independent manner, some additional solution information is shared between adjacent blocks having common interfaces. This information is stored in an additional two layers of overlapping "ghost" cells associated with each block as shown in Figure 1. At interfaces between blocks of equal resolution, these ghost cells are simply assigned the solution values associated with the appropriate interior cells of the adjacent blocks. At resolution changes, restriction and prolongation operators, similar to those used in block coarsening and division, are employed to evaluate the ghost cell solution values. After each stage of the multistage time-stepping algorithm, ghost cell values are reevaluated to reflect the updated solution values of neighboring blocks. With the AMR approach, additional interblock communication is also required at interfaces with resolution changes to strictly enforce the flux conservation properties of the finite-volume scheme [17, 18, 19]. In particular, the interface fluxes computed on more refined blocks are used to correct the interface fluxes computed on coarser neighboring blocks so as to ensure that the fluxes are conserved across block interfaces.

PARALLEL IMPLEMENTATION

The parallel block-based AMR solver was designed from the ground up with a view to achieving very high performance on massively parallel architectures. The underlying upwind finite-volume solution algorithm, with explicit time stepping, has a very compact stencil and is therefore highly local in nature. The hierarchical data structure and self-similar blocks make domain decomposition of the problem almost trivial and readily enable good load-balancing, a crucial element for truly scalable computing. A natural load balancing is accomplished by simply distributing the blocks equally amongst the processors. Additional optimization is achieved by ordering the blocks using the Peano-Hilbert space filling curve to minimize inter-processor communication. The self-similar nature of the solution blocks also means that serial performance enhancements apply to all blocks and that fine grain parallelization of the algorithm is possible. The parallel implementation of the algorithm has been carried out to such an extent, that even the grid adaptation is performed in parallel.

Other features of the parallel implementation include the use of FORTRAN 90 as the programming language

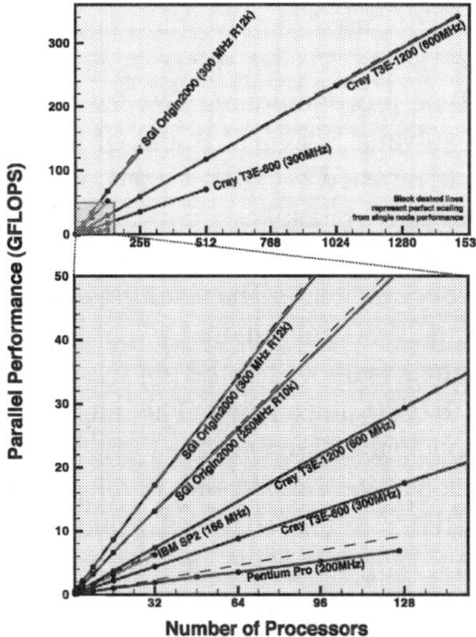

FIGURE 2. Parallel speedup of BATSRUS on various architectures. Black dashed lines represent perfect scaling from single node performance.

and the message passing interface (MPI) library for performing the interprocessor communication. Use of these standards greatly enhances the portability of the code and leads to very good serial and parallel performance. The message passing is performed in an asynchronous fashion with gathered wait states and message consolidation.

Implementation of the algorithm has been carried out on Cray T3E supercomputers, SGI and Sun workstations, on Beowulf type PC clusters, on SGI shared-memory machines, on a Cray T3D, and on several IBM SP2s. BATSRUS nearly perfectly scales to 1,500 processors and a sustained speed of 342 GFlops has been attained on a Cray T3E-1200 using 1,490 PEs. For each target architecture, simple single-processor measurements are used to set the size of the adaptive blocks. The scaling of BATSRUS on various architectures is shown in Figure 2.

Implicit Time-Stepping

In BATSRUS we have a number of time stepping algorithms implemented. The simplest and least expensive scheme is a multistage explicit time stepping, for which the time step is limited by the CFL stability condition. We have also implemented an unconditionally stable fully implicit time stepping scheme [27, 28]. The second order implicit time discretization (BDF2) requires the solution of a non-linear system of equations for all the flow variables. This can be achived by the Newton-

Krylov-Schwarz approach: a Newton iteration is applied to the non-linear equations; a parallel Krylov type iterative scheme is used to solve the linear systems; the convergence of the Krylov solver is accelerated with a Schwarz type preconditioning. We have implemented two Krylov solvers: BiCGSTAB and GMRES. A modified block incomplete LU (MBILU) preconditioner is applied on a block by block basis. Since every block has a simple Cartesian geometry, the preconditioner can be implemented very efficiently. The resulting implicit scheme requires about 20-30 times more CPU time per time step than the explicit method, but the physical time step can be 1,000 to 10,000 times larger. This implicit algorithm has a very good parallel scaling due to the Krylov scheme and the block by block application of the preconditioner.

In BATSRUS, we can combine explicit and implicit time stepping. Magnetosphere simulations include large volumes where the Alfvén speed is quite low (tens of km/s) and the local CFL number allows large explicit time steps (tens of seconds to several minutes). In these regions implicit time stepping is a waste of computational resources. Since the parallel implicit technique we use is fundamentally block based we only treat those blocks implicitly where the CFL condition would limit the explicit time step to less than the selected time step (typically ~ 10 s). Needless to say, this combined explicit-implicit time stepping represents more computational challenges (such as separate load balancing of explicit and implicit blocks). Overall, this solution seems to be a very promising option, but other potential avenues need to explored before one makes a final decision about the most efficient time-stepping algorithm for space MHD simulations. These questions will be discussed in an upcoming paper [29].

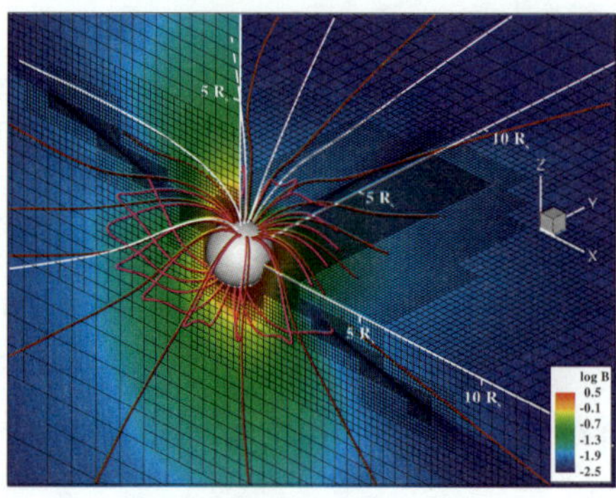

FIGURE 3. 3D representation of the steady-state solar wind solution. The shading represents $\log|\mathbf{B}|$ in the (x,z)- and (x,y)-planes. The thin black lines are the computational mesh and the thick solid lines are magnetic field lines: grey denotes the last closed field lines, black is open field lines expanding to the interplanetary medium just above the heliospheric current sheet, and finally, white lines show open magnetic field lines in the (y,z)-plane.

APPLICATIONS

BATSRUS has been extensively applied to global numerical simulations of the inner heliosphere including CME propagation [30, 31], the coupled terrestrial magnetosphere-ionosphere [32, 33, 34], and the interaction of the heliosphere with the interstellar medium [35]. In addition, it has also been successfully applied to a host of planetary problems ranging from comets [36, 37], to Mercury [38], Venus [39], Mars [40], Saturn [41], to planetary satellites [42, 43].

In this section we briefly summarize our most ambitious space weather simulation so far, in which we used BATSRUS to simulate an entire space weather event, from its generation at the Sun through the formation and evolution of a CME, to its interaction with the magnetosphere-ionosphere system [30, 31]. In this simulation we resolved multiple spatial and temporal scales and took advantage of frequent grid refinements and coarsening to follow the CME through interplanetary space. The total number of cells varied between 800,000 and 2 million as the solution evolved. The simulation used 13 levels of grid refinement. The simulation ran faster than real time on a 512 node Cray T3E-600 supercomputer. This simulation demonstrates that we have the necessary experience to undertake the research outlined in this proposal.

Here we only show a few highlights of this simulation. The detailed results have been published in JGR-Space Physics [31].

A steady state solar wind was obtained in the corotating frame for a tilted rotating Sun. The intrinsic magnetic field was approximated by the superposition of a tilted (with respect to the rotation axis) octupole and dipole. Figure 3 depicts a three-dimensional representation of the predicted pre-event steady-state solar wind solution in the vicinity of the Sun. The narrow dark region shown in Figure 3, which also coincides with regions of higher mesh refinement, corresponds to the beginning of the heliospheric current sheet. Due to the combined effects of magnetic tilt and solar rotation, the current sheet is tilted with respect to the rotation axis, and deformed, and resembles a "ballerina skirt."

Figure 4 shows a 3D representation of the magnetic field configuration 9 hours after the initiation of the CME. The density enhancement first leads to the "filling" of the closed magnetic field lines with additional plasma and subsequent expansion of the closed field line region. One can see that the closed field lines become greatly stretched by the outward moving plasma. This is due to the fact that the plasma β (the ratio of the kinetic and magnetic pressures) is quite large and the magnetic field

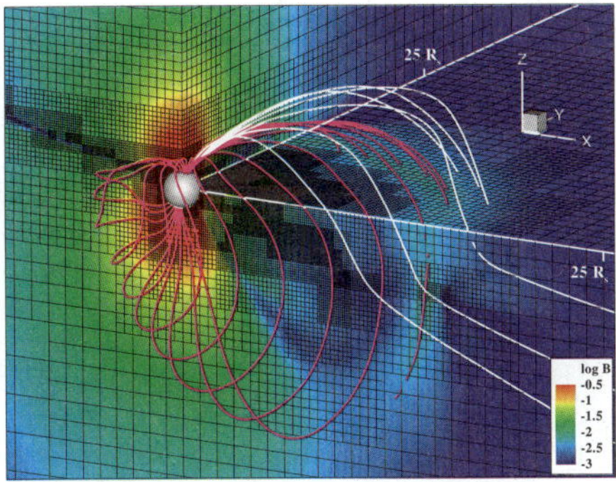

FIGURE 4. 3D representation of magnetic field lines 9 hours after the initiation of a CME. Grayscale represents $\log(B)$, white lines are open magnetic field lines, grey lines represent magnetic field lines with both ends connected to the Sun.

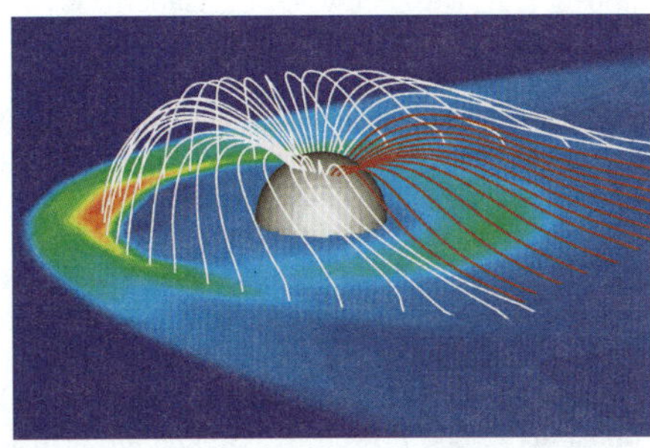

FIGURE 5. 3D representation of the last closed terrestrial field lines for southward IMF conditions. White field lines form the dayside magnetopause, while black ones map to the magnetotail. The greyscale represents normalized thermal pressure.

is "carried" by the outward moving plasma. We also note the decrease of magnetic field strength behind the leading edge of the outward moving disturbance.

The dynamic response of the global magnetosphere to the changing solar wind conditions produced by the density-driven CME was also computed as part of this simulation. The global magnetospheric configuration for quiet-time southward IMF conditions is shown in Figure 5. During the event the solar wind velocity remained nearly radial with the speed gradually decreasing from about 550 km/s to about 450 km/s. The solar wind dynamic pressure increased from its pre-CME value of 2.25 nP (at $t = 72$ hrs) to 4.6 nP at the peak of the event.

The ionospheric potential and convection patterns also change during the CME event. The ionospheric convection shows the two-cell pattern of ionospheric convection typical for southward-type IMF conditions. The convection pattern is also "twisted" due to the presence of a nonzero IMF B_y component. The most important change in the ionosphere is the doubling of the cross-cap potential drop from 30 kV at 70.5h to 60 kV some 27 hours later.

Overall, this simulated space weather event was not very geoeffective. It is expected that we will be able to generate more geoeffective CMEs with the help of more realistic explosive event generation modules. This simulation, however, demonstrates the present capabilities of BATSRUS.

CONCLUDING REMARKS

With the combination of adaptive mesh refinement, domain-decomposition parallelization, and robust finite-volume solvers, methods for solving the ideal MHD equations are developing into powerful tools for a number of applications. With attention to some issues particular to solar-wind modeling (high Alfvén speeds, strong embedded magnetic fields, pressure positivity and divergence control), these tools are becoming quite sophisticated. Much of the work to be done in improving these tools is in coupling them to solvers for regions in which semi-relativistic ideal MHD is not a sufficient model. The results presented in this paper, while preliminary, hint at the new abilities and insights that can be gained from this approach.

ACKNOWLEDGEMENTS

This work was supported by DoD MURI grant F49620-01-1-0359, NSF KDI grant NSF ATM-9980078, NSF CISE grant ACI-9876943, and NASA AISRP grant NAG5-9406. G. Tóth is partially supported by the Education Ministry of Hungary (grant No. FKFP-0242-2000). We also acknowledge the contributions of the developers of TIEGCM and RCM to the coupled magnetosphere model. In particular, the contributions of Ray Roble, Stanislav Sazykin and Richard Wolf are acknowledged and appreciated.

REFERENCES

1. Lax, P. D., and Wendroff, B., *Communications on Pure and Applied Mathematics*, **13**, 217–237 (1960).
2. Godunov, S. K., "Symmetric form of the equations of magnetohydrodynamics (in Russian)," in *Numerical Methods for Mechanics of Continuum Medium*, Siberian Branch of USSR Acad. of Sci., Novosibirsk, 1972, vol. 1, pp. 26–34.
3. Harten, A., *J. Comput. Phys.*, **49**, 357–393 (1983).
4. Powell, K. G., An approximate Riemann solver for magnetohydrodynamics (that works in more than one dimension), Tech. Rep. 94-24, Inst. for Comput. Appl. in Sci. and Eng., NASA Langley Space Flight Center, Hampton, Va. (1994).
5. Roe, P. L., and Balsara, D. S., *SIAM J. Appl. Math.*, **56**, 57–67 (1996).

6. Gombosi, T. I., Tóth, G., De Zeeuw, D. L., Hansen, K. C., Kabin, K., and Powell, K. G., *J. Comput. Phys.*, **177**, 176–205 (2002).
7. Boris, J. P., A physically motivated solution of the Alfvén problem, Tech. Rep. NRL Memorandum Report 2167, Naval Research Laboratory, Washington, D.C. (1970).
8. Powell, K. G., Roe, P. L., Linde, T. J., Gombosi, T. I., and Zeeuw, D. L. D., *J. Comput. Phys.*, **154**, 284–309 (1999).
9. van Leer, B., *J. Comput. Phys.*, **32**, 101–136 (1979).
10. Roe, P. L., *J. Comput. Phys.*, **43**, 357–372 (1981).
11. Linde, T. J., *A Three-Dimensional Adaptive Multifluid MHD Model of the Heliosphere*, Ph.D. thesis, Univ. of Mich., Ann Arbor (1998).
12. Brackbill, J., and Barnes, D., *J. Comput. Phys.*, **35**, 426–430 (1980).
13. Dai, W., and Woodward, P. R., *J. Comput. Phys.*, **142**, 331 (1998).
14. Londrillo, P., and Aanna, L. D., *Astrophys. J.*, **530**, 508–524 (2000).
15. Dedner, A., Kemm, F., Kröner, D., Munz, C., Schnitzer, T., and Wesenberg, M., *J. Comput. Phys.*, **00**, 00–00 (2001), submitted.
16. Tóth, G., *J. Comput. Phys.*, **161**, 605–652 (2000).
17. Berger, M. J., *Adaptive Mesh Refinement for Hyperbolic Partial Differential Equations*, Ph.D. thesis, Stanford Univ., Stanford, Calif. (1982).
18. Berger, M. J., *J. Comput. Phys.*, **53**, 484–512 (1984).
19. Berger, M. J., and Colella, P., *J. Comput. Phys.*, **82**, 67–84 (1989).
20. Berger, M. J., and LeVeque, R. J., "An adaptive Cartesian mesh algorithm for the Euler equations in arbitrary geometries," in *Proc. 9th AIAA Computational Fluid Dynamics Conference*, AIAA Paper No. 89-1930, Buffalo, NY, 1989.
21. Berger, M. J., and Saltzman, S., *Appl. Numer. Math.*, **14**, 239–253 (1994).
22. Quirk, J. J., *An Adaptive Grid Algorithm for Computational Shock Hydrodynamics*, Ph.D. thesis, Cranfield Inst. of Technol., Cranfield, England (1991).
23. Quirk, J. J., and Hanebutte, U. R., A parallel adaptive mesh refinement algorithm, Tech. Rep. 93-63, ICASE (1993).
24. Paillère, H., Powell, K. G., and De Zeeuw, D. L., "A wave-model based refinement criterion for adaptive-grid computation of compressible flows," in *30th AIAA Aerospace Sciences Meeting*, AIAA-92-0322, Reno, Nevada, 1992.
25. Powell, K. G., Roe, P. L., and Quirk, J., "Adaptive-mesh algorithms for computational fluid dynamics," in *Algorithmic Trends in Computational Fluid Dynmaics*, edited by M. Y. Hussaini, A. Kumar, and M. D. Salas, Springer-Verlag, New York, 1993, pp. 303–337.
26. De Zeeuw, D. L., and Powell, K. G., *J. Comput. Phys.*, **104**, 55–68 (1993).
27. Tóth, G., Keppens, R., and Botchev, M. A., *Astron. Astrophys.*, **332**, 1159–1170 (1998).
28. Keppens, R., Tóth, G., Botchev, M. A., and van der Ploeg, A., *Int. J. for Num. Meth. in Fluids*, **30**, 335–352 (1999).
29. Tóth, G., De Zeeuw, D. L., Gombosi, T. I., Hansen, K. C., Powell, K. G., Ridley, A. J., Roe, P. L., and Sokolov, I. V., *J. Geophys. Res.*, **00**, 00–00 (2002), in preparation.
30. Gombosi, T. I., De Zeeuw, D. L., Groth, C. P. T., Powell, K. G., and Stout, Q. F., *J. Atmos. Solar-Terr. Phys.*, **62**, 1515–1525 (2000).
31. Groth, C. P. T., De Zeeuw, D. L., Gombosi, T. I., and Powell, K. G., *J. Geophys. Res.*, **105**, 25,053 – 25,078 (2000).
32. Gombosi, T. I., De Zeeuw, D. L., Groth, C. P. T., Powell, K. G., and Song, P., "The length of the magnetotail for northward IMF: Results of 3D MHD simulations," in *Physics of Space Plasmas*, edited by T. Chang and J. R. Jasperse, MIT Press, Cambridge, Mass., 1998, vol. 15, pp. 121–128.
33. Song, P., De Zeeuw, D. L., Gombosi, T. I., Groth, C. P. T., and Powell, K. G., *J. Geophys. Res.*, **104**, 28,361–28,378 (1999).
34. Song, P., Gombosi, T., De Zeeuw, D., Powell, K., and Groth, C. P. T., *Planet. Space Sci.*, **48**, 29–39 (2000).
35. Linde, T. J., Gombosi, T. I., Roe, P. L., Powell, K. G., and De Zeeuw, D. L., *J. Geophys. Res.*, **103**, 1889–1904 (1998).
36. Gombosi, T. I., De Zeeuw, D. L., Häberli, R. M., and Powell, K. G., *J. Geophys. Res.*, **101**, 15233–15253 (1996).
37. Häberli, R. M., Gombosi, T. I., DeZeuuw, D. L., Combi, M. R., and Powell, K. G., *Science*, **276**, 939–942 (1997).
38. Kabin, K., Gombosi, T. I., De Zeeuw, D. L., and Powell, K. G., *Icarus*, **143**, 397–406 (2000).
39. Bauske, R., Nagy, A. F., Gombosi, T. I., De Zeeuw, D. L., Powell, K. G., and Luhmann, J. G., *J. Geophys. Res.*, **103**, 23,625–23,638 (1998).
40. Liu, Y., Nagy, A. F., Groth, C. P. T., De Zeeuw, D. L., Gombosi, T. I., and Powell, K. G., *Geophys. Res. Lett.*, **26**, 2689–2692 (1999).
41. Hansen, K. C., Gombosi, T. I., DeZeeuw, D. L., Groth, C. P. T., and Powell, K. G., *Adv. Space Res.*, **26**, 1681–1690 (2000).
42. Kabin, K., Gombosi, T. I., DeZeeuw, D. L., Powell, K. G., and Israelevich, P. L., *J. Geophys. Res.*, **104**, 2451–2458 (1999).
43. Kabin, K., Combi, M. R., Gombosi, T. I., De Zeeuw, D. L., Hansen, K. C., and Powell, K. G., *PSS*, **49**, 337–344 (2001).

Temporal and Spatial Variations of Heliospheric X-Ray Emissions Associated with Charge Transfer of the Solar Wind with Interstellar Neutrals

I. P. Robertson[1], T. E. Cravens[1], and S. Snowden[2]

[1]*Dept. of Physics and Astronomy, 1251 Wescoe Hall Dr., University of Kansas, Lawrence, KS 66045, U.S.A.*
[2]*NASA Goddard Spaceflight Center, Greenbelt, MD 20771, U.S.A.*

Abstract. X-rays should be generated throughout the heliosphere as a consequence of charge transfer collisions between heavy solar wind ions and interstellar neutrals. The high charge state solar wind ions resulting from these collisions are left in highly excited states and emit extreme ultraviolet or soft X-ray photons. X-rays should also be generated because of charge transfer collisions with neutral hydrogen in the Earth's geocorona. Originally a simple model was developed in which both the solar wind and the interstellar neutrals were assumed to be spherically symmetric and time independent. In our updated results, the hot model of Fahr [1] was used to model spatial variations of interstellar helium and hydrogen. At the same time a simple model was created to simulate X-ray radiation due to the Earth's geocorona. With the updated information, time independent maps of the heliospheric X-ray emission across the sky were created. Measured time histories of the solar wind proton flux were used in this updated model and the results were compared with "long term enhancements" in the soft X-ray background measured by the Röentgen satellite (ROSAT) for the same time period.

INTRODUCTION

In 1996 X-ray emission from the comet Hyakutake was discovered [2]. Cravens [3] proposed that this emission could be explained by charge exchange collisions between heavy solar wind ions and cometary neutrals. The product ion is left in an excited state and eventually emits a photon in the X-ray or EUV region of the spectrum [4]. Cox [5] suggested that this same process could be applied to interaction between solar wind ions and interstellar neutrals, or neutrals in the Earth's geocorona, and might be able to explain some of the temporal variation in the observed soft X-ray background (SXRB). Dennerl et al. [6] even suggested that this charge exchange mechanism might be able to explain the Long Term Enhancement (LTE) part of the SXRB. Cravens [7] consequently created a simple model of the heliospheric X-ray emissions and concluded that about half of the SXRB could be explained by this mechanism. Robertson et al. [8] and Cravens [9] slowly increased the complexity of the model and found a significant correlation between solar wind proton fluxes and LTE X-ray intensities.

Originally, the interstellar neutral density was approximated with a simple mathematical formula. The model has been improved upon by using Fahr's [1] hot model for interstellar neutrals, as described in the current paper. Time independent maps were created and time dependent behavior was also studied.

MODEL OF X-RAY PRODUCTION BY SOLAR WIND CHARGE EXCHANGE

The following equation was used by Cravens [2, 3] and Cravens et al. [9] to calculate the X-ray and EUV power density:

$$P_{X\text{-ray}} = \alpha n_n n_{sw} u_{sw} (\mathrm{eV\ cm^{-3} s^{-1}}) \qquad (1)$$

where α contains all the atomic cross sections, the transition information and solar wind heavy ion composition. α is different for interstellar helium and hydrogen and also varies with solar wind speed. For the time being we have used the same slow solar wind value of α for both helium and hydrogen, although the helium value should probably be somewhat less than the hydrogen value. The value

CP679, *Solar Wind Ten: Proceedings of the Tenth International Solar Wind Conference*, edited by M. Velli, R. Bruno, and F. Malara

used is: $\alpha \approx 6\times10^{-16}$ eV cm^2. The solar wind density is denoted as n_{sw}, and the solar wind speed as u_{sw}. The solar wind density is presently considered to be spherically symmetric, and its dependence on radial distance (r) is as follows: $n_{sw} = n_{sw0}(r_o/r)^2$, where r_o is 1 AU and n_{sw0} is the solar wind density at 1 AU.

The interstellar neutral density (a combination of interstellar helium and hydrogen) is denoted as n_n. Originally the helium and hydrogen densities were approximated by a simple mathematical formula. Fahr's hot model has been adopted to better calculate these densities. In this model gravitational focusing, ionization losses, and radiation pressure are taken into account. Due to these factors the neutral hydrogen density is lower in the downwind direction from the sun than in the upwind direction. For helium, however, gravitational focusing produces a cone of enhanced helium abundance downwind from the sun near 1 AU. The unperturbed interstellar hydrogen density is taken to be 0.1 cm^{-3}, and that of helium to be 0.02 cm^{-3}. Fahr [1] noted that even though the helium density is initially much smaller than that of interstellar hydrogen, its density reaches the same order of magnitude as the hydrogen density 1 AU on the upwind side. On the downwind side the density can be more than an order of magnitude larger than the hydrogen density.

The production rate, integrated over a path length s, starting at Earth and going out to 200 AU, yields the X-ray intensity. Solar wind ions can also charge transfer with neutral hydrogen in the Earth's geocorona outside the magnetopause and produce X-rays. For the time being, a simple mathematical formula is used to estimate this geocoronal intensity:

$$4\pi I_{geo} = 5\alpha n_{sw} u_{sw} n_{H0} R_E (10\, R_E/R_{mp})^2 \qquad (2)$$

where $R_{mp} \approx 15\, R_E$ is the magnetopause distance from the Earth in the flanks and n_{H0} = 25 cm^{-3} is a reference value of the exospheric hydrogen density at 10 R_E. The results are in agreement with an estimate of Cox [5].

TIME INDEPENDENT RESULTS

Figure 1 shows the heliospheric X-ray intensities in the equatorial plane for a solar wind speed of 400 km/s, a solar wind density with n_o = 7/cm^3, and neutral densities from Fahr's hot model. The location of the Earth is at the vernal equinox, as indicated on the small insert in the figure. The direction of the interstellar wind is also indicated with the large arrow.

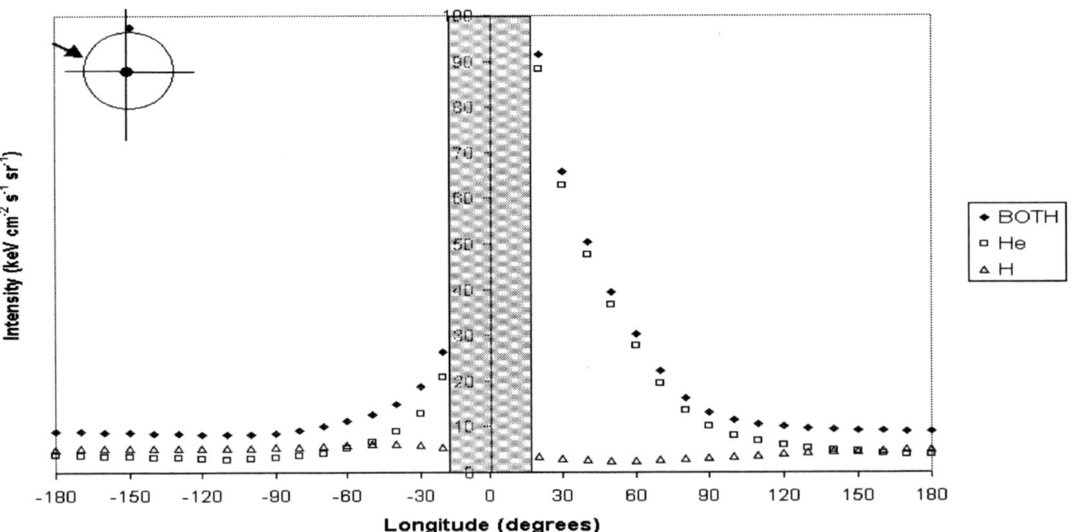

FIGURE 1. Time independent variation with ecliptic longitude of heliospheric X-ray intensities for look directions in the ecliptic plane. See text for details.

The coordinates used are Earth-centered solar ecliptic; consequently, zero degrees points towards the sun (an area blocked out in the graph), -90° points into the interstellar wind and +90° points towards the tail. There is little variation in the X-ray intensity due to charge exchange with interstellar hydrogen. The contribution due to helium varies much more, with a significant enhancement near 30° when the look direction intersects with the "helium cone" due to gravitational focusing. Note, though, that a more careful determination of α for helium could reduce these intensities by as much as a factor of 2.

TIME DEPENDENT RESULTS

Time dependent X-ray intensities were calculated for fixed look directions. We used solar wind proton fluxes measured by the IMP-8 spacecraft for the time period 1996-1998. The day number is the number of days after January 1, 1996. Once again the solar wind proton flux is assumed to be spherically symmetric and the flux decreases as $1/r^2$. Again Fahr's hot model is used to model the interstellar neutral density. Figures 2 and 3 show the X-ray intensity due to helium, hydrogen and the Earth's geocorona. The total intensity is also plotted. Cravens et al. [9] show a plot for the same time period, but it was produced without the use of Fahr's model. Two cases were considered: the Earth located in the upwind direction (summer) and the Earth located in the downwind direction (winter). In both cases depicted, the look direction is northward from the ecliptic plane. In Figure 2, the Earth is in the downwind direction (winter solstice). Because the Earth is immersed in the helium cone, the dominant contribution, both in variation as well as intensity, is from helium. The variation in hydrogen is not noticeable at this scale. Even though the geocoronal contribution is minimal, it does contribute to the variation in total intensity. In Figure 3, the Earth is in the upwind direction (summer solstice). The hydrogen contribution is dominant but has very little variation. More variation can be seen in the helium contribution and even more in the contribution from the geocorona. It should be noted that these plots were generated using identical alphas for the calculations of the helium as well as the hydrogen contributions. As noted earlier, that is probably not the case. It can be noticed, however, that even if the alpha for helium would be smaller by a factor of 2, the helium continues to be dominant in the X-ray variability.

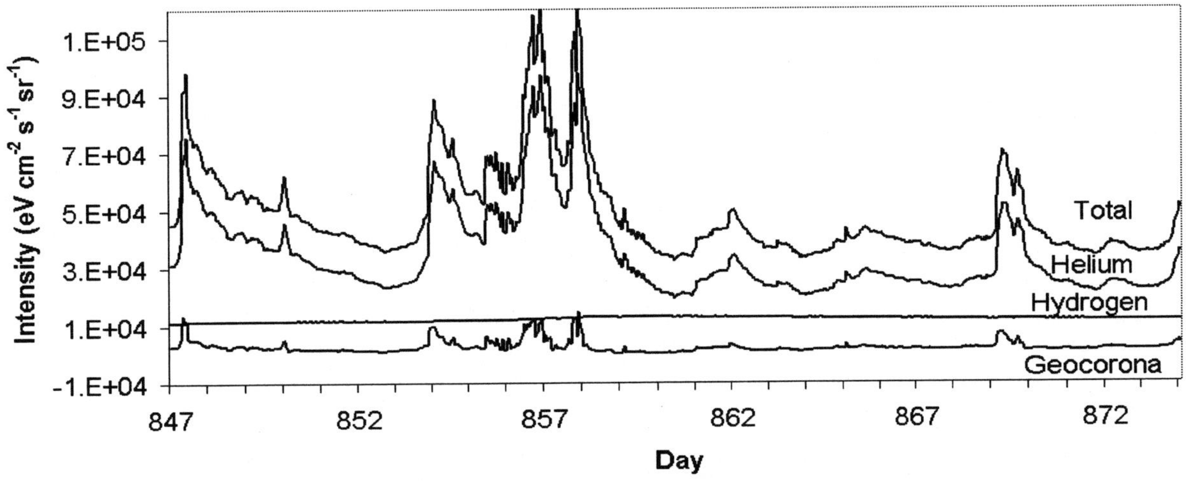

FIGURE 2. X-ray intensity at Winter Solstice versus time. Earth is located in the downwind direction.

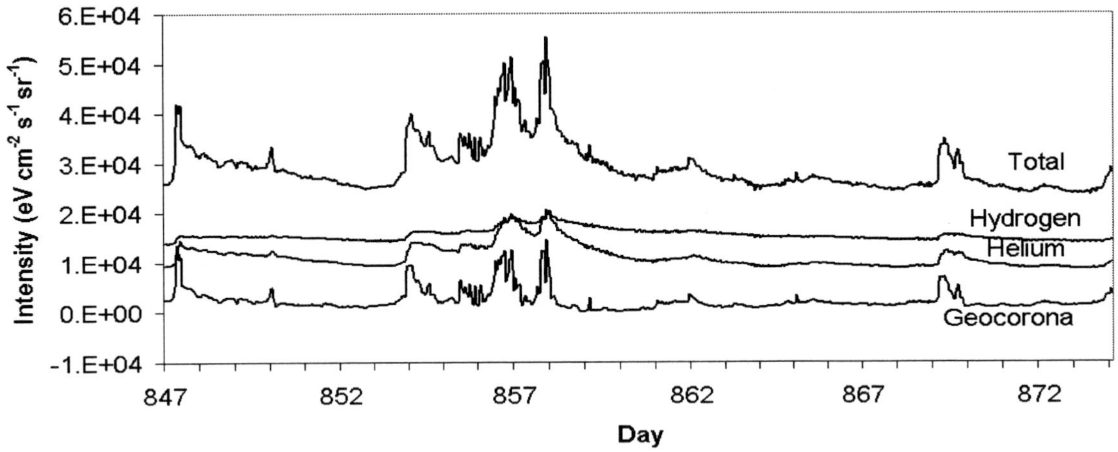

FIGURE 3. X-ray intensity at Summer Solstice versus time. Earth is located in the upwind direction.

ROSAT LONG-TERM ENHANCEMENT DATA

Cravens et al. [9] directly compared ROSAT LTE 1/4 keV data with measured solar wind proton fluxes for the same time period. Even though no selection was made with respect to look-direction, a significant correlation was found with a linear regression coefficient of $R = 0.7$. Cravens next plotted the total, helium and geocoronal intensities versus ROSAT LTE data and found a strong correspondence between the modeled X-ray intensity and the LTE data. The downwind plot with Fahr's densities shows considerable more variation than the X-ray intensities Cravens used and, consequently, would show an even stronger correspondence between X-ray intensity and LTE data.

CONCLUSIONS

Our improved model of soft X-ray intensities observed at Earth further supports the suggestion that LTEs, in the observed soft X-ray background, can be explained by charge exchange between heavy solar wind ions and interstellar neutrals. Further improvements in the modeling of X-ray intensities should include:

1. A more accurate calculation of α for different species;
2. A solar wind model which includes latitudinal variations;
3. An improved model of the geocoronal X-ray emission.

ACKNOWLEDGMENTS

NASA Planetary Atmospheres grant NAG5-4358 and NSF grant ATM-9815574 at the University of Kansas are acknowledged.

REFERENCES

1. Fahr, H. J., *Astron. Astrophys.* **14**, 263-274 (1971).
2. Lisse, C. M. , et al., *Science* **274**, 205 (1996).
3. Cravens, T. E., *Geophys. Res. Lett.* **24**, 105 (1997).
4. Cravens, T. E., *Science* **296**, 1042-1045 (2002).
5. Cox, D. P., "Modeling the local bubble," in *The Local Bubble and Beyond*, edited by D. Breitschwerdt, M. J. Freyberg, and J. Trümper, New York, Springer-Verlag, 1998, p. 121.
6. Dennerl, K., Englhauser, J., and Trümper, J., *Science* **277**, 1625 (1997).
7. Cravens, T. E., *Astrophys. J.* **532**, L153 (2000).
8. Robertson, I. P., Cravens, T. E., Snowden, S., and Linde, T., *Space Science Reviews* **97**, 401-405 (2001).
9. Cravens, T. E., Robertson, I. P., and Snowden, S. L., *J. Geophys. Res.* **106**, 24,883-24,896 (2001).

Solar Probe - The First Flight Into the Sun's Corona

Kenneth A. Potocki and Peter D. Bedini

Johns Hopkins University Applied Physics Laboratory

Abstract. The NASA Solar Probe mission to the inner frontier of the heliosphere is a part of the Sun-Earth Connection theme within the Office of Space Science. A NASA-appointed Science Definition Team has defined a Solar Probe mission and its scientific objectives. These include making measurements to understand the processes that heat the solar corona and produce the solar wind, subjects of continuing scientific debate. The Solar Probe mission will accomplish these objectives with a combination of *in situ* measurements designed to characterize the local heating and acceleration of plasma near the Sun and high resolution images to detect small-scale, transient magnetic structures at and around the Sun. In order to sample the solar corona acceleration region, Solar Probe will fly to four solar radii from the center of the Sun in an orbit inclined 90° to the plane of the ecliptic. Engineering solutions to design a probe that can withstand the near-Sun environment have been proposed for several decades. Current status of the Solar Probe concept is provided in this paper.

INTRODUCTION

NASA's Sun-Earth Connection Theme seeks to understand our changing Sun, and its effects on the solar system, life, and society. Solar activity, including CMEs, couples into interplanetary space. Only by exploring the outer coronal regions can we fully understand this coupling. The solar wind forms planetary magnetospheres, which shield the Earth and other planets from interplanetary and galactic particles, energy, and mass. Solar wind bursts cause magnetic storms that can damage Earth-orbiting spacecraft, cause ground power system outages, and create radiation hazards to man in space. The variability of the solar wind and embedded magnetic fields controls magnetospheric/ionospheric dynamics and the aurora at Earth. The solar wind is the origin of magnetospheres and of the aurora on Earth, and Solar Probe will study the origin of the solar wind.

SCIENCE OBJECTIVES

The present Solar Probe study is an engineering study, and relies solely on the 1999 Science Definition Team (SDT) findings for any science direction. The SDT report[1] and the associated Mission and Project Description document[2] are the governing documents. The SDT categorized the Solar Probe science objectives into three groups. Listed in priority order, they are:

Group 1 Objectives

- Determine the acceleration processes and find the source regions of the fast and slow solar wind at maximum and minimum solar activity;

- Locate the source and trace the flow of energy that heats the corona;

- Construct the three-dimensional coronal density configuration from pole to pole and determine the subsurface flow pattern, the structure of the polar magnetic field, and their relationship with the overlying corona;

- Identify the acceleration mechanisms and locate the source regions of energetic particles, and determine the role of plasma waves and turbulence in the production of solar wind and energetic particles.

Group 2 Objectives

- Investigate dust rings and particulates in the near-Sun environment;

- Determine the outflow of atoms from the Sun and their relationship to the solar wind;

CP679, *Solar Wind Ten: Proceedings of the Tenth International Solar Wind Conference*,
edited by M. Velli, R. Bruno, and F. Malara

• Establish the relationship between remote sensing, near-Earth observations at 1 AU and plasma structures near the Sun.

Group 3 Objectives

• Determine the role of x-ray microflares in the dynamics of the corona; and

• Probe nuclear processes near the solar surface from measurements of solar gamma rays and slow neutrons.

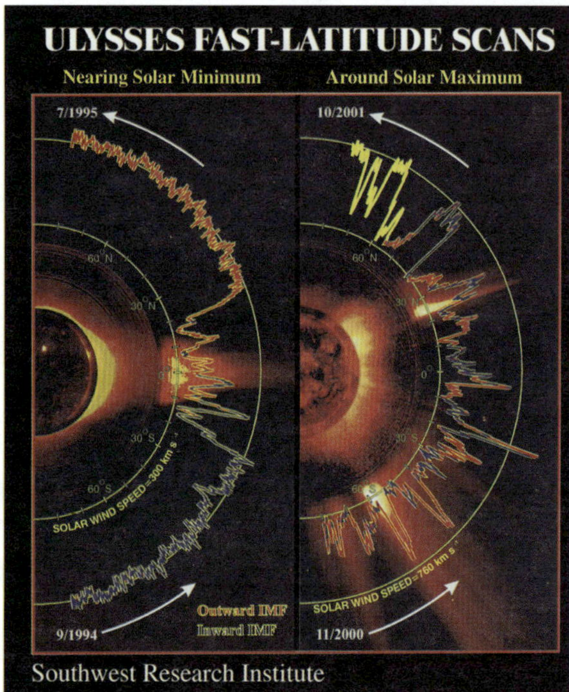

FIGURE 1. Ulysses/SWOOPS data overlaid on solar images provided by the EIT and LASCO/C2 instrument teams on SOHO and the Mauna Loa coronagraph team from NCAR's High Altitude Observatory.

CURRENT STATUS

Solar Probe is now a mission study in the Sun-Earth Connection Theme.

The Johns Hopkins University Applied Physics Laboratory (JHU/APL) has been tasked by the Goddard Space Flight Center to perform a new conceptual mission design study of the Solar Probe Mission with support from the Jet Propulsion Laboratory.

Key Technical Challenges

- Survive 3000-Sun intensity at perihelion (four solar radii)
- Maintain reliable power over mission life
- Protect against hostile near-sun dust environments
- Control attitude during flyby for survival imaging
- Minimize coronal scintillation effects of communications
- Accommodate in-situ and imaging instruments

New technology is being considered for its capability to solve these challenges. Solar Probe will be a mission of totally new discovery, not incremental science.

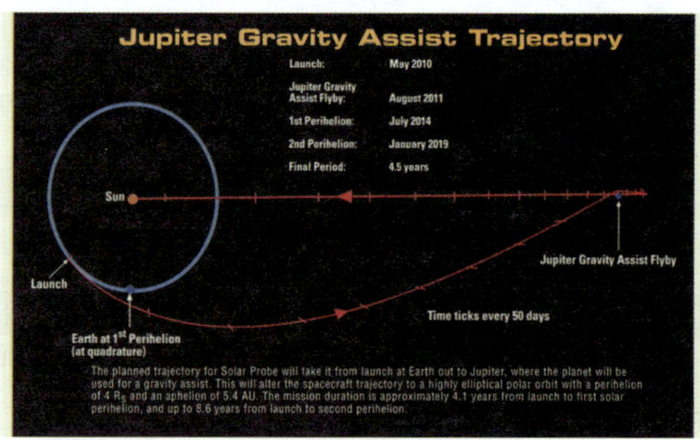

FIGURE 2. Solar Probe Mission trajectory showing the use of the Jupiter Gravity Assist to achieve a polar solar orbit.

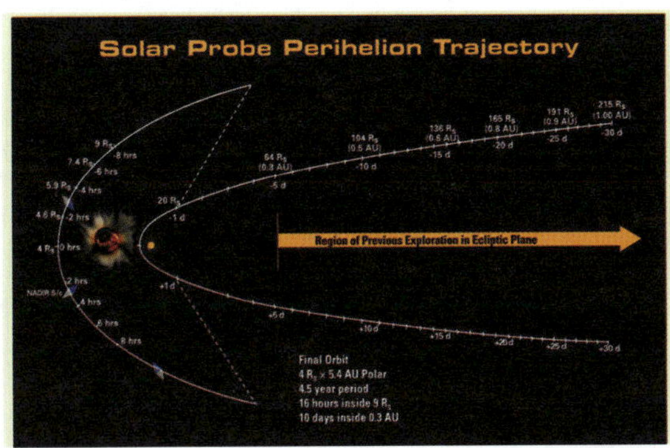

FIGURE 3. Solar Probe Perihelion Trajectory with an inset view of the 16-hour period around perihelion.

ACKNOWLEDGMENTS

Work on this engineering study is supported under NASA Contract NAS5-01072.

REFERENCES

1. *Solar Probe: First Mission to the Nearest Star*, Report of the NASA Science Definition Team for the Solar Probe Mission, (1999).

2. *Solar Probe - Mission and Project Description*, NASA OSS Announcement of Opportunity AO 99-OSS-04 Europa Orbiter/Pluto-Kuiper Express/Solar Probe, (1999).

Solar Wind Plasma Experiment on Solar Orbiter: dealing with the need for a sufficient phase-space resolution

R. D'Amicis*, R. Bruno*, MB. Cattaneo*, B. Bavassano*, G. Pallocchia* and J.A. Sauvaud†

*Istituto di Fisica dello Spazio Interplanetario (CNR) - Via Fosso del Cavaliere, 100 - 00133 Rome - Italy
†CESR (CNRS/Université de Toulouse), BP 4346 - 31029, Toulouse Cedex, France

Abstract. Solar Orbiter is proposed as a space mission dedicated to study the solar surface, the corona and the solar wind by means of remote sensing and in-situ measurements, respectively. At 0.21 AU, closest approach to the Sun, the full 3-D particle velocity distribution should be sampled as fast as a few tens of msec in order to study the growth phase of the instabilities. This implies some restrictions on the maximum phase space resolution given the limited allowed bit-rate for data transmission. In this paper we evaluate consequences of this limitation for solar wind distributions with different parameters.

INTRODUCTION

Our present knowledge of the solar corona, the solar wind and the 3-D heliosphere is based on the results of missions such as Helios, Ulysses, Yohkoh, Soho and Trace. Solar Orbiter will provide us with new insights in the plasma kinetic processes that act in the Sun's atmosphere and in the extended corona as it will explore the inner regions of the Solar System (the perihelion will be at 0.21 AU). The study of the kinetics of the solar wind is fundamental to fully understand the mechanism acting within the plasma's micro state. In particular, it helps us to understand the role of microinstabilities generated by non-Maxwellian characteristics and anisotropies in phase-space of the velocity distribution function. In order to study these microinstabilities it would be extremely important to study them during their growth phase. To do so, we should be able to sample the whole 3-D velocity distribution function fast enough, i.e. with a time resolution of the order of a few tens of msec, which is the time taken by the s/c to go across a scale length of the order of the typical Larmor gyroradius at 0.2 AU. Taken into account the need to follow the instability for about 100 gyroperiods, typical time duration of the growth phase, and the allowed bit rate of 5 kb/s, it follows that the maximum possible resolution would be 10 energies, 10 azimuths and 10 polar angles (see the Assessment Study Report, 2000).

In this report we show that this phase-space resolution is generally not sufficient to fully resolve the ion distribution and to compute correctly its moments.

NUMERICAL SIMULATIONS

The solar wind distribution can be represented by a bi-Maxwellian, defined in the following way:

$$f = n\left(\frac{m}{2\pi kT_\perp}\right)\left(\frac{m}{2\pi kT_\parallel}\right)^{1/2} exp\left\{-\frac{m}{2kT_\perp}\left[(v_y-V_y)^2 + (v_z-V_z)^2\right] - \frac{m}{2kT_\parallel}(v_x-V_x)^2\right\}$$

where m is the proton mass, k is the Boltzmann's constant and $V_{x,y,z}$ are the plasma flow velocity components. This distribution represents the most general equilibrium status for a magnetized plasma and it is written in the magnetic field's frame, i.e. chosen in such a way that the x axis is parallel to **B**. It is characterized by the following 6 variables: the number density, n; the parallel and perpendicular components of the temperature, $T_\parallel$ and $T_\perp$, with respect to the magnetic field's direction; the particle velocity components v_x,v_y,v_z. We assume that the wind flow is parallel to the x axis.

Taking into account only the major ions, we have extrapolated the proton parameters to 0.21 AU, closest approach to the Sun for Solar Orbiter, using the radial trends evaluated from Helios 2 observations. In particular, we chose day 100th (Marsch, 1982) as a reference for the slow solar wind and day 107th (Marsch, 1982; Marsch, 1991) for the fast solar wind as shown in Table 1. In Table 2, we show the temperature radial gradients and their indexes for both slow and fast wind. From the

CP679, *Solar Wind Ten: Proceedings of the Tenth International Solar Wind Conference*,
edited by M. Velli, R. Bruno, and F. Malara

radial trends shown in Table 2 and adopting r^{-2} for the density radial dependence, we obtained the parameters of the distribution as shown in Table 3. We chose proton velocity values slightly different from those of Helios (especially for what concerns the slow wind) because we intend to study solar wind in quasi-extreme conditions. The parameters characterizing the alpha particles, as obtained by Helios 2 (Marsch, 1991), are shown in Table 4.

TABLE 1. Solar Wind Parameters from Helios 2.

Day 1976	R [AU]	v_p [km/s]	n_p [cm^{-3}]	$T_{\parallel p}$ [eV]	$T_{\perp p}$ [eV]
100	0.333	429	71.9	18.6	15.7
107	0.291	781	28.3	48.8	83.4

TABLE 2. Temperature radial gradients.

Radial Gradient	Index	Slow wind	Fast wind
$T_{\parallel} \propto r^{-\beta}$	$\beta_{\parallel}$	1.0	0.72
$T_{\perp} \propto r^{-\beta}$.	$\beta_{\perp}$	0.9	1.12

TABLE 3. Solar Wind Parameters extrapolated at 0.21 AU.

R [AU]	v_p [km/s]	n_p [cm^{-3}]	$T_{\parallel p}$ [eV]	$T_{\perp p}$ [eV]	T_{tot} [eV]
0.21	250	181	29.5	23.8	25.7
0.21	800	54.3	61.7	120	101

Computing the moments of the distribution

Given a distribution function f, one can define the moment of order n of such distribution:

$$M_n \equiv \int f(\mathbf{v})\mathbf{v}^n d^3$$

from which one can evaluate:

- the number density: $N = \int f(\mathbf{v})d^3v$,
- the number flux density vector: $N\mathbf{V} = \int f(\mathbf{v})\mathbf{v}d^3v$ (from which one can compute the velocity dividing by N),
- the moment flux density tensor: $\Pi = m\int f(\mathbf{v})\mathbf{v}\mathbf{v}d^3v$,
- the energy flux density vector: $\mathbf{Q} = \frac{m}{2}\int f(\mathbf{v})v^2\mathbf{v}d^3v$.

Converting the momentum flux tensor and the energy flux vector to the plasma's ref. frame, one obtains the pressure tensor, $P = \Pi - \rho\mathbf{V}\,\mathbf{V}$ and the heat flux vector, $\mathbf{H} = \mathbf{Q} - \mathbf{V}P - \frac{1}{2}\mathbf{V}Tr(P)$. Using the definition $P \equiv NkT$, one can convert the pressure tensor into a temperature tensor. These quantities can be written by means of the

TABLE 4. Alpha particles parameters.

Parameters	n	$T_{\parallel}$	$T_{\perp}$	T_{tot}
Protons	n_p	$T_{p\parallel}$	$T_{p\perp}$	T_{ptot}
α particles	0.05 n_p	2 $T_{p\parallel}$	2 $T_{p\perp}$	2 T_{ptot}

counts registered by the instrument, defined as: $C = \int\int\int Svfd^3v$ where S is the effective cross-section and $d^3v = v^2\cos\theta\, d\theta\, d\phi\, dv$, θ is the polar angle and ϕ is the azimuthal angle. As the incidence is perpendicular, $\theta \sim 0$, so $cos\ \theta \sim 1$. Dealing with discrete quantities, one can rewrite the preceding as: $C_{ijk} = f_{ijk}Sv^3\Delta v_i\Delta\phi_j\Delta\theta_k$ or in a more compact way as $C_{ijk} = f_{ijk}Gv^4$ where the geometric factor, $G = S\ cos\theta\ \Delta\phi\Delta\theta\Delta v/v$ has been introduced. From the preceding equation, the distribution function can be evaluated: $f_{ijk} = C_{ijk}/Gv^4$ (for further reading see Paschmann et al., 1998). As an example, we will obtain the number density expression. Coming back to its definition and substituting the preceding expression for f, one obtains:

$$N = \int\int\int \frac{C}{Gv^4}v^2 cos\theta\, d\theta\, d\phi\, dv$$

that becomes:

$$N = \frac{\Delta\theta\Delta\phi}{G}\sum_{v_i}\frac{\Delta v_i}{<v_i^2>}\sum_{\phi_j}\sum_{\theta_k} cos\theta_k C_{ijk}$$

where $\Delta\theta$ and $\Delta\phi$ are the amplitudes of the polar and azimuthal sectors, respectively, while Δv_i is the amplitude of the energy channel. The average over v_i^2 is an average over the four adjacent energy sub-channels that constitute an energy channel while $cos\theta_k$ is computed using the central angle of the polar sector. Expressions for the other moments are obtained in a similar way.

We assume a field of view varying from -45° to +45° both in azimuthal and polar direction (D'Amicis et al., 2001). The energy variation law is assumed exponential and the energy limits vary from 15000 eV to 10 eV in order to cover the entire energy distribution.

We computed the principal moments as a function of some possible combinations of the number of azimuths, polar angles and energy steps, i.e. varying $\Delta\theta$, $\Delta\phi$ and ΔE, starting from the resolution suggested in the Assessment Study Report of 9° and 10 energy steps, down to 5° and 32 energy steps (Helios resolution). Our model distribution is characterized by a proton bi-Maxwellian distribution, whose parameters are shown in Table 3.

Results shown in Table 5 suggest that the lowest resolution causes large uncertainties in the computation of the principal moments. On the contrary, the other resolutions reproduce quite well the model. It is evident that, in general, the higher the resolution the closer to the

TABLE 5. Moments of the distribution as a function of the number of energy steps, of polar and azimuthal angles for a fast wind.

Resolution			Moments					
n^o ϕ steps	n^o θ steps	n^o E steps	n [cm^{-3}]	V_x [km/s]	$T_{\parallel}$ [eV]	$T_{\perp 1}$ [eV]	$T_{\perp 2}$ [eV]	T_{tot} [eV]
10	10	10	25.9	731.	48.5	103.	114.	88.0
10	10	20	52.0	803.	54.0	121.	134.	103.
10	10	32	53.4	800.	54.9	120.	133.	103.
12	12	20	52.0	803.	54.0	121.	130.	102.
12	12	32	53.4	800.	54.9	120.	129.	101.
18	18	20	52.0	803.	53.9	121.	125.	99.8
18	18	32	53.4	801.	54.8	120.	124.	99.6

model are the computed moments. Moreover, the number of energy steps plays a very important role. Actually, the computation of n, V_x and $T_{\parallel}$ is more sensitive to this parameter. The transverse velocity components are very small, as expected, of the order of unity. As an example, we can compare the $10 \times 10 \times 32$ resolution to that of $18 \times 18 \times 20$. The previous parameters are closer to the model when we use 32 energy steps instead of 20 energy steps, although the total resolution is lower than in the other case. We also notice that n, V_x and $T_{\parallel}$ are not very sensitive to the resolution in θ and ϕ provided that the number of energy steps is kept constant. On the other side, increasing the angular resolution allows to determine the perpendicular temperature more accurately. Finally, the highest resolution does not show very different results from the previous cases. An analogous computation for a slow wind (see table 3) showed better results even at the lowest resolution because slow wind is sampled in energy using narrower energy channels, given that the wind is peaked at low energies and $\Delta E/E$ is kept constant.

Sampling the distribution function

The next step is to study how the instrument samples a distribution function, obtained as the sum of two bi-Maxwellians, one for protons (see Table 3) and the other one for alpha particles (see Table 4), varying the number of energy steps, polar and azimuthal angles, both for slow and fast solar wind. We simulated all the cases described in Table 5 but we noticed that the alpha peak emerges when we use 24 energy steps, at least. The left panels of Fig. 1 which refer to the lowest resolution, show the counts (top panel) as a function of velocity and the contour plots (bottom panel) computed integrating over the polar angle. It is clear that this resolution is not sufficient to fully resolve the distributions of the major ions since the alpha particle peak does not show up. In the right panels the same fast wind is resolved using a higher resolution, i. e. 12 ϕ, 12 θ and 32 E and the alpha peak is now clearly identifiable. We also simulated the sampling for the maximum resolution ($18 \times 18 \times 32$), but we did not obtain a better spatial definition of our distribution function.

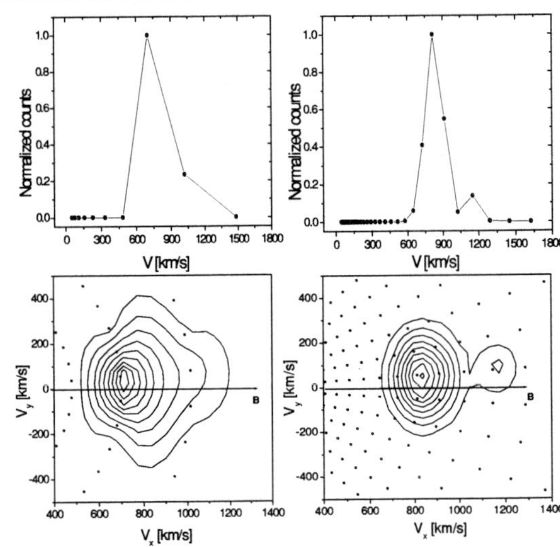

FIGURE 1. Fast wind. In the upper panels: counts vs. velocity. In the lower panels: contour plots integrated over θ. On the left-hand side panels: 10 ϕ, 10 θ and 10 E. On the right-hand side panels: 12 ϕ, 12 θ and 32 E. The first contours correspond to 10 % of the maximum phase-space density.

Finally, we simulated a slow solar wind as shown in Fig. 2 for the highest resolution. We represented three distributions: protons, alpha particles and protons+alpha particles. We observe that the proton and alpha particle distributions can not be separated. We notice that the proton distribution is dominant and the alpha particles are entirely masked.

Co-rotating phase

Solar Orbiter will correlate in-situ and remote-sensing measurements at 45 solar radii from a co-rotational vantage point. It is then very important to simulate this phase of the orbit which will identify the links between the activity on the Sun's surface and the resulting evolution of the corona and inner heliosphere. To do this, we have to add a V_y component, which can be estimated of the order of $\sim$ 100 km/s, to the bulk velocity. The aim of this study is to verify whether the field of view of the analyzer can cover the whole distribution function for both slow and fast wind. Fig. 3 shows that this is the case for a fast wind for both low and high resolution. On the contrary, the slow wind is at the limits of the field of view of the analyzer. This problem can be simply solved by shifting one angular sector downwards, so that we are able to observe the entire distribution during the whole orbit.

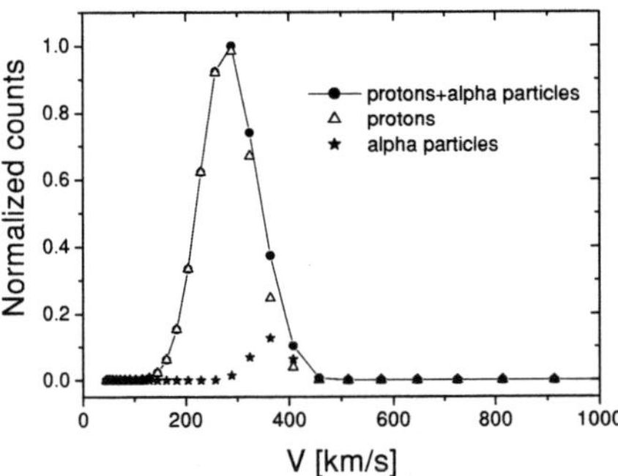

FIGURE 2. Slow wind. The picture shows three distributions: protons, alpha particles and protons+alpha particles with a velocity resolution of 18 ϕ, 18 θ and 32 E.

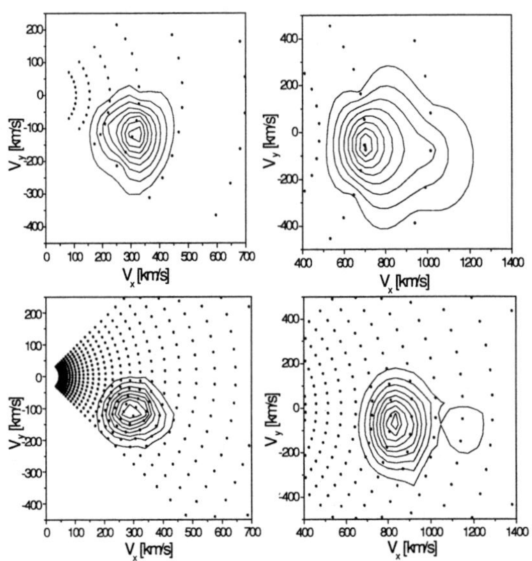

FIGURE 3. Co-rotating phase. The upper panels show the distribution with the lowest resolution ($10 \times 10 \times 10$) for slow (left) and fast wind (right). The bottom ones show the highest resolution ($18 \times 18 \times 32$): slow wind (left) and fast wind(right). The first contours correspond to 10 % of the maximum phase-space density.

CONCLUSIONS

We studied the moments computation as a function of the resolution used for energy, polar and azimuthal angles. For a fast wind, the resolution firstly suggested in the Assessment Study Report seems to be inadequate. It is important to emphasize how the computation of n, V_x and $T_{||}$ mainly depends on the number of energy steps. Increasing the angular resolution allows only to determine the perpendicular temperature values more accurately. The sampling of the distribution functions shows that, for a fast wind, the lowest resolution does not show any evidence for the alpha particles peak. By increasing the energy steps and leaving almost unchanged the angular resolution, we observe that the alpha peak emerges. For a slow wind, even with the highest resolution, we are not able to separate alpha particles from protons. Finally we find that during the co-rotating phase, a field of view from -45° to $+45^o$ centered around 0° seems not to be appropriate to observe the slow wind distribution. A simple solution can be found by shifting the azimuthal field of view adequately.

To better resolve the major ions' distribution, we propose a higher phase-space resolution than the one firstly suggested in the Assessment Study Report. From our simulations, it emerges that an angular resolution of 10 ϕ and 10 θ is sufficient but, a minimum of, at least, 24 energy steps is needed.

REFERENCES

1. D'Amicis, R., Bruno, R., Bavassano, B., Cattaneo, M. B., Baldetti, P., Pallocchia, G., "Numerical Study of a quasi 3D Top-Hat Analyzer", in *Solar Encounter: The First Solar Orbiter Workshop*, ESA SP-493, Tenerife, Spain, 2001, pp. 205–209.
2. Marsch, E., Mühlhäuser, K.-H., Schwenn, R., Rosenbauer, H., Pilipp, W. and Neubauer, F. M., *J. Geophys. Res.*, **87**, 52-72, (1982).
3. Marsch, E., "Kinetics Physics of the Solar Wind Plasma", in *Physics of the Inner Heliosphere II*, edited by R. Schwenn, E. Marsch, Springer-Verlag, Berlin Heidelberg, 1991, Vol. **21**, pp. 45–133.
4. Paschmann, G., Fazakerley, A. N. and Schwartz, S. J., "Moments of Plasma Velocity Distributions" in *Analysis Methods for Multi-Spacecraft Data*, edited by Paschmann G. and Daly P. W., SR-001, International Space Science Institute, 1998, pp. 125–158.
5. *Solar Orbiter: A High-Resolution Mission to the Sun and the Inner Heliosphere*, Assessment Study Report, SCI(2000)6, July 2000.

Multi-Angle Viewing of the Sun and the Inner Heliosphere

Alexander Ruzmaikin and MASSÉ Science Team[1]

Jet Propulsion Laboratory, California Institute of Technology, 4800 Oak Grove Drive, Pasadena, CA 91109, USA, E-mail: aruzmaik@pop.jpl.nasa.gov

Abstract. We describe the concept of a proposed mission, called Multi-Angle Solar Sources Explorer (MASSÉ), that would observe the Sun and the inner Heliosphere from an orbit at 0.72 AU over all solar longitudes. It would, in co-ordination with observations from Earth's side, investigate the sources of solar activity from their origin deep within the Sun, their emergence onto the photosphere, and their ejection into the Heliosphere. It carries a Doppler-magnetic imager, and in situ energetic particle, solar wind, and magnetic field detectors. Three-dimensional views of the convection zone, where solar activity originates, are reconstructed by correlating MASSÉ and earth-side Doppler signals from acoustic wave packets traversing deep solar layers. Magnetic images reveal the evolution of active regions over their life-time and allow the study of emerging fields from deep layers. Particle, plasma, and magnetic field data provide information on the sites and mechanisms of acceleration of hazardous high-energy particles produced by coronal mass ejections.

INTRODUCTION

Viewing only one side of the Sun at a time has handicapped solar observers. Half a solar rotation is needed before the farside becomes fully visible. Hence the processes developing on time scales of weeks, for example the full evolution of active regions, cannot be directly observed. Another critical limitation is that we cannot observe the solar deep interior.

Helioseismology that analyzes acoustic p-waves traveling within the solar interior has made it possible to partially overcome both obstacles [1]. Helioseismic studies have established the thermal structure and distribution of differential rotation inside the solar convection zone. Large active regions on the solar farside can be detected by analysis of "double-bounced" waves [2].

However, current helioseismology utilizes viewing from a single point and thus has serious limitations in probing the 3-D structures of flows and magnetic fields inside the Sun. For example, the solar core studies are very limited. The dynamo region located near the bottom of the convection zone cannot be well investigated with a narrow field of view or bounced waves. An effective way to overcome these limitations is simultaneous observation of the Sun from multiple longitudes including observations from the earth side.

Multiviewing of the inner Heliosphere uncovers the development of space weather related activity and helps to understand mechanisms of acceleration of energetic particles by coronal mass ejections.

Here we briefly discuss science benefits of multiviewing and a potential mission to carry it out.

THE SUN: INTERIOR AND SURFACE

The solar interior can be probed through correlation of Doppler images taken from two positions (Fig. 1). This correlation gives travel times (and phases) of waves traversing different depths of the Sun. The travel times are influenced by inhomogeneities, flows and magnetic fields thus carrying information about the interior structure and dynamics [1].

Solar activity originates in interplay between magnetic fields and fluid motions within the solar convection zone [3]. Fields change on time scales from minutes to the 11-year solar cycle and longer. They emerge at the solar surface, diffuse, reconnect, and are transported by the solar wind into the Heliosphere. Conducting plasma moving in the convection zone is a generator (dynamo) for the solar magnetic field. Thus the dynamo is the key to understanding the origins of solar activity.

[1] The MASSE science team includes A. Ruzmaikin (PI), E. Stone (particle lead), J. Harvey (solar magnetic lead), R. Ulrich (helioseismology lead), K. Ogilvie (in-situ plasma/magnetic lead), M. Acuna, A. Cummings, J. Feynman, B. Goldstein, D. Gough, K. Harvey, A. Kosovichev, A. Lazarus, C. Lindsey, D. Mewaldt, C. Ng, D. Reames, P. Scherrer, S. Tomczyk, Y. Toomre, T. von Rosenvinge, M. Wiedenbeck, G. Zank.

CP679, *Solar Wind Ten: Proceedings of the Tenth International Solar Wind Conference*, edited by M. Velli, R. Bruno, and F. Malara

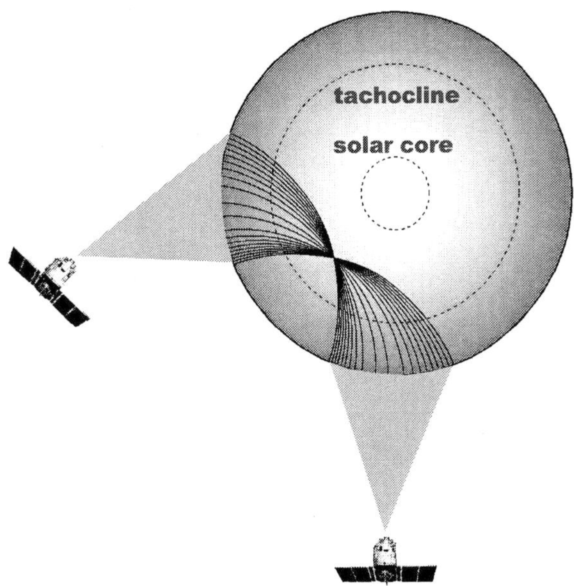

FIGURE 1. Imaging from two positions allows the probing all depths of the Sun.

One type of motion that is poorly known is called giant cells. Giant cells are expected to have lifetimes comparable to the Sun's rotation period; the rotation and convection at this scale are interrelated and interact with magnetic fields. Another type of dynamo related motion is meridional circulation, which is directly measured on the solar surface. First attempts have been made to follow it below the photosphere [4] but available data (SOHO and GONG) do not provide sufficient accuracy to definitively establish the pattern of meridional flows deep within the convection zone.

Current modeling of the dynamo [5-7] emphasizes the importance of a region deep inside the Sun, called the "tachocline,,, a thin layer where differential rotation of the convective zone, which stretches field lines into strong toroidal belts, ends and a nearly uniform rotation of the deeper interior begins. It is here where the dynamo is believed to work. In each solar cycle some critical changes occur in the dynamo as the field switches from a dominantly axisymmetric at solar minimum to a dominantly non-axisymmetric at solar maximum. The non-axisymmetic fields are associated with preferred longitudes of solar activity [8]. The tachocline can be probed by the Doppler viewing from different solar longitudes.

The structure and rotation of the Sun's energy-generating core remain enigmatic. Because magnetic fields decay very slowly in the core, it can have kept memory of initial stages of the Sun's formation in a field now frozen into the highly-conductive core. Current models of rotation based on data taken from a single observation point are limited in depth to >0.4 solar radii. With a spacecraft at the farside, the core rotation can be measured by using wave packets that bracket the solar center, passing both to the east and to the west of the core center.

Magnetic imaging of the solar surface from multiple longitudes allows the study of:

Active region evolution. Magnetic field evolution of a specific area on the Sun can only be observed for about 10 days before it rotates out of view from the Earth. Active regions systematically evolve through significant portions of their life while they are on the unseen side of the Sun, so their evolution from one configuration to another can only be described statistically. There are unanswered questions such as: What makes some active regions so energetic, while others are placid? Why do some active regions last for a long time while others, seemingly similar, quickly vanish? What is the magnetic flux budget history of an active region? The uninterrupted view of solar active regions development allows photospheric magnetic fields to be studied without confusing its spatial and temporal changes.

Magnetic clustering. During the rising phase of the solar cycle magnetic activity clusters at a few solar longitudes for many months [9,10]. Clusters of activity are often the sources of fast CMEs and solar energetic particles (SEPs) and should be monitored carefully. The cause of this clustering observed in the distribution of active regions [9] and associated magnetic fields on the Sun [11, 8] remains unknown; it may be a manifestation of the low-wavenumber, non-axisymmetric modes of the dynamo [8]. The preferred longitudes manifest themselves in the Heliosphere. It has been found that the radial component of the interplanetary field and the non-axisymmetric solar field rotate with a period of 27.03 days [12, 13]. This period does not coincide with the period of the surface or core rotation [8]. Multi-angle imaging to probe different depths of the convection zone is needed in search for the cause of this periodicity associated with preferred longitudes.

It has long been known that CMEs tend to occur in sheared magnetic fields. However, physical interpretation of line-of-sight earth-side observations is ambiguous (variations of field strength, direction, or location?) because they provide only the component of the vector magnetic field in the line of sight. Vector field measurements suffer from a 180°-direction ambiguity, and current vector instrumentation is less sensitive than required to isolate the cause of changes in weak fields. Much of the ambiguity will be resolved by making line-of-sight observations of the same event from different directions. Earth-Sun-spacccraft angles of ~20-100° are especially useful for this.

Two-sided magnetic imaging provides almost full-Sun boundary conditions, replacing current synoptic maps that confuse time and space, and thus greatly advancing MHD modeling of the Heliosphere [14].

INNER HELIOSPHERE

Multiviewing enhances the scientific understanding needed to forecast space weather. Technological developments and continuous human presence in space sharpen the need to accurately predict CME-related hazardous Solar Energetic Particle (SEP) events on time scales of days to weeks. This requires observation of the solar farside [15]. Several indications of impending CMEs are seen in photospheric magnetic fields such as development of large, long-lived magnetic clusters fed by newly emerging magnetic flux. Many of the most hazardous SEP events in the past 30 years were associated with CMEs erupting from these clusters.

Previously the two Helios' measured high-energy particles inside 1 AU but only protons, electrons, and helium ions. Elemental abundances for SEP events were determined from Helios 1 in only 2 events (21 June 1980 and 3 July 1982). ACE and WIND data show that composition and spectral signatures are key to understanding SEP acceleration and transport.

Two main classes of SEP events have been recognized: impulsive and gradual [16]. It is generally agreed that these involve separate acceleration mechanisms, resulting in different compositions and other signatures. The SEP events that represent a significant hazard to astronauts and space hardware, are gradual. Important questions must be answered to verify models and provide bases for SEP event forecasting:

What is the role of self-amplified waves in controlling particle transport? Fast CME-driven shocks may continuously accelerate particles. According to the quasilinear theory of wave-particle interaction and models of SEP transport [20,17] streaming particles produce waves that tend to trap them near the shock, producing more streaming particles and waves. These waves can be investigated by observing of wave intensity and cross (speed-magnetic field) helicity at 0.7 AU, where shocks, wave and particle intensities are much higher than at 1 AU.

How do CME-driven shocks accelerate particles? A vantage point inside 1 AU will help to find the answer. As the shock surface expands past this point to other spacecraft at 1 AU, correlated observations will allow evaluation of spatial gradients of shock and energetic particle characteristics. Models [16-18] will then be applied to study the evolution of SEP abundances, energy spectra, and maximum particle energies (Fig. 2). Simultaneous measurements of particles from spacecraft at different longitudes and radii relative to the Sun will lead to the site of the acceleration (e.g., at the nose of the shock or distributed across the shock front).

How do the early plateau and maximum SEP intensities scale with radius? Observations at 1 AU indicate that particle intensity in a given energy range sometimes rise by more than an order of magnitude to maximum intensity at or near shock crossing. A study of plateau intensity versus heliocentric distance has been made [21] but scant observations of the radial dependence are available. The spatial dependence of the early plateau is important because of an impact on astronauts and spacecraft hardware. It is also important to study maximum intensities of heavy ions because of their greater ionizing power.

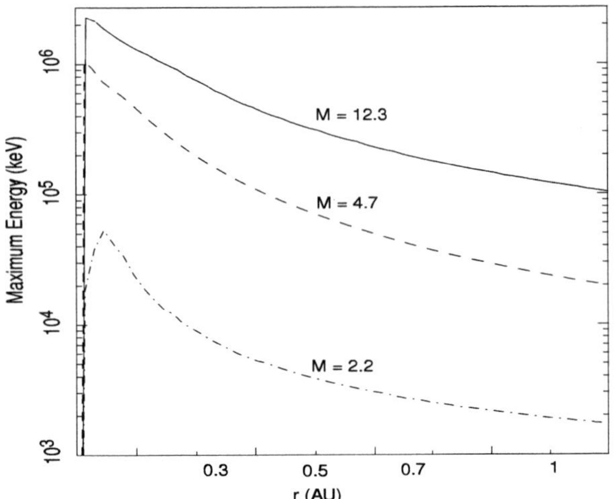

FIGURE 2. Maximum energy of particles accelerated at interplanetary shocks. Mach number at 0.5 AU is shown [18]. As shocks propagate into a weaker magnetic field, maximum energies drop, although particles accelerated earlier can be trapped behind the shock and then gradually leak out ahead of the shock.

Are there hybrid SEP events? Advanced Composition Explorer measurements [22] have shown that a significant fraction of events have a mixture of impulsive and gradual characteristics. These events typically are associated with both a large X-ray flare and a CME. Joint observations from two or more spacecraft at different viewing angles can distinguish whether material in these events is confined to a narrow cone, as expected for flare-accelerated material, or whether radial and latitudinal composition variations are consistent with a CME shock-acceleration model.

MISSION CONCEPT

To carry out the science objectives described above, a mission, called Multi-Angle Solar Sources Explorer (MASSE), was proposed to NASA. MASSE travels to a circular orbit 0.72 AU from the Sun, using a Venus gravity assist (Fig. 3). With a 584-day as seen from the Earth, MASSE traverses Earth-Sun-spacecraft angles at the rate of 225° per year passing the Sun's far side 8 months after launch. (The STEREO drift relative to Earth is only 22.5°/year.) Data are returned via weekly downlinks to the Deep Space Network.

A Doppler-Magnetograph has been designed to perform simultaneous Doppler/magnetic imaging of the Sun. Doppler data from MASSE and GONG (SDO) would permit the detection of strong magnetic fields in the interior, while magnetic measurements reveal their manifestation at the surface. MASSE-GONG observations provide unique two-component magnetic field. Near the limb (90° ± 30°), MASSE observes magnetic fields at sites of the initiation of CMEs seen from earth-side.

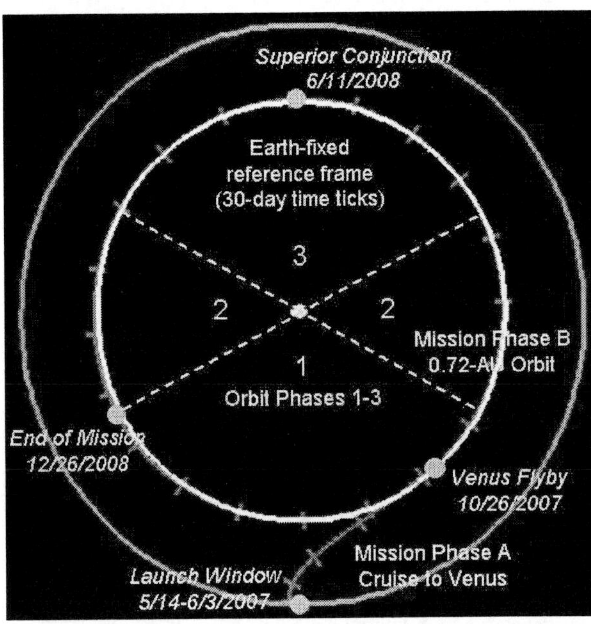

FIGURE 3. A sample of trajectory and mission phases.

Particle composition measurements and energy spectra can be made with Low- and High-Energy Telescopes as flown on ACE and designed for STEREO. These instruments provides spectra from ~2 to >150 MeV/nucleon for ions from H to Ni, as well as ~0.2-5 MeV electrons. Coordinated measurements with an earthside spacecraft give the gradient; three spacecraft would give parabolic estimates of the amplitude and location of the SEP peak; four spacecraft would see inhomogeneities in SEP and in shocks and CMEs.

Solar wind plasma and magnetic measurements can be made with ion and electron spectrometers, and a magnetometer that identify CMEs, measure their speed, and detect interplanetary shocks and waves associated with high-energy particle acceleration.

ACKNOWLEDGEMENTS

This research was conducted in part at the Jet Propulsion Laboratory, California Institute of Technology, under contract with NASA. We thank S. Stephens for managing the preparation of the proposal.

REFERENCES

1. Duvall, T. et al., eds., *Helioseismic Diagnostics of Solar Convection and Activity*, Kluwer Publs., 2001.

2. Lindsey, C. and Braun, D., *Science*, **287**, 1799-1801 (2000).

3. Parker, E., *Cosmic Magnetic Fields*, Oxford Univ. Press, 1978.

4. Giles, P., Duvall, T., Scherrer, P., and Bogart, R., *Nature*, **390**, 52-54 (1997).

5. Caligari, P., Moreno-Insertis, F. and Schussler, M., *Astrophys. J.*, **441**, 886-902 (1995).

6. Fisher, G. H. et al., *Solar Phys.*, **192**, 119-139 (2000).

7. Dikpati, M. and Gilman, P., *Astrophys. J.*, **552**, 348-353 (2001).

8. Ruzmaikin, A., et al., *J. Geophys. Res.*, **106**, 8363-8370 (2001).

9. Gaizauskas, V., Harvey, K., Harvey, J. and Zwaan C, *Astrophys. J.*, **265**, 1056-1065 (1983).

10. Benevolenskaya, E., Hoeksema, T., Kosovichev, A. and Scherrer, P., *Astrophys. J.*, **517**, L163-L166 (1999).

11. Hoeksema, T., and Scherrer, P., *Astrophys. J.*, **318**, 428-436 (1987).

12. Neugebauer, M., et al., *J. Geophys. Res.*, **105**, 2315-2324 (2000).

13. Henney, C. and Harvey, J., *Solar Phys.*, in press (2002).

14. Mikic, Z., Linker, J., Schnack, D., Lionello, R. and Tarditi, A., *Phys. Plasmas*, **6**, 2217-2224 (1999).

15. Feynman, J. and Gabriel, S., *J. Geophys. Res.*, **105**, 10,543-10,564 (2000).

16. Reames, D., *Space Sci. Rev.*, **90**, 413-491 (1999).

17. Ng, C. K., Reames, D. V. and Tylka, A. J., *Geophys. Res. Lett.*, **26**, 2145-2148 (1999).

18. Zank, G., Rice, W. K. M. and Wu, C. C., *J. Geophys. Res.*. **105**, 25,079 (2000).

20. Lee, M. A., *J. Geophys. Res.*, **88**, 6109-6119 (1983).

21. Ng, C. and Reames, D., *Astrophys. J.*, **424**, 1032-1048 (1994).

22. Cohen C. et al., *Geophys. Res Lett.*, **26**, 26,907-26,911 (1999).

A Realistic Interstellar Explorer

Ralph L. McNutt, Jr. and the Realistic Interstellar Explorer Team

G. B. Andrews, R.E. Gold, A. G. Santo, R. S. Bokulic, B. G. Boone, D. R. Haley, J. V. McAdams, M. E. Fraeman, B. D. Williams, M. P. Boyle, (JHU/APL)
D. Lester, R. Lyman, M. Ewing, R. Krishnan (ATK-Thiokol)
D. Read, L. Naes, (Lockheed-Martin ATC)
M. McPherson, R. Deters (Ball Aerospace)

The Johns Hopkins University Applied Physics Laboratory
Laurel, MD, U.S.A.

Abstract. From observations and theory we know that the unshocked solar wind extends at least 80 AU from the Sun but likely no more than ~100 AU in the region from which the local interstellar wind blows. The much larger region of the shocked solar wind and heliosheath extend out to at least several hundred AU, and fast neutrals from charge-exchanged supersonic solar wind protons disturb the very local interstellar medium to ~500 AU or more. Thus to really understand the interaction of the solar wind with the local external medium, a properly-instrumented, in situ probe to this region of space is required. For more than 20 years, an "Interstellar Precursor Mission" has been discussed as a high priority for multiple scientific objectives. The chief difficulty with actually carrying out such a mission is the need for reaching significant penetration into the interstellar medium (~1000 Astronomical Units (AU)) within the working lifetime of the initiators (<50 years). We have revisited an old idea for implementing such a mission. The probe and its perihelion carrier are launched initially to Jupiter as a combined package and then fall to the Sun where a large propulsive □V maneuver propels the package on a high-energy, ballistic escape trajectory from the solar system. Outbound in deep space, the two separate, and the probe takes data with its onboard instruments and autonomously downlinks the data to Earth at regular intervals. The implementation requires a low-mass, highly-integrated spacecraft to make use of available expendable launch vehicles. We provide a first-order cut at many of the engineering realities associated with such a mission. These separate into (1) the systems constraints imposed on the perihelion package by the combination of the propulsion system, carrying the needed propellant into perihelion, and the associated thermal and mechanical constraints, and (2) the requirements of power, autonomous operations, and data downlink from the probe itself. We find that many of the requirements for a low-mass probe that operates autonomously for this mission are common for either this propulsion concept or more advanced low-thrust concepts, e.g., solar sails and ion propulsion. We describe an implementation that could make such a mission into reality in the next 10 to 20 years.

INTRODUCTION

A mission past the boundary of the heliosphere has been discussed for more than 25 years and would yield a rich scientific harvest[1-4]. To the best of our current knowledge, the external ionized interstellar wind flow is supersonic with respect to the Sun. Hence, the heliosphere will set up a shock wave in the local interstellar medium, and the external shock may be as much as ~300 AU away at its closest. So 1000 AU is "clear" of the influence of the Sun on its surroundings and is a reasonable distance goal for a probe mission to the very local interstellar medium[5-10].

MISSION CONCEPT

The goal for the mission is to reach this distance within the working lifetime of the probe developers (<50 years) using a solar gravity assist (due to Oberth, 1929)[11]. In addition, launching toward a star enables comparison of local properties of the interstellar medium with integrated properties determined by detailed measurements of the target-star spectrum. Salient mission features include: (1) launch to Jupiter

CP679, *Solar Wind Ten: Proceedings of the Tenth International Solar Wind Conference*,
edited by M. Velli, R. Bruno, and F. Malara

and use a retrograde trajectory to eliminate angular momentum, (2) fall into 4 solar radii from the center of the Sun at perihelion, and (3) use an advanced-propulsion system ΔV maneuver to increase probe energy[6,7,9,10]. The star ε Eridani was selected as the target (Fig. 1).

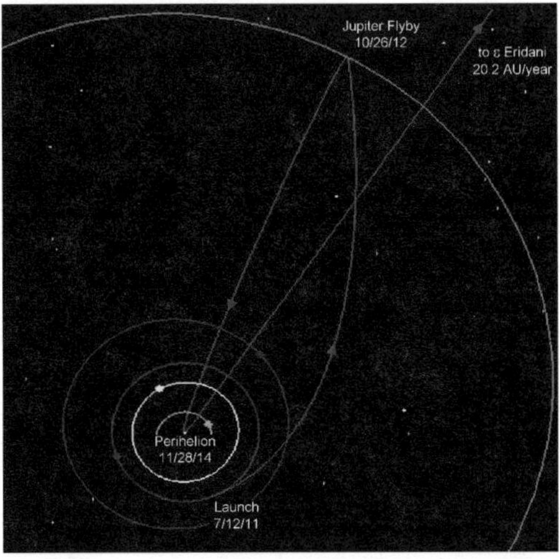

FIGURE 1. Trajectory and mission event-dates.

ENABLING TECHNOLOGIES

The enabling technologies link science, instruments, the spacecraft engineering, and reality[6,7,9,10]. To keep the mission based in a realizable infrastructure, we use Evolved Expendable Launch Vehicle (EELV) capabilities. Maximum capability is provided by the Delta IV 4050H + Star 48B that we have selected.

Innovations that are needed include: high I_{sp}, high-thrust propulsion (for perihelion maneuver, ~15 minutes), a carbon-carbon thermal shield, a technology that can be proven with the Solar Probe mission, and long-range, low-mass telecommunications. The extreme distance and low available power requires an optical downlink.

For power we need an efficient Radioisotope Thermoelectric Generator (RTG). For stable operation and a long extended mission life we can use proven, safe technology with Pu-238 plus extended operations with Am-241 in the same RTG container.

To enable the fifty-year design lifetime and keep options open for significantly longer operations, we need low-temperature (<150K) electronics that inherently provide long life with minimal heaters/insulation as well as fully autonomous operational capability. The latter implies open-loop control, software autonomy, and autonomous safing and recovery. These features could allow possible extension to multi-century flight times while maintaining data taking and downlink operations. Appropriate use of redundancies could extend probe lifetime to >1000 years (~20,000 AU). The model here is the nominal five-year design life of Voyager 1 and 2 that has now been exceeded by a factor of five.

To reach ~20 AU yr^{-1}, the probe needs to be accelerated by ~10 to 15 km/s during about 15 minutes around perihelion[6,11]. Such acceleration was provided to Ulysses in Earth orbit with a 20-metric-ton, 3-stage solid-fuel rocket. To accomplish the same thing near perihelion a much lighter, i.e., high specific impulse, system is required.

We considered several propulsion options[5-7,9,10]. Nuclear pulse propulsion ("Orion") offers very good performance but does not scale to small systems. Nuclear thermal propulsion (NTP) for small systems is limited to solid-core reactors and the I_{sp} is limited by fuel pellet coefficient of thermal expansion (CTE) and chemical reactivity. Solar thermal propulsion (STP) is being investigated for use in orbital transfer vehicles at 1 AU, although use at 1 AU requires concentrators. For this mission, the location of the "burn" eliminates the need for concentrators (albeit replacing it with a need for a very-high-temperature thermal shield/heat exchanger) while potentially avoiding the inherent CTE problems in a solid-core NTP system. Hence, STP is the propulsion option we have implemented [12].

Trade Studies

We have concentrated on an STP system[12], but most results also apply to NTP. To obtain a sufficiently large I_{sp} to provide ΔV, we have examined LH_2, CH_4, and NH_3 and maximized the propellant temperature (up to structural failure). Trades include pressure versus flow rate, heating, and recombination. LH_2 was selected over NH_3 as primary STP system propellant. The higher NH_3 density allows for smaller thermal shield mass, but the higher I_{sp} for LH_2 allows more packaging room for probe and room for "growth" with lighter materials. The requirements for heat exchanger coating are driven due to erosion of carbon-carbon by all of the studied propellants. We sized the propellant tank for propellant requirements, including considerations of storage for cruise and the need for a cryostat required for long-term LH_2 storage, as well as requirements on pressure and expulsion during the propulsive maneuver.

Spacecraft Systems

The spacecraft consists of a pre-perihelion carrier that is dominated by the thermal shield, cryostat, STP system, and the instrumented probe (Fig.2). Following the perihelion "burn" the carrier is discarded and the instrumented probe deploys instruments and the main communications optic (Fig. 3), verifies its status, established the optical communications downlink[13], and begins its long cruise/data-taking mission.

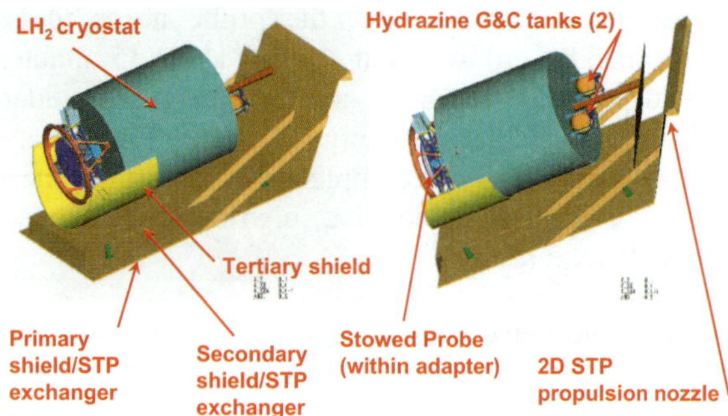

FIGURE 2. Probe and thermal-shield carrier in pre-perihelion configuration.

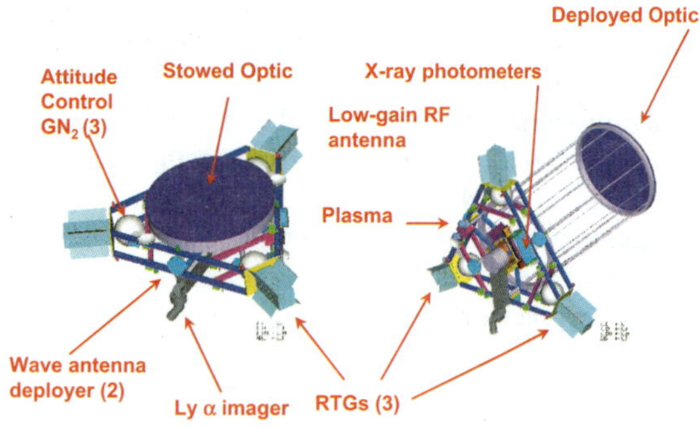

FIGURE 3. Probe configuration during cruise

The mass estimate of the payload is 12.16 kg using Be structure. The power estimate of payload is 1.87 W using ultra-low power (ULP) electronics operating at cryogenic temperatures (Table 1).

TABLE 1. Model Science Instrument Payload.

Instrument	Mass (kg)	Power (W)
Totals	12.16	1.87
Magnetometer	1.89	0.18
Plasma wave	1.48	0.08
Plasma spectrometer	0.97	0.53
Lyman alpha imager	3.43	0.13
Cosmic ray spectrometer	0.84	0.16
Energetic particles	0.80	0.33
Dust experiment	0.70	0.13
X-ray spectrometer	2.05	0.33

The mass and power of the probe have been estimated in some detail. The mass estimate by subsystem is 147.15 kg using a Be structure. The power estimate by subsystem is 19.81 W using ULP (Table 2). Special consideration has been given to lightweighting the structure and all other components, minimizing the power, and maximizing the probe lifetime such that the mission requirements can be met with the specified EELV.

For example, with respect to avionics we eliminate signal harness and maximize autonomy by using multiple ULP processors, each configured with two wireless communication transceivers.

We use RTGs based upon multicouples with four General Purpose Heat Sources (GPHS) and direct voltage (no DC/DC converters). Am-241 is identified as available RTG fuel: 1 Pu-238 RTG + 2 Am-241 RTGs provide for 51.3 W remaining after 50 years and ~6.27W after 500 years (all three RTGs).

An integrated optical communications, attitude, guidance, and control approach ensures required burst-mode performance of 500 bps from 1000 AU (over 5 light-days). Already at 100 AU, electromagnetic waves take 13.9 hours to travel from the spacecraft to Earth, so full open-loop autonomy is required.[13].

TABLE 2. Probe Mass and Power

Component	Mass (kg)	Power (W)	Notes
Science instruments	12.16	1.87	Allocated total is 10.0 kg; power is nominal, not peak
Integrated avionics	1.44	1.38	Includes command and data handling, guidance and control
Power system	45.60	2.12	Three RTGs: 1 with Pu-238, 2 with Am-241
Telecommunications	16.17	11.28	890 nm laser downlink plus RF for inner solar system
Structure	18.32	0.00	Beryllium truss
Attitude control	5.40	1.69	Attitude fine control integrated with laser downlink
Thermal	0.00	0.00	All thermal input via passive waste heat from RTGs
Propulsion	41.49	0.16	Cold-gas N2 system warmed by RTG waste heat; 56W max
Dry mass/power total	140.58	18.50	Subtotals of components before harness
Total with harness	147.15	18.87	Signal harness is virtual with RF; power harness use wires
Power reserve	0.00	0.94	Add 5% power reserve
Total for probe including reserve	147.15	19.81	Probe totals including reserves

Thermal requirements include: survive cruise prior to perihelion pass, protect the propellant, heat the propellant for the perihelion "burn, " and use waste heat from the RTG to eliminate heater-power. Analysis with Thermal Synthesis System (TSS) software shows the carbon-carbon (CC) primary shield can be limited to a temperature (2924K) if we use ~100 kg of CC aerogel backing on shield while the LH_2 remains thermally isolated until needed.

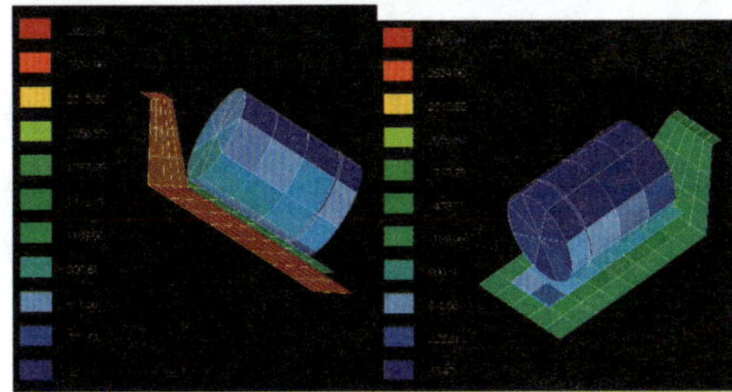

FIGURE 4. Peak temperatures (degC) at perihelion.

SCHEDULE

Table 3 gives a schedule for implementation.

TABLE 3. Schedule.

Date(s)	Activity
2000-2002	Advanced technology development studies
2002-2003	Update NASA strategic plan
2003-2007	Development of small probe technologies
2004-2007	Development of solar-sail demo mission
2004-2007	Development of Solar Probe mission
2007-2010	Focused development of Interstellar Probe
2009-2012	Design and launch of solar-sail probe
2012	Test Solar Probe performance at perihelion
2012-2015	Design and launch perihelion STP probe
2015-2065	Data return from out to 1000 AU
35850	Probe reaches vicinity of ε Eridani

SUMMARY AND CONCLUSIONS

The initial concept[5-7] for a robust, long-lived 50-kg probe has proven difficult to design. A robust probe is more likely to have a mass of ~150 kg, even with new technologies and materials. Novel propulsion is key. The continuing problem is the mass of the cryostat. Due to the system dry mass, a 20 AU/yr goal may not be realizable, although 12 AU/yr (60 km/s) may be. More technical definition and engineering work is required, but there is a road to actually implementing a mission that has remained elusive for at least 25 years. Ad Astra!

ACKNOWLEDGMENTS

This work was supported under Task 7600-039 from the NASA Institute for Advanced Concepts (NIAC) under NASA Contract NAS5-98051.

REFERENCES

1. Jaffe, L. D., and Ivie, C. V., *Icarus,* **39**, 486-494, (1979).

2. Holzer, T. E., et al. The Interstellar Probe: Scientific objectives for a Frontier mission to the heliospheric boundary and interstellar space, NASA Pub., 1990.

3. Mewaldt, R. A., Kangas, J., Kerridge, S. J., and Neugebauer, M., *Acta Astron.,* **35**, Suppl., 267-276 (1995).

4. McNutt, R. L., Jr., Gold, R. E., Roelof, E. C., Zanetti, L. J., Reynolds, E. L., Farquhar, R. W., Gurnett, D. A., and Kurth, W. S., *J. Brit. Int. Soc.,* **50**, 463-474 (1997).

5. McNutt, R. L., Jr., "A realistic interstellar explorer", Proc. Workshop on Interstellar Exploration in the Next Century, 1998, California Institute of Technology.

6. McNutt, R.L., A Realistic Interstellar Explorer, Phase I Final Report, NIAC CP98-01, May 1999.

7. McNutt, R. L., Jr., Andrews, G. B., McAdams, J., Gold, R. E., Santo, A., Oursler, D., Heeres, K., Fraeman, M., B. Williams, B., "A realistic interstellar explorer", STAIF-2000 Proc. (2000).

8. Liewer, P. C., Mewaldt, R. A.,. Ayon, J. A., and Wallace, R. A., "NASA's interstellar probe mission", STAIF-2000 Proc. (2000).

9. McNutt, R. L. Jr., et al., "Low-Cost Interstellar Probe" Paper IAA-L-0608, Fourth IAA International Conference on Low-Cost Planetary Missions, Laurel, MD, May 2000; Acta Astronautica, in press, 2002.

10. McNutt, R. L., Jr., et al., "A Realistic Interstellar Probe" COSPAR Colloquium on The Outer Heliosphere: New Frontiers, Potsdam, Germany, July 24–28, 2000; COSPAR Colloquia Series, 11, 431-434 (2001).

11. McAdams, J. V., and McNutt, R. L., Jr., "Ballistic Jupiter gravity-assist, perihelion-ΔV trajectories for a realistic interstellar explorer," Paper AAS 02-158, 2002.

12. Lyman, R. W., et al., "Solar Thermal Propulsion for an Interstellar Probe" 37th Joint Propulsion Conference, July 8-11, 2001, Salt Lake City, Utah.

13. Boone, B. G., R.S. Bokulic, R. S., Andrews, G. B., and McNutt, R. L., Jr., SPIE Paper 4821-26, in press, 2000.

Interstellar Pathfinder – A Mission to the Inner Edge of the Interstellar Medium

D.J. McComas[1], P.A. Bochsler[2], L.A. Fisk[3], H.O. Funsten[4], J. Geiss[5], G. Gloeckler[3,6], M. Gruntman[7], D.L. Judge[7], S.M. Krimigis[8], R.P. Lin[9], S. Livi[8], D.G. Mitchell[8], E. Möbius[10], E.C. Roelof[8], N.A. Schwadron[1], M. Witte[11], J. Woch[11], P. Wurz[2], T.H. Zurbuchen[3]

[1]*Southwest Research Institute, San Antonio, TX USA*
[2]*University of Bern, Bern, Switzerland*
[3]*University of Michigan, Ann Arbor, MI USA*
[4]*Los Alamos National Laboratory, Los Alamos NM USA*
[5]*International Space Science Institute, Bern, Switzerland*
[6]*University of Maryland, College Park, MD USA*
[7]*University of Southern California, Los Angeles, CA USA*
[8]*The Johns Hopkins University, Laurel, MD USA*
[9]*University of California Berkeley, Berkeley, CA USA*
[10]*University of New Hampshire, Durham, NH USA*
[11]*Max Planck Institut für Aeronomie, Lindau, Germany*

Abstract. Interstellar Pathfinder (ISP), our first step into the interstellar medium, is a scientific investigation to study the outer boundary of our heliosphere and the interstellar matter that flows into it. A wind of interstellar neutral gas penetrates to within several astronomical units (AU) of the Sun, giving us a direct sample of present-day galactic matter. ISP is a mission to this inner edge of the interstellar medium. Using highly sensitive instrumentation, ISP will determine the composition of our local interstellar environment. It will also take the first global images of the boundary region of the heliosphere at 100 to 150 AU. These measurements will allow ISP to answer fundamental questions about the origin of the solar system and the stars, about the evolution of our galaxy and of the universe, and about the characteristics of our local galactic environment and its influence on the heliosphere.

Scientific Background and Goals

As the heliosphere moves through the local interstellar medium, it is thought to form several plasma structures including an upstream bow shock, a heliopause that separates the solar wind from the ionized portion of the interstellar medium at about 150 astronomical units (AU), and a termination shock of the supersonic flow of the solar wind at ~100 AU. This interaction is shown schematically in the left panel of Figure 1. Because these structures are so distant, the common conception is that only distant Voyager, or eventually Interstellar Probe, has any expectation of reaching interstellar space. However, the inner edge of the interstellar medium is far closer. The solar wind does not stop neutral particles from the interstellar medium from penetrating relatively undisturbed to within 3 AU of the Sun. Further, energetic neutral atoms generated from ions accelerated in and around the various structures in the outer heliosphere also propagate freely into the inner heliosphere.

CP679, *Solar Wind Ten: Proceedings of the Tenth International Solar Wind Conference*, edited by M. Velli, R. Bruno, and F. Malara

Interstellar Pathfinder (ISP) is a mission to this inner edge of interstellar space. The right side of Figure 1 shows a blowup of the inner heliosphere, the ISP orbit, and the various sources observed by this mission. ISP will carry instruments that are much more capable than anything previously flown, allowing us to determine the properties and composition of the neutral interstellar gas, including its isotopic composition, and the structure of the outer heliosphere, for the first time. These measurements will reveal vital new data on the physical state and composition of the interstellar medium and our heliosphere's interaction with it.

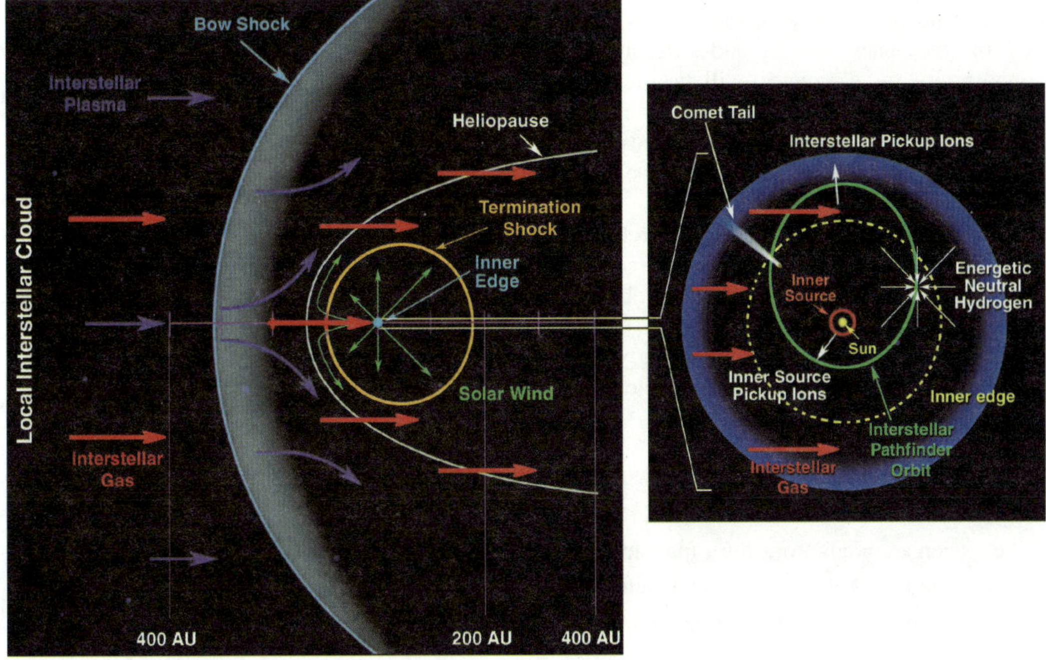

Figure 1. Schematic representation of the heliospheric structure formed by its interaction with the local interstellar medium (left) and an expanded view of the processes observable within the region of space sampled by Interstellar Pathfinder (right).

Interstellar Pathfinder was designed to answer five fundamental questions:

1. What is the composition of the interstellar medium, and what does this tell us about the evolution of the universe and galaxy and about the birthplace of the Sun? The elemental and isotopic composition of cosmic matter gives crucial information needed to determine the origin and evolution of the universe. The Pickup Ion Composition Spectrometer (PICSPEC) will gather this information for the local interstellar cloud, the only accessible sample of the present-day galaxy. By comparing with the known composition of our solar system, which formed from galactic matter 4.6 billion years ago, we can see how galactic material evolved chemically over time, extrapolate what it was like in the early universe, and unravel the dynamical processes occurring in our galaxy, such as the rate of in falling matter and stellar nucleosynthesis, and place limits on the location where our Sun was born.

2. What is the physical state of the local interstellar cloud, and how does the heliosphere interact with it? The Neutral Interstellar Helium Detector (NIHED) will directly determine the flow speed, temperature, and density of interstellar helium neutrals, while PICSPEC provides the complementary information for hydrogen from pickup ions. This combination reveals the physical characteristics of the surrounding interstellar cloud and its dynamic interaction with the heliosphere.

3. What are the characteristics of the heliospheric termination shock? The Shock and Heliosphere Energetic Neutral Atom (SHENA) imager will provide a full-sky map of the energetic processes at the termination shock where the solar wind is slowed to subsonic speed. These images will show us the nature and spatial variation of this shock. When we combine them with data from the Voyager Spacecraft, which will pass through this region soon, we will be able to formulate a complete global picture of this important, distant boundary.

4. What is the nature of the inner source of pickup ions and the composition of comets? The inner source of pickup ions is created when the solar wind is embedded in dust grains and released as neutrals to be ionized and picked up by the flow of the solar wind. Superimposed on the inner source, and offering a significant opportunity for discovery, will be pickup ions in the distant tails of comets. If a sufficient number of comets are detected, a systematic study of the volatile elements in comets will be possible, particularly in small comets.

5. What is the nature of the interaction of pickup ions with the solar wind? The interaction of solar wind with the interstellar neutrals is brought about through pickup ions. The PICSPEC instrument is the first of its kind, designed specifically for the study of pickup ions. The dynamic behavior of these pickup ions will be studied in unprecedented detail.

IMPLEMENTATION

All of the science goals of the Interstellar Pathfinder mission can be accomplished with a suite of three primary instruments and two supporting "off-the-shelf" monitors:

- The **Pickup Ion Composition Spectrometer (PICSPEC)** is a large-collecting-power ion mass spectrometer, with a collecting power of over 300 times that of the Ulysses SWICS instrument. SWICS provided data that enabled the discovery of most of the pickup ion species and led to the dramatic increase in our present understanding of the LIC. For PICSPEC, we achieve this high sensitivity by eliminating the need to step in voltage, as was done on SWICS. Rather, the deflection is measured by sensing the positions of the particles. This simple but powerful change permits us to measure with unprecedented sensitivity the velocity distributions and abundance of all pickup ions, including the rare isotopes D, ^{3}He, ^{18}O, ^{22}Ne and ^{38}Ar, as well as molecular pickup ions up to mass 50.
- The **Shock and Heliosphere Energetic Neutral Atom (SHENA)** imager is a full sky imager of energetic ENAs coming from the termination shock. SHENA is optimized for measuring fluxes of heliospheric energetic H atoms and has the required large geometric factor and angular resolution to accumulate 7° pixel resolution images of the heliospheric termination shock in five energy bands. SHENA measures neutral H atoms in the 0.3–6.0 keV energy range by first converting them to H^{+}, using foil stripping, and then using energy and coincidence analysis of the post-accelerated ions. This technique combines four simple, flight-proven elements into a new configuration for detecting and imaging ENAs.
- The **Neutral Interstellar Helium Detector** is an imager of interstellar neutral He, as well as of O and Ne during favorable portions of the orbit. NIHED is a high-sensitivity, high-angular-resolution version of Ulysses/GAS instrument, with 400 times larger collecting area and a far better signal-to-noise than GAS. NIHED has the necessary sensitivity, angular resolution, and background rejection to measure precisely the interstellar neutral He bulk flow parameters (flow vector, temperature, density), detect nonthermal tails in the He distributions, and determine the flow speed of interstellar O and Ne atoms.
- The **Solar Wind Monitor** comprises the solar wind ion and 3D electron spectrometers.
- The **Solar Ultraviolet** sensor provides a monitor of solar extreme ultraviolet.

The principal requirement for ISP's orbit is to reach the inner edge of interstellar space at ~3 AU from the Sun. Thus, a 3-year, in-ecliptic, 1 x 3 AU heliocentric orbit is used. This orbit places the spacecraft within the inner edge of the neutral LIC for more than 1 year, which is necessary for collecting the very rare interstellar isotopes. It is also ideal for accumulating detailed ENA images of the termination shock and the vast region beyond. Additional 3-year orbits are possible, but are not required to fulfill the primary objectives of the ISP mission.

STATUS AND CONCLUSIONS

Interstellar Pathfinder was rated Category 1 (the top science rating) and low risk in the 2001 Medium Explorer (MIDEX) competition, but for some unknown reason was not selected for Phase A study and mission development. Since then, the Sun-Earth Connection community has been carrying out two major strategic planning exercises. Both of these, NASA's Sun-Earth Connection Roadmap and the National Research Council's Decadal Survey, have rated the ISP science as extremely high priority, further solidifying the case for this critical mission. The Interstellar Pathfinder team will continue to optimize our mission design and implementation plan and, with the support of the heliospheric community, expect to be an outstanding contender for finally bringing a MIDEX mission to heliospheric science.

ACKNOWLEDGMENTS

We gratefully acknowledge all of the important contributions made to the Interstellar Pathfinder mission proposals by the host of scientists, engineers, and technical and administrative support personnel, who have worked on this effort over the years.

The Energetic Particles Spectrometers (EPS) on MESSENGER and New Horizons

S. A. Livi, R. McNutt, G. B. Andrews, E. Keath, D. Mitchell, G. Ho

Johns Hopkins University - Applied Physics Laboratory
11100 Johns Hopkins Road
Laurel, MD 20723 - USA

Abstract. In the course of this decade, two NASA deep space mission to the inner and outer heliosphere, MESSENGER to the planet Mercury and New Horizons to the planet Pluto, will carry onboard energetic particle spectrometers. The combination of measurements near the Sun (0.3 AU), and from the outer heliosphere (up to almost 40 AU), will ideally complement the information available from Ulysses and from near-Earth orbiting spacecraft, yielding boundary conditions on the processes that accelerate energetic particles. EPS is a hockey-puck-size Time-of-Flight (ToF) spectrometer that measures ions and electrons over a broad range of energies and pitch angles. Particle composition and energy spectra will be measured for H to Fe from 15 keV/nucleon to 3 MeV/nucleon and for electrons from 15 keV to 1 MeV. The ion section of EPS is a compact ToF telescope with two main components: a ToF section and a Solid State Detector (SSD) array to measure separately velocity and total energy of the incoming particles. Electrons are identified in EPS by the presence of an energy signal and by the absence of start or stop pulses, since energetic electrons have low efficiency for production of secondary electrons when passing through thin foils. For both ions and electrons the angle of arrival is determined by the position of the solid-state detector that collects the particle.

INTRODUCTION

Both the Sun and the dynamic heliosphere are prodigious accelerators of energetic ions and electrons. The evolution of the particles' intensities, angular distributions, energy spectra, and composition can yield incisive diagnostics of the acceleration process, if the particle's propagation from the acceleration site can be understood. EPS can reveal remarkably different scientific results in the vicinity of Mercury and of Pluto, because the physics of energetic particles takes on extreme opposite characteristics in the inner and outer heliosphere.

The orbit of Mercury at a mean distance of 0.39 AU minimizes the particle propagation effects from the Sun and transient shocks in the inner heliosphere. Therefore observations by EPS can directly address:

- Where and when prompt solar particles are accelerated and released from the Sun in association with solar flares and coronal mass ejections

- Under what conditions energetic particles are accelerated by fast shocks in the inner heliosphere

- How energetic particles interact with large-scale magnetic structures in the solar wind (like interplanetary coronal mass ejections), both inside and outside the orbit of Mercury

- Detection at the orbit of Pluto at a mean distance of 40 AU maximizes the effects of particle propagation and in situ energy changes. EPS observations can reveal the acceleration mechanisms of plasma/particle interactions in the outer heliosphere because:

- Great outbursts of solar activity can create populations of solar energetic particles that take ~100 days to pass by on their way out of the solar system

- Distances are so immense that field-aligned propagation becomes negligible compared to convection with magnetic field lines in the solar wind, so solar wind structure molds the energetic particle structures.

- Energetic particle populations can survive only if there is a balance between acceleration at shocks and/or magnetic field compressions on the one hand, and adiabatic deceleration in the expanding solar wind on the other.

CP679, *Solar Wind Ten: Proceedings of the Tenth International Solar Wind Conference*,
edited by M. Velli, R. Bruno, and F. Malara

EPS DESCRIPTION

The Energetic Particle Spectrometer (EPS) is a hockey-puck-size Time-of-Flight (ToF) spectrometer that measures ions and electrons over a broad range of energies and pitch angles. Particle composition and energy spectra will be measured for H to Fe from ~15 keV/nucleon to ~3 MeV/nucleon and for electrons from 15 keV to 1 MeV. EPS was developed on a NASA Planetary Instrument Definition and Development (PIDDP) grant focused on providing a low-mass, low-power instrument[1,2] that could measure the energetic pickup ions produced in the vicinity of Pluto[3,4] in a cometary-type interaction[5-8] An engineering model has operated well at the accelerator facility at GSFC. The instrument combines minimal spacecraft resource requirements with versatile measurement characteristics for exploratory missions.

The EPS was chosen for and will fly on the MESSENGER (NASA Discovery Program) mission to Mercury that launches in March 2004[9]. Only minor changes will be made to this instrument for the Pluto mission: (1) more open collimator to increase the geometric factor, (2) slower microprocessor clock to decrease secondary power by 375 mW, (3) low-current microchannel plates (MCPs) to decrease secondary power another 250 mW, and (4) a thinner front foil (6.6 μg C (300 Å)) to lower the energy threshold in the much lower UV background present.

Ion Measurement

EPS is a compact ToF telescope with two main components: a ToF section and a Solid State Detectors (SSD) array. The SSD comprises six ion implanted planar silicon detectors, each with four pixel, two dedicated to ion measurements and two to electron measurements.

Particles enter the system through a mechanical collimator that delimits the look direction and hit a thin (4 mg/cm^2) polyimide foil. A layer of 500 Å of Aluminum covers the entrance foil, to enhance Ly-α rejection properties. A deployable cover protects the EPS front foil during launch.

Electrons are released from the inner surface of the start foil and focused to a well-defined region on a microchannelplate (MCP) to generate the start signal in a dedicated anode. The incident ions are not affected by the electric fields of the focusing optics. After a 6-cm flight path ions traverse the stop foil, which is a 4 mg/cm^2 polyimide foil, covered by 500 Å of palladium to reduce light (both visible and UV) transmission toward the SSDs. The secondary electrons released by the stop foil are steered to the MCP and generate the stop signal. Electron trajectory simulations show that there is less then 400ps dispersion in the transit time of the secondary electrons from the foil to the MCP. Sub-nanosecond dispersion is required so as not to misidentify ion species.

The difference in time between start and stop (1ns to 200ns), together with the distance between the two foils, give a measure of the particle's velocity. The total charge released in the SSD is proportional to the energy of the parent particle.

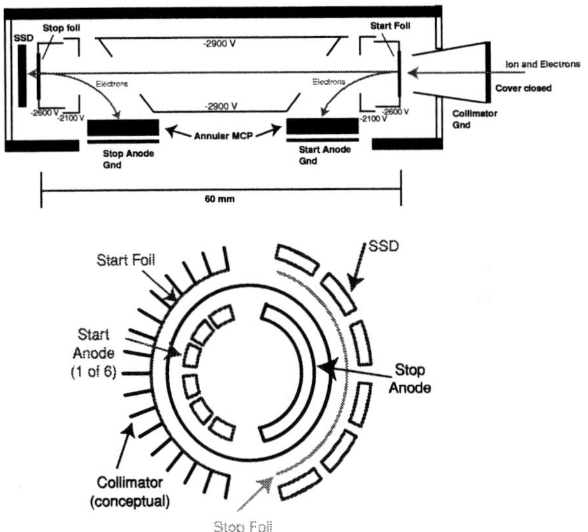

FIGURE 1. Principle of operation of EPS.

Electron Measurement

Energetic electrons have low efficiency for production of secondary electrons when passing through thin foils. A zero ToF technique for measuring energetic electrons would suffer from low foreground-to-background ratio, in a radiation environment like the Earth radiation belt. A thin layer (flashing) of 1mm of aluminum covers half of the SSD in EPS. This dead layer stops protons with energy less then 200 keV; on the other hand electrons lose less then 10 keV energy by the interaction with this dead layer. Electrons are identified in EPS by the presence of an energy signal and by the absence of start and stop pulses. Under such conditions, a signal of 50keV in the detector can be generated either by a 250keV proton that did not generate either a start or a stop pulse (a low probability event), or by a 60 keV electron. The ToF spectra collected in the adjacent SSD (without flashing) will be used during ground data analysis for checking and correcting for the proton contamination.

For both ions and electrons the elevation angle of arrival is determined by the position of the solid-state detector that collected the particle. This type of instrument is best suited to a spinning spacecraft, but these spacecraft are three-axis stabilized during data taking, so the azimuthal angle of arrival is set by the instantaneous orientation of the spacecraft..

The main instrument parameters of EPS are summarized in Table 1.

Table 1. EPD Sensor Characteristics

Geometric Factor (G-total instrument) =		0.03 cm^2 sr
Angular Coverage (Instantaneous) =		160° x 12°
Angular Pixels (elevation x azimuth) =		13.5° x 22.5°
Energy Coverage		See Table Below
Energy Resolution (Spectral Mode) =		8 keV
$\Delta E/E$ Capability at 50 keV =		16%
$\Delta E/E$ Capability at >100 keV =		< 8 %
Mass Species Separation		See Table Below
Species	Energy Range	Logic
Ions	> 5 keV	ToF only
Protons	15 keV - 3.0 MeV	ToF x E
Alphas	25 keV - 3.0 MeV	ToF x E
CNO	60 keV - 30 MeV	ToF x E
Electron Spectra	15 keV - 1.0 MeV	Singles/Foil Technique
Relative Electron Rates	1.0 to 2.0 MeV	Singles

CONCLUSIONS

The EPS onboard MESSENGER and the EPS part of the New Horizon payload ideally complement each other, the first flying in the inner solar system, and the second in the outer. The two instruments are designed and tailored for planetary magnetospheric measurements, yet they will make important contributions to our understanding of the processes that generate energetic particles in the Sun's atmosphere and accelerate particles in interplanetary space.

REFERENCES

1. McNutt, R. L., Jr., Mitchell, D. G., Keath, E. P. Paschalidis, N., Gold, R. E. and McEntire, R. W., *SPIE Proceedings*, **2804**, 217-226 (1996).

2. Andrews, G. B., Gold, R. E., Keath, E. P., Mitchell, D. G., McEntire, R. W., McNutt, R. L., Jr., and Paschilidis, N., *SPIE Proceedings*, **3442**, 105-114, (1998).

3. Bagenal., F., and McNutt, R. L., *Geophys. Res. Lett.* **16**, 1229-1232 (1989).

4. Bagenal, F., Cravens, T., Luhmann, J., McNutt, R., and Cheng, A.F., "Pluto's interaction with the solar wind", in Pluto and Charon, edited by S. A. Stern and D. J. Tholen, Tucson: Univ. of Arizona Press, 1997, pp. 523-555.

5. Gloeckler, G., Hovestadt, D., Ipavich, F., Scholer, M., Klecker, B. and Galvin, A., *Geophys. Res. Lett.*, **13**, 251-254 (1986).

6. Sanderson, T. R., Wenzel, K.-P., Daly, P., Cowley, S. W. H., Hynds, R. J., Smith, E J., Bame, S. J., and Zwickl, R. D., *Geophys. Res. Lett.*, **13**, 411-414 (1986).

7. Richardson, I. G., Cowley, S. W. H., Hynds, R. J., Sanderson, T. R., Wenzel, K.-P., and Daly, P. W., *Geophys. Res. Lett.*, **13**, 415-418 (1986)

8. Gloeckler, G., Geiss, J., Schwadron, N. A., Fisk, L. A., Zurbuchen, T. H., Ipavich, F. M., von Steiger, R., Balsiger, R. H., and Wilken, B., *Nature*, **404**, 576-578 (2000).

9. Gold, R. E., Solomon, S. C., McNutt, R. L., Jr., Santo, A. G., Abshire, J. B., Acuña, M. H., Afzal, R. S., Anderson, B. J., Andrews, G. B., Bedini, P. D., Cain, J., Cheng, A. F., Evans, L. G., Follas, R. B., Gloeckler, G., Goldsten, J. O., Hawkins, S. E.,III, Izenberg, N. R., Jaskulek, S. E., Ketchum, E. A., Lankton, M. R., Lohr, D. A., Mauk, B. H., McClintock, W. E., Murchie, S. L., Schlemm, C. E., II, Smith, D. E., Starr, R. D., and Zurbuchen, T. H., *Planet. Space Sci.*, **49**, 1467-1479 (2001).

Heliospheric Constellation: Understanding the Structure and Evolution of the Solar Wind

M. B. Moldwin[1], P. C. Liewer[2], N. Crooker[3], J. F. Fennell[4], J. Feynman[2], H. O. Funsten[5], B. E. Goldstein[2], J. T. Gosling[5], J. E. Mazur[4], V. J. Pizzo[6], C. T. Russell[1], J. Weygand[1]

1 Institute of Geophysics and Planetary Physics and Department of Earth and Space Sciences, UCLA, Los Angeles, CA 90095-1567
2Jet Propulsion Laboratory, California Institute of Technology, Pasadena, CA 91109
3 Center for Space Physics, Boston University, Boston MA 02215
4 The Aerospace Corporation, El Segundo CA, 90245-4691
5 Los Alamos National Laboratory, Los Alamos, NM 87545
6 Space Environment Center, NOAA, Boulder CO

Abstract. The Heliospheric Constellation (HELICON) mission concept calls for the first constellation of spacecraft to make coordinated measurements of the solar wind magnetic field, plasma and energetic particle distributions and composition in order to determine scale-lengths of solar wind structures and to resolve ambiguities in temporal and spatial variability. Specifically, HELICON enables the resolution of a wide array of critical questions of solar wind structure and dynamics. The HELICON's Science Objectives are as follows: (1) Determine the structure and evolution of expanding interplanetary coronal mass ejections, (2) Use supra-thermal and energetic particles to determine the source populations of solar events and the scale sizes of Interplanetary Coronal Mass Ejections (ICMEs) and Corotating Interaction Regions (CIRs) (3) Determine the structure and nature of the heliospheric current sheet, and (4) Examine the causes of variability in the solar wind. This brief report describes the mission concept and scientific rationale for such a mission.

INTRODUCTION

While we have learned much about solar wind structure and dynamics from single spacecraft observations, we have been stymied in our attempts to unambiguously test our understanding of the physical processes controlling the solar wind because with only one or two uncoordinated measurements we cannot separate spatial from temporal variability. The four major HELICON science objectives answer several of the fundamental problems in heliospheric physics that can only be solved with coordinated, simultaneous, multi-spacecraft observations. The next sections describe the mission concept, instrument suite, and conclude with how the mission answers the four major HELICON science objectives.

MISSION DESCRIPTION

HELICON consists of six identical spacecraft (Figure 1) in heliocentric orbit. One triad leads the Earth at 1AU with increasing inter-spacecraft spacing throughout the course of the mission. The other triad is in a one-year period elliptical orbit with perihelion at 0.8AU and aphelion at 1.2AU. Figure 2 shows how the orbits evolve over a course of a year in the middle of the prime mission. The orbital positions (and hence spacecraft separations) have been determined using The Aerospace Corporation's Satellite Operations Analysis Program (SOAP). One year after launch, the spacecraft sets are radially aligned within a 25° longitudinal band with the elliptical spacecraft between 1.15 and 1.2AU. Three months later, the spacecraft are longitudinally spread across 60° with the spacecraft ranging in distance from 0.9 to 1.1AU. Every six months the spacecraft "weave" back and forth with this same pattern.

Each spacecraft includes an identical instrument suite that continuously samples the solar wind with a 1-minute resolution. The instrument suite consists of a UCLA-built magnetometer with a range of +/- 1000 nT and a precision of +/- 0.05 nT; a Los Alamos

CP679, *Solar Wind Ten: Proceedings of the Tenth International Solar Wind Conference*, edited by M. Velli, R. Bruno, and F. Malara

National Laboratory-built solar wind plasma analyzer that can measure electrons (from 3 eV to 8 keV) and

Figure 1. Each HELCION micro-spacecraft carries an identical instrument suite. The six spacecraft are stacked and fit inside a Delta II shroud.

ion composition (from 0.4 to 10 keV/q) with an energy resolution of 5% and a mass resolution of 20%; and an Aerospace Corporation-built energetic particle instrument that can measure electrons (from 15 keV to 2 MeV) and ion composition (from 0.7 to 10 MeV for heavy ions and 0.08 to 10 MeV for protons) with a mass resolution of 10%.

SCIENCE OBJECTIVES

Structure and Evolution of ICMEs: The magnetic and density structure of Interplanetary Coronal Mass Ejections (ICMEs) can only be well determined with multi-spacecraft in situ observations. The current paradigm of the three-dimensional structure of ICMEs can be tested in three ways with HELICON: (1) At multiple longitudes at constant radial distance to determine the azimuthal stretching of the structure (Figure 3); (2) At multiple locations along the axis of the structure to measure the bending of the axis, and (3) At multiple radial distances to determine the radial evolution of the structure. The IMF and solar wind observations can be used at each location independently and the data can be inverted to give local structure and orientation or they can be combined in a model that allows stretching, bending, and expansion.

HELICON determines the relationship between the plasma composition structure and the magnetic topology at each location within the ICME to relate the structure to its coronal source. HELICON measures the presence or absence of bi-directional suprathermal electrons throughout the structure to determine its connectivity to the Sun and its magnetic topology. By examining this relationship at six different locations in space, HELICON provides a very rigorous test of our understanding of the 3D geometry of the structures and their relationship to the coronal magnetic field. The HELICON spacecraft separations span the ICME size estimates of 10° to 60° in longitude and from 0.1 to 0.25AU in the radial direction. Studies with WIND and NEAR show that over this range of longitude scales concurrently observed ICMEs have structure that is nearly identical at the shortest distances, to being nearly opposite at the greatest distances. Similarly WIND and NEAR revealed significant radial evolution over a distance of 0.2AU [*Mulligan et al.,* 1999]. Because of the 15 unique spacecraft pairs, nearly the full range of these separations is made simultaneously during much of the mission.

Scale-Size of Solar/Interplanetary Events: HELICON tests our understanding of the acceleration process, energetic particle transport, and scale-size of events by making simultaneous measurements of the energetic particle flux, composition, and IMF at a wide range of scale-lengths. In particular, HELICON answers many of the questions regarding ICME-driven shocks and CIRs.

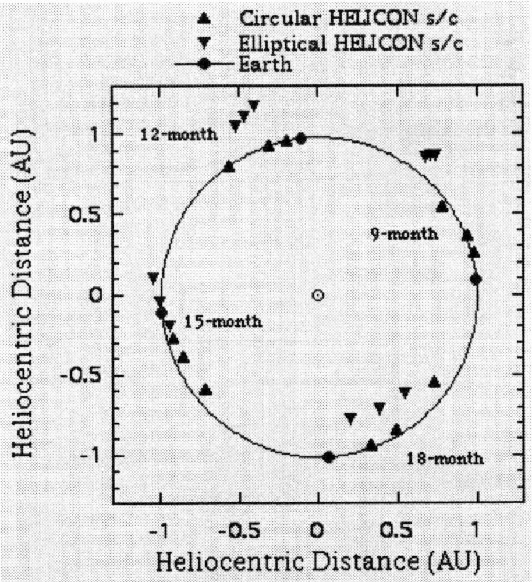

Figure 2: The HELICON satellite configuration evolves with time and is shown at four intervals during the prime mission. The constellation configuration alternates between having the two sets of spacecraft radially aligned (at month 12 and 18) to longitudinally spread (at month 9 and 15) every 3 months as the elliptical orbit spacecraft "weave" back and forth around the 1AU spacecraft.

ICME events: HELICON samples energetic particles from interplanetary shocks associated with fast coronal mass ejections. The simultaneous measurements, at different rigidities and at different locations along the shock front, determine how the source strength depends on particle rigidity and heliolongitude. The only existing multi-spacecraft measurements of ICME-related events have shown dramatic differences in the time-intensity profiles across longitudes as small as ~25° [e.g. *Reames,* 1999].

Therefore, HELICON is well configured in heliolongitude to probe the acceleration and transport along the shock.

Particle acceleration and transport in CIRs: HELICON addresses key questions of Corotating Interaction Regions (CIR)-related composition by taking a snapshot of the particle population at different longitudes at the same time, thus separating the effects of corotation from time dependence in the corotating frame. Since some CIRs populate the IMF with energetic ions and electrons for many tens of degrees of heliolongitude, only a multiple-platform constellation such as HELICON can sample the CIR acceleration sites from different locations along the reverse shock at the same time.

The only previous radial survey of CIR-related energetic ions near 1AU used the Helios 1 & 2 spacecraft [*Van Hollebeke et al.* 1978]. Based on the Helios results, it is expected a radial gradient in particle intensity within a *single* CIR as large as a factor of 5 over the radial sampling of HELICON (0.2AU).

Therefore, to achieve our science objectives with respect to the source and scale-size of solar energetic particle events, HELICON needs to identify solar wind transients such as shocks and CIRs at a range of scale-lengths from these structures. Making measurements of the magnetic field, bulk plasma moments (n, T, V), suprathermal electron flux and energetic particle flux at a wide range of energies (20 keV to several MeV) will accomplish our objectives.

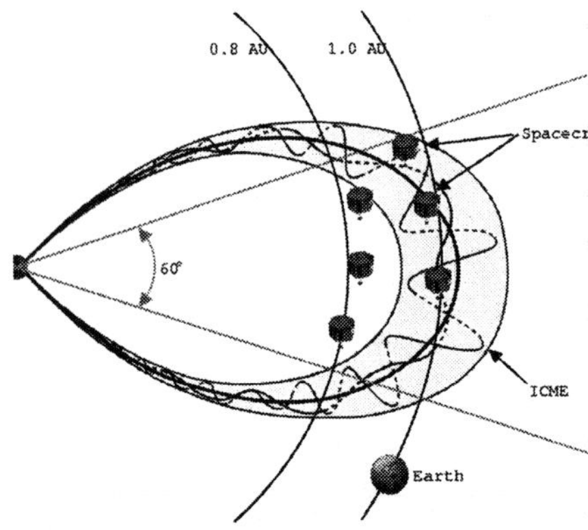

Figure 3: HELICON will be able to make six measurements within an ICME in order to determine structure and evolution. The spacecraft are shown at their positions two years after launch.

Nature and Structure of the Heliospheric Current Sheet: HELICON makes simultaneous measurements above, below, and within the HCS at a wide range of scale-lengths (See Figure 4) in order to address the three-dimensional structure of the HCS. The combined IMF measurements, plasma composition, and suprathermal electron direction information determines the magnetic topology and connection to the Sun to test models of the HCS structure such as those of Crooker et al. [1996].

The perturbation of the HCS due to transients in the solar wind are also examined by making field and particle measurements both within the transients, at the HCS, and across the boundary between them. By making simultaneous observations across the HCS over a range of azimuthal and radial scale lengths, HELICON unambiguously separates spatial from temporal dynamics and better understands the relative role of forcing at the Sun compared with dynamics caused by the interaction of streams in the interplanetary medium.

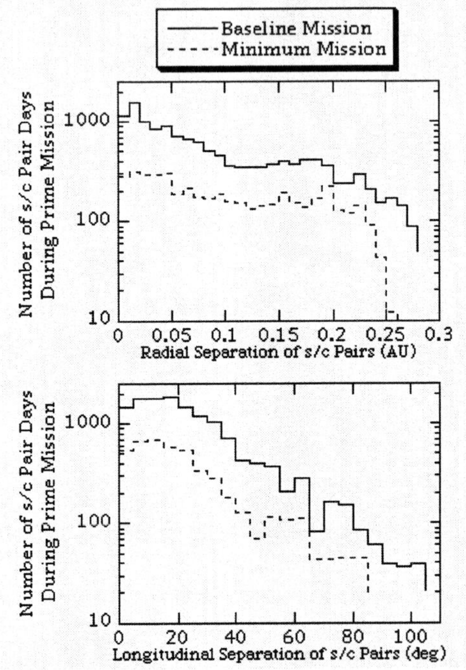

Figure 4: The distribution of the number of days of HELICON spacecraft pair separations. Top panel shows the radial separation and the bottom panel shows the azimuthal separation. The solid line represents the coverage for the Baseline Mission; the dashed line shows the coverage for the Minimum Mission.

Source of Solar Wind Variability: HELICON samples a wide range of spatial/temporal scales due to the wide range of spacecraft separations. Presently, we do not know if solar wind variability on the 0.1-10 hour time scales is caused by changes in the wind from a single source (time variability) or from sampling wind from different sources (spatial variability) or both. Assuming a ballistic mapping, the difference between the spacecraft longitude and the source longitude is given by $\Delta\varphi=\Omega\Delta t$, where Δt ($=R_{sc}/V_{sc}$) is the travel time of the solar wind parcel from the Sun to the spacecraft. Thus, two spacecraft separated in longitude and/or radius sample solar wind from different locations at the same time and sample solar wind from the same location at different times. For a solar wind velocity of 400 km/sec, spacecraft at the same longitude but separated radially by 0.1AU measure solar wind from sources separated by about 6° at the same time or from the same source region with a time difference of 10 hours; spacecraft separated by 6° but at the same distance from the Sun provide similar sampling. With spacecraft separations ranging from 0.0001 to 0.3AU HELICON samples time scales ranging from 1-200 hours and source size (at the source surface) of 0.6° to 100° and resolves the ambiguity.

HELICON, by separating temporal and spatial changes in the composition, can give information about the scales of the physical processes involved in this intermittent release; e.g., the time scales over which the composition remains constant can be related to the loop size. Active region time scales are typically minutes to hours and thus one would expect large temporal variation at these scales in solar wind from active regions. If slow wind near the current sheet is in blobs [*Wang et al.,* 1998], expected temporal variations in density are on the order of several hours. As another example, plasma sheets at the heliospheric current sheets typically take 1-10 hours to pass by a spacecraft, corresponding to an angular size of 0.6° -6°. HELICON is able to determine the temporal and spatial scales of solar wind.

CONCLUSIONS

Heliospheric Constellation (HELICON) is the first mission to make coordinated, simultaneous, multi-point measurements to study the radial and longitudinal evolution of a single parcel of the solar wind. It measures both spatial and temporal variations of solar wind streams in order to sample scale-sizes never before observed. HELICON consists of six identical spacecraft in two sets of three, each having distinct one-year period heliocentric orbits. One triad orbits the Sun at 1AU and the other triad in an elliptical orbit from 0.8 to 1.2AU. The spacing of the six HELICON spacecraft evolves through the mission, allowing measurements of multiple scale-lengths of the solar wind ranging from <1° to 90° in longitude and 0.0001 to 0.3AU in the radial extent. Joint studies, such as with the WIND and NEAR spacecraft, have shown that these separations are precisely those needed to measure the structure of interplanetary coronal mass ejections, corotating interaction regions, and the heliospheric current sheet.

HELICON enables the resolution of a wide array of critical questions of solar wind structure and dynamics. The HELICON's Science Objectives are as follows:

-Determine the structure and evolution of expanding interplanetary coronal mass ejections

-Use supra-thermal and energetic particles to determine the source populations of solar events and the scale sizes of ICMEs and CIRs

-Determine the structure and nature of the heliospheric current sheet

-Examine the causes of variability in the solar wind.

HELICON examines the evolution of stream structure as a function of distance from the Sun and time, thereby enabling the examination of both the interaction of solar wind streams and the effects on stream structure by variations at its source. HELICON provides for the first time the ability to study energetic particles simultaneously at many points along interplanetary shocks to unambiguously determine the particle sources and their transport. Finally, HELICON provides the information needed to develop predictive capabilities such as correlation lengths, correlation times, and the radial scale length of shock evolution. These studies are essential for the use of remotely obtained heliospheric data to predict conditions at the Earth with the advance warning desired by many users of space weather data.

The HELICON mission concept and satellite configuration are evolving to take advantage of advances in satellite design and instrument miniaturization. The HELICON mission is designed to fit under a NASA MIDEX cost-cap. The HELICON mission is a partnership between UCLA, NASA JPL, Los Alamos National Laboratory, and The Aerospace Corporation.

ACKNOWLEDGMENTS

We thank the HELICON engineering teams for their hard work and continued support to make HELICON a reality. The engineering team was lead by A. Gerber (JPL) and J. Cantrell (SDL).

REFERENCES

1. Crooker, N. U., et al., Heliospheric plasma sheets as small-scale transients, *J. Geophys. Res., 101*, 2467-2474, 1996.
2 Mulligan,T., et al., Inter-comparison of NEAR and Wind Interplanetary Coronal Mass Ejection Observations *J. Geophys. Res., 104*, 28217-28223, 1999.
3. Reames, D. V., *Space Sci. Rev. 90,* 413-489, 1999
4. Van Hollebeke et al. *J. Geophys. Res. 83*, 4723-4731, 1978.
5. Wang, Y.-M., et al., Origin of streamer material in the outer corona, *Astrophys. J., 498*, L165-L168, 1998.

The Ultraviolet and Visible-light Coronagraph of the HERSCHEL experiment

M. Romoli[a], E. Antonucci[b], S. Fineschi[b], D. Gardiol[b], L. Zangrilli[b], M.A. Malvezzi[c], E. Pace[a], L. Gori[a], F. Landini[a], A. Gherardi[a], V. Da Deppo[d], G. Naletto[e], P. Nicolosi[e], M.G. Pelizzo[d], J.D. Moses[f], J. Newmark[f], R. Howard[f], F. Auchere[f], J.P. Delaboudinière[g]

[a]*Dip. di Astronomia e Scienza dello Spazio - Univ. di Firenze, Italy*
[b]*Osservatorio Astronomico di Torino - INAF, Italy*
[c]*Dip. di Elettronica - Univ. di Pavia - INFM, Italy*
[d]*Dip. di Elettronica e Informatica - Univ. di Padova, Italy*
[e]*Dip. di Elettronica e Informatica - Univ. di Padova – INFM, Italy*
[f]*Naval Research Laboratory, Washington DC, USA*
[g]*Institut d'Astrophysique Spatiale, Université Paris XI, Orsay, France*

Abstract. The Herschel (HElium Resonant Scattering in the Corona and HELiosphere) experiment, to be flown on a sounding rocket, will investigate the helium coronal abundance and the solar wind acceleration from a range of solar source structures by obtaining the first simultaneous observations of the electron, proton and helium solar corona. The HERSCHEL payload consists of the EUV Imaging Telescope (EIT), that resembles the SOHO/EIT instrument, and the Ultraviolet and Visible Coronagraph (UVC).UVC is an imaging coronagraph that will image the solar corona from 1.4 to 4 solar radii in the EUV lines of HI 121.6 nm and the HeII 30.4 nm and in the visible broadband polarized brightness. The UVC coronagraph is externally occulted with a novel design as far as the stray light rejection is concerned. Therefore, HERSCHEL will also establish proof-of-principle for the Ultraviolet Coronagraph, which is in the ESA Solar Orbiter Mission baseline.The scientific objectives of the experiment will be discussed, togetherwith a description of the UVC coronagraph.

INTRODUCTION

This investigation is named after John F.W. Herschel (1792-1871), son of William Herschel. John Herschel's observation of Halley's Comet in 1835 led him to hypothesize a *repulsive force* from the Sun opposing gravity. This repulsive force is known today as the Solar Wind. HERSCHEL is conceived as a NASA Sounding Rocket Program providing new EUV and visible-light coronal observations to directly measure and to characterize in detail the properties of the two most abundant elements, Hydrogen and Helium. In particular, HERSCHEL will be able to:

- provide the first global images of the HeII corona;
- provide the first global EUV images of the corona for the two most abundant elements, H and He;
- provide the first maps of He abundance in the corona;
- provide the first global maps of the solar wind outflow (H^0 and He^+ outflow).

In addition to the above science objectives, HERSCHEL will establish a proof-of-principle for the UVC, which is in the ESA Solar Orbiter Mission[1,2]. Table 1 summarizes HERSCHEL objectives and the way they are accomplished. The HERSCHEL instrument package consists of the Extreme Ultraviolet Imaging Telescope (EIT)[3], which will provide HeII (30.4 nm) images from disk to 1.5 $R_\odot$, and FeXI, FeXII, and FeXV coronal images from disk to 1.5 $R_\odot$ for Si XI extraction; and the Ultraviolet and Visible-light Coronagraph (UVC), which will provide HI Ly-α (121.6 nm) images, HeII Ly-α (30.4 nm) images, and polarized brightness (pB) images from 1.4 to 4.0 $R_\odot$.

UVC OPTICAL DESIGN

UVC consists of two coronagraphs with identical optical design: coronagraph A for the imaging of HI121.6nm and visible-light corona; coronagraph B for the imaging of HeII 30.4nm and visible-light corona. The two coronagraphs differ mainly in the mirror coatings: coronagraph A has Al+MgF_2 and coronagraph B has Mo/Si multilayer. The optical layout of UVC (see Fig.1 and Fig.2) consists of two sections: the boom assembly and the telescope assembly. The boom assembly accomplishes the principal action required by a coronagraph: the rejection of the solar disk radiation. The telescope assembly is the imaging optics where the image of the solar corona is formed and detected. The telescope design is an off-axis gregorian.

CP679, *Solar Wind Ten: Proceedings of the Tenth International Solar Wind Conference*, edited by M. Velli, R. Bruno, and F. Malara

TABLE 1. HERSCHEL investigation summary.

Science Objectives	Approach	Physical Parameter	Observations
1. Investigate origin of the **slow solar wind**	Establish relative roles of **gravitational settling** and **Coulomb friction** in slow solar wind acceleration regions (e.g. streamers, coronal hole boundaries, fine structures)	• **He^+ density** map < 3 $R_\odot$ • **H^0 density** map < 3 $R_\odot$ • **e^- density** map (to account for He II doppler dimming comp.)	
2. Investigate the acceleration mechanism of the **fast solar wind**	Measure **outflow velocities** of different ions in coronal holes via doppler dimming	• **He^+ density** and **doppler dimm. wind speed** map < 2 $R_\odot$ • **H^0 density** and **doppler dimm. wind speed** map < 2 $R_\odot$ • **e^- density** map < 3 $R_\odot$	• **Ly-α HeII** intensity (UVC) • **Ly-α HI** intensity (UVC) • **FeX,FeXII,FeXV** int. to derive collisionally excited emission contribution to 304nm bandpass, e.g., Si XI (EIT)[1] • **Ly-α HeII** low corona (EIT) • **K corona pB** (UVC) • **Stray light** profiles (UVC)
3. Investigate variation of **helium abundance** in coronal structures: a) departures from primordial composition b)fractionation region for helium in the solar atmosphere	Measure **Helium abundance** profile in the extended corona (1.25 → 3.5 $R_\odot$)	• **He^+ density** map < 3 $R_\odot$ • **H^0 density** map < 3 $R_\odot$	
4. Facilitate **future investigation** of CMEs, kinematics, and solar cycle evolution of the electron, proton, and helium corona	Establish a proof of concept via suborbital investigation that will lead to the develop ment of the UVC strawman for **Solar Orbiter**	• **He^+ density** and **doppler dimm. wind speed** map < 2 $R_\odot$ • **H^0 density** and **doppler dimm. wind speed** map< 2 $R_\odot$ • **e- density** map < 3 $R_\odot$	

[1] Technique explained in Delaboudinière[4]

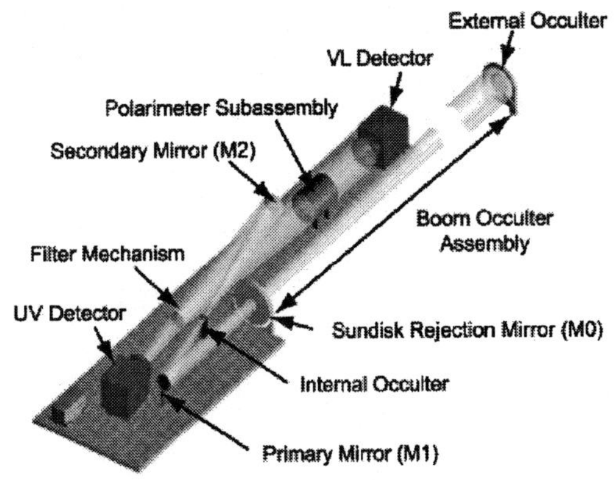

FIGURE 1. Coronagraph design showing a single coronagraph. The second coronagraph will be mounted on the opposite side of the optical bench.

Boom assembly. The solar disk-light is blocked on the front aperture by the external occulter (EO). The disk-light passing through the annular front aperture of EO is reflected back through it by a sun-disk rejection mirror (M0) behind the external occulter. This "solar shield" has an entrance aperture (A1) which is completely in the shadow of the external occulter. A1 allows the telescope primary mirror (M1) to collect the coronal light going through EO. M0 is a slightly tipped spherical mirror whose focal length is equal to the EO-M0 distance. In this way, the mirror collects all direct sunlight entering the instrument, focuses the solar image on one side of the front aperture and rejects it outside the instrument.

Telescope assembly. The primary mirror, M1, creates real images of the diffracting edges of EO and A1. The image of the EO edges is blocked by the internal occulter (IO), while that of A1 is dumped in the Lyot trap (A3) placed around and behind the secondary mirror (M2). In this way, M2 focuses only the coronal light on the detector plane. Both M1 and M2 are ellipsoids. Coronagraph A is equipped with a narrow band interference filter that transmit the HI line. The visible-light back-reflection of the HI Ly-α filter is used for the measurement of the polarized brightness (pB) of the K-corona. The HeII coronagraph is equipped with a two slot filter wheel that accomodates a HeII band pass filter and a mirror for the polarized brightness. The selection of the HeII 30.4 nm bandpass is accomplished by the multilayer coating of the mirrors.

UVC OPTICAL PERFORMANCES

Predicted coronal intensities. Fig.3 shows the predicted coronal intensities. The HI 121.6 nm curves are actual data from UVCS/SOHO[4] (streamers[6], coronal holes[7]). Also shown is the HI geocoronal emission. Visible light intensities are derived from Allen[8]. The helium intensities inferred from EIT data[4] show a dominance of stray light near 1.5$R_\odot$, showing the need for coronagraph observations. The predicted streamer HeII intensities are derived from electron densities and temperatures given by Gibson et al.[9].

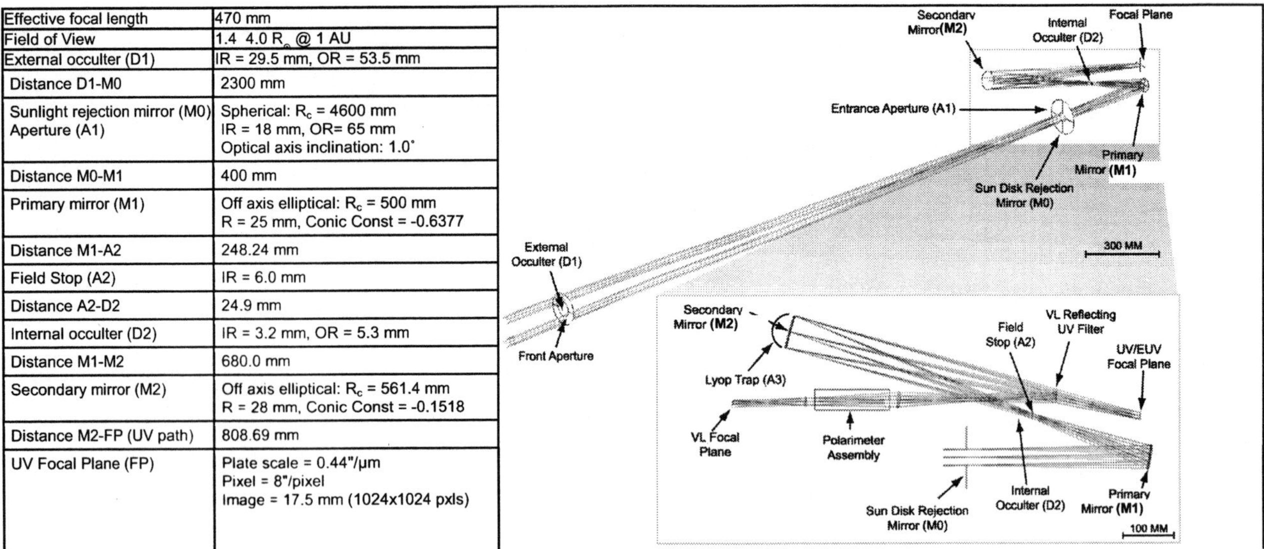

Effective focal length	470 mm
Field of View	1.4 4.0 $R_{\odot}$ @ 1 AU
External occulter (D1)	IR = 29.5 mm, OR = 53.5 mm
Distance D1-M0	2300 mm
Sunlight rejection mirror (M0) Aperture (A1)	Spherical: R_c = 4600 mm IR = 18 mm, OR= 65 mm Optical axis inclination: 1.0°
Distance M0-M1	400 mm
Primary mirror (M1)	Off axis elliptical: R_c = 500 mm R = 25 mm, Conic Const = -0.6377
Distance M1-A2	248.24 mm
Field Stop (A2)	IR = 6.0 mm
Distance A2-D2	24.9 mm
Internal occulter (D2)	IR = 3.2 mm, OR = 5.3 mm
Distance M1-M2	680.0 mm
Secondary mirror (M2)	Off axis elliptical: R_c = 561.4 mm R = 28 mm, Conic Const = -0.1518
Distance M2-FP (UV path)	808.69 mm
UV Focal Plane (FP)	Plate scale = 0.44"/μm Pixel = 8"/pixel Image = 17.5 mm (1024x1024 pxls)

FIGURE 2. UVC optical layout and optical parameters.

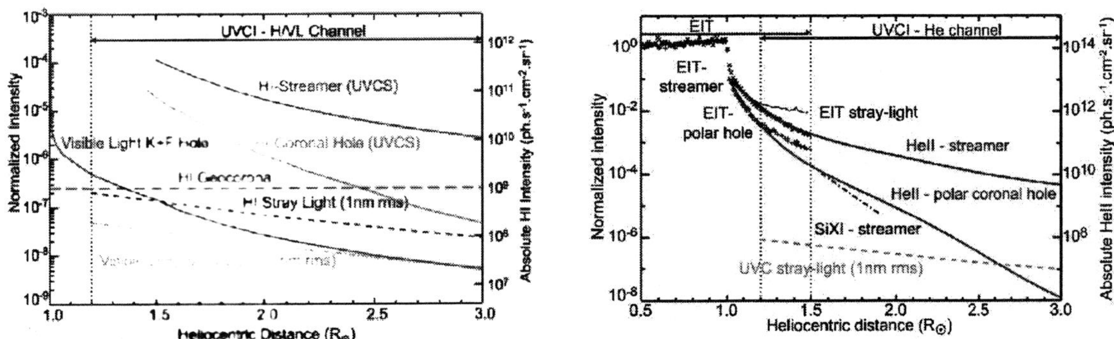

FIGURE 3 Predicted intensities and stray light in the UVC coronagraph A (left, HI and Visiblelight channels) and B (right, HeII channel).

The predicted coronal hole HeII intensities are derived using electron densities from Guhathakurta et al.[10], electron temperatures extrapolated from CDS/SUMER measurements[11] to interplanetary value[12], and outflow velocities from UVCS[12]. The modeled intensity of the SiXI 30.32 nm, in a streamer is shown, starting at 1.5$R_{\odot}$. Above 1.75$R_{\odot}$, the SiXI contribution to the signal in the UVCI HeII channel is negligible. This can be easily understood as the intensity profile of the collisionally produced SiXI line drops with the square of the electron density, whereas the intensity of the resonant HeII Ly-a drops only linearly with the electron density.

Predicted count rates. Fig.4 shows the predicted count rates. Helium will be detected up to at least 2$R_{\odot}$ in coronal holes. Statistically significant measurements will therefore be possible up to 3$R_{\odot}$ in streamers. In coronal holes, using the CCD in binning mode, we will reliably detect hydrogen up to 3$R_{\odot}$ and helium up to 2.6$R_{\odot}$. The fast solar wind acceleration occurs within these heights in coronal holes. Note that the real helium coronal hole count rate may be higher because the model includes significant Doppler dimming and does not account for the possible helium enhancement detected by Delaboudinière[4]. This shows an important feature of the HERSCHEL experiment: the FOVs of EIT and UVCI overlap. This will provide an important in-flight confirmation of the laboratory intercalibration.

Stray light rejection. Stray light inside the coronagraphs is produced by two sources:

1. disk light entering directly the external occulter;
2. disk light diffracted by the edges of the external occulter.

Disk light entering directly the external occulter (D1) is reflected back outside D1 by M0. The major source of stray-light is the disk-light diffracted by the edges of the external occulter, D1, and by the edge of the entrance aperture, A1, in M0.

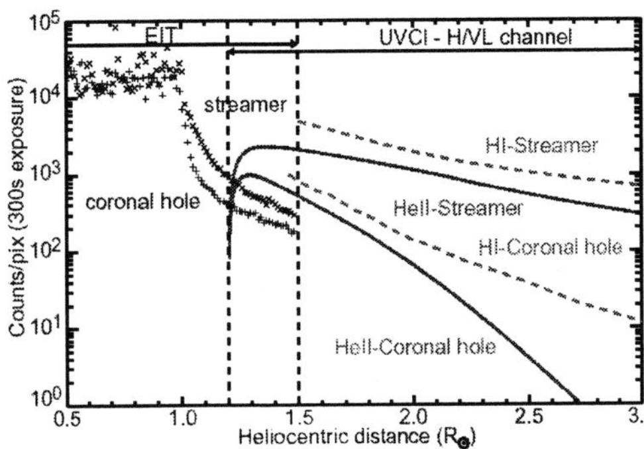

FIGURE 4. Predicted UV count rates.

This diffracted light is specularly reflected by the primary mirror, M1, which forms two images of D1 and A1. The internal occulter (D2) is placed where M1 creates the real image of D1. The real image of A1 is formed by M1 on the plane containing M2, just outside M2 edge. This image is dumped in the Lyot trap (A3) placed around and behind M2. Estimates of the stray light for both coronagraphs are given in Fig.3. The stray light estimates are based on the assumption of a 1 nm rms surface roughness[13]. One of HERSCHEL objectives is the test of the stray light in this novel coronagraphic design.

Multilayer coating performances Single optics can be used for all 3 UVC wavelengths. Fig.5 gives the reflectivity in EUV and visible wavelength range for the Mo/Si multilayer with a 30.4nm period. The key element in the Solar Orbiter UVCI instrument concept is that the mirrors with coatings optimized for 30.4 nm still have good reflectivity at 121.6 nm and in the visible. In the case of the HI 121.6 nm line, the off band rejection of the visible light is provided by Al/MgF2 interference filter such as the commercially available Acton 120 nm filter. The choice of two coronagraphs on HERSCHEL is explained with the short duration of the observation and with the importance of comparing visible light stray light (the most critical requirement) produced by multilayer optics to that of a more traditional Al/MgF_2 coated optics.

Optical performances. Fig.6 shows the optical performances of the EUV and visible-light paths. The optical performances of the VL path are limited at low solar radii (<1.8$R_\odot$) by the diffraction and at high solar radii by the pixel size. The optical performances of the EUV paths are essentially pixel limited. Diffraction and aberration for 30.4 nm bands are less than 1 pixel everywhere.

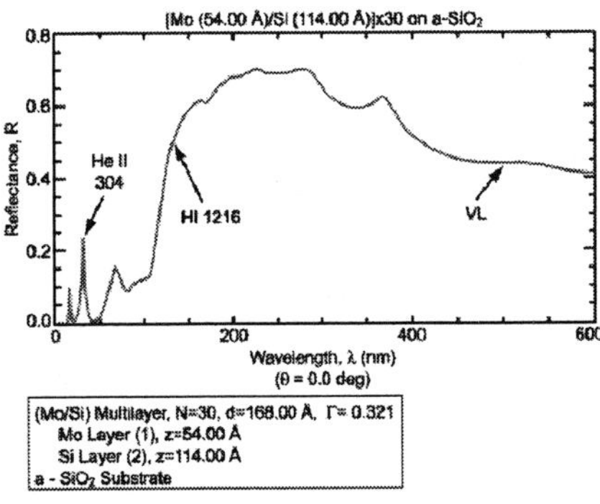

FIGURE 5. Reflectance of Mo/Si multilayer optimized for 30.4nm.

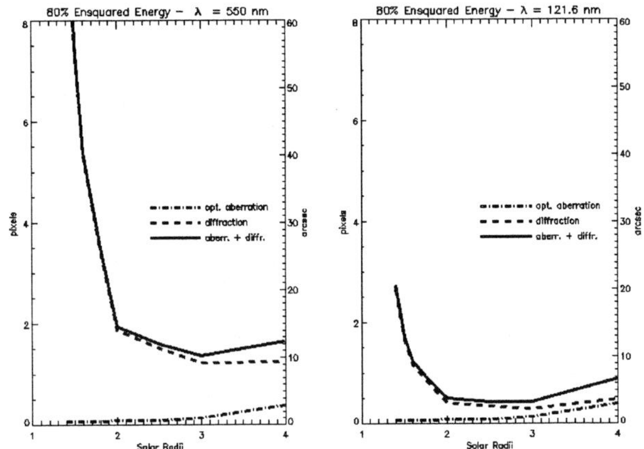

FIGURE 6. UVC optical performances: Visible-light channel (left), and HI channel (right).

REFERENCES

1. *Solar Orbiter: A High Resolution Mission to the Sun and the Inner Heliosphere,* ESA-SCI(2000)6, July 2000.
2. Antonucci, E., et al., Proc. SPIE 4139,378 (2000).
3. Delaboudinière, J.P., et al., *Sol. Phys.* 162,291 (1995).
4. Delaboudinière, J.P., *Sol. Phys.* 188,259 (1999).
5. Kohl, J.L., et al., *Sol. Phys.* 162,313 (1995).
6. Strachan, L., Suleiman, R., Panasyuk, A.V., Biesecker, D.A., Kohl, J.L., *ApJ* 571,1008 (2002).
7. Cranmer, S.R., et al., *ApJ* 511,481 (1999).
8. Allen, C.W., *Astrophysical Quantities,* The Athlone Press, London (1985).
9. Gibson, S., et al., *JGR* 104,96 (1999).
10. Guhathakurta, M., et al., *JGR* 104,9801 (1999).
11. David, C., et al., *A&A* 336,L90 (1998).
12. Antonucci, E., Dodero, M.A., Giordano, S., *Sol. Phys.* 197,115 (2000).
13. Romoli, M., et al., Proc. SPIE 4498,27 (2001).

An Interstellar Neutral Atom Detector (INAD)

Stefano Livi[1], Eberhard Möbius[2], Dennis Haggerty[1], Manfred Witte[3], and Peter Wurz[4]

1) Johns Hopkins University – Applied Physics Laboratory
2) University of New Hampshire
3) Max Planck Institut für Aeronomie
4) Physikalisches Institut – University of Bern

Abstract. Direct detection of interstellar neutrals is a powerful technique for enlarging our knowledge about the media surrounding our solar system. We present in this paper a combination of two telescopes and a pointing device that would enable precise and detailed measurements of the density, velocity, temperature, and composition of the neutral particles that penetrate through the heliospheric bow shock and the heliopause.

INTRODUCTION

The interstellar medium is commonly thought to be far away, beyond the termination shock of the supersonic solar wind at perhaps 80–100 astronomical units (AU) from the Sun, or beyond the heliopause, where the solar wind encounters the interstellar medium, at about 150 AU. Common conception has it that only distant Voyager, or eventually Interstellar Probe, has any expectation of reaching interstellar space. However, the inner edge of the interstellar medium is far closer. The solar wind does not affect neutral particles from the interstellar medium; they penetrate relatively undisturbed to within 3 AU of the Sun. Measurements made with an instrument such as INAD would reveal vital new data on the physical state and composition of the interstellar medium. The capability of INAD to observe the velocity distributions of He, H, and O will also allow to determine the deceleration and heating of inflowing interstellar gas through charge exchange interaction in the heliospheric interface region [Möbius et al., 2001].

INAD comprises three elements: the Neutral Interstellar Helium Telescope (NIHeT), dedicated to detecting and measuring the properties of interstellar Helium, the Neutral Atom Telescope (NAT) focused on measurements of Hydrogen, Oxygen and Carbon, and the Actuating Platform (AP) that orients the two telescopes in space, and permits, combined with spacecraft rotation, an almost complete coverage of the sky.

NEUTRAL INTERSTELLAR HELIUM TELESCOPE (NIHET)

NIHeT is a high-sensitivity, high-angular-resolution version of Ulysses/GAS, which observed the local distribution function of interstellar He for the first time [Witte et al., 1992]. NIHeT has the necessary sensitivity, angular resolution, and background rejection to precisely measure the interstellar neutral He bulk flow parameters (flow vector, temperature, density), and to detect non-thermal tails in the He distributions.

Principle of Operation.

Figure 1 illustrates NIHeT's principle of operation: particles enter the NIHeT analyzer through a multi-slit mechanical collimator [Taylor, 1993, Möbius et al., 1998] of 2° FWHM transmission and 65% transparency.

Following the first mechanical collimator, there is a set of cylindrical concentric plates, held at different potentials. The electric field between the plates changes the trajectory of charged particles, which are then absorbed by the second mechanical collimator, which is physically identical to the first. However neutral particles pass directly through both sets of collimators and hit a conversion surface covered with a thin layer of lithium fluoride (LiF).

CP679, *Solar Wind Ten: Proceedings of the Tenth International Solar Wind Conference,*
edited by M. Velli, R. Bruno, and F. Malara

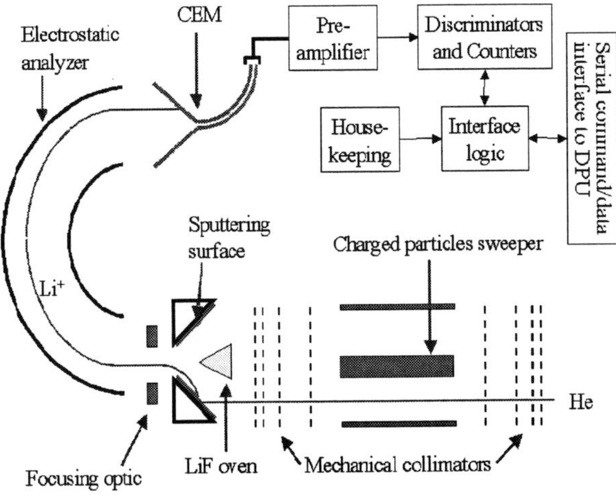

FIGURE 1. NIHeT principle of operation.

Upon impact, secondary positively charged ions are sputtered from the LiF surface [Witte et al. 1992], accelerated, and guided by ion optics to a channel electron multiplier (CEM), where they are counted by conventional electronics. The ion optical system eliminates most of the background photoelectrons and efficiently traps UV photons scattered or reflected from the LiF surfaces, which contribute noticeably to the background in Ulysses/GAS. The sputtering efficiency of positive ions from impacting atoms is energy dependent (Fig. 2), about 1% at 60 eV and $<10^{-4}$ below 20 eV. This $<10^{-4}$ efficiency prevents detection of the <16 eV interstellar H atoms.

Because sputtering efficiency degrades slightly with contamination of the LiF surface, new clean layers can be deposited in flight by evaporation of pure

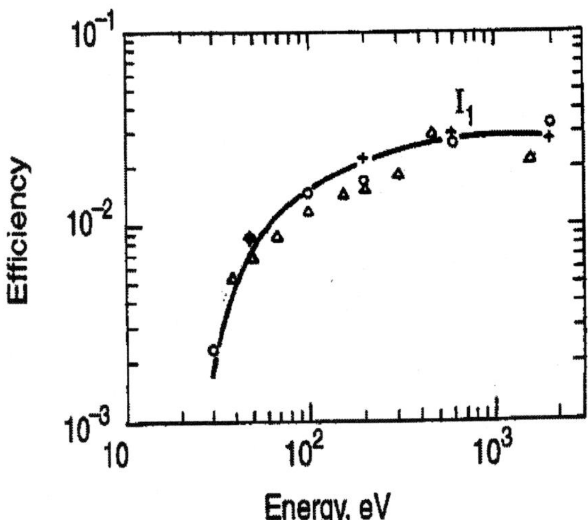

FIGURE 2. NIHeD Sputtering efficiency of secondary, positively charged ions (I_1) from a LiF surface after impact of a helium atom [Witte et al. 1992].

LiF from a heated supply, a technique used successfully by Ulysses/GAS.

Two modifications significantly improve the performance of NIHeT over Ulysses/GAS: first, the collection area will be increased by a factor of 400, second, a significant reduction of the background level. The capability of the GAS-instrument to observe low particle fluxes was limited by a signal-to-noise ratio (SNR) not better than 10, due to background counts from the g-radiation of the RTG, from penetrating cosmic ray particles, and predominantly from UV-photons, which could reach the CEM after just one specular reflection in the collimator system. In NIHeT, however, only the cosmic ray background in the somewhat larger CEM will matter, as UV-photons as well as photo-electrons will be efficiently suppressed by the subsequent ion-optics, including a spherical electro-static analyzer (ESA). In addition, as the cosmic ray background is essentially omni-directional while the signal appears in a peaked, angular distribution (<15° FWHM), the background contribution can be easily determined and removed, producing a very low SNR. Finally, the conical Field of View (FoV) is collimated to 1º (FWHM from the combined collimating effect of the two multi slit mechanical collimators) to improve angular resolution.

NEUTRAL ATOM TELESCOPE (NAT)

NAT is a mass-resolving neutral gas sensor that will produce images of the flow pattern of interstellar neutral O and H as well as neutral C. NAT has the necessary sensitivity, angular resolution, and background rejection (Table 1: NAT performance characteristics) to measure the interstellar neutral O and H bulk flow parameters (flow vector, temperature, density), detect non-thermal tails in the O distributions, and to search for neutral C from a dust or cometary origin. The necessary performance for interstellar gas detection was recently demonstrated with a prototype instrument [Wieser et al., 2001].

NAT basic design is the same as the zero-dimensional camera concept of NIHeT. The major difference is that after the collimator NAT uses the surface charge exchange technology that is necessary for multi species determination [Wurz 2000], and was successfully used in IMAGE/LENA [Ghielmetti et al., 1994; Wurz et al., 1995]. It also makes use of subsystems successfully flown on ACE and Cluster.

Principle of Operation.

Figure 3 illustrates NAT's principle of operation: particles first traverse a collimator identical in form and function to the one foreseen for NIHeT. Neutral

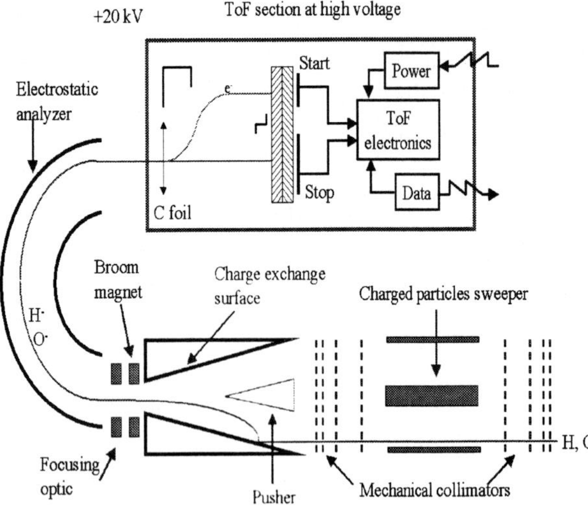

FIGURE 3. NAT principle of operation

particles continue their trajectory, hit a conversion surface at shallow angle of ~10°, and are scattered approximately in specular direction. Upon reflection a fraction of the atoms become negatively ionized. Diamond (natural and synthetic), $BaZrO_3$, AlN, or MgO on a highly polished tungsten substrate have been shown to exhibit very good conversion efficiencies [Wurz et al., 1997, 1998, Jans et al., 2000, 2001, Wieser et al., 2002]. The newly born negative ions are accelerated and guided by ion optics, identical to NIHeT, into a time-of-flight (TOF) section after a post-acceleration to 15–20 kV. The ion optical system eliminates most of the background photoelectrons and efficiently traps UV photons scattered or reflected from the conversion surfaces. In addition, a magnet is added to remove the low energy electrons produced from the conversion surface.

The negative ions enter the TOF section through a thin (~2–3 μg/cm^2) C-foil and proceed to hit the Stop portion of a microchannelplate (MCP) detector. Secondary electrons emitted from the C-foil are accelerated and deflected on to the Start portion of the MCP. The TOF is derived from the timing difference between both signals. The mass of the incoming negative ion can be calculated from the time of flight knowing the length of the TOF section and the post acceleration potential. The TOF section has been adapted from the Cluster/CODIF, FAST/TEAMS and Equator-S/ESIC sensors [Möbius et al., 1998; Rème et al., 1997]. Because of the TOF technique, this sensor is inherently insensitive to Ly-α and secondary electron background.

ACTUATED PLATFORM

The two instruments (NAT and NIHET) will be mounted on a common actuated platform that allows pointing in any direction, from parallel to anti-parallel to the spin axis of the spacecraft. Combined with the spacecraft spin, this movement points the two instruments through the full 4π celestial sphere. Actuated platform mechanisms have been flown successfully on numerous missions in the past, mostly as a telescope subsystem. A similar system (that uses the same motor, encoder, bearings, and flexible feedthroughs) is part of the TIMED/SEE instrument and is being developed for the telescope MDIS on the Messenger spacecraft.

The mechanism will be positioned so that the two instruments scan through the direction of the incoming interstellar neutral wind on every spacecraft rotation. The typical mode of operation will be a step of 1° every 15 min, fully controlled by the DPU. This operational mode results in 100% coverage of the sky (4π steradian) during each half of the orbit.

INSTRUMENT PARAMETERS

The main characteristics of the INAD package are summarized in Table 1. Note the geometric factor is increased by a factor of 50 compared to Ulysses/GAS, because of the much larger entrance area. Note also the increased angular resolution by factor of 8, resulting from the new collimator design; the extreme improvement in signal-to-noise ratio, owning to the electrostatic analyzer; and the capability of measuring atoms other then Helium, possible because of the charge exchange surface coupled with time-of-flight.

The INAD package has been optimized to achieve the largest geometric factor at the smallest possible mass and power consumption, while using as much as possible of existing subsystems, for reliability and cost optimization. Total mass is estimated to be around 6.5 kg, maximum power 5W.

TABLE 1. INAD performance characteristics

	NIHeT	NAT	Ulysses/GAS
Atoms measured	He, O and Ne	O, H, C	He
Energy range	30–130 eV [up to > 1 keV]	5–600 eV	30–130 eV
Angular Coverage	4π	4π	4π
Angular resolution	1°×1°	1°×1°	2°×4°
Active area	38 cm^2	38 cm^2	0.1 cm^2
Signal/background	40,000	$>10^5$	10
Geometric factor	1×10^{-2} cm^2 sr.	1×10^{-2} cm^2 sr.	2×10^{-4} cm^2 ster
M/ΔM [FWHM]	N/A	10	N/A

CONCLUSIONS

The neutral component of interstellar matter penetrates within 3 AU, and can therefore be reached and analyzed with a relatively contained effort. We described an instrument that has two hundred times more sensitivity than it's progenitor, and can measure for the first time the velocity distribution of interstellar Hydrogen and Oxygen. The required resources (6.5 kg, 5W) would be compatible with those available onboard a small satellite.

REFERENCES

1. Ghielmetti, A., E.G. Shelley, S. Fuselier, P. Wurz, P. Bochsler, F. Herrero, M.F. Smith, and T. Stephen, Optical Eng., 33 (1994) 362–370.
2. Jans, S., P. Wurz, R. Schletti, K. Brüning, K. Sekar, and W. Heiland, "Scattering of Atoms and Molecules from Barium Zirconate Surfaces," Nucl. Instr. Meth. B 173(4), (2001), 503–515.
3. Jans, S., P. Wurz, R. Schletti, T. Fröhlich, E. Hertzberg, and S. Fuselier, "Negative Ion Production by Surface Ionization Using Aluminium-Nitride Surfaces," J. Appl. Phys. 85(1) (2000), 2587–2592.
4. Möbius, E., L. M. Kistler, M. Popecki, K. Crocker, M. Granoff, Y. Jiang, E. Sartori, V. Ye, H. Rème, J.A. Sauvaud, A. Cros, C. Aoustin, T. Camus, J. L. Médale, J. Rouzaud, C.W. Carlson, J.P. McFadden, D.W. Curtis, H. Heetderks, J. Croyle, C. Ingraham, E.G. Shelley, D. Klumpar, E. Hertzberg, B. Klecker, M. Ertl, F. Eberl, H. Kästle, E. Künneth, P. Laeverenz, E. Seidenschwang, G.K. Parks, M. McCarthy, A. Korth, B. Gräwe, H. Balsiger, U. Schwab, M. Steinacher, The 3-D Plasma Distribution Function Analyzers With Time-of-Flight Mass Discrimination for CLUSTER, FAST and Equator-S, Measurement Techniques in Space Plasmas, R. Pfaff, J. Borowski, D. Young eds., Geophys. Monograph 102, 243, 1998.
5. Möbius, E., Y. Litvinenko, L. Saul, M. Bzowski, D. Rucinski, Interstellar gas flow into the heliopshere, in: The Outer Heliosphere: The Next Frontiers, K. Scherer, H. Fichtner, H.-J. Fahr and E. Marsch eds., COSPAR Coll Series, Vol. 11, p. 109–120, 2001.
6. Rème, H., J.M. Bosqued, J.A. Sauvaud, A. Cros, J. Dandouras, C. Aoustin, Ch. Martz, J. L. Médale, J. Rouzaud, E. Möbius, K. Crocker, M. Granoff, L. M. Kistler, D. Hovestadt, B. Klecker, G. Paschmann, M. Ertl, E. Künneth, C.W. Carlson, D.W. Curtis, R.P. Lin, J.P. McFadden, J. Croyle, V. Formisano, M. DiLellis, R. Bruno, M.B. Bavassano-Cattaneo, B. baldetti, G. Chionchio, E.G. Shelley, A.G. Ghielmetti, W. Lennartson, A. Korth, H. Rosenbauer, I. Szemerey, R. Lundin, S. Olson, G.K. Parks, M. McCarthy, H. Balsiger, The CLUSTER Ion Spectrometry Experiment, Space Science Reviews, 79, 303, 1997.
7. Taylor, S. C., Throughput Capabilities of a High Resolution Collimator, M.S. Thesis, Univ. New Hampshire,(1996)
8. Wieser, M., P. Wurz, P. Bochsler, E. Möbius, J. Quinn, S. Fuselier, Test of Neutral to Negative Ion Conversion Surfaces in a Prototype Sensor for Interstellar Neutral Gas Measurement, AGU Spring Meeting, Boston, May, 2001.
9. Witte, M., H. Rosenbauer, E. Keppler, H.-J. Fahr, P. Hemmerich, H. Lauche, A. Loydl, and R. Zwick, The Interstellar Neutral Gas Experiment on Ulysses, Atron Atrophys. Suppl., 92 (1992)
10. Witte, M., H. Rosenbauer, M. Banaszkiewicz, and H.-J. Fahr, The Ulysses Neutral Gas Experiment,: Determination of the Velocity and Temperature of the Interstellar Neutral Helium, Adv. Space Res., 13 (1993)
11. Wurz, P., M.R. Aellig, P. Bochsler, A.G. Ghielmetti, E.G. Shelley, S. Fuselier, F. Herrero, M.F. Smith, T. Stephen, Opt. Eng., 34 (1995) 2365.
12. Brown, M. P., and Austin, K., The New Physique, Publisher City: Publisher Name, 1997, pp. 25–30.
13. Wieser, M., P. Wurz, K. Brüning, and W. Heiland, "Scattering of Atoms and Molecules off a Magnesium Oxide Surface," Nucl. Instr. Meth. B 192 (2002), 370–380.
14. Wurz, P., R. Schletti, and M.R. Aellig,, "Hydrogen and Oxygen Negative Ion Production by Surface Ionization Using Diamond Surfaces," Surface Science 373 (1997), 56–66.
15. Wurz, P., T. Fröhlich, K. Brüning, J. Scheer, W. Heiland, E. Hertzberg, and S.A. Fuselier "Formation of Negative Ions by Scattering from a Diamond (111) Surface," proceedings of the Week of Doctoral Students 1998, (eds. J. Safránková and A. Kanka), Charles University, Prague, Czech Republic (1998), 257–262.
16. Wurz, P., "Detection of Energetic Neutral Particles," The Outer Heliosphere: Beyond the Planets, (eds. K. Scherer, H. Fichtner, and E. Marsch), Copernicus Gesellschaft e.V., Katlenburg-Lindau, Germany, (2000), 251–288.

Author Index

A

B

C

D

E

F

G

H

I

J

K

L

M

N

O

P

Q

R

S

T

U

V

W

X

Y

Z